OXFORD MEDICAL PUBLICATIONS

Samson Wright's applied physiology

D1344814

Samson Wright's
applied physiology

Thirteenth edition

CYRIL A. KEELE

*Formerly Director of Rheumatology Research Department
at The Middlesex Hospital Medical School;
Emeritus Professor of Pharmacology and Therapeutics,
University of London*

ERIC NEIL

*John Astor Professor of Physiology,
University of London,
at The Middlesex Hospital Medical School*

and

NORMAN JOELS

*Professor of Physiology.
The Medical College of St Bartholomew's Hospital,
London*

OXFORD
OXFORD UNIVERSITY PRESS
NEW YORK   TORONTO
1982

Oxford University Press, Walton Street, Oxford OX2 6DP
London   Glasgow   New York   Toronto
Delhi   Bombay   Calcutta   Madras   Karachi
Kuala Lumpur   Singapore   Hong Kong   Tokyo
Nairobi   Dar es Salaam   Cape Town
Melbourne   Auckland
and associated companies in
Beirut   Berlin   Ibadan   Mexico City

© Oxford University Press, 1961, 1965, 1971, 1982
First published 1926
Thirteenth edition 1982

British Library Cataloguing in Publication Data

Wright, Samson
Samson Wright's applied physiology.—13th ed.—
(Oxford medical publications)
1. Human physiology
I. Title   II. Keele, Cyril A.
III. Neil, Eric   IV. Joels, Norman
612   QP34.5
ISBN 0–19–263211–6
ISBN 0–19–263210–8 Pbk

Set by Western Printing Services Limited, Bristol
Printed in Great Britain by Butler & Tanner, Frome, Somerset

Contents

PART XII Physiology of reproduction

Preface

Our preface to the twelfth edition (1971) referred to the extensive revision needed to update the book since its predecessor of 1965.

Eleven years on, the incorporation of the results of morphological and functional studies of ever increasing sophistication has necessitated an almost entirely new book. Text and figures have been radically altered.

We have preserved the format of twelve parts dealing with the major 'systems'. The *interrelation* of the several systems, which characterizes mammalian and, of course, human physiology, has been documented by numerous cross-references to the relevant pages, but our purpose has been to underline the integration of physiological functions which characterizes the performance of the body as a whole in health. Disruption of such integration causes symptoms and signs revealed by Nature's own experiments in dysgenesis and in disease in general; indeed, such have often shown the pathway for appropriate experiments to define the sequence of such a dissolution of the corporate society of the body tissues.

The inclusion of modern data and their current interpretation has required a pruning of accounts of pioneer experiments which provided the foundation on which modern work continues to build. The present day student whether medical, dental, paraclinical, or a so-called pure scientist is beset by university courses which now embrace a host of subjects which occupy time, energy, and thought. Some of these subjects have physiology as a parent; none of them is irrelevant to the development of an educated and sentient student. Nevertheless our awareness of such preoccupations have led us to present references mainly to reviews and books rather than to original papers. Reviews in any case refer to original papers and the interested and devoted student can read them and discuss them with his or her teacher.

As before we thank our colleagues for allowing us to use illustrations and for help and comments on the manuscript. Publishers of books and journals have been similarly generous and their help is appropriately acknowledged in the legends of the figures.

We have appreciated the sterling secretarial help provided by Miss Jennifer Williams, Mrs Maureen Goodwin, and Mrs Jean Speller.

Lastly, we are most grateful to the editorial staff of the Oxford University Press for their unfailing assistance with the galleys and page proofs, and for their advice about the final format of the book.

C.A.K.
E.N.
N.J.

London
April 1982

PART I
The internal environment

The cell and body fluids

THE CELL

Structure of the cell

All living organisms are composed of cells, just as molecules are composed of atoms. The cell was recognized as the unit of structure in plants by Schleiden and in animal tissue by Schwann in 1839. Virchow, in 1859, confirmed the cellular hypothesis showing that all cells must necessarily be derived from pre-existing cells: *omnis cellula e cellula*.

The development of light microscopy showed that the cell possesses a membrane, a cytoplasm which seemed to possess a vague internal organization or cytoskeleton and which contained a nucleus and various inclusion bodies—such as the centrosome, the Golgi apparatus, and various other ill-defined structures of granular nature [FIG. I.1].

Two developments during the last 40 years have enormously advanced our knowledge. One was the use of the electron microscope. The resolution of the best light microscope is about 1 μm (one thousandth of a millimetre). The cell boundary is of the order of 10 nm ($\frac{1}{100}$ μm) and hence appeared only as a thin dense line at the outer boundary of the cytoplasm, forming a marginal zone between the cytoplasm and the extracellular region. The electron microscope has a thousand times the resolving power of the light microscope and its use has revealed the complexity of the cell membrane. Similarly, for the cell inclusions—details of the nucleus and organelles now can be clearly defined. Just as the atoms were once thought to be the ultimate unit in chemistry and have since been shown to contain a positive nucleus with electron shells, so the cell can be redescribed in terms of its constituents.

The second development has lain in the elegant and sophisticated use of chemical and biochemical techniques in the study of the functional activities of the cell as a whole, and even more so of the working mechanisms of its different constituents. Thus, lysis of the cell, which destroys its membrane, causes the extrusion of its several constituents. 'Homogenization' of a tissue by mechanical means, such as grinding the tissue or disrupting its components ultrasonically, yields a broth of previously intracellular material. In this broth are suspended particles of different density and the use of ultracentrifuges with ever increasing gravitational power allows differential centrifugation of these various components of the cell. These can then be examined structurally by the electron microscopist and also chemically by the biochemist. As Brachet (1961) has said, the cell biologist now seeks to explain in molecular terms what he can 'see' with his electron microscope, whereas the biochemist has become a biochemical cytologist, as interested in the structure of the cell particles as in their biochemical activity. Neither discipline can ignore the other—the combination of efforts is required.

Even before these remarkable advances occurred and accelerated, there were nevertheless two clear points of understanding about cell characteristics. First, it was realized that cell function was epitomized by the ability of the cell to harness energy and transform it, and that this transformation of energy was necessary for the

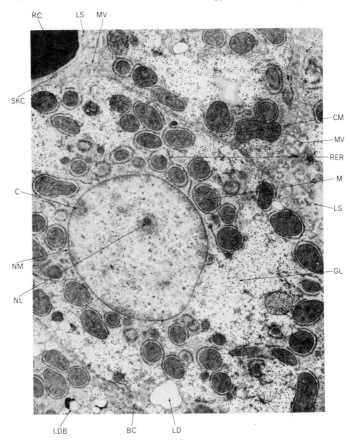

FIG. I.1. Electron microscope picture of liver cell. (Kindly supplied by Mr R. P. Gould.)

BC = Bile canaliculus
C   = Chromatin
CM = Cell membrane
GL = Glycogen
LD = Lipid droplet
LDB = Lysosomal dense body
LS = Liver sinusoid
M   = Mitochondrion
MV = Microvilli of liver cell projecting into the space of Disse
NL = Nucleolus
NM = Nuclear membrane
RC = Red cell in liver sinusoid
RER = Rough endoplasmic reticulum
SKC = Sinusoidal Kupffer cell

maintenance of the integrity of cellular structure and of the intracellular environment. Such energy transformations include the use of energy of the sunlight to produce chemical bond energy, as in plant cells. Other examples include the transformation of chemical bond energy into mechanical work, as in muscle cells, or into electrical energy, as in nerve cells, or even into visual light, as in the firefly tail.

Secondly, the cell interior was understood to contain macromolecules synthesized by energy mechanisms characterizing the cell. No macromolecule occurs in the non-living environment unless it represents the previous activity of cells now dead. Primitive life began with the spontaneous synthesis of complicated macromolecules at the expense of smaller molecules. It is the supreme ability of the cell to synthesize these larger molecules—deoxyribonucleic acid, ribonucleic acid, proteins, etc.

The cell membrane

Viewed with the electron microscope after fixation with osmium tetroxide, the cell membrane, or plasma membrane, measures approximately 7.5 nm (75 Å) in thickness. It appears to consist of three layers, each approximately 2.5 nm thick, in which internal and external electron-dense layers are separated by a central layer of low density [FIG. I.2].

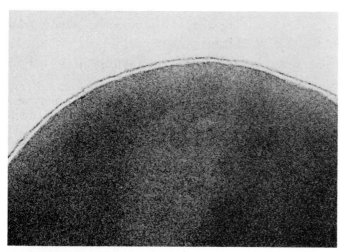

FIG. I.2. Electron micrograph of the human red blood cell plasma membrane. (With permission from J. David Robertson.)

The Davson–Danielli membrane model

This layered structure seen with the electron microscope accords well with the molecular model of membrane structure based on physicochemical evidence proposed by Davson and Danielli. In this model the membrane consists of a bimolecular layer of lipid molecules, corresponding to the clear zone in electron micrographs, with associated protein layers of monomolecular thickness on the inner and outer surfaces, corresponding to the electron-dense layers. The bimolecular lipid layer consists largely of phospholipid molecules arranged as shown in FIG. I.3. A certain amount of cholesterol is also present in this layer and is probably closely packed with the phospholipid, increasing the structural stability of the cell membrane. In FIG. I.3 each black dot represents the polar end of the phospholipid molecule which is hydrophilic (attracted to water), and which consists of the phosphorus–carbon–nitrogen part of the phospholipid complex. This end abuts on the outer protein layer. The remainder of the molecule (a long-chain hydrocarbon fatty acid) is hydrophobic (insoluble in water). This long-chain part of the molecule is arranged perpendicular to the membrane. To this

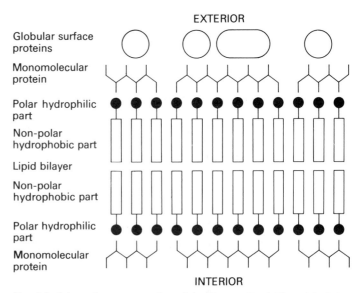

FIG. I.3. Schematic representation of the Davson–Danielli model of the molecular organization of a cell membrane or so-called 'unit membrane'. (See text.)

non-polar chain is opposed another non-polar part of a second phospholipid molecule—oriented in the opposite direction and leading to the polar 'phosphate' group which abuts on the inner protein layer. Danielli and his collaborators proposed that these layers of surface protein are present in the form of extended chains, which would explain the low surface tension of the cell membrane, but other protein components of the membrane structure may exist in a globular, unextended configuration.

The Singer membrane model

About ten years ago a modification of the Davson–Danielli model was put forward by Singer, Lenard, and their co-workers. In this lipid–globular protein mosaic model, which was derived from thermodynamic considerations, instead of a continuous layer of protein on the surfaces of the membrane there is a discontinuous mosaic of globular proteins contained in a phospholipid bilayer matrix [FIG. I.4]. Singer suggested that while a small fraction of the lipid

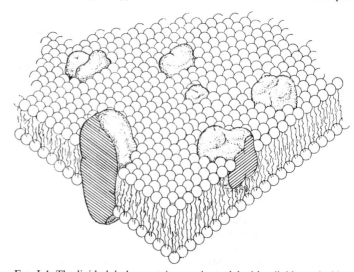

FIG. I.4. The lipid-globular protein mosaic model with a lipid matrix (the fluid mosaic model); schematic three-dimensional and cross-sectional views. The solid bodies with stippled surfaces represent the globular integral proteins, which at long range are randomly distributed in the plane of the membrane. At short range, some may form specific aggregates, as shown. (In cross section and in other details, the underline of FIG. I.5 applies.) (From Singer, S. J. and Nicolson, G. L. (1972). *Science, N.Y.* **175**, 720.)

may interact specifically with certain of the membrane proteins most of the lipid is in a fluid form. Thus, the concept is of an essentially dynamic or fluid membrane in which the proteins are randomly distributed and relatively free to move in two dimensions in the plane of the membrane, though it has been suggested recently that the mobility of some of the membrane proteins may be impeded by membrane-associated components, such as microfilaments and microtubules at the inner membrane surface.

The proteins are partially embedded in the phospholipid bilayer and partially protrude from it. The uncharged, hydrophobic groups of the proteins are largely within the lipid, while the hydrophilic ionic groups protrude at the surface [FIG. I.5]. Some of the proteins

FIG. I.5. The lipid–globular protein mosaic model of membrane structure: schematic cross-sectional view. The phospholipids are depicted as in FIG. I.4, and are arranged as a discontinuous bilayer with their ionic and polar heads in contact with water. Some lipid may be structurally differentiated from the bulk of the lipid (see text), but this is not explicitly shown in the figure. The integral proteins, with the heavy lines representing the folded polypeptide chains, are shown as globular molecules partially embedded in, and partially protruding from, the membrane. The protruding parts have on their surfaces the ionic residues (− and +) of the protein, while the non-polar residues are largely in the embedded parts; accordingly, the protein molecules are amphoteric. The degree to which the integral proteins are embedded and, in particular, whether they span the entire membrane thickness depend on the size and structure of the molecules. The arrow marks the plane of cleavage to be expected in freeze-etching experiments. (See text.) (From Singer, S. J. and Nicolson, G. L. (1972). *Science, N.Y.* **175**, 720.)

may span the entire thickness of the membrane; a glycoprotein present in large amounts in red blood cell membranes has been shown to be orientated with its carbohydrate-rich region on the outer surface and the region with a high ratio of polar amino-acids on the inner surface of the membrane. Evidence suggesting that substantial amounts of protein are deeply embedded in many membranes has been provided by freeze-fracture experiments. In this technique a frozen section is fractured with a microtome knife. A plane of weakness exists between the lipid bilayers of cell membranes along which cleavage occurs. The preparation may then be 'etched' by removing some of the frozen water by sublimation; the surface is then shadow-cast with metal and the replica examined in the electron microscope. In such replicas the intercalated protein molecules are represented as randomly distributed particles lying in a smooth matrix [FIG. I.6].

Chemical analyses of cell membrane preparations indicate that in addition to the main constituents, protein and lipid, which account for 60–70 per cent and 20–40 per cent of the dry weight respectively, there is also a small amount of carbohydrate constituting 1–5 per cent of the dry weight. The carbohydrate is present in mucoproteins such as those which constitute the ABO blood group antigens [p. 46]. Most of these mucoproteins lie in the outer surface of the membrane, but as mentioned above, some may extend through the

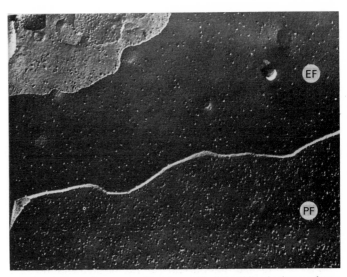

FIG. I.6. A freeze-fracture replica of an epithelial cell line in tissue culture. The fracture face EF shows few particulate entities, while the fracture plane PF shows a greater number of intramembranous particles; the latter are probably associated with intercalated proteins having diameters ranging from 4 to 10 nm. (With permission from E. L Benedetti and I. Dunia.)

entire thickness of the membrane with their carbohydrate-rich regions on the outer surface.

Passage of substances through the cell membrane

The arrangement of the cell membrane certainly accounts for the lower permeability of the membrane to water-soluble substances and allows us to understand that the lipid-soluble gases oxygen and carbon dioxide can pass across the membrane with great ease. On the other hand, the explanation of how water-soluble substances do penetrate (as they undoubtedly do) is difficult. To circumvent the difficulty the hypothesis was proposed that there were *pores* in the membrane. Needless to say, such pores were understood to be necessarily small because of the outstanding difference in concentration between the intracellular concentration of sodium compared with that in the extracellular phase. [Na$^+$] inside is less than one-tenth of [Na$^+$] outside. The intracellular concentration of potassium, on the other hand, is some thirty times that of the extracellular phase. The diameter of the hydrated sodium ion is only 5.12×10^{-1} nm and that of potassium even less, 3.96×10^{-1} nm. Any pore in the cell membrane must therefore be of this order of size, and here of course the light microscope with its limited resolving power was quite useless as a tool. Even the advent of the electron microscope was of little use for electron microscopic techniques do not allow resolution of holes of $5 – 10 \times 10^{-1}$ nm. Indirect techniques based on the measurement of the speed of transit of substances across the cell membrane (Solomon 1960) have, however, yielded evidence of the existence of pores of $7 – 8 \times 10^{-1}$ nm diameter in the membrane. These pores in the otherwise continuous lipid bilayer are probably lined with protein molecules with their polar groups orientated towards the aqueous phase. Since it is now believed that most of the protein molecules are not rigidly fixed within the cell membrane the pores are not necessarily permanent gaps in the membrane, but may be essentially short-lived routes of entry. It has also been suggested that the formation of pores may be associated with a transient change in a localized region of the lipid layer from a continuous lamellar form to a micellar structure of blocks of lipid. The newly created gaps between these lipid blocks would thus form fresh pores.

While the presence of pores may account for the passage of ions and small hydrophilic molecules into and out of the cell when an appropriate concentration gradient exists, i.e. by diffusion, it cannot explain the movement of larger hydrophilic molecules across the cell membrane or of ions and molecules against a concentration gradient. For these processes a *transport mechanism* is required, the process being one of *active transport* when it is dependent on a supply of metabolic energy. One of the most important and widespread transport mechanisms of the cell membrane is the *ionic pump* which is responsible for the capacity of the cell to extrude Na^+ and accumulate K^+ in each instance against an existing concentration gradient. The dependence of this ionic pump on metabolic energy is revealed by the rise in the intracellular Na^+ concentration and the fall in the intracellular K^+ concentration when slices of brain or kidney tissue are deprived of their energy supply by cooling, exclusion of oxygen or glucose, or by the addition of metabolic poisons such as cyanide or dinitrophenol. The energy source for the ionic pump is adenosine triphosphate (ATP) and an enzyme system capable of rapidly breaking down ATP into ADP and phosphate with the liberation of energy was shown to be present in crab nerve membrane by Skou (1957). This ATP-ase, which is much more active in the presence of Na^+ and K^+ ions, has subsequently been found to be a protein component of the cell membrane in every tissue with an ion pump. The probable mode of operation of the pump is as follows:

At the inner border of the membrane a specific site on a carrier molecule, probably a phosphoprotein, is phosphorylated by ATP (a process which is stimulated by the presence of Na^+) and binds sodium ions. The phosphoprotein–sodium ion complex is then transferred to the external surface of the membrane, where the sodium is released. The mechanism by which the carrier transports the ion across the membrane is unknown. Danielli has suggested various possibilities—the carrier may diffuse across the membrane, it may rotate so that the sodium ion combining site is presented to the outer surface of the membrane, or it may transfer the ion to the next in a chain of similar carriers. Release of the sodium ion to the external medium is followed by the binding of a potassium ion. This stimulates dephosphorylation and the ion-binding site returns to the inner surface of the membrane where potassium is released and the cycle begins again. This basic pattern may be modified in some cells—for example, a phospholipid, which could diffuse freely in the lipid bilayer, may take the place of the phosphoprotein as the phosphorylated intermediate. Pumping mechanisms exist for other ions besides Na^+ and K^+. For example, the parietal cells of the gastric mucosa transfer both H^+ and Cl^- from the intracellular fluid to the lumen of the gastric glands against concentration gradients. In the case of the H^+ ions the concentration gradient may be as great as one million-fold. The cells of the thyroid gland have a remarkable ability to concentrate inorganic iodide, so that the concentration of iodide within the thyroid gland may be fifty times that in the plasma. Amino acids and glucose are examples of other substances whose concentration within the cell may considerably exceed that in the surrounding extracellular fluid, indicating the existence of appropriate transport mechanisms. Sometimes two transport mechanisms may be linked. Thus, in the presence of substances which inhibit sodium transport, the accumulation of glucose by the cells of the intestinal mucosa and the accumulation of iodide by the cells of the thyroid are also blocked.

Insulating properties of the cell membrane

Long before the structural details of the cell boundary were known electrical measurements had shown that the interior of the cell was some 60–70 mV negative to the outside. Measurements of the capacity of the cell membrane showed this to be 1 microfarad cm^{-2}. The membrane may be considered as acting as the dielectric material of a charged condenser. The difference in charge across the membrane, 100×10^{-10} metres thick (100×10^{-1} nm), results in a

potential difference of, say 70 mV or 0.07 V. The electric field (stated in volts per metre) within the membrane is thus:

$$\frac{0.07}{100 \times 10^{-10}} = 7 \text{ million volts metre}^{-1}.$$

This example, based on one given by Engleberg (1966), shows that the cell membrane has a very high insulating value indeed, for the highest dielectric strength (volts per metre) which a good commercial insulator (such as rubber) can stand without breaking down is about one million volts per metre.

Intracellular organelles

The endoplasmic reticulum, ribosomes, and Golgi bodies. The 'interior of the cell' or cytoplasm is interlaced by a highly complicated arrangement of internal membranes forming tubules and vesicles. This branching network, or reticulum, is more concentrated in the inner, endoplasmic region of the cell than in its more peripheral, ectoplasmic region—hence its name, the endoplasmic reticulum. Through this network of canaliculi (each bounded by unit membrane) substances may be delivered from the outer membrane of the cell proper to the membrane of the nucleus or to other inclusion bodies of the cells, such as the mitochondria. Some authorities claim that the internal tubular system is continuous with the external cell membrane proper and is formed by the folding of this membrane. Robertson (1973), who has done much to clarify our thinking on the structural details of the cell, provides the accompanying illuminating diagram [FIG. I.7]. He points out that such an arrangement would cause us to think of the cell as a three-phase system: (i) a cytoplasmic phase; (ii) a phase constituted by the contents of the endoplasmic reticular tubules; and (iii) a membrane phase which separates the first two. This concept would imply that any piece of membrane in the cell must have been formed from pre-existent membranes. The so-called nuclear membrane [see below] has indeed been shown to be simply a system of sacs of the endoplasmic reticular system arranged around the spherical bit of cytoplasm which contains the main genetic material (DNA and RNA) of the cell, i.e. the nucleus.

Robertson's views would suggest that the endoplasmic reticulum would provide an 'intracellular circulatory system'. It would also

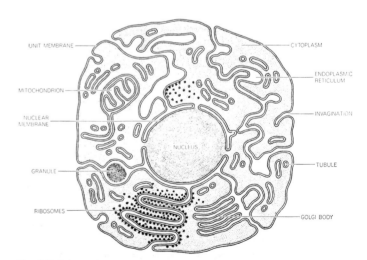

FIG. I.7. Schematic diagram based on the author's concept of cell structure, showing the extensive distribution of membrane within the cell and some of the many membranous organelles so far identified. Since the membranes within cells always have the unit membrane pattern, it may be that most of these organelles are formed from the unit membrane structure bounding all cells. (Robertson, J. D. (1962). *Scient. Am.* Repr. 151. Copyright by Scientific American, Inc. All rights reserved.)

ensure a far greater surface of exposure of the cytoplasm to the 'extracellular' environment. In general the membranes of the endoplasmic reticular components are subdivided from their electron microscope appearance into 'rough' and 'smooth'. Rough membranes appear so because they are studded with granules on their 'cytoplasmic' surface. These granules are the ribosomes, so-called because they are rich in ribonucleic acid (RNA). As will be seen [p. 475], their function lies in the synthesis of protein in which RNA plays a major role. In keeping with these findings, cells which produce large amounts of protein are packed with ribosomes. Thus, a liver cell may contain as many as 100 million ribosomes of about 15 nm diameter. The ribosome is *not* bounded by membrane.

'Rough' endoplasmic reticulum, in which ribosomes are closely applied to the endoplasmic reticulum, is characteristic of cells which produce protein 'for export'. Examples of such protein-secreting cells are the cells of the liver which produce plasma protein, pancreatic cells which produce digestive enzymes, fibroblasts which produce collagen, and plasma cells which produce antibodies. The proteins synthesized on the ribosomes somehow penetrate the unit membrane and are stored, segregated from the rest of the cytoplasm within the distended cisternae of the endoplasmic reticulum. On the other hand, in cells in which the proteins synthesized are incorporated into or used by the cell itself (for example, the contractile proteins synthesized by muscle cells or the haemoglobin synthesized by the precursors of the red blood cells), the endoplasmic reticulum is smooth and scanty but free ribosomes are scattered throughout the cytoplasm.

The Golgi body (or complex) was first described by Golgi in nerve cells at the end of the nineteenth century and was subsequently found in all animal cells. Found always in the vicinity of the nucleus, electron microscope studies now reveal that it is part of the endoplasmic reticulum, consisting of flattened parallel, smooth-surfaced unit membranes together with numerous smaller vesicles.

Palade and his co-workers have obtained evidence that in the exocrine cells of the pancreas the proteolytic enzymes trypsin and chymotrypsin, produced in an inactive form on the rough endoplasmic reticulum and secreted into the cisternae, pass to the Golgi body. The enzyme precursors then become enclosed in vesicles produced by budding of the smooth endoplasmic reticulum of the Golgi body. These vesicles or zymogen granules then pass to the cell surface, fuse with the plasma membrane, and discharge their contents into the lumen of the acinus [FIG. I.8]. A similar process may take place in other protein-exporting cells.

Mitochondria. These occur in variable numbers from a few hundred to a few thousand in different cells, related often to the energy required for the functional activity of the cell. Rapidly acting skeletal muscle, which requires speedy replenishment of its energy, is rich in mitochondria and in its endoplasmic reticular apparatus. Heart muscle, which does not require such rapid replenishment, has only a moderately developed sarcoplasmic reticulum, but abounds in mitochondria 1–2 μm in length and 0.3–1.0 μm in diameter distributed primarily between, and in close approximation to, the myofibrils and comprising 40–50 per cent of the cell volume.

Mitochondria are variable in size, varying from 0.5 μm to as much as 12 μm in length, and in shape; many are filamentous, some are globular. They are highly structured organelles delimited from the cytoplasm by a double membrane [FIG. I.9]. The inner aspect of this membrane gives rise to numerous infoldings, which form cristae composed of membrane pairs which project into the interior of the mitochondrion [FIG. I.10].

Mitochondria play a vital role in cellular metabolism providing the fundamental respiration energy transformations by which, through oxidative phosphorylation, the energy contained in the nutritive material brought to the cell is converted to adenosine triphosphate (ATP), the prime medium of energy supply. Thus, the

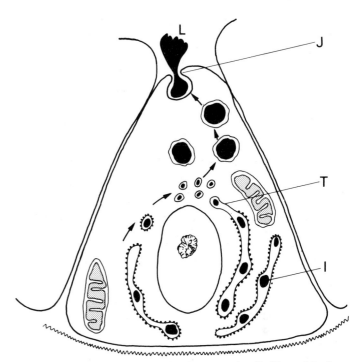

FIG. I.8. Possible secretion mechanism in pancreatic cells. The vesicles fuse with the plasma membrane and liberate their contents into the lumen of the acinus.

J = Plasma membrane
L = Lumen of acinus
I = Small zymogen granules in the cavities of the endoplasmic reticulum
T = Transition from rough endoplasmic reticulum to smooth membranes bounding vesicles in the Golgi region

(Redrawn from Ambrose, E. J. and Easty, D. M. (1977). *Cell biology*, p. 182. Nelson, Sunbury-on-Thames. After Harris, G. H. (ed.) (1964). *Introduction to molecular biology*. Longmans Green, London.

mitochondria, which serve as the 'power plants' of the cell, contain the enzymes of the tricarboxylic acid cycle [p. 454], enzymes for the catalysis of the oxidation of fatty acids and amino acids and the enzymes needed for coupling electron transport with the tricarboxylic acid cycle. The enzymes exist in a sterically ordered array along the cristae and surface membranes, thereby assuring stepwise aerobic biochemical reactions which result in the production of ATP. The evidence for this last statement is provided by the fact that a mitochondrial suspension prepared from cells can be broken up into fragments of the individual mitochondria and each fragment can conduct only a part of the complete reaction sequence.

Lysosomes. Also bounded by a membrane, but of entirely different function. 'The lysosome comes in a bewildering assortment of shapes and sizes even in a single type of cell; they cannot be identified solely on the basis of their appearance' (de Duve 1963). Their function has been elucidated clearly by de Duve. Four types of lysosomes may be distinguished—'storage granules', 'digestive vacuoles', 'residual bodies', and 'autophagic vacuoles'.

The original form of the lysosome is the storage granule which consists of enzymic granules (formed by ribosomes) and wrapped up in a lipoprotein membrane (formed possibly in a specialized region of the Golgi complex) to form a roughly spherical organelle, some 250–750 nm in diameter. The granules (about 60×10^{-1} nm diameter) consist of protein and are of a variety of types of hydrolytic enzymes. These include ribonuclease and deoxyribonuclease, phosphatases, proteolytic enzymes, glycosidases and sulphatases. Clearly such a conglomeration can have only one general function—a lytic or digestive one.

The membrane which surrounds this wasps' nest clearly has a

vital function for if such a swarm were to escape indiscriminately, cell death would ensue quickly. Indeed lysosomes are responsible for the rapid post-mortem autolysis of, for example, the intestinal mucosa subsequent to the death of the host individual.

However, the discriminate function of the lysosomes is of great importance to the cell. When the cell ingests substances by phagocytosis a phagosome or food vacuole is formed. Several phagosomes may fuse together to form a single vacuole. A lysosome now fuses with the phagosome forming a digestive vacuole. The products of digestion diffuse through the lysosome membrane into the cell cytoplasm. The digestive vacuole continues its digestive activity until only indigestible material remains, whereupon it shows the final stage of a 'residual body'. In some cells, such as the amoeba, this may be extruded from the cell itself, but most cells of the organized mammal cannot effect this and the residual body is used again and again in digestive activity [see FIG. I.29, p. 53].

The autophagic vacuole is a lysosome, in which can be identified remnants of cell debris (mitochondria or endoplastic reticulum) formed in the host cell.

White corpuscles in the blood live only for a short period. Their function is that of scavenging and they contain many lysosomes,

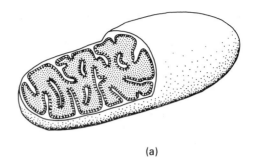

FIG. I.10. (a) Cut-away drawing of a typical mitochondrion showing the two membrane layers separated by a fluid-filled space called the intrastructure space. The space within the inner membrane is known as the interstructure space. (b) Diagrammatic representation of the structure of a mitochondrion. (Redrawn from Ambrose, E. J. and Easty, D. M. (1977). *Cell biology*, p. 203. Nelson, Sunbury-on-Thames. (a) from Green, D. E. in Brachet, J. (ed.) (1964). Copyright Scientific American, Inc. All rights reserved. (b) from Keeton, W. T. (1967). *Biological science*. Norton, New York. Copyright W. W. Norton and Company, Inc.)

which, at the time the cell is discharged from the bone marrow into the bloodstream, are full of granules. When the white cell engulfs a bacterium for instance, the granular content of its lysosomes steadily disappears as they are used up in the digestive processes.

Cell damage due to oxygen lack or to cell poisons causes disruption of the lysosomes which not only wrecks the cell itself, but leads to escape of the digestive enzymes into the extracellular fluid and thence by diffusion to neighbouring structures.

Centrosomes (or centrioles). These paired structures become plainly visible under the light microscope only when the cell approaches its hour of division, a time at which these organelles function as the poles of the spindle apparatus that divides the chromosomes. Under the electron microscope a pair of centrosomes is then seen clearly at each pole of the nucleus and all show a cylindrical structure made up of eleven fibres with two in the centre and nine arranged 'round the clock'.

Each pair of centrosomes gives rise to another when the cells divide. Their exact function during mitosis is still unknown. It is interesting that ciliated cells possess a structure called a kinetosome at the base of the cilium. These are undoubtedly concerned with the motility of the cilium and have been described as 'monomolecular muscles'. They have a structure identical with that of the centrosome and also replicate when the cell divides. It has been suggested that the centrosomes pull the chromatin material apart on cell division.

The nucleus. The cell nucleus is contained in or bounded by a nuclear membrane. This is composed of two unit membranes and shows large pores of 100 nm diameter. It is probable that the nuclear membrane is composed of the terminal parts of the endoplasmic reticulum [FIG. I.7] and that the pores which provide direct continuity between the nucleus and the cytoplasmic sap represents gaps between the 'end feet' of the endoplasmic reticulum.

The nucleus itself is spherical and approximately 10 μm in diameter; 80 per cent of it is water, but 80 per cent of its dry weight is protein. The remainder of the dry weight is made up by 18 per cent deoxyribonucleic acid (DNA) and 2 per cent ribonucleic acid (RNA). The nucleic acids have an affinity for the basic dyes toluidine blue and methyl green, and this has led to the location of nucleic acids in material examined by the light microscope. Additionally, the nucleic acids show intense absorption of light of wavelength 260 nm (ultraviolet). Thus, microphotos taken in the

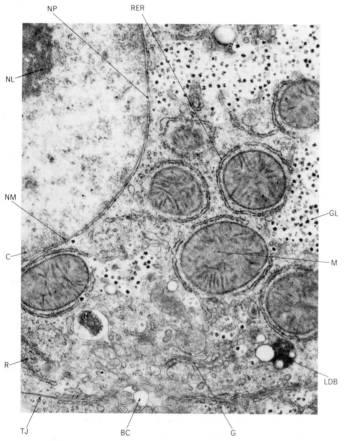

FIG. I.9. Liver cell (Kindly supplied by Mr R. P. Gould)

BC = Bile canaliculus
C = Chromatin
G = Golgi complex
GL = Glycogen granule
LDB = Lysosomal dense body
M = Mitochondrion
NL = Nucleolus
NM = Nuclear membrane
NP = Nuclear pore
R = Ribosomes
RER = Rough-surfaced endoplasmic reticulum
TJ = Tight junction

ultraviolet microscope show the location of the nucleic acids as dark areas.

The dense staining network of the nucleus consists of DNA. This network is called chromatin and it gives rise to the chromosomes which can be identified prior to and during mitotic division of the cell. Densest of all the nuclear material is the nucleolus. Composed of a network of helically-coiled fibres it is rich in RNA and is probably the sole site of RNA in the nucleus. The nucleolus is most prominent when the cell is synthesizing protein. The relations of DNA and RNA to cell reproduction and protein are discussed on pages 473–7.

Linkages between cells

In tissues, cells are not usually in close contact throughout their borders; there is a space of about 10–20 nm between them filled with interstitial fluid. This enables water, ions, and other molecules in the extracellular fluid to gain access to the greater part of the cell surface.

Nevertheless, the structural integrity of organs and tissues requires that there are linkages between cells so that their position relative to one another is maintained. Three types of linkage have been described. These are illustrated in Fig. I.11 and comprise:

1. Desmosomes. The opposed membranes of two adjacent cells become thickened, the thickened region being circular in plan view. Beneath each thickening lies a feltwork of fine filaments embedded in a condensation of the cytoplasm. No filaments or other structures connecting the two thickened regions have been described and the nature of the actual linkage remains unknown.

2. Intermediate junctions. These resemble desmosomes but are more extensive and the thickening is less well-marked.

3. Tight junctions or zona occludentes. Here the outer layers of the adjacent plasma membranes fuse forming a barrier to diffusion. These junctions are typical of epithelia and enable the tissue to form a water-tight barrier, e.g. in the bladder.

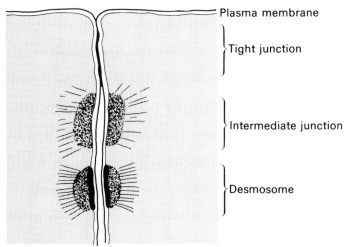

Fig. I.11. Diagram showing types of cell junction. (See text.)

REFERENCES

BRACHET, J. (1961). The living cell. *Scient. Am.*, September.
CAPALDI, R. A. (1974). A dynamic model of cell membranes. *Scient. Am.*, March.
DE DUVE, C. (1963). The lysosome. *Scient. Am.*, May.
ENGLEBERG. J. (1966). In *The physiological basis of medical practice* (ed. C. H. Best and N. B. Taylor) Chap. 1. Churchill Livingstone, Edinburgh.
FOX, F. C. (1972). The structure of cell membranes. *Scient. Am.*, February.
GLYNN, I. M. and KARLISH, S. J. D. (1975). The sodium pump. *A. Rev. Physiol.* **37**, 13.
NEUTRA, M. and LEBLOND, C. P. (1969). The Golgi apparatus. *Scient. Am.*, February.
NOMARA, M. (1969). Ribosomes. *Scient. Am.*, October.
RACKER, E. (1968). The membrane of the mitochondrion. *Scient. Am.*, February.
ROBERTSON, J. D. (1973). The organization of cell membranes. In *Cell biology in medicine* (ed. E. E. Bittar). Wiley, New York.
SINGER, S. J. and NICOLSON, G. L. (1972). The fluid mosaic model of the structure of cell membrane. *Science, N.Y.* **175**, 720.
SOLOMON, A. K. (1960). Pores in the cell membrane. *Scient. Am.*, March.
STAEHELIN, L. A. and HULL, B. (1978). Junctions between living cells. *Scient. Am.*, May.

Membrane transport

TRANSFER OF SUBSTANCES ACROSS CELL MEMBRANES

The physiological activity of a cell is very dependent on the ease with which it can gain materials from, or lose materials to, its surrounding environment. A cell can only grow and function if its requirements for oxygen, ions, amino acids, vitamins, and a score of other chemical substances are met, and the waste or secretory products of its metabolism are removed. Even transport to or from the cell is not sufficient, because substances must pass through the cell boundary into or out of the cell. These transport processes range from relatively simple *passive* processes such as diffusion to extremely complex *active transport* mechanisms that depend on the presence of special molecules within the cell membrane and require the expenditure of chemical energy derived from cellular metabolism. In discussing these processes it will be convenient to discuss first the transport of non-electrolytes, such as water, glucose, and amino acids, and then the transport of ions such as sodium, potassium, and chloride, the charged nature of which presents an additional complexity.

Transport of non-electrolytes

Diffusion

In a solution the molecules of both solute and solvent are in constant, rapid, random thermal movement and frequent collisions occur between molecules. This results in the process of diffusion— the movement of solute which occurs when a membrane or boundary separates two fluids at the same hydrostatic pressure but differing in their concentration of solutes, i.e. when there is a concentration gradient across the membrane (or within a single solution). If all the solutes are freely diffusible through the membrane, an equilibrium position is reached with equal concentrations on each side, i.e. the gradient has disappeared. If one fluid also contains material which is not diffusible through the membrane, then passive diffusion to equilibrium may result in an *unequal* division of the freely-diffusible solutes as well (the Donnan equilibrium with ions, see p. 16).

If the hydrostatic pressures on the two sides of the membrane are unequal, a flow of water and dissolved solutes will occur by *filtration*.

The time for equilibration by diffusion is proportional to the square of the diffusion distance; over a distance of a few micrometres only seconds are required, whereas days are needed where the diffusion distance is a few centimetres. This aspect of diffusion has important biological consequences. It limits the size of individual cells, since cellular metabolism depends upon the rapid diffusion of oxygen and substrates from the membrane to metabolic sites within the cell; in practice, few mammalian cells exceed 20 μm in diameter. It also means that every cell must have ready access to a transport system from which its requirements may be obtained; in the human body no metabolically active cell is more than 20 μm from a capillary.

The rate of diffusion of molecules down a concentration gradient is given by the *Fick equation*:

$$\mathrm{d}v/\mathrm{d}t = J = -\,DA(\mathrm{d}c/\mathrm{d}x),$$

where $\mathrm{d}v/\mathrm{d}t$ is the rate of diffusion, in $mol\ s^{-1}$; $\mathrm{d}c/\mathrm{d}x$ the concentration gradient down which diffusion is occurring, in $mol\ cm^{-3}\ cm^{-1}$ (the minus sign indicates that diffusion is taking place in the direction of decreasing concentration); and A is the area of the plane of solution at right angles to the movement. The proportionality constant, D, is the *diffusion coefficient*, and is the number of molecules diffusing in 1 second down a concentration gradient of 1 $mol\ cm^{-3}\ cm^{-1}$ when the area of the plane is 1 cm^2. The actual value of D depends on the size of the molecule and the viscosity of the solution. The greater the size of the molecule and the greater the viscosity of the solution, the smaller will be the diffusion coefficient. Thus, while the diffusion coefficient in water at 25 °C of haemoglobin, a large molecule (MW 68 000), is $6 \times 10^{-7}\ cm^2\ s^{-1}$, that of sucrose (MW 342) is nearly ten times greater, $5.2 \times 10^{-6}\ cm^2\ s^{-1}$, and the self-diffusion coefficient of water in water, 2.4×10^{-5} $cm^2\ s^{-1}$, is forty times that of haemoglobin.

The importance of the area of the surface across which diffusion takes place (A) in determining the rate of diffusion is illustrated by the enormous surface area of the capillary beds; the capillary surface area in 1 g of brain tissue is about 250 cm^2. Individual cells may have their surface area increased by the development of microvilli. These are a conspicuous feature of epithelial cells in the small intestine and kidney tubules and at both of these sites there is a rapid passage of substances between the cell and its surroundings. The rate of diffusion may also be regulated by varying the area of surface available. At rest only a fraction of the total lung surface is used for gas exchange. During exercise, however, the total surface area of 75 m^2 becomes available to meet the need for increased diffusion of oxygen and carbon dioxide.

Diffusion across cell membranes

The Fick equation can be used to describe the passage of substances across the cell membrane by passive diffusion. Here the concentration gradient will be the difference between the concentrations of the substance just within its outer and inner borders, i.e. $(C_{o_m} - C_{i_m})$, divided by the thickness of the membrane, x_m.

Thus $J = D_m A_m (C_{m_o} - C_{m_i})/x_m$ where A_m is the area of the cell membrane and D_m the diffusion coefficient for the molecule in the material of the membrane. The viscosities of cell membranes are 100–1000 times as great as that of water so that the values of D are two or three orders of magnitude less than the corresponding diffusion coefficients in water.

The cell membrane is lipid in nature, thus the concentrations of a solute just within the outer and inner borders of the membrane, C_{o_m} and C_{i_m} respectively, will not be the same as the concentrations in the immediately adjacent aqueous phases of the extracellular fluid and cytoplasm. The effect of different levels of lipid solubility in determining the concentration gradient for the solute *within the membrane*, and hence the rate of diffusion across the membrane, is illustrated in Fig. I.12. The relation between the concentration of the solute in the membrane lipids and in aqueous solution is given by the partition coefficient K_m, where $K_m = C_m/C_{water}$, i.e. the ratio of the concentrations when the solute is distributed at equilibrium between water and the material of the cell membrane. Using this value of K_m the Fick equation for the rate of diffusion across the cell membrane can be re-written in terms of the concentrations of solute in the extracellular and intracellular fluids (C_o and C_i respectively),

i.e. $J = D_m A_m K_m (C_o - C_i)/x_m$.

The terms D_m, K_m, and x_m are generally combined and replaced by P, known as the *permeability coefficient* (i.e. $P = D_m K_m/x_m$) and the equation simplified to

$$J = P A_m (C_o - C_i).$$

From this equation it can be seen that where transport across the cell membrane is dependent solely on the *passive transport* mechanism of solubility-diffusion, the rate of transport is directly proportional to the difference in concentration of the solute on either side of the membrane. However, the size of the proportionality constant, the permeability coefficient, varies greatly both with the membrane and with the molecule concerned. For example, the water permeability of the human red blood cell membrane is 100 times that of the membrane of the frog egg, while the permeability coefficient of the red cell membrane for water is 10^8 times the permeability coefficient for sucrose. As noted above, permeability coefficients are determined by the diffusion coefficient for the solute in the material of the membrane (D_m), the partition coefficient between the membrane and extracellular and intracellular fluid (K_m) and membrane thickness (x_m). For any given cell membrane thickness must be the same for the passage of all solutes, and diffusion coefficients vary by less than tenfold. Thus, the enormous range of permeability coefficients is due to a correspondingly large variation in the size of the partition coefficients. The most important determinant of the partition coefficient is the extent of hydrogen bonding between the molecule and water; as the number and strength of hydrogen bonds between the molecule and water increase, more energy is required to transfer the molecule from the aqueous to the lipid phase and the partition coefficient diminishes. The addition of a single hydroxyl group to a molecule can reduce its partition coefficient more than 100-fold. A second important factor is the number of methylene ($= CH_2$) groups in the molecule. In this case an additional methylene group may increase the partition coefficient up to ten times, the mechanism being very complex.

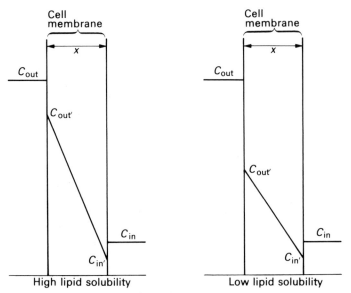

Fig. I.12. Effect of lipid solubility (oil/water partition coefficient) on the concentration gradient of a substance within the cell membrane. C_{out} and C_{in} are the concentrations of the substance in extracellular and intracellular fluid; $C_{out'}$ and $C_{in'}$ the corresponding concentrations just within the membrane at its outer and inner borders respectively. The higher the lipid solubility, the more closely $C_{out'}$ approaches C_{out} and the steeper is the concentration gradient within the membrane (and thus the greater will be the driving force for the diffusion of the substance across the membrane). (Redrawn from Dawson, H. and Segal, M. B. (1975). *Introduction to physiology*, Vol. 1. Academic Press, London.)

Partition and hence permeability coefficients may also differ between isomers of the same molecule. Isomeric differences in molecular conformation lead to differences in intramolecular hydrogen bonding and since an increase in intramolecular hydrogen bonding reduces the ability of the molecule to form intermolecular hydrogen bonds with water, differences in partition coefficient will result.

Having discussed the factors concerned in the transfer of molecules across the cell membrane by passive diffusion, we can now examine the extent to which this passive process of solubility-diffusion contributes to the total membrane transport. However, before we do so, it will be appropriate to consider the problem of the passage of water through the cell membrane.

Transport of water across the cell membrane

In pure water or in an aqueous solution the molecules of water are in continuous random movement just as are the molecules or ions of a solute. Water molecules are capable therefore of 'diffusing' as do the molecules or ions of a solute. If there exists a difference in the concentrations of water molecules on either side of a membrane there will be a net movement of water through the membrane, at a rate determined by the water concentration gradient and the permeability coefficient of the membrane for water. Differences in water concentration may arise, that is a water chemical potential gradient may be established for two reasons:

1. There may be a difference in *hydrostatic pressure* on either side of the membrane. In this situation the increased concentration of water molecules results from their tighter packing in the compartment at the higher hydrostatic pressure.

2. There may be a difference in *solute concentration* on either side of the membrane. This leads to the movement of water by *osmosis* and is a major determinant in the transfer of water across the membranes of living cells.

Osmosis

In an aqueous solution the concentration of water molecules is less than in pure water, since the addition of solute to pure water results in a solution which occupies a greater volume than the original pure water. Thus, a chemical potential gradient for water will also be present when pure water and an aqueous solution, or two aqueous solutions containing different concentrations of the same solute, are separated by a membrane permeable to water but impermeable to the solute. Such an *ideal* membrane, i.e. permeable to water but impermeable to solutes such as glucose, is afforded by a porous pot with copper ferrocyanide deposited across its pores. If a glucose solution is placed inside such a pot and the pot is placed upright in water, pure water diffuses from the exterior to the interior of the pot, and the pressure and volume within rise. Alternatively, the movement of water can be prevented by applying pressure to the glucose solution. Isolated cell membranes are not ideal membranes, permeable only to water, but are *partially permeable*, i.e. they are fully permeable to water, electrolytes, and small crystalloid molecules but hold back the larger colloidal particles (macromolecules) such as those of proteins and polysaccharides. They are like molecular sieves, with penetration governed by molecular size and shape and, to a large extent, by solubility in the membrane lipids. Thus, when a bag formed by a partially permeable membrane, e.g. a cellophane sac, containing a non-diffusible colloid is immersed in water, the effect is similar to that seen when a copper ferrocyanide pot containing a glucose solution is immersed in water. Unless sufficient pressure is applied from within the sac to establish equilibrium without net movement of water, water will diffuse in to dilute the colloidal solution.

This phenomenon is called *osmosis*. The osmotic pressure of a solution is defined as the hydrostatic pressure increment that would

have to be imposed on that solution to prevent entry of pure water through a boundary permeable only to water.

There is much misunderstanding about this, but the reader should by now be aware that it is not strictly correct to talk of the 'osmotic pressure' exerted by glucose or plasma proteins; solute molecules do *not* attract water to themselves: they do *not* exert a 'negative pressure' across the membrane. As has already been emphasized, osmosis concerns a property of water and the effect of dissolved substances such as glucose or proteins on that property. As a consequence of this 'osmotic effect' of solutes, the chemical potential of water in an aqueous solution is lower than that of pure water. Thus, if a solution is separated from water by a membrane there is a 'potential difference' between the two phases, and equilibrium can only be established by:

1. Free distribution of solute on both sides of the membrane; this is only possible with diffusible solute and permeable membrane.

2. Passage of water through the membrane which eliminates the potential difference by diluting the solution—osmosis with an indiffusible solute.

3. Application of pressure to the solution to increase the potential back to that of pure water and thus prevent net movement of water—osmotic pressure.

If diffusible and non-diffusible solutes are present, only the non-diffusible ones will contribute permanently to the pressure increment required, since the other solutes will be equally distributed at equilibrium (if the Gibbs–Donnan effect on ionic distribution is not applicable—see p. 16).

Osmolar concentration and osmotic pressure

The osmotic pressure of a solution depends solely on the number of particles (undissociated molecules, ions, colloidal micelles) in solution, and not on their size or weight. A mole of any substance represents the same (very large) number of molecules; it can be calculated that the aqueous solution of any undissociated substance at molar concentration (1 mol l^{-1}) would exert an osmotic pressure of 22.4 atmospheres under ideal conditions and with an ideal membrane. For example, a molar solution of glucose (MW = 180) exerts an ideal osmotic pressure of 22.4 atmos., while a glucose solution of 5 mM l^{-1} (= 0.9 g l^{-1} as in plasma) has an ideal osmotic pressure equivalent to $22.4 \times 0.005 = 0.112$ atmospheres (= 85 mm Hg). In general terms, the pressure required for equilibrium with respect to water is proportional to the concentration of solute which cannot penetrate the membrane. Its value is given by the Van't Hoff relationship:

$$\Delta\pi = RT\Delta C,$$

where $\Delta\pi$ and ΔC are the difference of osmotic pressure and the difference of solute concentration respectively on the two sides of the membrane, R is the gas constant, and T the absolute temperature. When actually measured, osmotic pressures, though linearly related to the concentration of solutes over the range of relatively low values encountered in most physiological situations, are usually less than those computed on the basis of the Van't Hoff equation. This is because mutual interaction between the molecules of solute leads to their failure to behave as individual units, and the net result is that their effective concentration is reduced. A more precise value of osmotic pressure is therefore given by the equation

$$\Delta\pi_{\text{actual}} = RT\gamma\Delta C,$$

where γ is the *activity coefficient* for the particular solute.

In view of the dependence on numbers of particles the concentration of osmotically-significant particles is often best expressed as an osmolarity (osmoles, or milli-osmoles, per litre). One osmole (= 1000 milli-osmoles) of a substance is that quantity of it which would have a calculated ideal osmotic effect equivalent to 22.4

atmos. when present in 1 litre of solution. For an undissociated non-electrolyte, the molar and osmolar concentrations are identical. In the case of substances which dissociate completely into ions, each ion has the same osmotic effect as an undissociated molecule; thus, a molar solution of sodium chloride ($Na^+ Cl^-$) exerts an ideal osmotic effect of 2×22.4 atmos. and is therefore 2 osmolar (2 osmol l^{-1}). If substances dissociate only partially the appropriate multiplication factor must be found by experiment. If molecules associate, as with macromolecules in colloidal particles, the osmotic effect will be lower than that from the separate molecules.

The osmotic effect of a mixture of solutes is the sum of the effects from each component separately, and we refer to osmolarity, instead of molarity, when we wish to show the additive osmotic effect of individual ions.

Osmosis and cell membranes

Osmotic gradients are of great importance in determining the flow of water across cell membranes. In animal cells their importance is much greater than that of hydrostatic pressure gradients, though hydrostatic pressure gradients are of considerable significance in fluid exchange across the walls of the capillaries and in filtration at the renal glomerulus. Hydrostatic pressure gradients across the cell membrane are generally very small, while even small concentration gradients generate large osmotic pressure gradients, a concentration gradient of only 10 mM of a non-electrolyte to which the membrane is impermeable producing an osmotic gradient of 0.22 atmos. (= 167 mm Hg).

Despite this great physiological importance of osmosis and the very extensive investigations to which this has led, our understanding of the osmotic forces which determine the flow of water across the membranes of living cells is still incomplete. Some of the problems are exemplified by the distinction between *isosmotic* solutions and *isotonic* solutions. Solutions with the same ideal osmotic pressure are termed *isosmotic*. If separated by an ideal membrane (permeable only to water) no net movement of water would occur. The situation will be different with the 'real' membranes of cells. *Isolated* cell membranes are *partially-permeable* and allow not only water but also certain small molecules and ions to pass. However, there is no predetermined relationship between the partial-permeability of an isolated membrane and its apparent permeability when forming the boundary of a living cell. The cell boundary is *selectively-permeable* and what is selected at any time depends upon the metabolic condition and activity of the cell. An isolated cell membrane may be fully permeable to small ions, e.g. Na^+, yet the working cell may specifically maintain them at internal concentrations quite different from the external concentrations. If the state of the cell is changed, e.g. by cooling, poisoning, or altered composition, then the selective-permeability of its boundary may change also. If the cell metabolism ceases, the ions may penetrate freely at the sole dictate of the concentration gradients. As a result of these partially-permeable and selectively-permeable properties of the living cell membrane a net movement of water may take place across the cell membrane even though the intracellular fluid and the surrounding fluid are isosmotic (as defined above). However, if the two solutions separated by the cell membrane come to equilibrium without net transfer of water, they are termed *isotonic* solutions. For example, the intracellular fluid of human red blood cells is isotonic across the red cell membrane with 0.92 per cent sodium chloride; this means that when 'living' red cells are suspended in this strength of sodium chloride solution, they neither gain nor lose water, though they will lose some diffusible glucose and urea to the solution. This situation, and the obvious isotonicity of red cells with plasma, is only maintained by active transport [p. 15].

An accepted physiological convention terms a fluid isotonic (without further qualification) when it is isotonic across the red cell membrane with cell fluid; hypertonic or hypotonic fluids have osmotic effects greater or less than this, and would remove water from, or add water to, red cells suspended in them. Since we know so little about the permeability of membranes and the concentration gradients maintained by living cells, it is impossible to predict the actual 'tonicity' of any given solution; this can only be found by experiment.

Relation between permeability coefficient and partition coefficient

The pore hypothesis. We can now return to the question of whether a passive solubility-diffusion process can adequately account for the measured rates of transport of molecules across the cell membrane. From a knowledge of the rate of entry of a non-electrolyte into the cell (J), the area of the cell membrane (A_m), and the concentrations of the non-electrolyte outside and inside the cell (C_o and C_i respectively), the permeability coefficient (P) can be obtained using the relationship

$$J = P A_m (C_o - C_i).$$

As discussed earlier, if the passage of the non-electrolyte through the cell membrane is dependent solely on solubility-diffusion, the permeability coefficient should be directly proportional to K_m, the partition coefficient for the non-electrolyte in the membrane lipid. For a great many molecules this is the case. For example, ethyl alcohol, with a partition coefficient of 0.03, representing a relatively high lipid-solubility, penetrates cell membranes readily; while glycerol, with a partition coefficient of 0.0001 is relatively lipid-insoluble, and its rate of entry into cells is slow. However, many small molecules, including those of water, urea, D-glucose, and amino acids, do not appear to behave according to this relationship. Their permeability coefficients are very much higher than would be predicted from their oil/water partition coefficients; the permeability of the cell membrane to water may be 1000 times its expected value. For these small molecules the rate of transfer across the membrane is related primarily to molecular size, the permeability coefficient being inversely proportional to the cube root of the volume of the molecule.

Glucose appears to be the largest molecule for which molecular volume rather than lipid solubility is the dominant factor in determining the rate of membrane permeation. The glucose molecule has a radius of 0.42 nm and this has led to the suggestion that a number of *water-filled pores* are present in the membrane and that small molecules cross the membrane through these aqueous channels rather than by dissolving in the membrane lipids. If these pores were of 0.4 nm radius, only 0.02 per cent of the area of the cell membrane would have to be occupied by pores to account for the high permeabilities of small molecules. However, it has been questioned whether such water-filled pores need exist as distinct entities. Their postulated dimensions are only about twice those of a water molecule, and spaces of this order may exist between the highly orientated hydrocarbon tails of the quasi-crystalline array of membrane lipids. Small molecules could squeeze through these spaces, but larger molecules could not pass through the membrane without distortion of the lipid array. Thus, more energy would be required for the diffusion of larger molecules through the membrane.

Facilitated diffusion

As described above, certain substances, notably glucose and amino-acids which are essential for cell metabolism, cross the cell membrane in response to a concentration gradient much more rapidly than can be explained on the basis of their lipid solubility. Nor can this rapid permeation be satisfactorily explained by passage through water-filled pores; the size of the molecules is too large.

The mechanism responsible for this rapid transport has been termed *facilitated diffusion*. A good example of this mechanism is the transport of glucose into the red cell and this 'model' has been widely used in studies of the process. Facilitated diffusion is believed to depend upon the presence in the cell membrane of a relatively small number of '*carrier*' *molecules*. These ferry the glucose across the membrane by first binding to the sugar at the border of the membrane at which the glucose concentration is higher (generally the outside of the cell). The sugar–carrier complex then transfers the glucose to the other border of the membrane, where it dissociates to deliver the sugar into the fluid on that side. The process is illustrated in FIG. I.13.

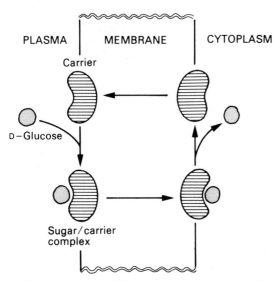

FIG. I.13. Simple carrier model for sugar transport across red cell membrane. (Redrawn from Ross, G. (ed.) (1978). *Essentials of human physiology*. Year Book Medical, London.)

There is considerable evidence in support of this carrier-mediated transport hypothesis as an explanation for the rapid transport of sugars and amino acids.

1. Kinetics of facilitated diffusion. If the rate of glucose entry is studied as a function of the glucose concentration gradient across the cell membrane, it can be seen that the rate of glucose entry is always much greater than predicted from the partition coefficient for glucose, i.e. glucose enters much faster than can be explained by a simple solubility-diffusion mechanism [FIG. I.14]. As the external concentration of glucose increases, this faster rate of entry of glucose into the cell continues to increase, at first linearly. However, it does not continue to increase linearly, as might be expected if glucose were entering through 'pores', but tails off to reach a maximal value which is unaffected by further increase in the external glucose concentration. This is taken to indicate saturation of the carrier system which is responsible for the active uptake. The substrate S and the carrier C react together to form a complex CS and from the Law of Mass Action:

$$K_m = \frac{[S] \times [C]}{[CS]},$$

where K_m is the dissociation constant for the complex CS. K_m is determined from the equation:

$$\frac{V}{V_{max}} = \frac{[S]}{K_m + [S]},$$

where V is the rate of transport at a substrate concentration [S] and V_{max} is the maximal transport rate. When $V = \frac{1}{2}V_{max}$, K_m equals the substrate concentration. Obviously $1/K_m$ measures the affinity of the carrier for the substrate. For glucose transport by the red blood cell V_{max} is 500 μM ml cells^{-1} min^{-1} and K_m, the concentration of glucose that produces 50 per cent of the maximal transport, is 5 mM. These features of carrier-mediated transport are shown in FIG. I.14 (a).

Now as

$$V = \frac{V_{max} [S]}{K_m + [S]}$$

$$\frac{1}{V} = \frac{K_m + [S]}{V_{max} [S]}.$$

If we fractionalize

$$\frac{1}{V} = \frac{K_m}{V_{max}} \times \frac{1}{[S]} + \frac{[S]}{V_{max} [S]}$$

and

$$\frac{1}{V} = \frac{K_m}{V_{max}} \times \frac{1}{[S]} + \frac{1}{V_{max}}.$$

This is the equation for a straight line ($y = ax + b$) where, if $1/V$ is plotted as a function of $1/[S]$, the intercept on the y axis is $1/V_{max}$ and the slope is K_m/V_{max}. The negative intercept on the x axis gives $-1/K_m$.

FIGURE I.14(b) shows the results of FIG. I.14(a) plotted in this double reciprocal manner. K_m can be estimated from the negative x intercept. As S is expressed in moles per litre, the dimension of K_m is in terms of molarity.

These equations for carrier-active transport are identical with those which define enzyme kinetics. The equation

$$V = \frac{V_{max} [S]}{K_m + [S]}$$

is the so-called Michaelis–Menten equation. The double reciprocal graph is known as the Lineweaver–Burk plot.

2. Substrate specificity. Facilitated diffusion displays considerable substrate specificity. Thus, glucose entry is a stereospecific process. Though the physiological isomer D-glucose is rapidly transported into the cell, the optical isomer L-glucose is not, K_m for L-glucose being 600 times larger than K_m for D-glucose. This degree of substrate specificity implies the matching of a highly specific structure on the carrier molecule by an equally specific structure on the part of the substrate. The D-glucose system also transports other sugars. These include D-mannose (K_m 20 mM), D-galactose (K_m 30 mM), D-xylose (K_m 60 mM), and D-arabinose (K_m 100 mM). All are hexoses or pentoses with a six-membered (pyranose) ring in the same chair conformation (C1 conformation) as in D-glucose.

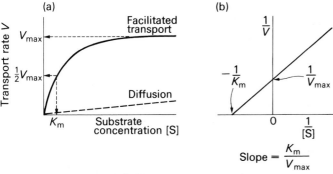

FIG. I.14. (a) Rate of substrate transport as a function of substrate concentration for simple diffusion and facilitated (carrier) transport. (b) Lineweaver–Burk plot; graphical representation of equation

$$\frac{1}{V} = \frac{K_m}{V_{max}} \times \frac{1}{[S]} + \frac{1}{V_{max}}.$$

(See text for further details.)

However, as the K_m values show, their relative affinities for the sugar transport system are all less than that of D-glucose. This appears to be related to the number of OH groups in the equatorial plane of the pyranose ring; the greater the number of such OH groups, the lower the K_m.

Substrate specificity also leads to competition when two structurally similar substrates compete for the available sites on the carrier molecule. An example of this is the blocking of galactose entry into the red cell by an increase in the glucose concentration in the plasma. The kinetics of this inhibition are of the truly competitive type in that the addition of D-glucose causes an apparent increase in the K_m for D-galactose but does not alter V_{max}. In such cases, the relative rates of active transport of the two substrates will depend on their relative affinity for the carrier and on the relative concentrations of each in the neighbourhood of the carrier sites. The reduction in the permeability coefficient for glucose as the concentration is raised, which was illustrated in Fig. I.14(a), can be regarded as an example of substrate competition in which the given sugar 'competes with itself', the competition being between identical, rather than similar, molecules.

3. Effects of poisons. In view of the similarity between enzyme kinetics and the kinetics of facilitated diffusion, it is not surprising that poisons such as $HgCl_2$ and dinitrofluorobenzene, which react with proteins in the cell membrane, should markedly inhibit facilitated diffusion. However, poisons such as cyanide and dinitrophenol, which interfere with the production of metabolic energy, do not interfere with facilitated diffusion, suggesting that metabolic energy is not required for this process. This is consistent with the fact that facilitated diffusion is always in the direction of the glucose concentration gradient. By contrast, other transport mechanisms, which can carry glucose and other substrates across the cell membrane against a concentration gradient [see below] and for which a supply of metabolic energy is required, are inhibited by these non-specific metabolic poisons.

Facilitated diffusion has been demonstrated in cells of a variety of tissues. Facilitated uptake of sugars has been described in muscle, adipose tissue, and the choroid plexus and capillaries of the brain, as well as in red and white blood cells. Facilitated transport of amino acids has been found to occur in the intestinal epithelium and red and white blood cells. Where facilitated diffusion mechanisms exist in the same cell for more than one type of molecule, i.e. sugars, fatty acids, and amino acids, the cell behaves as if it possessed several types of carrier, one for transporting sugars, one for fatty acids, and one for amino acids. Thus, there is no substrate competition between these classes of transported substances. The amino acids, moreover, appear to be grouped according to whether they are neutral, basic, or acidic, so that while substrate competition occurs between members of a given group, it is not seen between the groups. Thus there is mutual interaction between neutral amino acids such as glycine and alanine, but the transport of neither is affected by basic amino acids like lysine, or acidic amino acids like glutamic acid.

Little is known of the identity of the membrane carriers responsible for facilitated diffusion. Using radioactively labelled sugars, a protein material to which the radioactive sugar is attached has been isolated from the red blood cell membrane. This may be a membrane protein–sugar complex representing the intermediate state in which the sugar is held during the transport process.

Sodium co-transport

The mechanism of facilitated diffusion provides for the uptake of substances by the cell much more rapidly than would be possible by solubility–diffusion. Nevertheless, this accelerated transport can only take place down a pre-existing concentration gradient for the

particular substrate. However, in many cells and tissues there is a membrane mechanism which is capable of transporting substances from the surrounding fluid to the interior of the cell *against* a concentration gradient. A notable example is the brush-border membrane of the epithelial cells lining the small intestine. When 3-O-methyl glucose, a poorly metabolized sugar, is being absorbed the concentration within the cell may reach ten times that in the extracellular fluid. The transport mechanism resembles facilitated diffusion in that (i) the rate of sugar accumulation displays saturation kinetics; (ii) it is a stereospecific process, D-glucose but not L-glucose being accumulated in the intestinal epithelium; (iii) several related sugars share the same transport system though it is more highly specific than the facilitated diffusion carrier system—only hexoses, and only hexoses with an equatorial OH group on carbon number 2, are transported; and (iv), as might be expected from this specificity, there is mutual competition between the sugars that are transported.

The outstanding differences between this mechanism for membrane transport and facilitated diffusion are firstly, that, as might be expected from a process that must involve metabolic work in transporting substrate against a concentration gradient, it is inhibited by metabolic poisons such as cyanide and dinitrophenol, which block the provision of metabolic energy; and secondly, that it is strongly related to the concentration of *sodium* in the medium. The transport of sugar across the brush-border membrane of the intestinal epithelial cells is markedly reduced by replacing the sodium in the lumen of the gut by lithium or potassium. The stimulant effect of increasing the sodium concentration is shown in the Lineweaver–Burk plots of Fig. I.15. Increasing the sodium concentration up to 145 mM l^{-1} does not affect V_{max}, the maximum rate of uptake (the intercepts on the $1/V$ axis, which give the values of $1/V_{max}$, are identical at all levels of sodium concentration) but dramatically reduces K_m, the sugar concentration at which $V = \frac{1}{2}V_{max}$ (the values of $-1/K_m$ are given by the intercepts of the lines on the $1/[S]$ axis). There is other evidence linking sodium with this transport mechanism. (i) Not only does an increase in luminal sodium concentration stimulate the transport of sugars, but conversely, an increase in luminal glucose concentration stimulates the uptake of sodium ions across the brush-border membrane of the intestinal epithelial cell. At physiological sodium concentrations one sodium ion enters the cell for each glucose molecule transported across the

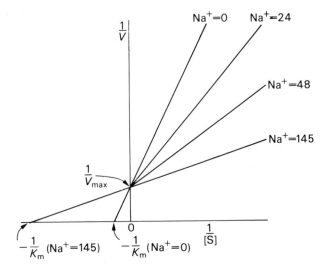

Fig. I.15. Lineweaver–Burk plots showing the effect of increasing luminal sodium concentration on the rate of sugar transport by the intestinal epithelium. Increasing [Na^+] from 0 to 145 mM l^{-1} does not alter V_{max} (all the lines intersect on the $1/V$ axis) but greatly increases $-1/K_m$ (the intercept of the lines on the $1/[S]$ axis moves to the left), i.e. the affinity of the carrier for the sugar is increased.

membrane. (ii) The transport of sugars is blocked by ouabain, a potent inhibitor of the membrane-bound sodium and potassium-dependent ATPase which is responsible for active sodium transport across the cell membrane [p. 15].

The association of this form of substrate transport with the transport of sodium by the cell has led to the use of the term *sodium co-transport* to describe the process. FIGURE I.16 illustrates a model

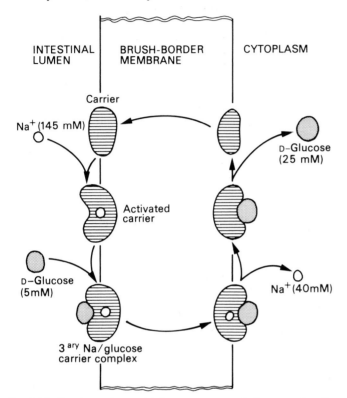

FIG. I.16. Simple carrier model for co-transport of glucose and sodium across the brush-border membrane of the intestinal epithelium. (Modified from Ross, G. (ed.) (1978). *Essentials of human physiology.* Year Book Medical, London.)

of the sodium co-transport system for glucose in the brush-border membrane of the intestinal epithelial cell. The characteristic feature of this model is a carrier which in the absence of sodium ions has a low affinity for glucose. However, in the presence of sodium ions the carrier undergoes allosteric modification so that its affinity for glucose is increased. Thus, the extent of binding of glucose to the carrier will depend on the concentrations of both glucose and sodium ions. At the luminal border of the brush-border membrane the sodium concentration is high, so that, even though the glucose concentration is relatively modest, the high (sodium-activated) affinity of the carrier for glucose results in a high concentration of the carrier–sugar–sodium complex. At the inner, cytoplasmic, border of the membrane, the sodium ion concentration is kept at a much lower level by the activity of the Na^+/K^+ exchange pump (see below). Sodium, therefore, leaves the carrier–sugar–sodium complex; the affinity of the carrier for glucose is thereby much reduced, and glucose is released into the cytoplasm even though the glucose concentration is higher than in the lumen. In this way glucose is accumulated within the cell against a concentration gradient. Formerly it was thought that this transport of glucose was the process requiring metabolic energy. It is now appreciated that the carrier–sugar–sodium complex simply moves across the membrane along a concentration gradient, the concentration being high at the luminal border, where it is being formed, and low at the cytoplasmic border, where it dissociates. (However, this concept of the carrier as a moving molecule should only be regarded as a useful

abstraction; the membrane is too thin for any appreciable movement to take place. It seems more likely that the carrier does not move but that the transported molecule, in this case glucose, moves along it to be released at the appropriate end when the local conditions dictate a change in affinity.) The energy-requiring step in the process, which is poisoned by cyanide or ouabain leading to the cessation of glucose accumulation, is the continual pumping of sodium back out of the cell. If this ceases, sodium concentration within the cytoplasm rises so that sodium remains attached to the carrier–sugar–sodium complex, which thus retains its high affinity for glucose and fails to release the sugar.

Sodium co-transport is a widely-distributed cellular mechanism:

1. Co-transport of sodium and sugars has been described in the proximal renal tubule and gall-bladder as well as in the intestine.

2. Co-transport of sodium and amino acids has also been described in the intestine and proximal renal tubule and in addition in brain cells, choroid plexus, leucocytes, and ascites tumour cells. There appear to be at least four separate carriers for amino acids; for neutral amino acids, dibasic amino acids, dicarboxylic amino acids, and imino acids.

3. Sodium co-transport may be responsible for the transport of catecholamines by cells of the nervous system and may be involved in the transport of choline and para-aminohippuric acid by the kidney.

4. Certain hereditary diseases have now been identified as due to disorders of sugar and amino acid co-transport systems. These include glucose–galactose malabsorption by the intestine, malabsorption of neutral amino acids by the intestine (Hartnup's disease), and cystinuria.

Transport of ions

The ionic compositions of extracellular and intracellular fluids are compared in TABLE I.1. It is clear that there are considerable differences in the concentrations of some of these ions on either side of the membrane, notably sodium, chloride, and bicarbonate, which are present in much higher concentration in extracellular fluid, whereas the potassium concentration in the intracellular fluid is much the higher. The higher concentrations in the intracellular fluid of protein anions and of phosphate anions (which are in the main, part of larger organic ions) can be disregarded in the present context since the membrane is generally impermeable to particles of this size, but the concentration gradients of the smaller, permeant, ions will tend to drive Na^+, Cl^-, and HCO_3^- into, and K^+ out of the cell, as would be the case with non-electrolytes. However, ion transport is more complex, since ions are charged particles and there is normally a potential difference across the cell membrane. The intracellular potential is negative to that registered in the

TABLE I.1 *Typical values for the ionic composition of mammalian intracellular and extracellular fluids. The units are mM l⁻¹*

Ion	Intracellular fluid		Extracellular fluid	
	Cations	Anions	Cations	Anions
K^+	155		5	
Na^+	12		145	
Mg^{2+}	15		2	
Ca^{2+}	2		2	
Cl^-		4		110
HCO_3^-		8		27
Protein⁻		64		15
PO_4^{2-}		90		2
Others		18		
Total	184	184	154	154

extracellular phase and ranges from −50 mV in liver cells to −90 mV in muscle. It can be demonstrated by impaling the cell with a microelectrode of 0.5 μm diameter. The cell membrane is hardly damaged and 'seals' the breach of its continuity so that, with suitable amplification, this voltage difference between the intracellular and extracellular phases may be demonstrated.

The effect of this potential difference across the cell membrane is to attract the positively charged Na^+ and K^+ cations into the cell, and to drive the negatively charged Cl^- and HCO_3^- anions out of the cell. The net movement of any ion across the cell membrane will therefore reflect the resultant of the combined concentration (chemical potential) and electric potential gradients. In solution, the rate of ion diffusion (J) under the influence of these combined forces is described by the equation:

$$J = -DA[dc/dx + (zCF/RT)\, dV/dx],$$

where D is the diffusion coefficient, A the area of the plane through which ion movement is taking place, dc/dx the concentration gradient, z the valency of the ion, C the ion concentration, F the Faraday constant (96 500 coulombs per mol of ion), R the gas constant (8.316 joules per degree), T the absolute temperature, and dV/dx the electrical potential gradient. This is, of course, the diffusion equation described earlier [p. 8] with an added term taking account of the electrical potential gradient. When applying this equation to the transport of ions across the cell membrane, D, the diffusion coefficient for the ion in aqueous solution, may be replaced by P, the permeability coefficient, where $P = K_m D_m/x$. (K_m is the oil/water partition coefficient for the ion, and D_m the diffusion coefficient for the ion in the membrane lipid; x is the membrane thickness.) Using a derivation of the ion flux equation above, values for ion permeabilities can be obtained from measurements of the membrane potential (E_m), and ion movements estimated with radioactive tracers.

Cell membrane permeability to ions

Measurements made as just described have indicated that the permeabilities of the frog muscle cell membrane to potassium and sodium ions are 6×10^{-7} and 8×10^{-9} cm s^{-1} respectively. Thus, the cell membrane is about 100 times as permeable to potassium as to sodium ions. When measurements of potassium and sodium ion permeabilities are made using a phosphatidylcholine bilayer as a model of the cell membrane, two striking differences emerge; (i) the permeabilities to both ions are very much lower than in the muscle cell membrane ($P_K = 3.4 \times 10^{-12}$ cm s^{-1}; $P_{Na} = 1 \times 10^{-12}$ cm s^{-1}), and (ii) potassium permeability is only 3–4 times that of sodium. Two explanations have been proposed for the higher permeabilities of both ions in the cell membrane as compared with the lipid bilayer model; (i) that a carrier is present in the cell membrane and serves to ferry ions across it, and (ii) that there are small water-filled pores in the cell membrane through which these monovalent ions can pass. This latter seems the more likely on current evidence.

The greater permeability of the cell membrane to potassium ions as compared to sodium ions has been explained in terms of the forces governing the transfer of the ions from water to the membrane sites controlling permeation. There is an electrostatic attraction between the cations and negative charges at these sites, which promotes permeation, while permeation is opposed by the attractive forces between the cations and water, i.e. the ion hydration energies. Normally, there are relatively few charges on the membrane sites, so the dominant force is the attraction between the ion and the water of the aqueous solution. Since less energy is required to tear a larger ion free from water, potassium, which is a larger ion than sodium, generally possesses the higher permeability. However, under certain circumstances, e.g. during the action potential, there is an increase in the negative charge at the membrane sites

controlling permeability, so that the electrostatic attraction becomes the dominant factor. Since this attractive force varies inversely with the square of the distance between the binding site and the centre of the ion, it favours the smaller, sodium, ion. Thus, the permeability of the cell membrane to sodium may increase greatly, exceeding that to potassium. It seems probable that there are separate pores or 'channels' for sodium and potassium permeation and that only the channels responsible for sodium permeation exhibit the increased negative charge.

The membrane potential

As previously mentioned, there is a potential difference across the membrane of all living cells, the interior of nerve and muscle cells being 70 to 90 mV negative with respect to the extracellular fluid. This potential difference stems from the differences in the permeability of the membrane to the various ion species which it separates. The membrane is relatively highly permeable to the principal intracellular cation, potassium, and to the principal extracellular anion, chloride, but is poorly permeable to the principal extracellular cation, sodium, and is impermeable to the protein and organic phosphate anions which constitute the bulk of the intracellular anion [TABLE I.1].

To appreciate why this should lead to the establishment of a membrane potential, let us first simplify the situation and assume that the membrane is permeable only to one ion, say potassium. Because the potassium concentration in the intracellular fluid (155 mM l^{-1}) is higher than that in the extracellular fluid (5 mM l^{-1}), potassium ions will tend to diffuse out of the cell. This flux ($J_{K_{(i \to o)}}$) will be determined by the membrane permeability to potassium and the concentration difference so that

$$J_{K_{(i \to o)}} = P_K\, \Delta C_K.$$

However, since the cations moving out of the cell are unaccompanied by anions (to which the membrane is impermeable) an excess of anion remains within the cell which becomes negatively charged. Thus, the diffusion of potassium ions out of the cell has set up a potential difference across the membrane, the *diffusion potential*. This potential will tend to draw potassium ions back into the cell by electrostatic attraction and this flux ($J_{K_{(o \to i)}}$) will be determined by the membrane conductance for potassium ions, G_K, (conductance is the reciprocal of resistance) and the potassium diffusion potential, E_K, so that

$$J_{K_{(o \to i)}} = G_K E_K.$$

As more potassium ions leave the cell down their concentration gradient, so the diffusion potential increases and with it this inward flux of potassium, until eventually the two fluxes are the same,

$$\text{i.e. } G_K E_K = P_K\, \Delta C_K$$

(It is worth noting that the actual number of potassium ions which must cross the membrane to set up this diffusion potential is remarkably small and the change in the internal potassium concentration is for all other purposes negligible.) The value of the diffusion potential, or membrane potential, E_m, at which this equilibrium is established is given by the Nernst equation:

$$E_m = \frac{RT}{zF} \ln \frac{[K^+]_o}{[K^+]_i}$$

ln represents the natural logarithm; the other symbols have already been defined. At a temperature of 30 °C this equation simplifies to:

$$E_m = \frac{8.316 \times 303}{96\,500}\, 2.3 \log_{10} \frac{[K^+]_o}{[K^+]_i} \text{ volts}$$

whence

$$E_m = 60 \log_{10} \frac{[K^+]_o}{[K^+]_i} \text{ mV.}$$

Thus, the *equilibrium potential* for potassium ions, i.e. the membrane potential at which there would be no net movement of potassium across a membrane permeable only to potassium ions, would be:

$$E_m = 60 \log_{10} \frac{[K^+]_o}{[K^+]_i} = 60 \log_{10} \frac{5}{155} = 60 \log_{10} 0.032$$

$$= 60 \times (\bar{2}.505) = 60 \times (-1.49) = -89.4 \text{ mV}.$$

We can similarly calculate the values for the equilibrium potentials assuming that the membrane was permeable only to sodium ions, or only to chloride ions. Using the concentrations of these ions shown in TABLE I.1, for sodium;

$$E_m = 60 \log_{10} \frac{[Na^+]_o}{[Na^+]_i} = 60 \log_{10} \frac{145}{12} = 60 \log_{10} 12.1$$

$$= 60 \times 1.08 = +65 \text{ mV};$$

and for chloride:

$$E_m = \frac{60}{-1} \log_{10} \frac{[Cl^-]_o}{[Cl^-]_i} = -60 \log_{10} \frac{110}{4} = -60 \log_{10} 27.5$$

$$= -60 \times 1.44 = -86.4 \text{ mV}$$

(the term -1 represents the valency of the negatively charged chloride anion). The interested reader may note that the E_m values for potassium and chloride ions are very similar. This results from the similarity of the ratios $[K^+]_o/[K^+]_i$ and $[Cl^-]_i/[Cl^-]_o$, which suggests that, in this example, the distribution of both the potassium and chloride ions is a passive consequence of the Donnan effect [see p. 16].

Comparison of these calculated equilibrium potentials with the membrane potential of approximately -90 mV as measured with an intracellular microelectrode, shows that the intracellular potassium and chloride concentrations are close to their equilibrium values. On the other hand, the membrane potential is far from the equilibrium value for sodium, and would have to be $+65$ mV to account for the low intracellular sodium concentration. With the inside of the cell 90 mV negative to the extracellular fluid both concentration and electrical gradients are present to promote a leak of sodium ions into the cell. Thus, maintainance of a low sodium concentration within the cell must depend on the continual extrusion by a *sodium pump* of sodium ions entering the cell. The situation has been compared to a leaking boat in which the water can be kept at an acceptable level and the boat kept afloat only by vigorous operation of the bilge pump.

Though the sodium pump ensures that the cell membrane behaves as if it was effectively impermeable to sodium, the membrane is, in fact, permeable to sodium (though much less so than to potassium) and the resultant passive sodium influx, even though associated with an equivalent active sodium efflux, can modify the membrane potential. Thus, the membrane potential differs from that predicted on the basis of absolute sodium impermeability. This is taken account of in the so-called 'constant-field equation' of Goldman, which includes terms for the ion permeabilities as well as for their intracellular and extracellular concentrations. This equation is:

$$E_m = 60 \log_{10} \frac{P_K[K^+]_o + P_{Na}[Na^+]_o + P_{Cl}[Cl^-]_i}{P_K[K^+]_i + P_{Na}[Na^+]_i + P_{Cl}[Cl^-]_o}.$$

If chloride is regarded as being distributed passively, the equation simplifies to:

$$E_m = 60 \log_{10} \frac{P_K[K^+]_o + P_{Na}[Na^+]_o}{P_K[K^+]_i + P_{Na}[Na^+]_i}.$$

Using established estimates for the permeability coefficients and concentrations of sodium and potassium in frog muscle, the membrane potential may be predicted from this equation:

$$E_m = 60 \log_{10} \frac{(6 \times 10^{-7} \times 2.5) + (8 \times 10^{-9} \times 120)}{(6 \times 10^{-7} \times 140) + (8 \times 10^{-9} \times 9)}$$

$$= -92 \text{ mV}.$$

This agrees closely with the measured valued of -90 mV. The Goldman equation also shows that the membrane potential is reduced (i.e. becomes less negative) when sodium permeability rises. During the action potential in nerve and muscle sodium permeability rises sufficiently to reverse the polarity of the membrane potential which briefly approaches the sodium equilibrium potential [p. 270].

Active sodium transport. The sodium pump

Active transport of sodium out of the cell against the electrical and chemical gradients accounts for between 20 and 45 per cent of the total metabolic energy expended by the cell. The process has been studied in nerve, muscle, and red blood cells and the following features have emerged.

1. The rate of active sodium transport increases with internal sodium concentration, but displays saturation kinetics. The K_m is about 20 mM, which is similar to the intracellular sodium concentration. Saturation kinetics have been seen previously to be a characteristic of membrane carrier systems in facilitated diffusion and sodium co-transport.

2. Glynn and his colleagues have shown that there is a tight obligatory coupling of sodium efflux to potassium influx in red blood cells. Active sodium transport out of the cell only occurs if potassium is present in the extracellular fluid, and is accompanied by the simultaneous transport of potassium into the cell. This inward transport of potassium also displays saturation kinetics with a K_m of about 2 mM potassium, which is of the same order as the normal extracellular potassium concentration. The inward potassium transport ceases if sodium is absent from the intracellular fluid, indicating that these two active fluxes are coupled together. It appears that three sodium ions are pumped out of the cell for every two potassium ions pumped in. The mechanism is thus an *electrogenic* pump; by removing three cations for every two returned, it increases the negative potential within the cell. However, the contribution of the sodium pump to the membrane potential is small; the membrane potential is less than 3 mV in excess of the value predicted by the constant field equation.

3. The presence of Mg-ATP *inside* the cell is essential for active sodium and potassium transport. For every three sodium ions pumped out, one molecule of Mg-ATP is hydrolysed. Metabolic poisons, such as cyanide or dinitrophenol, which inhibit the production of ATP, also inhibit the sodium/potassium pump.

4. Linked sodium and potassium transport is inhibited when cardiac glycosides, such as ouabain, are added to the *external* medium. Studies using radioactively labelled ouabain to estimate the amount of ouabain binding, suggest that there are 10^3 to 10^6 pumping sites on each cell.

A carrier model based on these observations is illustrated in FIG. I.17. The pump is represented by a carrier which shuttles from one side of the cell membrane to the other. With each cycle three sodium ions are transported from the cytoplasm to the external medium, and two potassium ions are carried back into the cytoplasm. Energy for the carrier is provided by the hydrolysis of ATP to ADP and inorganic phosphate at the inner border of the membrane. Ouabain inhibits the pump at the outer border.

Sodium/potassium ATPase

A membrane-bound enzyme which hydrolyses ATP to ADP and inorganic phosphate provided Na^+ and Mg^{2+} are present, was discovered by Skou in 1957. The following observations, among many others, have established that this enzyme is very closely related

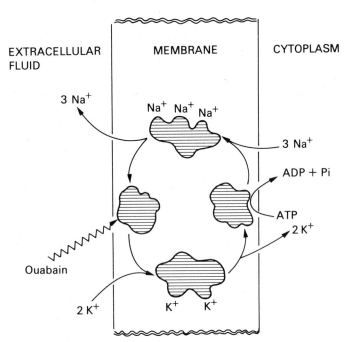

FIG. I.17. Simple carrier model of Na/K exchange pump in cell membranes. (Redrawn from Ross. G. (ed.) (1978). *Essentials of human physiology*. Year Book Medical, London.)

to the active transport of sodium and potassium across the cell membrane:

1. In the wide range of species which have been examined, there is a good correlation between the presence of the enzyme and the presence of sodium/potassium membrane pumps.

2. In order for ATP to be hydrolysed by the enzyme potassium must be present in the external medium and sodium must be present within the cell. The K_m values of sodium and potassium for the hydrolysis of ATP by the enzyme are similar to the K_m values for sodium and potassium transport.

3. Mg-ATP is only hydrolysed when it is inside the cell. The enzyme has no action on extracellular Mg-ATP. The rate of hydrolysis of Mg-ATP shows a close relationship to the rate of ion transport.

4. ATP hydrolysis, like ion transport, is inhibited by the presence of ouabain in the external medium. A given concentration of ouabain produces a similar inhibition of ATP hydrolysis and ion pumping.

Recent attempts to purify the enzyme have shown it to consist of two proteins. The larger has a molecular weight of 95 000 and possesses binding sites for sodium, potassium, ATP, and ouabain. The smaller, a glycoprotein of molecular weight 55 000 has no known function. The purified enzyme has been incorporated into artificial lipid bilayer membranes which, in the presence of ATP, have been shown to transport sodium and potassium ions. Thus, there is much evidence suggesting that the enzyme is indeed the 'ion pump'.

Donnan membrane equilibrium

The presence within the cell of non-diffusible anions, principally protein and organic phosphate anions, has important effects on the distribution of the diffusible (penetrating) ions.

Consider two ionized solutions *a* and *b* filling compartments of constant volume and separated by a partially-permeable membrane. It was shown theoretically (Gibbs) and confirmed experimentally (Donnan), that *at equilibrium*:

1. Each solution will be electrically neutral—its total charges on cations will equal those on anions:

$$[cations]_a = [anions]_a \text{ and } [cations]_b = [anions]_b.$$

2. The product of the diffusible ions on one side of the membrane will equal the product of the diffusible ions on the other

$$[\text{diffusible cations}]_a \times [\text{diffusible anions}]_a = [\text{diffusible cations}]_b \times [\text{diffusible anions}]_b$$

from which it follows that

$$\frac{[\text{diffusible cations}]_a}{[\text{diffusible cations}]_b} = \frac{[\text{diffusible anions}]_b}{[\text{diffusible anions}]_a}.$$

For example, with the freely diffusible ions of sodium chloride solutions, the equilibrium position will be reached when *a* and *b* have the same concentration of both ions (symmetrical distribution). But if one or more indiffusible ions are also present, the ionic distribution of the diffusible ions at equilibrium will be asymmetrical. Consider that in our simple system sodium chloride ($Na^+ Cl^-$) is present in solutions *a* and *b* but that only *a* contains a salt $Na^+ X^-$, where X is an indiffusible anion unable to cross the membrane (as is the case with many protein and organic phosphate anions), thus:

	Na^+	Na^+	
Solution *a*	Cl^-	Cl^-	Solution *b*
	X^-		

The penetrating ions (Na^+ and Cl^-) diffuse until equilibrium is attained. The two criteria established above will hold, namely

$$\left.\begin{array}{l} [Na^+]_a = [Cl^-]_a + [X^-]_a \\ [Na^+]_b = [Cl^-]_b \end{array}\right\} \text{ electrical neutrality}$$

and

$$\left.\begin{array}{l} [Na^+]_a \times [Cl^-]_a = [Na^+]_b \times [Cl^-]_b \\ \dfrac{[Na^+]_a}{[Na^+]_b} = \dfrac{[Cl^-]_b}{[Cl^-]_a} \end{array}\right\} \begin{array}{l} \text{products and ratios} \\ \text{of } \textit{diffusible} \text{ ions.} \end{array}$$

From these relationships it follows that

$$[Na^+]_a > [Cl^-]_a$$

and therefore that

$$[Na^+]_a > [Na^+]_b$$
$$[Cl^-]_a < [Cl^-]_b$$

and $[Na^+]_a + [Cl^-]_a > [Na^+]_b + [Cl^-]_b.$

Hence, *at equilibrium*, the cation (Na^+) concentration on the side of the membrane containing the non-penetrating anion X^- is greater than the cation concentration on the other side; the opposite is true of the diffusible anion (Cl^-) concentration which will be greater on the side without the non-diffusible anion. This is known as the Gibbs–Donnan Membrane Equilibrium, and the ratios $[Na^+]_a / [Na^+]_b$ and $[Cl^-]_b / [Cl^-]_a$ are termed the Donnan ratios. The necessary equality of these ratios for all species of diffusible ion can be understood by reference to the Nernst equation. At equilibrium

$$E_m = E_{Na} = 60 \log_{10} \frac{[Na^+]_a}{[Na^+]_b}$$

but likewise

$$E_m = E_{Cl} = 60 \log_{10} \frac{[Cl^-]_b}{[Cl^-]_a}$$

thus

$$\frac{[Na^+]_a}{[Na^+]_b} = \frac{[Cl^-]_b}{[Cl^-]_a}.$$

A further consequence of the Gibbs–Donnan Effect is that there is a greater number of ions in *a* than in *b*. The osmotic consequences of this are considered below.

It must be realized that the Gibbs-Donnan effect is brought about by differential permeability and passive transport, but of course the effect can be magnified or opposed by active transport, which can maintain an intracellular ion at a constant concentration *as though it were a non-penetrating ion*.

Living cells contain an excess of non-penetrating anions, mainly organic phosphates and proteins. The effects of this intracellular anion can be summarized thus:

1. The distribution of the diffusible anions and cations between cells and extracellular fluids is unequal, even when active transport does not apply. This is responsible for the permanent electrical difference which exists across cell membranes.

2. The concentration of cation inside a cell is greater than outside. In the case of the red cell the excess is about 10 mM per litre of cell fluid, but the situation is complicated by the active intake of K^+ and the expulsion of Na^+. The intracellular $[H^+]$ is somewhat greater than that of the external medium, e.g. the pH of red cells is 7.2 in plasma of pH 7.4.

3. The diffusible anion concentration is lower inside cells than outside. In the case of human red cells, the ratio of intracellular chloride to plasma chloride, i.e. the Donnan ratio, is 0.7 (see p. 188 for the 'chloride shift'). The much lower ratios found in nerve are due to a high membrane potential arising from other causes.

Donnan effect, osmosis, and active transport

When a membrane separates two compartments, in one of which there is a solution containing a non-diffusible anion, an equilibrium is established in which the final concentration of *diffusible* ions (cations + anions) in the solution containing the non-diffusible ion, exceeds that in the other solution. This was demonstrated in the example on p. 16, i.e.:

$$[Na^+]_a + [Cl^-]_a > [Na^+]_b + [Cl^-]_b.$$

If the concentration of diffusible ions in the solution containing the non-diffusible anion is greater then the total ion concentration (diffusible + non-diffusible ions) must be greater still. Thus the solution in this compartment exerts a greater osmotic effect.

It was assumed in the diagram on p. 16 that solutions *a* and *b* are confined in volume by rigid walls, as would be the case with vegetable cells. If the compartment walls are not rigid, then water will flow from *b* to *a*, *a* having the greater osmotic pressure. This will dilute *a* and concentrate *b* and disturb the Donnan equilibrium which can only be re-established by diffusible ion passing with the water from *b* to *a*, and so on. It is clear, that with no limitation on volume, true equilibrium cannot be established until all the water and ions on side *b* have passed to side *a*.

For animal cells, whose elastic walls are unable to resist an influx of water brought about by osmotic effects, rupture might seem inevitable. Indeed this would be the case were it not for the occurrence of active transport by the sodium/potassium membrane pump, which is adjusted so as to oppose the Donnan effect and prevent excess of cation, and therefore of water, entering the cell. Even a large difference in tonicity across the natural membrane of a metabolizing cell does not normally cause a large difference in hydrostatic pressure to develop, because the system is not in true equilibrium and metabolic work must be performed continuously to maintain it. When isolated tissue slices of kidney or brain are incubated at 37 °C in physiological saline solutions (containing the appropriate amounts of cations and anions in the proportions found in extracellular fluid) and some metabolic substrate such as α-ketoglutarate is added, the cells only gradually swell and their loss of K^+ and gain of Na^+ is small. However, when cellular metabolism is reduced—by lowering the temperature to zero, by adding meta-

bolic poisons such as cyanide or dinitrophenol, or by excluding metabolic substrate from their suspension medium—then the cells swell rapidly and they lose K^+ and gain Na^+. (Such changes occur *in vivo* in pathological states and are described by the term 'cloudy' swelling used by the histologist.) As should be clear by now, the cellular swelling results *not* from the failure of a hypothetical pump responsible for the active transport of water out of the cell, but from the failure of a metabolic pump which the cell uses to extrude sodium. All evidence so far obtained indicates that water always diffuses passively; its movement in and out of the cell is dictated by the active transport of solute—usually sodium.

The work done to maintain or increase concentration differences across a membrane by water transfer is called 'osmotic' work. For example, the osmotic work done by the kidneys in concentrating a urea solution from 5 mM l^{-1} (as in blood filtrate) to 330 mM l^{-1} (as in urine) is about 10.5 kJ mol^{-1}, regardless of whether this is a direct transfer of water or a transfer secondary to the movement of Na^+.

Permeability to protein

Although it is generally true that cell membranes are relatively impermeable to protein, this does not hold for specific proteins and specific types of cell, as the following examples show: placental membranes are permeable to maternal protein antibodies; foreign protein antigens have access to antibody-producing cells; protein enzymes are secreted into the lumen of the gastro-intestinal tract from specific cells; the liver both takes up protein and releases protein into the lymph and bloodstream; the enzyme ribonuclease can penetrate certain cells and destroy their RNA.

Transport across epithelia

Epithelia are continuous sheets of cells that cover the surfaces of organs and line body cavities. The most important function of many organs is the transport of water and solutes across these epithelia, as in the formation of urine by the kidney and the digestion and absorption of foodstuffs in the gastro-intestinal tract.

The principles involved in the passage of substances by passive diffusion through the cells of an epithelial layer have already been discussed, though it must be remembered that this involves passage from the extracellular environment through the cell membrane into the cytoplasm, at one face of the cell; and from the cytoplasm through the cell membrane into a possibly different extracellular environment, at the opposite face.

However, it is not always necessary for substances to traverse the full depth of the cell from its apical surface to its basal surface in order to cross the epithelial layer. In many epithelia the membranes of adjacent cells come into close contact over a short length of their lateral borders, to form a seal between the cells, which extends around their circumference. These ring-like seals are known as '*tight junctions*' or '*zona occludentes*'. Between the remainder of the lateral borders of the adjacent cells there are narrow clefts of variable dimensions, the lateral intercellular spaces, which form extensions of the extracellular fluid [Fig. I.18]. In some epithelia, such as those of the gall-bladder, proximal renal tubule, small intestine, and choroid plexus, most of the passive flux of sodium and chloride is across the tight junctions and through the lateral intercellular spaces, rather than through the epithelial cells themselves, and the rate of permeation is correspondingly greater. In other epithelia, known as 'tight' epithelia, e.g. distal renal tubule and urinary bladder, virtually all the transport is through the cell.

Moreover, in some epithelia, the cell membranes do not come into contact to form tight junctions completely around the cells but aqueous channels or pores are left between them. Such pores, which have been described in muscle capillaries and ependyma, enable solute to by-pass the cells completely, and explain the high permeability of such epithelia to polar molecules such as sucrose, inulin, and haemoglobin. The pores in the muscle capillaries are 3–5 nm in diameter and occupy about 0.2 per cent of the area of the capillary wall.

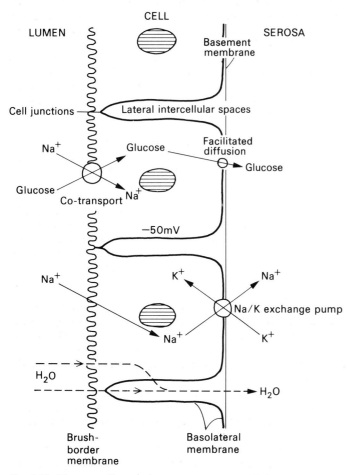

FIG. I.18. Glucose, sodium, and water transport across the intestinal epithelium. (Redrawn from Ross, G. (ed.) (1978). *Essentials of human physiology.* Year Book Medical, London.)

Active transport across epithelia

The transport of water, and of solutes such as ions, sugar, and amino acids, occurs across epithelia such as those of the stomach, gall-bladder, intestine, renal tubule, and choroid plexus, in the absence of external driving forces, e.g. gradients of hydrostatic pressure, osmotic pressure, or solute concentration. In some cases transport may even take place against uphill gradients of electric potential or solute concentration. Some of the mechanisms involved can be appreciated by considering the transport processes for glucose, sodium, and water by the intestinal epithelium. These are represented diagrammatically in FIG. I.18.

Glucose. Glucose is transported from the intestinal lumen into the cell across its brush border by the sodium co-transport system. This raises the local concentration at the luminal side of the cell above that at the serosal border, creating a concentration gradient along which glucose diffuses to the basal and lateral margins of the cell. The accumulation of sugar by the cell raises the overall internal glucose concentration above that of the extracellular fluid at the serosal surface, so that glucose diffuses out of the cell across the basolateral membrane. This transport, though passive, is speeded by facilitated diffusion, i.e. there is a carrier for glucose in these

regions of the cell membrane. Thus, the overall rate of glucose transport depends on both the kinetics of the sodium co-transport glucose carrier in the brush-border membrane, and the kinetics of the facilitated diffusion carrier in the basolateral membrane. The process of glucose transport is dependent on metabolic energy, since this is required for the continued operation of the sodium/potassium exchange pump. If this pump fails the gradient of sodium concentration across the brush-border membrane, which is essential for sodium co-transport, disappears and glucose is no longer accumulated. Thus, in the presence of metabolic inhibitors, there is failure both of glucose uptake into the cell from the lumen, and of glucose transport across the epithelium into the extracellular fluid on the serosal aspect and thence into the blood draining the intestine.

Sodium. As is the case with other mammalian cells, sodium is not in equilibrium across the membrane of the cells of the intestinal epithelium. Sodium ions therefore diffuse continuously into the cell along their electrical and chemical potential gradients. More enters across the brush-border membrane since the sodium permeability is higher than that of the basolateral membrane. Sodium also enters in association with glucose and amino acids brought into the cell by the co-transport system. A steady state of low sodium concentration within the cell is maintained in the face of this entry of sodium by the activity of the sodium/potassium exchange pump. All the pump sites appear to be on the basolateral portion of the cell membrane. This has been established (i) by biochemical analysis of the brush border and basolateral membranes, which has shown all the membrane-bound Na/K ATPase to be located in the basolateral membrane, and (ii) when sodium pumping has been abolished by radioactively labelled ouabain, subsequent autoradiography has shown the ouabain to be present at the basolateral border.

Thus, the transport of sodium across the intestinal epithelium is a two-stage process. Sodium first enters the cell at the luminal border under the influence of passive forces, and is then actively pumped out across the serosal border. Such two-stage transport of sodium was first described by Ussing in the frog skin, which has the capacity to absorb sodium from its freshwater surroundings, in which the sodium concentration is much lower than in frog extracellular fluid. In this situation the concentration gradient against which sodium is being transported may be as great as 100 to one. Two-stage sodium transport also occurs across the epithelia of the gall-bladder, renal tubule, and choroid plexus.

Water. The passage of water across epithelia is governed by the primary passage of sodium. The diffusion of sodium into the intestinal cell and its subsequent active extrusion into the extracellular fluid at the basolateral border creates an osmotic effect leading to a flow of water in the same direction. There is, however, a complication to be considered. If the flow of water depends simply on the presence of an osmotic gradient, we should expect the osmolarity of the fluid coming away from the transporting epithelium to be higher than that on the opposite side. Thus, the osmolarity of the cerebrospinal fluid in the ventricle should be higher than that of the extracellular fluid of the choroid plexus and of the plasma in the capillaries of the plexus. Yet this is not the case. There is no detectable difference in osmolarity and the process is described as *iso-osmolar secretion.* The same phenomenon is true of other epithelia, such as the intestine, where water transport follows that of sodium. This raises the question—how can water be transported across the epithelium when no effective osmotic gradient apparently exists? It could be argued that the water permeability of the cells is so high that water movement follows almost instantaneously on the transport of sodium by the Na/K pump. However, the measured permeability to water, though high, is not high enough to account for this apparently iso-osmotic flow, and other explanations must be sought.

The current view of water transport across epithelia stems from the work of Diamond and Bossert. Though their observations were made on the gall-bladder, their conclusions are applicable to the intestine. They noted that the filling of the intercellular clefts varied with the flow of fluid across the epithelium. Thus inhibition of active sodium transport, and hence of fluid flow, was associated with narrowing of the clefts. They concluded from this that the lateral margins of the cells, bordering the clefts, were the site of the active sodium transport. They further suggested that sodium transport was restricted largely to the portion of the cleft close to the tight junction, creating a region of hyperosmolality in this part of the cleft, into which water would therefore pass both from the cell and more directly from the lumen across the tight junction. The resulting fluid accumulation in this portion of the cleft would in turn create a hydrostatic pressure, promoting the flow of fluid down the cleft towards its serosal end. As the fluid flows along the cleft, further water enters, so that by the time it emerges from the cleft its osmolarity is little different from that of the fluid on the inside of the epithelium. In epithelia which transport large amounts of fluid the lateral membranes of cells bordering the intercellular clefts show much infolding and interdigitation. Presumably this increases the area of membrane exposed to the hyperosmolar fluid in the intercellular cleft and thus increases the rate of diffusion of water through the membrane.

Finally, it should be noted that in all these examples of active transport across epithelia, the sodium/potassium exchange pump plays a key role. This serves to emphasize the fundamental importance of this pump in nearly all physiological processes. This is reflected in the fact that the pump consumes between 20 and 45 per cent of the total metabolic energy used by the body.

Body water and body fluid

Total body water

Deuterium oxide (D_2O) or tritium oxide is injected intravenously. Each diffuses rapidly and evenly through the entire body water including the transcellular component. The degree of dilution which the substances suffer allows the calculation of total body water.

The average value in men is 62 per cent of the body weight (range 54–70 per cent). In women total body water is 51 per cent (range 45–60 per cent) of the body weight. In infants the range is 65–75 per cent of the body weight.

Total body water bears an almost constant relation to the *lean body mass* (i.e. the total weight of fat-free tissue). TBW = 73.2 per cent of the lean body mass. Lean body mass can be calculated from the specific gravity of the subject which is determined by weighing the subject in air and again when totally immersed in water (making a correction for the volume of air in his lungs). The difference in weights represents the volume of water displaced. The weight in air divided by the weight of water displaced gives the specific gravity. Spare individuals have a specific gravity of 1.10 whereas grossly fat subjects may show a specific gravity of 1.02. Fat has a much lower density (0.9) than other tissues (1.06). Bone has a specific gravity of 1.56.

Stored fat is water-free. Thus a total body water of 44 litres in a young man of 73 kg represents 60 per cent of his body weight. If he increased his weight to 80 kg by laying down fat his body water would represent only 55 per cent of his body weight.

Different tissues have different water contents as can be determined by their separate analysis. Bone has only 20 per cent water; skin 70 per cent; and heart, brain, and lung 75–80 per cent.

Extracellular fluid volume. This is measured by determining the dilution of a substance (injected in solution intravenously) which remains extracellular in its distribution. Probably sodium thiosulphate most nearly fulfils this requirement. The extracellular fluid volume measured by thiosulphate injection is 17 per cent of the body weight (range 15.4–19 per cent) in men and is slightly above 17 per cent in women. Thus a man of average build, 70 kg weight and with a total body water of 43.5 litres would have an extracellular fluid volume of 12 litres.

Other substances—inulin, mannitol, ^{36}Cl, ^{38}Cl, ^{24}Na, etc.—have been used to measure extracellular fluid, but give misleading figures. However, if extracellular fluid is measured by thiosulphate dilution the figure obtained should be qualified by terming it 'thiosulphate space'.

Intracellular fluid volume. This cannot be measured directly, but is estimated as the difference between the volumes of total body water and extracellular fluid. Thus, if total body water = 43.5 litres and extracellular fluid = 12 litres, intracellular fluid volume = 31.5 litres.

The extracellular fluid volume consists of (i) plasma, (ii) interstitial fluid, and (iii) *transcellular fluids* (which are separated from the plasma by both the vascular endothelium and by another epithelium). These transcellular fluids include the cerebrospinal fluid (150 ml), joint fluids, urine in the urinary tract, and digestive juices in the alimentary canal.

Plasma volume. This may be measured by determining the degree of dilution of a dye which escapes only slowly from the bloodstream after intravenous injection. Evans' blue (T 1824) binds with plasma albumin when injected and the dilution of the dye after injection of a known amount can readily be determined colorimetrically.

Alternatively, 50 ml of plasma are treated with radioiodine (^{131}I) *in vitro*; the iodine combines with albumin. The treated plasma is injected intravenously and the degree of dilution of its radioactivity determined.

The plasma volume is 2.8–3.0 litres in men and 2.4 litres in women. Plasma accounts for 4.3 per cent of the total body weight in both men and women.

Red cell volume. If a sample of blood is removed from a superficial vein it can (after heparinization to prevent clotting) be placed in a graduated tube (haematocrit) and centrifuged. The red cells sediment and the packed cell volume is read. In 100 ml of blood there are about 45 ml cells and 55 ml plasma. Hence, if the plasma volume is 2750 ml, the red cell volume is 2250, and the approximate blood volume is 5000 ml. However, blood from superficial veins usually has a higher haematocrit value than that elsewhere in the vascular system, so for accurate measurements of the total red cell mass and blood volume it is better to measure red cell mass directly.

Direct determinations of red cell mass entail a preliminary treatment of blood (5 ml withdrawn from the subject) with an isotonic solution containing radioactive phosphate ($^{32}PO_4$) in a siliconed tube at 37 °C for 2 hours. Inorganic $^{32}PO_4$ ions enter the cells and are conjugated to form organic phosphate ions which are indiffusible. The incubated blood is centrifuged and the separated cells are resuspended in isotonic saline. The sample is injected intravenously and samples of blood are withdrawn, at intervals of 10 minutes, from the vein. The degree of dilution of the radioactive cells is determined and hence the total red cell volume.

Electrolyte composition of cells

Biopsy specimens can be used for direct determination of electrolyte concentrations. Total exchangeable sodium, potassium, and calcium can be measured in life by determining the degree of dilution of radioactive isotopes injected intravenously.

Sodium. A known amount of ^{24}NaCl injected intravenously penetrates wherever sodium has access and the final dilution of ^{24}Na gives the measure of the mass of body sodium with which it can exchange—the total exchangeable sodium. This value is not identical with that of the total body sodium for about half the sodium content of bone is not available for exchange. In adults the total exchangeable sodium is 41.4 mM per kg body weight, so in a man of 70 kg this would give 2950 mM Na exchangeable. Total body sodium is 3700 mM. Of this 1500 mM is in bone and only half of bone sodium is available for exchange. Thus, 3700 minus 750 mM = 2950 mM are exchangeable. The average concentration of sodium in extracellular fluid is 143 mM l^{-1} and as there are 12 litres of extracellular fluid they contain 1720 mM. Hence 31.5 litres of intracellular fluid contain 2950–1720 mM = 1230 mM Na or on average just over 40 mM l^{-1}. (Lower figures than this are found in muscle, nerve, and liver.)

Potassium. Similar experiments reveal that 95 per cent of body potassium is exchangeable. The total average exchangeable K is 3200 mM in men and 2300 mM in women (46 mM kg^{-1} and 40.5 mM kg^{-1} respectively). The average concentration of potassium in extracellular fluid is 4–4.5 mM l^{-1} and in intracellular fluid is 145 mM l^{-1}.

Intracellular fluid thus differs radically from extracellular fluid in its electrolyte pattern. Intracellular fluid has a high K concentration and a low Na concentration; extracellular fluid has a low [K$^+$] and a high [Na$^+$] [TABLE I.1].

The blood

Blood consists of plasma and cells (red cells, white cells, and platelets). It is primarily a medium for the carriage of O$_2$, nutrient materials, hormones, and anti-infective agents (e.g. antibodies) to the tissues, and for the removal of CO$_2$ and other waste products from the tissues and their elimination from the body. The almost ubiquitous distribution of blood in the body and its unique chemical characteristics make it a most efficient transport system.

The properties of haemoglobin allow the carriage of the large amounts of O$_2$ needed for metabolic activities [p. 183]. The buffering power of haemoglobin is also an important factor in helping to maintain constancy of blood pH [p. 191].

Plasma proteins exert an osmotic pressure which influences the exchange of fluid between blood and tissues [p. 82]. Plasma proteins also combine with many substances, e.g. iron, thyroxine, and steroid hormones to form transportable complexes from which the active components are released at the appropriate sites.

Plasma and platelets contain all the factors required for clotting [pp. 20 and 30]. Thus loss of blood from injury is reduced by inherent properties of blood itself.

Antibodies belong to the γ-globulins. They are essential for the development of resistance to infection [p. 54].

Haemagglutinins and agglutinogens are important genetically and also in relation to blood transfusion [p. 47].

Leucocytes and plasma take part in the reactions of inflammation [p. 62].

REFERENCES

MACFARLANE, R. G. and ROBB-SMITH, A. H. T. (1961). *Functions of the blood*. Oxford.

THOMPSON, R. B. (1977). *The disorders of the blood*. London.
WILLIAMS, W. J., BEUTLER, E., ERSLEY, A. J., and RUNDLES, R. W. (1977). *Hematology*, 2nd edn. New York.
WINTROBE, M. M. (1974). *Clinical hematology*, 7th edn. Lee and Febiger, London.

THE PLASMA PROTEINS

The total plasma protein concentration is 64–83 g per litre. Two principal groups of plasma proteins are conventionally recognized: albumin and the globulins. The globulin fraction is subdivided into α_1-, α_2-, β-, and γ-globulins and fibrinogen. The average normal concentrations of the main plasma proteins in g per litre are:

Albumin, 48; Globulins, 23; Fibrinogen, 3.

A number of proteins with specific physiological functions have been partially isolated from the globulin fraction by electrophoresis. Among these proteins are prothrombin, plasma thromboplastin, isohaemagglutinins, angiotensinogen, immune globulins, and anterior pituitary hormones.

Properties of plasma proteins

Only those properties which need to be known to understand the methods of separating the proteins of the plasma, and their physiological properties, will be considered here.

Precipitation by salts. Different proteins are precipitated from solution by addition of different concentrations of salts. Thus albumin is precipitated by saturation with $(NH_4)_2SO_4$, globulin by half-saturation with $(NH_4)_2SO_4$. This method of separation gives a mean normal plasma albumin/globulin ratio of 1.7.

Fractional precipitation. In order to isolate individual plasma proteins in quantity, E. J. Cohn has devised methods based on fractionation with low salt concentrations at low temperatures, varying the pH and modifying conditions by the addition of alcohol.

Six main functional protein fractions have been obtained in this way:

  I. Fibrinogen + antihaemophilic globulin [p. 23].
 II. Immunoglobulins (= γ-globulins). Antibodies.
III. (i) Isohaemagglutinins (= β- and γ-globulins); (ii) prothrombin, fibrinolysin, complement (= α-, β-, and γ-globulins).
 IV. Angiotensinogen, alkaline phosphatase and some lipoproteins (= α- and β-globulins).
  V. Albumin.
 VI. The mother liquor: albumin and β-globulin; follicle stimulating hormone of the anterior pituitary.

Many of the proteins mentioned above have been isolated in a high degree of purity, e.g. albumin (in a form suitable for intravenous injection clinically); isohaemagglutinins (in 16 times the concentration found in pooled plasma); immune globulins (active against diphtheria, influenza virus, measles, mumps, typhoid bacillus) with 15–20 times the immune potency of pooled plasma; proteins concerned in blood clotting [p. 23].

Sedimentation in ultracentrifuge. The different proteins sediment at different rates when solutions of them are spun at very high speeds in the Svedberg ultracentrifuge: separation can thus be effected.

Isoelectric point. Proteins can ionize either as acids or as bases owing to the fact that the side chains of their constituent amino acids contain a selection of amino groups (NH$_2$) and carboxyl groups (−COOH). In alkaline solution, e.g. plasma, pH 7.4, the proteins ionize as acids, and free protein anions, negatively charged, are formed. In acidic solutions, the protein amino groups act as bases, taking up H^+, while the ionization of the carboxyl groups is suppressed. The protein then carries a net positive charge.

$$^+H_3N - \quad \quad - COO^-$$
$$^+H_3N - \quad PROTEIN \quad - COO^- \quad \quad \text{At isoelectric point}$$
$$^+H_3N - \quad \quad - COO^- \quad \quad \text{net charge} = 0.$$

(After Davenport, H. W. (1958). *ABC of acid base chemistry*, 4th edn. Chicago.)

At an intermediate pH (specific for each protein) the protein molecule carries equal numbers of positive and negative charges and hence has a net charge of zero. This pH value for electrical neutrality of the molecule is known as the *isoelectric point*. The actual value of the isoelectric point for a given protein depends on the relative number and relative dissociation strengths of its acidic and basic groups.

Buffer action. Plasma proteins act as buffers by virtue of their powers of H^+ acceptance, but they account for less than one-sixth of the total buffering power of the blood [see p. 192].

Molecular weight and shape. The molecular weight of plasma albumin is 69 000, and that of fibrinogen about 330 000. Penetration through the capillary wall depends not only on the size but also on the shape of the molecules. Albumin passes more readily than the others, but all plasma proteins pass through in small amounts and appear in the lymph. When capillary permeability is increased (e.g. in anoxia, urticaria, inflammation) all the proteins escape much more readily than normal.

Osmotic effects. The plasma proteins normally have an osmotic effect of 25 mm Hg and thus influence the exchange of fluid between blood and tissue spaces [see p. 82].

Viscosity. The resistance to the flow of fluid (at constant velocity) through a capillary (of constant bore) depends almost entirely on the viscosity of the fluid. The viscosity of the blood is thus a factor in maintaining the peripheral resistance and thereby, the arterial blood pressure. The viscosity of a protein solution depends far more on the shape of the protein solution molecule than on its size; the less symmetrical the molecule the greater is its viscosity. For this reason the following solutions have equal viscosities: 250 g l^{-1} albumin, 150 g l^{-1} γ-globulin, 20 g l^{-1} fibrinogen; each of these solutions has a viscosity equal to that of twice concentrated plasma (i.e. a plasma with 150 g of 'mixed' proteins per litre). Surprisingly enough the viscosity of whole blood (i.e. plasma plus suspended corpuscles) is also only that of twice concentrated plasma; this means that the corpuscles and the plasma contribute equally to the total viscosity of the blood.

Electrophoretic mobility. As already explained, proteins form *cations* in solutions which are *acid* with respect to their isoelectric points and form *anions* in solutions which are *alkaline* with respect to their isoelectric points. If the force of an external electrical field is applied to protein molecules dissolved in a suitable buffered electrolyte, they are caused to move; in acid solutions the protein ions (being positively charged) move towards the cathode; in alkaline solutions the protein ions (being negatively charged) move towards the anode. Each protein moves with a characteristic mobility which varies with the protein (also with pH, viscosity of the solvent, and the nature and concentration of the dissolved salts); therefore when a solution of mixed proteins (in a tube) is placed in an electrical field with the electrical poles at the ends, the contained proteins move at different rates (*electrophoresis*).

Electrophoretic separation of plasma proteins

Since the various plasma proteins have different surface charges they therefore migrate at different rates; prealbumin and albumin move fastest and γ-globulins slowest, with the other plasma proteins at intermediate rates. Although the separation is not complete numerous fractions can be detected and their concentrations measured by densitometric scan and staining [TABLE I.2 and FIG. I.19].

Transport function. Plasma proteins combine loosely with many chemical agents, including hormones, e.g. thyroxine and cortisol; metals, e.g. iron and copper; and numerous drugs. The bonds are

TABLE I.2. Summary of main features of plasma proteins

Fraction	Molecular weight	Plasma concentration (g l^{-1})	Function
Prealbumin	6×10^4	0.3	Binds thyroxine and triodothyronine
Albumin	6.9×10^4	40	Colloid osmotic pressure: binds hormones, fatty acids, bilirubin, drugs
α$_1$-Globulins	4.5×10^4	4.0	Includes antiprotease
α$_2$-Globulins			
Ceruloplasmin	1.6×10^4	0.4	Copper transport
Haptoglobins	9×10^4	1.2	Binds haemoglobin
β-Globulins			
Transferrin	9×10^4	2.4	Iron transport
Components of Complement	2×10^5	1.6	
Plasminogen	1.4×10^5	0.7	Fibrinolysis
Fibrinogen	3.5×10^5	3.0	Blood clotting
Prothrombin	6.8×10^4	1.0	Blood clotting
γ-Globulins			
IgG	1.5×10^5	10.0	All Igs are
IgA	1.7×10^5	1.6	antibodies
IgM	1×10^6	1.0	(Immunoglobulins)
IgE	2×10^5	0.1	[FIG. I.19]

firm enough to limit free diffusion of the small molecular substance and thus to reduce its biological activity. The bound form may serve as a reservoir from which the free hormone, metal or drug is slowly released.

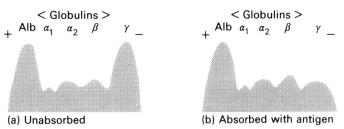

Fig. I.19. Association of antibody activity with γ-globulin serum fraction. Hyperimmune serum is separated into major fractions by electrophoresis before (a) and after (b) absorption with antigen. Only the γ-globulin fraction is reduced. (Redrawn from Roitt, I. M. (1980). *Essential immunology*, 4th edn. Blackwell, Oxford.)

RELATION OF DIET TO PLASMA PROTEINS

This may be studied in the *standard plasma-depleted dog*, as described by Whipple. Whole blood is withdrawn and the corpuscles reinjected suspended in Ringer–Locke solution (i.e. a protein-free fluid); this procedure (plasmapheresis), if repeated daily, leads to a progressive diminution in the concentration of plasma protein, as the rate of protein withdrawal exceeds the rate of regeneration. Depletion is continued for some weeks after the plasma-protein concentration has fallen to 40 g l⁻¹, in order to exhaust the protein reserves. Thereafter, on a standard diet, the rate of plasma-protein formation is constant.

The results show, as might be expected, that plasma proteins are normally formed from food proteins, but that in protein starvation they may be formed from tissue protein. The efficacy of a food protein depends on the degree of its chemical resemblance in amino-acid pattern to the plasma protein which it is going to form; very naturally plasma proteins are the most efficient raw materials. Plasma proteins can also be satisfactorily synthesized from amino acids if the ten essential ones are present, i.e. leucine, isoleucine, methionine, phenylalanine, histidine, arginine, lysine, tryptophan, valine, threonine. As albumin and globulin have distinctly different amino-acid patterns, some proteins (e.g. those from muscle and viscera) favour albumin formation, while others (e.g. plant and grain proteins) favour globulin formation. The presence of infection depresses protein regeneration.

ORIGIN OF PLASMA PROTEINS

Albumin and the proteins concerned in blood clotting (fibrinogen, prothrombin) are probably formed exclusively by the liver. In disease of the liver the concentration of these constituents in the plasma may fall markedly.

Immunoglobulins are formed in lymphoid tissue by the *plasma cells* [p. 57].

When the liver cells are extensively damaged, though the plasma albumin falls, the plasma immunoglobulin fraction frequently rises, probably as a result of plasma cell hyperplasia.

After haemorrhage, fibrinogen, globulin, and albumin are regenerated in that order, complete restoration being effected in a few days.

In nephrosis albumin may be lost in the urine at the rate of 25 g daily for several months before its concentration in the plasma is finally lowered—further evidence that the body can form this substance when necessary on a large scale.

In infancy, low total plasma-protein concentrations are found (e.g. 50–55 g l⁻¹) owing to the low albumin content. Albumin and globulin decrease in the first 6 months of pregnancy, but the fibrinogen increases.

The relationship between plasma proteins and tissue proteins is probably an intimate one. Whipple suggests that the proteins of the cells can be divided into three categories:

1. Fixed cell protein which is indispensable for cell life or activity.

2. Dispensable reserve protein which can be called upon for energy and other purposes in starvation.

3. Labile reserve protein which can be readily turned out into the bloodstream to maintain the plasma-protein concentration.

In haemorrhage or in protein starvation such outflow from the tissues into the plasma takes place. If plasma proteins are given intravenously they can supply all the tissue needs for protein, and food protein can be temporarily dispensed with; this observation suggests that plasma protein can be readily incorporated into the tissues. Proteins taken by mouth are hydrolysed to amino acids in the intestine and then readily built up into plasma or tissue protein. One must conceive of rapid interchanges between the proteins in the liver, plasma, and the tissues generally. As far as is known these interchanges always involve complete protein breakdown to amino acids, followed by synthesis or resynthesis. This view is confirmed by tracer studies using ¹⁵N, ³⁵S, and heavy hydrogen (deuterium) introduced into appropriate amino acids. Upwards of 30 per cent of such labelled amino acids are incorporated into the body proteins within three days, exchanging with the existing unlabelled amino acids, so that the amount and composition of the protein remains unchanged. Such studies indicate that 10 per cent of liver protein and 5 per cent of plasma protein are destroyed and regenerated in three days; the turnover of muscle and skin proteins is about one-quarter that of the liver, with that of other visceral organs intermediate.

REFERENCES

Desgrez, P. and Traverse, P. M. (1966). *Transport function of plasma proteins*. Amsterdam.

Laki, K. (1968). *Fibrinogen*. Dekker, London.

Macfarlane, A. S. (1964). Metabolism of plasma proteins. In *Mammalian protein metabolism* (ed. H. N. Munro and J. B. Allison) Vol. 1, pp. 297–341. New York.

Putnam, F. W. (1965). Structure and function of the plasma proteins. In *The proteins* (ed. H. Neurath) Vol. 3, pp. 153–267. New York.

Sandor, G. and Kawerau, E. (1966). *Serum proteins in health and disease*. London.

Turner, M. W. and Hulme, B. (1971). *The plasma proteins: an introduction*. Pitman, London.

Whipple, G. H. (1956). *The dynamic equilibrium of body proteins*. Thomas, Springfield, Ill.

Haemostasis: blood coagulation

When the wall of a blood vessel is breached blood escapes into the surroundings, provided that the blood pressure within the vessel exceeds the pressure outside it. Natural *haemostasis*, or the spontaneous arrest of bleeding by physiological processes, can be effective in stopping bleeding from small vessels but incised wounds of large arteries cause bleeding which cannot be controlled by natural haemostasis. From a mechanical point of view bleeding from an injured vessel will cease:

1. If the internal and external hydrostatic pressures become equal. This can result from a rise in external pressure, caused for example by accumulation of blood in the surrounding tissues, or from a fall in intravascular pressure due to local vasoconstriction or a fall in general blood pressure.
2. If the hole in the vessel becomes blocked by solid material, e.g. by deposition and aggregation of platelets.

The formation of a solid plug derived from the blood is the most important mechanism in natural haemostasis and if this is defective bleeding is very difficult to control, as in haemophilia. The *haemostatic plug* is formed initially by adherence of platelets to subendothelial collagen; platelet aggregation then occurs and fibrin deposition promotes this process; the fully formed blood clot contains also red cells and leucocytes within the fibrin meshwork. In vessels which are injured but not disrupted, platelet thrombi ('white bodies') form and break down rapidly without fibrin formation. The haemostatic plug is relatively stable, except in a condition such as haemophilia in which fibrin formation is deficient. The platelets and fibrin are equally important for haemostasis (Thomas 1977).

BLOOD COAGULATION

The essential reaction in coagulation of the blood is the conversion of the soluble protein fibrinogen into the insoluble protein fibrin by means of an enzyme, thrombin. Fibrinogen exists in the circulating blood as such; thrombin does not, but is formed from an inactive circulating precursor, prothrombin, when the blood is shed. The activation of prothrombin depends on the presence of Ca^{2+} and of factors which are derived from damaged tissues, disintegrating platelets, and from the plasma itself. The formation of prothrombin (in the liver) depends on the absorption from the bowel of adequate amounts of vitamin K.

Blood is normally fluid while circulating in the blood vessels, but clots when shed. The fluidity of the blood in the body depends on the special physical properties of the intact vascular endothelium, on the rate of blood flow, and on the presence in the blood of natural anticoagulants. Clinical states of excessive bleeding (haemorrhagic states) may be due to:

1. Impaired coagulability of the blood due to some abnormality in the complex physicochemical system concerned in clotting.
2. Alterations in the vessel walls preventing them contracting down after local injury.

Abnormal coagulation of the blood within the blood vessels (intravascular thrombosis) is generally due to alterations in the vascular endothelium combined with slowing of blood flow.

The process of blood clotting has been most frequently studied in glass tubes which activate the clotting mechanisms much more quickly than paraffin, silicone, or polythene surfaces. Negatively charged surfaces promote clotting and other silicates such as kaolin or colloidal silica are even more active than glass, probably because they provide a large surface of contact with blood. Crystals of a normal body constituent, monosodium urate, can also promote surface activation of clotting, and the natural triglyceride of stearic acid acts in a similar way.

Blood-clotting factors

Biochemical techniques of isolation and purification, immunological methods of identification, and specific clinical disorders of blood coagulation have together produced evidence for the existence of thirteen factors concerned in the reactions which lead to blood clotting. The factors are designated by roman numerals according to the recommendations of an International Committee. They will first be dealt with in numerical order, which indicates the historical sequence of their discovery, and then their interactions which promote clotting will be discussed.

Factor I (fibrinogen). Fibrinogen is a soluble plasma protein (MW 330 000) which is acted upon by thrombin to form insoluble fibrin clot. In the absence of fibrinogen (afibrinogenaemia) clotting does not occur. The plasma contains 2.5–4.0 g fibrinogen per litre.

Factor II (prothrombin). This inactive precursor of thrombin is formed in the liver and is decreased in hepatic disease. Its formation depends on the presence of vitamin K. Inert prothrombin (MW 69 000) is converted to the proteolytic enzyme *thrombin* (MW 33 000) by *prothrombin activator*.

Factor III (tissue factor, tissue extract, thromboplastin). This converts prothrombin to thrombin in the presence of factors V, VII, and X, Ca^{2+}, and phospholipid.

Factor IV (calcium). Ionic calcium is essential for clotting. It is required for the formation of prothrombin activator, for conversion of prothrombin to thrombin, and for the formation of insoluble fibrin clot.

Removal of Ca ions by K oxalate, Na citrate, or sodium edetate (EDTA, Versene) prevents clotting *in vitro*.

Factor V (labile factor). Factor V deficiency produces a rare haemorrhagic state first described by Owren (1947). This factor is required for conversion of prothrombin to thrombin by tissue extract and plasma factors. Factor V is consumed during clotting and is therefore absent from serum.

Factor VII (stable factor, autoprothrombin I). Factor VII deficiency is a rare natural occurrence but is frequently induced by oral anticoagulant drugs of the coumarin type, overdosage of which may lead to bleeding. Factor VII is required for the formation of prothrombin activator by tissue extract. It is not consumed during clotting and therefore is present in serum as well as plasma.

Factor VIII (antihaemophilic globulin [AHG], antihaemophilic factor). Classical haemophilia is due to the congenital absence of this factor. Factor VIII is required for the formation of prothrombin activator from blood constituents; it is consumed during clotting and is therefore absent from serum. *In vivo* the half life of factor VIII is 10–20 h.

Factor IX (Christmas factor, plasma thromboplastin component, autoprothrombin II). Lack of factor IX is associated with a congenital haemorrhagic state resembling haemophilia (Christmas disease). It is needed for the formation of prothrombin activator from blood constituents. Factor IX occurs in plasma and it is activated during clotting so that the activity in serum is much greater than in plasma.

Factor X (Stuart–Prower factor). Congenital absence of factor X produces a haemorrhagic state. Factor X is present in plasma and serum.

Factor XI (plasma thromboplastin antecedent). Congenital absence of factor XI causes a haemorrhagic state. It is required for formation of prothrombin activator from blood constituents and occurs in plasma and serum.

Factor XII (Hageman factor, contact factor). In factor-XII deficiency blood clots very slowly in glass but there is no haemorrhagic state. Factor XII is activated by glass and similar surfaces and it takes part in the formation of prothrombin activator from blood constituents. It is present in plasma and serum.

Factor XIII (fibrin-stabilizing factor). This is a plasma protein which causes polymerization of soluble fibrin to produce insoluble fibrin. Hereditary deficiency of factor XIII leads to a haemorrhagic state.

Platelets contain phospholipids which are essential for clotting in the absence of tissue extract.

Physiology of clotting process

Thrombin–fibrinogen reaction. The formation of fibrin clot is the only visible and measurable part of the clotting process and our knowledge of the earlier stages is ultimately based on their effects on the speed of conversion of fibrinogen to fibrin. This last stage is the best understood biochemically. Thrombin, acting as an enzyme, disrupts arginylglycine bonds in fibrinogen to split off two fibrino-peptides A and B. The changes may be represented diagrammatically as follows:

Type of Reaction

1. Proteolysis. Fibrinogen $\xrightarrow{\text{Thrombin}}$ Fibrin monomer + peptides
2. Polymerization. Fibrin monomer → Fibrin polymer (soluble fibrin clot)
3. Clotting. Fibrin polymer $\xrightarrow{\text{Factor XIII}}$ Insoluble fibrin clot.

Factor XIII acts as a transamidase in the presence of Ca ions.

Conversion of prothrombin to thrombin. There is no circulating thrombin in normal blood but its precursor prothrombin, an α_2-globulin, is present in plasma in a concentration of about 100 mg l^{-1}. Prothrombin is converted to thrombin by means of *prothrombin activator*. This is also called thromboplastin.

Prothrombin activator. This is formed in two main ways, one as the result of tissue damage (extrinsic system), the other by activation of an intrinsic system consisting entirely of blood constituents. In both cases the key reaction is the conversion of factor X to its active form Xa (a = active). Factor Xa then interacts with factor V and phospholipid to form prothrombin activator. In this stage factor V acts as a co-factor and phospholipid provides a surface on which the reagents are concentrated. We now have to consider how factor X is converted to Xa.

In the *extrinsic* system the reactions are as follows:

Factor X $\xrightarrow[\text{Ca ions}]{\begin{array}{c}\text{Tissue extract}\\+\\\text{Factor VII}\\+\end{array}}$ Xa $\xrightarrow[\text{Phospholipid}]{\text{Factor V}}$ Prothrombin activator

Russell's viper venom in the presence of Ca ions activates factor X but does not require factor VII.

In the *intrinsic* system the reactions leading to the formation of Xa are more complicated and take several minutes for completion. The surface-mediated reactions which activate factor XII to initiate clotting also activate the fibrinolytic and kinin-forming systems [p. 25]. Macfarlane (1964) has proposed an ingenious hypothesis to explain the sequence of reactions involved in blood clotting.

Enzyme–cascade hypothesis. Macfarlane suggests that surface contact induces a sequence of changes in which an inactive precursor is converted to an active enzyme which then acts on the next precursor to form the next active enzyme and so on as set out in the accompanying diagram:

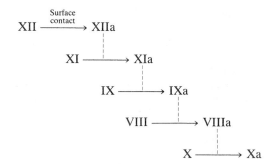

Thus the clotting mechanism of the intrinsic system is regarded as a series a series of enzyme–substrate reactions leading to the formation of Xa which then acts in the same way as described for the extrinsic system. Once Xa is formed clotting takes place within a few seconds, but contact with a foreign surface, e.g. glass or urate crystals, produces clotting only after 4–8 minutes. During this time the above reactions are presumed to be taking place. One very important aspect of the enzyme cascade is that it is an amplifying system so that minute amounts of the earlier factors can lead to final rapid conversion of large amounts of fibrinogen to fibrin. For example, Esnouf and Macfarlane (1968) quote the following figures:

The clotting of 1 ml of blood involves 3 mg of fibrinogen, 100 µg of factor II, 10 µg of factor X, 0.1 µg of factor VIII, and probably less of the earlier factors. Moreover, to be effective in stopping bleeding fibrin formation, once it begins, must be completed suddenly. In many cases of haemophilia fibrin formation begins within the normal time but the subsequent rate of formation is slow and haemostasis ineffective. In normal blood the first appearance of fibrin may not occur for some minutes, but once started the process is complete within seconds. This is due to the ultimate high speed of formation of thrombin.

A complete diagrammatic representation of the extrinsic and intrinsic reactions which promote clotting is shown in FIG. I.20.

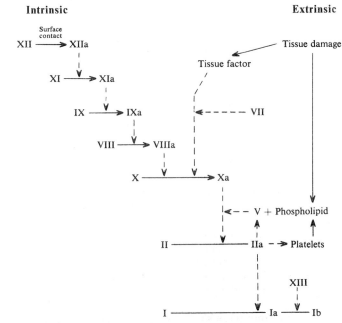

FIG. I.20. Scheme showing the sequence of changes which lead to blood clotting by activation of the intrinsic and extrinsic systems.
II = prothrombin; IIa = thrombin
I = fibrinogen;   Ia = fibrin;   Ib = stabilized fibrin.
———→ Transformation
- - - → Action
(Macfarlane, R. G. (1967). *Br. J. Haemat.* **13**, 437.)

Two comments on this scheme must be made. First, it is an oversimplification of an extremely complicated situation. It has not yet been shown that all the reactions in the cascade are enzyme–substrate in type.

Secondly, there are additional reactions not shown in Fig. I.20. For example, thrombin exerts positive feedback effects on factors concerned in the formation of prothrombin activator; it 'activates' factor VIII, perhaps by releasing it from its attachment to fibrinogen; it also increases the activity of factors V and XIII; it promotes aggregation of platelets and thus increases the amounts of available phospholipid. However, after activating these mechanisms thrombin soon inactivates them. Thus thrombin will intensify the reactions promoting local clotting, but only for a short time, i.e. until a firm clot has been formed. The spread of clot formation from a site of injury does not normally occur, partly because of the reactions just mentioned which cut it short, and partly because blood contains inhibitors of probably all the factors known to be concerned with clotting. Occasionally these restraining processes fail and widespread intravascular coagulation occurs.

Seegers (1967) has proposed an apparently different scheme for conversion of prothrombin to thrombin. The postulated sequence of events is set out below:

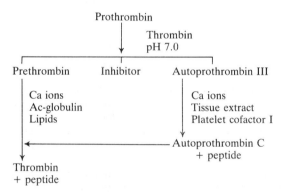

Thus prothrombin yields three derivatives before thrombin is formed from it. However, Esnouf and Macfarlane (1968) point out that autoprothrombin C is probably identical with Xa. Prethrombin may be identical with factor II of other workers, and Seegers' 'prothrombin' is probably a complex protein containing subunits equivalent to factors VII, IX, X, and II of the sequential hypothesis.

Tests for defects in blood clotting

With increasing knowledge of the factors involved in blood clotting tests have been devised to demonstrate various defects in the process. The simple measurement of clotting time is the basis from which the complex tests have been developed.

Clotting time. This is measured by delivering 1 ml venous blood into a small test tube, rocking gently in a water bath at 37 °C, and noting the time when the blood stops flowing. The normal clotting time is 6–12 minutes in glass and 20–60 minutes in siliconed tubes. Prolongation of clotting time occurs with marked deficiencies of the intrinsic prothrombin activator (thromboplastin) system, i.e. factors VIII, IX, X, XI, and XII, as well as deficiencies of factor V, prothrombin, and fibrinogen.

Prothrombin time (Quick's one-stage test). Oxalate plasma is prepared and more than enough ionized calcium salt + tissue thromboplastin (powdered dried brain) are added to produce optimum conditions for activation of prothrombin to thrombin. It is assumed that the concentration of fibrinogen is normal. The normal prothrombin clotting time is 11–16 s. A glance at Fig. I.20 will show that this test measures not only the concentration of prothrombin

but also the concentrations of factors V, VII, and X, all of which participate in the extrinsic activation of prothrombin. In clinical use the prothrombin time of a patient's plasma is always compared with that of a normal plasma. The prothrombin time is normal in haemophilia and Christmas disease since tissue factor does not require factors VIII and IX for activation of prothrombin; it is prolonged with deficiencies of prothrombin and factors V, VII, and X, such as occur with deficiency of vitamin K. Anticoagulant therapy with vitamin-K antagonists (e.g. warfarin) is controlled by measurement of prothrombin time which should be prolonged to about twice the normal value, i.e. to 25–30 s, to help prevent intravascular thrombosis.

Thromboplastin (prothrombin activator) generation test. This is a two-step method for obtaining fuller information about the production of thromboplastin (prothrombin activator) by the intrinsic pathway. The process requires normal concentrations of platelet phospholipid and factors V, VIII, IX, X, XI, and XII. These factors normally occur in a platelet suspension + serum (factors IX, X, XI, and XII) + Al(OH)$_3$-absorbed plasma (factors V, VII, XI, and XII). A mixture of these three preparations with Ca^{2+} sets thromboplastin formation going and the amount formed after varying intervals of time can be estimated by the ability to clot normal or substrate plasma. This second step is analogous to the one-stage prothrombin test except that the thromboplastin is intrinsic and not extrinsic in origin.

Intrinsic thromboplastin formation is defective in thrombocytopenia, haemophilia, Christmas disease, and in hypoprothrombinaemia.

REFERENCES

BIGGS, R. (ed.) (1976). *Human blood coagulation, haemostasis and thrombosis*, 2nd edn. Blackwell, Oxford.
THOMAS, D. (ed.) (1977). Haemostasis. *Br. med. Bull.* **33**, No. 3.
—— (ed.) (1978). Thrombosis. *Br. med. Bull.* **34**, No. 2.

Fibrin and fibrinolysis

As blood clots, fibrin is laid down as a network of fine threads which entangle the blood cells. The freshly formed threads are extremely adhesive, sticking to each other, to the blood cells, to the tissues, and to certain foreign surfaces; this adhesiveness makes the clot an effective haemostatic agent. Freshly shed blood sets in a soft jelly-like mass; gradually this clot contracts down (retracts) to about 40 per cent of its original volume, squeezing out serum (*serum = plasma minus fibrinogen and other factors used up in clotting*). The final clot is tougher and more solid and elastic, and is presumably a more efficient bung for damaged vessels. 'Clot retraction' is impaired if the platelets have been artificially removed, and also in disease conditions which have a low platelet count, although the coagulation time is not prolonged. Clots formed in the tissues have ultimately to be disposed of as healing takes place; the dissolution of the clot—fibrinolysis—is due to the action of the proteolytic enzyme called *fibrinolysin* or *plasmin*.

The process of fibrinolysis is clearly opposed to that of blood clotting and since both processes are activated by injury to blood and tissues it is important to compare the components of the fibrinolytic system with those of the blood-clotting system.

In the clotting system there are elaborate processes involved in the conversion of inactive prothrombin to the active enzyme thrombin which immediately changes fibrinogen to fibrin. In the fibrinolytic system too there is no free plasmin but blood plasma contains its inactive precursor *plasminogen*. Plasminogen is converted to plasmin by means of plasminogen activators which may be *intrinsic* or *extrinsic*.

1. Plasminogen $\xrightarrow{\text{Extrinsic or intrinsic plasminogen activator}}$ Plasmin

2. Fibrin $\xrightarrow{\text{Plasmin}}$ Small peptides (fibrinogen degradation products)

If blood from a healthy person is incubated the clot remains solid for weeks. Various forms of bodily or mental stress, such as surgical operations, violent exercise, or injection of adrenaline promote fibrinolytic activity with dissolution of the clot within a few hours, through activation of the intrinsic plasminogen activator in plasma. In people who die a violent sudden death the blood is fluid and incoagulable as the result of fibrinolysis. Intrinsic plasminogen activator becomes absorbed, together with plasminogen, on to fibrin clot so that when the system is activated the proteolytic activity of the plasmin will be wholly exerted on the clot itself.

Extrinsic activators of plasminogen are widely distributed throughout the cells and fluids of the body. The tissue activators probably occur in the microsomes. Urine contains a plasminogen activator called *urokinase*.

By analogy with the clotting system, there are inhibitors in plasma which can inhibit plasmin (antiplasmin) or prevent the activation of plasminogen. Fibrinolytic activity can be influenced by drugs and hormones. Adrenal corticosteroids and the antidiabetic drug phenformin enhance fibrinolytic activity, whereas ε-aminocaproic acid (EACA) and a trypsin inhibitor called aprotinin (*Trasylol*) inhibit it.

Physiological role of fibrinolysis. It has been suggested that in physiological conditions the clotting system of plasma is continually forming small amounts of fibrin which are deposited to form a thin layer on vascular endothelium, and that the fibrinolytic system is constantly in action to prevent excessive fibrin formation. If the clotting system predominates intravascular thrombosis tends to occur; if fibrinolysis predominates there might be a tendency to bleeding. All this is speculative but if free plasmin circulates in large amounts in the blood it can cause bleeding by its capacity to digest several clotting factors, and it also interferes with fibrin polymerization.

Tissue injury promotes blood clotting by release of tissue factor, and fibrinolysis by release of tissue plasminogen activator. Removal of fibrin may be part of the normal process of healing. Fibrin formation occurs quickly to promote haemostasis; fibrinolysis is a much slower process.

In addition to its fibrinolytic activity plasmin can form plasma kinins (bradykinin, kallidin) and thus contribute to the vascular and sensory features (pain) of the inflammatory response to injury [p. 63].

REFERENCE

FEARNLEY, G. R. (1969). Fibrinolysis. In *Recent advances in blood coagulation* (ed. L. Poller) p. 229. London.

INTRAVASCULAR THROMBOSIS

A *thrombus* is a solid mass formed in the living heart or blood vessels from constituents of the blood. The process called *thrombosis* is distinguished from extravascular clotting or clotting in wounds, and also from clotting which occurs in blood vessels after death.

Thrombosis always begins by deposition on the vascular endothelium of masses of platelets which grow by adhesion of other platelets as they flow by; the laminae of platelets (which fuse together and lose their identity) stand out as layers running transversely to the bloodstream; passing leucocytes adhere to their borders ('like flies on sheets of sticky flypaper'). The platelets liberate thromboplastins 'so that filaments of fibrin spread out from them on all sides and, meeting with filaments from the next lamella, hang in festoons between them'. The lamellae of platelets thus 'braced together by fibrin' entangle masses of red cells so that finally a solid mass of peculiarly constructed 'clot'—in fact, the 'thrombus'—is formed (McCallum). The red cells disintegrate and lose their haemoglobin; the initial red thrombus as it ages becomes yellowish grey; but newly formed thrombi added to it will be red.

Thrombi form most readily where there is local damage to the vascular endothelium, and slowing down of the bloodstream, e.g. in small leg veins, on atheromatous patches in small arteries, on damaged valves in the heart, or in the atrial appendage in atrial fibrillation. It is surprising, however, to find thrombi forming on the damaged wall of the aorta where 'it might seem that the pulsating torrent of blood would allow no chance for the deposition of pioneer platelets'.

Postoperative thrombosis

After surgical operations (especially those involving the abdomen) or childbirth, and in patients confined to bed for long periods, thrombosis may occur in leg veins; the condition is called postoperative thrombosis, decubitus thrombosis (because the recumbent position is an important causal factor), or thrombophlebitis (because the thrombosis is accompanied by inflammatory changes in the vein wall). The thrombus frequently arises in valve-cusp pockets in the deep veins of the calf. Thrombosis can be detected as follows: ^{125}I-labelled fibrinogen is injected intravenously. At a site of thrombosis the deposition of fibrin can be recorded by a local accumulation of ^{125}I detected by a suitable counter [FIG. I.21]. This method has shown that thrombosis may actually occur in the calf muscles during the course of an operation, but clinically thrombosis may not be diagnosable for several days, and in many cases not at all. The thrombosis which begins in the veins of the calf muscles may spread upwards, in some instances ascending to the popliteal, femoral, or even the iliac veins. In a small proportion of cases fragments of clot become detached (embolus) and are carried to the pulmonary arteries where they produce pulmonary infarction. This type of pulmonary embolism is a well-recognized cause of sudden death in the postoperative period.

Mechanism

This condition is due partly to changes in the circulation and in the properties of the blood and partly to local injuries to the leg veins.

1. The circulation in the veins of the legs and trunk (but not in the arms) is considerably slowed down after operations. The normal venous return depends on muscular contraction and respiratory movements; after abdominal operations the legs are moved very little, and the movements of the diaphragm may be hampered by a tight abdominal bandage or by flatus, or inhibited by the pain of the abdominal incision. Thrombosis is much rarer after operations on the upper part of the body.

2. There are changes in the composition of the blood owing to the general tissue response to the trauma: (i) the plasma fibrinogen concentration is raised; this may increase rouleaux formation of red cells. (ii) The platelet count is raised; the platelets also become more 'sticky' and so more liable to adhere to the lining of the blood vessels. There is a direct relationship between the extent of these blood changes and the incidence of intravascular thrombosis.

3. Sepsis may be a factor; thus most strains of *Staphylococcus aureus* produce a toxin which rapidly clots human blood.

4. The calf veins may be damaged owing to the limbs lying limply on the operating table or in bed; thromboplastins are liberated locally promoting thrombin formation.

Prevention

Postoperative venous thrombosis may be prevented in the following ways: (1) intermittent compression or electrical stimulation of

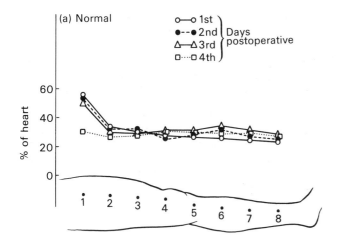

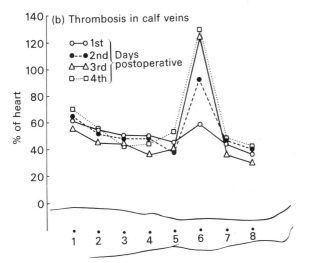

Fig. I.21. Postoperative ¹²⁵I-fibrinogen studies in (a) a patient who recovered without complications, and (b) a patient who developed a deep venous thrombosis, probably in soleal sinusoids. In both patients the percentage uptake of ¹²⁵I, compared with the precordial count, was measured at eight points in the lower limb, as shown diagrammatically. In (a) the percentage ¹²⁵I uptake is constant from point to point during the first four days after the operation (point number 1 is higher than the others owing to the proximity of the pelvic vessels). In (b) point number 6 shows increased uptake of ¹²⁵I, detectable on the first postoperative day and reaching a peak on the third day after the operation. There is no evidence of spread of the thrombosis to other points of the limb. (Reproduced by kind permission of Professor Leslie Le Quesne of the Middlesex Hospital, London, and of the Editor of the *British Journal of Surgery*.)

the calf muscles during surgical operations prevents stagnation of venous blood; (ii) administration of dextran, aspirin, or dipyridamole decreases platelet adhesiveness; (iii) low doses of heparin or ancrod reduce the tendency of the blood to clot.

If deep venous thrombosis has occurred, the clot may be dissolved (thrombolysis) by promoting fibrinolysis with streptokinase; this is only effective if given within 72 h of thrombus formation. Anticoagulant drugs prevent spread of the thrombosis.

REFERENCES

Browse, N. L. (1974). *Br. med. J.* **iv**, 96.
Poller, L. (ed.) (1973). *Recent advances in thrombosis*. Churchill Livingstone, London.

Vitamin K and blood clotting

Vitamin K is a complex naphthoquinone derivative which is required for the synthesis of prothrombin and factors VII, IX, and X

in the liver. In the absence of the vitamin the blood-clotting time is prolonged and serious haemorrhages may occur.

REFERENCE

Martius, C. (1967). Chemistry and function of vitamin K. In *Blood clotting enzymology* (ed. W. H. Seegers). New York.

Distribution and chemistry

Vitamin K is widely distributed in nature; thus it is found in traces in green vegetables, cereals, and animal tissues generally. It can be synthesized by many bacteria including those normally present in the human intestine (e.g. *E. coli*); the bacterial flora probably provide an adequate supply of vitamin K in man. Vitamin-K deficiency from dietary abnormalities is very rare in man.

The formula of 1:4 naphthoquinone and the method of numbering the positions in the ring are indicated below:

1:4 naphthoquinone

There are at least two fractions present in natural vitamin K, of which K_1, 2-methyl-3-phytyl-1:4 naphthoquinone is the more active. It is soluble in fat solvents but insoluble in water; it can therefore only be given by mouth or intramuscularly, unless it is specially prepared for intravenous administration by being very finely emulsified.

Absorption in intestine. Absorption of vitamin K from the small intestine only occurs in the presence of adequate amounts of bile salts (which are also necessary for the absorption of other fat-soluble substances like vitamin D and the fats of the food).

The vitamin is absorbed into the lacteals and passes into the thoracic duct; on reaching the liver it participates in the processes leading to the synthesis of prothrombin and factor VII; the vitamin is *not* incorporated in the prothrombin molecule which is a protein containing no naphthoquinone. If bile is excluded from the bowel, vitamin-K absorption does not occur; if the liver is damaged or extirpated, prothrombin formation is decreased or stops altogether.

Clinical conditions associated with vitamin-K deficiency

Obstructive jaundice. In this condition bile is excluded from the bowel owing to obstruction somewhere in the biliary passages. The absence of bile salts from the bowel prevents vitamin-K absorption with the result that prothrombin and factor VII are decreased, and the prothrombin time [p. 25] is prolonged. In severe cases the clotting time may also be prolonged and haemorrhages may occur. In untreated cases operative procedures may involve death from uncontrollable bleeding. The state of the blood may be restored to normal by giving bile salts alone, bile salts plus vitamin K by mouth, or more rapidly by injecting vitamin K intramuscularly.

Chronic diarrhoea. The syndrome of vitamin-K deficiency has been reported in chronic diarrhoea. The cause may be failure of absorption as a result of the altered state of the intestinal wall and the associated diarrhoea, or an abnormal state of the intestinal bacteria leading to defective synthesis of the vitamin. The syndrome may also occur with defective fat absorption in sprue.

Liver disease. Severe liver damage may abolish prothrombin formation: vitamin-K administration then has no effect, even when injected. This lack of prothrombin response has been used as a test of liver efficiency [p. 443].

Haemorrhagic states in infants. Newborn babies commonly have plasma prothrombin concentrations which are as low as one-third to one-sixth of normal; the deficiency is even more marked in premature babies. Usually the plasma prothrombin returns to normal during the second week after birth. A grave haemorrhagic state has been described in infants in whom for some unexplained reason the plasma prothrombin falls to less than 1 per cent of normal. If vitamin K is given complete recovery of the blood may occur in less than 48 hours.

REFERENCES

BIGGS, R. (ed.) (1976). *Human blood coagulation, haemostasis and thrombosis*, 2nd edn. Blackwell, Oxford.
THOMAS, D. (ed.) (1977). Haemostasis. *Br. med. Bull.* **33**, No. 3.
—— (ed.) (1978). Thrombosis. *Br. med. Bull.* **34**, No. 2.

Bleeding due to defects in coagulation

Since clotting plays an important role in haemostasis, marked deficiencies of the blood-clotting factors may lead to bleeding. Indeed, the occurrence of such bleeding provides strong evidence for the separate existence of each of these various factors. Thus haemorrhagic disorders may be associated with lack of fibrinogen (factor 1); prothrombin (factor II); and factors V, VII, VIII, IX, X, XI, and XIII. The concentration of calcium (factor IV) would have to fall to below 1.25 mmol l^{-1} (5 mg per 100 ml) to delay clotting and this would produce profound tetany. Factor-XII deficiency does not produce bleeding.

Significant deficiencies of blood-clotting factors can arise from genetic defects in their formation. To prevent the tendency to intravascular thrombosis drugs can be given to antagonize the actions of clotting factors or to reduce their concentration in the blood; in this situation the aim is primarily to inhibit thrombosis, but overdosage will produce bleeding.

Haemophilia

Haemophilia is an inherited sex-linked anomaly invariably transmitted by females, who themselves show no symptoms, to males who manifest signs of the disease [FIG I.22]. The condition is characterized by a marked increase in the coagulation time. The bleeding time is not prolonged because minute breaches in the skin are sealed by contraction of the capillaries. Blood should be collected for examination by venepuncture; normal blood under these conditions clots in 5–10 minutes, while haemophilic blood may take from 1–12 hours. Severe bleeding occurs after injuries of any kind, even of the most trivial character.

The abnormality responsible for haemophilia is deficiency of factor VIII, and the closely related Christmas disease is due to lack of factor IX. These factors can be prepared in concentrated form from deep frozen fresh plasma (cryoprecipitates). They are of particular value in preventing bleeding after dental extractions. Factor VIII administration can prevent the crippling complications of bleeding into muscles and joints of haemophilic patients.

REFERENCES

BIGGS, R. (1974). *Lancet* **i**, 1339.
—— and MACFARLANE, R. G. (1966). *Treatment of haemophilia and other coagulation disorders*. Blackwell, Oxford.

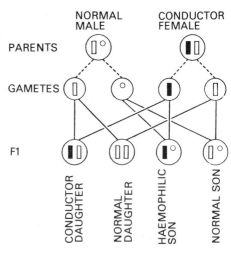

FIG. I.22. Transmission of haemophilia by conductor female to haemophilic son and conductor daughter. □ = normal X chromosome. ■ = X chromosome responsible for haemophilia. O = Y chromosome. The sex chromosomes in the *female* are similar in appearance and are labelled X and X; in the male the sex chromosomes are X, combined with a minute and neglible chromosome labelled Y. The gene responsible for haemophilia is present in the X chromosome; in the presence of another normal X chromosome the gene acts as a *recessive*, i.e. the individual has no signs of haemophilia (but can transmit the disease); certain constituents of the normal X chromosome may be responsible for this. When the ovum of a conductor female undergoes its reduction division, the two resulting cells differ: one contains a normal X chromosome □, the other a haemophilic X chromosome ■. At fertilization, if X of the ovum unites with X of a sperm, the offspring is a female; if X of the ovum unites with Y of a sperm the result is a male. If a haemophilic X ■ (of a female) unites with a Y (of a male) the haemophilic gene on X ■ is *unantagonized* by a normal X □ and the resultant is a *haemophilic son*. If a haemophilic X ■ unites with a normal X the result is a conductor daughter (who can transmit the disease but has no signs of it). Offspring not containing haemophilic X ■ are normal and their progeny are normal.

Anticoagulant drugs

In vitro blood clotting can be prevented by substances which sequester calcium, e.g. sodium citrate or oxalate, sodium edetate (EDTA).

In vivo the tendency to thrombosis can be inhibited by antagonizing clotting factors, by destruction of the key substance fibrinogen, or by inhibiting the synthesis of factors II, VII, IX, and X.

Heparin

Heparin was first isolated from the liver (hence its name) and shown to be a powerful anticoagulant substance; it was subsequently demonstrated in extracts of many other organs, e.g. lungs. Heparin is probably normally secreted by a scattered widely distributed system of connective tissue cells called the *mast cells*; it may help to maintain the normal fluidity of the blood.

Heparin inhibits blood coagulation both *in vitro* and *in vivo*; it acts slightly by preventing the activation of prothrombin to thrombin but mainly by neutralizing the action of thrombin on fibrinogen. It combines with a plasma co-factor antithrombin III (At III) and the conjugate is a more potent thrombin inhibitor than At III itself. It prolongs clotting time but has little effect on bleeding time. Heparin also activates lipoprotein lipase and thus promotes clearing of lipaemic plasma *in vivo*. The physiological role of heparin is not established. Chemically heparin is a polysaccharide derived from glucosamine (amino-glucose) and glucuronic acid, and containing many sulphate groups; it is thus related to mucoitin sulphate (of mucus) and chondroitin (of cartilage). Heparin owes its anticoagulant action to its strong electronegative charge, due to the sulphuric acid groups. Electropositive substances, such as toluidine blue and protamine, neutralize the negative charge of heparin and completely antagonize its anticoagulant action.

Certain dyes containing the grouping =NH change colour when added to a heparin solution in a test tube; this property is called metachromasia; toluidine blue, for example, changes in colour to purple in the presence of heparin. (The reaction is specific for sulphuric acid esters ($R-OSO_3H$) and their salts if they have a high molecular weight.) Heparin is quantitatively precipitated from watery solution by toluidine blue and thus rendered inert; this reaction has been developed as a quantitative test for heparin.

Mast cells

These cells, first described by Ehrlich, are widely distributed in many organs of all species from fish to man. They are found singly or in clumps; characteristically, they are arranged in close proximity to the walls of small blood vessels, and may even replace the lining endothelium. The cells contain numerous heparin granules which give a typical metachromatic purple reaction with toluidine blue. The heparin may be combined with the histamine which is also present in these cells.

1. There is a correlation between the number of mast cells in a tissue, its SO_4^{2-} content, and the amount of heparin that can be extracted from it. Thus the liver of sheep and oxen contains many mast cells and yields much heparin; the reverse is the case with the liver of the rat. On the other hand, the subcutaneous tissue in the rat is rich both in mast cells and heparin content. Considerable amounts of heparin are found in the walls of the large vessels, e.g. aorta and vena cava, where, too, the yield is related to the number of mast cells.

2. In conditions in which the heparin content of the blood is raised (e.g. anaphylactic shock) the mast cells show loss of both granules and of metachromatic reaction.

3. Metachromatic extracellular substance is also found under the intima and especially in the media of the aorta round the elastic fibres; it is also present in the substantia propria of the cornea. These findings, too, are related to a high heparin content.

Anaphylactic shock

Anaphylactic shock in dogs produces a profound fall of blood pressure, due to release of histamine, and incoagulability of the blood owing to an increase in heparin content; the mast cells appear exhausted from loss of granules. The blood platelet count falls.

Normal role of anticoagulants

Several antithrombins, including At III, occur in plasma and three of these play a physiological role (Lane and Biggs 1977).

Use of heparin

Heparin can be used to keep blood fluid *in vitro*. It can also be injected into patients to prevent the development or spread of venous thrombosis. Its action begins with a few minutes and lasts for hours.

REFERENCES

BARROWCLIFFE, T. W., JOHNSON, E. A., and THOMAS, D. (1978). *Br. med. Bull.* **34**, 143.
LANE, J. L. and BIGGS, R. (1977). In *Recent advances in blood coagulation*, Vol. 2 (ed. L. Poller) p. 123. Churchill Livingstone, Edinburgh.
RILEY, J. F. (1959). *The mast cells*. Churchill Livingstone, Edinburgh.

Therapeutic defibrination

Systemic poisoning by the venom of the Malaysian pit viper (*Agkistrodon rhodostoma*) is characterized by incoagulable blood, though spontaneous bleeding is trivial and there are no toxic effects on the red cells, liver, or nervous system. *In vitro* the venom has a direct coagulant effect on fibrinogen, probably by formation of an imperfect fibrin polymer. *In vivo* this polymer is diffusely distributed throughout the circulation and deposited on vascular endothelium where local activation of plasminogen to plasmin produces fibrinolysis. Thus Malaysian pit viper venom renders blood incoagulable by defibrination and a purified preparation *ancrod* (arvin) has been used therapeutically. Ancrod is a glycoprotein and is administered by injection. Large doses of ancrod are liable to produce haemorrhage, probably by excessive depletion of fibrinogen, but administration of antivenom quickly stops the action of ancrod.

Fibrinogenopenia can occur as a *congenital defect* and also occasionally in *pregnancy*, probably due to embolism of small tissue fragments (from prematurely separated placenta) rich in tissue activator which lead to diffuse foci of fibrin clots. This reduces blood fibrinogen level in the same way as ancrod and if the reduction is sufficiently severe haemorrhages will occur.

REFERENCES

REID, H. A. and CHAN, K. E. (1968). *Lancet* **i**, 487.

Vitamin K antagonists

Dicoumarol. This substance is a coumarin derivative related chemically to the naphthoquinone derivatives with vitamin K activity. Because of this chemical resemblance, dicoumarol acts as an anti-vitamin K by the process of substrate competition. It is thought that in the liver dicoumarol replaces vitamin K at the latter's normal site of action and thus prevents the vitamin from carrying out its normal physiological function. As dicoumarol cannot be utilized by the liver in the synthesis of prothrombin, the syndrome of vitamin K deficiency develops, i.e. the plasma prothrombin level fails, factors VII, IX, and X activity also decreases and blood coagulability is depressed. In this way dicoumarol act as an anticoagulant, but only *in vivo*. It is active by mouth and one dose may produce an effect lasting for several days. Dicoumarol has been used clinically, e.g. in venous thrombosis and coronary thrombosis.

Other anticoagulants which act as vitamin K antagonists include phenindione, warfarin, and nicoumalone, which differ from dicoumarol chiefly in their speed of onset and duration of action.

REFERENCES

DOUGLAS, A. S. (1962). *Anticoagulant therapy*. London.
INGRAM, G. I. C. (1961). *Pharmac. Rev.* **13**, 279.
MACKIE, M. J. and DOUGLAS, A. S. (1978). *Br. med. Bull.* **34**, 177.

BLOOD PLATELETS

Origin and structure

The bone marrow contains giant cells called *megakaryocytes* with a diameter of 35–160 μm. These are formed from bone-marrow stem cells via megakaryoblasts. Megakaryocytes contain an irregular ring of lobed nuclei. Blood platelets are formed within the cytoplasm of the granular megakaryocyte and are released into the circulation when the cell dies. Electron microscopy shows that platelet formation begins by the formation of microvesicles which coalesce to form a demarcation membrane for the platelets. Platelets are colourless, spherical, oval, or rod-shaped bodies 2–4 μm in diameter. Microcinematography under phase-contrast microscopy shows contractile vacuoles and vacuoles of pinocytosis. Leishman's stain shows a faint blue cytoplasm with distinct reddish-purple granules.

Electron microscopy reveals a cell membrane 6 nm thick which surrounds a cytoplasmic matrix containing Golgi apparatus, endoplasmic reticulum, 50–100 very dense granules, a few small mitochondria, microvesicles, microtubules, filaments, and granules with clear interiors. The platelet plasma-membrane consists of phospholipids, proteins, and, on the surface, glycoproteins in which there are receptors for ADP and thrombin.

Chemistry and metabolism

The platelet possesses numerous enzymes, is capable of a considerable expenditure of energy, and contains an actomyosin-like contractile protein, thrombosthenin, which is responsible for clot retraction.

The platelet contains ATP, ADP, histamine, adrenalin, and 5-hydroxytryptamine; clot-promoting phospholipids occur in the granules and in the cell membrane (platelet factors 1, 2, 3, and 4).

Platelets also contain the adenyl cyclase–cAMP and prostaglandin-forming systems.

Platelet functions

The functions of the normal platelet are concerned with *haemostasis, blood coagulation, phagocytosis*, and *storage* and *transport of substances*.

Haemostasis (arrest of bleeding). Perhaps the most important function of the platelets is the formation of the haemostatic plug to maintain the integrity of the vascular tree. Two properties of platelets help in the formation of plugs—*adhesiveness* to the damaged lining of blood vessels and *aggregation* of platelets so that they stick to each other [Fig. I.23].

Minor injury to blood vessels causes platelets to adhere to the exposed collagen in areas denuded of endothelium. Platelets adhere to other damaged cells and to foreign surfaces such as glass. The adhesiveness or stickiness of platelets is promoted by Ca ions and ADP.

Aggregation of platelets leads to the formation of 'white bodies' or microthrombi which may grow until they almost fill the lumen of a small vessel. At first the aggregation is reversible and the

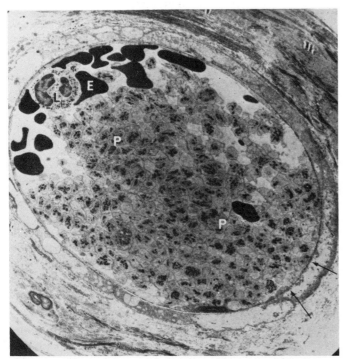

Fig. I.23. Electron micrograph of a section showing a mass of platelets (P) in an injured artery in the cheek pouch of a hamster is attached to the wall at the point (arrows) where the endothelium (E) has been destroyed by a microelectrode. (French, J. E. (1967). *Br. J. Haemat.* **13**, 595.)

platelets may be freed from each other by the circulating blood, only for fresh platelets to aggregate once more. If the vascular damage is more severe the aggregation becomes irreversible, platelet granules are discharged, leucocytes begin to adhere to the degranulated platelets, and fibrin is deposited. It must be emphasized that platelet adhesion and aggregation can occur without any evidence for activation of clot-promoting mechanisms. Some process other than fibrin formation must account for these phenomena.

Born and his colleagues (1963) have shown that platelet aggregation produced by ADP in concentrations to 10^{-7} mol l^{-1}; adrenalin, noradrenalin, and 5 HT are also active. The suggested sequence of events is as follows: platelets adhere to collagen in a damaged vessel wall; in the sticking platelets ATP is converted to ADP by ATPase; ADP is released and promotes aggregation of passing platelets which form a haemostatic plug in small blood vessels. Release of tissue factor from the damaged vessel and phospholipid from platelets will promote thrombin formation (which also causes platelet aggregation) and a firm clot will seal the vessel permanently. The actions of ADP in causing platelet aggregation are antagonized by ATP, AMP, and adenosine.

Although the arrest of haemorrhage due to vascular injury is effected mainly by the formation of platelet aggregates, other reactions contribute to haemostasis. After initial dilatation, damaged small blood vessels constrict for about 20 min, 5 HT perhaps being concerned in this process. During this period blood clotting occurs and when the capillary circulation returns to normal a firm clot has usually formed and prevents further bleeding.

In addition to ADP the prostaglandin-forming system is involved in platelet aggregation. Arachidonic acid is released from the platelet cell membrane by phospholipase and then rapidly oxidized by the enzyme cyclo-oxygenase (also called prostaglandin synthetase) to unstable cyclic endoperoxides PGG$_2$ and PGH$_2$. Cyclo-oxygenase is strongly inhibited by aspirin, indomethacin, and other non-steroidal anti-inflammatory drugs. In platelets the prostaglandin endoperoxide PGG$_2$ is further metabolized into the even more unstable compound *thromboxane A$_2$* which is a highly potent platelet aggregating agent and a constrictor of arterial muscle (Moncada and Vane 1978). However, in the arterial endothelium the endoperoxides PGG$_2$ and PGH$_2$ are converted to *prostacyclin* (PGI$_2$) which is a very potent inhibitor of platelet aggregation and also a vasodilator. Thus platelets and vascular endothelium may have opposing effects on platelet aggregation and vascular tone, and thromboxane A$_2$ and prostacyclin may be more important physiologically than the stable prostaglandins PGE$_1$ and PGE$_2$ [p. 559].

Blood clotting. In blood which has been deprived of platelets the coagulation time in glass tubes is prolonged, the activation of prothrombin is incomplete, and the formed clots do not retract. Platelets are necessary for the intrinsic clotting process which they help to promote by release of 'platelet factor 3', a phospholipid which participates in the conversion of prothrombin to thrombin by factors X and V.

Clot retraction is ascribed to shortening of the fibrin fibres produced by contraction of attached platelet pseudopodia, which contain the actomyosin-like protein. Agents which inhibit cell metabolism or enzyme activity also inhibit clot retraction.

Blood which is allowed to clot in a glass tube at 37 °C usually begins to show clot retraction after 30 min.

The physiological importance of clot retraction is not established but it may be that a compacted clot is a more effective haemostatic plug.

Phagocytosis. Carbon particles, immune complexes, and virus particles undergo phagocytosis by platelets.

Storage and transport. Platelets contain stores of 5 HT and histamine which are released when platelets disintegrate; these amines will then act on blood vessels. Platelets can take up 5 HT against a concentration gradient.

Platelet production and survival. The normal blood platelet count by direct methods is $150-400 \times 10^9 \, l^{-1}$.

The measurement of platelet survival has been carried out by transfusing platelets labelled with ^{51}Cr or ^{32}P (in DFP). The survival time is 8–12 days. Platelets are destroyed mainly in the spleen and in conditions of overactivity of the spleen (hypersplenism, p. 54) the platelets may almost disappear from the circulation. The platelet count is increased after trauma.

The platelet count is normally very constant. Platelet production is depressed by transfusion of platelets and enhanced by removal of platelets from the circulating blood (thrombocytophoresis). There must be some regulatory feedback mechanism but there is no firm evidence for the existence of a platelet-regulating hormone analogous to erythropoietin [p. 37].

HAEMORRHAGIC STATES

The preliminary discussion [above] suggests that clinical haemorrhagic states may be due to:

1. Defective blood clotting.
2. Defective capillary contractility.
3. The combined defects.

Defective blood coagulation

In cases of this disorder a firm clot is not formed following an injury during the period of capillary contraction. When the capillaries finally open up once more, oozing will recommence, and can only be controlled by measures, either general or local, that restore blood coagulability. In these clinical states haemorrhages may occur anywhere in the body 'spontaneously'; it must be supposed that the capillaries in many regions are constantly being exposed to trivial and unnoticed trauma and that the resulting tiny blood-leaks are normally effectively sealed off. But when there is decreased blood coagulability from any cause the normal harmless leaks become noticeable and may even lead to dangerous haemorrhage.

In haemorrhagic states coagulating agents applied sufficiently firmly to superficial bleeding areas may be useful temporarily; e.g. factor VIII in haemophilia [p. 28] or active preparations of thrombin. The dressing may consist of fibrin sheets or foam which do not need removal and are absorbed during healing.

Defective capillary contractility

The clinical condition in which the capillary abnormality results in bleeding is known as purpura, which must now be considered.

Purpura

This is a condition in which there is a tendency to 'spontaneous' haemorrhages, usually beneath the skin, from the various mucous membranes, and in internal organs. Purpura may be *symptomatic* or *primary*. Symptomatic purpura may be allergic or it may result from various infections (e.g. infective endocarditis, typhus), from many drugs in susceptible subjects (e.g. iodine, bismuth, ergot, quinine, sedormid), and in cachectic states (e.g. cancer). Primary (idiopathic) purpura occurs most often in children, has no constant associations with other maladies, and is occasionally congenital or hereditary. Severe purpura with haemorrhages in the skin and from mucous membranes is called purpura haemorrhagica.

In purpura the blood coagulation time is normal; but there is clear evidence of an abnormal state of the capillary wall. The following tests are used.

Capillary resistance. If firm pressure (e.g. by inflating a blood-pressure cuff at 60 mm Hg for 2 minutes) or a suction force (e.g. by the negative pressure used in 'cupping') is applied to the skin, the local (and sometimes the distal) capillaries leak blood, leading to the appearance of a crop of minute haemorrhages (petechiae); this abnormal response is evidence of diminished resistance (increased 'fragility') of the capillary endothelium.

Bleeding time. The ear is pricked and the escaping blood is dried every 15 seconds on the edge of a circle of filter paper; normally bleeding ceases after 2–6 minutes. Under these conditions no blood clot can form locally and arrest of bleeding depends exclusively on capillary contraction. In purpura the damaged capillaries fail to close and thus bleeding may continue for 10–20 minutes or even longer. The blood-clotting time is normal, because although the clot formed is too pulpy to stop effectively the flow of blood from a puncture, it suffices to stop the free flowing of blood in a test tube. The prothrombin concentration test is normal, because here the clotting does not depend on natural thromboplastin in the platelets, but on the addition of tissue thromboplastin (brain extract).

Skin microscopy. Examination in cases of primary purpura reveals that the skin capillaries are very irregular and distorted in form, sometimes branching; after puncture these vessels remain patent with the result that free bleeding proceeds from the needle track for several minutes. In symptomatic purpura the capillaries are anatomically normal but (because of the presence of a toxic agent or other cause) they do not contract effectively in response to injury.

The facts just presented indicate the importance of capillary defect in producing the haemorrhages of purpura. Probably the capillary changes are localized, thus accounting for bleeding from restricted parts (e.g. skin and mucous membranes).

Relation of blood platelets to purpura

In many cases of purpura there is a reduction in the platelet count (thrombocytopenic purpura) which may be down to 50 or even $10 \times 10^9 \, l^{-1}$; in some cases, however, the platelet count remains normal (athrombocytopenic purpura). With low platelet counts, though the coagulation time is normal, the clot that forms is soft and friable, does not retract well, and is doubtless a less satisfactory bung for damaged capillaries. Electron microscopic studies show that platelets are incorporated into endothelial cell cytoplasm and this is thought to strengthen the endothelium. It is clear, however, that a low platelet count alone cannot be responsible for the initial occurrence of haemorrhages. The injection of agar-serum into an animal greatly reduces the platelet count, but does not produce haemorrhages. The bleeding may cease in a severe case of purpura several days before the platelet count rises, and the platelets have been known to disappear clinically from the blood without haemorrhages occurring. It is possible that in some cases of purpura enough platelets 'stick' on to the wall of damaged capillaries (as they 'stick' on to any injured endothelium) to lower the count in the peripheral blood; alternatively the factors responsible for the capillary abnormality may be independently destroying the platelets. Thus in 'hypersplenism' there is clear evidence of excessive platelet destruction, often associated with purpura [p. 54].

A number of cases of primary thrombocytopenic purpura recover spontaneously after a time; resistant cases may be improved or cured by cortisone or ACTH, which are said to decrease capillary fragility. If all else fails splenectomy will cure about 70 per cent of severe primary thrombocytopenic purpuras. The platelet count usually rises, but does not run parallel with the arrest of the haemorrhage. It has been suggested that the spleen in some unknown way may be responsible for the capillary abnormality as well as for the thrombocytopenia.

REFERENCES

BIGGS, R. (ed.) (1972). *Human blood coagulation, haemostasis and thrombosis*. Blackwell, Oxford.
BORN, G. V. R. and CROSS, M. J. (1963). *J. Physiol., Lond.* **168**, 178.
DAVEY, M. G. (1966). *The survival and destruction of human platelets*. Basel.
JOHNSON, S. A. (1967). Platelets in haemostasis. In *Blood clotting enzymology* (ed. W. H. Seegers) pp. 379–420. New York.
MARCUS, A. J. (1969). *New Engl. J. Med.* **280**, 1278.
MONCADA, S. and VANE, J. R. (1978). *Br. med. Bull.* **34**, 129.
TURPIE, A. G. G., McNICOL, G. P., and DOUGLAS, A. S. (1971). In *Recent advances in haematology* (ed. A. Goldberg and M. C. Brain) p. 249. Churchill Livingstone, London.

HAEMOPOIESIS [Fig. I.24]

The development of red blood corpuscles, white blood corpuscles, and platelets is discussed in the appropriate sections. A few general points concerning their development will now be considered.

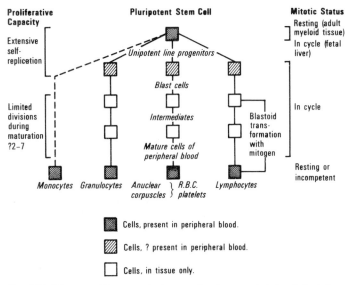

FIG. I.24. Diagram of haemopoiesis according to current state of physiological knowledge derived from observations with cell markers supplementing the concepts of classic morphology. (Barnes, D. W. H. and Loutit, J. F. (1967). *Lancet* **ii**, 1138.)

There has been much argument in the past as to whether the different types of blood cell arise from a single pluripotent stem cell (monophyletic theory) or whether there are separate stem cells for each main variety of blood cell, i.e. granulocyte, monocyte, lymphocyte, erythrocyte, and platelet (polyphyletic theory). There is now evidence that bone marrow, and even peripheral blood, contain a very small proportion of cells which are capable of considerable self-replication, and also differentiation into precursors of each of the main types of blood cell. Barnes and Loutit (1967) suggest the following terminology:

The *pluripotent stem cell* (*haemocytoblast*), 18–23 μm in diameter, with large nucleus and small rim of cytoplasm is an ancestral cell which undergoes extensive replication and can differentiate into the *unipotent line-progenitor cell*, which also undergoes extensive replication but is committed to a particular line of development. These line-progenitive cells (also called stem cells) seem, by functional tests with markers, to be in the long term a wasting

population which needs to be 'topped up' from pluripotent cells.

The pluripotent stem cells have been identified by their capacity to form colonies of cells in the spleen and bone marrow when they are injected into mice which have been given a large enough dose of X-rays to destroy the intrinsic haemopoietic tissue. A single colony, itself originating from one cell, reproduces further colony-forming units, which may also form colonies of the erythropoietic, granulopoietic, or other types. The pluripotent stem cells and the unipotent line progenitors have not yet been identified morphologically; in view of their minute numbers in proportion to other cells, this is hardly surprising.

The blast cells and intermediate cell types undergo a limited number of divisions (2–7) during their maturation, but they are not capable of the extensive replications of the unipotent line progenitors. The end cells of all types in the peripheral blood, except for the lymphocytes which can show blastoid transformation, are incapable of further division.

There is no fundamental conflict between the mono- and polyphyletic theories of haemopoiesis. Under physiological conditions the replacement of effete mature blood cells may take place largely by multiplication within 'compartments' of already differentiated unipotent precursor cells, but there may be some supplementation of each line from a 'compartment' of undifferentiated pluripotent cells.

In health the numbers of erythrocytes, leucocytes, and platelets in peripheral blood are kept constant within quite narrow limits. This implies that the rates of formation, release, and destruction are balanced in some way and there must be feedback control mechanisms to achieve this homeostatic regulation. The rate of erythrocyte production is enhanced by hypoxia, which acts mainly by stimulating the secretion of *erythropoietin* from the kidney, but no such definite mechanisms are known for the control of leucocyte and platelet productions. These matters are discussed more fully in the relevant sections.

REFERENCES

BARNES, D. W. H. and LOUTIT, J. F. (1967). *Lancet* **ii**, 1138.
BOGGS, D. R. (1966). *A. Rev. Physiol.* **28**, 39.
PORTER, R. and FITZSIMONS, D. W. (eds.) (1976). *Congenital disorders of erythropoiesis*. Amsterdam.

The red blood corpuscles (erythrocytes)

DEVELOPMENT

In the early embryo, blood formation takes place first in the mesoderm of the yolk sac (the area vasculosa) and later in the body of the fetus (mesoblastic stage). The mesoderm consists originally of a syncytium or nucleated mass of protoplasm without cell outlines. This syncytium then gives rise to a network of capillary vessels lined by endothelium and containing plasma (which is formed by liquefaction of the cytoplasm). Erythropoiesis takes place intravascularly; in places the endothelial cells proliferate and differentiate to form masses of nucleated haemoglobin-bearing cells which fill and distend the capillary lumen. These cells become free and circulate in the bloodstream and finally lose their nuclei to give rise to non-nucleated discs. According to Gilmour this early mesoblastic blood formation is the only example of intravascular haemopoiesis observed in the *human* embryo; later, the formation of red and white cells is extravascular. After the third month of fetal life, the spleen and especially the liver are the most important sites of blood formation (hepatic stage). Nucleated red cells develop from the mesenchyme between the blood vessels and the tissue cells.

About the middle of fetal life the bone marrow begins to act as a blood-forming organ (myeloid stage), the function becoming progressively more important as erythropoiesis in the liver decreases, so that the marrow is normally the sole region where red cells are formed after birth. During the second half of intra-uterine life very few nucleated red cells are found in the circulating blood, though reticulocytes are still very numerous. Occasionally in adult life when the marrow cavity is nearly obliterated by sclerosis or other disease, the spleen and liver again become important sites of blood formation.

Bone marrow

Distribution

Bone marrow may be yellow or red. Yellow marrow consists of fat cells, blood vessels, and a minimal framework of reticulum cells and fibres; in red marrow numerous blood cells of all kinds and their precursors (erythroid and myeloid) are present too. At birth all the bones are filled throughout their length with highly cellular red marrow. With increasing age the marrow becomes more fatty, the process setting in first in the distal bones of the limbs (tarsus and carpus), then in the intermediate (tibia, fibula, radius, ulna), and finally in the proximal bones (femur, humerus). At the age of twenty all the marrow of the long bones is yellow except for the upper end of the femur and humerus. In the adult, red marrow persists mainly in the vertebrae, sternum, ribs, and bones of the skull and pelvis. There is obviously ample room in the long bones of the adult for considerable expansion of the red marrow. Children, in relation to their weight, have relatively more red marrow than adults but have little reserve space on which to draw if needed; for example, in times of stress, or after severe haemorrhage or haemolysis, blood formation may take place in other regions, e.g. liver and spleen. In adults, examination of samples of marrow aspirated from the sternum or iliac crest during life gives much information about the state of marrow activity. *Post mortem* the sternum, a rib, or a vertebra is examined when diminished marrow activity is suspected; the shaft of a long bone is generally studied when evidence of extension of haemopoiesis is sought.

Vascular arrangement. The nutrient artery of the bone breaks up into smaller branches, which lead to a network of intercommunicating sinusoids. These vessels are lined by a thin endothelium (like capillaries elsewhere) but when dilated have the capacity of large veins.

Functions of the red marrow

These are:

1. Formation of red blood corpuscles.

2. Formation of granulocytes and to a less extent of monocytes and lymphocytes.

3. Formation of blood platelets.

4. Destruction of red cells by macrophages which constitute part of the lining of the blood sinuses.

Cytological methods

1. Smears of living marrow and blood cells may be stained (vital staining) with Janus green and neutral red; the former stains the mitochondria and the latter the cell granules. In these wet preparations, motility if present is retained.

2. Fixed marrow sections, marrow smears, or blood films are stained with eosinate of methylene blue (Leishman's stain), to show nuclear details and the basophilia or eosinophilia of the cytoplasm.

3. Living erythroid cells may also be stained with cresyl blue to demonstrate the presence of a reticulum in the cytoplasm.

STAGES OF ERYTHROPOIESIS

The nucleated precursors of the mature erythrocyte, which normally occur only in the bone marrow, have been given various names by different haematologists. The simplest terminology of these precursors formed from the haemocytoblast is shown in the TABLE I.3.

The nucleated cells and reticulocytes in the whole body constitute the *erythron*.

Normal erythropoiesis is called *normoblastic*; in anaemia due to deficiency of vitamin B_{12} or of folic acid erythropoiesis is *megaloblastic* [p. 39]. TABLE I.3 shows that most of the nucleated erythroid cells in the marrow are the more mature forms which are normally being steadily transformed into erythrocytes. About 2×10^{11} erythrocytes leave the bone marrow every day to enter the bloodstream.

Cytology of erythroid cells

In general, developing marrow cells (myeloid or erythroid) show the following changes: the cells decrease in size and the cytoplasm becomes relatively more extensive; nucleoli disappear and the chromatin becomes progressively coarser; the cytoplasm becomes less basophilic owing to decrease of ribonucleic acid, and the pigment or granules characteristic of maturity appear. Developing red cells have the characteristics shown in TABLE I.3 and FIG. I.25.

Stage I. Pronormoblast (proerythroblast)

This early cell, which is derived from stem cells, is large (diameter 15–20 μm); the cytoplasm is a deep violet-blue and there is a small crescent, showing paler staining, round the nucleus; the cell is devoid of haemoglobin. The nucleus is large (12 μm) occupying about three-quarters of the cell volume and the chromatin forms a fine stippled reticulum and contains several nucleoli; it develops

TABLE I.3 *Terminology of the precursors from the haemocytoblast*

Cell stages	Terminology	Cytoplasmic staining	Mitosis	Number per 100 nucleated cells in bone marrow
I	Pronormoblast (Proerythroblast)	Basophil	+ in states of stress	1–3
II	Early normoblast	Basophil	+	1–3
III	Intermediate normoblast	Polychromatophil	+	4–8
IV	Late normoblast Reticulocytes Erythrocytes	Eosinophil	−	8–16

into the early normoblast. The pronormoblast only shows mitosis in states of stress.

Stage II. Early normoblast (early erythroblast)

This cell is somewhat smaller and shows active mitosis; the nucleoli have disappeared and the chromatin network is fine and shows a few nodes of condensation. Haemoglobin is formed from stage II to stage IV.

Stage III. Intermediate normoblast (late erythroblast)

The cell is still smaller (diameter 10–14 μm) and shows active mitosis; the resting nucleus shows further condensation of the chromatin; haemoglobin increases, its eosinophil staining giving the cytoplasm a polychromatic appearance.

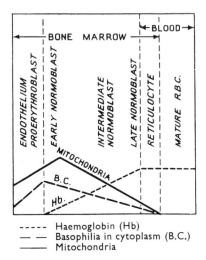

----- Haemoglobin (Hb)
— — Basophilia in cytoplasm (B.C.)
——— Mitochondria

Fig. I.25. Chart showing stages in development of red blood corpuscles.

Stage IV. Late normoblast (normoblast)

This cell represents a maturation of the previous stage; mitosis has now ceased; the cell diameter is 7–10 μm; the nucleus is small, the condensed chromatin assuming a 'cartwheel' appearance and finally becoming uniformly deeply stained (pyknotic). The haemoglobin has increased and in well washed preparations the cytoplasm gives an eosinophil reaction. Pyknosis is a stage in the degeneration of the nucleus which breaks up and finally disappears owing to extrusion or lysis; a young red cell (reticulocyte) is thus formed.

Maturation of the erythroblasts thus involves a decrease in the size of the cell, increased condensation and finally pyknosis of the nucleus, accumulation of haemoglobin, and a change in staining

reaction of the cytoplasm from basophil via polychromatophil to eosinophil.

Reticulocyte

The young red cell is so called because on vital staining with cresyl blue a network of reticulum is apparent in the cytoplasm in the form of a heavy wreath, or as clumps of small dots, or as a faint thread connecting two small nodes. (All the nucleated precursors of the reticulocyte (normoblasts) also give this staining reaction.) The reticulum probably consists of remnants of the basophil cytoplasm of the immature cell (chemically the reticulum is made up of ribonucleic acid). If red cells are stained with eosin and methylene blue the presence of the reticulum in the young cells (reticulocytes) leads to a diffuse mauve staining of the cell—polychromatophilia. In pathological states this stained basophil material is sometimes present in clumps which appear as discrete blue particles. This finding, known as basophil punctation (or punctate basophilia), is especially obvious after poisoning with lead. As the red cell ages, the reticulum disappears. In the newborn, 2–6 per cent of the red cells in the circulation are reticulated; the number falls during the first week to less than 1 per cent, at which level it remains throughout life. Their number is increased whenever red cells are being rapidly manufactured. For example, if haemolytic poisons are injected intravenously to destroy the circulating red cells, the marrow proliferates, and numerous young red cells pass into the bloodstream; it is then found that 25–35 per cent of the circulating red cells are reticulocytes. An increase in the reticulocyte count (reticulocytosis) is the first blood change noted when pernicious anaemia is treated with vitamin B_{12} [p. 39].

NORMAL RED CELL (ERYTHROCYTE)

The human red cell is normally a circular, non-nucleated, biconcave disc [Fig. I.26]; it is very elastic and can undergo astonishing deformation when passing through narrow capillaries. The normal diameter is 6.5–8.8 μm (mean 7.3 μm). The surface area of the red cell is about 140 μm², much greater than if its volume were contained in a sphere. Thus O_2 and CO_2 exchange are maximal with the biconcave configuration.

Red-cell indices

The old inaccurate counting-chamber methods for red-cell counts, haemoglobin concentration, and mean corpuscular volume have been replaced by electronic machines which may also be automated (though all such apparatus requires frequent checking). Table I.4 contains mean values which have been obtained from normal persons.

Table I.4

Direct measures	Adult men	Adult women	Children (1 year)
Red-cell count (RCC)			
per litre of blood	5.5×10^{12}	4.8×10^{12}	4.4×10^{12}
Haemoglobin (Hb)			
measured as cyanmethaemoglobin in photoelectric absorptiometer			
g per decilitre (100 ml) of blood	15.5	14	12
Mean corpuscular volume (MCV)			
in femtolitres (fl)	85	85	85
1 fl = 10^{-15} litre			
Packed (red) cell volume (PCV haematocrit)	0.47	0.42	0.40
= (MCV × RCC) centrifugation of heparinized blood gives volume of packed red cells as a proportion of whole blood [p. 19]			
Derived measures			
Mean corpuscular Hb concentration (MCHC) (Hb/MCV × RCC)	33 g dl⁻¹	33 g dl⁻¹	33 g dl⁻¹
Mean corpuscular haemoglobin (MCH) (Hb/RCC)	29.5 pg	29.5 pg	27.0 pg

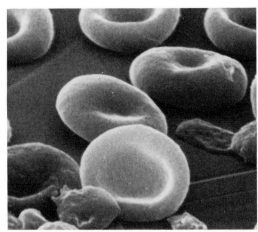

FIG. I. 26. Scanning electron micrograph of human red blood cells. These cells show the normal appearance of biconcave discs. The smaller rugged cells present are platelets.

Kindly supplied by Dr P. M. Rowles, Middlesex Hospital Medical School.

The above values, and modes of expression are those recommended by the British Committee for Standards in Haematology, incorporating SI units.

Note the following points of comparison with older modes of expression:

1. Red-cell count at 5.5×10^{12} $1^{-} = 5.5 \times 10^6$ mm^{-3} or 5.5×10^6 $\mu 1^{-1}$.
2. Hb concentration at 15.5 g dl^{-1} = 15.5 g per 100 ml.
3. Mean corpuscular volume at 85 fl = 85 μ^3 (μm^3).
4. Packed cell volume at 0.47 = 47 per cent. (PCV is now given as a proportion and not a percentage.)

Erythrocyte membrane [see p. 2]. The viability of the red cell depends on the integrity of its membrane which probably consists of a bimolecular layer of phospholipids covered on both sides by a protein layer. Though the membrane is freely permeable to water, Na$^+$, K$^+$, and Cl$^-$, a cation-pump mechanism keeps the intracellular [Na$^+$] low and the [K$^+$] high. The energy for this pump is provided by membrane ATPase which requires Mg^{2+}, and Na$^+$ and K$^+$ for full activation; ATP is formed during glycolysis and its hydrolysis provides the ultimate source of energy for the sodium pump. When red-cell metabolism ceases, as in cold-stored blood, the ions move between plasma and cells according to their concentration gradients.

The red-cell membrane can be broken by certain physical stimuli:
Mechanical fragility. When red cells are shaken with glass beads for one hour, 2–5 per cent of the cells are lysed (haemolysis). In some haemolytic anaemias the proportion is higher.
Autohaemolysis. Normal blood (+ anticoagulant) when kept at 37 °C for 24 h shows <0.5 per cent haemolysis. The value is higher in some haemolytic anaemias.
Osmotic fragility. Red cells placed in 'physiological saline' (a solution of NaCl in water, containing 150 mmol l^{-1} of Na$^+$ and of Cl$^-$, or 0.9 g NaCl per 100 ml) remain intact for several hours. Red cells placed in distilled water are quickly lysed (*haemolysis*) because the intracellular ions cause osmotic movement of water into the cells which swell until they burst. As the concentration of NaCl is reduced from the physiological level, haemolysis of normal cells begins at [Na$^+$] of 83 mmol l^{-1} (0.5 per cent saline) and is complete at [Na$^+$] of 50 mmol l (0.3 per cent saline). Red cells which are lysed by saline in concentrations of Na$^+$ >85 mmol l^{-1} are said to show increased fragility. This occurs, for example, in red cells from patients with hereditary spherocytosis, in whom haemolysis may commence in saline with [Na$^+$] = 116 mmol l^{-1} and be complete at

[Na$^+$] = 75 mmol l^{-1}. The red cells of venous blood are slightly more fragile than those of arterial blood in normal persons.

In the performance of osmotic fragility tests the red cells are placed in different concentrations of NaCl solutions and the amount of haemolysis is measured at each concentration. A curve may be drawn relating the saline concentration to the degree of haemolysis (measured by haemoglobin concentration after centrifugation). Osmotic fragility is related to the shape of the red cell; the more spherical it is the greater the fragility, i.e. the higher the concentration of saline at which haemolysis occurs.

Erythrocyte sedimentation rate (ESR)

If blood containing an anticoagulant is allowed to stand in a narrow vertical tube, the erythrocytes settle to the bottom half of the tube. The rate at which this occurs is called the erythrocyte sedimentation rate (ESR). The red cells sediment because their density is greater than that of plasma; this is particularly so when the cells aggregate to form rouleaux (that is the cells are piled on top of each other). Fibrinogen favours rouleaux formation and the ESR is enhanced by proteins which enter the plasma in inflammatory and neoplastic diseases; the ESR is increased in pregnancy after 3 months.

The Westergren technique is the most sensitive. Four volumes of freshly drawn blood are mixed with one volume of 38 g l^{-1} of sodium citrate and allowed to stand in a 2-mm diameter glass tube for one hour at 22–27 °C. The upper level of red cells is recorded and the column of clear plasma is measured, the mean normal value for males being 4 mm and for females 8 mm. Any ESR value exceeding 12 mm is regarded as abnormal.

HAEMOGLOBIN

Erythrocyte precursors synthesize *haemoglobin* which is essential for transport of oxygen from lungs to tissues [p. 183], for transport of CO$_2$ in the opposite direction [p. 187], and for the regulation of blood pH. The inclusion of haemoglobin within erythrocytes is most effective for functional purposes since it avoids the consequences of the presence of high plasma haemoglobin concentrations, viz. increased blood viscosity, raised osmotic pressure, rapid destruction of haemoglobin by the reticulo-endothelial system [p. 44], and excretion by the kidney (haemoglobinuria).

The synthesis of haemoglobin requires the provision of nutrients such as protein, vitamins, and minerals (especially iron) and only takes place in the cells of the erythroid series (normoblasts).

Haemoglobin consists of the protein globin united with the pigment haem.

Haem is an iron-containing porphyrin known as iron-protoporphyrin IX. The porphyrin nucleus consists essentially of four pyrrole rings joined together by four methine (=CH−) 'bridges'; the porphyrins are thus tetrapyrroles. The 'skeleton' of the formula of haem is shown in FIG. I.27; the pyrrole rings are numbered I, II, III, IV; the carbon atoms of the methine bridges are labelled α, β, γ, δ; the positions to which side chains are attached are numbered 1–8. The side chains at the respective positions are: 1, methyl (−CH$_3$); 2, vinyl (−CH = CH$_2$); 3, methyl; 4, vinyl; 5, methyl; 6, propionic acid (−CH$_2$.CH$_2$.COOH); 7, propionic acid; 8, methyl. Thus side chains 1, 3, 5, and 8 are methyl; 2 and 4 are vinyl; 6 and 7 are propionic acid.

The iron in haem is in the *ferrous* (Fe^{2+}) form. The iron is attached to the N of each pyrrole ring and to the N of the iminazole group in the associated globin; a 'bond' is available for loose union with O$_2$ (in *oxyhaemoglobin*) or CO (in *carboxyhaemoglobin* better called *carbonmonoxyhaemoglobin*). In oxidized haemoglobin this place is occupied by an OH group on the *ferric* (Fe^{3+}) atom.

When reduced or oxygenated haemoglobin is treated with an oxidizing agent, the Fe^{2+} is oxidized to ferric iron (Fe^{3+}); the sixth bond is attached to OH. The compound is called *methaemoglobin*; it cannot unite reversibly with gaseous oxygen; the O of the attached OH is not given off in a vacuum. Reduced haemoglobin is commonly represented as Hb, oxyhaemoglobin as HbO_2; methaemoglobin may be represented as HbOH.

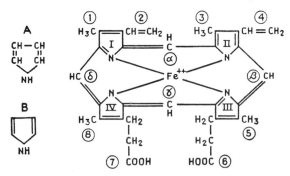

FIG. I.27. Chemistry of haem (iron-protoporphyrin IX). A = Pyrrole ring in full. B = Pyrrole ring, conventional outline. Haem: Pyrrole rings are numbered I, II, III, IV. C atoms of methine bridge are labelled α, β, γ, δ. Side chains are numbered 1–8.

Synthesis of haem

The starting substances are succinic acid and glycine, the probable sequence being as follows:

α–Amino–β–keto adipic acid (unstable)

δ-Amino laevulinic acid

These reactions are brought about by the enzyme δ-*aminolaevulinic acid synthetase* (*ALA synthetase*) with pyridoxal or pyridoxal phosphate as coenzyme.

Two molecules of δ-ALA condense to form porphobilinogen:

2 molecules of δ laevulinic acid

ALA dehydrase

Porphobilinogen

Other enzymes promote the formation of protoporphyrin IX and finally ferrous iron is introduced into the protoporphyrin molecule by the enzyme *haem synthetase* (Goldberg 1971).

All the enzymes concerned with porphyrin synthesis are under genetic control. Abnormalities in porphyrin metabolism may be inherited or acquired but they are all due to disorders of enzyme control of the processes of synthesis.

Porphyrins. Excessive production of porphyrins or their precursors may occur in the bone marrow or the liver. The commonest

disorder of porphyrin metabolism is *acute intermittent porphyria* of *hepatic* origin. This is characterized by episodes of abdominal pain, vomiting, constipation, tachycardia, hypertension, signs of peripheral neuropathy, and psychological disturbances. The disease is due to a marked increase in hepatic ALA-synthetase activity in the liver, and large amounts of δ-aminolaevulnic acid and porphobilinogen can be detected in the urine. The disease is inherited as a Mendelian dominant character but in many cases the attacks are provoked by drugs such as barbiturates, sulphonamides, and steroids.

Normal blood haemoglobin

In males the mean blood haemoglobin concentration is 15.5 g dl^{-1}; in 90 per cent of cases the range is 14–18 g dl^{-1}. In females the mean Hb concentration is 14.0 g dl^{-1} with a range of 12–15.5 g dl^{-1}. At birth the Hb concentration is 23 g dl^{-1}, falling to 10.5 g dl^{-1} at the end of the third month. The concentration then rises gradually to reach 12 g dl^{-1} at 1 year. It is most important to remember that *one gram of haemoglobin when fully saturated combines with 1.34 ml oxygen*. Thus Hb concentration is an index of the oxygen-carrying capacity of the blood.

Factors controlling haemoglobin formation

Quantitative technique. Whipple's standard anaemic dog has been employed. A normal dog weighing 10 kg has about 200 g of haemoglobin in the circulation. The animal is bled weekly to reduce the total circulating haemoglobin each time to 60 g. During the first 6–8 weeks the dog seems to be able to mobilize unidentified reserves of haemoglobin to help make good the deficiency, but these are finally exhausted. Subsequently the amount of haemoglobin newly formed each 14 days is readily measured by determining the amount of haemoglobin that has to be removed to bring the total in the blood down to the basal level of 60 g. These studies show that in addition to iron, dietary protein is of great importance in the treatment of haemorrhagic anaemia. Porphyrin is readily synthesized.

Role of protein. A low protein intake, e.g. in kwashiorkor [p. 485] retards haemoglobin regeneration even in the presence of excess iron; the limiting factor here is lack of globin. Globin itself (in the food) is the protein which is used most economically in haemoglobin formation. Some food proteins are less effective than others; thus bread and other cereals, dairy products, most vegetables and fruits, and salmon are relatively inert. Liver, kidney, spleen, and heart are most potent, muscle occupying an intermediate position. Though liver is of great value in the treatment of haemorrhagic anaemia in dogs, it seems to have no such exceptional efficacy in haemorrhagic anaemia in man. Studies with amino-acid mixtures show that the usual essential amino acids are needed for haemoglobin formation.

Role of iron. This is discussed on pp. 41–3.

Haemoglobin structure and function

Haemoglobin is built from four polypeptide chains of two types each in duplicate. Normal adult haemoglobin (HbA) contains four polypeptide chains, two called α and two called β, each chain being associated with one haem group. Thus there are four haems to the molecule (i.e. a tetramer of MW 68 000). HbA is written in brief HbA $(\alpha_2\beta_2)$.

Hb combines loosely and reversibly with oxygen, through an attachment of the latter to Fe^{2+} of haem. Combination of oxygen with one haem molecule facilitates combination of oxygen with the other three haem molecules; this accounts for the physiological oxygen-dissociation curve of human blood [p. 184].

The affinity of Hb for oxygen is influenced by the presence of 2,3-diphosphoglycerate (2,3-DPG) in the red blood corpuscles. This is a product of the metabolism of glucose via the Embden–Meyerhof pathway, and an enzyme in red cells (2,3-DPG mutase) causes 2,3-DPG to become the major part of the organic phosphate in red cells. As the concentration of 2,3-DPG rises the affinity of Hb for oxygen falls and the oxygen dissociation curve is shifted to the right [p. 185]. At high altitudes (e.g. 5000 m) 2,3-DPG concentration in red cells increases by 50 per cent and thus makes more oxygen available to the tissues. As 2,3-DPG accumulates, it depresses hexokinase and 2,3-DPG-mutase activity by negative feedback control. Glucose metabolism is thus related to oxygen transport via intracorpuscular 2,3-DPG concentration. Stored blood loses its 2,3-DPG and oxygen affinity increases. The addition of inosine and adenosine to stored blood increases 2,3-DPG formation and restores the normal affinity of haemoglobin for oxygen.

Varieties of haemoglobin

Apart from species differences, several varieties of haemoglobin occur in man; in all the *haem* moiety is the same, physical and chemical differences being due to variations in the composition of the peptides of the globin fraction.

Physiological haemoglobins. These can be separated by electrophoresis. Adult haemoglobin is of two types:

Haemoglobin A $(\alpha_2\beta_2)$ is the preponderant form. It is a spheroidal molecule with a molecular weight of 68 000 and has already been discussed.
Haemoglobin A₂ $(\alpha_2\delta_2)$ is a minor component in normal adults. Delta chains have a slightly different amino-acid composition compared with β chains.
Haemoglobin F $(\alpha_2\gamma_2)$ occurs in fetal red cells and has usually disappeared by 2–3 months after birth [p. 588]. It differs from HbA in having γ chains in place of β chains and in having a greater affinity for oxygen. HbF is much more resistant to alkali than HbA and this property is made use of in a photoelectric colorimetric method to estimate HbF in the presence of HbA.

Haemoglobinopathies. Haem synthesis is normal in abnormal haemoglobins.

Disorders of globin synthesis are of two main types:

1. Abnormal polypeptide chains are formed due to substitution of an abnormal amino acid in a chain of HbA.
2. The synthesis of a polypeptide chain is suppressed.

Haemoglobin S (HbS) is the most important pathological haemoglobin. It occurs in 10–20 per cent of negroes and is due to substitution of valine for glutamic acid at position 6 in the β chain of HbA. When HbS is reduced it becomes much less soluble and gelling of the haemoglobin inside the cells leads to changes in shape of the cells, i.e. 'sickling'. Sickle-shaped cells greatly increase blood viscosity and such cells are very liable to undergo haemolysis. *Sickle-cell anaemia* is a serious condition which is often fatal before middle age. Sickle-cell trait is inherited as a Mendelian dominant but the full-blown disease is autosomally recessive. *Haemoglobin C* is similar to HbS but is not associated with sickling. There are many other abnormal haemoglobins.

Thalassaemia (Mediterranean anaemia) occurs in the homozygous offspring of parents each of whom carries the relevant gene (heterozygotes). There is a defect in synthesis of the polypeptide chains, α and β, of HbA. Hence there occurs α- or β-thalassaemia in which either α or β chains are not synthesized; β-thalassaemia is the commoner. The red cells are abnormal in having reduced amounts of HbA, to compensate for which there are increased amounts of HbA₂ $(\alpha_2\delta_2)$ and HbF $(\alpha_2\gamma_2)$. The red cells are rapidly haemolysed

in vivo and a hypochromic anaemia occurs. Children with β-thalassaemia fail to thrive and die young.

REFERENCES

GOLDMAN, H (1971). In *Recent advances in haematology*. Churchill Livingstone, Edinburgh.
LEHMANN, H. and HUNTSMAN, R. G. (1974). *Man's haemoglobins*. Amsterdam.
WEATHERALL, D. J. (1971). The abnormal haemoglobins. In *Recent advances in haematology* (ed. A. Goldberg and M. C. Brain) p. 194. Churchill Livingstone, London.

REGULATION OF ERYTHROPOIESIS

Anaemia and erythrocytosis

In normal circumstances erythropoiesis is regulated so as to maintain the number of erythrocytes and haemoglobin content within a narrow range. In *anaemia* there is a reduction in number of circulating erythrocytes or a decrease in their content of haemoglobin. In *erythrocytosis* there is an increase in the number of circulating red cells and an increase in haemoglobin concentration in the blood.

Anaemia occurs when the erythropoietic tissues cannot supply enough normal erythrocytes to the circulation. The normal balance between production and destruction is maintained by a daily output of 2×10^{11} erythrocytes from the bone marrow, the cells surviving for about 120 days. The balance is upset if there is excessive loss of blood, e.g. by haemorrhage, or if there is some defect in the production of erythrocytes by the bone marrow.

In some cases of anaemia there is merely a reduction in the number of erythrocytes, MCV, MCHC, and MCH being normal. This is called a *normocytic, normochronic anaemia*.

In *normocytic, hypochromic* anaemia MCV is normal but MCHC and MCH are reduced. In *microcytic, hypochromic* anaemia MCV is also reduced.

In *macrocytic* anaemias the MCV may exceed 100 fl and MCH is raised, but MCHC is normal.

Erythrocytosis may be *relative*, i.e. due to reduced plasma volume (e.g. in dehydration) which increases red-cell count and PCV, and also leucocytes and platelets. This is a *polycythaemia* in which the numbers of all blood cells per litre are increased, but the total mass of cells is not increased.

In *true* erythrocytosis there is an *absolute* increase of red-cell mass. In many cases this is due to increased stimulation of the bone marrow by *erythropoietin*, hypoxia being the commonest underlying cause.

In *polycythaemia vera* there are absolute increases in the numbers of red cells, white cells, and platelets. Erythropoietin concentration in the blood is normal and the cause of the generalized hyperplasia of the bone marrow is unknown.

Factors promoting erythropoiesis

Hypoxia. Oxygen lack, when sufficiently severe, stimulates red-cell formation in the bone marrow and produces erythrocytosis. Hypoxia is responsible for the enhanced erythropoiesis which occurs at high altitudes [p. 204] in congenital heart disease, after haemorrhage, and during the course of many anaemias. Conversely, a higher than normal oxygen pressure depresses erythropoiesis. The hypoxic stimulus to erythropoiesis is not a direct action on the bone marrow but is mediated by the very important regulating hormone erythropoietin.

Erythropoietin (Ep)

Erythropoietin is now an established hormone which may be called erythropoiesis-stimulating hormone (ESH) or more simply Ep

(Gordon 1973). It is a circulating glycoprotein which in higher organisms is the prime regulator of erythropoiesis. Plasma Ep has a molecular weight of 46 000 and is made up of 74 per cent protein and 26 per cent carbohydrate (including 10 per cent sialic acid). Its activity is destroyed by trypsin and sialidase. Sialic acid may serve as the attachment site of Ep to carrier substances. Ep has been isolated and purified from the blood of sheep made severely anaemic by phenylhydrazine, a haemolytic agent, and from the urine of patients with anaemia due to hookworm infestation. Ep occurs in normal human plasma and urine and its concentration in body fluids of experimental animals and in man is increased by diverse *hypoxic* stimuli. Thus it is probable that Ep is concerned with normal erythropoiesis as well as that which follows emergency states of hypoxia.

Ep increases the numbers of nucleated RBC in the erythropoietic organs and the numbers of reticulocytes and RBC in peripheral blood; the simplest method of assay is to measure the incorporation of ^{59}Fe into the haemoglobin of newly formed RBC. The best *in vivo* assay method is to measure the increase in ^{59}Fe incorporation in mice which have been exposed to reduced P_{O_2} and then returned to normal P_{O_2} for several days. The exhypoxic mouse can detect very small amounts of Ep. *Immunochemical* methods of assay of Ep are even more sensitive and correlate well with the bioassay.

Sites and modes of action of Ep. Ep stimulates erythropoiesis but has no effect on leucocyte or platelet formation. It promotes erythropoiesis not only in adult bone marrow but also in fetal yolk sac, liver, and spleen.

When Ep is added to cultures of rat bone-marrow cells it induces specific differentiation of a haematopoietic precursor stem cell, the Ep-responsive cell, to form the earliest recognizable members of the nucleated erythroid cell line, leading to the appearance in succession of pronormoblasts, normoblasts, and reticulocytes. Synthesis of RNAs, DNA, globin, and ferritin and stroma formation precede the development of haemoglobin (which occurs after 6 hours). The stimulant action of Ep on haem synthesis is abolished by actinomycin D. Ep enhances the activity of δ-aminolaevulinic-acid synthetase [p. 36]. The subcellular site(s) of Ep action is not certain.

Site(s) of formation of Ep. The relation of the kidney to erythropoiesis is well established from both clinical and experimental studies. Thus anaemia often accompanies renal deficiency in man, and hypernephroma may produce erythrocytosis. Experimental bilateral nephrectomy in rats inhibits their erythropoietic response to hypoxia (e.g. after acute haemorrhage or exposure to low P_{O_2}).

An extract of kidney from hypoxic rats only becomes erythropoietically active after incubation with plasma. The substance in the kidney has been called the renal erythropoietic factor (REF) or *Erythrogenin* (Eg). Erythrogenin is extracted from the light mitochondrial and microsomal cell fractions, and acts on a plasma substrate, *erythropoietinogen*, which is an α-globulin, to form Ep which circulates in the bloodstream. This system is analogous to, but distinct from, the renin–angiotensinogen–angiotensin-I system.

$$\text{Erythropoietinogen} \xrightarrow[\text{Erythrogenin}]{\text{Hypoxia}} \text{Erythropoietin}$$

Erythrogenin, erythropoietinogen, and Ep are antigenically distinct from each other. Anti-Ep serum inhibits erythropoiesis *in vivo*.

The Eg–Ep system may be activated within the kidney since erythrogenin and its substrate are both available for interaction therein. Erythrogenin occurs in the juxtaglomerular cells but probably also in other renal cells in the cortex and medulla.

Extrarenal production of Ep in man. Although the kidney is the main site of erythrogenin formation, Ep can be detected in the plasma of anephric man and increases after haemorrhage. However, the hypoxic stimulus has to be much greater than in normal persons to trigger extrarenal Ep production. Experiments in rats suggest that the reticulo-endothelial system of the liver and spleen may be responsible for extrarenal erythrogenin production.

Factors influencing Ep production

Oxygen deficiency is the fundamental stimulus for erythropoiesis, whether produced by hypobaric hypoxia, bleeding, cardiorespiratory disturbances, or by haemoglobins that do not release O_2 easily. *Vasoconstrictors* can promote Ep formation by inducing renal hypoxia. Anti-Ep serum blocks the erythropoiesis produced by *in vivo* infusion of angiotensin, 5-HT, and prostaglandin E_1. Noradrenaline and vasopressin can also stimulate erythropoiesis. It is not known whether all these responses are physiological.
Nucleotides. cAMP, NAD, and NADP all stimulate erythropoiesis; the nicotinamides are completely antagonized by anti-Ep serum.
Products of red-cell destruction (*haemolysates*) enhance erythropoiesis, their effects being antagonized by anti-Ep. This could well be a physiological feedback process.

Hormones

Androgens stimulate erythropoiesis by a process blocked by anti-Ep serum. This effect may explain the higher haemoglobin and RBC values in males.

Adrenal cortical steroids and ACTH. Small, physiological doses of adrenal cortical steroids and ACTH stimulate erythropoiesis via Ep production; large doses of steroids are inhibitory.

TSH, thyroid hormones, pituitary growth hormone, prolactin, and parathyroid hormone all enhance erythropoiesis and, except PTH, they act by promoting Ep formation.
Ovariectomy enhances erythropoiesis, and oestrogens in moderate amounts depress the erythropoietic response to hypoxia.

Fetal Ep

Anaemia in either the fetus or the mother stimulates the production of fetal Ep. The fetus appears to be more sensitive than the mother to reduction in maternal RBC count. This may be due to the fact that the O_2 dissociation curve for fetal blood is shifted to the left of that for the adult [p. 587]. The fetus is thus on the brink of hypoxia and maternal hypoxia may lead rapidly to increased Ep formation by the fetus, with much smaller effects on the mother.

Inactivation and excretion of Ep. There are inhibitors of Ep activity in many organs, including the kidney, but the liver appears to be the most important site of inactivation of Ep. Normal persons excrete Ep in the urine.

Possible neural control of Ep production. There is some suggestive evidence that erythropoiesis is enhanced by electrical stimulation of certain regions of the hypothalamus. The effect might be mediated by release of pituitary hormones or by autonomic nervous stimulation, leading to renal vasoconstriction and hypoxia.

Clinical implications and applications

Polycythaemias. Ep is detectable in normal plasma and urine (immunochemical assay) and the amounts are greatly increased in patients with secondary *polycythaemia* (erythrocytosis) arising from respiratory or cardiac abnormalities which lead to renal hypoxia. Patients with haemoglobins which have a greater than normal affinity for O_2 also have raised plasma and urinary Ep levels.

Certain tumours of the kidney, adrenal, liver, or cerebellum may form Ep as an abnormal metabolic product and are thus associated with erythrocytosis.

By contrast, patients with primary polycythaemia (*polycythaemia rubra vera*, PV) have little or no detectable Ep in plasma or urine. In mice certain viral filtrates induce polycythaemia and it is possible that a virus acts like Ep to produce PV in man.

Anaemias. In anaemias due to blood loss, haemolysis, or decreased RBC production, plasma Ep content is increased and abnormally large amounts of Ep are excreted in the urine.

In other cases anaemia may be due to decreased production of Ep, such as occurs in chronic renal disease, protein deficiency, cirrhosis of the liver, rheumatoid arthritis, and chronic inflammation due to infections. In these latter conditions therapeutic administration of purified Ep may be beneficial, especially in patients with anaemia due to chronic renal disease.

Other endocrine influences. In *thyroid* deficiency (myxoedema, cretinism) anaemia may occur, due to reduced Ep formation. Administration of thyroid hormones restores red-cell production to normal. In thyrotoxicosis the red cells appear to be normal but increased Ep formation causes erythroid hyperplasia in the bone marrow and increased concentration of 2, 3-diphosphoglycerate in red cells enhances the dissociation of O_2 from oxyhaemoglobin. However, a small percentage of patients with Graves' disease have pernicious anaemia due to auto-immune atrophy of the gastric mucosa.

In adenohypophysial deficiency (e.g. Simmonds' disease, p. 514] there is anaemia, while Cushing's syndrome is associated with polycythaemia due to oversecretion of either ACTH or adrenocortical hormones.

REFERENCES

ADAMSON, J. W. and FINCH, C. A. (1974). In *Textbook of endocrinology* (ed. R. H. Williams) p. 963. Saunders, Philadelphia.
GORDON, A. S. (1973). *Vitams Horm.* **31**, 105.

Deficiency anaemias

Vitamins

Ascorbic acid is necessary for normal folate metabolism; it also aids absorption of iron by reducing ferric (Fe^{3+}) to ferrous (Fe^{2+}) compounds. Anaemia in scurvy is usually due to bleeding. Pyridoxine [p. 492], riboflavine [p. 492], and vitamin-E deficiencies may occasionally be associated with anaemia.

Maturation factors

In *aplastic anaemia* the red bone marrow is often absent and replaced by fatty tissue. In such cases the production of red cells and white cells is deficient. In a few cases the marrow is hyperplastic and pronormoblasts are formed which, however, fail to mature. In either case the total volume of red cells is reduced and there are also leucopenia and thrombocytopenia. The reduction in white cells predisposes to infection and the low platelet count may lead to bleeding, which enhances the anaemia.

In many instances the cause(s) of aplastic anaemia is not known. However, cytotoxic drugs used in the treatment of malignant disease inevitably depress bone-marrow production of blood cells and in a few people a hypersensitivity reaction to certain drugs, e.g.

chloramphenicol and sulphonamides, may produce aplastic anaemia, agranulocytosis etc.

Maturation of nucleated red cells (normoblasts)

For this stage of development special maturating substances are necessary. These include vitamin B_{12} (cobalamin) and folic acid, in the absence of which the red bone marrow becomes hyperplastic and spreads throughout the shafts of the long bones (femur, tibia, fibula, humerus, radius, ulna). Derangement of DNA synthesis leads to proliferation of abnormal nucleated erythroid cells, called early, intermediate, and late *megaloblasts*, in the bone marrow. These megaloblasts are larger than the corresponding normoblasts and the cytoplasm becomes prematurely filled with haemoglobin. The late megaloblast loses its nucleus to become a macrocyte (mean diameter = 8.2 μm; MCV = 95–160 fl; MCH = 50 pg; but MCHC is normal since the increased amount of haemoglobin is distributed throughout a larger cell). There is a marked reduction in the number of circulating red cells, e.g. to $1 - 2 \times 10^{12}$ per litre. Nucleated red cells are sometimes seen in blood smears. Owing to excessive red-cell destruction, a mild haemolytic jaundice is often present. Granulocytopenia and thrombocytopenia are common.

Megaloblastic haemopoiesis occurs typically in *pernicious anaemia* and in anaemia due to *folate deficiency*. Megaloblastosis occurs because DNA formation is limited while RNA formation is not. Both B_{12} and folate are necessary for the synthesis of thymidylate, the nucleotide of thymine which is found in DNA but not RNA. *Pernicious anaemia* is due to reduced absorption of vitamin B_{12} (cobalamin) from the ileum. The primary lesion is atrophy of the gastric mucosa, with reduced secretion of intrinsic factor (IF) by the parietal cells. Normally IF forms a complex with B_{12} and promotes its absorption from the ileum. Gastrectomy and various intestinal lesions may diminish the absorption in the ileum of the B_{12}–IF complex.

Clinical features of pernicious anaemia. The onset is usually insidious with weakness, lassitude, and shortness of breath on exertion. Soreness of the tongue; tingling and numbness in the hands and feet are common and diarrhoea sometimes occurs. In advanced cases peripheral neuropathy and demyelination of the nerve tracts in the spinal cord may lead to *subacute combined degeneration of the cord*, affecting both afferent and efferent pathways. Severe mental disturbances may also occur.

In untreated patients, pernicious anaemia shows periods of remission followed by relapse and the disease is ultimately fatal. Early treatment by injection of cobalamin rapidly converts megaloblastic haemopoiesis in the bone marrow to normal and within a few hours the patient feels stronger and the appetite quickly improves. Vitamin B_{12} eventually restores the peripheral blood picture to normal, the earliest objective sign of response being a reticulocytosis, which reaches a peak of 15–40 per cent of the total red cell count at 7–10 days after the commencement of B_{12} treatment. After this the reticulocyte count returns to the normal level of 0.5–2.0 per cent, while the total red-cell count and haemoglobin concentration rise steadily to normal. The administration of B_{12} prevents the development of subacute combined degeneration of the cord; if signs of demyelination already exist B_{12} promotes improvement of the neurological state.

Administration of therapeutic doses of folic acid to patients with pernicious anaemia relieves many of the symptoms by restoring red-cell production to normal, but folate has no beneficial effect on the neurological complications, which may indeed get worse. Folate deficiency does not itself cause neurological disturbances.

Neither B_{12} or folic acid have any beneficial effect on the gastric atrophy of pernicious anaemia. Administration of adrenal corticosteroids may promote some regeneration of both chief and parietal cells in the gastric mucosa and thereby increase the secretion of IF, enhance the absorption of B_{12}, and so lead to haematological remission. Steroids may also enhance the ileal uptake of the B_{12}–IF complex. These effects are of theoretical interest only, since injected cobalamin is vastly more effective in overcoming vitamin B_{12} deficiency.

Vitamin B_{12}

The molecule of vitamin B_{12} consists of two major portions.

1. A nucleotide (5,6-dimethyl benziminazole).
2. A corrin ring, which is porphyrin-like but has a cobalt atom at its centre.

There are four well-known forms of vitamin B_{12}, with different ligands attached to the Co atom: (i) Deoxyadenosylcobalamin (coenzyme B_{12}). This occurs in liver and other cells and produces isomerization of methylmalonyl CoA. It is probably the main dietary form of B_{12} and is present in animal tissues and bacteria (but not in plants). (ii) Methylcobalamin is found in blood plasma. It methylates homocysteine to form methionine. (iii) Hydroxocobalamin is formed from the other cobalamins when they are exposed to light. It is bound more firmly than cyanocobalamin to plasma globulins and is the best form of B_{12} for the treatment of pernicious anaemia. (iv) Cyanocobalamin is the form in which B_{12} was first crystallized. It is slowly converted to hydroxocobalamin on exposure to light. The body converts cyanocobalamin slowly to its physiological forms.

Absorption of vitamin B_{12}

Sources. Human beings are entirely dependent on dietary B_{12}, which is synthesized by micro-organisms in water, soil, or the intestine. B_{12} occurs in nearly all animal tissues but not in plants (unless contaminated with bacteria). The minimal daily requirement of B_{12} in normal adults is about 2.0 μg and a normal mixed diet contains 5–30 μg daily.

The daily loss of B_{12} is about 0.1 per cent of the total pool of B_{12} (3–4 mg). When the size of the pool decreases, as occurs after partial gastrectomy or in vegans (people who eat no foodstuffs of animal origin), the daily loss of B_{12} is reduced, in absolute terms. Thus the amount required for normal health may be obtained from a diminished total store of B_{12}. One mechanism which helps to achieve this state is the entero-hepatic circulation of B_{12}. B_{12} is excreted in bile to the extent of 0.5–5.0 μg daily and in deficiency states B_{12} absorption is enhanced.

Intrinsic factor (IF)

Vitamin B_{12} is absorbed by two processes:

1. A passive, inefficient process which occurs throughout the jejunum and ileum. This does not require IF and is quantitatively unimportant.
2. A second highly efficient process involves combination with IF to form an IF–B_{12} complex which is largely absorbed in the ileum in man. This is an active process which is most effective for physiological amounts of B_{12} (e.g. 23 μg).

Human IF from gastric juice is a glycoprotein of molecular weight 50 000 to 60 000. One molecule of IF binds one molecule of B_{12} to form an IF–B_{12} complex. This binding is prevented by 'blocking' IF antibody. IF bound to B_{12} is relatively resistant to proteolytic enzyme degradation, but free IF is rapidly destroyed by proteolytic enzymes in the gut. The IF–B_{12} complex passes into the ileum where absorption of B_{12} occurs as follows:

1. The IF–B_{12} complex is attached to receptors in the ileal mucosa. This process does not require energy but is calcium dependent and is optimal at pH 7.0. The attachment is inhibited by specific antibodies against the ileal brush-border membrane or by human 'binding' IF antibody.

2. The entry of B_{12} into the ileal cell is a slow, energy-dependent process. During this state IF is split off from the IF–B_{12} complex and B_{12} may be converted to the coenzyme form which is stored in mitochondria.

3. B_{12} is released into the portal blood freed from IF.

B_{12} transport. Studies with radioactive cobalt have shown that B_{12} begins to enter portal blood 3–4 hours after oral administration and its peak concentration occurs after 8–12 hours. In the blood, B_{12} becomes bound to two plasma globulins (transcobalamins), which have a binding capacity of 1 μg B_{12} per litre. Peripheral blood plasma usually contains only 20 ng B_{12} per litre, since B_{12} is rapidly taken up by the tissues, especially the liver. The B_{12}-binding globulins are concerned with transport and with storage of B_{12}.

Diagnosis of vitamin-B_{12} deficiency. The symptoms, physical signs, megaloblastic anaemia with macrocytosis have already been mentioned, together with leucocytopenia and thrombocytopenia. Deficiency of B_{12} is confirmed when these abnormalities are corrected by a physiological dose (e.g. 1 μg) of B_{12}. Other confirmatory procedures include:

Demonstration of reduced absorption from the ileum of cyanocobalamin incorporating radioactive Co. When B_{12} is administered together with pure IF, or with gastric juice, normal absorption is restored.

Estimation of serum vitamin B_{12} deficiency by microbiological or isotope dilution assay.

Methylmalonic acid excretion. Deoxyadenosyl B_{12} acts as a coenzyme for isomerization of methylmalonic acid and its final conversion to succinyl-CoA. It is therefore not surprising that in severe B_{12} deficiency urinary excretion of methylmalonic acid is increased. Administration of B_{12}, but not folate, restores methylmalonic-acid metabolism to normal.

Since methylmalonic acid accumulates in the body in B_{12} deficiency, and since folate administration does not affect this accumulation, it has been suggested that the increased concentration of methylmalonic acid in body fluids might be responsible for the neuropathy associated with B_{12} deficiency. However, in the rare congenital condition of methylmalonic aciduria the accumulation of methylmalonic acid does not cause nervous lesions. The cause of nerve damage in B_{12} deficiency is still unkown.

Immune phenomena in pernicious anaemia

Parietal cell antibodies can be detected by immunofluorescent and complement fixation techniques in 90 per cent of adult patients with pernicious anaemia. It has been suggested that pernicious anaemia is an auto-immune disease, demonstrated by the atrophic gastritis which destroys both parietal and chief cells. This accounts for the complete absence of secretion of HCl, IF, and pepsin in response to food, histamine, pentagastrin, or cholinergic drugs.

Intrinsic factor antibodies of two types may be found in the sera of patients with pernicious anaemia:

(i) A *blocking* antibody inhibits the combination of IF with B_{12}.
(ii) A *binding* antibody becomes attached to the ileal receptors to which the IF–B_{12} complex is normally bound.

Both these types of antibody reduce B_{12} absorption from the gut, but injected cobalamin can bypass these immune phenomena and restore normal health.

Folic acid (folate)

Folic acid consists of three components: pteridine, para-aminobenzoic acid, and glutamic acid. It is called *pteroylglutamic acid* and is usually found in natural sources as pteroyl polyglutamates. Folate is reduced in the body by the enzyme *dihydrofolate reductase* to form dihydrofolate (DHF) and tetrahydrofolate (THF) the latter becoming methylated to 5-methyl-THF.

Dietary sources. The richest source is *liver* which contains folate mostly as 5-methyl-THF. It also occurs in oysters, spinach, Brussels sprouts, pulses, cabbage, and orange juice; however, since some of the dietary folates occur as polyglutamates (containing more than three glutamic acid residues), which are irregularly absorbed from the gut, it is difficult to estimate the effective folate content in the diet. Food folate is rapidly destroyed by heating and much is lost by discarding cooking water. The daily requirement of folate is 100–200 μg; an average Western diet contains 500–800 μg daily which allows for losses by cooking and poor absorption of polyglutamates.

Absorption. Pteroylpolyglutamates are largely hydrolysed to pteroyl-monoglutamate by a carboxypeptidase in the jejunum, where rapid folate absorption occurs to give peak plasma folate concentration at about one hour after ingestion. The jejunal mucosa can convert pteroylmonoglutamate to 5-methyl-THF, which is the chief form of folate in plasma and liver. The liver contains most of the folate stored in the body (5–15 mg kg^{-1}). In the blood, folate is concentrated in erythrocytes and leucocytes. Some folate is metabolized in the body and small amounts are excreted in urine and faeces. The concentration of 5-methyl-THF in bile is much higher than in plasma and an enterohepatic circulation is important in maintaining normal blood folate levels. However, when folate intake is low, the body stores of folate become completely depleted in 3–4 months as the result of catabolism and excretion from the body.

Function of folates

Folates function as coenzymes in the THF form in mammalian cells in which they accept and transfer one carbon units, e.g. $\equiv$CH in purine synthesis and –CHO in pyrimidine synthesis. They are thus essential for DNA synthesis. THF also converts serine to glycine. The metabolic conversion of histidine to glutamic acid depends on the presence of THF as shown below:

Histidine —→ Urocanic acid —→ FormiminoGlutamic acid (FIGLU)

THF
5–Formimino–THF ←

Glutamic acid
Purine precursors

In folate deficiency the administration of histidine (15 g by mouth) leads to an abnormally high excretion of FIGLU in the urine. This test is, however, not specific; FIGLU excretion is also increased in B$_{12}$ deficiency.

Interaction between folate and vitamin B$_{12}$. Both B$_{12}$ and 5-methyl-THF are concerned with the conversion of homocysteine to methionine. Both B$_{12}$ and folate promote thymidylate and DNA synthesis.

Folate deficiency

This produces the same symptoms, signs, and megaloblastic anaemia as seen with B$_{12}$ deficiency, except that neuropathy occurs only in the latter, and with folate deficiency the gastric parietal cells may secrete acid and intrinsic factor. Deficiency is due to:
Reduced folate intake may occur in poor people, the elderly after gastrectomy, in kwashiorkor, in chronic alcoholics, etc.
Malabsorption of folate occurs in tropical sprue, coeliac disease, and after jejunal resection.
Increased utilization of folate occurs particularly in pregnancy

(when 300–400 μg folate daily supplement is required), in premature infants, in thyrotoxicosis, and in haemolytic anaemia. *Antifolate drugs* (e.g. methotrexate).

Diagnostic tests. If administration of 200 μg of folic acid restores the blood picture to normal there must have been folate deficiency.

REFERENCES

BABIOR, B. B. (ed.) (1975). *Cobalamin, biochemistry and pathophysiology.* New York.
CHANARIN, I. (1969). *The megaloblastic anaemias.* Blackwell, Oxford.
—— (1973). *Lancet* ii, 538.
DAVIDSON, S., PASSMORE, R., BROCK, J. F., and TRUSWELL, A. S. (1973). *Human nutrition and dietetics*, 6th edn, p. 172. Churchill, Edinburgh.
HOFFBRAND, A. V. (1971). The megaloblastic anaemias. In *Recent advances in haematology* (ed. A. Goldberg and M. C. Brain) p. 1. Churchill Livingstone, London.
KASS, L. (1976). *Pernicious anemia.* Philadelphia.
MATTHEWS, D. M. and LINNELL, J. C. (1979). *Br. med. J.* iii, 533.

IRON METABOLISM AND IRON DEFICIENCY ANAEMIA

The body of a healthy adult contains 4–5 g (75–90 mmol) of iron in the following forms:

1. Blood haemoglobin contains about 2.5 g (45 mmol).
2. Storage iron (2/3 as ferritin, 1/3 as haemosiderin) amounts to 1–1.5 g (18–27 mmol).
3. Myoglobin (myohaemoglobin), in red muscle [p. 187]. Contains 0.2 g (3.5 mmol) of iron.
4. Intracellular enzymes containing iron-protoporphyrins, including cytochrome oxidase, catalase, peroxidase, and cytochrome, account for less than 0.1 g of iron.

Myoglobin, peroxidase, and catalase consist of distinctive proteins bound with the same porphyrin as is found in the haem of haemoglobin, namely iron-protoporphyrin IX; their molecular weights are respectively 17 000, 44 000, and 225 000 (cf. haemoglobin, 68 000). The striking differences in the properties of these compounds must depend on the protein part of the molecule. Cytochrome and cytochrome oxidase contain an iron-protoporphyrin nucleus which differs from iron-protoporphyrin IX in the side-chains which are attached to the pyrrole rings; the attached proteins are also different.

The functions of the tissue iron-porphyrins (i.e. 3 and 4 above) can be briefly summarized thus: (i) myoglobin unites loosely and reversibly with molecular oxygen. It therefore serves as a small tissue O$_2$ reserve for the needs of very vigorous muscular activity [p. 187]. (ii) Peroxidase 'activates' H$_2$O$_2$ to oxidize suitable substrates. (iii) Catalase decomposes H$_2$O$_2$ to form water and molecular oxygen. Its cellular significance is unknown. (iv) Cytochrome and cytochrome oxidase are concerned with oxidation processes and electron transfer in the tissues, and not, like haemoglobin, with the transport of molecular oxygen.

Iron balance

Adult men

The intracellular iron-containing enzymes and myohaemoglobin are stable substances the iron of which cannot be called upon for other purposes; blood haemoglobin on the other hand is continually undergoing destruction. The iron content of haemoglobin is 0.33 per cent; thus 100 ml of blood containing 15 g of haemoglobin contain 50 mg of iron. As the red cells live about 120 days, 0.8 per cent of the total blood haemoglobin, i.e. that contained in 50 ml of blood, is destroyed daily, releasing about 25 mg of iron. It is essential to remember that for all practical purposes iron that has been absorbed from the intestine or some parenteral route is not subsequently excreted from the body, or only in small amounts. Thus an adult man on an adequate iron intake only excretes about 0.4 mg daily in the urine and about 0.8 mg in the bile, while very little is lost from the mucosa of the alimentary canal. It follows therefore, that the iron which is released from the destruction of haemoglobin, after being temporarily stored in the reticulo-endothelial system (presumably as ferritin) is used again for fresh haemoglobin synthesis. It will be pointed out later that iron is absorbed from the food in the intestine with considerable difficulty, i.e. only a fraction of the food iron is actually absorbed into the body, the rest being passed out in the faeces. It is, therefore, hard to say what iron intake is needed to provide any particular bodily iron need. Whenever iron loss exceeds iron absorption the blood haemoglobin falls, i.e. anaemia develops. The food iron requirement of a healthy adult male (who is not suffering from any form of blood loss) is negligibly small; a normal diet containing 5–10 mg iron daily is more than adequate to maintain a normal state of the blood in such healthy men.

Adult women

The iron loss, and in consequence the need for food iron, is greater in women than in men because of:

1. The blood loss during the monthly *menstrual period*. This is on average about 50 ml and if it regularly exceeds 80 ml iron-deficiency anaemia is liable to develop. To combat this the iron intake should be increased to 20 mg daily, of which about 2 mg will be absorbed into the body.

2. *Child-bearing*. Although menstrual blood loss is in abeyance, considerable iron loss occurs during the course of pregnancy and lactation as the following data show:

> Iron content of fetus at term = 400 mg
> Iron content of placenta and uterus = 150 mg
> Iron content of blood lost at delivery = 170 mg
> Iron content of milk during lactation = 180 mg
> (6 months)
> Total iron loss in 15 months (460 days) = 900 mg
> Average daily iron loss = 2 mg

The main iron demand on the mother occurs during the last two months of pregnancy when the fetus and placenta are growing rapidly; this is followed by the blood lost during labour. Iron-deficiency anaemia is common during pregnancy and supplementary iron administration is required. However, it should be remembered that decreased haemoglobin concentration in the blood during pregnancy is partly due to the fact that plasma volume increases relatively more than red-cell volume, so that estimation of haemoglobin concentration may give an exaggerated indication of anaemia [p. 583].

Growing children

During the period of growth the blood must 'grow' to keep pace with the rest of the body; a positive iron balance must therefore be maintained to enable the necessary additional haemoglobin to be formed. The following data indicate the size of the problem. At birth a baby weighs 3 kg and contains 300 ml of blood; the total iron content of the body is 400 mg. At one year it weighs 10 kg and contains about 1000 ml of blood. To produce the necessary increase in haemoglobin and body iron a positive iron balance of 500 mg of iron is required (=1.3 mg daily) over and above the iron needed to compensate for normal wastage. Even when mother and infant are on a satisfactory diet the infant's haemoglobin level falls from a birth value exceeding 100 per cent (of normal adult level) to 75 per cent at 3 months; between 6 and 12 months it is 85 per cent. Haemoglobin formation cannot, therefore, even under good cir-

cumstances keep pace with the increase in plasma volume. In families on a low iron intake the infant is probably born with poor iron reserves on which to draw, and it gets less iron in its diet; as a result the haemoglobin falls progressively, reaching a minimum at the end of the first year. Subsequently, as the rate of growth slows down, the blood picture improves steadily during childhood. A fresh strain develops in girls with the onset of puberty and menstruation.

If the anaemic children are given adequate amounts of iron, rapid recovery of the haemoglobin occurs.

Blood iron

Iron occurs in blood in three forms:

1. As *haemoglobin* in the red blood cells; 15 g of haemoglobin per dl of blood contains 50 mg of iron.

2. In combination with a plasma β-globulin called *transferrin* which is present in a concentration of 2.5 g per litre. Each molecule of transferrin can combine with 2 Fe^{3+} ions reversibly. The transfer of iron to the haemoglobin-forming cells (normoblasts) only occurs from the iron–transferrin complex. During the transfer ferric iron is reduced to ferrous iron. The *total iron-binding capacity of plasma* is 45–72 μmol l^{-1} (250–400 μg per 100 ml).

The *plasma iron concentration* is 13–32 μmol l^{-1} (70–175 μg per 100 ml). Normally the iron binding capacity of plasma is about 30 per cent saturated. Plasma iron is increased after administration of iron salts for treatment of iron-deficiency anaemia, and also when red-cell formation is depressed (aplastic anaemia, pernicious anaemia). Plasma iron is decreased when iron absorption is decreased or when red-cell formation is rapid, as after haemorrhage. With normal dietary intake of iron plasma iron level is quite constant.

3. As *ferritin*, which can be measured in serum by radioimmunoassay. Although transferrin iron level may convey some information about the body needs for iron, serum ferritin is the best indicator of the iron stores in the body.

Intake of iron

Normally, adequate amounts of iron occur in a mixed diet. Foodstuffs vary both in their content of iron and the availability of iron for absorption into the body. The iron in foods of animal origin is better absorbed than iron in foods of vegetable origin. About 10 per cent of dietary iron is absorbed and converted to haemoglobin.

Many factors influence the *absorption of iron* from the gut:

1. Absorption occurs mainly in the duodenum and upper jejunum, via the brush border of the intestinal mucosa into the bloodstream (not lymph).

2. Ferrous iron (Fe^{2+}) is better absorbed than ferric iron (Fe^{3+}). Reducing substances, such as ascorbic acid, enhance iron absorption by converting Fe^{3+} to Fe^{2+}.

3. The presence of phosphate or phytate in foods may reduce ionic iron absorption by forming insoluble iron salts. However, it seems that phytic acid in cereals inhibits iron absorption less than was formerly supposed because cereals mostly contain an enzyme, phytase, which destroys phytic acid.

4. Haem is absorbed direct, i.e. without splitting off of iron from protoporphyrin. Non-haem iron absorption is favoured by the presence of acid (HCl from stomach).

5. Decrease in iron stores in the body, i.e. iron-deficiency anaemia, enhances iron absorption. Exposure to hypoxia at high altitudes increases erythropoiesis and increases iron absorption. Conversely, an increase in iron storage in the body reduces iron absorption by the gut mucosa.

Control of iron absorption by the intestinal mucosa

The percentage absorption of iron is influenced by the above-mentioned factors. On an average mixed Western diet, 6 per cent of dietary iron is absorbed in males, 14 per cent in females during the years of menstrual bleeding, and 20 per cent in iron-deficient persons.

FIGURE I.28 shows diagrammatically the ways by which the intestinal mucosal cell controls iron absorption. Normally, only a small proportion of dietary iron passes through the cell into the portal blood. Some iron is stored as ferritin within the mucosal cell; this may come from absorbed dietary iron, but iron also enters the mucosal cell from the blood, perhaps directly from transferrin or perhaps by transport of plasma ferritin. When the mucosal cell is shed, its content of iron will be excreted via the faeces. FIGURE I.28 shows too how with iron deficiency iron from the intestinal lumen is more completely absorbed by the mucosal cell, cell ferritin content is reduced, and more iron passes into the blood. Correspondingly, less iron is excreted via the faeces, less iron passes from blood into the mucosal cell, and when the cell is sloughed less iron is lost from the body. With iron-overloading the mucosal cell accumulates a large store of ferritin, partly from the diet (at first) and partly from other iron stores via the blood [FIG. I.28]; iron absorption is decreased but more iron than usual would be lost by desquamation.

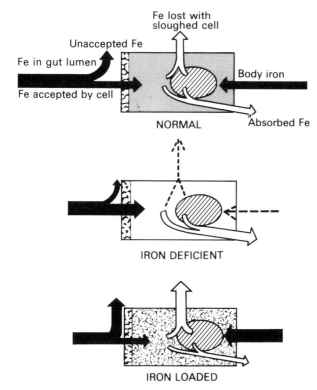

FIG. I.28. Control of iron absorption by the intestinal mucosa. After the iron has entered the cell, it may proceed into the body according to metabolic requirements. Alternatively, some of the iron may become fixed in epithelial ferritin to be lost when the cell is sloughed at the end of its life span. In iron-deficient subjects there appears to be little or no mechanism to retain absorbed iron within the villous epithelial cells and dietary iron thus readily passes into the body. Conversely in iron-loaded subjects, the large quantities of mucosal ferritin cause increased trapping of absorbed iron within the cell; this provides a mechanism for preventing further increase in body iron content. (Redrawn from Dagg, J. H., Cumming, R. L. C., and Goldberg, A. (1971). Disorders of iron metabolism. In *Recent advances in haematology* (ed. A. Goldberg and M. C. Brain) p. 77. London.)

Iron-deficiency anaemia

Iron-deficiency anaemia is of the *microcytic hypochromic* type [p. 37]. Both red-cell count and haemoglobin content are lowered, the latter more than the former. The bone marrow may show normoblastic hyperplasia. In this type of anaemia the daily requirement for maximal regeneration of haemoglobin may be 25–50 mg

of iron. This cannot be supplied in the diet and therapeutic iron preparations are essential.

Storage forms of iron

Iron is stored in tissues such as the liver, spleen, bone-marrow, lymph nodes, and reticulo-endothelial cells in two forms, ferritin and haemosiderin:

Ferritin is water-soluble and consists of a protein called *apoferritin* (MW 450 000) combined with *ferric* hydrophosphate to give a total molecular weight of 900 000.

Haemosiderin is granular, water-insoluble, and is a conglomerate of ferritin molecules. It gives a blue colour (Prussian blue) when ferricyanide is added to it. Most storage iron occurs as ferritin (2/3) and the remaining 1/3 as haemosiderin.

Increased storage of iron (siderosis) is caused by:

1. Excessive iron intake. This occurs among the Bantu, who cook their maize and other cereals in iron pots. Kaffir beer is also rich in iron and the total daily intake of iron may be 100 mg. Many cheap wines and cider contain enough iron to increase iron stores in the body.

2. Excessive destruction of erythrocytes, e.g. in haemolytic anaemias, especially after repeated blood transfusion, may also produce siderosis.

3. Failure to control iron absorption (with normal intake of iron) may lead to *haemochromatosis*. This may be hereditary and is associated with cirrhosis of the liver and chronic pancreatitis with diabetes mellitus. Since iron is not excreted to any significant extent, the only way to remove it is by repeated venesection.

Other metals in haemopoiesis

Copper. Copper is essential for haemoglobin synthesis in mammals, probably by promoting the absorption, mobilization, and utilization of iron. There is no evidence that anaemia from copper deficiency occurs in man, except perhaps in nutritional anaemia of infants. Very little copper is required; adequate amounts occur in the diet and most iron preparations contain traces of copper. There is no indication for copper administration in the treatment of anaemia.

Cobalt. Sheep and cattle develop anaemia when fed on a diet deficient in cobalt. This occurs because ruminants manufacture their vitamin B_{12} by bacterial action in the rumen. There is no evidence that cobalt deficiency anaemia occurs in other animals or man.

In many species large doses of cobalt produce polycythaemia, perhaps by increasing the production of erythropoietin.

REFERENCES

GROSS, F. (ed.) (1964). *Iron metabolism*. Springer, Berlin.
JACOBS, A. and WORMWOOD, M. (eds.) (1974). *Iron in biochemistry and medicine*. Academic Press, London.
MOORE, C. V. (1961). Iron metabolism and nutrition. *Harvey Lect.* **55**, 67.

FATE OF THE RED BLOOD CORPUSCLES. JAUNDICE

Duration of the life of red cells

Radioactive iron and heavy nitrogen after injection *in vivo* are quickly incorporated into red cell precursors. By the end of the reticulocyte stage, however, the power to take up these tracers is lost. The method suggests that it takes about 7 days for complete maturation.

Transfusion of reticulocytes and also studies of maturation *in vitro* indicate that these cells mature in about 4 days; hence development from the pronormoblast to the reticulocyte takes about 3 days. Multiplication of red cells occurs by mitosis of nucleated forms up to, but not including the late normoblasts. The rate and sites of blood formation may be gauged by injecting radioiron and taking serial counts at intervals over the bones, using a scintillation counter to measure radiation: thus a high count over the spleen of an adult would indicate extramedullary blood formation, a condition which may occur after widespread destruction of the normal bone marrow.

The life span of a mature erythrocyte is about 120 days, and since a reticulocyte takes 4 days to mature, it follows that if entry of reticulocytes into the peripheral circulation precedes final maturation there should normally be about 3 per cent of reticulocytes in the blood; since the actual figure is less than 1 per cent, it must be presumed that the majority of red cells are already mature when they leave the marrow.

Life span of the mature erythrocyte

This was originally studied by differential agglutination: a large volume of Group O blood is transfused into an anaemic Group A recipient. Counts of surviving Group O donor cells are made at intervals by diluting samples of the recipient's blood with Group B serum (containing α or anti-A agglutinins) which clumps the Group A recipient's cells and leaves the O donor cells free. Counts of the latter show that some of the transfused cells are still surviving after 100 days.

A simpler method depends on 'tagging' red cells with radioactive chromium. A few ml of the subject's blood are withdrawn, $Na_2{}^{51}CrO_4$ is added and the cells are washed. They are then returned to the circulation and the radiations are counted at intervals. The radioactive salt is so strongly adsorbed on the treated red cells that little elution occurs after injection. If the curve for loss of radioactivity is suitably corrected, it is found to be almost linear, falling to zero at about 120 days, the half life of the cell (50 per cent destruction) being about 60 days.

It must be realized that after the reticulocyte stage has been passed there are no means of determining how old any particular erythrocyte is: some will be young and capable of surviving 4 months; others will be senescent and ripe for destruction. Curves for survival times will give the maximum survival time of the youngest cells, the mean survival time, and the average rate of destruction of the cells in a heterogeneous sample.

Red-cell survival time in disease

Haemolytic anaemia. In haemolytic disorders the normal red-cell survival time of 120 days is much reduced. Anaemia does not usually develop until the survival time has fallen to about 20 days, as the bone marrow can produce a compensatory increase in red-cell production before it finally fails to keep pace with the rate of destruction. The enhanced erythropoiesis in the bone marrow is due to hypoxic stimulation of erythropoietin formation [p. 37].

Increase in red-cell destruction may reduce red-cell count and blood haemoglobin level; the serum unconjugated bilirubin concentration is raised, producing jaundice, and urinary urobilinogen excretion is increased; in severe cases haemoglobin and methaemoglobin are present in plasma and haemoglobinuria occurs. Selective uptake of radioactive chromium by the spleen may be detected by scanning the body surface after injection of ^{51}Cr-labelled red cells into the bloodstream. Evidence of increased erythropoiesis is provided by erythroid hyperplasia of the bone marrow, more rapid incorporation of radioiron into red cells, and by an increase in the proportion of reticulocytes in peripheral blood (normally < 1 per cent, in haemolytic anaemia 5–20 per cent of red cells).

Haemolytic anaemia may be due to corpuscular defects or to extracorpuscular abnormalities.

Corpuscular defects. *Hereditary spherocytosis* (congenital haemolytic anaemia, familial acholuric jaundice) is inherited as an autosomal dominant character. The red cells are mainly small and rounded (microspherocytes); they show increased osmotic fragility [p. 35] so that haemolysis may commence at 0.7 per cent saline and be complete at 0.4 per cent saline. Microspherocytes survive for only 15–20 days *in vivo*, both in patients with this disorder and when transfused into normal persons with the same ABO blood group. The spleen is enlarged and serum bilirubin level may be raised to 20–50 μmol 1^{-1}. The nature of the biochemical lesion is unknown but microspherocytes are more vulnerable than normal erythrocytes even in a physiological environment.

In *haemoglobinopathies* and *thalassaemias* [p. 37] there is increased haemolysis and sometimes severe anaemia.

Deficiency of certain *red-cell enzymes* render them more susceptible to haemolysis. For example, *glucose 6-phosphate-dehydrogenase* deficiency, inherited as a sex-linked character, decreases the reducing power of the cell and makes it more vulnerable to damage by oxidizing agents such as primaquine and other antimalarial drugs. *Pyruvate-kinase* deficiency may also produce haemolytic anaemia.

In the rare *paroxysmal nocturnal haemoglobinuria* the red cells are hypersensitive to lysis by complement [p. 59].

Extracorpuscular defects. Transfusions of blood with *ABO incompatibility* produces haemolysis [p. 49] and *Rh-D incompatibility* may produce haemolytic disease of the newborn [p. 48].

In *auto-immune haemolytic disorders* antibodies are formed against the patient's own red cells. These autoantibodies may be of IgG or IgM class and they may be bound to the red-cell membrane or occur in serum. Some react maximally at 37 °C (warm type) others at 20 °C (cold type). Spherocytosis and increased red-cell osmotic fragility may accompany the haemolytic anaemia.

Hypersplenism, whatever the cause, may enhance destruction of erythrocytes, leucocytes, and platelets.

REFERENCES

DACIE, J. V. (1960; 1962; 1967). *The haemolytic anaemias*, 2nd edn, Part I (1960); Part II (1962); Part III (1967); Part IV (1967). Dekker, London.
BRAIN. M. C. (1971). The red cell and haemolytic anaemia. In *Recent advances in haematology* (ed. A. Goldberg and M. C. Brain) p. 146. Churchill Livingstone, London.

Site of destruction of red cells

The red cells are destroyed by the reticulo-endothelial system. Haemoglobin is released and eventually, as explained below, both iron and globin are split off and bilirubin is formed. The released iron is most carefully stored in the body probably by becoming bound with a special tissue protein called *apoferritin* to form the iron-containing protein, *ferritin* [p. 43]. This 'reserve iron' is presumably released into the circulation for haemoglobin formation in the red marrow as and when required.

Macrophage (reticulo-endothelial) system

The term reticulo-endothelial system is used in this book to refer to certain phagocytic cells found mainly in the bone marrow, liver, lymph nodes, spleen (and subcutaneous tissues). In the marrow the cells form part of the lining of the blood sinuses (littoral cells); in the liver they lie at intervals along the vascular capillaries (Kupffer cells); in the lymph nodes they line the lymphatic paths; in the spleen they are found in the pulp. The characteristic feature of the cells of the reticulo-endothelial system is their power to ingest foreign colloidal particles; thus if carmine or Indian ink is injected intravenously into living animals the cells take up the dye, are deeply stained by it, and so are easily recognized. Cells of similar histological appearance which do not take up these dyes are not included in the reticulo-endothelial system. Because they ingest large particles the cells are called macrophages. The functions of these cells are as follows:

1. They ingest and destroy erythrocytes and form and release bilirubin. They also destroy leucocytes and platelets.

2. They ingest bacteria and are thus concerned with the defence of the body against infection. They rapidly increase in number under these conditions with resulting enlargement of the organs which are rich in these cells, e.g. spleen, lymph nodes.

3. They ingest and 'process' antigen which then stimulates antibody formation in plasma cells.

The reticulo-endothelial system functions as a physiological unity; if any part of it is put out of action the rest of the system undergoes compensatory hypertrophy and makes good the deficiency.

Origin of bilirubin

1. In the normal animal, and particularly when blood destruction is actively proceeding, the macrophages (reticulo-endothelial system) contain fragments of red cells, free haemoglobin, or iron in an inorganic form which gives the Prussian-blue reaction.

2. There is a trace of bilirubin in the circulating blood; but the blood leaving the spleen and the bone marrow contains significantly higher concentrations of bilirubin than the arterial blood, proving that bilirubin is formed in these organs.

3. Although the Kupffer cells of the liver also make bilirubin, the liver is *not* indispensable for bilirubin formation. After total extirpation of the liver, bilirubin rapidly accumulates in the body; the plasma and body fat are coloured yellow, and jaundice is present in animals which survive for more than six hours. Jaundice develops almost as rapidly when the spleen and the other abdominal viscera are removed as well, proving that the bone marrow macrophages are important sites of bilirubin formation.

4. After splenectomy (in the intact animal), compensatory hypertrophy of the other macrophages takes place, and the rate of bilirubin formation is not diminished. In bruises, after the passage of time the haemoglobin of the extravasated blood is converted progressively into bilirubin owing to the activity of the local macrophages; these, however, are not concerned with the normal destruction of circulating blood cells.

Chemistry of bilirubin formation

Haemoglobin contains haem (=iron protoporphyrin IX) attached to globin. In the reticulo-endothelial cells the haem part of the molecule is altered by the oxidation of the C of one of its methine (= CH) bridges. The tetrapyrrole ring structure is thus broken and the four pyrrole groups become arranged as a straight chain. As a result of this chemical change the green iron-containing compound choleglobin is formed; as its name implies the molecule still contains the original globin. Next both iron and globin are split off and bilirubin (sometimes called haemobilirubin) is formed.

This compound is only soluble in lipid solvents but is extruded into the plasma in a colloid form bound with α-globulin. It is this protein conjugation of bilirubin which is responsible for the solubility of the bilirubin complex in the plasma and which prevents its excretion by the kidney. On reaching the liver, bilirubin (or bilirubin-globulin) enters the hepatic cells and therein undergoes conjugation with glucuronic acid to form water-soluble bilirubin monoglucuronide and bilirubin diglucuronide (these compounds are sometimes called cholebilirubin). These compounds pass by the bile ducts to the intestine, where by bacterial degradation, mainly in the colon, stercobilinogen (= urobilinogen) is formed. Some urobilinogen is reabsorbed and goes via the portal system to the liver, whence some escapes into the general bloodstream and some is re-excreted in the bile. Urobilinogen, unlike bilirubin itself, is filtered off by the kidney and is excreted in the urine. If the urine is allowed to stand urobilinogen is oxidized to urobilin. That amount of stercobilinogen which is not reabsorbed from the intestine is excreted in amounts of 20–250 mg per day in the faeces. Some of this stercobilinogen is oxidized to stercobilin. These compounds are responsible for the brown colour of the faeces.

Jaundice

Jaundice is a yellow colour of the skin, conjunctivae, and other tissues caused by the presence of an excess of bilirubin in the plasma and tissue fluids. The normal adult range of plasma bilirubin is 5–18 µmol l^{-1} (0.3–1.0 mg per 100 ml); the mean is 10 µmol l^{-1} (0.5 mg per 100 ml).

An excess of bilirubin in the blood can result from three causes:

1. Excessive breakdown of red blood cells—haemolytic jaundice.

2. Infective or toxic damage to the liver cells—hepatic or hepatocellular jaundice.

3. Obstruction of the bile ducts—obstructive jaundice.

A useful classification of these three types is: (i) pre-hepatic; (ii) hepatic; (iii) post-hepatic jaundice.

In differentiating these three types of jaundice it is important to remember that the bile contains bile salts secreted by the liver which are responsible for the emulsification and absorption of fat from the intestine. Correspondingly, hepatic damage or biliary obstruction leads to an abnormally high fat content of the faeces. Thus the faeces are bulky, greasy, and often foul-smelling due to chemical degradation of the fats by intestinal bacteria. If biliary obstruction is complete, stercobilin must be absent from the faeces which are correspondingly pale, and there is no urobilin in the urine.

On the other hand, haemolytic jaundice is not associated with any abnormalities of fat digestion and absorption; the faeces do not contain fat but may contain increased amounts of stercobilin. The blood picture, normal in other types of jaundice, shows anaemia in haemolytic jaundice with the presence of reticulocytes and abnormal red cell forms.

As has been said, bilirubin formed by the reticulo-endothelial cells is lipophilic, non-water-soluble, and exists in the plasma as a protein conjugate. It is not excreted in the urine, hence bilirubinuria is not found in haemolytic jaundice (hence the term acholuric), although urobilinuria may be very obvious. Bilirubin mono- and diglucuronides however are water-soluble. In hepatic and post-hepatic jaundice the bilirubin glucuronides formed by the liver cells are 'regurgitated' back into the bloodstream. There these glucuronides are conjugated with plasma protein but their greater water solubility allows them to dissociate from the protein and to escape through the glomerular filter, bilirubinuria occurring when the plasma bilirubin-glucuronide level exceeds 35 µmol l^{-1} (2 mg per 100 ml). The bile salts also escape into the bloodstream in hepatic and post-hepatic jaundice and are excreted in the urine; there is no excretion of bile salts in pre-hepatic jaundice.

The van den Bergh test. When a mixture of sulphanilic acid, hydrochloric acid, and sodium nitrite (diazo reagent) is added to serum containing an excess of bilirubin glucuronide a reddish-violet colour results, the maximum colour intensity being reached within 30 seconds. This is the so-called *direct* reaction. When the reagents are mixed with serum containing an excess of bilirubin itself or bilirubin–protein complex, as formed by the reticulo-endothelial system no colour develops until alcohol is added, whereupon the reddish-violet colour makes its appearance. This is the so-called *indirect* reaction. In this indirect reaction, the addition of alcohol solvent provides the means of solution for the water-insoluble bilirubin, which is thus enabled to react with the diazo reagent.

From the above discussion it should be clear that in haemolytic (= pre-hepatic) jaundice the plasma gives only an indirect van den Bergh reaction whereas in hepatic or post-hepatic jaundice most of the excess pigment present in the plasma will give a direct reaction.

The liver forms *plasma proteins* and in infective hepatitis there is often evidence of abnormal changes in the plasma protein level. The plasma albumin level is often below normal: the globulin level may however be increased and electrophoresis reveals that this increase is due to excessive production of γ-globulins. Even if the total globulin level is not above normal there is usually a decrease in the A/G ratio.

Various empirical tests show positive results when serum containing an excessive amount of γ-globulin is treated with suitable reagents. One commonly employed is the *thymol turbidity* test in which the serum is added to a thymol-barbitone solution of low ionic strength; normally this results in faint opalescence but if there is an excess of γ-globulin the solution becomes turbid.

Alkaline phosphatase present in the plasma is derived from osteoblasts in the bone. Alkaline phosphatase is excreted in the bile and if there is biliary obstruction or if infective hepatitis is associated with obstructive features, as is often the case when the swollen liver cells occlude the minute bile canaliculi, the alkaline–phosphatase level in the plasma rises. The accompanying Table presents some of the more usual chemical and pathological changes in the three types of jaundice.

	PRE-HEPATIC	HEPATIC	POST-HEPATIC
Urine urobilinogen	++	+	− (if complete obstruction)
Urine—bile salts	absent	present	present
Urine bilirubin	absent	present	present
Faecal urobilinogen	increased	reduced	reduced or absent
Faecal fat	absent	increased	increased
Liver function tests	normal	impaired	normal or impaired
Alkaline phosphatase	normal	high	very high
Thymol turbidity	slight	marked	usually slight
Plasma albumin	normal	reduced	normal or reduced
Plasma γ-globulin	normal	increased	normal
van den Bergh	indirect+	mainly direct+	mainly direct+

Jaundice (hyperbilirubinaemia) of the newborn

It is unusual for the plasma bilirubin concentration in normal adults to exceed 18 μmol l^{-1} (1 mg per 100 ml); higher values are, however, very common in normal young infants. In 68 of 110 normal newborn infants the plasma bilirubin in the umbilical cord blood exceeded 18 μmol l^{-1} (1 mg per 100 ml). In many infants the bilirubin level continues to rise after birth to a peak which is generally reached during the first week, and then declines; if the plasma bilirubin exceeds 80 μmol l^{-1} (5 mg per 100 ml) clinical jaundice is always present. The level of serum bilirubin is higher in premature infants than in those born at term, and shows an inverse relationship to the body weight. In any case it rarely exceeds 200 μmol l^{-1} (12 mg per 100 ml). The cause of this 'physiological' jaundice of the newborn has been much discussed:

1. The jaundice has been attributed to excessive haemolysis. In support of this view it is emphasized that a rapid fall of red-cell and haemoglobin concentration normally occurs after birth; the fall is maximal during the second week and may continue into the third month; as a result the blood picture changes from one of polycythaemia (by adult standards) to one of anaemia. But the usual evidence of excessive haemolysis is absent; thus no haemolysins

have been demonstrated and there is no raised saline fragility of the red cells; there is no relationship between the time of onset of the fall or the rate of fall of the red-cell count and the onset or the severity of the jaundice. The fall in the red-cell count immediately after birth is probably due to decreased activity of the red marrow and not to increased haemolysis.

2. The jaundice of the newborn is probably due to hepatic immaturity. Many physiological functions are imperfectly carried out in the newborn, e.g. temperature regulation, voluntary movements. *In utero* the bilirubin formed is mainly eliminated via the placenta; it is secreted to a lesser extent by the liver into the bile and reaches the intestine to form the green meconium. Immediately after birth the liver has to eliminate all the bilirubin formed; it would seem that frequently the liver is unable to deal adequately with this task during the first 10 days of life; jaundice therefore develops. Bilirubin excretion in the faeces in these jaundiced infants is always decreased.

One of the *congenital* forms of hyperbilirubinaemia is due to a defect in the glucuronide-conjugating machinery.

Excretion of bile pigments in urine

Bilirubin (haemobilirubin) is not excreted by the kidney, but bilirubin glucuronide (cholebilirubin) is excreted when the plasma level exceeds 35 μmol l^{-1} (2 mg per 100 ml). In general then, haemolytic jaundice tends to be acholuric, i.e. there is no bile pigment in the urine; in obstructive jaundice, however, bile pigment (and bile salts) appear in the urine in amounts which are proportional to the concentration of these substances in the blood.

REFERENCES

GRAY, C. H. (1970). The bile pigments. *Br. med. Bull.* **13**, 94.
LATHE, G. H., CLAIREAUX, A. E., and NORMAN, A. P. (1958). In *Recent advances in paediatrics*, 2nd edn (ed. D. Gairdner) p. 87. Churchill Livingstone, London.
MACLAGAN, N. F. (1970). Diseases of the liver and bilary tract. In *Biochemical disorders in human disease* (ed. R. H. S. Thompson and I. D. P. Wootton) 3rd edn, p. 129. Churchill Livingstone, London.

BLOOD GROUPS

Three main groups of 'factors' are present in human blood cells which enable the cells of different individuals to be differentiated; these factors have great clinical, medicolegal, and genetical interest. The chief groups are:

1. A, B, and O.
2. Rh (Rhesus) factors.
3. M and N.

Classical (A, B, O) blood groups

Human beings can be divided into four main groups according to the presence (or absence) in their red cells (and in certain tissue cells) of the substances called A, B, and O: 42 per cent contain substance A; 9 per cent contain substance B; 3 per cent contain substances A and B; 46 per cent contain substance O. The percentage distribution given above is for Western European peoples. Some of the Eastern European peoples show a higher proportion (up to 40 per cent) of Group B. Pure American Indians belong almost exclusively to Group O.

The groups are correspondingly called group A, group B, group AB, and group O. More refined analysis shows that substance A can be subdivided into two sub-groups called A_1 and A_2; A_1 includes 75 per cent of all group A, A_2 forms 25 per cent. Group AB is similarly divided into A_1B and A_2B.

A and B are called group specific substances and chemically are polysaccharides; they are *agglutinogens*, i.e. in the presence of a suitable antibody called an *agglutinin*, agglutination or clumping of the red cells occurs. The agglutinin acting on agglutinogen A is called α or anti-A; the agglutinin acting on agglutinogen B is called β or anti-B. (These agglutinins are also called isohaemaglutinins.) Group specific substance O does not normally act as an agglutinogen and there is no corresponding agglutinin. Very rarely, after repeated transfusions, group O cells evoke an irregular agglutinin (anti-O) in susceptible recipients. Group O cells are not agglutinated by agglutinins α or β.

The so-called Landsteiner's law states the following: if an agglutinogen is present in the red cells of a blood, the corresponding agglutinin must be absent from the plasma; if the agglutinogen is absent the corresponding agglutinin must be present. The first part of this law is a logical outcome of the situation, for if both agglutinogen and agglutinin were present the cells would be agglutinated. The second part is a fact but not a necessary sequence, for when the agglutinogen is absent from blood the agglutinin might well be absent too. In fact absence of Rh agglutinogen is not normally accompanied by presence in the plasma of the anti-Rh agglutinin; similarly absence of M or N substance is *not* accompanied by the presence of anti-M or anti-N in the plasma. It is only in the case of the A, B, O blood groups that absence of A is associated with the presence of anti-A (α) and the absence of B with the presence of anti-B (β). The finding is strange and at present inexplicable.

If the serum of the four classical groups is examined, the agglutinins are found to be distributed as follows: Group A contains β; Group B contains α; Group AB contains no agglutinin; Group O contains α and β. The full description of the four groups taking into account both agglutinogens and agglutinins would be, therefore: Aβ; Bα; AB; Oαβ. The agglutinin α is not a single substance but can be subdivided into α_1 and α proper: α_1 only agglutinates A_1; α proper agglutinates both A_1 and A_2. The relative amounts of α proper and α_1 vary in the sera of Groups B and O.

To determine an individual's classical blood group an isotonic saline suspension of his red cells is mixed on a slide:

1. With a test serum containing agglutinin α.
2. With a test serum containing agglutinin β.

The results are diagnostic as shown by the following Table. When no agglutination occurs the red cells remain separate and evenly distributed; when agglutination occurs the cells are massed together in clumps and lose their outline. With high-titre (i.e. powerful) agglutinins the cells are massed into a few large clumps; with weaker agglutinins more numerous but smaller clumps are formed.

Cells	Serum from Group A subject Agglutinin β [anti-B]	Serum from Group B subject Agglutinin α [anti-A]
A	−	+
B	+	−
AB	+	+
O	−	−

+ = Agglutination.   − = No agglutination.

The agglutinogens A and B first appear in the sixth week of fetal life; their concentration at birth is one-fifth the adult level and it progressively rises during puberty and adolescence. Group-specific substances A and B are not limited to the red cells but are found in many organs: salivary glands and pancreas++; kidney, liver, and lungs+; testis. They are water-soluble and in about 80 per cent of people (secretors) they appear in the body fluids in the following

relative concentrations: saliva and semen, 600; amniotic fluid, 175; red cells, 8–32; tears, 5; urine, 3; cerebrospinal fluid, 0.

Only 50 per cent of newborn infants have demonstrable agglutinin, and this has simply filtered across the placenta from the mother. The specific agglutinins appear at 10 days, rise to a peak at 10 years, and then decline. At all ages there are marked variations in agglutinin content in different individuals: the range for the 'titre' (i.e. concentration) of agglutinin α is 8–2048 (most at 128); for β, 2–1024 (most at 32). The figures quoted indicate the *extent to which the serum can be diluted before losing its agglutinating potency*; a high titre represents high agglutinin activity. The specific agglutinins act best at low temperatures and against well-diluted cells; with weak sera and high cell concentrations the cells may 'mop up' the agglutinin without being agglutinated.

Non-specific agglutinins may occur clinically which act in the 'cold' (at 0–35 °C) and not usually at body temperature; a person's cold agglutinins may agglutinate his own red cells.

Relation of A, B, O blood groups to blood transfusions

The effects of transfusing the blood of any group into the circulation of a member of another group can readily be worked out if the following additional points are borne in mind. The serum agglutinins of the donor can usually be ignored (if their potency is not too high) as they are sufficiently diluted by the much larger volume of the plasma of the recipient to produce no ill effects and they are also neutralized by soluble agglutinogens which are found free in the recipient's body fluids. It follows, therefore, that the agglutinogens of the cells of the recipient need not usually be considered. Account need only be taken of the effect of the *serum agglutinins of the recipient on the cells (agglutinogens) of the donor*. There are exceptions to this rule, for sometimes the titre of agglutinins (anti-α or anti-β) in group O plasma is exceptionally high and may agglutinate the recipient's cells after transfusion. Except in emergency, intra-group transfusion should be used. The following Table indicates the effects that would be produced by transfusing cells of any group into a recipient of any group (the sign + indicates agglutination of the cells and incompatibility; the sign − indicates no agglutination, and therefore compatibility):

Recipient (agglutinins in serum in italics)	Red cells of donor				Percentage of individuals in each group
	AB	A	B	O	
AB + no agglutinin	−	−	−	−	AB 3
A + agglutinin β	+	−	+	−	A 42
B + agglutinin α	+	+	−	−	B 9
O + agglutinins αβ	+	+	+	−	O 46

The following important conclusions can be drawn from the preceding Table. Group A and Group B can only safely receive red cells from Group O and their own group. Group AB contains no agglutinins in the serum and have therefore been called the *universal recipients*. The cells of Group O contain no agglutinogen and are not agglutinated by the members of any group; Group O have therefore been called *universal donors*. The classical terms, universal donor and recipient, are however no longer valid as they ignore the complications produced by the existence of the Rh factors. It should be emphasized that the only safe method of determining compatibility is to test *directly the serum of the recipient against the donor's corpuscles*; this should, whenever possible, be carried out in practice. If carried out as described on p. 49, it will safeguard against Rh as well as against ABO incompatibility.

Rh (Rhesus) blood groups

The Rh groups are of outstanding clinical importance. The original discovery was made as follows: red cells of the rhesus monkey were injected into rabbits; the rabbit responds to the presence of an antigen in these cells by forming an antibody which agglutinates rhesus red cells. The surprising observation was then made that if the immunized rabbit's serum is tested against *human* red cells, agglutination occurs in 85 per cent of white men; these people are called Rh+ (positive) and their serum contains no Rh antibody. No agglutination occurs in 15 per cent; these are called Rh− (negative) and their serum also contains no Rh antibody except in the circumstances explained below.

There are several varieties of Rh antigen and of Rh antibody; the commonest Rh antigen is called D, and its antibody is called anti-D. Blood group antigens are the result of the action of genes which are present in the chromosomes. The gene corresponding to the antigen D is also called D; when D is absent from the chromosome, its place is occupied by the alternate form (allelomorph) called d. A Rh gene is inherited from both the father and the mother. If gene D is carried by both sperm and ovum the resulting gene composition (genotype) of the offspring is DD; if the gametes carry D and d respectively the result is Dd; if both gametes carry d, the result is dd. DD (called homozygous) and Dd (called heterozygous) are both Rh+; dd (homozygous) is Rh−. Of the 85 per cent Rh+ English people examined, 35 per cent were DD and 48 per cent were Dd; the remaining 2 per cent of the Rh+ people had some other genotype containing D. In the case of a homozygous father of genotype DD, all the sperms contain D; with a heterozygous father of genotype Dd, half the sperms contain D and half d. These factors are of importance in relation to the genotype of the child and the likelihood of development of haemolytic disease.

As noted previously, in the case of individuals whose red cells contain no D agglutinogens, anti-D agglutinins are not naturally present in the plasma [cf. the ABO groups, p. 46], but the production of anti-D may be evoked by:

1. Transfusion of a Rh− individual with Rh+ blood (experimentally, 0.5 ml may suffice).
2. By the presence of a Rh+ fetus in a Rh− mother; in the latter case the titre of anti-D is not likely to be high, unless the woman has undergone one or more pregnancies.

A serum containing anti-D acting *in vitro* may agglutinate Rh+ cells suspended in saline (complete agglutinins). In other cases no such agglutination is demonstrable *in vitro*, although the transfused Rh+ cells may produce a severe reaction *in vivo*. In such cases, the presence of agglutinins may be made manifest by allowing them to act *in vitro* on Rh+ cells suspended not in saline but in ox albumin (20 per cent). Such anti-D is designated as an incomplete antibody.

Rh factor and haemolytic disease

The child of a Rh− mother (genotype dd) and a Rh+ father (genotype DD) must be Rh+ (Dd); if the Rh+ father is Dd the offspring may be Rh+ (Dd) or Rh− (dd). If the mother is Rh− and the fetus is Rh+, serious complications may occur: cells containing D may pass across the placenta from the fetus to the mother; the latter responding by forming anti-D which returns to the fetal circulation and tends to destroy the fetal red cells. The degree of damage done to the fetus depends on the magnitude of the maternal anti-D response and the ability of maternal Rh agglutinins to cross the placenta. Generally no harm is done during the first pregnancy; but serious results may occur in the second or later pregnancies depending on the degree of sensitivity of the mother. If the mother has been immunized previously by a Rh+ transfusion at any time, *even in childhood*, a dangerously high response may occur during a first pregnancy. But it should be emphasized that in most cases, agglutinins are *not* formed and the great majority of matings between a Rh− mother and a Rh+ father result in normal offspring; presumably fetal red cells do not usually cross the placental barrier, and so maternal agglutinins are not evoked.

Effects of anti-D on the fetus

The changes in the fetus may be termed *haemolytic disease* because they are due to the destruction of the red cells by maternal anti-D. The following are the chief clinical syndromes:

Hydrops fetalis. The fetus is grossly oedematous; it either dies *in utero*, or if born prematurely or at term, it dies within a few hours.

Icterus gravis neonatorum (haemolytic disease of the new born). The infant is born at term; it is jaundiced [haemolytic jaundice, p. 45] or becomes so within 24 hours. There may be no anaemia at birth because the excessive red-cell destruction is more or less compensated for by an intense normoblastic response of the marrow, associated with a high reticulocyte count and the presence of many nucleated red cells in the circulating blood (erythroblastaemia; erythroblastosis fetalis). The rate of red-cell destruction by anti-D is maximal at birth and so anaemia may develop in the first few days. There may be severe neurological lesions involving the basal ganglia especially; they secondarily become stained bright yellow with bile pigment (kernicterus) [p. 361]; the liver may also be severely damaged and death may occur from liver failure. Free anti-D (derived from the mother) is present in the infant's blood for at least one week after birth and continues to destroy the infant's cells, though at a diminishing rate, all this time.

The best treatment for severe haemolytic disease is an *exchange transfusion* carried out soon after birth. A polythene catheter is passed along the umbilical vein into the inferior vena cava; small quantities of the infant's blood are successively withdrawn and replaced by an equal volume of compatible *Rh negative blood*. The infant's Rh+ red cells which were doomed to destruction are thus removed from the circulation and the infant's organs do not have to deal with the products of their disintegration; the infant is given an adequate supply of Rh− red cells which will survive in the circulation for the normal length of time. In some cases with persistent elevation of the serum bilirubin, a second exchange transfusion may be needed. The danger level for bilirubin is about 18 mg per 100 ml plasma: above this value there is a risk of kernicterus.

Prevention of Rh-haemolytic disease. It has been known for some time that ABO incompatibility between mother and baby nearly always prevents rhesus immunization of the mother. The most probable explanation is that any Rh+ fetal cells which cross into the maternal circulation are destroyed by the mother's naturally occurring anti-A or anti-B before they have had time to stimulate the production of Rh antibodies.

A technique for demonstrating fetal cells in maternal blood shows that transplacental haemorrhage is most likely to occur near or just after delivery and that the greater the number of Rh+ fetal cells in the maternal circulation just after delivery the more likely the mother is to produce immune Rh antibodies. In the more usual case where mother and baby are ABO compatible the destruction of Rh+ fetal cells in the maternal blood can be brought about by injection of anti-D (in the form of gamma-globulin containing a high titre of anti-D) soon after childbirth. In this way maternal Rh immunization can be prevented (Clarke 1975).

Rh factor and blood transfusion

The following rules should be observed:

1. To avoid sensitization it is essential that Rh as well as ABO

testing should be carried out whenever possible before a blood transfusion is performed, whatever the sex of the recipient.

2. No Rh− female at any age before the menopause should ever be given a Rh+ blood transfusion if it can possibly be avoided. If she is Rh−, she becomes sensitized by the injected Rh+ blood and forms anti-D; she is likely to destroy, subsequently, any Rh+ fetus with which she becomes pregnant. In other words the transfusion may make her permanently childless.

3. Any Rh− woman who has given birth to a child suffering from haemolytic disease is undoubtedly immunized to D and her serum contains anti-D. Likewise any Rh− person who is given Rh+ blood may become sensitized and the serum may contain anti-D. Anyone whose serum contains anti-D will show all the changes attributable to mismatched transfusion if they are transfused with a blood, otherwise compatible, but containing D.

Direct cross-matching. The only sure safeguard against transfusion complications is to match the serum of the recipient directly against the cells of the donor. The donor's cells should be diluted in normal saline, which will reveal the presence of anti-ABO agglutinins in the recipient's serum and also complete anti-D. A second suspension of donor cells in 20 per cent albumin is also tested against the recipient's serum, to reveal incomplete anti-D if this be present. The latter test may be supplemented by the direct Coombs test carried out on washed donor cells after exposure to the recipient's serum. Agglutination in any of these tests indicates incompatibility.

Numerous other agglutinogens commonly occur, which are usually named after the recipients in whom they occasionally evoke the formation of agglutinins (e.g. Kell, Luther, Duffy, etc.). These agglutinogens do not obey Landsteiner's law in that absence of the agglutinogen from the red cell is not associated with the presence of the corresponding agglutinin in the plasma. Further they very rarely evoke the formation of agglutinins even after transfusion, though where such formation does occur, typical reactions may follow subsequent transfusions. So, too, these agglutinogens are very occasional causes of haemolytic disease in the newborn. However, for practical purposes, Kell, Luther, Duffy (and also M and N agglutinogens) may usually be disregarded.

Use of plasma for transfusion

When comparatively small volumes of plasma are used its agglutinin content can be ignored because its titre is reduced to harmless levels by dilution in the recipient's plasma. If large volumes of plasma are employed the injected agglutinins may be harmful. It is then wiser to use plasma from which the agglutinins have been removed by contact with the appropriate red cells (conditioned plasma). If a blood bank is used in which blood of all four groups is pooled, the α and β agglutinins are absorbed by the A, B, and AB cells and thus removed.

Effects of incompatible (mismatched) blood transfusions

When whole blood is injected intravenously into patients (blood transfusion), serious symptoms and even death can occur if the recipient's serum contains antibodies (α, β, or anti-D) which agglutinate the donor's red cells. The red cells are first agglutinated and then undergo haemolysis. The following types of clinical reaction occur:

Inapparent haemolysis

The injected red cells are rapidly destroyed, the recipient's blood returning within a week or less to its pre-transfusion state. No other symptoms are observed.

Post-transfusion jaundice

The injected red cells are destroyed and the bilirubin formed from the released haemoglobin accumulates in the blood in sufficient amounts to produce jaundice (haemolytic jaundice). The amount of urobilinogen in the urine is increased.

Severe reaction with haemoglobinuria and renal failure

Soon after beginning the transfusion, perhaps after a few ml of blood have been introduced, the patient complains of violent pain in the back or elsewhere, and tightness in the chest; these symptoms are attributed to the agglutinated red cells forming clumps which block capillaries. The masses are then haemolysed; the released haemoglobin colours the plasma red; some of the haemoglobin is converted into a brown pigment. With rapid destruction of the haemoglobin, bilirubin accumulates in the plasma and stains the tissues (jaundice).

The volume of urine is greatly decreased. This is due in part to a fall of arterial blood pressure (e.g. down to 70 mm Hg) and perhaps to local vascular disturbances in the kidneys, leading to grave reduction of the glomerular filtrate volume. If the urine is acid and glomerular filtration is slow, the haemoglobin which passes through the glomeruli is precipitated in the tubules, possibly as acid haematin, owing to the filtrate becoming acid as it passes down the tubule. The lumen of the tubule is thus obstructed and this may interfere with the flow of urine. (Intravenous injection of haemoglobin solution into man and animals has caused transient oliguria, but precipitation of pigments does not occur unless there are coincident vascular changes.) However, the anuria which develops, and which may become almost complete, is probably due chiefly to vascular disturbance involving the glomeruli. Owing to renal failure the nitrogenous constituents and the potassium content of the blood rise, and a condition of latent uraemia results. In severe cases the patient may die in 8–10 days after developing lethargy and coma; marked hyperkalaemia is present before death. It is important to note that after incompatible transfusion, tubular damage occurs (lower nephron nephrosis), and that during recovery there is a stage when potassium is being filtered from the plasma by the glomeruli, but is not resorbed by the tubules: under these conditions there is a real danger of death from hypokalaemia.

In a series of cases in which incompatible blood of the ABO groups was given, no death occurred when less than 350 ml of blood were transfused.

Inheritance of classical blood groups

The four classical blood groups depend on three genes named after the corresponding factor, A, B, and O. Each person's blood group is determined by the two genes which he receives from each parent. Genes A and B, whether present in the red cell separately or together, may be demonstrated by the use of anti-A or anti-B serum. The presence of the O gene is not easily demonstrated, and to anti-A serum AA and AO cells react alike, both serologically being Group A.

Accordingly:

If a person receives genes	$\begin{cases} A + A \\ or \\ A + O \end{cases}$	$\begin{matrix} B + B \\ or \\ B + O \end{matrix}$	A + B	O + O
His blood group is	A	B	AB	O
His genotype is	AA or AO	BB or BO	AB	OO

This information can be used in investigating cases of disputed paternity. A baby must receive one of three possible genes (A, B, or O) from each parent. Further, each parent transfers one or two genes to the child: an A parent (genotype AA or AO) can give A or O, a B parent (genotype BB or BO) B or O, an AB parent A or B, and an O parent (OO) O only. It must be remembered however that the child's blood may not be set in its true ABO type until as late as one year after birth.

It follows that:

If the baby's group is	Parents must have given it	So if mother was	Father could not have been
O	O + O	No matter which	AB
AB	A + B	No matter which	O
A	A + O or A + A	B or O	B or O
B	B + O or B + B	A or O	A or O

The medicolegal value of blood grouping tests is greatly enhanced if the MN and the various Rhesus factors are also studied.

The M and N factors depend on two minor blood genes. It is very rare indeed for these to elicit an output of agglutinins when injected into man and they may therefore be ignored in carrying out transfusions. They are, however, antigenic to rabbits and agglutinating sera can be prepared by injecting human M or N cells into rabbits.

Each person carries two of the genes of the M and N group, i.e. M + M (=M), N + N (=N), or M + N (=MN). As a result:

If a baby's supplementary group is	Parents must have given it	So if mother's supplementary group was	Father could not have been
M	M + M	No matter which	N
N	N + N	No matter which	M
MN	M + N	N	N
MN	M + N	M	M

It should always be remembered that blood grouping tests can never prove that any suspected person *is* the actual father; they can only show that he could not possibly have been the father or that he (like many others) might have been.

Use of stored blood

Blood transfusion is the ideal treatment for severe haemorrhage when given promptly and in adequate amounts. Modern medical and surgical procedures have greatly increased the number of transfusions given and the amount of blood used for each transfusion; hence it has become necessary to institute blood banks, where blood stored at 4 °C is always available. Ordinarily such blood will be grouped and cross-matched, intragroup transfusion being used. If time does not permit the grouping and cross-matching of the recipient, O Rh− blood should be used. In cases of extreme urgency (war casualties; train accidents) it may be necessary to give O Rh+ blood, though a proportion of recipients will develop anti-D agglutinins.

Red cells undergo rapid changes during storage in simple citrate solutions even at 4 °C. They are preserved much longer in the presence of glucose. This acts partly by liberating lactic acid, the resulting fall of pH favouring survival both *in vitro* and *in vivo*; this effect is usually reinforced by using disodium hydrogen citrate instead of trisodium citrate as an anticoagulant. The chief effect of glucose, however, is to provide a substrate for the metabolism, which even at 4 °C is still important and contributes to cell survival. During cold-storage, reduction of metabolism decreases active

transport and cations move with the concentration gradient so that cell K^+ falls, and plasma K^+ rises from the normal 4–5 mM 1^{-1} to 20 or 30 mM 1^{-1} in 2 weeks, while cell Na^+, normally about 12 mM 1^{-1}, rises to 30 or 40 mM 1^{-1}. The changes may be summarized as follows:

1. Increase in cell Na^+, decrease in cell K^+, with a net increase in cell total base and water.

2. As a result the cells swell and become shorter and fatter, i.e. more spherocytic [p. 35]: in consequence they undergo haemolysis more readily in hypotonic solution and may rupture *in vitro* in salt concentrations as high as 0.8 per cent.

3. Spontaneous haemolysis of the cells takes place to an increasing degree while in contact with their own plasma in the blood bank.

4. The balance of phosphorylation and dephosphorylation is disturbed, phosphoric esters breaking down to liberate inorganic phosphate, which may rise in the cells from 2 to 30 mg per 100 ml, while adenosine triphosphate (ATP) decreases, the fall expressed as P being from 10 to 2 mg per 100 ml cells. It is probable that the immediate link between glucose and glycolysis on the one hand and cell nutrition and active cation transport on the other is ATP.

Changes in stored blood after transfusion

It is found that if abnormal cold-stored red cells are incubated with glucose *in vitro* at 37 °C, their metabolism greatly increases; consequently the Na^+ is extruded from the cells against the concentration gradient, the K^+ is drawn back and the ionic pattern of the red cells returns to normal. Similarly if the stored cells are transfused, they become normal ('reconditioned') in less than 48 hours with respect to Na^+ and K^+ content, volume, shape, and saline fragility.

The survival of stored red cells after transfusion is a better test of the value of a blood preservative than are *in vitro* tests and, as a result of such tests, the standard acid citrate glucose (dextrose) blood diluent (ACD) has been introduced: it contains 100 ml 2 per cent disodium hydrogen citrate and 20 ml 15 per cent glucose to which are added 420 ml blood. After 14 days at 4 °C the red cells of blood so treated show 80 per cent survival 24 hours after transfusion and thereafter the surviving cells are destroyed at the rate of about 1 per cent per day. Stored blood is not a suitable medium for conveying leucocytes or platelets to a recipient, for the former disintegrate *in vitro* after 2 or 3 days; the latter may survive longer but both leucocytes and platelets disappear within a few days of transfusion.

Dangers of blood transfusion

Apart from the dangers of incompatibility, there is a risk of mechanical overloading of the circulation, particularly in patients with cardiac damage. There are also certain 'chemical' risks:

1. It has been seen that stored cells lose potassium to the external plasma and that about 48 hours elapse before the K returns to the red cells; in the earlier part of this period there is in the case of massive transfusions a real risk of the patient dying from hyperkalaemia; this has occurred in replacement transfusion for erythroblastosis fetalis.

2. With massive transfusions the normal conversion of citrate to bicarbonate may be delayed and the patient may suffer from lack of ionized calcium; this is more likely to occur in patients with liver disease or in induced hypothermia; in the case of massive transfusions with citrated blood it is usual to 'cover' the citrate by giving calcium gluconate.

3. Citrate is normally oxidized to bicarbonate by the tissue cells [citric-acid cycle, p. 454]; with defective kidney function alkalosis might result.

Stored plasma

Plasma can be stored in the liquid form for many months: if dried it can be kept for years and under all conditions of temperature; it can be reconstituted by adding sterile water. Plasma is of great value in traumatic shock, and burns. Conditioned plasma [p. 49] can be given in very large volumes without fear of agglutinating the recipient's red cells. The plasma was originally prepared from large pools representing many donors and some batches carried the virus of infective hepatitis. Present methods employ only small pools, but even so the risk of hepatitis remains. For this reason transfusion with the polysaccharide dextran (6 per cent in saline), is preferred by many. The large molecules of this substance are well retained in the circulation and help to maintain the blood pressure.

REFERENCES

CLARKE, C. A. (1975). *Rhesus haemolytic disease*. MTP, Lancaster, England.

DACIE, J. V. (1967). *The haemolytic anaemias*. Part IV. *Haemolytic disease of the newborn*, pp. 1261–345. Churchill Livingstone, London.

MOLLISON, P. L. (1972). *Blood transfusion in clinical practice*, 5th edn. Blackwell, Oxford.

RACE, R. R. and SANGER, R. (1975). *Blood groups in man*, 6th edn. Blackwell, Oxford.

The white blood corpuscles (leucocytes)

The white blood corpuscles (leucocytes) are divided into three groups:

1. *Granulocytes* (10–14 μm): characterized by the presence of granules in the cytoplasm and a lobed nucleus. Using Leishman's stain, three types of cell can be recognized by the character of their granules: neutrophil (or polymorphonuclear leucocytes) with rather fine red-brown granules; eosinophil, crammed with large red granules; and basophil containing purple-blue granules. The nuclear chromatin of this group of leucocytes is coarse and 'ropy'.

2. *Lymphocytes* (small, 7–10 μm; large, 10–14 μm): these are round non-granular cells with large round nuclei. They are divided into large and small lymphocytes. The nuclear chromatin is coarser and 'lumpy'.

3. *Monocytes* (10–18 μm): this is a convenient term used to describe a group of cells which are ill-understood and suffer from a multitude of labels. The group includes the mononuclear cells, hyaline, and transitional cells among others. They are large pale cells with a pale-staining round or indented eccentric nucleus, the chromatin of which is finely reticular. The cytoplasm is usually pale blue and clear, but sometimes it contains fine purple dust-like granules (azur granules) which may be few or numerous.

White cells contain a variety of substances, which in the case of the granulocytes, include histamine. Most of this is found in the eosinophils and basophils. Lymphocytes, monocytes, and red cells contain no histamine. The basophil leucocytes contain heparin, slow-reacting substance-A (SRS-A), and eosinophil chemotactic factor [p. 60]. Lymphocytes contain biologically active proteins called lymphokines [p. 61].

Normal count

The normal range of the white-cell count is 4–11 × 10⁹ per litre. Considerable variations between these limits may occur in the same individual from day to day, from hour to hour, and even from minute to minute. The count is made up approximately as follows: granulocytes 70 per cent (neutrophils 50–70, eosinophils 1–4, basophils 0–1), lymphocytes 20–40, monocytes 2–8 per cent. (In children neutrophils are about 20 per cent less and lymphocytes about 20 per cent more than in adults.) Representative *absolute* numbers of the different cells per litre are roughly as follows: neutrophils 3000–6000 × 10⁶, eosinophils 150–300 × 10⁶, basophils 0–100 × 10⁶, lymphocytes 1500–2700 × 10⁶, monocytes 300–600 × 10⁶.

In newborn infants the white count is about 20 × 10⁹ per litre; after the second week it begins to decline, reaching normal adult levels at 5–10 years. During infancy the lymphocyte is the predominant blood cell, constituting 40–50 per cent of the count.

Changes in white-cell count in disease

An increase in the total circulating leucocytes above 11 × 10⁹ per litre is known as *leucocytosis*; a decrease below 4 × 10⁹ per litre is a *leucopenia*. A differential leucocyte count enables the percentage and absolute numbers of the different varieties of white cell to be determined. The absolute figures are far more important than the alterations in the relative proportions. According to the type of cell involved, a leucocytosis may be described as neutrophil, eosinophil, or basophil leucocytosis, lymphocytosis, or monocytosis. Leucopenia is generally due to a neutropenia, i.e. a decrease in the neutrophil cells.

Development

In the embryo the white corpuscles develop in the mesoderm and migrate secondarily into the blood vessels. In extra-uterine life the granulocytes normally develop exclusively in the red marrow; the lymphocytes and monocytes also develop from stem cells in the bone marrow. T-lymphocytes mature in the thymus.

Granulopoiesis

The sequence of cells which give rise to the granulocytes in the marrow is as follows:

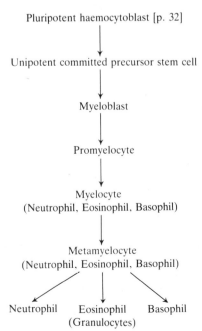

Pluripotent haemocytoblast [p. 32]

↓

Unipotent committed precursor stem cell

↓

Myeloblast

↓

Promyelocyte

↓

Myelocyte
(Neutrophil, Eosinophil, Basophil)

↓

Metamyelocyte
(Neutrophil, Eosinophil, Basophil)

Neutrophil   Eosinophil   Basophil
(Granulocytes)

The cells from the myeloblast stage onwards can be identified in stained bone-marrow smears in man. The characteristics of the myeloblast, myelocyte, and mature granulocyte are described below.

Myeloblast (12–18 μm diameter). This cell develops from precursor stem cells; the nucleus is pale, purple-blue, large, and round with finely stippled chromatin and several nucleoli. The cytoplasm consists of a narrow blue rim without granules. Protein synthesis is very active, as shown by a highly developed endoplasmic reticulum.

Myelocyte (10–15 μm diameter). These cells are characterized by the appearance of granules in the cytoplasm. Using Leishman's stain, the myelocytes may be classified according to the colour of their granules into *neutrophil* (most), *eosinophil*, and *basophil* (very scanty). The cytoplasm of the myelocytes as a whole is more extensive and less basophilic; the nucleus is smaller and more basophilic, the nucleoli have disappeared and the chromatin is coarser.

Other methods of identifying leucocytes include supravital staining of wet preparations. Here Janus green stains mitochondrial structures, and neutral red the vacuoles. Motility may be studied in these preparations and in unstained cells under phase-contrast microscopy: it is marked in neutrophils and the later myelocytes, present in eosinophils, basophils, and monocytes, absent in myeloblasts and lymphoblasts, and of different character in lymphocytes.

Leucocytes

Each type of metamyelocyte gives rise to the corresponding leucocyte (neutrophil, eosinophil, and basophil); the nucleus indents, and then becomes lobed; the granules, in fresh preparations (on a warm stage) instead of being motionless, show dancing or streaming movements, and the cytoplasm becomes more liquid, so that amoeboid movements occur. The more mature white cells are found lying just external to the sinusoids, and pass actively through the intact endothelial lining of these vessels into the circulation. Senile leucocytes lose their motility, the granules are still and no longer stain with neutral red; the cells break up readily in films, and the nucleus can be seen lying free, surrounded by faintly staining granules. As a leucocyte ages the complexity of the nuclear lobulation increases; thus a very young leucocyte has a horseshoe-shaped nucleus, while an old cell may show four or five lobes joined together by very faint strands of chromatin.

Neutrophils (neutrophil granulocytes)

Morphology. The living neutrophil, as seen by phase-contrast microscopy when it is moving on a glass slide, is 10–15 μm in diameter, but the shape of the cell is constantly changing as it undergoes amoeboid movements. The cytoplasm contains 50–200 dense granules [lysosomes, p. 5] which in human neutrophils are about 200 nm in diameter. Under the electron microscope the granules are seen to consist of a finely granular matrix bounded by a typical membrane. The cytoplasm of the mature neutrophil contains little endoplasmic reticulum, a small Golgi apparatus, and few mitochondria. The neutrophil is an end cell incapable of division.

Life history. Radioactive labelling has shown that the entire maturation process from myeloblast to neutrophil takes about 3 days.

In a healthy man 2–3 × 10^10 neutrophils are circulating in the blood at any one time and an equal number are marginated on vessel walls or sequestered in closed capillaries. Neutrophils have a half life of about 6 hours in the circulation; after emigration into the tissues they never return to the bloodstream and survive in the tissues for a few days. For every circulating neutrophil 50–100 mature cells are held in the bone-marrow reserve. The relative constancy of the blood neutrophil count suggests that there must be an efficient feedback mechanism to control the release of mature cells from the bone marrow. Dead leucocytes release a granulocyte-inducing factor which causes mobilization of reserve cells in the marrow and also stimulates new granulocyte formation. ACTH and cortisol produce a neutrophil leucocytosis. These factors may participate in the regulation of the numbers of neutrophils in circulating blood.

The turnover of neutrophils is very high. In man 50–100 ml of packed neutrophils are eliminated daily, probably mainly into the intestine and out via the faeces or into respiratory secretions. Neutrophils which die in the tissues are taken up by macrophages.

Increases in the blood neutrophil count are due to:

1. Mobilization of marginated or sequestered neutrophils from blood vessels. Adrenalin and exercise produce transient neutrophilia in this way.
2. Release of stored neutrophils from the bone marrow. This produces only a transient neutrophilia.
3. A lasting neutrophilia can only be maintained by increased neutrophil production in the bone marrow, e.g. in infections.

A decrease in blood neutrophil count (neutropenia) is brought about by the same factors acting in the opposite way.

Metabolism. Neutrophils have glycogen in their cytoplasm. Most of the energy for phagocytosis and motility comes from glycolysis which leads to lactic acid formation. The capacity of neutrophils to function in anaerobic conditions is of great value in the removal of bacteria or cell debris from necrotic tissue. Neutrophils also show active metabolism of neutral lipids and phospholipids during phagocytosis.

Neutrophil granules. Neutrophil granules contain large amounts of protein and traces of lipids and nucleic acids. Enzymes released from ruptured granules include cathepsins, phosphatases, nucleases, nucleotidase, and β-glucuronidase. The granules are thus regarded as *lysosomes* [p. 5].

Physiology of neutrophils. Neutrophils play an important role in inflammation, particularly when this is of bacterial origin. At a site of injury or inflammation the vascular endothelium becomes sticky and neutrophils adhere both to the altered endothelium and also to each other, forming clumps. Divalent cations are required for adhesion, and plasma proteins, especially fibrinogen, promote it. Neutrophils are motile cells and after they have become adherent to the endothelium they emigrate from the bloodstream into the tissues. Electron-microscopic studies suggest that leucocytes emigrate by passing through the junctions between endothelial cells.

Chemotaxis. Chemotaxis means directed movement and is most easily demonstrated *in vitro*, though chemotaxis of neutrophils and macrophages probably occurs also *in vivo* during an inflammatory response.

In vitro chemotaxis has been observed in a two-compartment chamber (Boyden chamber). The compartments are separated from one another by a Millipore filter containing micropores about 3 μm in diameter. The pore diameter is less than that of the neutrophil or macrophage cells placed in the upper compartment; the cells can therefore enter the medium in the lower compartment only by active motility such as enables these cells to squeeze between endothelial cells *in vivo*. The substances to be tested for chemotactic activity are placed in the lower compartment and the number of cells on the lower surface of the filter is counted after 2–4 h at 37 °C. Some chemotactic substances act directly to promote directional movement of neutrophils; such substances are called *cytotaxins*. Other chemotactic substances act indirectly, most commonly only in the presence of plasma or serum in which activated chemotactic components of complement, C3a, C5a, and C567 are formed. The substances which act indirectly are called *cytotaxigens*.

Examples of cytotaxins are casein-culture filtrates of certain bacteria, e.g. *E. coli* and anaerobic corynebacteria and the supernatant fraction of virus-infected cells (plus the endogenous activated components of complement). The cytotaxigens include antigen–antibody complexes and plasmin.

It is very probable that chemotaxis occurs *in vivo* but conditions make accurate observations much more difficult than *in vitro* (Wilkinson 1974).

When a neutrophil comes into contact with a bacterium *phagocytosis* occurs [FIG. I.29]. In this process of cell membrane of the neutrophil becomes invaginated until the bacterium is completely engulfed in a digestive pouch. Phagocytosis (cell eating) is clearly analogous to pinocytosis (cell drinking). Phagocytosis of bacteria is aided by the presence of substances in plasma called *opsonins*; it requires the presence of divalent cations and is prevented by calcium chelating agents; phagocytosis occurs in the pH range 6.0–8.0 and is not dependent on oxygen. When a bacterium has become engulfed in the digestive pouch bactericidal substances (e.g. lysozyme) and enzymes are released from the lysosomal granules into the pouch where they kill and digest the bacterium. The enzyme peroxidase forms H_2O_2 which is strongly bactericidal.

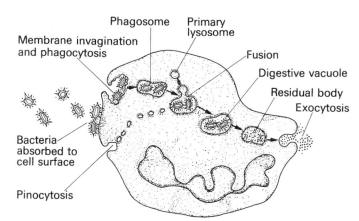

FIG. I.29. Schematic diagram of endocytosis: both phagocytosis of immunoglobulin-coated bacteria and pinocytosis are shown. The fusion of primary lysosomes with the phagosome to form the digestive vacuole, the subsequent degradation of the bacteria leading to the formation of a residual body, and the expulsion of indigestible components are also depicted. (Redrawn from Wintrobe, M. M. *et al.* (1974). *Clinical hematology*, 7th edn. Philadelphia.)

Phagocytosis is concerned not only with the elimination of foreign bacteria but also with the removal of inanimate particles such as antigen–antibody complexes, which may help to prevent the damaging effects produced by such complexes. Many foreign inanimate substances, e.g. carbon particles, sodium urate crystals, are taken up by neutrophils, which in this sense act as scavengers.

Evidence of role of neutrophils in resistance to infection.

Bacterial infections usually produce a neutrophil leucocytosis, up to $20-40 \times 10^9$ per litre. Great multiplication of the myelocytes occurs in the marrow, and leucocytes are discharged into the blood stream in enormous numbers, with an increase in proportion of young neutrophils in the circulation. The stimulus to increased neutrophil production in many cases is provided by bacterial endotoxin and the response seems an appropriate way of dealing with

an invasion by pathogenic micro-organisms. However, neutrophilia also results from the local inflammatory reaction evoked by the presence of dead or dying tissue in the body, e.g. in myocardial infarction, so neutrophil leucocytosis is not conclusive evidence for the role of neutrophils in resistance to bacterial infection.

Resistance to microbial infection is reduced by any factors which interfere with the formation, maturation, release into the circulation or phagocytic activity of neutrophils.

In agranulocytosis (reduction or absence of granulocytes) there is a greatly increased susceptibility to bacterial infection. Patients with granulocytic leukaemia are also susceptible to infection, in this case because the very high number of immature neutrophils do not possess full phagocytic activity. Cortisone and related substances increase the number of circulating mature normally functioning neutrophils but host resistance to infection is reduced because neutrophil emigration into the tissues is impaired.

Primary intrinsic neutrophil defects. In some persons lack of resistance to infection may be due to impairment of chemotaxis, phagocytosis, or of intracellular killing of bacteria. For example, in chronic granulomatous disease of childhood there is recurrent infection in early life with *Staphylococcus aureus* and Gram-negative pathogens. Neutrophils are normal or increased in number, phagocytosis occurs normally, but the ingested bacteria are not killed This is due to lack of formation of H_2O_2.

Injury produced by neutrophils. In certain types of experimental inflammation, e.g. Arthus reaction, the degree of damage to the tissues is proportional to the number of neutrophils which accumulate. In neutropenic animals the reaction is minimal. Tissue damage is due to release of lysosomal contents.

Endogenous pyrogen. Neutrophils contain a fever-producing substance (*endogenous pyrogen*) which is an important mediator of the febrile response to bacterial pyrogens [p. 355].

Eosinophils. The eosinophils are phagocytic, but less motile than neutrophils.

The eosinophil granules are lysosomal in nature and contain most of the enzymes found in neutrophil granules. They have a very high peroxidase content which partly accounts for their parasiticidal action, e.g. versus schistosomes.

Eosinophils collect at sites of allergic reactions. It has been suggested that they limit the effects of mediators (e.g. histamine, bradykinin) of some types of antigen–antibody reaction.

The level of circulating eosinophils is reduced by adrenal corticosteroids and hence by secretion of ACTH. The eosinopenia is caused by sequestration of eosinophils in the lungs and spleen and by their destruction in the circulating blood.

Basophils. The granules of the basophils contain histamine and heparin, so these cells are analogous to tissue mast cells [p. 29].

Monocytes are larger than granulocytes and have a horseshoe-shaped nucleus. They are best described as mononuclear phagocytes or *macrophages* as compared with neutrophil granulocytes which are termed *microphages*. Monocytes, which are formed in the bone marrow, circulate in the bloodstream and leave the circulation to form the *reticulo-endothelial system* in spleen, liver, lymph nodes, etc.

REFERENCES

BIGGAR, W. D. (1975). *Lancet* i, 991.
BREWER, D. B. (1972). *Br. med. J.* ii, 396.
WILKINSON, P. C. (1974). *Chemotaxis and inflammation*. Churchill, Edinburgh.

The spleen

The spleen is an organ whose functions are mostly shared with other organs and tissues. It is not essential to life, and good health can be maintained after splenectomy. There is no evidence that the spleen is an endocrine organ.

Blood formation

During the second half of fetal life the spleen forms red blood corpuscles. In anaemia due to destruction of the bone marrow this function may be resumed in adult life.

The spleen contains lymphoid tissue which like that elsewhere forms lymphocytes and plasma cells.

In leukaemias, the normal cells of the splenic pulp are replaced by lymphoid cells in lymphatic leukaemia, and myeloblasts and myelocytes in myeloid leukaemia.

Blood destruction

The spleen is part of the reticulo-endothelial system and destroys aged red cells, platelets, and perhaps leucocytes. After splenectomy nucleated red cells appear in the circulating blood and the percentage of reticulocytes rises. Leucocytosis and thrombocytosis occur.

Defence reactions

The lymphoid tissue and reticulo-endothelial cells (macrophages), like these tissues elsewhere, participate in defence reactions against toxins (diphtheria, tetanus), bacteria, and larger parasites by the formation of antibodies and by phagocytosis.

Reservoir of red corpuscles

In some mammals, e.g. cat and goat, the spleen acts as a reservoir of red blood cells which can be discharged into the general circulation by sympathetic stimulation or by noradrenalin. Anoxia can increase the oxygen-carrying power of the blood by stimulating the sympathetic nerves to the spleen and the adrenal medulla.

In man there is normally no reservoir of blood in the spleen. After peripheral intravenous injection of ^{51}Cr-labelled red cells the uptake of radioactivity recorded by an external counter over the spleen indicates rapid equilibration, which suggests a high rate of blood flow through the spleen without pooling. The same technique in the cat shows slow equilibration and evidence for pooling.

Although normal red cells do not form a pool in the normal human spleen, experimentally injected spherocytes and other abnormally shaped red cells do and the trapped cells can be discharged into the circulation by noradrenaline. A splenic pool is also found when the red cells are normal but the spleen is enlarged as the result of disease (e.g. leukaemia, Hodgkin's disease). Normally blood flows rapidly through the human spleen via the sinusoids. When pooling occurs there is extrasinusoidal accumulation of red cells among the lymphatic cords. Prolonged pooling in the spleen can promote destruction of red cells, leading to anaemia.

Splenectomy in man is of therapeutic value in cases where the spleen is the predominant organ of blood cell destruction.

In *hereditary spherocytosis* anaemia results from the abnormal tendency for the red cells to pool in the spleen; prolonged pooling causes excessive destruction and a haemolytic anaemia. When cells from a patient with this condition are transfused into a normal person they survive for 14 days (normal survival is about 100 days). Thus the abnormality resides primarily in the spherocytic red cells.

Splenectomy is also of value in auto-immune haemolytic states, in cases of hypersplenism causing excessive destruction of red cells, granulocytes, and platelets, and in some cases of purpura haemorrhagica due to excessive destruction of platelets [p. 31].

REFERENCES

PRANKERD, T. A. J. (1963). The spleen and anaemia. *Br. med. J.* **ii,** 517.
WINTROBE, M. M. *et al.* (1974). Functions of the spleen. In *Clinical hematology*, 7th edn, pp. 354–70. Saunders, Philadelphia.

Immunity and inflammation

Immunity as a branch of medical science grew historically from the study of bodily resistance to infections and has been naturally associated with bacteriology. During this century, and particularly during the past 25 years, it has become clear that immune phenomena have a much wider application than to protection against pathogenic micro-organisms. This has led to the rise of *immunology* as a separate discipline, with its special techniques, concepts, and terminology. From the physiologist's point of view there are certain aspects of immunology which must be discussed in relation to the normal properties and functions of the plasma immunoglobulins, the blood cells, lymphoid tissues, and the thymus.

Inflammation is usually regarded as a pathological state but when it is considered as the response of living tissue to injury it is perhaps better dealt with as a disturbance of physiological functions, with particular involvement of blood vessels, leucocytes, plasma, and sensory nerve endings, associated with release of normal constituents of cells. Minor degrees of injury occur commonly enough to be regarded as 'normal' events, and inflammation contains within its disorder the seeds of repair and restoration of normal function. Many immune reactions are inflammatory in nature so it is appropriate to discuss immunity and inflammation in the same section.

IMMUNITY

It was known in ancient times that no person who had had an attack of smallpox would have a second attack, even during a severe epidemic. It was also realized that this immunity was specific in that smallpox did not confer protection against other diseases. Attempts were made in China and India to protect children against this 'most terrible of all the ministers of death' by inoculation with dried crusts from the skin of infected patients, in order to induce a mild attack. In 1798 Jenner showed that inoculation with material from cowpox vesicles protected against subsequent exposure to smallpox. This process of vaccination was a most important landmark in the history of immunity. With the rise of bacteriology during the latter half of the nineteenth century came advances in understanding of the processes by which immunity is achieved. It was shown that animals can form protective *antibodies* against pathogenic parasites and their toxins. These antibodies circulate in the blood plasma and thus are carried to all parts of the infected host so that they can kill invading microorganisms or neutralize toxins. In addition Metchnikoff showed that phagocytes ('eating cells') played a protective role by attacking and digesting many types of pathogenic bacteria. Thus protection against infection is attained by humoral and cellular mechanisms. The cellular mechanism can act very quickly but antibodies are only formed after an interval of 1–2 weeks from exposure to infection. In the 1890s it was shown that injection of animals with diphtheria or tetanus toxin provoked the formation of specific antitoxic sera. These sera, by providing ready-made antibodies, were effective in the treatment of diphtheria and in the prevention of tetanus (passive immunization). During this century a much more effective means of protection has been developed in which modified diphtheria and tetanus toxins (toxoids) are administered to people to stimulate their own antibody formation and create an active immunity. Active immunization can be induced also against bacterial infections, e.g. typhoid, tuberculosis, and against viral infections, e.g. smallpox, yellow fever, poliomyelitis.

Up till about 1900 antibody formation was regarded solely as a means of protecting the body against noxious foreign materials, living or dead, which had penetrated into the blood or tissues. However, it was then found that antibody formation could be induced by injection of bland proteins, as well as harmful microbes or their toxins. It thus became clear that antibody formation was a basic physiological response to all types of foreign proteins which entered the body without being broken down by the proteolytic enzymes of the alimentary tract. Substances which promoted antibody formation were called *antigens* and it was shown that antibody formation in response to a particular antigen was highly specific.

In 1900 Landsteiner showed that blood serum from some persons caused clumping (agglutination) of red cells of some others and thus revealed the different antigenic components of A, B, O blood groups in human erythrocytes and the isoagglutinins in serum. It has since been shown that red cells contain many other antigens, such as the D antigen in Rh+ persons. When Rh+ red cells are administered to Rh− persons antibody (agglutinin) production is initiated and may have serious consequences [p. 48].

In 1902 Portier and Richet injected an extract of the tentacles of the sea anemone into dogs. The first injection had little or no effect, but a second injection three weeks later caused rapid collapse and death in 25 minutes. They called this phenomenon *anaphylaxis* (without protection) in contrast to *prophylaxis* (promoting protection) and implied that the first injection had removed the natural protection against the toxin whose noxious properties were revealed by the second lethal injection. Arthus (1903) described local anaphylaxis, an oedematous, haemorrhagic reaction produced by repeated injection of horse serum into rabbits, and von Pirquet and Schick (1905) showed that anaphylactic shock occurred in man after repeated injections of horse serum and that even a single injection produced *serum sickness* after a delay of 7–10 days. It was found that in all these cases the anaphylactic phenomena were due to specific interactions between reinjected or persisting antigen and antibody formed in response to a previous injection of antigen; anaphylactic shock occurs when antigen combines with antibody fixed on cells, the interaction leading to release of histamine and other substances which are mainly responsible for the manifestations of shock.

Anaphylaxis is a state of *hypersensitivity* which is induced by exposure to certain kinds of antigenic stimuli. The word *allergy*, with which hypersensitivity is often equated, strictly means an *altered capacity to react* and when introduced by von Pirquet in 1906 it was applied to specifically induced reactions which were either enhanced or depressed. The word allergy is nowadays mostly used to mean altered reactivity which is *increased above normal*, and the term *immunity* implies a *decreased reactivity* to an external noxious antigenic agent.

Allograft (homograft) and xenograft (heterograft) reactions

The transplantation of tissues or organs from one animal to another has been studied for many years and the knowledge gained has been of great value in relation to transplants in man. It has long been known that in human beings a skin-graft from one person to another (allograft) does not take, although skin from one part of a person is readily accepted at other sites of this person (autograft). An organ or tissue transplant from one animal species to another (xenograft) is also rejected. Exceptions to these generalizations

show that genetic factors are involved. In lines of mice with genetic constitutions that are identical, or very nearly so, in all members, grafts of skin or other organs are readily accepted between different individual animals. The same applies to identical twins. Individuals with the same genetic constitution are termed *syngeneic*, those with different genetic constitutions are called *allogeneic*. Genetic differences are associated with differences in chemical composition of the tissues in different individuals so that one animal's tissues act as antigens in another animal.

The allograft reaction is an immunological response produced by the recipient against the transplant antigens of the graft. For the first few days after transplantation an autograft and an allograft are indistinguishable. They both heal into place and rapidly acquire a new blood supply and lymphatic drainage. After a week or so the circulation to an allograft diminishes, the graft becomes necrotic and after 2–3 weeks is sloughed off. By this time an autograft has healed firmly in position and remains intact for the rest of the animal's life. When the blood supply to an allograft begins to fail the underlying bed of tissue is infiltrated with mononuclear cells, mainly lymphocytes and histiocytes. If after a first graft has been rejected a second graft from the same donor is applied the allograft rejection is greatly speeded up and the graft is sloughed in 3–4 days. This 'second set reaction' is due to previous immunization and is associated with rapid invasion by lymphocytes, plasma cells, and polymorphs.

Rejection of a first skin allograft is brought about by mechanisms like those involved in delayed hypersensitivity reactions. The reaction is thus mediated by lymphocytes, serum antibodies being of little importance. In the second set reaction, in addition to lymphocytes, serum antibodies are involved and produce the Arthus-type features with polymorph infiltration, intravascular thrombosis, and necrosis.

It has been suggested that the biological significance of the allograft reaction lies in its ability to reject occasional mutant cells formed during the normal course of cell division in the body. Mutant cells may carry different antigens from those present in normal cells and rejection of mutant cells may be an important process for the preservation of an unchanged genetic constitution. This 'immunological surveillance' may help to prevent the development of cancer. If so it is a process which all too frequently fails.

Immunological tolerance

Ehrlich was the first to emphasize that animals do not usually make any immunological response to their own plasma proteins or tissue cells although these are excellent antigens in other species. He described this state as 'horror autotoxicus'. Burnet and Fenner in 1949 proposed the following explanation. The capacity to make an immunological response to foreign antigen develops late in fetal life or even after birth; all potential antigens with which the cells are in contact during the period of immunological immaturity are recognized as 'self', while materials with which first contact is made after this period are recognized as 'not self' and will evoke an immunological response. Burnet predicted that the ability to recognize 'self' would be due to the fact that when the precursors of antibody-producing cells in fetal life encounter potentially antigenic materials in the tissues, they are subsequently unable to make a specific immune response to these materials.

This prediction was soon confirmed by Medawar's skin-grafting experiments in mice. He showed that if a mouse of a pure-line strain was inoculated during fetal life or just after birth with tissue cells from a mouse of another strain it would subsequently accept an allograft of skin from this second strain as readily as an autograft of its own skin. Skin from a third unrelated strain was rejected. The change induced by the foreign cells in the recipient was called *acquired tolerance*. When a tolerant recipient, bearing a foreign skin-graft was injected with lymph-node cells from a mouse of the recipient's strain which had previously been immunized against donor-strain tissue, the hitherto tolerated graft was rapidly rejected.

Immunological tolerance is thus an induced specific unresponsiveness to substances which can induce sensitization in animals from soon after birth. It is natural to argue from this that not only do foreign antigens evoke tolerance but that the natural non-antigenicity of body constituents is due to a similarly induced tolerance. All potentially antigenic material encountered in fetal life, whether 'self' or 'not self', elicits no response either then or subsequently, at least as long as it persists in the tissues. In normal conditions the antigenic material to which the fetus is exposed is that from its own body components. This is the 'self-recognition' process postulated by Burnet.

Antibody production

An older theory concerning antibody production was that the antigen molecule acted as a *template* at the site of production and thereby impressed a new pattern on the antibody molecule so that it would combine specifically with the antigen molecule. This theory, however, did not explain either immunological tolerance or the persistence of antibody production long after the presumed disappearance of antigen from the body.

Clonal-selection theory (Burnet). This theory proposes that certain mesenchymal cells (probably lymphocytes) which are called 'immunologically competent', are each genetically endowed with the capacity to respond to one, or at most very few, molecules of specific antigenic pattern by making antibodies against this pattern. A particular antigen activates only the few cells bearing the appropriate complementary reactive sites, but the whole population of cells together covers a wide range of specific reactivities. Antigenic stimulation promotes the proliferation of cells of the appropriate reactivity, producing a 'clone' of cells, all of similar specific reactivity (a clone is the population of cells descended by asexual reproduction from a single cell). Antigen thus acts as a trigger for proliferation and not as a template for specificity. In fetal life antigen inhibits cell activation instead of stimulating it.

Auto-immunity and human disease

So far we have emphasized that immune processes are directed against antigenic components of micro-organisms and toxins and proteins of innocuous character, the common feature being that these antigens are foreign to the reacting animal. The body does not normally make antibodies, or develop other kinds of immune response, against constituents of its own tissues. However, no processes are perfect and it is not surprising that the mechanisms maintaining immunological tolerance should sometimes fail and allow the production of antibodies against, or sensitization by, products of the animal's own tissues. The terms used to describe these events are *auto-immunization* and *autosensitization*. The antibodies are called *auto-antibodies* and the term *auto-immune disease* refers to conditions in which auto-immunity is thought to play a pathogenic role.

Auto-immunization might occur when new antigenic materials are formed at any time after the period of immunological immaturity. This applies to spermatozoa and to newly evolved mutant cells. Some potential antigens are *anatomically segregated* so that there is normally a barrier between them and immunologically competent cells. A breakdown of this barrier at any time after early infancy leads to auto-antibody formation. The formation of auto-antibodies to spermatozoa, brain constituents, heart, pancreas, and thyroglobulin illustrate this process. Simple organic chemicals (e.g. penicillin) can combine with normal body proteins to make a complex antigen whose specificity is determined by the small molecular drug. Finally, it is to be expected that acquired tolerance should occasionally break down and permit the appearance of immunologically reactive cells (what Burnet calls 'forbidden clones') which promote auto-antibody formation.

The presence of auto-antibodies, or even of sensitization to an auto-antigen, does not establish the auto-immune nature of a disease process. It is necessary to show in addition that administration to normal animals of serum antibodies or immunologically potent cells from lymph nodes, etc., can reproduce the disease. In man, the best established example of auto-immune disease is *acquired haemolytic anaemia* in which the patient forms a haemolysing antibody against an antigen in his own red cells. Other diseases ascribed to auto-immune mechanism include Hashimoto's disease and other forms of lymphocytic thyroiditis, rheumatic fever, rheumatoid arthritis, systemic lupus erythematosus, and myasthenia gravis. In some instances auto-immune disease may result from tissue damage by a virus which thereby allows the escape of normally sequestered antigen. It has been suggested that the orchitis associated with mumps is due to initial damage to the testis by the virus and subsequent auto-immune reaction evoked by absorption of seminal fluid into the blood and lymph.

Nature of antibodies, immunoglobulins

Electrophoretic techniques, ion-exchange chromatography, and ultracentrifugal analysis have shown that antibodies are located in the serum γ-*globulins* which are therefore called *immunoglobulins*. These are the slowest moving serum proteins on electrophoresis and they form a family of protein molecules which are heterogeneous, yet related. They can be further distinguished by using them as *antigens* to form antisera in animals. In this way human immunoglobulins have been characterized [TABLE I.5].

Immunoglobulin structure. Serum antibodies or immunoglobulins are all made up from units consisting of four peptide chains covalently linked by disulphide bonds. Within this framework the total immunoglobulin population in any individual shows an amazing degree of heterogeneity related to diversity of function of cellular origin. The peptide chains are of two types, 'heavy' and 'light' [FIG. I.30]. The chains can be separated from each other by chemical reduction of the S–S bonds. In IgG the heavy chain has a molecular weight of 50 000 and the light chain a molecular weight of 20 000. The IgG molecule, with 2 heavy + 2 light chains, therefore has a molecular weight of 140 000. High molecular weight antibodies (e.g. IgM) consist of polymers of the four chain units.

TABLE I.5 *Properties of major human immunoglobulin classes* (Roitt 1980)

WHO nomenclature	Molecular weight	Serum level	Percentage of total immunoglobulins
IgG	140 000	8000–16 000 mg 1^{-1}	80
IgA	160 000	1400–4200 mg 1^{-1}	13
IgM	900 000	500–2000 mg 1^{-1}	6
IgE	200 000	17–50 µg 1^{-1}	0.002

All classes of antibodies have similar light chains, associated with properties common to all immunoglobulins. Heavy chains of different classes have distinct physicochemical properties. The notation IgG, IgA, IgM, IgD distinguishes immunoglobulins according to the antigenic determinants of their heavy chains. It is not yet known what determines the specific combining activity of antibodies but it must be related to the amino-acid sequences in the Fab fragment.

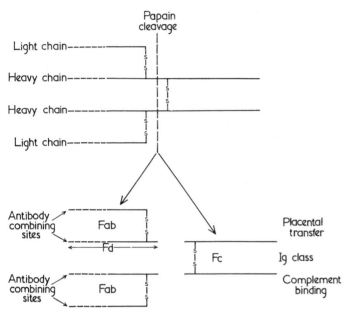

FIG. I.30. Immunoglobulin structure (IgG). Papain splits IgG into two Fab (antibody fragments) and one Fc (crystallizable fragment). The properties conferred by the Fab and Fc fragments are indicated.

Biological properties of immunoglobulins. IgG is the most abundant immunoglobulin class which includes the antibodies stimulated by natural infection or artificial immunization with viruses or bacteria and their products. These antibodies are distributed equally between blood and tissue fluids. They cross the placenta from mother to fetus (by active transport) and are found in milk, saliva, nasal and bronchial secretions but not in cerebrospinal fluid. When combined with antigen they activate complement and initiate complement fixation. This leads to immune adherence and facilitates phagocytosis (opsonin activity); it also promotes chemotactic activity for polymorphonuclear leucocytes. Thus IgG antibodies are well adapted to neutralize exotoxins or viruses and to promote phagocytosis.

IgA antibodies are unusual in that they not only occur in plasma but are secreted into saliva, intestinal juice, respiratory secretions, and colostrum. These antibodies lyse bacteria in the presence of lysozyme which also occurs in the secretions containing IgA. Thus IgA antibodies help to protect mucous surfaces.

IgM antibodies are regular concomitants of almost all specific antibody responses and because of their large molecular size they are predominantly intravascular. Each IgM molecule possesses at least five identical antigen-combining sites so these antibodies are particularly suitable for attachment to surfaces of cells with large numbers of similar antigenic sites. IgM activates complement, promotes phagocytosis, and causes cell lysis by digesting holes in the cell membrane at the sites of antibody attachment. IgM antibodies are much more effective, molecule for molecule, than IgG antibodies in lysing cells, and are well suited for dealing with particular antigens, such as bacteria, in the bloodstream. IgM antibodies appear early in response to infection.

Other groups of antibodies include IgD and IgE. The latter are the heat-labile skin-sensitizing antibodies (also called reagins) re-sponsible for the anaphylactic type of immediate hypersensitivity. IgE antibodies become fixed to tissues and when they have become attached to mast cells contact with antigen leads to rupture of the cell membrane, disintegration of the granules, and release of histamine and SRS-A [p. 60]. The concentration of IgE antibodies in serum is normally very low.

Sites of antibody formation. Immunoglobulins are synthesized in all the lymphoid tissues of the body, except the thymus. The main antibody-producing cells are the *plasma cells* which are formed from B lymphocytes.

REFERENCES

British Medical Bulletin (1967). Delayed hypersensitivity. *Br. med. Bull.* **23**, No. 1.
BURNET, F. M. (1969). *Self and non-self.* Cambridge.
HERBERT, W. J. and WILKINSON, P. C. (1977). *A dictionary of immunology.* Blackwell, Oxford.
HOLBOROW, E. J. (1967). An ABC of modern immunology. *Lancet* **i**, 833, 890, 942, 995, 1049, 1098, 1148.
ROITT, I. M. (1980). *Essential immunology*, 4th edn. Blackwell, Oxford.
WEIR, D. M. (1977). *Immunology*, 4th edn. Churchill Livingstone, Edinburgh.

Lymphoid tissue

Functionally, lymphoid tissue, through its content of lymphocytes, plasma cells, and macrophages, is responsible for the development of immunity and for states of immunological hypersensitivity. Formerly, lymphocytes were classified only according to their size, i.e. small, medium, or large. However, it has been shown that there are two main types of immunologically competent lymphocytes, T (thymus-dependent) lymphocytes and B (bursa or bone-marrow processed) lymphocytes. T lymphocytes are concerned with cell-mediated responses, and B lymphocytes, which give rise to plasma cells, promote humoral antibody formation. B cells have membrane-bound immunoglobulins on their surface; T cells do not.

T lymphocytes are responsible for *cell-mediated immunity*, which is directed against intracellular pathogenic micro-organisms (many bacteria, viruses, and fungi). T lymphocytes are also involved in the rejection of foreign transplants. B lymphocytes are chiefly involved in the production of *humoral immunity* in which plasma cells secrete antibodies against encapsulated pyogenic bacteria (e.g. pneumococcus, streptococcus). Both types of lymphocyte can also promote characteristic types of hypersensitivity (immunologically sensitized states).

Embryological development

Lymphopoiesis first appears in the liver and thymus of the human fetus at about three months. The fetal liver (and later the bone marrow) forms stem cells which migrate to the thymus, where some of them are activated to become immunologically competent T lymphocytes. At birth, the thymus weighs 10–12 g and cell-mediated immunity is well developed so that graft rejection can occur.

At birth, the development of peripheral lymphoid tissue (lymph nodes, spleen, gut-associated lymphatic tissue) is very slight and immunoglobulin secretion very small. IgA and IgE occur in serum in trace amounts; IgM is formed before birth and is present in significant amounts in neonatal serum; IgG occurs in neonatal serum in high concentration but this is maternal IgG which has diffused across the placenta. After birth, serum IgG concentration falls steadily for 2–3 months, when the child's own IgG production takes over, and IgG serum level subsequently rises gradually over a period of years.

Location of T and B lymphocytes in adult

Peripheral blood lymphocytes comprise 60–80 per cent T cells, 20–30 per cent B cells.
Thoracic duct lymphocytes comprise 85–90 per cent T cells, 10–15 per cent B cells.
Lymph nodes contain T cells in the paracortical region and B cells in subcapsular region, in germinal centres and in medullary cords.
The spleen contains T cells in periarteriolar sheaths, and B cells in germinal centres, red pulp, and around periarteriolar sheaths.

Role of bone marrow in lymphopoiesis

In adult life the bone marrow forms *stem* cells which are capable of maturing into competent B or T cells. Congenital deficiency of stem cells leads to marked reduction in numbers of both B and T cells, and hence severe impairment of humoral and cell-mediated immune responses.

Thymus

Maturation of stem cells in thymus. Stem cells migrate from bone marrow to thymus, where some of them become immunologically competent T cells. These enter the bloodstream from which they are taken up and stored in the paracortical region of lymph nodes. The thymus is active in fetal life, the organ grows in adolescence and atrophies in old age.

Neonatal thymectomy in mice has profound effects:

Lymphopaenia (reduction in number of circulating lymphocytes).
Reduction in number of T lymphocytes in paracortical region of lymph nodes.
Deficiency in cell-mediated immune responses, e.g. inability to reject a foreign skin graft.
Depressed humoral antibody formation in response to many antigens (i.e. those which need 'co-operation' between T and B cells).
Wasting, apparently due to infection, since it does not occur in germ-free mice. Infection is often fatal.

In older animals, with a fully developed lymphoid system, thymectomy produces only mild immunological defects after a long latent period. However, an intact thymus is required to restore the immune system after depletion of T cells by total body irradiation or by antilymphocyte serum.

Maturation of peripheral lymphoid tissue

Antigenic stimulus. If normal animals (thymus intact) are reared under germ-free conditions, peripheral lymphoid tissue (lymph nodes, gut, spleen) does not develop and formation of immunologically competent B lymphocytes is much impaired. Thus antigenic stimulation by natural bacterial flora appears to be essential for B lymphocyte replication and plasma cell formation leading to humoral antibody secretion.

In birds, there is an organ called the bursa of Fabricius situated in the posterior end of the alimentary tract. This bursa is essential for development of immunologically competent B lymphocytes. Bursectomy during embryonic life leads to failure of development of germinal centres and plasma cells, and hence to agammaglobulinaemia, though T lymphocyte development is unimpaired. The term B lymphocyte was used as an abbreviation of bursa. However, in mammals there is no bursa, though it has been suggested that the gut-associated lymphoid tissue as a whole (tonsils, Peyer's patches, appendix, etc.) acts like the bursa of Fabricius. Alternatively, the bone marrow itself may promote maturation of B lymphocytes.

Antibody formation. The formation of specific antibodies in response to an injected antigen can be followed by measuring the serum antibody levels. When a protein antigen such as diphtheria toxoid is injected into a rabbit there is a long interval (12–14 days) before antibody appears in the blood; this *primary response* is small and transient. If the toxoid is later reinjected a *secondary response* is produced in which the antibody rises to a high level within 3–4 days. If a secondary response is aroused by injection of antigen into the foot of a rabbit, the efferent lymph from the draining lymph nodes contains a much higher concentration of antibody than the afferent lymph from the injection site. This shows that lymph nodes can form antibodies. Antigenic stimulation promotes antibody formation also in the bone marrow, spleen, and lymphoid tissue of the gut, but not in the thymus.

During a secondary response in a lymph node there occurs enlargement of lymphoid follicles (germinal centres) with an increase of their large and medium lymphocytes, an increase in numbers of small lymphocytes in the perifollicular zone, and proliferation of plasma cells in the medullary cords. The use of immunofluorescence techniques has shown that antibody formation takes place chiefly in the plasma cells, which are themselves formed from B lymphocytes.

Agammaglobulinaemia (hypogammaglobulinaemia). In human congenital agammaglobulinaemia there is a deficiency of peripheral lymphoid tissue and of B lymphocytes, absence of plasma cells, and marked reduction of serum IgG levels. After maternal IgG has disappeared from the blood (by 3 months) there is a great susceptibility to certain bacterial infections (e.g. pneumonia, meningitis). On the other hand, cell-mediated immunity, brought about by T lymphocytes, is normal. Hence the common virus infections of childhood, such as measles, mumps, and chicken pox, pursue a normal course leading to subsequent immunity, and vaccination against smallpox is successful.

Cellular co-operation in the immune response

The large mononuclear cells of the *monocyte–macrophage* series are essential for full antigen stimulation of both B cell and T cell mediated responses. Macrophages probably modify antigen structure and also help to concentrate antigen at the lymphocyte cell surface.

Co-operation between T and B cells is also important [Fig. I.31]. Although T cells do not secrete antibody their presence greatly enhances B cell responses to certain antigens, and after neonatal thymectomy the antibody response to such antigens is much depressed. The processes involved in this co-operation are complex.

The responses of small lymphocytes to antigen depend on whether they have been previously primed or not. The 'memory' cells (primed) respond much more quickly than unprimed T or B small lymphocytes.

Humoral immunity

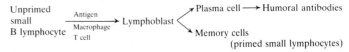

Cellular immunity

Lymphokines are soluble factors (MW 20 000–80 000) which appear in the fluid surrounding stimulated (activated) T lymphocytes. They inhibit migration of macrophages, are chemotactic for monocytes, initiate exudation of cells, and may increase vascular permeability [p. 6].

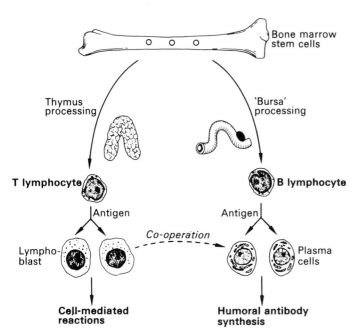

Fig. I.31. Processing of bone-marrow cells by thymus and gut-associated central lymphoid tissue to become immunocompetent T and B lymphocytes respectively. Proliferation and transformation to cells of the lymphoblast and plasma-cell series occurs on antigenic stimulation. (Redrawn from Roitt, I. M. (1980). *Essential immunology*, 4th edn. Blackwell, Oxford.)

Lymphocyte life span

Recirculation of small lymphocytes. Gowans showed that small lymphocytes labelled with radioisotope circulate in both blood and lymphatic tissues. This *recirculation* occurs as follows:
The small lymphocytes in peripheral blood (mostly T cells) have a special affinity for the endothelium of the post-capillary venule in lymphoid tissue. They pass through this endothelium to be stored in the appropriate part of the lymph node, i.e. T cells in the paracortical region and B cells in the subcapsular region or germinal centres. From these regions small lymphocytes can enter efferent lymphatics and finally the thoracic duct, which conveys them into the bloodstream where the recirculation is completed. The time spent by lymphocytes in the blood is very short compared with the time spent in lymphoid tissue. Both T and B lymphocytes can have a long or short life span, but T cells generally have a long life span, e.g. 2–4 years, while most B cells are thought to be short-lived (days or weeks).

In the bone marrow and thymus the majority of lymphoid cells are short-lived (days) and undergo rapid renewal. The proportion of thymocytes which become immunologically competent T cells is small. Thymic lymphopoiesis is not dependent on immunological stimulation; it is influenced by hormones, especially by adrenal corticosteroids which inhibit lymphopoiesis. The thymus involutes gradually after puberty.

Complement

We have already seen that plasma contains complex systems of proteins and enzymes which participate in such phenomena as blood clotting and fibrinolysis [pp. 23–6] and the plasma kinin-forming system is discussed on page 561. However, the most complex system of all is probably that known as *complement*, which plays an important part in many types of immunological response. Complement has been described as a 'multimolecular self-assembling biological system which constitutes the primary humoral mediator of antigen–antibody reactions' (Muller-Eberhard 1975). The biological consequences of its activation are:

1. Irreversible structural and functional alterations of cell membranes leading to cell death. This helps to kill off many types of pathogenic micro-organisms and thus increases host resistance to infection.

2. Complement activation is also involved in some types of hypersensitivity reaction, which cause release of vasoactive amines (e.g. histamine from mast cells), chemotaxis of neutrophils, and contraction of smooth muscle. Activation of the full complement system may kill body cells which possess complement receptors on their surface (i.e. in nephrotoxic nephritis and auto-immune haemolytic anaemia).

As in the case of blood clotting, complement may be activated in two ways: (i) The so-called classical mode of activation by IgG or IgM complexes involves eleven proteins in which three components called C1q, C1r, and C1s comprise the *recognition unit* (C1q forms the initial attachment to antigen–antibody complex); components C2, C3, C4 form an *activation unit* which stimulates components C5, C6, C7, C8, C9 to become the *membrane-attack system*, which causes cell lysis. Thus full activation of the complement system leads to cell death, and, as with the blood-clotting system, the sequential activation of the components of complement may be described as an amplifying cascade.

In some instances the activation may not go all the way and then the activation of C3 to C3a and C5 to C5a can be seen to exert effects of their own, such as formation of *anaphylatoxin*, which promotes release of histamine and SRS-A from mast cells, and a *chemotactic factor* for polymorphonuclear leucocytes; at the same time immune adherence to phagocytes is promoted. (ii) An alternative (properdin) pathway for complement activation is stimulated by endotoxin or by IgA aggregates; this involves only five plasma proteins.

The complement system is virtually independent of other plasma systems and tentative suggestions of its relationship to blood-clotting, fibrinolytic, and kinin-forming systems only emphasize the considerable complexity of all these systems. Finally, it must be stressed that plasma contains numerous inhibitors of all these systems. Lack of an inhibitor of active C1 is associated with 'hereditary angioneurotic oedema', in which acute episodes of non-inflammatory oedema may occur in skin and mucous membranes.

REFERENCE

MULLER–EBERHARD, H. J. (1975). *A. Rev. Biochem.* **44**, 697.

Immunological hypersensitivity reactions

These reactions are of different types (Roitt 1980):

I. Anaphylactic type (immediate)

Introduction of specific antigens into appropriately sensitized persons can produce immediate reactions (within a few minutes) in skin and mucous membranes. In skin the reaction appears as wheal and flare, in the nasal mucosa swelling and irritation lead to sneezing (hay fever, allergic rhinitis), and in the bronchi and bronchioles mucosal swelling and increased smooth muscle tone produce 'bronchial asthma', characterized by great difficulty in breathing (especially expiration) and signs of hypoxia (cyanosis, etc.). (See page 557 for anaphylaxis in experimental animals.)

Autoradiographic and electron-microscopic studies have shown that in this type of hypersensitivity IgE immunoglobulins (reagins) are specifically bound to mast cells, circulating basophils, and in the lung to other unidentified cells; raised IgE levels in serum are found in asthmatic patients. However, although IgE is the most important immunoglobulin involved, IgG may also contribute to the reactions.

The presence of IgE homocytotropic antibodies in serum can be demonstrated by injecting the serum into the skin of a normal person (or better into a monkey) to induce passive sensitization in the recipient. At the site of injection IgE antibodies become fixed on the surface of nearby mast cells and subsequent injection of specific antigen causes histamine release, with wheal and flare reaching a peak after 30 min (Prausnitz–Küstner test).

Chemical mediators of immediate hypersensitivity reactions.

When specific antigen is brought into contact with IgE-coated mast cells or blood basophils, the following chemical mediators are released:

Histamine [p. 556].

Slow-reacting substance-A (*SRS-A*, where A means anaphylaxis).

Eosinophil chemotactic factor (*ECF-A*), which attracts eosinophils into the region of antigen–IgE interaction.

Phase-contrast microscopic studies show that while the mediators are being rapidly released the mast cells undergo degranulation in response to antigen challenge. However, in the lung, SRS-A is not only released but also synthesized by antigen–IgE interaction, and these processes take place in cells other than mast cells (Brocklehurst 1973). In lung tissue removed from an asthmatic patient antigen interacts with fixed IgE to release all the mediators mentioned above. It has long been known that antihistamine drugs are ineffective in asthma and it seems very probable that SRS-A is the most important mediator in this type of hypersensitivity response.

SRS-A is chemically a *leukotriene* and is formed by the action of the enzyme lipogenase on arachidonic acid (Lewis 1981).

Actions of SRS-A. It produces slow, prolonged contraction of a few types of smooth muscle. In the guinea-pig it acts strongly on the ileum, only slightly on the bronchioles, and not at all on the uterus. SRS-A acts powerfully on *human* bronchiolar muscle, producing prolonged contraction with no signs of tachyphylaxis. SRS-A also sensitizes this muscle to the contractile effects of acetylcholine, histamine, and plasma kinins [p. 556].

Influence of cyclic AMP on anaphylactic type of mediators. An *increase* in intracellular concentration of cAMP reduces the amounts of histamine, SRS-A, and ECF-A released by antigen-IgE interaction. A raised cAMP level also reduces the formation of SRS-A.

Drugs which raise intracellular cAMP level will antagonize the effects of antigen–IgE interaction. For example, *isoprenaline* (isoproterenol) and salbutamol activate adenyl cyclase and therefore increase cAMP formation; *theophylline*, which inhibits the phosphodiesterase which destroys cAMP will also tend to increase the intracellular concentration of cAMP. Both these drugs are beneficial in asthma.

Conversely, substances which reduce intracellular cAMP concentration will enhance mediator release by antigen–IgE interaction. Thus α-adrenergic receptor stimulation by phenylephrine or β-adrenergic receptor blockade by propranolol will tend to make asthma worse.

Cholinergic (parasympathetic) nerve stimulation releases acetylcholine, which enhances antigen-induced release of histamine, SRS-A, and ECF-A. These effects are antagonized by atropine. Cholinergic nerve stimulation does not alter intracellular cAMP concentration but it does stimulate guanylate cyclase and thus increases intracellular cyclic guanosine monophosphate (cGMP) concentration. Cyclic AMP and cGMP have opposite effects on antigen–IgE interactions.

The sequence of biochemical events which lead to release of chemical mediators may be as follows:

1. Antigen combines with two adjacent molecules of IgE on the surface of the target cell.
2. Activation of serine esterase (Ca^{2+} required).
3. Energy must be provided by glycolysis in presence of Ca^{2+}; cGMP enhances (and cAMP inhibits) movement of mast-cell granules along microtubules, fusion of granule membrane with plasma membrane, and secretion of mediators by exocytosis.

REFERENCES

Brocklehurst, W. E. (1973). *Proc. R. Soc. Med.* **66**, 1198.
Lewis, G. P. (1981). *Trends in Pharmacological Sciences*, Vol. 2, No. 1, p. vi.

II. Cytotoxic-type hypersensitivity

When antibodies bind to antigen on a cell surface they promote contact with phagocytes:

1. By reducing electric charge on the cell surface.
2. By opsonic adherence. This happens with Rhesus incompatibility when IgG antibodies from the Rh− mother cross the placenta to react with the D-antigen on fetal red cells through opsonic adherence. This leads to red-cell destruction (*haemolytic disease of the newborn*).
3. By immune adherence in which the C3 component of complement promotes phagocytosis as happens in *auto-immune haemolytic anaemia*.

In some cases the whole complement system may be activated to produce cell lysis by direct damage to cell membranes.

III. Complex-mediated hypersensitivity

The union of soluble antigens and antibodies in the body may produce an acute inflammatory reaction. The immune complex may fix complement and thus release anaphylatoxins and chemotactic factors from C3 and C5. Anaphylatoxins release histamine and thus increase vascular permeability. The chemotactic factors promote an influx of polymorphonuclear leucocytes (especially neutrophils) which remove some of the antigen–antibody complexes by phagocytosis. Phagocytosis results in release of proteolytic enzymes, kinin-forming enzymes, and cationic proteins, which further increase vascular permeability by direct action, release of kinins, and release of more mast-cell histamine. Platelet aggregation is promoted, leading to the formation of microthrombi, and activation of the full complement system may lead to cell lysis. Thus soluble immune complexes may cause considerable damage when they become localized in any part of the body.

The results of immune complex formation largely depend on the relative proportion of antigen and antibody.

With gross *antibody excess* the complexes are rapidly precipitated and localized at the site of introduction of antigen. This happens in the Arthus type of reaction, which occurs in hyperimmunized animals after injection of antigen. An intense inflammatory reaction reaches a peak within 3–8 h and then subsides gradually, though sometimes with necrosis. There is massive polymorphonuclear-leucocyte invasion with the consequences just described.

In man, an Arthus-type reaction is seen in 'Farmer's lung'. In this condition there is sensitization to thermophilic actinomycetes growing in mould hay. Inhalation of spores to these organisms evokes severe respiratory difficulty within 6–8 h due to a complex-mediated hypersensitivity reaction in the mucosa of the respiratory tract.

Antigen excess (serum sickness). Formerly, when large amounts of horse serum, containing antibodies against diphtheria or tetanus toxin, were frequently used to produce passive immunity, a condition called serum sickness sometimes developed. Pyrexia, generalized urticaria, joint swelling, and enlargement of lymph nodes was seen 7–10 days after injection. The horse serum proteins acted as antigens, and when antibody formation (primary response) reached a certain level soluble immune complexes produced signs of hypersensitivity. Since antigen concentration was continuously declining, the duration of serum sickness was brief (about 7 days). In persons previously sensitized to horse serum, re-injection produced a much quicker antibody response with severe and sometimes fatal reactions.

IV. Cell-mediated (delayed-type) hypersensitivity

This type of hypersensitivity is seen in allergic reactions to bacteria, viruses, and fungi; in contact dermatitis evoked by some simple chemicals; and in the rejection of foreign transplants (allo- or xenografts). The best example is the delayed reaction which follows intradermal injection of tuberculin in persons previously infected with *Mycobacterium tuberculosis*. Erythema and induration develop after several hours and the reaction is maximal at 24–48 h. Histologically, there is initial perivascular cuffing with mononuclear cells, then a more extensive exudation of mononuclear and polymorphonuclear cells. The latter soon move away from the lesion leaving behind mononuclear cells (lymphocytes and cells of the mononuclear-macrophage series). This type of cellular infiltrate differs markedly from the polymorphonuclear infiltrate of the Arthus reaction.

Cellular basis. Delayed-type hypersensitivity differs from Types I to III in that the *serum* of the sensitized animal or person does not contain antibodies capable of transferring the sensitivity to normal animals or persons. However, it has been clearly shown that this type of hypersensitivity can be transferred to normal animals by injecting cells from blood or peritoneal exudate of sensitized animals. There is much evidence that T lymphocytes are the key cells involved in delayed-type hypersensitivity:

1. In children with thymic deficiency, and in thymectomized chickens, delayed hypersensitivity responses are defective. In neonatally thymectomized mice skin allografts are not rejected.
2. In children with primary immunoglobulin deficiency, and in bursectomized chickens, delayed hypersensitivity reactions are relatively unimpaired.
3. When guinea-pig skin is sensitized by contact with a chemical (e.g. chlorodinitrobenzene) which evokes delayed-type hypersensitivity, only the paracortical thymus dependent regions of the relevant lymph nodes show early proliferation and differentiation to blast forms of lymphocytes. In this instance, the small molecular chemical is called a *hapten*; the chemical does not become antigenic until it combines with some large molecular body constituent (e.g. a protein) but the specificity of the hypersensitivity is determined by the hapten.

The cell-mediated hypersensitivity reaction is initiated by antigen, probably after processing by macrophages. The modified antigen reacts with receptors on the surface of appropriate T lymphocytes and activates these cells to become large blast cells, which undergo mitosis and at the same time release soluble protein substances called *lymphokines*. Among these are macrophage-migration-inhibition factor (MIF) which inhibits the movement of macrophages but increases their phagocytic activity; monocyte-chemotactic factor; skin-reactive factor, which increases vascular permeability and promotes exudation of cells; a platelet-aggregation factor; a factor which promotes mitosis in non-sensitized lymphocytes. At the site of injection of antigen sensitized lymphocytes release lymphokines, which cause large numbers of additional mononuclear cells and non-sensitized lymphocytes to enter the area, and macrophages will be prevented from leaving it.

Rejection of skin allografts is almost entirely effected by T lymphocytes which produce cellular infiltration, as just described, and also kill off the transplant cells. Humoral antibodies may contribute to rejection of kidney allografts.

V. Stimulatory hypersensitivity

The thyroid cell receptors normally combine with pituitary TSH which activates membrane adenyl cyclase and thus increases cyclic AMP concentration within the thyroid cells. This promotes the synthesis and secretion of thyroid hormones. Human thyroid-stimulating antibodies (TSAb) also combine with thyroid Graves' disease [p. 545].

Prevention or suppression of the allograft reaction. Avascular tissues such as the cornea and cartilage do not arouse the allograft reaction, but grafts of tissues which possess a fibrous stroma and a blood supply are all liable to rejection. Some organs are better tolerated than others; a transplanted kidney arouses less reaction than an allograft of skin. Successful organ transplantation can only be achieved by reducing the allograft reaction to a minimum.

A transplanted organ can survive indefinitely if the donor and recipient are identical (monozygotic) twins. With any other relationship the chance of finding a completely compatible donor is very remote, but for any given recipient it is possible to find some donors whose organs will evoke milder allograft reactions than those from other donors. Attempts have been made to detect *histocompatibility antigens* (HLA) so that the most suitable donor can be selected. For example, Brent and Medawar (1963) in experiments on guinea-pigs injected allogeneic lymphocytes of the potential recipient into the skin of potential tissue donors. A local inflammatory reaction, like the tuberculin reaction, developed after 34–48 hours at the site of intradermal injection. The degree of this reaction varied from one donor to another but in all cases was closely correlated with the degree of reaction in a skin-graft transplanted from the injected animal back to the lymphocyte donor. Similar reactions have been obtained by intradermal injection of homologous lymphocytes in man, the reaction being greater the bigger the difference in genetic constitution; autologous lymphocytes cause no response and killed homologous lymphocytes are also ineffective. Detection of white-cell antigens by agglutination and other tests has also proved helpful.

Histocompatibility tests can show which donor's tissue will cause least reaction in the recipient. This is helpful because other measures to reduce the allograft reaction will then be more effective. *Immunosuppressive agents* were developed after it was found that total body X-irradiation prolonged the survival of allografts. Cortisone and related steroids, and numerous drugs used to inhibit the growth of malignant cells, can suppress the allograft reaction in various ways. An antilymphocyte globulin (ALG), prepared by injection of lymphocytes of one species into an animal of another species, prevents lymphocytes from participating in immune reactions. ALG has been used to suppress reactions to human allografts, e.g. kidney transplant (*British Medical Journal* 1975).

REFERENCES

ALLEN, L. (1967). Lymphatics and lymphoid tissue. *A. Rev. Physiol.* **29**, 216.
British Medical Journal (1975). Editorial. *Br. med. J.* **i**, 644.
ELVES, M. W. (1966). *The lymphocytes.* Lloyd-Luke, London.

GESNER, B. M. (1965). In *The inflammatory process* (ed. B. W. Zweifach, L. Grant, and R. T. McCluskey) pp. 281–322. New York.

GOWANS, J. L. (1959). *J. Physiol., Lond.* **146**, 54.

YOFFEY, J. M. (ed.) (1967). *The lymphocyte in immunology and haemopoiesis*. London.

INFLAMMATION

The word 'inflammation' comes from the Latin *inflammare*, to burn. No definitions of inflammation are entirely satisfactory but the important points are that it is a process aroused by sub-lethal injury to tissues; that it involves reactions in tissue cells, blood vessels, and blood cells; and that it consists of a sequence of changes which usually lead to healing. Celsus (1st century AD) summarized the clinical signs of inflammation as *redness, swelling, heat,* and *pain,* to which Galen (2nd century AD) added *loss of function.*

The causes of inflammation are many. *Infections* by bacteria, viruses, and toxins form an important group which may be illustrated by boils and carbuncles due to staphylococci, pneumonia due to pneumococci, measles and mumps due to viruses. In most of these cases there is also fever.

Allergic reactions such as those of delayed hypersensitivity are inflammatory in nature. *Trauma* of different kinds and degrees evokes inflammatory responses such as those which follow mechanical injury, excessive heat or cold (heat burn, frostbite).

Macroscopic observations

Minor damage to the skin occurs often enough to be regarded as a normal or physiological event. A firm stroke of the skin evokes the 'triple response' described by Lewis in the 1920s [p. 141]. Redness, swelling, and heat suggest that this rapidly developing reaction is a mild form of inflammation. Sunburn, or ultraviolet erythema, shows all the classical features of inflammation. In some cases, as with a heat burn leading to blister formation, an inflammatory pleural effusion or an effusion into a joint in a patient with rheumatoid arthritis, swelling is an outstanding feature.

The redness and heat of inflamed skin are due to vasodilatation and increased blood flow through cutaneous blood vessels. Skin temperature at rest does not normally exceed 33 °C but since the blood temperature is 37 °C an increased flow can raise the skin temperature to a value nearer to 37 °C and the affected skin will then feel hotter than uninflamed skin. The swelling of inflamed skin is due to escape of protein-rich fluid (approaching the composition of plasma) from the bloodstream into the extracellular spaces. In a blister or an inflammatory joint effusion the exudate often contains 4–5 g per 100 ml of proteins with the same proportions of albumin and globulins as occur in plasma. This shows that vascular permeability to proteins has vastly increased. This may be confirmed by intravenous injection of dyes such as Evans' blue or pontamine sky blue, which are firmly bound to plasma proteins; the dye accumulates in the area of inflammation, causing blueing. Pain and tenderness in inflammation are discussed elsewhere [p. 396]. Increased tissue tension due to exudation into extracellular fluid is one factor in the production of pain, but sensitization of sensory nerve endings by released chemicals is perhaps even more important. Loss of function in inflammation is due to reflex inhibition of muscular movements by pain or to mechanical restrictions imposed by swelling.

Nervous system and inflammation. In certain types of inflammation the nervous system is involved in the vascular responses as well as in the sensory effects. The well-known counter-irritant substances, mustard oil and capsaicin from pepper, cause cutaneous vasodilatation and swelling as well as pain. The vascular effects (and of course the sensory) are not seen in skin whose nerves have been cut and allowed to degenerate. It is probable that both vasodilatation and enhanced vascular permeability are mediated by an axon reflex. This is analogous to the axon reflex-mediated flare induced by histamine, and indeed histamine release may contribute to the actions of counter-irritants.

However, most types of inflammation can occur in denervated tissues so the nervous system is not essential for the response.

Microscopic observations

The vascular and cellular responses of inflammation have been studied with the light microscope and the electron microscope. With the light microscope observations have been made not only on stained sections of dead tissue but also on living tissues such as the rat mesentery or the rabbit's ear provided with a transparent window.

After injury there is transient contraction of arterioles, followed by dilatation of arterioles, capillaries, and venules. Vasodilatation may be initiated by an axon reflex but is maintained by the actions of substances released by the injury. During the vasodilatation blood flow is at first increased but later slows progressively. This stasis is due to loss of fluid from the venules, partly as a result of increased hydrostatic pressure and probably more to increased permeability of the vascular wall, allowing the escape of plasma proteins from venules and/or capillaries. At first the blood cells maintain their normal position in the central axial stream but soon the distinction between axial stream and clear outer zone in the small vessels disappears and leucocytes begin to stick to the damaged endothelium. If they stick long enough they penetrate the microvascular wall and enter the tissue spaces. Both plasma proteins and leucocytes pass between and not through the endothelial cells, though the processes involved are independent of each other. In most cases injury causes leakage of plasma proteins long before leucocyte emigration occurs. The escape of plasma proteins is a purely passive affair, and this is partly true for leucocytes, which, however, also show active amoeboid movements (*diapedesis*). A leucocyte can pass through a vessel wall in 2–13 minutes. Substances liberated from bacteria and damaged tissues attract leucocytes and give direction to their movements (*chemotaxis*).

Injury also causes platelets to adhere to the endothelium and to each other, thus forming platelet clumps or thrombi, ADP being involved in this process [p. 30]. It is not known what causes leucocytes to adhere to endothelium.

Protein-rich exudate contains all the plasma clotting factors and injury will release tissue factor and thus promote the formation of fibrin clot. Complement participates in many types of inflammatory response. The other plasma proteins are removed from the extracellular fluid by the lymphatics, there being no evidence for reabsorption via small blood vessels.

Cells in inflammatory exudates. In the early stages of inflammation (after a few hours) polymorphonuclear leucocytes are the predominant cells in an exudate. At a later stage mononuclear cells predominate. The early dominance of polymorphs is due to the fact that they move faster than mononuclear cells; the latter live the longer and so become the dominant cells in chronic inflammation. Most of the mononuclear cells in exudates are monocytes from the blood, the remainder being lymphocytes. Monocytes are converted in the tissues into macrophages, giant cells, and histiocytes, all of which are phagocytic.

Polymorphs are essential for the production of the Arthus reaction in rabbits in which the antigen–antibody interaction causes an acute vasculitis with a heavy accumulation of polymorphs. If circulating polymorphs are suppressed by administration of nitrogen mustard the Arthus reaction is strongly inhibited. On the other hand, the increased vascular permeability produced by a heat burn, or by an irritant such as turpentine, is not diminished by suppression of circulating polymorphs.

Endogenous mediators of inflammation

It is possible that the inflammatory response is due to the direct effects of noxious stimulation, but there is much evidence to suggest that inflammation is brought about by endogenous mediators, released from cells of the blood, blood vessels, or extravascular tissues, or formed in plasma and extracellular fluid. There are several natural constituents which are capable of producing the vascular and sensory phenomena of inflammation and we shall deal with those which have been detected in inflammatory exudates, natural or experimentally induced, or in perfusates of inflamed tissue.

Histamine. Histamine is dealt with on page 556, where its role in anaphylactic shock and in other allergic states is discussed. When turpentine is injected into the pleural cavity of a rat exudate is formed rapidly and during the first half-hour histamine and 5-hydroxytryptamine can be isolated from the fluid. This initial phase of effusion is suppressed by prior administration of an antihistamine drug (e.g. mepyramine maleate). The histamine and 5HT involved in this reaction are released from mast cells. Histamine enhances vascular permeability by an action on venules, causes vasodilatation by axon reflexes and by direct action on small vessels and excites sensory nerve endings to cause itch or pain.

5-Hydroxytryptamine (5HT, serotonin). This substance occurs in mast cells of the mouse and rat, but not in other species. It contributes to the early vascular responses of inflammation in the mouse and rat, and in all mammals its release from disintegrating blood platelets may arouse pain, particularly in the presence of bradykinin [p. 398].

Permeability factors and plasma kinins. Plasma and exudates can be activated by contact with foreign surfaces to form a *globulin permeability factor* which may increase vascular permeability directly or through the formation of *plasma kinins* [p. 000], both acting on the venules. A turpentine-induced pleural effusion in the rat contains globulin permeability factor for several hours after the histamine and 5HT have disappeared. This factor is antagonized by salicylate. Mepyramine and salicylate together almost completely suppress the turpentine effusion. The kinin system is probably involved in the vascular responses to thermal burns, ultraviolet irradiation, and bacterial infection. Bradykinin is found in perfusates of animal and human skin heated to 45 °C or above. Plasma kinins can also be detected in joint effusions of patients with gouty or rheumatoid arthritis.
Prostaglandins contribute to the vascular, cellular, and sensory phenomena of inflammation [pp. 339 and 559].

Mediators of delayed-hypersensitivity reactions

Histamine, 5HT, and the plasma kinin system are probably involved as mediators in immediate or early responses to injury and in the initial phases of anaphylactic and other hypersensitivity reactions. The possible mediators of delayed hypersensitivity reactions (e.g. tuberculin reaction) will now be considered.

The tuberculin reaction generally shows the same features as non-immune acute inflammatory reactions, viz., increased vascular permeability, early polymorph emigration, and later predominance of mononuclear cells, probably derived from monocytes and lymphocytes. The difference is in the time scale; in delayed hypersensitivity the increase in vascular permeability takes 24 hours to reach its peak and 4–5 days to subside and the vasodilation follows a similar course.

It has been suggested that histamine is responsible for the vascular phenomena of delayed hypersensitivity. If this is so it must be through some process other than the explosive release from mast cells, because in immediate hypersensitivity (anaphylactic) reactions in the skin the flare and weal last only 30–45 minutes. Schayer and Kahlson have suggested that the enzyme histidine decarboxylase continuously synthesizes histamine at the site of injury and that such locally produced histamine would not be antagonized by antihistamine drugs. These drugs certainly do not prevent the delayed-hypersensitivity response, and the evidence supporting the role of histamine is inconclusive. Other possible mediators include permeability globulins, plasma kinins, lymphokines, and substances released from polymorphonuclear-cell lysosomes, e.g. proteases and a cationic protein. RNA is also a potent enhancer of vascular permeability.

HEALING

The immediate reactions of tissue to trauma or bacterial invasion may be regarded as a first line of defence which also sets in motion the processes which promote repair. The vascular and cellular responses have been studied using the transparent-chamber technique in the ear of the rabbit. After insertion of the chamber the healing process can be observed in all its stages under the microscope.

Plasma and some blood cells exude from the damaged periphery and a fibrin clot is formed. Macrophages, formed from blood monocytes, invade the fibrin clot and remove the breakdown products of cells, as well as fibrin itself. New blood vessels then grow into the area, commencing as capillary sprouts from pre-existing vessels around the chamber; the new capillaries later develop into arterioles, true capillaries, and venules and vasomotor nerves grow into the smooth muscle. Lymphatic vessels grow in the same way as blood capillaries but are always independent of them. The cells and vessels in healing wounds are immersed in an amorphous matrix or 'ground' substance which during the first few days of healing has a relatively high content of sulphated mucopolysaccharide. The mucopolysaccharide concentration falls when collagen fibrils appear. Healing finally depends on the formation of collagen in the area which has been cleared of debris and revascularized. Collagen, the main fibrillary component of connective tissue, is formed from fibroblasts which grow into the healing area at the same time as the new blood vessels (6–7 days). Contraction of wounds is probably due to shortening of newly formed fibrous tissue.

Healing of epithelium is a special process which is best seen in the skin. When an incised wound of the skin heals by first intention the epithelium at the edges rapidly proliferates to bridge the gap before the formation of new connective tissue; in a granulating wound epithelium grows in from the periphery and also from hair follicles and sebaceous glands in the exposed area. With very superficial skin lesions, e.g. blister formation, the epithelial barrier in human skin is re-formed in 5–7 days. The spread of epithelium is due mainly to rapid mitosis of the basal layer and also perhaps to movement of cells across the denuded area. It has been suggested that normally mitosis is actively inhibited by a substance released by neighbouring epithelial cells (chalone). Destruction of an area of epithelium thus allows the marginal epithelial cells to proliferate in an uninhibited way until the damaged area is completely covered.

Factors which influence healing

Many factors influence healing. Locally, infection and a poor blood supply impair or delay healing. General factors include nutrition and hormone action. Healing of wounds is impaired in scurvy, due to vitamin-C (ascorbic acid) deficiency, in which collagen formation is defective [p. 495]. Protein lack, particularly deficiency of methionine, acts likewise. Cortisol and related compounds, in large doses, interfere with collagen formation and wound healing, but physiological or small therapeutic doses do not impair healing in man. When applied topically to the skin corticosteroid derivatives may promote healing in many types of dermatitis, by suppressing the inflammatory response.

It must be emphasized that no substances are known which accelerate the rate of healing above the natural rate, but where any factor is deficient or abnormal, correction of the abnormality will restore the rate to normal.

The processes of healing vary in different organs. The changes which lead to regeneration in the liver are of particular interest because, after removal of a substantial part of the organ, or severe damage by disease, the remaining liver cells proliferate until the total liver weight has returned to normal. This is a remarkable illustration of a long-term homeostatic process. The stimulus to enhanced mitosis is conveyed by the blood, since partial hepatectomy in one of a pair of parabiotic rats induces mitosis in the intact liver of the non-hepatectomized partner within 2–3 days of the operation. The nature of the substance is not known. In the process of regeneration the normal proportion between parenchymal cells and stroma is maintained. Anterior pituitary growth hormone is necessary for liver regeneration and the process is facilitated by thyroxine and cortisol.

REFERENCES

FLOREY, H. W. (ed.) (1970). *General pathology*, 4th edn. London.

SLOME, D. (ed.) (1961). *Wound healing*. London.

SPECTOR, W. G. and WILLOUGHBY, D. A. (1968). *The pharmacology of inflammation*. London.

ZWEIFACH, B. W., GRANT, L., and McCLUSKEY, R. T. (1965). *The inflammatory process*. New York.

PART II
The heart and circulation

General considerations

An average adult has a blood volume of some 5–6 litres and a resting cardiac output of about 5½ litres. In heavy exercise the cardiac output may reach 25 litres. The main purpose of the circulation is to supply the tissues of the various organs with nutrient substances (oxygen, carbohydrate, amino acids, fats, hormones, and immunological agents) and to remove waste products of their metabolism. Subsidiary functions exist—thus, skin blood flow subserves the thermoregulatory functions of the organism providing a flow which occasionally far exceeds the metabolic requirements of the skin when the body is subjected to thermal stress.

During the three score years and ten biblically allotted to man a little pump weighing some 300 g, 'the size of a clenched fist' is responsible for the circulation. No engineer has yet devised a pump with the long-term performance of the heart. In 70 years the output of the two ventricles exceeds 400 million litres—even if man remained in the resting state. The heart possesses four chambers—two atria, which though contractile also possesses reservoir properties and two ventricles which acts as the pumps proper. The atrium and the ventricle on the right side of the organ are respectively separated from their counterparts on the left by an interatrial septum and an interventricular septum [p. 94]. The right atrium and ventricle receive blood which has returned from the tissues via the veins and which reaches the right atrium at a pressure *only slightly above zero*. The blood is successively pumped from the right atrium and the right ventricle into the pulmonary artery which divides into many arterioles and then into pulmonary capillaries. These capillaries are only one cell thick and provide an enormous surface area for the diffusion inwards of oxygen from the air sacs of the lungs and the diffusion outwards of the waste product, carbon dioxide, into the air sacs from which the carbon dioxide is finally expelled from the body by rhythmic respiratory mechanisms. The oxygenated blood is then collected from the pulmonary capillaries by the pulmonary venules and veins and is transported thereby to the left atrium and left ventricle. The left ventricle has the task of pumping blood into the systemic circulation via the main distributing artery—the aorta. The aorta divides into arteries which supply the various regional circuits. These arteries themselves undergo successive divisions and the smaller branch arteries finally deliver the blood to vessels between 500 and 100 μm diameter—the arterioles. These in turn subdivide and lead via precapillary sphincters to capillaries which, in enormous numbers, permeate the tissues and because of their tremendous surface area and extreme thinness of wall (one cell thick) allow diffusion of nutrient substances to the immediately adjacent tissue cells. Capillary blood likewise by diffusion takes up carbon dioxide and waste products from the tissue cells. The capillaries distribute the blood finally to venules and thence to the veins which serve (among other functions) as conduits for the return of blood to the heart. A massaging action by the skeletal muscles which surround the veins considerably assists venous return.

The heart, then, provides the energy for the circulation. Its phasic ejection of blood raises the pressure in the aorta and may be regarded as the source of current which provides potential. The analogy of Ohm's law enables an understanding of the factors required

$$\underset{\text{(Current)}}{I} = \frac{E\,\text{(Potential)}}{R\,\text{(Resistance)}}$$

E is the pressure head for flow provided by *I*. *R* is the resistance offered by the smaller vessels of the arterial tree. It must be clearly understood that blood flows from the heart to the periphery and thence via the veins back to the heart only because of an *energy* (or *pressure*) *gradient* [see p. 66]. Because of the resistance to flow exhibited in different parts of the periphery this pressure gradient must be adequate though variable. As the heart output is intermittent (some 70–75 beats a minute) the pressure in and the flow into the arteries from the heart is pulsatile. This means that the arterial system should possess a low input impedance. The tissues themselves require desirably a steady flow through the capillaries so as to benefit maximally from the diffusion exchanges between the blood and tissues. Lastly, the 'requirements' of tissues vary with their metabolic activity at any one time—as exemplified by the situation in heavy muscular exercise where some 20 litres of blood per minute are required by the skeletal muscle from a total cardiac output of, say, 25 litres per minute.

Clearly, cardiac work could be lessened by increasing the radius of the vessels into which the blood is delivered, thereby reducing the pressure required to provide an adequate flow. Unfortunately the blood volume then needed to fill these larger tubes would represent a large fraction of the body weight—it would also entail a much greater provision of red cells by the bone marrow. Alternatively, the work of the heart could be reduced by considerably increasing the distensibility of the vascular system for this would decrease the pulsatile pressure and hence the impedance. However, such an increase in distensibility of the tree would cause a forfeiture of the effects of sudden increases in cardiac output, which occurring in biological emergencies may be vitally important to the survival of the animal. If the system took long to 'inflate', the urgent requirement for blood supply to muscles, brain, and heart itself in emergency circumstances could not be achieved, for the pressure head available to such organs would rise too slowly for it to be of much use.

Some compromise then has been arrived at. The *arteries* are *distensible* and contribute to the conversion of the pulsatile ejection of the heart into a steady flow to the tissue capillaries. The *arterioles* offer a *resistance* to the flow towards the capillaries and step down the hydrostatic pressure within these capillaries. If this were not so the loss of blood volume by transudation of fluid across the capillary walls could only be offset by considerably increasing the plasma protein concentration and hence the colloid osmotic pressure of the plasma. Lastly, the *veins* have been equipped to serve as *capacity* vessels in which, by appropriate variations of their diameter, the mobilization of blood to the heart or the minimization of acute increases in forward flow across the circulation are achieved in various circumstances.

Blood enters the right or left atrium at a pressure near zero. The left ventricle pumps the blood into the aorta where it reaches a peak value during the cardiac contraction phase (systole) of 120 mm Hg or so [Fig. II.1]. During diastole the aortic pressure subsides to some 80 mm Hg, due to the elastic recoil of the arterial system and to the resistance to outflow offered by the peripheral arterioles. *This combination of elasticity and resistance converts the pulsatile ejection of the heart into a steady outflow.* An important experiment of Borelli performed three hundred years ago illustrates this point.

A tube containing a valve is allowed to dip into water. The tube is connected to a rubber bulb from which an outlet tube debouches. Rhythmic squeezing of the bulb causes fluid to be sucked into the rubber bulb, whence it is discharged through the outlet. If the outlet tube is rigid, the emission of the fluid occurs in spurts whether the outlet tube be widely patent or constricted. If the outlet tube is elastic, flow is still intermittent unless the end is constricted, whereupon emission becomes steady.

The vascular system may be conveniently regarded as comprising: 1. Windkessel vessels; 2. precapillary resistance vessels; 3. precapillary sphincters; 4. capillary exchange vessels; 5. postcapillary resistance vessels and capacity vessels (veins); and in some circuits 6. shunt vessels.

1. *Windkessel vessels* are represented by the aorta and its large branches—vessels which are highly elastic. Systolic ejection distends these and, subsequent to the closure of the aortic valve at the

termination of systole, the elastic recoil of the vessels sustains the pressure head better and renders the blood flow to the periphery steadier than it would otherwise be. Potential energy, stored during cardiac contraction by the elastic tissues of the aorta and its branches, is reconverted into kinetic energy for the circulation during the diastolic phase. Degenerative changes in the media of the large vessels cause a loss of arterial elasticity and a high pulse pressure (systolic pressure minus diastolic pressure) results owing to the lack of the Windkessel effect.

2. *Precapillary resistance vessels (arterioles)* provide the great majority of the peripheral resistance. Changes in their radius moreover determine the blood supply of the different regional circuits. Precapillary resistance vessels usually exhibit an efficient local myogenic control of their own vascular radius and on this myogenic tone is superimposed an extrinsic neural control effected by sympathetic constrictor nerves. These nerves discharge at an impulse frequency of say 1 imp s^{-1} but the rate of discharge may be increased to 10–16 imp s^{-1} in appropriate circumstances (e.g. haemorrhage) or entirely suspended (e.g. in the skin vessels during heat stress).

3. *Precapillary sphincters*, themselves part of the precapillary resistance vessels, are particularly important in determining the size of the capillary exchange area which is perfused at any one moment in the tissue, for increase in the patency of the sphincters causes an increase in the number of capillaries open. The radius of the precapillary sphincters is controlled both by neurogenic factors and by the local concentration of tissue metabolites.

4. *The capillary exchange vessels* are tubes consisting of a single layer of endothelial cells. Solutes pass to and from the tissues across the capillary walls. The capillaries represent the key section of the circulatory system, but they themselves are not controlled by either nervous or metabolic factors. It is the alteration of precapillary sphincteric tone which determines the number of capillaries patent

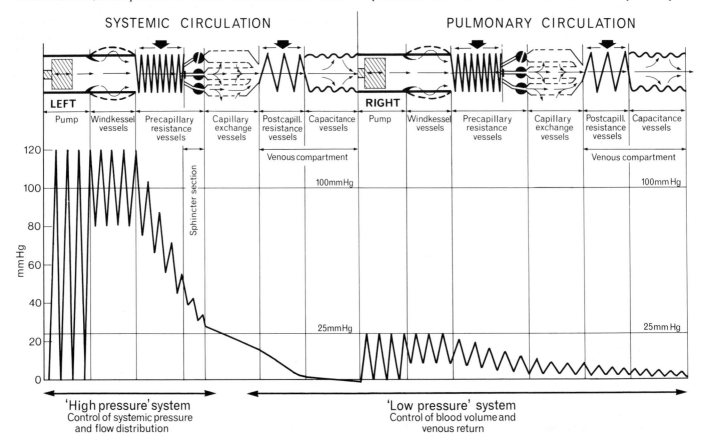

Fig. II.1. Schematic diagram illustrating the various series-coupled sections and the pressure drop along them. (Folkow, B. and Neil, E. (1971). *Circulation*. Oxford University Press, New York.)

and hence the surface area available for exchange between blood and interstitial fluid. In resting tissue only some 20–25 per cent of the capillaries are patent; the onset of tissue activity is attended by relaxation of the sphincters and perhaps by maximal opening of the relevant capillary exchange bed.

5. *Capacity vessels* are represented by the venular and venous compartments. These, though contributing little to the over-all resistance of the circuit are important sites of change in the *capacity* of the vascular system. Changes in luminal figuration (from elliptical to circular cross-sectional profiles) and changes in the myogenic tone of the veins induced by sympathetic constrictor nerves are of great importance in adjusting the capacity of the system, particularly in postural changes. When man assumed the upright posture a considerable strain on the circulatory system arose, for veins are distensible and the hydrostatic pressure of the column of blood in veins below the level of the heart tended to 'pool' the blood in such parts as the feet and legs. Both sympathetic nervous discharge on the myogenic tissue of the veins and the provision of extravascular compressive forces by skeletal muscle contractions offset these hydrostatic effects.

6. *Shunt vessels* occur only in a few tissues—most notably in the skin. Such vessels by-pass the capillaries and, if patent, permit a rapid flow of blood which serves no nutritive purpose to the tissue concerned. Their patency is controlled entirely by sympathetic vasoconstrictor discharge. Thermal stress causes, via the central nervous system, abolition of such discharge and the tremendous increase of cutaneous blood flow which results allows dissipation of heat from the body surface.

THE DIMENSIONS OF THE VASCULAR SYSTEM

In resting man a cardiac output of 5½ litres per min suffices to supply oxygen and nutriments to the whole body. The husbandry of the circulation must be careful, for regional measurements of the flow through the various parallel circuits show that the brain receives 750 ml min^{-1}, the liver 1500 ml min^{-1}, the kidneys 1200 ml min^{-1}, the muscles 600–900 ml min^{-1}, and the heart itself 200 ml min^{-1}. These flows total 4½ l min^{-1} and we have not allowed for skin and bone flow. Any one regional circuit receives a blood flow which depends on a pressure head (mean blood pressure), which is the same for each, and on its local vascular resistance, which varies from tissue to tissue.

Resistance can only be expressed quantitatively from a knowledge of flow and pressure head. For such a calculation flow must be expressed in unit or quantitatively comparable tissue weight. Conventionally, flow is described as ml per 100 g per min. TABLE II.1 provides information about the blood flow, resistance, and oxygen usage of the more important organs.

Muscle in resting man, though using 20 per cent of the total oxygen required by the body receives some 15 per cent of the cardiac output, although its flow is only 2.7 ml per 100 g per min. In maximal exercise the flow may reach values of 50–70 ml per 100 g per min, which represents a flow through 30 kg muscle of some 20 l min^{-1}. (Muscle bulk is 40–45 per cent of the body weight.) This enormous increase in muscle blood flow is achieved by the vasodilatation of the arterioles and precapillary sphincters caused by local metabolites produced by the active tissue. It is *not* due to nervous factors. In resting muscle, however, the influence of sympathetic constrictor nerves on the precapillary resistance vessels is partly responsible for keeping the muscle blood flow at the low values recorded. When one considers that even a doubling of flow from 2.7 ml per 100 g per min to 5.4 ml per 100 g per min would require a bulk muscle blood flow increase from 840 ml min^{-1} to 1680 and that this must be provided by the pump, it is clear that the muscle blood vessels represent by far the most important site of the peripheral resistance. Interestingly, the precapillary resistance vessels which supply the muscle have themselves a high degree of basal myogenic tone, for even sympathetic denervation only increases resting muscle blood flow some two or three times. As has been seen, this basal tone can be much more profoundly reduced when the metabolic requirements of the muscle are increased by exercise.

Skin vessels also manifest a high regional resistance in resting circumstances, mainly due to the sympathetic constrictor control of the precapillary resistance vessels and the A-V shunts. Thermal stress, however, which reflexly inhibits this tonic constrictor discharge, may cause the skin blood flow to achieve values of 150–200 ml per 100 g per min—a bulk skin flow of as much as 4 l min^{-1}, indicating how much extra load is put on the heart when the subject exercises in torrid climates.

Salivary glands may sporadically achieve a blood flow of 600 ml

TABLE II.1. Blood flow, resistance, and oxygen usage of the more important organs

Organ	Weight (kg)	Blood flow during rest (Max. vasodil. = [1])			Oxygen usage during rest			
		ml min^{-1}	ml per 100 g per min	% total card. output	A-V O$_2$ difference, ml per 100 ml blood	ml min^{-1}	ml per 100 g per min	% total O$_2$ usage
Brain	1.4	750 [1500]	55	14	6	45	3	18
Heart	0.3	250 [1200]	80	5	10	25	8	10
Liver	1.5	1300 [5000]	85 }	23	6	75	2	30
GI. tract	2.5	1000 [4000]	40 }					
Kidneys	0.3	1200 [1500]	400	22	1.3	15	5	6
Muscle	35	1000 [20 000]	3	18	5	50	0.15	20
Skin	2	200 [4000]	10	4	2.5	5	0.2	2
Remainder skeleton, bone marrow, fat connective tissue, etc.)	27	800 [4000]	3	14	4.4	35	0.15	14
TOTAL	70 kg	5500		100		250		100

The values in the table are 'rounded' figures and roughly describe the situation in average man during rest. Figures within brackets give in very approximate terms organ blood flows at maximal vasodilatation of the respective circuits.

(Folkow, B. and Neil, E. (1971). *Circulation*. Oxford University Press, New York.)

per 100 g per min and the intestinal mucosa during digestive episodes may receive as much as 500 ml per 100 g per min.

The kidneys with a blood flow of more than 400 ml per 100 g per min have the important function of filtering the blood and to the requirements of the filtration process have to be added those for reabsorption of the sodium which is simultaneously filtered. It is not surprising that the oxygen usage of renal tissue is linearly related to the filtered load of sodium.

Finally, the most important regional circulations of all, those of the heart and brain are conspicuous in manifesting a control of vascular resistance which is effected rather by chemical than by nervous influences. In the resting body the left ventricle receives via the coronary circulation some 85 ml per 100 g per min, the total coronary flow being say 250 ml min^{-1}. In heavy exercise, left ventricular coronary flow reaches 300–400 ml per 100 g per min in a heart beating 180 min^{-1} with a stroke volume perhaps as much as 200 ml. The coronary vessels are extremely sensitive to the metabolic effects of oxygen lack and the resistance of the circuit is correspondingly adjusted to the local metabolic needs of the cardiac tissue. Normally, myocardial tissue extracts some 65–70 per cent of the oxygen supplied by the arterial blood. Thus, if the arterial oxygen content is 20 ml per 100 ml, blood withdrawn from the coronary sinus, which drains the left ventricle, contains only 6–7 ml per 100 ml even when the individual is at rest. It follows that the huge increase in oxygen usage necessitated by the cardiac work in heavy exercise can only be met by maximal coronary vasodilatation of a degree proportional to the metabolic activity of the heart.

The cerebral circulation provides a flow of some 55 ml per 100 g per min and this varies little unless the chemical composition of the blood or the cerebral tissues changes markedly. In the brain the local [H$^+$] is the main determinant of cerebral vascular resistance—hyperventilation causes cerebral vasoconstriction and conversely a rise in [H$^+$] dilates the cerebral vessels. There is reason to believe that regional activity within the brain varies the local blood flow so that the increased local [H$^+$] resulting from, say activity in the occipital cortex, is followed by vasodilatation in that region. Elsewhere in the body, as will be seen, reflex neural adjustments of sympathetic constrictor tone ensure that the main blood pressure is kept fairly stable and hence that the pressure head for perfusion does not vary much. The cerebral vessels possess only a sparse innervation by vasoconstrictor nerves and the cerebral vascular pressure head is thus maintained and stabilized by extracerebral adjustments.

CARDIOVASCULAR INNERVATION

The heart receives an innervation both from the sympathetic nervous system, and a parasympathetic supply via the two vagus nerves. The sympathetic nerve cells lie in the intermediolateral cell horn of the upper five thoracic segments of the spinal cord. Their axons (preganglionic fibres) pass into the sympathetic trunk to the superior, middle, and inferior cardiac ganglia. Here they synapse and the postganglionic fibres pass via the superior, middle, and inferior cardiac sympathetic nerves to supply nodal tissue and the muscle of atria and ventricles. Sympathetic stimulation causes increase in the rate of the heart by increasing the rhythmicity of the sinu-atrial node and of much more importance it increases the force and speed of the myocardial contraction [inotropic action, p. 99]. It is dubious whether there is much tonic influence of the cardiac sympathetic nerves in circumstances extant in the circulation of the individual at rest. It is certain that an increase in sympathetic discharge, such as is seen in biological emergencies, such as flight, fright or fight, or in haemorrhage, are of vital importance to the organism in securing maximal mobilization of the pumping mechanisms of the heart.

The vagus nerves which supply the heart have their cell bodies in the medulla in the nucleus ambiguus of the vagus [p. 131]. Preganglionic fibres are distributed in the vagi to synapse with ganglion cells which are found in the vicinity of the sinu-atrial and atrioventricular nodes and in the atria. No vagal motor fibres are distributed to the ventricles.

Stimulation of the vagus profoundly reduces the rate of impulse generation by the sinu-atrial node [negative chronotropic action, p. 120]. It also reduces the rate of propagation of the cardiac impulse and diminishes the force of atrial contraction. Vagal stimulation does *not* influence contraction of the ventricular myocardium directly because the vagi do not innervate the ventricular muscle. Nevertheless, electrical stimulation of the cardiac vagi does slow the heart—it may even stop it—and reduce the cardiac output, but these effects are due to the vagi stopping or slowing the generation of the cardiac excitatory process by the s-a node.

In the circulation of the resting man or laboratory animal the heart beats under what is called vagal tone. Commonly, in man the resting heart rate is 68–75 beats min^{-1}. Atropine, which abolishes the transmission of the vagal effects upon the s-a node [p. 120], causes the heart rate to increase to 160–180 min^{-1}. The difference between the resting heart rate before and after atropine gives a measure of the degree of vagal tone.

The activity of the medullary centres respectively responsible for vagal and sympathetic effects on the heart is subservient to the effects of afferent impulses coming from nerve endings in the cardiovascular system itself and also to the sporadic influence of nerve centres in the higher parts of the neuraxis (cortex and hypothalamus). The tachycardia of emotion is an example of 'higher centre influence' [p. 131]. The medullary reflex influences are less well known and require a little more attention.

Situated in the adventitia of the wall of the first part of the internal carotid arteries (carotid sinus) and in the adventitia of the wall of the transverse part of the aortic arch are stretch receptors [p. 124]. Those in the carotid sinuses are nerve endings of the glossopharyngeal (IXth) nerve, whereas the aortic nerve endings are of vagal origin. There are also similar vagal nerve endings at the junction of the venae cavae with the right atrium and at the junction of the pulmonary veins with the left atrium, although these endings are subendocardial or subintimal in distribution. Ventricular vagal receptors likewise exist, though more sparsely. Lastly, the pulmonary trunk and its main branches possess adventitial vagal nerve endings of stretch receptor function. All these afferent nerve endings are responsive to mechanical distortion and in the main serve as stretch receptors although the names 'baroreceptors' and 'mechanoreceptors' are often accorded them. These receptors provide a negative feedback mechanism for the cardiovascular system [p. 124]. The 'stretch' which they signal, by way of impulse activity in the fibres which ascend as afferents to the medullary cardiovascular centres, has the reflex effects of reducing sympathetic discharge to the heart and blood vessels and increasing vagal activity destined mainly for the heart. Each systolic ejection of the heart causes stretch of the blood vessels and correspondingly, a phasic burst of impulses. If the mean blood pressure rises or if the pulsatile ejection of the heart be excessive, more impulses ascend to cause reflex sequelae which tend to redress the situation by appropriate cardiac and vascular responses. Conversely (and of great importance), a fall in systolic ejection of the heart which would tend to lower the systemic blood pressure, results in fewer baroreceptor impulses which as a consequence cause less efferent vagal activity and more sympathetic activity, which again helps to stabilize the cardiovascular system. These nerves are often referred to as the 'buffer nerves' for these reasons.

PERIPHERAL VASCULAR INNERVATION

Overwhelmingly preponderant in their distribution to the vascular tree are the sympathetic vasoconstrictor nerves. These arise like those of the heart from cell bodies in the intermediolateral horns of the spinal segments, but, whereas the heart receives a supply from only the upper five thoracic segments, the systemic vascular tree receives fibres from cells of origin situated in each of the twelve thoracic segments and in the upper two segments of the lumbar part of the spinal cord. This *thoracolumbar outflow* exerts a tonic influence on virtually all of the vascular tree with the exception of the cerebral regional circulation, as noted previously, and with the notable exception of the true capillaries of the system as a whole. Although the main arteries and veins do receive sympathetic constrictor innervation, the main density of such innervation occurs in the arterioles and precapillary resistance sections and in the post-capillary venules. The influence of this sympathetic discharge on the peripheral resistance and on the precapillary/postcapillary resistance ratio is profound [p. 72]. Adjustments of sympathetic tonic activity can again be produced by both higher centre influences and by mechanoreceptor reflexes just as was the case with the heart. A rise of mean blood pressure or of pulse pressure in the systemic arteries is reflexly smoothed out or offset by the attendant increase of negative feedback which, by reducing the sympathetic tonic discharge, lowers the peripheral resistance. Conversely, a fall in circulating blood volume (haemorrhage) is accompanied by a reflex increase in arteriolar and venular tone which helps to sustain the mean level of blood pressure by increasing the peripheral resistance and by reducing venular capacity.

Section of the spinal cord in the lower cervical region causes a fall of blood presure from its resting mean value of c. 100 mm Hg to 40 mm Hg. Conversely, stimulation of the lateral funiculi of the intact spinal cord produces a striking rise of blood pressure, for the 'excitor' pathways between the medullary centre units and the relevant thoracolumbar sympathetic cell bodies traverse both sides of the spinal cord with this lateral distribution.

Destruction of the central parts of the medulla between the level of the facial colliculi [p. 124] and the obex causes a fall of blood pressure similar to that ensuing on spinal transection. Stimulation of this intact region on the other hand causes a rise of blood pressure. It is for such reasons that this region was designated the *vasomotor centre* [p. 124].

Vasodilatation of parts of the vascular system can be achieved in different ways—*nervous* and *chemical. Nervous* mechanisms of vasodilatation can be:

1. Reduction of sympathetic constrictor tone. Examples of this are provided by the flushed skin of heat stress, or the experimentally induced vasodilatation of the muscles, viscera and skin caused by artificial stimulation of the carotid sinus or aortic vagal afferent nerves.
2. Specific activation of vasodilator nerves. Such nerves may be subdivided into: (i) Sympathetic cholinergic vasodilators. These supply only skeletal blood vessels. They are far outweighed in importance by the sympathetic noradrenergic constrictor supply to the skeletal muscle blood vessels. The constrictor supply to skeletal muscle blood vessels is tonically active—the vasodilator supply is not. Vasodilatation of the skeletal muscle blood vessels by the cholinergic sympathetic fibres is seen only in biological emergencies, where, in the life of wild animals, it may have importance in the survival of the potential victim. The sudden onset of sympathetic vasodilatation consequent solely upon higher centres which participate in the 'alarm reaction' may help to initiate the conversion of the animal into a head-heart-lung-muscle pump system. Within a few seconds, however, metabolite-induced vasodilatation in the muscles far exceeds the potential of the sympathetic vasodilators.

(ii) Parasympathetic vasodilator nerves. These are best represented by the nervi erigentes which supply sexual erectile tissue. The vasodilator supply to the salivary glands provides another example. In each case the activity of such nerves, though contributing to pleasure and fulfilling important biological functions, plays little part in the overall economy of the circulation.

(iii) Dorsal root vasodilators [axon reflex, p. 141].

Chemical vasodilatation is of overwhelming importance in conditions of increased metabolic activity of the tissue. The best examples are provided by the skeletal and cardiac muscle circulations, but as has been stated, a regional distribution of the intracerebral blood flow may be mainly dictated by such mechanisms.

REFERENCE

Folkow, B. and Neil, E. (1971). *Circulation*. Oxford University Press, New York.

DIMENSIONS OF AND FLOW VELOCITY IN THE COMPONENTS OF THE VASCULAR SYSTEM

The aorta in man has a cross-sectional area of about 4 cm^2. With a cardiac output of 5.4 l min^{-1}, 90 ml of blood will pass through the aortic lumen each second at a *mean* velocity of 22.5 cm s^{-1}. As the arteries subdivide, the radius of the individual branches of course decreases, but the total cross-sectional area increases. Thus, there are approximately 8000 small arteries with an internal radius of 0.5 mm and each has a cross-sectional area of $(22/7 \times 0.05)^2$ cm^2. Their total cross-sectional area is $22/7 \times (0.05^2 \times 8000)$ cm^2 = 63 cm^2. This is 16 times that of the aorta. The rate of flow through each of these arteries is $22.5/16 = 1.4$ cm s^{-1}.

There are about forty-thousand million capillaries in the systemic circuit of a man weighing 70 kg, but only about a quarter of these are patent at any one time in resting conditions. The average capillary has a radius of 3 μm (and a length of approximately 750 μm). The linear velocity of blood flow through them is of the order of 0.3 mm s^{-1} which is only 0.3/225 or roughly one-seven hundredth of the mean velocity of flow in the aorta. Hence, the total cross-sectional area of the capillary bed in resting conditions, when only 25 per cent of the capillaries are patent, is 700 times that of the aorta. If they were all patent at any one time, which must be rarely if ever the case, their total cross-sectional area would be 2800 times that of the aorta.

Such a *total* (patent) capillary exchange surface would be $2\pi r l \times$ *40 000 000 000* μm^2 = $44/7 \times 3 \times 750 \times 0.4$ square metres = 560 square metres. Of this normally 25 per cent is available, i.e. 140 square metres of surface for the exchange of solutes and gases between the blood and the tissues. It can be calculated that the muscles (30 kg of them) would account for some 250 square metres of capillary surface when their vessels were fully patent.

On the venous side, the progressive confluence of the postcapillary venules, etc., leads finally to the two venae cavae. The cross-sectional area of each vena cava is approximately 50 per cent bigger than that of the aorta and as the blood flow per second through the two of them is the same as that through the aorta, then the linear velocity through each of them, if they each carried equal flows, would be one-third of that in the aorta—7–8 cm s^{-1}. Actually the inferior vena cava returns about two-thirds of the cardiac output so its flow velocity would be higher and that of the superior vena cava lower. However, the important point is that flow is rapid in the big veins. The relatively sluggish ooze which occurs when a medium-sized arm vein is cut does not indicate that flow in the vein is slow—the bleeding is sluggish because the pressure within the vein is only 2–3 mm Hg above atmospheric and hence the pressure head causing the bleeding is very small. If any remains unconvinced, let him try the Harvey experiment of clearing a venous segment of a superficial arm vein by firmly stroking in a centrifugal direction, finally occluding the vein distally by pressure with the fingers; on releasing the finger pressure the vein fills very promptly indeed owing to the high flow velocity.

Pressure and flow

Fluid flows from one site (a) to another (b) if the total fluid energy at (a) exceeds that at (b) (Burton 1965).

Total fluid energy comprises three items: 1. Pressure energy (*P*)—potential energy; 2. Gravitational potential energy (σgh)—the capability of doing work because of differences in fluid level; 3. Kinetic energy ($\frac{1}{2}pv^2$) (where v = velocity and p = density).

This third term, the *kinetic energy per unit volume*, is important. Bernoulli showed that when fluid flows along a tube, its pressure measured by a side tube is lower when its velocity is higher. In a fluid flowing at one level where we can ignore the gravitational energy factor.

$$E = P + \frac{1}{2}pv^2$$

(if negligible dissipation of energy occurs).

In FIG. II.2, the flow velocity increases where the tube narrows and the lateral pressure decreases; on reaching the wider part of the tube again, part of the kinetic energy is now transformed into pressure energy.

The numerals in the diagram show clearly that flow is down a gradient of energy, *not* of pressure. As Burton points out, this kinetic factor has to be borne in mind when pressures are measured by catheterizing blood vessels.

In the arterial system the kinetic energy factor is negligible, except in the aorta in heavy exercise.

In the vena cava, kinetic energy is important only during exercise.

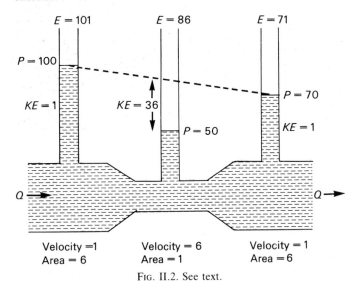

Fig. II.2. See text.

In the atria and pulmonary artery, kinetic energy is not inconsiderable at rest and is very important when the cardiac output is increased. Thus, it can be calculated (Burton 1965) that the kinetic energy factor is 50–55 per cent of the total in the venae cavae, atria, and pulmonary artery.

When measuring pressure, unless the tip of the catheter has its opening at right angles to the stream (measuring 'side pressure') the pressure recorded is not accurately that in the blood at that point.

If the catheter faces the oncoming bloodstream or if an endcan-

nula is tied into the artery the flow of blood is prevented and the manometer registers a pressure higher than that of the blood by the factor $\frac{1}{2}pv^2$ [FIG II. 3].

Downstream

Fig. II.3. See text.

If the catheter opening faces down stream the pressure recorded is lower than the pressure in the fluid by slightly less than $\frac{1}{2}pv^2$ (flow is distorted by streaming round the catheter tip and this reduces the correction required). Such a catheter placement is the accepted technique for the recording of pulmonary arterial pressure [see p. 145].

In the circulation at rest, the kinetic energy factor is of little importance on the systemic arterial side and of only moderate significance in the systemic veins. Flow therefore can be regarded as occurring between the root of the aorta and the right atrium by virtue of a pressure head ($P_1 - P_2$).

In a rigid tube system, it would not matter to the flow produced if both P_1 and P_2 were raised by the same amount, for $P_1 - P_2$ would be exactly the same. Only the transmural pressures (registered by side tubes) would all be increased by the same amount.

When man stands, nothing happens to the pressure head ($P_1 - P_2$). Only the *transmural pressure* is affected. Here, however, it is necessary to understand that the dependent arteries, and particularly the capacitance vessels (venules and veins), are distensible and the capillaries are porous. The physical consequences of this would be serious if physiological adjustments did not compensate [see p. 91].

REFERENCE

BURTON, A. C. (1965). *Physiology and biophysics of the circulation.* Chicago.

THE MYOGENIC ACTIVITY OF THE SMALL VESSELS AND THE FACTORS WHICH INFLUENCE IT

Basal myogenic tone

Basal myogenic tone is produced by the inherent myogenic activity of the smooth muscles of the arterioles, metarterioles and precapillary sphincters. The arterioles possess a continuous smooth muscle coat which shows a fairly synchronized myogenic activity, due to the activity of pacemaker cells (whose site changes from time to time), which is propagated in a cell-to-cell manner. This pacemaker activity is accentuated by a stretching force, so that when the transmural pressure is raised an arteriolar constrictor response occurs to reduce the pressure transmitted downstream.

The precapillary sphincters and metarterioles also possess 'independent' pacemaker muscle cells which show asynchronous contractile activity. The activity of these pacemaker cells is responsible for the 'tone' of these small vessels; metabolites released locally can suppress the pacemaker activity, whereupon the sphincters relax and the capillary bed governed by them opens. Thus, if a precapillary sphincter is initially closed by contraction of its muscular elements, its capillary bed is not perfused and the cylinder of tissue cells supplied by this microcirculation continues to produce metabolites which accumulate outside the sphincter causing it to relax; the dissipation of these metabolites by the blood flow then allows the inherent pacemaker activity to redevelop, whereupon the sphincter tends to close again.

The basal myogenic tone produced by these mechanisms can be assessed by comparing the flow through the vascular circuit, when the tissues are at rest with the sympathetic nerves cut, with that recorded when the vessels are maximally dilated by, say, the intra-arterial injection of chloral hydrate. The pressure head and trans-mural pressure are kept constant in these two circumstances. Such studies can be made by the Mellander technique [p. 73] which also allows an investigation of capillary filtration coefficient changes as well.

Muscle vessels possess a high degree of basal myogenic tone equivalent to 10–15 *peripheral resistance units* (PRU_{100}); chloral hydrate (which causes maximal vasodilatation) lowers this resistance to 1.5–2 PRU. The flow and pressure head measurements in denervated muscle vessels which allow these calculations are respectively: in basal conditions, $P_1 - P_2 = 100$ mm Hg, flow = 7–10 ml per 100 g per min; maximal vasodilatation $P_1 - P_2 = 100$ mm Hg; flow = 70–100 ml per 100 g per min.

These precapillary resistance vessels of the muscle vascular circuit are very susceptible to the effect of natural metabolites produced by local tissue activity. The postcapillary resistance vessels are *not*, however, thus the precapillary: postcapillary ratio always falls from say 4–5:1 to 2–3:1 when the circuit is dilated by chemical factors.

TABLE II.2. Capillary filtration coefficient (CFC: ml per mm Hg × 100 g × min) as related to blood flow in ml per min × 100 g.

Tissue	Species	Resting conditions		Maximal vasodilatation	
		CFC	Blood flow	CFC	Blood flow
Muscle (limb)	man	0.006	3–5	0.03–0.04	50–60
Muscle	cat	0.01–0.015	6–10	0.04–0.05	50–60
Jejunum	cat	0.1–0.15	30–50	0.4–0.45	250–275
Jejunal mucosa–submucosa	cat	0.2–0.25	50–80	0.9–1	400–500
Kidney	man	1–1.2	300–400		

Relationship between filtration coefficient and blood flow in different tissues and species.
(See also *Br. med. Bull.* (1963). **19**, 155–60.)

Other vascular circuits like those of the salivary glands, which indulge only sporadically in high metabolic activity, show a high basal myogenic tone which is strikingly reduced when the tissue does work, just as is the case with the muscle vessels. The Table shows the relationship between resting blood flow and that during maximal vasodilatation in several tissues.

Neurogenic influences

These are superimposed on the basal myogenic mechanism. The 'constrictor' fibres of the sympathetic terminate on the *outer* sheath of the smooth muscle of the media. The arterioles show a dense innervation which is revealed by a histochemical fluorescence technique (Falck 1962) specific for catecholamines which permits visualization of the noradrenalin content of the axon terminals. The metarterioles and precapillary sphincters are less generously provided. The venules (surprisingly) are only feebly endowed.

Increased sympathetic discharge causes three effects: 1. a rise in total vascular resistance; 2. an increase in the precapillary/post-capillary resistance ratio which reduces the mean capillary hydrostatic pressure and hence favours the osmotic uptake of fluid from the interstitium [see p. 82]; 3. a reduction of the capacity of the postcapillary venules and veins.

The resting sympathetic discharge to the vessels of the microcirculation has never been measured directly, but it has been inferred from experimental results obtained by artificially stimulating the sympathetic trunks supplying the vascular circuits of the tissue under study (Mellander 1960).

FIGURE II.4 shows the response of the vessels to sympathetic stimulation frequencies from one impulse every 4 seconds to 16 imp s^{-1}. Even the initial stimulation at a very slow impulse rate shows a brisk response, particularly of the capacity vessels (drop in limb volume) and a more moderate reduction in flow due to an increase in resistance. The whole gamut of responses of which the vascular elements is capable occurs over a range of sympathetic discharge which does not exceed 16 imp s^{-1}. Sympathetic stimulation alters the distribution of fluid between the interstitium and the vascular compartment. After the initial sharp drop of volume there is a slower fall of volume and the tangent drawn to this gives an indication of the rate of this slower fall of volume. This slow shrinkage is due to the fall in hydrostatic pressure in the capillary which occurs as a consequence of an increase in the precapillary: postcapillary resistance ratio. Sympathetic vasoconstriction thus helps to mobilize extravascular fluid from the tissues and this effect, coupled with that of reducing the venous capacity, is of course of enormous importance in haemorrhage.

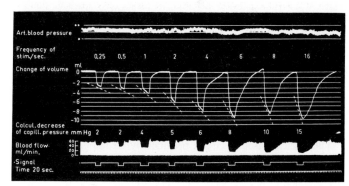

FIG. II.4. Mellander preparation. Effects of sympathetic stimulation at different frequencies on volume and blood flow of the hind quarters. Note that the effects on volume are maximal with a stimulation frequency of 6 imp s^{-1}. The effects of sympathetic stimulation on blood flow are maximal only when the stimulation frequency reaches 16 imp s^{-1}. (Mellander, S. (1960). *Acta physiol. scand.* Suppl. 176.)

Mellander constructed FIG. II.5 from the results of experiments similar to that of FIG. II.4. The percentage response of the capacity vessels to sympathetic discharge develops more quickly at the low rates of induced sympathetic activity. This result may at first sight seem incompatible with the histological demonstration of a notably sparse innervation of the skeletal muscle venules made by Fuxe and Sedvall (1965). However, it can be argued that even a slight shortening of the smooth muscles of the veins where the wall:lumen ratio is very low causes a relatively great decrease in their volume. Moreover, as the venous pressure is low, even a few muscle cells and noradrenergic terminals may well be enough to cause a considerable constriction of the venous vessels. Thirdly, some of the reduction in volume of the capacity vessels may be due to their tendency to assume an elliptical cross-sectional area when their transmural pressure drops and thus to diminish their capacity. This geometric change may indeed have much to do with the striking changes of 'capacity' which occur when very slow rates of sympathetic stimulation are employed.

In natural circumstances, variations in sympathetic discharge are caused either by reflex alterations of mean blood pressure affecting the sino-aortic mechanoreceptors and therefore the activity of the medullary cardiovascular centre, or by corticohypothalamic discharges which occur in circumstances of conscious emotional involvement. The regional vascular circuits are not equally affected by reflexly-induced changes involving activity of the medullary cardiovascular centres. The muscle circuit, for instance, is briskly responsive to sino-aortic mechanoreceptor reflexes—the kidney vessels are not.

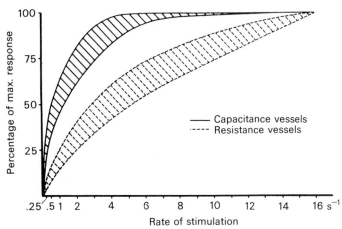

Fig. II.5. The response of resistance vessels (dashed lines) and of capacitance vessels (solid lines) to increasing frequency of sympathetic stimulation. (Mellander, S. (1960). *Acta physiol. scand.* Suppl. 176.)

At rest, muscle blood flow is 3–5 ml per 100 g per min (20–30 PRU_{100}). Sympathetic blockade increases flow to 7–10 ml per 100 g per min (10–15 PRU_{100}). Maximal sympathetic vasoconstriction will reduce the flow to 1 ml per 100 g per min even when the pressure head is 100 mm Hg (. ·. to 100 PRU_{100}). The blood flow through the skeletal muscle bed must be small indeed in haemorrhagic hypotension (when the pressure head may only be 50 mm Hg).

The muscle vascular circuit provides by far the greatest site of peripheral vascular resistance, for not only is it very sensitive to an increase in sympathetic discharge, but also its sheer bulk (30 kg) is the overriding factor. The splanchnic circuit, often erroneously cited as the most important site of peripheral vascular resistance, is quantitatively far less significant. The importance credited to the splanchnic bed stems from the experimental results of stimulating the cut peripheral ends of the splanchnic nerve; such stimulation does, of course, raise the blood pressure very strikingly. More recent work has shown, however, that this experimental result is more likely due to a constriction of the capacitance venules and veins, which secures the expulsion of perhaps as much as 40 per cent of the regional intestinal volume (which itself is 7–9 ml per 100g tissue in the gut in the absence of sympathetic discharge). Thus, the rise of pressure is more due to a 'transfusion' of blood thereby increasing cardiac output, than to the primary increase in resistance itself. The gut vessels show a high 'resting' flow when denervated—40–60 ml per 100 g per min and this is halved initially by strong sympathetic stimulation. However, if the stimulation continues the flow returns to about 80–90 per cent of the control level. The mechanism of this 'autoregulatory escape' in intestinal vessels is not yet fully understood. Sympathetic stimulation causes a reduction of 30–50 per cent in CFC in the intestinal vessels and this lower CFC value is maintained throughout stimulation despite the return of the

blood flow to near normal values. Thus, there is a decrease in the capillary exchange surface to about two-thirds to a half of its control value even when the resistance to flow has returned to near normal during continued sympathetic stimulation, and this suggests that there may be a redistribution of blood flow from say the mucosa to submucosal vessels (Mellander and Johansson 1968).

Pressure, resistance, and the precapillary: postcapillary resistance ratio

Figure II.6 shows graphically the pressure changes in the different compartments of the systemic circuit. The pressure head $P_1 - P_2$ remains fairly stable. Even in heavy exercise the mean systemic pressure is only increased by some 25–30 per cent. However, the *profile* of the pressure drop varies with different circumstances. Characteristically in the resting animal or man the greatest pressure drop occurs as the blood traverses the precapillary resistance vessels. If a widespread state of vasoconstriction occurs—say in response to a slow haemorrhage—the precapillary resistance rises so that, although the mean pressure in the arteries is well maintained, the pressure drop across these small vessels is steeper and bigger. Conversely, in heavy exercise there is a marked dilatation of the muscle precapillary resistance vessels and, moreover, turbulence in the larger vessels contributes to there being a greater drop in pressure as the blood traverses the arteries themselves. Capillary hydrostatic pressure is raised because of precapillary vasodilatation and the profile of pressure drop becomes somewhat more linear.

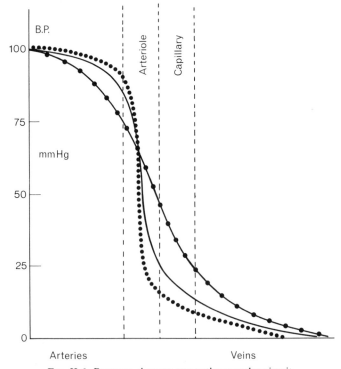

Fig. II.6. Pressure changes across the vascular circuit.
——— = Rest
—●— = Vasodilatation
●●●● = Vasoconstriction

Capillary pressure—the precapillary: postcapillary resistance ratio

Fluid can only be lost by or gained from the intact circulation in the capillary section. Two opposing forces influence fluid transfer—the hydrostatic pressure P_{cap} pushing fluid out and the osmotic pressure of the plasma proteins pulling it in. The osmotic forces are dealt with on page 82. Here, our concern is with the mechanisms that adjust P_{cap}.

The capillaries are simple single cell-walled endothelial tubes and have no power whatever of independent contraction.

Thus, if we consider the flow through a capillary:

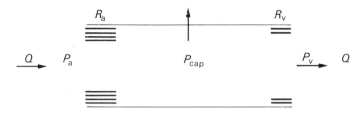

Q is the volume which enters from the arterioles and leaves via the venules each second.

P_a is the pressure on the arterial side of the capillary, say 100 mm Hg, and P_v is the venular pressure. R_a and R_v are resistances.

Suppose the mean capillary pressure at the middle part of the capillary (c) is P_{cap}.

Then:

$$Q = \frac{P_a - P_{cap}}{R_a} \qquad = \frac{P_{cap} - P_v}{R_v}$$

$$\therefore \quad P_a.R_v - P_{cap}.R_v = P_{cap}.R_a - P_v.R_a$$

$$\therefore \quad P_{cap}(R_a + R_v) = P_a.R_v + P_v.R_a$$

divide through by R_a

$$\therefore \quad P_{cap}(1 + R_v/R_a) = P_a R_v/R_a + P_v$$

$$\therefore \quad P_{cap} = \frac{P_a R_v/R_a + P_v}{(1 + R_v/R_a)}$$

Thus, the mean hydrostatic capillary pressure is governed by a ratio of precapillary:postcapillary resistance.

In the vascular circuit of muscle, this ratio R_a/R_v is about 5/1.

Substituting values for P_a and P_v and for R_a/R_v

$$P_{cap} = \frac{(100 \times 1/5) + 5}{(1 + 1/5)} = 21 \text{ mm Hg.}$$

The colloid osmotic pressure due to the plasma proteins is some 25 mm Hg. The interstitial fluid contains some protein and may exert an osmotic pressure from outside of perhaps 3–5 mm Hg. It can be seen then that the resistance ratio of 5/1 allows a mean capillary pressure which about balances the osmotic pressure. As a whole, this ideal systemic capillary and its hundreds of thousands of companions will show no net loss or gain of fluid and the total volume of tissue perfused by such capillaries will remain the same—isovolumic. Of course, at the 'arteriolar' end of the capillary the hydrostatic pressure will be higher than the mean and some fluid will be forced out of the capillary bed into the interstitium. However, at the venous end of the capillary, fluid will be pulled back by the osmotic forces as the hydrostatic pressure is lower. There is thus a tissue fluid circulation, first recognized by Starling (1896). It has been calculated that in the course of a day some 24 litres of fluid escape from the capillaries and that about 20–22 litres of this returns directly to the capillaries—the remainder (2 or more litres) being transported back to the systemic veins via the lymphatic system [see p. 147].

Pappenheimer and Soto-Rivera (1948) perfused the blood vessels of an isolated hind leg and determined the net gain or loss of weight when 'arterial' and 'venous' perfusion pressures were adjusted. An infinite number of pairs of values of arterial and venous pressures at which neither net gain nor loss of weight occurred were found. In such circumstances the preparation was described as isogravimetric and it was postulated that the mean capillary pressure was such as to balance the osmotic pressure influence. Further details are given on page 82 *et seq.*

Mellander (1960) used a principle similar to that of Pappenheimer and Soto-Rivera (1948), but measured the volume of the hind limb plethysmographically. The hind limbs and pelvis of a cat are almost separated by dissection; only the aorta and inferior vena cava preserve continuity between the 'preparation' and the remainder of the animal [FIG. II.7]. The hind limbs are enclosed in a water-filled plethysmograph, and changes in their volume are recorded. The arterial inflow pressure is recorded from the inferior mesenteric branch of the aorta. Adjustment of a screw clamp placed around the aorta proximal to the site of recording allows the maintenance of a constant arterial inflow pressure if so desired. The inferior vena cava is cannulated and venous return from the hind limbs is recorded by leading the entire caval flow to a Gaddum flow meter connected to a piston recorder. The height of the flow meter can be adjusted and correspondingly the venous outflow pressure from the hind limbs can be varied. Blood returns from the flow meter via a funnel into the upper part of the abdominal vena cava.

Both lumbar sympathetic chains containing the postganglionic sympathetic fibres destined for the hind quarters are isolated; they can either be preserved in continuity (and thus act as the motor pathways of reflex effects relayed over the vasomotor sympathetic mechanism) or alternatively they are sectioned and their distal ends placed on electrodes for electrical stimulation when required.

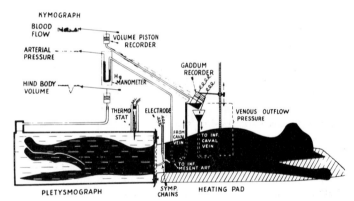

FIG. II.7. Illustration of the Mellander preparation. (Mellander, S. (1960). *Acta physiol. scand.* Suppl. 176.)

At a steady arterial pressure and venous outflow pressure the blood flow to the hind limbs is, of course, constant and the volume recorder likewise shows only pulsatile variations about a steady mean volume. A slight increase in venous outflow pressure such as induces an increase in capillary hydrostatic pressure of 5 mm Hg causes an increase in limb volume which occurs in two stages: first a sudden and rapid increase due to distension of capacity vessels (venules) and second a slower increase which is due to transcapillary filtration occurring at a greater rate than in the control circumstances. Conversely a reduction in venous outflow pressure causes a two-stage reduction in volume of the limbs.

FIGURE II.8, shows the effects of stimulating the sympathetic vasoconstrictor fibres to the hind limbs. Three events are worthy of comment:

(a) The blood flow decreases abruptly due to vasoconstriction of resistance vessels, particularly the arterioles.

(b) Almost simultaneously the volume of the hind limbs decreases rapidly, due to the vasoconstriction of capacity vessels (venules and veins).

(c) Slightly later there occurs a slow reduction in limb volume due to the passage of fluid from the tissues into the capillaries. This influx is caused by the reduction of hydrostatic pressure within the

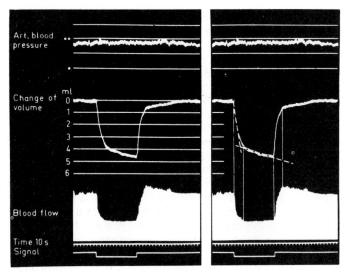

Fig. II.8. Effects of sympathetic stimulation on volume and blood flow of the cat's hind limbs. Stimulation frequency 2 imp s⁻¹ (signal). Constant inflow and outflow pressures. The second tracing (identical with the first) shows the initial rapid fall of volume followed by the slow fall of volume due to transcapillary influx of interstitial fluid. Note accompanying changes of blood flow. (Mellander, S. (1960). *Acta physiol. scand.* Suppl. 176.)

capillaries consequent upon the increase in arteriolar resistance. Mellander has shown that this slow reduction in limb volume in these circumstances can be offset, despite the continuation of sympathetic vasoconstrictor stimulation, by raising the venous outflow pressure by some 6 mm Hg. This then gives a measure of the fall in capillary hydrostatic pressure induced by the sympathetic stimulation.

The results of such studies indicate that variations of P_{cap} do result from the effects of sympathetic stimulation. An increase in sympathetic activity always lowers the mean capillary pressure because the pre/postcapillary resistance is increased. Only in long-term hypotension does this response alter [p. 151].

It is well known that sympathetic blockade reduces the arterial pressure; high spinal anaesthesia produces the same effect because the anaesthetic paralyses the intermediolateral horn cells of the thoracolumbar outflow. The fall in pressure is due partly to a reduction in cardiac output, mainly to reduction of the precapillary and postcapillary resistance and partly to the venodilatation of the venous capacity vessels. The mean capillary pressure, however, may not fall—it may rise because the pressure transmission across the arterioles is more complete.

Just as sympathetic vasoconstriction favours a fall in P_{cap}, exercise or tissue activity causes a rise of P_{cap}. The precapillary resistance vessels, particularly the precapillary sphincters, relax in the local presence of metabolites formed by the active-tissue cells and P_{cap} rises strikingly. This favours fluid transudation into the interstitium and the raised osmolality of the interstitial fluid further promotes this fluid transference. The postcapillary resistance venules are not very susceptible to the influence of local chemical factors, so the sum effect of all these changes is to lower the precapillary:postcapillary resistance ratio.

One way of assessing the change in P_{cap} which is caused, say, by the effect of sympathetic stimulation, is to raise the postcapillary pressure by hoisting the venous outflow level by a known height. If raising the height of the venous outflow pressure by 5 mm Hg just prevents the slow uptake of fluid from the limbs caused during the sympathetic stimulation, then it can be inferred that P_{cap} had been reduced by approximately that amount. This technique is useful in another way—in measuring the *capillary filtration coefficient*. The capillary filtration coefficient (CFC) is the fluid filtered per mm Hg rise of pressure per 100 g tissue per min.

Venous outflow pressure is raised by, say, 5 mm Hg by adjusting the height of the collecting device. The volume record, previously steady (isovolumic conditions), shows a prompt rise due to venous capacity changes and then a slow upward slope. The tangent of the slope gives the rate of fluid filtration and as the time and the weight of tissue in the plethysmograph are known, the CFC can be calculated. Resting innervated muscles shows a CFC of 0.010–0.0150 ml per min per 100 g per mm Hg in the cat. CFC is increased fourfold by maximal dilatation which indicates that only about a quarter of the skeletal muscle capillary bed is patent in resting conditions. The CFC in intestinal vessels at rest is about 10 times that in muscle (0.1 ml per min per 100 g per mm Hg) and in maximal vasodilatation 0.4 ml per min per 100 g per min per mm Hg. This reflects the enormous density and perhaps the greater permeability of the intestinal capillary bed. Skin shows a CFC about three times that of muscle when measured in conditions of thermal comfort.

P_{cap} is of course not the same in all capillaries. It is obvious, for instance, that P_{cap} could not possibly be 20 mm Hg or more in the pulmonary capillaries, for this is higher than the mean pulmonary arterial pressure. Indeed, if the P_{cap} in the pulmonary circuit is raised (by left heart failure for instance) pulmonary oedema occurs very promptly and may threaten life of itself.

Even in the systemic capillaries P_{cap} varies—for instance P_{cap} in the feet cannot be only 20 mm Hg when man stands quietly. If the pressure in the dorsal vein of the foot is directly recorded it is found to be 80–90 mm Hg and as the blood does not flow backwards, the capillary pressure must be higher than this.

The liver sinusoids likewise must show a P_{cap} which is much lower than 20 mm Hg—the sinusoids receive the bulk of their blood supply from the portal vein and the portal venous pressure is about 7–10 mm Hg. The pressure in the sinusoids cannot be above 6 mm Hg. Does this mean that the sinusoids are constantly 'picking up' fluid from the hepatic interstitium? Again, obviously not, because the liver would look like a wrung out sponge in next to no time. The answer, of course, lies in the extraordinary permeability of the liver sinusoids to protein. It would be of little avail for the liver to manufacture protein if its access to the blood were only by way of the lymph, with its slow rate of flow. Protein can and does pass freely across the sinusoidal wall, hence the osmotic pressure exerted by the proteins already *in* the plasma is offset to a great extent by the osmotic force exerted by proteins in the interstitium. Thus, a P_{cap} of 6 mm Hg is adequate to balance this reduced plasma osmotic factor.

REFERENCES

FOLKOW, B. and NEIL, E. (1971). *Circulation*. Oxford University Press, New York.
MELLANDER, S. (1960). *Acta physiol. scand.* **50**, Suppl. 176.
—— and JOHANSSON, B. (1968). *Pharmac. Rev.* **20**, 117.
PAPPENHEIMER, J. R. and SOTO-RIVERA, A. (1948). *Am. J. Physiol.* **152**, 471.
STARLING, E. H. (1896). *J. Physiol., Lond.* **19**, 312.

STRUCTURAL AND FUNCTIONAL FEATURES OF COMPONENTS OF THE VASCULAR CIRCUIT

All blood vessels, except the capillaries, have walls which are constituted by three coats, the innermost tunica intima, the tunica media, and the tunica adventitia (outermost).

The *tunica intima* consists of a single continuous layer of endothelial cells. In larger blood vessels this is separated from the media by the acellular internal elastic lamina. Vessels of less than 100 μm in diameter show breaks in the elastic lamina and as they get smaller this elastic lamina disappears and short processes of the endothelial cell pierce the basement membrane to come into close relationship with the smooth muscle of the media—the so-called myoendothelial junction.

The *tunica media* consists of smooth muscles. The *adventitia*, which surrounds the media, is made up of collagen and fibroblasts.

Anatomical features of different parts of the vascular circuit

1. *The arterial vessels* comprise (a) the large elastic vessels, such as the aorta and its immediate branches; (b) large muscular distributing arteries; and (c) small arteries and arterioles.

Rhodin (1967) has classified the arterioles and their subdivisions as follows on the basis of electron microscope studies.

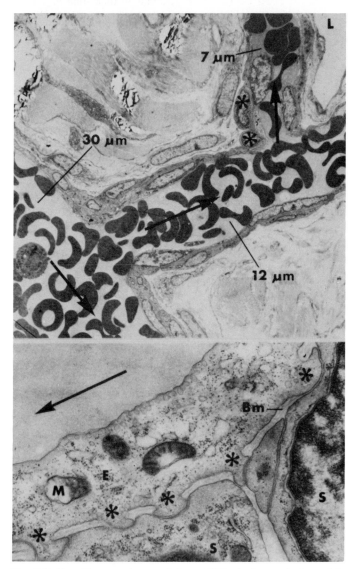

FIG. II.9. Top. Electronmicrograph (mag. 1750 ×) of a terminal arteriole with an inner diameter of 30 μm, at the point where a smaller 12 μm terminal arteriole is given off. The direction of the blood flow marked by arrows. Smooth muscle cells of a precapillary sphincter (**) indicate the beginning of an arterial capillary (7 μm), with a near-by lymphatic capillary (L). Nuclei of the smooth muscle cells forming the sphincter are marked (**). Preparation from the subdermal layer of the rabbit hind leg.

Bottom. Enlargement of part of the precapillary sphincter area seen in the top figure. The lumen and direction of blood flow indicated by arrow. The endothelial cell (E), containing some mitochondria (M) is separated from the smooth muscle cells (S) by a basement membrane (Bm). Short processes (*) of the endothelial cell pierce the basement membrane and make membranous contacts (myo-endothelial junctions) with the smooth muscle cells (S). The two smooth muscle cells seen in this figure correspond to those marked by (**) in the top figure. Magnification 25 000 ×. (Rhodin, J. G. (1967). The ultrastructure of mammalian arterioles and precapillary sphincters. *J. Ultrastruct. Res.* **18**, 181.)

(i) Arterioles: 100–50 μm diameter with a fragmented sheath of connective tissue elements instead of the internal elastic lamina. The tunica media contains 2–4 layers of smooth muscle.

(ii) Terminal arterioles: less than 50 μm in diameter with no internal lamina and a single layer of helically arranged smooth muscle cells. Myo-endothelial connections are seen between the intimal endothelium and the smooth muscle cells [FIG. II.9].

(iii) Precapillary sphincters and metarterioles: side branches debouching at right angles from the terminal arterioles; 10–15 μm diameter with a single layer of circularly arranged smooth muscle cells at their junction with the terminal arteriole (precapillary sphincter) and irregularly placed smooth muscle cells extending for about 20 μm from this junction (metarterioles). Both these types show many myo-endothelial junctions [FIG. II.10]. Regions of close contact between neighbouring smooth muscle cells ('tight junctions—with a low electrical resistance) are particularly prominent. Lateral contacts between adjacent muscle cells in which a process of one cell appears to invaginate the wall of its neighbour are frequently seen and are likely to be associated with extensive intercellular electrical coupling. Rhodin has suggested that the myo-endothelial junctions may form a means whereby metabolites can be exchanged between the lumen of the vessel and the smooth muscle cells.

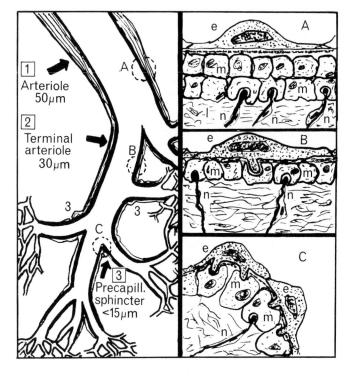

FIG. II.10. Diagrams summarizing the anatomical features (on the left) and the electron microscopic characteristic of the precapillary vessels. A, B, and C are respectively arteriole, terminal arteriole, and precapillary sphincter. Note the myo-endothelial junctions in B and C.
Key to lettering in the right hand figures:
e = endothelial cell, m = smooth muscle cell, n = nerve fibre.
(Rearranged and redrawn from Rhodin, J. G. (1967). *J. Ultrastruct. Res.* **18**, 181.)

The small vessels are innervated by terminal axons which are partly surrounded by Schwann cells; these axons run with the arterioles towards their terminal branches. Terminal axons are distributed to the precapillary sphincters. These terminal axons do not penetrate the scanty media but finish 80–500 nm (800–5000 Å) from the muscle cells.

2. *The capillaries* consist of a single layer of non-contractile

endothelial cells. They are not innervated. Details of their structural characteristics are described on page 82.

3. *Veins.* (a) Smaller veins or venules are distinguishable from arterioles by their thinner wall in relation to their lumen. Even the smallest veins contain an internal elastic lamina. Their innervation is sparse compared with that of the arterioles but, as they possess no external elastic lamina, it is possible that the terminal axons which supply them may effect a closer contact with the smooth muscle cell membranes of their wall.

(b) Large veins vary in structure according to their site in the body. In general, however, their internal elastic lamina is well developed. They contain two layers of smooth muscle in their media separated by a layer of connective tissue. The inner layer is disposed circularly and the outer layer forms a spiral of large pitch. Only the outer layer is innervated.

Functional characteristics of vascular smooth muscle

Bozler (1948) defined two types of smooth muscle cells—a *visceral* or *single unit* type and a multiunit type.

Visceral single unit cells exhibit automatic myogenic activity. This inherent activity is enhanced by mild stretch. Characteristically, structures which contain such muscle cells evince cell-to-cell propagation of electrical and mechanical activity. 'Pacemaker' cells exist whose activity spreads to and drives neighbouring smooth muscle cells. The ureter, the intestine and the uterus abound with smooth muscles of the 'visceral' type. As will be seen, the smooth muscle cells of the precapillary sphincters behave in this manner.

Multiunit smooth muscle cells show little or no evidence of inherent activity or of cell-to-cell propagation of excitation. They are dominated by motor nerves. They are not excited by mild stretch.

The vascular tree contains both these types of smooth muscle. Large arteries and veins contain muscle cells which show no automaticity and which are not excited by stretch. These muscle cells are supplied by sympathetic and some of them by parasympathetic nerves and their activity is in the main dominated by these extrinsic nerves. These large vessels then are characterized by the presence of multiunit muscle cells in their media. Multiunit muscle cells are also found in the cutaneous A–V shunts.

Visceral single unit muscle cells abound in the precapillary sphincters. Microcirculation studies of the small vessels in the living anaesthetized animal provide ample evidence of the existence of 'pacemaker' activity of the muscle elements of these structures. 'Vasomotion' and rhythmic closure of the sphincters can easily be seen. Mild increases of stretch such as produced by a rise in the transmural pressure across the vessel evoke a powerful vasoconstrictor response. Funaki (1961) was the first to record intracellularly from vascular smooth muscle cells and FIG. II.11 shows typical pacemaker activity with low oscillating membrane potentials from which rhythmically developed local potentials and excitation propagation ensue. Mild stretch increases the rate of 'spontaneous discharge'.

This type of vascular smooth muscle cell is peculiarly well suited for local control of the luminal diameter of the precapillary sphincter and for adjusting the resistance to perfusion of the capillary bed which the sphincter guards. The accumulation of local metabolites relaxes the sphincter whereupon its capillary bed opens up; dissipation of the metabolites is then followed by the resumption of inherent myogenic tone and the sphincter shuts until the local chemical situation again causes relaxation of the smooth muscle elements.

Bohr (1965) has suggested that the sodium permeability in these visceral smooth muscle cells is relatively high but also fluctuates. The increased sodium permeability may be referable to a feebler degree of Ca^{2+} fixation which the visceral smooth muscle cell exhibits compared with that of skeletal muscle. Gentle stretch of

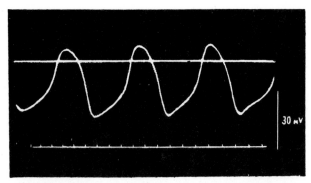

FIG. II.11. A series of spike discharges recorded intracellularly from a vascular smooth muscle fibre (venule). In this record the resting potential is −26. The peak of the action potential is +10 mV. Note the initial slow potential (pre-potential) from the summit of which the spike arises. The horizontal line indicates zero potential. Calibration, 30 mV; time mark, 30 ms; temperature, 29 °C. (Funaki, S. (1961). *Nature, Lond.* **191**, 1102.)

the cell reduces the membrane potential and increases the rate of 'spontaneous discharges and contractions' perhaps by further increasing sodium permeability. Hyperpolarization of the cell membrane stabilizes it and abolishes the 'spontaneous spikes and contractions'.

In arterial vessels with more than one layer of muscle cells in the media (arteries and arterioles, but not terminal arterioles) the noradrenergic postganglionic sympathetic fibres end in varicosities which are in very close approximation to the *outermost* layer of smooth muscle cells only. This differentiation implies that NA release affects only a few of the muscle cells of the media which, on contracting, unload the inner layers. In the arterioles, cell-to-cell spread of myogenic activity may secure a contractile response of the adjacent cells when NA excites the outermost muscle cells.

The smooth muscle cells of the pre- and postcapillary vessels are influenced by chemical agents which reach them directly from the blood stream and also indirectly from the interstitial fluid. Blood borne agents can of course gain access to the interstitium by transcapillary diffusion; tissue metabolites mainly act from outside by virtue of their concentration in the interstitium. The larger vessels have vasa vasorum which distribute blood-borne agents directly to the adventitia and outer media.

Noradrenalin released by the postganglionic sympathetic terminals is bound to α-receptors in the vascular smooth muscle cell membrane and evokes depolarization and contraction of the muscle, perhaps by releasing Ca^{2+} from the membrane itself. These α-receptors show a fairly general distribution.

REFERENCES

BOHR, D. F. (1965). Individualities among vascular smooth muscles. In *Electrolytes and cardiovascular diseases* (ed. E. Bajusz) pp. 342–55. Basel.
BOZLER, E. (1948). *Experientia* **4**, 213.
FALCK, B. (1962). *Acta physiol. scand.* **56**, Suppl. 197.
FUNAKI, S. (1961). *Nature, Lond.* **191**, 1102.
RHODIN, J. G. (1967). *J. Ultastruct. Res.* **18**, 181.

PHYSICAL FACTORS WHICH INFLUENCE VASCULAR FLOW

Hagen (1839), and particularly Poiseuille (1842, 1846), showed experimentally the factors which governed the volume flow of fluid flowing steadily through *rigid* tubes of capillary diameter *in vitro*.

The volume flow (Q) per unit time was proportional to the

pressure head $P_1 - P_2$, the fourth power of the radius (r) and inversely proportional to the tube length (L).

$$Q = K\frac{Pr^4}{L}$$

K is a constant for the liquid concerned at any one temperature; it is extremely temperature dependent, being larger as the temperature rises.

The flow of treacle through a tube under the influence of a given pressure head is much slower than that of water. If the treacle is warmed it flows a good deal faster than when it is cold, but it still flows at a much slower rate than does water. The term *viscosity* (meaning 'a lack of slipperiness' of the fluid) describes the behaviour of these different fluids. We have already understood that temperature affects viscosity.

The constant in the equation above is the reciprocal of the viscosity (η) and the expression may be rewritten to give the Hagen–Poiseuille equation:

$$Q = \frac{(P_1 - P_2)\pi r^4}{8L\eta}$$

The term $\pi/8$ derives from the mathematical deduction of the volume passing per unit time. This was first made (independently) by Wiedermann (1856) and Hagenbach (1860) (see Folkow and Neil 1971).

From the law of Ohm ($I = E/R$); as I is represented by Q and E by $(P_1 - P_2)$, the resistance R is represented by $8L\eta/\pi r^4$. *The longer the tube or the higher the viscosity or the smaller the radius, then the higher is the resistance to flow.*

The surprise perhaps is that it is the *fourth* power of the radius of the tube which influences resistance. Thus, if the radius be halved the resistance to flow is increased sixteen times, so, to force the same fluid volume through the tube per unit time requires sixteen times the pressure head initially required.

Poiseuille's results (and their mathematical interpretation) were based on measurements of the flow of a Newtonian fluid, i.e. a fluid which showed a viscosity which was unaffected by flow rate, through *rigid* tubes under the influence of a *steady head* of pressure. As the heart beats rhythmically, ejecting blood phasically into a system of elastic tubes, the situation 'in life' is rather far removed from the conditions which Poiseuille analysed. However, some of the factors which would seem to be immediately complicating, act in opposite directions and actually contribute to the provision of an approximately steady flow.

Thus, the large elastic vessels store part of the energy produced by the phasic cardiac ejection as potential energy—their walls are distended during systole (the Windkessel function). In diastole, the recoil of the walls of the Windkessel vessels *converts this stored potential energy into kinetic energy and, together with the resistance offered by the arterioles*, these two influences secure a *fairly steady flow* through the systemic capillaries [cf. Borelli, p. 66]. The distension of the smaller resistance vessels means that, *in the absence of other influences*, a given pressure head would increase their internal radius and would correspondingly cause a raised flow at that pressure. However, these same arteriolar vessels are themselves characterized by a generous content of smooth muscle in their media. This muscle, although itself 'elastic' in character, i.e. distensible— also reacts actively to stretch by contracting, as does cardiac muscle, and this basic myogenic property limits, and indeed may offset or even perhaps exceed, the initial expansile response of an 'elastic' vessel to transmural pressure. FIGURE II.12 shows four graphs of flow against pressure. In (A) the relationship is linear, as would be defined by a Poiseuille formula (viscosity and length changes being ignored). In (B) the increase in flow as a result of a rising pressure head is greater than in (A)—the initially collapsed vessels are distensible. Obviously the vessels cannot be unlimitedly

distensible because the adventitia, etc., contain inextensible fibro-collagenous elements, so at high pressures the flow would again be linear.

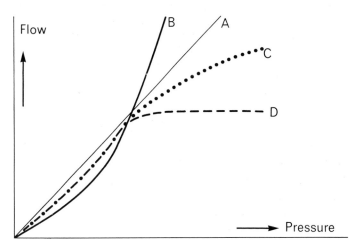

FIG. II.12. Pressure-flow curves.
A = Rigid tube.
B = Distensible vessel.
C = Distensible vessel containing active myogenic elements whose contraction offsets the distensible effects of raised pressure.
D = Distensible vessel containing myogenic contractile elements which serve to stabilize flow over a wide range of pressure (say 80–200 mm Hg). Such vessels would show 'autoregulation'.

In (C) the myogenic contractile response to stretch is depicted as offsetting the 'elastic' effects exerted by the pressure rise; the curve is concave to the pressure axis. Curve (D) shows the result when the myogenic elements of the wall even exceed the 'elastic' effects of a raised pressure.

Superimposed on these physical features are the additional effects caused by an increased sympathetic vasoconstrictor discharge to the resistance vessels. Such discharge constricts the lumen of the innervated vessels (mainly the arterioles) and thereby decreases the flow for a given pressure head. Even when higher pressures are *artificially produced* they do not increase the flow much when maximal sympathetic vasoconstriction is being exerted, for the arteriolar wall when thus constricted becomes thicker and therefore less easily distended.

In natural conditions, a rise of arterial pressure (caused, for example, by a sudden increase of cardiac output, as in the initial stages of muscular exercise) provokes a reflex reduction of sympathetic vasoconstrictor discharge via the arterial baroreceptors [p. 124] and this induces relaxation of the smooth muscle and hence vasodilatation with an increase in r. The stretching force provided by the raised pressure which would otherwise increase the lumen, is thus minimized by the reflex arteriolar vasodilatation, which itself lowers the resistance.

Peripheral resistance units (PRU)

Using Ohm's law, it is convenient to have a simple term for resistance. The term employed is the peripheral resistance unit (PRU). This expresses the resistance to flow through 100 g tissue, in ml per min for a given pressure gradient (in mm Hg).

$$PRU_{100} = \frac{(P_1 - P_2)\ \text{mm Hg}}{Q\ (\text{ml per 100 g per min})}$$

In resting muscle for example, a flow of 3 ml per 100 g per min is provided by a pressure head of 100 mm Hg when the resistance is 33.3 PRU. In maximal dilatation (exercise) the resistance falls to perhaps $100/70 = 1.4$ PRU.

Viscosity

Viscosity is an important factor in the resistance term of the Poiseuille equation.

Isaac Newton (1713) described the 'internal friction' or 'defectus lubricatitis = lack of slipperiness' which we now call viscosity. (The word viscosity is derived from 'viscum', which means mistletoe. The Romans used a sticky concoction from mistletoe as bird-lime to daub on the branches of trees in order to snare small birds.) The term internal friction, or lack of slipperiness emphasizes that when fluid moves along a tube, laminae in the fluid slip on one another and move at different speeds thereby causing a velocity gradient in a direction perpendicular to the wall of the tube. This velocity gradient is called the *rate of shear. A simple viscous liquid is known as a Newtonian fluid and is defined as one whose viscosity does not vary with the rate of shear and remains constant at different rates of laminar flow.*

It can be shown experimentally (Osborne Reynolds 1883) and proved mathematically, that a Newtonian fluid moving parallel to the axis of a tube with a constant velocity possesses a paraboloid profile [FIG. II.13]. Reynolds used a long cylindrical tube through which fluid flowed at a pressure head provided by a raised reservoir. He introduced a thin stream of dye into the axial stream and noted that the dye remained in the centre of the stream showing this paraboloid profile. Only if the velocity of flow became high did flow change from laminar or streamlined to turbulent [FIG. II.14].

In laminar flow, the central axis of flow moves with the highest velocity (V_{max}) and the successive cylindrical laminae move progressively more slowly as one moves from the central axis to the wall; at the wall the velocity of flow of this (infinitely thin) 'shell' of fluid is zero. *The resistance met by fluid moving in streamlined fashion through the blood vessels has nothing to do with friction between the fluid and the vessel wall, but is due to the friction between the adjacent laminae of the fluid.*

The bigger this internal friction, the greater is the difference of velocity between the laminae and the greater the coefficient of viscosity.

The fundamental equation developed by Newton, which allowed the definition of the coefficient of viscosity is:

$$F = A\eta \times \frac{\Delta V}{\Delta x}$$

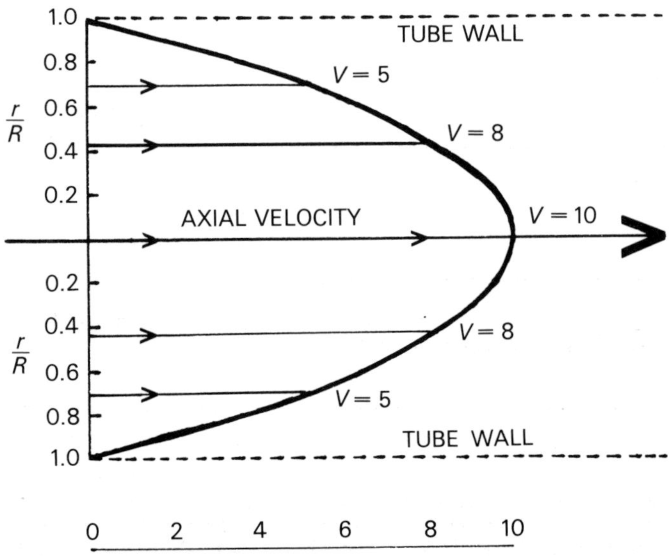

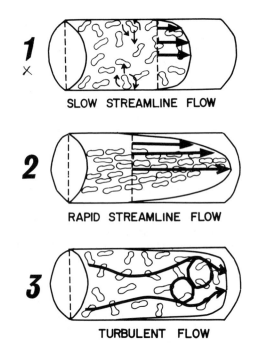

FIG. II.14. Diagram of different features of streamline and turbulent flow.

Where F is the tangential force or 'drag' exerted between the fluid laminae, A is the area of contact between them, and $\Delta V/\Delta X$ is the velocity gradient—i.e. the increase in velocity (cm s⁻¹) for each centimetre distance at right angles to the direction of fluid flow.

The unit of viscosity introduced was called the *poise* (after Poiseuille). A fluid of 1 poise viscosity has a force of 1 dyne cm⁻² of contact between layers when flowing with a velocity gradient of 1 cm s⁻¹ cm⁻¹.

Water has a viscosity of 0.01 poise at 21 °C, so the practical unit has become the centipoise—one hundredth the value of the poise.

Viscosimeters (or viscometers), measure viscosity *in vitro*. The simplest is that devised by Ostwald [FIG. II.15]. The bulb is filled with the test fluid. The liquid is then sucked up to above the mark 1

FIG. II.13. Laminar flow. (After McDonald, D. A. (1960). *Blood flow in arteries*. Arnold, London.)

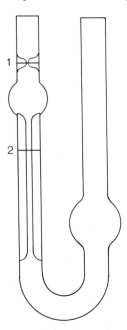

FIG. II.15. Viscometer (Ostwald). The right-hand bulb is filled with the test liquid via the wide vertical tube. The test fluid is then sucked up into the narrow limb to above point 1. The time taken to fall from 1 to 2 is measured both for the test liquid and for water.

and the time taken for the fluid to fall under its hydrostatic head from 1 to 2 is measured and compared with that taken for water at the same temperature. In another type of viscometer the fluid in a container is exposed to two smooth surfaces (e.g. rotating coaxial cylinders) which can be moved relative to each other at measured rates. The force required to produce a given rate of shear is measured and is then related to that for water.

Measurements of viscosity show that the temperature of the fluid is very important. Water has a viscosity of 1 cP at 20.3 °C, but at 37 °C it is only 0.695, whereas at 0 °C it is approximately 1.8. The viscosity of plasma and blood is even more sensitive to temperature change. This is obviously important when considering the cutaneous circulation, subjected as it is to physical changes in the thermal environment and the physiological responses thereto.

The term *relative viscosity* is often used; the viscosity of the fluid is expressed relative to that of water. Water has a viscosity of 0.695 cP at 37 °C, so plasma which has a viscosity of 1.2 cP at 37 °C has a relative viscosity of 1.7.

Blood viscosity

As blood consists of corpuscles suspended in plasma, one would hardly expect it to behave like a homogeneous Newtonian fluid. *In vitro*, blood has a *relative* viscosity of 4–5 when measures at moderate or high shear rates. When the *in vitro* measurements are made at very low shear rates the viscosity is greatly increased [Fig. II.16]. These facts indicate that blood *in vitro* shows *anomalous* viscosity in that it does not manifest the properties of a Newtonian fluid (the viscosity of which is independent of the rate of shear).

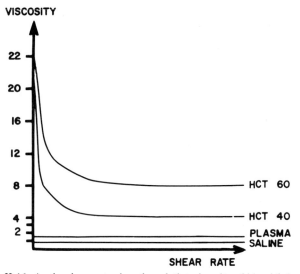

Fig. II.16. As the shear rate rises the relative viscosity of blood (whether with a haematocrit of 40 or of 60) becomes constant with increasing flow and blood behaves like a Newtonian fluid such as plasma or saline. (From Burton, A. C. (1965). *Physiology and biophysics of the circulation.* Year Book Medical, Chicago. Used by permission.)

Poiseuille began his investigations on the flow of liquids through capillary tubes seeking to define the factors controlling the flow of blood in the capillaries in life. He showed *in vitro*, as has been seen, that the volume flow is: 1. proportional to the pressure head; 2. to r^4; and 3. inversely proportional to the length, L. The constant of proportionality between Q and the quantities 1, 2, and 3 is inversely proportional to the coefficient of viscosity and is constant for any given fluid at any given temperature. As the late Leonard Bayliss (1952) wrote, in one of the best accounts of the rheology of blood available, 'it is one of the ironies of scientific progress, however, that the rate of flow of blood through a glass tube—and indeed, through the vessels of an animal—is not proportional to the applied pressure (i.e. to the shearing stress) except at high rates of shear; is not proportional to the fourth power of the radius when this is less than 0.01 cm (as it is in most of the vessels of the living animal) and is not in all circumstances inversely proportional to the length of the tube'.

Anomalous viscosity

Factors which influence the situation and which contribute to the anomalous viscosity of blood can be discussed under four headings: 1. shear rate; 2. haematocrit; 3. tube radius and length; 4. temperature.

Shear rate. The viscosity of blood rises at low shear rate. When shear rate is high, the erythrocytes occupy the central axis of the tube (Bayliss 1959), moving with their long axes parallel to the direction of flow where the flow rate is fastest [Fig. II.14] and where the intermolecular differences of shear rate and hence the friction between cells and plasma is least. At high rates of flow blood behaves almost as a Newtonian fluid with constant viscosity [Fig. II.16]. When the flow rate and hence the shear rate is low, this tendency of the red cells to occupy the axis of the stream is minimal and is moreover offset by collisions between the suspended particles—both these influences increase the internal friction and viscosity.

The tendency to axial streaming as the flow rate rises is responsible for the phenomenon of plasma skimming. The cell-free zone of plasma at the periphery of the tube is preferentially directed into a branch vessel which branches at a large angle from the supply vessel and the haematocrit value of the blood passing through this branch is correspondingly low. However, the cell-poor zone is only 5–20 μm wide and the cell-free zone 5 μm wide. Plasma skimming is thus of no importance whatever when *large* vessels branch, but may be of appreciable significance in small vessel branching.

Haematocrit. The relative viscosity of blood *in vitro* increases as the haematocrit value rises above its normal figure of 45 per cent. However, this effect is much less marked when the blood flows through the tubes of capillary diameter (6 μm) [Fig. II.17]. The red cells are very deformable and themselves, with a diameter of 7 μm, can be squeezed through the lumen in single file with little extra force required (Prothero and Burton 1962). Thus, whether a few, or a lot, of red cells pass through the capillary per second has little effect on blood viscosity *in the capillaries—this approximates to that of plasma* (see following pages).

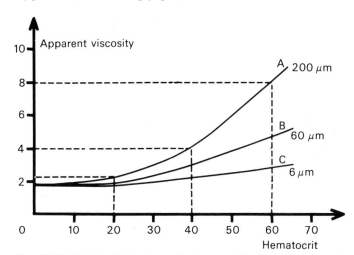

Fig. II.17. Relationship between the haematocrit and relative viscosity when blood flows through tubes of diameter: A = 200 μm; B = 60 μm; C = 6 μm. Flow rate is high. (After Haynes, R. H. (1961). *Trans. Soc. Rheol.* **5,** 85.) (From Burton, A. C. (1965). *Physiology and biophysics of the circulation.* Year Book Medical, Chicago. Used by permission.)

The Fåhreus–Lindqvist effect—tube diameter and blood viscosity. Fåhreus and Lindqvist (1931) measured the apparent viscosity of human blood in tubes varying between 575 and 40 μm diameter using a constant pressure head of 100 mm Hg; they arranged that the length of each capillary tube was proportional to its diameter. Thus, the *shearing stress at the wall* was much the same in all the tubes. They showed that the relative viscosity of blood fell when the tube diameter was less than 300 μm. In a tube of 40 μm diameter the relative viscosity of the blood was only 70 per cent of that through tubes of 150 μm or greater diameter. Bayliss (1952) confirmed and extended these observations [Fig. II.18] making measurements of flows through tubes of only 14 μm diameter at the tip. He found that the relative viscosity of the blood passing through a tube of 14 μm was only 47 per cent of that through tubes larger than 150 μm diameter.

As neither water, serum, nor plasma altered its relative viscosity when flowing through tubes of different sizes, the Fåhreus–Lindqvist effect is obviously due to the presence of the erythrocytes. Two hypotheses have been advanced—one suggests that the width of the cell-free zone at the periphery is constant whatever the tube diameter and correspondingly occupies a bigger proportion of the whole in narrow tubes—as a result the total relative viscosity would be less. The other suggestion is, that in calculating Poiseuille's formulation, it is impermissible to integrate as if the laminae of fluid were infinitely thin, when the tube diameter is not much different from that of the red cells. Instead a summation process should be employed, in which it is supposed that the velocity gradient is discontinuous, there being alternate layers of uniform velocity (occupied by the corpuscles) and layers in which shear is occurring (occupied by the plasma).

Laminar flow has a paraboloid front and the volume, Q, is given by the equation:

$$Q = 2\pi \frac{P_1 - P_2}{4L\eta} \int_\varrho^R (R^2 - r^2) r.dr,$$

where R is the radius of the tube and r the radius of a lamina of fluid intermediate between the central axis and the wall of the tube. The integration of R^3 gives $R^4/4$ so if $R = 5$, its integral is $5^4/4 = 156$. However, if the five separate laminae are *summed* instead of *integrated*, we get: $1^3 + 2^3 + 3^3 + 4^3 + 5^3 = 1 + 8 + 27 + 64 + 125 = 225$.

In the integration formula we get:

$$Q = \frac{\pi P_1 - P_2}{2L\eta} \times 156$$

and in the summed form we get:

$$Q = \frac{\pi P_1 - P_2}{2L\eta} \times 225$$

as Q and P are measured, the value of η calculated from the measurements will be 156/225 (in the example given), when the integration method is used, times that deduced from the summation technique. Sigma (Σ) is the Greek symbol used in mathematics for 'summation of' and this explanation of the reason for the low viscosity in capillary tubes, given first by Dix and Scott Blair (1940), is referred to as the 'Sigma phenomenon'.

If the apparent relative viscosity of blood in 'wide' tubes (large enough to render any variation in their dimensions unimportant), be taken as η_∞^* and if we assume that the shear takes place in discontinuous layers between unsheared layers of thickness, δ, we find that the corresponding apparent relative viscosity η_a^* in a capillary tube of radius a is given by:

$$\frac{\eta_\infty^*}{\eta_a^*} = 1 + \frac{2\delta}{a} + \frac{\delta^2}{a^2}$$

(Bayliss 1952)

Figure II.18 shows experimental values listed in the literature plotted as points and the theoretical values of η_a^*/η_∞^* drawn as continuous lines. The agreement is quite satisfactory.

Copley (1960) has additionally suggested that the vascular endothelium of the capillaries may secret a substance (? a mucopolysaccharide) which further helps to lower the viscosity of the blood in the capillaries. Such a property, together with that resulting from the readily deformable nature of the red cells, may explain why *the viscosity of blood traversing the capillaries differs little from that of plasma*.

Temperature. The cutaneous and subcutaneous vessels and even those of the deeper regions in the limbs are subjected to considerable alterations of temperature. The most common effect is that of cooling, which raises the viscosity. Thus, when the hand is kept in ice water regional viscosity of the blood shows a threefold increase.

To summarize—there are differences between the results of measuring blood viscosity *in vivo* and those obtained *in vitro*. In general viscosity *in vivo* is lower and shows a relative value (compared with water) of 3.5. At 37 °C water has a viscosity of 0.695 cP, so blood *in vivo* has an apparent viscosity of 2.5 (at 37 °C). However, in small tubes (less than 25 μm radius) the apparent or relative viscosity is about halved—1.2 cP (identical with that of plasma at 37 °C). Hence in capillaries unaffected by temperature changes, blood viscosity is similar to that of plasma and the blood behaves approximately as a Newtonian fluid. Poiseuille's formula therefore provides an adequate condensation of the behaviour of capillary blood flow with η approximately constant at constant temperature.

Turbulence

Laminar or streamlined flow which is characteristic of most parts of the vascular system is silent and shows a linear relationship with pressure. Above a certain flow velocity, the flow ceases to rise

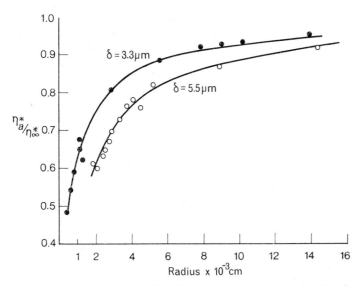

Fig. II.18. The effect of the radius of the tube on the apparent viscosity of blood.

Ordinate: Ratio of $\dfrac{\text{apparent viscosity in tube of radius } a}{\text{apparent viscosity in tube of radius greater than 0.05 cm}}$

Abscissa: Radius of tube (a) in cm × 10^{-3}
Collected data from the literature. (Redrawn from Bayliss, L. E. (1952). In *Deformation and flow in biological systems* (ed. A. Frey-Wissling) North Holland, Amsterdam.)

linearly with the pressure and *turbulence* develops, at first gradually. When turbulence is fully established the fluid flow is approximately proportional to the square root of the pressure head.

Osborne Reynolds (1883) made a classic investigation of this phenomenon, which Poiseuille himself had indeed noted but which he himself did not analyse rigorously. Reynolds showed that a thin stream of dye, which he injected into fluid flowing under a high pressure head through a long glass tube, ceased to have an axial distribution when the velocity of flow became high and showed vortex-like eddies. The dye then filled the tube. The critical average velocity of flow at which the motion became turbulent was dependent on the radius of the tube and on the viscosity and density of the fluid.

$$V = \frac{R_e \eta}{p_r}$$

where V is the critical *mean* velocity, R_e is Reynolds' number, p is the density, η the viscosity of the fluid, and r is the radius of the tube.

Hence
$$R_e = \frac{Vpr}{\eta}.$$

If V is measured in cm s^{-1}, r in cm, density is unity, and viscosity in *poises* (*not* centipoises) then, in a long straight tube, turbulence develops when R_e exceeds 1000. (If the diameter is used in the formula than R_e would be 2000.)

Turbulent flow entails a greater energy loss than does laminar flow because more energy is dissipated in creating the kinetic energy of the eddies.

None of the small 'resistance' vessels of the vascular system shows turbulent flow—only streamline flow. The ventricles and the aorta, however, are normal sites of turbulence.

The critical value of 1000 for turbulent flow applies *only* to long straight tubes. If a small area of narrowing occurs in a tube or in a vessel affected say by the development of an atheromatous plaque, the mean velocity through this reduced lumen is greatly increased and turbulence develops at and beyond the site of constriction. This causes a vibration which can be felt and a murmur or bruit which can be heard with the stethoscope. Turbulence due to a stenosis occurs at much lower Reynolds' numbers than 1000.

The heart sounds are due to closure of the valves and do not necessarily involve turbulence. However, if flow becomes very high, as in heavy exercise, *quite normal* systolic murmurs can be heard with the stethoscope.

Aortic turbulence occurs during systolic ejection. The *mean* velocity of the blood in the aorta during the whole cardiac cycle is about 20 cm s^{-1} in a man at rest with a cardiac output of 5.6 litres and a heart rate of 70. The stroke volume (80 ml) is expelled through the aortic cross-sectional area of 4 cm^2 ($r = 1.1$ cm). Thus, the Reynolds' number for *mean* flow is

$$R_e = \frac{20 \times 1.1}{0.03} = 714$$

(the blood has a viscosity of 0.03 poise at 37 °C).

During systolic ejection alone, however, the Reynold's number exceeds 1000 and turbulence results. In heavy exercise, aortic turbulence becomes very marked and extends into the larger arterial branches.

The Korotkow sounds heard in auscultatory sphygmomanometry originate from turbulence as the systolic blood pressure pushes blood through the semi-collapsed vessel beyond the partially occluded brachial artery. Once the cuff pressure is below that of the diastolic pressure the downstream vessel is wide open and the sound disappears.

Turbulent flow itself need not cause murmurs, but when it occurs in a region where wall resonance is prominent then vibrations of the wall are aroused and these are responsible for the murmur heard. Such vibrations characteristically occur where blood speeds through a narrowed region into a wider lumen (e.g. the ductus arteriosus, traumatic a-v anastomoses, or in valvular stenosis).

REFERENCES

BAYLISS, L. E. (1952). In *Deformation and flow in biological systems* (ed. A. Frey-Wyssling). Amsterdam.
—— (1962). The rheology of blood. In *Handbook of physiology*, Section 2, Circulation, Vol. I, p. 137. American Physiological Society, Washington.
BURTON, A. C. (1965). *Physiology and biophysics of the circulation.* Chicago.
CARO, C. G., PEDLEY, T. J., SCHROTER, R. C., and SEED, W. A. (1978). *The mechanics of the circulation.* Oxford University Press.
DIX, F. J. and SCOTT-BLAIR, G. W. (1940). *J. appl. Phys.* **11**, 574.
FÅHREUS, R. and LINDQVIST, T. (1931). *Am. J. Physiol.* **96**, 562.
HAGEN, G. H. L. (1839). *Ann. d. Phys. u. Chem.* **46**, 423.
HAGENBACH, E. (1860). *Ann. d. Physik.* **109**, 385.
HAYNES, R. H. (1961). *Trans. Soc. Rheol.* **5**, 85.
MCDONALD, D. A. (1960). *Blood flow in arteries*. London.
POISEUILLE, J. L. M. (1846). Paris. Mém. Savants Étrangers, **9**, 433.
REYNOLDS, O. (1883). *Phil. Trans.* **174**, 935.
WIEDEMANN, G. (1856). *Ann. d. Physik.* **99**, 221.

The capillary circulation

The capillary bed offers an extensive surface area for exchange between blood and tissues, with a minimal distance required for such diffusion. It comprises a dense network of narrow (3 μm radius), short (750 μm) tubes with total cross-sectional and wall surface areas which are huge in relation to the length of the constituent tubes. Each 'ideal', capillary 750 μm long and 3 μm radius, will have a cross-sectional area of 30 μm^2 and a wall surface area of 15 000 μm^2. The total cross-sectional area of the capillary bed *if fully patent* is some 2800 times that of the aorta. There are various computations of the size of the total capillary exchange surface in man of 70 kg weight, and the figure generally agreed on is 550–600 square metres. This however implies that the total number of systemic capillaries is between 40 and 50 thousand million and that they were all patent, a situation which is never seen in life. At rest, probably only 25 per cent of this total number of capillaries are patent—perhaps 10 thousand million with a combined total surface area of 150–200 square metres and a transverse cross-sectional area about 700 times that of the aorta.

In experimental studies of flow in animals [p. 86] the total capillary surface area in skeletal muscles has been estimated as 7000 cm^2 per 100 g. In man with say, 30–35 kg muscle this would be equivalent to an exchange surface area of 200–250 square metres with some four to five hundred capillaries per cubic millimetre—a figure which is comparable with that obtained from actual counts of capillaries. Some tissues show a much smaller density than this, such as bone, connective tissue and smooth muscles. Others, like the heart, brain, kidneys, liver, and gastro-intestinal mucosa, have maximal flows which may be far greater and which possess a much more dense capillary network.

In resting man, the *capillary flow rate* is of the order of 0.3 to 0.5 mm s^{-1}. Again there must be a considerable variation in this value—between say, the flow in fat depot tissues, which is much slower and that in metabolically active tissues.

FIGURE II.19 shows the approximate relationship between the transverse cross-sectional area and the linear velocity across the vascular system.

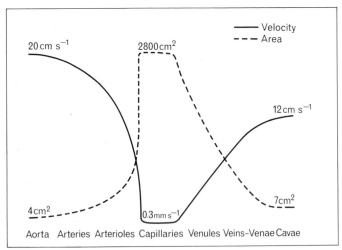

Fig. II.19. Velocity of flow and cross-sectional area at different stages of the vascular circuit.

As Poiseuille showed: Regional pressure drop $= \dfrac{QL}{nr^4}$

where Q is the flow, L the length, and n is the number of parallel vessels.

Clearly *the flow resistance in the capillary section is small because* L *is so short* (750 μm).

THE STRUCTURE OF THE CAPILLARIES

On the basis of electron microscope studies three types of capillary [Fig. II.20] may be classified:

1. Continuous (non-fenestrated) capillaries in which the single layer of endothelial cells is continuous except at the intercellular region. This is about 10 nm wide and is partly occupied by a homogeneous material. Recently it has been shown (Karnovsky 1967) that some of these intercellular spaces are traversed by channels some 4 nm wide. These channels are the routes traversed by fluid filtered from or absorbed by the capillary and by water-soluble substances which exchange with the interstitial fluid by diffusion [p. 84].

2. Fenestrated capillaries occur in the renal glomeruli, in glands of both exocrine and endocrine type, in the choroid plexuses and in the intestinal villi. They are prominent in the vasa recta of the renal

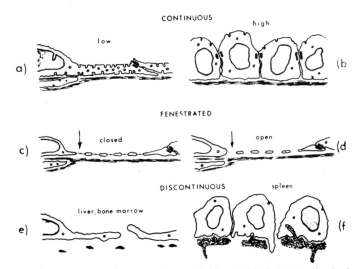

Fig. II.20. Classification of capillaries. (Majno, G. (1965). In *Handbook of physiology*, Vol. 2 Circulation III, p. 2293. American Physiological Society, Washington.)

medulla where they may act as counter-current exchangers. The fenestrations seen are intracellular openings which presumably serve the function of rapid and large transudations of fluid such as occur in the renal glomeruli (filtering 170 litres of fluid per 24 hours).

3. Discontinuous capillaries ('sinusoids'). These possess a thin endothelial layer with large gaps between the individual cells. Bone marrow, liver and spleen are characterized by such sinusoids which seem adapted not only for the exchange of large protein molecules but even of erythrocytes themselves. A blood transfusion may be given via a cannula inserted into the bone marrow.

TRANSCAPILLARY EXCHANGE

Three types of transport occur: 1. Filtration–absorption which governs plasma volume; 2. Diffusion which is responsible for supply of nutrients to and the removal of waste from the tissue cells; 3. Micropinocytosis.

Filtration–absorption

The forces concerned in this exchange [Fig. II.21] are the hydrostatic pressure exerted across the wall of the capillary, which favours filtration through the pores and the effective osmotic pressure of the plasma which favours absorption of interstitial fluid through the pores (Starling 1896). If filtration and absorption are balanced in any organ, conditions are 'isovolumetric' or 'isogravimetric'. The hydrostatic pressure causing filtration is obviously the difference in hydrostatic pressure *inside* the capillary P_{cap} minus that outside in the interstitial fluid P_{if}. Both vary—particularly P_{cap}. It was not until Landis (1930) measured P_{cap} directly that much headway was made in investigating Starling's hypothesis. Landis showed that the pressure at the arteriolar end of the capillary loop in the human finger held at heart level was about 32 mm Hg and at the venous end of the loop was some 12 mm Hg. Mean capillary pressure is usually taken as 24–25 mm Hg. Clearly if the colloid osmotic pressure be 25 mm Hg, the arteriolar end of the capillary loop would lose fluid,

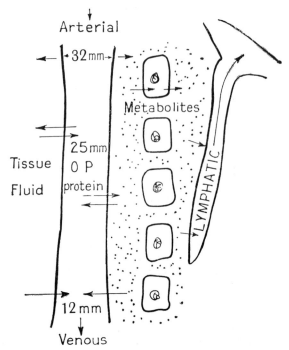

Fig. II.21. Fluid interchanges between plasma, tissue spaces, and lymphatics.

whereas the venous end of the capillary would be the site of reabsorption [FIG. II.21].

P_{cap} is not of course constant in any one tissue for it depends on the state of tone of the resistance vessels. Moreover, the resting level of P_{cap} characteristically differs in certain tissues. The glomeruli (70 mm Hg), the lungs (8 mm Hg), and the liver (6 mm Hg) illustrate this statement.

For any one tissue, P_{cap} is determined by the ratio between the precapillary and postcapillary resistances.

$$P_{cap} = \frac{P_A R_v/R_a + P_V}{1 + R_v/R_a}$$

where P_A and P_V are the central arterial and venous pressures respectively [p. 73].

P_{if} does not vary much in most tissues—and is within one or two mm Hg of atmospheric pressure.

Hydrostatic load, as in the legs in the erect position, must obviously raise the regional P_{cap} but does not primarily increase P_{if}, except where dense fascia surrounds both the tissue and its contained vessels, thus transmitting the increased pressure in the distensible veins to the interstitial space. In such circumstances as these, the situation then approximates to that of the truly encapsulated organs such as the brain and the abdominal viscera. Such organs 'float' in rigid or semi-rigid fluid compartments and changes in their position relative to the heart level produce nearly equal changes in both P_{cap} and P_{if} so that the over-all transmural pressure is little affected.

The filtration process. The hydrodynamic flow of pore-restricted filtration seems essentially to obey Poiseuille's law, providing that the radius of solvent and solute particles is less than 1/20th of that of the pores. The plasma proteins are not filtered to a great extent and thus exert an effective osmotic pressure which restricts the fluid bulk filtered. Plasma contains 300 milli-osmoles per litre of crystalloids, and as each 1000 milli-osmoles per litre will exert an osmotic pressure of 22.4 atmospheres the total osmotic pressure of plasma is about 6.7 atmospheres. The crystalloids can and do pass back and forth across the capillary wall with great ease so that they do *not* contribute to the *effective* osmotic pressure. Effective osmotic pressure results from the plasma proteins, which constitute 7 g per 100 ml of plasma and exert an osmotic pressure of 25 mm Hg. Sixty-five to eighty per cent of this osmotic pressure is due to albumin which not only exceeds the globulins in concentration (A/G = 1.8) but which possesses a molecular weight of only 70 000 compared with those of the globulins which range from 100 000 to 450 000. It should be remembered that it is the *number* of molecules of solute which matters. Thus, 1 g albumin in 100 ml exerts an osmotic effect equivalent to 6 mm Hg, whereas the same concentration of globulin gives an osmotic pressure of 1.5 mm Hg.

Protein content of the intestinal fluid. Some protein traverses the capillary wall, however, and this means that the interstitial fluid contains some protein and exerts an osmotic pressure, π_{if}. Even though this is low it offsets the full osmotic pressure effect of the plasma protein concentration. The liver with its discontinuous sinusoids may have an interstitial fluid protein concentration of 80–90 per cent of the plasma protein concentration. The intestine (fenestrated capillaries) (40–60 per cent), and the skeletal muscle (continuous capillaries) (10–30 per cent) provide other examples.

It appears that the *relative* concentrations of the proteins in the interstitial fluid are the same as those in the plasma. This result is surprising—it would seem more probable that molecular sieving would restrict the transfer of globulins rather than the smaller albumin molecules. It has been postulated that most of such protein transfer as does occur, takes place through relatively few wide-bore 'capillary leaks' [see p. 90]. Additionally, it has been suggested that

micropinocytosis or cytopempsis [p. 90] contributes to this situation by adding certain globulins to the interstitial fluid by active transport.

Quantitative studies of the filtration–absorption process

Pappenheimer and Soto-Rivera (1948) used an experimental preparation in which the weight of an isolated perfused hind limb of the cat was continuously measured during changes of perfusion inflow pressure and changes of venous outflow pressure. The plasma protein concentration of the perfusing blood could be varied.

The arterial pressure and venous outflow pressure were initially adjusted to maintain the limb at constant weight. A sudden rise of arterial pressure caused an initial prompt increase in limb weight (due to increased venous volume as the venules and small veins changed their cross-sectional profile from elliptical to circular under the influence of the increased pressure transmitted via the capillaries). There followed a slow rise of limb weight due to the filtration of fluid from the capillaries.

Conversely, absorption of fluid could be measured when the arterial pressure was lowered below that required to maintain the isogravimetric state.

Similar effects could be produced by maintaining the arterial pressure constant and changing the venous pressure. Quantitatively it was found that changes of venous pressure required to produce rates of filtration or absorption comparable with those aroused by alterations of arterial pressure were only one-fifth to one-tenth as great.

It follows that there are an infinite number of pairs of values of arterial and venous pressure at which the leg will remain at constant weight. Thus the tendency to absorb fluid as a result of a fall in arterial pressure can be counterbalanced by raising the venous pressure.

Considering the capillary, if Q is the flow and R_v is the resistance to flow from the effective mid-point of the capillary to the vein and pV is the venous pressure.

$$R_v = \frac{(pC - pV)}{Q}.$$

In the isogravimetric state, the mean isogravimetric pressure pC_i is given by

$$pC_i = Q_iR_v + pV_i.$$

The technique then, consists of altering the blood flow by varying the arterial pressure and adjusting the venous pressure to maintain constant weight. If the values of Q_i are plotted against those of pV_i the value of pV_i at zero flow equals the isogravimetric mean capillary pressure [FIG. II.22]. The slope of the line is of course R_v; it is found to be constant and independent of flow in these conditions (note: the vessels of the perfused limb are denervated).

The isogravimetric pressure determined in this fashion is equal to the sum of all pressures which oppose the filtration which this capillary hydrostatic pressure would otherwise produce. These opposing forces would clearly include the extravascular interstitial fluid pressure (which is *ordinarily* negligible) and the effective colloid osmotic pressure of the plasma. Here again the effective colloid osmotic pressure across the capillary membrane will be determined by the concentration of proteins in the plasma minus that of proteins in the interstitial fluid.

Pappenheimer and Soto-Rivera determined the values of isogravimetric pC_i required when different plasma protein concentrations were used in the perfusing blood. They found that the mean capillary pressure required to secure isogravimetric conditions was slightly less (2 mm Hg less) than the osmotic pressure of the plasma proteins over the range of protein concentrations employed. They ascribed this slight difference to there being a small concentration of protein (0.7 g per 100 ml) in the interstitial fluid.

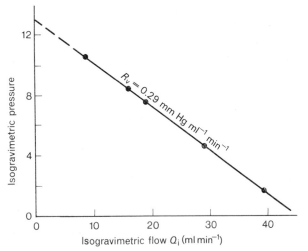

Fig. II.22. See text. (Pappenheimer, J. R. and Soto-Rivera, A. (1948). *Am. J. Physiol.* **152,** 471.)

These results provided for the first time quantitative proof of the Starling hypothesis. The rate of fluid exchange across capillaries is simply proportional to the difference between the mean hydrostatic pressure in the capillary ($pC = QR_v + pV$) and the sum of all pressures opposing filtration (isogravimetric pressure pC_i). It is independent of the absolute values of these quantities [Fig. II.23].

The slope of the line gives the capillary filtration coefficient (0.014 ml per min per 100 g tissue per mm Hg) in this preparation. The CFC is a measure of the *hydrodynamic conductivity* of the capillary wall.

The range of CFC values for different tissues is a wide one—that of the renal glomeruli is 200 times the CFC of muscle capillaries.

In the human forearm the filtration coefficient is about 0.006 ml per min per 100 g per mm Hg. If this value were taken as representative of the whole body, this would be 6 litres per 24 hours in a man of 70 kg per mm Hg filtering pressure. If the filtering pressure

averaged 4 mm Hg (osmotic pressure equals, say, 25 mm Hg and the mean resting filtering pressure, say, 29 mm Hg) then not less than 24 litres of fluid is filtered from the capillaries each day.

Of this volume all but 2–4 litres is reabsorbed by the capillaries. The remainder is absorbed by the lymphatics.

During the day the heart pumps, say, 8000 litres of blood. The 'filtration fraction' of the capillaries is 24/8000 = 0.3 per cent.

The extravascular circulation resulting from the capillary filtration and absorption is very important indeed for the regulation of blood volume but is of minimal significance in providing for the exchange of nutrient and waste materials between the blood and the tissues. This metabolic exchange occurs by diffusion mechanisms. Diffusion of a substance from the capillaries is determined by its lipid-solubility. Oxygen and carbon dioxide, being lipid soluble, can utilize the entire surface area of the capillary wall, diffusing with ease through the plasma membranes of the endothelial cells themselves. Lipid-insoluble molecules, on the other hand, can only diffuse through intercellular pores which provide aqueous channels for such diffusion.

Diffusion

The fundamental law of free diffusion is that of Fick (1855).

$$\frac{dn}{dt} = DA \frac{dc}{dx}$$

where dn/dt is the rate of linear diffusion (quantity n, time t) in direction x, D is the diffusion coefficient (cm² s⁻¹), A is the cross-sectional area, and dc/dx is the concentration gradient.

In a steady state dc/dt is constant and

$$n = DA \frac{\Delta c}{\Delta x}.$$

The driving force for diffusion is provided from the random kinetic movements of the diffusing molecules.

However, as stated previously, we cannot consider the total surface area of the capillary wall as available for the diffusion of lipid-insoluble substances. Such molecules can only pass back and forth via the intercellular pores which necessarily occupy only a fraction of the wall area [Fig. II.24].

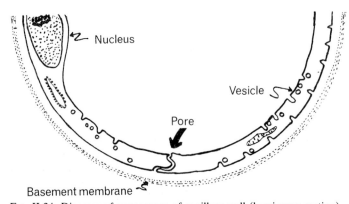

Fig. II.24. Diagram of appearance of capillary wall (hemi-cross section). Note pore and the presence of numerous infoldings of the capillary surface with associated vesicles. These are supposed to be involved in micropinocytosis (cytopempsis). An amorphous basal membrane surrounds the capillary.

However, if the lipid-insoluble molecules are small in relation to the size of the pores the law of Fick will still suffice to describe the situation.

Naturally, the area for such diffusion is now determined by the

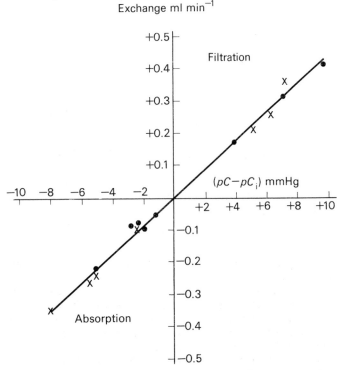

Fig. II.23. Capillary filtration–absorption exchange. (Redrawn from Pappenheimer, J. R. and Soto-Rivera, A. (1948). *Am. J. Physiol.* **152,** 471.)

size of the individual pores and by their number per unit area. The Fick equation can be rewritten:

$$A_p = \dot{n} \times \frac{\Delta x}{D \Delta c}$$

where A_p is the total *pore* area.

As the path length Δx is rarely known in practice, it is more useful to solve for the pore area per unit path length $(A_p/\Delta x)$ as follows:

$$\frac{A_p}{\Delta x} = \frac{\dot{n}}{D \Delta c}.$$

Once the pore area per unit path length $(A_p/\Delta x)$ has been estimated for any membrane, that membrane may then be used to determine the free diffusion coefficients of any test molecule.

Renkin (1954) studied the diffusion of different test substances through an artificial cellulose membrane which had pores of 1.6 nm. The *apparent* pore area (A_s) per unit path length for free diffusion $A_s/\Delta x$ diminished as a function of the molecular radius of the test molecule [FIG. II.25]. Now as the true pore area of the membrane remained constant, this reduction in $A_s/\Delta x$ must mean *restriction* to free diffusion as the molecular size increased. This in turn can be expressed by using the term 'restricted diffusion coefficient' (D') instead of the value 'diffusion constant' (D).

$$D' = D \frac{A_s}{A_p}$$

where A_s is the apparent pore area for the solute and A_p is the true pore area.

The theory of restricted diffusion was worked out by Pappenheimer, Renkin, and Borrero (1951). In brief, two factors exist which impede the flow of molecules through pores which possess dimensions *similar* to that of the diffusing molecules.

1. Friction between the molecule and the pore walls.
2. Steric hindrance—it is assumed that a molecule must pass through the pore opening without striking the edge of the pore.

The theoretical restriction to diffusion arising from these sources and which alter D to D' can be calculated and an equation derived which can be tested experimentally:

$$\frac{A_s}{A_p} = \frac{D'}{D} = \left[1 - \left(\frac{a}{r}\right)\right]^2 \left[1 - 2.1\left(\frac{a}{r}\right) + 2.09\left(\frac{a}{r}\right)^3 - 0.95\left(\frac{a}{r}\right)^5\right]$$

(where a is the molecular radius of the molecule under study and r is the pore radius.)

FIGURE II.26 shows that the experimental points fit the smooth curve drawn from this equation.

The figure also shows that membranes with pore size sufficient to allow even a slow penetration of plasma proteins (pore radius, say, 4 nm) nevertheless do cause some restriction to the diffusion of far smaller molecules. Thus, glucose, with a molecular radius of 0.37 nm, shows a diffusion through pores of 4 nm radius which is only approximately two-thirds of its free diffusion. The diffusion of water (molecular radius 0.15 nm) through the same pores is only slowed by 14 per cent. *This differential restriction to the diffusion of small solute molecules and water is the essential factor underlying transcapillary fluid shifts caused by transient changes in the concentration of small molecules in either plasma or tissue fluids.*

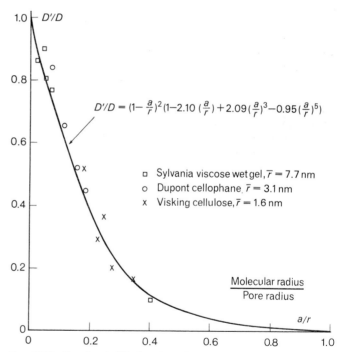

FIG. II.26. Restricted diffusion through artificial porous membranes of various pore sizes. The smooth curve is drawn from the theory of restricted diffusion. (Adapted from Renkin.) (Landis, E. M. and Pappenheimer, J. R. (1963). *Handbook of physiology*, Vol. 2, Circulation II, p. 961. American Physiological Society, Washington.)

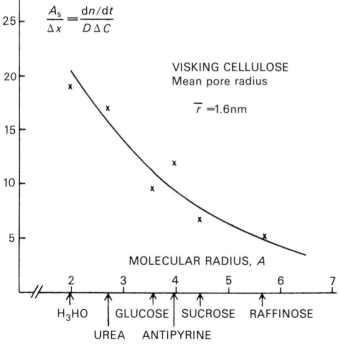

FIG. II.25. Apparent pore areas per unit path length as a function of molecular size. The smooth curve is constructed from the theory of restricted diffusion, assuming a mean pore radius of 1.6 nm. Mean pore radius determined on the same membrane from combination of diffusion and filtration was 1.9 nm. Similar data for diffusion of lipid insoluble molecules through the walls of muscle capillaries are shown in FIGURE. II.28. (Adapted from Renkin.) (Landis, E. M. and Pappenheimer, J. R. (1963). In *Handbook of physiology*, Circulation II, p. 961. American Physiological Society, Washington.)

When we turn to the problem of the living capillary wall, we are confronted by the experimental fact that a molecule as large as inulin (radius = 1.5 nm) can penetrate the wall rapidly. Albumin, on the other hand, (radius ≃ 4 nm) cannot. This gives some preliminary idea of the size of the pores to be expected. However, the quantitative determination of the approximate size of the intercellular pores in capillaries required carefully controlled experimental conditions. Pappenheimer, Renkin, and Borrero (1951) employed the isolated perfused hind limb preparation already described. If an inert lipid-insoluble substance was added to the inflow blood it caused a transient osmotic disturbance consisting of two related processes: 1. the added molecules tend to diffuse out into the interstitial fluid; 2. interstitial fluid tends to be drawn into the capillary.

Both processes continue until the concentration difference across the capillary wall approaches zero. In the isolated perfused hind limb the osmotic transient can be prevented by continually adjusting the mean capillary hydrostatic pressure so as to maintain an isogravimetric condition. The increment of mean capillary hydrostatic pressure required to do this is of course equal to the partial osmotic pressure (or diffusion pressure) exerted by the added molecules.

Now, if flow be maintained constant and arterial and venous concentrations of the fluid are known, the net rate of diffusion of the test molecule during the osmotic transient can be determined.

$$\dot{n} = Qb(C_a - C_v)$$

where Qb is the amount delivered by blood flow and C_a and C_v are the concentrations in arterial and venous blood. But as already stated above

$$\dot{n} = \frac{A_p}{\Delta x} \times D\Delta c$$

and from Van't Hoff's law
$\Delta\pi = \Delta cRT$ (where $\Delta\pi$ is the rise in osmotic pressure).

Hence $\quad \dfrac{A_p}{\Delta x} = \dfrac{Qb(C_a - C_v)}{D.\Delta\pi} RT.$

$\Delta\pi$—the rise in osmotic pressure—is given by the rise in mean capillary hydrostatic pressure which has to be adjusted to preserve the isogravimetric state, so all the factors needed to determine the pore area per unit path length available for diffusion of the test molecule can be obtained experimentally.

The results of an experiment in which raffinose was added to the perfusing fluid are shown in Fig. II.27.

It can be seen that the pore area for diffusion (per unit path length) remained fairly constant. Indeed the capillary diffusion area calculated was unaffected by mechanically induced changes of blood flow produced in other experiments. This independence of $A_s/\Delta x$ with respect to blood flow indicates that the rate of filtration and absorption of the test molecule plays a negligible role in determining the rate of transfer of the test molecule.

From Fig. II.27 it can be seen, for example, that 10 minutes after the addition of raffinose the net flux was 0.1 mg per 100 g per s and the partial osmotic pressure was 15 mm Hg.

$$\frac{A_s}{\Delta x} = \frac{RT}{D} \times \frac{\dot{n}}{\Delta\pi}$$

Where A_s is the pore area available for raffinose.

$\qquad RT = 25 \times 10^9$ dyne-cm-mol^{-1}
$\qquad \Delta\pi = 15$ mm Hg $= 15 \times 1.333 \times 10^3$ dyne cm^{-2}
$\qquad\qquad = 20 \times 10^3$ dyne cm^{-2}
$\qquad D_{\text{raffinose}} = 0.59 \times 10^{-5}$ cm^2 s^{-1}
$\qquad\qquad \dot{n} = 0.1$ mg s^{-1}. Molecular weight of raffinose = 594
$\qquad\qquad = 1.67 \times 10^{-7}$ mol s^{-1}

$\therefore \quad A_s/\Delta x = \dfrac{25 \times 10^9}{0.59 \times 10^{-5}} \times \dfrac{1.67 \times 10^{-7}}{20 \times 10^3}$

$\qquad A_s/\Delta x = \dfrac{0.44 \times 10^4}{11.8 \times 10^{-2}}$

$\qquad\qquad = 0.38 \times 10^5$ cm.

Figure II.28 shows the results obtained from the measurements of the diffusion of several lipid-insoluble test molecules. Just as in artificial porous membranes, the restricted pore area decreases as a

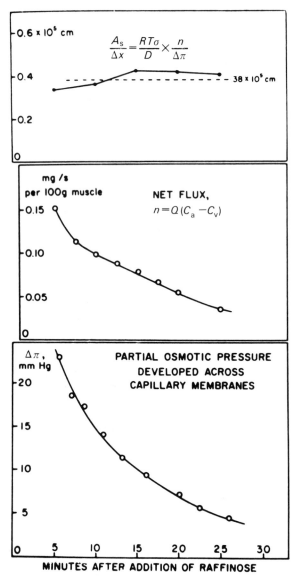

Fig. II.27. Diffusion of raffinose from the capillaries of a perfused cat hind limb. At zero time 20 mm l^{-1} raffinose was added to the perfusion reservoir. The final distribution volume of raffinose in perfused tissue was 19 per cent of limb volume. The capillary diffusion area per unit path length calculated for raffinose was 0.38×10^5 cm; this value was independent of time, extravascular fluid volume, or of mechanically induced changes of blood flow. (Adapted from Pappenheimer *et al.* 1951.) (Landis, E. M. and Pappenheimer, J. R. (1963). In *Handbook of physiology*, Vol. 2, Circulation II, p. 961. American Physiological Society, Washington.)

function of the molecular radius of the test substance. Extrapolation to zero molecular radius indicates that the true pore area per unit path length in muscle capillaries is, say, about 0.7×10^5 cm.

If $A_p/\Delta x$ is $0.6 - 0.7 \times 10^5$ when Δx is taken as unity, then by giving an approximate value for the true dimensions of Δx—the path length—the total pore area can be calculated. The average thickness of the capillary wall is, say, 1 µm $= 1 \times 10^{-4}$ cm. Hence, $A_p = 0.6 - 0.7 \times 10^5 \times 10^{-4} = 6-7$ cm^2.

Histological estimations of the total capillary surface area in muscle made by Pappenheimer and his colleagues give a value of about 7000 cm^2. Hence, the pore area constitutes no more than one-thousandth of the total capillary surface area in muscle. The figure of 7000 cm^2 relates to the fully dilated capillary bed. When, owing to the activity of the precapillary sphincters, the capillary bed is partly shut down, the pore area for the diffusion of lipid-insoluble molecules is correspondingly reduced.

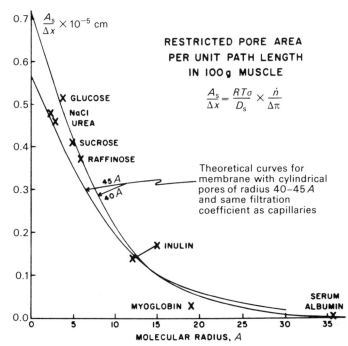

FIG. II.28. Restricted diffusion of lipid-insoluble molecules from the capillaries of perfused cat hind limbs. Each point represents the mean value of data from several experiments. The curves are constructed from the theory of restricted diffusion and filtration. The data fit theoretical restricted diffusion through pores of radius 4.0–4.5 nm in a membrane having the same filtration coefficient as the capillaries in the hind limb. (Recalculated from the data of Pappenheimer *et al.* 1951.) (Landis, E. M. and Pappenheimer, J. R. (1963). In *Handbook of physiology*, Vol. 2, Circulation II. p. 961. American Physiological Society, Washington.)

The size of the individual pores can also be calculated. It is assumed that pore flow is laminar and obeys Poiseuille's equation:

$$Q_f = \frac{n \pi r^4}{8\eta} \times \frac{\Delta P}{\Delta x}$$

$$Q_f = \frac{A_p r^2}{8\eta} \times \frac{\Delta P}{\Delta x}$$

$$r = \sqrt{\left(\frac{8\eta}{\Delta P/\Delta x} \times \frac{Q_f}{A_p} \right)}.$$

When quantities are corrected for osmotic reflection, etc. (see Landis and Pappenheimer 1963) the pore radius calculated is of the order of 4 nm.

Each square centimetre of capillary membrane contains about a thousand million pores.

As the pore area per unit path length in 100 g muscle is, say, 0.6 × 10⁵ cm and the concentration of water available for diffusion in either direction is about 55 molar (0.99 g ml⁻¹) the Fick diffusion equation can be used to calculate the diffusion rate of water across the capillary. Substituting these values and that of the diffusion coefficient of water (3.4×10^{-5} cm² s⁻¹) in the equation gives a calculated diffusion rate of 2 g s⁻¹. The total volume of plasma within the capillaries of 100 g muscle is only 1 ml, so plasma water exchanges 120 times per minute with that of the interstitial fluid.

The plasma flow per minute is, say, 3 ml min⁻¹, so the diffusion of water back and forth across the capillary wall is *forty times* the rate of plasma flow. It will be remembered that filtration and absorption is only 0.3 per cent of the plasma flow.

NaCl, urea and glucose pass back and forth across the membrane at rates which are respectively 20, 18, and 10 times that at which they are delivered by the bloodstream.

These high rates of exchange occur in spite of the small pore area

because the path length is short. Although the exchange rate is high it is a good deal less than would be the case if diffusion were free. Diffusion is restricted and the 'apparent pore area' for a substance decreases as its molecular radius increases. For large lipid-insoluble molecules the restriction to diffusion becomes so great that the degree of molecular sieving is determined largely by the rate of filtration.

Lipid-soluble molecules

Lipid-soluble molecules can diffuse through the plasma membranes of the endothelial cells themselves and are not restricted to pore-passage. Thus, urethane, paraldehyde, or triacetin injected into the blood perfusing the isolated hind limb preparation caused no osmotic transient at all (Renkin 1952).

Similarly, O_2 and CO_2, which both have high lipid solubility, diffuse far more rapidly than water across the capillary walls.

Measurement of the PS product

Diffusion of a solute from the blood flowing through the capillaries depends on the permeability of the vessel wall to the solute, on the wall surface area and on the velocity of flow of the blood. If flow becomes slow, this limits the rate of solute delivered by the vessel to the interstitial fluid. As flow velocity increases, a level is eventually reached at which the time of transit is insufficient for complete equilibration across the vessel wall and solute transport becomes limited by the permeability and surface area of the vessel. To describe the condition of incomplete equilibration it is convenient to define the quantity capillary clearance (*C*) of a solute as the imaginary volume of blood which, if completely equilibrated with tissue in a given time would have transported the same quantity of solute.

$$\text{Capillary clearance } C = QE = Q\frac{(A - V)}{(A - T)}$$

where *A*, *V*, and *T* refer to the solute concentrations in arterial blood, venous blood, and tissue fluid respectively.

Renkin has used the isolated hind limb preparation to study the clearance of ⁸⁶rubidium and/or ⁴²K injected into the arterial inflow. Such substances, gaining access to the interstitial fluid by diffusion from the capillaries, are taken up very rapidly by the muscle cells and the interstitial fluid acts as an effective sink providing that the arterial concentration of ⁸⁶Rb or ⁴²K is kept low. Thus, their concentrations in the interstitial fluid can be regarded as zero.

The ideal situation in the capillary, where a substance is passing out of the capillary blood into the interstitial fluid by diffusion into a sink in which the extravascular fluid concentration is kept effectively zero, is analogous to the loss of heat from a hot body or to the loss of charge from a charged condenser or to the decay of a radioactive element. The equations governing these processes are:

$$\begin{aligned} \theta_t &= \theta_o e^{-at} & \text{(cooling)} \\ Q_t &= Q_o e^{-at} & \text{(condenser)} \\ A_t &= A_o e^{-kt} & \text{(radioactive decay)} \end{aligned}$$

where θ_t, Q_t, and A_t respectively represent the temperature, charge, and radioactivity after a time '*t*' following their reduction from initial values of θ_o, Q_o, and A_o.

Now, from general principles, if

$$y = b e^{ax} \quad \text{then } ax = \log_e \frac{y}{b} = 2.3 \log_{10} \frac{y}{b}$$

and if $\quad y = b e^{-ax}$ then $ax = \log_e \frac{b}{y} = 2.3 \log_{10} \frac{b}{y}.$

So if $\quad \theta_t = \theta_o e^{-at}$

$$at = 2.3 \log_{10} \frac{\theta_o}{\theta_t}.$$

In the case of the capillary, if A is the original concentration of the radioactive tracer (^{42}K or ^{86}Rb) and V is the final concentration in the venous effluent, the general equation above can be put in the form:

$$V = A.e^{-at}.$$

If the blood flow Q is measured, then Q is obviously reciprocally related to 't'; moreover, the factor 'a' can be replaced by the PS product to yield the equation:

$$V = A.e^{-PS/Q}$$

It follows that
$$\frac{PS}{Q} = 2.3 \log_{10} \frac{A}{V}$$

or
$$PS = 2.3Q \log_{10} \frac{A}{V}.$$

The *extraction* (E) of the substance is $\dfrac{A-V}{A}$ or $1 - \dfrac{V}{A}$ as a fraction.

$$\left(1 - \frac{V}{A}\right) = 1 - e^{-PS/Q}.$$

Clearance (C) of the substance is QE, hence

$$\text{Clearance } (C) = Q(1 - e^{-PS/Q})$$

and
$$PS = -Q \log_e (1 - E).$$

If Q and E are known, PS and clearance can be calculated.

Renkin used hind limb preparations perfused with blood supplied from a reservoir; the volume of the venous effluent was measured. Radioactive tracer, ^{42}potassium or ^{86}rubidium, was added to the reservoir and arterial and venous concentrations were accurately determined. From the knowledge of Q and E thus obtained clearance and PS values were calculated.

The dimensions of P are moles diffusing per unit time, per unit concentration difference and per unit surface area. S represents the total capillary surface in 100 g muscle. The PS product has the dimensions of cm³ per 100 g per min of tissue (or ml per 100 g per min).

Figure II.29 shows a family of curves drawn according to the clearance equation showing theoretical clearance-flow relations for several arbitrary values of the PS product. On the same graph are some of Renkin's experimental results. At low blood flow, the clearance of ^{86}Rb approaches equality with flow (line with slope of 1). At high flow, clearance is limited by the PS product which is therefore a measure of the diffusion capacity of the capillary bed for the substance.

In resting muscle, with a blood flow of 6–10 ml per 100 g per min through denervated vessels and with a third of the total capillary surface area perfused, PS is about 5.5 ml per 100 g per min. In denervated intestinal vessels PS is 30–50, probably indicating the greater size and density of the pores in fenestrated capillaries. In the coronary circuit, PS values of 80 or more are obtained.

The most important feature of the relation between capillary clearance and blood flow in skeletal muscle is the effect of sympathetic vasomotor stimulation which decreases clearance at every flow rate. Maximal stimulation reduced PS from 5.5 to 2.5 [Fig. II.29]. On the other hand, metabolic vasodilatation induced either by stimulation of the somatic nerve supplying the skeletal muscle or after interruption of blood flow for some minutes caused an increase of PS at every flow rate and in Figure II.29 raised PS to 9.0.

These results underline the view that at rest the capillary surface area is determined by the tone of the precapillary sphincters. This sphincteric tone can be increased by sympathetic vasoconstrictor nerve stimulation and decreased by the action of local metabolites and the effects of such changes of sphincteric tone are reflected in the PS values obtained in the different circumstances.

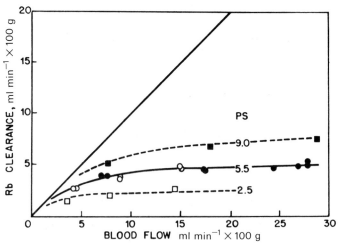

Fig. II.29. Influence of vasomotion on clearance of ^{86}rubidium in skeletal muscle. (Renkin, E. M. (1966). In *Coronary circulation and energetics of the myocardium* (ed. G. Marchetti and B. Taccardi) p. 18. Karger, Basel.)

Quite unlike the sympathetic vasoconstrictor effects, the stimulation of a sympathetic vasodilator nerve on PS is nil. Flow of course increases for a given perfusion pressure head, but PS remains unaltered. It follows that sympathic vasodilator nerves cause relaxation of the arterioles but *not* of the precapillary sphincters. As a result more blood goes by 'thoroughfare' channels from arterioles to venules.

Permeability to large molecules

The key to the accurate measurement of the permeability to substances with molecular weights above 10 000 is found in the lymph. Lymph formation is initiated by ultrafiltration through the capillary wall. *The quantity of capillary fluid filtered is tiny in comparison with the plasma flow and only a small fraction of this becomes the lymph.* Drinker showed that the protein content of lymph in leg lymphatics was as much as 40 per cent of that in the plasma—elsewhere it may be higher (e.g. liver). (This protein has escaped initially from the plasma through the capillary walls.) Drinker claimed that the relatively high plasma protein content of lymph was due to two successive ultrafiltration processes, one from the arteriolar end of the capillary into the interstitial space, the other from the interstitial space to the venular end of the capillary.

If C_1 is the plasma protein concentration in the capillary plasma, at the arterial end of the capillary, ultrafiltration occurs at F ml s^{-1}. The quantity of protein solute transport per unit time in the ultrafiltrate is:

$$M_F = \frac{C_1 F}{K}$$

where K is the ratio of the effective membrane diffusion area for the solvent water (A_w) to that for the protein solute (A_s).

At the venous end of the capillary fluid returns to the blood stream at $(F - L)$ ml s^{-1}—the difference between filtration rate and lymph flow rate.

The quantity of protein carried back into the blood per unit time, M_{-F} is:

$$M_{-F} = \frac{C_2(F - L)}{K}$$

where K is the same molecular 'sieving constant' and C_2 is the concentration of protein solute in the interstitial fluid or lymph.

As C_2 is less than C_1 there is a net outward *diffusion* of solute protein. The quantity transported per unit time by Fick's law is:

$$M_D = (C_1 - C_2)P$$

where P is the capillary membrane permeability to the solute, equal to $D_s(A_s)/(\Delta x)$.

The total transport of high molecular weight solute from blood to lymph, M_L, is the algebraic sum of these three processes:

$$M_L = M_F - M_{-F} + M_D.$$

M_L is also given by the equation:

$$M_L = C_2 L$$

$$\therefore C_2 L = \frac{C_1 F}{K} - \frac{C_2(F - L)}{K} + (C_1 - C_2)P.$$

The lymph/plasma ratio R

$$R = \frac{C_2}{C_1} = \frac{F + PK}{L(K - 1) + (F + PK)}.$$

R is thus a function of four variables; lymph flow, filtration rate, permeability, and the reciprocal of the sieve coefficient.

This equation can be simplified: for a large molecule such as protein, A_s is far smaller than A_w and therefore K is much greater than one. Thus, $(K - 1)$ can be replaced by K. The product PK is far greater than F, so in the denominator of the equation $(F + PK)$ can be replaced by PK alone.

$$R \cong \frac{PK}{LK + PK} = \frac{P}{L + P}.$$

Thus a solution for P gives a reasonable approximation for permeability in terms of two measurable quantities—the lymph flow (L) and the lymph/plasma solute concentration ratio (R).

$$P \cong L\left(\frac{R}{1 - R}\right).$$

FIGURE II.30 shows the results of two sets of experiments by Grotte (1956) as replotted by Renkin (1964).

Dogs were given intravenous injections of dextrans (of molecular weights in the 10 000–20 000 range). The kidneys were tied off to prevent the excretion of the dextrans. Lymph was collected from superficial lymphatics of the hind limb. Lymph and plasma samples were removed at hourly intervals and analysed for dextran concentration. Within 4 hours or so of injection the lymph/plasma dextran concentration ratio (R) became constant. FIGURE II.30(A) shows the relationship between R and the molecular weight of the dextran, and FIGURE. II. 30(B) shows P (the permeability coefficient), calculated from the data, plotted against the molecular weight.

The data from another set of experiments are shown on the same graphs. In this series, dextrans of molecular weights up to 300 000 were injected 24 hours before the analyses were begun. The kidneys were not tied off, so molecules of molecular weight <60 000 were excreted in the urine during this period. Then the procedure of the first experiment was repeated. The data are shown on the right-hand side of each graph.

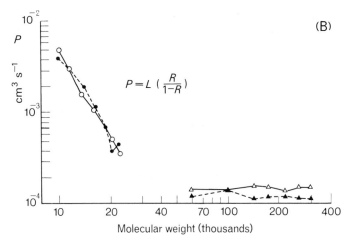

FIG. II.30. Calculation of a spectrum of capillary permeabilities from data on lymph flow and composition.

A. Concentration ratios, lymph/plasma of dextran molecules of graded sizes in leg lymph of dogs. Data of Grotte (1956).

B. Permeability of capillary walls to dextrans of graded sizes calculated from data above. See text for explanation. (Renkin, E. M. (1964). *Physiologist* **7**, 13.)

In each type of experiment, control values of R were established from the samples obtained during these conditions and then venous congestion was produced by obstructing venous return with a femoral cuff. Lymph flow doubled and steady state values of R fell. FIGURE II.30(B) shows that despite the large difference in R, the calculated permeabilities in the two conditions are nearly the same. Permeability falls rapidly with increasing molecular weight in the 10 000 to 20 000 range—from 5×10^{-3} cm^{-1} s^{-1} at 10 000 to 4×10^{-4} cm^{-1}s^{-1} at 22 000. However, in the 60 000–300 000 range it becomes constant at approximately 1.3×10^{-4} cm^{-1} s^{-1}.

FIGURE II.31 shows the capillary permeability of a wide range of inert lipid insoluble molecules (Renkin 1964). Both abscissa and ordinate are logarithmic. The data are derived from the results of various experimental groups. The pecked lines on the graph are

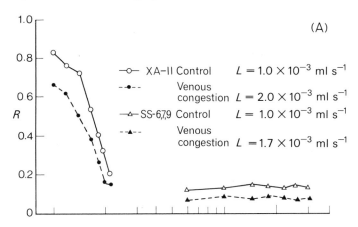

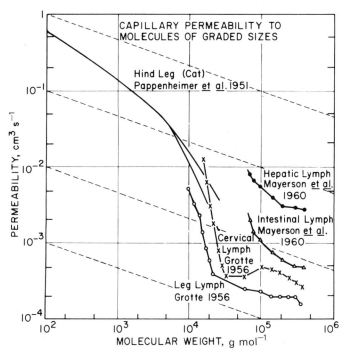

FIG. II.31. Capillary permeability to lipid-insoluble molecules of graded sizes. Collected data from the literature. (Renkin, E. M. (1964). *Physiologist* **7**, 13.)

plotted with a slope of $-\frac{1}{2}$ and represent the dimunition in transport to be expected if free diffusion were the means of transport. [The free diffusion coefficient is inversely proportional to the square root of the molecular weight of the substances (Graham's law). Any slope which declines more steeply indicates that there is a restriction to free diffusion.]

The figure shows clearly that, whereas the transport of solutes of molecular weight 100–10 000 is only moderately restricted, there is a sharp inflection of the curve, indicating greater restriction to diffusion of above 10 000 MW.

Above 60 000, however, there is little further fall other than can be accounted for by free diffusion—the slope of the line is in no case greater than $-\frac{1}{2}$.

In view of the fact that the rapid fall-off of R (lymph/plasma concentration ratio) is limited to the smaller of the large molecules, whereas molecules above, say, 60 000 show no further reduction of R, led Mayerson *et al.* and Grotte independently to propose that two transport systems existed in the capillary wall: 1. A small pore system of 3–4.5 nm which would account for the transport of molecules up to 60 000 or so. This would explain the rapid fall of permeability in this molecular weight range as the molecular size increased. 2. *A less extensive* system of 'leaks' or large openings which would account for the high molecular weight 'tail' of the lymph/plasma concentration curve. Mayerson *et al.* (1960) added the suggestion that the relatively high concentration of very large molecules in the lymph might be due to transport by vesicles (micropinocytosis or cytopempsis). It can be estimated from the data available that a *large pore system would have a total surface area of less than 1 per cent of that of the small pores* (Renkin 1964). (It should be remembered in this context that the small pore system itself represents only some 0.1 per cent of the total surface area of the capillary wall. Hence the large pore system has a total area only one-hundred thousandth of the wall.) Similarly, it can be calculated that micropinocytosis plays no role of importance in the transport of molecules which have molecular weights of less than 10 000. Vesicular transport should be featured by a zero slope, for it represents an active transport by the capillary endothelium of large molecules which is entirely independent of diffusion, free or restricted. Further work is required to provide functional evidence of the complete flatness of the tail of the permeability/molecular weight curve. The present data suggest that this may be so, but they are not entirely conclusive. Micropinocytosis, like a large pore system, would be valuable in the transport of the large molecules of immunological proteins, etc., which can and do gain access to the tissues.

Micropinocytosis

Micropinocytosis involves active transport. It is very slow and can hardly be envisaged to contribute much to the *total* transcapillary exchange. Electron microscope studies reveal 'vacuoles' which *seem* to have been in the process of traversing the endothelial cells of the capillary wall. As noted above, micropinocytosis may provide an active transport route for macromolecules (such as γ-globulins) which have limited powers of access to the tissues by diffusion. It may provide a route for the transfer of molecules by active expulsion against a concentration gradient—even proteins from the interstitial space could be transferred to the blood in this fashion.

REFERENCES

FICK, A. (1855). *Ann Physik.* **94**, 59.
GROTTE, G. (1956). *Acta chir. scand.* Suppl. **211**, 1.
KARNOVSKY, M. J. (1967). *J. cell Biol.* **35**, 213.
LANDIS, E. M. (1930). *Am. J. Physiol.* **93**, 353.
—— and PAPPENHEIMER, J. R. (1963). In *Handbook of physiology*, Vol. 2, Circulation II, 961–1034. American Physiological Society, Washington.
MAJNO, G. (1965). In *Handbook of physiology*, Vol. 2, Circulation III, 2293–375. American Physiological Society, Washington.
MICHEL, C. C. (1972). In *Cardiovascular fluid dynamics*, Vol. II (ed. D. H. Berger). Academic Press, New York.
PAPPENHEIMER, J. R. (1953). *Physiol. Rev.* **33**, 387.
—— and SOTO-RIVERA, A. (1948). *Am. J. Physiol.* **152**, 471.
—— RENKIN, E. M., and BORRERO, L. M. (1951). *Am. J. Physiol.* **167**, 13.
RENKIN, E. M. (1959). *Am. J. Physiol.* **197**, 1205.
—— (1964). *Physiologist* **7**, 13.
—— (1966). In *Coronary circulation and energetics of the myocardium* (ed. G. Marchetti and B. Taccardi) pp. 18–30. Basel.
STARLING, E. H. (1896). *J. Physiol., Lond.* **19**, 312.

Veins and venous return

The venules and veins which collect the blood returning to the heart from the capillaries are characterized by thin walls which are collapsible when the transmural pressure falls below about 6 cm H_2O. These vessels then assume an elliptical profile; this geometrical change of their cross-sectional profile is important in modifying both their capacity and their resistance to flow.

The venules leading from the capillaries are first invested by a connective tissue sheath which surrounds the endothelial cells and muscular elements develop in the wall as the vessels increase in size. The veins proper are arbitrarily accorded a luminal diameter of at least 500 μm and at this juncture the endothelial lining first shows the internal folds which are the venous valves. These valves permit flow only towards the heart. Not all veins possess them, but they are prominent in the venous channels of the legs (particularly) and in the arms. The venae cavae, portal and cerebral veins do not possess functional valves. The veins of the limbs can be divided into superficial and deep vessels. The superficial veins run in the subcutaneous tissue; they themselves possess numerous valves and moreover are connected with the deep veins by a wealth of communicating channels which themselves are characterized by an abundance of valves permitting flow only from the superficial to the deep vessels. These communicating channels allow the superficial veins to empty some of their blood into the deep veins during rhythmic muscular contractions in the legs and arms.

As might be expected the leg veins are thicker walled and possess a greater development of adventitial connective tissue than those elsewhere. Such structural modifications seem to be related to the gravitational stresses which these leg veins have to endure.

THE DISTENSIBILITY OF VEINS

The pressure–volume relationship of a vein is quite different from that of an artery of comparable size [FIG. II.32]. Thus the volume of the vein increases strikingly when the transmural pressure is raised. However, this phenomenon is not so much due to any stretching of the wall as to the geometric change in the cross-sectional profile as the transmural pressure rises. Thus, if an ellipse has the dimensions of 0.7 cm major axis (*a*) and 0.1 cm minor axis (*b*) its perimeter is $\pi(a + b) = 0.8 \pi$ cm. A circle with a radius of 0.4 cm (*r*) has the same perimeter—$2\pi r = 0.8\pi$ cm. However, the cross-sectional area will influence the pressure–volume response of the vessel so that simply changing the profile from elliptical to circular by slightly increasing the transmural pressure will considerably raise the volume accommodated per unit length. Only when the cross-sectional profile is circular will a further increase of pressure cause stretch of the distensible elements of the venous wall. It can be seen

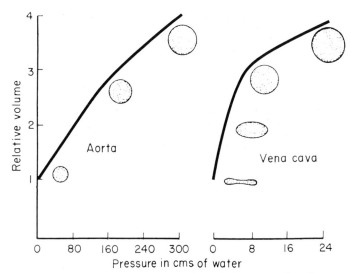

FIG. II.32. Comparison of the distensibility of the aorta and of the vena cava. The way in which the cross-section of the vessels changes in the two cases is also indicated. (From Burton, A. C. (1965). *Physiology and biophysics of the circulation.* Year Book Medical, Chicago. Used by permission.)

from FIG. II.32 that the distensibility of the vein in this range is less than that of an artery of comparable size.

This 'collapsibility' of the thin-walled veins is of the utmost importance in allowing striking changes of venous capacity in response to quite slight changes of transmural pressure over a certain range. Moreover, the resistance which the vein offers to blood flow is likewise modified by this geometrical alteration in the venous profile. If the transmural pressure becomes zero or even negative (extravascular pressure exceeds intravascular pressure) the vein collapses completely and resistance to flow increases sharply. Such changes occur in veins above heart level—the neck veins are collapsed in normal individuals (when sitting or standing) above a level which is 5–10 cm higher than that of the heart. In the figure shown [FIG. II.33] a fluid-filled thin-wall tube connected to a reservoir shows collapse above the level of the outflow tube, for the pressure outside the tube is equal to that inside the tube. Below the outflow tube the pressure inside exceeds that outside and the tube is distended by transmural pressures which increase as the depth of the tubing increases. Quiet standing then considerably increases the volume of blood in the leg veins. On the other hand, the *resistance* offered by the leg veins to venous return is less, for the veins are all of circular profile.

In an ellipse with major axis a and minor axis b the equation determining flow for a given pressure head is:

$$Q \propto \frac{2a^3b^3}{a^2 + b^2}.$$

In the case of a circle, $a = b = r$ and the 'resistance factor' per unit length is:

$$\frac{2r^6}{2r^2} = r^4.$$

If $r^2 = ab$
and $b = a\alpha$
when the elliptical and circular cross-sectional areas are the same then the ratio of flows through the two tubes will be:

$$\frac{ellipse}{circle} = \frac{2\alpha}{1 + \alpha^2}$$

which is less than unity.

FIGURE II.34 shows that as a venous segment with a circular profile changes to one of an increasingly elliptical configuration the

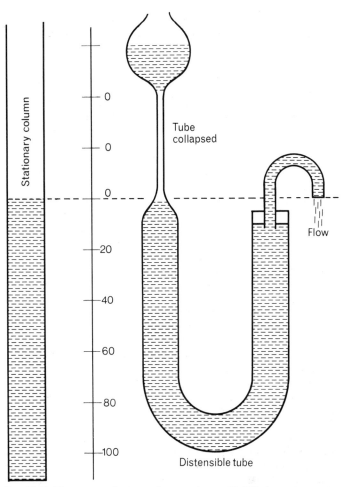

FIG. II.33. The pressure in the stationary column of fluid is dependent on its specific gravity and on the vertical distance from the point of measurement to its meniscus.

A collapsible tube is distended only as long as the internal pressure exceeds the external pressure. These two pressures are exactly equal in the segment of tube which is collapsed. (After Rushmer, R. F. (1961). *Cardiovascular dynamics.* Saunders, Philadelphia.)

flow decreases in the manner shown. On the same graph is shown the effect of developing the elliptical shape from that of a circle while maintaining the perimeter constant.

These changes of resistance due to alterations of 'luminal geometry' complicate calculations of 'pressure gradient' made from simultaneous measurements of peripheral venous pressure and right atrial pressure. If partial collapse occurs between these sites of measurement the peripheral venous pressure measured becomes independent of the right atrial pressure.

Blood flows from the periphery to the right heart because of an energy gradient. The energy concerned is a sum of potential energy (pressure) and kinetic energy. As the veins anastomose to form fewer but bigger channels the flow velocity increases as we approach the heart, until the linear velocity is of the order of 15 cm s^{-1}. However, the kinetic energy factor is not so important in the venous system of resting man, but becomes very significant in circumstances of exercise, accounting for as much as 50 per cent of the total.

POSTURE AND THE VEINS

When man lies horizontally the mean pressure in his dorsalis pedis artery is about 100 mm Hg; that in the corresponding vein is, say, 15 mm Hg and that in the right atrium is sensibly zero. If he lies

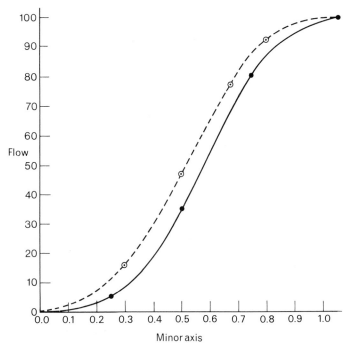

Fig. II.34. Volume flow through elliptical tubes (ordinate) as a function of the minor semi-axis of the elliptic cross-sectional area (abscissa). Broken line = same cross-sectional area as a circle. Unbroken line = same perimeter as circle. When minor axis = 1 then this radius is the same as that of the circle and 100 per cent flow is that through a circular tube. (After Brecher, G. A. (1956). *Venous return*. Grune and Stratton, New York.)

quietly, avoiding muscle movements, *all the valves in his veins are open* permitting a continuous forward flow to the heart due to the pressure gradient. The mean capillary pressure in such circumstances is some 20 mm Hg.

When he stands (or is tilted into the upright position) the hydrostatic pressure in his dorsalis pedis artery is increased by some 85–90 mm Hg (the difference between heart level and that of the foot) and so is that in the capillaries and in the dorsal vein of the foot—but his right atrial pressure remains about the same. If the blood vessels of the leg were rigid tubes nothing would have altered, because the pressure head from the aorta to the right atrium is entirely unaffected by this postural adjustment. Only the transmural pressure is increased—for instance in the foot—by 85–90 mm Hg.

However, the vessels of the legs are not rigid tubes and particularly the veins are susceptible to increases of transmural pressure both 'geometrically' and 'distensibly'. The thin-walled veins become circular and yield also to stretching. The result is 'venous pooling'—a descriptive, though in a way unfortunate term, suggesting as it does, stagnation. The veins simply accommodate more blood and correspondingly less is available for return to the heart. The situation is similar to that induced by haemorrhage. The reduction of venous return in turn induces a fall in cardiac output and this tends to drop the arterial pressure head.

Meanwhile the effective pressure head to the brain has decreased by some 50 mm Hg (for the same hydrostatic reasons), but so has that in the cerebral venous sinuses in which the pressure may fall to −40 mm Hg. Within the cranium the cerebrospinal fluid pressure also falls by a similar amount so the transmural pressure of the intracranial vessels does not alter and, as we have seen, the pressure head from the intracranial arteries across the capillaries to the intracranial veins remains the same. The cerebral venous sinuses are in any case held open by extravascular tissues and this may cause disaster if a breach of their walls is made inadvertently during an intracranial operation when the patient is in a semivertical

position. As the intraluminal pressure of the cerebral venous sinus is then well below atmospheric, then on severing the wall of the vein by accident, air is sucked into the vessel and may cause a fatal air embolism.

The extracranial cerebral veins and those of the neck are collapsed or semicollapsed in the upright position and this contributes an increase of resistance offered to venous return from the head.

The subject tilted passively into the upright position suffers transient dizziness. This is not due to any loss of the arteriovenous pressure head supplying his brain caused by hydrostatic factors in the cerebral circuit, but is due to the fall in the cardiac output (due to diminished over-all venous return) and the consequent fall of arterial pressure and the rise in extracerebral venous resistance. However, the reduction of aortic pressure is evanescent because of compensatory vasoconstriction, reflexly induced both from the carotid sinus and the aortic areas. Particularly the carotid sinus suffers (for hydrostatic reasons) a fall in its mean arterial blood pressure but both reflexogenic areas are influenced by the drop in aortic pressure consequent upon the decrease in cardiac output. Their afferent impulse activity diminishes and this causes reflex sympathetic vasoconstriction, both of the precapillary resistance vessels and of the veins. The arteriolar vasoconstriction restores the aortic pressure essentially to its normal level and the venular and venous constriction effectively reduces the amount of venous pooling. Moreover, the increase in the precapillary-postcapillary resistance ratio reduces the hydrostatic pressure and this reduction, particularly in the capillaries of the feet, minimizes the transudation of fluid from these capillaries.

The importance of these sympathetic constrictor responses reflexly evoked by the postural change can be realized by examining the effects of tilting in a subject who has suffered blockade of the sympathetic effect or responses by the administration of hexamethonium. When tilted the subject faints promptly. However, if a pressure suit (g-suit) has been previously fitted from the waist downwards then quite moderate inflation of the suit (to a pressure of, say, 20–25 mm Hg) prevents his faint when he is tilted. This small extravascular pressure prevents venous pooling and correspondingly offsets the serious fall of venous return, cardiac output and aortic pressure which otherwise occurs.

Normal man, when standing quietly or when tilted into the head-up position, does show a fall of cardiac output from 5.5 to, say, 4.0 l min⁻¹, although his aortic blood pressure may be maintained sensibly normal. Thus the sympathetic venoconstriction and arteriolar constriction only minimize the problem offered by the postural influences on the veins.

Mellander, Öberg, and Odelram (1964) have further shown that postural changes may induce important changes in the precapillary resistance vessels by a 'myogenic' influence of the change in transmural pressure in the dependent vessels. The increase in transmural pressure in the foot, for instance, reduces the capillary filtration coefficient from 0.0077 ml per mm Hg per 100 g per min to 0.0017 ml per mm Hg per 100 g per min due to a myogenic response of the precapillary sphincters to the increased stretching force. Thus, the fluid transuded from these foot capillaries is much less than otherwise would be the case because the capillary surface area perfused is so much reduced. Oedema, which does occur, is nevertheless formed at a much slower rate.

Abdominal veins and posture

Venous distension in the abdomen in response to postural alterations is minimized by the accompanying rise in intra-abdominal pressure. The abdominal contents behave as if the abdomen were fluid filled. The pressure in the abdominal veins exceeds that of the intra-abdominal contents of only 5–10 cm H_2O whatever the position of the body (Rushmer 1961). Even when the individual voluntarily raises his intra-abdominal pressure, such as when straining to defecate, although his abdominal veins are compressed blood cannot pass backwards into his legs because retrograde flow is prevented by valves in the iliac and femoral veins. Similarly, venous backflow into the brain is prevented by the simultaneous rise of cerebrospinal fluid pressure.

Venous return to the thorax

The intrathoracic pressure varies in eupnoea between 5 cm H_2O subatmospheric in expiration and 10 cm H_2O subatmospheric in inspiration. As the pressure falls during inspiration the abdominal pressure rises and the two factors combine to favour venous return to the thorax in inspiration. Brecher and his colleagues (e.g. Hubay *et al.* 1954) have shown clearly how a negative endotracheal pressure increases superior vena caval flow in the closed chest [FIG. II.35] and conversely, how positive pressure respiration decreases venous return.

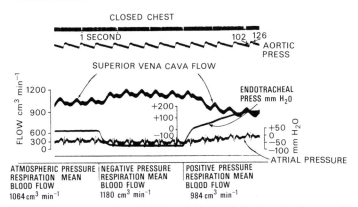

FIG. II.35. Segment of an original optical record showing the changes of mean blood flow in the superior vena cava at different endotracheal pressures in the closed chest. Tracings from top to bottom: time; aortic blood pressure in millimetres of mercury; mean blood flow in superior vena cava in millilitres per minute; endotracheal pressure in millimetres of water; right atrial pressure in millimetres of water. (Hubay, C. A., Waltz, R. C., Brecher, G. A., Praglin, P., and Hingson, R. A. (1954). *Anesthesiology* 15, 445.)

Finally, venous return, which is mostly dependent on the forward push from behind (*vis a tergo*) provided initially by the heart and transmitted in the form of a positive pressure from the arteries via the arterioles and capillaries, can nevertheless be assisted phasically by the suction force on the atrium (*vis a fronte*) exerted by the contraction of the ventricle. Blood flows from the atrium into the ventricle because the atrial pressure exceeds that in the ventricle. This pressure gradient may steepen (and hence the blood flow increases) either because the atrial pressure increases or because the ventricular pressure decreases (suction). Brecher (1958) has produced cogent evidence of ventricular suction in showing that subatmospheric ventricular pressures may be recorded when the a–v orifice is temporarily clamped. Venous return through the superior vena cava is increased in more normal circumstances with each ventricular contraction [FIG. II.36].

THE MUSCLE PUMP AND VENOUS RETURN

As stated before, the pressure recorded in a vein of the dorsum of the foot is about 120 cm H_2O (90 mm Hg) in man standing quietly. The conditions are artificial, however, for normally the subject will be making slight muscle movements and these lower the pressure. Pollack and Wood (1949) showed that even one step lowered this pressure to a level of some 60 mm Hg, whereupon the pressure gradually returned to 85 mm Hg at a rate which was determined by the arterial blood flow to the foot. Repeated steps lowered the

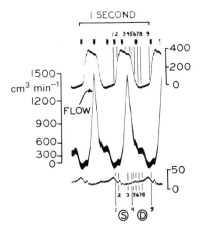

FIG. II.36. Venous return during different parts of the cardiac cycle (open chest). Tracings from top to bottom: time, right intraventricular pressure in mm water, superior vena cava flow in ml min⁻¹, superior vena cava pressure in mm water. Numbers 1 to 9 denote the phases of the cardiac cycle. S = systole, D = diastole. (Brecher, G. (1956). *C.r. II Congrès International d'Angéiologie* 306.)

pressure to about 30 mm Hg [FIG. II.37]. Such muscular contractions could not lower the pressure in this manner unless the continuity of the blood column between heart and foot were temporarily but repeatedly interrupted during the steps. When the man stands at rest, the valves in the leg veins are all open, permitting continuous forward flow of blood towards the heart but likewise permitting the full hydrostatic effect of the continuous column of blood. When the muscle contracts the venous segments are squeezed and the rise of pressure forces blood towards the heart out of each successive segment; the valves prevent back flow. As soon as the muscles relax the depleted segments are promptly refilled from the more peripheral venous channels and also from the superficial veins via their valved communicating channels. The more frequent and powerful such rhythmic movements are the more efficient this 'muscle pumping'. The communication between the superficial and the deep veins by channels which only allow the flow from the superficial to the deep veins (because these channels are valved) is of great importance. These superficial veins, which are unsupported themselves by muscle actions are thus 'milked' into the deep veins and much of their blood is drained off by the deep veins.

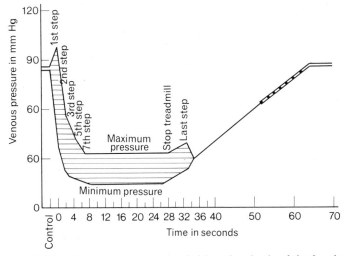

FIG. II.37. The venous pressure recorded in a dorsal vein of the foot is 85 mm Hg when the man stands quietly. Walking lowers the pressure to approximately 30 mm Hg. On stopping the venous pressure rises gradually to 85 mm Hg again. (Pollack, A. A. and Wood, E. H. (1949). *J. appl. Physiol.* **1**, 649.)

Otherwise the superficial veins would sustain the full effect of the hydrostatic column of blood. As Greenfield (1962) has written, drainage of blood from the superficial to the deep veins ceases to be an apparent anomaly when it is remembered that the mean pressure in the heart ventricles greatly exceeds that in the atria and veins whence they receive their inflow.

REFERENCE

Caro, C. G., Pedley, T. J., Schroter, R. C., and Seed, W. A. (1978). Chapter 14 in *The mechanics of the circulation*. Oxford University Press.

Structure and properties of the heart

STRUCTURE OF THE CARDIAC CHAMBERS

The heart contains four chambers—two thin-walled atria separated from each other by an interatrial septum and two thicker-walled ventricles which possess a common wall in the interventricular septum. Atria and ventricles are connected by a fibrous A–V ring. This ring is penetrated on the right side by the tricuspid valve and on the left by the mitral (or bicuspid) valve [Fig. II.38].

The two valves consist of flaps or cusps which are attached at the periphery of the valve ring. Chordae tendinae (which take origin from papillary muscles arising from the inner border of the ventricle) are attached to the free edges of the valve cusps and act like guy ropes.

On the right side the pulmonary orifice (which leads from the

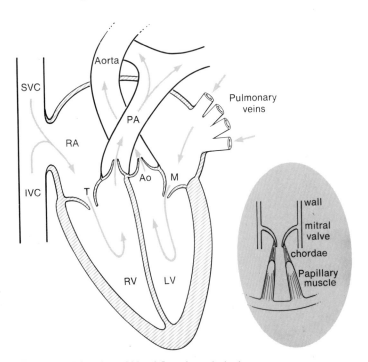

Fig. II.38. Direction of blood flow through the heart.

SVC = Superior vena cava PA = Pulmonary artery
IVC = Inferior vena cava M = Mitral valve
RA = Right atrium LV = Left ventricle
T = Tricuspid valve Ao = Aorta
RV = Right ventricle

Inset figure shows attachment of papillary muscles and chordae tendineae to the cusps of the mitral valve. (Folkow, B. and Neil, E. (1971). *Circulation*. Oxford University Press, New York.)

ventricle to the pulmonary artery) is guarded by the pulmonary or semilunar valve which consists of three flaps. A similarly constructed valve (aortic) is situated at the aortic orifice (which leads from the left ventricle to the aorta). These valves open at the onset of ventricular ejection and close when the relevant arterial pressure exceeds that of the corresponding ventricle when it begins to relax.

Closure of both atrioventricular valves causes the first heart sound and closure of the semilunar valves causes the second heart sound.

Special junctional tissues of the heart

This term is employed to describe certain tissues in the heart which are concerned with the initiation and propagation of the heart beat. They include the sinu-atrial node and the junctional tissues—the latter consisting of the atrioventricular node; the atrioventricular bundle, or bundle of His; the right and left divisions of the bundle, and their arborizations under the endocardium and the terminal fibres which penetrate the ventricular substance [Fig. II.39]. The bundle of His and all its ramifications may be called Purkinje fibres.

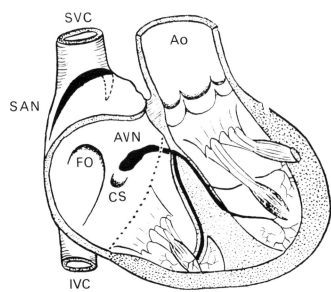

Fig. II.39. The specialized tissue of the mammalian heart. The pulmonary trunk and part of the right ventricle have been removed. Ao = aorta. AVN = atrioventricular node giving origin to the a.v. bundle. Note how the bundle passes beneath the attachment of the tricuspid valve (the septal cusp of which has been removed) and then, after a short course below the membranous part of the interventricular septum, divides straddlewise over the fleshy septum. CS = opening of coronary sinus. FO = fossa ovalis. IVC = inferior vena cava. SAN = sinu-atrial node. Note its horseshoe shape. SVC = superior vena cava. (Drawn by Professor E. W. Walls.)

1. The sinu-atrial node is situated at the junction of the superior vena cava and the free border of the right atrial appendix, and extends down along the sulcus terminalis for a distance of 2 cm. It is 2 mm in width and has a rich capillary blood supply. It consists essentially of thin, elongated muscle fibres (about one-third the size of heart-muscle fibres), fusiform in shape and longitudinally striated, which interlace with one another in a plexiform manner. These fibres normally initiate the heart beat; for this reason the sinu-atrial node is called the *pacemaker*. Nerve cells and fibres forming the excitor relay of the vagus nerve and excitor (postganglionic) fibres of the sympathetic are also present.

2. The atrioventricular node is situated at the posterior and right border of the interatrial septum near the mouth of the coronary sinus. Atrial muscle fibres from the region of the coronary sinus collect fanwise, interlace, and unite with the atrioventricular node.

In structure this node is identical with the sinu-atrial node. The sinu-atrial node is a right-sided structure developing from tissue which lies at the entrance of the primitive right great vein (which later becomes the superior vena cava). This explains the position of the node and its supply by the right vagus nerve. The atrioventricular node is a left-sided structure, supplied therefore by the left vagus nerve, and developed from tissue in the vicinity of the entrance of the left great veins which becomes the coronary sinus. The anatomical position of the two nodes is thus also accounted for [Keith].

3. The bundle of His runs upwards to the posterior margin of the membranous part of the interventricular septum and then forwards below it, ensheathed and isolated in a canal. At the anterior part of the membranous septum, in front of the attachment of the septal cusp of the tricuspid valve to the a.v. ring, the bundle forks. The left division pierces the membrane and then lies on the upper border of the muscular septum to enter the subendocardial space of the left ventricle beneath the union of the anterior and right posterior cusps of the aortic valve. The right division passes down the right side of the septum and is mainly transmitted in the moderator band. Both branches are continued as an arborization of fibres lying under the endocardium of both ventricles from which terminal fibres penetrate the ventricular wall.

Purkinje tissue (bundle of His and its branches) differs histologically from cardiac muscle. In man the Purkinje fibres are somewhat larger (range 10–46 μm, mean 16μm); the cell outlines are indistinct; the central cytoplasm is granular and contains several nuclei; the peripheral cytoplasm contains myofibrillae but these are separated by more sarcoplasm; the glycogen content is greater.

Arrangement of muscle in the atria and ventricles

The atria are thin walled and subserve a capacity function as well as that of contraction. The thin atrial wall consists of two main muscular systems—one encircles both atria and the other, arranged at right angles to this, is independent for each atrium. The excitation process responsible for contraction arises at the SA node and spreads like ripples over a pond through the atrial wall to the A–V node at a rate of 1 metre per second. The ventricles serve as the pumps. The right ventricle supplies the lung circuit and the left the systemic circuit.

The ventricles contain much more muscle than do the atria; the left ventricle, which has to do the larger amount of work, is thicker than the right.

It is customary to describe four groups of fibres all arranged spirally (Sands Robb 1942):

1. Superficial bulbospiral system which, arising from the left side of the ventricular-aortic ring and the mitral ring, terminates in the interventricular septum after following a posterior and obliquely downward course.

2. Superficial sinospiral fibres originating from the back of the tricuspid valve pass obliquely over the front of the right ventricle to the cardiac apex and thence turn inward to terminate in the anterior papillary muscles of the left ventricle.

The arrangement of these two spiral bundles ensures that on ventricular contraction blood is virtually wrung out of the heart, but it must be clearly understood that the ventricles do not empty themselves completely. After systole there is a residual volume of blood left (end-systolic volume) which varies in amount.

3. Deep sinospiral fibres encircle the bases of both ventricles and contribute importantly to the ejection of blood into the arteries.

4. Deep bulbospiral fibres form a thick cuff round the mitral and aortic orifices and are restricted to the left ventricle. They form a 'canal' for left ventricular ejection and perhaps contribute to the closure of the mitral valve.

When the ventricles contract, the base (the A–V ring) is pulled downwards and the heart rotates to the right so that the apex is pushed forwards in closer approximation to the chest wall in the region of the fifth intercostal space, causing the 'apex beat', which can be palpated.

PRESSURE PULSES IN THE CARDIAC CHAMBERS AND RELATED VESSELS: ASSOCIATED PHENOMENA

Using catheters and high fidelity manometers, optical or electrical, the sequence of events in the various chambers which occur during cardiac contraction can be clarified [FIG. II.40].

At the beginning of the cardiac cycle, both atria and ventricles are relaxed and filled with blood by venous return achieved by the *vis a tergo*. Atrium and ventricle on each side are in continuity as the A–V valves are open and the pressure in each cavity is almost identical. Following the impulse generation of the SA node, the atrial muscle contracts and P_A rises with P_V following. As P_A is greater than P_V atrial contraction adds to the diastolic volume of the ventricle. The atrial contraction wave lasts about 0.1 s in a cardiac cycle of 0.8 s (rate = 75 min^{-1}). As it passes off, the pressure in both atrium and ventricle falls and the ventricle has meanwhile been invaded by the excitation process which has spread from the SA node across the atrial muscle to the A–V node and thence via the bundle of His and the Purkinje tissue. Ventricular contraction begins and promptly P_V exceeds P_A and the A–V valves close (causing the first heart sound). The ventricle is now a closed cham-

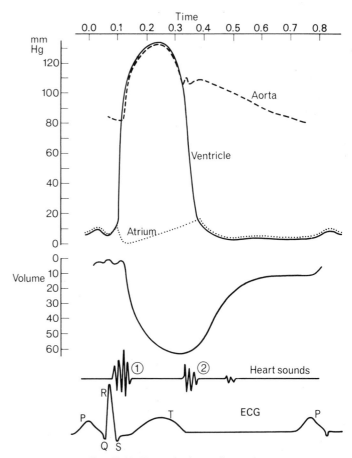

FIG. II.40. Events in the cardiac cycle.

ber and the pressure rises promptly during the 'isometric' phase which lasts about 0.05 s. The ventricular pressure then rapidly exceeds that in the artery and the semilunar valve opens. There follows the ejection phase and during this arterial and ventricular pressures follow each other closely. Initially in this phase ejection is rapid, as evidenced by the ventricular volume tracing, and the arterial pressure rises because blood enters the vessel faster than it can escape via the peripheral arteriolar branches. The summit of the pressure curve is reached when aortic entry and run-off become equal and subsequently the pressures decline as the ventricular contraction begins to subside while flow from the artery to its peripheral branches continues to be high. The total period of ventricular systole is 0.3 s and when this ends the ventricular pressure drops sharply. The arterial pressure is better sustained however owing to elastic recoil of the vessel wall and almost immediately the arterial pressure exceeds that in the ventricle, thereby causing closure of the semilunar valve and the sharp second heart sound. The initial part of ventricular diastole which follows is isometric, lasting some 0.08 s ending in opening of the A–V valve because the atrial pressure exceeds that in the ventricle. Rapid filling of the ventricle then occurs for 0.1–0.12 s, although the pressure in both chambers still falls owing to the continued rapid relaxation of the ventricle (see volume curve). Finally, the phase of slow filling or diastasis lasting 0.19 s terminates the cardiac cycle. This slow filling is due to the continued venous return filling both atrium and ventricle and readjusting the end diastolic volume of the ventricle.

Ventricular diastole lasts 0.5 s (in a cardiac cycle of 0.8 s) made up as follows: 0.08 s (isometric relaxation) + 0.12 s (rapid filling) + 0.2 s (diastasis) + 0.1 s (during *atrial* systole).

Obviously when the heart beats at 180 min^{-1} with a total cardiac cycle of only 0.33 s these periods of systole and diastole given for a heart rate of 75 min^{-1} are shortened. Diastole suffers particularly and herein lies the danger of pathological tachycardia—the heart is inadequately filled and its output rapidly declines.

The atrial pressure curve shows three well marked waves:

1. Is due to atrial systole and has already been described.
2. With the onset of ventricular systole there is a rise of pressure due to bulging of the A–V valve into the atrium. This is soon succeeded by a sharp fall of atrial pressure beginning at the time of onset of ventricular ejection. The abrupt fall of pressure is attributed to the passive lengthening of the atrium caused by the downward movement of the A–V ring as the ventricular contraction proceeds. The atrial pressure is now exceeded by the venous pressure and atrial filling begins.
3. The rising phase of this wave is due to venous return. The atrial pressure eventually exceeds that in the ventricle and the opening of the A–V valve, which follows as a simple physical consequence, terminates the phase of ventricular isometric relaxation. As the atrial blood passes into the relaxing ventricle atrial pressure falls to rise again during the phase of diastasis.

Heart sounds

These may be heard by placing the ear on the chest wall or by using a stethoscope. They may be 'displayed' by using a recording microphone on the chest wall connected with suitable recording equipment (phonocardiogram) [FIG. II.41].

The first heart sound is due partly to vibrations of the A–V valves on closure and in the adjacent cardiac wall and partly to turbulence.

The second heart sound is due mainly to coaptation of the valves. If the pressure in the aorta or pulmonary artery is unusually high the corresponding sound is excessively loud, as the rebound of the valves is pronounced.

These are the classical heart sounds. The first sound has a fre-

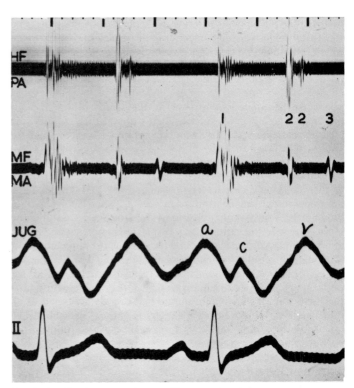

FIG. II.41. Simultaneous phonocardiograms, jugular venous pressure tracing and e.c.g. from a normal subject. Records from above downwards: time 1/25 s and 1/5 s, high-frequency phonocardiogram recorded from the pulmonary area, low-frequency phonocardiogram recorded from the mitral area, jugular venous pressure tracing and e.c.g. Lead II.

In the phonocardiograms, 1 represents the first heart sound, 2.2 the aortic and pulmonary components of the second heart sound, and 3 the third heart sound. The aortic component of the second sound precedes the pulmonary component and is well recorded at the mitral area. The third heart sound is detected at the mitral area only. A soft systolic murmur is present. (By courtesy of Dr Aubrey Leatham.)

quency of 30–80 s^{-1} and a duration of 0.05 s and is commonly described as 'lubb'. That due to closure of the tricuspid valve is most audible at the right sternal border in the fourth intercostal space—and of the mitral is best heart at the apex of the heart. The second sound has a higher pitch 'dup' (frequency 150–200 s^{-1}) and a shorter duration (0.025 s). The pulmonary second sound is best heard in the parasternal line in the second left intercostal space and that of the aortic valve in the second intercostal space near the right side of the sternum. The third and fourth 'sounds' are less important, being usually inaudible—they can be detected on the phonocardiogram. The third sound is due to vibrations of the cardiac walls produced by the rapid filling phase of the ventricles. The fourth sound, of low frequency and amplitude, occurs during atrial systole.

The heart sounds signal the duration of ventricular systole [FIG. II.40]. The characteristics of the sounds are of course clinically important.

A split first sound is not uncommon and merely indicates that, as is the case physiologically, the mitral valve shuts after the tricuspid. A split second sound on the other hand is suggestive of delayed conduction in the right or left branch of the bundle of His.

Murmurs are heard whenever turbulence becomes excessive. Any increase in the velocity of blood flow favours turbulence [see p. 80] and consequently such sounds are heard in strenuous exercise, in anaemia, in thyrotoxicosis, etc. If the valve aperture is narrowed (stenosis) the velocity of flow and hence its turbulence increase. Mitral stenosis causes a diastolic (or presystolic) murmur which converts the first mitral sound into *rrrubb*. Aortic stenosis causes a systolic aortic murmur, *rrup*. Incompetence of the valves produced by disease allows regurgitation of blood and mitral incompetence thus causes a systolic mitral murmur which is described as 'lush' in sound. Aortic valvular incompetence causes a softening and prolongation of the second sound during early diastole. In disease both stenosis and incompetence may coexist.

Ventricular volume

A glass container is fitted with a rubber diaphragm perforated by a central hole to permit the fitting of the ventricles—the rubber diaphragm fits snugly round the A–V ring to provide an air-tight seal. The glass container is connected by tubing to a piston recorder and the changes in ventricular volume during the cardiac cycle are recorded graphically. With the onset of atrial systole, ventricular volume increases. At the beginning of ventricular systole and closure of the A–V valve there follows the period of isometric contraction during which ventricular volume does not of course change. The onset of ejection is accompanied by quite a minor change of ventricular volume—this initial expulsion is mainly having the effect of expanding the ascending aortic wall. Subsequently, however, ventricular volume falls more sharply and then the rate of ejection subsides [FIG. II.40]. The end of ejection occurs simultaneously with the incisura but the volume remains unaltered indicating the isometric relaxation phase, which is terminated by rapid filling of the ventricle. Subsequently, slow filling restores the ventricular volume.

From the pressure and volume curves the mechanical work of the heart [see p. 109] can be computed [FIG. II.42].

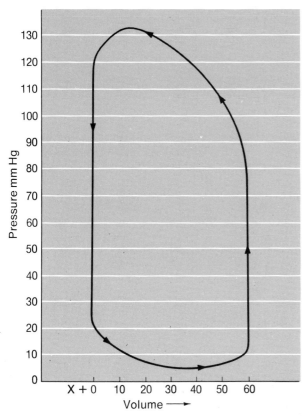

FIG. II.42. Work diagram of the heart, constructed from the pressure and volume curves shown in FIG. II.40. (Folkow, B. and Neil, E. (1971). *Circulation*. Oxford University Press, New York.)

Electrocardiogram

Though discussed separately [p. 112] it is important to correlate the ECG with the mechanical events described above.

The P wave is due to atrial systole and its duration is 0.1 s approximately. It precedes the first wave of the atrial pressure curve. The QRS which represents the invasion of the ventricle by the excitation process consists of Q (which is probably associated with the excitation of the upper interventricular septum), R (excitation of the apical septum and apices of the ventricular walls) and S (the excitation of the more basal parts of the ventricle). The QRS complex is complete just *before* the semilunar valves open. The T wave is complete by the time that the semilunar valves close [FIG. II.40]. The P–R interval lasting 0.13–0.16 s gives an indication of the conduction velocity from atrium to ventricle including that in the bundle of His. A P–R interval of >0.2 s indicates delayed conduction which usually occurs in the bundle of His or its branches.

HISTOLOGICAL STRUCTURE OF HEART MUSCLE

Light microscopy. The myocardium consists of columns of striated muscle fibres 100 µm long and 15 µm broad. The fibres are surrounded by a very rich capillary network. The muscle fibres are arranged in a syncytial fashion. Each fibre consists of an outer membrane, the sarcolemma, surrounding numerous striated myofibrils 1 µm wide which are arranged longitudinally. Numerous mitochondria lie between the myofibrils. Each myofibril, which contains the contractile proteins actin and myosin, is interrupted at intervals of 1.2–2.5 µm by dark lines known as the Z lines. Two Z lines limit longitudinally the functional unit of the myocardium—the sarcomere.

Electron microscopy. This extends our understanding of the structural pattern. The sarcomere consists of alternate light and dark bands. The central band, 1.5 µm long, is dark and contains *myosin* filaments 10 nm in diameter [FIG. II.43]. This central band, known as the A band, is flanked by an I band on each side which appears light and which consists of filaments of the protein actin (5 nm broad).

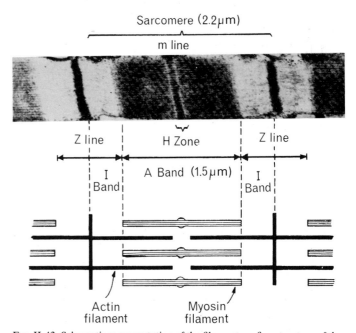

FIG. II.43. Schematic representation of the filamentous fine structure of the sarcomere (below) in relation to the sarcomere band pattern (above). (Spiro, D. and Sonnenblick, E. H. (1964). *Cyclopedia of medicine. Surgery and specialties.* F. A. Davis, Philadelphia.)

Myosin and actin can combine reversibly to form actomyosin. Their interaction is the fundamental basis of muscle contraction. The evidence for this statement is given at length in the section on skeletal muscle.

The sarcolemma is invaginated by a system of transverse tubules which penetrate the sarcomere at the Z line. The lumina of these tubules is continuous with the extracellular fluid. Additionally there is a reticulum of interconnected longitudinal tubules which lie in close contiguity to the myofibrils. At each Z line these longitudinal tubules approximate to but do not anastomose with the transversely disposed invaginations of the sarcolemma. The term 'triad' is used to describe the arrangement of the terminal dilatations of two longitudinal tubules (known as cisternae), one each side of a transverse tubule [FIG. II.44].

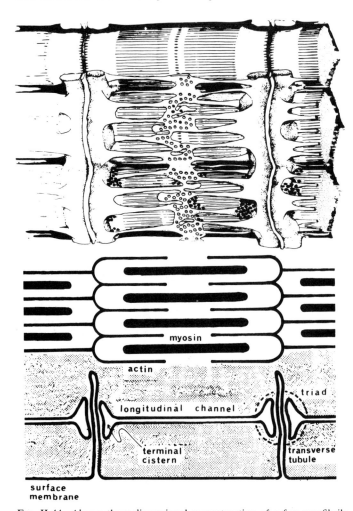

FIG. II.44. *Above*: three-dimensional reconstruction of a few myofibrils running from left to right with endoplasmic reticulum in close contact. Two triads are seen at the upper edge of the picture each with a central T-tubule and two adjacent cisternae (from Peachey, L. (1965)).
Lower: two-dimensional diagram showing connection between fibre surface and the endoplasmic reticulum. (Modified from Weidmann, S. (1967); from Folkow, B. and Neil, E. (1971). *Circulation*. Oxford University Press, New York.)

Depolarization of the sarcolemma causes the development of an electrotonic potential which is conducted through the transverse tubule and then causes depolarization of the adjacent cisternae. This depolarization causes a release of Ca^{2+} and the calcium diffuses into the myofibrils and activates myosin, which then splits ATP, yielding the energy necessary for the formation of the actomyosin complex and for the contraction of the muscle. The calcium ions so

released are then actively transported back into the longitudinal tubules of the sarcoplasmic reticulum and relaxation occurs.

Although the cardiac muscle fibres show a syncytial arrangement when viewed through the light microscope, electron microscope studies show that the membranes bordering each cardiac muscle cell, though apposed, are nevertheless separated by narrow interposed intercellular spaces. This close apposition of the limiting membranes of adjacent muscle cells is responsible for the appearance of the intercalated disc which runs transversely across the cardiac muscle bundles. Heart muscle is *not* a structural syncytium but it behaves as if it were, the reason being that the electrical resistance offered by the intercalated disc is very low. Thus, when one cell is excited the excitation process spreads easily across the intercalated disc to its neighbours.

BIOPHYSICAL ASPECTS OF THE CONTRACTILE FEATURES OF ISOLATED CARDIAC MUSCLE

Muscle may be described in terms of a model which consists of a contractile element, CE, arranged in series with an elastic element, SE, and in parallel with another elastic element—PE [FIG. II.45].

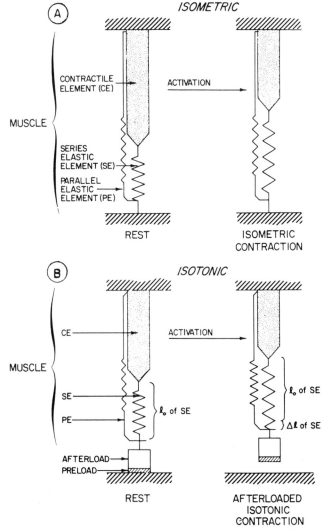

FIG. II.45. Three-component model for muscle (after Hill). 1, Contractile component (CE); 2, elastic component in series with contractile component (SE); and 3, elastic component (PE) in parallel with 1 and 2. *A*: isometric contraction; *B*: isotonic contraction. (Sonnenblick, E. H. (1962). *Fedn Proc.* **21**, 975.)

When the muscle is excited, CE shortens and the force developed thereby stretches SE.

Abbott and Mommaerts (1959) used the papillary muscle from the ventricle as a convenient cardiac muscle preparation in which the muscle fibres were longitudinally disposed in parallel fashion. Their findings on the biophysical features of contractile performance of this preparation have been confirmed and extended by Sonnenblick (1962) and his colleagues.

The muscle preparation is attached to a lever at one end and secured at the other to a tension transducer [FIG. II.46]. Multiple electrodes are placed along the muscle so that electrical stimuli may be simultaneously delivered to almost all the muscle fibres. The initial length of the muscle is adjusted by having it support various small preloads; a micrometer stop above the lever then fixes this resting length. Once this stop has been adjusted, any additional load placed at the lower end on top of the preload weight can have no effect on the resting muscle length. However, the muscle encounters this additional load when it contracts—hence the term 'after load' for this extra weight. The total load suffered by the muscle when it shortens is the preload plus the after load.

When the isotonic lever is fixed, external shortening of the muscle cannot occur and an isometric tension response is recorded. Fundamental results are obtained by studying the isometric force of contraction developed on stimulation of the muscle preparation, starting from different resting lengths.

Resting length—active tension relationships of isometric contraction

FIGURE II.47 shows that the isometric twitch tension developed during stimulation is increased by increasing the resting length of the muscle. This is the basis of Starling's Law of the Heart [p. 101]—that the initial diastolic length of the ventricular fibres determines their force of contraction.

FIGURE II.48 shows that the contractile force developed at any one resting length is increased when noradrenaline is added to the fluid surrounding the muscle in the bath. This effect of the catecholamine is known as a *positive inotropic effect* and as will be seen

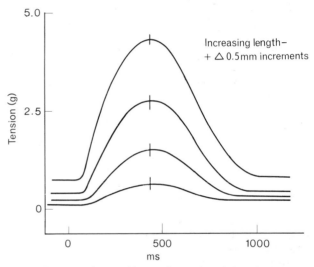

FIG. II.47. Four superimposed isometric muscle twitches obtained from a cat papillary muscle. From below upwards the four twitches were obtained at initial lengths of 8.5, 9.0, 9.5, and 10.0 mm. (Sonnenblick, E. H. (1962). *Am. J. Physiol.* **202**, 931.)

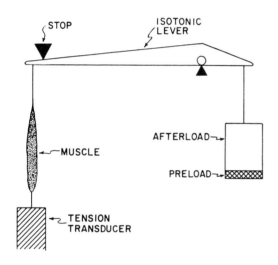

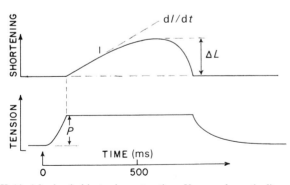

FIG. II.46. Afterloaded isotonic contraction. *Upper*: schematic diagram of apparatus. Cat papillary muscle is attached below to a tension transducer. Free upper end of the muscle is connected to an isotonic lever. Initial length of the muscle is established by a small preload. A stop is then set which keeps initial length of the muscle constant prior to onset of contraction. Loads added to preload are only encountered by the muscle with onset of contraction. Loads added to preload are only encountered by the muscle with onset of contraction. *Lower*: recording of force (tension) and shortening for an afterloaded contraction. With stimulation, force development begins. When force (*P*) equals load, shortening begins at a maximal rate (d*l*/d*t* or *V*) and is maintained for a short period of time. Δ*L* is net shortening of the muscle which equals distance the load is moved. (Sonnenblick, E. H. (1962). *Fedn Proc.* **21**, 975.)

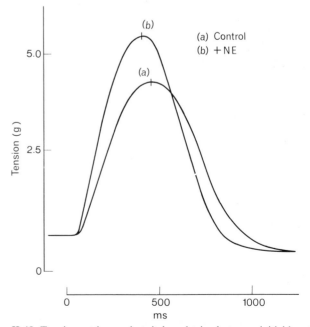

FIG. II.48. Two isometric muscle twitches obtained at same initial length. 'a', control twitch; 'b', twitch obtained when noradrenaline was added to the bathing solution to a concentration of 0.5 μg ml⁻¹. (Sonnenblick, E. H. (1962). *Am. J. Physiol.* **202**, 931.)

[p. 102] is of the greatest importance in influencing the contractile force of the ventricle (and hence the stroke volume) of the heart *in situ* in the intact heart. Until recently the effect of increasing the resting length of cardiac muscle fibres in improving their contractile force and that of positive inotropic influences (e.g. noradranalin) in increasing contractile force developed from the same resting length have been regarded as distinct. This proposition is no longer tenable as will be discussed later [p. 105]. Nevertheless, it is most convenient to discuss the mechanical consequences of these two effects separately to provide a clear understanding of their 'individual' contribution to the performance of cardiac muscle and that of the working heart.

Isotonic recording of papillary muscle

In isotonic recording, the muscle is free to undergo external shortening and our interest is directed to the velocity with which it manages to shorten.

In an isotonic contraction the shortening of CE develops a force which first matches the load and then exceeds it, so the load moves and SE is stretched by a constant force equal to the load and the length of SE henceforth remains constant throughout the contraction.

Hence the velocity of shortening of the muscle is an expression of the performance of CE, independent of SE.

The force–velocity relation of active CE is as shown in Fig. II.49. When the load is zero, velocity is maximal and this is termed V_o. When the load is increased progressively, the velocity of shortening finally becomes zero and the force developed is the maximal isometric force referred to as P_o.

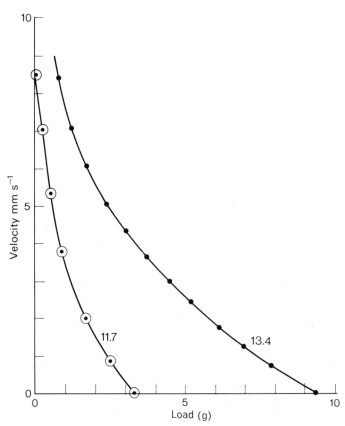

FIG. II.49. Force–velocity curves (papillary cardiac muscle).
⊙—⊙ Initial length of muscle = 11.7 mm
●—● Initial length of muscle = 13.4 mm

Note that when initial length is greater muscle develops a greater P_o but not a greater V_{max}. (After Sonnenblick, E. H. (1962). *Am. J. Physiol.* **202**, 931.)

When the initial length is varied the force–velocity relationship alters; however, although—as might be expected from the previous section—an increase in initial length increases P_o it has no effect on V_{max} [Fig. II.49].

Now, when catecholamines are added to the bath solution the muscle force—velocity curve shows an increase in *both P_o and V_{max}* [Fig. II.50]. Similar effects are produced by raising the calcium ion concentration of the bath fluid.

Thus, cardiac muscle can alter its work and power (rate of working) at any one load and muscle length by nature of its changing force–velocity relationships in different chemical environments.

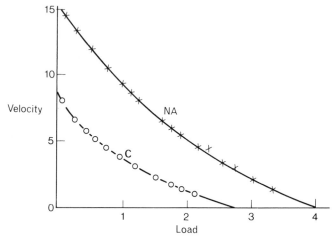

FIG. II.50. Force–velocity curves (cat cardiac papillary muscle). Velocity in mm s^{-1}; load in g.

C = control.
NA = after addition of noradrenalin (0.05 μg ml^{-1}) to the muscle bath. Noradrenalin increases both P_o and V_{max} (positive inotropic action). (After Sonnenblick, E. H. (1962). *Am. J. Physiol.* **202**, 931.)

These results obtained on isolated muscle preparations are of basic importance in understanding the main features of performance shown by the intact heart. It is clear that diastolic filling of the ventricle will affect the initial length of the ventricular fibres and that the stroke volume or beat output of the ventricle will be related to the vigour and velocity of ventricular contraction. The heart–lung preparation allows a rigorous study of the performance of ventricular muscle in circumstances of altered diastolic volume and of altered chemical composition of the blood.

The heart–lung preparation

Introduced by Starling, between 1910–1914, the use of this preparation clarified the important features of ventricular performance.

The chest of the animal is opened during artificial ventilation. Loose ligatures are placed around both venae cavae, and the brachiocephalic artery. The left subclavian artery is tied. A reservoir of blood leads to a cannula which is then tied into the superior vena cava (the inferior vena cava is cannulated and a manometer attached records atrial pressure). A cannula is inserted into the brachiocephalic artery and the ligature tightened. Blood pumped out by the left ventricle is directed via an elasticity chamber to a resistance *R* which consists of a rubber finger-stall with the end cut off, linking the inlet and outlet tubes of a cylindrical glass tube. A side arm of the cylindrical tube allows the pressure outside the finger-stall to be varied by a pump, thus tending to collapse the finger-stall. Blood passes through the lumen of the finger-stall if its pressure is high enough to overcome that outside the finger-stall. The outlet tube leads to a warming coil and thence back to the reservoir [FIG. II.51]. The lungs are artificially ventilated and oxygenate the pulmonary arterial blood. The preparation is effectively 'denervated' because the blood supply to the animal's brain is cut off by tying the cannula into the brachiocephalic artery, so the central neurones quickly perish. However, the peripheral vagal and sympathetic axons remain viable for an hour or two and though not discharging any impulses naturally, they can be artificially excited by electrical stimulation.

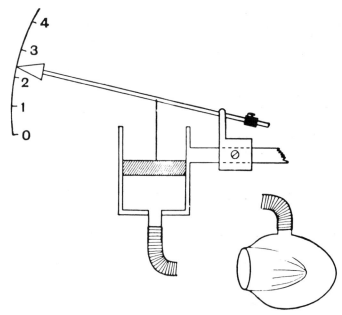

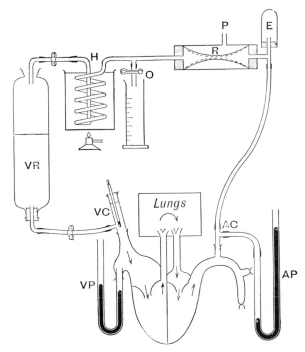

FIG. II.51. Knowlton–Starling heart–lung apparatus (after Hemingway). AC, arterial cannula; E, air-chamber, to produce elasticity; H, heating apparatus; O, outlet for determination of output (when determining output this clip is opened for a given time and outlet to venous reservoir closed); P, to pressure-bottle; R, peripheral resistance (dotted line shows position during increased resistance); VC, venous cannula, VR, venous reservoir; VP, venous pressure (regulated by screw clip from reservoir). (McDowall, R. J. S. (1951). *Handbook of physiology*, 41st edn. John Murray, London.)

Stroke volume of the heart–lung preparation is registered by a cardiometer attached to a piston recorder. The pericardium surrounding the heart is slit and peeled back from the ventricles. The cardiometer [FIG. II.52] is fitted over the heart and the perforated rubber diaphragm is of such size as to fit snugly in the atrioventricular ring, forming an air-tight fitting which is not too restrictive to embarrass the filling of the ventricles from the atria. The pericardium is then drawn over the cardiometer itself and ligated at its apex round the outlet tube. When the ventricles (*both* ventricles) contract, blood is displaced into the pulmonary artery, and the aorta and the artificial circuit and an equal volume of air enters the cardiometer from the piston recorder, with the result that the piston moves down as registered by the lever. Conversely, diastole attended by venous filling of the ventricles displaces air to the piston recorder and the lever moves up. (Note changes in volume registered by the piston–lever system are changes in volume of the blood in the *ventricles* and *not* in the atria, which are separated from the cardiometer by the rubber diaphragm.)

Linden (1968) has published three figures obtained from a heart–lung preparation which summarize some important features of ventricular performance in response to changes of muscle length produced by altering venous filling and in response to peripheral sympathetic stimulation.

FIGURE II.53 shows the cardiometer recording of the ventricular

FIG. II.52. Diagram of cardiometer (bottom right) and piston recorder used to indicate changes in volumes of the ventricles in the heart–lung preparations. (McDowall, R. J. S. (1951). *Handbook of physiology*, 41st edn. John Murray, London.)

stroke volume at a right atrial pressure of some 4 cm H_2O—the *change of volume* cause by the stroke output and therefore the stroke volume is about 8 ml per beat. The rhythmic variations of the cardiometer trace are (1) fast, due to the rate of ventricular beat and (2) slow, due to the variations induced by artificial ventilation of the lungs by a positive pressure pump.

The right atrial pressure is raised (by raising the height of the venous reservoir) at the signal from 4 cm H_2O to 10 cm H_2O. This increased atrial pressure causes an increased inflow into the ventricles which are filled more completely and the end-diastolic volume (upper border of the cardiometer trace) increases, as indeed does the volume after systole (the end-systolic volume). However, within 10–15 seconds it is clear that a new steady state has been established in which, at a greater diastolic volume, the stroke volume (the vertical distance on the arc of the lever trace between the top and bottom of the cardiometer record) is greater than before. Thus, an increase in the end-diastolic volume of the ventricles and therefore of the initial resting length of their fibres is accompanied by an

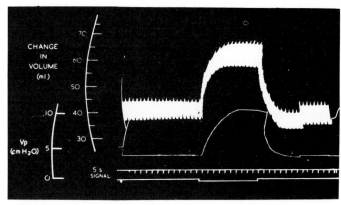

FIG. II.53. Example of the Starling mechanism. From above downwards: cardiometer trace, indicating change in volume in both ventricles; venous pressure, from right atrium; time marker, 5 seconds; signal marker. Both vagus nerves and both ansae subclaviae were sectioned. During the signal the inflow to the right atrium was increased suddenly by a set amount. (Linden, R. J. (1968). *Anaesthesia* **23**, 566.)

increased stroke volume. This is Starling's Law—*that the force of contraction of the ventricular muscle fibre is proportional to its initial resting length*. This result is in harmony with that already described on the response to stimulation of isolated muscle strips to an increase in their initial resting lengths.

FIGURE II.54 shows the behaviour of the *same preparation* when the peripheral cardiac sympathetic nerves were stimulated (by applying the stimuli to the ansa subclavia from which these nerves are distributed to the heart). The position of the cardiometer level was adjusted on the drum, but, as can be seen, the stroke volume at a mean right atrial pressure of 4 cm H_2O was as before, about 8 ml. On stimulating the sympathetic supply to the heart and thereby provoking the release of noradrenalin from the postganglionic nerve endings, there results a striking fall of end-diastolic and end-systolic volume. However, within 10–15 seconds a more steady situation is reached, in which although the end-diastolic volume is reduced, that of the end-systolic volume is reduced even more and thus the difference between these two, which is the stroke volume, is greater than during the control conditions. Thus, the stroke volume is slightly greater during sympathetic stimulation, although the end-diastolic volume, and hence the initial length of the ventricular muscle fibres, is less. This effect of sympathetic stimulation is due to a positive inotropic influence of noradrenalin on the cardiac muscle fibres.

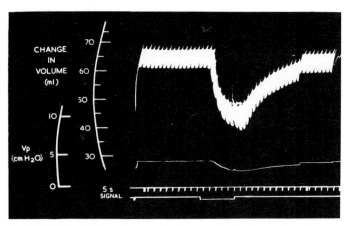

FIG. II.54. Stimulation of sympathetic nerves to the heart. Records as in previous figure. During the signal the left ansa subclavia was stimulated supramaximally at 4 pulses per s. (Linden, R. J. (1968). *Anaesthesia* **23**, 566.)

(It is interesting to note that the right atrial pressure falls during sympathetic stimulation, to return to control values when stimulation ceases. This fall of atrial pressure is *not* responsible for the reduced end-diastolic volume. It is at first sight paradoxical, for the venous reservoir filling the atria is held at a constant height. It must be due to an increased suction of blood from the atria by the vigorously contracting ventricles—the oft-disbelieved force of *vis a fronte* [see p. 93].)

To summarize then, the decrease in end-diastolic volume coupled with a (slightly) increased stroke volume must mean that the ventricles are emptying themselves more completely as they beat and the reserve blood in the heart has been drawn upon. At this juncture it must be stressed that the *normal* heart never expels the whole of end-diastolic blood content when it beats. Just as there is a functional residual capacity in the lungs, consisting of a residual volume of alveolar gas which cannot be expelled, plus a volume of one litre or so which can be expelled by a vigorous expiration after the normal quiet expiration has been completed [p. 157], so has the heart a residual reserve of blood. The only difference is that the volume of blood which *cannot* be expelled by the most vigorous systolic contraction is very small so that a heart which is stimulated

to beat more vigorously by noradrenalin infusion may in these *special circumstances* following haemorrhage almost completely empty itself and the forcible coaptation of the ventricular walls towards the end of systole may cause subendocardial petechiae.

In more natural circumstances, however, the heart beating in the circulation of resting man contains an appreciable volume of blood when systole is complete and the stroke volume has been expelled. A more vigorous systole draws upon and reduces this reserve end-systolic volume.

FIGURE II.55 shows the combined effects of sympathetic stimulation and an increase in the filling pressure of the ventricles on the ventricular stroke volume. After a control period (beat output = 8 ml, atrial pressure 4 cm H_2O) sympathetic stimulation commences at signal A and *is continued* until the end of signal C. As before the end-diastolic and end-systolic volumes decrease to reach an almost steady state at which the stroke volume is slightly increased (the reduction of the end-systolic volume is greater than that of the end-diastolic volume) though the atrial pressure has fallen. At this point (signal B) the venous reservoir is quickly raised to restore the atrial pressure to its control value and this increased reservoir height is maintained until signal C. The ventricular volume increases in response to the increased filling which this higher venous pressure secures and the stroke volume is obviously increased compared with that seen in the control period. Thus, the ventricles beating under the combined influences of increased venous filling (Starling effect) and sympathetic stimulation put out much more blood per stroke at the same (or similar) end-diastolic volume and atrial pressure.

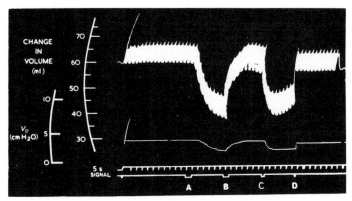

FIG. II.55. Combined effect of stimulation of sympathetic nerves and Starling mechanism. Records as in previous figures.
At signal A, the left ansa subclavia was stimulated and this stimulation continued until the signal mark C.
At signal B the right atrial inflow was increased (by hoisting the reservoir).
At signal C both extra inflow and sympathetic stimulation were stopped.
At signal D the drum was stopped for 2 minutes. (Linden, R. J. (1968). *Anaesthesia* **23**, 566.)

At signal C sympathetic stimulation is stopped and the reservoir is lowered to its control position. The ventricular volume falls not to its control value, but to a much smaller value and the atrial pressure also falls. During the ensuing 20 seconds it can be seen that the end-diastolic volume is slowly rising and that the right atrial pressure is also climbing, albeit at a much slower rate. These 'after-effects' are probably due to there being some residual catecholamine in the ventricular tissue which is slowly removed by a combination of uptake by the nerve endings and metabolism.

At signal D the drum is stopped for 2 minutes and on restarting the record shows that both the ventricular volume and the atrial pressure have more nearly returned to the values of the initial control period.

These elegant experiments show quite simply that ventricular performance per beat can be strikingly influenced either by increasing the initial length of the cardiac fibres or by increasing the vigour of contraction of the fibres at any given length by exciting the sympathetic nerves supplying the ventricles. The combination of increased venous filling and the positive inotropic effect of sympathetic stimulation may secure a greatly increased stroke volume from the same end-diastolic volume—the increased 'wringing out' of the blood from the ventricles caused by the sympathetic inotropic influence on ventricular contraction would lower the end-diastolic volume unless an increase of venous filling restored the *status quo*. Without a clear appreciation of these factors, one might consider that the demonstration of an increased stroke volume (as occurs in exercise in the intact animal or man), without an increase in right atrial pressure or an increase in cardiac volume, would seem to abnegate Starling's Law. However, Starling's Law is basic to the performance of the heart muscle and the effect of added positive inotropic influences merely modifies the response curve of the ventricular force developed by contraction from any one initial ventricular muscle length. Such a conclusion is supported from experimental results obtained which are detailed below.

Families of Starling curves

Starling initially expressed the ventricular performance of the heart–lung preparation by plotting stroke volume against right atrial pressure. More recently Sarnoff and his colleagues (see Sarnoff and Mitchell 1962) have used left ventricular stroke work and the left ventricular end-diastolic pressure as co-ordinates and this graphic mode of presentation will be used here.

FIGURE II.56 shows (a) the experimental points obtained in control conditions of measuring the stroke work against the LVEDP and (b) those determined during sympathetic stimulation. The figure shows in quantitative terms the same features described by the actual experimental record of FIG. II.55. At a given LVEDP the stroke work of the left ventricle is greatly increased by sympathetic stimulation. (We infer that the end-diastolic pressure in the ventricle is related to the stretch and therefore the initial length of the ventricular muscle fibres.) It will be noted that the relationship between stroke work and LVEDP is only linear over part of the range. Once the 'filling pressure' becomes high (10–12 cm H₂O) there is further increase in stroke work which reaches a plateau; this plateau is at a much lower level in the absence of positive inotropic effects.

The Starling curve obtained during sympathetic stimulation was that recorded during *maximal* sympathetic excitation. When a series of progressively less intense stimuli were employed, each particular degree of sympathetic stimulation influenced the position of the Starling curve plotted from the relevant experimental points. Thus, depending on the degree of sympathetic engagement, a *family* of Starling curves could be constructed from the experimental results (Sarnoff 1955).

One of the great problems of investigating the biophysical performance of the ventricles more rigorously is to obtain a satisfactory measurement of the length of the ventricular muscle fibres. Starling fully realized that cardiac muscle, just as skeletal muscle, contracted with a force proportional to its initial length, but the geometric complexity of the arrangement of the ventricular muscle fibres led him to use the mean right atrial pressure as the reference point, arguing that this gave a measure of the 'filling pressure' which stretched the ventricle to its initial length in diastole. The modern use of LVEDP is more accurate, but even this remains one step removed from the measurement of length which is really required. For instance, how are we to decide whether or not the *distensibility* of the ventricle is increased by sympathetic stimulation, for if it *were* increased any given LVEDP would cause greater ventricular filling and hence a greater end-diastolic length. To surmount this difficulty, Linden and Mitchell (1960) developed a lever–potentiometer

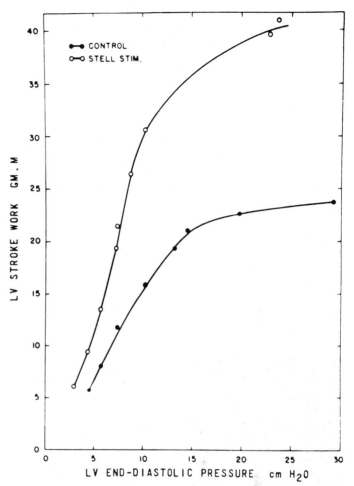

FIG. II.56. Relationship between left ventricular stroke work and LVEDP before (closed circles) and during (open circles) stimulation of the left stellate ganglion. Both vagi cut; heart rate constant at 171 min⁻¹. (Sarnoff, S. J. *et al.* (1960) *Circulation Res.* **8**, 1100. By permission of the American Heart Association, Inc.)

system to record the changes in length of a segment of the myocardium. (The ventricular wall is impaled by two needles and the lever-potentiometer system records the distance between them.) The MSL (myocardial segmental length) is taken as representative of that of the remainder of the ventricular myocardium.

Alteration of the height of the venous reservoir changes the diastolic MSL as shown in FIG. II.57. Stimulation of the cardiac sympathetic did not affect the relationship between MSL and LVEDP obtained in control conditions and we can conclude that excitation of the sympathetic nerves does not increase the distensibility of the muscle. However, as Linden (1963) points out, when the ventricle is filling rapidly the motive force of the ventricular wall may be described in terms not only of the static elastic factor of the ventricular muscle, but additionally in those of viscous damping (in which the velocity is a component) and inertia (mass × acceleration). Ventricular relaxation is more rapid in early diastole with increased sympathetic stimulation [FIG. II.58] which indicates that although such stimulation does not increase the static factor of 'stretchability' it lowers the inertial and viscous factors.

Other positive inotropic effects on ventricular performance

Although sympathetic stimulation and similarly, though of less quantitative importance in the intact body, the concentration of catecholamines (released by the adrenal medulla) in the blood provides far and away the most important positive inotropic effect on the ventricles, other influences do contribute. Calcium ions can

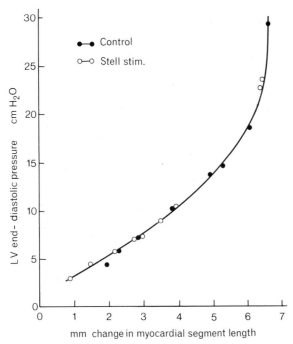

FIG. II.57. Relationship between LVEDP and the changes in length of a segment of the left ventricular myocardium before (closed circles) and during (open circles) stimulation of the left stellate ganglion. (Sarnoff, S. J. *et al.* (1960). *Circulation Res.* **8**, 1100. By permission of the American Heart Association, Inc.)

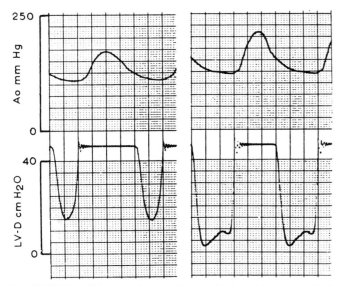

FIG. II.58. The effect of an increased sympathetic activity on ventricular 'filling time'. From above down: aortic blood pressure; pressure in the ventricle recorded only during diastole. Left panel, control. Right panel, during stimulation of the left stellate ganglion. The heart rate was held constant throughout at 180 beats per minute by stimulating the left atrium. (Linden, R. J. (1965). *Sci. Basis Med.*, p. 164. London.)

be shown to cause a positive inotropic effect but the effect is pharmacological rather than physiological for, thanks to humoral control, the $[Ca^{2+}]$ of the blood is kept remarkably constant. An increase in the rate of the heart causes an increased contractility of the ventricle which develops progressively over a few beats. This 'staircase phenomenon' or *treppe* is of little quantitative importance. Finally a sudden increase of aortic pressure in the heart–lung preparation was shown by Anrep to increase the contractility of the ventricle, but again this 'Anrep effect' is of minor importance. Linden (1968) has investigated these several influences of positive

inotropic nature on the performance of the ventricle in the intact dog. He and his colleagues have shown that the maximal rate of change of pressure in the ventricle during its isometric contraction gives an excellent indication of the ventricular inotropic response (Furnival, Linden, and Snow 1968). FIGURE II.59 shows how notably stimulation of the left ansa subclavia (supplying most of the ventricular sympathetic fibres) increases dp/dt_{max}.

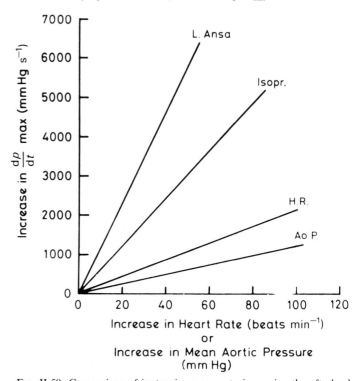

FIG. II.59. Comparison of inotropic response to increasing the afterload (mean aortic pressure, Ao.P.) or the frequency of contraction (H.R.) with the inotropic response to increasing stimulation of the left ansa subclavia or infusion of isoprenaline. (Linden, R. J. (1968). *Anaesthesia* **23**, 566.)

When the separate influences of the various factors causing positive inotropic effects are assessed [FIG. II.59] it can be seen that the 'treppe' effect and the Anrep effect are much less important than that of sympathetic stimulation.

The influence of the vagi on ventricular performance

The vagi do not supply motor fibres to the ventricles in the mammal, although they do of course to the atria and the s.a. node. The dramatic effects of vagal stimulation on cardiac performance are exercised on the pacemaker and on the force of atrial contraction and are *not* due to a direct effect on the ventricles.

The influence of the vagi on ventricular contractility defined by the ventricular function curves (i.e. the graph of stroke work against LVEDP) can only be assessed by artificially pacing the heart via electrodes implanted into the wall of the atrium, for otherwise the heart may stop if subjected to strong vagal stimulation.

FIGURE II.60 shows that vagal stimulation of the *paced* heart has no effect on the ventricular function curve. MSL/LVEDP curves also show no effect of the vagi on the distensibility of the ventricle.

The depression of atrial contractility caused by the vagi may slightly diminish that phase of ventricular filling which is due to atrial contraction, but the reduction of stroke volume due to this factor is of trivial degree.

Heart rate and cardiac output in the heart–lung preparation

Tachycardia *per se*, unaccompanied by positive inotropic effects does not increase the output per minute of the heart in the heart–

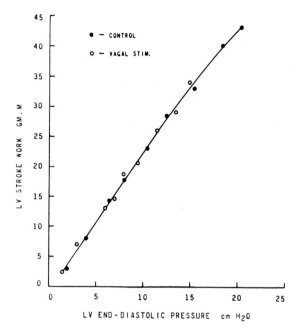

FIG. II.60. Heart rate held constant at 187 min⁻¹. Vagal stimulation causes no change in ventricular contractility. (Sarnoff, S. J. *et al.* (1960). *Circulation Res.* **8**, 1100. By permission of the American Heart Association, Inc.)

lung preparation. An increase in heart rate inevitably means a reduction in the diastolic period and it is in this phase only that the ventricle is filled. Hence, really fast hearts in resting man, such as seen in acute attacks of paroxysmal tachycardia [p. 118], where the heart rate rises above 200 min⁻¹, usually show the signs of incipient failure, for the right atrial pressure backs up, the blood being unable to get into the ventricle in adequate amounts to sustain a forward activity of ventricular stroke volume. Heart rate increases of lesser order if unaccompanied by an improvement in the force of each beat do not much improve the total output/minute, for although more beat outputs are ejected per minute, because of the infringement of the diastolic filling period, the stroke volume delivered per beat is less.

It is an entirely different matter when the increase in rate is accompanied by a positive inotropic influence. Thus sympathetic stimulation causes both an increase in rate and force of ventricular contraction, but unless the venous return to the heart is increased the increase of cardiac output cannot be maintained because the heart 'uses' up its end-systolic reserve. During muscular exercise the massaging action of the muscles on the veins secures a rapid return of blood to the heart beating rapidly and vigorously due to sympathetic stimulation.

Sympathetic stimulation in a 'paced' heart decreases the LVEDP and doubles the diastolic period because systole is shortened [FIG. II.58]. When the heart rate is allowed to increase sympathetic stimulation causes a shorter systole, with relatively longer proportion of the total length of the cardiac cycle available for diastolic filling than if the tachycardia induced had been unaccompanied by the positive inotropic influence of the catecholamine liberated.

This increased vigour of ventricular contraction which accompanies the tachycardia of sympathetic stimulation causes suction of blood into the atria and ventricles—*vis a fronte*. The best proof of the existence of this suction force was provided by Bloom (1955) using an *excised* rat heart beating in a beaker containing warm Ringer solution. Each beat caused an expulsion of fluid through the aortic stump and the heart moved through the solution as if jet-propelled. The atrial wall was drawn into the mitral orifice during each diastole. The energy for ventricular filling must have been derived from the elastic recoil of ventricular relaxation for no hydrostatic pressure gradient existed between the interior and exterior of the heart.

Excitation–contraction coupling

Until recently the two factors of 'diastolic' length and extrinsic inotropic influences which modify myocardial performance were regarded as independent regulatory mechanisms. Modern work (see Jewell (1978) for references) has shown that they are not. An appreciation of this view requires an account of the mechanism of excitation contraction coupling in cardiac muscle.

Depolarization of the sarcolemma by the cardiac action potential is conducted radially along the transverse tubules to the centre of the muscle fibre. There it causes the release of calcium from the terminal portions of the longitudinal tubules of the sarcoplasmic reticulum; (these terminal parts contain a high concentration of calcium in the resting state). The 'resting' calcium concentration of the myofibril sarcoplasm is normally only about 10^{-7}M and the release of Ca^{2+} from the longitudinal SR raises it approximately one-hundred-fold. In the resting state troponin exerts an inhibitory effect on actin via tropomyosin but troponin has an avidity for calcium ions and binds calcium to its saturation point when the sarcoplasmic Ca^{2+} concentration rises to 10^{-5}M. As a result the inhibitory influence of the troponin–tropomyosin complex on actin is removed. Simultaneously the rise in calcium concentration in the sarcoplasm activates prosthetic groups of the myosin fibrils which act as an enzyme (ATPase) catalysing the breakdown of ATP. This provides the energy for contraction of the actomyosin complex and for the muscle fibrils as a whole by activating the sliding filament system. There follows an active re-uptake of calcium ions by the sarcoplasmic reticulum; as the myofibrillar Ca^{2+} concentration falls, relaxation of the myofibril occurs. FIGURE II.61 summarizes these changes.

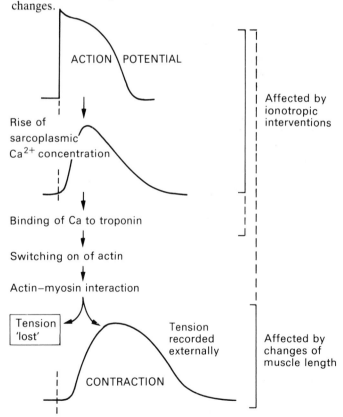

FIG. II.61. Schematic diagram to show the main events in excitation–contraction coupling in cardiac muscle. The solid brackets on the right indicate events that are generally considered to be influenced by inotropic interventions and changes of muscle length. These have been extended with broken lines to take account of the recent discoveries described in the text. (Redrawn from Jewell, B. R. (1978). In *Developments in cardiovascular medicine* (ed. C. J. Dickinson and J. Marks) pp. 129–44. MTP Press, Lancaster.)

Using the calcium ion-sensitive bioluminescent protein aequorin (injected into cells of frog's atrial trabeculae) and simultaneously recording tension developed of the thin unbranched trabecular strips, Allen and Blinks (1978) showed conclusively that the calcium involved in activation enters the muscle cell from the extracellular space preceding the onset of tension by the muscle [Fig II. 62].

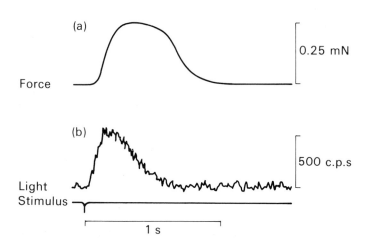

Fig. II.62. Mechanical (a) and luminescent responses (b) from an aequorin-injected atrial trabecula of *R. pipiens*. Isometric contractions at 0.5 Hz; 21 °C (128 sweeps averaged). Light signals are calibrated in photon counts per second (c.p.s.). (Redrawn from Allen, D. G. and Blinks, J. R. (1978). *Nature, Lond.* **273**, 509–13.)

The removal or disruption of the sarcolemma either by microdissection or by chemical techniques (using detergents such as Triton X or chelating agents such as EDTA) produces the so-called 'skinned fibre'. Such preparations can be induced to contract by raising the extracellular concentration of Ca^{2+} from a value of $10^{-7}M$ to $10^{-5}M$. The relative tension produced can be conveniently plotted against pCa where pCa is the negative logarithm to the base 10 of $[Ca^{2+}]$. The curve obtained is sigmoid in form [Fig. II. 63] and supposedly indicates the binding of calcium ions by troponin which reaches saturation at $10^{-5}M$ (pCa 5).

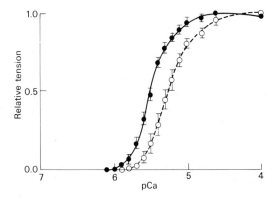

Fig. II.63. Tension-pCa curve showing how the tensions produced in tonic contractions of 'skinned' preparations of cat ventricular muscle varies with the bathing Ca^{2+} concentration. Tensions have been expressed as a percentage of the maximum value observed, and the Ca^{2+} concentration is given as pCa where pCa = $Log_{10} Ca^{2+}$. (Redrawn from Brandt, P. W. and Hibberd, M. G. (1976). *J. Physiol., Lond.* **258**, 76P.)

Hibberd and Jewell (1979) have shown that the contractile mechanism of skinned muscle is more sensitive to changes of pCa when the muscle is stretched. Measuring sarcomeric length they showed that the curve represented by Figure II.63 moved to the left when the resting length of the sarcomere was increased from 1.9 μm to 2.4 μm.

Ter Keurs *et al.* (1980) using intact thin ventricular laminae have studied the effects of different calcium concentrations on the relations between tension and sarcomere length derived from contractions at constant muscle length. Their results are consonant with the proposition that cardiac muscle length influences contractile performance by affecting excitation–contractile coupling.

If then the degree of activation of the contractile complex is dependent upon the amount of calcium bound by troponin then length-dependence of activation could be due to effects of muscle length on either the affinity of troponin for calcium or the extent of rise in the sarcoplasmic calcium concentration induced by excitation of the cell. Fabiato and Fabiato (1975) using skinned ventricular fibres obtained results suggesting that changes of length influenced the amount of calcium released from the sarcoplasmic reticular stores when contraction was induced. Allen and Blinks (1978) noted that the amplitude of the aequorin signal increased with extracellular Ca^{2+} but decreased with stretch and suggested that the prime influence of stretch on cardiac contractility might be attributable to a change in the sensitivity of the contractile apparatus to Ca^{2+} as has been described by Endo (1972) in skeletal muscle, and by Fabiato and Fabiato in skinned ventricular fibres. Allen and Kurihara (1979) noted that there was no difference in peak aequorin light emission when the resting length of ventricular fibres was altered and isometric contractions were provoked but stressed that the falling phase of light emission was shorter during a contraction occurring from a greater resting length. They concluded that changes of muscle length affected the calcium sequestering mechanism. Allen and Kurihara (1981) later obtained results which indicated the binding constant of troponin for calcium ions increased with the number of detached cross bridges between actin and myosin.

The two factors—amount of Ca^{2+} released and the efficacy of the sequestration mechanism for mopping up the calcium ions released—may operate in concert in inducing the changes of contractile force provoked by alterations of 'diastolic length'. Whatever be the final conclusion, the evidence discussed shows clearly that length–tension and 'inotropic' effects such as those induced by catecholamines may have a common origin. There is a wealth of evidence that the catecholamines strikingly increase the inward current carried by calcium ions during contraction of cardiac muscle.

REFERENCES

Abbot, B. C. and Mommaerts, W. F. H. M. (1959). *J. gen. Physiol.* **42**, 533.
Allen, D. G. and Blinks, J. R. (1978). *Nature, Lond.* **273**, 509.
—— and Kurihara, S. (1979). *J. Physiol., Lond.* **292**, 68P.
—— (1981). *J. Physiol., Lond.* **310**, 75P.
Baker, P. F. (1978). *Prog. Biophys. molec. Biol.* **24**, 177.
Blinks, J. R., Rudel, R., and Taylor, S. R. (1978). *J. Physiol., Lond.* **277**, 291.
Endo, M. (1972). *Nature New Biol.* **237**, 211.
—— (1977). *Physiol. Rev.* **57**, 71.
Fabiato, A. and Fabiato, F. (1975). *Nature, Lond.* **256**, 54.
—— —— (1977). *Circulation Res.* **40**, 119.
Hibberd, M. G. and Jewell, B. R. (1979). *J. Physiol., Lond.* **290**, 30P.
Jewell, B. R. (1977). *Circulation Res.* **40**, 119.
—— (1978). In *Developments in cardiovascular medicine* (ed. C. J. Dickinson and J. Marks) pp. 129–44. MTP, Lancaster.
Lakatta, E. G. and Jewell, B. R. (1977). *Circulation Res.* **40**, 251.
Linder, R. J. (1965). *Scientific basis of medicine. Annual Reviews*, p. 64.
—— and Snow, H. M. (1974). In *Recent advances in physiology*, 9th edn (ed. R. J. Linden). Churchill Livingstone, Edinburgh.

SONNENBLICK, E. H. and SKELTON, C. L. (1974). *Circulation Res.* **35**, 517.
STARLING, E. H. (1918). *The Linacre lectures on the law of the heart.* London.
TER KEURS, H. E. D. J., RIJNSBURGER, W. H., VAN HEUNINGEN. R., and NAGELSMIT, M. J. (1980). *Circulation Res.* **46**, 703.

Cardiac performance during exercise

An ordinarily fit young man can increase his cardiac output in maximal exercise from 5.5 l min^{-1} to 25 l min^{-1}. His heart rate rises from the resting level of 50–75 min^{-1} to 180 min^{-1}. His initial stroke volume rises from, say, 80 ml (HR = 70) to about 140 ml. How are these changes achieved?

They are achieved by a combination of sympathetic inotropic and chronotropic influences on contractile force of the ventricle and the rate of ejection respectively, coupled with a Starling mechanism of increased venous return (secured by the muscle pump and by venoconstriction which reduces the venous capacity) which *tends* to increase the LVEDP and hence to increase the initial length of the ventricular fibres. The two mechanisms act in concert and whether the ventricular volume at end-diastole increases, decreases or remains the same respectively depends on: 1. whether the sympathetic inotropic and chronotropic effects at first lag behind the influence exerted on ventricular filling by the increased venous return; 2. whether the sympathetic effects at first 'get ahead' of the influence of the boosted venous return; or 3. whether they exactly balance throughout. It should be obvious that the two factors must reach a steady-state equilibrium during exercise of some minutes' duration, otherwise the heart will progressively dilate or will progressively wring itself out. The interesting phase is that which occurs initially.

Starling (1918) first stated 'if a man starts to run his muscular movements pump more blood into the heart. As a result the heart is overfilled. Its volume, both in systole and diastole, enlarge progressively until by the lengthening of the muscle fibres so much more active surfaces are brought into play within the fibres that the energy of the contraction becomes sufficient to drive on into the aorta during each systole, the largely increased volume of blood entering the heart from the veins during diastole.' Such a statement *by itself* pays no attention to the influence of the sympathetic innervation, but a later quotation from Starling's lecture to The Royal Army Medical Corps (1920) indicates that he was aware that other factors could influence cardiac size, although the reference to them is rather vague: 'In studying the reactions of the isolated heart, dilatation of the heart seems to be the only mechanism of the unfailing response of this organ to any increase in the demands made upon it. But the effort of throwing this organ into the circle of control by the central nervous system is that it is kept in rest or activity in an equable condition and the dilatation, which was so marked a condition of its reaction when isolated, is reduced to such small dimensions in the heart, reined in and controlled by the cardiac centres and helped by the correlated changes in other organs, that it becomes imperceptible in the intact animal and is not revealed, for instance, by any radiographic study of the heart during exercise.'

The measurement of heart size at rest or in exercise is extremely difficult. Liljestrand, Lysholm, and Nylin (1938) used X-ray techniques and took anteroposterior and lateral photographs simultaneously of subjects at rest and exercise. They calculated stroke volumes at rest and in exercise and claimed that the stroke volume rose from 44 ml to 78 ml. Both these figures are probably ridiculously low, judging by more direct measurements made with established techniques (dye output measurements, etc.). They found that exercise increased the volume of the heart from 581 ml to 673 ml during hard work (1260 kg m min^{-1} on the bicycle ergometer). This is an increase of 15 per cent, but of course, which part

of the heart—right side, left side, atria or ventricles—was quite unknown. As Linden (1963) has said, a doubling of stroke volume would only require a 25 per cent increase in diameter and this degree of accuracy cannot be obtained with a single X-ray photograph. It is doubtful whether it could be obtained with two simultaneous X-rays at right angles.

A whole gamut of methods has been used in dogs. Rushmer (1961) particularly has introduced a whole series of new techniques into the study of cardiac performance. Inductance gauges placed within the ventricle to measure a diameter of the ventricle, and strain gauges wound round the ventricle to measure circumference have been chronically implanted during thoracotomy (and if necessary cardiotomy) followed by surgical restitution of the heart and chest and recovery. Electronic recording techniques then allow beat to beat measurements of the parameter chosen as the change from rest to exercise occurs. The sonar technique has been employed—transducers of barium titanate crystals are sewn to opposite walls of the ventricle to serve respectively as transmitter and receiver of pulsed ultrasonic (3 megacycles s^{-1}) sound waves and the distance apart is recorded electronically.

Finally, electromagnetic flowmeters have been implanted round the aorta and give a beat to beat record from which the cardiac output per minute is computed.

It is not uncommon to find that the exercising dog increases its cardiac output mainly by increasing its rate and to a very minor extent by increasing its stroke volume. Thus, in one experiment the heart rate increased from 95 (control) to 285 approximately and the stroke volume rose only from 36.2 ml to approximately 40 ml during heavy exercise. The rise of cardiac output from approximately 3.5 l min^{-1} at rest to about 11 l min^{-1} was almost secured by an increase in rate, stroke volume increasing little. These findings, however, perhaps obscure the *extraordinary* fact that the stroke volume is even maintained, never mind increased slightly, when we consider the enormous shortening of the diastolic phase which such heart rates entail. When one considers the additional fact that the coronary flow to the left ventricle is mechanically impeded during systole it is remarkable that the pump can perform so adequately at these very high rates without any sign of heart failure. It is worth while remembering that though these dogs are 'fit' in the veterinary medical sense of the word, their physical performance and cardiac responses are unlikely to be improved by thoracotomy, the implantation of flow meters hither and thither—in short the trauma of a major operation some few months previously.

The ability of these dog hearts to maintain and even to increase their stroke volume is not attended by any rise of their central venous pressure. As maintenance of stroke volume for 3–4 minutes must entail a balance of venous return and stroke output, one must infer that the influence of the sympathetic discharge, besides improving the inotropic performance of the contracting ventricle must also have improved the priming of the ventricle during the truncated diastolic filling period. Sympathetic discharge not only increases the force of atrial contraction, but also lessens the inertial and viscous resistance to diastolic filling of the heart. The heart fills much more quickly, but *not* because of a raised central venous pressure. *Vis a fronte* presumably contributes more markedly to ventricular filling in exercise than it does at rest.

The inductance gauge has provided beat to beat evidence of the changes in *diameter* of the heart which occur when exercise is commenced on a treadmill. The diameter decreases sharply with the beginning of exercise and then returns to something similar to its control value. Two uncertainties exist: 1. the sudden commencement of the treadmill often causes a scrabbling of feet before the dog matches the speed of the track and this may cause a 'startle' reaction with an unduly heavy barrage of sympathetic impulses; 2. the 'diameter' measured is in only one plane and inferences of the changes in stroke volume from the change in the diameter are not necessarily valid.

The wealth of data accumulated about the stroke volume changes in the dog during exercise is not matched by that provided by studies in man; moreover the changes in the size of the heart are even less documented. There are a few series of measurements of cardiac output changes in exercising man and from the total output and the heart rate the stroke volume can of course be calculated, but there have been objections (e.g. Linden 1963) to the conditions of the measurements. Thus, Asmussen and Nielsen determined the cardiac output of men first at rest and then exercising at 1260 kg m min⁻¹. However, the measurements at rest were made with the subjects in the supine position—in which heart size and stroke volume are greater than those of man quietly standing, whereas the exercise measurements were made with them in the upright position. Thus, their finding that there was no increase in the stroke volume in exercise (during which the cardiac output increased from 6.4 l min⁻¹ (rest) to 20.9 l min⁻¹ is not above reproach (see Linden 1963). Similarly, the studies of Donald *et al.* (1955) were conducted on subjects lying flat on a couch throughout, first at rest and then pedalling a cycle wheel. As stated, the end-diastolic volume of the ventricles in a supine subject is larger, as is the stroke volume, than in subjects standing quietly.

Reviewing the problem, Linden (1965) gave sample figures for the situation of man at rest and exercising maximally, standing in both cases:

	Resting	*Exercising*
Oxygen uptake ml min⁻¹	400	3000–3500
Cardiac output l min⁻¹	5	25
Heart rate	70	180
Stroke volume	70	140
Residual (end-systolic volume) (ml)	75	40
End-diastolic volume (ml)	145	180
Cycle time (s)	0.85	0.33
Ventricular systole (s)	0.3	0.2
Ventricular diastole (s)	0.55	0.13

In resting man the heart spends 21.5 s in systole and 38.5 s in diastole every minute and puts out 5 l min⁻¹. In maximal exercise the heart spends 36 seconds in systole and only 24 seconds in diastole during each minute and yet puts out 25 litres of blood. Though the ejection performance is impressive the transfering of 25 litres of blood from the venous side in this shortened diastolic period is indeed remarkable.

Linden (1965) has concluded that the increased stroke volume of maximal exercise is derived partly from an encroachment upon the end-systolic reserve volume (due to sympathetic inotropic effects) and partly from a 'Starling effect' of an increased end-diastolic volume secured by increased venous return. Both mechanisms discussed in the sections on isolated cardiac muscle and on the heart–lung preparation contribute to the performance of the heart in intact man.

The figures given in the table are of course only sample values. In maximal exercise, physique and training can increase cardiac performance to levels of 35 or even 40 l min⁻¹. Such maximal performances, however, are only achieved by international class athletes. Such trained athletes have much larger end-systolic reserves than do ordinarily fit young adults—professional cyclists for instance, who probably call upon their end-systolic reserve in a sprint to a greater extent than any other performers, have end-systolic volumes of the order of 180 ml and their estimated heart weight is 500 g, instead of 300 g. Characteristic of these highly trained individuals is a heart rate at rest which is far slower (perhaps 40 min⁻¹) than that of the untrained but moderately fit young adult. Training, then, invariably lowers the heart rate, but the reason for this is obscure. The maximal steady heart rate which can be achieved in exercise by trained or untrained individuals is much the same—say 180 min⁻¹. The trained man's ability to reach a cardiac output of,

say, 36 l min⁻¹ depends on his ability to eject a stroke volume of 200 ml—a 'tidal' volume which exceeds the total *end-diastolic volume* (180 ml) of the man who can only summon up a cardiac output of 25 l min⁻¹. The professional cyclist has drawn upon his huge end-systolic reserve and his more muscular left ventricle can eject 200 ml of blood in 0.2 s!

Moderate exercise can be achieved by the trained athlete with quite trivial increases of heart rate—the untrained man with much less end-systolic reserve begins to encroach upon his heart rate reserve much sooner and may show little evidence of an increased stroke volume in moderate exercise.

It is obvious that the output of the right ventricle must be balanced, within a beat or two, with that of the left ventricle. Hamilton (1955) pointed out that the Starling mechanism forms the basis for preserving this balance which 'must be exact and since the two ventricles are subject to the same hormonal and nervous influences they each act as a control for the other . . . only a delicate adjustment of strength of contraction to degree of filling serves as an hypothesis to explain their maintained balanced output in face of the fact that left ventricular pressure load and coronary supply are much more variable than right'. As Sarnoff (1955) has written— 'this statement can hardly be improved upon'.

The mechanism of sympathetic stimulation in exercise is discussed under the heading 'cortico-hypothalamic influences on the circulation' on page 131 *et seq.*

Cardiac output and oxygen usage of the body in exercise

There is a roughly linear relationship between cardiac output and total oxygen uptake in exercise, providing the latter does not exceed approximately 2 l min⁻¹; at greater rates of working with higher oxygen usages the increase of cardiac output is less than will satisfy a linear relationship [Fig. II.64].

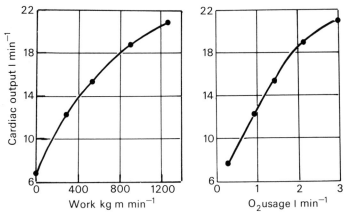

Fig. II.64. Exercise. (Drawn from the results of Asmussen, E. and Nielsen, M. (1952) *Acta physiol. scand.* **27**, 217.)

Measurement of cardiac output in man

Dye injection method

In Hamilton's dye method a known amount of the dye, Evans' blue (T1824), is injected into a vein. By its passage through the heart and pulmonary circulation it will be evenly distributed in the bloodstream and its mean concentration during the first passage through an artery can be determined from successive samples of blood taken from the artery. The blood flow in litres per second (*F*) is given by the formula

$$F = \frac{I}{c.t}$$

where I is the total amount of dye injected, c is the mean concentration of the dye, and t is the duration in seconds of the first passage of the dye through the artery.

Before the injection of dye, 10 ml of venous blood are withdrawn from the basilic vein through the cannula later to be used for injection. This venous sample is divided into two samples of 5 ml; to one of these is added sufficient dye to give a concentration of 0.5 mg per 100 ml (standard). The other sample is used as a blank. One ml of dye solution containing 5 mg is then injected rapidly into the basilic vein and immediately after the injection the arm is raised to a vertical position and gently stroked to further the inflow. Samples of arterial blood are taken at intervals of 0.5–2.0 seconds into a series of tubes. These are later centrifuged together with the blank and standard tubes and the concentrations of dye are determined photocolorimetrically. The concentrations of successive samples are plotted on semi-logarithmic paper. From the resulting curves the mean concentration of the dye can be calculated. FIGURE II.65 shows two results obtained respectively during an experiment in which the subject was at rest and during an experiment in which he performed work at an intensity of 1260 kg m min^{-1}. Each curve shows that the dye concentration reaches a peak and then steadily declines only to rise again owing to re-circulation of the blood containing dye. Naturally, in the work experiment the re-circulation occurs earlier. However, if the early descent from the initial peak is continued as a straight line to cut the abscissa, the point on the time scale at which this occurs gives the duration of the first passage of the dye through the artery (t). In the *rest* experiment 5 mg of dye was injected, the mean concentration of the dye during its first passage through the artery was 1.6 mg l^{-1} during the time t of 39 seconds. Therefore in 39 seconds 5 mg of dye must have been diluted by 5/1.6 = 3.1 litres. In 1 minute the flow (= cardiac output) would have been 3.1 × 60/39 = 4.77 l min^{-1}.

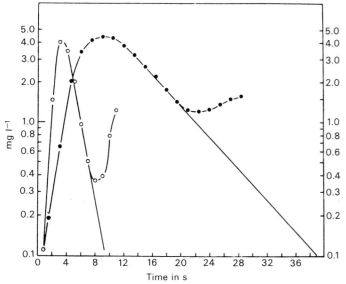

FIG. II.65. Dye concentration curves. ●—● = rest experiment. Cardiac output 4.73 l min^{-1}. ○—○ = work experiment. 1260 kg m min^{-1} Cardiac output 21.9 l min^{-1}. (Asmussen, E. and Nielsen, M. (1952). *Acta physiol. scand.* **27**, 217.)

Similarly in the *work* experiment, t was 9 seconds, c was 1.51 mg l^{-1}, and the cardiac output was 21.9 l min^{-1}. Dye dilution curves are used nowadays in the investigation of congenital and acquired heart disease. Unlike the direct Fick method, catheterization of the heart is not required, and this offers a very real advantage in the study of ill patients. In heart failure the appearance time of the dye is

prolonged and the dye concentration in the blood climbs slowly to a low peak from which it slowly falls.

Dye curves are useful in the investigation of cardiac septal defects. If a septal defect is associated with a shunt of blood from the left to the right heart some of the dye entering the left heart from the lungs passes back into the right heart and finally reappears early on the down stroke of the 'primary' curve. If the shunt is from right to left, some of the dye gains access to the left heart across the shunt and causes a hump on the ascending limb of the dilution curve; the peak of this curve (due to the appearance of dye which has traversed the whole of the pulmonary circuit in the normal manner) of course occurs later.

Thermal dilution method

This is in effect a variant of the dye dilution technique using heat, or more generally 'negative heat' (i.e. cold), as the indicator. A bolus of cold fluid is rapidly injected through a catheter inserted into a peripheral vein and advanced to the right atrium. The quantity of thermal indicator delivered to the blood is derived from the volume of injectate and its temperature as recorded by a fast response thermistor placed in the lumen of the catheter near the injection orifice. The resultant change in blood temperature is measured by a second thermistor bead mounted in the end of a cardiac catheter which is passed retrogradely from a peripheral artery to the base of the aorta, or advanced from a peripheral vein through the right heart into the pulmonary artery. The shape and time-course of the change in aortic or pulmonary artery blood temperature is similar, though not identical, to that of the dye concentration curve following dye injection. The formula for the calculation of cardiac output is also analogous to that for the dye technique, and is

$$F = \frac{V_I \times (T_B - T_I)}{\Delta T_B \times t}$$

where F is the blood flow in l s^{-1}, V_I the volume of injectate, T_B and T_I the blood and injectate temperatures respectively, ΔT_B the mean change in aortic (or pulmonary arterial) blood temperature, and t the duration in seconds of the first passage of thermal indicator through the artery (corrections for the different specific gravities and specific heats of injectate and blood have been neglected).

Though more care is required to avoid errors when using thermal dilution rather than dye dilution, the fact that blood need not be withdrawn for sampling or calibration makes the method applicable to infants and small children in whom the blood volume is limited. Moreover the rapid dissipation of the indicator (cold) allows measurements to be repeated in rapid succession.

REFERENCE

GANZ, W. and SWAN, H. J. C. (1972). Measurement of blood flow by thermodilution in man. *Am. J. Cardiol.* **29**, 241–6.

Direct application of the Fick principle

$$\text{The pulmonary blood flow} = \frac{O_2 \text{ usage ml per minute}}{(A - V) \ O_2 \text{ difference ml per 100 ml}} \times 100.$$

The pulmonary blood flow is equal to the cardiac output. Arterial blood is obtained by arterial puncture in man and the O_2 content is determined. Mixed venous blood is withdrawn by a catheter inserted via the antecubital vein from the right ventricle and its O_2 content determined. Ideally, the O_2 usage is determined simultaneously by means of closed circuit spirometry [see p. 212].

Arterial blood contains 19 ml per 100 ml and the mixed venous blood contains 14 ml per 100 ml, the $(A - V) \ O_2$ difference is 5 ml per 100 ml. If the resting O_2 usage is 250 ml the cardiac output is 5000 ml min^{-1}.

This method suffers from various disadvantages:

1. The subject may well be alarmed by the ritual which attends its performance and the cardiac output may therefore be somewhat higher than normal.

2. It is dangerous in the case of subjects doing heavy exercise for an indwelling atrial or ventricular catheter may precipitate ventricular fibrillation which is fatal.

Clearly, if a subject performing severe exercise has an oxygen usage of 4 l min⁻¹, his cardiac output cannot be less than 4000 × 100/19 = 21 litres (for his mixed venous blood cannot contain less than 0 ml per 100 ml of O_2) and is likely to be more than 21 litres. Highly trained athletic subjects have exceeded 5.3 litres O_2 usage per minute and it is likely that their cardiac output is of the order of 35 l min⁻¹.

The inapplicability of the direct Fick method in heavy exercise conditions led to the use of the dye method.

Cardiac work and oxygen usage

Mechanical work is defined as the product of force and the distance moved by the point of application of that force. When fluid moves as the result of applied pressure the volume of fluid displaced times the pressure gives the external work done.

The systolic ejection of blood into the aorta can be described first by defining the amount of work performed in any small fraction of time

$$\Delta W = P_v \times \Delta V$$

which integrated over the period of systolic ejection is

$$W = \int P \mathrm{d}v.$$

The net mechanical (= external) work of the ventricular cycle is given by plotting the pressure-volume relationships, which are experimentally recorded [FIG. II.42]. The area enclosed gives the total mechanical work done per beat.

A complete description of the factors involved during the ventricular cycle is very complicated and, though provided by Otto Frank (1920), entails the consideration of some influences which to date have not been satisfactorily investigated. Most experimentalists have agreed on a simple formula which is satisfactory for practical purposes:

$$\text{Mechanical work per beat} = QR + \frac{mv^2}{2g}$$

where Q = stroke volume in ml,
  m = mass of blood moved in grams,
  v = mean velocity in the aorta, and
  R = mean arterial pressure.

Suppose the stroke volume of resting man is 80 ml and the arterial mean pressure is 100 mm Hg (= 0.1 m × 13.6 = 1.36 m H_2O)

$$QR = 80 \times 1.36 = \text{approximately 109 g metres}$$

This QR product is a measure of the potential energy developed by systolic ejection—this potential energy is used to expand the Windkessel vessels—the aorta and its large vessels.

The term $mv^2/2g$ gives the kinetic energy developed. The *mean* velocity v is about 0.5 m s⁻¹.

$$\frac{mv^2}{2g} = \frac{80 \times \frac14}{2 \times 9.8} \text{ g metres}$$

$$= \text{approximately 1 g metre.}$$

The kinetic factor is of trivial importance in considering the work per beat done by the left ventricle in resting man. It becomes much more important in exercise conditions when the mean velocity reaches values of 2.5 m s⁻¹. In such circumstances the stroke output may be 180 ml and the mean blood pressure 120 mm Hg or 1.65 m H_2O.

$$QR + \frac{mv^2}{2g} = 180 \times 1.65 + \frac{180 \times (2.5)^2}{2 \times 9.8}$$

$$= 298 + 50$$

$$= 350 \text{ g metres approximately.}$$

The kinetic energy factor now accounts for about one-seventh of the total.

The right heart of course does less work because the mean pulmonary arterial pressure is only about one-fifth of that in the aorta. However, the kinetic energy factor is the same as that in the systemic circuit, so:

$$QR + \frac{mv^2}{2g} = 22 + 1 = 23 \text{ g metres per beat in resting man}$$

and $QR + \dfrac{mv^2}{2g} = 60 + 50 = 110$ g metres per beat in heavy exercise

Thus, in heavy exercise the total external work done by the two ventricles is 350 + 110 g metres per beat = 460 g metres. Per minute, at heart rate of 180 min⁻¹, the total mechanical work is:

$$460 \times 180 = 82\ 800 \text{ g metres min⁻¹}$$
$$= 83 \text{ kg metres min⁻¹ approximately.}$$

Now, 1 ml oxygen is equivalent to 2 kg metres so, if the process were 100 per cent efficient and if the oxygen usage of the ventricular myocardium were devoted solely to the performance of external work, 83 kg metres min⁻¹ would require 42 ml O_2. The oxygen usage of the human heart in these conditions is not known for certain, for obvious reasons—it would be criminally foolish to have a coronary sinus catheter in place, thereby inviting ventricular arrhythmia and death. However, calculations, based on values obtained in man in moderate exercise and on values obtained in direct coronary sinus samples and flow measurements in more strenuously exercising dogs, suggest that myocardial oxygen usage may reach 250 ml min⁻¹.

Hence, the mechanical efficiency, calculated from these figures, is (42/250) × 100 = 16–17 per cent.

Two points should be clearly understood. (a) The external work of the heart is only a small proportion of the total energy transformed by the cardiac cells. (b) The physicist's definition of work implies that if a force, no matter how strong, does not move its point of application or if a pressure developed does not move a volume of blood, then no work is done. A remark credited to the late J. L. Lilienthal Jr may be quoted: 'The guy who thought up that definition never stood holding the suitcases while his wife searched in her handbag for the airtickets.'

Burton (1965) has provided a simple and useful analogy of the situation. FIGURE II.66 shows a crane with a hook lifting scrap iron into a truck. The external work done by the motor of the crane is the weight times the height it is lifted. Modern cranes use an electromagnet to develop the force on the scrap iron. Just to maintain this force, energy must be continuously supplied by the current of the electromagnet and this extra energy consumption must be added to that needed to perform the external work of lifting the iron into the truck. In heart muscle similarly, the total energy turnover is mechanical work *plus* the energy cost of maintaining the force (tension) of the muscle. Hill showed in skeletal muscle that the extra rate of energy turnover is proportional to the tension and hence the energy cost will be the mean tension and the time it is maintained—the so-called tension–time integral.

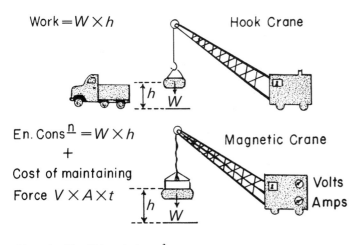

$$\text{Work} = W \times h \qquad \text{Hook Crane}$$

$$\text{En. Cons}^{\underline{n}} = W \times h$$
$$+$$
$$\text{Cost of maintaining}$$
$$\text{Force } V \times A \times t \qquad \text{Magnetic Crane}$$

Volts
Amps

<u>Muscle</u> $E = W \times h + \alpha \int T. dt$

<u>Heart</u> $E = \int P. dV + \alpha \int T. dt$

Fig. II.66. Illustrating the analogy of an electromagnetic crane (cf. *hook crane*) to maintenance heat of the heart muscle. (Burton, A. C. (1957). *Am. Heart J.* **54**, 808.)

External cardiac work is governed by the *product* of mean pressure and stroke volume.

If one disregards the kinetic work, then if the oxygen usage of the heart is devoted only to external mechanical work it should not matter whether volume or pressure in the *QR* product is varied provided that the product remains constant.

However, studies of the performance of the isolated supported heart show clearly that myocardial total energy turnover, as expressed in myocardial oxygen usage, rises much more when the heart beats against an increased pressure (*R*) than when it increases its stroke volume (*Q*) against a steady pressure (Sarnoff *et al.* 1958).

When the left ventricular external work is increased by raising the pressure and keeping heart rate and stroke volume constant myocardial oxygen usage rises *pari passu* with the increase of work and the myocardial efficiency (external work/O_2 usage) is unaltered [Fig. II.67].

In contrast, increasing the stroke volume against a constant pressure (heart rate kept constant) does not greatly increase myocardial O_2 usage, hence efficiency rises in a striking fashion [Fig. II.67].

Finally, the mean arterial pressure was held constant and the heart output was raised in three series of experimental runs—at a heart rate of 120 min⁻¹, at a heart rate of 160 min⁻¹ and at a heart rate of 200 min⁻¹. The higher the heart rate, the greater was the myocardial oxygen usage for any given cardiac output.

Myocardial oxygen usage thus bears little relationship to external cardiac work *per se*. Sarnoff's results indicated that oxygen usage was most closely related to the total tension developed by the ventricular muscle as reflected by the total area beneath the systolic portion of the aortic pressure pulse. The term tension time index (TTI) is used to express this parameter. TTI is calculated as the product of the mean systolic pressure, the duration of systole and the heart rate.

[Note that although this interpretation of tension has been used for more than 25 years, tension (*T*) is the product of the pressure and the existing radius. At any given mean pressure tension *per se* is governed by the ventricular radius at that time.]

The main point is that the external work of the heart is a much smaller item in its energy turnover than is the TTI and the efficiency calculated is correspondingly small because the numerator of the equation is small compared with the denominator.

$$\text{ME} = \frac{\text{External work}}{\text{External work} + \text{TTI}} \times 100.$$

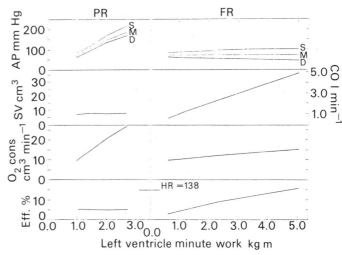

Fig. II.67. Contrasting effects on myocardial oxygen consumption of increasing work by increasing mean aortic pressure (pressure run, PR) and increasing work by increasing cardiac output (flow run, FR). Heart rate was held constant at 138 min⁻¹ throughout. AP aortic pressure; SV, stroke volume; Eff, per cent efficiency. (Sarnoff, S. J. *et al.* (1958). *Am. J. Physiol.* **192**, 141.)

An increase of external work by increasing stroke volume, but not arterial resistance, has a marked effect on the numerator but little effect on the denominator—hence the efficiency rises. An increase of external work by raising the resistance increases the mean systolic pressure developed and increases the TTI as evidenced by oxygen usage so efficiency shows little change. As Burton (1965) has said, these facts are of great clinical importance in the management of cardiac patients. It is much more dangerous for the patient to have an angry argument, which increases heart rate and blood pressure and hence the TTI, than to engage in moderate exercise where the increased stroke volume required is of little import in influencing the total work of the heart.

Braunwald *et al.* have questioned that the TTI is the best criterion or determinant of myocardial oxygen usage. They point out that the level of contractile state (as reflected in measurements of V_{max}) is of great importance. Myocardial oxygen usage may be increased by positive inotropic effects even when TTI may be reduced. For the meantime it is sufficient to regard TTI as a good yardstick of the requirements of myocardial metabolism.

REFERENCES

ABBOT, B. C. and MOMMAERTS, W. F. H. M. (1959). *J. gen. Physiol.* **42**, 533.
ALLEN, D. E. and BLINKS, J. R. (1978). *Nature, Lond.* **273**, 509.
BAKER, P. F. (1978). *Prog. Biophys. molec. Biol.* **24**, 177–223.
BLINKS, J. R., RUDEL, R., and TAYLOR, S. R. (1978). *J. Physiol., Lond.* **277**, 291.
BRAUNWALD, E., ROSS, J. and SONNENBLICK, E. H. (1976). *Mechanisms of contraction in the normal and failing heart.* Little Brown, Boston.
ENDO, M. (1972). *Nature New Biol.* **237**, 211.
—— (1977). *Physiol. Rev.* **57**, 71.
FABIATO, A. and FABIATO, F. (1975). *Nature, Lond.* **256**, 54.
—— —— (1977). *Circulation Res.* **40**, 119.
JEWELL, B. R. (1977). *Circulation Res.* **40**, 221.
—— (1978). In *Developments in cardiovascular medicine* (ed. C. J. Dickinson and J. Marks) pp. 129–44. MTP, Lancaster.
LAKATTA, E. G. and JEWELL, B. R. (1977). *Circulation Res.* **40**, 251.
LINDEN, R. J. (1965). *Scientific basis of medicine annual reviews*, p. 164.
—— and SNOW, H. M. (1974). In *Recent advances in physiology*, 9th edn (ed. R. J. Linden). Churchill Livingstone, Edinburgh.
SONNENBLICK, E. H. and SKELTON, C. L. (1974). *Circulation Res.* **35**, 517.
STARLING, E. H. (1918). The Linacre lecture on the law of the heart. London.
—— (1920). *J. roy. Army med. Cps* **34**, 258.

Origin and spread of the cardiac impulse

Rhythmicity of the cardiac chambers

If the frog heart be removed from the body and perfused in retrograde fashion (via a cannula tied in the truncus arteriosus) by a solution of electrolyte composition and glucose content similar to that of blood it will beat for hours. The heart beat, like that of a mammal, is inherently rhythmic.

In situ the beat of the frog heart may be recorded by a lever attached by thread to a hook inserted in the apex of the ventricle. The excursions of the lever can be recorded on a kymograph. If the recording lever system is sensitive enough the graphic record shows each beat to comprise the sequential contraction of three parts of the heart which can by inspection be identified as (a) initial contraction of the sinus venosus, (b) contraction of the atria followed by (c) contraction of the ventricle.

Stannius tied a tight ligature between the sinus venosus and the atria. The rhythm of the sinus venosus continued virtually unimpaired; contraction of the atria followed by that of the ventricle briefly ceased. Shortly, however, atrial rhythm reasserted itself and caused atrial contraction accompanied by ventricular contraction; atrial rhythm (which dictated ventricular rhythm) was slower than that of the sinus venosus. Then a second ligature was tied in the atrioventricular ring. Ventricular contraction now ceased although atrial rhythm was maintained. After some time ventricular rhythmic contractions began themselves, but this *idioventricular rhythm* was slower than before. It may be concluded that in the frog heart (i) the inherent rhythm is initiated in the sinus venosus which acts as a pacemaker, (ii) that the cardiac impulse spreads to and excites sequentially the atria and the ventricles, (iii) that separated (by crushing) from the sinus venosus the atrial muscle cells contain potential pacemakers which can assume the role of initiating cardiac impulses which in turn can dictate the ventricular rhythm, (iv) that even in the absence of pacemaker control from the sinus and the atria the ventricle can develop its own slower rhythm due to its content of pacemaker cells.

Normally putative atrial and ventricular pacemakers are held in check by the faster rhythm developed in the sinus venosus.

The Stannius experiments cannot be performed on the mammalian heart for the sinu-atrial node (the homologue of the sinus venosus) is incorporated into the wall of the right atrium, which precludes the tying of ligature I and the mammalian ventricles are entirely dependent on a coronary blood supply. Thus, the second Stannius ligature occludes the coronary arteries and causes ventricular fibrillation.

Using electrocardiographic (ECG) recordings Lewis showed that the same sequence of spread of the contractile process over the mammalian heart occurred as in the frog. Thus, heating or cooling of the region of the sinu-atrial node caused an acceleration or slowing of the heart beat, the ECG complexes being normal. The same thermal changes applied elsewhere were either ineffective or caused changes of rate with alterations of the normal ECG complexes. Electrophysiological studies showed that depolarization of the sinu-atrial node always preceded that elsewhere in the heart. Destruction of the sinu-atrial node by cautery or crushing led to the heart beat slowing and electrical recordings revealed that such cardiac cycles were initiated from the a.v. node. The ECG complexes showed disturbances of the P wave identifiable with abnormalities of conduction through the atrial muscle. The application of an electrical stimulus to a normal heart causes a premature beat which manifests a normal ECG complex *only* if the stimulus is applied to the sinu-atrial node—elsewhere the ECG complex of the extrasystole is abnormal. Paired surface electrodes placed on the surface of the heart show that the excitation process initiated from

the sinu-atrial node is conducted across the atrial walls like the ripples on a pond at a speed of 1 m s^{-1}. At the a.v. node the speed of conduction slows to 0.2 m s^{-1}, but then speeds up on reaching the bundle of His to reach 4 m s^{-1} in the specialized Purkinje tissue. The bundle is the sole conducting strand between atria and ventricles. Its two branches pass the impulse to the cardiac apex (endocardial surface within 13–15 ms after it leaves the a.v. node). The excitation process is propagated by the Purkinje arborization along the endocardial surface and reaches the base of both ventricles within 40 ms. Within 55 ms the epicardial surface of the right ventricle and in 65 ms that of the canine left ventricle have been excited [FIG. II.68].

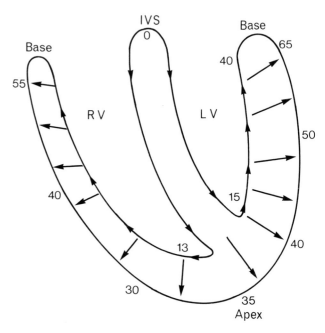

FIG. II.68. Diagram to illustrate the time relations of the spread of the excitation process to the ventricles of the dog. Figures show time in milliseconds. O represents the arrival of the impulse at the top of the interventricular septum (IVS).

Bundle of His

Experimental evidence proves that the bundle of His is the sole connecting strand between the atria and ventricles. If the bundle of His is compressed, a proportion of the impulses transmitted from the atria by the a.v. node may fail to reach the ventricles which thus beat at a slower rate than that of the atria. If the bundle is completely divided in animals or destroyed by disease in man, the atria and ventricles are completely 'dissociated', and beat quite independently one of the other; the atria then beat at their usual rate in response to impulses generated in the s.a. node, while the ventricles beat much more slowly—thirty to forty times per minute—in response to a new rhythm centre in their own substance, probably situated in the part of the bundle below the point of section.

Clinical electrocardiography

As has been seen, when the heart beats, an electrical process which precedes the sequence of associated mechanical changes spreads from the s.a. node. The electrical currents which develop can most easily be demonstrated in the thoracotomized experimental animal by slitting the pericardium and placing the distal segment of the cut phrenic nerve on the surface of the ventricle. The diaphragm then contracts at the rate of the heart, owing to the electrical excitation of the distal phrenic nerve fibres.

Initially the string galvanometer of Einthoven was used and Lewis in particular made great contributions to our understanding of the normal and abnormal rhythms of cardiac action employing this instrument. These workers used bipolar leads placed on the limbs and connected to the galvanometer—the so-called *classical limb leads*. From such studies they obtained characteristic records [Fig. II.69] which they labelled as shown.

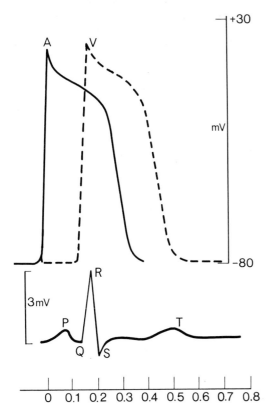

Fig. II.69. (Diagrammatic). Above: intracellular potential changes occurring during one cardiac cycle in an atrial fibre (A) and a ventricular fibre (V).
Below: PQRST complex of electrocardiogram (surface leads). Note the *intra*cellular potential changes by 100–105 mV while the QRS complex registers only 3 mV.
Note also that the isoelectric record during the S–T interval disguises the fact that electrical activity is present in the ventricular muscle during this period, as it is apparent from the action potential record of the individual ventricular muscle fibre shown above. The ECG recorded at the surface of the body represents the *resultant* of activity in the individual fibres and the ECG record is isoelectric during this period simply because the complex pathways taken by the electric currents in the heart result in these individual potential changes cancelling one another out when recordings are made using surface leads.

Name of unipolar lead	Position of exploring electrode
VR . .	Right arm
VL . .	Left arm
VF . .	Left leg (F = foot)
VC . .	Chest
	Six chest positions are described:
V_1 . .	4th intercostal space to right of sternum
V_2 . .	4th intercostal space to left of sternum
V_3 . .	Midway between left sternal border and mid-clavicular line on a line joining positions 2 and 4
V_4 . .	5th intercostal space in midclavicular line
V_5 . .	5th intercostal space in left anterior axillary line
V_6 . .	5th intercostal space in left midaxillary line
V_7 . .	5th intercostal space in left postaxillary line

They proved that P was caused by atrial excitation, QRS by the invasion of the ventricular muscle, and that T represented changes caused by the terminal repolarization of the ventricular muscle.

Nowadays unipolar precordial leads and unipolar limb leads are additionally employed and these will be described first.

Unipolar leads

One electrode (the exploring electrode) is placed on an area of the body surface. The other (indifferent) electrode is kept at approximately zero potential by connecting electrodes placed respectively on the right arm, left arm, and left leg to a central terminal through a 5000 ohm resistance. (The currents from the three limbs neutralize one another.) The indifferent electrode undergoes no significant change during the cardiac cycle. When a unipolar lead is used the ECG records the potential changes which affect the exploring electrode only.

The following unipolar leads are used; they are labelled V, followed by a letter or by a number describing the position of the exploring electrode.

The wiring employed is, by convention, such that when the exploring electrode is negative relative to the central zero terminal, the record is deflected downwards: when the exploring electrode is positive, the record is deflected upwards.

Unipolar precordial leads. These are influenced by electrical activity throughout the heart but particularly by that part of the heart nearest to the electrode. Thus leads V_1 and V_2 reflect right ventricular activity and leads V_5 and V_6 reflect left ventricular activity. Spread of the excitation process towards the electrode gives an upward deflection and spread away from it causes a downward deflection. The first part of the ventricle to be excited is the upper part of the interventricular septum. This is supplied by a twig from the left bundle branch so that the spread of excitation here is from left to right. Hence the initial deflection in lead V_1 is upward and that in V_6 is downward. The excitation process subsequently involves the remainder of the septum and the ventricular walls (from endocardial to epicardial surface) rapidly. As the left ventricle is much thicker-walled than the right the electrical changes occurring in the left ventricle dominate the precordial ECG. Thus in V_1 the main QRS deflection is negative (i.e. downward) whereas that in V_6 is positive (i.e. upward) [Fig. II.70].

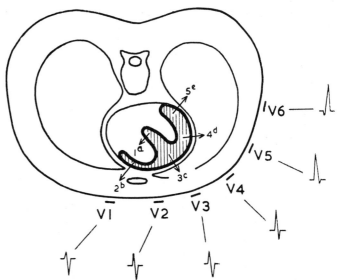

Fig. II.70. Cross-section through the chest to show the precordial leads and their relation to the heart; a, b, c, d, and e, show the order in which the electrical impulse spreads through the ventricle. Note the alteration in the configuration of the QRS complex between V_1 and V_6. (Diagram by Dr D. S. Short.)

Unipolar limb leads. These reflect the electrical activity of that part of the heart which faces the electrode. Thus VF reflects the electrical activity of the inferior surface of the heart, VL that of the left upper side of the heart and VR the cavity of the ventricles [FIG. II.71].

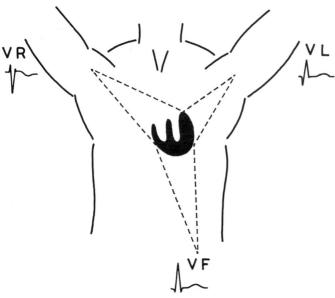

FIG. II.71. The relation of the unipolar limb leads to the heart. The right arm lead faces the cavity of the ventricle. The foot lead faces the inferior surface of the heart; this may be formed by the right or the left ventricle or by both, depending on the position of the heart. The left arm lead may face the cavity of the ventricles or the outside of the left ventricle, depending on the position of the heart. (Diagram by Dr D. S. Short.)

The configuration of VL and VF depends on the position of the heart within the chest and thus varies greatly with the phase of respiration. When the heart is vertical, as in deep inspiration, the part of the heart which faces VF is the left ventricle and the QRS pattern in the record obtained correspondingly resembles that recorded over the left ventricle by the precordial lead V_6. In expiration, the heart lies more obliquely and the inferior surface may be formed entirely by the right ventricle so that the QRS pattern recorded by VF resembles that of the precordial lead V_1 (which is placed over the right ventricle). In expiration, the left ventricle faces VL so that the record from this lead resembles that of V_6.

Bipolar leads

Classical limb leads. These are the leads which have been most extensively used in man; a vast amount of information has consequently accumulated correlating the electrical with the clinical and post-mortem findings. These leads were also used in the earlier fundamental studies of the origin and spread of the cardiac impulse and in the analysis of experimentally induced abnormalities of the heart's action. The classical limb leads are:

Lead I: from right arm and left arm.
Lead II: from right arm and left leg.
Lead III: from left arm and left leg.

The deflection recorded in a bipolar limb lead represents the algebraic sum, at any moment, of the potentials of the two constituent leads, as follows:

Lead I   = Lead VL–VR
Lead II = Lead VF–VR
Lead III = Lead VF–VL

Since the position of the heart in the chest affects the unipolar limb leads it will also affect the bipolar limb leads. Thus when the heart is

vertical there is a predominant R in Lead III and a predominant S in Lead I. This pattern, which is known as right axis deviation, is also seen in right ventricular hypertrophy (e.g. due to pulmonary valvular stenosis). When the heart is more horizontal there is a predominant R wave in Lead I and a predominant S in Lead III. This pattern (left axis deviation) is also seen in left ventricular hypertrophy (as occurs in aortic stenosis or in arterial hypertension).

Bipolar precordial leads. These have been largely superseded by unipolar precordial leads. CR leads, however, are sometimes used. In these the exploring electrode is placed on the chest in one of the seven chest positions called 1–7. The other electrode is placed on the right arm. Since the arm electrode is far from the heart, the ECG mainly reflects the activity of the surface of the heart so that the resulting record closely resembles that of the unipolar precordial leads.

Interpretation of electrocardiogram

Atrial complex (P wave)

Limb Leads I, II, and III [FIG. II.72].—The P wave is upright and has a rounded or pointed summit; its duration is 0.1 second. It

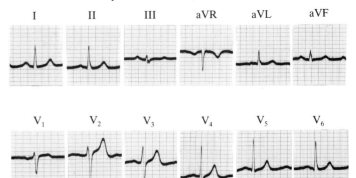

FIG. II.72. Electrocardiographic records of a normal man (Leads I, II, and III; unipolar limb Leads VR, VL, and VF; and unipolar precordial leads 1–6). (By courtesy of Department of Cardiology, The Middlesex Hospital.)

represents the passage of the impulse from the s.a. node over the atria; the a.v. node is reached at about the summit of P. The magnitude of the P wave is some guide to the functional activity of the atrial muscle. In mitral stenosis the left atrium is hypertrophied, and correspondingly the P wave is prominent, or bifurcate. In atrial fibrillation the P wave disappears and is replaced by a series of fine irregular oscillations, corresponding to the rapid irregular excitation of the atria [FIG. II.78]. If the cardiac impulse arises in an abnormal focus and spreads in other than the usual direction, the P wave is altered or even inverted, becoming a downward deflection [FIG. II.77].

Conduction time of bundle of His (P–R interval)

The time interval from the beginning of P to the commencement of R is a guide to the conduction time of the bundle of His. When the classical limb leads are used the true conduction time is the interval from the top of P (when the a.v. node is excited) to the beginning of Q (when the invasion of the ventricles commences). It is easier, however, to measure the P–R interval (the Q wave is inconstant and of small size), and undoubtedly it is mainly taken up by the passage of the impulse along the bundle and its branches. The P–R interval in health varies between 0.13 and 0.16 s, the extreme limit of the normal being 0.2 s. When it exceeds 0.2 s there is delayed conductivity in the bundle of His. When the P–R interval is shorter than normal, the impulse has probably arisen in the a.v. node and has therefore excited the ventricles sooner than is normally the case.

Ventricular complex (QRST waves)

Following P there is a brief isoelectric period; a succession of deflections then appears, namely: Q, a small (often inconspicuous) downward one; R a prominent upward one; and S, another downward one; the record then returns once more to the base line. The duration of QRS in man is about 0.08 s; the upper limit of the normal is 0.1–0.12 s. The final T wave is a broad upward deflection, with an average duration of 0.27 s. The duration of QRST is thus about 0.4 s. The upstroke R just coincides with the onset of ventricular systole; the end of the T wave coincides approximately with the end of ventricular systole or may outlast it slightly.

QRS represents the stage of ventricular depolarization. It is affected by structural changes in the ventricular muscle and by delay in conduction in the bundle and its branches. Thus left ventricular hypertrophy (as in hypertension) gives an abnormally tall R wave in V_6 and a deep S wave in V_1. Conversely right ventricular hypertrophy gives a tall R wave in V_1 and an unusually large S wave in V_6. Loss of left ventricular muscle bulk (as in extensive infarction) gives a stunted R wave in V_6. Damage to one or both bundle branches leads to prolongation of ventricular depolarization and thus to a QRS complex which exceeds 0.12 s in duration.

The S–T segment and T wave represent the stage of repolarization. Deviations from normal are caused by ischaemia and overdosage with digitalis.

These changes are of great importance clinically and will now be considered in more detail.

Electrocardiographic changes in myocardial lesions

Lesions of the ventricular myocardium may be produced experimentally:

1. By a local application of a 0.2M KCl solution which depolarizes the surface membrane of the affected muscle fibres.
2. By tying various blood vessels to produce patches of ischaemia and consequent death of the affected areas. Clinically, localized cardiac ischaemia may result from the occlusion of coronary vessels by a thrombus. These myocardial lesions give rise to distinctive electrocardiographic patterns.

FIGURE II.73 shows the classical evolution of the form of the ECG after anterior cardiac infarction. Within a few hours the most conspicuous abnormality is the appearance of the ST deviation (a). Normally the ECG record is isoelectric between the end of S and the beginning of T, but shortly after myocardial injury the ST segment becomes elevated in some leads and depressed in others. After some days the ST deviation decreases and T wave abnormalities appear (b). Usually, if the ST segment is elevated the T wave becomes inverted and vice versa. After several weeks the ST segment becomes normal and the T wave changes alone remain (c). After months or years, they too may return to normal. When the injury of infarction affects the posterior (or, more accurately, the inferior) part of the heart, the ECG changes in Leads I and III are the *opposite* of those shown in FIG. II.73. The ST elevation is seen in Lead III and the ST depression in Lead I; the T wave, upright in Lead I, is inverted in Lead III. The mode of production of these changes is not fully understood, but as is the case with skeletal muscle a damaged area generates an injury potential.

Right and left axis deviation

Clinically one frequently obtains curves characterized by a tall R wave in Lead I and a big S wave in Lead III. These are called examples of deviation of the main electrical axis of the heart to the left, or briefly left axis deviation [FIG. II.74, c]. Conversely a big S in Lead I and a big R in Lead III is called right axis deviation. Such axis deviation is noted with mechanical displacement of the heart or with the movements of the heart which occur with the phases of respiration. Hypertrophy of the left ventricle, such as occurs as a result of chronic arterial hypertension or aortic incompetence, is associated with left axis deviation [FIG. II.74, c]. Conversely hypertrophy of the right ventricle, e.g. from mitral stenosis or emphysema is associated with right axis deviation [FIG. II.74 a].

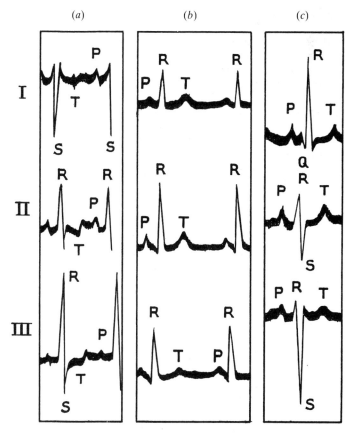

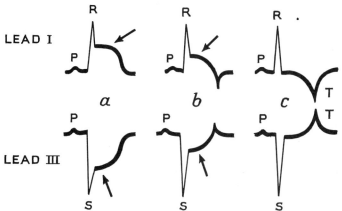

FIG. II.73. Diagram illustrating the changes in the ventricular complexes in Leads I and III after coronary thrombosis affecting anterior surfaces of ventricles; a = elevated ST segment in Lead I and depressed ST segment in Lead III; b = T waves becoming evident; c = ST displacement disappeared. The T wave finally points in the opposite direction to the original ST displacement in each lead. (Parkinson and Bedford (1928). *Heart* **14**.)

FIG. II.74. Electrocardiogram in right and left axis deviation; *a* = right axis deviation (in a woman with mitral stenosis). The T wave is inverted in all leads because of treatment with digitalis; *b* = normal record; *c* = left axis deviation (in a woman aged 52 with hypertension); BP 245 mm Hg systolic, 145 mm diastolic; enlarged left ventricle.

Abnormalities of cardiac rhythm

Heart block

By heart block is meant a condition in which there is defective conduction in some part of the heart. This might theoretically occur:

1. At the s.a. node.
2. At the a.v. node or in the bundle of His.
3. In one of the branches of the bundle of His. Clinically, the first is rare but the other two fairly common.

Sinu-atrial nodal block. There is block within the substance of the node and the impulse occasionally fails to escape to activate the remainder of the heart. On such occasions a whole heart beat is lost. During such a pause neither atrial nor ventricular contraction occurs nor is there any electrical variation. After an interval which is usually less than two complete cardiac cycles the heart resumes its normal action. The condition is at once unmasked by exercise, when the 'missing' beats reappear and the heart suddenly doubles its rate. It then accelerates further like a normal heart. Sinu-atrial heart block may be produced by vagal stimulation and is correspondingly relieved by atropine which blocks the effect of acetylcholine on the s.a. node.

Atrioventricular block. There is block within the a.v. node. There may be any grade of block from a very slight delay, so that the P–R interval may slightly exceed normal (0.2 s), to complete block in which atrioventricular conduction ceases entirely. Any defect in a.v. conduction short of complete block is termed partial heart block.

1. *Partial heart block* may appear in two forms: (i) delayed conduction, in which the P–R interval exceeds 0.2 s; (ii) dropped beats. At first the P–R interval may lengthen with successive beats until one ventricular beat is lost, the cycle being repeated indefinitely.
2. *Complete heart block* [FIG. II.75]. The ventricles beat with a slower and independent rhythm dictated by a portion of the conducting tissue below the site of the block. The atria and ventricles beat at entirely different rhythms. Correspondingly the P waves and the QRST complexes bear no relationship to each other. Providing that the bundle branches are intact the QRST complexes are themselves normal in character.

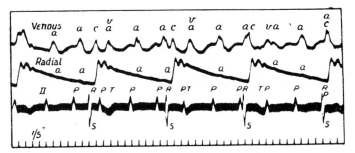

FIG. II.75. Complete heart block in man. Records from above downwards are: jugular venous pulse, radial pulse, and electrocardiogram (Lead II). Note the regular relationship between the jugular *c* wave, the radial primary wave, and the R wave (all indicative of ventricular activity). There is a regular relationship also between the jugular *a* waves and the P wave (indicating atrial activity). In their respective records the *a* waves and P waves follow regularly. There is no relationship between atrial and ventricular events.

If complete heart block develops suddenly, there may be a delay before the ventricles start beating at their own rate. Naturally, in this period the systemic blood pressure falls to a very low level and the blood supply to the brain becomes inadequate If the ventricular

standstill lasts for only a few seconds it may give rise merely to dizziness and faintness, but if it is more prolonged it may cause loss of consciousness, convulsions, and even death. An attack of complete heart block accompanied by fainting constitutes the *Stokes–Adams* syndrome.

Permanent atrioventricular block is usually due to obstruction of the coronary artery supplying the bundle tissue.

Bundle branch block. In this condition there is a block of one of the branches of the bundle of His. Thus the excitation process cannot be propagated directly to one of the ventricles although it is satisfactorily conducted to the other. The excitation process has to make a detour through ventricular muscle on the affected side to regain the bundle tissue below the site of the block. Consequently there is a delay in the activation of one ventricle. The QRS complex is correspondingly prolonged (exceeding 0.12 s) and is abnormal in form [FIG. II.76]. Repolarization is likewise affected so that the ST segment and the T wave show abnormal features. Bundle branch block does not affect the heart rate. It does, however, cause delay in the contraction of one ventricle and this may give rise to a detectable widening of the interval between the aortic and pulmonary

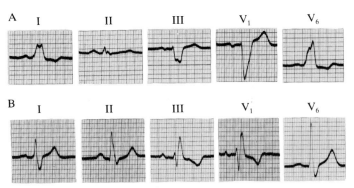

FIG. II.76. Electrocardiograph Leads I, II, and III, and unipolar precordial Leads V_1 and V_6. Record A = patient with left bundle branch block. Record B = patient with right bundle branch block. Compare these records with those of the normal subject [FIG. II.72]. (By courtesy of Department of Cardiology, the Middlesex Hospital.)

components of the second sound. This is more easily noticeable in right bundle branch block because the right ventricle normally contracts before the left. Bundle branch block is usually due to obstruction of the coronary arterial branch supplying the bundle tissue.

New rhythm centres

Normally the s.a. node acts as pacemaker:

1. If the node is destroyed or cooled, while the a.v. node is warmed, the a.v. node becomes the pacemaker. Thus the impulse passes from this node at the same time to both atria and ventricles and these contract simultaneously. The P–R interval is much shortened; the P wave is inverted or buried in the R wave [FIG. II. 77]. In clinical records, shortening of the P–R interval strongly suggests that the impulse is arising in or near the a.v. node.
2. If the ventricles are suddenly dissociated from the atria by bundle block they assume a new idoventricular rhythm slower than that normally dictated by the s.a. node.
3. Occasional spontaneous beats may arise anywhere in the substance of the atria or ventricles. True rhythmic function—the ability to initiate and maintain the heart beat over prolonged periods—is only present in the nodes and junctional tissues. The occasional spontaneous beats are known as ectopic beats, premature beats, or as *extrasystoles*. Sometimes an ectopic focus may initiate the heart beat for several seconds or for hours. This is known as ectopic rhythm or tachycardia. This abnormal heart

action which usually appears and declines abruptly is also called paroxysmal tachycardia. If the ectopic focus is in the atria it is termed paroxysmal atrial tachycardia and, if in the ventricles, paroxysmal ventricular tachycardia.

Single ectopic beats (extrasystoles)

1. Atrial. If the atrium is stimulated during diastole after its refractory period has passed, it responds with a premature contraction. An impulse is transmitted to the ventricles, which contract too. The rate of recovery of the bundle is slower than that of the other parts of the heart, so that if an impulse reaches it prematurely, conduction along it is delayed; the P–R interval, therefore, is prolonged. The P wave is abnormal. The ventricular complex, QRST, is normal. The next atrial impulse arising in the s.a. node appears after a pause equal to the normal diastolic period or a little in excess of it.

2. Ventricular. If the ventricle is stimulated after the refractory period has passed, i.e. after the end of systole, it responds by contracting. As the ventricle obeys the 'all-or-none' law, it responds to the maximum of its ability if the stimulus is adequate. The actual force of the contraction depends on the extent to which the ventricles have recovered from their previous contraction, and the degree of filling which has taken place. If the ventricle is stimulated early in diastole, the contraction it gives is feeble and may be insufficient to open the semilunar valves. The premature beat then only gives rise to the heart sound and is not accompanied by arterial pulsation at the wrist. If the premature contraction occurs later in diastole, the ventricle may contract sufficiently forcibly to discharge its contents. As the ventricle has not had time to become completely filled, the output during this premature beat is less than normal, and the pulsation felt at the wrist is small [FIG. II.77].

Following the extrasystole there is a long pause (compensatory pause). The duration of the 'extrasystolic' cycle and the 'returning' cycle (i.e. the cycle following on the premature contraction) is equal to two normal cycles. The reason for this compensatory pause is simple. The normal atrial impulse which follows the extrasystole finds the ventricles in a refractory condition, and no response is

obtained. The ventricles must therefore wait for the succeeding atrial impulse before contracting [FIG. II.77].

Electrocardiographic features. The electrical record of the ectopic beat shows an abnormal ventricular complex which is not preceded by a P wave; the next P wave is generally 'buried' within this ventricular complex. The QRS is prolonged and abnormal in appearance.

Paroxysmal tachycardia

In this condition the heart may suddenly accelerate to 150 or 200 min^{-1}. The new 'pacemaker' may be situated in the atrium or in the ventricle (giving rise to atrial or ventricular tachycardia respectively). In one case studied, the output per minute fell from the normal level in that subject of 5.6 litres to 2.5 litres, and the output per beat from 77 ml to about 15 ml. This example illustrates how a very rapid heart rate at rest may cripple circulation. Diastole is too brief to allow proper filling of the heart, so that each beat only discharges a small amount of blood. The rest period is very short, the heart fatigues, and signs of heart failure ultimately appear.

Ectopic atrial tachycardias: flutter and fibrillation

Response of atria to rapid stimulation. As the frequency of stimulation of the atria (dog) is increased the following sequence of events takes place:

1. With rates up to 290 min^{-1} the atria respond regularly with contractions of uniform size; a simple atrial tachycardia is thus set up.

2. When the atria are stimulated at 290–380 min^{-1}, a condition called the 'partially refractory state' develops. By this is meant that at any given moment some atrial muscle fibres are responsive, while adjacent fibres are refractory. In response to each stimulus at this rate some groups of fibres react and others fail to respond. If a large part of the atria responds, a big beat is obtained; when a small area reacts, a small beat results. The fibres which respond at one beat are not sufficiently recovered and fail to react to the succeeding stimulus, so that alternate large and small contractions result.

As the refractory period of the ventricles is longer than that of the atria, they develop a 2:1 response when stimulation occurs at 350 min^{-1}. The a.v. node cannot transmit more than 270 impulses per minute.

3. With still more rapid rates of stimulation, e.g. 380–450 min^{-1}, or over, the length of the refractory period of the atrial muscle is increased, and is longer than the interval between succeeding stimuli. A stage of 2:1 response of the atria then results, i.e. every other stimulus finds the atria in a completely refractory state, and so a contraction results from alternate stimuli only.

Efficient conduction in the atria depends on the excitation process meeting responsive tissue which can transmit the impulse farther. When the atria are stimulated at high rates, conduction in the atria is slowed; considerable areas of the atrial wall may be in a refractory state when the excitation process reaches them. These form areas of obstruction, and the impulse instead of flowing smoothly must wind about irregularly to seek out portions of tissue which have recovered. The rate of conduction in the atria when they are stimulated at high rates may fall to 500 mm s^{-1} (normal = 1000 mm).

Following brief stimulation at these very high rates, atrial flutter or atrial fibrillation may set in.

Lewis thought that at certain rates the excitation process made a circuit around a ring of atrial muscle, e.g. the tissue joining the mouths of the inferior and superior venae cavae. He believed that such a circus movement accounted for the clinical conditions of atrial flutter and atrial fibrillation.

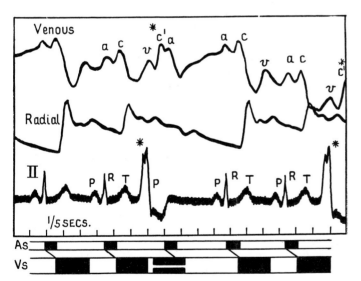

FIG. II.77. Ventricular extrasystole. Simultaneous venous, radial, and electrocardiographic curves from a patient, showing an extrasystole arising in the ventricle. The diagram placed below the figure illustrates the mechanism of the heart during the period of the disturbance. As, Vs = atrial and ventricular systole. The premature ventricular complex is abnormal in character; the P wave is buried in, instead of preceding, the ventricular complex. Similarly the c wave is premature and precedes the a wave. The pause following the premature beat is longer than normal (compensatory pause); the atrial *rhythm* is undisturbed. (Lewis, T. (1925). *Mechanism and graphic registration of the heart beat.*)

In atrial flutter the circus movement follows a regular pathway. If the length of the pathway is sufficient the excitation process returns to the atrial ectopic focus to find it has recovered from its refractory period and the impulse is re-propagated round the circle. If the speed of conduction of the impulse round the circle is slowed, the same state of affairs occurs. Thirdly, if the refractory period of the atrial muscle is shortened, the returning impulse will be repropagated. Any of these features may be seen in diseased hearts. Thus the atria may be dilated—a long pathway. Ischaemia may slow conduction and drugs may shorten the refractory period. Flutter occurs clinically at a rate of 200–350 min^{-1} in diseased hearts, particularly those manifesting atrial dilatatation. At atrial rates above 200 min^{-1} the a.v. node and Purkinje system cannot transmit every impulse; as a result the ventricles respond to every alternate atrial impulse (2:1 block) or, at higher flutter rates, occasionally to every third atrial impulse (3:1 block).

In atrial fibrillation the atria exhibit no co-ordinated contraction; inspection of the heart shows merely an irregular tremulousness of the surface. Most of the atrial muscle fibres are completely or partially refractory at any instant and the excitation wave takes an irregular sinuous path, which varies. The AV node is stimulated at irregular intervals and the ventricular beats become irregular. Fibrillation commonly supervenes on flutter in diseased hearts and the atria may be excited as fast as 450 min^{-1}. Atrial fibrillation, which abnegates the chance of any effective atrial contraction, is not of itself fatal, because atrial contraction is only responsible for 20–25 per cent of ventricular filling. However, the irregularity which the condition superimposes on the ventricular rhythm is dangerous and this must be prevented either (1) by direct electrical shock delivered to the chest wall, timed to fire at the peak of the R wave (when the ventricle is completely refractory) causing depolarization of the atria and the abolition of ectopic foci, thereby (hopefully) allowing the normal sinus rhythm to be restored; (2) drug therapy such as digitalis. The action of digitalis on the ventricular rate is complex— the drug lengthens the refractory period of the transmission system of the a.v. node by reflexly increasing vagal activity (perhaps by promoting the activity of ventricular receptors—page 127) and by a direct action on the a.v. node. The ventricular muscle thereby receives a more regular transmission of impulses and can beat more slowly and powerfully. It is interesting that digitalis actually increases the rate of atrial fibrillation by shortening the refractory period of atrial muscle as a consequence of the increased vagal activity which the drug induces.

Electrocardiographic features. In atrial arrhythmias the pattern of the QRST complex is normal but the ventricular rate is more rapid than normal—usually between 130 and 200 min^{-1}. In paroxysmal atrial tachycardia the P wave is abnormal, usually inverted, and bears a constant relation to the QRS complex. In paroxysmal atrial flutter the atrial rate is so rapid (200–400 min^{-1}) that atrial repolarization is affected. As a result each P wave tends to be followed by an atrial T wave pointing in the opposite direction thus giving an undulating or saw-toothed appearance in some leads. As a rule the QRS complex regularly follows alternate P waves. In atrial fibrillation there is no co-ordinated atrial contraction and therefore no P wave. There are irregular electrical oscillations called fibrillation waves (f) [Fig. II.78] which occur at a rate of 300–500 min^{-1}. The bundle of His cannot conduct impulses at such high frequencies. The QRS complex is irregular in time with a rate of about 150 min^{-1}.

Physiological effects. Tachycardia of any abnormal variety leads to a fall in cardiac output. Diastole is too brief to allow proper filling of the heart so that each beat discharges only a small amount of blood. The rest period is very short, the heart fatigues and signs of heart failure ultimately appear.

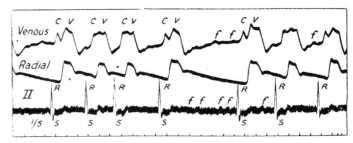

Fig. II.78. Atrial fibrillation in man. Simultaneous records of jugular venous tracing, radial curve, and electrocardiogram (Lead II). The P waves are replaced by small irregular oscillations (f, f). The ventricular complexes are of supraventricular origin and occur at completely irregular intervals (note interval between R waves).

The radial curve displays the gross pulse irregularity. No a waves are present in the jugular curve. (Lewis, T. (1925). *Mechanism and graphic registration of the heart beat*. Shaw, London.)

Causes. Paroxysmal atrial tachycardia is frequently found in otherwise normal subjects. Atrial flutter and fibrillation are normally associated with mitral stenosis, severe hyperthyroidism, or with coronary arterial disease.

Vagal stimulation has a striking effect on atrial arrhythmias. If the atrial rate is not too rapid vagal stimulation may stop the arrhythmia and lead to a restoration of sinus rhythm. For this reason pressure on the carotid sinus is often successful in paroxysmal atrial tachycardia.

Ectopic ventricular tachycardia

Ventricular tachycardia or fibrillation may be produced experimentally by ligating a coronary artery or occasionally by inhaling chloroform. Clinically ventricular tachycardia and ventricular fibrillation are frequently observed and recorded during operations on the heart.

Ventricular fibrillation is fatal unless treated promptly. It is the cause of sudden death in coronary occlusion. If it supervenes in patients in hospital in intensive care units it may be aborted by applying electrodes directly either side of the heart and passing a high voltage (500 volts) direct current for a few milliseconds across the heart. Whereas this can be employed during cardiac operations, it can hardly be used in patients with intact chests. In such cases, the heart *may* be defibrillated by electrodes applied to the chest wall which deliver higher direct current voltages. If repeated electric shocks prove ineffective, then the chest should be opened and the heart massaged manually to promote some coronary flow and some improvement of potential cardiac contractility before delivering direct current shocks. If thoracotomy and direct cardiac shocks prove impracticable in the circumstances external cardiac massage consisting of rhythmic powerful thrusts to the chest wall may promote the requisite coronary flow which renders the cardiac muscle susceptible to the influence of external direct current shocks.

Vectorcardiography

The voltage from Lead I is connected to an oscilloscope in such a manner than this voltage determines the horizontal deflection of the beam. Vertical deflection is governed by the Lead III voltage. The cathode-ray beam describes three loops successively—one for the P wave, one for the QRS complex, and one for the T wave. Between each loop the beam pauses because there is zero voltage in each of the leads.

The main or QRS loop gives the simultaneous mean electrical axis [Fig. II.79].

Vectorcardiography has never achieved the popularity of the standard 12-lead ECG.

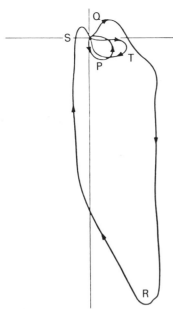

Fig. II.79. Tracing from a normal planar vectorcardiogram.

Echocardiography

A piezoelectrical crystal is excited intermittently by electronic means and transmits bursts of ultrasound. Usually, ultrasound frequencies of 2¼ million cycles per second are used in bursts of 1000 s⁻¹. The ultrasound waves penetrate tissues and are reflected back to the crystal which is arranged to receive the 'echo' between the transmitting periods. The time for the echo to be received depends on the distance travelled. Bone and air absorb ultrasound but the heart muscle and valves are good reflectors and it is the motion of these structures that is detected. At 2.25 million Hertz frequencies, structures 15–20 cm from the chest wall can be examined.

A variety of systems are used: in A mode scanning the intensity of the echo is shown on the *x* axis of the oscilloscope whereas the time of its arrival back at the receiver is displayed on the *y* axis. This technique has been superseded by M mode echocardiography with time on the *x* axis, distance on the *y* axis, and intensity (brightness) on the *z* axis. The electrocardiogram is recorded simultaneously for timing of the cardiac cycle events; the movement of, say, mitral valve cusps can be readily seen (see Sokolow and McIlroy (1978) who give details of these and more sophisticated methods).

REFERENCES

Chou, T. and Helm, R. A. (1974). *Clinical vectorcardiography*, 2nd edn. New York.
Feigenbaum, H. (1976). *Echocardiography*, 2nd edn. Saunders, Philadelphia.
Goldman, M. J. (1976). *Principles of clinical electrocardiography*, 9th edn. Los Altos, California.
Leatham, A. (1975). *Auscultation of the heart and phonocardiography*, 2nd edn. Churchill, Edinburgh.
Lewis, T. (1928). *Mechanism and graphic registration of the heart beat.* Shaw, London.
Sokolow, M. and McIlroy, M. B. (1977). *Clinical cardiology*. Los Altos, California.

THE ELECTROPHYSIOLOGY OF CARDIAC MUSCLE CELLS

The advent of the technique of microelectrode penetration of cells has led to an enormous advance in our understanding of the processes underlying the sequential excitation of the cardiac chambers. In the 1950s the impalement of Purkinje fibres by Draper and Weidmann (1951) showed the general features of the cardiac poten-

tial. At rest most Purkinje (and ventricular fibres) were not inherently rhythmic and showed a resting (polarized) membrane potential of approximately 90 mV negative inside with respect to outside—similar to that recorded in resting nerve [p. 267]. This resting potential was steadily maintained until the cell was depolarized by a cathodal electrical stimulus whereupon the potential changed fulminantly to values of 20 to 30 mV positive inside to outside. This rapid alteration of potential on depolarization differed little from that seen in nerve. However, repolarization in the Purkinje cell requires some 400 ms [Fig. II. 80]. This is entirely different from the situation in skeletal muscle or nerve in which the action potential is complete within a few milliseconds [p. 269]. Because of this long action potential the cardiac cells do not regain their excitability until the mechanical contraction is almost complete [Fig. II.81]. Thus the heart cannot be tetanized as can skeletal muscle [p. 255].

Moreover repolarization occurs in several steps—a rapid initial fall from +30 mV to −10 mV followed by a 'plateau phase' in which the membrane potential only falls to −40 mV and a last stage

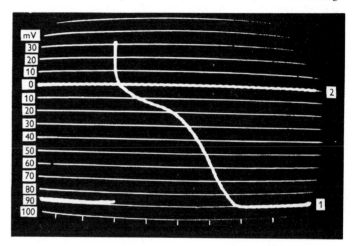

Fig. II.80. Action potential of Purkinje fibre (dog) recorded by an internal electrode. Time intervals 100 ms. Ordinate: inside potential relative to outside. Potentials above zero are positive inside; those below zero are negative inside. Resting potential = −90 mV. Peak of action potential = +31 mV. (Draper, M. and Weidmann, S. (1951). *J. Physiol., Lond.* **115,** 74.)

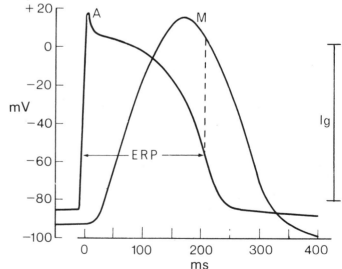

Fig. II.81. Action potential (A) and isometric tension (M) of cardiac muscle.
Peak of isometric tension curve precedes the end of the effective refractory period (ERP) (After Brooks, C. M. *et al.* (1955). *Excitability of the heart.* Grune and Stratton, New York.)

in which a relatively more rapid drop to the resting value of −90 mV is achieved.

Hutter and Trautwein (1956) succeeded in recording intracellularly the rhythmic action potentials of cells in the sinus venosus of the frog. These differed from the Purkinje and ventricle cells in showing (a) that the final polarized potential was rarely more negative than −50 mV (b) that during the diastolic period the membrane potential progressively changed from −50 mV to approximately −40 mV, whereupon the fulminant upstroke of the action potential occurred to achieve a positive potential of some +10 mV.

Hutter and Trautwein (1956) examined the influence of vagal and of sympathetic stimulation on the action potentials of these rhythmically beating sinus venosus cells [Fig. II.82(a, b)]. Vagal stimulation abolished the slow diastolic change from −50 mV to −40 mV and indeed hyperpolarized the cell (i.e. rendered the inside of the cell more negative with respect to the outside) and abolished the spontaneous rhythm. Sympathetic stimulation increased the rate of diastolic depolarization and increased the rate of beat [Fig. II.82(c)]. It is notable that it additionally hyperpolarized (rendered the cell more negative inside) at the end of the action potential and increased the height of the action potential.

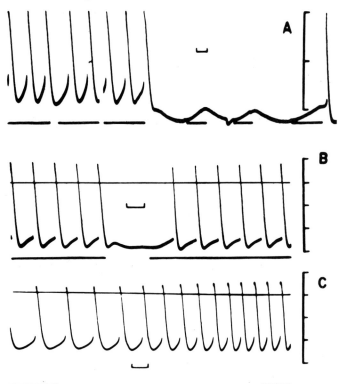

FIG. II.82. Effect of vagus and sympathetic stimulation on pacemaker potentials in the sinus venosus of the frog heart: (A and B) vagi stimulated during break in lower trace; and (C) vagosympathetic stimulation in an atropinized heart. Voltage calibration in 20 mV steps. Time: 1 second. (Weidmann, S. (1957). *Ann. N.Y. Acad. Sci.* **65**, 663.) (From results of Hutter and Trautwein 1956.)

They showed that ACh increased the extrusion rate of ^{42}K from the sinus venosus. First the sinus venosus was loaded with radioactive potassium by allowing it to beat in a solution containing ^{42}K. On transferring the tissue to Ringer they then measured the rate of loss of the isotope [Fig. II.83]. Acetylcholine caused a threefold increase in the rate of loss of the positively charged K⁺, thereby hyperpolarizing the cells. Tissue loss of isotope reverted to its control rate when atropine (which prevents the membrane action of ACh) was added even though ACh was still present.

These important results revealed the characteristics of pace-

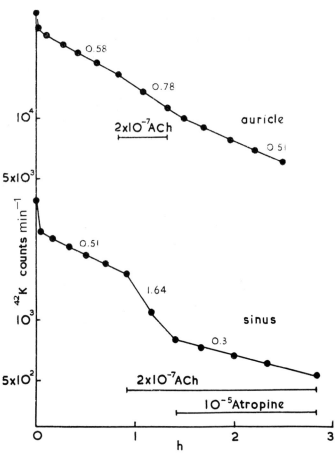

FIG. II.83. An experiment on the isolated sinus venosus (lower graph) and right atrium (upper graph) of a tortoise heart. The tissues were loaded with ^{42}K and the rate of loss of the isotope was studied. The small figures give the rate constants in h⁻¹ at different stages of the experiment. In the sinus venosus the action of acetylcholine was stopped by atropine. Abscissae: time (h). Ordinates: isotope content of tissue in counts min⁻¹ plotted on a logarithmic scale. (Hutter, O. F. (1957). *Br. med. Bull.* **13**, 176.)

maker tissue, i.e. cells which show a progressive depolarization during diastole. Such cells are found in nodal tissue and in some of the Purkinje tissue. Normally the rate of rise of diastolic depolarization is 15 to 60 mV s⁻¹ in sinus nodal tissue whereas that of a.v. nodal tissue and such Purkinje cells as do show pacemaker activity is appreciable slower. Moreover, sinus tissue cells register only some −50 mV at the end of repolarization compared with much more negative values (e.g. −90 mV) registered in atrial, Purkinje, and ventricular cells. Once the membrane potential achieves a threshold value of −40 mV there occurs a fulminant rise of inward current carried by sodium ions and the action potential 'takes off'. It follows that the threshold excitation is achieved earliest in the sinus and that the action potential generated there is conducted to the other cardiac cells (some of which are latent pacemakers) and arrives there before their own diastolic depolarization to threshold is reached. Thus the normal sequential rhythmic beat of the various parts of the heart is achieved. If the natural pacemaker of the sinus node is destroyed then the fastest latent pacemaker (often the a.v. node) takes over.

It is important to note that the conduction velocity in the AV node is low (only 0.2 m s⁻¹), which ensures an appreciable delay between atrial and ventricular contraction. This slow conduction is partly referable to the small size (radius 7 μm) of the AV nodal tissue (compared with that of Purkinje fibres—radius 50 μm) which conduct at 4 m s⁻¹ and partly to the fact that the ionic currents generated by the nodal fibres are far less than those developed by the Purkinje tissue (Noble 1979).

Voltage clamping and ionic current studies

The large size (1000 μm long and 50 μm radius) of Purkinje cells favours micro-electrode penetration and voltage clamp studies [p. 271]. Some Purkinje cells moreover evince pacemaker properties and this has led to their usage in experiments designed to analyse such features. Unfortunately, there are differences between the nature of the ionic currents responsible for pacemaker and action potentials in Purkinje cells compared with those now recognized in atrial cells and those assumed to occur in s.a. nodal tissue. Nevertheless, the study of Purkinje cells may be summarized first.

1. On depolarizing the cell membrane, the first ionic current to be recorded is an inward flow of sodium current (similar to that in nerve) due to an increase of at least one hundred-fold in sodium conductance. Just as in nerve, this sodium current is selectively blocked by tetrodotoxin [see p. 275]. The inward sodium current provides most of the upstroke of the action potential. Sodium conductance then falls but not completely to its resting level.

2. A slower and smaller inward flow of current is carried by Ca^{2+} (Beeler and Reuter 1970) which can be experimentally distinguished from the inward sodium current. Thus the threshold membrane potential for activating i_{Na^+} in ventricular fibres is -60 mV whereas that for activating $i_{Ca^{2+}}$ is -30 mV.

The inward calcium current is abolished when a Purkinje fibre is bathed in a calcium-free medium [FIG. II.84].

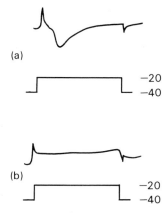

(a)

(b)

FIG. II.84. Evidence for Ca^{2+} current in the dog ventricle. (a) Response of membrane current (top) to a depolarization (maintained for 600 ms) from -40 mV to -25 mV in Tyrode solution containing 1.8 mM Ca^{2+}. Note the presence of a transient inward current. (b) The response to a similar depolarization which occurred while the preparation was bathed in Tyrode which contained no Ca^{2+}; no inward current occurs. (When calcium ions were readmitted to the bathing solution the inward current reappeared.) (Redrawn from Beeler, G. W. and Reuter, H. (1970). *J. Physiol., Lond.* **207**, 165, 191, 211.)

By clamping the membrane voltage at -40 mV, $i_{Ca^{2+}}$ can be studied by itself as a response to graded steps of further depolarization. Thus when the voltage is raised to -25 mV as in FIG. II.84(a) a calcium inward current occurs for some 200 ms. This $i_{Ca^{2+}}$ is unaffected by tetrodotoxin (unlike i_{Na^+}) but is blocked by Mn^{2+} (which does not alter the initial inward sodium current).

The inactivation of the inward $i_{Ca^{2+}}$ is far slower than that of the initial sodium current when the membrane potential is near zero and this perhaps explains the long-lasting plateau of the cardiac action potential. The entry of calcium ions during this plateau provides an important link between the excitation and contraction in cardiac muscle.

3. Quite unlike the situation in nerve, threshold excitation causes an initial *fall* in potassium conductance. This was first shown by Hutter and Noble in 1960 and by Noble (1962). The decrease in

potassium conductance with depolarization is known as anomalous rectification. In Purkinje fibres K^+ conductance shows two components. The first ($g_{K^+_1}$) is time-independent and shows anomalous or inward-going rectification—i.e. the channels pass outward current in response to positive potential changes less readily than they pass inward current in response to negative potential changes. This existence of inward-going rectification has the result of reducing the total ion movements during the action potential and is important in determining the effects of variations of extracellular potassium concentration on cardiac rhythmicity in clinical conditions.

The second potassium component is a small slower time-dependent conductance $g_{K^+_2}$. In Purkinje tissue the activation of $g_{K^+_2}$ is thought to be important in terminating the plateau phase of the action potential. The sharp drop in the membrane potential is accompanied by a fall in the two inward currents carried by sodium and calcium ions [FIG. II.85].

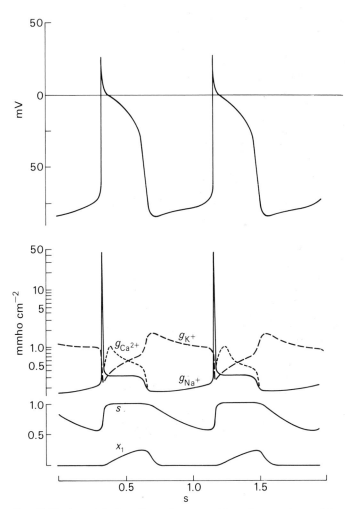

FIG. II.85. A model of cardiac action potentials and pacemaker activity. Changes of g_{Na^+}, g_{K^+}, and $g_{Ca^{2+}}$ are shown in relation to the action potential. For an account of the gating variables s and x_1, which influence the time-dependent changes of g_{K^+}, see Noble (1979) Chapters 6 and 7. (Redrawn from Noble, D. (1979). *The initiation of the heart beat*, 2nd edn. Clarendon Press, Oxford.)

On reaching its most negative value after repolarization, the membrane potential of a pacemaker Purkinje cell then shows the steady ascent to more positive values which characterizes the 'pacemaker potential'. E_m is very negative and $i_{K^+_2}$ is at its highest. Even though g_{Na^+} is low the large difference between E_m (-80 mV) and E_{Na^+} ($+61$ mV) produces an appreciable inward sodium current. Initially this sodium current only slightly exceeds the outward current of potassium ions ($i_{K^+_2}$) but $g_{K^+_2}$ decays and $i_{K^+_2}$ drops continuously, allowing the inward sodium current to produce a progressively more positive membrane potential. The outward potassium current sharply declines towards the end of the pacemaker potential and most of the rapid increase in the rate of depolarization is attributable to this. Once the sodium threshold of approximately -60 mV is achieved, the spike potential 'takes off' [FIG. II.86].

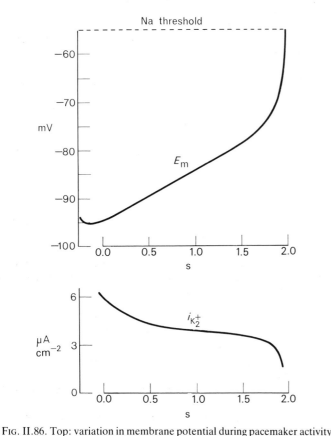

FIG. II.86. Top: variation in membrane potential during pacemaker activity in a Purkinje cell.
Bottom: $i_{K^+_2}$ changes (calculated from voltage clamp results). (Modified from Noble, D. and Tsien, R. W. (1968). *J. Physiol., Lond.* **195**, 185.)

As Noble (1979) points out, however, the difference between the pacemaker range of potentials in the Purkinje fibres (-90 to -60 mV) and that in sinus and atrial tissue (-60 mV to -40 mV) 'suggests caution in applying analysis of one tissue to the other'. However, sinu-atrial cells have not yet been submitted to voltage-clamp studies owing to the technical difficulties; atrial fibres are themselves normally quiescent but can be induced to display pacemaker properties and rhythmic beats by applying a steady depolarizing current (Brown, Clark, and Noble 1972). Analysis of the features of such activity reveals that its mechanism differs from that of Purkinje fibres in that no activation of K⁺ currents can be found in the range of -90 to -60 mV in atrial fibres whereas these currents are substantial in Purkinje cells. The threshold for activation of K⁺ current in atrial cells is about -40 mV.

Acetylcholine causes a large increase in K⁺ membrane conductance in sinus and atrial tissue; it hyperpolarizes the cell, reduces the action potential duration, and may block conduction to neighbouring atrial fibres and to the AV node.

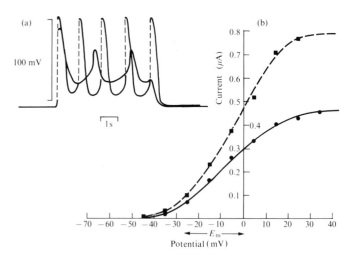

FIG. II.87(a) Pacemaker activity induced in an atrial trabeculum by the *same* magnitude of depolarizing current before (full line) and after (broken line) addition of adrenalin (10^{-6} g ml⁻¹). Note that after adrenalin addition the early diastolic potential is about 20 mV lower. (b) The activation curve for the outward K⁺ currents before (●) and after (■) addition of adrenalin. [Modified from Brown, H. F. and Noble, S. J. (1974). *J. Physiol., Lond.* **238**, 51P.]

Adrenalin causes an increase in outward potassium current in atrial cells—a change which is the opposite of that required to accelerate the beat. The increased beat rate induced by adrenalin is due to its augmenting effect on the inward calcium current $i_{Ca^{2+}}$ [FIG. II.87(c)].

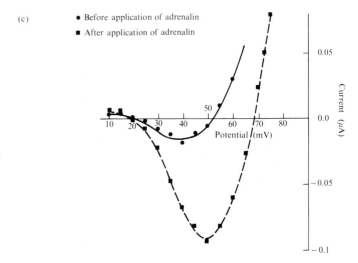

FIG. II.87(c). The initial current–voltage relation in TTX Ringer (sodium current abolished) before and after the addition of adrenalin (10^{-6} g ml⁻¹). The inward Ca²⁺ current is markedly increased and attains a peak value when a depolarization of 50 mV is induced. [Modified from Brown, H. F. and Noble, S. J. (1974). *J. Physiol., Lond.* **238**, 51P.]

Vassort (1973) using frog atrial trabeculae about 100 μm in diameter measured both ionic currents and mechanical responses during stepwise depolarizations. Addition of adrenalin (5×10^{-6}M) increased the slow inward current of calcium and the tension developed by the muscle, peak effects on both being seen with depolarizations of about 90 mV [FIG. II.88]. Thus the primary effect of adrenalin is to increase the phasic tension by increasing calcium conductance.

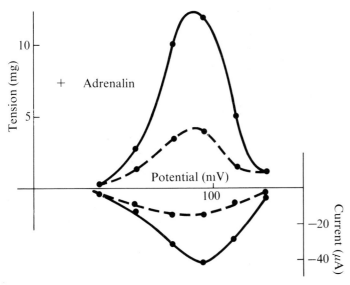

FIG. II.88. Peak tensions (above the abscissa) and maximal slow inward calcium current (below) plotted against depolarizing voltage values in Ringer solution (broken lines) and after the addition of adrenalin (5×10^{-6}M), shown by solid lines. (Vassort, G. (1973). *Eur. J. Cardiol.* **1**, 163.)

More recently, Brown *et al.* (1979) have shown that adrenalin increases the rate of rise of the up stroke of the action potential of atrial fibres (immersed in tetrodotoxin Ringer) along with an increase in slow inward calcium current. Finally, Brown *et al.* (1980) have succeeded in voltage clamping of isolated tiny pieces of sino-atrial nodal tissue of rabbit hearts and have shown that the slow inward current $i_{s.i.}$, important for the upstroke of the action potential, is increased by adrenalin.

REFERENCES

ALLEN, D. G. and BLINKS, J. R. (1978). *Nature, Lond.* **273**, 509.
BEELER, G. W. and REUTER, H. (1970). *J. Physiol., Lond.* **207**, 165, 191, 211.
BROWN, H. F. and NOBLE, D. (1974). *J. Physiol., Lond.* **238**, 51.
—— CLARK, A., and NOBLE, S. J. (1972). *Nature New Biol.* **235**, 30.
—— —— —— (1976). *J. Physiol., Lond.* **258**, 521, 574.
—— DI FRANCESCO, D., KIMURA, J., and NOBLE, S. J. (1980). *J. Physiol., Lond.* **307**, 12P.
—— —— and NOBLE, S. J. (1979). *J. Physiol., Lond.* **290**, 31P.
KATZ, A. M. and MESSINEO, F. C. (1981). *Am. Heart J.* **102**, 491.
NOBLE. D. (1979). *The initiation of the heartbeat*, 2nd edn. Clarendon Press, Oxford.
VASSORT, G. (1973). *Eur. J. Cardiol.* **1**, 163.

Neural control of the cardiovascular system

HISTORICAL INTRODUCTION

Important discoveries were made in the nineteenth and early twentieth centuries:

1. Smooth muscle in blood vessels (Henle 1840).
2. The resistance to fluid flow through narrow tubes is inversely proportional to the fourth power of the radius and proportional to the tube length (Poiseuille 1846).
3. Vagal stimulation slows and sympathetic stimulation accelerates the heart (Weber and Weber 1845).
4. The kymograph (Ludwig 1946) allows continuous measurement of heart rate and blood pressure.
5. Sympathetic nerve section causes vasodilatation; stimulation induces vasoconstriction. Section of the cervical cord causes hypotension (Bernard 1851–1858).
6. Reflex inhibition of heart rate and hypotension result from stimulation of the central end of the 'depressor' (aortic) branch of the vagus (Ludwig and Cyon 1867).
7. Paralysis of the solar sympathetic plexus results in venodilatation and a failure of venous return to the heart (Goltz 1864).
8. Transection of the medulla below the acoustic striae causes hypotension. Stimulation of structures in the floor of the lower part of the IVth ventricle induces hypertension. The concept of a vasomotor centre is introduced (Ludwig's laboratory 1870–1873).
9. Pulmonary blood flow can be calculated if the oxygen usage of the body be simultaneously measured together with the difference in arterial and mixed venous oxygen content of the blood (Fick 1870). Direct measurements made by Zuntz and Hagemann (1898) (using horses) furnish the first quantitative values of cardiac output and stroke volume.
10. The force of isometric ventricular contraction of the frog's heart varies with increased venous filling (Frank 1895).
11. Electrical stimulation of the hypothalamus promotes hypertension (Karplus and Kreidl 1909).
12. Stroke volume of the dog's heart is determined by its end diastolic filling (Starling 1910–1915).
13. The carotid sinus contains stretch receptors whose afferent fibres pass via the IXth nerve to the medulla inducing reflex vagal stimulation and sympathetic vasomotor inhibition (Hering 1920–1927).
14. Dye dilution methods of determining cardiac output are introduced by Hamilton (1928).
15. High-fidelity optical manometers allow phasic measurements of pressure changes in the heart and blood vessels (Hamilton 1928; Wigger 1928).
16. Electromagnetic flow meters chronically implanted allow phasic changes of regional blood flow to be recorded (Kolin 1936).

Such developments are fully documented elsewhere (see Fishman and Richards 1964; Folkow and Neil 1971).

REFERENCES
FISHMAN, A. P. and RICHARDS, D. W. (eds) (1964). *Circulation of the blood: men and ideas*. Oxford University Press, New York.
FOLKOW, B. and NEIL, E. (1971). *Circulation*. Oxford University Press, New York.

MEDULLARY CARDIOVASCULAR CENTRES

Alexander (1946) studied the responses of blood pressure, heart rate, and sympathetic impulse activity to stimulation or ablation of discrete areas of the bulbopontine region. Stimulation prevoked predominantly depressor effects when applied to the more medial and slightly more caudal parts of the floor of the IVth ventricle. Such depressor effects were most likely caused by an inhibition of cells, otherwise tonically discharging impulses exciting the thoracolumbar sympathetic vasoconstrictor fibres.

Pressor effects were evoked by stimulating the rostral and lateral regions of the bulbopontine area.

When the brainstem was cut across at the level of the facial genu, the blood pressure fell and this fall was presumably due to the reduction in sympathetic discharge which the section caused. The stimulation of sensory components in the sciatic nerve which had previously caused striking reflex pressor effects now evoked only a feeble pressor response. Depressor responses to electrical stimulation of the medial 'depressor area' were even more striking than before.

When the brainstem was sectioned at the level of the obex blood pressure fell to 40 mm Hg and sympathetic activity disappeared. No further pressor effects could be induced by electrical stimulation of the bulb.

Previous experiments of this type (Dittmar, Ranson, and Billingsley, etc.) had led to the use of the term vasomotor centre. The *concept* is useful—that this region of the medullary-pontine area contains neurons which exert a predominantly excitatory effect on spinal sympathetic vasoconstrictor activity. However, the term is too restrictive, for the 'pressor' areas contain neurons which excite positive chronotropic and inotropic responses from the heart. Moreover, there is no clear anatomical separation between the pressor and depressor areas in the intermediate region of the upper medulla—they overlap. Hence, it is appropriate to use the term 'medullary cardiovascular' centre recognizing that (though even this is an oversimplification) the area contains both neurons which excite thoracolumbar sympathetic fibres to the heart and blood vessels and neurons which inhibit these sympathetic fibres.

On this region, play both the tonically active afferent fibres from the baroreceptors of the carotid sinus and aortic regions and those from the atriovenous junctions and ventricles of the heart itself, as well as corticohypothalamic descending pathways which may sporadically induce striking excitatory or inhibitory influences upon them. The proprioceptors of the cardiovascular system (arterial baroreceptors and cardiac mechanoreceptors) exert some tonic restraint upon the medullary neurones responsible for sympathetic vasoconstrictor tone. Chemoreceptors from the carotid and aortic bodies exert little tonic influence on these medullary neurones in eupnoeic conditions, but in anoxia or asphyxia they excite them and thereby provoke powerful reflex sympathetic effects.

The nucleus ambiguus of the vagus and not as formerly believed the dorsal motor nucleus is the so-called cardio-inhibitory centre of the older literature. It lies lateral to the medullary reticular neurones which modify spinal sympathetic discharge. It receives via the nucleus of the tractus solitarius connections from the baroreceptors, cardiac mechanoreceptors, and chemoreceptors, all of which can increase its tonic activity. In the resting animal vagal tone secures a slower heart rate than would otherwise be the case. The vagal tone is reflexly aroused—thus section of the carotid sinus and aortic nerves conspicuously raises the heart rate. The nucleus ambiguus also receives fibres from the corticohypothalamic pathways. These may excite or inhibit it [p. 131].

The fact that sino-aortic denervation abolishes vagal tone indicates that these motor vagal nuclear cells are not tonically active unless played upon by such afferent nerves. In this respect these vagal nuclear cells differ from those of the cardiovascular centre which excite the sympathetic. The activity of these latter cells is inherently tonic and is increased by sino-aortic deafferentation.

Spinal autonomic centres

After spinal transection, sympathetic discharge is reduced due to the loss of medullary 'drive' and the blood pressure falls. The level to which it falls initially depends on the site of transection. A section at T_1 cuts off all the thoracolumbar sympathetic neurones from the 'cardiovascular centre' and the fall of blood pressure is more profound than that caused by a transection at say T_{10}. However, within days or weeks the spinal sympathetic cell bodies appear to recover some tonic discharge and moreover are capable of responding by increased activity to nocuous stimulation.

REFLEX EFFECTS ON THE CARDIOVASCULAR SYSTEM

Mechanoreceptors

These are found in: (a) the systemic arterial tree at specific sites; (b) the pulmonary trunk and its division into the right and left pulmonary arteries. Both the systemic and pulmonary mechanoreceptors are situated in the adventitia. (c) The atrio-caval junctions and the pulmonary veno-atrial junctions. These are subendocardial. (d) The atria and ventricles.

With the exception of the carotid sinus, which is supplied by the glossopharyngeal nerve, all the other mechanoreceptors are vagal.

The systemic arterial mechanoreceptors

These are found in the carotid sinus, the aortic arch, the root of the right subclavian artery and in the junction of the thyroid artery with the common carotid artery [FIG. II.89]. They are known as baroreceptors—they are sensitive to *stretch* which is usually caused by a rise of pressure expanding the arterial wall. If the stretch is prevented by surrounding the wall with a closely applied rigid plaster of Paris cast the receptors do not respond when the intraluminal pressure is raised.

The carotid sinus region lends itself to experimental isolation and the bulk of the work on the properties of the arterial baroreceptors has been done on this region.

The carotid sinus. The initial part of the internal carotid artery is dilated and it is this segment which is called the sinus [see FIG. II.89]. The muscle of the media in this region is sparse and eccentrically distributed in the circumference of the wall.

Afferent nerve endings of the sinus nerve which itself is a branch of the glossopharyngeal nerve are distributed lavishly in the adventitia but few penetrate as far as the media.

The afferent fibres from the baroreceptors have their bipolar cell bodies in the petrous ganglion and their central processes enter the medulla and synapse with the neurones of the medullary cardiovascular centre and the nucleus tractus solitarius of the vagus.

Bilateral carotid occlusion causes a rise of systemic blood pressure. The effect is abolished if the carotid sinus nerves have been previously cut [FIG. II.90]. Hence, the response is not due to cerebral ischaemia. Similarly, if all the efferent branches of the carotid artery are clamped in a normal animal (thus causing cerebral ischaemia) the systemic blood pressure does not rise.

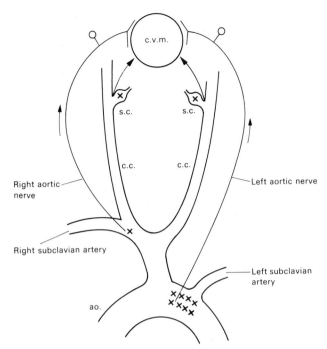

FIG. II.89. Diagram of arterial baroreceptor sites in the dog (shown by crosses). c.v.m. = cardiovascular medullary centre; s.c. = carotid sinus; ao. = aorta; c.c. = common carotid artery.

The effect of common carotid occlusion is reflex. The fall of pressure in the carotid sinus reduces the stretch on its wall and the stretch receptors lessen their discharge. As the normal tonic effect of this discharge on the medullary cardiovascular centre is inhibitory, these medullary neurones now escape and increase their drive to the sympathetic vasoconstrictors and to the cardiac sympathetic fibres. Blood pressure increases, partly because of an increase in peripheral arteriolar resistance and partly because of an increase in the force and rate of the heart beat.

If the aortic nerves are cut the rise of blood pressure, caused by temporary common carotid occlusion, is greater, because the vagal proprioceptors from the aortic areas normally limit the rise of blood pressure.

Section of both carotid sinus nerves causes a rise of systemic blood pressure [FIG. II.90]. The loss of inhibitory impulses from the carotid sinus nerves allows 'escape' of the cardiomedullary neurones and they increase their drive to the sympathetic vasoconstrictors and to the cardiac sympathetic fibres.

Perfusion of the carotid sinuses at high pressure causes a fall of systemic blood pressure [FIG. II.91]; when the perfusion pressure is reduced from normal levels the systemic pressure rises. The reflex rise of systemic pressure caused by sinus hypotension is due to the lessening of tonic inhibitory impulse activity which allows reflex escape of the sympathetic vasoconstrictor and cardiac sympathetic fibres.

Carotid sinus perfusion entails tying all efferent branches near the bifurcation except the external carotid artery. A pump supplies blood to a cannula tied into the common carotid artery and passing through the arterial segment the blood returns through a cannula tied into the external carotid artery, via a resistance to the reservoir which feeds the pump. The perfusion pressure in the carotid sinus is adjusted by altering the pump output or more commonly by changing the resistance of the outflow from the segment. Alternatively the static pressure in the sinus segment can be raised by attaching a pressure reservoir to the common carotid cannula and tying all efferent branches of the carotid bifurcation [e.g. FIG. II.91].

A mean blood pressure of some 60–70 mm Hg is required in the sinus segment before any reflex inhibition is exercised upon the

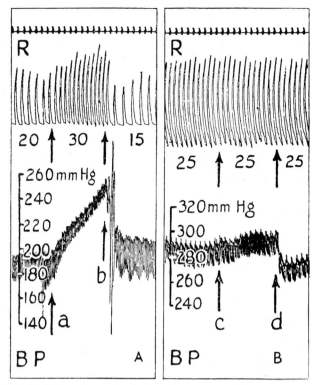

FIG. II.90. Reflex effects of occlusion of common carotid arteries on blood pressure. Experiment on dog: chloralose. Aortic nerves cut. Records from above downwards: time in 3 seconds and arterial blood pressure.

A. Between a and b occlude common carotid arteries: fall of pressure in carotid sinuses produces a rise of arterial blood pressure. At b, release arteries: sudden distension of carotid sinuses, reflexly causes temporary slowing of heart and fall of blood pressure.

Both carotid sinus nerves cut between A and B. Blood pressure rises from 190 to 280 mm Hg; note readjustment of blood pressure scale.

B. Between c and d repeat occlusion of common carotid arteries: slight mechanical increase of blood pressure. (Heymans, C. and Bouckaert, J. (1931). *J. Physiol., Lond.*)

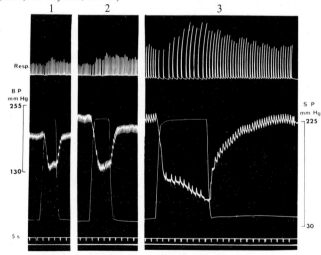

FIG. II.91. Dog, 13.7 kg. Bilateral preparation of carotid sinuses. Moissejeff 'blind sac' technique. Records from above downwards: Respiration, systemic blood pressure, static sinus pressure.

1. Rise of pressure in both sinuses from 40 mm Hg to 230 mm Hg. Between 1 and 2 left vagus cut. Systemic pressure increases.

2. Repeat rise of pressure in both sinuses—fall of systemic blood pressure is greater. Between 2 and 3 right vagus cut. Systemic blood pressure again rises.

3. Repeat rise of pressure in both sinuses. Note profound fall of blood pressure with no evidence of 'compensation' during sustained sinus hypertension. Note also marked slowing of breathing during sinus hypertension. (Eric Neil 1968.)

medullary cardiovascular centres. This is sometimes spoken of as the threshold of the sinus reflex. It is not accurate to say that a mean pressure of 70 mm Hg causes no *discharge* of the afferent mechanoreceptor nerves—it does—but the discharge is not adequate to modify the behaviour of the medullary neurones and therefore exerts no detectable reflex influence. At a normal mean pressure of 140 mm Hg in a dog, afferent impulse activity already causes tonic reflex restraint of sympathetic activity and tonic reflex stimulation of cardiovagal discharge. When the pressure drops below this, carotid and aortic baroreceptor discharge lessens [FIG. II.92(a) and (b)], sympathetic discharge is reflexly increased, and vagal activity is reflexly lessened. Both these reflex responses serve to minimize the fall of blood pressure. Conversely, if some primary influence raises

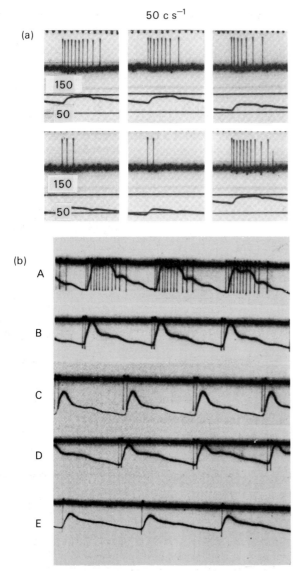

FIG. II.92. (a) Six successive records of the impulse activity of a single carotid baroreceptor fibre during one cardiac cycle at different systemic arterial pressures. Each record shows from above downwards time (50 Hz), impulse activity, and carotid arterial pressure. Calibration lines of 150 mm Hg and 50 mm Hg refer to arterial pressure. The first record is obtained at normal blood pressure. The subsequent four records were obtained as the arterial pressure was successively reduced by bleeding. The final record shows the restoration of impulse activity when the blood lost was returned to the animal. (b) Each record shows from above downwards, impulse activity in a single fibre of the left aortic nerve and blood pressure recorded from the common carotid artery. Records taken at different mean pressures: A 125 mm Hg; B 80 mm Hg; C 62 mm Hg; D 55 mm Hg; E 42 mm Hg.

the systemic pressure the increased impulse discharge in the sinus nerve afferents tends to reduce sympathetic activity and to increase vagal activity which again serves to minimize the rise of blood pressure. The sinus nerves (and the aortic nerves) act as 'blutdruckregler' (Kahn 1930) or blood pressure regulators. Samson Wright (1930) called them 'buffer nerves' which more than any other term describes their mode of action.

The effect of vagal section on carotid sinus reflex responses. A rise of pressure in both vascularly isolated innervated carotid sinuses causes a fall of systemic blood pressure but the fall is limited by the influence of the other mechanoreceptors still present and functional in the cardiovascular system. Thus, in FIGURE II.91 (1) the rise of sinus pressure causes a fall of systemic pressure from 200 mm Hg to 140 mm Hg, but having reached its nadir the systemic pressure begins to creep upwards. The reason is that aortic mechanoreceptors and the cardiopulmonary receptors are now exposed to this much lower systemic pressure and their own impulse discharge lessens, which itself tends to reduce their inhibitory effect on the medullary cardiovascular centres and allows some escape of these which expresses itself as an increased sympathetic vasoconstriction. This secondary effect serves to hoist the peripheral resistance somewhat and thereby to increase the blood pressure. Between (1) and (2) the left vagus was cut, whereupon the systemic pressure rose to 240 mm Hg (due to the loss of some cardio-aorto-pulmonary mechanoreceptor tonic discharge) and the same rise of sinus pressure now lowered the systemic pressure by 95 mm Hg—buffering by the surviving mechanoreceptor fibres was less complete but still the blood pressure rose slightly during sinus hypertension indicating that these surviving buffer nerves were operating. Between (2) and (3) the right vagus was cut and, with it, all known surviving mechanoreceptor afferents. Sinus hypertension, as before, now caused a fall of systemic blood pressure from 240 mm Hg to 80 mm Hg, and there was no evidence whatever of 'compensation' during the period of raised sinus pressure for there were no buffer nerves left to operate in this manner.

Aortic arch. The stretch receptors are situated in the adventitia of the wall of the transverse part of the arch, adjacent to the root of the left subclavian artery. The aorta represents the surviving part of the fourth left branchial arch and the root of the right subclavian artery represents the surviving remnant of the fourth right branchial arch. The right aortic nerve though distributed in minor degree to the arch of the aorta mainly stems from the root of the right subclavian artery. The two aortic nerves run parallel to the vagus enclosed in the vagus sheath—they may be separate from the vagal trunk, as in the rabbit. When the vagus sheath is opened, under magnification the aortic nerves may often be seen to lie on the main vagal trunk (cat and dog). The nerves join the superior laryngeal branch of the vagus near its junction with the vagal trunk. Their fibres course with the superior laryngeal afferent fibres into the trunk and have their bipolar cell bodies in the nodose ganglion. The central processes run on into the medulla to synapse with the cell bodies of the nucleus of the tractus solitarius of the vagus and with neurones which influence thoracolumbar sympathetic discharge. The conduction velocity of the aortic myelinated fibres is of the order of 20–45 metres per second.

Vagal cardiac mechanoreceptors

1. Myelinated atrio-caval and pulmonary veno-atrial receptors

Paintal (1953) initially classified these vagal afferents as *A* and *B* type. *A* type receptors discharge during atrial systole only and their impulse burst occurs in the P–R interval of the ECG [FIG. II.93]. *B* type receptors (also vagal) do not discharge in the P–R interval but

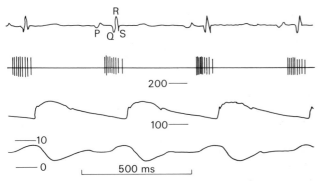

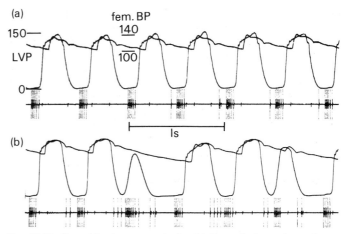

FIG. II.93. Type A discharge pattern of a right atrial receptor. Records from above downwards; ECG, right atrial vagal afferent discharge, arterial blood pressure (mm Hg), and right atrial pressure (cm H_2O). The fibre discharges only during atrial systole.

FIG. II.95. Cat, chloralose anaesthesia. (a) Records from above downwards: femoral arterial pressure, left ventricular pressure, and electroneurogram of left atrial vagal afferent fibre. Note (i) discharge precedes ventricular systole; (ii) femoral arterial pressure rise occurs later than the rise of ventricular pressure (pulse delay). (b) Same preparation showing two ventricular extrasystoles, the first of which does not open the aortic valve (compensatory pause in pulse record). Each ventricular extrasystole alters the pattern of afferent vagal discharge.

show activity which precedes the onset of atrial contraction which varies in its intensity with the atrial venous filling (V wave) and which reaches its peak after the T wave of the ECG [FIG. II.94]. The bigger the venous return the greater the discharge of these *B* type fibres. They would seem to be suited to provide information about circulating blood volume—'volume receptors'—see page 149 *et seq.*

Many objections to a rigorous division of the myelinated cardiac atrial receptors into *A* and *B* types have been raised. Certainly there exist fibres whose discharge is characterized by an '*A*'-type burst followed by activity during venous filling of the atrium. Such receptors have been designated 'intermediate type receptors'. In any case, a fibre showing '*A*'-type discharge rhythm can alter its characteristics when, for instance, ventricular extrasystoles occur [FIG. II.95(a) and (b)]. A full discussion of these problems is provided by articles in Hainsworth, Kidd, and Linden (1979).

Cardiac myelinated afferents subserve two roles: 1. Their stimulation provokes tachycardia [FIG. II.96], as shown by Linden and his colleagues, who proved that distension of the pulmonary veno-atrial orifice or a pouch of the left atrium (which excites the left atrial receptors) induces reflex tachycardia. Similarly distension of the superior vena caval–right atrial junction induces tachycardia (Linden *et al.* 1970). Such findings suggest that the atrial receptors of both sides of the heart may be responsible for the Bainbridge reflex.

2. The left atrial mechanoreceptors have also been credited with a role in modifying the output of ADH (Koizumi and Yamashita 1978). Increased distension of the left atrium causes a moderate diuresis, but this is unlikely to be solely due to a reduction of ADH output as was originally claimed. Thus Linden and his co-workers have claimed that the diuresis is partly due to the reflex secretion of a diuretic hormone which is, however, unidentified chemically as yet (see Hainsworth *et al.* 1979).

2. Non-myelinated atrial afferents (C fibres)

First described by Coleridge *et al.* (1973) these, unlike the receptors

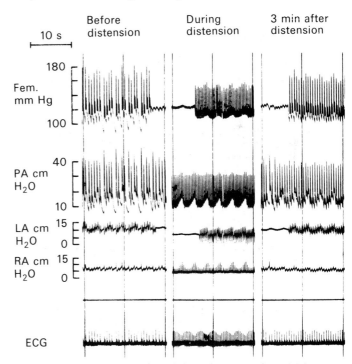

FIG. II.96. Effects of distension of the pulmonary vein–atrial junction. (Ledsome, J. R. and Linden, R. J. (1964). *J. Physiol., Lond.* **170**, 456.)

of myelinated vagal fibres which are located mainly at the vein-atrial junctions, are scattered throughout the atria and the interatrial septum. Their discharge is sparse (1 s^{-1}) if present at all, and is irregular. Increase in the atrial pressure increases their impulse activity. Thorén (1979) considers that an increase in their discharge provokes reflex vasodilatation, particularly in the renal vascular circuit.

3. Ventricular C fibres

Coleridge, Coleridge, and Kidd (1964) showed that such fibres discharged with an irregular rhythm at a low rate (1 s^{-1}) and that their site was in the left ventricle. Probing the epicardium increased their activity as did injections of veratridine. Muers and Sleight (1972) showed that inotropic stimulation by adrenaline markedly

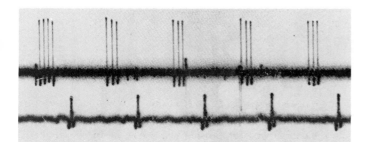

FIG. II.94. B type discharge pattern of a right atrial vagal receptor. Records from above downwards; impulse activity and ECG record of five cardiac cycles (Neil, E. and Joels, N. (1961). *Arch. exp. Path. Pharmak.* **240**, 453.)

stimulated the receptors, as did partial occlusion of the aorta [e.g. Fig. II.97] or the coronary sinus. Such fibres are far more numerous than those of atrial receptors of unmyelinated fibres and their reflex effects include profound bradycardia, reflex inhibition of sympathetic discharge (particularly in the renal circuit), and correspondingly systemic hypotension.

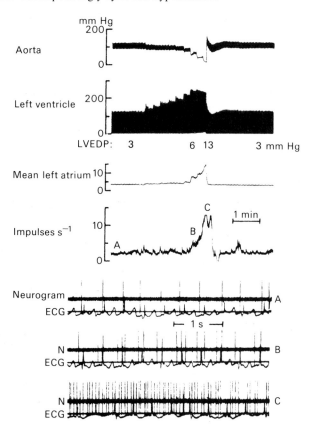

Fig. II.97. Effect of a graded aortic occlusion on aortic blood pressure, left ventricular pressure, left ventricular end-diastolic pressure (LVEDP), mean left atrial pressure, and spike frequency in a single left ventricular C fibre. Letters in spike frequency recording (fourth panel) correspond to neurograms below. During graded aortic occlusion this receptor does not respond to a change in left ventricular systolic pressure from 120 to 220 mm Hg. However, when aortic occlusion is further accentuated, receptors increase discharge in parallel with increase in LVEDP. (Thorén, P. (1977). *Circulation Res.* **40**, 415.)

Patients subjected to coronary angiography often show bradycardia when the contrast medium is injected and Sleight (1975) suggested that this may be due to the chemical excitation of such ventricular receptors. This was proved to be the case by Eckberg *et al.* (1975) and by Coleridge *et al.* (1979). Thorén (1979) considers that vasovagal syncope and the bradycardia of myocardial infarction may be ascribed to the reflexes aroused by an increase in ventricular C fibre activity.

REFERENCES

COLERIDGE, H. M., COLERIDGE, J. C. G., and KIDD, C. (1964). *J. Physiol., Lond.* **174**, 323–9.
—— —— DANGEL, A., KIDD, C., LUCK, J. C., and SLEIGHT, P. (1973). *Circulation Res.* **33**, 87–97.
ECKBERG, D. L., WHITE, C. W., KIOSCHOS, J. M., and ABBOUD, F. M. (1974). *J. clin. Invest.* **54**, 1445–61.
HAINSWORTH, R., KIDD, C., and LINDEN, R. J. (eds.) (1979). *Cardiac receptors.* Cambridge University Press.
JOHANSSON, B. (1962). *Acta physiol. scand.* **57**, Suppl. 198, 1–191.
KOIZUMI, K. and YAMASHITA, H. (1978). *J. Physiol., Lond.* **385**, 341–58.
—— NISHINO, H., and BROOKS, CHANDLER McC. (1977). *Proc natn. Acad. Sci. U.S.A.* **74**, 2177.

KOLLAI, M., KOIZUMI, K., YAMASHITA, H., and BROOKS, CHANDLER McC. (1978). *Brain Res.* **150**, 519–32.
MUERS, M. F. and SLEIGHT, P. (1972). *J. Physiol., Lond.* **221**, 283–309.
PAINTAL, A. S. (1953). *J. Physiol., Lond.* **120**, 596–610.
—— (1973). *Physiol. Rev.* **53**, 159–227.
SLEIGHT, P. (1975). Neural control of the cardiovascular system. In *Modern trends in cardiology* (ed. M. F. Oliver) pp. 1–43. Butterworths, London.
THORÉN, P. (1977). *Circulation Res.* **40**, 415–21.
—— (1979). *Rev. Physiol. Biochem. Pharmacol.* **86**, 1–94.

THE HEART RATE

The resting heart rate—65–75 beats per minutes—can be much lower in athletes (40 min⁻¹). The heart normally beats under the influence of the cardiac vagal nerves. Atropinization, which blocks postganglionic vagal transmission to the s.a. node, will raise the heart rate to 150 min⁻¹. In maximal athletic activity the heart rate may be as high as 180–200 min⁻¹ due to sympathetic chronotropic discharge which in turn is 'driven' by corticohypothalamic activity which presumably simultaneously reduces the efficacy of the mechanisms responsible for resting vagal tone.

1. Baroreceptors and the heart rate

Cardiac vagal impulse activity is reflex in origin and is aroused by impulses from the carotid and aortic baroreceptors. The sinus and aortic afferents send impulses via the IXth and Xth nerves respectively into the tractus solitarius (TS) in the medulla where they relay at the nucleus of the tractus (NTS). Second-order neurones then excite the cell bodies of the nucleus ambiguus (NA). These are the cell bodies of the cardiac vagal motor neurones which course down the vagus to effect ganglionic relays at various sites—s.a. node, atria, and a.v. node—in the heart itself. Complete proof that the NA *is* the so-called cardiac vagal centre was provided by McAllen and Spyer (1977). Impaling cell bodies of the NA with a microelectrode they showed that antidromic stimulation of the central end of the peripheral cardiac vagal branches evoked depolarization of NA cells.

Impulse activity of cardiac vagal motor neurones depends primarily on baroreceptor input which in turn depends on the height of the mean arterial blood pressure and on the pulse pressure. In normal circumstances this fairly constant input is phasically interrupted by the activity of the 'inspiratory centre'. Each time inspiration occurs it is due to neuronal activity of cells in the medulla. These, besides initiating inspiration, discharge to the nucleus of the tractus solitarius and the nucleus ambiguus and inhibit both the relay of the baroreceptor–NTS–NA pathway and the activity of cardiac cell bodies themselves thus switching off cardiac vagal motor discharge. The result is that the heart quickens with inspiration and slows during expiration—the phenomenon of *sinus arrhythmia*.

Cell bodies in the NA discharge also to the intrinsic muscles of the larynx via the recurrent laryngeal nerves. The abductor muscles of the larynx contract widening the laryngeal aperture as inspiration occurs, owing to impulses firing from the NA. It is clear from Fig. II.98 that each burst of 'inspiratory' neuronal activity is associated with a silencing of the cardiac vagal motor neurone. Even when baroreceptor traffic is increased by raising the pressure in an isolated innervated carotid sinus each inspiratory movement is accompanied by a silencing of cardiac vagal discharge and cardiac acceleration [Fig. II.99].

The quickening of the heart beat with natural inspiration is not wholly attributable to interruptions of cardiac vagal discharge. The cardiac sympathetic branches show a rhythmic activity too. However, recordings from these branches can only be obtained by thoracotomy which in turn presupposes artificial ventilation by a

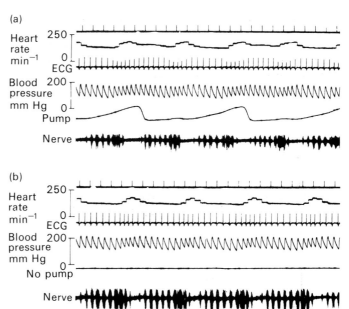

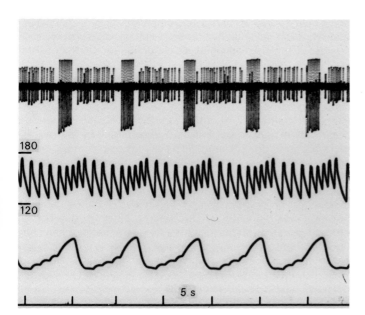

FIG. II.98. Dog 16.2 kg. Chloralose anaesthesia. Spontaneous breathing. Records from above downwards: electroneurogram of an efferent slip of the cervical vagus, containing a single fibre supplying the abductor (inspiratory) muscle of the larynx and a single fibre of a cardiac vagal efferent; blood pressure; and respiratory excursions (inspiration upwards) recorded by a pneumotachometer. Time marker shows 5-s intervals. *Note* that with each burst of 'inspiratory' laryngeal neurone activity the smaller cardiac vagal impulses cease and the heart quickens. During the expiratory periods vagal activity is resumed and the heart slows. (Eric Neil 1976.)

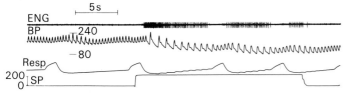

FIG. II.99. Dog 17.3 kg. Morphine–chloralose–urethane anaesthesia. Spontaneous respiration. Right carotid sinus isolated but innervated. Records from above downwards: time signal interval 5 s; electroneurogram (ENG) of single cardiac vagal efferent; systemic blood pressure (BP); respiration (inspiration upwards); and pressure in the right carotid sinus (SP) recorded from the right lingual artery. Left carotid artery clamped.

Initially, zero pressure in right carotid sinus. No cardiac vagal discharge visible. On raising sinus pressure during inspiration, no vagal discharge is seen and no bradycardia occurs until expiration occurs. Each subsequent inspiration abolishes vagal discharge and is accompanied by tachycardia. (Eric Neil 1973.)

pump. FIGURE II.100 shows that in such circumstances the variations in heart rate and in sympathetic activity do *not* follow the respiratory excursions induced by the pump but exhibit a rhythm which is presumably that determined by the cyclical activity of the pontomedullary neuronal complex 'respiratory centre'. Even when the pump is disconnected and the animal is allowed to suffer mild asphyxia the cardiac sympathetic discharges and the heart rate continue to show this same rhythmic variation, proving that in these circumstances afferent discharges of pulmonary stretch receptors of the Hering–Breuer type [see p. 170] cannot be responsible for the rhythm.

The importance of the role of the *central* respiratory rhythm compared with that of the pulmonary vagal afferents also applies to the behaviour of the cardiac vagal efferents. FIGURE II.101 shows records obtained in a spontaneously breathing dog where the rhythm of cardiac vagal firing is clearly related to the natural expiratory periods. However, when respiratory movements are paralysed by administration of gallamine (Flaxedil) the rhythmic discharge of the cardiac vagus persists unaltered although natural

FIG. II.100. Cat. Chloralose–urethane anaesthesia. Thoracotomized and artificially ventilated. Each record (a and b) shows from above downwards: time (1 s intervals), heart rate min^{-1}, ECG, systemic blood pressure, respiratory excursions, and efferent activity in right cardiac sympathetic. (a) Pump ventilation: discharge occurs rhythmically but at a *central* rhythm—not that of the pump. (b) Pump disconnected; sympathetic discharge still shows central rhythm. (F. Habibollahi and P. M. I. Sutton 1979.)

respiratory movements cease. When pump ventilation is instituted the rhythm continues as before, bearing no relation to that of the pump. Note that the striking tachycardia which occurs after Flaxedil administration is due to the blocking of vagal transmission (vagolytic action) by the drug. Although the cardiac efferents discharge as before, their effect is negligible because of the atropine-like influence of Flaxedil.

2. Chemoreceptors and the heart rate

The carotid and aortic chemoreceptors [see p. 204] lie in the carotid bodies, in the left and right carotid bifurcations, and scattered around the aortic arch and its immediate branches—the so-called aortic bodies. They are innervated by branches of the IXth nerve and the vagus respectively and their afferent fibres ascend to relay in the tractus solitarius of the medulla. These chemoreceptors are responsive to oxygen lack—hypoxic, anaemic, stagnant, or histotoxic [see p. 208]; to hypercapnia; or to acidaemia; and above all to asphyxia. Thus, carotid occlusion provokes their discharge in the experimental animal—clamping the common carotid arteries lowers the blood flow through the carotid chemoreceptors and produces a degree of tissue asphyxia [FIG. II.102].

Normally, resting chemoreceptor activity is sparse in animals breathing air and makes little, if any, contribution to cardiovascular features, such as heart rate, inotropic myocardial performance, or to overall sympathetic vasomotor activity. However, when the carotid (or aortic) bodies are separately perfused, cardiovascular and respiratory responses are aroused. These include reflex vasoconstriction and a rise in systemic blood pressure, hyperpnoea but somewhat protean changes in heart rate which may increase, decrease or remain little affected. Until the work of Daly and Scott (1963) these reflex effects of carotid body stimulation on heart rate were poorly documented. Daly and Scott convincingly showed that carotid chemoreceptor excitation evoked alterations of heart rate that were very dependent on the accompanying reflex hyperpnoea which occurred. They showed that if the animal were artificially ventilated, thereby preventing the reflex response of hyperpnoea to chemoreceptor excitation, such carotid body stimulation induced

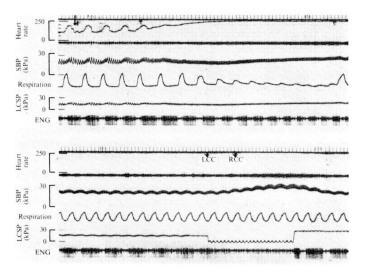

FIG. II.101. Dog 14.6 kg. Morphine–chloralose–urethane anaesthesia. Initially breathing spontaneously. The left carotid bifurcation was prepared either for natural blood flow through the common carotid artery or for carotid sinus-body perfusion. The right external carotid and occipital arteries were ligated. Each record shows from above downwards: time in seconds, heart rate, systemic blood pressure (SBP), respiration, left carotid sinus blood pressure (LCSP) (recorded via a lingual catheter), and the electroneurogram (ENG) of a cervical vagal strand. The upper record shows the animal initially breathing spontaneously and manifesting sinus arrhythmia, accompanied by rhythmic bursts of cardiac vagal discharge occurring during expiration. At the arrow (F) Flaxedil infusion is commenced and is suspended at the second arrow. Note that the heart rate accelerates prior to any failure of natural respiration reaching a value of more than 250 min⁻¹. Natural breathing fails but the rhythmic discharge of the cardiac vagal unit continues. Pump respiration is commenced at the third arrow. The lower record (taken about 1.25 min later) shows the heart rate to be above 250 min⁻¹ (for the recording is identical with that of the seconds time-scale) and the central rhythmic vagal discharge continuing without relation to the rhythm of pump ventilation. In this lower record two arrows ↓ show clamping of first the left and then the right carotid arteries, respectively (LCC and RCC). Note that the pressure in the left carotid artery is now submitted to a pulsatile stimulus slightly in excess of 0 and that the cardiac vagal unit falls silent. On raising the left carotid sinus pressure to 30 kPa during the inspiratory phase of pump ventilation, cardiac vagal discharge reappears immediately and then shows a rhythm unrelated to that of the pump, but virtually identical with that of the 'respiratory centre'. (Neil, E. (1979). In *Cardiac receptors* (ed. R. Hainsworth, C. Kidd, and R. J. Linden). Cambridge University Press.)

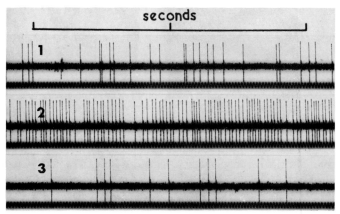

FIG. II.102. Electroneurogram of single chemoreceptor unit of the carotid sinus nerve.
Record 1. Common carotid artery patent.
     2. Common carotid artery clamped.
     3. Common carotid artery patent.
  BP unchanged by these manoeuvres because the carotid sinus nerve is cut.
  Time records below each neurogram = 50 c s⁻¹. (Joels, N. and Neil, E. 1963.)

reflex bradycardia. Moreover, in animals in which the pulmonary branches from the lungs had been cut and the chest wall reconstituted so as to permit spontaneous breathing, carotid body excitation always caused bradycardia.

The situation can be clarified by recording from single or few-fibre preparations of the cardiac vagus dissected from the otherwise intact cervical vagal trunk in spontaneously breathing dogs. As already described [p. 128] cardiac vagal activity in the dog shows a respiratory rhythm, vagal discharge accompanied by bradycardia occurring only during expiration. When the isolated carotid bodies are perfused such as by Ringer–Locke solution equilibrated with high CO_2 tensions or by solutions which contain minute concentrations of sodium cyanide (which provokes histotoxic anoxia) the reflex effect of hyperpnoea virtually suppresses the resting vagal discharge and the heart rate accelerates. However, the normal central respiratory rhythm can itself be suppressed by stimulating the central end of the superior laryngeal nerve; both inspiratory and expiratory muscle electromyogram activity are abolished. FIGURE II.103 (a) and (b) show the responses of the breathing, cardiac vagal discharge, and the heart rate to superior laryngeal stimulation on which is superimposed the effect of the injection of either CO_2/Ringer or Ringer containing cyanide into the isolated innervated carotid body. The onset of superior laryngeal stimulation can be seen to suppress the diaphragmatic EMG throughout the period of such stimulation. During this period, chemoreceptor stimulation induces an enormous discharge of cardiac vagal activity with consequent asystole.

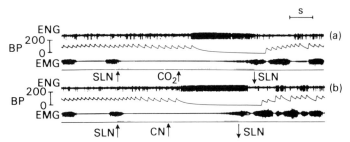

FIG. II.103. Dog. Morphine–chloralose anaesthesia. Spontaneous respiration. Left external, common carotid, and occipital arteries tied. Left lingual artery cannulated. Records (a) and (b) both show from above downwards: ENG of cardiac efferent vagus, blood pressure, diaphragmatic EMG, and signal marker. Between the arrows (SLN), stimulation of the central end of the right superior laryngeal nerve, which inhibits EMG activity. At the arrow (CO_2), 0.5 ml of saline equilibrated with 100 per cent CO_2 was injected via the lingual artery (record a). In record (b) 50 µg of NaCN in 0.5 ml Ringer solution was similarly injected (CN). Note vagal discharge and cardiac asystole. (Eric Neil 1974.)

Similar results can be obtained when nasal reflexes (with afferent fibres coursing in the trigeminal nerves) are provoked and chemoreceptor excitation is induced against this background. Profound bradycardia and/or asystole supervenes; these effects produced in dogs (Angell-James and Daly 1972) are even more dramatic in naturally diving animals such as the seal (Elsner *et al.* 1977). Indeed, the striking 'diving reflex', in which marked cardiac slowing is accompanied by profound vasoconstriction in the muscles, is due to this combination of nasal and chemoreceptor stimulation. The seal is converted virtually into a head–heart–lung preparation and the heart beats slowly, adequately maintaining the pressure required to supply the vital central nervous system. The muscles work anaerobically and the degree of vasoconstriction is so striking that the lactic acid concentration of the blood does not rise until the animal surfaces—whereupon the lactacidaemia attains concentrations of 15 mmol per litre (Scholander 1940, 1962).

These profound changes in heart rate caused by a combination of chemoreceptor stimulation and reflexes from the nose or larynx are of considerable clinical importance. In the past, trivial 'minor operations' have been carried out under nitrous oxide/oxygen anaesthesia. Nitrous oxide rarely induces surgical anaesthesia unless the concentration of oxygen supplied with it is reduced below that in room air. If the patient, already subjected to a degree of hypoxia, stops breathing there is a natural tendency to intubate the larynx and this may well provoke cardiac asystole. Such minor operations should never be carried out with this anaesthetic unless the patient has been premedicated with atropine.

The situation relating to the interaction of cardiac chronotropic reflexes and respiratory reflexes may be summed up in Fig. II.104.

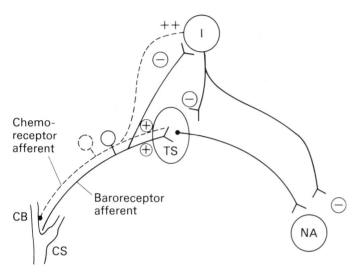

Fig. II.104. For explanation see text. I = 'inspiratory centre'; TS = tractus solitarius; NA = nucleus ambiguus; CB = carotid body; CS = carotid sinus.

The baroreceptors are solely responsible for resting vagal tone in the normally breathing animal. Their reflex pathway via the sino-aortic nerves includes synapses in the tractus solitarius whence axons from the nucleus of the tractus pass by monosynaptic or polysynaptic relays to the nucleus ambiguus and thence via the vagi to the heart. Collateral branches of the sino-aortic nerves induce a feeble suppression of neurons of the 'inspiratory centre'. This baroreceptor influence on respiration can for practical purposes be ignored. However, the chemoreceptor components of the sino-aortic nerves also relay in the nucleus of the tractus solitarius but, unlike the baroreceptors, exert a striking stimulatory effect on the inspiratory components of the central respiratory complex. In eupnoea the chemoreceptor input is of trivial import but in hypoxia the afferent traffic induces marked hyperpnoea. The central inspiratory discharge normally 'gates' the baroreceptor input and relay to the nucleus ambiguus in the eupnoeic animal—whether the 'gating' is at the tractus solitarius or at the nucleus ambiguus (or both) is not yet certain. However, the effect is to prevent cardiac vagal discharge during inspiration even in such eupnoeic animals. When chemoreceptor excitation is induced, the reflex stimulation of the central 'inspiratory' neurones is powerful enough to override chemoreceptor reflex vagal bradycardia. Only if the activity of the central respiratory neurones is abolished by nasal or laryngeal stimulation or by artificial ventilation causing hypocapnia can the primary chemoreceptor reflex bradycardia be unmasked.

3. Cardiac receptors and the heart rate

As already described, atrial receptors cause reflex tachycardia [p. 127] and ventricular receptors, if sufficiently stimulated provoke bradycardia.

4. Nociceptive stimuli and the heart rate

The afferent stimulation of somatic nerves may evoke either pressor or depressor reflex effects. Unmyelinated C-fibre stimulation causes hypertension (due to sympathetic vasoconstriction) and tachycardia the C fibres synapse with pressor regions of the medullary centre and provoke sympathetic stimulation (Johansson 1962). The stimulation of thin myelinated fibres from deep tissues on the other hand causes bradycardia and hypotension. These fibres synapse with ventral regions of the medullary depressor area and induce over-all sympathetic inhibition.

5. Corticohypothalamic effects on the heart rate

Everyone knows that many emotional effects can modify the heart rate. Excitement or fear causes tachycardia, whereas a sudden shock may induce a dropped beat or bradycardia.

Many of such changes can be provoked by stimulation of the limbic system. The limbic system includes cortical and subcortical structures which lie beneath the neocortex and which form a ring around the brain stem [Fig. II.105] (limbus = a ring). The paleocortical components of the limbic system are the hippocampus (dentate gyrus + Ammon's horn), pyriform lobe and the olfactory bulb. Intermediate between the position of the paleo and neocortex lies the cingulate gyrus (sometimes itself known as the limbic cortex). Subcortical components of the limbic system include the amygdaloid nuclei, the septal nuclei, the hypothalamic and thalamic nuclei, the caudate nucleus and the mesencephalic reticulum.

The plentiful connections of the hypothalamus with limbic lobe structures are shown in diagrammatic form in Figure II.105.

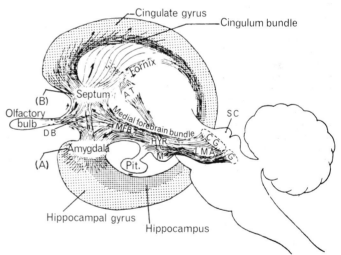

Fig. II.105. Schema of the limbic system with some of its major connections via the median forebrain bundle (MFB) to the hypothalamus and midbrain. (MacLean, P. D. (1958). *Am. J. Med.* **25**, 611.)

Löfving (1961) studied cardiovascular changes induced by stimulation of the rostral end of the cingular gyrus near the genu of the corpus callosum. Dorsal to the callosal genu, cingulate stimulation provoked bradycardia and hypotension—due to a combination of vagal hyperactivity and a reduction of sympathetic drive to the heart and particularly the muscle vessels—renal and cutaneous circulations were barely affected. These cortically induced effects were abolished by elective lesions of the anterior hypothalamus or by electrolytic destruction of the medial bulbar reticular formation. Such results presumably indicate the neuronal pathway from the cingulate region.

Complicating the issue, however, were the effects provoked by cingulate stimulation ventral to the genu of the corpus callosum, which were quite the opposite—hypertension with pronounced vasoconstriction of the splanchnic and muscle circulatory beds and tachycardia. Nevertheless, behavioural responses of conscious animals confronted with stressful or even dangerous situations have indeed been divided into 'defence reaction' and 'playing dead' responses. Playing dead is behaviourly most pronounced in the opossum ('playing possum'). The creature when confronted by danger shams death-displaying flaccidity, hypopnoea, bradycardia, and hypotension and dropping to the ground. The would-be predator is presumably misled and turns away. The defence reaction is just the opposite—piloerection, hypertension, and tachycardia accompanied by spitting, snarling, and hyperactivity.

In man, the behavioural counterpart of the playing dead reaction may be the faint. Some individuals respond to stress in this manner—others with the 'defence' reaction. Fainting is associated with a loss of muscle tone, with bradycardia, and hypotension, yet with an increased muscle blood flow.

Abrahams, Hilton, and Zbrozyna (1960) mapped the responses of lightly anaesthetized cats to stimulation of the hypothalamus and showed that the region of the hypothalamus from which muscle vasodilatation could be provoked was topographically similar to, if not identical with, that which Hess described as inducing the 'defence' reaction in the conscious animal. This region in the ventral hypothalamus near the midline lies ventral to the preoptic nucleus and dorsal successively to the optic chiasma and mamillary bodies [FIG. II.106]. The tract from this region runs on into the midbrain in the substantia nigra, dorsal to the cerebral peduncles above the origin of the IIIrd nerve and then into the bulbar region dorsal to the pyramid. It establishes connections with medullary reticular neurones and these synapses are not influenced by baroreceptor afferent discharge. Stimulation of this hypothalamic region causes, apart from skeletal muscle vasodilatation, retraction of the nictitating membrane, dilatation of the pupil, piloerection, and respiratory stimulation.

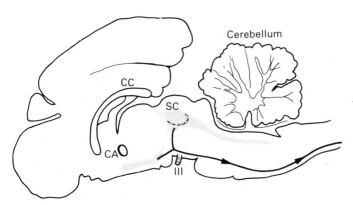

FIG. II.106. Paramedian sagittal section of cat's brain. Shaded areas in hypothalamus, central grey matter, mesencephalic tegmentum and medulla are areas integrating the defence reaction. Solid line with arrows shows pathways dorsal to the cerebral peduncle for cardiovascular pattern of 'defence' response. CC = corpus callosum; CA = anterior commissure; SC = superior colliculus. (Redrawn from Hilton, S. M. (1975). *Brain Res.* **87**, 213–19.)

A similar series of responses could be provoked by stimulation of the tegmentum of the midbrain and of the central grey matter surrounding the cerebral aqueduct and Abrahams *et al.* concluded that the three areas were so adjacent anatomically that no functional distinction between them could be made from their results.

In conscious cats, mild stimuli delivered by chronically implanted electrodes, whose tips lay in this hypothalamic 'defence' area, caused pricking of the ears, pupillary dilatation, raising of the head and skeletal muscle vasodilatation. In its earliest manifestations this could be described as an 'alerting' reaction and similar signs could be provoked by physiological stimuli, such as sounds, flashes of light, or mild cutaneous stimulation.

Tachycardia and an increased force of heart beat, hypertension and venoconstriction accompany the skeletal muscle vasodilatation as responses to stimulation of the 'defence' area and these cardiovascular changes seem appropriate as a preparatory response for exercise whether it be expended in fight or flight.

Hilton and Spyer (1971) showed that electrical stimulation of a localized area in the anterior hypothalamus and preoptic region induced vagal hyperactivity, revealed by marked bradycardia, sympathetic inhibition, evidenced by a fall in peripheral resistance and blood pressure and apnoea. The area defined is small and circumscribed and lies ventral and caudal to the anterior commissure whence it projects dorsal to the fornix into the dorsal hypothalamus to pass caudally to the tractus solitarius and to the ventrolateral medulla. FIGURE II.107(a) and (b) show respectively the site of the hypothalamic 'defence' and 'depressor' areas and their projection to the medulla.

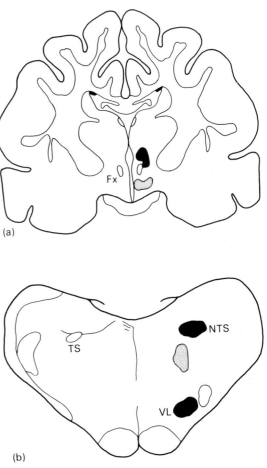

FIG. II.107(a). Diagrammatic coronal section at level of anterior hypothalamus showing areas integrating the full patterns of defence (hatched) and baroreceptor (solid black) responses. Fx = fornix. (b) Diagrammatic coronal section at level of medulla just rostral to obex, showing area of primary afferent relay for baroreceptor afferents (NTS, nucleus of the tractus solitarius), ventrolateral area initiating sympathetic inhibition (VL), area integrating defence reaction (hatched), and area of efferent pathway for cardiovascular components of defence reaction (open 'circle'). TS = solitary tract. (Modified from Hilton, S. M. (1975). *Brain Res.* **87**, 213–19.)

Hilton (1975) regards this pathway as part of the efferent arm of the baroreceptor reflex. Spyer (1972) has indeed shown that this anterior hypothalamic 'depressor' area contains neurones which respond to rises of pressure in a vascularly isolated carotid sinus; two-thirds of the neurons which did respond showed enhanced activity but the remainder were inhibited. Hilton regards the hypothalamic pathway of ascending fibres excited by baroreceptor stimulation as playing an important role in the normal baroreceptor reflex. However, a decerebrate animal shows brisk baroreceptor cardiovascular reflex responses. Nevertheless, this may be due to the removal of hypothalamic 'defence area' impulse activity which in the anaesthetized cat exerts a depressant effect on baroreceptor reflex bradycardia (Lopes and Palmer 1976) and Hilton's concept of a functional longitudinal integration of cardiovascular reflexes evoked by baroreceptor stimulation has merit.

REFERENCES

ANGELL-JAMES, J. and DALY, M. DE BURGH (1972). *Symposia of the Society for Experimental Biology*, pp. 313–41. Cambridge University Press.
CALARESU, F. R., FAIERS, A. A., and MOGENSON, G. J. (1975). *Prog. Neurobiol.* **5**, 1–35.
DALY, M. DE BURGH and SCOTT, M. J. (1963). *J. Physiol., Lond.* **165**, 179–97.
ELSNER, R., ANGELL-JAMES, J. E., and DALY, M DE BURGH (1977). *Am. J. Physiol.* **232**, H517.
HAINSWORTH, R., KIDD, C., and LINDEN, R. J. (1979). *Cardiac receptors.* Cambridge University Press.
HILTON, S. M. and SPYER, K. M. (1971). *J. Physiol., London.* **218**, 271.
KOEPCHEN, H. P., LUX, H. D., and WAGNER, P. H. (1961). *Pflügers Archiv.* **273**, 443–61.
KOIZUMI, K. and BROOKS, C. M. (1972). *Ergebn, Physiol.* **67**, 1–68.
KORDY, M. T., NEIL, E., and PALMER, J. F. (1975). *J. Physiol., Lond.* **247**, 24–25P.
LIPSKI, J. McALLEN, R. M., and SPYER, K. M. (1975). *J. Physiol., Lond.* **251**, 61–78.
LOPES, O. U. and PALMER, J. F. (1976). *Nature, Lond.* **264**, 454.
McALLEN, R. M. and SPYER, K. M. (1976). *J. Physiol., Lond.* **258**, 187–204.
—— —— (1978). *J. Physiol., Lond.* **282**, 353–64.
NEIL, E. (1974). *Proc. XXVI Int. Cong. Physiol.*
SCHOLANDER, P. F. (1940). *Hvalrådets Skrifter Norske Videnskaps-Akad.* Oslo, No. 22.
—— (1962). *Harvey Lect.* **57**, 93–110.
SPYER, K. M. (1972). *J. Physiol., Lond.* **224**, 245–57.
—— (1979). In *Integrative functions of the autonomic nervous system* (ed. C. M. Brooks, K. Koizumi, and A. Sato) pp. 283–92. Elsevier, Amsterdam.
THOREN, P. (1979). *Rev. Physiol. Biochem. Pharmacol.* **86**, 1–94.

Circulation through special regions

CORONARY CIRCULATION

The right and left coronary arteries supply the myocardium, each arising from the aorta immediately above the semilunar valves. The left coronary artery after a very short course divides into a left circumflex and an anterior descending branch. The circumflex branch traverses to the left in the A–V groove ending in a posterior descending branch. The anterior descending branch courses the interventricular groove to reach the apex, yielding some septal branches in transit.

The right coronary artery passes along the right atrioventricular sulcus towards the back of the heart where it gives off several descending branches to both ventricles. Each of the coronary rami during their superficial course to the apex give off branches which enter the myocardium. At the apex terminal divisions of the various arteries pass inwards to supply the inner layers of the myocardium and the papillary muscles from which arise myocardial capillaries which lead to veins. Eighty per cent of the left coronary artery inflow drains into the coronary sinus; some (15 per cent) enters the right ventricle by deep venous channels. Eighty to ninety per cent of the right coronary inflow drains via the anterior cardiac veins into the right atrium—a minor fraction drains into the coronary sinus. The great majority of coronary sinus flow is drained from the left

ventricle and coronary sinus catheterization samples have yielded valuable information of blood flow in the left ventricle and of left ventricular metabolism.

It is unfortunate that the dog, widely used for studies of coronary flow and myocardial metabolism, shows a pattern of coronary artery distribution which differs from that in man. In the dog there is a striking preponderance of left coronary arterial supply—some 85 per cent of the myocardium receiving blood therefrom. Only 20 per cent of men have a left coronary artery preponderance. Half, indeed, show right preponderance and in about 30 per cent there is a balance of myocardial supply via the two arteries.

Nevertheless, the basic data agree fairly well for dog and man under resting conditions. Each species shows a left coronary flow of 70–85 ml per 100 g per min and it is assumed that this flow is supplied to left ventricular tissue. The left ventricular oxygen usage is 8–10 ml per 100 g per min—three-quarters of this oxygen usage is achieved during systole. The coronary artery–coronary sinus A–V O_2 difference is about 13–14 ml, a desaturation of 70 per cent. In severe exercise in trained dogs coronary flows as high as 500–600 ml per 100 g left ventricle per minute have been recorded.

Measurements of coronary flow have been made:

1. In man, by the nitrous oxide technique [see p. 138] in which the subject breathes 15 per cent N_2O in an oxygen–nitrogen mixture for 10 minutes and samples of arterial and coronary sinus venous blood are removed either continuously (the two samples giving the integrated A–V difference, a final venous sample being taken for V_{10}) or simultaneously at intervals during the 10 minute period.

$$\text{Coronary flow per 100 g per min} = \frac{100\,V_t.S}{\int_0^t (A-V)\,dt}$$

where A = arterial concentration
V = coronary sinus venous concentration
S = partition coefficient for N_2O between blood and myocardial tissue.
V_t = venous concentration of N_2O after equilibrium is reached between blood and myocardium in time t.

The integrated A–V difference is determined by measuring the arterial and venous blood contents of N_2O in the two syringes which both draw blood from artery and coronary sinus respectively over the 10 minute period which is sufficient for equilibrium. V_t is the coronary sinus venous blood N_2O content taken separately at 10 minutes. S, the partition coefficient, is unity.

Gregg and co-workers compared the results of the N_2O method with direct measurements of flow, made with a rotameter, in animal experiments and found reasonable agreement in resting animals.

2. In animals, electromagnetic flowmeters have been implanted round the main left coronary artery and in some cases round its circumflex branch (Gregg 1966). The left coronary artery supplies almost the whole of the blood which drains from the coronary sinus, so catheterization of the coronary sinus allows sampling for oxygen content of blood mainly from the left ventricle. From the flow measured and the A–V O_2 difference myocardial oxygen usage can be directly determined. Similarly, substrate uptake can be studied. Flowmeter measurements have the huge advantage of showing phasic flow as well as the flow per minute. Moreover, thanks to the devoted work of Gregg and his pupils over the course of 25 years, the technique can now be employed in survival animals trained to undergo exercise, etc.

Coronary flow in resting man

The blood flow to the left ventricle is 70–85 ml per 100 g per min. The coronary A–V oxygen difference is about 12 ml per 100 ml blood and the myocardial oxygen usage at rest is 7–9 ml per 100 g per min. It will be noted that coronary sinus blood is very desaturated even when the heart beats in the 'resting' circulation. If the arterial blood contains 19 ml O_2 per 100 ml then the venous blood contains 7 ml and the desaturation suffered in 12/19 × 100—about 63 per cent—so the coronary sinus blood is only 35 per cent saturated and has a pO_2 of less than 20 mm Hg.

Mechanical effects of ventricular contraction

Whereas the pressure head provides flow as in other organs, the coronary resistance is of course influenced by the extravascular pressure which rhythmically varies with systole. Consequently, the flowmeter record shows a strong phasic variation of coronary flow [Fig. II.108].

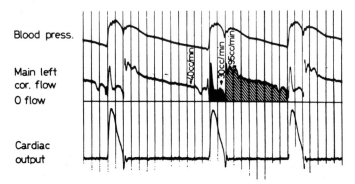

Fig. II.108. Record showing phasic aortic blood pressure and phasic flow in the main left coronary artery and ascending aorta obtained by means of a strain gauge and electromagnetic flow metres. Dog at rest. Black area is systolic flow; line shaded area is diastolic flow. Time, 0.1 s (vertical lines).

Blood press. (mm Hg)	90	Oxygen usage (cm³ min⁻¹)	6.7
Stroke cor. flow (cm³)	0.98	Cor. A–V oxygen (ml)	13.2
Stroke syst. cor. flow (cm³)	0.18	Cardiac output (ml min⁻¹)	2189
Stroke diast. cor. flow (cm³)	0.80	Stroke volume (ml)	42

(Gregg, D. E., Khouri, E. M., and Rayford, C. R. (1965). *Circulation Res.* **16**, 102. By permission of American Heart Association, Inc.)

In the dog at rest, left coronary arterial flow here is 51 ml min⁻¹ with a stroke flow of 0.98 ml. This stroke flow is made up of 0.18 ml during systole and 0.80 ml during diastole. With the onset of isometric contraction, flow declines sharply because the myocardial tissue pressure is rising steeply and the aortic pressure head is minimal. When the ejection phase of systole occurs, the improvement of the aortic pressure causes a sharp peak in the flow which quickly subsides due to the throttling effect of the high intramural myocardial pressure in the contracting ventricle. Flow then reaches a low value and maintains this until the advent of aortic valve closure; flow then rises to a peak and thereafter the flow values follow the contour of the aortic pressure.

Some of the flow registered during the initial part of systole may be into the extramural part of the left coronary artery and its branches, expanding these (because of the throttling effect of the ventricular contraction on the intramural vessels); some may be distributed to atrial or right ventricular myocardium which do not raise their intramural pressure to anything like the same extent as does the left ventricle. Nevertheless, as in this example, about 80 per cent of the total coronary flow occurs in diastole during which the mean pressure head is of course reduced, so the true vascular resistance caused by the 'tone' of the vessels is even lower.

Kirk and Honig (1965a) investigated the intramyocardial pressure in the beating heart [Fig. II.109a and b] utilizing the principle that fluid flow through a collapsible segment will cease when the external pressure on the segment exceeds the distending pressure. A curved needle was inserted as shown and a collapsible segment was located at the point marked 'hole'. (This hole had been made

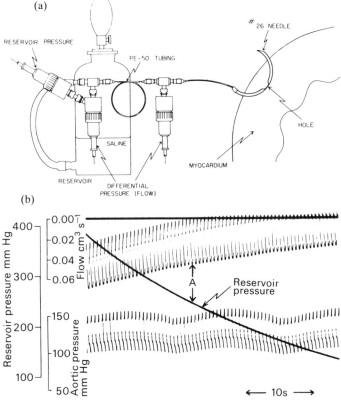

Fig. II.109. (a) Schematic representation of apparatus used to measure myocardial tissue pressure.

(b) Representative recording showing method for determining systolic tissue pressure. Aortic and reservoir pressures have separate co-ordinates and zero flow through the collapsible segment is the upper base line. The segment was located halfway across the wall. As the reservoir pressure falls systolic flow decreases until at 'A' it just ceases. Reservoir pressure of 240 mm Hg at this point is taken as peak systolic tissue pressure. Note that peak aortic pressure is but 150 mm Hg. (Kirk, E. S. and Honig, C. R. (1964). *Am J. Physiol.* **207**, 361.)

before insertion by filing through most of the wall of the needle leaving only a thin connecting strip at this site.)

A reservoir pressure of saline well in excess of aortic pressure caused flow through the needle; flow and aortic pressure were recorded simultaneously. Flow was maximal in diastole, minimal in systole and diminished as the reservoir pressure fell. At a reservoir pressure of 240 mm Hg, flow just ceased in systole and this pressure was taken as the *peak systolic tissue pressure*. In this record the 'collapsible segment' was located equidistant from the epicardial and endocardial surface and the peak aortic systolic pressure was only 150 mm Hg. Further reduction of the reservoir pressure reduced diastolic flow and finally at a reservoir pressure which was approximately zero, flow ceased entirely.

Examination of the peak systolic tissue pressure recorded at different depths of the myocardium showed that the systolic tissue pressures in the inner half of the myocardial wall uniformly exceeded the peak ventricular blood pressure [Fig. II.109b]. In the outer half of the myocardium the reverse was the case and peak systolic tissue pressures near the epicardium were much lower than the peak systolic aortic pressure.

A gradient in coronary blood flow during systole was thus shown to exist, the outer layers of the ventricular wall receiving a larger share of the flow and the innermost wall presumably receiving no flow at all at the peak of systole. A theoretical analysis of the situation, though too detailed for inclusion in this text, should be consulted by those interested (see Kirk and Honig 1964a).

Kirk and Honig (1964a, b) then investigated the capillary

density, myocardial pO_2 and flow distribution within the different layers of the myocardium. Capillary density was 750 capillaries per mm² in superficial layers and 1100 capillaries per mm² in the deeper layers. Half-intercapillary distance in the deep layers was 16.5 μm and in the superficial layers 20.5 μm. The minimum diffusion distance was thus 20 per cent shorter in the deeper layers and this is of great importance in determining the myocardial pO_2. Flow measurements were made by determining the rate of disappearance of depots of radioactive iodine (Na¹³¹I) injected at various depths into the wall. The faster the flow, the greater was the disappearance rate. Transmural flow was higher in the superficial layers and in those nearest the epicardium the flow proceeded throughout the cycle at a rate equal to that of the diastolic flow in the deeper layers. Flow in the deepest layers ceased during systole. If the left coronary artery was perfused at a constant rate, vagal arrest of the heart caused an increased flow in the vessels of the deeper layer at the expense of the superficial layers, indicating that the effect of the heart's contraction is indeed to limit flow in the subendocardial layers.

Myocardial pO_2 measurements showed that pO_2 decreased as the probe (platinum cathode) was moved inwards from the epicardial layer. Mean oxygen tension was some 40 mm Hg in the superficial sites and approximately 20–25 mm Hg in the deep layers. The average mean pO_2 in the myocardium was 31 mm Hg, but as is seen a gradient of 15–20 mm Hg pO_2 exists from outside inwards. Such results, obtained in a heart beating under resting conditions, indicate that the inner myocardium may well have to rely on some anaerobic metabolism under conditions of stress. In this respect it is interesting to note that the myoglobin content of the cells in the deeper layers is higher than that in the superficial myocardium. Mgb acts as an oxygen store which is charged up during the diastolic flow of blood and which yields its oxygen in the stage of systolic throttling of blood flow. Nevertheless, Lochner and Nasseri (1959) found coronary venous pO_2 to average 10.5 mm Hg in dogs performing mild treadmill exercise. The mean tissue pO_2 in the subendocardial layers must have been less than this and that in the epicardial layers more. If the pO_2 were less than 10 mm Hg then this is insufficient to charge the Mgb fully, again suggesting that the inner myocardium is forced towards anaerobic metabolism in stress conditions. Lastly, the ratio of lactate to pyruvate in the deep myocardium is higher [indicating anaerobic metabolism; see p. 217] than in the epicardial muscle.

These several facts may help to explain the great incidence and severity of necrosis in the inner half of the myocardial wall following coronary occlusion and also the predominance of a subendocardial distribution of necrosis found *post mortem* in subjects with no obvious history of coronary disease or of anginal attacks. It also explains the remarkable predilection of the hypertrophied heart to subendocardial ischaemia and injury. As the muscle fibres increase in bulk, the intercapillary distance inevitably increases and the efficacy of the supply of oxygen by diffusion correspondingly diminishes.

Capillary density in the myocardium

Myers and Honig (1964) calculated that there were 750–1100 capillaries per mm² in the superficial and deep myocardium respectively from their results on myocardial tissue content of blood. Wearn (1928), using an injection technique designed to identify all capillaries in the tissue, counted 5000 capillaries per mm². Though many think this figure is too high, this would indicate that in the heart beating in resting conditions, only about one-fifth of the capillary bed is patent and that the opening of precapillary sphincters increases the number of patent capillaries.

Renkin (1967) has pointed out that cardiac muscle, just as skeletal muscle, has one capillary per muscle fibre, but the skeletal muscle fibres in gracilis for instance are much bigger than the

cardiac fibres—50 μm diameter as opposed to 20 μm. FIGURE II.110 shows that there will be many more capillaries per unit mass of heart. There are 400 capillaries per mm² in gracilis and 2500 per mm² in the heart model.

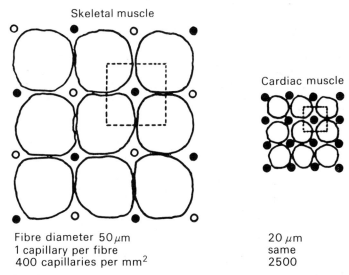

Skeletal muscle

Cardiac muscle

Fibre diameter 50 μm
1 capillary per fibre
400 capillaries per mm²

20 μm
same
2500

FIG. II.110. Diagram illustrating relative capillary densities in skeletal and cardiac muscle. (Renkin, E. M. (1967). In *Coronary circulation and energetics of the myocardium* (ed. G. Marchetti and B. Taccardi). Karger, Basel.)

⁸⁶Rb clearance studies have been performed in hearts by Winbury and his colleagues (Renkin 1967). PS measurements [see p. 87] show that PS increases with flow representing a graded metabolic vasodilatation with metabolic rate. The mean PS of the lowest flow measurements [FIG. II.111] is 16 ml per 100 g per min; at a mean flow of 55 ml per 100 g. PS is 80 ml per 100 g per min. PS shows no further increase with flow.

Maximal PS represents the diffusion capacity of the capillary surface area so the results indicate that about 75 per cent (60/80) of the coronary capillaries are patent at even the lowest flow. Though there is some discrepancy between this proposition and the findings of Kirk and Honig, the main point is that the PS measurements of 60–80 ml per 100 g per min are 8–10 times those found in skeletal muscle indicating the huge capillary surface area for diffusion provided by the coronary vascular bed.

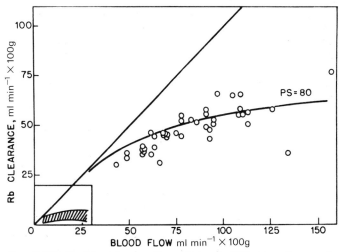

FIG. II.111. Capillary clearance-flow relation for ⁸⁶Rb in cardiac muscle. Data of M. Winbury *et al.* The inset compares the range for skeletal muscle. (Renkin, E. M. (1967). In *Coronary circulation and energetics of the myocardium* (ed. G. Marchetti and B. Taccardi). Karger, Basel.)

Coronary flow during sympathetic stimulation and in exercise

Stimulation of the left ansa subclavia or stellate ganglion increases the mean flow in both right and left coronary arteries. The hyperaemia outlasts the stimulation and usually outlasts the increased heart rate which the period of sympathetic excitation simultaneously induces.

There are no cholinergic sympathetic vasodilator fibres to the coronary vessels for even after α blockade and β blockade sympathetic stimulation never causes vasodilatation. The vagi are not proven to supply the coronary vessels and though vagal stimulation sufficient to cause cardiac arrest does increase coronary flow (Sabiston and Gregg 1957) this effect is probably due to the lessened intramural tissue pressure and extravascular resistance caused by asystole.

Adrenalin and noradrenalin infused into the circulation also provoke an increased coronary flow due to their stimulant effect on cardiac contractile force and metabolism. Hence the release of these hormones in stress circumstances does presumably contribute to an increase in coronary flow (secondary to metabolic changes in the heart), but this effect is of far less quantitative significance than that caused by the increased impulse discharge in the sympathetic nerves, which occurs simultaneously.

Exercise

Gregg and his colleagues (see Gregg 1966) have studied phasic coronary flow in conscious dogs at rest and during exercise, using electromagnetic flow meters chronically implanted on the main left coronary artery or on its circumflex branch.

One example will suffice to show an important point. After 40 seconds of heavy (treadmill) exercise the following changes had occurred:

1. Cardiac output increased three-and-a-half-fold.
2. Heart rate increased about threefold—hence the stroke volume was slightly increased.
3. Mean blood pressure increased about 30 per cent.
4. Coronary flow increased fourfold.
5. The ratio of systolic coronary flow to diastolic coronary flow which was approximately 0.22 at rest increased to 0.9 during exercise.
6. Myocardial oxygen usage increased fivefold, because a greater coefficient of utilization (A–V/A) of oxygen accompanied the fourfold increase in coronary flow.

Where is this increased flow during systole going? One possibility is that the flowmeter records extramural vessel expansion produced during systole when the arterial pressure is high and the myocardial resistance of tissue pressure is higher. The blood then surges into the intramyocardial tissues as diastole supervenes. The other is that this increased flow during systole is distributed to the more superficial layers of myocardial muscle (where the ratio of peak systolic aortic pressure/peak myocardial tissue pressure is higher than unity (Kirk and Honig 1964)), but *not* to the deeper myocardial layers, where the reverse obtains and flow must be cut off during systole. Lastly, some of the blood during systole may go to the atrial regions and some to the right ventricle as well as to the left ventricle; the extravascular tissue pressure in the right ventricle and atria is much lower and the extravascular resistance offered is correspondingly less.

In man, exercise may increase the cardiac output fivefold or more, and cardiac metabolic demands must increase by at least this factor and probably more, for the heart beats faster against a higher mean pressure.

If the resting coronary flow increases five times then flows of 300–400 ml per 100 g per min will be distributed to the heart muscle in heavy exercise and as the (A–V) difference probably increases as well from 12 ml per 100 ml blood to perhaps 18 ml, the oxygen usage of the heart could well be as much as 70 ml per 100 g per min. In top class athletes capable of even greater cardiac outputs in maximal performance these figures must be exceeded.

Emotional excitement. The patterns of coronary flow response to various types of excitement such as fright, olfactory and auditory stimuli are similar to those of muscular activity. The duration of the response is short however, and the circulatory response transitory. Sympathetic discharge, which increases in the 'startle' reaction, causes tachycardia and an increased ventricular performance against a higher mean systemic pressure. As the left ventricular work increases so does the myocardial oxygen usage and metabolism and the coronary hyperaemia which ensues is again of 'chemical' causation.

The chemical background of coronary vasodilatation

Hilton and Eichholtz (1925) showed that oxygen lack caused a huge increase in the coronary blood flow recorded in a heart–lung preparation. The preponderance of evidence since then has favoured a causal relationship between cardiac metabolic activity (as reflected by oxygen usage) and CBF. Berne has suggested a possible metabolic regulation of CBF as follows (see Berne and Levy 1977).

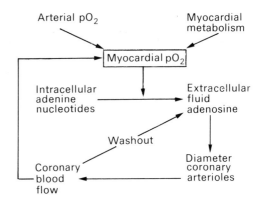

ATP and ADP are very potent vasodilator substances, being four times as active as AMP and adenosine itself. However, it is dubious whether these phosphorylated compounds can cross cell membranes. The nucleoside adenosine, however, does traverse myocardial cell membranes and if liberated by the myocardial cell could gain access to the resistance vessels, including the precapillary sphincters of the coronary system. Berne has found that inosine and hypoxanthine, which are both metabolic oxidation products of adenosine without vasodilator properties, are released into the coronary venous blood in hypoxia and presumably arise from the myocardial adenine nucleotides which are disrupted by hypoxia.

For Berne's hypothesis to be valid it is necessary to show that adenosine formed in hypoxia is not deaminated (to form inosine) before it escapes from the myocardial cell. To date adenosine has not been demonstrated in coronary sinus blood, but this may well be due either to its destruction by adenosine deaminase in the blood or to its entrance into the red blood cells.

Other influences claimed to cause coronary vasodilatation in exercise and in hypoxia are of course [K⁺] and hyperosmolality. Neither has proved a satisfactory candidate for election yet.

Myocardial metabolism

Bing (1965) pioneered the study of myocardial metabolism in man and animals by coronary sinus blood sampling more than 25 years ago. He showed that the heart can use many foodstuffs as substrates.

After a meal or after glucose infusion the heart uses mainly carbohydrate (glucose, lactate, pyruvate) and the myocardial RQ exceeds 0.9. Fasting lowers the RQ and if prolonged this falls to 0.7, which characterizes that of fat metabolism. Fatty acids and ketones infused into the blood are then avidly extracted by the heart. The human myocardium can supply itself with 67 per cent of its energy requirements by burning fatty acids which it obtains from the plasma albumin-bound non-esterified fatty acid (NEFA) fraction.

The substrates of oxidative metabolism provide the energy for oxidative phosphorylation—the resynthesis of ATP (the energy coinage of the cell).

When cardiac muscle contracts ATP is broken down to ADP and inorganic phosphate, but the ATP is rapidly resynthesized by the Lohmann reaction from creatine phosphate. Only by inhibiting the enzyme creatine phosphotransferase can we be sure whether ATP *does* break down when the muscle contracts.

$$\text{ATP} \xrightarrow{\text{actomyosin}} \text{ADP} + \text{P}_{\text{in}}$$

Lohmann

$$\text{ADP} + \text{CP} \xrightarrow[\substack{\text{creatine phospho-} \\ \text{transferase}}]{} \text{ATP} + \text{C}$$

The enzyme can be inhibited by fluorodinitrobenzene (FDNB), and skeletal and cardiac muscle, when treated with FDNB, will contract normally a few times during which ATP is broken down (but CP is not).

When the heart is subjected to hypoxia the creatine phosphate concentration drops strikingly and the ATP/ADP quotient of the myocardium falls precipitously.

The myocardial cell contains the enzyme phosphorylase in an inactive form, phosphorylase b. This form can be activated by the addition of cyclic adenylic acid (cAMP) to active phosphorylase a.

Cyclic adenylic acid is formed from ATP by the enzyme adenylate cyclase which itself is powerfully activated by adrenalin:

$$\text{Adrenalin}$$
$$\searrow$$
$$\text{ATP} \xrightarrow[\text{cyclase}]{\text{adenylate}} 3'5' \text{ AMP}$$
$$\downarrow$$
$$\text{Phosphorylase b} \longrightarrow \text{Phosphorylase a}$$

Phosphorylase a breaks down glycogen to yield glucose 1-phosphate, which is converted to glucose 6-phosphate. This compound is then broken down in muscle by glycolysis to give pyruvic acid and thence via the respiratory transfer chain to yield energy and the resynthesis of ATP.

It would be attractive to suppose that the inotropic effect of adrenaline on cardiac muscle is due to its promotion of these changes, but this remains speculative.

Metabolism in the failing heart

Fleckenstein (1967) has classified heart failure into two types:

1. The reduction of contractility is due to a deficient concentration of ATP. Such a situation is seen in hypoxia or ischaemia.
2. The reduction of contractility is due to a deficient utilization of energy-rich phosphate compounds, although the concentration and supply of these substances is adequate. Interference with the effect of Ca^{2+} in excitation-contraction coupling is the best example of this; such heart failure is responsive to Ca^{2+} or to cardiac glycosides.

Clinical heart failure is more closely related to type 2.

Bing (1965) has stressed that the failing heart shows the same power of extraction of metabolic substrates, making it improbable that a disturbance in energy production is responsible for failure. Even in decompensated heart failure, where the heart is dilated, there is little abnormal in the coronary flow picture at rest or in the response of myocardial oxygen usage to exercise.

To date we do not know the fundamental biochemical cause of heart failure.

Angina pectoris

The rhythmic clenching of the fist when the inflow of arterial blood is prevented by a sphygmomanometer cuff placed on the upper arm and inflated to 200 mm Hg will soon cause the development of an intense 'ischaemic pain' which develops to become intolerable.

Ischaemia of the heart muscle produces the same subjective effects. When the oxygen supply of the rhythmically contracting myocardium fails to keep pace with its oxygen requirements, pain results. The pain is substernal and radiates to the left shoulder, inner side of the left arm and to the angle of the jaw. The subject knows from experience that this pain can be alleviated by coming to a dead halt if he has been exercising. If the attack is precipitated by emotional excitement, as in an outburst of rage, the subject's preoccupation with the pain will cut short his acrimonious outbursts promptly.

Anginal pain is relayed by non-myelinated afferents in the arterial adventitia and myocardium whose endings are excited by (unknown) anaerobic metabolites. Once the anaerobic situation is relieved, coronary flow will wash these substances away from the nerve endings which then cease to discharge. The afferent fibres course up the cardiac sympathetic branches reaching the dorsal roots of segments T_1–T_4 and thence relay in the substantia gelatinosa Rolandi; lateral spinothalamic neurones are synaptically excited by this discharge and this causes pain. The typical referred pain of the shoulder and inner arm is due to the posterior roots involved.

The underlying cause of angina is a narrowing of the coronary arteries by atherosclerosis.

Vasodilator drugs, such as glyceryl trinitrate or sodium nitrite, provide relief. However, much of their therapeutic effect derives from their promotion of splanchnic venodilation which lessens venous return; end-diastolic volume decreases and the tension which the smaller ventricle has to develop is reduced ($P = T/R$).

Coronary thrombosis

This results from the detachment of a thrombus from a coronary artery, which then occludes an arterial branch lower down. The area supplied by this branch undergoes ischaemic necrosis (myocardial infarction). Pain is intense and enduring and a shock-like syndrome ensues with hypotension, nausea, vasoconstriction of the skin. The affected myocardial area fibroses if the attack is not fatal. If the artery occluded is too large, death ensues because of ventricular fibrillation—such deaths are remarkably rapid, the subject literally slumping and dying on the spot.

REFERENCES

BERNE, R. M. and LEVY, M. N. (1977). *Cardiovascular physiology*. Mosby, St. Louis.

BING, R. J. (1965). *Physiol. Rev.* **45**, 171.

FLECKENSTEIN, A., DÖRING, H. J., and KAMMERMEIER, H. (1967). In *Coronary circulation and energetics of the myocardium* (ed. G. Marchetti and B. Taccardi) pp. 220–36. Basel.

GRANATA, L., OLSSON, R. A., HUVOS, A., and GREGG, D. E. (1965). *Circulation Res.* **16**, 114.

GREGG, D. E. (1966). In *The physiological basis of medical practice*, 8th edn (ed. C. H. Best and N. B. Taylor) Chap. 44, pp. 795–812. Edinburgh.

—— KHOURI, E. M., and RAYFORD, C. R. (1965). *Circulation Res* **16**, 102.

HILTON, R. and EICHHOLTZ, F. (1925). *J. Physiol., Lond.* **59**, 413.

KATZ, A. M. and MESSINEO, F. C. (1981). *Am. Heart J.* **102**, 491.

KHOURI, E. M., RAYFORD, C. R., and GREGG, D. E. (1965). *Circulation Res.* **17**, 427.

KIRK, E. S. and HONIG, C. R. (1964). *Am. J. Physiol* **207**, 361 and 661.

MYERS, W. W. and HONIG, C. R. (1964). *Am. J. Physiol.* **207**, 653.

RAYFORD, C. R., KHOURI, E. M., and GREGG, D. E. (1965). *Am. J. Physiol.* **209**, 680.

RENKIN, E. M. (1967). In *Coronary circulation and energetics of the myocardium* (ed. G. Marchetti and B. Taccardi) p. 18. Basel.

CEREBRAL CIRCULATION

The adult brain weighing some 1400 g receives a total flow of approximately 750 ml min^{-1} (50–60 ml per 100 g per min). The oxygen usage is 3.3 ml per 100 g per min or a total of 45 ml min^{-1} —which is 20 per cent of that of the whole body at rest.

Anatomical features

In man the internal carotid and vertebral arteries (forming the basilar artery) provide the entire blood supply. United in the circle of Willis (lying at the base of the brain) this arterial anastomosis provides six cerebral branches for distribution to the cortex and subcortex and brainstem. The basilar artery supplies the occipital lobes, cerebellum, pons, and medulla. The internal carotid arteries supply the upper brain stem and the remainder of the cerebral hemispheres.

As the cranium is rigid (except in the young child) and as the brain is virtually incompressible the combined volume of brain tissue, cerebrospinal fluid and intracranial blood is nearly constant. The Monro–Kellie hypothesis states that the blood volume in the cranial cavity is approximately constant (Monro 1783; Kellie 1824).

Superficial and deep veins, which anastomose, open into the venous sinuses which lie between the dura mater and the bone. These veins have no valves and are kept open by the structure of the dura around their orifices. The blood drains from the brain mainly via the internal jugular vein but also by channels which join the vertebral venous plexus and by anastomoses with the orbital and pterygoid plexuses. These vertebral anastomoses can indeed deal with the entire venous return from the brain. Thus, Batson reported a case of complete obstruction of the upper part of the superior vena cava; the patient survived).

Measurement of cerebral blood flow

Pioneered by Schmidt, important quantitative measurements of cerebral blood flow were made using the bubble flowmeter in monkeys and these direct values were compared with those obtained with the nitrous oxide technique showing the latter to be accurate and acceptable for use in man. The subject breathes 15 per cent N$_2$O in an oxygen–nitrogen mixture for 10 minutes. During this period blood samples are removed from the internal jugular bulb (which drains the deep cerebral vascular bed) and simultaneously from a peripheral artery. Five arterial and five venous samples are withdrawn simultaneously and their N$_2$O concentrations analysed and plotted graphically [FIG. II.112]. The integrated A–V nitrous oxide difference can be obtained graphically or by calculation. More recently, some favour continuous slow withdrawal of arterial and cerebral venous blood throughout the 10 minute breathing period. These two total samples give the overall A–V difference. Two additional samples are taken to give the arterial and venous N$_2$O concentrations at the end of the 10 minute period. By the Fick principle, in a steady state:

$$\text{CBF per 100 g per min} = 100 \frac{\text{Amount of N}_2\text{O taken up by brain}}{(\text{A–V})\text{N}_2\text{O concentration}}$$

$$\therefore \text{ CBF per 100 g per min} = \frac{100 V_U . S}{\int_0^U (\text{A–V})\mathrm{d}t}.$$

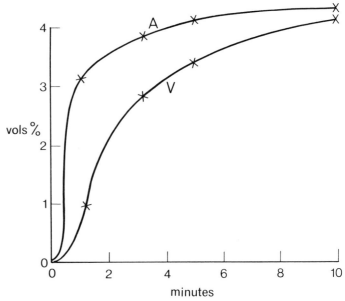

FIG. II.112. Arterial (A) and cerebral venous blood (V) concentrations of N$_2$O while inhaling 15 per cent N$_2$O in an oxygen–nitrogen mixture. Ordinate shows concentration of N$_2$O (ml per 100 ml of blood). (After Kety, S. S. and Schmidt, C. F. (1948). *J. clin. Invest.* **27**, 476.)

Where V_U = venous N$_2$O concentration after equilibrium is reached in the brain tissue during time U.
S = partition coefficient for N$_2$O between blood and brain tissue.

As the partition coefficient of N$_2$O between brain and blood is unity, the amount of N$_2$O taken up by the brain can be calculated from V_U at equilibrium and the integrated (A–V) difference.

The N$_2$O technique can only be used in steady states and not for the measurement of rapidly changing blood flows. It has been replaced by the use of intra-arterial injection of ^{133}Xe (a gamma emitter). 2 mCi of ^{133}Xe dissolved in 2 ml saline are injected within 1–2 s via a thin catheter in the internal carotid artery. The clearance is recorded for 10 minutes.

^{133}Xenon thus injected intra-arterially is carried to the brain and enters and exits at a rate solely dependent on diffusion and solubility. As the solubility of xenon is far higher in air than in blood or brain tissue, the great majority of it will be excreted by the lungs and there is little or no effective arterial recirculation. Thus, as the isotope equilibrates rapidly with the brain tissue during the injection and when the injection stops, the arterial blood now containing no isotope washes the gas out of the brain at a rate which depends solely on the blood flow.

The gamma emissions of ^{133}Xe are recorded by scintillation crystals mounted externally; an array of crystals (collimated with lead so that each 'looks at' a defined volume of brain) is mounted over various points of the patient's scalp. The signals from the crystals and their associated photomultipliers are fed through pulse height analysers into a multichannel tape-recorder, which in turn feeds the information through a scaler and a chart recorder. These give the shape of the clearance curve.

The calculation of the mean blood flow through the volume of brain 'seen' by each crystal depends on the formula

$$\text{Flow (ml g}^{-1}\text{ min}^{-1}) = \frac{\lambda b \times (H_{max} - H_{10})}{A_{10}}$$

Where λb = brain–blood partition coefficient; H_{max} = maximal height of clearance curve; H_{10} = height at 10 min; A_{10} = area under clearance curve [FIG. II.113].

The clearance curve shows two components—the more rapid

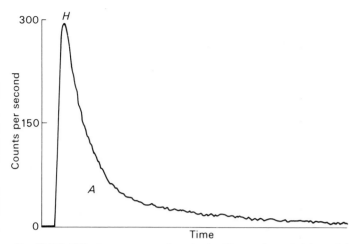

FIG. II.113. ^{133}Xe clearance curve from brain. H = maximum height and A = area under the curve. (Redrawn from Harper, A. M. (1970). In *Cerebral circulation* (ed. G. D. McDowall). Little Brown, Boston.)

flow through grey matter and another through white matter. When plotted on semilog paper a straight line drawn through the 'tail' of the curve is subtracted from the primary curve [FIG. II.114] and the '$T_{\frac{1}{2}}$' (the time taken for the radioactivity to decline to half its initial value) of the resulting exponential components is calculated. Blood flows for each exponential can be calculated separately.

Flow (ml g^{-1} min^{-1}) for fast (grey) or slow (white) matter is

$$F = \frac{\lambda \times \log_e 2}{T_{\frac{1}{2}}} = \frac{\lambda \times 0.693}{T_{\frac{1}{2}}} .$$

Using these various techniques, it has been confirmed that *total* cerebral blood flow is 50–55 ml per 100 g per min (Kety and Schmidt 1948; Lassen *et al.* 1964; Ingvar *et al.* 1965; Harper 1966). In normal man this is remarkably constant, although this should not disguise the fact that *local* increases of CBF occur when there is an increase in regional function. Total cerebral flow is not even much affected by sleep, and this is believed to be due to the maintained requirement for a supply of oxygen and glucose and for the removal of metabolites. The brain is particularly vulnerable to oxygen lack and it is not surprising that a flow of only 25 ml per 100 g per min

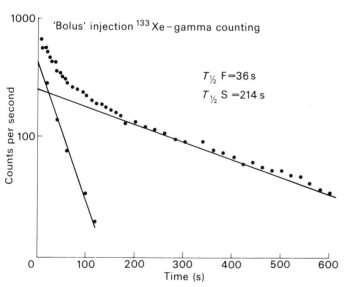

FIG. II.114. Semilog plot of idealized clearance curve showing exponential stripping in fast (F) and slow (S) components. The intercepts of the two slopes at time 0 gives a measure of the relative weights of the two tissue components. (Redrawn from Harper, A. M. (1970). In *Cerebral circulation* (ed. G. D. McDowall). Little Brown, Boston.)

may cause loss of consciousness. Murray Harper (1976) quotes a 'critical flow level' for failure of neuronal function (as indicated by EEG monitoring) in man undergoing carotid surgery, as approximately 18 ml per 100 g per min.

As elsewhere in the vascular system the blood flow through the brain depends on the perfusion pressure (mean arterial pressure minus the cerebral venous pressure, which is effectively the intracranial pressure) and the resistance. When recumbent, the mean arterial pressure is 100 mm Hg and the internal jugular pressure is less than 10 mm Hg. If upright, the mean cerebral arterial pressure falls some 20–30 mm Hg, but the internal jugular pressure and that of the c.s.f. (intracranial pressure) falls similarly owing to the gravitational forces. The cerebral vessels and the attached non-collapsible channels of the venous sinuses act like a siphon holding the perfusion pressure fairly constant. There is, however, an element of danger here for patients submitted to intracranial operations when tilted into the head-up position. Inadvertent breach of a venous sinus (in which the pressure is subatmospheric) causes air to be sucked into the lumen with perhaps fatal results of air embolism. Normally, as stated above, the sinuses are held open by fibrous attachments to their walls.

When the mean systemic pressure is lowered by repeated haemorrhages, CBF remains steady until the pressure falls below 65 mm Hg [FIG. II.115]. Conversely, until the mean pressure rises from 100 mm Hg to perhaps 140 mm Hg, CBF is reasonably constant. This autoregulation of CBF is seen only when the arterial P_{CO_2} and P_{O_2} are maintained at or near normal values; when arterial CO$_2$ tension is raised the pressure flow relationship is linear [FIG. II.116].

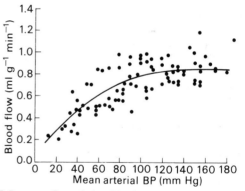

FIG. II.115. Pressure–flow relationship of the cerebral cortex in dogs studied under normocapnia. (A. M. Harper (1966). *J. Neurol. Neurosurg. Psychiat.* **29**, 398–403.)

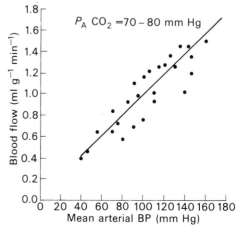

FIG. II.116. Pressure–flow relationship of the cerebral cortex in dogs studied under hypercapnia. (A. M. Harper (1966). *J. Neurol. Neurosurg. Psychiat.* **29**, 398–403.)

In man, a maintained rise of arterial CO_2 tension of 1 mm Hg causes an increase of about 3 ml per 100 g per min in the CBF. The CBF decreases about 1.5 ml per 100 g per min per 1 mm Hg fall of arterial CO_2 tension below its normal range. At 20 mm Hg tension little further vasoconstriction occurs because such low levels of flow tend to induce cerebral hypoxia with the production of lactic acid (a cerebral vasodilator when it diffuses into the cerebral interstitium); above 80 mm Hg CO_2 tension CBF does not increase further probably indicating that the vessels are maximally dilated.

The mechanism of cerebral autoregulation is believed to be chiefly metabolic—as the perfusion pressure and therefore the driving force for flow decreases, CO_2 accumulates to produce cerebral vasodilatation by virtue of its conversion to H^+. In more serious conditions of reduced perfusion, cerebral hypoxia causes the release of pyruvic and lactic acids, which diffusing into the interstitial fluid further raise its H^+.

Some (e.g. Ekstrom-Jodal 1970; Eklöf *et al.* 1971) believe that myogenic responses of the small precapillary blood vessels (which contract in the face of a rise in distending pressure according to the Bayliss–Folkow hypothesis) [p. 70] contributed at least in part to autoregulation of CBF. Thus, when saline was quickly injected into a branch of the middle cerebral artery causing a sudden rise in transmural pressure, flow transiently increased and returned to normal in a matter of two seconds—such a rapid reactivity is likely to be myogenic.

Purves (1974) has furnished evidence that vasodilator nerves supplying the cerebral vessels are responsible for autoregulation, showing that cervical sympathectomy abolishes it, as does atropine—indicating that the nerve supply concerned is cholinergic sympathetic vasodilator. However, many groups of workers have claimed that sympathectomy does not influence autoregulation of the deep cerebral vessels; they attribute Purves' results to effects on the extraparenchymal resistance—i.e. major arteries which though intracranial are outside the brain itself. Such would include the pial vessels (Harper 1975).

Some anaesthetic agents undoubtedly influence cerebral blood flow and metabolism. Thus even light thiopentone anaesthesia reduces CBF and metabolism by 30 per cent.

Halothane, trichloroethylene, cyclopropane, ether, and chloroform all cause a reduction in cerebral blood flow and a depression of metabolism. Two per cent halothane in air decreases CBF by 25 per cent and reduces metabolism by 30 per cent. Such effects may stem from the influence of these agents on vascular sensitivity.

Vigorous hyperventilation for a minute or so will lower the arterial pCO_2 to 15 mm Hg in some subjects. Dizziness results, partly due to the acapnia and alkalosis but also due to oxygen lack of the cortical neurons. The cerebral vasoconstriction is extreme—the retinal vessels in this respect serve as a 'mirror' of the cerebral vasculature and a common complaint is a narrowing and blurring of the field of vision. In addition to reducing the overall blood flow, the local effect of reduction of the $[H^+]$ of the interstitial fluid presumably causes increased tone of the precapillary sphincters which diminishes the surface area of the capillary beds in the grey matter. Finally, the low pCO_2 exerts a 'reverse Bohr' effect and causes the O_2 dissociation curve to move over to the left so that the pressure head of oxygen released from the oxyhaemoglobin is much lower [see p. 185].

Hyperbaric oxygen (oxygen under more than 1 atmosphere pressure) constricts the cerebral vessels. Lambertsen and his colleagues (1955) showed that men breathing O_2 at 3.5 atmospheres pressure (arterial $pO_2 = 2300$ mm Hg) nevertheless had a jugular pO_2 of only 75 mm Hg because of the reduction in cerebral flow caused by vasoconstriction. If 2 per cent CO_2 were added to the gas inspired its effects overcame the influence of high pO_2 and relaxed the cerebral vessels so that the jugular venous pO_2 rose to the startling figure of

1000 mm Hg. A *high* cerebral pO_2 is dangerous, for neuronal metabolism is disrupted and convulsions, coma and death occur [see p. 210]. The 'protective' effect of increased precapillary resistance 'steps down' the pO_2 to which the neurones are eventually exposed. The slightest vitiation by CO_2 of the gas mixture breathed is dangerous because the vasodilatation so induced exposes the neurones to a much higher local pO_2.

The rapid oxygen usage of the brain is essentially due to the grey matter. White matter consists of axons and (if their O_2 usage is similar to that of peripheral axons) their O_2 usage is only 0.3 ml per 100 g per min. White matter comprises about 60 per cent of the brain weight so, per 100 g of brain, white matter would use 0.2 ml min^{-1} and as the brain uses 3.3 ml per 100 g per min the grey matter uses 3 ml min^{-1}. In a brain weighing say 1500 g, 40 per cent of this being grey matter, 600 g of grey matter use 40–45 ml min^{-1} of oxygen, and 900 g of white matter use only about 3 ml O_2 min^{-1}. *The grey matter, less than 9 per cent of the body weight uses nearly 20 per cent of the oxygen consumed by the whole body at rest.* In keeping with this high rate of oxygen usage, the grey matter has a capillary density of nearly 4000 capillaries per mm^2.

Central venous pressure is raised considerably when straining at stool or coughing but cerebral blood flow is unaffected—the intracranial c.s.f. pressure of course rises simultaneously, which minimizes the change of pressure in the transmural venules and capillaries and perhaps prevents their rupture; simultaneously there is a rise of transmitted arterial pressure.

Intracranial pressure. Normally the c.s.f. occupies about 10 per cent of the intracranial volume and is at a pressure between 0 and 7 mm Hg. Large volumes of c.s.f. may be displaced into the spinal canal, where volume is made available by compression of the epidural venous plexuses. This mechanism provides about 65 per cent of the compensatory capacity of the rigid skull. Once this 'reserve' has been used up small extra additions cause striking rises in intracranial pressure, which may correspondingly approach the mean arterial pressure. Cushing was the first to note that an acute rise of intracranial pressure led to the development of arterial hypertension which served to maintain the pressure head for cerebral blood flow. Kety investigated the effects of increased intracranial pressure (caused by tumours) in man on CBF measured by the N_2O technique. When lying down the jugular venous pressure is about 10 cm H_2O. Kety found that rises of cerebrospinal fluid pressure from 10 cm to 45 cm H_2O caused a linear rise of arterial blood pressure and the cerebral blood flow remained constant despite the increase in extravascular resistance. When cerebrospinal fluid pressure exceeded 45 cm H_2O, arterial pressure fell progressively and the subjects became comatose. It is believed that the rise of cerebrospinal fluid pressure from its normal levels initially tends to produce bulbar asphyxia and the effect of this on the vasomotor centre is to excite it thereby evoking extracerebral vasoconstriction and the rise of the mean arterial blood pressure. When the rise of cerebrospinal fluid pressure becomes too severe the bulbar centres become depressed and the 'compensatory' mechanism fails.

REFERENCES

EKLÖF, B., INGVAR, D. H., HAGSTRÖM, E., and OLIN, T. (1971). *Acta physiol. scand.* **82**, 172–6.
EKSTRÖM–JODAL, B., HÄGGENDAL, E., LINDER, L. E., and NILSSON, N. J. (1971). *Eur. Neurol.* **6**, 6–10.
HARPER, A. M. (1966). *J. Neurol. Neurosurg. Psychiat.* **29**, 398–403.
—— (1975). In *Cerebral vascular diseases* (ed. J. P. Whisnant and B. A. Samdok) pp. 27–47. Grune & Stratton, New York.
INGVAR, D. H., CRONQUIST, S., EKBERG, R., RISEBERG, J. and HØEDT-RASMUSSEN, K. (1965). *Acta neurol. scand.* Suppl. **14**, 72–8.
—— and LASSEN, N. A. (1974). (eds.) *Brainwork: the coupling of function metabolism and blood flow through the brain.* Academic Press, New York.

KETY, S. S. and SCHMIDT, C. F. (1948). *J. clin. Invest.* **27**, 484–92.
LASSEN, N. A. and HØEDT-RASMUSSEN, K. (1966). *Circulation Res.* **19**, 681–8.
PURVES, M. J. (1972). *The physiology of the cerebral circulation.* Cambridge University Press.
——(ed.) (1978). *Cerebral vascular smooth muscle and its control.* Elsevier, Amsterdam.
STANDGAARD, S., MACKENZIE, E. T., SEMGUPTA, D., ROWAN, J. O., LASSEN N. A. and HARPER, A. M. (1974). *Circulation Res.* **34**, 435–40.

CUTANEOUS CIRCULATION

Skin weighs about 2 kg in an adult. Its blood flow depends not so much on the metabolic activity of the skin itself as on the requirements for the maintenance of body temperature. Under maximal heat load, the skin flow may total as much as 3–4 l min⁻¹; conversely, when the body is exposed to cold stress, total skin blood flow falls (as a result of sympathetic vasconstriction) to less than 50 ml min⁻¹, so that almost the full insulating power of the skin and subcutaneous fat is then realized.

A nude man's comfortable environmental temperature at rest is about 27 °C. At lower ambient temperatures his skin vessels are strongly constricted and blood flow is directed to the deeper tissues. In the limbs, the arteries supplying the extremities have accompanying veins (venae comites) which serve as a counter-current exchange mechanism to favour heat exchange. In cold environments, the cooled blood returning from the surface helps to lower the temperature of the arterial blood distributed to the surface.

The average skin flow of a naked man in resting thermal equilibrium (at a surrounding temperature of 25–30 °C) is 10–15 ml per 100 g per min, but there are considerable regional variations. This blood flow is only doubled by blocking the sympathetic vasoconstrictor nerves; normally the skin vessels are tonically under the influence of a sparse sympathetic discharge. However, whereas this is true for the skin as a whole, it is not correct in the case of the skin of the hands and feet. The vascular circuit of skin of these areas shows *numerous* A–V anastomoses, which are much less frequent in the skin of the rest of the body. These A–V anastomoses are normally dominated by sympathetic vasoconstrictor activity and are closed. When heat stress abolishes the sympathetic discharge to these structures (by a hypothalamic mechanism, p. 131) the A–V anastomoses dilate widely and hand and foot skin flow increase strikingly. Thus, the hand blood flow reaches 30 ml per 100 g per min after acute sympathetic blockade (environmental temperature = 27 °C). Seventy per cent of the hand consists of muscles, tendons, and bone which have a blood flow below 10 ml per 100 g per min, so this total flow recorded represents a skin flow (largely through the patent A–V anastomoses) of 70–80 ml per 100 g per min. These A–V anastomoses are most numerous in the fingers.

Skin which has few or no A–V shunts does not show the same degree of hyperaemia on sympathetic blockade and flows rarely exceed 25 ml per 100 g per min in such circumstances. Far higher flows than this can be achieved by exposing the body to heat stress, values of 150 ml per 100 g per min being recorded by Edholm, Fox, and Macpherson (1956) in forearm skin in such circumstances. It follows that the ordinary precapillary resistance vessels and precapillary sphincters of the skin evince a considerable *basal* tone. On this background of basal myogenic tone operate the postganglionic sympathetic fibres. At comfortable environmental temperatures, their discharge is only sufficiently effective to halve the flow from that allowed by the basal myogenic tone itself. However, exposure to cold causes a hypothalamic stimulation to this sympathetic discharge and flow is reduced to as little as 1 ml per 100 g per min.

There are no specific vasodilator fibres to skin, which at first sight seems surprising when one considers the well-known phenomenon of blushing. Blushing may be due to bradykinin release secondary to a brief corticohypothalamically controlled discharge of sympathetic cholinergic fibres to the sweat glands. Bradykinin is the most potent vasodilator known.

In heat stress maximal activation of the sweating mechanism may secure a total sweat rate of 2 l h⁻¹. Bradykinin release during this process doubtless contributes to an enormous hyperaemia which may exceed 3 l min⁻¹.

The skin flow requirements for homoeothermia obviously throw a big load on cardiovascular function, for as much as 3–4 l min⁻¹ has to be supplied by the heart to this circuit. Moreover, this requirement is *dominant* and men working under maximal heat loads may simply collapse with circulatory failure unless supervised adequately.

Cold vasodilatation

Although exposure of the skin to temperatures of 0 °C for considerable periods may not cause damage, lower temperatures or prolonged exposure, particularly in damp conditions, produce lesions (such as 'trench foot').

'Cold' vasoconstriction reduces the supply of nutriment to the skin, but at the same time the direct effect of temperature and the reduction of the flow of warm blood lower the metabolic rate of the tissue and metabolite accumulation is very slow. Unlike the muscle precapillary sphincter response to vasomotor discharge (which 'escapes' because of the local accumulation of metabolites) the response of the precapillary resistance vessels of the skin to maximal vasoconstrictor discharge is well sustained and does not 'escape' to any great extent.

When the skin temperature drops below 10 °C, pain may be experienced. In such circumstances cold vasodilatation takes place. This phenomenon is largely due to axon reflexes operating particularly on A–V anastomoses. Tissue injury causes the liberation of histamine which excites the sensory terminals and produces long-lasting vasodilatation by an axon reflex pathway [see below]. In addition low temperatures promote the formation of plasma kinins.

Tissue trauma

Lewis (1927) described the triple response to cutaneous trauma. A firm strong stroke across the skin using, say a pencil point, evokes a series of responses classified as: 1. red reaction; 2. flare; and 3. wheal.

The red reaction. This is a dilatation of precapillary sphincters directly due to the presence of histamine and/or polypeptides released from the damaged skin. Characteristically it outlines the stroke and is sometimes known as the red line. It is not mediated by nerves for the capillary vasodilatation which is responsible for the appearance is a purely passive phenomenon consequent upon the relaxation of the precapillary sphincters. Local anaesthetization of the skin does not prevent the red reaction.

Flare. Flare is due to dilatation of the arterioles, terminal arterioles and precapillary sphincters. It causes an irregular erythematous area which surrounds the site of the red line. The skin temperature overlying this area is raised because the drop in arteriolar resistance permits an increased blood flow. Local anaesthetization abolishes the reaction so it is mediated by nerves. However, if the relevant nerve trunk—say the ulnar nerve is blocked at the elbow in an otherwise normal subject and the pencil stroke is made on the otherwise normal skin supplied by the ulnar nerve the flare is still produced. This indicates that though the mechanism is nervous it does not involve CNS connections. It transpires that the response of arteriolar dilatation is mediated by a branching of the terminal axons of C fibres in the manner shown in FIGURE II.117—the axon reflex. The phenomenon of posterior root vasodilators is simply due to an artificial method of stimulating these terminal branched C fibres. The substance liberated by the nerve endings around the arterioles is not known; it produces a long-lasting vasodilatation which is unlike that of any of the known vasodilator substances.

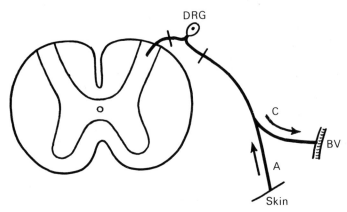

FIG. II.117. Antidromic vasodilators in dorsal nerve roots, axon reflex in the skin and mechanism of flare. DRG = dorsal root ganglion; BV = cutaneous blood vessel.

Wheal. Eventually, if the stroke stimulus has been strong enough, a blister-like appearance develops spreading from the margins of the red line within the flare area. This is due to increased capillary permeability due to damage, coupled with the rise of capillary pressure occasioned by dilatation of the precapillary resistance vessels. The fluid formed contains substantial amounts of protein.

The features of the triple reaction are due to the liberation of histamine. The intradermal injection of histamine in known amounts can simulate the triple response closely in appearance and time course. Polypeptides, such as bradykinin, are also released in response to injury.

Some individuals are unusually prone to develop striking triple response reactions (dermatographia). Skin injury caused by a wide variety of physical and chemical factors give rise to these vascular responses. Their common origin is in the release of histamine (and polypeptides) from the damaged area.

Features of nervous control of the skin circulation

Normally the skin vessels are subjected to a low rate of sympathetic constrictor discharge which effectively limits the flow through them. Vasodilatation of this vascular circuit is never 'neurogenically' active, but is due to the reduction of vasoconstrictor impulse activity. Vasodilatory effects can additionally result from the action of bradykinin released by sympathetic discharge to the sweat glands (e.g. blushing).

Most of the control of vasoconstrictor activity is exercised from the hypothalamus, in response to temperature changes which are recorded centrally and which also cause the excitation of lateral spinothalamic tracts excited by temperature receptors in the skin. As the heat load rises gradually, first the A–V anastomoses of the hands, ears and feet dilate due to the reduction of their regional sympathetic discharge. Later the remainder of the skin vessels dilate, again due to the progressive withdrawal of sympathetic vasoconstrictor activity. Sweat gland activation follows later if the maximal dilatation of A–V anastomoses and the already moderate dilatation of the skin resistance vessels are inadequate themselves to effect the restoration of thermal equilibrium. The evaporation of

sweat yields 2.4 kJ g^{-1}; additionally bradykinin release secures a maximal vasodilatation of the skin vessels in the vicinity.

Emotional effects on cutaneous skin vessels are relayed from corticohypothalamic 'centres' to the thoracolumbar sympathetic cell bodies and thence to the skin vessels. Blanching of the skin—'white with fear', 'livid with passion', 'the blush of shame', etc., keep the novelists occupied in describing the pantomime of emotional exteriorization.

REFERENCES

BURTON, A. C. and EDHOLM, O. G. (1955). *Man in a cold environment.* London.
EDHOLM, O. G., FOX, R. H., and MACPHERSON, R. K. (1956). *J. Physiol., Lond.* **134**, 612.
GREENFIELD, A. D. M. (1963). In *Handbook of physiology*, Section 2, Vol. II, p. 1324. American Physiological Society, Washington.
LEWIS, T. (1927). *The blood vessels of the human skin and their responses.* London.
STRÖM, G. (1960). In *Handbook of physiology*, Section 1, Neurophysiology, Vol. II, p. 1173. American Physiological Society, Washington.

MUSCLE CIRCULATION

About 40 per cent of the body weight is muscle—approximately 30 kg. The total blood flow through resting muscle is some 800 ml, for the flow per 100 g per min is only 2–3 ml. In exercise, with a cardiac output of 25 l min^{-1}, the muscle blood flow is probably 20 l min^{-1} and thus reaches about 70 ml per 100 g per min. This profound vasodilatation indicates that the vascular resistance is high in resting muscle and indeed the muscle vascular circuit provides by far the most important contribution to the total peripheral resistance.

Some of the vascular resistance offered by the muscle circuit is due to tonic sympathetic discharge. The precapillary resistance vessels are generously endowed with sympathetic vasoconstrictor nerves. These discharge at a rate of 1 impulse per s when the animal or man is recumbent and increase their rate of impulse activity to 2 or 3 s^{-1} in the upright position. This discharge increases the pre/postcapillary resistance ratio and lowers the mean capillary pressure, which in turn helps to increase the uptake of tissue fluid. Sympathetic discharge also causes constriction of the postcapillary venules which helps to mobilize blood towards the heart.

Barcroft (1963), using venous occlusion plethysmography, has provided us with the quantitative details of resting muscle blood flow under the influence of sympathetic resting discharge and after sympathetic blockade. Local anaesthetization of the ulnar, radial, and median nerves at the elbow doubles the resting blood flow through the forearm. These somatic nerves contain the sympathetic fibres destined for the forearm muscle vessels. Naturally the local anaesthesia of the somatic fibres concomitantly reduces skeletal muscle tone, but this effect of itself does not underly the increase in muscle blood flow for the local blockade of the somatic nerves at the elbow has no influence on the resting blood flow in a sympathectomized subject although of course the loss of skeletal muscle tone is as great.

Sympathetic vasoconstriction can be temporarily abolished by heating the subject. Thus, forearm blood flow is doubled when the legs of the subject are immersed in water at 45 °C for 30 minutes. Hypothalamic responses to the rise in temperature cause an inhibition of the normal bulbar vasomotor drive to the thoracolumbar sympathetic neurones.

Sympathetic vasoconstriction is increased after haemorrhage and the muscle blood flow is profoundly reduced. These effects in man can be analysed with more precision in animal experiments such as those using the Mellander preparation [p. 73]. The vasoconstriction of hypotension is largely reflex in nature. Both reduced baroreceptor inhibition and increased chemoreceptor excitation of the vasomotor centre resulting from haemorrhagic hypotension contribute to the increased impulse traffic in the sympathetic vasoconstrictor nerves.

When the sympathetic fibres supplying the hind limbs are stimulated synchronously at, say, 10 imp per s, in the Mellander preparation muscle blood flow falls to some 15 per cent of its normal value. This is the maximal reduction of flow which sympathetic activity can effect. The flow drops to values of 0.3–0.5 ml per 100 g per min.

The temporary blockade or the permanent loss (by surgical removal) of the sympathetic vasoconstrictor supply to the skeletal muscle vessels only doubles the resting muscle blood flow. Yet the exercise hyperaemia induced by muscle activity in a sympathectomized subject will increase the flow just as in a normally sympathetic-innervated limb to 70 ml per 100 g per min and as has been stressed, this increased flow is due to chemical effects on the precapillary resistance vessels. It follows that the basal myogenic tone of the precapillary resistance vessels is high. These vessels are among the best examples of the 'Bayliss–Folkow response' to a raised transmural pressure. Such a rise of transmural pressure induces a *myogenic* contraction which by raising the precapillary vessel tone 'protects' the capillaries from an undue rise of capillary pressure. These responses are important in the leg muscle circuits in the erect position for they reduce the rate of oedema formation.

Sympathetic vasodilator nerves

From the point of view of the teacher it is in some ways unfortunate that the sympathetic vasodilator nerves to muscle vessels were ever discovered. Once the junior student hears that there are sympathetic vasodilator fibre nerves, he tends to jump to two totally erroneous conclusions: (a) that the sympathetic supply to the muscle vessels is solely vasodilator and (b) that the sympathetic vasodilator nerves are responsible for exercise hyperaemia.

Both these conclusions are resoundingly wrong.

The great preponderance of sympathetic nerves which supply muscle vessels are vasoconstrictor. Exercise hyperaemia, when established, is entirely independent of the local sympathetic supply to the muscle vessels. It is just as quantitatively impressive in sympathectomized limbs as in normal limbs and is due to chemical factors.

Even when these facts are pounded home, the more mediocre student clings tenaciously to a third and equally wrong belief—viz., that the sympathetic vasodilator fibres are important in securing an improved nutritive flow to the muscles in the premonitory and initial stages of the exercise. They are *not*. Their effects are solely to reduce the vascular peripheral resistance of the muscle bed by dilating arterioles (but *not* precapillary sphincters) and bypassing the muscle capillaries thus providing a thoroughfare channel across to the venules. In this respect they are probably important in preventing too fulminant a rise in the blood pressure at the beginning of exercise, when a doubling of the stroke volume and an increased heart rate would produce a staggering degree of hypertension unless the total peripheral resistance were reduced.

The sympathetic vasodilator nerves are activated by cortico-hypothalamic-reticulo-spinal pathways, which are quite separate from the vasomotor centre—thoracolumbar spinal paths. The thoracolumbar sympathetic cells which operate the vasodilator effects are not influenced by medullary afferents such as baroreceptor and chemoreceptor fibres and play no part in the changes of peripheral resistance which such afferents secure.

The evidence for these statements was provided by Lindgren and Uvnäs (1954), who mapped out the brainstem route which the corticohypothalamic descending pathway followed.

Barcroft *et al.* (1944) proved the presence of this cholinergic innervation in man. Volunteers were first submitted to venous occlusion of both legs (which traps some 700 ml of blood therein) followed by venesection and bleeding. Fainting ensued and as the faint occurred forearm blood flow increased despite the fact that the arterial blood pressure was falling. When this phenomenon was analysed by studying the blood flow in both forearms, one of which was normal and the other with the radial, median, and ulnar nerves blocked ('sympathectomized'), fainting caused a rise of flow in the normal arm and a passive fall of flow in the denervated forearm. The rise of flow in the normal forearm must have been due to active vasodilatation.

Sympathetic vasodilator excitation (by hypothalamic stimulation) may increase the muscle blood flow from 2–3 ml to say, 30 ml per 100 g per min (but not more) as shown in animal experiments.

The role of this mechanism is still disputed. There is evidence from experiments in man that mental stress and emotion cause muscle vasodilatation, and animal experiments have shown that stimulation of the 'defence area' in the anterior hypothalamus p. 132] evokes muscle vasodilatation along with the somatomotor features of the alerting reaction—ears pricking, muscle tensing, etc. Hence, the system is believed to operate only in emergencies and is credited with importance in helping to secure biological survival when the animal is threatened by a predator. The oft-quoted hypothetical scene in the jungle gives the essence of our ideas—the gazelle grazing, hears the crackle of a twig, which alerts him to the proximity of a predator. His alarm reaction provokes flight together with a substantial muscle vasodilatation. As stated, however, the muscle vasodilatation does *not* improve his nutritive capillary muscle flow, for Renkin has shown that the capillary permeability surface area product *PS* is not increased by sympathetic vasodilator stimulation [p. 88]. The increased flow is due to arteriolar dilatation but the precapillary sphincters are not supplied by the sympathetic vasodilators and do not open in response to these nervous influences. As the capillary flow and the surface area of the capillary bed is determined by precapillary sphincteric tone, the huge increase of capillary blood flow in exercise has to await the relaxation of these sphincters caused by the local chemical effects of metabolites and hyperosmolality which rapidly develop as a consequence of the fulminant increase of muscle metabolism. The sympathetic vasodilators thus ensure that the peripheral resistance of the muscle vessels is reduced immediately the increased cardiac output takes place and thereby provide a 'safety valve mechanism' against an undue rise of arterial pressure. Once exercise is established, the local chemical effects in the muscle *completely* oversway sympathetic vasodilator influences.

Exercise hyperaemia

In exercising muscle flows of 50–70 ml per 100 g per min indicate a profound reduction of local vascular resistance. Of more importance, this fall of resistance is due not only to arteriolar vasodilatation but also to the dilatation of the precapillary sphincters. This enormously increases the size of the capillary bed and hence its surface area, thereby facilitating diffusion of oxygen to and CO_2 from the active tissues. The transmitted hydrostatic pressure rises sharply in the capillaries [p. 73] and fluid transfer from capillaries to the interstitium increases. Even a brief spell of intense muscular exercise reduces the plasma volume by 10 per cent and it is this filtration mechanism which is responsible for the haemoconcentration seen in man after exercise, not the expulsion of red cells following contraction of the spleen because the splenic capsule in man has no muscle with which to contract. Splenectomized man shows the same haemoconcentration after exercise as does a normal individual.

With the proof that exercise hyperaemia in man was independent of the presence of sympathetic nerves and was therefore chemical in origin, some 40 years have been spent in seeking which chemicals might be responsible. The obvious candidates of CO_2 excess, O_2 lack, increased $[H^+]$, lactic acid, ATP, ADP, AMP, etc., were tested and though each caused *some* vasodilatation none was satisfactory in the starring role. At the moment two factors are under discussion: hyperosmolality of the interstitial fluid and a rise of $[K^+]$ in the vicinity of the precapillary sphincters.

Mellander *et al.* (see Mellander and Johansson 1968) found that strenuous forearm or leg exercise caused an increase of regional venous osmolality of up to 30 mOsmol kg^{-1}. The infusion of hypertonic solutions into the brachial artery in amounts which raised regional venous osmolality by 15 mOsmol kg^{-1}, lowered forearm flow resistance to 30 per cent of its control value, increasing flow from some 3 ml to 8–12 ml per 100 g per min. Hyperosmolality inhibits myogenic pacemaker activity and does indeed cause relaxation of vascular smooth muscle cells of isolated portal vein preparations (see Mellander and Johansson 1968) but there are no data about the responses of vascular smooth muscle cells of the precapillary sphincters to such osmotic effects. However, the influence of hyperosmolality on these sphincters will bear a closer scrutiny, for there is no question that the increased delivery of osmotically active small molecules from the active tissue cells always occurs in exercise.

The proposition that an increased $[K^+]$ of the interstitium might be partly responsible for the hyperaemia of exercise was advanced by Dawes in 1947 and has received some support from more recent work of Kjellmer. Contracting muscle cells do liberate potassium and during maximum exercise vasodilatation the venous effluent from the contracting muscle may double in K$^+$ (despite the hyperaemia). A raised $[K^+]$ in the medium surrounding isolated strips of small arteries causes relaxation.

Although the detailed cause(s) of exercise hyperaemia remains obscure these three factors, possibly acting necessarily in concert—$[K^+]$, hyperosmolality and O_2 lack—are being further investigated in the hope that a solution of the problem may be forthcoming.

Mechanical interference of muscle contractions

Arterial flow is impeded and may even cease during intense phasic muscle contractions, but the flow during the relaxation period occurs uninhibited through a maximally dilated bed. Myoglobin [p. 187] acts as an oxygen acceptor during this relaxation period and when contraction once more interrupts capillary flow the myoglobin yields its oxygen to the myofibrils. If the muscle is engaged in a sustained contraction of more than say, 10 seconds' duration, the myoglobin supply of oxygen is exhausted and anaerobic metabolites accumulate. These cause fatigue and eventually ischaemic pain. Forearm muscles are of 'phasic character' and engaged in their normal phasic activity can sustain quite heavy work loads because the relaxation periods allow recharging of their myoglobin 'accumulator' system. It is common experience that carrying a heavy suitcase is more tiring although the metabolic demand of the arm muscles is far less than that of maximal intermittent activity; the sustained contraction 'runs down' the myoglobin accumulator.

Following heavy phasic exercise the blood flow does not subside immediately, but falls exponentially from its high level during the exercise to resting values. Just as the oxygen debt of exercise can be measured, by determining the excess oxygen usage over and above the resting O_2 consumption in the post-exercise period, so can the 'blood debt' be computed.

Reactive hyperaemia

If the arterial supply to a limb is cut off for 5 minutes by pumping up a sphygmomanometer cuff to 200 mm Hg then on releasing the cuff pressure the limb shows obvious cutaneous hyperaemia and this is accompanied by muscle hyperaemia. Venous occlusion plethys-

mograph measurements show that blood flow may transiently reach values of as much as 35–40 ml per 100 g per min, a hyperaemia due to the combined dilatation of arterioles and precapillary sphincters caused by the local accumulation of metabolites. These metabolites are quickly dissipated by the flushing effect of the increased blood flow which their presence initially secures. Reactive hyperaemia is more transient than is post-exercise hyperaemia.

Tonic and phasic muscles and their blood supply

The account give above relates in essence to the features of the circulation through phasic muscle, which constitutes some three quarters of the total muscle mass. Phasic muscle—synonymous with 'white' muscle—shows rapid phasic contractions. Its vascular features have already been described—a resting flow of 3 ml per 100 g per min which may increase to 60–70 ml per 100 g per min in maximal exercise.

Tonic muscle contracts more slowly ('red' muscle) and, being concerned with the maintenance of posture, its activity is rather of the steady prolonged type achieved by asynchronous contraction which requires a relatively low O_2 usage. Such muscles comprise some 25 per cent of the total muscle mass and show features which differ in many respects from those of the phasic type. The resting flow far exceeds that of phasic muscle, being 20–30 ml per 100 g per min; this indicates a relatively low basal myogenic tone. Their vascular bed is as much as three times the size of that of phasic muscle and their flow capacity is as high as 150 ml per 100 g per min. The greater surface area of the capillary bed and their lower oxygen requirements render them unlikely to be exposed to oxygen debt as occurs in phasic muscle.

REFERENCES

BARCROFT, H. (1963). In *Handbook of physiology*, Section 2, Circulation, Vol. II, p. 1353. American Physiological Society, Washington.
FOLKOW, B. and NEIL, E. (1971). *Circulation*. Oxford University Press, New York.
MELLANDER, S. and JOHANSSON, B. (1968). *Pharmac. Rev.* **20**, 117.

THE SPLANCHNIC CIRCULATION

The combined vascular beds of the liver, spleen and the gut are called the splanchnic circulation.

At rest this vascular circuit receives some 1500 ml per min^{-1}. The hepatic artery, which supplies the liver only, furnishes about 25 per cent of this. *All* the splanchnic circulation passes through the liver, which receives 75 per cent of its blood flow from the portal vein, which in turn drains the blood form the gut and spleen. When parts of the gastro-intestinal tract and pancreas (which weigh about 2.5 kg in all) become active then blood flow increases—if they all became active at once the blood flow would exceed 4 litres. However, activity is sequential in the course of digestion and absorption of food, so the actual hyperaemia (which increases the portal blood flow to the liver) is only some 50 per cent above that at rest.

Generally speaking, the various sections of the gastro-intestinal tract show an inner mucosa and a muscular wall. The smooth muscle layer shows a flow of 40 ml per 100 g per min during maximal vasodilatation and about 10 ml per 100 g per min in resting conditions. The mucosal layer receives about 50–60 ml per 100 g per min at rest and as much as 300–400 ml per 100 g per min when maximally dilated.

During activity the blood flow increases from that of rest owing to vagal activity (in the stomach) and either humoral activity (?) or the local release of bradykinin from the mucosal glands and metabolites in the intestinal tract itself. Measurements of CFC show that this is high in the mucosal capillaries—ten times that for skeletal muscle. The mucosal vessels have an enormous capillary surface available for absorption and pore-bound secretion.

There is countercurrent system of villous blood vessels in the small intestine (Lundgren 1968). Substances which are lipid-soluble when carried in the 'venous' descending limbs of the vascular hairpin loop pass by diffusion across into the ascending 'arterial' limb because of a concentration gradient. High concentrations of absorbed substances are thus reached in the outer parts of the villi and these substances leave relatively slowly via the venous drainage. Such a system automatically slows down the entrance of rapidly absorbed solutes into the blood.

Oxygen tends to leave the arterial ascending limbs of the villi and diffuses across into the venous descending limb. The pO_2 at the tip of the villi is lower than that at their base (Lundgren 1968).

Sympathetic vasoconstriction. The rich supply of sympathetic fibres to the gut is tonically active but provides only a moderate neurogenic contribution to the splanchnic vascular resistance. Thus, regional flow in the intestine is only increased by 25 per cent when this sympathetic discharge is abolished by section or by blockade. The rise of systemic blood pressure on stimulating the splanchnic nerve is due more to venoconstriction, which is a 'capacity effect' serving as a 'transfusion' into the vascular bed, than to a resistance effect.

Liver blood flow

Receiving 100 ml per 100 g per min, of which the hepatic artery provides 25–30 ml per 100 g per min, the liver blood flow is high compared with that of most tissues. As 30 ml of hepatic arterial blood pass through the sinusoids at a pressure head of 100 mm, the resistance to arterial flow is 3.3 PRU_{100}. At maximal arterial dilatation flow may reach 100 ml per 100 g per min or more and the resistance falls to 1 PRU_{100} or less.

Myogenic arterial tone is offset by vasodilator metabolites produced locally and the balance between these two factors causes an *autoregulation* of *arterial* hepatic flow. Portal blood flow is not autoregulated and increases after a meal due to functional hyperaemia in the gut. Usually hepatic arterial inflow increases when portal flow is reduced—probably because the local metabolite concentration relaxes the hepatic arterial precapillary sphincters.

Both the artery and the portal venous distribution are supplied by vasoconstrictor sympathetic nerves. Sympathetic stimulation causes a marked reduction of the capacitance branches of the portal system and helps to 'mobilize' blood towards the heart.

The difference between hepatic arterial (systemic) pressure, say 100 mm Hg and that of the portal vein (7–12 mm Hg) indicates that as both systems supply the hepatic sinusoids the precapillary resistance in the arterial system must be high, whereas that in the portal vein must be feeble. Both streams coalesce to yield a sinusoidal pressure of 6–8 mm Hg. The hepatic sinusoids are perforated and allow a free exchange of macromolecules across their wall. Liver lymph contains almost as much protein as does the plasma.

The entire splanchnic circuit contains about 20 per cent of the blood volume and the liver contains a third of this. Thus, about 400 ml of blood are in the liver at any instant which represents an intrahepatic depot to be drawn upon. The blood content per unit mass is high—

$$\frac{400}{1500} \times 100 = 27 \text{ per cent.}$$

REFERENCES

LUNDGREN, O. (1968). *Acta physiol. scand.* **72**, Suppl. 303, 1–42.

THE PULMONARY CIRCULATION

Five centimetres long in man, the pulmonary trunk divides into a (longer and wider) right pulmonary artery and a left pulmonary artery. These arteries rapidly subdivide into terminal branches which are thinner walled and wider than their systemic counterparts. Small muscular arteries, 1000–100 μm in diameter, are adjacent to the respiratory bronchioles and alveolar ducts. The pulmonary arterioles, 30 μm in diameter, contain only a thin rim of muscle. They subdivide to yield a network of capillaries which are about 4 μm radius and about 350 μm long (both wider and shorter than are systemic capillaries); this vascular reticulum surrounds the alveoli sandwiched between their walls. They have an effective wall surface area of about 60 square metres at rest. In heavy exercise this increases to 90 square metres. The transit time across the capillaries is about 1 s at rest and only 0.3 s in heavy exercise. Venules and veins receive the oxygenated blood from the capillaries and the four main pulmonary veins distribute it to the left atrium. The volume of blood in the pulmonary circuit at rest is about 600 ml. As the heart itself contains about 400 ml there is a litre of blood in the thorax. The pulmonary blood volume is higher in recumbent man—when he stands some of the pulmonary reservoir is used to supply the heart with blood for pumping and this partly offsets the postural reduction of systemic venous return.

The pressures in the pulmonary artery are much lower than those in the systemic arteries. In man the mean pulmonary arterial pressure is about 10–15 mm Hg; the systolic pressure is about 20–25 mm Hg and the diastolic is 6–12 mm Hg. The pulse pressure is 12–15 mm Hg. The left atrial pressure is about 5 mm Hg and the pulmonary capillary pressure is usually taken as 8 mm Hg—and it will be noted this is far below the plasma colloid osmotic pressure of 25 mm Hg.

The pulmonary artery is thus very distensible (low pulse pressure) and offers a low resistance to flow; the work of the right heart is much less than that of the left.

In heavy exercise, the cardiac output of 25 l min^{-1} has to cross the pulmonary circuit. It does so without there being any demonstrable increase in the pressure head, which remains much the same as that rest $(15 - 5 = 10$ mm Hg). It would appear that the resistance offered by the pulmonary vascular circuit falls *pari passu* with the increase in pulmonary blood flow. FIGURE II.118 shows the resistance (mm Hg ml per min^{-1}) changes with perfusion pressure in an isolated blood perfused lobe of a dog lung.

However, in the intact circulation there is a complicating factor to be considered when the velocity of blood flow through the pulmonary circuit increases. As stated on page 70, the driving force of blood flow is *not* the pressure head, but the total energy. If no gravitational factor is involved (the subject being recumbent):

$$E = \text{Pressure} + \tfrac{1}{2}\varrho v^2.$$

It can be calculated that at resting cardiac output, where the velocity of pulmonary arterial blood flow is 90 cm s^{-1}, the kinetic energy accounts for 13 per cent of the total energy and at a threefold cardiac output (270 cm s^{-1}), kinetic energy is now nine times as great (3^2) and accounts for 52 per cent of the total energy (Burton 1965). Thus, the pressure measurements are misleading in exercise, indicating an unduly striking ability of the pulmonary arterioles to distend.

The transmural pressure in the capillaries is well below that of the colloid osmotic pressure of the plasma. Thus, the pulmonary capillaries are unlikely to lose fluid, which is as well, because the alveoli themselves do not possess lymphatics. If P_{cap} in the pulmonary circuit does rise above 24 mm Hg or so, then pulmonary oedema results, as in left ventricular heart failure.

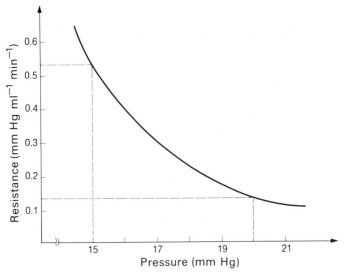

FIG. II.118. Approximate relationship between perfusion pressure (i.e. transmural pressure) and flow resistance in an isolated, blood-perfused lung lobe of the dog. Note the marked reduction of resistance for such moderate rises (from 15 to 20 mm Hg) in pressure as may occur during exercise. (Modified from Edwards, W. S. (1951). *Am. J. Physiol.* **167**, 756; from Folkow, B. and Neil, E. (1971). *Circulation.* Oxford University Press, New York.)

Pulmonary capillary flow

The magnitude of pulmonary capillary flow has been established by using the body plethysmograph. The subject sits in the chamber, and the chamber pressure is continuously measured. If he rebreathes the air of the chamber for a short while the chamber pressure falls only slowly because he produces about as much carbon dioxide as he uses oxygen. However, if he rebreathes from a bag containing 80 per cent nitrous oxide and 20 per cent oxygen (the bag is in the chamber with him as shown in FIGURE II.119) then the

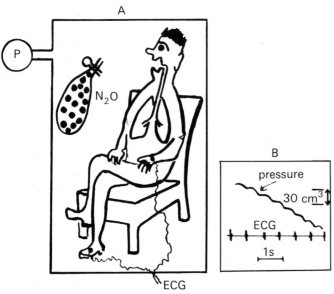

FIG. II.119. Body plethysmograph. Ⓟ = pressure recorder; ECG simultaneous recorded. N$_2$O is a bag containing 20 per cent O$_2$ and 80 per cent N$_2$O from which the subject can rebreathe.

When the subject holds his breath the record shows that N$_2$O uptake is pulsatile with the heart beat. The record is one of *change of pressure* in the plethysmograph but can be converted into a volume change by calibrating the pressure rise caused by the injection of a known volume of air into the chamber. (After Lee, G. de J. and Du Bois, A. F. (1955). *J. clin. Invest.* **34**, 1380.)

chamber pressure falls more quickly owing to the absorption of the highly soluble nitrous oxide in his lungs. If he holds his breath for a few seconds the slope of the pressure record then obtained shows variations with the phases of the cardiac cycle [FIG. II.119]. During systole the slope is higher, whereas in diastole it is barely evident. The instantaneous rate of blood flow can be calculated from a knowledge of the solubility coefficient of nitrous oxide and its mean alveolar concentration. The results obtained by Lee and Du Bois (1955) show that capillary flow through the lungs is pulsatile even in resting man. During systole, flow reaches a peak of more than 10 l min^{-1} whereas it falls to a level of 2–3 l min^{-1} during diastole.

The regional distribution of blood flow to the lungs is described on page 166.

Pulmonary vasomotor nerves

The stimulation of sympathetic fibres distributed to the lung vessels increases the vascular resistance. This has been unequivocally demonstrated by I. de Burgh Daly and his school (see Daly and Hebb 1967). Failure to demonstrate pulmonary vasomotor effects in the past stemmed from the fact that the bronchial circulation was not separately perfused in the 'isolated lungs' preparation artificially perfused by a pump. The viability of the sympathetic nerves to the pulmonary vessels depends on their getting a blood supply from the bronchial vessels. Daly perfused the bronchial circulation separately and, if necessary, interrupted this perfusion only for the brief period required to obtain control values of pulmonary flow before and after a short burst of stimuli delivered to the stellate ganglion (from which course the pulmonary sympathetic fibres). The lungs were perfused by a pump and pulmonary arterial and left atrial pressures were measured, thereby giving the pressure head of the perfused circuit. Left atrial outflow was recorded and invariably this flow diminished sharply during sympathetic stimulation [FIG. II.120]. Pulmonary arterial pressure rose, but left atrial pressure showed little if any change—proof of a pulmonary vasoconstriction. No information at present exists of the site of this resistance change whether arteriolar, venular or even of pulmonary A–V shunts. Again, there is no guarantee that the vasomotor nerves do not exert a more important effect on the capacity of the pulmonary circuit than on its resistance.

Pulmonary vascular reflexes

Reflexes affecting the pulmonary vascular resistance have been demonstrated (Daly and Daly 1959). Baroreceptor stimulation (aortic or carotid sinus) induces reflex dilatation of the pulmonary vessels, whereas chemoreceptor stimulation provokes reflex pulmonary vasoconstriction. There is evidence that the sympathetic nerves provide a slight tonic vasoconstriction of the pulmonary vessels. The reflex effects described above are abolished by stellatectomy and are therefore mediated by the sympathetic nerves.

The pulmonary trunk and the right and left pulmonary arteries are the site of adventitially placed vagal mechanoreceptors similar in appearance to those in the carotid sinus and aortic arch. An increase in pressure in the pulmonary artery provokes bradycardia and hypotension and in general the systemic cardiovascular reflex effects produced by stimulation of these pulmonary baroreceptors are qualitatively similar to those which may be provoked from the sino-aortic areas.

Ledsome and Linden (1964) showed that the stimulation of vagal mechanoreceptors at the junction of the pulmonary veins with the left atrium provokes tachycardia. These vagal receptors could be the fingers of the afferent arm of the Bainbridge reflex. Vagal receptors in a similar site—maybe the same ones—when stimulated, induce diuresis and would seem to be concerned in the regulation of blood volume [see p. 149].

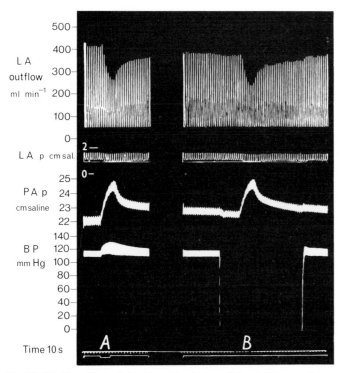

FIG. II.120. The effect of electrical stimulation of the stellate ganglion on pulmonary vascular resistance. The innervated isolated left lung of a dog was perfused with blood through the pulmonary artery at a constant head of pressure and through the bronchial arteries. No ventilation of the lung. Two stimulations (A and B) of the left stellate ganglion caused a fall in blood flow through the lungs (top record). Note the second response (B) was obtained while there was no blood flowing through the bronchial circulation, i.e. at zero bronchial arterial blood pressure (BP). LA, left atrial; PA, pulmonary artery; p, pressure. (Daly, I. de B. (1958). *Q. Jl exp. Physiol.* **43**, 2.)

A dramatic reflex response is seen on producing multiple micro-emboli in the pulmonary small vessels. The sequel to such micro-emboli (which are produced by the intravenous injection of starch grains) is an extraordinary tachypnoea. The afferent fibres are vagal and they must be small for they resist cooling to 3 °C. Phenyl diguanide injected intravenously will excite these fibres, as will 5HT (Paintal 1955) [p. 175]. It has been suggested, but not proved, that these nerve endings are situated immediately adjacent to, but not in, the smallest pulmonary arterioles (in which the starch grains lodge).

Chemical effects on the lung vasculature

Hypoxia induces pulmonary vasoconstriction (Duke 1957). This effect may well be mediated by the systemic chemoreceptors which induce reflex sympathetic stimulation, but the persistence of hypoxic pulmonary vasoconstriction after ergotamine and atropine favours a direct effect of oxygen lack on the lung vessels. The isolated perfused lung shows pulmonary vaosconstriction when nitrogen is substituted for air as the ventilatory gas mixture.

Chronic hypoxia is associated with a marked increase in pulmonary arterial pressure and with a later development of right ventricular hypertrophy. Cattle which live at high altitudes (2500 or more metres) develop brisket disease in which pulmonary hypertension, severe right ventricular failure, and oedema of the brisket are the outstanding features. Thick pulmonary precapillary vessels develop in high altitude dwellers and even children born and raised at altitude show pulmonary hypertension.

Acute hypercapnia has no effect on the pulmonary flow and resistance in unanesthetized man and animals, but in anaesthetized animals passively ventilated at constant ventilation volume, CO_2

added to the mixture evokes pulmonary vasoconstriction. In the latter case the degree of acidaemia induced by the CO_2 is greater than that in the spontaneously breathing animal which can increase its ventilation under the CO_2 stimulus. Acidosis of any type causes pulmonary vasoconstriction.

REFERENCES

BURTON, A. C. (1965). *Physiology and biophysics of the circulation*. Chicago.
DALY, I. DE BURGH and DALY, M. DE BURGH (1959). *J. Physiol., Lond.* **148**, 220.
—— and HEBB, C. O. (1966). *Pulmonary and bronchial vascular systems*. Edward Arnold, London.
DAWES, G. S. and COMROE, J. H. (1954). *Physiol. Rev.* **34**, 167.
DUKE, H. N. (1957). *J. Physiol., Lond.* **135**, 35.
FISHMAN, A. L. (1963). In *Handbook of physiology*, Secton 2, Circulation, Vol. II, p. 1667. American Physiological Society, Washington.
LEDSOME, J. R. and LINDEN, R. J. (1964). *J. Physiol., Lond.* **170**, 456.
LEE, G. DE J. and DU BOIS, A. F. (1955). *J. clin. Invest.* **34**, 1380.
PAINTAL, A. S. (1955). *Q. Jl exp. Physiol.* **40**, 348.

LYMPHATIC FLOW

Lymphatic capillaries, composed of single cell-walled endothelial tubes (which have closed ends), are distributed lavishly among the tissue cells forming a network similar to that of the vascular capillaries. Some 95 per cent of the protein lost from the vascular system per day is returned by the lymphatic vessels. The lymphatic capillaries, unlike their vascular counterparts, are freely permeable to protein.

The capillaries debouch upon larger vessels which contain a sparse amount of smooth muscle in their walls and whose endothelial linings feature valves which permit only unidirectional flow towards the central veins. Lymph glands are situated at various points along the course of the larger lymphatics. Reaching these glands the lymph vessels subdivide forming narrower vessels which open into the sinus of the gland. Lymphocytes, other cellular elements and γ-globulins enter the small lymph vessels which drain the gland and these vessels once more join up to form larger vessels. The two large terminal channels—the right and left thoracic ducts, finally empty into the right and left subclavian veins respectively at their junction with the jugular veins. The flow of lymph is very slow and only 2–4 litres a day drain into the vascular system. However, this trivial daily fluid volume contains *some 195 g of protein*, which is of great importance in terms of preserving the protein content of the plasma.

Protein transport by the lymphatic system

Mayerson (1963) investigated the transport of proteins by the lymphatics in two ways; (a) by analysing samples of thoracic duct lymph and plasma for radioactive iodinated albumin which he injected into a cannulated leg lymphatic vessel; (b) by isolating and catheterizing the afferent and efferent vessels of the popliteal lymph node and infusing and collecting test substances.

Following the commencement of peripheral infusion of the radioprotein, radioactive material appeared in lymph sampled from the thoracic duct and its concentration increased sharply to a level which was maintained during the infusion period of 50 minutes and for another 10 minutes during an ensuing infusion of saline. Then there was a sharp fall of concentration [FIG. II.121]. Plasma radioactivity reached a maximum value (0.2 per cent of that in the thoracic duct) in an hour or so and then maintained this value.

It is obvious then that *macromolecules* which have gained access to the lymphatic system can only escape by entering the blood-stream. However, urea and dextran molecules can escape freely and any substance with a molecular weight less than 5000 was found to do so.

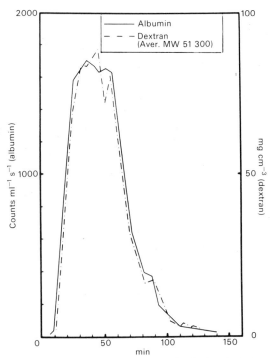

FIG. II.121. Concentration of dextran and [131I]-albumin in lymph and plasma. Dextran and [131I]-albumin solutions infused centrally into leg lymphatic of anaesthetized dog at zero time at rate of 0.5 ml min⁻¹. Infusions of dextran and albumin stopped after 50 min and 0.9 per cent saline infusion started at same rate for next 100 min. All values are corrected for free iodine. (Mayerson, H. S. (1963). In *Handbook of physiology*, Section 2, Circulation, Vol. II, p. 1035. Physiological Society. Washington.)

The question arises—how do the big molecules get in if they cannot get out? Two opinions are expressed as to the mode of entrance. Some think that micropinocytosis is responsible for entrance and others that the lymphatic capillaries possess gaps between the cells which constitute their wall. Even particulate matter, such as the carbon particles of Indian ink, can penetrate the lymphatic capillaries and indeed this is the basis of the well-known method of delineating cutaneous capillaries, by the intradermal injection of Indian ink or dye solutions.

Once in, the extralymphatic forces of compression push the lymph forwards into the larger vessels and the lymphatic valves prevent retrograde movement. The lymphatics larger than capillary size are not permeable to macromolecules which have therefore no means of escape except via the thoracic duct.

The protein content of lymph varies with the different tissues which the regional lymphatics drain. In the liver lymph the protein concentration is some 6 g per 100 ml—about 80–85 per cent that of the plasma. In the resting leg the lymphatic protein content is only 1–1.5 g per 100 ml. It is decreased during exercise because the fluid transfer into the lymph increases more than does the protein transfer.

The lymphatics of the intestinal villi (lacteals) absorb and transport lipids—indeed the milky appearance of these vessels following the digestion of a fatty meal led to their discovery by Aselli (1627) (although he himself had little idea of their nature and it was left to Bartholin (1651) and Rudbeck (1653) to discover their function and to give them the name lymphatics).

Lymph flow

As only 2–4 litres of lymph seep back into the bloodstream in 24 hours and, as the capacity and dimensions of the lymphatic system are not notably different from that of the capillaries and veins, it follows that lymph flow is very slow. Even a fistula in the thoracic duct only drips lymph at 0.5–1.0 ml min⁻¹.

In regional lymphatics, flow is notably accelerated (but still remains fairly slow) by venous obstruction, which may increase the flow by a factor of five or ten and by vasodilatation of the neighbouring arterioles and precapillary sphincters. Muscular exercise adds a mechanical massaging action to promote further the rate of lymph flow in muscle lymphatics. The secretory action of glands is accompanied by increased lymph flow due to vasodilatation of the local blood vessels.

REFERENCES

MAYERSON, H. S. (1963). In *Handbook of physiology*, Section 2, Circulation, Vol. II, Chap. 30, pp. 1035–73. American Physiological Society, Washington.

REGULATION OF BLOOD VOLUME

Extravascular factors—osmolarity and sodium concentration of extracellular fluid

How is the blood volume kept fairly constant at some 75–80 ml kg⁻¹ body weight in an adult? Obviously there must initially be a balance between fluid intake and fluid output by urine, sweat, and the insensible loss of fluid through the skin and the lungs. Such a balance is achieved by an intake which is influenced by thirst if it is inadequate, and an output which is adjusted with respect to its various routes. Thus, heat load, causing sweating, sharply reduces urine output as does an undue loss of fluid from the gastro-intestinal tract because of diarrhoea or incessant vomiting.

A rise in the crystalloid osmolar concentration of the interstitial fluid in the vicinity of the paraventricular and supraoptic nuclei deliberately induced by the injection of a microvolume of hypertonic saline via an indwelling cannula causes a goat to drink copiously. The output of ADH from the posterior pituitary is increased when the osmolar concentration of the interstitial fluid surrounding these hypothalamic nuclei rises. ADH minimizes fluid loss by the kidney, acting on the permeability of the collecting tubules [p. 227].

Angiotensin, present in excess in the plasma or when injected in minute amounts into the hypothalamus (Fitzsimons 1969), causes copious drinking in conscious rats. Angiotensin is formed when the output of renin from the kidney is increased and this in turn occurs in circumstances of reduced blood flow to the kidney. This mechanism is responsible for a big output of renin during hypotension following haemorrhage and the insatiable thirst of individuals after severe blood loss is likely to be due to this action of angiotensin on the hypothalamic centres.

Changes of the water intake or the water loss themselves alone are insufficient to restore salt/water balance. Sodium provides the skeleton of the interstitial fluid and the water the clothing. There can be no long-standing situation of water retention alone (with the exception of the unfortunate and unnatural iatrogenic situation of water intoxication which was unwittingly produced in labour wards where the expectant mothers, given posterior pituitary hormone containing ADH repeatedly for the medical induction of the labour, were fiercely and repeatedly exhorted by the nursing sisters—as is their wont—to drink copious flagons of water and developed water intoxication for obvious reasons).

Water needs sodium to accompany it in order to stay in the body fluids, so sodium balance is just as important to consider when discussing the regulation of plasma volume or 'body water' as is water itself.

Sodium concentration in the body fluids again obviously depends on the intake and the output. The intake is usually adequate, but the output can vary enormously either in physiological circumstances, such as heat load, where large amounts of sodium are lost in the sweat or in pathological conditions, such as excessive diarrhoea or in renal tubular damage. The sodium concentration of the extracellular fluid is adjusted by the hormone aldosterone and to a lesser extent by cortisol, both of which are elaborated by the adrenal cortex. Aldosterone liberation is itself influenced and perhaps governed by the concentration of angiotensin in the blood. Angiotensin, which powerfully increases aldosterone output, is liberated from the α_2 globulins of the plasma by renin which itself is poured out during renal ischaemia, such as occurs in hypotension. Thus, the sodium framework of the ECF is preserved as far as possible by reducing Na loss in the urine as secured by the renal tubular action of the increased secretion of aldosterone.

Certain diseases—among them congestive heart failure, hepatic cirrhosis, and nephrosis—are characterized by a larger plasma volume than normal. Each of these conditions is characterized by secondary hyperaldosteronism with sodium retention. The retention of sodium causes a secondary retention of water and in addition to circulatory plethora such patients manifest oedema which may be gross.

As detailed on page 4, although water can diffuse readily from the extracellular fluid into the cells, sodium is effectively precluded from doing so by the cellular sodium pumps. If 1 litre of water is drunk the fluid is distributed throughout the 50 litres (approximately) of body water so the effective rise of plasma volume is relatively trivial (2 per cent) even if the renal excretion of water did not increase, as however it does because of the inhibition of ADH liberation induced by the hypotonicity of the body fluids. When sodium chloride is injected intravenously in hypertonic solution, however, the plasma volume and that of the extracellular fluid rises and the adjustment of the situation by renal excretion is slow.

Intracellular factors—intrathoracic blood volume and cardiopulmonary receptors

Gauer and Henry (1951) suggested that blood volume is monitored by stretch receptors in the *intrathoracic* circulation; substantial evidence now supports their claims. They noted that the low pressure system (systemic veins, right heart, pulmonary circulation, and left ventricle in diastole) contains 85 per cent of the blood volume and that the compliance of this low pressure system is about 3 ml per mm Hg per kg body weight). This is highly distensible compared with the arterial system (0.015 ml per mm Hg per kg body weight). Measurements of compliance reveal that the distensibility of the low-pressure system is constant, provided that changes of blood volume are less than say 10 per cent. Consequently, this system registers changes in blood volume as changes in the tension of the walls of the vessels and cardiac chambers which comprise it. Although the compliance of the extrathoracic component is similar to that of the intrathoracic compartment, as the latter has a much smaller volume (700–900 ml) than that in the systemic veins (3–3.5 1), it is more greatly distended by a given volume increment and correspondingly vascular receptors in the walls of the intrathoracic system might well be suited to report changes of blood volume. The atrial receptors have been implicated in serving just this purpose.

The first clue that intrathoracic volume changes could cause striking changes of plasma volume was provided by H. C. Bazett *et al.* in 1924. Immersion of standing subjects up to the neck in thermoneutral water for 3–4 hours caused such a profuse diuresis that the blood haematocrit was significantly increased. Later work has shown that such whole body immersion provides a very useful method for investigating the control of plasma volume. When man stands in a bath which is progressively filled with water, blood which initially is pooled in the dependent regions [FIG. II.122 (a)] is displaced by the rising hydrostatic pressure of the water towards the thorax. When the bath water reaches the level of the diaphragm, the situation is as shown in FIGURE II.122 (b)—the distribution of blood volume is similar to that seen when man is prone on a couch.

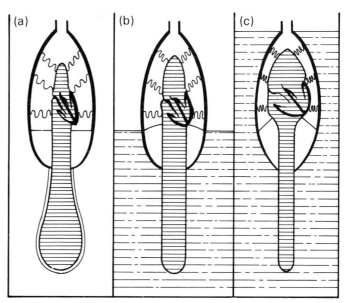

FIG. II.122. Effect of postural changes and immersion to the neck on the distribution of blood volume in the low-pressure system. (a) Standing. (b) Immersion to the diaphragm. (c) Immersion to the neck. The difference of intrathoracic blood volume between (a) and (c) is about 700 ml. (Redrawn from Gauer, O. H. and Thron, H. L. (1965). Postural changes in the circulation. In *Handbook of physiology*, Section 2, Circulation, Vol. 3 (ed. W. F. Hamilton and P. Dow) pp. 2409–39. Physiological Society, Washington.)

Further filling of the bath now displaces more of the blood volume into the thorax [FIG. II.122 (c)]. The central venous pressure rises perhaps 15 mm Hg and as the intrathoracic pressure only increases by about 2 mm Hg (owing to compression of the thoracic wall) the transmural pressure is about 13 mm Hg greater in the intrathoracic veins, cardiac chambers, and pulmonary circuit (and also in the systemic arteries). Biplanar X-rays (Lange *et al.* 1974) showed that the heart volume of man standing in air increased by as much as 300 ml when he stood immersed to neck level in water, and that atrial distension accounted for the majority of this increase. Cardiac stroke volume increased by 30 per cent. Epstein *et al.* (1975) found that whole body immersion induced an increase of sodium excretion ($U_{Na}V$) and potassium excretion (U_KV) almost identical with that induced by a saline infusion of two litres (over the course of two hours) as shown in FIGURE II.123.

In normal circumstances, man relies on drinking in response to the perception of thirst. Two factors provoke him to drink: (i) hyperosmolarity of cells of the paraventricular nuclei of the hypothalamus; thus the subcutaneous injection of polyethylene glycol causes a rise in osmolarity in ECF and provokes thirst; (ii) the degree of distension in the intrathoracic vessels and cardiac chambers. Over-distension of this part of the cardiovascular system prevents thirst. Conversely, a reduction of thoracic blood volume caused by ligating the inferior vena cava provokes thirst.

Astronauts do not suffer thirst; their intrathoracic blood volume is higher in the situation of weightlessness than in normal circumstances. Recent experiments cited in a Skylab report show a fluid loss from the body of the order of a litre or more; blood volume was reduced by 10 per cent. The fluid loss is not so much due to excess excretion as to a lessened water intake of some 300 ml per day. It is supposed that this 'loss of thirst' can be attributed to the increased barrage of impulses coming from the stretch receptors of the intrathoracic low pressure system. Fitzsimons (1979) has stated that a deficiency in information at diencephalic level from the cardiac receptors is interpreted as thirst.

Evidence of the role of the intrathoracic receptors in defending against a reduction of central blood volume has been provided by

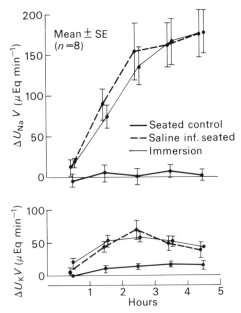

FIG. II.123. Comparison of the effects of immersion and acute saline infusion (2 litres over 120 minutes) in the seated posture on the rates of sodium and potassium excretion in eight normal subjects. Data are expressed as the absolute changes from the preimmersion hour ($\Delta U_{Na}V$ and $\Delta U_{K}V$) Immersion resulted in significant increases in $\Delta U_{Na}V$ (hours 2–5) and $\Delta U_{K}V$ (hours 2–4), which were not different from the increases induced by saline infusion. (Redrawn from Epstein, M., Pins, D. S., Arrington, R., Denunzio, A. E., and Engström, R. (1975). *J. appl. Physiol.* **39**, 66.)

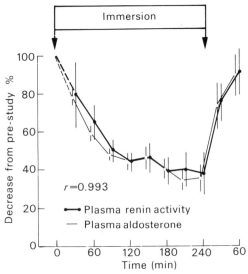

FIG. II.124. Comparison of the effects of immersion on plasma renin activity (PRA) and plasma aldosterone in subjects in balance on a 10-mM Na diet. Data are expressed in terms of per cent change from the preimmersion hour. The suppression of plasma aldosterone paralleled the suppression of PRA throughout the immersion period. Similarly, both PRA and plasma aldosterone recovered in parallel following cessation of immersion. (Redrawn from Epstein, M., Pins, D. S., Sancho, T., and Haber, E. (1975). *J. clin. Endocr. Metab.* **41**, 618–25.)

Henry *et al.* (1968) in dogs subjected to non-hypotensive bleeding. Central venous pressure and ADH titre of the blood were measured and in other experiments aortic baroreceptor and atrial mechanoreceptor discharge were recorded. A haemorrhage sufficient to lower the blood volume by 10 per cent (which did not lower the systemic blood pressure) caused a fall of CVP by 2–3 cm H_2O. ADH titre rose sharply and atrial receptor discharge fell by 50 per cent; arterial baroreceptor discharge was much less affected. The authors concluded that the fall in atrial distension was the main cause of the rise in ADH titre. Similar conclusions were reached by Share (1975) who studied the effects of haemorrhage in patients undergoing haemodialysis; a loss of 10 per cent of blood volume, which did not lower the mean arterial pressure or the arterial pulse pressure, caused a 70 per cent increase in ADH titre.

Blood volume control must involve control of sodium excretion and Gauer and Henry review the evidence for implicating cardiac receptors as afferents in a reflex regulation of sympathetic discharge to the renal vessels and juxtaglomerular apparatus (JGA). Oberg and Thoren (1973) found that distension of the left ventricle (which promotes the discharge of C-fibre receptors) reflexly provokes renal vasodilatation by inhibiting sympathetic discharge to the kidney. Some of the renal sympathetic nerves supply the JGA and act thereon by influencing β-adrenergic receptors. Renal glomerular filtration rate and hence renal tubular sodium load can be modified by this cardiac reflex pathway and the sodium concentration at the macula densa may well be affected. Epstein *et al.* (1976) have shown that immersion-induced increase in central volume produced a profound fall in both plasma renin concentration and plasma aldosterone concentration in man [FIG. II.124]. The subjects' hormonal concentrations returned to normal within an hour of the end of the immersion period.

Integration of osmocontrol and volume control. In water deprivation, volume and osmocontrol mechanisms are believed to work using the ADH mechanism to restore both volume and osmotic pressure. However, McCance (1936) provoked experimental

salt deficiency in man by exposing his subjects to thermal stress, (which caused profuse sweating) and allowing them only water to drink. After a weight loss of two kg, he noted that they retained water and abandoned the homeostatic control of osmolarity. Volume control took precedence.

Arndt (1965) infused water into a carotid loop of trained dogs in amounts sufficient to change the local osmotic pressure in the diencephalon. When the osmotic pressure fell by 2 per cent, a water diuresis occurred. However, if the dog was bled 8 per cent of its blood volume during the water infusion, no diuresis occurred; only when the blood was reinfused did a profuse diuresis result.

Red cell component

The red cell component of the blood volume is adjusted in an entirely different manner and the details of the process are obscure. However, it is known that the red cell output of the bone marrow (providing that the raw material of iron and proteins are freely available and that the various haemopoietic factors, such as vitamin B_{12}, copper, manganese, vitamin C, and thyroxine are not deficient) depends on the influence of the hormone erythropoietin [p. 37]. Hypoxia increases the output of erythropoietin from the kidney and causes polycythaemia. The blood volume increases because of the rise in red cell mass. Conversely, in some forms of chronic renal disease anaemia occurs as a result of deficient output of erythropoietin.

Control of fluid exchange between plasma and the extracellular fluid

The balance between the effective hydrostatic pressure across the capillary wall and the effective osmotic pressure exerted in the opposite direction across the capillary endothelium must obviously be fairly exact over the course of a day. Twenty litres of fluid escape a day from the capillaries to the interstitial fluid. Of this fluid 16–18 litres are reabsorbed into the capillaries themselves and the remaining 2–4 litres are returned via the lymph.

The hydrostatic pressure in the capillary is governed by the ratio of precapillary/postcapillary resistance [p. 72]. Dilatation of the precapillary resistance vessels raises P_{cap} and correspondingly increases fluid loss. Conversely, sympathetic stimulation increases the recapillary/postcapillary resistance ratio and lowers P_{cap} so that the uptake of fluid is favoured. After haemorrhage, the anhydraemia is restored at least partly by reflex sympathetic vasoconstriction which, by lowering P_{cap}, favours fluid uptake from the interstitium. P_{cap} is also influenced by both local physical and chemical influences on the precapillary resistance vessels. A rise of transmural pressure in the arteries causes a myogenic response which itself reduces the transmitted pressure to the capillaries. The local accumulation of metabolites produced by increased tissue activity causes an increased osmolality of the interstitial fluid and meanwhile reduces the constrictor response of the precapillary resistance vessels to sympathetic stimulation and indeed reduces basal myogenic tone. Thus, in exercise not only is the mean capillary pressure raised in the active muscles (and the size of the capillary bed increased), but also the osmolality of the interstitium rises sharply and increases the osmotic attraction of fluid from the capillaries into the interstitium. The haemoconcentration which results from exercise is due to these factors.

The osmotic pressure exerted by the plasma proteins is only fully effective if the osmotic pressure of the interstitium is very low. In tissue injury proteins released from the damaged cells are rapidly broken down by enzymes released from the injured lysosomes and the osmotic pressure of the interstitial fluid increases considerably; concomitant damage to the capillary wall allows the escape of proteins and the end result is inflammatory oedema, or in the case of burns, a copious weeping of protein-rich fluid from the surface which quickly causes haemoconcentration.

The factors which determine the concentration of plasma protein are not clearly understood—the relative constancy of plasma protein concentration suggests that there must be some mechanism, but where it is established and what is its nature are questions still unanswered.

Lastly, the neurogenic mechanisms of regulating the precapillary/ postcapillary resistance ratio and venous capacity are those of systemic arterial and cardiovascular mechanoreceptors whose actions are detailed elsewhere. It has been claimed that the atrial receptors may be concerned with the regulation of ADH and aldosterone output but the evidence is equivocal.

Volume regulation in pathophysiological conditions

1. Congestive heart failure. The plethora associated with CHF is ascribed in part to a reduction in the density of the afferent receptors in the subendocardium of the atria and ventricles. Animals with chronic right-heart failure (induced by surgical lesions) show a greatly reduced atrial impulse discharge in response to artificially induced rises of atrial pressure. Venous back pressure affects the liver—disordered hepatic metabolic functions (e.g. hormonal metabolism) and ascites may result.

2. Paroxysmal tachycardia. This causes a rise of the atrial mean pressure. Diuresis is characteristically seen as a result.

3. Vagal dysfunction in mediastinal tumours and lung cancer.
Some cases have been reported who develop orthostatic hypotension. Immersion in the water bath did not cause the usual diuresis nor the fall in ADH levels seen in normal subjects. It may be supposed that the cardiac afferent vagal pathways had been damaged by the tumours.

4. Cirrhosis is characterized by plethora and may often be associated with ascites. The condition is associated with secondary hyperaldosteronism, probably because the damaged liver cannot metabolize the hormone. Sodium retention (and therefore water retention) results.

5. Nephrosis too manifests hyperaldosteronism with sodium retention.

REFERENCES

FITZSIMONS, J. T. (1979). *The physiology of thirst and sodium appetite.* Monograph No. 35 of the Physiological Society. Cambridge University Press.
GAUER, O. P. and HENRY, J. P. (1963). *Physiol. Rev.* **43,** 423.
—— —— (1976). In *International review of physiology. Cardiovascular physiology* II, Vol. 9 (ed. A. C. Guyton and A. W. Cowley). Butterworth, London.
MCCANCE, R. A. (1936). *Lancet,* **i,** 643, 765, 823.

Traumatic shock

Shock is a syndrome featured by a cardiac output which is inadequate to maintain tissue nutrition. Haemorrhage, trauma, burns, and dehydrating conditions, such as heat stress, prolonged vomiting, or diarrhoea, may all cause shock.

Clinical picture

Following injury and haemorrhage the patient looks pale, grey, or cyanotic, breathes rapidly and shallowly, is restless and fidgety, but dull mentally, and shows a lessened sensibility. His skin is cold but moist, his pulse thready and rapid. Insatiable thirst is a characteristic symptom and on drinking, vomiting is particularly prone to occur. His blood pressure is low.

Severe haemorrhage of course causes a fall of blood pressure despite the 'compensatory' sympathetic discharge which results from the reduction of baroreceptor activity and the increase in chemoreceptor discharge. Arterioles, precapillary sphincters and venules and veins are alike constricted. The pulse is rapid and the feeble pulse pressure is manifest as a thready pulse. The increase in the precapillary/postcapillary resistance causes an uptake of tissue fluid and this is the reason for the pinched features of shock. Renal blood flow is profoundly reduced by sympathetic constriction and the patient may be anuric.

The increased sympathetic discharge thus contributes to survival in the immediate stages following bleeding. Animals which have been sympathectomized may die if 30 per cent of their blood volume is removed, whereas those with intact sympathetic nervous systems can tolerate a greater blood loss. This sympathetic discharge results from the escape of the medullary cardiovascular centres from the tonic inhibitory influence of the baroreceptors, but it can be abolished by exposing the patient to heat load. Thus, hot-water bottles and warm environments do not favour the patient's chances of survival because hypothalamic responses to warming abolish sympathetic vasoconstriction.

Although the immediate effects of sympathetic vasoconstriction may allow the individual to survive, the long-sustained throttling of the regional circulations—particularly those of muscle and kidney may give rise to untoward effects. If the kidneys remain ischaemic too long then functional powers will be irreparably damaged as evidenced by a loss of their ability to concentrate the urine, documented by Howard (1962) on the basis of surgical experience with war casualties in Korea.

The situation in the muscle vascular circuit has been analysed by Mellander and Lewis (1963), who found that long-sustained hypotension in the muscle vascular bed caused a failure of the precapillary sphincters to respond to sympathetic stimulation by constriction. The constrictor response of the postcapillary venules to sympathetic stimulation was better sustained. Hence, although the normal response (and that soon after haemorrhage) of the microcirculation to sympathetic stimulation is an increase of the precapillary/postcapillary resistance ratio which leads to a fall in capillary pressure and to an uptake of tissue fluid this response is entirely altered after sustained hypotensive ischaemia.

Sympathetic stimulation now decreases the precapillary/post-capillary resistance ratio and increases P_{cap} so that fluid loss from the vascular system occurs. This is probably only one of many of the untoward changes which contribute to the cause of death in irreversible shock.

The failure of the precapillary sphincters and arterioles to respond adequately to sympathetic stimulation in sustained ischaemia is most likely due to the influence of anaerobic metabolites and hyperosmolality of the smooth muscle in their walls.

Burns

The only type of traumatic shock which is associated with haemoconcentration is that following severe burns. The burnt areas weep protein-rich fluid through the capillary walls. The breakdown products of cellular damage raise the osmotic pressure of the tissue fluid and this together with the increase of permeability in the injured capillaries leads to a rapid loss of fluid. *Plasma* transfusion is urgently required to offset this loss and to prevent the rise of viscosity which haemoconcentration causes.

The insatiable thirst of the conscious shocked patient is interesting and characteristic. Its cause *might* be due to the effect of angiotensin, which is probably present in excess in the plasma in shock. Renal ischaemia causes the release of renin which forms angiotensin. Fitzsimons and Simons (1969) have shown that angiotensin injected intravenously promotes drinking in rats and Epstein, Fitzsimons, and Rolls (1970) have shown that as little as 0.2 μg of angiotensin provokes drinking when administered via a microcannula chronically implanted in the anterior lateral regions of the hypothalamus.

Other causes of shock

It must be admitted that some injuries which are unaccompanied by bleeding will still cause severe shock. Bone fractures are an example. Pain may be severe and experiments on animals have shown that the sensory nociceptive impulses aroused serve as contributory factors. Thus, local anaesthetization of the part or spinal anaesthesia are alike in their efficacy in restoring the blood pressure and cardiac performance to more nearly normal levels. Indeed, in injuries in conscious man or animals nociceptive impulses must play a considerable role in inducing the reactions of shock—sustained pain without blood loss may provoke a faint. Finally, the conscious appreciation of a comminuted fracture not only by the pain it causes, but by visual inspection, can hardly improve 'morale' and the sight of one's own blood spraying from a cut artery may well induce vagal hyperactivity and fainting.

Infection supervening on the injury again will hasten the deterioration of cardiovascular performance. Thus, strangulated bowel or perforated duodenum release endotoxins into the peritoneum which, absorbed by peritoneal capillaries or lymphatics, cause peripheral arteriolar paralysis and prejudice myocardial contractile activity.

The end result of the untreated shock syndrome may be death due to myocardial failure. When shock does not respond to transfusion and other therapeutic measures it is said to be irreversible.

Treatment

Treatment of shock is to give blood in cases of injury other than burns, and to give plasma to burned patients as soon as possible. Prolonged diarrhoea, such as occurs in cholera or in the summer diarrhoea of infants, necessitates saline transfusions together with appropriate antibiotics.

Some in the past have given vasoconstrictor agents to shocked patients; the rationale is false. Others have urged the use of sympathetic blocking drugs. Unfortunately it is impossible to block the sympathetic effects on the arterioles without blocking those on the veins, and pooling of the blood in the veins when their capacity increases as a result of the sympathetic blockade is not desirable in a circulation which is already depleted. The main precept of treatment is to transfuse as soon as possible and to avoid heating the patient.

REFERENCES

Bock, K. D. (ed.) (1962). *Shock, pathogenesis and treatment*. Basel.
Epstein, A. N., Fitzsimons, J. T., and Rolls, B. J. (1970). *J. Physiol., Lond.* **210**, 457.
Fitzsimons, J. T. and Simons, B. J. (1969). *J. Physiol., Lond.* **203**, 45.
Folkow, B. and Neil, E. (1971). *Circulation*. Oxford University Press, New York.
Howard, J. (1961). In *Shock* (ed. K. D. Bock). Basel.
Mellander, S. and Lewis, D. H. (1963). *Circulation Res.* **13**, 105.

High blood pressure

High blood pressure is usually referred to as hypertension. Clinically there are two main types:

1. Primary or essential hypertension, of unknown causation.
2. Secondary hypertension, due to phaeochromocytoma or to renal or endocrine disease.

ESSENTIAL HYPERTENSION

Essential hypertension is defined, quite arbitrarily, as being present when the casual arterial blood pressure persistently exceeds 150/90 or 160/100 mm Hg. However, when the effect of age is taken into account, there may be no sharp dividing line between normal and hypertensive blood pressure levels.

Essential hypertension may be benign or malignant.

Benign form

In the early stages the hypertension is moderate, e.g. 210/110 mm Hg. The blood pressure, especially the systolic, fluctuates considerably: during sleep or emotional and physical rest, the pressure may be normal; in states of stress the pressure rises to excessive levels. Later the hypertension becomes 'fixed' in the abnormal range and cannot be reduced to normal by rest or sedatives like the barbiturates. There is compensatory cardiac hypertrophy; the walls of the small arteries and arterioles become thickened; renal changes appear, e.g. an increased volume of night urine, albuminuria, or slight haematuria. After a period which may vary from a few years to 20 years death occurs from heart failure, vascular accidents (haemorrhage or thrombosis), or renal failure.

Malignant form

The condition is so named because death occurs within 6 months to 2 years of its first recognition. The blood pressure is much higher than in the benign form, e.g. 260/150 mm Hg. The peripheral vascular changes include acute arteriolar necroses, which are readily seen in the retinal vessels. There is papilloedema. Renal failure is common.

Malignant hypertension occurs as a complication of both essential and secondary types of hypertension. It is probably due to the great rise in blood pressure *per se* and the condition can revert to the benign form if the blood pressure is sufficiently reduced by treatment with reserpine and ganglion-blocking drugs (Pickering 1968).

Mechanism of development

In well established hypertension cardiac output is normal and as the viscosity of the blood is normal it follows that the peripheral resistance must be raised. The precapillary resistance vessels, notably the arterioles, offer a greater resistance to flow.

For years arguments have raged about the mechanism of development of hypertension, and a whole classroom-full of candidates have been examined for the causative agent. The increased peripheral resistance has been variously ascribed to sympathetic overactivity, or to the influence of blood-borne agents, such as angiotensin and catecholamines.

Thus, sympathetic overactivity has been invoked because of the marked influence of emotional tension in causing hypertension even in normal individuals and many have claimed that the reduction of blood pressure effected by the administration of ganglioplegic drugs or adrenergic blocking drugs supports their proposition that the hypertension is due to excessive sympathetic discharge. This seems about as logical as claiming that essential hypertension is due to an increased blood volume because bleeding reduces the blood pressure. 'Total' sympathectomy may alleviate hypertension for some time, but the operation is not performed so frequently nowadays as was the case 45 years ago because its palliative effect is temporary.

Blood-borne angiotensin is unlikely to be the cause of established hypertension—there is no rise in the plasma content of angiotensin.

Folkow (see Folkow and Neil 1971) has suggested that the establishment of increased arteriolar resistance in enduring hypertension is due to a hypertrophy of the vascular wall in response to a long-standing increase in load. Just as other tissues increase in bulk with repeated load, such as skeletal muscle, heart muscle, ureteric muscle, when there is an intermittent obstruction below, so do the arteriolar muscles. Folkow, Grimby, and Thulesius (1958) proved that even with maximal vasodilatation (and therefore complete relaxation of the muscles of their media) the regional flow resistance was much higher in hypertensive subjects than in normal individuals. The ratio between resting flow resistance and that during maximal vasodilatation was normal. The conclusion follows that the resistance vessels (arterioles) of the hypertensive patient are structurally narrowed owing to the development of morphological changes induced by the pressure load. Now for a given shortening of the muscle elements, the resistance to flow will increase to a greater degree in hypertensive than in normal subjects because of the increase in the wall/lumen ratio. Contraction of the outer layer of the smooth muscle of the media produced by stimulation of the sympathetic vasoconstrictor nerves will more effectively displace the entire mass of the wall into the lumen. The increase in wall/lumen ratio enhances the vasoconstrictive response caused by a given contraction induced by either neurogenic or chemical agencies. Although the thickened arteriolar walls still respond to vasoconstrictor or to vasodilator agents these adjustments are operated from a different base line. Folkow argues that these structural changes of arteries and arterioles will become widespread and will reduce the distensibility of the wall. Thus, the carotid sinus baroreceptors, which above all are sensitive to phasic stretch, discharge less vigorously than before in animals that have endured hypertension for months (McCubbin, Green, and Page 1956). Although the baroreceptors in these hypertensive dogs increased their discharge when the blood pressure was further increased and reduced when the blood pressure temporarily fell, the baroreceptor reflex mechanisms were 'reset' at a higher pressure level.

This hypothesis of the mechanism of development of the increased arterial and arteriolar resistance of established hypertension is superficially satisfactory but there remains the question as to how the increasing load (which causes the medial hypertrophy) develops in the first place. Here the arguments become speculative, but are based on premises which are reasonably tenable. Thus it is asserted and indeed borne out by experience, that normal individuals themselves sporadically display bouts of hypertension in everyday episodes. It is claimed that if such episodes are frequent and if the individual is perhaps prone by reason of heredity (and there *is* a strong hereditary element in essential hypertension), intermittent bouts of corticohypothalamic discharge which indeed provoke sympathetic vasoconstriction and the outpouring of chemical agents—catecholamines, cortisol, ADH, etc.—lead more inexorably to muscle hypertrophy in the media of the resistance vessels. This in turn permanently increases the arterial resistance and the work of the heart. The heart hypertrophies and the enlargement of the muscle fibres increases the intercapillary distance and lowers the chances of the cardiac muscle being adequately supplied with oxygen. The end-result may be attacks of myocardial ischaemia or left ventricular failure or cerebral haemorrhage.

Thus, hereditary make up *and* environmental factors are combined to set the stage for the fully fledged situation of essential hypertension. The environmental factor is presumably an important variable providing the whole gamut of experience from 'peace, perfect peace' through the 'slings and arrows of outrageous fortune' to the Sten guns of modern urban living. South Sea Islanders show little, if any, increase in their mean blood pressure level with age; would-be inhabitants of developing townships, scratching both for a tenement and for a means of livelihood, show a remarkably increased tendency to hypertension.

Essential hypertension can, of course, be treated by hypotensive drugs. Adrenergic blocking drugs, e.g. guanethidine and methyldopa, interrupt the sympathetic pathways and are frequently used.

Reserpine, which reduces vasoconstrictor tone both by blocking peripheral sympathetic action and by depleting both CNS and peripheral structures of their noradrenalin content, is often employed together with ganglion-blocking drugs.

SECONDARY HYPERTENSION

Secondary hypertension is most commonly due to renal disease. It may also be caused by a phaeochromocytoma, by excess secretion of glucocorticoids or of aldosterone or finally by coarctation of the aorta.

Renal hypertension results as a consequent of nephritis or cystic disease or pyelonephritis or of renal arterial stenosis. Primarily, the hypertension results from an increased output of renin which forms angiotensin. Angiotensin not only increases precapillary resistance but also induces increased aldosterone output which, in turn, causes salt and water retention. Frequently surgical removal of the kidney, if this be affected on one side only, will eliminate the hypertension. The fact that such is not always the case indicates that the long-standing changes of medial hypertrophy provide the important background of hypertension.

Phaeochromocytoma is a tumour of the adrenal medulla which liberates noradrenalin. Surgical removal is indicated and the hypertension thereupon subsides.

Glucocorticoid hypersecretion occurs in Cushing's disease due to a tumour of the zona fasciculata and reticulosa of the adrenal cortex. The hypertension is mainly due to the salt-retaining properties of the excessive amounts of cortisol secreted. The hypertension is of mild degree and disappears if surgical removal is successful.

Primary aldosteronism (Conn's syndrome) is due to a tumour of the zona glomerulosa of the adrenal cortex, which produces excessive quantities of the salt-retaining hormone aldosterone. Surgical removal cures the hypertension.

Coarctation of the aorta is associated with a hypertension above the aortic constriction and a hypotension below, which by depleting the normal renal flow, arouses the secretion of renin and the formation of angiotensin. Nowadays surgical repair is effected in childhood.

REFERENCES

FOLKOW, B. and NEIL, E. (1971). *Circulation*. Oxford University Press, New York.

—— GRIMBY, G., and THULESIUS, O. (1958). *Acta physiol. scand.* **44**, 255.

HENRY, J. P., MEEHAN, J. P., and STEPHENS, P. M. (1967). *Psychosomat. Med.* **27**, 408.

McCUBBIN, J. W., GREEN, J. H., and PAGE, I. H. (1956). *Circulation Res.* **4**, 205.

PAGE, I. H. and McCUBBIN, J. W. (1963). In *Handbook of physiology*, Section 2, Circulation, Vol. III, p. 2163. American Physiological Society, Washington.

PICKERING, G. W. (1968). *High blood pressure*, 2nd edn. Churchill Livingstone, London.

SILVERTSSON, R. (1970). *Acta physiol. scand.* Suppl. 343.

PART III
Respiration

The rhythmic act of quiet breathing (eupnoea) entails an active inspiratory movement in which the diaphragm descends and the chest wall is pulled outwards by contractions of the external intercostal muscles. The volume of the thoracic rhythmic respiration is under the control of 'centres' in the brainstem which secure successively inspiration (active) and expiration (passive in quiet breathing). In inspiration the size of the thoracic cavity is increased by descent of the diaphragm due to its contraction and by an outward movement of the ribs due to contraction of the external intercostal muscles. The diaphragm is supplied by motoneurons of C3, 4, and 5, and the intercostal muscles by motoneurons of T1–T12. Both these sets of motoneurons are excited by discharge of the 'inspiratory centre' of the respiratory centre complex. The *elastic lungs* follow this thoracic expansion passively and room air (fresh air) is thereby drawn into the depths of the lung. At the end of inspiration the external intercostal muscles and the diaphragm relax and the elastic recoil of the thoracic wall and lungs causes passive expiration.

During breathing, air enters by the nose and mouth and passes through the glottis to the trachea and thence by two main bronchi and their branches, the bronchioles, to the terminal or respiratory bronchioles. The respiratory bronchioles give rise to alveolar ducts, which lead via the atria to the pulmonary alveoli [Fig. III.1]. The pulmonary alveoli consist of large flat, thin epithelial cells which lie in immediate proximity to the numerous pulmonary capillaries.

Fig. III.1. Diagram of termination of respiratory bronchiole. b = Respiratory bronchiole; v = Vestibule; a = Atrium; s = Air sac; c = Alveolus. (After Miller.)

In quiet respiration, an adult breathes 6–7 l min^{-1}; his breathing rate is about 12–14 min^{-1} and the amount of air inspired or expired per breath (tidal air) is approximately 500 ml. At rest an adult uses about 250 ml O_2 min^{-1} and expires about 200 ml CO_2 min^{-1}. In heavy exercise the pulmonary ventilation volume may exceed 80 l min^{-1} and the oxygen usage may rise above 3½ l min^{-1}.

THE MECHANISM OF BREATHING

During inspiration the thorax is enlarged: (1) by movements of the ribs outwards and upwards; these differ according to the position of the individual ribs (2) by descent of the diaphragm.

Rib movements

1. The first pair of ribs is jointed between the spinal column and the manubrium sterni. In quiet inspiration this part of the thorax moves but little; in hyperpnoea the manubrium sterni moves upwards and the upper thorax is increased in its anteroposterior diameter.

2. The 2nd to 6th ribs inclusive slope obliquely downwards and forwards from their joints with the spinal column, each being successively longer and more oblique than its rostral neighbour. On inspiration the ribs move upwards to assume a more horizontal position owing to contraction of the external intercostal muscles which are excited by impulse activity in the external intercostal branches of the relevant thoracic segmental nerves. These external intercostal muscles pass obliquely forwards and downwards from their origin at the lower border of the rib near the tubercle to an insertion into the upper border of the rib below.

When the muscle contracts it tends to pull the lower rib upwards and the upper rib downwards. However, the first rib is secured in its position by the scalene muscles which contract so the external intercostals can only raise the ribs. In causing such an effect they are assisted by the mechanical advantage to the upward movement resulting from the attachment of the muscle fibres below to the anterior end of the long arm of a lever system and above to the posterior end of the long arm [Fig. III.2(a)]. In the case of the internal intercostal muscles the arrangement is the opposite [Fig. III.2(b)]. These muscles show little if any activity in quiet breathing but become active in hyperpnoea, serving as expiratory muscles. Their attachments are, as shown in Figure. III.2, such as to produce a mechanical advantage for their expiratory function.

The movements of these ribs (2–6) cause an increase in the anteroposterior diameter of the chest and, by virtue of their bowed mid-part, an increase in the transverse diameter of the thorax.

The lower ribs (7–10) also swing outwards and upwards in inspiration and widen the transverse thoracic diameter.

Diaphragmatic movements

The diaphragm consists of muscle fibres and a central tendinous portion. The muscle fibres arise from the xiphisternum, from the inner surfaces of the lower six ribs and as crura from the lumbar vertebrae and arcuate ligaments to converge on the domed central tendon. The muscle fibres sweep upwards from their origin along the inner part of the thoracic cage and then arch towards the midline. During inspiration, as a result of discharge in the phrenic neurons (C3, 4, 5) the muscle fibres contract and draw the central tendon downwards. In eupnoea the diaphragm shows an excursion

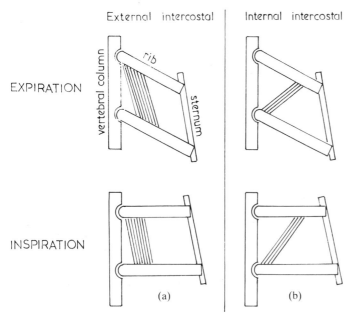

External intercostal | Internal intercostal

vertebral column

rib

sternum

EXPIRATION

INSPIRATION

(a) | (b)

FIG. III.2. Movements of the ribs induced by contraction of the external (a) and the internal (b) intercostal muscles.

of some 1.5 cm; in deep breathing the diaphragm may show as much as a 7 cm excursion. These movements can be examined using radiographic techniques.

The diaphragmatic movements account for as much as 75 per cent of the tidal volume in eupnoea. Nevertheless, they are not *essential* for respiration because external intercostal muscle activity alone (after bilateral anaesthetization of the phrenic nerves) can still produce the respiratory excursions necessitated by moderate activity of the individual.

The abdominal muscles (obliqui, transversus, and rectus) on contraction raise the intra-abdominal pressure and draw the lower ribs down and medially. Though probably inactive in quiet respiration they contract vigorously in all voluntary expiratory manoeuvres. They are the most important muscles involved in forced expiration.

The scaleni and sternomastoids play little if any part in quiet breathing but become active in voluntary static inspiratory efforts and expiratory efforts.

The intrinsic muscles of the larynx supplied by the recurrent laryngeal nerve; the abductor muscles (posterior crico-arytenoid) contract early in the inspiratory phase. Their paralysis causes inspiratory stridor. The adductor muscles begin to contract early in expiration but their contraction is not complete and it seems likely that the main respiratory function of these adductors is protective. Thus they prevent aspiration of foreign bodies and by contracting during the early compressive stage of the cough reflex enable a high intratracheal pressure to be developed.

In the fetal mammal the airless lungs completely fill the thoracic cage. They are invested by visceral pleura which is continuous with the parietal pleura which lines the chest wall. These two pleural layers are separated by only a very thin layer of fluid which is adhesive and unexpansile.

With the initiation of air breathing after birth the first inspiration causes enlargement of the chest: (a) by descent of the diaphragm which increases the vertical dimension of the chest; and (b) by contraction of the external intercostal muscles. This thoracic expansion causes in turn expansion of the lungs which are virtually dragged after the chest wall because of the adhesive and inexpansile properties of the intrapleural fluid. This intrapleural fluid can with-

stand a pull of some 3500 mm cm^{-2}. The expansion of the lungs is naturally attended by a fall in the *intrapulmonary* pressure so that air is pulled into the lungs via the tracheobronchial tree. [Indeed, if the body of the baby is born first (breech delivery), then precautions must be taken against the first breath of the babe occurring before the head has cleared the birth canal, for otherwise the child aspirates the fluid detritus in the vagina. As the initiation of breathing is assisted by the exteroceptive stimulation provided by the relatively cold environment of room air, the limbs and torso of the breech-delivered infant should always be wrapped in warm towels to prevent or at least to minimize the chances of premature respiratory activity before the head of the child is extracted.]

Once inspiration thus expands the lungs their elastic recoil tendency causes a slight subatmospheric pressure to develop between the visceral and parietal pleural layers and this 'negative' intrapleural pressure develops to a degree which is of course dependent upon the amount of thoracic expansion which occurs. The intrapleural pressure can be recorded by inserting a needle into the intrapleural 'space'—the tip lies in the fluid between the pleural layers. A quiet inspiration causes the intrapleural pressure to drop 5 mm Hg or so below atmospheric. A really deep inspiration will cause the development of as much as 30 mm Hg pressure below that of the atmosphere.

The pressure in the lungs—intrapulmonary pressure—varies. At end-expiration in quiet breathing the pressure is atmospheric [FIG. III.3]. With the development of the next inspiration the intrapulmonary pressure drops to about 3 mm Hg below atmospheric, but regains the full atmospheric value at end inspiration. When expiration supervenes passively, the elastic recoil of the lung causes the intrapulmonary pressure to swing slightly to the positive side (3 mm Hg) but it regains the atmospheric value by the time that the quiet expiration is being completed.

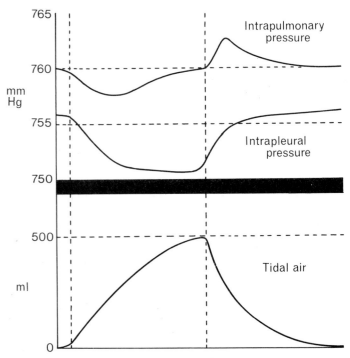

FIG. III.3. Changes in intrapulmonary and intrapleural pressures during the respiratory cycle.

Intrapulmonary pressure can be greatly changed by artificial respiratory 'gymnastics' such as by voluntary inspiratory or expiratory efforts against a closed glottis. Forced expiration against a closed glottis (Valsalva's manoeuvre) or against a column of mercury may produce a positive intrapulmonary pressure of 100 mm

Hg or more. Forced inspiration against a closed glottis (Müller's manoeuvre) can reduce the intrapulmonary pressure to 80 mm Hg below the atmospheric value.

LUNG VOLUMES AND CAPACITIES

Volumes

Tidal volume is the volume of air breathed in or out during quiet respiration (about 500 ml).

Inspiratory reserve volume is the maximal volume of air which can be inspired after completing a normal tidal inspiration—i.e. inspired from the end-inspiratory position—(2000–3200 ml).

Expiratory reserve-volume is the maximal volume of air which can be expired after a normal tidal expiration, i.e. expired from the end-expiratory position (750–1000 ml).

Residual volume is the volume of gas which remains in the lungs after a maximal expiration (1200 ml).

Capacities

Vital capacity is the maximal volume of air which can be expelled from the lungs by forceful effort following a maximal inspiration (4.8 litres in the male and 3.2 litres in the female.) [See FIG. III.4.]

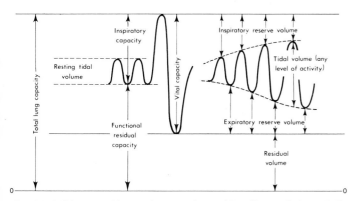

FIG. III.4. Diagram of lung volumes and capacities. (Pappenheimer, J. R. et al. (1950). *Fedn Proc. Fedn Am. Socs exp. Biol.* **9**, 602.)

Clearly the vital capacity is related to the size and development of the subject. It is usually 2.6 l m^{-2} surface area in the male and 2.1 l m^{-2} in the female. It is increased in swimmers and divers. It is decreased in older people and in diseases of the respiratory apparatus, e.g. poliomyelitis, respiratory obstruction, pleural effusion, pneumothorax, pulmonary fibrosis, emphysema, and pulmonary oedema. It is also reduced in pregnancy and in ascites, etc.

The vital capacity is altered by posture, being greater when measured in the upright position owing to the decreased pulmonary blood volume in the standing subject.

If the vital capacity is recorded on a kymograph (spirogram) the volume of air expelled can be timed [FIG. III.5]. Normally 80 per cent of the vital capacity should be expired in the first second (FEV$_1$). This measurement is a much more sensitive index of the severity of obstructive disease than is provided by the vital capacity itself, but it does not allow for the differentiation of the various causes of obstruction.

The total lung capacity is the volume of gas contained in the lungs after a maximal inspiration. The total lung capacity is thus equal to the sum of the vital capacity plus the residual volume and is of the order of 6000 ml in the adult male.

Inspiratory capacity, the maximal volume of gas which can be inspired from the resting expiratory level.

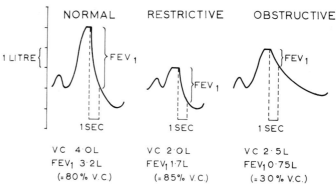

VC 4·0L
FEV$_1$ 3·2L
(≈ 80% V.C.)

VC 2·0L
FEV$_1$ 1·7L
(≈ 85% V.C.)

VC 2·5L
FEV$_1$ 0·75L
(≈ 3·0% V.C.)

FIG. III.5. The three records are taken from spirometer tracings. Inspiration up, and expiration down, and the records read from left to right. After a deep inspiration the subjects breathed out as forcefully and as rapidly as they could. The 'restrictive' record was obtained from a subject with kyphoscoliosis; the 'obstructive' from a patient with emphysema. Although the vital capacity is gravely reduced in the 'restrictive' patient, the proportion of the vital capacity expired in the first second is normal (unlike that in the 'obstructive' patient). (Campbell. E. J. M. and Dickinson, C. J. (1960). *Clinical physiology*. Blackwell, Oxford.)

Functional residual capacity (FRC), the volume of gas remaining in the lungs at the resting expiratory level. FRC is the sum of residual volume plus expiratory reserve volume.

With the exception of the FRC all other lung volumes and capacities can be measured by the use of a simple spirometer. The FRC can be determined by an open-circuit method in which the subject inspires pure oxygen for a period of 5 minutes and expires into a large spirometer previously washed out with oxygen and therefore nitrogen free. The expired gas contains nitrogen from the subject's lungs and its volume and concentration are measured. As it is known that the alveolar gas normally contains about 80 per cent nitrogen (a sample can be analysed) and as this nitrogen is eventually 'washed out' of the lungs into the spirometer a simple calculation gives the residual volume. Thus if 40 000 ml of expired gas in the spirometer contains 5 per cent nitrogen than 2000 ml of nitrogen must have been washed out of the lungs and as the lungs contain 80 per cent N$_2$ the FRC must be 2000 × 100/80 = 2500 ml.

The FRC is increased in conditions of hyperinflation of the lung which may result from emphysema or asthma.

Maximum breathing capacity (MBC), nowadays called the *maximum ventilation volume* (MVV), represents the greatest volume of air that can be ventilated on command during a given interval. This index of ventilatory function depends on the complete co-operation of the subject, who is asked to breathe as rapidly and as deeply as he can for a fifteen-second interval. The volume of air moved is either recorded by a spirometer fitted with a writing point or is collected in a Douglas bag; the result is expressed as litres per minute. Normal subjects can attain a maximal ventilation volume of 100 l min^{-1} or more. Males in the age group of 16–34 years show a range of 82–169 l min^{-1} (mean value = 126 l min^{-1} ± 28.6 l min^{-1}). Older subjects are not capable of such high ventilation volumes. The ability to reach a high MVV depends upon the muscular forces available, on the compliance of the thoracic walls and lungs, and on the airway resistances set up. MVV is profoundly reduced in patients with emphysema or in patients with airway obstruction.

REFERENCES

BOUHUYS, A. (1977). *The physiology of breathing*. Grune & Stratton, New York.

COMROE, J. H. (1965). *Physiology of respiration*. Year Book Medical, Chicago.

Work of breathing

ELASTIC FORCES—COMPLIANCE AND RELAXATION PRESSURES

The contraction of the respiratory muscles causes the expansion of the thoracic cage and thereby secures a fall in the intrapulmonary pressure which allows the ambient atmospheric pressure to push air into the lungs. The work done by the respiratory muscles can be expressed as the force of muscular contraction multiplied by the distance of muscle movement. More conveniently it can be described in terms of *pressure* multiplied by *volume*. The volume inspired for a given change of intrapulmonary pressure is almost entirely related to the elastic properties of the lungs and chest wall when a fit man breathes quietly at rest. The *static* elastic resistance can be measured by determining the airway pressure associated with a given lung volume in a subject *who has completely relaxed his respiratory muscles and his glottis*. In such circumstances his airway pressure is identical with his intrapulmonary (or intra-alveolar) pressure. The airway pressure is recorded by a manometer connected to a tube which is inserted via one nostril into the nasopharynx. The subject inspires from a spirometer; after the inspiration is completed, the spirometer lead is closed and the subject relaxes his respiratory muscles and glottis. The airway pressure and the volume which has been inspired from the spirometer are measured. From a series of inspirations and expirations with appropriate measurements of pressure and volume the pressure–volume diagram of the lungs and thorax is constructed. FIGURE III.6 shows that the airway pressure is zero (with respect to atmospheric) at the normal end-expiratory volume. In such a situation the normal end-expiratory volume is that of the functional residual capacity. The term 'relaxation volume' is often used as a synonym, and the 'relaxation-volume point' is correspondingly that at which the airway pressure is zero in the circumstances defined. On either side of the relaxation volume point the pressure–volume relationship is approximately linear over a limited range and the slope of the line $\Delta V/\Delta P$ gives a measure of the compliance of the lungs and thorax combined. In normal young adults $\Delta V/\Delta P$ is approximately 0.13–0.2 l cm H_2O^{-1}.

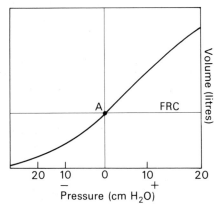

FIG. III.6. Static compliance of the lungs and chest wall. Point A represents the situation at the functional residual capacity (where pressure is zero with respect to atmospheric) at the normal end-expiratory volume.

The reciprocal of *compliance* is *elastance* ($\Delta P/\Delta V$). Thus the elastance of the lungs and chest wall is approximately 5–8 cm H_2O per litre.

The *relaxation pressure curve* can be analysed into its two components—the elasticity of the lungs and the elasticity of the chest wall and diaphragm.

Compliance of the lungs alone can be measured by recording the static *transpulmonary* pressure. At constant volume and with the airways unobstructed the pressure in the alveoli (*intrapulmonary pressure*) is identical with that measured by the nasal catheter. The *transpulmonary pressure* is given by the *intrapulmonary pressure* minus the *intrapleural pressure*. This intrapleural pressure can be recorded by inserting a needle into the intrapleural space and by injecting a tiny bubble of air therein, whereupon suitable manometric recordings can be made. Of more convenience is the technique which employs an air-containing latex balloon sealed over a catheter which is passed via the nostril *into the lower part of the thoracic oesophagus*. The intra-oesophageal (intrapleural) pressure is then measured at various end-inspiratory or end-expiratory volumes, the subject relaxing his respiratory muscles and glottis prior to and during the measurements. FIGURE III.7(a) shows the lung compliance curve and FIG. III.7(b) compares this curve with that of the lungs and chest wall and with that of the chest wall itself. Two points should be noted:

(i) The lung compliance is greater than that of the lungs plus thorax, being 0.22 litres per cm H_2O.
(ii) The relaxation pressure curve of the lungs lies to the right of that of the chest wall, indicating that the lungs are always tending to collapse whereas the thoracic cage is always 'deflated' below its relaxation volume.

Lung compliance is usually measured over the approximately linear part of the pressure–volume curve—that which expresses the volumes of the lung during normal breathing. Clinically, the term *specific compliance* is often used.

$$\text{Specific compliance} = \frac{\text{Compliance}}{\text{Functional residual capacity}}$$

(in l. cm $H_2O^{-1}l^{-1}$). Lung compliance itself is less for small lungs than large lungs and is less the more the lungs in a given subject are inflated. It is seriously reduced in emphysema, in pulmonary fibrosis, and in pulmonary congestion.

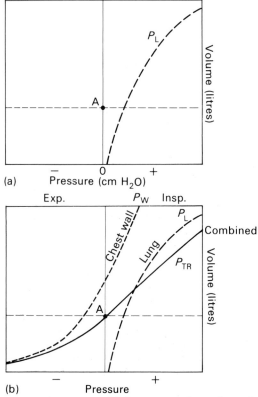

FIG. III.7(a). Lung compliance curve. (b) Components of compliance. Point A = functional residual capacity; P_W = chest wall; P_L = lung; P_{TR} = total (lung + chest wall).

MAXIMAL INSPIRATORY AND EXPIRATORY PRESSURES

Measurements are made with a recording manometer connected to a one-hole rubber stopper cut to fit one nostril. The other nostril is held closed while the subject expires or inspires with maximum force. Initially the subject first exhales maximally and then inserts in his mouth a tube connected to a spirometer from which he inhales any desired percentage of the vital capacity. Without changing the volume of air in the lungs he closes his mouth and inserts the nasal tube for pressure recording. FIGURE III.8 shows the results obtained. It is evident that the expiratory pressures are larger when the chest is inflated and that the inspiratory pressures are larger when it is deflated. Several mechanical factors are involved, but two stand out in importance. First, the inspiratory muscles are at their most favourable part of their length–tension relationship when the lung volume is small, whereas the muscles responsible for forceful expiration are stretched to a length at which they can develop their greatest tension when the lung volume is large. Second, the tendency of the chest wall to recoil towards a greater chest volume is most when the chest is deflated whereas the recoil of the lung–chest system towards a smaller volume is greatest when this system is expanded.

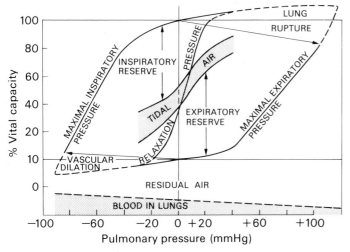

FIG. III.8. Pressure–volume diagram of the human chest. (After Fenn, W. O. (1951). *Am. J. Med.* **10**, 77.)

These curves for maximum inspiratory and expiratory pressures outline the total range of pressures and volumes which can be concerned in any manoeuvre (Fenn 1951). The right side of the figure at relatively low pressures is the area concerned in continuous pressure breathing. At higher pressures of 40–50 mm Hg air begins to leak through the bronchiolar tissues and may produce interstitial emphysema in the neck. When the pressure exceeds 80 mm Hg the lungs may rupture. This has occurred in sailors practising submarine escape techniques. If the breath is held during their ascent the pressure may rise to such levels that air is forced into the blood vessels causing cerebral emboli. Similarly, in explosive decompression, such as might occur if cabin pressure suddenly dropped in a stratosphere plane, the chest is maximally expanded until the excess air can escape.

During defecation the pulmonary pressure may be very high but there is little strain on the lungs because they are not expanded much.

The left-hand side of the figure shows that the maximal inspiratory pressure which can be developed is approximately 80 mm Hg. The inspiratory pressure curve indicates the maximal depth of

submersion which could be tolerated by a subject breathing ambient pressure through a tube reaching to the surface. It has been shown that at five feet (1.5 m) depth of water (110 mm Hg) inspiration is impossible. In such an experiment the pressure in the lungs is not negative relative to the atmospheric pressure but is negative relative to that in the water surrounding the body, so that the differential effect is identical to that produced by the application of an equal negative pressure to the lungs.

The relative portion of the total pressure attributable to muscle contraction and elastic recoil respectively at each chest volume is represented by the relaxation pressure (recoil) and maximum pressure (recoil plus muscle contraction).

Airway resistance and lung viscosity

Christie likened the lungs to a pair of bellows with an elastic recoil mechanism furnished by a spring [FIG. III.9]. To expand the bellows, force must be expended on their handles sufficient to extend the spring, to deform the fabric of the bellows and to overcome the air-flow resistance offered by the nozzle. Clearly the extension of the spring will determine how much force is required—how much air is drawn into the bellows. Deformation and air-flow resistance will require a force proportional to the rate at which air is drawn in. The spring represents an *elastic* resistance which by analogy must be overcome by the respiratory muscles when the lungs are expanded during inspiration. The lung tissue and the flow of air through the tracheobronchial system represent a *viscous* resistance.

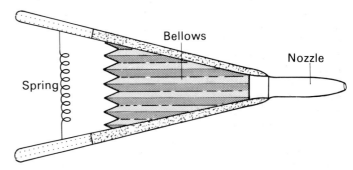

FIG. III.9. Representation of the lungs as a pair of bellows with an elastic recoil mechanism furnished by a spring.

Viscous resistance to breathing is responsible for the fact that the changes in intrapleural pressure and lung volume are out of phase with each other during breathing [FIG. III.10]. The intrapleural pressure change precedes that of the lung volume in each respiratory cycle. As inspiration begins, the intrapleural pressure decreases more quickly than expected because the development of force which it represents is required not only to overcome elastic resistance but also viscous resistance and viscous resistance itself increases as the rate of lung expansion increases. As the rate of expansion of the lung lessens towards the end of inspiration, viscous resistance diminishes to vanish at the end of inspiration with the lungs temporarily stationary. The rate of change of lung volume

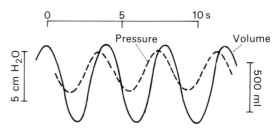

FIG. III.10. Intrapleural pressure and tidal volume. If the elastic pressure is substracted from the total pressure this gives the pressure exerted against non-elastic resistance.

during expiration is reversed and the pressure required to overcome viscous resistance is also reversed and is subtracted from the pressure necessary to overcome the elastic resistance.

The pressure–volume relation of the respiratory cycle is thus a closed loop [Fig. III.11]. In the diagram the dashed line shows the relationship which would be furnished by a solely elastic resistance system. The solid line features the relationship actually determined by appropriate measurements.

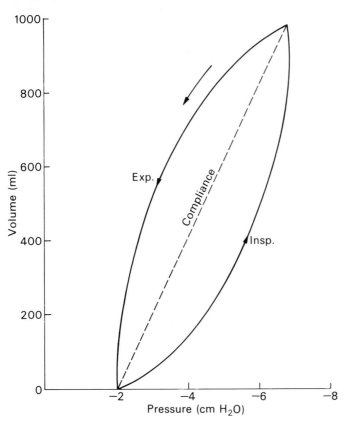

Fig. III.11. The area to the right of the compliance curve (dashed) represents the flow-*resistive* (non-elastic) work done on the lungs and air during inspiration. The area to the left of the compliance curve (within the loop) shows the *resistive work* done during passive expiration. The width of the loop is increased when there is an increased airway resistance. Bronchial asthma notably increases airway resistance. In emphysema the loop is much wider than normal. Note: at faster ventilation rates flow-resistive work increases, but elastic work does not.

Air flow in the respiratory tract is chiefly streamlined and partly turbulent. Turbulence increases as the rate of flow increases. The resistance offered to turbulent flow is much greater than that to streamlined flow. By measuring the pressure difference between the mouth and the alveoli and expressing it in terms of unit rate of the air flow which is measured simultaneously (i.e. in cm H_2O l^{-1} s^{-1}) the 'non-elastic resistance' (which includes viscous resistance of the tissues as well as airway resistance) is obtained. The 'non-elastic resistance' is usually expressed as the pressure difference between the mouth and the alveoli at a given rate of air flow (e.g. 1.5 cm H_2O at 0.5 l s^{-1}; 3.5 cm H_2O at 1.0 l s^{-1}). In emphysematous subjects and in asthmatic subjects the 'non-elastic resistance' is greatly increased.

REFERENCES

Bouhuys, A. (1977). *The physiology of breathing*. Grune & Stratton, New York.
Milic-Emili, J. (1974). Chapter 4. In *Respiratory physiology* (ed. J. G. Widdicombe) Series 1, Vol. 2, pp. 105–38. MTP International Review of Science. Butterworths, London.

MECHANICS OF THE PLEURAL SPACE

At resting volume the opposed recoil of the lung inwards and the chest wall outwards tends to separate the visceral from the parietal pleura with a pressure of some 5 cm H_2O. As the pleurae are permeable both to gases and to liquid it might be supposed that one or other or both would collect in the intrapleural space. That this does not happen is due to the sigmoid shape of the oxygen dissociation curve [p. 184] and to the difference in the colloid osmotic pressures in blood and tissue fluid [p. 82]. The fall in oxygen pressure caused by the removal of a given volume of oxygen (5 ml per 100 ml) from the arterial blood greatly exceeds the rise of carbon dioxide pressure which results from the addition of an approximately equal amount of carbon dioxide (4 ml per 100 ml). Thus the sum of the tensions of dissolved gases in the venous blood and tissue fluids is substantially less than atmospheric. The intrapleural gas pressure is some 70 cm H_2O less than that in air or blood. Therefore, no gas could be extracted from the pleural liquid or the blood to produce a gas bubble unless the gas pressure in the pleural space became more than 70 cm H_2O subatmospheric. Actually the gas pressure in a closed pneumothorax is only a few centimetres of water below atmospheric—say 3 cm H_2O. So the gas diffuses into the blood under a pressure of $70 - 3 = 67$ cm H_2O. As the gas diffuses out of the pleural space the space contracts, keeping the pressure of the gaseous phase constant until all the gas has been absorbed.

Formation or absorption of pleural fluid is determined, as elsewhere, by the balance of hydrostatic and colloid osmotic forces in the capillaries of the pleural surfaces. Visceral pleural capillaries are supplied by the pulmonary circulation and their hydrostatic pressure, ignoring gravitational forces, is some 10 cm H_2O. As the pressure on the visceral pleural surface attributable to the elastic recoil of the lung is about -5 cm H_2O, the transcapillary pressure is 15 cm H_2O tending to filter fluid into the pleural space. Pleural fluid itself, however, has a low protein content—only 1.5 g per 100 ml which would give a maximal pleural fluid osmotic pressure of 8 cm H_2O. Plasma colloid osmotic pressure is 25 mm Hg or 34 cm H_2O so the colloid OP difference is $34 - 8 = 26$ cm H_2O. Thus the total driving force causing fluid absorption by the visceral pleural capillaries is $26 - 15$ cm $H_2O = 11$ cm H_2O.

The parietal pleural capillaries are part of the systemic circulation and manifest a mean capillary pressure of some $30 - 35$ cm H_2O. As the colloid osmotic factor is the same, there exists a driving force of about 5–9 cm H_2O filtering fluid into the pleural space. The balance between filtration from the parietal capillaries and absorption into the visceral capillaries maintains the presence of a thin layer of fluid between the pleural surfaces with the two advantages of free sliding of the lung on the chest wall and the synchronous transmission of changes of chest volume to the lung. The total volume of intrapleural fluid is of the order of 2 ml.

REFERENCES

Agostoni, E. (1972). Mechanics of the pleural space. *Physiol. Rev.* **52**. 57–128.

SURFACE TENSION OF THE LUNG LINING

Von Neergard (1929) found that gas-free fluid-filled lungs could be distended by pressure almost twice as easily as when the lungs were filled with air. Pattle (1956) showed that pulmonary oedema fluid possessed an unduly low surface tension and concluded that some substance was secreted into the alveoli which produced a lowering of surface tension.

Figure III.12 shows the pressure–volume relations in the excised lungs of a cat. With expansion caused by gas-free saline solution the lungs fill evenly, with no evidence of an opening pressure which must be overcome before the fluid begins to enter. The pressure–volume curve obtained in this manner indicates the true elastic behaviour of lung tissue. It will be noted that the pressure–volume curve obtained during inflation of the lungs is virtually identical with that determined during the subsequent deflation.

The pressure–volume curves obtained when the lung is successively inflated and deflated with air are entirely different. First, there is no filling (increase in volume) until a pressure of more than 8 cm H₂O is reached and subsequent inflation is irregular until complete expansion occurs at a pressure of 20 cm H₂O. When the lung is allowed to deflate its pressure volume curve is displaced far to the left of that derived from progressive inflations, so that the pressure at any volume is less on expiration than on inspiration. As this phenomenon of hysteresis is not manifest in the fluid-filled lung it has been ascribed to the effects of surface tension present at the interface of air and the fluid lining the alveolar walls.

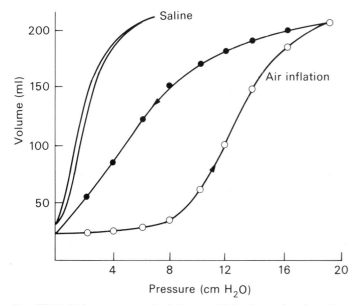

FIG. III.12. Volumes expressed relative to initial volume after degassing. Note absence of hysteresis in saline-filled lungs and marked hysteresis in air-filled lung. (After Radford, E. Jr (1957). In *Recent studies of mechanical properties of mammalian lungs* (ed. J. W. Remington) pp. 177–90. American Physiological Society, Washington, DC.)

Surface tension is defined as the tangential force in the surface acting perpendicularly per unit length across any line in the surface. It arises owing to the tendency of any surface to decrease to a minimum. The molecules at the surface are inwardly attracted to adjacent molecules of their liquid phase which is not counterbalanced because they lie at the surface. The surface layer has thus an excess of energy proportional to the surface area; the constant of proportionality between surface area and surface energy is the surface tension coefficient which is expressed in dynes per centimetre.

The effect of surface tension in a thin-walled sphere such as a soap bubble or an alveolus is to tend to make the sphere smaller. To oppose this there must be an equal and opposite force which is created by the pressure within the sphere. In the soap bubble the tension in the wall is equal to that exerted by the pressure inside the bubble. According to the law of Laplace

$$P = \frac{2T}{r},$$

where r is the radius of the sphere. If the surface tension in the soap bubble remains constant then the pressure required to inflate it

increases as the bubble diminishes in size. For this reason, when two soap bubbles of different size are put in communication with each other the small bubble empties into the larger.

Pattle was the first to study the properties of bubbles expressed from lung tissue and emphasized the physiological importance of a low surface tension on lung mechanics. Clements showed that normal lung tissue extracts contained a substance which lowers surface tension. Clements measured the surface tension of films, made from alveolar lining substance, on a Wilhelmy balance. In the expanded film the surface tension was some 45 dyne cm⁻¹, but when the surface area was compressed the surface tension fell to low values such as 5 dyne cm⁻¹. If slow cyclical changes in the area of the surface were produced, the surface tension plotted against surface area formed a loop demonstrating hysteresis [FIG. III.13].

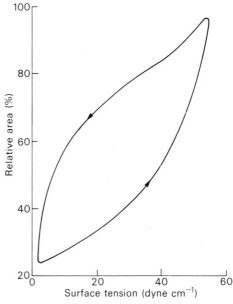

FIG. III.13. Surface tension of aqueous lung extract as a function of surface area.

Now on theoretical grounds one would predict that a system of alveolar 'bubble' surfaces connected to a single tube and possessing different dimensions and radii would be unstable. If their surface tension remained constant and independent of their surface area the smaller alveoli would empty first as the lung deflated, eventually leaving only the larger alveolar sacs gas filled. However, as FIG. III.12 shows, pressure can be lowered markedly during deflation from a maximal volume before a significant fraction of the volume is expelled. Moreover, direct observation of the lung surface reveals the stability of alveoli of various sizes as the lung deflates. Thus the stability of the alveoli is dependent on the fact that surface tension is generally reduced and because additionally it adjusts itself to the surface area.

The substance or substances responsible for lowering the surface tension of the fluid lining the alveoli has not yet been identified with certainty. Known as *pulmonary surfactant* it has been prepared by ultracentrifugation from endobronchial lavage fluid. Such preparations contain 74 per cent phospholipids among which lecithin comprises 56 per cent. Dipalmitoyl lecithin is the predominant molecule and makes the principal contribution to the surfactant. Ten per cent of the surfactant material is protein of which one fraction contains a protein in which hydrophobic amino acids outnumber hydrophilic amino acids two-to-one. King and Clements (1972) consider this protein to be well suited to a role as the apoprotein of pulmonary surfactant. It has considerable ability to bind phospholipids.

Surfactant is claimed to be produced by large epithelial cells lining the alveoli. These cells (known as pneumocytes or Type II alveolar cells) contain osmophilic inclusion bodies which are believed to be secretion products. The cells show microvilli on their free alveolar surface—a feature of cells with known transport functions (Heinemann and Fishman 1969).

Atelectasis of the new-born—the so-called hyaline membrane disease—is associated with a lack of surfactant material and a correspondingly high lung surface tension. Extracts from the lung tissue of such infants show reduced surfactant action permitting only a reduction of the surface tension of saline films to some 20–25 dyne cm^{-1} instead of 4–8 dyne cm^{-1}.

REFERENCES

CLEMENTS, J. A. (1971). *Archs intern. Med.* **127**, 387.
HEINEMANN, H. O. and FISHMAN, A. P. (1969). *Physiol. Rev.* **49**, 1.
KING, R. J. and CLEMENTS, J. A. (1972). *Am. J. Physiol.* **223**, 715.
NEERGAARD, R. VON (1929). *Z. ges. Med.* **66**, 373.
PATTLE, R. E. (1956). *J. Path. Bact.* **72**, 203.
—— (1965). *Physiol. Rev.* **45**, 48.
STRANG, L. B. (1974). Chapter 2. In *Respiratory physiology* (ed. J. E. Widdicombe) pp. 31–66. Butterworths, London.

ALVEOLAR VENTILATION AND PERFUSION. DIFFUSING CAPACITY

Respiration involves the gaseous exchange of oxygen and carbon dioxide by diffusion between the alveoli and the pulmonary capillary blood. This exchange depends primarily on the volume of alveolar ventilation and the volume of blood flow through the pulmonary capillaries which are in contact with the ventilated alveoli. The surface area available for gaseous exchange is 90 square metres.

Alveolar ventilation

In quiet breathing a healthy man inspires and expires about 500 ml of air. Some of this inspired air (about 150 ml), merely fills the conducting air passages between the mouth and nose and the respiratory bronchioles. The remainder effectively ventilates the alveoli thereby lowering the pCO_2 and raising the pO_2 of the alveolar air. As the alveoli still contain some 2–2.5 litres of air after a quiet expiration (functional residual capacity), the over-all reduction of pCO_2 and rise of pO_2 by one subsequent quiet inspiration of 350 ml is small; nevertheless the expulsion of about 350 ml of this 'diluted' alveolar air containing 6 per cent CO_2 and 14 per cent O_2 by the following expiration ensures that the alveolar gas pressures of oxygen and carbon dioxide are kept relatively constant, despite the continuous delivery of CO_2 and removal of O_2 by the pulmonary capillary blood. Clearly, the volume of *alveolar* ventilation is of paramount importance in this respect. The total pulmonary ventilation per minute (ordinarily some 6–7 l min^{-1}) at rest is usually achieved by a tidal air of 500 ml and a respiratory frequency of 12–14 min^{-1}. If the ventilatory volume were kept steady then the breathing can be modified by altering the frequency. Thus at a rate of 36 min^{-1} a pulmonary ventilation volume of 6 l min^{-1} would require a tidal air of only 167 ml; conversely, at a rate of 6 min^{-1} the same pulmonary ventilation volume would require a tidal air of 1000 ml. Although the total ventilation would be the same the respective alveolar ventilations would be vividly different, for it must be remembered that 150 ml of air in each breath merely fill the dead space. Hence in rapid shallow breathing the alveolar ventilation is grossly inadequate, whereas in deep slow breathing it may be excessive; correspondingly these alterations exert profound effects on the oxygenation of the pulmonary

capillary blood and on the rate of removal of carbon dioxide therefrom.

The total alveolar ventilation can only be deduced by measuring the total ventilation and subtracting the volume required for ventilation of the dead space.

Dead space may be defined as:

1. Anatomical dead space.
2. Physiological dead space.

In health the two are identical. The *anatomical dead space* is simply the internal volume of the airway between the mouth and nose and the alveoli. Physiological dead space includes this volume and two additional volumes: (i) the volume of inspired gas which ventilates alveoli which receive no pulmonary capillary blood flow and (ii) the volume of inspired gas which ventilates alveoli in excess of that volume which is required to arterialize the blood in their local pulmonary capillaries. These two additional volumes are ordinarily negligible in the healthy subject. The anatomical dead space can be measured with the aid of a nitrogen meter, which allows the rapid and continuous measurement of the nitrogen concentration of the inspired and expired air. From a mouthpiece fitted to the subject a sample of air is continuously drawn off, via a narrow side tube, by a vacuum pump; the sample passes at low pressure (2 mm Hg) through a glass tube across which a high voltage is applied. As a result of this the gas emits radiation and glows brightly. Nitrogen glows with an orange–pink colour and this radiation is picked up by a filter which passes the N_2 bands in the range 310–480 nm to a photoelectric cell, which analyses and records the N_2 concentration. When the subject is breathing air, the N_2 concentration of the expired air is about 80 per cent. After a quiet expiration the subject takes a deep breath of pure oxygen and then breathes out slowly and evenly. During the inspiration of oxygen and the early part of the subsequent expiration the N_2 meter registers 0 per cent N_2 but subsequently the N_2 content of the remainder of the expiration rises rapidly to reach a plateau at say 60 per cent N_2 [FIG. III.14].

If a square front were preserved between the dead space gas and the alveolar gas, the expired nitrogen concentration would remain zero until a volume equal to that of the anatomical dead space was expired and would then rise sharply to the final plateau concentration. Needless to say this is not the case, as there is some mixing between the dead space and alveolar gases during the expiration. However, by placing a vertical line on the record in such a position that the shaded area A is equal to the non-shaded area B [FIG. III.14], the volume expired up to this point is the anatomical dead space.

Physiological dead space

All the CO_2 expired is derived from the alveoli, which alone indulge in gaseous exchange.

Hence:

Volume of tidal air × pCO_2 in the expired air = volume of air from the alveoli per breath × mean pCO_2 in the alveoli.

The arterial pCO_2 gives the most satisfactory value for the mean alveolar pCO_2.

Let tidal volume = 500 ml.
and x = volume of dead space air.
Then $(500 - x)$ = volume from the alveoli per breath.
Suppose pCO_2 in expired air = 28 mm Hg
and arterial pCO_2 = 40 mm Hg
$$(500 \times 28) = (500 - x)40 + (x \times 0)$$
$$\therefore x = 150 \text{ ml.}$$

As has been said, in health the anatomical and physiological dead space volumes are equal. In disease, where there may be excessive ventilation/perfusion ratios in parts of the lung the physiological dead space volume may greatly exceed that of the anatomical dead space.

RESPIRATORY DEAD SPACE
(SINGLE BREATH ANALYSIS)

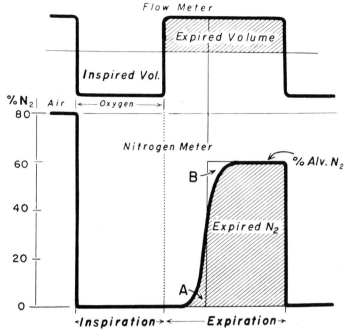

FIG. III.14. Single breath analysis of anatomical dead space. *Above:* Volume flow of inspired and expired gas. A constant flow rate is pictured for ease of measurement, though this would not be obtained in actual practice. *Below:* Nitrogen concentration of inspired and expired gas following a single breath of oxygen. (Comroe, J. H., Forster, R. E., DuBois, A. B., Briscoe, W. A., and Carlsen, E. (1962). *The lung.* Year Book Publishers, Chicago.)

The sampling of alveolar air

Haldane–Priestley method

The subject with the nose clipped makes a rapid maximal expiration down a narrow tube about three feet long and of one inch internal bore and then occludes the mouthpiece with his tongue. A sample of the air contained near the mouthpiece of the tube is withdrawn via a side tube. This, the last to be expelled from the lungs, is taken to be alveolar air.

Continuous sampling method

The subject breathes through a mouthpiece fitted with inspiratory and expiratory valves. About 10 ml of gas are removed from the last part of each expiration via a side tube situated just beyond the expiratory valve. The alveolar air thus obtained is passed through two gas analysers placed in parallel, which simultaneously analyse oxygen (Pauling tensimeter) and carbon dioxide (infrared gas analyser). The apparatus dead space must be minimal. On making a normal expiration the tidal air escapes through the expiratory valve into a 300 ml rubber tube. The air just beyond the valve is thus trapped and is ready for sampling. In the following inspiration, the negative mouthpiece pressure activates an aneroid device which closes a thermionic valve relay, which in turn operates and opens the sampler valve. At the end of inspiration the mouthpiece air pressure returns to atmospheric level and the sampling valve correspondingly closes.

Comparison of the results obtained by this method reveals that the pCO_2 so measured is about 2–2.5 mm Hg less than that calculated from Haldane–Priestley samples delivered after the end of a normal expiration. The pO_2 is correspondingly slightly higher as

determined in the continuous sampling method than in the H–P method.

Indirect method

Riley and his associates have devised an indirect method of determining the 'effective' alveolar gas pressures (i.e. those which, if present continuously and uniformly throughout the lung in all functioning alveoli, would permit CO_2 and O_2 exchange between alveoli and blood during a series of breaths in amounts equal to the gaseous exchanges which can be actually measured by analyses of inspired and expired air). The formula used (it can be derived arithmetically) is:

$$\text{Effective alveolar } pO_2 = \text{inspired } pO_2 \times \frac{\% \text{ N}_2 \text{ expired air}}{\% \text{ N}_2 \text{ inspired air}}$$

$$- \frac{\text{arterial } pCO_2}{\text{Expired air RQ}}$$

where the expired air RQ [respiratory quotient, p. 211] is the ratio of:

$$\frac{CO_2 \text{ expired per minute}}{O_2 \text{ used per minute}}.$$

The arterial pCO_2 is measured directly by arterial sampling. The expired air is collected and its volume and composition are measured.

From these various results it can be stated that the mean alveolar pCO_2 is about 40 mm Hg. The mean alveolar pO_2 is about 100 mm Hg.

Uniformity of alveolar ventilation

In the healthy resting subject alveolar ventilation is reasonably uniform and averages 2.0–2.5 l m^{-2} min^{-1}. The very complexity of the lung structure however renders it unlikely that alveolar ventilation is absolutely uniform and in the diseased lung there may be obvious unevenness of ventilation:

1. *The single-breath technique*, using the N$_2$ meter as already described, yields valuable information of irregular alveolar ventilation. If the subject, initially breathing room air, makes a single deep inspiration of O_2 and then expires slowly and evenly, the N$_2$ concentration of his expired air can be continuously analysed. In the healthy subject the N$_2$ concentration of the air expired rises very little (+ 1 per cent) if measured between an expired volume of 750 ml and 1250 ml [FIG. III.15]. In the diseased lung, however, such as in asthma, emphysema, fibrosis, or congestive heart failure, the N$_2$ concentration rises appreciably as the expired air volume increases from 750 to 1250 ml.

2. *Nitrogen wash-out by multiple breath technique.* The mean concentration of N$_2$ in the expired air is measured, breath by breath, in a subject respiring pure oxygen and the values are plotted on semi-log paper against the number of breaths. Uneven ventilation can be readily detected.

Ventilation–perfusion ratio

Hitherto we have concentrated on the ventilatory properties of the lungs. Nevertheless efficient gaseous exchange between alveoli and pulmonary capillary blood necessitates perfusion of the alveoli by the pulmonary capillary blood stream. Approximately four litres of air ventilate the alveoli each minute and five litres of blood pass through the pulmonary capillaries in the same time. Hence the mean ventilation/perfusion ration is 4/5 or 0.8. Each lung normally displays the same ratio, receiving 2 litres of ventilation and 2.5 litres of blood respectively. It may be asked whether it is necessary to go further than to determine the alveolar ventilation volume and the

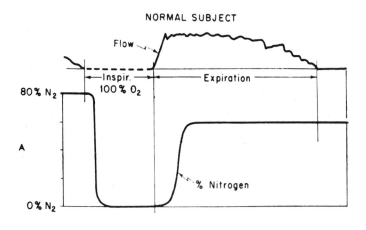

NORMAL SUBJECT

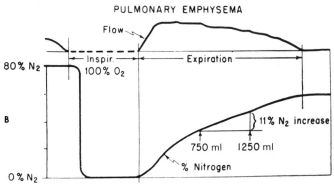

PULMONARY EMPHYSEMA

FIG. III.15. Uniformity of alveolar ventilation. A. Nitrogen meter record that would be obtained from normal subject with even distribution. B. Nitrogen meter record in patient with uneven distribution of air. The method of measuring the single breath is shown on the lower record: the increase in nitrogen concentration is measured for a 500 ml sample of alveolar gas (between 750 and 1250 ml of expired gas). (Comroe, J. H. *et al.* (1962). *The lung.* Year Book Publishers, Chicago.)

cardiac output, for these values will give the mean ventilation/perfusion ratio. The answer to this question is in the affirmative and can be illustrated by an example given by Comroe. Suppose the total alveolar ventilation and the pulmonary blood flow were normal, but that all the blood went to one lung which was not ventilated and that all the alveolar ventilation occurred in the remaining lung which received no blood flow. Asphyxial death would rapidly ensue. This example, deliberately extreme, indicates however that non-uniformity of the ventilation/perfusion ratio may cause abnormalities in the oxygenation of the arterial blood. The causes of uneven ventilation have been mentioned—asthma, pneumothorax, emphysema, pulmonary fibrosis; non-uniform blood flow may occur due to anatomical shunts, regional reduction in the pulmonary vascular bed (such as in emphysema), pulmonary embolism and in cases of raised pulmonary resistance (pneumothorax, fibrosis, heart failure). Obviously, uneven ventilation may exist without any abnormalities of pulmonary blood flow (and vice versa), or lastly, both alveolar ventilation and blood flow may be non-uniform (as in emphysema). A detailed survey of the problem would be beyond the scope of this text but one example may be briefly mentioned:

Let us suppose that the pulmonary blood flow is normal (5 l min^{-1}) and uniform but that the lungs are unevenly ventilated, one receiving much more of the tidal air than the other. Because of the poor ventilation of one lung, blood passing through it is inadequately oxygenated and its haemoglobin saturation may be only 90 per cent (instead of the normal 97 per cent) corresponding to an arterial pO$_2$ of only 65 mm Hg [see p. 184]. The blood passing through the hyperventilated lung will be slightly 'over-arterialized' but as the alveolar pO$_2$ cannot rise to 150 mm Hg due to the

presence of CO$_2$ in the alveoli, the saturation of the haemoglobin in the blood will be only raised say from 97 per cent to 98 per cent. The saturation of the mixed arterial blood will therefore be $(0.5 \times 98 + 0.5 \times 90) = 94$ per cent corresponding to an arterial pO$_2$ of only 75 mm Hg, which is seriously below the usual arterial pO$_2$ of approximately 95 mm Hg. The uniformity of ventilation/perfusion may be assessed as follows:

1. The physiological dead space and the anatomical dead space can be measured. If the physiological dead space greatly exceeds the anatomical dead space there must be alveoli which have a high ventilation/perfusion ratio.

2. By continuous measurement of the CO$_2$ content of the expired alveolar air by a rapid analyser; normally the expired air from the alveoli changes little in its CO$_2$ content during the completion of the breath. If the CO$_2$ content of the last part of the alveolar gas is much higher than the initial value the ventilation/perfusion ratio is not uniform, being high initially and low in the alveoli contributing to the last part of the tidal air.

3. In cases with obvious anoxaemia, there may be 'shunting'. This may be due to the presence of an anatomical by-pass channel through which some of the blood flows or may be due to the perfusion of poorly ventilated alveoli. If the patient breathes oxygen the blood remains anoxaemic in the case of the anatomical shunt, but the blood becomes fully saturated if the case is one with blood passing through poorly ventilated alveoli.

Diffusing capacity of the lungs

O$_2$ and CO$_2$ are transferred across the alveolar and pulmonary capillary membranes by diffusion. The diffusion coefficient of a gas is the volume measured in millilitres which diffuses through one square centimetre of membrane per minute when there is a pressure difference of one mm Hg across the membrane.

The diffusion capacity of the lungs for oxygen is given by the equation:

$$D_{O_2} = \frac{A_L \times dO_2}{t_L}$$

where D_{O_2} is the diffusion capacity in ml, A_L is the total area of the diffusing surface and t_L is the distance through which the gas must diffuse from the alveoli into the plasma; dO$_2$ is the diffusion coefficient.

By Fick's law of diffusion the rate of diffusion is proportional to the pressure gradient. The rate of diffusion of oxygen per minute across the alveolar and capillary membranes of the lung is obviously the rate of oxygen uptake per minute.

Hence $V_{O_2} = D_{O_2} \times \bar{P}_{(A-B)}$
where V_{O_2} is the O$_2$ uptake per minute and $\bar{P}_{(A-B)}$ is the mean alveolar-capillary gradient of oxygen tension. The measurement of D_{O_2} therefore requires the determination of V_{O_2} which is easy, and an estimate of $\bar{P}_{(A-B)}$ which is hard to obtain. Only if the mean capillary pO$_2$ be known (and the mean alveolar pO$_2$), can $\bar{P}_{(A-B)}$ be calculated; the difficulty in measuring mean capillary pO$_2$ is responsible for there still being some uncertainty as to the numerical value of D_{O_2}.

Let us consider the mixed venous blood entering the arterial end of the pulmonary capillary—its oxygen tension will be about 40 mm Hg and its haemoglobin saturation will be 70–75 per cent. The alveoli contain oxygen at a partial pressure of about 100 mm Hg and the arterialized blood leaves the capillary at an oxygen tension of 95 mm Hg. Owing to the high tension gradient to which the blood is exposed at the beginning of the capillary and owing to the shape of the oxygen dissociation curve, the blood oxygen tension will rise rapidly in the first part of the capillary and more slowly in the latter portion of the capillary. Even if the blood pO$_2$ at the end of the capillary were accurately known, the mean capillary oxygen tension would not be

$$\frac{\text{'arterialized' pO}_2 + \text{mixed venous pO}_2}{2}$$

for the oxygen tension does not rise linearly.

It may be asked why the arterial pO$_2$ (measured directly) cannot be taken as identical with that at the end of the pulmonary capillary. The reason is, that blood leaving the pulmonary capillaries with an oxygen tension almost the same as that in the alveoli is joined by venous blood from the bronchial circulation; by the time the blood reaches the systemic arteries, further contributions of venous blood have been made by the anterior cardiac and Thebesian veins which empty into the cavities of the left heart. Although this venous admixture is small and causes only slight effects on the saturation of the haemoglobin, it reduces the pO$_2$ in the arterial blood below that which exists in the pulmonary end-capillary blood. Thus suppose 5 parts of venous blood (saturation 70 per cent; pO$_2$ 38 mm) mix with 95 parts of pulmonary end-capillary blood (saturation 97 per cent; pO$_2$ 100 mm). Then the mixed blood has a final saturation of (0.05 × 70 + 0.95 × 97) = 95.6. The corresponding pO$_2$ is 92 mm hence the pO$_2$ has fallen 8 mm below that of the end-capillary level. Whereas this is of trivial import in terms of gas carriage, it makes the measurement of D_{O_2} very inaccurate.

The difficulty can be resolved to some extent by having the subject breathe 12 per cent O$_2$. In such circumstances the blood may leave the pulmonary capillaries only 70 per cent saturated at a pO$_2$ of 38 mm and the effect of venous admixture on this anoxaemic end-capillary blood is correspondingly very small.

Thus, suppose 95 parts of blood (70 per cent saturated) mix with 5 parts of blood from the bronchial and anterior cardiac veins (50 per cent saturated at a pO$_2$ of 27 mm) the final mixture (arterial blood) will be (0.05 × 50 + 0.95 × 70) = 69 per cent saturated and the pO$_2$ is 37.5 mm Hg, i.e. a drop of only 0.5 mm Hg from the end-capillary value.

Suppose then, that in a subject breathing 12 per cent O$_2$, analysis of his mixed venous blood, withdrawn by a cardiac catheter from the pulmonary artery, reveals that the haemoglobin saturation is 50 per cent and the pO$_2$ is 27.5 mm Hg. Arterial blood sampled directly has a pO$_2$ of 38.0 mm Hg. Let us assume that the arterial pO$_2$ is identical with that of the end-capillary blood. We wish to determine the mean tension gradient of oxygen between the alveoli and the pulmonary capillary blood, i.e. that gradient which if maintained along the entire length of the capillary would cause the same diffusion of oxygen as actually occurs. The following method is adopted:

Haemoglobin saturation (per cent	pO$_2$	$P_a - P$	$\dfrac{1}{P_a - P}$	$\Sigma \dfrac{1}{P_a - P}$	$\dfrac{\Sigma \dfrac{1}{P_a - P}}{\text{Total } \Sigma \dfrac{1}{P_a - P}}$
50	27.5	19.5	—	—	—
52	28.5	18.5	0.054	0.054	0.071
54	29.5	17.5	0.057	0.111	0.146
56	30.5	16.5	0.061	0.172	0.226
58	31.5	15.5	0.0645	0.2365	0.309
60	32.5	14.5	0.069	0.3055	0.401
62	33.5	13.5	0.74	9.3795	0.498
64	34.5	12.5	0.080	0.4595	0.603
66	36.0	11.0	0.091	0.5505	0.723
68	37.0	10.0	0.100	0.650	0.853
70	38.0	9.0	0.111	0.761	1.000

Mean alveolar pO$_2$ = 47 mm Hg.

A Table is constructed in which the first two columns respectively show each 2 per cent increase in haemoglobin saturation between 50 per cent and 70 per cent and the corresponding pO$_2$ (read from an oxygen dissociation curve). The third column of the Table gives the tension gradient which exists at each of these 'points'. As the rate of diffusion from alveoli to blood is proportional to the tension gradient, the time taken for each successive increment of oxygen uptake will be proportional to the reciprocal of the tension gradient (fourth column). The sum of the reciprocals (fifth column) will be proportional to the total time spent by the blood in the capillary. The sum of the reciprocals at any point can be expressed as a fraction of the total sum of the reciprocals and therefore as a fraction of the length of the capillary traversed by the blood at that point (sixth column).

By plotting the corresponding pO$_2$ against the fraction of the capillary length traversed FIGURE III.16 is obtained. The mean capillary pO$_2$ can then be solved graphically—the shaded areas in FIGURE III.16 must be equal. In the particular example given the mean capillary pO$_2$ is 33.5 mm Hg and the mean tension gradient is 47 − 33.5 = 13.5 mm Hg.

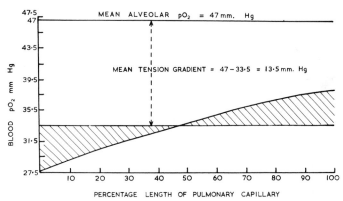

FIG. III.16. Graphic representation of the rise in pO$_2$ of blood traversing the pulmonary capillaries.

Hence in this example if the oxygen usage at rest (V_{O_2}) was 250 ml min^{-1}:

$$V_{O_2} = D_{O_2} \times \bar{P}_{(A-B)}$$
$$250 = D_{O_2} \times 13.5$$
$$D_{O_2} = 18.5 \text{ ml}$$

Values for D_{O_2} determined at rest range between 15 and 35 ml although the average figure is close to 20 ml. The uncertainty of accurate measurement of the end-capillary oxygen tension is responsible for the variation of the D_{O_2} figure. It is assumed in the application of the above method that the breathing of low oxygen mixtures does not alter the diffusing capacity of the lung—an assumption that may not be entirely justified.

The difference between alveolar pO$_2$ and the arterial pO$_2$ depends on two factors:

1. A membrane component which is responsible for there being some difference between mean alveolar pO$_2$ and end-capillary pO$_2$.

2. A physiological shunt component which is responsible for the difference between end-capillary pO$_2$ and arterial pO$_2$. This includes the effect of the admixture of bronchial and cardiac venous blood and the effect of passage of some blood through poorly ventilated alveoli. The breathing of 12 per cent O$_2$ has two effects (i) it increases the 'membrane component', so that the alveolar-end capillary pO$_2$ difference becomes more considerable and the mean alveolar-mean capillary oxygen gradient becomes measurable, and (ii) it greatly reduces the effect of the 'shunt component'.

D_{O_2} is much greater in exercise, rising to values of 65 ml or more—a threefold increase which is presumably due to dilatation of capillaries normally patent in the lung bed and to the opening up of additional pulmonary capillaries.

In lung disease patients may show a reduction of D_{O_2}. Thus a thickening of the alveolar-capillary membrane, such as occurs in sarcoidosis and berylliosis, or a reduction of the pulmonary capillary bed (fibrosis or emphysema) may markedly diminish the diffusing capacity. As a result such patients may show serious arterial anoxaemia during exercise. The inhalation of 100 per cent O_2 during exercise raises arterial saturation to normal in these patients which allows a differential diagnosis between impaired diffusion and an anatomical shunt.

Carbon dioxide diffuses 20 times more readily than oxygen and hence impaired diffusion rarely causes CO_2 retention.

REFERENCES

BOUHUYS, A. (1974). *The physiology of breathing*. Grune & Stratton, New York.

COMROE, J. H. (1965). *Physiology of respiration*. Year Book Medical, Chicago.

ROUGHTON, F. J. W. (1954). Chapter 5. In *Respiratory physiology in aviation* (ed. W. M. Boothby) p. 51. USAAF School of Aviation Medicine.

DISTRIBUTION OF BLOOD FLOW THROUGH THE LUNGS

West (1965) has shown the inequality of blood flow in the lungs using carbon dioxide labelled with radioactive oxygen $C^{15}O_2$ [FIG. III.17].

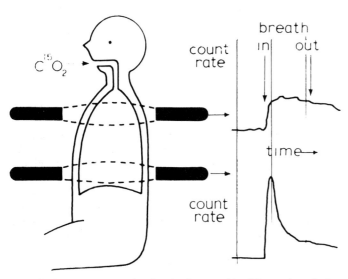

FIG. III.17. Measurement of regional pulmonary blood flow and ventilation using radioactive CO_2. Pairs of scintillation counters examine the anteroposterior cores of the lung. The counting rate at the end of inspiration is proportional to the ventilation of the lung in the counting field, and its volume; the slope of the tracing during breath-holding (clearance rate) measures the regional blood flow. (West, J. B. (1965). *Ventilation/blood flow and gas exchange*. Blackwell, Oxford.)

Pairs of scintillation counters are sited over anteroposterior cores of the lung of a seated subject. The subject inhales a single breath of $C^{15}O_2$ and holds his breath for 15 seconds. The initial rise in counting rate is determined by the ventilation of the lung in the counting field and its volume. The decline of the counting rate (clearance rate) is due to removal of $C^{15}O_2$ by the regional pulmonary flow and the slope of the tracing is a measure of this flow.

Similar results can be obtained using reactor-produced xenon-133 (half-life 5 days). The radioactive xenon is made up in saline solution and is injected via a catheter into the superior vena cava. As the labelled blood enters the pulmonary capillaries the xenon is liberated into the alveoli because of its low solubility. It remains

there during the breath-holding period and during this time the lung is scanned from the base upwards by a pair of scintillation counters. The xenon is then distributed evenly throughout the alveoli by rebreathing; this produces the situation which would have resulted if all the alveoli had been evenly perfused in relation to their volume. A second scan is now made and the comparison of the two scans allows the calculation of the blood flow per unit alveolar volume.

These techniques show that there is a steady fall in blood flow per unit alveolar volume with the rib number from the bottom to the apex of the lung when the subject is sitting or standing. In the upright lung the apical blood flow is almost nil. Posture and exercise modify this distribution. When supine, apical flow increases, basal flow remains similar and the two become more equal. When supine, the posterior regions of the lung receive more flow than do the anterior. Exercise causes, of course, an increase in blood flow in all regions, but flow becomes more even throughout the lung.

FIGURE III.18 shows a simple model which depicts the regional distribution of pulmonary blood flow.

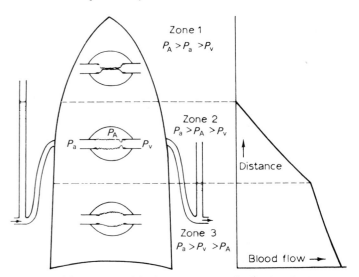

FIG. III.18. Diagram to explain the topographical distribution of pulmonary blood flow. The lung is divided into three zones by the relative sizes of the pulmonary arterial (P_a), pulmonary venous (P_v) and alveolar (P_A) pressure. For details see text. (West, J. B. (1965). *Ventilation/blood flow and gas exchange*. Blackwell, Oxford.)

In the upper zone 1 the alveolar pressure P_A exceeds the arterial pressure and there is no flow. Presumably this is because thin-walled collapsible vessels, such as the capillaries, are directly exposed to alveolar pressure.

In zone 2 arterial pressure exceeds alveolar pressure, but alveolar pressure in turn exceeds venous pressure. Here each vessel behaves like the resistor in a Starling heart–lung preparation—that is, a collapsible tube surrounded by a pressure chamber.

When the chamber (alveolar) pressure exceeds the downstream (venous) pressure, flow is determined not by the arterial–venous pressure difference, but by the arterial–alveolar pressure. The reason for this is that the thin tube offers no resistance to the collapsing pressure so that the pressure inside the tube is the same as the pressure outside. The result is that the pressure gradient responsible for flow is the perfusing pressure minus chamber pressure. The collapsible tube actually develops a constriction in its downstream end (where the pressure inside the tube is least). Under ordinary flow conditions there will be a linear relation between volume flow rate and the arterial-alveolar pressure difference. As the arterial pressure increases steadily down the lung due to the hydrostatic effect and the alveolar pressure remains the same at each level blood flow increases steadily from the top to the bottom of zone 2.

In zone 3 as venous pressure (below heart level) is increased and now exceeds alveolar pressure the collapsible tubes are held open and the ordinary arteriovenous pressure difference determines flow. However, the flow increases steadily as we move down the lung because the alveolar pressure remains the same at each successive level, whereas the transmural pressure steadily increases and distends the vessels, decreasing their flow resistance down the zone.

Distribution of ventilation; alveolar ventilation–perfusion ratio

Unlike the regional blood flow through the various parts of the lung the change in ventilation from the base to the apex of the lung is not marked. The ventilation per unit volume does increase linearly from apex to base, but the rate of change of ventilation measured is only about one-third of that of blood flow.

If we assume a pulmonary blood flow of 6 l min^{-1} and an alveolar ventilation of 5.1 l min^{-1} then the *alveolar ventilation–perfusion ratio* is

$$\frac{0.51}{0.6} = 0.85 \text{ overall.}$$

Although the overall figure is 0.85 there are considerable differences in the regional ventilation–perfusion ratio values. At the base of the lung the ratio is 0.63. At the apex of the lung it is 3.3 [FIG. III.19]. In other words the alveoli at the base are slightly over-perfused in relation to their ventilation, whereas the apical alveoli are grossly under-perfused in relation to their ventilation.

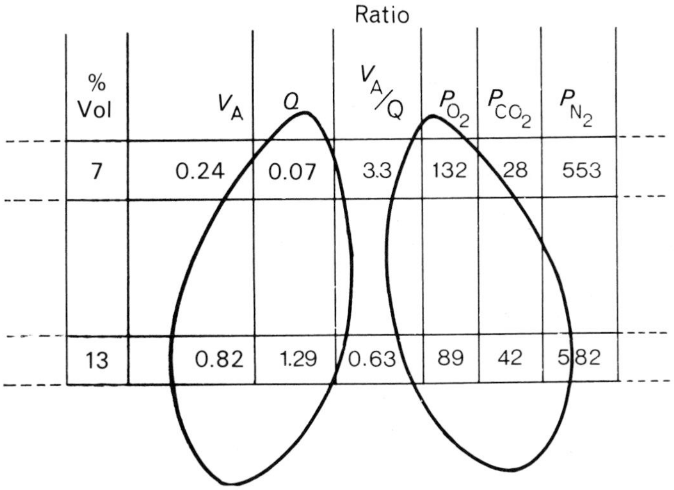
Ratio
% Vol
V_A
Q
V_A/Q
P_{O_2}
P_{CO_2}
P_{N_2}
7
0.24
0.07
3.3
132
28
553
13
0.82
1.29
0.63
89
42
582

FIG. III.19. Ventilation (V_A), blood flow (Q), ventilation perfusion ratio ($V_{A/Q}$), and gas tensions at different zones of the lungs. (After West, J. B. (1965). *Ventilation/blood flow and gas exchange.* Blackwell, Oxford.)

The oxygen–carbon dioxide diagram [FIG. III.20] is useful in allowing a depiction of the pO_2 and pCO_2 values of gas and blood samples in their relation to one another. pO_2 in the inspired air is 150 mm Hg and pCO_2 is zero. If a lung unit were ventilated but not perfused such would be its alveolar gas composition (Point I). A 'Normal' lung unit, adequately ventilated and perfused would have a normal alveolar composition of pO_2 100 mm Hg and pCO_2 of 40 mm Hg. The capillary blood leaving this unit would have the same composition (point A).

Mixed venous blood has a pO_2 of approximately 40 mm Hg and an alveolar pCO_2 of 46 mm Hg. If an alveolar unit were not ventilated but were supplied by such mixed venous blood its alveolar gas composition would be the same (point $\bar{V}$).

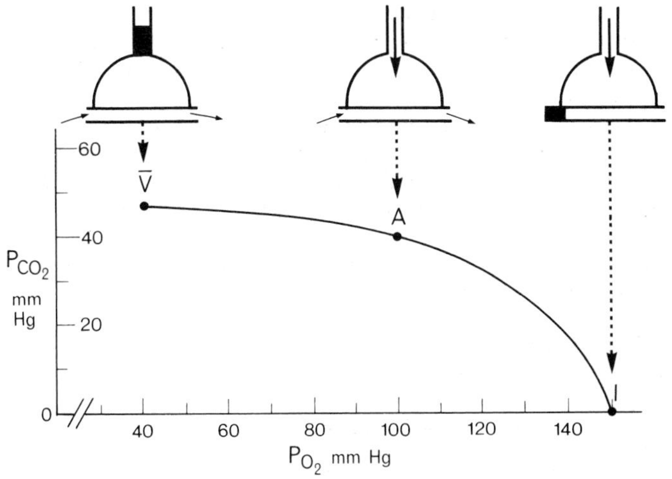
P_{CO_2} mm Hg
60
40
20
0
$\bar{V}$
A
I
40
60
80
100
120
140
P_{O_2} mm Hg

FIG. III.20. The oxygen–carbon dioxide diagram. The three figures above represent: perfused non-ventilated lung, normally perfused and ventilated lung, and ventilated non-perfused lung. (After West, J. B. (1965). *Ventilation/blood flow and gas exchange.* Blackwell, Oxford.)

The V – I line has a shape which is determined by the nature of the blood dissociation curves for oxygen and CO_2. The ends of the line are fixed by the composition of inspired air (I) and mixed venous blood (V) respectively.

From what has been said the V – I line (or ventilation–perfusion line) defines the gaseous tensions of either alveolar air or capillary blood at various levels of the lung. At the apices $\dot{V}_{A/Q}$ is high. Hence pO_2 is high (130–135 mm Hg) and pCO_2 is low (pCO_2 = 25–30 mm Hg). At the base of the lungs pCO_2 is 42 mm Hg and pO_2 is about 90 mm Hg.

The CO_2/O_2 exchange ratio can be superimposed on this diagram. Thus, if the CO_2 produced is equal to the oxygen used (the exchange ratio R = 1) the gas exchange R line has a slope of 45 degrees, if R is less than 1 the slope is flatter.

Similarly, exchange (R) lines for capillary blood can be superimposed on the CO_2/O_2 diagram, but with more difficulty. The definition of R = 1 is that equal *volumes* of oxygen and carbon dioxide are exchanged. However, in the case of blood, owing to the shapes of the O_2 dissociation and CO_2 dissociation curves, the exchange line for blood is not linear. Nevertheless, the lines can be drawn from a knowledge of the dissociation curves and can be depicted as fanning out from the point V which itself depicts the 'mixed venous blood' point.

Any gas R line must meet its corresponding blood R line on the ventilation–perfusion line.

REFERENCES

WEST, J. B. (1970). *Ventilation/blood flow and gas exchange*, 2nd edn. Blackwell, Oxford.
—— (1975). *Anesth. Analg.* **54,** 409.

Neural control of respiration

THE 'RESPIRATORY CENTRE'

The so-called 'respiratory centre' comprises interconnected neurons situated in the medulla, pons, and in the lower part of the midbrain. If the brainstem is transected below the calamus scriptorius breathing ceases.

Rhythmic respiration persists even when the brainstem is transected at the level of the upper part of the medulla although its pattern is less smooth than that of animals with an intact neuraxis. Such an experimental finding led earlier workers to define 'inspiratory' centres and 'expiratory' centres in the medulla by demonstrating effects of *electrical stimulation* delivered to different parts of the medulla. Such investigations showed that in general stimuli delivered to the more medial parts of the medulla evoked prolonged inspiratory activity (during the maintenance of the stimulation) when the electrode tip was inserted into the more ventral region. These responses could be evoked between the levels of the rostral medulla and the calamus scriptorius. Conversely, stimulation of medullary sites lying more dorsally provoked expiration. FIGURES III.21 and III.22 show the localization of these 'centres'—sometimes termed the 'medullary respiratory centre'.

Hering (1868) and Breuer (1868) showed that breathing became deeper and slower when the vagi were cut and conversely that inflation of the lungs inhibited respiration. This is not the case in animals whose brainstem has been transected at the level of the rostral medulla. Thus the influence of the vagal pulmonary stretch receptors [p. 169] must be exerted at a higher level of the neuraxis.

Midpontine transection yields a preparation in which rhythmic respiration more normally resembles that of an anaesthetized animal *providing that the vagi are intact*. Vagotomy alters this pattern and the 'midpontine' preparation develops *apneustic* breathing which, though rhythmic, shows a preponderant proportion of the respiratory cycle to be in the inspiratory phase, expiration only briefly intervening in each cycle. Coupled with the fact that stimulation of the lateral-most part of the reticular formation of the middle and lower pons induced marked inspiratory activity even in intact animals the evidence suggested the presence of an 'apneustic centre'. This apneustic centre was supposed to be inhibited by vagal afferents from the pulmonary stretch receptors aroused by inspiration, so that its stimulatory effect on the medullary 'inspiratory' neurons was switched off allowing expiration to supervene.

Lastly, it was shown that even in a vagotomized animal, midcollicular transection did not prevent rhythmic breathing of a reasonably normal pattern and did not provoke apneusis. However, when a further transection was made through the midpons, apneusis developed. It was concluded that a 'pneumotaxic centre' existed in the upper pons and lower mesencephalon which served normally to

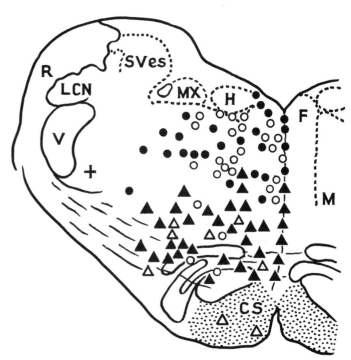

FIG. III.22. Distribution of inspiratory and expiratory centres as defined by stimulation experiments. Section through medulla at level B–B, FIGURE III.21. ▲, Inspiratory centre, ○ ●, Expiratory centre. CS, Corticospinal (pyramidal) tract; F. Medial longitudinal fasciculus (bundle); H, Hypoglossal nucleus; LCN, Lateral cuneate nucleus; MX, Motor nucleus of vagus; M, Medial lemniscus; R, Restiform body; SVes, Vestibular nucleus; V, Spinal tract of V. (Pitts (1941). *Am. J. Physiol.* **134**, 192.)

check the activity of the apneustic centre even when the vagal pulmonary afferents were severed.

FIGURE III.23 displays this idealized system. It was supposed that the inspiratory neurons were activated directly by the arterial CO_2 tension and discharged over spinal pathways to cervical 3, 4, 5, motor neurons which form the phrenic nerves to the diaphragm and to the 12 thoracic segments innervating the external intercostal muscles. As a result of inspiration, pulmonary afferent vagal traffic increased and inhibited the apneustic centre removing its drive on the medullary inspiratory neurons; expiration was normally believed to occur passively. The pneumotaxic centre was supposed to relay cortical messages which could promote voluntary expiration or inspiration.

The chemical stimulus provided by the arterial CO_2 tension was considered to influence the inspiratory and expiratory 'centres' themselves. This was entirely disproved when Loeschcke, Mitchell, and others proved that central chemical effects were due to an influence on neurons situated near the surface of the *ventral* medulla [p. 196]—geographically remote from the classical 'inspiratory' centre.

Modern studies have shown that the discharge of neurons in the vicinity of the tractus solitarius occurs immediately before (and during) that of phrenic motoneurons. Von Baumgarten and Kanzow (1958) designated these R_α and R_β neurons: each discharged prior to and during phrenic activity but whereas R_α neurons were silenced by lung inflation, R_β neurons increased their activity during this manoeuvre. These neurons are situated just rostral to the obex and lie mainly in the ventrolateral part of the tractus solitarius complex. The axons of these inspiratory units pass by bulbospinal pathways to the cervical and thoracic motor neurons.

In such anaesthetized animals it is possible to distinguish a group of neurons placed more ventrally in the medulla in the nucleus ambiguus and the nucleus retroambigualis. Some of these discharge in expiration—others during inspiration. However,

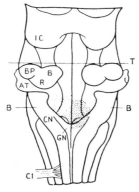

FIG. III.21. Identification of 'inspiratory' and 'expiratory' centres. Dorsal view of brainstem (cat) with cerebellum removed. The respiratory centres are projected on the floor of the fourth ventricle. To avoid overlapping, the *expiratory* centre is shown only on the *left*, and the *inspiratory* centre only on the right.

FIGURE III.22 is a cross-section at the level shown by the lines B–B. Section at level T, combined with double vagotomy, produces apneusis. The pneumotaxic centre thus lies in the pons above level T.

IC, inferior colliculus. BP, brachium pontis (middle peduncle). B, brachium conjunctivum (superior peduncle). C1, first cervical root. R, restiform body (inferior peduncle). AT, acoustic tubercle. CN, cuneate nucleus. GN, gracile nucleus. (Pitts *et al.* (1939). *Am. J. Physiol.* **126**, 673.)

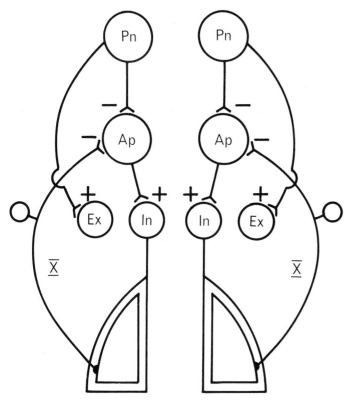

FIG. III.23. Diagram of the organization of the respiratory centres.

Pn = Pneumotaxic centre
Ap = Apneustic centre
In = Inspiratory centre
Ex = Expiratory centre
X = Vagus afferent from the lung parenchyma

All centres are bilaterally represented and there are generous interconnections across the midline which are omitted in this diagram only to preserve pictorial clarity.

The apneustic centre (ordinarily under some degree of inhibition exercised by the pneumotaxic centre) nevertheless excites the inspiratory centre which in turn excites the spinal motor neurons promoting diaphragmatic and chest movement. Expansion of the lungs excites vagal stretch receptors which reflexly inhibit the apneustic centre discharge.

The pneumotaxic centre can cause excitation of the respiratory centre in hyperpnoea.

studies made in paralysed and artificially ventilated cats show no concentration or clusters of inspiratory or expiratory neurons (see Karczewski (1974) for references).

'Respiratory' neurons are numerous in the pons and can be located by microelectrodes in the lateral upper pons (nucleus parabrachialis medialis (NPBM), and locus caeruleus) in the middle pons (nucleus pontis caudalis) and in the lower pons in the rostral part of the nucleus gigantocellularis. The densest concentration of neurons with respiratory activity is in the NPBM and these are active even in cats with spinal transection at C_7 which are paralysed and artificially ventilated. Nowadays NPBM is regarded as the equivalent of the 'pneumotaxic centre'. Bilateral destruction of NPBM, followed some months later by bilateral vagotomy produces apneustic breathing, but only during anaesthesia. When the animal regains consciousness apneusis is replaced by normal rhythmic breathing. Subsequent anaesthesia once more yields apneusis which reverts to the normal pattern when the anesthesia wears off.

Seemingly, apneusis results from an abnormal situation of the bulbopontine complex and cannot be identified simply as due to an unopposed activity of an 'apneustic centre' which, when the central neuraxis is deprived of inputs from the 'pneumotaxic centre' (and higher centres) and vagal afferent input exerts the dominant influence on respiration. The fact that the upper brainstem can com-

pensate for the loss of the pneumotaxic centre and vagal afferent input in the awake animal indicates that these 'higher centres' play a more important role than that which they have hitherto been accorded. In short, the term 'respiratory centre' should be discarded—one might refer to a bulbopontine respiratory neuronal complex, laborious though the term may be, even then realizing that the basic activity of this complex is profoundly modified by the influence of suprapontine mechanisms in the conscious animal—an influence which is modified or suppressed by anaesthesia.

Recent results from several laboratories suggest that the mechanism for terminating inspiration and switching to expiration involves a slowly increasing centrally generated inspiratory activity (CIA) which, combined with vagal pulmonary stretch receptor fibre discharge, suddenly activates a set of 'switch neurons'. These 'switch neurons' terminate inspiration and maintain inhibition of inspiratory neuronal discharge during the subsequent expiratory phase. It is possible that these neurons are the R_β neurons described by von Baumgarten and Kanzow. After vagotomy, CIA alone attains the threshold for the 'inspiratory off switch' (IOS); inspiratory duration is longer and CIA reaches a higher peak level. An interaction between CIA and vagal input occurs in a group of neurons in the ventrolateral nucleus of the tractus solitarius (such as described by von Baumgarten and Kanzow 1958). Von Euler and Trippenbach (1976) suggest that the CIA input to these solitary tract neurons is mediated by a recurrent feedback from those inspiratory neurons in the tractus solitarius which project to the spinal segmental motor neurons.

The IOS mechanism receives excitatory contributions from three sources—CIA, vagal stretch receptors, and a tonic input from NPBM. In addition to these excitatory inputs, IOS receives an inhibitory input related to the level of arterial CO_2 tension. The effect of CO_2 on the threshold is unaffected by lesions in the inspiratory inhibiting neurons of NPBM. Moreover, such NPBM lesions do not influence the initial rate of rise of CIA or its dependence on CO_2. Von Euler and Trippenbach conclude that although NPBM does not form the basic central pattern *generator* for respiratory rhythmicity it exerts an important modulating influence on the IOS mechanism during inspiration.

REFERENCES

BAUMGARTEN, R. VON and KANZOW, E. (1958). *Arch. ital. Biol.* **96**, 361.
BREUER, E. (1868). *S. Ber. Akad. Wiss. Wien* **58**, 909.
EULER, C. VON and TRIPPENBACH, T. (1976a). *Acta physiol. scand.* **96**, 338–50.
—— (1976b). *Acta physiol. scand.* **97**, 175.
HERING, E. (1868). *S. Ber. Akad. Wiss. Wien* **57**, 672.
KARCZEWSKI, W. A. (1974). Chapter 7. In *Respiratory physiology* (ed. J. G. Widdicombe) MTP International Review of Science, pp. 197–219. Butterworths, London.

Role of pulmonary stretch receptors

The *pulmonary stretch receptors*, more precisely described by Adrian (1933) as 'slowly adapting pulmonary receptors stimulated by lung inflation', are present in the smooth muscle of the bronchial wall down to the small cartilaginous bronchi. The receptors are localized chiefly at the points of bronchial branching, where the smooth muscle is thickest. Whether they also exist in bronchiolar smooth muscle is less certain. It has been claimed that pulmonary stretch receptors lie in the visceral pleura, but there are few receptors in this tissue and removing the visceral pleura does not abolish pulmonary stretch receptor activity. However, there are similar slowly adapting stretch receptors in the tracheal and bronchial segments of the extrapulmonary airways. Histologically, the endings appear as terminal arborizations similar to the baroreceptor endings in the carotid sinus and aortic arch, and under the electron microscope the terminal axons are seen to be crowded with mitochondria and to contain glycogen and a few vesicles of widely ranging diameters. The afferent fibres run in the vagus, are myelinated and of large diameter with conduction velocities in the range 14 to 59 m s⁻¹. Fibres from pulmonary stretch receptors are the most common myelinated afferents from the lung found in the vagus.

Electrophysiological studies of the action potential discharge in vagal afferent fibres have shown that the lung volume threshold for activity of the pulmonary stretch receptors is low, usually within the tidal volume range of eupnoeic breathing. The threshold for some receptors may be below the functional residual capacity so that they discharge during the expiratory pause as well as during inflation and are only inhibited by deflation of the lungs. Other endings, particularly those in the trachea and extrapulmonary bronchi, are stimulated by deflation as well as by inflation. During spontaneous eupnoeic breathing vagal impulse activity from pulmonary stretch receptors increases with the onset of inspiration, dying down as expiration begins [Fig. III.24]. In the artificially ventilated cat maximum impulse frequencies of 100 to 300 s⁻¹ may be attained at inflation pressures up to 3 kPa (30 cm H₂O). If the inflation is maintained there is a slow adaptation of the receptors [Fig. III.25(a)]. The discharge of the receptors is altered by changes in lung compliance. A large inflation of the lung, which increases compliance by opening up collapsed alveoli, raises the threshold, and decreases the discharge of the receptors. Conversely the receptors are sensitized in pulmonary congestion, oedema, and bronchoconstriction, conditions in which a decrease in compliance leads to a greater mechanical pull on the airways. The sensitivity of the receptors may also be influenced by the tone in the smooth muscle of the airways. Injection of histamine, which contracts airway smooth muscle, usually increases pulmonary stretch receptor discharge but this may be secondary to the increased distension of the airways due to the lung collapse which this drug also produces. It is possible that the discharge of these receptors may be reduced by contraction of the smooth muscle in the absence of this indirect mechanical effect.

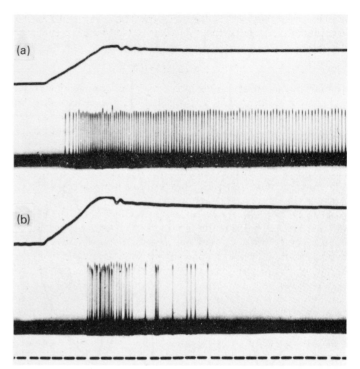

FIG. III.25. Responses to inflation of the lungs from two kinds of afferent fibre in the vagus of the cat. The upper record (a) is from a pulmonary stretch receptor which adapts slowly to maintained inflation; the lower record (b) is from a lung irritant receptor which adapts rapidly. Upper traces, intratracheal pressure; lower traces, action potentials; time in 0.1 s. (From Knowlton, G. C. and Larrabee, M. G. (1946). *Am. J. Physiol.* **151,** 547.)

Reflex effects

Artificial inflation of the lungs of anaesthetized animals inhibits the inspiratory muscles. Both the diaphragm and the inspiratory intercostal muscles are affected. This reflex, the *Hering–Breuer inflation reflex*, described by Breuer and Hering in 1868, has its afferent pathway in the vagus and is abolished by vagotomy, following which breathing usually becomes slower and deeper. The pulmonary stretch receptors appear to be the receptors responsible for the reflex, since the reflex is blocked by cooling the vagus to a temperature which blocks conduction in pulmonary stretch afferents but which leaves intact conduction in non-myelinated afferents and their reflex responses. The Hering–Breuer inflation reflex is similarly abolished by selective blockade of the myelinated fibres by the application of direct current to the vagus. Conversely, respiration can be inhibited by stimulation of the central end of the cut vagus using stimulus intensities which excite only large myelinated fibres.

In man, pulmonary stretch receptors have been identified histologically in the lungs and pulmonary stretch fibre activity has been seen during eupnoea in recordings from single vagal fibres. However, the response to lung inflation in eupnoea is weak in anaesthetized human subjects, while in healthy conscious subjects Guz and his colleagues (1970) have shown that local anaesthetization of both vagi does not alter the pattern of breathing, minute volume, or P_{CO_2}. In unanaesthetized laboratory animals, the reflex, though still present, is similarly weaker than in the anaesthetized state.

In addition to inhibiting inspiratory muscle activity, stimulation of pulmonary stretch receptors by inflation of the lungs reflexly relaxes tracheobronchial smooth muscle. Possibly this may modulate airway calibre during individual breaths since the discharge in efferent vagal fibres to airway smooth muscle usually ceases late in inspiration, when the pulmonary stretch receptors are most strongly stimulated. The motoneurons to the adductor (constrictor)

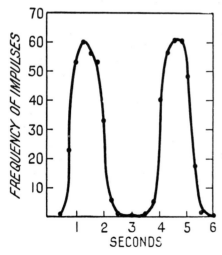

FIG. III.24. Vagal action potentials during two normal respiratory cycles. Action potentials are recorded in a single fibre of the vagus nerve in the decerebrate cat. The frequency of the impulses reaches a maximum of 60 s⁻¹ at the height of inspiration and falls to a minimum at the end of expiration. (Adrian (1933). *J. Physiol., Lond.*)

muscles of the larynx are also inhibited. Thus in general, during inspiration, the airways from the larynx to the bronchi are reflexly dilated.

Physiological role of the inflation reflex

The main action of the reflex is to inhibit the inspiratory muscles and thereby to limit tidal volume. If the reflex is abolished by vagotomy in experimental animals then breathing becomes slower and deeper, though minute volume and alveolar ventilation are not greatly altered, suggesting that the reflex has little effect on the relationship between gas tensions and respiratory minute volume. However, as already mentioned, in unanaesthetized healthy man bilateral vagal blockade does not alter the pattern of breathing or end-tidal P_{CO_2}.

A second, striking effect of vagotomy is seen when respiration is stimulated by asphyxia or inhalation of CO_2. After vagotomy the increased ventilation in response to these stimuli results almost entirely from an increase in tidal volume without any change in the frequency of breathing. This effect occurs both in anaesthetized and unanaesthetized animals and in unanaesthetized man.

Euler and his colleagues have studied the way in which the inflation reflex adjusts the pattern of breathing and have represented their results by a diagram in which tidal volume is plotted against the duration of inspiration [Fig. III.26]. During stimulation of breathing by hypercapnia the increase in tidal volume was accompanied by a decrease in the duration of inspiration. (This inverse relationship between tidal volume and the duration of inspiration is complementary to the direct relationship between tidal volume and respiratory frequency, both of which increase during hypercapnic hyperpnoea.) Presumably the duration of inspiration is reduced because the more vigorous inspiratory drive leads to a more rapid build-up of tidal volume, so that pulmonary stretch receptor discharge more quickly attains an intensity sufficient to cut short inspiration. In addition to the increase in the frequency of breathing as a result of shortening the duration of inspiration, Clark and Euler (1972) found that in hypercapnic hyperpnoea there is also a direct relationship between the duration of inspiration and the duration of expiration. Thus the rise in respiratory rate is due to a shortening of expiratory as well as inspiratory time. That the lung inflation reflex is responsible for the hyperbolic relationship between tidal volume and inspiratory duration was demonstrated by the abolition of this relationship by vagotomy, after which hypercapnia increased tidal volume but not the frequency of breathing [Fig. III.26—cat vagotomy].

These findings may help to explain why the inflation reflex appears to be absent in healthy eupnoeic man. If, during quiet breathing the rhythm of respiration as determined by the intrinsic activity of the medullary and pneumotaxic centres of the brainstem respiratory complex is such as to cut short inspiration before the pulmonary stretch discharge reaches a level adequate to stop breathing, then vagal blockade will not alter the pattern of breathing and the inflation reflex will appear to be absent. Initially, as the depth of breathing increases there will be no reduction in the duration of inspiration. As is illustrated by Figure III.26, it is only when tidal volumes increase to values at which pulmonary stretch receptor activity is sufficient to inhibit further inspiration that the inflation reflex becomes manifest. This threshold level of tidal volume seems to be about 1.0 to 1.5 l in man.

The physiological importance of the inflation reflex thus appears to be that, in situations in which an increased central respiratory drive increases the rate of build-up of tidal volume, it produces an increase in respiratory frequency in addition to the increase in tidal volume, thereby magnifying the increase in minute ventilation.

There may be a second physiological advantage derived from the inflation reflex. For any given ventilation there is an optimal combination of respiratory frequency and tidal volume which minimizes the work of the respiratory muscles. This optimal pattern depends on the mechanical conditions of the lungs, in particular lung compliance and resistance. Since pulmonary stretch receptors may 'sense' mechanical conditions in the lungs they should contribute to the establishment of this optimal pattern.

Since pulmonary stretch receptors can influence airway smooth muscle tone, it is also possible, though as yet uncertain, that they may be important in reflexly adjusting airway calibre to an optimal value. This optimal value would represent a balance between the advantages of a dilated airway with a smaller resistance to airflow, and a constricted airway with a smaller dead-space.

In summary, therefore, whilst the Hering–Breuer vagal inhibitory inspiratory reflex may exert a considerable influence in determining the pattern of breathing when respiration is increased, it is of minor importance during eupnoea in adult conscious man. Attention has correspondingly been directed in the past two decades to the role of proprioceptors situated in the thoracic parietes.

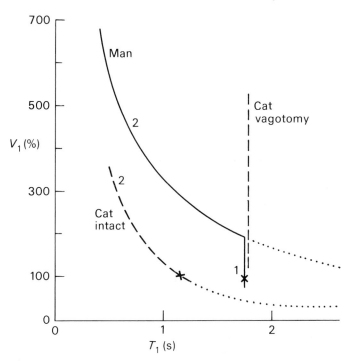

FIG. III.26. The relationship between volume of inspiration (ordinate, V_I, as a percentage of eupnoeic tidal volume) and time of inspiration (abscissa T_I). The lines labelled (2) show the hyperbolic relationship between the two variables when the Hering–Breuer reflex limits inspiratory duration; larger inspiratory volumes correspond to shorter inspiratory times and therefore more rapid breathing. The vertical lines (1) correspond to constant-frequency breathing when the time of inspiration is limited not by the Hering-Breuer reflex but by the intrinsic activity of the respiratory complex of the brainstem. The crosses show eupnoeic points. For the cat (interrupted lines) only vagotomy produces constant time of inspiration and frequency. For man (continuous line) the eupnoeic point is on the vertical line and the Hering–Breuer reflex has not reached its central threshold. (From Widdicombe, J. G. (1974). In *Recent advances in physiology*, 9th edn (ed. R. J. Linden). Churchill Livingston, Edinburgh. Modified from Clark and Euler (1972). *J. Physiol., Lond.* **222**, 267.)

REFERENCES

ADRIAN, E. D. (1933). *J. Physiol., Lond.* **79**, 332.
CLARK, F. J. and EULER, C. VON (1972). *J. Physiol., Lond.* **222**, 267.
GUZ, A., NOBLE, M. I. M., EISELE, J. H., and TRENCHARD, D. (1970). In *Breathing*. Hering–Breuer Centenary Symposium (ed. R. Porter) pp. 17–40. Churchill, London.

Proprioceptive afferents of respiratory muscles and their control

The intercostal muscles contain numerous muscle spindles as well as tendon end-organs and Pacinian corpuscles (Barker 1962). The diaphragm on the other hand is only sparsely supplied with spindles which are indeed less numerous here than are tendon end-organs (Corda *et al.* 1965).

The observation that section of the cervical and thoracic dorsal spinal roots decreases respiratory movements in animals was made in patients by Nathan and Sears (1960). Patients in whom these sensory roots had been severed to alleviate the pain caused by inoperable tumours, showed less intercostal and/or diaphragmatic activity depending on whether the thoracic or appropriate cervical roots were cut. These findings can be interpreted satisfactorily by considering the activity of muscle spindles during the phases of the respiratory cycle.

The role of muscle spindles in influencing reflexly the tone of skeletal muscle is considered at length on pp. 297 *et seq.* and the account given here presents only a summary. The spindle lies parallel to the intercostal muscle fibres. Its central part contains annulospiral nerve endings of IA afferents which ascend to gain access via the appropriate dorsal spinal root to the spinal cord. Here these afferents establish monosynaptic connection with the α-motoneuron which supplies and excites the intercostal muscle wherein the spindle lies. Stretch of the intercostal muscle elongates the spindles and causes IA afferent discharge which in turn reflexly excites the α-motoneuron thereby causing contraction of the intercostal which thus 'resists' the stretch. Within the spindle, however, there are also muscle fibres designated as intrafusal muscle fibres to differentiate them from the ordinary skeletal muscle elements which in turn may be called extrafusal fibres. The intrafusal fibres of the spindle themselves are innervated by motor fibres of small calibre—the so-called gamma motoneurons [p. 297] or better called fusimotor fibres. Stimulation of these fusimotor neurons causes contraction of the intrafusal fibres and produces a distortion of the central annulospiral endings of the spindle. This arouses an increased afferent discharge of the IA sensory fibres and this in turn reflexly excites the α-motoneuron discharge to the extrafusal fibres of the intercostal muscle, which as a result contract. This sequence of events caused by excitation of the gamma motoneurons produces then a reflex contraction of the intercostal muscles which may be described as 'gamma-led'.

When the chest is inflated passively by positive pressure the intercostal muscles are stretched and electroneurographic recordings of the IA afferents show an increased discharge. If the appropriate dorsal root is preserved in the main and the impulse activity of a fine twig containing only a few IA afferents is recorded, the impulse activity of these increases slightly before the reflex response (to the intact IA afferent fibre discharge) occurs; this reflex response is a contraction of the homologous intercostal muscle, which can be recorded electromyographically. When the whole dorsal root is cut the electromyographic response of that intercostal muscle to inflation of the chest is abolished.

Critchlow and von Euler (1963), who obtained these findings, further showed that the intercostal spindle afferent discharge during natural respiration was *increased* in rate during inspiration [FIG. III.27]. As inspiration 'takes the load off' the spindle (for the extrafusal muscle is *contracting*) the increased afferent discharge in this phase must be caused by an increased fusimotor activity. Additional information was provided by selective blockade of the motoneurons supplying the intercostal muscle. It is possible to block the small fusimotor neurons in the motor nerve by the local application of lignocaine [FIG. III.27] leaving the larger α-

motoneurons still, in the main, capable of conduction. Such blockade led to a completely different picture of afferent discharge; the IA afferents now showed an increased activity during expiration (due to stretch of the intercostal muscle) and a decreased activity during inspiration (owing to contraction of the muscle taking the load off the spindle).

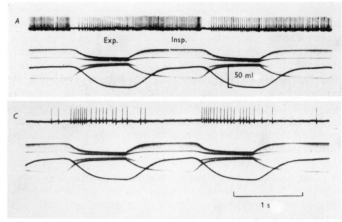

FIG. III.27. Afferent discharge from an inspiratory muscle spindle.

A: control. *C*: 3.5 min after lignocaine to the intercostal nerve (which paralyses the efferent γ fibres). In each record, from above downwards: electroneurogram, changes in intercostal width (the amplitude modulated high-frequency trace) and tidal volume. Note 'reversal' of the discharge pattern caused by lignocaine. (Critchlow, V. and von Euler, C. (1963). *J. Physiol., Lond.* **168**, 820.)

In harmony with this explanation, recordings of the activity in the motoneurons of the external intercostal nerve normally showed evidence of both fusimotor activity and α-motoneuron activity occurring with each inspiration. Additionally, however, some fusimotor fibres discharged tonically. Lignocaine application blocked the fusimotor discharge.

Motoneuron preparations of *internal* intercostal nerves (supplying the internal intercostal muscles) showed rhythmic firing of gamma motoneurons in the expiratory phase usually unaccompanied by α-motoneuron discharge [FIG. III.28]—in keeping with the knowledge that in eupnoea, expiration is achieved passively and is unaccompanied by contraction of the internal intercostal muscles.

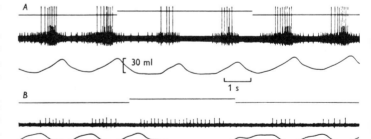

FIG. III.28. The reflex effects of weak tetanic stimulation of the left vagus nerve on (*A*) the inspiratory γ and α afferent activity recorded from a branch to the external intercostal muscle, and (*B*) the expiratory γ efferent activity of a branch to the internal intercostal muscle. Stimulation at a frequency of 47 s⁻¹ during the periods marked. (Eklund, G., von Euler, C., and Rutkowski, S. (1964). *J. Physiol., Lond.* **171**, 139.)

Eklund, von Euler, and Rutkowski (1964) found that artificial over-ventilation could completely abolish inspiratory α-activity though leaving gamma discharge with an inherent respiratory rhythm. Hypoxia caused a rapid exacerbation of both alpha and gamma type discharge [FIG. III.29], as did hypercapnia, and neither of these effects was abolished by paralysing the respiratory muscles

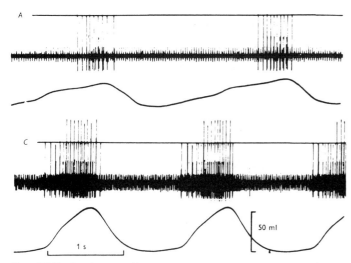

FIG. III.29. Inspiratory efferent activity in a nerve branch to external intercostal muscle to show the effect of hypoxia. *A*, spontaneous breathing air. *C*, after 3 min of breathing 8 per cent O_2 in N_2. The small spikes verified with lignocaine as belonging to γ fibres. (Eklund *et al.* (1964). *J. Physiol., Lond.* **171**, 139.)

with gallamine and administering the gas mixtures by artificial ventilation.

Vagal section in the spontaneously breathing animal, though slowing the respiratory rate, did not affect the relation between inspiratory alpha and gamma activity. Weak vagal stimulation diminished both α- and γ-activity in the external intercostal nerve but slightly increased γ-activity in the internal intercostal nerve [FIG. III.28].

Succinylcholine is a profound stimulant of muscle spindles and, when injected intravenously, considerably exacerbated not only alpha discharge (as might be expected) but often γ-activity.

Passive stretching of the intercostal muscle, by pulling on the ribs, notably increased the gamma discharge and this, too, indicates that proprioceptive effects on the gamma discharge to the respiratory muscle stretched do exist in the intercostal muscle as opposed to the situation in the limb musculature.

From these findings von Euler and his colleagues have evolved the following concepts:

Inspiratory and expiratory alpha and gamma motoneurons are driven from the bulbar respiratory integrating mechanisms in a reciprocal manner. During eupnoea, expiration is passive and the majority of the 'expiratory' alpha motoneurons remain silent. The descending rhythmic signals to the spinal alpha and gamma neurons are related to the demand for a required tidal volume, which is of course achieved by an appropriate change of length of the inspiratory muscles. Intrafusal and extrafusal muscles contract together and if the extrafusal shortening equals that of the intrafusal fibres the spindle afferent discharge will not alter. However, the extrafusal shortening will depend on the loading of the muscle and will be affected by imposing a resistance on the respiratory mechanism. A greater resistance prevents the inspiratory extrafusal muscles from shortening so much, but will not affect the contraction of the intrafusal fibres. As a consequence the afferent discharge from the spindle will increase and this in turn will reflexly excite the alpha motoneurons more powerfully with the result that the 'misalignment' of length change between intra- and extrafusal muscles is reduced and the requisite tidal volume will be again achieved. Corda *et al.* (1965) showed that the IA afferent discharge from external intercostal spindles did indeed increase when the inspiratory load was heightened by occlusion of the trachea. Similarly, Campbell *et al.* (1961) showed that any increase of the respiratory load exacerbated the force of contraction of the respiratory mus-

cles. Euler and his colleagues found that this response was mediated by dorsal roots and was unaffected by vagotomy [FIG. III.30].

It would seem that during eupnoea the 'gamma drive' to the system is primary, for the intercostal afferents rhythmically increase their discharge during contraction of the inspiratory muscles. This spindle discharge exerts a drive on the alpha neuron system to achieve the requisite change of length of the inspiratory muscles.

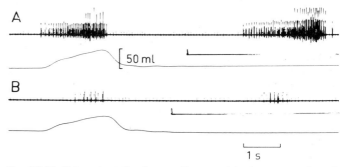

FIG. III.30. Cat: vagi cut. Inspiratory efferent activity from a nerve branch to the external intercostal muscle to show the effect of tracheal occlusion (mark). Lower tracings represent tidal volume. *A*: dorsal roots intact. *B*: After severing dorsal roots T4–8 ipsilaterally and T5–8 contralaterally. (Corda, Eklund, and von Euler (1965). *Acta physiol. scand.* **63**, 391.)

To summarize, monitoring of tidal volume in man seems rather to depend upon the 'load detecting reflex' of the thoracic muscle spindles and their spinal connections than on the vagal stretch receptors of the lung parenchyma which are of greater importance in this respect in animals lower in the evolutionary scale. This does not imply that the vagal receptor discharge plays no part in man but only that it is subservient in the importance of its influence to that of the parietal receptor reflexes. The vagal Hering–Breuer type reflex can be demonstrated in amphibia such as the frog—animals in which the mechanism of respiration is entirely different. The frog takes air into the mouth and then closes the mouth and nares and raises the floor of the mouth. This force-pumps air into the paired single-cavity air sacs which are unprotected against such inflation by any bony chest wall. A generous innervation by pulmonary vagal stretch receptors exists and afferent discharge increases in these vagal fibres as inflation proceeds and this inhibits the buccal movements. The reflex obviously serves as a protective mechanism. In the mammal the chest wall moves primarily in respiration and appropriately the monitoring system has its receptor elements situated in the thoracic parietes.

REFERENCES

BARKER, D. (ed.) (1962). *Symposium on muscle receptors.* Hong Kong.
CAMPBELL, E. J. M., DICKENSON, C. J., DINNICK, O. P., and HOWELL, J. B. L. (1961). *Clin. Sci.* **21**, 309.
CORDA, M., EKLUND, G., and EULER, C. VON (1965). *Acta physiol. scand.* **63**, 391.
CRITCHLOW, V. and EULER, C. VON (1963). *J. Physiol., Lond.* **168**, 820.
EKLUND, G., EULER, C. VON and RUTKOWSKI, S. (1964). *J. Physiol., Lond.* **171**, 139.
NATHAN, P. W. and SEARS, T. A. (1960). *J. Neurol. Neurosurg. Psychiat.* **23**, 10.

Role of other afferent nerves

1. Pulmonary vagus

In addition to the pulmonary stretch receptors, there are two other types of receptor end-organ with afferent pathways in the vagus which can reflexly influence ventilation and the pattern of breathing. These are lung-irritant receptors and type-J receptors.

(i)Lung-irritant receptors

Many histological studies have shown that throughout the airways from the trachea to the respiratory bronchioles there are afferent end-organs, the nerve fibres of which ramify under and between the columnar cells of the epithelium. In the trachea and extrapulmonary bronchi, where they respond to inhaled mechanical and chemical irritants and reflexly cause coughing, they have been termed 'cough receptors', but since stimulation of the histologically identical endings situated in the intrapulmonary bronchi and the bronchioles does not cause coughing the receptors in these sites have been called 'lung-irritant receptors'.

The afferent fibres from the lung irritant receptors run in the vagus, are myelinated, and belong to the Aδ group with conduction velocities in the range 4 to 26 m s^{-1}. They do not usually discharge during eupnoea in anaesthetized animals but if the depth of breathing is increased there may be a few impulses with each breath. Mechanical or chemical stimulation leads to a rapidly adapting response consisting of a brief burst of impulses at a high frequency [FIG. III.25(b)]. Five main types of stimulation have been found to excite lung-irritant receptors:

(a) Inhalation of chemical or mechanical irritant gases and aerosols, such as ammonia and ether vapour, cigarette smoke, and 'inert' carbon dust which presumably act directly on the receptor terminals [FIG. III.31].

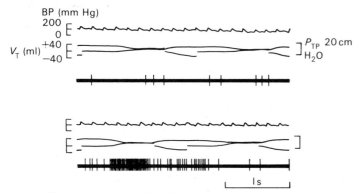

FIG. III.31. Response of a lung irritant receptor to inhalation of carbon dust. Traces from above down: systemic arterial blood pressure (BP), transpulmonary pressure (P_{tp}), tidal volume (V_T) zeroing at points of zero airflow, and action potentials in a single vagal nerve fibre from a lung irritant receptor. Upper record, control showing slow spontaneous discharge; lower record, 20 s after the start of inhalation of dust, showing maximum stimulation of the receptor. Rabbit was anaesthetized, paralysed, and artificially ventilated and vagotomized. (Redrawn from Sellick, H. and Widdicombe, J. G. (1971). *J. appl. Physiol.* **31**, 15.)

(b) Contraction of airway smooth muscle induced by histamine injections or aerosols. This response is secondary to the muscle contraction and not due to a direct action of the drug, since no response is obtained if the histamine is administered after paralysing the smooth muscle with isoprenaline.

(c) Large inflations and deflations of the lungs above the eupnoeic tidal volume or below the functional residual capacity, e.g. collapse of the lung following a pneumothorax. Presumably these volume changes distort the epithelium in which the receptors lie.

(d) Conditions which decrease lung compliance, such as pulmonary congestion due to left heart failure or atelectasis. These conditions lead to increased mechanical pull on the airways.

(e) A miscellaneous group including pulmonary microembolism, anaphylactic reactions, and injections of drugs such as 5-

hydroxytryptamine and phenyl diguanide. These agents may have a mechanical or chemical irritant action, or both.

Thus in general lung irritant receptors respond to any sudden change in the mechanical or chemical environment of the airway epithelium. Their ability to signal such changes is greatly enhanced by their rapid adaptation in response to a maintained change in their environment. In this respect their behaviour is comparable to that of Pacinian corpuscles and the touch receptors of the skin.

Reflex effects. The characteristic reflex response of the breathing to stimulation of the lung-irritant receptors is *hyperpnoea*, both in man and experimental animals. A vagally mediated hyperpnoea occurs in lung deflation, pneumothorax, inhalation of chemical and mechanical irritant gases and aerosols, lung congestion, anaphylaxis, microembolism, and bronchoconstriction due to drugs such as histamine. Though cough receptors, pulmonary stretch receptors, and J-receptors are also excited by many of these stimuli, only the lung-irritant receptors respond consistently to all of them.

The effect of the lung-irritant receptors on breathing may be of particular significance in three circumstances.

(a) Since they are stimulated by lung deflation and by pneumothorax, the lung-irritant receptors must contribute to the vagally mediated increase in minute volume induced by lung deflation or collapse—the *Hering–Breuer deflation reflex*. In animals, lessening of pulmonary stretch receptor discharge, which will augment ventilation by removing a restraint on inspiration, will also contribute to this reflex response. However, in healthy man the pulmonary stretch receptors seem to have no effect on breathing during eupnoea, and thus the hyperventilation following a pneumothorax is due probably only to the activation of lung irritant receptors.

(b) Experimental animals and human subjects can be observed during normal breathing to take occasional spontaneous deep 'augmented' breaths. The response is vagally mediated and is associated with a decreased compliance due to mild progressive collapse of alveoli which are opened up by the manoeuvre. A decrease in lung compliance lowers the threshold and increases the discharge of lung irritant receptors and since their reflex action is to augment inspiration, they are the most likely mechanism for this phenomenon. Activity of lung-irritant receptors is similarly likely to be responsible for the gasping inspirations of newborn babies in whom lung compliance is low until the lungs have been fully expanded, and for the 'paradoxical reflex' described by Head (1889) in which vigorous inflation of the lung reflexly induces a contraction of the diaphragm further augmenting the size of the breath.

(c) In rabbits lung-irritant receptors are stimulated during the hypernoea caused by carbon dioxide or asphyxia, presumably because of the large lung inflations. It is therefore likely that a 'positive-feedback' from the lung irritant receptors contributes to the hyperpnoea, though the reduction in the ventilatory response to hypercapnia after vagotomy or vagal blockade in man is probably due mainly to the abolition of pulmonary stretch receptor discharge.

As well as affecting the breathing, the lung-irritant receptors also cause reflex vagal bronchoconstriction in response to the inhalation of irritant gases and aerosols, microembolism of the lungs, the administration of histamine, and possibly pulmonary congestion, though the physiological advantage of this reflex is not yet clear. Stimulation of lung-irritant receptors also causes reflex laryngeal constriction due to increased activity of laryngeal adductor motoneurones, leading to raised resistance to expiratory airflow.

(ii) J-receptors

Studies with the light microscope have revealed occasional non-myelinated nerve fibres in the lung parenchyma, usually arising from the network of nerve fibres surrounding small arterioles, but the techniques used do not distinguish between motor and afferent fibres or between nerve endings and nerve fibres. However, with the electron microscope, axon bundles can be seen in intimate relation to the endothelial cells of the pulmonary capillaries, and while some of these axons contain vesicles characteristic of cholinergic and noradrenergic autonomic motor fibres, other fibres do not contain vesicles and could be afferent.

The possibility that there might be receptors in the lung parenchyma arose with the finding by Dawes, Mott, and Widdicombe, in 1951, that intravascular injections of certain non-physiological drugs, in particular phenyl diguanide, but also including 5-hydroxytryptamine, nicotine, and capsaicin, caused a vagal reflex characterized by apnoea and rapid shallow breathing, bradycardia, and hypotension. Paintal in 1955 identified the afferent nerves as non-myelinated vagal fibres with conduction velocities mainly below 3 m s⁻¹. Impulses frequencies during stimulation reach 20 to 50 impulses s⁻¹ with average values about 7.5 impulses s⁻¹. The fact that the endings respond within 2.5 s of the injection of phenyl diguanide into the right heart, and within less than 1 s when halothane is insufflated into the lungs [Fig. III.32] is strong evidence that these receptors lie in the lung parenchyma, presumably the alveolar wall. They have therefore been termed juxtapulmonary capillary receptors, or J-receptors.

The receptors are mechanosensitive and can be localized by probing the lungs in open-chest animals. However the weak response to large inflations [Fig. III.32], large deflations or pneumothorax suggest that they play little part in the response to physiological lung volume changes.

The receptors are strongly stimulated in four pathological conditions: (a) pulmonary congestion, caused by occlusion of the aorta; (b) pulmonary oedema, caused by occlusion of the aorta or injection of oedema-producing drugs as alloxan; (c) pulmonary microembolism; and (d) inhalation of strong irritants such as chlorine gas. The first three conditions have in common that they produce an increase in alveolar interstitial fluid, or even overt pulmonary oedema. This led Paintal (1970) to conclude that the receptors are primarily sensitive to the content of interstitial fluid between the capillary endothelium and alveolar epithelium [Fig. III.33]. It is not clear whether irritant gases act directly on the endings or by the production of interstitial oedema. A further significant observation is that the changes in breathing during pneumonia induced experimentally by carragenin or inhalation of steam are mediated chiefly by non-myelinated vagal fibres originating almost certainly from J-receptors.

Reflex effects. Injection of phenyl diguanide causes an expiratory apnoea followed by rapid shallow breathing in the cat, and an inspiratory apnoea followed by rapid breathing in the rabbit. There is reflex hypotension and bradycardia in both species. Though phenyl diguanide can also excite lung-irritant receptors and peripheral chemoreceptors, the main action is on the J-receptors as evidenced by the short latency of the response when the drug is injected into the right ventricle and by its persistence after conduction in myelinated fibres of the vagus has been blocked by cooling. Stimulation of J-receptors may also reflexly cause bronchoconstriction and has recently been shown to produce a powerful reflex contraction of the adductor (constrictor) muscles of the larynx. Stimulation of J-receptors by phenyl diguanide has additionally been found to inhibit the spinal monosynaptic reflex response to excitation of skeletal muscle stretch receptors in the cat hind-limb.

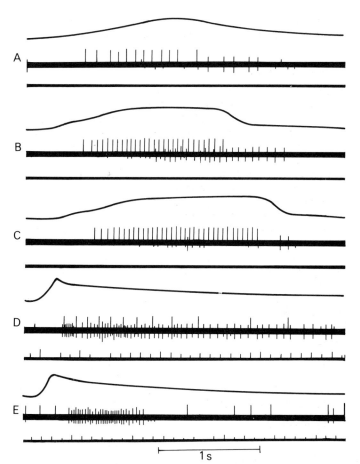

Fig. III.32. Responses of type J receptor (small diphasic spike) to inflation of the lung (A, B, and C) with open chest and insufflation with halothane (60 ml) in D and E. The large monophasic spikes in A, B, C, and D are those of a pulmonary stretch fibre and these show the consistent response to inflation of the lung with 60 ml in A and about 150 ml in B and C in contrast to the variable response in the type J fibre which is excited during the deflation phase in A, during inflation in B (but after a significant delay); there is comparatively little effect in C (again about 150 ml inflation). Insufflation of halothane had no excitatory effect on pulmonary stretch fibre in D (normal circulation) in contrast to the marked excitation of the type J receptor which is excited with a similar latency in E after cutting the great vessels and removing the ventricles. From above downwards in each record, intratracheal pressure, impulses in filament and 0.1 s time marks. Gain of amplifier for intratracheal pressure in A, D, and E is twice that in B and C. (From Paintal, A. S. (1969). *J. Physiol., Lond.* **203**, 511.)

This last effect has led Paintal to suggest that the J-receptors may have a physiological role in severe exercise. He postulates that severe exercise causes an increase in interstitial fluid in the lungs, which excites J-receptors, which reflexly inhibit spinal monosynaptic reflexes and thereby limit the power of contraction of skeletal muscle. This sequence, which he terms the J-reflex, raises interesting possibilities, but remains untested and highly speculative.

Table III.1 summarizes the reflex responses to stimulation of the three types of lung receptor.

2. Other vagal afferents

(a) Afferent fibres from the laryngeal mucous membrane are concerned with the cough reflex which guards the respiratory passages against the entrance of foreign bodies and which assists in their expulsion. In coughing there is a deep inspiration followed by closure of the vocal cords (adductor contraction) and a vigorous expiratory movement towards the end of which the vocal cords

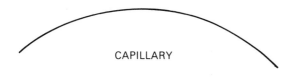

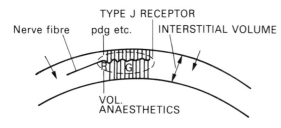

FIG. III.33. Schematic representation of the likely location of the type J receptor lying in the interstitial tissue, perhaps connected to collagen fibrils. The ending is stimulated by an increase in interstitial volume (or pressure) produced when the inflow of fluid into the interstitial tissue (which acts like a sponge) exceeds removal of the fluid. Volatile anaesthetics and other chemical substances (for example phenyl diguanide—pdg) act on the regenerative region, R, of the ending, while the rise in interstitial volume acts on the generator region, G, of the ending. (From Paintal, A. S. (1964). *Pharmac. Rev.* **16**, 341–80.)

abduct thus permitting explosive expiration, during which the linear air velocity may achieve a peak speed of 960 km h⁻¹.

(b) Afferent fibres in the superior laryngeal nerve (together with sensory fibres of the trigeminal nerve) are concerned with the swallowing reflex by exciting the deglutition centre in the medulla. During a swallowing movement, respiration is inhibited in whatever phase of the cycle the swallowing was initiated. This reflex has an obvious protective value against the aspiration of foodstuffs into the respiratory passages.

(c) Chemoreceptor afferents from the aortic body stimulate respiration in circumstances of anoxia or acidaemia.

(d) Aortic baroreceptors and atrial and ventricular cardiac mechanoreceptors exert a feeble and unimportant restraining effect on the respiratory centre in normal circumstances.

(e) Vena-caval receptors situated near the atriocaval junction cause reflex hyperpnoea when stimulated by central venous distension.

TABLE III.1. *Reflex responses to stimulation of types of lung receptor*

Reflex effector	Responses of:		
	Pulmonary stretch receptors	Lung-irritant receptors	J-receptors
Breathing	Inhibition	Hyperpnoea	Apnoea and rapid shallow
Bronchial muscle	Relaxation	Contraction	? Contraction
Laryngeal calibre	Dilation	Constriction	Strong constriction
Heart	? Tachycardia	Unknown	Bradycardia
Vascular resistance	? Increase	Unknown	Decrease
Airway mucus	Unknown	Secretion	Unknown
Spinal reflexes	Unknown	Unknown	Depression
Sensation	? No action	Unpleasant	? Unpleasant

From Widdicombe J. G. (1974). In *Recent advances in physiology*, No. 9, (ed. R. J. Linden). Churchill Livingstone, Edinburgh.

3. Non-vagal afferents

(i) *Glossopharyngeal baroreceptor afferents and chemoreceptors* exert effects respectively similar to those of the vagal baroreceptors

and chemoreceptors. These afferent nerves run in the carotid sinus nerve, a branch of IX.

(ii) *Sensory endings of the Vth cranial nerve and olfactory nerves* reflexly cause sneezing. A sneeze is preceded by a deep inspiration; closure of the glottis then occurs during the early development of the subsequent violent expiratory movement, so that when the glottis is finally opened the expiration is explosive.

(iii) *Nociceptive afferents* reflexly stimulate the respiratory centre—hence the use of slapping in attempting to initiate respiration in a newborn infant suffering from asphyxia livida.

(iv) *Afferent nerves from joints* reflexly stimulate breathing. If the muscles of a limb (which is isolated from the trunk except for nervous connections) are stimulated electrically, the joint movements thereby produced cause hyperpnoea; this is, of course, abolished when the connecting nerves are cut. Such reflexes may contribute to the hyperpnoea of exercise in the intact animal.

REFERENCES

CLARK, F. J. and EULER, C. VON (1972). On the regulation of depth and rate of breathing. *J. Physiol., Lond.* **222**, 267–95.

DAWES, G. S., MOTT, J. C., and WIDDICOMBE, J. G. (1951). *J. Physiol., Lond.* **115**, 258.

GUZ, A., NOBLE, M. I. M., EISELE, J. H., and TRENCHARD, D. (1970). The role of vagal inflation reflexes in man and other animals. In *Breathing; Hering–Breuer Centenary Symposium* (ed. R. Porter) pp. 17–40. Churchill, London.

HEAD, H. (1889). *J. Physiol., Lond.* **10**, 1–70; 279–90.

MILLS, J. E., SELLICK, H., and WIDDICOMBE, J. G. (1970). Epithelial irritant receptors in the lungs. In *Breathing: Hering–Breuer Centenary Symposium* (ed. R. Porter) pp. 77–92. Churchill, London.

PAINTAL, A. S. (1955). *Q. Jl exp. Physiol.* **40**, 89.

—— (1970). The mechanism of excitation of type J receptors, and the J reflex. In *Breathing: Hering–Breuer Centenary Symposium* (ed. R. Porter) pp. 59–71. Churchill, London.

WIDDICOMBE, J. G. (1974). Reflexes from the lungs in the control of breathing. In *Recent advances in physiology*, No. 9 (ed. R. J. Linden). Churchill Livingstone, Edinburgh.

Dyspnoea

Normally, breathing goes on without intruding on consciousness. Dyspnoea literally means difficult breathing—'a consciousness of the necessity for increased respiratory effort'. When the breathing enters consciousness unpleasantly and produces discomfort, it is called dyspnoea. This definition is not entirely satisfactory because it excludes the grave disturbances of breathing which may occur in unconscious subjects (e.g. in diabetic or uraemic coma). Hyperpnoea simply means increased breathing; for a time it does not impinge on consciousness, and so represents a stage preceding the onset of dyspnoea. An ordinary person is not aware of any increase in the breathing until the pulmonary ventilation is doubled. Real discomfort develops when the ventilation is increased four- or fivefold; this level of ventilation is called the *dyspnoea point*. Dyspnoea is not wholly a pathological phenomenon, for it develops in normal subjects during strenuous exertion.

The following factors require further consideration:

1. Breathing reserve (BR)

The maximal ventilation volume (MVV) is the maximal voluntary pulmonary ventilation in litres per minute, determined during a 15-second period. The pulmonary ventilation in litres per minute at rest or under any other specified conditions is designated PV. The breathing reserve (BR) under the specified conditions is MVV–PV. The percentage breathing reserve (per cent BR) is (MVV–PV)/

(MVV) × 100. It is also called the *dyspnoeic index*. If its value falls below 60 per cent (range 60–70 per cent) dyspnoea is generally present. The per cent BR may be lowered owing to a decrease in maximal ventilation volume (MVV) or a rise in pulmonary ventilation (PV).

Example: Normal resting person, MVV = 100 l min⁻¹; PV = 8 l min⁻¹; BR = 92 per cent. When owing to exertion PV increases to 40 l min⁻¹ the per cent BR falls to 60 per cent and dyspnoea is present.

2. Vital capacity

A decrease in vital capacity decreases the maximal ventilation volume and thus the percentage breathing reserve; it therefore predisposes to dyspnoea. As the depth of breathing approaches the vital capacity the sense of discomfort increases.

3. Mechanical efficiency

A person with a low mechanical efficiency uses more energy than a normal person to do a given amount of work: his O_2 consumption and pulmonary ventilation are correspondingly greater and he thus developes dyspnoea earlier.

PATHOLOGICAL DYSPNOEA

Pathological dyspnoea may be classified as follows:

1. Due to mechanical or nervous hindrance to the respiratory movements.
2. Cardiac dyspnoea—e.g. left ventricular failure or decompensated mitral stenosis; both these conditions are associated with pulmonary congestion.
3. Metabolic acidaemia in the terminal stages of chronic nephritis or in diabetic ketosis. The raised [H⁺] of the cerebral interstitial fluids 'drives' the respiratory centre excessively.
4. Chronic anoxia.
5. Thyrotoxicosis—there is little evidence of dyspnoea at rest, but it develops during work.

Of these various causes only the first two groups will be further discussed.

Mechanical and nervous hindrance to pulmonary ventilation

Generally speaking there is a decrease in the maximal ventilation volume and in the vital capacity of this group. An excellent example is provided by the condition emphysema.

Emphysema

In emphysema, the lungs are voluminous and inelastic; there is widespread destruction of both the alveolar walls and the pulmonary capillaries. In a normal subject of say 60 years of age, the lung volume is only about three litres when the chest wall is relaxed at the end-tidal expiratory position. FIGURE III.34 shows that the lung volume of an emphysematous subject 60 years of age was eight litres, when the chest wall was relaxed. The lungs have lost their elasticity. Measurement of the compliance gave a value of 0.9 litre per cm H_2O for the emphysematous and 0.25 litre per cm H_2O for the normal subject of this age [FIG. III.34]. FIGURE III.35 shows that there is:

1. An increase in the FRC of the lungs.
2. A notable decrease in the amount of air which can be expelled within the first second of a forced expiration (FEV₁) following a maximal inspiration. FEV₁ is only 0.75 litre compared with FEV₁ for a normal subject (3.2 litres). The normal subject expels his vital capacity in less than 4 seconds; the emphysematous subject does

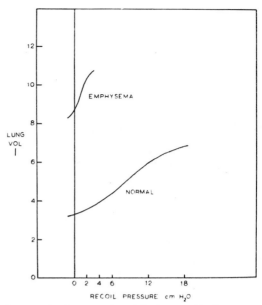

FIG. III.34. The elastic properties of the lungs. This diagram relates the volume of the lungs to their elastic recoil pressure as measured in the pleura or oesophagus. (Campbell, E. J. M. (1958). *Post-grad. med. J.* **34**, 30.)

not succeed in doing so even in 8 seconds and has to take another breath because of an asphyxial stimulus.

Expiration during natural breathing at rest is achieved by the passive recoil of the lungs. The recoil pressure of the lungs maintains the pressure in the airways above that in the intrapleural space and drives air up to the mouth. In emphysema this elastic recoil is mostly lost. If an attempt is made to accelerate the rate of airflow by using the expiratory muscles, then the positive intrapleural pressure thus developed narrows the airways in addition to raising the alveolar pressure. The flow resistance ('non-elastic resistance') is thereby considerably increased. These factors which hinder tidal expiration or forced expiration, also impair the efficiency of coughing. The linear velocity of airflow in the trachea may reach 960 km h⁻¹ in coughing in a normal subject but attains perhaps only one-sixth of this value in the severely emphysematous patient.

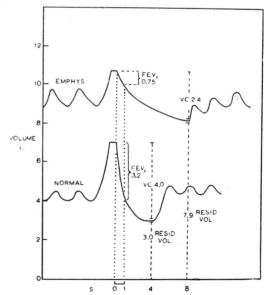

FIG. III.35. The lung volumes and the forced vital capacity. VC, vital capacity; FEV₁, forced expired volume in 1 s. (Campbell, E. J. M. (1958). *Post-grad. med. J.* **34**, 30.)

Further tests of the emphysematous subject reveal that:

1. The physiological dead space is perhaps doubled.

2. Large parts of the emphysematous lungs are poorly ventilated (N_2 wash-out is slow on breathing O_2).

3. The resting diffusing capacity for O_2 (ml O_2 per minute per mm Hg) is low—say one-fifth of normal.

4. The arterial pO_2 is lowered perhaps to 50 mm Hg—the saturation of haemoglobin with oxygen in the arterial blood is correspondingly lowered from 98 per cent to 80–85 per cent.

5. The arterial and alveolar pCO_2 may be very high—e.g. 60 mm Hg. Correspondingly the plasma bicarbonate concentration is raised thus helping to compensate for this respiratory acidosis. The pulmonary ventilation is partly maintained by an anoxic stimulus. Hence if the patient is placed in an oxygen tent or is made to breathe oxygen for therapeutic purposes his breathing is depressed and the arterial and alveolar pCO_2 may exceed 100 mm Hg CO_2; narcosis develops with muscle twitchings, and cerebral vasodilatation occurs giving a throbbing headache, papilloedema, and a raised c.s.f. pressure.

Asthma

Bronchial constriction is present which diminishes the vital capacity and the maximal ventilation volume and cause hyperinflation of the lung. During an acute attack the breathing capacity is so limited by bronchospasm that intense dyspnoea is present. Treatment with bronchodilator drugs (e.g. adrenalin) ameliorates the symptoms of an acute attack. A marked improvement of the vital capacity and maximal ventilation volume are seen as a result of such therapy when the drugs are administered to a subject who is not at the time suffering from an acute attack. The emphysematous subject is not aided by bronchodilator drugs.

Other causes of mechanical hindrance to lung movements

These include collapse or consolidation of the lungs, obstruction to the main airways, and pneumothorax. Whereas asphyxia may always be an important contributory factor some of the dyspnoea experienced in these conditions may be evoked reflexly via the vagi. Thus the hyperpnoea caused by blocking the bronchus in an experimental animal may disappear when the vagus nerve is cut on the affected side.

Cardiac dyspnoea

This has as its general cause, pulmonary congestion. This, besides diminishing the vital capacity, also reflexly stimulates the breathing by a variety of mechanisms:

1. Anoxic anoxia.
2. Sensitization of vagal J-receptors (see above).

MECHANISMS OF RESPIRATORY SENSATION

The physiology of respiratory sensation is controversial and undoubtedly complex: it is likely that both normal awareness of respiratory position and movement, and also the distress of breathing in extreme conditions, both physiological and pathological, are due to the interaction of many inputs. Moreover, as with other forms of conscious sensation, the quality and intensity of respiratory sensation is undoubtedly influenced by the cerebral cortex, e.g. the degree of attention or anxiety evinced by the individual with regard to his breathing. However, two sources of impulses which have been established as making an important contribution to respiratory sensation are the respiratory proprioceptors, i.e. mechanoceptors in the respiratory muscles and joints, and the lung receptors with a vagal afferent pathway.

Respiratory proprioceptors and respiratory sensation

We are normally aware of the degree of lung distension though we do not generally take cognisance of this information unless we specifically direct our attention to it. The intercostal muscles contain many muscle spindles and the costovertebral joints possess mechanoreceptors sensitive to rib displacement. Though the diaphragm, the main muscle of breathing, has few muscle spindles its tendon has Golgi tendon organs. These proprioceptors are presumably the principal sites of origin of the impulses giving rise to the conscious sensation of lung distension, since this sensation is still present in subjects with bilateral vagal blockade.

The activity of mechanoreceptors in the respiratory muscles and joints has also been held to be responsible for dyspnoea, the discomfort in breathing experienced in extreme physiological or pathological conditions. It has been suggested by Howell and Campbell that dyspnoea results when there is '*length–tension inappropriateness*'. That is to say, when as a result of some hindrance to breathing, the change in the volume of the chest, gauged from the displacement of the joints (signalled by the joint receptors) or from the change in muscle length (signalled by the muscle spindles) differs from the 'expected' change. This 'expected' change could be represented by the tension generated in the respiratory muscles or by the intensity of the motorneuron discharge causing the muscle contraction. According to this hypothesis, dyspnoea would be experienced when there was 'inappropriateness' between this input to the respiratory musculature and the output as signalled by the respiratory proprioceptors. There is some experimental support for this hypothesis. For example, the bursting sensation associated with breath-holding is greatly alleviated and the time for which the breath can be held, is prolonged by phrenic nerve block, which abolishes the diaphragmatic contractions which occur during breath-holding. On the other hand, inspiratory efforts against a closed glottis are not in themselves unpleasant, although the degree of 'length–tension inappropriateness' is extreme, since there is muscular tension without volume displacement. Moreover, as described below, dyspnoea may also be associated with the activity of lung receptors. Thus it seems likely that a sense of 'length–tension inappropriateness', based on information from respiratory proprioceptors, is but one element in a multifactorial mechanism, and that 'length–tension inappropriateness' will only give rise to dyspnoea if these other variables, such as the degree of excitation of lung receptors and the tensions of the blood gases, also attain certain values.

Lung receptors and respiratory sensation

There is ample evidence indicating that stimulation of lung receptors can give rise to unpleasant respiratory sensations. Examples include:

(i) The inhalation of irritant gases or aerosols usually causes a distressing or painful sensation localized to the thorax or a sense of tightness in the chest. Since lung irritant receptors are stimulated by ether, ammonia, and cigarette smoke [p. 174], and J-receptors are stimulated by chlorine, halothane, and ether [p. 175], activation of lung receptors is the most likely origin of this sensation.

(ii) Passage of an endobronchial or endotracheal tube, which has been shown to excite only tracheal cough and lung-irritant receptors, leads to an immediate sensation usually described as painful, burning or irritating.

(iii) Reflation of a collapsed lung causes an instantaneous painful tearing sensation. In experimental animals reinflation of a collapsed lung evokes a marked discharge from lung-irritant receptors.

There is also much evidence that lung receptors play a major role in producing the sensation of dyspnoea. TABLE III.2 lists the responses of the various types of lung receptors in a variety of conditions causing dyspnoea in man. In every instance the stimulus

TABLE III.2 *Vagal reflexes initiated from the lungs, and responses of lung receptors*

Stimulus	Vagal reflex response of			Response of		
	Hyperpnoea	Broncho-constriction	Dyspnoea in man	Pulmonary stretch receptors	Lung irritant receptors	Type J receptors
Atelectasis	+	?	+	−	+	(+)
Pneumothorax	+	?	+	−	+	○
Ammonia	+	+	+	(+)	+	(+)
Cigarette smoke	+	+	+	?	+	?
Histamine aerosol	+	+	+	?	+	○
Lung congestion	+	?	+	(+)	+	+
Micro-embolism	+	+	+	○	+	+
Anaphylaxis	+	+	+	?	+	?

+, positive response; ○, no response; −, negative response; ?, unknown; brackets indicate weak or variable response.

From Widdicombe J. G. (1971). *Scientific Basis of Medicine, Annual Reviews*, **148**.

causing dyspnoea in man also promotes activity of lung irritant receptors in experimental animals, strongly suggesting a causal relationship. As indicated in TABLE III.2, J-receptor activation [see p. 175] may additionally contribute to the sensation of dyspnoea in lung congestion and micro-embolism. Guz and his colleagues have shown that bilateral anaesthesia of the vagus nerves prolongs breath-holding time in man and ameliorates the dyspnoea of lung diseases such as sarcoidosis and fibrosis, and the breathlessness produced by breathing a hypercapnic gas mixture. Furthermore, reflex hyperventilation results from stimulation of lung-irritant receptors in experimental animals, and in man vagal anaesthesia reduces both the hyperpnoea of hypercapnia and the hyperventilation which is often associated with the dyspnoea of lung disease. Conversely, patients with functional block of the spinal cord at C2–3, in whom the only likely afferent pathway is that from the lungs, still experience the respiratory distress of hypercapnia, breath-holding, and pulmonary congestion.

Thus it seems that the proprioceptors of the muscles and joints of the respiratory apparatus are responsible for the awareness of lung volume or thoracic shape in healthy quiet breathing. On the other hand lung pain and dyspnoea are due primarily to stimulation of lung-irritant receptors and, perhaps to a lesser extent, of J-receptors, though it is possible that a sense of 'length-tension inappropriateness', generated by the respiratory proprioceptors, may be an additional factor of smaller importance.

REFERENCES

GUZ, A., NOBLE, M. I. M., EISELE, J. H., and TRENCHARD, D. (1970). Experimental results of vagal block in cardiopulmonary disease. In *Breathing: Hering–Breuer Centenary Symposium* (ed. R. Porter) pp. 315–29. Churchill, London.

HOWELL, J. B. L. and CAMPBELL, E. J. M. (1966). *Breathlessness*. Blackwell, Oxford.

NOBLE, M. I. M., EISELE, J. H., TRENCHARD, D., and GUZ, A. (1970). Effect of selective peripheral nerve blocks on respiratory sensations. In *Breathing: Hering–Breuer Centenary Symposium* (ed. R. Porter) pp. 233–46. Churchill, London.

SEARS, T. A. (1971). Breathing: a sensori-motor act. In *The scientific basis of medicine Annual Reviews 1971* (ed. I. Gilliland and J. Francis). Athlone Press, London.

WIDDICOMBE, J. G. (1971). Breathing and breathlessness in lung diseases. In *The scientific basis of medicine Annual Reviews 1971* (ed. I. Gilliland and J. Francis). Athlone Press, London.

—— (1974). Reflexes from the lungs in the control of breathing. In *Recent advances in physiology*, No. 9 (ed. R. J. Linden). Churchill Livingstone, Edinburgh.

Artificial respiration

Artificial respiration is called for in man in two types of respiratory failure:

1. When breathing fails due to drowning, inhalation of irrespirable or poisonous gases, suicidal and accidental overdoses with narcotics, overdosage with anaesthetics, or electrocution.

2. In gradually progressive respiratory failure due to paralysis of respiratory muscles, e.g. in poliomyelitis or diphtheria.

In the chronic group there is usually ample warning of the impending disaster, and arrangements can be made to use a breathing machine. In the acute group no moment must be lost and the treatment must be capable of being carried out by the first instructed person on the scene and requires the use of no more equipment than can be readily improvised almost anywhere. There is another fundamental difference between the chronic and acute forms of respiratory failure. In the former, machine-breathing is instituted while the circulation is functioning normally; if the mechanical respiration is efficient no circulatory derangement develops and the blood flow to all the organs remains normal. In acute asphyxia, on the other hand, failure of the circulation and of the central nervous system follows rapidly in the train of respiratory arrest. The body is unfortunately quite unadapted for dealing with complete or even very severe oxygen lack, which 'stops the machine and wrecks the machinery' in a matter of minutes. Let us consider the sequence of events in drowning, for example. After one minute or so of complete lack of oxygen consciousness is lost; within another minute or two the respiratory centre, which initially was stimulated, ceases to function. The vasomotor centre is more resistant and may maintain vasoconstriction for a little longer (asphyxia livida), but soon it too fails, and full peripheral vasodilatation sets in, with a resulting fall of blood pressure to about 40 mm Hg (asphyxia pallida). Most important of all, the heart—unlike skeletal or smooth muscle—can function normally for only a short time in the absence of oxygen; it has no type of anaerobic metabolism to fall back on. The force of the heart beat in severe anoxia rapidly weakens and the chambers greatly dilate; the output into the blood vessels is reduced to a trickle, and the blood flow to the organs almost ceases. When fibrillation of the ventricles develops the chances of recovery are of the slenderest. Acute asphyxia thus presents a combination of respiratory, circulatory, and nervous system failure with which the treatment employed must cope effectively if it is to succeed.

Mouth to mouth breathing

Mouth to mouth breathing is superior to all other methods (Safer and McMahon 1959). The operator kneels on the left of the supine patient. The neck of the patient is first extended. The lower jaw is held upwards by placing the left thumb in the mouth and grasping the symphysis mentis with the corresponding fingers. After clearing water or vomitus from the mouth, the patient's nose is closed by the fingers of the right hand. The operator then 'enfolds' the patient's mouth between his own lips, ensuring an air-tight connection; he then exhales forcefully. The chest movement obtained is observed and the force of expiration modified accordingly. The elastic recoil of the patient's chest secures passive expiration during which phase the operator withdraws his mouth. The procedure is repeated 12 to 20 times a minute.

In some cases some of the air blown into the patient's pharynx gains access to his stomach. Such an eventuality can be checked from time to time by palpation. The air can be expressed by pressure on the upper abdomen.

REFERENCE

SAFAR, P. and MCMAHON, M. C. (1959) *A manual for emergency artificial respiration*. Baltimore.

Mechanical methods

Drinker's method. This involves the use of an airtight tank into which the patient is placed, with the head outside. Alternative negative and positive pressures are obtained in the tank by means of electrically driven pumps, and the effect is to produce movements of the chest wall resembling those of normal inspiration and expiration; the negative pressure pulls on the chest wall and produces inspiration; the positive tank pressure compresses the chest and produces expiration. Patients with respiratory paralysis following poliomyelitis have been kept alive for years in this way.

Anaesthetic machine. In respiratory failure in the operating theatre artificial respiration can be given from an anaesthetic machine, from a simple pump like the Oxford inflator, or by blowing periodically down an inserted endotracheal tube.

Whatever method is used it is essential to ensure that the airway is free.

REFERENCES

GARLAND, T. L. (1955). *Artificial respiration*. London.
WHITTENBERGER, J. L. (1955). *Physiol. Rev.* **35**, 611.

THE ANALYSIS OF THE RESPIRATORY GASES (OXYGEN AND CARBON DIOXIDE)

The Haldane apparatus was devised for this purpose [FIG. III.36]. A sample of inspired (or expired or alveolar air) is drawn into a calibrated burette at atmospheric pressure and its volume measured. It is then exposed to a solution of caustic potash (KOH) which absorbs its CO_2 content, and the residual volume is read again at atmospheric pressure. Subsequently, it is exposed to chromous chloride solution, which absorbs oxygen and the final volume (nitrogen) is determined. If say 10 ml of air was initially taken into the burette and 9.60 ml remain after CO_2 absorption, then the CO_2 percentage in the sample is 4 per cent. If after absorption of oxygen the final volume is 8.0 ml, the oxygen percentage is 16 per cent.

The Haldane apparatus is used for calibrating various types of analytical apparatus which more quickly provide the results of respiratory gas analysis. These include:

1. The mass spectrometer

A gas sample is introduced into the apparatus and the gas is reduced to a pressure of 10^{-6} mm Hg, upon which it is ionized by a beam of electrons, which produces positively charged ions of oxygen, CO_2, nitrogen, and argon. These are deflected by a magnetic field as shown in FIGURE III.37. The heavier the ionic mass of the gas, the longer path it takes to reach the collecting electrode. [$O_2 = 32$, Argon $= 40$, $CO_2 = 44$, and $N_2 = 28$]. If one collector electrode is used, the path taken by the ions depends on the magnitude of the electric voltage of the collector. By varying this negative voltage

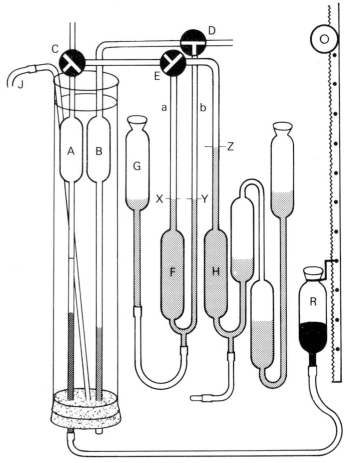

FIG. III.36. Haldane air gas apparatus. A: measuring burette; B: temperature compensator; F–G: potash solution; H: chromous chloride solution. (From Douglas, C. E. and Priestley, J. G. (1937). *Human physiology*, 2nd edn. Oxford University Press.)

between limits 25 times a second, the ions of the component gases may be sequentially collected and electronic amplification allows a continuous recording of the changes in oxygen and carbon dioxide composition during breathing [FIG. III.38].

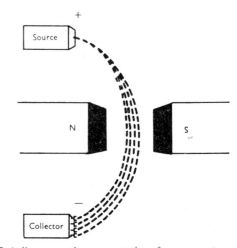

FIG. III.37. A diagrammatic representation of a mass spectrometer. The gas to be analysed is reduced to a very low pressure (10^{-6} mm Hg) and ionized by bombarding it with electrons. The positively charged ions of nitrogen, oxygen, carbon dioxide, and argon take separate paths through the magnetic field to the collector electrodes. By using one electrode for each gas and measuring the electrode current the percentage composition of the mixture can be determined. (From Green, J. H. (1976). *An introduction to human physiology*, 4th edn. Oxford University Press.)

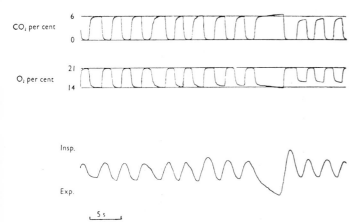

FIG. III.38. Recordings of the CO_2 and O_2 levels in the inspired and expired air using the mass spectrometer (upper traces) and respiratory movements using a stethograph (lower trace) during quiet breathing and following a prolonged expiration. The end expiratory level of CO_2 and O_2 gives an indication of the alveolar air composition. (From Green, J. H. (1976). *An introduction to human physiology*, 4th edn. Oxford University Press.)

2. The paramagnetic oxygen analyser

All substances when placed in a magnetic field are either attracted or repelled by one pole of a magnet. Strongly magnetic substances are called *ferromagnetic* (iron, cobalt, nickel). Substances that are feebly attracted by a magnet are termed *paramagnetic*; those that are slightly repelled by a magnet are called *diamagnetic*. Very powerful magnets are required to detect paramagnetic or diamagnetic properties. The Pauling oxygen analyser measures oxygen partial pressure in a gas mixture, operating by virtue of the paramagnetic properties of oxygen; the other gaseous constituents are diamagnetic. A small glass sphere filled with nitrogen is suspended by a vertically stretched silica fibre between the poles of a magnet. If the sphere moves in the magnetic field it rotates around the silica fibre. When oxygen is passed through the space between the magnetic poles and therefore surrounds the spherule, the sphere is subjected to a magnetic force depending on the difference between its own magnetic susceptibility and that of the oxygen and rotates, causing torsion of the silica fibre. Equilibrium is reached when the magnetic rotational force is balanced by the restoring force of the twisted silica fibre. By fixing a mirror to the fibre the reflection of a light source on to a scale calibrated in units of oxygen tension permits the measurement of P_{O_2}. The instrument requires some 30–40 seconds to reach equilibrium and is therefore too slowly responsive for continuous measurement of oxygen partial pressure during breathing. However, it is particularly suitable for measurements of oxygen pressure in collected samples.

3. The infrared CO_2 analyser

CO_2 absorbs radiant heat. Infrared radiation is passed from a source through a sampling chamber to a detector; when the sampling chamber is filled with a gas mixture containing CO_2 the radiation is lessened. The analyser can be used for continuous recording of CO_2 concentration in respiratory gases. The respiratory gases have the following percentage composition. Those for expired and alveolar air are of man at rest.

	Inspired air (per cent)	Expired air (per cent)	Alveolar air (per cent)
O_2	21	16	14
CO_2	0	4	6
N_2	79	80	80

These are 'idealized' figures. It will be noted that the percentage of nitrogen (obtained by subtracting the combined percentages of oxygen and CO_2 from 100) is higher in alveolar and expired air than in inspired air. This is *not* because nitrogen is secreted by the lungs but is due to the fact that the expired air volume is less than that inspired. Not all the oxygen absorbed is converted to expired carbon dioxide. The inspired air volume is 80/79 times the expired volume.

If the subject (with nose clipped) breathes through a mouthpiece fitted with inspiratory and expiratory valves, the expired air can be collected in a rubber bag (Douglas bag) over a known time. The air expired is saturated with water vapour at body temperature, but cools in the bag to room temperature; water also condenses in the bag and the gas is finally saturated with water at room temperature. The air collected in the bag is first submitted to sampling for analysis and then its volume is measured by expelling the contents of the bag through a gas meter. The volume read on the meter is converted to its NTP (dry) equivalent—i.e. 760 mm Hg (101.3 kPa) and 0 °C using the formula

$$V_{NTP} = V \times \frac{273}{273 + t} \times \frac{(B - p)}{760},$$

where V is the volume collected, t the room temperature, B the barometric pressure, and p the saturated water vapour pressure at room temperature. Suppose the corrected volume of expired air per minute is 6 litres. Then the subject whose expired air composition is 16 per cent O_2, 4 per cent CO_2 as shown in the table, will have an oxygen uptake and carbon dioxide output as follows:

O_2 uptake/min = Vol. O_2 inspired minus Vol. O_2 expired

$$= \left[\left(\frac{80/79 \times 21}{100} \right) \text{ minus } \left(\frac{16}{100} \right) \right] \times 6000$$

$$= [21.3 - 16] \times 60 = 318 \text{ ml.}$$

CO_2 output/min

$\qquad$ = Vol. CO_2 expired minus Vol. CO_2 inspired (zero)

$\qquad$ = 6000 × 4/100 = 240 ml.

BLOOD GAS MEASUREMENTS

1. Blood gas content

This can be determined most satisfactorily by means of the Van Slyke–Neill manometric apparatus [FIG. III.39]. Samples of arterial or mixed venous blood are withdrawn using a well-greased syringe containing a drop of heparin solution. Scrupulous care must be taken to avoid contact with the air. On removing the syringe from the needle used for puncture of the vessel a polythene cap is placed over the nozzle. The blood is then displaced into an Ostwald pipette, which delivers one ml into the manometric apparatus. The details of the technique are beyond the scope of this text and can be found in textbooks of practical physiology. However, the principle of the method is that the gases contained in the blood are liberated by a combination of chemical displacement and exposure to a vacuum. After such treatment the volume of gases liberated is reduced to 2 ml and the pressure exerted by the gases is read on a mercury manometric column.

If CO_2 is alone to be determined, the blood is admitted to the chamber together with gas-free lactic acid; by lowering the mercury reservoir, the mercury and blood/acid mixture is exposed to a vacuum and CO_2 is displaced, together with some oxygen and all the nitrogen which the blood contained. Mechanical shaking assists this displacement. The mercury and aqueous solution are then *slowly* returned to the chamber (to avoid reabsorption of some of the CO_2) and cautiously brought up to the 2 ml mark. The manometric pressure is read on the left-hand column. The mercury reservoir is then lowered slightly to produce a lowered chamber pressure and 1 ml of sodium hydroxide is slowly admitted to the chamber from the cup, permitting absorption of the CO_2. The mercury reservoir is then raised and once more the fluid meniscus is brought back to the 2 ml mark; the manometric pressure is again read. The difference in the two pressure readings before and after CO_2 absorption allows the calculation of the CO_2 content of the blood sample.

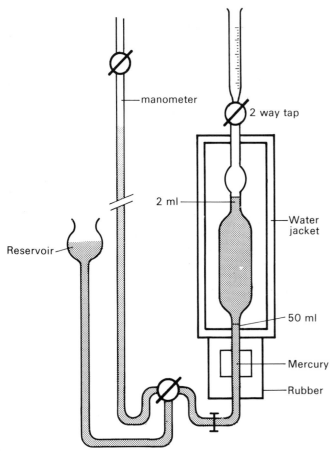

manometer

2 way tap

2 ml

Reservoir

Water jacket

50 ml

Mercury

Rubber

FIG. III.39. The Van Slyke–Neill manometric apparatus.

The analysis of blood oxygen content requires the lysis of the red cells by saponin and the conversion of the haemoglobin to methaemoglobin by potassium ferricyanide. The blood saponin–ferricyanide mixture is exposed to a vacuum and mechanical agitation as described above and the volume of gases reduced to 2 ml as before. The pressure exerted may be read and then the CO_2 is absorbed and the pressure of the oxygen and nitrogen determined at 2 ml volume. The oxygen is absorbed by admitting a sodium hydrosulphite solution containing anthraquinone-β-sulphonate (which catalyses the absorption by the hydrosulphite) and the final volume noted. The difference between the pressures registered at constant volume before and after O_2 absorption yields the oxygen content of the blood.

It is possible by the use of acidified saponin–ferricyanide to do both CO_2 and O_2 determinations on a single sample, but the acidification of the ferricyanide causes a somewhat viscous and sludgy mixture with blood in the chamber, and this technique does not yield such high accuracy as when the gas contents are determined separately.

Using tonometers of 250 ml capacity, small samples of blood can be introduced and the tonometer filled with gas mixtures of known composition. The tonometer is rotated for ten or more minutes in a waterbath at 37 °C or any desired temperature; the blood spreads in

a thin film over the inside of the vessel and equilibrates with the gas mixture. The tonometer is then held upright in the bath and the blood drains into the narrow region above the sampling tap, to be removed and submitted to analysis of O_2 and/or CO_2 content by the Van Slyke apparatus. The gaseous pressure and composition is determined and in this way the so-called oxygen or carbon dioxide dissociation curves are constructed from an appropriate series of measurements [see pp. 185–7].

The Van Slyke apparatus remains the most accurate method of determining blood gas content. Several methods have more recently been developed which allow estimations of gas contents using micro-samples (0.03 ml blood). These include those of Natelson and of Roughton and Scholander; the techniques are described in practical textbooks.

2. Blood gas tensions

(a) **Bubble method** introduced by Krogh for the determination of gaseous tensions in blood. A small bubble of air is exposed to blood flowing from an artery into a micro-capillary from which the blood returns to the artery below. The bubble equilibrates with the blood and can then be withdrawn into a micro-gas analysis apparatus, such as that described by Roughton and Scholander. The gas tensions thus determined are identical with those of the arterial blood.

(b) **CO_2 electrode.** A glass electrode (sensitive to changes of pH) surrounded by a thin layer of dilute bicarbonate solution is confined in a Teflon plastic container which separates it from the blood sample. The plastic membrane is permeable to CO_2 which, equilibrating with the bicarbonate solution, causes a change of pH registered by the electrode. The instrument is calibrated repeatedly against solutions of known CO_2 tension.

(c) **O_2 polarography.** This technique measures the electrical current produced by the electrochemical reduction of oxygen:

$$O_2 + 4e = 2 \times O^{2-}.$$

At the same concentration of oxygen in the solution, the amount of reduced oxygen (and therefore the current) depends on the voltage applied to a polarizable (platinum) electrode connected with a non-polarizable Ag–AgCl electrode both immersed in the solution under study [Fig. III.40].

If the voltage is raised from 100–200 mV, as it rises so does the current, but within the range of 400–800 mV, current is little affected; at such voltages each O_2 molecule is almost instantaneously reduced and the 'reduction current' is proportional to the amount of oxygen which reaches the electrode. As Lübbers (1966) has stressed, the 'reduction current' does *not* measure the oxygen tension of the solution—it measures the oxygen availability of the sampling electrode. The main condition for measuring oxygen tension is that oxygen is transported to the platinum electrode only by diffusion. If the solution is homogeneous, then the arrival of oxygen at the surface of the platinum depends on the gradient between it and the sample. The platinum electrode surface oxygen tension is zero, so a constant diffusion zone can be established by a membrane permeable to oxygen interpolated between the electrode system and the solution (or blood). O_2 is then transported only by diffusion and the current developed will be linearly related to the oxygen tension of the sample. Lübbers (1966), using a thin Teflon–cellophane membrane 15 μm thick, found a response time of 50 per cent in 2 s and 100 per cent in 15 s, and achieved a sensitivity of 0.8 nA per mm Hg oxygen tension. He discusses various modifications of the design of the apparatus and the problems which they themselves may pose. He also deals with the theory and practice of the polarographic method used for measuring oxygen *content* of the blood—a concept first introduced by Percy Baumberger of Stanford University in 1940. If the chemically

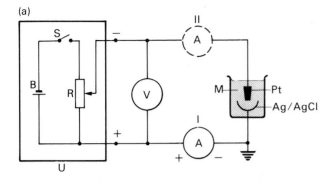

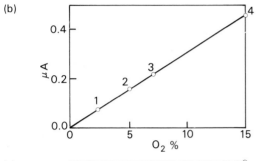

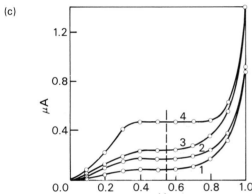

FIG. III.40(a). Polarographic circuit. By U a negative voltage can be applied to the platinum electrode. The current from the reduction of oxygen is measured by A. In position I the instrument A can be grounded. B = battery; R = resistance; S = switch; V = voltmeter; A = ammeter; Pt and Ag/AgCl are platinum and silver–silver chloride electrodes. (b). Calibration curve drawn from the polarograms (c) with different gas mixtures (1–4). Zero P_{O_2} = zero current. (c). Polarograms with four different O_2–N_2–CO_2 gas mixtures (1–4). The polarograms show a well-formed plateau. (Redrawn from Lübbers, D. W. (1966). In *Oxygen measurement in blood and tissue* (ed. J. P. Payne and D. W. Hill). Churchill, London.)

bound oxygen of HbO_2 is converted to physically dissolved oxygen, the resultant oxygen tension can be measured as described above. Briefly, the oxygen tension of the blood sample is first measured in the usual fashion by a micro Pt-electrode and the blood is then transferred to a chamber in which sodium ferricyanide releases the chemically bound oxygen, so that it enters physical solution. This solution is redisplaced to the vicinity of the electrode chamber, which again measures oxygen tension. From the difference in oxygen tensions the O_2 content is calculated.

(d) **Gas chromatography.** Two columns each containing a thermal conductivity detector are connected in parallel. A stream of helium is directed through both columns. In one column, CO_2 is separated from the other gases and the difference in thermal conductivity between this column and its parallel companion is registered as a peak, whose height is proportional to the percentage of CO_2 in the sample. O_2 and N_2 are separated and measured in the

other column. This technique can be modified to measure blood gas contents, by mixing the sample with the appropriate releasing agents before displacing the gases evolved with helium and passing the resulting mixture through the chromatographic apparatus.

REFERENCE

LÜBBERS, D. W. (1966). In *Oxygen measurement in blood and tissue* (ed. J. P. Payne and D. W. Hill). Churchill, London.

THE CARRIAGE OF OXYGEN AND CARBON DIOXIDE BY THE BLOOD

The function of the circulation is to distribute oxygen from the lungs to the tissues and to return carbon dioxide produced by the tissues to the lungs, whence it is expired. Arterial blood leaves approximately in equilibrium with the gaseous phase of the alveolar air. The oxygen partial pressure of arterial blood is about 100 mm Hg (13.3 kPa); the carbon dioxide partial pressure (or tension) is about 40 mm Hg (5.3 kPa). Gases dissolve in a solvent according to the temperature—in this case 37 °C—and, by *Henry's Law* according to their partial pressure. At 100 mm Hg and 37 °C only 0.3 ml oxygen dissolve in 100 ml blood. At 40 mm Hg only 2.8 ml carbon dioxide enter solution per 100 ml.

Arterial blood *content* of oxygen is, however, normally some 19.0–19.5 ml oxygen per 100 ml. Arterial content of carbon dioxide is approximately 48–49 ml per 100 ml. Obviously the two gases must exist in combination with constituents of the blood in addition to their dissolved component. Haemoglobin plays an important role in the carriage of the blood gases.

It is convenient to consider oxygen carriage first.

Oxygen carriage

Haemoglobin

X-ray crystallography has shown that haemoglobin consists of four polypeptide chains each carrying a haem molecule (Perutz 1964). Haem is an iron-containing porphyrin consisting of four pyrrole rings joined by methine ($-CH=$) groups containing ferrous iron in its centre [FIG. III.41].

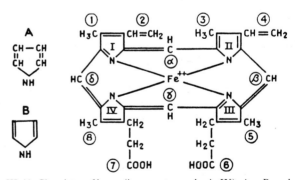

FIG. III.41. Chemistry of haem (iron-protoporphyrin IX). A = Pyrrole ring in full. B = Pyrrole ring, conventional outline. Haem: Pyrrole rings are numbered 1, II, III, IV. C atoms of methine bridge are labelled $\alpha, \beta, \gamma, \delta$. Side chains are numbered 1–8.

The four polypeptide chains consist, in adult haemoglobin [p. 36], of two α-chains each containing 141 amino-acids connected by peptide ($-CO.NH-$) links and two β-chains consisting of 146 amino-acids in peptide combination. Perutz and co-workers have succeeded in establishing the amino-acid sequence of these α- and β-chains. The important feature of the polypeptide chains is the attachment of the iron atom of the haem between histidine residues of the chain at positions 58 and 87 on the α polypeptide complex and between 63 and 92 on the β complex. The link between Fe^{2+} and histidine 87 of the α polypeptide and histidine 92 of the β chain is direct. The link between histidine 58 (α) and histidine 63 (β) and the relevant Fe^{2+} of their respective haems *can be interrupted by the introduction of an oxygen atom* [Fig. III.42].

The complete molecule of haemoglobin when deoxygenated shows close contact of the alpha-chains with the beta-chains, but a fair degree of separation of the two alpha-chains from each other; similarly the two beta-chains have little contact with each other. When the haemoglobin takes up oxygen the amino end of an α-chain moves towards the carboxyl end of the other α complex and a similar arrangement occurs with the two β-chains.

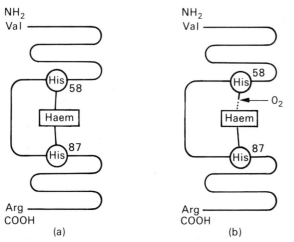

FIG. III.42. The histidine linkages in the *alpha*-chain of Hb. (a) In reduced haemoglobin. (b). In oxyhaemoglobin.

When oxygenated haemoglobin loses its oxygen the molecule expands exposing the terminal NH_2 groups and some of the imidazole groups. This is believed to account for the greater power of reduced haemoglobin to mop up H^+ (and CO_2 as carbamino-haemoglobin) compared with that of oxyhaemoglobin.

Haemoglobin contains 0.334 per cent of its weight as Fe^{2+}. As the atomic weight of iron is 56, the minimal molecular weight of haemoglobin would thus be 56×334—approximately 16 700.

Until the 1920s this was indeed believed to be the molecular weight of haemoglobin. Two classical investigations disproved this: (1) Adair, measuring the depression of the freezing point was able to determine the osmotic pressure and hence the molecular weight. He obtained a value of approximately 67 000. (2) Svedberg (1926) introduced the method of ultracentrifugation in which the solution containing the protein is exposed to gravitational forces over 100 000 g by centrifuging at 60 000 revolutions per minute. Again a value of approximately 67 000 was obtained.

It followed that there must be four iron atoms in each molecule of haemoglobin and hence four haem groups each of which must combine with a molecule of oxygen. It should be noted that 1 gram of haemoglobin has been experimentally shown to combine with 1.36 ml O_2. Thus 16 700 g of haemoglobin will combine with 22 400 ml O_2 which is 1 mole of oxygen.

The equation defining the situation is therefore:

$$Hb_4 + 4O_2 \rightleftharpoons Hb_4 O_8$$

Obviously it is unlikely that four oxygen molecules collide with one molecule of haemoglobin simultaneously. Adair therefore suggested that the reaction took place in stages:

$$Hb_4 \quad + O_2 \longrightarrow Hb_4 O_2$$
$$Hb_4 O_2 + O_2 \longrightarrow Hb_4 O_4$$
$$Hb_4 O_4 + O_2 \longrightarrow Hb_4 O_6$$
$$Hb_4 O_6 + O_2 \longrightarrow Hb_4 O_8$$

Each of these reactions would have its appropriate equilibrium constant (K_1, K_2, K_3, and K_4) and Adair was able to express the overall equilibrium between haemoglobin and oxygen. However, the equation which he developed is only applicable, as he himself pointed out, to solutions of haemoglobin in which the osmotic pressure, Π, is related to the molar concentration of protein, c, by the equation

$$\Pi = cRT.$$

Such an equation is only valid in considering dilute solutions (less than 5 g per 100 ml) of haemoglobin. In the corpuscle, haemoglobin exists at 30–35 per cent of the cell water and, as Roughton (1964) has stated, these conditions cause a wide deviation of the experimental results from those predictable from the Adair equation.

Margaria (1963) concluded that the first three constants K_1, K_2 and K_3 were related in such a way as to suggest: (a) that no interaction occurs in the first three oxygenation reactions and (b) that the affinity of the iron for oxygen in the fourth reaction is some 125 times that shown in the first three reactions. He derived a simplified equation replacing the four constants, K_1, K_2, K_3, and K_4 by a single constant K together with a factor 'm' which expresses the increased affinity of the fourth haem group for oxygen in the final progress towards saturation. His equation expresses the situation as

$$\% \text{ saturation} = \frac{\left(\dfrac{1 + KP_{O_2}}{KP_{O_2}}\right)^3 + m - 1}{\left(\dfrac{1 + KP_{O_2}}{KP_{O_2}}\right)^4 + m - 1}$$

If K is given a value of 0.0144 and m a value of 100, a curve may be drawn [Fig. III.43] which accurately describes the oxygen dissociation curve of man (see Lambertsen 1968).

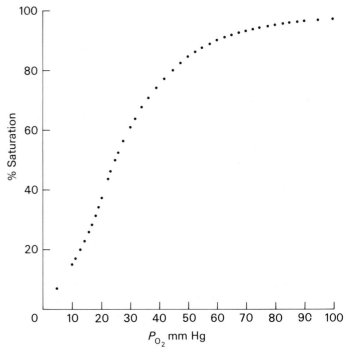

FIG. III.43. Oxygen dissociation of human blood at pCO_2 of 40 mm Hg and 37 °C.

REFERENCES

ADAIR, G. S. (1925). *Proc. R. Soc.* **A108**, 628.
LAMBERTSEN, C. J. (1974). In *Medical physiology*, 13th edn, Vol. 2 (ed. V. B. Mountcastle) Chap. 60, pp. 1399–422. Mosby, St. Louis.

MARGARIA, R. (1963). *Clin. Chem.* **9**, 745.
PERUTZ, M. F. (1967). *Harvey Lect.* **63**, 217.
ROUGHTON, F. J. W. (1964). In Respiration, Vol. 1 (ed. W. O. Fenn and H. Rahn) Chap. 31, pp. 767–827. American Physiological Society, Washington, DC.
SVEDBERG, T. (1926). *J. Am. chem. Soc.* **68**, 430.

DETERMINATION OF THE OXYGEN DISSOCIATION CURVE OF BLOOD

5 ml samples of blood are placed in a series of cylindrical glass vessels (tonometers) of 250–300 ml capacity. The tonometers are filled with a gas mixture of known composition of oxygen and carbon dioxide and nitrogen. They are rotated in a waterbath at 37 °C and the blood spreads in a thin film on the walls of the tonometer reaching equilibrium with the gaseous phase. The blood is then removed and analysed for its oxygen content and the gaseous phase is again analysed for its composition. From such results an oxygen dissociation curve is constructed in which oxygen content (in ml per 100 ml or in mM l^{-1}) is plotted as the ordinate against the partial pressure of oxygen (in mm Hg or in kPa) as the abscissa. Alternatively the 'percentage saturation' of haemoglobin with oxygen may be plotted on the ordinate. When blood is equilibrated at oxygen pressures above 100 mm Hg (13.3 kPa) the haemoglobin is fully saturated with oxygen. The oxygen content of such blood is made up from combined oxygen and dissolved oxygen. At 150 mm Hg (20 kPa) at 37 °C 0.5 ml O_2 are dissolved. Hence if the oxygen content of such blood is 20.5 ml per 100 ml, 20.0 ml per 100 ml represent 100 per cent saturation of the blood haemoglobin and this figure can be used correspondingly. Thus 10.1 ml O_2 per 100 ml will be 10 ml per 100 ml combined oxygen plus 0.1 ml dissolved (approximately) and this can be expressed as 50 per cent saturated (at a P_{O_2} of c. 30 mm Hg).

Nowadays with the advent of SI units the ordinate can be expressed as millimolar content of oxygen. A millimole of oxygen is 22.4 ml so if blood contains 205 ml when 'fully saturated' it will contain $205/22.4 = 9.15$ mM l^{-1} of which approximately 0.2 mM l^{-1} (0.5 ml per 100 ml) will be dissolved.

The term *oxygen capacity* is used to indicate the quantity of oxygen contained by 100 ml blood when it is fully saturated by equilibration at 150 mm Hg (20 kPa) at 37 °C. Then and only then can the *dissolved* oxygen content be allowed for. If the blood were equilibrated at this temperature with pure oxygen at say 700 mm Hg dry gas pressure its *dissolved* oxygen would be of the order of 2.1 ml per 100 ml or 0.9 mM l^{-1} (Henry's law).

On studying the O_2 dissociation curves of blood they are seen to be sigmoid in shape [FIG. III.44]. This sigmoid shape is largely due to the increased affinity of the fourth oxygenation reaction as already stated. The 'flat top' of the curve means that the alveolar (and hence arterial) tension of oxygen may drop from 100 mm Hg to 60 mm Hg without much decreasing the degree of saturation of the haemoglobin. Thus at moderate altitude subjects suffer little impairment in their uptake of oxygen by the blood.

The 'steep part' of the curve shows that large amounts of oxygen can be liberated from the blood with relatively minor falls of oxygen tension. This means that the pressure gradient of oxygen between blood and tissues (on which diffusion to the tissue cells depends) is kept high. Mixed venous blood returning to the right heart in a man at rest may be 70 per cent saturated and may have an oxygen tension of about 35–40 mm Hg. When the body tissues become active (e.g. muscles or glands) their oxygen usage rises and their own oxygen tension falls. This increases the pressure gradient between capillary blood and tissues and their supply of oxygen correspondingly improves by increased desaturation of the capillary blood. Two factors aid in this improvement of supply of oxygen,

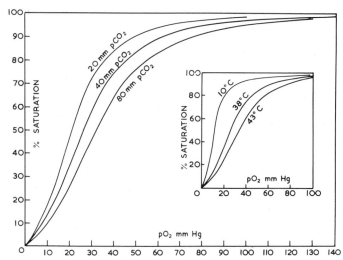

FIG. III.44. The large graph shows the effect of CO_2 on the oxygen dissociation curve of human blood at 38 °C. The inset figure shows the effect of temperature on the human blood oxygen dissociation curve determined at a pCO_2 of 40 mm Hg in each case.

which can be understood by inspection of the oxygen dissociation curves.

(a) The effect of CO_2 and pH. As pCO_2 increases (or as pH decreases, the steep part of the blood oxygen dissociation curve is displaced to the right, the so-called Bohr effect after its discoverer [FIG. III.44]. It should be noted that the effect of CO_2 in modifying the *saturation* of haemoglobin at oxygen pressures above 80 mm Hg is trivial. It is the *desaturation* which is aided by a raised CO_2 pressure, thereby increasing the pressure gradient between the capillary blood and the peripheral tissues. It must be understood that when tissues become more active their own CO_2 (and perhaps other sources of hydrogen ion) production increases and this facilitates the unloading of oxygen to them. At the same time the important role of CO_2 in relaxing the tone of the precapillary sphincters causes a large increase in the capillary surface area and a smaller intercapillary distance, with correspondingly improved diffusion of oxygen to the tissues.

(b) The effect of temperature. The curve is displaced to the right by a rise in temperature. Tissue activity again exerts its own influence on peripheral oxygen supply by promoting vasodilatation and increased capillary surface area and by aiding desaturation of the haemoglobin.

The result of increasing tissue activity is an increased coefficient of utilization. At rest the coefficient of utilization is:

$$\frac{O_2 \text{ taken up by the tissues}}{O_2 \text{ content of arterial blood}} \quad \frac{5}{20} = 25 \text{ per cent.}$$

In heavy muscular work, venous blood may leave the muscles with less than 4 ml O_2 per 100 ml—a coefficient of utilization of 80 per cent. In such circumstances the muscle blood flow also increases considerably so the total oxygen abstracted per minute is greatly raised.

The effect of diphosphoglycerate concentration in the red cell

The mature mammalian erythrocyte is non-nucleated and maintains its intracellular composition by glycolysis. The cell is directly permeable to glucose which on entry is converted by enzymes of the glycolytic pathway successively to glucose 6-phosphate,

3-phosphoglyceraldehyde, 1,3-diphosphoglycerate, and thence by a mutase to 2,3-diphosphoglycerate (DPG).

$$
\begin{array}{l}
COO^- \\
| \\
HC\!-\!O\!-\!PO_4 \quad \text{2,3-Diphosphoglycerate} \\
| \\
H_2C\!-\!O\!-\!PO_4
\end{array}
$$

2,3-Diphosphoglycerate exists in high concentration in erythrocytes. It has an important physiological function, acting as a highly charged polyanion which influences the affinity of haemoglobin for oxygen by binding to the β-chains of *reduced* haemoglobin (but not to those of oxygenated haemoglobin).

$$2,3DPG + HbO_2 \rightleftharpoons Hb\!-\!2,3DPG + O_2.$$

Consequently the presence of DPG favours the dissociation of oxygen from HbO_2 and shifts the oxygen dissociation curve to the right. Acidaemia hinders glycolysis and reduces the 2,3DPG concentration in the cell. Hypoxia on the other hand causes a rise in [DPG] and at high altitude this increases the unloading of oxygen to the tissues. This increased DPG concentration at altitude is secondary to the hypocapnia which occurs and with it the fall in intracellular [H⁺] of the erythrocytes.

Fetal haemoglobin contains 2 α-chains and 2 γ-chains, unlike adult Hb which possesses 2 α-chains and 2 β-chains. The γ-chains bind DPG much more feebly than do the β-chains which largely accounts for the greater affinity of fetal Hb for oxygen compared with that of its adult counterpart [see p. 587].

Kinetics of the reaction between oxygen and haemoglobin

Hartridge and Roughton (1924) rapidly mixed a solution of reduced haemoglobin with oxygenated Ringer solution and using a reversion spectroscope observed the flow of the effluent mixture passing down the glass tube. The instrument detects the slight alteration of the absorption bands when reduced haemoglobin is converted into oxyhaemoglobin. The combination of Hb and O_2 occurs in less than 10 ms. When a red cell suspension is used the combination takes longer because the oxygen has to diffuse into the cells but still occurs in 15–20 ms. Blood passes through the pulmonary capillaries in about 0.75 s in man at rest and in as little as 0.3 s in heavy exercise, so the 'safety factor' afforded by the rate of combination is a large one.

The dissociation of oxyhaemoglobin occurs at a similar speed and (unlike that of the combination of Hb and O_2) is accelerated by a rise of [H⁺], CO_2 pressure, and temperature.

REFERENCE

HARTRIDGE, H. and ROUGHTON, F. J. W. (1923). *Proc. R. Soc.* **B94**, 336.

OXYGEN CARRIAGE IN THE BODY

1. Each 100 ml of arterial blood passes to the tissues carrying about 0.3 ml of oxygen in solution, and about 19 ml in combination with haemoglobin. The tension of oxygen is about 100 mm Hg (13.3 kPa); it must be remembered that tension is a property solely of the gas in solution. The oxygen tension in resting tissues is probably just a little lower than that found in the venous blood, e.g. about 35 mm Hg. Owing to the great difference of oxygen pressure, oxygen rapidly passes out of the plasma through the capillary wall and tissue fluid to reach the tissue cells. The oxygen tension in the blood falls to about 40 mm Hg; complete equilibrium with the tissues is not achieved. The oxyhaemoglobin in the corpuscles is now exposed to an O_2 tension of 40 mm Hg in the plasma around it,

and therefore cannot retain all the oxygen it has hitherto held in combination. Dissociation occurs, and about 30 per cent of the oxygen present (e.g. 5–6 ml per 100 ml) is liberated. This volume of gas cannot remain in solution in the plasma, which already holds as much O_2 as it can; the oxygen liberated from the corpuscles must therefore diffuse out into the tissue fluid.

As a result the venous blood leaves with an oxygen tension of 40 mm Hg and an oxygen content of about 14 ml per 100 ml. There is slightly less oxygen in solution, and considerably less oxygen in combination with haemoglobin. The arteriovenous oxygen difference in oxygen content of arterial and venous blood is 19 − 14 = 5 ml per 100 ml.

In this instance the coefficient of utilization, which is

$$\frac{O_2 \text{ taken up by tissues}}{O_2 \text{ content of arterial blood}} = \frac{5}{19} = 0.26, \text{ or } 26 \text{ per cent.}$$

2. When a tissue is active, e.g. the skeletal muscles during vigorous exercise, the venous blood becomes far more extensively reduced. Very active muscles may abstract almost all the oxygen brought to them in the arterial blood, i.e. the coefficient of utilization is extremely high. The mixed venous blood (that in the right atrium) may have an oxygen content of 7–8 ml per 100 ml in hard work, or 3–4 ml per 100 ml in extremely violent exercise, corresponding to coefficients of utilization of 65 per cent and 80 per cent respectively; as the mixed venous blood consists partly of blood from skin and viscera, which is only slightly reduced, it is clear that the blood from the active muscles must be almost completely free from oxygen. This more extensive reduction in active tissues is brought about as follows:

(i) The number of patent capillaries becomes greatly increased [see p. 216]. Owing to the greatly increased total cross-section of the vascular bed locally, the linear velocity of the blood through the tissue is slowed down and thus more time is available for dissociation and diffusion of oxygen. The total blood flow is of course greatly increased because of local arteriolar dilatation.

(ii) A large surface of blood is in contact at any one time with the tissues, and the gaseous interchange is further facilitated. These circulatory effects, due to CO_2 and related acid metabolites, far outweigh in importance those effects due to a shift of the oxygen dissociation curve in ensuring a better supply of oxygen to active tissues.

(iii) As the tissue is consuming O_2 at a great rate, the oxygen tension within it probably falls to zero. A very steep oxygen pressure gradient exists between plasma and tissues, permitting rapid diffusion.

The oxygen tension in the blood falls, e.g. to 30 mm Hg; the oxyhaemoglobin dissociates and gives off about 60 per cent of the oxygen combined with it. In an active tissue the temperature rises, larger amounts of CO_2 are evolved, and the H⁺ ion concentration tends to go up. The rise of temperature and of CO_2 tension increase the rate of dissociation of HbO_2. The increased CO_2 tension means that HbO_2 gives off more oxygen at any given oxygen tension.

In the ways described, activity of a tissue, by altering the calibre of the precapillary resistance vessels and changing the local O_2 and CO_2 tension and the temperature, may enable three times the volume of oxygen to be abstracted from the same volume of blood flow compared with the resting utilization. As the blood flow is also increased the total oxygen abstracted in unit time is correspondingly raised.

3. When the mixed venous blood (O_2 tension 40 mm Hg) passes through the lungs it is exposed to an oxygen tension in the alveoli of 100 mm Hg. Owing to the great difference of oxygen pressure, the gas rapidly diffuses from the alveoli through the thin pulmonary and capillary epithelium into the plasma. The arterial blood finally leaves the lungs almost fully saturated with oxygen (97 per cent saturated), at an oxygen tension of 100 mm Hg. The oxygen con-

tent is then about 19 ml (the exact figure depends on the haemoglobin content); only 0.3 ml is in solution.

The peculiarities of fetal haemoglobin and the mechanism of O_2 transport in the placenta between mother and fetus are discussed on page 587.

Myoglobin

This is found in muscles, particularly those which are engaged in slow repeated contractions, e.g. leg muscles and in the hearts of large mammals. Myoglobin has a molecular weight of 16 800, contains one three-hundredth of its weight of iron and therefore contains one atom of iron per molecule. Myoglobin combines reversibly with oxygen according to the equation:

$$Mgb + O_2 \rightleftharpoons MgbO_2$$

At equilibrium

$$K = \frac{[MgbO_2]}{pO_2[Mgb]} \text{ where } K = \text{ the equilibrium constant.}$$

The percentage saturation 'y' is given by the equation:

$$y = 100 \frac{[MgbO_2]}{[Mgb] + [MgbO_2]}$$

$$= 100 \times \frac{KpO_2}{1 + KpO_2}$$

$$\text{therefore } \frac{y}{100} = \frac{KpO_2}{1 + KpO_2}.$$

This is the equation of a rectangular hyperbola. The oxygen dissociation curve of myoglobin has this form [FIG. III.45]. It can be seen that myoglobin takes up oxygen at low pressures much more readily than does blood. Hence when blood reaches the muscles the myoglobin can extract oxygen from the blood haemoglobin. The rate of association of myoglobin with oxygen is very fast. Blood haemoglobin shows the Bohr effect, i.e. the dissociation curve is moved to the right when the pCO_2 increases. The unimolecular myoglobin does not show the Bohr effect; hence in the muscles the increased unloading of O_2 fom blood is not impaired by a decreased power of loading up O_2 on the part of myoglobin. Even at a tension

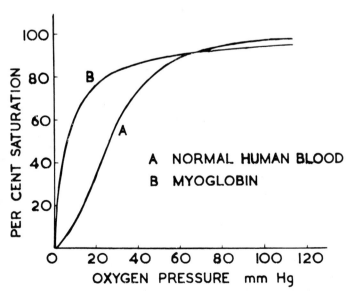

FIG. III.45. A. Oxyhaemoglobin dissociation curve of human blood at 38 °C. pH = 7.40. B. Oxygen dissociation curve of myoglobin under similar conditions. (Roughton, F. J. W. (1954). *Handbook of respiratory physiology*. USAAF Aviation School of Medicine.)

of O_2 of 40 mm Hg the myoglobin is 95 per cent saturated with oxygen. It is not until the pO_2 falls below 5 mm that the myoglobin becomes less than 60 per cent saturated. However, the tissue oxidases can operate satisfactorily even at this low oxygen tension.

Myoglobin presumably acts as a temporary oxygen store in the muscles. Thus during a muscular contraction the blood flow may be sharply reduced to increase again when the muscle relaxes. During the period of ischaemia the muscle uses oxygen supplied from its myoglobin which was charged up prior to the contraction.

CARBON DIOXIDE CARRIAGE

As already stated, plasma equilibrated at 37 °C with a gas mixture containing CO_2 at a partial pressure of 40 mm Hg (approximately 5.3 kPa), which is the alveolar or arterial partial pressure of CO_2, dissolves only 2.8 ml of the gas per 100 ml.

Arterial blood at pCO_2 of 40 mm Hg and at 37 °C contains some 48 ml of CO_2 per 100 ml. (As 1 mmol of CO_2 = 22.26 ml this can be re-expressed as 21.56 mmol i.e. *c.* 22 mmol CO_2 per litre.) Thus CO_2 must exist in blood mainly in the combined form.

Mixed venous blood in resting man contains about 52 ml CO_2 per 100 ml (23.36 mmol l^{-1}) at a partial pressure of 46 mm Hg (6.1 kPa). As the cardiac output is approximately 5.5 l min^{-1} in man at rest $(52 - 48) \times (5500/100) = 220$ ml CO_2 are given up per minute by the body—about 10 millimoles.

Just as it is customary to consider the oxygenation of the venous blood in the lungs first, so it is conventional to start with the uptake of CO_2 by the blood in the peripheral tissues. *Owing to a pressure gradient* CO_2 diffuses rapidly from tissues into the capillary blood in solution, first into the plasma and then into the red cells. CO_2 forms carbonic acid in solution, but only slowly and to a very limited extent if the reaction is uninfluenced by other factors.

$$CO_2 + H_2O \rightleftharpoons H_2CO_3.$$

Equilibrium is reached when no more than 0.2 per cent of the carbon dioxide is converted into H_2CO_3.

H_2CO_3 itself ionizes:

$$H_2CO_3 \rightleftharpoons H^+ + HCO_3^-.$$

The actual sequence would be trivial if it were not for two important facts.

(a) the *corpuscles* contain the enzyme carbonic anhydrase, which catalyses the reaction $CO_2 + H_2O \rightleftharpoons H_2CO_3$. An enzyme cannot itself alter the equilibrium point of a reaction, but if the second reaction: $H_2CO_3 \rightleftharpoons H^+ + HCO_3^-$ could be 'persuaded' to proceed to the right, then and only then would the kinetic influence of carbonic anhydrase be useful. Fortunately this 'secondary' influence is to hand.
(b) The histidine residues of the α- and β-chains of haemoglobin act as hydrogen acceptors [see p. 193] which 'mop up' the hydrogen produced in the corpuscle by the formation of H_2CO_3 therein from the CO_2 entering. Moreover, it transpires that, just as CO_2 reduces the amount of oxygen which can be carried by haemoglobin at any given pO_2, so does the release of oxygen from oxyhaemoglobin increase the hydrogen acceptance powers of the haemoglobin molecule and thereby permit a greater carrying power of the blood for carbon dioxide.

CO_2 dissociation curves of blood *in vitro* are constructed in the manner detailed for oxygen dissociation curves. FIGURES III.46 and III.47 show the dissociation curves for oxygenated and reduced blood. The influence of oxygen in reducing the carrying power of the blood for CO_2 is usually known as the C–D–H effect after Christiansen, Douglas, and Haldane who first described it in 1914. Not all of this 'extra' carbon dioxide carried by reduced blood at a given pCO_2 can be attributed to a difference in the hydrogen

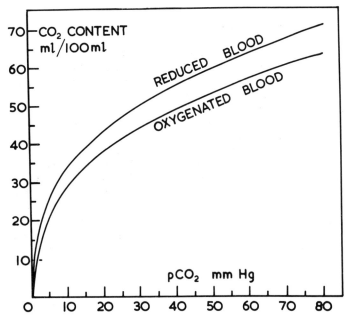

FIG. III.46. CO_2 dissociation curves of human blood. Temperature = 37 °C.

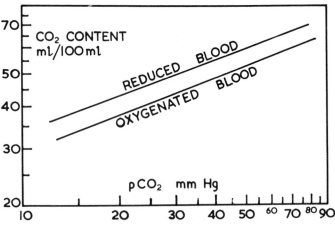

FIG. III.47. The same experimental results as shown in FIGURE III.46 plotted as log CO_2 content/log pCO_2.

acceptance powers between reduced and oxygenated haemoglobin. Some of it accrues from the fact that CO_2 entering the corpuscles combines directly (and rapidly) with haemoglobin itself to form a carbamino compound

$$CO_2 + Hb\,NH_2 \leftrightharpoons Hb\,NH\,COOH.$$

Reduced haemoglobin has a greater power to form this direct compound [see p. 189] and indeed about 20 per cent of the carbon dioxide entering the peripheral capillaries is accommodated as carbaminohaemoglobin. Consideration of the evidence for carbaminohaemoglobin formation will be deferred until we have dealt with the sequence of reactions which result from the hydration of CO_2 to carbonic acid and its subsequent ionization. As stated, the hydrogen ions have mostly been 'tucked away' in the histidine components of the haemoglobin. However, the bicarbonate ion concentration of the corpuscle has risen and this creates an imbalance between corpuscle and plasma. Bicarbonate ions diffuse out into the plasma but the corpuscular membrane is impermeable to the haemoglobin anion, which imposes a Gibbs–Donnan equilibrium on the system. In such a system [p. 16] the relative concentrations of the diffusible anions is given by

$$\frac{[HCO_3^-]\ \text{cells}}{[HCO_3^-]\ \text{plasma}} = \frac{[Cl^-]\ \text{cells}}{[Cl^-]\ \text{plasma}} = \frac{[OH^-]\ \text{cells}}{[OH^-]\ \text{plasma}} = r \simeq 0.7$$

As bicarbonate anions diffuse out, chloride ions move into the cell—the so-called chloride shift. (The metallic cations cannot leave the corpuscle owing to the Na^+–K^+ pump.)

The result of this chloride shift is that the bulk of the bicarbonate ions formed in the corpuscle as a result of CO_2 entry are finally accommodated in the plasma joining the 26 mmol l^{-1} of HCO_3 already present. The chloride shift mechanism takes about 750 ms, which considering the blood only spends 750 ms in the pulmonary capillaries in man at rest might render this reaction rate-limiting in CO_2 evolution. However, the carbaminohaemoglobin disruption in the pulmonary capillaries is rapid.

In the lungs the situation is reversed. Oxygenation of the haemoglobin occurs simultaneously with the expulsion of H^+ from the histidine residues. The chloride shift operates in the opposite direction as the reactions

$$H^+ + HCO_3^- \rightarrow H_2CO_3 \rightarrow CO_2 + H_2O$$

proceed as shown.

KINETICS OF TRANSPORT

Before 1933 little thought was given to the rate of uptake or evolution of CO_2. The uptake of CO_2 was understood to depend on the formation of bicarbonate ions derived from the reactions

$$CO_2 + H_2O \rightleftharpoons H_2CO_3$$
$$H_2CO_3 \rightleftharpoons H^+ + HCO_3^-$$

The formation of H_2CO_3 from CO_2 and water is, however, slow. As stated previously the blood spends only 0.75 s in the pulmonary capillaries and perhaps only 0.33 s in these capillaries during exercise. Roughton and his colleagues therefore examined the kinetics of CO_2 transport and found that:

1. A catalyst, which Eggleton named carbonic anhydrase, is present in the red cells; this enzyme accelerates the reaction: $H_2O + CO_2 \rightleftharpoons H_2CO_3$.
2. The rapid formation of carbaminohaemoglobin from CO_2 and haemoglobin assists the speedy uptake of CO_2 by the blood.

Carbonic anhydrase

Roughton used a manometric technique for examining the rate of uptake or evolution of CO_2. A specially designed vessel ('boat') was used in which a bicarbonate or phosphate solution, plasma or blood could be placed in one compartment and a suitable acid reagent in the other [FIG. III.48]. On shaking the vessel the two solutions were mixed and CO_2 was evolved at a rate which could be followed manometrically. Alternatively the boat and its connections with the manometer were filled with a CO_2/air mixture and a bicarbonate solution, plasma or blood was placed in the boat itself. On agitating the boat the solution took up CO_2 at a rate which could be measured. When phosphate solution, normal blood, and blood treated with cyanide were separately shaken with CO_2 and the rates of uptake of CO_2 measured, the uptake of CO_2 by normal blood was much more rapid than that of the phosphate. Some, but not all of this difference was due to the presence of carbonic anhydrase in the blood. If cyanide were added to the blood the enzyme was poisoned but still there was a rapid uptake of CO_2 in excess of that ascribable to the passage of CO_2 into solution. This in turn was found to be due to the formation of carbaminohaemoglobin [FIG. III.49].

Further analysis showed that carbonic anhydrase was present in the erythrocytes. In purified form carbonic anhydrase is a protein (molecular weight 30 000) containing one atom of zinc per molecule.

Carbonic anhydrase can be inhibited by cyanide, azide, and sulphide in concentrations of 10^{-3}–10^{-4} molar. It can also be in-

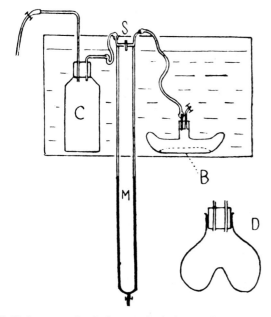

FIG. III.48. Apparatus for the boat method of measuring reaction velocity. D shows the end elevation of the boat, B, and the manner in which the bottom is divided into two separate compartments. M is the manometer, and C a jar compensating for temperature and pressure changes. The two limbs of the manometer can be thrown into communication by opening the clip S. The whole is immersed in a waterbath. (Brinkman, R., Margaria, R., and Roughton, F. J. W. (1938). *Phil. Trans. R. Soc.* **B232**, 65.)

hibited by sulphonamides in concentrations of 10^{-5}–10^{-8} molar. The most notable inhibitor of the enzyme is acetazolamide (*Diamox*). Much use of acetazolamide has been made in investigating the functional importance of carbonic anhydrase, which occurs not only in the erythrocytes but also in the gastric mucosa, in pancreatic tissue and in the renal tubular cells. The results obtained are discussed in the appropriate sections.

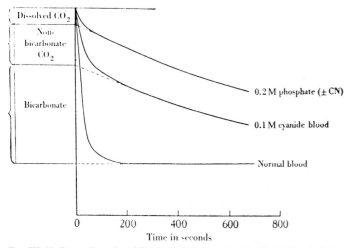

FIG. III.49. Rate of uptake of CO_2 by 0.2 M phosphate buffer (pH = 7.4), by blood poisoned with 0.1 M cyanide, and by normal blood. The ordinates represent the pressure of CO_2 in the gas phase in each case.

For phosphate buffer the initial rapid uptake is due simply to dissolved CO_2, the slow delayed uptake is due to the formation of bicarbonate from CO_2.

For cyanide blood, the rapid initial uptake is due to dissolved CO_2 together with some rapid, non-bicarbonate, chemical compound of CO_2. The slow delayed uptake is due to bicarbonate formation.

For normal blood, the dissolved CO_2, the non-bicarbonate-bound CO_2, and the bicarbonate-bound CO_2 are all taken up in the rapid initial phase. (Meldrum, N. V. and Roughton, F. J. W. (1933). *J. Physiol., Lond.* **80**, 113.)

Carbaminohaemoglobin

As just described, there is still a rapid uptake of CO_2 in excess of that taken up in solution when blood is exposed to CO_2/air mixtures even after the carbonic anhydrase has been inactivated by cyanide. This is due to the formation of carbaminohaemoglobin, i.e. the direct combination of CO_2 with amine ($-NH_2$) groups of the molecule. Ferguson and Roughton studied the effects of rapidly mixing haemoglobin solutions (equilibrated with CO_2) with separate solutions of alkaline barium chloride and sodium bicarbonate. The solutions were rapidly mixed in the special chamber devised by Hartridge and Roughton for the study of the equilibrium between haemoglobin and oxygen, and the composite solution was speedily delivered into a centrifuge tube surrounded by ice. The barium salt of a carbamino compound is soluble and fairly stable at 0 °C if the pH is sufficiently high, whereas in such conditions CO_2, H_2CO_3, and HCO_3^- are all converted to CO_3 and are precipitated as $BaCO_3$. Hence the carbon dioxide content of the supernatant fluid gave the concentration of carbamino-bound CO_2 in the original solution.

From such studies it was found that the carbamino-bound CO_2 content of a haemoglobin solution varied only slightly with CO_2 pressures above 10 mm Hg, at which tension the carbamino-binding power was almost saturated.

The reason for this is that although the reaction: $CO_2 + HbNH_2 \rightleftharpoons HbNHCOOH$ goes to the right with increasing CO_2 tension the increase in CO_2 concentration also causes more formation of H_2CO_3 and hence more H^+ from its dissociation. H^+ reacts with the NH_2 groups to form $-NH_3^+$ so there is less $-NH_2$ to combine with CO_2.

Of some importance, however, is the finding that reduced haemoglobin has a much greater power of forming carbamino-bound CO_2 at any given pCO_2 than has oxyhaemoglobin. It is for this reason that the carbamino-binding power of the respiratory pigment plays a significant role in CO_2 turnover in the body, for the entrance of CO_2 to the capillary blood from the tissues is accompanied by a coincidental deoxygenation of the oxyhaemoglobin. The difference between the CO_2 carrying power of oxygenated blood and reduced blood is partly due to the difference in the carbamino-combining power of oxyhaemoglobin and reduced haemoglobin and partly due to the greater power of hydrogen ion acceptance of reduced haemoglobin compared with that of oxygenated haemoglobin.

However, these findings were obtained in haemoglobin solutions which contained no 2,3-diphosphoglycerate (DPG). It is now known that CO_2 and DPG compete for the same binding sites of the Hb molecule. When DPG is added to Hb solutions carbamate formation is less, particularly when the Hb solution is reduced. Nowadays it is thought probable that carbamino binding accounts for only 10 per cent of the total transport of CO_2.

Carbamino binding occurs only at the terminal α-amino groups of the four globin chains and can be prevented by 'blocking' these groups by combining them with cyanate. The terminal groups on the two β-chains react with DPG and it is these which have the highest affinity for CO_2.

REFERENCES

KILMARTIN, J. V. and ROSSI-BERNARDI, L. (1973). *Physiol. Rev.* **53**, 836.
MICHEL, C. C. (1974). Chapter 3. In *Respiratory physiology* (ed. J. G. Widdicombe) pp. 67–104. Butterworths, London.
ROUGHTON, F. J. W. (1964). Chapter 31. In *Handbook of physiology*, Section 3 *Respiration*, Vol. 1 (ed. W. O. Fenn and H. Rahn) pp. 767–826. American Physiological Association, Washington.

Hydrogen ion concentration of blood

The term pH means $\log \frac{1}{[\text{H}^+]}$ and was introduced to obviate the need for expressing hydrogen ion concentrations (which are very small indeed in blood) in terms of 10^{-7}, etc. However, the new use of SI units has introduced terms which are easily applicable to the situation.

Remembering that the dissociation constant of water K_w has the value 10^{-14} at 23 °C then as $[\text{H}^+] = [\text{OH}]$ both must have the value of 10^{-7} mol l⁻¹. pH = pOH = 7.0. Water is regarded as neutral.

A solution with a hydrogen ion concentration of 1×10^{-7} moles per litre is the same as one with $[\text{H}^+] = 100$ nanomoles per litre. Thus pH = 7.0 = 100 nM hydrogen ion per litre.

The addition of 0.3 to such a pH value is equivalent to halving the hydrogen ion concentration so it follows that pH 7.30 is the same as $[\text{H}^+] = 50$ nM l⁻¹. Using this information the table given below is easily understandable

pH	[H⁺] (nanomolar)
7.0	100
7.1	80
7.2	62.5
7.3	50
7.4	40
7.5	31.25
7.6	25
7.7	20
7.8	15.6
7.9	12.5
8.0	10

The conversion of a pH value into [H+] and vice versa requires some explanation and the following examples should prove helpful.

To convert a pH value, say 7.45, into the hydrogen ion concentration in nM l⁻¹ is done as follows:

$$\text{pH} = -\log[\text{H}^+] = 7.45$$
$$\log[\text{H}^+] = -7 - 0.45 = -8 + 0.55 = \bar{8}.55.$$

Therefore $[\text{H}^+] = 10^{-8} \times 3.55$ (3.55 is antilog of 0.55)
$[\text{H}^+] = 35.5$ nM l⁻¹.

To convert a hydrogen ion concentration, say of 107 nM l⁻¹ into the corresponding pH

$$-\log[\text{H}^+] = \text{pH}$$
$$[\text{H}^+] = 107 \times 10^{-9}$$
$$\log[\text{H}^+] = -9 + 2.029 \text{ (where 2.029 is log of 107)}$$
$$-\log[\text{H}^+] = +9 - 2.03$$
$$\text{pH} = 6.97.$$

It will be noted that in common logarithms the *mantissa* (numbers to the right of the decimal point) is always positive. The *characteristic* (number to the left of the decimal point) may be positive or negative. A bar is placed above the figure of the characteristic if it is negative.

When the pH of arterial blood is measured at 37 °C using a glass electrode, the sample must be treated anaerobically, i.e. precautions must be taken to prevent any loss of CO_2. The value obtained (approximately 7.40) is the pH of 'true' plasma—the plasma phase in gaseous and electrochemical equilibrium with the corpuscles suspended in it. This value of 7.40 is characteristic of the normal individual. In acidaemia it may fall to as low as 7.00 and in severe alkalaemia it may rise to as much as 7.80. These values roughly correspond to the range of true plasma pH compatible with survival.

The relative constancy of 'true' plasma pH is remarkable considering that hydrogen ions are continuously entering the blood.

Fortunately the major potential contributor, CO_2, is removed by alveolar ventilation (no less than 330 litres of CO_2 (equal to 15 000 millimoles of H^+) being expired daily). $H_2PO_4^-$ and HSO_4^- radicles are excreted in the urine (approximately 160 mmol day⁻¹). They are derived from the oxidation of phosphorus- and sulphur-containing proteins ingested in the diet. Obviously these components must gain access to the blood before they can be removed by the kidney yet the pH of the blood plasma phase remains stable. The stability of hydrogen ion concentration is due to the presence of *buffer systems* in the blood, extracellular fluid, and intracellular fluid.

Buffers and buffer systems

Buffers are substances which confer the ability of a solution to resist a change of pH when acid (a proton donor, such as hydrogen ion) or base (a proton acceptor) is added to it. Most buffers of biological fluids are composed of a weak acid and its conjugate base.

The dissociation reactions for members of a buffer pair are:

Weak acid HA: $\quad \text{HA} \rightleftharpoons \text{H}^+ + \text{A}^-$
Conjugate base NaA: $\text{NaA} \rightarrow \text{Na}^+ + \text{A}^-$

In such a mixture, nearly all the anions A^- result from the dissociation of the conjugate base; the equation derived from the Law of Mass Action (Guldberg and Waage) is:

$$K^1_A = \frac{[\text{H}^+][\text{A}^-]}{[\text{HA}]}$$

which may be rewritten as:

$$[\text{H}^+] = K^1_A \times \frac{[\text{HA}]}{[\text{A}^-]} \text{ (Henderson)}$$

or in its negative logarithmic form (Hasselbalch)

$$\text{pH} = pK^1_A + \log \frac{[\text{A}^-]}{[\text{HA}]}.$$

On studying the *titration curve* of a buffer solution it shows the general form of FIGURE III.50.

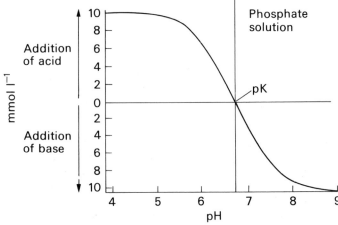

FIG. III.50. Titration curve of a 20 mM solution of NaH_2PO_4–Na_2HPO_4. Buffer value is maximal at 6.8 which is the pK for phosphate.

The ability of the system to resist changes of pH is maximal when pH of the solution is identical with that of the pK value. When pH = pK, then from the Henderson–Hasselbalch equation:

$$\log \frac{[\text{A}^-]}{[\text{HA}]} \text{ is 0 thus } [\text{A}^-] = [\text{HA}] \text{ for the log of unity is 0.}$$

The *buffer value* of the solution is obviously greatest in the middle range of the titration curve (where the slope is virtually linear) and this gives a measure of the *buffering power of the system*—the number of moles of acid or base which must be added to a molar solution in order to change the pH by one unit. Van Slyke (1922) proved that the maximal value for any *single* buffer system is 0.575 per pH unit—i.e. that 0.575 moles of H^+ must be added to a molar solution of the buffer to induce a change of pH of one unit.

From what has been stated, the buffer value of any one system in the body fluids must be related to its pK value compared with the pH value which characterizes the body fluids. Its *overall* contribution to buffering power *must of course also depend on its molar concentration*.

In the body, molar concentrations can be conveniently replaced by defining buffer value in terms of millimoles of hydrogen ion which must be added per millimole of the buffer system to induce a change of unit pH or, alternatively, the number of millimoles of anions associated with a change of unit pH when hydrogen ions are added to, or taken from, the solution.

Considering the inorganic phosphates $[H_2PO_4^-]$ and $[HPO_4^{2-}]$ in plasma, despite the fact that the pK of the phosphate system is 6.8 and therefore close to the pH of plasma, the inorganic phosphates in plasma are of trivial importance as buffers *because their concentration (3–5 mg per 100 ml) is so small*. Nevertheless, it is possible to predict from the Henderson–Hasselbalch equation that the two components exist in a ratio of

4 of $[HPO_4^{2-}]$ to 1 of $[H_2PO_4^-]$ at a plasma pH of 7.40

for: $7.4 = 6.8 + \log \dfrac{[HPO_4^{2-}]}{[H_2PO_4^-]}$

$= 6.8 + 0.6$, and 0.6 is the log of 4.

Only in the urine, where the phosphates are highly concentrated, do they exert an important role in buffering.

This example serves to show, however, that the ratio of components of a buffer system at a given pH can be predicted from a knowledge of the pK of the system.

In the case of the bicarbonate/carbonic acid system the equation expresses the situation as:

$$pH = pK + \log \frac{[HCO_3^-]}{[H_2CO_3]}$$

or more usually as:

$$pH = pK + \log \frac{[HCO_3^-]}{\alpha pCO_2}$$

where α is the solubility coefficient of CO_2 and *all* the CO_2 present is supposed to exist as H_2CO_3. As previously stated, the equilibrium of the reaction $H_2O + CO_2 \rightleftharpoons H_2CO_3$ is such that only 0.1–0.2 per cent of the CO_2 is actually converted into carbonic acid. The pK of the above equation is usually given as 6.1 but if this last consideration is taken into account, the 'true' pK is 3.1. However, the important point lies in the fact that CO_2 itself is volatile and is expelled by the lungs by alveolar ventilation, which is in turn determined by the hydrogen ion concentration of the cerebral interstitial fluid [p. 196].

The cerebral interstitial fluid and the c.s.f. normally contain so little protein that the protein buffering power of these fluids is negligible. $[H^+]$ of the cerebral interstitial fluid is determined in essence by the ratio of $\alpha pCO_2/[HCO_3^-]$. CO_2 is freely diffusible; HCO_3^- much more slowly so. Consequently any rise in pCO_2 is reflected within a few seconds as a rise in $[H^+]$ of the cerebral interstitial fluid and this stimulus excites the 'respiratory centre'. Hyperpnoea 'blows off' the extra CO_2 until the *status quo* is regained.

In other words the bicarbonate/carbonic acid system is heterogeneous—acid entering the blood and converting HCO_3^- to H_2CO_3 merely causes an increase in CO_2 tension which in turn promotes

the removal of CO_2 from the body—pCO_2 remains remarkably constant as a result. The buffer value of the bicarbonate/carbonic acid system is independent of pH within the range seen in intracellular and extracellular fluid and depends only on the concentration of bicarbonate ion as long as pCO_2 is maintained constant or nearly so, by appropriate modifications of alveolar ventilation.

When non-carbonic acids (e.g. lactic acid, aceto-acetic acid) enter the blood they cause not only a decrease in $[HCO_3^-]$ but also a decrease in the anionic concentrations of proteins. When CO_2 is formed in the tissues it titrates only the non-carbonic buffers, which mainly consist of proteins—haemoglobin, plasma proteins and intracellular proteins of the tissues and organic phosphates in the cells).

Buffering by proteins with special reference to haemoglobin

Proteins are composed of amino-acids which possess amino and carboxyl groups:

Amino-acids can ionize in two ways. The NH_2 group can serve as a hydrogen ion acceptor, ionizing as NH_3^+, and the carboxyl group can serve as a hydrogen ion donor ionizing as COO^-. There is a characteristic $[H^+]$ for each amino-acid at which its net charge is zero—the sum of its positive charges equalling the sum of its negative charges. This hydrogen ion concentration is defined as the *isoelectric point* of the amino acid—the corresponding pH is termed pI. At its isoelectric point the amino-acid behaves as an amphoteric electrolyte (Zwitterion). For an acidic amino acid such as glutamic acid $(COOH—CHNH_2 (CH_2)_2 COOH)$ pI is far below 7—in this case pI = 3.2. For a basic amino-acid such as lysine $(COOH—CH_2NH_2(CH_2)_3—CH_2—NH_2)$ the isoelectric point is 9.7.

However, proteins are composed of peptide links (—CO.NH—) of the amino acids and this linkage in most cases 'uses up' their ionizing propensities so the great majority of these combinations play no part in buffering. In the special case of haemoglobin, although the α-chains have 141 amino-acid residues, the terminal amino-acids are valine, with its NH_2 group and arginine with its COOH group at the other end. The β-chains with 146 amino-acid residues have valine NH_2 and histidine COOH at their ends.

In all then, there are four terminal valine NH_2 groups, two arginine COOH groups and two histidine COOH groups. The pK value of these terminal COOH groups is approximately 4.0 —hence they exist almost entirely as COO^- at a plasma pH of 7.4 or at an erythrocyte pH of roughly 7.2. They can therefore be ignored as potential hydrogen acceptors or donors. The four valine NH_2 groups have a pK of 7.7 and cannot be totally ignored as potential hydrogen acceptors but there are *only* four of them. Haemoglobin has a buffer value of about 2.85 millimoles of hydrogen ion per unit change of pH per mM of haemoglobin per litre. The maximal buffer value of a *single* buffer system is 0.575 millimoles of hydrogen ion per unit change of pH per mM concentration of buffer (Van Slyke 1922). It follows that we must look elsewhere in the haemoglobin molecule than at the terminal NH_2 groups of the α and β polypeptide chains to account for this extraordinarily high buffering power of oxygenated haemoglobin.

Thanks to the work of Perutz, Kendrew, and their colleagues the exact composition of the α- and β-chains has been determined and we can concentrate on those amino-acid components which, despite their peptide linkages, might conceivably play an important role in buffering hydrogen ion changes in blood. Lysine

(COOH—CH—NH$_2$—(CH$_2$)$_3$—CH$_2$—NH$_2$) obviously forms peptide links still leaving one NH$_2$ group free to accept hydrogen ion as—NH$_3^+$. There are indeed 44 lysine molecules in the four polypeptide chains but as the isoelectric point of lysine is 9.7, at a plasma pH of 7.4 or an erythrocyte pH of 7.2, lysine must exist almost exclusively in the ionized form of —NH$_3^+$ and cannot play any role.

There are in all 56 dicarboxylic acids (aspartic and glutamic) in the four polypeptide chains but their 'side chain' COOH groups have pK values of approximately 4.0. Thus at plasma pH they exist as COO$^-$ and play no part in buffering.

Attention concentrates on the histidine components of the haemoglobin molecule of which there are 38, 20 in the two α-chains and 18 in the two β-chains. Histidine has the formula:

$$NH_2-CH.COOH$$
$$|$$
$$CH_2$$
$$|$$
$$HC = C$$
$$| \quad |$$
$$HN \quad N$$
$$\backslash C \nearrow$$
$$|$$
$$H$$

Even when combined in the polypeptide chain, histidine remains a potential hydrogen ion acceptor as shown:

$$-NH-CH.CO- \quad \xrightarrow{H^+} \quad -NH.CH.CO-$$

UNIONIZED   IONIZED

The $\equiv$ N in the imidazole ring accepts hydrogen to become $\equiv$ NH$^+$. The pK value of histidine is approximately 7.0 which is very near the operative pH of the erythrocyte (7.2) and the plasma 7.4. There is little doubt that the buffering properties of haemoglobin per se can mainly be attributed to its histidine components.

We have so far referred to the millimole of haemoglobin as being the unit of 16.7 g l^{-1}. In these terms the buffer value of haemoglobin is 2.85—that is 2.85 mM of hydrogen ion must be added to a solution containing 16.7 g l^{-1} of haemoglobin in order to change its pH by one unit [see FIG. III.54, p. 193]. As the maximal millimolar buffer value of a single ionizable group is 0.575 this value of 2.85 must imply that a *minimum* of 2.85/0.575 = approximately 5 ionizable units must be required. As each of the haem-polypeptide units of 16.7 g l^{-1} contains either ten (α-chain) or nine (β-chain) histidine units it is probable that all of them are concerned in the buffering properties of the molecule because *some* of them may be affected by neighbouring amino acids linked by the polypeptide connections and may exert less than their maximal influence.

The preponderant contribution of haemoglobin to the buffering power of blood as compared with that of the plasma proteins can best be appreciated by the following numerical examples.

When 5 mmol of hydrochloric acid are added to neutral water (pH 7.0) the hydrogen ion concentration of the resultant solution is, of course, 5 mmolar. As the pH must be the negative logarithm of [H$^+$] = 5×10^{-3} it will be $(+0.7 - 3) = 2.30$.

Now let us consider blood which contains erythrocytes and plasma. The following equation approximately defines the quantitative buffer powers of the plasma proteins.

$$[Prot^-] = 0.1 \times [Prot] \times (pH - 5.08),$$
where [Prot$^-$] = millimolar concentration of protein anions and [Prot] = grams of protein per litre.

The plasma protein concentration is approximately 70 grams per litre. As the haematocrit value is 45/55 the plasma protein concentration per litre of blood is $55/100 \times 70 = 38.5$ grams per litre. At pH 7.4:

$$[Prot^-] = 0.1 \times 38.5 \times (7.40 - 5.08) = 8.84 \text{ millimols.}$$

At pH 7.2: [Prot$^-$] falls to 8.16 millimols which is a reduction of approximately 0.7 mM or 700 000 nM which represents the amount of hydrogen ion 'mopped up' while the free hydrogen ion change has only been from 40 nM to 62.5 nM l^{-1}.

As stated before *the buffer value of a system is defined as the number of millimoles of hydrogen ion which must be added to a system to change its pH by one unit*. In this case 0.7 mmol hydrogen ion changed the pH by 0.2, so the contribution of the plasma proteins to the buffer value of blood is 3.5 millimoles of hydrogen ion per unit-pH change.

Far more important than the plasma proteins is haemoglobin. One litre of *red cells* contain about 335 grams of haemoglobin, i.e. at a concentration of some 33.5 per cent. If the millimole of haemoglobin is taken as 16.7 grams (the amount that binds one millimole of oxygen) the litre of red cells contain 20 millimoles of haemoglobin. However, a litre of blood contains only 450 ml of red cells so blood has a millimolar concentration of 9 millimoles of haemoglobin (another way of stating that $9 \times 16.7 = 150$ is the blood concentration of Hb in grams per litre).

When the ionization of oxygenated haemoglobin in solution is studied, the following approximate equation expresses the experimental facts.

$$[HbO_2^-] = 2.85 \times [HbO_2] \times (pH - 6.63)$$
where [HbO$_2^-$] = millimoles per litre of haemoglobin ions and [HbO$_2$] = concentration of haemoglobin in millimoles per litre (regarding 16.7 g as the millimole) [see FIG. III.54, p. 193].

If 5 millimoles of hydrogen ion are added in the form of a strong acid to a solution containing 9 millimoles of haemoglobin at a pH of 7.40 the change in pH can be calculated.

Clearly, 2.85 millimoles of hydrogen ion added to 1 millimolar haemoglobin solution will cause a drop of pH by 1 unit. Therefore 5 millimoles of hydrogen ion will reduce the pH by $5/(9 \times 2.85) = 0.194$ units. Let us say that the pH falls by 0.2 units—from its initial value of 7.40 to 7.20. The increase in free hydrogen ion concentration has only been from 40 nanomoles per litre to 62.5 nanomoles per litre, despite the fact that 5 *million* nanomoles of hydrogen ion have been added.

The total buffer value of haemoglobin in a litre of blood is slightly less than that (2.85) given from the studies on haemoglobin in a litre of water. A litre of blood contains only 0.84 litre of water, the remainder being occupied by haemoglobin and plasma proteins. Acid added to a litre of blood is thus 1.0/0.84 more concentrated than is indicated by the change in the anion concentration. Thus the actual buffer value of haemoglobin per litre of blood is $9 \times 2.85 \times 0.84 = 21.5$ millimoles [H$^+$] per unit pH and that of the plasma proteins is 3.25 mM [H$^+$] per unit pH.

The sum of the contributions of haemoglobin and the plasma proteins yields approximately the slope of the CO$_2$ titration curve of whole blood—viz. $21.5 + 3.25 = 24.15$ mM of hydrogen ion per unit pH.

As these figures are approximate in the case of the plasma proteins it is reasonable to say that the slope of the titration line of *whole blood* is about 25 mM of hydrogen ion per unit pH change [FIG. III.51].

TABLE III.3 shows the approximate concentrations of bicarbonate ions, oxyhaemoglobin ions, and plasma protein ions at four different pH values produced *by altering the CO$_2$ tension alone* and FIGURE III.52 plotted from these results shows how as the CO$_2$

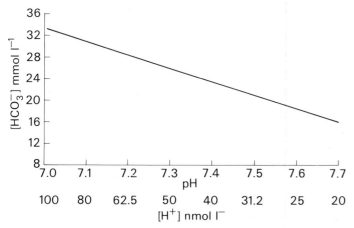

FIG. III.51. The titration curve of oxygenated blood in terms of the alterations of $[HCO_3^-]$ caused by increments of $[H^+]$ produced by increasing CO_2 tension.

TABLE III.3

pH	$[HCO_3^-]$	$[P.Prot^-]$	$[HbO_2^-]$	$\begin{bmatrix} Total \\ Prot^- \end{bmatrix}$	Total anion
7.03	33.7	7.6	8.7	16.3	50.0
7.23	28.6	8.3	13.0	21.3	49.9
7.43	23.3	9.0	17.3	26.4	49.7
7.63	18.1	9.7	21.8	31.5	49.6

tension falls the $[HCO_3^-]$ diminishes and the protein ionization rises, mainly because of the change in oxyhaemoglobin ionization. Plasma protein ionization changes only trivially (from 7.6 mmol l^{-1} to 9.7 mmol l^{-1}).

It should be noted that the sum of the protein anions and the bicarbonate anions equals approximately 50 mmol l^{-1} [FIG. III.53]. This figure yields the value accorded to the *buffer anion concentration*—sometimes misleadingly referred to as its stoichiometric equivalent in cations, i.e. *buffer base*. There is no merit in this

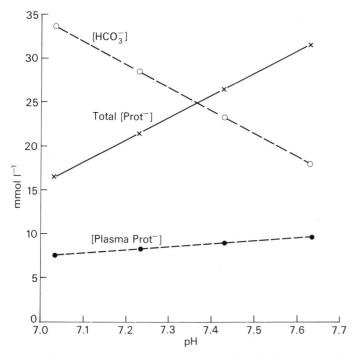

FIG. III.52. Changes in bicarbonate and protein ionization with pH is oxygenated blood exposed to varying CO_2 tensions as in FIGURE III.51.

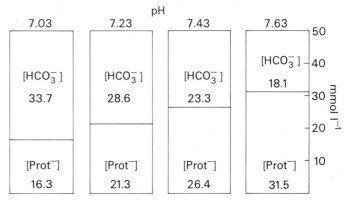

FIG. III.53. Note that the total buffer anion concentration is unchanged at approximately 50 mmol l^{-1} when the pH is altered by varying the CO_2 tension although the parcellation of the $[HCO_3^-]$ and protein anion [Plasma prot$^-$ and HbO_2^-] changes.

term—there is no electrochemical difference between the metallic cations which 'cover' the buffer anions (whose charge and concentration varies with pH) or the so-called fixed anions (e.g. Cl$^-$).

It should now be obvious that any change in the ionization of haemoglobin will result in an alteration in the number of bicarbonate anions which can be 'covered' by the cations present in the blood (approximately 150 mmol l^{-1} *total* of which only about 50 mmol l^{-1} are *available*. Two examples can be given, one normally far more important than the other:

1. Deoxygenation of oxyhaemoglobin in the tissues with the formation of reduced haemoglobin.
2. Alteration of the temperature at which the blood is equilibrated.

Reduction of oxyhaemoglobin lessens its ionization and there are correspondingly a greater number of cations 'free' to cover bicarbonate ions. Thus, at any given CO_2 tension reduced blood carries more CO_2 as bicarbonate than does oxygenated blood [see FIG. III.54]. Indeed the reduction of 1 mmol of HbO_2 allows the uptake of 0.7 mmol extra bicarbonate without there being any change in the hydrogen ion concentration [FIG. III.54]. This is, of course, advantageous in the tissues where deoxygenation of oxyhaemoglobin and the formation and uptake of CO_2 proceed simultaneously.

The alteration of temperature is not of the same importance in normal circumstances, but has to be borne in mind when the body is artificially cooled as in cardiac operations performed on hypothermic patients. Such patients are cooled to temperatures of 25 °C and at such levels the protein ionization is strikingly reduced so the

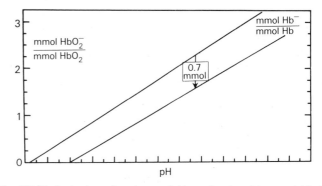

FIG. III.54. Ionization of oxyhaemoglobin and reduced haemoglobin in solutions at 37 °C. Upper line = ionization of oxyhaemoglobin. Lower line = ionization of reduced haemoglobin. Note that reduction of oxyhaemoglobin at pH 7.40 lowers the concentration of haemoglobin ions by 0.7 mmol mmol^{-1} (Hb). (Data of Nashat, F. S. and Neil, E. (1956). *Proc. XXth Int. Congr. Physiol.*)

carrying power of the blood for CO_2 as $[HCO_3^-]$ is markedly *increased*.

So far we have considered the situation when the haemoglobin, bicarbonate, and (unimportant in this instance) the plasma protein ionic concentrations total approximately 50 mmol l^{-1}. However, when any acid (source of H^+) stronger than H_2CO_3 enters the blood the extra supply of hydrogen ions promptly reduces the total concentration of buffer anion

$$HbO_2^- + H^+ = H\,Hb$$
$$Prot^- + H^+ = H\,Prot$$
$$HCO_3^- + H^+ = H_2CO_3$$

As a result most of the influx of H^+ is temporarily masked by buffering although in the fullness of time the extra H^+ *must* be excreted so that the buffer anionic concentration can eventually be restored. Such excretion of hydrogen ions is carried out by the kidney.

Of the immediate reactions, however, that of $HCO_3^- + H^+ = H_2CO_3$ is unique in that H_2CO_3 is potentially volatile. Thus

$$H_2CO_3 \rightleftharpoons CO_2 + H_2O$$

and the carbon dioxide formed is blown off in the lungs. Moreover, the activity of the respiratory central neuronal complex is markedly increased by the $[H^+]$ of the cerebral interstitial fluid (which contains a negligible concentration of protein):

$$[H^+] \propto \frac{\alpha pCO_2}{[HCO_3^-]}$$

and the fall in $[HCO_3^-]$ coupled with the rise in the CO_2 tension stimulates alveolar ventilation.

The equation:

$$[H^+] \propto \frac{\alpha pCO_2}{[HCO_3^-]}$$

has led to the classification of acidosis (or alkalosis) under two headings—primary 'non-gaseous' and primary 'gaseous'. The following examples may be given:

1. Non-gaseous acidosis: a decrease in $[HCO_3^-]$ which is, however, primarily caused by an increase in $[H^+]$ of the plasma:
   (i) Diabetic ketosis—entrance of β-hydroxybutyric acid into the blood.
   (ii) Renal failure—in which the kidney fails to excrete its normal quota of hydrogen ions.
   (iii) Muscular exercise (severe)—in which lactic acid may be produced.
   (iv) Starvation—in which keto acids are produced.
   (v) Infantile diarrhoea (loss of $NaHCO_3$).
The appropriate compensatory reactions are (a) increased breathing which lowers the alveolar pCO_2; (b) increased renal excretion of H^+. Obviously this latter compensation is impossible in renal failure which has caused the primary condition.

2. Gaseous acidosis: primary increase in pCO_2 and $[H_2CO_3]$ caused by:
   (i) Emphysema—the respiratory function of the lungs is inadequate.
   (ii) Depression of the respiratory centre—e.g. by morphine.
The appropriate compensation is the renal excretion of hydrogen ions which raises the buffer anionic concentration of HCO_3^-. The kidney forms a highly acid urine with a high concentration of ammonium ions [see p. 229].

3. Non-gaseous alkalosis: primary increase in $[HCO_3^-]$ (and also protein and haemoglobin ionization) caused by:
   (i) Ingestion of bicarbonate or citrate (e.g. for peptic ulcer therapy or to produce an alkaline urine).

(ii) Pyloric obstruction and high intestinal obstruction in which H^+ are lost from the body.
The breathing compensatory reactions are (a) decreased breathing which raises pCO_2 and hence $[H_2CO_3]$; (b) increased renal excretion of bicarbonate.

4. Gaseous alkalosis: primary decrease in pCO_2 and hence $[H_2CO_3]$ caused by:
   (i) Voluntary overbreathing.
   (ii) Chronic oxygen lack which causes hyperpnoea.
In both cases the appropriate compensation is effected by the kidney which secretes an alkaline urine containing bicarbonate. Whereas the Henderson–Hasselbalch formulation has thus been of value in understanding the above changes, it has perhaps concentrated the attention of students too much on the buffer properties of the bicarbonate/carbonic acid system itself. Even allowing for the fact that the bicarbonate/carbonic acid system contains the volatile component CO_2 it is necessary to remember that the buffer properties of blood are mainly dependent on the presence of haemoglobin. The bicarbonate/carbonic acid system acts as a go-between, playing an important role in that (a) CO_2 can be evolved in the lungs; (b) bicarbonate can be excreted in the urine. Haemoglobin enjoys neither of these advantages.

REFERENCE

Van Slyke, D. D. (1922). *J. Biol. Chem.* **52**, 525.

THE LOG pCO_2/pH GRAPH

When blood samples are equilibrated in tonometers at various CO_2 tensions and the corresponding pH values measured the points lie on a straight line if the CO_2 tensions are expressed logarithmically. If the pH values are obtained from the corresponding samples of true plasma [p. 190] the points lie on the same line [Fig. III.55]. The so-called pH of blood is thus that of its true plasma phase.

The effect of increasing or decreasing the buffer anion concentration, on the position of the log pCO_2/pH line

The position of the log pCO_2/pH line of blood depends on the concentration of buffer anion available. Normally blood contains about 48–50 mM buffer anion per litre and the corresponding log pCO_2/pH line is shown as [A] in Figure III.56.

1. Addition of H^+. After addition of sufficient lactic acid to reduce the buffer anion concentration to 35 mM l^{-1} the position of the log pCO_2/pH line is [B] in Figure III.56.
2. Addition of HCO_3^-. By adding sodium bicarbonate the concentration of buffer anion is increased to 65 mM l^{-1} and the corresponding log pCO_2/pH line is shown as [C] in Figure III.56.

Table III.4 shows the pH values of these different bloods at $pCO_2 = 40$ mm Hg.

Table III.4

	Acidaemia	Normal	Alkalaemia
Buffer anion concentration mM l^{-1}	35.0	49.0	65.0
pH at $pCO_2 = 40$ mm Hg	7.17	7.37	7.55

From the Fig. III.56 it can be appreciated that compensation can be effected:

1. By altering the pCO_2.
2. By renal adjustments which correct the change in buffer anion concentration.

In the 'acidaemic' blood, if the pCO_2 were lowered to approximately 15 mm Hg the pH of the blood would be restored to 7.37.

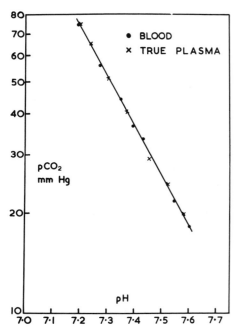

Fig. III.55. Log pCO_2/pH lines (at 37 °C) of blood and true plasma. (Nashat, F. S. and Neil, E.)

This theoretical value could be achieved in the body by vigorous hyperpnoea, occasioned by the acidaemia. In actual fact, the stimulus to breathing caused by acidaemia is never so intense as to offset completely the influx of H^+. Hyperpnoea does develop but what might be called the 'respiratory compensation' is only partial, thus merely reducing the displacement of blood pH. The compensation is completed by renal excretion of H^+—as a result the log pCO_2/pH line becomes normal in position.

In the 'alkalaemic' blood the pCO_2 would have to rise to 65 mm in order for the pH to return to 7.40. This theoretical value could be reached by reducing alveolar ventilation. Certainly the respiratory centre is depressed when the H^+ concentration of the blood falls, but the respiratory compensation is never complete. Again the

compensation is completed by the kidney which excretes bicarbonate, thus restoring the log pCO_2/pH line to its normal position.

Obviously the acid–base picture can also be looked at from the viewpoint of bicarbonate concentration of the true plasma. As has been stated, the relationship between true plasma $[HCO_3^-]$ and pH is approximately linear for any given blood composition [Fig. III.51, p. 193] when the blood is titrated with CO_2.

At pH 7.40, 40 mm Hg pCO_2 and 37 °C

$$as\ pH = pK + \log \frac{[HCO_3^-]}{[\alpha pCO_2]}.$$

and as pK = 6.10, $\log \frac{[HCO_3^-]}{\alpha pCO_2} = 1.30$

1.30 is the $\log_{10}$ of 20 and α (the solubility coefficient of CO_2) is 0.031 mmol per mm Hg per litre, so the term αpCO_2 in this case equals 1.25 mmol and $[HCO_3^-]$ must be 20 × 1.25 = 25 mmol.

Let such a situation be represented as point Ⓐ on the true plasma bicarbonate titration line N–N [Fig. III.57]. If acid (e.g. lactic) enters the blood, its initial effect is to increase $[H^+]$, (diminish pH), to displace bicarbonate as CO_2 and to reduce protein ionization. The result is artificially represented by a movement from point Ⓐ to point Ⓑ situated on another titration line (Z–Z) which shows a reduction of buffer anionic concentration. Compensation could theoretically be achieved by respiratory stimulation which lowers pCO_2 and hence raises pH towards point Ⓒ which would restore the pH to its initial value. Such respiratory compensation is never complete and what actually occurs, given time, is a return to point Ⓐ secured by the excretion of H^+ by the kidneys (as shown by the dotted line in Fig. III.57).

These simple considerations should not be obscured by too many quantitative data based sometimes obsessively on considerations of the acid–base characteristics of blood alone. In the body other buffering mechanisms, notably those of the intracellular proteins and organic phosphates, modify the 'arithmetical' alterations of $[H^+]$ predicted from considerations of the behaviour of blood *in vitro*. As the intracellular body water containing these buffering mechanisms is 35 litres in bulk, whereas blood itself represents only 5 litres, it should be simple to understand that 'whole-body buffering' is more effective than that surmized from studies of acid–base equilibrium of blood itself.

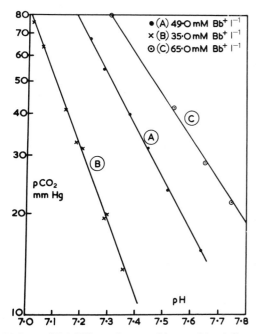

Fig. III.56. Log pCO_2/pH lines. A, normal human blood; B, acidosis; C, alkalosis. (Redrawn from data of Brewin, E. G. Gould, R. P. Nashat, F. S., and Neil, E. (1955). *Guy's Hosp. Rep.* **104**, 177.)

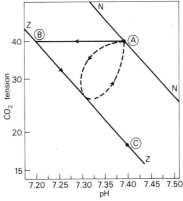

Fig. III.57. Diagrammatic representation of the course of events following the entrance of lactic acid into the blood. Abscissa: pH; ordinate: CO_2 tension (log scale). Point A shows the position of a normal arterial sample on the normal log P_{CO_2}/pH line N–N. Addition of lactic acid depletes the buffer anions and tends to displace the arterial point towards point Ⓑ on the log P_{CO_2}/pH line Z–Z. Respiratory compensation could be fully achieved by lowering the P_{CO_2} to point Ⓒ which would restore the original arterial pH value. In actual fact the respiratory stimulation caused by the fall in pH lowers the arterial P_{CO_2} and the renal excretion of hydrogen ion restores the status quo in a manner depicted by the dotted elliptical line shown in the diagram.

The chemical regulation of respiration

[H⁺] AND BREATHING

Immediately before and after the turn of the last century Haldane and his colleagues showed that the ventilation volume of man at rest was very sensitive to a change in the pCO_2 of the air breathed, but was comparatively insensitive to the pO_2 of the inspired air. They showed that the inspiration of gas mixtures, containing an increasing amount of CO_2 but the same oxygen content, caused a hyperpnoeic response which seemed to secure a relatively stable pCO_2 of the alveolar air and therefore of the arterial blood [FIG. III.58].

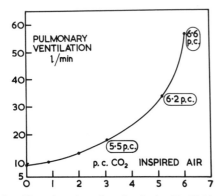

FIG. III.58. Graphic registration of results obtained by Haldane and Priestley. The inset figures show the alveolar CO_2 concentration of the subject during inhalation of the various CO_2 mixtures. The normal resting alveolar CO_2 concentration of the subject was 5.5 per cent. (Haldane, J. S. and Priestley, J. G. (1935). *Respiration*. New Haven, Yale University Press.)

As has been seen, the alveolar pCO_2 is normally of the order of 40 mm Hg, which approximately corresponds to an alveolar percentage of 5.6 per cent CO_2. This stabilization of the pCO_2 is achieved by a balance between the metabolic delivery of some 200–250 ml CO_2 by the tissues of the resting body and an alveolar ventilation of some 4 l min⁻¹ by the room air which contains a *negligible* concentration (0.03 per cent) of CO_2. If CO_2 is added to the gas mixture inspired, the alveolar and arterial pCO_2 are brought back to normal. As FIGURE III.58 shows this compensatory mechanism has its limits. Clearly, if the mixture inspired exceeds 6 per cent no amount of alveolar ventilation will bring the alveolar pCO_2 back to normal and indeed, although the pulmonary ventilation reaches some 60 l min⁻¹ when mixtures containing 6–6.5 per cent CO_2 are inspired, the alveolar pCO_2 exceeds 46 mm Hg (6.6 per cent CO_2).

Increased acidity of the blood, cerebrospinal fluid, and other body fluids also increases breathing. Leusen (1954) showed unequivocally that artificial perfusion of the ventriculo-cisternal system (normally occupied by cerebrospinal fluid) with fluids containing an abnormally low [HCO_3^-] would cause hyperventilation with an associated reduction in the arterial pCO_2.

Leusen showed that either a rise in the pCO_2 or of [H⁺] in the cerebrospinal fluid would stimulate breathing. Further work by Loeschcke, Mitchell, Severinghaus and others indicated that the chemosensitive areas of the medulla were situated superficially on the ventral surface of the brain stem and not in the medullary respiratory centres. Thus, irrigation of the floor of the fourth ventricle itself by bicarbonate solutions equilibrated with high pCO_2 mixtures (and therefore acid) did not stimulate the breathing and the application of pledgets soaked in artificial solutions saturated with CO_2 to the floor of the fourth ventricle was alike ineffective in promoting increased respiratory activity. However, the application of the solutions or the soaked pledgets to the ventrolateral surface of the medulla caused prompt respiratory stimulation [FIG. III.59].

These medullary chemosensitive areas were shown to be irresponsive to hypoxic solutions; they were depressed by local application of cold solutions or of solutions containing procaine.

Agreement has been generally reached that the effective stimulus of these chemosensitive areas is the [H⁺] of their extracellular fluid.

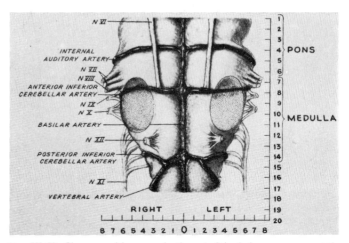

FIG. III.59. Chemosensitive zone in the cat. Stippled areas represent the region of respiratory chemosensitivity to high CO_2 and H⁺, nicotine, or acetylcholine. Numbers on bottom and on the right indicate stereotaxic co-ordinates in millimetres. (Mitchell, R. A., Loeschcke, H. H., Severinghaus, J. W., Richardson, B. W., and Massion, W. H. (1963). *Ann. N.Y. Acad. Sci.* **109**, 661.)

The results of Pappenheimer (1966) and his colleagues showed for the first time the sensitivity of the central respiratory mechanism to changes in the composition of the cerebrospinal fluid in chronic unanaesthetized animals. Goats were used and guide tubes were chronically implanted so that their ends lay respectively over the lateral ventricles or over the dura of the cisterna magna. Sharp trocars could sporadically be inserted through the guide tubes to pierce the dura. Perfusion of sterile artificial cerebrospinal fluid could thus be achieved and the fluid collected and analysed from the unanaesthetized animal. After training, the goats tolerated the application of a respiratory mask, which allowed the administration of various gas mixtures and the collection of the expired air for measurements of composition and volume. Chronic carotid van Leersum loops were made so that arterial sampling could be achieved by superficial puncture through the skin.

The respiratory characteristics of a 40 kg goat standing quietly are similar to those of a 70 kg man in the basal state. The oxygen usage is approximately 240 ml min⁻¹ and the alveolar ventilation about 5–5.5 l min⁻¹. The average increment of alveolar ventilation per 40 kg body weight is 0.9 litres per min per mm Hg increase of arterial pCO_2. This response is, however, considerably diminished by even light barbiturate anaesthesia. Perfusion of artificial 'control' solutions of synthetic cerebrospinal fluid altered neither the resting pulmonary ventilation volume nor the ventilatory response to inspired CO_2 in goats whose cerebrospinal fluid was formed and circulated naturally.

The pCO_2 in cisternal outflow fluid was independent of that in the inflow at the perfusion rates employed (approximately 2 ml min⁻¹). Even an increase of 50 mm Hg in the inflow fluid pCO_2 did not alter that in the effluent by more than 1 mm Hg, indicating that CO_2 exchanges rapidly between the perfusion fluid and the cerebral tissue and that equilibration is achieved completely by the time the fluid reaches the cisterna. The pCO_2 in the cisternal outflow approximates to that in the jugular venous blood and is closely similar to the pCO₂ in the cerebral tissue interposed between the blood capillaries and the cerebrospinal fluid.

Perfusion of the ventriculo-cisternal system was carried out using synthetic solutions containing normal, low and high bicarbonate concentrations and during each perfusion the respiratory responses to inhaled CO₂ were recorded. At any given pCO₂ in the cerebrospinal fluid (and hence in the arterial blood), the alveolar ventilation varied with the [HCO₃⁻] in the artificial cerebrospinal fluid. At high [HCO₃⁻] in the perfused 'cerebrospinal' fluid (e.g. 30 mM l⁻¹) breathing was depressed compared with that when the perfused 'cerebrospinal' fluid [HCO₃⁻] was normal (22 mM l⁻¹); conversely, low [HCO₃⁻] in the perfused fluid (16 mM l⁻¹) caused an increase in ventilation for the same pCO₂.

In further experiments, goats were rendered *chronically* alkalotic by the administration of bicarbonate, or *chronically* acidotic by giving them ammonium chloride. Samples of (naturally formed) cerebrospinal fluid were taken from the cisternal outflow together with samples of arterial blood while the goat inspired various CO₂ mixtures and its ventilation was measured in each case.

When the alveolar ventilation, measured in the normal, acidotic or alkalotic goat during these CO₂ inhalations, was plotted against the [H⁺] of the cisternal fluid outflow sampled, it was clear that the respiratory response to CO₂ is a single exponential function of [H⁺] in the cerebrospinal fluid (i.e. that alveolar ventilation plotted on a log scale is linearly related to the concentration of [H⁺] expressed in nM per kg H₂O [FIG. III.60].

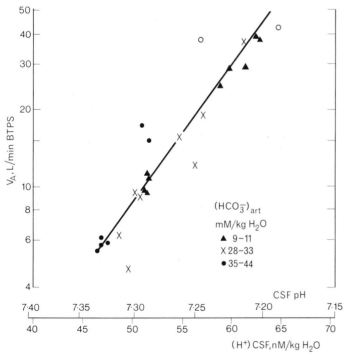

FIG. III.60. Ventilatory response to CO₂ inhalation in normal, in chronic metabolic acidosis and in chronic metabolic alkalosis (goat). Abscissa [H⁺] in c.s.f. (linear scale) Ordinate: alveolar ventilation (log scale). (Fencl, V., Miller, S., and Pappenheimer, J. R. (1966). *Am. J. Physiol.* **210**, 459.)

The graph shows that a fall of pH of only 0.15 increases alveolar ventilation from 5 to 45 l min⁻¹. The ventilation of a chronically alkalotic goat breathing 10 per cent CO₂ is identical with that of a chronically acidotic goat breathing air, providing that the [H⁺] of the ventricular fluid in each case is the same. Pappenheimer (1967) points out that the maximum range of cerebrospinal fluid pH seen in *chronic* circumstances which is compatible with life is only 0.2 pH, viz, from 7.38 breathing air during extreme alkalosis to 7.18

breathing CO₂. If the cerebrospinal fluid pH falls below 7.18 the animals reach their maximal breathing capacity, whereas above 7.38 the animals die from hypoventilation.

We can now reconsider the experimental results obtained when the ventriculo-cisternal system is *perfused acutely* with artificial cerebrospinal fluid solutions of different [HCO₃⁻]. The ventilation is measured in each circumstance while the animal breathes CO₂ mixtures and the cerebrospinal fluid cisternal fluid [H⁺] is determined simultaneously; it transpires that it is not the [H⁺] in the perfusate which by itself determines the ventilation. FIGURE III.61 shows that there are an infinite number of values of alveolar ventilation for any one [H⁺] in these *acute* circumstances. Hence the nervous elements sensitive to [H⁺] cannot be in direct contact with fluid in the large cavities.

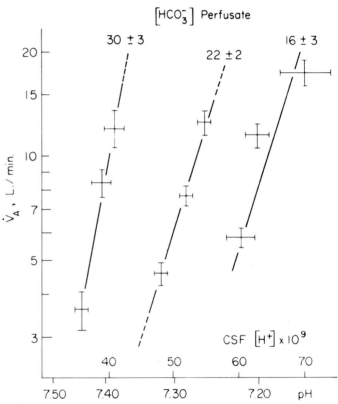

FIG. III.61. Ventilatory responses to CO₂ inhalation for normal goats during ventriculo-cisternal perfusions with 30, 22, and 16 mEq [HCO₃⁻] per litre. Crosses represent means and standard error from 75 steady-state periods in six goats. (Pappenheimer, J. R. (1967). *Harvey Lect.* 71.)

FIGURE III.62 shows that a goat in *chronic* severe acidosis (plasma [HCO₃⁻] = 10 mM per kg H₂O) has a cerebrospinal fluid [HCO₃⁻] of some 15 mM per kg H₂O. A goat in chronic severe alkalosis with a plasma [HCO₃⁻] of 45 mM per kg H₂O has a cerebrospinal fluid [HCO₃⁻] of 25 mM per kg H₂O. Thus, the cerebrospinal fluid [HCO₃⁻] alters only 10 mM per kg H₂O, while the plasma [HCO₃⁻] changes by 35 mM per kg H₂O in the difference between *chronic* metabolic alkalosis and *chronic* metabolic acidosis. This suggests that there is an exchange of [HCO₃⁻] between the plasma and the cerebrospinal fluid and further experiments show this to be an active transport. The exchange of materials between the perfusate and the brain can be determined from the volume and composition of the fluid entering the lateral ventricles and leaving the cisterna (Pappenheimer *et al.* 1964). Substances enter the ventricles in the perfusion fluid in newly formed cerebrospinal fluid and by exchange across the ependymal linings of the ventricular system. Substances leave in the outflow, in bulk absorbate from the arachnoid villi and by transependymal exchange. If the rate of formation of newly

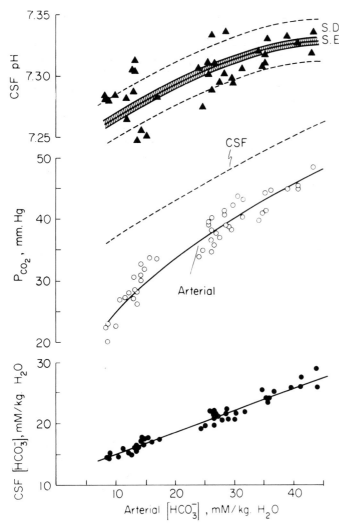

FIG. III.62. $[HCO_3^-]$, pH, and pCO_2 in c.s.f. and blood during chronic variations of blood $[HCO_3^-]$ induced by the administration of NH_4Cl or $NaHCO_3$. (Fencl, V. *et al.* (1966). *Am. J. Physiol.* **210**, 459.)

formed fluid and the bulk absorption are known, these components of the exchange can be subtracted from the total, leaving the measured transependymal flux.

In the ensuing account, the following symbols are used (Pappenheimer *et al.* 1964):

$\dot{V}$ = rate of flow of perfusion fluid ml min^{-1}

c = concentration μM per g H_2O

i, o, v, p, f, a = subscripts referring respectively to inflow, outflow, ventricular fluid, capillary plasma, freshly formed c.s.f., and fluid absorbed in bulk

$\bar{c}$ = mean concentration in ventricular fluid

C_{Inulin} = clearance of inulin = $\dot{V}_a$ [see p. 235]

$\dot{n}$ = net transependymal flux rate (μM min^{-1})

k_{out}, k_{in} = transependymal flux coefficients out of or into the ventricles, ml min^{-1}.

The net transependymal flux rate, $\dot{n}$ is given by:

$$\dot{n} = k_{out}c_v - k_{in}c_p.$$

If we consider the exchange of HCO_3^- between cerebrospinal fluid and plasma, then:

The total rate of entrance of HCO_3^- into the ventricles is equal to: perfusion inflow + bulk secretion + transependymal exchange:

$$\dot{V}_ic_i + \dot{V}_fc_f + k_{in}c_p.$$

The total rate of exit of HCO_3^- is equal to perfusion outflow + absorption in bulk + transependymal exchange:

$$\dot{V}_oc_o + \dot{V}_ac_o + k_{out}c_v.$$

In the steady state the total entrance rate *must* equal the total exit rate of HCO_3^-:

$$\dot{n} = \dot{V}_ic_i - \dot{V}_oc_o + \dot{V}_fc_f - \dot{V}_ac_o.$$

It can be shown that the rate of formation of new fluid can be measured from the dilution of inulin added to the perfusion fluid and that bulk absorption can be determined from the clearance of inulin and that

$$\dot{V}_f = \dot{V}_o - \dot{V}_i + C_{Inulin}$$

and

$$\dot{V}_ac_o = C_{Inulin}c_o.$$

Substituting,

$$\dot{n} = \dot{V}_i(c_i - c_f) - (\dot{V}_o + C_{Inulin})(c_o - c_f).$$

The only quantity on the right hand side of this equation which cannot be measured is c_f—the concentration of HCO_3^- in freshly formed cerebrospinal fluid. However, this is believed to be closely similar to, and may be assumed to be identical with, that in plasma, so that

$$\dot{n} = \dot{V}_i(c_i - c_p) - (\dot{V}_o + C_{Inulin})(c_o - c_p)$$

which allows the measurement of the transependymal flux of bicarbonate.

The question is whether there is net movement of bicarbonate ion from plasma in the cerebral capillaries across the ependyma into the ventricular cerebrospinal fluid when there is a steady-state concentration difference of 20 mM per kg H_2O during metabolic alkalosis or conversely, whether a net flux occurs from cerebrospinal fluid to plasma in acidosis when the cerebrospinal fluid $[HCO_3^-]$ exceeds that in the capillaries by 5 mMolal. To solve it, Pappenheimer *et al.* perfused the ventriculo-cisternal system with artificial solutions and measured the flux of HCO_3^-. During chronic alkalosis, when the natural plasma $[HCO_3^-]$ exceeded that of the cerebrospinal fluid by 18 mM per kg H_2O, net flux was only zero when the solution perfused through the ventriculo-cisternal system contained 18 mM per kg H_2O less than did the plasma.

In the chronic acidotic goat, in which the *natural* cerebrospinal fluid $[HCO_3^-]$ exceeded that in the plasma by 5 mM per kg H_2O, net transependymal flux was only zero when the artificial solution perfused contained 5 mM per kg H_2O of $[HCO_3^-]$ more than that of the plasma. In each case, zero net transependymal flux occurred when the perfusate $[HCO_3^-]$ was equal to that in the natural c.s.f. in alkalosis or acidosis [FIG. III.63].

The slope of the flux relationship is unaffected by the acidity of the blood, which indicates that the passive permeability of tissues interpolated between plasma and cerebrospinal fluid remains the same. The ion pump which maintains the concentration difference of HCO_3^- between plasma and ventricular fluid hence works in such a manner that flux due to active transport is balanced exactly by passive flux in the opposite direction so that in *chronic* conditions, net flux is zero.

Net flux occurs only when steady-state conditions are disturbed, as for example happens when a non-equilibrium $[HCO_3^-]$ solution is perfused through the ventricles. During net flux there must be a concentration gradient for diffusion of HCO_3^- through the interstitial fluid as shown in FIGURE III.64.

The pH at any point in the tissue between cerebrospinal fluid and capillary will be determined by the $[HCO_3^-]$ at that point in equilibrium with the pCO_2 of cisternal fluid, for molecular CO_2 diffuses rapidly and is distributed uniformly throughout the tissue.

By trial and error Pappenheimer *et al.* found that alveolar ventilation became a single function of pH during perfusion of the ventricular system with varying $[HCO_3^-]$ providing that the pH is

referred to a point three-quarters of the distance along the concentration gradient of HCO_3^- between the cerebrospinal fluid and the ion-pump to blood [FIG. III.64]. The respiratory effects of inhaled CO_2 and of perfusion of the ventriculo-cisternal system with artificial solutions containing various bicarbonate ion concentrations can both be expressed as a single function of the $[H^+]$ of cerebral interstitial fluid close to the site of HCO_3^- exchange with the cerebral capillary plasma. Moreover, the relationship between the alveolar ventilation and the interstitial fluid in these perfusion experiments is identical with that shown between alveolar ventilation and the pH of large cavity fluid in *chronic* acidosis or alkalosis where the net flux of HCO_3^- between cerebrospinal fluid and plasma is zero [FIG. III.65].

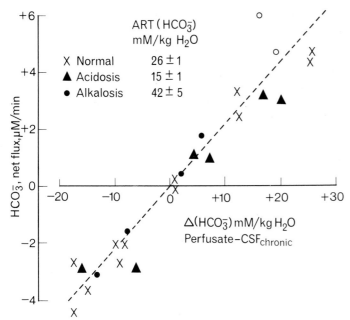

FIG. III.63. Transependymal flux of HCO_3^- plotted as function of concentration difference between $[HCO_3^-]$ of perfusate and the $[HCO_3^-]$ which occurs naturally in c.s.f. at each acid–base condition. ($CSF_{chronic}$). (Fencl, V. *et al.* (1966). *Am. J. Physiol.* **210**, 459.)

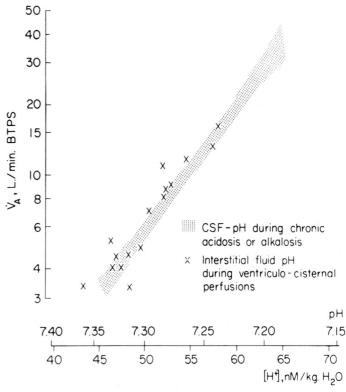

FIG. III.65. Ventilatory responses to CO_2 inhalation during chronic disturbances of acid–base balance and during perfusion of the ventriculo-cisternal system with various bicarbonate solutions. Shaded area is the result of a statistical analysis of results from 81 steady-state observations on five goats during metabolic alkalosis or acidosis. The abscissa for these data refer to $[H^+]$ in large cavity fluid. The crosses refer to measurements made during net flux of HCO_3^- caused by perfusions of the ventricles and cistern and the abscissa for these data refers to interstitial fluid. (Fencl, V. *et al.* (1966). *Am. J. Physiol.* **210**, 459.)

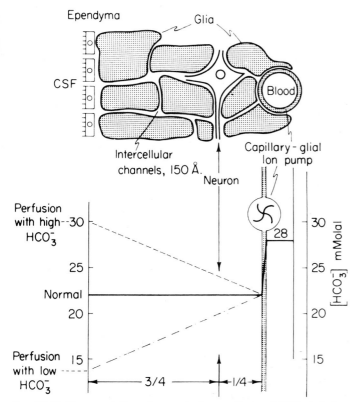

FIG. III.64. Representation of concentration gradients of HCO_3^-] during zero net flux (normal) and during net flux through interstitial fluid caused by perfusion of the ventricular system with high or low $[HCO_3^-]$.

The neuron is depicted at a point three-quarters of the distance along the concentration gradient; at this point the pH of the interstitial fluid is always such as to be a unique determinant of ventilation while CO_2 mixtures are breathed and during ventricular perfusion with various $[HCO_3^-]$. (Pappenheimer, J. R. (1967). *Harvey Lect.* 71.)

Thus, the neural elements sensitive to $[H^+]$ are located in interstitial fluid at a site nearer to that of ion exchange with the plasma than to that of the large cavity fluids. Additionally, it seems that the hydrogen ion concentration of the cerebrospinal fluid in the large cavities is identical with that of the interstitial fluids in *steady-state chronic conditions*. Hence, as there is no appreciable gradient of CO_2 pressure between cerebral capillary blood and cisternal fluid the $[HCO_3^-]$ of interstitial fluid must be identical with that of the cerebrospinal fluid in the large cavities.

Effectively then, CO_2 acts as a respiratory stimulus by providing a rapid means of altering the $[H^+]$ in the interstitial fluid bathing the bulbar neurons which are part of the respiratory centre complex. The rise of $[H^+]$ produced stimulates the breathing.

OXYGEN LACK

However, respiration is affected also by oxygen lack, in a manner which depends not only on the degree of anoxia but also on the rate of onset of anoxia.

When anoxia is severe and fulminant, such as occurs when a miner enters a shaft containing methane and/or CO, or an airman at great altitude is cut off from his oxygen supply, the individual loses consciousness in a matter of seconds and will die unless help is obtained. The miner rescued from such a predicament may believe he has been knocked down by his mate and act accordingly.

A more gradual development of anoxia causes a dulling of the intellect and the senses. The subject's performance deteriorates, although he himself may be quite unaware of what is happening. The symptoms of this less fulminant anoxia include emotional lability—he may be apathetic or may manifest an excited and often belligerent mien. Self-confidence tends to be excessive and this further impairs his chances of survival, as he undertakes tasks which he would avoid as dangerous when in his normal rational state. As painful sensation is obtunded in the anoxic state, this, coupled with a deterioration of motor performance, may lead him to injure himself; thus the miner suffering from carbon monoxide poisoning may burn himself with his lamp. 'Anoxia not only stops the machine but wrecks the machinery'; when Haldane said this, he was stressing the profound deleterious effect of oxygen lack on the nervous system. Protracted oxygen lack causes *irreversible* damage of the CNS neurons. Thus, the patient who suffers too long a period of cardiac arrest, before the heart is artificially caused to resume beating, thence providing an effective circulation, may recover his respiratory rhythm, etc., but will remain a 'vegetable' and spend the rest of his existence as an imbecile owing to the loss of his higher centres. The CNS neurons cannot sustain an oxygen debt.

Respiration in acute anoxia

Haldane showed that a subject rebreathing from a spirometer in a closed circuit, so that his expired CO_2 was absorbed by soda lime, manifested no hyperpnoea until the inspired gas contained less than 14 per cent O_2, which is equivalent to a pO_2 of 100 mm Hg. In such circumstances the alveolar pO_2 would be approximately 60 mm Hg for the alveolar gas would contain CO_2 at a partial pressure of 40 mm Hg, as, by definition, the breathing is unchanged and as the metabolic production of CO_2 has not altered. Clearly the respiratory response to oxygen lack is not a very sensitive one if this anoxic stimulus occurs in the *acute* state. If the O_2 content of the spirometer gas rebreathed fell below 14 per cent then the subject did develop hyperpnoea. Thus, the inhalation of 10 per cent O_2 in N_2 increases the breathing from say 7 l min^{-1} to 8 l min^{-1} and studies subsequent to those of Haldane have recorded that men breathing 8 per cent O_2 in N_2 may double their ventilation volume. Schizophrenic patients were exposed to 4.2 per cent O_2 in N_2 and increased their breathing from 7 l min^{-1} to 30 l min^{-1} before becoming unconscious. Anoxia than *can* increase the breathing, but is far less effective than a raised pCO_2 as a stimulus.

The results of Pappenheimer and others before have shown that in ordinary circumstances the pulmonary ventilation volume is governed by the [H$^+$] of the cerebral interstitial fluid surrounding the 'respiratory neurons'. This [H$^+$] in turn depends essentially on the ratio of pCO_2/[HCO$_3^-$] and the pCO_2 is itself modified by the degree of alveolar ventilation. Hence it can be seen that if anoxia does stimulate the breathing, the increased alveolar wash-out of CO_2 will lower the pCO_2 and will reduce the central chemical drive provided by the cerebral interstitial fluid [H$^+$]. It has been known since 1930 that anoxia excites a discharge in afferent nerves whose endings (chemoreceptors) are situated in the carotid and aortic

bodies [p. 204] and that the increase in discharge in these nerves is progressive as the arterial pO_2 falls. Artificial stimulation of these nerves excites the respiratory mechanism reflexly. With moderate hypoxia, however, the increase in chemoreceptor discharge is insufficient to drive the breathing in the resting subject, presumably because any initial stimulation of ventilation lowers the central pCO_2 and [H$^+$], which offsets the effect of the increased afferent activity. Only if the anoxia is sufficiently intense to cause marked chemoreceptor stimulation does the central respiratory complex become subservient to its influence. Breathing then increases, despite the fall in pCO_2 and [H$^+$] which this hyperpnoea causes.

Acute anoxia would thus be a much more effective stimulus of respiration if it did not cause acapnia and alkalosis of the cerebral interstitial fluid. Asphyxia, in which both oxygen lack and CO_2 excess occur simultaneously, causes enormous hyperpnoea. Again this was first clearly demonstrated by Haldane who studied the respiratory response of his subjects when they rebreathed from a spirometer which initially contained air. No soda lime tower was interposed between subject and spirometer, so as the subject used up the oxygen from the spirometer, he added his own metabolic CO_2 to the increasingly anoxic mixture which he rebreathed.

Interaction of oxygen lack and carbon dioxide excess

Now the question arises—are the two stimuli of oxygen lack and CO_2 excess simply additive or is the response to CO_2 amplified or sensitized by hypoxia? The answer to this problem has been provided by Nielsen and Smith (1951) and by Lloyd, Jukes, and Cunningham (1958). Both teams examined the effect of increasing the alveolar pCO_2 on the breathing while keeping the alveolar pO_2 steady at different values. FIGURE III.66(a) shows that an increase of pCO_2 causes effects on the breathing which are markedly influenced by the alveolar pO_2. The response to CO_2 is strikingly greater when the alveolar pO_2 is maintained throughout at 40 mm Hg than when it is kept at the normal value of 100 mm Hg. The slope of the ventilation volume/pCO_2 line increases as the alveolar pO_2 is lowered [FIG. III.66b]. Anoxia and hypercapnia (raised pCO_2) *interact*; anoxia sensitizes the response of the respiratory

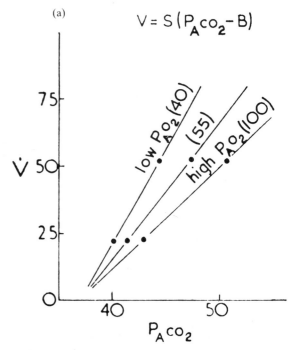

(a)

$$V = S(P_A co_2 - B)$$

$\dot{V}$

low P_{O_2} (40)

(55)

high P_{O_2} (100)

$P_A co_2$

FIG. III.66(a). Shows the effect of pO_2 on the relation between pulmonary ventilation and alveolar pCO_2.

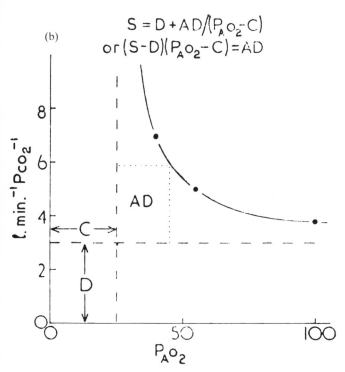

(b)

$$S = D + AD/(P_{A}O_{2} - C)$$
$$\text{or } (S-D)(P_{A}O_{2} - C) = AD$$

FIG. III.66 (b) The slope of the lines shown in (a) is plotted against alveolar pO_2. (Cunningham, D. J. C. and Lloyd, B. B. (eds.) (1963). *The regulation of human respiration.* Blackwell, Oxford.)

mechanism to excess CO_2 (and H^+). The two stimuli are *not* simply additive.

At this juncture we may consider the use which has been made of the $\dot{V}/pCO_2$ lines in providing a more quantitative description of the influence of various factors—chemical and physical—on respiration.

Ventilation /pCO₂ graphs at different pO₂ values

The ventilation volume/pCO_2 line at any one alveolar pO_2 may be extrapolated to a 'zero ventilation' which corresponds to an alveolar pCO_2 value on the abscissa. This theoretical value of pCO_2 may be regarded as that at which CO_2 (or $[H^+]$) causes no respiratory stimulation of itself. Nielsen and Smith (1951), who first noted this, called the point of intersection the 'apnoea point'.

All the $\dot{V}/pCO_2$ lines, though differing in slope according to the pO_2, when extrapolated meet at this common point and the $\dot{V}/pCO_2$ relationship at any given constant pO_2 may be expressed by the equation (Lloyd and Cunningham 1963):

$$\dot{V} = S(pCO_2 - B)$$

where $\dot{V}$ = ventilation volume and S is the slope of the line for a given pO_2. FIGURE III.66(b) shows the results of FIGURE III.66(a) plotted graphically.

The essentially hyperbolic relationship between S and alveolar pO_2 has been described by Lloyd and Cunningham and their co-workers in the following equation:

$$S = D\left[1 + \frac{A}{pO_2 - C}\right]$$

and substituting this value of S in the equation above gives:

$$\dot{V} = D(P_{CO_2} - B)\left[1 + \frac{A}{pO_2 - C}\right].$$

This equation has four parameters, A, B, C, and D. Of these B has already been defined—the alveolar pCO_2 value at which CO_2 itself exerts no effect on the breathing. FIGURE III.66(b) shows the parameters A, C, and D added to the graph.

It can be seen that D is the minimum slope of the $\dot{V}/pCO_2$ line, i.e. the value of S when pO_2 is infinity and there is no hypoxic drive. C, on the other hand, is the critical alveolar pO_2 value at which the slope becomes infinite. A is a parameter which describes the sensitivity to hypoxia. It is obtained by drawing a line stepped off from the C ordinate at a height of $2D$ to meet the experimentally determined S/pO_2 plot.

[Note: Students become confused about this and I have found the anology of the rheobase and the chronaxie helpful. The excitability of a tissue, such as nerve, muscle, etc., to electrical stimulation may be determined by finding the time taken for a current of known strength to excite—the so-called strength–duration curve. This is of hyperbolic form as shown [FIG. III.67]. The weakest current strength which can just excite the tissue if allowed to pass for an adequate (and long) time is called the rheobase. Obviously this is an inconvenient and laborious yardstick (or milestick) of excitability. It is therefore conventional to express excitability in terms of the much shorter time which a current of strength twice that of the rheobase must pass in order to excite the tissue. This duration is called the chronaxie.

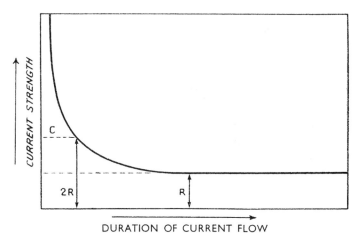

FIG. III.67. Strength–duration curve. Rheobase and chronaxie. The curve shows the relationship between the *strength* of stimulation (current strength) and the minimum *duration* of stimulation (i.e. duration of current flow) necessary to set up an impulse. R = Rheobase. 2R = 2 × R. When 2R is the strength of stimulus, the minimum duration of current flow which can produce excitation is called the chronaxie (C).

Now if the Lloyd–Cunningham picture be examined and considered in the light of this analogy D could be regarded as the rheobase and A as the chronaxie.]

Respiratory regulation in a variety of circumstances, such as acidosis, alkalosis, hyperthermia, catecholamine infusion, chronic anoxia, etc., may be described in terms of the four parameters, A, B, C, and D, determined from the experimental results obtained.

Thus, FIGURE III.68 (a and b) shows the effect of acidosis on the respiratory response of a normal subject. After his control values had been determined the subject ingested ammonium chloride for a week and the regimen was repeated. FIGURE III.68(a) shows that B, the intercept parameter, was much lower than before. Chronic acidosis then affects the response to CO_2. However, none of the other parameters, A, C, or D showed any change, so acidosis does not affect the response to acute hypoxia FIGURE III.68(b).

Noradrenalin infusion is known to stimulate the breathing and an analysis made in the terms described, reveals that this effect is due to its increasing the parameter A (sensitivity to hypoxia). B, D, and C are unaffected by noradrenalin. This influence on the hypoxic sensitivity is probably due to an action of the catecholamine on the arterial chemoreceptor mechanisms.

Chronic anoxia is, as has been stated, attended by a resting pulmonary ventilation in excess of that seen at sea level. Michel and Milledge (1963) analysed the respiratory regulation of four men acclimatized at a height of 5800 m where the barometric pressure is only half an atmosphere. They compared the respiratory responses at sea level with those at altitude, using the equations already described. They found that parameters A and C were little affected by several months at high altitude, C being unchanged and A being increased in three subjects, but reduced in the fourth. B, however, was approximately half and D was approximately double the sea level values. The reduction in B is due to the reduction of cerebrospinal fluid $[HCO_3^-]$ and (more slowly) the plasma $[HCO_3^-]$ which occurs during acclimatization—as the subjects were at altitude for several months, these changes were complete.

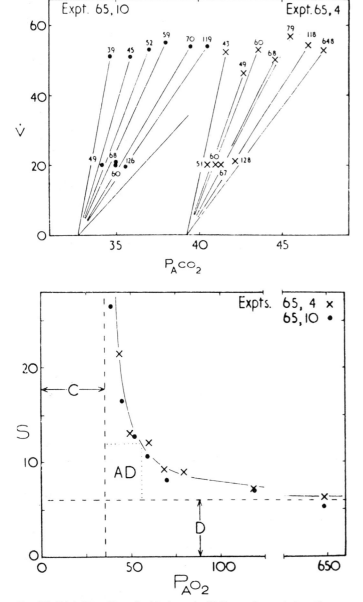

FIG. III.68(a). The effect of acidosis on the V, $P_{A CO_2}$, $P_{A O_2}$ relation. Crosses ×, experiment before, closed circles ●, experiment during acidosis. Numbers near points indicate $P_{A O_2}$. The fan of V, $P_{A CO_2}$ lines is displaced 7 mm to the left during acidosis, but is otherwise largely unchanged. (b) The effect of acidosis on the S, P_{O_2} relation. The slope S of the V, $P_{C O_2}$ lines is plotted against $P_{A O_2}$. Crosses ×, experiment before, closed circles ●, experiment during acidosis. The hyperbola of the equation has been fitted to the crosses ×. The proximity of the closed circles ● to the hyperbola indicates that acidosis has little effect on parameters A, C, and D. (Cunningham, D. J. C. *et al.* (1961). *Q. Jl exp. Physiol.* **46**, 323.)

The rise in parameter D has been previously observed, although it was simply described before as a rise in sensitivity to CO_2 shown by acclimatized individuals even when hypoxia was abolished by the inhalation of oxygen-rich mixtures.

Acclimatization to anoxia

As stated previously, the subject exposed to acute anoxia develops no hyperpnoea until the alveolar pO_2 falls below an approximate value of 60 mm Hg. The situation can be graphically depicted by the alveolar air diagram [FIG. III.69]. In this graph the alveolar pCO_2 is plotted on the ordinate and the alveolar pO_2 on the abscissa. The normal subject at sea level has an alveolar pO_2 of say 100 mm Hg and an alveolar pCO_2 of say 39 mm Hg and his respiratory situation can be defined on the diagram as point (1).

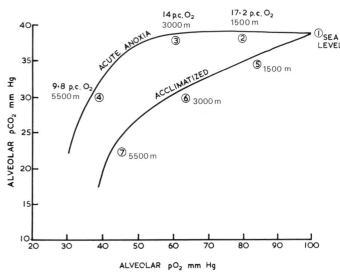

FIG. III.69. The alveolar gas tensions of subjects exposed to acute anoxia and to chronic anoxia (see text). (Modified from a figure of Rahn, H. and Otis, A. B. (1949). *Am. J. Physiol.* **157**, 145.)

When he is exposed to various degrees of acute hypoxia his alveolar pO_2 will fall, but as his metabolic production of CO_2 does not alter, his alveolar pCO_2 will not change unless the degree of hypoxia is sufficient to increase his breathing, whereupon his alveolar pCO_2 will fall. The degree of oxygen lack can be adjusted either by having him breathe gas mixtures of lower oxygen *content* at sea level, or by exposing him in a low-pressure chamber to breathing room air. The partial pressure of oxygen which he inspires is determined by the percentage composition times the total dry gas pressure and can be adjusted either by lowering the percentage composition or by lowering the total barometric pressure.

Point (2) thus represents the alveolar gas composition of a man acutely exposed *either* to breathing 17.2 per cent O_2 at sea level (inspired pO_2 = 128 mm Hg) *or* to breathing air at a simulated (or genuine) altitude of 1500 m (barometric pressure = 630 mm Hg; pO_2 = 128 mm Hg). Similarly point (3) represents the alveolar gas situation of a man acutely exposed to either an altitude of 3000 m (barometric pressure = 520 mm Hg) or to breathing 14 per cent O_2; in each case the pO_2 of the inspired gas mixture is 104 mm Hg. In neither case is there any fall of the alveolar pCO_2, which shows that his breathing has not been stimulated. His alveolar pO_2 is progressively reduced of course, *pari passu* with the fall of pO_2 in the inspired air. Only when the inspired pO_2 decreases further does the hyperpnoeic response to acute hypoxia develop. Point (4) shows the situation of an individual acutely exposed to a simulated altitude of 5500 m (barometric pressure = 400 mm Hg or alternatively, to breathing 9.8 per cent O_2 at sea level; in each case the inspired pO_2 is of the order of 80 mm Hg. His respiratory response to such circumstances is evidenced by the fall in his alveolar pCO_2 which has dropped from 40 mm Hg to approximately 30 mm Hg. It is clear that even this degree of hypoxia, though admittedly causing some hyperpnoea, is not an impressive respiratory stimulant. Assuming that the metabolic production of CO_2 remains the same, breathing has only increased 4/3 times—some 33 per cent.

When the situation is re-examined during circumstances of chronic anoxia, however, a striking difference in respiratory behaviour is revealed. Points (5), (6), and (7) graphically depict data obtained on subjects chronically exposed to altitutdes of 1500, 3000, and 5500 m respectively. Each of these 'alveolar points' shows that breathing is increased progressively with *chronic* anoxia—the 'acclimatized' individual develops a greater response to chronic anoxia compared with that of the subject acutely exposed. The biological value of such a response is obvious, for it means that for any given inspired pO_2 the acclimatized subject can secure by increased breathing a higher alveolar pO_2 than does the individual who is unacclimatized. Hence, the haemoglobin in his blood will be more fully saturated.

These facts have been recognized since 1914 but the underlying explanation of them has remained the subject of protracted argument. The essential basis of acclimatization was recognized by 1920 to reside in the re-adjustment of the $[H^+]$ of the plasma following a primary reduction of the arterial pCO_2 which occurs when hypoxic stimulation of the breathing increases the alveolar ventilation.

As

$$[H^+] \propto \frac{pCO_2}{[HCO_3^-]}$$

and as pCO_2 falls, clearly the appropriate compensation could be achieved by reducing $[HCO_3^-]$. Until the last decade, the restitution of the situation was considered to be the function of the kidney—the production of a more alkaline urine was repeatedly documented in investigations of acclimatized subjects. The sequence was interpreted as follows: hypoxic hyperpnoea → lowered arterial pCO_2 → lowered pCO_2 in the renal tubular cells. The renal tubular cells form H_2CO_3 from CO_2 and H_2O and this is accelerated by carbonic anhydrase. As stated on page 188 the enzyme cannot change the equilibrium point of this reaction *per se*, but the H_2CO_3 ionizes and the H^+ is secreted by active transport on the part of the tubular cells into the tubular fluid. Here H^+ combines with HCO_3^- which has been filtered and H_2CO_3 is formed. In the absence of any other hydrogen ion acceptors in the tubular fluid this H_2CO_3 mainly (99.9 per cent) gives rise to CO_2, for the equilibrium point of the reaction:

$$H_2O + CO_2 = H_2CO_3$$

is such that 999/1000 exists as CO_2. The CO_2 produced in the tubular fluid diffuses from the tubular fluid across the tubular cells into the capillaries. This is the mechanism of absorption of bicarbonate from the glomerular filtrate. When the arterial pCO_2 is low, the provision of H^+ by the tubular cells is inadequate and as a consequence bicarbonate is not absorbed so completely and eventually appears in the urine which becomes more alkaline than normally. The depletion of bicarbonate from the body which results helps to restore the $pCO_2/[HCO_3^-]$ ratio and hence the $[H^+]$ of the blood.

Though these changes undoubtedly do occur, they are not nowadays regarded as the primary reason for acclimatization hyperpnoea. The renal response is too sluggish to account for this and is moreover, unattended by a complete restoration of the plasma pH. Mitchell, Severinghaus and their colleagues have provided the more likely explanation in important studies of the cerebrospinal fluid compensation of man during the development of acclimatization.

Four subjects were studied in the Barcroft Laboratory (University of California) on White Mountain at 3800 m. Hypoxia drove

the breathing only slightly at first, but within 24 hours hyperpnoea developed progressively and arterial pCO_2 fell from 40 to 33 mm Hg and blood pH rose from 7.43 to 7.48. The pCO_2 of the cerebrospinal fluid fell from 49 to 39 mm Hg.

The reduction of pCO_2 was such that the cerebrospinal fluid pH could be calculated to have increased from its normal sea level value of c. 7.32 to 7.43 [FIG. III.70]. However, the c.s.f. pH measured by spinal tap showed little change. This indicates that the bicarbonate concentration of the c.s.f. must have been substantially reduced and indeed measurements of c.s.f. $[HCO_3^-]$ showed that it had fallen by as much as 5 mM l^{-1}. This reduction in c.s.f. $[HCO_3^-]$ was *not* due to renal excretion of plasma bicarbonate because the plasma $[HCO_3^-]$ fell only 1–2 mM l^{-1} in the course of a week. After discarding other possibilities Severinghaus *et al.* (1963) concluded that active transport of bicarbonate was the underlying cause of the rapid stabilization of c.s.f. pH and that the restoration of the $[H^+]$ of the milieu of the central chemoreceptors was due to this. These findings, which preceded the studies of goats referred to on page 196, stress the role of the active pump between the c.s.f. and the cerebral capillaries in adjusting the $[H^+]$ of the cerebral interstitial fluid milieu of the bulbar respiratory neurons.

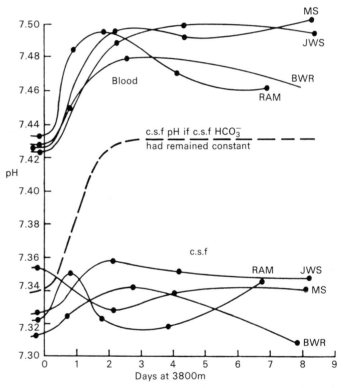

FIG. III.70. Acclimatization of four subjects (RAM, JWS, MS, and BWR) to altitude, breathing air.
  Solid lines show pH values obtained from blood and c.s.f. samples.
  Pecked line shows the change in c.s.f. pH there would have been if the c.s.f. $[HCO_3^-]$ had remained constant. (Severinghaus, J. W. (1965). In *Cerebrospinal fluid and the regulation of ventilation* (ed. C. Brooks, F. Kao, and B. B. Lloyd). Blackwell, Oxford.)

The mechanism of acclimatization of 'respiratory control' can thus be envisaged as a sequence: anoxic hyperpnoea → lowered arterial, c.s.f. and interstitial fluid pCO_2 → decreased $[H^+]$ of the cerebral interstitial fluid → reduction of effect of hypoxic 'drive', so that respiration though increased is not greatly increased. Reduced $[H^+]$ of cerebral interstitial fluid → increased transport of bicarbonate out of c.s.f. and interstitial fluid so that $pCO_2/[HCO_3^-]$ and $[H^+]$ are brought back to near normal. This adjustment takes time and hence the response of the subjects' ventilation to an altitude of 3800 m takes a day or so to develop.

Long before these sophisticated findings were available it was known that when men, acclimatized at altitude, returned swiftly to the lowlands their breathing remained increased. This provided a puzzle, for the consensus of opinion was that the breathing of the acclimatized man was determined by oxygen lack acting through the peripheral arterial chemoreceptors. Though acid–base equilibrium adjustments of the blood were known to occur in acclimatization they were recognized to be incomplete and were regarded as simply 'allowing' the anoxic drive to exert its full effect. Later experiments showed that the breathing of a fully 'acclimatized hypoxic' subject was little affected when he inhaled pure oxygen; moreover, the advent of the helicopter allowed the rapid transport of men from high altitude to lowland laboratories where the maintenance of hyperpnoea was more fully documented and confirmed.

The situation here is now much better understood. Once the c.s.f. and cerebral interstitial fluids are credited with the role of primarily influencing the respiratory neurone complex by virtue of their [H+], then it is clear that as these fluids contain only a negligible amount of protein their [H+] at any given [HCO₃⁻] will vary more wildly with a change in pCO₂, because these fluids contain no other buffer anion. Even if there were no bicarbonate pump between the c.s.f. and the blood and if the c.s.f. [HCO₃⁻] *only reflected* the plasma [HCO₃⁻] then the depletion of plasma and c.s.f. bicarbonate would render particularly the c.s.f. susceptible to an increment of CO₂ pressure. However, as has been seen, active processes secure a c.s.f. [HCO₃⁻] in acclimatization which is as much as 5 mM l⁻¹ less than the sea-level value, so the [H+] is even more affected by a rise of pCO₂. Hence, when the subject acclimatized to oxygen lack is suddenly exposed to a normal pO₂ of 150 mm Hg, any reduction in his breathing which may result from this oxygenation will raise his arterial and cerebral interstitial fluid pCO₂ and will cause a striking rise of [H+] which by stimulating the medullary neurons will in turn cause increased breathing. The subject has to 'unacclimatize' himself, and just as acclimatization takes time, so does the restoration process required.

It is wrong to suppose that the breathing of acclimatized man is unaffected by the low pO₂ for this is only the case in man at rest. If he indulges in any physical activity, which after all is usually his ordinary purpose, then his breathing is less if his oxygen supply can be increased. Thus, Åstrand (1954) showed that the substitution of oxygen for room air diminished the ventilation volume of an exercising subject acclimatized for 5 days at an altitude of 4000 m and the heavier the exercise the greater the reduction of ventilation that oxygen effected.

It is likewise wrong to imagine that the chemoreceptors become less responsive to hypoxia with the development of acclimatization. Direct measurements of their afferent activity in cats (Åstrand 1954) show that this is maintained after four days at a simulated altitude of 4000 m. It is the restoration of a more normal [H+] of the milieu surrounding the medullary neurons which renders these structures *less responsive* to the afferent input from the chemoreceptors—just as in the case in man at rest in a normobaric environment.

Other adaptations seen in hypoxic acclimatization

These include 1. raised cardiac output; 2. polycythaemia. Each of these changes improves the mean capillary pO₂ and hence the pressure gradient between capillary and tissue cell. Clearly, if a tissue extracts say 10 ml O₂ from each 100 ml of blood which supplies it then the desaturation which the blood suffers is approximately 50 per cent and the blood leaves the tissues with a pO₂ of say 26 mm. If the blood flow be doubled, the percentage desaturation of the blood is halved and the venous blood leaves the tissues at a saturation of 75 per cent and a pO₂ of say 40 mm. The pressure gradient to the tissues is thereby improved. Barcroft reported a

doubling of the cardiac output even at 4250 m and doubtless the cardiovascular response at higher levels is even greater.

Polycythaemia due to red cell proliferation increases the concentration of haemoglobin in the blood and hence its O₂ carrying power. The haemoglobin figures given for Sherpas and the acclimatized lowlanders at 4875 m were 19.0–19.3 g per 100 ml—equivalent to an oxygen carrying power of some 25.5–26.0 ml O₂ in combined form. The greater haemoglobin concentration ensures that the tissue oxygen requirements can be satisfied by a smaller desaturation of the oxyhaemoglobin present in the blood and hence a smaller drop in the capillary pO₂. To take the example already given, if the blood containing 20 ml O₂ per 100 ml (Hb = approximately 15 g per 100 ml) gives off 10 ml then the desaturation is 50 per cent and the pO₂ is 26 mm Hg. But if the [Hb] increases to say 20 g per 100 ml, then its 100 per cent saturation O₂ content is 20 × 1.34 = 26.8 ml O₂ and 10 ml from this leaves it with 16.8 ml O₂ and the desaturation is 10/26.8 × 100 = approximately 30 per cent. Hence the blood exits from the tissues at 70 per cent saturation with a pO₂ of approximately 35 mm. Again, the pressure head of O₂ between blood and tissues has been improved.

Polycythaemia is *not* due to a direct effect of low pO₂ on the bone marrow. Tissue culture experiments have shown that division of the littoral cells is not encouraged by low pO₂. Hypoxia acts indirectly causing the liberation of a hormone erythropoietin—a glycoprotein—from the kidney and perhaps elsewhere besides and it is this substance which stimulates the proliferation of red cells [p. 37].

The increase in red cells seen in dwellers at high altitudes is such that r.b.c. counts of 7½–8 million per mm³ are well documented. Such polycythaemia greatly increases blood viscosity [p. 79] and this in turn throws a great load on the heart which is itself pumping at least twice as much blood. Somehow a compromise is effected and counts in excess of 8 million are rare except in disease.

CAROTID AND AORTIC BODIES. CHEMORECEPTORS AND CHEMORECEPTOR RESPIRATORY REFLEXES

The carotid bodies are found above and near the bifurcation of the common carotid artery on both sides of the neck. Weighing only 2 mg and of dimensions little bigger than a pin head, they are difficult to see at all with the naked eye. In most animals they receive their blood supply from the occipital and ascending pharyngeal arteries and they are most commonly found closely adherent to the common stem of the occipital and ascending pharyngeal arteries which branch from the external carotid artery just cephalad to the carotid bifurcation [FIG. III.71]. Afferent nerves, whose endings serve as chemoreceptors, traverse medially and slightly dorsally in the carotid sinus nerve (which also contains the baroreceptor affe-

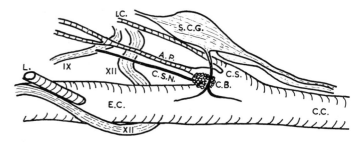

FIG. III.71. Carotid bifurcation of cat. C.C., E.C., and I.C. = common, external, and internal carotid arteries. C.B. = carotid body. C.S.N. = sinus nerve. XII = hypoglossal nerve. IX = glossopharyngeal nerve. A.P. = ascending pharyngeal artery. C.S. = carotid sinus. L = lingual artery. S.C.G. = superior cervical ganglion. (Heymans, C. and Neil, E. (1958). *Reflexogenic areas of the cardiovascular system*. Churchill, London.)

rents from the carotid sinus) and this sinus nerve joins the glosso-pharyngeal nerve. Chemoreceptor afferents thus gain access to the medulla with the remainder of those from the IXth nerve.

The aortic bodies are scattered conglomerations of epithelioid tissue similar in structure to those of the carotid body. They lie in the main between the concavity of the aortic arch and the convexity of the pulmonary trunk and the origin of the right and left pulmonary arteries. These bodies are innervated by the vagi and chemoreceptor afferent fibres can be found in both vagi and both aortic nerves. Again these afferent fibres (whose cell bodies lie in the nodose ganglion of the vagus) are distributed to the medulla and effect synaptic connections with the cardiovascular and respiratory 'centres'.

The carotid body has been intensively investigated in terms of function, blood flow and its histological characteristics.

The classical proof of the importance of the carotid body as a chemoreceptor was provided by Heymans in 1930. He showed that sodium cyanide injected into the common carotid artery stimulated respiration only if the carotid sinus nerve on that side were intact. Hence, sodium cyanide stimulates respiration only reflexly. Heymans then showed that anoxic anoxia caused hyperpnoea only if the carotid chemoreceptor nerves and the aortic chemoreceptors (in the vago-aortic nerves) were intact [FIG. III.72]. The inhalation of carbon dioxide mixtures, on the other hand, stimulated the breathing almost equally well whether the chemoreceptors were intact or not. This is true of the effects of acidaemia too, so it can be said that, providing the animal is not hypoxic, the stimulation of the breathing which CO_2 or H^+ may cause is *not* essentially due to the peripheral arterial chemoreceptors, but is due to a central effect—now known to be due to a modification of the $[H^+]$ in the interstitial fluid which surrounds the medullary neurons [see p. 199].

(a)   (b)

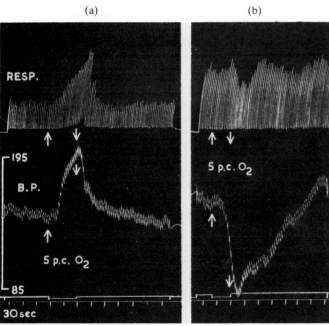

FIG. III.72. Cat. Effect of breathing 5 per cent O_2 spontaneously. (a) Before, (b) after cutting the sinus and vagus nerves. (Glaister, D. H. and Hearnshaw, J.)

Reflex responses to carotid body perfusion

Artificial perfusion of the carotid segment containing the carotid body is relatively easy using a technique similar in essence to that which has been described in the section on carotid baroreceptors [p. 125]. The perfusion medium—blood, plasma, or Ringer–Locke solution, is pumped through the segment via the common carotid artery. A cannula in the external carotid artery returns the blood to

a reservoir from which it is pumped once more through the segment. All arterial branches except those which supply the carotid body are tied. The venous drainage of the carotid body is carefully preserved and the blood flow through the organ drains into the internal jugular vein and thence into the general circulation. Injections of chemical substances can be made into the entrance limb of the system and the efficacy of these in arousing reflex chemical effects on the respiration is tested.

Using such techniques, Heymans showed that:

1. Hypoxia of the perfused carotid segment caused reflex hyperpnoea if the relevant sinus nerve were intact.
2. Perfusion of the segment with blood at high pCO_2 or H^+ caused reflex hyperpnoea.
3. Sodium cyanide added to the perfusing blood caused reflex hyperpnoea.

Electroneurography of the chemoreceptor afferents

Just as thin slips of the sinus nerve containing a few units or even a single unit can be made to examine the impulse activity of baroreceptor fibres so can the technique be used to identify and study chemoreceptor responses. Desirably the pO_2 and pH of the blood, which naturally supplies or is artificially perfused through the carotid body, should simultaneously be known and, as will be seen later, it is of some importance to know the pressure of perfusion and the blood flow through the carotid body. However, the important features can first be summarized before going into the details of blood flow, etc.

1. In an anaesthetized animal, breathing air, the chemoreceptors show a resting discharge which is characteristically irregular in character. This discharge is intensified by hypoxic hypoxia as might be expected from the results of the reflex respiratory studies [FIG. III.73]. The chemoreceptor preparations are made from a sinus nerve which is first cut near its junction with the glossopharyngeal nerve; the fine twig consisting of pure chemoreceptor fibres is made by trial and error by peeling back delicate bundles in the direction of the carotid body, from the sinus nerve trunk. Clearly, the impulse activity in these chemoreceptor afferents sampled can have no reflex effect on the medullary centres because the whole

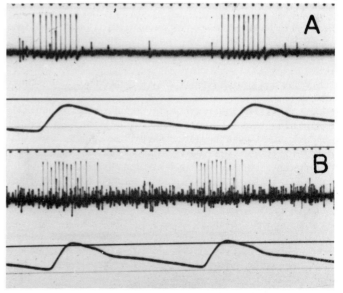

FIG. III.73. Afferent impulses in sinus nerve twig. A, breathing air. B, breathing 5 per cent O_2 in N_2. Single baroreceptor unit is little affected by hypoxia. Chemoreceptors fire briskly. Femoral blood pressure (calibration lines 150 and 100 mm). Time 50 c s^{-1}.

of the corresponding sinus nerve has been cut. However, the afferent activity recorded is taken as representative of that which occurs in the remaining chemoreceptor fibres (in the opposite sinus nerve and in both vagi) which are intact and which exert reflex effects on the respiration.

When the impulse activity of a single chemoreceptor fibre is studied at different arterial oxygen tensions its impulse frequency at an arterial pO_2 of 100 mm Hg is sparse, approximately 1–2 impulses per second providing the arterial CO_2 tension is in the normal range (30–35 mm Hg in an anaesthetized cat). On lowering the arterial oxygen tension by substituting hypoxic gas mixtures for the animal to breathe, discharge increases but may only attain 6–12 impulses per second [FIG. III.74]. The discharge shows little adaptation.

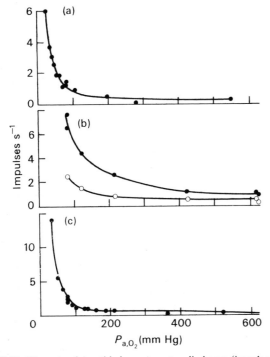

FIG. III.74. The rate of carotid chemoreceptor discharge (impulses s^{-1}) in single fibres, plotted against the arterial O_2 tension (mm Hg) all from the same cat. (b) Shows two fibres (●, ○) from the same strand, (a) and (c) are from single fibres. Arterial pCO_2 was 28–31 mm Hg for (a) and (b) and 31–34 mm Hg for (c). Mean arterial pressure was 95 ± 7 mm Hg in (a) and (b) and was 123 ± 6 mm Hg in (c). (Redrawn from Biscoe, T. J., Purves, M. J., and Sampson, S. R. (1970). *J. Physiol., Lond.* **208**, 121.)

2. When the arterial tension of CO_2 is increased on the other hand there is a prompt response of chemoreceptor discharge which quickly adapts (McCloskey 1968) if the stimulus is maintained. This prompt initial response to hypercapnia is not however seen if the animal has been given the carbonic anhydrase inhibitor acetazolamide [FIG. III.75]. This suggests that the efficacy of CO_2 as a chemoreceptor stimulant is dependent upon the formation of $[H^+]$ from CO_2. The carotid body contains carbonic anhydrase which presumably normally catalyses the hydration of CO_2 to H_2CO_3; subsequent ionization raises the extracellular $[H^+]$ in the vicinity of the chemoreceptor neurons.

FIGURE III.76 shows that the increase of discharge of a single chemoreceptor fibre is almost linear over a CO_2 tension range of 25–50 mm Hg, providing that the arterial oxygen tension does not vary.

When the carotid body is perfused by solutions equilibrated at a low oxygen tension and a high CO_2 tension the discharge is greater than additive. Seemingly the 'interaction' between hypercapnia and hypoxia on the respiratory response of the intact animal or man

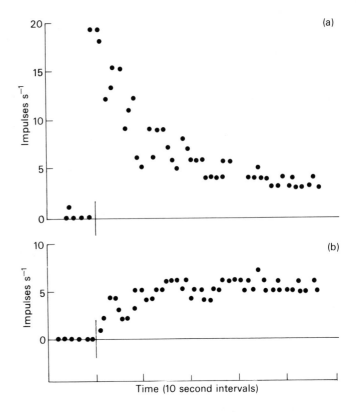

FIG. III.75. Cat. Single chemoreceptor fibre. Changes in discharge frequency on sudden application of a hypercapnic stimulus before (a) and after (b) inhibiting the carbonic anhydrase of the carotid body with acetazolamide. (Redrawn from results of McCloskey, D. I. (1968). In *Arterial chemoreceptors* (ed. R. W. Torrance). Blackwell, Oxford.)

[p. 200] is at least partly referable to the increase in peripheral chemoreceptor drive.

3. Stagnant hypoxia (which is equivalent to asphyxia) powerfully stimulates the chemoreceptors. This is most simply shown by recording discharge before, during and after occlusion of the common carotid artery supplying the carotid body under study [FIG. III.77]. As stated on page 207 haemorrhagic hypotension causes chemoreceptor asphyxia.

Activity of the chemoreceptor afferents in haemorrhagic hypotension helps to sustain the reflex vasoconstriction which

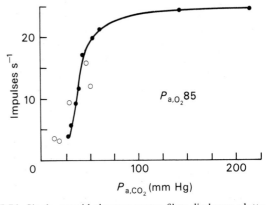

FIG. III.76. Single carotid chemoreceptor fibre discharge plotted against arterial CO_2 tension. Rate and volume of artificial ventilation were varied to change arterial CO_2 tension and so also the pH. Arterial oxygen tension was 85 mm Hg throughout. ● Results after the CO_2 tension was increased; ○ results after the CO_2 tension was decreased. (Redrawn from Biscoe, T. J., Purves, M. J., and Sampson, S. R. (1970). *J. Physiol., Lond.* **208**, 121.)

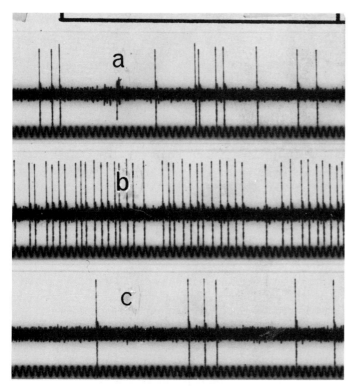

FIG. III.77. Chemoreceptor preparation (sinus nerve). BP 120 mm Hg. Spontaneous breathing (air).

    a. Before clipping corresponding common carotid artery.
    b. Common carotid clipped.
    c. After releasing clip.
    No change in blood pressure (nerve cut).

    Above: signal shows 1 second between marks.
    Below: 50 c s^{-1} time trace.

(Neil, E. and Joels, N. (1963). In *Regulation of human respiration* (ed. D. J. C. Cunningham and B. B. Lloyd). Blackwell, Oxford.)

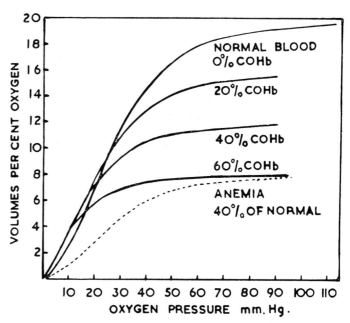

FIG. III.78. Oxygen dissociation curves of (1) human blood, at pH 7.40 and 38 °C, containing various percentages of carboxyhaemoglobin, and of (2) anaemic human blood containing only 40 per cent of normal haemoglobin content. (Roughton, F. J. W. (1954). *Handbook of respiratory physiology.* USAAF School of Aviation Medicine.)

occurs [p. 151] and is responsible for the increased breathing which is one of the features of haemorrhagic hypotension.

4. Anaemic hypoxia is chronically present in the severely anaemic patient. It also occurs acutely in carbon monoxide poisoning. In neither case is the breathing stimulated at rest. Carbon monoxide is therefore trebly dangerous as a poisonous gas. 1. It uses up the available haemoglobin rendering it unavailable for oxygen combination. 2. The presence of COHb in the blood causes an alteration of the HbO$_2$ dissociation curve of such haemoglobin as is available for combination with O$_2$. The HbO$_2$ curve is pushed over towards the left and is more convex to the *y* axis [FIG. III.78]; as a consequence a given degree of desaturation of HbO$_2$ requires a greater fall of pO$_2$ than ordinarily is the case. This lowers the pressure head available for oxygen diffusion between the capillary blood and the tissue cells. 3. CO does not stimulate the breathing and hence the alveolar pO$_2$ is not raised when the subject breathes air vitiated by CO.

Edwards and Mills (1968) found that CO undoubtedly does cause some chemoreceptor stimulation. The irresponsiveness of the resting ventilatory volume to this increasing chemoreceptor traffic remains unexplained—but so does that to mild hypoxic hypoxia for that matter. Admittedly, we know now that hypoxic drive only becomes effective if the c.s.f. [H$^+$] can be adjusted, as in acclimatization, and this takes time, but the details are still obscure.

Structure and blood supply of the carotid body

The carotid body is sometimes called the carotid glomus (glomus = a ball of yarn; plural = glomera). This term arose from an investigation of its histological structure using the light microscope. Groups of large epithelioid cells or glomera are seen which are separated from each other by connective tissue. The whole organ is virtually interlaced by sinusoids and presents an unusually rich vascular appearance. This adjacence of the epithelioid cells (which themselves receive a generous terminal innervation by sinus nerve fibres) to the capillary blood seems appropriate for an organ which 'tastes' the blood composition.

The blood flow through the carotid body was first measured by Daly, Lambertsen, and Schweitzer (1954) by collecting the venous drainage. It is at first sight very small—only some 40 μl min^{-1}. However, the carotid body weighs only 2 mg and if the blood flow is expressed in the conventional terms it is 2000 ml per 100 g per min, which is four times that of the thyroid gland weight for weight. Likewise the oxygen usage of the carotid body was determined by Daly *et al.* and can be expressed as 9.0 ml per 100 g per min. This very high oxygen usage is similar to that of nerve cells. The vigorous oxygen usage of the carotid body is in a way overshadowed by the enormous blood flow which it enjoys, so that the venous effluent from the organ still contains a high percentage saturation of haemoglobin. On inspection through the microscope the carotid body vein(s) can always be identified by the bright red appearance which is in sharp contrast to the purplish colour of other veins in the vicinity.

The vascular resistance of the organ appears to be very low but stimulation of postganglionic sympathetic nerves, which supply it from the adjacent superior cervical ganglion, increases this vascular resistance and lowers the flow through the organ. Such facts have to be taken into account when considering the reasons for chemoreceptor discharge in haemorrhagic hypotension—a situation always accompanied by sympathetic vasoconstriction. Nevertheless, although such vasoconstriction undoubtedly contributes to a lower flow through the carotid body in haemorrhagic hypotension and thereby causes some increase in chemoreceptor discharge of itself, it is only contributory; even after denervation of the local sympathetic supply, haemorrhagic hypotension still causes a striking excitation of impulse activity in the chemoreceptor nerves.

The intimate details of chemoreceptor excitation are beyond the scope of this book. More specialized texts, such as those of Torrance (1968, 1974), discuss these matters in great detail.

Aortic bodies

The ease of isolation and perfusion of the carotid body, which moreover causes little disturbance of the systemic circulation or of the normal respiratory mechanism, has been responsible for the wealth of information about the carotid glomera compared with the paucity of our detailed knowledge of the aortic bodies. Paradoxically, the aortic chemosensory areas were defined before the chemoreceptor function of their carotid homologues was even guessed at. Corneille Heymans and his father found in 1927 that if the head of a dog (connected with its trunk only by the vagi) were separately perfused from a donor animal then changes in the chemical composition in the trunk of the recipient caused changes of movements of the larynx of the 'head' section even though the composition of the blood supplying the head remained constant. These laryngeal movements were recorded as representative of the activity of the central respiratory apparatus. Suitable exclusion experiments showed that the chemical effects exerted from the trunk were stemming from the aortic area. Naturally, the reflex laryngeal movements were no longer aroused after the vagi were cut.

It seems likely that the aortic bodies and the carotid bodies by virtue of their respective afferent connections with the neuraxis act as a functional entity. Nevertheless, Daly and Ungar (1966) have shown that the carotid bodies in the dog subjected to perfusion with hypoxic blood cause a far greater reflex stimulation of breathing than do the aortic bodies when similarly stimulated.

REFERENCES

ACKER, H., FIDONE, S., PALLOT, D., EYZAGUIRRE, C., LÜBBERS, D. W., and TORRANCE, R. W. (1977). *Chemoreception in the carotid body.* Springer, Berlin.

ÅSTRAND, P.-O. (1954). *Acta physiol. scand.* **30**, 335 and 343.

BAND, D. M., CAMERON, I. R., and SEMPLE, S. J. G. (1969). *J. appl. Physiol.* **26**, 261.

BISCOE, T. J. (1971). *Physiol. Rev.* **51**, 437.

—— PURVES, M. J., and SAMPSON, S. R. (1970). *J. Physiol., Lond.* **208**, 121.

BROOKS, C., KAO, F., and LLOYD, B. B. (eds.) (1965). *Cerebrospinal fluid and the regulation of ventilation.* Blackwell, Oxford.

CUNNINGHAM, D. J. C. (1974*a*). *Q. Rev. Biophys.* **6**, 433.

—— (1974*b*). Integrative aspects of the regulation of breathing. In *Respiratory physiology*, Vol. 2 (ed. J. G. Widdicombe) pp. 303–69. MTP International Review of Science. Butterworths, London.

—— SHAW, D. G., LAHIRI, S., and LLOYD, B. B. (1961). *Q. Jl exp. Physiol.* **46**, 323.

DALY, M. DE B. LAMBERTSEN, C. J., and SCHWEITZER, A. (1954). *J. Physiol., Lond.* **125**, 67.

—— and UNGAR, A. (1966). *J. Physiol., Lond.* **182**, 379.

EDWARDS, MCIVER W., and MILLS, E. (1969). *J. appl. Physiol.* **27**, 291.

FENCL, V., MILLER, S., and PAPPENHEIMER, J. R. (1966). *Am. J. Physiol.* **210**, 459.

HEYMANS, C. and NEIL, E. (1958). *Reflexogenic areas of the cardiovascular system.* Churchill, London.

HOWE, A. and NEIL, E. (1972). Enteroreceptors, Arterial chemoreceptors. In *Handbook of sensory physiology* III/1 (ed. E. Neil). Springer, Berlin.

LAHIRI, S. and MILLEDGE, J. S. (1967). *Resp. Physiol.* **2**, 310.

LEUSEN, I. (1954). *Am. J. Physiol.* **176**, 39 and 45.

LLOYD, B. B. and CUNNINGHAM, D. J. C. (eds.) (1963). *The regulation of human respiration*, pp. 331–49. Blackwell, Oxford.

MCCLOSKEY, D. I. (1968). In *Arterial chemoreceptors* (ed. R. W. Torrance). Blackwell, Oxford.

MICHEL, C. C. and MILLEDGE, J. S. (1963). *J. Physiol., Lond.* **168**, 631.

MITCHELL, R. A., LOESCHHKE, H. H., MASSION, W. H., and SEVERINGHAUS, J. W. (1963). *J. appl. Physiol.* **18**, 523.

NEIL, E. and JOELS, N. (1963). In *Regulation of human respiration.* Blackwell, Oxford.

NIELSEN, M. and SMITH, H. (1951). *Acta physiol. scand.* **24**, 293.

PAPPENHEIMER, J. R. (1967). *Harvey Lect.* 71–94.

—— FENCL. V., HEISEY, S. R., and HELD, D. (1965). *Am. J. Physiol.* **208**, 436.

SEVERINGHAUS, J. W., MITCHELL, R. A., RICHARDSON, B. W., and SINGER, M. M. (1963). *J. appl. Physiol.* **18**, 1155.

TORRANCE, R. W. (1968). *Arterial chemoreceptors.* Blackwell, Oxford.

—— (1974). Arterial chemoreceptors. In *Respiratory physiology*, Vol. 2 (ed. J. G. Widdicombe) pp. 247–71. MTP International Review of Science. Butterworths, London.

WINTERSTEIN, H. (1956). *New Engl. J. Med.* **255**, 331.

TYPES OF HYPOXIA

1. Hypoxic type

This is characterized by a low arterial pO_2 which is inadequate to saturate the haemoglobin fully.

Such hypoxic hypoxia may be due to:

(i) Low pressure of O_2 in the air—e.g. in mines *or* at altitude where although the percentage O_2 content of the air is normal the total barometric pressure is low.

(ii) Bulbar poliomyelitis or cervical transection.

(iii) Obstruction of the respiratory passages.

(iv) Inadequate lung ventilation or pulmonary oedema.

(v) Thickening of the alveolar membrane.

(vi) Congenital heart disease with a right to left cardiac shunt.

2. Anaemic type

This is less serious in its effects than the hypoxic form. As the oxygen tension in the blood is normal, the rate of tissue oxidation is maintained at its usual level. No increase in the pulmonary ventilation occurs at first; i.e. the breathing does not respond readily to a decrease in the volume of oxygen in the arterial blood, so long as the appropriate tension is maintained. At rest, the prejudicial effect on the tissues is relatively slight. Of the 19 ml per 100 ml of oxygen normally present in arterial blood, only 5 ml are used up under resting conditions. As a person with 50 per cent haemoglobin carries 9.5 ml oxygen per 100 ml blood, his resting requirements are readily satisfied. But his capacity to do work is greatly diminished because he has not the normal reserves of oxygen in the blood to call upon.

(i) Lack of haemoglobin. In severe anaemias the venous blood is very reduced. In exercise, the chief method which is available in these subjects for increasing the oxygen supplies to the tissues is to increase the cardiac output. Acceleration of the pulse occurs with slight exertion.

(ii) Altered haemoglobin. In poisoning with the nitrates, nitric-oxide–haemoglobin and some methaemoglobin are formed. Some sulphonamide derivatives produce methaemoglobin; marked cyanosis may be present, owing to the presence in the blood of this coffee-coloured compound. The patient may, however, feel well so long as enough normally functioning haemoglobin (i.e. capable of transporting oxygen) is available, because the arterial oxygen tension is normal.

(iii) Carbon monoxide poisoning causes not only anaemic hypoxia but renders such oxyhaemoglobin as exists in the blood less capable of yielding its oxygen to the tissues. CO has about 250 times as great an affinity for Hb as has O_2. Hence with a normal alveolar pO_2 of 100 mm Hg a CO tension of 0.4 mm Hg in the alveoli would cause progressive carboxyhaemoglobinaemia. CO combines with Hb in the same manner as does oxygen, occupying the free linkage of the iron atom in the haem.

Such oxyhaemoglobin as is present in the arterial blood of a person suffering from CO poisoning dissociates in a manner quite different from that normally seen; the shape of the dissociation curve of this residual oxyhaemoglobin is no longer sigmoid and the curve itself is shifted to the left [FIG. III.78]. Thus the oxyhaemoglobin gives off its oxygen much less readily.

People suffering from CO poisoning should be lifted from the vitiated atmosphere and adequately ventilated with oxygen mechanically if necessary. Care should obviously be taken to avoid acapnia during mechanical ventilation.

REFERENCE

JOELS, N. and PUGH, L. G. C. (1958). *J. Physiol., Lond.* **142**, 63.

3. Stagnant type

This occurs when the cardiac output and blood flow to the organs are diminished because of heart failure [p. 137], impaired venous return, haemorrhage, or shock. The tension of oxygen in the blood is normal, but the amount reaching the tissues is inadequate. The rate of tissue oxidation is normal, because oxygen is supplied at a high pressure head. As the blood circulates more slowly in the tissues there is more time available for reduction of oxyhaemoglobin. Furthermore, the impaired circulation causes CO_2 accumulation in the tissues which facilitates the giving-off of oxygen. Thus the tissues make the most effective use of what oxygen does reach them in the blood.

4. Histotoxic type

This occurs in poisoning with cyanide, which interferes with tissue oxidation, by paralysing cytochrome oxidase. Narcotics also depress tissue oxidation by interfering with dehydrogenase systems.

Cyanosis

Cyanosis is a blue colour of the skin and/or the mucous membranes. It is caused by the presence of blood in the minute vessels which contains more than 5 g per 100 ml of haemoglobin in reduced (or altered) form.

The two classic causes of cyanosis are hypoxic and stagnant hypoxia. Of these stagnant anoxia is the more common cause, particularly when it is of the local type (cold blue hands, etc.).

Individuals who suffer from methaemoglobinaemia show cyanosis. Sufferers from carbon monoxide poisoning do not show cyanosis: 1. COHb is cherry pink, 2. HbO_2 in the presence of COHb does not readily become reduced.

Anaemic individuals are less likely to be cyanosed because if they had more than 5 g per 100 ml of such haemoglobin as they possess in the reduced form they would be dead.

Voluntary hyperpnoea

If the subject breathes very deeply and quickly for 2–3 minutes and then allows his breathing to act independently of voluntary control, the bout of overventilation is followed by depressed breathing or by apnoea (cessation of rhythmic breathing). A phase of periodic breathing commonly sets in, after which respiration gradually becomes normal. Examination of the alveolar air at the end of the overventilation shows that the CO_2 tension is greatly reduced, e.g. to 15 mm Hg (normal 40 mm) and the O_2 tension is raised, e.g. to 140 mm Hg (normal 100 mm) [FIG. III.79].

The apnoea is due solely to the CO_2 lack (i.e. CO_2 tension below the minimum necessary for rhythmic breathing) and is not the result of the excess O_2 in the alveolar air. 1. If the overventilation is carried out with air containing 5 per cent CO_2 so that the alveolar CO_2 concentration is not lowered, no apnoea occurs; in fact hyperpnoea persists after discontinuing the experiment although the alveolar O_2 concentration is raised to about 19 per cent. 2. The inhalation of pure O_2 very markedly raises alveolar O_2 concentra-

tion, e.g. to 80 per cent; but it does not alter alveolar CO_2 concentration and has no effect whatever on breathing.

During the period of apnoea, oxygen is being steadily removed from the alveoli, so that two minutes later the alveolar oxygen tension is only 30 mm Hg; severe oxygen lack is thus present. At the same time CO_2 has been passing continuously from the tissues into the venous blood, diffusing into the alveoli. Thus the alveolar and arterial CO_2 tension slowly rises, reaching a level of about 36 mm Hg [FIG. III.79]. Though this tension is still below the threshold level, 'air hunger' is experienced and breathing is resumed. This resumption of the breathing is due to a combination of factors (i) the rise of CO_2 tension almost to threshold, (ii) the stimulating action of the severe anoxia. It can be shown that if either factor is lacking breathing does not recommence.

If pure oxygen instead of air is taken in during the period of forced breathing, the apnoea is greatly prolonged, because there is no anoxic stimulus to interact with the hypercapnic (raised [H^+]) factor.

Under normal conditions variations in the CO_2 tension alone control breathing; but during the period of apnoea following voluntary overventilation, the O_2 lack which develops enables breathing to be resumed before the CO_2 tension has risen fully to the normal threshold level. When breathing begins again, fresh oxygen supplies are taken in, the alveolar oxygen rises, and the oxygen want is temporarily relieved; at the same time too, CO_2 is washed out from the alveoli and the arterial CO_2 tension falls. Thus both factors in the coalition of forces which brought about respiratory activity have been weakened. The breathing therefore ceases again. Similar events occur during the next period of apnoea: the alveolar oxygen is used up and increasing oxygen want develops; CO_2 accumulates and the alveolar and arterial CO_2 rise. Breathing resumes again. This cycle is repeated several times [FIG. III.79]. It may be wondered why the process does not continue indefinitely. The reason is that the brief bouts of breathing do not succeed in washing out CO_2 from the alveoli as rapidly as it is being turned out from the tissues and blood. The alveolar (and arterial) CO_2 tension is a little higher at the end of each period of apnoea than it was at the end of the preceding one, thus gradually rising to threshold level. Regular breathing is then finally established.

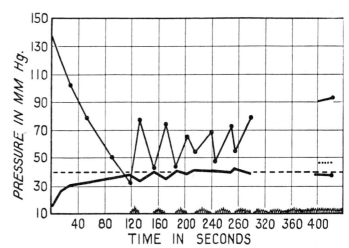

FIG. III.79. Changes in the breathing and in the tension of gases in aveolar air *after* forced respiration for two minutes. Upper curve, pressure of oxygen in alveolar air. Middle curve, pressure of CO_2 in alveolar air. Lower curve, respiration. Straight interrupted line = normal CO_2 tension (40 mm Hg). (Douglas and Haldane.)

The results described are by no means invariably obtained. In some normal subjects apnoea does not follow the overventilation, but there is some decrease in the pulmonary ventilation; or the apnoea may pass smoothly into regular breathing of increasing amplitude. Sometimes the overventilation continues in spite of a marked fall of alveolar CO_2 tension and seems to be outside the subject's voluntary control.

The general effects of overbreathing are best studied when the pulmonary ventilation is increased to a more moderate extent (e.g. two- to threefold) and kept up for a longer time, e.g. 5 to as long as 30 minutes. They include:

1. *Respiratory alkalaemia.* The blood pH rises to 7.55 or even 7.60. Voluntary hyperventilation never lowers the alveolar CO_2 concentration below 2 per cent (pCO_2 = 14–15 mm Hg) because the 'exercise' of hyperventilation itself causes an increase in CO_2 metabolically produced. Hence the CO_2 expired is not a measure of the CO_2 'washed out' of the body. The O_2 usage must be determined and the CO_2 metabolically produced at an RQ of, say, 1.0 is then equivalent to this value. If this CO_2 volume metabolically produced is then subtracted from the total CO_2 expired, the volume of CO_2 washed out per minute can then be stated.

Some subjects use so much energy overbreathing that the overall volume of CO_2 actually washed out is small, and the fall in their alveolar CO_2 is correspondingly slight.

2. *Renal compensation for the alkalaemia.* The kidneys excrete an alkaline urine containing bicarbonate. The reason for this lies in the failure of the renal tubule cells to provide sufficient H^+ owing to the low pCO_2 in the blood and tissues. As HCO_3^- can only be absorbed from the tubule lumen if H^+ is secreted therein, HCO_3^- and Na^+ are lost in the urine. The NH_4^+ concentration of the urine becomes vanishingly small, again due to the failure of the renal cells to manufacture H^+ which would combine with NH_3 formed by the distal tubule cells. The urine contains keto acids.

3. *Cardiovascular changes.* The cardiac output is raised by voluntary hyperventilation. Kety and Schmidt (1948) found cardiac outputs of 8–8.5 l min^{-1} in subjects breathing 25 l min^{-1}. In voluntary hyperventilation the blood pressure is never reduced, largely owing to this increase of cardiac output. The commonest effect of voluntary hyperventilation on the blood pressure is a slight rise—largely referable to the increase in cardiac output.

4. *Neurological changes.* The subject becomes dizzy and lightheaded; consciousness is dulled and may even be lost. Paraesthesiae are common—e.g. numbness and tingling of the extremities. Tetany often develops with characteristic changes—stiffness of the face and lips, carpopedal spasm and increased excitability of the motor nerves. Carpal spasm causes the *main d'accoucheur* phenomenon. Chvostek's sign, a spasm of the facial muscles on tapping the facial nerve, is positive. Tetany is usually ascribed to a lowering of the calcium ion concentration in the blood and tissue fluids. Thus, parathyroprivia, which causes a fall in the total blood calcium, produces tetany. In hyperventilation tetany, however, the total blood calcium is normal but the calcium ionic fraction is reduced owing to the increase in calcium proteinate formation caused by the alkalosis. The ionization of the proteins: H Prot $\rightleftharpoons$ Prot$^-$ + H^+ is favoured by alkalosis and the increased concentration of plasma protein ions 'covers' more of the calcium ions. The increased susceptibility of the motor nerves to mechanical stimulation can also be ascribed to the reduction in calcium ion concentration.

The cerebral symptoms are largely due to acapnic anoxia. Acapnia causes cerebral vasoconstriction (the cerebral blood flow may fall from 55 ml per 100 g per min to 35 ml per 100 g per min (Kety and Schmidt 1948)) and a shift of the oxygen dissociation curve to the left so that the unloading of O_2 from the oxyhaemoglobin occurs with greater difficulty. Many of the cerebral symptoms (such as dizziness) can be relieved by substituting 100 per cent O_2 for air as the inspired gas.

It should be noted that positive pressure mechanical hyperventilation (particularly in an anaesthetized subject) can lower the blood pressure considerably, for such artificial ventilation not only reduces the cardiac output but also causes a more serious fall of alveolar pCO_2. In these circumstances there is no excess in CO_2 production over the resting level and the alveolar pCO_2 can hence be reduced to much lower levels.

High oxygen pressure

Pure oxygen (at atmospheric pressure) can be breathed in man without ill effect for short periods, e.g. for several hours. There are few records of the effects of such inhalations for longer periods owing to fear of producing the pulmonary damage which develops in small animals breathing pure oxygen for several days; 100 per cent O_2 has been breathed in a few instances for two days without ill effect in man; 60 per cent O_2 mixtures can be breathed in man indefinitely with perfect safety.

In man, breathing oxygen at 3.5 atmospheres pressure, Lambertsen and his colleagues (1955) found an arterial pO_2 of some 2100 mm Hg. However, the cerebral venous pO_2 was approximately normal indicating that the cerebral vessels had constricted, partly due to the influence of the high arterial pO_2. Lambertsen calculated that the mean cerebral capillary pO_2 was of the order of 850 mm Hg. When 2 per cent CO_2 was added to the inspired air the cerebral arterial pO_2 was little affected, but the cerebral venous pO_2 rose in startling fashion to about 1000 mm Hg owing to the cerebral vasodilatation which the inspired CO_2/O_2 mixture induced. The greater incidence and earlier onset of nervous symptoms evoked by hyperbaric oxygen mixtures which contain CO_2 is explained by the higher cerebral tissue pO_2 which occurs owing to the cerebral vasodilatation. Such symptoms and signs include dazzle, giddiness, nausea and twitching of lips and limbs as early features; more prolonged exposure causes convulsions and unconsciousness. High oxygen pressures cause inactivation of sulphydryl groups of enzymes and coenzymes in the Krebs cycle. Two particular reactions are implicated (a) the conversion of pyruvic acid to acetyl coenzyme A and (b) the conversion of α-ketoglutaric acid to succinyl coenzyme A. Both these reactions require as a co-factor the substance lipoic acid (6–8-dimercapto-octanoic acid)

$$CH_2-CH_2-\underset{\underset{S}{\displaystyle |}}{CH}-\underset{\underset{S}{\displaystyle |}}{CH_2}.\ (CH_2)_4\ COOH)$$

which can exist in the reduced form with sulphydryl groups at sites 6 and 8. If the pO_2 is high these groups are oxidized and the two reactions described above cannot be completed. As a result the Krebs cycle 'stops turning' and as this cycle is responsible for the resynthesis of 15 mols ATP per mol pyruvate, the supply of the 'energy-rich' ATP dies off and the metabolic functions of the cells correspondingly suffer. The provision of $-SH$ groups in the form of reduced glutathione offsets the toxic action of high oxygen pressures on cellular metabolism. Similarly, the chelation of metallic ions, such as Cu^{2+}, by the addition of EDTA (ethylenediamine tetra-acetic acid) or other chelating agents also protects cellular metabolism against the effects of high pO_2 (see Haugaard 1968).

These effects of high oxygen pressure have to be borne in mind now that hyperbaric oxygenation chambers are in use (e.g. in Amsterdam and Glasgow) in which pressures of 2–3.5 atmospheres are employed to aid surgical operations on the heart and lungs. The patient breathes oxygen and his oxygen stores are correspondingly increased. At a pO_2 of 2100 mm Hg some 6.3 ml O_2 per 100 ml is present in the dissolved form. Such chambers are also employed in the treatment of CO poisoning and of infections by anaerobic organisms.

REFERENCES

DICKENS, F. and NEIL, E. (eds.) (1964). *Oxygen in the animal organism.* Pergamon, Oxford.

HAUGAARD, N. (1968). *Physiol. Rev.* **48**, 311.

KETY, S. and SCHMIDT, C. F. (1948). *J. clin. Invest.* **27**, 484.

Breath-holding

If a subject holds his breath at the end of a quiet expiration he can maintain voluntary apnoea for a period of 45–55 seconds before reaching a 'breaking point' at which the urgent desire to breathe becomes dominant. During breath-holding the alveolar pO_2 falls and the alveolar CO_2 rises providing two obvious reasons for the breaking point. As we have seen the breathing is more likely to be stimulated by a given increase in pCO_2 if the alveolar pO_2 is reduced—there is interaction between the two stimuli. If the subject repeats the experiment having 'washed out' his lungs with a few breaths of 100 per cent O_2 prior to holding his breath, his breath-holding time is prolonged some 15–20 seconds. When the alveolar pO_2 is high (c. 650 mm Hg) to begin with:

1. There is no oxygen-lack stimulus during the breath-holding time.
2. The sensitivity of the respiratory centre to the rise of alveolar pCO_2 is less in the absence of oxygen lack.

A third experiment in which the subject overbreathes room air for one minute before holding his breath reveals that the subsequent breath-holding time is further prolonged to perhaps two minutes or more. By overbreathing, the alveolar pCO_2 is reduced to 15–20 mm Hg before breath-holding begins; hence it is likely that quite severe oxygen lack may occur in the extended breath-holding period before the pCO_2 has even reached normal figures. Lastly, if the subject overbreathes pure oxygen for one minute before breath-holding the period of voluntary apnoea may be extended to five minutes.

If the previously eupnoeic subject holds his breath at the end of a quiet inspiration, his alveoli will contain some 3 litres of gas of which about 14 per cent is oxygen. He thus begins breath-holding with some 420 ml O_2 and his body at rest uses oxygen at 300 ml min^{-1} and produces CO_2 at 250 ml min^{-1}. His lung volume is thus bound to shrink not only because of the greater metabolic usage of oxygen from the lungs than is provided by the metabolic production of CO_2, but also because of the much greater solubility of CO_2 in the body fluids and the differences between the oxygen and carbon dioxide dissociation curves.

At breaking point, typical figures for alveolar gas are 8 per cent O_2 (56 mm Hg pO_2) and 7 per cent CO_2 (49 mm Hg pCO_2). It has long been known that subjects can go on rebreathing at an unrestricted lung volume longer than they can hold their breath and reach a high pCO_2 in so doing. Fowler (1954) showed that if subjects who had reached the break-point of voluntary breath-holding and whose alveolar air when expelled contained approximately 8 per cent O_2 and 7 per cent CO_2 then took six to eight breaths from a bag containing gas of a similar composition, the unpleasant 'desire to breathe' sensation was relieved within two breaths and they could hold their breath for a further 20 seconds. These important results indicate that although changes in blood gas composition obviously play a part in generating the distressing symptoms which limit our power to hold the breath, they are not directly responsible for these symptoms. They also show that the actual movements of the chest wall and lungs alleviate the unpleasant sensation. This accords with the common observation that breath-holding time can be extended by making respiratory movements behind a closed glottis.

Guz *et al.* (1966) showed that bilateral local block of the vagi and glossopharyngeal nerves in conscious normal men prolonged the breath-holding time at all lung volumes and prevented the mounting unpleasantness associated with breath-holding in normal circumstances. Such findings indicate that the 'drive to breathe' which develops during breath-holding is mediated by the vagi and by the glossopharyngeal nerves. They might also be taken to imply that the consciousness of the afferent activity in these nerves (e.g. chemoreceptor afferents, lung deflation receptor fibres, etc.) is the cause of the unpleasant sensation. Campbell *et al.* (1967) suggested an alternative explanation—that the sensation arises because the chest and lungs are not permitted to move and this normal response to the increased vagal and glossopharyngeal afferent activity is prevented. They therefore administered tubocurarine chloride intravenously to two normal unanaesthetized male subjects, to secure temporary paralysis of the respiratory muscles, and examined the effects of 'breath-holding' time and the distressing sensation caused by the apnoea. They reasoned that if the afferent activity in IXth and Xth nerves caused the unpleasant sensation, paralysis of the respiratory muscles would not affect it and the 'breath-holding' time would be the same.

The two volunteers first breathed a mixture of 68 per cent O_2 and then received 1.2 mg atropine intravenously. Holding their breaths at the functional residual capacity one subject held out for 80 seconds and the other for about 100 seconds. After tubocurarine, apnoea at resting lung volume was tolerated for at least 4 minutes by both subjects and the distressing sensation normally provoked by breath-holding was entirely absent. They point out that their results must be reconciled with those of Fowler who showed that the unpleasant sensations of breath-holding could be relieved by chest movements, whereas they had shown that a complete absence of chest movement was unassociated with a feeling of distress. They suggested (as did Fowler) that the unpleasant sensation is engendered by afferent discharges from receptors in muscles, joints, or tendons which are stimulated when the respiratory muscles contract but when chest movement is prevented. The sensation is thus absent if on the one hand such muscle contraction is prevented, as in their experiments, or on the other if movement is permitted. The question why vagal block abolishes the unpleasant sensation if the afferent pathway of the sensation begins in the receptors of the chest wall is answered by the suggestion that such vagal block interrupts afferents from the lung which cause reflex contraction of the respiratory muscles during breath-holding in normal circumstances. In the absence of this reflex drive the respiratory muscles do not contract and no sensation is aroused.

The somatic afferents responsible for the eventual sensation of distress cannot yet be identified with certainty. Joint receptors would seem most likely implicated for afferents from joint receptors in the limbs, unlike those from muscle spindles, are responsible at least for conscious proprioceptive sensation.

Campbell *et al.* (1969) report a study by Newsom Davis and Semple, who investigated a patient with spinal transection at C2. This patient, in spite of possessing intact vagi, was quite unaware of the movement or volume of his lungs. These results, too, suggest that the sensation of breath-holding, however aroused, results from contractions of the respiratory muscles and not from the stimulation of parenchymatous vagal afferents.

REFERENCES

Campbell, E. J. M., Freedman, S., Clark, T. J. H., Robson, J. G., and Norman, J. (1967). *Clin. Sci.* **32**, 425.
—— Godfrey, S., Clark, T. J. H., Freedman, S., and Norman, J. (1969). *Clin. Sci.* **36**, 323.
Fowler, W. S. (1954). *J. appl. Physiol.* **6**, 539.
Guz, A., Noble, M. I. M., Widdicombe, J. G., Trenchard, D., Mushin, W. W., and Makey, A. R. (1966). *Clin. Sci.* **30**, 161.

RESPIRATORY QUOTIENT

The Respiratory Quotient (RQ) is the ratio of the volume of CO_2 evolved from the lungs over the volume of O_2 absorbed from the lungs in one minute.

The respiratory quotient has been intensively studied in many conditions in health and disease and the main results are summarized below; but it must be confessed that the respiratory quotient frequently gives no information about metabolic processes in the body; on the contrary, a knowledge of the metabolic processes taking place is generally necessary to interpret the RQ. Furthermore the RQ throws no light on the stages of intermediate metabolism.

1. Effect of combustion of foodstuffs, intermediate metabolism, and diet

If pure foodstuffs are burnt in a bomb calorimeter their 'respiratory quotient' can be measured directly.

(i) With pure carbohydrate it is 1.

$$C_6H_{12}O_6 + 6O_2 = 6CO_2 + 6H_2O$$

i.e. the volume of CO_2 evolved is equal to the volume of oxygen used.

(ii) With the fatty acids derived from food fat the RQ is about 0.7.

$$C_{17}H_{35}COOH + 26O_2 = 18CO_2 + 18H_2O.$$
$$RQ = 18/26 = 0.693.$$

Fat contains very little oxygen which must consequently be provided in sufficient amount to oxidize both the hydrogen and the carbon of the fat molecule.

(iii) With pure protein it is about 0.8 (indirectly calculated).

When fat or carbohydrate is completely combusted in the body the RQ is always 0.7 or 1.0 respectively, whatever the intermediate stages through which the foodstuff passes. This fact is illustrated by the following example. Let us suppose that food fat or reserve fat is converted in the body to carbohydrate which is then combusted. The reactions involved in these transformations can be crudely summarized as follows:

(a) $C_{18}H_{36}O_2 + 8O_2 = 3C_6H_{12}O_6$.
    stearic acid   glucose
(b) $3C_6H_{12}O_6 + 18O_2 = 18CO_2 + 18H_2O$.

Adding the two reactions to get the RQ of the whole process:

(c) $C_{18}H_{36}O_2 + 26O_2 = 18CO_2 + 18H_2O$,

i.e. RQ = 18/26 = 0.693, which is the RQ of the direct complete combustion of fat.

When carbohydrate is converted into fat which is stored and not combusted the reaction can be summarized as follows:

(d) $3C_6H_{12}O_6 = C_{18}H_{36}O_2 + 8O_2$.

O_2 thus spared is used by the cells for oxidation processes, so reducing the O_2 intake from the lungs without decreasing tissue oxygen consumption or CO_2 formation or elimination. The RQ is the ratio of CO_2 output from the lungs/O_2 intake from the lungs; as the denominator (the O_2 intake in the lungs) has fallen owing to reaction (d) the RQ value for the whole metabolism of the body is raised.

Conversely reaction (a) (above) while taking place alone involves additional uptake of O_2 in the lungs over and above that needed for oxidation of the foodstuffs; in the ratio CO_2/O_2 the denominator is increased and so the RQ of the whole metabolism of the body is lowered.

It follows therefore that if the RQ is determined from measurements extending over long continuous periods, the value obtained will approximately indicate the mixture of foodstuffs undergoing oxidation, e.g. if carbohydrate predominantly is being oxidized the RQ will approximate to 1, if fat the RQ will be about 0.7. There is a tendency also for the tissues normally to utilize preferentially the foodstuff most freely available so the RQ will tend to vary with the composition of the food. The non-nitrogenous residues of protein enter the common metabolic pool and are then treated essentially as intermediates from the oxidation of carbohydrate and fat.

The RQ determined over a short period will depend on the RQ of

the intermediate processes through which all the foodstuffs are passing during the time. Thus rats were given the whole of their daily diet in one meal; the RQ subsequently determined at intervals varied between 0.27 and 1.6.

2. Alterations in pulmonary ventilation

The RQ is affected (independently of any change in the nature of foodstuffs oxidized) when, in conditions of stable metabolism, the pulmonary ventilation is altered either voluntarily or secondarily to a rise of body temperature or to fluctuations in the H^+ ion concentration of the blood. Thus:

(i) Voluntary hyperpnoea washes out excessive quantities of CO_2 (without increase in O_2 consumption); the RQ may rise considerably above unity.

(ii) In acidaemia from any cause, e.g. ingestion of NH_4Cl, or secretion of the alkaline intestinal juices, there is hyperpnoea with increase in CO_2 output but again without corresponding rise in O_2 consumption; the RQ therefore rises. The same applies to the hyperpnoea of raised body temperature.

(iii) Conversely in alkalaemia, e.g. from ingestion of $NaHCO_3$, the breathing is depressed, CO_2 is retained in the body, and the RQ falls.

3. Violent exercise

During very violent exercise, lactic acid enters the blood, forms H_2CO_3 with the HCO_3^- of the plasma, and liberates large additional volumes of CO_2 which are eliminated from the lungs; the RQ may then exceed 2. During recovery from exercise, CO_2, which is derived from oxidative processes in the muscles, is retained in the blood in large amounts to reform the bicarbonate, and the RQ falls to a very low value. The RQ is not affected in a constant manner by moderate exercise because there is no acidaemia and roughly the resting mixture of foodstuffs is being metabolized [cf. p. 216].

4. Diabetes mellitus

The diabetic patient shows a low RQ due to the increased dissimilation of fats and the decreased dissimilation of carbohydrate [p. 508].

Metabolic rate

If different proportions of carbohydrate and fat are burnt in a calorimeter with 1 litre of oxygen, it is found that pure carbohydrate (RQ = 1) yields about 5 kcal, pure fat (RQ = 0.7) yields about 4.8 kcal, and various mixtures of carbohydrate and fat (RQ between 0.7 and 1) give between 4.8 and 5 kcal (20.1 and 20.9 kJ).

It is assumed, probably unjustifiably that the RQ values obtained in man have a similar significance. The RQ being known, the heat value of 1 litre of oxygen is deduced from a table; if the oxygen consumption is also known the heat production can be determined. Thus, if the RQ = 0.8 (calorific value per 1 litre O_2 = 4.875), and the O_2 consumption per minute is 0.25 litre, then the heat production per minute is 0.25 × 4.875 = 1.22 kcal (5.1 kJ).

Clinically the metabolic rate under conditions of complete rest and fasting (basal metabolic rate) is calculated from the oxygen consumption alone. This is determined most readily by means of the Benedict–Roth apparatus which consists essentially of an accurately constructed tank filled with oxygen and suspended in water. It is connected by means of tubing through a soda-lime tower to the patient's mouth (the nose is clipped). The patient re-breathes from the tank; the CO_2 formed is removed by means of the soda lime, and the decrease in the volume of oxygen in the tank is a direct measure of the oxygen consumption. The CO_2 output is not determined, and the respiratory quotient is assumed to be 0.75 (as the subject is fasting), corresponding to an output of 4.82 kcal l^{-1} of oxygen consumed [Figs. III.80 and III.81].

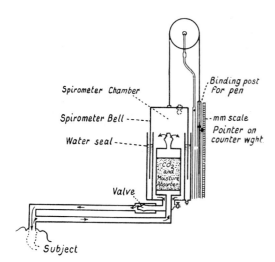

Fig. III.80. Benedict–Roth apparatus for determining the metabolic rate. The arrows in the diagram show the direction of the flow of air during respiration. (Beaumont, G. E. and Dodds, E. C. (1947). *Recent advances in medicine*, 12th edn. London.)

Basal metabolic rate (basal metabolism)

By this is meant the energy output of an individual under standardized resting conditions, i.e. at complete bodily and psychical rest, 12–18 hours after a meal (post-absorptive period) and in an equable environmental temperature. It can be determined by the methods just described. Under such conditions a proportion of the energy liberated is used to maintain the activities of vital organs like the heart, brain, or glands, but the greater part is converted into heat so as to maintain body temperature and prevent it from falling below the normal level.

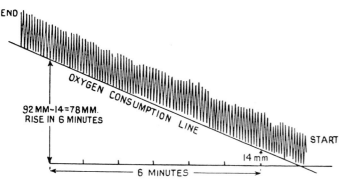

Fig. III.81. Graphic record of determination of metabolic rate. The record (which reads from *right to left*) is obtained using the apparatus shown in the preceding figure. This particular instrument is so constructed that each 1 mm fall of the spirometer bell (and corresponding rise in the record) represents 20.73 ml oxygen used (this value is not corrected to NTP). If the calorific value of oxygen is assumed to be 4.82 kcal l⁻¹, then 20.73 ml of oxygen represent the evolution of 0.1 kcal. In this experiment the rise in the record in 6 minutes is 78 mm corresponding to an uncorrected heat output of 78 × 0.1 kcal = 7.8 kcal, or 1.3 kcal min⁻¹. (Beaumont, G. E. and Dodds, E. C. (1947). *Recent advances in medicine*, 12th edn. London.)

Clinically the basal metabolic rate (BMR) is expressed as a percentage above or below the accepted normal standard for the individual taking into account his age, height, weight, etc. Thus a BMR of +50 means one which is 50 per cent above the normal average for that person.

Factors influencing metabolic rate

1. Surface area. The basal metabolism is most closely related to the surface area and is less directly related to height or weight. The surface area can be calculated from the following formula if the height and weight are known (Du Bois): $A = W^{0.425} \times H^{0.725} \times 71.84$, where A = surface area in cm², W = weight in kg, and H = height in cm.

In the male 40 kcal (165 kJ), and in the female about 37 kcal (155 kJ), are given off every hour per square metre of body surface (the surface area of an average adult is about 1.8 m²); or, expressed in terms of body weight, the basal metabolism amounts to about 1 kcal (4.2 kJ) kg⁻¹ h⁻¹. The values quoted for metabolic rate are average normal values; an average implies that higher and lower values occur.

2. Age. The basal metabolism is considerably greater per sq metre of surface in children than in adults; there is a further gradual fall in the metabolism during adult life as age advances. These facts are well shown in TABLE III.5 below.

TABLE III.5

Age (years)	BMR (kcal m⁻² h⁻¹)	
	Male	Female
2	57.0	52.5
6	53.0	50.6
8	51.8	47.0
10	48.5	45.9
16	45.7	38.8
20	41.4	36.1
30	39.3	35.7
40	38.0	35.7
50	36.7	34.0
60	35.5	32.6

Note 1 kcal = 4.2 kJ.

3. Starvation or prolonged undernutrition damps down the metabolic rate. For example, in a man who had fasted 31 days, the daily basal metabolism diminished from 958 to 737 kcal m⁻² of surface area—a fall of over 20 per cent. In poorly nourished patients the reduced body weight can finally be maintained on considerably less than the standard basal caloric requirements.

4. Body temperature. For every rise of 0.5 °C in the internal temperature of the body, the basal metabolism increases by 7 per cent. The chemical reactions of the body, like those occurring in a test-tube, are speeded up by a rise of temperature. Thus, a patient suffering from pneumonia with a temperature of 42 °C (about 4 °C above the normal) would have an increase of 50 per cent in his metabolism (and in his pulmonary ventilation) because of this fever alone.

5. External temperature. Exposure to cold increases the metabolism; there is consequently increased heat production which helps to maintain the normal body temperature. Exposure to external heat of brief duration has little effect on metabolism, as compensation is effected mainly by increasing heat loss; if the exposure is prolonged, a gradual fall in the metabolic rate takes place.

6. Ductless glands. (i) The active principle of the thyroid gland—thyroxine—acts as a general catalyst, speeding up the metabolic activities of the tissues [p. 539].

Thus in thyrotoxicosis, in which there is increased secretion of the active principle of the gland, the basal metabolism may increase in a severe case up to double the normal (i.e. the BMR is up to +100). In myxoedema, in which there is lessened secretion of thyroxine, metabolic activity may be depressed to 70 per cent or even 60 per cent of the normal (i.e. the BMR is −30 or −40;

the last value is about the lowest ever observed clinically and represents the minimum metabolic level).

(ii) Adrenalin increases the metabolic rate, but to a less extent than thyroxine. The injection of 1 mg adrenalin in man increases heat production up to about 20 per cent for a few hours only.

(iii) The anterior pituitary influences the metabolic rate indirectly through its thyrotrophic hormone.

7. Effect of food. The taking of food stimulates metabolism. This effect is not equally marked with all classes of foodstuffs, being least with carbohydrate and fat, and greatest with protein. If 125 g of protein in the form of meat are eaten at one meal, the metabolism rises to reach a maximum after 3–5 hours and then slowly declines; the peak percentage increase is 10–35 per cent; the total increase in metabolism may average 20 per cent over a period of 4–6 hours. If carbohydrate and fat are eaten in amounts of equal calorific value, the metabolism increases by only 5–10 per cent. The more striking effect of protein is termed its *specific dynamic action*. The increase in metabolism is attributed to a stimulating action on cellular metabolism of the fatty acid residues which are left after the NH_2 groupings have been removed from the amino-acids. The stimulating action of a carbohydrate meal is attributed to extra carbohydrate being burnt to provide the energy necessary to convert some of the glucose into a glycogen store; this energy is really being stored and not expended because the resulting glycogen has a higher chemical energy level than the glucose from which it is derived. If an ordinary mixed diet is taken, the metabolism is increased by 50–150 kcal daily (200–600 kJ).

8. Exercise. Lastly, and most important, there is an increase in the metabolism with muscular work. During very violent exercise the oxygen consumption per minute may rise from 250 ml to 4 litres, or even more, i.e. the metabolism may increase over 16 times.

HAEMO-RESPIRATORY CHANGES OF EXERCISE

Ventilation in exercise

In steady-state exercise the pulmonary ventilation per minute shows a roughly linear relationship to the oxygen usage occasioned by the exercise. Approximately 25 litres of pulmonary ventilation are required per litre of O_2 used [FIG. III.82].

Varieties of exercise

1. The term 'very severe exercise' is used with reference to muscular activity which by reason of its severity can only be kept up for a very short time; examples of this are a 100 metres or 400 metres race at top speed; at the end of such a race the runner is completely exhausted. From the physiological point of view the characteristic feature of this form of exercise is the inability of the haemo-respiratory systems to supply the muscles during the period of exercise with all the oxygen they require for their tremendous level of activity. A so-called *oxygen debt* is incurred [p. 217], i.e. a large volume of oxygen has to be absorbed after the exercise is over to dispose of metabolites which accumulated in the muscles during activity because of their relatively anoxic state.

2. 'Moderate exercise' is of a kind that can be kept up for long periods, e.g. vigorous walking at 8 km h⁻¹ or steady running. Such exercise may involve a considerable measure of exertion and necessitate extensive haemo-respiratory adjustments. Its outstanding physiological feature is the ability of the body to supply the active muscles with practically all the oxygen that they require immediately it is wanted and to a degree which is proportional to the level of activity. These points are further discussed below.

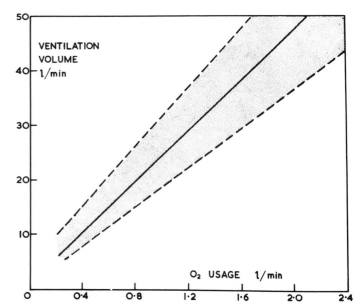

FIG. III.82. Pulmonary ventilation of man in exercise showing scatter of results obtained by many different experimentalists. The solid line is drawn showing a PV of 25 litres per litre of O_2 used per minute. (Redrawn from a graph in *Handbook of respiratory data in aviation* (1944). Committee on Aviation Medicine, Washington, DC.)

Metabolic studies on exercise in man

The resting oxygen consumption and the CO_2 output are first determined by means of the Douglas bag technique. The expired air is then collected during and after exercise, and the gaseous exchanges again determined. The metabolism over and above the resting level can thus be calculated, and is termed the excess metabolism of exercise.

Moderate exercise

1. Pulmonary ventilation. This is increased in any one form of exercise to a degree which is proportional to the intensity of the exercise. Thus FIGURE III.83 shows the increase in pulmonary ventilation and oxygen usage shown by a subject walking at different speeds. Walking at 8 km h⁻¹ required an oxygen usage of 2.5 l min⁻¹ and a pulmonary ventilation of 60.9 l min⁻¹.

FIGURE III.84 on the other hand shows the oxygen usage of a man running at different speeds. Although the oxygen usage increases with greater speeds of running it can be seen that at 10.9 km h⁻¹ the oxygen usage is only 2.2 l min⁻¹. Applying the simple approximation of 25 litres pulmonary ventilation per litre O_2 used, this corres-

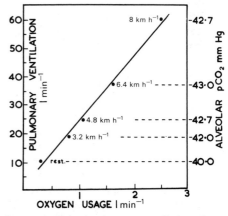

FIG. III.83. Oxygen usage and pulmonary ventilation of a man walking at different speeds as indicated. Alveolar pCO_2 values obtained from the subjects are shown on the right.

ponds to a pulmonary ventilation of 55 l min⁻¹ for running at 10.9 km h⁻¹. Hence the statement made above—that in any one form of exercise the pulmonary ventilation increases proportionally to the intensity of that exercise. A man walking at 8 km h⁻¹ is obviously less efficient than a man running at 10.9 km h⁻¹ and one has only to carry out the experiment to realize the subjective sensation of effort associated with very brisk walking.

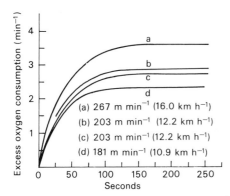

FIG. III.84. Relation of oxygen consumption to severity of exercise (speed of running) in man. Ordinate: excess oxygen consumption above resting level (l min⁻¹). Abscissa: time in seconds after onset of running. More oxygen is consumed at higher speeds; the oxygen consumption rises gradually and reaches its maximum after about 2 min. (Hill, A. V. (1927). *Muscular movement in man.* McGraw Hill, New York.)

The increased pulmonary ventilation of moderate exercise when a steady state has been reached is not easy to explain:

(i) FIGURE III.83 shows that the alveolar pCO_2 of the subject rose from 40 mm to about 43 mm during walking at 8 km h⁻¹. At first sight this might seem a trivial stimulus, but how important it is, is shown by holding the breath during such exercise, even if the subject has previously been breathing oxygen during the exercise itself. The breaking point of breath-holding is soon reached, for the CO_2 production in a man walking at 8 km h⁻¹ is about 2.4 l min⁻¹—i.e. a ninefold increase over the resting value. The eight- or ninefold increase in the breathing during walking at 8 km h⁻¹ removes the CO_2 almost as fast as it is delivered to the lungs, but the equilibrium reached is one in which the alveolar (and arterial) pCO_2 maintains a steady positive stimulus to the breathing. Other factors probably contribute to the hyperpnoea of exercise. These include:

(ii) A rise of body temperature which besides causing stimulation of the respiration itself also sensitizes the response of the respiratory mechanism to the arterial pCO_2 (Lloyd and Cunningham 1963).

(iii) A slight fall in the arterial pO_2, which itself is capable of stimulating the breathing, also sensitizes the respiratory response to pCO_2 [see p. 200]. At first sight it may seem strange that the arterial pO_2 may fall when the breathing is increased sixfold. It must be remembered (a) that the blood spends much less time in the pulmonary capillaries during exercise owing to the greatly increased capillary blood flow per minute; (b) that the mixed venous blood reaching the pulmonary capillaries has a much lower pO_2; (c) that the flow of blood in the pulmonary capillaries is pulsatile and this is exaggerated in conditions of exercise. Thus with systole the mixed venous blood is ejected quickly into the pulmonary bed and correspondingly the initial length of the capillary is traversed at a speed which does not favour equilibration of the blood with the alveolar air. (d) It has been shown directly that the arterial pO_2 may fall slightly in subjects working at only moderate intensity. The 'oxygen lack' stimulus is more important still as the exercise increases in severity as is shown by the effect of breathing oxygen instead of air during exercise at different levels of work [FIG. III.85].

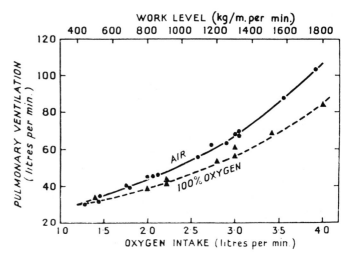

FIG. III.85. Effect of breathing 100 per cent O_2 on pulmonary ventilation during exercise on the bicycle ergometer. Note that the ventilation when breathing air at rest is little different from that while breathing oxygen. (Åstrand, P. O. (1954). *Acta physiol. scand.* **30**, 343.)

(iv) In addition to these important effects breathing is also stimulated by (a) reflexes from the moving joints and (b) impulses from the higher centres produced by emotional factors. The latter also contribute to the increase in breathing which may precede the exercise (see Åstrand and Rodahl 1970).

In moderate exercise of this intensity there is no respiratory stimulus from lactacidaemia. The respiratory quotient remains almost unchanged.

REFERENCES

ÅSTRAND, P. O. (1956). *Physiol. Rev.* **36**, 307.
—— and RODAHL, K. (1977). *Textbook of work physiology*, 2nd edn. McGraw Hill, New York.
BANNISTER, R. G. and CUNNINGHAM, D. J. C. (1954). *J. Physiol., Lond.* **125**, 118.
—— —— and DOUGLAS, C. G. (1954). *J. Physiol., Lond.* **125**, 90.
CUNNINGHAM, D. J. C. and LLOYD, B. B. (1963). *The regulation of human respiration.* Blackwell, Oxford.

2. Oxygen usage. This is increased in any one form of exercise proportional to the intensity of the exercise. The 'severity' of exercise which can be performed steadily is a subjective sensation which appears to be related rather to the oxygen usage which such exercise requires, than to the actual work done. This point is made clear by FIGURE III.86. Three forms of exercise, weight lifting, running on a treadmill, and cycling on a bicycle ergometer, all at specified rates, are compared in terms of their oxygen usage and work done. The subjective sensation of the 'heaviness' of the work load (shown at the right of the figure) is related to the oxygen usage and not to the work load.

At 100 per cent efficiency one litre of oxygen burnt is equivalent to 19.62 kilojoules. Thus if a subject on a bicycle ergometer was doing 600 kg metres (= 600 × 9.81 or 5886 Nm or 5.9 kJ) work and used 1.5 litres of oxygen (= 29.43 kJ) his efficiency would be

$$\frac{5.9}{29.4} \times 100 = 20 \text{ per cent.}$$

It will be noted that 1 ml O_2 is equivalent (at 100 per cent efficiency) to 2 kg metres work.

FIGURE III.84 shows that during running at 16 km h⁻¹ the O_2 usage was 3.6 l min⁻¹. It is convenient to consider here the means by which most of this enormous amount of oxygen is supplied to the active muscles. The main reactions are as follows:

(i) Breathing. The increase in pulmonary ventilation already

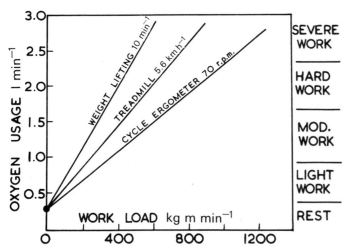

FIG. III.86. Comparison of work on bicycle ergometer (efficiency = 22 per cent), treadmill (efficiency 15 per cent) and in weight lifting (efficiency 10 per cent). Note that subjective sensation of work is related to the O_2 usage and *not* to the work done. Arbitrary scales of 'work sensation' as follows: light: 0.4–0.8 litres O_2 per min; moderate: 0.8–1.6 l min^{-1}; hard: 1.6–2.4 l min^{-1}; and severe: > 2.4 litres O_2 per min. (Redrawn from a graph in *Handbook of respiratory data in aviation*. Committee on Aviation Medicine (1944). Washington DC.)

described introduces large amounts of fresh air into the lungs (and drives out CO_2).

(ii) O_2 Uptake in the lungs. Large amounts of oxygen are taken up from the lungs by the blood.

(a) The mixed venous blood arrives at the lungs in a more reduced form. Normally at rest the venous oxygen content is 14 ml per 100 ml, the amount taken up from the alveolar air is 5 ml per 100 ml, and the arterial oxygen content is 19 ml per 100 ml. In hard work the venous blood may contain 7 or even as little as 3 ml per 100 ml; in becoming normally arterialized it takes up 12–16 ml per 100 ml. The oxygen intake may by this means alone increase more than threefold.

(b) The cardiac output is also increased about fourfold or more (e.g. up to 25 l min^{-1}; in international standard athletes perhaps as high as 35 l min^{-1}).

The oxygen uptake in the lungs may in these ways be increased to the requisite amount, e.g. sixteenfold (from 0.25 to 4 l min^{-1}).

(iii) Supply of oxygen to the tissues. A great blood supply (and therefore oxygen supply) to the muscle results from the large left heart output and the redistribution of the blood in the systemic circulation.

(iv) Removal of oxygen by the tissues. This is effected as follows:

(a) Dilatation and increase in the number of patent capillaries in the muscles slows the rate of the blood flow and allows more time for gaseous exchanges. (b) Low oxygen tension in the tissues allows oxygen to diffuse more readily and to a greater extent out of the blood. (c) High CO_2 tension and raised temperature increase the extent and rate of dissociation of oxyhaemoglobin.

The blood which leaves the tissues is thus very extensively re-duced, and the mixed venous blood may contain as little as 3 ml per 100 ml oxygen instead of the usual 14 ml per 100 ml. Experimen-tally, the O_2 consumption of active muscle may rise thirtyfold. This could be achieved by an increased blood flow (e.g. × 10) and an increased utilization of oxygen (e.g. × 3); it is known that the muscle blood flow in exercise may increase even up to thirtyfold, so that an ample margin is available.

It will be noticed again how every part of the body makes its contribution to the general effort. The pulmonary 'bellows' supply the oxygen. The heart and vasomotor mechanism send the blood

mainly to the parts which need it. The tissues abstract as much oxygen as they can from the blood which is supplied to them.

The O_2 consumption falls rapidly to its resting level when the exercise is ended.

3. CO_2 output. The elimination of the large amounts of CO_2 formed in the body is effected in a manner analogous to that described for oxygen usage.

4. Respiratory quotient. Special attention has been paid to the RQ of the total metabolism or of the excess metabolism of exercise. The RQ of moderate exercise proves to be about the same as the pre-exercise value (e.g. 0.85). Thus when walking at 3.2, 4.8, 6.4, and 8 km h^{-1} the RQ varied from 0.85 to 0.9; during the previous resting period it was 0.8 to 0.88. If these results mean anything they suggest that the body uses the various foodstuffs in exercise in roughly the same proportions as at rest.

5. Cardiac output. In running at the rate of 14.5 km h^{-1}, an oxygen intake of about 4 l min^{-1} may be attained and may be kept up for half an hour. This involves a cardiac output per minute of 25 to 30 litres. If oxygen is breathed during the period of exercise, the oxygen consumption may be as high as 5.9 l min^{-1} in highly trained athletes, and the cardiac output is then some 40 l min^{-1}. This latter figure probably represents the limit of haemo-respiratory perfor-mance.

6. Fatigue. In steady prolonged exercise, fatigue is due to a number of ill-understood factors; in the main it is attributed to changes in the brain resulting from slight anoxia and increased H$^+$ ion concen-tration. Afferent impulses set up in the active muscles (in part perhaps by the local physico-chemical changes) give rise to discom-fort and contribute to the sense of weariness. Some of the stiffness may be due to swelling of the muscle from accumulation of exuded fluid from the blood.

Severe exercise

1. Pulmonary ventilation. The characteristic feature of this type of exertion, which is necessarily of relatively brief duration, is that the breathing remains much above the resting level for a prolonged period after the exertion is over. In the case of a man who ran 200 m in 23.4 seconds, the pulmonary ventilation returned to normal in 27 minutes; after a 400 m race followed by severe gymnastics, in 44 minutes; after 'standing-running' for 4 minutes (breathing oxygen), in 87 minutes. In the case of a 100 m sprint, the subject may scarcely draw breath during the race, but marked dyspnoea develops later.

2. Lactic acid formation. During such very violent exercise, owing to relative anoxia of the muscles, lactic acid accumulates therein and diffuses out into the bloodstream and throughout the body fluids.

The resting level of blood lactate is 9–18 mg per 100 ml, i.e. 1–2 mmol l^{-1}. As a result of violent exercise the level may rise to 100 or even to 200 mg per 100 ml (11–22 mmol l^{-1}). As lactate is freely diffusible, the blood lactate level represents the concentration in the muscles, the interstitial fluid, and possibly in the intracellular fluid generally. If the total volume of body fluid contains a lactate concentration of 200 mg per 100 ml (i.e. equal to that in the blood), the total lactate accumulation is 90 g. After the exercise is over this lactate is disposed of; the blood lactate concentration steadily diminishes and reaches the normal level after a variable period of time, sometimes after as long as 60 minutes.

3. Respiratory quotient. The respiratory quotient of excess metabolism during the period of exertion first rises above 1 and may reach 1.5 or 2; the maximum figure is usually attained shortly after the end of exercise. During the recovery period the RQ falls below normal, e.g. to 0.5.

These changes can be readily accounted for. The lactic acid which, as we have just noted, accumulates in the plasma during violent exercise, is buffered as usual by the bicarbonate.

$$H^+ + HCO_3^- \rightarrow H_2CO_3 \rightarrow CO_2 + H_2O.$$

This reaction results in the liberation of large amounts of CO_2, without any equivalent utilization of oxygen. In addition, various foodstuffs are burnt in the muscles, oxygen being used and CO_2 evolved with an RQ of 1 or less. When the RQ of excess metabolism is 2, for each 1 molecule of CO_2 resulting from oxidation processes in the muscles, at least another 1 molecule of CO_2 is evolved from the $NaHCO_2$ of the plasma. The extent to which the RQ exceeds 1 is some index of the intensity of the exertion. The low RQ following the exercise is due to CO_2 being retained in the blood to re-form the bicarbonate.

4. Recovery after severe exercise.
After severe exercise the oxygen consumption (like the pulmonary ventilation) remains initially far above the resting level; thus following violent standing-running the O_2 intake per minute was at the rate of 1800 ml at 30 seconds, 1250 ml at 50 seconds, 750 ml at 100 seconds, and 500 ml at 140 seconds (the resting level was 350 ml min^{-1}). The O_2 intake declines slowly further but may not return to resting level for 30–120 minutes.

The volume of O_2 used after the exercise is over, in excess of the resting O_2 consumption for the same length of time, is called the 'oxygen debt'. It can be measured as follows:

(i) The resting O_2 consumption is determined. (ii) The post-exercise oxygen consumption is measured until it has fallen to its pre-exercise value; as a rule this occurs in 30 minutes. The resting oxygen consumption in (say) 30 minutes is then deducted from the post-exercise O_2 consumption for the same time. The difference is the O_2 debt. Oxygen debt figures as high as 15–18 litres have been observed. These results can be readily accounted for. During very violent exercise all the tremendous circulatory and respiratory reactions prove inadequate to supply the active muscles with their full O_2 requirements. As has been repeatedly emphasized, the active muscles, in spite of their large O_2 uptake are contracting in a sense anaerobically. Much lactate accumulates in the body and probably other products of muscular metabolism as well. The recovery oxygen is used to dispose of the various waste products which have accumulated during the bout of violent activity. (iii) During the first few minutes of the recovery process there is no decrease in the amount of lactate in the plasma nor, presumably, in the muscles. It seems, therefore, that metabolites other than lactate are disposed of first. The oxygen used for this purpose is termed the 'alactic acid debt'. (iv) When this has been dealt with the lactate is removed as follows:

(a) The lactate of the muscles is first disposed of; i.e. it is re-oxidized to pyruvate and then dissimilated to CO_2 and water or perhaps partially reconverted into glycogen. (b) The sodium lactate of the blood is in the ionized form as Na^+ and lactate$^-$; some of this lactate is now taken up by the liver and is there rebuilt into glycogen. (c) The rest diffuses into the muscles (as the local lactate concentration falls) and other organs to be oxidized and then dissimilated or converted into glycogen. Such lactate as is oxidized in the tissues yields CO_2 (lactate cannot be oxidized in the blood stream). (d) The CO_2 thus formed, on entering the blood, is retained there to a considerable extent (instead of being blown off in the lungs) to re-form the bicarbonate which was depleted during the period of exercise. (e) The CO_2 output from the lungs is thus considerably less than the oxygen consumed, and in this way the low RQ which follows violent exercise is accounted for.

5. Blood reaction.
Lactic acid enters the plasma and causes an increase in the H^+ ion concentration. This acidaemia stimulates respiration and increases the pulmonary ventilation to an extent which is sufficient to give rise to subjective symptoms of distress or dyspnoea. The acidaemia is still present, of course, at the end of this severe type of exercise. The ventilation, therefore, does not return to normal for a prolonged period, e.g. for half an hour or more after the exercise is over; as already pointed out throughout this time the oxygen consumption (and the cardiac output and blood flow to the muscles) remains above the resting level.

After the exercise is over, the normal blood reaction is restored (i) by blowing off the excess CO_2; (ii) by restoring the bicarbonate of the plasma to its original level, as described above.

It must be emphasized that strenous exercise does not depend on concurrent oxidation; lactic acid and other waste products are formed in amounts far greater than the maximum oxygen intake can cope with. The chemical processes (partly aerobic, but to a varying extent also anaerobic) going on in the muscles yield energy which enables activity to be continued. The excess lactic acid in man is temporarily buffered in the tissues and in the bloodstream, accumulates throughout the body fluids, and is disposed of after the exercise is over. We are thus permitted to do work far in excess of the greatest oxygen supplies which the heart and lungs can provide at the time, and we accumulate lactic acid and other substances. We metaphorically overdraw our oxygen account during the exercise and repay the debt later. The size of the repayment during the recovery period is, of course, equal to the oxygen debt incurred during activity. This acidaemia which is still present at the end of exercise serves, so to say, as a guarantee that the ventilation will be adequate to supply all the oxygen necessary for the recovery process.

PART IV
The kidney and the regulation of body fluids

Structure and functions of the kidney

Introduction

The two kidneys, each weighing about 150 g in the adult, filter the blood and reject its waste products in dissolved form in the urine. Bilateral nephrectomy in animals (or renal failure in man) causes a progressive rise in the plasma concentration of urea, potassium ion, and hydrogen ion and convulsions, coma, and death ensue within a few days.

In 24 hours some 1700 litres of blood are filtered by the kidneys and about 170 litres of cell and protein-free filtrate are formed during the day. By active transport on the part of the renal tubule cells, the essential constituents of this filtrate are restored to the circulation, together with water which is absorbed *passively* as an osmotic consequence of this active solute transport. Waste products such as urea, K^+ and H^+ are only partly absorbed and indeed the tubule cells actively secrete K^+ and H^+ into the tubular fluid. These rejected substances osmotically attract some water and in the ordinary course of events approximately 1.5 litres of hypertonic urine is finally passed by the kidney into the ureter and thence to the bladder wherefrom it is voided periodically.

Common experience reveals that the daily volume of urine may vary widely according to whether the fluid intake is high (beer drinking, etc.) or whether the fluid loss by extrarenal routes—notably sweating—is excessive. The comment that 'the urinary output in Baghdad is a puff of dust', though pithy and indeed giving the essence of the fluid balance situation, is inaccurate because solutes cannot be excreted in solid form, but osmotically attract a minimal or obligatory volume of fluid—some 500–600 ml per day. In man the kidneys cannot concentrate urine to more than approximately five times the osmolar concentration of the plasma.

STRUCTURE OF THE KIDNEY

FIGURE IV.1 shows the shape of the kidney cut vertically. It consists of an outer reddish cortex and an inner medulla which is paler in hue. The inner border of the medulla features ten calyces into which project the medullary papillae. The calyces debouch upon the widened end of the ureter which exits from the hilum—the indentation on the medial border—and passes to the bladder. The hilum is also the site of entrance of the renal artery and the site of exit of the renal vein.

The medulla contains ten to fifteen pyramids which terminate medially in the renal papillae.

The functional unit of the kidney is the nephron, of which the human kidney contains about a million. These million nephrons drain into the renal pelvis. FIGURE IV.2 shows the different parts of the nephron.

Bowman's capsule is the initial dilated part which contains an invaginated tuft of capillary vessels—the glomerulus. Bowman's capsule and the glomerulus together constitute the Malpighian—*not* the Pygmalion—corpuscle. These 'corpuscles' were first described by Marcello Malpighi in 1666, who considered them to be glandular in nature and did not recognize that they contained tufts of the very capillaries which he had been the first to describe (in the frog air sac) in 1660.

The investment of the epithelial structure which forms Bowman's capsule by the glomerular capillary tuft results in the tuft being clothed in a visceral layer of epithelium which is continuous at the site of entrance of the afferent and efferent arterioles with the parietal layer of the epithelial capsule proper. The visceral epithelial cells (known as podocytes), covering the capillaries, are separated from the capillary cells by a basement membrane which is the only intact structure in the filtering surface of the glomerulus. The visceral epithelial cell layer itself is not continuous, being connected with the basement membrane only by a series of pedicels

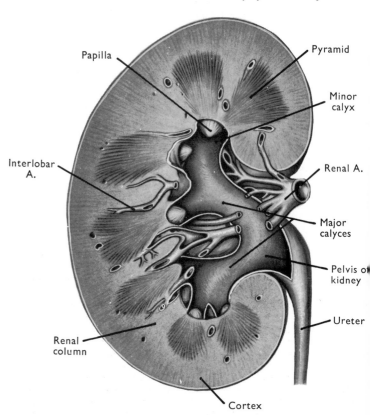

FIG. IV.1. Coronal section of the kidney.

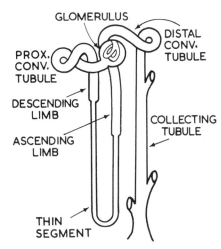

GLOMERULUS

DISTAL
CONV.
TUBULE

PROX.
CONV.
TUBULE

DESCENDING
LIMB

COLLECTING
TUBULE

ASCENDING
LIMB

THIN
SEGMENT

Fig. IV.2. Diagram of nephron.

or 'feet'. These minor processes of the epithelial cell themselves arise in turn from major extrusions of the epithelial cell bodies which run parallel to the capillary wall; the 'feet' implant in an interdigitating manner upon the basement membrane leaving small clefts between each tiny process [Fig. IV.3].

The major part of each glomerular endothelial cell is a thin layer of cytoplasm which forms the capillary lining. Over the site of the nucleus the endothelial cell bulges into the capillary lumen. The thin parts of the endothelial cells are riddled with pores 100 nm in diameter [Fig. IV.3]. These pores are too large to restrain the plasma constituents but serve to expose the basement membrane (sometimes termed the lamina densa) to the free flow of plasma by removing the endothelial cytoplasmic barrier (Selkurt 1963).

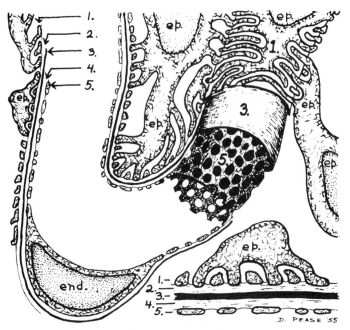

Fig. IV.3. A schematic illustration of a glomerular capillary. At the upper right, parts of three epithelial cells (ep.) are shown with terminal processes interdigitating upon the capillary surface (layer 1). The appearance of these feet in cross-section is indicated at the left, and in the insert at the lower right. The epithelial feet are slightly embedded in a cement layer (2) which in turn rests upon the dense structural portion of the basement membrane (layer 3). An inner cement layer (4) provides a bed for the endothelium (layer 5). The very attenuated endothelial sheet is perforated by closely spaced holes a little over 0.1 μm in diameter, as may be seen in surface view to the right of the figure, and in the transverse section to the left, and in the insert. (Pease, D. C. (1955). *J. Histochem. Cytochem.* **3**, 295.)

The glomerular capillaries are not simple loops, but form a freely branching anastomotic network. Each glomerulus contains some six lobules and each of these consists of 3–6 capillary loops—20–40 loops in all. Many anastomoses occur between the capillaries within any one lobule. The glomerulus is supplied with blood by an afferent arteriole and the blood leaves from the tuft by an efferent arteriole. This arrangement allows the maintenance of a much higher pressure (60 mm Hg) in the glomerular minute vessels than in capillaries elsewhere [see p. 74]. This high capillary pressure is well adapted for the filtration function which the glomeruli subserve.

The *proximal convoluted tubules* (PCT), whose lumen is continuous with that of Bowman's capsule, consist of cells with scalloped outline and brush border. This brush border is formed by numerous microvilli which enormously increase the surface available for absorption. Mitochondria abound particularly in the basal cytoplasm, which is in keeping with the fact that these proximal tubular cells are responsible for the active transport of some 80 per cent of the sodium filtered out of the tubular fluid into the peritubular capillary blood and correspondingly have a high rate of oxygen usage.

The loop of Henle consists of a descending limb which arises in continuity with the terminal part of the proximal tubule; this descending limb is continued into the thin segment where the epithelium is of pavement-type; from this segment arises the ascending limb which is formed by low cuboidal epithelium.

The *distal convoluted tubule* (DCT), is characterized by cuboidal epithelium, which on its luminal border shows only few scattered microvilli. The DCT begins near the pole of the glomerulus and establishes a close proximity to the afferent arteriole of its parent glomerulus. At this site the cells of the tubule are columnar rather than cuboidal and are huddled close together. It is for this reason that this part of the DCT is called the macula densa (= crowded spot) [see Fig IV.6, p. 220]. The macula densa and the adjacent juxtaglomerular part of the afferent arteriolar wall are believed to be functionally associated [see p. 220].

The *collecting tubules* receive the tubular fluid from the DCT. Lined by clear cuboidal epithelium they converge into papillary ducts which lead to the renal papillae and thence to the pelvis. The renal papillae—those portions of the medullary pyramids which project into the renal pelvis—contain not only the papillary ducts but also the hairpin loops of the juxtamedullary nephrons and the vasa recta [see below].

The glomeruli all lie in the cortex, as do the DCT, and all but the terminal part of the PCT. The actual depth of the glomeruli in the cortex, however, is of some importance in terms of their corresponding loops of Henle. Glomeruli which lie adjacent to the cortical surface have nephric loops which penetrate only a short way into the medulla before turning back to the cortex. Some of their loops indeed are restricted to the cortical substance. Glomeruli near the cortico-medullary junction—the so-called *juxtamedullary* glomeruli—possess long tubular loops which penetrate deeply to near the tip of the renal papillae before turning back towards the cortex. These comprise about 14 per cent of the nephrons of the kidney and their Henle loops manifest a vascular supply which is strikingly different from those of the 'cortical glomerular' nephrons.

BLOOD SUPPLY OF THE KIDNEY

From the renal artery branch interlobar arteries which ascend between the medullary pyramids to give rise to the descriptively named arcuate arteries which course between cortex and medulla parallel to the cortical surface. Interlobular arteries arise from these arcuate vessels and run towards the surface. These are the parent arteries of the thick-walled afferent arterioles which supply the glomerular capillaries [FIG. IV.4]. The short efferent arterioles have a luminal diameter similar to that of the afferent vessel, but possess a thinner wall. Formed from the confluence of the glomerular capillaries, they promptly subdivide to provide a peritubular capillary plexus which embraces the convoluted tubules in the cortex. The venous drainage of these nephrons shows a marked species variation, but in man, stellate veins drain into interlobular veins which in turn pass to the arcuate, then the interlobular, and finally the renal vein.

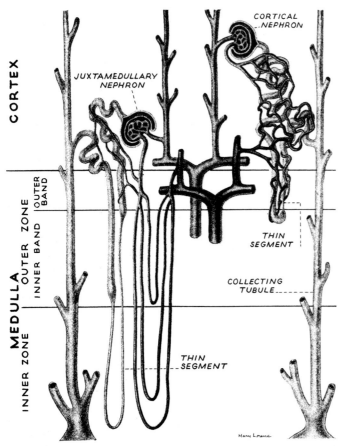

FIG. IV.4. Diagram showing the significant differences between a cortical nephron (right) and one located in the juxtamedullary region. (Smith, H. W. (1951). *The kidney*. New York.)

The juxtamedullary glomeruli are drained by short straight efferent arterioles which quickly subdivide into a leash of straight vessels (vasa recta) which run into the medulla. From the efferent arteriole itself and from the proximal part of the vasa recta, numerous capillary branches supply a subcortical plexus. The vasa recta, grouped into vascular bundles, traverse and supply the outer medullary zone and enter the inner medullary zone where the bundles break up into separate vessels. A variable number of vasa recta supply the capillary plexus of the inner medullary zone but many continue without branching to reach the papilla tip where they form a capillary network [FIG. IV.5]. Venous drainage of the subcortical and medullary regions is by long ascending vasa recta (from the innermost medulla) or by short vessels draining the outer medulla and subcortical zone. These finally open into the arcuate veins. The arterial and venous vasa recta are in close proximity to both the loops of Henle and the collecting tubules.

REFERENCE

SELKURT, E. (1963). In *Handbook of physiology*, Section 2, Vol. II. Chap. 43, p. 1457. American Physiological Society, Washington, DC.

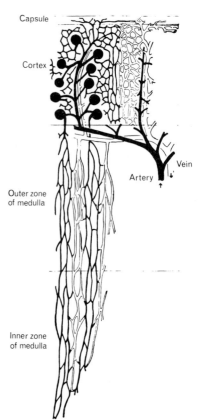

FIG. IV.5. The vasa recta system of the kidney. (Maximov, A. A. and Bloom, W. (1957). *Textbook of histology*. Saunders, Philadelphia.)

THE JUXTAGLOMERULAR APPARATUS

Adjacent to its site of division to form the glomerular capillaries the afferent arteriole shows an asymmetrical thickening of its wall in the media. This *Polkissen* (polar cushion), comprising myoepithelioid cells forms part of the juxtaglomerular apparatus (JGA) or the juxtaglomerular complex as some prefer to call it. The juxtaglomerular complex also contains an elliptical plate of tubular epithelial cells from the initial part of the DCT—the macula densa—and Mesangial or Lacis cells. The lacis cells fill the triangle described by the afferent, and the efferent arterioles and the distal tubular cells [FIG. IV.6].

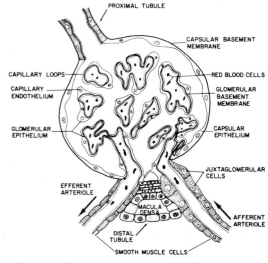

FIG. IV.6. Diagram of juxtaglomerular complex. (Ham, A. W. and Leeson, T. S. (1961). *Histology*, 4th edn. Lippincott, Philadelphia.)

The myoepithelioid cells have well-developed Golgi apparatus and endoplasmic reticulum, abundant mitochondria and ribosomes. They often contain secretory granules. These have been shown by Cook (1967) to contain renin. The granulation increases as a consequence of sustained hypotension in the afferent arteriole, in sodium deficiency and following adrenalectomy. The lacis cells also show granulation in conditions of extreme 'hyperactivity'.

The macula densa (MD) cells are specialized distal tubular epithelial cells. They are not well adapted for reabsorption and do not show signs of secretory activity. But characteristically they have their Golgi apparatus between their nuclei and their outer borders contiguous to the efferent arteriolar wall. In other tubular cells the Golgi apparatus is between the nucleus and the luminal border of the cell. The basement membrane of the macula densa cell is discontinuous where it comes in contact with the Polkissen cells. The relevance of these observations is not readily understood, but will be referred to when the subject of autoregulation of renal blood flow and glomerular filtration rate is considered in detail.

FUNCTIONS OF THE KIDNEY

Glomerular filtration

Large quantities of blood, amounting to 20–25 per cent of the resting cardiac output, flow through the kidneys each minute. About a tenth of this amount is filtered off into Bowman's capsule as the blood passes through the glomeruli. The filtrate is actually an 'ultrafiltrate' of plasma. It contains virtually no protein and no cells. Otherwise its composition is identical to that of plasma. Samples of the filtrate obtained from Bowman's capsule by direct puncture in newts, rats, and some monkeys shows that the filtrate is identical with plasma in respect of osmolality, electrical conductivity, concentration of electrolytes, and of smaller organic molecules like glucose, urea, and creatinine.

The rate of glomerular filtration determined directly, i.e. the single-nephron glomerular filtration rate (SNGFR), in rats is 20–60 nl min^{-1}. The glomerular filtration rate measured indirectly in intact animals is about 0.35–0.60 ml g^{-1} kidney weight min^{-1}. In normal man it is around 125 ml min^{-1}.

The mechanism of glomerular filtration is like that responsible for tissue-fluid formation. The glomerular blood pressure constitutes a filtering force driving fluid out of the blood vessels into Bowman's capsule; it is opposed by the osmotic pressure of the plasma proteins (= 25 mm Hg) which tends to hold water in the blood vessels. The net filtering force is the glomerular blood pressure minus the plasma protein osmotic pressure (e.g. 50 − 25 = 25 mm Hg). The volume of glomerular filtrate depends on several factors: (i) directly on the magnitude of the net filtering force, i.e. it is increased by a rise in glomerular pressure or by a fall in plasma protein osmotic pressure. (ii) Within certain limits the volume of glomerular filtrate varies directly with the renal blood flow. (iii) The hydrostatic pressure in Bowman's capsule is normally zero (i.e. atmospheric pressure). Should it rise, e.g. as a result of obstruction to the outflow of urine, it resists ultrafiltration.

In normal circumstances the glomerular membranes are absolutely impermeable to molecules 4 nm (40 Å) in diameter or just over 70 000 in molecular weight. Whereas the ratio of the concentration of crystalloids and inulin in the filtrate to their concentration in plasma is unity, the same ratio for haemoglobin is only 0.03 and for plasma albumin, of about 70 000 MW, less than 0.01. In disease the permeability of the glomerular membrane may increase so that greater amounts of albumin and of other large molecules are filtered.

Reabsorption in the tubules

As the glomeruli form a filtrate (at a rate of 170–180 litres per day in man) it follows that the tubules must modify this filtrate, for the urine differs widely in volume and composition from a filtrate of plasma [see TABLE IV.1].

TABLE IV.1

Constituent	Concentration in plasma; mg per 100 ml	Total in 170 litres glomerular filtrate		Total excreted in 24 h urine	Total re-absorbed in tubules	
Water		170	l	1.5 l	168.5 l	
Glucose	100	170	g	none	170	g
HCO$_3^-$	150	255	g	0.1 g	255	g
Na$^+$	330	560	g	5 g	555	g
Cl$^-$	365	620	g	9 g	611	g
K$^+$	17	29	g	2.2 g	26.8 g	
Phosphate (as P)	3	5.1 g		1.2 g	3.9 g	
Ca^{2+}	10	17	g	0.2 g	16.8 g	
Urea	30	51	g	30 g	21	g
HSO$_4^-$ (as S)	2	3.4 g		2.7 g	0.7 g	

Glucose is reabsorbed completely and of, say, 170 litres per day of water which pass into the tubules from the glomeruli 168.5 litres are reabsorbed. Only 1.2 per cent of the sodium, calcium, and chloride filtered is finally voided in the urine. In ordinary circumstances the urine is acid (pH 6.0–6.6) and contains very little bicarbonate; when an alkaline urine is passed large amounts of bicarbonate are excreted.

Methods of study of tubular function

The action of the tubular cells on the filtrate can be inferred by comparing the excreted load of a substance X ($U_x V$) to its filtered load (P_xGFR). U_x and P_x represent the concentrations of the substance X in urine and plasma respectively while V stands for the volume of urine formed per minute. Such indirect calculations do not indicate the site within the tubule where the effect is produced, nor can they differentiate between the relative contributions of reabsorption and secretion should a substance be treated both ways in different parts of the tubule.

It is nowadays possible to insert small capillary pipettes into different segments of the renal tubules in experimental animals and to study tubular function more directly.

The pipettes can be used to draw samples of the tubular fluid or alternatively used to introduce fluids of known composition into an isolated segment of the nephron. The fluid can then be collected through the same pipette and the changes in its composition noted. Finally, two pipettes can be introduced at a distance and the segment of the tubule between them perfused *in situ*. Castor oil injected into the tubule serves to isolate a segment and prevent its content from mixing with that of other parts of the tubule.

Tubular segments have been isolated and perfused *in vitro* to study transport across their wall.

It is possible to measure electrical conductivity, solute concentrations, and osmolality in minute nanolitre samples of fluid.

THE PROXIMAL TUBULE

In the proximal tubule the volume of the glomerular filtrate is reduced by 75–80 per cent. Glucose, sodium chloride, and sodium bicarbonate are actively reabsorbed accompanied by the passive flow of water. The reabsorption is iso-osmotic so that the osmolality of the fluid that remains in the tubule is unchanged at about 300 mOsm.

The proximal tubule is highly permeable to water and to small solute molecules. It is about 90 times more permeable than the gall-bladder. The high permeability is, at least in part, due to the enormous cell membrane surface area afforded by the microvilli in the 'brush border'. However, the proximal tubular epithelium is also permeable to slightly larger molecules like sorbose and raffinose which cannot actually traverse cell membranes. The finding suggests that the permeability of the tubular epithelium is also due to the presence in the proximal tubule of paracellular pathways; an idea which is supported by electrical resistance measurements. The resistance across the tubular epithelium as a whole is less than it is across the cellular membrane. The extracellular shunting paradoxically occurs through what are often described as 'tight junctions'.

Water reabsorption in the tubule is evidenced by the increase in the concentration of a marker (like inulin) which is filtered into the tubule but once there it is neither reabsorbed nor added to by tubular secretion. For all practical purposes, creatinine, a naturally occurring substance, also satisfies these requirements and is commonly used as the indicator for water reabsorption. The ratio of the concentration of creatinine in the tubular fluid (TF_{Cr}) to its concentration in plasma (P_{Cr}) is 1 in Bowman's capsule, rises to 3 in the middle of the proximal tubule [FIG. IV.7], and is estimated to reach 5 by the end of the tubule. The end of the proximal tubule is not accessible for direct micropuncture and the concentration there of various substances is predicted by linear extrapolation of data obtained from puncturing the tubule in its middle as a function of tubular length.

Glucose is completely reabsorbed in the proximal tubule provided its concentration in the plasma is normal. FIGURE IV.7 shows that the TF/P ratio for glucose falls to less than 0.2 halfway through the length of the proximal tubule while the ratio for creatinine rises to nearly 3.

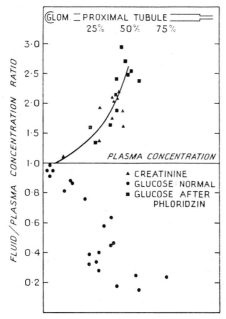

FIG. IV.7. Absorption of glucose and water in the renal tubules. Experiments on guinea-pigs and rats. Ordinate: concentration ratios between the fluid collected (by direct puncture) from Bowman's capsule or the proximal tubule, and plasma. 1.0 means that the fluid and the plasma have the same concentration. 25 per cent, 50 per cent, 75 per cent represent points one-quarter, half, and three-quarters the distance along the proximal tubule. (Redrawn from Walker *et al.* (1941). *Am. J. Physiol.* **134**, 587.)

Glucose reabsorption is inhibited by the systemic administration of the glucoside phloridzin. It is thought that the glucose moiety of the phloridzin molecule engages the membrane carrier which is involved in the active transport of glucose.

Glucose that is not reabsorbed in the proximal tubule passes out in the urine—glucosuria.

When the plasma concentration of glucose exceeds 10–12 mM l^{-1} the reabsorption of glucose is not complete and sugar appears in the urine. This level is often referred to as the glucose threshold. The amount of glucose that passes in the urine increases linearly with plasma glucose once the reabsorptive limit of the proximal tubule is exceeded [see FIG. IV.18(a)]. The limit is referred to as the tubular maximum (Tm) and the reabsorption of glucose as a Tm-limited process. The Tm is set by the saturation of the transport mechanism. The tubular maximum for glucose (Tm$_G$) is about 2 mM per minute. Note that this is a measure of a quantity per unit time and not a concentration, as it is in 'threshold'.

Sodium reabsorption is the major operation performed by the proximal tubule. About 18 moles of sodium are filtered each day. Of these some 13–14 moles are reabsorbed in the proximal tubule. The reabsorption is an active process and requires the expenditure of energy for it occurs against an electrochemical gradient. The lumen of the tubule is 4 mV negative to the interstitial fluid; this transcellular potential difference suggests that any passive movement of cations could only occur towards the lumen. The O$_2$ consumption of the kidney is directly related to its sodium reabsorption.

The active reabsorption of sodium entrains a series of passive transport processes: to maintain electrical balance the sodium ion reabsorbed is either accompanied by an anion or it is replaced by a cation transported into the tubular lumen.

The reabsorption of salt, furthermore, necessitates the reabsorption of water to maintain the osmotic equilibrium.

That the sodium reabsorption is primary is shown by the following experiment:

Dogs are infused intravenously with considerable amounts of the inert sugar alcohol mannitol. Mannitol, though filtered, is not reabsorbed; in the tubular fluid it exerts a large osmotic effect which greatly limits the reabsorption of water. Hence a large volume of urine is passed; an osmotic diuresis is produced. The volume of urine formed may amount to 67 per cent of the glomerular filtrate. In the circumstance sodium excretion also rises but this never exceeds 25 per cent of the filtered load of sodium. Thus the reabsorption of sodium is primary compared with that of water. Other experiments using sophisticated micropuncture techniques confirm the same point: it can be shown that exchanging one permanent anion in the tubule with another does not influence sodium reabsorption and that sodium reabsorption can proceed even when adverse osmotic gradients are set-up and augmented by this reabsorption.

At present it is considered that there are two sodium pumps. (i) An electrogenic pump, which actively ejects Na$^+$ into the renal interstitium; Cl$^-$ moves passively with Na$^+$. This pump is inhibited by ethacrynic acid or by replacing the Cl$^-$ of the tubular fluid by an anion which has limited powers of penetrating the tubular cells. (ii) A coupled Na$^+$–K$^+$ exchange pump which ejects Na$^+$ from the cell in exchange for K$^+$, which is transported into the cell from the interstitial fluid. This mechanism is inhibited by ouabain and is rendered less effective if the extracellular [K$^+$] is lowered [FIG. IV.8].

The chief Na–K exchange pump resides in the more distal parts of the tubule, but in the proximal tubule a similar exchange between Na$^+$ and H$^+$ occurs. (In severe acidosis this mechanism may account for 25–30 per cent of the sodium reabsorption.)

Sodium reabsorption in the proximal tubule is thought to be modified by changes in the flow, hydrostatic, and colloid osmotic pressures in the peritubular capillaries. Either an increase in hydrostatic pressure or a fall in colloid osmotic pressure inhibits reabsorption, while the reverse may increase it.

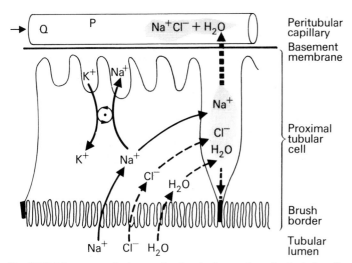

FIG. IV.8. Movement of anions and cations in the renal proximal tubule cell. Dashed lines show passive movement; solid lines show active transport.

Bicarbonate

About five moles of bicarbonate ions are filtered a day. Nearly 90 per cent of this load is usually reabsorbed in the proximal tubule. The TF/P concentration ratio for the HCO_3^- falls to 0.4 by the middle of the tubule. The pH of the filtrate is little altered. That is if the pH of the filtrate is measured by a glass electrode introduced into the tubule.

The mechanism of the bicarbonate reabsorption is believed to be as follows:

The CO_2 within the tubular cell, which is derived from the metabolism of the cell but also from the circulating CO_2 and CO_2 reabsorbed from the tubular lumen, is hydrated to form carbonic acid. This ionizes to produce hydrogen ions which diffuse into the tubular lumen in exchange for sodium ions. In the tubular lumen the H_2CO_3 formed is dehydrated to give CO_2 which passes back into the cell and water which is excreted. The Na^+ and HCO_3 ions that accumulate in the cells are transferred to the extracellular fluid. The HCO_3^- absorbed into the blood stream has thus not actually been absorbed from the tubular fluid, but the end result is the same, for the HCO_3^- filtered is absorbed as CO_2 [FIG. IV.9].

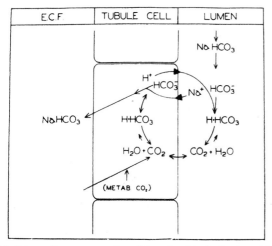

FIG. IV.9. Schematic representation of the role of the hydration of CO_2 and H^+–Na^+ exchange in reabsorption of sodium bicarbonate. (Brazeau, P. and Gilman, A. (1953). *Am. J. Med.* **15**, 765.)

Whether the exchange of H^+ for Na^+ is 'carrier linked' (i.e. the two ions share the same carrier, so that they could be transported in opposite directions) is not certain. For though hydrogen ion extrusion from the cell is undoubtedly active, the movement of sodium into the tubular cell may very well be passive down a concentration gradient. The transport of sodium at the *basal* membrane is always active.

The hydration and dehydration of CO_2 in the cell and in the tubular fluid is catalysed by the enzyme carbonic anhydrase which is abundantly found in the tubular cell and also in the infoldings of the 'brush border'.

Bicarbonate reabsorption varies significantly with the hydrogen ion concentration of the extracellular fluid [see p. 229].

Urea

Large amounts of urea are reabsorbed in the proximal tubule. The TF/P ratio for urea in the middle of the proximal tubule reaches 2 suggesting that some 30 per cent of the filtered urea is reabsorbed. (TF/P of creatinine at the same point is 3.) The reabsorption of urea in this segment is passive and depends on the concentration gradient achieved as a result of the active reabsorption of salt and water. The more salt and water reabsorbed, the greater is the gradient that favours the movement of urea from the tubular lumen into the bloodstream.

Urea reabsorption also depends on the rate at which the tubular fluid flows past the absorbing surface. This suggests some limitation to the diffusion of the molecule. Indeed the different segments of the tubule have different permeabilities to urea. The excretion of urea, quite understandably, increases in diuresis.

Phosphate

The phosphate concentration of the glomerular filtrate is approximately 1.5–2 mM l^{-1}. Hence, some 300 mM of phosphate are filtered daily; of this filtered load some 30–40 mM l^{-1} are excreted in the urine. Phosphate reabsorption is restricted to the proximal tubules.

As the pH of the proximal tubule is unaltered one must assume that HPO_4^{2-} and $H_2PO_4^-$ are reabsorbed in the same ratio as they exist in the filtrate.

Phosphate reabsorption is active and is Tm limited. The Tm_{PO_4} is very low at 0.1–0.15 mM min^{-1} (unlike the Tm for glucose which is set at about 2 mM). The excreted load of phosphates parallels the filtered load once the filtrate exceeds 2 mM min^{-1}. The plasma level of phosphate is consequently about 2 mM l^{-1}.

The capacity of the tubular cell to reabsorb phosphates is independent of sodium, H^+, Cl, water, and urea reabsorption. It is inversely related to the plasma level of glucose and the administration of phloridzin increases the Tm_{PO_4} but much more significantly the Tm_{PO_4} is modified by the level of the circulating parathyroid hormone. Prolonged administration of the hormone in hypoparathyroid patients depresses phosphate reabsorption due to lowering of the Tm_{PO_4}.

Sulphate, amino acids, and proteins

These are all absorbed through discrete reabsorptive processes characterized by very low Tm values. They are therefore excreted freely when their filtered load increases.

There are at least three specific and distinct mechanisms for the reabsorption of amino acids. Amino acids reabsorbed by the same mechanism compete for transport but those that are reabsorbed by different mechanisms influence each other's reabsorption only inasmuch as the availability of energy is concerned.

Potassium

Virtually all the filtered potassium is reabsorbed in the proximal tubule. The TF/P ratio in the middle of the tubule is as low as it is for glucose. The reappearance of K ions in the distal tubular fluid and the urine produced in higher TF/P ratios indicates that the ion is secreted later on in the tubule.

Potassium reabsorption is active and is not Tm limited.

THE LOOP OF HENLE

Sodium, chloride, and water are reabsorbed in different parts of the loop. At the beginning of the loop and in its descending limb water is reabsorbed freely. The reabsorption is passive and is imposed by the hypertonic interstitial environment. It is accompanied by the diffusion of sodium ions from the interstitial fluid into the tubular lumen.

In the ascending limb of the loop, and for the first time in the renal tubule, the reabsorption of sodium chloride is dissociated from that of water. The epithelium in this region is impermeable to water, consequently the fluid that leaves the ascending limb is invariably hypotonic relative to plasma.

The reabsorption of sodium chloride is active in the thick part of the ascending limb. The tubular epithelium here can establish transepithelial concentration gradients which are considerably greater than those established in the proximal tubule (200 mM l^{-1} difference in $[Na^+]$). Recent evidence suggests that the chloride ion is the primary absorbate and that its transport is active. Sodium ions follow the chloride ions passively. The lumen of the ascending limb of the loop is 3–9 mV positive relative to the peritubular fluid.

The mechanism of the movement of salt across the thin part of the ascending limb is not clear. It is considered to be passive and that it occurs down a concentration gradient.

Urea may be passively reabsorbed in the loop.

The rate of salt reabsorption in the loop varies directly with the load of salt delivered into it from the proximal tubule. In general some 50 per cent of the load is reabsorbed.

Frusemide and other 'loop diuretics' produce an osmotic diuresis by inhibiting the active reabsorption of chloride in the thick part of the ascending limb of the loop of Henle.

THE DISTAL TUBULE

In this segment too, the reabsorption of salt and water are independent of each other. They are furthermore governed by different control mechanisms.

At the beginning of the distal tubule the tubular fluid is hypotonic relative to plasma but varying amounts of water are reabsorbed as the fluid passes along the tubule. The amount of water reabsorbed depends on the circulating level of the antidiuretic hormone vasopressin. The maximal osmolality likely to be achieved in the distal tubule is that of plasma. The TF/P osmolality can never exceed unity in this section.

Sodium is actively transported across the tubular epithelium by a powerful pump which can operate against considerable concentration gradients. The transtubular sodium concentration gradient at the end of the distal tubule may be 130 mM l^{-1}. A potential difference in excess of 100 mV, lumen negative, is observed in these circumstances. Normally chloride follows passively. There is, however, some evidence to suggest that there may be some active chloride reabsorption.

The main transport mechanism in this segment is one where sodium and potassium ions are exchanged at the basal membrane of the tubular cell. The way in which chloride ions move into the tubular cell is still in debate. It can, however, be assumed that the movement of both sodium and chloride ions into the tubular cell from the lumen is down a concentration gradient and passive.

Sodium reabsorption in this segment is modified by the plasma concentration of suprarenal corticoids, especially aldosterone.

THE COLLECTING DUCT

This is the final common pathway for the filtrate produced in a number of nephrons. It invariably lies in the renal medulla and is surrounded by the medullary interstitium so that the filtrate, whether formed in cortical or in juxtamedullary glomeruli, always passes through the collecting duct and is exposed to the medullary interstitium before it is produced as urine.

In the collecting duct salt and water are reabsorbed independently of each other. The reabsorption is usually hypo-osmotic; the water movement is motivated by the hypertonicity of the medullary interstitium, but the rate of the reabsorption is regulated by the water permeability of the collecting duct epithelium as modified by the plasma level of antidiuretic hormone.

Some sodium chloride is actively reabsorbed against a considerable gradient. The reabsorption is promoted by the circulating mineralocorticoids.

Substantial amounts of urea are reabsorbed in the collecting duct. Urea could be reabsorbed passively as its concentration builds up through the reabsorption in the tubules of water and solutes. There is, however, some evidence that urea transport could occur uphill against a concentration gradient in protein depleted rats (Clapp 1966). In dogs, furthermore, the apparent transport of urea is inhibited by iodoacetate, a known inhibitor of anaerobic glycolysis (Goldberg, Wojtczak, and Ramirez 1967).

REFERENCES

Clapp, J. R. (1966). *Am. J. Physiol.* **210**, 1304.
Goldberg, M., Wojtczak, A. M., and Ramirez, M. A. (1967). *J. clin. Invest.* **46**, 388.

TUBULAR SECRETION

Two natural constituents of plasma are secreted by the tubular cells in quantity. These are the potassium and hydrogen ions. However, tubular secretion plays a major role in ridding the organism of foreign substances, for plasma could only be completely cleared of a substance in one passage through the kidney if the substance is actively secreted by the tubular cell. The tubular secretion of a drug plays an important role in maintaining its blood level and hence its therapeutic efficiency. For all intents and purposes tubular secretion is a mirror image of tubular reabsorption. It occurs in different parts of the tubule, it could be Tm limited or it could depend on the time the peritubular blood spends in contact with the tubular lumen. It is mainly active but again conditions to favour passive secretion could be set by the active transport.

Potassium secretion

Almost all the filtered potassium is reabsorbed in the proximal tubule. No more than 10 per cent of the filtered load is ever found in the earliest part of the distal tubule. The unsettled debate about whether potassium is reabsorbed, secreted, or recirculated in the loop of Henle is not worth considering at this stage. It suffices to state that most of the potassium produced in the urine is due to secretion of the ion in the distal tubule. Distal tubular secretion is estimated to contribute 75 per cent of the excreted potassium in rats kept on normal control diets. Tubular secretion of potassium is greatly reduced in sodium and potassium depletion (Malnic, Klose, and Giebisch 1966). The TF/P ratio for potassium often exceeds 5 and may reach 10 by the end of the distal tubule.

The secretion of potassium is most probably passive and occurs down an electrochemical gradient; the distal tubule being up to 100 mV negative to the interstitial fluid. The distal tubular cell permeability to potassium is high enough to allow free movement of the ion.

Potassium secretion is modified by changes in sodium and hydrogen ion excretion. It is often suggested that in the distal tubule potassium and hydrogen ions are both exchanged for the sodium which is reabsorbed and that the two ions compete for the exchange mechanism. Recent evidence suggests that the competition between the potassium and hydrogen ions is not for a common carrier but depends on the intracellular content of the two ions. The entrance of potassium ions into the cell is known to reduce the cellular hydrogen ion content and vice versa.

Potassium secretion is modified by the plasma level of mineralo-corticoids.

Secretion of hydrogen ions is active and occurs throughout the tubule. The most significant secretion occurs in the distal tubule. Hydrogen ions are secreted against a gradient which increases along the kidney tubule. The magnitude of the gradient is obviously determined by the acid–base status of the animal. Pitts 'quotes' a gradient of a thousand to one at the end of the distal tubule when the animal produces urine with a pH of 4.4. The hydrogen ion concentration of this urine is 40 000 nM l^{-1} compared to 40 nM l^{-1} for plasma.

Secretion of drugs

Certain substances are actively secreted from the plasma into the proximal tubular fluid. Stop–flow techniques have revealed that para-aminohippuric acid (PAH), which is a foreign substance used for the estimation of renal plasma flow [p. 235], is secreted by the proximal tubule cells. Diodone is similarly secreted, as are the mercurial diuretics. Penicillin is also secreted by the proximal tubular cells.

MECHANISMS OF URINARY CONCENTRATION

The mammalian kidney elaborates urine which varies widely in its solute concentration. Man can produce urine of 50 mOsm l^{-1} when over hydrated or concentrated urine containing 1200 mOsm l^{-1} when dehydrated. Certain mammals adapted to living in arid environments produce more concentrated urine; the chinchilla's urine, for example, is 7.5 Osm l^{-1}. The following observations made independently over the years helped in clarifying the mechanisms by which the versatility is achieved:

1. It has long been known that only creatures that have loops of Henle in their nephrons can concentrate urine (i.e. birds and mammals) and that the concentrating ability of the mammalian kidney is directly related to the length of Henle's loop (O'Dell and Schmidt-Nielsen 1960).
2. Micropuncture studies showed that tubular fluid is isotonic in the proximal segment and hypotonic in the early part of the distal segment. Wirz (1956) has furthermore demonstrated that in rats producing urine of low osmolality the hypotonicity of the tubular fluid persisted throughout the length of the distal segment while in animals that produced isotonic or hypertonic urine the tubular fluid regained isotonicity by the last third of the distal segment. The tubular fluid at the end of the distal tubule was thus either hypotonic or isotonic, but it was never hypertonic [FIG. IV.10].
3. Tissue slices from different parts of the kidney had different osmolalities. Slices from the cortex or the outer zone of the medulla were iso-osmotic with plasma while sections from the inner zones of the medulla were hypertonic. The osmolality was greater the farther the section analysed was from the corticomedullary junction.

Slices from the papilla showed the greatest osmolality (see Ullrich, Kramer, and Boylan 1961).

The hyperosmolality of the medulla was due to increased concentrations of sodium chloride and urea in the tissue (Atherton, Hai, and Thomas 1968).

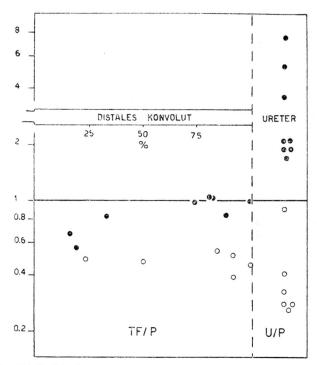

FIG. IV.10. Distal tubular fluid/plasma ratios (*TF/P*) and ureteral urine/plasma ratios (*U/P*) of total molecular concentration in the concentrating (●) and the diluting (○) rat kidney. The site of micropuncture is given as per cent of the length of the distal convoluted tubule. (Wirz, H. (1956). *Helv. physiol. pharmac. Acta* **14**, 353.)

4. The medullary osmotic gradient was reduced in water diuresis and increased when vasopressin was infused in rats.
5. The osmolality of the urine produced, usually equalled the osmolality of the tissue at the tip of the renal papilla.
6. The elaboration of concentrated urine depended on the relatively low renal medullary blood flow. An increase in blood flow was found to precede the production of dilute urine.
7. The study of the anatomy of the renal medulla showed the intricate relationship between the constituent structures. The ascending and descending limbs of the loop of Henle lay very close to each other, and were surrounded by the ascending and descending segments of the vasa recta. The collecting ducts lay in the vicinity. The long loops of Henle, belonging to the juxtamedullary nephrons were seen to make 'hair-pin' bends at or near the tip of the renal papilla.

Countercurrent flow of blood and tubular fluid (haarnadel gegenström—Wirz 1956) is evident when the tip of the papilla is observed through a stereoscopic dissecting microscope (Jamieson 1974).

From these observations it is argued that 'dilution' of the urine occurs before the distal convoluted tubule is reached while its 'concentration' is achieved by the reabsorption of water from the tubular fluid as it traverses the collecting duct.

The amount of water reabsorbed in the collecting duct depends (i) on the magnitude of the force available for translocating the water, and (ii) on the water permeability of the collecting duct.

There is no evidence that water is actively transported. On the other hand, since the duct lies in the hypertonic environment of the renal medulla it follows that water can be reabsorbed if the collecting duct were permeable to water but not as permeable to the solutes producing the medullary hyperosmolarity. This it is.

The amount of water abstracted from the tubular fluid in the collecting duct and the volume and concentration of the urine formed could thus be related to the osmolarity of the medullary interstitium. When other conditions remain constant the concentration of the urine varies with medullary osmolarity. There is clearly an inverse relationship between the volume of urine produced and the medullary osmolarity.

The water permeability of the duct is a direct function of the circulating levels of plasma antidiuretic activity (ADA). The higher the plasma ADA the greater would be the concentration of the urine formed and the less its volume—assuming other variables remain unchanged.

REFERENCES

ATHERTON, J. C., HAI, M. A., and THOMAS, S. (1968). *J. Physiol., Lond.* **197,** 429.
JAMIESON, R. (1974). In *MTP International Review of Science*, Vol. 6 (ed. K. Thurau) p. 199. Butterworth, London.
MALNIC, G., KLOSE, R. M., and GIEBISCH, G. (1960). *Am. J. Physiol.* **211,** 529.
O'DELL, R. and SCHMIDT-NIELSEN, B. (1960). *Fedn Proc. Fedn Am. Socs exp. Biol.* **19,** 366.
ULLRICH, K. J., KRAMER, R., and BOYLAN, J. W. (1961). *Prog. cardiovasc. Dis.* **3,** 395.
WIRZ, H. (1956). *Helv. physiol. pharmac. Acta* **14,** 353.

THE MEDULLARY OSMOLARITY

The medullary osmolarity depends on the balance between the rate at which solute is introduced into the region and the rate at which it is removed from the region.

The generation of the medullary osmolarity

The exact mechanism by which the medullary hyperosmolarity is generated is not known. Currently it is thought to be produced by a system of 'countercurrent multiplication'.

Sodium or chloride ions are actively transported out of the ascending limb of the loop of Henle into the surrounding insterstitium [FIG. IV.11]. This movement of salt is not accompanied by water reabsorption so that the concentration of salt in the ascending limb decreases, but its concentration in the interstitial fluid increases. In the renal medulla all the tubular structures, excluding the ascending limb, are in osmotic equilibrium. The descending limb thus acquires the increased osmolarity of the surrounding interstitium. This effect is multiplied as new iso-osmolar filtrate arrives at the descending limb and forces the concentrated tubular content towards the tip of the loop of Henle and the 'hairpin' bend. Here the forceful active transport of salt from the adjacent ascending limb increases the concentration of the fluid in this segment further. The active transport mechanism in the thick ascending limb is assumed to be very potent and to operate against considerable concentration gradients in excess of 200 mOsm l^{-1}; whether the thin part of the ascending limb transports sodium actively or not is still debated. The passage of urea into the medullary interstitium is thought to be mainly passive and to occur in the collecting duct down a concentration gradient. The collecting duct is assumed to be particularly permeable to urea.

The operation of the system depends on the anatomical 'hairpin' configuration of the loop of Henle and was first described by Kuhn in 1942. Wirz, Hargitay, and Kuhn (1951) provided the experimental evidence.

There are few conditions that are known to change the operation of the countercurrent multiplication system or the rate of

generation of the medullary osmolarity. Vasopressin probably augments the rate of deposition of solute in the medullary interstitium.

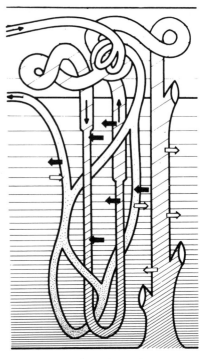

FIG. IV.11. Scheme of a nephron—flow in the tubules and blood vessels. The density of the transverse shading indicates the relative osmotic pressure at various levels of the kidney substance from outer cortex to inner medulla. White arrows = transport of water (passive). Black arrows = transport of crystalloids (active or passive). (Wirz, H. (1956). *Helv. physiol. pharmac. Acta* **14,** 353.)

The dissipation of the medullary osmolarity

The maintenance of the medullary osmolarity critically depends on the blood flow through the region. If this were large the rate of removal of solute from the medullary interstitium would equal the rate at which solute is deposited so that the region would continue to be in osmotic equilibrium with the blood. This is probably what happens in the renal cortex. However, medullary flow is relatively small and accounts for only a small fraction of the total renal blood flow. This permits the creation and maintenance of an area of hyperosmolarity. Besides, the arrangement of the descending and ascending limbs of the vasa recta in close proximity to each other permits the operation of a counter-current exchanger, where water in the blood effectively short-circuits the vasculature that dips into the medullary tissue. The faster the flow the less effective is the bypass. This is analogous to the heat exchange that occurs in the human limb. The finger tips are cooler than the core of the body when the arm blood flow is low because the blood heat bypasses the peripheral vessels and is shunted away from the artery going into the limb to a vein at the same level carrying blood in the opposite direction. As blood flow to the limb increases the shunting is reduced and fingers warm up.

Medullary blood flow varies in a variety of conditions. It is increased in haemodilution and volume expansion.

For practical purposes it is therefore reasonable to conclude that the medullary osmolarity is an inverse function of medullary blood flow. When other conditions are unchanged an increase in flow dissipates medullary osmolarity, reduces water reabsorption, and leads to the production of large volumes of dilute urine [FIG. IV.12].

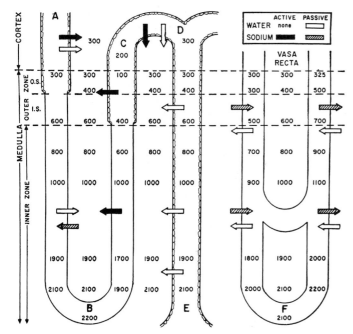

FIG. IV.12. The loop of Henle is a countercurrent osmotic multiplier, basically as conceived by Gottschalk and Mylle. Sodium chloride is actively reabsorbed in the ascending limb of the loop of Henle; the hyperosmotic interstitium abstracts water from the water-permeable descending limb (and possibly sodium chloride diffuses into this limb), thus increasing the concentration of sodium in the tubular urine delivered to the ascending limb and converting the system to a countercurrent multiplier. Since an osmotically dilute fluid is delivered to the distal convoluted tubule, the ascending limb must be relatively impermeable to water. (Reprinted by permission, *American Journal of Physiology*.)

The system operates to establish an osmotic concentration gradient in the interstitium, increasing from the corticomedullary junction to the tip of the loop, and to concentrate the urine by the passive abstraction of water through the water-permeable collecting ducts. The *vasa recta* (right) serve as *countercurrent exchangers*, promoting the over-all efficiency and carrying away in the ascending capillaries the water abstracted from the urine.

Sites for which data on the osmotic pressure of the tubular urine are now available: A, proximal convoluted tubule (cortex); B, tip of loop of Henle in papilla; C, distal convoluted tubule near macula densa, and D, near entry to collecting duct; E, papillary urine; and F. blood in the capillaries near tip of the papilla. O.S. outer stripe, I.S. inner stripe of the outer zone of the medulla. The figures are nominal estimates of osmotic pressure (in mOsm per kg water) given by Gottschalk and Mylle. (Smith, H. W. (1959). *Bull. N.Y. Acad. Med.* **35**, 293.)

THE PERMEABILITY OF THE COLLECTING DUCT AND THE DISTAL CONVOLUTED TUBULE TO WATER

In the absence of vasopressin (the antidiuretic hormone) the epithelial cells that form the walls of the distal convoluted tubules and collecting ducts are relatively impermeable to water. Their permeability is markedly increased in the presence of vasopressin. The increased permeability is sufficient to allow the tubular fluid to equilibrate with the surrounding interstitial fluid during the time the two fluids spend in apposition. Thus the tubular fluid in the distal convoluted tubule acquires the osmolarity of the cortical interstitial fluid, i.e. becomes iso-osmotic and the fluid in the collecting duct becomes hyperosmotic like the medullary interstitium. Equilibration may not be achieved if the tubular flow rate is excessive despite adequate levels of ADH in the plasma.

Vasopressin acts only when placed on the blood side of the tubular epithelium though its action is mainly on the luminal side. Like most other polypeptide hormones its action is mediated through the intracellular formation of cyclic 3,5-adenosine monophosphate (cAMP). Its action on the permeability of the luminal membrane is thought to be mediated by the production of hyaluronidase.

The normal level of vasopressin is 0.4–0.8 microunits per ml. Maximal water reabsorption obtains when the level rises to 1–2 microunits. The plasma level of vasopressin is labile since vasopressin is released in response to a large number of stimuli, some being more sustained than others. Once released into the circulation vasopressin is rapidly metabolized. The half time in man for an injected dose is around 7–8 minutes.

So of the factors that influence the concentrating ability of the kidney the permeability of the collecting ducts is the most fickle. It is therefore the most important minute to minute determinant of urinary concentration and volume in natural circumstances. It must, however, be stressed that the reabsorption of water ultimately depends on the presence of the physical force capable of translocating water across the tubular wall (i.e. the medullary hyperosmolarity). Vasopressin can only facilitate the movement of water by exposing the tubular contents to the full impact of the physical force surrounding them.

REFERENCES

KUHN, W. and REIFFEL, K. (1942). *Z. Physiol. Chem.* **276**, 145.
WIRZ, H., HARGITAY, B., and KUHN, W. (1951). *Helv. physiol. pharmac. Acta* **9**, 196.

H+ AND ACIDIFICATION OF THE URINE

The word oxygen means generator of acid, so it is understandable that metabolism produces a preponderance of acid substances. The chief product, CO_2, is volatile and is expelled in volumes exceeding 300 litres per day from the lungs—an amount equivalent to 15 litres of normal HCl. The anions formed daily from the oxidation of P and S of protein exceed the amount of metallic cation (Na, K, and Ca) ingested in the average diet by 50–100 mM per day. Under abnormal conditions organic acids like acetoacetic and β-hydroxybutyric acid also accumulate; in strenuous exercise lactic acid may temporarily appear from the muscles. The excess of anion, whether normal or pathological, can only be excreted in the urine. In all body fluids, electrical neutrality must be maintained, i.e. anions like phosphate or sulphate must be 'covered' by an equivalent amount of cation. If these anions which are excreted in the urine were fully covered by metallic cations an equivalent amount of cation (chiefly Na+) would be drained out of the body with disastrous results. The problem is solved by the manufacture by the kidney of two other cations, H+ and NH4+, which 'cover' the excreted anions and hence conserve an equivalent amount of metallic cations. The kidneys cannot secrete a urine which is more acid than pH 4.5. Hence there is a limit to the amount of acid excreted in the free form, defined by the equation:

$$pH = pK + \log \frac{[A^-]}{[HA]}$$

and if pK is equal to the pH of the urine, half the excreted acid must be covered by metallic cation. If only the sodium salts of strong acids were present in the tubular fluid, H+–Na+ exchange would result in the formation of a strong acid and the manufacture and secretion of H+ by the tubular cells would soon stop. The presence of buffer salts (particularly the phosphates) in the tubular fluid is therefore vital for the excretion of the maximal amounts of H+ and, correspondingly, for the reabsorption and hence conservation of the maximal amount of Na+.

The two most important buffer systems present in the tubular fluid are $HPO_4^{2-}/H_2PO_4^-$ and HCO_3^-/H_2CO_3.

At a pH of 7.4 (approximately that of the glomerular filtrate), since $pK_{H_2CO_3} = 6.1$, the ratio $\dfrac{[HCO_3^-]}{[H_2CO_3]}$ is 20/1 for

$$pH = pK + \log \frac{[HCO_3^-]}{[H_2CO_3]}$$

$$7.4 = 6.1 + 1.3$$
$$1.3 = \log 20.$$

At a urinary pH of 6.1, $\log \dfrac{[HCO_3^-]}{[H_2CO_3]} = 0$ and hence

$$[HCO_3^-] = [H_2CO_3].$$

Correspondingly the phosphate equilibrium in plasma is such that four parts of the so-called 'basic phosphate' (HPO_4^{2-}) exist to one part of 'acidic phosphate' ($H_2PO_4^-$) at a pH of 7.4, for $pK_{phos} = 6.8$

$$7.4 = 6.8 + \log \frac{[HPO_4^{2-}]}{[H_2PO_4^-]}$$

$$7.4 = 6.8 + 0.6$$
$$0.6 = \log 4.$$

At a urinary pH of 5.8 ten parts of phosphate are in the 'acidic' form and one part in the 'basic' form ($5.8 = 6.8 + \bar{1}$); for each phosphate ion excreted as $H_2PO_4^-$ one sodium ion is saved and one hydrogen ion is excreted.

The phosphates are by far the most important buffers in the urine, although in acid urine the formation of NH_4 from NH_3 and H^+ helps to keep the $[H^+]$ lower than it would otherwise be.

A change of pH in the tubular fluid could be theoretically achieved by a tubular transport system which operated either by:

1. Causing selective reabsorption of HPO_4^{2-} or
2. Causing selective reabsorption of HCO_3^- or
3. Secreting H^+ into the tubular fluid. Pitts showed that the H^+ excreted in the urine, by subjects in ammonium chloride acidosis, far exceeded that filtered. Hence the hydrogen ion concentration of the urine is provided by the excretion of H^+ into the tubular fluid.

The tubule cells form H_2CO_3 from CO_2 (derived from the blood and from their own metabolism) and H_2O under the influence of carbonic anhydrase. H_2CO_3 ionizes to yield H^+ and HCO_3^-. H^+ formed thus in the cell exchanges with Na^+ in the tubular fluid; effectively $NaHCO_3$ is reabsorbed into the blood and HPO_4^{2-} accepts the H^+ to form $H_2PO_4^-$ [Fig. IV.13].

It might be asked why this mechanism produces acidification of the distal tubular fluid, when a similar mechanism has been described to effect bicarbonate absorption in the proximal tubule, in which little change of pH occurs in the fluid. The reason of course lies in the amount of HCO_3^- available as H^+ acceptor in the distal tubule. In the proximal tubule abundant HCO_3^- is present to accept H^+ and form H_2CO_3 which yields CO_2 and H_2O. CO_2 passes across the tubule cell into the blood and the reactions are almost *isohydric*. In the distal tubule the secretion of H^+ occurs *pari passu* with the absorption of Na^+ and the amounts gaining access to the distal fluid are in excess of the HCO_3^- which remains in the tubule. We can consider the process to consist of two (artificial) stages.

Let us assume that the distal tubule contains initially say 100 mM Na^+ per litre (0.7 isosmotic) and also say 14 mM HCO_3^- per litre. Stage 1 would consist of H^+ passing into the tubule forming H_2CO_3 and hence CO_2 in the tubular fluid. The CO_2 is absorbed as in the proximal fluid. Stage 1 would represent isohydric absorption of HCO_3^-, for, as in the proximal tubule, Na^+ absorbed would enter the blood in partnership with HCO_3^- formed in the tubule cell. However, more H^+ is excreted than there is HCO_3^- to act as a hydrogen acceptor—correspondingly the HCO_3^- is 'used up', and the remainder of the mechanism (stage 2) is the use of HPO_4^{2-} as a hydrogen acceptor as shown in Figure IV.13. Even the presence of HPO_4^{2-} does not prevent the tubular fluid becoming more acid, but it permits a far greater H^+ secretion by virtue of its buffer properties than would otherwise be the case, for the pH of the tubular fluid does not drop too low.

It must also be remembered that the fluid passes far more slowly through the distal tubule than through the proximal tubule. This presumably aids the equilibrium processes required. The lowest urinary pH is between pH 4.4 and 4.5, i.e. the maximal H^+ gradient which can be achieved is about 1000.

By inhibiting the enzyme carbonic anhydrase the formation of H_2CO_3 and hence H^+ by the tubule cells is slowed. Acetazolamide (*Diamox*) is the most effective inhibitor.

$$\underset{S}{\overset{\displaystyle \overset{N{=\!=}N}{\underset{\displaystyle}{CH_3CONH.C \qquad C.SO_2NH_2}}}{}}$$

Acetazolamide (2-acetamido-thiadiazole-5-sulphonamide)

On administering acetazolamide, the urine becomes alkaline and an abundant excretion of sodium bicarbonate occurs. The excretion of sodium is necessitated by the failure of reabsorption of HCO_3^-.

Time	Urine flow ml min⁻¹	Urine pH	Excr. titratable acid	Excr. HCO₃⁻ µEq min⁻¹	Excr. Na⁺ µEq min⁻¹	Excr. K⁺ µEq min⁻¹
Control	0.9	5.4	164	0.3	116	52
110 min after acetazolamide 10 mg kg⁻¹	6.2	7.8	0	750	1315	248

In this particular case the bicarbonate excretion rose over two thousand times. Hence as only ⅛ of the filtered bicarbonate gains access to the distal tubule, some of the effects of carbonic anhydrase inhibition of H^+ secretion and HCO_3^- absorption must have been exerted in the proximal segment.

At the same time as inhibiting the secretion of H^+, acetazolamide increases the excretion of K^+ by the distal tubule. K^+ and H^+ competitively occupy a common transport mechanism; the relationship is not one-to-one because H^+ has a much greater affinity

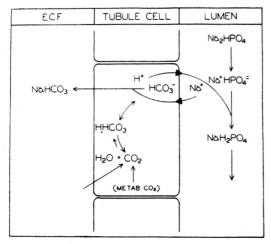

Fig. IV.13. Schematic representation of the role of the hydration of CO_2 and of H^+–Na^+ exchange in the acidification of urinary buffers. (Brazeau, P. and Gilman, A. (1953). *Am. J. Med.* **15**, 765.) (After Berliner, Kennedy, and Orloff (1954). *Ciba Symposium on The Kidney*, p. 147. London.)

for the transport mechanism than has K⁺. The increase in K⁺ excretion is genuinely due to an increase in secretion of K⁺ by the distal segment and is not due to a diminished reabsorption of K⁺ by the proximal segment.

In normal dogs the administration of KCl decreases H⁺ excretion and causes systemic acidosis; K⁺ depletion of the body causes alkalosis, for H⁺ excretion becomes excessive.

Bicarbonate reabsorption

Some 170–180 litres of filtrate, containing 25 mM l^{-1} of bicarbonate, are filtered daily in man—a total load of 4250–4500 mM. At a pH of 6.1, 1½ litres of urine formed per day contain 2 mM HCO_3^- = 1.25 mM l^{-1}. Bicarbonate reabsorption is thus normally almost complete.

By altering the alveolar pCO_2 and hence the tissue tension of CO_2 throughout the body the amount of bicarbonate reabsorbed by the renal tubules can be markedly changed. Thus, by *overbreathing*, the bicarbonate reabsorption is reduced and bicarbonate is excreted (accompanied inevitably by sodium) in the urine. Conversely, the inhalation of CO_2 rich mixtures enhances the reabsorption of HCO_3^- by the tubular cells. Lastly, if tubular H⁺ excretion and bicarbonate reabsorption are inhibited by acetazolamide, the excretion of sodium bicarbonate which results can be minimized by the inhalation of CO_2 rich mixtures which raise the tissue tension of CO_2. Such a procedure makes available a sufficient amount of CO_2 for the renal tubular cell to form H_2CO_3 even although its carbonic anhydrase activity is greatly reduced by the acetazolamide.

FIGURE IV.14 shows the relationship between the alveolar pCO_2 and the renal reabsorption of bicarbonate. CO_2 tension of course exerts its effect by forming intracellular H_2CO_3 and hence H⁺—only by secreting H⁺ can the tubule cell achieve the reabsorption of HCO_3^-.

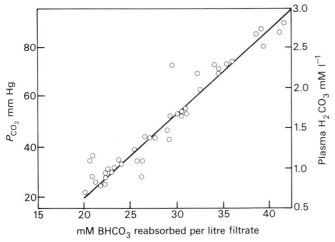

FIG. IV.14. The effects of induced variations of plasma CO_2 tension on the rate of reabsorption of sodium bicarbonate.

(Redrawn from Brazeau, P. and Gilman, A. (1953). *Am. J. Physiol.* **175**, 33.)

If the CO_2 tension is chronically raised, as for instance in emphysema, bicarbonate reabsorption is complete and there results a chronic high level of HCO_3^- in the plasma. Conversely if the bicarbonate plasma concentration is artificially raised, $[HCO_3^-]$ in the glomerular filtrate is so high that complete reabsorption is impossible. At an arterial pCO_2 of c. 40 mm Hg reabsorption per 100 ml of glomerular filtrate normally stabilizes at a value between 2.6 and 2.8 mM, and above this concentration bicarbonate is excreted in the urine. There is, so to speak, too much HCO_3^- for the H⁺ to partner; any excess of HCO_3^- over and above the amount which H⁺ can convert into H_2CO_3, and thus absorb from the tubular fluid as CO_2, must correspondingly be excreted in the urine and

must take Na⁺ with it as an ionic partner. The osmotic effect of this bicarbonate loss is seen as a diuresis. This type of diuresis is seen in:

1. Voluntary hyperpnoea.
2. Systemic alkalosis.
3. Administration of potassium salts (which reduces H⁺ transport and hence promotes HCO_3^- excretion).

It is interesting to consider whether active sodium reabsorption is the primary event in Na⁺/H⁺ exchange by the renal tubular cells. Overbreathing normally causes the excretion of an alkaline urine containing sodium bicarbonate; it might be argued that because H⁺ formation and excretion by the tubule cells is so deficient, then Na⁺ absorption is reduced and therefore must be secondary. However, if HCO_3^- remains in the urine it must take Na⁺ with it as an ionic partner and the excretion of some $NaHCO_3$ is simply a consequence of this. Moreover, a very different state of affairs occurs when a *sodium-deficient subject* overbreathes. He cannot produce an alkaline urine because his sodium absorption is complete; as a result most of the bicarbonate ion, as well as the chloride ion of the tubular fluid, is absorbed passively accompanying the sodium, for there is little H⁺ formed by the acapnic tubular cells to exchange with the sodium. 'Alkalosis superimposed on sodium deficiency leads to a conflict between two regulatory functions of the kidney and the regulation of acid–base balance is sacrificed to the conservation of the body's stores of sodium' (Robinson).

Ammonium mechanism

The chief circumstance which provokes increased ammonium excretion is metabolic acidosis. The tubular epithelium forms NH_3 from the amide nitrogen of glutamine (glutamic acid amide) which circulates in the blood. Sixty per cent of the NH_3 formed is derived from this source by the action of the enzyme glutaminase present in the tubule cells. The remainder of the NH_3 produced is derived from the α-amino nitrogen of other circulating amino-acids (glycine, alanine, leucine, and aspartic acid) by the action of amino-acid oxidase. Dissolved NH_3 is a proton acceptor, appropriating H⁺ from an aqueous environment to become NH_4^+ (NH_3H^+).

$$NH_3H^+ \leftarrow NH_3 + H^+.$$

Correspondingly $H^+ = K_{am} \dfrac{[NH_3H^+]}{[NH_3]}$.

Cells are permeable to NH_3 but not to NH_3H^+. Hence NH_3 can diffuse from the tubule cell into the tubular fluid where it captures H⁺ according to the above equation. The higher the hydrogen ion concentration, the bigger the formation of non-diffusible NH_4^+ (or NH_3H^+) which is correspondingly excreted. At the same time the formation and excretion of NH_4^+ allows the excretion of anions which would otherwise require the accompaniment of metallic cations. The ammonium mechanism thus assists the conservation of sodium by the body. The ammonium content of the urine is negligible until the pH falls below 6.0. Ammonium excretion then increases linearly as the urinary pH falls below this value. In chronic acidosis more ammonium is excreted at any given urinary pH.

Titratable acidity. The amount of alkali in mM per litre which is required to adjust the pH of a sample of urine to that of plasma (7.4) is described as a measure of titratable acidity. The titratable acidity (mM l^{-1}) multiplied by the daily volume of urine (in litres) gives the amount of sodium conserved by the renal mechanism of H⁺ secretion. In a healthy man this is normally 20–30 mM per day but it may increase to 150 mM per day in severe diabetic acidosis. The ability of the kidney to excrete such amounts of acid depends on the concentration of buffer substances available—notably on the phosphates. In diabetic ketosis the urine contains large quantities of the free acids, β-hydroxybutyric (pK 4.7) and acetoacetic.

Actually the total secretion of H$^+$ by the kidney is equal to the titratable acid of the urine plus the ammonium excreted daily, for NH$_3$ accepts H$^+$; the whole amount of Na$^+$ conserved by the kidneys is given by the sum of these two.

Chloride reabsorption in segments distal to the proximal convoluted tubule. It is probable that the chloride ion is absorbed passively with Na$^+$. The absorption of Na$^+$ is 'linked' with the secretion of H$^+$ in the proximal and the distal tubules. However, the H$^+$ which gains entrance to the tubular fluid forms H$_2$CO$_3$ and thence H$_2$O and CO$_2$ in the proximal tubule and H$_2$PO$_4^{2-}$ in the distal tubule and in each case 'disappears'. Hence, despite this exchange, a passive absorption of Cl$^-$ probably occurs to maintain the balance of cations and anions in the tubular fluid.

RENAL BLOOD FLOW

Renal blood flow exhibits a number of characteristics:

1. It is very high compared to flow through other organs; 300–400 ml per 100 gm per minute is an average figure. The renal vascular circuit has very little basal tone. The PRU$_{100}$ [p. 71] is 0.25 compared to 2.0 in the cerebral vascular circuit. Furthermore, the normal resting flow through the kidney amounts to 80 per cent or more of the maximal flow possible. This obviously contrasts with muscle where the resting flow is only 1/30th of the maximal.

2. The oxygen extraction across the kidney is always low. The arteriovenous (A–V) difference is about 1.5 ml of oxygen per 100 ml of blood. This is low compared to the overall extraction in the body as a whole where the A–V difference is 4–5 ml and is extremely low compared to the extraction of 10–12 ml in the coronary circulation.

This finding should, however, not be construed as indicating low oxygen consumption by the kidney. An oxygen consumption of 19–20 ml min^{-1} for the two kidneys (weighing 300 g in man) represents about 8 per cent of that of the whole body at rest.

3. In most organs the oxygen arteriovenous difference is inversely related to blood flow. This is not the case in the kidney. The A–V difference does not change despite massive alterations in blood flow indicating that the oxygen consumption varies with blood flow.

4. The renal circulation shows a remarkable constancy in face of blood pressure changes. In most organs an increase in perfusion pressure is accompanied by a reduction in the vascular resistance across the organ so that blood flow increases. In the kidney there is an increase in resistance which parallels the increase in pressure and blood flow is unchanged (autoregulation).

5. Blood flow within the kidney is not homogenous; the cortex is better perfused than the medulla.

The intrarenal distribution of blood flow

The measurement of transit times of injected dye by Kramer and his colleagues was the first successful attempt at measuring differentially the blood flow through various regions of the kidney. In this investigation photoelectric reflectometers were used. One was placed on the surface of the kidney while another was inserted via

the ureter to lie close to the renal papilla. A cold light source was pierced through the cortex into the renal medulla. This arrangement permitted the direct measurement of the passage time of a dose of Evans blue through the different parts of the kidney [Fig. IV.15]. Blood-flow values were calculated from mean passage times and known volumes of the vascular system. The passage time of the dye between the light source and the extrarenal reflectometer indicated cortical blood flow while the passage time between the light source and the papillary reflectometer indicated medullary flow.

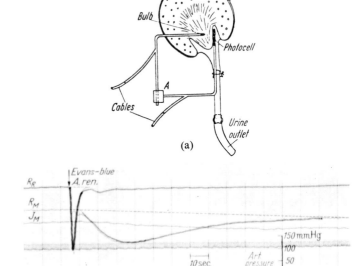

Fig. IV.15(a). Determination of medullary and cortical circulation times from which corresponding blood flows are calculated. (b) Original tracings from the devices shown above. Note the widely different passage times of Evans blue injected into the renal artery in cortical and medullary tissue. (Kramer, K., Thurau, K., and Deetjen, P. (1960). *Pflügers Arch. ges. Physiol.* **270**, 251.)

These studies on dogs by Kramer and his colleagues (see Thurau 1964) have clearly shown that the perfusion rate in the cortex far exceeds that in the medulla—indeed only 1–2 per cent of the renal blood flow traverses the *inner* medulla. This low flow in the inner medulla is due to a low velocity of the blood traversing the vasa recta which in turn is due to the following three factors (i) the vasa recta in the dog may be as long as 40 mm—more than 40 times the length of an 'ideal systemic capillary' and this length factor of course raises the resistance as calculated by the Poiseuille equation [p. 76]. (ii) Additionally, the dehydration suffered by the blood as the vasa recta near their hairpin bend in the innermost part of the medulla is considerable, for the osmotic pressure of the interstitium may be as much as 2000 mOsm l^{-1}. Hence, the blood becomes polycythaemic and its viscosity rises sharply [p. 78]. (iii) The hydrostatic pressure head is not high.

TABLE IV.2

Source	Weight % of total kidney	Intrarenal circulation times	Vascular volume		Blood-flow rates min^{-1}		
			ml per 100 g tissue	ml per 100 g kidney	ml per 100 g tissue	ml per 100 g kidney	% of total renal blood flow
Cortex	70	0.02	19.2	13.5	458	321	92.5
Outer medulla	20	0.086	19.2	3.9	112	22.4	6.5
Inner medulla	10	0.75	22.0	2.2	29	2.9	1.0

TABLE IV.2 (Thurau 1964), from results of the Göttingen laboratory, shows that the cortex receives 450–500 ml per 100 g per min. About three-quarters of the kidney is cortex so that the total cortical weight in *both* kidneys in man would be ¾(2 × 150) = 225 g. Thus 225 g cortex receive 500 × 225 = 1125 ml min⁻¹. To put it another way, the renal cortices, representing 0.3 per cent of the total body weight, receive 20 per cent of the resting cardiac output.

The whole of the renal medulla receives 7–10 per cent of the total renal flow and the inner medulla itself gets only 1 per cent. As might be expected, if the inner medullary blood flow is artificially increased, osmotic stratification of the inner medullary interstitium is greatly reduced and in such circumstances urine osmolality can be lowered to or even below that of the plasma. A relatively slow medullary flow is of great importance for the development of hyperosmolarity of the inner medulla and hence for the production of hypertonic urine. It should be noted that the term 'relatively slow' is employed. Inner medullary flow at 30 ml per 100 g per min is fifteen times that in resting muscle and is half that of the brain.

The original method of Kramer can be accused of damaging the kidney. More recently, studies of the clearance of radioactive isotopes of inert lipid-soluble gases from the renal blood have been employed. Tracer doses of ⁸⁵Kr or ¹³³Xe are injected into the renal artery and their clearance monitored by scintillation counters on the body surface directed towards the kidney. Washout of the isotope follows a complex curve [FIG. IV.16] containing three or four exponentials. The slopes of these exponentials have been shown by autoradiographic and anatomical studies respectively to represent: (i) blood flow in the outer cortex; (ii) flow in the juxtamedullary cortex and outer medulla; (iii) flow in the inner medulla; and (iv) in the perinephric fat.

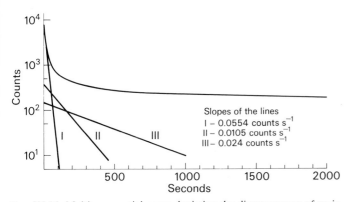

FIG. IV.16. Multiexponential curve depicting the disappearance of an injected dose of ¹³³Xe from the kidney. This is mathematically analysed into three component exponentials, the slopes of which are also shown. The flow rates that these show are respectively 233, 44, and 10 ml per 100 g tissue per minute.

Another method employs the injection of radioactively labelled plastic microspheres of a size that permits them to pass freely along afferent arterioles, but not the glomerular capillaries. When the experiment is terminated the kidney is removed and the radioactivity in its various parts is assessed. Repeated measurements in the same experiment under different conditions is rendered possible by the use of variously labelled spheres. Nashat (1974) discusses the pros and cons of these methods.

Anatomically, the regional distribution of flow may be based on the difference in the resistance offered by cortical and juxtamedullary nephrons. The regional distribution is relevant to the operation of the countercurrent exchange system in the renal medulla.

Autoregulation of blood flow and glomerular filtration rate

Renal blood flow and glomerular filtration rate remain remarkably constant at blood pressures between 90 and 200 mm Hg. Two facts suggest that the changes in vascular resistance responsible for this constancy of blood flow despite the changing arterial pressures, are essentially brought about by alterations in the calibre of the preglomerular resistance section—the *afferent arterioles*.

(i) Direct micropuncture results show that the peritubular capillary pressure remains much the same whether the blood pressure is 90 or 200 mm Hg.

(ii) GFR is constant, which strongly suggests that the varying arterial pressure is prevented, by changes of afferent arteriolar tone, from affecting the glomerular capillary pressure.

This *autoregulation* of renal blood flow is not due to the influence of renal nerves, for it persists after denervation of the kidney and the pharmacological blockade of the intrarenal ganglia. It is dependent on the intrinsic myogenic tone of the renal vessels, for it is abolished by the addition of papaverine, chloral hydrate or sodium cyanide to the renal blood. After papaverine the renal blood flow varies linearly with the arterial blood pressure in the range 90–200 mm Hg.

The probable explanation of autoregulation of blood flow is that it is due to alterations in the myogenic tone of the afferent arteriolar vessels in response to changes in the perfusing pressure. As discussed on page 70 this mechanism was first adumbrated by Bayliss in 1902, and Folkow, during the last decade, has supported this hypothesis with many experimental results although these were obtained in other regional circulations. The stretch of the afferent arterioles caused by a rise of pressure causes a contraction response of the smooth muscle of the arteriolar media and this reduction in calibre raises the resistance to blood flow.

The Bayliss–Folkow mechanism explains the time course of the response of the RBF to sudden changes of arterial perfusion pressure. Thus a sudden increase of perfusion pressure causes, in the initial 4 or 5 seconds, a passive rise of flow which then subsides sharply to a value well below normal by 15 seconds and finally increases again to its 'control value' by 20–30 seconds (Thurau and Kramer 1959). This latency and the phasic change of resistance are consonant with the contractile features of smooth muscle.

There is evidence that the autoregulation of renal blood flow and glomerular filtration rate may be related to the filtered load of sodium chloride.

Thurau and Schneerman have demonstrated a negative feedback relating the delivery of salt to the macula densa to afferent arteriolar resistance. An increase in sodium chloride delivered to the macula densa region of a single nephron was shown to be accompanied by a reduction in the filtration rate of the same nephron.

The mechanism of this 'tubuloglomerular feedback', as this relation is now known, is not clear, but it is thought that it might depend on a local interaction of renin and angiotensin in the interstitium of the afferent arteriole (see Schneerman *et al.* 1970).

Autoregulation is a feature only of the cortical blood flow. Medulary blood flow does not show autoregulation, but as only 8 per cent of the total RBF is supplied to the medulla the effects of pressure on the RBF due to changes of medullary flow are masked.

Renal blood flow, renal oxygen consumption, and the tubular reabsorption of sodium

By far the largest chemical energy expenditure of the kidneys is required to reabsorb the filtered load of sodium. Some 550 g of sodium are filtered every 24 hours and all but 5 g or so are reabsorbed. This entails a high oxygen usage. The rapid metabolic rate of renal tubular tissue is in a way disguised by the bright red colour of renal venous blood. Renal venous blood is 80–85 per cent saturated and has a correspondingly high pO₂. The renal A–V O₂ difference is only 1.5 ml per 100 ml, but of course the RBF is so high at 1300 ml min⁻¹ that the renal oxygen usage is approximately 20 ml min⁻¹. The total weight of both kidneys is 300 g so the renal oxygen usage is nearly 7 ml per 100 g per min, which compares with that of the heart beating in man at rest. Again, it is the cortex, which contains the proximal and distal convoluted tubules, which is characterized by this very high oxygen usage—that of the medulla is much lower.

When the oxygen consumption of the kidney is measured at different rates of RBF obtained by varying the blood pressure over a wide range the relationship is found to be linear [Fig. IV.17(a)] providing that the perfusion pressure exceeds 50 mm Hg at which RBF is some 200 ml per 100 g per min. (Below such perfusion pressures the kidney does not filter. Presumably the increasing pressure then opens up the capillaries and filtration proceeds progressively as the pressure rises.)

This linear relationship between O_2 usage and blood flow above pressures of 50 mm Hg also holds for GFR. As the filtered sodium load at any one [Na^+] in the plasma is determined by the GFR it follows that O_2 usage is linearly related to the Na load and this is seen to be the case in Figure IV.17(b). The scatter of values is very small and the relationship of Na reabsorption to oxygen usage (Q_{Na}/O_2) is about 7 when each is expressed in mM per 100 g tissue weight per min. As about 10 per cent of tubular oxygen usage is concerned with the active transport of other substances (K^+, Ca^{2+}, Mg^{2+}, and organic substances) this estimation of Q_{Na}/O_2 is too low and the actual value is more nearly 8.

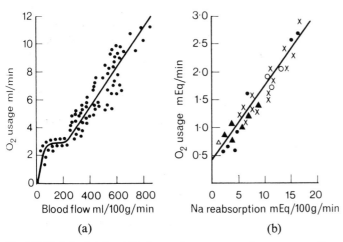

FIG. IV.17. In both (a) and (b) the renal oxygen usage figures are per 100 g kidney. (After Kramer, K. and Deetjen, P. (1964) in *Oxygen in the animal organism* (ed. F. Dickens and E. Neil). Pergamon Press, Oxford.)

The deliberately wide variations of blood pressure required to yield these differing renal blood flows do not cause any material change in the A–V O_2 difference *provided* that the kidney is filtering and therefore its tubules are reabsorbing most of the Na filtered. If the perfusion pressure is too low (below 50 mm Hg, corresponding to a renal blood flow of as little as 200 ml min⁻¹) this constancy disappears [Fig. IV.17(a)].

In skeletal or cardiac muscle, metabolism determines the blood flow. The kidney is unique in that blood flow, hence glomerular filtration rate and hence the filtered sodium load, determines renal metabolism.

THE RENIN ANGIOTENSIN SYSTEM

Tigerstedt (1898) showed that extracts of the renal cortex caused hypertension when injected intravenously. Braun-Menendez *et al.* and Page and Helmer separately showed in 1940 that the rise of

blood pressure was not a direct effect of renin but was due to renin acting as a proteolytic enzyme on a plasma *a*-2 globulin substrate *angiotensinogen* to form an inactive decapeptide Angiotensin I. A further enzyme then forms octapeptide Angiotensin II. Angiotensin II is the most potent pressor substance known, causing generalized arteriolar constriction. Additionally, Angiotensin II stimulates the secretion of aldosterone from the zona glomerulosa of the adrenal gland and thereby promotes sodium reabsorption by the kidney. Angiotensin is rapidly destroyed by angiotensinase.

Plasma renin levels are usually expressed in ng Angiotensin II formed per 100 ml plasma in 3 hours' incubation at 37 °C. Normally the concentration is some 200 ng per 100 ml.

Plasma renin activity depends on the rate of release of renin into the systemic circulation. This in turn is directly related to the degree of granulation observed in the JGA on histological examination.

Renin is released into the systemic circulation in response to:

(i) Sympathetic stimulation of JGA.
(ii) A sustained fall in the systemic blood pressure or more precisely a fall in the renal perfusion pressure at the afferent arteriole.
(iii) Hyponatraemia.

Thurau and Mason (1974) suggest that the release of renin into the systemic circulation is only one of two routes through which renin leaves the JGA, the other being 'local' where renin is released into the interstitial tissue of the JGA complex and the vessels it contains. It is thought that it is through this release route that the renin angiotension system affects tubuloglomerular feedback and plays a part in the autoregulation of GFR and RBF.

Local renin release is stimulated by an increase in the delivery of sodium chloride to the macula densa region of the tubule.

Nervous control of renal blood flow

As has been stated, autoregulation is not dependent on nerves. Although the kidney is richly supplied with sympathetic nerves (derived from segments T4–L2 of which T10–T12 are the most important) which are distributed from the splanchnic nerves and renal plexuses along the renal vessels, their influence is minimal in resting conditions. Transplanted kidneys show renal blood flows and GFR values within the normal range. However, neurogenic constriction of the kidney vessels can occur in hypoxia of anoxic or hypotensive stagnant type and is mediated by carotido-aortic chemoreflexes.

Profound neurogenic renal vasoconstriction is well documented in states of sustained hypotension and shock. Cortical blood flow is grossly reduced and indeed the one organ which may be irreparably damaged after the otherwise successful treatment of wound shock by transfusion, is the kidney. Howard (1962) reported his results based on the observation of five thousand battle casualties in Korea, where the average evacuation time to the forward hospital for transfusion was 3½ hours. Some casualties developed a fatal oliguria, others a modest oliguria, but a striking azotaemia. Others maintained a normal urinary volume, but both GFR and RBF were reduced and the concentrating power of the kidney tubules was reduced.

Kramer (1964) has pointed out that circulatory failure (sustained hypotension) causes renal failure because the persistent vasoconstriction of the afferent arteriole, coupled with the hypotension, stops glomerular filtration. At mean pressures of 50–60 mm Hg the RBF is only 30 per cent of its control value. The glomerular capillary pressure is insufficient to produce filtration and the renal tubules have no (or little) sodium to absorb. The oxygen usage of the kidney falls but there is no evidence of renal hypoxia until the blood flow falls below 10 per cent of normal, whereupon ischaemic structural changes occur.

The medullary blood vessels still carry about 25 per cent of their normal blood flow and in the absence of tubular fluid flow through the loops of Henle this medullary blood flow 'washes away' solutes from the interstitial fluid and dissipates the osmotic stratification which is the normal feature of the medullary interstitium. During recovery from circulatory failure following transfusion, restoration of the GFR provides tubular flow to the loops of Henle and the osmotic stratification is once more built up. Until hyperosmolarity of the interstitium of the inner medulla is restored the kidneys *cannot* form a concentrated urine.

REFERENCES

BRAUN-MENENDEZ, E., FASCIOLO, J. C., LELOIR, F., and MUÑOZ, J. M. (1940). *J. Physiol., Lond.* **98**, 283.

BRITTON, K. (1968). *Lancet* **i**, 335.

DAVENPORT, H. W. (1969). *The ABC of acid–base chemistry*, 5th edn. Chicago University Press.

GIEBISCH, G. (1958). *J. cell. comp. Physiol.* **51**, 221.

GOTTSCHALK, C. and MYLLE, M. (1959). *Am. J. Physiol.* **196**, 927.

HOWARD, J. M. (1962). In *Shock* (ed. K. D. Bock). Springer, Berlin.

JOHNSON, P. C. (ed.) (1964). *Autoregulation of blood flow*. American Heart Monograph, No. 8. New York.

KOEFORD-JOHNSEN, V. and USSING, H. H. (1953). *Acta physiol. scand.* **28**, 60.

KRAMER, K. A. (1964). In *Oxygen in the animal organism* (ed. F. Dickens and E. Neil). Pergamon Press, Oxford.

MALNIC, G., KLOSE, R. M. and GIEBISCH G. (1960). *Am. J. Physiol.* **211**, 529.

NASHAT, F. S. (1974). In *Recent advances in physiology*, 9th edn (ed. R. J. Linden) Chap. 5. Churchill Livingstone, Edinburgh.

PAGE, I. H. and HELMER, O. M. (1940). *J. exp. Med.* **71**, 495.

PITTS, R. F. (1969). *Physiology of the kidney and body fluids*, 2nd edn. Chicago University Press.

SCHNEERMAN, J. WRIGHT, F. S., DAVIS, J. M., STECKELBERG, W. V., and GRILL, G. (1970). *Pflügers Arch. ges. Physiol.* **318**, 147.

SELKURT, E. (1963). In *Handbook of physiology*, Section 2, Vol. II, Chap. 43, p. 1457. American Physiological Society, Washington.

SMITH, H. W. (1951). *The kidney: structure and function in health and disease*. New York.

THURAU, K. (1964). *Am. J. Med.* **36**, 698.

—— and KRAMER, K. (1959). *Pflügers Arch. ges. Physiol.* **269**, 77.

—— and MASON, J. (1974). In *MTP international review of science*, Vol. 6 (ed. K. Thurau) p. 367. Butterworth, London.

TIGERSTEDT, R. and BERGMAN, P. G. (1898). *Skand. Arch. Physiol.* **8**, 223.

WINTON, F. R. (1956). *Modern views on the secretion of urine*. Churchill, London.

WIRZ, H. (1956). *Helv. physiol. pharmac. Acta* **14**, 353.

—— and BOTT, P. A. (1954). *Proc. Soc. exp. Biol. N.Y.* **87**, 405.

—— HARGITAY, B., and KUHN, W. (1951). *Helv. physiol. pharmac. Acta* **9**, 196.

ANTIDIURETIC HORMONE (ADH)

The antidiuretic hormone (ADH) released from the posterior lobe of the pituitary [p. 522] exerts a tonic effect on water reabsorption in the collecting tubule which results in the formation of perhaps 1 ml of urine from the delivery of some 15 ml tubular fluid to the collecting duct each minute. Hypothalamic ADH secretion is responsive under ordinary conditions to the tonicity of the plasma and extracellular fluid. Thus the ingestion of water causes dilution of the body fluids and reduces or abolishes the secretion of ADH; as a result a water diuresis develops owing to the escape of large volumes of tubular fluid into the urine. The diuretic response of a subject or of a trained conscious dog to a given water load is very predictable and this forms a convenient way of 'assaying' the amounts of ADH released by superimposed stimuli.

ADH secretion is regulated by the following factors:

1. The crystalloid osmotic pressure of the plasma.
2. Possibly by the volume of the plasma.
3. Afferent impulses from various parts of the body, and emotional states.

Effect of changes in plasma crystalloid osmotic pressure

Injection into the carotid artery of a hypertonic solution of NaCl induces an immediate antidiuretic response due to the secretion of ADH. An exactly similar effect can be provoked by injecting an appropriate dose of ADH; this amount equals the quantity released by the gland during the experimental procedure. The antidiuretic response to hypertonic saline is reduced to 10 per cent of its former value by removal of the posterior lobe; this small residuum is due to the fact that some secreting tissue in the stalk and median lobe still remains.

When the hypertonic solution is infused intravenously at a constant rate into a conscious animal or man ADH release begins when the osmolarity of the plasma exceeds a critical level. In man this is approximately 290 mOsmol l^{-1}.

The effective stimulus to ADH secretion is the increase in the crystalloid osmotic pressure in the carotid arterial blood. Thus:

1. Solutions of NaCl, Na_2SO_4, or sucrose of equal osmolar concentration produce effects of equal magnitude.
2. The amount of ADH released is directly proportional to the degree of hypertonicity induced in the carotid blood.

These results are believed to be due to stimulation of specific osmoreceptors which respond to the changes in crystalloid osmotic pressure. These osmoreceptors set up nerve impulses which reflexly stimulate the hypothalamus, and so the neurohypophysis. It is supposed that at normal blood tonicity the receptors discharge steadily and so maintain reflexly a steady secretion of ADH; if the plasma becomes hypertonic both the receptor discharge and the secretion of ADH are increased and vice versa if the tonicity of the plasma falls. In the dog the receptors lie in or close to the supraoptic nucleus, being supplied by deep cerebral branches of the internal carotid artery.

It is interesting to note that the osmoreceptors respond with maximal intensity to changes in concentration of the two principal ions of extracellular fluid, i.e. Na^+ and Cl^-. The osmoreceptors do not respond to the intracarotid injection of hypertonic solutions of urea or glucose, for urea and glucose can pass across their membranes with ease.

Physiological role of osmoreceptors

Effects of raised crystalloid osmotic pressure. Normally this may be produced by water deprivation; by an excess of simple salt (NaCl); or, finally, by an excess of other solutes (e.g. urea, glucose).

1. Water deprivation: release of ADH reduces urine flow to an obligatory minimum thus conserving water.
2. Salt excess: though the NaCl output in the urine increases greatly, the urine flow increases only moderately. The osmotic diuresis due to the increased tubular fluid concentration of salt is offset by the increase in ADH secretion. Any water drunk subsequently is initially retained (as the result of this excessive secretion of ADH). Although thereby it increases the volume of body fluids, ADH thus helps, however, to lower the crystalloid osmotic pressure towards normal.
3. Increase of blood urea: this normally leads to increased urea excretion which is facilitated by an increased urine volume but hampered by oliguria. The non-responsiveness of the osmoreceptors to raised blood urea prevents there being a release of ADH which would hamper urea excretion by reducing the urinary volume. Similarly the insensitivity of the osmoreceptors to hyperglycaemia means that ADH does not interfere with the polyuria of diabetes mellitus.

Effects of lowered crystalloid osmotic pressure. 1. Water drinking promotes a transient polyuria or diuresis due to inhibition of ADH secretion caused by plasma hypotonicity.

2. Simple salt deprivation which produces hypotonicity of the ECF causes an initial diuresis due to lessened secretion of ADH. This reduces the ECF volume but helps to maintain the tonicity of the body fluids. Subsequently, complex changes occur due to the interaction of several factors, of which altered secretion of ADH is only one. It is possible, too, that if plasma hypotonicity develops very slowly over a long period the behaviour of the osmoreceptors is altered. Thus if water is drunk by a person suffering from chronic salt lack the resulting water diuresis is delayed, prolonged, and incomplete.

Volume receptors and ADH secretion

It is probable that cardiac atrial and venous vagal receptors act as volume receptors which, on being overstimulated in circumstances of plethora or raised atrial and venous pressure, cause reflex inhibition of ADH secretion which in turn leads to diuresis [see p. 149; Gauer and Henry 1976].

Effects of afferent impulses and emotional states

1. In experimental animals, exercise produces secretion of ADH, not directly, but because of its emotional concomitant; if the animal becomes accustomed to the procedure from constant repetition, the effect of exercise becomes negligibly small. The amount of ADH secreted in response to the emotional component of exercise is measured as usual by the extent to which a previously induced water diuresis is diminished. As the antidiuretic action of exercise is still obtained after renal denervation this 'exercise effect' is not mediated by nerves. Emotion acts by increasing hypothalamic activity and hence ADH secretion.

2. Similar effects are produced clinically by anger, fear (e.g. an impending venepuncture), a faint from any cause (e.g. haemorrhage, prolonged standing or as part of a vasovagal syndrome); by afferent impulses from various sources, ranging from operative trauma to suckling; by drugs which act directly on the supraoptic nucleus, e.g. nicotine (and hence smoking), anticholinesterases (note that acetylcholine applied directly to the supraoptic nucleus causes ADH secretion), and morphine. Clinically the stimulus may cause a massive discharge of ADH in such concentrations that general circulatory effects result, including pallor due to contraction of the precapillary sphincters in the skin.

REFERENCES

Gauer, O. H. and Henry, J. P. (1976). *Int. Rev. Physiol.* **9**, 145.
Jewell, P. A. and Verney, E. B. (1957). *Phil. Trans. R. Soc.* **B240**, 197.
Thorn, N. A. (1958). *Physiol. Rev.* **38**, 169.

THE REGULATION OF SALT EXCRETION AND OF THE EXTRACELLULAR FLUID VOLUME

Sodium chloride is the most abundant solute in the extracellular fluid and forms the skeleton around which the body water is held; the efficiency of osmoregulation is such that deviations from a constant solute concentration in the extracellular fluid are not tolerated for any length of time. Volume distortions can therefore occur only as a result of alterations in the solute content of the extracellular fluid and the regulation of volume is achieved by the regulation of salt balance.

Some mammals, like sheep, have special salt-satiety centres; their salt intake is controlled. This is not true of man whose salt intake depends on taste and social habit. On the other hand, salt is lost in sweat, it is secreted in the gastro-intestinal tract and is excreted by the kidney, but of the three channels only renal excretion is regulated to subserve volume regulation.

The regulation of sodium excretion by the kidney is highly complex. Sodium chloride is filtered at the glomerulus but is also reabsorbed throughout the tubular system.

There is plenty of evidence to confirm that sodium excretion cannot be regulated by changes in the GFR. This does not mean that sodium excretion is not influenced by changes in the GFR when these are substantial; all it implies is that these influences are not employed by the organism to regulate sodium excretion.

Ideas that different nephrons have different propensities as far as salt handling is concerned, though attractive, have not as yet been confirmed. One is therefore left to concede that the regulation of sodium excretion rests primarily, if not entirely, on the regulation of its tubular reabsorption.

There is no single factor to which tubular sodium reabsorption can be solely related and which accounts for all the patterns of sodium excretion seen in different circumstances. The following factors are, however, known to influence sodium excretion singly or in combination:

1. An increase in the plasma concentration of sodium promotes sodium excretion. The farther the level is deviated from the 'normal' the greater is the effect. A decrease in P_{Na} inhibits sodium excretion. Disturbances in P_{Na} are, however, not commonly encountered in clinical practice. Hyponatraemia is seen infrequently, hypernatraemia hardly ever.

But the action of P_{Na} depends on the concentration of protein in the peritubular capillary system so that the same P_{Na} exerts a greater effect if it is accompanied by a reduced post-glomerular protein concentration.

2. A fall in the concentration of plasma proteins promotes natriuresis. The effect is often masked; for plasma proteins are likely to exert their effect in the peritubular environment where the protein concentration is $[P_{pr}] \times 100/100$-FF [p. 236]. This suggests that the GFR, the renal blood flow and the haematocrit of blood all play a part in deciding the final effect of plasma proteins on sodium excretion.

3. The plasma levels of aldosterone and other glucocorticoids have an important role in regulating sodium excretion. Aldosterone is mainly responsible for the salt retention seen when salt intake is chronically curtailed while its absence leads to death as a result of salt loss and potassium retention in Addison's disease.

Aldosterone is secreted in response to hyponatraemia, hyperkalaemia and to increases in the plasma levels of angiotensin II.

4. Stimulation of the renal nerves may increase sodium reabsorption. A direct effect on the tubular cell is claimed but this is very difficult to distinguish from any indirect action which the nerves might exert, such as reducing GFR and RBF and increasing systemic renin release.

5. The infusion of large volumes of saline to produce an 'expansion of volume' in experimental animals leads to increased salt and water excretion. The effect is powerful enough to overcome the influence of salt-retaining hormones such as aldosterone. As the effect is not always explicable by distortions in the known factors a 'natriuretic hormone' which could act directly and specifically on the renal tubules has been postulated. Considerable research, however, has so far failed to confirm the existence of such a substance or to indicate the site and mode of its production.

'Volume expansion' may produce its action through changes in the known factors interacting in such a way as to produce an augmented and not readily predictable effect.

6. An increase in left atrial pressure has been shown to increase salt and water excretion in experimental animals. Part of this effect is due to a reduction in the release of vasopressin from the neurohypophysis, but there is no doubt that it also occurs when the neurohypophysis and its connections are experimentally ablated.

Whatever causes the effect is blood-borne and its production depends on the integrity of the vagus nerve, and of certain regions in the midbrain.

This section clearly shows that our knowledge of the regulation of volume is not as clear as it is of osmoregulation. It also emphasizes that this regulation is multifactorial and could not be truthfully simplified.

Even more difficult to answer is the question of how the organism is appraised of the distortions in volume when they occur. For the time being it suffices to appreciate that there are no specific receptors that measure volume and that volume distortions may be represented as pressure changes in the cardiovascular system or as changes in the chemical composition of blood and the extracellular fluid.

USE OF CLEARANCE VALUES AND OTHER SPECIAL METHODS IN STUDY OF RENAL ACTIVITY

Clearance value

The clearance value [C] of a plasma constituent is the volume (in ml) of plasma which contains the amount of the constituent which is excreted in the urine in one minute. Consider the clearance value for urea (C_{urea}): plasma concentration of urea [P_{urea}] is 0.3 mg ml^{-1}; the amount of urea excreted in the urine in one minute [$U_{urea}V$] is 20 mg.

$$C_{urea} = \frac{U_{urea}}{P_{urea}} = \frac{20}{0.3} = 67 \text{ ml},$$

where U_{urea} stands for the urinary concentration of urea and V stands for the urine flow in one minute.

Thus 67 ml of plasma contain the amount of urea which is excreted in the urine in one minute. The term clearance value is somewhat misleading because no part of the plasma is completely cleared of urea. Of 700 ml of plasma containing 210 mg urea which flow through the kidney in one minute the volume filtered out through the glomeruli is only 120 ml containing 36 mg of urea of which 20 mg escape in the urine. The clearance value is thus a so-called 'virtual volume'; it is the result of an arithmetical calculation. For all that, the determination of the clearance value for certain substances provides a measure of the volume of the glomerular filtrate and of renal efficiency.

Inulin clearance—glomerular filtration rate (GFR).

The soluble polysaccharide inulin is filtered out from the glomeruli in the same concentration as in plasma; in the tubules the inulin is neither reabsorbed not secreted. If these statements are true then the inulin clearance value is equal to the volume of the glomerular filtrate.

Suppose the plasma inulin concentration [P_{in}] is 1 mg ml^{-1} and the inulin excretion in the urine per minute [$U_{in}V$] is 127 mg. Then

$$C_{in} = \frac{U_{in}V}{P_{in}} = \frac{127}{1} = 127 \text{ ml},$$

i.e. 127 ml of plasma have been 'virtually' cleared of inulin.

As inulin is assumed to be neither reabsorbed nor secreted in the tubules, the only way in which 127 ml of plasma could have been cleared of inulin is by the filtration of 127 ml of (protein-free) plasma through the glomeruli. In other words, the inulin clearance value (127 ml) is, as stated above, the glomerular filtrate volume.

Inulin is a carbohydrate polymer with a molecular weight of 5200 corresponding to 32 hexose molecules. Non-toxic and physiologically inert, it has an even lower diffusibility than might be expected from its large molecular weight because it is an elongated molecule. It has been shown that inulin passes into the capsular fluid of amphibia. Inulin is not excreted by the kidneys of aglomerular fishes. In man, alterations in the plasma inulin concentration

(under conditions of constant glomerular filtration rate) do not affect the inulin clearance value. Thus if the plasma inulin concentration [P] is doubled, twice the amount of inulin is filtered out from the glomeruli, and twice the amount is passed out in the urine; as P and UV rise to the same extent the clearance value is unaltered. (NB the *excretion rate* of inulin rises linearly with the plasma concentration [FIG. IV.18].

The clearance values for all substances eliminated solely by glomerular filtration should be the same under constant conditions. In the dog, the clearance values for inulin, creatinine, and sodium ferrocyanide are identical, suggesting that they are treated in the same way. In view of the highly selective character of both tubular reabsorption and secretion, it would be most improbable that these three widely different chemical compounds should be treated similarly by the tubules. The more likely proposition is that all three are removed by the physical process of filtration and that the tubules exert no influence on their subsequent excretion.

Phloridzin abolishes the power of the tubule cells to reabsorb glucose. In the phloridzinized dog the glucose clearance becomes identical with that of inulin and creatinine. In man, after phloridzin the glucose/inulin clearance ratio is 0.9 and, if a larger dose could be safely given, the ratio would presumably reach unity as in the dog.

To determine the glomerular filtration rate in man a large initial dose of inulin is injected, which is followed by a constant inulin infusion at a rate which compensates for its loss in the urine. A reasonably constant plasma level is thereby maintained. The inulin concentration in the plasma of venous blood [P], the urinary concentration [U], and the volume of urine excreted per minute are determined. The inulin clearance in a male of surface area of 1.73 m^2 is 127 ml min^{-1}. The range is ± 25.8 ml. Females show a slightly lower GFR (expressed at the same surface area of 1.73 m^2)—109 ± 13.5 ml min^{-1}.

Measurement of renal blood flow by clearance methods

Certain organic iodine compounds, e.g. *diodrast*, are completely or almost completely extracted from the blood during each passage through the kidney. Let us assume a plasma diodrast concentration of 1 mg per 100 ml and a renal plasma flow of 700 ml min^{-1}: 120 ml of glomerular filtrate are formed, containing 1.2 mg of diodrast; hence 580 ml of plasma pass to the tubular blood vessels. This amount of plasma contains 5.8 mg of diodrast which is actively secreted by the tubules, from the blood into their lumina. Hence the whole of the diodrast originally present in the 700 ml of plasma is cleared from the blood. Thus diodrast clearance gives a measure of the renal plasma flow. At low plasma concentrations of diodrast the extraction ratio by the kidney exceeds 0.91. The deficit in the extraction ratio below 1.0 may be attributable to there being some blood flow through the kidney which does not pass through peritubular blood vessels; hence the diodrast clearance may be taken as representing the volume of plasma presented for clearance to the functioning renal parenchyma.

Organic iodine compounds, such as diodrast, often have unpleasant side-effects when injected and their use has been largely superseded by that of *para-aminohippuric acid* (PAH). This is treated similarly by the kidney, giving a clearance value which is identical with that of diodrast. It is interesting to note that both diodrast and PAH depress the secretion of penicillin by the renal tubules. Plasma levels of penicillin are obviously more easily maintained if secretion of the antibiotic by the kidney tubules can be minimized. A search for a suitable method of doing this led to the use of PAH which safely competes with penicillin to occupy the secretory transport mechanism of the tubule cells.

The clearance rate of PAH is 654 ± 163 ml min^{-1} in a man of 1.73 m^2 and 592 ± 153 in a woman of the same surface area. The practical procedure in outline is as follows: after a large priming dose of PAH intravenously, the compound is continuously administered by drip so as to maintain the plasma concentration despite loss in the urine. Plasma venous concentration is thus equivalent to the arterial concentration; the urinary concentration and volume are simultaneously determined.

Filtration fraction. The inulin/PAH clearance ratio ×100 is described as the filtration fraction [FF]. This is 19.2 ± 3.5 per cent.

Since it is now thought that the colloid osmotic pressure in the peritubular environment might influence the tubular reabsorption of salt and water, a knowledge of the filtration fraction has acquired special significance. The post-glomerular protein concentration is equal to

$$\frac{[P_{pr}] \times 100}{100 - FF},$$

where $[P_{pr}]$ is the concentration of plasma proteins. The filtration fraction varies directly with the haematocrit of the blood perfusing the kidney and expresses a divergence between renal blood flow and the glomerular filtrate which could not be attributed to changes in afferent arteriolar resistance.

Osmotic and free-water clearances

These can be calculated as follows:

$$C_{osm} = \frac{U_{osm} V}{P_{osm}}$$

$$C_{H_2O} = V - C_{osm}.$$

(The subscript 'osm' stands for osmolarity.)

They measure the rate at which plasma is cleared of osmotic particles and of 'free water' that is 'not bound' to the osmotic particles respectively.

In man C_{osm} is just under 3 ml min^{-1} and is increased in osmotic diuresis as is seen when diuretics that enhance salt excretion are administered.

The free-water clearance is usually negative. This means that the volume of urine excreted is less than the osmolar clearance or, in other words, that the urine is more concentrated than plasma. In water diuresis, seen following the ingestion of water, free-water clearance becomes positive. TABLE IV.3 shows the approximate relationship between the concentration of total solutes in the urine expressed as osmoles per litre and its specific gravity (which is the easiest and one of the most useful bedside determinations of renal function). Specific gravity (SG) is compared with water which is 1000.

TABLE IV.3

SG	1005	1007	1010	1015	1020	1025	1030	1035
osmol l^{-1}	0.2	0.3	0.4	0.6	0.8	1.0	1.2	1.4

The concentration of plasma and glomerular filtrate is 0.3 osmol l^{-1}.

The minimum volume of urine that is necessary to contain the excreted solutes at maximal concentration is termed the '*volume obligatoire*'.

The amount of nitrogenous waste (mainly urea) excreted varies with the protein intake. Minimal nitrogenous excretion occurs when the protein intake is nil and the full calorific requirements of the body are provided by non-protein foodstuffs—fat and carbohydrate. TABLE IV.4 shows how the *volume obligatoire* can thus be reduced to 150 ml per day.

Note that the osmolar clearance falls from 2.7 ml min^{-1} on a mixed diet to about 0.5 ml min^{-1} in fasting but with glucose intake.

Tm$_{PAH}$—the maximal rate of tubular excretion. Clearly if the plasma concentration of PAH be progressively raised a point is

TABLE IV.4

Diet	Milliosmoles of solute to be excreted in 24 h	Volume of Urine (ml)		
		SG 1015 0.6 osmol	SG 1020 0.8 osmol	SG 1035 1.4 osmol (maximal concentration)
Usual mixed intake	1200	2000	1500	850
Fasting	800	1300	1000	550
Fasting+sufficient glucose to meet calorific requirements	200	350	250	150

reached at which the tubules are secreting a maximal amount of PAH and yet complete extraction of the PAH from the blood can no longer be achieved.

The PAH 'secretory' load to the tubules is given approximately by the equation:

Tubular load (PAH) = P_{PAH} × (RPF − GFR)
(where P_{PAH} is the plasma concentration of PAH).

With ordinary levels of P_{PAH} the secretory power of the tubules is such that PAH is completely extracted during its passage through the kidney. With increasing concentrations of PAH in the plasma the secretory powers of the tubules reach a maximum, and at still higher P_{PAH} levels the extraction ratio falls below unity. As the normal high clearance value of PAH is dependent upon complete extraction by the tubules, the clearance rate of PAH falls when the maximal tubular secretory power is exceeded. Correspondingly, if the amount of PAH excreted in the urine and the amount of PAH filtered are plotted against P_{PAH}, the excretion of PAH in the urine eventually increases linearly with P_{PAH}, parallel to the graph showing filtration of PAH. The two lines are separated by an amount of PAH excretion which represents the maximal secretory powers of the tubules for PAH [FIG. IV.18(b)]. This value (Tm$_{PAH}$) gives a useful quantitative definition of the functional secretory powers of the tubular mass.

In man (1.73 m^2 surface area) Tm$_{PAH}$ = 79.8 ± 16.7 mg min^{-1}.
In woman (1.73 m^2 surface area) Tm$_{PAH}$ = 77.2 ± 10.8 mg min^{-1}.

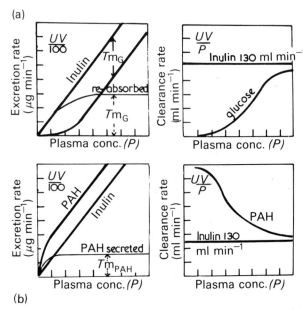

FIG. IV.18. Excretion of clearance graphs. (a) = Inulin and glucose. (b) = Inulin and PAH.

Maximal tubular reabsorption of glucose (Tm_G). The functional reabsorptive power of the tubular mass can be similarly expressed quantitatively in terms of the ability of the tubule to reabsorb glucose. At ordinary plasma levels glucose is not excreted in the urine. As the plasma concentration is raised successively, the filtered load becomes too great for the tubular reabsorptive powers and glycosuria occurs. The clearance of glucose (normally zero) becomes positive and rises progressively. The excretion graph [Fig. IV.18(a)] shows that at high plasma concentrations the excretion of glucose rises linearly with the plasma concentration, parallel to the graph showing 'filtered glucose', but below it by an amount equivalent to Tm_G.

Tm_G in a man of 1.73 m^2 surface area is 300–350 mg min^{-1}.

Energy mechanisms involved in tubular reabsorption and secretion

Clear evidence exists that the secretion of PAH, diodrast, penicillin, and probably probenecid is effected by the same enzyme systems of the tubular cell. The Krebs tricarboxylic acid cycle and energy-rich phosphate from ATP are involved. Thus 2:4 dinitrophenol (DNP), which uncouples oxidative phosphorylation from the respiratory processes of the cell, profoundly diminishes the secretion of the above-mentioned compounds. Dehydroacetic acid (DHA), which inhibits succinic dehydrogenase and therefore blocks the Krebs cycle, has a similar effect. On the other hand acetate increases the secretion of PAH and diodrast; acetate is known to react with coenzyme A and ATP to form acetyl coenzyme A which is probably active acetate [see p. 452].

Glucose reabsorption is unaffected by DHA or DNP and is not increased by acetate; presumably the carrier mechanism involved differs from that required for the secretory processes. Both glucose reabsorption and secretion of PAH, etc., are reduced or abolished by cyanide (which interferes with cytochrome systems).

REFERENCE

Pitts, R. F. (1974). *Physiology of the kidney and body fluids*, 3rd edn. Chicago University Press.

Physiology of micturition

FUNCTIONS OF THE URINARY BLADDER

Storage of urine

The mammalian urinary bladder stores urine which has been formed by the kidneys and transported to the bladder via the ureters. Normally the bladder is emptied at intervals of up to several hours. The urine stored in the bladder remains unchanged in chemical composition although its osmolarity is much greater than that of the plasma in capillaries only 12 nm away from the luminal surface of the transitional epithelium (urothelium). The urothelium differs radically from other types of epithelium, e.g. that of the small intestine. Urothelial cells grow very slowly and have a life-span of 200 days (compared with that of human ileal enterocytes which is about three days). Urothelium forms a complete barrier to the passage of water and solutes, probably because it contains a cerebroside which is chemically related to sphingomyelin (which occurs in myelinated nerve fibres).

REFERENCES

Hicks, R. M. (1975). *Biol. Rev.* **50**, 215.
The Lancet (1975). Editorial. *Lancet* **ii**, 913.

Efferent nerve supply of bladder

The efferent fibres to the bladder come both from the sympathetic and the sacral autonomic.

The detailed anatomy of these nerves has been carefully studied in man [Fig. IV.19]. The sympathetic connector cells lie in the grey matter of the first and second lumbar segments of the spinal cord.

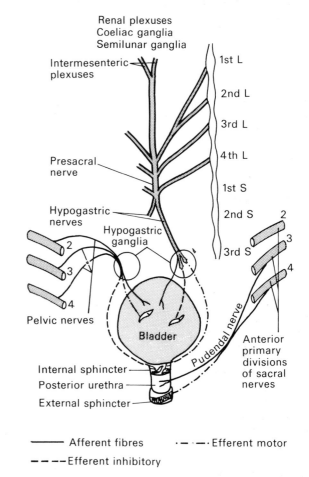

FIG. IV.19. Diagram of innervation of the bladder and sphincters in man. The pelvic nerves (nervi erigentes) are parasympathetic. The hypogastric nerves are sympathetic. The pudendal nerves are somatic. (Redrawn from Learmonth, J. R. (1932). *Am. J. Surg.* **16**, 270.)

The connector fibres take various routes—through the lateral sympathetic chain, and through the coeliac and superior mesenteric ganglia—to form a nerve lying in front of the sacrum—the presacral nerve. This divides at the level of the first piece of the sacrum into two hypogastric nerves which end in the hypogastric ganglia on the lateral aspects of the rectum. In the lower animals the sympathetic relay station is in the inferior mesenteric ganglia (at the origin of the inferior mesenteric artery) from which hypogastric (postganglionic) nerves arise. In man, however, the inferior mesenteric ganglia are rarely present. The sacral autonomic fibres arise from cells in the second sacral, and less constantly from the third sacral segments, and pass in the nervi erigentes, likewise to end in the hypogastric ganglia which serve as the ganglionic relay station for both sets of fibres. From the anterior border of these ganglia, sympathetic and autonomic postganglionic fibres arise and pass to the bladder, both the body (detrusor) and internal sphincter.

The prostatic urethra and the external sphincter receive efferent somatic fibres from the pudic nerves.

Effects of sympathetic stimulation

In animals the body of the bladder relaxes and the sphincter vesicae contracts. Learmonth obtained the following results on stimulating the presacral (sympathetic) nerve in man: closure of the ureteric orifices, contraction of the internal sphincter, increase in the tone of the trigone and vasoconstriction in this region. In addition, there was contraction of the muscle of the seminal vesicles, the ejaculatory ducts, and the prostate, and the contained secretions were squeezed out. No effect was observed on the dome or lateral walls of the bladder. Intravenous injection of adrenaline in man lowers the pressure in the comfortably full bladder; it is therefore probable that the sympathetic nerves can exert some inhibitory influence on the body of the bladder in man as in lower animals.

Effects of sacral autonomic stimulation

The internal sphincter is relaxed, the detrusor is stimulated, and the bladder is emptied.

Afferent supply of bladder

The afferent fibres probably take a double route:

1. Along the sympathetic into the dorsal nerve roots of L1 and 2, and the lower thoracic segments.
2. Along the sacral autonomic into the sacral dorsal nerve roots.

They subserve two functions: (i) they indicate the degree of distension of the bladder; (ii) convey pain sensibility. Traction on the presacral (sympathetic) nerve gives rise to a crushing kind of pain which is felt in the bladder itself, i.e. it is localized surprisingly accurately and is not referred to the skin segments. The pain of bladder disease is diminished to a considerable extent by section of the presacral (sympathetic) nerve; it is abolished totally, however, when the sacral parasympathetic fibres are also cut.

Cortical control

The path of the afferent fibres within the central nervous system is not known. A higher centre for control of the bladder is described at the top of the motor area of the cerebral cortex on the medial aspect of the hemisphere in association with the centres for the perineal structures. The path in the human spinal cord for these motor fibres from the cortex lies lateral to the pyramidal tracts intermingled with the spinocerebellar fibres.

Postural activity in the bladder

The bladder is an organ in which the phenomena of tone have been extensively studied:

1. If fluid is injected into a bladder 24 hours after death, no rise of internal pressure initially takes place. The fluid serves first of all to straighten all the folds that are present; then, as more fluid is pumped in the pressure rises with increasing rapidity for every fresh rise in bladder content. On reaching the limit of elasticity of the bladder wall the gradient of ascending internal pressure becomes very steep before the final bursting of the organ.
2. In the initial stage of shock following complete transection of the spinal cord, similar results are obtained, i.e. the bladder responds purely passively to distension with urine.
3. The normal bladder, however, responds differently. If fluid is introduced into the bladder through a catheter, e.g. in quantities of 50 ml at a time, the bladder pressure rises immediately on each occasion; but if an interval is allowed to elapse after each bout of filling, the pressure frequently falls to some extent, though the volume, of course, remains unchanged. These points are well

brought out in FIGURE IV.20. The intact bladder thus responds initially to distension by a contractile resistance which leads to a rise of pressure; subsequently the bladder wall actively adjusts itself to the new conditions, its fibres presumably elongate and the contents are gripped at a lower pressure than that which obtained previously. When one talks of tone in the bladder or other hollow viscera, one has in mind these active adjustments of the muscle coat to variations in internal volume; as a result the contents are subjected to pressures which vary to some extent, but not directly, with their volume. When the bladder volume exceeds 500 ml the pressure tends to rise more sharply and may exceed 10–15 cm of water.

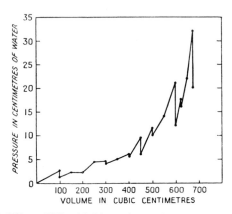

FIG. IV.20. Effect of filling bladder on internal pressure. Ordinate = bladder pressure in cm H_2O; abscissa = bladder volume in ml. 50 ml of fluid were introduced into the bladder via a catheter and the resulting pressure recorded. A short interval was then allowed. The vertical drops of pressure indicate the adaptation that then takes place. Further quantities of 50 ml were then successively introduced and the procedure repeated. Note the much more rapid rise of pressure when the bladder volume exceeds 500 ml. (Denny-Brown and Robertson (1933). *Brain* **56**.)

4. As the bladder is filled progressively there is first the gradual rise of internal pressure just described; later transient pressure waves appear which are at first small and unaccompanied by any sensation. With higher bladder volumes (e.g. 700 ml), these contractions may cause pain and are associated with a sharp rise of pressure. By an effort of the will, i.e. a decision to hold the urine, the contractions may pass off wholly or partially and the discomfort disappears [FIG. IV.21]. This, however, is only temporary, for the contractions recur, a pressure of 100 cm of water or more is attained, discomfort is acute and the subject begs to be allowed to empty his bladder. Conversely with a moderate degree of filling (e.g. 200 ml volume), a decision to micturate leads to a sharp rise of

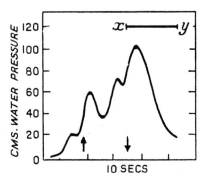

FIG. IV.21. Effect of consciousness on bladder pressure. Ordinate = bladder pressure in cm H_2O; abscissa = time in 10 s. In the interval between the first and second arrows there was conscious 'preparedness for micturition', which was associated with the initial rapid rise of pressure; during *x–y* there was a strong desire to hold water, associated with cessation of bladder contraction.

vesical pressure before any actual effort is made, and independently of any contraction of the abdominal wall [Fig. IV.21].

5. The sphincter muscles, too, maintain an attitude or possess tone, but their function is to maintain a constant position and keep the urethral orifice closed, except, of course, during micturition, when their tonic activity is reciprocally temporarily inhibited.

Voluntary micturition

From these observations the mechanism of voluntary micturition can be deduced. As the bladder fills, its internal pressure slowly rises as already indicated. Contraction waves appear which stimulate the pressure receptors in the muscle coat and send afferent impulses to the spinal cord and thence up to the brain, which give rise to a 'desire to micturate'. If it is inconvenient to micturate, impulses from the cerebral cortex cause inhibition and elongation of the bladder wall, mainly apparently by inhibiting sacral parasympathetic excitor activity, but also possibly via the sympathetic nerves which relax the bladder musculature. The pressure in the organ falls and the desire to micturate temporarily passes off [Fig. IV.21]. By constant practice the bladder can be accustomed to accommodate very large volumes of urine before an uncontrollable and unbearable rise of intravesical pressure occurs. It is well known that many sedentary workers, and women especially, can hold their urine for very long periods and without discomfort.

On the other hand, the bladder may be voluntarily emptied in response to the 'desire to micturate'. Impulses from the cortex then pass down to the sacral segments and along the nervi erigentes to stimulate the bladder wall and to relax the internal sphincter; the centre controlling the external sphincter is reflexly inhibited. Certain accessory muscles are also involved: the perineum is relaxed, the abdominal wall contracts, the diaphragm descends, and the breath is held with the glottis closed. The resulting rise of intra-abdominal pressure compresses the bladder from without. The intravesical pressure rises steeply and the bladder is evacuated. Micturition can, of course, be carried out voluntarily before urgent afferent impulses have been received.

In young children the process is different. Postural activity in the bladder is less perfect, and quite small quantities of urine may raise the pressure sufficiently to send afferent impulses up to the cord. The higher centres are not involved and micturition occurs entirely reflexly (perhaps at the spinal level) from stimulation of the nervi erigentes.

The postural behaviour of the bladder can be influenced by a variety of circumstances. In certain forms of emotional stress, as in the anxious moments preceding an examination, the power of the bladder to elongate its fibres as its contents increase, appears to be in abeyance. The well-known result is that there is a constant desire to micturate, with the passage of only small quantities of urine on each occasion. Frequency is produced in a similar way when the bladder wall (or mucous membrane) is irritated by inflammation or stone. The nature of the disturbance in the so-called 'enuresis of children' is not clear. Many authorities believe the trouble is psychological and that it is analogous to the 'emotional' frequency described above.

Effects of interference with nervous control of bladder

Section of sympathetic supply

In man, the immediate effects, as would be expected, are relaxation of the ureteric orifices, the trigone, and internal sphincter. After complete sympathetic denervation, ejaculation can no longer take place, though psychical orgasm and erection occur; sterility therefore results. Later, the internal sphincter may recover and close completely, though it gives way easily when a catheter is passed. After an initial and inconstant period of frequency of micturition, bladder function is re-established in a comparatively normal way.

Injury to sacral nerve supply

In severe cases there is complete loss of voluntary micturition; the external sphincter is flaccid. The bladder responds peculiarly to distension: with a small volume, e.g. 200 ml, the pressure may rise rapidly to 50 cm of water; the bladder then adapts itself quickly leading to a sharp fall of pressure to say 30 cm of water. After a time the bladder, though deprived of its motor innervation, empties itself periodically automatically through the intervention of the local peripheral neuromuscular mechanisms. The 'isolated' bladder, if otherwise healthy, can respond to adequate internal distension by contraction of the detrusor muscle; these contractions, however, are not as powerful or as well co-ordinated as normally so that micturition is incompletely performed and has to be aided by the compression effect of abdominal contraction. Large quantities of residual urine (e.g. 300 ml) may be left in the organ.

Injury to afferent supply

The afferent impulses from the bladder may be lost or diminished from lesions of the lumbosacral dorsal nerve roots, as in tabes. The patient is unaware of the state of distension of the bladder. He can still micturate at will, but if he does not do so at regular intervals the accumulation of urine precipitates involuntary automatic evacuation; or else the pressure in the bladder may rise till the resistance of the sphincter is passively overcome and dribbling occurs.

Injury to cortical control

Efferent control of the bladder is disturbed in lesions in the vicinity of the pyramidal tracts, to the lateral side of which the fibres for the voluntary regulation of the bladder are probably found. The patient is completely conscious of all events occurring in his bladder, but voluntary control of micturition is interfered with. The patient has difficulty in initiating the act, and, further, he cannot hold his water at will: when the desire comes on he responds to it immediately. In other words, voluntary control of both motor and inhibitory fibres to the bladder is impaired. Such symptoms occur in any spinal lesions involving both pyramidal tract regions, e.g. disseminated sclerosis, early compression of the cord, or syringomyelia.

Acute transection of the cord

If the spinal cord is completely transected by some acutely acting injury, voluntary micturition is completely abolished. The activity of the detrusor muscle remains in abeyance for a long period, but sphincter tone returns very soon. At this stage the bladder responds to filling in the same way as the dead organ or like an elastic bag. Retention of urine is therefore complete from an early stage. If no catheter is passed, the bladder becomes increasingly overstretched. The sphincter is finally forced open by the high intravesical pressure and small quantities of urine escape at frequent intervals—a condition of 'retention with overflow'. Owing to the excessive stretching of the bladder wall, its nutrition suffers and it becomes very prone to infection. Cystitis may occur, and death results from the usual complications of ascending urinary infection. If the inflammation is of a lower grade, the bladder shrinks and the musculature is so damaged in consequence that it becomes incapable of ever again responding normally to internal stimuli; it contracts at irregular intervals and evacuates small amounts of urine.

If the bladder is catheterized within 24 hours of the spinal injury, fitful evacuation of urine may occur at intervals, but most of the urine has to be drawn off. As recovery occurs, the bladder responds with increasing force to distension with urine so that the pressure rises sharply; but if a constant volume is left in the bladder the pressure rapidly falls again. In the course of time, reflex micturition becomes more perfectly established. The volume of urine present in the bladder before reflex evacuation occurs shows wide individual variations. If the subject takes a deep breath his bladder, which usually empties at a content of e.g. 380 ml, may empty when it contains no more than e.g. 80 ml. Similar results follow when the flexor reflex is elicited or the glans penis is stimulated [p. 363]. Reflex evacuation can often be produced when a catheter is in position, though it may be impossible otherwise. Presumably in such cases the co-ordinated inhibition of the sphincter has not yet been established, and the detrusor muscle is unable to overcome the resistance of the sphincter.

When the functional activity of the isolated region of the spinal cord becomes depressed through toxaemia, reflex bladder activity passes off in the reverse order in which it has returned; the detrusor weakens first (with retention and overflow), and finally sphincter tone is lost, and dribbling results. A similar state of affairs develops in any condition associated with extensive destruction or inflammation of the distal end of the spinal cord.

Effects of obstruction

Obstruction to the outflow of urine from the bladder may be due to anatomical obstacles (e.g. enlarged prostate or urethral stricture) or to functional derangements (e.g. a tonic sphincter with a relatively weak detrusor); retention of urine consequently tends to occur, and the intravesical pressure rises till it causes expulsion of the urine. When the obstruction is mechanical, the stretching of the bladder wall acts at first as a growth stimulus resulting in hypertrophy of its fibres and increase in their expulsive power, which for a time enable the obstacle to be overcome. But when, finally, the bladder wall is overstretched, it becomes paralysed, and normal evacuation no longer takes place. At first, small rhythmical contractions may occur and account for some dribbling, but in the main the escape of urine is due to overflow from the overfilled organ. This is referred to as 'retention with overflow', or 'passive incontinence'.

Regulation of water balance and composition of body fluids by the kidney

WATER DIURESIS

If 1–2 litres of water are drunk, absorption takes place rapidly from the intestine. The passage of water into the blood slightly dilutes the plasma and decreases its crystalloid o.p. by about 3 per cent (i.e. corresponding to a decrease of concentration of about 10 milliosmoles per litre) [FIG. IV.22]; the plasma volume increases slightly. These changes are so small because the excess water is distributed throughout 50 litres of body water, and the kidneys increase the loss of fluid in the urine.

Body fluid changes

The trivial dilution of the plasma proteins decreases their osmotic pressure to a negligible extent (e.g. from 25 to 24 mm Hg); likewise no significant rise occurs in the capillary blood pressure. The flow of water out of the blood vessels is the result of the decreased crystalloid o.p. of the plasma.

The crystalloid o.p. of the interstitial fluid is higher than the reduced o.p. of the plasma; water therefore passes from the plasma into the interstitial spaces. (Electrolytes simultaneously diffuse in the other direction, i.e. from the interstitial fluid into the plasma,

but the rate of movement of the water in the opposite direction is faster and more important. At equilibrium the volume of both plasma and interstitial fluid is increased and both fluids which together comprise the ECF have an identical but lower crystalloid o.p.).

The balance that is normally maintained between intra- and extracellular fluid is disturbed and owing to the fall in crystalloid o.p. of the ECF, fluid enters the cells. Finally the ingested water is distributed equally throughout all the body water and all the body fluids have a slightly lower crystalloid o.p. [FIG. IV.22]. If there were no renal response and if the 2 litres of water ingested were to be stored in the 45 litres of body water the crystalloid o.p. would fall by only 4.4 per cent.

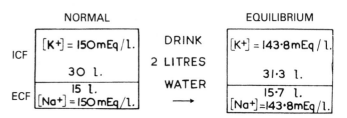

FIG. IV.22. Changes in ICF and ECF produced by drinking 2 litres of water.

Renal changes

A renal response occurs after a latent period of 15–30 minutes [FIG. IV.23]; the flow of urine rises (from the 'resting' value of 50 ml h^{-1}) to its peak usually within an hour and a half when a maximum excretory rate of 10 ml min^{-1} or more may be attained; the diuresis declines and is usually over in 3 hours by which time the excess urinary output has about equalled the excess fluid intake.

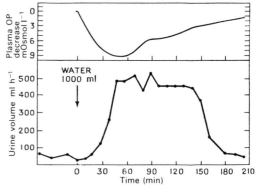

FIG. IV.23. Changes in urine volume and plasma *crystalloid* osmotic pressure after drinking 1000 ml of water in man. Urine volume in ml per hour. Change in plasma crystalloid osmotic pressure in milliosmoles per litre. The maximal fall in plasma crystalloid osmotic pressure is about 10 milliosmoles per litre. The dilution of the plasma precedes by about 15 minutes the onset of diuresis. (After Baldes and Smirk (1934). *J. Physiol., Lond.* **82.**)

As the volume of urine increases, its specific gravity falls, e.g. to 1001; there may be a slight total increased excretion of NaCl and urea during the diuresis although the percentage concentration of these substances is, of course, very low. This initial washing out of solids is compensated for by a lessened rate of excretion after the diuresis is over. The point to emphasize, however, is that the kidney responds selectively by an enormous increase in water output with little associated loss of solids. There are no changes in renal dynamics. There is no increase in renal blood flow; there is no increase in GFR unless the urinary volume exceeds 900 ml h^{-1}; it is quite clear therefore that the diuresis is due to decreased reabsorption of water by the renal tubules. A urinary output of 900 ml h^{-1} (= 15 ml min^{-1})

means that, of every 130 ml of glomerular filtrate although 16 ml, as usual, reach the distal tubule per minute, 15 ml are allowed to escape in the urine (compared with 1 ml as normally).

The level of diuresis is of the same order of magnitude as that seen in severe clinical diabetes insipidus and is likewise due to lack of ADH. The dilution of the plasma crystalloids 'switches off' the hypothalamico-hypophysial mechanism responsible for ADH secretion. It will be noted that the dilution of the plasma precedes the onset of diuresis. During this period it may be supposed that the circulating ADH is being destroyed and that no fresh ADH is being secreted; as the blood ADH level falls diuresis sets in.

Water intoxication

This is uncommon in clinical cases. It was at one time noted in women receiving posterior-pituitary extract for a medical induction of labour (the oxytocin content causes the initiation of uterine contractions). When they were exhorted at the same time to drink water the danger of water intoxication became correspondingly acute, for the ADH content of the posterior-pituitary injections prevented the normal tubular rejection of the excess water as usually occurs in a water diuresis. Headache, nausea, and incoordination of movements are the main features of water intoxication; the changes in the crystalloid o.p. in all the tissues, plus the direct effect of the raised intracranial pressure on the oedematous brain due to the great increase in the volume of all the body fluids, account for the symptoms.

EFFECTS OF EXCESS SALT

Changes in body fluids

If a strong NaCl solution is drunk or injected the plasma crystalloid o.p. rises; water moves from the interstitial spaces into the plasma, initially increasing its volume, but at the same time the salt diffuses out into the interstitial spaces. The outflow of salt causes a secondary outflow of water by osmotic action. Ignoring the intermediate changes just described the net result is a uniform increase in the NaCl content of the extracellular fluid without change in the relative or absolute water content of plasma or interstitial fluid. The raised osmotic pressure of the ECF leads to a flow of water from the cells into the ECF. The final result is an *increase* in *crystalloid concentration* and osmotic pressure throughout the body fluids but there is a *decrease* in the *volume* of the ICF and an increase in the volume of the ECF [FIG. IV.24].

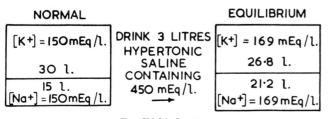

FIG. IV.24. See text.

Renal changes

In an experiment in which 28 g NaCl, i.e. 484 mEq, were ingested the urinary flow was increased from 30 to 120 ml h⁻¹ and the rate of chloride excretion rose from 0.2 to 1.2 g h⁻¹ (the amount filtered per hour is about 24 g). Reabsorption of chloride (and inferentially sodium), though less complete than usual, was still proceeding on a very large (and seemingly unnecessary) scale. The maximal urinary NaCl concentration occurred between the third and twelfth hour when it was about 0.55 molar (= 3.4 per cent NaCl, compared with the normal average of 1 per cent). The rate of excretion of salt is thus comparatively slow.

These points are not simply of academic interest for the survivor of a shipwreck stranded on a raft may be confronted with the problem of whether to drink *sea water* (which is hypertonic) or not. The 'average' salt content of sea water is 450–500 mEq l⁻¹ and the kidney has great difficulty in elaborating a urine containing as much salt as this. If the urine formed over a long period contains less than 450–500 mEq l⁻¹ then the drinking of sea water occasioned by thirst leads inexorably to cellular dehydration and death.

RESULTS OF SIMPLE SODIUM CHLORIDE DEPLETION

Simple sodium chloride deficiency can be produced experimentally in various ways:

1. Salt is excluded as completely as possible from the diet. Severe sweating is induced to cause loss of salt; plain water is given to replace water lost in the sweat.

2. In animal experiments an isotonic glucose solution is introduced into the peritoneal cavity. Glucose diffuses into the blood; Na⁺ and Cl⁻ diffuse out. The fluid is periodically replaced by fresh glucose solution; NaCl is thus withdrawn from the body.

Body fluids

There is a uniform decrease in the Na⁺ and Cl⁻ concentration in the ECF; owing to the fall in its crystalloid o.p. water moves from the interstitial fluid into the cells. Initially therefore the volume of the ECF falls, that of the ICF rises and the cells swell. The decrease in extracellular water necessarily involves a decrease in plasma volume; but as this raises the plasma protein osmotic pressure, plasma volume is relatively better preserved than is the volume of the interstitial fluid. The concentration of Na⁺ and Cl⁻ in the plasma is decreased.

Renal changes

Almost complete reabsorption of Na⁺ and Cl⁻ is accomplished finally by the distal nephron; a process which is stimulated by aldosterone; correspondingly Na⁺ and Cl⁻ may almost disappear from the urine. As the plasma becomes progressively hypotonic, despite Na⁺ and Cl⁻ conservation by the kidney, ADH secretion is suppressed via the osmoreceptors and a moderate polyuria develops. During the first 3–4 days of salt lack the body weight falls owing to water loss. This loss of fluid, however, helps to keep up ECF tonicity which appears to be maintained at the expense of ECF volume. NaCl depletion always leads to a decrease in ECF volume and therefore of plasma volume which may be severe enough to induce circulatory failure. The decrease in plasma volume is accompanied by a decrease in cardiac output. The usual compensatory vasoconstriction occurs and apparently affects the renal vessels, for although arterial pressure is for a time maintained, the renal plasma flow and the glomerular filtration rate are decreased, the latter by as much as 30 per cent. There is a general impairment of tubular activity for several reasons:

1. Decreased renal blood flow.

2. A normal Na⁺ concentration in the *milieu intérieur* is necessary for the correct functioning of all 'excitable' tissues (nerve, skeletal and cardiac muscle) and perhaps of all cells.

3. The swelling of the cells may disturb their metabolism. Renal insufficiency is shown by the incomplete excretion of nitrogenous constituents; the blood urea level may rise from 30 to 80 mg per 100 ml; the urea clearance is decreased to 40–50 per cent of the control value. A contributory factor may be the breakdown of tissue proteins. Water drinking does not induce the usual striking diuresis [FIG. IV.25]. The plasma [Na⁺] may fall finally by as much as 20 per

cent, evidence that 'tonicity' is ultimately sacrificed to preserve volume.

Overventilation in a normal subject causes the excretion of an alkaline urine with the excretion of sodium bicarbonate. When a sodium-depleted subject overbreathes, the urine cannot be made alkaline for the kidney continues to reabsorb and conserve sodium. This inability of the sodium-deficient patient to excrete an alkaline urine may lead to errors in diagnosis, for a patient who has suffered prolonged vomiting or repeated gastric aspirations is alkalotic and also sodium deficient. Sodium deficiency causes him to pass an acid urine and his condition may be mistaken for acidosis. If he is supplied with sodium in the form of neutral salt then his urine becomes alkaline (McCance 1936). As Robinson (1954) has pointed out, alkalosis superimposed upon a deficiency of sodium leads to a conflict between two regulatory functions of the kidney, and the regulation of acid base balance is sacrificed to the conservation of the body's stores of Na^+.

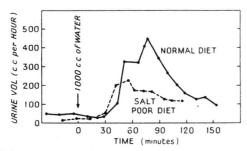

Fig. IV.25. Effect of salt-poor diet on renal response to water drinking. In each case 1000 ml of water were drunk (arrow). On a salt-poor diet the diuretic response is smaller than on a normal diet. (After McCance and Widdowson.)

It should be stressed that Na^+ is a physiologically important ion; Cl^- however is merely an 'accompanying anion', maintaining electrical neutrality and acid–base equilibrium, and has no specific action of its own. Salt deficiency is physiologically equivalent to Na^+ lack.

Clinical syndromes of salt loss

Heat (stoker's) cramp

Men working very hard in hot moist atmospheres (stokers, miners) sweat profusely; if they replace the water lost but not the salt, NaCl deficiency is produced as already explained [Fig. IV.26]. A common symptom is widespread, intense and exceedingly painful cramps of the muscles probably due to the harmful effect of the low Na^+ content of the interstitial fluids. The cramps are relieved by drinking saline (0.5 per cent) or by taking salt tablets. Nowadays such workers are given salt tablets prophylactically.

Grave salt-deficiency syndrome

In severe cases (shown by increasing pallor) compensatory vasoconstriction can no longer make up for the low cardiac output; the

blood pressure falls, causing giddiness, fainting, and cold sweating; the pulse rate rises. There is lassitude, stupor, headache, asthenia as well as cramps, nausea, and vomiting and a 'shock-like' state results with signs of renal failure. The clinical picture is one of general grave illness which may be variously diagnosed. A ready clue to the diagnosis is obtained by demonstrating the absence of Na^+ and Cl^- in the urine, and a low plasma Na^+ level. Final confirmation is obtained by the rapid improvement resulting from the administration of saline solution.

Certain other points should be stressed. The normal fullness (turgor) of the skin and mucous membranes depends on the volume of interstitial water. In salt deprivation the skin feels flabby and mucous membranes are dry; there is a 'putty-like' feel to the muscles. The eyeball tension is low. *There is no sensation of thirst,* which is in sharp contradistinction to water deficiency.

EFFECT OF WATER DEPRIVATION

As water is constantly being lost from the body in urine, expired air, and insensible perspiration, the irreducible minimum water loss (in the absence of sweating) is about 1200 ml daily, representing 2.5 per cent of the total body water. If corresponding amounts of water are not drunk, water depletion develops leading to changes in the body fluids which are the reverse of those described for water drinking.

Body fluids. The extracellular fluid is reduced in volume but its crystalloid concentration and osmotic pressure rise. Water is consequently drawn out of the tissue cells. All the fluid compartments ultimately lose water and gain in crystalloid content and all end up with the same (i.e. slightly more than normal) crystalloid osmotic pressure. Initially the water loss is borne chiefly by the ECF water; later proportionally larger contributions are made by ICF. The metabolism of the shrunken cells is disturbed; there is protein breakdown and loss of K^+. The volume of urine is decreased to a minimum by increased secretion of ADH, the excretion of electrolyte is increased, which though helping to preserve the tonicity of the ECF results in electrolyte depletion. It should be emphasized that in so-called simple water deprivation there is not only obvious water loss but also additional electrolyte loss for the reason just given. These facts should be borne in mind when restorative measures are planned; chronic water loss is really water + electrolyte loss and must be dealt with by giving water + electrolytes.

As the plasma volume falls, the venous and capillary pressures fall while the osmotic pressure of the plasma proteins rises. These two factors tend to keep up plasma volume at the expense of the interstitial fluid. Consequently determinations of plasma volume or of haemoglobin or of plasma protein concentration are not a reliable guide to the degree of water deprivation because the plasma volume changes last and least (thus helping to maintain the cardiac output); its relative constancy masks the larger water changes which are occurring extravascularly. When changes in plasma volume or composition are obvious, the clinical state is grave.

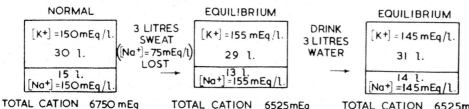

Fig. IV.26. Theoretical changes in body water compartments following sweating and replacement of water lost. (The kidney is assumed to play no part in the reactions so as to simplify the presentation.) (See text.)

Renal changes in man. The effects of complete deprivation of water in man for four days have been studied; the diet was otherwise adequate in all constituents. The renal blood flow remained unchanged: the volume of glomerular filtrate was reduced at most by 20 per cent. There was no decrease in plasma volume and no haemoconcentration, presumably because the interstitial fluid reserves were called upon. The body weight decreased by about 3.5 kg. The volume of urine was reduced to 30–40 ml h^{-1}. At urinary outputs of 30 ml h^{-1} or under, urea, creatinine, phosphate, total nitrogenous and total non-nitrogenous solids become maximally concentrated.

Water deprivation in infants. These effects of water deprivation are especially important in infants owing to the peculiarities of their renal function. *In utero*, about half the work of the kidney is done by the placenta; at birth the kidney is perhaps not fully developed functionally. The urine in infants is hypotonic (not hypertonic as it is normally in adults), i.e. an excessive volume of urine is needed (by adult standards) to eliminate a fixed amount of solid. The glomerular filtration rate is also low. Water deprivation or excess fluid loss rapidly induces renal failure and leads to retention of urea and electrolytes. In illness in children associated with sweating, diarrhoea, or vomiting, the fluid intake should be kept high and the protein intake restricted to minimize formation of nitrogenous waste products.

Problems of thirst

Thirst is not a simple sensation but a 'feeling' or affect. Thus we always say 'I am thirsty' just as we say 'I am hungry, happy, miserable, or tired'. This group, then, represents emotional states, and it may be unwise simply to look for receptors, afferent paths, centres, and the rest of the physiological apparatus that we invoke to account for single sensations. Thirst is of course experienced in water deprivation. An unpleasant feature of thirst is the dryness of the mouth, which persists after anaesthetizing the mouth, though it is at once abolished by increasing the water content of the body by any route. Dryness of the mouth and throat is produced by atropine which arrests salivary secretion; an urgent desire is aroused to moisten the mouth but the condition is probably not identical with the more complex phenomenon of thirst. In simple salt deficiency when, as explained, the cells are swollen (their water content is increased and the volume of the interstitial fluids is depleted), thirst is absent, both in man and animals. Animals given hypertonic saline suffer severe thirst and drink copiously; in this condition the *cells are shrunken* and hypertonic from loss of water, but the interstitial fluid volume is greatly increased. The 'affect' perhaps depends on the state of the body cells as a whole; how the brain is informed of this state is unknown but the altered state of the brain itself may contribute to the feeling of thirst. It is important, however, to recognize that thirst is not a guide to the total water content of the body. Though thirst is marked in simple water deprivation it is absent in the state of dehydration which accompanies salt deficiency.

There is evidence that a 'thirst centre' is located in the hypothalamus [see also p. 345].

1. Andersson and McCann (1956) injected minute amounts of hypertonic saline directly into the dorsal and ventral hypothalamus of unanaesthetized goats. The goats drank large amounts of water (*polydipsia*).

2. Electrical stimulation of an area of hypothalamus reaching anteriorly to the anterior commissure and posteriorly to the bundle of Vicq d'Azyr again caused polydipsia.

3. Electrocoagulation of this area caused permanent hypodipsia; the goats, which would not drink water, nevertheless drank milk.

REFERENCES

Andersson, B. and McCann, S. M. (1956). *Acta physiol. Scand.* **35**, 312.
Gamble, J. L. (1954). *Extracellular fluid.* Cambridge, Mass.
McCance, R. A. (1936). *Lancet* i, 643, 765, 823.
Robinson, J. R. (1954). *Reflections on renal function.* Blackwell Scientific, Oxford.

Kidney function in disease

RENAL FUNCTION TESTS

Urine volume. A normal person produces widely variable amounts of urine depending on a host of factors. These include water intake, solute excretion but also disturbances in renal function. Normally some 1–2 litres are produced per day. Oliguria is a reduction in the volume voided and is pathological if it continues despite variations in salt and water intake. Polyuria is the state where more urine is voided than normal and is significant if persistent. A normal person excretes 70 per cent of a litre of water he drinks rapidly within five hours.

The *specific gravity* of the urine varies between 1001 and 1034 depending on its osmolal concentration. After 12 hours abstinence from water or liquid food a subject should produce a urine having 1000 mOsm l^{-1} and whose specific gravity is about 1025. Failure to do that indicates abnormal renal function. The persistent production of urine which is isosmolar with plasma (300 mOsm l^{-1} = SG 1007) despite variation in water intake, is often spoken of as isosthenuria.

Urinary sediment

When freshly voided urine is centrifuged, for about 5 min at 3000 rev min^{-1}, the formed elements it contains settle to form a sediment. This normally is made up of a few epithelial cells, one or two red cells, and one or two white cells per high-power field of the microscope ($\times 400$). An occasional clean and colourless cast of the tubule is also seen (hyaline cast). These casts are made of high molecular weight mucoprotein produced in the distal tubular epithelium. Casts may entrap cells to become granular casts. The increase in the number of cells or of casts indicates renal destruction of glomeruli or tubular cells.

Proteinuria. Some low molecular weight proteins (e.g. albumin) are normally filtered in the glomerulus. These are usually reabsorbed in the proximal tubule. Albuminuria is diagnosed when a patient excretes more than 150 mg albumin per day. It could be induced by changes in the permeability of the glomerular capillaries or by failure of the reabsorptive process in the tubule. Orthostatic proteinuria is an example of the former mechanism.

Blood levels of urea and of creatinine are useful indices of glomerular filtration. The normal blood level of urea ranges between 3 and 7 mM l^{-1}. The level is nearly doubled when the GFR is halved. The normal plasma creatinine level is between 45 and 110 µmol l^{-1} and is also increased with failing glomerular filtration.

Glomerular filtration rate is best measured as the clearance of inulin, or of a number of other markers such as EDTA, potassium ferrocyanide, and vitamin B$_{12}$. Clinically, however, the clearance of endogenous creatinine is the one that is commonly used. It is determined over 24 hours. The urinary excretion of creatinine [$U_{cr}V$] in 24 hours is measured and related to the creatinine concentration in one plasma sample obtained within this period [P_{cr}]. The normal C_{cr} is 100–140 ml min^{-1} in men and is slightly less for women.

The creatinine clearance is used as an index to evaluate the excretion of other substances. The U/P of a substance is divided by the U/P of creatinine and the ratio is taken to indicate the net excretion of *X*. In more refined analyses renal plasma flow is measured by the clearance of PAH or radioactively labelled hippuran. The transport maxima for glucose and PAH could be estimated and indicate tubular mass. Tubular function is also indicated by the excretion of phenolsulphophthalein (PSP). At least 50 per cent and usually 70 per cent of the injected dye is excreted in two hours.

EFFECTS OF RENAL ISCHAEMIA AND INJURY

The kidney is prone to damage by ischaemia or by noxious agents:

1. If the blood supply is arrested even for a few minutes proteinuria occurs due to an increased glomerular capillary permeability.

Postural proteinuria is due to renal vasoconstriction; it occurs in susceptible subjects when standing, particularly in a hot environment; in these circumstances venous pooling takes place in the legs, venous return is reduced, and cardiac output falls. Compensatory reflex vasoconstriction occurs which affects (among others) the renal vessels and the consequent decrease in renal blood flow may be sufficient to damage the glomerular membranes and thus to cause proteinuria.

2. Complete renal ischaemia for eight hours causes severe tubular damage and it is questionable whether the changes produced are reversible. If the ischaemia is partial, the severity of the effects produced are proportional to its duration. Renal blood flow may be reduced by vasoconstriction in many conditions, e.g. haemorrhage, trauma, dehydration. Tubular injury may result from poisoning (e.g. CCl_4 or Hg^{2+}), from bacterial toxins, from the products released from injured muscle (*crush syndrome*), or from back pressure effects resulting from the formation of acid haematin in the tubules following mismatched blood transfusions or in haemolytic disease. The effects of ischaemia and toxic agents are additive. The lesion may consist of 'massive necrosis' of the proximal tubule or there may be more patchy lesions throughout the nephron. There may indeed be tubulorhexis in which damage to the basement membrane and tubule cells permits fluid to escape from the tubule into the peritubular tissue spaces. Oedema of the kidney occurs which obstructs venous return from the renal parenchyma, thereby aggravating the ischaemia. Oliguria or anuria develops.

Acute renal failure

In cases that recover several stages are recognized: 1. phase of anuria (or oliguria); 2. return of urine flow ('early diuretic phase'); 3. increasing polyuria ('late diuretic phase'); 4. progressive recovery. The changes are due to the reduced GFR, and varying grades of tubular failure.

Phase of anuria

The chances of recovery in patients with reversible tubular lesions depend greatly on how skilfully the case is handled. The following points should be noted:

1. Renal plasma flow. This may be as low as 100 ml min^{-1}; normal 700 ml min^{-1}.

2. Water balance. Though only a small volume is lost in the urine (0–300 ml day^{-1}), the unavoidable water loss in insensible perspiration, sweat (if any), and the expired air continues; some water is produced in the tissues as a result of oxidation of the foodstuffs. In the absence of vomiting or diarrhoea it is desirable that the water intake be limited to 1 litre daily. Anuric patients in the past were given large volumes of fluid by mouth or intravenously in the

mistaken belief that the kidneys in this condition resembled a blocked pipe which could be forced open by raising the pressure. But no degree of overhydration can reopen the renal vessels or the renal tubules; the excess fluid may kill the patient by producing water intoxication or pulmonary and cerebral oedema.

3. Electrolyte balance. As no significance electrolyte loss is occurring no electrolytes need be given; the diet should be electrolyte-free till diuresis sets in. If vomiting occurs, the vomit should be filtered and the ejected fluid and electrolytes returned.

4. Nitrogen balance. On a normal diet containing 100 g of protein, about 16 g of nitrogenous waste (chiefly urea) are formed; in complete starvation about 10 g; and on a protein-free diet providing the full calorific requirements, about 3–5 g are formed. In the last two cases the nitrogenous products are derived from breakdown of tissue protein. During anuria, inevitably the blood nonprotein nitrogen (NPN) including urea, uric acid, and creatinine must rise progressively, producing literally uraemia. The rate of rise is obviously least if a high-calorie protein-free solution is given. This can be achieved by the intravenous infusion of glucose.

Haemodialysis may be required. The patient's radial artery is connected to a cellophane tube immersed in an isotonic solution similar to that of plasma but urea-free. The patient's blood is returned from the system to a peripheral vein. Dialysis removes urea and other non-protein nitrogen constituents.

Diuretic phase

As the circulation is restored, glomerular filtration is resumed and fluid enters the tubules; but, as their damaged epithelium is relatively functionless, the filtrate escapes largely unchanged in the urine. The renal blood flow returns to normal after some months; full tubular function is not restored until much later. Consequently, as the GFR rises the urine flow increases because of inadequate tubular reabsorption of water, and marked polyuria develops.

The effects of this state of affairs can be worked out from TABLE IV.5. In the case of Na$^+$ and Cl$^-$ if the filtration load were reduced to one-fiftieth of normal and no tubular reabsorption occurred, the normal amount of Na$^+$ and Cl$^-$ would still be excreted in the urine. In the case of K$^+$, the filtration load could be reduced to one-tenth with a similar result. If the volume of fluid filtered were reduced to one-hundredth of normal and no tubular reabsorption occurred, 1.7 litres of urine would be passed daily. It is obvious that if the glomerular filtrate volume were larger (say over 10 per cent of normal) and the tubules relatively functionless, a severe loss of water and electrolytes would occur. During the diuretic phase, it is essential to replace the lost electrolytes (Na$^+$, Cl$^-$, K$^+$) as well as water; otherwise the blood level falls, with resulting extracellular fluid depletion in spite of the adequate water intake.

TABLE IV.5

	Na$^+$	Cl$^-$	K$^+$	Urea
Plasma concentration (mg per 100 ml)	330	365	17	30
Total filtered in 170 litres of glomerular filtrate in g (24 hours)	560	620	20	51
Amount normally excreted in urine in g (24 hours)	5	9	2.2	30

The degree of tubular inefficiency can be gauged in several ways:

1. The concentration ratio of creatinine in urine and plasma (U/P) is a measure of water reabsorption. Normally U/P for creatinine is 100/1; in the diuretic phase it may be 1/1 indicating no water reabsorption.

2. The power of tubular transport can be assessed by determining

the Tm_G and Tm_{PAH}, or the PAH extraction rate (E); the last may fall from 90 per cent to 0.40 per cent at various stages in the disease.

The position with the waste products is quite different. Consider a rate of urea formation of say 8 g per day (normally, on an average diet, 30 g per day). Normally, one half the urea filtered is returned to the blood from the tubular fluid; when the tubules are damaged their relative impermeability to urea is lessened and more urea correspondingly returns to the blood. Hence, during the early stage of polyuria, the clearance rate of urea remains low and the blood urea concentration is correspondingly raised. Ultimately however renal function is more fully restored and urea clearance becomes normal.

Syndromes of renal failure

The detailed processes which are carried out by the kidney are numerous; it is unlikely that in disease all these functions will be disturbed simultaneously or to the same degree. Each individual failure contributes to the clinical picture. Renal failure thus presents clinically as a series of syndromes—a combination of symptoms and signs that can be conveniently termed 'the uraemias'. The main constituent features are described below as headings which relate to the failure of different renal functions:

1. Failure to excrete nitrogenous waste products.
2. Acidaemia.
3. Raised serum K^+.
4. Cellular dehydration.

REFERENCE

PITTS, R. F. (1974). *Physiology of the kidney and body fluids*, 3rd edn. Year Book Medical, Chicago.

Effects on body fluids of derangements of the alimentary canal

GENERAL CONSIDERATIONS

In *intestinal obstruction* there is loss of water and electrolytes in varying proportions, thus disturbing both the volume and composition of the body fluids; the fluid which is lost from the body may be neutral, acid, or alkaline in reaction so leading to alkalaemia or acidaemia. If the obstruction is accompanied by changes in the state of the intestinal blood vessels there may be loss of plasma or even of whole blood from these vessels into the lumen of the gut increasing the difficulties of plasma volume regulation. The clinical condition is aggravated by pain, vomiting, and the development of infection.

TABLE IV.6 indicates the daily average volumes of the digestive juices which enter the bowel from the blood and which are normally practically completely reabsorbed into the blood.

TABLE IV.6

Secretion	Volume (ml)	Total NaCl (g)
Saliva	1500	7
Gastric juice	2000–3000	15–30
Bile	500–1000	7
Pancreatic juice	500–800	8
Succus entericus	? 3000	? 18

FIGURE IV.27 sets out the electrolyte composition of the main gastro-intestinal secretions and compares it with that of plasma.

Complete loss of the gastro-intestinal secretions for 24 hours would deprive the body of about 8 litres of water or nearly one-fifth of the total body water, a loss sufficient in itself to produce a grave clinical condition. This water loss would be accompanied by the loss of the electrolytes in the juices, the net result being the loss of an approximately isotonic solution of varying acid–base balance. The total NaCl loss in a day might be 50 g compared with a total NaCl content in all the body fluids of the order of 100 g (and a daily salt intake of say 10 g). Over and above the obvious loss by vomiting unavoidable fluid losses are taking place: in the lungs (400 ml), by insensible perspiration (600 ml), and in the urine (the minimum volume necessary to eliminate the waste products from the body is about 500 ml), or a total of at least 1500 ml. If sweating is taking place the water loss is even greater. If none of the water drunk is retained then the water lost is the volume vomited plus the loss by the other routes just indicated (minus the water resulting from oxidation of the foodstuffs). The loss of the body fluid in intestinal obstruction is not a question of simple dehydration, i.e. simple loss of water; it is a loss of water accompanied by loss of electrolytes.

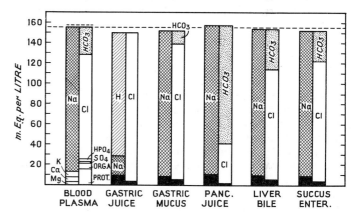

FIG. IV.27. Electrolyte composition of plasma, gastric juice, gastric mucus, pancreatic juice, liver bile, and succus entericus. (After Gamble (1949). *Extracellular fluid*. Cambridge, Mass.)

RESULTS OF LOSS OF GASTRIC SECRETION

The main gastric glands secrete a fluid which, apart from its enzyme content, is practically a 0.1 N solution of HCl; gastric mucus and the secretion of the pyloric region is alkaline, the contained electrolytes being Na^+, Cl^-, and HCO_3^-. The volume of fluid secreted by the main gastric glands normally far exceeds that formed in the pyloric glands. Loss of gastric secretion (by vomiting or aspiration) leads to loss of water and of Cl^-, a relatively smaller loss of Na^+ and a variable loss of H^+ (acid).

Clinical pyloric obstruction

The effects depend on the relative loss of water and electrolytes. In practice the net result clinically is a relatively greater loss of electrolytes than of water. The extracellular fluid thus becomes hypotonic as well as reduced in volume; the plasma shows a marked fall of Cl^-, a less marked fall of Na^+, and an increase of HCO_3^- [FIG. IV.28]. Because the crystalloid osmotic pressure of the extracellular fluid is reduced, fluid flows into the cells, thus further decreasing the extracellular fluid volume. The volume of the plasma is, however, partly protected by the rise of plasma protein osmotic pressure which helps to retain fluid in the blood [note the increase in R in FIG. IV.28] which is due mainly to increased protein concentration.

As the plasma volume decreases, the red-cell count and the haemoglobin concentration correspondingly rise. The kidney responds as expected by a reduction in the volume of urine to the minimal value, by the maximal degree of reabsorption of electrolytes and by the secretion of an alkaline urine. Ultimately the kidney fails because of impaired blood supply and the harmful effects of the electrolyte imbalance (chiefly lack of Na^+).

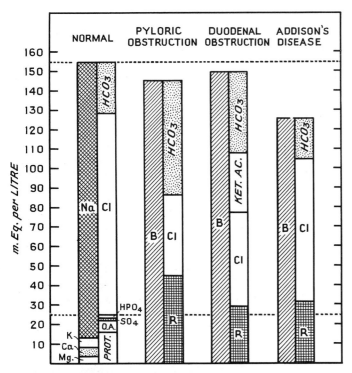

FIG. IV.28. Electrolyte composition of normal plasma and the plasma in pyloric obstruction. O.A. = organic acids; B = total base; R = combined concentration of organic acid, phosphate, sulphate, and protein. Note increase of R. (After Gamble (1949). *Extracellular fluid*. Cambridge, Mass.)

Gastric aspiration plus water replacement

FIGURE IV.29 illustrates an experiment performed in man. Continuous aspiration of the gastric juice was carried out for several days, water and energy requirements being made good by drinking water and a glucose solution. The loss is thus of electrolytes only: of Cl^-, relatively less Na^+, and of H^+. The extracellular fluid became hypotonic (though unchanged in volume) and water doubtless flowed into the cells, increasing their volume. The plasma showed the expected decrease in Cl^- and an increase in HCO_3^- (the plasma bicarbonate rose from 65 to 140 ml of CO_2 per 100 ml); there was no haemoconcentration, showing that the water intake was adequate. The kidney showed the characteristic failure which results; although the urine was alkaline (as would be expected in alkalaemia) the plasma bicarbonate continued to rise from Na^+ depletion: the volume of urine was reduced to only 500 ml daily in spite of a plentiful water intake. Nitrogenous excretion was inadequate (e.g. the total urea output was decreased) leading to nitrogenous retention (azotaemia): the blood urea rose from 30 mg to 130 mg per 100 ml. On administration of saline the plasma Cl^- rose to normal, replacing HCO_3^-; the plasma bicarbonate fell and plasma Na^+ doubtless rose. Renal activity was restored and the retained nitrogenous constituents were soon eliminated, blood urea returning to normal.

Similarly, if a salt-free solution (e.g. a glucose solution) is in-

jected in clinical cases of pyloric obstruction in sufficient volume to make good the water loss, the plasma volume can be fully restored and the anhydraemia relieved, but the alkalaemia and (to a smaller extent) the renal failure are unaffected.

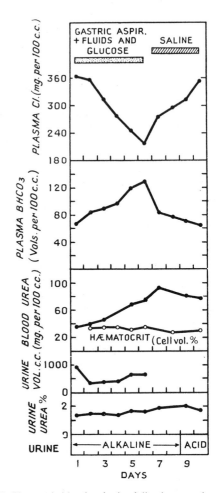

FIG. IV.29. Changes in blood and urine following experimental aspiration of gastric contents. Adult in clinically normal state (apart from slight alkalaemia due to alkali therapy). Aspirate 1½–2 litres of gastric contents daily for five days; salt intake restricted; fluid and glucose given freely. Note fall in plasma Cl^-, rise in blood urea and plasma bicarbonate, and decreased urine volume. On sixth day give saline injections: blood composition rapidly restored to normal. (Nicol (1940). *Q. Jl Med.* **9**, 98.)

Acid urine in alkalaemia from salt loss due to vomiting

In some patients who have suffered severe salt loss as a result of prolonged vomiting, an acid urine is passed in spite of the alkalaemia as explained on page 229.

If NaCl is given to such cases, a highly alkaline urine containing large amounts of HCO_3^- and HPO_4^{2-} is passed and the blood reaction rapidly returns to normal.

SIMPLE HIGH SMALL-INTESTINE OBSTRUCTION

With simple obstruction high up in the small intestine, absorption of saliva, gastric juice, bile, and pancreatic juice (and ingested food and water) is interfered with; these fluids accumulate above the block and distend the bowel, setting up powerful, colicky peristaltic movements and profuse vomiting. As is evident from the data in TABLE IV.6, in a very short time an enormous loss of water may occur which is probably the main factor responsible for death. It should be borne in mind that the usual additional fluid is being lost from the body in the urine, from the skin (insensible perspiration 600 ml and sweat), in the breath (400 ml), and in the faeces (if any are passed).

The fluid lost is approximately isosmotic and neutral; as water is being lost at the same time by the uncontrollable routes there may be relatively more loss of water than of salt; on the other hand, if some ingested water is being retained the loss of salt may be the greater.

The intestinal distension referred to above aggravates the condition in various ways: 1. intestinal secretion is stimulated; 2. vomiting is worsened; 3. there is interference with the circulation in the gut wall leading to venous and capillary engorgement. The rise of capillary pressure and the altered permeability of the capillary endothelium leads to the escape of large quantities of protein-rich fluid into the gut wall. When half the length of the small intestine is involved a volume of fluid equal to half the plasma volume may be lost in this way. All the factors enumerated above intensify the anhydraemia. Signs of dehydration make their appearance when the net fluid loss is equal to 6 per cent of the body weight, i.e. to about 4 litres; it is obvious that such a loss can occur with startling rapidity.

The swollen, overstretched intestinal wall loses tone and contractility, and complete paralysis develops above the level of the obstruction; haemorrhages may occur in the gut wall, patches of necrosis develop, and finally perforation may take place. At any stage infection may complicate the issue. In advanced cases respiration is depressed owing to the alkalaemia and periods of apnoea may occur; there may be signs of tetany. The plasma volume is as usual initially well maintained because the fluid loss is in large measure borne by the interstitial space; but as the anhydraemia becomes worse the plasma volume is substantially decreased, and circulatory failure occurs, producing anoxia and cyanosis and aggravating the renal failure.

LOW SMALL-INTESTINE OBSTRUCTION

Symptoms develop more gradually and are less severe as absorption of water and solutes can take place more satisfactorily above the level of the block. On the other hand, as intestinal distension becomes more marked, local circulatory disturbances develop; loss of fluid into the gut wall itself then becomes a serious factor, intensifying the anhydraemia.

PRINCIPLES OF TREATMENT OF WATER AND ELECTROLYTE LACK

The essential principle is to determine accurately the nature of the disturbance and to take necessary measures to correct it promptly. In clinical conditions in which varying degrees of depletion of water, Na^+ and pH changes have produced a complex alteration of the electrolyte pattern and the fluid volume and distribution, it is found that as soon as enough water and salt (sodium chloride) have been provided to restore renal activity the kidneys can be relied upon to restore the electrolyte pattern of the body fluids accurately to normal by their own specialized activities.

Simple water loss

When practicable give plenty of water to drink; in addition inject 5 per cent glucose solution intravenously to produce rapid effects: the sugar is metabolized (yielding energy) leaving the water of the solution to make good the water deficiency. Salt (or weak saline) should also be given to make good the associated salt lack.

Water and salt loss

Saline (0.9 per cent) is given by all convenient routes until Cl^- appears in the urine, indicating that the Cl^- content of the extracellular fluid is restored to normal. Water or a glucose solution can then be given to make good any remaining shortage of body water.

Simple intestinal obstruction

Saline (0.9 per cent) must be given on the lines indicated above. Tetany can be relieved by injection of 10 ml of 10 per cent calcium gluconate. Great relief is obtained clinically by continuous aspiration of the gastric and duodenal contents to relieve the vomiting and the intestinal distension. However, this continuous aspiration, e.g. for six days, may withdraw over 90 g of sodium chloride from the body and a very large volume of fluid. The water and salt loss so produced is just as real as when it results from vomiting, though less spectacular and less distressing.

A salt-lactate solution (containing NaCl and Na lactate) is of value in states of low plasma bicarbonate; the addition of KCl may be useful as in Butler's solution which contains the following constituents in mEq per litre: Na^+, 30; K^+, 15; Cl^-, 22; lactate$^-$, 20; also HPO_4^{2-}, 3; all in 5 per cent glucose. The lactate ion is metabolized, freeing the Na^+ which re-form $NaHCO_3$ (as occurs during the recovery phase after severe muscular exercise).

Isotonic (1.5 per cent) sodium bicarbonate solution is of value as an emergency measure in dangerous degrees of acidosis due to $NaHCO_3$ depletion until salt or salt-lactate solutions can be given.

OBSTRUCTION ASSOCIATED WITH INTERFERENCE WITH LOCAL BLOOD SUPPLY

Usually the veins are obstructed. Capillary engorgement becomes intense; the peritoneum is filled with an exudate resembling plasma; the gut lumen is filled with a thick red exudate. Organisms proliferate rapidly in the contents of the bowel and in its wall liberating toxins which are probably the cause of death. The clinical condition is aggravated by serious loss of plasma or of blood into the bowel, and the presence of infection. After a time the fluid in the bowel becomes darker and finally black and exceedingly toxic: if injected intraperitoneally into another animal it proves rapidly fatal.

If the terminal part of the ileum is anastomosed to the rectum and the intervening isolated portion of the colon is closed, no evidence of intoxication appears. The lower end of the closed colon is later found filled with a sausage like mass having the appearance and consistency of normal faeces. No toxins are produced and no symptoms develop.

PART V
Muscle, and the nervous system

Skeletal muscle

MICROSCOPIC STRUCTURE

Skeletal muscle consists of long cylindrical muscle fibres which vary in length from 1–40 mm and in thickness from 50–100 μm or more. The muscle fibres are enclosed in a structureless membrane called the sarcolemma. Each muscle fibre contains many myofibrils (1–2 μm diameter) which lie parallel to one another among sarcoplasm, both being enclosed by the sarcolemma. The myofibrils themselves are striated and are aligned within the sarcolemma so that corresponding points of their striation pattern lie at the same level giving the appearance of discs crossing the whole thickness of the muscle fibre.

When viewed under magnification with ordinary transmitted light the striations of the muscle fibre are not seen unless the condenser is stopped down, upon which they become visible, presenting an appearance which depends on the focusing of the microscope. Thus, if the microscope is focused exactly on a thin muscle fibre the striations are barely visible; if the microscope barrel is then raised, the highly refractive regions become bright, but if the barrel is lowered beyond the exact focus the highly refractive bands become dark. It is essential to understand these points, for the term 'dark' and 'light' bands are meaningless unless the focusing position of the microscope is specified. Using the deep focusing position, the muscle fibre is seen to show alternate dark and light crossbands. The dark band contains highly refractile material which is also *birefringent*. (Birefringence means that the refractive index of the tissue, when the electrical oscillations of the light used to measure it are parallel to the axis, is different from that measured when they are perpendicular.) Another term for birefringent is *anisotropic*, so it is this property which has led to this band being called the anisotropic or A band. The alternate light bands are isotropic, and are called the I bands [Fig. V.1]. In the centre of each A band is found a slightly less refractile region named the H band (after Hensen who discovered it; some authorities believe that H is derived from the German word *hell* for light). Lastly, in the

centre of the I band is found a narrow line of highly refractile material which therefore looks dark with this method of focusing and which is called the Z line (from the German, *Zwischenscheibe* = between-disc).

The unit of muscle has long been regarded as the substance included between two Z lines—this unit is called a *sarcomere*.

Great advances have been made in our knowledge of the functional units of muscle by the application of: 1. interference microscopy; 2. electron microscopy and X-ray studies of prepared muscle fibrils.

Interference microscopy

Briefly, the interference microscope employs the principle that varying thickness and refractive index of an object cause varying degrees of retardation in the phase of the light transmitted through them. The light transmitted through the object is allowed to interfere with a reference beam, which has followed a parallel path to the specimen beam but is displaced sideways from it, so that it passes only through the fluid surrounding the specimen. The two beams are coherent. The resultant intensity depends on the phase difference between the two beams, i.e. the extra retardation suffered in passing through the specimen instead of through the same thickness of fluid alone. Hence the intensity of the image at each point is determined by the retardation of the light along its path through the corresponding points in the object.

When a living isolated frog-muscle fibre is photographed with an interference microscope the A band is seen to remain of constant width whether the fibre is stretched or electrically stimulated to contract. When a negative electric pulse is conducted, via a fluid-filled micropipette of 2 μm diameter, to the membrane of a muscle fibre, the potential difference across the membrane in contact with the pipette tip is reduced while the rest of the membrane is unaffected. Only when the tip of the pipette is placed opposite a Z line does any local contraction of the muscle occur. The A bands on each side of the affected Z line themselves show no contraction but they come closer together due to narrowing of the I bands. These observations by A. F. Huxley strongly suggest that the reduction of the membrane potential which is a necessary preparatory stage for muscle fibre contraction, can be achieved only by an inward conduction of activation along the Z line.

By splitting muscle in a suitable medium, separate myofibrils (1–2 μm in diameter) can be obtained which reveal the same striations as the parent fibre which is itself some 50–100 μm in diameter. Such myofibrils cannot be made to contract by electrical stimulation, for the muscle membrane has been destroyed, but shortening of the myofibril can be induced by adding ATP to the medium in which they are suspended. Alternatively the myofibril can be stretched. In these various conditions the width of the A band remains constant and only the I band changes in length. However the H band is noted to change in width *pari passu* with the degree of separation of the Z lines; when the Z lines become more

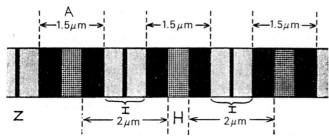

FIG. V.1. Diagram to show the dimensions that remain constant as a fibre is stretched. Microscopic appearance of muscle sarcomeres. Refractile material is dark. (Huxley, A. F. (1956). *Br. med. Bull.* **12**, 167.)

closely approximated the H line is narrow and conversely the H band becomes wider as the inter-Z distance is increased.

Interference microscopy has been used to measure the relative amounts of protein present in the A and I bands of the myofibril. It is known that muscle fibres and myofibrils contain two main proteins, *myosin* and *actin*. Myosin can be dissolved out of the myofibril by pyrophosphate solution; after such treatment a 'ghost' fibril is left consisting of material stretching from the Z lines to the position of the edge of the former H line; the A band has disappeared. The myosin-free fibre can then be further treated with potassium iodide solution which extracts actin. When observed in the phase contrast microscope the 'ghost' fibrils are now seen to have lost all their visible material between the Z lines and the H zones, leaving behind the Z lines themselves, still joined together by some backbone material which provides continuity of structure. Hence it would seem that myosin is confined to the A band whereas actin filaments stretch from the Z line through the I band into the A band, terminating at the H zone.

Electron microscopy

These conclusions have been confirmed by electron microscopy. The structural proteins are arranged in longitudinal filaments spaced out a few tens of nanometres apart. In the A band lie the myosin filaments 10–11 nm thick and about 45 nm apart, extending from one end of the A band to the other. The actin filaments are thinner 4–5 nm in diameter) and stretch from the Z line to the edge of the H zone. In the A band, each of the myosin filaments (which are themselves arranged hexagonally) is surrounded by six actin filaments; it shares these with its six nearest neighbour myosin filaments, each of which also has a total complement of six actin filaments. These observations suggest that striped muscle is built up of two overlapping sets of filaments [Fig. V.2].

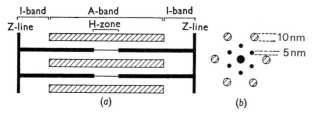

FIG. V.2. Diagrammatic representation of arrangement of filaments in striated muscle: (a) longitudinal, (b) sectional. (Huxley, H. E. (1956). *Endeavour* **15**, 177.)

Electron microscope pictures of muscle fibrils either at resting length, contracted, or stretched confirm the evidence that:

1. The A band remains the same width.
2. The H zone changes in width according to the alteration of the distance between the Z lines.

Hence the I filaments (actin) remain at constant length as do the A filaments and the change in length of the sarcomere during contraction or during applied stretch is achieved by the two filaments sliding over each other. Further work has shown that there are cross-bridges between the filaments, each actin filament being connected to the myosin at intervals of about 40 nm. These cross-links are the structural expression of the biochemical interaction between actin and myosin; as ATP can affect that interaction it may enable the two types of filament to slide past each other.

Protein constituents of muscle

Muscle contains 80 per cent water, but the great majority of its dry weight consists of protein. The muscle proteins, actin, myosin, tropomyosin, and troponin, associated with the contractile

mechanism comprise about 60 per cent of this and the remaining two-fifths of muscle protein is shared by protein enzymes (as in other cells) and by stroma proteins, which serve as an inert scaffolding, holding the remaining structures in place.

Myosin can be extracted, together with actin, by soaking minced muscle in salt solutions of ionic strength 0.4–0.5 M. Its separation in a fairly pure and uncontaminated form is best achieved by special solutions, such as Hasselbalch–Schneider solution which consists of 0.47 M potassium phosphate, 0.1 M potassium phosphate buffer pH 6.5, and 0.01 M sodium pyrophosphate.

Myosin has a characteristically-shaped molecule—like a golf-club—with a short compact head and a long shaft. The length of the molecule is 150 nm and its diameter is 2–4 nm. Brief digestion of the molecules by trypsin solutions yields two fragments named heavy meromyosin (HMM) and light meromyosin. HMM has a molecular weight of 350 000 and LMM 150 000. Heavy meromyosin (30 nm long) molecules have a large globular head (4 nm diameter) with a short tail, whereas light meromyosin is a simple linear strand 2nm in diameter and 100 nm long. LMM seems to possess only structural function—it is composed of two α-helices, like a two-stranded rope. HMM, however, has two important actions exerted, possibly at one, but probably two, chemically active sites in its globular head. At one of these binding to actin occurs and at the other, ATP is hydrolysed. As Wilkie (1968) puts it lucidly, 'heavy meromyosin appears to be the very kernel of the contractile machine, for within this small volume (about 4 nm diameter by 30 nm long) there are the means for producing breakdown of the chemical fuel (ATP-ase site) and also for producing a mechanical effect (binding site)'.

Individual myosin molecules do not join to each other when immersed in solutions of high ionic strength, but if the solution has an ionic strength of 0.2 or so, aggregation of the myosin molecules occurs to form rods or filaments, visible even by light microscopy. The centre of the filament is composed only of 'tails'. The remainder of the filament shows 'heads' projecting from the surface. Myosin filaments grow by the molecules arranging themselves, *in one of two directions depending on which end of the filament a given molecule is joining*.

Actin is structurally attached to the Z membrane. It can be dissolved by 0.6 M potassium iodide and can then be purified. It can exist in two forms:

1. G-actin (globular actin) has a spherical molecule of 5.5 nm diameter; its molecular weight is 60 000. Each molecule binds one molecule of ATP firmly.
2. F-actin (fibrous actin) represents polymerization of G-actin molecules to form a two-stranded helical chain and these bead-like double helices (5.5 nm in diameter) are found in the living thin filaments. Actin can bind to myosin to form actomyosin.

Cross-bridges and the sliding filament hypothesis

The head of the myosin molecule has the actin-binding properties and the enzymatic properties which the cross-bridges are assumed to possess.

The myosin molecules aggregate with their heads pointed in one direction along half of the filament and in the opposite direction along the other half [Fig. V.3]. The heads serve as the cross-bridge and the inherent directionality of the myosin aggregates is a crucial feature of the sliding filament hypothesis. In the muscle fibril the thin fibres move towards each other in the centre of the A bands so that it is required that all the elements of force generated by the cross-bridges in one half of the A band be oriented in the same direction and that the direction of the force be reversed in the other half. The direction of the force developed can be explained by the arrangement of the myosin molecules themselves—pointing in the same direction in half of each thick filament and in the opposite direction in the other half.

When F-actin chains are treated with heavy meromyosin, electron microscope pictures show that the resulting compound has an arrow head pattern. The arrows always point in the same direction over the length of a given filament—seemingly the underlying structure of the actin imposes the pattern. All the actin molecules in a given thin filament are oriented in the same sense, all can interact in an identical manner with the given myosin cross-bridge.

FIG. V.3. The molecular structure of myosin makes it aggregate as shown. The head of the molecule is shown as a 'club-head' and the tail as a straight line. Tails join in the centre; heads extend as projections at the ends, oppositely pointed at each end.

From such studies, H. E. Huxley proposed a schematic model of the actin–myosin arrangement [FIG. V.4].

According to the sliding filament theory, proposed simultaneously by A. F. Huxley and H. E. Huxley (1964), the force of contraction is developed by the cross-bridges in the 'overlap' region between actin and myosin (in the A band) and active shortening is caused by movement of the cross-bridges, causing one filament to slide over the other. The cross-bridges, however, are about 45 nm apart, which is only 5 per cent of the length of the half-sarcomere. Skeletal muscle, however, can shorten by some 30 per cent so each individual cross-bridge must detach itself from one site on the actin and reattach itself to another site further along, repeating the process five or six times, 'with an action similar to a man pulling in a rope hand over hand' (Wilkie 1968).

BRIDGE-FREE REGION

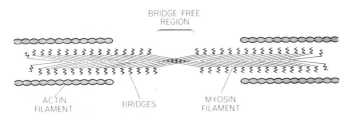

ACTIN FILAMENT   BRIDGES   MYOSIN FILAMENT

FIG. V.4. Contact of actin and myosin in muscle might be made in the manner schematically illustrated here. The thin actin filaments at top and bottom are so shaped that certain sites are closest to thick myosin filament in the middle. The heads of individual myosin molecules (*zigzag lines*) extend as cross-bridges to the actin filament at these close sites. (Huxley, H. E. (1965). *Scient. Am.* **213**, 18. Copyright by Scientific American, Inc. All rights reserved.)

One of the earliest predictions from the sliding filament theory was that when the muscle was stretched, the overlap between the myosin and actin filaments would diminish and the tension developed by the muscle when it was caused to contract would likewise decrease. In order to determine that such behaviour—which is known to be a feature of intact muscle with its thousands of myofibrils—could indeed be attributed to actin/myosin overlap in the individual sarcomeres very special experimental techniques were required. These have been provided by A. F. Huxley and his colleagues.

Gordon, Huxley, and Julian (1966) first developed a technique which enabled the study of a single isolated muscle fibre (freed from connective tissue) undergoing contractions while a selected part of its total length was held constant. The selected part was defined by two 'markers'—tiny pieces of gold leaf stuck to the fibre with tap

grease. These were placed at the extremities of a segment of the fibre which showed uniform spacing under microscopic examination. The position of these two markers was followed continuously by a photo-electric spot-follower device from which was derived an electric signal proportional to the distance between them. The length of the whole fibre was controlled by a high-speed motor connected by a lever to the tendon at one end of the fibre and the segment of the fibre between the markers was held at constant length by driving this motor from the length signal as a negative feedback arrangement. Tension was measured using an RCA-5734 (movable anode) transducer connected to the tendon at the opposite end from the motor.

Gordon, Huxley, and Julian (1966) found that only small amounts of tension were developed by fibres in which the sarcomere length was greater than 3.65 μm. As the sarcomere length was increased progressively by stretching from 'slack' (2.05 μm) to 3.65 μm (where overlap ceases) the developed isometric tension remained constant up to 2.2 μm and then fell linearly to zero at 3.65 μm [FIG. V.5]. Such behaviour conforms exactly to the change in the number of 'bridges' (H. E. Huxley 1963) on the thick filaments which are overlapped by thin filaments, since, although the bridges are distributed uniformly along most of the length of each thick

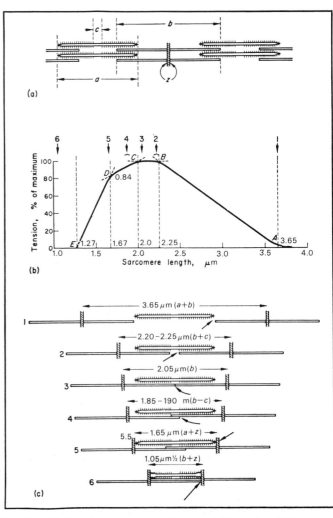

FIG. V.5 (a). Standard filament lengths. $a = 1.60$ μm; $b = 2.05$ μm; $c = 0.15$–2 μm; $z = 0.05$ μm. (b) Tension-length curve from part of a single muscle fibre (schematic summary of results). The arrows along the top show the various critical stages of overlap that are portrayed in (c). (c) Critical stages in the increase of overlap between thick and thin filaments as a sarcomere shortens. (Gordon, A. M., Huxley, A. F., and Julian, F. J. (1966). *J. Physiol., Lond.* **184**, 170.)

filament there is a gap of 0.2 µm at the middle of each thick filament where no bridges are seen. Hence, all the bridges are overlapped at a sarcomere length equal to the length of this gap plus the length of a thin filament (2.05 µm) or 2.20–2.25 µm. Shortening below this length cannot increase the number of bridges overlapping, so tension developed does not change. In summary, isometric tension was found to be closely proportional to the number of bridges on each thick filament which were overlapped by a thin filament, over the whole range from the slack length up to lengths where overlap no longer occurred.

Studying isometric tension development on stimulation of the muscle fibre when it was below 'slack length' at rest, Huxley *et al.* found that the isometric tension produced decreased as the sarcomeric resting length was reduced, and reached zero at a sarcomere length of 1.3 µm. It appears that extensive overlap of the filaments interferes with the formation of cross-bridges (Wilkie 1968).

Excitation–contraction coupling

Muscle contracts normally in response to a stimulus, natural or artificial, delivered down its motor nerve [p. 255]. At the neuromuscular junction acetylcholine is released, which effectively depolarizes the membrane (sarcolemma) by increasing its permeability to sodium, and this sets off an action potential which is propagated along the muscle fibre. The action potential is followed by a wave of contraction. *The link between excitation and contraction in skeletal muscle is provided by a highly specialized system of internal conduction within the muscle fibre.*

Transverse tubules (the T system) pass inwards along the Z line (in the frog) [FIG. V.6] or along the A-I boundary (in the mammal) from openings in the sarcolemma. They are inwardly directed extensions of the sarcolemma and their lumina (30 nm wide) are thus in continuity with the interstitial fluid (or extracellular fluid) which surrounds the muscle fibrils. Such continuity provides a simple means of conveying an electrical signal derived from the depolarization of the cell membrane.

H. E. Huxley demonstrated that the whole of the *transverse tubular system* was continuous with the extracellular spaces, by immersing live frog sartorius muscle for an hour at room temperature in a solution containing the protein ferritin. The muscle was then fixed in glutaraldehyde, post-fixed in osmium tetroxide and after embedding in Araldite, examined under the electron microscope. The transverse tubules contained large numbers of ferritin molecules (recognizable by their dense iron-containing core), as did the extracellular space of the myofibrils. No ferritin molecules were found elsewhere in the muscle fibres. The uptake of ferritin by the muscle was freely reversible for, if after an hour of immersion of a fresh muscle in ferritin solution, the muscle was soaked in Ringer for a further hour, subsequent fixation and sectioning yielded specimens which contained little if any ferritin in the T system.

In addition to the T system, muscle fibrils possess a well marked longitudinal sarcoplasmic reticulum which runs parallel to the fibrils in the intrafibrillar spaces and which gives off numerous interconnecting side branches.

At certain parts of the fibril (H band and Z line) these tubules fuse to form a sac that girdles each fibril. Electron microscopic studies reveal that the Z line in frog muscle is the site of contiguity of three elements and these are called *triads*. FIGURE V.6 shows details of a triad in a frog myofibril cut longitudinally. The central part is the transverse tubule (T system) cross-section. Longitudinally on either side are found the vesicles or sacs of the dilated longitudinal sarcoplasmic reticulum. These sacs are named terminal cisternae. They are often accompanied by deposits of glycogen granules.

The close proximity of the terminal cisternae of the longitudinal reticulum and the transverse tubular system would at first sight

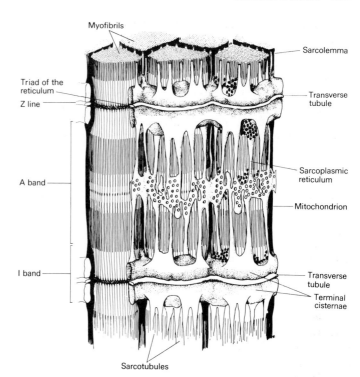

FIG. V.6. Schematic representation of the distribution of the sarcoplasmic reticulum around the myofibrils of skeletal muscle. The longitudinal sarcotubules are confluent with transverse elements called the terminal cisternae. A slender transverse tubule (T tubule) extending inward from the sarcolemma is flanked by two terminal cisternae to form the so-called triads of the reticulum. The location of these with respect to the cross-banded pattern of the myofibrils varies from species to species. In frog muscle, depicted here, the triads are at the Z line. In mammalian muscle there are two to each sarcomere, *located at the A-I junctions.*

(Modified after L. Peachey, from Fawcett, D. W. and McNutt, S. (1965). *J. Cell Biol.* **25**, 209. Drawn by Sylvia Colard Keene.)

suggest that there was continuity between the structures. Such is not the case—the soaking of muscle in ferritin solutions yields preparations which, though showing abundant evidence of ferritin in the T system, evince no sign of the protein in the terminal cisternae. Contiguity, not continuity is as clear a description in this case as in that of the synapse or neuromuscular junction. Nevertheless, the close association of the two systems in the triad structure is most extensive in muscles which contract and relax very rapidly; in such muscles the triads are found at the junctions of the A and I bands and hence there are two triads per sarcomere. Such morphological features are suggestive of there being some connection between the site of the triad and the triggering of contraction of the myofibrils.

It appears that the electrical excitation process, arriving by way of the T system depolarizes the cisternal membranes and triggers a sudden release of calcium ions within the myofibril from the sarcoplasmic reticulum.

It was previously thought that the Ca^{2+} released acted simply as a co-factor for the myosin ATPase sites which were actin and Ca^{2+} activated. However, the discovery of the two proteins tropomyosin and troponin which are found in association with actin has modified this viewpoint. The arrangement of actin, tropomyosin, and troponin is shown in FIGURE V.7. F-actin is a double stranded helix. Tropomyosin (molecular weight 64 000) is also a coiled strand (two α-helices), which fits 'in the groove' of actin. Troponin is bound to tropomyosin (not to actin) and is distributed along the thin filament at intervals of 40–41 nm forming reactive sites which influence the contractile process [FIG. V.7]. When the local calcium concentration is low (10^{-7} M) as in muscle at rest, the troponin molecules inhibit the neighbouring actin molecules and prevent activation of

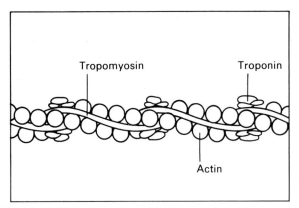

FIG V.7. The helical arrangement of the regulatory proteins tropomyosin and troponin on the thin filament of actin. Tropomyosin exists as filaments of which one is present in each of the two grooves of the actin molecule. (After Ebashi *et al.* (1969).)

the myosin ATPase sites. When the sarcoplasmic [Ca²⁺] rises, the troponin molecules avidly take up the extra calcium ions—two or three Ca²⁺ per molecule of troponin. This calcium uptake results in a conformational change and in a reduction of the inhibitory effect of troponin on actin [FIG. V.8]. The myosin ATPase sites are then 'freed' to split ATP and the actin molecules are rendered available for attachment to the HMM projections of the thick myosin filaments.

In the resting muscle, calcium is sequestered in the sarcoplasmic reticulum, which surrounds the myofibrils and correspondingly the interstitial fluid, which immediately bathes them, has a low [Ca²⁺]. The sudden rise of [Ca²⁺] leads to their diffusion into the myofibrils and initiates the contraction process as described above. The protein aequorin [p. 106] which emits light when activated by Ca²⁺ can be injected into the sarcoplasm (Taylor *et al.* 1975). Excitation of the contractile process causes a brief luminescence, which precedes the tension developed by the contraction.

The calcium ions are then recaptured by active pumping on the part of the membranes of the sarcoplasmic reticulum and the fibres relax. Meanwhile the ADP is regenerated to ATP. The action potential lasts only 1 ms, while the muscle twitch may be ten to hundred times as long in duration. Hence, the contractile system can be reactivated long before the contraction has begun to subside,

and the muscle can be effectively tetanized by appropriately frequent stimuli.

The ability of the sarcoplasmic reticulum to sequester Ca²⁺ has been proved directly. A muscle fraction comprising mainly sarcoplasmic reticulum has been proved to be capable of 'picking up' sufficient Ca²⁺ to lower the concentration of the medium surrounding the muscle to 10^{-7} mol l⁻¹. At such a concentration of Ca²⁺ actomyosin preparations *relax* in the presence of ATP. The sequestration of Ca²⁺ by the SR in resting muscle helps to account for the relaxed state of the myofibrils. When the [Ca²⁺] *rises* the effect of calcium is to trigger the hydrolysis of ATP by myosin ATPase.

Isotonic and isometric contraction

Muscle may be depicted [FIG. V.9] as consisting of a contractile component (CC) in parallel with an elastic component (PEC) and in series with another elastic component (SEC). PEC represents the elasticity of the structural elements other than the contractile proteins—connective tissue sheaths and such proteins of the muscle as provide only scaffolding. The tendons feature as the series elastic component.

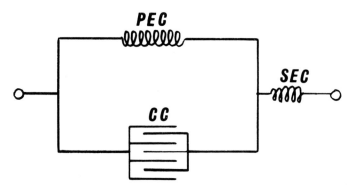

FIG. V.9. Equivalent mechanical components in muscle:
CC = contractile component.
PEC = parallel elastic component.
SEC = series elastic component.

If a muscle is suspended vertically from an attachment to a spring it can be stretched by attaching weights to the lower end and the passive tension developed can be measured. The passive or resting tension-length relationship is not linear—muscle does not obey Hooke's law. The resistance to stretch resides mainly in the series elastic component but also to some extent in the parallel elastic component. The myofibrils themselves (CC) act only as a passive viscous element and have very little resistance to stretch. FIGURE V.11 shows the nature of the resting tension/length curve (P).

On stimulation, the contractile component is transformed into an active structure which converts chemical energy into mechanical work. The characteristics of PEC and SEC are unchanged during contraction. When the contracting muscle is allowed to shorten as a whole, its speed of contraction depends on the force which opposes its shortening. This force can obviously be equated with the tension that it has to develop—for instance, if one attempts to pick up a motor car, the speed of contraction measured is nil, because one cannot lift it. On the other hand, maximal velocity registered by a commissionaire in picking off a pound note tip may be impressive. The force–velocity curve of muscle is shown in FIGURE V.10. The curve fits an equation (Hill):

$$V = \frac{(P_0 + P)b}{(P + a)}$$

where *V* is the initial speed of shortening, *P* is the force acting on the muscle, P_0 is the maximal tension the muscle can develop, *a* is a

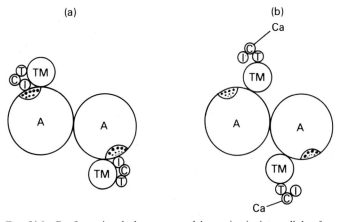

FIG. V.8. Configurational change caused by a rise in intracellular free calcium ions. (a) Before activation the actin site capable of binding myosin (shown by the shaded area) is covered by the troponin–tropomyosin complex. (b) When intracellular [Ca²⁺] rises above a critical level, calcium is bound by TnC and this causes a conformational change which uncovers the binding site for myosin. Myosin can now interact with actin. T, C, and I are subunits of troponin; A = actin; TM = tropomyosin; Ca = calcium ions. (After Potter, D. and Gergely, J. (1974). *Biochem. J.* **13**, 2697.)

constant with the dimensions of a force, and *b* is a constant with the dimensions of a velocity.

When the load is such that no external shortening can occur, the velocity is termed V_0 and the maximum tension is P_0. On the other hand, when the load is negligible the muscle contracts at maximal speed (V_{max}).

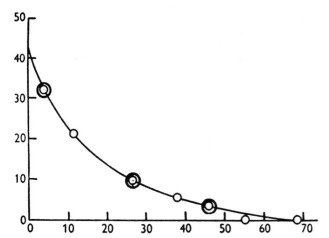

FIG. V.10. Force–velocity curve of tetanized muscle at 0 °C. Abscissae: force (g wt). Ordinates: velocity (μm s^{-1}). Small circles: experimental points. Large circles: points 'used up' in fitting the theoretical curve. Agreements between theory and experiment is significant only at other points on the curve. Curve drawn from Hill's equation (see text). (Wilkie, D. R. (1956). *Br. med. Bull.* **12**, 177.)

Isometric contraction. In isometric contraction the muscles do no external work—such contractions occur in life when we simply maintain a posture against gravity. The myofibrils themselves of course shorten and in doing so, stretch the series elastic component (SEC)—the overall tension developed can be recorded.

Now, if the events during an isometric contraction are considered, they are as follows:

At the onset there is no load opposing the contraction of the myofibrils ($P = 0$) and CC therefore shortens at V_{max} which however, stretches SEC, increasing its tension which confronts the further performance of CC, thereby reducing its speed of shortening. The tension which the muscle can develop rises to its maximum (P_0), at which point the speed of shortening of CC is now zero.

The isometric tension developed by the muscle depends on its initial length [FIG. V.11]. Beyond a certain length, which approximately coincides with its resting length under natural conditions in the body, tension developed declines and eventually reaches zero. The reason for this is that the actin and myosin filaments are disengaged by the artificial elongation of the sarcomeres and no development of active forces by the cross-bridges between the two proteins is possible [p. 250].

The tension recorded, however, is the 'total tension' T. Some of this is passive tension, as seen in FIGURE V.11. In order to determine the active tension we must subtract passive tension, yielding the curve A in FIGURE V.11. The form of the *total* tension/length curve developed by contraction of a muscle from different initial lengths depends on how much connective tissue (elastic component) that muscle contains. The form of the *active* tension/length curve of different muscles is the same because the properties of the contractile component are the same.

If a single maximal stimulus be delivered to a muscle it responds with a single twitch. If the stimuli be delivered more rapidly (100–200 s^{-1}) the mechanical response during the period of stimulation shows a rapid development of tension to a level which is well sustained as a plateau, until stimulation ceases. This tetanic tension is usually double that of the twitch. There is no difference whatever

in the active performance of CC in the two situations—the smaller twitch tension is simply due to the fact that the contraction of CC induced by one stimulus lasts only 20–40 ms and during this period there is insufficient time for the tension to climb to its full value because an appreciable delay occurs before CC can stretch SEC sufficiently. Thus a rise of temperature which diminishes the duration of the twitch further reduces the twitch/tetanus tension ratio, although it has no effect whatever on the tension developed during tetanic contraction.

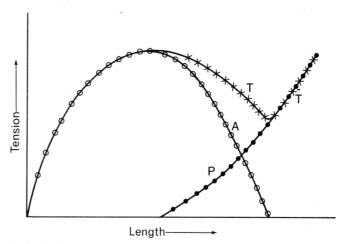

FIG. V.11. Tension–length curves in muscle (tension developed on stimulation). T shows total tension; A = active tension; P = passive tension.

The active muscle tension developed on stimulation is maximal when the initial length is that of the resting muscle in the body.

If a second maximal stimulus be delivered to the muscle shortly after the first, one of two things may happen:

1. The muscle does not respond at all, or responds more feebly. This shows the absolute or relative refractory period of the muscle—2–3 ms, but in skeletal muscle, quite unlike cardiac muscle [p. 119], this period is very short compared with the duration of the mechanical response.

2. Usually it is possible to excite the muscle with a second stimulus, producing a mechanical response which is superimposed on the first. This is summation and as the speed of the repetition of the stimuli burst increases the mechanical response becomes fused (tetanus). All muscle contractions in the body are tetanic.

Isotonic contraction. In isotonic contraction, external work is done—the muscle is allowed to contract and move a weight. This form of muscular work is carried out when walking or when lifting a load.

To study the force of isotonic contraction developed with various loads, it is best to arrange that the muscle is subjected to 'after load', as in FIGURE V.12. The stop is adjusted so that the initial muscle length is the same whatever weight is hung from the lever. On stimulation the muscle contracts, CC shortens and stretches SEC, so that muscle tension rises, but in this phase the tension rises just as it does in isometric contraction. When the force developed by the muscle just exceeds the effect of the weight, the muscle as a whole begins to shorten; thereafter the tension in the muscle remains constant (isotonic) throughout the remainder of shortening and the contraction recorded is due to the shortening of CC alone.

The work done by the muscle is the force times the distance moved.

Heat production and energetics of contraction

When the muscle contracts, its heat production increases, before either the rise in tension evoked in isometric contraction or the

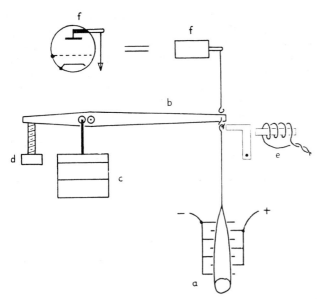

Fig. V.12. Diagram of apparatus for measuring the mechanical properties of muscle. (a) Muscle (frog's sartorius) lying on stimulating electrodes. (b) Duralumin lever. (c) Muscle load. (d) Movable stop. (e) Electromagnetic stop. (f) Transducer.

shortening recorded during isotonic contraction. This precedence can be shown experimentally only by slowing the mechanical events, such as by cooling the muscle (frog) to 0 °C; a further trick to achieve this slowing of the mechanical process is to soak the muscle in hypertonic Ringer solution. These manoeuvres allow a convincing demonstration of the precedence of heat production over the mechanical contractile processes [Fig. V.13].

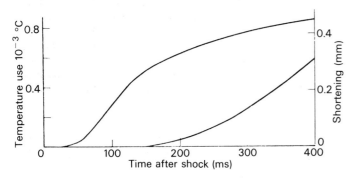

Fig. V.13. Early stages of heat production (upper curve) and shortening (lower curve) in twitches of toad's sartorius at 0 °C, after soaking in concentrated Ringer (2.28R). Muscle length 30 mm, load 1.6 g. The maximum shortening was 1.24 mm: the total temperatures rise 0.97×10^{-3} °C. (Hill, A. V. (1958). *Proc. R. Soc. B.* **148**, 397.)

Chemical changes during contraction

Muscle contains ATP, which itself is usually described as the energy coinage of the cell. For many years the hydrolysis of ATP to yield ADP, inorganic phosphate and *energy* has been regarded as the initiatory mechanism of muscular contraction but the experimental proof that this is so has not been furnished until recently. With only 2–3 μmoles of ATP per gram of muscle at rest stimulation even to exhaustion causes no detectable change in the concentration of ATP. Muscle contains another 'energy rich' compound—phosphocreatine—which rapidly rephosphorylates the ADP formed by hydrolysis of ATP and this obscures any change which initially occurs in [ATP].

$$\text{ADP} + \text{CP} \xrightleftharpoons[]{\text{creatine phosphotransferase}} \text{ATP} + \text{C} \qquad \begin{array}{l}\text{(Lohmann}\\ \text{reaction).}\end{array}$$
$$[0.03] \quad [22] \qquad\qquad\qquad [3] \quad +[4]$$

The bracketed numbers show the equilibrium concentration of the substances (in μmol g^{-1}) in resting muscle at 0 °C.

When ATP is hydrolysed, the [ADP] increases and the enzyme creatine phosphotransferase catalyses the restoration of equilibrium achieved as the reaction goes from left to right. As a result, ATP is resynthesized and one cannot detect any change of [ATP] on stimulation of the muscle to exhaustion. As the equilibrium constant of ([ATP] [C])/([ADP] [CP]) is 20, about 99 per cent of any ATP that is split is restored.

However, Wilkie (1968) has used fluorodinitrobenzene to inhibit creatine phosphotransferase and has shown (a) that frog muscles at 0 °C treated with this compound will contract normally for a hundred twitches, and (b) that during this time ATP is broken down while CP is not.

In normal circumstances the depleted store of creatine phosphate which is the result of muscle activity has to be made good by recovery processes. If oxygen is available, each two-carbon substrate source (when oxidized via the Krebs cycle and the respiratory chain) secures the rephosphorylation of 15 molecules of ADP. When the ATP concentration is raised by such means, the 'Lohmann' reaction proceeds from right to left and the small rise in [ATP] causes a big increase in [CP].

In the absence of sufficient oxygen, the concentration of NAD [p. 448] can only be maintained at the level necessary by means of the reaction:

$$\underset{\text{Pyruvic acid}}{\text{CH}_3\text{CO.COOH} + \text{NADH}_2} \quad \rightarrow \quad \underset{\text{Lactic acid}}{\text{CH}_3\text{CHOH.COOH} + \text{NAD}}$$

This of course restores the concentration of NAD needed for the conversion of 2-phosphoglyceraldehyde to 2-phosphoglyceric acid, but at the same time liberates lactic acid (which is a *nuisance*, causing acidaemia). When oxygen *is* available, lactate (transported by the circulation) is oxidized in other organs, such as the heart; some is converted to glycogen, *not* in the muscle, but in the liver [see p. 440].

Muscular fatigue

Sustained contractions of the muscles or repeated muscular contractions lead to voluntary fatigue. This fatigue is due to changes in the muscles themselves. Merton (1954) recorded the action potential and the tension developed by the adductor pollicis in response to repetitive maximal shocks to its motor nerve (ulnar). After a control period during which the normal response of action potential and tension developed to motor nerve stimulation was recorded, the subject made a maximal voluntary contraction which exerted an initial tension of 7 kg but which over the course of two minutes weakened from fatigue. On relaxing the voluntary effort the twitch response to ulnar nerve stimulation was negligibly small, although the action potential of the muscle was unimpaired. The twitch response recovered rapidly. The experiment was repeated with the circulation arrested; the muscle became fatigued much earlier and on suspending the voluntary effort the twitch response to ulnar nerve stimulation had again disappeared and remained in abeyance until circulation was restored. Throughout the experiment the action potential evoked by motor nerve stimulation was unaltered, indicating that neuromuscular transmission was unimpaired.

There is no evidence that muscular fatigue is due to a failure of the action potential-contraction coupling. Presumably fatigue occurs owing to changes in the muscle induced by (1) anoxia, (2) accumulation of metabolites. Both these changes are offset by blood flow, but there are mechanical problems in maintaining blood flow during sustained contraction, for the rise in intramuscular tension tends to prevent the passage of blood.

REFERENCES

CARLSON, F. D. and WILKIE, D. R. (1974). *Muscle physiology*. Prentice-Hall, Englewood Cliffs, New Jersey.
EBASHI, S., ENDO, M., and OHTSUKI, I. (1969). *Q. Rev. Biophys.* **2**, 351–84.
GORDON, A. M., HUXLEY, A. F., and JULIAN, F. J. (1966). *J. Physiol., Lond.* **184**, 170.
HILL, A. V. (1958). *Proc. R. Soc.* **B148**, 397.
HUXLEY, A. F. (1974). *J. Physiol., Lond.* **243**, 1–43.
—— (1977). *The pursuit of knowledge*. Cambridge University Press.
—— and HUXLEY, H. E. (1964). Organizers of a discussion of the physical and chemical basis of muscular contraction. *Proc. R. Soc.* **B160**, 433–542.
HUXLEY, H. E. (1957). *J. biophys. biochem. Cytol.* **3**, 631.
—— (1963). *J. molec. Biol.* **7**, 281.
—— (1971). *Proc. R. Soc.* **178**, 131.
JUNGE, D. (1976). *Nerves and muscle excitation*. Sinauer, Sunderland, Maryland.
MERTON, P. A. (1954). *J. Physiol., Lond.* **123**, 553.
TAYLOR, S. R., RÜDEL, R., and BLINKS, J. R. (1975). *Fedn Proc. Fedn Am. Socs exp. Biol.* **34**, 1379–81.
WILKIE, D. R. (1968). *Muscle*. Edward Arnold, London.
—— (1968). *J. Physiol., Lond.* **195**, 157.

PROPERTIES OF THE MOTOR UNIT

The motor unit

Skeletal muscles receive their motor nerve supply from the ventral horn cells of the spinal cord or from corresponding cells in the motor cranial nuclei. Each ventral horn cell supplies from 5 to 2000 muscle fibres, varying with the individual muscle. A ventral horn cell and its efferent fibre is a motor neuron; it is called by clinicians the lower motor neuron. A motor neuron together with the group of muscle fibres which it innervates is called a motor unit. The smallest group of muscle fibres that can ever be employed naturally in the body, either in reflex or voluntary activity, is obviously that supplied by a single motor neuron. The size of unit varies inversely with the precision of the movements performed by the part; e.g. in the limb muscles the unit may contain up to 2000 muscle fibres, in the extrinsic eye muscles less than 5.

Kugelberg and Edström (1968) induced prolonged contraction of the muscle fibres innervated by a single motor axon (i.e. a motor unit). This depleted the glycogen content of the relevant muscle fibres. When the tissue was stained for glycogen these depleted fibres appeared pale against the normally pink-stained fibres.

All the efferent fibres passing to *skeletal* muscle are excitatory. There are no efferent nerves which on stimulation produce relaxation of the muscle, *i.e. there are no inhibitory somatic efferents*. (In the case of smooth and cardiac muscle the efferent supply is both excitatory and inhibitory.) Skeletal muscle contraction under natural conditions always results from discharge of the motor neurons; muscular relaxation is the result of a decrease or cessation of discharge of the motor neurons.

Response of muscle to motor nerve stimulation

1. A single electrical shock of adequate strength applied to the motor nerve gives rise to a simple muscle twitch [FIG. V.14]. After a short latent period the muscle contracts and then relaxes; the twitch is over in about 0.1 second. If the stimulus applied is of threshold strength it can (theoretically) excite one motor nerve fibre and its related group of muscle fibres, i.e. one motor unit. As the strength of the stimulus applied to the nerve is increased, more nerve fibres, and therefore more groups of muscle fibres respond, and the strength of the resulting contraction is correspondingly greater. With maximal stimulation, all the nerve fibres are excited and consequently all the muscle fibres supplied by the motor nerve contract.

2. If a second maximal stimulus is applied at varying intervals after the first, the following results are obtained. If it falls during the latent period of the muscle (i.e. the first few milliseconds), the arrival of the nervous impulse produces no additional response—the muscle is said to be completely refractory. If applied later, a second muscular response results, leading to further development of tension (summation of effects) irrespective of the phase of the muscle cycle in which the stimulus is applied, i.e. whether during rising tension, the height of contraction or relaxation [FIG. V.14]. The tension resulting from two maximal stimuli applied successively at suitable intervals may thus be considerably greater than that from a single stimulus of the same strength.

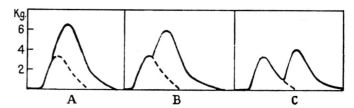

FIG. V.14. Summation of responses in skeletal muscle. Isometric contraction curves of mammalian nerve–muscle preparation. Ordinate = tension in kg. In each record the lower curve represents the response of the muscle to a single maximal stimulus to the motor nerve. The continuous thick line represents the response to the initial stimulus followed by a second stimulus. The second stimulus in A was applied during the rise of tension, in B at the height of contraction, and in C during relaxation. Further development of tension occurred in each case. (After Cooper, S. and Eccles, J. C. (1930). *J. Physiol., Lond.* 69.)

3. If a series of maximal stimuli are applied at increasingly short intervals, increasingly complete degrees of summation take place. At low rates (in mammalian muscle, 10–20 stimuli s^{-1}) the mechanical fusion is incomplete and the muscle gives a tremulous response (partial or sub-tetanus). At higher rates (about 60 s^{-1}) the mechanical fusion is complete, and full tetanus results [FIG. V.15]. It is essential to appreciate that the more complete the tetanus the greater is the tension exerted by the fibres, and the steadier their pull; in fact, with full tetanus it may be almost impossible to detect the slightest flicker in the mechanical record, though the electrical record shows a series of discrete waves corresponding to the arrival of each nervous impulse [FIG. V.16, B]. As maximal stimuli were

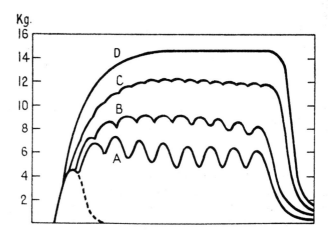

FIG. V.15. Genesis of tetanus. Response of mammalian nerve–muscle preparation. Isometric contraction records. Tension (ordinate) in kg. Lowest curve—response of muscle to single maximal stimulus to motor nerve (simple twitch). A, B, C, D—responses to rapidly interrupted maximal repetitive stimuli: A at 19, B at 24, C at 35, and D at 115 stimuli per second. Curves A, B, C show partial tetanus; curve D shows full tetanus. Note as the frequency of stimulation rises the tension developed becomes greater and is sustained more steadily. (After Cooper, S. and Eccles, J. C. (1930). *J. Physiol., Lond.* 69.)

employed in the experiments described in 2 and 3 above, the increase in tension produced by repetitive stimulation cannot be ascribed to more nerve and muscle fibres coming into action. The greater tension of a tetanus compared with a twitch is thus due to each muscle fibre generating a greater tension when repetitively stimulated.

These results are important. In both reflex acts and voluntary movements the behaviour of the motor units must depend on the character of the discharge from the motor neurons. Their rate of discharge may vary from 5–10 to 100–150 s⁻¹; the degree of tetanus resulting (i.e. whether partial or complete) and the consequent nature and strength of the contraction will vary correspondingly.

The degree of activity of each motor unit can thus be finely graded from the centre. The number of ventral horn cells activated during any reflex or other kind of act may be varied; the number of motor units in action at any moment is thus regulated and obviously the larger the number of active motor units the greater the tension resulting.

The central discharge is asynchronous, i.e. the cells in what is called the motor neuron pool, which innervates the muscle, do not fire off impulses simultaneously. Thus, the different muscle units are at any one moment in different phases of activity; when one group is contracting another is relaxing, and vice versa. Algebraic summation occurs, the individual variations are evened out and the muscle gives a steady pull.

Electrical changes in skeletal muscle

1. Stimulation of a muscle through its nerve causes a localized potential to appear at the motor end plate. When this end plate potential attains a critical magnitude it generates a propagated muscle action potential which travels simultaneously to both ends of the muscle fibre [Fig. V.16]. This single stimulus produces a single muscle action potential; it is completed during the phase of rising mechanical tension of the twitch [Fig. V.16, A].

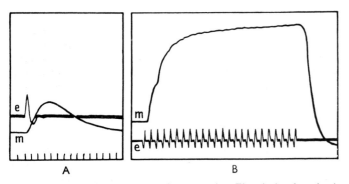

Fig. V.16. Mammalian nerve–muscle preparation. Electrical and mechanical changes in skeletal muscle in response to motor nerve stimulation. m = Mechanical record. e = Electrical record of whole muscle. One recording electrode is placed on the belly of the muscle, the other on the tendon (belly-tendon lead).

A. Response to single stimulus. Diphasic action potential which is completed in the early part of the contraction phase. Time in 0.01 second.

B. Response to stimulation at the rate of 67 per second. There is almost complete fusion of the mechanical contraction waves, but the action potential waves are distinct and discrete and follow the stimulation rate. (Sherrington, C. S. *et al.* (1932). *Reflex activity of spinal cord.* Oxford University Press, London).

If the nerve is stimulated repetitively a muscle action potential develops in response to each stimulus in the series even when the mechanical record shows complete fusion of the contraction waves [Fig. V.16, B]. The electrical record indicates the rate at which the nerve is being stimulated.

When an electrical stimulus is applied directly to a normal muscle, the response is due to stimulation either of the nerve fibres

within the muscle or of the motor end plates; the stimulation employed is thus still 'indirect'.

2. Direct stimulation of the muscle fibres can only be achieved after treatment of the muscle with curare which blocks neuromuscular transmission.

(i) When a curarized muscle is stimulated directly the sequence of events resembles that occurring in a stimulated nerve fibre. A local catelectrotonus is first produced; when this reaches a critical magnitude a propagated action potential is generated (cf. nerve spike potential) which travels at a low velocity. The interior of a resting muscle fibre is negative compared with the surface; during activity the surface becomes negative compared with the interior, i.e. the resting polarization is reversed. The muscle action potential is due to the movement of ions between the intracellular and extracellular fluids.

3. When a muscle is in a state of tone or is contracting voluntarily, the motor neurons discharge asynchronously. The electrical record obtained under these conditions from the muscle as a whole shows rapid irregular variations, quite unlike the rhythmic, repetitive pattern of the electrical record in Figure V.16, B, which is the result of synchronous stimulation of all the motor nerve fibres. The surface electrode leads record the sum of all the motor unit activity present at any moment in the muscle. To observe the rate of firing of individual motor units and of individual ventral horn cells, it is necessary to use a concentric needle electrode, i.e. a hypodermic needle down which a fine insulated wire is inserted so that the bare tip of the wire just shows at the point of the needle. This limits the electrical pick-up to the muscle fibres in the vicinity of the needle tip.

4. Although the motor unit is the physiological unit of muscle action, it is possible in cases of injury or disease of the lower motor neuron to record the action potentials resulting from the spontaneous random discharge of single muscle fibres or of groups of fibres constituting a fraction of a motor unit (fibrillation potentials, Fig. V.17, A, B). The normal motor unit action potential is the sum of the action potentials of all the individual muscle fibres of which the unit is composed. Thus the motor unit action potential is of greater amplitude than the single fibre action potential; it is also of longer duration because the muscle fibres of a motor unit fire off with a temporal dispersion of some 5–10 milliseconds [Fig. V.17].

Effects of section of motor nerve. Lower motor neuron lesion

Injury or destruction of the ventral horn cells (or the cells in the motor cranial nuclei) or of the motor fibres supplying the muscles produces a characteristic series of changes.

Results of section of a motor nerve. 1. The nerve fibres distal to the point of section undergo degeneration; this applies to both the efferent fibres (and ultimately to the motor end plates, below) and to the afferent fibres (and the muscle sense-organs from which they arise).

2. The ventral horn cells (and to a minor extent the cells of the dorsal ganglia) undergo chromatolysis.

3. The muscle fibres which have been deprived of their efferent nerve supply become completely paralysed; all reflexes inducing reflex tone are abolished and so the muscles are flaccid.

4. After three months the motor end plates become distorted or disappear. The denervated muscle fibres progressively shrink, presumably from disuse, for a period up to three years. Later, if re-innervation has not taken place, 'disruptive' changes occur. The muscle fibres split longitudinally into individual fibrils and also fragment transversely; later they are converted into tubes filled with deeply-staining nuclei and granular material and finally disappear. Within the first three years re-innervation of the muscle may restore it to a varying extent structurally to normal; after three years useful recovery is improbable.

5. Varying degrees of functional recovery may occur, depending

on the success with which regeneration of the motor fibres takes place. The new outgrowths take place from the central cut ends of the nerve fibres. The difficulties in the way of satisfactory reinnervation of the muscles are of the same kind as those described for regeneration of sensory nerves.

6. Fibrillation potentials [Fig. V.17, A, B] are found in muscles three weeks or more after nerve injury. They are due to increased sensitivity of muscle fibres to acetylcholine (ACh). In innervated muscle ACh produces membrane depolarization *only* when applied to the motor end plate. After denervation the entire muscle membrane becomes as sensitive as the end plate to the depolarizing action of ACh. Fibrillation disappears when the end plates disappear or the muscle fibres become irresponsive.

The amount of fibrillation activity and the frequency of discharge are greatly increased by drugs like neostigmine, which potentiates the action of acetylcholine [Fig. V.17, C]; neostigmine may induce fibrillation which was not detectable previously.

7. Changes in strength–duration curve. The method used in man is as follows: (i) a small stimulating electrode is placed over the 'motor point', i.e. the point on the skin which is nearest to where the motor nerve enters the muscle; (ii) another large (or 'indifferent') electrode is applied to the skin of some distant region. Square pulses of current of duration 0.01, 0.1, 1.0, 10, and 100 milliseconds are applied through the electrodes; the minimal voltage needed to produce muscular contraction is noted and the results plotted as a strength–duration curve for the most excitable fibres in the motor nerve.

Following a nerve injury, the affected nerve fibres become inexcitable after a few days. If only the most excitable fibres are injured, the strength–duration curve subsequently obtained represents the response of the intact less excitable fibres. If the entire motor nerve is severed and all its fibres degenerate, the denervated muscle will respond ultimately only to adequate direct stimulation of its constituent fibres. As the chronaxie of muscle fibres is much longer than that of nerve fibres, the strength–duration curve is correspondingly altered. Stronger stimuli or currents of longer duration must be employed to elicit a muscular contraction. Should the motor nerve fibres regenerate later, the curve gradually returns to the normal pattern [Fig. V.18]. When the denervated muscle fibres degenerate completely no response is obtained to electrical stimulation of any duration or (tolerated) intensity.

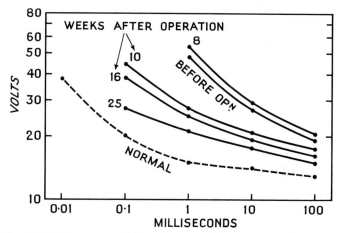

FIG. V.18. Effect on strength–duration curve of human muscle of section of motor nerve followed by resuturing and regeneration of motor nerve fibres. Ordinate: strength of stimulus in volts. Abscissa: duration of current flow in milliseconds. (Ritchie (1944). *Brain* **67**, 322.)

If the ventral horn cells are destroyed, no regeneration, and therefore, no recovery can occur. The functional loss is purely motor (and not sensory, too).

Neuromuscular transmission

The motor nerves to skeletal muscle are cholinergic in nature. As the motor nerve approaches a striated muscle fibre it loses its medullary sheath. The naked axis cylinder breaks up into terminal ramifications which make close and extensive contact with a specialized part of the muscle sarcoplasm known as the motor end-plate [Fig. V.19].

The sequence of events at the neuromuscular junction is as follows:

1. Arrival of an impulse at the motor nerve terminals.
2. Release of ACh from the presynaptic nerve terminal vesicles in quanta, each containing less than 10^4 molecules of ACh.
3. Binding of ACh to end-plate receptors and production of local endplate potential (depolarization).
4. When the end-plate potential reaches a certain critical magnitude it depolarizes the surface membrane of the muscle fibre and sets up a propagated muscle action potential
5. The muscle spike potential precedes the development of mechanical tension. The relationship between these two events is described as excitation–contraction coupling.

The process by which a nerve impulse releases ACh from synaptic vesicles at the terminal is promoted by calcium ions and inhibited by magnesium ions.

Castillo and Katz (1956) showed that when a micro-electrode is placed within the end-plate of a frog's muscle numerous miniature end-plate potentials (mepps) are recorded at rest, i.e. without nerve stimulation. These potentials are too small to produce contraction of the muscle. They are attributed to random release of ACh from synaptic vesicles when the latter collide with the membrane of the nerve terminal. This collision must open two barriers,

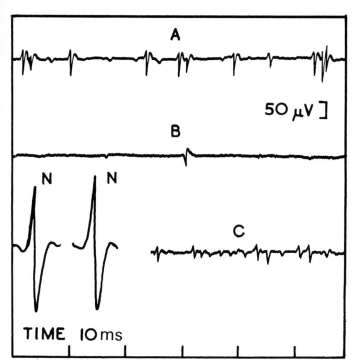

FIG. V.17. Normal muscle action potentials and fibrillation potentials. Records taken with concentric needle electrode in skeletal muscle in man. N, N. Representative action potentials of motor unit of normal muscle. Potentials are mainly diphasic. Note their magnitude and duration. A. Fibrillation potentials in denervated muscle. The deflections are small and very brief. B. Occasional fibrillation potentials in another denervated muscle. C. Same muscle as in B after administering neostigmine (an anticholinesterase). The frequency of the fibrillation potentials is greatly increased. Time in 10 ms. (Weddell *et al.* (1944). *Brain* **67**.)

the membrane of the vesicle, and the nerve membrane, so that ACh is suddenly discharged by exocytosis. In addition ACh is also released by a nonquantal process which does not produce mepps.

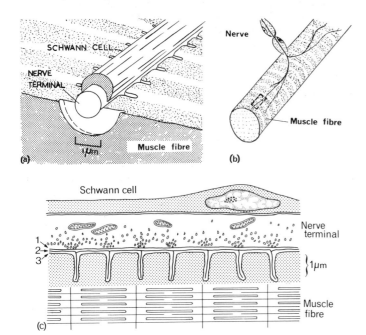

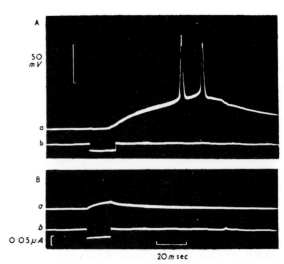

FIG. V.20. External and intracellular application of acetylcholine to a motor end-plate. In A, an ACh-filled micropipette was placed on the outside of an end-plate, and a quantity of ACh was released by passing a brief outward-directed current pulse through the pipette (registered in trace *b*). It produced the effect shown in trace *a*, a depolarization, developing after a diffusion delay and culminating in two spikes. Between records A and B, the ACh-pipette entered the muscle fibre. An outward pulse produces now a small catelectronic potential, but no spike potential. (del Castillo, J. and Katz, B. (1956). *Prog. Biophys.* **6**, 122.)

FIG. V.19. Diagram of neuromuscular junction of frog.
(a) One portion of the junction.
(b) General position of endings of motor axon on muscle fibre, showing portion (a) as small rectangle.
(c) Schematic drawing from electron micrographs of a longitudinal section through the muscle fibre.
1. Terminal axon membrane.
2. 'Basement membrane' partitioning the gap between nerve and muscle fibre.
3. Folded postsynaptic membranes.
(Katz, B. (1966). *Nerve, muscle and synapse*. McGraw-Hill, Maidenhead.)

The motor nerve terminal is in intimate contact with the deeply folded postsynaptic end-plate membrane [FIG. V.19]. The tip of the fold is never more than 1 μm from the presynaptic membrane. Thus ACh has only to diffuse through this short distance to reach the end-plate receptors, and the calculated time for this diffusion is a few hundred microseconds. ACh then combines very briefly with the receptor molecules on the surface of the postsynaptic membrane and opens up channels to cations, chiefly Na^+ and Ca^{2+}. The inward flow of Na^+ causes the end-plate potential and when this exceeds 30–40 mV the propagated spike potential in the muscle is initiated and muscular contraction occurs. Normally the end-plate is very much more sensitive to ACh than the rest of the muscular fibre, but after degenerative motor nerve section the whole muscle fibre responds to ACh. *Intracellular* application of ACh to the motor end-plate is ineffective [FIG. V.20].

The mammalian nerve–muscle junction resembles that of the frog. An intracellular electrode in the motor end-plate in the rat records both spontaneous miniature end-plate potentials and 'giant' potentials, up to 12 mV in height, of presynaptic origin. The latter are due to synchronous release of many quanta of ACh (without motor nerve impulses). These giant potentials are responsible for fibrillation (localized twitches) in muscles treated with anticholinesterases.

Microphysiological studies by Kuffler and his colleagues (1975), using iontophoretic application of ACh as well as motor nerve stimulation, have shown that a quantum of ACh acts over an area of 1–2 μm² on the post synaptic membrane. The presynaptic release sites are separated by about 2μm and when acetylcholinesterase

(AChE) is fully active the released ACh is destroyed so quickly that the quanta act independently of each other. However, when AChE is inhibited by anticholinesterases (e.g. neostigmine) ACh is destroyed more slowly and can diffuse further so that a threefold potentiation of the effects of the released quanta can occur. Micro-iontophoretic application of ACh produces potentials very similar to miniature end-plate potentials [FIG. V.21]. Kuffler and Yoshikami (1975) have calculated that a quantum contains less than 10^4 molecules of ACh, and that the conductance change produced by one molecule of ACh results in a net flux of over 3×10^3 univalent ions through the membrane. The concentration of ACh in the synaptic cleft can rise to 300 μmol l⁻¹ A presynaptic vesicle with an inner diameter of 50 nm would contain ACh in a concentration of about 250 mmol l⁻¹, a very high concentration indeed.

The work of Katz, Kuffler, and others provides the most elegant and detailed confirmation of the original studies of Dale and his colleagues in 1936 which suggested that ACh is the chemical trans-

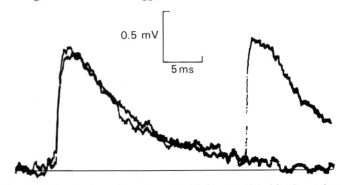

FIG. V.21. Comparison of pipette-evoked ACh potential with min epp in a snake muscle fibre. One ms pulses were passed through an ACh pipette whose tip was at the edge of a synaptic bouton. Two iontophoretically produced responses (left) are superimposed; during one of them a spontaneous min epp occurred (right). Rise time (10–90 per cent) of the responses to the pipette-applied ACh is about 1.1 ms, and of the min epp about 0.75 ms. The preparation had been lightly treated with collagenase. This treatment removes some, but not all of the AChE. This accounts for the slightly slower than normal time course of both potentials, although no anticholinesterase was used. (Kuffler, S. W. and Yoshikami, D. (1975). *J. Physiol.*, *Lond.* **251**, 465.)

mitter at the motor nerve–skeletal muscle junction. The modern techniques also show how the weak electrical signals in the presynaptic nerve terminals are amplified by the release of ACh.

Acetylcholinesterase (AChE)

The action of ACh liberated by a single nerve impulse is very brief so that the muscle end-plate membrane can be rapidly repolarized and become sensitive to a subsequent release of ACh. Only in this way can graded repetitive muscular contractions occur in response to different frequencies of nerve stimulation.

ACh is removed from its site of release partly by diffusion but mainly by destruction by AChE. Histochemical studies have revealed a high concentration of AChE in the palisade structure of the motor end-plate. After administration of an anticholinesterase drug a single stimulus to the motor nerve produces a repetitive response in the muscle, with increase in tension, due to accumulation of ACh.

AChE is present not only at the postsynaptic membrane but also at the presynaptic membrane on motor nerve terminals. Inhibition of AChE at both sites may not only enhance the action of ACh on the postsynaptic membrane at the end-plate but may also increase the local concentration of ACh sufficiently to stimulate motor nerve terminals antidromically and promote release of ACh at neighbouring nerve terminals.

Block of neuromuscular transmission

Neuromuscular transmission may be blocked either by inhibiting release of ACh from the presynaptic motor nerve terminals or by preventing the actions of ACh on the postsynaptic membrane of the motor end-plate.

1. *Presynaptic block* is produced by botulinum toxin which combines irreversibly with the cholinergic motor nerve terminals and interferes with synthesis or release of ACh. In this way it produces muscular paralysis.

2. *Post-synaptic block* can be produced by several substances.

α-Bungarotoxin is a polypeptide, containing 74 amino acid residues, which has been isolated from the venom of the krait, a very poisonous snake which causes muscular paralysis. α-Bungarotoxin produces postsynaptic neuromuscular blockade by combining irreversibly with ACh receptors on the motor end-plate. Radioactive α-bungarotoxin has been used to reveal the number of ACh receptors in the end-plate.

Neuromuscular blocking drugs of clinical value may act in two main ways: by *competition* or by *depolarization*.

The most important competitive blocking drug is *tubocurarine* (the active alkaloid of tube curare). Claude Bernard was the first to show that curare abolishes the response of a muscle to indirect stimulation (via the motor nerve), without preventing the contraction evoked by direct stimulation of the muscle. Bernard concluded that it acted at the neuromuscular junction. Curare does not prevent release of acetylcholine but merely stops the latter acting on the motor end-plate. Tubocurarine acts competitively with acetylcholine to combine with the receptors on the motor end-plate. Tubocurarine does not itself depolarize the membrane, and when attached to receptors it prevents acetylcholine from exerting its full stimulant action. Tubocurarine reduces the size of the end-plate potential in proportion to the number of receptors occupied by this alkaloid. Paralysis occurs when the end-plate potential is reduced below the threshold required to set off a propagated action potential in the muscle fibre.

The action of tubocurarine at the motor end-plate resembles that of atropine towards the muscarinic actions, and that of hexamethonium towards the ganglionic actions, of acetylcholine.

By preserving acetylcholine, *anticholinesterase* drugs [p. 413] overcome the competitive type of block.

The best known depolarizing blocking drugs are *decamethonium* (C10) and *suxamethonium* (succinylcholine). These drugs act like acetylcholine to depolarize the motor end-plate; they therefore produce initial muscular twitching. The paralysis is due to persistent depolarization since these drugs are removed from the motor end-plate much more slowly than acetylcholine.

Clinically, neuromuscular blocking drugs are used to produce muscular relaxation during surgical operations and to reduce movements during electroconvulsion treatment of psychotic patients.

REFERENCES

BLINKS, J. R., RÜDEL, R., and TAYLOR, S. R. (1977). *J. Physiol., Lond.* **227**, 291–323.

CASTILLO, J. DEL and KATZ, B. (1956). *J Physiol., Lond.* **132**, 630.

DALE, H. H. (1953). *Adventures in physiology*, p. 530. Wellcome Institute for the History of Medicine, London.

HOBBIGER, F. (1976). Pharmacology of anticholinesterase drugs. In *Neuromuscular junction. Handb. exp. Pharmak.*, Vol. 42, p. 487 (ed. E. Zaimis). Springer, Berlin.

KATZ, B. (1967). *Nerve, muscle and synapse*. McGraw, London.

KOELLE, G. B. (ed.) (1963) Cholinesterase and anticholinesterase agents. *Handb. exp. Pharmak.* Suppl. 15. Berlin.

KUFFLER, S. W. and NICHOLLS, J. G. (1977). *From neurone to brain*. Sinauer, Sunderland, Mass.

——and YOSHIKAMI, D. (1975). *J. Physiol., Lond.* **251**, 465.

KUGELBERG, E. and EDSTRÖM, L. (1968). *J. Neurol. Neurosurg. Psychiat.* **31**, 424–33.

STEVENS, C. F. (1979). The neuron. In *The brain*, pp. 15–28. Scientific American, New York.

Myasthenia gravis

Myasthenia gravis is a rare disease characterized by great skeletal muscular weakness and rapid onset of fatigue. The muscles first and most affected are those supplied by the cranial nerves, but in severe cases general muscular weakness may cause the patient to become bedridden, and death may occur from paralysis of the respiratory muscles. The muscular weakness resembles that produced by injection of tubocurarine, the defect being located at the neuromuscular junction. Electrical stimulation (at 40 Hz) of a motor nerve to a striated muscle normally produces a sustained muscular contraction during the period of stimulation; in myasthenia gravis the muscular tension rapidly declines [FIG. V.22], though, as with tubocurarine, normal neuromuscular transmission is restored by administration of neostigmine or other anticholinesterase drugs.

Histological studies of muscle from myasthenic patients show presynaptic and postsynaptic abnormalities at the neuromuscular junction. Functionally, miniature end-plate potentials in myasthenic muscle are much smaller than normal, though the density of synaptic vesicles in nerve terminals is not reduced. The evidence suggests that in myasthenia gravis an important abnormality is a marked reduction in the ACh content of the synaptic vesicles. This

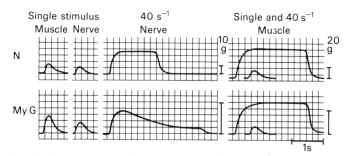

FIG. V.22. Responses of isolated external intercostal muscles removed from a control subject (N) and a patient with myasthenia gravis (MyG) to indirect (nerve) and direct (muscle) stimulation, at 32 to 33 °C. Muscle tension was recorded with a strain gauge. (Redrawn from Lambert, E. H. and Elmqvist, D. (1971). *Ann. N.Y. Acad. Sci.* **183**, 183–99.)

is responsible for the reduced amplitude of both miniature end-plate potentials and the end-plate potentials induced by motor nerve stimulation. However, there is also evidence that in myasthenic muscle the motor end-plates are less sensitive than normal to the stimulant action of ACh. Patients with myasthenia gravis are hypersensitive to the paralysing action of tubocurarine, and anticholinesterases, by reducing the rate of destruction of ACh by AChE, are therapeutically beneficial.

Neostigmine produces very striking symptomatic improvement, each dose acting for several hours: after administration of this drug patients who have pronounced facial weakness and cannot swallow or get out of bed, rapidly develop a great increase in muscular power so that they can eat a hearty meal, walk or run about the room, and even perform light manual work.

Role of thymus

In many cases of myasthenia gravis the thymus is enlarged and thymectomy cures or alleviates the condition in 80 per cent of patients. Myasthenia gravis is probably an autoimmune disease which may arise as follows:

1. The serum frequently contains an IgG autoantibody which combines with the receptor protein for ACh in striated muscle. This combination reduces the number of muscle end-plates which can respond to ACh released from motor nerve terminals.

2. There may be an autoimmune thymitis associated with the release of a hormone called thymopoietin (formerly known as thymin). Thymopoietin is a polypeptide (molecular weight 5562) which causes neuromuscular block in experimental animals; neostigmine reverses this block (Goldstein and Schlesinger 1975).

Neonatal myasthenia gravis in children born of mothers with adult myasthenia gravis could be due to placental transfer of either autoantibodies or thymopoetin from maternal to fetal circulation. Likewise the beneficial effects of cortisol administration in myasthenia gravis could be due to reduction of autoantibody formation or inhibition of thymopoietin release from the inflamed thymus.

REFERENCES

AHARONOV, A., ABRAMSKY, O., TARRAB-HAZDAI, R., and FUCHS, S. (1975). *Lancet* **ii**, 340.
GOLDSTEIN, G. and SCHLESINGER, D. H. (1975). *Lancet* **ii**, 256.

The Nervous System

ELEMENTARY FEATURES OF THE ANATOMY OF THE CENTRAL AND PERIPHERAL NERVOUS SYSTEMS

The central nervous system (CNS) comprises the brain and spinal cord. The brain comprises all structures which are intracranial; cerebrum, midbrains, pons and medulla, and the cerebellum [FIG. V.23].

The cerebrum is derived from the primitive forebrain which can be subdivided into the telencephalon (consisting of the two cerebral hemispheres and their interconnections) and the diencephalon containing the two thalami. The midbrain (mesencephalon) is interposed between the diencephalon and the rhombencephalon (comprising the pons, cerebellum, and medulla).

At the foramen magnum at the base of the skull the neural tube continues caudally as the spinal cord to terminate at the lower border of the first lumbar vertebra. The cord is about 45–50 cm long, thus much shorter than the vertebral canal which contains it. Below the first lumbar vertebra the canal contains the lumbar and sacral roots in leashes—known as the cauda equina.

In cross-section the spinal cord is about 2 cm in diameter. Anteriorly, a prominent fissure and posteriorly a less obvious sulcus

can be seen; between these fissures lies the central canal. The so-called grey matter consisting of neuronal cell bodies (perikarya) forms an H-shaped figure with longer and narrower posterior horns and blunt anterior horns. The horizontal stem of the H constitutes the grey commissure, which surrounds the central canal.

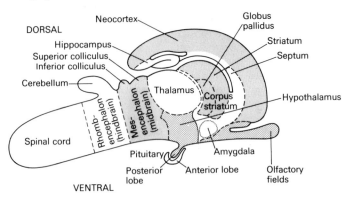

FIG. V.23. Schematic view of a mammalian brain and spinal cord. The hindbrain consists of pons, medulla, and cerebellum. The midbrain shows two dorsal elevations—the colliculi. The forebrain contains the cerebral cortex (outermost) which overlies the basal ganglia (corpus striatum and globus pallidus). The remainder of the forebrain is the diencephalon: the upper two-thirds comprises the thalamus and the lower third the hypothalamus which connects with the pituitary complex. (After Nauta, W. J. H. and Feirtag, M. (1979). In *The brain*, Chap. IV. Scientific American, New York.)

The ventral or anterior horn contains the cell bodies of axons which pass out in the ventral (or anterior) roots. These axons are purely motor in function. The dorsal (posterior) horn receives the fibres of the posterior roots which are entirely sensory in function. Their cell bodies are found in the posterior root ganglion which is a swelling on the posterior root. These ganglion cells are bipolar—their axons connect both with peripheral structures and with structures in the dorsal horn.

There are thirty pairs of motor and sensory roots—eight cervical, twelve thoracic, five lumbar, and five sacral. The arm, for example, is supplied by C5–T1—the so-called brachial plexus. The lumbar plexus is formed from the upper four lumbar nerves. The sciatic nerve contains fibres from L4 and 5, and S1, 2, and 3.

FIGURE V.24 shows a simplified picture of the spinal cord in cross-section. Surrounding the grey matter is the white matter, consisting of myriads of ascending and descending axons cut across. The white matter can be conveniently divided into three white columns (funiculi)—anterior, lying between the median fissure and the anterior roots; lateral between the anterior and posterior roots; and dorsal between posterior roots and the posterior fissure.

The figure also shows a *monosynaptic* reflex arc. The afferent nerve fibre from a muscle spindle reflexly excites the motor neuron which supplies the same muscle. The two types of fibre are shown

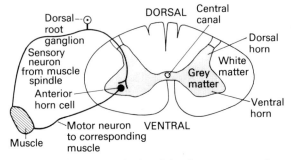

FIG. V.24. Cross-section of the spinal cord showing a monosynaptic connection between a dorsal root (sensory) neuron with its bipolar body and an anterior horn cell (motor).

widely separate in the diagram for clarity. However, in actuality the sensory and motor fibres soon join to form a common trunk—the so-called *mixed nerve* which issues from the vertebral canal (as a peripheral spinal nerve) via the intervertebral foramen.

STRUCTURE OF THE NERVOUS TISSUE

The term *neuron* is used to describe the nerve cell (cell body, soma, perikaryon) and its processes, the dendrites and the axon (axis cylinder, nerve fibre) [FIG. V.25]. The nutrition of the axon and the preservation of its structure depend on its intact connection with its cell body.

Neurons vary considerably in shape and size in different parts of the body. Granule cells from the cerebellum are 5 μm in diameter, large motor cells of the anterior horn of the spinal cord are up to 120 μm in diameter. Axons vary from a few micrometres in length up to 90 cm long.

Under the phase-contrast microscope, a living neuron has a granular cytoplasm, differentiated under the cell membrane to form a superficial gel layer enveloping a relatively fluid core. The contractile property of the gel layer results in there being a continual flow of axoplasm from the cell body to the periphery. The large vesicular nucleus contains a single prominent nucleolus adjacent to which, in the female, can often be seen a large granule representing the sex chromatin (Jordan 1978).

After fixation, special stains reveal the presence in the cytoplasm of basiphil masses or Nissl granules (bodies), numerous rod-like mitochondria, a Golgi apparatus, and fine, long filaments of neurofibrillae. The mitochondria and neurofibrillae extend into the axon hillock and axon, but no Nissl granules are found here.

Nissl substance. The granules are stained with basic dyes, e.g. methylene blue, thionine, or cresyl violet. Their size and number vary with the physiological condition of the cell. Fatigue, the action of certain poisons, and section of the axon cause the Nissl granules to disintegrate into a fine dust which eventually disappears (chromatolysis).

Electron microscope (EM) studies reveal that these basiphil masses are composed of many thin, parallel arranged, membrane-bounded cavities, or cisternae [FIG. V.25]. These are similar to the (less numerous) rough-surfaced membranes found in the cytoplasm of most cells which collectively make up the endoplasmic reticulum or ergastoplasm of the cell. Covering the surface of the membranes and giving them their rough appearance are many minute particles. This granular component consists largely of ribose nucleoproteins (RNP); these substances stain with basic dyes and account for the basiphilic staining properties of the Nissl granules.

The generous distribution of these small RNP granules is one of the most striking morphological features of the neuron. The same kind of association between RNP granules and endoplasmic reticulum is found in the cytoplasm of most active gland cells (e.g. acinar cells of pancreas)—that is in cells which sustain an intense protein production.

Golgi apparatus. EM studies reveal a number of agranular or smooth-surfaced membranes and a vesicular component which correspond to similar structures described in other cells forming the basis of the classical Golgi apparatus as seen with the light microscope.

Mitochondria. These are many in number and may be rod-like or spherical in form as in other cells.

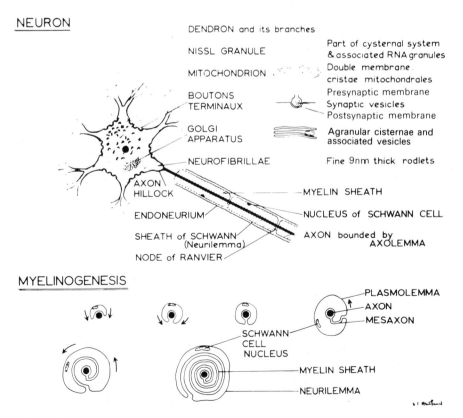

FIG. V.25. The upper figure shows (diagrammatically) the structures found in the nerve cell by examination, with the light microscope (on the left); the electron microscopic details of these structures are shown on the right.

The lower part of the figure shows the process of myelinogenesis in peripheral nerve.

Neurofibrillae. In the cytoplasm, threads, 6–10 nm in diameter and of variable length, traverse the cytoplasmic matrix, forming a loose feltwork of fibrils in the cytoplasm; they are morphologically similar to those described in the axoplasm of peripheral nerves. Presumably they represent the structures, which, after aggregation and suitable staining, result in the neurofibrillae visible with the light microscope.

Dendrites. These branch repeatedly immediately they leave the cell and also contain Nissl granules, mitochondria and neurofibrillae.

The axon. The axon or nerve fibre arises from a part of the cell (axon-hillock) in which there are no Nissl granules.

Medullated nerves

A medullated (myelinated) nerve fibre consists of the following structures from within outwards:

1. A central core of semifluid axoplasm which flows from the cell body to the periphery; if the axon is sectioned this axoplasm pours from the cut end; if the axon is ligated the axoplasm accumulates proximal to the constriction and causes swelling.

The axoplasm contains a fibrillar component 9 nm in diameter which runs parallel to the axis of the fibre and which is similar to that forming the feltwork of fibrils in the cell body.

Mitochondria are present and also elements of the endoplasmic reticulum.

The axolemma, only detectable with electron microscope techniques, separates the axoplasm from the surrounding structures.

2. The axon may be surrounded by a myelin (medullary) sheath which is a specialized set of Schwann cells arranged in a linear manner.

Myelinogenesis. Myelinogenesis begins with Schwann cells growing round the axon and completely enveloping it along its length [FIG. V.25]. The cell membrane of the Schwann cell surrounding the axon is connected to the outer cell membrane of the Schwann cell by the double mesaxon. The Schwann cells rotate and wrap round the axon many closely packed, helically arranged layers of double membranes. Each membrane is composed of two lipid layers sandwiched between layers of protein and they form the myelin sheath of the nerve fibre. The thickness of the myelin sheath is determined by the number of membrane layers wrapped round the axon.

The nodes of Ranvier which interrupt the myelin sheath at intervals indicate the junction between adjacent non-syncytial Schwann cells.

3. The neurilemma (sheath of Schwann) is the outermost cell membrane of the Schwann cell; under it lies a thin layer of Schwann cell cytoplasm and its peripheral nucleus.

Unmyelinated nerves

In unmyelinated nerve fibres, the axons are enclosed only by a Schwann cell that has not spun a myelin sheath around them. All the postganglionic fibres of the autonomic nervous systems are non-medullated (unmyelinated), as are those fibres in the somatic nervous system which are less than 1 μm in diameter; larger somatic fibres and the preganglionic autonomic fibres are medullated. Non-medullated nerves abound in the CNS and in the dorsal nerve roots.

Myelination of nerve fibres in the CNS itself occurs by a different process. The important cells are the oligodendroglia, *not* the Schwann cells [see p. 369]. Myelin formation occurs intracellularly not by the proliferation of a surface membrane, the mesaxon.

4. Surrounding the neurilemma of medullated nerve fibres in peripheral nerve trunks is a thin layer of fine reticular fibres which form the endoneurium.

Bundles (fascicles) of nerve fibres are enclosed in a connective tissue capsule, the *perineurium*, and a number of such fascicles are bound together by connective tissue fibres called the *epineurium*.

Axonal transport. Far from being an inert mechanical support for the neurilemma the axoplasm is the roadway for busy traffic moving in both directions. Centrifugal transport is both slow (1 mm day^{-1})—carrying protein components essential for the growth or regeneration of the axon and fast (10–20 cm day^{-1}) carrying enzymes required for the manufacture of transmitters. Centripetal transport is also fast. These features of axoplasmic streaming can be established either by following the transport of radioactively labelled amino acids, e.g. [^{3}H]-leucine, outwards from the cell body or by mapping retrograde (centripetal) transport of horse-radish peroxidase, which accumulates in the cell bodies. Probably viruses and tetanus toxin pass up the nerve in this manner.

Axoplasmic transport is abolished by DNP, azide, cyanide or prolonged anoxia. All these block oxidative phosphorylation.

THE SYNAPSE

The brilliant histological work of Cajal (1906) demonstrated the invariable separation between the axon terminals of one nerve fibre and the dendritic processes of an adjacent neuron. This site of contiguity (never continuity) was named the synapse by Sherrington after a suggestion by Verrall (a classics don of Trinity College, Cambridge) that the Greek verb synapsein', meaning 'to clasp' would adequately describe the contiguity which Cajal had demonstrated histologically and which Sherrington had inferred from his functional studies.

Synapses may be found where an axon terminal is contiguous with (i) a dendrite, (ii) a dendritic spine (a branch of a dendrite), or (iii) the cell body surface itself (soma or perikaryon)—FIGURE V.26. Only the electron microscope can define the fine structure of the surfaces contiguous at the synapse. The EM shows that a gap of 20–30 nm separates the pre- and postsynaptic membranes. The

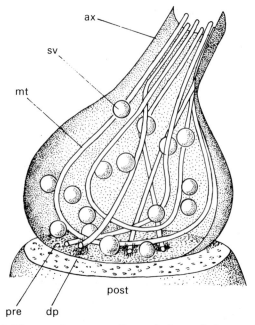

FIG. V.26. Diagram of a synapse. Preterminal axon (ax), synaptic vesicle (sv), microtubules (mt), presynaptic membrane (pre), postsynaptic process (spine or dendrite) (post), point where the dense projection is thought to anchor the microtubules to the presynaptic membrane (dp). (After Gray, E. G. (1977). *The synapse*, 2nd edn. Carolina Biology Readers, Burlington, N. Carolina.)

presynaptic cytoplasm contains vesicles some 50 nm in diameter clustered near the membrane. Mitochondria containing ATP abound. The terminal part of the presynaptic axon tends to be bulbous but contains no protein threads (neurofilaments) which are a feature of the axon itself.

The synaptic vesicles contain the chemical transmitter responsible for excitation (or inhibition as the case may be) of the neuron next in the chain. These vesicles are transported down the axon along microtubules in the axon [Fig. V.27].

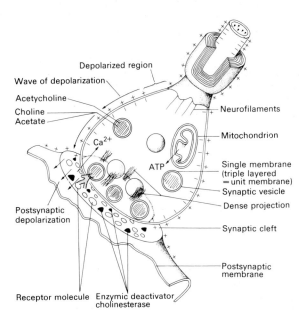

Fig. V.27. Diagram of a synapse where acetylcholine is the transmitter. The dark-staining dense projections possibly play a role in guiding the vesicles to specific regions of the subsynaptic membrane. (After Gray, E. G. (1977). *The synapse*, 2nd edn. Carolina Biology Readers, Burlington. N. Carolina.)

Synapses on dendritic spines tend to have a pronounced post-synaptic thickening and a wider synaptic cleft (type 1) compared with those on the soma (type 2). After fixation with aldehyde followed by osmium tetroxide, synaptic vesicles at excitatory synapses appear circular whereas those at inhibitory synapses look flat or elongated (Uchizono 1965). The reason for this difference is not known. In the cortex most of the excitatory synapses are situated on the tips of the dendritic spines but in the spinal cord the dendrites have no spines and the synapses are found directly on the dendritic shafts (Gray 1977).

REFERENCES

Cajal, Santiago Ramón y (1906). *Histologie du système nerveux de l'homme et des vertébrés*, Vols. I and II.Maloine, Paris.

Aidley, D. J. (1971). *The physiology of excitable cells.* Cambridge University Press.

Cold Spring Harbor Symposia on Quantitative Biology (1976). Vol. 40, *The synapse.*

Gray, E. G. (1969). *Prog. Brain Res.* **31**, 141.

—— (1977). *The synapse*, 2nd edn. Carolina Biology Readers, Burlington, N. Carolina.

Jordan, E. G. (1978). *The nucleolus*, 2nd edn. Carolina Biology Readers, Burlington, N. Carolina.

Peters, A., Palay, S. L., and Webster, H. de F. (1976). *The fine structure of the nervous system*, 2nd edn. Harper & Row, New York.

Uchizono, K. (1965). *Nature, Lond.* **207**, 642.

Peripheral nerve—elementary considerations of functional characteristics

1. Conduction velocity

Few discoveries have rivalled that of the nerve-muscle preparation in the wealth of data and interpretation which can be derived from it. When the sciatic nerve supplying gastrocnemius is stimulated electrically using sufficient stimulus strength, the muscle contracts. If the origin of the muscle be immobilized and the tendon freed from its insertion and attached to a lever system, the tension developed by the muscle can be recorded on a kymograph. Using this preparation Helmholtz (1850) was the first to disprove his great teacher Johannes Muller who had asserted that nervous conduction was instantaneous. Helmoltz prepared a long length of the sciatic nerve of a frog with its associated gastrocnemius arranged for recording of the muscle response to a single maximal stimulation of the nerve delivered as far peripherally as possible [Site Ⓐ Fig. V.28]. The muscle response was a twitch recorded on the kymograph paper. When he moved the stimulating electrodes nearer the muscle [Site Ⓑ] and repeated the stimulation the muscle response. though identical in size, occurred earlier. By measuring the distance between the stimulation Site Ⓐ and that of Site Ⓑ and that between the peaks of the two muscle twitches recorded on the kymograph paper accompanied by a time tracing Helmholtz determined the conduction velocity of the sciatic nerve fibres. Thus, the distance between stimulus Site Ⓐ and stimulus Site Ⓑ was, say, 3 cm. The delay between the peaks of the respective muscle twitches was 1.5 ms. Thus, as the time taken for the 'nerve impulse' to travel 3 cm between A and B was 1.5 ms the conduction velocity was 20 metres per second. Helmholtz also measured conduction velocities in man recording the difference in latency of contraction of the thenar (thumb) muscles in response to stimulation of the ulnar nerve at the axilla and at the elbow. He found higher conduction velocities (50–60 ms^{-1}), which he showed to be related to the higher temperature of the human arm.

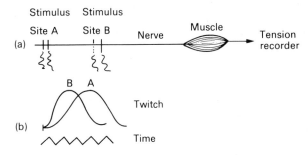

Fig. V.28. Principle of measurement of conduction velocity of nerve.

2. Cathodal excitation

When electrodes attached to a battery are placed on a nerve, current flows from the anode into the nerve and traversing the length of nerve flows out at the cathode. If points of application of the electrodes are separated by a few centimetres the nerve-muscle preparation can be used to show that the muscle twitch occurs earlier when the cathodal electrode is applied to the nerve nearer to the muscle than is the case when the anodal electrode is the more proximal. Alternatively if two pairs of electrodes are applied to a nerve and each pair of electrodes is arranged to give a barely threshold shock, only when the cathode of each pair are applied simultaneously does a large muscle contraction result [FIG. V.29]. Moreover, if the nerve is anaesthetized between the anodal and cathodal electrode sites no difference in muscle response occurs when the cathode is the more proximal, whereas the stimulus required to excite the muscle twitch must be considerably increased when the anodal electrode is proximal to the cathodal site.

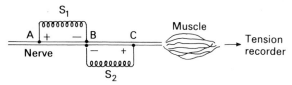

FIG. V.29. Nerve–muscle preparation. Shock 1 and Shock 2 are separately arranged to cause individually a barely detectable contraction of the muscle. When the two shocks are applied simultaneously only if the cathodes are approximated as shown does a large muscle contraction result.

3. Threshold

Again the nerve-muscle preparation can be used to demonstrate that a certain strength of stimulus must be applied to provoke a twitch. Such a strength may be defined as threshold—but the threshold so determined depends on the rate of rise of the current and on the duration of its application. If the current duration is less than, say, 10 ms, the strength of threshold current required is greater than that needed when a longer period of current application is employed. The 'strength–duration curve' of muscle has proved valuable in monitoring the clinical improvement of patients who have suffered nerve injury [p. 257]. The weakest current which will excite a tissue if allowed to flow for an adequate time is called the *rheobase*. The duration which a current twice the strength of the rheobase must flow to excite the tissue is termed the *chronaxie*.

4. Accommodation

If the rate of rise of current application is slowed, as for example is the case when a condenser is connected between the electrodes, then the current required to provoke muscle contraction is larger. Indeed, if the condenser be of sufficiently large capacity no contraction is produced whatever the strength of current, for this is employed in charging up the condenser. This effect of slowing down the rate of rise of current strength on the ability of the nerve to respond to such excitation is known as accommodation.

5. Injury potential

When electrodes placed on an intact nerve are connected through a galvanometer, no voltage difference can be detected. However, if the nerve is transected and one electrode is applied to the cut surface while the other remains in contact with the intact nerve trunk, the galvanometer registers a negative potential difference of some tens of millivolts at the cut surface. This phenomenon is described as the injury potential. It is indicative of the fact that the axoplasm of the intact resting nerve fibre is at a potential negative to that of the exterior—the nerve fibre is said to be 'polarized'. The injury potential is never as large as the true 'resting potential' which can only be recorded by measuring the potential difference between an extracellular electrode and a microelectrode inserted into the axoplasm itself. Such a technique was impossible until the late 1930s when using the giant axons of the squid (discovered by Young in 1936) which are 0.5–1 mm diameter, Curtis and Cole and Hodgkin and Huxley employed electrodes of 50 μm for insertion down the axoplasmic interior of the axon. The resting potential so measured was −60 to −90 mV (inside negative to outside).

6. Diphasic and monophasic action potentials

Before the advent of intracellular electrodes it was possible to demonstrate the electrical changes associated with the passage of a nerve impulse aroused by electrical stimulation using extracellular electrodes. The impulse consists of a wave of negativity of short length which traverses the axon at a velocity which varies with the diameter of the fibre (and the temperature).

As first one and then the other electrode 'samples' this wave of negativity there is a potential difference between them which changes its sign as the impulses pass from one to the other [FIG. V.30].

It should be understood that the demonstration of a diphasic action potential by extracellular electrodes presupposes that the electrodes be sufficiently separated. For instance, if the action potential recorded in a frog nerve is, say, 1 ms in duration and the conduction velocity is 25 metres per second, the length of the active 'depolarized' region is 2.5 cm which may be similar to that of the permissible separation of the electrodes.

The monophasic action potential is recorded by placing one electrode on the cut or crushed end of the nerve and the other placed on the intact nerve trunk, nearer the stimulating electrode.

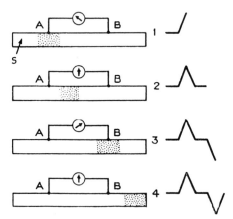

FIG. V.30. Production of diphasic action potential in nerve fibre. Stippled part of nerve fibre = area of activity.
Electrical changes. Stimulate fibre at S. 1. Activity develops at A. 2. Activity passes away from A. 3. Activity develops at B. 4. Activity passes away from B.

7. All or none law

A nerve trunk supplying a muscle is stimulated and the muscle tension is recorded. The muscle contraction developed requires a certain strength of current—threshold. However, further increase in stimulus strength produces larger contractions until a maximal response is obtained.

This, however, is simply attributable to the fact that the nerve trunk contains many fibres of different diameter and threshold. A classic demonstration by Keith Lucas (1909) showed that nerve fibres did display an all-or-none behaviour. He used the dorsocutaneous muscle of the frog which is supplied by less than ten motor neurons. Steadily increasing the stimulus did not result in a smooth increase in contraction—the size of the muscle contraction recorded showed a series of abrupt steps and the number of such steps never exceeded the number of motor neurons supplying the muscle.

The all-or-none law is proved by the use of intracellular electrodes—the action potential when aroused (threshold) is of constant size even if the strength of stimulation is further increased. It follows that the intensity of a natural stimulus adequate to excite a receptor of a single afferent nerve fibre must be signalled by the frequency of the impulses in the fibre and not by their amplitude.

8. Compound action potentials of the nerve trunk

The advent of the cathode ray oscillograph in the early 1930s allowed the display of action potentials recorded extracellularly from a nerve trunk, such as the sciatic, which like all other peripheral nerve trunks contains both motor and sensory fibres. When the sciatic nerve is removed from the body and both ends of the nerve preparation are secured, the preparation is placed in a trough and immersed in paraffin oil. Stimulating electrodes are sited at one end of the nerve and recording electrodes (connected via an amplifier to a cathode ray oscilloscope) at the other end several centimetres from the stimulating electrodes. The delivery of a shock of sufficient strength to excite *some* of the nerve fibres causes propagation of the impulse in both directions but as the recording electrodes are 'downstream' only, this need not concern us at the moment. The largest single fibre components of the nerve trunk are the most excitable and are, moreover, those which are the fastest conducting. In a mammalian nerve the largest fibres are about 20 μm in diameter; they are myelinated and there is an approximate correlation between fibre diameter and conduction velocity in myelinated nerve—the conduction velocity in metres per second is approximately six times the fibre diameter in micrometres. Somatic myelinated fibres in peripheral nerve range from 20 μm to 2 μm and correspondingly possess conduction velocities of 120 m s^{-1} to 12 m s^{-1}. When the stimulus strength is adequate to excite all of these, the compound action potential shows the so-called A peak which itself can be subdivided into Aα, Aβ, and Aγ fibre peaks [FIG. V.31]. Knowing the distance between the stimulating and recording electrodes and the time taken for the action potential to reach the latter, conduction velocities can be calculated.

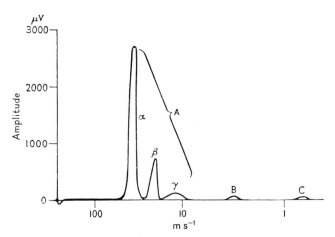

FIG. V.31. Mixed nerve of frog. Compound action potential recorded at a site distant from that of stimulation. (Walsh, E. G. (1957). *Physiology of the nervous System*. Longmans, Green, London.) (Data from Erlanger and Gasser, modified from Bell, Davidson, and Scarborough 1950.)

Increasing the stimulus strength above that which excites all the fibres contributing to the A peak, brings in a second fibre component of the nerve trunk which can be recorded and designated as the B peak. This component of the compound action potential is due to excitation of thinly myelinated fibres belonging *only* to the preganglionic autonomic nerves. Their conduction velocity is about 4–15 m s^{-1}; even when maximally excited their contribution in peak size to the compound action potential is small. Lastly, the least responsive to electrical stimulation are the unmyelinated nerves. Each fibre is small in diameter and shows slow conduction (0.5–2.5 m s^{-1}); combination of their electrical response yields the C peak which is very small in amplitude compared with that of the A peak. Such results of extracellular recording should not disguise the fact that the C fibre components of the sensory fibre content far outnumber the myelinated sensory fibres.

All peripheral nerves contain both motor and sensory fibres and no distinction is possible when an isolated section of nerve is stimulated and oscilloscope recordings are made from it. However,

if the mixed nerve is stimulated, say, in the limb itself and recordings are made from the appropriate segmental dorsal roots the compound action potential shows A and C peaks but not a B peak for there are no preganglionic autonomic fibres in the dorsal (sensory) root. The *histological* analysis of the afferent fibre components of a mixed nerve such as the sciatic or the ulnar can be made by sectioning the relevant segmental ventral (motor) roots and allowing the animal to recover for some six weeks. The motor fibres degenerate and the remaining fibres must all be afferent and can be counted after sacrifice of the animal.

Reflex action

If a frog is decapitated and suspended vertically and one of its dependent legs is pinched, the leg is quickly withdrawn from the nocuous stimulus (flexion reflex). However, if the spinal cord (already severed from the brainstem and higher 'centres' such as the cerebrum) is itself destroyed by passing a wire down the vertebral canal then nocuous stimulation of the foot no longer evokes flexor withdrawal. This is clear proof that the central nervous system must be involved in the flexion response. Indeed the word 'reflex' indicated that sensory messages relayed by afferent nerves were 'reflected' in the CNS to exit via motor nerves. Magendie (1822) proved that the dorsal roots of the spinal cord subserved a sensory function and that the ventral roots carried nerve fibres of purely motor function. Histology at that time did not permit the demonstration of the anatomical features of the terminations of the sensory axons and their relations with the cell bodies and their processes (dendrites of the motor neurons).

The reflex arc forms the functional unit of the CNS. An afferent (sensory) nerve rarely directly engages the effector or efferent (motor) neuron although it does indeed do so in the case of the stretch reflex in muscles which provides the basis for muscle tone [p. 297]. Much more often the incoming afferent axons synaptically excite and (in some cases inhibit) intercalary neurons which may not necessarily be in the same segment of the cord. These intercalary (internuncial) neurons in turn synaptically influence others, which finally modify the activity of the motor neurons concerned in the reflex. Such reflex arcs are termed polysynaptic.

General features of reflex excitation

1. Synaptic delay. Nowadays this can be accurately measured. If a shock is given to the axons which synaptically engage a neuron then the excitation of the said neuron can be recorded by an intracellular electrode. Synaptic delay is of the order of 0.2 ms or even less.

2a. Spatial summation. If fibres from an afferent nerve A and another afferent nerve B are stimulated separately neither may cause a reflex response, i.e. each stimulation has been subliminal. If, however, both are stimulated simultaneously a reflex response is evoked. Each afferent nerve on being excited liberates sufficient chemical transmitter to cause some excitatory postsynaptic potential (EPSP [see p. 281]); neither alone produces sufficient EPSP to induce the motoneurons to discharge a spike potential but when both afferents are stimulated the development of EPSP by the motoneuron is adequate for the initiation of a spike. Spatial summation is an important feature of central synaptic transmission.

2b. Temporal summation. This is of less importance. It is illustrated by the effect of repetitive stimulation of an afferent fibre in evoking a reflex response although individual stimuli of the same strength are ineffective.

3. Occlusion. When two afferent excitatory nerves (*a* and *b*)—each of which can evoke the flexor reflex—are simultaneously stimulated, it is sometimes found that the tension developed by the flexor muscle under observation is less than the sum of the tension produced by each afferent stimulated separately; thus if *a* produces in the muscle a tension of value 9 (in arbitrary units), and *b* also a tension of 9, stimulation of *a* and *b* together may only yield a tension of 12 (instead of the expected 18). This phenomenon is referred to as occlusion; it is due to the fact that some of the spinal motor neurons (in this case producing 6 units of tension in the muscle) are common to both *a* and *b*. As these motor neurons are maximally excited when *a* or *b* is stimulated separately, they naturally give no greater response when *a* and *b* are stimulated together. In brief, occlusion is due to, and is convincing evidence of, afferent fibres overlapping in their central distribution [FIG. V.32, A].

4. Subliminal fringe. Sometimes, however, the tension yielded by *a* and *b* combined is greater than the sum of the two reflex responses taken singly [FIG. V.32, B]. This result indicates that each afferent while fully activating a certain number of motor neurons acts also on a further number subliminally, and that some of these subliminally influenced motor neurons are common to *a* and *b*. Concurrent subliminal excitations can thus sum to produce liminal stimulation. This type of result is another illustration of spatial summation.

Subliminal fringe is of great importance in reflex co-ordination. It enables one level in the nervous system to reinforce the action of another. Thus, sometimes, feeble stretch of an extensor muscle produces a weak reflex contraction; rotation of the head alone [p. 302] may also produce little increase in the activity of the muscle. But the combination of the two procedures may give rise to a considerable reflex contraction of the muscle—again an example of summation of subliminal fringes.

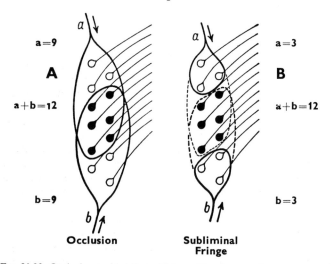

FIG. V.32. Occlusion and subliminal fringe. The diagram shows a group of motor neurons. The clear cells are exclusively influenced by the afferent fibres *a* or *b*; the dark cells are common to both *a* and *b*.
A: Stimulation of *a* or *b* effectively excites 9 motor neurons. Stimulation of *a* and *b* together excites only 12 motor neurons, because 6 are common to both afferents. Occlusion takes place.
B: Stimulation of *a* or *b* effectively excites 3 motor neurons (enclosed by continuous line) and produces a subliminal effect on another 6 motor neurons (enclosed by dotted line). Stimulation of *a* and *b* stimulates 12 motor neurons as the two subliminal fringes effectively sum.

5. Recruitment and after-discharge. These features are well brought out when the crossed extensor reflex (e.g. contraction of the quadriceps as a result of stimulation of an afferent nerve in the opposite limb) is contrasted with the motor tetanus of the same muscle [FIG. V.33].

1. The motor nerve is stimulated for a few seconds at a frequency sufficiently high to produce complete tetanus. After a brief latency

the tension developed by the muscle rises sharply to a maximum; when the stimulus is discontinued, the tension diminishes rapidly as the muscle relaxes [FIG. V.33, B].

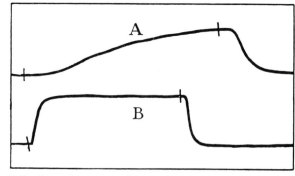

FIG. V.33. Crossed extensor reflex compared with motor tetanus. Quadriceps muscle. A: Crossed extensor reflex produced by contralateral afferent stimulation, frequency 38 s^{-1}; B: Tetanus of muscle produced by peripheral motor nerve stimulation, frequency 38 s^{-1} (isometric records). The vertical lines indicate the beginning and the end of stimulation. (Liddell and Sherrington (1923). *Proc. R. Soc. B.*)

2. When the crossed extensor reflex is elicited there is a longer latency, which is characteristic of reflex activity; under continued afferent stimulation the tension in the quadriceps rises relatively gradually to its maximum. When stimulation is stopped the tension is maintained for some time and then slowly declines [FIG. V.33, A]. The explanation of these differences is as follows:

As the muscle fibres are under the same mechanical and nutritive conditions in both experiments and the same tension develops on both occasions, the same number of muscle fibres may be presumed to be contracting in both reactions. The rapid rise of tension in the motor tetanus indicates that all the muscle fibres involved contract practically synchronously; the slower development of tension in the reflex contraction suggests that first a few motor units are affected, then gradually more and more become involved, until the full quota is in action. In other words, with continued excitation of an afferent nerve, an increasing number of motor neurons is brought under excitation; this is called excitatory recruitment. Recruitment is due to '*pseudo*-temporal summation', i.e. with repetitive afferent stimulation there is a progressive increase in the 'background' activity of the internuncials. This leads to an increase in the excitability of more and more motor neurons until spatial summation raises the local synaptic potential to threshold, causing discharge [p. 267]. If afferent stimulation is continued for a long time, effective summation takes place at a steadily increasing number of motor neurons. Indefinite prolongation of the afferent stimulus does not produce unlimited recruitment; a stimulation plateau is reached [FIG. V.33, A]. There is thus a limit to the number of motor neurons that can be recruited.

On discontinuance of afferent stimulation the tension may remain unaltered for several seconds; this is called the after-discharge plateau. As it is equal in height to the stimulation plateau, *all* the motor neurons which were ultimately excited during afferent stimulation must still be discharging. When relaxation of the muscle sets in, it proceeds more gradually than in the motor tetanus. In the latter all the muscle fibres go out of action together; in the case of the reflex it must be supposed that the motor neurons stop discharging successively and as this happens the corresponding muscle fibres relax. Recruitment thus gives inertia, and after-discharge provides momentum to reflex movements, making these movements smoother in onset and termination.

FIGURE V.34 shows the after-discharge in the flexor reflex elicited by a single afferent stimulus. Action potentials (representing the results of motor neuron discharge) are present right to the end of

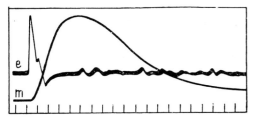

FIG. V.34. Reflex after-discharge. Mechanical (m) and electrical response (e) to single shock to afferent nerve producing reflex response. A large number of action potentials are visible in the electrical record right up to the end of muscular activity. These represent the after-discharge of the spinal ventral horn cells. Relaxation is consequently more gradual than in a motor nerve twitch. Time (on base line) in 0.01 second. (Creed, R. S. Denny-Brown, D., Eccles, J. C. Liddell, E. G. T. and Sherrington, C. S. (1932). *Reflex activity of the spinal cord*. Oxford University Press, London.)

muscular relaxation. The motor neurons thus continue to discharge long after afferent stimulation has ceased. After-discharge is attributed to persistent stimulation of the motor neurons from the internuncial background. Impulses go on wandering in these paths for varying periods after afferent stimulation ceases, continue to bombard the motor neurons, and so maintain the after-discharge [FIG. V.35].

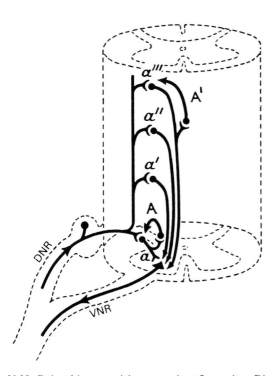

FIG. V.35. Role of internuncial neurons in reflex action. Diagrammatic section of spinal cord. DNR = dorsal (posterior) nerve root; VNR = ventral (anterior) nerve root: *a*, *a'*, *a''*, *a'''*, A, A' = internuncial neurons.

The shortest reflex arc is via DNR, a, VNR. Longer reflex arcs involving delay paths are shown via *a'*, *a''*, or *'''*. A, A', are 'reverberators'. Note how an impulse passing from DNR to VNR along *a* may branch to excite A, which in turn re-excites *a*; similarly an impulse along *a'''* may branch to excite A', which in turn re-excites *a'''*.

Nerve biophysics

Much of our understanding of the biophysics of nerve conduction has stemmed from studies of giant axons in invertebrates such as *Sepia* (cuttlefish) and *Loligo* (squid). These giant nerve fibres are c. 500 μm in diameter and this allows their longitudinal penetration by an internal wire electrode itself approximately 100 μm in diameter. When the internal electrode is connected across a voltmeter with an external electrode in the sea water bathing the axon a membrane potential of about 50–70 mV (inside negative to outside) is registered.

The axoplasm inside the membrane can be squeezed out from the cut end and chemically analysed. Squid axoplasm contains far more potassium ions than does squid blood or the sea water which is commonly used as the bathing medium in such experiments. Conversely, axoplasm has a much lower concentration of sodium than that of the extracellular fluid. TABLE V.1 shows the discrepancy in these and other ionic concentrations.

TABLE V.1. *Concentrations of ions and other substances in squid axons*

Substance	Concentration (mmol kg H_2O^{-1})	
	Axoplasm	Sea water
K^+	400	10
Na^+	50	450
Cl	40–150	540
Ca^{2+}	0.4	10
Mg^{2+}	10	55
Organic anions	365	—
Water	865 g kg^{-1}	966 g kg^{-1}

In the external fluid about 90 per cent of the osmotic pressure is due to sodium and chloride ions whereas in the axoplasm these ions contribute less than 10 per cent of the total osmotic balance, potassium and organic anions replacing sodium and chloride respectively.

The potassium ions inside the fibre are not bound to proteins. Osmotic balance could not be accounted for unless the main ionic constituents of the axoplasm were free; moreover, axoplasm has an electrical conductivity at least three-quarters of that of sea water.

In the resting state the nerve fibre is said to be *polarized*. The resting membrane is assumed to be more permeable to potassium than to sodium ions. As the potassium ions are more concentrated within the axoplasm they tend to set up a potential difference with the inside negative and the outside positive, for the negative organic anions which cannot permeate the membrane hold the potassium ions from diffusing down the concentration gradient from axoplasm to the external solution. If the membrane were permeable to potassium ions alone the transmembrane potential difference would approach the equilibrium potential for a potassium electrode as defined by the Nernst equation

$$E_K = \frac{RT}{nF} \log_e \frac{[K]_o}{[K]_i}$$

where E_K is the equilibrium potential of the potassium ion defined in the sense internal potential minus external potential; R is the gas constant (8.316 joules per degree); T is the absolute temperature, n is the valency (unity), and F the Faraday (96 500 coulombs per mole). $[K]_o$ and $[K]_i$ are the external and internal concentrations of potassium respectively.

The development of the Nernst equation is as follows: As there is a negative charge within the cell and as there is a thirtyfold intracellular concentration of potassium within the cell, then although the concentration gradient favours the diffusion of K^+ outwards the negative intracellular charge impedes this. At equilibrium electrical forces balance the tendency of K^+ to move out and inward and outward fluxes are equal.

The work done per mole in bringing the potassium ions (or any ion) to a potential E_1 from a potential of E_0 is:

$$W = nF (E_1 - E_0)$$

where W is work done and therefore difference of potential energy; F is the Faraday (coulombs per mole); and n is the valency.

The difference of potential energy per mole between two solutions of different concentrations of the ion is identical with the work done in changing the volume containing one mole of the ion, for the solvent molecules only separate the ions. The situation is analogous to that of compressing a gas:

From the gas law ($PV = RT$)

$$W = - \int_{E_0}^{E_1} P dV$$

Hence $W = - \int_{E_0}^{E_1} RT \frac{dV}{V} = RT \log_e \left[\frac{E_0}{E_i} \right]$

$$= -RT \log_e \left[\frac{c_0}{c_i} \right]$$

where c_0 and c_i are the ionic concentrations outside and inside.

At equilibrium

$$nF (E_1 - E_0) - RT \log_e \frac{c_0}{c_i} = 0$$

$$\therefore E_1 - E_0 = \frac{RT}{nF} \log_e \frac{c_0}{c_i} .$$

$(E_1 - E_0)$ is the electrochemical potential at equilibrium E at which there is no net flux of the ion. The Nernst equation describes this situation at equilibrium for any ion.

At a temperature of 18 °C this equation simplifies to:

$$E_K \text{ (volts)} = \frac{8.316 \times 291}{96\,500} \times 2.3 \log_{10} \frac{[K^+]_o}{[K^+]_i}$$

$$= 0.058 \log_{10} \frac{[K^+]_o}{[K^+]_i} \text{ volts.}$$

$$E_K = 58 \log_{10} \frac{[K^+]}{[K^+]} \text{ millivolts.}$$

In squid blood (20 mmol $[K^+]$) and with an axoplasmic potassium concentration of 400 mmol, the value for E_K would be

$$E_K = 58 \log_{10} \frac{20}{400} = 58 \log_{10} \frac{1}{20} = 58 \times (-1.3)$$

$$= -75 \text{ mV.}$$

Resting membrane potentials of −70 mV have been recorded in

intact axons with natural circulation. In isolated axons immersed in sea water the potential differences recorded are rather smaller. However, the dependence of the *resting* potential on the disparity between $[K^+]_o$ and $[K^+]_i$ is strikingly shown by varying the potassium concentration in the external medium [FIG. V.36].

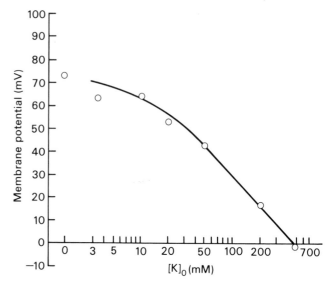

FIG. V.36. Effect of external potassium concentration on the resting membrane potential of *Sepia* axons. Temperature 17–19 °C. (Redrawn from Hodgkin, A. L. and Keynes, R. D. (1955). *J. Physiol., Lond.* **128**, 61–88.)

When the sodium concentration of the external medium is altered, however (choline replacing sodium to preserve the electrochemical and osmotic situation) this has no effect on the *resting* membrane potential. The resting cell membrane thus appears to be *relatively* impermeable to sodium. However, when labelled ions ^{24}Na and ^{42}K are used it is found that the flux rates of sodium and potassium ions are of the same order of magnitude—i.e. that the amounts of each species which exchange across the fibre surface in any given time are approximately equal. But if we consider the transfer of ions *from the external fluid into the fibre*, we must bear in mind that $[Na^+]_o$ exceeds $[K^+]_o$ some forty times so that equal inward flux rates must mean that potassium can penetrate the membrane far more easily than sodium. In other words, the *permeability* or conductance of the membrane (g) (which for any ion may be defined as the ratio of its inward flux to its *outside* pressure) is much greater for potassium than for sodium. In resting *Sepia* axons Keynes (1949) found a ratio of 15:1 for $g_K : g_{Na}$.

Cable properties of nerve fibres. As has been seen the nerve fibre is provided with an electrochemical battery which is charged at some 70–100 mV negative to its surroundings and which maintains large differences of ion concentrations. *It is the nerve membrane itself which possesses this property*. Thus, if the inserted microelectrode scrapes the inner surface of the axonal membrane irreversible damage results associated with a permanent local *depolarization*—i.e. the transmembrane potential falls to zero and nerve impulse conduction is blocked. The potential energy stored in the membrane is normally utilized during the passage of an impulse so as to ensure its forward conduction over considerable distances without loss of signal strength. The presence of the insulating membrane containing capacitance and resistant elements endows the axon with cable-like properties [FIG. V.37]. However, if we compare the electric constants of nerve with those of ordinary communication cables, the leakage of the axon membrane per square centimetre is a hundred million times higher and its capacity about a million times higher, while its core conductivity is a hundred million times lower than that of a commercial cable. As a result any *subthreshold* signals in a nerve fibre (depending as in an inert cable entirely upon energy supplied from outside) only travel 1–2 mm along the fibres, being rapidly attenuated. However, the cable properties of the axon play a very important part in impulse transmission. They provide an electric link between adjacent zones of the nerve and ensure that currents which have been generated by the electric activity of one region will spread into the adjacent resting region sufficiently to excite it and to advance the signal further.

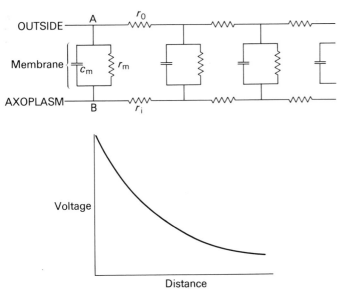

FIG. V.37. Diagram of the passive electric properties of a cable-like length of axon containing capacity and resistance elements. Current applied between points A and B produces transmembrane potentials which decay exponentially with distance.

When an electric current passes through the surface membrane the *permeability* of the membrane to ions is altered and if this alteration be of sufficient degree (threshold) the electric charge of the membrane becomes unstable and a self-regenerating and propagating electrical change develops. *This process of excitation occurs at the region where electric current passes outwards through the fibre membrane—i.e. at the cathode or in front of an action potential.* Threshold excitation occurs *when the transmembrane potential is sufficiently reduced* and manifests itself in a rapid reversal of the membrane potential so that the axoplasm becomes electrically positive to the extracellular fluid. This is followed within a matter of 1–2 ms by a return to the resting level. This action potential or spike potential [FIG. V.38] is of an explosive all-or-none character which acts as a travelling cathode, generating local eddy currents which excite neighbouring regions of the nerve. It travels at a constant strength and speed in any one nerve fibre (if the temperature is constant), leaving the fibre ready to respond again identically after a brief refractory period of 1–2 ms. The action potential relies for its transmission on the local energy resources of the successive nerve segments which it traverses and not upon that supplied at its point of origin, providing that the initial stimulus itself is of threshold strength.

Alterations in membrane permeability during the passage of a nerve impulse

In 1938 Curtis and Cole showed that a threshold excitation of a single axon increased the membrane conductance fortyfold [FIG. V.38] and in 1939 Hodgkin and Huxley proved that the transmembrane potential difference changed from one that was some 60 mV inside negative to outside to a value which was approximately 40 mV inside positive to outside. The height of the spike potential

was approximately 100 mV; it lasted slightly longer than 1 ms, the membrane potential returning to approximately its resting value.

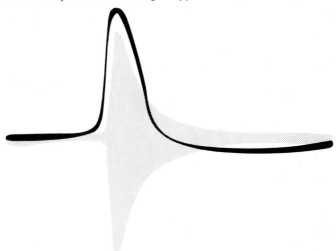

FIG. V.38. The impedance change during an action potential. Two superimposed records are shown. The thin line is an action potential of a squid giant axon and the continuous curve is a measure of the ease with which ions can pass across the membrane (impedance). Time scale logarithmic. Widening of the impedance trace indicates a reduction of resistance of the axon membrane and is not due to a change in membrane capacity. Resistance falls from 1000 Ω cm^{-2} to 25 Ω cm^{-2} during the action potential. (Redrawn from Cole, K. S. and Curtis, H. J. (1938). *J. gen. Physiol.* **22**, 649.)

The changes thus described can be satisfactorily explained by assuming that the selective permeability of the membrane is momentarily reversed, sodium now being able to penetrate much faster than can potassium. The inrush of sodium ions therefore briefly causes the axoplasm to become electrically positive with respect to the bathing medium. Consonant with this proposition the height of the spike is smaller when the extracellular [Na$^+$] is lowered [FIG. V.39]; in sodium free ECF no action potential develops at all and the nerve fibre becomes inexcitable.

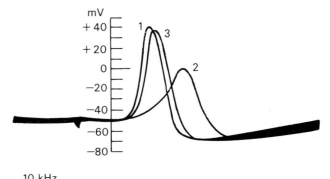

FIG. V.39. Effect of sodium-deficient extracellular fluid on the action potential. Records 1 and 3: axon in sea water. Record 2: axon in one-third sea water and two-thirds isotonic dextrose. (Record 3 was obtained after Record 2 and shows the recovery of the spike height and rate of development of the spike almost to that shown in Record 1. (Redrawn from Hodgkin, A. L. and Katz, B. (1949). *J. Physiol., Lond.* **108**, 37.)

The resistance (reciprocal of conductance) of the squid axon falls from its resting value of 1000 ohm cm^{-2} to 25 ohm cm^{-2} during the upstroke of the action potential but neither the membrane electrical capacity nor the resistance of the axoplasm itself alters appreciably. *These facts indicate that the spike potential arises from permeability changes in the membrane itself* and this has been proved to be so in studies made by Hodgkin, Baker, and Shaw and their colleagues. They submitted the squid axon to a series of sweeps of a roller (which extruded the axoplasm) and either reinflated or perfused the axon with isotonic solutions of varying composition. These internal perfusates necessarily contained potassium but the anions were unimportant providing that the pH of the solution was approximately buffered to pH 7.5. Such preparations yielded action potentials which closely resembled those of the normal intact axon [FIG. V.40].

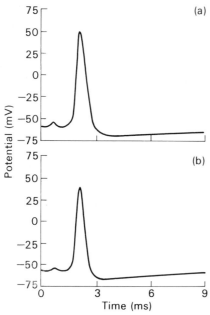

FIG. V.40. Action potentials from an axon perfused with potassium sulphate (a) and from an intact axon (b) are quite similar. (Redrawn from Baker, P. F., Hodgkin, A. L., and Shaw, W. I. (1962). *J. Physiol., Lond.* **164**, 330.)

If the internal solution which has replaced the axoplasm contains more sodium than does normal axoplasm the action potential is smaller and slower in its development [FIG. V.41]. Replacement of the Na-rich solution with isotonic potassium sulphate restores the normal spike potential. It can be inferred that the difference in

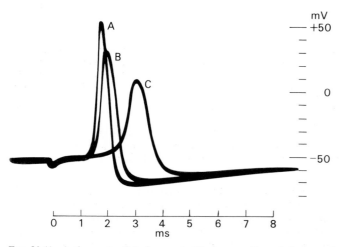

FIG. V.41. Action potentials from a 'refilled' axon. Record A: isotonic potassium sulphate has replaced the axoplasm. Record B: one-quarter of potassium replaced by sodium. Record C: one-half of potassium replaced by sodium. (Redrawn from Baker, P. F., Hodgkin, A. L. and Shaw, T. I. (1962). *J. Physiol., Lond.* **164**, 330.)

sodium concentration provides the electromotive force which generates the action potential.

Hodgkin and Huxley (1952) in a classic series of papers proposed that the nerve membrane be represented by an electrical circuit shown in FIGURE V.42. The electrical capacity of the membrane (1 μF cm^{-2}) is shown as C_m; conducting pathways are represented by parallel channels each possessing a battery and a resistance. At rest R_{Na} is high (g_{Na} is low) compared with R_K and the membrane potential is near E_K.

Normally when the axon is excited an impulse propagates along the nerve fibre upon which the transmembrane potential changes with time and distance and currents, varying in time, flow through all the elements in the cable. To simplify the experimental situation for analysis the voltage clamp technique is used. This consists in measuring the flow of current through a definite area of membrane of the giant axon when the membrane potential is kept uniform over this area or is changed in a stepwise manner by a feedback amplifier.

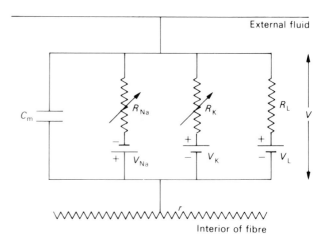

FIG. V. 42. Diagram of an element of the excitable membrane of a nerve fibre. (Redrawn from Hodgkin, A. L. (1964). *The conduction of the nervous impulse*. Liverpool University Press.)

The analysis of the electrical changes in the axon during activity

The resting membrane of the axon is not solely permeable to K$^+$; the membrane is not totally impermeable to Na$^+$ and it is also permeable to Cl$^-$. The potential developed across a membrane with such properties has been expressed in the Goldman–Hodgkin–Katz 'constant-field' equation which assumes a uniform voltage throughout the axonal membrane.

$$E_m = \frac{RT}{F} \log_e \frac{P_K[K]_o + P_{Na}[Na]_o + P_{Cl}[Cl]_i}{P_K[K]_i + P_{Na}[Na]_i + P_{Cl}[Cl]_o}$$

where P is the permeability coefficient of each respective ion. Obviously, if P_K is far greater than P_{Na} or P_{Cl} then the equation reduces to that of Nernst.

As stated previously [p. 263], when the membrane is depolarized by an outward flow of current such as produced by an applied cathode or by an adjacent active region invaded by an action potential, sodium permeability g_{Na} rises immediately and sodium ions rush in down the concentration gradient. Sodium enters more quickly than potassium leaves and this further lowers the transmembrane potential contributing to a further depolarization which explosively accelerates the entrance of sodium. This causes the rising phase of the action potential and the inside of the cell rapidly becomes positive reaching values not far short of the equilibrium potential for sodium E_{Na} (approximately 60 mV positive), at which net inward sodium movement is zero. Additionally, the rise in g_{Na} ceases and sodium permeability falls (inactivation). This occurs at the peak of the spike. Approximately simultaneous with the peak of sodium conductance potassium conductance begins to rise, reaching its peak about 0.5 ms later. g_{Na} falls more swiftly to its resting value (in 1–1.5 ms) than g_K which returns to normal in about 3 ms [see FIG. V.48, p. 273].

The electric currents which flow during the propagation of the nerve impulse are energized by the resting ionic gradients for sodium and potassium.

Voltage clamp studies

Two fine silver wire electrodes are inserted down the axis of the giant fibre for a distance of 30 mm or so [FIG. V.43]. The potential difference across the membrane is measured between one of these wires (B) and the electrode C immediately outside the axon in the sea water which surrounds it. Wire A is used for passing current through the membrane to the external electrode D. The voltage wire B is connected to the input of an amplifier whose output goes to the current wire A, the direction of the connections being such that any accidental change of membrane potential is almost completely offset by the current that the amplifier sends through the membrane. A second input to the amplifier allows rectangular pulses to be delivered. When such a pulse is fed in, the amplifier automatically sends through the current wire A whatever current is required to make the membrane potential undergo stepwise changes propor-

tional to those which are applied through the second input. This current is displayed on a cathode-ray oscilloscope and photographed.

When the two inputs to the amplifier are the same, no current flows—i.e. in the resting axon with an internal potential of −65 mV and the setting on the amplifier source at the same potential. When the setting of the variable potential is changed from −65 mV to −15 mV the difference in potential between the two inputs to the amplifier evokes a passage of current in such a direction as to change the axoplasmic potential to −15 mV and to hold it there (negative feedback). In normal circumstances an action potential would be initiated and sodium ions would enter, making the axoplasm more positive. The feedback system ensures that the potential inside the fibre is stabilized at −15 mV by passing a compensating current in the opposite direction. This current (which is measured) compensates for the ionic currents through the membrane.

When the membrane potential is suddenly artificially reduced to zero there is an almost instantaneous discharge of *capacity* current from the capacitative element of the membrane [FIG. V.44] lasting a few *microseconds*. Analysis of this initial pulse confirms that the capacity of the axon membrane is approximately 1 microfarad per square centimetre. Subsequent to this ultrashort discharge the remaining current flow is *ionic* in nature and flows for a few milliseconds; this current can be analysed during the period in which the membrane potential is held constant by the feedback system.

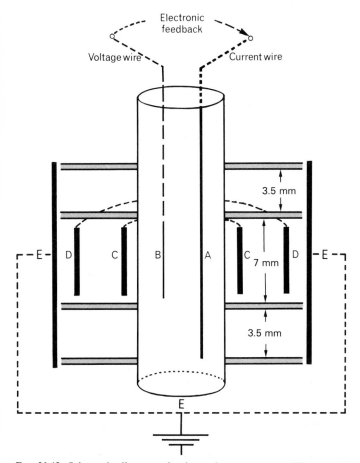

FIG. V.43. Schematic diagram of voltage-clamp apparatus. The axon is immersed in sea water and the stippled horizontal lines represent partitions in a box made of insulating material which guided the current flow. Potential difference across membrane was measured between wires B and C; current passed from wire A to electrode E. Current through the middle section of the nerve was measured as potential drop in sea water between wires C and D. (After Huxley, A. F. (1964). *Science, N.Y.* **145**, 1154. Copyright by the American Association for the Advancement of Science.)

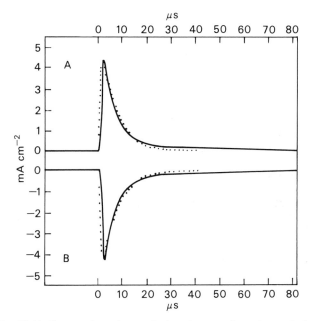

FIG. V.44. Current through capacitative element of membrane during a voltage clamp. Abscissa: time in microseconds. Ordinate: membrane current density (mA cm^{-2}) with inward current taken as positive. At $t = 0$ the potential difference between external and internal electrodes was displaced +40 mV in curve A or −40 mV in curve B. The continuous curves were traced from experimental records. The dotted curves were calculated according to the equation

$$I^* = 6.8 \left[\exp(-0.159t) - \exp(-t) \right]$$

where I^* is the current in mA cm^{-2} and t is time in μs. (Hodgkin, A. L., Huxley, A. F., and Katz, B. (1952). *J. Physiol., Lond.* **116**, 424.)

When the normal potential difference across the membrane is *increased* (i.e. the inside of the fibre being made more negative, say from a resting value of −65 mV to −100 mV) the very small current which results is always inward, as it should be from Ohm's law. When the inside of the fibre is made more positive by an equal amount, however, the currents are larger and show a marked *early* phase in which the current direction is against that of the change in membrane potential. If it were not for the feedback this current would drive the inside of the fibre still more positive and would produce the rising phase of an action potential. This transient initial inward current is due to a movement of sodium ions down their concentration gradient and is succeeded by an outward current carried by potassium ions.

By changing the extracellular concentration of Na and K their respective contributions to the ionic current aroused by depolarization can be varied. Thus FIG. V.45 shows the ionic currents obtained when the axon was in sea water and was submitted to a sustained depolarization of 56 mV [FIG V.45A]. When the resting axon was immersed in sea water in which 90 per cent of the sodium chloride was replaced by choline chloride a similar depolarization of 56 mV yielded FIG. V.45B. The inward current has disappeared

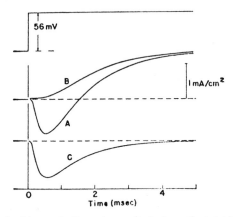

FIG. V.45. Squid axon (voltage clamped). A shows the total ionic current which flows during a sustained depolarization of 56 mV when the axon is in sea water. B shows the ionic current during a depolarization of 56 mV when the axon is in a solution in which sodium is replaced by choline, but which in other respects is similar to sea water. The ionic current carried by potassium ions is solely outwards (upwards in the figure). C is obtained by subtracting B from A and shows the inward current carried by sodium. (Hodgkin, A. L. and Huxley, A. F. (1952). *J. Physiol., Lond.* **117**, 500.)

and only an outward current remains; this current is carried by potassium ions and can be seen to be slightly delayed (c. 1 ms) compared with the onset of current development occurring when the fibre was immersed in normal sea water. By subtracting curve B from curve A, curve C shows the sodium current which is wholly inward in these circumstances and which, moreover, disappears in about 4 ms despite the maintenance of the clamped depolarizing potential. As will be noted later, the outward potassium current shows no such disappearance. The identification of this outward current with potassium efflux is validated by measuring membrane current density simultaneously with potassium efflux using radioactive potassium (^{42}K). The increase in potassium ion efflux is linearly proportional to the current density with a slope equal to Faraday's constant so the outward current is indeed carried by potassium ions [FIG. V.46]. If the membrane is depolarized so that the total potential difference equals that of a sodium concentration cell ($E_{Na} = +55$ mV) there is no sodium current—only the delayed outward potassium current and for displacements beyond this value of V_{Na} the sodium current is outward [FIG. V.47].

The separation of the ionic current into its two components allows the estimation of the membrane conductivity for sodium and for potassium [FIG. V.47]. By dividing the ionic currents by the respective electrochemical gradients g_{Na} and g_K can be calculated

$$I_{Na} = g_{Na} (E-E_{Na})$$

$$I_K = g_K (E-E_K)$$

where $V = E - E_r$, E_r being the resting potential. Thus V, V_{Na}, and

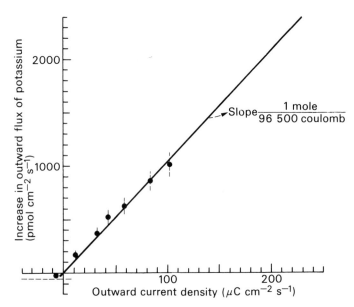

FIG. V.46. Abscissa: mean outward current density. Ordinate: mean increment in ^{42}K outflux associated with flow of current in μC cm^{-2} s^{-1} (Hodgkin, A. L. and Huxley, A. F. (1953). *J. Physiol., Lond.* **121**, 403–14.)

V_K can be measured directly as displacements from the resting potential.

FIGURE V.48 shows the changes in conductance produced when the interior of the axon, at rest some 50–60 mV negative to the outside, is suddenly made more positive by 56 mV—a change with corresponds to short-circuiting the membrane. Sodium conductance increases rapidly from a very small value to 25 mmho cm^{-2} and thereafter declines exponentially *despite* the maintenance of depolarization. Potassium conductance, again starting from a low level, after about 0.5 ms responds to the depolarization, rising in an S-shaped curve to about 20 mmho cm^{-2} and maintaining this value *despite* continuance of the depolarizing voltage. Both sodium and potassium conductances decline exponentially when repolarization of the membrane is permitted but the speed of decline of g_{Na} is ten times faster than that of g_K.

It is important to understand that sodium conductance can be reduced in two different ways. If the resting potential, displaced by a depolarizing stimulus, is rapidly restored the system controlling sodium permeability quickly reverts to its resting value. But if the depolarization is maintained the sodium conductance is reduced

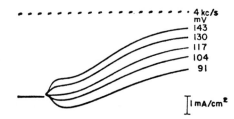

FIG. V.47. Membrane currents when the internal potential is raised from −62 mV to values comparable to the peak of an action potential. Axon in sea water; temperature 3.5 °C; *outward current upwards*. The records for 91− and 104−millivolt displacement of membrane potential show a phase of *inward current*, while those for 130 and 143 millivolts show an early hump in the outward current. The record at 117 millivolts depolarization shows neither, and it is therefore taken to be very close to the sodium equilibrium potential, at which the current carried by sodium is zero. *Note*—if the resting potential is −62 mV, a depolarization of 117 mV would change it to +55 mV ($\equiv E_{Na}$). The outward currents are solely due to potassium unless the depolarization exceeds 117 mV. (From Hodgkin, A. L., Huxley, A. F., and Katz, B. (1952). *J. Physiol., Lond.* **116**, 424.)

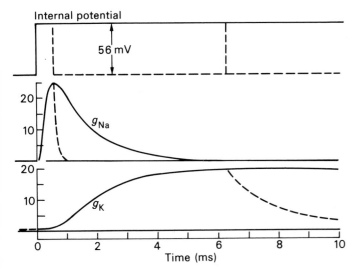

Internal potential

g_{Na}

g_K

FIG. V.48. Time course of sodium conductance (g_{Na}) and potassium conductance (g_K) associated with depolarization of 56 mV; vertical scale in mmho cm^{-2}. The continuous curves are for a maintained depolarization; broken curves give the effect of repolarizing the membrane after 0.6 or 6.3 ms. (From Hodgkin (1958), based on Hodgkin and Huxley (1952). *J. Physiol., Lond.* **116**, 473–96.)

more slowly by a process termed *inactivation* (Hodgkin and Huxley 1952). Once inactivation has occurred the membrane must remain in a repolarized state for several milliseconds before a second stimulating pulse is effective once more. (This is the essential basis of the refractory state and the features of the absolute and relative refractory period.) Potassium conductance, however, shows no inactivation during sustained depolarization.

FIGURE V.49 shows families of curves for changes in sodium and potassium conductances caused by different voltage displacements (which are sustained). The experimental points (circles) fit smooth curves which are solutions of equations developed by Hodgkin and

Huxley (1952). The details of the development can also be found in Hodgkin's Sherrington Lectures (1964).

Although sodium conductance changes precede those of potassium those governing g_K can be considered as more simple and will be discussed first. Hodgkin and Huxley defined a quantity n which varied with first-order kinetics—i.e. for each value of membrane potential there was a corresponding equilibrium value of n which was approached exponentially with a time constant which was also a function of membrane potential. Examining changes of g_K caused by a depolarization of 25 mV they decided that although during repolarization the changes of g_K could be fitted by a first order equation, those shown during depolarization required a fourth-order equation. In such a case the rise of g_K from zero to its finite value is described by $(1 - \exp(-t))^4$ while the fall is given by $\exp(-4t)$.

Thus they wrote:

$$g_K = g_K n^4$$

where g_K is a constant with the dimensions of conductance cm^{-2} and n is a dimensionless variable which can vary between 0 and 1.

They used rate constants α_n and β_n which varied with voltage but not with time, having the dimensions of [time]$^{-1}$

$$dn/dt = \alpha_n (1-n) - \beta_n n.$$

Their interpretation assumed that potassium ions can cross the membrane only when four similar particles occupy a certain region of the membrane. n represents the proportion of the particles in a certain position (say, inside the membrane) and $1-n$ the proportion elsewhere (e.g. outside the membrane). α_n determines the rate of transfer from outside to inside and β_n the transfer in the opposite direction. Thus, in this case β_n would decrease when the membrane is suddenly depolarized.

The time constant is given by

$$\tau_n = \frac{1}{(\alpha_n + \beta_n)}.$$

Hille (1970) has presented graphically changes in ionic conductance of potassium and the time constant τ_n with alterations of membrane potential [FIG. V.50].

At the resting potential n is small and the potassium conductance or permeability correspondingly low. With depolarization, potassium permeability is activated but relatively slowly because of the fairly high value of the time constant n compared with that which governs activation of sodium permeability.

In the case of sodium, Hodgkin and Huxley concluded that sodium movement depended on the distribution of charged particles which only when they occupied particular sites in the membrane allowed sodium transference. They tested a hypothesis which considered that sodium conductance was proportional to the number of sites on the *inside* of the membrane which were occupied simultaneously by three activating molecules but which were not blocked by an inactivating molecule. The proportion of *activating* molecules on the inside was designated m and $1-m$ refers to the proportion of such molecules on the outside. As the features of inactivation of sodium conductance already referred to had to be taken into account, they proposed that h should designate the proportion of *inactivating* molecules on the outside and $1-h$ the proportion on the inside of the membrane. α_m or β_h and β_m or α_h represented transfer rate constants in the two directions. Time constants were respectively:

$$\tau_m = \frac{1}{(\alpha_m + \beta_m)}$$

$$\tau_h = \frac{1}{(\alpha_h + \beta_h)}$$

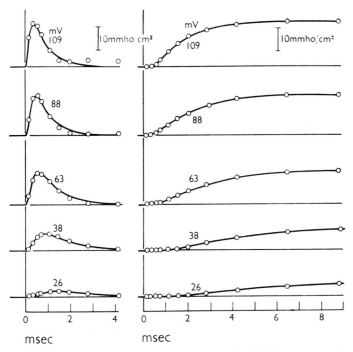

msec   msec

FIG. V.49. Time course of sodium (left) and potassium (right) conductances for different sustained displacements of membrane potential at 6 °C; the numbers give the depolarization used. The circles are experimental estimates and fit smooth curves obtained from equations. (Hodgkin, A. L. and Huxley, A. F. (1952). *J. Physiol., Lond.* **117**, 500.)

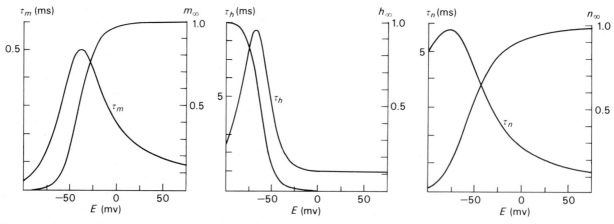

FIG. V.50. The steady-state values and time constants for the parameters *h*, *m*, and *n* in the Hodgkin–Huxley model. Note different time scales. The resting potential is $E = -65$ mV. (Redrawn from Hille, B. (1970). *Prog. Biophys.* **21**, 1–32.)

The relevant equations were:

$$g_{Na} = m^3 h g_{Na}$$
$$dm/dt = \alpha_m (1-m) - \beta_m m$$
$$dh/dt = \alpha_h (1-h) - \beta_h h.$$

From such studies Hodgkin and Huxley were able to calculate the variations of potassium and sodium permeabilities which would account for the propagated action potential. Their theoretical model agreed remarkably faithfully with that actually recorded [FIG. V.51].

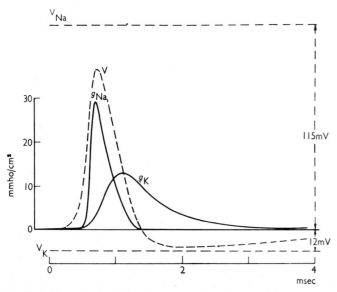

FIG. V.51. Theoretical solution for propagated action potential and conductances at 18.5 °C. Total entry of sodium = 4.33 pmol cm^{-2}; total exit of potassium = 4.26 pmol cm^{-2}. The impulse is passing from right to left. (Hodgkin, A. L. and Huxley, A. F. (1952). *J. Physiol., Lond.* **117**, 500.)

As the impulse advances along the nerve fibre the potential difference across the membrane immediately in front of the active region is altered by electric currents flowing in a local circuit through the axoplasm and the surrounding fluid medium. This increases g_{Na} and sodium ions which enter render the inside of the axon positive and provide the current needed to excite the next segment. At the crest of the impulse the slower changes which result from depolarization take effect; g_{Na} falls and g_K rises so that K$^+$ leaves the fibre at a greater rate than that which Na$^+$ enter. As a consequence the potential swings back towards the equilibrium potential of the potassium ion [FIG. V.52].

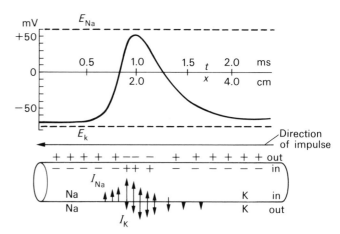

FIG. V.52. Membrane potentials in active nerve, either at the same place at different times (*t* scale) or at different places at the same time (*x* scale). Lower diagram shows ion flow, with the length of the arrows indicating rate of flow. Note that Na$^+$ inflow precedes K$^+$ outflow. The nerve impulse is propagating to the left. (Modified from Aidley, D. J. (1971). *The physiology of excitable cells*. Cambridge University Press. After K. S. Cole.)

Saltatory conduction in myelinated nerve fibres

As previously described, many vertebrate nerve fibres are surrounded by a fatty sheath known as myelin. This sheath is interrupted every millimetre or so to form the nodes of Ranvier. Myelin is formed from numerous layers of Schwann cell membrane caused by the mesaxon being wrapped repeatedly round the axon itself. Contact between the axon membrane proper and the extracellular fluid is established only at the nodes.

The myelin sheath itself has a much higher resistance and a much lower transverse capacitance than the axon membrane. Transverse resistance is some 160 000 Ω cm^2; capacitance is only 0.0025 μf cm^{-2}. At the node of Ranvier membrane resistance is 20 Ω cm^2 and capacitance 3 μF cm^{-2}. Thus it is likely that when current is passed across the membrane its flow through the nodes will exceed that in the internodal segments. Moreover one might expect the myelinated fibre to manifest conduction from node to node—i.e. discontinuous or saltatory (*saltare* being Latin for 'to jump') in nature.

Such has proved to be the case. Using single myelinated fibres Tasaki and his colleagues showed that threshold stimulus intensity was lowest at the nodes [FIG. V.53] and that conduction-blocking agents such as cocaine were only effective when applied to the nodes. Tasaki and Takeuchi (1942) placed a nerve fibre in three pools of Ringer's solution which were insulated from each other by

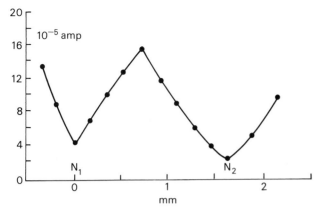

FIG. V.53. The variation in threshold intensity of stimulation along the length of a single myelinated fibre. N_1 and N_2 mark the position of two nodes of Ranvier. (Tasaki, I. (1953). *Nervous transmission.* Thomas, Springfield, Ill.)

air gaps. The two outer pools were earthed and the middle pool was connected to earth through a resistance. When the fibre was electrically excited all the radial current flowing across that part of the fibre which was in the middle pool of Ringer flowed through the resistance and could be measured by the potential of the middle pool with respect to earth. The results were clear-cut; inward currents occurred only when there was a node in the middle pool—inward currents must therefore be confined to the nodal regions [FIG. V.54].

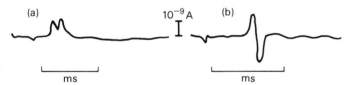

FIG. V.54. Radial currents in a short length of a myelinated fibre during the passage of an action potential (see text). (a) shows no inward current when the middle pool of Ringer does not contain a node of Ranvier. (b) Inward current (downwards) when the middle pool of Ringer contains a node of Ranvier. (Tasaki, I. and Takeuchi, T. (1942). *Pflügers Arch. ges. Physiol.* **245**, 274.)

Huxley and Stämpfli (1949) extended these observations. They drew a single myelinated fibre through a short length (0.7 mm) of a capillary 40 μm diameter filled with Ringer's solution [FIG. V.55(a)]. They measured the distribution of current during an impulse. A record of the potential difference across the capillary

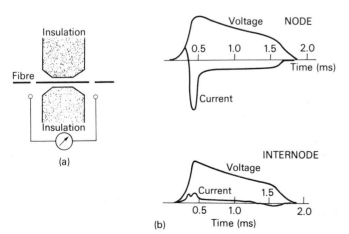

FIG. V.55. Potential and current changes during a propagated impulse. Outward current in each case is plotted upwards. (Huxley, A. F. and Stämpli, R. (1949). *J. Physiol., Lond.* **108**, 315.)

gave the longitudinal current in the external fluid as a function of time. Using a micromanipulator to slide the fibre through the capillary and taking a series of records the longitudinal current was found as a function of time and distance. The radial current through the node or through the myelin was obtained by differentiating the longitudinal current with respect to distance and the potential difference across the surface by integrating with respect to distance. FIGURE V.55(b) shows the radial current in the node or internode of the fibre. As in Tasaki and Takeuchi's experiments inward current during excitation is confined to the node: the myelinated segments show only a smaller, outward, component [FIG. V.55(b)].

If current enters or leaves the axoplasm only through the nodes, then the flow of current through any internodal part of the axon during the approach of a nerve impulse should be the same at all points along the internode. Hence the recorded action currents should have the same time relation at all sites along any single internode. This is actually the case [FIG. V.56].

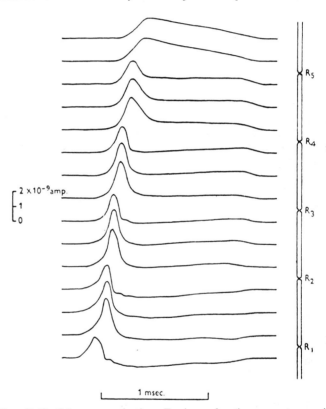

FIG. V.56. Saltatory conduction. Tracings of action current records obtained at a series of positions along a single medullated fibre. Diagram of fibre of right-hand side shows positions where each record was taken. The three records from any one internode are practically synchronous, while records from different internodes are displaced in time. (Huxley, A. F. and Stämpfli, R. (1949). *J. Physiol., Lond.* **108**, 315.)

There is no doubt that saltatory conduction is the mode of transmission in myelinated nerve. It would seem to have the advantage of being more rapid. [FIG. V.57] shows a comparison between the 'local circuit' in unmyelinated and myelinated fibres.

However, as Rushton (1951) pointed out myelinated fibres only manifest an increase in conduction velocity when their size exceeds 1 μm [FIG. V.58].

Pharmacological methods of separating sodium and potassium currents

Tetrodotoxin (TTX) obtained from the ovaries of the Japanese puffer fish is a deadly poison which selectively abolishes the voltage-sensitive sodium permeability. TTX acts only on the outside of

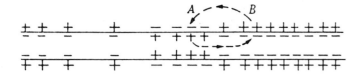

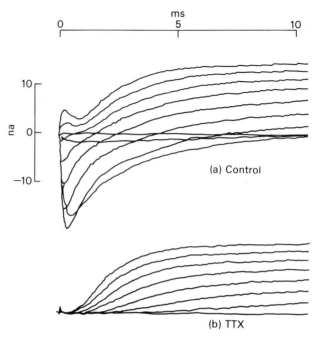

FIG. V.57. Diagrams illustrating the local circuit theory; the upper sketch represents an unmyelinated nerve fibre, the lower a myelinated nerve fibre. (Hodgkin, A. L. (1958). *Proc. R. Soc.* **B148**, 1.)

the neuronal membrane; if TTX is perfused through the squid axon it is ineffective in blocking the sodium current aroused by depolarization. However, if an axon poisoned by TTX applied externally is depolarized, inward sodium current is prevented, although the delayed outward potassium current is unaffected [FIG. V.59].

Tetraethyl ammonium (TEA), on the other hand, blocks the potassium permeability mechanism selectively when applied [FIG. V.60].

Pronase (a proteolytic enzyme) when perfused through a squid axon has a curiously selective influence. It abolishes the *inactivation* of sodium conductance of the axonal membrane without modifying the *activation* of the conductance of sodium or of potassium. When applied to the exterior of the axon it has no effect at all so it would seem that the molecules involved in or responsible for the process of sodium inactivation are situated only on the inner surface of the axonal membrane (Armstrong, Bezanilla, and Rojas 1973). Such molecules which cause the inactivation of g_{Na} fulfil the role of the h factor in the Hodgkin–Huxley equations.

Channels and gating particles

It is now customary to use the term 'channels' for the pathways for the movements of sodium and potassium. Such a term tentatively denotes the concept of pathway without requiring that different ions have different pathways or even that pathways be discrete localized structures (Hille 1970).

A prepulse to −50 mV inactivates the Na channels. A prepulse to −80 mV removes inactivation. A prepulse to −200 mV strongly delays the opening of potassium channels. In all cases the time course of the opening of sodium channels bears no simple relation to the time course of the opening of K channels—the different channels have independent kinetics. Moreover, the channels differ

FIG. V.59. Voltage clamp currents in a node of Ranvier (*Rana pipiens*). TTX concentration 300 nM. Note that TTX abolishes the inward sodium current; only the outward potassium current remains. (Hille, B. (1966). *Nature, Lond.* **210**, 1220–2.)

in their structure. A sodium channel is selective for Na and Li ions and is less permeable to K ions. It may frequently be unoccupied and may be asymmetrical with the highest energy barrier at its outer end. A potassium channel which accepts Rb as well as K is less

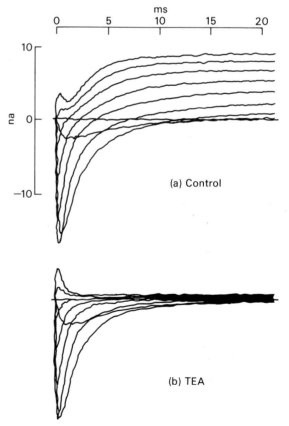

FIG. V.60. Voltage clamp currents in a node of Ranvier (*Rana pipiens*). Note that TEA blocks the outward potassium current; only the inward sodium current remains. (Hille, B. (1966). *Nature, Lond.* **210**, 1220–2.)

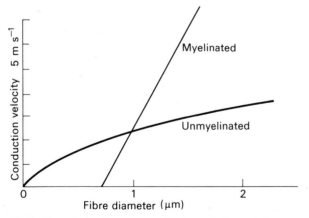

FIG. V.58. The relation between conduction velocity and fibre size in myelinated and unmyelinated nerve. Myelinated fibres conduct faster only when greater than 1 μm in diameter. (Rushton, W. A. H. (1951). *J. Physiol., Lond.* **115**, 101–22.)

permeable to Na. It may retain several K ions at one time and may be asymmetrical with the highest energy barrier at its inner end.

Hille (1971, 1973) has used voltage clamp methods on single frog axons to test the ability of various substances to permeate and carry current through sodium and potassium channels. His data suggest that the sodium channel is 0.3 − 0.5 nm wide and the potassium channel is 0.3 nm wide and is shorter than that of sodium. In the squid axon the sodium channel is twelve times more permeable than that to potassium.

Using tritiated TTX, Rang and Ritchie (1968) have counted the 'pores' for sodium transport and find several hundred such pores per square micrometre in the squid axon. This is only a small proportion of the membrane lipoprotein molecules (which number four million per square micrometre).

Hodgkin and Huxley (1952) who showed the dependence of permeability changes of Na and K on the transmembrane potential suggested that such dependence of g_{Na} and g_K arose from the effect of the electrical field on the orientation of molecules with a charge or dipole moment. They termed these *activating or gating particles*, which occupying particular sites allow, for example, the passage of sodium ions— an increase in g_{Na}. The fall of g_{Na} with inactivation could be attributed to a relatively slow movement of another particle which blocked the sodium channel. They envisaged another system of charged particles that moved relatively slowly under the electric field influence to open the potassium gates. They suggested then that specific charged activating and inactivating particles move in the membrane to open or to close the sodium and potassium channel gates.

Unfortunately, the gating currents within the membrane are, as Hodgkin and Huxley surmised, very small indeed compared with the ionic currents carried by sodium ions, and moreover, tend to be swamped by capacitative currents flowing at the same time. However, in 1973 Armstrong and Bezanilla and Keynes and Rojas independently measured these gating currents, using a technique of alternately depolarising and hyperpolarizing the membrane by equal amounts, so as to produce identical but opposite capacitative currents. The gating current could thus be identified and measured (see Kuffler and Nicholls (1977) for discussion).

REFERENCES

General reading

AIDLEY, D. J. (1971). *The physiology of excitable cells.* Cambridge University Press.
COLE, K. S. (1968). *Membranes, ions and impulses.* University of California Press, Berkeley.
HODGKIN, A. L. (1958). *Proc. R. Soc.* **B148**, 1.
—— (1964). *The conduction of the nervous impulse.* Liverpool University Press.
KUFFLER, S. W. and NICHOLLS, J. G. (1977). *From neuron to brain.* Sinauer, Sunderland, Mass.

Selected papers

ARMSTRONG, C. M. and BENZANILLA, F. (1973). *Nature, Lond.* **242**, 459.
—— —— (1974). *J. gen. Physiol.* **63**, 533
—— —— and ROJAS, E. (1973). *J. gen. Physiol.* **62**, 375.
BAKER, P. F. (1966). *Endeavour* **25**, 166.
—— HODGKIN, A. L., and SHAW, T. I. (1962). *J. Physiol., Lond.* **164**, 330; 355.
CURTIS, H. J. AND COLE, K. S. (1938). *Nature, Lond.* **142**, 209.
—— —— (1939). *J. gen. Physiol.* **22**, 649.
—— —— (1940). *J. Cell comp. Physiol.* **15**, 147.
GOLDMAN, D. E. (1943). *J. gen. Physiol.* **27**, 37.
HILLE, B. (1966). *Nature, Lond.* **210**, 1220.
—— (1967). *J. gen. Physiol.* **50**, 1287.
—— (1970). *Prog. Biophys.* **21**, 1.
—— (1971). *J. gen. Physiol.* **58**, 599.
—— (1973a). *J. gen. Physiol.* **61**, 669.
—— (1973b). *J. gen Physiol.* **61**, 699.

HODGKIN, A. L. (1958). *Proc. R. Soc.* **148**, 1.
—— and HUXLEY, A. F. (1952a). *J. Physiol., Lond.* **116**, 449; 473; 497.
—— —— (1952b). *J. Physiol., Lond.* **117**, 500.
—— —— (1953). *J. Physiol., Lond.* **121**, 403.
—— —— and KATZ, B. (1952). *J. Physiol., Lond.* **116**, 424.
—— and KATZ, B. (1949). *J. Physiol., Lond.* **108**, 37.
—— and KEYNES, R. D. (1955). *J. Physiol., Lond.* **128**, 28; 61.
HUXLEY, A. F. and STÄMPFLI, R. (1949). *J. Physiol., Lond.* **108**, 315.
KATZ, B. (1966). *Nerve, muscle and synapse.* McGraw Hill, New York.
KEYNES, R. D. and ROJAS, E. (1973). *J. Physiol., Lond.* **233**, 28p.
—— —— (1974). *J. Physiol., Lond.* **239**, 393.
MARMONT, G. (1949). *J. Cell comp. Physiol.* **34**, 151.
RANG, H. P. and RITCHIE, J. M. (1968). *J. Physiol., Lond.* **196**, 183.
RUSHTON, W. A. H. (1951). *J. Physiol., Lond.* **115**, 101.
TASAKI, I. (1953). *Nervous transmission.* Thomas, Springfield, Ill.
—— and TAKEUCHI, T. (1942). *Pflügers Arch. ges. Physiol.* **245**, 274.

RADIOACTIVE ISOTOPE STUDIES

The conclusion that with the passage of a train of impulses the axon gains a small amount of sodium and loses a similar quantity of potassium has been demonstrated by Hodgkin and Keynes (1955) and their colleagues using radioactive sodium and potassium isotopes. Two techniques can be employed: (1) the axon is immersed in either artificial sea water or in artificial sea water containing ^{24}Na. The axon can be 'loaded' by stimulation during its immersion in radioactive sea water and is then washed in a steady flow of artificial sea water containing ordinary sodium ions. The fluid is collected and its radioactivity measured at suitable intervals. Radioactive sodium is slowly extruded. (2) Radioactive Na or K can be injected via a microsyringe into the axoplasm uniformly over a known distance (some 20 mm) and its extrusion studied.

Each type of experiment has contributed to our understanding of the factors involved during the resting and excitation phases relating to sodium uptake and potassium extrusion.

FIGURE V.61 shows the ^{24}*Na content* of a giant axon which, first at zero level, rises (slightly) when the axon is immersed in ^{24}Na sea

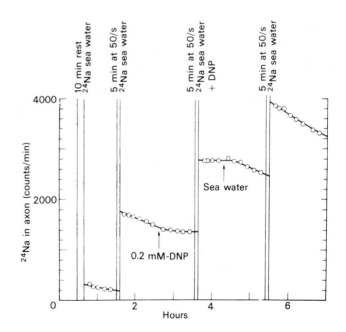

FIG. V.61. Effect of 0.2 mMol.DNP on sodium entry during stimulation of a squid axon. Temperature 17 °C. (From Hodgkin and Keynes 1955.) The abscissa is time, and the ordinate gives the amount of ^{24}Na *inside* the fibre; external ^{24}Na was washed away with a stream of sea water. (Hodgkin, A. L. and Keynes, R. D. (1955). *J. Physiol., Lond.* **128**, 28.)

water. Reimmersed in normal sea water the radioactive sodium is slowly extruded during the subsequent 40 minutes. Repetitive stimulation of the axon in ²⁴Na sea water for 5 minutes notably raises its internal concentration which then falls exponentially when the axon is again exposed to sea water *until* 0.2 mM DNP (which uncouples respiration and phosphorylation processes) is added to the bathing fluid. Extrusion of ²⁴Na stops. DNP does not hinder the *uptake* of ²⁴Na, for a further stimulation for 5 minutes of the axon in ²⁴Na sea water containing DNP raises its internal ²⁴Na content as much as the previous stimulation did—only the *extrusion* of ²⁴Na is abolished until normal sea water is readmitted, whereupon active extrusion recommences. Finally, stimulation when the axon is again immersed in the radioactive solution once more raises its ²⁴Na content and on re-exposure to normal sea water active extrusion is pronounced.

FIGURE V.62 shows another method of following ²⁴Na extrusion by measuring the *efflux* in fluid superfusing the axon. Initially, the axon has been 'loaded' by stimulation in ²⁴Na sea water whereafter normal sea water superfuses the fibre. An exponential loss of ²⁴Na is reflected by the radioactivity of the effluent fluid which is plotted on a log scale. This active extrusion promptly ceases when DNP is added to the superfusate but this metabolic inhibition is completely reversible, for on removing the DNP, extrusion soon recovers and regains the same slope as that prior to DNP. Exactly the same results can be obtained by microinjection of ²⁴NaCl into the axon and following the efflux of sodium in normal sea water and in sea water containing DNP. Cyanide and azide act similarly to DNP in inhibiting active extrusion of sodium.

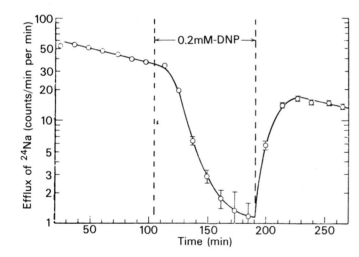

FIG. V.62. Action of 2:4 dinitrophenol on the outward movement of sodium in a *Sepia* axon which had been recovering from the effect of stimulation. Temperature 18 °C. Abscissa, time after end of stimulation in solution containing ²⁴Na. Ordinate, rate at which ²⁴Na leaves fibre. (Hodgkin, A. L. and Keynes, R. D. (1955). *J. Physiol., Lond.* **128**, 28.)

²⁴²K movements can be followed in the resting axon and during the period subsequent to stimulation in ⁴²K sea water. FIGURE V.63 shows that after immersion for 15 minutes in ⁴²K sea water the axon at rest has taken up ⁴²K, for on readmitting sea water the resting axon loses ⁴²K at an initially high rate which falls exponentially. When stimulated in the ⁴²K solution it takes up ⁴²K and subsequently extrudes it when again superfused by sea water. A brief period of stimulation in this normal sea water increases the rate of extrusion of ⁴²K.

From the data obtained, inward and outward fluxes of potassium and sodium could be quantitatively determined. Inward fluxes were determined directly and outward fluxes were calculated from the rate constant for loss of radioactive ions in inactive sea water. At rest the outward flux of K⁺ was 3–4 times greater than the inward

flux because the resting membrane potential was well below the theoretical potential calculated from the potassium concentration ratio. Resting inward flux was approximately 16 picomoles per square centimetre (Keynes 1951).

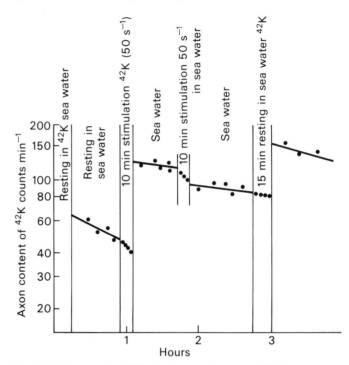

FIG. V.63. Movement of ⁴²K in *Sepia* axon. For entry of ⁴²K, 1 count min⁻¹ was equal to 1.17×10^{-11} mol K cm axon⁻¹. Diameter of axon 168 μm. (Keynes, R. D. (1951). *J. Physiol., Lond.* **114**, 119.)

The resting inward flux for sodium was 60 picomoles per square centimetre and the outward flux about half this. Now most of this outward flux cannot be due to the diffusion of free ions because the concentrations of sodium inside and outside would give a flux ratio for free diffusion of Na⁺ ions of 45—not 2. Most of the outward flux of sodium must be due to sodium carried in non-ionic form, probably by an active sodium pump.

Stimulation of the axon at 100 Hz increased the inward flux of potassium more than threefold and the outward flux ninefold. Inward flux and outward flux of sodium were both increased some twenty times by stimulation, yielding proof of a notable increase of sodium turnover during activity.

The immediate effect, therefore, of the passage of a train of impulses is a small increase in axoplasmic sodium concentration and a similar fall of potassium concentration. Keynes and Lewis (1951) found that each impulse caused a net gain of 3.8 picomoles per square centimetre of Na⁺ and a very similar loss of K⁺. These experimental results validate the proposition that changes in membrane permeability to Na⁺ and K⁺ can adequately account for the amount of charge which must be transferred across the membrane in order to alter the potential by the amount recorded. The action potential is approximately 100–120 mV in height. The capacitance of the membrane is 1μF cm⁻². As $Q = CV$ the amount of current required is (1×10^{-6}) farads $\times 0.12$ (volts) = 1.2×10^{-7} coulombs. 1 mole of a univalent ion carries 96 500 coulombs so the transference of 1.2×10^{-12} mol cm⁻² of sodium (or 1.2 pmol cm⁻²) of sodium inward would be minimally required to account for the spike potential. The values measured by Keynes and Lewis (1951) exceed these minimal requirements.

Returning to the experimental results obtained with the use of inhibitors such as DNP, the important point is that DNP, azide, or cyanide (all of which inhibit active sodium extrusion) do *not* prevent the nerve fibre from conducting a large number of impulses and do *not* appreciably alter either the resting membrane potential or the size of the action potential. Probably, in the giant axon, concentration differences can be dissipated slowly by leakage and downhill movements of Na⁺ and K⁺ provide the immediate source of energy for propagating spike potentials. Even when the axoplasm is replaced by K_2SO_4 solutions the axon will still conduct hundreds of thousands of nerve impulses.

From such experiments it is postulated that there are two systems in the axon membrane—one which is driven by metabolism and is responsible for building up concentration differences, the other relatively independent of metabolism and responsible for controlling downhill movements of sodium and potassium during the action potential [Fig. V.64].

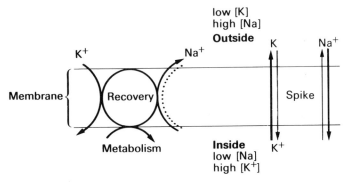

Fig. V.64. Diagram illustrating the movement of ions through the membrane. The downhill movements which occur during the impulse are shown on the right; uphill movements during recovery are shown on the left. The broken lines represent the component of the sodium efflux which is not abolished by removing external potassium ions. (Hodgkin, A. L. and Keynes, R. D. (1955). *J. Physiol., Lond.* **128**, 28.)

These two systems differ in their properties:

1. Ouabain, which inhibits active transport by combining with adenosine triphosphatase (ATPase), has no effect on the action potential.

2. Changes in the extracellular concentration of divalent ions such as Ca^{2+} and Mg^{2+}, which profoundly affect excitability of the axon, have little effect on the pumping system.

3. Lithium which can replace sodium in the extracellular fluid without influencing the nature and size of the action potential is not actively extruded by the metabolically driven sodium pump.

4. The maximum rate at which the secretory system can transport ions is 50 pmol cm⁻² s⁻¹ whereas movements as high as 10 000 pmol cm⁻² s⁻¹ occur during the action potential.

The extrusion of sodium by the pump mechanism is loosely coupled with the absorption of potassium in the manner showed by Figure V.64. Thus a chemical substance which reduces sodium extrusion also lowers potassium uptake without there being any change in the membrane potential. Moreover, if the axon is placed in potassium-free fluid, sodium extrusion falls to a third of its normal value, to be rapidly restored when potassium is added to the bathing medium [Fig. V.65].

Energy-rich phosphate compounds

Caldwell (1956, 1960) showed the effect of metabolic inhibitors on the energy-rich phosphate compounds in the axoplasm. ATP is present and arginine phosphate in the squid axon takes the place of creatine phosphate in mammals acting as a phosphagen.

When cyanide (or DNP) is applied, sodium extrusion is inhibited; ATP is dissimilated to ADP and P_i and arginine phosphate to arginine and phosphate. When the metabolic poisons are removed, ATP and phosphagen are resynthesized and sodium extrusion is restored. When cyanide is used, arginine phosphate breaks down first and [ATP] only falls when [arginine phosphate] is very low. The sodium extrusion declines with a time course similar to that of ATP, but the active transport mechanism is only normally operative when both arginine phosphate and ATP are present.

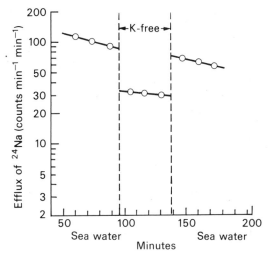

Fig. V.65. Effects of potassium-free sea water on sodium efflux from a *Sepia* axon. When not in a test solution the axon was in artificial sea water containing 10.4 mM K.

Caldwell, Hodgkin, Keynes, and Shaw (1960) studied the effects of intra-axonal micro-injections of ATP, arginine phosphate, and creatine phosphate on the extrusion of sodium by axons poisoned with cyanide. Creatine phosphate had no effect when injected inside (or applied outside) the axon. Arginine phosphate and ATP, though utterly ineffective when added to the extracellular fluid, caused a marked restoration of sodium extrusion when injected *inside* the axon [Figs. V.66 and V.67]. Their action was transient and the restoration of sodium extrusion was approximately proportional to the quantity injected [see Fig. V.67].

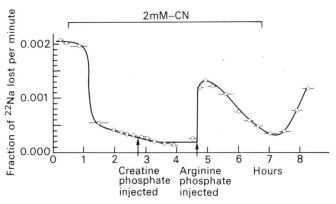

Fig. V.66. The effect on the outflow of sodium from an axon poisoned with cyanide of injecting first creatine phosphate and then arginine phosphate. The mean concentrations of the axon immediately after the injections were 15.3 mM—creatine phosphate and 15.8 mM—arginine phosphate. (From Caldwell, P. C., Hodgkin, A. L., Keynes, R. D., and Shaw, T. C. (1960). *J. Physiol., Lond.* **152**, 545.)

The injection of arginine phosphate into a cyanide-poisoned axon raises the sodium outflow [Fig. V.68]. During the period of two hours in which this effect had reached and maintained its plateau, brief substitutions of K⁺-free ECF sharply reduced sodium efflux; this suggests a coupled movement of Na⁺ out and K⁺ in. As the rate of Na⁺ efflux progressively falls away (cyanide still being present in the ECF) substitution of K⁺-free sea water no longer alters progressively the rate of sodium extrusion. During this phase the arginine phosphate is being progressively hydrolysed to arginine.

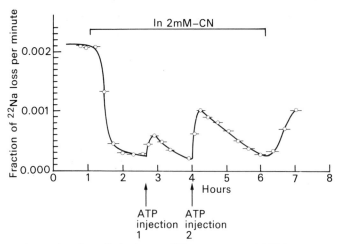

FIG. V.67. The effect of injecting two different amounts of ATP *into* an axon poisoned with cyanide; injection 1 raised the ATP concentration in the axon by 1.2 mM and injection 2 by 6.2 mM. (From Caldwell *et al.* (1960). *J. Physiol., Lond.* **152**, 545.)

When axons are poisoned with cyanide their normal influx of potassium is reduced. The injection of arginine phosphate restores the potassium influx to the same level as in the normal resting axon.

The energy provided by the hydrolysis of the energy-rich phosphates thus drives a pump which extrudes sodium and takes in potassium. Vital to the performance of this pump is the presence of an ATPase which requires the presence of both potassium and sodium to activate it. Thus, if the ECF is potassium-free the hydrolysis of ATP by this enzyme is inhibited and the pump stops. The enzyme is inhibited by ouabain which abolishes Na extrusion. Skou (1965), who discovered the Na/K ATPase, pointed out that it fulfilled the minimal requirements for the Na/K transport system of the membrane. It should be (1) located in the membrane, (2) have an affinity for Na^+ that is higher than that for K^+ at a site located on the inside of the membrane, (3) have an affinity for K^+ that is higher than for Na^+ at a site located on the outside of the membrane, (4) contain an enzyme system that can catalyse ATP hydrolysis thus converting the energy from ATP (13 kcal mol⁻¹) into a movement of

cations, and (5) be capable of hydrolysing ATP at a rate dependent on concentration of Na^+ inside and also of K^+ outside the cell.

Na/K ATPase is a lipoprotein and has a molecular weight of 250 000 daltons. The use of almost pure ATPase from kidney tissue has allowed the preparation of an antiserum which inhibits Na^+ transport by erythrocyte membranes.

Radioactive labelling of ATP (AT³²P) has shown that when the ATPase system is treated with AT³²P the amount of labelling depends on the ionic composition of the medium. Post, Sen, and Rosenthal (1965) showed that the labelling is low in the presence of Mg^{2+} alone but is markedly increased when Na^+ is added. The inclusion of sodium and potassium ions in the medium results in a large reduction in labelling compared with that seen with sodium ions alone. This reduction is offset however by adding ouabain. Their results can be explained in terms of alternate phosphorylation and dephosphorylation of the membrane during pumping. Phosphorylation requires sodium. The addition of potassium favours dephosphorylation by a mechanism which is blocked by ouabain.

Baker (1966) has suggested a model which incorporates the major properties of the sodium pump [FIG. V.69].

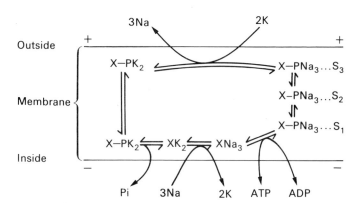

FIG. V.69. A hypothetical model of the sodium pump. It is based on the idea that the affinity for sodium and potassium ions of a cycling carrier is modified by combination with phosphate. (Baker, P. F. (1966). *Endeavour* **25**, 166.)

To explain the high affinity for potassium outside the cell and for sodium inside it is proposed that metabolic energy is used to convert a carrier molecule that itself has a high affinity for Na^+ and is loaded with sodium, into a carrier which has a high affinity for potassium. If this molecule is passed to a site within the membrane the exchange of sodium on the carrier with internal potassium might be avoided. According to this model the exchange of internal sodium with external potassium must occur through alternate breakdown and resynthesis of ATP. At the outer face of the membrane when potassium is present, the carrier will react with K^+ and return to the inside of the cell where it is dephosphorylated and the potassium exchanged for sodium. If the ECF is K^+ free and if Na^+ is available external Na^+ will tend to exchange with Na^+ on the carrier before the carrier returns to the inside surface of the membrane. When both external Na^+ and K^+ are absent, sodium might dissociate from the carrier and the free carrier return to the inside.

The properties of exchange diffusion can be explained if the carrier has to cross the membrane by means of a series of sites (S_1, S_2, S_3). For exchange diffusion to occur the carrier must traverse these sites freely. If more than one site is occupied, then exchange is impossible and exchange diffusion will be seriously reduced. Baker points out that a combination of ouabain with site S_3 would block coupled pumping, exchange diffusion and the loss of sodium accompanied by an anion. Electrogenic properties of the sodium pump might be explained if sodium passes out as a positively charged complex and potassium returns in a neutral form.

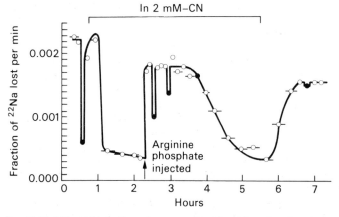

FIG. V.68. Effect of injecting a large quantity of arginine phosphate into a cyanide-poisoned fibre on Na efflux into K-free solution (●) and 10 mM K (○). Immediately after the injection the mean concentration of arginine phosphate in the fibre was 33 mM. (From Caldwell *et al.* (1960). *J. Physiol., Lond.* **152**, 545.)

REFERENCES

BAKER, P. F. (1966). *Endeavour* XXV, 166.
CALDWELL, P. C. (1956). *J. Physiol., Lond.* **132**, 35P.
—— (1960). *J. Physiol., Lond.* **152**, 545.
—— HODGKIN, A. L., KEYNES, R. D., and SHAW, T. I. (1960). *J. Physiol., Lond.* **152**, 545.
—— and KEYNES, R. D. (1960). *J. Physiol., Lond.* **154**, 177.
HODGKIN, A. L. (1964). *The conduction of the nervous impulse.* Liverpool University Press.
—— and HUXLEY, A. F. (1953). *J. Physiol., Lond.* **121**, 403.
—— and KEYNES, R. D. (1955). *J. Physiol., Lond.* **128**, 28.
—— and SHAW, T. I. (1961). *Nature, Lond.* **190**, 885.
KEYNES, R. D. (1951*a*). *J. Physiol., Lond.* **113**, 99.
—— (1951*b*). *J. Physiol., Lond.* **114**, 119.
—— and LEWIS, P. R. (1951). *J. Physiol., Lond.* **114**, 151.
POST, R. L., SEN, A.K., AND ROSENTHAL, A. (1965). *J. biol. Chem.* **240**, 14–37.

MICROELECTRODE STUDIES OF RESPONSES OF MOTONEURONS TO SYNAPTIC TRANSMISSION

A motoneuron supplying a given muscle can be monosynaptically excited by stimulation of the afferent nerves from a synergistic muscle. On inserting the microelectrode into the motoneuron the resting membrane potential of about -70 mV is registered. The motoneuron can be identified by stimulating its axon electrically in the appropriate muscle nerve. This causes the discharge of the cell body in response to the antidromic conduction of the stimulus. An electric stimulus to the corresponding muscle *afferent* nerves (Group IA) causes a presynaptic volley to be conducted to the synapse, and thus initiates the production of the depolarizing potential—the excitatory postsynaptic potential (EPSP)—within 0.5 ms. The size of the EPSP depends to some extent on the intensity of the afferent stimulus [FIG. V.70]. If this is adequate the EPSP rises by some 8 mV, upon which the motoneuron generates a

spike potential and discharges a nerve impulse. These changes deserve closer analysis.

Generation of excitatory postsynaptic potential

Let us suppose that the synapse between afferent terminals and part of the soma-dendritic membrane of the cell body is as represented in FIGURE V.71, B.

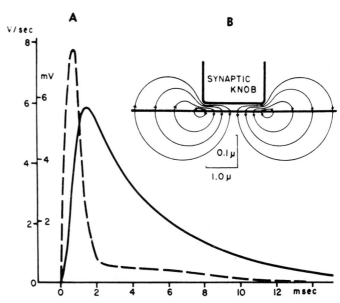

FIG. V.71. A. The continuous line is the mean of several monosynaptic EPSPs, while the broken line shows the time course of the subsynaptic current required to generate this potential change. B. Diagram showing an activated excitatory synaptic knob and the postsynaptic membrane. As indicated by the scales for distance, the synaptic cleft is shown at 10 times the scale for width as against length. The current generating the EPSP passes in through the cleft and inward across the activated subsynaptic membrane, but outward across the remainder of the postsynaptic membrane. (Eccles, J. C. (1957). *The physiology of nerve cells.* Johns Hopkins University Press, Baltimore.)

The surface of the cell membrane involved in the synapse is called the *subsynaptic membrane* and the remainder of the motoneuron cell membrane may be referred to as the *postsynaptic membrane*. When the afferent fibre is stimulated electrically a propagated nerve impulse eventually arrives at the afferent presynaptic terminals (synaptic knob) [FIG. V.71, B]. The question arises whether this nerve impulse itself causes a flow of electric current in the synaptic cleft which excites the subsynaptic membrane and thereby effects synaptic transmission. The answer is no—for during the rising phase of the presynaptic spike potential there is no development of EPSP. Synaptic excitation is effected by the liberation of a chemical transmitter from the presynaptic terminals by the propagated afferent volley. This transmitter causes a temporary change in the subsynaptic membrane so that it becomes equally permeable to all ions; effectively this means that the subsynaptic membrane becomes a short circuit, passing a current inwards which tends to bring its potential to 0 mV. The membrane potential of the postsynaptic membrane thereupon provides a voltage which causes current to pass outwards through the postsynaptic membrane and inwards through the subsynaptic membrane, thus generating the EPSP [FIG. V.71, B].

In keeping with this explanation are the effects of changing the ionic composition of the motoneuron by injecting ions electrophoretically through the microelectrode into its interior. Such changes of ionic composition exert a profound effect on the membrane potential and spike potential but have virtually no influence on the EPSP.

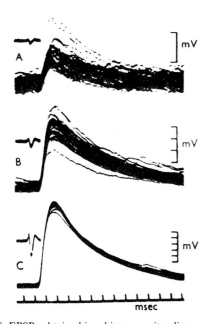

FIG. V.70. A–C. EPSPs obtained in a biceps-semitendinosus motoneuron with afferent volleys of different size. Inset records at the left of main records show afferent volley recorded near entry of dorsal nerve roots into spinal cord. They are taken with negativity downward and at a constant amplification for which no scale is given. Records of EPSP are taken at an amplification that decreases in steps from A to C as the response increases. Separate vertical scales are given for each record, formed by superposition of about forty faint traces. (Eccles, J. C. (1957). *The physiology of nerve cells.* Johns Hopkins University Press, Baltimore.)

By grading the size of the afferent volley, the size and time course of EPSP may be closely studied. Beginning within 0.4 ms of the arrival of the presynaptic impulse, the EPSP rises to its summit 1 ms later and, if its height is insufficient to cause the generation of a spike, then declines exponentially with a time-constant of 4 ms. If its size is sufficient (about 6–10 mV) to cause the generation of a spike, the impulse so generated destroys the EPSP. Two successive afferent volleys, neither of which alone can cause a spike discharge from the motoneuron, each produce some EPSP and, providing that the interval between them is not too great, may cause temporal summation of EPSP to the critical level (6–10 mV) necessary for the initiation of a spike [Fig. V.72]. Similarly, the simultaneous excitation of two separate afferent fibres both of which effect synaptic connections with the motoneuron under study may cause summation of EPSP to the critical level and may thus produce a spike discharge from the motoneuron. This is an example of spatial summation; the excitation of either fibre alone does not produce synaptic transmission of a spike potential.

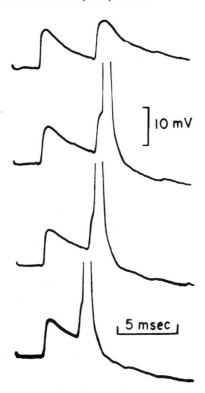

Fig. V.72. Intracellular potentials set up in a biceps-semitendinosus motoneuron by two afferent volleys in the biceps-semitendinosus nerve. The volley interval is progressively decreased from above downward. Note the generation of a spike potential (truncated) at all but the longest interval. The latency of the spike decreases as the interval shortens. (Eccles, J. C. (1957). *The physiology of nerve cells*. Johns Hopkins University Press, Baltimore.)

Generation of the spike potential

A spike potential can be evoked from the motoneuron either by synaptic excitation or by direct stimulation of the cell itself. In the latter case a special double-barrelled microelectrode is employed; the current is applied through one barrel of the electrode which is inserted into the cell and the membrane potential is recorded through the other barrel [Fig. V.73]. Using this technique, it can be shown that a recorded depolarization of slightly less than 10 mV can generate an impulse. This threshold depolarization level is similar to the level achieved by EPSP during natural synaptic excitation. In each case then, depolarization initiates the spike. A given afferent volley becomes more effective in exciting the synaptic propagation

of a spike potential if the motoneuron is meanwhile submitted to progressive depolarization by means of directly applying current to its interior; conversely hyperpolarization of the membrane by such current application renders the motoneuron less susceptible to synaptic activation.

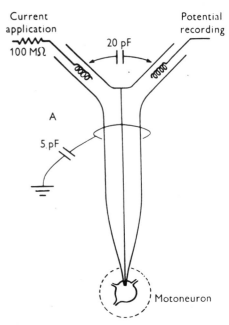

Fig. V.73. Double-barrelled microelectrode and its immediate connections. Typical values are given for the several electrical characteristics which are significant in the use of the electrode. (Eccles, J. C. (1957). *The physiology of nerve cells*. Johns Hopkins University Press, Baltimore.)

Analysis shows that when EPSP depolarization, or depolarization by the direct application of current, reaches the critical level it evokes a spike potential first in the initial segment (IS) of the axon and the axon hillock—the so-called IS spike. This potential is 30–40 mV in height and 'takes off' from a threshold depolarization of 6–10 mV which has previously been achieved by the EPSP. The IS spike therefore requires a relatively low degree of depolarization for its own initiation but, once initiated, itself produces a further depolarization of 30–40 mV which is apparently additionally required to secure the generation of the SD spike from the soma-dendritic membrane. The SD spike is an additional 50–60 mV in height [Fig. V.74]. The entire duration of the IS–SD potential is only 1 ms. The neuronal spike potential is caused, as in the axon, by a sequential alteration of membrane permeability to first sodium and then potassium ions. Thus the rising phase of the spike can be notably

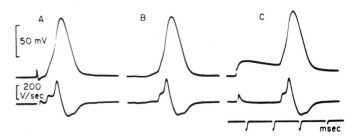

Fig. V.74. A, B, and C, respectively, show the intracellular potentials produced when a spike response of a motoneuron is generated antidromically, synaptically (monosynaptic), and by a directly applied current (by a double electrode). An electrically differentiated record of each potential wave lies immediately below it. Note that there is an inflection at approximately the same level (about 30 mV) on the rising phases of the three spikes. (From results of Fatt, P. (1956). *J. Neurophysiol.*, and Eccles, J. C. (1957). *The physiology of nerve cells*. Johns Hopkins University Press, Baltimore.)

slowed and reduced in amplitude by the intracellular injection of sodium ions.

Central inhibition as studied in spinal motoneurons

The outlines of the mechanism involved in central inhibition have been clarified by intracellular studies. As an example let us consider the intracellular responses of a motoneuron which supplies the biceps semitendinosus muscle (flexor) to an afferent volley of impulses in the sensory nerves derived from the muscle spindles of the quadriceps (extensor). On receipt of the afferent volley (conducted synaptically) the membrane potential of the motoneuron changes from -66 mV to -68 mV (i.e. the cell is hyperpolarized [Fig. V.75]). This hyperpolarization potential can be called the IPSP (inhibitory postsynaptic potential) and is approximately a mirror image of the EPSP. The development of IPSP begins about 1–1.5 ms after the entry of the afferent volley into the spinal cord, i.e. appreciably later than the development of EPSP, which begins in 0.5 ms, rises to a summit in 1.5–2 ms after its onset, and decays exponentially. Analysis of IPSP can be carried out as was described in the case of EPSP. The results suggest that IPSP is due to the influence of a synaptically liberated inhibitory chemical transmitter on the subsynaptic membrane, whereby an outward current develops in this region and consequently a hyperpolarization of the whole postsynaptic membrane. In these circumstances the subsynaptic membrane becomes highly permeable to potassium and chloride ions but retains its high degree of impermeability to sodium ions. IPSP can show temporal and spatial summation just as does EPSP.

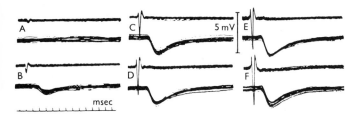

Fig. V.75. *Lower* records give *intracellular* responses of a biceps-semi-tendinosus motoneurone to a quadriceps volley of progressively increasing size, as is shown by the *upper* records which are recorded from the L6 dorsal root by a *surface* electrode (*downward deflections signalling negativity*). Note three gradations in the size of the IPSP; from A to B, from B to C, and from D to E. All records are formed by the superposition of about forty faint traces. Voltage scale gives 5 mV for intracellular records, downward deflections indicating membrane hyperpolarization. (Coombs, J. S., Eccles, J. C., and Fatt, P. (1955). *J. Physiol., Lond.* **130**, 326.)

So-called direct inhibition. If the muscle spindles (afferent nerve endings) of an extensor muscle are stimulated by stretching the muscle there results reflex contraction of the muscle itself and almost simultaneously reflex relaxation of its antagonist (flexor).

Intracellular recordings from the motoneurons have shown that the afferent impulses in the nerve fibres from the muscle spindles can pass along collaterals which synaptically excite special neurons of the neighbouring *intermediate nucleus of Cajal*. These neurons in turn discharge impulses within 1 ms of the time of entry of the afferent volley into the spinal cord which traverse very short axons and liberate an inhibitory transmitter at their synaptic connection with the antagonist motoneurons. Thus the pathway of 'direct inhibition' is not monosynaptic. The excitation of the agonist motoneurons by the spindle afferent fibres is monosynaptic, but the concomitant inhibition of the antagonist motoneurons is disynaptic.

Renshaw cells and motoneuron inhibition. If the motoneuron axon is stimulated electrically, an antidromic impulse is conducted into the cell body and, after discharging the motoneuron, produces

inhibition of this and some other motoneurons at that segmental level. Motor nerve axons give off some collateral branches as they traverse the cord towards the ventral root, and these collaterals make excitatory synaptic connections with a special group of interneurons situated in the ventromedial part of the ventral horn Fig. V.76]. These cells (described by Renshaw) in turn send axons which

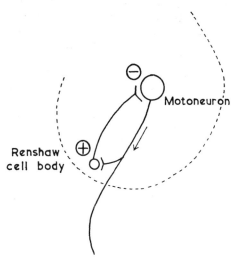

Fig. V.76. The figure shows the outline of the ventral (or anterior) horn of the grey matter of the spinal cord. Collaterals of the motoneurons effect synaptic connections with the Renshaw cells which in turn send inhibitory axons to the motoneurons. (After Eccles, J. C., Fatt, P., and Koketsu (1954). *J. Physiol., Lond.* **126**, 524.)

effect inhibitory synaptic connections with the motoneurons. Antidromic electrical stimulation is of course an abnormal way of exciting Renshaw cell activity; in natural circumstances the Renshaw cells are activated when the motoneurons are excited synaptically and thereupon 'feed back' on the motoneurons, curbing their discharge. The Renshaw cells are stimulated by acetylcholine and there is every reason to believe that the motor axon collateral endings liberate ACh as the natural excitatory synaptic transmitter for the Renshaw cells, just as do the main motor axon terminals at the muscle end plate. The intra-arterial injection of ACh directly excites the Renshaw cells and this excitatory action is increased by anticholinesterases such as eserine. Conversely dihydro-β-erythroidine which blocks cholinergic transmission depresses the responses of Renshaw cells to synaptic stimulation.

Golgi tendon organ afferents and inhibition of interneurons. Stimulation of the Golgi tendon organs may cause reflex inhibition of their corresponding muscles. This synaptic inhibition of the motoneurons is effected by a pathway which involves the intercalation of an interneuron whose cell body lies in the ipsilateral intermediate nucleus. The Golgi tendon organ afferents (Group I B) produce excitatory postsynaptic potentials only in the region of these interneuron cells. The interneuron cells then discharge over a short axonal pathway releasing an inhibitory transmitter on the agonist motoneuron. Note: Stimulation of the Golgi tendon endorgans by more natural means than tugging on the tendon as a whole does not cause inhibition of the tone of the muscle—see page 333.

Summing up these various types of central inhibition two main points emerge:

1. Any given afferent nerve releases only one type of transmitter from its terminal branches. In the case of the dorsal root ganglion cells with their primary afferent fibres, the nerve terminals liberate an excitatory transmitter. The Renshaw cells and some intermediate neurons liberate an inhibitory transmitter.

2. It seems that inhibitory cells are short axon neurons lying in the grey matter, while all transmission pathways, in and out of the cord and possibly up and down the cord, are formed by the axons of excitatory cells (Eccles).

These examples of inhibition can be classified as *postsynaptic*. They depend on the delivery of an inhibitory chemical by the incoming axons which hyperpolarizes the subsynaptic membrane. There is another form of inhibition which is exercised *presynaptically* which is analysed below.

Presynaptic inhibition. Barron and Matthews (1938) showed that stimulation of the dorsal root fibres caused, after some delay, a long-lasting depolarization within the relevant segment and indeed in adjacent segments. This depolarization which spreads electrotonically into the dorsal roots was termed the *dorsal root potential* and is now termed primary afferent depolarization (PAD). The afferent terminals which were thus partially depolarized then exert less vigorous synaptic excitatory effects on subsequent stimulation and this phenomenon is called presynaptic inhibition.

Frank and Fuortes (1957) showed that some motoneurons subjected to inhibitory afferent stimulation did not reveal any IPSP changes or any change in their threshold to direct stimulation through the microelectrode (situated in their cell body), but did display a marked reduction in the EPSP response to stimulation of a volley delivered by excitatory afferent stimulation. Eccles and his colleagues (see Eccles 1964) proved that such responses were caused by *presynaptic inhibition*. The general features of presynaptic inhibition can be explained by considering the response of an extensor motoneuron, such as supplies the calf muscle plantaris, to a conditioning volley delivered by stimulating the afferent nerves from the flexor muscles (biceps-semitendinosus) or to its own monosynaptic excitatory afferent stimulation. FIGURE V.77 shows the control EPSP evoked by stimulation of the IA afferents from plantaris recorded intracellularly in the appropriate motoneuron; following a conditioning volley initiated by 22 volleys to the flexor nerve the EPSP response to excitatory afferent stimulation is profoundly reduced in amplitude, although its time course is unchanged. This reduction in EPSP amplitude is accompanied by a lessened response of the monosynaptic reflex response recorded from the corresponding ventral root. Presynaptic inhibition may last for 200 ms or more. It is ascribed to the depolarizing effect induced by axo-axonic synapses—primary afferent depolarization (PAD). Afferent fibres can influence each other by means of col-

laterals which innervate interneurons. These then establish synaptic contacts with the presynaptic terminals of other afferent fibres. For example, a volley in a IA afferent fibre from a flexor muscle produces depolarization of the presynaptic terminal supplying an extensor motor neuron through the action of interneurons. This depolarization of the presynaptic terminal reduces the amplitude of the action potential and hence the amount of chemical transmitter released by the terminals of the IA afferent fibre from the extensor muscle [FIG. V.77].

Eccles classified the interneurons which caused presynaptic inhibition as D cells. Most of such cells lie at the base of the dorsal horn about 1.7–2.5 mm from the dorsal surface. FIGURE V.78 shows some of the simplest pathways of PAD.

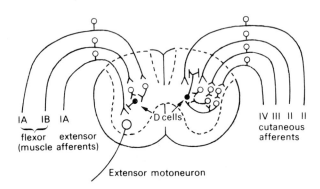

FIG. V.78. Schematic diagram illustrating interneuronal pathways of PAD. The left-hand side illustrates the convergence of flexor IA and IB fibres onto a D type interneuron which in turn makes contacts on the terminals of an afferent IA fibre from extensor muscle. On the right-hand side pathways for presynaptic inhibitory actions on cutaneous primary afferents are shown. (Schmidt, R. F. (1974). In *Handbook of sensory physiology*, Vol. II (ed. A. Iggo) pp. 151–206. Springer, Berlin.)

Presynaptic inhibition is antagonized by the convulsant drug picrotoxin but is entirely irresponsive to strychnine. The convulsant effect of strychnine is due to its profound depression of postsynaptic inhibition. Barbiturates greatly accentuate presynaptic inhibition.

The nature of the inhibitory transmitter at the axo-axonic synapse is still not known for certain. γ-Aminobutyric acid (GABA) is a strong candidate. Presynaptic transmission is depressed by GABA and its influence is blocked by the drug's antagonists picrotoxin and bicuculline. Depletion of GABA by semicarbazide results in reduced presynaptic inhibition of monosynaptic reflexes [see Curtis (1969) and Schmidt (1974) for references].

Volleys in myelinated cutaneous afferents are particularly powerful in inducing presynaptic inhibition of the cutaneous afferents themselves. Muscle afferents receive predominantly presynaptic inhibitory influences from the other muscle afferents.

It can be assumed that in the vertebrate CNS all types of myelinated somaesthetic primary afferent fibres are subjected to presynaptic inhibition. Very little is known about presynaptic inhibition of axon terminals stemming from other than primary afferent axons. A general opinion that presynaptic inhibition is restricted to the primary terminals of the first-order neurons may only reflect a lack of evidence. There are indications that axons of the second order in somaesthetic pathways, axon terminals in the Vth cranial nucleus, and in the lateral geniculate nucleus are subjected to presynaptic inhibition (see Schmidt 1974).

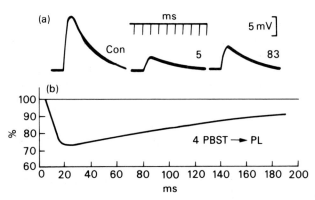

FIG. V.77. Reduction of the amplitude of EPSP in motoneurons by presynaptic inhibition. In (a) control EPSP evoked by a Ia afferent volley, and its reduction by a conditioning tetanus of 22 group 1 volleys taken at 5 ms and at 83 ms after the end of the tetanus. (b) Plotting of the time course of the EPSP depression as a percentage of the control. Abscissa, time in ms. Ordinate, percentage of control. The records were taken from a plantaris motoneuron, the inhibition being evoked by four group 1 conditioning volleys from the posterior biceps-semitendinosus muscle nerve. (Eccles, J. C., Eccles, R. M., and Magni, F. (1961). *J. Physiol., Lond.* **159**, 147.)

REFERENCES

BARRON, D. H. and MATTHEWS, B. H. C. (1938). *J. Physiol., Lond.* **92**, 276.
CURTIS, D. R. (1969). *Prog. Brain Res.* **31**, 171.

Eccles, J. C. (1964). *The physiology of synapses.* Springer, Berlin.
Schmidt, R. F. (1974). In *Handbook of sensory physiology*, Vol. II (ed. A. Iggo) pp. 151–206. Springer, Berlin.

Receptors

Impressions of the outside world are conveyed by sensory nerve impulses which are generated in the peripheral nerve endings of the afferent fibres. Cutaneous receptors respond to touch, temperature changes, and injurious stimuli. The eyes respond to changes of light intensity and wavelength, the ear to sound waves and the olfactory and gustatory nerve endings to changes of chemical composition. Such receptors are called *exteroceptors*. The *interoceptors* respond to changes within the body itself and comprise the *proprioceptors* which are stimulated by changes of position of the body (i.e. muscle spindles, joint receptors, vestibular receptors), the *visceroceptors*, e.g. baroreceptors and nociceptors, and the *chemoceptors*.

Cutaneous receptors

Some of the receptors subserving sensations of touch, light pressure, heat, and cold are constructed on a uniform plan. They consist of a lamellated connective tissue capsule which surrounds a soft cellular core in which the axon ends after losing its medullary sheath. They vary only in the complexity of this general design.

1. Tactile corpuscles of Meissner are ellipsoidal structures which occur in groups in the cutaneous papillae. The density of these structures is high in the skin of the finger tips, lips, and nipples and the orifices of the body. Other tactile corpuscles form basket-like arborizations round the base of the hair follicles; they are stimulated by mechanical displacement of the hair. The tactile sensibility of an area of skin bearing hairs, e.g. the external surface of the forearm, is considerably reduced after it is shaved. The hairs act as levers amplifying the mechanical displacement of the sensory endings round the follicles. Tactile receptors are nerve endings of A fibres (β, γ, δ group). They are rapidly adapting; hence we do not feel our clothes once they have been donned.

2. Pacinian corpuscles, which resemble an onion in shape and lamellation, are large receptors found in large numbers in the subcutaneous tissues and in the neighbourhood of tendons and joints. They respond to deformation caused by firm pressures and are quickly adapting. Their afferent fibres are of the A group.

3. End bulbs of Krause occur in the conjunctiva, in the papillae of the lips and tongue, in the skin of the genitalia and in the sheaths of nerves. They are spherical but otherwise resemble the tactile corpuscles. They are mechanoreceptors; their fibres belong to the A group.

4. The organ of Ruffini in the dermis is supplied by a large myelinated axon. It is probably a mechanoreceptor.

5. Free unencapsulated nerve endings serve as receptors which respond to nocuous stimuli. These peripheral endings may be terminal branches of thin myelinated (Aδ) or unmyelinated fibres (C group). Free nerve endings can subserve sensations of touch and temperature, as well as pain.

Proprioceptors

These include:

1. Muscle spindles.
2. Tendon end organs of Golgi.
3. Joint receptors.

4. Vestibular receptors in the maculae of the saccule and utricle and in the cupulae of the semicircular canals.

Muscle spindles [p. 298], tendon end-organs [p. 333], and vestibular receptors [p. 304] are discussed elsewhere.

Joint receptors

These are mainly of the Pacinian and Golgi type which are situated in the ligaments of the joint and form the endings of afferent nerves of Group I (12–15 µm); some Group II fibres (7–10 µm) arise from Ruffini end organs which are found in the capsule. Three types of response were observed by Boyd and Roberts (1953) and by Skoglund (1956) in investigating the activity of joint receptors.

1. Golgi end organs in the ligaments show impulse discharge which varies with the exact position of the joint. The discharge is slowly adapting and is relatively insensitive to movement.

2. The end organs of Ruffini on the other hand, situated in the capsule are very responsive to movements of the joint, but adapting quickly do not signal much information about the exact position of the joint.

3. Pacinian corpuscles in the ligaments are very sensitive to quick movement or to vibration and according to Skoglund might serve as 'acceleration detectors'.

Afferents from the joint receptors relay in the cord, and second order fibres travelling in the posterior columns transmit impulses relaying in the thalamus to the contralateral sensory cortical areas I and II, and to the ipsilateral sensory area II, within fifteen milliseconds.

REFERENCES

Boyd, I. A. and Roberts, T. D. M. (1953). *J. Physiol., Lond.* **122**, 38.
Gordon, G. (ed.) (1977). *Br. med. Bull.* **33**, 89–177.
Hunt, C. C. (1974). *Muscle receptors. Handbook of sensory physiology*, Vol. II, No. 2. Springer, Berlin.
Iggo, A. (ed.) (1973). *Somatosensory system. Handbook of sensory physiology*, Vol. II. Springer, Berlin.
Skoglund, S. (1973). In *Handbook of sensory physiology*, Vol. II (ed. A. Iggo) pp. 111–36. Springer, Berlin.

General properties of receptors

1. Methods of study. The action potential may be recorded in single nerve fibres laid on non-polarizable electrodes. The Pacinian corpuscle is sufficiently large to be isolated and stimulated directly [p. 286].

2. Specificity of response. The receptors of specialized structure respond only when an appropriate, specific stimulus is applied; other kinds of stimuli are ineffective. Thus the tactile corpuscles or Pacinian corpuscles respond to deformation; the tendon organs to stretch; the otolith organ to the pull of gravity; the retina to light; the cochlea to vibrations of the basement membrane set up by sound. The receptors are thus peripheral analysers which respond to a specific environmental change by generating nerve impulses.

3. Adaptation. The appropriate natural stimulus is applied and maintained. The receptor responds to a constant stimulus not with a single discharge, but with a repetitive burst of nerve impulses; *adaptation* is not immediate. The duration of the burst varies greatly among the different receptors.

(i) In the case of stretch receptors, e.g. the muscle spindles, adaptation is slow; the discharge continuing for as long as the muscle is stretched, though the frequency of the discharge declines from its initial peak.

(ii) In touch receptors adaptation is rapid. Thus if a hair is bent, impulses are set up only during the movement, but not after it has ceased, although the hair is kept in the abnormal position.

(iii) The rate of adaptation in temperature receptors is intermediate between that of muscular and tactile receptors.

(iv) Pain receptors show little adaptation.

4. Effect of strength (intensity) of stimulus. As the strength of stimulation is increased the frequency of the discharge rate rises from low levels, e.g. 5–10 s⁻¹ to maximal levels of 300 s⁻¹ or higher. Thus in FIGURE V.79a pressures of 250 g and 500 g applied to the cat's toe-pad, produced peak discharge rates of about 250 and 400 per second respectively.

5. Effect of extent of stimulation. An extensive stimulus activates more receptors and consequently impulses pass back along a larger number of nerve fibres.

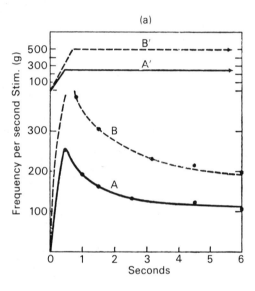

(a)

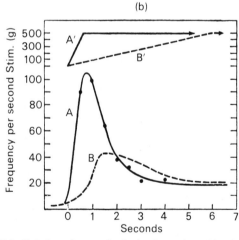

(b)

FIG. V.79(a). Relation of strength of stimulus and resulting frequency of afferent impulses. Stimulus A': weight of 250 g applied to cat's toe-pad. Resulting maximal frequency (A) is 250 s⁻¹. Stimulus B': weight of 500 g. Resulting maximal frequency (B) exceeds 400 s⁻¹. Note rapid adaptation in both cases as stimulus is maintained. (b) Relationship between rate of stretch of muscle and frequency of impulses set up in receptors. Vertical axis—*Above*: Weight attached to muscle to produce stretch and stimulate receptors. *Below*: Frequency of afferent nerve impulses per sec set up by stretching muscle and recorded in distal end of cut muscle nerve.

In A', the stretch reached maximum in less than 1 s; in B' the same maximum stretch was attained after 6 s. Note that the maximal frequency of the nerve impulses (A) in response to A' is over 100 s⁻¹ compared with 40 s⁻¹ (B) in response to B'. (Modified from Adrian, E. D. (1928). *Basis of sensation.*)

6. Action potential. The action potential recorded varies in spike width with the diameter and type of conducting afferent fibre, but not with the nature of the receptor. The modality of a sensation (e.g. whether it is touch, heat, cold) this in no way depends on the characteristic of the impulse in an afferent fibre.

Temperature receptors

Temperature receptors are found on the chest, nose, nipples, anterior surface of the arm and forearm, and on the abdomen. There are far more 'cold' spots than 'warm' spots.

'Warm' receptors respond with a fairly steady discharge at any one tissue temperature between 35 and 45 °C. The *maximum* frequency of the steady discharge is generally at a tissue temperature of 38–43 °C but even then seldom exceeds 10 impulses per second. Sudden heating of the skin produces a rapid volley of impulses but this is of phasic character.

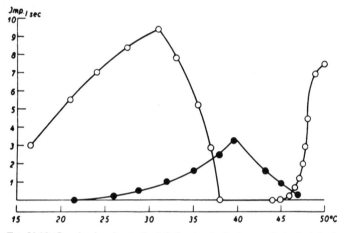

FIG. V.80. Graphs showing to the left the steady discharge of a typical single cold fibre (open circles), in the middle a typical single warm fibre (filled circles), and to the right the paradoxical cold fibre discharge (open circles) as a function of the temperature. (Dodt, E. and Zotterman, Y. (1952). *Acta physiol. scand.* **26,** 345.)

'Cold' receptors in the cat's tongue [FIG. V.80] were localized within 0.2 mm to a site in the vicinity of the papillae. Single fibre preparations were made from the lingual nerve. Cold receptors fire with a steady discharge at any one tissue temperature between 10 and 35 °C. The maximal frequency of steady discharge is at a tissue temperature of 25–30 °C; such a maximum is of the order of 10 impulses per second. Sudden cooling of the skin causes a rapid volley of impulses. If the tissue temperature is raised beyond 45 °C the warm receptors do not respond but the cold receptors discharge briskly.

Both the warm receptors and the cold receptors respond primarily to the temperature of the tissues which immediately surround them and *not* to the gradient of temperature between the deep subcutaneous tissue and the surface. Cold receptor fibres are thin myelinated fibres 1.5–2 µm in diameter. Warm receptor fibres are slightly larger.

Electrophysiological basis of natural excitation of afferent nerve endings

Gray and Sato studied the mechanism of natural excitation of a single Pacinian corpuscle, which is a mechanoreceptor. The corpuscle was stimulated mechanically by a rod and the potential change between the receptor and its sensory nerve beyond the first node of Ranvier was recorded.

The results showed that the application of a mechanical stimulus caused a local potential change which increased in magnitude with increase of the stimulus up to a definite maximum upon which a propagated response was initiated in the nerve fibre terminal within the corpuscle [Fig. V.81]. The characteristics of the local response

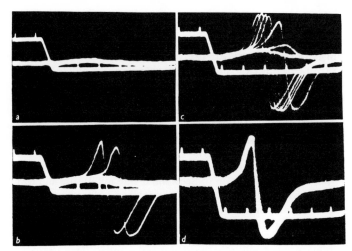

Fig. V.81. Potentials from a fresh preparation with four different strengths of stimulus. Top beam (at left) signals time course and amplitude of stimulus and time 1 ms. Bottom beam: recorded potential. (Gray, J. A. B. and Sato, M. (1953). *J. Physiol., Lond.* **122**, 610.

(or generator potential) were similar to those of electrotonic potentials.

The action potential which they recorded from the sensory nerve was made up of three components:

1. The receptor or generator potential.
2. The potential from the sensory nerve terminal.
3. The activity produced by the first node of Ranvier.

By local application of procaine they were able to abolish 2 and 3, leaving the receptor potential itself. A certain minimal stimulus was required in order to excite the receptor potential; greater stimuli than this increased the receptor potential to a maximum value [Fig. V.81]. Similarly the rate of application of the stimulus proved important, for below a certain rate stimuli did not excite a potential, whereas above this rate increasing speeds of application evoked more generous responses. Lastly, successive stimuli caused summation of the receptor potentials [Fig. V.82].

It would seem probable that the membrane of a receptor is polarized as is the case with the axon. Mechanical deformation or current application depolarizes the membrane and allows a flow of ions causing potential changes which, reaching a threshold value, initiate a similar change in the sensory fibre and hence evoke the propagated spike potential.

Thus in the normal preparation a generator potential aroused by mechanical stimulation arouses the initiation of a spike potential in the sensory nerve fibre terminals [Fig. V.81].

Katz obtained similar results in a study of the frog's muscle spindle. When the spindle was stretched its sensory endings were partially depolarized as could be recorded from the sensory axons near the spindle itself. The size of the potential evoked depended on the rate and extent of the mechanical stimulus.

REFERENCES

GRAY, J. A. B. (1959). *Handbook of physiology*, Vol. 1 *Neurophysiology*, p. 123. American Physiological Society, Washington, DC.
IGGO, A. (1977). *Br. med. Bull.* **33**, 97–102.
KUFFLER, S. W. and NICHOLLS, J. G. (1977). *From neuron to brain*. Sinauer, Sunderland.

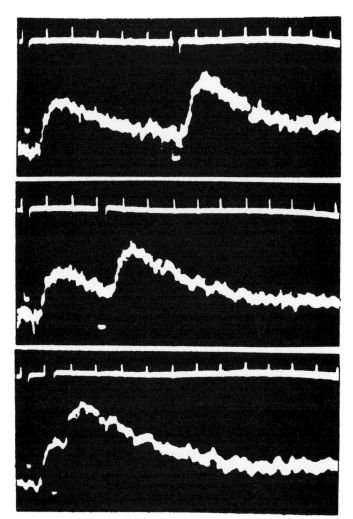

Fig. V.82. Summation of receptor potentials. Time marker in 1 ms intervals also signals the stimuli. (Gray, J. A. B. and Sato, M. (1953). *J. Physiol., Lond.* **122**, 610.)

Peripheral nerves

The peripheral nerves can be subdivided into:

1. The cutaneous nerves which supply the skin and which carry afferent impulses which may be consciously interpreted as 'touch', 'pain', 'itch', 'warm', and 'cold' sensations. These sensations constitute cutaneous sensibility Cutaneous nerves contain four times as many non-myelinated fibres as myelinated axons. Nociception is conveyed by non-myelinated and small myelinated fibres.

2. The nerves to muscles, joints, and other deep structures. These carry afferent impulses from the joints which are *consciously* interpreted as joint movement or position and afferent impulses from the muscle spindles which provide proprioceptive information relayed to the cerebellum (*non-conscious* proprioception). In addition afferent impulses from Pacinian corpuscles relay to conscious levels signalling deep pressure, and impulses in thin myelinated and unmyelinated fibres give rise to conscious sensations of muscle pain. The sensations of deep pressure and proprioception are sometimes referred to as deep sensibility.

DEGENERATION AND REGENERATION IN PERIPHERAL NERVES

The changes which constitute (Wallerian) degeneration in nerve cells and their processes may be initiated by transection, or crushing of the nerve fibre, local injection of toxic substances, or interference with their blood supply.

Injuries to nervous tissues are of different orders of severity. Sunderland classifies them as follows:

First-degree injuries. These are the most common and are caused by:

1. Pressure being applied to a nerve for a limited time thus occluding its blood supply and resulting in a local anoxia sufficient to impair the nerve's function.

2. Direct effect of pressure on axons causing local injury. Damage of this kind will be repaired so that function returns within a few hours to a few weeks. The axon is not destroyed but merely loses its functional properties for a short time.

Second-degree injuries. These are the result of prolonged and/or severe pressure being exerted on some part of the nerve, such as was formerly carried out deliberately, in crushing the phrenic nerve so as to paralyse one half of the diaphragm in tuberculosis.

Death of the axon at the site of pressure is followed by death of the axon distal to it because the axon is separated from the cell body on which it depends for its nutrition and, therefore, its existence. Chromatolysis occurs in the cell body.

Regeneration of the axon is facilitated by the presence of uninterrupted endoneurial tubes, down which the regenerating axons grow to effect their former connections. The axoplasm grows down the endoneurial tubes at the rate of 2–3 mm per day. Remyelination of the axon closely follows its growth down the endoneurial tubes.

Third-, fourth-, and fifth-degree injuries. Third-degree injuries are characterized by the endoneurial tubes becoming interrupted; fourth-degree injuries by the fascicles becoming disorganized; and fifth-degree injuries by the nerve trunk being severed.

Consideration will now be given to the processes of degeneration and regeneration in a fifth-degree injury.

Changes in the nerve cell body. Within 48 hours of section of the nerve the Nissl substance begins to disintegrate into a fine dust (*chromatolysis*), which may lose its staining reaction by 15–20 days. The dissolution of the Nissl substance (RNA) occurs in order that the protein manufacturing processes can be mobilized to help the neuron to survive. The Golgi apparatus fragments and dwindles in amount. The cells swell from increased fluid content and become rounded. The neurofibrils disappear; the nucleus becomes displaced to the cell margin and may even be completely extruded, in which case the cell atrophies and completely disappears.

The degree of chromatolysis depends on the variety of neuron affected, the nature of the injury, and the proximity of the site of injury to the nerve cell. In cases of severe injury, death of the cell often results.

Repair begins about 20 days after nerve section and is complete in 80 days. The Nissl substance and Golgi apparatus gradually reappear; the cell regains its normal size and the nucleus returns to its central position. Cell repair may occur even if the axon does not regenerate.

Afferent neurons. If the peripheral axon of a posterior root ganglion is divided the changes described above occur but pass away more quickly. If the central axon is cut, the changes in the ganglion cell are slight (Similarly, in tabes, where the central axons of the posterior nerve roots are damaged by syphilis, the ganglion cells show very slight changes.) The synaptic terminals in the spinal cord show characteristic changes; they swell after 24 hours, begin to disintegrate after 3 days, and disappear within 6 days.

Central nervous system. Most of the affected cells atrophy completely. As usual the atrophy is more intense when the fibres are cut close to their parent cells.

Autonomic ganglia. Here changes are difficult to demonstrate owing to the normal sparseness of the Nissl substance.

CHANGES IN THE NERVE FIBRE

Stage of degeneration. Within 24 hours after nerve section the following histological and chemical changes occur simultaneously along the whole length of that part of the nerve *distal to the site of injury*.

The axis cylinder breaks up into short lengths and within a few days only a little debris is left in the space formerly occupied by the axon. The myelin sheath breaks down, rather more slowly than the axon, into oily droplets. At this stage, 10–60 days (at the most), the degenerating fibres can be revealed by the Marchi technique.

The cells of the sheath of Schwann divide mitotically and form cords of cells lying within the endoneurial tubes. Macrophages from the endoneurium invade the degenerating myelin sheath and axis cylinder and remove the debris forming the remains of these structures by phagocytosis. As the debris is removed the Schwann cell cytoplasm gradually fills the endoneurial tubes. This process is completed by 3 months. The physical destruction of myelin occupies 8–10 days after nerve injury; chemical destruction of myelin takes 8–32 days after nerve section.

Degeneration in the *central stump* may occur for a variable (usually short) distance from the point of section.

At the *site of injury* the Schwann cell tissue differentiates into thin elongated cells which grow in all directions from the cut end of the distal stump at the rate of up to 1 mm per day. 'Useful' growth is towards the central end of the cut nerve. Some growth of Schwann tissue occurs from the cut surface of the central part of the fibre but activity in the distal stump is much greater. Eventually the gap between the cut surfaces of the injured nerve fibre is bridged by Schwann cells. Gaps of up to 3 cm may be bridged by Schwann cells; but if the two cut ends of the nerve are stitched together the Schwann cells accomplish their task much more easily. Fibroblasts also play an important part by forming scar tissue.

Stage of regeneration. While the above processes are occurring the central axon elongates and then grows out in all directions by extension of pseudopod-like structures of fibrils, up to 100 in number. The regenerating fibrils appear to be guided by the strands of Schwann cells into the ends of the peripheral endoneurial tubes. Two to three weeks after nerve section the peripheral tubes contain varying numbers of developing fibrils, some none, some as many as 25. Eventually all the fibrils re-innervating one tube degenerate except a single successful one which thereupon progressively enlarges to fill the tube. Fibrils from one fibre may enter several tubes but probably only one fibril will survive.

The daily rate of growth is 0.25 mm in the junctional area of scar tissue, 3–4 mm in the peripheral stump, and rather faster in the distal end of a crushed nerve where the mechanical conditions for regeneration are more suitable than those in a cut nerve.

Medullary sheaths begin to develop in about 15 days and follow the course of the growing fibrils. The Schwann cells filling the endoneurial tubes form the medullary sheath round the successful fibril in the same way as in the developing nerve fibre. Completion of the medullary sheath occurs within one year.

Increase in fibre diameter takes place very slowly. The diameter finally attained is limited by the diameter of the peripheral tube and the size of the parent nerve cell. Regenerated fibres rarely attain a fibre diameter more than 80 per cent of normal.

The frameworks of the sensory and motor endings can persist for months. When the growing axon tip ultimately reaches the nerve ending it establishes a connection with it which at first may be atypical in appearance but later becomes normal. Many fibres will establish connections with new kinds of endings in new situations. Functional complications may occur as a result of this. It is possible that when a number of nerve fibrils grow down one tube to the same ending the appropriate one alone survives to establish effective connections. It is also conceivable that when a new fibre reaches skin or muscle it may call forth locally the appearance of a suitable ending. (In tissue culture, when a nerve fibre meets a myoblast, a motor end plate is formed.) As is explained below, functional recovery lags considerably behind anatomical regeneration.

Regeneration of neurons never takes place in the central nervous system; regeneration of the central axon of a dorsal nerve root may occur as far as the pia, but there is no penetration into the spinal cord. The proliferation of glial cells forms scar tissue which heals the lesion.

Effects of cutaneous nerve section

Immediate effects

In Head's classical experiments, the radial and external cutaneous nerves were exposed in the neighbourhood of the elbow and small portions were excised; the cut ends were at once united to facilitate regeneration of the fibres. All forms of cutaneous, i.e. superficial sensibility were abolished over the radial half of the forearm and the back of the hand. Deep sensibility, however, was retained as it is mediated by afferent fibres in the nerves arising from subcutaneous tissue, muscle, and joints. Deep pressure was recognized. Excessive pressure gave rise to aching pain. There was awareness of the position and of movements of the part.

Recovery of skin sensibility

1. Anatomical considerations. The histological changes of degeneration and regeneration following nerve section have been described; broader anatomical factors must now be discussed. Under ideal conditions each sprouting central axon would grow down its old neurilemmal sheath to its own type of receptor in its original locality. Such perfect anatomical recovery never takes place even when the two cut ends of the nerve are carefully stitched together immediately after the section, because the two cut ends of any one fibre will never be placed in apposition; much better anatomical repair may occur, for obvious reasons, when a nerve is crushed and not cut. When the two ends of a cut nerve are separated by a gap, many of the sprouting central axons get lost, never reach the peripheral sheaths, and so never function. Such fibres as do reach the distal end of the nerve will establish faulty connections either with a receptor which is of their own type but which lies in a different region of the skin, e.g. a touch fibre originally supplying the base of the thumb may connect with a receptor at the tip of the thumb, or with a different type of receptor, e.g. a touch fibre may connect with a temperature receptor, or both faults may occur.

2. Functional results of imperfections of regeneration. Functional recovery is likewise imperfect.

(i) If many skin receptors fail to establish central connections, skin sensitivity is reduced, and areas of anaesthesia may be present.

(ii) The ability to localize a stimulus is impaired when abnormal connections are established between the periphery and the cerebral cortex. Thus, suppose that point *a* at the base of the thumb was connected originally with area A in the sensory cortex; 'experience' has 'taught' us that cortical activity at A represents stimulation of point *a*. If new peripheral connections are established, cortical area A may become connected with, say, point *b* at the tip of the thumb.

A stimulus at *b* will then be interpreted centrally as coming from *a*, i.e. there will be false localization which can only be overcome, if at all, by prolonged practice. Or again, cortical touch area A may become connected wtih a skin temperature receptor; temperature stimuli either will not be recognized or will be misinterpreted as touch stimuli.

3. Recovery in the Head type of experiment. In Head's experiment, in which the anatomical conditions were almost ideal, recovery occurred in two fairly well defined stages. The first phase of crude recovery began after 8 weeks and was maximal in 30 weeks. The returning sensation was punctate, being limited to some pain spots, hot spots, and cold spots, the intervening areas of skin being insensitive. The findings at this stage are as follows:

(i) Pain. A pin-prick cannot be located accurately, and the pain radiates widely and is not infrequently referred to some part at a distance from the point actually stimulated. A stronger nocuous stimulus than normal must be applied before pain is felt, but the sensation when it does occur has a most unpleasant quality.

(ii) Temperature. A sensation of cold is produced by temperatures below 24 °C; similarly, a sensation of heat results from temperatures above that of the body—the lower limit being between 38 and 45 °C. It is impossible to recognize any temperature quality between 24 and 38 °C.

During the next year or longer, the finer and more discriminative aspects of sensibility return. Intermediate grades of temperature can now be recognized.

The two stages described by Head often overlap to a varying extent; functional recovery is less satisfactory in most patients with nerve injuries than in his case.

Dorsal nerve roots

All the afferent impulses from the periphery (including the viscera) pass into the dorsal nerve roots and enter the spinal cord. The medullated A fibres in the dorsal roots vary in diameter from 1–20 µm; in addition, there are numerous non-medullated C fibres present, constituting about 40 per cent of the total number. The dorsal roots on entering the cord divide into a *medial* bundle which contains all the segmental large myelinated *afferents* (Group I) and some smaller ones (Groups II and III), and a *lateral* bundle which carries only the small *afferent* fibres (Groups III and IV). After section of the lateral division of the dorsal root, pain sensibility is abolished in the peripheral distribution of that root. These Group III and IV fibres relay in the substantia gelatinosa Rolandi; second-order fibres ascend in the lateral spinothalamic tract.

Group I fibres in the medial division (a) relay with cells in Clarke's column to form second-order fibres which run in the dorsal spinocerebellar tract and (b) form mono- and disynaptic connections with the adjacent motoneurons. Group II and III fibres carrying 'touch', 'deep pressure', and 'joint proprioceptive' impulses form diffuse connections with the dorsal interneurones and relay to the brain stem, thalamus, and cortex via the dorsal columns or via the ventral spinothalamic tract.

Skin distribution of the dorsal roots. Dermatomes

The skin area supplied by a dorsal nerve root (dermatome) cannot be mapped out simply by determining the extent of the anaesthetic zone resulting from section of the particular root, owing to the extensive degree of overlap between adjacent dorsal roots. The methods which have been employed in man are:

1. Cut three dorsal roots (in operations carried out for the relief of intractable pain) above and three below the root investigated and determine the area of residual sensibility; this area gives the full extent of the intact dermatome.

2. Stimulate the peripheral end of the cut dorsal root (exposed at operation) and map out the area of resulting cutaneous vasodilation or mark out the area of herpetic eruption in which the dorsal root ganglia are inflamed as a result of virus infection.

Section of dorsal nerve roots

Section of the dorsal nerve roots produces the following results in the corresponding segments of the body:

1. Loss of all forms of sensation—pain, temperature, touch, muscle and visceral sensibility; trophic changes appear.
2. Loss of all reflexes, superficial and deep; loss of muscle tone.
3. Great clumsiness in the movement of the part, which makes its use almost impossible. This disturbance develops because all parts of the central nervous system—the cerebral cortex and the centres concerned with the reflex control of posture—are deprived of afferent impulses from the joints and muscles.

METHODS OF MAPPING NERVE FIBRE PATHWAYS

1. Staining techniques

The older methods made use of the fact that normal myelinated nerve fibres stain black with osmic acid which the myelin reduces to the black osmium oxide. The best known techniques are:

(a) Weigert–Pal. The principle of the method is as follows: The cerebrosides and lecithins of myelin, treated with dichromate normally reduce the latter to chromium dioxide (CrO_2). CrO_2 acts as a mordant towards haematoxylin. If the myelin sheaths have degenerated, with phagocytosis of the lecithins, no reduction of dichromate occurs and the scar area appears colourless when haematoxylin is used to stain the tissue. This technique is useful for investigating the topography in peripheral nerve and in the brain stem and cord of *myelinated* axons only.

(b) Marchi method. Normal myelin, since it is readily oxidized by $K_2Cr_2O_7$, does not stain with osmium tetroxide after treatment with dichromate. During the *early* stages of myelin degeneration, however, oleic acid is liberated; this is not oxidized by dichromate and this stains readily with osmium tetroxide. The useful limits of this method are from ten days to one month after the trauma has been inflicted.

After, say, destruction of the motor cortex (Area 4) on one side, the Marchi method shows the degenerating fibres of the pyramidal tract as black dots in serial sections, allowing its course to be plotted.

Neither of these methods demonstrates the nerve fibres themselves, for each is related to the delineation of the myelin. Nonmyelinated tracts, which are overwhelmingly preponderant, require techniques which make use of the affinity of nerve fibres for silver salts (which they reduce to a deep brown silver proteinate). Chemically, the *silver methods* are much like those of photography. The tissue is soaked in silver nitrate, then the silver is precipitated in those parts which have absorbed it most. Golgi, Cajal, and Bielschowski pioneered these techniques for the study of central nervous structures.

Degenerating axons can be stained by the *Nauta* method—a silver method which selectively stains dying axons as far as their terminal ramifications; this technique does *not* stain intact axons.

The Alzeimer–Mann–Haggquist method uses methyl blue and eosin. Axons stain blue and myelin sheaths stain red.

2. Electrical methods

These can be retrograde or anterograde. Thus a neuronal cell body can be impaled by a microelectrode which can either be used to record depolarization when its axon is stimulated antidromically or to stimulate the cell body which discharges orthodromically down its own axon.

3. Several new methods have proved useful in living neurons

(i) Tritiated amino-acids are injected into a cluster of neuronal cell bodies. These enter the perikarya and are transported by axoplasmic flow to the axon terminals. Autoradiography displays the pathway.

(ii) Horseradish peroxidase is injected among a group of axon terminals. After incubation with an appropriate substrate, granules appear in the cell bodies. This technique has proved particularly useful in showing rapid retrograde axonal transport.

(iii) Transmitter substances (e.g. noradrenalin, serotonin, or dopamine) can be converted into fluorescent derivatives by treating the nervous tissue with formaldehyde or glyoxylic acid and exposing the tissue to ultraviolet light.

(iv) Similarly, an enzyme involved in the manufacture of a particular transmitter is injected into the animal and provokes the formation of antibodies that combine with it specifically. These purified antibodies are then labelled with a fluorescent dye and utilized to stain selectively neurons which contain the relevant enzyme.

Afferent paths in the spinal cord

FIGURE V.83 shows a traditional view of some of the main afferent tracts. On the basis of more recent work, Webster (1977) discusses the somaesthetic pathways under the headings anterolateral quadrant, dorsolateral quadrant and those of the dorsal funiculi and nuclei.

FIGURES V.84 and V.85 show a cross-section of the spinal cord delineating roughly the laminar arrangement of the grey matter as first described by Rexed (1952).

Lamina I contains small fusiform cells which receive A δ (small myelinated) and C (unmyelinated) fibres.

Laminae II and III receive similar fibre terminations. They contain small cells which are also 'supplied' by dendrites from cell bodies in the deeper layers. These two laminae represent what was earlier described as the substantia gelatinosa Rolandi. Their neurons are mainly short and effect synaptic connections with neurons from lamina IV. Some of the neurons of lamina III, however, do supply axons which course in the gracile and cuneate tracts of the dorsal columns.

Lamina IV contains large cell bodies; many of their dendrites pass dorsally. Many hair cell fibres and fibres whose endings respond to very light pressure end in lamina IV. Lamina IV cells contribute axons to the ipsilateral dorsal funiculi and to the spinocervical pathway; others send axons across the midline to ascend in the lateral spinothalamic tract.

Lamina V again contains large cells which are synaptically stimulated by fibres subserving either light touch or nociceptor excitation. Their axons contribute substantially to the contralateral spinothalamic tract and the ipsilateral spinocervical tract.

Lamina VI is only clearly featured in the *cervical* and *lumbar* regions. It represents the site of synapse of the muscle and joint afferents. The IA afferents from muscle spindles establish monosynaptic connections here. Some of the cells give rise to the ventral spinocerebellar tract (VSCT).

Laminae VII and VIII also receive muscle and joint afferents together with some from the skin. Each of these laminae receives fibres from the reticulospinal tract stemming from cells in the pontomedullary reticular region [FIG. V.86]. Lamina VIII receives predominantly pontine reticulospinal fibres (which course in the ventral quadrant of the cord) whereas VII receives mostly medullary reticulospinal fibres whose tract is in the lateral quadrant of the cord.

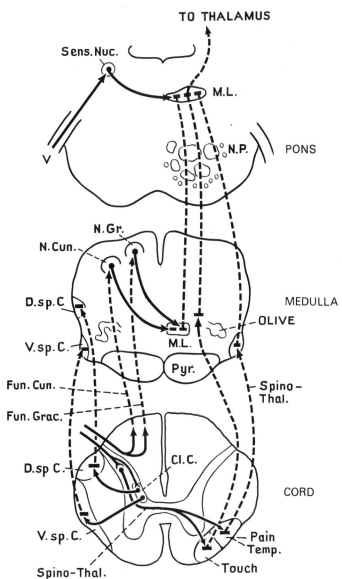

FIG. V.83. Diagram of course of ascending tracts. V = Afferent fibres of fifth nerve. Sens.Nuc. = Sensory nucleus of fifth nerve. N.P. = Nuclei pontis. Pyr. = Pyramidal tract. N.Gr. = Nucleus gracilis. N.Cun. = Nucleus cuneatus. D.Sp.C. = Dorsal spinocerebellar tract. V.Sp.C. = Ventral spinocerebellar tract. M.L. = Medial lemniscus. Spino-Thal = Spinothalamic tract. Fun.Cun., Fun.Grac. = Funiculus cuneatus and gracilis. Cl.C. = Clarke's column of cells.

The anterolateral quadrant besides containing spinothalamic tracts additionally contains ascending tracts to the brain stem reticular formation, the inferior olive, and the superior colliculus. The majority of the cell bodies of the spinothalamic tract lie in lamina V—some are found in lamina I and others in the nucleus of the lateral funiculus [FIG. V.84]. Spinoreticular fibres arise from cells in contralateral laminae VII and VIII.

In the dorsolateral quadrant, fibres other than those of the dorsal spino-cerebellar tract pass from cells of laminae IV and V via the spinocervical tract to the lateral cervical nucleus which lies immediately lateral to the dorsal horn of segments C_1 and C_2. Here, after relay, they decussate and travel with the medial lemniscus. They are important in the transference of information relating to cutaneous sensibility.

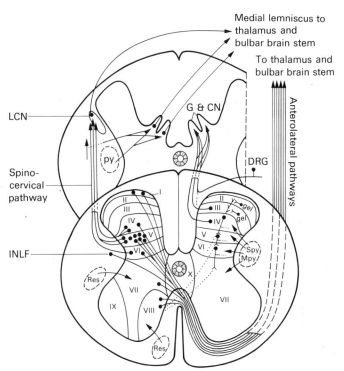

FIG. V.84. Scheme of somaesthetic pathways in the spinal cord. DRG = Dorsal root ganglion. INLF = Interstitial nucleus of lateral funiculus. py = Pyramidal tract. gel = Substantia gelatinosa. LCN × Lateral cervical nucleus. Res = Reticulospinal tract. G and CN = Gracile and cuneate nuclei. Mpy = Motor cortex pyramidal tract fibres. Spy = Somatosensory cortex pyramidal tract fibres. Roman numerals I to X indicate Rexed's laminae. (Webster, K. E. (1977). In *Somatic and visceral sensory mechanisms* (ed. G. Gordon) p. 114. *Br. med. Bull.* **33**, No.2.)

The dorsal columns, very well developed in man, are *wholly myelinated*. Only 25 per cent of these fibres arrive at the gracile and cuneate nuclei without relay. The medullary gracile and cuneate nuclei contain two cytoarchitecturally distinct zones—a caudal half where the cells are grouped in clusters and a more rostral part which presents a reticular appearance with a mixture of large and small multipolar neurons. Primary afferents are mostly distributed to the

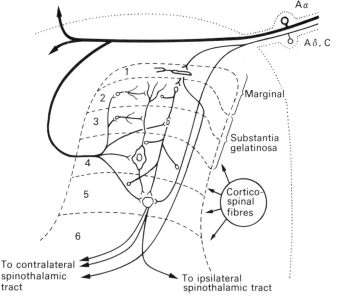

FIG. V.85. Diagrammatic presentation of spinal cord dorsal horn structure and primary afferent projection into the cord. Numerals indicate Rexed's laminae.

RETICULOSPINAL FIBRES

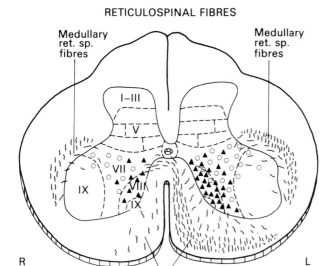

▲ Sites of termination of pontine ret. sp fibres
○ Sites of termination of medullary ret. sp fibres

FIG. V.86. Diagram showing the sites of termination and the course within the spinal cord of the cat of reticulospinal fibres coming from the pontine and medullary region of origin, respectively. Note difference in distribution of the two components. Roman numerals refer to Rexed's laminae of the cord. (Nyberg-Hansen, R. (1965). *J. comp. Neurol.* **124**, 71.)

'cluster' areas; non-primary fibres pass to the reticular region of the nucleus. The reticular zone only of the gracile and cuneate neurons receives contralateral fibres from the somatosensory cortex.

Interpolated between the head end of the gracile nucleus and the descending vestibular nucleus is nucleus Z [FIG. V.87] which receives most of its input from the ipsilateral hindlimb via the DSCT.

The classical somaesthetic region of the thalamus is the ventroposterior (or posteroventral) nucleus—a medial (arcuate) part receiving trigeminal afferents from the face and a more lateral part which is supplied from the medial lemniscus [p. 293]. Between the ventroposterior and ventrolateral nuclei of the thalamus there is an intermediate nucleus which receives afferents (proprioceptive from nucleus Z) and which projects to cortical area 3a [see p. 321]. The posteroventral nucleus fibres project to the sensory cortex area 3, 1, 2 [p. 320].

REFERENCES

GORDON, G. (ed.) (1977). *Somatic and visceral sensory mechanisms. Br. med. Bull.* **33**, No. 2.
HEAVNER, J. E. (1975). *Pain* **1**, 239.
REXED, B. (1952). *J. comp. Neurol.* **96**, 415.
SCHEIBEL, M. E. and SCHEIBEL, A. B. (1969). *Brain Res.* **13**, 417.
WEBSTER, K. E. (1977). In *Somatic and visceral sensory mechanisms* (ed. G. Gordon) pp. 113–20. *Br. med. Bull.* **33**, No. 2.
WILLIS, W. D. and COGGESHALL, R. E. (1978). *Sensory mechanisms of the spinal cord.* Wiley, Chichester.

As a generalization it has been customary to identify the main afferent tracts clinically with various sensory functions. Though an over-simplification this practice has proved useful in helping to define the sites of clinical lesions and therefore justifies a full description.

1. Cutaneous sensibility (a) Dorsal root fibres subserving *pain* and *temperature* enter the cord in the vicinity of the cells of the dorsal horn of grey matter. They are very small myelinated or unmyelinated fibres. The sensory endings of nociceptive fibres are unmyelinated naked nerve terminals. On entering the cord the

fibres branch extensively and run several segments up and down the cord in the most lateral part of the posterior grey horn (the so-called Lissauer's tract) before effecting a synapse in the apical zone of the posterior horn (which is termed the substantia gelatinosa Rolandi after its discoverer). Second-order axons cross the midline anterior to the central canal and reaching the lateral funiculus, ascend as the lateral spinothalamic tract. As this tract ascends the cord it itself shows evidence of lamination; its constituents derived from the sacral segments lie most laterally and posteriorly within the lateral spinothalamic fasciculus, whereas those from the cervical segments lie more anteriorly and medially.

Patients suffering from severe intractable pain (e.g. from carcinomatous metastases) have been treated by anterolateral chordotomy—an operation in which the lateral spinothalamic tract, lying near the lateral surface can be severed by an appropriate section using a special knife.

As its name implies, the tract carrying these impulses is destined for the thalamus. In the pons the spinothalamic fibres are joined by those from the gracile and cuneate nuclei to form the medial lemniscus [see p. 293].

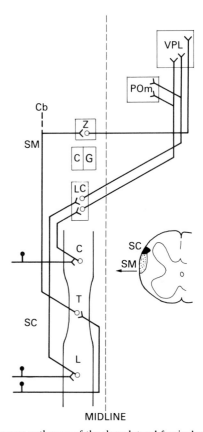

FIG. V.87. Sensory pathways of the dorsolateral fasciculus. The organization of the spinocervical and spinomedullothalamic pathways is shown. Primary afferent fibres end upon cells of the dorsal horn in the cervical (C) and lumbar (L) enlargements. The second-order neurons project through the spinocervical tract (SC) to the lateral cervical nucleus (LC). The axons of third-order cells decussate and project through the medial lemniscus to the medial part of the posterior nuclear complex (PO$_m$) and the ventral posterior lateral (VPL) nucleus of the thalamus. Other primary afferent fibres ascend in the dorsal column to the thoracic cord (T), where they synapse upon cells in Clarke's column (and also on cells caudal to Clarke's column), which in turn project in the spinomedullary tract (SM) to nucleus Z. Some of these axons also project to the cerebellum (Cb). Nucleus Z projects contralaterally to the VPL and adjacent parts of the ventral lateral nucleus. The transverse section of the spinal cord indicates the locations of the spinocervical and spinomedullary pathways. (Willis, W. D. and Coggeshall, R. E. (1978). *Sensory mechanisms of the spinal cord*, p. 262. Wiley, Chichester.)

(b) Some of the fibres subserving touch after entering the cord and branching also relay in the dorsal horn to give rise to second-order fibres, which in the main cross to the opposite side of the cord, finally to ascend in the lateral boundary of the anterior funiculus (near the exit of the ventral roots) as the ventral or anterior spinothalamic tracts. This tract too becomes incorporated in the medial lemniscus of the brain stem. This tract is of less importance because tactile sensation is also conveyed by tracts in the dorsal columns. However, these latter fibres ascend in the *ipsilateral* dorsal columns.

2. Position sense (a) Conscious sense of position, vibration, and deep pressure is carried by axons from joint receptors and Pacinian corpuscles. These, entering via the posterior root, branch and enter the dorsal column of the same side. Those from the legs ascend the cord as the fasciculus gracilis which lies medially in the dorsal column to terminate in the gracile nucleus in the medulla. Those from the thoracic and cervical segments lie more laterally in the cord and terminate in the cuneate nucleus of the medulla. Second-order fibres from these two nuclei which lie in the dorsal medulla [FIG. V.83] sweep vertically across the midline (internal arcuate fibres) to form the medial lemniscus. In addition to conscious proprioceptive information, the medial lemniscus thus formed carries tactile and vibration sensations, destined for the posteroventral nucleus of the thalamus [p. 313] and after relay, finally by thalamo-cortical fibres to the post-central gyrus (sensory cortex).

(b) *Unconscious proprioception* is subserved by the spinocerebellar tracts. Afferent fibres from the muscle spindles and tendon end-organs enter the posterior root. The dorsal spinocerebellar tract (DSCT) is composed of large axons whose cell bodies (synaptically excited by these afferent fibres) are found in Clarke's nucleus or column which is located in the dorsal and medial region of the grey matter in the segments T_6 to C_7. The pathway of the dorsal spinocerebellar tract is entirely ipsilateral. Proprioceptive information is transferred via the Group IA and IB afferents and inferior cerebellar peduncle to the anterior lobe of the cerebellum very quickly indeed for the conduction velocities of the IA and IB fibres—approximately 100 m s⁻¹—is matched by that in the dorsal spinocerebellar tract itself which transmits impulses at 110 m s⁻¹.

The ventral spinocerebellar tract (VSCT) receives fibres predominantly from the lumbosacral regions. Most of the cell bodies of the VSCT are found in the third to sixth lumbar segments. The axons of these cells immediately cross the cord and ascend ventral to the DSCT to enter the cerebellum via the superior cerebellar peduncle. All VCST neurons receive strong polysynaptic influences from the ipselateral flexor reflex afferents, i.e. low and high threshold cutaneous afferents and groups II and III muscle afferents. These exert a predominantly inhibitory influence on VCST activity.

The brain stem

The previous section has described the spinal course of the major sensory pathways. Their passage up the brain stem can be conveniently studied in brain stem sections at successive levels using as additional landmarks the sites of both motor and sensory nuclei of the cranial nerves and some of the major descending (motor) pathways.

A section through the lower third of the medulla shows the decussation of the pyramidal tracts [FIG. V.88]. These are corticospinal fibres. They pass from the ventral aspect of the medulla

through the base of the ventral horn of grey matter, and come to lie in the lateral columns where they descend into the spinal cord as the lateral corticospinal tract (LCST). The dorsal columns are wider, and therefore push the dorsal horns of grey matter farther apart from the pons to the upper cervical region. The central canal of the spinal cord is still situated centrally, but is now approaching the dorsal surface. On the surface in the ventrolateral region, on each side, are the rubroreticulospinal tract, the dorsal and ventral spinocerebellar tracts, and the spinothalamic tract.

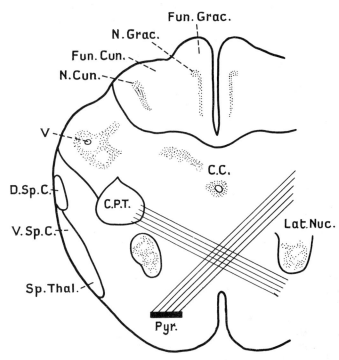

FIG. V.88. Section through medulla at level of decussation of pyramids (diagrammatic). Fun.Grac., Fun.Cun. = Funiculus gracilis and cuneatus; N.Grac., N.Cun. = Nucleus gracilis, Nucleus cuneatus; V. = Descending root of fifth nerve; C.C. = Central canal; Pyr. = Pyramidal tract; C.P.T. = Crossed pyramidal tract; D.Sp.C., V.Sp.C. = Dorsal and ventral spinocerebellar tracts; Sp. Thal. = Spinothalamic tract; Lat. Nuc. = Lateral nucleus.

At the level of the olive [FIG. V.89], the central canal is now approaching the dorsal surface of the medulla, and is about to open out at the calamus scriptorius into the fourth ventricle. The dorsal columns have been replaced by masses of grey matter—the funiculus gracilis ends in the nucleus gracilis, and the funiculus cuneatus in the nucleus cuneatus. From these nuclear cells a new relay of fibres arises. The majority cross the middle line (internal arcuate fibres) to lie dorsal to the pyramid as the medial lemniscus. Some fibres pass to the inferior cerebellar penduncle (restiform body) of both sides.

In the floor of the fourth ventricle at the level of the calamus scriptorius and in close relation to the dorsal nucleus of the vagus, are situated certain important cell groups subserving cardiovascular and respiratory activity and those acting as vomiting and deglutition centres. The respiratory centres lie medially and paramedially in the reticular formation of the pons and medulla.

The inferior olivary nucleus, dorsal to the pyramid, consists of a wavy layer of grey matter. Fibres pass to the restiform bodies on both sides. These are the climbing fibres which excite the cerebellar Purkinje cells. The olivary cells also provide descending axons of the bulbospinal tract which lies in the ventral part of the spinal cord. The fibres in the ventral spinothalamic tract now migrate medially to lie just dorsolaterally to the medial lemniscus, while the lateral spinothalamic fibres carrying pain and temperature remain lateral to the olive.

The nuclei of the twelfth, eleventh, tenth, ninth, and seventh cranial nerves may conveniently be described here.

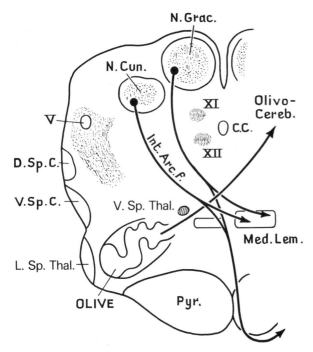

FIG. V.89. Section through medulla at level of olivary nucleus (diagrammatic). Int.Arc.f. = Internal arcuate fibres; Med.Lem. = medial lemniscus; Olivo-Cereb. = Olivocerebellar fibres; XI, XII = Nuclei of eleventh and twelfth nerves; N.Grac., N.Cun. = Nucleus gracilis and cuneatus; Pyr. = Pyramid; C.C. = Central canal; V. = Descending root and nucleus of fifth nerve; D.Sp. C., V.Sp.C. = Dorsal and ventral spinocerebellar tracts; L.Sp.Thal. = Lateral spinothalamic tract; V.Sp.Thal. = Ventral spinothalamic tract.

XII (hypoglossal nerve). The nucleus of the hypoglossal nerve extends through the lower two-thirds of the medulla. It first lies ventrolateral to the central canal, and when the fourth ventricle appears it lies in its floor, close to the middle line. The hypoglossal nerve fibres pass out ventrally between the olive and pyramid.

If the hypoglossal nerve is stimulated, the tip of the tongue is pushed over to the opposite side. Conversely, if the twelfth nerve is paralysed, the tongue when projected deviates to the same side (by the unopposed action of the other nerve). The affected side of the tongue shows the signs of a lower motor neuron paralysis–wasting of the muscle fibres (with consequent wrinkling of the mucous membrane) and fibrillation.

XI (accessory nerve). The nucleus of origin of the spinal part of this nerve consists of cells lying in the lateral part of the ventral horn of grey matter in C1–5. The fibres pass dorsally and then bend outwards to emerge at the side of the cord and medulla as lateral roots. These contain large medullated efferent (and some afferent) fibres for the supply of the sternomastoid and trapezius. (The bulbar part of the nucleus [shown in FIG. V.89] is best regarded as part of the vagus nucleus.)

X (vagus nerve). The afferent fibres of the vagus arise in the jugular and nodose ganglia from unipolar cells and convey impulses from somatic and visceral structures. Some fibres ascend to end in the dorsal nucleus, which lies in the floor of the fourth ventricle lateral to the hypoglossal nucleus; other fibres descend to form the tractus solitarius and end in adjacent nerve cells (the nucleus of the

tractus). These two nuclei give rise to the autonomic (involuntary) fibres of the vagus and also connect up with the nucleus ambiguus.

The nucleus ambiguus is the somatic ('voluntary') motor nucleus of the vagus, and lies ventrolateral to the dorsal nucleus. The fibres run first dorsally and then curve round to emerge dorsal to the olive and supply such structures as the muscles of the larynx [FIG. V.90] and the sino-atrial and atrioventricular nodes and the atria.

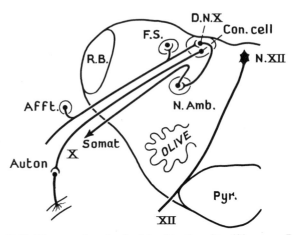

FIG. V.90. Diagram of mode of origin of tenth and twelfth nerves. R.B. = Restiform body = inferior cerebellar peduncle; D.N.X. = Dorsal nucleus X; F.S. = Fasciculus solitarius; Con.Cell = Connector cell which gives rise to some of the autonomic fibres in vagus; N.Amb. = Nucleus ambiguus which gives rise to vagal motor fibres to the heart and larynx.

IX (glossopharyngeal nerve). The ninth (glossopharyngeal) nerve is arranged in exactly the same way as the tenth. The afferent fibres have their cell bodies in the petrosal ganglion. They end in nuclei which form the upward continuation of the dorsal nucleus of the tenth (ascending fibres) and of the tractus solitarius (descending fibres). The motor nucleus which supplies the pharynx is in line with the nucleus ambiguus. The autonomic fibres arise in the dorsal nucleus and in the nucleus ambiguus.

VII (facial nerve). Though the seventh (facial) nucleus is situated in the lower pons, it must be considered here because the arrangement is like that already described. The afferent fibres have their cell bodies in the geniculate ganglion and pass in the nervus intermedius to end in nuclei in line with the dorsal nucleus of the ninth and tenth and the fasciculus solitarius. The autonomic fibres arise in the dorsal nucleus. The motor nucleus is in line with the nucleus ambiguus. The fibres run dorsally and form a loop round the sixth nerve nucleus and then turn ventrally and laterally.

In lesions of the pyramidal tract (supranuclear lesion) voluntary movements in the face are lost, but emotional movements (frowning, smiling) are retained because of the separate motor pathway for emotional exteriorization. If the seventh nerve is injured in any part of its course (infranuclear lesion) all types of movements of the face are equally affected.

We may summarize these facts thus:

1. The motor somatic nucleus of XII is a column of cells lying near the middle line.

2. The afferent fibres of VII, IX, and X divide into:

(i) Ascending fibres which end in a column of cells lying lateral to the nucleus of XII = column of dorsal nuclei of VII, IX, and X.

(ii) Descending fibres which end in grey matter lying just lateral to the column of dorsal nuclei = fasciculus (tractus) solitarius and its nucleus.

3. Many of the autonomic efferent fibres of VII, IX, and X arise in the column of dorsal nuclei.

4. The somatic efferent fibres of VII, IX, X, and XI arise from a column of cells which extend from the lower border of the pons to C5, and are situated lateral and ventral to the column of dorsal nuclei.

A section of the junction of pons and medulla shows the restiform body and the entrance of the eighth nerve [FIG. V.91].

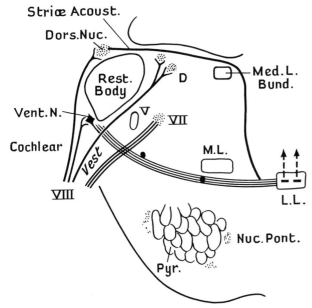

FIG. V.91. Section through the lower pons (diagrammatic). Striae Acoust. = Striae acousticae; Dors.Nuc., Vent.N. = Dorsal and ventral cochlear nucleus; Cochlear = Cochlear division of VIII; Vest. = Vestibular division; D. = Deiters' (lateral vestibular) nucleus; M.L. = Medial lemniscus; L.L. = Lateral lemniscus; Trap. = Trapezoid body; Med.L.Bund. = Medial longitudinal bundle; Nuc.Pont. = Nuclei pontis; Pyr. = Pyramidal tract; VII = Dorsal nucleus of VII; Rest.body = Restiform body; V = Descending root and nucleus of fifth.

In FIGURE V.91 the cochlear division of the eighth nerve is seen to pass dorsal to the restiform body (inferior cerebellar peduncle) and the vestibular division ventral. The ventral cochlear nucleus lies between the two divisions of the nerve, and the dorsal nucleus on the dorsolateral aspect of the restiform body. From the latter the striae acousticae cross in the floor of the fourth ventricle, and from the former the trapezoid body traverses the substance of the pons. The two groups of fibres turn up as the lateral lemniscus which is lateral to the medical lemniscus. The vestibular division ends in the vestibular nuclei which form an important reflex centre, coordinating the position of the eyes and limbs with that of the head and helping to maintain tone in the extensor or antigravity muscles; when these nuclei are destroyed in the decerebrate cat decerebrate rigidity is greatly reduced.

Other features are:

1. The pyramidal tracts beginning to break up into bundles.
2. The medial lemniscus, which is now joined by the spinothalamic tract conveying the fibres for pain, temperature, and touch.
3. The restiform body (inferior cerebellar peduncle).

THE PONS

The appearance of the pons is modified by the presence of numerous transversely crossing bundles of the brachium pontis (middle peduncle of the cerebellum) which are running from the pons to the opposite cerebellar hemisphere and vice versa. These break up the pyramidal tract into scattered groups of fibres, between which lie small masses of grey matter, the nuclei pontis. The medial lemniscus is now joined by central fibres from the sensory cranial nuclei, e.g. X, IX, VIII, V.

At the upper border of the pons, the fourth ventricle narrows gradually into the aqueductus cerebri (Sylvii). Above it on each side appear two masses of longitudinally running fibres, the brachia conjunctiva (superior cerebellar peduncles). These arise mainly from the dentate nucleus of the cerebellum, and as they pass forward approach the middle line and decussate to reach the opposite red nucleus and, further headwards, the thalamus. The ventral spinocerebellar tract turns over the lateral aspect of this peduncle to enter the vermis of the cerebellum [FIG. V.92].

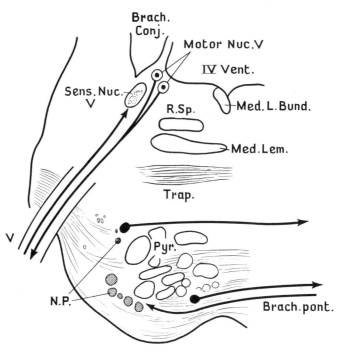

FIG. V.92. Section through upper part of pons (diagrammatic). Brach.Conj. = Brachium conjunctivum; V. = Fibres of fifth nerve; IV Vent. = Fourth ventricle; R.Sp. = Rubrospinal tract; Med.Lem. = Medial lemniscus; Trap. = Trapezoid body; Med.L.Bund. = Medial longitudinal bundle; N.P. = Nuclei pontis; Pyr. = Pyramidal tract; Brach.pont. = Brachium pontis.

THE MIDBRAIN (MESENCEPHALON)

The midbrain consists essentially of two structures: the superior and inferior colliculi dorsally, and the cerebral peduncles ventrally. Between them is the aqueductus cerebri (Sylvii) which is surrounded ventrally and laterally by the nuclei of origin of the oculomotor nerves [FIGS. V.93 and V.94].

The cerebral peduncles are great masses, chiefly of white matter, uniting the pons with the thalamic region of the cerebrum. They consist of three parts from before backwards:

1. Basis pedunculi, containing fibres from the cerebral cortex to the pons and cord. The middle three-fifths are occupied by the pyramidal tracts; the medial one-fifth by frontopontine and corticonuclear fibres, and the lateral one-fifth by temporopontine fibres.

2. Substantia nigra. A mass of deeply pigmented cells [see p. 359].

3. Tegmentum (Latin: a cover). A region where longitudinally and transversely running fibres are intermingled in a complex manner. Three decussations take place within it which from below upwards are:

(i) Of the dentatothalamic fibres.
(ii) Of the rubroreticulospinal tracts.
(iii) Of the tectospinal tracts.

The *red nucleus* extends from the hypothalamus to the caudal border of the superior colliculi [FIG. V.94]. It consists of two groups of cells: 1. n. magnocellularis—composing the caudal third of the nucleus and made up of large nerve cells which give rise to the rubrospinal tract; 2. n. parvocellularis—which consists of small cells forming the cranial two-thirds of the whole nucleus. In man, the large cells are few and the rubrospinal tract is small; most of the cells of the nucleus give rise to a rubroreticular tract which ends in the reticular grey matter in the brain stem from which reticulospinal fibres transmit the impulses to the spinal cord. By means of its afferents from cerebellum, vestibule, and muscles, and its efferent rubrospinal and rubroreticular fibres, the red nucleus plays an important part in mammals in helping to maintain normal body posture and normal muscle tone. In animals, decerebrate rigidity results from a midcollicular transection of the brain stem.

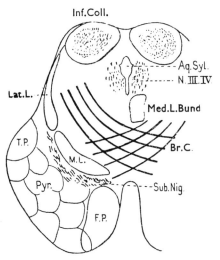

FIG. V.93. Section through midbrain at level of inferior colliculi (diagrammatic). Inf.Coll. = Inferior colliculi; Lat.L. = Lateral lemniscus; M.L. = Medial lemniscus; Aq.Syl. = Aqueductus cerebri (Sylvii); Med.L.Bund. = Medial longitudinal bundle; Br.C. = Brachia conjunctiva; N.III.IV. = Nuclei of third and fourth nerves; Sub.Nig. = Substantia nigra; T.P., F.P. = Temporopontine, frontopontine fibres; Pyr. = Pyramidal tract.

The superior colliculi are an important centre for visual reflexes. By means of the tectospinal tract they reflexly alter the position of the eyes, head, trunk, and limbs in response to retinal impulses. Colliculonuclear fibres pass to the third nerve nucleus to cause constriction of the pupil during the light reflex. The tectospinal tract connects the superior colliculi with the pupil-dilator centre in the thoracic cord (T1 and 2).

V (trigeminal nerve). The efferent fibres arise from the motor nucleus lying at the side of the grey matter bounding the aqueductus cerebri and fourth ventricle in the lower midbrain and upper pons [FIG. V.92].

The afferent fibres are derived from the semilunar (Gasserian) ganglion and end in a mass of grey matter lying lateral to the motor nucleus—the principal sensory nucleus. Long descending fibres are given off which end round adjacent grey matter. This column of cells and fibres—descending root (spinal tract) of the fifth nerve, descends to the upper cervical region of the cord. From these sensory nuclei a new relay of fibres passes across the middle line to join the medial lemniscus.

III, IV, and VI. Though we speak of the nuclei of the third, fourth, and sixth nerves as if they were isolated structures, they simply constitute one long column of grey matter controlling the movements of the eyes. They are linked by fibres of the medial longitudinal bundle which itself receives axons from the vestibular nuclei [p. 304]. This column extends from the upper part of the midbrain surrounding the ventral and lateral part of the aqueductus cerebri (III, IV), down to the upper part of the pons in the floor of the fourth ventricle (VI). The fibres of the third nerve take a curved course in the tegmentum to emerge on the medial side of the crus

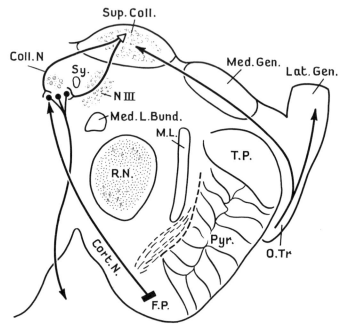

FIG. V.94. Section through the midbrain at the level of the superior colliculi (diagrammatic). Sup.Coll. = Superior colliculi; Coll.N. = Colliculonuclear fibres; Sy. = Aqueductus cerebri (Sylvii); Med.Gen., Lat.Gen. = Medial and lateral geniculate bodies; N.III. = Nucleus of third nerve; O.Tr. = Optic tract; Med.L.Bund. = Medial longitudinal bundle; R.N. = Red nucleus; M.L. = Medial lemniscus; T.P. = Temporopontine; Pyr. = Pyramidal tract; F.P. = Frontopontine; Cort.N. = Corticonuclear fibres.

[FIG. V.94]. The fourth nerves decussate in the roof of the aqueductus cerebri. The sixth nerve passes between the pyramid bundles to emerge at the lower margin of the pons.

The ascending pathways can now be summarized:

ASCENDING PATHS IN THE BRAIN STEM

1. Medial lemniscus. From the gracile and cuneate nuclei a fresh relay of fibres sweeps ventrally across the midline (internal arcuate fibres) to form the medial lemniscus which lies near the midline dorsal to the pyramid. The medial lemniscus at this stage conveys only the impulses carried in the dorsal columns, i.e. those from joint receptors, from touch receptors, from Pacinian corpuscles, and from other sources.

2. Spinothalamic tract. The fibres which carry impulses from pain and temperature receptors [p. 291] remain on the surface of the medulla lateral to the olivary nucleus, where they mingle with the ventral spinocerebellar tract. The fibres from touch receptors diverge medially from these to rejoin the rest of the touch fibres (which are lying in the medial lemniscus). In the pons, the pain and temperature fibres of the spinothalamic tract have also passed medially to join the main sensory path. The medial lemniscus thus enlarged is now carrying all the impulses from the skin and muscles of the opposite side of the body below the head [FIG. V.83].

3. Spinocerebellar tracts.

These convey impulses from muscle and tendon receptors but also contain cutaneous afferent fibres. The dorsal tract enters via the inferior cerebellar peduncle and the ventral via the superior peduncle.

4. Cranial nuclei.

Afferent impulses from the face enter in the fifth nerve, and end in the principal sensory nucleus in the pons, and in the long descending root and adjacent grey matter which extend down to the upper cervical cord. From this long column of grey matter a fresh relay arises which crosses the middle line to enter the medial lemniscus. The fibres of the ophthalmic division of the fifth nerve end in the lowest part of the spinal part of the descending root; the maxillary division ends in the middle part of the root, while the superior part of the nucleus receives the inferior or mandibular division of the nerve. In other words, the face is represented 'upside-down' in the nucleus. The medial lemniscus also receives accessions of fibres from other sensory cranial nuclei, particularly the vagus (from the respiratory tract), the dorsal nuclei of the ninth and seventh nerves (taste fibres) [see p. 392] and from the vestibular division of the eighth nerve (from the labyrinth) [p. 304].

In the midbrain the medial lemniscus lies in the tegmentum dorsal to the substantia nigra and close to the middle line; it passes through the subthalamic region to end in the thalamus [FIG. V.83]. From the thalamus another neuron relay projects to the sensory cortex. Fibres in the medial lemniscus ascend the brain stem and give off many collaterals to the adjacent neurons of the reticular system. This sensory influx to the reticular system is of great importance in maintaining and varying behavioural alertness.

Muscle tone

Muscle tone is a state of partial tetanus of the muscle maintained by an asynchronous discharge of impulses in the motor nerves supplying the muscle. Tone is reflexly engendered by the impulse activity of afferent nerves whose endings lie in the muscle spindles. These afferent nerves reflexly excite α-motoneurons which supply the identical muscle which contains the spindle afferents. The reflex arc is monosynaptic and both the afferent and efferent limbs of the arc are composed of large (12–20 μm diameter) fibres. Destruction of the afferent limb (e.g. by tabes) or the efferent limb (e.g. poliomyelitis, trauma) of the arc abolishes tone. Although the reflex arc is spinal, supraspinal nervous pathways modify the reflex in the intact animal. Thus transection of the brain stem in the midcollicular region causes exaggerated muscle tone and Sherrington used such preparations (decerebrate) to investigate the features of muscle tone. Conversely spinal transection above the level of the spinal segmental region under study notably diminishes the briskness of reflex response [p. 362].

Sherrington showed that when an innervated muscle is stretched, it responds by contracting and if the stretch is maintained so is the contraction [FIG. V.95]. If the nerves supplying the muscle are cut, the paralysed muscle behaves like any elastic structure. Sherrington proved also that section of the dorsal roots, which carried afferent fibres from the muscle under study, likewise abolished not only resting tone but the active contractile response which normally occurred on stretching the muscle.

The stretch reflex is the contractile response of all innervated muscles to stretching, but is most prominently featured by the antigravity (extensor) muscles.

The afferent nerve endings in the muscle which are excited by stretch are the muscle spindles [p. 298]. These specialized receptors

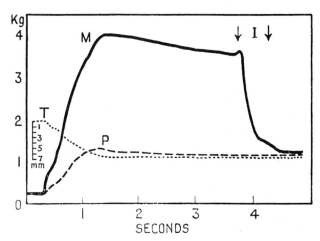

FIG. V.95. Stretch reflexes. Response of intact and paralysed muscle to stretch. Ordinates: Muscle tension in kg, and stretch of muscle in mm. Response of knee extensor muscle. T = Stretch of 6.5 mm applied; M = Innervated muscle; P = Paralysed muscle; I = Inhibitory afferent nerve stimulated, which abolishes the stretch reflex, and tension in M falls to level of P. Myograph multiplies tendon movement 62 times. (Liddell, E. G. T. and Sherrington, C. S. (1924). *Proc. R. Soc.* **B96**, 212–42.)

lie in parallel with the skeletal muscle fibres and are elongated when the skeletal muscle is stretched; this excites the nerve endings of large afferent nerve fibres and provokes contraction by a monosynaptic reflex excitation of the α-motoneurons supplying the muscle.

One might think that such a receptor organ situated in *parallel* with the muscle fibres would cease to signal information about the length of these muscle fibres if the muscle were excited to contract either by voluntary effort or by supraspinal nervous discharges of involuntary origin. Such is not the case, however, because the spindle itself is supplied by *intrafusal* motor nerve fibres—small motoneurons (fusimotor or gamma neurons—1–5 μm in diameter) which by causing contraction of *intrafusal* muscle elements influence the deformation of the central part of the spindle in which lie the large afferent nerve endings. The function of these large afferents is to signal such distortion of the central part of the spindle and to cause reflex alterations of α-motoneuron discharge to the *extrafusal* skeletal muscle.

Eccles and Sherrington (1930) showed that the pure motor nerve to a muscle consisted of two main groups of fibres—large fibres of 12–20 μm diameter (α motoneurons) and small fibres of 2–6 μm diameter (γ motoneurons). To study a pure motor nerve, it is first necessary to cut all the dorsal sensory roots which carry afferent fibres from the muscle under study distal to the posterior root ganglia. The sensory components of the muscle nerve degenerate within weeks and the myelinated nerves which remain are solely motor.

The significance of this bimodal distribution was not understood until the mid-1940s when it was shown that the small motor neurons did not supply the extrafusal skeletal muscle fibres, but instead supplied the intrafusal fibres of the muscle spindles.

The ventral root supplying motor nerves to a muscle was isolated and the tension of the relevant muscle was recorded. Stimulation of the ventral root caused excitation of the motoneurons, which was recorded distally, and contraction of the muscle. A weak stimulus caused excitation only of the α motoneurons and a contraction of the muscle. Increasing the strength of the stimulus led to increasing muscle contraction in response and to an increase to a maximum of the α motoneuron component of the compound action potential. Further increase of stimulus strength caused no greater development of contractile tension, but produced a second component in the compound action potential recording [FIG. V.96] in the form of a slower peak and from conduction velocity measurements this second peak was attributed to the excitation of γ-motoneurons with a fibre size of 4 μm (conduction velocity c. 27 m s^{-1}). Finally, an important result was obtained by compressing the nerve bundle distal to the site of electrical stimulation but proximal to that of recording of the compound action potential. By grading the pressure exerted, it was possible to abolish conduction in the larger fibres without seriously prejudicing that in the γ motoneurons, as evinced by the continued response of the γ peak to electrical stimulation although the α peak response to such stimulation had disappeared. In these circumstances, although γ conduction was intact, no muscle contraction occurred on stimulation of the compressed nerve root; only when the pressure was released was stimulation once more effective in arousing muscle contraction. It follows that the γ motoneurons do not themselves innervate the skeletal muscle fibres.

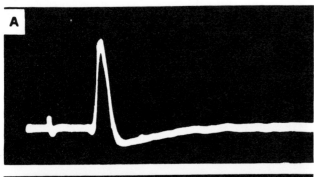

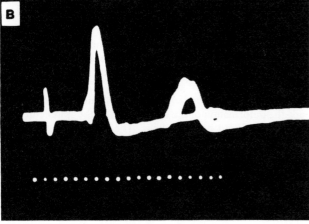

Fig. V.96. Recording from motor fibres to soleus. Stimulation of ventral root of S2, multiple sweeps. A: Stimulus strength just maximal for fast-conducting fibres. B: Stimulus strength increased during exposure. A high threshold, relatively slow-conducting nerve component appears while the large spike height remains unchanged. Entire potential complex shifts slightly due to latency shortening during increase in stimulus strength. Peak conduction velocity for fast fibres 76 m s^{-1} and 27 m s^{-1} for slow fibres. This particular ventral root contained an unusually large proportion of small nerve fibres. Time base 2000 Hz. (Kuffler, S. W., Hunt, C. C., and Quilliam, J. P. (1951). *J. Neurophysiol.* **14**, 29.)

If the experiment is repeated and recordings are additionally made from Group I afferents from the muscle spindles (peeled off the otherwise intact dorsal nerve root [see Fig. V.101] then stimulation of the γ motoneurons alone causes increased discharge in these Group I afferents which arise from the central or equatorial region of the muscle spindle. Contraction of the intrafusal fibres of the spindles, caused by γ motoneuron excitation, distorts the equatorial region and thus arouses afferent impulse activity in the Group I fibres.

The muscle spindle

The mammalian muscle spindle, as its name implies, is a fusiform structure several millimetres long situated in parallel with the skeletal muscle fibres and attached to the endomysium of these muscle fibres at both its ends [Fig. V.97]. The spindle contains 2–12 intrafusal muscle fibres, the diameter of which may vary between 6 and 28 μm. The central third or equatorial region of the spindle is 80–200 μm wide and in this region the intrafusal fibres are surrounded by fluid contained in a capsule. The presence of the fluid and the capsule here gives the spindle its characteristic shape. In this central region the intrafusal fibres possess many nuclei and the fibres show only sparse striations; at their polar ends, striations are well marked.

Barker (1948) introduced the term 'nuclear bag' for the dense aggregation of nuclei in the central region of the spindle and the term 'myotube region' for the parts of the intrafusal fibres on either side of the nuclear bag. It should be understood that the nuclei in the central region are not aggregated in any bag other than the intrafusal muscle fibre itself.

Boyd (1962) described two kinds of intrafusal fibre on the basis of the structure which they respectively displayed in their equatorial region. He divided the intrafusal fibres into:

(a) Nuclear bag fibres—showing an aggregation of nuclei in the equatorial region.

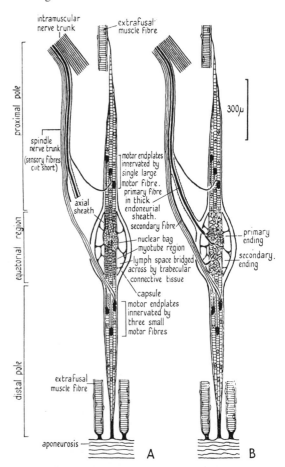

Fig. V.97. Diagram of an idealized rabbit's muscle spindle; polar regions shortened to about half their length. In A the efferent innervation by fine nerve fibres is shown, the afferent innervation being omitted to expose structure of the equatorial region. The small motor end plates are shown as black discs. B shows the same spindle complete with afferent innervation comprising one primary and one secondary ending. (Barker, D. (1948). *Q. Jl micros. Sci.* **89**, 143.)

(b) Nuclear chain fibres which show only a single line of nuclei lying in a chain [Fig. V.98].

The nuclear bag fibres receive a *motor* innervation from γ_1 nerve fibres which end in discrete end plates near the poles of the fibre. The chain fibres receive a *motor* innervation from finer γ_2 motor fibres, which form a diffuse terminal reticulum along much of the polar regions of the chain fibre.

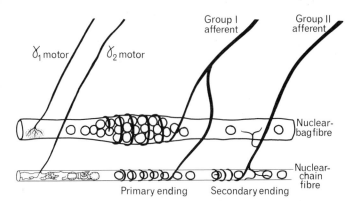

Fig. V.98. Greatly simplified diagram of the central region of the muscle spindle (after Boyd). (Matthews, P. B. C. (1964). *Physiol. Rev.* **44**, 219.)

Both types of intrafusal fibre receive a primary annulo-spiral *afferent* innervation in their equatorial region. These *primary* afferent nerves are of large diameter (12–20 µm) with correspondingly fast conduction velocities and are the Group IA fibres.

In addition there is a secondary innervation by smaller afferent fibres which is always polar and *not* equatorial in its distribution and which though often spirally arranged round the nuclear chain fibres is never so arranged on the nuclear bag fibres. These secondary afferent fibres belong to the Group II afferents.

Electrophysiological studies have shown that the responses of primary and secondary afferent nerve endings to mechanical stimulation differ. When a skeletal muscle is stretched, the muscle spindle primary endings signal both the instantaneous length of the muscle and the velocity at which it is being stretched while the secondary endings signal rather the instantaneous length [Fig. V.99].

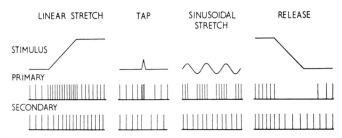

Fig. V.99. Diagrammatic comparison of the responses of 'typical' primary and secondary endings to various stimuli. The responses are drawn as if the muscle were under moderate initial stretch and as if there were no fusimotor activity. (Matthews, P. B. C. (1964). *Physiol. Rev.* **44**, 219.)

Under static conditions, the differences in behaviour between the primary and secondary nerve endings are slight—each shows a similar relationship between the frequency of its discharge and the extension applied to the muscle. Matthews (1964) has shown that the large dynamic response of the primary ending is caused by its situation in the relatively non-viscous equatorial region. The feeble dynamic response of the secondary ending is attributed to its site being in a region of the intrafusal fibre which has similar visco-elastic properties to those of the rest of the fibre.

Boyd (1971) in functional studies of isolated innervated mammalian spindles has shown that γ_1 fibres which terminate in discrete polar end plates on the nuclear bag fibres act as dynamic fusimotor fibres, whose primary action is to alter the dynamic sensitivity of the primary ending during muscle stretch. The γ_2 fibres, which end in 'gamma trails' on nuclear chain fibres, increase the static sensitivity of the primary ending to maintained changes in muscle length.

Matthews (1962) showed that the stimulation of both types of fibre increased the discharge of the primary ending when the muscle was at constant length. However, on stretching the muscle during stimulation of static γ fibres the normal pronounced dynamic response of the primary afferent ending was absent so that it behaved like a secondary ending. In contrast, stimulation of dynamic fusimotor fibres increased the response of the primary ending to the dynamic stimulus of stretching [Fig. V.100]. The functional effects of the static and of the dynamic fusimotor fibres were so different as to make it seem *probable* that they were in keeping with the histological classification of γ_1 and γ_2 fibres.

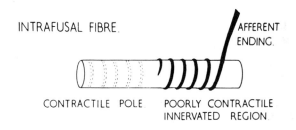

NATURE OF CONTRACTION.	RESPONSE OF ENDING.
A. SHORTENING WITH INCREASE OF 'VISCOSITY.'	DYNAMIC TYPE.
B. SHORTENING WITH 'VISCOSITY' UNCHANGED.	STATIC TYPE.

Fig. V.100. Diagram of part of intrafusal muscle fibre. (Brown M. C. and Matthews, P. B. C. (1966). In *Control and innervation of skeletal muscle* (ed. B. L. Andrew). Livingstone, Edinburgh.)

The role of γ efferents in induced muscle contraction

Direct recordings of activity in γ fibres are made by dissecting filaments from the otherwise intact ventral root. Discharge in such fibres is tonic in an anaesthetized 'intact' animal. Spinal transection above the segmental level of recording profoundly depresses such discharge, and accompanying this there is a decrease of tone in the relevant muscle under study. Conversely, intercollicular transection of an anaesthetized animal (Sherringtonian decerebration) markedly increases both γ discharge and muscle tone.

The exaggerated muscle tone of an intercollicular decerebrate animal is locally reduced by the application of a weak procaine solution to a muscle nerve. This local loss of muscle tone occurs at a stage when only the *gamma* fibres of the nerve are blocked. Local anaesthesia of a nerve *first* blocks the *small* fibres (quite the opposite of applied pressure which first blocks the large fibres).

Granit, Merton and co-workers argued that the degree of muscle tone is largely determined by the state of activity in the γ efferents which, causing contraction of the intrafusal fibres, arouse Group IA afferent discharge and thereby reflex excitation of the α moto-neurons supplying the extrafusal skeletal muscle fibres.

Their evidence was provided by studies of the activity of Group

IA afferents arising from spindles of a muscle whose tension was recorded simultaneously. The technique of 'sampling' IA activity in a single or few-fibre filament from an otherwise intact dorsal root was employed [Fig. V.101].

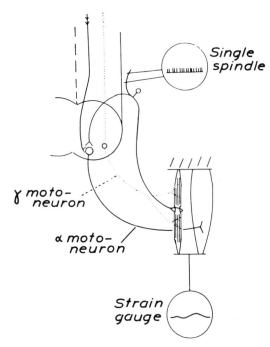

FIG. V.101. Diagram illustrating arrangement of experiment for reflex work with simultaneous spindle control. Muscle with parallel intrafusal fibre containing spindle connected to strain gauge. Afferent fibre discharge from spindle isolated in dorsal root and projected on oscilloscope. Destination of α and γ fibres from ventral root of spinal cord shown. (Granit, R., Holmgren, B., and Merton, P. A. (1955). *J. Physiol., Lond.* **130**, 213.)

Contractions of limb extensor muscles can be induced by either brain stem stimulation or, conveniently by neck reflexes. The tone in the limb muscles can be reflexly influenced by movements of the head on the neck ('neck reflexes') [see p. 302]. Ventroflexion of the head causes a contraction of the hind limb extensors whereas dorsiflexion of the head causes a relaxation of the hind limb extensors. Correspondingly, if myographic and Group IA afferent recordings are made from soleus (a hind limb extensor) changes in myographic tension and IA afferent activity can be simultaneously recorded and their responses to such neck reflexes assessed.

When the neck was rapidly ventroflexed and held there the muscle contraction was rapid and well maintained. The contraction, was however, slightly preceded and then accompanied by a striking increase in Group IA afferent discharge. This increase in IA afferent discharge was not *due* to the contraction of the muscle for that would itself have reduced it, but was ascribed to increased fusimotor discharge. The increase in IA discharge promoted the contraction of the muscle by exciting its α motoneurons. After cutting the dorsal roots carrying the relevant IA afferents, ventroflexion no longer induced any contraction of the muscle although the spindle afferent activity increased as before. This proved that the neck reflex did not excite the α motoneurons in the intercollicular decerebrate animal, but only the γ fibres which of course could not influence skeletal muscle contraction in the absence of the afferent limb of the reflex arc.

The term 'follow-up length servo' was introduced by Merton to describe the *modus operandi* of the γ loop mechanism. He considered the spindles to serve as misalignment receptors, recording the discrepancy between the muscle and the spindle lengths after the appropriate bias was exerted on the spindle by the gamma

discharge. The atonia of spinal transection could be interpreted then as partly due to a loss of γ bias due to cutting off supraspinal facilitation. Conversely, 'Sherringtonian' (intercollicular) decerebrate rigidity could be attributed to 'spindle cramp' caused by an excessive supraspinal drive on the γ motoneurons, exerted by the remaining preponderant 'rump' of extrapyramidal facilitatory neurons in the hindbrain.

However, exaggerated muscle tone can be produced by direct α motor neuron drive as revealed by the technique of ischaemic decerebration.

Intercollicular decerebration and ischaemic decerebration

Transection of the brain stem at intercollicular level which leads to decerebrate rigidity of 'Sherringtonian' type is a traumatic procedure which before the advent of blood transfusion and antibiotics hardly favoured the survival of the animal for more than a few hours. The operation is performed under ether anaesthesia. The common carotid arteries are ligated and the parietal bone trephined. The vertebral arteries are compressed as they exit from their canals to course across the atlas vertebra and during this period a knife or spatula inserted through the trephine hole is directed obliquely to slice through the hemispheres along a plane immediately anterior to the tentorium cerebelli which is bony in the cat. This transection of the brain stem occurs at the intercollicular level. Cerebral tissue rostral to the section is scooped out and vertebral artery compression meanwhile released. Packing with swabs bathed in thrombin solution limits the haemorrhage which occurs but blood loss is inevitable and is sometimes severe. In any case, the chances of infection of the extensive wound site usually precluded there being much chance of obtaining a chronic decerebrate preparation.

Only when the animal has 'blown off' the ether does decerebrate rigidity develop being particularly evident in the extensor antigravity limb muscles. Characteristically this hypertonia is abolished by dorsal root section of the appropriate segments receiving afferent nerves from the limbs.

It was a search for a method of producing decerebrate rigidity which was less traumatic and which rendered the animal less prone to infection and more likely to survive as a chronic preparation that led Pollock and Davis (e.g. 1930) to develop the technique of *ischaemic decerebration*. They tied both carotid arteries and the basilar artery at the junction of the pons and the medulla. Such an operation required only two slits in the neck for carotid ligation plus trephining a hole in the ventral basiocciput through the roof of the mouth, the jaws being widely separated by retractors. On removing the small circle of bone, the basilar artery could be seen between the medullary pyramids and suitably ligated. The circular bone fragment was replaced and secured and the buccal mucosa sutured. The small neckwounds were closed and the chances of chronic survival of the preparation were good. Again, as the anaesthetic wore off, striking spasticity of the limbs developed; superficially the decerebrate preparation closely resembled the features of that resulting from intercollicular transection. However, there was one striking difference—the ischaemic spasticity was in no way reduced by deafferentation of the limbs by appropriate dorsal root section indicating that in this preparation hypertonia was induced by an excessive drive of α motoneurons from the surviving rump of the rhombencephalon.

The difference between the effects of deafferentation remained puzzling until the role of fusimotor (γ) fibres in provoking IA afferent discharge in turn reflexly exciting α motoneurons had been elucidated. Meanwhile, however, it was already known that cerebellectomy strongly increased the spasticity of a Sherringtonian decerebrate animal and histology of the surviving rump of rhombencephalic tissue, including the cerebellum, showed that about half of the cerebellum had necrosed in the ischaemic preparation. Such knowledge led to an important experimental investigation by Granit, Holmgren, and Merton (see Granit 1955) in which they recorded spindle IA afferent activity in a precollicular decerebrate cat together with records of electromyographic activity and tension of the relevant muscle. The cerebellum was exposed by removing the left side of the tentorium. Initially spindle IA afferent discharge was tonic and muscle tension and e.m.g. activity of moderate degree. On cooling the culmen (anterior lobe of the cerebellum [p. 311] IA afferent activity ceased altogether, but muscle tension increased and, of course, e.m.g. activity as well. Rewarming the anterior lobe restored the *status quo* (IA afferent activity and a more moderate degree of muscle tension). Thus ischaemic decerebrate rigidity is due to a direct drive of the α motor neurons and is quite independent of IA afferent discharge. Moreover, the disappearance of IA afferent discharge when the anterior lobe of the cerebellum is inactivated suggests that this afferent discharge is itself aroused by fusimotor activity when the cerebellum was intact.

Additional evidence of the role of α and γ motoneurons in the two types of decerebrate preparation has been provided:

1. Matthews and Rushworth (1957) showed that the local injection of procaine into nerve trunks supplying the spastic muscles of an ischaemic decerebrate cat did not abolish muscle tone until the α motor nerves were themselves paralysed. Procaine paralyses the thinner fusimotor fibres first. In a Sherringtonian decerebrate preparation, as might be expected, procaine rapidly reduces the spasticity.

2. Henatsch and Ingvar (1956) showed that chlorpromazine given systemically abolishes fusimotor activity. Correspondingly, the spasticity of a Sherringtonian decerebrate cat disappeared whereas that of an ischaemic decerebrate preparation was unaffected by the administration of the drug.

3. Neck reflexes [p. 302] provoke alterations of tone in the extensor muscles of the leg. As already stated, forcible ventroflexion of the neck provokes a lively fusimotor response in the leg extensors in the Sherringtonian decerebrate cat but has no effect in the ischaemic preparation. Ventroflexion in each type of decerebrate does provoke contraction of the leg extensors but this is induced primarily by α activity in the ischaemic preparation.

The fact that cerebellectomy increases the spasticity of an intercollicular preparation suggests that the cerebellum exerts a tonic inhibitory influence on nuclei of the rhombencephalon whose axons passing spinally themselves exert a facilitatory effect on the spinal α motor neurons. Such nuclei may be those supplying the vestibulospinal and some of the reticulospinal tracts, for the hypertonia of the ischaemic decerebrate is abolished by section of the vestibular component of both VIIIth nerves, or by bilateral labyrinthectomy after cerebellectomy in an otherwise normal animal.

Normally it would seem that muscle tone in the intact animal though affected by myotatic (stretch reflex) and non-myotatic influences is primarily determined by the myotatic component. When the cerebellum is intact most influences on muscle tone seem to be relayed by the fusimotor route. The cerebellum is an important site of α–γ linkage; after cerebellectomy muscle tone seems to be subservient to a vestibulospinal drive which operates on α motor-neurons.

The clinical assessment of tone

The clinician assesses the resistance offered to passive displacement. He uses his experience in deciding whether such resistance is 'normal', low, or high. In essence he is using the same criterion as does the physiologist with an experimental animal—each is sampling the response to stretch.

REFERENCES

BARKER, D. (1948). *Q. Jl microsc. Sci.* **89**, 143.
BOYD, I. A. (1958). In *Electroencephalography and clinical neurophysiology*, p. 406. Amsterdam.
—— (1962). *Phil. Trans. R. Soc. B.* **245**, 81.
—— (1971). *J. Physiol., Lond.* **214**, 30–31P.
BROWN, M. C. and MATTHEWS, P. B. C. (1966). In *Control and innervation of skeletal muscle* (ed. B. L. Andrew) p. 18. Livingstone, Edinburgh.
ECCLES, J. C. (1957). *Proc. 1st Int. Cong. Neurol. Sci.*, p. 81.
—— ECCLES, R. M., and LUNDBERG, A. (1957). *J. Physiol., Lond.* **137**, 22.
—— and SHERRINGTON, C. S. (1930). *Proc. R. Soc. B.* **106**, 326.
ELDRED, E., GRANIT, R., and MERTON, P. A. (1953). *J. Physiol., Lond.* **122**, 498.
GRANIT, R. (1955). *Receptors and sensory perception.* Yale University Press, New Haven, Conn.
HAMMOND, P. H., MERTON, P. A., and SUTTON, G. G. (1956). *Br. med. Bull.* **12**, 214.
HENATSCH, H. D. and INGVAR, D. (1956). *Arch. Psychiat. Z. Neurol.* **195**, 77–93.
HUNT, C. C. (ed.) (1974). Muscle receptors. In *Handbook of sensory physiology*, Vol. III/2. Springer, Berlin.
KOEZE, T. H., PHILLIPS, C. G., and SHERIDAN, J. D. (1968). *J. Physiol., Lond.* **195**, 419.
KUFFLER, S., HUNT, C. C., and QUILLIAM, J. P. (1951). *J. Neurophysiol.* **14**, 29.
LEKSELL, L. (1945) *Acta physiol. scand.* **10**, Suppl., 31.
MATTHEWS, P. B. C. (1962). *Q. Jl exp. Physiol.* **47**, 324.
—— (1964) *Physiol. Rev.* **44**, 219.
—— (1972). *Mammalian muscle receptors and their central actions.* Arnold, London.
—— (1981). *J. Physiol., Lond.* **320**, 1.
—— and RUSHWORTH, G. (1957). *J. Physiol., Lond.* **135**, 245.
—— and STEIN, R. B. (1969). *J. Physiol., Lond.* **200**, 723.
MERTON, P. A. (1953). In *The spinal cord* (ed. G. E. W. Wolstenholme). Churchill, London.
POLLOCK, L. I. and DAVIS, L. (1930). *J. comp. Neurol.* **50**, 377.
—— —— (1931). *Am. J. Physiol.* **98**, 47.
SHERRINGTON, C. S. (1898). *J. Physiol., Lond.* **22**, 319.

Postural reflexes

Some idea of the part played by different regions of the nervous system in the regulation of posture may be gained by a study of animals (or clinical cases) in which parts of the neuraxis have been removed.

Spinal preparation

Transection of the spinal cord in man or in animals, e.g. in the mid-thoracic region, causes an initial stage of profound flaccidity and an absence of reflexes for a period of days (animals) or weeks (man) [see p. 362].

Later, the isolated cord recovers some of its reflex function; tone returns to the somatic musculature, but the reflex contraction in the affected limb muscles is quite insufficient to support the weight of the animal. *Spinal man cannot stand unsupported.*

Decerebrate preparation

Transection of the midbrain between the colliculi causes decerebrate rigidity. The limbs are hyperextended, The tail and the head are dorsiflexed, and the back is concave owing to extreme hyperextension of the spine (opisthotonos). The animal can be carefully balanced to show a caricature of standing on its four legs, but the slightest displacement causes the decerebrate animal to topple over—*it has no righting reflexes.* Nevertheless the decerebrate preparation does evince:

1. Stretch reflexes.
2. Positive supporting reaction.
3. Crossed extensor reflexes.
4. Tonic neck reflexes.
5. Tonic labyrinthine reflexes.

The stretch reflex has already been described [p. 297].

Local and segmental static postural reflexes

Positive supporting reaction. Decerebrate rigidity basically depends on a harmoniously operating group of stretch reflexes. These highly localized reflexes produce contraction of the antigravity muscles and reciprocal inhibition of the antagonistic muscles. The resulting posture is reinforced and modified by the positive supporting reaction; this is a remarkable irradiating reflex which produces simultaneous contraction of extensors and flexors of a limb converting it into a solid rigid pillar, well adapted to maintaining the body weight. The positive supporting reaction is most easily elicited after removal of the anterior lobe of the cerebellum. The procedure is as follows: press against the pads of the fingers or toes and dorsiflex and hand or the foot; the afferent impulses arise from the stimulated skin and from the muscles (mainly the interossei) which are stretched. All the muscles of the limb reflexly contract, i.e. both the protagonists and the antagonists.

Segmental static reactions. The decerebrate animal also possesses reflex mechanisms for adjusting the position of one limb in relation to alterations in the state of another. The *crossed extensor* reflex is a case in point; impulses from one leg reflexly produce extension of the opposite limb. Another example is the *shifting reaction*: flex (say) the right limb and allow the body to veer to the right; owing to stretch of the adductors of the left limb, the right limb is reflexly caused to extend.

Attitudinal reflexes

The postural activities of the decerebrate preparation exceed those found in the spinal preparation. The posture is now a co-ordinated one of the whole body instead of being limited, as in a spinal animal, to the hind limbs (or to all four limbs in a high spinal transection), and the degree of tone is adequate to maintain, but not to adopt, the upright position. In the decerebrate animal the posture of the trunk and limbs can be adjusted:

1. In accordance with alterations in the position of the head in space.
2. By changing the position of the head relative to the trunk.

In the first case the afferent impulses arise solely from the otolith organ of the vestibule (tonic labyrinthine reflexes); in the second case additional afferent impulses come from the neck muscles (tonic neck reflexes). The new position reflexly imposed on the body persists for as long as the new position of the head is maintained.

Tonic neck reflexes. To study these separately, bilateral extirpation of the labyrinths is carried out. The neck reflexes are set up by alterations of the position of the head relative to the body.

1. Ventroflex the head: the fore limbs flex and the hind limbs become more extended.
2. Dorsiflex the head: the fore limbs extend and the hind limbs flex. The purpose of these responses seems obvious: the position of the body is being adapted, e.g. 'for looking under a shelf, or looking up to a shelf'.
3. Press ventralwards on the lower part of the cervical vertebral column: all four limbs flex (vertebra prominens reflex), as 'in an animal crawling into a hole'.
4. Rotate or incline the head in various directions: to simplify

description, the limbs on the side to which the jaw is turned are called 'jaw limbs'; the limbs to which the vertex is turned are called 'skull (or vertex) limbs'. In general it may be said that the jaw limbs extend (to support the weight of the head) and the skull limbs flex.

The receptors for the neck reflexes are probably Pacinian corpuscles in the ligaments of the cervical vertebral joints, particularly in the atlanto-occipital joint and also from neck muscle spindles.

The centres for reflexes 1, 2, and 4 is in the upper cervical region of the spindal cord; the afferent impulses pass in the dorsal roots of C1–3 and come chiefly from the muscles of the back of the neck. The descending paths are the long propriospinal tracts. The centre for 3 is in the lower cervical region.

Tonic labyrinthine reflexes are studied after section of the dorsal nerve roots of C1, 2, 3 or after immobilizing the head, neck, and upper thorax by means of a plaster jacket (to prevent neck reflexes from coming into play). The labyrinthine reflexes are due to alterations in the position of the head relative to the horizontal plane.

1. If the animal is placed in the supine position, maximum tone is present in the antigravity muscles.
2. In the prone position, with the snout 45 degrees below the horizontal plane, tone in the extensor muscles is reduced to a minimum; in intervening positions intermediate grades of tone are present.

The purpose of these reactions is not very clear; they disappear after section of the vestibular nerves. The receptors are in the otolith organ as shown by the following experiment. If an anaesthetized guinea-pig is centrifuged at high speed, the otolithic membranes are detached (as proved by microscopic examination), but the ampullae of the semicircular canals are unharmed: the tonic labyrinthine reactions are, however, abolished. Alterations in the position of the otoliths thus reflexly modify tone in the muscles of the limbs. The centres for the labyrinthine reactions are the vestibular and reticular nuclei; the descending tracts employed are the vestibulospinal and reticulospinal.

After unilateral labyrinthine extirpation [p. 306], the unopposed activity of the intact otolith organ results in flexion and rotation of the head to the side of the lesion; this in turn evokes secondary modifications in the position of the limbs through the operation of the neck reflex. As a result the jaw limbs extend and the skull limbs show diminished extensor tone. When tonic labyrinthine and neck reflexes are simultaneously evoked, they produce the algebraic sum of the separate responses. Thus, if the head is dorsiflexed in a decerebrate animal (with intact labyrinths), labyrinthine impulses produce increased tone in all four limbs, while the proprioceptors from the neck tend to extend the fore limbs and flex the hind limbs. The actual result observed is extension of the fore limbs (as both reflexes tend to increase extensor tone) and little change in the hind limbs (because the two reflexes are exerting antagonistic influences).

The attitudinal reflexes can be easily demonstrated in normal animals and man where they are carried out more smoothly because the flexor muscles also participate in the reactions. A cat sees a piece of meat in the air; the head extends, the fore limbs extend, the hind limbs are unaffected or flex, but can be suddenly extended when the animal wants to spring.

Posture in the decorticate preparation. The whole cerebral cortex is removed but the basal ganglia and the brain stem are left intact. The posture adopted by such a decorticate preparation varies with the species and the conditions under which the animal is examined.

In the dog or cat the posture when the animal is on its feet is a normal one and walking movements can be reflexly performed. In the decorticate monkey, however, tone is gravely disturbed; there is full extension of the hind limbs and a semiflexed posture of the fore limbs; walking movements do not occur. The findings in decorticate man are very similar. The legs are fully extended; the arms lie across the chest, semiflexed at the elbow, the forearms slightly pronated, and the wrists and fingers flexed.

In decorticate man typical neck reflexes can be elicited. When the head is rotated to the right, the right arm extends at the elbow and becomes abducted; the left arm becomes fully flexed with the hand touching the neck; the right leg is extended and the left leg is flexed.

Righting reflexes. By means of the righting reflexes the decorticate cat or rabbit can bring its head right way up and get the body into the erect position under all circumstances. If the animal is laid on its side or on its back, the head at once rights itself, the body follows suit, and finally the animal resumes the upright posture. The decerebrate animal, though it can remain insecurely in the upright position if put there, can never actively assume that position; it has no righting reflexes. The righting reflex consists of a chain of reactions following one another in an orderly sequence as each reaction produces its successor.

1. Labyrinthine righting reflex. With the animal's head in the lateral position impulses arise in the saccules which lead reflexly to righting of the head.

2. Body righting reflex. With the animal on its side, the side of the trunk in contact with the bench is undergoing constant stimulation, while the other side in contact with air is not. This asymmetric stimulation of the deep structures in the body wall also reflexly rights the head. The head can thus be righted even after double labyrinthectomy.

If a labyrinthless animal is laid on its side and a weighted board is placed on the upper side of the animal so that streams of impulses pass up from both sides of the trunk, the head falls back into the lateral position.

3. Neck righting reflex. The reflexes just described act primarily on the neck muscles and right the head. The trunk, however, remains as before in the lateral position, so that the neck is twisted. This evokes a further reaction—the neck righting reflex—which brings the thorax and lumbar region successively into the upright position. If righting of the head is prevented, impulses from the body surface may cause righting of the body directly.

4. Limbs. The appropriate posture of the limbs is largely attained by impulses arising in the limb muscles themselves. The righting reflex can be well demonstrated in the intact cat if it is blindfolded and dropped with the legs pointing upwards. The cat turns itself round with great speed and alights gently on all fours.

The chief centre for this group of righting reactions is in the neighbourhood of the red nucleus [FIG. V.102].

5. Optical righting reflexes. In animals with the visual cortex intact, righting of the head is also brought about reflexly by means of optical impulses. In the intact cat, dog, or monkey, after denervation of the labyrinths and the neck muscles, righting cannot take place if the animal is blindfolded, but is successfully carried out if the eyes are open. The centre for the optical righting reflex is in the visual (calcarine) cortex, whence impulses pass ultimately to the neck muscles to right the head. In man the optical righting reflexes are far more important than those of the labyrinth.

Righting reflexes still occur after cerebellectomy but they are imperfectly executed. The cerebellum is essential for all precise movements.

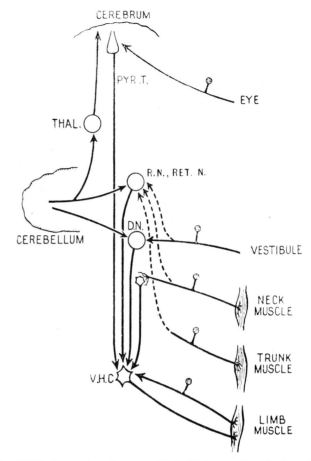

FIG. V.102. Regulation of posture. Thal., Thalamus. Pyr. T., Pyramidal tract. R.N., Red nucleus. Ret.N., Nuclei of reticular formation of brain stem. D.N., Deiters' nucleus (Lateral vestibular nucleus). V.H.C., Ventral horn cell. This figure neglects to show the gamma-loop mechanism. The reader can help to memorize this concept by modifying the diagram correspondingly.

REFERENCES

GRANIT, R. (1955). *Receptors and sensory perception*. Yale University Press, New Haven, Conn.
MAGNUS, R. (1924). *Körperstellung*. Springer, Berlin.
—— (1925). *Proc. R. Soc.* **B98**, 339.
—— (1926). *Lancet* **ii,** 531, 585.

Mechanism of standing in man

The comfortable stance of a normal man is quite unlike the imperfect caricature of standing displayed by a decerebrate cat; it is hardly surprising to find that the detailed mechanisms employed differ, though fundamental similarities persist. When a man is comfortably balanced in the upright position there is remarkably little e.m.g. activity in the muscles of the trunk and thighs. The explanation of this surprising finding is that the disposition of the skeleton, the ligaments, and the soft parts is such that a momentary insecure balance can be maintained passively. This passive erect posture in the absence of all muscular activity is, as stated, momentary and insecure, and the person would immediately fall if muscular activity did not develop. (A person whose muscles are paralysed cannot stand.) Once he begins to fall, reflex compensatory muscular reactions set in which restore the state of balance: the muscular contraction then ceases till the next deviation from the erect position occurs. A standing man can fall in any direction: forwards, backwards, or sideways. The muscles which oppose the fall are acting as antigravity muscles; depending on the direction of the fall, any of the muscles of the trunk or legs act as antigravity muscles.

Thus when the body sways forwards the extensors of the trunk and the flexors of the leg contract sufficiently to restore the balance; when the sway is backwards the recti abdominis and the leg extensors contract; when the sway is sideways the contralateral external oblique muscle responds. These responses are reflexly produced, partly as a result of impulses from stretch receptors in the trunk and legs and partly from the receptors in the head, mainly the eyes. With the eyes closed the upright stance is less steady and more swaying of the trunk occurs. This observation shows that visual afferents are concerned in the reflex maintenance of the upright stance in man.

THE VESTIBULAR APPARATUS

The vestibular apparatus (labyrinth) is a complex sense organ which is stimulated by 1. gravity and 2. rotation movements. It plays an important role in postural activity; it gives rise to afferent impulses which reflexly adapt the position of the trunk and limbs to that of the head and enable the erect position of the head and the normal attitude of the body to be maintained. Impulses from the vestibule also reach the cerebral cortex and subserve the recognition of the position and movements of the head.

Anatomy

The vestibular apparatus consists of the three semicircular canals and the otolith organ (the saccule and utricle) [FIG. V.103].

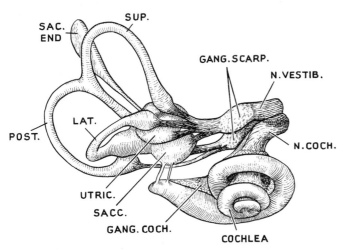

FIG. V.103. Anatomy of vestibular apparatus. N.Vestib. = vestibular nerve. N.Coch. = cochlear nerve. Gang.Scarp. = Scarpa's (vestibular) ganglion. Gang.Coch. = cochlear ganglion. Utric. = utricle. Sacc. = saccule. Sup.,Post.,Lat. = superior, posterior, and lateral semicircular canals. Sac.end. = sacculus endolymphaticus. (Modified from Hardy (1934). *Anat. Rec.* **59.**)

1. The canals are the lateral (horizontal), superior, and posterior, each being in a different plane at right angles to the others. The left superior canal is in the same plane as the right posterior canal, and vice versa. The membranous canals contain the endolymph and are enclosed in bony canals, from which they are separated by the perilymph; each canal commences as a dilation or ampulla, containing a projecting ridge, the crista. The canals open into the utricle by means of five apertures, one being common to the superior and posterior canals.
2. The utricle communicates with the saccule by means of the ductus endolymphaticus. Both the saccule and utricle contain a projecting ridge, the macula. The canalis reuniens unites the saccule and the duct of the cochlea.

Structure of crista and macula

The crista and macula are the specific receptors of the vestibular apparatus and have a similar structure. Covering the ridge is a tall columnar epithelium (hair cells) giving attachment to long stiff hairs which project into a firm gelatinous material, the cupula terminalis [FIG. V.104]. Between the hair cells lie the fibres of origin of the vestibular division of the eighth nerve.

1. In the canals the cupula rises to the roof of the ampulla, acting as a movable partition which divides the ampulla into two compartments [FIG. V.106].
2. In the saccule and utricle the cupula contains many chalky particles, the otoliths, hence the name the otolith organ. When the head in man is in the normal erect position, the macula of each utricle is approximately in the horizontal plane, with the cupula, hairs, and otoliths rising vertically from the macular epithelium; the macula in each saccule then lies in the vertical plane, with the hairs and otoliths projecting horizontally sideways into the cupula.

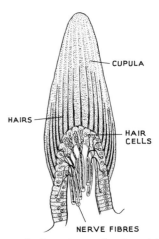

FIG. V.104. Structure of crista of ampulla of semicircular canal. (After Camis.)

Central vestibular connections

The nerve endings in the maculae and cristae continue as nerve fibres which have their cell bodies in the bipolar cells of the vestibular ganglion; the central axons of the vestibular nerve enter the medulla ventral to the inferior peduncle and dorsal to the descending root of the fifth nerve. The axons divide into ascending and descending branches which end in four nuclear masses:

1. The large medial (principal) nucleus in the pons and medulla.
2. The descending nucleus associated with the descending vestibular fibres.
3. The superior nucleus (of Bechterew) at the level of the sixth nucleus.
4. The lateral nucleus (of Deiters) in the lower pons.

For all practical purposes these four nuclei can be treated as a single functioning entity.

Fibres from the vestibular nuclei pass
(i) to the paleocerebellum of both sides via the restiform body;
(ii) directly and via the cerebellum to the red nucleus and the nuclei of the reticular formation in the brain stem;
(iii) in the median longitudinal bundle to the oculomotor nuclei of both sides;
(iv) via the medial lemniscus to the opposite thalamus and thence to the opposite temporal lobe;
(v) down the vestibulospinal tracts to the ventral columns of the spinal cord to end directly (and also via short interneurons) round ventral horn cells.

Mode of action of otolith organ (saccule and utricle)

The maculae of the saccule and utricle are stretch receptors, the effective stimulus being the pull of gravity on the cupula and contained otoliths and hairlets; the hair cells are thus deformed with resulting stimulation of the nerve filaments which lie between them. As might be expected the saccules are affected by a lateral tilt of the head: thus if the head is tilted laterally to the right (to rest on the shoulder) the cupula of the right saccule hangs downwards and pulls on its macula which is maximally stimulated; the cupula of the left saccule points upwards and 'rests' on the macula, this being the position of minimal stimulation of the nerve endings. Ventral or dorsal flexion of the head (fore and aft tilt) affects the utricular maculae; with the head erect the cupulae in the utricles point upwards providing a minimal stimulus; when the head is bent well forward, or back, the cupulae are pendent, pulling on the maculae and so stimulating them maximally.

Action potentials can be recorded in the appropriate branches of the vestibular nerve or in its nucleus in the medulla. A tilt of as little as 2.5 degrees stimulates the appropriate maculae; as the tilt increases the frequency of the discharge progressively rises [Fig. V.105]. The general law of the receptors thus applies here, that the frequency of the impulse stream is directly related to the strength of stimulation. If the head is kept in any particular position, the impulse discharge pattern persists for as long as the position is maintained, except for some slight reduction in the discharge rate; the receptors thus show little adaptation.

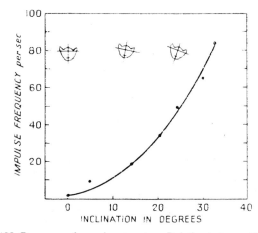

Fig. V.105. Response of saccular receptors. Relation between tilt of head and frequency of impulses in vestibular nerve. Decerebrate cat. The degree of lateral tilt of the head is shown in the upper diagrams. The impulses were recorded from the right nerve while the head was being tilted to the right (i.e. right cheek down). As the tilt was increased the impulse frequency rose correspondingly. When the head was tilted to the left, the discharge in the right nerve ceased. (After Adrian (1943). *J. Physiol., Lond.* **101**, 393.)

Mode of action of the semicircular canals

Direct observations have been made on the exposed semicircular canals in fish; a drop of oil is introduced into the canal. On rotation the behaviour of the cupula is photographed. As the rotation begins the endolymph in the canal moves, as shown by the shift in the position of the drop of oil; the cupula, which rises up as a septum completely dividing the ampulla in two, becomes bent over in the direction of the endolymph movement to an angle of up to 30 degrees [Fig. V.106].

The effective stimulus to each ampulla is rotation of the head in the plane of its canal. Consider the case of rotation of the head in the horizontal plane, in the direction of the arrows as shown in Figure V.107; the left ampulla is 'leading' its canal while the right ampulla is 'trailing' behind its canal. As the endolymph possesses

inertia it does not move as fast, initially, as the canal in which it is contained; thus in a shortlived rotational movement of the head (e.g. one or two turns) the endolymph lags behind; this is equivalent to a flow in the reverse direction from that of the head movement. In the experiment illustrated by Figure V.107 the initial endolymph movement is thus towards the right ampulla and away from the left ampulla; both cupulae presumably swing to the left. Vestibular action potentials show that the frequency of the impulses from the right ampulla is increased while that from the left ampulla is decreased; i.e. in the case of the lateral (horizontal) canals, the 'trailing' ampulla is stimulated while the 'leading' ampulla is depressed.

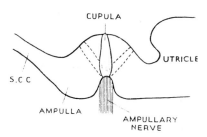

Fig. V.106. Mode of stimulation of semicircular canals. S.C.C., semicircular canal. The cupula, situated on the top of the crista, completely blocks the ampulla of the membranous canal. The cupula is caused to swing by movements of the endolymph.

The stimulus to the cristae is due to the swing of the cupula set up by the endolymph; it seems that a swing in a certain direction in any canal increases the stimulus to the nerve endings, while a swing in the opposite direction in that canal decreases the stimulus to the nerve endings. The combination of increased impulse discharge from one ampulla and decreased impulse discharge from the other, may form the basis of the interpretation of the direction of the movement. The degree of alteration of the frequency of the impulse discharge is directly related to the rate of acceleration of the rotational movement.

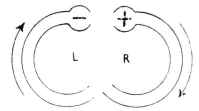

Fig. V.107. Diagram to illustrate mode of action of lateral (horizontal) semicircular canals. The arrow shows the direction of movement of the head. The right ampulla is stimulated; the nervous discharge from the left ampulla is decreased.

As the rotation is continued, the endolymph takes up the same rate of movement as its canal; the cupula, by reason of its own elasticity, then returns (in about 30 seconds) to its original resting position, and the resting nervous discharge is resumed (i.e. the discharge in the 'active' ampulla decreases and that in the 'depressed' ampulla increases). On cessation (or deceleration) of the movement, changes occur which are the reverse of the initial ones. The endolymph, by reason of its momentum, continues to move after the canal has come to rest; thus in Figure V.107, on cessation of head movement, the endolymph will continue to flow in the direction shown by the arrows, i.e. to the right; the cupulae will swing to the right. As this is the opposite direction to that at the beginning of the movement the subjective impression will be of a movement in the reverse direction to the previous rotation. Finally, the cupulae, by reason of their elasticity, regain their resting position, and the sense of movement ceases.

The vertical posterior canal acts similarly except that for some unknown reason the ampulla which is leading is stimulated [Fig. V.108, B]; the details for the vertical superior canal are uncertain [Fig. V.108, C]. The left superior and the right posterior canals act as a functional pair, as do the right superior and the left posterior canals. With the head at rest there is a steady 'spontaneous' discharge of impulses from all the six ampullae. Characteristic modifications of this discharge pattern are set up by rotary movements in any direction.

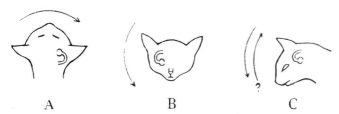

Fig. V.108. Direction of rotation which stimulates the three semicircular canals (cat). A: horizontal canal; stimulating direction is with ampulla trailing. B: posterior canal (rotation in transverse plane); stimulating direction with ampulla leading. C: superior (anterior) canal (rotation in median plane); results uncertain. (Adrian, E. D. (1943). *J. Physiol., Lond.* **101**, 397.)

The semicircular canals give information about *movements*, the otolith organ about the *position* of the head.

Extirpation experiments. Bilateral extirpation of the labyrinths in birds produces grave disturbances; the animals subsequently cannot stand or fly. In the cat or monkey the initial symptoms soon disappear and the animal behaves fairly normally if allowed the use of its eyes; it fails, however, to orientate itself under water and would soon drown ; it cannot right itself when falling blindfold through the air. Muscle tone is not permanently decreased.

Unilateral extirpation gives rise to complex derangements of postural activity.

The immediate effects are skew deviation of the eyes (i.e. one eye rolls upwards and outwards and the other downwards and outwards), nystagmus, rotation, and lateral flexion of the head (so that the *occiput* is turned to the *side of the lesion*); these changes are due to the unopposed action of the intact labyrinth. The altered position of the head sets up neck reflexes which, secondarily, modify the posture of the trunk and limbs. The limbs on the side of the lesion (the side to which the vertex is pointing, or skull (vertex) limbs) flex, and the limbs on the opposite side (jaw limbs) extend. There is spiral rotation of the trunk.

The permanent effects are:

1. Nystagmus: there is a slow swaying movement towards the side of the lesion and a quick return towards the midline.
2. The reciprocal changes in tone and the head rotation persist.
3. The rotation of the trunk diminishes.

If one canal, e.g. the horizontal, is removed, spontaneous movements are set up in the plane of that canal, as no impulses are sent up to give information about the movement and thus check it.

Relation of vestibule to regulation of posture. The results of extirpation show that the vestibule normally plays an important part in the regulation of posture. Studies on the tonic labyrinthine reflexes and on the righting reflexes, which have been fully considered, indicate how the otolith organ helps reflexly to maintain the upright position of the head and to adjust the position of the body to that of the head in space or relative to the trunk.

Barany's caloric test. The semicircular canals in man can be readily stimulated. By throwing the head backwards 60 degrees and

looking to the opposite side at an angle of 50 degrees from the middle line and up to the ceiling, the lateral (horizontal) canal is placed vertical. The right ear, for example, is then syringed with cold air, which causes (by convection currents) a downward movement of the endolymph; this is equivalent to moving the head to the opposite side (to the left). The patient complains of giddiness, and if allowed to stand tends to fall to the right; nausea and vomiting may occur. The following results can be noted:

1. Nystagmus. The short jerk is to the opposite side (the left), and the slow movement is to the same side (the right).

The slow deviation to the same side is due to impulses which reach the eye nuclei from the vestibule via the vestibular nuclei and medial longitudinal bundle. The short jerk is probably a compensatory movement initiated by the cerebral cortex. The nystagmus illustrates the law of reciprocal innervation. De Kleijn removed the (right) eyeball in the rabbit and connected the internal and external rectus muscles to levers. Nystagmus was produced by syringing the (right) external meatus. The slow phase of the movement was to the right and while the external rectus slowly contracted the internal rectus relaxed correspondingly slowly; in the quick phase the internal rectus contracted suddenly and the external rectus as rapidly relaxed.

The clinical terminology of vestibular nystagmus is confusing, as attention is mainly paid to the 'quick' or correcting component. Thus a so-called 'right horizontal nystagmus' is one in which the quick movement is to the right, and the slow to the left. For completeness it may be mentioned that nystagmus may also be vertical or rotary.

2. Past pointing. On attempting to raise and lower the arms and touch a given point on a tape (when the eyes are shut), the limbs deviate out to the stimulated side.

3. Spontaneous deviation of the limbs occurs towards the stimulated side. The tendency to fall is also towards the syringed side.

Abnormal vestibular stimulation also produces complex autonomic disturbances such as alterations in blood pressure, heart rate, respiration, and bowel tone and movements; these changes are well seen in the phenomenon of sea-sickness.

To summarize, the function of the vestibular apparatus in man is to indicate head movement and to transmit impulses to the cranial nuclei III, IV and VI, which supply the eye muscles (via the median longitudinal bundle) which compensate for this head movement. As a result the position of the eyes is held approximately constant allowing visual fixation on moving objects. Thus, when the head is nodded about a bitemporal axis, there occur normally reflex contractions of the superior and inferior recti, which serve to keep the visual axis horizontal. After bilateral labyrinthectomy patients suffer from 'jumbled vision' when subjected to rapid movement. Thus, such a patient wheeled in a bath chair over uneven ground cannot recognize people. Reading newsprint on a bus must tax the vestibulo-visual 'fixation mechanism' of the normal subject to the utmost.

REFERENCES

Kornhuber, H. H. (ed.) (1974). Vestibular system. In *Handbook of sensory physiology*, Vol. VI/1, Part 1. Springer, Heidelberg.

THE RETICULAR FORMATION

The term 'reticular formation' is used both in an anatomical and a physiological sense. Anatomically the reticular formation is poorly defined; the name has been applied to those parts of the brain stem which are characterized by an interlacing network of fibre bundles. The reticular formation is not a morphological unit but is composed of many nuclei of very different structure. These nuclei are scattered throughout the central part of the brain stem; the term 'reticular nucleus' should be taken to exclude cranial nerve nuclei and relay nuclei of the cerebellar system and the relay nuclei of the lemniscal systems. The tegmentum of the medulla, pons, and midbrain contains the bulk of the nuclei of the reticular formation [FIG. V.109].

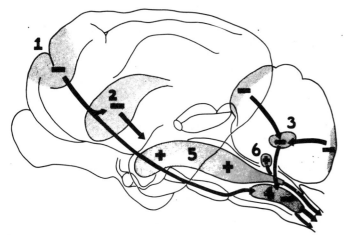

FIG. V.109. Diagrammatic reconstruction of cat's brain showing suppressor and facilitatory systems concerned in spasticity. Suppressor pathways are: 1, cortico-bulbo-reticular; 2, caudatospinal; 3, cerebello-bulbo-reticular; 4, reticulospinal. Facilitatory pathways are: 5, reticulospinal; 6, vestibulospinal. (Lindsley, D. B., Schreiner, L. H., and Magoun, H. W. (1949). *J. Neurophysiol.* **12**, 197.)

Cytoarchitectural studies have shown the presence of more than 50 reticular nuclear masses in the brain stem (Olzewski and Baxter 1952).

Magoun and Rhines (1946) showed that stimulation of the ventromedial part of the medullary reticular formation caused an inhibition of movement, induced either reflexly or by cortical stimulation. This focused attention on the contribution of this region to the extrapyramidal inhibitory projection to the spinal cord. The same workers found that facilitatory effects could be produced on cortically or reflexly evoked movements, by electrical stimulation of the tegmentum of the midbrain and pons, or of the hypothalamus and subthalamus. Hence the reticular formation also sends and relays facilitatory extrapyramidal fibres to the spinal neurons.

These two systems comprise the descending reticular pathway which projects over reticulospinal tracts.

Lastly, it was shown that lesions of the reticular formation in the region of the midbrain caused profound somnolence in chronic survival animals; this change in behaviour was accompanied by marked alterations in the EEG pattern. Correspondingly the electrical stimulation of these areas modified the EEG causing desynchronization. It was recognized that *ascending* tracts in the reticular formation conveyed sensory information to the thalamus which maintained thalamocortical circuits responsible for wakefulness [p. 317]. The ascending reticular system is anatomically co-existent with the so-called descending facilitatory reticular system; no doubt the one influences the other.

Some of the evidence for the statements presented above may now be considered in more detail.

The descending inhibitory reticular projection

FIGURE V.109 shows the site of the bulbar 'inhibitory reticular region'. Electrical stimulation of this area causes the abolition or reduction of movements induced reflexly by pyramidal stimulation, or by stimulation of the motor cortex. Stimulation of the inhibitory areas of the cortex or of the caudate nucleus, evokes an outburst of potentials in the bulbar reticular region as recorded by electrodes inserted at this site, accompanied by an inhibition of movements artificially induced by stimulation of the motor cortex. The bulbar inhibitory reticular region itself appears normally to be under the influence of the inhibitory areas of the cortex and the caudate nucleus. The corticospinal inhibitory extrapyramidal pathway projects over this route to the bulbar reticulum, which in turn relays the projection by reticulospinal pathways which course in the anterolateral funiculi of the cord. In addition to the influence of the cortex, the anterior lobe and paramedian lobules of the cerebellum project via the fastigial nucleus to the pontomedullary region and further serve to reinforce the inhibitory influence of this bulbar region upon the spinal neurons.

The descending facilitatory reticular projection

Experiments similar to those described above show that a surprisingly large area of the brain stem [as outlined in FIG. V.109] causes facilitation of cortically or reflexly induced movements. These effects are abolished by the destruction of the lower extremity of the area, as might be expected. They are independent of the vestibular nuclei and vestibulospinal tracts and are mediated by bilateral reticulospinal pathways which run in the lateral funiculi of the cord [FIG. V.109].

Decerebrate rigidity is a release phenomenon; the mid-collicular transection cuts off the extrapyramidal and pyramidal fibres from the cortex and diencephalon. The 'rump' of lower nervous system which is released by decerebration from the modifying influence of the higher levels, produces reflexly the abnormal pattern of tone which constitutes decerebrate rigidity. Mid-collicular transection interrupts the corticobulbar pathway responsible for extrapyramidal inhibition. The bulbar reticular inhibitory area is now less active, depending solely upon the cerebellobulbar connections for its inhibitory drive on the spinal motoneurons. On the other hand the mid-collicular transection has also interrupted the descending facilitatory reticular projections. If the cut surface of the brain stem of a cat displaying rigidity following intercollicular transection is stimulated in the region of the tegmentum the rigidity is further exaggerated and the stretch reflexes become even more vivid. The intercollicular section has caused some loss of facilitatory drive from higher levels. The question arises as to how the intact remainder of the facilitatory system maintains its drive. One contributory factor may partially explain this. The ascending reticular system receives collaterals from the great sensory pathways as they ascend the brain stem; ordinarily this sensory influx activates the ascending reticular system which in turn activates the cortex. Some of this sensory inflow is relayed by the neurons of the 'ascending system' to neighbouring neurons of the descending reticular system and the residual activity of the facilitatory reticulospinal pathway is dependent on such an influx.

To summarize: decerebrate rigidity is a state of release in which exaggerated proprioceptive reflexes are responsible for a state of heightened muscle tone. The exaggeration of the muscle reflexes is in turn dependent upon a release of the spinal neurons from an inhibitory extrapyramidal barrage; a residual facilitatory barrage probably contributes to the state of 'functional exaltation' of the stretch reflexes.

Although the above account of decerebrate rigidity has been presented in terms of the effects of interruption of reticulospinal projections it must not be forgotten that the vestibular nuclei also influence the state of the spinal motor neurons via vestibulospinal pathways. Thus decerebrate rigidity is abolished by cutting the vestibulospinal tracts in the ventral funiculi of the cord. Similarly a unilateral and isolated lesion of the lateral vestibular nucleus abolishes decerebrate rigidity in the ipsilateral limbs. It would seem therefore that the vestibular nuclei themselves are more active following the interruption of the extrapyramidal corticobulbar tracts. The marked influence of the position of the head on the state of tonus in the limbs [see p. 302] is a manifestation of labyrinthine attitudinal reflexes relayed through the vestibular nuclei and thence via vestibulospinal tracts to the spinal motor neurons.

Granit and Kaada (1952) found that stimulation of the facilitatory reticular neurons increases the rate of discharge of the γ efferents, as indicated by recording from the Group IA afferent fibres from the muscles. Conversely stimulation of the inhibitory reticular neurons reduces or abolishes γ efferent activity.

REFERENCES

Granit, R. and Kaada, B. R. (1952) *Acta physiol. scand.* **27**, 130.
Magoun, H. W. (1954). In *Brain mechanisms and consciousness* (ed. J. F. Delafresnaye). Blackwell, Oxford.
Massion, J. (1967). *Physiol. Rev.* **47**, 383.
Olszewski, J. and Baxter, D. (1954). *Cytoarchitecture of the human brain stem*. Philadelphia.
Pompeiano, O. (1973). In *Handbook of sensory physiology*, Vol. II (ed. A. Iggo) Chap. 12, pp. 381–488. Springer, Berlin.

The cerebellum

The cerebellum lies dorsal to the medulla and pons in the posterior (occipital) fossa and three peduncles (superior, middle, and inferior) connect it with the brain stem.

Anatomically, it is customary to divide the cerebellum into:

1. Two large, laterally placed cerebellar hemispheres.
2. A small central portion, the vermis, so called 'because it resembles a worm bent on itself to form almost a complete circle'.

Phylogenetically the initial formation of a primitive cerebellum occurred in the fish and was responsible for the co-ordination of swimming movements, receiving information from the lateral line sensory organs which subserved a function similar to that of the vestibular apparatus in mammals. This early 'vestibular' part of the organ is termed the archicerebellum and is represented in mammals by the flocculus and nodulus (the so-called flocculonodular lobe).

Further evolutionary development yielded the corpus cerebelli and in the higher forms shows the conspicuous increase in size of the cerebellar hemispheres. Phylogenetically this can be subdivided into an anterior lobe consisting of lingula, lobulus centralis, and culmen, separated by the fissura prima from a posterior lobe. Functionally, the anatomical anterior lobe is related to the lobulus simplex which, however, lies on the other side of the fissura prima. This anterior lobe receives spinocerebellar afferent fibres and is referred to as the paleocerebellum. The posterior lobe consists predominantly of the ansiform and paramedian lateral lobules [Fig. V.110]. The median (vermal) and paramedian (paravermal) parts of the posterior lobe contain the declive, folium and tuber of the vermis and the pyramis and uvula; their neurons project to the fastigial and interpositus nuclei respectively. The lateral ansiform and paramedian lobules are described as neocerebellar and project to the dentate nucleus. The neocerebellum functionally acts to co-ordinate phasic movements of the limbs, trunk, neck, and head including those of the eyes.

Unlike the cerebral cortex the whole of the cerebellar cortex shows the same cytoarchitectural pattern. Three layers are recognized, an outer molecular layer, a middle Purkinje cell layer, and an inner nuclear layer [Fig. V.111]. This external cerebellar cortex which is extensively folded constituting the *folia* overlies the white matter (which consists of afferent and efferent axons [Fig. V.112]).

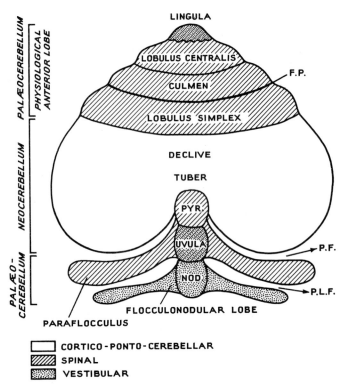

Fig. V.110. Diagram of primate cerebellar cortex laid out flat and looked at from the dorsal surface to show principal subdivisions and afferent connections. F.P. = primary fissure, separating anatomical anterior lobe from posterior lobe. (Physiological anterior lobe includes lobulus simplex.) P.F. = prepyramidal fissure. P.L.F. = posterolateral fissure = posterior border of posterior lobe. Anatomical posterior lobe extends from P.F. to P.L.F. (Modified from Dow and Fulton.)

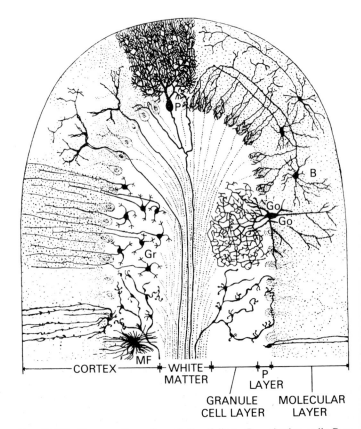

Fig. V.111. Cross-section of cerebellar folium. B = basket cell; P = Purkinje cell; Gr = granule cell; Go = Golgi cell. MF = mossy fibre.

The deep cerebellar nuclei (fastigial, interpositus, and dentate) lie internal to the white matter.

Two types of afferent fibres enter the cerebellum—mossy fibres and climbing fibres [FIG. V.112]. The mossy fibres are so called because they terminate in a series of moss-like glomeruli making axodendritic synaptic connections with the important granule cells which together with Golgi cells constitute the main cellular component of the inner nuclear layer. The mossy fibres comprise spinocerebellar, cuneocerebellar, vestibulocerebellar, reticulocerebellar, and corticopontocerebellar afferents. Each mossy fibre establishes synaptic connections with many granule cells.

The Purkinje cells are large and flasked shaped and lie in a single layer. There are approximately fifteen million in man. *Their axons form the sole efferent projection of the whole cerebellar cortex* and pass to form synaptic connections in the deep cerebellar nuclei. The axons of the deep cerebellar nuclei project to the vestibular and reticular nuclei, to the nuclei of VI, IV, and III controlling the extrinsic eye muscles and to the contralateral red nucleus. The phylogenetically most recent part of the dentate nucleus, termed the neodentate, projects to the contralateral ventrolateral thalamic nucleus.

The Purkinje cell dendrites traverse the outer molecular layer reaching the outer surface. Their branching is conspicuous, forming a two-dimensional tree which is flat in the plane at right angles to the long axis of the folium. Purkinje dendrites provide a huge surface area for axodendritic synapses [FIG. V.112].

The extracerebellar afferents which synapse with the Purkinje cell are the climbing fibres [FIG. V.112]. These originate mainly from cells in the inferior olivary nucleus. They establish a one-to-one connection with the Purkinje cell and excite it to discharge.

The granule cells—hundreds of which are excited by each mossy fibre—are small and very numerous. Each granule cell axon

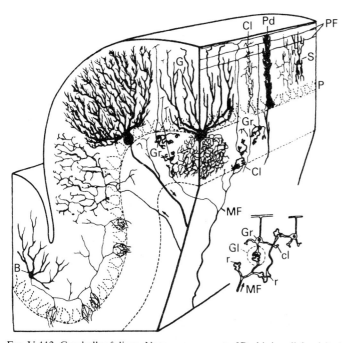

FIG. V.112. Cerebellar folium. Note arrangement of Purkinje cell dendrites (Pd) and basket cell axons in the transverse plane of the folium and the longitudinal arrangement of the parallel fibres (PF). B = basket cell; Cl = climbing fibre; Gr = granule cell; MF = mossy fibre; P = Purkinje cell; Pd = Purkinje cell dendrites; PF = parallel fibre; S = stellate cell (Jansen and Brodal 1958). The inset figure on the right illustrates the relation between the rosettes (r) of the mossy fibres (MF) with the dendritic claws (cl) of the granule cells (Gr) in a glomerulus (Gl). Other elements of the glomerulus are not shown. (From Hámori, J. and Szentágothai, J. (1966). *Acta biol. Acad. Sci. hung.* **15**, 95–117.)

ascends to the outer molecular layer and then bifurcates to form a T [FIG. V.113]. The two branches of the T run along the long axis of the folium and are designated parallel fibres. They pass at right angles through the branches of the Purkinje cells and establish synaptic contacts with the Purkinje cell dendrites. The synaptic transmission from the parallel fibres to the Purkinje cell dendrites is excitatory. Thus the sequence mossy fibre–granule cell–parallel fibre–Purkinje cell provides excitation of the P cell.

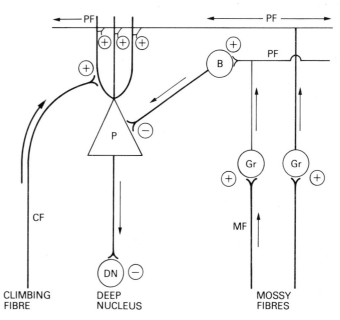

FIG. V.113. Diagrammatic representation of a granule cell axon. PF = parallel fibre effecting excitatory connections with Purkinje cell dendrites; B = basket cell; Gr = granule cell; MF = mossy fibre; P = Purkinje cell.

However, an inhibitory input system is provided by the basket cells and the Golgi cells. The basket cells in the deepest part of the outer molecular layer possess a relatively sparse dendritic tree which, situated in the same plane as that of the P cell dendrites is similarly traversed by the parallel fibres which provide the basket dendrites with an excitatory input. Basket cell axons then pass at right angles to the parallel fibres immediately superficial to the Purkinje cell bodies to which they give off collaterals which terminate around the Purkinje cell body and provide a marked inhibitory input to the P cells. Stellate cells, situated superficially in the molecular layer are also excited by the parallel fibre input and their axons also inhibit Purkinje cells.

Lastly, the large Golgi cells which lie below the P cell layer are excited by the collaterals of climbing fibres and their dendrites which pass outwards into the molecular layer are excited by the parallel fibres. Golgi cell axons which branch extensively in the inner nuclear layer establish inhibitory synaptic contacts with the granule cell dendrites [FIG. V.114].

Cerebellar connections and functions

I. Afferent

1. Dorsospinocerebellar tract arising from Clarke's column (T1–L2) carries mainly proprioceptive impulses from IA and IB afferents (but also cutaneous afferents) from the trunk and leg. It enters the ipsilateral inferior cerebellar peduncle and is distributed to the anterior lobe, pyramis, uvula, and the median part of the paramedian lobe.

2. Ventrospinocerebellar tract arises from border cells of the anterior horn and is predominantly a crossed tract which ascends the cord in the ventrolateral funiculus to enter the cerebellum via

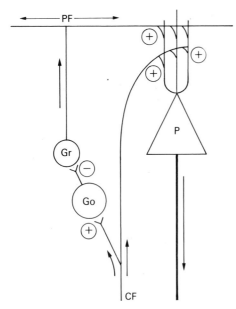

FIG. V.114. Diagrammatic representation of a Golgi cell axon. CF = climbing fibre; Go = Golgi cell; Gr = granule cell; PF = parallel fibre of granule cell; P = Purkinje cell.

the superior peduncle. It carries a large proportion of exteroceptive (cutaneous) fibres and proprioceptive fibres from Group IB and joint receptors. It is distributed mainly to the vermis and the anterior lobe.

3. Olivocerebellar tract is mainly a crossed pathway via the superior peduncle which supplies all parts of the cerebellar cortex and the deep cerebellar nuclei. The inferior olivary nucleus itself receives fibres from all levels of the spinal cord, from brain stem nuclei and from the opposite cerebral cortex.

4. Vestibulocerebellar tract arises in the medial and descending vestibular nuclei, enters via the inferior peduncle and supplies the flocculonodular lobe and uvula. It is ipsilateral.

5. Cuneocerebellar tract carrying proprioceptive impulses from the arm and neck muscles arises in the external arcuate nucleus and passes via the inferior peduncle to the ipsilateral anterior lobe and the pyramis and uvula.

6. Tectocerebellar tract from the superior and inferior colliculi relays fibres respectively from the eye and ear and entering via the superior peduncle passes to the lobulus simplex and the declive and tuber.

7. Pontocerebellar tract which arises from the pontine grey matter occupies the middle peduncle and is distributed to all parts of the cerebellar cortex except the flocculonodular lobe. It is chiefly crossed. This tract carrries impulses from the cerebral cortex, from the spinal cord and from many brain stem nuclei.

8. Rubrocerebellar tract from the caudal two-thirds of the red nucleus is both crossed and uncrossed, entering via the superior peduncle. It is distributed mainly to the dentate nucleus. It probably transmits impulses which have originated from the motor cortex, relaying in the red nucleus.

9. Reticulocerebellar tract arises in the lateral reticular nucleus, is uncrossed, and is distributed via the inferior peduncle to the whole of the cerebellar cortex.

II. Efferent

Purkinje cell axons pass to the deep cerebellar nuclei in an orderly manner. Their influence on these nuclei is solely inhibitory. They have been proved to synthesize GABA (gamma aminobutyric acid) which is released at their terminals. Purkinje cells have a high GABA content. The vermis projects to the medial or fastigial nucleus. The lateral part of the cortex, the ansiform lobule, projects

to the dentate nucleus and paramedian parts of the cerebellum project to the nucleus interpositus.

The fastigial nucleus distributes axons to the vestibular nuclei and to the medullary reticular formation; vestibulospinal and reticulospinal activity is modified accordingly. Similarly, this nucleus projects to the medial longitudinal bundle which integrates the activity of the nuclei of VI, IV, and III supplying the extrinsic eye muscles and to the ascending reticular formation.

The dentate and interpositus nuclei send fibres via the superior peduncle to the red nucleus and the thalamic ventrolateral nucleus of the opposite side. Thalamocortical fibres distribute the information to areas 4 and 6 of the motor cortex. Rubroreticulospinal fibres transmit to the spinal segments. As the dentatothalamocortical path is crossed and as the corticopontocerebellar pathway is also crossed [FIG. V.115] each cerebellar hemisphere provides and receives information which assists its 'comparator' function in modifying movements on its own side of the body.

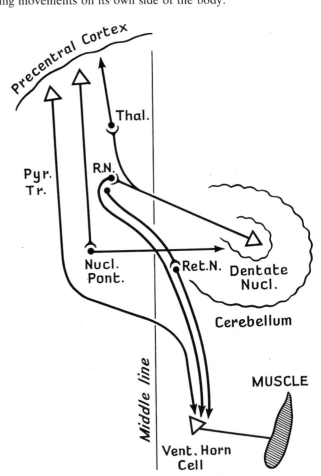

FIG. V.115. Some connections of neocerebellum. Thal., = Thalamus. Pyr. Tr., = Pyramidal tract. R.N., = Red nucleus. Nucl.Pont., = Nuclei pontis. Ret.N., = Reticular nuclei.

Archicerebellum-flocculonodular (FN) lobe connections

Its connections, afferent and efferent, are with the vestibular nuclei. Vestibular afferents (from saccule, utricle, and semicircular canals) pass directly or after relays in the vestibular nuclei via the inferior peduncle to the FN lobe; efferents therefrom return via the inferior peduncle to the vestibular nuclei. From these nuclei the vestibulospinal tract connects with the spinal motoneurons. The FN lobe is a long relay superimposed on the vestibulospinal tract for controlling body posture.

Effects of stimulation of the FN lobe. Stimulation of the FN lobe

produces electrical responses in the anterior and middle ectosylvian gyri of the cortex cerebri (this indicates the extent of cortical projection of the vestibular receptors).

Effects of extirpation. The main disturbance is of equilibrium shown by an inability to maintain an erect posture. A monkey so afflicted cannot stand without swaying. Children are not uncommonly affected by a tumour (medulloblastoma) which develops from cell rests in the nodulus. This produces similar unsteadiness in standing and walking (trunk ataxy). Section of the inferior peduncle causes identical symptoms in animals. Excision of the nodulus in the dog abolishes motion sickness.

Anterior lobe functions. 1. Receiving area for tactile, proprioceptive, corticopontine, and auditory and visual impulses. (i) By stimulating tactile receptors in the skin and recording electrical responses in the cerebellum it has been shown that there is a localized projection to the anterior lobe (and also to the paramedian lobules). Thus stimulation of the tail induces electrical activity in the lingula, that of the hind limb, activity in the lobulus centralis; responses were obtained in the culmen on stimulating the forelimb and in the lobulus simplex on exciting tactile receptors of the face and head [FIG. V.116].

(ii) Proprioceptive stimulation evokes responses in local areas of the anterior lobe which show the same topographical relationship as found in the case of tactile stimulation.

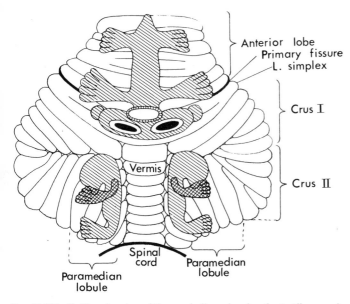

FIG. V.116. Outline drawing of the cerebellum showing the tactile areas in schematized form. The anterior area encompasses the lobulus simplex and anterior lobe and is an ipsilateral projection. The posterior area is located primarily in the paramedian lobules bilaterally but may extend into crus I and II and medially into pyramis. Note body plan in each tactile area; it is less definite in the contralateral paramedian lobule than in other areas. (Snider, R. S. (1950). *Archs Neurol. Psychiat., Chicago* **64**, 196.)

2. Modification of muscle tone and stretch reflexes. (i) Stimulation of the anterior lobe may inhibit tone, mainly in the extensor muscles of the same side of the body. Again, topographical relationships can be observed, for stimulation of the culmen, centralis and simplex produces effects in the hindlimbs, forelimbs, and neck respectively. It is probable that these effects are mediated via the rostrolateral part of the fastigial nucleus to the ipsilateral vestibular nuclei and to the bulbar reticular formation; these nuclei in turn relay the effects to the spinal motoneurons. After damage to the rostrolateral fastigial nuclear cells, anterior lobe stimulation causes an accentuation of decerebrate rigidity, via an ipsilateral fastigial

projection which relays through the rostromedial cells of the fastigial nucleus.

In intercollicular decerebrate animals stimulation of the anterior lobe may cause not only inhibition of extensor muscle contraction but also active flexion of the limbs. Granit and Kaada (1952) showed that stimulation of the medial part of the anterior lobe inhibits the I A afferent discharge from ipsilateral muscle spindles, whereas electrical excitation of the lateral parts of the anterior lobe increases activity in the muscle spindles of the extensors. Such results strongly suggest that these anterior lobe effects are exercised via the γ efferents to the muscle spindles. Granit and his colleagues have further shown that temporary suppression of anterior lobe activity (by surface cooling) abolishes γ discharge to the intrafusal fibres of the muscle spindle; this discharge reappears on restoring anterior lobe function (by rewarming).

At first sight it may seem that the anterior lobe exercises its effects solely on the γ system and that the changes in muscle tone are simply manifestations of altered γ activity consequent upon stimulation of the anterior lobe. However, the anterior lobe must form an important link between the α and γ systems responsible for muscle tone, for ablation of the anterior lobe, which increases intercollicular decerebrate rigidity, abolishes all signs of γ activity on the spindles. Attitudinal reflexes tested after anterior lobe ablation, show effects which are independent of γ discharge and one is forced to the conclusion that the exaggerated muscle tone is due to excessive discharge of the α neurons which is no longer reflexly modified by γ effects. This excessive α firing is due to vestibulospinal projections which have escaped from the inhibitory influence of the cerebellum.

(ii) Ablation of the anterior lobe enhances extensor tone in the intercollicular decerebrate animal particularly in subprimates; the stretch reflexes of such a preparation are exaggerated. As stated above, modification of γ activity can no longer be induced. If an anterior lobe ablation is performed on an otherwise normal animal there is likewise an increase in extensor tone and the positive supporting reaction (magnet reaction) is particularly marked. Effects are restricted to the hind limb if the lobulus centralis is alone ablated and correspondingly changes of tone occur only in the forelimb if the culmen is removed.

Besides these postural changes, there is some ataxia which interferes with the execution of voluntary movement. Movement is less precise in range and force and in speed and direction.

(iii) Stimulation of the cerebral motor cortex elicits electrical responses in the cerebellum; stimulation of the 'face' area of the cortex causes electrical activity in the lobulus simplex and responses in the culmen occur on stimulating the 'arm' area of the motor cortex. Again there is the same topographical picture as is found for peripheral sensory stimulation.

(iv) Auditory and visual stimulation evoke electrical activity in an identical area of the cerebellum. This includes the lobulus simplex [see FIG. V.117]. These effects are abolished by destruction of the colliculi.

Neocerebellum (most of the cerebellar hemispheres).

Connections. The main connections are with the opposite cerebral hemisphere and the upper brain stem.

1. Corticopontine fibres arise in the frontal lobes (areas 4 and 6) and the temporal lobes and end in the nuclei pontis which also receive fibres from the pyramidal tracts. From the nuclei pontis the pontocerebellar fibres cross to the opposite cerebellar hemisphere in the middle peduncle.

2. From the dentate nucleus the superior cerebellar peduncle arises. Dentatothalamic fibres pass to the opposite side to reach the ventrolateral thalamic nuclei, where they are relayed to the cerebral cortex (areas 4 and 6). [FIG. V.115]. By means of these

cerebral–cerebellar–cerebral interconnections the cerebrum and cerebellum mutually influence one another's activities. Each cerebellar hemisphere influences the opposite cerebral cortex; in its turn the excitomotor cortex, via the corticospinal tracts, controls the movements of the opposite side of the body. Because of the double decussation (i.e. of the superior peduncles and of the pyramidal tracts) each cerebellar hemisphere controls voluntary movements on its own side of the body.

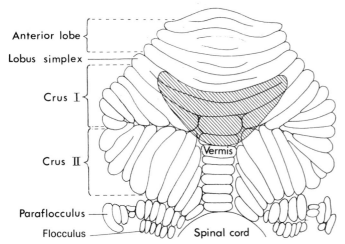

FIG. V.117. Outline drawing of the cerebellum showing that the auditory and visual areas, as determined by click and photic stimulation, are coextensive. Note that this so-called audiovisual area lies primarily in the lobulus simplex, folium, and tuber vermis but extends into crus I and II. (Snider, R. S. (1950). *Archs Neurol. Psychiat., Chicago* **64**, 196.)

In addition to these well-documented connections the neocerebellum also receives spinal afferents. Thus the pyramis and paramedian lobule show face, arm, and leg 'areas' arranged craniocaudally [FIG. V.116], in which appropriate electrical responses occur when tactile receptors in the corresponding parts of the periphery are stimulated. Similar responses can be evoked by proprioceptive stimulation. Likewise electrical stimulation of the paramedian lobule evokes responses in the contralateral sensorimotor cortex and, vice versa, the entire paramedian lobule can be excited from the anterior ectosylvian gyrus. The ansiform lobule consists of two crura (crus I and crus II) of which crus II can be activated more readily from the cortex. Stimulation of the 'leg' area of the motor cortex causes electrical responses in the lateral part of crus II whereas the medial part of crus II can be more obviously activated by stimulation of the face area of the cortex.

The cerebellum responds not only to proprioceptive but also to tactile, visual and auditory stimulation. Almost every structure which projects to the cerebellum also receives a projection from it (seldom direct). The new concept of cerebellar function is that the organ acts as a comparator of a servo-mechanism. Thus it receives both a representation of the corticospinal activity which is transmitted to the muscles, and almost simultaneously a representation of the result in terms of the muscle movement from the proprioceptors of the muscles. In addition to peripheral proprioceptive impulses it receives from tactile receptors and from the eye and ear further information which allows a comparison of the 'true' corticospinal input and the proprioceptive indication of the position of the limb. Then by cerebellocortical relays, the motor activity caused by the corticospinal discharge can be modified appropriately. Normal muscular co-ordination presupposes that the α and γ systems are linked. Movements in which the γ system plays no part are movements which occur without adequate cerebellar information from the muscle spindle; it is not surprising that such movements should be ill controlled. Nevertheless there is to date no evidence that stimulation of the neocerebellum itself causes any modification of γ

efferent activity—it is the anterior lobe from which these effects can be induced.

Damage to the neocerebellum causes most of the dysfunction which has been noted as the result of cerebellectomy in animals, or of clinical cerebellar disease in man. This is as expected, for the so-called neocerebellum forms an overwhelming preponderance of the total mass of the organ.

Results of lesions. There are characteristic disturbances of posture and movement. In unilateral lesions the changes are predominantly found on the same side of the body.

1. DISTURBANCES OF POSTURE. (i). Atonia. There is homolateral atonia. The muscles feel flabby and the limb swings about like a flail.

(ii) Attitude. The face is rotated towards the opposite side. The homolateral shoulder is lower than its fellow. The leg is abducted and rotated outwards; the weight is thrown on the sound leg so the trunk is bent with the concavity towards the affected side.

(iii) Spontaneous deviation. If the eyes are closed and the arms are held straight out in front of the body the homolateral arm sways laterally.

(iv) Nystagmus is common in cerebellar lesions.

(v) Deep reflexes. The knee jerk is characteristically pendular, i.e. after the initial reflex response, the leg, on falling continues to swing freely to and fro.

2. DISTURBANCES OF VOLUNTARY MOVEMENT. (i) There is feebleness (*asthenia*) of movement; the muscles tire readily. Voluntary movements are carried out slowly.

(ii) *Ataxia* (inco-ordination of movement) is marked. It is unaffected by closing the eyes (unlike that of tabes) because conscious proprioception (which is mediated via the dorsal columns, lemniscus, and thalamus to the cortex) is unaffected. There is *decomposition* of the movement—movement seems to occur in obvious stages; *asynergia*—lack of co-ordination between protagonists, antagonists, and synergists; and *dysmetria*—the movement is ill executed in direction, range, and force.

(iii) *Intention tremor* is a conspicuous feature of neocerebellar lesions. This coarse tremor ($4–6$ s^{-1}), as its name implies, is most conspicuous when the part is used in a voluntary movement; it becomes progressively exacerbated as the movement develops. Intention tremor can be induced in experimental animals by an isolated lesion of the dentatothalamic pathway (which carries the efferent projection of the neocerebellum to the ventrolateral nucleus of the thalamus, whence it is relayed to areas 4 and 6).

These disturbances can be demonstrated by:

(a) Finger–nose test: the patient attempts to place the tip of the finger of the outstretched hand on the tip of the nose.

(b) Adiadochokinesis: the patient is unable to carry out rapid pronation and supination movements.

(iv) Gait. The patient tends to deviate to the affected side and brings himself back to the original line—a zigzag path results.

(v) Speech is slow and lalling owing to the imperfection of execution of the movements of the laryngeal muscles and tongue.

It must be noted that the three classical signs which occur *late* in the disease of disseminated (multiple) sclerosis and are grouped as Charcot's triad: viz. nystagmus, intention tremor, and lalling speech, are all referable to disturbances of the cerebellar connections with the brain stem. Disseminated sclerosis, as its name implies, is a widespread demyelinating disease of the central nervous system, which produces both sensory and motor disturbances, but these late signs are due to involvement of the cerebellar pathways.

Other clinical conditions associated with cerebellar dysfunction include:

1. In acute irritative lesions (e.g. vascular lesions) of the cerebellum, giddiness is severe and forced movements occur, which turn the patient so that the face on the side of the lesion is in contact with the pillow.

2. Tumours of the neocerebellum produce signs closely resembling those already detailed. The nystagmus, as stated, consists of a slow to-and-fro movement on looking to the affected side, and a rapid to-and-fro movement on looking to the opposite side. If the flocculonodular lobe is involved, the signs are bilateral and mainly involve the trunk.

3. Tumours growing from the sheath of the eighth nerve usually involve the cerebellum later in their course.

4. In a group of diseases called hereditary ataxy, of which Friedreich's disease is the best known, the spinocerebellar tracts or other cerebellar connections tend to degenerate early, producing characteristic signs.

REFERENCES

ECCLES, J. C. (1966). In *Muscular afferents and motor control* (ed. R. Granit) pp. 19–36. Wiley, Chichester.
—— ITO, M., and SZENTAGOTHAI, J. (1967). *The cerebellum as a neuronal machine*. Springer, Berlin.
GRANIT, R. and KAADA, B. (1952). *Acta physiol. scand.* **27**, 130.
HOLMES, G. (1939). Cerebellum of man. *Brain* **62**, 1–30.
JANSEN, J. (1957). *Acta physiol. scand.* **41**, Suppl. 143.
—— and BRODAL, A. (1954). *Aspects of cerebellar anatomy*. Tanum, Oslo.
OSCARSSON, O. (1957). *Acta physiol. scand.* **42**, Suppls. 145, 146.
—— (1973). In *Handbook of sensory physiology* (ed. A. Iggo) Chap. 11, pp. 339–80. Springer, Berlin.
PUPILLI, G. and FADIGA, R. (1964). *Physiol. Rev.* **44**, 373.

The thalamus

Each thalamus is a large ovoid diencephalic mass of grey matter which lies obliquely across the path of the corresponding cerebral peduncle. The two thalami lie close together in their rostral two-thirds separated only by the third ventricle [FIGS. V.118–V.120]. The two thalami are joined across the midline by the mass intermedia. Their caudal thirds are more widely divergent; the corpora quadrigemina lie between them.

The thalamus is a great sensory relay station. Ascending sensory fibres synapse here and are projected in turn to the primary cortical sensory areas. The thalamus also receives important afferent impulses from the ascending reticular formation which it relays to widespread areas of the cerebral cortex, receiving in turn impulses from corticothalamic fibres. By means of these reticulothalamocortical and corticothalamoreticular connections the state of 'alertness' of the cortex, in response to afferent sensory information, can be varied.

The external medullary lamina consisting of thalamocortical and corticothalamic fibres covers the lateral surface of the thalamus. An attenuated layer of nerve cells (the reticular nucleus) separates the internal capsule from the external medullary lamina.

The internal medullary lamina [FIG. V.120] consisting mainly of internuclear thalamic connections divides the thalamus into lateral, medial and anterior nuclear masses. The lateral nuclear mass in turn can be divided into ventral and dorsal groups of nuclei. The ventral nuclear group contains the ventral anterior, ventral lateral, ventral posterior nuclear groups and most posteriorly the medial and lateral geniculate bodies. The dorsal nuclear group is comprised by the pulvinar, lateral posterior and lateral dorsal nuclei [FIG. V.122].

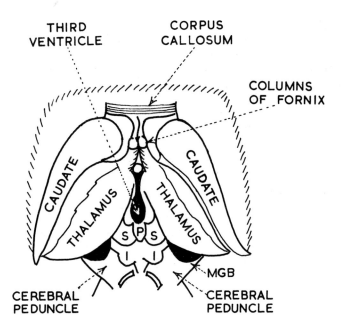

FIG. V.118. S and I = superior and inferior colliculi. P = pineal body. MGB = medial geniculate body.

The medial nuclear mass contains the intralaminar nuclei, the centromedian nuclei, the medial nucleus, and the nuclei of the midline.

The anterior nuclear mass is enclosed by the bifurcation of the internal medullary lamina.

Functionally it is most convenient to consider the thalamus as containing: 1. Extrinsic nuclei; 2. Intrinsic nuclei.

1. Extrinsic nuclei are also known as cortical relay nuclei. Afferent fibres from extrathalamic sources synapse here. The axons of the nuclear cells are distributed to the primary cortical areas—those of the pre- and postcentral cortices and those of the visual and auditory cortical receiving areas. The extrinsic nuclei comprise:

(i) *Posteroventral nucleus* (PLV, PMV). This receives the medial lemniscus and the trigeminal lemniscus. The medial lemniscus carries afferent fibres from the gracile and cuneate nuclei and afferent

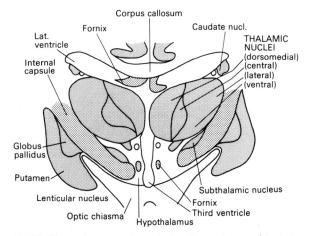

FIG. V.119. Coronal section through the diencephalon with thalamus, hypothalamus, subthalamus, basal nuclei, and related tracts identified. (From Curtis, B., Jacobson, S., and Marcus, E. M. (1972). *An introduction to the neurosciences*. Saunders, Philadelphia.)

LAT.

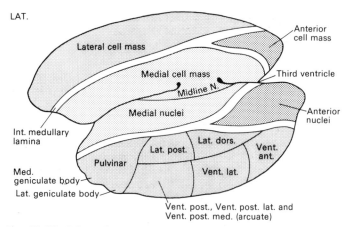

FIG. V.120. Schematic representation bilaterally of the location of the thalamic nuclei. (From Curtis, B., Jacobson, S., and Marcus, E. M. (1972). *An introduction to the neurosciences.* Saunders, Philadelphia.)

fibres from the ventral and lateral spinothalamic tracts. Hence impulses from touch receptors, pressure receptors, joint receptors, temperature receptors, and pain receptors end in synaptic relation with these nuclear cells. With the possible exception of the 'pain' impulses they are all relayed to the postcentral cortex. The trigeminal lemniscus carries the same types of afferents from the face, together with 'taste' fibres. Muscle afferents from the limbs which have relayed in nucleus Z [p. 292] yield fibres which synapse here to be forwarded to area 3a of the cortex [p. 321].

The afferent fibres which come respectively from the face, arm, and leg end in the posteroventral nucleus so that the 'face-receiving area' lies most medially, the 'arm-receiving area' lies immediately lateral to it, and the 'leg'receiving area' lies, in turn, most laterally [FIG. V.121]. Thus transection in the lumbar spinal region causes cell degeneration only in the most lateral part of the posteroventral nucleus, whereas transection in the cervical spinal cord leads to the subsequent degeneration of all the posteroventral nucleus except its most medial division (known as the arcuate nucleus) which receives only trigeminal afferents from the face. If cortical ablations

are made it is found that removal of the 'face area' of the postcentral sensory cortex causes degeneration of the arcuate nucleus; ablation of the 'leg area', which lies most medially in the postcentral gyrus, leads to degeneration of the lateral part of the posteroventral nucleus.

Confirmatory evidence of this topographical relationship is provided by electrophysiological studies. Tactile stimulation of the foot evokes potentials which can be suitably recorded by an electrode whose tip lies in the lateral part of the posteroventral nucleus; another electrode implanted in the most medial part of the postcentral gyrus records potentials shortly after their receipt by the thalamic electrode.

(ii) *Lateroventral nucleus* (LV) [FIG. V.120]. This receives the dentatothalamic fibres from the cerebellum, carrying proprioceptive information. The lateroventral nuclear cells project in turn to the precentral motor cortex—areas 4 and 6.

(iii) *Anterior nucleus*. This, the most rostral mass, is surrounded by the divisions of the internal medullary lamina and bulges into the lateral ventricle. It receives the anatomically distinct mammillothalamic tract (the bundle of Vicq d'Azyr) which carries impulses relayed in the mammillary bodies from the hippocampus. The anterior nucleus sends fibres to the cingular gyrus (area 24). The functional significance of the whole of this pathway is considered elsewhere.

(iv) The *medial geniculate bodies* receive a topically organized projection of auditory fibres from the cochlear nuclei and inferior colliculi which they relay to the auditory areas of the cortex. Destruction of small areas within the medial geniculate body reduces the hearing of restricted bands of tone frequency.

(v) The *lateral geniculate bodies* receive the primary visual neurons. After destruction of retinal areas the cells in the l.g.b. show transneuronal degeneration; it can be shown that the macula of the retina projects to the caudal two-thirds of the l.g.b. whereas

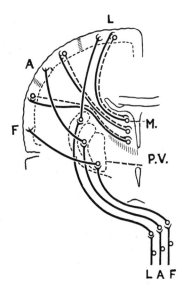

FIG. V.121. Connections of posteroventral nucleus and medial nucleus of thalamus, L, A. F = afferent fibres from leg, arm, and face which relay in the brain stem and cross over to the opposite side in medial lemniscus to end in the posteroventral nucleus of thalamus (P.V.). They relay there to pass to the leg, arm, and face areas of the sensory cortex. M. = the medial thalamus, which is connected with the cortex by a 'private' thalamocortical and corticothalamic pathway. This to-and-fro path is the circuit which is responsible for the resting electroencephalogram.

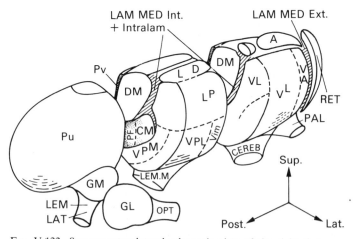

FIG. V.122. Superoposterolateral schematic view of the right thalamus showing the respective positions of the different nuclei (orientation is indicated by arrows). Sections have been made in order to make apparent the relative position of the main structures mentioned in this paper. A = anterior; CEREB = brachium conjunctivum (cerebellar afferents); CM= nucleus 'centre median'; DM = dorsalis medialis (or medialis); GL = nucleus geniculatus lateralis; GM = nucleus geniculatus medialis; Intralam. = nuclei intralaminares; LAM. MED. Ext. = lamina medullaris externa; LAM. MED. Int. = lamina medullaris interna; LD = nucleus lateralis dorsalis; LEM. M. = lemniscus medialis; LEM. LAT. = lemniscus lateralis; LP = nucleus lateralis posterior; OPT = tractus opticus; PAL = pallidal afferents; Pf=nucleus parafascicularis; Pu=nucleus pulvinaris; Pv = nuclei paraventriculares; RET = nucleus reticularis; VA = nucleus ventralis anterior; Vim = nucleus ventralis intermedius; VL = nucleus ventralis lateralis; VPL = nucleus ventralis posterior lateralis; VPM = nucleus ventralis posterior medialis. (From Albe-Fessard, D. and Besson, J. M. (1972). In *Handbook of sensory physiology* (ed. A. Iggo) Chap. 13, pp. 489–560. Springer, Berlin.)

the remainder of the retinal neurons end more rostrally. The lateral geniculate cells project to the calcarine cortex (geniculocalcarine tract).

2. *Intrinsic* nuclei receive afferents mainly from other structures in the thalamus. They comprise the midline and intralaminar nuclei (which though well developed in the lower mammalia are less so in the primates), the dorsomedial and dorsolateral nuclei, and the pulvinar whose development is marked in the primates. The dorsomedial and intralaminar nuclei have generous connections with the frontal lobes and the hypothalamus. The midline and intralaminar nuclei project also to the neostriatum. The pulvinar projects to the inferior parietal lobes. The dorsolateral nuclei receive fibres from and project to the precuneate gyrus.

These nuclei degenerate after appropriate cortical lesions. The intralaminar nuclei include the centrum medianum, parafascicular, limitans, central, paracentral, and central lateral nuclear groups. They receive the reticulothalamic, tegmentothalamic, and tectothalamic tracts; they relay to the reticular complex or reticular nucleus of the thalamus.

The *reticular* nucleus is a shell of nerve cells that surrounds the dorsal thalamus on its lateral and anterior sides [FIG. V.122]. It is situated between the internal capsule and the external medullary lamina.

Heavily myelinated fibres of the thalamic radiations split the curved ribbon of cells of the reticular nucleus into clusters and thereby give the nucleus its reticular appearance.

The reticular nucleus receives fibres from all the intralaminar nuclei, which in turn are the sites of synapses of afferent tracts from the ascending reticular formation. These tracts (tegmentothalamic and reticulothalamic) carry impulses aroused in the reticular formation of the brain stem by sensory collaterals. Impulses which are relayed from the reticular core of the brain stem via intralaminar nuclei and the thalamic reticular nucleus to the cortex are described as 'non-specific'. This term 'non-specific' is used because the reticulothalamic path exerts great effects on the electrical activity of widespread areas of the cortex. This influence of the intrinsic and the reticular nuclei on cortical activity can only be appreciated from studies of the electroencephalogram.

Thalamic syndrome

The posteroventral nucleus relays skin, muscle, and taste afferents to the postcentral cortex. The posterolateral nucleus relays cerebellar impulses to the excitomotor cortex areas 4 and 6.

A lesion of these nuclear masses sometimes results from thrombosis of the local artery of supply. As might be expected, there occur profound muscular weakness, decreased muscle tone, and ataxia (due to damage to the cerebellar afferents). Accompanying these symptoms there is a loss of discriminative aspects of sensation such as occurs in lesions of the sensory cortex itself. The loss includes the sensation of light touch, tactile localization, and discrimination, intermediate grades of touch and the appreciation of small movements at joints. These symptoms of sensory loss can be referred to destruction of the posteroventral nucleus. All these symptoms and signs occur on the opposite side of the body. They may be accompanied by altered emotional effects.

REFERENCES

ALBE-FESSARD, D. and BESSON, J. M. (1972). In *Handbook of sensory physiology* (ed. A. Iggo) Chap. 13, pp. 489–560. Springer, Berlin.
GORDON, G. (ed.) (1977). Somatic and visceral sensory mechanisms. *Br. med. Bull.* **33**, No. 2.
JACOBSON, S. (1972). In *An introduction to the neurosciences*, Chap. 12, pp. 294–300. Saunders, Philadelphia.

The electroencephalogram

By placing two electrodes on the scalp and leading via suitable amplifiers to a cathode ray oscillograph or to an ink-writing device a record is obtained of the electrical activity of the cerebrum. A similar record which resembles in a general way that of the electroencephalogram (EEG) may be obtained by placing the electrodes directly on the exposed surface of the cerebral hemispheres—this is termed the *electrocorticogram*.

If an EEG record is taken in a normal subject who refrains from mental activity and who keeps his eyes closed, the usual pattern of electrical activity consists of a sequence of waves which recur at a frequency of 8–12 Hz. These alpha waves occur in bursts or 'spindles'; they gradually build up and then recede as shown in FIGURE V.123. Their average amplitude is 50 microvolts, but this varies from cycle to cycle. If the eyes are opened or if the subject indulges in mental arithmetic or any other purposive activity, the large alpha waves disappear ('blocking') and the record shows only small irregular oscillations. Seemingly the cortical cells when not engaged in any specific task discharge synchronously about 10 times per second: their pattern of 'inactivity' is a mark-time beat. Such activity is highly characteristic for any given individual.

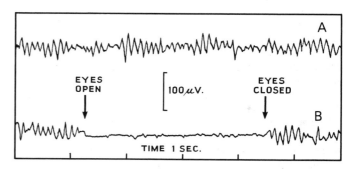

FIG. V.123. Normal human electroencephalogram (EEG). A. Normal record taken with eyes closed, showing alpha waves. B. Normal record. On opening the eyes the regular alpha rhythm is replaced by small irregular oscillations. The normal waves return on closing the eyes. Time in seconds.

The EEG is extensively used as an adjunct to clinical diagnosis. The objects of clinical recording are:

1. To determine the distribution of electrical activity over wide areas of the cortex.

2. To observe activity arising simultaneously in different areas of the brain.

To attain these objects six simultaneous tracings are taken. Many electrodes are placed on the scalp and are connected to the amplifiers in pairs. A typical distribution of the electrodes is shown in FIGURE V.124.

Each tracing is a complex rhythmic wave which never repeats itself precisely. An instrumental analyser automatically plots a histogram of the wave components underneath the wave complex [FIG. V.124] giving the relative amplitudes of the sine wave components present in the natural EEG. The sine waves may be classified according to their frequency as follows:

Frequency Hz	Name	
1–3.5	delta	(δ)
4–7	theta	(θ)
8–13	alpha	(α)
14–30	beta	(β)

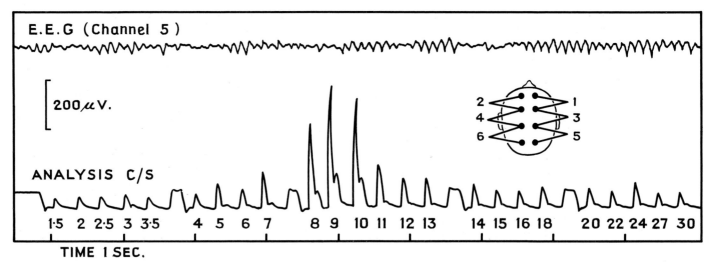

FIG. V.124. Normal human electroencephalogram and its analysis. Upper record: EEG, Channel 5. The disposition of the surface electrodes on the head is shown in inset (top = front, bottom = back of head). Lower record: Analysis of Channel 5. The height of each wave is proportional to the magnitude of the contribution of the particular wave frequency.

Alpha rhythm is present maximally in the occipital and parieto-occipital areas and is usually diminished by visual or mental activity. Theta rhythm is often found over the parietal and temporal areas; the waves have a low amplitude (10 microvolts). The largest waves are the delta group which usually appear during sleep, attaining an amplitude of 100 microvolts. Delta activity may often be evoked by overbreathing. The presence of an intracranial tumour may cause mechanical stress in neighbouring parts of the brain which in turn act as a source of delta wave activity. Correspondingly a tumour can be localized by EEG tracings which show evidence of a focus of delta wave production.

The neurophysiological basis of the EEG remains poorly understood. Such evidence as is available does not suggest that the slow rhythmic waves are formed by the summation of individual spikes. By recording from the brain at different depths it has been shown that the waves originate in grey rather than white matter and that the nuclear masses of the thalamus make an important contribution to the EEG record. Thermocoagulation of the outer four layers of cortical grey matter does not alter the pattern of the EEG. If a slab of cortex is isolated from all its nervous connections it becomes quiescent. If an area of cortex is isolated from its cortical connections by deep circumcision its activity persists until undercutting destroys its thalamic connections.

In certain states of anaesthesia, the electrocorticogram shows waves occurring in groups separated by periods of inactivity. Stimulation of a sensory nerve or a thalamic relay nucleus causes a short latency (1–5 ms) potential wave in a focal part of the cortex. This occurs irrespective of the presence or absence of spontaneous activity recorded therefrom. Thus peripheral stimulation of a nerve from the foot evokes a potential only in the 'foot' area of the sensory cortex, i.e. the most medial part of the postcentral gyrus. (It is in this way that the distribution of axons from the thalamic relay nuclei has been elucidated).

If on the other hand the intralaminar thalamic nuclei are stimulated the cortical response is (i) bilateral and widespread throughout the cortical areas, (ii) of longer latency (15–60 ms) and of longer duration. In response to repetitive stimuli delivered at a rate of 8–12 s⁻¹ the cortical record shows potential waves which at first increase in size progressively ('recruiting response') and which, if the repetitive stimulation is maintained, wax and wane as do the naturally occurring waves of the electrocorticogram [FIG. V.125]. The widespread occurrence of these waves is not simply due to their intracortical conduction from primary receptive fields of the cortex. Thus they appear in areas of cortex isolated except for their thalamocor-

tical connections. The long latency of the cortical responses normally evoked by stimulation of the intralaminar or medial nuclei is not due to subcortical delay but rather to the setting up of reverberating activity in cortical circuits by the widespread arrival of the thalamocortical impulses.

The diffuse character of the cortical recruiting response to stimulation of the intralaminar and medial nuclei, has led to their thalamocortical projection being called the 'diffuse' system. Likewise the term 'non-specific' system is employed to distinguish this projection from that of the cortical relay nuclei. There is no doubt that the recruiting response is independent of the cortical relay nuclei. Thus isolated destruction of the posteroventral nucleus abolishes the potential response of the postcentral gyrus to electrical stimulation of the medial lemniscus, but does not affect the appearance of spontaneous waves of electrical activity which characterize the electrocorticogram of this area; nor does it affect the development of the recruiting response to stimulation of the intralaminar nuclei. Although these terms 'non-specific' and 'diffuse' are widely used, they are not particularly felicitous since they suggest that the thalamic regions from which the recruiting responses can be obtained are themselves diffuse and ill-defined. On the contrary, the thalamic regions from which these recruiting responses can be obtained by electrical stimulation comprise only

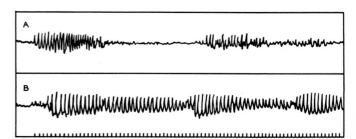

FIG. V.125. Stimulation of medial thalamus produces cortical potentials resembling the normal resting electroencephalogram or electrocorticogram. Lightly anaesthetized cat: electrocorticogram from middle suprasylvian gyrus. A. Spontaneous electrical waves at frequency of 8–12 s⁻¹, appearing in bursts and alternating with periods of feeble electrical activity. B. During the period indicated by the signal (lowest line) the medial thalamus was stimulated at the frequency shown by the signal. Electrical waves develop in the cortex, resembling those of the normal electroencephalogram; the waves wax and wane in magnitude, and show a uniform frequency. These waves appear all over the cerebral cortex; they appear in areas of cortex isolated except for their thalamocortical connections. (Morison, R. S. and Dempsey, E. W. (1942). *Am. J. Physiol.* **138**, 283–96.)

the intralaminar, the centromedian, dorsomedial, and antero-medial nuclei, the anterocentral nucleus and the reticular nucleus.

The ascending reticular system and the electroencephalogram

The EEG of an unanaesthetized animal or man in the wakeful state shows an incessant low-voltage high-frequency discharge [Fig. V.126]. In sleep or under barbiturate anaesthesia this record changes to one characterized by high voltage slow waves (δ waves). The change from sleep to wakefulness is characterized by a reversion of the EEG record to the rapid low-voltage discharge.

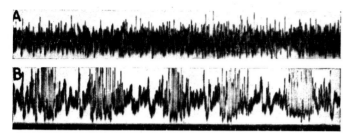

Fig. V.126. Effect of the transection of the brain stem on the cortical activity of the cat. A, intact brain, irregular fast activity characterizing the waking state of the brain (temporal lead, suprasylvian gyrus); B, 'cerveau isolé' preparation, same lead: regularly spaced spindles of alpha waves, a typical aspect of the sleep condition of the cortex resulting from the transection of the brain stem at the collicular level. (Bremer, F. (1953). *Some problems in neurophysiology*. The Athlone Press, London.)

Bremer introduced the *encéphale isolé* preparation in order to study the EEG changes in the unanaesthetized cat. Under ether anaesthesia the atlanto-occipital membrane is opened and the medulla is transected just above the first cervical spinal segment; thereafter the cat is maintained on artificial respiration without anaesthesia. In such a preparation the brain is active and the record obtained resembles that of the wakeful animal; the cranial nerves continue to conduct visual, auditory, and exteroceptive stimuli to the brain stem. Thus the visual receiving areas of the cortex shows fast low-voltage waves until both optic nerves are cut, whereupon the discharge changes to one of rhythmic bursts of large amplitude waves which are separated by periods of 10–15 seconds in which no activity is manifest. This latter record resembles that seen in quiet sleep.

In further experiments Bremer studied the effects of a mesencephalic transection of the brain stem at the collicular level immediately caudal to the oculomotor nuclei—the so-called *cerveau isolé* preparation. The rapid irregular electrical activity ('desynchronized activity'), shown in the electrocorticogram of the intact animal under light ether anaesthesia, changed to one in which a monotonous succession of bursts of alpha waves occurred, separated by pauses of a fairly constant duration [Fig. V.126]. This record showed all the electroencephalographic features of sleep. Visual and olfactory pathways alone remained intact and when suitably stimulated caused desynchronization with resultant fast low-voltage activity.

Bremer concluded that EEG signs of somnolence were due to section of the great sensory pathways running headwards to the thalamus to activate the cerebral sensorium. However, later work by Moruzzi and Magoun and their respective collaborators has shown that the 'sleep pattern' EEG of the *cerveau isolé* is due not to the severance of the lemnisci themselves, but rather to the interruption of the ascending reticular system which normally showers impulses on the non-specific nuclei of the thalamus. Thus:

1. Discrete electrolytic lesions of the classical sensory pathways themselves in the mesencephalon do not cause any fundamental alteration in the electrocorticogram. When the behavioural effects of these lesions are studied in chronic survival animals it is found that such animals do not show drowsiness or coma [Fig. V.127].

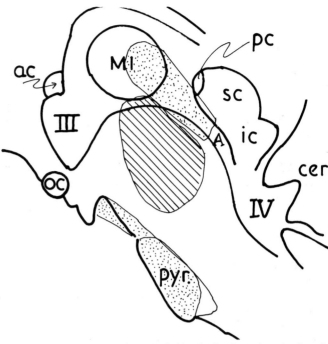

Fig. V.127. Outline of brain stem (midsagittal) of monkey. Lesions in dotted areas do not cause coma. Lesion in central (lined) area causes coma; sc = superior colliculus; ic = inferior colliculus; pc = posterior commissure; ac = anterior commissure; cer = cerebellum; MI = massa intermedia; oc = optic chiasma; pyr = pyramid; III = third ventricle; IV = fourth ventricle; A = cerebral aqueduct. (Redrawn from French, J. D. and Magoun, H. W. (1952). *Archs Neurol. Psychiat., Chicago* **68**, 591.)

2. Lesions of the mesencephalic tegmentum, subthalamus, hypothalamus, and the medial part of the thalamus cause 'synchronization' of the electrocorticogram. Correspondingly such lesions caused somnolence or coma in survival animals [Fig. V.127]. The extent to which the ascending reticular system was anatomically involved by the lesion directly influenced the degree of disturbance of behaviour and of the resting electrocorticogram.

3. Single-shock stimulation of peripheral limb nerves, single flashes of light on the retina, or single clicks delivered to excite the cochlear nerve endings, all evoke potentials in the nervous elements of the reticular system and cause transient desynchronization of the electrocorticogram.

4. Even in animals in which the lemnisci and spinothalamic tracts have been destroyed, stimulation of the brain stem still evokes desynchronization of the electrocorticogram. The ascending reticular system receives throughout its course (from the lower pons to the level of the thalamus) afferent collaterals from the long auditory and somatic sensory pathways and afferents from the visceral and visual systems [Fig. V.128]. By relaying such afferent impulses to the cortex the ascending reticular system exercises a strong desynchronizing action on the EEG and according to Magoun and others is responsible for maintaining a state of wakefulness or alertness.

As might be expected, a single peripheral stimulus gives rise to impulses which ascend to the cortex in both classical sensory pathways and in the ascending reticular system; the impulses travel slowly by multisynaptic relays in the reticulum but are conducted finally to widespread areas of the cortex, whereas those in the great sensory pathways are propagated with great speed to the primary receiving areas. French, Verzeano, and Magoun conclude that whereas the great sensory pathways subserve discriminative sensory perception, the reticular system functions by arousing consciousness or alertness 'without which the above-mentioned sensory discrimination and effective response would be impossible'.

Some of the thalamic connections of the ascending reticular system have been described by Gastaut (1959) and Papez. They relay in the nuclei of the midline and in the intralaminar nuclei before passing on to the thalamic reticular nucleus where there is a further relay. From this nucleus fibres are widely distributed to all parts of the cortex (although each part of the nucleus supplies only one cortical area (Rose 1952).

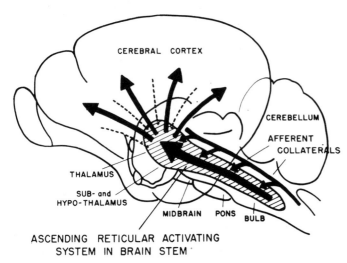

FIG. V.128. Outline of brain of cat, showing distribution of afferent collaterals to ascending reticular activating system in brain stem. (Starzl, T. E., Taylor, G. W., and Magoun, H. W. (1951). *J. Neurophysiol.* **14**, 479.)

It would appear from such a description that the thalamic nuclei which receive fibres from the ascending reticular system are those which are concerned with the recruiting response of the EEG. The effect of these ascending reticular impulses is to produce desynchronization of the EEG. On the other hand stimulation of the intralaminar and midline nuclei at a frequency of 8–12 Hz produces the recruiting response which is identical with 'synchronization' of the EEG. However, Dempsey and Morison (1942), who originally described these results, found that a more rapid stimulation of the intralaminar or midline nuclei caused desynchronization of the EEG. Whether the ascending reticular impulses 'break' a resting mark-time cortical rhythm which is a function of the non-specific thalamocortical pathway in this way, or whether they exert their effects on the EEG by other pathways remains unknown at the present. Left to themselves the units of the thalamocortical system adopt a resting rhythm, characterized by the phasic waxing and waning of the alpha rhythm. Influx from the ascending sensory and reticular pathways abolishes this rhythm and causes 'arousal'.

Dell, Bonvallet, and Hiebel (1954) have shown that adrenalin activates the ascending reticular system, thus contributing to EEG arousal–the electrical record changes to one of the fast low-voltage wave activity. Carotid sinus distension on the other hand exercises a depressant effect on the mesencephalic reticular system and produces slow wave activity on the EEG.

REFERENCES

BREMER, F. (1953). *Some problems in neurophysiology*. Athlone Press, London.
DELL, P., BONVALLET, M. and HIEBEL, M. (1954). *Electroenceph. clin. Neurophysiol.* **6**, 599.
DEMPSEY, E. W. and MORRISON, R. S. (1942). *Am. J. Physiol.* **135**, 293.
GASTAUT, H. (1954). In *Brain mechanisms and consciousness* (ed. J. F. Delafresnaye) p. 117. Blackwell, Oxford.
HAZEMANN, P. and MASSON, M. (1976). *ABC d'electro-encephalographie*. Masson, Paris.
PAPEZ J. W. (1959). *Electroenceph. clin. Neurophysiol.* **8**, 117.
ROSE J. E. (1952). *Res. Publ. Ass. nerv. ment. Dis.* **30**, 454.

Structure of cerebral cortex

Histological studies carried out over many years have led to the cerebral cortex being subdivided into many areas each with its distinctive cellular arrangement. The detailed cell structure of any cortical area is called its *cytoarchitectonics* or its cytoarchitecture.

According to Economo, in typical regions of the cortex, six cell layers can be recognized which are numbered I to VI from without inwards [FIG. V.129].

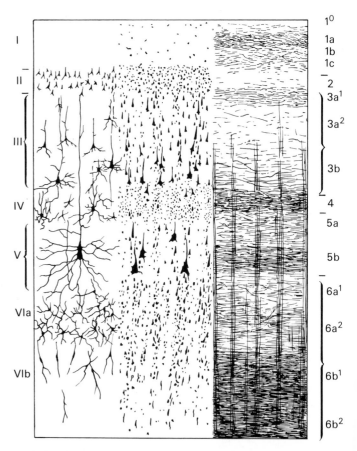

FIG. V.129. Cerebral cortical structure (diagrammatic). I molecular layer; II external granular layer; III outer pyramidal layer; IV internal granular layer; V inner pyramidal layer; VI fusiform layer; 3a¹ band of Bechterew; 4 outer band of Baillarger; 5b inner band of Baillarger. Left-hand picture shows cells; right-hand picture shows fibres.

I Molecular or plexiform layer: Apical dendrites from pyramidal cells in deeper layers ascend and ramify horizontally. Scattered horizontal cells (Cajal) and axon terminals of Martinotti cells are found here. Perhaps this layer is a site of a horizontal spread of neuronal activity—either excitatory or inhibitory.

II External granular layer: characterized by closely packed small fusiform cells whose dendrites either ascend to layer I or spread laterally within this layer. Some of the axons of these cells pass to layers V and VI and synapse with large pyramidal cells (layer V) or spindle pyramidal cells (layer VI). The afferent fibres which establish synaptic connections with layer II cells are derived from Martinotti cells and from granular cells of layer IV.

III Outer pyramidal layer: contains larger cells whose apical dendrites ascend to layer I and whose basilar dendrites pass horizontally in the same layer. Their efferent axons pass mainly to layers V and VI. The afferent fibres to the layer III cells are supplied by axons of granular cells from layer IV and from Martinotti cells of layers V and VI.

IV Internal granular layer: densely packed by small star-shaped (stellate cells). Their dendrites are mostly distributed within this layer and receive a dense supply of specific thalamocortical afferents. This characterizes the so-called 'sensory' cortical areas 3, 1, and 2. The axons of the stellate cells mostly terminate in layers V and VI although some do ascend to the outer layers.

V Inner pyramidal layer: contains large pyramidal cells, some of which (area 4) are very large and have been designated the giant cells of Betz (60–120 μm by 30–80 μm). Their apical dendrites pass to the outer layers whereas their basilar dendrites are confined to this layer. The axons of the large pyramidal cells project to the white matter (with collaterals to the same and more external cortical layers) and terminate on the cells of the brain stem (e.g. the gracile and cuneate nuclei) and spinal cord.

VI Fusiform layer: contains many spindle-shaped cells whose dendrites arborize in the outer layers and whose axons project to subcortical nuclear groups.

(The cells of Martinotti, pyramidal in shape, *characteristically* possess axons which pass *outwards* to the cortical surface and which never enter the white matter. Their axons, like those of the granule and horizontal cells are short.)

There are four principal bands of transversely running nerve fibres in layers I, III, IV (outer line of Baillarger), and Vb (inner line of Baillarger), respectively. The longitudinally running fibres penetrate outwards as far as layer II.

Apart from the typical cortex just described (or isocortex) there is the allocortex which is contructed on an entirely different plan. The allocortex includes the uncus, hippocampus, and the gyrus dentatus; in man it constitutes only about one-twelfth of the cerebral surface.

Five fundamental types of isocortex may be differentiated; types 2, 3, and 4 are essentially alike and differ one from the other in details; types 1 and 5, on the other hand, contain very obvious distinctive features [FIG. V.130].

Types 2, and 3, 4. These have the six typical laminae previously described.

Type 2. Frontal type (anterior two-thirds of the frontal lobe, superior parietal lobule, and part of the temporal lobe). The granule cells are triangular.

Type 3. Parietal type (parietal lobe and junctional region of parietal, occipital, and temporal lobes). There is an increase in the depth and density of the two granule layers II and IV, and these cells are round in shape; the pyramidal cells are smaller, slender, and more numerous.

Type 4. Polar type (only at the frontal and occipital poles). The cortex is narrow, and all the layers are reduced in depth though the cells are more densely packed.

Type 1. Agranular cortex. This is characteristic of the excitomotor regions of the cortex and is thus found in the posterior third of the frontal lobe anterior to the fissure of Rolando. It is, however, also found on the convexity and medial surface of this region, in

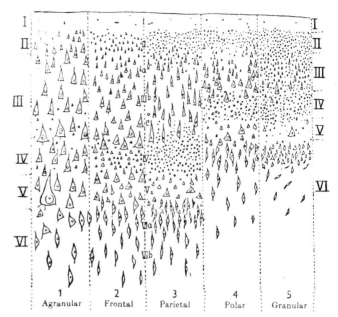

FIG. V.130. Diagram of the five fundamental types of structure found in the cerebral cortex. (Economo, C. von (1929). *Cytoarchitectonics of the human cerebral cortex.* Springer, Berlin.)

Broca's area, and the anterior part of the island of Reil. As the name implies, granule cells are completely absent; the cells in layer II and IV have become pyramidalized, i.e. replaced by pyramidal cells.

Type 5. Granular cortex. This is characteristic of the sensory zones, e.g. the postcentral gyrus (general body sensation), calcarine region (vision), Heschl's gyrus (hearing). The granules have largely replaced the pyramidal cells in layers III and V, i.e. the cells have become granulized. In the visual cortex the internal granule layer IV is divided into two parts by transversely running fibres (the line of Gennari).

A map illustrating some of the cytoarchitectonically discrete areas is shown in FIGURE V.131. Such maps are useful when stimulation or extirpation experiments are planned to determine more precisely the functional attributes of different cortical regions.

Afferent fibres to the cortex can be classified as:

(a) **Specific thalamocortical:** these terminate in layer IV and layer

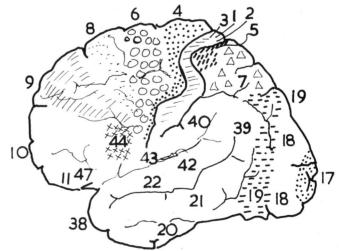

FIG. V.131. Cytoarchitectonic areas of the human cortex (lateral surface). Note position of areas 3, 1, and 2 (postcentral sensory cortex) and areas 5 and 7 (parietal lobe). (After Brodmann, K. (1914).) *Vergleichende Lokalisationslehre der Grosshirnrinde.* Barth, Leipzig.)

III but give collaterals to the apical dendrites of the pyramidal cells of V (which are passing outwards to the pial surface). Such afferents are exemplified by the VP thalamocortical fibres to area 1 and those from the lateral geniculate body to area 17.

(b) **Non-specific thalamocortical:** (whose exact origin from the thalamus is not known) which ascend to layer I *but* which give collaterals to each of the deeper layers II-VI.

(c) **Association fibres** (connecting different areas of the cortex) and commissural fibres (connecting the cortices of the two hemispheres). These terminate mainly in layers I–IV.

Cortical efferent fibres can be classified as:

(a) They pyramidal cell axons of layers V and VI which provide the preponderance of projection, commissural, and association fibres.

(b) Pyramidal and granule cell axons of layers II–IV are predominantly distributed within these same cortical layers.

(c) Martinotti cells (with ascending axons), horizontal cells (layer I) and granule cells, all have short axons which are distributed within the cortex itself.

Somatic sensory cortex and parietal lobe

The functions of that part of the parietal cortex which subserves conscious skin and muscle sensibility have been studied in a number of ways.

Cortical potentials in animals

The points of immediate termination of afferent impulses in the sensory areas (receiving areas) of the cortex can be mapped out by electrical means. The cortex is exposed in an anaesthetized animal; stimulation of skin or muscle sets up a burst of action potentials in a restricted region of the cortex which represents the arrival point of the impulses. The map thus determined when the body wall (skin and muscles) is stimulated in the monkey is shown in FIGURE V.132.

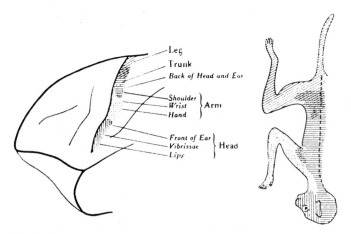

FIG. V.132. Projection of body wall of monkey on sensory cortex. (Adrian, E. D. (1941). *J. Physiol., Lond.* **100**, 159–91.)

The receiving area in all species is mainly concerned with those parts of the body which are most closely related to the outside world; thus in the cat there is a large area for the claws, in the rabbit for the mouth region, in the dog for the face, and in the monkey for the hands and face. The impulse patterns set up by the peripheral

receptors become modified as they pass through the various relays in the central nervous system as a result of central summation, irradiation, and after-discharge; the patterns arriving in the sensory area thus differ significantly from those set up by the sensory endings. Strangely enough, no cortical responses have yet been detected in the animal species examined, following the application of thermal or painful stimuli.

Further analysis shows that light tactile stimulation of the various parts of the body evokes potentials in the appropriate parts of the postcentral gyrus closely resembling the 'map' of the cortical responses to electrical stimulation of the periphery. In addition to the inverted representation of the body in the postcentral gyrus a second somatic sensory area has been found in the ectosylvian gyrus. Here potentials arise in the anterior (rostral) end during tactile stimulation of the head and potentials are evoked in the posterior part during tactile stimulation of the hind limb. There is good evidence of somatotopic localization. It is not clear what role this second somatic sensory area fulfils. It has been suggested that it receives impulses from the contralateral paramedian lobule of the cerebellum. Certainly the stimulation of this sensory cortical area causes electrical responses in the paramedian lobule.

Thus the body is 'represented' *twice* in the somatic sensory cortex, in areas designated SI and SII [FIG V.133].

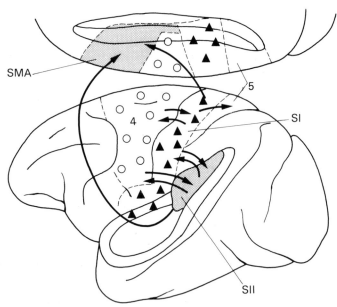

FIG. V.133. Schematic diagram to show the sites in the monkey of the four areas of the cortex considered to be importantly involved in somatic sensation and the ipsilateral association fibres between them. Reciprocal connections join SI and SII to each other and to area 4, and small projections pass from SI and SII to the SMA, but only SI sends fibres to area 5 of the parietal cortex. SI = first somatic sensory area; SII = second somatic sensory area; SMA = supplementary motor area. (Redrawn from Jones, E. G. and Powell, T. P. S. (1969). *Brain* **92**, 477–502.)

SI, lying in the post-central gyrus contains the cytoarchitectural fields 3, 1, and 2. Anteriorly, SI merges with a transitional field 3a which separates area 3b from area 4 [FIG. V.134].

In area 3b two-thirds of the units are excited by light touch; in areas 1 and 2 most cells respond to 'deep' stimuli such as pressure and joint movements. None of the SI cells is influenced by nociceptive stimuli.

Areas 3a and 3b receive a dense input from the thalamus; these afferents terminate in layer IV and the deep part of layer III. Areas 1 and 2 receive fewer thalamic fibres [FIG. V.135].

The cortex of SI is supplied by afferents from the opposite side of the body but from both sides of the face. The body is represented upside down (as can be shown by recording cortical potentials in response to peripheral stimulation) and the extent of its representation depends on the animal species. The pig's snout, the cat's vibrissae and the monkey and man's hand and face have a large area of representation.

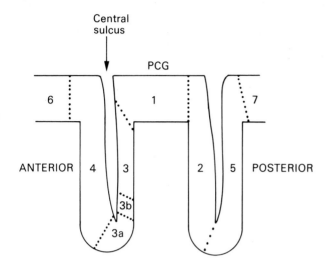

FIG. V.134. Fronto-occipital sagittal section of the 'hand area'. PCG = post-central gyrus.

Microelectrode penetrations of cells in SI show that a cell responds to physiological stimulation of one type of receptor only. When the penetrations are made perpendicular to the cortical surface most cells respond to the same type of stimulus. Mountcastle (1957) concluded that SI was organized functionally in a columnar fashion; this is consonant with the predominantly radial arrangement of axons and dendrites shown in Golgi impregnated preparations, and is similar in arrangement to that in area 4 (Asanuma and Rosén 1972). SII occupies the parietal cortex and is mostly buried in the superior bank of the Sylvian fissure. Unlike SI

it is supplied by afferents from both sides of the body. Like SI it manifests a dermatomal sequence of representation (although there is more overlap). The posterior region of SII receives afferents from the legs whereas the anterior part receives facial afferents; the 'face' area of SII is immediately adjacent to that of SI.

Recently, SII has been subdivided into an anterior and posterior division. Anteriorly, fibres from the ventroposterior thalamic nucleus (many of which are collaterals of fibres destined for SI) are prominent. Thus, when recording from VP thalamic cells antidromic responses can be elicited by stimulation of SI. About half these units can be excited by SII stimulation. Such VP units themselves can be shown to be naturally (orthodromically) responsive to peripheral tactile stimulation of appropriate parts of the body surface.

Cells in the posterior part of SII are responsive not only to tactile stimulation but additionally can be excited by auditory, visual, and nociceptive stimuli. These SII cells are supplied by the posterior group of thalamic nuclei whose cells receive lemniscal and spinothalamic inputs.

Area 5 of the parietal cortex contains neurons which have been recently studied in unanaesthetized monkeys trained to perform movements in response to sensory cues. Such neurons impaled by microelectrodes react to passive or active rotation of a joint or joints. Few respond to tactile stimuli. The major input to these area 5 cells seems to be from cells of areas 1 and 2 themselves excited by joint and deep tissue receptors (via thalamic relays). Like the other areas in SI and SII, area 5 displays a columnar organization.

Lastly, area 4—long regarded as the classical 'motor' area—contains cells which are activated by joint movements and muscle palpation. Some of these cells may receive an input from muscle spindles whose thalamic relays are as yet ill-defined (Jones and Powell 1976).

Association fibres interlink these areas (SI, SII, area 5, and area 4) involved in somatic sensation. SI and SII are reciprocally connected with each other and with area 4. SI projects to area 5 which in turn projects to area 7 and to area 6. Each of these areas receives thalamic and association fibres. Areas 3a and 3b are predominantly

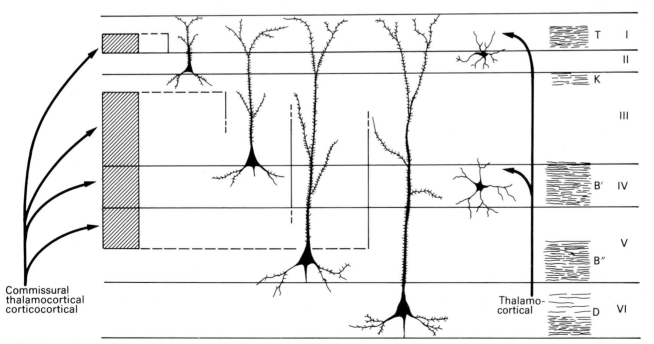

FIG. V.135. Schema showing laminar pattern, termination, and interpretation of sites of termination upon post-synaptic neurons of thalamic, commissural, and corticocortical fibres in area SI. Seventy-five per cent of thalamic afferents end on dendritic spines of pyramidal cells (left); 25 per cent end on stellate cells (right). On the extreme right are shown the horizontally disposed bands of intracortical association fibres. T=tangential plexus of layer 1; K=stria of Kaes; B′, B″ are the inner and outer bands of Baillarger. D=deep plexus of layer VI. (Redrawn from Jones, E. G. and Powell, T. P. S. (1972). In *Handbook of sensory physiology* (ed. A. Iggo) p. 605. Springer, Berlin.)

excited by thalamic fibres. Areas 1, 2, and 5 and SII receive more association fibres than from thalamic nuclei.

Commissural fibres connect the corresponding somatosensory areas with those of the opposite hemisphere. These commissural fibres, like the association fibres, are mostly axons of pyramidal cells situated in layer III. SI projects to the contralateral areas SI and SII, but SII only supplies its opposite counterpart.

All the somatosensory areas—particularly SI and area 4 send fibres to the caudate and putamen. SI sends fibres back to its own thalamic projection nuclei and additionally to the tectum, pons (and via pontine connections to the cerebellum). Furthermore, SI efferents project to the spinal segments, terminating in the dorsal part of the dorsal horn and interestingly to the gracile and cuneate nuclei. Labelling of the cells of origin in SI by retrograde axonal transport of horseradish peroxidase has helped to establish these pathways.

It is probable that each area of the somatosensory cortex is effecting important relays on the ascending somatic pathways thereby contributing to the cortical control of movement.

Parietal lobe ablation

Removal of the parietal cortex on one side causes a severe perceptual imbalance—there is a defective response to all stimuli that arise in the contralateral visual or somatic fields. The spatial extent of the stimulus delivered to the affected field has to be greater to enable it to compete for the perceptual process (Denny-Brown 1966). The significance of the defect in motor behaviour caused by such unilateral lesions can only be clearly assessed by making symmetrical bilateral lobe lesions. Woolsey and Bard (1936) noted that proprioceptive placing, which had been lost after unilateral parietal lobectomy, returned after removal of the other parietal lobe.

Lateral parietal lobe ablation profoundly disorders exploratory movements. Tactile and visual placing on an approaching surface are imperfect; the monkey cannot pick up small objects and feeds rather by lowering the mouth to the food. If the post-central gyrus (areas 3, 1, 2, 5) is bilaterally removed, visual placing is retained but tactile placing is lost, initially completely, slowly returning to an imperfect order of attainment. When blindfolded, the creature's hand and foot are inactive to tactile stimulation. Righting, tilting, and hopping reactions remain undisturbed, as are optic righting reactions, and visual placing is precise. In the monkey at least, visually directed movements of the hands and limbs and visual avoiding responses are not dependent on the integrity of the post-central gyrus. If the post-central gyrus lesion is extended to include the upper lip of the Sylvian fissure, avoiding responses to touch are exaggerated and explosive, and violent withdrawal occurs in response to a pinprick. The removal of this small piece of the most lateral part of the post-central gyrus has an effect similar to that of removal of area 7, whereas removal of the cortex posterior to the central fissure results in effects resembling those seen after removal of area 4.

Ablation of the posterolateral parietal cortex (particularly that part of area 7 which lies deep in the intraparietal sulcus) causes gross impairment of visual responses. A unilateral lesion results in pronounced neglect of the opposite perceptive field both visual and somatic. Bilateral ablation abolishes visual placing—coarse tactile placing is retained. Optic righting reactions are lost. There is an inability to make use of visual information. Two examples are constructional apraxia (inability to copy designs or objects) and spatial disorientation in which the patient cannot find his way about even in familiar surroundings.

Electrical stimulation in man.
Electrical stimulation of the exposed surface of the post-central gyrus in a conscious man leads to hallucinations of tactile stimulation, feelings of numbness or tingling or a sense of movement or pressure which are referred to the opposite side of the body; but there is never any complaint of pain. Irritation arising from disease usually produces similar symptoms. Stronger electrical stimulation may set up movements owing to impulses passing forward to the motor areas. A high degree of topographical representation of the different parts of the body can be demonstrated by electrical methods in the postcentral gyrus in man corresponding roughly to the motor localization in the precentral cortex, though more overlapping takes place.

Clinical evidence.
The post-central gyrus, whcih directly receives the afferent paths, is presumably mainly concerned with mediating the elementary sensations of touch, pressure, heat, cold. The more posterior part of the parietal lobe (superior parietal lobule, part of the supramarginal and angular gyri) is thought to be associated with the more elaborate processes of discrimination between stimuli and the recognition of common objects placed in the hand (*stereognosis*) without looking at them. Study of patients with disease or injury to the parietal lobe shows that the parietal cortex is concerned with the following aspects of sensation:

1. Differences in the relative intensity of different stimuli: heat is not merely distinguished from cold, but warm objects are distinguished from warmer, cold from colder, rough from rougher, and so forth.

2. Recognition of spatial relationships: (i) Tactile localization: the precise point stimulated is accurately located. (ii) Tactile (two-point) discrimination: two points of a compass placed close together are recognized as two and not as one. (iii) The extent and direction of small joint displacements can be estimated accurately. Thus the relations of a stimulus in one-, two-, or three-dimensional space are clearly defined.

3. Appreciation of similarity and difference in external objects brought in contact with the surface of the body (without the aid of visual impressions): differences and similarity of size, weight, form, and texture are thus recognized.

4. Stereognosis: this is the most elaborate function subserved by the parietal cortex. It necessitates perfect reception of the impulses set up by the stimuli from the object. The sensations produced are 'synthesized' in the cortex and compared with previous similar sensory 'memories'. We thus recognize an object as a fifty pence coin and distinguish it from a ten pence coin by the presence of the heptagonal edge (for which we deliberately feel).

It should be emphasized that the forms of sensation enumerated in paragraphs 1–4 above are lost or impaired in lesions of the parietal lobe. (The face area of the post-central gyrus is closely associated with the taste area.)

As gross forms of anaesthesia are rare in clinical lesions of the sensory cortex, casual examination of the sensory nervous system may reveal no abnormality. The patient must always be tested for stereognosis (which is almost constantly defective) and in the other ways indicated above. Prolonged examination at one time must be avoided, as the sensory cortex fatigues readily.

Motor changes in lesions of sensory cortex.
In lesions of the sensory cortex the patient's complaint is commonly of clumsiness (ataxy) and 'uselessness' of the affected side (which is on the side opposite to that of the diseased cortex). This finding again emphasizes the overwhelming importance of proper afferent guidance of the motor areas of the brain.

REFERENCES

ADRIAN, E. D. (1947). *The physical background of perception*. Oxford.
ASANUMA, H. and ROSÉN, I. (1972). *Expl Brain Res.* **14**, 243.
BURTON, H. and JONES, E. G. (1976). *J. comp. Neurol.* **168**, 249.
DENNY-BROWN, D. (1966). *The cerebral control of movement*. Liverpool University Press.

JONES, E. G. and POWELL. T. P. S. (1970). *Phil. Trans R. Soc.* **257**, 45.
—— —— (1972). In *Handbook of sensory physiology*, Vol. II (ed. A. Iggo) pp. 579–620. Springer, Berlin.
MOUNTCASTLE, V. B. (1957). *J. Neurophysiol.* **20**, 408.
—— (1978). *Proc. R. Soc. Med.* **71**, 14–28.
POWELL, T. P. S. (1977). In *Somatic and visceral sensory mechanisms* (ed. G. Gordon) *Br. med. Bull.* **33**, No. 2.
WERNER, G. and WHITSEL, B. L. (1973). In *Handbook of sensory physiology*, Vol. 2 (ed. A. Iggo) pp. 621–700. Springer, Berlin.
WOOLSEY, C. N. and BARD, P. (1936). *Am. J. Physiol.* **116**, 165.

Excitomotor areas, pyramidal tracts. Clinical hemiplegia

Excitomotor areas

The term 'excitomotor area' is generally applied to that part of the cortex of the frontal lobes which on stimulation gives rise to skeletal muscle responses. As this region lies anterior to the central sulcus it is commonly called the *precentral motor cortex*. It is divisible into several cytoarchitectonically distinct zones [FIG. V.136] with fairly distinctive functions.

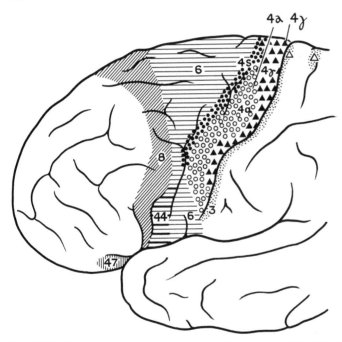

FIG. V.136. Diagram of precentral motor cortex in man. Area 4 is divided into two zones: 4γ, contains 'giant' Betz cells. 4α, contains no giant Betz cells. 47 = area 13 (Walker). △ = Betz cells in postcentral area. 4s = suppressor band. 8, contains frontal eyefield. (After Bonin, G. von, in Bucy, P. S. (1944). *Precentral motor cortex*. Thomas, Springfield, Ill.)

Area 4. This occupies almost the whole length of the precentral gyrus; it extends medially not only to the margin of the central sulcus but also over on to the medial surface of the hemisphere.

Area 4 is one of the main regions of origin of the pyramidal tracts. Stimulation of area 4 produces co-ordinated movements of the opposite side of the body. Histologically area 4 is divisible into two zones:

1. Area 4γ which lies most caudally [FIG. V.136] which is also known as the area giganto-pyramidalis owing to the presence of the giant cells of Betz in the fifth layer of this region (Betz cells are *not* found elsewhere).

2. Area 4a which constitutes the bulk of area 4 and lies most rostrally.

Area 6. This area contributes:

1. Descending fibres to the pyramidal tract.
2. Horizontal fibres which pass caudally to excite area 4 neurons.
3. Descending motor fibres which do not run in the pyramidal tract and are therefore, by definition, extrapyramidal.

Many of the pyramidal axons pass without relay to the spinal segmental levels where they form synapses with either internuncials in the dorsal horn or with the motoneurons themselves. The extrapyramidal fibres on the other hand have many synapses in their descending path with cells of the nuclear masses *en route*. The nuclei of the striatum (caudate and putamen), the globus pallidus, the hypothalamus, and the nuclei of the reticular formation are all the site of extrapyramidal fibre synapses. Just as is the case with the pyramidal fibres the extrapyramidal projections cause effects on the motoneurones of the opposite side of the spinal cord.

Some of the extrapyramidal projections from the cortex are excitatory, others are inhibitory to the motoneurons of the cord.

Stimulation of area 6 produces predominantly excitatory effects; complex co-ordinated movements of the opposite side of the body occur even when the pyramidal tracts have been destroyed, so these must be mediated by extrapyramidal pathways.

In addition to these descending fibres which influence the spinal motoneurones, areas 4 and 6 give origin to corticopontine fibres which relay via the nuclei pontis to the opposite neocerebellum, and to corticothalamic fibres which form synapses with the thalamic nuclei; from the thalamic nuclei axons pass back to the cortex.

Area 8 lies in the middle frontal convolution [FIG. V.136]. Stimulation causes movements of the eyes and the area is correspondingly termed the frontal eye field. Stimulation causes a conjugate deviation of the eyes to the opposite side. Corticonuclear fibres from area 8 descend first in the anterior limb of the internal capsule, then in the medial fifth of the pes pedunculi, to pass finally dorsally to the eye nuclei of the opposite side. Thus corticonuclear fibres pass from the left cortex to supply the right sixth nucleus which innervates the external rectus of the right eye.

If the right sixth nerve nucleus is itself stimulated both eyes deviate to the right; there is simultaneous contraction of both the right external rectus and the left internal rectus, although the latter muscle is innervated by the third nerve. Internuncial fibres establish connections between the homolateral sixth and third nuclei, whence fibres pass to the opposite third nerve nucleus and thence to the internal (medial) rectus.

Pyramidal tracts

These pass from the motor area to the spinal ventral horn cells and to all the motor cranial nuclei except those supplying the external eye muscles. The pyramidal cells and tracts constitute the upper motor neurons; the spinal and cranial motor neurons constitute the lower motor neurons. The pyramidal tract fibres to the spinal ventral horn cells constitute the corticospinal tract; the fibres to the motor cranial nuclei constitute the corticobulbar tract.

There are only 35 000 Betz cells in area 4 of the human motor cortex on one side. There are 1 000 000 fibres in the pyramidal tract, of which about 600 000 are myelinated. About 6 per cent of these are above 10 μm in diameter (36 000); these presumably represent the axons of the Betz cells.

The pyramidal tract does not arise exclusively from the areas 4 and 6. If areas 4 and 6 are ablated in the monkey many pyramidal fibres do not degenerate. Only if large frontal and parietal lobe ablations are additionally made do the remaining pyramidal fibres degenerate. These histological results have been confirmed by electrophysiological studies in which it was shown that antidromic stimulation of the medullary pyramids evoked potentials in both frontal and parietal lobes (in monkeys).

Course of the pyramidal tracts [FIG. V.137]

The pyramidal tracts converge from the precentral cortex through the corona radiata to reach the internal capsule. This is a mass of white fibres lying between the basal ganglia, limited laterally by the lenticular nucleus (putamen + globus pallidus) and medially by the caudate nucleus and thalamus. In horizontal section the internal capsule is V-shaped, the point of the V looking medially. The pyramidal tracts lie in the bend (the genu), and the anterior two-thirds of the posterior limb (occipital part). The fibres from before backwards are concerned with the control of head, shoulder, elbow, wrist, finger, trunk, hip, knee, and toe movements in the order named [FIG. V.138]. (It should be noted that extrapyramidal fibres, both excitatory and inhibitory are intermingled here with the pyramidal fibres.) Immediately behind the pyramidal tracts lies the condensed sensory path, and the visual path a little farther back. More posteriorly still lie the auditory fibres and the temporopontine tract. In the anterior limb (frontal part) are found fibres from the frontal lobes mainly (and also from the parietal and temporal lobes), to the basal ganglia, the eye nuclei, and other regions of the brain stem (especially the pons). Even a small injury here may produce widespread motor and some sensory disturbance.

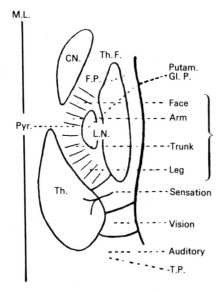

FIG. V.137. Horizontal section through the base of the brain to show the basal ganglia and the internal capsule (diagrammatic). M.L. = Middle line; Pyr. = Pyramidal tract; C.N. = Caudate nucleus; Th. = Thalamus; Th.F. = Thalamofrontal tract; F.P. = Frontopontine and Corticonuclear fibres; Putam. = Putamen; Gl.P. = Globus pallidus; L.N. = Lenticular nucleus; T.P. = Temporopontine fibres.

In the crus the pyramidal fibres lie ventral to the substantia nigra, occupying the middle three-fifths of this region. (The medial fifth carries the corticonuclear and frontopontine, and the lateral fifth the temporopontine, fibres.) In the pons the pyramidal fibres are broken up into a series of scattered bundles by the nuclei pontis and the crossing fibres of the brachium pontis. Stimulation of the cut surface of the crus and pons in the ape shows that there is well-marked localization of the pyramidal fibres for different parts of the body. From without inwards the order of the fibres is foot, hip and knees, abdomen and chest, fingers and wrist, face and tongue. Throughout the brain stem, the corticobulbar fibres are crossing to reach the motor cranial nuclei of the opposite side. In the medulla the corticospinal fibres reunite to form a compact ventrally projecting mass, the pyramid. The pyramidal tracts were so named because they were recognized as the tract in the pyramid (and not because they arise in pyramidal-shaped cells in the cerebral cortex).

In the lower part of the medulla, the main pyramidal decussation takes place. Most of the fibres cross over to the opposite side and dorsally, to come to lie in the lateral columns of the spinal cord as the crossed pyramidal (lateral corticospinal) tract. Some (about 15 per cent of the fibres) stay on their own side and in their original position and continue into the cord close to the anterior median fissure as the direct pyramidal (anterior corticospinal) tract. The crossed fibres ultimately connect directly and via dorsal inter-neurons with ventral horn cells.

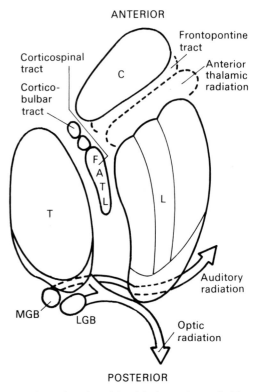

FIG. V.138. Horizontal section of the right internal capsule. T = thalamus; C = caudate; L = lenticular nucleus; MGB and LGB = medial and lateral geniculate bodies; F = face; A = arm; T = trunk; L = leg.

The direct pyramidal fibres do not as a rule extend beyond the lower cervical or midthoracic region.

Effects of stimulating motor area

Results in apes

In apes the motor area occupies the whole length of the precentral convolution; it extends backwards into the fissure of Rolando (sulcus centralis) (about one-third of the excitomotor area is 'buried' in the fissure), but never on to the postcentral convolution.

Faradic stimulation of this region causes movements on the opposite side of the body. The body is represented upside down in the cortex. Thus, stimulation of the most medial part of the precentral gyrus causes movements of the opposite leg, whereas stimulation in the most lateral region causes contractions of the facial muscles of the opposite side. The responses of the motor cortex to punctate faradic stimulation can be 'mapped out' as a 'simiusculus' in the ape. Such figurines have huge fingers and toes and huge jaws and tongues—very similar to the area of representation of these parts in the corresponding 'sensory cortex' in the postcentral gyrus.

A second somatotopic representation of the body musculature is found lying almost wholly on the medial wall of the hemisphere. This has been termed the 'supplementary motor area'. The face, neck, and tail are represented rostrocaudally along the cingular sulcus.

In view of the relative crudity of the experimental conditions it is remarkable that the stimulation of motor points in the cortex produces movements which are so suggestive of 'fragments' of voluntary movement. Stimulation of a motor point is not identical with stimulating the cells of origin of the pyramidal tract, for these lie in the fifth layer, separated by a huge mass of internuncial neurons from the surface.

Results in man

In patients under local anaesthesia, faradic stimulation of the motor area elicits responses closely resembling those described for the ape, i.e. discrete isolated movements (on the opposite side) of a single segment of an extremity or a single part of the trunk or head. The cortical representation is arranged as in the ape, except that separate foci exist for each of the fingers, and these occupy a relatively large area. The focus for the thumb is most inferior, and that for the fifth finger most superior [FIG. V.139]. Stimulation of the points controlling the upper part of the face, the pharynx, the vocal cords, and the muscles for closing the jaws, usually gives bilateral reactions. Electrical stimulation of the most inferior part of the precentral gyrus produces rhythmic co-ordinated movements of the lips, tongue, mandible, larynx, and pharynx. Epileptic attacks beginning in this area commence with the same type of movement—chewing, licking, swallowing, and grunting.

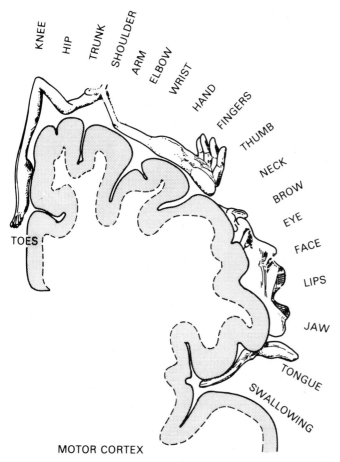

FIG. V.139. Coronal section through the cerebral hemisphere showing the 'motor homunculus'. Note the large area devoted to the thumb and fingers and to the face. (After Penfield, W. and Rasmussen, T. (1955). *The cerebral cortex of man*. Macmillan, New York.)

These results are obtained by *faradic* stimulation of the surface of the cortex. Though they differ from those obtained more recently using more discriminative techniques [p. 328] they nevertheless provide a sufficient basis of understanding for the interpretation of the effects of experimental and clinical lesions of the motor area and pyramidal tracts.

Results of destruction of motor area or pyramidal tracts

1. Extirpation of motor areas. In apes the left motor cortex was exposed and the arm area demarcated by electrical stimulation. The region yielding primary movements of fingers, wrists, and elbow was excised. A few hours later the opposite arm showed drooping of the wrist, weakness of the elbow, and to a less extent of the shoulder; no movement was possible in the fingers. One month later good recovery of most arm movements had occurred, but there was lasting decrease in the skill with which the fingers could be used. The operation area was re-exposed, but on electrical stimulation it yielded no response.

2. Section of pyramidal tracts. Removal of the 'arm' area destroys the cells of origin not only of the corticospinal fibres but of many extrapyramidal fibres also. To determine the role of the pyramidal tracts alone, the tracts must be cut at a level where they are not commingled with other descending tracts which subserve other functions. The medullary pyramid consists almost exclusively of corticospinal fibres.

Section of the medullary pyramid produces a relatively uncomplicated picture of the results of loss of pyramidal tract function.

(i) *Effects on movement.* In rhesus monkeys section of one medullary pyramid produces disturbances of movement closely resembling those resulting from extirpation of the motor area. There is marked weakness of the opposite arm and leg; the hand is most gravely affected, discrete movements of the fingers being lost. Simpler movements persist, e.g. striking, kicking, scratching, reaching, grasping, walking, but are less easily and accurately performed.

(ii) *Effects on tone and reflexes.* There is a decrease in muscle tone in the limbs (tone in the back, neck, and thorax is decreased very little). The deep reflexes are slow but not decreased. The superficial reflexes (e.g. abdominal, cremasteric) are abolished at first; later they return but are sluggish and weak.

A sharp contrast exists between the decreased muscle tone and reflexes following allegedly pure section of the pyramidal tract and the spasticity and exaggerated reflexes following clinical lesions which involve both pyramidal and extrapyramidal fibres.

In the chimpanzee, two 'release' phenomena develop after section of the medullary pyramid: (a) the Babinski plantar response makes its appearance [cf. p. 326]; (b) the proprioceptive grasp reflex can now be obtained: if an object is firmly pressed into the palm, it is reflexly grasped.

Role of pyramidal tracts in voluntary movement

The two outstanding facts described:

1. That artificial stimulation of 'motor' points produces movements which resemble fragments of a voluntary movement.

2. That impairment or loss of voluntary movements results from extirpation of the motor cortex or section of the pyramidal tracts, indicate that the pyramidal tract is a path which is employed in voluntary movement and in fact is indispensable for performance of certain types of movement. This statement does *not* mean that the motor cortex *initiates* voluntary movements. The pyramidal tracts may be regarded as a long internuncial pathway, linking the great afferent pathways which end in the cerebral cortex (especially those from the 'distance receptors' (eyes, ears), and from the muscles

(directly and via the cerebellum) with the lower motor neurons (cranial and spinal).

Clinical lesions of the pyramidal tracts

Throughout most of its course the pyramidal tract is commingled with other descending tracts. In man, injury to the pyramidal tract in the brain gives rise to weakness or paralysis on the opposite side of the body; the condition is called *hemiplegia*. The symptoms and signs in any individual case depend on the severity of the lesion, whether it is acute or chronic, and on the extent to which the other descending (extrapyramidal) pathways are involved. In general there is both weakness and loss of voluntary movement associated with *spasticity*. There is no constant relationship between the severity of loss of willed movement and the degree of hypertonus present.

Hemiplegia

Acute lesions in man. An acute lesion in the internal capsule, where the pyramidal tracts are condensed in a small area, produces *initially* a stage of shock in which all the muscles of the opposite side are toneless and in which no reflex movements can be elicited; in these respects, the condition resembles spinal shock [p. 362]. These findings again emphasize the degree to which the higher levels of the nervous system participate in so-called spinal and brain stem reflexes. Some two or three weeks later reflex activities return to the affected side of the body, showing that such reflexes can be mediated by the isolated lower levels.

The paralysis on the affected side affects voluntary movements of the face, leg, and arm. Movements which usually involve both sides of the body, e.g. those of respiration, movements of the back, and abdominal wall, are generally retained; it would seem that one pyramidal tract can control the lower motor nuclei supplying the muscles concerned on both sides. If the corticonuclear fibres escape injury, eyeball movements persist. Emotional movements remain intact; frequently they are elicited very readily and are exaggerated in character. The patient cannot voluntarily wrinkle up his forehead, frown, or whistle, but the muscles concerned in these acts are employed perfectly efficiently when the patient's face exteriorizes pleasure, surprise, or annoyance.

Stage of recovery. Chronic lesions. As the stage of shock passes away, or in chronic lesions, the clinical picture as explained is that of paralysis and disturbances of tone and reflexes. The common clinical syndrome which is described below is attributable to a combined lesion of the pyramidal tract and the descending inhibitory extrapyramidal pathways.

1. MUSCLE TONE AND POSTURE. Muscle tone is abnormal in distribution and excessive in degree in certain muscles, i.e. the limbs are placed in an abnormal position and tend to be fixed there; to this latter feature the term *spasticity* is applied. The upper limb, instead of lying at the side in the customary way, is adducted at the shoulder, the elbow is semi-flexed, the forearm is pronated, and the wrist and fingers are flexed. The limb is involuntarily maintained in the unnatural position indefinitely without fatigue; it can only be moved passively with difficulty. The leg is adducted, extended at the knee, and physiologically extended at the ankle (anatomical plantar flexion). The muscles do not waste greatly, because though not used in voluntary movement they are continuously in action to maintain the posture described.

2. ASSOCIATED REACTIONS. This term is used to describe certain movements which can be reflexly aroused on the affected side by such 'semi-involuntary' movements as yawning and stretching and by any forceful sustained voluntary muscular contraction on the normal side; the responses are modified by the position of the head

relative to the trunk, i.e. by neck reflexes. Thus in the standing hemiplegic subject, looking straight ahead, when the normal fist is clenched, the spastic arm moves slowly into increased flexion at elbow, wrist, and digits. If the procedure is repeated with the head rotated towards the hemiplegic side, the spastic arm moves into extension and abduction, the fist remaining closed.

3. DEEP REFLEXES. The deep (i.e. tendon) reflexes are modified in a characteristic way. The ones commonly elicited are the knee jerk and the ankle jerk in the lower limb, and the triceps and supinator jerk in the upper limb. The deep reflexes are simply fractionated stretch reflexes, and their general characteristics can be studied in the case of the human knee jerk.

Knee jerk. With the knee supported in the flexed position, the patellar tendon is sharply tapped; the muscle is stretched, the muscle spindles are stimulated, and impulses pass up into the third and fourth lumbar dorsal nerve roots to enter the spinal cord and end directly round ventral horn cells. Efferent impulses pass out in the lumbar ventral nerve roots to produce contraction of the quadriceps extensor and at the same time reciprocal inhibition of the antagonistic hamstrings takes place.

The deep reflexes, like the stretch reflexes, *usually* vary with the degree of existing muscle tone. In hemiplegia there is commonly increased extensor tone. The knee jerk (like the other deep reflexes) is exaggerated, and what is even more characteristic, it is more sustained, i.e. relaxation is still further prolonged.

Ankle clonus is often present, and is also due to the state of heightened muscle tone. When the ankle is forcibly dorsiflexed, rhythmic movements of dorsiflexion and plantar flexion take place at the ankle joint. Owing to the increased tone of the posterior calf muscles, stretching of the fibres sets up not a simple reflex contraction or ankle jerk, but a rhythmic series of contractions or clonus. This is doubtless due to a synchronous (instead of the more usual asynchronous) discharge of the ventral horn cells, leading to a tremulous partial tetanus. Sharp displacement of the patella may induce clonic contractions of the stretched quadriceps (patellar clonus).

4. SUPERFICIAL REFLEXES. The superficial reflexes commonly tested are:

(i) Abdominal: stroking the skin of the abdomen produces contraction of the underlying muscle.

(ii) Cremasteric: stroking the inner side of the thigh results in the testis being retracted towards the inguinal canal.

(iii) Plantar: stroking the sole of the foot produces a downward movement (plantar-flexion) of the great toe and the small toes.

All the superficial reflexes are lost on the affected side in cases of hemiplegia. As these reflexes are also lost when the medullary pyramid is cut, they are mediated by the pyramidal tract; thus their loss in hemiplegia is due to the injury to the pyramidal tract.

(iv) Babinski response: in hemiplegia an abnormal reflex can be elicited from the sole of the foot, *different in every respect from the usual plantar response*. The stimulus employed must be nocuous or painful, e.g. firm scratching with the finger-nail. The reflex can be elicited from a fairly wide area, from the calf, the thigh, and even as high as the groin; it consists first of an upward movement (dorsiflexion) of the great toe and fanning out of the small toes. The anatomists misleadingly call dorsiflexion of the great toe 'extension'. The abnormal toe response is therefore sometimes called the 'extensor response'; this term should not be used and the response should be described (after its discoverer) as the Babinski sign or the Babinski response. The Babinski response is a 'fraction' of a released spinal flexor or withdrawal reflex. Thus, in some cases of hemiplegia, nocuous stimulation of the sole of the foot produces a reaction extending far beyond the toes; there may be dorsiflexion (upward movement) of the ankle, flexion of the knee, and even flexion of the hip. The more extensive reaction is identical in every way with the spinal flexor reflex. The Babinski response is a release phenomenon; as it appears after isolated destruction of the medullary pyramid, its occurrence in man must be attributed to a lesion of the pyramidal tract.

5. DISTURBANCES IN VOLUNTARY MOVEMENT. In a recovering case of hemiplegia, considerable improvement occurs in the leg and the patient may walk about with little more disturbance than a slight limp; a good deal of power returns to the arm and face. The incidence of permanent loss of movement, in order of severity, is as follows: movements of the hands and especially the digits; extension of wrist; supination of the forearm; abduction and elevation of the upper arm; dorsiflexion of the foot and toes; flexion of the proximal leg joints. The recovery may be attributed to restoration of function to temporarily damaged pyramidal fibres or to the more effective use of extrapyramidal pathways arising in the cortex.

Walshe has analysed the loss of movement in the arm and hand in progressively developing hemiplegia. Paresis first appears in the movements of the hand and digits and spreads so as to involve the proximal part of the limb. The first movements to disappear are those in which the interossei, lumbricales, and the flexors and opposers of the thumb are involved. Abduction and extension of the fingers are affected before adduction. Certain combinations of movements disappear, as is well shown by asking the patient to touch each finger-tip in succession with the tip of the thumb; there is a stage in the development of the hemiplegia in which, while each digit can be flexed, the necessary movements of adduction and apposition are lost and the thumb and finger-tips are not effectively approximated 'but flex futilely into the palm'. The limb is progressively 'denuded of movements' and therefore of movement combinations.

Modern concepts of pyramidal function

The classical somatotopic representation of the body 'upside down' in the motor cortex is only obtained in experiments in which faradic stimulation is used. If the motor cortex is stimulated with single surface-cathodal pulses of 5 ms duration and a weak current strength, it is commonly found that only three regular responses may be obtained, i.e. movements of the thumb and index finger, movements of the big toe, and movements of the angle of the mouth. Of these responses, that of the thumb can be evoked at the lowest threshold of stimulation [FIG. V.140]. If the current strength is increased, this response can be induced by the stimulation of an increasingly wide area of the motor cortex. The movements of the big toe and angle of the mouth which can be elicited from discrete areas, separate from that responding to low-threshold stimulation to produce thumb movement, can in turn be evoked from widespread areas if the current strength is increased. The areas for thumb, toe, and mouth response overlap when the current strength is increased and in these circumstances all three movements may result from a single stimulus of one cortical point (Liddell and Phillips 1950). Many years ago Hughlings Jackson pointed out that an epileptic fit usually begins either in the hand, the angle of the mouth, or in the foot.

EFFERENT INFLUENCES ON THE ANTERIOR HORN CELLS FROM SUPRASPINAL STRUCTURES

Corticofugal fibres. Fetz (1968) showed that *corticospinal fibres* were distributed to the spinal laminae IV, V, VI, and VII. Effects of such cortical projections were predominantly inhibitory in lamina IV and predominantly excitatory in lamina VI. It will be remembered that lamina IV cells provide the main contribution to the dorsolateral ascending tracts [see FIG. V.84]. There is an important difference between the distribution of corticospinal neurons from the somatosensory area of the cortex and those from the precentral motor cortex. The somatosensory corticospinal fibres end dorsomedially in laminae IV to VI where they could influence the transmission of afferents from muscles, skin, and joints, not only by modifying their segmental effects but also by influencing the transmission of impulses in these cells whose axons travel via the spinocerebellar and dorsolateral tracts. The 'motor' area corticospinal neurons are distributed predominantly to the ventrolateral parts of laminae V to VII, close to the interneurons which send their short axons to the anterior horn cells. Such corticospinal fibres could influence exteroceptive and proprioceptive reflex transmission and motor output.

Phillips and Porter (1964) recorded intracellularly from a motorneuron of the median nerve of the baboon. They showed that single surface-anodal pulses of 0.2 ms duration delivered to the lowest threshold (thumb–index finger) area of the opposite precentral gyrus induced α motor neuron EPSP responses of a graded nature as the stimulus strength increased [FIG. V.141]. This growth of the EPSP induced by increasing the stimulus strength indicates that the α motoneuron itself receives synapses from more than one corticospinal cell; the attainment of maximal EPSP indicates that all the cells connected monosynaptically to it are being stimulated. Phillips and his colleagues have designated such a collection of cells as a 'colony'. This *direct* innervation of spinal motor neurons by pyramidal neurons is *not* seen in the cat; one or more interneurons are intercalated.

FIGURE V.142 shows the mean amplitudes of maximal monosynaptic EPSPs evoked by cortical stimulation in motoneurons (all of which were identified by antidromic stimulation of the relevant nerves). EPSPs evoked from motoneurons of the small muscles of the fingers and those from the long extensors of the fingers were greater than the others. This is consonant with our clinical experience that damage to corticospinal neurons profoundly impairs isolated movements of the fingers; *extension* of the fingers is almost as severely affected. Similarly, pyramidotomy in the baboon destroys the precision patterns of digital performance.

Landgren *et al.* (1962) recorded from motoneurons supplying the muscles of the hand and then stimulated the motor cortex with single surface-anodal stimuli. Pure EPSPs which resulted were identical in shape and time course with those aroused by monosynaptic excitation of the appropriate group IA afferents [FIG. V.143]. Thus, some corticospinal neurons can directly influence the background on which the muscle spindle afferents exert their

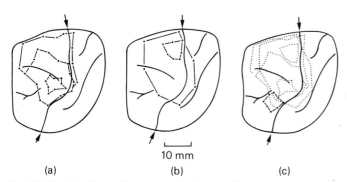

10 mm

(a)   (b)   (c)

FIG. V.140. Left Rolandic region of baboon, with arrows to mark the Rolandic fissures. Precentral gyrus to left of fissure. Stimulation by single surface-cathodal rectangular pulses, duration 4.5 ms. (a) Extent of area for flick movements of thumb, index, and little finger at strengths 1.2, 1.6, and 2.7 mA. (b) Area for hallux and middle toe, strengths 2.05 and 4.7 mA. (c) Area for angle of mouth, 2.35 mA. Areas of (a) and (b) shown by dotted lines. (Redrawn from Liddell, E. G. T. and Phillips, C. G. (1950). *Brain* **73**, 125–40.)

effect. By moving the focal anode to different sites of the cortex, Landgren *et al.* could alter the proportions of EPSP and IPSP in the same α motor neurons; IPSP was attributed to the excitation of inhibitory interneurons interposed between the corticospinal axon terminals and the motor neuron impaled.

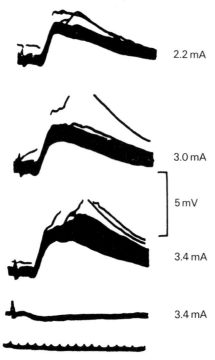

FIG. V.141. Superimposed intracellular records of monosynaptic EPSPs evoked in motoneuron of median nerve (membrane potential −74 mV) by 0.2 ms, surface-anodal pulses to lowest-threshold area of opposite precentral gyrus. Strength (mA) noted beside each record. Note that the EPSP increases as current strength is raised. Lowest record—extracellular (electrode withdrawn from cell). Calibrations: amplitude, see 5.0 mV scale; time, 1000 s⁻¹. (From Phillips, C. G. and Porter, R. (1964). In *Physiology of spinal neurons* (ed. J. C. Eccles and P. Schade) *Progress in Brain Research*, Vol. 12, pp. 222–42. Elsevier, Amsterdam.)

In addition to these influences of the corticospinal tract (CST) on α-motoneurons, Fidone and Preston (1969) showed that fusimotor (γ) neurons [p. 298] responded to CST stimulation. CST stimulation usually inhibited extensor fusimotors but excited most of the flexor fusimotor neurons.

Besides corticospinal influences, the segmental anterior horn cells receive fibres—probably relayed by segmental internuncial neurons from reticulospinal tracts and from tracts, though labelled rubrospinal and tectospinal, can be regarded as components of the descending reticular system [p. 307]. Vestibulospinal pathways cause a preponderant facilitatory effect on spinal motor neuron activity. Note that the cerebellar nuclei [see p. 309] which *inhibit* the activity of the vestibular nuclei normally serve to check this vestibulospinal excitation of the spinal motor neurons.

Intracortical microstimulation

Asanuma and his colleagues have developed the use of brief shocks to intracortical neurons delivered by glass-coated tungsten micro-electrodes. They used a pair of such electrodes, one for stimulation and the other for extracellular recording of pyramidal tract neurons. In each of these electrodes the sharpened tungsten wire projected approximately 10 μm from the glass. The electrodes were parallel and were micromanipulated independently along parallel axes within the cat's motor cortex. A pyramidal tract neuron (PTN) was identified by antidromic stimulation of the contralateral medullary pyramid and the recording microelectrode was then left *in*

situ. The stimulating electrode (microcathode) was then used to measure the threshold for single pulses of 0.2 ms duration delivered at different depths from the surface of the cortex. The success of direct stimulation was verified when the impulse was obscured by the stimulus artefact, by collision between the orthodromic impulse and a suitably timed antidromic impulse. The latency of responses to direct stimulation was less than half a millisecond.

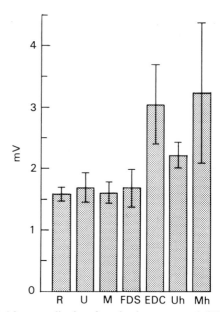

FIG. V.142. Mean amplitudes of maximal monosynaptic EPSPs evoked by corticospinal volleys in motoneurons identified by antidromic stimulation of whole interosseous nerve ('radial', R), whole ulnar (U), whole median (M) (including four palmaris longus motoneurons) and nerves to flexor digitorum sublimis (FDS), extensor digitorum communis (EDC), ulnar nerve to intrinsic muscles of hand (Uh), and median nerve to intrinsic muscles of hand (Mh). (From Clough, J. F. M., Kernell, D., and Phillips, C. G. (1968). *J. Physiol., Lond.* **198**, 145–66.)

When the threshold was less than 10 μA, histology showed the tip of the microelectrode to be within the PTN layers of the cortex. Phillips and Porter (1977) provide details of the techniques and the problems of interpretation associated with these methods.

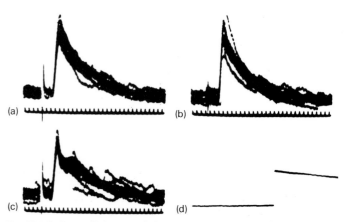

FIG. V.143. Intracellular records from α motoneuron of deep radial nerve of baboon. (a) EPSP evoked by unifocal surface-anodal pulses, strength 1.25 mA, applied to the 'hand' area of the cortex. Time, 1000 Hz in (a), (b), and (c). (b) Group 1A monosynaptic EPSP. (c) Same pulses applied to cortex at a point 3 mm distant from point stimulated in (a). Note smaller EPSP and inhibitory erosion of its decaying phase. (d) Calibrating 1.5 mV rectangular voltage step to amplifier input. (Landgren, S., Phillips, C. G., and Porter, R. (1962). *J. Physiol., Lond.* **161**, 91–111.)

Asanuma and Rosén (1972) mapped the buried 'arm area' of the motor cortex in monkeys conscious and capable of spontaneous movement but tranquillized by small doses of pentobarbitone. The cortex was first mapped with trains of 0.5 mA pulses delivered to the surface to find the thumb and finger area. This was then probed with the tungsten–glass microcathode and stimulated at 333 Hz, 0.5 ms duration. Motor thresholds at different points ranged from 2–10 µA. Responses of the thumb and fingers were recorded by stroboscopic photography. FIGURE V.144 shows the points that were stimulated in one parasagittal plane and the thumb movements that were obtained from the effective points. The figure shows the general direction of the radially oriented architecture of the cortex and suggests that the effective points for thumb extension (for instance) are disposed in a columnar arrangement along the axis of the radial bundles.

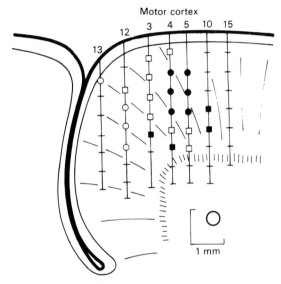

FIG. V.144. Microstimulation of effective spots within the depth of the cortex, strength 10 µA. Points marked by horizontal bars along the electrode tracks evoked no responses. Open circles, flexion of thumb; open squares, extension of thumb; filled circles, adduction of thumb; filled squares, abduction of thumb. Circle within millimetre scales has been added to indicate maximum probable radius of physical spread of 10 µA ($k = 272$ µA mm^{-2}). (Redrawn from Phillips, C. G. and Porter, R. (1977). *Corticospinal neurons*. Academic Press, London.)

Phillips and his colleagues—e.g. Andersen *et al.* (1975)—investigated the arm area of a baboon using a combination of surface-anodal and microcathodal stimulation; the baboons were lightly anaesthetized and fine EMG electrodes were placed in the first dorsal interosseous muscle, in a thumb muscle, and in extensor digitorum communis. They were able to determine the threshold for single motor units of these muscles at different track depths [FIG. V.145].

Recordings of PTN activity during movement

A major advance in understanding the role of the cortex and the pyramidal tract in movement derived from the study of recordings of the activity of PTN themselves (Hardin 1965; Evarts 1965, 1966, 1968, 1969); these are summarized in Evarts (1979). Only a few of these important experimental results will be quoted here. In brief, Evarts recorded the nerve impulse discharges produced by identified PTN cells in conscious monkeys which had been trained to carry out a number of movement tasks. Identification of PTN neurons was achieved by recording responses to antidromic stimulation of the medullary pyramid. The stimulating electrodes were implanted under anaesthesia and a bony defect in the cranium was

created. This allowed the subsequent penetration, weeks after the recovery of the animal from the initial operation, of the motor cortex by extremely fine microelectrodes micromanipulated through the dura. Once the electrode tip was sufficiently adjacent to a cortical cell to record its discharges selectively, medullary pyramidal (antidromic) stimulation tested whether this was a PT

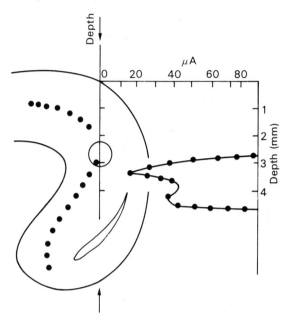

FIG. V.145. Tracing of parasagittal section of right arm area of baboon showing track of stimulating microcathode. Position of some Betz cells are marked by dots. The figure shows on the right a plot of the threshold against depth for a single motor unit of extensor digitorum communis. (Note that the threshold was lowest (10 µA, six 0.2 ms pulses at 500 Hz) at tip point 3.5 mm depth.) (Redrawn from Phillips, C. G. and Porter, R. (1977). *Corticospinal neurons*. Academic Press, London.)

cell or not. The antidromic responses occurred with latencies of less than 5 ms; axonal conduction velocities were usually greater than 10 m s^{-1}, most commonly about 60 m s^{-1} and some as high as 80 m s^{-1}. Obviously, the technique used renders successful recording from an impaled cell more likely if the cell be large, with a correspondingly faster conducting axon. Phillips and Porter (1979) remind us that in the macaque monkey 85 per cent of the PTN are 1 µm or less in diameter and that only 2–3 per cent of the tract fibres exceed 6 µm.

Evarts did identify two populations of PTN. Cells with antidromic latencies of <1 ms with conduction velocities of 50 m s^{-1} or more were either silent in the resting animal or showed only sparse discharge. Cells with slower conduction velocities (longer response latencies) showed more tonic discharge in the resting animal. When the trained animal moved the right arm, a PTN cell in the left 'arm area' previously silent emitted a burst of impulses some 100 ms *before the movement began* [FIG. V.146].

For most PTN studied the discharge frequency of the cell was related primarily to the force and the changes of force with time required to effect the movement.

Further experiments examined Betz cell activity in monkeys that had been trained to react to involuntary movement of their arm. The monkey was first taught to set and maintain a given position of a handle (or to move it with precision). Then a coloured lamp was switched on and the monkey took the colour as a cue as to how to react to an impending displacement of the handle. A red light meant that the handle should be pulled back and a green light required that the monkey pushed the handle forward. After training, some 40 ms sufficed for the monkey to respond correctly. This is an interesting example of the role of sensory cortical structures in relaying signals which control the Betz cell output. Another striking example is provided by the recording of activity in the ventrolateral thalamic neurons which occurs prior to movement (Evarts 1970).

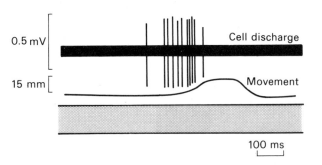

FIG. V.146. Left cortical PTN cell discharge. The monkey grasped a horizontal lever and pulled it forwards 15 mm with its right hand. The movement is recorded on the second trace. The top trace shows that the neuron begins firing about 100 ms before the movement. (Redrawn from Porter, R. (1973). *Prog. Neurobiol.* **1**, 1–51.)

Influences of pyramidal tract activity

Antidromic stimulation delivered to the medullary pyramid and thus of the PT axons reveals that there are cells with 'fast' (20–70 m s⁻¹) and 'slow' (< 20 m s⁻¹) axons. 'Slow' cells can exert a direct recurrent facilitatory action. These slow cells outnumber the fast cells.

Cutaneous, auditory, and visual inputs converge on the PT cells. Low threshold muscle afferent volleys from the forelimb evoke both EPSP and IPSP responses from non-pyramidal neurons adjacent to the PT cells. It is possible that these low threshold responses (from the muscle spindles of the forelimb) may serve as an input to the sensorimotor cortex probably via area 3b, which controls PT output.

The dentatothalamic tract, which excites neurons of the ventrolateral nucleus of the thalamus, causes excitation of the 'fast' PT cells by monosynaptic relays of this thalamocortical connection and excitation of the 'slow' PT cells by polysynaptic relays.

In addition to exciting α motor neurons directly, the PT excites the interneurons of the polysynaptic reflex arcs, such as the inhibitory pathway from the spindle primary afferents of one muscle to the motoneurons of its antagonists and the interneurons in flexor arcs originating from skin afferents, group III and IV muscle afferents and joint afferents. When anterior horn motoneurons are impaled by the microelectrode, it can be shown that the eventual polysynaptic influence of PT stimulation is predominantly to cause EPSP responses in flexor motoneurons and IPSP responses in extensor motoneurons.

Pyramidal tract stimulation can produce presynaptic inhibition by depolarizing cutaneous afferent nerve endings, Golgi tendon organ afferents, and Group II muscle afferents. It cannot inhibit transmission from the primary muscle spindle afferents in this manner.

The large cell neurons of the red nucleus seem to be a meeting-point between cortical, cerebellar, and spinal cord neurons. The collaterals of 'fast' PT axons induce IPSP responses in RN cells by polysynaptic pathways whereas the 'slow' PT fibres induce monosynaptic EPSP responses. (Note that PT recurrent collaterals induce similar actions on the PT cells themselves.) These same RN cells can be excited from the nucleus interpositus of the cerebellum.

'Slow' PT axons excite the *climbing* fibres which, arising in the ventral lamella of the inferior olive, act as an excitatory input to the Purkinje cells of the cerebellum; (the Purkinje cells in turn act as an inhibitory output to the dentate nucleus and hence the dentatothalamocortical pathway). As stated above, this last pathway causes excitation of the 'fast' PT cells by monosynaptic thalamo-

cortical connections and of the 'slow' PT cells by polysynaptic relays.

'Fast' PT axons on the other hand excite the pontine nuclei, which give rise to some of the mossy fibre afferents to the cerebellar cortex.

These generous connections render the 'comparator' function of the cerebellum available to the sensorimotor cortex during the sequence of a movement (Phillips 1971).

PT axons provide collaterals to the gracile and cuneate nuclei, trigeminal nuclei, and the VPL nucleus of the thalamus [FIG. V. 147]. These collaterals presumably can adjust the gain and discrimination of somaesthetic pathways destined for the cortex and provide for comparison of afferent input with efferent 'commands'.

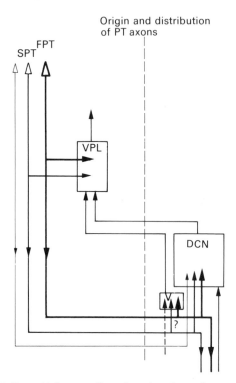

FIG. V.147. Pyramidal axon collaterals and corticonuclear axons to somaesthetic relay nuclei. (From Phillips, C. G. and Porter, R. (1977). *Corticospinal neurones. Their role in movement.* Academic Press, London.)

REFERENCES

ANDERSEN, P., HAGAN, P. J., PHILLIPS, C. G., and POWELL, T. P. S. (1975). *Proc. R. Soc.* **B 188**, 31–60.
ASANUMA, H. and ROSÉN, I. (1972). *Expl Brain Res.* **14**, 243–56.
CLOUGH, J. F. M., KERNELL, D., and PHILLIPS, C. G. (1968). *J. Physiol., Lond.* **198**, 145–66.
—— PHILLIPS, C. G., and SHERIDAN, J. D. (1971). *J. Physiol., Lond.* **216**, 257–79.
EVARTS, E. V. (1965). *J. Neurophysiol.* **28**, 216–28.
—— (1966). *J. Neurophysiol.* **29**, 1011–27.
—— (1968). *J. Neurophysiol.* **31**, 14–27.
—— (1969). *J. Neurophysiol.* **32**, 375–85.
—— (1970). *Physiologist* **13**, 191–8.
—— (1979). In *The brain*, Chap. VIII, pp. 98–107. Scientific American.
FETZ, E. E. (1968). *J. Neurophysiol.* **31**, 69–80.
FIDONE, S. J. and PRESTON, J. B. (1969). *J. Neurophysiol.* **32**, 103–15.
HARDIN, W. B. (1965). *Archs Neurol. Psychiat., Chicago* **13**, 501–12.
LANDGREN, S., PHILIPPS, C. G., and PORTER, R. (1962). *J. Physiol., Lond.* **161**, 112–15.
LIDDELL, E. G. T. and PHILLIPS, C. G. (1950). *Brain* **73**, 125–40.
PHILLIPS, C. G. (1969). *Proc. R. Soc.* **B173**, 141–74.
—— (1971). *Proc. 3rd Int Cong. Primat.*, Zurich, Vol. 2, pp. 2–23. Karger, Basel.
—— (1973). *Proc. R. Soc. Med. V* **66**, 987–1002.

—— (1975). *Can J. neurol. Sci.* **2**, 209–18.

—— and PORTER, R. (1964). In *Physiology of spinal neurons* (ed. J. C. Eccles and J. P. Schadé) *Progress in Brain Research*, Vol. 12, pp. 222–42. Elsevier, Amsterdam.

—— —— (1977). *Corticospinal neurones. Their role in movement*. Academic Press, London.

PORTER, R. (1972). *Brain Res.* **40**, 39–43.

—— (1973). *Prog. Neurobiol.* **1**, 1–51.

SEESLE, B. J. and WIESENDANGER, M. (1982). *J. Physiol., Lond.* **323**, 245.

CORTICAL PARTICIPATION IN THE MUSCLE SERVO LOOP

Vallbo (1973) recorded discharge in primary spindle afferents (from a finger flexor) in the median nerve of man. When the finger was flexed the afferents increased their discharge at approximately the same time as the muscle began to contract [FIG. V.148]. This discharge was sustained when the muscle was allowed to shorten freely in a ramp movement, suggesting that there was a continuous increase in fusimotor outflow [FIG. V.149]. When the isotonic movement was rapid (ballistic movement) spindle unloading occurred and the primary ending did reduce its discharge.

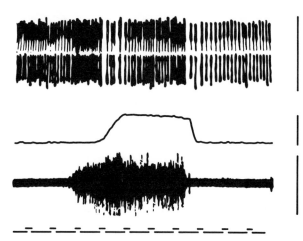

FIG. V.148. Response of a spindle primary ending to an isotonic contraction. From the top shown the single unit impulses, the angle at the metacarpophalangeal joint and the EMG activity when the subject flexed his ring finger. The flexion was opposed by a load corresponding to a torque of 0.1 Nm at the metacarpophalangeal joint. Calibrations: 100 µV, 150° (bottom) and 140° (top), 0.2 mV. Time signal: 1 s. (From Vallbo, Å. B. (1973). In *New developments in electromyography and clinical neurophysiology* (ed. J. E. Desmedt) Vol. 3, pp. 251–62. Karger, Basle. By courtesy of the publishers.)

The original 'follow-up length' servo proposed by Merton [p. 300] cannot account for these results in that spindle afferent discharge accelerated during *but not in advance of the* α *motorneuron activity which initiated the movement*.

Marsden, Merton, and Morton (1972) investigated the role of stretch reflexes during voluntary contractions of flexor pollicis longus made by human subjects. The muscle supplies only the terminal phalanx of the thumb. It arises from the upper two-thirds of the front of the radius—its fleshy belly (containing its spindles) lies deeply in the forearm. From this muscular belly a rounded tendon issues which, traversing the carpal tunnel and the palm, runs laterally to the thumb to be inserted into its terminal phalanx. Marsden *et al.* arranged that the proximal phalanx was held in a clamp so that the axis of rotation of the top joint coincided with the axis of a low-inertia electric motor. The spindle of the motor carried a potentiometer at one end which recorded thumb position and at the other end an arm against which the pad of the thumb pressed to move or to be moved by the motor. This arm incorporated a strain

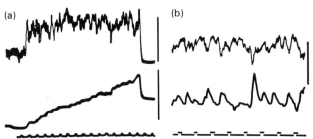

FIG. V.149. Response of a spindle primary ending during an isotonic contraction. Same unit as in FIGURE V.148. (a) Relation to joint angle. The top trace represents the impulse frequency of the single unit and the lower trace the angle at the metacarpophalangeal joint when the subject slowly flexed his ring finger. The events associated with the second half of this contraction are also illustrated in (b). Calibrations: 0 and 25 imp s⁻¹, 155° (bottom) and 145° (top). Time signal: 1 s. (b) Relation to the speed of joint movement. The upper trace shows the single unit impulse frequency and the lower trace the time derivative of the joint angle signal and, hence, it represents the speed of the joint movement. Calibrations: 0 and 25 imp s⁻¹. Time signal: 1 s. (From Vallbo, Å. B. (1973). In *New developments in electromyography and clinical neurophysiology* (ed. J. E. Desmedt) Vol. 3, pp. 251–62. Karger, Basle. By courtesy of the publishers.)

gauge which measured the force exerted by the thumb. The degree of contraction of the muscle was recorded by a pair of surface electrodes placed over the muscle belly in the forearm; e.m.g. activity aroused was amplified and integrated and the degree of contraction of the flexor was recorded by the slope of this integrated record. A standing current in the motor held the motor back against a stop and thus defined the starting position of the thumb. The subject made standardized flexion movements by being given a tracking task on a cathode ray tube. By keeping two spots together he executed a 20° movement in 1.2 s. All movements were begun against a constant force offered by the standing current in the motor but about half-way through the movement perturbations were suddenly introduced. Four situations were selected: the movement continued to the end against a constant load (control—*C*) or abruptly halted by an increase in load (*H*) or driven back by imposing a sudden force which elicited a stretch reflex (*S*) or allowed to accelerate by suddenly reducing the load (release response *R*). FIGURE V.150(a) shows the results—each of the four situations was recorded 16 times and the record traces show the average. Recording lasted for 250 ms during the mid-range of the movement and the various perturbation were applied 50 ms after beginning recording. Such perturbations all caused an alteration of muscle activity after approximately only 50 ms. A reduction of the load is followed by a decrease in activity compared with that during control whereas halting or reversing the movement (*H* and *R*) increased muscle force. This latency of 50 ms is far too short for these responses to be voluntary attempts at compensation—such voluntary compensatory efforts require 140–150 ms—and they are therefore indicative of a true automatic servo mechanism which is based on the stretch reflex—a reflex which manifests just such a latency.

When the magnitude of the initial load is increased the size of these servo responses increases *pari passu* and vice versa. Thus, a sudden extension of a relaxed thumb causes little if any stretch response—zero gain. The force of the corrective responses is adjusted to match the delicacy of the movement required—i.e. the 'gain' of the servo is proportional to the load [FIG. V.151].

As the number of α motoneurons active does increase with force and as all α motoneurons receive a spindle input, Marsden *et al.* (1972) inferred that the spindle input does not itself lead to excitation of α motoneurons unless the gain input is active at the same time. They pointed out that an input to the servo, which caused the motoneuron to become accessible or more accessible to spindle

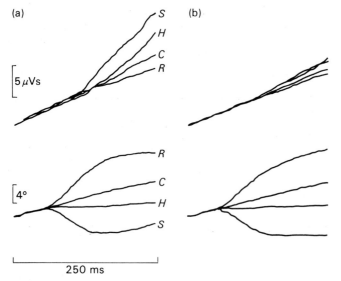

(a)   (b)

FIG. V.150. Servo-type responses from the muscle flexing the top joint of the thumb. During a course of a tracking movement against an initially constant resistance the movement may either be reversed (*S*) eliciting a stretch reflex, or halted (*H*), or allowed to accelerate by a reduction in the opposing resistance (*R*). In the control trials (*C*) none of these things happens. The top records are the integrated electromyogram of the flexor muscle. The bottom records give the angular position of the terminal phalanx of the thumb. Each trace is the average of 16 trials. (a) Records taken with a normal thumb. (b) Records taken in the same experiment after anaethesia of the thumb induced by occluding the blood supply for 90 min with a cuff at the wrist. Subject P.A.M. (From Marsden, C. D., Merton, P. A., and Morton, H. B. (1972). *Nature, Lond.* **238**, 140–3.)

excitation would provide a 'focusing effect' permitting the higher centres to use servo action for discrete contractions of one small part of a large muscle despite the fact that excitation from a single spindle input is widely distributed among the motoneurons of a large composite muscle. The need for such a 'focusing effect' in movements of individual fingers had already been emphasized by Phillips (1969).

Anaesthetization of the thumb by a ring block with xylocaine at the base of the proximal phalanx depresses the stretch reflex and servo action is lost (FIG. V.150(b) shows that an effect similar to

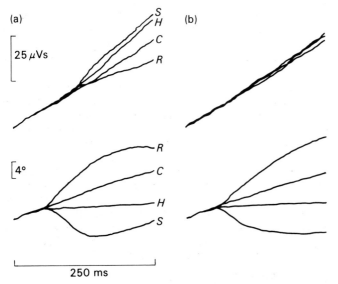

(a)   (b)

FIG. V.151. As FIG. V.150, but with the movements against a ten times greater initial resistance. Note that the scale of the integrated e.m.g. is five times that of FIG. V.150. Subject P.A.M. (From Marsden, C. D., Merton, P. A., and Morton, H. B. (1972). *Nature, Lond.* **238**, 140–3.)

that evoked by such a ring block can be produced by inflating a narrow cuff round the wrist for some 90 minutes). The subjects reported that it required more conscious effort to start the movement when the thumb was anaesthetic. Thus, anaesthesia caused by xylocaine or ischaemia provides both objective and subjective evidence of the advantage of servo action.

If the muscle is fatigued by working it during ischaemia produced by inflating a pressure cuff on the upper arm its contractility is lowered and it has to be more strongly activated to achieve the same force. This is exemplified by an increase in the e.m.g. activity which entails a reduction in the recording gain [FIG. V.152]. This makes intelligible the well-documented experience that in piano playing during ischaemia of the arm there is little deterioration in performance until the muscles are so weak that they cannot execute the required movements at all. Muscle fatigue is compensated for by an increase of gain elsewhere in the servo of the same type as occurs with increased load.

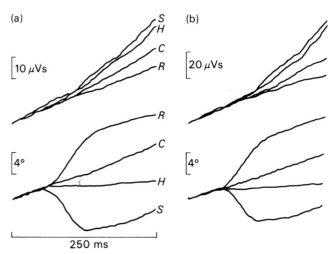

(a)   (b)

FIG. V.152. Gain compensation during fatigue. The initial force in this experiment was low, the same as for FIG. V.150. (a) Fresh muscle. The muscle was then fatigued by applying maximal shocks to the median nerve at the elbow at 50 s⁻¹ for 60 s. Recovery was prevented by a blood pressure cuff on the upper arm, arresting the circulation. (b) Muscle fatigued. The electrical responses are of the same form as before fatigue, but scaled up by a factor of roughly 2, as shown by the calibrations. Force records (not illustrated) showed that the forces developed in the halt and in the other responses were the same as before fatigue. In these experiments each trace is the average of eight trials. The period of circulatory arrest was too short to cause ischaemic nerve block. Subject P.A.M. (From Marsden, C. D., Merton, P. A., and Morton, H. B. (1972). *Nature, Lond.* **238**, 140–3.)

The latency of servo action in the thumb is some 50 ms whereas tendon jerk latencies (e.g. brachioradialis reflex) are only 20–25 ms. Merton (1979) has suggested that the explanation of this extra delay is that the stretch reflex arc for the thumb runs to the cerebral cortex and back. Phillips (1969, see also Phillips and Porter 1977) has shown that the rapidly conducting pyramidal tract fibres end monosynaptically on the motoneurons of the baboon's hand. It is also known that spindle afferent impulses are conducted very rapidly to cortical area 3a [see p. 321]. Phillips suggested that in the primate's hand the spinal stretch reflex arc 'had been overlaid in the course of evolution by a transcortical circuit'.

Phillips (1969) presented a speculative diagram [FIG. V.153] of the working of a system based on the co-ordinated distribution of suprasegmental motor output between α and γ motoneurons and their co-activation in movements. The diagram is designed to focus on the force and velocity of movements. The intended movement is from *a* to *c* and is supposed to be achieved by a graded recruitment of α motoneurons. The movement would progressively reduce the total feedback from the spindles of the prime movers (FIG. V.153

unloading effect) but the diagram illustrates the situation which would arise if the fusimotor neurons are also recruited (FIG. V.153 fusimotor effect) and that this fusimotor effect exactly balances that of the unloading effect so that there would be a null influence on the Group IA feedback as long as the movement continued at the prescribed velocity towards its target c. FIGURE V.153 (a) shows the effect on Group IA feedback of slowing of the movement caused by an 'unexpected' resistance encountered at b. This reduces the unloading effect [FIG. V. 153 (a), full line] but the fusimotor effect is unchanged so the total spindle feedback must increase. The primary endings [FIG. V.153 (a)] under the influence of both dynamic and static fusimotor neurons will increase their discharge abruptly at time b and because they are sensitive to velocity as well as length the further increase in their discharge will be the greater the faster the actual unloading effect [FIG. V.153 (a), interrupted line]. Since the feedback from the primary spindle endings excites the α motoneurons its abrupt increase at b will tend to evoke a prompt increase in force at the very instant of deviation and its further increase will only be checked if the movement is brought back to its course towards target c. (These compensated responses are not shown in the diagram.) The secondary endings [FIG. V.153 (a), s] are also supposed to be under static fusimotor influence and would signal the increasing discrepancy between the 'intended' and actual movements. Although these secondary endings do not stimulate the α motoneurons their discharge as well as that of the primary endings may well be important in 'informing' the cerebral and cerebellar cortices.

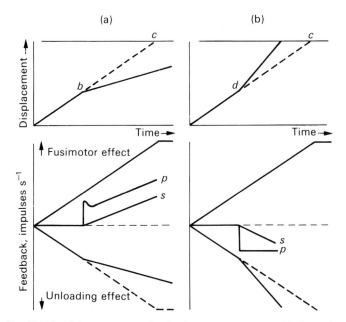

FIG. V.153. Alpha-gamma coactivation in a ramp movement. (a) Shows the effect of an *increase* in the resistance to the movement superimposed at point b. (b) shows the effect of a *decrease* in the resistance to the movement applied at point d. For further explanation see text. (From Phillips, C. G. (1969). *Proc. R. Soc.* **B173**, 141–74.)

FIGURE V.153 (b) shows the effects of an 'unexpected' reduction of resistance at time d with a consequent speeding up of the movement. Feedback from the primary endings under dynamic fusimotor influence would be abruptly reduced so that the movement would tend to slow down and resume its approach to target c. The behaviour of primaries under static fusimotor influence would resemble that of the secondaries [FIG. V.153 (b), s].

Operation of such a system in movements of the hand of a baboon requires (i) that the neurons of the cortical hand area project to fusimotor neurons besides their powerful monosynaptic projection to α motoneurons; (ii) that the hand muscles contain

spindles which are unloaded by movement and made to discharge by fusimotor activation; and (iii) that the spindles supply excitatory feedback to the γ motoneurons. Modern investigations have furnished evidence for the existence of such pathways.

REFERENCES

MARSDEN, C. D., MERTON, P. A., and MORTON, H. B. (1972). *Nature, Lond.* **238**, 140–3.
—— —— —— (1976). *J. Physiol., Lond.* **257**, 1–44.
MERTON, P. A. (1979). In *Human physiology* (ed. O.C.J. Lippold and F. R. Winton) Chap. 59, pp. 392–405. Churchill Livingstone, Edinburgh.
PHILLIPS, C. G. (1969). *Proc. R. Soc.* **B173**, 141–74.
VALLBO, A. B. (1973). In *New developments in electromyography and clinical neurophysiology* (ed. J. E. Desmedt) Vol. 3, pp. 251–62. Karger, Basle.

CONSCIOUS APPRECIATION OF POSITION AND MOVEMENT

Hitherto joint receptors have been accorded primary importance in providing 'conscious proprioceptive sensation'. Matthews (1977) however stresses (a) that patients who have had an entire replacement of the hip-joint (ball and socket) show minimal loss in their inability to detect externally imposed displacement; normally about 0.5° can be detected and after prosthesis the threshold is raised only to 1.2° displacement. (b) In subjects in whom a finger is 'ring-blocked' by local infiltration of anaesthetic and the hand rendered insensient by anoxia (induced by an occlusive pressure cuff at the wrist) awareness of passive flexion-extension movements of the digit still persisted. (c) Most joint receptors discharge maximally at either extreme of movement, discharging only sparsely in the mid-range, yielding a somewhat capricious source of kinaesthetic information.

None of this evidence, of course, denies a *contribution* of joint receptor afferents to kinaesthesia—there is no question that in normal circumstances they play some role. Nevertheless rather than being the sole determinants of our conscious kinaesthetic sense they only contribute to it.

The muscle spindles and their afferent connections undoubtedly contribute to conscious kinaesthesia (Matthews 1972, 1977). Their ascending afferent pathways reach the gracile and cuneate nuclei or nucleus Z and after relay there pass to the ventroposterior nucleus of the thalamus. Electrical recordings in area 3a of the cortex show responses to group IA excitation. (Area 3a lies immediately posterior to area 4 'motor area' and is anterior to areas 1 and 2 (sensory area).) The input from 3a passes by cortico-cortical connections to area 4 by numerous short association fibres.

Longitudinal vibration of a tendon is a highly selective stimulus of the spindle primaries of that muscle. Transverse percutaneous vibration of intact tendons in man arouses illusions of movement. Thus, if both forearms are suspended in a horizontal plane so that gravity does not act on the elbow joint, vibration of the biceps tendon at a frequency and amplitude insufficient to arouse reflex contraction of the muscle arouses the illusion that the elbow is being extended (i.e. as if the biceps were being lengthened). The direction of the illusory movement can be demonstrated by the blindfolded subject by tracking it with the opposite arm which is similarly suspended [FIG. V.154]. Whether Golgi tendon-end organs also contribute to such illusions is dubious, although their discharge too is indeed affected by such vibrations.

Golgi tendon-end organs and their function

Their function was once believed to be only that of inducing the 'lengthening reaction'. Serving as high-threshold receptors to passive stretch their discharge was shown to cause disynaptic inhibition

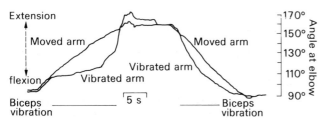

FIG. V.154. The effect of vibration applied to an arm that the subject was using to make a tracking movement. The left arm was moved by the experimenter and the subject was asked to track it with his right arm. During the periods indicated vibration was applied to the biceps of the right arm which was the one which was being moved voluntarily. The arm was moving in the vertical plane with the upper arm lying horizontal so that the biceps muscle will have been contracting throughout. (From Goodwin, E. M., McCloskey, D. I., and Matthews, P. B. C. (1972). *Brain* **95**, 705–48.)

of the motoneurons which supplied the specific muscle concerned.

It is now accepted that Golgi organs function essentially as contractile tension receptors. Their natural transducer properties have been difficult to appreciate, simply because the tension sensed by such a receptor is not easily related to the tension measured at the muscle tendon *as a whole*. Most Golgi organs are found at musculo-aponeurotic junctions and this complicates the issue. Morphological and electrophysiological studies indicate that the mechanical events leading to tendon end-organ excitation occur in two steps: (i) contraction of muscle fibres inserted on the receptor increases the tensile force of the tendon end-organ collagen bundles; this in turn (ii) squeezes the afferent terminals interwoven among them.

The threshold of tendon end-organs to their natural stimulation is very low as they can and do respond to the contraction of a single motor unit. They reflect local conditions in the muscle, relaying the information via IB afferents (see Barker (1974) for references). Golgi tendon end-organ discharge is also influenced by vibration when the relevant muscle is contracting.

Phillips and Porter (1977) have provided an excellent diagram indicating a possible scheme of the neural organization of movement performance incorporating many of the features described in the preceding pages [FIG. V.155].

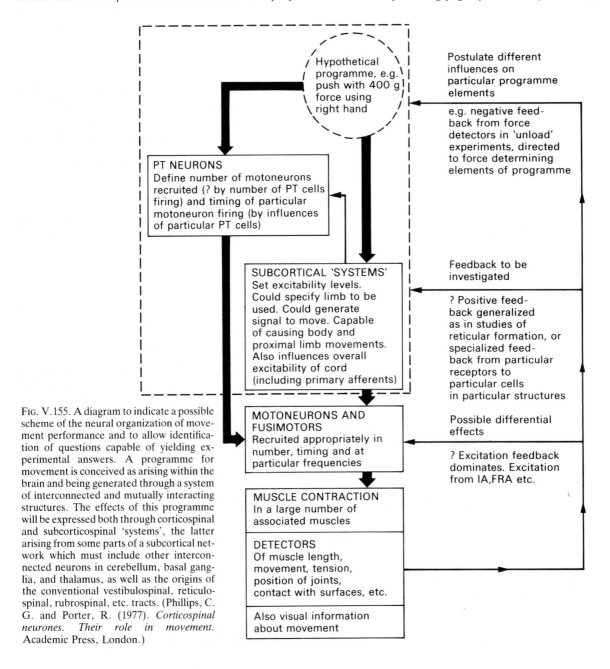

FIG. V.155. A diagram to indicate a possible scheme of the neural organization of movement performance and to allow identification of questions capable of yielding experimental answers. A programme for movement is conceived as arising within the brain and being generated through a system of interconnected and mutually interacting structures. The effects of this programme will be expressed both through corticospinal and subcorticospinal 'systems', the latter arising from some parts of a subcortical network which must include other interconnected neurons in cerebellum, basal ganglia, and thalamus, as well as the origins of the conventional vestibulospinal, reticulospinal, rubrospinal, etc. tracts. (Phillips, C. G. and Porter, R. (1977). *Corticospinal neurones. Their role in movement.* Academic Press, London.)

VOLUNTARY MOVEMENTS

Despite the huge advances made in our knowledge of central nervous control of movement described above, our understanding of the processes involved remains incomplete. The process of learning, by which movements which are initially executed clumsily and with difficulty become ultimately skilful and easy, has not yet been accounted for in satisfactorily physiological terms.

Voluntary movements are voluntary in their aim—i.e. to do something. For simple experimental and clinical tests, the aim is usually a displacement of joints or of segments of the body in a certain direction and to a certain extent; the subject is told to clench the fist, flex the elbow, extend the knee, and the like. He judge by inspection, how well he has succeeded in carrying out the instructions. The subject will 'bend the arm', and it happens; visual and proprioceptive impulses tell him that it has happened. We cut the bread, open the door, kick a ball, and judge our success.

An attempt is often made to draw a distinction between 'voluntary' and 'automatic' movements, the movement of walking frequently being placed in the latter category. No really hard-and-fast distinctions can be made. Walking is learnt by the child with difficulty, and at first requires constant attention. With time and practice it becomes almost automatic in character, so that we walk while thinking of something quite different. A simple movement involves groups of muscles. Shaking hands we use: 1. Prime movers (protagonists)—the flexors of the fingers; 2. Antagonists—the extensors of the fingers which relax; 3. Synergists—muscles which though not essential for the prime movement nevertheless facilitate its execution. The synergists are the extensors of the wrist—it is almost impossible to 'shake hands' unless the wrist is extended.

Voluntary movements are graded in two ways—by varying the discharge rate of any individual motoneuron and by varying the number of motoneurons and therefore the number of motor units in action [FIG. V.156]. The component parts of the movement must

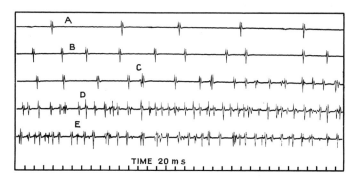

FIG. V.156. Muscle action potentials during voluntary movement. Action potentials (recorded with concentric needle electrodes) in the first dorsal interosseous muscle. The motor unit potentials are triphasic as is not uncommonly the case. Time, 20 ms. A. Commencing contraction potentials are at low frequency. B. Frequency increasing. C. Additional motor units come into action. D, E. Discharge frequency of active units rises. (W. F. Floyd.)

also follow one another in appropriate sequence, and the whole movement must be correctly related to the stimuli which arouse it. Further, the movement is superimposed on an appropriate basis of postural activity.

Disturbances of voluntary movement

Voluntary movements may be disturbed by abnormal function of parts of the cortex other than the sensory receptor areas and the executive precentral cortex.

1. If the nature of the object is not recognized—*agnosia* is present. A man shown a pencil does not recognize it as such, but thinks it is a cigar and uses it accordingly. The movement is planned and executed correctly, but the effect is ludicrous, because the idea in the patient's mind of the nature of the object was incorrect.

2. An object may be recognized, but the main idea of the movement cannot be analysed into its component parts. A man was supplied with a cigar and matchbox, which he recognized. Asked to use them, he opened the matchbox and stuck the cigar in and tried to shut the box as though it were a cigar-cutter. Then, taking the cigar out, he rubbed it on the side of the box as though it were a match. The main idea of the action was to light the cigar, but the components of the idea were not directed to the proper object nor followed in the right order.

3. Lastly, the nature of the object may be recognized, the general idea of action is normal, but the appropriate impulse patterns do not reach the precentral cortex intact—a normally worked-out idea is not exteriorized into the corresponding pattern of movements. Such a condition is termed *motor apraxia*. Liepmann described a case of right-sided motor apraxia, the left side being normal. There was no paralysis on the right side. Asked to brush the examiner's coat, he took the lower corner of the coat correctly with the left hand and held it; but with the right hand he picked up the brush and made a rhythmical series of movements in the air above his right ear. At the telephone, he put the receiver to his ear with the left hand, but with the right hand he took the mouthpiece and put it to his forehead, making nodding and puffing movements all the while. Another patient who showed similar difficulties said, 'Je comprends bien ce que vous voulez, mais je ne parviens pas à le faire.' Motor apraxia has been described following lesions of the prefrontal cortex or subcortex and of the corpus callosum.

SPEECH

Spoken and written speech are important examples of skilled voluntary movements. Forms of speech depend primarily on the integrity of the afferent mechanisms. For spoken speech we must first be able to hear sounds—that necessitates an intact auditory pathway from the ears to the auditory centres; secondly, we must be able to understand them—this process is believed to be related to the activity of the adjacent 'auditory-psychic' areas; the term 'auditory speech centre' is used to describe the special cortical region concerned. Similarly, written speech requires an intact pathway from the eyes to the visual cortex to enable us to see; the written symbols must be correctly interpreted. This process is believed to be related to the activity of the 'visual speech centre' (possibly part of the general 'visuo-psychic' area). The sensory speech centres are usually found on the left side of the brain in right-handed individuals. The exteriorization of speech demands the skilled use of many muscles, such as those of the tongue, larynx, or hand. The excitomotor areas of the brain, their descending tracts, and the related lower motor neurons, must be functioning normally, and appropriately guided by the cerebellar muscular, labyrinthine, and other afferents. It is further supposed that, as for other voluntary movements, some higher centre mediates the 'planning' of the details, sequence, and duration of the various movement patterns employed. It is suggested that there is a 'motor speech centre' for both spoken and written speech in the left prefrontal lobe, in the neighbourhood of the alleged higher centre for voluntary movements in general. The 'higher' motor centre for spoken speech is often called Broca's centre, the left inferior frontal convolution. Excisions in man in the vicinity of area 44 produce motor aphasia of varying degrees of severity and duration. Stimulation of this area in conscious man disturbs its function and produces arrest of speech; one patient said later, 'I knew what I wanted to say but could not' [cf. the comment of the patient with apraxia [see above] 'je ne parviens pas à le faire']. Stimulation of the posterior part of the parietal lobe or of the upper temporal lobe also produces 'aphasic arrest' (sudden speechlessness) suggesting that these regions are also involved in speech.

The sensory speech centres are undoubtedly linked by important pathways with the motor centres and guide them in their activities.

The above schema can be summarized as follows:

EARS→Auditory Centres—→'Auditory Speech Centre' →'Motor Speech Centres'

EYES→Visual Centres—→'Visual Speech Centre'

—→Excitomotor Areas→Lower Motor Neurons→Muscles

This schema is too rigid, and underestimates the closeness of co-ordination that exists between various regions of the cerebral cortex. It is probably unjustified to define and locate speech centres with any precision or to assign to them relatively autonomous functions. The left side of the brain is far more important than the right in right-handed individuals for the carrying out of speech functions in general. Speech functions should be considered as a whole, both on the receptive and the expressive side, and both for spoken and written speech. It is best to talk of a 'cortical speech area' in a vague way rather than to use the terms 'speech centres', which give a misleading idea of non-existent precision of localization. As speech is dependent on visual, auditory, and proprioceptive impulses, the speech area must be in the closest functional and anatomical connections with those regions of the cortex which primarily receive these impulses. The temporosphenoidal lobe, the hinder part of the parietal lobe (including the supramarginal and angular gyri), and the island of Reil, probably represent the principal receptive side of the speech area [Fig. V.157]. The area is continued forward to include that part of the prefrontal lobe from which the efferent pathway may possibly emerge. The whole of this vast area is correlated with the highest intellectual activities; and injury to this region depresses not only speech but also other intellectual activities. Injuries to the incoming pathways (auditory, visual, or proprioceptive), to the outgoing pathway to the excitomotor areas, or to the cerebral grey matter itself, cause fairly characteristic disturbances of speech, which are discussed more fully below. In general, it may be said that injury to the path to the visual speech centre (*a* in schema) may produce word-blindness (inability to understand written words). Injury to the outgoing pathway to the excitomotor areas (*b* in schema) may produce pure motor aphasia (inability to speak articulately), or inability to write (agraphia), or both, without much mental disturbance. Otherwise most cases of aphasia which result from injuries to the cerebral grey matter itself, usually show interference with both the expressive and the appreciative side of speech and deterioration of the intel-

lectual faculties. Thus, simple sentences may be understood, while those involving more complex ideas cannot be followed. The patient may obey when told to raise his hand, but does not know what to do if he is asked to touch his right hand with his left forefinger.

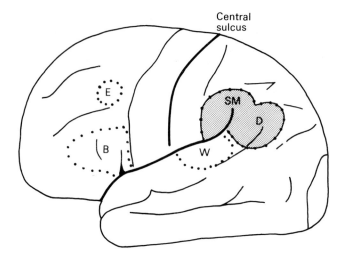

FIG. V.157. Speech areas. B = Broca's area (area 44); W = Wernicke's area (area 22); D = Dejerine's area; E = Exner's area (motor writing centre). SM=supramarginal gyrus. Note: Wernicke's area and Dejerine's area are both 'sensory speech centres'. See text. (After Penfield, W. and Roberts, G. (1959).)

Clinical conditions

Pure motor aphasia. This is a condition in which there is loss of articulate speech without mental confusion or deterioration. The patient is dumb, though the excitomotor cortex and its efferent paths are intact. The condition can be regarded as a special example of apraxia; a certain highly complicated group of movements, i.e. those of speech, are lost. It is due to a subcortical lesion, which cuts off the speech area from the excitomotor cortex.

Pure word-blindness. This is the inability to recognize the meaning of written or printed words, which appear as hieroglyphics. The patient is unable to read aloud or copy print into writing. Speech otherwise is not disturbed. Word-blindness (*alexia*) is a special form of visual agnosia; there is an inability to read. It is due to a lesion, subcortical in position, which cuts off the visual centres from the speech area. It is usually situated deep in the substance of the occipital lobe behind the angular gyrus, and often involves the optic radiation, which is close by and is passing to the occipital cortex. Word-blindness is thus often associated with right hemianopia.

Dysarthria. Dysarthria is a disturbance of speech resulting from defects of the excitomotor areas in the cortex and their connections, i.e. the pyramidal tracts, cranial nuclei, cranial nerves, muscles; from disturbances of tone, like rigidity; or from lack of cerebellar control of postural and voluntary activity. Any of these disorders or a combination of them disturb speech to some extent; it should be noted that the muscles cannot be used effectively for any purpose, including that of speech.

Connections and functions of the prefrontal lobes

The prefrontal lobes of the cerebral cortex lie anterior to the excitomotor cortex (areas 4 and 6 and the frontal eyefield region of area 8) and also extend on to the medial aspect of the hemisphere as far back as the anterior end of the corpus callosum, thus including the precallosal part of the cingular gyrus (area 24). The prefrontal lobes increase markedly in size as one ascends the phylogenetic scale; but it should be emphasized that the parietal and temporal lobes enlarge even more conspicuously.

Connections

These are mainly to and from the thalamus and hypothalamus; connections are also established with other regions of the cerebral cortex.

Afferent connections [Fig. V.158]

1. Many fibres pass from the dorsomedial nucleus of the thalamus to most of the areas of the prefrontal lobes (areas 8, 9, 10, 11, 12 on the lateral and adjacent medial surface; areas 44–47 in the inferior frontal convolution). Groups of cells in this thalamic nucleus project on the circumscribed areas of the cortex. As the medial nucleus of the thalamus receives afferents from the posterior hypothalamus it follows that the impulses that reach the prefrontal lobes via the medial nucleus represent a 'resultant' of hypothalamic as well as of thalamic activity.

2. Fibres from the anterior nucleus of the thalamus project on to the precallosal part of the cingular gyrus (area 24). The hippocampus sends efferent fibres via the fornix to the mammillary bodies of the hypothalamus whence a new relay transmits the impulses to the anterior thalamic nucleus. The hippocampus is thus ultimately projected to inhibitory area 24.

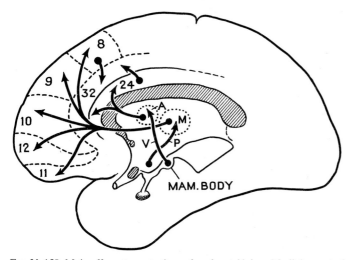

Fig. V. 158. Main afferent connections of prefrontal lobes. Medial aspect of right cerebral hemisphere. A = Anterior nucleus of thalamus. M = Medial (dorsomedial) nucleus of thalamus. P = Fibres from hypothalamus to M. V = Mammillothalamic tract. (Le Gros Clark, W. (1948). *Lancet* **i**, 354.)

Closed-circuit connections with the thalamus

Like most regions of the cerebral cortex the prefrontal lobes establish to-and-fro connections with the anterior, medial, and the adjacent so-called 'intralaminar' thalamic nuclei; this type of closed circuit is responsible for the 'resting' electrocorticogram or electro-encephalogram.

Intercortical connections

1. Area 32 [Fig. V.158] receives afferents from the frontal inhibitory areas 8s and 24s and also from the other inhibitory areas (4s, 2s, 19s). No other connections of area 32 are known. The significance of this intensive subjection of area 32 to such widespread inhibitory influences is not understood.

2. A long tract runs back from the frontal eyefield in area 8 to area 18 in the occipital lobe (the parastriate area, which surrounds the visual cortex and receives afferents from it). The frontal lobes are thus linked with the visual system. Ablation of area 8 causes visual agnosia (object vision hemianopsia). Proprioceptive data from the eye muscles are integrated in area 8 and through the fronto-occipital projection (area 8 to area 18) these data can be correlated with retinal impressions. When this connection is broken, object vision hemianopsia results.

3. Fibres from the prefrontal areas 44–47 (and from the occipital area 18) pass into the temporal lobes. The prefrontal lobes thus establish connections, both direct and roundabout (i.e. via area 18), with the temporal lobes which in their turn receive numerous association fibres from most parts of the cerebral cortex [Fig. V.159].

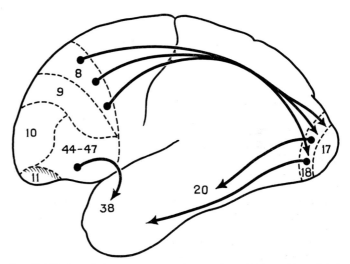

Fig. V.159. Association tracts connecting perefrontal lobes with occipital and temporal lobes. Lateral aspect of cerebral hemisphere. (Le Gros Clark, W. (1948). *Lancet* **i**, 355.)

Efferent connections [Fig. V.160]

1. The inhibitory areas 8s and 24s discharge to the caudate nucleus.

2. Area 10 provides many of the fibres of the frontopontine tract which passes between the head of the caudate nucleus and the putamen in the anterior limb of the internal capsule to the pontine nuclei and thence to the cerebellum.

3. Area 8 has an important projection to the tegmentum of the midbrain (corticotegmental fibres).

4. Areas 9 and 10 send fibres to the ventral and medial thalamic nuclei and to the tegmental reticular formation.

5. The hippocampus, uncus, area 13 and the amygdala project via the fornix to the mammillary bodies (hypothalamus).

As the prefrontal lobes are linked up with the thalamus and the hypothalamus by so many fibres passing in both directions, the entire system may be considered as functioning as an integrated whole. The prefrontal lobes are linked up with the cortical visual areas and the temporal lobes; they also affect the autonomic nervous system via the hypothalamus (and also more directly via the brain stem). The activities of the nervous 'complex' consisting of prefrontal lobes, thalamus, and hypothalamus are correlated with some of the higher intellectual activities, the personality, emotional affects, and forms of behaviour of social significance.

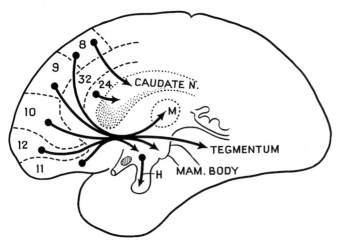

FIG. V.160. Main efferent connections of prefrontal lobes. Medial aspect of right cerebral hemisphere. M = Medial (dorsomedial) nucleus of thalamus. H = Tract from hypothalamus to posterior pituitary. Mam. body = Mammillary body. (Le Gros Clark, W. (1948). *Lancet* i, 355.)

Experimental studies

Results of ablation

Some of the more striking consequences of ablation of the prefrontal areas in monkeys are summarized below.

1. Alterations of activity. In the monkey, prefrontal ablations, especially those involving area 13 (on the orbital surface) produce initially a state of apathy: the animal sits with a blank expression on its face, the head is drooped, and it stares into space ignoring the approach of human beings; such movements as are carried out are sluggish. After some days or weeks the animal passes into a state of hyperactivity which persists unchanged for months. The animal is constantly on the move, incessantly walking or pacing about 'like a caged lion'; the movements are without aim and the animal seems unable to control or check them. Sometimes the movements become almost maniacal in their violence; they cease during the hours of darkness.

Similar hyperactivity can be provoked by lesions in the septal-preoptic regions: it is believed that hyperactivity results from the interruption of projections from area 13 to the hypothalamus. Lesions of the hypothalamus itself abolish the responses obtained by stimulating area 13. Area 13 has connections with both the hypothalamus and the caudate nucleus.

2. Alterations in emotional exteriorization. Monkeys were trained to discriminate between weights; when they chose the heavier weight correctly they were suitably rewarded. The postcentral gyrus was then extirpated, with the result that discrimination was grossly impaired. The animal, finding itself in difficulties about choosing correctly, often flew into a tantrum of rage. These tantrums were abolished by lesions experimentally placed in the prefrontal lobes in area 24 [FIG. V.160]. Cases have been recorded in which transverse lesions of the posterior margins of areas 13 and 14 have produced a clinical state resembling the 'sham rage' of the decorticate animal.·

3. Alterations in social behaviour. Unilateral or bilateral ablation of the rostral part of the cingular gyrus was immediately followed by marked changes in social behaviour The monkey lost its preoperative shyness and fear of man. It would approach the experimenter and examine his fingers with curiosity, instead of cowering (as it would normally) in the far corner of the cage. In a large cage, with other monkeys, it showed no grooming or acts of affection towards its companions; in fact, it behaved as though they were inanimate. It would walk over them, walk on them if they happened to be in the way, and would even sit on them. It would openly take food from its companions and appeared suprised when they retaliated; it was not aggressive, but seemed merely to have lost its 'social conscience'.

4. Impairment of memory. The monkey was shown two inverted cups, and food was placed under one of them; the monkey was trained to raise the cup covering the food; if it chose correctly it was given the food to eat. To test the memory a screen was placed between the monkey and the cups immediately after one had been filled. The normal monkey could still make a correct choice after the cups had been out of sight for 90 seconds; after prefrontal ablation it might choose wrongly after an interval of only 5 seconds.

5. Impairment of learning capacity. Despite earlier claims to the contrary, it can now be asserted that frontal lobe ablation in monkeys and man is followed by an impairment of learning capacity and of other more purely intellectual functions. This loss of learning power is proportional to the extent to which the neopallium, or its projections, are involved.

6. Results of stimulation. The prefrontal areas by their connections with the hypothalamus and also by their direct brain stem connections can influence autonomic activities. Thus stimulation of area 13 produces changes in heart rate, blood pressure, and breathing. Stimulation of the central end of the vagus fibres from the lungs produces circumscribed activity in area 13 and nowhere else in the cerebral cortex. Stimulation of area 13 affects the motility and secretory activity of the gut.

Prefrontal lobectomy

This operation was introduced as a method of treating involutional depression and manic depressive psychosis by Moniz (1936). Jacobsen (1935) (see Fulton 1951), studying the behaviour of a chimpanzee during the performance of discrimination tests, noted that the animal became neurotic. Bilateral frontal lobectomy relieved the neurosis. Moniz interrupted the subcortical connections of the orbitofrontal area (prefrontal leucotomy) in psychotic patients and reported favourable results.

Effects of excision in man

The results in man are very variable and depend in part on the extent and site of the operation and whether it is unilateral or bilateral; graver disturbances are usually noted after bilateral excisions. If the excitomotor areas are spared there is no impairment of volitional movement or alteration in muscle tone or reflexes. Sometimes there is surprisingly little change in the personality—particularly if the prefrontal excision has been unilateral as, for example, for removal of a tumour.

In cases involving bilateral excision there may be a striking deterioration in social behaviour—a loss of emotional restraint being a common feature.

Prefrontal leucotomy

Deliberate lesions of the prefrontal areas are practised for certain types of mental disorder. The operation of prefrontal leucotomy was introduced by Moniz; a burr hole is made on each side of the skull above the zygoma and behind the orbital margin. A special needle is introduced into the brain and carried through an arc upwards to cut the subcortical white fibres in the plane of the coronal suture just anterior to the tip of the lateral ventricle. The aim of the operation is to sever the connections between the thalamus and the prefrontal lobes. It should be remembered that when the deep thalamocortical connections are cut most of the prefrontal cortex is put out of action, as the cortical association fibres are relatively few.

To an increasing extent therapeutic leucotomies are being carried out by an operation which enables the surgeon to see exactly what he is doing. The newer operations consist of section of selected parts of the corticothalamic connections or of excision of limited areas of the prefrontal cortex (topectomy).

The type of mental disorder for which prefrontal leucotomy has been most helpful is that showing the clinical picture of 'mental tension'. This is described as a 'persistent emotional charge' sustaining and to some extent determining the clinical picture; this 'charge' is 'always of an unpleasant quality, invariably distressing and sometimes intolerable to the patient. Its presence is shown by irritability, rage, fear, or other forms of emotional excitation, insomnia, and on the motor side, restlessness, aggressiveness, destructiveness, or impulsive behaviour'.

Generally only chronic cases that have not responded to any other form of therapy have been operated on. Following the operation there is usually an initial state of confusion in which the patient has impaired memory, may not know who or where he is and generally shows marked depression of intellectual activity; there may also be urinary (and occasionally faecal) incontinence. Some cases pass through a phase of severe irritability in which they are restless, tear at their dressings, and become abusive and violent on the slightest provocation; they are usually ravenously hungry. These symptoms pass away gradually; a proportion of the patients (e.g. 30 per cent in one carefully selected series) recover sufficiently to leave the mental hospital and live at home or to do useful work; but they rarely fully regain their pre-illness personality and ability.

The principal mental changes after leucotomy are:

1. The patient's mood changes to one of cheerfulness.
2. The patient becomes less self-critical than a normal person, holding an exalted opinion of his abilities though his higher intellectual faculties are impaired.
3. Concentration is less evident; his attention is easily distracted.
4. Mental drive is reduced; there may be a lack of initiative.

Fulton, reviewing the subject of frontal lobe function concluded that the orbito-insulo-cingulate areas of the frontal lobe are concerned with emotion and that the more lateral elements of the frontal cortex (neopallium) are concerned with 'intellectual capacity'. Correspondingly he has suggested that this functional differentiation justifies a more discriminative approach to the problem of surgical alleviation of psychoses and severe neuroses. Space does not permit other than a crystallization of the evidence derived from post-operative studies.

(i) Bilateral removal of the cingulate gyrus has notably improved obsessional states and aggressive psychotics but has proved of no avail in the treatment of schizophrenia.

(ii) Bilateral removal of area 13 has proved effective in the alleviation of schizophrenia and depression. Neither of these operations causes any impairment of intellectual capacity.

(iii) Bilateral removal of areas 9 and 10 has produced improvement of aggressive and overactive psychotics and its effects seem similar to those of cingulectomy.

If the cortical-subcortical connections in the median ventral quadrant are destroyed by electrocoagulation, intractable pain and major psychotic symptoms can be relieved without impairing intellectual capacity. Lesions which extend to the lateral surfaces of the frontal lobe, or to the more lateral projections to the underlying structures, cause a serious reduction in intellectual capacity.

REFERENCES

BARKER, D. (1974). In *Muscle receptors* (ed. C. C. Hunt). *Handbook of sensory physiology*, Vol. III/2. Springer, Berlin.

FULTON, J. F. (1951). *Frontal lobotomy and affective behaviour*. London.

GAINOTTI, G. (1972). *Cortex* **8**, 41–55.

GREENBLATT, M. and SOLOMON, H. C. (1958). *Res. Publ. Ass. nerv. ment. Dis.* **36**, 19–36.

HUBEL, D. H. (ed.) (1979). *The brain*. Scientific American, San Francisco.

KETY, S. (1978). *Harvey Lect.* **71**, 1–22.

MONIZ, E. (1936). *Tentatives opératoires dans le traitement de certaines psychoses*. Masson, Paris.

MATTHEWS, P. C. B. (1972). *Mammalian muscle receptors and their central actions*. Arnold, London.

—— (1977). In *Somatic and visceral sensory mechanisms* (ed. G. Gordon) *Br. med. Bull.* **33**(2), 137–42.

PHILLIPS, C. G. and PORTER, R. (1977). *Corticospinal neurones. Their role in movement*. Academic Press, London.

Sleep

According to Bremer (1954) the waking state, in mammalia at least, is an expression of 'a dynamic equilibrium between the activation of cerebral neuronal networks maintained by the incessant impact of innumerable ascendant and associative impulses and the cumulative functional depression resulting from the very continuity of this state of excitation'. The ascending reticular system integrates and amplifies corticopetal impulses travelling by the classical sensory pathways; the sum of these ascendant impulses sustains the cortical excitatory state and thereby maintains 'wakeful activity'.

These views are based on the effects of mesencephalic transection (*cerveau isolé*) and on the results of electrolytic lesions of the ascending reticular system. Either of these procedures produces behavioural sleep and somnolence and simultaneously causes the EEG to assume a wave pattern which resembles that of sleep.

It is common knowledge that procedures which minimize sensory stimulation favour the onset of natural sleep. Thus the room is darkened, the body musculature is relaxed, the temperature of the body's surroundings is made 'comfortably warm'; silence is a useful adjunct to the process of falling asleep. Anxiety and emotion make sleep more difficult; it is known that adrenaline causes activation of the ascending reticular system. There is much in favour of the hypothesis that sleep results from a reduction in the sensory afflux. However, the problem is much more complex than this. Sleep is more likely when the subject is tired even though the surroundings themselves do not predispose to sleep.

A sleep cycle comprises a period in which (i) there is a progressive reduction of muscle activity (recorded by the EMG) which is accompanied by a gradual slowing of EEG waves which increase in amplitude, followed by (ii) a phase of rapid irregular movements of the eyes accompanied by low voltage fast and irregular EEG activity. Each cycle has therefore been designated to consist of non-rapid eye movement sleep NREM or slow-wave sleep (SWS) and rapid eye movement sleep (REM), sometimes called 'paradoxical sleep'.

Typically, four or five of such cycles occur during a night's sleep in an adult. Each cycle lasts about an hour and a half and comprises a period of approximately 75 minutes of NREM and 15 minutes of REM. Towards morning the NREM periods lessen and the REM periods increase.

NREM sleep itself has been arbitrarily divided into four stages. Stage 1 is characterized by fast low amplitude EEG activity accompanied by EMG evidence of skeletal muscle discharge not noticeably different from that of the individual when resting and awake. Stage 2 evidences the appearance of the so-called sleep spindles and is associated with a marked decrease in EMG activity. Sleep spindles consists of runs of a few seconds duration of regular waves at 14–15 per second superimposed on a low voltage background. In either Stage 1 or 2 the sleeper is readily arousable. In Stage 3 sleep deepens and the EEG shows sleep spindles now superimposed on a background of waves of delta type (frequency 1–2 per second and of 100 microvolts amplitude). Stage 4 is associated with a high threshold for awakening and manifests an EEG record consisting solely of the slow high-voltage delta waves.

This sequence of NREM stages is now followed by the REM phase, in which the EEG changes to rapid low voltage irregular waves, the heart and respiratory rates quicken and in which, besides the rapid eye movements, gross body movements occur occasionally against a background of *decreased* muscle tone. Far from being associated with a lower threshold for arousal, sleep is deeper in this REM stage and this disparity between the fast low voltage EEG—normally associated with wakefulness—and the behavioural observations has led to the term 'paradoxical' sleep in describing this phase. REM sleep then merges into NREM as the next cycle begins. Dement and Kleitman (1957) found that the REM stage is associated with dreaming. When their subjects were awakened in the REM phase 80 per cent of them reported that their dreams had been interrupted! When similarly aroused during NREM sleep, only 7 per cent reported the interruption of a dream.

If subjects are aroused repeatedly in the REM phase they become irritable—one can hardly blame them—but on subsequent nights they manifest many more periods of REM sleep. It has been argued that REM sleep is necessary for mental well-being and indeed barbiturates and monoamine oxidase inhibitors which reduce REM sleep do cause untoward psychological effects in some individuals.

NREM sleep is believed to be partly due to the reduction of ascending reticular activity which, in the conscious animal, 'alerts' the higher centres and which causes desynchronization of the EEG. However, a slow rhythmic stimulation of the central regions of the pons and medulla promotes synchrony and sleep in experimental animals (see Moruzzi 1972) whereas transection at mid-pontine level causes desynchronization of the EEG.

The neurons of the midline raphé nuclei of the pons and medulla are serotoninergic. Selective destruction of these nuclei produces a marked increase in the total duration of wakefulness, both from a behavioural and an electrical standpoint (Jouvet 1972). In intact animals the inhibition of the synthesis of serotonin at the level of tryptophan hydroxylase by parachlorophenylalanine (PCPA) de-

pletes the serotonin contents of these neurons and causes insomnia. This insomnia is immediately reversible by the injection of 2–5 mg kg⁻¹ of 5-hydroxytryptophane (5HTP), which is rapidly converted to 5HT in the raphe nuclei. If the rostral raphe nuclear cell bodies are destroyed surgically, the 5HT content of their axons and terminals degenerates and insomnia results, which is not alleviated by 5-hydroxytryptophane, presumably because the degenerated neurons cannot synthesize 5HT.

Jouvet and his colleagues (see Jouvet 1969, 1972, 1974) have found that the control of the sleep-waking cycle by 5HT is regulated by other monoaminergic systems which act directly or indirectly. Thus the bilateral destruction of the dorsal noradrenergic bundle at the level of the isthmus causes profound hypersomnia, which Jouvet ascribes to an increase in 5HT turnover because:

(i) this hypersomnia is prevented by pretreatment with *p*-chlorophenylalanine;

(ii) 24 hours after the lesion there is a very marked increase in the conversion of [³H]-tryptophane to [³H]-5HT in the cortex, although no alteration of endogenous 5HT could be detected.

These results support the suggestion that some noradrenergic neurons may control the activity of the 5HT neurons of the raphé system (Jouvet 1969).

Sleep promoting factor

In 1913 Legendre and Pieron transfused cerebrospinal fluid from dogs, sleep-deprived for several days, to the cisterna magna of normal dogs; the recipients slept for some three hours longer than normal. Pappenheimer (1979) and his colleagues used his method of perfusing the ventriculo-cisternal system in goats chronically prepared [see p. 366] which allows the collection of relatively large volumes of cerebrospinal fluid effluent without causing discomfort to the goat. Such c.s.f. effluent was infused into the ventricular system of rats, cats, or monkeys. Rats sleep normally some two-thirds of the time during daylight and only about a third of the time during darkness. Pappenheimer used a regular 12-hour alternation of light and dark cycles. When the rats were infused with 0.2 ml of c.s.f. (collected from a sleep-deprived goat) during the last hour of a daylight period they showed more slow-wave sleep in the ensuing six hours than rats which had been submitted to the transfusion of 'control' c.s.f. These same effects were obtained when the donor samples of the sleep-deprived goats had been passed through a

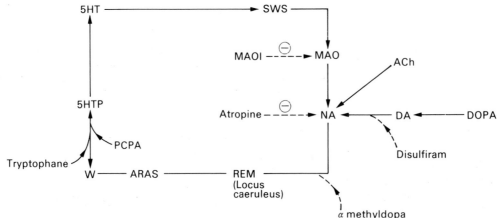

W = wake; ARAS = ascending reticular activating system
PCPA inhibits tryptophane hydroxylase (TRY) ⟶ decreases brain 5HT ⟶ TOTAL INSOMNIA
MAOI (prevents 5HT catabolism) ≫ SWS (total suppression of REM)
Disulfiram inhibits dopamine β hydroxylase
Atropine suppresses final steps of desynchronized sleep
W ⇌ SWS ⇌ REM
NEVER   W ⟶ REM

membrane filter that prevented the passage of molecules of molecular weights greater than 500 daltons. The sleep-promoting factor (factor S) is a polypeptide (350–500 daltons) and has been extracted from the brains of sleep-deprived animals. The purified factor S transfused via the cerebral ventricles in doses of 100 pmol does not cause an immediate effect—probably because it requires an hour for the substance to reach brain stem neurons. Thereafter such infusions increased the sleeping period of rabbits from 40 per cent per hour to 60–70 per cent per hour. The effect lasted for at least six hours. Not only the duration but also the amplitude of the slow waves of the EEG rhythm was increased by factor S. Indeed artificially transfused factor S causes behavioural and EEG effects which closely simulate those manifest following a period of sleep deprivation when the animal is then permitted to sleep.

The duration of action of factor S indicates that it is stable for hours, which led Pappenheimer to infer that some of it produced in sleep-deprived animals would be absorbed via the sagittal sinus and thence pass from the blood to be excreted by the kidney. Purified extracts of urine from men submitted to sleep deprivation do indeed promote the duration of SWS in rabbits when administered via the cerebral ventricular route.

Further work showed that factor S must reach the brain stem neurons in order to promote sleep. If the cerebral aqueduct is blocked, infusion of factor S is ineffective, which suggests that either the periaqueductal tissue (which contains the ascending reticular activating system) or neurons of the pontomedullary rhombencephalon is the site of action of factor S.

REFERENCES

Bremer, F. (1954). In *Brain mechanisms and consciousness* (ed. J. F. Delafresnaye). Blackwell, Oxford.
Dement, W. and Kleitman, N. (1957). *Electroenceph. clin. Neurophysiol.* **9**, 673.
Jouvet, M. (1969). *Science, N.Y.* **163**, 32.
—— (1972). *Ergebn Physiol.* **64**, 166–307.
—— (1974). *Proc. Int. Cong. Physiol. Sci.*, New Delhi, Vol. 10, p. 192.
Kales, A. and Kales, M. (1974). *New Engl. J. Med.* **290**, 487.
Legendre, R and Piéron, H. (1913). *Z. allg. Physiol.* **14**, 235–62.
Moruzzi, G. (1972). *Ergebn Physiol.* **64**, 1–165.
Pappenheimer, J. R. (1979). *Johns Hopkins med. J.* **145**, 49–56.

CONDITIONED REFLEXES

Pavlov recognized two distinct classes of reflexes.

1. The inborn or unconditioned reflex which is present in all normal members of a species: i.e. what is generally called a reflex, such as the knee jerk, light reflex, secretion of saliva when food is introduced into the mouth, and the flexor reflex.

2. The acquired or conditioned reflex which depends for its appearance on the formation of new functional connections in the central nervous system, and is therefore peculiar to the individual. The term 'conditioned' refers to the fact that certain conditions must be present if this class of response is to develop. It would help clear thinking if these reactions were called 'conditioned responses'.

A simple example will illustrate how a conditioned reflex is established. The introduction of food into the mouth is a stimulus which sets up reflexly the unconditioned response of salivary secretion, and is therefore termed an unconditioned stimulus. If a neutral stimulus, e.g. the ringing of a bell, is applied so as to coincide with the unconditional stimulus (the taking of food) and if the procedure is repeated several times, the initially neutral stimulus finally acquires fresh properties (and nervous connections) and can now of itself elicit a secretion of saliva. The flow of saliva in response to ringing the bell is an example of a conditioned reflex; the procedure of ringing the bell has become for the individual under experiment a conditioned stimulus (i.e. one which elicits a conditioned reflex).

Conditioned reflexes are always built up primarily on the basis of inborn reflexes. The salivary reflex is commonly employed because the response of the glands can be expressed quantitatively in terms of the volume of saliva secreted. The salivary duct is usually brought up to the surface, the saliva is collected, and the volume automatically recorded; to prevent extraneous factors influencing the animal it is usual to have the observer and the recording apparatus in a separate room from that in which the experimental animal is placed.

Establishment of positive or excitatory conditioned reflexes

1. The animal must be alert and in good health, and there must be complete freedom from all simultaneously operating nervous influences.

2. The conditioned stimulus (or rather the external stimulus which is to become the conditioned stimulus, e.g. for eliciting the salivary flow) must begin to operate before the unconditioned stimulus is applied; e.g. the bell must begin to sound before any food is put into the mouth. The conditioned stimulus must also be allowed to continue to act so as to overlap the unconditioned, i.e. the bell continues to ring while the animal is being fed.

3. Almost any stimulus if suitably employed may become a conditioned stimulus; it may be one to which the animal was previously indifferent or even one which is mildly nocuous in character.

4. Necessity for reinforcement. For a conditioned stimulus to retain its new properties it is essential that it should always be followed by the unconditioned stimulus. Thus, in the example previously mentioned, if ringing of the bell (conditioned stimulus) is carried out several times alone and is not followed by placing food in the mouth (unconditioned stimulus), it soon ceases to elicit a salivary flow—in other words, the signal has become misleading, it no longer heralds the administration of food and so is ignored. The process of following up a conditioned stimulus with the basic unconditioned stimulus is termed reinforcement.

5. The disappearance of a natural agency may become an effective conditioned stimulus. Thus a bell is rung continuously while the dog is brought into the experimental room. The sound is then cut out and food is administered. After several repetitions the cessation of the sound becomes an effective conditioned stimulus (as happened to Londoners during the Second World War with the 'cut out' of flying bombs).

Conditioned reflexes can be built up with difficulty in decorticate animals but such responses as are painstakingly developed are of little precision. Pavlov considered that conditioned reflexes were mediated solely by the cerebral cortex.

REFERENCE

Pavlov, I. P. (1927). *Conditioned reflexes* (trans. G. V. Anrep). Oxford University Press, London.

The limbic system—limbic lobe and hypothalamus

The limbic system consists of the limbic lobe (gyrus fornicatus = gyrus cinguli, isthmus, hippocampal gyrus, and uncus) and the related subcortical nuclei, amygdala, septal nuclei, hypothalamus, and anterior thalamic nuclei.

The fornix is the main projection of the hippocampus, uncus and amygdala to the hypothalamus (mammillary bodies). By means of the bundle of Vicq d'Azyr the anterior nuclei of the thalamus and, in turn the cingular gyrus, can be excited by this cortico-hypothalamico-thalamo-cortical circuit.

Kaada (1951) has shown that the limbic system represents the primary area of control of autonomic function in the forebrain. Vivid changes in heart rate, blood pressure, gut movements, pupillary reactions, etc., can be induced by stimulation of the limbic system. The hypothalamus which has long been regarded as the important region of central autonomic control can be most profitably considered as a part of the limbic system.

EMOTION

'Of points where physiology and psychology touch, the place of one lies at "emotion"'. A discussion of the physiology of the emotions must begin by recognizing that man is body and mind, a 'psycho-biological whole', and that emotion has a mental and a physical side. The mental side consists of cognitive, affective, and conative changes. The following example illustrates the meaning of these terms: thus, I hear a noise which I recognize as that of an exploding bomb (cognition); I feel frightened (affect), and I want to take shelter (conation). The physical side of an emotion consists of changes in viscera and skeletal muscles; these are often widespread and involve the co-ordinated activity of both the autonomic and somatic nervous systems.

Fear is accompanied by tachycardia, tachypnoea, vasoconstriction of the skin, sweating ('cold sweat') pilo-erection, pupillary dilatation, and dryness of the mouth. Muscular tremors occur. Thus both sympathetic and somatic activity take place.

Grief is accompanied by tears, excess nasal secretion, and skin pallor. Muscular tone is reduced and movements are slow and feeble. Again, both somatic and autonomic functions are disturbed.

The complex patterns of emotional exteriorization are achieved in the main by the prefrontal-hypothalamic-thalamic complex. The orbito-insulo-temporo-cingulate areas of the cerebral cortex in particular are intimately concerned in the production of autonomic features of emotion. These areas project:

1. To the hypothalamus which in turn sends fibres to the bulbar autonomic centres.
2. To the reticular formation of the brain stem which modifies somatic motoneuron activity appropriately.

There is no doubt that the mental and viscerosomatic sides of an emotion *can* be dissociated—a patient with a vascular lesion of the internal capsule may manifest the viscerosomatic changes which ordinarily accompany emotion but in this case the 'affect' is lacking—he feels nothing. However, viscerosomatic changes though not mainly responsible for affect may modify it. More important than the feed-back of viscerosomatic information from the body is the complex sustained activity of the higher levels.

Experimental production of emotional changes. Transection of the brain stem immediately rostral to the thalamus causes the remarkable reaction of 'sham rage'. The cat on recovery from the anaesthetic may display vivid somatic and autonomic changes which are the counterpart of intense fury in the normal animal, but which are in these circumstances precipitated by the most trivial stimuli.

Similar outbursts of rage can be evoked in the conscious animal by electrical stimulation (Hess technique) if the tips of the implanted electrodes lie in the posterolateral hypothalamus.

Bard showed that the sham rage reaction was dependent upon the integrity of the posterior hypothalamic areas. Bard and

Mountcastle (1948) found that removal of the *whole of the neocortex*, far from causing 'rage' attacks in the cat, rendered the animal unduly placid. Subsequent bilateral destruction of the amygdala or limbic cortex rendered the cat subject to the rage attacks. Such results suggest:

1. That the neocortex lowers the threshold of rage reactions.
2. The amygdala and limbic cortex exert inhibitory effects on the hypothalamic mechanism associated with the rage reaction.

Unfortunately for the clarity of this particular theme, amygdalectomy, which of itself causes the otherwise intact cat (or dog) to become subject to rage reactions has the reverse effect in the macacus monkey and in the wild Norway rat. The wild Norway rat, which in normal circumstances apparently fully justifies its name, becomes gentle and placid after bilateral amygdalectomy. Similarly, the vicious and intractable macacus monkey becomes much more tolerable as a companion after amygdalectomy. As Bard (1959) points out 'no explanation is at hand for these striking species differences but it is evident that the result obtained bears a relation to the natural temperament or the degree of domestication of the animal'.

Further experimentation has indicated that the limbic system not only participates in the elaboration and integration of the activity whereby emotion is expressed, but is also involved in the ascending pathways leading to the, as yet unidentified, regions of the brain where emotion is experienced. This has been demonstrated using the technique of *self-stimulation*. An electrode is implanted in a stereotaxically defined region of the limbic system and the animal, generally a rat, placed in a box containing a bar which when pressed delivers a shock through the electrode. If after the bar has been accidentally pressed a few times, the animal carefully avoids it, the experience caused by the shock is presumably unpleasant. However, if the electrode is placed in certain regions, notably the posterior hypothalamus just anterior to the mamillary bodies, but also, though less effectively, parts of the anterior hypothalamus, midbrain tegmentum, pre-optic region, and median forebrain bundle, the animal will stimulate itself repeatedly, as often as 5000 times an hour. If permitted, the animal will continue to stimulate itself even to the point of exhaustion. Stimulation at these sites obviously yields pleasurable sensations suggesting that these are either the sites at which such sensations are produced, or more probably, that they lie on the ascending pathway to the regions ultimately responsible.

THE HYPOTHALAMUS

The hypothalamus is a diencephalic structure which, separated from the thalamus by the hypothalamic sulcus, forms the antero-inferior wall and floor of the third ventricle. It extends forwards from the mesencephalic tegmentum to the pre-optic area. It receives a more generous blood supply than does any other structure in the brain. The hypothalamus contains many nuclear masses which may be conveniently grouped as follows:

1. Pre-optic area (medial and lateral pre-optic nuclei)
2. Supra-optic area (supra-optic, suprachiasmatic, paraventricular, and anterior nuclei)
3. Tuberal area (ventromedial, dorsomedial, arcuate, lateral, and posterior nuclei)
4. Mammillary area (medial and lateral mammillary, pre- and supramammillary nuclei) [FIG. V.161].

Except for the lateral pre-optic, lateral, and lateral mammillary nuclei, these nuclei lie in the highly cellular medial region of the hypothalamus in which the nuclear masses are distinct. In the lateral region the cells are more diffusely arranged and dispersed

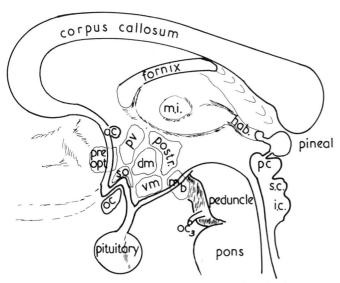

FIG. V.161. The hypothalamic nuclei. Midsagittal section of brain stem: a.c. = anterior commissure; p.c. = posterior commissure; s.c. = superior colliculus; i.c. = inferior colliculus; hab. = habenula; oc_3 = oculomotor nerve; oc = optic chiasma; m.i. = massa intermedia. Hypothalamic nuclei: preopt = pre-optic nucleus; so = supra-optic nucleus; pv = paraventricular nucleus; dm = dorsomedial nucleus; vm = ventromedial nucleus; mb = mammillary body; postr = posterior nucleus.

among the fibres of the *medial forebrain bundle*, which extends throughout the entire lateral hypothalamus and provides the pathway through which the majority of the hypothalamic connections with other regions of the brain are established.

Connections

The principal nervous connections of the hypothalamus are with the limbic system and the midbrain tegmentum [FIGS. V.162 and V.163].

Afferent

1. From other parts of the *limbic* system.
    (i) The pyriform cortex and the associated subcortical amygdaloid nuclei send a massive contribution, the so-called 'ventral pathway', to the medial forebrain bundle.
    (ii) A supplementary pathway from the amygdaloid nucleus, the stria terminalis, runs to the ventromedial nucleus of the hypothalamus.
    (iii) From the hippocampus the postcommissural fornix runs to

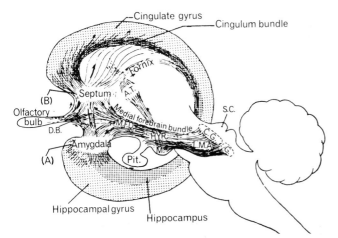

FIG. V.162. Schema of the limbic system with some of its major connections via the median forebrain bundle (MFB) to the hypothalamus and midbrain. (MacLean, P. D. (1958). *Am. J. Med.* **25**, 611.)

the mammillary bodies. Other hippocampal fibres are carried to the medial forebrain bundle in the precommissural fornix which runs through and partly relays in the septal nuclei.
    (iv) From a restricted region of the hippocampus, the medial cortico-hypothalamic tract runs to the arcuate nucleus. This pathway and the stria terminalis are the only two major afferent pathways running directly to the medial hypothalamus. Other afferent fibres take an indirect route via the medial forebrain bundle and lateral hypothalamus to reach the medial hypothalamus.
    (v) The cingulate gyrus establishes connections with the hypothalamus through its projections to the hippocampus.

2. From the *midbrain tegmentum* there is a massive projection of catecholamine- and 5-hydroxytryptamine-containing fibres to the mammillary nuclei and the medial forebrain bundle. It is through this route that the ascending sensory pathways project to the hypothalamus since they do not establish direct connections with it.

3. The *frontal cortex* and the dorsomedial nucleus of the thalamus may send fibres to the medial forebrain bundle. However, recent studies suggest that these pathways, if they exist, are less extensive than formerly believed (Raisman 1966). Nevertheless, considerable areas of the neocortex can project to the hypothalamus indirectly, through their connections with the hippocampus.

Efferent

1. The hypothalamus sends fibres to the amygdaloid nuclei and thence, after relaying, to the pyriform cortex, both in the 'ventral pathway' from the lateral hypothalamus and the stria terminalis from the ventromedial nucleus.

2. Fibres in the medial forebrain bundle run from the hypothalamus to the septal nuclei and after relaying, to the hippocampus.

3. The mammillo-thalamic tract (of Vicq d'Azyr) which mostly arises from the medial mammillary nucleus, passes to the anterior thalamic nucleus and thence to the cingulate gyrus.

4. Descending fibres (which arise mainly in the lateral hypothalamic nuclei lying amidst the medial forebrain bundle) pass to the reticular formation of the tegmentum and thence to the motor nuclei of the bulb and to the spinal motor neurons, thereby contributing to the extrapyramidal facilitatory pathway. These fibres also constitute a potential route over which the hypothalamus can exert its effects upon the autonomic nervous system since direct monosynaptic connections to regions containing either sympathetic or parasympathetic neurons have not been demonstrated.

5. The hypothalamus was formerly believed to project periventricular fibres direct to the dorsomedial nucleus of the thalamus, but like the corresponding afferent connections, it is now thought that comparatively few fibres follow this route.

6. The hypothalamicohypophyseal tract originates from the supra-optic and paraventricular nuclei and runs to the posterior pituitary.

REFERENCES

BARD, P. (1959). *Medical physiology*, 11th edn. Mosby, St Louis.
—— and MOUNTCASTLE, V. B. (1948). *Res. Publ. Ass. nerv. ment. Dis.* **27**, 362.
ECCLES, J. C. (1980). *Bull. Mem. Acad. Roy. Med. Belg.* **135**, 697.
KAADA, B. R. (1951). *Acta physiol. scand.* **24**, Suppl. 83.
RAISMAN, G. (1966). Neural connexions of the hypothalamus. *Br. med. Bull.* **22**, 197.

FUNCTIONS OF THE HYPOTHALAMUS

These include:

1. Participation in the elaboration of emotional behaviour and the experience of emotion, as described above.

2. Regulation of the activity of the anterior pituitary gland through the release of releasing factors and release-inhibiting factors into the hypothalamico-hypophysial portal system [p. 000].

3. Formation of the posterior pituitary hormones and the regulation of their release [p. 523].

4. Participation in the integrated control of the cardiovascular system [p. 131].

5. Regulation of body temperature [p. 349].

6. Control of hunger and feeding.

7. Control of water intake and the sensation of thirst.

These last two functions will now be discussed in detail.

The control of hunger and feeding

Bilateral lesions in the ventromedial nuclei cause hyperphagia (excessive eating); the animals become grossly obese. Subsequent lesions made in the lateral hypothalamic area produce anorexia and the animals die of starvation even though food is plentiful. The suggestion has been made that the ventromedial nuclei act as a 'satiety' centre owing to their functioning as 'glucoreceptors'. There is evidence that these nuclear cells take up radioactive glucose more actively from the blood than do adjacent parts of the brain (Larsson 1954). If gold thioglucose is injected into rats, the preferential uptake of the glucose by the ventromedial hypothalamic nuclei leads to cellular degeneration therein, owing to the toxic action of the gold which accompanies the glucose. As a result the rats become hyperphagic and correspondingly obese.

Larsson (1954) showed that stimulation of the lateral hypothalamic region (caudolateral to the mammillary bodies) caused conscious goats to eat voraciously during the period of stimulation. He believed that he was stimulating hypothalamic projections destined to influence the activity of the dorsal motor nucleus of the vagus.

It has been suggested that there are two hypothalamic centres concerned with hunger and eating

1. The ventromedial satiety centre.

2. The lateral feeding centre.

The satiety centre if inadequately supplied with glucose is supposed to activate the 'feeding centre'; if charged up with glucose, the satiety centre either ceases to stimulate the feeding centre or may actively inhibit it. Cannon showed that hunger pangs are sensations derived from contractions of the empty stomach. However, patients after total gastrectomy manifest hunger. Similarly patients who have suffered vagotomy may be hungry. It would seem that hunger is aroused by changes in the neuraxis; these changes cause 'food-drive' and also cause autonomic effects such as hunger contractions and gastric secretion. The injection of insulin, in doses sufficient to lower the blood glucose level, causes hunger contractions and gastric secretion; both these effects are mediated by vagal motor activity and may be aroused by the lowered glucose content of the ventromedial hypothalamic nuclei.

REFERENCE

LARSSON, S. (1954). *Acta physiol. scand.* **32**, Suppl. 115.

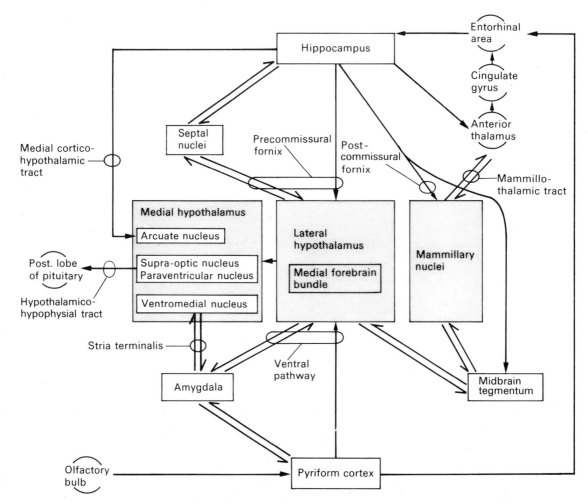

FIG. V.163. Schematic diagram of principal fibre connections between the hypothalamus and other regions of the brain. (Modified from Raisman, G. (1966). *Br. med. Bull.* **22**, 197.)

The control of water intake: thirst

The hypothalamus is involved in the maintenance of fluid balance by participating in the control of water intake as well as in the control of water loss by the body. However, before considering the specific involvement of the hypothalamus, a brief review of the overall control of water intake and its relation to the sensation of thirst will be helpful.

One of the most important mechanisms for securing an adequate level of fluid intake is the sensation of thirst. Thirst is not merely a simple sensation aroused by dryness of the mouth, but a complex 'internal' or 'general' sensation with a considerable emotional component or 'affect'. Thus we say 'I am thirsty' just as we say 'I am hungry, happy, miserable, or tired'. That thirst is more than a feeling of dryness in the mouth has been known since the early nineteenth century, when it was shown that the thirst of dogs running around in the sun and of patients suffering from rabies or cholera could be relieved by the intravenous administration of fluids. Conversely, Claude Bernard demonstrated in 1856 that moistening the mouth and throat without restoring the body fluids does not relieve thirst, while other workers found that thirst persisted after cutting the sensory nerves to the tongue, mouth, and pharynx. A thirsty dog with a gastric fistula would drink copiously but if the water escaped through the fistula its thirst was not assuaged. However if water was introduced through the fistula directly into the stomach and the fistula temporarily occluded to prevent its escape the dog's thirst was relieved. Thus a reduction in body water appears to be the essential factor leading to the 'general' sensation of thirst. This is not, of course, to deny that the dryness of the mouth which accompanies the depletion of body water is a powerful signal of a pressing need to drink. Moreover dryness of the mouth due to local factors such as speaking, smoking, breathing through the mouth, or eating dry or highly flavoured food can lead to drinking even though there is no reduction in the body water. Indeed in normal circumstances the greater part of our fluid intake is initiated by such local oropharyngeal factors or associated with social habits and can be regarded therefore as anticipatory of future needs rather than as restoration of a water deficit.

True thirst, as opposed to simple dryness of the mouth, is always associated either with cellular dehydration leading to shrinkage of the cells or with a diminution in extracellular fluid volume. *Cellular dehydration* may be provoked by water deprivation, the administration of hypertonic saline, or potassium depletion. Holmes and Gregersen demonstrated in 1950 the role of cellular dehydration in stimulating drinking by showing that after intravenous injection of hypertonic saline, sodium sulphate, sodium bicarbonate, or sorbitol—all substances which cause water to leave the cells—dogs drank more water than when given hypertonic urea, isomannide, or glucose—substances which penetrate cells easily. The thirst which follows ingestion of alcohol may similarly be due to osmotic withdrawal of water from the cells, though alcohol is believed to inhibit the release of ADH from the posterior pituitary so that a loss of body water due to increased urine output may also be a factor.

The threshold for the increase in intracellular osmolarity at which drinking is stimulated has been calculated for man and dog by Wolf and found to be 1–2 per cent. In view of the close similarity between these values and the increase in osmolarity found by Verney to cause release of ADH, Wolf suggested that there are thirst receptors in the hypothalamus, a suggestion for which a good deal of direct experimental support has now accumulated [see below].

Thirst is also a prominent feature of *extracellular dehydration* which may occur clinically as a result of haemorrhage, vomiting, or diarrhoea. Experimentally the extracellular fluid volume can be reduced secondarily to a fall in sodium concentration brought about by restricting the dietary intake of sodium or by peritoneal dialysis with isotonic glucose. The extracellular fluid volume can be de-

pleted without affecting the crystalloid concentration by injecting saline with an hyperoncotic colloid concentration into the peritoneal cavity. This withdraws extracellular fluid by a Starling mechanism. Though thirst is a prominent symptom when haemorrhage occurs clinically, bleeding is not a satisfactory way of producing an extracellular fluid deficit experimentally since it is difficult to remove sufficient fluid without debilitating the animal through acute anaemia.

The receptors which initiate the sensation of extracellularly induced thirst have not been positively identified. However, bearing in mind that the most important adverse effect of extracellular fluid loss is to impair the circulation through a fall in blood volume, that 80 per cent of the blood volume is held in the low-pressure capacitance vessels near the heart and that changes in extracellular fluid volume are reflected in the filling of these vessels, it is likely that the receptors concerned are the stretch receptors in the walls of the capacitance vessels and the atria with their afferent fibres running in the vagus nerves. Moreover, these receptors are known to be on the afferent side of reflex arcs that influence vasopressin and aldosterone secretion and therefore the excretion of water and sodium in response to changes in extracellular fluid volume.

Recent studies by Fitzsimons have indicated that a *renal factor* may contribute to extracellularly induced thirst. Constriction of the abdominal aorta above the renal arteries in rats produces drinking, whereas constriction of the aorta below the renal arteries, or above the renal arteries after bilateral nephrectomy, is ineffective. Drinking can be induced by injection of extracts of the renal cortex and this renal factor causing drinking is probably renin. Renin acts on plasma angiotensinogen to produce angiotensin I which is converted to angiotensin II by a converting enzyme found in the pulmonary vessels. Intravenous infusion of renin or angiotensin II causes the water-replete rat to drink. Drinking also follows injection of renin, angiotensin I, or angiotensin II into the anterior diencephalon in rat, monkey, goat, and rabbit, the regions of greatest sensitivity being the septum, the medial preoptic nucleus, and the anterior hypothalamic area. The minimum effective dose of angiotensin II given by intracerebral injection, 0.5 ng in the rat, is at least 1000 times smaller than the minimum intravenous dose. Though the normal circulating level of angiotensin II in the rat is very much lower still, 30 pg ml^{-1} plasma, it must be remembered that normally the stimulus to renin release will be a thirst stimulus in its own right and may be expected to augment any central effect of angiotensin.

Other factors which stimulate the thirst mechanism include hypokalaemia and hypercalcaemia. In hypokalaemia a direct effect has been postulated additional to that attributable to cellular dehydration resulting from loss of potassium. Thirst neurons also appear to be directly stimulated by elevated plasma calcium levels, severe thirst being a marked symptom in vitamin D intoxication and hyperparathyroidism.

Satiety. The mechanisms that terminate drinking following a fluid deficit are also complex. Where cellular dehydration has occurred it can be shown that satiety is reached when sufficient water has been drunk to restore the cells to normal size. However when water is drunk to relieve thirst caused by extracellular dehydration, the ingested fluid is distributed throughout all the body fluids and overhydration of the cells may halt drinking well before the plasma deficit is made up. The extracellular volume is eventually restored because extracellular thirst stimuli, after a delay of some hours, also increase sodium appetite and the resulting increase in salt intake enables the ingested water to be retained in the extracellular fluid compartment. Obviously the aim of drinking is achieved when the fluid deficit is replaced. However, absorption of fluid from the alimentary tract, though rapid, is not immediate, and mechanisms appear to exist which prevent drinking from overshooting the needs of the animal. Thus an animal will cease drinking when an amount of water equal to the deficit has reached the stomach even though absorption may hardly have started and certainly will not have been completed. In several species this inhibition seems to be due to gastric distension since drinking ceases if an amount of water equal to the deficit is placed directly in the stomach or if the stomach is distended with an inflated balloon. However, this is not the case in the dog, suggesting that additional satiety mechanisms may be involved, initiated by the motor acts of drinking and swallowing and the passage of fluid through the oropharynx. Thus a dog with an oesophageal fistula will cease drinking after an amount of water about twice that of the deficit has been drunk, even though none of this water will have reached the stomach or been absorbed.

The hypothalamus. Participation of the hypothalamus in the regulation of water intake has been demonstrated using a variety of techniques including destruction or electrical stimulation of localized areas and intracerebral micro-injection.

During the 1930s W. R. Hess developed methods for studying the effects of electrical stimulation of the brain in conscious animals and found that stimulation of certain parts of the hypothalamus in the cat caused eating and drinking. Andersson and McCann in the mid 1950s elicited drinking in the unanaesthetized goat by localized injections of hypertonic saline as well as electrical stimulation. Injections of 0.01 ml or less of hypertonic saline into the hypothalamus near the paraventricular nucleus caused the goat to drink 2–8 litres of water within 5 minutes. Similar polydipsia (excessive drinking) followed electrical stimulation of the region between the anterior columns of the fornix and the mamillothalamic tracts. Electrocoagulation of this area caused permanent hypodipsia; the goats would not drink water but nevertheless continued to drink milk and take soup. Stimulation of the lateral hypothalamus also has been found to produce drinking in the rat while bilateral destruction of the area leads to a failure to drink where there is a need for water.

There is evidence that the neural systems for drinking in response to thirst induced by cellular and extracellular dehydration are anatomically separate. Damage to the lateral pre-optic area has been shown in the dog and rat to abolish drinking in response to cellular dehydration produced by intravenous administration of hypertonic saline, but did not affect drinking following water deprivation (in which there is both cellular and extracellular dehydration) or intraperitoneal injection of a solution of hyperoncotic colloid in saline (which leads to extracellular dehydration only).

The osmoreceptors which initiate drinking in response to cellular dehydration have been localized just anterior to the hypothalamus and are therefore probably separate from the osmoreceptors involved in ADH release. Injection of 2 μl of hypertonic saline or hypertonic sucrose into the lateral preoptic area induced drinking in the rabbit and rat, but neither isotonic saline, distilled water, nor hypertonic urea caused any drinking. Drinking could not be produced by injection of hypertonic sucrose elsewhere in the hypothalamus and the drinking that normally follows intravenous hypertonic saline was abolished by electrolytic destruction of the lateral preoptic area.

While the neurons that mediate drinking due to cellular dehydration have been localized in the lateral pre-optic area where they can be selectively destroyed without affecting the drinking response to extracellular depletion, the neurons that mediate extracellularly induced thirst are more diffusely spread and similar localization has not been possible. However the loss of response to thirst of either origin when lesions are placed in the lateral hypothalamus suggests that both neural systems converge in this region.

Though the hypothalamus appears to be of primary importance in the control of drinking, experiments involving selective destruction or stimulation indicate that other regions including the septal nuclei, amygdala, and hippocampus can influence water intake. In the monkey the anterior cingulate area seems of particular importance in drinking behaviour.

Neurotransmitters and drinking. The injection of acetylcholine or carbachol into the lateral hypothalamus of the rat causes drinking. Atropine blocks this response and also abolishes drinking in the water deficient rat. Injection of carbachol into many parts of the limbic system induces drinking which is eliminated after destruction of the lateral hypothalamus. Thus in the rat there appears to be a system of cholinergic neurons subserving drinking which converge on the lateral hypothalamus. However, in other species including the cat, monkey, and rabbit, carbachol inhibits drinking. Intrahypothalamic injections of adrenalin and noradrenalin have variable effects on drinking, but subcutaneous injections of the β adrenergic agonist isoprenaline induces copious drinking in the rat, the response being prevented by the β adrenergic blocker propranolol. This action of isoprenaline does not appear to represent a direct effect on the hypothalamus since it is abolished by nephrectomy, suggesting that the renin–angiotensin system is concerned in drinking caused by β adrenergic stimulation.

REFERENCES

ANDERSSON, B. and LARSSON, S. (1961). Physiological and pharmacological aspects of the control of hunger and thirst. *Pharmac. Rev.* **13**, 1–16.
ECCLES, J. C. (1980). *Bull. Mem. Acad. Roy. Med. Belg.* **135**, 697.
FITZSIMONS, J. T. (1972). Thirst. *Physiol. Rev.* **52**, 468–561.
—— (1979). *The physiology of thirst and sodium appetite.* Monographs of the Physiological Society No. 35. Cambridge University Press.

Regulation of body temperature in man

Man is homeothermal, i.e. he maintains his body temperature constant in spite of wide variations in environmental temperature. The term 'body temperature' refers to the temperature of the deeper structures (e.g. viscera, liver, brain). The skin usually has a lower temperature than the depths of the body. This provides a temperature gradient which leads to heat loss from the depths to the surface of the body and thence from the surface to the environment. In hot countries these temperature gradients may be reversed.

The normal deep body temperature in man at rest is 36–37.5 °C. In health it is always kept fairly close to this level by maintaining a balance between heat gain and heat loss [FIG. V.164].

HEAT GAIN

Heat gain is due to:

1. Heat produced in the body.
2. Heat taken up under certain circumstances from the environment.

1. Heat production (thermogenesis)

Heat is produced by the metabolic activities of the body. In some organs, such as the liver and the heart, heat production is relatively constant. On the other hand, skeletal muscle makes a very variable contribution to heat production, at rest very little, in exercise a great deal. The muscular contractions of shivering play an important part in preventing a fall of body temperature in a cold environment.

Heat production under standard resting (basal) conditions is 4 kJ (one kcal) per kg of body weight per hour, or 155–170 kJ (37–40 kcal) (depending on sex) per square metre per hour. This output works out at about 7100 kJ (1700 kcal) day^{-1} in an average man and 6300 kJ (1500 kcal) in an average woman. Moderate physical activity increases heat production to a total of 10 500–12 500 kJ (2500–3000 kcal) day^{-1}; if very heavy work is done the total heat output may rise to 25 000 kJ (6000 kcal) or more, per day. Short bursts of extremely severe exercise may increase heat production temporarily to 10–16 times the basal level. The 'specific heat' of water is 1, i.e. one (small) calorie raises the temperature of 1 g of water by 1 °C. The specific heat of physiological saline, and therefore of the body (which is 62 per cent water) is also, approximately, 1. Therefore, if there were no heat loss, the temperature of the body under basal conditions would rise by 1 °C h^{-1}, and under conditions of normal activity it would rise by 2 °C h^{-1}. But so efficient are the mechanisms for bringing about heat loss that only when work is heavy or when the environmental conditions interfere with the heat loss mechanisms does the body temperature rise well above the normal range.

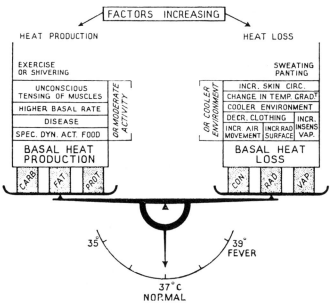

FIG. V.164. Balance between factors increasing heat production and heat loss. (Du Bois, E. F. (1938). *Ann. intern. Med.* **12.**)

2. Heat gained from the environment

The body can take in heat from objects hotter than itself: (i) by direct radiation from the sun or heated ground, or (ii) by reflected radiation from the sky. This type of heat intake is independent of the temperature of the air. The amount of heat gained by radiation can be reduced by wearing garments which reflect the radiations or by making use of any available shade. In the hot desert the body takes up more heat when naked than when covered by thin white clothes.

When, however, the air temperature exceeds that of the skin, the body surface takes up in addition heat from its immediate surroundings; this last sort of heat gain is a great burden to people living in hot climates.

HEAT LOSS

Heat is lost from the body in several ways:

1. By radiation from the body to cooler objects at a distance.
2. By conduction and convection to the surrounding atmosphere if its temperature is lower than that of the body (or rather, that of

the skin). The air in immediate contact with the skin is warmed; the heated molecules move away and cooler ones come in to take their place; these in turn are warmed and so the process goes on. These air movements constitute convection currents.

3. By evaporation of water. The essential fact to remember is that when 1 g of water is converted into water vapour 2.43 kJ (0.58 kcal) of heat is needed and has to be taken up from the environment; this heat is known as the latent heat of vaporization and is measured in kJ per gram. Thus when 1 kg (approximately 1 litre) of water evaporates, 2430 kJ (580 kcal) are taken from the immediate surroundings. Evaporation of water takes place from the lungs, and from the skin. Evaporation of sweat is the principal means of heat loss when the body temperature tends to rise.

The different methods and routes of heat loss are considered more fully below.

Radiation

The magnitude of heat loss (or gain) by radiation depends on the size of the body surface and on the average temperature difference between the skin and the surrounding objects. The part played by radiation, as already indicated, varies widely with climatic conditions. In a temperate climate, a resting person (wearing ordinary clothes) loses about 60 per cent of his heat production by radiation: if he is working, although heat production is increased, there is no great increase in the absolute amount of heat lost by radiation because the skin temperature rises only slightly. (During work the percentage of the total heat loss which is due to radiation obviously becomes smaller.)

Lungs

The expired air leaving the lungs is saturated with water vapour at body temperature; this water is derived by vaporization from the moist mucous membranes of the respiratory passages. The amount of water vapour taken up depends on the initial state of the inspired air; when dry it takes up a good deal but when saturated with water vapour it takes up none at all. On an average (for what such a figure is worth, considering the range of atmospheric conditions) the water loss from the lungs is about 300 ml day^{-1}, equivalent to a heat loss of nearly 730 kJ (200 kcal), i.e. 300 × 2.42 kJ.

Some body heat is also lost via the lungs by raising the temperature of the inspired air to body temperature; therefore an increase in pulmonary ventilation (especially when the air is dry and cool) increases heat loss.

Skin

Heat loss from the skin occurs by:

1. Conduction–convection to a degree which varies with the temperature gradient between the skin and the surrounding atmosphere. Heat loss from the bare head in an environmental temperature of −4 °C may amount to half the total resting heat production in man. Bald men and babies should certainly wear fur or woollen hats in winter! Skin temperature varies directly with its blood flow; the calibre of the skin vessels, especially those in the hands, feet, and face which are under intense vasomotor control can be nicely adjusted to varying body needs. External cold produces cutaneous vasoconstriction and consequently reduces skin blood flow and decreases heat loss; external heat has the opposite effect. The flow of heat from the body depths to the skin is mainly due to conduction within the body, i.e. the carriage of heat from the depths to the surface by the warm blood.

2. Evaporation of water. Water is lost from the skin (a) by insensible perspiration; (b) by sweating.

(a) Insensible perspiration. This consists of the passage of water by diffusion through the epidermis (it is called 'insensible' because it cannot be seen or felt); the fluid lost is not formed by sweat glands. It amounts to 600–800 ml per 24 hours, equivalent to a heat loss by evaporation of about 1700 kJ (400 kcal). It is produced over the whole body surface at a fairly uniform rate and is largely independent of environmental conditions.

(b) Sweating. Because of its outstanding importance in temperature regulation this mode of heat (and water) loss requires detailed discussion.

Sweat glands

There are two kinds of sweat glands in man.

1. *Eccrine*, which are distributed generally over the body surface and secrete a dilute solution containing sodium chloride, urea, and lactic acid. The NaCl concentration is variable, i.e. 0.1–0.37 per cent, and depends on the level of adrenocorticoid activity. Eccrine glands are densest on the palms and soles, next dense on the head, and much less dense on the trunk and extremities. Eccrine glands are supplied by cholinergic fibres present in sympathetic nerves. Atropine inhibits eccrine sweating.

2. *Apocrine* glands develop from hair follicles. They are found mainly in the axilla (though eccrine glands also occur here), round the nipples and, in the female, in the labia majora and mons pubis.

In the horse all sweat glands are apocrine in type, though they resemble most human eccrine glands in their role in temperature regulation.

Apocrine glands are not supplied by secretory nerves, but are stimulated by adrenalin carried in the bloodstream (Lovatt Evans 1957). Atropine does not inhibit this secretion.

In the human axilla apocrine sweat is a milky odourless fluid. The typical axillary odour develops as the result of bacterial action.

Secretion of sweat

Secretion is produced by direct or reflex stimulation of the centres in the spinal cord, medulla, hypothalamus, or cerebral cortex. Eccrine sweat secretion is increased in the following conditions:

1. With rise of external or of body temperature; this so-called *thermal sweating* is produced in two ways: (i) by the rise of body temperature directly affecting the hypothalamic centres; and (ii) reflexly from the stimulated 'warm' nerve endings in the skin.

2. In emotional states (*mental sweating*); this type is limited as a rule to the palms, soles, and axillae, though in extreme cases it may become more generalized. Mental sweating is due to impulses sent out from the higher centres. The term *hyperhidrosis* is used to describe excessive sweating in the regions usually involved in mental sweating. The condition is due to overactivity of centres controlling sweat secretion and not to any abnormal behaviour of the glands themselves (Chalmers and Keele 1952). Hyperhidrosis is reduced by atropine or by ganglion-blocking drugs, and is of course abolished by sympathectomy.

3. In exercise; both thermal and mental factors play a part.

4. Sweating also occurs in nausea and vomiting, in fainting, in hypoglycaemia and in asphyxia. This is due to sympathetic activity.

5. Gustatory sweating occurs in hot climates when spicy foods such as curry or capsicum are eaten. Pain nerve endings in the mouth are stimulated to produce reflex sweating in the head and neck.

Thermal sweating. The maximum rate of thermal sweat secretion in one hour may be as high as 1.7 litres; 8–11 litres may be lost in 5–8 hours. More usually the 24-hour maximal secretion is 12 litres. In the dry, torrid heat of Boulder City, Nevada (maximal shade temperature 34–38 °C, humidity 6–30 per cent), the sweat output was (in one subject) 79 litres in 14 days (i.e. 5.5 litres day⁻¹). Every litre of sweat which is evaporated from the skin leads to the loss of

2430 kJ (580 kcal) of heat from the body. The evaporation of 1.7 litres (the maximum value for 1 hour) results in a heat loss of 4100 kJ (1000 kcal); that of 12.0 litres (the maximum for one day) gives a heat loss of 29 000 kJ (7000 kcal); that of 5.5 litres (the maximum average for a day, recorded over a period of days) gives a heat loss of 7300 kJ (3000 kcal). If the sweat does not evaporate from the skin but is wiped away or merely runs down the body, no heat loss occurs at all. Sweating under such conditions is just a useless or harmful form of fluid loss.

In similar conditions of controlled experimental hyperthermia men sweat more profusely than women. By varying the amount of sweat secreted, the amount of heat which the body can lose can extend over an immense range. When the external temperature exceeds that of the body, evaporation (which in practice means evaporation of sweat) is the only method of heat loss available. Heavy sweating involves a rapid loss of water together with salt from the body; dehydration and also salt deprivation will occur, with the usual consequences, unless enough water is drunk and adequate amounts of salt are taken.

When the external air is not only hotter than the body but is also saturated with water vapour heat loss becomes impossible because the sweat cannot evaporate. Thus, in a dry room at a temperature of 115–126 °C, no rise of body temperature occurs; on the other hand, a stay of 15 minutes in a moist room at 55 °C may raise the body temperature to 37.7 °C.

Temperature regulation has priority over the maintenance of water and salt balance; sweating will thus continue even though it produces severe dehydration and marked salt loss; it is arrested only by circulatory failure.

As sweat comes from the blood, rapid sweating demands a large cutaneous blood flow, and therefore dilatation of the skin blood vessels. This vasodilatation is brought about by:

1. External heat acting directly on the vessels.

2. Reflexly from the cutaneous 'warm endings'.

3. By the rise of blood temperature acting directly on the hypothalamic centre.

The neurogenic vasodilatation may be due to the action of a vasodilator bradykinin-like polypeptide formed as the result of sweat gland activity (Fox and Hilton, 1958). The cutaneous vasodilatation decreases the peripheral resistance, lowering the diastolic pressure, but is accompanied by an increase in the cardiac output.

REFERENCES

Chalmers, T. M. and Keele, C. A. (1952). The nervous and chemical control of sweating. *Br. J. Derm.* **64**, 43.

Fox, R. H. and Hilton, S. M. (1958). *J. Physiol., Lond.* **142**, 219.

Lovatt Evans, C. (1957). Sweating in relation to sympathetic innervation. *Br. med. Bull.* **13**, 197.

NORMAL BODY TEMPERATURE

The temperature of the surface of the body varies with the environment, and the body influences skin temperature in such a way as to help maintain the deep or central body temperature constant. There are, however, temperature differences between different deep structures within the body, and it is not justified to take the temperature of any one structure as representative of deep body temperature as a whole. The temperature in a central organ is determined by its own heat production, its insulation and by the temperature and rate of flow of blood through it. The rectal temperature has been widely assumed to be an accurate index of the temperature of the blood to which the hypothalamic thermoregulatory receptors are exposed. However, when heat is introduced into the body, e.g. by immersing a limb for 3 minutes in warm water, the mouth temperature rises significantly, with little or no change in rectal temperature. The rise in mouth temperature is closely correlated with the temperature rise in the subclavian artery [FIG. V.165]. Oesophageal temperature and the temperature of the tympanic membrane or external auditory meatus respond in the same way as the mouth temperature. Thus the rectal temperature gives a poor reflection of rapid changes in blood temperature but for slower changes, as in fever, the rectal temperature behaves similarly to mouth temperature and is less subject to technical errors.

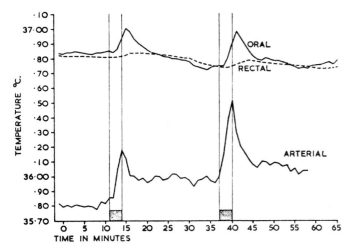

FIG. V.165. Temperature measured in the mouth, rectum, and subclavian artery in a normal subject during immersion of one forearm in a bath of warm water (shaded areas indicate duration of immersion). (Cranston, W. I. (1966). *Br. med. J.* **ii**, 69.)

The clinical thermometer takes 3 or 4 minutes to reach a constant maximum reading in the mouth. A misleadingly high reading is obtained after a hot drink. A misleadingly low reading is obtained after a cold drink, or if the nose is blocked, preventing the mouth from being closed.

Diurnal variation

FIGURE V.166 shows typical fluctuations of body temperature in a normal person during three days. A diurnal variation of 1.5 °C may occur in any normal person. The possible extent of the 'tolerated' diurnal variation is not always fully appreciated, because it is very unusual to take records at frequent intervals during the night in normal persons, and it is commonly at the inconvenient hours of 2–6 a.m. that the minimum temperatures are observed.

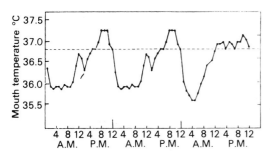

FIG. V.166. Diurnal variation of body temperature. (Samson Wright.)

As the mouth temperature rises, the heart rate increases, and vice versa.

The normal temperature rise during the day is due to increased heat production from muscular activity and metabolism of food.

During sleep the temperature falls, partly because of bodily inactivity but also because of less perfect temperature regulation. When a healthy man is kept in bed the normal type of diurnal temperature variation takes place, but the evening maximum is not so high.

A reversal of the ordinary daily routine, e.g. on going on to night work, ultimately reverses the temperature curve. This does not occur during the first night, for at the hour of the usual minimum (e.g. 3–4 a.m.) there is an imperative desire to sleep. After a few days, acclimatization takes place and the sleepiness passes off.

In the tropics the body temperature is about 0.5 °C above the normal range found in temperate zones. In nervous people, and especially in children, elevations of temperature occur for which there is no adequate explanation.

Age

The temperature in infants is irregular at first, but periodicity gradually sets in with the development of regular periods of activity and rest. Temperature regulation is imperfect: a fit of screaming causes a rise; a cold bath may lower the temperature by 4 °C. In the aged the temperature is subnormal: the body is less active, the circulation is feeble and there is less power of compensation for changes in external temperature. Old people are intolerant of extremes of external temperature.

Menstruation

During menstruation the average temperature is at a minimum. It rises slightly during the next 14 days; the waking temperature (recorded in bed) shows a distinct rise at about the time of ovulation. During the second half of the menstrual cycle the waking temperature is 36.7–37.2 °C. This rise may be due to progesterone [p. 573].

Exercise

Bodily activity greatly increases heat production; only 25 per cent of the energy liberated by the chemical changes in muscle is converted into work, the balance being evolved as heat. The effect on body temperature depends in part on the extent to which heat loss can be increased to make up for heat production; this in turn depends greatly on environmental conditions. However, other factors may also contribute to the rise in temperature. After a three-mile race temperatures as high as 40–41 °C have been recorded in athletes.

Emotion

Emotional stimuli can raise the body temperature by as much as 2 °C and may account for some unexplained fevers.

THE HYPOTHALAMUS AND TEMPERATURE REGULATION

The hypothalamus is the principal 'centre' for the integration of the mechanisms concerned in the regulation of body temperature though the human behavioural responses to thermal stress, such as wearing clothes and building houses, indicate that in man the cerebral cortex also plays an important role. Through its close connections with the thalamus the hypothalamus receives information concerning the temperature of the surface of the body via afferents from thermal receptors in the skin, while receptors in the hypothalamus itself detect changes in the temperature of the blood perfusing this region. On the efferent side the hypothalamus has access to both the somatic and the autonomic nervous systems and can thus modify muscular and glandular activity, cutaneous circulation, sweat secretion, and pulmonary ventilation. Several lines of evidence have established this central role of the hypothalamus in temperature regulation.

Ablation or selective destruction

After removal of the cerebrum in a dog, normal temperature regulation is preserved provided the thermal stress is not too great. The animal responds appropriately to external heat and cold, but cools down if left for a long time in a cold room. However if the midbrain is transected, so that the afferent impulses cannot reach the hypothalamus along the sensory pathways nor can efferent impulses from the hypothalamus reach the muscles and blood vessels, the animal becomes poikilothermic, i.e. its temperature passively follows that of its surroundings. In man a similar result may follow injuries to the pons and medulla which cut off the greater part of the body from the influence of the temperature-regulating centres. If the weather is warm, the room overheated, and the patient well wrapped up and surrounded with hot bottles, the body temperature rises. Similarly, if the room is cold and the patient only lightly covered, the body temperature falls.

Localized destruction of the *anterior hypothalamus* in the region of the pre-optic nucleus leads to a rise in body temperature which may culminate in the death of the animal from hyperthermia. This region of the hypothalamus appeared to exert a continuous inhibition on the mechanisms responsible for heat production as well as activating the somatic and autonomic reactions of panting, sweating, and vasodilatation which are normally engaged when the animal is exposed to a raised environmental temperature. Thus the anterior hypothalamus came to be regarded as a 'heat-responsive' or 'heat loss' centre.

Localized destruction of the *posterior hypothalamus* abolishes the responses of the animal to cold; thus shivering and piloerection and vasoconstriction do not occur and the body temperature falls progressively so that the posterior hypothalamus came to be regarded as a 'heat-maintenance' centre. If the caudal hypothalamic lesion is sufficiently extensive, the descending pathways from the more rostrally situated 'heat-responsive' centre, which pass caudally through the lateral hypothalamic areas, may be interrupted as well. The animal then becomes poikilothermic and can neither prevent its body temperature falling when exposed to cold nor protect itself against a rising body temperature caused by increasing the environmental temperature.

A temporary disturbance of temperature regulation may occur in man after trauma or neurosurgical operations in the vicinity of the hypothalamus. Generally this results in fever which is treated with ice-packs, cold water enemas, or the administration of barbiturates which selectively depress the mechanisms responsible for heat production. Less frequently, the more caudal region of the hypothalamus may be affected resulting in hypothermia which must be corrected by artificial heating. Because general anaesthetics depress the heat-producing mechanisms and lower body temperature, operating theatres must be kept warm and the patient's temperature kept under observation during surgery. The body temperature may also fall if other parts of the temperature-regulating mechanism are depressed, e.g. by chlorpromazine which reduces the activity of the ascending reticular system, or by tubocurarine which paralyses skeletal muscle.

Electrical stimulation

Electrical stimulation of the anterior hypothalamus causes panting, sweating, and vasodilatation—all mechanisms which play a part in heat loss. Stimulation of the posterior hypothalamus increases sympathetic activity causing hypertension, tachycardia, vasoconstriction, shivering, and pupillary dilatation, all of which may feature to a greater or lesser extent in the normal response to cold. Thus the effects of electrical stimulation are also consistent with the view that the anterior hypothalamus is the site of the 'heat-loss' centre and the posterior hypothalamus that of the 'heat-maintenance' centre. However localized destruction and electrical stimulation are relatively crude techniques when applied to such a complicated neuronal network as the hypothalamus, and may result in the simultaneous destruction or stimulation of several distinct and possibly opposing mechanisms. With the introduction of more sophisticated techniques it has become increasingly apparent that the classical concept of anterior 'heat-loss' and posterior 'heat-maintenance' hypothalamic centres may be tenable no longer and, as described below, an alternative hypothesis is taking its place.

Local heating and cooling

Thermodes consisting of fine tubes through which warmed or cooled water is passed have been implanted in the hypothalamus. Local warming can also be produced by passing a 'diathermy' type of high-frequency current through electrodes precisely positioned in the hypothalamus with the aid of a stereotaxic apparatus. The pre-optic region has been shown to be sensitive to such local temperature changes, warming of this region leading to panting and sweating while cooling led to heat conservation and increased heat production. Thus the preoptic region of the hypothalamus appears to function as a *'thermostat'*, regulating the activity of the more posterior hypothalamic regions which integrate the mechanisms responsible for heat loss or heat conservation and production, so as to correct any deviation from a *'set-point'* value of hypothalamic temperature. This 'thermostat' has also been described as a *'proportional controller'* because the magnitude of the thermoregulatory response that is induced is directly related to the extent to which hypothalamic temperature varies from its 'preset' value. The work of Hammel and his colleagues has suggested that this 'set-point' value of hypothalamic temperature is not fixed but may change, as for example in fever or exercise [see below]. The concept that hypothalamic control of heat loss and heat conservation and production is directed towards the maintenance of an adjustable 'set-point' temperature, also helps to explain the relationship between skin temperature and deep body temperature in temperature regulation. Under certain conditions, shivering may occur in cold environments even though the hypothalamic temperature is higher than in hot environments when heat dissipation is taking place. FIGURE V.167 illustrates the findings from an experiment in which a 'thermal clamp' device was used to set the local hypothalamic temperature to any desired level while deep body temperature remained constant. The effect of changing hypothalamic temperature was examined when the dog was exposed to different environmental temperatures. As environmental temperature rose, presumably diminishing the discharge from cutaneous cold receptors while increasing that from the warm receptors, there was a large fall in the level of hypothalamic temperature below which heat production began to increase and a smaller fall in the hypothalamic temperature above which evaporative heat loss became considerable. Use of this 'thermal clamp' technique has also enabled deep body temperature to be varied independently of hypothalamic temperature and has revealed that the 'set-point' levels of hypothalamic temperature at which heat dissipation or heat conservation and production are initiated depend on deep-body (core) temperature as well as on the temperature of the environment. This suggests that there are temperature receptors in the core which project to the hypothalamus as well as the temperature receptors in the skin. Various reports have indicated that such *extrahypothalamic deep-body temperature receptors* may be present in the spinal cord (dog, cat, guinea-pig), in or close to the walls of the great veins (dog, sheep), and in the viscera (sheep).

Unit activity in hypothalamic neurons

The electrical activity of individual hypothalamic neurons has been investigated with microelectrodes. Though most units studied did not alter their activity with changes in hypothalamic temperature, in the anterior hypothalamus cells have been found which increased

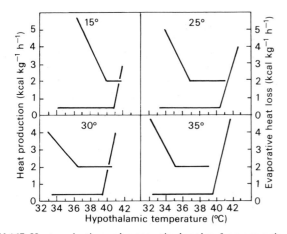

FIG. V.167. Heat production and evaporative heat loss from a conscious dog during manipulation of hypothalamic temperature at air temperatures of 15, 25, 30, and 35 °C. Curves with negative slopes show heat production; curves with positive slopes show evaporative heat loss.
(Adapted from Hellstrom, B. and Hammel, H. T. (1967). *Am. J. Physiol.* **213**, 547. From Hellon, R. F. (1971). Central thermoregulation. In *Handbook of sensory physiology.* Springer, Berlin.)

their firing in response to a rise in hypothalamic temperature [FIG. V.168] while others increased their firing when local hypothalamic temperature fell. It is tempting to speculate that, if the warm-sensitive neurons are responsible for activating the heat dissipating pathways from the hypothalamus, and if the cold-sensitive neurons are responsible for activating the heat-conservation and heat-

production pathways, the combination of heat-sensitive and cold-sensitive neurons could subserve the role of the hypothalamic thermostat. It has further been suggested that these temperature-sensitive anterior hypothalamic neurons could provide the basis of a 'set-point' mechanism. Over the physiological range of hypothalamic temperatures both warm and cold sensors are active, but a temperature change increases the discharge of one group while simultaneously decreasing that of the other. Thus there will be a value of hypothalamic temperature at which the activity of the two sets of sensors will be balanced in terms of the body temperature regulating responses which they engage. This would be the 'set-point' value of hypothalamic temperature [FIG. V.169].

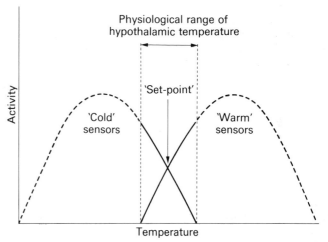

FIG. V.169. Determination of hypothalamic 'set-point' temperature by the balance of activity in warm- and cold-sensitive neurons. The response characteristics of these temperature sensors have been represented by bell-shaped curves. The interrupted portions of these curves are a hypothetical projection based on the suggestion that the temperature-sensitive characteristics of these sensors may be similar to those of the warm and cold receptors in the cat's tongue as described by Hensel and Zotterman (see p. 286).

Other neurons have been found in the hypothalamus which altered their firing rates in response to peripheral warming and cooling]FIG. V.170] while a few neurons have been found to increase their activity in response to both a rise in hypothalamic and peripheral temperature. This would be consistent with the concept of a temperature-regulating system sensitive to the interaction of thermal information from both sources. An alternative hypothesis for the 'set-point' mechanism suggests that some of the hypothalamic neurons with firing rates unaltered by changes in hypothalamic

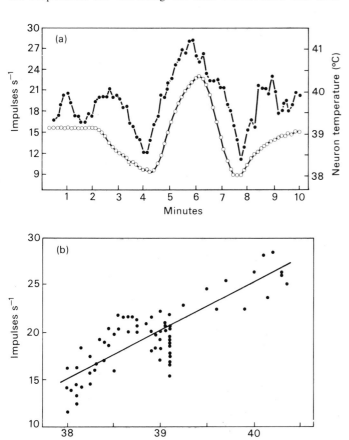

FIG. V.168. Response of a hypothalamic neuron sensitive to warming. (a) Time plot of firing rate (●) at 8-s intervals and neuron temperature (○). (b) Correlation of firing rate and temperature, each point derived from corresponding pairs of points in (a); line drawn from regression equation.
(From Hellon, R. F. (1967). *J. Physiol.*, *Lond.* **193**, 381.)

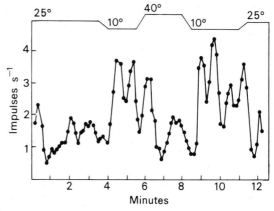

FIG. V.170. Responses of a hypothalamic neuron in a rabbit (urethane anaesthesia) during changes in ambient temperature. The top of the frame shows ambient temperature (°C).
(From Hellon, R. F. (1970). *Pflügers Arch. ges. Physiol.* **321**, 56.)

temperature may also constitute part of the temperature-controlling system. Their discharge could provide a 'reference' signal which might represent the 'set-point' temperature, while some function of the difference between the firing rates of these temperature-insensitive neurons and of the neurons sensitive to warming or cooling would provide an 'error' signal. The extent to which heat-dissipating or heat-conserving and producing mechanisms were brought into action to correct any deviation from the 'set-point' temperature would then depend on the direction and magnitude of this error signal.

Hypothalamic neurotransmitters and body temperature regulation

The hypothalamus contains high concentrations of adrenalin, noradrenalin, and 5-hydroxytryptamine (5HT or serotonin). Feldberg and Myers (1973) have shown that injection of 5HT into the lateral cerebral ventricle of the unanaesthetized cat produces a large increase in body temperature, whereas adrenalin and noradrenalin cause a fall in body temperature [FIG. V.171]. As the same effects followed injections of minute amounts of these amines into the anterior pre-optic region of the hypothalamus, it seemed that temperature regulation might be achieved by the activity of two sets of hypothalamic neurons; one responsible for initiating heat conservation and production and utilizing 5HT as its transmitter, and the other responsible for initiating heat dissipation and utilizing a catecholamine transmitter. Unfortunately for this simple and attractive hypothesis, though responses similar to those in the cat are seen on intraventricular injection in the dog, ox, and monkey, 5HT in the rabbit, sheep, and rat causes a fall and catecholamines a rise in body temperature. It is not known what effects intraventricular injections of these amines have in man. However some of this species variability may be due to the use of excessive doses leading to synaptic blockade rather than excitation or may

arise from differences in the thermoregulatory state of the animal at the time of the injection.

Myers and Sharpe (1968) have adduced more direct evidence for the release of specific transmitter substances in the hypothalamus during thermoregulation. Sterile saline was injected into the anterior hypothalamus of a donor monkey, then withdrawn and injected into the anterior hypothalamus of a recipient animal. When the donor animal was in thermally neutral conditions this had no effect on the recipient monkey, but when the donor was cooled this 'push–pull' procedure caused the recipient to shiver and its temperature rose. Heating the donor led to a fall in the body temperature of the recipient. More recently, the saline withdrawn from the donor monkey during cooling has been shown to contain an increased concentration of 5HT, while the liberation of noradrenalin was found to be augmented by peripheral warming.

Myers and his colleagues have shown that in the unanaesthetized monkey cholinergic mechanisms are also involved in the hypothalamic regulation of body temperature. Micro-injection of acetylcholine or carbachol at sites scattered diffusely throughout both the anterior and posterior hypothalamus led to an intense rise in body temperature accompanied by vigorous shivering. However, in a circumscribed area, at the junction of the posterior hypothalamus and mesencephalon, the injections led to a marked fall in temperature. In complementary experiments the 'push–pull' technique of saline injection and withdrawal was used to study acetylcholine release in the hypothalamus. In the anterior hypothalamus enhancement of acetylcholine release was found only in response to cooling, but in the posterior hypothalamus and mesencephalon, while cooling enhanced acetylcholine release at the majority of sites, in a proportion of cannula positions acetylcholine release was increased by heating the monkey. These findings suggested that two cholinergic pathways are involved in temperature regulation; a heat conservation and production pathway originating in the anterior hypothalamus and passing through the posterior hypothalamus, and a heat dissipation pathway originating in the posterior hypothalamus.

FIGURE V.172 illustrates a scheme proposed by Myers to explain the inter-relationship in the control of body temperature of these neurochemical systems in the hypothalamus of the monkey. The cells in the preoptic region of the hypothalamus which contain 5HT are possibly the cells which increase their firing in response to cooling, and when these cells are stimulated by cooling or a

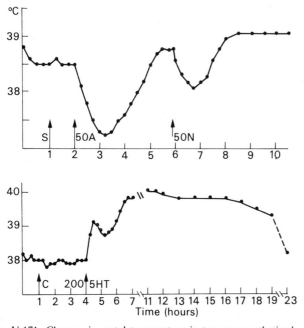

FIG. V.171. Changes in rectal temperature in two unanaesthetized cats. Each arrow indicates an injection of 0.1 ml into a lateral ventricle. In the upper record, the successive injections contained 0.9 per cent NaCl(S), 50 µg adrenalin (50 A), and 50 µg noradrenalin (50 N). In the lower record, the injections were of 100 µg creatinine sulphate (C), and 200 µg 5HT (200 5HT).
(From Fox, R. H. (1974). Temperature regulation with special reference to man. In *Recent advances in physiology*, No.9. Adapted from Feldberg, W. and Myers, R. D. (1964). *J. Physiol., Lond.* **173**, 226.)

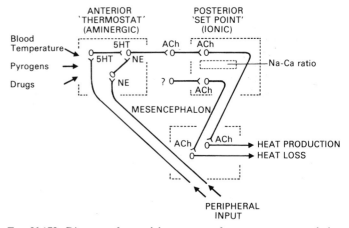

FIG. V.172. Diagram of a model to account for temperature regulation under normal conditions as well as during a pyrogen-induced fever. Factors which affect the aminergic 'thermostat' in the anterior hypothalamus are given, and the outflow from the posterior hypothalamic 'set-point' is mediated by a cholinergic system which passes through the mesencephalon. 5HT = 5-hydroxytryptamine; NE = noradrenalin; ACh = acetylcholine. (From Myers, R. D. (1971). *Pyrogens and Fever*. CIBA Foundation Symposium. Elsevier, Amsterdam.)

pyrogen, a cholinergic pathway to the posterior hypothalamus, initiating heat production, is activated. On the other hand, when the noradrenalin (NA)-containing cells, which may be the cells whose discharge is increased by heating, are stimulated by warming, the 5HT-cholinergic heat production pathway is inhibited by noradrenergic blockade of the synapse of the 5HT cell on the acetylcholine-containing cell. The suppression of the heat production pathway permits the second cholinergic system in the posterior hypothalamus to activate the efferent heat loss pathway, which may also be stimulated from extrahypothalamic sources. Note that this scheme proposes that the noradrenergic anterior hypothalamic neurons initiate heat loss by inhibiting the 5HT-cholinergic heat production pathway, rather than by direct activation of the cholinergic heat loss pathway in the posterior hypothalamus. This is because no evidence has been obtained in the monkey for a chemically mediated pathway responsible for heat loss running from the anterior to the posterior hypothalamus.

The scheme illustrated in FIGURE V.172 is also consistent with the hypothesis that body temperature is determined by a 'set-point' temperature mechanism in the posterior hypothalamus, the particular set-point value representing the resultant of the activities of the two cholinergic outflows in a given set of circumstances, while regulation about this set-point value is achieved by the activity of an aminergic 'thermostat' in the pre-optic region of the anterior hypothalamus.

A major factor determining the set-point temperature appears to be the *ratio of sodium to calcium ions in the posterior hypothalamus*. In several species including the cat, rabbit, rat, and monkey, an increase in the ratio of sodium to calcium ions has been shown to lead to a sustained hyperthermia and a decrease in this ratio to a sustained hypothermia. These temperature changes were produced only when the infusions of excess sodium or calcium ions were made into the cerebral ventricles or the mammillary region of the posterior hypothalamus. Infusion into other hypothalamic regions was ineffective as were infusions of potassium or magnesium ions. After the body temperature had been reset at a new level by a maintained disturbance of the balance of sodium and calcium ions in the posterior hypothalamus the animal thermoregulated to restore this new set-point level when attempts were made to alter its body temperature by instilling hot or cold water into the stomach.

REFERENCES

BENZINGER, T. H. (1969). Heat regulation: homeostasis of central temperature in man. *Physiol. Rev.* **49**, 671–759.
BLIGH, J. (1972). *Essays on temperature regulation* (ed. J. Bligh and R. Moore). North Holland, Amsterdam.
CABANAC, M. (1975). Temperature regulation. *A. Rev. Physiol.* **37**, 415–39.
COOPER, K. E. (1966). Temperature regulation and hypothalamus *Br. med. Bull.* **22**, 238.
FELDBERG, W. S. and MYERS, G. D. (1963). *Nature, Lond.* **200**, 1325.
FOX, R. H. (1974). Temperature regulation with special reference to man. In *Recent advances in physiology*. No. 9 (ed. R. J. Linden). Churchill Livingstone, Edinburgh.
HELLON, R. F. (1971). Central thermoreceptors and thermoregulation. In *Handbook of sensory physiology*, Vol. III/i. *Enteroceptors* (ed. E. Neil). Springer, Berlin.
HENSEL, H. (1981). *Thermoreception and temperature regulation*. Academic Press, London.
MYERS, R. D. and SHARPE, L. G. (1968). Temperature in the monkey: transmitter factors released from the brain during thermoregulation. *Science, N.Y.* **161**, 572–3.

Role of ductless glands

On cooling a dog, the metabolic rate increases by 7 per cent before any indication of increased muscle activity sets in; this is probably a measure of the extent to which the ductless glands help in the 'fine regulation' of body temperature.

1. Adrenal medulla. Exposure to external cold reflexly stimulates secretion of adrenalin which stimulates metabolism and decreases heat loss.

2. Thyroid. External cold (via hypothalamic thyrotrophin releasing hormone and pituitary thyrotrophin) stimulates thyroid secretion which increases heat production, mobilizes glycogen, and stimulates gluconeogenesis.

(i) When an animal (rat) is transferred to a cold environment there are histological changes in the thyroid indicative of functional activity. The gland becomes intensely congested; the amount of colloid decreases and loses its affinity for haematoxylin; the lining cells become columnar and the mitochondria enlarge and become more distinct. On exposure to an external temperature of 37 °C the gland passes into a 'resting' state: the alveoli of the gland are distended with deeply staining colloid, and the mitochondria become indistinct.

(ii) Thyroidectomized animals show impaired temperature control. Patients with myxoedema tend to have a subnormal temperature; the febrile response to injection of bacterial vaccines is reduced.

3. Adrenal cortex. Exposure to external heat or cold stimulates secretion of adrenal corticoids.

RESPONSES TO HEAT AND COLD IN MAN

Man's ability to survive and work in hot or cold climates depends on *physiological* mechanisms for temperature regulation and on *behavioural* responses by which man alters the environment so as to warm or cool the body to the zone of comfort. In the maintenance of homeothermy the physiological response is the more important versus heat and the behavioural response versus cold. Heat is continuously generated inside the body and to maintain constant body temperature there must be a net flow of heat from the body to the environment in all climates. In the cold, heat loss may be easily modulated by varying the insulation to impede heat flow; in hot climates heat must be extracted against the natural gradient by some form of heat pump. An overcoat is easier to design and wear than a refrigerator; since hyperthermia is a more serious problem than hypothermia, man has evolved a greater reserve of heat-eliminating than heat-conserving capacity (Fox 1965). Man is structurally and functionally best fitted to live in a hot, wet environment. The naked human body has little hair as insulation and the two million sweat glands serve admirably to promote cooling by evaporation of sweat. Anthropologists have shown that man evolved in a tropical climate, and migrations to colder regions of the world were made possible only by the development of behavioural responses to reduce heat loss from the body by the wearing of clothes, building of shelter and by the use of fire for warming the micro-climate.

Bodily response to heat

Effect of exercise

When more heat is produced as a result of physical exertion there is a compensatory increase in heat loss.

1. The blood flow through the skin is greatly increased, leading to a rise of skin temperature, and therefore to a greater temperature gradient between the body surface and the environment. Heat loss by radiation only increases by 60 kJ (15 kcal) per hour for each 1 °C rise of skin temperature. Total heat loss by conduction–convection and radiation, though greater than at rest (e.g. rising from 300 to 600 kJh^{-1} (75 to 150 kcal h^{-1}) now constitutes only 20–30 per cent, instead of 80 per cent of the total heat loss from the body.

2. The main heat loss in exercise is due to increased secretion and vaporization of sweat. In one experiment, maximal heat production

was at the rate of 2900 kJ h⁻¹ (700 kcal h⁻¹); heat loss also greatly increased but never exceeded 2500 kJ h⁻¹ (600 kcal h⁻¹); body temperature, therefore, rose temporarily. Sweat secretion in exercise is partly 'thermal' and partly 'mental' in origin; it subsides slowly after the exercise is over.

Though deep-body temperature rises during exercise this is not simply a consequence of the increased heat production, since the rise in temperature is independent of environmental temperature over a wide range, and is determined solely by the metabolic rate [Fig. V.173]. The simplest explanation for this would be an elevation of the hypothalamic 'set-point' temperature, the rise being related to the rate of working, and a direct influence from the motor cortex has been proposed to account for this. However, this relatively simple explanation has proved to be inadequate to explain all the effects of exercise on thermoregulation and an elevation of set-point is probably only one factor in what has now been recognized as a situation which is extremely complicated and at present incompletely understood.

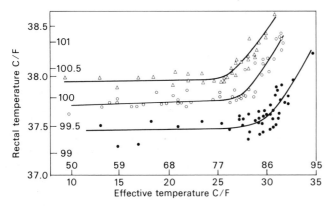

Fig. V.173. Equilibrium rectal temperatures of a subject working at energy expenditures of 180 (●), 300 (○) and 420 (△) kcal h⁻¹ in a wide range of climatic conditions. (4.2 kJ = 1 kcal.)
(From Lind, A. R. (1963). *J. appl. Physiol.* **18**, 51.)

Effect of a hot bath

The subject gets into a hot bath and stays in it for about five or six minutes; observations are made while in the bath and after getting out of the bath until the control levels are regained. In a bath at 43.5 °C the mouth temperature may rise from about 36.5 °C to over 38.0 °C; the heart rate may increase from 75 to 115 and the respiration rate from 12 to 28 min⁻¹. Mouth temperature may subsequently take some 30 minutes to return to normal, though respiration and heart rate recover rather more rapidly. This experiment shows clearly the limitations of normal temperature regulation, and how it can break down quickly under conditions of stress. In a bath at 43.5 °C heat is being taken up by the immersed parts of the body, and heat loss depends mainly on evaporation of sweat from the exposed parts of the body. As the atmosphere in the bathroom is very humid, evaporation of sweat is probably minimal. Body temperature is therefore raised by the heat released by metabolism as well as by the heat taken up from the surrounding medium. This experiment is of clinical interest in showing that the body temperature may be raised for a considerable time after a hot bath.

Effects of raised air temperature

The main automatic compensatory reactions consist (as in exercise) of cutaneous vasodilatation and sweating; adrenalin and thyroxine secretion are inhibited; adrenal corticoid secretion is increased. If the external temperature is lower than body temperature, the increased blood flow through the skin leads to increased heat loss by conduction, convection, and radiation. If the external temperature

exceeds body temperature, heat loss can only result from evaporation of sweat. Some of the changes are considered in detail below.

1. Sweating. When the room temperature is raised slowly, e.g. from 26 to 41 °C, there is a long latent period (e.g. 30 minutes) before sweating sets in. Sweating usually commences quite suddenly and coincides with and is due to a rise of internal temperature (acting centrally) and some rise of skin temperature (of the order of 1–1.5 °C) (acting reflexly). Sweating progressively increases in intensity and may continue to increase even when the room temperature declines; the skin temperature may later fall owing to loss of heat produced by vaporization of sweat. In hot weather sweating may be produced immediately on exposure to greater external heat.

The importance of sweating in the response to heat is well illustrated by the findings in a patient with *congenital absence of sweat glands*. In the winter his temperature regulation was normal; in the summer, however, it broke down. Thus in July and August his morning temperature was 36.0 °C and his evening temperature 39.2 °C; pulse rate and respiration followed the temperature changes. The patient and a normal subject were exposed naked in a hot, moist room for 30 minutes with the following results:

	Skin temperature (°C)	Oral temperature (°C)	Weight loss skin and lungs (g)	Urine volume (ml)
Normal atmosphere:				
Patient	34	37		
Control	33	37		
Hot moist atmosphere:				
Patient	40	38.5	22	270
Control	37.5	37	262	10

The patient's skin remained dry and velvety; he felt sick and ill and 'panted like a dog'; there was considerable diuresis; pyrexia developed (the mouth temperature rose to 38.5 °C). The control sweated profusely, passed very little urine, and maintained his normal body temperature.

2. Changes in blood and circulation. There is an increase in cardiac output but a decrease in diastolic pressure, since the peripheral resistance is reduced by cutaneous vasodilatation. There is initially an increase in plasma volume. The effect on systolic pressure is variable, depending on the extent to which the increased cardiac output compensates for the decreased peripheral resistance. The pulse rate always increases even if there is no significant pyrexia.

The circulation under these conditions is adversely affected by standing. Normally, on passing from the horizontal to the erect position the blood vessels in the legs are reflexly constricted which prevents accumulation of blood in the dependent parts, and maintains an adequate blood supply to the brain; during heat exposure vascular tone is dominated by temperature requirements and the normal postural adjustments do not occur. Thus, in hot environments, a subject who is strapped to a board soon faints if he is passively tilted from the horizontal towards the erect position and kept there for some time: in addition his pulse rate further increases.

All these effects are enhanced by accompanying anhydraemia and salt loss. When these are severe the plasma volume, the cardiac output, and the blood supply to the skin decrease; as the cutaneous blood supply becomes inadequate, sweat secretion diminishes and may cease. With the main avenue of heat loss thus closed, body temperature rises 'explosively' and the general condition becomes precarious.

3. Heat cramps. If the Na^+ and Cl^- lost in the sweat are not made good, muscular cramps develop.

4. Hyperpnoea develops, the alveolar CO_2 falls and there is, consequently, an alkalaemia which is compensated for by the passage of an alkaline urine, and decreased NH_3 formation by the kidney.

5. Heat exhaustion. This is due to hyperpyrexia, to salt loss, and to dehydration.

6. Acclimatization. Prolonged exposure to external heat leads to 'exhaustion' of the sweat glands; the volume of sweat secreted falls and its Na^+ and Cl^- content rises, i.e. the concentration of these ions resembles more closely that of a filtrate from the plasma. On the other hand after repeated bouts of exposure to external heat useful adaptations develop: (i) sweating commences at a lower level of body temperature; (ii) the Na^+ and Cl^- content of the sweat falls with resulting better preservation of the electrolyte content (and crystalloid o.p.) of the extracellular fluids. This latter adaptation is due to increased secretion of the pituitary hormone ACTH which in turn causes a discharge of adrenal corticoids. The latter promote the reabsorption of Na^+ and Cl^- from the sweat back into the blood.

Role of clothing and wet-bulb thermometer level in response to heat. About 20–30 litres of air are entangled between the clothes. In hot weather the clothes retain moisture and the cooling effect of sweating is diminished. Evaporation takes place from the surface of the clothes and only affects the skin through a layer of damp undergarments. Opening the jacket, waistcoat, and shirt gives a feeling of greater comfort, and more moisture can evaporate from the skin surface.

Haldane, attired in flannel trunks, canvas trousers, boots, and stockings, exposed himself to a hot, humid atmosphere (high wet-bulb thermometer temperature). If the wet-bulb temperature was 29.5 °C, the rectal temperature remained normal when the subject was at rest in still air. If the wet-bulb rose above this level, the rectal temperature rose markedly. At 31.5 °C the rise of body temperature was 0.5–0.8 °C h⁻¹; at 34.5 °C, 1 °C h⁻¹; at 36.7 °C, 2.2 °C h⁻¹. In moving air, e.g. with an air current of 50m min⁻¹, a wet-bulb temperature of 35 °C could be borne without rise of internal temperature. If muscular work was done in still air, the limit of wet-bulb temperature was still lower.

Heat stroke

This grave syndrome is most commonly seen in the tropics and during heat waves in subtropical countries. It also occurs as a complication of diseases causing hyperpyrexia. The symptoms are due to hyperpyrexia, salt loss, and dehydration. If the environmental temperature is high heat cannot be lost by radiation and convection, and if, in addition, the air is moist and still, evaporation of sweat cannot take place. In some cases, but by no means all, the onset of hyperpyrexia is preceded by sudden cessation of sweating. In healthy people the precipitating cause is strenuous physical activity during hot weather, which leads to sudden overloading of the thermoregulatory mechanisms and a rise of body temperature to 40–42 °C or even higher. In most cases the onset is acute with sudden loss of consciousness, and in severe or fatal cases convulsions occur. A vicious circle is established since the rise of body temperature increases body metabolism and heat production. Hypotension and circulatory failure reduce blood flow and heat transport to the skin and thus reduce heat loss. The circulatory shock impairs renal and hepatic function and damages the central nervous system.

In the treatment of heat stroke reduction of the hyperpyrexia is the first essential. Immersion of the patient in ice-cold water, or application of ice-filled rubber bottles helps to lower body temperature and the induced cutaneous vasoconstriction can be prevented by chlorpromazine which also inhibits shivering and convulsions.

FEVER

This term is defined as a raised central temperature. It results from a disturbance of the temperature regulating centre. Fever is most commonly caused by:

1. Infections due to bacteria (e.g. pneumonia), viruses (e.g. influenza), and protozoa (e.g. malaria).
2. Tissue destruction, as for example in cardiac infarction, uninfected neoplasms, serum sickness, and rheumatic fever.

In fever there is increased heat production, but this alone does not account for the pyrexia, because the changes are of an order of magnitude which could be easily compensated for under normal conditions; but heat loss is decreased simultaneously, fully accounting for the rise of body temperature. At febrile temperatures thermoregulation in response to heating or cooling the body is just as precise as in the normal state, and the usual diurnal variation in body temperature is present. Fever is thus attributed to a resetting of the set point of the central thermostat.

In the initial stage a rigor or shivering often occurs and increases heat production. The skin vessels are constricted, minimizing heat loss; a rapid rise of temperature therefore takes place and the blood pressure rises. During the fastigium, when the temperature is at its height, the cutaneous vessels are relaxed, the skin is flushed and the blood pressure falls, especially in hypertensive patients; the sweat glands are usually inactive. The rise of temperature increases the metabolic rate by accelerating the oxidative activities of the body; but the heat loss is relatively inadequate to cope with this increased production. During the defervescence, when the temperature falls, marked sweating occurs, and heat loss is now greater than heat production.

Some of these phenomena are well demonstrated during a malarial rigor. During the shivering attack there is a sudden marked increase in heat production, e.g. from 330 to 950 kJ h⁻¹ (80 to 230 kcal h⁻¹). Heat loss, on the other hand, is unchanged or decreased because of cutaneous vasoconstriction so that all the extra heat liberated is stored in the body. The body temperature may rise to 41 °C as temperature regulation is in abeyance. After an interval the thermostatic control is readjusted to 37 °C; sweating sets in, and with increased vaporization there is additional heat loss, bringing the body temperature down to normal once more.

The general manifestations of fever are due to the pyrexia, dehydration, disturbed electrolyte balance in the blood (e.g. Na^+ and Cl^- loss, alkalaemia), and the action of toxins.

Pyrogens

The word 'pyrogen' was introduced by Burdon Sanderson (1876) to denote fever-producing substances extracted from putrefying meat. It is now used to describe all substances which produce fever on parenteral injection.

Pyrogens occur in bacteria and in mammalian tissues. Bacterial pyrogens occur mostly in Gram-negative bacteria and are also known as bacterial endotoxins. They are not destroyed by boiling or autoclaving and may therefore occur in parenterally-administered infusions of saline, etc., unless precautions are taken to prevent bacterial contamination during preparation of the solutions. Bacterial pyrogens have been identified as lipopolysaccharides and are highly potent agents. In man intravenous injection of 0.1 μg can produce fever.

Bacterial pyrogens cause fever when injected into the cerebral ventricle or directly into the hypothalamus in cats and monkeys. However, there is some doubt as to whether bacterial pyrogens normally cross the blood–brain barrier and it is likely that they induce fever by interacting with polymorphonuclear leucocytes to produce a polypeptide *leucocyte pyrogen* (endogenous pyrogen). After intravenous injection, bacterial pyrogen alone produces a rise of temperature after a latent period of 70–75 minutes; after incubation with blood for 3 hours at 37 °C the same amount of bacterial pyrogen produces a rise of temperature in 20 minutes. Leucocyte pyrogen appears to be much more potent and specific in its action. Injections of leucocyte pyrogen into the preoptic and anterior regions of the rabbit hypothalamus led to fever after about 8 minutes, whereas the latent period following injections of bacterial pyrogen into the same sites was about 25 minutes (Cooper *et al.* 1967). Moreover the amount of leucocyte pyrogen required to produce fever when injected into the anterior hypothalamus was only one hundredth of the amount required when given intravenously, whereas with bacterial pyrogen the amounts required by either route were the same. There is no evidence that leucocyte pyrogen acts by release of 5HT or noradrenalin.

Fever can also be produced experimentally by the injection of *prostaglandin E₁* into the cerebral ventricles in cats, rabbits, sheep, rats, and monkeys, and the prostaglandin content of the cerebrospinal fluid increases during fever induced by bacterial pyrogens. As with leucocyte pyrogen, microinjection of prostaglandin E₁ into the anterior hypothalamus produces fever, while injection into the posterior hypothalamus is ineffective. In prostaglandin-induced fever, as in pyrogen-induced fever, normal thermoregulation takes place at the elevated body temperature level. These similarities between the effects of pyrogen and prostaglandin have led to the suggestion that prostaglandin E₁ may be a mediator of pyrogen-induced fever. There are minor differences between the effects of pyrogen and prostaglandin E₁. On injection into the anterior hypothalamus leucocyte pyrogen produces a febrile response only after a latency of several minutes; this latency is absent after prostaglandin E₁ injection. However, such a difference is to be expected if prostaglandin E₁ mediates the effects of leucocyte pyrogen. Another difference is that 4-acetamidophenol, an aspirin-like drug, blocks the fever induced by pyrogens but not that induced by prostaglandin E₁. Again this is consistent with the role of mediator suggested for prostaglandin since Vane has found that aspirin-like drugs block prostaglandin synthesis.

Both pyrogens and prostaglandin E₁ increased the activity of cold-sensitive neurons in the hypothalamus and depressed that of warm-sensitive neurons while the discharge of temperature-insensitive hypothalamic neurons was largely unaltered. 'Biasing' the response of the temperature-sensitive neurons in this way might be expected to lead to an elevation of body temperature until a new hypothalamic temperature was attained at which the previous balance of activity between cold-sensitive and warm-sensitive neurons was restored. Studies of hypothalamic unit activity have also suggested that antipyretics act by blocking the effect of pyrogens on the temperature-sensitive neurons. Wit and Wang (1968) reported that the depression of the discharge of warm-sensitive neurons in the pre-optic region of the hypothalamus which followed administration of bacterial pyrogen, was reduced by the injection of the sodium salt of acetylsalicyclic acid. On the other hand, salicylates have no effect on the temperature of the normal animal or on the hyperthermia produced by hypothalamic cooling, indicating that such antipyretics probably do not affect the temperature-sensitive cells directly nor do they directly influence the activity of the effector pathways for heat loss or heat production and conservation.

Antipyretic drugs

Drugs such as sodium salicylate and aspirin (acetylsalicylic acid)

reduce body temperature in fever due to infections; they do not reduce normal temperature, nor the temperature in hyperthermia due to high environmental temperature. They probably act by antagonizing the actions of pyrogens on the hypothalamus. They 'reset the thermostat' and the temperature returns towards normal as the result of increased heat loss brought about by cutaneous vasodilatation, hydraemia and sweating. The drugs are still effective when sweating is prevented by atropine.

Cortisone and related compounds are also antipyretics. Their mode of action is unknown.

REFERENCES

Cooper, K. E., Cranston, W. I., and Honour, A. J. (1967). Observations on the site and mode of action of pyrogen on the rabbit brain. *J. Physiol., Lond.* **191**, 325–7.

Moncada, S. and Vane, J. R. (1978). *Br. med. Bull.* **34**, 129.

Wit, A. and Wang, S. C. (1968). Temperature-sensitive neurons in preoptic/anterior hypothalamic region: actions of pyrogen and acetyl salicylate. *Am. J. Physiol.* **215**, 1160–9.

Bodily responses to cold

In most parts of the world the environmental temperature is well below that of deep body structures. Hence there must be active heat production to maintain the normal deep body temperature at a constant level. If the environmental temperature falls greatly the body can help to keep its internal temperature constant by increasing heat production or by reducing heat loss.

1. Heat production is increased by raising the metabolic rate. Cold promotes hunger which increases food intake. Secretions of adrenalin and of thyroid hormones produce small increases in metabolic rate. Shivering and voluntary muscular activity release large amounts of heat in the muscles. In the newborn infant *non-shivering thermogenesis* in brown fat is promoted by sympathetic activity.

2. Heat loss is reduced by cutaneous vasoconstriction. This reduces the amount of heat transferred by the blood from the body core to the body surface. In the limbs a counter-current heat exchange between the warm arterial blood and the cooled blood in the venae comitantes further helps to conserve body heat. The cutaneous vasoconstriction may raise the blood pressure.

Although most cutaneous vessels constrict in response to cold, the vessels of the hand, foot, ear, and face may exhibit 'cold vasodilatation'. This occurs on exposure of these parts to moderate cold (e.g. 5–10 °C) when the body is generally warm. It may be regarded as a desirable response which will prevent the development of frostbite. The cold vasodilatation response increases as a person becomes acclimatized to cold.

Heat loss is also reduced by a thick layer of subcutaneous fat which provides excellent insulation. This may explain the endurance by long-distance swimmers of temperatures of 15 °C for several hours. An extra 1 mm of subcutaneous fat is equivalent to raising the temperature of the water by 1.5 °C.

Clothing in cold climates

The purpose of clothing as a protection against cold is to provide a layer of thermal insulation so as to reduce the amount of heat lost from the body surface to the environment. In fact the insulation is provided by the 'dead' air trapped in the clothing. Animal furs are excellent in this respect, not only for their original owners, but also for human beings. This has long been recognized by Eskimos who wear two layers of fur, one facing outwards and the other facing inwards. The maximum thickness of dead air should be maintained even on movement. For example, gloves should be made with curved fingers since the fingers are more commonly bent than straight. When the fingers are bent inside ordinary straight-fingered gloves the dead air layer at the flexures is much narrowed and thermal insulation is reduced.

When exercise is undertaken in a cold climate the movements of air inside the clothing should increase heat loss from the body surface. If heat loss is inadequate sweating occurs and is very uncomfortable in these circumstances. The ideal clothing for cold climates incorporates materials which are permeable to water and impermeable to air movements. Such features are possible because movement of air through a material is a different process from diffusion of water vapour.

Adaptation to cold

Without protection by clothing, shelter and heating man could not survive in the polar regions; no physiological adaptation could enable a nude man to withstand temperatures of $-70\,°C$ to $-100\,°C$ or blizzards of 60 km h^{-1}! If artificial insulation is really effective physiological adaptation does not occur.

The Eskimos' igloo is made of hard snow which is an excellent insulator by virtue of myriads of small air cells. The oil lamp used for heating melts the inner layer of snow which refreezes to form a smooth reflecting surface which conserves radiant heat; the temperature rises to about 30 °C. The modern polar hut has an outer weatherproof skin, walls made of material layered to trap pockets of air and an internal lining of aluminium foil to reflect radiant heat from diesel stoves. Men at polar bases are out of doors for less than 10 per cent of the time and no general adaptation to cold takes place. Eskimos and Lapps have a higher blood flow through the hands than white men living under the same conditions. This may be related to the raised metabolic rate brought about by a high protein diet.

True adaptation to cold occurs among the Australian aborigines and Kalahari bushmen. They can sleep almost naked, without shivering, at air temperature down to 4 °C and with a skin temperature of 27–28 °C. In these conditions there is some fall in body temperature but this is limited by marked cutaneous vasoconstriction. This process is called insulative hypothermic acclimatization and is particularly well suited for those parts of the world in which people are undernourished and exposed to high temperatures during the daytime.

By contrast a naked European shivers when the air temperature drops below 27 °C or the skin temperature below 33 °C. The onset of shivering prevents a fall in body temperature but the discomfort prevents sleep.

Hypothermia

Hypothermia occurs naturally in hibernating animals in which, with the onset of hibernation, the body temperature varies with that of the environment over the range 5–15 °C. In man hypothermia may be *accidental* or it may be used deliberately in *surgery of the heart or brain*.

Accidental hypothermia may occur in healthy people who are exposed to excessive cooling. Immersion in the cold sea after shipwreck or exposure of inadequately clothed persons on mountains can overcome the capacity of the thermoregulatory mechanisms so that the deep body temperature falls. At 33 °C temperature regulation fails, and at 29–31 °C consciousness is lost. Below 30 °C glucose is metabolized slowly or not at all; breathing is depressed and the heart rate very slow. Death occurs at body temperatures below 25 °C.

Accidental hypothermia also occurs in elderly persons living in inadequately heated rooms. Old people when exposed to the cold show a smaller increase in O_2 consumption, a greater fall of rectal temperature and less cutaneous vasoconstriction than young people. Many geriatric patients die from hypothermia every winter.

The development of hypothermia may be accelerated by drugs such as chlorpromazine and sedatives. A few patients suffer from chronic hypothermia in which the defect is a low setting of the hypothalamic thermostat. Temperature regulation occurs but the usual level is abnormally low (Cranston 1966). In these cases there is no evidence of endocrine disease. Hypothermia occurs in myxoedema because the patients cannot increase their metabolic rate on exposure to cold.

Hypothermia is deliberately induced in some patients undergoing operations on the heart and brain. The body temperature is lowered to 21–25 °C by cooling the blood or the surface of the body, and at these temperatures the O_2 usage by the tissues falls to about a quarter of the normal. Cardiac arrest for 10 minutes under these circumstances does not cause permanent damage to the heart or brain and allows the performance of operations on these organs which would be impossible without cooling. Certain drugs have been used to facilitate the lowering of body temperature. For example, chlorpromazine (*Largactil*) increases heat loss through cutaneous vasodilation, brought about by central inhibition of vasomotor tone and by peripheral antagonism of noradrenalin and adrenalin.

Resuscitation of partially frozen mammals

Some mammals can survive reduction of deep body temperatures to the range $+1$ to $-5\,°C$. Rats can withstand cooling to $0–1\,°C$ for 1 hour and golden hamsters can be revived after several hours of cardiac and respiratory arrest at this temperature. Some hamsters survive deep body temperature down to $-5\,°C$ but if extensive ice crystallization occurs recovery is exceptional. Ice formation probably damages tissues by dehydration. Individual cells such as spermatozoa and red blood cells can be preserved by freezing at $-79\,°C$ provided that dehydration is prevented by the incorporation of glycerol in the suspending fluid.

It is unlikely that frozen animals could be kept in suspended animation for long periods, and the procedure is not yet applicable to man. The lowest deep body temperature yet tolerated by a human being is 9 °C for one hour. The patient, who had an inoperable pelvic carcinoma, recovered consciousness and lived for 38 weeks without apparent alteration in neurological or intellectual functions (Niazi and Lewis, quoted by Parkes and Smith, 1959.)

REFERENCES

Adaptation to the environment. In *Handbook of physiology*, Section 4 (ed. D. B. Dill). American Physiological Society, Washington, DC (1964). Numerous articles on adaption to heat and cold.

ATKINS, E. and SNELL, E. S. (1965). Fever. In *The inflammatory process* (ed. B. W. Zweifach, L. Grant, and R. T. McCluskey) pp. 495–534. Academic Press, New York.

BENZINGER, T. H. (1969). *Physiol. Rev.* **49,** 671.

BORISON, H. H. and CLARK, W. G. (1967). Drug actions on thermoregulatory mechanisms. In *Advances in pharmacology* (ed. S. Garattini and P. A. Shore) Vol. 5, pp. 129–212. Academic Press, New York.

BURTON, A. C. and EDHOLM, O. G. (1955). *Man in a cold environment.* Arnold, London.

EDHOLM, O. G. and BACHARACH, A. L. (eds.) (1965). *The physiology of human survival.* Academic Press, New York.

Cold, Carlson, L. D. and Hsieh, A. C., pp. 15–22.

Heat, Fox, R. H., pp. 53–79.

FELBERG, W. (1965). A new concept of temperature regulation in the hypothalamus. *Proc. R. Soc. Med.* **58,** 395.

—— and MYERS, G. D. (1963). *Nature, Lond.* **200,** 1,325.

HAMMEL, H. T. (1968). Regulation of internal body temperature. *A. Rev. Physiol.* **30,** 641.

HENSEL, H. (1981). *Thermoreception and temperature regulation.* Academic Press, London.

NEWBURGH, L. H. (ed.) (1968). *Physiology of heat regulation and the science of clothing.* Hafner, Philadelphia.

PARKES, A. S. and SMITH, A. U. (1959). *Br. med. J.* **i,** 1295.

SMITH, A. U. (1957). Experimental hypothermia in animals. In *Lectures on the scientific basis of medicine*, Vol. 5, p. 19. London.

The basal ganglia

The basal ganglia include (i) the corpus striatum (consisting of the caudate nucleus and the lenticular nucleus; (ii) the subthalamic nucleus; (iii) the substantia nigra of the mesencephalon.

The corpus striatum consists of the phylogenetically more recent caudate nucleus and the putamen. The lenticular (bean-shaped) nucleus contains the putamen (latin for husk or shell) and which lies laterally to the phylogenetically older globus pallidus. The globus pallidus is so called because it looks pale in fresh sections of the brain, although when it is stained by the Weigert–Pal method (which displays myelinated axons) it appears darker than the putamen and caudate.

The caudate nucleus, as its name implies has a tail (terminating at the amygdaloid nucleus), it also possesses a head [FIG. V.174].

The anterior limb of the internal capsule intervenes between the head of the caudate nucleus and the putamen of the lenticular nucleus. However, the cortical efferent fibres traversing the internal capsule do not separate the caudate and putamen entirely as the head of the caudate is in continuity through a bridge of grey matter (beneath the anterior limb of the internal capsule) with the putamen. Moreover, numerous grey strands extend between the two nuclei passing across the internal capsule; hence the name 'corpus striatum'.

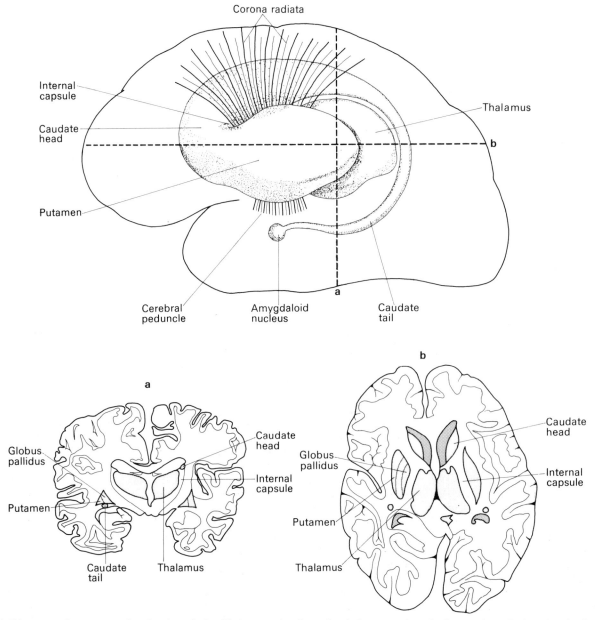

FIG. V.174. Diagrammatic representation showing relationship between basal ganglia, thalamus, and cerebral cortex in sagittal section (top) and coronal (a) and horizontal (b) sections.
(From Curtis, B., Jacobson, S., and Marcus, E. M. (1972). *An introduction to the neurosciences*. Saunders, Philadephia.)

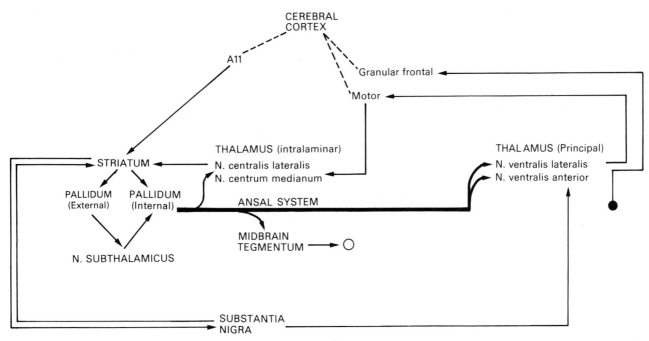

FIG V.175. A scheme of the principal connections of the basal ganglia. At (●) the thalamic nucleus ventralis anterior projects on to several subcortical regions, including the thalamic nucleus centralis lateralis. The midbrain tegmentum at (○) probably establishes polysynaptic relationships with the spinal cord and the thalamic intralaminar nuclei.
(From Webster, K. E. (1975). *Proc. R. Soc. Med.* **68**, 203.)

The course taken by striatal efferents appears at first sight to be extraordinarily complicated, but knowledge of the evolution of the region which they traverse shows a simple plan underlying the arrangement of these efferents. Thus they form an ancient evolutionary system of fibres travelling in a caudo-dorso-medial direction, but their pathways are necessarily altered by the huge development of the corticospinal fibres which pass caudally from the cerebrum at right angles to the striatal efferents, which correspondingly either thread their way through the peduncular fibres or curve round the peduncle. The German expression *Kammsystem* means 'comb system' and perfectly describes the direct course through the internal capsule of fibres from the putamen and the dorsal pallidus, for these fibres resemble the teeth of a comb in a hank of hair (corticofugal fibres). The less descriptive term *fasciculus lenticularis* is used for this striatal system.

The *ansa* ('loop') *lenticularis* is comprised by fibres which have curved round the cerebral peduncle and which are destined for the nucleus ventralis anterior thalami to the anterior and midline nuclei of the thalamus, the red nucleus, and the substantia nigra.

Nowadays the term neostriatum has largely replaced the older expression 'corpus striatum' whereas the expression paleostriatum (= pallidum = globus pallidus) is commonly employed. The globus pallidus itself is subdivided into external and internal parts by a medial medullary lamina. In the main the external part is efferent to the substantia nigra, subthalamic nuclei, and reticular nuclei, whereas the internal division receives afferent fibres from these structures and projects via the *ansal* system to the ventrolateral and ventroanterior nuclei of the thalamus with other projections to the centromedian and centrolateral nuclei of the thalamus.

The principal afferents received by the globus pallidus stem from the caudate and putamen. Its chief efferents form two main bundles—the ansal fasciculus (field H_2 of Forel) which passes to the thalamic ventroanterior and ventrolateral thalamic nuclei and the ansa lenticularis which passes to the reticular nuclei of the mesencephalon and to the red nucleus, substantia nigra, and subthalamic nuclei.

The neostriatum receives afferents from the cortex, thalamus, and substantia nigra. Cortical fibres come from all four lobes but mainly from the frontal and parietal areas. Thalamostriate fibres stem from the intralaminar, medial and ventral anterior nuclei. Nigrostriate fibres (which are dopaminergic) stem from the dorsomedial pars compacta of the substantia nigra and are distributed in a topically ordered manner. (The ventrolateral pars reticulata of the substantia nigra projects to the ventrolateral and ventroanterior nuclei of the thalamus and thence, after relay, to the cerebral cortex [see FIGS.V.175, V.176, and V. 177].)

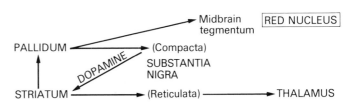

FIG. V.176. The connections of the substantia nigra. The midbrain tegmentum and red nucleus may be involved in the genesis of postural tremor. (From Webster, K. E. (1975). *Proc. R. Soc. Med.* **68**, 203.)

In summary, the basal ganglia, like the cerebellum, receive information from all parts of the cerebral cortex and project back to the motor cortex via the ventrolateral and ventroanterior thalamus. Both basal ganglia and cerebellum exert an influence on motor performance. However, the cerebellum receives large oligosynaptic inputs from the spinal cord and the basal ganglia do not. Moreover, the cerebellum has direct access to major motor pathways (such as the vestibulospinal and rubrospinal tracts) other than those in the corticospinal tracts whereas the basal ganglia has only indirect influence on such extrapyramidal pathways (Webster 1975).

Functional studies

The relative inaccessibility of the basal ganglia has hindered our understanding of their function. The three classical methods—electrical stimulation, ablation, and the recording of electrical potentials are difficult to apply and their results hard to interpret owing to the adjacence of other structures and nervous pathways

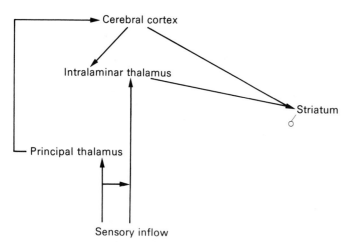

FIG. V.177. A scheme of forebrain relationships based partly on the work of Jones, E. G. and Powell, T. P. S. (1971) *Brain* **92**, 477–502. At (○) the striatum projects on to various subcortical structures shown in FIG. V.175. (From Webster, K. E. (1975). *Proc. R. Soc. Med.* **68**, 203.)

which are functionally unrelated to the basal ganglia. Our knowledge remains somewhat fragmentary and may be listed: (1) Stimulation of the neostriatum inhibits neurons in the motor cortex but excites inferior olivary cells. Such neostriatal stimulation applied to the ventromedial part of the caudate nucleus causes inhibition of spinal reflexes whereas dorsolateral caudate stimulation induces an exaggeration of spinal reflexes. Low-frequency stimulation of the caudate nucleus may interfere with the acquisition of a visual discrimination task or may induce a catatonic state. In unanaesthetized animals stimulation of the head of the caudate nucleus causes contraversive turning of the animal's head. The movement is sustained during the stimulation. Phasic movements are never induced by stimulation of either component of the corpus striatum.

(2) Stimulation of the pallidum, on the other hand, converts a cortically induced phasic movement into a state of tone which has been described as 'plastic'. Thus, during an extensor movement the limb can be arrested in a state of partial extension and 'frozen' there; although passive manipulation can alter the position this is in turn maintained. This 'freezing' effect is not associated with any undue rigidity. In conscious animals learned barpressing behaviour (rewarded with milk) ceases during pallidal stimulation but even though the stimulation continues, on the direct presentation of the milk the animal will drink it.

Even widespread lesions of the striopallidum do not cause rigidity, tremor or choreo-athetosis. The range of muscle movements is only slightly reduced.

In short, physiological investigations of the types described cast remarkably little light on the interpretation of symptoms and signs manifested by patients who have been identified as showing signs of basal ganglial disorders.

Clinical evidence

Parkinson's disease (paralysis agitans)

Usually occurring in the middle-aged patient, paralysis agitans is associated with degenerative changes in the striopallidum and the substantia nigra. The concentration of dopamine in these regions is much reduced. Characteristically there occur: rigidity, weakness of movements, and tremor.

1. Rigidity affects mainly the big muscle masses of the proximal parts of the limbs. Though it affects both protagonists and antagonists it is usually most marked in the biceps, in the knee flexors and in sternomastoids. The back is flexed, the arms adducted and flexed,

and the knees are bent. In advanced cases the rigidity is so marked that the patient can be carried about like a statue. Voluntary or involuntary movement becomes progressively more difficult as the rigidity progresses. Walshe (1929) showed that movement could be temporarily restored to normal by the local injection of 1 per cent procaine solution into the appropriate muscles. Thus, when biceps and triceps were injected with procaine, rhythmic flexion and extension of the elbow were notably improved. It seems likely that this result is due to paralysis of the γ efferents supplying the muscle spindles, and not, as was originally suggested, to blocking of the afferent nerves from the muscles. The striopallidum conveys inhibitory extrapyramidal fibres to the reticular formation; the exaggeration of γ fibre activity in Parkinson's disease might well be due to the striopallidal lesions which are its main pathological features. Chlorpromazine, which decreases the activity of cells in the ascending reticular formation (and which decreases γ discharge in experimental animals) reduces the rigidity of Parkinson's disease.

2. The weakness and poverty of movement which characterize Parkinson's disease result in early defects in fine movements of the fingers, in the fine variations in speech, and in movements of expression. Though rigidity when present must be a contributory factor in these motor defects it is not an essential accompaniment of the poverty of movement. The face becomes mask-like and emotional movements of the facial muscles occur seldom. Tendon jerks become progressively more difficult to elicit as the rigidity increases.

3. Tremor consists of regular rhythmically alternating contractions of a muscle group or groups and their antagonists, about six times a second. Most commonly it involves the fingers, hands, lips, or tongue, and often it causes movements of pronation-supination. Tremor becomes characteristically increased in emotion or when the patient is aware that someone is watching him. It ceases during sleep. Adrenalin, which is secreted from the adrenal medulla in emotional states excites the mesencephalic components of the ascending reticular system and notably increases Parkinsonian tremor. Sleep, which damps down the activity of the ascending reticular units, abolishes the tremor. Since tremor cannot be produced in animals by experimental lesions of the striopallidum but only by lesions of the reticular system, it seems likely that its frequent presence in paralysis agitans may be ascribed to the associated degenerative lesions in the reticular formation and its ascending connections.

It is now considered that the essential cause of Parkinson's disease is degeneration of nerve cells which normally produce dopamine in the substantia nigra. The resulting lack of dopaminergic activity upsets the balance which normally exists between the excitatory effect of cholinergic fibres ending in the corpus striatum and the inhibitory influence of dopaminergic nerves which have neighbouring terminations there. Thus, reserpine (which depletes dopamine stores in the appropriate nerve endings) and phenothioazines (which block dopamine receptors in the striatum) exacerbate the signs and symptoms of the disease. Physostigmine which enhances the action of acetylcholine likewise increases tremor and rigidity.

Dopamine cannot cross the blood–brain barrier but its precursor laevo-*dopa* (dihydroxyphenylalanine) can.

Dihydroxyphenylalanine

In Parkinson's disease the neostriatum contains perhaps only half its usual content of dopamine (which is normally very high—

3.5 µg g^{-1}—compared with elsewhere). The administration of dopa presumably allows the dopaminergic cells to decarboxylate the compound to form dopamine:

$$\text{Dopamine}$$

Benefit is relatively slow and some three months are required before maximal improvement is achieved.

In addition to Parkinson's disease which occurs as a degenerative condition in the middle-aged and elderly the striopallidum is involved in:

(i) *Encephalitis lethargica* in which there are striopallidal lesions as a sequel to encephalitis in youth.

(ii) *Wilson's disease* (progressive lenticular degeneration), a bilateral affection of the lenticular nuclei (putamen + globus pallidus) associated with multilobular hepatic cirrhosis.

(iii) *Kernicterus*: the pallidum is damaged and stained yellow in haemolytic disease of the new-born which results from the noxious action of Rhesus antibodies. If the child survives, it may show rigidity, chorea, and athetosis (and often mental deficiency).

(iv) *Chorea and athetosis*. These two disorders are characterized by spontaneous movements; lesions are usually found in the caudate nucleus, putamen, and sometimes in the ventrolateral nucleus of the thalamus.

In chorea the spontaneous movements are irregular, brief, and rapid. If the arm is affected there is, in addition, decreased tone and muscular weakness, especially of the hand (there is complaint of 'uselessness' of the hand); the arm hangs limply. Voluntary movements of the arm are not carried out smoothly but tend to be abrupt and sudden.

In athetosis the spontaneous movements are slow and confluent—one phase blending with the next; the limb thus undergoes a series of complex writhings and contortions like those of a hula-hula girl.

The pattern of the involuntary movements of chorea and athetosis suggests that they are caused by impulse discharges which arise in the cerebral cortex; the movements only occur when the descending paths from the excitomotor cortex are intact.

Experimentally it has been shown that an isolated lesion of the caudate nucleus may produce choreo-athetosis. On the basis of these results it has been suggested that the spontaneous movements are due to interruption of the inhibitory pathway from area 4s via the caudate nucleus to the thalamus, thus blocking the thalamocortical circuit and depriving the cortical excitomotor area of an afferent control from the subcortical ganglia. Wilson suggested that a lesion of the thalamus might deprive the cerebral cortex of cerebellar control or lead to disordered cerebellocortical influence and thus give rise to the spontaneous movements; he drew attention to the other signs suggestive of cerebellar dysfunction which are found in chorea, e.g. decreased muscle tone, weakness, and inco-ordination of willed movements.

It is claimed that the movements of athetosis may be controlled by removal of area 6, or by section of the ventral column of the spinal cord, without encroaching on the crossed pyramidal tracts. Both these operations were devised to reduce the activity of some of the corticospinal extrapyramidal pathways. The movements persist or are made even worse by division of the posterior roots supplying the affected limb.

It is clear from the discussion that it is premature to assign a precise pathology to chorea and athetosis.

The subthalamic nucleus (corpus Luysii) is sometimes included in the basal ganglia. Little is known of its functions. It is occasionally the site of a lesion (usually vascular) in man; the unfortunate patient is afflicted by spontaneous attacks of inco-ordinated movement of the whole of the contralateral side of the body (hemiballismus). These movements are mediated by the corticospinal pathway and are so drastic in their effects that the patient may quickly succumb. It has been suggested that surgical removal of the appropriate motor cortex may be justifiable in such cases.

Spinal lesions

Tabes dorsalis

The essential lesion in the syphilitic disease is a degeneration of the dorsal nerve roots central to their ganglia, affecting especially the fibres which ascend in the dorsal columns (i.e. those concerned with deep sensibility) and those which subserve pain.

The phenomena are subjective and objective:

Subjective phenomena. Lightning pains: these come in attacks, with intervals of freedom. The first effect of the disease seems to be to stimulate pain fibres in the dorsal nerve roots. The pain may be referred to skin, muscle or bone, and may vary in intensity from slight discomfort to intolerable agony.

Objective findings. The fibres conveying pain impulses and those from the deep structures are affected first and to the greatest extent.

(i) Pain sensibility. Pain sensibility may be lost, its appreciation is delayed, or poorly localized. As in syringomyelia, loss of pain sensibility may be responsible for trophic disturbances. The perforating ulcers found under the ball of the foot may arise from the neglect of a corn.

As the dorsal roots mainly affected in tabes are those which supply the lumbosacral and the cervicothoracic regions of the cord, the common areas of anaesthesia are round the anus, over the legs, upper chest, and the ulnar borders of the hands. The involvement of the fifth nerve accounts for the anaesthesia of the central part of the face.

The joint deformity known as the Charcot joint may be produced thus: an injury is inflicted on a joint, or a mild sprain or subluxation may occur; as no pain is aroused, the condition is neglected, receives no effective treatment, and becomes progressively worse until considerable damage to the articular surfaces of the bones results.

(ii) Deep sensibility. As the fibres that ascend the dorsal columns are damaged there is loss of sense of position, passive movement, and vibration sense.

(a) Loss of sense of position: with the eyes closed the patient is unaware of the exact position of the various parts of his body.

(b) Loss of sense of passive movement; while the eyes are closed, the great toe, for example, may be flexed or extended without the patient's knowledge of the alteration of position imposed on the toe.

(c) Loss of vibration sense: after striking a tuning-fork, the handle is placed on some bony prominence. Normally, a sensation is felt which is varyingly described as a 'buzzing' or 'electrical' sensation. When dorsal column sensibility is lost, the patient states that he merely feels a cold object in contact with his skin.

(iii) Reflexes. The muscles are flabby and excessive movement is permitted at the joints without producing discomfort.

The knee jerk and ankle jerk which depend on the integrity of 'stretch afferents' in the dorsal nerve roots, are also lost.

(iv) Voluntary movement. There is considerable disturbance of voluntary movement, illustrating the important general principle that the indispensable basis of all purposeful and effective motor activity is accurate information.

Syringomyelia

Syringomyelia is a rare condition of excessive overgrowth of neuroglial tissue, accompanied by cavity formation involving the grey matter round the central canal of the spinal cord. The crossing fibres subserving pain and temperature, which decussate in the grey commissure, are damaged, with resulting loss of these sensations. The touch fibres which cross in this region are likewise destroyed, but touch has a double path; the fibres which ascend in the dorsal columns escape. We thus obtain the characteristic *dissociated anaesthesia* of this disease, i.e. there is loss of pain and temperature sensibility, while the sense of touch is retained.

The symptoms and signs are usually referred to and noticed in the hand and arms, owing to the predilection of syringomyelia for the *cervical* enlargement of the spinal cord. Accompanying the sensory changes are alterations in muscle tone at the level of the lesion and below the level of the lesion. At the level of the lesion the gliosis and cavitation may spread to involve the anterior horn cells, thus causing flaccid paralysis of the muscles (usually of the hand). At a later stage, the pyramidal and extrapyramidal tracts are gradually destroyed as they lie in the adjacent white matter and there develops a progressive spastic diplegia of the legs (below the level of the lesion).

Owing to loss of the nociceptive reflexes and of pain and temperature sense the affected part is not withdrawn, reflexly or consciously, from a nocuous stimulus and correspondingly may suffer serious damage. Often a patient with syringomyelia may show signs of previous whitlows in the fingers (Morvan's deformities).

Complete transection of the spinal cord

Spinal transection causes immediate and permanent loss of sensation and voluntary movement below the level of the lesion.

In acute transection of the cord the patient feels himself cut in two. The higher centres are unaffected and the mind remains clear, but the whole of the body below the level of the lesion is deprived of all activity.

Stage of flaccidity

The muscles are completely paralysed; all the reflexes are abolished and muscle tone is lost; the muscles lie in any position imposed on them by gravity. There is complete loss of all sensation; cramplike pains, however, are present at the level of the lesion. The bladder and rectum are generally paralysed. The sphincter vesicae, however, frequently retains its functions, or recovers very rapidly, with consequent retention of urine. The penis is flaccid, and erection is impossible. As the vasoconstrictor fibres leave the cord between the first thoracic and second lumbar segments, a transection below the level of the second lumbar produces very little fall of blood pressure, while section at the first thoracic level causes a fall of blood pressure equal to that resulting from destruction of the vasomotor centre, i.e. to a level of 40 mm Hg.

The venous return from the limbs depends largely on the contractions of the skeletal muscles. The amount of blood reaching a muscle depends, too, on the functional activity of the part. As the legs are immobile this blood flow and venous return is reduced; they are usually cold and blue. The skin is dry and is very liable to be affected by serious sloughing bed-sores.

If the lesion is at the level of the sixth thoracic segment, all impulses coming in from the abdominal viscera are cut off from the brain. Griping sensations or distension of the viscera are not appreciated.

The above phenomena belong to the *stage of flaccidity*; the isolated segments of the spinal cord have lost their power of mediating reflex functions. To this temporary state, the term *spinal shock* is applied. In looking for the cause of this condition the following facts must be noted.

Spinal shock. 1. The shock affects the distal segments of the cord only and not the segments headward to the injury. The monkey with its cord divided in the thoracic region, goes on looking out of the window and catching flies. The fall of blood pressure is not responsible for shock, because the fall is equally marked in the headward part of the animal, which shows no flaccidity after transection.

2. Operative shock plays no essential part. The method of transection is unimportant; cutting the cord across quickly or tearing it deliberately makes no difference. Furthermore, if the animal is allowed to recover from the shock and a second transection is then made a few segments lower down, the reflex activities of the distal end of the cord are quite unaffected by the second operation.

3. The higher the animal is in the scale of development, the more profound and the more lasting is the condition of spinal shock. In the cat it lasts a few moments, in the monkey a few days, while in man it persists for about three weeks. In the higher animals the spinal cord has few truly autonomous activities and is correspondingly profoundly affected by separation from the brain stem and cortex. This supports the view that the simple spinal reflex is a fiction and that nearly all so-called spinal reflexes employ long supraspinal arcs involving the higher levels of the nervous system.

The general principle which emerges from this discussion is very important. The nervous system does not consist of a series of isolated units but is a closely-knit and integrated whole. Damage to any part of the nervous system disturbs its smoothness of working and, until compensation has been established, the functional failure is more severe than can be accounted for by the anatomical lesions. To this depression of function in distant parts of the nervous system the term *diaschisis* (Monakow) is applied.

Stage of reflex activity

As the stage of shock passes off, functional activity returns first in smooth muscle. The sphincter vesicae (if affected at all) recovers very soon, but the detrusor of the bladder regains its powers more slowly. The consequent retention of urine must be dealt with by catheterization. Tone next returns to the hitherto paralysed blood vessels, as the connector cells in the cord begin to act independently of the vasomotor centre. The blood pressure is thus restored to about its normal level. The isolated segments of the cord can also mediate as centres for vasomotor reflexes.

1. Muscle tone. Tone in skeletal muscle returns after two or three weeks. The flexor muscles of the lower limbs now become less flabby and offer some resistance to the fingers. This returning tone is, of course, reflex in character and is produced by impulses entering the cord from the muscles. It is worth noting that the isolated cord 'favours' the flexor neurons and muscles, and this fact will make clear many of the findings; thus the extensor muscles remain flabby for a longer period and do not attain the same degree of tone as the flexors. *All* the muscles, however, are hypotonic, even the flexors themselves, because the stretch reflexes (which are mainly responsible for muscle tone) are feeble when mediated by the spinal cord alone especially in man, and need reinforcement from the brain stem centres. The limbs tend to adopt a position of slight flexion, and the paralysis is therefore referred to as *paraplegia in flexion*; the limbs cannot support the body weight. The muscles undergo no wasting, because, though paralysed for voluntary movements, they are in constant reflex activity.

2. Reflex movements. Spontaneous involuntary flexor movements of the limbs occur. The small toes are separated and raised; the anterior tibials, the hamstrings, recti abdominis, and adductors of the thigh can be felt to harden, but their load is initially too great and they fail to move the limbs. Contraction of these flexor groups of muscles is accompanied by reciprocal inhibition of the extensor muscles of the limbs.

(i) *Flexor reflex*. The reflex movement that returns first is the flexor reflex. To elicit this reflex a nocuous stimulus is necessary, i.e. one which tends to injure the part and which causes pain in the intact organism. The reflex is obtained most easily by stimulating the skin or deep structures of the sole of the foot, but the maximum receptive field is much wider than this, so that when the cord has recovered stimulation of the leg as high as the groin or perineum is effective. From the latter regions a stronger stimulus is necessary, and the response is less constant in its appearance. The movement consists of dorsiflexion ('upward movement') of the big toe and abduction of the other toes; these reflex toe movements constitute the Babinski response; as the reflex spreads there is dorsiflexion of the foot, flexion of the knee and hip, and abduction of the thigh. The antagonistic muscles are inhibited.

The flexor reflex is a withdrawal or defence reflex which removes the limb from an injurious agency. All the muscles which are contracted in the flexor reflex are called physiological flexors; the antagonists (which contract in the extensor thrust or the crossed extensor reflex) are physiological extensors. It should particularly be noted that the dorsiflexors of the ankle are the physiological flexors in that region.

The normal plantar reflex in man in response to stimulation of the sole of the foot consists of a downward movement, i.e. plantar flexion of the toes; it only appears with the development of the pyramidal tracts and replaces the more primitive reflex. In the normal plantar reflex the tensor fasciae femoris contracts; in the abnormal Babinski response the hamstring muscles generally harden.

(ii) *Mass reflex*. In some cases a very widespread reaction is readily elicited by scratching any point on the lower limbs or the anterior abdominal wall below the level of the lesion. The response obtained consists of:

(a) Flexor spasms of both lower extremities and contraction of the anterior abdominal wall.

(b) Evacuation of the bladder even when its contents may only be half the amount which must normally be present before reflex emptying occurs; this may be partly due to the abdominal compression raising intravesical pressure to threshold level.

(c) Profuse sweating below the level of the lesion. To understand the distribution of the sweating, it must be remembered that the sweat fibres to the head and neck arise from T1, 2, and those to the arm from T5–9. With a lesion at the level of T1, the whole body sweats when the mass reflex is obtained, as all the sympathetic fibres leave the cord below the level of the lesion.

The mass reflex is less commonly observed in modern cases of spinal transection. In any case it does not manifest itself until several months after the original lesion.

(iii) *Coitus reflex*. This is produced by stimulation of the glans penis, or the skin round the genitals, anterior abdominal wall, or anterior and inner surface of the thighs. The response consists of swelling and stiffening of the penis, withdrawal of the testes because of contraction of the cremaster muscles and curling up of the scrotal skin from the action of the dartos. The recti abdominis, flexors of the hip, and adductors of the thighs also contract. Seminal emission may occur. On ceasing stimulation, the penis and lower limbs relax.

(iv) *Deep reflexes*. The knee jerk returns about one to five weeks later than the flexor responses. It consists at first merely of a tightening of the slack quadriceps muscle, but later actual extension of the knee may occur. The knee jerk is a 'fractionated stretch reflex' and stretch reflexes are generally feeble in the spinal animal.

It is therefore found that though the quadriceps may contract fairly briskly it relaxes immediately, and the limb drops quite limply and not gradually as in the normal person. The ankle jerk may return later still. If ankle clonus is present it consists of no more than a few irregular and unequal jerks.

(v) *Observations on cases with increased extensor activity*. The ultimate clinical picture in patients studied during the Second World War differed in certain interesting respects from those described above; these differences are probably due to the better general health of the patients, which permitted a maximal degree of functional recovery of the isolated spinal cord. Generally about six months after the occurrence of the transection marked activity appeared in the extensor arcs, resulting in heightened extensor reflexes and the appearance of extensor spasms. The detailed findings are summarized below.

(a) The ankle jerk and knee jerk became exaggerated; quadriceps clonus and ankle clonus were sometimes noted.

(b) If the limb muscles were passively stretched abruptly, e.g. if the flexed thigh was suddenly extended, a reflex extension of the same or both limbs occurred. The contraction often involved both extensor and flexor muscles, converting the limb into a 'solid pillar'. Relaxation subsequently developed slowly. A few patients in whom this response was well marked could stand for a long time in a bath of warm water without support.

(c) The mass reflex as described above was not obtained. Either mass flexion or mass extension of the limbs occurred, which was not accompanied by sweating or emptying of the bladder.

(d) Stimulation of the glans penis produced the genital response described above, but it was not accompanied by seminal emission or limb movements.

The spinal transection syndrome of the First World War thus seems to represent the functional activity of an incompletely recovered spinal cord.

3. Autonomic reflexes. (i) Reflex evacuation of the bladder is gradually established; reflex defaecation also occurs.

(ii) The skin, which hitherto has been dry and scaly, now shows sweating again; it becomes more healthy, and ulcers heal up rapidly. Because of the improved vascular tone and the return of reflex activity to the skeletal muscles, the circulation through the limbs is greatly improved, and they become warm and of good colour.

Stage of failure of reflex activity

If general infection or toxaemia occurs, failure of reflex function develops. The reflexes become increasingly difficult to elicit; the receptive fields become narrowed down to the optimum areas from which the reflexes can be obtained. The mass reflex disappears. The threshold for all reflexes is raised, and fewer groups of muscles are involved in the motor responses. The muscles waste and become flaccid, and bed-sores develop, which still further lower the general state of the patient.

Incomplete transection of the spinal cord

If the spinal cord is gravely injured, but does not suffer complete division, a state of spinal shock develops identical with that already described. When the stage of reflex activity returns, certain striking differences present themselves. In cases of incomplete transection some of the descending fibres in the ventrolateral columns of the cord (especially the vestibulospinal and reticulospinal tracts) may have escaped injury and so some connections persist between the brain stem and spinal cord. These tracts mainly reinforce the activity of the extensor motoneurons—correspondingly cases of incomplete transection manifest extensor hypertonia.

1. Reflex tone returns to the extensor muscles, and so the legs lie extended at hip and knee, with the toes pointing slightly downwards. The condition is therefore called *paraplegia in extension*.

2. Involuntary movements are relatively infrequent, but when they occur involve an increase of extensor tone, producing downward movements of the feet and toes.

3. Reflex movements. (i) Extensor thrust reflex. The reflex is elicited as follows: the lower limb is passively flexed and allowed to rest on the bed; the patient's foot is then pressed up with the palm of the hand. Active contraction of the quadriceps and posterior calf muscles (the physiological extensors) occurs, and the limb straightens out. This reflex is often absent when the cord is completely divided in man.

(ii) The flexor reflex can be obtained by nocuous stimulation of the sole of the foot. The flexion movement, however, is small, and the receptive field only extends to the knee. It is usually accompanied by active and forcible extension of the opposite limb (crossed extensor reflex).

(iii) Gentle flexion of one limb produces extension of the opposite limb (Phillipson's reflex). The flexed limb then becomes extended and the opposite one flexed; the responses alternate in each limb, producing a steppage movement.

The range of reflex response is greater now that more of the nervous system is available. It is clear also that movements of locomotion can be carried out to some extent (reflexly) by the lower levels of the central nervous system—but of course walking as such is impossible.

Tendon jerks and stretch reflexes in disease

Merton (1956, 1979) has pointed out that although tendon jerks and stretch reflexes were supposed to have the same reflex pathway their excitability sometimes differs. Thus, tendon jerks are of normal briskness in Parkinson's disease and in cerebellar lesions although in Parkinson's disease the muscles are rigid and in cerebellar disease they are flabby and hypotonic. As the stretch reflex itself has more than one component (primary and secondary afferents from the spindle) and probably possesses a cortical circuit too it is possible to explain this difference between the two types of response. Nevertheless, the tendon jerks are usually exaggerated in cases of spastic hemiplegia and the knee and ankle jerks are, of course, lost in syphilis which destroys the dorsal roots of the lumbar enlargement. In many respects the clinical term 'tendon reflex' is unfortunate because although the jerks are reflexly elicited by tapping the tendon of a slightly stretched muscle, it is from the muscle spindles of the relevant structure that the afferent impulses ascend in Group I afferents to effect monosynaptic connections with the appropriate α motoneurons and to cause contraction of the muscle stretched. Even when the overlying skin and the muscle's tendon are anaesthetized the so-called tendon jerk is unimpaired.

The clinical tendon jerks are usually tested at the ankle (S1, 2), the knee (L3, 4), the biceps (C5, 6), triceps (C7, 8). Additionally, brachioradialis (C7, 8) and the jaw jerk (Vth cranial nerve) can be tested.

The cerebrospinal fluid

ANATOMY

The central nervous system is enveloped by the meninges; from without inwards these are termed dura, arachnoid, and pia. The dura consists of fibrous tissue lined by endothelium. Separating it from the arachnoid is the subdural space, which contains a small amount of fluid resembling lymph. The dura ends at the lower border of the second sacral, the spinal cord at the lower border of the first lumbar vertebra. The spinal theca can therefore be punctured in the lower lumbar region without fear of injury to the cord.

The arachnoid is separated from the pia by the subarachnoid space, which contains the cerebrospinal fluid. The pia invests the nervous substance very closely. The arachnoid, however, does not dip into the sulci or fissures (with the exception of the longitudinal sulcus), and invests the spinal cord quite loosely. There are also definite dilatations of the subarachnoid space called cisternae. The cisterna magna is found in the interval between the medulla and the undersurface of the cerebellum. [FIG. V. 178]. The cisterna pontis lies on the ventral aspect of the pons and contains the basilar artery. The cisterna basalis (interpeduncularis) is formed by the arachnoid bridging across the interval between the tips of the temporal lobes, and contains the circle of Willis. Prolongations of the subarachnoid space extend along the sheaths of the spinal and cranial nerves, particularly the optic. The rest of the cerebrospinal fluid lies in the ventricles of the brain. The ventricles establish a connection with the extraventricular fluid through the foramen of Magendie (medial aperture) in the middle line in the inferior part of the roof of the fourth ventricle, and the foramina of Luschka (lateral apertures) at the extremities of the lateral recesses of this ventricle. As the arteries and veins enter and leave the brain substance they are surrounded by the perivascular spaces, which are continuous at one end with the subarachnoid space and at the other with the fine spaces which surround the nerve cells. The flow along these perivascular spaces (which correspond to the spaces containing interstitial fluid in other organs) is normally outwards to the subarachnoid space, and they serve to remove waste products resulting from cell activity.

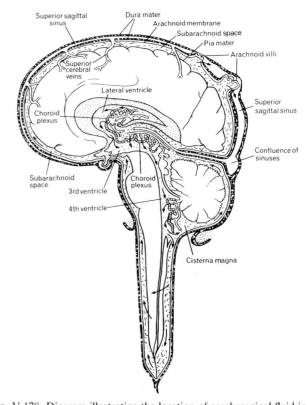

FIG. V.178. Diagram illustrating the location of cerebrospinal fluid in the ventricles and subarachnoid spaces (after Rasmussen).
(Davson, H. (1964). *A textbook of general physiology*, 3rd edn, p. 231. Churchill, London.)

Choroid plexuses. Some arteries (accompanied by a covering of pia) pass through the brain substance to reach the lining ependymal layer in the lateral, third and fourth ventricles. They then break up into complex capillary networks which project into the ventricular cavities and become lined by the now much folded ependymal cells. The blood vessels and their lining epithelium constitute the choroid plexuses. The epithelium becomes differentiated into cubical cells containing mitochondria, granules, and vacuoles, evidence that the cells are the seat of active metabolic processes.

Composition

Cerebrospinal fluid is a clear colourless fluid with a specific gravity of about 1005. It is almost protein-free (20–30 mg per 100 ml) and almost cell-free (lymphocytes c. 5 mm^{-3}). It contains less glucose (c. 4 mM l^{-1}) than plasma.

Its approximate crystalloid composition in the mammal is shown in the TABLE (Pappenheimer 1967).

Ionic composition of c.s.f. and plasma ultrafiltrate
(in mM per kg H₂O)

Constituent	Plasma Ultrafiltrate	c.s.f.
Na$^+$	145	150
K$^+$	4.5	2.8
Ca^{2+}	3	2
Mg^{2+}	1.5	2.3
H$^+$	40×10^{-6}	50×10^{-6}
Cl$^-$	115	130
HCO$_3^-$	28	22

The [H$^+$] of c.s.f. is $50 = 10^{-6}$ mM per kg H₂O, which equals 50×10^{-9} g kg^{-1} H₂O or 50 ng l^{-1}. Hence the pH of c.s.f. is approximately 7.3.

There are significant differences between the concentrations of some of the ionic constituents in the plasma ultrafiltrate and the c.s.f.

FORMATION AND ABSORPTION OF CEREBROSPINAL FLUID

Site of formation

It is formed by the choroid plexuses, especially by the large plexuses which are found in lateral ventricles. The cerebro-spinal fluid passes from the lateral ventricles through the foramina of Monro (interventricular foramina) into the third ventricle, aqueductus Sylvii (cerebral aqueduct), and fourth ventricle, and out through the foramina of Luschka and of Magendie into the subarachnoid space, both cerebral and spinal [FIG. V.179]. The evidence is as follows:

1. If the interior of the lateral ventricle is exposed at operation in man, drops of clear fluid can be seen exuding from the surface of the choroid plexus.
2. Experimentally, if a catheter is introduced into the third ventricle, a continuous flow of fluid is obtained. On the other hand, if the aqueductus Sylvii is occluded, the lateral ventricles become greatly distended by the retained fluid, the condition being called internal hydrocephalus.
3. When one foramen of Monro is occluded, unilateral hydrocephalus (of the corresponding lateral ventricle.) results. If, however, the corresponding choroid plexus is coincidentally extirpated the ventricle remains collapsed.
4. Obstruction of the foramina of Magendie and Luschka results in distension of the whole ventricular system.

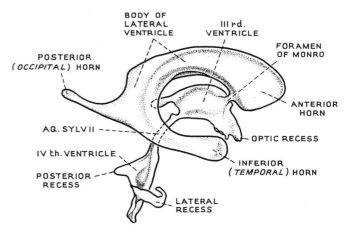

FIG. V.179. Lateral view of cast of human ventricles. (After Retzius, from Tilney and Riley.)

Route of absorption

Cerebrospinal fluid is absorbed mainly via the arachnoid villi into the dural sinuses and the spinal veins; to a minor degree fluid may pass along the sheaths of the cranial nerves into the cervical lymphatics and also into the perivascular spaces. Roughly four-fifths of the fluid is absorbed via the cerebral arachnoid villi and most of the rest via the spinal villi. The arachnoid villi are small finger-like projections (lined by the usual flat epithelial cells of the subarachnoid spaces) which project into the venous sinuses as shown in FIG. V.180. The Pacchionian bodies (arachnoidal granulations) are simply exceptionally large villi; they are few in number and present only in the adult.

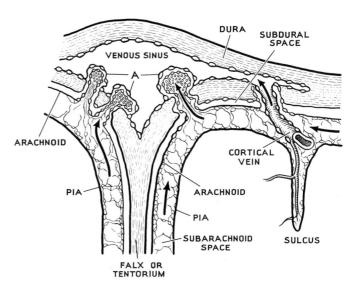

FIG. V.180. Diagram to show relations of pia arachnoid, arachnoid villi, and cortical veins to dural sinuses. Note the blood vessel coming from the brain substance surrounded by a continuation of the subarachnoid space. The arrows show the direction of flow of the cerebrospinal fluid. A = arachnoid villi. (Modified from Weed.)
(Cushing, H. (1926). *Studies in intracranial physiology and surgery*.)

Mechanism of formation and absorption

The fluid is formed by secretory activity of the epithelial cells of the choroid plexuses of the intraventricular system. The capillaries of the choroid plexuses are covered by a thick, highly differentiated epithelium which features all the intracellular organelles in the profusion which is required for active transport activity. This active secretion of c.s.f. by the choroid plexuses involves active transport of Na^+ via the Na^+/K^+ activated ATPase system and Cl^- by the carbonic anhydrase system. A potential difference exists between c.s.f. and the extracellular fluid of the body, such that the c.s.f. is usually positive to the ECF. This potential difference varies between -3 mV at pH 7.6 and $+12$ mV at pH 7.0. If the ions in the c.s.f. and plasma were distributed passively, then at a pH of 7.4 with a potential difference of $+5$ mV (c.s.f. $-$plasma) the sodium distribution ratio should be 0.8, whereas in actual fact the c.s.f./plasma ratio for Na^+ is 1.04. Thus Na^+ must be actively transported into the c.s.f. Potassium, on the other hand, if distributed passively between plasma and c.s.f., would also have a ratio of distribution of 0.8, whereas that found $(2.8/4.5 = 0.6)$ indicates that K^+ is actively transported out from the c.s.f. into the plasma. Ca^{2+} like K^+ is actively transported from c.s.f. to plasma, whereas Mg^{2+} is actively transported into the cerebrospinal fluid.

Inhibitors such as ouabain and dinitrophenol which block active Na^+ transport decrease c.s.f. production.

Pappenheimer and his colleagues (e.g. Heisey, Held, and Pappenheimer 1962) have made a penetrating study of the problems of c.s.f. formation and absorption in chronically prepared unanaesthetized goats.

Ventriculocisternal perfusions allowed the measurement of inflow rate into the lateral ventricle and outflow rate from the cisterna magna; measurements were simultaneously made of the concentrations of test substances (such as inulin, labelled water, etc.) in the ingoing and outflowing fluid. From such data the clearance of a substance from c.s.f. could be determined.

Each test involved the perfusion over the period of an hour and a half of, say, 100 ml of fluid similar to natural c.s.f. but containing additionally substances under study, such as inulin. Inflow rate was kept constant at, say, 1.5 ml min^{-1} and was maintained appreciably in excess of that at which the natural c.s.f. was formed by the choroid plexuses (c. 0.16 ml min^{-1} in the goat). The rate of fluid outflow from the cistern was likewise measured continuously and the concentration of the test substance in this outflow fluid determined.

In each case, steady state measurements were made at a particular perfusion pressure and a series of such determinations were made over a range of perfusion pressures (from -10 cm H_2O to 30 cm H_2O).

Suppose that the perfusing solution is introduced into the ventriculocisternal system at a rate $\dot{V}_i$ and that it contains inulin at a concentration c_i. During its passage through the ventricular system its concentration of inulin will be slightly diluted by the formation of naturally produced c.s.f. From the cisternal exit cannula issues fluid at a rate V_o with an inulin concentration of c_o. The amount of inulin thus 'recaptured' from that introduced is $\dot{V}_o c_o$. Meanwhile, some of the perfusion fluid containing inulin is reabsorbed from the subarachnoidal space via the arachnoidal villi. Suppose that the rate of absorption of fluid is $\dot{V}_a$ and that its inulin concentration is also c_o. Then the rate of disappearance of inulin from the fluid into the blood is $\dot{V}_a c_o$.

This can be expressed in terms of *clearance* of inulin from c.s.f. to blood. The clearance rate of inulin in this context is expressed in terms of the minimal volume (in ml) of c.s.f. which are required to supply the amount of inulin which disappears from the subarachnoid fluid into the plasma per minute, and is identical with the rate of absorption $\dot{V}_a$.

Now as
$$\dot{V}_a c_o = \dot{V}_i c_i - \dot{V}_o c_o$$

Inulin clearance $(C_{inulin}) = \dot{V}_a = \dfrac{\dot{V}_i c_i - \dot{V}_o c_o}{c_o}.$

As $\dot{V}_i$ and $\dot{V}_o$ are recorded and as $\dot{V}_a$ can thus be determined from

inulin clearance measurements, the rate of formation $\dot{V}_f$ of naturally produced cerebrospinal fluid can in turn be calculated:

$$\dot{V}_i + \dot{V}_f = \dot{V}_o + \dot{V}_a.$$
$$\therefore \dot{V}_f = \dot{V}_o + \dot{V}_a - \dot{V}_i$$
$$\dot{V}_f = \dot{V}_o + C_{inulin} - \dot{V}_i.$$

FIGURE V.181 shows steady-state measurements of $(\dot{V}_o - \dot{V}_i)$ and calculated clearances of inulin and $\dot{V}_f$ at different perfusion pressures.

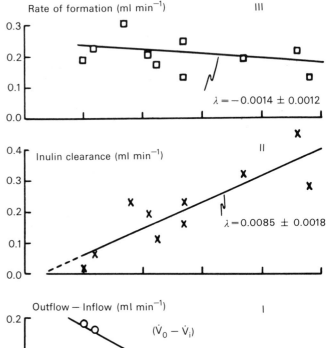

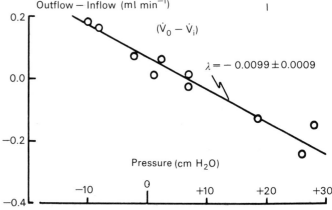

FIG. V.181. Rate of formation of cerebrospinal fluid and its relation to hydrostatic pressure in a goat. Rate of formation (panel III) calculated from the sum of inulin clearance (panel II) and outflow minus inflow rates (panel I).

(Heisey, S. R., Held, D., and Pappenheimer, J. R. (1962). *Am. J. Physiol.* **203**, 775.)

At 30 cm H_2O pressure (high) inulin clearance was as much as 0.4 ml min^{-1}. Inulin clearance fell virtually to zero when the perfusion pressure was reduced to -10 cm H_2O so almost all the inulin introduced by perfusion was recaptured in the cisternal outflow. However, the inulin concentration within the ventricular system was the same whatever the perfusion pressure so it follows that loss of inulin by diffusion through the ventricular walls was negligible. At pressures higher than -10 cm H_2O the clearance of inulin increased linearly with pressure suggesting that inulin is indeed cleared by bulk absorption through the villi.

From such results, it can be concluded that the rate of formation of c.s.f. in the normal conscious goat is some 0.16 ml min^{-1}. In keeping with the evidence that c.s.f. is secreted and that its formation requires active chemical work is the finding that such substances as ouabain (which inhibits Na^+/K^+ ATPase), dinitrophenol (which uncouples oxidation from phosphorylation) and acetazolamide (which inhibits carbonic anhydrase) all diminish the rate of formation of c.s.f.

In man the rate of formation is of the same order—perhaps 0.2 ml min^{-1}. The content of c.s.f. in man is 130–150 ml, of which some 30 ml is in the ventricular system and the remainder in the subarachnoid space.

Fluid pressure

The normal pressure of the fluid measured by lumbar puncture depends on a balance between its rate of secretion and of absorption. The value in man in the lateral recumbent position varies between 100–200 mm H_2O. The pressure in the sitting position is 200 mm H_2O higher than in the recumbent position. A rise of venous pressure such as follows coughing or crying hinders absorption and so raises the cerebrospinal fluid pressure. Compression of the internal jugular vein has a similar effect (Queckenstedt's sign).

FUNCTIONS OF CEREBROSPINAL FLUID

1. It serves as a fluid buffer. 2. It acts as a reservoir to regulate the contents of the cranium; if the volume of the brain or of the blood increases, cerebrospinal fluid drains away; if the brain shrinks, more fluid is retained. 3. It may serve as a medium for nutrient exchanges in the nervous system; in the main, however, the brain carries out its metabolic exchanges directly with the blood.

Effects of cerebral tumour

If there is any increase in the intracranial contents, as by a tumour of the cerebrum, additional room is made by expulsion of some cerebrospinal fluid. Then the blood vessels are compressed, the gyri flattened, and gradually there is a general rise of pressure above the tentorium which is transmitted to the prolongations of the subarachnoid space round the optic nerves. The first structures in the nerve to suffer from compression are the veins: blood can still flow along the arteries and reach the optic disc, but its return is interfered with. In consequence, the minute vessels at the nerve head become engorged and swollen; the fluid exudes from them at this point of least resistance. The resulting appearance, on ophthalmoscopic examination, is termed *papilloedema*.

Pressure is then exerted on the posterior fossa of the skull. The cerebellum is gradually driven into and through the foramen magnum, fills this aperture like a cork, and thus impedes the escape of fluid into the spinal canal, whence about one-fifth of the total absorption of fluid normally occurs. In addition the fourth ventricle foramina (of Magendie and Luschka) are probably distorted and partially blocked. A vicious circle is thus established: as cerebrospinal fluid cannot escape from the ventricles and is not absorbed, hydrocephalus results. This further raises the intracranial pressure and wedges the cerebellum still more firmly into the foramen magnum; death results from medullary anaemia.

Hydrocephalus

By this term is meant a pathological accumulation of cerebrospinal fluid; the hydrocephalus may be internal or external, according as to whether the excess fluid is in the ventricular system or in the subarachnoid space. Hydrocephalus is due to the obstruction to the outflow of the fluid. The obstruction may be:

(i) Intraventricular, blocking the foramen of Monro, the cerebral aqueduct, or the fourth ventricle foramina. The fluid in the occluded ventricles cannot escape and cannot undergo absorption locally; its volume progressively increases because of continued formation of fluid by the choroid plexuses.

(ii) Extraventricular: (a) preventing the free flow of the fluid throughout the subarachnoid space and so diminishing the total surface of arachnoid villi available for absorption. Thus a block at the foramen magnum prevents the fluid from entering the spinal arachnoid and thus cuts off one-fifth of the absorbing surface; a block at the tentorial opening (cisterna ambiens) prevents the fluid from passing from the posterior fossa of the skull into the supratentorial subarachnoid space where most of the absorption normally occurs; (b) inflammatory changes in the leptomeninges may occlude the arachnoid villi; (c) thrombosis of the dural sinuses prevents the escape of the fluid from the subarachnoid space into the veins.

Communicating and non-communicating hydrocephalus. The site of the obstruction can often be determined by injecting phenolsulphonphthalein into the lateral ventricle. Normally it appears in fluid obtained by lumbar puncture in 2–3 minutes, and in the urine in 10–12 minutes. If it does not appear in the spinal fluid or only after a long delay the hydrocephalus is non-communicating, i.e. the site of the block is not distal to the fourth ventricle foramina. If it appears in the spinal fluid normally but in the urine after an undue delay, the hydrocephalus is communicating, i.e. the block is in the meninges distal to the fourth ventricle foramina, in the arachnoid villi, or in the dural venous sinuses.

LUMBAR PUNCTURE

This is carried out by introducing a needle into the subarachnoid space below the termination of the spinal cord, usually between the spines of the fourth and fifth lumbar vertebrae.

It is performed for: diagnostic purposes, to relieve raised intracranial pressure, e.g. in meningitis or uraemia, the introduction of drugs (e.g. streptomycin and spinal anaesthetics).

Use of lipiodol

Lipiodol is a very heavy liquid, opaque to X-rays, which can be injected into the cisterna magna. Normally it drops by reason of its weight to the bottom of the spinal theca. If the subarachnoid space is obstructed by tumour or by adhesions the lipiodol is held up at that point and its position can be demonstrated by X-rays.

The effect of intravenous injection of hypertonic solutions on the c.s.f. pressure

When 50 ml of 10 per cent NaCl solution is injected intravenously c.s.f. pressure falls for 2 hours due to absorption of fluid from the c.s.f. into the plasma. However, the effect is temporary because Na^+ and Cl^- eventually move into the c.s.f. themselves and equilibrium is re-established. Nevertheless, the temporary benefit secured may be of value in conditions of raised intracranial pressure (e.g. as caused by a cerebral tumour). Papilloedema may be relieved, consciousness restored and intracranial operations made easier as bulging of the brain is prevented.

Hypertonic glucose injections (i.v.) likewise exert but a transient benefit as the glucose is metabolized. Hypertonic urea exerts a more prolonged effect because of the low rate of penetration of urea into the c.s.f. and the slowness of renal excretion of urea.

REFERENCE

HEISEY, S. R., HELD, D., and PAPPENHEIMER, J. R. (1962). *Am. J. Physiol.* **203**, 775.

The blood–brain barrier

Paul Ehrlich was the first to demonstrate that when the dye trypan blue was injected intravenously subsequent examination of the tissues after sacrifice of the animal revealed that all organs were stained except the brain. This led to the concept of the blood–brain barrier. Later work has established that only water, CO_2, and O_2 cross the cerebral capillaries easily, whereas the passage of many 'physiological' substances is slow.

Two barriers in fact exist—one located at the choroid plexus–cerebrospinal fluid interface and the other between the cerebrospinal fluid and the brain capillaries elsewhere than the plexuses. The choroid plexus is slowly permeable to small ions and lipid-insoluble substances and is impermeable to large lipid-insoluble compounds.

The most spectacular progress in our understanding of the blood–brain barrier has been provided by Brightman, Karnovsky and their colleagues. The capillaries in brain differ from those in other organs. The endothelial cells are surrounded by a continuous belt of tight junctions which are 'tight' for molecules down to a molecular weight of 2000. The endothelial cells are covered by foot processes from astrocytes. The covering is, however, incomplete, the spaces between these glial processes being large enough to permit the passage of protein markers with a molecular weight of 40 000. The structural basis for the homeostatic regulation of the interstitial fluid of the brain may be depicted diagrammatically [FIG. V.182].

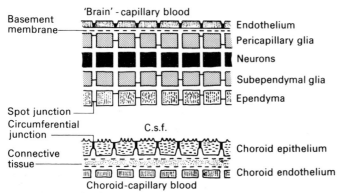

FIG. V.182. Schematic representation of the structures involved in the exchange of materials between the blood, cerebrospinal fluid, and cells within the vertebrate brain. The intercellular spaces between endothelial cells of 'brain' capillaries are occluded by circumferential junctions (*zonulae occludentes*) which prevent the escape of proteins, such as peroxidase, from the blood. Circumferential junctions between choroid epithelial cells also act as a physical barrier for the escape of materials from the blood into the cerebrospinal fluid.
(Reproduced from Cohen, M. W., Gerschenfeld, H., and Kuffler, S. W. (1968). *J. Physiol., Lond.* **197**, 363–80.)

Crone and colleagues injected a mixture of albumin-bound dye and various test substances into the carotid artery, immediately sampling superior sagittal sinus outflow. An analysis of the dilution curves [FIG. V.183] reveals the rate of transfer across the blood–brain barrier. Such studies show that the blood–brain barrier permeability even to small polar substances is very limited. The presence of lipophilic and hydrophilic groups in a molecule is decisive in determining whether a substance passes quickly or not into the brain. Thus the introduction of two hydroxyl groups into the molecule of propanol ($CH_3.CH_2.CH_2 OH$), thereby changing it to the highly polar D-glycerol, strikingly reduces the transcapillary loss from 95 to 3 per cent. Similar techniques have revealed that D-glucose passes the blood–brain barrier by facilitated diffusion and that the blood–brain barrier distinguishes between dextro- and laevo-glucose.

When the venous outflow concentration of test molecules is compared in (say) brain and muscle circulations, it can be shown that the ratio in the effluent between [^{14}C]-D-glucose and ^{42}K is less than unity in the brain and above unity in muscle. Muscle possesses passive porous capillaries and potassium escapes faster than glucose by diffusion. Brain capillaries behave quite differently. Similarly, glucose transference from blood to brain is faster than that of mannitol, whereas in muscle capillaries with their larger porosity the two molecules are transferred at a similar rate.

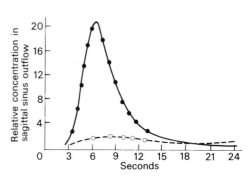

FIG. V.183. Relative concentrations of glycerol and propanol in superior sagittal sinus blood following intracarotid injection: note that because of the rapid transcapillary loss of propanol (peak 'extraction' 90 per cent) the relative venous concentration of propanol is very low. (From Crone, C. (1971). In *Ion homeostasis of the brain* (ed. Siesjö, B. K. and Sørensen, S. C.) p. 56. Munksgaard, Copenhagen.

When tritiated mannitol or tritiated glucose are injected rapidly, immediate sacrifice and freezing of the brain followed by autoradiography reveals that whereas mannitol is localized only intravascularly, glucose permeates the entire tissue. There is no preferential localization of glucose and no evidence is obtained that glial cells preferentially take up glucose.

In suggesting answers to the oft-posed question 'Why a blood–brain barrier?' Crone (1971) points out that the barrier enables close control of an interstitial fluid composition strikingly different from that of plasma. Even wide ranges of potassium concentration do not appreciably modify [K^+] in the interstitial fluid (Bradbury and Davson 1965). As the low [K^+] of the extracellular phase of neurones and glial cells is highly important for their function, this alone would provide an argument for the barrier. Furthermore, the continuous release of transmitters such as serotonin, noradrenalin, and dopamine from the central neuronal terminals might be less effective if such substances were to be dissipated via the immediately adjacent capillary blood flow. On the other hand, the neuronal excitant L-glutamic acid which stimulates cortical neurons cannot reach the synaptic structures involved, from the glutamate sources in the plasma.

REFERENCES

BRADBURY, M. W. B. and DAVSON, H. (1956). *J. Physiol., Lond.* **181**, 151.
CRONE, C. (1970). In *Capillary permeability* (ed. C. Crone and N. A. Lassen). (Benzon Symposium II.) Munksgaard, Copenhagen.
KUFFLER, S. W. and NICHOLLS, J. G. (1977). *From neuron to brain.* Sinauer, Sunderland, Mass.
PAPPENHEIMER, J. R. (1967). *Harvey Lect.* 71–94.
SIESJÖ B. K. and SØRENSEN, S. C. (eds.) (1971). *Ion homeostasis of the brain.* (Benzon Symposium III.) Munksgaard, Copenhagen.
WALSHE, F. M. R. (1929). *Lancet* **i**, 963.
WEBSTER, K. E. (1975). *Proc.R.Soc.Med.* **68**, 203–10.

GLIA

Supporting cells

Although there are one hundred thousand million neurons in the CNS, supporting cells exceed even this number by some six times. These supporting cells form the structural matrix of the CNS but additionally play a vital role in transporting gas, water, electrolytes, and metabolites from the capillaries to the neurons and they remove waste products from the neurons. Unlike the neurons these supporting cells normally undergo mitotic division. They can be classified as astrocytes, oligodendrocytes, microglia, and ependyma. Generically they comprise the neuroglia [FIG. V.184].

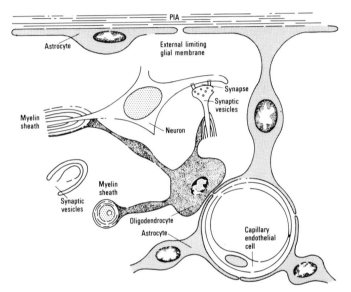

FIG. V.184. Diagrammatic representation of relationship between neuron, astrocyte, and oligodendrocyte.
(From Curtis, B. A., Jacobson, S. and Marcus, E. M. (1972). *An introduction to the neurosciences*. Saunders, Philadelphia.)

Astrocytes are large. They form the essential skeleton of the CNS. They segregate synapses and form a barrier at the inner and outer surface of the brain and around the capillaries and blood vessels. They can proliferate and form scars after neuronal damage. Their processes at the outer surface fuse with those of the pia and astrocytic processes fuse with ependymal processes to form the internal glial membrane.

Oligodendrocytes as their name implies, have fewer dendrites. They have dense clumps of rough endoplasmic reticulum and have a much darker cytoplasm and nucleus than astrocytes. They form and maintain myelin—they may also digest the myelin.

Microglia cells are of mesodermal origin unlike astrocytes and oligodendrocytes which arise from ectoderm. Ovoid and small they seem to be multipotential, able to act as phagocytes or to become astrocytes.

Mononuclear cells—macrophages only appear in abundance after myelin destruction.

Ependymal cells line all parts of the ventricular system. Cuboid and ciliated, their cell processes fuse with astrocytic processes to form the inner limiting glial membrane, In modified form, ependymal cells are found attached to the capillaries in the lateral ventricles and third and fourth ventricles where they form the choroid plexus.

Glial cell properties

Glial cells exhibit a resting membrane potential of −85 to −90 mV—appreciably higher than that of neurons (−75 mV). The glial membrane behaves like a perfect potassium electrode and accurately follows the Nernst equation. Changes in sodium and chloride concentrations do not modify this membrane potential.

This dependence of the glial membrane potential on external potassium concentration underlies the finding that the stimulation of cortical neurons in the vicinity of an impaled glial cell causes a predictable depolarization of the glial cell due to the leakage of potassium from the neurons into the interstitial fluid.

When plasma potassium concentrations are altered from 1.5 mmol to 99 mmol, cerebrospinal fluid concentration of potassium changes only from 1.7 mmol to 3.7 mmol. Glial activity is evidenced by the potential across the glial cell membrane which alters by 59 mV per tenfold increase or decrease in the ratio $[K_o]/[K_i]$ [FIG. V.185]. These experiments of Kuffler *et al.* (1966) were performed on the optic nerve of *Necturus* (newt) at 20 °C. When, on the other hand, glial membrane potentials are measured at different *plasma* concentrations the change of their potential is only 22 mV per tenfold increase in [K+] of the plasma. Thus, the glial membrane potential responds to changes of interstitial fluid [K+], which reflect changes in the c.s.f. concentration of K+, rather than that of blood.

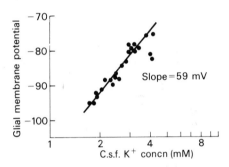

FIG. V.185. Glial cell membrane potentials measured in mud puppies with intact circulations. Each point represents data from one animal maintained in a high potassium environment. The solid line gives the slope predicted by the Nernst equation for a tenfold change in potassium concentration. (After Cohen, M. W., Gerschenfeld, H. M., and Kuffler, S. W. (1968). *J. Physiol., Lond.* **197**, 363–80.)

Cerebrospinal fluid is formed by the choroid plexuses. The choroid capillary wall is of fenestrated endothelium—protein molecules such as peroxidase (MW = 43 000 daltons) can leak out—unlike the brain capillaries, where tight junctions are a feature of the intercellular sites.

The underlying choroidal epithelial cells structurally resemble secretory cells and tight junctions are a feature of the intercellular regions of this choroid epithelial belt [FIG. V.182]. This epithelial layer is credited with preventing proteins from escaping and regulating the c.s.f. formed by active transport.

REFERENCES

COHEN, M. W., GERSCHENFELD, H. M., and KUFFLER, S. W. (1968). *J. Physiol., Lond.* **197**, 363.

CSERR, H. F. (1971). *Physiol. Rev.* **51**, 273.

KUFFLER, S. W. and NICHOLLS, J. G. (1966). *Ergeb. Physiol.* **57**, 1.

—— —— and ORKAND, R. (1966). *J. Neurophysiol.* **29**, 768.

PART VI
The special senses

The ear

THE OUTER AND MIDDLE EAR

The external auditory meatus, about 2.5 cm long, leads inwards and forwards to the ear drum (tympanic membrane). The tympanic membrane consists of connective tissue covered with skin on the outside and mucous membrane on the inside. The ear drum is the shape of a shallow funnel with the apex of the funnel (umbo) pointing inwards. It is probably aperiodic. Movements of the ear drum set up by the pressure variations caused by sound waves are relayed to the three ossicles consisting of the malleus, incus, and stapes. The handle of the malleus is attached to the inner side of the tympanic membrane, its tip being situated at the umbo. The head of the malleus is firmly bound by ligaments to the incus. The long process of the incus passes downwards and parallel to the handle of the malleus and articulates with the head of the stapes. The oval base of the stapes fits into the oval window of the cochlea, being surrounded and attached to the margins of the oval window by the annular ligament. The ossicular bones increase the efficiency of transmission of the sound waves to the inner ear. The cross-sectional area of the lightly damped vibrating ear drum is about 50 mm^2; that of the heavily damped oval window is 3 mm^2. The foot of the stapes rocks on a fulcrum provided by the lower margin of the oval window to which it is more firmly attached by the dense annular ligament; elsewhere round its circumference the annular ligament is less firm.

Two muscles, stapedius and tensor tympani, modify the transmission of sound waves. Tensor tympani, as its name implies, keeps the ear drum taut. Attached to the neck of the malleus, the muscle on contraction pulls the handle of the malleus and the ear drum inwards and increases its tension. Tensor tympani can be reflexly excited by loud sounds and this reflexly diminishes the amplitude of vibration of the ear drum. Stapedius is inserted into the neck of the stapes and tends to pull the stapes out from the oval window. It, too, serves a protective function and can be reflexly excited by loud sounds.

The Eustachian tube connects the middle-ear cavity with the pharynx. It serves to equalize the pressures on either side of the ear drum. Normally its pharyngeal opening is closed, but it can be opened by swallowing or yawning which contracts the tensor palati muscle. If an inequality of pressure develops between air in the middle ear and that in the external environment, swallowing thus allows an equalization of the pressures. When catarrhal infections cause closure of the pharyngotympanic tube this swallowing mechanism is no longer effective. The air trapped in the middle ear is partially absorbed; as a result the ear drum bulges inwards causing much discomfort and loss of hearing.

THE INNER EAR

The cochlea is a tube, coiled (2½ turns) round a central bony pillar, the modiolus. At the lower end, the cochlea contains two apertures, sealed respectively by the oval window and the round window. The cochlear lumen is divided by two membranes into three compartments. The basilar membrane, on which is situated the specialized organ of Corti, is attached medially to the osseous spiral lamina and laterally to the fibrous spiral ligament. Reissner's membrane is attached to the medial wall in the region of the limbus and laterally to the upper margin of the stria vascularis. Between the basilar membrane and Reissner's membrane is the scala media which is filled with endolymph. Above Reissner's membrane is a compartment filled with perilymph named the scala vestibuli and below the basilar membrane another compartment containing perilymph is named the scala tympani [Fig. VI.1]. At the apex of the cochlea a small opening (the helicotrema) permits continuity between the scala vestibuli and the scala tympani.

If the coiling of the cochlea is ignored for diagrammatic purposes the appearance is as shown in Fig. VI.1.

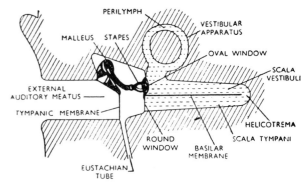

Fig. VI.1. Diagrammatic arrangement of the organ of hearing. The sound pressure fluctuations move the tympanic membrane, activating the lever system of small bones (malleus and stapes shown). The oscillations of the stapes in the oval window set up vibrations in the fluid (perilymph) of the scala vestibuli, in the basilar membrane itself, and in the scala tympani, eventually causing movement of the flexible round window membrane. The vestibular apparatus is concerned with balance of the body. The cochlea is shown straight, not coiled, for simplicity, and Reissner's membrane and scala media have been omitted from this diagram. (von Békésy, G. and Rosenblith, W. A. (1951). In *Handbook of experimental psychology* (ed. S. S. Stevens). Wiley, New York.)

When the stapes is pressed into the oval window the pressure is transferred to the perilymph in the scala vestibuli and thence to the scala media, causing a downward movement of the basilar membrane. Finally, the pressure is transmitted via the scala tympani to the round window, causing it to bulge outwards into the middle ear. Conversely, an outward movement of the stapes and oval window causes an upward movement of the basilar membrane.

The basilar membrane and organ of Corti

The basilar membrane is composed of fibres which run laterally from the osseous spiral lamina to the spiral ligament. At the base of the cochlea the basilar fibres are only 0.15 mm long but at the apex of the cochlea they are 0.4 mm long. The basilar membrane is not under tension, contrary to earlier beliefs. Thus when fine cuts are made (by a microdissection technique) in the membrane the cut edges do not gape as does tissue under tension. Moreover when a hair probe is used to exert a pressure on the exposed basilar membrane the deformation pattern produced is quite unlike that of a membrane under tension (Békésy). The basilar membrane can best be likened to a gelatinous sheet covered by a thin homogenous fibre layer. There are marked differences in stiffness in the basilar membrane at the apex and that at the base. The volume displacement of the cochlear partition for a given hydrostatic pressure at the apex is 200 times that near the stapes. Békésy scattered small crystals of silver on the surface of the taut basilar membrane and used stroboscopic observation to show the movement of the membrane, in response to vibrations of known frequency and amplitude imparted to the oval window by means of a mechanically controlled piston. The higher the frequency of vibration imparted to the oval window the nearer to the stapes was the region of basilar membrane which showed the maximal amplitude of vibration. Low frequency vibration was associated with a maximal amplitude of vibration of the basilar membrane near the apex of the cochlea. However, the amplitude of vibration did not show a sharp maximum. The vibrations of the membrane are heavily but not critically damped. The basic pattern of movement of the basilar membrane is that of a travelling wave—like that of the arterial pulse wave. Such a wave initiated by a 200-cycle tone is shown in Fig VI.2. In the case of a complicated travelling wave initiated by a sound consisting of several tones and their harmonics, the high frequency components are eliminated nearer the stapes than those of longer wavelengths. Thus the apex of the cochlea is affected only by low frequency tones while the base of the cochlea, though responding to low frequencies, is mainly affected by high frequencies.

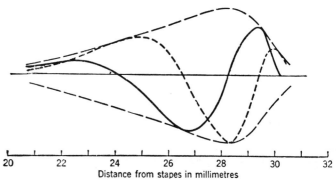

Fig. VI.2 Diagram illustrating travelling wave theory of basilar membrane movement. Solid and short dash lines represent same sound wave at two instances of time. Long dash line is described by connecting the peaks at successive instants of time. Scale at bottom represents distance along basilar membrane. (von Békésy, G. (1947). *J. acoust. Soc. Am.* **19**, 452.)

The organ of Corti

At the junction of the basilar membrane with the bony spiral lamina, and projecting into the scala media, stand the two rods of Corti. Internal to the inner rod is a single hair cell; external to the outer rod are three or four hair cells. There are in all 3 500 inner hair cells and 20 000 outer hair cells. They are innervated by nerve fibres of the cochlear division of the eighth nerve. These fibres have their cell bodies (27 000 in all) in the spiral ganglion which is the counterpart of the dorsal root ganglion in the case of the spinal nerves. From the upper surface of the hair cells, which are 8–12 µm in diameter, project tiny hairs 4 µm long and 0.1 µm thick which pass through a thin dense granular reticular lamina and embed themselves in the tectorial membrane. The tectorial membrane is a spiral ribbon attached at one end of the limbus and by one surface and its outer edge to the organ of Corti in the region of Hensen's cells which lie outside the external hair cells [Fig. VI.3]. The tectorial membrane is a stiff gelatinous structure composed of glycoprotein material.

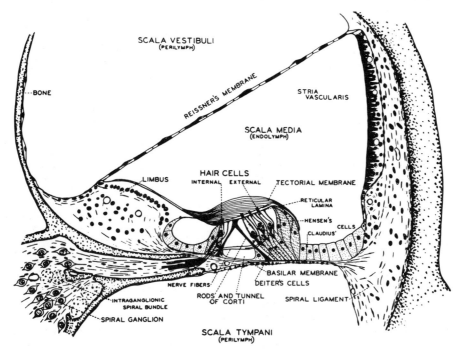

Fig. VI.3. Diagram, based on a camera lucida drawing of a cross-section of the second turn of the cochlea in the guinea-pig. (Davis, H. (1953). Acoustic trauma in the guinea-pig. *J. acoust. Soc. Am.* **25**, 1180.)

Between the rods of Corti lies the tunnel of Corti, which is filled not with endolymph but with perilymph.

The mechanism of stimulation of the cochlea has been elucidated by the use of electrophysiological methods.

Endolymphatic potential. Békésy introduced a microelectrode into the cochlea. By micromanipulation, the electrode was made to traverse the scala vestibuli to puncture Reissner's membrane and thus to penetrate the scala media. He found that the scala media showed a steady DC potential, being 50–100 mV positive to the scala vestibuli. The interior of the cells of Reissner's membrane and of the cells of the organ of Corti and the stria vascularis was about 30 mV negative to the perilymph in the scala vestibuli. On penetrating into the scala tympani the perilymph therein was virtually identical with the potential of the scala vestibuli. Thus Reissner's membranes separates two fluids of widely different composition and across the membrane, providing that it is not damaged grossly, exists a potential difference of 50–100 mV. Endolymph has an electrolyte concentration very similar to that of intracellular fluid $K^+ = 137.7$ mM l^{-1}; $Na^+ = 15.2$ mM l^{-1}; Cl^-; $= 107.7$ mM l^{-1}). Perilymph on the other hand is very similar in composition to extracellular fluid ($K^+ = 4.8$ mM l^{-1}; $Na^+ = 153.7$ mM l^{-1}; $Cl^- = 120.6$ mM l^{-1}). Obviously the endolymphatic potential cannot be explained on the basis of a sodium or potassium ion concentration difference, for the endolymph potential is positive with respect to that of the perilymph unlike, for instance, the interior of a cell which is negative with reference to its surroundings. The endolymphatic potential is directly dependent upon an adequate oxygen supply; it can be abolished by introducing sodium cyanide into any one of the three scalae. It can be increased by a downward movement of the basilar membrane and, conversely, an upward displacement of the membrane reduces the DC potential. It has recently been suggested that the stria vascularis is the source of the potential and that movements of the basilar membrane affect the DC potential value by altering the forces on the hair cells which are embedded in the tectorial membrane. Thus a radial displacement of the tectorial membrane also modifies the potential.

The injection of Ringer solution (which resembles extracellular fluid) into the scala media abolishes the DC potential although, as might be expected, it has no effect when injected into the scala tympani. Conversely, the injection of a potassium-rich, sodium-poor solution into the scala media does not alter the DC potential, but abolishes it if the injection is made into the scala tympani. Injection of the fluid into the scala vestibuli has, however, no effect. Thus Reissner's membrane is impermeable to sodium and potassium whereas the basilar membrane is freely permeable to these ions. It is the reticular lamina [FIG. VI.3] which is a barrier to diffusion; correspondingly the tunnel of Corti must be filled by a fluid like perilymph. As the nerve fibres from the organ of Corti traverse this tunnel they would be unable to conduct impulses if it were filled with endolymph (which has the same electrolyte composition as intracellular fluid).

The cochlear microphonic potentials. Wever and Bray (1930) found that when the eighth nerve (cat) was placed on electrodes and the action potentials amplified in the usual manner the loudspeaker recorded faithfully pure tones fed into the ears as sound waves up to frequencies of 20 000 Hz. These potentials were mistaken at first for nerve action currents, but subsequent investigation showed that they were microphonic potentials originating in the cóchlea and spreading into adjacent tissues. The microphonic potentials are resistant to ischaemia and anaesthesia and show no latency or refractory period; all these properties differentiate them from nerve impulses. They are produced by the transformation of mechanical energy into electric energy in much the same way as pressure on a quartz crystal induces a piezo-electric potential. By introducing microelectrodes into different turns of the cochlea it has been possible to establish that the source of the cochlear microphonic potentials is in the external hair cells at the level of the reticular lamina. The microphonic potentials can be recorded optimally by placing one electrode in the scala media and one electrode in the scala tympani. Békésy has concluded that they are produced by a 'generator' located between the tectorial membrane and the basilar membrane. They are probably developed as a modulation of the DC endolymphatic potential. As has been stated above, the DC potential is altered by movements of the basilar membrane (which bears the hair cells and which thus moves them in relation to the tectorial membrane); if the movement of the membrane is slow and periodic the corresponding changes in the DC potential are indistinguishable from the cochlear microphonics (CM). Just as the DC potential is abolished by introducing a high-potassium solution into the scala tympani so likewise the cochlear microphonic in response to a sound-stimulus is eliminated.

By recording CM at various points in the cochlea of a guinea-pig, Tasaki (using microelectrodes) has established that the basal turn of the cochlea responds to all frequencies of sound stimuli. Only low frequencies cause a microphonic response from the apical turn [FIG. VI.4]. The recording of CM has simplified and, at the same time, has enormously improved our knowledge of the 'tonal localization' of the cochlea. Thus a local degeneration of the organ of Corti can be caused in animals (and man) by prolonged exposure to a loud tone; the cochlear microphonic generated by this particular band of frequency is abolished. Histological confirmation of the degeneration *post mortem* allows us to compile a map of tonal localization in the cochlea. Deaf albino cats and waltzing guinea-pigs, in which the organ of Corti is congenitally underdeveloped, show no cochlear microphonics. In the pouch-young opossum, the development of the organ of Corti begins in the upper part of the basal turn; correspondingly the cochlear microphonics at this stage of the animal's development can only be evoked by a restricted band of medium frequencies.

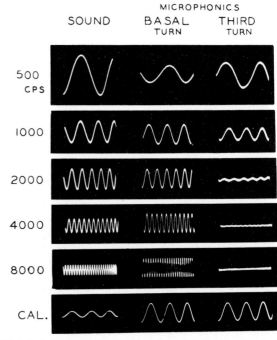

FIG. VI.4. Relation between intensity of sound (left column) and microphonic responses recorded from basal turn (middle) and from third turn (right) of cochlea. Three records at each frequency were photographed simultaneously. Compare relative amplitudes and phase differences. Calibration at bottom indicates sound pressure of approximately 57dB above 0.0002 microbar at 1000 Hz (left) and 1/3 mV peak to peak (middle and right). (Tasaki, I. (1954). *J. Neurophysiol.* **17**, 97.)

At present, the available evidence is in favour of the theory that the increase in current flow associated with the cochlear microphonic stimulates the terminal non-medullated auditory nerve endings. The excitatory phase of CM (i.e. increasing negativity in the scala media) is associated with a current flow outwards across the membrane of the nerve fibres.

Action potentials of the eighth nerve. The response of single first-order neurons of the eighth nerve to sounds of different frequencies and intensities has been reported by Tasaki [FIG. VI.5]. At threshold intensity the nerve endings respond only to a narrow range of frequencies. As the intensities are increased, the nerve endings respond to an increasingly wide range of sound frequencies which lie below that to which they maximally respond, but there is little if any impulse discharge to frequencies above that of maximal responsiveness.

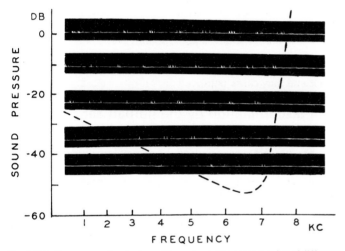

FIG. VI.5. Responses of a single auditory nerve fibre to tone pips of different frequencies and intensities. Dotted line shows boundary of response area of this fibre. (Tasaki, I. (1954). *J. Neurophysiol.* **17**, 97.)

Recordings from second-order neurons are technically easier. Again, a given single fibre shows an obvious peak of response to a particular sound frequency. By increasing the sound intensity the fibre shows responses to sounds of a wider range of frequency, but there is a sharper 'cut-off' of impulse discharge to frequencies below that causing the maximal response than there is in the case of the first-order neuron [FIG. VI.6]. It seems clear, however, that the auditory nerve endings do not respond to only one frequency except at threshold intensity of the sound stimulus. We can assume that at such threshold intensity for a given pure tone, a specific region of the basilar membrane vibrates and consequently a specific or individual nerve ending may signal this. With greater intensities of sound, a considerable length of the basilar membrane is caused to vibrate and, correspondingly , other nerve endings are activated. Nevertheless, the maximal vibration is at the site affected by the threshold intensity and the nerve endings in this region will fire at maximal impulse rate. The 'place' theory of hearing, which supposes that different frequencies of sound activate different areas of the basilar membrane and hence different auditory neurons, gains some support from these findings. However, the remarkable power of the human auditory mechanism to discriminate between sounds in the 64–2000 Hz range, can hardly be explained by this analysis of the cochlear microphonics and auditory nerve impulses simply in terms of peripheral localization in the cochlea. All recent findings suggest that the apical turn of the cochlea contains the units which respond to the lower five octaves (1024–32 Hz). It is difficult to believe that place discrimination could possibly be so selective.

Wever has therefore proposed that a 'volley principle' may

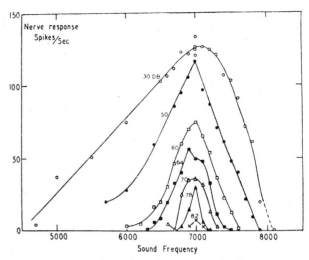

FIG. VI.6. The relation of intensity and frequency to the discharge of a single auditory neuron. As the intensity level is increased (numbers attached to the curve get smaller), a wider range of frequencies excite the fibre, and each frequency excites more discharges. Distance along abscissa is probably a function of distance along the basilar membrane. Intensity level equals dB below a reference level. (Galambos, R. and Davis H. (1943). *J. Neurophysiol.* **6**, 39.)

account for frequency discrimination for tones up to 2000 Hz. By rotation of activity, a group of nerve fibres from the apical turn of the cochlea, reproduces in its total output of nerve impulses the frequency of the stimulating sound waves. Thus pure tones up to 2000 Hz yield clear synchronous volleys of action potentials. The duplex theory of hearing which combines the place principle and the volley principle is gaining favour. This suggests that place 'coding' of sensory information is responsible for the discrimination of sound frequencies above 2000–3000 Hz (upper limit = 16 000–20 000 Hz) and the volley principle accounts for the 'coding' of sound frequencies in the lower register.

Effect of efferent activity in the auditory nerve on the afferent output from the cochlea. Galambos (1956) has shown that the impulse activity in the afferent fibres of the eighth nerve, in response to any given sound stimulus, is greatly reduced by the electrical stimulation of the efferent olivocochlear bundle which courses in the acoustic nerve. It is of considerable interest that such efferent stimulation did not affect the size of CM recorded simultaneously with the action potentials, although the latter might even be completely inhibited. It seems that the sensitivity of the auditory mechanism can be varied by efferent discharge along the acoustic nerve—in the same manner as the sensitivity of the muscle spindle discharge can be altered by γ efferent discharge.

PHYSICAL PROPERTIES OF SOUND. SOUND PERCEPTION

The simplest sound wave is a sine wave such as is produced from a tuning fork. The vibrating fork causes air pressure changes in its vicinity and these travel through the air as waves which are perceived by the auditory mechanism as sound. The wave frequency determines the pitch of the sound. If two tuning forks are simultaneously set in vibration and one emits sound waves of two or three times the frequency of the other the resultant wave form set up by the two forks will be different from that of either fork vibrating alone. The wave form of any periodic vibration can be resolved however (by Fourier's theorem) into a series of simple waves, of which that with the lowest frequency is called the *fundamental*, and the others, whose frequencies are multiples of this fundamental, are called *harmonics*. The quality of a sound depends on its wave form; many sounds are of very complicated wave form but can be resolved by mathematical analysis into fundamental and numerous harmonics, providing that they are periodic. If a sound contains many components which have frequencies which are not simple multiples of the fundamental, it is designated 'noise'. The auditory mechanism is much more sensitive to sounds of 1000–3000 Hz than to sounds outside this range [FIG. VI.7].

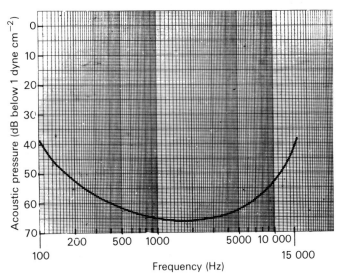

FIG. VI.7. Audibility curve for a normal subject.

The intensity of a sound is measured in terms of the energy or power (in microwatts, or ergs per second) passing through an area of 1 cm² placed at right angles to the direction of sound propagation. Alternatively it can be expressed, in dynes cm⁻² in terms of pressure. Loudness is not synonymous with intensity—for loudness is a psychological term. To take a simple example, the human auditory mechanism cannot perceive a sound of frequency above 20 000 Hz although that of a dog can—hence the development of whistles which emit a sound of frequency above that of 20 000 Hz for summoning dogs. Whatever the intensity of the blast (even though it may cause the dog to put its paws over its ears) the owner of the whistle cannot hear it at all, and so suffers no sensation of loudness. Even in the audible range of sound frequencies for man, there is a great difference in the sensitivity; thus the sound energy sufficient to produce threshold hearing is only 10^{-9} to $10^{-16}\,\mu\text{W cm}^{-2}$ for a note of 2000 Hz. The pressure variation caused by such an intensity of sound is only 1/12 000 bar or 1/12 000 dyne cm⁻². At the lowest frequency audible to the human ear (16 Hz) the pressure required is 100 bars and at the upper limit (20 000 Hz) the pressure required for threshold audibility is 500 bars.

In view of the huge range of intensities involved, a scale of change of intensity is employed to describe them. The unit is the bel—if the intensity of a sound E_1 increases ten times (to E_2) the intensity level is raised by one bel—the scale is thus logarithmic.

$$N_b = \log \frac{E_2}{E_1} \text{ (where } N_B = \text{ the number of bels).}$$

The decibel which is one-tenth of a bel is more convenient.

$$N_{dB} = 10 \log \frac{E_2}{E_1}.$$

where E_1 is the reference sound. The energy of a sound wave is the square of the pressure exerted by the sound and is also the square of the voltage applied to a loudspeaker. Hence

$$N_{dB} = 20 \log \frac{P_2}{P_1} = 20 \log \frac{V_2}{V_1}.$$

The decibel is about the least change in intensity which can be detected. The standard reference intensity is defined in physical units as 0.0002 dyne cm⁻² (10^{-16} W cm⁻²); such a sound pressure is approximately equal to that required in playing a 1000 Hz note so as to be just at the threshold of audibility in young people. The accompanying TABLE shows how the full range of sound energy from the threshold to the loudest tolerable sound can be expressed on the scale between 0 and 120 dB. The sound energy at the top of the scale is one million million times that required for threshold hearing.

TABLE (AFTER BEATTY)

Decibels	
120	Pneumatic drill at operator's position.
100	Loud motor-horn—7 m.
80	Very heavy street traffic, New York.
50	Conversational voice—4 m.
30	Quite suburban street—London.
20	Whisper—2m.
10	Rustle of leaves in gentle breeze.
0	Faintest audible sound.

When the sound is very loud indeed (intensity over 120 dB) it can be felt and may cause pain.

Pitch discrimination by the human ear

The ear is most sensitive to sound energy and to pitch variations in the 1000–3000 Hz range—indeed the minimal fractional difference in frequency which is perceptible is 0.3 per cent. Thus a change of 1000 to 1003, 2000 to 2006, and 3000 to 3009 can be detected. Expert musicians can even improve on this. At very low frequencies (32–64 Hz) the minimal fractional difference in frequency perceptible is 1 per cent and in the upper frequency range (16 000 Hz to 20 000 Hz) auditory discrimination is very poor. Altogether there are about 2000 gradations in pitch detectable by the human auditory mechanism. As the basilar membrane is just over 30 mm in length the place theory would require an average shift of 0.015 mm for each pitch change! The range 1000–3000 Hz is the most important for the auditory perception of human speech. If a normal subject listens to a loudspeaker transmitting the voice of a reader, he can detect about 85 per cent of the words used, even if the loudspeaker transmits only the frequencies of above 1000 Hz, although the energy of the original speech sounds is thereby reduced by 80 per cent. Conversely, if the voice is heard through a loudspeaker which transmits only frequencies below 1000 Hz, although 85 per cent of the energy is still transmitted, the intelligibility of the speech is greatly reduced, being only 40 per cent of normal.

An audibility curve can be plotted by determining the energy required to produce threshold hearing at each of the octave frequencies. A valve oscillator is connected via an amplifier to headphones and the subject signals when each pure tone is just audible when sounded for a short duration. FIGURE VI.7 shows an audibility curve in which the sound energy (expressed in dB below 1 volt input to the loudspeaker) is plotted against frequency. The curve shows how much more sensitive is the hearing in the 1000–3000 Hz range of frequency.

Clinical audiometry

Tests are conducted similarly, but the instrument is calibrated so that the threshold intensity for each tone (as determined in a large group of normal subjects) is produced by adjusting the intensity dial to 0 dB. Hearing loss is determined by increasing the intensity until threshold audibility is achieved for each tone tested; the corresponding dB increase on the dial is noted. The test is conducted in a sound-proof room; one ear at a time is tested and the subject wears a headphone. He flashes a light whenever he hears the sounds, which are produced sporadically.

The importance of clinical audiometry lies in the information yielded not only of the degree of deafness but also of the frequency range in which it is most manifest [e.g. FIG. VI.8]. As a result, hearing aids can be designed to overcome some of the acoustic problems of the individual patient.

Speech audiometry is the most practical way of assessing hearing. A list of phonetically balanced words at fixed levels of intensity is reproduced on gramophone records and a class of children wearing headphones can be tested simultaneously.

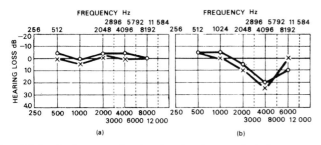

FIG. VI.8. Audiograms showing hearing loss against frequency. A hearing loss of 0 dB implies normality according to the British Standard used in the calibration of the audiometer; a negative loss (above the zero line) indicates an auditory threshold lower than normal. ○ = Right ear × = Left ear. (*a*) Normal hearing. (*b*) Moderate loss of hearing at 4000 Hz in both ears. Subjectively this loss would be unnoticed. (Burns, W. (1959). *Symposium on Engine Noise and Noise Suppression*. Institution of Mechanical Engineers, London.)

Masking. It is common knowledge that the voice must be raised when attempting to converse in noisy surroundings. Quiet conversation conducted at an intensity level of 60 dB will be inaudible if the noise background exceeds this intensity—it has been masked.

Masking represents the inability of the auditory mechanism to separate the total stimulation into the separate components.

The degree to which one component of a sound is masked by the remainder can be measured by finding two thresholds. If one tone is played alone, the audibility threshold can be determined—the absolute threshold for this tone. If the audibility threshold for the tone is then redetermined during the sounding of a separate component, this, the 'masked threshold', can be expressed, as a ratio of the absolute threshold, in decibels.

Masking tends to be greater for tones of approximately similar frequency than for tones widely different in frequency. Low-frequency tones mask high-frequency tones more easily than vice versa; this must be taken into account in the design of deaf aids.

The effects of noise on hearing

The structural and functional integrity of the ear may be seriously affected by continued noise. Broadly speaking the more intense the noise the greater is the hearing loss; the degree of hearing loss and its position in the tonal spectrum is related to the frequency value of the noise. The duration of the period to which the subject is exposed to the noise influences the hearing loss.

Continued daily exposure to industrial noises usually causes a loss of sensitivity in the 4000 Hz range of hearing which gradually progresses and is later accompanied by a loss of sensitivity of hearing in the lower frequency ranges. When the deterioration of hearing affects the 300–3000 Hz range, social handicap results owing to the impairment of the subject's understanding of conversation.

AUDITORY PATHWAYS

The auditory nerve after a short course enters the medulla and divides into components, which pass respectively to the ventral and dorsal cochlear nuclei [FIG. VI.9]. These are the sites of the first synapse. Second order neurons from the cochlear nuclei pass by different pathways to the neighbouring nuclei (superior olive, trapezoid body) on both sides of the brain stem and third-order neurons pass up in the lateral lemniscus to the inferior colliculi of both sides; some of these fibres send collaterals to the medial geniculate bodies. From the inferior colliculi many fibres project to and relay in the medial geniculate bodies. Finally, the medial geniculate body neurons project to the auditory cortex. The auditory cortex in both the cat and the dog lies in the ectosylvian gyrus; it receives an orderly 'tonotopic' representation—the anterior part receiving impulses which originated from the cochlear apex, and the posterior part of the gyrus receiving impulses which arose from the base of the cochlea [FIG. VI.9]. Throughout the auditory pathway tonotopic localization is preserved; thus selective destruction of minute portions of the medial geniculate body causes deafness for restricted bands of frequencies.

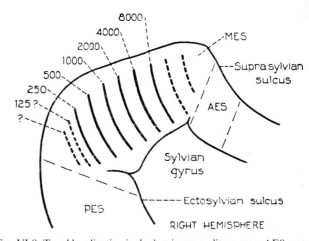

FIG. VI.9. Tonal localization in dog's primary auditory area. AES, anterior ectosylvian area; PES, posterior ectosylvian area; MES, middle ectosylvian area. Bands indicate point at which strychnine spikes occurred with lowest intensity. Bands indicated by dashes were not determined experimentally. (Tunturi, A. R. (1950). *Am. J. Physiol.* **162**, 493.)

Each ear is bilaterally represented in the auditory pathway from the medulla upwards and projects about equally to the two cerebral hemispheres. Thus the removal of one auditory cortex has only a slight effect on auditory acuity. Deafness is hardly ever produced by cortical lesions.

Deafness

The two main types of deafness are conductive deafness and nerve deafness.

Conductive deafness

This may be due to foreign bodies, wax, otitis media, which damages the drum and/or the ossicles, or pathological fixation of the stapes in the oval window (ostosclerosis). The hearing loss is fairly uniform throughout the frequency range.

The skull bones themselves conduct sound to the cochlea, though less effectively, and the basilar membrane can be set in vibration even if the sound energy is not delivered preferentially to the bones of the skull, so hearing loss is never complete.

If a vibrating tuning fork is held near the ear the air-conducted sound can be heard much longer in a normal subject than if the base of the vibrating fork is placed on the mastoid process. The bone-conducted sound becomes inaudible because of the masking effect of sound in the room which is carried by air-conduction. Thus if a finger be placed in the external meatus the bone-conducted sound remains audible for a longer time. In patients with conductive deafness air-conducted sounds are less well perceived and bone-conducted sounds may persist longer. This is the basis of Rinne's test. In normal subjects the air-conducted sound from the tuning fork is heard longer than that heard by bone conduction when the base of the fork is held on the mastoid (Rinne's test positive). In conductive deafness Rinne's test may be negative (bone conduction greater than air conduction).

Weber's test. In this test the base of a vibrating fork is placed on the vertex in the midline. Normally both ears 'hear' the sound equally well. In unilateral conductive deafness the sound is better 'heard' in the affected ear. This is again because masking due to air-conducted sounds is less obtrusive on the affected side.

Nerve deafness

This may be 1. degenerative; 2. hereditary; 3. toxic (quinine, measles); 4. acoustic trauma (boilermakers); 5. meningitic; 6. acoustic neuroma.

In nerve deafness both air-conducted and bone-conducted sounds are less well perceived (Rinne's test). Weber's test reveals that the 'deaf ear' remains deaf to bone-conducted sound and the sound is thus lateralized and perceived by the normal ear.

REFERENCES

BURNS, W. (1973). *Noise and man*, 2nd edn. Murray, London.
DAVIS, H. (1957). *Physiol. Rev.* **37**, 1.
DAVIS, H., DEATHERAGE, B. H., ELDREDGE, D. H., and SMITH, C. A. (1958). *Am. J. Physiol.* **195**, 251.
GALAMBOS, R. (1956). *J. Neurophysiol.* **19**, 424.
JERGER, J. F. (1973). *Modern developments in audiology.* Academic Press, New York.
KEIDEL, W. and NEFF, W. D. (1976). In *Handbook of sensory physiology*, Vol. V/3. Springer, Berlin.
LICKLIDES, J. C. R. (1951). In *Handbook of experimental psychology* (ed. S. S. Stevens) p. 985. Wiley, New York.
MISRAHY, G. A. HILDRETH, K. M., SHINABERGER, E. W., CLARK, L. C., and RICE, E. A. (1958). *J. acoust. Soc. Am.* **30**, 247.
TASAKI, I. (1954). *J. Neurophysiol.* **17**, 97.
—— (1957). *A. Rev. Physiol.* **19**, 417.
TAYLOR, W. (1973). *Disorders of auditory function.* Academic Press, New York.
VON BEKESY, G. (1956). *Science, N. Y.* **123**, 779.

The eye

Anatomical description

The adult human eyeball is a nearly spherical structure one inch in diameter. Only the anterior segment of the eye, the cornea, is transparent, the remainder being opaque. The wall of the eyeball contains three layers 1. external supporting, 2. middle, 3. retina, from without inwards. The external layer consists of dense connective tissue which is white in colour except for that covering the central anterior portion of the eye where it forms the transparent cornea. Except in the cornea the external layer forms the sclera. The middle layer, bluish in colour (hence the name uveal), contains numerous blood vessels. It is thin over the posterior two-thirds of the eyeball but thickens to form the ciliary body more anteriorly.

This circular structure gives rise to ciliary processes which extend inwards and itself continues forwards to form the iris [FIG. VI.10].

The pupil, which is the aperture surrounded by the iris, can be altered in diameter by change of tone of the smooth muscle in the iris.

The middle layer behind the ciliary body is called the choroid. Anteriorly the choroid forms the ciliary body as a ring on the inner side of the sclera.

The retinal layer contains an outer pigmented layer which lines the whole of the inner surface of the middle layer of the eye and an inner nervous layer which contains the photo-receptors (rods and cones). The crystalline lens is attached to the ciliary body by means of the suspensory ligament or zonule. The zonule has a broad area of attachment to both the capsule of the lens and to the ciliary body.

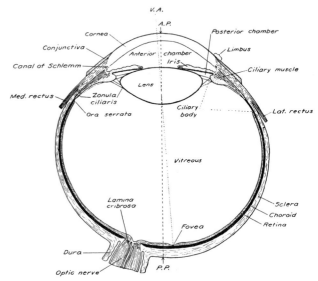

FIG. VI.10. Schematic horizontal meridional section of right eye. × 3. *A.P.*, anterior pole; *P.P.*, posterior pole; *V.A.*, visual axis. (Redrawn and modified from Salzmann.)

The space between the cornea and the anterior lens surface is called the anterior chamber. The narrow circular space between the iris, the lens, and the ciliary body is termed the posterior chamber. Both anterior and posterior chambers contain aqueous humour, a thin watery fluid containing crystalloids in concentrations similar to that of plasma. Aqueous humour contains also a great deal of hyaluronic acid which is kept in the depolymerized state by hyaluronidase present in the ciliary body. Hence the viscosity of the aqueous humour is low. It is probable that most of the aqueous humour is formed by dialysis of crystalloid-containing fluid from the capillaries of the ciliary processes. As the usual intra-ocular tension is 25 mm Hg the hydrostatic pressure within these capillaries must be substantially greater in order that the oncotic pressure of the plasma proteins may be overcome. Once formed, aqueous humour passes from the posterior chamber, forwards between lens and iris to enter the spaces of Fontana which, communicating with the anterior chamber, are found in the angle of the iris. From the spaces of Fontana, the fluid passes across the endothelium lining the canal of Schlemm, which drains the fluid into the ocular veins. Aqueous humour is formed at a rate of 2 mm³ per minute. In inflammatory conditions the iris may adhere to the anterior surface of the lens at the pupillary margin—as a result the normal flow of fluid through the pupil is obstructed and the iris balloons forwards into the anterior chamber—the clinical condition of *iris bombé*. Blockage of the canal of Schlemm leads to an abnormally high intra-ocular pressure— the eyeball tension rises so that the eyeball on palpation feels as hard as a stone. The condition is known as *glaucoma*. Glaucoma can be precipitated in people over 40 years of age by the injudicious local use of atropine. Atropine blocks the response of the sphincteric muscle of the iris and the ciliary muscle to third nerve impulses thereby causing mydriasis (dilatation of the pupil) and paralysis of accommodation [cycloplegia—see p. 387]. As a result the iris falls back into the ciliary angle and interferes with drainage by the canal of Schlemm.

The vitreous body is an amorphous transparent gel which occupies the intra-ocular space between the lens and posterior zonular surface anteriorly and the internal layer of the retina. It is not known whence it is formed. It contains albumin and hyaluronic acid; the latter is responsible for the high viscosity of the vitreous.

The eye and light refraction

Approximately speaking, the refractive index of cornea, aqueous humour, and vitreous humour is 1.33. The crystalline lens, however, has a refractive index of 1.42. For practical purposes then, the light rays suffer refraction at the cornea and the surfaces of the lens. The greatest refraction, as might be expected, occurs at the air-cornea surface; the lens contributes less, but variations in the lens curvature, such as occur with accommodation, are of great importance in vision.

The path of light through the eye is complicated in that, having suffered refraction at the cornea, it is further refracted towards the normal at the anterior surface of the lens and then refracted away from the normal at its posterior surface. However, as an approximation it can be shown that the optic properties of the compound eye can be simplified as in the 'reduced' eye of Listing. This has a single spherical surface of radius 5.6 mm separating two media of refractives indices 1 and 1.336, situated in the aqueous humour 1.4 mm behind the cornea. The nodal point or optical centre of the schematic eye is 7 mm behind the anterior surface of the cornea. (A ray passing through the nodal point does not suffer refraction.) The eye is about 24 mm in length, so the distance of the nodal point from the retina (which is the focal length) is 24 − 7 = 17 mm. The refracting power of a lens is expressed in terms of dioptres.

$$\text{Dioptric power} = \frac{1}{\text{focal length in metres}}.$$

Hence the refractive power of the schematic eye is $\frac{1000}{17}$ = about 59 D. Actual measurements of the total refractive power of the normal human eye using X-ray beams gives a value almost identical with this figure. FIGURE VI.11 shows how the path of the light can be drawn and the image formed on the retina can be constructed. The light rays from a distant point (more than 6 metres) are approximately parallel and will therefore be brought to a focus on the retina (which is the posterior principal focus).

Those rays which pass through the nodal point suffer no refraction. FIGURE VI.11 shows that the retinal image is inverted. By

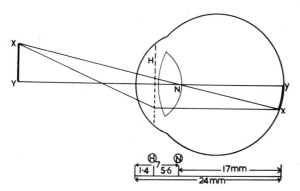

FIG. VI.11. The reduced eye. H = principal point. N = nodal point.

pressing on the eye the reader can convince himself that the stimulation of the retinal elements of the nasal part of the retina causes a sensation of vision in the temporal field. The bright ring of light (phosphene) thus obtained always occurs in the 'visual field' opposite to the part of the retina stimulated. Similarly pressure on the upper surface of the eyeball through the closed lid causes a phosphene in the lower field of vision and vice versa.

The dioptric power of the eye (59 D) is reduced by some 16 D after removal of the lens. The cornea is thus responsible for 43 D of the refractive power of the eye and this is borne out by direct determinations of its dioptric power. Obviously the refractive power of the cornea will be lost when the head is immersed in water (which has the same refractive index as has the corneal surface).

The size of the retinal image can be calculated by similar triangles. Let us suppose that the object is 1 metre high and is 10 metres distant.
Then:

$$\frac{\text{size of object}}{\text{size of image}} = \frac{\text{distance of object from nodal point}}{\text{distance of image from nodal point}}$$

$$\frac{1000}{X} = \frac{10\,000}{17}$$

$$\therefore X = 1.7 \text{ mm}.$$

The angle subtended at the nodal point by the lines XN and YN [FIG. VI.11] is the 'visual angle'. Visual acuity may be expressed in terms of the visual angle. The average subject can resolve two points and recognize their duality when the angle which they subtend is 1 minute. The space on the retina separating the two images can thus be calculated to be 4.5 μm.

Accommodation

The dioptric power of the cornea is invariant in mammals in an air medium. The ability of the eye to focus on objects at varying distances (accommodation) is due to a mechanism which allows changes of curvature of the anterior surface of the lens. This can be simply shown by the following experiment described by both Purkinje and Sanson. The subject 'relaxes' his eye by gazing into the distance. An observer holds a lighted candle to one side of the subject's eye and sees three images reflected therefrom—two upright and one inverted. The brighter of the two upright images is reflected from the cornea and the other which is larger is reflected from the anterior surface of the lens. The inverted image which is faint and small is formed by reflection from the posterior surface of the lens.

(That these images are indeed formed from the surfaces named is shown by holding a candle in front of three watch glasses placed upright on an optical bench, the two proximal glasses being placed convex forwards and the distal glass concave forwards. If a black matt surface be placed behind the third watch glass the experimenter sees the three images of his candle in the proximal glass and can show how the brightness of each can be varied by moving the watch glasses along the bench. Correspondingly, the size of any of these images can be diminished by increasing the curvature of the appropriate watch-glass.)

The subject then gazes at a near object; the brighter upright image and the inverted image show respectively no change and a very slight change. The second image becomes smaller and brighter, indicating that the anterior surface of the lens has both increased its curvature and has moved forwards during accommodation. By increasing its convexity the lens increases its power. The radius of curvature of the anterior surface of the lens is 11–12 mm when the eye is adapted for distant vision (resting position) and 6–7 mm when the eye is accommodating maximally for near vision. The posterior surface of the lens hardly alters its radius of curvature, which is 6 mm at rest and 5.5 mm in full accommodation.

The bulging forward of the anterior lens surface is due to the relaxation of tension in the lens capsule which permits the elastic lens to assume a more spherical form. In order to appreciate these changes it is necessary to understand some of the details of the anatomy of the region [FIG. VI.10].

The crystalline lens is surrounded by an elastic capsule which in turn is attached to the zonula. The zonula, composed of delicate connective tissue fibres, is split medially into anterior and posterior laminae which enfold the lens capsule and blend with it near the equator. Laterally the two laminae fuse and the zonula is attached to the inner surface of the ciliary body. The ciliary body is a circular zone of tissue which extends forwards from the ora serrata towards the circumference of the lens. The ciliary body contains three structures, the smooth ciliary muscle, the ciliary processes, and the orbiculus ciliaris. The ciliary muscle fibres in turn consist of:

1. Radial fibres which originate near the corneosclerotic junction and are inserted into the choroid near the posterior margin of the ciliary body.

2. Sphincteric or circular muscle fibres which lie more centrally.

When the ciliary muscle contracts, it pulls the ciliary body forwards and inwards towards the lens; correspondingly, the tension exerted by the zonula on the lens capsule is reduced [FIG. VI.12]. As a result, the elastic lens bulges forwards. The accommodation reflex includes both this change of the dioptric power of the lens and pupillary constriction which shuts off the more peripheral parts of the lens and this allows the light to fall only on the centre of the lens, in which the accommodative changes are most pronounced. The motor pathway responsible for both pupillary constriction and ciliary muscle contraction is the third nerve (autonomic). Accommodation for near vision also entails convergence of the visual axes, which again involves the third nerve (which supplies the internal rectus muscles). Accommodation can be paralysed by atropine and homatropine (cycloplegia).

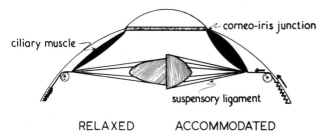

RELAXED   ACCOMMODATED

FIG. VI.12. Diagram of the mechanism of accommodation. The junction formed by the cornea and iris is indistensible. Hence contraction of the ciliary muscle pulls the ciliary body forwards and inwards, thus allowing relaxation of the suspensory ligament.

The crystalline lens is a biconvex structure, enclosed in a highly elastic capsule which is thinner over the posterior than the anterior surface and thinner in its central part than at the periphery. The lens is composed of ribbon-like elongated epithelial cells which are arranged in concentric laminae. The central core of the lens posesses a higher refractive index than does the remainder. The lens has no blood supply but satisfies its low metabolic requirements by taking up substances from the aqueous humour. Glucose is taken up by the lens substance and is metabolized (mainly anaerobically) to lactic acid (glycolysis); the lactic acid diffuses into the aqueous humour. Some aerobic metabolism is achieved, despite the absence of cytochrome oxidase, by using the hydrogen acceptor glutathione. Thus if glutathione is removed from the lens by dialysis, aerobic metabolism is abolished, only to return when glutathione is again added to the medium. In senile cataract there is no glutathione in the lens.

Presbyopia (loss of accommodation) is due to sclerosis of the lens

substance, which thus fails to adapt itself to a more spherical shape when the zonule is relaxed in the accommodation reflex. The lens contains 30 per cent protein by weight which is denatured by ultraviolet light; as a result of such irradiation the proteins tend to agglutinate and coagulate in the presence of Ca^{2+}. The changes which occur in cataract are more commonly seen in countries where the ultraviolet radiations are intense, e.g. Egypt and the Arctic. Cataract is associated with a swelling and loss of elasticity of the lens, coagulation of the lens proteins and, of course, lens opacity. The high incidence of cataract in diabetics has been attributed to an action of glucose which renders the proteins more readily coagulable by light.

Amplitude of accommodation. This is the difference in refractive power of the eye in the two states of complete relaxation and maximal accommodation. As has been stated, the normal eye at rest brings parallel rays to a focus on the retina and possesses a dioptric power of 59 D. In addition to this 'resting' power, however, the eye can be accommodated for near vision so that an extra refractive power is brought into play when a young adult focuses clearly on a near object, say 10 cm away. The amplitude of accommodation can be expressed by the number of dioptres of refractive power which are additionally obtained by adapting the eye for near vision. Thus, if the resting power of the eye to bring parallel rays to a focus is taken as zero, then the extra refractive power conferred by accommodation which permits an object at

10 cm distance to be focused clearly is 10 D $\left(\dfrac{1}{0.1} \right)$. The nearest

point at which an object can be clearly seen is called, logically enough, the *near point*. The ability of the eye to accommodate decreases throughout life—hence the near point recedes, slowly during the first 35–40 years and much more quickly afterwards. This loss of accommodative power is due to changes in the lens substance and not to changes in the power of the ciliary muscles. A man of 70 years of age may have a near point as much as 100 cm away and would thus possess only 1 D of accommodative power. The Table shows typical figures for the near point and amplitude of accommodation at different ages.

Age (yrs)	Near point (cm)	Amplitude of accommodation
10	9	11
20	10	10
30	12.5	8
40	18	5.5
50	50	2
70	100	1

Spherical and chromatic aberration

Even the *emmetropic* (i.e. normally formed) eye is subject to some defects of this nature although they are minimized. Spherical aberration by a glass lens results from the fact that the peripheral part of the lens possesses a greater refractive power than does its centre. In the crystalline lens, however, the central part possesses a greater refractive power than does the remainder of the lens substance. Moreover, the iris, which covers the outer part of the lens, also functions to reduce any spherical aberration.

Chromatic aberration is a manifestation of the different refraction suffered by the colours comprising white light, which depends on their wavelength. Red light with a longer wavelength is refracted the least and blue light conversely the most. If the eye looks at bright light the middle wavelengths are focused, the blue rays meet in front of the retina and the red rays behind and neither is brought to a point focus. Although the eye is subject to chromatic aberration, this is of no physiological importance.

Myopia and hypermetropia

These are examples of *ametropia* in which the refractive state of the eye differs from that of emmetropia in which parallel rays are focused on the retina without accommodation. Both myopia and hypermetropia are due to an incongruity between the refractive power and the axial length of the eye. In myopia the eye is too long and in hypermetropia the eye is too short [Fig. VI.13].

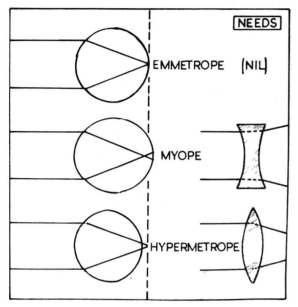

Fig. VI.13. Diagram of the normal, myopic, and hypermetropic eye, and the optical corrections required in myopia and hypermetropia.

Myopia. As the eye is too long relative to its refractive power, parallel light rays are brought to a focus in front of the retina. Only if the object is brought more closely towards the eye can its image be focused, for the incident rays on the cornea then become increasingly divergent. Thus the far point, unlike that of the emmetropic eye, is at a finite distance and may be only a few inches from the eye in severe myopia. The range of accommodation of the myopic eye is much less than usual, for the far point (at which accommodation is completely relaxed) and the near point are closely approximated. Myopia can be corrected by concave glasses which cause divergence of the incident rays [Fig. VI.13] and the power of the lens required gives a measure of the degree of myopia. As the myope ages, his near point retreats but the aged myopic subject may never need glasses for reading fine print, as his near point in youth is so close to the corneal surface that even in old age his near point may be at a distance which is normal for a young subject.

Hypermetropia. The parallel rays are brought to a focus behind the retina as the axial length of the eye is too short. Thus even distant objects can only be focused by using some accommodation. Correspondingly the hypermetropic eye shows a hypertrophy of the ciliary muscle. His near point is more distant than that of the emmetrope. He requires glasses for comfortable reading at an earlier age than does the normal subject.

Hypermetropia can be corrected by a convex lens which increases the convergence of light rays incident upon the cornea [Fig. VI.13].

Astigmatism

As the name suggests this is an error of vision in which the light rays are not brought to a point focus on the retina. The focus for horizontal rays differs from that for vertical rays. Hence an astigmatic subject looking at a piece of graph paper may focus on the vertical lines and may fail to focus the horizontal lines and vice versa. The defect is most commonly due to a difference in the horizontal and vertical curvatures of the cornea; occasionally the same abnormality affects the lens. If the curvature is greater along the vertical meridian, astigmatism is said to be 'with the rule'; if the horizontal curvature is greater, the astigmatism is 'against the rule'.

Retinal layers; photoreceptors

The retina comprises several layers of which the light-sensitive receptors are outermost. The retina possesses three layers of neurones, i.e. from inwards outwards, ganglion cells, bipolar cells, and light-sensitive rods and cones [Fig. VI.14]. Beyond the receptor elements lie the pigment epithelium and lamina vitrea. The light traverses the ganglion cells and bipolar cells before finally reaching the rods and cones. At the posterior pole of the eye lateral to the optic disc a central pit (fovea centralis) is found in which the path of the light rays to the photoreceptors is uninterrupted by the ganglion and bipolar cells and their respective fibres. The fovea centralis is occupied solely by cones. It is on this part of the retina that the image of small objects falls when the eye is used for discriminate vision. In the fovea, each cone cell is connected to one ganglion cell; elsewhere, several cones may converge upon one ganglion cell. The cone density in the fovea centralis is $150\,000$ mm^{-2}. Beyond the fovea there are both rods and cones. In the human retina there are $125\,000\,000$ rods and $7\,000\,000$ cones. As the optic nerve contains only $800\,000$ nerve fibres there is considerable convergence; in the outer parts of the retina where rods are much more densely distributed than cones, one ganglion cell may be connected to as many as 300 rods; similarly 10 cones may converge upon one ganglion cell.

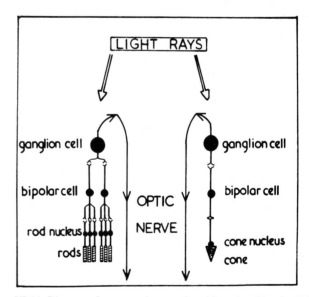

Fig. VI.14. Diagram of nervous elements found in retina. For simplicity, one optic nerve fibre is shown as serving several rod units (on the left) and one optic nerve fibre is shown serving a single cone (on the right). Note that the light rays transmitted through the cornea have to traverse the ganglion and bipolar layers before reaching the photoreceptors.

The ganglion cells and their fibres cannot themselves function as light receptors, for the point of entrance of the optic nerve (optic disc) is quite insensitive to light and constitutes the 'blind spot'. The presence of the blind spot can be shown by a simple experiment and its position can be determined by the method of similar triangles.

Thus if a piece of paper is marked with a cross and, 6 cm to the right of this, a circle, on fixating on the cross with the right eye only, the circle is not seen if the paper is moved to a distance of about 24 cm, because its image falls on the blind spot. The retinal distance between the fovea centralis (on which the image of the cross is formed) and the blind spot can be calculated, for (by similar triangles):

$$\frac{\text{Retinal distance between fovea and blind spot}}{17 \text{ mm}} = \frac{60}{240 \text{ mm}}$$

∴ Retinal distance = 4 mm. Hence the optic disc lies 4 mm to the nasal side of the fovea in each eye.

The electron microscope has clarified the details of organization of the visual receptor—bipolar—ganglion cell connection. FIGURE VI.15 shows the schematic organization of the retina. The central ends of the rods (spherules) and the cones (pedicles) are invaginated by dendrites of the bipolar cells. Horizontal cells establish further synaptic connections with rods and cones in the manner shown in the figure. Often a bipolar dendrite making synaptic contract with a cone pedicle or a rod spherule is accompanied by two dendrites from a horizontal cell—an arrangement which is known as a *triad*.

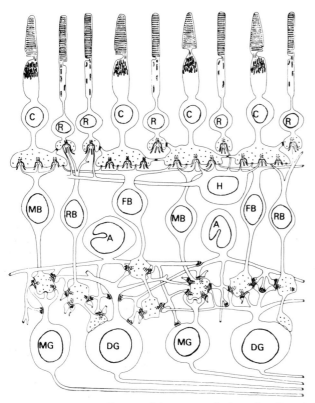

FIG. VI.15. A summary diagram of the contacts in the retina. R, rod; C, cone; MB, midget bipolar; RB, rod bipolar; FB, flat bipolar; H, horizontal cell; A, amacrine cell; MG, midget ganglion; DG, diffuse ganglion. (Redrawn from Dowling, J. E. and Boycott, B. B. (1966). *Proc. R. Soc.* **B166**, 80–111.)

A single foveal cone pedicle may exhibit as many as twelve such triads and it is believed that the horizontal cells are in a very strategic position for regulating receptor—bipolar synaptic transmission.

The bipolar cells themselves can be divided into (a) rod bipolar (RB) which make both axodendritic and axosomatic contacts with diffuse ganglion cells— so called because they establish synapses with many bipolar cells, thus providing 'retinal convergence', (b) two types of cone bipolar cells— *midget* which though connected to single foveal cones have many synaptic contacts with the cone pedicle and *flat* bipolar cells, which connect with groups of cones. The midget bipolar cells in turn make contact with midget

ganglion cells—one to one, whereas the diffuse ganglion cells connect with all types of bipolar cell. The synapses between the processes of the bipolar and ganglion cells occur in the inner plexiform layer. Amacrine cell processes make synaptic contacts with both ganglion cell dendrites and bipolar cell dendrites. The adjacent bipolar cell axon and amacrine process synapses with the dendrites of the ganglion cells constitute a *dyad*. The bipolar cell process effecting the synapse with the ganglion cell is characterized by a ribbon-like accumulation of dense material and such synapses are designated ribbon synapses.

These second synaptic connections seen in the inner plexiform layer are believed to be the site of major processing of the visual image. Centre-surround interaction and selective responsiveness to motion or direction of motion is considered to be a function of the amacrine cell system. Their horizontal interconnections are supposed to cause a sharpening of the edges of any stimulated field on the retina.

Visual acuity

Visual acuity is the reciprocal of the angle subtending two objects which are recognized as discrete. It can be measured by finding the distance at which two points or parallel lines are no longer recognized as separate. The visual angle then subtended is calculated and the acuity of vision is expressed as its reciprocal. The angle usually obtained is one minute, which corresponds with a distance between retinal images of 4.5 μm. As the diameter of a foveal cone is just less than 3 μm this suggests that one unstimulated cone separates those stimulated by the two light sources. Some people, however, can see separate images even when the visual angle is only 25 seconds—in which case the retinal distance between the images is only 2 μm which casts doubt on the above interpretation. Moreover, it is surely difficult to hold the head and eyes so steadily that the movement is limited to less than 5 μm. Visual acuity is tested by the ability of the subject to recognize test-letters on a chart which because of their size and distance subtend known visual angles.

The letters which are black on a white background illuminated suitably, possess details (spaces and breadth of stroke) which subtend an angle of one minute, at a requisite distance. Each line of letters may thus have a figure of 60 metres, 36, 24, 18, 12, 9, 6, and 5 metres noted beside it. If the subject, who stands at 6 metres distance, views the chart with one eye at a time and can read no further than the '18 metres' line, his visual acuity is 6/18, etc.

Various factors affect the visual acuity:

1. Obviously the visual acuity depends on the type of stimulus; the illumination of the surface, the time of exposure, and the brightness contrast must be considered.

2. Normal and abnormal errors of the refractive powers of the eye play an important part. Thus chromatic aberration tends to reduce the visual acuity whereas the use of monochromatic light increases visual acuity. Spherical aberration is reduced by pupillary constriction. Astigmatism, myopia, and hypermetropia are the usual causes of low visual acuity.

3. Visual acuity is maximal at the fovea centralis where the cones are closely packed and where they each have connection with a single ganglion cell. The periphery of the retina has a visual acuity of less than 1/30th of that of the fovea. Foreign bodies in the eye such as occur in industry (e.g. lathe workers) are far more often found in the periphery of the cornea than in the centre, the flying object being perceived much less easily at the periphery of the field of vision.

Photopic and scotopic vision

Daylight (photopic) vision is a function of the cones and twilight (scotopic) vision is a function of the rods. The eye responds to light of wavelengths between 400 nm and 750 nm. Below 400 nm (the ultraviolet rays) and above 750 nm (the infrared rays) the receptors are not stimulated.

In general, photopic vision (due to cone receptor mechanisms) applies to brightness levels above 1 millilambert and scotopic (rod) vision is used if the brightness levels are less than 10^{-3} millilamberts. Between these values both mechanisms contribute.

Photopigments. The rods contain a purplish red protein (MW = 270 000) *rhodopsin* or *visual purple*, which is bleached by light. This pigment can be prepared in pure solution and its bleaching sensitivity to various wavelengths of the spectrum can be determined. Similarly, its absorption curve can be likewise determined at different wavelengths.

Rods are composed of a pile of discs normal to the rod axis. Rhodopsin molecules lie with their resonating axes in the planes of these discs and can move freely in these planes but not out of them.

If rods are activated by light absorbed by rhodopsin and if every quantum so absorbed is equivalent in its contribution to vision, it follows that the rhodopsin absorption spectrum should coincide with the scotopic or twilight sensitivity function. Crescitelli and Dartnall (1953) measured the density spectrum of human rhodopsin extracted from an excised human eye and found it to coincide with the scotopic spectral sensitivity determined by Crawford (1949) [Fig. VI.16]. Clearly, flashes that are equally absorbed are equally well seen.

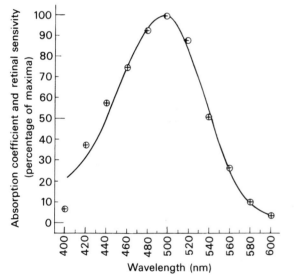

Fig. VI.16. Absorption spectrum of visual purple (curve) and spectral sensitivity of the dark-adapted eye (points). (Redrawn from Crescitelli, F. and Dartnall, H. J. R. (1953). *Nature, Lond.* **172**, 195.)

The pinkish substance rhodopsin consists of a protein, opsin, combined with retinene, which is the aldehyde of vitamin A. Hubbard and Wald (1951) synthesized rhodopsin thus:

$$\text{vitamin A} \underset{\text{NAD}}{\overset{\text{NADH}_2}{\underset{\text{alc. dehyd.}}{\rightleftharpoons}}} \text{retinene} \xrightarrow{+ \text{opsin}} \text{rhodopsin}$$

They used vitamin A extracted from cod liver, alcohol dehydrogenase from horse liver, and opsin from animal retinae. However, when pure crystalline vitamin A was employed, no synthesis occurred. This led to the investigation of the structure of retinene and retinol (vitamin A). It transpired that pure vitamin A is an alcohol with a *trans* configuration and its oxidation yields the compounds *trans* retinaldehyde. Only the 11-*cis* isomer of retinaldehyde, however, combines readily with opsin to form rhodopsin.

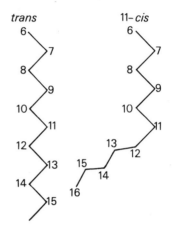

The structure of the 11-*cis* compound:

shows it to be curled up—to yield a shape which fits snugly into the opsin molecule.

The long *trans* molecule does not fit and will not attach to opsin.

When a quantum of light is absorbed by rhodopsin, the initial result is an isomerization of the *cis*-retinene part of the molecule to yield the *trans* form which detaches from the rod and migrates to the pigment epithelial layer, leaving the opsin in the rods. It is envisaged that the 11-*cis* isomer of vitamin A is formed and stored in the pigment epithelium whence it migrates into the retina to be oxidized to retinene, thence to combine with the free opsin to reform rhodopsin.

At the threshold of vision a single quantum isomerizes one molecule of retinene and this generates an effect that operates within a fraction of a second 0.1 mm distant.

After light adaptation the visual threshold is raised, so that entering a dimly lit room one finds it initially difficult to see. However, visual perception improves quickly at first and then more slowly reaching near maximal sensitivity in half an hour [Fig. VI. 17]. The process is termed 'dark adaptation'—with its completion, visual sensitivity is increased 10 000-fold.

Initially it was proposed that the rise of threshold caused by light was due to the bleaching of rhodopsin. As Rushton (1963) has pointed out, this explanation is of itself inadequate, for each rod contains 18 million molecules and catching 1 quantum per second per rod, months would be required for appreciable bleaching of the retinal rod population. We know that if the background field is suddenly brightened the new raised increment threshold is established within a couple of seconds.

Dowling and Wald (1960) employed a new method which clarified the matter. Using albino rats, they determined the threshold of the ERG (electroretinogram) at various stages of dark adaptation and graphed their results against the rhodopsin found in the rods by

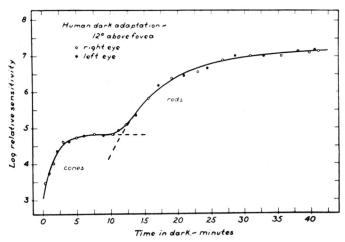

Fig. VI.17. Dark adaptation of the human eye, measured in a test patch fixated 12 degrees above the fovea. The ordinates show the logarithm of the relative visual sensitivity, the reciprocal of the visual threshold. The first rise of sensitivity is due to cones, and is virtually completed within from four to five minutes. The later rise of sensitivity is due to rods and occupies about 45 minutes. (Wald, G. (1954). *Transactions of the 4th Conference of Nerve Impulse.* Macey, New York.)

extraction from the retina of the killed animal (expressed as a percentage of the total content when dark adaption was complete). FIGURE VI.18 shows some of their findings. The abscissa shows the log threshold for the ERG of the animal immediately before death. The ordinate shows the percentage content of rhodopsin. The white circles show the result of a total bleach of rhodopsin followed by various periods in the dark. As rhodopsin regenerates, the log threshold falls and the relationship is linear. If threshold is expressed in units of the fully dark adapted value the relationship is simply: *log threshold is proportional to free opsin.*

FIGURE VI.18 also shows that vitamin A alcohol starvation of the rats yields the same relationship between rhodopsin content expressed as a percentage of the rhodopsin content of the normal animal and log threshold. In these circumstances, the lack of provision of 11-*trans* retinol necessarily results in a lack of 11-*cis* retinene which is required for the formation of rhodopsin—yielding the symptoms of night blindness. As the filled circles in FIG. VI.18 lie on the same line as before, it follows that the *log threshold is proportional to the free opsin content of the rods irrespective of whether the opsin is freed by light or lessened by starvation.*

Further advances in our understanding of the features of rhodopsin breakdown and regeneration derive from the ingenious studies of Rushton and of Weale and their colleagues (see Rushton 1963, 1972) using the principle of densitometry.

Retinal densitometry

When the eye is focused at infinity, a parallel pencil of light rays entering the pupil is focused on the retina and passes through it. Any light reflected from behind which falls upon the pupil will exit as a beam parallel to that entering. This reflected light represents a relatively small proportion of the incident light because the black melanin-pigmented layer behind the retina absorbs a great deal of it, whereas the visual pigments of the retina such as rhodopsin have absorbed some of it during its passage through them going in and then exiting. However, strong light bleaches rhodopsin but does not bleach melanin. Thus the *change* in reflected light intensity on bleaching is due to visual pigments only and in the peripheral part of the retina this is ascribable only to rhodopsin.

FIGURE VI.19 shows a diagram of the retinal densitometer. The light source A emits two beams which are united at the mixing cube B. A red filter F_1 is interposed and a polaroid P_1 is orientated so as to promote transmission through B. A green filter F_2 is placed in the other light pathway and P_2 is crossed with P_1 and hence promotes reflection. The combined beam passes through a rotating polaroid P which allows red and green light to pass alternately. The flickering image I, is reflected back into the photomultiplier PC and produces a flickering output. This output fluctuation may be reduced to zero by adjusting the wedge W until the red signal and the green signal are equal as judged by the photocell. Any increase in rhodopsin will now absorb more green light but not more red light, for rhodopsin does not absorb red (650 nm)—as depicted in FIG. VI.22. Thus an increase in rhodopsin will cause a flickering output which will require a shift in the wedge W to remove it. The wedge shift required will be that added wedge density needed to offset the rhodopsin double density which has accumulated in the green. (The measuring light traverses the rod layer twice.) Rushton was able to measure the change in double density of rhodopsin in the human retina by noting the wedge shift required to keep the photocell free from output fluctuations; such measurements required only about 5 seconds to do.

When a bright light of 1 unit (= 20 000 trolands) is shone on the dark adapted eye, rhodopsin bleaching occurred initially quickly, then more slowly, to level after 5 minutes. Increasing the bleaching light to 5 units or 100 units caused greater and faster bleaching of

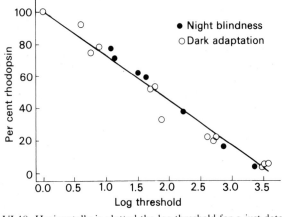

Fig. VI.18. Horizontally is plotted the log threshold for a just detectable ERG in the rat at various stages of dark adaptation following exposure to bright light (white circles). Vertically is plotted the per cent rhodopsin in the rods at that stage of dark adaptation. Black circles show the same relation when rhodopsin is removed not by light but by deprivation of vitamin A. (Redrawn from Dowling, J. E. and Wald, G. (1960). *Proc. natn. Acad. Sci. U.S.A.* **46**, 587.)

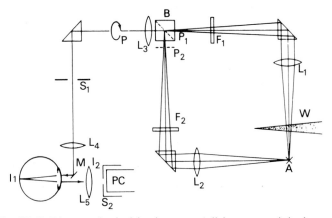

Fig. VI.19. Diagram of retinal densitometer. *A*, light source emitting beams 1 and 2 united at mixing cube B where images of the filament A coincide. This is focused by L_4 upon the cornea after reflection in the opthalmoscopic mirror M. The stop S_1 is uniformly illuminated and focused sharply upon the retina at I_1. Light reflected from behind the retina at I_1 is focused by L_5 upon the stop S_2 through which it passes to fall upon the photomultiplier cell PC. (Rushton, W. A. H. (1961). *J. Physiol., Lond.* **156**, 166–78.)

rhodopsin and further brightening of the light beyond 100 units caused no increase in bleaching. These results indicate that the 'levelling off' of bleaching seen with 1 unit or 5 units is due to equilibria being established under these respective conditions between the breakdown of rhodopsin and its regeneration.

When rhodopsin has been fully bleached by the 100 unit illumination, the light is switched off and brief measurements in weak light are made every minute during dark adaptation; the closed circles plot the course of rhodopsin regeneration, which is 50 per cent complete in 5 minutes and 90 per cent complete within 15 minutes [FIG. VI.20].

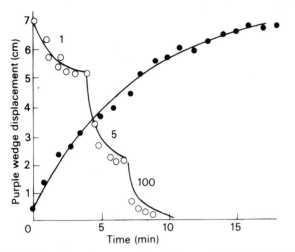

FIG. VI.20. Bleaching and regeneration in the eye of W. A. H. Rushton. Black circles: regeneration in the dark; white circles: bleaching under steady illumination of 1, 5, and 100 units of 20 000 trolands. (Campbell, F. W. and Rushton, W. A. H. (1955). *J. Physiol., Lond.* **130**, 131.)

Rhodopsin regeneration after full bleach takes about 18 minutes before the rod threshold falls below that of the cones. Thus rod threshold cannot be measured *earlier* than 18 minutes, because of the cones, or later, because regeneration is complete. However, there is a rare congenital condition where the eye contains few if any cones—*photanopia*, and Rushton studied the dark adaptation curve of such a subject and obtained results depicted in FIG. VI.21. The dotted curve depicts the dark adaptation of a normal subject;

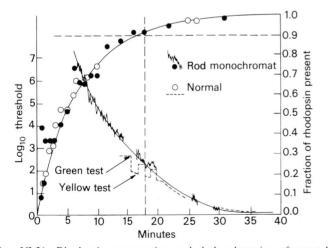

FIG. VI.21. Rhodopsin regeneration and dark adaptation after total bleaching in normal and rod monochromat. White circles normal, black circles monochromat shows that monochromat's rhodopsin regenerates normally. Irregular line traces dark adaptation in monochromat over a range of 6 log units. Dotted line shows dark adaptation of normal: cones for first 18 mins, after 20 mins pure rods which coincide with monochromat's curve. (Rushton, W. A. H. (1961). *J. Physiol., Lond.* **156**, 193.)

the wavy line starting after six minutes is the threshold of the rod monochromat starting at level 7 log units. The open circles show the regeneration of rhodopsin in the normal and the black circles those in the rod monochromat. The course of pigment regeneration is the same in both subjects. In man, each 5 per cent of rhodopsin bleached raises the threshold 1 log unit. In further experiments, Rushton showed that protanopes, deuteranopes and normals all have the same foveal dark-adaptation curves. Foveal dark-adaptation curves reveal the features of *cone* pigment regeneration.

Cones and colour vision

In man, cone vision sensitivity is 1/100th–1/1000th that of rod vision. Clerk Maxwell showed in 1860 that three colours, red, green, and blue could be used to match any known colour, thus supporting the theory advanced by Thomas Young and Von Helmholtz in the nineteenth century that there were three types of cone, each possessing its own characteristic photosensitive substance—one responding to red, another to green, and yet another to violet light. The different colour sensations were supposed to be caused by stimulation of various combinations of these three types of receptor.

Gallinaceous birds (fowls) possess only one pigment, iodopsin, which is contained in the retinal cones. The spectral absorption of iodopsin peaks at 562 nm. The pigment possesses the same chromophore as rhodopsin—i.e. 11-*cis* retinal—and iodopsin and rhodopsin differ only in their opsin.

The sensitivity of cone vision as measured in daylight at brightness levels of about 1 millilambert can be determined by plotting the wavelength against the reciprocal of the energy required to produce visual sensation at that wavelength, and the resultant curve is described as that for photopic vision [FIG. VI.22]. When the measurements of the absorption spectrum of iodopsin are added (as open circles) there is obviously close agreement between the spectral absorption and the photopic visual sensitivity. The figure also shows the similarity between scotopic visual sensitivity and rhodopsin absorption. The peak sensitivity of scotopic vision is approximately 500 nm, whereas that of photopic vision lies at about 560 nm. This difference accounts for the so-called *Purkinje shift*—a difference in the luminosity of colours in lights of different intensity first noted by J. E. Purkinje, the Czech physiologist. As dusk gathers, red and blue flowers, of similar luminosity during the day, alter their 'appearance' so that the blues appear to glow, whereas the reds become nearly black.

It should be noted that the spectral sensitivities shown in FIG. VI.22 are expressed in terms of fractions of unity. If the curves were plotted on an absolute scale of energy intensity, the ordinates would be enormously different because rod sensitivity far exceeds cone sensitivity. FIGURE VI.22 shows that the range of wavelengths of light to which the visual mechanism responds is between 395 and 700 nm. Light of less than 390 nm wavelength is termed ultraviolet (a greater frequency than violet) and above 750 nm wavelength is called infrared (less frequency than red). The wavelengths of spectral colours are approximately

400 nm—violet	550 nm—greenish yellow
450 nm—blue	600 nm—orange
500 nm—blue–green	650–700 nm—red

The phenomenon of trichromacy (three-colour matching) has generally been expressed in terms of red, green, and blue inputs and colour has been analysed in terms of R, G, and B spectral sensitivity curves; R, G, and B cones; and R, G, and B pigments. Human cone pigments have been named erythrolabe, chlorolabe, and cyanolabe (i.e. red-, green-, and blue-catching).

Colour blindness was first clearly documented in 1798 by John Dalton of atomic theory fame. He noted that he could not detect any difference between the hue of 'a laurel leaf and a stick of red sealing

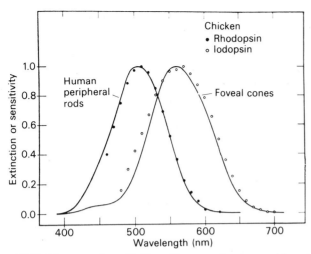

FIG. VI.22. The spectral sensitivities of human rod and cone vision compared with the absorption spectra of rhodopsin and iodopsin extracted from the chicken retina. The spectral sensitivity data are from Wald, G. (1945). *Science, N.Y.* **101** 653; and (1949) *Documenta Ophthal.* **3**, 94. (Wald, G. (1954). *Transactions of the 4th Conference of Nerve Impulse.* Macey, New York.)

wax'. Later, von Kries suggested the classification of colour blindness:

I. Trichomats
  a. Protanomaly
  b. Deuteranomaly

II. Dichromats
  a. Protanopia
  b. Deuteranopia
  c. Tritanopia

III. Monochromats

Normal and anomalous trichromats both require the three primary colours to match the whole variety of spectral colours but the anomalous employ unusual ratios of the primaries. Protanomalous subjects are less sensitive to red and need more red in matching, whereas deuteranomalous people need more green. Protanopes are insensitive to red and deuteranopes insensitive to green. Tritanopes, insensitive to blue, and monochromats are rare.

Protanomaly (PA), protanopia (P), deuteranomaly (DA), and deuteranopia are examples of X-linked inheritance. The gene for red–green deficiency is localized in the X-chromosome. It behaves recessively; the normal allele is dominant. Since males with the XY constellation carry only one X-chromosome, they are colour-blind if the X-chromosome is affected (hemizygous). Females with the XX constellation are colour-blind only if both X-chromosomes carry the gene for red–green deficiency, i.e. if the woman is homozygous for that gene. If only one X-chromosome is affected, the female is phenotypically normal but a conductor for colour blindness (heterozygous carrier of that gene). In European males, red–green defects occur in about 8 per cent. Coloured races show a lower incidence—between 2.5 and 5 per cent Deuteranomaly accounts for half the cases of red–green defects. Only 0.4 per cent of women ever show colour defects (Jaeger 1972).

Defects in colour vision are tested by

(i) *Holmgren's skeins* of coloured wool. The subject is asked to match a piece of wool from the assortment of skeins of various colours.

(ii) *Ishihara charts* which are printed with figures or designs in coloured circles on a background of circles coloured to look similar (to a colour-blind subject) to those of the figure itself.

(iii) *The Edridge Green lantern* in which the subject has to identify the colour of a small illuminated area, the size of which can be varied. This device is used by the Board of Trade to test would-be engine or lorry drivers for evidence of colour-defective vision.

Rushton (see 1972) has measured the foveal difference spectra of a protanope and a deuteranope. The fovea of the protanope was

bleached by a red light and the transmissivities for lights throughout the spectrum were measured first in the dark-adapted and then in equilibrium under this bleaching light. After recovery, the experiment was repeated using a blue–green bleaching light. The light intensities employed were such as to produce 50 per cent bleaching in each case. At every wavelength throughout the spectrum the transmissivity changes were identical in both the dark-adapted and the bleached conditions [FIG. VI.23]. This result shows unequivocally that there is only one measurable pigment in the foveal cones of the protanope—named *chlorolabe*.

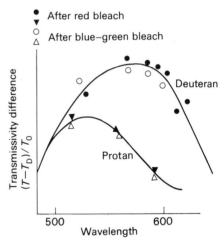

FIG. VI.23. Transmissivity changes in a protanope and a deuteranope. (Rushton, W. A. H. (1965). *Nature, Lond.* **206**, 1087.)

When Rushton studied a deuteranope, he again obtained coincidence of the transmissivities whether half bleaching was achieved using red light or the blue–green light. The deuteranope possesses only one pigment in the foveal cones—*erythrolabe*. The deuteranope transmissivity absorbs much further into the red region of the spectrum which corresponds with the fact that he can see much further into the red part of the spectrum than the protanope.

Later experiments of Rushton and his colleagues showed that the chlorolabe of the protanope and the erythrolabe of the deuteranope were identical with the relevant pigments in normal eyes. Kinetic studies of the cone pigments using the densitometer have revealed that pigment is bleached at a rate proportional to the 'quantum catch' and that regeneration proceeds at a rate proportional to the fraction of the pigment in the bleached state. These two processes are independent and hence are simply additive.

Little has been said about 'blue-sensitive' cones. It is known that they differ from other cones and resemble rods in three respects: they have poor optimum spatial and temporal resolution, they have a higher value for their lower limit for $\Delta I/I$ and they may be absent from the very centre of the fovea. Barlow (1972) suggests that if these properties are attributable to the nervous pathway it is possible that 'blue' cones either have the same type of pathway as the rods, or use the rod pathway in the light adapted state, when the rods are saturated. It has been noted that the Purkinje shift occurs abnormally in tritanopes, who presumably lack 'blue' cones.

REFERENCES

BARLOW, H. B. (1972). Single units and sensation: a neuron doctrine for perceptual psychology. *Perception* **1**, 371–94.

CRAWFORD, B. H. (1949). *Proc. R. Soc.* **B62**, 321–34.

CRESCITELLI, F. and DARTNALL, H. J. A. (1953). *Nature, Lond.* **172**, 195.

DOWLING, J. E. and WALD, G. (1960). *Proc. natn. Acad. Sci. U.S.A.* **46**, 587.

HUBBARD, R. and WALD, G. (1951). *Proc. natn. Acad. Sci. U.S.A.* **37**, 69–79.

JAEGER, J. F. (1972). In *Visual psychophysics* (ed. D. Jameson and L. M.

Hurlich) Vol. VII/4 *Handbook of sensory physiology*, Chapter 24, pp. 625–42. Springer, Berlin.

Rushton, W. A. H. (1963). In *Recent advances in physiology*, 8th edn. (ed. R. Creese) Chapter 5, pp. 140–77. Churchill, London.

—— (1972). In *Photochemistry of vision* (ed. H. J. A. Dartnall) Vol. VII/1 *Handbook of sensory physiology*, Chapter 9, pp. 364–94. Springer, Berlin.

The visual path

The visual fibres arise in the layer of nerve cells in the retina and pass backwards along the optic nerve to the optic chiasma. Partial decussation takes place; the fibres from the nasal sides of the retinae cross, and those from the temporal sides of the retinae remain uncrossed. The left optic tract therefore conveys fibres from the left halves of both retinae. Each half of the retina receives light rays from the opposite half of the field of vision. The left optic tract corresponds, therefore, to the right or opposite half of the field of vision. The fibres from the macula lutea (yellow spot), or region of most precise vision, behave in exactly the same way. The fibres from the nasal sides of both maculae cross, and those from the temporal sides remain uncrossed [Fig. VI.24].

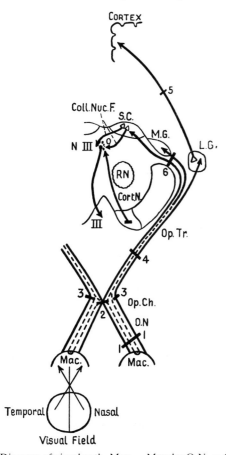

Fig. VI.24. Diagram of visual path. Mac. = Macula; O.N. = Optic nerve; Op. Ch. = Optic chiasma; Op. Tr. = Optic tract; N.III. = Nucleus of third nerve; III = Issuing fibres of third nerve; R.N. = Red nucleus; S. C. = Superior colliculi; Coll. Nuc. F. = Colliculonuclear fibres; M.G. = Medial geniculate body; L.G. = Lateral geniculate body; Cort. N. = Corticonuclear fibres (from frontal lobes to third nerve nucleus).

The optic tracts thus constituted wind round the outer side of the crura cerebri and end in two main areas:

1. The *superior colliculi* (or in the near-by pretectal area), which are not concerned with conscious vision but serve as a centre for visual reflexes (e.g. the light reflex, p. 387). Stimulation of the superior colliculi usually causes the eyes to move to the opposite side accompanied by turning of the head in the same direction, elevation of the eyebrows, opening of the palpebral fissures, and changes in pupil diameter (constriction or dilatation).

2. In the *lateral geniculate body*, from which a fresh relay arises which passes back in the optic radiation to the occipital cortex. The fibres pass through the internal capsule behind those for 'common sensation', and then pass deep in the substance of the temporal lobe round the outer surface of the lateral ventricle to reach the 'half-vision centre' in the occipital lobe; in man this centre is situated in the cuneus and lingual gyrus above and below the calcarine fissure on the medial aspect of the lobe (area 17) [Fig. VI.24]. The 'half-centre' is so called because, like the optic tract, each occipital centre represents the opposite half of the field of vision.

Central projection of visual fibres

About one-third of the fibres in the optic tract are derived from the maculae, and a similar proportion of the surface of the lateral geniculate bodies is devoted to their reception. Both anatomical and electrical studies have shown that there is an orderly point-to-point (topographical) projection of the retina, firstly in the lateral geniculate body, and secondly in the occipital cortex.

Lateral geniculate bodies. The grey matter of the lateral geniculate bodies shows six distinct layers, numbered 1 to 6. The fibres from the retina of the contralateral side end in layers 1, 4, and 6; the fibres from equivalent spots of the ipsilateral retina end in the same region but in layers 2, 3, 5. There is thus a fusion of equivalent spots in the lateral geniculate body whence fibres pass to the cortex [Fig. VI.25]. Thus after enucleation of one eye in man, layers 2, 3, and 5, in the lateral geniculate body of the same side show transneuronal degeneration, while layers 1, 4, and 6 remain intact.

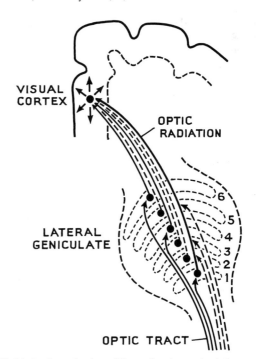

Fig. VI.25. Mode of termination of fibres of optic tract in six layers of lateral geniculate body and further course to visual cortex. Continuous lines = crossed fibres; interrupted lines = uncrossed fibres. (After Le Gros Clark, W. (1941). *J. Anat., Lond.* **75**, 232.)

Cortical representation. In man the peripheral part of the retina is represented well forward on the medial surface of the occipital lobe above and below the calcarine fissure. The macula has a very much larger central representation and occupies mainly the posterior part of the medial surface (though there is also a forward running tongue). In man the macular representation stops at the occipital pole [Fig. VI.26].

Each occipital half-vision centre (e.g. the left) represents the homolateral (in this case the left) halves of the two retinae and therefore (as stated) the opposite (right) halves of the field of vision. Furthermore, above the calcarine fissure (cuneus), the upper halves of the retinae (lower halves of the field of vision) are represented; below the calcarine fissure (lingual gyrus), the lower halves of the retinae (upper halves of the field of vision) are represented.

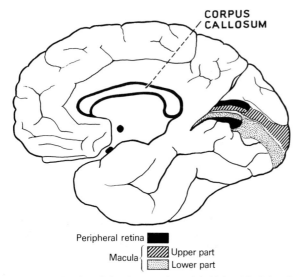

Peripheral retina ▮

Macula { ▨ Upper part
        { ▦ Lower part

FIG. VI.26. Localization of visual centres in occipital lobe. Medial surface of the brain in man. The representation of the macula is far greater than that of the peripheral part of the retina. (After Brouwer and Van Heuven (1934). *Res. Publ. Ass. nerv. ment. Dis.* **13**.)

Both the region of the occipital lobe which surrounds the visual receptive area (e.g. area 18) and the posterior parietal region are believed to function as a visuopsychic area. Activity in this region enables the nature of objects seen to be recognized, e.g. a pencil, paper, or ball is recognized as such.

Reference is made on page 337 to the fact that fibres from the frontal eyefield (area 8) pass back to area 18, and that the latter connects with the temporal cortex. Responses similar to those resulting from stimulating the frontal eyefields can be obtained on stimulating the occipital eyefields (in areas 17 and 18).

Results have more recently been obtained by studying the evoked potential in the occipital cortex resulting from stimulation of discrete parts of the retina. 1 mm of foveal retina is represented by some 16 mm in the cortex, whereas more peripheral parts of the retina receive only about 1/60th of this representation per mm.

ELECTROPHYSIOLOGICAL STUDIES

The retinal ganglion cells discharge fairly steadily even when they are not subjected to any input from the cones or rods. Kuffler (1953) showed that this resting discharge could be increased or decreased by shining a light on a small circular region of the retina. The response depended on where in this 'visual field' the light fell. An 'on' response resulted in an increased discharge during the light stimulus and an 'off' response resulted in a decreased discharge. Kuffler found that there were two distinct cell types. In one the receptive field comprised a central 'on' area surrounded by a circular zone which caused 'off' responses. In the other light falling on the centre of the receptive field caused 'off' responses and on the surrounding zone induced 'on' discharge [Fig. VI.27]. When the

light stimulus to the centre of the receptive field ceased, such a cell responded with a burst of impulses.

Two spots of light shone on separate parts of an 'on' area induced a bigger 'on' response than did either alone; if one part of an 'on' area was stimulated by light simultaneously with a point in the 'off' area the ganglion cell response was very weak, the two effects tending to neutralize each other.

The ganglion cell response to diffuse light affecting the whole retina was much less impressive than when a small circular spot covered the precise receptive field centre.

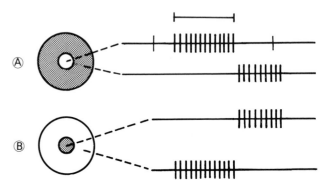

FIG. VI.27. The two main classes of receptive fields of ganglion cells: (A) 'On' centre cells respond to illumination of the centre part of the receptive field (shown by the bar); illumination of the surrounding zone inhibits discharge, but a burst of impulses follows when the light is switched off. (B) 'Off' centre cells – the responses are reversed.

The receptive field centres of the ganglion cells connected with the cones of the fovea were tiny whereas those of the ganglion cells connected with the photoreceptors of the peripheral retina were much larger. Such cells would receive messages from huge numbers of photoreceptors and might be regarded as functioning in dim lighting conditions.

Hubel and Wiesel (1962) studied geniculate cells in the same manner and found many of their properties were similar to those of the retinal gangliar units. [Fig. VI.28]. However, they manifested a greater capacity of the periphery of a geniculate cell's receptive field to cancel the effects of the centre of the field. Thus the geniculate body units increase the disparity of the visual responses to diffuse lighting compared with those to small centred spots of light.

The visual cortex itself receives many millions of fibres from the lateral geniculate cells which primarily synapse with cortical cells of the internal granular layer [p. 318], whence interconnections are made with cells of layers 3 and 5. These cells in turn project to neighbouring cortical areas and also to the superior colliculi and cerebellum. Single visual cortical cells respond to lines and edges in their receptive fields rather than to circular spots. Hubel and Wiesel classified them into two main groups—'*simple*' and '*complex*'. '*Simple*' cells which are stellate in shape respond to line stimuli such as a light line on a dark background (slit) or a dark line on a light background (bars). The response of such a cell depends on the orientation of the shape and its position on the cell's receptive field. Thus a given cell may respond to a vertical slit of light and fail to respond when the bar is displaced sideways or inclined away from the vertical [Fig. VI.28].

'*Complex*' are less discriminating as to the exact position of the stimulus but unlike the simple cells they respond with sustained firing to moving lines. Thus such a cell might respond with an increasing discharge to movement of a dark horizontal bar moved downwards in the receptive field, less vigorously when the bar is moved upwards again and show no response whatever to a vertical bar whether it is stationary in or moved across the receptive field. As Hubel (1963) points out such cells may be involved in the perception of form and movement.

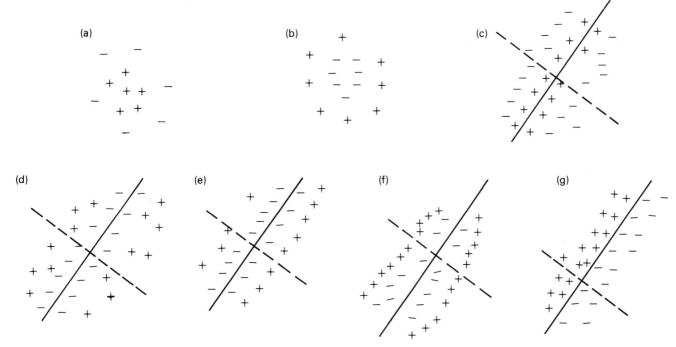

Fig. VI.28. (*a*) and (*b*) Concentric non-orientated fields of cells in the lateral geniculate body (cat) (*a*) 'on' centre (*b*) 'off' centre. (*c*)–(*g*) Simple fields of the visual cortex with specific axis orientation. (*c*) shows an orientation-selective field the centre of which responds to bright bars. (*d*) and (*f*) show direction-specific simple neurons which signal edges moving only in one direction perpendicular to their axis (dotted line). + signifies excitation by light; – signifies inhibition by light (and activation by darkness) for spot stimulation. The solid line marks axis orientation.

The simple cells are found in area 17. Though some of the complex cells can be found here, most of them are located in areas 18 and 19. The organization of the cells in these cortical areas is columnar, the columns of cells orientated perpendicular to the cortical surface comprising cell units which all possess similar receptive field orientations. It is believed that the complex cells integrate the discharges of the simple cells and that their output determines visual sensation.

It is curious that only about 25 per cent of the simple and 7 per cent of the complex cells show any difference in their responses to the linear orientation of the light stimulus when the colour of the light is changed. The lateral geniculate body on the other hand has been shown to contain both cells which manifest spectral responses similar to that of the intact visual system and cells some of which are excited by red light and inhibited by green light, and others which respond to green light but are inhibited by red.

REFERENCES

HUBEL, D. H. (1963). *Scient. Am.* **168**, 1–10.
—— and WIESEL, (1963). *J. Physiol., Lond.* **165**, 559–68.
KUFFLER, S. W. (1953). *J. Neurophysiol.* **16**, 37–68.

The light reflex

The nervous arc employed by the light reflex is probably as follows: the afferent fibres enter the superior colliculi or the adjacent pretectal region from the optic tracts. Here a new relay arises—the colliculonuclear fibres, which cross both in front and behind the aqueductus Sylvii to reach the most anterior part of the third nerve nucleus. As the fibres from each retina reach both optic tracts and both superior colliculi, it is found that shining light on one eye causes constriction of both pupils (consensual light reflex). The fibres of the third nerve relay in the ciliary ganglion and pass in the short ciliary nerves to the eye.

Convergence–accommodation reaction

During accommodation for near vision the ciliary muscle (on both sides) contracts, the suspensory ligament of the lens is relaxed, and the anterior surface of the lens becomes more convex. At the same time, the eyes converge owing to contraction of both medial rectus muscles, and the pupils are constricted. Convergence–accommodation is to some extent a willed movement, as the object must be definitely looked at before it occurs. It is suggested that visual impulses pass to the occipital cortex and are relayed to the frontal lobes. Fibres arise there which descend in the anterior limb of the internal capsule to reach the medial part of the pes pedunculi. These corticonuclear fibres then turn abruptly dorsally through the medial lemniscus to the opposite side, to end in the third nerve nucleus which supplies all three muscles mentioned.

Stimulation of area 19 in the occipital cortex leads to constriction of the pupil, the fibres concerned running first to the pretectal area; there is thus another pathway from the cerebral cortex that may be employed in pupilloconstrictor reactions.

Argyll Robertson pupil

This term is applied to a condition in which the pupillary constriction in response to a light stimulus is absent or notably diminished, while the sphincter pupillae still contracts during convergence–accommodation. From the description given above, it is clear that the only part of the reflex pathway for pupillary constriction which is 'private' to the light reflex and is not shared by the convergence–accommodation reaction is the part of the optic tract which enters the superior colliculi, the superior colliculi themselves, and the colliculonuclear fibres (the afferent fibres subserving accommodation pass up to the cortex); the oculomotor nerve is the final common path for both reactions. The Argyll Robertson pupil is often associated clinically with lesions in the vicinity of the aqueductus Sylvii and the superior colliculi which would interrupt the 'private' light reflex path. As syphilis of the nervous system commonly affects this region, this sign is very frequently found in this disorder.

Effects of Injury

The following definitions may help us to understand the effects of the commoner lesions of the visual tracts. [The numbers in brackets refer to Fig. VI.24.]

Hemianopia: blindness of half the visual field from causes other than retinal. This may be:

1. Bitemporal or binasal, i.e. loss of both temporal or both nasal fields of vision.
2. Homonymous: loss of the right or the left halves of both fields of vision.
3. Quadrantic: blindness of one quadrant only.

A lesion of the central part of the optic chiasma (2) (e.g. from pituitary tumours) where the fibres from the nasal side of both retinae cross causes bitemporal hemianopia [Fig. VI.29]. A lesion of the outer margins of the chiasma (3, 3) may damage the fibres from the temporal sides of the retinae, and cause binasal hemianopia. Any lesion of the optic nerve, chiasma, or tracts up to the point where the fibres for the superior colliculi leave (1–4), produces loss of the light reflex from the blind side of the retina. A lesion of the lateral geniculate body, optic radiation (5), or occipital cortex produces loss of sight, but the light reflex from the blind side of the retina is retained. A lesion of the optic tracts (4), or their continuation to the cortex (5), causes homonymous hemianopia: i.e. a lesion of the left tract or the left visual centre causes loss of the right halves of the field of vision in both eyes.

Incomplete lesions of the visual cortex lead to loss of colour vision; white objects are seen indistinctly, or sensation may be excited only by the more potent type of stimuli such as abruptly moving objects. Lesions of the lateral surface of the brain (areas 18 or 19, i.e. in the posterior parietal region [Fig. V.131], leave visual sensibility intact but cause disturbance of higher visual functions, such as loss of visual orientation of localization in space, impaired perception of depth and distance, loss of visual attention and inability to recognize visually the nature of common everyday objects.

Extrinsic muscles of the eye

The eyeball can be displaced laterally, medially, upwards, or downwards by six extrinsic muscles which consist of:

1. Lateral and medial rectus muscles. These rotate the eye laterally or medially about a vertical axis. Paralysis of the lateral rectus (which is supplied by the VIth cranial (abducens) nerve with its long cranial course) is prone to occur as a result of brain stem lesions or cavernous sinus thrombosis. Such paralysis results in the affected eye turning medially, due to the unopposed action of the medial rectus. There is an obvious strabismus (squint) and homonymous diplopia (double vision).

Paralysis of the medial rectus (supplied by the IIIrd cranial nerve causes lateral deviation of the eye (due to the unopposed action of the lateral rectus) and crossed (heteronymous) diplopia.

2. Superior and inferior oblique muscles. The superior oblique muscle supplied by the IVth (trochlear) nerve pulls the eyeball downwards and outwards and rotates it inwards. The major action of superior oblique is exerted when the eye is adducted upon which the muscle becomes a simple depressor. When superior oblique is paralysed, the patient with the eye adducted cannot move his eye downwards and suffers diplopia. The inferior oblique rolls the eye outwards and pulls it upwards. It is innervated by the IIIrd nerve.

3. Superior and inferior rectus muscles respectively cause upward and downward movements of the eye. Each has a subsidiary action of adduction. They are supplied by the IIIrd cranial nerve.

The fictitious formula $ER_6(SO_4)_3$ is helpful in remembering the innervation of the extrinsic muscles. ER is the external rectus (supplied by VI), SO_4 is the superior oblique (supplied by IV) and all the remaining muscles are innervated by the IIIrd nerve.

As the visual field is essentially binocular, it is of the utmost importance that the co-ordination of the movements of the two eyes is exact if visual images are to be registered on corresponding parts of their retinae. Otherwise diplopia results.

Eye movements

It is customary to describe four types of eye movement:
1. saccadic; 2. smooth pursuit; 3. vergence; 4. vestibular.

Saccadic movements are very rapid—up to $400°\,s^{-1}$ and occur with a reaction time of 200 ms. The eyes move in a conjugate fashion. Such movements characterize our inspection of stationary objects when the head is held steady. Though normally voluntary, saccades may be involuntarily aroused by peripheral visual or auditory stimuli. Saccades can be produced by stimulation of the frontal eye fields (area 8), but it is doubtful whether they are initiated therefrom, because electrical recordings from area 8 neurons show that they discharge during the movement rather than prefatory to its occurrence. Nevertheless, the control of saccades seems to be exercised by area 8; corticobulbar fibres pass to the pontine reticular formation whence they are relayed to the cranial nuclei of III, IV, and VI.

During steady fixation of a target, the saccades are small (a few minutes of arc) and the observer is unconscious of them. Their

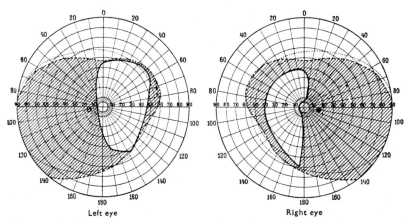

Left eye   Right eye

Fig. VI.29. Bitemporal hemianopia. Loss of both temporal halves of the visual fields in patient with pituitary tumour. The outer dotted line indicates the normal extent of the field of vision; the inner thick continuous line shows the field of vision in the patient. The shaded parts of the chart indicate the degree of visual loss. (H. Zondek, *Krankheiten der Endokrinen Drusen*.)

function might be to prevent sensory adaptation to the visual image. When the subject reads, there may be some five saccadic movements observable as he scans a single line of text.

Smooth pursuit movements occur voluntarily when the eyes track moving objects but take place involuntarily if a repetitive visual pattern is displayed continuously. Like saccades, these movements involve conjugate displacements of the two eyes, and allow us to follow moving targets accurately even at displacements of 30° s⁻¹. Pursuit movements depend on visual information from areas 17, 18, and 19 and from the superior colliculi. Control of the movements seems to be determined by the velocity of the moving object. Cortical and collicular neurones project by the corticotectal tracts to the cranial nuclei of III, IV, and VI.

Vergence movements allow focusing of an object which moves away from or towards the observer or when visual fixation shifts from one object to another at a different distance. Divergence or convergence occurs as is appropriate. In monkeys, electrical stimulation of areas 19 or 22 provoke vergence movements via the corticotectal tracts and the relevant cranial nuclei.

Vestibular movements are usually effective in compensating for the effects of head movements in disturbing visual fixation. For instance, when the head is tilted sideways the eyes rotate around their anteroposterior axes, so that their vertical axes are kept in alinement with the direction of gravity. The afferent nerves responsible for this have their receptors in the vestibular saccule. On the other hand, angular accelerations are signalled by the cristae of the semicircular canals [p. 305]. In each case the afferent impulses travel via the vestibular nuclei and the median longitudinal bundle to the nuclei of III, IV, and VI. Most rotations of the head do not involve angular rotations as fast as 300° s⁻¹ and the vestibular system can compensate for these. However, when the body is rotated at great speeds round a vertical axis (e.g. a skater performing a spin) eye movements show the so-called oculovestibular nystagmus, with a slow motion of the eyes in the opposite direction to that of the rotation—this is initiated by the vestibular mechanism—followed by a quick jerky binocular 'return' movement in the direction of rotation. This sequence is repeated as long as the angular *acceleration* lasts. It is likely that the fast component of nystagmus is mediated by mechanisms similar to those responsible for saccades.

REFERENCES

Alper, M. (1972). In *Handbook of sensory physiology* VII/4 *Visual psychophysics* (ed. D. Jameson and L. M. Hurvich) Chap. 12, pp. 303–30. Springer, Berlin.
Robinson, D.A. (1968). *Science, N.Y.* **161**, 1219–24.
—— (1970). *J. Neurophysiol.* **33**, 393–404.

The chemical senses

Chemical sensibility occurs even in the simplest living organisms. Unicellular protozoa, such as the amoeba or paramoecium react to nutrient particles by assimilating them, and to chemical irritants by avoidance movements. There must be different receptor mechanisms to initiate these entirely different motor responses. The chemical senses are well developed in all aquatic animals and in fishes there is differentiation of surface chemoreception into three types. Olfaction is confined to the olfactory pits and is mediated by the first or olfactory nerve. 'Taste' is widely distributed over the skin of the flanks as well as the mouth, and is mediated by the seventh or gustatory nerve, and chemical irritants act on receptors present in the skin over the whole body which is supplied by the fifth nerve and the spinal nerves. In amphibia, such as the frog, the sense of taste is confined to the mouth but receptors for chemical irritants are still found all over the skin. In mammals, including man, high sensitivity to chemical irritants is confined to the exposed mucous membranes of the eyes, nose, mouth, upper respiratory tract, and the anogenital apertures. This capacity to respond to surface application of chemical irritants was called by Parker (1922) the *common chemical sense* which has been regarded phylogenetically as the most primitive of the chemical senses from which the more specialized senses of taste and smell have evolved. Its purpose appears to be to promote rejection of, or withdrawal from, noxious chemicals in the immediate environment. The common chemical sense is thus comparable to the sense of pain produced by noxious physical stimuli.

In the nose and mouth common chemical sensibility is closely associated with the senses of smell and taste respectively. For example, the smells of the vapours of onions or ammonia consist of two components, one caused by stimulation of the olfactory apparatus, the other by stimulation of fifth nerve endings subserving the common chemical sense. After destruction of the olfactory apparatus or its nervous connections the typical odours of onions and ammonia are no longer detectable but the irritant properties of their vapours can still be appreciated. In rabbits nerve impulses can be recorded from the fifth nerve as well as the olfactory nerve when odours of certain subtances, which are not obnoxious in man, are introduced into the nose. Thus the common chemical sense contributes to the appreciation of odours, even in non-irritant concentrations.

Similarly, taste perception is often due to the blending of fifth nerve sensibility with the primary gustatory function of the chorda tympani and vidian nerves. Pungent spices and curries act largely by fifth nerve stimulation and large amounts can be truly painful.

The common chemical sense can thus play a physiological role, though excessive stimulation of this sense gives rise to pain. In terms of threshold concentrations of substances required for excitation smell is the most sensitive of the chemical senses, taste is intermediate and common chemical sensibility is the least sensitive.

The senses of smell and taste are mainly concerned with nutrition and are closely related to each other. If the nose is held, smell is abolished and the flavours of foods are largely lost. A cold in the head acts in the same way. Because odorous substances are dispersed in the air the sense of smell can give information about distant objects whereas taste informs only about substances in solution inside the mouth.

THE SENSE OF SMELL (OLFACTION)

The olfactory sense is highly developed in some mammals, such as the rabbit and dog, which are described as macrosmatic, but is much reduced in primates, including man, which are called microsmatic. In man smell is of obvious importance for nutrition, but it can also warn of serious environmental hazards such as the presence of hydrogen sulphide or hydrocyanic acid. The sense of smell contributes greatly to the joy of life, and much pleasure is derived from scented flowers and from perfumes.

Olfactory receptor area

The sense of smell arises mainly from stimulation of receptors in the yellowish-brown olfactory mucosa which lines the surface of the superior turbinate and the upper third of the nasal septum [Fig. VI. 30]. In man the total area of the olfactory mucosa on both sides is about 500 mm²; in the dog, which is macrosmatic, the area is much larger. The olfactory mucosa is above the main respiratory stream and sniffing is often required to stimulate the receptors. The olfactory mucosa contains special sensory cells (10–20 million in man),

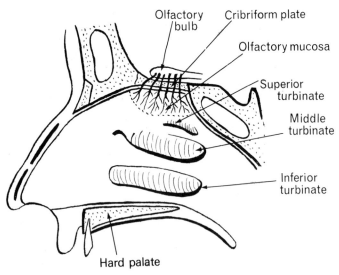

FIG. VI.30. Anatomy of olfactory mucous membrane. (Kindly supplied by Professor E. W. Walls.)

each of which is a bipolar neuron. These receptor cells lie between supporting cells [FIG. VI. 31], and their dendrites extend as naked processes (olfactory rods) which end in fine cilia (6–12 to each rod) which lie in the mucus covering the olfactory mucosa. The supporting cells end in microvilli which secrete mucus. The pigment is secreted by Bowman's glands and supporting cells; it consists of phospholipids, lecithin and their auto-oxidation products. It appears to be necessary for the sense of smell.

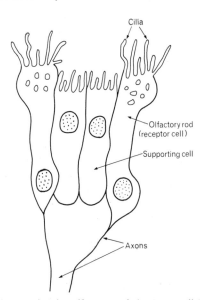

FIG. VI.31. Diagram showing olfactory rods (receptor cells) and supporting cells in the olfactory mucosa. (Kindly supplied by Professor E. W. Walls.)

The axons from the olfactory receptor cells are fine unmyelinated fibres, 0.2 µm in diameter, which run in bundles or fascicles containing 20–100 fibres; each fascicle is enclosed in one mesaxon or Schwann cell process. This arrangement differs from that for all other kinds of unmyelinated fibres in which there is an individual mesaxon for each individual unmyelinated nerve fibre. The fascicles run together in the fila olfactoria which pierce the cribriform plate and enter the olfactory bulb. Within the outer layers of the bulb the axons of the olfactory nerves enter the glomeruli and there form synapses with dendrites from mitral and tufted cells. These cells form the second-order neurons on the olfactory pathway. The arrangement is excellent for spatial summation since, in the rabbit,

each glomerulus receives impulses from 26 000 receptors and passes this information through 24 mitral cells and 68 tufted cells. The axons of the mitral cells form most of the lateral olfactory stria and run to the ipsilateral olfactory cortex which flanks the anterior perforated substance and includes the uncus [FIG. VI. 32]. Other fibres run via the intermediate olfactory stria to connect with the olfactory tubercle, and hence with the limbic system. The axons of the tufted cells run in the medial olfactory stria and cross the midline in the anterior commissure to form synapses with deeply located granule cells in the opposite bulb. The significance of this arrangement is not clear. In experimental animals section of the anterior commissure greatly impairs the sense of smell. Destruction of one olfactory bulb produces loss of smell only on the same side.

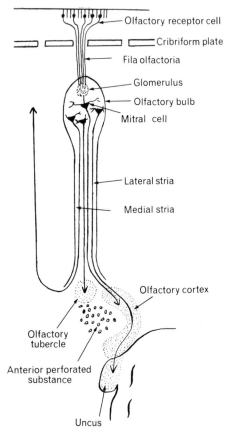

FIG. VI.32. Diagram of olfactory pathway. (Kindly supplied by Professor E. W. Walls.)

There is a regional projection pattern from the olfactory mucosa to the bulb. The upper part of the olfactory epithelium projects mainly to the upper part of the bulb and the lower part of the olfactory epithelium to the lower part of the bulb.

Inhibitory pathways. Efferent fibres in the olfactory striae can depress electrical activity in the bulb. These effects are mediated by the granule cells.

Physiology of olfaction

In dealing with any of the senses we have to consider the nature of the stimulus, how the stimulus is conveyed to the receptor cells, how the interaction between the stimulating agency and receptive mechanism is converted into generator potentials and nerve impulses and finally how the impulse traffic is sorted and coded by the central nervous system to provide the information concerning the nature of the stimulus.

Olfactory stimuli cannot be rigidly defined in the way that light or sound can be measured with respect to wave length, and intensity of stimulation. The allied chemical sense of taste has four primary qualities of sweet, sour, salt and bitter from which the numerous varieties of tastes of individual substances are compounded. There are no agreed corresponding primary qualities for the sense of smell.

To arouse the sense of smell molecules of the odoriferous substance must come into contact with the receptor cells of the olfactory epithelium. To achieve this the substance must take the high road through the nasal cavity, which is reached only by eddies from the low road during normal quiet breathing. Deep breathing or the act of sniffing more effectively brings the substance to the olfactory mucosa. The substance must then penetrate the mucus layer covering the olfactory epithelium, into which the cilia of the receptor cells project, probably in constant motion.

The odorous substance must now combine with receptors on the surface of the cilia and we are at once faced with the same sort of problems which arise when dealing with enzyme–substrate or drug–receptor interactions. The different regions of the olfactory mucosa may selectively absorb some substances more than others and in any small area there seem to be receptors for many different substances. Ultimately, the pattern of induced activity depends on the physical and chemical properties of the stimulant molecules and of the molecules of the receptive substance.

When odorous substances become adsorbed on to the olfactory mucosa electrical changes are set up. First, from the olfactory mucosa there develops a monophasic negative potential which lasts 4–6 seconds. This response, the *electro-olfactogram*, is probably the generator potential of the olfactory organ and is comparable to such potentials in other organs. Action potentials are then set up in the olfactory receptors and are conducted along the axons to the olfactory bulb where the high rate of spontaneous discharge is modified.

The receptors in the olfactory mucosa are relatively non-specific and odorants elicit complex spatiotemporal patterns of excitation. Recognition of odours depends on simultaneous activity in an ensemble of receptor neurons (Moulton 1976).

Numerous classifications of odours, and theories concerning the relation between chemical constitution of odorous molecules and character of sensation, have been proposed. None will be satisfactory until much more is known about the physical and chemical properties of both odorous molecules and receptor sites.

Individual variations in response.

Human beings vary in their sensitivity to odorous substances. In an Australian study nearly 20 per cent of adult males, but less than 5 per cent of females, could not distinguish the odour of 20 per cent potassium cyanide in water from that of water alone (Kirk and Stenhouse 1954). The same people detected other smells normally.

Belavoine (1941) studied the responses of 686 subjects (mostly male) who were asked to identify the following materials by the sense of smell: 1 per cent aqueous ammonia; 90 per cent alcohol; camphor; vinegar; 10 per cent sodium hypochlorite (chlorine odour). About two-thirds of the subjects identified the substances correctly; 30 per cent were inexact or uncertain and 2 per cent of otherwise normal people apparently had no sense of smell at all. In 5 subjects with anosmia (complete absence of the sense of smell) Henkin (1967) recorded no subjective or objective responses to the highly irritant vapours of pure pyridine or concentrated ammonium hydroxide. Not only was there absence of the primary olfactory sense but also of the common chemical sense from the nasal mucosa and pharynx, although the sense of taste was normal. The absence of the common chemical sense is probably a manifestation of a general indifference to noxious stimuli (innate insensitivity to pain). In people with apparently normal olfactory sense there are wide differences in sensitivity to different odorous chemicals.

Adaptation. Everyone has experienced adaptation to the smell of odorous substances (olfactory fatigue). An odour at first quite distinct soon fades and becomes imperceptible. It occurs to unpleasant stenches as well as to fragrant perfumes. Adaptation can develop within seconds or minutes, depending on the nature of the substance. The degree of adaptation can be measured by the rise in threshold concentration required to excite the sense of smell. Adaptation is selective, so that when fatigue has developed to one substance another odorous substance produces a normal sensation.

Adaptation also occurs to irritants which excite the common chemical sense. Workers in the paprika mills in Hungary adapt to the presence of the irritant capsaicin in the atmosphere and tolerate concentrations which cause pronounced sneezing and watering of the eyes in previously unexposed persons.

Influence of disease on olfaction. Damage to the olfactory mucosa or the olfactory path by trauma or disease can abolish the sense of smell (anosmia) or alter its character (parosmia).

General diseases can also modify the sense of smell. Patients with adrenal insufficiency have greatly enhanced sensitivity for taste and smell, particularly the latter. Reduction of the sense of smell (hyposmia) occurs in hypogonadism and vitamin A deficiency.

TASTE (GUSTATION)

What we commonly call 'tastes' are due not only to sensations aroused by stimulation of taste buds but also to sensations of common chemical sense, heat, cold, touch and especially olfaction. If the nose is held tastes may be remarkably altered so that boiled turnips, apples and onions are almost the same, ham and lamb cannot be distinguished, port tastes like sugar and claret like weak vinegar. Much of the discrimination attributed to the palate is really due to the sense of smell. On the other hand the 'odour' of chloroform is largely due to stimulation of sweet taste receptors (Moncrieff 1967). Flavour is the best word to describe the complex sensation comprising taste, odour, roughness or smoothness, hotness or coldness, pungency or blandness.

Taste receptors (taste buds)

The receptors for taste are chemoreceptors which are stimulated by substances dissolved in the oral fluids which bathe them. The taste buds are located on the edges and dorsum of the tongue, and on the epiglottis, soft palate and pharynx. On the *tongue*, taste buds lie on the surface of the fungiform papillae, in the grooves of the foliate papillae and the prominent vallate papillae arranged in a V at the back of the tongue. The small filiform papillae over most of the dorsum do not contain taste buds. In man the total number of taste buds is about 10 000. The number decreases in old age.

Taste buds are ovoid clusters of cells oriented vertically in the epithelial layer with a small pore opening on the surface. Human taste buds measure 60–80 μm in length and 40 μm in diameter at the thickest part. The cells within the taste bud were formerly classified in two types, gustatory and sustentacular (supporting), on the basis of differences in size, shape and staining properties. However, the electron microscope shows that many of the cells are transitional in character, and degenerating cells are also seen. It is probable that all the cells in the taste bud are sensory but in different stages of development. The taste cells are formed from the epithelial cells around the taste bud and migrate towards the centre as they mature and finally degenerate. In the rat the average life span for taste cells is 3–5 days. Each taste cell ends in microvilli at the tip near the pore.

The afferent nerves from the taste cells begin as minute fibres down to 50 nm in diameter, within the cell. These fibres form a nerve plexus near the basement membrane and become myelinated in the connective tissue underlying each taste bud. There are two or three large fibres to each taste bud and each fibre connects with one or more taste cells. Taste buds degenerate and disappear within a week after the taste nerve is cut. As the taste nerve regenerates to the periphery, the taste buds also regenerate. It is not known how the sensory nerve controls the activity of the taste-cell-forming epithelial cells.

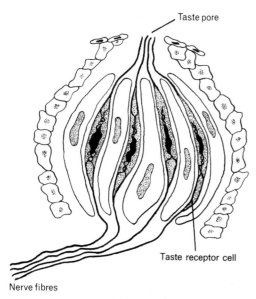

FIG. VI.33. Taste bud.

Taste nerve pathways

The sensory fibres from the taste buds in the anterior two thirds of the tongue run at first with the fibres subserving touch, temperature and pain in the lingual nerve. They leave this nerve to join the chorda tympani nerve which enters the brain as part of the seventh cranial nerve. The chorda tympani also contains the efferent fibres for salivation. Taste fibres from the posterior third of the tongue run in the glossopharyngeal (ninth) nerve and those from the epiglottis and pharynx in the vagus. The small myelinated taste fibres of all three nerves run into the *tractus solitarius* which has its nucleus in the medulla. The cell bodies of the second-order neurons are located in this tract and their axons cross the midline to join the medial lemniscus and terminate with fifth nerve fibres subserving touch, temperature and pain in the posteroventral nucleus of the thalamus. The third relay arises here and ends in the inferior part of the postcentral gyrus (sensory cortex), together with afferents from the face [FIG. VI. 34]. All afferents from the tongue, both of taste and common sensibility, travel along the same pathway to the thalamus and sensory cortex. When a taste-evoking substance, e.g. quinine, is placed on the tongue it produces changes in the electro-corticogram of the lower part of the postcentral gyrus. Electrical stimulation of this area can produce hallucinations of taste, and destructive lesions reduce gustatory and tactile sensibility of the tongue.

Receptor mechanisms. Taste-producing substances, dissolved in the oral fluid, act by forming a weak attachment to receptors on the microvilli of the gustatory cells. The nature of this combination and the way in which electrical impulses are generated in the gustatory nerves are unknown. The binding of substances to the receptors must be weak because the taste produced by any substance can be abolished by washing the tongue with water.

Modalities of taste

In marked contrast to the sense of smell there are only four basic modalities of taste, sweet, sour, bitter, and salt. In the tongue sweet sensitivity is greatest at the tip, sour at the sides, bitter at the back, while salt sensitivity is more homogeneous but greatest at the tip [FIG. VI.35]. The mid-dorsum is insensitive to all tastes. Individual taste buds can respond exclusively to salt, sweet or sour substances, or to some combination of two, three or four of the basic taste stimuli. The taste buds selectively sensitive for these four modalities of taste show no histological differences to account for their functional properties.

Electrophysiological studies have shown that taste receptor cells do not always belong to basic receptor types corresponding to the basic taste qualities. Individual sensory cells are differentially sensitive to chemicals; for instance, a nerve discharge from a given receptor cell may be produced by low concentrations of sodium chloride or by high concentrations of sucrose, but in another cell the reverse will hold. Thus it seems that each sensory cell possesses a cluster of receptors, the sensitivities of which vary among different sensory cells. Any one cell reacts to a varying degree to several different chemical stimuli, many of which fall in two or more of the four classical basic taste categories.

Diamant and Zotterman (1959) have recorded the electrical responses of the exposed chorda tympani nerve in anaesthetized man, during operations undertaken to mobilize the stapes. The integrated responses to touch, and to applications to the tongue of sodium chloride, sucrose, saccharine, quinine and acetic acid are shown in FIG. VI. 36. These were as expected, but the depression of spontaneous activity by water was of particular interest. In the cat, dog, pig, and rhesus monkey there are taste receptors which respond to distilled water, but in the rat and man there are no such receptors. The significance of this difference is not clear.

Substances producing basic taste sensations

Sour. Except in the case of hydrogen ions and sour taste there is no

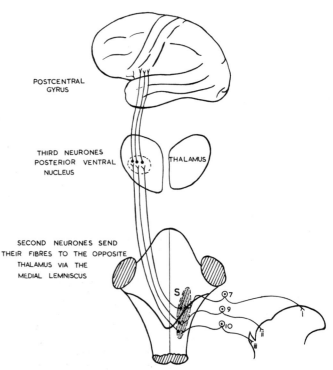

FIG. VI.34. Diagram of taste pathway.

7, 9, 10 = The first neurons in ganglia of facial, glossopharyngeal, and vagus nerves. Their peripheral processes innervate taste buds in anterior two-thirds of tongue (i), posterior one-third of tongue (ii), and region of epiglottis (iii).

S = Nucleus of tractus solitarius.

(Kindly supplied by Professor E. W. Walls, Middlesex Hospital Medical School.)

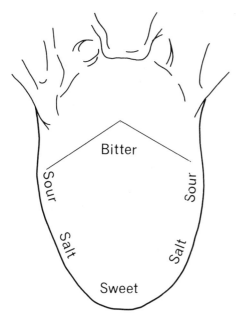

FIG. VI.35. Diagram to show parts of the tongue where the four tastes are best appreciated.

clear relationship between chemical constitution and basic or primary taste sensation. The nature of the combination between taste-producing substance and specific receptor substance is unknown though speculations have been proposed.

The sour taste is due almost entirely to hydrogen ions and the degree of sourness is roughly related to the degree of dissociation. Some acids have other tastes as well; for example, citric acid is sweet as well as sour and picric acid is bitter and sour. Amino acids may taste sour, bitter, sweet or salty depending on concentration, and varying among different subjects. Weak organic acids, such as acetic acid which occurs in vinegar, are more sour than would be expected from their degree of dissociation. The threshold pH for HCl is 3.5 and for acetic acid 3.8. The pH threshold for sour taste with acetic acid-sodium acetate buffer is 5.9; this high value may be due to enhancement of the effect of H^+ by acetate ions.

Salty. Sodium chloride is the reference substance for pure salty taste, though oddly enough at threshold concentration (0.02 M) it tastes sweet. The anion contributes most to the tastes of salts but the cation can modify the anionic effect. Potassium salts tend to be bitter as well as salty and potassium iodide is only bitter. Salts of heavy metals such as mercury have a metallic taste and lead acetate and beryllium salts a sweet taste.

Bitter. Bitter tastes are produced by many different types of chemical substance. Quinine sulphate is the classical bitter substance with a threshold of 0.000 008 M. Strychnine hydrochloride is even more potent with a threshold at 0.000 0016 M. Other alkaloids such as morphine and nicotine are bitter, as also are caffeine, urea, phenyl-thiourea, and certain salts such as magnesium sulphate. Many sweet substances have a concomitant bitter taste or aftertaste, e.g. saccharin. This double taste is most apparent as the substance moves from the front of the tongue, where sweet tastes are appreciated, to the back where bitter sensitivity is particularly developed. (A sweet toffee is kept in the front of the mouth and bitter beer is tossed to the back.)

Sweet. The sweet taste is associated primarily with organic compounds except for certain inorganic salts of lead and beryllium. Sucrose is the standard reference substance for sweetness, with a threshold concentration of 0.01 M. Glucose has a threshold of 0.08

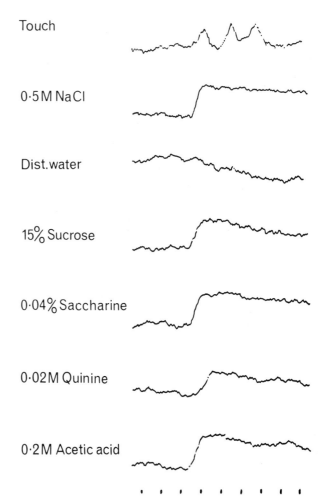

FIG. VI.36. Integrated electrical responses of the whole chorda tympani of man to sapid solutions poured on the tongue. Note that water depressed the spontaneous afferent activity. Time in seconds. (Diamant *et. al.* (1963). In *Olfaction and taste* (ed. Y. Zotterman). Pergamon Press, Oxford.)

M. Fructose is sweeter than sucrose, and maltose, galactose and lactose are less sweet than glucose. However, the taste sensation is not the same for all these sugars. Other sweet tasting substances include alcohols, glycerol, aldehydes, ketones, amides, esters, and chloroform. Synthetic sweeteners, such as saccharin (threshold concentration 0.000 023 M), dulcin and cyclamate are used as substitutes for sucrose in diabetics and obese persons in whom sugar intake must be reduced.

A substance called gymnemic acid, extracted from the leaves of an Indian plant called *Gymnema sylvestre*, selectively abolishes the sensation of sweet, leaving bitter, salt, and sour unaffected.

Miracle fruit. After the berries of the West African plant *Synsepalum dulcificum* have been chewed, lemons taste like oranges and grapefruit tastes sweet. This 'miracle fruit' gives to sour maize bread and to sour palm wine a pleasant sweet taste. The active constituent of the fruit is a 'taste-modifying protein' which itself is tasteless and does not affect the sensations of bitter or salt. Kurihara and Beidler (1969) suggest that the protein becomes attached through part of its molecule to the cell membrane near a sweet receptor. When the $[H^+]$ of the surrounding medium is raised by acids, the sweet receptor membrane undergoes a conformational change, so that the sugar residues which form part of the taste-modifying protein can fit into the sweet receptor site.

Factors influencing gustatory sensations

Area. Stimulation of a single papilla or of a limited area of the tongue by one drop of solution produces weaker sensations than does tasting of the same solution by the whole mouth. The threshold decreases as the area of stimulation increases up to 60–90 mm².

Temperature. The maximum sensitivity to taste-producing solutions occurs when their temperature is within the range 30–40 °C.

Individual variation. Variations in sensitivity to taste-producing solutions are well known, and there is a general reduction in sensitivity in older people. Borg *et al.* (1967) found it of little value to compare the chorda tympani nerve responses of individual patients with the subjective responses from other subjects, because of the variations in individual response.

There are two conditions in which there are clearly identifiable taste defects:

Familial dysautonomia. This is a rare congenital disorder characterized by an inability to recognize by taste even saturated solutions of sodium chloride, sucrose or urea, and the threshold for identification of HCl is much higher than normal (Henkin 1967). This condition is also accompanied by orthostatic hypotension, absence of lacrimation, hyporeflexia, and relative insensitivity to temperature and to noxious stimulation. Thus the sensory defects are widespread.

Selective taste blindness. There are no instances of complete loss of only one modality of taste. However, there is a condition inherited as a Mendelian recessive trait, in which there is a very marked rise in threshold to the bitter taste of phenylthiourea (phenylthiocarbamide, PTC). The normal threshold is about 0.00002 M but in non-tasters it is 0.008 M. Different studies have shown that between 3 and 4 per cent of Caucasians are non-tasters. The defect is highly selective since there is no taste blindness to other bitter substances nor to substances which taste sweet, salty or sour.

Adaptation. It is common knowledge that sapid (taste-producing) substances quickly produce adaptation if kept in one place in the mouth. To keep the taste going sweets are continually moved about in the mouth. In experimental studies sapid solutions are made to flow continuously over the tongue; taste intensity rapidly falls and the threshold for the particular substance rises considerably. Borg *et al.* (1967) have shown that in man the adaptation is peripheral [Fig. VI.37]. There is a reduction in the summated chorda tympani nerve response to 0.2 M NaCl solution which parallels the reduction

in subjective estimation of saltiness. The nerve response in the rat shows much less adaptation which emphasizes the difficulty in arguing from animal experiments to man.

Adaptation to one acid produces adaptation to other acids, presumably because H⁺ is the stimulus in all cases. There is no cross adaptation to salts, and bitter and sweet substances are intermediate.

Interaction between sapid substances is a well known phenomenon; the reduction of the sour taste of fruits by sucrose is a good example. In electrophysiological studies on the response to a mixture of 10 per cent sucrose + acid at pH 2.5 there was no evidence of mutual peripheral inhibition, so the sour-sweet interaction must be presumed to be central.

Acceptance and rejection of foods. Taste sensation is most directly related to the taking of food and to the rejection or avoidance of some (but by no means all) noxious substances. Thus taste leads to one of two reactions, acceptance or rejection. Among the four basic tastes sweet is the most generally acceptable (though in the long run purified sucrose promotes dental decay and perhaps atherosclerosis), bitter quickly becomes unpleasant beyond a small degree, and acid and salt are pleasant at first but moderate concentrations become unpleasant.

Acceptance or rejection of substances may be related to the metabolic state. It is claimed that the sense of taste may guide animals and human beings to select what is good for them; adrenalectomized animals and patients with adrenal cortex insufficiency seek out salt and patients with hypoglycaemia find strong sugar solutions more palatable than when the blood sugar is normal. But since diabetics may also have a craving for sweet it is clear that the hedonic value of a taste stimulus may not be in the best interest of the patient.

REFERENCES

Beidler, L. M. (ed.) (1971). *Handbook of sensory physiology*, Vol. IV. *Chemical senses*, Part 1. *Olfaction*; Part 2. *Taste*. Springer, Berlin.
Diamant, H. and Zotterman, Y. (1959). *Ann. N.Y. Acad. Sci.* **81**, 358.
Hayashi, T. (ed.) (1967). *Olfaction and taste II*. Pergamon Press, London.
Henkin, R. I. (1967). In *Olfaction and taste II*, pp. 321–35. Pergamon Press, London.
Katsuki, Y., Sato, M., Takagi, S. F., and Oomura, Y. (eds.) (1980). *Food intake and the chemical senses*. Japan Scientific Societies Press.
Moncrieff, R. W. (1967). *The chemical senses*, 3rd edn. Hill, London.
Moulton, D. G. (1976). *Physiol. Rev.* **56**, 578.
Schneider, R. A. (1967). The sense of smell in man. *New Engl. J. Med.* **277**, 299.
Wolstenholme, G. E. W. and Knight, J. (eds.) (1970). *Taste and smell in vertebrates*. Churchill Livingstone, London.

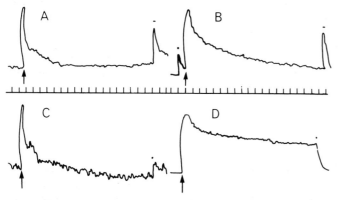

Fig. VI.37. The summated chorda tympani response to a continuous 3 min flow of 0.2 M NaCl. A, B. C are human responses for patients Nos. 1, 2, 3 respectively and D is a rat response. Dots indicate response during application of distilled water; arrows, onset of salt. Tape recorded data processed under identical conditions with rise and fall time constants of 1.5 s. The tape recorder was off at beginning of B. Time base in 10 s intervals. (Diamant *et. al.* (1965). *Acta physiol. scand.* **64**, 67.)

Pain

Pain is an unpleasant sensory experience distinct from other sensory modalities such as touch, warmth, and cold. It can be elicited by noxious stimulation in normal persons and it is also the outstanding symptom in many diseases. Pain is usually associated with tissue or cell damage and serves as a warning that such damage is taking place. Pain gives information about states of the body, but, unlike other sensations, not about the nature of the stimulus. If the conducting nerve pathway is uniformly and seriously damaged the total information conveyed is reduced and numbness or local anaesthesia develop. If, however, the large nerve fibres are selectively destroyed, e.g. in herpes zoster, pain may be aroused by stimuli which on normal skin cause no pain. Thus with normal afferent nerve pathways pain results from tissue damage, but with hypersensitive nerve pathways non-noxious stimuli can also evoke pain.

Pain cannot be defined in words which would mean anything to a person who had not experienced it. It is a subjective affair though it may be accompanied by measurable physiological responses such as reflex withdrawal movements, changes in vasomotor tone, blood pressure, heart rate, and breathing and sweating. Sherrington defined pain as 'the psychical adjunct of an imperative protective reflex'. This draws attention to its value to the organism as a whole. The rapid reflex withdrawal response evoked by a pin-prick, a blow, or contact with a hot object prevents more serious or widespread injury. Thus it limits the damage produced by noxious stimuli. The development of conditioned reflex responses based on the association of pain with other stimuli, may lead to protective movements which result in the avoidance of injury altogether. Pain certainly gives protection from harmful agencies acting on the surface of the body. (Persons who lack the sense of pain are very prone to burns and other injuries.) Pain may also lead to appropriate action to deal with certain pathological states. Patients with obstructive coronary artery disease get angina pectoris on exertion, due to the relatively poor blood supply to the heart. The pain makes them stop and rest, and thus protects the diseased heart from the serious consequences of over-activity. The pain of acute inflammation is relieved by resting the affected part. This also appears to be beneficial. Pain is also often helpful in the diagnosis of disease, e.g. acute appendicitis, renal colic, myocardial infarction, etc.

However, pain has its limitations as a protective warning system. For example, some types of acute severe pain after injury, such as post-operative pain, serve no useful purpose, and *chronic* pain such as that due to untreatable malignant disease, persistent inflammation as in rheumatoid arthritis, or nerve lesions such as that producing post-herpetic neuralgia are of no value to the suffering patient. From another viewpoint, the value of pain is much reduced by the fact that some noxious agencies such as ultraviolet light and ionizing radiation do not arouse pain at the time of exposure.

Affective component of pain

Pain is an unpleasant experience and therefore has a large emotional or affective accompaniment. By contrast, some kinds of sensation are almost devoid of affect, e.g. muscle sense; moderate temperature may be associated with a pleasurable affect; touch may be associated with every degree of affect, according to what is being touched (and by whom).

EXPERIMENTAL AND PATHOLOGICAL PAIN

Whilst pain may be produced experimentally in normal subjects without much emotional accompaniment, the pain of disease is frequently enhanced by apprehension concerning its significance. A student who experiences the pain induced by ischaemic muscular contractions is not unduly alarmed, partly because he knows that the pain can be very rapidly relieved by restoring the circulation and also because the pain has no significance for his future. Conditions are different in a patient with ischaemic muscular pain (e.g. intermittent claudication or angina pectoris). Even though the peripheral stimulus to pain might be of the same intensity as in the experimental subject the patient suffers much more, mainly because he is anxious about his future health. Conversely, emotional factors can block pain. Beecher (1956) noted that many badly

wounded soldiers tolerated their injuries remarkably well, presumably because of emotional relief, amounting sometimes to euphoria, at escaping from the horrors of war.

Pain sensitivity varies greatly amongst different people. A few persons are apparently insensitive to noxious stimuli (in spite of a histologically intact nerve apparatus) which produce pain in the vast majority; others may be hypersensitive. The reaction of the patient to the pain of disease will depend on the cause of the pain, its location, its severity, his pain-sensitivity (Keele, K. D. 1968), and his psychological disposition. One of the most important things every doctor has to learn is how to assess the significance of a patient's description of his pain.

Qualities of pain

Pain varies greatly in quality according to the provoking stimulus and the site of the stimulation. Most people describe pain in terms of the external agency which produces it, or in words which express picturesquely how they feel their pain might have been produced. Terms such as 'pricking', 'burning', 'tearing', 'cutting', 'stabbing', 'crushing', and 'aching' are commonly used.

Those who have been subjects in experimental studies on pain would readily distinguish cutaneous pain from deep pain. Indeed, Lewis almost regarded them as different sensations. In the skin, pricking and burning types of pain have been distinguished by Hardy, Wolff, and Goodell (1952). Ischaemic muscle pain is distinct from cutaneous pain, and different from both are colicky pains in the hollow viscera. It is difficult to know how many different qualities of pain can be distinguished but it can safely be said that distinctive qualities are associated with the following structures: skin, subcutaneous tissue, muscle, and hollow viscera. All types of pain are readily distinguished from itch. Superficial pain leads to reflex withdrawal movements, increase in heart rate, rise in blood pressure, and changes in breathing, whereas deep pain characteristically produces faintness, nausea, sweating, bradycardia, and a fall in blood pressure.

Thus different types of unpleasant experience are described by the word 'pain'. It is not known whether they can all be explained in the same physiological terms.

TISSUES FROM WHICH PAIN CAN BE ELICITED

Pain can be elicited from many different tissues in the body using physical or chemical stimuli. Certain tissues are much more sensitive or respond to a wider range of stimuli than others. It is very important to remember that an inflamed tissue is hyperalgesic, i.e. its pain-threshold is lowered. This means that some tissues which are not pain-sensitive in their normal state may become so when they are inflamed.

STIMULI OF PAIN

Pain results from stimulation of 'nociceptive' nerve endings or nerve fibres by physical or chemical agencies. The sensory experience may be greatly modified by other changes within the central nervous system but the origin of nearly all types of pain is to be sought in peripheral structures.

Physical stimuli

Pain is produced in the skin by many kinds of physical stimuli, thermal, mechanical, and electrical, which have the common property of being potentially or actually harmful. Pain and the accompanying imperative protective reflex and voluntary responses minimize the amount of damage inflicted by the noxious stimulus.

In man, raising the *skin temperature* to 45 °C or above evokes pain. Exposure to cold (0 °C) is also painful. Noxious heating of the skin in animals promotes reflex withdrawal movements.

Pain due to *excessive pressure* or *tension* occurs commonly on the surface of the body. A blow on the shin, a knock on the head, or the pulling of hair are examples. Visceral pain, too, is often due to excessive tension on nerve endings in smooth muscle, e.g. the pain of childbirth which is associated with powerful uterine contractions and the colics of the alimentary, biliary, and urinary tracts. Stretching of the walls of blood vessels by excessive dilatation or by traction can also produce pain. The throbbing headache which follows injection of histamine is due to distension of intracranial arteries; the headache associated with fever is of similar origin and so, possibly, is the well-known 'hang-over' headache of the 'morning after'.

Changes in intracranial pressure may produce headache by stretching the walls of intracranial arteries or venous sinuses. Reduction of intracranial pressure by removal of cerebrospinal fluid produces headache in this way; headache following lumbar puncture is probably due to leakage of cerebrospinal fluid.

In *migraine* the typical unilateral headache is sometimes associated with distension and increased amplitude of pulsation of the branches of the external carotid artery on the affected side (Wolff 1963). In such cases ergotamine gives relief by reducing the amplitude of arterial pulsation. Noradrenalin acts similarly.

Another example of pain due to tension is the headache produced by sustained contraction of the muscles of the neck and scalp. The contractions may be initiated by a trivial primary focus of irritation. The spasmodic contraction of the sternomastoid—the 'stiff-neck'—is very painful.

It is often said that the pain of inflammation is due to increased tension in the affected tissue. There may certainly be some increase of tension but the important factor is the reduction of pain threshold, i.e. hyperalgesia, probably due to biochemical changes in the tissues. The hyperalgesia may so enchance the effect of physical stimuli that arterial pulsation may be painful and some trivial contact, which is barely perceived in normal tissues, can produce agonizing pain. Anyone who has had a 'boil' will be able to confirm this. Tissues which are not normally pain-sensitive may become so when inflamed.

Pain due to compression of nerves, e.g. by a tumour, a prolapsed intervertebral disc, or an aneurysm, probably arises from an interference with their blood supply. Complete ischaemia would soon block conduction of all nerve impulses and lead to loss of sensation in the innervated region. However, lesser degrees of compression, or intermittent compression may increase the irritability of nerves and lead to paraesthesiae and hyperalgesia. Thus normally subthreshold stimuli become painful. Ischaemia probably produces metabolic changes in the affected nerves but the nature of these changes is not known.

Causalgia

The word 'causalgia' means 'heat pain' and is applied to a condition in which there is persistent burning pain following nerve injuries. There is a reduction of threshold to 'burning' pain but not to 'pricking' pain. Causalgia is accompanied by sweating and vasomotor changes in the affected area. Causalgia has been variously attributed to:

1. The effect of surrounding tissues on uninsulated C-nerve fibres.

2. To the absence of other sensory modes, such as touch, during regeneration.

3. To the formation of artificial synapses between postganglionic sympathetic nerve fibres and C-fibres owing to the breakdown of insulation. The continuous discharge of nerve impulses along efferent sympathetic nerve fibres would thus set up afferent impulses in adjacent pain nerve fibres. This view is supported by the recorded beneficial effects of sympathectomy. In patients who have benefitted from sympathectomy, injection of noradrenalin into the affected region produces reappearance of the causalgic pain.

Chemical stimuli

It is possible that physical pain stimuli act via chemical mediators. This is at present impossible to establish for rapidly acting transient pain stimuli such as pin-prick or momentary contact with a hot object. However other physical stimuli such as a severe blow which produces a bruise, or a burn which leads to blistering, produce not only immediate pain but pain which persists for some time after the physical stimulation has ceased; in these cases it may be that chemical factors are responsible for the lasting pain and hyperalgesia.

Inflammatory pain

Inflammation consists essentially of the cellular and humoral responses to tissue injury produced by irradiation, heat, mechanical trauma, or infective and allergic agencies. The vascular and cellular responses of inflammation have been much studied; less attention has been given to the causes of the accompanying pain and hyperalgesia.

Let us consider the pain of a *heat burn*, which may be studied experimentally in normal subjects by application of a heated brass rod to skin. FIGURE VI.38 is a graphic record of the pain produced in

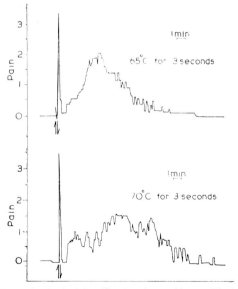

FIG. VI.38. Records of pain following experimental burns. The burns were induced in the skin of the forearm by application of a heated brass rod 1 cm in diameter, at 65 °C in one subject and 70 °C in the other, the duration of application being 3 seconds in both cases. Pain was recorded as described by Armstrong *et al.* (1953). Ordinate = Pain Scale, where 1 = slight pain, 2 = moderate pain, 3 = severe pain, and 4 = very severe pain. Note the peak of pain during application of heat, followed by latent period of 27 and 21 seconds respectively when no pain was felt, and then a slowly waxing and waning of what Lewis called the 'delayed' pain of a burn. (Keele, C. A. (1958). Causes of pain. In *Lectures on the scientific basis of medicine*, Vol. 6, p. 143. The Athlone Press, London.)

two subjects by 3 seconds' exposure to temperatures of 65 and 70 °C respectively. The initial high peak of pain coincided with the period of heating, after which the pain rapidly subsided to zero. Then after 21–27 seconds pain returned and rose more slowly to moderate intensity and then decayed again during a period of about 10 minutes, after which the area was hyperalgesic but not spontaneously painful. It is the delayed pain which is attributed to chemical factors. The hyperalgesia is similar to that seen following other forms of injury. Lewis suggested that the chemical agents which produce hyperalgesia in an inflamed tissue produce spontaneous pain when they reach a high enough concentration. This agrees with Lewis' findings that spontaneous pain can be induced in an area of hyperalgesic inflamed skin by occluding the circulation, or by warming the skin to 30–35 °C; both these procedures would increase the local concentration of pain-producing metabolites, the raised temperature increasing their formation, the ischaemia preventing their removal. Ischaemia of the nerves would increase the sensitivity to pain-producing metabolites.

Acute gouty arthritis

This is an excellent example of chemically induced pain. It was originally suggested by Garrod (1876) that an acute attack of gout was due to deposition of sodium urate in the affected joint. This has been amply confirmed in recent years by demonstrations of microcrystals of monosodium urate in synovial fluid during the phase of acute inflammation. Moreover, injection of a suspension of urate crystals into a normal joint provokes an acute inflammatory response comparable to a spontaneous attack. FIGURE VI.39 summarizes experiments performed by Faires and McCarty (1962) on themselves. In each a control injection of saline was given into one knee joint and an injection of urate crystals into the other. The acute inflammatory reaction to the urates was clearly greater than the experimenters bargained for and drastic measures were required to relieve the pain.

Ischaemic muscular pain

Lewis studied the causation of pain which develops in a limb when the circulation is entirely occluded. Exercise is carried out by performing a gripping movement with the hand at the rate of once per second. Pain sets in after 30 seconds and is intolerable in 70 seconds. Though the pain is diffusely felt, it is most marked in the muscles. Various possible causes of the pain may be considered.

1. It is not due to vascular spasm, because after occlusion the blood vessels of a limb lose their tone.
2. The pain is not due to muscular tension because it is continuous and not accentuated during contraction.
3. It is the result of activity because it is related to the amount of exercise which is performed.

Lewis suggested that muscular activity releases a pain-producing factor (P) which passes out into the tissue spaces and is normally removed by the bloodstream. If exercise is carried out with the circulation occluded this substance accumulates, and when it reaches a certain concentration, pain develops. When the circulation is restored, the pain disappears within a few (2–4) seconds. Lewis' P factor may consist of more than one substance, e.g. K^+, adenine nucleotides, and lactic acid.

Intermittent claudication. The above experiments explain the recurrent pain (intermittent claudication) produced in the legs during exertion in patients with narrowed limb blood vessels, e.g. in atherosclerosis. In this disease, the blood supply to the muscles is adequate for their needs during periods of rest; but during activity the blood flow cannot be increased sufficiently to cope with the additional requirements (i.e. there is *relative* ischaemia). The P factor consequently accumulates, giving rise to pain which in-

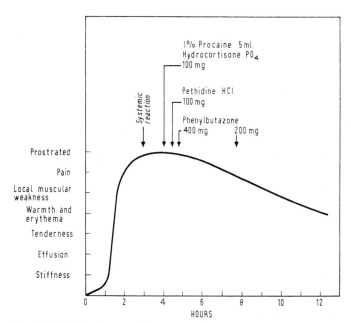

FIG. VI.39. Clinical response in two normal men to intrasynovial injection into knee-joint of sodium urate crystals. One line for similar reaction in both subjects. Moderate pain, large effusion after 24 hours. Slight tenderness, erythema, and moderate effusion after 48 hours. No subjective symptoms and barely detectable effusion after 72 hours. (Faires, J. S. and McCarty, D. J. (1962). *Lancet* ii, 683.)

creases in intensity until the patient is compelled to stop. During the period of rest the P factor is washed away and the pain disappears.

Pain of coronary occlusion. The pain of coronary occlusion, like that in an ischaemic limb, is presumably produced by ischaemia of the affected area of heart muscle. In an acute infarct there will also be release of 5-hydroxytryptamine and prostaglandins from platelets and of potassium and adenine nucleotides from muscle, and probably formation of plasma kinins. The pain persists as long as the sensory nerve endings in the ischaemic patch of heart remain alive.

Angina of effort. Attacks of anginal pain may be brought on by effort, without evidence of coronary occlusion. Attacks occur under circumstances that would be expected to throw an extra burden on the heart and so cause relative ischaemia. The blood supply of the heart is thus insufficient for its increased needs, the P factor accumulates and gives rise to pain.

Myonecrosis. The condition of myonecrosis is characterized by segmental hyaline necrosis of skeletal muscle fibres, as the result of which their contents, including the easily identifiable pigment myoglobin, are released into the extracellular fluid. The outstanding symptom of myonecrosis is skeletal muscular pain and the characteristic sign is myoglobinuria. The most interesting cause of myonecrosis is poisoning by certain sea-snakes (e.g. *Enhydrina schistosa*) which live in the sea off the coast of Malaya. The bite of the sea-snake is felt as a prick, but after this there is no pain at this site at all. After 1–2 hours pain develops in muscles throughout the body and after 3–6 hours myoglobin appears in the urine.

The toxin in sea-snake venom is not itself painful but by its disruptive effect on skeletal muscle cell membranes it releases the cell contents and these act on the appropriate afferent nerve endings to arouse pain. The pain of myocardial infarction might in part be due to necrosis of heart muscle.

Pain of peptic ulcer. The pain of ulcer on the stomach or duodenum is probably due mainly to the action of acid or other irritants on a hypersensitive region of the mucous membrane. The evidence is as follows:

1. The pain tends to follow a meal after an interval which varies with the position of the ulcer; the interval is about an hour with an ulcer near the incisura and three hours or more in duodenal ulcer. Pain generally sets in when gastric acidity reaches a characteristic threshold level, e.g. pH 1.5 in some, or lower levels of acidity in other cases.

2. The ingestion of food, which decreases gastric acidity, always relieves and usually abolishes duodenal ulcer pain. Food often relieves gastric ulcer pain, but usually the pain has disappeared before the next meal is taken.

3. Vomiting, which removes acid fluid from the stomach, relieves pain generally in gastric and sometimes in duodenal ulcer.

4. In both types of peptic ulcer pain is relieved by antacids.

5. In patients who are experiencing frequent bouts of spontaneous pain, the introduction of 200 ml of 0.5 per cent HCl into the stomach (when pain is absent) produces pain. If the acid is introduced into the stomach when the patient is free from spontaneous pain, no pain is produced. Similar results are produced by other acids, e.g. H_2SO_4.

In the crater of an active ulcer sensory nerve endings are exposed and their sensitivity may be enhanced by inflammatory products or changes in the local blood supply. In these circumstances the acid which is normally secreted by the gastric mucosa, or acid, or other irritants which are introduced into the stomach, may set up a sufficiently vigorous discharge from the nerve endings to give rise to a pain sensation. If the inflammation or congestion subsides the discharge set up by the less sensitive endings in response to the same stimuli may be inadequate to produce a pain sensation. Similarly during healing of the ulcer the nerve endings become covered with mucus or scab, blood clot, or granulation tissue that protects them from nocuous stimuli.

Pain may, however, occur in peptic ulcer cases with achylia or at low levels of intragastric acidity; in these it is probable that inflammation has so greatly sensitized the nerve endings that muscle spasm may arouse pain.

Nature of chemical factors which produce pain

Many substances can arouse pain when applied to the skin or mucous membranes or when injected into the body.

Extrinsic algogenic substances include strong irritants such as acids and alkalies, organic solvents, war gases and liquids which penetrate skin and mucous membranes. Injection of hypo- or hypertonic solutions or of certain drugs, e.g. thiopentone can also produce pain in skin and deeper structures. Other extrinsic algogenic agents occur in plant and animal stings and venoms which will be discussed later.

Intrinsic algogenic substances

It is, however, more important to consider pain production by intrinsic substances derived from living body cells and fluids. The role of HCl in the production of pain in patients with peptic ulcer has been discussed. The possible roles of other substances must now be considered (see Keele and Armstrong 1964).

Chemical stimulation of pain receptors may be studied experimentally by injecting the materials into the skin or deeper structures, but a more accurate and reproducible method is to apply the substances to the nerve endings in an exposed blister base. A blister is made by application of a cantharidin plaster; the fluid is aspirated and the raised epidermis is removed. An isotonic Ringer–Locke solution at pH 7.5 is applied to keep the blister base in a pain-free state. At intervals this bathing fluid is removed and the test fluids are applied. The intensity of pain is recorded graphically as described in FIGURE VI.38.

With this technique it has been shown that many substances normally present inside body cells can cause pain when released into the extracellular fluid. For example, when blood platelets disintegrate they release 5-hydroxytryptamine (5HT, serotonin) which produces pain on the blister base at 10^{-7} mol l^{-1}. ATP and prostaglandins are released at the same time and are also algogenic. Erythrocytes are cells which can be readily collected and separated in an intact state; when they are suspended in physiological saline and applied to a blister base they cause no pain but after lysis they evoke marked pain. This is partly due to their high content of K^+ (>100 mM l^{-1}) and partly to the presence of adenosine phosphates, AMP, ADP, and ATP. Leucocytes have a cationic protein in their lysosomes, which also produces pain. Mast cells contain histamine, which in high concentrations (>10 mg l^{-1} or 100 μmol l^{-1}) produces pain, but in lower concentrations (100 μg–10 mg l^{-1} or 1–100 μmol l^{-1}) arouses itch. Cellular secretions may also cause pain. Gastric juice is algogenic not only by virtue of its HCl content but perhaps also through the action of pepsin. Pancreatic juice contains two pain-producing agents, pancreatic kallikrein and trypsin. It is not in the least surprising that when gastric juice or pancreatic juice escape into the highly sensitive peritoneum, after perforation of a peptic ulcer or in pancreatitis, severe pain is produced. Bile is also algogenic. The pain associated with myonecrosis could be due to K^+, ATP, and perhaps larger molecular substances such as enzymes. There is no doubt that many intracellular substances can arouse pain when they escape into the extracellular fluid and make contact with the appropriate afferent nerve receptors.

Plasma kinins. In addition, plasma and extracellular fluid contain a protein system from which very active pain-producing *plasma kinins* can be formed. The best known of these plasma polypeptides are *bradykinin* which is a nonapeptide, and *kallidin* which is a decapeptide. These kinins are formed from a substrate called kininogen (an α-2-globulin) by the action of enzymes called *kininogenases*. These kininogenases are either extrinsic (e.g. salivary or pancreatic kallikrein) or intrinsic (plasma kallikrein). The best known procedure for activating the intrinsic plasma kininogenase is contact with a foreign surface, such as glass. Clotting factor XII (Hageman factor) is first activated and then initiates two separate chains of reactions which lead to blood clotting and kinin formation respectively. It is possible that kininogenases act directly as well as through kinin formation. Inflammatory exudates such as blister fluid and rheumatoid arthritic effusions also contain the full plasma kinin-forming system. Plasma kinins are rapidly destroyed by kininases present in plasma and certain tissues.

The plasma kinins have the following actions: they cause vasodilatation, increased vascular permeability and pain, all these being characteristic features of inflammation. Pain production by bradykinin is enhanced by 5HT.

The kinin-forming system is activated by crystals of monosodium urate in the same way as by glass, i.e. through initial activation of factor XII. This process may contribute to the pain and inflammation of acute gouty arthritis. Tissue injury also probably promotes plasma kinin formation.

Prostaglandins (PGs). These ubiquitous substances [p. 559] are capable of producing pain when present in high concentrations (e.g. 10 μmol l^{-1}) but lower concentrations produce hyperalgesia. PGs are formed from arachidonic acid, and since anti-inflammatory drugs such as aspirin and indomethacin inhibit PG synthesis, this may account for the relief of inflammatory pain and hyperalgesia induced by these drugs in rheumatoid arthritis.

Stings and venoms

The algogenic agents present in plant and animal stings and venoms, include substances identical with or very similar to those which occur in body cells or fluids. For example, *nettle stings* contain pain-producing concentrations of acetylcholine, histamine, and 5HT (dock leaves contain a 5HT antagonist), but the prolonged and recurrent burning pain and itch which may last for hours are probably due to a large molecular substance, perhaps a proteolytic enzyme. It must be stated clearly that formic acid does not occur in any stings or venoms except those of certain types of ant. When 10–20 per cent formic acid is applied to intact skin it penetrates to cause pain. The red wood-ant forms a secretion containing 20–70 per cent formic acid which it introduces after biting; this is indeed painful.

The venom of the common wasp contains high concentrations of 5HT, histamine, and a bradykinin-like polypeptide; the venom of the European hornet has a similar composition, but in addition carries the rapier thrust of 5 per cent acetylcholine. Bee venom is rich in histamine but its main pain-producing substances are larger molecules, a basic polypeptide (MW 3000) and an enzyme, phospholipase, which lyses cell membranes. Many painful venoms are polypeptides or proteins without enzymic activity. Such are the toxins in the stings of jellyfish, the venom of the deadly stonefish which arouses the most intense pain, and the venom of the scorpion. Snake venoms are formed in modified salivary glands, and delivered via the poisonous fangs into the victim's body. They vary in composition but the chief noxious agents are enzymes and toxins which cause local destruction, paralyse nervous tissue, damage the heart, and affect blood clotting. Tissue destruction is itself painful, but some snake venoms contain proteases and esterases which release bradykinin from plasma. It was indeed the action of *Bothrops jararaca* venom on plasma which led Rocha e Silva to the discovery of bradykinin.

Thus the stings and venoms of the plant and animal world cause pain by means of small molecular agents such as histamine, 5HT and acetylcholine, proteolytic and other enzymes, or toxins of protein or polypeptide structure. These are nature's weapons of chemical warfare.

NEURAL PATHWAYS CONCERNED WITH PAIN

Pain must ultimately be considered in several terms, anatomical, physiological, psychological, and sociological. When pain is produced by a pin-prick or a hot object it is clear that afferent nerves are stimulated, nerve impulses pass into the central nervous system and in conscious persons pain is experienced. Various attempts have been made to explain the production of pain in anatomical and physiological terms, though how physicochemical activities in the brain are translated into this particular sensory experience is beyond the scope of science. However, it should be possible to define the neural activities which accompany pain as distinct from other sensations.

Receptors for afferent pathways concerned with pain

Pain and reflex withdrawal movements are aroused in man by acute noxious stimulation of the skin. Electrophysiological studies in laboratory animals have shown that noxious stimulation of the skin excites receptors at the endings of afferent A-delta nerve fibres (Group III, conduction rate 10–30 m s⁻¹ and C fibres (Group IV, conduction rate 1.0 m s⁻¹ or less). The unmyelinated afferent C fibres are more numerous than the myelinated A-delta fibres;

however, some C fibres conduct impulses in response to non-noxious thermal and mechanical stimulation.

Brief noxious stimulation of *human skin* of the extremities, e.g. by pin-prick or sudden exposure to temperatures of 55–60 °C, produces a *double pain* sensation, the first (fast) component being attributed to activation of A-delta nerve fibres and the second to activation of slow-conducting C fibres; the two components may be separated by 1.0–1.5 s, according to the site of stimulation. The sensory responses associated with A-delta and C fibre nerve impulses have been studied in human skin by Van Hees and Gybels (1972), and by Torebjörk and Hallin (1974). They inserted tungsten microelectrodes (tip diameter 1–5 µm) percutaneously into fascicles of cutaneous nerves of normal persons and were thus able to record impulses in A-delta fibres and C fibres, in some cases from single fibres of either type. Electrical stimulation (at 0.2–1.0 s⁻¹) of the sensory nerves peripheral to the sites of the recording microelectrodes produced impulses in intact nerve fibres and these were correlated with the sensations experienced by the experimental subject. The main findings were as follows:

Weak electrical stimulation of the sensory nerve produced a sensation of 'slight touch' or 'tapping' associated with a few impulses in A-delta fibres after a short latent period. There were no impulses in C fibres. With increasing strength of electrical stimulation a sensation of stronger tapping or 'throbbing' developed, with more A-delta fibre impulses; when the stimulation was increased sufficiently to produce a sensation of 'pricking', nerve impulses were also produced in C fibres after a longer latent period. Finally, strong electrical stimulation of the nerve produced augmented and prolonged pain, sometimes burning in character, associated with numerous early impulses in A-delta fibres and later impulses in C fibres. (The possibility that efferent postganglionic sympathetic C fibres, which run centrifugally in cutaneous nerves, might contribute to the recorded C fibre impulses was excluded by nerve block central to the recording site.)

The effects of selective block of A and C fibres on sensation and nerve impulses induced by electric stimulation at the periphery have supported the above observations. Pressure on the cutaneous nerve blocked A fibre responses and the sensation of touch; injection of lignocaine (lidocaine) blocked C fibre responses and the prolonged burning pain; A fibre response was reduced but touch and prick were still felt.

Torebjörk (1974) recorded C fibre activity in the peroneal nerve on the dorsum of the human foot, using tungsten microelectrodes and applying mechanical, thermal and chemical stimuli to the appropriate area of skin. Gentle mechanical stimuli such as puffs of air, light touch and bending hairs produced no C fibre activity. Firm stroking of the skin, needleprick, and noxious heat (a glowing match) evoked C fibre impulses; itch powder (cowhage, *Mucuna pruriens*) and nettle leaves (*Urtica urens*) acted similarly and intradermal injection of 5 per cent KCl or topical application of acetic acid to the skin produced severe burning pain and marked C fibre activity, lasting for a few minutes. C fibre nociceptors are therefore called 'polymodal' since they respond to different modes of noxious stimulation.

Visceral nociceptors

Numerous mechanosensitive and chemosensitive receptors have been described in experimental animals. Many are concerned with physiological regulation of the *milieu interne* and their activation produces no conscious experience. However, pain of great intensity can arise from inner organs, accompanied by nocifensor reactions in skeletal muscle and autonomically innervated organs. Excessive distension of hollow viscera, such as the intestine or bladder, activates mechanoreceptors of C fibres but correlation between neural responses and pain cannot be studied in man.

In *skeletal muscle* afferent C fibre activity has been recorded in nerves of muscle contracting under ischaemic conditions. No such activity was recorded with ischaemia alone or when muscle contracted with intact blood flow. In unanaesthetized dogs experimental *occlusion of a coronary artery* produces behavioural manifestations of pain. In dogs, and in human patients with *ischaemic cardiac pain*, section of T1 and T4 dorsal roots, or corresponding sympathetic ganglia, abolishes signs of discomfort and pain. Recording of single units in T2 and T3 rami communicantes showed that coronary occlusion caused discharge of impulses in A-delta and C fibres. The A-delta discharge was correlated with the cardiac cycle, whereas that of C fibres was irregular. The A-delta fibres may be mechanosensitive and the C fibres sensitive to stimulation by chemical factors such as K+, lactic acid, adenine nucleotides, bradykinin and hyperosmolarity [cf. P factor, p. 397].

Electrophysiological studies suggest that nociceptor stimulation by *algogenic substances* is not specific since these substances may also excite large myelinated fibres not concerned with pain. Perhaps the most important effect of algogenic substances is to sensitize nociceptors to mechanical and thermal stimuli. This may occur particularly in *inflammation* in which primary hyperalgesia is so marked [pp. 397 and 403].

Spinal neurons in nociception

In peripheral nerves, nociceptors of A-delta and C fibre types have been clearly demonstrated. *C fibre nociceptors are much the more important in producing pathological pain.* The central processes of these fibres end around cells in the dorsal horn of grey matter of the spinal cord, especially in those of Rexed laminae I, IV, and V [p. 290].

Within the dorsal horn there are two main types of second-order nociceptive neuron:

1. Exclusively nociceptive neurons.
2. Neurons which receive a convergent input from both low-threshold mechanoreceptors (group II fibres) and C-fibre nociceptors.

Type 2 neurons outnumber type 1. They show certain features which correlate well with the pain evoked by noxious stimulation of the skin.

(i) The neurons of type 2 fire at a much higher frequency than those of C fibres stimulated by noxious heat. This suggests convergence of C-fibre input, which also accounts for temporal summation during radiant heat stimulation.

(ii) With heat-evoked activity these type 2 neurons show an after-discharge, lasting 20 s or more, after the stimulus is stopped. This contrasts with the almost immediate cessation of firing in peripheral C-fibre heat nociceptors, but agrees well with the after-sensation felt when noxious heat stimulation ceases.

The nociceptive neurons in the dorsal horn give rise to fibres which mostly cross to course upwards in the contralateral *ascending anterolateral (spinothalamic) tract*. All fibres in the spinothalamic tract are myelinated, and therefore fast-conducting. The nociceptive neurons pursue a polysynaptic pathway with relays in the *brainstem reticular system* on the way to the medial intralaminar and posterior nuclei of the *thalamus* and the *sensory cerebral cortex (S II)*. The pathway also connects with the *limbic lobe* and the *hypothalamus* and thus can evoke emotional responses and reactions involving the autonomic nervous system.

In considering the neuronal pathway concerned with pain one can see that there is specificity in response to noxious stimuli in the first neurons (A-delta and C nociceptor fibres), and in the second neurons, with cell bodies in the dorsal horn and fibres projecting to the brain in the *spinothalamic tract*. Section of this tract usually abolishes pain from the lower parts of the body, at least for 12 months or so. However, from the brainstem upwards the proportion of selectively nociceptive cells gets progressively smaller and the pattern of neuronal activity which underlies the experience of the many varieties of pain is not yet understood.

Neural pathways which suppress pain

Within the central nervous system are pathways which inhibit the activity of the central neurons excited by peripheral noxious stimulation. These inhibitory effects have been studied mainly in the nociceptive neurons in the dorsal horn of grey matter in the spinal cord. Two main anatomical pathways are involved:

1. *Segmental*. Stimulation of nerves in the same segment in which pain is felt can relieve such pain in many cases. Innocuous or noxious stimulation can be effective.
2. *Supraspinal*. A descending inhibitory pathway from the brain stem to Rexed laminae I and V can also promote pain relief. This pathway can be activated by direct electrical stimulation of extrasegmental afferent nerves.

Segmental inhibition is more potent than supraspinal inhibition of pain.

Segmental inhibition may be achieved by activation of group A afferent nerve fibres, as when a mechanical vibrator relieves pain of post-herpetic neuralgia. Noxious electrical stimulation of segmental afferents, e.g. *transcutaneous electrical stimulation* of cutaneous afferent nerves, may produce dramatic pain relief which sometimes persists long after nerve stimulation ceases. *Stimulation of the dorsal column of the spinal cord* to activate segmental collaterals also relieves pain.

Mesencephalic pain inhibitory system

A most important neural system for inhibition of pain is that which arises from the *mesencephalic grey matter* and descends to the dorsal horn cells in the spinal cord. This system can be activated by electrical stimulation to produce marked analgesia; it contains 'opiate receptors' to which morphine becomes bound before it produces analgesia. (Opiate receptors are also found in the dorsal horn of grey matter in the spinal cord.)

Anatomy. The system is situated in the following regions in laboratory animals (rat, cat, monkey) and in man: structures immediately surrounding the third ventricle; periaqueductal grey matter (PAG); substantia nigra; nucleus raphe magnus (dorsal raphe nucleus) in the medulla.

From the dorsal raphe nucleus arises a descending pathway, which is located in the *dorsolateral funiculus* of the spinal cord and terminates in the dorsal horn of grey matter, probably around nociceptive cells in Rexed laminae IV and V.

The periventricular and periaqueductal neurons may form synapses with the cells of the dorsal raphe nucleus. The central nervous pain inhibitory system receives connections from many other regions of the brain which can influence the development of analgesia.

Physiology. *Electrical stimulation* of the periventricular–periaqueductal–dorsal raphe nucleus system produces a very potent analgesia, and inhibition of withdrawal spinal reflexes evoked by noxious stimulation with electric shock, pin-prick, heat (temperatures > 50 °C), and chemical irritants.

These effects are specific and other sensory modalities, such as responsiveness to tactile and proprioceptive stimuli, are not affected even when analgesia is complete.

In experimental animals the analgesia often persists beyond the period of electrical stimulation. In patients with chronic pain analgesia may last for 20 hours after stimulation has ceased.

Stimulation of this system selectively inhibits the responses in second-order nociceptive neurons in the dorsal horn of grey matter of the spinal cord. The analgesia is abolished by section of the dorsolateral funiculus.

Morphine appears to activate the same system. Microinjection of morphine into periaqueductal grey matter inhibits spinal cord nociceptive reflexes, but not after section of the descending dorsolateral funiculus. After systemic administration, morphine increases neuronal activity in the dorsal raphe nucleus. It is not surprising that the specific morphine antagonist *naloxone* prevents the analgesia produced by stimulation of the mesencephalic pain inhibitory system, as well as that produced by morphine.

Endogenous peptides and pain

Endogenous peptides in the nervous system may be either algogenic or analgesic. *Substance P* is a peptide containing 11 amino acid residues which can be demonstrated in primary afferent nerve fibres by immunohistochemical techniques. The substance P content of such fibres, which terminate in the substantia gelatinosa of the dorsal horn of grey matter in the spinal cord, disappears after dorsal root section. Release of substance P from the dorsal root afferents is brought about by noxious stimulation of afferent A-delta and C fibres in peripheral nerves. Now the terminals of substance-P-containing afferents possess opiate receptors on the membrane surface and morphine analgesia may be partly due to receptor binding of morphine which reduces the release of substance P in the dorsal horn of grey matter. Opioid peptides, considered below, act similarly. Thus, in addition to its possible actions in other parts of the central nervous system and in the nerve plexuses of the alimentary tract, substance P might be a mediator of pain. (The name 'substance P', therefore, may be more appropriate than was realized when Gaddum and Schild first used it in 1934. Substance P must not be confused with Lewis' P factor where P meant Pain induced by ischaemic muscular contractions [p. 397].)

Enkephalins and endorphins

When opiate receptors were identified in various parts of the central nervous system by demonstration of binding sites of tritiated exogenous ligands such as dihydromorphine, it was natural to search for endogenous ligands for the same receptors. In 1975 Hughes, Kosterlitz and their colleagues isolated and later identified and synthesized two pentapeptides, which they called Met-enkephalin and Leu-enkephalin respectively. These peptides were found in the nervous system in the same regions where opiate receptors had been located and it was shown that their actions were virtually identical with those of morphine, and that naloxone was a specific antagonist of the peptides as well as of morphine (Hughes and Kosterlitz 1977). Later came the discovery that higher molecular weight peptides, e.g. β-endorphin, also acted like the enkephalins by combining with opiate receptors. However, whilst the enkephalins are unstable in brain tissue, having a half-life of seconds or a most one minute due to breakdown by peptidases, β-endorphin is much more stable. Incidentally, β-endorphin is one of several peptides which can be split off from a pituitary hormone called β-lipotropin; β-endorphin contains 30 amino acid residues and one end of its chain consists of the amino acid sequence in the pentapeptide, met-enkephalin. The exact mode of synthesis and release of enkephalins and β-endorphin in the brain is not yet known but it is probably justified to speak of enkephalinergic and endorphinergic neurones, and their most important locations and modes of action will now be considered.

As far as pain relief is concerned the opioid peptides appear to act in two main regions of the central nervous system, enkephalins being more abundantly distributed than endorphins (which are confined to the hypothalamus). Enkephalin-containing nerve terminals have been identified by an indirect fluorescence technique in the dorsal horn of the spinal cord, the spinal trigeminal nucleus, the periaqueductal grey matter, and the raphe nuclei [p. 400]. Electrical stimulation of periaqueductal (or periventricular) grey matter or raphe nuclei in experimental animals, and in a few human patients with chronic pain, relieves pain at least partly by activating a 5HT (serotonin)-containing nervous pathway which courses caudally in the dorsolateral funiculus of the spinal cord to inhibit nociceptive neurons in Rexed laminae I, IV, and V. This descending inhibitory pathway is activated by morphine, and the specific morphine antagonist naloxone blocks the actions of both morphine and the enkephalins released by electrical stimulation of the above mentioned sites. Since this pathway is serotoninergic depletion of 5HT reduces the supraspinal actions of morphine.

Morphine also stimulates an area in the vicinity of the lateral reticular nucleus in the brain stem extending up through the substantia nigra which activates the raphe nucleus via a dopaminergic pathway [Fig. VI.41].

In addition to this *supraspinal* pain-inhibitory pathway there is also a *spinal* level of inhibition by morphine and endogenous opioid substances. The terminals of afferent nociceptive substance-P-

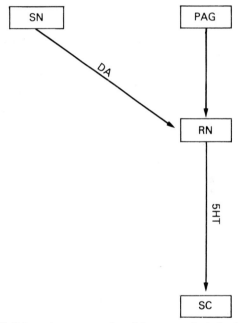

Fig. VI.40. Schematic representation of the proposed relationships of three brain sites at which electrical stimulation (or morphine microinjection) produces analgesia. SN = substantia nigra: PAG = periaqueductal grey; RN = raphe nucleus; SC = spinal cord. DA indicates that the SN–RN pathway is dopaminergic. (After Barnes, C. D., Fung, S. J., and Adams, W. L. (1979). *Pain* **6**, 207.)

containing nerve fibres in the substantia gelatinosa possess opiate receptors and there is evidence for the existence of enkephalinergic interneurons in this region. Morphine and opioid peptides may act *segmentally* to inhibit release of substance P.

The question arises as to whether or how this opioid peptide system is activated, especially in man. If enkephalin is continuously released one would expect that the administration of naloxone would either produce pain or reduce the threshold for pain aroused by noxious stimulation. Most investigators have reported that in normal persons naloxone does not affect pain threshold but in persons with a naturally high pain threshold naloxone may enhance the pain produced by noxious stimulation. In patients with chronic pain naloxone does not make the pain worse. This suggests that either the endogenous opioids have no influence on chronic pain or that these substances have been used up at a rate exceeding that of their synthesis and release. A reduction in the concentration of enkephalin-like peptides in the cerebrospinal fluid of patients with chronic pain has been reported.

In human beings many experimental studies cannot be carried out for obvious ethical reasons. However, electrical stimulation of periventricular grey matter has relieved chronic pain in a few patients, and it is suggested that pain relief after acupuncture, transcutaneous electrical stimulation, or counter irritation may also be due to activation of the descending inhibitory pathway from the dorsal raphe nucleus.

Opioid peptides (enkephalins, β-endorphin) which occur naturally in certain parts of the brain mimic very closely all the actions of morphine. They may act beneficially to relieve pain, but they also produce the undesirable effects of morphine.

It is important to remember that there is evidence for pain-inhibitory nerve pathways quite independent of that activated by opioid peptides.

Acupuncture has been used in China for many thousand years for the relief of pain. Recently, its efficacy has been generally confirmed and its mode of action widely studied. Nowadays, sharp, easily sterilizable needles are inserted into the skin usually at points selected according to traditional Chinese practice. The needle is rotated manually or an electric current is passed through it, so as to produce some local pain; this produces relief of pain in deeper structures especially those of the same spinal cord segment. To relieve pain acupuncture must excite A-delta fibres at a frequency not exceeding 3 Hz. Similar results are obtained by *transcutaneous* electrical current, which relieves chronic pain and raises pain threshold to various types of noxious stimuli. Because the effect of acupuncture can be antagonized by naloxone it has been suggested that the procedure acts by stimulating the dorsal raphe nucleus and hence the inhibitory pathway to the nociceptive cells in the dorsal horn of grey matter in the spinal cord (Mayer and Price 1976).

Hypnotic analgesia and *analgesia accompanying stress* are not antagonized by naloxone. It is therefore assumed that pain suppression in these cases is different from that produced by acupuncture.

Theories about pain

Physiologists would like to know what types of neural activity underlie the various forms of pain. No final answers can be given but certain basic concepts deserve consideration.

1. The intensity hypothesis. This proposes that excessive stimulation of any sensory receptors, e.g. tactile receptors, would cause pain. Electrophysiological recording of afferent impulses from low-threshold touch receptors has shown that maximal stimulation never produces pain or other manifestations of nociceptor excitation. However, pungent spices which have a pleasing taste in small concentrations can be really painful in high concentrations.

2. Specificity hypothesis. This proposes that there is a distinct, specific nociceptive nerve pathway from sensory endings in skin, muscle, viscera, etc. through the spinal cord, brain stem, and thalamus to the sensory cortex. Although the first two neurons in the series, starting at the periphery, can be shown by electrophysiological recording to respond to noxious stimulation, the evidence for a specific nerve pathway above the spinal cord is less clear (Zimmermann 1976).

3. Pattern hypothesis. Pain must be correlated with some patterns of activity in the neurons of the central nervous system, but the exact patterns are not yet known. Sensory nerve fibres of the A group (including A-delta fibres) will help to localize the site of stimulation, while excitation of C-fibre nociceptors produces pain with its typical unpleasant character and diffuse distribution.

Gate-control hypothesis (Melzack and Wall 1965). This proposes that large myelinated afferent A nerve fibres interact with small unmyelinated C fibres via inhibitory cells of the substantia gelatinosa (SG) in the dorsal horn of grey matter in the spinal cord. Stimulation of the large fibres enhances SG cell activity and blocks transmission of impulses to the nerve cells concerned with pain. C-fibre stimulation inhibits SG cells and thus promotes the passage of impulses along the pain-producing pathway in the spinal cord. A supraspinal descending inhibitory pathway is also recognized. The details of the proposed interactions between large and small fibres are no longer acceptable but the gate-control hypothesis has provoked much valuable experimental work which shows that stimulation of large myelinated nerve fibres can indeed inhibit the responses to C-fibre nociceptor activation, especially when the interactions occur in the same segment of the spinal cord.

These theories or hypotheses deal mainly with the first two neurons in the 'pain' pathway. They are incomplete and must be extended (as is done in a prophetic way in the gate-control theory) to take full account of the inhibitory influence of the periventricular-periaqueductal-dorsal raphe nucleus-descending dorsolateral funicular system. Indeed, for pain to be experienced, and reflex responses aroused, the activity of this endogenous system must either be suppressed or bypassed. It is not yet known how this comes about but it is well established that C fibres have a high threshold for noxious stimulation. Excitation of other nerve fibres might facilitate passage of C fibres impulses, or summation of C fibre activity (spatial or temporal) might force a passage to higher levels of the central nervous system.

LOCALIZATION OF PAIN

Localization of the source of pain is one of the most important aids to diagnosis of disease. Pain is accurately localized in the skin, but accuracy is lost as the source of the pain sinks deeper into the body. In infants it would seem, from the lack of local sign in such conditions as otitis media, that there is no localization of pain.

It may be said that pain is primarily localized to the segment corresponding to the stimulated nerves, and that accuracy of localization is superimposed on this segmental pattern. In the skin, accuracy of localization of pain appears to be based on the richness of its innervation, the multiple innervation of pain spots, and on the associated nervous mechanism for accurate localization of touch. In deeper structures, innervation is much sparser and there is no other sensory mode to aid in localization of pain. In these circumstances pain may be felt in any part of the affected segment.

True visceral and deep somatic pain is sometimes felt at the site of primary stimulation and may or may not be associated with pain referred to some other part of the segment. In such cases pain is relieved by injection of local anaesthetic into the site of noxious stimulation, but not by injection into other structures of the segment.

Referred pain from deep structures is that pain which occurs in addition to or in the absence of true visceral or deep somatic pain. It is felt at a site other than that of stimulation, in deep or superficial structures supplied by the same or adjacent neural segments. The pain is most frequently referred to other parts of the same segment, but it may be spread to adjacent segments higher in the spinal cord, then to lower adjacent segments and finally to corresponding segments on the opposite side of the cord. Pain is more commonly referred to the anterior than to the posterior half of the body, e.g. in peptic ulcer, angina pectoris, and ureteric colic. This is attributed to the fact that we are normally more conscious of the front than of the back half of our body. The pain of diaphragmatic pleurisy is often felt in the shoulder tip. This is due to the fact that both regions are innervated by cervical segments 3–5.

Localization of pain is believed to be a function of the cerebral cortex. Reference of pain is due to spread of excitation from the stimulated neurons of the same and adjacent segments. This spread takes place within the central nervous system and might occur at any level from the spinal cord up to the cortex. In the spinal cord there are neurons which are excited by both cutaneous and visceral A-delta fibres (Zimmermann 1976).

Hyperalgesia

The word 'hyperalgesia' is used to describe a state in which originally non-noxious stimuli produce pain, or in which noxious stimuli produce more pain than they normally do. Hyperalgesia may occur as a consequence of tissue damage or as a result of nerve lesions. Hardy, Wolff, and Goodell (1952) distinguished two main types of hyperalgesia associated with tissue damage.

Primary hyperalgesia

This type of hyperalgesia occurs within an area of tissue damage due, for example, to a burn, or bacterial infection. Pain threshold is lowered so that a non-noxious stimulus becomes painful. Primary hyperalgesia is characteristic of inflammatory lesions, e.g. the marked tenderness associated with a boil. As healing proceeds the hyperalgesia gradually disappears.

The neurophysiological basis of primary hyperalgesia has been studied in conscious man using tungsten microelectrodes to record single-unit activity from C nociceptors in intact cutaneous nerves (Torebjörk and Hallin 1978). Parts of the receptive fields of single C units were heated to 50 °C for 1–2 min. In these areas there was initial hypoalgesia and reduced discharge of C nociceptor units in response to mechanical and thermal stimuli; after 5–10 min hyperalgesia developed in the damaged areas so that the mechanical and thermal stimuli were more painful than in normal skin and C nociceptor fibre impulses were correspondingly increased in frequency. The hyperalgesia which lasted for up to two hours was associated with a reduced threshold for noxious mechanical and thermal stimulation. The sensitization was attributed to release of algogenic substances from damaged skin [cf. p. 397].

Secondary hyperalgesia

This is associated with tissue damage but occurs in undamaged tissue adjacent to, and sometimes extending some distance from, the site of injury. There is no lowering of pain threshold but with a given noxious stimulus the pain is felt more intensely in an area of secondary hyperalgesia than in normal skin.

Mechanisms producing hyperalgesia

Primary hyperalgesia is produced by local chemical changes in the damaged tissues. These changes may be sufficient to induce pain at rest, especially when the circulation to the affected part is occluded. Both hyperalgesia and pain have been attributed to substances such as histamine, 5-hydroxytryptamine, plasma kinins, and PGs.

Secondary hyperalgesia is probably due to spread of excitation within the central nervous system. It may be associated with spasm of skeletal muscle, local vasomotor changes, and secretion of glands controlled by the affected neural segments. The convergence of cutaneous and visceral afferents on to common dorsal horn cells means that subliminal activation by one type of afferent lowers the threshold to stimulation of the other.

Referred pain may or may not be accompanied by secondary hyperalgesia. If there is such hyperalgesia, injection of procaine into the superficial or deep hyperalgesic structures will reduce the amount of pain. When cutaneous hyperalgesia is marked, pain originating from deep structures may be greatly relieved by spraying ethyl chloride (which acts by cooling) on the affected area of skin (Wolff and Wolf 1958).

In some cases the spread of excitation in the central nervous system produces painful contractions of skeletal muscle, which may be widespread and remote from the original site of noxious stimulation. Injection of procaine into the affected muscles abolishes this type of pain.

Hyperalgesia due to nerve lesions. Selective damage to large sensory fibres allows light tactile stimuli to produce severe pain even in the absence of peripheral tissue damage.

REFERENCES

BEECHER, H. K. (1956). *Pharmac. Rev.* **9**, 59.
British Medical Journal (1981). How does acupuncture work? *Br. med. J.* **283**, 746–8.
FAIRES, J. S. and McCARTY, D. J. (1962). *Lancet* **ii**, 683.
GARROD, A. B. (1876). *A treatise on gout and rheumatic gout (rheumatoid arthritis)* 3rd edn. Longman Green, London.
HARDY, J. D., WOLFF, H. G., and GOODELL, H. (1952). *Pain sensations and reactions*. Baillière, Tindall and Cox, London.
HEES, J. VAN and GYBELS, J. M. (1972). *Brain Res.* **48**, 397.
HILGARD, E. R. (1975). *Pain* **1**, 213.
KEELE, C. A. and ARMSTRONG, D. (1964). *Substances producing pain and itch*. Arnold, London.
—— and SMITH, R. (1962). *UFAW Symp. Assessment of Pain in Man and Animals*. London.
KEELE, K. D. (1957). *Anatomies of pain*. Blackwell, Oxford.
—— (1968). *Br. med. J.* **i**, 670.
MAYER, D. J. and PRICE, D. D. (1976). *Pain* **2**, 379.
MELZACK, R. and WALL, P. D. (1965). *Science, N.Y.* **150**, 971.
MERSKEY, H. and SPEAR, F. G. (1967). *Pain. Psychological and psychiatric aspects*. London.
NATHAN, P. W. (1977). *Br. med. Bull.* **33**, 149.
NOORDENBOS, W. (1959). *Pain*. Elsevier, Amsterdam.
PAYNE, J. P. and BURT, R. A. P. (eds.) (1972). *Pain*. Churchill Livingstone, London.
SINCLAIR, D. (1981). *Mechanisms of cutaneous sensation*. Oxford University Press.
TERENIUS, L. (1978). *A. Rev. Pharmac. Toxicol.* **18**, 189.
TOREBJÖRK, H. E. (1974). *Acta physiol. scand.* **92**, 374.
—— and HALLIN, R. G. (1974). *Brain Res.* **67**, 387.
—— —— (1978). In *Abstracts 2nd Wld Cong. Pain*, p. 236.
WOLFF, H. G. (1963). *Headache and other head pain*, 2nd edn. Oxford University Press, New York.
ZIMMERMANN, M. (1976). In *International review of physiology and neurophysiology II*, Vol. 10 (ed. R. Porter) p. 180. University Park Press, Baltimore.
See the journal *Pain* published by Elsevier for numerous articles and references on the subject.

ITCH (PRURITUS)

Itch is a characteristic sensation distinct from other sensory modalities (Rothman 1954; Shelley and Arthur 1957; Keele 1957). Pain and itch have much in common:

1. Both originate from stimulation of superficial free nerve endings in the skin. Itching is produced by excitation of endings immediately beneath the epidermis and perhaps from intraepidermal endings as well. Itching does not occur after the epidermis has been completely removed. Pain is elicited also from deeper nerve endings in the skin and of course from many structures other than skin.

2. The nerve pathway (C fibres) mediating itch accompanies fibres subserving pain in the ascending anterolateral tracts in the spinal cord. Cordotomy abolishes itch as well as pain.

3. Low-intensity heat, or electrical stimulation of the skin can produce itching: stronger stimulation produces pain. This finding has suggested that itch is subthreshold pain.

4. Around an area of spontaneous itching, e.g. due to gnat bite, there develops a much larger area of itchy skin in which itching is very readily set up by light friction. The itchy skin corresponds to secondary hyperalgesia and is likewise attributed to activity in the central nervous system.

Itch differs from pain in many ways:

(i) Both itch and pain have their own peculiar qualities. Each has a wide range of intensity from the just perceptible to the completely intolerable.

(ii) Itch provokes scratching, superficial pain leads to evasive movements. The pain produced by scratching suppresses itch.

(iii) Immersion of skin in water at 40–41 °C alleviates itching, but aggravates burning pain (Lewis).

(iv) Morphine relieves pain, but can make itching worse. It may be concluded that itch is not subthreshold pain.

Experimental itch

Itch powder (Cowhage, *Mucuna pruriens*). Itch powder consists of very fine sharp-pointed spicules or trichomes which cover the seed pods of the leguminous plant *Mucuna pruriens*. When rubbed into the skin the powder produces marked itch and sometimes pain. There is often erythema, weal, and flare, but itch can occur in their absence. The itching is not mechanical in origin since boiled spicules are inactive. The pruritogenic agents in itch powder are:

(i) A histamine liberator (Broadbent 1953).
(ii) Mucunain, a proteolytic enzyme (Shelley and Arthur 1955).

Shelley and Arthur (1957) have found that in addition to mucunain, many other proteinases, e.g. papain, trypsin, and plasmin produce itching on intradermal or intra-epidermal administration. The action occurs within a few seconds so it cannot be due to histamine liberation (histamine produces itching only after a delay of 50–60 seconds). It is suggested that the itching of skin diseases, e.g. atopic dermatitis, lichen planus, bacterial or fungal infection, is mainly due to release from cells of the epidermis, bacteria, or fungi of endopeptidases which act directly on the most superficial nerve endings in the skin.

It seems likely that in urticaria itching is due to the histamine released by the antigen–antibody reaction in the skin. In dermographia (whealing in response to firm pressure on the skin, see p. 142) histamine release occurs too deep in the dermis to stimulate the superficial itch nerve endings (Broadbent 1953).

Central itching

Administration of certain drugs, e.g. morphine, eserine, and di*iso*propylfluorophosphonate by intraventricular or intracisternal injection into cats produces marked scratching (Feldberg and Sherwood 1954). This suggests that itching may sometimes be of central nervous origin. The generalized pruritus associated with obstructive jaundice, Hodgkin's disease, chronic nephritis, and diabetes, in which there may be no primary pathological change in the skin, might arise from such central stimulation.

Tickle

Tickle may be defined as itch produced by a light external moving stimulus. Tickle resembles itch in being conveyed by the spinothalamic tract, in provoking scratching, and in being inhibited by pain.

REFERENCES

BROADBENT, J. L. (1953). *Br. J. Pharmac.* **8**, 263.
FELDBERG, W and SHERWOOD, S. L. (1954). *J. Physiol., Lond.* **123**, 148.
HARDY, J. D., WOLFF, H. G., and GOODELL, H. (1952). *Pain sensations and reactions*, p. 216. Baillère, Tindall and Cox, London.
KEELE, C. A. (1957). *Proc. R. Soc. Med.* **50**, 477.
—— (1970). *A. Rev. Med.* **21**, 67.
—— and ARMSTRONG, D. (1964). *Substances producing pain and itch.* Arnold, London.
ROTHMAN, S. (1954). *Physiology and biochemistry of the skin.* University of Chicago Press.
SHELLEY, W. B. and ARTHUR, R. P. (1955). *Archs Derm Syph., Chicago* **72**, 399.
—— —— (1957). *Archs Derm. Syph., Chicago* **76**, 296.
ZOTTERMAN, Y. (1972). In *Pain* (ed. J. P. Payne and R. A. P. Burt). Churchill Livingstone, London.

PART VII

The autonomic nervous system

Chemical transmission

Introduction

In this section the structure and functions of the autonomic nervous system will be discussed. The physiology of smooth muscle is included. Neurochemical transmission was first discovered in relation to the autonomic nervous system and this process has since been shown to occur at somatic motor nerve—skeletal muscle junctions and at synaptic sites in the central nervous system. The subject of neurochemical transmission in general is dealt with in this section, but there are many references to this topic in other parts of the text [pp. 257, 283, 360, and 499].

The autonomic nervous system was originally described by Langley as an *efferent* system of nerves which controls the activities of the heart, blood vessels, glands, and smooth muscles throughout the body. The word 'autonomic' was introduced for two reasons:

1. Many of the viscera innervated by this system, e.g. heart, intestine, and uterus show spontaneous rhythmic activity when suspended in a suitable medium after removal from the body. In the isolated intestine reflex responses to distension are mediated by the independent local myenteric nerve plexus.

2. The efferent neurons which control the viscera have their cell bodies in autonomic ganglia situated outside the central nervous system.

However, *in vivo*, the activities of the autonomic nervous system must be co-ordinated with those of the somatic nervous system and there are reflex pathways involving all levels in the central nervous system which ensure that visceral activities harmonize with those of skeletal muscle.

The autonomic nervous system has also been called the vegetative, visceral, or involuntary nervous system. The word 'vegetative' is of historical interest only and the word 'involuntary', which accurately describes the lack of voluntary control over most visceral functions, is becoming less true as evidence accumulates that blood pressure and heart rate can be influenced directly by persons skilled in the arts of yoga, meditation, and relaxation. The best name would be the *efferent visceral nervous system*, which arises from, and is controlled by, the central nervous system.

Visceral efferent nerves differ in one very important respect from somatic efferent nerves to skeletal muscle. Many visceral nerves, e.g. vagus supply to the heart and sympathetic nerves to the gut, are directly *inhibitory* to the innervated organs, whereas somatic nerves to skeletal muscle are solely *excitatory* and are therefore called motor nerves. Visceral nerves are either excitatory or inhibitory.

Visceral afferent nerves. Like the somatic nervous system the autonomic nervous system receives information via afferent nerves from both somatic and visceral structures. Visceral afferents have cell bodies in dorsal root ganglia or the comparable structures in some cranial nerves; their peripheral processes run mainly in the vagus, splanchnic, and pelvic nerves; their central processes enter the spinal cord or brain stem to relay with neurons which run to higher levels of the central nervous system and produce responses in somatic and visceral structures. Because the peripheral processes of visceral afferent nerves run with efferent visceral nerves, for example, the vagus contains more afferent than efferent nerve fibres, visceral afferents have been called 'autonomic afferents'. This is wrong because visceral activities may be influenced by impulses conveyed along both somatic and visceral afferent nerves (Newman 1974). It must again be emphasized that the autonomic nervous system comprises the neurons which form the efferent visceral nerves of the sympathetic and parasympathetic systems.

GENERAL ARRANGEMENT

The study of the autonomic nervous system is considerably simplified if attention is first devoted to the broad principles on which the system is constructed. In a somatic spinal reflex arc three neurons may be involved [see FIG. VII.1]:

1. The afferent or receptor neuron (A) with its cell body in the dorsal root ganglion.

2. An internuncial cell in the dorsal horn of grey matter which by means of its axon transmits the impulse to the ventral horn; it might be called the connector neuron (B).

3. The ventral horn cell and its axon—the excitor neuron (C) which transmits the efferent impulses to skeletal muscle [cf. FIG. VII.1].

In the nerve supply of the viscera three neurons can also be recognized [FIG. VII.1]:

(i) There is an *afferent* proceeding from an internal organ. The nutrient cell lies in the dorsal root ganglion (or its cranial equivalent), and a central process is sent into the grey matter.

(ii) The *connector cell* is situated not in the dorsal horn but in the adjacent grey matter, the exact position differing in the various regions of the nervous system. In the thoracic region, for example, the connector cells lie in the lateral (intermediolateral) horn of grey matter. The connector fibre must, of course, connect the afferent neuron with the excitor neuron which actually supplies a viscus. But the excitor cells are not found within the central nervous system; they have migrated outwards to form masses of cells situated peripherally. The connector fibres histologically are mostly medullated B fibres (white rami); they are called *preganglionic fibres*, as they pass to a peripheral ganglion.

(iii) The excitor cells lie peripherally, either as ganglia or as isolated groups of cells. They are best termed *ganglion cells*. Each preganglionic fibre branches to supply several ganglion cells and

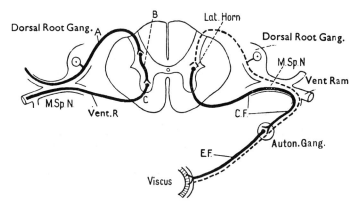

FIG. VII.1. General arrangement of autonomic nervous system (on right) contrasted with that of somatic nervous system (on left). A, B, C = Afferent, connector, and excitor neurons of somatic nervous system; Vent.R. = Ventral nerve root; M.Sp.N. = Mixed spinal nerve; Dotted line = Afferent visceral fibre with nutrient cell in dorsal root ganglion; Lat.Horn = Lateral horn of grey matter which gives rise to C.F. = Connector (preganglionic) fibre which passes in ventral root, mixed spinal nerve, and ventral ramus (Vent.Ram.) to end in Auton. Gang. = Autonomic ganglion; E.F. = Excitor (postganglionic) fibre ending in viscus; Dorsal Root Gang. = Dorsal root ganglion. (After Ranson.)

each ganglion cell is supplied by branches from several preganglionic fibres. From the ganglion cells fibres arise which, by devious routes, reach the various organs which they innervate. The fibres are mostly non-medullated C fibres (grey rami). They have been called excitor fibres but since some fibres are functionally inhibitors they are best described non-committally as *postganglionic fibres.*

Connector cells are not present uniformly in all the segments of the nervous system, but in certain regions only, namely [FIG. VII.2]:

(a) In the brain, in connection with the nuclei of certain cranial nerves: i.e. III, VII, IX, and X.

(b) In the whole of the thoracic region, and in the first two lumbar segments of the spinal cord.

(c) In the second and third sacral segments of the spinal cord.

Connector fibres therefore leave the central nervous system from these regions only, and constitute three great systems of outflowing fibres, which are termed the cranial (or bulbar), the thoracicolumbar, and sacral outflows respectively.

The autonomic nervous system may be usefully subdivided on the basis of the anatomical situation of the connector cells and fibres. Thus the connector cells in the brain, their axons in the cranial nerves specified, and their related ganglionic neurons constitute the cranial division of the system or the *cranial autonomic.* The connector cells in the thoracicolumbar region, their axons in the corresponding ventral roots and in their subsequent ramifications, and their ganglionic neurons constitute the *sympathetic nervous system.* The sacral connector cells, their axons in the sacral ventral roots, and all their related ganglionic neurons constitute the *sacral autonomic* [FIG. VII.2].

We can therefore make this classification:

AUTONOMIC NERVOUS SYSTEM—
(i) Cranial autonomic.
(ii) Sympathetic.
(iii) Sacral autonomic.

Another classification can be made from a functional standpoint. The cranial and sacral divisions have similar physiological actions and form the *parasympathetic system.* When stimulated, the sympathetic and parasympathetic nerves produce antagonistic effects on the organs which they both supply (e.g. heart, pupil). Under natural conditions, however, the two systems act synergistically.

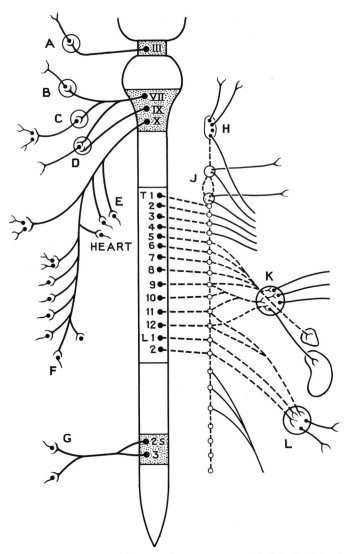

FIG. VII.2. General plan of autonomic nervous system. On *left*: Cranial and sacral autonomic (parasympathetic) system. Thick lines from III, VII, IX, X, and S2, 3 are preganglionic (connector) fibres. A, ciliary ganglion; B, sphenopalatine ganglion; C, submaxillary and sublingual ganglia. D, otic ganglion; E, vagus ganglion cells in nodes of heart; F, vagus ganglion cells in wall of bowel; G. sacral autonomic ganglion cells in pelvis; thin lines beyond = postganglionic (excitor) fibres to organs. On *right*: Sympathetic nervous system. Dotted lines from T1–12; L1, 2 are preganglionic fibres; H, superior cervical ganglion; J, middle and inferior cervical ganglia (the latter fused with the 1st thoracic ganglion to form the stellate ganglion); K, coeliac and other abdominal ganglia (note preganglionic fibres directly supplying the adrenal medulla); L, lower abdominal and pelvic sympathetic ganglia; continuous lines beyond = postganglionic fibres.

Some organs are, however, innervated by one division only (e.g. uterus, adrenal medulla, and most arterioles from the sympathetic only; glands of stomach and pancreas from the parasympathetic only). The ganglion cells of the sympathetic are situated, as a rule, at a distance from the organ innervated, though the pelvic organs of the urogenital tract have sympathetic ganglia close to or within their substance. Thus there are 'long' adrenergic neurons consisting of postganglionic fibres running from the sympathetic chain to the heart, blood vessels, glands, etc., and 'short' adrenergic neurons which supply the uterus, bladder, and vas deferens. Surgical removal of the sympathetic chain is therefore not a complete sympathectomy; the ganglia of the parasympathetic are usually in close proximity to the organs (the sphenopalatine and the otic ganglion are, however, obvious exceptions to the latter statement).

Cholinergic and adrenergic neurons

All preganglionic neurons (parasympathetic and sympathetic) are *cholinerg c*, as also are postganglionic fibres of the parasympathetic. Some anatomically sympathetic fibres, e.g. to sweat glands, are cholinergic. Postganglionic sympathetic neurons are generally *adrenergic*.

SYMPATHETIC NERVOUS SYSTEM

In describing the anatomical details of the autonomic nervous system it is necessary to note the situation of the connector cells, the path taken by the preganglionic fibres, the position of the ganglion cells, and the path of their fibres.

The connector cells of the sympathetic lie in the lateral horn of grey matter in the thoracic region and in the corresponding grey matter in the first and second lumbar segments; the preganglionic fibres pass out in the ventral root corresponding to the segment from which they arise and then enter the mixed spinal nerve. The latter divides into a small dorsal and a large ventral ramus; the preganglionic fibre is continued in the ventral ramus, but soon leaves it to form a branch which passes to the lateral sympathetic chain. This branch is the white ramus communicans which passes to the ganglia of the sympathetic; the white ramus is merely a portion of the preganglionic fibre. The preganglionic fibre may end in a ganglion of the lateral sympathetic chain, or pass on to more distantly situated ganglia like those in the neck or abdomen. The postganglionic fibres which arise from these ganglia take various routes to the periphery. The salient anatomical features of the sympathetic nervous system are indicated in TABLE VII.1 and those of the parasympathetic in TABLE VII.2

TABLE VII.1 *Sympathetic*

Organs supplied	Site of connector cells	Site of ganglion cells	Route of postganglionic fibres
Head and neck	T1, 2		
(1) Eye	T1, 2	Superior cervical ganglion	Along internal carotid artery
(2) Face	T1, 2		Along external carotid artery
(3) Skin of head and neck	T1, 2		With cervical plexus
(4) Cerebral vessels	T1, 2	Superior and inferior cervical ganglia	Along internal carotid and vertebral arteries
Thoracic viscera	T3, 4	Superior, middle, and inferior cervical ganglia (man) Stellate ganglion (animals)	Cardiac branches of sympathetic
Forelimb	T5–9 (sometimes also T2–4)	Middle and inferior cervical, first and second thoracic ganglia (man) Stellate ganglion (animals)	With brachial plexus
Hindlimb	T10–L2	Lumbar and sacral ganglia	With lumbosacral plexus
Abdomen	T6–L2		
(i) Viscera of abdomen proper	T6–12 (chiefly)	Upper abdominal ganglia (superior mesenteric, coeliac, etc.)	Along blood vessels
(ii) Pelvic viscera	L1, 2 (chiefly)	Inferior mesenteric ganglia (animals) Hypogastric ganglia (man)	Along blood vessels and in hypogastric nerves
Thoracic and abdominal parietes	T1–12	Ganglia of lateral sympathetic chain	With intercostal nerves

TABLE VII.2 *Parasympathetic*

Cranial nerve	Site of connector cells	Site of ganglion cells	Structures supplied
III	Cranial part of IIIrd nerve nucleus	Ciliary ganglion	Sphincter pupillae Ciliary muscle
VII	Dorsal nucleus of VIIth nerve (superior salivary nucleus)	Sphenopalatine ganglion In salivary glands	Lacrimal gland Submaxillary and sublingual glands
IX	Dorsal nucleus of IXth nerve (inferior salivary nucleus)	Otic ganglion	Parotid gland
X	Nucleus ambiguus of Xth nerve (vagus)	*Heart.* Sino-atrial and atrioventricular nodes	Atrial and junctional tissue
	Dorsal nucleus of Xth nerve (vagus)	*Bronchi.* Local	Smooth muscle Mucous glands
		Alimentary canal Myenteric (Auerbach's) plexus	Gastric and intestinal glands
		Submucous (Meissner's) plexus	Smooth muscle Pancreas: exocrine and endocrine cells
Sacral	Segments 2 and 3 of sacral cord *Nervi erigentes*	Hypogastric ganglia	Most of large intestine Bladder Prostate Blood vessels of penis Uterus

The effects of stimulation of autonomic adrenergic and cholinergic nerves are summarized in TABLE VII.3, and many of these effects are discussed in more detail in other sections of the text.

Central autonomic control

Smooth muscles, the heart, and secretory glands (exocrine, and some endocrines, such as the pancreatic islets and adrenal medulla) come under the immediate control of sympathetic and parasympathetic nerves. Autonomic ganglia are relay stations without independent activity, but the connector cells within the *spinal cord* show tonic activity even after the cord is severed from the brain. Autonomic reflexes demonstrable in spinal animals (including man) are those concerned with maintenance of vasoconstrictor tone (which keeps up a stable though low arterial blood pressure), sweating, and reflex emptying of the urinary bladder, rectum, and seminal vesicles.

Above the spinal cord there are many levels at which autonomic nerve activity is controlled, but it is important to note that there is considerable overlap between autonomic and somatic centres of integration. In the *medulla oblongata* and *pons* there are 'centres' which are concerned with reflex control of respiration and blood pressure [pp. 124 and 167] and higher levels of integrative control include particularly the *hypothalamus* [p. 342], the *limbic system*, the *thalamus* and *corpus striatum*, and *cerebral cortex*. The hypothalamus is not only the main site of integration of the whole autonomic nervous system, e.g. in the control of body temperature [p. 349], but is also the structure on which visceral afferents, and nerve pathways from higher levels of the central nervous system converge to influence numerous *endocrine glands* via the adeno-

TABLE VII.3 *Responses of effector organs to autonomic nerve impulses*

Organ	Adrenergic nerve impulses	Receptor type	Cholinergic nerve impulses
Eye			
Iris radial muscle	Contraction (mydriasis)	α	—
Iris circular muscle (sphincter pupillae)	—		Contraction (miosis)
Ciliary muscle	—		Contraction, accommodation for near vision
Lid smooth muscle	Lid retraction		—
Heart			
S–A node	Tachycardia	β_1	Bradycardia. Cardiac arrest
Atria	Increased contractility and conduction velocity	β_1	Decreased contractility Decreased conduction velocity
A–V node and conduction system	Increased conduction velocity	β_1	Decreased conduction velocity A–V block
Ventricles	Increased contractility and conduction velocity	β_1	—
	Increased irritability; extrasystoles	β_1	
Blood vessels			
Coronary	Dilatation	$\alpha_1 \beta$	?
Skin and mucosa	Constriction	α	—
Skeletal muscle	Constriction	$\alpha_1 \beta_2$	—
Cerebral	Slight constriction	α	—
Abdominal visceral	Constriction	$\alpha_1 \beta$	
Lung			
Bronchial muscle	Relaxation	β_2	Constriction
Bronchial glands	—		Stimulation of mucus secretion
Stomach			
Motility and tone	Decrease	β	Increase
Sphincters	Contraction	α	Relaxation
Secretion	Inhibition		Stimulation, especially enzymes
Intestine			
Motility and tone	Decrease	$\alpha_1 \beta$	Increase
Sphincters	Contraction	α	Relaxation
Secretion	?		Increase
Gall-bladder	Relaxation		Contraction
Urinary bladder			
Detrusor	Relaxation	β	Contraction
Trigone and internal sphincter	Contraction	α	Relaxation
Ureter			
Motility and tone	Usually increase		?Increase
Uterus	Variable (responses influenced by female sex hormones and by pregnancy)		Variable
Sex organs	Ejaculation in male		Vasodilatation and erection (penis, clitoris)
Skin			
Arrectores pili	Contraction	α	—
Sweat glands	—		Secretion
Adrenal medulla	—		Secretion of Ad and NA
Liver	Glycogenolysis Gluconeogenesis	β	—
Pancreas			
Exocrine glands	—		Secretion
Islets	Glucagon secretion ↑ Insulin secretion ↓	α	Insulin secretion [p. 506]
Salivary glands			
Parotid	—		Secretion
Submaxillary	Thick secretion		Watery secretion
Lacrimal glands	—		Secretion
Nasopharyngeal glands	—		Secretion
Autonomic ganglion cells	—		Stimulation
Adipose tissue	Release of free fatty acids (FFA)	β	—

hypophysis and water balance and milk ejection via the neuro-hypophysis [p. 523]. The role of *chemical transmitters* in various parts of the central nervous system is discussed on page 419.

REFERENCES

BACQ, Z. M. (1975). *Chemical transmission of nerve impulses. A historical sketch.* Pergamon, Oxford.

BURN, J. H. (1971). *The autonomic nervous system*, 4th edn. Blackwell, Oxford.

FELDBERG, W. (ed.) (1957). The autonomic nervous system. *Br. med. Bull.* **13**, 153.

JOHNSON, R. H. and SPALDING, J. M. K. (1974). *Disorders of the autonomic nervous system.* Blackwell, Oxford.

McLENNAN, H. (1970). *Synaptic transmission*, 2nd edn. Saunders, Philadelphia.

NEWMAN, P. P. (1974). *Visceral afferent functions of the nervous system.* Arnold, London.

PICK, J. (1970). *The autonomic nervous system.* Lippincott, Philadelphia.

SMOOTH MUSCLE

The physiology of skeletal muscle and neuromuscular transmission has been dealt with in relation to the somatic nervous system [p. 257]; cardiac muscle is discussed in the section on heart and circulation [p. 94]. It is appropriate to discuss smooth muscle in relation to the autonomic nervous system.

The term 'smooth', 'plain', or 'unstriated' muscle is applied to the contractile tissue present in the hollow viscera, blood vessels, the bronchi, exocrine glandular ducts, and certain structures in the eye and skin. It possesses no visible cross-striations. The electron microscope shows that the individual myofibrils are striated but the striations do not form a regular pattern. As in skeletal and cardiac muscle the contractile proteins are actin and myosin. Smooth muscle cells are spindle-shaped, the dimensions varying from one type to another. In the gut they are 30–40 μm long and 5 μm in diameter at the level of the centrally placed nucleus. Uterine smooth muscle cells are up to 300 μm long and 10 μm wide, whereas vascular smooth muscle cells are 15–20 μm long and 2–3 μm in diameter.

The speed of contraction in smooth muscles is very slow compared with skeletal muscle, and the duration is often prolonged. Tone, or persistent contraction of smooth muscle may be myogenic in origin or due to the continuous arrival of nerve impulses (neurogenic).

The functional diversity of smooth muscle activity justifies the use of the term 'smooth muscles'. For example, the muscle in the iris or blood vessels maintains continuous tone, the uterus shows only occasional activity, during late pregnancy and of course during labour; the urinary bladder contracts as a unit, while in the ureter and intestine waves of contraction pass from one end to the other. All types of smooth muscle have a membrane potential and in most smooth muscles depolarization leads to conducted action potentials which trigger contraction. However, in some smooth muscles, e.g. arteries and trachea, contraction occurs without action potentials and sometimes without local depolarization. The functional activity of all smooth muscles is controlled by autonomic nerves.

The smooth muscle of most organs consists of small bundles of individual smooth-muscle cells (often about 12 cells to a bundle). Except in the muscular wall of arterioles, in which the smooth muscle may only be one cell thick, the unit of smooth muscle is the bundle, in which individual cells are coupled to each other either by protrusion of one cell into another or by close apposition of smooth-muscle cells with a gap junction [FIG. VII.3]. This structural arrangement explains the electrical coupling between smooth-muscle cells since the junctions between them are areas of low resistance. In addition there is electrical coupling between the separate bundles of smooth-muscle cells, due to close apposition of

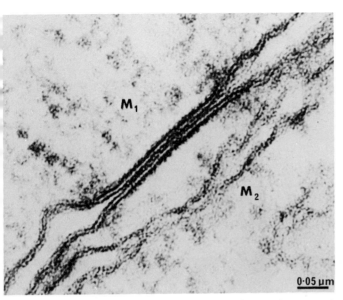

FIG. VII.3. Gap junction between apposing cell membranes of two smooth muscle cells in tissue culture; glutaraldehyde fixation. (Courtesy of Professor Geoffrey Burnstock.)

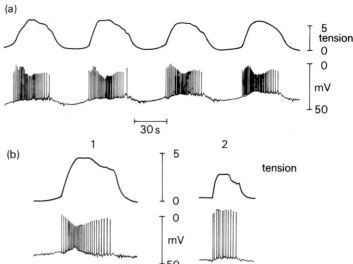

FIG. VII.4. Pacemaker activity and conducted action potentials in isolated segments of myometrium from parturient rats. Intracellular recordings of contractions and electrical activity (b, 1, and 2). (a) one type of pacemaker, which causes the slow, periodic fluctuations in membrane potential that culminate in the rhythmic discharge of action potentials and contractions. (b) 1: second type of pacemaker, which determines the frequency of action potential discharge during a single contraction. Pacemaker potentials are visible as the membrane depolarization preceding the individual spikes. (b) 2: conducted action potentials in another segment of muscle; note their abrupt rise from a flat baseline and their amplitude, which is larger than pacemaker spikes. Spike frequency and train duration are much less than in b, 1; this results in a smaller contraction. (Redrawn from Marshall, J. (1974). *Handbook of physiology*, Section 7, Vol. IV, Part 1, p. 473. American Physiological Society, Washington, DC.)

these bundles and to the presence of small branching bundles which pass from one main bundle to an adjacent bundle. Because of this structure action potentials propagate not only along a muscle bundle, but also from one bundle to others nearby, thus producing co-ordinated contraction (or relaxation) of a large amount of smooth muscle.

Pacemakers and driven activity in smooth muscle

In some organs, such as the gastrointestinal tract, ureter, and uterus, there are smooth-muscle cells which act as pacemakers (cf. sino-atrial node, p. 112). These cells show slow depolarization which leads to an action potential, followed by repolarization and repetition of the sequence at a frequency of about one impulse s^{-1} [FIG. VII.4]. The spread of this action potential may cause other smooth-muscle cells to fire repetitively at the same frequency but without any phase of slow depolarization (driven action potential).

Resting membrane potential of smooth muscle

The resting membrane potential (and action potentials) of smooth muscle can be measured by means of a very fine intracellular microelectrode. The resting membrane potential of visceral smooth muscle (e.g. taenia coli) is unstable but is usually about 50–55 mV (interior negative to exterior). This is a good deal less than the resting membrane potentials of ventricular muscle (80 mV) and skeletal muscle (90 mV). The difference is due to the relatively high intracellular concentrations of chloride and sodium and the relatively low potassium concentration in smooth muscle.

Action potentials

When the transmembrane potential falls (depolarization), at a certain critical value *action potentials* are set up and propagate, as already described, to produce contraction of the smooth muscle. During the rising phase of the action potential calcium ions probably carry most of the inward current, sodium ions playing only a minor role (tetrodotoxin, which selectively blocks sodium entry, does not inhibit spontaneous or evoked action potentials in smooth muscle). The concentration of calcium inside the smooth-muscle cell is raised to a level which is sufficient to initiate contraction. The action potential in smooth muscle rises more slowly and is more prolonged than in skeletal muscle; the development of tension in smooth muscle is directly proportional to the frequency of spike potentials

and since this frequency is less than in skeletal muscle the speed of contraction is much slower in smooth muscle [FIG. VII.4].

Hyperpolarization. When the transmembrane potential increases to say 75 mV, hyperpolarization is said to have occurred and this is accompanied by relaxation of smooth muscle.

Factors influencing smooth-muscle activity. Visceral smooth muscle is influenced by stretch and by the actions of the autonomic nerve chemical transmitters acetylcholine and noradrenalin [pp. 411 and 413]. In addition some inhibitory nerves may act by release of ATP (purinergic nerves, see Burnstock (1979)). The muscle of the taenia coli and ureter contract in response to stretch. When the intracellular concentration of Ca^{2+} is less than 10^{-7} M smooth muscle relaxes; maximal contractile activity occurs when intracellular $[Ca^{2+}]$ reaches 10^{-5}–10^{-4} M.

Nerve supply to smooth muscle

Smooth muscle is generally innervated by autonomic postganglionic nerves (the smooth muscle in fetal membranes and umbilical vessels has no nerve supply). In those types of smooth muscle which are innervated by both divisions of the autonomic nervous system the parasympathetic and sympathetic are antagonistic to each other in their effects. The nerves of either division can be excitatory or inhibitory at neuromuscular junctions.

Autonomic nerve terminals cannot be seen with the ordinary light microscope but with the introduction of the electron microscope and of specific histochemical techniques for identifying transmitter substances in nerves it is now possible to follow autonomic nerves to their terminals, to see details of their structure, and to show the relationship of their terminal axons to smooth-muscle cells. Adrenergic axon endings in smooth muscle can be visualized after partial denervation and the use of the fluorescence histochemical technique to reveal the remaining nerve terminals. Adrenergic nerve terminations differ from somatic nerve endings in skeletal muscle in the following ways:

1. The ramifications of the autonomic axon terminals are longer than the innervated smooth-muscle cells.

2. There is considerable overlap of the terminals of different axons, i.e. each smooth-muscle bundle is innervated by several axons.

3. The preterminal autonomic axons are non-myelinated and have a beaded appearance due to the presence of varicosities which become more numerous in the terminal region of the axon. The terminations of cholinergic neurons possess varicosities like those in adrenergic neurons.

FIGURE VII.5 shows a comparison between the termination of a somatic nerve which supplies individual skeletal-muscle fibres and the termination of an autonomic nerve which supplies smooth-muscle bundles.

Postganglionic axons enter smooth muscle in bundles of up to 100 axons, which divide to form smaller bundles which run parallel to the muscle bundles. The ultimate innervation of smooth-muscle bundles is through small axon bundles which never lose their Schwann cell covering, and through 'naked' axons which are embedded in smooth-muscle cells. In smooth muscle there is no specialized region corresponding to the end-plate in skeletal muscle. The transmitters in autonomic nerves are stored in vesicles in the varicosities, and on release by nerve impulses the transmitter has to cross a gap of 20 nm in the case of single naked axons, or 50–200 nm in the case of small axon bundles, to reach the adjacent smooth muscle.

Some smooth-muscle cells show irregular miniature potentials which may be analogous to miniature end-plate potentials in skeletal muscle. These small potentials are seen in smooth muscle containing 'close-contact' varicosities from which randomly released quanta of transmitter can easily traverse the short distance (20 nm) from neuron to muscle. The passage of nerve impulses causes the simultaneous release of many quanta of transmitter and causes excitatory or inhibitory junction potentials in the smooth muscle, leading to contraction or relaxation [FIG. VII.6].

The frequency of impulses in autonomic nerves is less than in motor nerves to skeletal muscle. Vasoconstrictor and vasodilator nerves discharge at 1–2 impulses s^{-1} under resting conditions and with intense reflex activity the rate may rise to 6–8 impulses s^{-1}. Artificial stimulation at 15–30 impulses s^{-1} gives maximal effects and above this rate fatigue sets in and the output of transmitter per impulse decreases. The physiological frequency of firing in most parasympathetic nerves is similar to that in sympathetic nerves; for example, the sacral nerves to the bladder discharge tonically at less than one impulse s^{-1}.

Smooth muscles also contain afferent nerve endings which are stimulated by pressure or tension. Excessive tension may excite sensory receptors which give rise to pain, such as that of biliary colic and intestinal colic or that associated with uterine contractions in childbirth.

REFERENCES

BENNETT, M. R. (1972). *Autonomic neuromuscular transmission*. Cambridge University Press.
BÜLBRING, E. and BOLTON, T. B. (eds.) (1979). *Smooth muscle. Br. med. Bull.* **35**, No. 3.
—— BRADING, A., JONES, A., and TOMITA, T. (eds.) (1970). *Smooth muscle*. Arnold, London.
BURNSTOCK, G. (1979). *Br. med. Bull.* **35**, 255.
COBB, J. L. S. and BENNETT, M. R. (1969). *J. cell Biol.* **41**, 287.
GRAY, E. G. (1956). *Proc. R. Soc.* **B146**, 416.

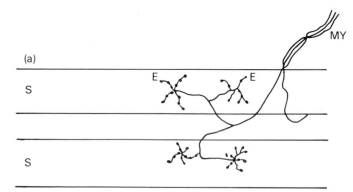

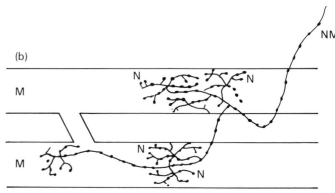

FIG. VII.5. Comparison between the termination of a somatic nerve in an amphibian slow striated muscle and the termination of an autonomic nerve in a mammalian smooth muscle. (a) A single small myelinated nerve fibre (MY) enters a slow striated muscle and branches, losing its myelination; each of these branches further subdivides and ends in two endplates (E) on a muscle fibre (S); the 'en grappe' endplates consist of varicose nerve twigs. (After Gray 1956.) (b) A single non-myelinated varicose nerve fibre (NM) enters a smooth muscle and branches, each branch entering a muscle bundle (M); the branches further subdivide within the muscle bundle to give three systems of varicose nerve terminals (N).

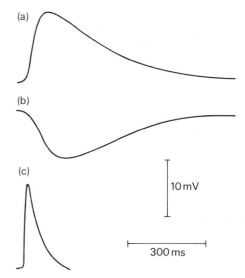

FIG. VII.6. Junction potentials in smooth muscle produced by autonomic nerve stimulation. In (a) there is an excitatory potential, in (b) an inhibitory potential, and in (c) an excitatory potential of quicker onset and briefer duration, due to release of transmitter from close-contact nerve varicosities.

NEUROCHEMICAL TRANSMISSION

All neurons convey information either from one neuron to others or from a neuron to muscular or glandular structures. Conduction along nerve fibres is electrical in nature but transmission of activity at synapses is almost exclusively chemical in nature.

Neurons are capable of synthesizing numerous chemical substances which can be released to act on other cells. Some neurons in the hypothalamus release substances into the bloodstream to act on structures at various distances from these neurons. For example, the hypothalamico-neurohypophysial neurons secrete substances (ADH, oxytocin) which are carried by the bloodstream to faraway organs, such as the kidney and uterus, on which they act [p. 523]. Such substances are called hormones [p. 499]. This can be regarded as *neurohormonal* secretion. The same applies to the substances released from hypothalamic neurons into the portal vessels which supply the adenohypophysis [p. 518] and to catecholamine release from the adrenal medulla [p. 413].

The term *neurochemical transmission* (or simply *chemical transmission*) is applied to substances formed by neurons and released from axonal terminals to traverse the very narrow gap (measured in nm) between presynaptic endings and postsynaptic membranes of neuronal, muscular or glandular cells. The presynaptic transmitter is stored mainly in vesicles (and perhaps also in the cytoplasm) located in axon terminals. The stored transmitter is released by nerve action potentials and diffuses to combine with specific postsynaptic membrane receptors. This in turn increases or decreases the activity of the postjunctional cell, which can be recorded by intracellular electrodes as excitatory postsynaptic potentials (EPSPs) or inhibitory postsynaptic potentials (IPSPs) [p. 281]. The synapse acts as an electrochemical transducer which serves as an amplifier so that weak electric currents in presynaptic nerve terminals lead to potent effects on postsynaptic cells. When a local EPSP reaches a critical size the depolarization initiates a conducted action potential in the postsynaptic cell. A local IPSP involves hyperpolarization of the postsynaptic cell membrane. The recorded electrical changes in the synaptic region are due to the opening or closing of ionic channels in the cell membrane [p. 281].

Historically the idea of chemical transmission was first proposed in 1905 by Elliott who suggested that postganglionic sympathetic nerves acted by release of adrenalin (Ad). In 1921 Loewi showed that vagal stimulation of a perfused frog's heart released a substance which slowed the rate of a second frog's heart. He also showed that stimulation of the sympathetic nerves to the first heart also increased the rate of beating of the second heart. Loewi called the substances 'Vagusstoff' and 'Acceleransstoff' respectively. Subsequently he identified vagusstoff as acetylcholine (ACh), and in 1946 von Euler showed that postganglionic sympathetic nerves released noradrenalin. During the 1930s Dale and his colleagues demonstrated that in the autonomic nervous system all preganglionic nerve fibres, and all postganglionic parasympathetic nerve fibres released acetylcholine. Cannon showed that postganglionic sympathetic fibres released an adrenalin-like substance which he called 'sympathin'. Dale introduced the terms 'cholinergic' and 'adrenergic' to define nerves which released ACh or Ad-like substance respectively (see Bacq 1975) for historical account).

The main evidence that a substance acts as a neurochemical transmitter is as follows:

1. The neuron must be capable of synthesizing and releasing the substance at its presynaptic terminals. Synthesis is effected by enzymes from appropriate substrates; release of the substance must occur in response to nerve impulses (action potentials in presynaptic terminals) and should be small or absent when there are no nerve impulses.

2. The released substance must be identified by biological and/or chemical tests. When applied directly to the innervated structure it must combine with the appropriate receptors and exactly mimic the effects of nerve stimulation.

3. There must be physiological processes for removal or destruc-

tion of the putative transmitter. Reuptake by presynaptic auto-inhibitory autoreceptors is important (Langer 1980).

4. Drugs which enhance release or reduce the rate of removal of the substance should increase the response of the innervated structure to nerve stimulation. Drugs which block synthesis or release of the supposed transmitter should reduce the response of the postsynaptic structure to nerve stimulation. Likewise, drugs which combine with postsynaptic receptors without eliciting physiological responses should prevent the responses to nerve stimulation.

New techniques have helped greatly in the more detailed study of the process of chemical transmission. Radio-isotopic labelling of precursors and of transmitters, immunohistochemical and immunofluorescence techniques, isolation of receptors, electron microscopic studies, and direct application of supposed transmitters by microiontophoresis are among the many procedures which have contributed to further understanding of neurochemical transmission. Their value will be discussed in relation to individual transmitters, e.g. ACh, NA, DA, amino acids, and peptides.

REFERENCES

Bacq, Z. M. (1975). *Chemical transmission of nerve impulses. A historical sketch*. Pergamon, Oxford.
Cotterell, G. A. and Usherwood, P. N. R. (eds.) (1977). *Synapses*. Blackie, Glasgow.
Elliott, T. R. (1905). *J. Physiol., Lond.* **32**, 401.
Euler, U. S. von (1946). *Acta physiol. scand.* **12**, 73.
Langer, S. Z. (1980). *Trends Neurosci.* **3**, 110.
Loewi, O. (1921). *Arch. ges. Physiol.* **189**, 239.

CHOLINERGIC TRANSMISSION

Acetylcholine (ACh)

$$CH_3 - \overset{\overset{\displaystyle CH_3}{|}}{\underset{\underset{\displaystyle CH_3}{|}}{N^+}} - CH_2 - CH_2 - O - \overset{\overset{\displaystyle}{}}{\underset{\underset{\displaystyle O}{\|}}{C}} - CH_3$$

Acetylcholine is the transmitter substance at cholinergic nerve terminals.

Synthesis. ACh is synthesized by a specific enzyme called *choline acetyltransferase* (choline acetylase) which occurs in all cholinergic neurons. The enzyme is a basic protein and its presence can be detected by immunohistochemical techniques. It is synthesized in the perikaryon and migrates along cholinergic neurons to the nerve terminals where ACh synthesis mainly occurs. Choline acetyltransferase requires the presence of choline, acetyl Co-A, ATP, and glucose to form ACh. Acetyl Co-A and ATP, formed in mitochondria, and glucose occur in the neuronal cytoplasm. Choline, a quaternary ammonium compound, does not cross membranes easily but cholinergic nerve terminals possess a specific carrier mechanism for the active transport of choline from extracellular fluid into neuronal cytoplasm. Choline transport across membranes is inhibited by a drug called hemicholinium. When all the required components are present choline acetyltransferase synthesizes ACh in the cytoplasm [FIG. VII. 7].

Storage. In cholinergic neurons ACh synthesis is a continuous process and the ACh is mostly stored in minute vesicles, 30–60 nm in diameter, which electron microscopy reveals in large numbers in the synaptic terminals. Each 'synaptic vesicle' contains a quantum or packet of a thousand or more ACh molecules. However, there is always some ACh present in the cytoplasm.

Release of ACh. When ACh is released from cholinergic presynaptic nerve terminals it speedily traverses the small (10–30 nm) gap between these terminals and the receptors on the postsynaptic membrane to which it becomes bound. It is thought that the released ACh comes from the synaptic vesicles when the latter fuse with the nerve terminal membrane. After the release of ACh by exocytosis [p. 514] the vesicular membrane is reabsorbed. The fusion of synaptic vesicles with presynaptic neuronal membrane occurs as the result of random Brownian movements and even when the nerve fibre is at rest there are minute end-plate potentials which are attributed to the release of a few quanta of ACh. The passage of an action potential along the nerve terminal produces the synchronous release of a great number of such quanta. The fusion of synaptic vesicle with neuronal membrane is promoted by Ca^{2+} and inhibited by Mg^{2+}. Depolarization of cell membranes enhances the entry of Na^+ and Ca^{2+}.

ACh may also be released directly from the cytoplasm of presynaptic cholinergic nerve terminals, without involvement of synaptic vesicles. With such a process quantal release of ACh and miniature end-plate potentials is difficult to explain.

In normal circumstances the rates of synthesis and release of ACh are sufficient to sustain muscular and glandular responses to high physiological frequencies of cholinergic nerve stimulation. If, however, ACh synthesis is impaired by reduction of choline entry into presynaptic nerve terminals (e.g. by hemicholinium) the store of ACh is rapidly depleted and eventually cholinergic transmission fails.

ACh is released continuously at postganglionic parasympathetic and at all preganglionic nerve terminals (both sympathetic and parasympathetic). This results from the constant passage of nerve impulses at a low frequency (e.g. one or a few impulses s^{-1}), which is sometimes called tonic activity. If autonomic cholinergic nerves are cut and time allowed for degeneration ACh release ceases.

Actions on the postsynaptic membrane

ACh has two main types of action on postsynaptic membranes (Dale 1914):

1. A *muscarinic* action, in which ACh acts like muscarine (an alkaloid from the poisonous mushroom *Amanita muscaria*).
2. A *nicotinic* action in which ACh acts like nicotine.

The muscarinic actions are typically those exerted by ACh released from postganglionic parasympathetic nerve terminals in the heart, smooth muscle, and exocrine glands [TABLE VII.3, p. 408]. The muscarinic actions are slow in onset and prolonged; they are antagonized by atropine which combines with ACh receptors at the sites of muscarinic actions. The nicotinic actions of ACh are those produced on autonomic ganglia and skeletal muscle; they are quick in onset and of brief duration and are antagonized by hexamethonium (autonomic ganglia) or by tubocurarine (skeletal muscle). The muscarinic and nicotinic receptors in postsynaptic membranes are activated by different regions of the ACh molecule. Some cholinoceptive neuronal membranes possess both nicotinic and muscarinic receptors, e.g. autonomic ganglion cells and the Renshaw cells in the central nervous system [p. 283]. In such cells ACh produces both a fast, brief excitatory postsynaptic potential (EPSP), due to activation of nicotinic receptors, and a slow, prolonged EPSP, due to activation of muscarinic receptors. However, most cholinoceptive cells in the central nervous system have muscarinic receptors.

The excitatory effects of ACh on postsynaptic membranes are due to increases in permeability to Na^+ and Ca^{2+}. Muscarinic inhibition, e.g. in the heart, is caused by a specific increase in K^+ and Cl^- permeability (and is associated with hyperpolarization).

Acetylcholine receptors.
Cholinergic nicotinic receptors have been studied in detail in vertebrate skeletal muscle [p. 257]. The number and distribution of ACh receptors have been mapped in the postsynaptic membrane by intracellular recording and microiontophoretic application of ACh. Receptor distribution has also been measured by application of radioactive α-bungarotoxin which combines irreversibly with nicotinic receptors on skeletal muscle fibres. In normal circumstances the ACh receptors are almost entirely restricted to the region of the motor endplate; after denervation there is a spread of receptors to other parts of the muscle membrane and this leads to *denervation hypersensitivity* to ACh.

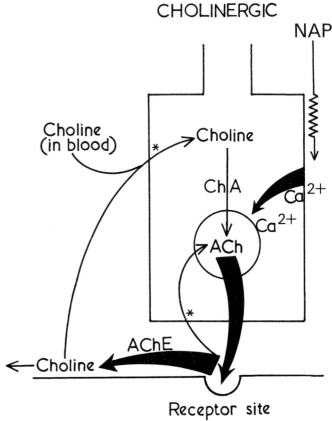

FIG. VII.7. Diagram showing processes involved in synthesis, release and disposal of acetylcholine at *cholinergic* nerve terminal and receptor site.
ACh = Acetylcholine   NAP = Nerve action potential
ChA = Choline acetyltransferase   *→ = Sites of active
AChe = Acetylcholinesterase   transport
(Kindly supplied by Dr P. F. Heffron, Middlesex Hospital Medical School.)

Fambrough, Devreotes, and Card (1977) have proposed the following explanation for these findings. Chemically, after isolation by detergents, the nicotinic ACh receptors have been shown to be hydrophobic glycoproteins. They are synthesized on polysomes bound to endoplasmic reticulum, glycosylated in the Golgi apparatus, and subsequently incorporated into the plasma membrane of which they form an integral part that can act as an ion-transporting structure. At the motor endplate regions the receptors are 'stabilized' by the presence of cholinergic nerve terminals. At extrajunctional sites ACh receptors are also synthesized and incorporated into the plasma membrane, but these receptors are rapidly reabsorbed into the cell interior and destroyed by lysosomal enzymes. After denervation the extrajunctional receptors become more stable and this accounts for the spread of sensitivity to ACh beyond the motor endplate region. Thus cholinergic innervation of skeletal muscle simultaneously stabilizes junctional endplate receptors on the surface of the folds of the postsynaptic membrane, and renders extrajunctional receptors more susceptible to intracellular destruction. The processes by which these effects are mediated are not fully understood but muscular activity *per se* is an important factor (Fambrough 1979).

The muscarinic ACh receptor has received much less attention.

REFERENCES

DALE, H. H. (1914). *J. Pharmac. exp. Ther.* **6**, 147.
FAMBROUGH, D. M. (1979). *Physiol. Rev.* **59**, 165.
—— DEVREOTES, P. N., and CARD, D. J. (1977). The synthesis and degradation of acetylcholine receptors. in *Synapses* (ed. G. A. Cottrell and P. N. R. Usherwood). Blackie, Glasgow.
THESLEFF, S. and SELLIN, L. C. (1980). *Trends Neurosci.* **3**, 122.

Removal of acetylcholine

Acetylcholine is rapidly removed from its site of action. This is brought about by:

1. Diffusion. In a few sites diffusion is almost rapid enough to account for the rate of decay in action of acetylcholine. However, in most sites there is a barrier to free diffusion of acetylcholine.

2. Acetylcholinesterase. A specific acetylcholinesterase (true cholinesterase) is found in high concentration in nearly all postsynaptic structures related to cholinergic neurons. The active site of this enzyme consists of an *anionic* (N$^+$-attracting) subsite and an *esteratic* (ester-binding) subsite [FIG. VII.8]. Hydrolysis of acetylcholine involves the formation of a reversible enzyme–substrate complex, followed by acetylation of the esteratic subsite and release of choline into solution:

$$E + ACh \rightarrow EACh \rightarrow EA \rightarrow E$$
$$\quad\quad\quad\quad\quad \downarrow \quad\ \downarrow$$
$$\quad\quad\quad\quad\quad Ch \quad A$$

E = Enzyme. ACh = Acetylcholine. Ch = Choline. A = Acetyl.

The acetylated cholinesterase (EA) is very unstable and reacts speedily with water to give free enzyme and acetic acid. Choline has similar actions to acetylcholine but is very much weaker, so the hydrolysis of acetylcholine is equivalent to its inactivation. The choline is available for resynthesis of acetylcholine. The significance of acetylcholinesterase in removing acetylcholine is revealed by the action of acetylcholinesterase inhibitors which enhance and prolong the effects of cholinergic nerve stimulation.

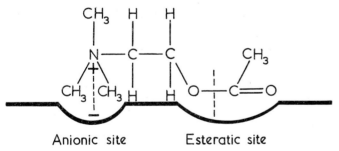

FIG. VII.8. Diagram to show the sites on acetylcholinesterase to which acetylcholine becomes bound. Anticholinesterases are also bound to these sites.

3. Reuptake of ACh is insignificant, but choline uptake is essential for ACh synthesis. In this respect cholinergic neurons differ from adrenergic neurons and GABA-forming neurons, both of which take up their respective transmitters very readily.

Anticholinesterases. These drugs produce their effects mainly by inhibiting acetylcholinesterase and butyryl cholinesterase (plasma or pseudocholinesterase). Preservation of ACh enhances the effects of cholinergic nerve activity [see TABLE VII.3, p. 408]. Anticholinesterases are of different types. Edrophonium (Tensi-

lon) combines reversibly with the anionic site and in man acts for only a few minutes. Carbamates, such as the alkaloid eserine (physostigmine) and the synthetic neostigmine, form an acylated enzyme which is more stable than acetylated cholinesterase, but the complex dissociates within a few hours. Organophosphates (e.g. diisopropyl phosphofluoridate, DFP) combine only with the esteratic site, but the phosphorylated enzyme is so stable that the action is virtually irreversible.

Transmission in autonomic ganglia

Preganglionic nerves of the whole autonomic nervous system are cholinergic. The ganglion cell receptors are mainly nicotinic in nature but there are also some muscarinic receptors. Autonomic ganglia also possess 'small intensely fluorescent' neurons containing dopamine (DA). Muscarinic receptors and DA neurons may modify the nicotinic actions of ACh.

Block of ganglionic transmission

1. Nicotine *depolarizes* ganglion cells and in large doses produces block of transmission after initial stimulation.

2. *Competitive inhibition* is produced by drugs such as hexamethonium which combines with ACh receptors on ganglion cells. These inhibitors block autonomic ganglionic transmission without initial stimulation.

Ganglion blockade interrupts both sympathetic and parasympathetic pathways.

Sympathetic blockade produces cutaneous vasodilation with a warm skin and, on standing or vertical tilting, a fall in arterial blood pressure. This postural hypotension is due partly to the reduced peripheral resistance in dilated arterioles, but mainly to the loss of venomotor tone. This leads to pooling of blood, under the influence of gravity, in the dependent parts of the body. Venous return is reduced and cardiac output falls. The compensatory vascular adjustments which prevent a fall in blood pressure when a normal person stands up are reflexly elicited via the sympathetic vasoconstrictor nerves to arterioles, capillaries, and veins. Ganglion blockade abolishes such reflexes. The hypotensive effect has been much used in the treatment of patients with essential and malignant hypertension.

Sympathetic blockade also prevents ejaculation by paralysing the musculature of the seminal tract.

Parasympathetic blockade, e.g. by atropine has the following effects:

Delay in gastric emptying; relaxation of intestinal muscle leading to constipation and even to paralytic ileus; difficulty in emptying the bladder; dilatation of the pupil and relaxation of the ciliary muscle, preventing accomodation for near vision; many secretions are inhibited; the mouth and skin become dry.

Ganglion blocking drugs were formerly used in the treatment of hypertension but have been replaced by drugs which act more selectively to inhibit sympathetic nervous activity [p. 417].

REFERENCE

VOLLE, R. L. (1966). *Pharmac. Rev.* **18**, 839.

Adrenergic transmission

Some years ago ACh was regarded as the 'typical' chemical transmitter and much was discovered about its synthesis, release, postsynaptic action, and extraneuronal disposal by cholinesterase. At that time adrenergic transmission was less well understood, but in recent years the introduction of new techniques has changed the picture radically so that, although much remains to be discovered, more is known about adrenergic than about cholinergic transmission. The scheme of biosynthesis of noradrenalin and adrenalin from tyrosine, proposed by Blaschko in 1939, has been fully confirmed and amplified by discovery of the enzymes responsible for each stage of the sequence. The use of subcellular fractionation and the scintillation counter, the discovery of the fluorescent histochemical technique, and analysis of the modes of action of many drugs have provided most valuable information on the sites of synthesis of *catecholamines* and of their distribution, release from adrenergic neurons, sites of action on postsynaptic structures, and the metabolic disposal of these potent substances.

The naturally occurring catecholamines are dopamine (DA), noradrenalin (NA, norepinephrine), and adrenalin (Ad, epinephrine). The central nervous system contains neurons which synthesize and release NA [p. 419] and in the basal ganglia and hypothalamus there are neurons which form and release dopamine as the neurotransmitter [p. 419]. The neurons which form and release NA and DA have local actions, but the catecholamines (NA and Ad) released from the adrenal medulla enter the circulation and produce widespread effects throughout the body.

Adrenergic nerve stimulation also releases prostaglandins (PGE$_1$ and PGE$_2$) which antagonize the actions of catecholamines and reduce the release of NA (Horton 1973).

Evidence for role of noradrenalin as adrenergic transmitter

1. Adrenergic nerves can synthesize noradrenalin from tyrosine. The presence of noradrenalin can be detected by biological or fluorimetric assays of extracts of sympathetic nerves, or by demonstration of highly fluorescent catecholamine derivates after treatment of freeze-dried tissue with formaldehyde vapour. Immunohistofluorescence techniques have helped to localize the sites of enzymes involved in catecholamine synthesis, and radioimmunoassay has been introduced for measurement of NA and Ad. Drugs which inhibit noradrenalin synthesis reduce the response to adrenergic nerve stimulation.

2. Release of noradrenalin can be detected after postganglionic sympathetic nerve stimulation. In perfused organs stimulation of sympathetic nerves to the perfused heart, limbs, or other organs leads to the appearance in the perfusate of a substance which, when separated from other substances by absorption and chromatographic techniques, can be identified as noradrenalin. In the intact animal noradrenalin can similarly be detected in the venous outflow.

After postganglionic sympathetic nerve section release of noradrenalin fails at the same time as the response to nerve stimulation.

3. The actions of noradrenalin on adrenergically innervated organs are indistinguishable from the effects of adrenergic nerve stimulation.

4. The inactivation of infused noradrenalin and of noradrenalin released by nerve stimulation is due mostly to uptake by adrenergic nerve terminals and to a small extent to enzymic inactivation. Drugs which interfere with noradrenalin uptake enhance the responses to adrenergic nerve stimulation.

Catecholamine biosynthesis

Noradrenalin is synthesized in peripheral and central adrenergic neurons and in the chromaffin cells of the adrenal medulla, where adrenalin is also synthesized. The chemical reactions involved are shown below, starting with the essential amino acid L-tyrosine.

The rate-limiting reaction is the oxidation of tyrosine to dopa by tyrosine hydroxylase. Inhibition of this enzyme greatly reduces noradrenalin synthesis; tyrosine hydroxylase in adrenergic nerves is inhibited by the end product, free NA, which exerts a negative feedback control over NA synthesis. This inhibition of tyrosine hydroxylase is due to combination of NA with the enzyme's cofactor tetrahydropteridine. DA acts in the same way. The enzyme which decarboxylates dopa also forms 5-hydroxytryptamine and

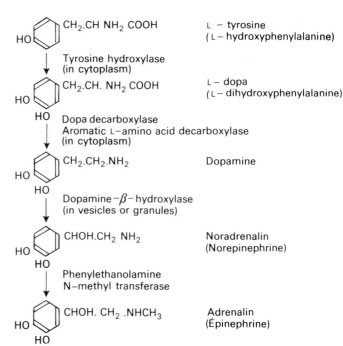

histamine from the corresponding amino acids; hence it is called aromatic L-amino acid decarboxylase. The compound dopamine is generally regarded as an intermediary in the biosynthesis of noradrenalin. However, in certain parts of the brain, e.g. caudate nucleus, it occurs in much higher concentrations than noradrenalin and appears to have actions of its own. The formation of adrenalin from noradrenalin in the adrenal medulla by the enzyme phenylethanolamine N-methyl transferase requires a methyl donor such as methionine.

Adrenalin, noradrenalin, and dopamine are described as *catecholamines*. A closely related catecholamine is the synthetic compound isopropylnoradrenaline (isoprenaline, isoproterenol) in which the N-methyl group of adrenalin is replaced by an isopropyl group.

Measurements of the rate of accumulation of both endogenous and radioactive NA in ligated nerves, and of the recovery of NA stores after depletion by reserpine (which reduces vesicle storage) suggest that NA-containing vesicles are transported from the cell bodies along adrenergic nerves, perhaps via microtubules, to the nerve terminals at a rate of 5–10 mm h^{-1}. The enzymes tyrosine hydroxylase, dopa decarboxylase, and dopamine β-hydroxylase are also conveyed by axoplasmic flow (Blaschko 1973).

Synthesis of noradrenalin in postganglionic sympathetic neurons ceases after section and degeneration of the postganglionic nerves but is unaffected by preganglionic nerve section.

Storage. Adrenalin, noradrenalin, and dopamine are stored in the form of granules, 0.1–0.5 μm in diameter, in the chromaffin cells of the adrenal medulla. These granules are distinct from other subcellular particles. Noradrenalin occurs in adrenergic nerve terminals in vesicles 40–50 nm in diameter, and some is probably free in the cytoplasm. The concentration of NA in vesicles may reach 1 mol l^{-1}. The storage granules in the adrenal medulla and in adrenergic neurons contain high concentrations of complexes of catecholamine and ATP in a molecular ratio of 4:1.

Release. The release of catecholamines from the cells of the adrenal medulla is brought about by the action of acetylcholine released by stimulation of the preganglionic fibres of the splanchnic nerve. Acetylcholine acts by promoting entry of calcium ions which then release catecholamines from the granules by exocytosis. Release of NA from adrenergic neurons is effected by nerve impulses in the presence of calcium ions. NA release is accompanied by release of dopamine β-hydroxylase which suggests that the whole contents of synaptic vesicles are released directly into extra-neuronal fluid (exocytosis) (Bennett 1972).

Removal and inactivation of catecholamines after release

Catecholamines are removed and inactivated by processes different from those which deal with ACh. The actions of released NA are terminated by two uptake processes:

1. *Uptake$_1$* (neuronal). Free NA in the synaptic cleft is taken up into presynaptic adrenergic nerve terminals by a process termed Uptake$_1$ which involves active transport into the neuronal cytoplasm. This specific uptake accounts for inactivation of about 85 per cent of released NA. Uptake$_1$ can work against a concentration gradient of 10 000 to 1 and deals with exogenous NA as well as with that released from adrenergic nerves. Uptake$_1$ is less specific for Ad than for NA.

2. *Uptake$_2$* (extraneuronal) is less important than uptake$_1$ for removing NA, accounting for only about 15 per cent of that released from adrenergic nerve endings. Uptake$_2$ is extraneuronal and is mediated by postsynaptic cells such as those of the heart and smooth muscle. This form of uptake is followed by intracellular metabolic inactivation by the enzymes *monoamine oxidase (MAO)* in mitochondria and *catechol-o-methyl transferase (COMT)*.

MAO is an oxidative deaminase which acts as follows:

3,4–Dihydroxymandelic acid

COMT transfers a methyl group as follows:

Normetanephrine

Adrenalin (epinephrine) is analogously metabolized to meta-nephrine.

The end product of both NA and Ad metabolism by the combined actions of MAO and COMT is the compound:

4- Hydroxy-3-Methoxymandelic acid

The presence of MAO and COMT in the liver largely accounts for inactivation of orally administered catecholamines (which pass via the portal vein to the liver). These enzymes also inactivate circulating catecholamines from the adrenal medulla. MAO also destroys 5-hydroxytryptamine.

The presence of MAO in mitochondria in adrenergic nerve terminals prevents the accumulation of free NA in axonal cytoplasm, but MAO cannot destroy NA stored in vesicles.

The processes of synthesis and reuptake provide enough NA for transmission at physiological rates of nerve stimulation. Exhaustion of transmitter stores would only occur at very high rates of stimulation or if the processes of synthesis are impaired.

Normally, only 2–3 per cent of catecholamines released into body fluids escape enzymic destruction: this amount is excreted in the urine, largely as the inactive glucuronide.

REFERENCES

BENNETT, M. R. (1972). *Autonomic neuromuscular transmission*. Cambridge University Press.
BLASCHKO, H. (1973). *Br. med. Bull.* **29**, 105.
HORTON, E. W. (1973). *Br. med. Bull.* **29**, 148.
IVERSEN, L. L. (1971). Review of transmitter uptake mechanisms. *Br. J. Pharmac.* **41**, 571.
—— (ed.) (1973). Catecholamines. *Br. med. Bull.* **29**, No. 2.
TRIGGLE, D. J. and TRIGGLE, C. R. (1976). *Chemical pharmacology of the synapse*. Academic Press, New York.

Action of catecholamines

Noradrenalin is released at postganglionic sympathetic nerve endings and exerts mainly local effects at its site of release. Adrenalin is the chief catecholamine released from the adrenal medulla and is distributed throughout the body by the bloodstream in circumstances described on page 417. Tumours of chromaffin tissue (phaeochromocytoma) contain large amounts of both NA and Ad which may be released to act on systems throughout the body [p. 418]. Dopamine has minor peripheral actions but has more important actions on the central nervous system [pp. 360 and 419].

The responses of adrenergically innervated organs are set out in TABLE VII.3 [p. 408]. These are mediated mostly by NA release at adrenergic nerve terminals, but in certain conditions the release of Ad (and to a minor extent NA) from the adrenal medulla may reinforce or modify the effects of adrenergic nerve activity.

Adrenalin generally acts like noradrenalin but differs in some respects.

1. The cardiovascular system in man. An *intravenous* infusion of Ad at a rate of 0.1–0.3 µg kg^{-1} min^{-1} raises systolic blood pressure, lowers diastolic blood pressure, increases heart rate, increases the cardiac output, and reduces the total peripheral resistance. In conditions of stress the adrenal medulla secretes Ad at about this rate. *Intravenous* infusion of NA raises both systolic and diastolic blood pressure, slows the heart rate, slightly reduces cardiac output, and increases the total peripheral resistance. The bradycardia is due to carotid sinus and aortic baroreceptor reflex vagal stimulation, resulting from the raised blood pressure.

Intra-arterial infusion of Ad produces vasodilatation in *skeletal muscle*, whereas similar infusion of NA leads to vasoconstriction. In normal circumstances at rest sympathetic tone, with continuous release of NA, produces a relatively small blood flow through skeletal muscle [p. 142].

In most regional vascular beds the sympathetic nervous control via liberated NA predominates over the effects of adrenomedullary secretion of Ad.

2. The central nervous system. The roles of DA and NA in the central nervous system are discussed on pages 419–20.

In man, intravenous infusion of Ad produces anxiety, apprehension, initial stimulation of breathing, and coarse tremors of the extremities. NA is much less active in these respects. Catecholamines do not cross the blood–brain barrier, but where this barrier is absent, e.g. in the area postrema, they may enter the brain freely.

Metabolic actions

Adrenalin and noradrenalin have actions on carbohydrate and lipid metabolism, adrenalin being more potent on the former and noradrenalin on the latter.

Carbohydrate metabolism. Adrenalin raises the *blood sugar*, occasionally to levels high enough to produce glycosuria. An intravenous infusion, e.g. for 8 hours, produces a long-lasting rise in blood sugar, even in a fasting animal. It also raises the blood *lactate* level and increases total O_2 consumption by up to 30 per cent (*calorigenic action*).

The *hyperglycaemia* is due mainly to actions on the liver and pancreas.

Liver. Adrenalin acts on the liver to promote glycogenolysis and enhanced gluconeogenesis from lactate.

Pancreas. Adrenalin inhibits glucose-induced secretion of insulin from the β-islet cells. Sudden stoppage of insulin secretion leads rapidly to hyperglycaemia.

Adrenalin causes the breakdown of glycogen to glucose in the following way. It first activates the enzyme adenyl cyclase which converts ATP to cyclic $3':5'$-adenosine monophosphate (cAMP). The mode of action of cAMP in promoting glycogenolysis is discussed on page 501. The enzyme phosphoglucomutase converts glucose 1-phosphate to glucose 6-phosphate. The *liver* possesses a phosphatase which hydrolyses the latter to glucose and inorganic phosphate, the glucose being released into the bloodstream. In *muscle* adrenalin promotes glycogenolysis as in the liver but since muscle lacks glucose 6-phosphatase lactic acid is formed.

Adrenalin increases blood lactate and pyruvate, partly by promoting glycogenolysis in muscle, but also as the result of simultaneous increases in blood glucose and free fatty acid (FFA) levels. Noradrenalin, which increases the plasma FFA level, does not raise the blood lactate because it does not cause hyperglycaemia.

Lipid metabolism. Adrenalin and noradrenalin activate a specific lipase in adipose tissue and muscle which breaks down triglycerides to FFA and glycerol. This lipolysis might be mediated by cyclic AMP. It is antagonized by insulin. In the liver some of the excess FFA is converted into ketone bodies. The catecholamines make available for active tissue more oxidizable structures, such as FFA, glycerol, and ketone bodies, and at the same time depress the oxidation of glucose. The lipolytic action of adrenalin is brief, that of noradrenalin prolonged.

The availability of the metabolic fuels, FFA and glucose, must be rapidly adjusted to the changing energy requirements of the body under varying conditions of activity and stress. The large fuel depots of triglyceride and glycogen can be drawn on in response to body demands. The sympathetic nervous system, including the adrenal medulla, can, through release of its catecholamine mediators, speedily activate the enzymes which catalyse the breakdown of these stores of energy.

Sites of action of catecholamines

NA, Ad, and DA act directly on the structures which they influence. The original proposal by Ahlquist (1948) that catecholamines acted on two types of membrane adrenoceptor, α and β, has been amply confirmed and extended by studies with agonists and antagonists. Generally NA acts on α-receptors, isoprenaline on β-receptors, and Ad on both α- and β-receptors.

α-receptors are of two kinds, α_1-receptors on postsynaptic membranes which are mainly excitatory, e.g. in blood vessels and the non-pregnant uterus; α_2-receptors are located on presynaptic nerve terminals of cholinergic and adrenergic nerves. Activation of neuronal α_2-receptors is inhibitory, and relaxation of intestinal muscle by Ad can be partly due to reduced release of ACh from cholinergic nerves. In the case of adrenergic nerves, high concentrations of NA have an autoinhibitory effect on transmitter release (Bülbring 1979).

β-receptors are also of two kinds, β_1 and β_2, but both are located mostly on postsynaptic membrane sites. β_2-receptors are typically those associated with relaxation of smooth muscle, e.g. in skeletal muscular blood vessels, intestine, and bronchioles; β_1-receptors occur in cardiac muscle and their activation increases heart rate and

contractility [p. 408]. However, not all β-receptors can be categorized as β_1 or β_2 [see Table VII.3, p. 408].

Stimulation of β-receptors activates adenyl cyclase which in turn promotes c-AMP formation etc. [p. 500]. It seems likely that α-receptor activation increases intracellular $[Ca^{2+}]$ and β-receptor activation decreases intracellular $[Ca^{2+}]$.

Summary of actions on α- and β-receptors:

	α-Receptors	β_1-Receptors	β_2-Receptors
Noradrenalin	++	+	±
Adrenalin	+	++	+±
Isoprenaline	−	++	++
Salbutamol	−	±	++

The actions of α- and β-receptor blocking drugs agree with differences in receptor properties [see below].

Sympathomimetic drugs

The term 'sympathomimetic' was introduced by Dale to describe those substances which closely mimic the effects of sympathetic nerve stimulation. Such substances may act *directly* on α- or β-receptors or they may act *indirectly* by displacing noradrenalin from its storage sites or by preventing its uptake by adrenergic nerve terminals.

The directly acting sympathomimetics include noradrenalin, adrenalin, and isoprenaline. When noradrenalin is introduced into a sympathetically innervated tissue part is taken up by the receptors of the effector cells (uptake$_2$) and most by the adrenergic terminals (uptake$_1$). For this reason when postganglionic sympathetic fibres undergo degenerative section *supersensitivity* to noradrenalin develops. When the uptake sites are eliminated all the added noradrenalin can combine with tissue receptors so that the effector response is enhanced.

Indirectly-acting sympathomimetics include tyramine which displaces noradrenalin from adrenergic neurons and cocaine, imipramine, and desipramine which inhibit neuronal uptake of noradrenalin.

REFERENCES

Acheson, G. H. (ed.) (1966). Second Symposium on Catecholamines. *Pharmac. Rev.* **18**, No. 1.
Ahlquist, R. D. (1948). *Am. J. Physiol.* **153**, 586.
Barcroft, H. and Swan, H. J. C. (1953). *Sympathetic control of human blood vessels*. Arnold, London.
Bennett, M. R. (1972). *Autonomic neuromuscular transmission*. Cambridge University Press.
Blaschko, H., Sayers, G., and Smith, A. D. (eds.) (1975). Adrenal gland. Adrenal medulla. In *Handbook of physiology*, Section 7, Vol. VI, p. 283. American Physiological Association, Washington, DC.
Bulbring, E. (1979). *Br. med. Bull.* **35**, 285.
Burn, J. H. (1971). *The autonomic nervous system*, 4th edn. Blackwell, Oxford.
Iversen, L. L. (ed.) (1973). Catecholamines. *Br. med. Bull.* **29**, 91.

ANTAGONISTS OF ADRENALIN AND NORADRENALIN

The term *adrenergic receptor blocking drug* is used to describe substances which combine with α- or β-receptors to prevent the corresponding actions of the catecholamines. Thus we have α- and β-receptor blocking drugs.

α-receptor blocking drugs include the *ergot* alkaloids (e.g. *ergotamine* in large doses), *phentolamine, piperoxan,* and *phenoxybenzamine*. The last is a haloalkylamine with a very prolonged action. All these drugs produce peripheral vasodilatation. They antagonize the effects of circulating amines more readily than those of the noradrenalin released at adrenergic nerve endings.

β-receptor blocking drugs include *propranolol*, which blocks β_1 and β_2 receptors, and *practolol*, which blocks β_1-receptors almost exclusively. Both antagonize the actions of catecholamines on the heart; they slow heart rate and reduce the force of contraction, and by thus decreasing the work of the heart they help to relieve angina pectoris. The β_2-receptor block by propranolol may lead to bronchoconstriction.

Adrenergic neuron blocking drugs are substances which inhibit the release of noradrenalin at adrenergic nerve terminals. They do not antagonize the actions of released noradrenalin or circulating adrenalin. Bretylium and guanethidine are hypotensive drugs which inhibit NA release.

Interference with adrenergic transmission can also be brought about by α-methyl dopa (methyl-dopa, *Aldomet*) which is converted to α-methyl noradrenalin in adrenergic nerve terminals. This substance combines with presynaptic α_2-receptors on adrenergic nerve terminals to inhibit release of NA.

Noradrenalin synthesis is directly inhibited by α-methyl-p-tyrosine and by 3, 5-L-tyrosine which both act by inhibiting tyrosine hydroxylase.

6-Hydroxydopamine destroys adrenergic neurons. It does not cross the blood–brain barrier, but after intracisternal administration it destroys local adrenergic and dopaminergic neurons. It does not destroy the cells of the adrenal medulla.

Reserpine. This Rauwolfia alkaloid depletes tissues of their stores of 5-hydroxytryptamine and of catecholamines. The sedative action of reserpine may be related to depletion of these substances from the hypothalamus.

Reserpine also removes catecholamines from the chromaffin cells of the adrenal medulla, and noradrenalin from adrenergic postganglionic sympathetic nerves. The latter effect probably contributes to the hypotensive action of reserpine (Muscholl and Vogt 1958). The depletion of noradrenalin in adrenergic nerve terminals is due to an inability to store this substance in the intraneuronal granules. Synthesis and uptake of exogenous noradrenalin are normal but if storage is impaired, this transmitter is rapidly destroyed by monoamine oxidase in the intraneuronal mitochondria.

REFERENCES

BOURA, A. L. A. and GREEN, A. F. (1965). Adrenergic neurone blocking agents. *A. Rev. Pharmac.* **5**, 183.
GREEN, A. F. (1962). Antihypertensive drugs. *Adv. Pharmac.* **1**, 161.
LANGER, S. Z. (1980). *Trends Neurosci.* **3**, 110.
MORAN, N. C. (1967). New adrenergic blocking drugs: their pharmacological, biochemical and clinical actions. *Ann. N.Y. Acad. Sci.* **139**, 541.
MUSCHOLL, E. and VOGT, M. (1958). *J. Physiol., Lond.* **141**, 132.

The adrenal medulla

STRUCTURE

The cells of the adrenal medulla are large, ovoid, and columnar in type, and are grouped in clumps around the blood vessels. Many of the cells contain fine granules which are coloured brown by chrome salts. This chromaffin reaction is due to oxidation of adrenalin and noradrenalin, or their precursors, in the granules. Electron microscopy and differential centrifugation have shown that the chromaffin granules are distinct from mitochondria. The granules

contain adrenalin or noradrenalin + ATP and dopamine β-hydroxylase.

The catecholamines, adrenalin and noradrenalin, are the active principles secreted by the adrenal medulla. They are synthesized as described on page 414. The enzyme phenylethanolamine N-methyl transferase, which converts NA to Ad, only acts when chromaffin cells are in contact with adrenal cortical tissue. ACTH promotes formation of tyrosine hydroxylase and dopamine β-hydroxylase.

The proportions of adrenalin and noradrenalin vary from species to species. In rabbits noradrenalin comprises 4.5 per cent of the total catecholamines, whereas in the domestic fowl it amounts to 70 per cent. In man the adrenal medulla contains about 80 per cent adrenalin + 20 per cent noradrenalin. In early fetal life the medulla and other chromaffin tissue, such as the organs of Zuckerkandl, contain only noradrenalin. The proportion of adrenalin increases steadily after birth.

Histochemical studies have shown that adrenalin and noradrenalin occur in separate kinds of chromaffin cells. Stimulation of different regions of the hypothalamus in cats releases different proportions of these two amines into the adrenal venous blood. This suggests that the adrenalin- and noradrenalin-containing cells have separate innervations. Reflex stimulation also acts selectively; carotid occlusion and asphyxia stimulate mainly noradrenalin secretion, whereas sciatic nerve stimulation and hypoglycaemia cause mainly adrenalin secretion.

Chromaffin cell tumours (*phaeochromocytomas*) may contain predominantly adrenalin or noradrenalin, some containing exclusively the latter.

FUNCTIONS OF THE ADRENAL MEDULLA

Unlike the adrenal cortex, the adrenal medulla is not essential for life. Nevertheless it contributes to the fight or flight reactions which occur in conditions of emergency (Cannon). Secretion of catecholamines from the adrenal medulla can be measured by determining their concentrations in adrenal or peripheral venous blood, using biological or fluorimetric assay.

Nervous control of adrenal medulla

In physiological conditions the secretion of catecholamines from the adrenal medulla is entirely controlled by the splanchnic nerves, whose fibres end round the medulla cells. The splanchnic fibres are preganglionic and release acetylcholine as their transmitter to act on the medullary chromaffin cells which correspond to autonomic ganglion cells. When the splanchnic nerve is stimulated the released acetylcholine excites the chromaffin cell by promoting an inward movement of calcium ions across the plasma membrane. This causes the chromaffin granules to discharge their content of catecholamines, ATP, and dopamine β-hydroxylase to the cell surface and hence into the adrenal venous blood.

Splanchnic nerve activity is controlled by centres in the reticular formation in the medulla oblongata and hypothalamus which exercise a higher control. The activity of these centres may be modified by afferent impulses from many parts of the body. Secretion from the adrenal medulla ceases after splanchnic nerve section.

Regulation of adrenal medullary secretion

Resting secretion. It has been estimated that in the 'resting' dog the output of adrenalin is 54 pmol kg^{-1} min^{-1} and that of noradrenalin is 11.5 pmol kg^{-1} min^{-1}, measured in adrenal venous plasma.

One type of chromaffin cell secretes adrenalin and another type noradrenalin. Each type of cell may be stimulated selectively or both may secrete together.

Secretion in conditions of emergency or stress. Cannon originally demonstrated secretion of adrenalin in cats whose hearts had been sensitized by denervation. In small animals the heart rate increased in response to as little as 2.5 µg l⁻¹ (13.5 pmol l⁻¹) of adrenalin. If any change in the state of the animal caused cardiac acceleration this must have been due either to catecholamine secretion by the adrenal medulla, or (less likely) to release of 'sympathin' (NA) from sympathetic nerves. If the response was abolished by splanchnic nerve section it must have been due to adrenal medullary secretion. Measurements of blood catecholamine concentrations have confirmed and extended Cannon's findings. They show that catecholamine secretion is stimulated by: (i) physical exertion and certain emotional states, (ii) exposure to cold, (iii) fall of arterial blood pressure, (iv) asphyxia and cerebral anaemia, (v) anaesthetic and convulsant drugs, (vi) stimulation of afferent nerves, (vii) hypoglycaemia.

The secreted catecholamines, especially adrenalin, may help in the following ways:

In muscular exercise there will be redistribution of blood to the active muscles, and increase in the force and rate of the heart will help to cope with increased venous return. Increased glycogenolysis in the liver and lipolysis in adipose tissue will provide more glucose and FFA as energy-producing fuels for skeletal and cardiac muscle.

In hypoglycaemia adrenalin helps to raise blood glucose by enhancing hepatic glycogenolysis, by promoting gluconeogenesis from lactate and by inhibiting insulin secretion.

On exposure to cold the calorigenic action of adrenalin will help to maintain body temperature by increasing heat production and cutaneous vasoconstriction will help to reduce heat loss. As Cannon put it, from the standpoint of temperature regulation adrenalin secretion is a fine adjustment, shivering a 'coarse' adjustment.

The value of adrenal medullary secretion in psychological stress is obscure.

Adrenal medulla and sympathetic nervous system

The adrenal medulla and sympathetic nervous system are both involved in the immediate bodily responses to emergency situations, but they can function independently of each other. It is therefore misleading to speak of a 'sympathico-adrenal system' as a physiological unit. The sympathetic nerves, through their liberated noradrenalin, are mostly concerned with regulation of vascular tone, blood flow, and blood pressure, while the adrenal medulla through its predominant secretion of adrenalin, has its most important actions on metabolism

All adrenomedullary secretion of catecholamines can be abolished by surgical demedullation of the adrenal glands. Adrenergic nerve function can be very largely but not completely eliminated in the following ways:

Surgical sympathectomy. It must be emphasized that surgical removal of the sympathetic chains does not eradicate the 'short' adrenergic neurons in pelvic viscera.

By drugs which block transmission through autonomic ganglia, prevent release of noradrenalin or antagonize the actions of noradrenalin on α- and β-receptors.

By administration to newborn animals of an antiserum against the sympathetic nerve growth factor. This *immunosympathectomy* greatly reduces development of sympathetic nervous tissue.

Adrenodemedullated, sympathectomized animals with nearly complete elimination of adrenergic neuronal activity maintain good health in a quiet non-stressful environment. In the female cat or dog reproduction occurs normally. The blood pressure shows an initial fall but full recovery takes place later. It seems that in the absence of catecholamines the walls of the arterioles can develop sufficient

tone to maintain the peripheral resistance. *Supersensitivity* to injected adrenalin or noradrenalin develops, as shown by greater pressor and positive inotropic responses. After acute suppression of sympathico-adrenal activity there may be temporary signs of parasympathetic overactivity but the heart rate and pupil size soon return to normal. Emotional excitement causes no rise in blood sugar or red-cell count and the normal rise in blood pressure does not occur. Responses to emergency situations are impaired. The animals are very sensitive to cold and at an environmental temperature of 4 °C they show no vasoconstriction, piloerection, or shivering and die from hypothermia after periods of exposure which can be withstood by normal animals. Muscular exercise leads more rapidly to fatigue and exhaustion, owing to lack of conversion of glycogen to glucose and fat to free fatty acids, both of which are necessary to provide energy for muscular work. The normal capacity to withstand these stressful situations is only present when the sympatho-adrenomedullary responses are reinforced by adrenocortical secretion.

REFERENCES

Blaschko, H., Sayers, G., and Smith, A. D. (eds.) (1975). Adrenal gland. Adrenal medulla. In *Handbook of Physiology*, Section 7, Vol. VI, p. 283. Washington, DC.

Brodie, B. B., Davies, J. I. Hynie, S., Krishna, G., and Weiss, B. (1966). Interrelationships of catecholamines with other endocrine systems. *Pharmac. Rev.* **18**, 273.

Lee, J. and Laycock, J. (1978). *Essential endocrinology*, p. 36. Oxford University Press.

Melmon, K. L. (1974). Catecholamines and the adrenal medulla. In *Textbook of endocrinology*, 5th edn (ed. R. H. Williams). Saunders, Philadelphia.

Zaimis, E. (1967). Immunological sympathectomy. In *Lectures on the scientific basis of medicine*, pp. 59–73. Athlone Press, London.

Adrenal medullary tumour (phaeochromocytoma)

A tumour consisting of adrenal medullary tissue is called a phaeochromocytoma. These tumours, usually benign, may occur in the adrenal glands or in para-aortic rests of adrenal medullary tissue which run down to the organ of Zuckerkandl. Phaeochrome tumours occur in less than 1 per cent of all cases of arterial hypertension, but their diagnosis is important because they produce a type of hypertension which can be cured by surgery.

The hypertension due to a phaeochrome tumour may be paroxysmal or sustained. The features of a paroxysm are: headache, severe palpitations, sweating, nausea, and vomiting, increased depth and rate of breathing, anxiety, weakness, dizziness, and substernal pain. On physical examination the patient looks anxious, the skin is pale, cool, and moist; pupils are dilated. Heart rate is usually increased, but in some patients there is bradycardia. Extrasystoles may occur. Arterial blood pressure rises considerably, occasionally to 300+/200 mm Hg. Body temperature is raised, there is hyperglycaemia and the basal metabolic rate is increased. The sweating, which may be profuse, is due to raised body temperature, central action on the hypothalamic temperature regulating centre, and possibly to direct stimulation of sweat glands by the catecholamines.

All these symptoms and signs can be produced by infusions of large doses of adrenalin or noradrenalin. With such doses the previously described differences between these two amines no longer occur, so that adrenalin raises diastolic blood pressure and noradrenalin increases metabolic rate. Phaeochrome tumours contain varying proportions of adrenalin and noradrenalin (and also their precursor, dopamine).

REFERENCE

ACHESON, G. H. (ed.) (1966). Second Symposium on Catecholamines. *Pharmac. Rev.* **18**, 645.

MELMON, K. L. (1974). Catecholamines and the adrenal medulla. In *Textbook of endocrinology*, 5th edn (ed. R H. Williams). Saunders, Philadelphia.

Chemical transmission in the central nervous system

The anatomical complexity of the central nervous system makes the study of chemical transmission very difficult since it is impossible to isolate central nervous neurons and detect transmitter release as in peripheral nerves. However, technical advances during the past 25 years have revealed not only much more detailed information about transmitter synthesis, storage, release, and mode of action on presynaptic and postsynaptic receptors in peripheral organs but have also provided methods for study of chemical transmission within the central nervous system.

Firstly, older methods of analysis permitted measurement of the amounts of acetylcholine, catecholamines and serotonin in various parts of the central nervous system and the presence of synthesizing enzymes for these substances was detected in brain homogenates or slices. Acetylcholinesterase was also found to be present. ACh may act on cholinoceptive neurons to cause convulsions when it is injected into the carotid artery or into the cerebral ventricles. ACh can be detected in small cups, filled with eserinized saline, placed over activated cerebral cortex and it is released from the neurons which terminate on Renshaw cells [p. 283]. These cells possess both rapidly responding nicotinic receptors and slowly responding muscarinic receptors. Most central nervous cholinoceptive receptors are muscarinic. However, although ACh was the first neurotransmitter to be recognized noradrenalin (NA), dopamine (DA), 5-hydroxytryptamine (5HT, serotonin), histamine, glutamic acid, gamma aminobutyric acid (GABA), glycine, and peptides must now also be considered as transmitters or modulators within the central nervous system.

The evidence for the presence of catecholamines within the central nervous system has been greatly extended by the formaldehyde fluorescence histochemical technique which reveals the presence of NA, DA, and also 5HT. Each of these substances can be distinguished from the others.

DA is normally a precursor of NA [p. 414], but the nigrostriatal fibres, which run from the substantia nigra to the caudate nucleus, and other neurons such as those supplying the median eminence are dopaminergic in nature. This means that DA is the end-product of the catecholamine-forming sequence, and that dopaminergic neurons do not contain dopamine β-hydroxylase. The marked reduction in DA concentration in the substantia nigra and striopallidum in Parkinson's disease is discussed on page 360; the role of DA in the control of adenohypophysial activity is discussed on page 518.

The isolation of the enzymes concerned with the conversion of tyrosine to catecholamines has permitted the use of immuno-chemical fluorescence techniques to detect these enzymes in adrenergic and dopaminergic neurons. Electron microscopy of adrenergically innervated tissues has revealed the presence of dense-core synaptic vesicles in the terminal varicosities of adrenergic nerves; these vesicles have been shown to contain NA and dopamine β-hydroxylase. In addition, since monoamines such as DA, NA, and 5HT undergo reuptake (uptake$_1$) by the nerve terminals from which they are released, autoradiography, using

[^{3}H]-labelled monoamines injected into the cerebral ventricles or directly into the brain, localizes the sites of monoaminergic nerve terminals. Finally, the use of the various methods for detecting monoaminergic neurons can be applied after experimental lesions of the brain have interrupted the nerve pathways. Such lesions can be produced surgically or by local administration of 6-hydroxy-dopamine which is selectively taken up by, and destroys, adrenergic neurons.

FIGURE VII.9 shows the distribution of adrenergic and dopaminergic neurons in the central nervous system of the rat (Livett 1973). There is an extensive distribution of adrenergic neurons with cell bodies mainly in the pons. There are both ascending and descending NA pathways. The distribution of 5HT-containing neurons is similar but FIGURE VII.9 shows the distribution of DA-containing neurons to be different; the cell bodies are more rostral and the fibres have a more restricted distribution.

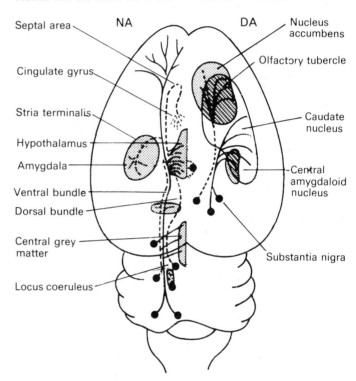

FIG. VII.9. Horizontal representation of ascending NA and DA pathways in the rat brain. (Redrawn from Livett, B. G. (1973). *Br. med. Bull.* **29**, 93.)

The functional aspects of catecholamines in the central nervous system have been discussed by Vogt (1973) and Hornykiewicz (1972, 1973). Phylogenetically DA may be the more primitive transmitter, but has been replaced by β-hydroxylated amines during evolution, except in such regions as the nigrostriatal system.

Since neither NA or DA crosses the blood–brain barrier their actions cannot be studied by injection into the bloodstream. When NA has been administered by *intraventricular* injection in cats it causes a stuporous state with a fall in body temperature. Injection of NA into into newly hatched chicks (which have a deficient blood–brain barrier) causes sleep. These findings suggest that NA directly depresses neuronal activity but it could have the same effects if it stimulated inhibitory neurons.

Application of NA by microiontophoresis to cells of the brain stem may produce excitation, inhibition, or no effect on nerve cell activity. Excitation might be due to vasoconstriction leading to ischaemic stimulation and inhibition could arise from NA action on α_2-presynaptic neuronal receptors or on α_1-postsynaptic receptors. NA is regarded by many investigators as an inhibitory transmitter in the thalamus, cerebral cortex, and cerebellar cortex. Application of NA to the supra-optic and paraventricular nuclei (which are stimulated by ACh) causes inhibition. Noradrenergic neurons suppress ACTH secretion by inhibiting the activity of the neurons which synthesize and secrete corticotrophin-releasing factor [p. 518]. It is thus probable, but not finally established, that NA generally inhibits neuronal activity. The same is true for DA with respect to its inhibitory action on the abundant cholinergic neurons in the caudate nucleus and putamen (corpus striatum). Stimulation of the substantia nigra releases DA in the corpus striatum; degeneration of the nigrostriatal tract, and loss of DA produces parkinsonism [p. 360]. The retina also contains some inhibitory dopaminergic neurons. However, the dopaminergic neurons in the hypothalamus stimulate the release of GnRH and DA is probably the prolactin-inhibiting factor in the median eminence. These substances are carried by the hypothalamico-hypophysial portal vessels to the adenohypophysis on which they act [p. 512].

In spite of their widespread distribution noradrenergic neurons comprise only about 1 per cent of all neurons in the central nervous system, and even in the hypothalamus which contains the brain's highest concentration of NA only 5 per cent of the nerve terminals do contain NA. The corpus striatum contains the highest concentration of DA in the brain, but only 15 per cent of the nerve terminals are dopaminergic. *Serotoninergic* (5HT-containing) neurons are less numerous than catecholamine-containing neurons. The presence of descending serotoninergic neurons in the brain stem and spinal cord is essential for the analgesic action of morphine [p. 401].

Cholinergic neurons comprise about 10 per cent of all central nervous neurons. It is, however, clear that monoaminergic and cholinergic neurons together constitute only a small proportion of the total number of neurons in the central nervous system. In quantitative terms the most important neurotransmitters are certain amino acids and also some polypeptides.

Amino acid neurotransmitters

There are two main types of amino acid neurotransmitter, which, unlike ACh and the catecholamines, appear to be active only in the central nervous system and not in peripheral synapses (Werman 1972; Krnjević 1974; de Feudis 1975).

1. *Excitatory amino acids.* These include glutamic and aspartic acids which depolarize most neurons in the mammalian CNS.
2. *Inhibitory amino acids.* These include gamma-aminobutyric acid (GABA) and glycine which hyperpolarize neuronal cell membranes.

The problem about amino acids is that they occur widely as residues which are joined together to make peptides and proteins throughout the body; GABA is the only potential neurotransmitter which is synthesized solely in the central nervous system. All the above-mentioned amino acids occur in higher concentrations in grey matter (containing synapses) than in white matter; there is evidence for their release from presynaptic nerve terminals and their actions mimic the effects of neuronal excitation. The evidence concerning inhibitory amino acids is much more convincing than that for the excitatory amino acids. Amino-acid transmitters are mainly inactivated by reuptake processes, but metabolic degradation also occurs.

Glycine is almost certainly an inhibitory transmitter in the spinal cord and brain stem and perhaps to a lesser degree at higher levels in the CNS. The evidence is as follows:

1. Glycine concentration in the ventral grey matter of the spinal cord is higher than that of any other amino acid. Most of the glycine in the CNS is synthesized *de novo* from glucose via serine.
2. Iontophoretic administration of glycine via a microcannula

mimics the effects of inhibitory neuron stimulation on anterior horn cells, which become hyperpolarized due to increased Cl⁻ conductance.

3. Strychnine, a spinal convulsant drug, antagonizes the inhibitory action of glycine. [³H]-strychnine is taken up by glycine receptors at separate but mutually interacting sites (Snyder 1975).

The evidence suggests that glycine is the neurotransmitter formed and released by inhibitory interneurons which act on motor neurons in the brain stem and spinal cord.

GABA is inhibitory to cells in polysynaptic pathways in the spinal cord and to cells in the cerebral cortex. The Purkinje cells of the cerebellum contain GABA which when released hyperpolarizes the cells of Deiter's nucleus and the other deep cerebellar nuclei. Strychnine does not antagonize the actions of GABA but picrotoxin does (picrotoxin does not block the action of glycine). GABA is probably the major inhibitory transmitter in the brain stem, cerebral cortex, and cerebellum. It has both presynaptic and postsynaptic actions.

Glutamic acid is the most abundant amino acid in the brain. It is concentrated in dorsal sensory nerve terminals. Glutamate depolarizes spinal motoneurons and cortical neurons. It increases Na⁺ conductance. It may be an important excitatory transmitter in dorsal root afferents and at other sites in the CNS. The role of aspartic acid as an excitatory transmitter is less certain.

Polypeptides as neurotransmitters

The concept of polypeptide-synthesizing neurons was first proposed in relation to the specialized cells in the anterior hypothalamus which form vasopressin and oxytocin. These octapeptides are transported along the neurons of the hypothalamico-neurohypophysial tract to the posterior pituitary. The passage of nerve impulses along this tract releases these peptide hormones from axonal nerve endings—a process called *neurosecretion* [p. 523]. In the 1950s anatomical and physiological evidence showed that the hypothalamus also controlled secretion of anterior pituitary hormones by synthesizing and releasing hormones which are carried along the hypothalamico-hypophysial vessels. In recent years these hormones have been shown to be polypeptides [p. 518]. The use of radioimmunoassay, immunofluorescence, and immunohistochemistry, together with the light microscope and electron microscope has revealed the presence of TRF, GnRH, and somatostatin not only in the hypothalamus but also in the rest of the central nervous system including the spinal cord (Guillemin 1976). These peptides are rapidly destroyed (in seconds or a few minutes) after release from their neurons so their actions must be very localized, as are those of neurotransmitters in general. The hypothalamic hormonal peptides in the minute concentrations which act upon the adenohypophysis, also have profound effects on spontaneous electrical activity of neurons when locally applied to various parts of the brain.

Locally synthesized peptides may act on neighbouring neurons to influence conditioned reflexes and learning; periaqueductal and periventricular regions of the brain contain receptors with which morphine and related drugs combine. These 'opiate receptors' occur in nerve structures concerned with pain perception, and there exist in the brain and pituitary peptides which bind to these receptors specifically, the binding being inhibited by the morphine antagonist naloxone [p. 401].

The physiological roles of chemical mediators or modulators are difficult to demonstrate conclusively, but it is now clear that the spontaneous activity of central nervous neurons, and the modifications of activity induced by afferent nervous stimulation, could be due to release of the many chemical agents discussed in this section. Chemical transmitters such as acetylcholine and noradrenalin act very briefly because they are very rapidly destroyed or inactivated. Peptides released by neurosecretion may act for seconds or minutes and thus account for longer lasting effects which sometimes result from very short periods of afferent nerve stimulation.

REFERENCES

GUILLEMIN, R. (1976). Physiological and clinical significance of hypothalamic and extrahypothalamic brain peptides. *Triangle* **15**, 1–7.
HORNYKIEWICZ, O. (1973). *Br. med. Bull.* **29**, 172.
HUGHES, J. (ed.) (1978). *Centrally acting peptides.* Macmillan, London.
LIVETT, B. G. (1973). *Br. med. Bull.* **29**, 93.
VOGT, M. (1973). *Br. med. Bull.* **29**, 168.
WERMAN, R. (1972). *Comp. Biochem. Physiol.* **18**, 745.

Other substances such as histamine, prostaglandins, and cyclic AMP have been proposed as mediators of neuronal activity in the CNS, but the evidence is inconclusive.

Chemical transmission at synapses is unidirectional, i.e. from presynaptic nerve terminals to postsynaptic cell membranes. It is selectively influenced by drugs which act on presynaptic or postsynaptic sites. It is not clear why there should be so many chemical transmitters in different parts of the CNS. Theoretically, transmission could be mediated by one excitatory and one inhibitory transmitter, or even more simply by one transmitter substance acting on many different kinds of receptors.

Possible role of glial cells [p. 369]. There is evidence that glial cells can take up various neurotransmitters and it has been suggested that it is a major function of neuroglia to keep the extraneuronal spaces of CNS free of transmitters (Krnjević 1974).

REFERENCES

Research Publications. Association for Research in Nervous and Mental Diseases (1972). Vol. 50 Neurotransmitters.
British Medical Bulletin (1973). Vol. 29, No. 2 Catecholamines.
BACQ, Z. M. (1975). *Chemical transmission of nerve impulses. A historical sketch.* Pergamon, Oxford.
CURTIS, D. R. and CRAWFORD, J. M. (1969). Central synaptic transmission. *A. Rev. Pharmac.* **9**, 209.
DALE, H. H. (1953). *Adventures in physiology.* Pergamon, London.
DE FEUDIS, F. V. (1975). Aminoacids as central neurotransmitters. *A. Rev. Pharmac.* **15**, 105.
GUILLEMIN, R. (1976). Hypothalamic and extrahypothalamic brain peptides. *Triangle* **15**, 1.
HUGHES, J. (ed.) (1978). *Centrally acting peptides.* Macmillan, London.
IVERSEN, L. L. (1967). *The uptake and storage of noradrenaline in sympathetic nerves.* Cambridge University Press.
—— (1977). In *Synapses* (ed. G. A. Cottrell and P. N. R. Usherwood) p. 137. Blackie, Glasgow.
KATZ, B. (1967) *Nerve, muscle and synapse.* McGraw Hill, London.
KOELLE, G. B. (ed.) (1963). Cholinesterase and anticholinesterase agents. *Handb. exp. Pharmak.* Suppl. 15.
KRNJEVIĆ, K. (1974). Chemical nature of synaptic transmission in vertebrates. *Physiol. Rev.* **54**, 418.
KUFFLER, S. W. and YOSHIKAMI, D. (1975). *J. Physiol., Lond.* **251**, 465.
McILWAIN, H. and BACHELARD, H. S. (1971). *Biochemistry and the central nervous system.* Churchill Livingstone, London.
McLENNAN, H. (1970). *Synaptic transmission*, 2nd edn. Saunders, Philadelphia.
TRIGGLE, D. J. and TRIGGLE, C. R. (1976). *Chemical pharmacology of the synapse.* Academic Press, London.

PART VIII
Digestion

Structure of gastrointestinal (GI) tract

The wall of the tube passing from the mouth to the anus has four coats [Fig. VIII. 1].

(a) *The serous coat*: The pharynx, oesophagus, and rectum are attached to surrounding structures by fibrous tissue. The remainder of the alimentary canal lies less firmly attached in the abdomen with the surface covered, except along its attached border, by a serous membrane (the visceral peritoneum).

(b) *The muscular coat*: There is an inner circular and an outer longitudinal layer of smooth muscle. At the sphincters the circular layer is thicker. Between these two layers of smooth muscle lies the *myenteric plexus of Auerbach*. This intrinsic nerve plexus is under extrinsic autonomic control by both parasympathetic and sympathetic nerve fibres. The activity of the muscles of the gastrointestinal tract promotes both local mixing and forward propulsion of the contents of the gut.

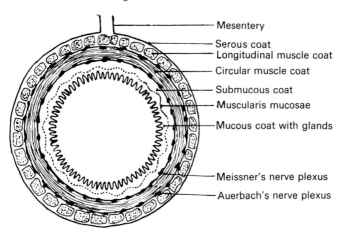

FIG. VIII.1. Diagram of cross-section of alimentary tract, showing the different coats of the wall.

(c) *The submucous coat*: this consists of loose connective tissue, blood vessels, and lymphatics. Between the submucous coat and the inner circular muscle lies the *submucous nerve plexus of Meissner* which receives connections from that of Auerbach and from the extrinsic autonomic nerves.

(d) *the mucous coat*: this is lined by epithelium and consists of a stroma containing glands and of the muscularis mucosae. The stroma consists of a loose connective tissue rich in lymphocytes. The muscularis mucosae consists of smooth muscle.

SALIVARY GLANDS

There are three pairs of salivary glands—parotid, submandibular, and sublingual. The parotid gland lies in front of the ear. Its secretion passes via Stensen's duct which opens into the mouth opposite the site of the second molar tooth. The duct may be easily cannulated. The submandibular gland lies medial to the mandible in the so-called submaxillary triangle. Its duct (*Wharton's duct*) opens into the floor of the mouth at the side of the frenulum linguae. The sublingual glands lie immediately subjacent to the mucosa of the floor of the mouth. Numerous ducts collect and discharge the sublingual secretion into the sublingual part of the mouth.

Salivary glands may contain either mucous cells or serous cells. Mucous cells contain large translucent granules (consisting of a precursor of mucin) and such cells appear pale or translucent in histological sections. Serous cells have opaque small zymogen granules (consisting of a precursor of ptyalin). The parotid gland is purely serous. The submandibular gland contains both types of cell, but is predominantly serous, whereas the sublingual gland, though also a mixed gland, is predominantly mucous. Mucous glands form a viscid secretion containing mucin; serous glands form a thin watery secretion containing ptyalin. Both sets of glands contain their secretory cells in acini in each of which the cells are arranged round a central lumen which leads to a duct. These ducts join successively to form intralobular and interlobar ducts which lead to the main duct.

Innervation of the salivary glands

All glands receive both parasympathetic and sympathetic nerve supplies; the parasympathetic innervation is by far the more important.

The parotid gland receives parasympathetic fibres. Cells of the inferior salivary nucleus (dorsal nucleus of IXth) provide preganglionic neurones which course via the tympanic nerve and the small superficial petrosal nerve to the otic ganglion to form a synapse with the otic ganglion cells. Postganglionic fibres from these cells join the auriculotemporal nerve to reach the parotid gland, where they are distributed to the secreting cells and to the blood vessels of the gland. As 'taste' fibres from the posterior third of the tongue pass by the glossopharyngeal nerve to end in the dorsal nucleus of IX afferent impulses from the mouth reflexly excite salivary secretion.

The submandibular and sublingual glands receive their parasympathetic innervation, again interrupted by a ganglionic relay, originating from the superior salivary nucleus (dorsal nucleus of VIIth). Preganglionic fibres course in the nervus intermedius, join the facial nerve and leave by its chorda tympani branch to reach the lingual nerve. These preganglionic fibres form synapses with ganglion cells scattered in the vicinity of the sublingual and submandibular glands. Postganglionic fibres supply both secreting cells and blood vessels of the glands. 'Taste' fibres from the anterior two-thirds of the tongue, conveyed by the nervus intermedius to the superior salivary nucleus provide another simple reflex arc for salivary secretion.

Salivary secretion

Stimulation of the chorda tympani causes vasodilatation and secretion of the submandibular gland. The secretory response is abolished by atropine, but the vasodilatation is not. Hilton and Lewis (1957) have shown that parasympathetic stimulation liberates a proteolytic enzyme kallikrein from the gland cells which acts on plasma α_2-globulins in the interstitial fluid to form the vasodilator nonapeptide bradykinin.

The formation of saliva entails secretory work on the part of the gland cells. Saliva is hypo-osmolar compared with plasma and the work required for the elaboration of such a solution far exceeds that which could be provided by the hydrostatic pressure of the blood supplying the gland. Secretory work of the gland cells entails an increased metabolic turnover and the oxygen consumption of the gland correspondingly shows a fivefold increase during activity compared with that at rest. There is a discrepancy between the initiation of the raised secretory rate and that of the raised oxygen usage. Moreover, the oxygen usage remains high for some minutes after secretion in response to nerve stimulation has ceased. It has been suggested that the rapid onset of secretion in response to stimulation is allowed by the breakdown of energy-rich phosphate compounds which require oxidative energy for their resynthesis during the recovery period after stimulation.

Burgen and Seeman (1958) showed that the acinar cells secrete K^+ and HCO_3^- into the acinar lumen, accompanied by sufficient Cl^- to preserve electrical neutrality. This primary secretion is approximately isotonic owing to the simultaneous passage of water into the acinar lumen. The ducts which drain the acini have a rich blood supply and possess, moreover, cells in their walls which are histologically more complex than those which ordinarily line simple conduits. In keeping with these facts the salivary duct cells *actively reabsorb K^+* and an accompanying anion and transfer some Na^+ into the saliva. The end result is hypotonic saliva, so the duct cells must be relatively impermeable to water.

Iodide and thiocyanate, which are excreted in saliva, are also actively transported from the plasma directly into the lumen of the ducts by the cells of the duct wall; they are not transported by the acinar cells.

Saliva contains K^+ at a concentration ($15–20$ mM l^{-1}) higher than that in the plasma. Na^+ concentration of saliva is always lower than that in the plasma, as is that of chloride. $[Na^+]$, however, rises to a plateau concentration of $80–90$ mM l^{-1} as salivary flow increases [Fig. VIII.2]. Bicarbonate concentration usually exceeds that in the plasma and $[HCO_3^-]$ rises as the secretory rate increases. The pH of saliva, which is below 7.0 at low secretory rates, therefore increases

as the rate of salivary formation steps up. Mucin secreted by the mucous cells is a useful lubricant. Ptyalin secreted by the serous cells is an amylase which initiates the digestion of starch. Given time it can digest starch to maltose. Such digestion ordinarily continues in the interior of the bolus of food formed by chewing and mixing with the saliva even when this bolus has reached the stomach. The pH of gastric juice is far below that for optimal activity of the salivary enzyme, so once the gastric juice penetrates the bolus, ptyalin action is terminated.

The functions of saliva are mainly mechanical. It assists mastication and swallowing and it aids speech by lubricating the mouth. Saliva also minimizes the risk of buccal infections and dental caries. Salivary secretion is provoked either by the taste of food (inborn reflex) or by the thought of food (psychic or conditioned reflex).

REFERENCES

BURGEN, A. S. V. and EMMELIN, N. (1961). *Salivary glands*. Arnold, London.
—— and SEEMAN, P. (1958). *Can. J. Biochem. Physiol.* **36**, 119.
HILTON, S. M. and LEWIS, G. P. (1957). *Br. med. Bull.* **13**, 189.
JENKINS, G. N. (1970). *The physiology of the mouth*, 3rd edn. Blackwell, Oxford.

Swallowing

Swallowing, as other movements of the alimentary canal, can be studied by:

1. X-ray techniques. 2. Measurement of intraluminal pressures; this has been done by balloons, but more accurate results are obtained by open-tipped tubes connected to transducers which register pressure.

Swallowing occurs in three stages: oral, pharyngeal, and oesophageal; the first stage is voluntary, the other two are reflexly produced.

1. After mastication the food is rolled into a bolus, which lies in the curve of the tongue. Swallowing commences by closing the mouth and voluntary contraction of the mylohyoid muscles, which throws the bolus back between the pillars of the fauces on to the post-pharyngeal wall. This region of the pharynx has a rich sensory innervation from the glossopharyngeal nerves; when the local nerve endings (and also those in the soft palate and epiglottis) are stimulated, afferent impulses are set up which reflexly (via the so-called deglutition centre in the medulla) produce the complex co-ordinated movements occurring in the involuntary phases of swallowing.

2. The soft palate is elevated and thrown against the post-pharyngeal wall to close off the nasal cavity. The larynx rises with the elevation of the hyoid, and the pharynx is practically obliterated. The vocal cords are approximated, and breathing is momentarily inhibited. The posterior pillars of the fauces approximate to shut off the mouth cavity. The pharynx reopens to permit the passage of the bolus; the epiglottis guards the laryngeal opening until the bolus reaches the oesophagus which simultaneously opens up to receive it. Aspiration of the food into the larynx is also prevented by the associated reflex apnoea. Cricopharyngeus briefly relaxes and the bolus enters the upper oesophagus. Then the cricopharyngeal muscle contracts and the vocal cords open to allow the resumption of rhythmic breathing.

3. The bolus is then propelled along the oesophagus by peristaltic waves in its muscle coat. Gravity plays little part in this process, as the rate of progress along the oesophagus is not affected by posture; it is as rapid in the supine as in the erect position.

The swallowing reflex is temporarily abolished by anaesthetizing the pharynx with cocaine; it is deranged by lesions of the medulla

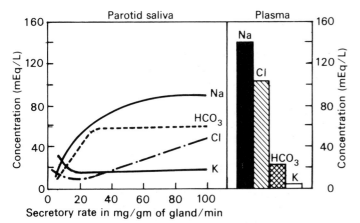

FIG. VIII.2. Electrolyte composition of human parotid saliva as a function of secretory rate.
[Redrawn from Bro-Rasmussen, F., Killmann, S., and Thaysen, J. H. (1956). *Acta physiol. scand.* **37**, 97.]

oblongata or of the ninth and tenth nerves. Food may then regurgitate into the nose or be aspirated into the larynx.

REFERENCES

Bosma, J. F. (1957). Deglutition: pharyngeal stage. *Physiol. Rev.* **37**, 275.
Code, C. F. (1968). *Handbook of physiology*, Section 6, Vol. IV. American Physiological Society, Washington, DC.

Upper oesophageal sphincter

The upper end of the oesophagus is normally shut off from the pharynx, and there is resistance to the passage of a gastroscope. The upper oesophageal sphincter can be relaxed voluntarily by sword swallowers or by those beer drinkers who can 'pour it down' without swallowing (Ingelfinger 1958). Normally this sphincter opens 0.2–0.3 seconds after the beginning of a swallow, remains open for 0.5–1.0 seconds, and then closes.

The oesophagus

At rest the oesophagus (25 cm in length) is relaxed and may contain air or other material. Peristalsis is usually initiated by swallowing but may also arise in response to local stimulation at various levels in the oesophagus. It consists of a lumen-obliterating contraction 4–8 cm in length which moves aborally at 2–4 cm s^{-1}. Once initiated, peristalsis travels down the oesophagus whether or not a bolus is present . Oesophageal peristalsis depends on the integrity of the vagal efferents and on local reflexes involving Auerbach's plexus. Afferent stimuli from various sources can modify peristalsis, but the existence of a supravagal swallowing centre is not established.

Antiperistalsis (or reverse peristalsis), presumably occurs in ruminants but has not been clearly demonstrated in man.

Lower oesophageal (cardiac) sphincter

The last 2–5 cm of the oesophagus is sphincteric in action although there is no anatomical sphincter. When swallowing is not occurring, the sphincter is in a state of tone and its walls are tightly in apposition. Within 1.5–2.5 seconds of swallowing the cardiac sphincter relaxes, the inhibition requiring intact vagi and Auerbach's plexus. As a result of this rapid relaxation even a swift-moving liquid bolus meets little resistance at the lower sphincter. When a peristaltic contraction reaches this region the sphincter closes and may undergo a strong, prolonged after-contraction. This prevents regurgitation of food, gastric juice, and air. If the intragastric pressure is excessively raised by rapid air-swallowing (aerophagy) or by the evolution of CO_2 from ingested $NaHCO_3$ the resistance of the cardiac spincter may be overcome, and gas is expelled from the mouth.

Gastrin [p. 427] increases the tone of the cardiac sphincter and may serve to keep it closed during gastric digestion.

The condition *achalasia* is characterized by a failure of the sphincter to relax on swallowing. It seems to be due to destruction of the local nerve plexuses. Such patients show an exceptional sensitivity of the cardia to circulating gastrin. They also show intense sphincteric spasm in response to very small doses of methacholine. This 'denervation sensitivity' has been employed diagnostically.

REFERENCE

Inglefinger, F. J. (1958). *Physiol. Rev.* **38**, 533.

NATURE OF PERISTALSIS

Bayliss and Starling defined true peristalsis as a co-ordinated reaction in which a wave of contraction preceded by a wave of relaxation passes down a hollow viscus; the contents of the viscus as they are propelled along would thus always enter a segment which had actively relaxed and enlarged to receive them. This type of movement was thought to be responsible for transferring the contents of the alimentary canal from the oesophagus through the stomach, small and large intestine, and finally to the anus.

The passage of peristaltic waves along the oesophagus depends on the continuity of the preganglionic vagal nerve supply but not on the integrity of the muscle coat. If the oesophageal wall is divided and the superficial nerve plexus is left intact, the peristaltic wave can still pass normally over the oesophagus.

In the stomach and intestines, however, peristalsis can occur in the absence of extrinsic nervous influences (preganglionic vagus or pre- and postganglionic sympathetic) though it is modified by the activity of these nerves. Thus gastric tone and motility are initially decreased by vagotomy in man [p. 429]. After section of the vagi and destruction of the abdominal ganglia of the sympathetic in animals, intestinal peristalsis continues normally for months. Normal gastric and intestinal peristalsis is attributed to a series of co-ordinated local nervous reflexes (involving possibly Auerbach's plexus) in response to the chemical and mechanical stimulation set up by the food.

The stomach

The stomach serves as a reservoir for the food ingested and by its secretory activity provides the enzymes and hydrochloric acid required for the initial digestion of protein. Its shape is shown in Figure VIII.3, which also indicates the terms applied to the various parts of the organ.

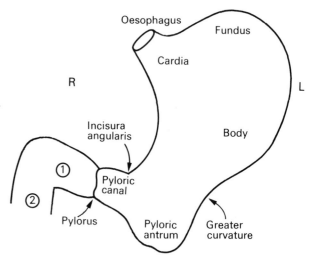

Fig. VIII.3. Gross anatomy of the stomach. Note that the pyloric canal ends at the pyloric sphincter. 1 and 2 are the first and second parts of the duodenum.

The part of the stomach to the left of the incisura angularis is the body, and that to the right the pyloric part: the part of the body above the level of the cardiac orifice is the fundus. The pyloric part is divided into the pyloric antrum or vestibule, and the pyloric canal. Functionally the first part of the duodenum (duodenal bulb or cap) is associated with the pyloric part.

Arrangement of musculature

The stomach has an outer longitudinal and an inner circular coat; between the mucous membrane and the circular coat is an additional incomplete but well-developed muscular layer which runs from the oesophagus down either side of the lesser curvature and then spreads out in a fan-like manner. These oblique fibres fuse finally with the circular; they are supposed to maintain the normal length of the lesser curvature.

At the pylorus the circular fibres are thickened; additional development of radially disposed fibres of the muscularis mucosae throws the mucous membrane into a well-marked fold—the pyloric sphincter.

When the stomach is empty its walls are firmly in apposition. When food enters the stomach, the muscle fibres are elongated to enlarge the size of the cavity uniformly in order to accommodate the new contents, without much change of internal pressure. With an opaque meal of average composition, the stomach assumes varying shapes. Most commonly it is **J**-shaped; the body forms a vertical tube and the pyloric part forms a horizontal or slightly ascending segment which is turned to the right. Foods pass through the stomach to the pyloric antrum roughly in the order in which they are swallowed; but if a heavier food succeeds a lighter it sinks in the stomach contents till it finds its own level.

SECRETIONS AND FUNCTIONS OF GASTRIC JUICE

Gastric mucous membrane

1. Main gastric glands. The bulk of the gastric mucosa contains the *main* gastric glands which possess short ducts and long alveoli [FIG. VIII.4]. The alveoli contain *chief* or *peptic* cells, which secrete pepsin, and less numerous ovoid *parietal* or *oxyntic* cells which secrete H⁺. The surface of the gastric mucosa consists of columnar cells which secrete mucus; this surface is pitted by the openings of the gastric glands—some 100 pits are found per mm², each pit being formed by the opening of the ducts of several glands.

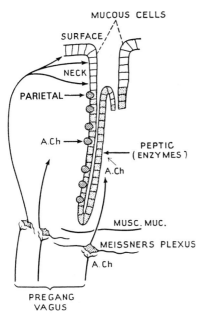

FIG. VIII.4. Diagram of nerve supply of gastric mucosa and of chemical transmitters concerned. Pregang. Vagus: preganglionic fibres of vagus end in Meissner's plexus. Transmitter is ACh (acetylcholine). Postganglionic fibres from Meissner's plexus end in gland cells.

Mucus is secreted by the surface cells as a gel which is alkaline; in the resting state the lining of the stomach is covered by an adherent mucoid alkaline layer. Any mechanical stimulation of the surface mucosa promotes the rate of formation of mucus; neither vagal stimulation nor histamine has any effect on mucus production by these cells. This layer of mucus serves a protective function preventing damage of the mucosa by acid or peptic digestion.

The numerous peptic cells contain granules of pepsinogen in the resting phase. Pepsinogen is a protein (MW = 42 500) which is converted by acid (optimally at pH = 2) to the active proteolytic enzyme pepsin (MW = 35 000) by the loss of polypeptide from the pepsinogen molecule. Pepsinogen extruded from the peptic cells in response to vagal stimulation is quite inactive and unless the pH of the gastric secretion is lower than 6.0 (as it almost invariably is) the formation of the active enzyme pepsin is very slow. Pepsin hydrolyses peptide bonds between phenylalanine and another amino acid and thus digests proteins to polypeptides of varying size and complexity.

The volume of secretion produced by the peptic cells cannot be determined directly, for it accompanies the greater volume of secretion by the parietal cells out of the ducts.

Atropine abolishes pepsinogen secretion, as might be expected.

Gastric HCl secretion is solely a property of the parietal cells found chiefly in the body of the stomach. Pure parietal secretion has been shown to be approximately 0.5 N HCl, but gastric juice collected from the body of the stomach is less acid than this owing to its contamination with non-acid secretions which dilute or partially neutralize it. The maximum acid secretory response has been measured before and after partial gastrectomy in patients with peptic ulcer and correlated with the parietal cell mass. The number of parietal cells in the excised mucosa is estimated and plotted against the reduction in maximum acid secretion following the operation. About one thousand million parietal cells can produce a maximal output of 20 mM H⁺ per hour in response to histamine stimulation.

The plasma concentration of H⁺ is 0.00004 mM l⁻¹ (40 ng l⁻¹) and the parietal cell raises the concentration in its secretion to 150 mM l⁻¹. This requires energy derived largely from oxidative metabolism of carbohydrate. Thus parietal secretion is abolished by such inhibitors as fluoroacetate (which stops the tricarboxylic acid metabolic cycle, p. 454) and dinitrophenol which though it permits oxidation to continue prevents the regeneration of energy-rich phosphate compounds such as ATP. According to Davenport (1967) the crucial reaction in H⁺ secretion is an oxidation in which a high energy intermediate compound is oxidized to a low-energy product and a hydrogen ion; the energy liberated is used to transfer the hydrogen ion into the gastric juice against the concentration gradient. As the high-energy compound is oxidized to generate the hydrogen ion it gives up an electron which must be carried down the electron-transporting system to be accepted by oxygen [FIG. VIII.5]; as a result one OH⁻ ion is formed for every H⁺

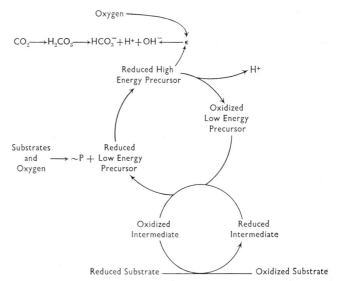

FIG. VIII.5. A scheme of the metabolic basis of gastric acid secretion. [From Davenport, H. W. (1971). *Physiology of the digestive tract*, 3rd edn. Year Book, Chicago.]

secreted and this OH⁻ must be neutralized if H⁺ secretion is to continue. Neutralization is accomplished by H⁺ provided from carbonic acid which is rapidly formed in the cell from metabolically produced CO_2 owing to the presence of carbonic anhydrase in the cytoplasm [Fig. VIII.6]. As a result each hydroxyl ion formed is replaced by a bicarbonate ion which is then discharged into the venous blood. Acetazolamide applied topically or given intravenously in large doses decreases H⁺ secretion by the parietal cells very considerably.

During gastric digestion, gastric venous blood contributes a greater amount of HCO_3^- to the systemic circulation and the plasma pH correspondingly rises; breathing is slightly depressed and the alveolar pCO_2 rises (alkaline tide).

The gastric mucosa also possesses a chloride pump. Chloride thus moves from the cell into the juice not only against a concentration gradient, but also against the electrical gradient, for the mucosal surface is 60 mV negative to the serosal surface of the stomach. The H⁺ and Cl⁻ pumps are closely coupled. Water moves passively from cell to juice and gastric juice is isotonic with the plasma.

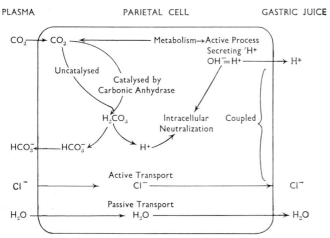

Fig. VIII.6. Role of carbon dioxide in intracellular neutralization during secretion of acid by the gastric mucosa.
[From Davenport, H. W. (1971). *Physiology of the digestive tract*, 3rd edn. Year Book, Chicago.]

In addition to pepsinogen (and its derivative pepsin) and HCl the main gastric glands secrete a heat-labile mucoprotein the *intrinsic factor* which combines very firmly with dietary vitamin B₁₂. (Gastric atrophy results in a failure to absorb vitamin B₁₂.) Rennin, an enzyme which curdles milk, acting on caseinogen which is converted into casein and thence into insoluble calcium caseinate, is not present in human gastric juice. It is found only in the gastric juice of young animals. Its function can be performed by pepsin.

2. Pyloric (antral) glands are found in the mucosa between the level of the incisura angularis on the lesser curvature and the pylorus and to a lesser extent on the greater curvature. These glands resemble the duodenal glands of Brunner in having long ducts and short alveoli; they secrete a mucus-rich alkaline viscid juice which is poor in enzyme content. Their rate of secretion is low (0.5–5 ml h⁻¹) and is unaffected by feeding or by vagal stimulation. The mucus secreted is supposed to lubricate the surface over which a large volume of chyme moves back and forth during digestion.

In the deeper portions of the pyloric glands are found the so-called 'G'-cells which secrete the important hormone *gastrin* [p. 427].

3. Cardiac tubular glands consisting of cells which secrete mucus

are found in the cuff of gastric mucosa which immediately surrounds the oesophagus.

Approximately 2.5–3 l of gastric juice are secreted daily.

Collection of pure gastric juice

In animals (dog) the following procedure is employed. A special stomach pouch is prepared as described by Pavlov (a Pavlov pouch [Fig. VIII.7]). A small pouch of the stomach is separated from the main body of the organ by a double layer of mucous membrane; the open end is brought up to the surface of the body. The nervous and vascular connections of the pouch are left intact. Pure gastric juice unmixed with food can be obtained from the pouch while the digestive processes are proceeding in the main stomach. The vagus nerve is exposed in the neck under anaesthesia and divided. A few days are allowed to elapse for the inhibitory fibres to the heart to degenerate; for some unknown reason the vagal fibres to the stomach (and pancreas) survive longer than those to the heart. The peripheral end of the vagus may then be stimulated in the unanaesthetized animal; a flow of gastric juice results after a short latent period, proving that the vagus is the secretory nerve to the stomach [Fig. VIII.7].

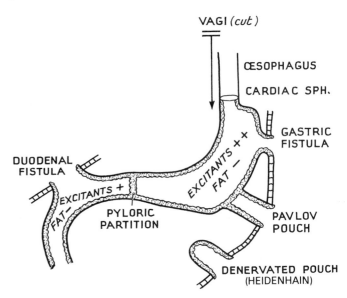

Fig. VIII.7. Diagram illustrating experimental analysis of humoral factors regulating gastric secretion. The pyloric partition is at the pyloric sphincter.

Note: a denervated stomach pouch (Heidenhain pouch) secretes in response to blood-borne gastrin.

In man the gastric juice is aspirated from the stomach after swallowing a fine rubber tube. The presence of blood or organic acids is suggestive of gastric ulcer or neoplasm. Endoscopy is used to check such a provisional diagnosis.

Regulation of secretion of gastric juice

Gastric secretion occurs in response to reflex stimulation of the vagi accompanied by the release of the excitatory hormone gastrin.

The taste of food reflexly provokes gastric secretion—the so-called *psychic* or *appetite juice*. In addition the sight, smell or thought of food causes gastric secretion by conditioned reflexes.

The sham feeding technique (Pavlov) proves the mechanism of production of appetite juice. The oesophagus is divided in the neck of a dog and the two ends are brought separately to the surface and sutured in position. The animal is allowed to recover from the operation; it is kept in good condition by placing food-stuffs in the lower oesophageal opening. When it feeds, food is collected from the upper oesophageal opening; despite the fact that the food never reaches the stomach, appetite juice begins after a latency of 5–7 minutes; the volume secreted reaches its peak within 1 hour and may persist for 3 hours. *The juice is highly acid and is rich in pepsin.* A similar secretion is provoked by insulin hypoglycaemia which causes increased vagal discharge [p. 428]. Appetite juice is abolished by atropinization or by section of the vagi.

The appetite juice formed as a result of vagal activity initiates production of gastrin and the digestion of the food which has entered the stomach. The presence of meat and its digestive products also causes the release of the hormone *gastrin* from the pyloric antral mucosa into the gastric venous blood, portal circulation and systemic circulation. Gastrin returns via the arterial blood to stimulate the parietal (oxyntic) cells of the body of the stomach. There is a latency of some 30–60 minutes between the introduction of food into the stomach via a gastric fistula and the secretion of gastric juice thus provoked humorally; the response lasts for 2 hours. The intravenous injection of pure gastrin provokes secretory activity within 5 minutes. Gastrin excites the liberation of a highly acid juice but it does not excite the production of pepsin.

It has been customary hitherto to regard gastric secretion as involving two separate phases—vagal and humoral. However, Uvnäs (1942) postulated that vagal stimulation caused a greater gastric secretion in the cat when the pyloric antral part of the stomach (known to produce the hormone gastrin) was intact. Nowadays it is recognized that vagal activity not only causes direct stimulation of the secretion of gastric juice by the parietal cells but also provokes the liberation of gastrin from the 'gastrin cells' of the pyloric antrum.

The gastrin cell is located in the deeper parts of the pyloric glands. It possesses electron-opaque granules which disappear after feeding, leaving vacuoles. Fluorescein-labelled antibodies to human gastrin have demonstrated these specifically reacting cells and their distribution in the mucosa corresponds with the areas of mucosa from which gastrin can be extracted. The gastrin cell is an example of the so-called APUD cells of Pearse. APUD means Amine Precursor Uptake and Decarboxylation. They can be demonstrated by formaldehyde-induced fluorescence. They are derived from the neural crest.

Gastrin is released both by vagal stimulation and by chemical stimuli within the gastric lumen. Gastrin release is also induced when a denervated transplanted *antral* pouch is distended as was proved by Grossman and his colleagues who studied acid secretion in a denervated (Heidenhain) pouch under these circumstances. Grossman also showed, however, that balloon distension of a denervated pouch of the *fundic gland area* induces acid secretion by exciting an intrinsic cholinergic mechanism within the walls of the fundus. Moreover, he proved that vago-vagal reflexes are also implicated, because the rate of acid secretion of a vagally innervated pouch is much greater than that of a denervated pouch, even when the pyloric antral area has been resected.

Clearly nervous and humoral stimulation of gastric secretion interact [Fig. VIII.8] and are not merely additive but are markedly synergistic. Thus sham-feeding causes a greatly reduced secretion from the parietal glands if the antrum (which produces gastrin) has been removed; the response can be restored by infusing gastrin during the sham-feeding, although the infused quantities of gastrin are by themselves inadequate to provoke secretion of juice.

Gastrin was identified by Gregory and Tracy and their colleagues in the 1960s as a polypeptide of 2000 molecular weight (17 amino acids) in two forms I and II, also known respectively as little gastrin I and little gastrin II. The two forms differ only by the presence of a sulphate group bound in ester form to the single tyrosyl in the molecule. The molecule is a heptadecapeptide whose biological activity resides in the C-terminal tetrapeptide amide: Tryp.Met. Asp.Phe–NH_2 (which is shared by the synthetic compound penta-

gastrin). As will be seen later, cholecystokinin which is identical with pancreozymin (CCK–PZ) has the same C-terminal tetrapeptide sequence as gastrin and indeed has *qualitatively* the same biological activity as gastrin. Their quantitative effects differ, however, in that whereas gastrin powerfully excites gastric secretion and only feebly influences contraction of the gall-bladder, CCK-PZ causes strong contractions of the gall-bladder and only moderately stimulates gastric secretion. CCK-PZ has 29 amino-acid residues in addition to the tetrapeptide moiety and these contain a sulphated tyrosyl group in the position of the seventh amino-acid residue from the C-terminal.

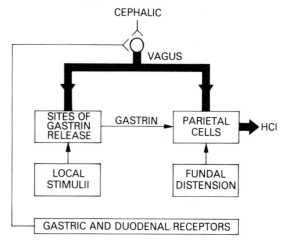

FIG. VIII.8. Interaction of vagal and humoral factors responsible for gastric secretion of HCl.
[Redrawn from Blair, E. L. (1974). In *Recent advances in physiology*, 9th edn (ed. R. J. Linden) Chapter 7. Churchill Livingstone, Edinburgh.]

Gastrin has been demonstrated in the blood by radioimmunoassay techniques. Protein ingestion increases the blood level in man.

Little gastrins have half-lives of about three minutes. They are inactivated by the liver and kidney cortex.

Pentagastrin is a synthetic compound which contains the four terminal amino acids of gastrin (Tryp-Met-Asp-Phen-NH_2) with β-alanine added at the N terminus. It is valuable for testing gastric secretory functions [p. 428].

Zollinger–Ellison syndrome. Some pancreatic tumours (of δ-cells) secrete large amounts of gastrin. The patient secretes excessive quantities of gastric HCl and is prone to peptic ulceration.

Inhibition of gastric acid secretion

Certain factors *inhibit* acid secretion. These are: (i) emotional—fear, grief, and sensations of nausea; (ii) high levels of [H^+] in the pyloric antrum or proximal duodenum; (iii) the presence of fat in the duodenum; (iv) the presence of hyperosmolar concentrations in the duodenum.

When the pH of the contents of antrum and duodenum falls below 1.5, the reduction of gastric acid secretion is absolute [Fig. VIII.9]. Hyperacidity of the antral contents prevents the *release* of gastrin; acidification of the antrum *does not* prevent the gastric acid secretory response to an *intravascular* infusion of gastrin. It is noteworthy incidentally that patients with achlorhydria (such as those with pernicious anaemia) have a very high concentration of gastrin in their blood; if they swallow a solution of hydrochloric acid of pH 2–3, the blood gastrin concentration falls.

The inhibitory effect of acid in the duodenum on the other hand is attributed to the release of an inhibitory hormone or *chalone* known as 'enterogastrone'. Thus acidification of the proximal duodenum stops the response of denervated fundic pouches to the intravenous injection of gastrin. Fat in the duodenal lumen also

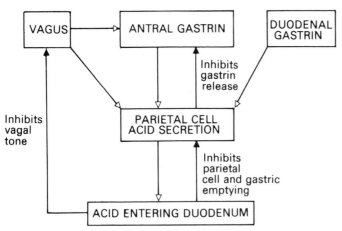

Fig. VIII.9. A summary of the inhibitory effects of acid in the antroduodenal region on the entry of acid into the duodenum.
[Redrawn from Blair, E. L. (1974). In *Recent advances in physiology*, 9th edn (ed. R. J. Linden) Chapter 7. Churchill Livingstone, Edinburgh.]

inhibits acid secretion in transplanted fundal pouches, but the fat must be partly hydrolysed to exert this effect—i.e. already partly digested by pancreatic juice and bile.

At present there is no certainty as to whether one or more chalones are responsible for the inhibitory effect of duodenal acid or fat on gastric acid secretion. Both secretin and CCK-PZ (pancreozymin) inhibit gastrin-stimulated acid production by the parietal cells in the dog, but not in the cat. A polypeptide of molecular weight approximately 5000 which contains 43 amino acids has been isolated from duodenal mucosa and has been shown to exert anti-acid secretory effects. It has been correspondingly named gastric inhibitory polypeptide (GIP). It is *not* structurally identical with either secretin or CCK-PZ. It is released by fat and digestion products present in the lumen of the duodenum. It inhibits gastric emptying and may be 'enterogastrone'.

Tests of gastric secretory function

Pentagastrin test. The synthetic compound pentagastrin simulates the effect of natural gastrin. Intramuscular injection (5 μg per kg) of pentagastrin induces a rapid increase of output of H$^+$ which reaches a peak of 25–40 mmol h^{-1} within 15 minutes in normal young men. This peak secretion rate is maintained for approximately 15 minutes more; H$^+$ secretion rate then slowly declines to the resting level during the course of the next half-hour.

Patients with duodenal ulcer often but not invariably show an increased response. Patients with pernicious anaemia show no response.

Insulin test. The intravenous injection of insulin causes hypoglycaemia which in turn provokes hypothalamic stimulation of the vagal nuclei. As a result gastric secretion occurs and the juice collected shows the high acidity and high pepsin content of 'vagal juice'.

If the hypoglycaemia caused by insulin be prevented by giving glucose, no gastric secretion occurs in response to the insulin injection.

The insulin test provides a check on the success of the operation of gastric vagotomy (which is performed for peptic ulceration).

MOVEMENTS OF THE STOMACH

The effects of extrinsic nerves on gastric motility

Blair (1974) states that at least three distinct neural mechanisms are involved in the effects of the vagus and sympathetic nerves. One of these induces contraction and two of them cause relaxation.

Jansson and Martinson (1965) showed that gastric relaxation is best recorded by studying alterations of gastric volume when the intragastric pressure is held constant below 8 cm H$_2$O. Blair and his colleagues used this technique to prove that the relaxation induced by vagal stimulation is not prevented by atropine or guanethidine. Intracellular recordings show that vagal and sympathetic stimulation may both cause hyperpolarization of the same cell. Gastric contraction may indeed result from appropriate vagal *or* sympathetic stimulation. The situation remains far from clear. Nevertheless, there is unequivocal evidence in man that gastric motility is decreased after vagotomy for peptic ulcer and that atropine stops gastric contractions for a long time. Wolf and Wolff (1947), who studied the exposed gastric mucosa of their subject 'Tom', noted that eserine (an anticholinesterase) induced powerful contractions of the stomach.

Although food intake increases the gastric peristaltic movements the gastric pressure remains low because of the law of Laplace $P = T/R$, where P is the pressure, T the wall tension, and R the radius. As gastric filling increases the radius, the tension in the wall rises, but the transmural pressure is little affected.

The ability of smooth muscle to develop active tension over a wide range of length aids the stomach to raise the intragastric pressure by contraction even when it is stretched by the food ingested.

The stomach shows a basic electrical rhythm which is initiated by 'pace-maker' cells situated near the fundus on the greater curvature and which propagates initially slowly—less than 1 cm s^{-1} but faster as the contraction wave approaches the antrum at a speed of 4 cm s^{-1}. In man these potentials (recorded extracellularly) occur about 3 per minute. This basal electrical rhythm is not necessarily followed by a contraction. When a fasting animal is fed, the number of potentials followed by contractions increases fivefold. The spread of electrical activity is myogenic involving direct electrical conduction from cell to cell by electrotonic spread through the low-resistance cell contacts provided by nexuses.

The stomach possess an intrinsic nervous system, but neither its precise organization nor its functional significance is known. Ganglion cells in the myenteric (Auerbach's) plexus and submucous (Meissner's) plexus are believed to be concerned with efferent stimulation of the muscle cells. They have not been identified by fluorescent techniques as containing catecholamines. However, degeneration studies following vagotomy and sympathectomy have shown that these autonomic nerves terminate in the two plexuses.

The basic electrical rhythm of the stomach is disrupted following vagotomy probably due to the unopposed action of the sympathetic nerves, for similar disturbances of the rhythm result from the injection of large doses of catecholamines into the bloodstream.

Gastric emptying

The food that enters the stomach is usually a mixture of liquids and solids, while the chyme that leaves the stomach is essentially liquid. Gastric emptying begins as soon as a large part of the gastric contents becomes fluid enough to pass the pylorus.

Radiological studies show that peristaltic waves arising in the neighbourhood of the cardia increase in force as they move towards the antrum. It was once believed that the rate of discharge of chyme (gastric fluid contents) was due to variations in tone of the pyloric sphincter but even if a tube is inserted in the pylorus, to maintain the patency of the connection between the stomach and the duodenum, the rate of gastric emptying is little affected. It is now accepted that it is the *force* of gastric peristalsis which determines emptying. Cineradiography reveals that most peristaltic waves reach the pyloric antrum where some of the gastric contents pass through the open pylorus to the duodenum and some are pushed back into the body of the stomach. As the contraction wave occludes the terminal antrum and pyloric canal the antral contents are almost completely returned to the body of the stomach to re-enter the antrum as relaxation supervenes. Gastric emptying results from a progressive wave of contraction which sequentially involves antrum, pylorus, and proximal duodenum.

The rate of emptying has been shown by Hopkins (1966) to be related to the square root of the volume of a liquid meal remaining in the stomach. This suggests that it is wall tension which is the controlling factor. The radius of a cylinder varies with the *square root* of its volume and from Laplace's law: $T \propto R$. How tension acts as the adequate stimulus to peristalsis is not known, but it may be by a nervous mechanism. Thus gastric receptors have been identified in the vagi by Paintal (1954) and by Iggo, showing a discharge rate which is proportional to wall tension. They may reflexly provoke motor vagal discharge which leads to gastric contractions.

The duodenum also modifies the rate of emptying. Thus whether food ingested is hyper- or hyposmolal, the chyme reaching the duodenum is very quickly rendered isosmolal and similarly the pH of the duodenum is raised to near 7.0 in the second part of the duodenum. It is possible that duodenal receptors serve to modify the rate of gastric emptying. One suggestion is that osmoreceptors in the duodenal mucosa exert such a controlling role. Hunt and colleagues have proposed that such an osmoreceptor would have membrane permeability characteristics similar to that of an erythrocyte. Thus if water alone is place in the stomach the duodenal osmoreceptor would distend when the duodenal lumen received the hypo-osmolar 'chyme' and this distension would mildly inhibit the rate of emptying [FIG. VIII.10]. If a solution of potassium chloride (non-penetrating) is ingested the receptor becomes smaller with a more pronounced inhibition of emptying rate. Conversely a solute which is actively transported into the receptor would cause a greater osmoreceptor cell volume even than water with a correspondingly faster rate of emptying [FIG. VIII.10]. FIGURE VIII.11 shows Hunt's (1961) results on ingesting solutions of KCl and NaCl at different concentrations. Weak concentrations of NaCl are more quickly emptied than water. However, when [NaCl] exceeds 250 milliosmol per litre (250 millimoles of NaCl molecules) emptying becomes slower, so presumably the membrane capacity for NaCl transport has been exceeded and entry of water into the osmoreceptor following facilitated transport of NaCl is now offset by a flux of water in the opposite direction. At 500 mosmol per litre, the rate of emptying equals that of water and further increases of [NaCl] progressively hinder emptying [FIG. VIII.11].

Hunt and others have also provided substantial evidence for their theory that there are 'acid' receptors in the duodenal wall. It is well known that duodenal acidity slows gastric emptying. Harding and

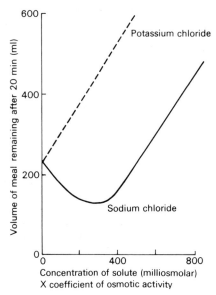

FIG. VIII.11. Diagram showing the relation between the volume of 750 ml 'test meals' of two solutes at varying concentrations and of water remaining in the stomach after 20 minutes and the effect of varying concentrations of solutes in the meals. The volume remaining at zero value on the abscissa represents the effect of water alone.
[Redrawn after Hunt, J. N. (1961). *Gastroenterology* **41**, 49.]

Leek (1972) showed that 'acid' receptors exist by directly recording the discharge of afferent fibres in the mesenteric nerves.

Results of vagotomy. Both vagi are severed, usually where they lie on the surface of the lower part of the oesophagus. When the resection is completely performed no gastric secretion occurs in response to injection of insulin, the sight or taste of food, or emotional disturbances. The chemical phase of gastric secretion and the response to pentagastrin are probably unaffected. In duodenal ulcer cases the volume and acidity of the night secretion are reduced; this indicates that reflex vagal secretion can occur even during sleep. Initially gastric tone and motility are greatly diminished; the emptying time of the stomach is increased from 2–3 hours to 12–24 hours or longer. The patient complains of fullness, occasionally of nausea, and belches up large volumes of gas, often foul-smelling. If slight pyloric narrowing is present before the operation the weakened stomach movements resulting from vagotomy may lead to prolonged retention of the gastric contents. The operation is usually combined with pyloroplasty to prevent retention of gastric contents by the increased pyloric tone caused by the vagal section.

After the operation the threshold for pain produced by distension (with a balloon) of oesophagus, stomach, duodenum, or small intestine is unchanged. The patient can vomit and experiences appetite, hunger, loss of appetite, fullness of the stomach, and pain if the ulcer fails to heal.

Vomiting

The phenomena of vomiting are as follows: nausea is first experienced, the secretion of saliva is increased, and the breathing becomes deep, rapid, and irregular. Retching may occur, which consists of simultaneous inco-ordinated spasmodic contractions of the respiratory muscles; the diaphragm, for example, descends when the expiratory muscles contract. The glottis closes and remains shut till the expulsion of the vomited material is effected. The pyloric part contracts firmly, and at the same time the body of the stomach relaxes so that the gastric contents are forced into it; antiperistalsis may sometimes take place in the stomach, but it is

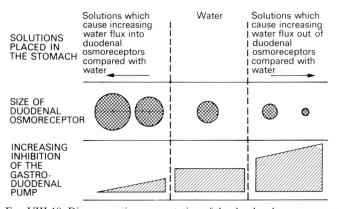

FIG. VIII.10. Diagrammatic representation of the duodenal osmoreceptor hypothesis in the control of gastric emptying.
[Redrawn from Blair, E. L. (1974). In *Recent advances in physiology*, 9th edn (ed. R. J. Linden) Chapter 7. Churchill Livingstone, Edinburgh. After Hunt.]

unimportant. The flaccid stomach is compressed by the raised intra-abdominal pressure resulting from the simultaneous descent of the diaphragm and the contraction of the abdominal wall. The cardiac sphincter is inhibited and the gastric contents are therefore driven into the dilated oesophagus. Some of this material is at once expelled from the mouth; some is moved up and down the oesophagus. Towards the end of the act of vomiting, the diaphragm relaxes, i.e. ascends, and all the expiratory muscles and the abdominal wall contract. As the glottis is closed the intrapulmonary pressure becomes positive. The oesophagus is thus compressed; it may also actively contract throughout its length or a wave of antiperistalsis may pass over it; its contents are thus emptied into the mouth. The palate is raised to shut off the nasal cavity from the throat.

The complex series of movements which occur during vomiting are controlled through a *vomiting centre* situated in the dorsal portion of the lateral reticular formation in the medulla. Afferent impulses to produce vomiting may arise in the stomach and other parts of the alimentary tract, the vestibular apparatus (e.g. in motion sickness), the heart, and other organs.

Apart from the vomiting centre there is a specialized *chemoreceptor trigger zone* on the medullary surface (Borison and Wang 1953). Drugs such as apomorphine, morphine, and digitalis, which are termed central emetics, actually stimulate the chemoreceptor trigger zone which is connected with the true vomiting centre. There is no evidence that chemical agents stimulate the vomiting centre directly.

REFERENCES

BLAIR, R. J. (1974) In *Recent advances in physiology*, 9th edn (ed. R. J. Linden) Chapter 7, pp. 279–339. Churchill Livingstone, London.
BORISON, H. L. and WANG, S. C. (1953). *Pharmac. Rev.* **5**, 193.
DAVENPORT, H. W. (1971). *Physiology of the digestive tract*, 3rd edn. Year Book, Chicago.
GREGORY, R. A. (1962). *Secretory mechanisms of the gastrointestinal tract.* Arnold, London.
HARDING, R. and LEEK, B. F. (1972). *J. Physiol., Lond.* **225**, 309.
HOPKINS, A. (1966). *J. Physiol., Lond.* **182**, 144.
HUNT, J. N. (1961). *Gastroenterology* **41**, 49.
JANSSON, G. and MARTINSON, J. (1965). *Acta physiol. scand.* **63**, 351.
PAINTAL, A. S. (1954). *J. Physiol., Lond.* **126**, 255.
UVNÄS, B. (1942). *Acta physiol. scand.* suppl. 13, 1.
WOLF, S. and WOLFF, H. G. (1947). *Human gastric function.* Oxford University Press, New York.

SECRETION AND FUNCTION OF PANCREATIC JUICE

The pancreas is a dual organ; the externally secreting alveolar tissue forms pancreatic juice; the islets of Langerhans form internal secretions, insulin [p. 503] and glucagon [p. 507].

The exocrine function of the pancreas is subserved by secretory acini and duct cells which form pancreatic juice. This juice, which contains enzymes and electrolytes, passes via intercalary and excretory ducts to be collected by two ducts which open into the second part of the duodenum. Ordinarily the major pancreatic duct joins the bile-duct and forms the ampulla of Vater [see FIG. VIII.14]. The accessory duct of Santorini enters the duodenum some 2 mm higher.

The pancreatic acini receive a vagal innervation. Preganglionic vagal fibres synapse with ganglion cells embedded in the pancreatic tissue; the postganglionic fibres innervate both the acinar cells and the smooth muscle of the ducts.

Collection of pancreatic juice

In animals a permanent duodenal fistula is made by inserting a metal tube into the second part of the duodenum opposite the opening of the main duct. The duct can be catheterized when required and the juice collected without interfering with the normal processes of digestion. The juice can be sampled for content of electrolytes and enzymes, its total volume measured, and, most important, the bulk of the juice can be returned (via a second tube in the metal cap) to the duodenum. The animal suffers a negligible loss of pancreatic juice and remains healthy.

In man, juice can be collected via a multilumen tube (passed via the nasopharynx) which is fitted with two inflatable balloons, one above the other below, the ampulla of Vater. The uncontaminated juice is aspirated through an orifice in the lumen of one of the tubes. Nowadays, with the aid of fibre optics a catheter can be introduced under direct vision into the duct.

Composition and volume of pancreatic juice

In man some 500–800 ml of juice are secreted daily. The *electrolyte composition of pancreatic juice* varies with the rate of secretion [FIG. VIII.12]. The concentration of bicarbonate is notably high (80 mmol l^{-1}) even at low rates of secretion and reaches values of approximately 120 mmol l^{-1} at the highest secretory rates. Chloride concentration falls as the bicarbonate concentration rises and the total concentration of these two anions remains constant; other anions are present in negligible concentration. The juice is isotonic with plasma. It is markedly alkaline (pH c. 8.40) owing to its high bicarbonate concentration. Na$^+$ and K$^+$ concentrations are almost identical with those of plasma and are invariant with changing rates of secretion.

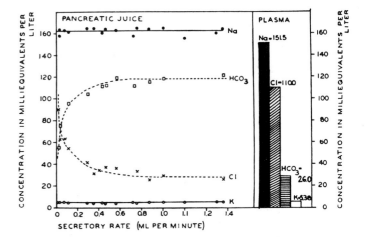

FIG. VIII.12. Relation between the rate of secretion and the concentrations of sodium, potassium, chloride, and bicarbonate in the pancreatic juice of the dog. The electrolyte content of the plasma is shown on the right for comparison.
[Bro. Rasmussen, F., Killman, S. A., and Thaysen, J. H. (1956). *Acta physiol. scand.* **37**, 97.]

Pancreatic secretion involves chemical work and is accompanied by an increased oxygen consumption. The duct cells of the pancreas contain carbonic anhydrase and the bicarbonate output by the gland is inhibited by acetazolamide in high concentration. It is believed that the duct cells form the bicarbonate and secrete it into the duct lumina; thus alloxan causes cellular vacuolization of the duct cells and causes a cessation of bicarbonate secretion without affecting either the histological picture of the acinar cells or their secretion of enzymes.

Enzyme content. Enzymes are secreted by the acinar cells; α-amylase, lipase, and proteolytic enzymes are the main constituents. Pancreatic α-amylase is stable in the pH range of 4–11 and has a molecular weight of some 45 000; it splits the α-1-4 glucosidic bond of starch and digests starch to maltose. Pancreatic lipase hydrolyses neutral fats (glycerol esters of fatty acids). Its pH range of activity is from 7–9. Its activity is greatly increased by bile salts. The proteolytic enzymes are secreted in inactive form from the acini as trypsinogen and chymotrypsinogen. Trypsinogen is converted to trypsin by enterokinase (an enzyme secreted by the duodenal mucosa) and by trypsin itself. Enterokinase and trypsin both split a bond between lysine and isoleucine in the trypsinogen molecule. Trypsin, thus liberated, is the most active proteolytic enzyme of pancreatic juice. It hydrolyses proteins by splitting bonds in the protein molecule which contain L-lysine and L-arginine in which the ε-amino or the guanidino group are free. As a result of trypsin activity proteins are digested mainly to small polypeptides; some amino acids are formed as well because trypsin can hydrolyse some dipeptides.

Chymotrypsinogen is converted into the active chymotrypsin by trypsin; chymotrypsin also digests proteins to small polypeptides.

Pancreatic juice also contains procarboxypeptidase which is converted to active carboxypeptidase by enterokinase. Carboxypeptidase splits peptide chains by the stepwise removal of amino acid residues from the free carboxyl group at the end of the chain. Ribonuclease and desoxyribonuclease are also found—both split nucleic acids (of ribose and and desoxyribose type respectively) into nucleotides.

Regulation of pancreatic secretion

Nervous phase

Within a few minutes of taking food, the flow of juice increases; this rapid response only occurs if the vagi are intact. Similarly stimulation of the vagus provokes a small volume flow of enzyme-rich juice of the consistency of glycerin. Gregory (1962) points out that this may be simply demonstrated in an anaesthetized dog by placing a gauze plug soaked in bicarbonate solution in the pyloric canal through a small incision in the stomach. Juice is collected from a cannula placed in the duct; the vagus, cut some 60 minutes previously, is stimulated. Unless the bicarbonate-soaked plug is inserted in the pyloric region, acid gastric juice reaches the duodenum and as this promotes secretin liberation there is no guarantee that the juice collected is not formed in response to secretin.

Vagal stimulation causes a disappearance of zymogen granules from the gland cells. During subsequent rest these are reformed. When radioactive-labelled amino acids are injected intravenously, pancreatic enzymes containing the radioactive compounds can be found in the gland within a minute or so. This indicates a rapid formation of enzyme protein; one site of such formation is in the ribonuclein particles attached to the cytoplasmic surface of the endoplasmic reticulum of the acinar cells. The enzyme is then transported across the boundary of the endoplasmic reticulum and is formed into granules within the intracisternal spaces of the reticulum; these granules move towards the Golgi apparatus and become invested with a membrane to become mature zymogen granules, which aggregate in the apical region of the acinar cell.

Vagal stimulation does not cause pancreatic secretion after the administration of atropine.

Hormonal phase

Two hormones, *secretin* and *pancreozymin*, are produced (by the duodenal and jejunal mucosa) which on liberation into the portal venous blood gain access to the systemic circulation and reaching the pancreatic tissue via the arterial blood stimulate the secretion of pancreatic juice. *Secretin*, a polypeptide (molecular weight = 5000) consists of 27 amino acids. It has been shown to be produced by argentaffin cells in the crypts of the duodenal mucosa. Its structure is known and strikingly resembles that of glucagon. It produces a flow of watery juice, rich in bicarbonate but free from enzymes. The flow of juice is closely related to the dose injected intravenously. Secretin potentiates the effect of CCK-PZ on the pancreas.

The original demonstration of the existence of a humoral factor

in the production of pancreatic juice was by Bayliss and Starling in 1902. In an anaesthetized dog a loop of jejunum was tied at both ends; its nerve supply was destroyed but its blood supply was preserved. Weak acid was introduced into the jejunal loop, whereupon a brisk secretion of pancreatic juice occurred. Starling at once recognized that a humoral factor was involved and proved this by cutting out a further piece of jejunum, rubbing its mucous membrane with sand and weak HCl and injecting the filtrate into the jugular vein. Within a few minutes a striking secretion of pancreatic juice took place.

Secretin causes contraction of the pyloric sphincter and delays gastric emptying.

Harper and Vass (1941) showed that an increase in enzyme output by the pancreas occurred when food substances, such as casein, were placed in the totally denervated small intestine. In their experiments a continuous flow of pancreatic juice was secured by repeated intravenous injections of secretin; the enzyme content of this juice was characteristically low unless the food substances were placed in the denervated intestine itself. An alcoholic extract of duodenal mucosa was made, and the alcohol removed, leaving a fat-free aqueous extract. Secretin was precipitated from this by adding bile salts; the supernatant fluid was saturated with sodium chloride and left in the dark for 48 hours. A protein precipitate formed, which after further purification could be dissolved and injected intravenously, whereupon it was found to stimulate enzyme output of the pancreas without increasing the volume flow of pancreatic juice. Its effects were almost identical with those of vagal stimulation with the notable exception that they were quite unaffected by the administration of atropine (Harper and Raper 1943). Harper and Mackay (1948) showed that pancreozymin caused depletion of the zymogen granules in the acinar cells. Final proof that pancreozymin is a hormone was provided by Wang and Grossman (1951) who showed that a transplant of the uncinate process of the pancreas into the mammary region responded by secreting enzyme-rich juice when acid, peptone, carbohydrates, amino acids and fat were introduced into the duodenum. All substances tested liberated both secretin and pancreozymin. Peptone proved most effective in liberating pancreozymin; acid or secretin itself was most effective in causing the production of secretin. Acid was only a feeble stimulant of pancreozymin liberation.

Pancreozymin is a polypeptide (33 amino acids). It is identical in structure with cholecystokinin. The substance is now designated CCK-PZ. Interestingly its five terminal amino acids (Gly-Tryp-Met-Asp-Phen-NH$_2$) are those of gastrin, which perhaps accounts for some stimulant properties that they have in common.

Immunofluorescence studies reveal that granular cells dispersed among the columnar cells of the duodenal and jejunal mucosal crypts are the source of CCK-PZ.

Pancreatic secretion is initiated by direct vagal stimulation of the exocrine secretory cells; vagal activity also releases gastrin from the pyloric antrum and this powerfully excites acid secretion from the gastric parietal (oxyntic) cells. Acid entering the duodenum in turn promotes the secretion of secretin and CCK-PZ. In addition, amino acids, some polypeptides and fatty acids further increase CCK-PZ output. Secretin amplifies the stimulation of enzyme and electrolyte rich pancreatic juice by a given dose of CCK-PZ; similarly CCK-PZ (liberated in response to amino acids in the duodenal lumen) amplifies the secretory response of the pancreas to secretin produced as a result of acidifying the contents of the intestinal lumen. Each hormone potentiates the action of the other. Both delay gastric emptying and stimulate the secretion of succus entericus.

The pure hormones secretin and CCK-PZ are now used to test pancreatic exocrine function, which is assessed by measuring bicarbonate output and trypsin production of the pancreas from samples aspirated from the duodenum.

Complete extirpation of the pancreas in man

Carcinoma of the pancreas may necessitate complete removal of the gland. The sequelae of pancreatectomy include diabetes mellitus [p. 508] and the development of digestive disturbances. Normally some 5 g of fat appear in the faeces daily; some 40–50 g fat can be so recovered in pancreatectomized subjects. The defect lies in the imperfect digestion of fat; as a result, absorption is poor. It is therefore surprising that most of the faecal fat is split, but this is due to the hydrolysis of the glycerides (which normally never reach the colon) by colonic bacteria; fatty acids thus formed cannot be absorbed from the colon. The faeces are of course bulky, pale and greasy.

After pancreatectomy the faecal nitrogen content (normally 1 g per day) increases four- to eightfold owing to incomplete proteolysis. There are no abnormalities of carbohydrate digestion and absorption.

These digestive disorders which lead to a loss of some 30 per cent of the calorific value of the food ingested may be minimized by feeding pancreatic digestive extracts.

Some of the abnormalities noted above are seen in children with congenital pancreatic fibrocystic disease.

REFERENCES

DAVENPORT, H. W. (1971). *Physiology of the digestive tract*, 3rd edn. Year Book, Chicago.
GREGORY, R. A. (1962). *Secretory mechanisms of the gastrointestinal tract*. Arnold, London.
—— and TRACY, H. J. (1961). *J. Physiol., Lond.* **156**, 523.
HARPER, A. A. and MACKAY, I. F. S. (1948). *J. Physiol., Lond.* **107**, 89.
—— and RAPER, H. S. (1943). Pancreozymin. *J. Physiol., Lond.* **102**, 115.
—— and VASS, C. C. N. (1941). *J. Physiol., Lond.* **99**, 415.
WANG, S. C. and GROSSMAN, M. (1951). *Am. J. Physiol.* **164**, 527.

THE BILE

Bile is secreted continuously by the hepatic cells into the bile capillaries [FIG. VIII.13] whence it is collected by the hepatic ducts which join to form the common bile-duct [FIG. VIII.14]. Between periods of digestion bile is diverted via the cystic duct into

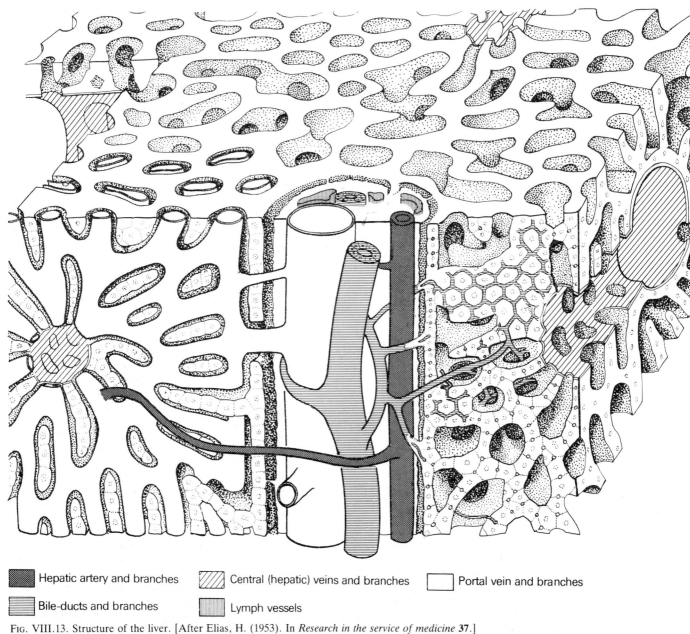

	Hepatic artery and branches		Central (hepatic) veins and branches		Portal vein and branches
	Bile-ducts and branches		Lymph vessels		

FIG. VIII.13. Structure of the liver. [After Elias, H. (1953). In *Research in the service of medicine* **37**.]

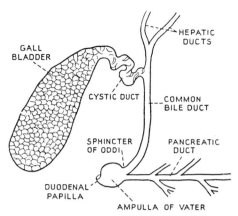

Fig. VIII.14. Anatomy of biliary tract. [Barclay, A. E. (1936).]

the gall-bladder where it is concentrated and stored. Some 500–1000 ml of bile are formed daily in man.

Constituents of the bile

1. Bile salts. These are synthesized by the hepatic cells. Cholic acid is formed from cholesterol and the acid side chain is conjugated either with taurine (aminoethylsulphonic acid, which in turn is derived from cystine) or with glycine to form taurocholic and glycocholic acids respectively [Fig. VIII.15]. At the pH of bile (7.3–7.7) taurocholate and glycocholate exist as anions, but they do not contribute to the osmotic pressure of the bile despite their concentration (10–20 mmol l^{-1}). Thus, the osmotic pressure of bile determined cryoscopically is about 300 mosmol l^{-1} (identical with plasma) but the sum of the electrolytes of the bile is far in excess of this figure. The bile salts, together with some cation, form aggregates which are osmotically inactive.

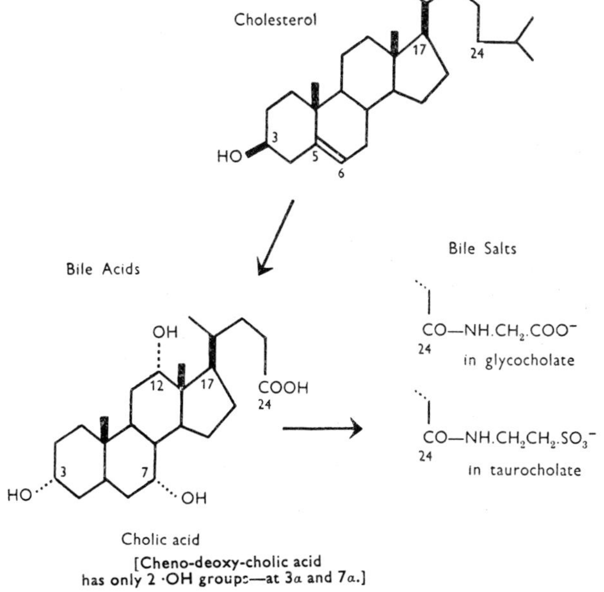

Cholesterol

Bile Acids

Bile Salts

CO—NH.CH$_2$.COO$^-$

in glycocholate

COOH

CO—NH.CH$_2$CH$_2$.SO$_3^-$

in taurocholate

Cholic acid

[Cheno-deoxy-cholic acid has only 2 ·OH groups—at 3α and 7α.]

Fig. VIII.15. See text.

Bile salts which enter the duodenum are reabsorbed in the portal vein and return to the liver—the so-called enterohepatic circulation. By giving ^{14}C-labelled bile salt the total pool of circulating bile salt can be measured by determining the radioactivity of bile collected by duodenal intubation. Some 3.5 g of bile salts comprises the total pool. As an ordinary meal causes a secretion of 6–8 g of bile salt there is a double circulation of the store with each meal.

Similar studies show that the half-life of the bile salts is some 3 days and the daily rate of synthesis is about 25 per cent. Bile salts injected intravenously are powerful stimulants of biliary secretion.

Hydrotropic action. The structure of a bile salt is such that one side of the molecule is hydrophilic ('water-attracting') and the opposite side is hydrophobic ('water-repelling' or 'lipide-attracting'). This is responsible for the characteristic property of the bile salts, that of lowering the surface tension of aqueous solutions, and therefore allowing the formation of stable solutions or emulsions of many fatty materials.

2. Electrolytes. Na$^+$ (180–220 mml^{-1}), K$^+$ (6–8 mM l^{-1}), and Ca^{2+} are the main cations, Cl$^-$ (60–70 mM l^{-1}) and HCO$_3^-$ up to 60–70 mM l^{-1} the important anions. An increase in the rate of secretion is accompanied by a rise in [HCO$_3^-$]. Acetazolamide given intravenously in high doses reduces [HCO$_3^-$] of bile and hence lowers its pH from 7.7 to 7.1.

3. Cholesterol. The biliary content of cholesterol is 0.6–1.7 g l^{-1}; the blood concentration of cholesterol is 2 g l^{-1} (5 mmol l^{-1}). The concentration of the bile which occurs in the gall-bladder may lead to a further rise in the cholesterol concentration and thus to its crystallization. Cholesterol crystals in turn form nuclei for the crystallization of bile pigments and calcium salts—gall stones.

4. Bile pigments are responsible for the colour of liver bile. They are solely excretory products and have no digestive function.

Control of bile secretion

Nervous and humoral mechanisms affect biliary secretion. Vagal stimulation increases flow and this effect is abolished by atropine. Secretin preparations increase biliary secretion and correspondingly all substances which cause secretin liberation, when introduced into the duodenum (acid, peptone, fatty acids), also evoke an increase of biliary flow. Following a meal, the biliary flow increases within an hour as gastric emptying occurs; it becomes maximal 3–5 hours after feeding. It is generally agreed that bile salts themselves are the most important choleretic (a substance which increases bile secretion) and the enterohepatic circulation of these substances ensures a continuance of biliary secretion during the digestive phase.

Functions of the gall-bladder—flow and storage of bile

During fasting the rate of secretion of bile is low, and the pressure in the ducts is correspondingly small. The sphincter of Oddi (which surrounds the bile duct as it enters the ampulla of Vater) is contracted and the sphincter can resist a pressure of some 30 cm of bile. When the pressure of bile in the common duct exceeds 7 cm or so bile passes without appreciable resistance along the cystic duct into the gall-bladder. The gall-bladder is thin walled and possesses a capacity of some 60 ml. However, its mucosa rapidly absorbs fluid and electrolytes, but not bile salts, bile pigments or cholesterol, which are correspondingly concentrated some 5–6 times. The absorption of bicarbonate is rapid, hence gall-bladder bile may have a pH value less than 6.0. Chloride is similarly quickly absorbed, as is sodium, but calcium and potassium are not and correspondingly reach concentrations of 6 mM l^{-1} and 40 mM l^{-1} respectively. The gall-bladder *secretes* mucin.

Within 30 minutes of feeding, the sphincter of Oddi relaxes concomitantly with the occurrence of contraction of the gall-bladder. Both these effects are mediated by the vagi. Additionally there is a humoral effect on gall-bladder contraction supplied by the liberation from the duodenal mucosa of CCK-PZ. The most effective stimuli for the liberation of CCK-PZ are fats and meat extracts. In cross-circulation experiments, fat placed in the duodenum of the donor dog causes contraction of the gall-bladder in the recipient.

As digestion proceeds to its completion, the tone of the sphincter of Oddi once more increases and bile is again prevented from access to the intestine. If this were not so, the choleretic effect of bile salts would be continuous.

Cholecystography. Organic iodine compounds which are opaque to X-rays are given orally or by intravenous injection. They are excreted in the bile and become concentrated in the gall-bladder. Iodipamide has a sufficiently high iodine content to be radiopaque even in the liver bile; hence the bile-ducts can be visualized on the X-ray screen. Tetraiodophenolphthalein on the other hand can only be seen in the gall-bladder after concentration of the bile; the shadow of the gall-bladder contents can be photographed—filling defects due to stone formation or absence of filling due to obstruction of the cystic duct can be identified. The organic iodine compounds are administered to the fasting subject and X-ray photographs taken some hours later. The subject then ingests a fatty meal (cream, olive oil, etc.) and photographs are taken at intervals to show whether the gall-bladder shadow disappears satisfactorily as its contraction expels its contents.

Removal of the gall-bladder

This operation gives rise in man to disadvantageous consequences:

1. The bile-ducts become dilated to accommodate to some extent the bile which is continuously secreted by the liver.

2. If the tone of the sphincter of Oddi is high, the pressure in the bilary passages rises until it may equal or exceed the secretory pressure of the liver cells and thus interfere with their activity.

3. If the tone of the sphincter is low (as it often is for a time after cholecystectomy), bile dribbles into the intestine when it is not needed and is consequently wasted.

The importance of the reservoir action of the gall-bladder is illustrated by the following experiment. If the common bile-duct is tied, and the gall-bladder removed, jaundice appears in 3–6 hours; but if the gall-bladder is left it can store so much bile pigment newly secreted by the liver that (after tying the bile-duct) jaundice does not develop for 36–48 hours. The rise in retained plasma bilirubin is similarly more marked and more rapid in the animal deprived of its gall-bladder.

Results of complete biliary obstruction

Complete obstruction of the bile-ducts produces results which are due to:

1. Absence of bile from the bowel. This leads to impaired digestion and reduced absorption of fats and corresponding changes in the faeces; reduced absorption of vitamin K leading to a fall of plasma prothrombin and haemorrhages [p. 443]; defective haemoglobin formation and anaemia of obscure origin; reduced absorption of fat-soluble vitamins A (and carotene) and D.

2. The effects on the body cells of retention of bile in the blood and tissue fluids.

The retention of bile leads to *jaundice* [p. 43], itching, and a profound loss of appetite. The blood level of all the organic bile constituents (bile pigments, bile salts, and cholesterol) is increased; bile pigments and bile salts appear in the urine.

3. The injury to the liver resulting from biliary obstruction.

The liver varies in size, is stained with bile pigment and is smooth and firm. The bile canaliculi are dilated, the hepatic cells are atrophied (mainly round the portal canals) and there is connective tissue overgrowth. The impairment of liver functions leads to the signs of parenchymal failure which are detailed on page 443.

The composition of the diet markedly influences the clinical state. In animals on a carbohydrate-rich diet, life may be prolonged for a year or more; an exclusively meat diet may, however, prove fatal within one week. A high glycogen content protects the liver cells against the harmful effects of various toxic agents.

Bile fistula

A complete bile fistula (in which all the bile is passed through an artificial opening to the exterior) results in loss of bile from the body; there is progressive impairment of bile secretion by the liver. There is no compression of the liver cells and therefore no parenchymal failure. The usual results of loss of water and electrolytes occur (if these are not replaced).

REFERENCES

DAVENPORT, H. W. (1971). *Physiology of the digestive tract*, 3rd edn. Year Book, Chicago.

SECRETION AND FUNCTIONS OF SMALL INTESTINE

Methods

To investigate the factors controlling the secretions of the small intestine in unanaesthetized animals, a loop of intestine is separated and both free ends are sutured into the abdominal wall; the continuity of the bowel is restored; the secretion of the loop can be collected and studied. In man, the small intestine can be blocked in two places by means of balloons; by suitable devices the intestinal contents above the proximal balloon or between the two balloons can be aspirated and examined.

Structure and secretory activity

The mucosal surface of the duodenum and the small intestine is adapted to provide a huge area for absorption. The mucosa shows finger-like projections of about 1 mm height called villi. The villi are covered by a layer of columnar cells which themselves possess a brush border consisting of microvilli 1 μm long and 0.1 μm broad. Each villus in its core contains a lymphatic vessel continuous with the lymphatic plexus of the submucosa, smooth muscle fibres, continuous with the muscularis mucosae, and an arteriole and a venule with their relevant capillary plexus. The core of the villus also contains a nerve net which has connections with the submucosal (Meissner's) plexus [FIG. VIII.16].

Between the villi are the intestinal glands which are also known as the *crypts of Lieberkühn*. Simple tubular glands, they do not penetrate the muscularis mucosae. They are lined by low columnar

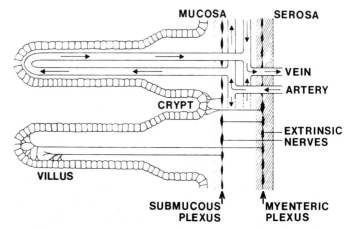

FIG. VIII.16. The arrangement of the vascular and nervous supply of the small intestine (diagrammatic). Note the crypts of Lieberkühn.

epithelium with goblet cells (which secrete mucus), argentaffin cells which synthesize secretin, and 5-hydroxytryptamine (a powerful stimulant of intestinal motility) and large acidophilic *Paneth* cells [FIG. VIII.17]. The epithelial and Paneth cells are zymogenic—that is they produce a great variety of enzymes capable of digesting proteins, carbohydrates, fats, and nucleic acids. They also produce *enterokinase* which activates trypsinogen, forming trypsin.

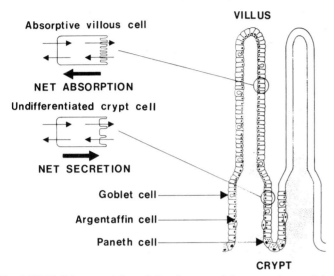

FIG. VIII.17. Diagram of the cellular elements of the mucosal part of the wall of the ileum. (See text for details.)

The epithelial cells at the bottom of the crypts shew evidence of active mitosis and there is a continuous process of replacement of cells shed from the tips of the villi by the migration upwards of the new cells formed in the crypts. The injection of tritiated thymidine into an animal leads to its incorporation into the DNA of cells of the crypts undergoing mitosis. Thereafter the migration of these cells towards the tips of the villi can be followed by serial autoradiography. Every three days the lining of the small intestine is replaced by this rapid turnover.

The *duodenum*, twelve finger-breadths in length (22 cm in life) possesses in addition to these general features of the small intestinal mucosa special submucosal mucous glands which resemble the gastric pyloric glands. Known as Brunner's glands they are tortuous, long, and penetrate the muscularis mucosae [FIG. VIII.18]. Their ducts empty into the crypts of Lieberkühn. Numerous in the first part of the duodenum, there are few below the common opening of the bile and pancreatic ducts. They show only a small basal secretion but the ingestion of fatty foods or the injection of secretin induces a large volume of alkaline mucous secretion from them as has been shown by studying the behaviour of isolated transplanted intestinal loops.

During digestion and absorption the villi contract quickly with an irregular rhythm, relaxing slowly. Their muscular fibres probably serve to pump lymph from the core of the villus towards the submucosal lacteals.

The *jejunum* was so christened because it was found to be empty at post mortem (Latin *jejunum* means fasting). The duodeno-jejunal junction is situated at the left side of the second lumbar vertebra. Arbitrarily the first 40 per cent of the small intestine is called jejunum—about 100 cm in life. The jejunum shows a progressive increase in numbers of goblet cells and lymphoid tissue. The lymph nodules lie immediately below the mucosa. The jejunal mucosa shows maximal folding, the folds being known as plicae circulares. The jejunum merges imperceptibly into the *ileum* (meaning roll or coil) which in living man is about 160 cm long. The goblet cells of the mucosa and the lymphoid tissues here reach their maximal density. The aggregated lymphatic follicles are known as Peyer's patches.

The crypts of Lieberkühn, which characterize the *whole* of the small intestinal length, continuously form the cells which migrate up the villus. These cells secrete the fluid and enzymes which form the *succus entericus* or *intestinal juice*. *Enterokinase* (enteropeptidase) is the most important—it activates trypsinogen. Other enzymes include maltase, invertase, dipeptidase, and alkaline phosphatase. These enzymes are located in the luminal (brush) border of the epithelial cell.

About three litres of fluid (similar in electrolyte composition to extracellular fluid with a pH however of 7.60) is secreted during the course of a day. The control of this secretory activity is not nervous nor humoral but is effected by local, mechanical, and chemical stimulation of the intestinal mucosa by the presence of chyme and the food particles which it contains.

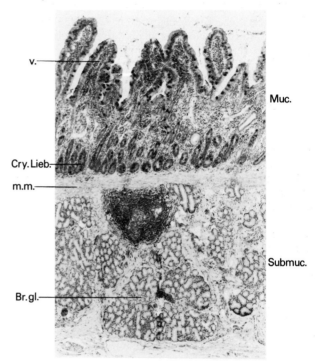

FIG. VIII.18. A light micrograph of the duodenum. The mucosa (Muc.) shows large, blunt villi (v) and numerous crypts of Lieberkühn (Cry.Lieb). Thick muscularis mucosae (m.m) separates the mucosa from the submucosa (Submuc). Numerous acini of Brünners glands (Br. gl.) fill the submucosa. No muscularis externis is shown.

As with the duodenal juice, a flow of intestinal juice follows a meal; it is slight during the first 2 hours, but shows a marked increase in the third hour; it is most obvious at the upper end of the gut.

Mechanical stimulation of the intestinal mucous membrane increases the volume and total enzyme output of the small intestine. Colicky intestinal contractions and ingestion of water, glucose, egg, and milk, produce similar results. Local irritants increase the volume of fluid and mucus secreted.

Digestion and absorption in the small intestine

1. 'Erepsin', a mixture of several specific enzymes, acts principally and rapidly on peptones and polypeptides, converting them into amino acids (it can also break down caseinogen and other proteins slowly); other enzymes complete the breakdown of nucleic acid via nucleotides and nucleosides to liberate purine and pyrimidine bases [p. 472].

2. Invertase converts sucrose into glucose and fructose; maltase breaks down maltose into two molecules of glucose; lactase converts lactose into glucose and galactose [p. 450].

Borgström *et al.* (1957) used an intubation technique in man to study the absorption of glucose and lactose; polyethylene glycol served as a non-absorbable indicator. After feeding, the two sugar samples were collected from different levels of the small intestine. Absorption was complete at 100 cm from the duodenum.

Optimal absorption of sugars requires the presence of both sodium and potassium. Replacement of sodium by lithium or mannitol abolishes active glucose transport. Ouabain which prevents active sodium absorption also inhibits glucose absorption. Crane and his colleagues have suggested that glucose transport occurs in two steps both sodium dependent. Step 1—glucose entry into the membrane of the villous cell requires no energy, whereas step 2—an accumulation step, requires energy to maintain a concentration gradient in the cell. It is possible that sodium influences the supply of energy from ATP by activating ATPase in the cell membrane.

It is probable that some disaccharides are absorbed as such into the epithelial cell, there to be hydrolysed by maltase and invertase present in the brush border of the villi. Nearly all the disaccharidase activity of the mucosa of the small intestine is detectable in the brush border.

3. Lipase is particularly concerned with the hydrolysis of the *primary* ester linkages. If it can hydrolyse the ester linkage at position 2 at all, it does it very slowly. It has been suggested that triglyceride digestion proceeds first by removing the terminal fatty acid to yield an α–β-diglyceride; the other terminal fatty acid is then removed leaving a β-monoglyceride. As this last fatty acid is linked by a secondary ester grouping, its removal requires isomerization to a primary ester linkage—a process which is very slow. Thus monoglycerides are the main end products of digestion; less than 25 per cent of the ingested fat is completely dissimilated to glycerol and fatty acids. With the aid of the bile salts, these products of digestion enter the mucosa of the villi, where within the epithelial cell α-monoglycerides are further hydrolysed to produce free glycerol and fatty acids, whereas β-monoglycerides are reconverted to triglycerides. This resynthesis requires 'activation' by formation of an acyl derivative of the fatty acid—a reaction which is catalysed by thiokinase and ATP.

$$R-COOH \xrightarrow[\substack{\nearrow \quad \searrow \\ ATP \quad ADP}]{\underset{\text{Thiokinase}}{\overset{Mg^{2+}}{\boxed{}}}} R-\overset{\overset{O}{\|}}{C}\sim S-CoA$$

The free glycerol formed in the lumen of the intestine, 20–25 per cent of the triglyceride ingested, passes into the capillaries of the villi and thence to the portal vein. Glycerol released within the epithelial cell combines with the fatty acids there and normally all free fatty acids formed in the wall are reincorporated into triglycerides, which enter the lacteals in the form of chylomicrons (1 μm diameter). These chylomicrons contain 0.5 per cent protein; puromycin, which stops protein synthesis, prevents their formation and leads to an accumulation of fat in the epithelial cells.

Fatty acids of more than 10 carbon atoms in length are found in the lymph of the thoracic duct as esterified fatty acids. If the carbon chain is shorter than 10, the acids are transported as unesterified or free fatty acids (NEFA or FFA) by the blood.

The bile salts are carried back to the liver and re-excreted in the bile—the so-called entero-hepatic circulation.

Cholesterol is absorbed directly into the lymphatics and recovered therefrom as cholesterol esters.

Almost complete absorption of the products of digestion and of other materials (e.g. water, salts, vitamins, vitamin B_{12}) normally occurs in the small intestine. The mechanism of absorption of the different substances is considered more fully on p. 458 (fat), p. 450 (carbohydrate), p. 464 (protein), p. 27 (vitamin K), p. 546 (calcium), p. 41 (iron).

Jejuno-ileal insufficiency

1. Excision of the small intestine must be very extensive before it interferes significantly with digestion or absorption. Thus in one patient almost the whole of the small intestine was resected (for regional ileitis), leaving only the duodenum, jejunum, and a small length of ileum. Carbohydrate absorption was 99 per cent complete, 70 per cent of the ingested protein was absorbed, but fat absorption showed very wide variations. Thus when 97 g were ingested only 5 g were absorbed; when 200 g were eaten, 80 g were absorbed; the respective losses in the faeces were thus 92 and 120 g. The lost fat was excreted mainly as fatty acid. Calcium absorption was greatly decreased owing to the formation of insoluble calcium soaps; the serum Ca fell with resulting attacks of tetany. The tetany was controlled by cutting down the fat intake and increasing the carbohydrate and giving large amounts of Ca and vitamin D.

2. More severe disturbances result from gastrocolic fistula, where much of the food completely short-circuits the small intestine by passing directly from the stomach into the transverse colon. In such patients, in addition to the changes described, there may also be failure of absorption of vitamin B_{12} (resulting in pernicious anaemia), and of vitamins (leading to complex multiple vitamin deficiencies).

3. Jejuno-ileal insufficiency also occurs in sprue and other obscure derangements of the small intestine.

Reduced absorption of foodstuffs from the small intestine may be due to (i) insufficient intake; (ii) inadequate digestion from lack of pancreatic or other juices; (iii) lack of materials necessary to promote absorption, e.g. bile salts; (iv) abnormal state of the wall of the small intestine; (v) inadequate length of available small intestine, e.g. after extensive resections or fistula formation.

Movements of the small intestine

Two different types of contraction occur; (a) rhythmic segmental contractions, (b) peristalsis.

Segmental contractions in the duodenum are of two types: (i) eccentric, (ii) concentric (Friedman *et al.* 1965). Eccentric movements involve localized contraction of segments of 1–2 cm which occurs in several regions at once or sequentially. They more markedly involve the outer longitudinal muscle layer and seem to subserve a mixing function. Concentric contractions involve the circular muscle layers over a segment longer than that described above and their end result, as observed radiographically during the sequelae of ingesting a barium meal, seems to be the emptying of the barium sulphate from the duodenum. The frequency of these segmental movements during digestion is highest in the duodenum —about 12 min^{-1}—in man and decreases as the distance from the pylorus increases. This gradient of rhythmicity has been observed in man and animals. The 'pacemaker' cells responsible for the initiation of segmental contractions are located in the second part of the duodenum in the neighbourhood of the point of entry of the bile and pancreatic ducts. A basic electrical rhythm of slow waves can be demonstrated which is conducted caudally by the longitudinal muscle layer. Contractions of the circular muscle cause spike potentials which occur irregularly and cause a local rise of intraluminal pressure.

The co-ordination of these rhythms is achieved by the myenteric plexus and results in propulsion of the intestinal contents down the ileum. It is possible that 5-hydroxytryptamine released during contraction renders the smooth muscle more sensitive to distention. Distention of a short segment induces proximal segmental contraction and distal relaxation.

Peristalsis consists of waves of contraction passing along the intestine, preceded by waves of relaxation. It is usually superimposed upon the rhythmic segmental contractions and the peristaltic contraction increases the local intraluminal pressure which the segmental contraction is itself causing. The peristaltic waves propel the intestinal contents towards the ileo-colic sphincter, demonstrating the so-called 'polarity' of the intestine. Thus, if a segment of intestine is cut out and sewn back in a reversed position to reestablish continuity of the ileum, the forward movement of the intestinal contents ceases even though the lumen of the intestine is patent. Each peristaltic wave lasts a second or two and propels the chyme forwards a few inches.

The polarity of the peristaltic wave has been described as the *myenteric reflex*. It is dependent upon the integrity of the myenteric plexuses and is abolished by applying nicotine or cocaine (which paralyse these plexuses) locally to the serosa.

Intestinal transit times can be measured by observing the passage of a radio opaque barium meal. The small intestine is about three metres in length and a barium meal may take about three hours to reach the ileo-colic valve after leaving the stomach. However, this period varies a great deal. It is affected by nervous and endocrine influences, by infections and by vitamin deficiencies. It is also modified by the contents of the lumen—such as saline cathartics. Diarrhoea is common in thyrotoxicosis whereas the myxoedematous patient manifests constipation. Grief may result in constipation and fear may cause diarrhoea. These influences may be mediated by the extrinsic autonomic nerves supplying the gut. Vagal stimulation accelerates and amplifies contraction of the small intestine, whereas sympathetic influences are preponderantly inhibitory.

The term *ileus* describes a loss of motility of the gut. Even the opening of the peritoneal cavity may induce immobility of the exposed intestine and this is particularly notable when the intestine is handled. Fortunately this untoward effect disappears soon after the operation is completed but, particularly when there has been peritoneal infection, a prolonged immobility known as *paralytic ileus* may supervene which is potentially dangerous. This postoperative complication is due to peritoneal receptor stimulation which leads to the excitation of discharge in afferent fibres which ascend in the splanchnic nerve trunks. These induce reflex inhibition (by sympathetic fibres) of the normal contractile responses of the gut to mechanical stimuli. Gaseous distention develops which exerts a further paralytic effect. The passage of a tube by mouth down to the ileum allows aspiration of gas and fluids and should be employed in dealing with this serious condition.

When the small intestine is obstructed by bands of adhesion or by other mechanical factors (such as the strangulation of a segment by a femoral or inguinal hernial ring) *colic*—a recurrent cramp-like pain—occurs and may be associated progressively with vomiting and circulatory collapse (shock). The severity of the symptoms and sequelae is greater the higher in the intestine the obstruction occurs. If the condition is not dealt with surgically, the vigorous peristaltic contractions which initially occur above the site of constriction subside and the gut loses tone, whereupon distention by gas and intestinal fluid ensues. Vomiting causes alkalosis and dehydration and the patient becomes gravely ill. In the special cases of volvulus, intussusception, or obstructed inguinal or femoral hernia, an additional hazard is the shutting off of the blood supply whereupon the loop of bowel involved becomes gangrenous. All efforts should be directed to the prompt alleviation of intestinal obstruction by surgery. In addition to the surgical measures required, the bowel must be decompressed by a tube passed from the mouth or nose to the area affected by post-operative ileus.

The ileo-caecal valve

At the ileo-caecal junction the ileum is invaginated into the caecal wall yielding an appearance similar to that of the cervix projecting into the vagina. This region serves as a valve which offers low resistance to the passage of ileal contents into the caecum but which strongly opposes reflux in the opposite direction. Direct observation in man (via a caecal fistula) shows the ileal opening to be 2–3 mm in diameter situated in the middle of a papilla. Rhythmic contractions of the circular muscle occur and, in the intervals between these, small jets of fluid escape into the caecum. Closure may be prolonged if no food has been ingested, but in a few minutes after a meal, rhythmic opening and closing every 30 seconds permits the escape of ileal fluid (Hightower 1968).

Gastrin causes relaxation of the ileo-caecal valve and secretin induces its contraction— effects which are the reverse of those exerted on the cardiac sphincter by these hormones.

REFERENCES

BORGSTRÖM, B., DAHLQVIST, A., LUNDH, G., and SJÖVALL, J. (1957). *J. clin. Invest.* **36, 1521.**
FRIEDMAN, G., WOLF, B. S., WAYE, J. D., and JANOWITZ, H. D. (1965). *Gastroenterology* **49**, 37.
HIGHTOWER, N. C. (1968). In *Handbook of physiology*, Section 6 *Alimentary canal* (ed. C. F. Code) Vol. IV, p. 2001. American Physiological Society, Washington, DC.

THE LARGE INTESTINE

Structure

The mucosal surface of the colon is smooth—there are no plicae circulares and no villi. Simple tubular glands are abundant, formed of simple columnar epithelium with huge numbers of goblet cells [FIG. VIII.19]. Lymph nodules are found in the ascending colon. Circular muscle is distributed as usual in the gut. The wall of the colon is sacculated and the mucous membrane is thrown into folds opposite the constrictions between the sacculations. The longitudinal muscle coat of the large intestine is not equally distributed throughout the wall but is collected into the three distinct bundles termed the *taenia coli* which can be seen through the serous coat.

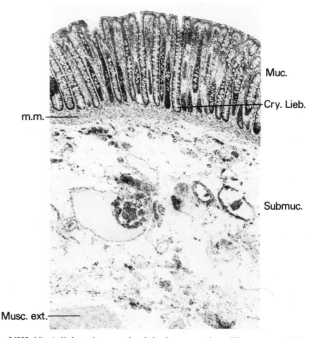

FIG. VIII.19. A light micrograph of the human colon. The mucosa (Muc.) consists of deep crypts of Lieberkühn (Cry.Lieb), containing numerous oval, light-staining goblet cells. A substantial muscularis mucosae (m.m) separates the colonic epithelium from the vascular submucosa (Submuc.). Smooth muscle of the muscularis externis (musc. ext.) is also present.

When these longitudinal bands which cause the puckerings of the wall (haustra) are cut, the mucosal folds can be smoothed out. Little fatty tags project from the colonic serosa; they are little peritoneal bags of fat called *appendices epiploicae*.

Movements of the large intestine

Direct pressure measurements have been made in man using water-filled balloons. Four types of waves of motility have been described.

Type I consists of small pressure variations lasting 5 seconds or so, which increase the local pressure by 5 cm of saline. Occurring about ten times a minute these waves presumably aid mixing of contents. Type II pressure waves again mainly contribute to mixing but last longer (30 s) and produce larger rises in pressure. Their frequency is slow—one or two per minute. Type III activity causes a prolonged but very minor rise in pressure. The important propulsive wave is that of *mass peristalsis* (type IV). In such contractions, which last 3–4 minutes, pressure rises steeply to a peak of as much as 100 cm saline, then declines more slowly. The result of this activity is to propel contents from the caecal region towards the rectum. These waves occur infrequently but are particularly evident following a meal, constituting the *gastrocolic reflex*. The proximal colon but *not* the distal colon is innervated by the vagus. The distal colon receives its parasympathetic innervation from the pelvic splanchnic nerves (S 2, 3, 4) via the hypogastric plexus. The parasympathetic nerves cause contraction of the colonic musculature. The sympathetic nerves supplying the colon leave the lumbar sympathetic chains to pass as the inferior mesenteric nerves to the colon. Sympathetic stimulation inhibits colonic motility; nevertheless bilateral removal of the lumbar sympathetic chains does not cause a lasting effect on colonic movements although the immediate post-operative sequelae do include colonic hyperactivity.

The time course of the passage of the luminal contents from the ileo-caecal junction to the junction of the colon with the rectum is variable. However, if the contents of a barium meal are followed by X-ray, the caecum is reached in say four to five hours, the hepatic flexure in six hours, splenic flexure in nine hours and the pelvic colon in about 12–18 hours [FIG. VIII.20]. Mass peristalsis does not drive the contents into the rectum itself until the accumulation of faeces at this juncture is sufficient. The pelvic-rectal flexure may act as a sphincter—there is some thickening of the circular muscle here. The rectum is normally empty. When a mass peristalsis wave drives the faeces into the rectum (and this is often associated with the gastro-colic reflex induced by a meal) the intraluminal rectal pressure rises to 25 cm saline and this usually provokes the desire to defaecate. In the adult this may be inhibited to a considerable extent by voluntary control. The distention produced by faeces or by gases stimulates receptors in the rectal wall which discharge afferent impulses via the pelvic nerves.

There are two anal sphincters—internal and external. The internal sphincter (consisting wholly of smooth muscle) and the external sphincter (consisting of somatic striated muscle innervated by the pudendal nerves) are normally kept closed even when the individual is under surgical anaesthesia. Garry and his colleagues (Bishop *et al.* 1956) have shown that the parasympathetic innervation of the internal sphincter via the pelvic splanchnic nerves is inhibitory though it is excitatory to the colonic musculature. It seems that distention of the rectum provokes pelvic afferent impulses arising from muscle receptors in the rectal wall. These induce reflex parasympathetic discharges (mainly from S2) over the pelvic splanchnic nerves which inhibit the internal sphincter and an inhibition of the discharge in somatic pudendal fibres which results in relaxation of the external sphincter.

The excitatory sympathetic innervation of the internal sphincter and rectal musculature is not involved in the defecation reflex—only the sacral segments of the spinal cord are concerned. If the desire to defecate is acceded to, a co-ordinated reflex results which

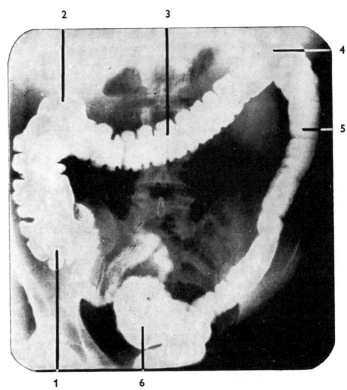

FIG. VIII.20. Normal radiograph of large intestine particularly well filled from end to end.

1. Caecum.
2. Hepatic flexure.
3. Transverse colon.
4. Splenic colon.
5. Descending colon.
6. Sigmoid colon.

empties the rectum and the contents of the descending colon. The diaphragm and abdominal muscles contract and the glottis is closed. A forced expiration against the closed glottis raises the intra-abdominal pressure—the so-called straining at stool. An important feature of the act of defecation is the contraction of the longitudinal muscle of the distal colon and rectum which shortens and straightens the pelvi-rectal passage. The levator ani muscles pull the anal canal upwards and this together with relaxation of the sphincter aids in expelling the faeces.

In infants, defecation occurs as a simple spinal reflex but social training brings control of the reflex by higher centres. Defecation then occurs only when the external sphincter is 'permitted' to relax. Spinal transection in the adult initially causes a retention of faeces but, unlike the case of micturition, the defecation reflex quickly reasserts itself, although of course no voluntary control is ever regained.

Disorders of large intestinal motility

1. *Hirschsprung's disease or megacolon*. This is seen in children and is due to degeneration of Auerbach's plexus. This blocks both peristaltic and mass contractions and accumulation of faeces in the large intestine causes megacolon. The condition usually involves only the pelvi-rectal junction and excision of this region followed by an anastomosis of the cut ends has proved satisfactory in such cases. Sympathectomy is fruitless because the condition is not due to excessive sympathetic motor discharge.

2. *Constipation* is most frequently the result of neglecting the call to defecate. If the sensation of fullness of the rectum is repeatedly ignored the sensory mechanism and its reflex effects become 'adapted'. The situation becomes progressively worse, whereupon the individual resorts to laxatives. The result is a dependence on laxatives.

Constipation was once believed to cause widespread toxic symptoms as a result of absorbing toxins from the bowel. However, the symptoms—including headache, restlessness, and irritability—and signs, such as furred tongue and foul breath, seem mostly to result from a prolonged distension and mechanical irritation of the rectum. Similar effects can be provoked by packing the rectum with cotton wool.

3. *Diarrhoea* is seen in infective diseases of the bowel such as typhoid and gastro-enteritis, in chronic ulcerative colitis and can be provoked by food poisoning or by the use of laxatives such as senna. The condition is of itself potentially serious, particularly in causing both dehydration and electrolyte loss (notably potassium).

Absorption and secretion in the large intestine

It is known that about a litre of chyme daily enters the colon from the ileum because direct evidence has been obtained in subjects with chronic ileostomies. Only about 100 ml of water reaches the rectum per day. The main sites of absorption of water are the caecum and ascending colon. The large intestine can absorb sodium, chloride and potassium ions, and glucose but cannot absorb protein, fat, or calcium.

The secretory activity of the colon is mainly confined to the secretion of mucus from the goblet cells. This lubricates the intestinal mass and serves to neutralize acids which are formed by bacteria in the large intestine. Colonic bacteria cause the production of CO_2, H_2S, hydrogen, and methane which contribute to flatus. Most of the flatus passed per rectum is, however, nitrogen derived from swallowing air. The bacteria of the ileum and colon produce the substances indole and skatole (methyl indole) from tryptophane.

The faeces

The faeces are derived partly from the ingested food, but mainly from the intestinal secretions. The faeces in starving animals are decreased in bulk but differ comparatively little in composition from those of normally fed animals. If vegetables and coarsely ground cereals are excluded from the diet, the faeces have a fairly constant composition, i.e. water, 65 per cent; solid material, 35 per cent (weight of *dried* faeces is about 30–50 g) made up approximately as follows: ash, 15 per cent; ether-soluble substances, 15 per cent; nitrogen, 5 per cent.

The *ash* consists mainly of compounds of calcium, phosphate, iron, and magnesium. The *ethereal* extract consists of fatty acids, neutral fat, a little lecithin, traces of cholic acid, its decomposition product dyslysin, and coprosterol, (derived from cholesterol). The faeces contain many *desquamated epithelial cells* and *bacteria*, most of which are dead. On a fat intake of 100 g daily not more than 5–7 g are normally lost in the faeces; on a protein intake of 100 g the faecal N content should not exceed 1.5 g (corresponding to 10 g of protein).

Ingested cellulose passes out unchanged, and substances which are enclosed in a cellulose wall escape digestion and absorption. The increased bulk of this undigested residue stimulates intestinal peristalsis. The passage of the food through the bowel is therefore quickened, and the digestive ferments have insufficient time to exert their full action [see TABLE VIII.1].

As the cellulose content of the food is increased by coarser milling of the flour, the bulk of the faeces is increased. They contain

TABLE VIII.1

Diet	Moist faeces (g)	Dried faeces (g)	Percentage of ingested food	Nitrogen lost (g)
Bread from fine flour	133	25	4.0	2.1
Bread from coarse flour	252	41	6.6	3.2
Brown bread	317	76	12.0	3.8

more water and solids; more of the ingested food and more nitrogen is lost to the body.

REFERENCES

BISHOP, B., GARRY, R. C., ROBERTS, T. D. S., and TODD, J. K. (1956). *J. Physiol., Lond.* **134**, 229.
TRUELOVE, K. G. (1972). *Physiol. Rev.* 46, 457.

The liver

STRUCTURE

The liver consists of lobes which are subdivided into lobules. The lobule is made up of ramifying columns of hepatic cells; the cell outlines are often indistinct so that the columns form a syncytium. The portal vein, hepatic artery, and bile-ducts, surrounded by a connective tissue capsule, enter the liver and branch repeatedly in the substance of the organ. The portal vein divides into branches, the interlobular veins, which surround the lobules; from these vessels blood passes between the liver cells in vascular capillaries to reach the centre of the lobule where it drains into the intralobular branches of the hepatic vein. The hepatic artery likewise divides into branches which accompany those of the portal vein between the lobules; ultimately the hepatic artery blood also enters the vascular capillaries, where it mixes with the blood from the portal vein. The vascular capillaries have no specific endothelial wall but ramify between the hepatic cells which constitute their boundaries. The intimate contact between the blood and the liver cell is well demonstrated by injection experiments; the injection material often penetrates from the vascular capillaries into the interior of the liver cells. This is an ideal arrangement, as the liver has to transform, or modify, many of the constituents of the blood. At intervals along the vascular capillaries are the stellate cells of Kupffer which are part of the macrophage or reticulo-endothelial system. They vary in number in different species; in man there are few.

Bile is formed in tiny vacuoles in the interior of the hepatic cells and is discharged through fine canaliculi into the bile capillaries. These (like the vascular capillaries) have no specific endothelium, but ramify between, and are lined by, the liver cells. FIGURE VIII.13 [p. 432] shows how there is always hepatic cell tissue between the fine bile capillaries and the much wider vascular capillaries, so that normally the blood and the bile are kept apart and never mix. The hepatic cells are well placed to transfer materials from the blood into the bile. At the periphery of the lobule the hepatic cells become continuous with, and transformed into, the cubical cells lining the bile-ducts.

Hepatic blood flow and oxygen usage

Hepatic blood flow can be measured approximately by infusing the dye bromsulphalein (BSP) intravenously at a constant rate. The dye is assumed to be excreted solely by the liver; hence if BSP is infused intravenously at such a rate as will maintain its concentration in the arterial blood constant, then the rate of infusion must equal the rate of excretion by the liver. The Fick principle states that if the concentrations of a substance X in the blood entering and leaving an organ are known and the rate of usage (or production) of the substance X by the organ be likewise known then the rate of blood flow through that organ can be calculated. In the present case, by sampling the hepatic venous blood by means of a catheter passed via the antecubital vein through the right heart and the inferior vena cava into the hepatic vein, and using the data described above, the hepatic blood flow can be calculated.

The estimated liver blood flow is 1500 ml min^{-1}, of which some 300 ml (or 20 per cent of the total) are delivered by the hepatic artery, and 1200 ml are derived from the portal venous supply.

The hepatic arterial blood contains some 19 ml O_2 per 100 ml (95 per cent saturated); hepatic venous blood contains about 13.4 ml O_2 per 100 ml; the hepatic arterio-venous difference is thus 5.6 ml per 100 ml. Portal venous blood contains perhaps 17 ml O_2 per 100 ml (85 per cent saturated) and the 'portal vein–hepatic vein' difference in oxygen content is 3.6 ml O_2 per 100 ml. Hence 300 ml of hepatic arterial blood supply $3 \times 5.6 = 16.8$ ml O_2 per minute of the requirements of the liver. Likewise, 1200 ml of portal venous blood supply $12 \times 3.6 = 43.2$ ml O_2 per minute of the liver's requirements. The total O_2 usage of the liver is (in this case) $43.2 + 16.8 = 60$ ml per minute of which about 70 per cent is supplied by the portal system (Sherlock 1964). Sherlock has pointed out that the portal venous blood is more desaturated during digestion and correspondingly represents 'an undependable source of oxygen, supplying least during digestion when hepatic activity is greatest'. (However this opinion disregards the fact that there is a marked increase of portal blood flow during digestion so that although the portal oxygen content per 100 ml of blood is reduced the total oxygen flow is increased owing to hyperaemia.)

Whereas the mean hepatic arterial pressure is about 100 mm Hg that in the portal vein is only 7 mm Hg. The pressure in the hepatic vein is only 5 mm Hg. As both the hepatic and portal systems converge on the sinusoids of the liver the arteriolar resistance in the hepatic arterial tree must be such that the blood enters the sinusoids at a pressure at least of similar order to that in the portal vein. Some authorities believe that the portal venous pressure may depend to some extent on the transmitted pressure from the hepatic arterioles.

Causes of hepatic anoxia

Centralobular damage or necrosis resulting from an inadequate O_2 supply to the liver may occur in the following conditions:

1. Diminished blood flow through the liver, e.g. (i) when the cardiac output is decreased (in 'shock', haemorrhage, heart failure); or (ii) in partial obstruction of the portal vein or of the hepatic artery.

2. Breathing air containing O_2 at a lowered tension (e.g. at high altitudes).

3. When the metabolism of the liver is raised, e.g. in hyperthyroidism; the O_2 usage of the liver is increased and the previously adequate O_2 supply becomes insufficient.

4. When the liver cells become swollen; as a result the vascular capillaries are compressed and obstructed so that the blood supply of the central centralobular cells is reduced. Swelling (oedema) of the hepatic cells is a regular initial effect of any injury to the liver; this explains why poisons absorbed from the intestine produce lesions which are usually most marked not at the periphery but in the centre of the lobule, in spite of the fact that the peripheral cells receive the highest concentration of the poison via the portal radicles. Swelling of the hepatic cells associated with centralobular necrosis also occurs when the cells become extensively laden with fat (so called fatty liver or fatty infiltration, p. 459).

Blood storage in the liver

The calibre of the hepatic vascular capillaries may vary considerably during life; the capillaries thus serve as a storage area for blood which can be discharged when necessary into the general circulation [cf. spleen, p. 54].

Portal obstruction

This may occur in a variety of circumstances:

1. (i) Ligature of one branch of the portal vein produces atrophy in the corresponding ischaemic area of the liver; the hepatic cells almost completely disappear, but the bile-ducts, blood vessels, Kupffer cells, and connective tissue are more resistant, and survive. The unaffected lobes (which are still normally vascularized) undergo compensatory hypertrophy, up to five times their original weight.

(ii) Ligature of the main portal vein in animals produces marked atrophy of the liver; if part of the liver is then excised no regeneration of the remaining tissue takes place.

(iii) Occlusion or narrowing of the main portal vein or of one of its branches may also occur clinically.

2. Obstruction to the portal flow is a common initial result of serious hepatic injury (hepatitis) owing to oedema of the cells. Later, when hepatic cells die, they are replaced by fibrous tissue which contracts down and obliterates the contained blood vessels; the more extensive the fibrosed area, the more severe is the resulting interference with the portal venous flow.

Clinical results of portal obstruction from any cause

The main results are as follows:

1. There is a rise of portal vein pressure; direct measurements at operation in normal man gives values of 15 cm H_2O (compared with 5–12 cm H_2O in the ankle veins); in hepatic fibrosis (so-called cirrhosis of the liver) the portal venous pressure may be as high as 40 cm H_2O.

2. *Ascites* (an abnormal accumulation of fluid in the peritoneal cavity) develops. The factors which contribute to the production of ascites are as follows:

(i) Owing to the raised pressure in the portal vein there is a corresponding rise of capillary blood pressure in the abdominal viscera leading to an excessive outflow of fluid into the peritoneal cavity.

(ii) In many patients with portal obstruction the hepatic cells are also damaged, leading to a fall of serum albumin concentration; this further facilitates the escape of fluid from the capillaries. The protein content of the ascitic fluid varies widely, e.g. from 0.1 to 3.8 g per 100 ml; this finding indicates that the permeability of the capillary wall to protein may be increased to a variable degree.

Ascites may develop in cases with disease of the liver parenchyma even without much rise of portal pressure if the plasma protein concentration is markedly reduced, e.g. to about half normal. Blood transfusion may cause removal of the ascitic fluid presumably because it raises the plasma protein level.

(iii) For some unknown reason ascites may be induced by a diet rich in meat; the fluid is absorbed when meat is excluded from the diet. The noxious agent in meat is not its protein content, because protein-free meat-extracts are even more effective in producing ascites than is an equivalent amount of whole meat.

(iv) In some cases of ascites an antidiuretic agent is present in the urine; this finding is interpreted as indicating that the pituitary antidiuretic hormone is present in the blood in excessive amounts, probably because the damaged liver cannot inactivate it in the normal way. As a result there is excessive reabsorption of water by the kidney leading to an increase in the volume of the body fluids; much of this fluid accumulates in the peritoneal cavity because of the abnormalities described in (i) and (ii) above.

(v) In some patients with chronic hepatitis (cirrhosis of the liver) the ascitic fluid accumulates very rapidly and may have to be withdrawn from the peritoneal cavity every few weeks. An important contributory factor is the failure of the kidney to excrete Na$^+$

ions; Na⁺ is retained, leading secondarily to water retention. On a liberal salt intake the amount of Na⁺ excreted in the urine may be less than 1 mM (0.023 g) daily. If the Na⁺ intake is drastically reduced the ascites tends to clear up. There is hypersecretion of adrenal corticoids which causes an excessive degree of reabsorption of Na⁺ from the lumen of the renal tubules into the blood. The Na⁺ content of the sweat and saliva is low in these patients. The situation is described as secondary hyperaldosteronism.

Thus in patients with portal obstruction ascites may be due to: increased capillary pressure in the portal bed; reduced serum albumin concentration; a toxic factor in meat; excess antidiuretic hormone: sodium retention resulting perhaps from excess adrenal corticoids. It is clear that hepatic and renal dysfunction are important contributory factors.

Ascites may also occur: (a) in cardiac failure: it is due to raised intra-abdominal venous pressure and renal dysfunction [cf. cardiac oedema]; (b) in nephrosis: it is due to lowered serum albumin concentration.

3. A collateral circulation is established between the portal and the systemic circulation, e.g. at the junction of oesophagus and stomach, in the mucosa of the anus, and in the abdominal wall. Haemorrhages may occur from the dilated venules of the bowel.

Cirrhosis is commonly associated with gastro-oesophageal varices whose rupture causes serious and potentially lethal bleeding.

FUNCTIONS OF THE LIVER

The functions are very numerous and are discussed in various sections of this book. They are summarized below:

1. Storage organ. The liver stores glycogen, fat, probably proteins, vitamins, e.g. vitamin A, vitamin B_{12}, other substances concerned in blood formation and regeneration, and blood.
2. Synthesis. The liver synthesizes the plasma proteins, fibrinogen, prothrombin, and (by virtue of its mast cells) heparin.
3. Bile secretion [p. 432].
4. Formation and destruction of red cells.
5. Detoxicating function [p. 442].
6. Metabolism. The liver is pre-eminently the central organ of metabolism. To discuss its role adequately would mean considering in detail the metabolism of carbohydrate, fat, and protein. It is more convenient, therefore, to consider the role of the liver when describing the metabolism of the individual foodstuffs. The following sections should be consulted:

(i) Carbohydrate metabolism [p. 450], especially role of liver glycogen and regulation of blood sugar [p. 456].

(ii) Fat metabolism [pp. 457–63].

(iii) Protein metabolism [pp. 463–73].

Complete extirpation of the liver

Experimental hepatectomy in animals has shown that the liver is essential to life; the animals survive only for a day or two. The principal results of complete hepatectomy are as follows:

1. There is a progressive fall of blood sugar (development of hypoglycaemia), e.g. to 40 mg per 100 ml (2.25 mM l⁻¹) or less, producing the characteristic symptoms for the species; in the dog there is marked muscular weakness, followed by convulsions, coma, and death. The aggravation of the clinical state runs parallel with the fall of the blood sugar. Within about 8–10 hours the dog lies in convulsions. *Only* if intravenous glucose is given to sustain blood glucose levels can the effects of hepatectomy on blood urea and bile pigment levels be studied at all.

The administration of glucose in doses of 0.25–0.5 g kg⁻¹ body weight to a flaccid and comatose animal produces astonishing effects: within 30 seconds the dog can walk and respond to a call; the heart beats more strongly. The blood sugar immediately following the injection reaches a high level; when it decreases again symptoms reappear. The development of fatal hypoglycaemia after hepatectomy proves that the liver is the organ responsible for the maintenance of the normal level of the blood sugar [cf. p. 456] on which the brain solely depends.

Light is thrown on the fate of certain sugars in the body by the following observations in the hypoglycaemic, liverless animal. Maltose, mannose, fructose, and glycogen act, on intravenous injection, like glucose; these four substances can thus be converted somewhere outside the liver into glucose. Galactose, however, is neither converted into glucose nor used in the liverless animal. This last observation is the basis of the galactose test of liver efficiency.

2. There is a progressive fall of the blood urea and rise of the blood amino acids. This proves that the liver is the chief site of deamination of amino acids and the only organ in which urea is formed.

3. Jaundice develops after total hepatectomy; bilirubin accumulates in the blood and tissues [FIG. VIII.21], and is excreted in the urine. Jaundice develops constantly in animals which survive longer than 6 hours; the plasma and fatty tissues become coloured. The blood gives a positive van den Bergh reaction for bilirubin [p. 45], first an indirect and then a biphasic reaction. The removal of the spleen and of all the abdominal viscera does not alter the rate of development of the jaundice. If haemoglobin is injected intravenously into the liverless animal considerable additional amounts of bilirubin are formed [FIG. VIII.21]. These observations prove that bilirubin can be formed extra-abdominally. The liver, however, is the only organ normally concerned with the excretion of bile pigment.

4. The coagulability of the blood is depressed owing mainly to a marked decrease in the concentration of plasma prothrombin and perhaps also to a decrease in plasma fibrinogen.

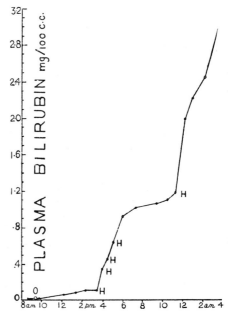

FIG. VIII.21. Accumulation of bilirubin in the blood following removal of the liver; conversion of haemoglobin into bilirubin. Experiment on dog. Ordinate: plasma bilirubin in mg per cent. At O, total removal of liver. Note gradual rise in bilirubin level. At each point marked H injection of haemoglobin intravenously. Note resulting marked increase in bilirubin level indicating extra-hepatic bilirubin formation. Considerable amounts of bilirubin were also excreted in the urine.
[Redrawn from Mann, F. C., Bollman, J. L., and Magath, T. B. (1924). *Am. J. Physiol.* **69**, 399.]

5. Hepatic insufficiency develops. If the blood sugar is kept normal by the continuous infusion of glucose the animal may survive for 18–24 hours. Finally, however, restlessness, dyspnoea, and vomiting occur; the animal becomes ataxic, and does not appear to hear or see. Coma and anuria develop, and death takes place quite suddenly. These symptoms are unrelieved by glucose and are not due to hypoglycaemia; these terminal manifestations are due to loss of some unknown liver functions (i.e. to some undefined form of hepatic failure).

Partial extirpation of the liver

The results produced depend on 1. the physiological reserve of liver tissue; 2. the regenerative power of the liver; 3. its nutritional state.

1. The physiological reserve is indicated by the fact that the body contains liver tissue far in excess of the minimal amount necessary for normal physiological function. Thus, in the dog, after removal of 80 per cent of the liver, bile salts and bile pigments are not retained in the blood or excreted in the urine. Even if 90 per cent of the bile-ducts are ligated the volume of bile secreted remains normal.

2. The regenerative power of the liver is illustrated by the following observations. In the dog, if three-fourths of the liver is removed, proliferation takes place in the remaining tissue as a result of active mitotic division of the cells; the original liver mass is restored in 6–8 weeks. The bile-ducts at the periphery of the lobules also sprout and bud off new clumps of cells; blood vessels and connective tissue soon invade the newly-formed areas. Excision can be repeated many times and is always followed by regeneration. Regenerative processes play an important part in the repair of the liver following the administration of hepatic poisons.

3. The importance of the nutritional state and the blood supply are dealt with on page 460.

Chronic hepatic insufficiency may be produced as follows: establish an Eck fistula, i.e. a lateral anastomosis is established between the portal vein and the inferior vena cava; the portal vein is tied headwards of the stoma, thus cutting off most of the blood supply to the liver. It becomes reduced to half or less of its original size; within 2 months remove 60 per cent of the liver. No regeneration now takes place, so that less than one-fifth of the original amount of liver tissue is left and very few of the cells which are present appear normal on histological examination. In spite of this, the animal maintains fairly normal health. The chief abnormalities present are as follows: (i) the blood sugar level tends to be slightly below normal; (ii) after injection of insulin the recovery of the blood sugar is greatly delayed; (iii) pancreatectomy produces only a slight hyperglycaemia as hepatic glucogenesis is impaired; (iv) adrenalin produces a less marked rise in blood sugar; (v) poisons are not well tolerated.

Relation of composition of diet to liver function

There are a number of important observations showing that the protein, carbohydrate, and fat intake are related to the efficient functioning of the liver.

1. Effects of high-fat diet and role of lipotropins [p. 459].

2. Protective action of methionine. As is pointed out on page 459, methionine, by donating CH_3 groups to form choline, is an indirect lipotropin. In addition it has been found that methionine (and to a smaller extent cysteine) protects the liver against the noxious action of certain poisons and especially against chloroform. A dog which has been kept on a low protein diet for a long time is highly susceptible to chloroform poisoning; light anaesthesia for 20 minutes kills the animal; post mortem, extensive necrosis of the liver cells is found. But if the dog is given methionine (or cysteine) a few hours previously, it can withstand deep anaesthesia for 40 minutes with little subsequent clinical disturbance and no signs of liver injury. A large meat meal (or plasma protein injected intravenously) has a similar protective action, doubtless because of its S-containing amino acid content. (It should be noted that choline is not protective against chloroform poisoning.)

3. Effects of grave protein insufficiency. (i) Under certain conditions, as yet imperfectly determined, grave protein insufficiency may, after a delay of some weeks, produce in rats, acute hepatic insufficiency due to massive necrosis of the liver. The necrosis involves large areas which are separated from each other by normal regions. In the affected areas the liver cells are dead and there is much congestion and many haemorrhages. If the animal recovers, post-necrotic scarring occurs; the normal regions hypertrophy and between them are found scars which interfere with the hepatic blood supply. Special stress has been laid on the fact that the cysteine content of these diets is deficient. The experiment makes it clear that in the absence of adequate supplies of amino acids (needed presumably for the maintenance of the liver proteins and enzymes), the liver cells may die.

(ii) A similar condition of massive necrosis occurs in animals fed on grain grown in soil rich in selenium. The selenium in the grain replaces the sulphur in the sulphur-containing amino acids, resulting in hepatic necrosis; the hepatic necrosis is prevented by administering methionine or cysteine.

4. In general, it may be emphasized that the liver resists many forms of stress best when its stores of carbohydrate and protein are ample; its efficiency is impaired when it is laden with fat. Thus a dose of carbon tetrachloride which is fatal (in the dog) within 24 hours if the liver is fatty at the time of administration is without effect if the liver is well filled with glycogen. In the case of tetrachlorethane, 1 ml produces severe intoxication and coma of 6–8 hours' duration if given 24 hours after the last meal; if given 12 hours after food, only mild symptoms result. The toxic action of ethyl alcohol on the liver varies inversely as its glycogen content; the toxic action of trinitrotoluene and arsphenamine is potentiated by protein deficiency and a high fat intake.

'Detoxicating' and protective action of the liver

The liver exerts its protective action in a variety of ways:

1. By *conjugation*, i.e. by combining the unwanted substance (or a derivative of it) with another molecule or chemical group, the resulting compound being excreted in the urine. It does not always follow that the excreted compound is less toxic than the original, as measured by the usual toxicity tests—it may even be more toxic. For this reason, and for the additional reason that many normal physiological compounds are chemically 'manipulated' by the liver in this way before being excreted, the term 'detoxication' commonly applied to conjugation in the liver is a misnomer, and would be better replaced by 'protective synthesis'. Examples are:

(i) Conjugation with sulphate. Many phenolic compounds are conjugated in the liver with sulphate, and excreted as 'ethereal sulphates' (sulphate esters).

(ii) Conjugation with glycine. Many aromatic acids that cannot be catabolized in the body are combined with glycine (or other amino acids in special cases) before excretion. Benzoic acid ($C_6H_5.COOH$) is transformed by the liver into hippuric acid (benzoylglycine, $C_6H_5.CO.NH.CH_2.COOH$); a similar detoxication of benzoic acid also occurs in the kidney. On the other hand, phenylacetic acid ($C_6H_5.CH_2.COOH$) is 'detoxicated' by combination with glutamine (the amide of glutamic acid).

(iii) Conjugation with glucuronic acid. Many drugs and hormones containing OH groups (either alcoholic or phenolic) combine with glucuronic acid to form glucuronides; thus pregnanediol (from progesterone) is excreted in the urine as pregnanediol glucuronide after conjugation in the liver, and similarly for the other urinary steroids.

(iv) Conjugation with acetic acid. Aromatic amino compounds react in the liver with acetic acid (as reactive '2C fragments') [p. 461] to form the corresponding acetyl derivatives, e.g. sulphanilamide forms acetylsulphanilamide, which is then excreted.

2. By complete *destruction*. Many compounds foreign to the body are destroyed in the liver by complete oxidation. Examples are nicotine and the short-acting barbiturates. Partial oxidation (or, less often, reduction) may precede the conjugation reactions described in 1 above.

Clinical hepatic failure

Disease of the liver in man may result in:

1. Excretory failure. Failure to excrete bile leads to retention of the constituents of bile and an increase in the concentrations of bile pigment, bile salts, cholesterol, and alkaline phosphatase in the blood. Bile salts and bile pigments appear in the urine; jaundice develops. The absence of bile from the intestine impairs digestion and, in addition decreases the absorption of fat and fat-soluble vitamins.
2. Portal obstruction. This condition is characterized by development of collaterals between the portal and systemic veins, and ascites.
3. Parenchymal failure. In this condition excretory failure may or may not be present; the term parenchymal failure is used to indicate that the liver cells are failing to carry out their other functions, to a greater or lesser degree.

A number of laboratory tests are available for assessing the form and extent of parenchymal hepatic failure.

Laboratory tests of liver function

The difficulty of producing experimental liver insufficiency, without completely removing the liver, should be borne in mind when considering means of testing liver function clinically. Extensive liver damage must be present in man before obvious signs of insufficiency present themselves. Some liver functions, are however, impaired sooner than others and these may serve as more sensitive tests. In disease of the liver parenchyma the excretion of bile is often not markedly affected and consequently there may be little jaundice and no increase in blood bilirubin; urea formation is another very resistant function. The following tests are of value.

Galactose tolerance test. When galactose is absorbed from the intestine it is normally converted by the liver into glycogen and subsequently into glucose which is dissimilated by the tissues. In the liverless animal, however, galactose is scarcely utilized at all and is only disposed of by excretion in the urine. The test consists of administering 40 g of galactose by mouth and determining the blood galactose at 0.5, 1.0, 1.5, and 2 hours. In normal subjects the blood galactose rises very slightly; in the presence of liver damage a greater rise takes place. The sum of the four galactose values (at the times stated) in mg per 100 ml gives the galactose index. The normal average is 68 and the maximum normal 160. Higher values indicate hepatic insufficiency. The galactose index is not raised in cases of diabetes mellitus, even when the glucose tolerance is markedly impaired.

Hippuric acid test. When sodium benzoate is taken by mouth it is conjugated in the liver and kidneys with glycine to form hippuric

acid. Clinically 6 g of Na benzoate are ingested; normally 3–3.5 g of benzoic acid are excreted in the urine as hippuric acid in the course of the next 4 hours; an excretion of 2.7 g is taken as the low limit of normality. In liver disease hippuric acid excretion is depressed owing to diminished conjugation.

Renal disease must, of course, be excluded, as impaired kidney functions gives the same result owing to diminished hippuric acid synthesis and diminished renal excretion of that which is formed.

Bromsulphalein excretion test. Normal liver cells secrete this dye (phenol and tetrabromphthalein disodium sulphonate) from the blood into the bile. 5 mg kg^{-1} of the dye are injected intravenously and specimens of blood collected after 5 and 45 minutes. Assume the initial concentration of the dye in the blood to be 100 per cent. After 5 minutes the blood concentration normally falls to 85 per cent and after 45 minutes to 5 per cent. A value at 45 minutes exceeding 10 per cent indicates liver damage.

Coagulability of the blood (prothrombin response to vitamin K). In liver disease the coagulability of the blood is decreased primarily owing to the lowered plasma prothrombin and Factor VII; if there is also a fall of plasma fibrinogen, it constitutes an additional contributory factor; as a result haemorrhages may occur. In such cases the administration of vitamin K by any route does not raise the abnormally low prothrombin level, i.e. the prolonged prothrombin clotting time, characteristic of liver disease, is unaffected.

Plasma albumin concentration. This is lowered in liver insufficiency because the liver is the sole site of albumin formation. For some unknown reason plasma globulin concentration is raised, often markedly. High protein feeding fails to restore the serum albumin level as the mechanisms concerned in the manufacture of albumin are impaired.

Enzymatic tests

Plasma alkaline phosphatase is derived from bone and liver tissues. Its normal concentration in adults is 3–13 King–Armstrong units per 100 ml. The enzyme is excreted in bile and if the plasma concentration exceeds 35 it indicates biliary obstruction. Liver cell damage, such as occurs in cell necrosis or cirrhosis, is associated with moderate increases in plasma concentration of the enzyme. Some difficulty exists in young adults in differentiating raised plasma phosphatase concentrations which may be due to osteoblastic activity associated with bone growth. Estimation of the related enzyme 5'-nucleotidase (normal plasma concentration 2–12 units per litre) may help. Its concentration is increased in liver disease but not in bone disease. Phosphatase derived from osteoblasts does not attack a nucleotide phosphate substrate.

Plasma transaminases. Liver cell destruction releases *alanine transaminase* from the cytoplasm of the hepatic cells. The enzyme catalyses the reaction

$$
\begin{array}{llll}
CH_3 & CH_2.COOH & CH_3 & CH_2.COOH \\
| & | & | & | \\
CHNH_2 \ + & CH_2 & \longrightarrow \quad CO \ + & CH_2 \\
| & | & | & | \\
COOH & CO.COOH & COOH & CHNH_2.COOH \\
\text{Alanine} & \text{α-ketoglutaric acid} & \text{Pyruvic acid} & \text{Glutamic acid}
\end{array}
$$

Another name for the enzyme is serum glutamic pyruvic transaminase—SGPT or GPT. A rise of GPT level in the plasma is more specific for liver disease than is a rise in *aspartate transaminase* (also known as serum glutamic oxaloacetic transaminase SGOT or GOT). This catalyses the reaction

$$CH_2.COOH \quad CH_2.COOH \qquad CH_2\ COOH \quad CH_2.COOH$$
$$| \qquad\qquad | \qquad\qquad\quad | \qquad\qquad |$$
$$CHNH_2 \ + \ CH_2 \quad\longrightarrow\quad CO \quad + \quad CH_2$$
$$| \qquad\qquad | \qquad\qquad\quad | \qquad\qquad |$$
$$COOH \qquad CO.COOH \qquad\quad COOH \quad CHNH_2.COOH$$

Aspartic acid	α-Ketoglutaric acid	Oxaloacetic acid	Glutamic acid

Aspartate transaminase is released in liver disease but is also raised in myocardial infarction.

Isocitrate dehydrogenase plasma concentration is increased in hepatocellular damage.

Hepatocellular failure in man

Failure of liver cell function can occur in almost all forms of liver disease, but is commonest in portal cirrhosis and acute virus hepatitis (Sherlock 1975). There is no characteristic histological picture in this condition.

The clinical features are: indigestion; muscular wasting and weakness; jaundice; low blood pressure with increased peripheral blood flow; tachycardia and increased cardiac output; fever; the breath is foetid 'like the smell of a freshly opened corpse'; the patient shows the neuropsychiatric syndrome of hepatic pre-coma or coma. The features of pre-coma are drowsiness, personality changes, intellectual deterioration, confusion, tremor, slow and slurred speech. Finally coma sets in. There are accompanying abnormalities in the electroencephalogram.

Hepatic coma has been attributed to failure of the damaged liver to detoxicate nitrogenous materials, chiefly ammonia, absorbed from the alimentary tract. Administration of a high protein diet, of ammonium salts or of methionine may precipitate coma, and conversely a protein-free diet can relieve it. Many intestinal bacteria produce ammonia, and in hepatic coma such bacteria are found unusually high up in the small intestine. Chemotherapy with oral neomycin has produced clinical benefit in many cases, but the improvement does not run strictly parallel to the fall in blood ammonia level or to the changes in bacterial flora in the gut. Other substances besides ammonia may contribute to the production of hepatic coma.

REFERENCE

SHERLOCK, S. (1975). *Diseases of the liver and biliary system*, 5th edn. Blackwell, Oxford.

PART IX
Metabolism

Chemical transformation and energy release

Metabolic pathways

During the oxidative breakdown of foodstuffs in the body, their complex molecular structures are not converted into the simpler discard products by a single chemical reaction, but undergo a well-defined sequence of small chemical manoeuvres, during some of which energy is liberated. (A similar difference exists between reducing your potential energy by jumping out of the window or by descending gradually down the stairs and landings.)

The foodstuff is said to be metabolized along a specific *pathway*: the pathway is defined by the sequence of intermediate chemical structures through which the foodstuff passes until the final waste products are reached. There may be dozens of intermediate stages for what appears, at first sight, to be a relatively simple change, and each step may be catalysed by its own highly specific enzyme. FIGURES IX.4 and IX.23 are diagrams of metabolic pathways.

Dissimilation

The terms *dissimilation* or *catabolism* are used to describe the total breakdown process undergone by a foodstuff down to the final end-products. In the case of carbohydrates and fats, the end-products are CO_2 and H_2O; with amino acids, end-products containing nitrogen (e.g. urea) also appear.

The opposite transformation, i.e. the building up of storage, structural, and functional materials from simple foodstuffs or intermediates is termed *assimilation* or *anabolism*.

Catabolic processes liberate energy because the energy content of a complex foodstuff is greater than that of its simpler degradation products. Conversely, anabolic processes will only proceed if energy from elsewhere is 'pumped into them'. The form in which metabolic energy is made manifest and usable is discussed below.

THREE PHASES OF CATABOLISM

Considering the variety of dietary proteins, carbohydrates, and fats to be metabolized, it might seem that the number of chemical changes required would be far beyond the capacity of a cell. But intermediary metabolism proceeds in three major phases, each of which introduces considerable simplification to the problem, so that the number of steps required to release the available energy from the multitude of substrates is unexpectedly small. In general terms, the three phases are as follows:

Phase I. This corresponds to the processes of intestinal digestion and absorption and the similar process in tissues when storage material is mobilized for dissimilation: polysaccharides are converted to simple hexose sugars; fats to glycerol and fatty acids; proteins to amino acids. From the thousands of complex foodstuff structures the problem is reduced to dealing with a few small soluble molecules—glucose and several closely-related isomers, glycerol, about ten fatty acids of varying chain length, and about twenty rather dissimilar amino acids.

The hydrolytic reactions of phase I themselves liberate relatively little energy but they prepare foodstuffs for metabolism proper.

Phase II. The various products from phase I are partially oxidized along converging pathways such that the products are CO_2, H_2O, nitrogenous discard materials, and one of three acids:

acetic acid (usually in the form of 'active acetate') [p. 454]
α-ketoglutaric acid
oxaloacetic acid
[see FIG. IX.8 for the structures of these]

All the C atoms of the fatty acids, two-thirds of the C atoms of carbohydrate and glycerol, and about half the amino acids yield acetic acid; the two keto acids arise from the other amino acids [see FIG. IX.18, p. 469]. The simplification is again evident.

During Phase II metabolism, about one-third of the available energy of complete dissimilation is released.

Phase III. The problem is the complete oxidation of the three acids remaining from phase II metabolism. For this, cells adopt a complicated metabolic 'cyclical' pathway common to all three acids. This pathway is called the 'citric acid cycle' after one of the chief intermediates, citric acid [FIG. IX.7, p. 454]. These transformations result in the acids being oxidized to CO_2 and H_2O with the release of the remaining two-thirds of the available chemical energy.

Phase III metabolism thus represents a final metabolic pathway common to the carbon structures of *all* the major foodstuffs, and provides a set of related and interchangeable intermediates through which the major metabolic materials can be transformed one into the other. All cells which require a supply of oxygen conduct phase II and III metabolism, with minor enzymic differences dependent on species.

The phased metabolic processes for glucose, fatty acids, and amino acids are considered in detail later.

ENERGY LIBERATION AND TRANSFER. HIGH ENERGY COMPOUNDS

All chemical reactions involve energy exchanges. The energy values are customarily expressed in heat units (kJ); a chemical reaction which liberates heat is said to be exothermic, and one which takes in heat is endothermic. However, the energy which can be made available for the performance of useful work by a chemical reaction is not always equivalent to the heat liberated, irrespective of the efficiency of the process. It is the available (metabolically usable) energy which is of importance in metabolic processes. Reactions releasing usable energy are termed *exergonic* and those requiring external energy to be supplied are termed *endergonic*. Most dissimilation (catabolic) reactions are exergonic. Thus the complete oxidation of 1 mole (180 g) of glucose to CO_2 and H_2O liberates energy equivalent to 2870 kJ of heat; in other words, the energy content of glucose is 2870 kJ mol^{-1} above the energy content of its oxidation products. To reverse the reaction and synthesize 1 mole of glucose from CO_2 and H_2O (as occurs in plants during photosynthesis), energy equivalent to 2870 kJ must be supplied to the system from an outside source.

A complex molecule has a higher energy content than the atoms or simpler molecules from which it is built because of the energy of formation of the chemical bonds which hold it together; this bond-energy is liberated when the bonds are broken. The bond-energy of the link between two given atoms of a molecule naturally depends on the structure of the *whole* molecule. This point is well illustrated by reference to the very important group of organic phosphates, which have the general formula $R \cdot O \cdot PO(OH)_2$, where R represents an organic radical derived, for example, from glucose, creatine, etc.

When most organic phosphates are hydrolysed (liberating phosphoric acid), energy equivalent to about 8–12 kJ mol^{-1} is made available as heat. In these cases, the compound is called a *low-energy* phosphate compound; the formulae of such phosphates are written R—ph, where—ph represents the phosphate group whose hydrolytic removal from the rest of the molecule results in the release of only a small amount of energy. But when some organic phosphates (of special structural types) are hydrolyzed, energy equivalent to about 42–50 kJ mol^{-1} is made available. These phosphates are termed *high-energy* phosphate compounds; their formulae are written R~ph, where ~ph represents a phosphate group whose hydrolytic removal from the molecule releases a surprisingly large amount of energy.

It must be made clear that high-energy compounds are *not* 'activated' forms of low-energy compounds. They have special structures which release unusually high amounts of useful energy (42–50 kJ mol^{-1}) on hydrolysis, because their energy content is that much greater than the energy content of their hydrolysis products. Equally well, high-energy compounds can only be synthesized if at least 42–63 kJ of energy per mole can be provided from some concurrent energy-yielding chemical reaction or other acceptable energy source. There is no justification for saying that a high-energy phosphate compound R~ph contains a high-energy phosphate 'bond'. The chemical energy is associated with the whole structure, the symbol ~ph merely showing which phosphate has to be hydrolyzed to cause the high-energy release.

Low-energy and high-energy phosphates

The metabolically-important phosphates of each type are:

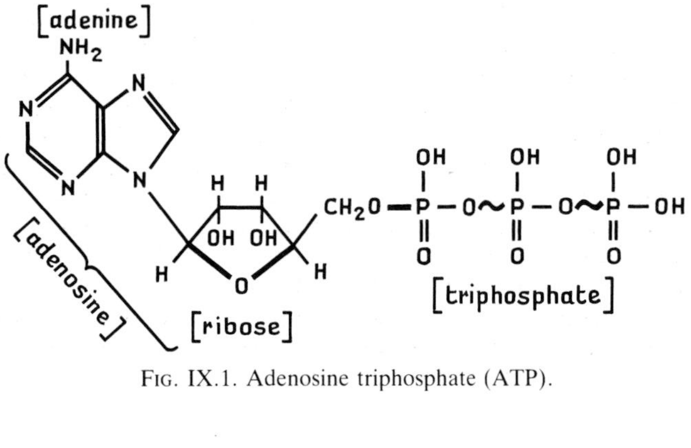

FIG. IX.1. Adenosine triphosphate (ATP).

1. Low-energy: glucose 1-phosphate and glucose 6-phosphate; fructose disphosphate; phosphoglyceraldehyde; monophosphoglyceric acid; adenosine monophosphate (AMP).
2. High-energy: adenosine disphosphate (ADP); adenosine triphosphate (ATP); creatine phosphate; diphosphoglyceric acid; phosphopyruvic acid; acetyl phosphate.
Thus:

Phosphate hydrolysis

phosphoric acid HO—$PO(OH)_2$	HO—ph	
creatine phosphate	creatine~ph [See FIG. IX.19]	1 high-energy
adenosine monophosphate (AMP)	adenosine—ph	1 low-energy
adenosine diphosphate (ADP)	adenosine—ph~ph	1 low-energy 1 high-energy
adenosine triphosphate (ATP)	adenosine—ph~ph~ph [full structure FIG. IX.1]	1 low-energy 2 high-energy

Role of high-energy phosphates

These high-energy phosphates are important in metabolism for the following reasons:

1. The phosphate group can be transferred directly to another organic molecule without much of the high energy of the reactant being dissipated as heat. The product is a phosphorylated molecule; it may or may not have a high-energy phosphate bond, depending on its structure, but it has a total energy content exceeding that of the non-phosphorylated molecule by 42–50 kJ mol^{-1}. Examples are the interaction of ATP with glucose:

$$\text{glucose} + \underset{\text{(ATP)}}{\text{adenosine}-\text{ph}\sim\text{ph}\sim\text{ph}}$$

$$\rightarrow \underset{\text{(glucose 6-phosphate)}}{\text{glucose}-\text{ph}} + \underset{\text{(ADP)}}{\text{adenosine}-\text{ph}\sim\text{ph}}$$

and with creatine

$$\text{creatine} + \text{adenosine}-\text{ph}\sim\text{ph}\sim\text{ph}$$

$$\rightleftharpoons \underset{\text{(creatine phosphate)}}{\text{creatine}\sim\text{ph}} + \underset{\text{(ADP)}}{\text{adenosine}-\text{ph}\sim\text{ph}}$$

It must again be emphasized that a molecule formed by phosphate transference from a high-energy phosphate compound is not necessarily a high-energy compound itself, though the whole molecule is raised to a higher energy level; only the special phosphates listed previously have high-energy phosphate structures. If the product is a high-energy phosphate it is *as though* the phosphate had taken a 'high-energy bonding' with it.

2. If the over-all dissimilation processes of metabolism are imitated in a test tube, the liberated energy is dissipated as heat. In the body, this energy of dissimilation, instead of immediately being lost as heat, is used to synthesize high-energy phosphate compounds (e.g. ATP); these compounds are stored, and the energy 'locked-up' in them is utilized as required [see 3 below]. The mechanism of the process is believed to be as follows: low-energy phosphate compounds undergo reactions (usually oxidations) that convert them into high-energy phosphate derivatives; the resulting phosphate groups, 'with their attendant high energy', are then used to phosphorylate ADP, giving ATP. The energy of the dissimilation has thus been channelled into the products of the reaction. For example, the direct oxidation *in vitro* of phosphoglyceraldehyde to phosphoglyceric acid liberates energy as heat; but in the body, this

stage of the dissimilation of carbohydrate proceeds thus [see also p. 453].

$$ph—G.CHO \xrightarrow{+HO-ph} ph—G.CH\begin{cases} OH \\ O—ph \end{cases}$$

(phosphoglyceraldehyde) (phosphoric acid)

$$\xrightarrow{\substack{oxidation \\ (-2H)}}$$

$$ph—G.COOH \xleftarrow{+ADP} ph—G.CO.O—ph$$

(phosphoglyceric acid) (diphosphoglyceric acid)
+ATP
(usable energy as~ph)

[ADP takes up the ~ph to form ATP]   [G = —OCH₂.CHOH—]

3. High-energy phosphate appears to be the sole source of energy that cells can use directly. It is used: (a) to effect chemical synthesis; (b) to perform work (muscular, osmotic, secretory); (c) to liberate heat (by hydrolysis) maintaining the body temperature but thereby dissipating the metabolic energy. The energy of muscle contraction is believed to be derived directly from ATP [p. 254].

High-energy esters. Coenzyme A

As well as the special phosphates described above, one other group of metabolic intermediates can be considered to be of the high-energy type. These are the acyl derivatives (R—CO—) of mercaptans (HS—R'). Such derivatives are thio-esters, R—CO—S—R'. On hydrolysis to the acid R—COOH and the mercaptan HS—R', there is a large energy release of 33–42 kJ g mol⁻¹, contrasting with only 8.4 kJ from the hydrolysis of oxygen esters R—CO—O—R'. Using the symbolism adopted for phosphates, all thio-esters have the structure

$$R—CO~S—R'.$$

The most important metabolic intermediates of this type are the acyl derivatives of coenzyme A (A for acylation). This coenzyme is a widely-distributed mercaptan with the chemical structure:

HS.CH₂CH₂.NH—pantothenic acid—diphosphate
—ribose phosphate—adenine

usually shortened to HS—Co A. Its esters are heavily involved in the dissimilation of the acids of Phase III metabolism and of the fatty acids.

For example, pyruvic acid, CH₃.CO.COOH, has a higher energy content than acetic acid, CH₃.COOH. When pyruvate is oxidized metabolically to acetate, the released energy is not wasted as heat. Instead, the actual product is the coenzyme-A ester of acetic acid (acetyl—Co A, CH₃—CO~S—Co A), a high-energy compound whose formation has 'mopped up' the energy released from the pyruvate. Acetyl—Co A is also known as 'active acetate', because its acetyl group constitutes acetic acid in a particularly reactive form. Obviously acetyl—Co A can readily undergo reactions which would require an outside supply of energy if free acetic acid was involved [p. 454]. Of course, the production of acetyl—Co A, or any acyl—Co A compound, is tantamount to the production of a molecule of ATP from ADP, at least as far as biological energetics are concerned:

$$CH_3.CO~S.Co\ A + ADP + HO—ph$$
$$\rightleftharpoons CH_3.COOH + HS.Co\ A + ATP$$

MECHANISMS OF BIOLOGICAL OXIDATION

Enzyme–coenzyme systems

All the reactions of metabolism depend upon the catalytic activity of enzyme systems. These consist of soluble protein (enzyme-protein, apoenzyme) together with various accessory substances. If the latter is a simple ion (e.g. Mg^{2+}, PO_4^{3-} it is called a cofactor and usually accelerates the enzyme action. If it is a complex organic, but non-protein, substance it is called a *coenzyme* and usually acts as an essential intermediate carrier of products of the enzyme-catalysed reaction. The enzyme-protein is generally specific for a particular chemical reaction or type of reaction; a particular coenzyme, however, may act as a carrier in a number of different systems (i.e. it is less specific).

In a sense, the coenzyme is one of the reacting substrates, being transformed chemically into some other product. This product can, however, undergo further change, regenerating the coenzyme for a repetition of the process; i.e. for the over-all reaction:

$$X—Y + Z \rightarrow X—Z + Y$$

a coenzyme may function thus:

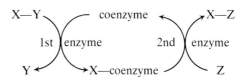

Examples are the hydrogen-carrying coenzymes discussed below, and coenzyme-A for the transference of acyl groups.

Certain proteins (the conjugated proteins) consist of a protein moiety chemically combined with an organic, but non-protein, unit called the prosthetic group of the protein. With enzyme-proteins of this type, the prosthetic group acts as an intermediate carrier of products of the reaction, i.e. the enzyme has a 'built-in' coenzyme; examples are the flavoproteins and cytochrome oxidase [p. 449].

Dehydrogenation

It is characteristic of many forms of tissue oxidation that the process does not involve the addition of atmospheric oxygen to the substrate, but rather the addition of water (or a related substance R—O—H) followed by dehydrogenation (removal of hydrogen). For a given reaction, a specific dehydrogenase enzyme catalyses the transference of the 2 H to the appropriate coenzyme acceptor, which is thereby reduced. Such oxidations are *anaerobic*, dependent only on a supply of coenzyme, not oxygen itself, though oxygen may be required in the final analysis for the regeneration of the coenzyme from its reduced form.

The commonest coenzyme acceptor for dehydrogenase reactions is a complex derivative of nicotinamide called nicotinamide adeninedinucleotide (NAD). Its constitution is shown in FIG. IX.2; in the reduced form, NAD.2H, the hydrogen is attached to the nicotinamide portion of the molecule. Some dehydrogenase enzymes cannot function with NAD; instead, they require the phosphate derivative of NAD, designated NADP as their hydrogen acceptor. Nicotinamide is one of the vitamins of the B-group [p. 491] because it participates in a fundamental metabolic process (the construction and function of dehydrogenase coenzymes) yet cannot be synthesized by humans in required amount and must be supplied as a micro-component of the diet.

The reduced coenzyme is ultimately reoxidized by passing its hydrogen along a chain of hydrogen carriers, finally to form water by reaction with molecular oxygen [p. 448].

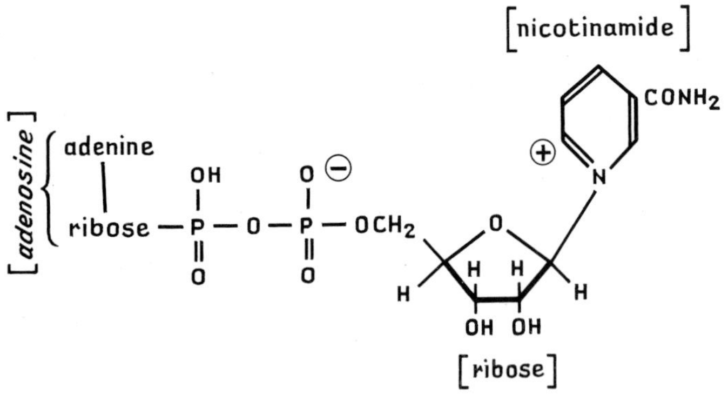

FIG. IX.2. Nicotinamide adeninedinucleotide (NAD).

Types of oxidative reaction

The following examples of biological dehydrogenation illustrate the general principles:

1. The oxidation of lactic acid to pyruvic acid, with NAD as hydrogen carrier (considered in detail on p. 452). This is a fully reversible reaction, without energy liberation provided that the reduced coenzyme, NAD.2H, is not reoxidized by some alternative process.

2. The oxidation of phosphoglyceraldehyde to phosphoglyceric acid, an important energy-liberating step along the pathway of glucose metabolism [p. 452]. The overall reaction is:

$$\text{ph-G-CHO + HO-ph + ADP + NAD} \rightarrow \text{ph-G-COOH + ATP + NAD.2H}$$

This is the classical example of energy-trapping by ATP formation. The conversion of each molecule of aldehyde to acid liberates usable energy, appearing as one ATP molecule synthesized from ADP, *plus* one hydrogenated NAD molecule which will generate three ATP molecules if subsequently reconverted to the coenzyme by oxygen and the appropriate enzyme chain [see below]. In the absence of oxygen, the main reaction will proceed only until all the coenzyme has been reduced, unless an alternative hydrogen acceptor is available. In muscle, this alternative can be pyruvic acid [see (1) above and FIG. IX.4].

3. The oxidative decarboxylation of pyruvic acid. An important energy-liberating stage of phase II metabolism is the oxidative conversion of the keto-acid pyruvic acid ($CH_3.CO.COOH$) to acetic acid ($CH_3.COOH$) and CO_2. Coenzyme is reduced, and the acetic acid appears as its high-energy ester with coenzyme A, $CH_3.CO{\sim}S.Co\ A$. The overall reaction

$$\text{CH}_3.\text{ CO.COOH + NAD + HS.Co A} \rightarrow \text{CH}_3.\text{CO}{\sim}\text{S.Co A + NAD.2H + CO}_2$$

represents an extremely complex series of steps, involving (as well as the coenzymes already mentioned) two other vitamins of the B group, thiamine and lipoic acid—another example of vitamins participating in a fundamental enzyme system.

The pyruvate to acetate conversion is *irreversible*. The energy liberated from one molecule of pyruvate appears as one high-energy acetyl-Co A *plus* one reduced NAD.

Other α-keto-acids (e.g. ketoglutaric acid) undergo a similar oxidative reaction.

Transmigration of hydrogen to unite with molecular oxygen to form water

The hydrogen released from various substrates by the action of dehydrogenases becomes attached to a coenzyme hydrogen acceptor, namely NAD (or, in certain cases, NADP or a flavin [see below]). It can remain there temporarily, but eventually the H has to be transferred elsewhere to regenerate the coenzyme. This subsequent transfer of hydrogen to molecular oxygen involves the mediator of a 'bucket chain' of carrier substances. The links in the chain are enzymes with prosthetic groups which carry hydrogen by being alternately reduced and reoxidized as the 'hydrogen load' is passed from one to the other down the line, finally to react with oxygen to give water (the 'respiratory chain' or aerobic sequence).

The overall reaction is:

$$\text{NAD.2H} + \tfrac{1}{2}\text{O}_2 \xrightarrow{\text{many steps}} \text{NAD} + \text{H}_2\text{O}$$

but this does not imply that the H atoms lost by the substrate are the same as those which finally combine with molecular oxygen; *transfer of electrons rather than of atoms is actually involved*. During the oxidation of one molecule of reduced coenzyme via this respiratory chain, metabolic energy equivalent to three ATP molecules is generated. The precise mechanism by which *oxidative phosphorylation* is coupled with hydrogen- and electron-transport is not known.

The chief links of the respiratory chain [FIG. IX.3] are flavoproteins, cytochromes, and cytochrome oxidase, substances shown by spectroscopy to be widely present in aerobic cells.

Flavoproteins. These are metal-containing enzymes which catalyse the transfer of hydrogen from reduced-coenzymes to the prosthetic group of the enzyme. This prosthetic group is a yellow pigment called flavin (or more precisely, flavin adenine dinucleotide, FADN); it is a derivative of riboflavin (vitamin B_2)—yet another example of the metabolic function of B-group vitamins [p. 492].

Cytochromes. The pigments are conjugated proteins carrying iron-porphyrin prosthetic groups allied to, but not identical with, the haem group of haemoglobin, and bound to proteins other than globin. Cells have their own sequence of cytochrome carriers which are alternately reduced and oxidized in appropriate order. The first

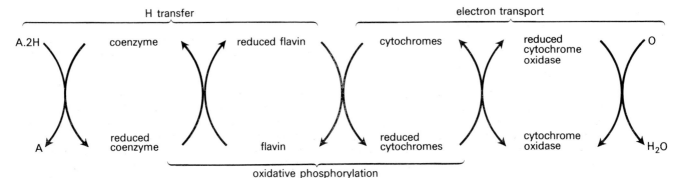

FIG. IX.3. Outline of respiratory chain oxidation.

cytochrome is reduced from the ferric iron state to the ferrous iron state by reduced flavins; it then, in turn, reduces the next cytochrome and itself returns to the oxidized state. The final reduction is of cytochrome oxidase.

Cytochrome oxidase. This enzyme is so called because it catalyses the reduction of its own iron-porphyrin prosthetic group by reduced cytochrome, thus regenerating the (oxidized) cytochrome. Reduced cytochrome oxidase, unlike the other carriers of the respiratory chain, is readily oxidized by molecular *oxygen*. In this way, cytochrome oxidase is restored to its oxidized form, and the hydrogen from the substrate has, in effect, reached its final resting place by combining with respiratory oxygen of the air to form water (recalling the notice President Harry S. Truman displayed on his desk—'The buck stops here').

Hydrogen transfer in terms of electron transfer

The coenzymes and flavins are reduced by addition of 2 H atoms, but the iron-containing cytochromes are reduced by electron transport (change of ionic charge).

A hydrogen atom removed during a dehydrogenation reaction undergoes 'acidic ionization' to give a proton (positively charged, denoted by H^+) and an electron (equally but oppositely charged, denoted by ε')

$$H \rightleftharpoons H^+ + \varepsilon'.$$
$$\text{atom} \quad \text{proton} \quad \text{electron}$$

The term oxidation is applied to a reaction in which a substance loses electrons (as well as, of course, to a reaction in which O is gained or H lost); similarly, a gain of electrons is termed reduction. Thus, considering the two forms of combined iron, ferrous (Fe^{2+}, reduced form) and ferric (Fe^{3+}, oxidized form), we have

$$\text{(reduced } Fe^{2+} \underset{+\varepsilon'}{\overset{-\varepsilon'}{\rightleftharpoons}} Fe^{3+} \text{ (oxidized)}$$

This is equally the case with iron-containing cytochrome (Cyto-Fe^{3+}) and cytochrome oxidase (CyOx-Fe^{3+}):

$$H \text{ (from substrate via Co and flavin)} \rightarrow H^+ + \varepsilon'$$
$$Cyto\text{-}Fe^{3+} + \varepsilon' \rightarrow Cyto\text{-}Fe^{2+}$$
$$Cyto\text{-}Fe^{2+} + CyOx\text{-}Fe^{3+} \rightarrow Cyto\text{-}Fe^{3+} + CyOx\text{-}Fe^{2+}.$$

Reduced cytochrome oxidase, CyOx-Fe^{2+}, has the unique property of transferring an electron to molecular oxygen:

$$2 CyOx\text{-}Fe^{2+} + \tfrac{1}{2} O_2 \rightarrow 2 CyOx\text{-}Fe^{3+} + O^{2-}$$

Protons from the initial ionization then react with the negatively-charged oxygen to give water:

$$2 H^+ + O^{2-} \rightarrow H_2O$$

In view of the close resemblance between haem and the porphyrin groups of the cytochromes and cytochrome oxidase, it is remarkable that the proteins derived from them differ so fundamentally. Haemoglobin combines reversibly with oxygen, but its iron is not oxidized; the iron of cytochromes is reversibly oxidized and reduced but oxygen is not directly involved except as the oxidizing agent of reduced cytochrome oxidase.

Cellular organization of respiration
Electron transport particle

Only mitochondria [p. 5] can effect the aerobic respiratory processes of the cell. They contain an organized repeating unit which is the smallest common denominator for both electron transfer and oxidative phosphorylation. The unit has been purified as a particulate entity—the electron transport particle, ETP—by application of an appropriately controlled amount of damage to isolated mitochondria. The ETP contains all the components of the respiratory chain, in strictly determined molecular proportions and (presumably) in organized spatial sequence. The particle 'accepts' hydrogen from reduced coenzymes, and uses molecular oxygen as terminal acceptor, without any other enzyme or factor being added. If the particle is slightly 'damaged' it may allow the aerobic oxidation sequence to proceed without the simultaneous generation of high-energy phosphate compounds; further damage gives smaller fragments which will conduct only portions of the electron transport sequence, but which are still recognizably particulate entities.

Relation of biological oxidation to energy release

From the above discussion, two distinct types of oxidative energy release will be apparent, each leading to the generation of high-energy compounds:

1. Anaerobic oxidation at substrate level: e.g. the generation of ATP from phosphoglyceraldehyde oxidation, or of acetyl-Co A from pyruvate oxidation;

2. Aerobic oxidation through the respiratory chain, whereby the reoxidation energy of each molecule of reduced coenzyme is channelled into the formation of three molecules of ATP from three molecules of ADP plus phosphate. This is often termed electron transport phosphorylation or respiratory phosphorylation. Its mechanism is obscure, but it differs from phosphorylation at substrate level in being 'uncoupled' by a number of substances like 2, 4 – dinitrophenol or some antibiotics. In the presence of these substances, there is continued oxidation of reduced coenzymes by oxygen but the energy is wasted as heat instead of being coupled to ATP synthesis. Dehydrogenations *as such* are independent of oxygen, but unless oxygen is available to reoxidize the H-carrier via the respiratory chain, the yield of energy will only be that of the phosphorylation at the substrate level.

REFERENCES

BALTSHEFFSKY, H. and BALTSHEFFSKY, M. (1974). Electron transport phosphorylation. *A. Rev. Biochem.* **43**, 871–98.
Ciba Foundation Symposia (1975). *Energy transformation in biological systems.* Associated Scientific Publishers, Amsterdam.
HAYAISHI, O (ed.) (1974) *Molecular mechanisms of oxygen activation.* Academic Press, New York.
LEHNINGER, A. L. (1972). *Bioenergetics*, 2nd edn. Benjamin, New York.
MEHLMAN, M. A. and HANSON, R. W. (eds.) (1972). *Energy metabolism and the regulation of metabolic processes in mitochondria.* Academic Press, New York.
WHITE, A., HANDLER, P., SMITH, E. L., HILL, R. L., and LEHMAN, I. R. (1978). *Principles of biochemistry*, 6th edn. pp. 265–422. McGraw-Hill, New York.

Carbohydrate metabolism

CARBOHYDRATES OF FOOD

The principal carbohydrates in the food are:

1. Polysaccharides, e.g. starch $(C_6H_{10}O_5)_n$ in vegetable foods. Cellulose and pectins cannot be digested by the enzymes of the human gut.

2. Disaccharides $(C_{12}H_{22}O_{11})$, e.g. sucrose (saccharose, cane- or beet-sugar), lactose (from milk), and maltose.

3. Monosaccharides: (i) Hexoses $(C_6H_{12}O_6)$, e.g. glucose (dextrose), and fructose (laevulose) in fruits and vegetables. (Galactose is not ingested as such but is split off from lactose.)

(ii) Pentoses $(C_5H_{10}O_5)$, not in the free form but in nucleic acid [p. 472] and in certain polysaccharides, e.g. the pentosans of fruits and gums.

The metabolic pathway of carbohydrates may be joined by certain food constituents which are not carbohydrates, e.g. glycerol.

Digestion

1. Amylase (pytalin) of the saliva can only digest starch after the natural plant granules have been burst, e.g. by cooking. It acts in a neutral or faintly acid medium (optimally at pH 6.5). Amylase digestion can thus continue in the stomach for about half an hour until it is arrested by the excessive acidity of the gastric contents. The (unimportant) digestive action of amylase may thus be summarized:

Starch $\longrightarrow$ erythro-dextrin $\longrightarrow$ achroo-dextrin $\longrightarrow$ maltose
(dextrins of decreasing complexity)   (to some extent)

2. The HCl of the gastric juice may hydrolyse some sucrose.

3. Amylase of the pancreatic juice rapidly converts all forms of starch and dextrins completely into maltose. It acts in an alkaline medium, and its digestive activity is increased by the presence of the bile salts.

4. Succus entericus contains three classes of enzymes, invertase, maltase, and lactase, which convert disaccharides into monosaccharides as follows:

$$\text{Sucrose} \xrightarrow{\text{Invertase}} \text{Glucose} + \text{Fructose}$$

$$\text{Maltose} \xrightarrow{\text{Maltase}} 2\ \text{Glucose}$$

$$\text{Lactose} \xrightarrow{\text{Lactase}} \text{Glucose} + \text{Galactose}$$

Digestion by these disaccharidases probably occurs in the luminal part (brush border) of the epithelial cells. Disaccharidase activity in man is maximal in the jejunum and proximal ileum. Hereditary defects of lactase or invertase lead to fermentative diarrhoea due to malabsorption of lactose or sucrose; in such cases mixtures of glucose with galactose or fructose are well absorbed.

Pentoses are liberated as an end-product of the digestion of nucleic acids and from the partial digestion of pentosans.

The end-products of carbohydrate digestion are, therefore, monosaccharides, by far the most important of which is glucose.

Bacteria in large intestine may convert some glucose into methane, CO_2, and other products.

Absorption

The absorption of sugars from the stomach and colon is normally negligible. Sugars (monosaccharides) are absorbed from the jejunum and upper ileum.

Whether monosaccharides are formed in the gut lumen or in the brush border of the epithelial cells, the process of absorption includes movement across the cell into the blood capillaries. Simple diffusion can play some part in absorption when the concentration of sugar in the gut exceeds that in the blood, but the main features of absorption can only be explained by more complex transport processes requiring energy. Some of the features of the transport process are:

1. *Specificity*. Glucose and galactose are absorbed at similar rates and much faster than mannose or pentoses, with fructose intermediate in rate of absorption.

2. *Active transport* is involved since sugars can move against their chemical potential. For example, glucose can be absorbed when its concentration in mesenteric blood is higher than that in the gut lumen. To concentrate the sugar, energy from cellular metabolism must be supplied [see p. 436].

3. *Ionic dependence*. The absorption of sugars depends on the presence of Na ions and probably also on the Na:K concentration ratio.

4. *Electrical activity*. In the absence of added substrate the serosal surface of the intestine is 1–5 mV positive relative to the luminal surface. When glucose is added to the luminal surface the transmural potential can rise to 12mV. Phloridzin, which is a highly specific inhibitor of glucose absorption, depresses the glucose stimulated potential without depressing endogenous metabolism of the absorptive cells.

Carrier hypothesis. To explain the features of the absorption of glucose and other sugars it is postulated that the sugar combines with a carrier, a mobile component of the cell membrane. The sugar–carrier complex moves the sugar across the lipid barrier of the cell membrane and releases the sugar inside the cell. To concentrate the sugar the carrier is coupled to a source of energy. Na affects the supply of energy from ATP by activating ATPase in the cell membrane.

Rate of absorption. The rate of absorption of sugars is influenced by the following factors:

1. The state of the mucous membrane and the length of time during which the carbohydrate is in contact with it. Absorption is thus depressed in diarrhoeal conditions (because of hurry), in enteritis, and in coeliac disease (the nature of the mucosal disturbance in this condition is unknown). The effects of widespread excision of the small intestine, and gastrocolic fistula are discussed on page 436.

2. Role of endocrines. (i) Thyroid: thyroxine acts directly on the intestinal mucosa stimulating absorption. The rate of glucose absorption is thus depressed in myxoedema, accounting for a flattened glucose tolerance curve, and stimulated in hyperthyroidism giving a 'diabetic-looking' glucose tolerance curve.

(ii) Anterior pituitary: affects absorption solely through its influence on the thyroid; hyperpituitarism induces thyroid overactivity and hypopituitarism induces thyroid atrophy, with the results to be expected from (i) above.

(iii) Adrenal cortex: glucose absorption is depressed in adrenal cortex deficiency; it can be restored to normal without the use of cortical steroids by a high salt diet which raises the Na^+ level in the blood to normal. Impaired glucose absorption can thus result from an abnormal state of the intestinal cells, secondary to the altered electrolyte content of the body fluids (which is characteristic of adrenal cortex deficiency) [see p. 535].

(iv) Insulin: has no effect on absorption.

3. Role of vitamins. Various members of the vitamin-B complex, e.g. thiamine, pantothenic acid, and pyridoxine, promote absorption.

Fate of hexoses. The hexoses are brought to the liver from the intestine in the portal blood. Galactose is converted exclusively in the liver into glucose (directly or via glycogen); fructose is converted both in the liver and in the muscles into glucose.

INTERMEDIARY METABOLISM OF CARBOHYDRATE

The following reactions occur in the body:

1. *Glycogenesis*, the synthesis of glycogen from glucose.

2. *Glycogenolysis*, the conversion of glycogen to glucose, mainly in the liver.

3. *Glycolysis*, the oxidation of glucose or glycogen to pyruvate and lactate by the Embden–Meyerhof pathway.

4. *The citric acid cycle* (Krebs cycle, tricarboxylic acid cycle). This is the final common pathway of oxidation of carbohydrate, fat and protein, through which acetyl-Co A is completely oxidized to CO_2 and H_2O.

5. *Hexose monophosphate shunt* (direct oxidative pathway, pentose phosphate cycle). This is an alternative to the Embden–Meyerhof pathway and the citric acid cycle, to CO_2 and H_2O.

6. *Gluconeogenesis*, the formation of glucose or glycogen from non-carbohydrate sources. The principal substrates for gluconeogenesis are glucogenic amino acids [p. 469], lactate, and glycerol.

Glycogen

Glucose is converted to glycogen in most tissues of the body, particularly in liver and muscle, which in man contain about 60 g and 150 g of glycogen respectively. The concentration of glycogen in the liver may reach 5 per cent after a high carbohydrate meal, or fall nearly to zero after 18 hours of fasting. The concentration in resting muscle is 0.7–1.0 per cent. *The brain cells contain little glycogen and depend on a continuous supply of glucose for their high metabolic activity.*

Glycogen is a polysaccharide consisting of many hundreds of glucose units joined together by glucosidic linkages (C—O—C) in a highly branched molecule with a molecular weight of up to 5 000 000. It is a suitable form in which to store carbohydrate because:

1. It is insoluble and so exerts no osmotic pressure. It cannot diffuse from its storage sites.

2. It has a higher energy level than a corresponding weight of glucose, though energy is indeed required to make it from glucose.

3. It is readily broken down to glucose in the liver to enter the bloodstream, and in many tissues (including the liver) it is degraded to lower intermediates which yield energy. The glycogen in muscle is consumed during muscular activity but is not easily re-converted to glucose even when hypoglycaemia is profound.

Glycogen synthesis.
Glucose is phosphorylated (by hexokinase + ATP) to form glucose 6-phosphate (in the liver the enzyme is glucokinase). This is an *irreversible* reaction. Glucose 6-phosphate is then converted to glucose 1-phosphate by phosphoglucomutase [Fig. IX.4].

Glucose 1-phosphate reacts with uridine triphosphate (UTP) to form *uridine diphosphate glucose* (UDPG) which serves as the source of glucose from which glycogen is formed by polymerization under the influence of the enzyme *glycogen synthetase*. The final reactions concerned with glycogen synthesis are not reversible. *Glycogen synthesis is promoted by insulin.*

Glycogenolysis.
Glycogen breakdown is brought about by phosphorylase as follows: the enzyme adenyl cyclase is first activated and catalyses the formation of cyclic 3'-5' AMP (cyclic AMP) from ATP; by transfer of its phosphate group cyclic AMP converts inactive phosphorylase *b* to active phosphorylase *a* which forms glucose 1-phosphate from glycogen [Fig. IX.5]. Glucose 1-phosphate is then converted to glucose 6-phosphate by phosphoglucomutase. *In the liver, but not in muscle, the specific enzyme glucose 6-phosphatase removes phosphate from the latter and promotes the entry of free glucose into the blood.* Adrenalin and glucagon promote hepatic glycogenolysis by stimulating adenyl cyclase. Adrenalin also activates adenyl cyclase in skeletal muscle but since this tissue contains no glucose 6-phosphatase the glucose 6-phosphate enters either the Embden–Meyerhof pathway or the hexose monophosphate shunt pathway and lactic acid is the final product. Glucagon does not stimulate muscle adenyl cyclase. ACTH activates adenyl cyclase in the adrenal cortex but not that in the liver or skeletal muscle.

Disease of glycogen storage or breakdown.
In *hepatorenal glycogenosis (von Gierke's disease)* liver and kidney cells are loaded with glycogen which, in the liver, is not broken down even by adrenalin and glucagon. There is congenital deficiency or even absence of glucose 6-phosphatase.

In *myophosphorylase deficiency glycogenosis (McArdle's syndrome)* there is marked muscular weakness on exercise. The muscle glycogen content is raised due to deficiency of phosphorylase; hence glycogen cannot be broken down to provide energy for muscular contraction.

Sources of liver glycogen.
Liver glycogen is formed from:

1. The hexose monosaccharide end-products of carbohydrate digestion, i.e. glucose, fructose, galactose.

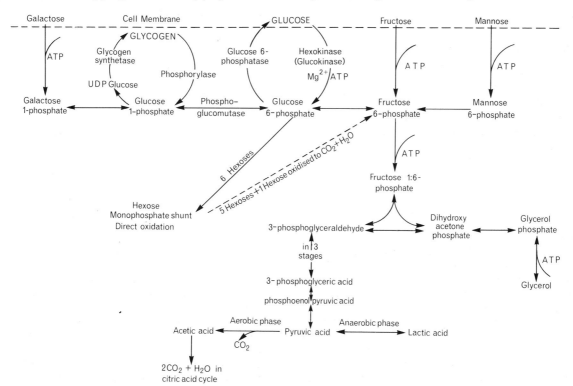

Fig. IX.4.

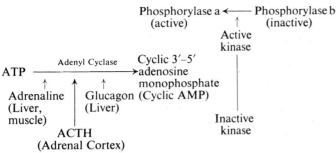

FIG. IX.5. Steps in formation of active phosphorylase.

2. 'Intermediates' of carbohydrate breakdown, especially lactic acid and pyruvic acid.

3. Glycerol derived from the hydrolysis of neutral fat.

4. Any 'intermediate' derived from the breakdown of the amino acids which enter into the so-called 'metabolic pool'. There is no doubt that in this way glycogen (and glucose) may be formed from many amino acids after they have been deaminated.

The quickest way of building up liver glycogen, however, is by raising the blood glucose level; this is obviously best done by the intravenous injection of glucose.

Role of liver glycogen. 1. It is the only immediately available reserve of blood glucose. 2. When liver glycogen is high the rate of deamination of amino acids in the liver is depressed and the amino acids remain available for protein synthesis. 3. A high level of liver glycogen depresses the rate of ketone formation from long-chain fatty acids. 4. Detoxication by acetylation or glucuronide formation of many substances is favoured by a high liver glycogen content. 5. A glycogen-rich liver is protected against the harmful effects of many poisons, e.g. carbon tetrachloride, ethyl alcohol, arsenic and bacterial toxins.

Glucose

The structural formulae of the three naturally occurring hexoses are shown in FIGURE IX.6.

Fate of glucose in the body

Glucose which is adsorbed from the gut or formed from other sugars may undergo the following changes:

1. Storage as glycogen in the liver and muscles.
2. Dissimilation in all tissues to yield energy. The complete oxidative breakdown of 1 g of glucose to CO_2 and H_2O yields energy equivalent to about 16.7 kJ.
3. Conversion to fat, and storage in fat depots.
4. Transamination of some intermediary products of glucose break-down, to form amino acids.

Glucose entry from the blood into most cells is facilitated by insulin. Glucose cannot be metabolized in the cell until it has been raised to a higher energy level by conversion to glucose 6-

phosphate. This phosphorylation process is accomplished by the enzyme *hexokinase*, and in the liver by an additional enzyme *glucokinase*; ATP acts as phosphate donor and Mg^{2+} are required:

$$\text{Glucose + ATP} \xrightarrow[\text{Mg}^{2+}]{\text{Hexokinase}} \text{Glucose 6-phosphate + ADP.}$$

Glucose 6-phosphate is a low-energy phosphate compound, but since the whole molecule is at a higher energy level than glucose, the hexokinase reaction is irreversible. In the liver free blood glucose is regenerated by glucose 6-phosphatase:

$$\text{Glucose 6-phosphate + H}_2\text{O} \xrightarrow{\text{glucose 6-phosphatase}} \text{Glucose + Phosphate + Heat.}$$

This reaction is also irreversible.

If ATP is to continue to function as an energy donor in the hexokinase reaction it must be steadily reformed by the phosphorylation of ADP. In muscle ATP is regenerated as follows:

$$\text{Creatine phosphate + ADP} \rightarrow \text{Creatine + ATP [see p. 254]}$$

Glucose 6-phosphate is a key compound from which many reactions can proceed:

1. It can lead via glucose 1-phosphate to the formation of glycogen, galactose, or glucuronic acid.
2. It can lead to the hexose monophosphate shunt, to glycolysis and to gluconeogenesis.

Glycolysis (Embden–Meyerhof pathway) [FIG. IX.4] Glucose 6-phosphate, whether formed by breakdown of glycogen or by phosphorylation of glucose, undergoes the following changes in its conversion to pyruvate:

I. *Glucose 6-phosphate* $\xrightarrow{\text{Phosphohexose isomerase}}$ *Fructose 6-phosphate*

Fructose and mannose enter the metabolic pathway here after phosphorylation by independent reactions.

II. *Fructose 6-phosphate* $\xrightarrow[\text{ATP}]{\text{Phosphofructokinase}}$ *Fructose 1, 6-diphosphate*

This raises the molecule to a higher energy level.

III. *Fructose 1, 6-diphosphate* $\xrightarrow{\text{Aldolase}}$ *Dihydroxy acetone phosphate (DHAP) + 3-phosphoglyceraldehyde.*

The latter are interconvertible. DHAP can be converted to glycerol phosphate which can be incorporated into triglycerides and phospholipids. If no glycerol is formed DHAP is converted to 3-phosphoglyceraldehyde so that *two* molecules of this triose are available for the next reaction.

IV. *3-phosphoglyceraldehyde* $\xrightarrow{\text{in 3 stages}}$ *3-phosphoglyceric acid*

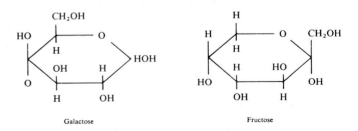

The numbers (1) and (6) show the positions at which phosphorylation occurs.

Glucose

Galactose

Fructose

FIG. IX.6.

(a) 3-phosphoglyceraldehyde + HO-ph → 1, 3-diphosphoglyceraldehyde

(b) 1, 3-diphosphoglyceraldehyde + NAD → 1, 3-diphosphoglyceric acid + NAD·2H

(c) 1, 3-diphosphoglyceric acid + ADP → 3-phosphoglyceric acid + ATP

Stage (a) involves the non-enzymic addition of phosphoric acid to phosphoglyceraldehyde. The pathways for glucose and glycerol (from fats) join here.

In stage (b), dehydrogenation of the addition product occurs (enzyme: triose phosphate dehydrogenase) to give diphosphoglyceric acid, which has one of its phosphates associated with a high-energy structure. The hydrogen from the dehydrogenation is passed directly to NAD [see Fig. IX.2]. It might be thought that the sequence of reactions described above would have to stop when all the NAD had been reduced to NAD·2H. However, it is permitted to continue because the reduced coenzyme is reoxidized by transferring its 2H either (a) directly to pyruvic acid (under anaerobic conditions only), or (b) through the respiratory chain of H acceptors until it is finally disposed of by uniting with molecular oxygen to form water [for details, see p. 449].

In stage (c) the phosphate of diphosphoglyceric acid ('together with its energy bond') is transferred to a molecule of ADP, raising it to the higher energy level of ATP. The energy liberated in passing from phosphoglyceraldehyde to phosphoglyceric acid has thus been retained within the system and used to synthesize ATP from ADP and phosphate.

V. 3-*phosphoglyceric acid* → 2-*phosphoglyceric acid* (lower energy phosphate) → *phosphoenolpyruvic acid* (high energy phosphate)

VI. *Phosphoenolpyruvic acid* + *ADP* ↔ *Pyruvic acid* + *ATP*

Energy yield

Each glucose unit of glycogen gives rise to two molecules of triose and subsequent products. Thus the dissimilation of one glucose unit of glycogen to two molecules of pyruvic acid generates energy sufficient to be bound as four newly-formed ATP molecules (two formed at step IVc and two at step VI). But one ~ph from ATP is used (in step II) to raise the energy content of the reactant to a level high enough to initiate the chain of dissimilation reactions. Therefore the net release of metabolically-available energy is equivalent to 3ATP = approx. 150 kJ mol^{-1} of glucose. If the process starts from free blood glucose, the net release is only two ATP, since the ~ph of one ATP is used in the hexokinase reaction [p. 451].

These figures only refer to phosphate energy capture at the level of substrate oxidation. In addition, 6 moles of ATP will be generated by the respiratory chain oxidation of the 2 moles of reduced coenzyme formed in step IVb. Thus if the dissimilation proceeds aerobically the energy yield from one free glucose to pyruvic acid is 8 ATP. But if the process is anaerobic (true anaerobic glycolysis) there will be no respiratory phosphorylation and only 2 ATP will be produced.

Fate of pyruvic acid

Pyruvic acid is a key substance in phase II metabolism. It is a metabolic stage reached by glycerol (from fat) and many amino acids (from protein) as well as all carbohydrates. Its further transformations include those to:

1. Lactic acid;
2. Glucose;
3. Oxaloacetic acid;
4. Acetyl-coenzyme A.

The latter two substances react together in the phase III 'common metabolic pathway' for final dissimilation.

1. Conversion of pyruvic acid to lactic acid

This conversion is important because it occurs in skeletal muscle working under conditions of relative oxygen lack, e.g. during maximal exercise. In the presence of adequate oxygen supplies, no lactic acid is formed; pyruvic acid is broken down through complicated intermediate steps to CO_2 and H_2O, and the reduced coenzyme (NAD.2H) is steadily reoxidized to NAD so that it can continue to function as hydrogen-carrier in stage IV (b). In the absence of adequate oxygen supplies all the NAD would soon be put out of action as it would all be converted to the reduced form, NAD.2H. But in these circumstances, pyruvic acid acts as a temporary H store (acting in place of the molecular oxygen which is not available), i.e. pyruvic acid dehydrogenates (oxidizes) the (reduced) NAD.2H back to (oxidized) NAD; the latter can then continue to act as hydrogen carrier in the energy-mobilizing reactions leading to the formation of more pyruvic acid. The pyruvic acid itself is reduced to lactic acid.

$$NAD.2H + \underset{\text{pyruvic acid}}{CH_3.CO.COOH} → NAD + \underset{\text{lactic acid}}{CH_3.CHOH.COOH}$$

Thus under anaerobic conditions, the muscle glycogen can be broken down into lactic acid to yield energy without the NAD being completely saturated with H (and so put out of action).

The reaction pyruvic acid ⇌ lactic acid is reversible: the same enzyme and coenzyme are involved whichever way the reaction is moving. If a muscle forming lactic acid is resupplied with enough oxygen, or the lactic acid circulates to a point of greater oxygen availability, the reaction is reversed and lactic acid is reconverted to pyruvic acid, thus:

$CH_3.CHOH.COOH+NAD→CH_3.CO.COOH+NAD. 2H$
$NAD.2H→NAD + 2H$
2H (*through many intermediate carriers*) + mol. O → H_2O + energy

The pyruvate ⇌ lactate conversion is more of a cul-de-sac than a pathway, for 2H must come out the way wherein it went.

The enzyme concerned is called muscle lactate dehydrogenase simply because it has been mainly studied in the lactic acid → pyruvic acid direction. When the enzyme is catalysing the reverse reaction it is catalysing the addition of H and would be more appropriately called a hydrogenase (but enzyme terminology must be taken as it is, with all its imperfections).

2. Reconversion of pyruvic acid to glucose

Several energy barriers prevent a simple reversal of glycolysis. The reaction phosphoenol-pyruvate → pyruvate is irreversible but can be achieved indirectly as follows: pyruvate → malate → oxaloacetate → phosphoenol-pyruvate. The existence of this, and the other energy barriers, means that glucogenesis from pyruvate can only occur if energy is provided from *without*, e.g. by ATP. The complete oxidation of 2 molecules of pyruvic acid or lactic acid provides more than enough energy to rebuild 10 molecules of pyruvic acid into 5 molecules of glucose.

3. Conversion of pyruvic acid to oxaloacetic acid. CO_2 assimilation

Some non-oxidative enzymic decarboxylations are reversible, i.e. the enzymes can also catalyse CO_2 uptake (assimilation). One such is the reaction of pyruvic acid with CO_2 to give *oxaloacetic acid* (OAA, ketosuccinic acid). The responsible enzyme is called *β-carboxylase*, present in most cells.

$$\underset{\text{pyruvic acid}}{CH_3.CO.COOH} + CO_2 ⇌ \underset{\text{oxaloacetic acid}}{HOOC.CH_2.CO.COOH}$$

The formation of OAA requires ATP energy, so that the tendency is rather for OAA to breakdown to pyruvic acid unless the pyruvic acid is being produced in bulk.

(This is an example of CO_2 being used as a molecular building material, as is also the case in the synthesis of urea [p. 468] and purines [p. 472]. CO_2 is not always a waste product.)

4. Conversion of pyruvic acid to acetyl-coenzyme A

The oxidative decarboxylation of α-keto-acids to esters of coenzyme A has already been described. Pyruvic acid gives acetyl-coenzyme A (acetyl-Co A). This high-energy compound is often called 'active acetate' because it takes part in enzymic reactions as though it were a hyperactive form of acetic acid. Indeed, the acetyl group of acetyl-Co A can react either as CH_3.CO— or as —CH_2COOH, i.e. *either* C atom can be 'active' in an appropriate enzymic reaction.

The pyruvate → active acetate conversion produces CO_2, ATP, and reduced NAD, and is *irreversible*.

A combination of reactions 4 and 3 described above results in the conversion of two molecules of pyruvic acid (2 × 3C) into one of oxaloacetic acid (4C) and one of acetyl-Co A (2C), the CO_2 produced in 4 being assimilated in 3.

Citric acid cycle. Phase III metabolism

The final dissimilation of pyruvic acid from Phase II glycolysis is not a simple matter of the successive loss of C atoms as CO_2. The normal route is one of quite unexpected complexity [see Fig. IX.7].

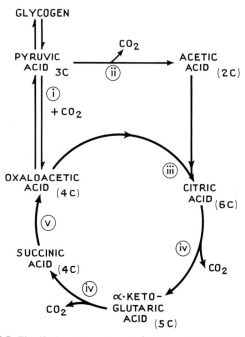

Fig. IX.7. The Krebs citric acid cycle [cf. Figs. IX.8, IX.12, IX.18].

(i) By CO_2 assimilation (carboxylase reaction) one molecule of pyruvic acid forms oxaloacetic acid (4C, two COOH groups).
(ii) Another molecule of pyruvic acid is oxidatively decarboxylated to active acetate.
(iii) The OAA and acetyl-Co A react together ('condensing enzyme') to form free Co A and *citric acid* (6C, three COOH groups = a tricarboxylic acid). This reaction is 'driven' by the energy-yielding split of the acetyl-Co A.
(iv) Through two further decarboxylations and related oxidations the citric acid (6C) is degraded through a 5C acid (ketoglutaric acid) to a 4C acid (succinic acid).

At this stage, the equivalent of the three C atoms of one of the original molecules of pyruvic acid has been converted into three molecules of CO_2 (though one of these three has been 'used' in the carboxylase reaction of the other pyruvic acid).

(v) The succinic acid (4C) undergoes further oxidation back into oxaloacetic acid (4C); this latter condenses with a further molecule of acetyl-Co A (from pyruvic acid or elsewhere) to reform the 6C citric acid, and so on.

The whole process is thus cyclical, with oxaloacetic acid acting 'catalytically'. (The above steps are indicated in Fig. IX.7 by the appropriate numbers). The system, which has been demonstrated in many tissues, is variously labelled the *tricarboxylic acid cycle*, the *citric acid* cycle, or the *Krebs cycle*, after H. A. Krebs, who propounded, and found experimental evidence for, the system. This investigator's name is also given to another cyclical enzyme system—the Krebs urea cycle [Fig. IX.17].

The structures of some of the known intermediates of the cycle are shown in Fig. IX.8. The citric acid cycle is, properly speaking, a system of enzymes, not of substrates; however, it is found convenient to think of it in terms of substrates, since some of the individual enzymes concerned have not yet been isolated.

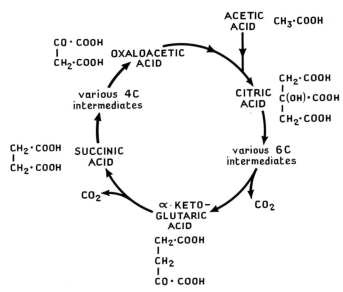

Fig. IX.8. Important intermediates of the citric acid cycle.

Energy yield. Each time the circuit is completed, one molecule of active acetate is 'drawn in' and dissimilated into 2 molecules of CO_2. The cycle only acts under aerobic conditions; each turn liberates 8 atoms of H, which are passed along the respiratory chain to molecular oxygen to form 4 H_2O and generate 12 molecules of ATP [p. 453].

The complete aerobic dissimilation of 2 molecules of pyruvic acid (i.e. one glucose unit) thus results in the generation of 38 ATP, made up as follows:

Glucose to 2 pyruvate = 2 + (2 × 3)
2 pyruvate to 2 Ac-Co A = 2 × 3
2 Ac-Co A to CO_2 = 2 × 12

This final yield of useful energy should be compared with the much smaller yield from the anaerobic expedient of lactate formation. The whole dissimilation process represents an efficiency of conversion of chemical energy into utilizable ATP energy of at least 50 per cent.

The cycle is probably unidirectional as shown; components of the cycle can 'escape' as pyruvic acid by reversal of the carboxylase reaction, step (i) [Fig. IX.7, and below].

Location of enzymes involved in citric acid cycle. Glycolysis is brought about by enzymes located outside mitochondria. Pyruvate passes through the mitochondrial membrane and is metabolized by the enzymes involved in the citric acid cycle which are arranged in the appropriate sequence along the shelves and inner walls of the mitochondria. Oxidative phosphorylation also occurs *only* inside mitochondria. Electron transport particles consisting of packets of NAD, flavoprotein, and cytochrome enzymes are arranged in an orderly sequence along the mitochondrial shelves [p. 5].

Citric acid cycle in protein and fat synthesis. Many amino acids after deamination are transformed directly or indirectly into acids which participate in the citric acid cycle. In this way they 'go round the cycle' to form oxaloacetic acid and are either dissimilated, or else give pyruvic acid and so may ultimately be built into glycogen or glucose.

(ii) Conversely, these constituent acids of the cycle (formed from carbohydrate via pyruvic acid) may be aminated (gain NH_2) to form amino acids which may then be used in protein synthesis, thus:

$$\text{pyruvic acid} \rightleftharpoons \underset{\substack{\text{oxaloacetic} \\ \text{acid}}}{\begin{array}{c}CH_2.COOH \\ | \\ CO.COOH\end{array}} \underset{\substack{\xrightarrow{\text{amination}} \\ \xleftarrow{\text{deamination}}}}{} \underset{\substack{\text{aspartic} \\ \text{acid}}}{\begin{array}{c}CH_2.COOH \\ | \\ CH(NH_2)COOH\end{array}} \rightleftharpoons \text{proteins}$$

(iii) Similarly, carbohydrate and fat metabolism have a common meeting point in the citric acid cycle [FIG. IX.7]. Acetyl-Co A is the chief product from fatty acid oxidation, bringing fat into the carbohydrate pathway. The reverse reaction will synthesize fatty acid by combination of acetyl-Co A obtained from pyruvic acid.

(iv) The glycerol for combination with fatty acids to give neutral fats can also be synthesized from carbohydrate by the reduction and hydrolysis of phosphoglyceraldehyde, an intermediate on the chain of reactions from glucose to pyruvic acid [p. 453].

The citric acid cycle is thus the meeting point for the metabolism and interconversion of carbohydrates, fats, and proteins—the final common pathway of metabolism.

Disorders of citric acid cycle. The cycle can be artificially disrupted by:

1. Cellular poisons, including arsenic, specifically inhibit the decarboxylation of keto-acids, and so block the entry of pyruvate to the cycle and the operation of the cycle itself.

2. Fluoroacetic acid, $F.CH_2.COOH$, and fluorine compounds giving rise to it, are highly lethal. The fluoroacetate enters the cycle as though it were 'active acetate', forming fluorocitric acid. Apparently this cannot undergo the subsequent reactions of the cycle and (like a spanner in the works) prevents citric acid itself from being handled. The cell has thus brought about its own destruction by catalysing a 'lethal synthesis' from fluoroacetate.

3. Thiamine (vitamin B_1) is an obligatory part of a coenzyme for the pyruvate $\rightarrow$ acetate conversion. In thiamine deficiency (beriberi p. 491) the cycle is slowed and pyruvate accumulates.

Substances like dinitrophenol which 'uncouple' respiratory oxidation from ATP formation do not interfere with the substrate conversions of the citric acid cycle, but do prevent them from generating energy as ATP. The cycle thus revolves 'in neutral gear', with disastrous results on the energy-requiring functions of any cell such as those responsible for secretion.

Hexose monophosphate shunt for aerobic dissimilation

The hexose monophosphate shunt occurs in some tissues such as the liver, lactating mammary gland, and adipose tissue and pro-

vides an alternative pathway to the Embden–Meyerhof reactions of glycolysis.

The initiating steps are the oxidation of glucose 6-phosphate to the corresponding acid, phosphogluconic acid, followed by oxidative decarboxylation to a series of 5C pentose phosphates. The H from these oxidations is passed to NADP. The pentose phosphates are either used for the synthesis of nucleotides [p. 472] or are transformed by an extremely complex cycle of reactions back to glucose phosphate

$$\text{6 pentose phosphate} \rightarrow \text{5 glucose phosphate}$$

The over-all dissimilation can be represented as

$$\text{glucose 6-ph} + 12 \text{ NADP} + 6 \text{ H}_2\text{O}$$
$$\rightarrow 6 \text{ CO}_2 + 12 \text{ NADP.2H} + \text{ph.OH}$$

Aerobic reoxidation of the reduced coenzyme generates $12 \times 3 = 36$ ATP, and the sequence will only proceed under these conditions.

It is not easy to estimate the amount of glucose catabolized by this alternative pathway. It may be more important as a source of pentose sugars for nucleic acid synthesis and of NADP.2H for fat synthesis, than as a route of energy liberation. Certainly the glycolytic route is more flexible in that it is able to continue under anaerobic conditions.

Metabolism of other hexoses

Fructose is released from sucrose in the intestine by the enzyme invertase. It is metabolized in the body as follows:

A small amount is phosphorylated by hexokinase to form fructose 6-phosphate. In liver and muscle another enzyme, *fructokinase*, effects the transfer of phosphate from ATP to form fructose 1-phosphate (unlike glucokinase, the activity of fructokinase is not affected by insulin). This is the major type of phosphorylation of fructose. Fructose 1-phosphate is split into dihydroxyacetone phosphate and glyceraldehyde. Glyceraldehyde is then phosphorylated and together with dihydroxyacetone phosphate it enters the Embden–Meyerhof pathway for glucose metabolism. Fructose 1-phosphate may also form fructose 1, 6-diphosphate [FIG. IX.4].

In liver, kidney and skeletal muscle fructose 1, 6-diphosphate can be converted to fructose 6-phosphate, and thence to glucose. As aldolase can convert the two triose phosphates to fructose 1, 6-diphosphate, much of the fructose in the liver is converted to glucose.

Fructose is metabolized actively by adipose tissue, and this metabolism is independent of that of glucose.

Galactose is liberated from lactose in the intestine by the enzyme lactase. It is readily converted in the liver to glucose, and the capacity to effect this conversion is a test of liver function (galactose tolerance test). The reactions involved are as follows:

$$\text{Galactose} \xrightarrow[\text{ATP}]{\text{Galactokinase}} \text{Galactose 1-phosphate}$$

$$\text{Galactose 1-phosphate} + \text{Uridine diphosphoglucose} \rightarrow$$
$$\text{(UDP Glucose)}$$

$$\text{Glucose 1-phosphate} + \text{Uridine diphosphogalactose}$$
$$\text{(UDP Galactose)}$$

$$\text{Uridine diphosphogalactose} \rightleftharpoons \text{Uridine diphosphoglucose}$$

The last reaction is reversible, so that if there is a reduced intake of galactose, glucose can be converted to form the galactose required in the formation of milk, glycolipids, and mucoproteins. Uridine diphosphoglucose is converted to glycogen and thence by glycogenolysis to glucose. The utilization of galactose, like that of glucose but unlike that of fructose, is dependent on insulin.

In the synthesis of lactose in the mammary gland galactose is converted to UDP-galactose which condenses with glucose 1-

phosphate to form lactose 1-phosphate. The lactose 1-phosphate is hydrolysed to form lactose.

There is an inherited metabolic disorder called *galactosaemia* in which there is a deficiency of the enzyme which converts galactose 1-phosphate to UDP-galactose. Ingested galactose accumulates in the blood and causes serious disturbances in growth and development. Galactose-free diets (no milk) allow normal growth, and bodily needs for galactose are provided by the formation of UDP-galactose from UDP-glucose.

BLOOD GLUCOSE REGULATION

The normal morning fasting level of blood glucose is 4.5 – 5.5 mmol l⁻¹, as measured by reduction methods. The more specific glucose oxidase method gives somewhat lower figures. After the ingestion of a meal containing carbohydrate the level rises temporarily to 6.5 – 7.5 mmol l⁻¹. After fasting for 24 hours or more the blood glucose level is maintained at 3.5 – 4.0 mmol l⁻¹.

The blood glucose level does not normally fall below 3.5 mmol l⁻¹, even with a very low carbohydrate intake. Hypoglycaemia is as harmful to the brain as is oxygen lack. Hyperglycaemia which leads to glycosuria is wasteful but otherwise not harmful.

Factors regulating blood glucose

The blood glucose concentration represents an equilibrium between the rate at which glucose is entering and leaving the bloodstream. Many factors contribute to the homeostatic processes which keep the blood glucose level constant within relatively narrow limits.

The carbohydrates in the diet tend to raise the blood glucose. The final products of digestion, glucose, fructose and galactose pass via the portal vein to the liver where fructose and galactose are con-

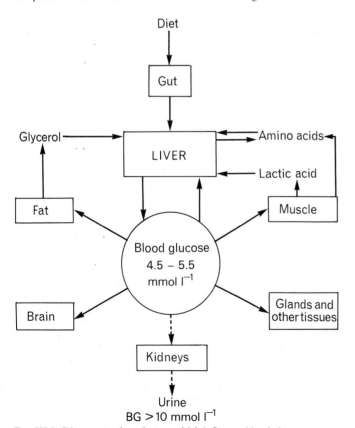

Fig. IX.9. Diagram to show factors which influence blood glucose concentration. See text.

verted to glucose. The liver serves as a receiving, manufacturing, storing and distributing centre for glucose which is then carried by the systemic bloodstream to all parts of the body; it is taken up and utilized by the brain, muscle, adipose tissue and secreting glands. The 'secretion' of glucose by the liver raises blood glucose, and the removal of glucose by actively metabolizing tissues lowers blood glucose [Fig. IX.9].

Liver as glucostat

The liver is the key organ in regulating the blood glucose. After hepatectomy in animals the blood glucose falls rapidly to lethal levels, showing that other organs contribute little to maintaining blood glucose. The liver responds sensitively to changes in blood glucose concentration. When the blood glucose is high the liver takes up glucose and stores it as glycogen. When the blood glucose is low there is a net loss of glucose from the liver to the bloodstream. Unlike other cells of the body, hepatic cells are freely permeable to glucose so the liver is capable of speedy responses to changes in blood glucose concentration. At normal blood glucose concentrations of 4.5 – 5.5 mmol l⁻¹ the liver is a net producer of glucose; at 8.5 mmol l⁻¹ the rates of uptake and output of glucose are equal. The reactions in liver cells to changes in blood glucose concentration are partly controlled by intermediary metabolites, ratios of oxidized to reduced coenzymes, availability of ATP or ADP, etc., these representing cellular processes which are independent of hormones. In addition, control by many hormones is superimposed on these more primitive control mechanisms.

Factors which tend to raise or lower blood glucose

Tend to raise	Tend to lower
Hunger	Satiety
Glucose absorption from gut	Glucose diffusion in ECF
Hepatic glycogenolysis	Muscular exercise
Adrenaline	Insulin
Glucagon	↑ Glucose oxidation
Gluconeogenesis	↑ Glycogen deposition
(in liver)	↑ Lipogenesis
Insulin antagonists	↓ Gluconeogenesis
Growth hormone	
Cortisol	[Glycosuria—in diabetes]
Insulin-destroying enzymes	

Factors tending to raise blood glucose

Hunger, a sensation aroused by a metabolic need, promotes eating, which helps to satisfy that need by raising the blood glucose. Satiety, on the other hand, stops food intake and tends to lower blood glucose.

Many of the factors which tend to raise blood glucose operate when the blood glucose level falls (hypoglycaemia). The first compensatory response to hypoglycaemia is glycogenolysis in the liver (within minutes), and the second response is gluconeogenesis, also mainly in the liver, which occurs over hours and even days.

Glycogenolysis is promoted by two hormones which are rapidly secreted in response to a fall in blood glucose—adrenalin and glucagon. In the liver both these cause glycogen breakdown by activating phosphorylase (via adenyl cyclase and cyclic AMP). In the liver glucose 6-phosphatase acts on glucose 6-phosphate to release glucose which enters the bloodstream. The activity of glucose 6-phosphatase is appropriately enhanced by a fall in blood glucose level, and at the same time glucokinase activity is depressed, thus decreasing glycogen synthesis. Adrenalin also promotes glycogenolysis in muscle, but owing to the lack of glucose 6-phosphatase in this tissue, glycolysis occurs and leads to the formation of lactate. The lactate enters the bloodstream and in the liver forms one of the substrates for gluconeogenesis. Glucagon does *not* cause glycogenolysis in muscle.

If the liver contains 60 g of glycogen (4 per cent by weight of a 1500 g liver) the complete oxidation of its derived glucose would only provide 1000 kJ of energy. Glycogenolysis is a suitably rapid response in an emergency but to maintain the blood glucose at a level (say 4.0 mmol l^{-1}) to satisfy the metabolic needs of the brain during prolonged starvation additional new sources of glucose are required. The substrates for gluconeogenesis are the non-nitrogenous portions of certain amino acids, lactate from muscle, and glycerol from split triglycerides in adipose tissue. Such sources are available for long periods of time. About 90 per cent of gluconeogenesis takes place in the liver. In normal man, during a fast of 2 or 3 days' duration, about 180 g (1 mole) of glucose are produced in 24 hours. Of this the central nervous systems uses about 140 g, and the red cells some 30 g, leaving very little for other tissues. Lipolysis can provide extra energy very rapidly through the release of FFA, which give 37.5 kJ g^{-1} when oxidized. In starvation muscle obtains up to 95 per cent of its energy from FFA; ketone bodies can also be utilized by muscle and other tissues. Fatty acids can very quickly modify carbohydrate metabolism in muscle. Thus in the isolated perfused heart fatty acids impair the ability to oxidize glucose to lactic acid and reduce the sensitivity to insulin. Glucose entry into the muscle cells is lessened; more glucose tends to remain in the blood. Lipolysis provides energy for heart muscle and other tissues and it also reduces the glucose usage of such tissues 'sparing it' for the brain which can metabolize *only* glucose.

Gluconeogenesis is under hormonal control. Glucagon, besides promoting glycogenolysis, also enhances gluconeogenesis from amino acids. Adrenalin enhances gluconeogenesis indirectly by inhibiting the secretion of insulin. Cortisol promotes gluconeogenesis partly by favouring the release of amino acids from proteins in muscle and bone, and partly by inducing the synthesis of gluconeogenic enzymes in the liver.

Cortisol and anterior pituitary growth hormone reduce the uptake of glucose by tissues.

Lipolysis is promoted by adrenalin, glucagon, cortisol, ACTH, and growth hormone. The FFA released not only provide an alternative fuel to glucose in muscle but they also reduce glucose uptake. These hormones help to maintain the blood glucose level partly by increasing glucose inflow into the blood and partly by reducing its outflow into the tissues.

Hypoglycaemia promotes the secretion of adrenalin, glucagon, cortisol, and growth hormone.

Factors tending to lower blood glucose

There are only two:

1. Starvation.
2. Insulin. The most important hypoglycaemic factor is insulin secreted by the β cells of the pancreatic islets. This hormone, whose detailed mode of action is discussed on page 507, has many actions which tend to lower blood glucose and its secretion in response to a rise in blood glucose level shows its importance as a physiological regulator of carbohydrate metabolism. Insulin antagonizes, in various ways, the cellular metabolic processes and the controlling influences of the hormones which tend to raise the blood glucose level. It lowers blood glucose mainly by promoting transport of glucose from extracellular fluid into cells. Inside cells glucose undergoes a variety of metabolic changes such as oxidation, deposition as glycogen and conversion to fat or amino acids. Insulin activates enzymes which produce all these changes.

REFERENCES

BROWN, D. H. and BROWN, B. I. (1975). Some inborn errors of carbohydrate metabolism. *MTP Int. Rev. Sci.* **5**, 391–426.

CLARK, M. G. and LARDY, H. A. (1975). Regulation of intermediary carbohydrate metabolism. *MTP Int. Rev. Sci* **5**, 223–66.

HANSON, R. W. and MEHLMAN, M. A. (eds.) (1976). *Gluconeogenesis*. Wiley-Interscience, New York.

HERS, H. G. (1976). The control of glycogen metabolism in the liver. *A. Rev. Biochem.* **45**, 167–89.

KREBS, E. G. and PREISS, J. (1975). Regulatory mechanisms in glycogen metabolism. *MTP Int. Rev. Sci.* **5**, 337–89.

WHITE, A., HANDLER, P., SMITH, E. L., HILL, R. L., and LEHMAN, I. R. (1978). *Principles of biochemistry*, 6th edn, pp. 423–531. McGraw-Hill, New York.

Fat metabolism

CHEMISTRY OF FATS (LIPIDES)

The lipides of physiological interest can be classified as follows:

1. Simple lipides. The most important are the neutral fats (triglycerides), i.e. glyceryl esters of fatty acids.

Triglycerides are formed from one molecule of glycerol and three molecules of fatty acid, thus:

$$
\begin{array}{lll}
CH_2OH & HOOC.R & CH_2O.OC.R \\
| & & | \\
CHOH & + HOOC.R' \longrightarrow & CHO.OC.R' \\
| & & | \\
CH_2OH & HOOC.R'' & CH_2O.OC.R'' \\
\text{Glycerol} & \text{Fatty acids} & \text{Triglyceride}
\end{array}
$$

where R, R′, and R″ represent three (possibly different) radicals of fatty acids. Diglycerides and monoglycerides are formed from one molecule of glycerol and two molecules or one molecule of fatty acid respectively. The most common of the many known fatty acids are:

(i) Palmitic acid, $CH_3(CH_2)_{14}COOH$ (a 16 C acid).
(ii) Stearic acid, $CH_3(CH_2)_{16}COOH$ (an 18 C acid).
(iii) Oleic acid, $CH_3(CH_2)_7CH = CH(CH_2)_7COOH$ (an 18 C unsaturated acid).

(i) and (ii) are fully saturated acids, (iii) is an unsaturated acid, containing one double bond, $—CH = CH—$. Practically all the natural fatty acids contain an even number of C atoms (as might be expected from their mode of synthesis, p. 462).

Waxes are esters of fatty acids with long-chain alcohols (instead of with glycerol, as in neutral fats).

2. Compound lipides. These are complex compounds formed from fatty acids, glycerol (or related substances), and various nitrogen-containing bases, and often containing phosphate groups. They are integral parts of the general cell structure; they are present in large amounts in nervous tissue; they are employed in fat transport [p. 433].

The chief types of compound lipide are:

(i) *Phospholipides* (phosphatides), containing glycerol, 2 molecules of fatty acid (generally unsaturated), phosphate, and a nitrogen-containing base (choline, $(CH_3)_3N^+CH_2CH_2OH$, in the case of the lecithins; ethanolamine (cholamine), $NH_2CH_2CH_2OH$, in the case of the cephalins).

(ii) *Sphingomyelins*, containing fatty acid, phosphate, choline, and a complex base (sphingosine), but no glycerol.

(iii) *Galactolipides* (cerebrosides), containing the monosaccharide galactose, fatty acid, and sphingosine, but no phosphate or glycerol.

3. Associated lipides. These are of two main types: (i) Those components (the split fats) that are obtained by the hydrolysis of lipides (i.e. glycerol, fatty acids, soaps). Soaps are salts of fatty acids, and are obtained by the hydrolysis of fats in alkali (saponification).

(ii) Those components that are associated with the lipides in tissue extracts simply because they are dissolved by the fat solvents: mainly (a) the steroids [cf. p. 529], e.g. hormones of the ovary, testis, adrenal cortex; cholesterol (esters of cholesterol with fatty acids are called cholesterides); (b) the fat-soluble vitamins (i.e. vitamins soluble in fats and in fat solvents) [p. 488].

The fat of food consists mainly of neutral fat, together with small amounts of free fatty acid, lecithin, and cholesterol esters. The nutritional importance of dietary fat is discussed on page 485.

DIGESTION AND ABSORPTION OF NEUTRAL FAT

Some hydrolysis of neutral fat takes place during cooking. None is achieved by the salivary secretions.

Stomach

Only in exceptional circumstances, may significant fat digestion occur in the stomach. A fat-splitting enzyme (gastric lipase) is present in the pure gastric juice from a Pavlov pouch; in addition, pancreatic lipase may regurgitate into the stomach from the duodenum. The activity of lipase in the stomach is obviously restricted, because the enzyme is sensitive to free acid and is destroyed by exposure to 0.02 per cent HCl for 15 minutes. Fat hydrolysis (lipolysis) may take place in the stomach in cases of achylia gastrica and in young suckling animals which ingest large quantities of milk: the fat of milk is present in an emulsified and therefore readily digested form, and moreover inhibits the secretion of gastric acid.

ABSORPTION OF FAT

Small intestine

Fat digestion begins in the small intestine owing to the action of pancreatic lipase and bile salts both of which enter the second part of the duodenum at the duodenal papilla. Pancreatic juice, which is alkaline, helps to adjust the pH of the initially highly acid chyme to one which is (at pH approximately 6.0) more favourable to the action of lipase (whose optimal pH is slightly above 7.0). Bile salts exert a detergent action, lowering surface tension and this promotes some emulsification of the fat thereby providing a greater surface area for the lipase to digest. Lipase acts at the interface between the fat particles and water and successively hydrolyzes the triglyceride into diglycerides, and monoglycerides with release of the associated fatty acids. Pancreatic electrolytes, monoglycerides, fatty acids, and bile salts thereupon form polymolecular aggregates (known as 'micelles') which are water soluble. These polymolecular aggregates can dissolve hydrophobic compounds in their interior and such constituents as cholesterol are thus made water soluble. The micelles themselves are passively absorbed into the luminal brush border of the villi where the bile salts are detached to return to the intestinal lumen finally to be actively reabsorbed more distally in the ileum. The 'fat' content of the micelle consists of monoglycerides and fatty acids. Once inside the intestinal cell they are dealt with in two ways'.

1. Monoglycerides and fatty acids containing more than 14 carbon atoms are re-esterified to tryglycerides; in this process the cell uses glycerol derived from α-glycerophosphate. The triglyceride globules are then 'coated' with β-lipoprotein (synthesized by the rough endoplasmic reticulum of the cell) and chylomicrons are formed. These enter the lymphatics which transport them to the thoracic duct and thence to the bloodstream. Chylomicrons are approximately one micrometre in diameter.

2. Short-chain fatty acids with less than 12–14 carbon atoms pass directly from the mucosal cells into the villous blood capillaries and are transported as free fatty acids (FFA)—a term which is synonymous with non-esterified fatty acid (NEFA). The fatty acids are bound to albumin in the bloodstream.

Movements of the villi compress both lacteals and villous capillaries and promote the mobilization of the lipids towards the thoracic duct and portal veins respectively.

Blood fat

Plasma lipid (triglycerides, free fatty acid, phospholipids, and steroids) are bound to plasma proteins to form *lipoprotein complexes* which vary in density according to their lipid content. As fat is lighter than water, the more lipid the protein contains, the less its density. Hence the lipoprotein complexes can be divided by centrifugation into very low density (VLD, <1.006), low density (LD), and high density (1.060–1.200). VLDLP is thus the shorthand designation of a very low density lipoprotein complex (which happens to be characteristic of triglycerides). Some 140–150 mg per 100 ml of triglycerides bound to lipoproteins are normally present in the plasma. Cholesterol concentration of plasma is approximately 5 mmol l^{-1} (200 mg per 100 ml). 75 per cent of this is esterified. Phospholipids (such as lecithin and cephalin) have approximately the same plasma concentration as cholesterol. They are bound with α-lipoproteins and form the HDLP constituent. The free fatty acids (FFA = NEFA) comprise 12 mg per 100 ml in resting conditions.

Fat stores

Fat is stored in the adipose cells of the fat depots as triglyceride. White adipose tissue represents the biggest store of energy in the body. Its main role is the maintenance of FFA (NEFA) concentration in the blood. White adipose tissue is found ubiquitously. It comprises large cells which contain a single fat droplet. According to the nutritional state of the body the cell content of cytoplasm is reduced as the neutral fat accumulates until the cell becomes only a thin cytoplasmic nucleated envelope containing the tryglyceride. Adipose tissue is highly specific in accumulating or releasing fat as the situation requires. Its oxygen usage is approximately 8 ml per 100 g min^{-1}—similar to that of the heart beating in resting man.

Brown adipose tissue is found only in the neonate. Its remarkable metabolic properties are described on page 596.

FATE OF FAT AFTER ABSORPTION

After absorption, fat is treated in various ways:

1. It undergoes complete oxidation in the tissues to yield energy, CO_2, and H_2O. When 1 g of mixed fat is completely burnt to CO_2 and H_2O, 38 kJ of heat are produced; during the metabolic dissimilation of fat, a large proportion of this amount of energy is made available to the body as high-energy phosphate ATP. Active acetate (Ac-Co A) is an intermediate and can be used in acetylation reactions and for the synthesis of acetoacetic acid and certain body components [p. 460].

2. It is stored (as neutral fat) in the fat depots. In contrast to the small carbohydrate reserves (0.5 kg), fat may be stored in the body in very large amounts. On an average, fat forms over 10 per cent of the body weight (i.e. about 7 kg in an adult, equivalent to an energy

reserve of 3800 kJ kg^{-1} of body weight, or more than a month's total food energy); in people who are over-weight the fat reserves are much bigger.

Adipose tissue should be regarded as a highly specific tissue, taking up fat differentially when fat is present in excess of the body's immediate metabolic needs, and releasing it when required. Adipose tissue takes up fat in the same selective manner as, for example, the thyroid takes up iodide.

Adipose tissue is under hormonal control (growth hormone, insulin, catecholamines) and the details of the regulation are considered under these headings.

3. Fat is built into the structure of all tissues. The structural lipid consist of the following:

(i) Lecithins (and the related cephalins).

(ii) Cholesterides (cholesterol esters of fatty acids).

The constituent fatty acids in these two groups are mainly unsaturated (unlike the fatty acids in depot fat, many of which are fully saturated). Lecithin and cholesterides *are essential* constituents of *all* cell membranes; lecithin is a component of the medullary sheath of nerve fibres.

(iii) Certain specialized lipids like the sphingomyelins and cerebrosides of the central nervous system.

Structural lipids are as integral a part of the cell architecture as are proteins; they constitute the *élément constant* of the total body lipid (in contrast to the '*élément variable*' or neutral fat of the fat depots). In starvation, the neutral fat in the depots is called upon and used for producing energy, but structural lipids (*élément constant*) are unaffected in amount.

Sources of depot fat

The neutral fat in adipose tissue is derived from two main sources:

(a) From food fat.

(b) From carbohydrate. The classical proof (Gilbert and Lawes) is as follows: young pigs fed up on barley deposit large amounts of fat, more than could be derived from the fat and protein of the food even if it were assumed that all the carbon of the ingested protein went to form fat; clearly the fat which has been deposited must have been partly derived from carbohydrate. Similarly, potatoes are rightly condemned as 'fattening', though the fat content of potatoes is almost nil (actually 0.1 per cent). The details of the carbohydrate → fat transformation are considered on page 462.

RELATION OF LIVER TO FAT METABOLISM

1. When fats are to be used in the body they are withdrawn from the fat reserve (the adipose tissue cells), and pass to the liver; the fat content of the liver may be little altered, as the fat is broken down as fast as it arrives.

2. The neutral fat content of the liver is increased in the following conditions:

(i) On a high-fat diet which is also deficient in the so-called lipotropic factors (i.e. choline and methionine; see below).

(ii) In starvation.

(iii) After pancreatectomy, in animals kept alive by adequate doses of insulin.

The fat-laden liver arising from these conditions is called a *fatty liver*.

3. The neutral fat in the liver is broken down by hydrolysis into glycerol and fatty acids (enzyme: liver lipase). (Phase I metabolism of fat.)

(i) The glycerol is utilized via the pathways of carbohydrate metabolism.

(ii) The fatty acids are oxidized to acetyl-Co A units containing 2 C each (Phase II); these fragments are either (a) completely oxidized

to CO_2 and H_2O with energy liberation (Phase III), or (b) recombined to give acetoacetic acid (a 4C compound, $CH_3.CO.CH_2.COOH$); this latter process is termed *ketogenesis*, since acetoacetic acid is a ketone.

It appears that the liver cannot further metabolize acetoacetic acid; any acetoacetic acid that the liver manufactures from fatty acids must be distributed to the tissues, where it is completely oxidized to yield CO_2, H_2O, and energy.

These reactions are more fully discussed on pages 461–3.

4. In circumstances of carbohydrate deficiency, the metabolism of liver fat can play a part in the maintenance of the blood glucose level. The liver can (i) increase the metabolism of fat (including the production of acetoacetic acid for energy utilization in tissues) and thus 'spare' the available carbohydrate, and (ii) possibly convert some small part of the fat into glucose (or glycogen).

A detailed discussion is given below.

Fatty liver. Lipotropins

In the condition of fatty liver, a grossly enlarged liver containing massive depositions of neutral fat is found. The causes of fatty liver enumerated in (2) above will now be considered in detail.

1. High fat diet. In animals given a high neutral fat diet, fatty liver develops. Under such circumstances, fat becomes the principal source of energy for the body and it is appropriate that large amounts of fat should be brought to the liver for complete or partial dissimilation; but one would expect that a suitable balance would be struck between the uptake of fat by the liver and its complete oxidation or its redistribution to the tissues as acetoacetic acid. It is surprising to find, therefore, that the uptake of fat by the liver should exceed, so markedly, its rate of despatch. The fat content of the fatty liver is, however, considerably decreased by administering: (i) methionine (or large amounts of proteins rich in methionine); or (ii) choline, or the related substance betaine; or (iii) lecithin (which contains choline).

Role of lipotropins. A substance which reduces the amount of liver fat is called a lipotropin (lipotropic factor). It is thought that the lipotropins other than choline itself are effective because they contain choline or because they promote choline synthesis. Considering the substances already mentioned we find:

$$H_2N.CH_2.COOH \longrightarrow H_2N.CH_2.CH_2OH$$
$$\text{glycine} \qquad\qquad \text{ethanolamine}$$

$$\xrightarrow[\substack{\text{from} \\ \text{methionine} \\ \text{or betaine}}]{+\ CH_3^-} (CH_3)_3N^+CH_2CH_2OH$$
$$OH^-$$
$$\text{choline (base)}$$

$$CH_3-S.CH_2.CH_2.CH(NH_2)COOH \qquad\qquad (CH_3)_3N^+.CH_2.COO^-$$
$$\text{methionine} \qquad\qquad\qquad\qquad\qquad \text{betaine}$$

(a) Methionine is a methyl donor; it supplies labile methyl (CH_3^-) groups to ethanolamine (cholamine, β-amino-ethanol) to form choline. The ethanolamine is synthesized in the body from glycine (probably via serine).

(b) Betaine (a methylated glycine derivative) is also a methyl donor, and its methyl group can be used to synthesize choline from ethanolamine. It is unlikely that betaine can be transformed directly into choline, though choline is known to be oxidized to betaine.

(c) Lecithin contains choline as part of the molecule and liberates choline on hydrolysis.

The mode of action of choline as lipotropin is unknown. It may promote the conversion of liver fat into choline-containing phospholipids (e.g. lecithins) which are more readily transferred from the liver into the blood. The important point, however, is that choline prevents and cures fatty liver.

Choline deficiency also gives rise to haemorrhagic necrosis of the kidneys; this disorder may likewise be due to insufficient lecithin formation in the renal cells.

Transport and role of choline. Choline is not transported in the blood in the free state but always in the combined form, i.e. as part of phospholipid molecules (lecithin, sphingomyelin). The normal level of blood choline in man (present as phospholipid) is 3–5 mmol l^{-1}; in states of choline deficiency giving rise to fatty liver, the blood choline is much less.

Choline is an indispensable constituent of the body; if the necessary groupings are not available for its synthesis, then sufficient choline must be provided, as such, in the food. It is not strictly speaking a vitamin (though occasionally put in this category) because it can be synthesized in adequate amounts by the body. As explained on page 467, choline promotes creatine synthesis because of its action as a methyl donor via methionine.

2. Starvation. In starvation, the stress of metabolism also falls on fat, i.e. the fat in the depots. As lipotropins are not available in adequate amounts, fat accumulates in the liver.

3. After pancreatectomy. If a pancreatectomized animal is kept alive with insulin, fatty liver develops even on a fat-free diet (e.g. a diet of lean meat and sugar). In pancreatectomized dogs, the liver weight may increase fourfold and the total lipid content increase thirtyfold over the normal values. The condition is cured by administering large amounts of choline or of methionine. The liability of the chronic diabetic patient to develop fatty liver should be noted.

Effects of fatty liver. When the liver cells are laden with fat their functional activity is directly depressed; in addition the swollen cells compress the vascular capillaries, and so diminish the blood supply, especially to the centralobular cells. With the development of fatty liver hepatic function is impaired, the most obvious symptom being reduced glucogenesis with the result that the fasting blood glucose level of an affected diabetic patient may fall to normal (or lower) even in the absence of insulin. The patient is also extremely sensitive to injected insulin.

End results of Fatty Liver: Diffuse Hepatic Fibrosis (Cirrhosis). Gross chronic fatty liver leads to the death of many liver cells and their replacement by scar tissue. The condition is then called diffuse hepatic fibrosis (cirrhosis of the liver). The scarring, by interfering with the hepatic blood supply, aggravates the condition, setting up a vicious circle leading to more necrosis and more scarring. Diffuse hepatic fibrosis secondary to fatty liver occurs clinically in the following states:

1. In severe malnutrition (especially in the tropics). The diet is deficient in fat and in protein; the lack of lipotropins is probably the main factor responsible for the condition. Kwashiorkor is characterized by an enlarged grossly fatty liver and other disturbances (steatorrhoea, macrocytic anaemia, and oedema).

2. In so-called 'alcoholic cirrhosis' the liver is initially 'fatty' and later shows fibrosis. The essential cause is a diet deficient in lipotropins aggravated by their defective absorption from the intestine. The drinking of excessive amounts of alcohol contributes only indirectly to the condition by leading to loss of appetite and decreased food intake and by setting up gastro-enteritis, thus interfering with absorption.

Relation of liver to ketogenesis

1. Ketone bodies. The name 'ketone bodies' (or 'acetone

bodies') is applied to the following three substances, which form a metabolically-related group:

$$acetoacetic\ acid\ (CH_3.CO.CH_2.COOH)$$
$$\beta\text{-hydroxybutyric acid }(CH_3.CHOH.CH_2.COOH)$$
$$acetone\ (CH_3.CO.CH_3)$$

(i) Acetoacetic acid is the parent substance of the group, and is formed during the dissimilation of (a) long-chain fatty acids from fat [p. 461], and (b) certain of the essential amino acids [p. 465]. In the body, acetoacetic acid is found associated with β-hydroxybutyric acid and small amounts of acetone.

(ii) β-Hydroxybutyric acid is the reduction product of acetoacetic acid (i.e. it is formed by the addition of 2H); the two acids are freely interconvertible.

(iii) Acetone only arises from acetoacetic acid by spontaneous and non-reversible decarboxylation (loss of CO_2), a reaction which occurs chiefly in the lungs and bladder.

It should be noted that one of the three 'ketone bodies', namely β-hydroxybutyric acid, is *not a ketone*, though it is metabolically derived from one (acetoacetic acid).

The interrelationships of the ketone bodies are summarized in FIG. IX.10.

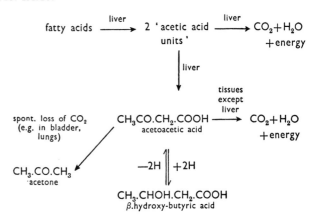

FIG. IX.10. Inter-relationships of the 'ketone bodies' [cf. FIG. IX.12].

The liver is the only organ which produces ketone bodies on any significant scale. A comparison of the ketone body content of the blood which enters and leaves the liver shows that the liver forms ketone bodies and delivers them into the hepatic vein for circulation to, and dissimilation by, the tissues. All the tissues, with the exception of the brain and the liver itself, can dissimilate acetoacetic acid to CO_2, H_2O, and usable energy.

2. Blood ketone level. Ketosis. The normal non-fasting blood ketone level is small (0.5–2 mg per 100 ml); even a short-term fast (of two or three days) increases this level as much as fiftyfold. The amount of circulating ketone depends upon the balance between (i) ketone formation by the liver and (ii) ketone dissimilation by the tissues. Little is known about the factors that determine ketone dissimilation; it is not apparently influenced by the hormones. There is, however, a maximum amount of fat which the tissue can use (mostly as acetoacetic acid), namely about 2.5 g of fat kg^{-1} day^{-1}, equivalent to 175 g of fat daily in a 70 kg man. The rate of ketogenesis in the liver varies greatly according to circumstances. If ketogenesis proceeds at an unduly high rate, exceeding the rate at which dissimilation can be carried on by the tissues, then ketones accumulate in the blood. This condition is called ketosis, and may lead to the excretion of ketone bodies in the urine. In extreme ketosis, the urinary ketone output may reach 100–120 g day^{-1}.

The responses of the body to the accumulation of ketones in the blood are discussed on page 510.

3. Mechanism of ketone body formation. β-Oxidation of fatty acids. The neutral fat in the liver is hydrolyzed as required, releasing glycerol and long-chain fatty acids, the majority containing 16 or 18 C atoms (e.g. palmitic, stearic, and oleic acids).

The next stage is the breakdown, mainly in the liver, of these long carbon chains into 'fragments' containing two carbon atoms each. This breakdown is an oxidative process, hydrogen being passed to NAD. Before this can occur, the acid has to be converted to its high-energy Co A derivative by reaction of the acid with ATP and Co A. The Co A ester undergoes dehydrogenation and hydration, followed by cleavage at the β-carbon atom by reaction with another molecule of Co A. This splits off acetyl-Co A, leaving the carbon chain with 2C less but still as a Co A ester, so that the whole process can be repeated, another acetyl-Co A splits off, and so on down the chain. The C chain of fatty acids is thus disassembled, 2C at a time, from the acid end [Fig. IX.11]:

$$R - - CH_2 - CH_2 \underline{\quad\quad} CH_2 - COOH$$
$$R - - CH_2 - CH_2 \underline{\quad\quad} CH_2 - CO\sim S\text{-}Co\text{-}A$$
$$R - - CH_2 - CH = CH - CO\sim S\text{-}Co\text{-}A \quad (+NAD.2H)$$
$$R - - CH_2 - CHOH - CH_2 - CO\sim S\text{-}Co\text{-}A$$
$$R - - CH_2 - CO\sim S\text{-}Co\text{-}A \quad + \quad CH_3 - CO\sim S\text{-}Co\text{-}A$$

$$\downarrow \quad \text{and so on}$$

Fig. IX.11. Sequence of intermediates for disassembly of fatty acid chain to acetyl-Co A.

Thus the 16 C chain of palmitic acid yields 8 fragments of acetyl-Co A (*active acetate*).

Isotope experiments using acids containing labelled carbon atoms in known positions have shown:

(a) The acetyl-Co A units formed in the liver can be completely dissimilated in the liver via the citric acid cycle if sufficient oxaloacetic acid is available from elsewhere. Coenzyme A is thus freed for further fatty acid cleavage.

(b) Instead of being oxidized, pairs of acetyl-Co A units can react together in random fashion, to form the 4C ketone body acetoacetic acid. Initially this condensation produces free Co A and aceto-acetyl-Co A, but the latter is rapidly converted to free acetoacetic acid:

$$2\ CH_3 - CO\sim S\text{-}Co\ A \rightarrow$$
$$CH_3 - CO - CH_2 - CO\sim S\text{-}Co\ A + HS\text{-}Co\ A$$
$$\underbrace{CH_3 - CO}_{2C} - \underbrace{CH_2 - COOH}_{2C}$$

The acetoacetic acid is distributed to the tissues, and is there oxidized to CO_2. It is usually assumed that the first step in the oxidative dissimilation of acetoacetic acid by the tissues, is its resplitting to acetyl-Co A.

The proportion of acetyl-Co A forming acetoacetic acid rather than being immediately oxidized to CO_2 is determined by the rate of the simultaneous carbohydrate dissimilation in the liver. If the rate of carbohydrate dissimilation is depressed (as in diabetes or starvation), then the proportion of acetoacetic acid formed rises, and ketosis may develop. A possible explanation for this effect is given on page 510.

UTILIZATION AND DISSIMILATION OF 'ACTIVE ACETATE'

'Active acetate' units are highly reactive, and immediately they are formed they undergo other changes, as follows:

1. Self-condensation to give acetoacetic acid. The random self-condensation of two acetyl-Co A forms the 4C compound, acetoacetic acid. This is carried in the circulation to other tissues which can dissimilate it [see 2 (i) below].

2. Dissimilation to CO_2 and H_2O. Acetyl-Co A and acetoacetic acid are completely dissimilated by entering the citric acid (Krebs) cycle [p. 454]. Acetyl-Co A (2C) and acetoacetic acid (4C) are the 'ready money' of fat metabolism. The long C chain of the fatty acids is appropriate for storage purposes only, and represents capital locked away in securities. This wealth is not negotiable directly, but must be converted into the ready money of 2C and 4C fragments (2C fragments for buying a paper or paying a bus fare: 4C 'notes' for putting the wealth into circulation). Just as ready money in sufficient quantity can be converted into securities, so can 2C fragments from any source be built up into the long C chain of the fatty acids.

(i) Acetyl-Co A from fatty acid is metabolized in exactly the same way as that formed by the oxidative decarboxylation of pyruvic acid (from carbohydrate). It combines with oxaloacetic acid (4C) to give the 6C tricarboxylic acid, citric acid; this then 'goes round the cycle,' losing two carbon atoms in the form of two molecules of CO_2 by two successive decarboxylations. As a result, energy is liberated which is made available to the body through the high-energy ATP; oxaloacetic acid is regenerated, and is ready for reaction with a further Ac-Co A [Figs. IX.7 and IX.12].

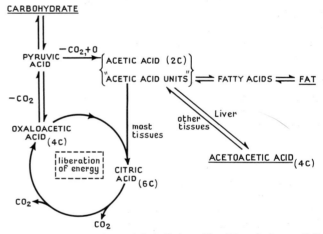

Fig. IX.12. The formation and dissimilation of fat. (Note the irreversibility of the pyruvic acid → acetic acid step.)

Energy yields. The energy of fatty acid dissimilation appears as ATP during the respiratory reoxidation of the coenzyme hydrogen carriers reduced during the aerobic chain cleavage and the aerobic citric acid cycle. Six carbon atoms of a chain dissimilated through 3 Ac-Co A to 6 CO_2 yields a total of 44 ATP, compared with the 38 for six carbon atoms as glucose. There is no counterpart in fat metabolism to the anaerobic glycolysis of carbohydrate.

(ii) It appears that the main step in the dissimilation of acetoacetic acid in the tissues is its re-splitting into two Ac-Co A by free Co A; these are then oxidized by the enzymes of the Krebs citric acid cycle present in the tissue concerned. Liver and brain seem unable to split acetoacetic acid, and thus cannot utilize it to any extent.

Acetoacetic acid may also be able to enter the citric acid cycle directly (say at a 4C stage) but this must constitute a minor pathway.

It can be seen from Figure. IX.12 that for the complete dissimilation of fats, oxaloacetic acid must be available from some source (e.g. from pyruvic acid or from aspartic acid); otherwise the Ac-Co A will form correspondingly larger quantities of acetoacetic acid in order to free the Co A, and ketosis will develop.

3. Acetylation reactions. The active Ac-Co A can be employed by the body in acetylating reactions, e.g. the synthesis of acetylcholine [p. 411] or the acetylation of amines during their 'detoxication' by the liver.

$$CH_3.CO-O.CH_2CH_2\overset{+}{N}(CH_3)_3OH^-$$
acetylcholine

$$CH_3.CO-NH.C_6H_4.SO_2NH_2$$
acetylsulphanilamide

4. Use as building unit for body components. Evidence from isotope experiments shows that labelled C atoms from acetic acid (or any other compounds giving rise to Ac-Co A) can be built into the molecules of (i) cholesterol (all its 27 C atoms come directly from acetate); (ii) haem porphyrins, and (iii) the intermediates of the citric acid cycle and their derivatives.

5. Resynthesis of fatty acid. 'Acetic acid units' (from any source) can be built up into long-chain fatty acids, and thence into fat.

This occurs by the stepwise addition of 2C acetyl-Co A units to the acid end of a carbon chain. The process requires a supply of H from reduced NADP, of ATP, and of CO_2 as a catalyst; its mechanism is *not* the reverse of fatty acid breakdown.

The cell microsomes are responsible for fatty acid synthesis, while oxidation only occurs in mitochondria.

Confirmation is provided by:

(i) All fats are built from fatty acids with an even number of C atoms.
(ii) Isotope experiments show that when acetic acid $CD_3.^{13}COOH$, labelled with ^{13}C and deuterium (**D**, heavy hydrogen), is fed to animals, the liver fats have an isotope distribution in the fatty acid chains thus:—$CD_2^{13}CH_2-CD_2.^{13}CH_2-CD_2.^{13}COOH$.

INTEGRATION OF FAT AND CARBOHYDRATE METABOLISM

Conversion of carbohydrate to fat

The well-known conversion of dietary carbohydrate into depot fat has been mentioned on page 459. It is obvious from FIGURE IX.12 that the pathways of carbohydrate and fat metabolism meet at the common intermediate acetic acid (acetyl-Co A). Since fatty acid is reversibly formed from acetyl-Co A the way is open for the transformation of carbohydrate via pyruvic acid and acetic acid into fatty acids. This transformation is promoted by insulin and depressed by anterior pituitary hormones.

Glycerol from the hydrolysis of fats is converted into a triose (probably phosphoglyceraldehyde) and joins the main pathway of carbohydrate metabolism [p. 453]. This reaction is reversible allowing carbohydrate to be used for the synthesis of glycerol when this is needed for the deposition of fatty acids as body fat.

carbohydrate ⇌ triose ⇌ glycerol

pyruvic acid

fat

acetic acid ⇌ fatty acids

Conversion of fat to carbohydrate

As stated above, the glycerol of fats can join the reversible pathway of carbohydrate metabolism and thus be built into glucose or glycogen. This would allow the formation of only 12 g of blood glucose from 100 g of fat, whereas it has been thought that much greater conversion than this could occur in the fasting animal. There is no doubt (from isotope experiments) that C *atoms* from acetic acid (and other molecules giving rise to active acetate, e.g. fatty acids) can be built into the molecule of glucose by the liver. A consideration of FIGURES IX.7 and IX.12, however shows that *no net increase* of glucose could arise in this way because each turn of the cycle takes in 2C as acetate and eliminates 2C as $2CO_2$. There is no clear pathway in animals by which the fatty acid → glucose conversion could be accomplished. Any net increase in the glucose of the body fluids that does come directly from fatty acid must be derived from the synthesis of, say, succinic acid (4C) which would directly enter into the Krebs cycle and could thus be converted into oxaloacetic acid and thence into pyruvic acid and glucose; this must be a minor pathway except possibly in gross ketosis when acetoacetate might form succinate in muscle. Under conditions of carbohydrate deficiency, the chief function of the liver is to provide acetoacetic acid for utilization by the tissues, thus 'sparing' the utilization of the more valuable blood glucose which is required for CNS metabolism.

Dependence of fat metabolism on carbohydrate

The complete dissimilation of fat via acetyl-Co A requires available supplies of oxaloacetic acid to 'catalyse' the Krebs cycle [FIG. IX.12]. This must be supplied from a source other than fat (since the pyruvic acid → acetic acid reaction is irreversible); the chief source of oxaloacetic acid is carbohydrate, via pyruvic acid and the carboxylase reaction [p. 454]. The complete dissimilation of fats in the liver thus requires the simultaneous oxidation of carbohydrate. This idea used to be expressed vividly in the phrase 'The fats burn in the flame of the carbohydrates.' Any change which reduces the rate of carbohydrate oxidation in the liver will reduce the rate of complete oxidation of acetyl-Co A locally, without reducing its rate of formation; indeed, its rate of formation may *increase* if fat is the only available energy source. However, as these highly reactive '2C fragments' cannot accumulate, they undergo self-condensation to form acetoacetic acid. The lower the rate of carbohydrate utilization in the liver relative to the fat utilization, the greater is the production of acetoacetic acid, and the amount which circulates in the blood correspondingly increases. If the tissues cannot dissimilate the increased supplies of acetoacetic acid which reach them then ketosis inevitably results. Starvation, diabetes, and the other circumstances that lead to ketosis all involve a decrease in the normal ratio of carbohydrate to fat utilization in the liver.

Causes of ketosis

Ketosis occurs clinically and experimentally in the following conditions:

1. With a low glycogen and a high fat content in the liver, e.g. on a high fat, low carbohydrate diet, or in starvation (when the liver glycogen reserves are exhausted and depot fat is being utilized). Under these conditions fat becomes the 'principal substrate of the liver cell' leading to an excessive ketone body formation. It should be noted that a small proportion of the ketones may be derived from ketogenic amino acids [p. 469]. The formation of glucose from glucogens may be simultaneously depressed. As the tissues cannot cope with all the ketones supplied to them, the blood ketone level rises. However, people differ greatly in their susceptibility to ketosis. Thus Eskimos can tolerate high fat diets that would cause gross ketosis in the average European. The body can apparently be 'trained' gradually in the utilization of very high fat diets.

2. From the injection of certain anterior pituitary extracts in

normal animals [p. 516]. Depot fat is mobilized and reaching the liver increases its lipid content. Glucose dissimilation is simultaneously depressed, and liver glycogen is decreased owing to its conversion into blood glucose. Anterior pituitary extracts thus promote ketone formation in the liver and ketosis.

3. Ketosis of Diabetes Mellitus. After removal of the islet tissue of the pancreas, the unantagonized action of the anterior pituitary leads to ketosis as described in 2. above. Injected insulin [p. 508] acts in the opposite manner; it promotes deposition of depot fat and decreases the flow of fat to the liver; it increases glucose dissimilation and glycogen deposition in the liver; it annuls the specific ketone-producing action of the anterior pituitary. Thus by its action on the fat depots and on the liver, insulin abolishes the ketosis of diabetes mellitus. It has no effect on ketone utilization by other tissues.

REFERENCES

Krebs, H. A. (1966). The regulation of release of ketone bodies by the liver. *Adv. Enzyme React.* **4**, 339–54.

Masoro, E. J. (1977). Lipids and lipid metabolism. *A. Rev. Physiol.* **39**, 301–21.

Rommel, K. and Bohmer, R. (eds.) (1976). *Lipid absorption: biochemical and clinical aspects.* University Park Press, Baltimore.

Volpe, J. J. and Vagelose, P. R. (1976). Mechanisms and regulation of biosynthesis of saturated fatty acids. *Physiol. Rev.* **56**, 339–417.

Wakil, S. (ed.) (1970). *Lipid metabolism.* Academic Press, New York.

White, A., Handler, P., Smith, E. L., Hill, R. L., and Lehman, I. R. (1978). *Principles of biochemistry*, 6th edn, pp. 568–633. McGraw-Hill, New York.

Protein metabolism

PROTEINS

The proteins are complex molecules built mainly from α-amino acids linked together in chains. The linkage between the amino acids is called a peptide bond; molecules built up from many amino acids are called polypeptides; proteins are types of polypeptide, but with large and complex structures often consisting of several polypeptide chains cross-linked between specific amino acid units. In spite of these complications, considerable progress has been made in elucidating the chemical structure of the simpler proteins. For example, the complete amino acid sequence (primary structure) of the insulin molecule is known [p. 504], but even this gives little indication of the spatial configuration (secondary structure) of the protein in its physiological environment. Hydrolysis (digestive or otherwise) converts a protein through stages of intermediate complexity (conventionally called metaproteins, proteoses, and peptones) to a mixture of peptides and then free amino acids. About twenty different amino acids have been found in the various proteins studied; most proteins contain a selection of 15–18 of these, though some proteins are built from only a few different amino acids.

The amino acids have the general formula

$$R.CH(NH_2).COOH$$

where **R** is any one of a variety of organic groupings as shown below. The peptide bonds of a polypeptide chain are —CO—NH— (amide) linkages of the following type:

$$\overset{\mathbf{R'}}{\underset{}{}} \qquad \overset{\mathbf{R''}}{\underset{}{}}$$
— — —CO—NH.CH.CO—NH.CH.CO—NH— — — —
peptide bonds

The properties of the proteins reflect the nature and mutual arrangement of the constituent amino acid units. Proteins with high and specific physiological activities (e.g. protein hormones, enzymes, antibodies) must owe these characteristics to unique amino acid arrangements at areas within the whole molecule; the slightest disturbance of these areas, even involving only one amino acid unit, may modify or completely abolish the physiological activity. Even small peptides with only a few constituent units may have intense physiological or pharmacological actions, but in general the importance of small peptides and free amino acids lies in their metabolic reactions.

Protein amino acids

The names and formulae of the principal amino acids obtained by the hydrolysis of proteins are set out below:

Amino acids with unsubstituted C chains

1. Glycine (α-aminoacetic acid)
   $NH_2.CH_2.COOH$ (R= H; the simplest amino acid)
2. Alanine (α-aminopropionic acid)
   $CH_3CH(NH_2)COOH$ (3C chain)
3. Valine
   $(CH_3)_2CH.CH(NH_2)COOH$ (5C branched chain)
4. Leucine
   $(CH_3)_2CHCH_2.CH(NH_2)COOH$ (6C branched chain)
5. Isoleucine
   $(CH_3)(C_2H_5)CH(NH_2)COOH$ (6C branched chain)

Hydroxyl-substituted amino acids

6. Serine (β-hydroxyalanine)
   $HO.CH_2.CH(NH_2)COOH$ (3C chain)
   $\beta \qquad \alpha$
7. Threonine
   $CH_3CH(OH).CH(NH_2)COOH$ (4C chain)

Sulphur-containing amino acids

8. Cysteine (β-mercaptoalanine)
   $HS.CH_2.CH(NH_2)COOH$ (3C chain—note the reactive HS. group)
9. Cystine (dicysteine, the oxidation product of cysteine)
   $S.CH_2.CH(NH_2)COOH$
   |
   $S.CH_2.CH(NH_2)COOH$
10. Methionine (α-amino-γ-methylthiobutyric acid)
    $CH_3S.CH_2CH_2.CH(NH_2)COOH$ (4C chain—note
    $\gamma \qquad \alpha$ the CH_3S. group)

Aromatic amino acids, derived from alanine

11. Phenylalanine (β-phenylalanine)

    ⬡—$CH_2.CH(NH_2)COOH$

12. Tyrosine (para-hydroxyphenylalanine)

    HO—⬡—$CH_2.CH(NH_2)COOH$ (a phenol)

13. Thyroxine and triiodothyronine, iodine-containing derivatives of tyrosine [see p. 538]

14. Tryptophan (β-indolylalanine)

    $CH_2.CH(NH_2)COOH$

The amino acids (1 to 14) set out above are neutral substances, since they contain one NH_2. (basic) group and one COOH (acidic) group which mutually neutralize each other.

Acidic amino acids

15. Aspartic acid (α-aminosuccinic acid)   (4C chain)

    CH₂—COOH
    |
    CH(NH₂).COOH

15 a. Asparagine (β-amide of aspartic acid)

    CH₂—CONH₂
    |
    CH(NH₂).COOH

16. Glutamic acid (α-aminoglutaric acid)   (5C chain)

    CH₂.CH₂—COOH
    |
    CH(NH₂).COOH

16 a. Glutamine (γ-amide of glutamic acid)

    CH₂.CH₂—CONH₂
    |
    CH(NH₂).COOH

Aspartic and glutamic acids (15 and 16 above) contain an additional COOH (acidic) group in the **R** side-chain (i.e. they are mono-amino-dicarboxylic acids) and are acidic substances. Asparagine and glutamine, with an amide (neutral) group in the sidechain are neutral substances, but readily give the corresponding acids on hydrolysis, or during the hydrolysis of proteins containing them.

Basic amino acids

17. Arginine (α-amino-δ-guanidinovaleric acid)

    NH
    ‖
    NH₂—C—NH.CH₂CH₂CH₂.CH(NH₂)COOH

                                                              (5C chain)

18. Lysine (α: ε-diaminocaproic acid)

    NH₂.CH₂CH₂CH₂CH₂.CH(NH₂)COOH   (6C chain)

19. Histidine (β-iminazolylalanine)

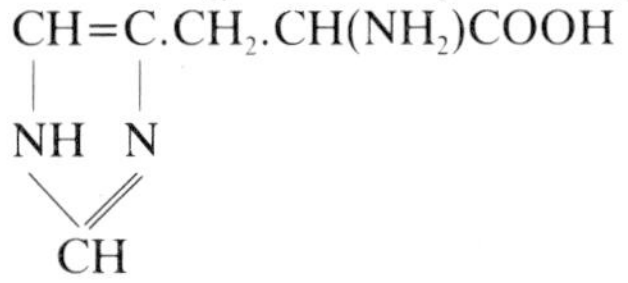

    CH=C.CH₂.CH(NH₂)COOH
    | |
    NH   N
     \\ ‖
      CH

The three amino acids (17, 18, 19) set out above have nitrogen-containing **R** groups and are basic substances.

Imino acids

20. Proline

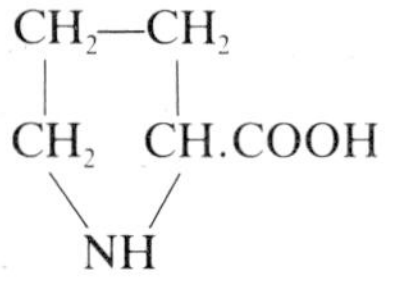

    CH₂—CH₂
    | |
    CH₂   CH.COOH
      \\ /
       NH

                                                              (5C chain)

21. Hydroxyproline

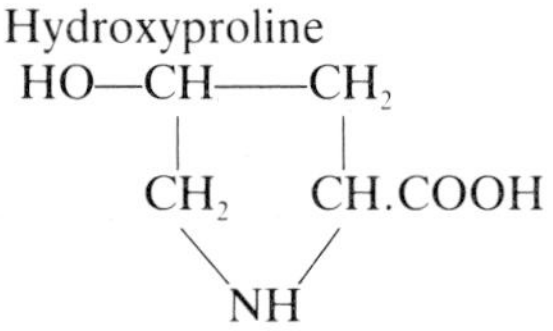

    HO—CH——CH₂
        | |
       CH₂   CH.COOH
         \\ /
          NH

Proline and hydroxyproline are imino acids, with an α. NH—as part of a ring system. They are neutral substances. Proline is a common constituent of most proteins, but hydroxyproline is only found in the collagen protein of connective tissue and bone.

Non-protein amino acids

Some amino acids have metabolic significance as intermediates in the degradation or interconversion of the protein amino acids though they are not themselves found in mammalian proteins. Examples are ornithine and citrulline of the Krebs urea cycle [p. 468, Fig. IX.17], homocysteine from methionine [p. 467, Fig. IX.15], and 5-hydroxytryptophan in the synthesis of serotonin from tryptophan [p. 558].

Peptides

Examples of smaller peptides of known physiological importance are:

1. Oxytocin and vasopressin, octapeptide hormones of the posterior pituitary [p. 523] hypothalamic releasing factors [p. 518], angiotensin [p. 561], endorphins [p. 401], etc.
2. Glutathione, a tripeptide of glycine, glutamic acid, and cysteine, which regulates the activity of oxidizing enzymes, particularly in brain.

DIGESTION OF PROTEINS

Mouth. None

Stomach. The pepsin–HCl mixture converts protein to proteoses and peptones (large polypeptides). Young mammals also secrete rennin, which partially hydrolyzes the caseinogen protein of milk, giving soluble casein; in the presence of calcium salts, casein forms a coagulum of calcium caseinate which (strangely) is readily digested enzymically. Even the existence of rennin in man is not established. Its role is discharged by pepsin.

Duodenum. Trypsin and chymotrypsin of pancreatic juice, acting in an alkaline medium, continue the process of hydrolytic disintegration begun by pepsin, liberating peptides of various chain lengths together with a few free amino acids.

Small intestine. Succus entericus contains a set of mixed peptidases ('erepsin') each of which specifically hydrolyzes terminal peptide bonds, and thus the breakdown of peptides to free amino acids is completed; this occurs partly in the lumen and partly in the wall of the intestine.

Digestion of nucleoprotein. The protein portion is removed from nucleoprotein by the acid of the stomach, and digested with the other food proteins. The freed nucleic acid is hydrolyzed enzymatically in the gut to nucleotides, nucleosides, and finally to the constituent pentoses, purines, and pyrimidines. These are absorbed and metabolized. The metabolism of purines, pyrimidines, and nucleic acids is discussed on page 478.

Absorption of amino acids

The liberated amino acids are absorbed into the intestinal capillaries and thence via the portal vein, to the liver and general circulation. Absorption is active. After the ingestion of protein, there is a transitory rise in the free amino acid content of the blood. Free amino acids can provide the whole of the body's requirements for protein.

AMINO ACID POOL. AMINO ACID TURNOVER. PROTEIN SYNTHESIS

The amino acids in the blood diffuse throughout the body fluids and reach all the tissue cells. At the same time, most of the tissue proteins (both 'structural' protein and 'functional' protein) are continually undergoing disintegration to release amino acids which likewise enter the circulation and thus become part of what is called the general 'amino-acid pool'. This steady and rapid tissue protein breakdown is taking place on a large scale. No functional distinction can be drawn between the fate of the amino acids derived from the food and those derived from the tissues. From the 'common amino acid pool', amino acids are taken up by the cells (each cell according to its specific needs) to be built into the cell structure as required. If a cell takes up as much amino acid as it loses, it is in a state of 'dynamic equilibrium'; if the loss is greater, the cell wastes; if the gain is greater the cell grows. [A general discussion of 'metabolic pools' is to be found on page 480.] From the rate of incorporation of isotopically labelled amino acids, the rate of synthesis of proteins can be calculated. In experimental animals this protein turnover is greatest in intestinal mucosa, followed by kidney, liver, brain, and muscle in that order. Each protein species is constantly lost and resynthesized at a characteristic rate; some part of this turnover reflects cell renewal and replacement of secreted protein (as in intestinal mucosa), but there is considerable true 'dynamic' degradation and resynthesis even within a cell. In man, the total protein turnover involves the breakdown and resynthesis of 80–100 g of tissue protein per day, about half of it occurring in the liver; on an average the plasma proteins are completely replaced every 15 days.

The amino-acid pool has no anatomical reality, but represents an availability of amino acid building units. Thus the pool 'contains' not only all the free amino acids of the blood and body fluids, but also the amino acids which would be freed by the net breakdown of a portion of the tissue proteins. The pool is constantly undergoing *depletion* because:

1. Large-scale deamination of presumably surplus amino acids takes place, the amino (NH_2) groups which are split off being transformed mainly into urea, leaving a 'non-nitrogenous residue'.

2. Amino acids and their derivatives (e.g. creatine) are lost in the urine and other excretions.

3. Amino acids are continually being built up into those proteins (e.g. hair) which are not part of the 'dynamic' systems.

On the other hand, the amino acid pool is always being *reestablished* by amino acids derived from the following:

1. Re-amination of certain non-nitrogenous residues.

2. Amination of appropriate 'fragments' which are present in the common metabolic pool (and therefore derived from carbohydrate and fat breakdown).

3. Amino acids split off from dietary protein and absorbed from the intestine into the blood.

This state of affairs is termed the 'continuing metabolism of the amino acids'.

The non-nitrogenous residues which are left after deamination of the amino acids are used in ways described below.

Dynamic equilibria

Not only the proteins of the body are in a state of 'dynamic equilibrium', i.e. a balance between simultaneous breakdown and synthesis; the same holds for practically all the materials of the body, even for such seemingly chemically inert material as the depot fat. Only a small proportion of the organic components of the body are 'highly dynamic'. In animals maintained from conception in an environment with a constant concentration of tritiated water (^3H_2O) (so that all the organic molecules became 'labelled' with tritium), it was shown that 50 per cent of the organic materials of the total animal have a half-life more than 100 days. Collagen protein particularly was uniquely inert and that present when maturity was reached was not broken down 'nor replaced' during the animals' lifetime.

EXAMPLE OF AMINO ACID UTILIZATION

The general principles of amino acid utilization (discussed in detail later) are illustrated by a study of the fate of ingested glycine, in which both C atoms and the N atom have been labelled with isotopes. After absorption, the labelled glycine mixes with the glycine already present in the body fluids. The changes undergone by the glycine show:

1. Some is incorporated into the tissue and plasma proteins, at rates varying with each protein.

2. Some is built into non-protein compounds which contain glycine as part of their structure, e.g. glutathione, glycocholic acid [p. 433], hippuric acid[p. 442].

3. Some is converted reversibly into the amino acid serine, and thence into cysteine.

4. Some is used in the synthesis of other body constituents for which glycine is a specific precursor, e.g. creatine [p. 470, Fig. IX.19], haem and bile pigments, purines, choline (probably via serine).

5. The remainder is deaminated. The NH_3 split off is (i) converted into and excreted as urea, or (ii) used to aminate various keto-acids to form other amino acids, or (iii) indirectly excreted as ammonium ions (NH_4^+) in the urine.

6. The non-nitrogenous residue which results from the deamination reaction (5) is either (a) dissimilated to CO_2 and H_2O, yielding energy, or (b) built up into glycogen or glucose.

NITROGENOUS EQUILIBRIUM

The body is in nitrogenous equilibrium when the nitrogen intake in the food over a given (long) period exactly equals the nitrogen lost in the excreta over the same period. The nitrogen is all in the chemically-combined form; 95 per cent of the nitrogen intake is in the form of dietary protein; the nitrogen loss is mainly via the urine (as urea, ammonia, uric acid, etc.) and, to a minor extent, in the faeces. Experiments measuring nitrogen intake and excretion under specified conditions are called *nitrogen balance* experiments. A subject in nitrogenous equilibrium is said to be in nitrogen balance; by a curious extension of this terminology, a subject whose intake of N is greater than the output (e.g. in growth) is said to have a *positive* nitrogen balance, and one whose intake of N is smaller than the output (e.g. in starvation) is said to have a *negative* nitrogen balance.

To establish nitrogenous balance, certain minimum amounts of protein (or equivalent amino acids) must be provided, to replace the inevitable losses from the dynamic equilibrium and metabolic utilization of amino acids. This minimum replacement requires amino acids of specific type, in adequate amount and in appropriate ratios. It is impossible to maintain N equilibrium on diets which are deficient in any one (or more) of these essential amino acids, no matter how much protein is consumed [see below].

Nitrogenous equilibrium in relation to diet is further discussed on pages 484–5.

Essential amino acids

The term *essential* is applied to those amino acids needed for replacement and growth, but which cannot be synthesized by the body in amounts sufficient to fulfil its normal requirements. They must therefore be supplied in the diet, usually combined in proteins.

The rest of the amino acids (the non-essential amino acids) can be synthesized in the body either:

1. By the amination of appropriate non-nitrogenous fragments derived from other sources.

2. In special cases directly from the essential amino acids. But it should be emphasized that the body is 'spared the trouble' of this synthesis if the non-essential amino acids are also available from the diet [see below].

Animals given a basal diet which contains no protein or amino acid, but which is otherwise complete in all respects, rapidly die. If the right type of protein supplements are added, then normal health and reproductive power are maintained in adult animals and satisfactory growth occurs in young animals [FIG. IX.13].

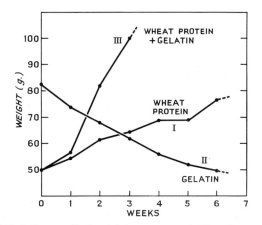

FIG. IX.13. Influence of lysine deficiency on growth. Growth curves for rats on non-nitrogenous basal diets supplemented with the various proteins indicated. Curve I: Wheat protein (deficient in lysine). Curve II: Gelatin (deficient in many amino acids, though high in lysine). Curve III: Wheat protein + gelatin (sufficient of all the essential amino acids). Normal growth curve.

[Drawn from data in papers by Hawk, and Osborne, and Mendel.]

In an extension of these experiments (and one of the great scientific achievements of this era), W. C. Rose demonstrated that a diet adequate for growth could be compounded from mixtures of relatively simple, chemically-defined substances, with pure synthetic amino acids completely replacing protein. By employing amino acid mixtures of varying composition, it was possible:

(i) To escape from the limitations imposed by using proteins of fixed composition, and thus to study the role of each amino acid separately and its interrelations with other amino acids; and (ii) to study the utilization of each amino acid quantitatively and to recommend a minimum level of intake.

For example, healthy men have lived for months on diets devoid of nitrogenous compounds except for known added amounts of pure synthetic amino acids. By measuring the nitrogen balance on various amino-acid supplements it has been found that the following eight amino acids are indispensable for human adults under normal conditions; exclusion of any one of these essential amino acids leads to a negative nitrogen balance [FIG. IX.14], fatigue, loss of appetite, and 'nervous irritability'; when the missing amino acid is added to the diet perfect health is promptly restored:

valine
leucine
isoleucine
threonine
methionine (which can also be converted to cysteine)
phenylalanine (which can also be converted to tyrosine)
tryptophan
lysine.

These synthetic diets must provide a high calorie intake if N balance is to be achieved, but the reason for this is not known.

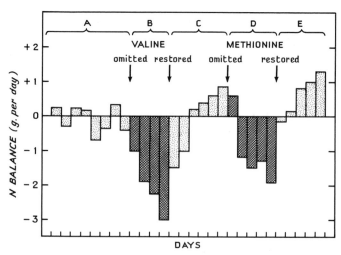

FIG. IX.14. The significance of essential amino acids in the maintenance of nitrogenous equilibrium in man. The subject was on a basal diet, total N content 7.04 g day^{-1}, 95 per cent of this N being furnished by a known mixture of 10 pure amino acids. On this diet, a fluctuating nitrogenous equilibrium is maintained (period A). Omission of valine (period B) or methionine (period D) immediately causes a negative balance to appear (e.g. total N loss over B = 9 g). Restoration of the missing amino acid is followed by the establishment of a positive balance (periods C and E).

[Redrawn from Rose, W. C. (1950). *J. biol. Chem.* **182**, 541.]

Quantitative aspects

For normal adults, the minimum amount of each essential amino acid which must be supplied per day when all other amino acids are present in abundance is of the order 0.3–1 g of the natural L-form. Rose suggests that a 'safe' intake would be double this. Naturally the requirement for essential amino acids varies with the amount of 'strain' placed on the protein-synthesizing capacity of the body, and during growth or lactation a higher intake (as mixed protein) is required [p. 599]. In addition, animal growth experiments indicate that dietary supplies of two other amino acids, *histidine* and *arginine*, may be required under conditions of growth or equivalent physiological stress; the capacity of the body to synthesize histidine and arginine, though adequate for protein maintenance, may not suffice for the more extensive calls of protein accumulation. Further it may be supposed that the amounts of essential amino acids required for specific physiological processes may differ significantly from those required for growth.

Non-essential amino acids

Within the body, the dietary 'non-essential' amino acids are just as important metabolically and structurally as the 'essential' ones. If they are not provided in the diet they must be synthesized from other N compounds. Certain amino acids or N sources can be converted to the full complement of non-essential amino acids more rapidly than others, reflecting the different metabolic pathways of each in the tissues. For this reason, maximum growth is only achieved when all the amino acids are presented to the tissues simultaneously and in appropriate ratios, as well as in sufficient amount. In animals, no amino acid is non-essential for maximum growth rates, because the task of synthesizing the non-essential amino acids presents too great a metabolic burden to permit optimum growth as well. But, under the conditions used by Rose, nitrogenous equilibrium can be maintained in adult men by supplementation of the 'safe' daily intake of essential amino acids (1.4 g of N total) with the surprisingly low amount of 2.5 g of N in the form of 'non-essential' glycine.

The nutritional aspects of amino-acid metabolism are further discussed on pages 484 *et seq*.

SPECIFIC METABOLIC ROLES OF AMINO ACIDS

In addition to being the building units of all the tissue proteins (including the enzymes and many of the hormones), amino acids have special roles in the formation of other components of the body. Some illustrative examples will be given:

1. Glycine is a fundamental building unit, and an inhibitory transmitter in the spinal cord [p. 283].

2. Arginine is part of the cycle which is responsible for urea formation [p. 468], and provides an imidine group for creatine synthesis [p. 471, FIG. IX.21].

3. Histidine is the precursor of histamine.

4. Aromatic amino acids. Phenylalanine (which is indispensable) can be irreversibly converted to tyrosine; the latter is therefore dispensable if sufficient phenylalanine is supplied. Tyrosine is the precursor for thyroxine [p. 538], noradrenalin, and thence adrenalin [p. 420], and the dark melanin pigments of the hair and skin. Tryptophan is essential for the formation of 5HT (serotonin) [p. 558].

5. Sulphur-containing amino acids. Methionine, cysteine, and cystine are the only important sources of sulphur that can be used for synthetic purposes in the body, e.g. for the formation of organic sulphates [p. 479] or taurine [p. 433].

Methionine (which is indispensable) can be irreversibly converted into cysteine (which is therefore dispensable).

Cysteine is readily converted to cystine under oxidative conditions, and the reaction is reversed under reductive conditions. The biological action of a protein containing a free SH. group (i.e. combined cysteine) is often completely altered upon oxidation to the corresponding disulphide (i.e. combined cystine).

6. Methionine (biological methylation). Certain compounds of the body, with structures containing methyl groups (CH_3—) attached to an atom other than C can take part in enzymic reactions whereby these methyl groups are transferred to suitable 'acceptors' which have no methyl group. Such reactions are termed *transmethylation* reactions, and the substrates (methyl donors) are said to possess *biologically labile methyl groups*. The most important compounds with biologically labile methyl groups are methionine (containing CH_3—S—, choline (containing $(CH_3)_3N^+$—), and the oxidation derivative of choline, betaine. The processes of transmethylation in the body have been investigated by the use of isotopes, and the results are summarized in FIGURE IX.15.

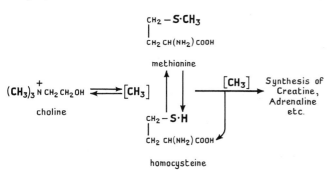

FIG. IX.15. Biological transmethylation.

The following points should be noted:

(i) Methionine enables choline to be synthesized in the body from ethanolamine ($NH_2.CH_2CH_2OH$) by donating the necessary three methyl groups.

(ii) Methionine can yield methyl groups to other suitable recipients, but these reactions appear to be irreversible Two examples are given:

(a) Methionine contributes a methyl group to guanidino-acetic acid, thus bringing about the synthesis of creatine [p. 471, FIG. IX.21]. This methyl group must be donated directly by methionine (but may come indirectly from choline donating a methyl group to homocysteine to form methionine).

(b) Methionine contributes a methyl group to noradrenalin for the synthesis of adrenalin [p. 420].

For these methylations, methionine has to be converted, by reaction with ATP, into an adenosine derivative of methionine which is the actual 'active' methylating agent.

Synthesis of labile methyl groups. 'Formate' metabolism

It was previously thought that the body tissues were unable to synthesize biologically-labile methyl groups; these had therefore to be supplied in the form of dietary methionine (an essential amino acid). However, later work using isotopes has shown that the tissues can synthesize substances containing labile methyl groups (e.g. choline) on a considerable scale if the normal dietary amounts of folic acid and vitamin B_{12} [p. 39] are present (as was not the case with the original experimental diets). It is probable that the newly synthesized methyl groups are used first to form methionine from homocysteine; the methionine then acts as a transmethylating agents as described above. Although the tissues can and indeed do synthesize methyl groups, dietary methionine is the most important source under normal conditions.

The precursor of synthetic methyl groups is a one-carbon unit called 'formate' which appears to bear the same kind of relationship to formic acid, H.COOH, that 'active acetate' bears to acetic acid [p. 468]. However, 'formate' is not a Co A ester; it is carried as formyl (H.CO-) or hydroxymethyl (HOCH$_2$—) group on the Na atoms of folic acid derivatives [p. 40]. 'Formate', and thence methyl groups, CH_3, arise chiefly from serine, $CH_2OH.CH(NH_2)COOH$.

Other metabolic reactions in which 'formate' plays a part are the interconversion of glycine and serine, and the synthesis of the purine nucleus.

DISSIMILATION OF AMINO ACIDS

Amino acids which are not used as such undergo oxidative deamination, predominantly in the liver; the amino (NH_2) group is released as ammonia (NH_3) which is then excreted as urea or used in the synthesis of other amino acids as described below. The non-nitrogenous residues, after undergoing various transformations, enter the citric acid cycle, and are either completely dissimilated or are converted into glucose, fat, ketone bodies, or other amino acids.

1. Oxidative deamination. The over-all reaction is the transformation of an amino acid, $R.CH(NH_2).COOH$, to the corresponding keto-acid, $R.CO.COOH$. This involves an oxidation (or more exactly, a dehydrogenation) (enzymes: amino-acid oxidases; hydrogen carriers: NAD or flavin) to give an imino acid, followed by a hydrolysis liberating ammonia, thus:

$$CH_3.CH(NH_2)COOH + NAD$$
$$\text{alanine} \qquad \qquad \text{coenzyme} \atop \text{(H-carrier)}$$

$$\longrightarrow CH_3.C(:NH)COOH + NAD.2H$$
$$\text{α-imino acid} \qquad \qquad \text{reduced} \atop \text{coenzyme}$$

$$CH_3C(:NH)COOH + H_2O \longrightarrow CH_3.CO.COOH + NH_3$$
$$\text{pyruvic acid}$$

2. Transamination. Deamination of an amino acid may be coupled with the simultaneous amination of a keto-acid. The process is called transamination (enzymes: transaminases), and the result is the transference of the NH_2 group from one amino acid and its use to synthesize another, thus:

$$
\begin{array}{c}
\text{CH}_3 \\
| \\
\text{CH(NH}_2\text{)COOH} \\
\text{alanine}
\end{array}
+
\begin{array}{c}
\text{CH}_2\text{CH}_2\text{COOH} \\
| \\
\text{CO.COOH} \\
\text{\textit{α}-keto-glutaric acid}
\end{array}
$$

$$
\rightleftharpoons
\begin{array}{c}
\text{CH}_3 \\
| \\
\text{CO.COOH} \\
\text{pyruvic acid}
\end{array}
+
\begin{array}{c}
\text{CH}_2\text{CH}_2\text{COOH} \\
| \\
\text{CH(NH}_2\text{)COOH} \\
\text{glutamic acid}
\end{array}
$$

Transamination is reversible and plays an important part in both the breakdown of amino acids and their synthesis from non-protein sources (e.g. from keto-acids of the citric acid cycle).

The serum levels of certain transaminases (especially of serum glutamic-oxaloacetate transaminase, S-GOT) are markedly raised in diseases and morbid conditions involving injury to large numbers of metabolically-active cells, e.g. in myocardial infarction.

3. Amination of non-nitrogenous residues. When ammonium salts or amino acids containing amino groups labelled with isotopic ^{15}N are administered, the ^{15}N label is later found incorporated into the amino groups of many different amino acids of the proteins.

Appropriate non-nitrogenous residues must therefore have been converted into amino acids by taking up NH_2 from NH_4^+ or other amino acids, i.e. by processes of direct amination or transamination. Amino acids from the amino-acid pool are continually being broken down by deamination, and the processes of direct amination or transamination are used to resynthesize some of these amino acids. Such synthetic processes are increased if a need arises for additional quantities of amino acid, e.g. the provision of glycine for 'detoxication' mechanisms [p. 442]. Products of deamination formed at one site can be reaminated elsewhere and so re-enter the 'amino-acid pool'.

4. Ammonia. Ammonia, formed by the kidney tubule cells, mainly from glutamine, diffuses into the lumen of the tubule and acts as a hydrogen ion acceptor.

Urea formation

The surplus ammonia which is formed by deamination and not used for reamination is converted into urea. Ammonia is very toxic to cells. *Urea, on the other hand, is harmless even in very high concentrations.*

Urea formation occurs *exclusively* in the liver. Removal of the liver produces the following results [FIG. IX.16].

1. When urinary secretion is maintained in the liverless animal supported by glucose infusion [p. 441], there is a steady decrease in the blood urea, e.g. from 5 to 1 mmol l^{-1}; the amount eliminated in the urine likewise shows a progressive and marked decrease. Practically all the urea which is excreted after hepatectomy can be accounted for by the decrease in the urea content of the body fluids.

2. If anuria follows the operation, the blood urea remains quite constant.

3. If the kidneys are removed in a normal animal there is a rise in the blood urea; if hepatectomy is now performed, the blood urea remains at the high level previously attained.

These results prove that the liver is the only site of urea formation, and that urea once formed is not destroyed in the body.

4. The liver is the chief site of deamination of amino acids. In the

liverless animal the blood amino-acid content progressively rises [FIG. IX.16]; injected amino acid is not deaminated to the normal degree and does not give rise to extra urea.

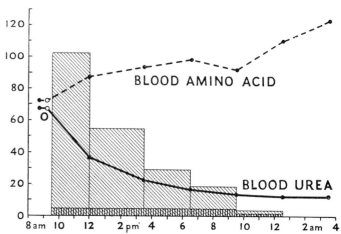

FIG. IX.16. Effect of removal of liver on blood urea, blood amino acid, and urea and amino acid excretion in urine. Experiments on dogs. At point marked O remove liver. Curve with broken line: blood amino-acid nitrogen in mg. per litre. Curve with continuous line: blood urea nitrogen in mg per litre. Rectangles with angle hatching: urea nitrogen + ammonia nitrogen excreted in urine in mg per hour. Rectangles with vertical hatching: amino-acid nitrogen in urine in mg per hour. Note rise of blood amino acid, fall of blood urea and decrease and ultimately disappearance of urea from urine.
[Redrawn from Mann, F. C. *et al.* (1926). *Am. J. Physiol.* **78**, 259.]

Essentially, urea formation in the liver consists in the union of NH_3 (2 mols) and CO_2 (1 mol) with the elimination of H_2O. This however, does not take place directly but through a cyclical system of enzyme reactions (called the Krebs urea cycle) in which the (non-protein) amino acid ornithine acts as 'catalyst' [FIG. IX.17]. Ornithine reacts with CO_2 and 1 mol of NH_3 to give citrulline, which itself reacts with a second mol of NH_3 to give arginine; the amidine group of arginine is split off as urea under the influence of the hydrolytic enzyme arginase, and ornithine is regenerated to continue the cycle. The interaction of amino acids is further complicated by the fact that the amino acids glutamic and aspartic are involved in the entry of NH_3 into the cycle. ATP energy is also consumed. By the use of isotopes it has been shown that the CO_2 used in urea formation comes solely and directly from blood HCO_3^-.

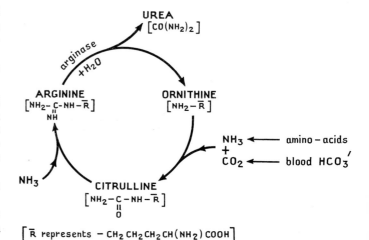

FIG. IX.17. Krebs urea cycle (ornithine cycle).

Fate of non-nitrogenous residues from amino acids

In general, the non-nitrogenous residues remaining after deamination enter the 'common metabolic pool' and are either completely dissimilated to CO_2 or (less often) are built up into other body constituents.

Some amino acids are *glucogenic*, i.e. they may give rise to glucose: some are *ketogenic* and may give rise to acetoacetic acid (and the other 'ketone bodies'). The experimental methods employed in the study of these reactions are:

1. Incubate liver slices with an amino acid, or perfuse the liver with an amino acid, and determine whether there is an increase in the glucose or ketone level in the medium or venous outflow.

2. Study the diabetic or phloridzinized animal either of which excretes large amounts of glucose in the urine. The administration of protein, like the administration of carbohydrate, increases the urinary glucose excretion, suggesting that food protein can give rise to glucose. If specific amino acids are administered, some (glucogenic) increase the glucose pool and the degree of glucosuria while others (ketogenic) increase the excretion of ketone bodies in the urine. *Glucogenic amino acids cannot counteract hypoglycaemia in hepatectomized animals; thus the liver is an indispensable factor for their transformation into blood glucose.*

3. Administer amino acids labelled with C isotopes, and investigate the isotope distribution (if any) in subsequently isolated glucose, glycogen, or ketones.

The results of such studies are as follows [FIG. IX.18].

(i) The ultimate fate of the non-nitrogenous residues derived from the essential amino acids, tryptophan, lysine, histidine, and methionine is unknown.

(ii) Leucine, isoleucine, and phenylalanine (three of the essential amino acids) and tyrosine are ketogenic. Their deamination and normal oxidation in the liver gives rise to acetoacetic acid, by known irreversible routes; this acetoacetic acid is subsequently dissimilated.

(iii) All the non-essential amino acids are glucogenic. This fact indicates that they give rise to an intermediary found on the reversible pathway of carbohydrate dissimilation. Since all the non-essential amino acids can be synthesized in the body from carbohydrate and a N source the route by which they enter the carbohydrate pathway (including the deamination stage) must be reversible. The identity of the intermediary has been determined in most cases:

(*a*) On oxidative deamination or transamination, alanine, and the amino acids which can be converted into it (glycine, serine, cysteine) give rise to pyruvic acid,

$$\underset{\text{(alanine)}}{CH_3.CH(NH_2)\,COOH} \overset{+O_2 - NH_3}{\rightleftharpoons} \underset{\text{(pyruvic acid)}}{CH_3.CO.COOH}$$

(*b*) Glutamic acid (and probably proline and arginine) give α-ketoglutaric acid.

$$\underset{\text{(glutamic acid)}}{COOH(CH_2)_2.CH(NH_2)COOH} \rightleftharpoons \underset{\text{(ketoglutaric acid)}}{COOH(CH_2)_2.CO.COOH}$$

(*c*) aspartic acid gives oxaloacetic acid,

$$\underset{\text{(aspartic acid)}}{COOH.CH_2.CH(NH_2)COOH} \rightleftharpoons \underset{\text{(oxaloacetic acid)}}{COOH.CH_2.CO.COOH}$$

These three products are all components of the citric acid cycle [FIG. IX.18, and p. 454 FIG. IX.8]; they can either be built into glucose, or dissimilated in the liver to yield energy, thus 'sparing', the dissimilation of glucose.

It is found by method (3) [above] that the carbohydrate apparently formed from glucogenic amino acids does not always arise by direct conversion of the amino acid to glucose. It can also come from the conversion to glucose of some other metabolite which has itself been 'spared' by the non-nitrogenous residue of the amino acid.

(iv) The remaining essential amino acids (threonine and valine) are glucogenic. The path by which they reach the carbohydrates is a non-reversible route as they are indispensable in the diet.

Fate of residues from aromatic amino acids

In certain *very rare* congenital disorders, the normal oxidative mechanisms of phenylalanine and tyrosine metabolism are deranged by 'blockage' at different points. This results in the urinary excretion of intermediate metabolites, and has given valuable information on the normal oxidation of these amino acids.

The following conditions are of particular interest:

1. *Phenylketonuria* (Fölling's disease), a form of idiocy in which there is an inability to convert phenylalanine to tyrosine. Phenylalanine accumulates, and phenylpyruvic acid (the deamination product of phenylalanine) appears in the urine.

2. *Alcaptonuria*, where homogentisic acid (2:5-dihydroxyphenylacetic acid) appears in the urine; such urine darkens considerably when alkaline and exposed to air, and may turn almost black. Homogentisic acid is apparently a normal metabolite from tyrosine, but alcaptonurics excrete it rather than oxidize it further to acetoacetic acid as its normal fate.

EXOGENOUS AND ENDOGENOUS METABOLISM

The metabolic processes occurring in the body were once divided into two fairly sharply defined compartments:

1. Metabolism of the foodstuffs (exogenous).
2. Metabolism of the tissues (endogenous).

The distinction drawn was analogous to that between the consumption of fuel (food) *by* an engine, and the wear and tear *of* the engine itself (tissues). But, as has been emphasized, the proteins and nucleic acids and most of the other constituents of the tissue cells are constantly undergoing breakdown processes similar to those that affect the foodstuffs in the gut; in general, the products of digestion and those of cell disintegration enter a common pool and

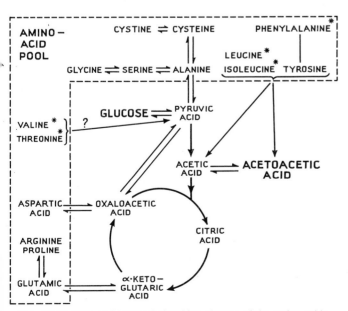

FIG. IX.18. Fate and inter-relationships of some of the amino acids.
* Indicates *essential* amino acids.

suffer a common fate. There was, however, a fundamental truth in the classical conception; some of the urinary constituents vary widely in amount in relation to the protein content of the food, while others are relatively independent of the diet. This distinction is brought out in TABLE IX.1 (from Folin), which shows the differences in the composition of the urine on two diets, both *adequate in calories*, but one *poor* in protein content.

TABLE IX.1

	Protein-rich diet	Protein-poor diet	
Volume of urine	1170 ml	385 ml	Greatly reduced
Total nitrogen	16.8 g	3.6 g	Greatly reduced
Urea nitrogen	14.7 g = 87%	2.2 g = 62%	Greatly reduced
Uric acid nitrogen	0.18 g	0.09 g	Halved
Ammonia nitrogen	0.49 g	0.42 g	Unchanged
Creatinine nitrogen	0.58 g	0.6 g	Unchanged
Inorganic SO_4	3.27 g = 90%	0.46 g = 60 %	Greatly reduced
Ethereal SO_4	0.19 g	0.10 g	
Neutral sulphur	0.18 g	0.20 g	Unchanged

Obviously the total nitrogen excretion in the urine is related to the protein intake. On low-protein diets the nitrogenous output falls from 16.8 to 3.6 g daily. The excretion of urea and inorganic sulphate is markedly dependent on the diet; the urinary output of these substances is an index of protein intake, or, in the (discarded) classical terminology, of *exogenous* metabolism. On the other hand, the excretion of creatinine, neutral sulphur and about half the uric acid is quite independent of the protein intake; in the classical terminology, the output of these substances is an index of *endogenous* metabolism. This term can be usefully retained, since the 'continuing metabolism' of the tissues is normally an equilibrium between simultaneous synthesis and breakdown and does not have to be taken into account when making comparison between dietary intake and urinary output. The end-products of endogenous metabolism appear to represent a slow, continuous, and wasteful 'seepage' of useful materials from the common metabolic pools.

Origin and physiological significance of nitrogen-containing constituents of urine

Mixed proteins contain an average of 16 per cent of N; thus the dissimilation of 100 g of protein (the recommended, though high, daily intake) yields 16 g of waste N, nearly all of which is excreted in the urine. The chief N-containing waste products in the urine are: urea; creatinine, and creatine; ammonium ions; uric acid. A small quantity of amino acids is also lost from the body in the urine.

Urea

1. Effect of Protein Intake. Urea is formed in the liver from ammonia derived from amino acids (or from ingested ammonium salts); the details of its mode of formation are given on page 468. As explained, the amino acids of the body constitute a 'pool' into which amino acids pass from food and tissues, and from which amino acids are taken, to be synthesized into protein or otherwise transformed as required by tissues. Normally, 'surplus' amino acids (whatever their origin) are deaminated and their ammonia is converted into urea; adult man has no means of storing amino acids except by a very limited increase in tissue protein. On a normal protein intake the amino-acid pool is 'overflowing' with amino acids from recently ingested food. On such a diet, the urinary urea is therefore mainly of food origin; in other words it is mainly (but by no means exclusively) exogenous. Within limits, then, the urea output varies directly with the recent protein intake. Urea is always the principal

nitrogenous constituent of the urine, constituting 60–90 per cent of the total urinary N. On a normal mixed diet, an adult's daily excretion of urea is 15–50 g (= 7–23 g urea-N).

2. Effect of Protein-free Diet (adequate in energy content). When no protein is eaten, amino acids are contributed to the 'pool' by the breakdown of tissue proteins only. One would perhaps expect that an equal amount of amino acid would be withdrawn from the 'pool' for restitution purposes and that urea formation would therefore cease. This, however, is not what happens. Instead, the 'pool' is slowly and continuously depleted of its essential amino acids, mainly through their conversion to high-priority expendable body components, such as creatine. The remaining amino acids not used for these special syntheses cannot be rebuilt into the proteins from which they come, because one or more of the required essential (non-synthesizable) amino acids needed for the protein synthesis has been utilized elsewhere; as they cannot be stored they are irreversibly broken down, yielding first ammonia and then urea. More tissue protein is broken down than is rebuilt, the net loss appearing partially as expendable metabolites (e.g. creatinine) but mainly as urinary urea. The urinary urea on a protein-free diet is, in the classical terminology, wholly endogenous, and its minimum level is about 4 g of urea per day. It should be emphasized that only under the conditions of this experiment (i.e. protein-free diet containing the full calorific requirements in the form of carbohydrate and fat), as opposed to conditions of starvation, is the minimum excretion of N and of urea seen. It is an 'experimental' situation!

3. Effect of Starvation (i.e. no protein and no calories). *In complete starvation, tissue protein (particularly muscle protein) is broken down to amino acids on a much larger scale than in 2*. Some of these amino acids are used for the restitution of high-priority expendable components as in 2. But, as no food energy is available, most of the liberated amino acids are deaminated and the residues utilized for energy purposes and to maintain the blood sugar level. Tissue protein is thus used in the same way as, and as a substitute for, food protein. In starvation, therefore, urea excretion is on a much larger scale than on a protein-free diet which is adequate in calories, because in the latter condition tissue protein is not being used for 'fuel'.

Creatine and creatinine

The structure and relationships of urea, guanidine, creatine, creatine phosphate (phosphocreatine), and creatinine are shown in FIGURE IX.19.

Distribution of creatine. Creatine occurs in greatest concentration in skeletal muscle, with lesser amounts in heart muscle, brain, and the uterus (especially during pregnancy). The creatine concen-

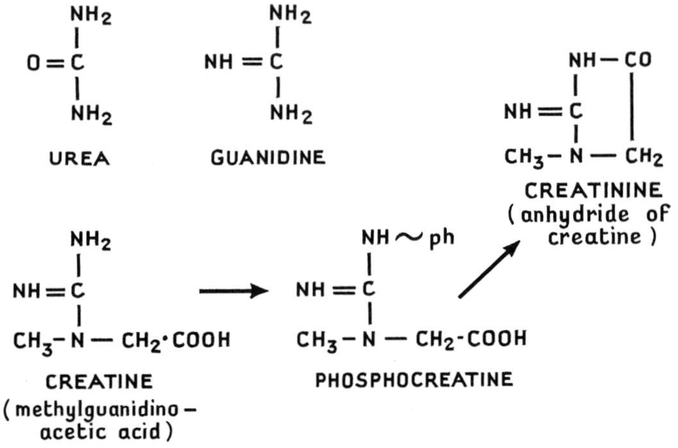

FIG. IX.19. The structures and relationships of creatine and creatinine.

tration of skeletal muscle increases steadily during fetal and post-natal growth until the adult level is finally reached.

In resting muscle, creatine exists largely as creatine phosphate (phosphocreatine), a high-energy compound with ~ph [p. 254]. Creatine phosphate is formed by the reaction of creatine with adenosine triphosphate (ATP); this reaction is reversible, creatine phosphate transferring its ~ph to adenosine diphosphate (ADP), and regenerating ATP as required. The energy of carbohydrate dissimilation in muscle is initially made available as ATP [p. 254], but it is stored as the high-energy phosphate groups of creatine phosphate. Creatine phosphate can be accumulated in quantity (unlike ATP), and is available for the rapid resynthesis from ADP of the ATP which is required for muscular work [Fig. IX.20].

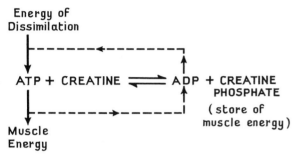

Fig. IX.20. Creatine phosphate as an energy store.

Formation of creatine and creatinine. Isotope experiments have shown that creatine is synthesized in the liver from three amino acids [Fig. IX.21]: arginine transfers its amidine group to glycine to form guanidinoacetic acid, which is irreversibly methylated by methionine.

The creatine is discharged into the blood (normal level = 180–720 μmol l^{-1} of whole blood and taken up by the muscles as required. Creatine in excess of the muscle storage capacity (e.g. after ingestion of large amounts of creatine) is disposed of by unknown means, though in special cases it is excreted in the urine [see below].

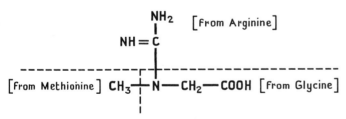

Fig. IX.21. Creatine synthesis. Note the three amino acids involved in the formation of creatine, and thence of creatine phosphate and creatinine.

The chief reaction of creatine is its phosphorylation by ATP to give creatine phosphate. Some of the store of creatine phosphate is continuously being lost to the body by a slow, spontaneous transformation to creatinine, by ring formation with loss of H_3PO_4; the creatinine is excreted in the urine. This is the major source of creatinine, little or none coming direct from creatine itself.

Excretion of creatine in urine (creatinuria). Creatine is not a normal component of the urine but may appear in the following conditions:

1. (i) It is constantly present in children of both sexes up to the age of puberty. This may be associated with the low storage power of the muscles for creatine at an early age.

(ii) It is found in an intermittent manner in the urine of women, but is not related to menstruation. There is a continuous creatine excretion during pregnancy. It rises to a maximum of 1.5 g daily after delivery, when it is probably derived from the involuting uterus.

(iii) Creatine is excreted irregularly by normal men.

2. In certain pathological states of muscle, e.g. the myopathies, creatine is excreted, because of the low storage power of the muscles. Even if only small amounts of creatine are ingested, 90 per cent or more appears unchanged in the urine.

3. In any condition in which unusual breakdown of the tissues (especially muscles) occurs, e.g. in starvation, diabetes, exophthalmic goitre, and fever (from the increased metabolic rate), creatine is excreted. In all states of undernutrition the muscles bear the brunt of the burden; as their substance is broken down for energy purposes, creatine is liberated, some of which is stored in the muscles which are still intact, and the rest is passed out in the urine as creatine.

Excretion of creatinine. Creatinine is formed in the body exclusively from creatine via creatine phosphate. Minute amounts of creatinine (so-called 'endogenous' creatinine) are present in blood (90 μmol l^{-1}); it is a non-threshold substance which is excreted in the urine by filtration from the glomeruli. If creatinine is ingested, the blood level is raised. This 'exogenous' creatinine is also secreted by the tubules as well as being filtered by the glomeruli.

The urinary output of creatinine during exercise is always greater than during inactivity; this is, however, immediately followed by a period of unusually low output, so that the 24-hour output is unaffected by work. The daily output of creatinine is thus reasonably constant and is independent of the protein intake or the total amount of nitrogen excreted. Creatinine output depends not so much on the muscle mass as on its creatine plus creatine phosphate store, though, of course, the two are related; the amount of creatinine excreted in 24 hours represents the conversion of about 2 per cent of the body creatine.

Urinary creatinine is a product of endogenous metabolism in the classical sense, representing the removal of the steadily expendable creatine. The excretion of creatinine means the loss of valuable methyl groups from the body; indeed of the normal dietary components only methionine can supply the methyl groups to replace them [p. 467], though tissue synthesis of methyl groups may make some contribution.

Ammonium ions. The origin and significance of urinary ammonium ions, NH_4^+, are fully discussed on page 229.

Amino acids. Traces of many amino acids and small peptides are found in normal urine, representing a mainly 'endogenous' excretion. An increase in the excretion of specific amino acids (probably due to faulty reabsorption by the renal tubules) may occur pathologically (e.g. in the Fanconi syndrome, which is a form of osteomalacia or rickets, resistant to vitamin D therapy, and characterized by urinary loss of phosphate, glucose, and amino acids). The investigation of urinary amino-acid patterns by paper chromatography may have diagnostic value.

Certain amino acids are used by the liver to 'detoxicate' unwanted compounds [p. 442]. These products, of which the chief is hippuric acid (from the condensation of benzoic acid with glycine), then appear in the urine.

Uric acid. This is the only end-product which appears in quantity from the metabolism of purines and nucleic acids in man.

REFERENCES

Bender, D. A. (1975). *Amino acid metabolism.* Wiley-Interscience, New York.

CANTONI, G. L. (1975). Biological methylation: selected aspects. *A. Rev. Biochem.* **44**, 435–51.

FELIG, P. (1975). Amino acid metabolism in man. *A. Rev. Biochem.* **44**, 933–55.

SCHIMKE, R. T. and KATUNUMA, N. (EDS.) (1975). *Intracellular protein turnover.* Academic Press, New York.

TRUFFA-BACHI, P. and COHEN, G. N. (1973). Amino acid metabolism. *A. Rev. Biochem.* **42**, 113–34.

WHITE, A., HANDLER, P., SMITH, E. L., HILL, R. L., and LEHMAN, I. R. (1978). *Principles of biochemistry*, 6th edn, pp. 677–755. McGraw-Hill, New York.

NUCLEIC ACID METABOLISM, URIC ACID. PURINES AND PYRIMIDINES

Purines

Purines are derived from a system of 2 fused rings each containing two N atoms, namely a 6-membered (pyrimidine) ring and a 5-membered (iminazole) ring [FIG. IX.25]. The chief purines found in nucleotides and nucleic acids are adenine and guanine. Uric acid is the final oxidation product (in man) of these purines; intermediate oxidation products are hypoxanthine and xanthine. Purines combine through their 9-nitrogen position with sugar residues.

Pyrimidines

Pyrimidines have a single 6-membered ring containing two N atoms. The major pyrimidines found in nucleotides and nucleic acids are cytosine, uracil, and thymine [FIG. IX.22]. All carry a hydroxy substituent at C-2 position and combine with sugar residues through their 3-nitrogen position.

Occurrence of purines and pyrimidines

Purines and pyrimidines are occasionally found as free bases, more usually as nucleosides and nucleotides, and as nucleic acids (polynucleotides).

Free purines. The diet contributes small amounts of free purines to the body, particularly hydroxypurines (xanthine and hypoxanthine) from meat extracts, and methylpurines (caffeine, theobromine) from tea, coffee, and cocoa. Normal urine contains very small amounts of a large selection of free purines derived both from purines synthesized in the body and from dietary purines; but in normal persons only uric acid, the major end-product of all purine metabolism, appears in quantity [p. 470]—if other purines accumulate then an abnormal block in the metabolic pathway to uric acid is indicated [FIG. IX.23].

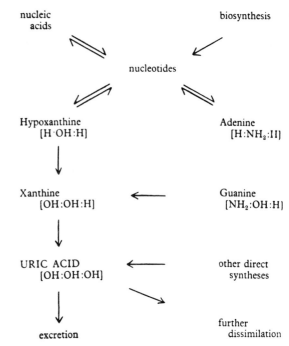

FIG. IX.23. Oxidative formation of uric acid. The substituents on positions 2:6:8 of the purine nuclei are shown in square brackets.

Nucleosides and nucleotides

When a purine or pyrimidine is linked to a sugar residue the resulting compound is called a *nucleoside*. If the sugar residue is also phosphorylated a *nucleotide* results. The biologically important nucleotides have ribose as their sugar residue (except in the case of the nucleotides composing deoxyribonucleic acids, which have 2-deoxyribose).

Examples:

adenine-ribose = adenosine, adenine nucleoside

adenine-ribose-phosphate = adenylic acid (adenosine monophosphate, AMP) adenine mono nucleotide

adenine-ribose-phosphate-phosphate = adenosine diphosphate (ADP)

adenine-ribose-phosphate-phosphate-phosphate = adenosine triphosphate (ATP) [FIG. IX.1]

hypoxanthine-ribose = inosine

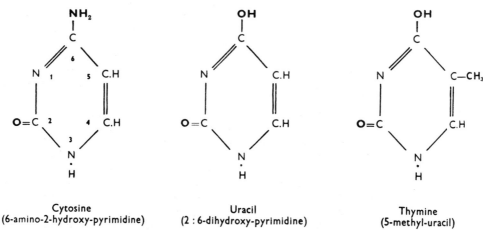

Cytosine
(6-amino-2-hydroxy-pyrimidine)

Uracil
(2 : 6-dihydroxy-pyrimidine)

Thymine
(5-methyl-uracil)

FIG. IX.22. Structures of some important pyrimidines.

hypoxanthine-ribose-phosphate = inosinic acid
uracil-ribose = uridine
uracil-ribose-phosphate = uridylic acid (uridine monophosphate)

Nucleotides have two distinct functions:

1. As coenzyme-like activators or carriers of biologically important substances.
2. As building units for the nucleic acids.

Nucleotide derivatives as coenzymes and carriers

1. Many enzymic processes only take place when the substrate has been 'activated' by combination with a specific nucleotide. The nucleotides are thus coenzymes, being regenerated at the end of the reaction [p. 448].

(i) Nucleotides of *adenine* serve to activate acids for biological reactions; the obvious case is adenosine triphosphate, ATP, whose terminal phosphate (from phosphoric acid) is activated for phosphorylation [p. 448]; in a similar way, amino acids, sulphuric acid, or carbonic acid can undergo their particular synthetic reactions only as the appropriate phosphate anhydride with adenosine monophosphate. On the other hand, most organic acids react as their derivatives with coenzyme A (A for acylation). Coenzyme A is built from an adenine phosphate and pantotheine, but the 'activated' acyl group of the organic acid is attached to the pantotheine end of the molecule, not to the phosphate. The metabolically-utilizable form of acetic acid is acetyl-coenzyme A [p. 447], and analogously for other acids of the tricarboxylic acid cycle [p. 454].

(ii) The nucleotides of *uracil* are concerned with the interconversion reactions of the sugars. For example, free glucose cannot be oxidized to glucuronic acid nor isomerized to galactose;

but it forms uridine-diphosphate-glucose =
uracil-ribose-phosphate-phosphate-α. glucose,

which is readily oxidized to uridine-diphosphate-glucuronic acid or isomerized to uridine-diphosphate-galactose, and from which glycoside derivatives of the new sugar can be generated.

(iii) The coenzymes required for reactions involving alcohols, especially for lipide syntheses from choline or glycerol, are nucleotides of *cytosine*.

2. The coenzymes of biological oxidation and hydrogen transfer [p. 448] are derivatives of adenosine phosphates containing a further ribose unit—the so-called 'di-nucleotides':

nicotinamide adeninedinucleotide (NAD, formerly called co-enzyme-I or DPN)
nicotinamide adeninedinucleotide phosphate (NADP, formerly called coenzyme-II or TPN) which bears a further phosphate on the adenosine ribose unit.

In each case, it is only the nicotinamide end which undergoes change—the reversible addition of 2H.

Another 'dinucleotide' component of the oxidation and reduction chain is flavin adenine dinucleotide FADN,

adenine-ribose-phosphate-phosphate-ribitol-flavin

3. The nucleotides of adenine are the prime movers for metabolic energy in the human body, and, as far as is known, in all living systems. All reactions producing, storing, or utilizing chemical energy involve ATP, or a compound at a high energy level whose existence has prerequired ATP.

Minute amounts of these highly-specific coenzyme and carrier nucleotides are found free in tissues and fluids; those in blood are predominantly associated with the cellular elements. The reactions which govern the synthesis of nucleotides must be among the most important regulatory mechanisms of the body.

NUCLEIC ACIDS AND GENETIC CODING

Miescher (1869) first isolated chemical products from cell nuclei using pus cells as a source. He found that the main chemical extracted ('nuclein') contained a lot of phosphorus, but he declared at once that this phosphorus was not in the form of lecithins which was the only organic phosphorus-containing product then known. He continued his work using the spermatozoa of salmon and found that nuclein constituted nearly 50 per cent of the organic material and that it itself contained 9.6 per cent of phosphorus and was of acid nature. Finally, he concluded that this material might be genetically active.

Nuclein was named nucleic acid in 1896 and by the end of the nineteenth century it had been isolated from yeast and from animal sources, such as the thymus. During the next 30 years it was shown that yeast nucleic acid contained phosphate, a pentose sugar identified as ribose and four nitrogenous bases—adenine, guanine, cytosine, and uracil. Thymonucleic acid was proved to contain phosphate, deoxyribose and four nitrogenous bases—adenine, guanine, cytosine, and thymine.

These two nucleic acids were finally named ribonucleic acid (RNA) and deoxyribonucleic acid (DNA). Cell fractionation studies revealed that DNA was restricted to the cell nuclei. RNA, however, was found both in the nuclei and in the cytoplasm. It was shown that the fundamental unit of the nucleic acids was a nucleotide. Each nucleotide consisted of the compound arranged thus:

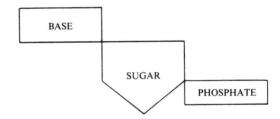

The nucleotides were joined by di-ester links between the phosphoric acid and the sugar molecule, viz:

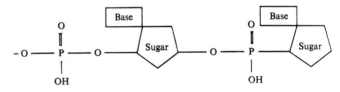

and such dinucleotides were first synthesized by Todd in 1955.

During this period it became clear from chemical analyses that both DNA and RNA were macromolecules with molecular weights of the order of 50 000 units.

X-ray crystallography showed that the nitrogenous bases were at right angles to the axis of the chain and that the planes of the sugar molecules and the bases were mutually at right angles. Pauling and Corey (1953) then postulated a helical structure for the molecule which was, however, of incorrect form.

Meanwhile, Chargaff (1950) had noted that in the DNA series the amount of guanine in gram moles always equalled that of cytosine and that the gram molar quantity of adenine always equalled that of thymine. This finding was a key factor in the solution of the structure of DNA, given correctly by Crick and Watson, based on their own findings and those of Wilkins (1953) obtained on X-ray diffraction patterns. X-ray diffraction studies of paracrystalline fibres from solutions of the sodium salt of DNA revealed that the molecule was in the form of a chain which possessed a repeating unit along its axis at intervals of 0.34 nm. Titration studies had shown that the phosphate groups were situated on the outside of the macromolecule.

Crick and Watson's model of DNA consists of two coaxial poly-nucleotide right-handed helices, each of which were of equal pitch and subtended an angle of 120 degrees at the axis [Fig. IX.24].

If we temporarily forget about the helical twisting of the molecule, we may use the analogy of a chain ladder. The rungs of the ladder are formed by the hydrogen bonding of two nitrogenous bases and the bases themselves. Such pairs of bases may occur in any sequence, but the two constituents of any one pair are invariant and thymine always pairs with adenine and guanine always with cystosine. These rungs are attached to the sides of the ladder by linkages between the bases and the sugar molecules. The sugar molecules in turn link with the phosphate molecules which complete that section of the side of the ladder [Fig. IX.24].

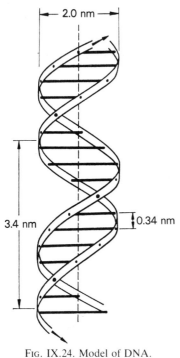

Fig. IX.24. Model of DNA.
[Cohen, D. (1965). *The biological role of the nucleic acids*. Arnold, London.]

The two helices 'run' in opposite directions differing in the direction of their sugar-phosphate linkages. Thus, if one chain has the linkage from the 3′ position of a sugar to the 5′ position of the next in a given direction the other will have the linkage 5′ to 3′ in that direction. To understand this more fully and indeed to follow some of the structural criteria which the Crick–Watson model must satisfy, it is necessary to consider the structural formula of the units.

Structure of the nitrogenous bases

Two of the bases are purines—adenine (6 aminopurine) and guanine (2 amino 6 oxypurine):

Adenine (6-amino-purine)   Guanine (2-amino-6-hydroxy-purine)

The remaining two, cytosine and thymine, are pyrimidines:

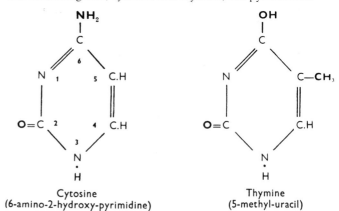

Cytosine
(6-amino-2-hydroxy-pyrimidine)

Thymine
(5-methyl-uracil)

Fig. IX.25.

The sugar 2-deoxyribose has the formula:

(1) CHO
(2) CH₂
(3) CHOH
(4) CHOH
(5) CH₂OH

(β ring form)

Linkage between a purine base and the sugar occurs between group 9 (base) and group 1 (sugar):

ADENOSINE

Pyrimidines link at group 3:

CYTIDINE

The resulting combinations of base and sugar are known as nucleosides and are designated adenosine, guanosine, cytidine and thymidine respectively.

The phosphate ester of a nucleoside is known as a nucleotide. Combination of the phosphate with the deoxyribose part of the nucleoside may occur at group 3′ or group 5′.

3′ nucleotide   5′ nucleotide

The macromolecule is joined together by diester links in which one phosphoric acid molecule forms bonds between the 3′ and 5′ positions of consecutive sugar molecules.

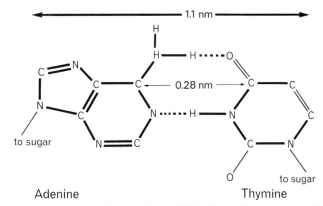

Fig. IX.26. Inter-chain base pairing in DNA. Figure shows 1.1 nm width. Dotted lines represent hydrogen bonds.

Adenine   Thymine

new DNA molecular chains are produced, each of which contains one strand from the original one; each new chain is identical with that of the parent.

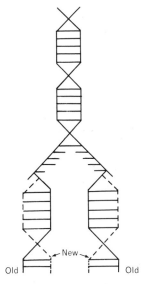

Fig. IX.27. Unzipping of DNA. Two complementary strands are reproduced alongside it from single components.

Dimensions of the double helix

The individual rungs of the ladder are 0.34 nm apart and one complete cycle of the helix is 3.4 nm [Fig. IX.24]. There are ten rungs (ten pairs of bases) in each complete turn.

The diameter of the helix is 2.0 nm, but the distance from chain to chain is about 1.1 nm. Only by fitting one purine to one pyrimidine can the distances be satisfied. Two pyrimidines would be too far apart and two purines could not be accommodated in this distance. Figure IX.26 shows an example of a base pair and the interbase distance.

As DNA is a double helix and as the complementary pairing is strictly defined it follows that given one strand of the two it is easy to predict the exact sequence of bases in the other. It would seem then that the double helix provides twice as much information as is necessary. The point is one of great importance, for the coding of genetic messages is believed to be determined by the order of sequence of the bases of the polynucleotide strand. The double helix is believed to split—on cell division for instance, and the two strands are, so to speak, unzipped [Fig. IX.27]. The information coded in each strand is used to control the production of a new strand which is complementary. The two new strands are constructed from new nucleotides provided by the cells. Hence, two

The coding of protein synthesis by the cell

The consideration that proteins are the basis of life (*proteuo*, occupy the first place) led to speculation as to how such complex molecules could be coded by messages which could be made up from an alphabet consisting of only four letters (the bases, A, C, G, T). Each protein contains all or most of 20 different types of amino acids [Table IX.2]. There may be 200 or so amino acid subunits linked together, containing these 20 amino acids in a specific sequence for a given protein.

Gamow (1954) pointed out that for the coding available from the four letter alphabet to be adequate to deal with the 20 word dictionary required, would necessitate code words of at least three letters. Clearly only four words can be formed if the words are only one letter in length. With two letter words, however, 4 × 4 code words could be possible and with three letter words 4 × 4 × 4 = 64 words—more than enough to deal with the twenty word dictionary [Table IX.2].

Before describing the modern views on such genetic messages it is necessary to consider the intermediate mechanisms provided by RNA.

DNA, though it must provide the *information* for protein synthesis to the ribosomes [p. 5] in the cytoplasm, does not itself leave the nucleus but uses a *messenger*—one form of RNA (mRNA). Only part of the DNA is required for transcription to mRNA and this transcription is provided by the two strands of the DNA which code for this particular message, unwinding and separating briefly. One strand of the DNA is copied on to the messenger and afterwards the original DNA structure is reformed.

TABLE IX.2

Amino acid	RNA code words	
Alanine	CCG	UCG
Arginine	CGC	AGA
Asparagine	ACA	AUA
Aspartic acid	GUA	
Cysteine	UUG	
Glutamic acid	GAA	
Glutamine	ACA	AGA
Glycine	UGG	AGG
Histidine	ACC	
Isoleucine	UAU	UAA
Leucine	UUG	UUC
Lysine	AAA	AAG
Methionine	UGA	
Phenylalanine	UUU	
Proline	CCC	CCU
Serine	UCU	UCC
Threonine	CAC	CAA
Tryptophan	CGU	
Tyrosine	AUU	
Valine	UGU	

RNA is like DNA except that the pyrimidine base, uracil, appears in the polynucleotide chain wherever the complementary base adenine appears at the complementary site on the DNA chain (in other words, uracil appears in RNA instead of thymine). RNA of course also contains ribose instead of deoxyribose. Message transmission from DNA, using for instance letters A, C, T, will be to RNA which uses letters UGA in the alphabet [FIG. IX.28].

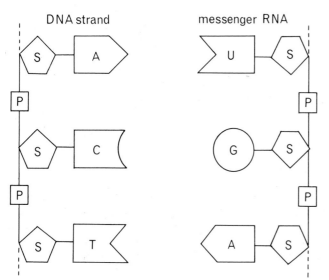

FIG. IX.28. Message transmission from DNA to mRNA.

Considering now the code letter combinations with the one, two and three letter words, and using U in the 'words' of RNA, they will be as shown in TABLE IX.3.

Messenger RNA (MW = 1 000 000) leaves the nucleus and passes to the ribosomes the sites of the protein synthesis. The sequence of bases in the messenger RNA specifies the sequence of amino acids required for the particular protein to be synthesized.

Another form of RNA (*transfer RNA or tRNA*) is now involved.

This is soluble in cytoplasm (and is sometimes known as sRNA) and is of a low molecular weight. Though really a single chain it has the shape of a twisted hairpin [FIG. IX.29]. One of the free ends always possesses the same last three nucleotidyl units (ACC). On this end an amino acid may be attached by an ester bond to the 3' carbon atom of the terminal adenosine. The other free end always possesses guanosine as its terminal group.

The head of the hairpin contains a sequence of three unpaired bases (UAC for example in FIG. IX.29).

These three unpaired bases at the head are believed to give the code for the particular amino acid that the tRNA transports. There is at least one tRNA for each of the 20 amino acids.

First, a given amino acid is activated at its carboxyl group by an activating enzyme in the presence of ATP and forms an enzyme-bound aminoacyl adenylate complex:

$$ATP + R.CH\, NH_2.COOH + enzyme$$
$$\rightarrow [R.CH\, NH_2.CO.AMP\text{–enzyme complex}]$$
$$+ \text{pyrophosphate}$$

The activated amino acid is then transferred to tRNA and the AMP and enzyme are liberated. tRNA transports it to the ribosome. Messenger RNA is already attached to the surface of the ribosome and moves along its surface in the manner of a tape

TABLE IX.3

Singlet code

A
G
C
U

Doublet code

A A	A G	A C	A U
G A	G G	G C	G U
C A	C G	C C	C U
U A	U G	U C	U U

Triplet code

A A A	A A G	A A C	A A U
A G A	A G G	A G C	A G U
A C A	A C G	A C C	A C U
A U A	A U G	A U C	A U U
G A A	G A G	G A C	G A U
G G A	G G G	G G C	G G U
G C A	G C G	G C C	G C U
G U A	G U G	G U C	G U U
C A A	C A G	C A C	C A U
C G A	C G G	C G C	C G U
C C A	C C G	C C C	C C U
C U A	C U G	C U C	C U U
U A A	U A G	U A C	U A U
U G A	U G G	U G C	U G U
U C A	U C G	U C C	U C U
U U A	U U G	U U C	U U U

engaging with the reading head in a tape recorder. At the ribosome surface the complementary tRNA molecules line up along the mRNA strand in consecutive groups of three nucleotides, each new tRNA molecule bringing the amino acid needed. As each tRNA 'delivers' its amino acid, the amino acids interact to form a peptide link—CONH with its preceding neighbour, thereby releasing its tRNA molecule for another job. This linking by the formation of a peptide bond requires a transfer enzyme, plus at least another enzyme and a cofactor, which is guanosine triphosphate. The polypeptides grow by this method as the messenger RNA moves along the ribosome.

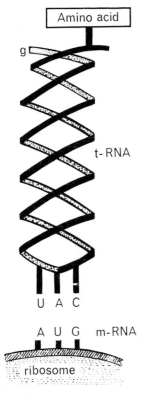

FIG. IX.29. Diagram of tRNA molecule lined up alongside the mRNA groups.

Different parts of the messenger RNA may be read simultaneously on different ribosomes, adding to the complexity, but also to the speed and efficiency of protein synthesis.

The triplet of consecutive bases in mRNA which is 'recognized' by the transfer RNA carrying the amino acid required is termed a *codon*. The sequence, three unpaired bases at the head of the loop of the appropriate tRNA–amino acid complex, is called an *anti-codon*.

It has been found that protein synthesis can occur in cell-free extracts of bacteria. The colon bacillus (*Escherichia coli*) is grown in a nutrient medium and the suspension of bacteria is centrifuged. The cells are gently broken up and DNA, RNA, ribosomes, enzymes and cofactors are thereby released from the cell sap. When sources of energy, such as adenosine triphosphate, are present the cell-free system will incorporate amino acids added and synthesize proteins. If the amino acids are radioactively labelled with ^{14}C the process of incorporation can be followed.

The addition of the specific enzyme deoxyribonuclease, which destroys DNA, halts the synthesis of protein in the cell-free media. However, if messenger RNA is then added the synthesis of protein is renewed.

The discovery of the enzyme polynucleotide phosphorylase (Ochoa 1955) has allowed the preparation of synthetic RNA

polymers by linking bases together in random order. In this way, Jones and Nirenberg (1962) synthesized an artificial RNA polymer containing only uracil, which they named poly-U. They then made a series of mixtures, each of which contained an active cell-free suspension and the 20 amino acids.

In each of the mixtures one amino acid was labelled radioactively and the remainder were non-radioactive. Poly-U was added to each mixture and the whole system was incubated for 90 minutes. Each mixture was then treated with trichloracetic acid (which precipitates such protein as may be formed), the precipitate was transferred to minute filter-paper discs and these were examined in a radiation counting unit to find out how well the poly-U had directed amino acids into the synthetic protein. It transpired that poly-U 'directed' phenylalanine.

Similar synthesis yielded poly-C (polycytidylic acid) and this dictated the synthesis of leucine.

Further experiments on such lines have shown that 18 of the 20 amino acids can be coded by 'words' containing only two different bases. Only methionine and aspartic acid seem to require a combination of U, G, and A. If the entire code does consist of triplets it is conceivable that correct coding may sometimes be achieved when only two out of the three bases read are recognized. Nirenberg suggests, however, that such imperfect recognition may occur more often with synthetic RNA polymers containing only one or two bases than it does with the mRNA of natural sources which always contains all four bases.

The manner in which a single amino acid is placed in its appropriate site has been elucidated by Chapeville and Lipmann and co-workers. They examined the effect of converting cysteine to alanine on the incorporation of either amino acid into the protein chain.

Cysteine (HS. CH_2 $CHNH_2$ COOH) and alanine (CH_3 $CHNH_2$ COOH) are each 3C chain amino acids which differ only in their terminal groups, alanine lacking the sulphur atom. Cysteine is known to be directed into protein by poly-UG. Alanine is directed into protein by poly-CG. Cysteine was radioactively labelled and was then enzymatically attached to its particular type of transfer RNA. This molecular complex was then treated with Raney nickel (made by dissolving out the aluminium from a nickel aluminium, using a strong alkali). Raney nickel removes sulphur from the cysteine molecule and leaves alanine without however detaching it from 'cysteine-transfer RNA'. This complex was then added to the artificial RNA made from uridine diphosphate and guanosine diphosphate (poly-UG) which should allow the incorporation of cysteine into a polypeptide, but not alanine, which requires poly-CG. However, the alanine-transfer RNA was incorporated into the polypeptide just as if it had been cysteine-transfer RNA. Clearly, there can be no 'recognition' of the amino acid once it has been attached to the transfer-RNA—the amino acid loses its identity and the tRNA carries it to the code word recognized.

As has been stated, the *codon* is given by the triplet of bases on the mRNA strand (the anticodon is formed by the three unpaired bases at the head of the transfer RNA complex). These bases must be complementary to those of the triplet forming the codon. Thus, if the codon is UUU the anticodon provided by the tRNA amino acid complex must be AAA (sometimes written aaa to avoid confusion).

TABLE IX.2 [p. 476] shows the various code triplets that have been defined for the different amino acids. The base sequence shown in the triplets is arbitrary. Some amino acids are coded by more than one triplet.

REFERENCES

CRICK, F. H. C. (1966). The genetic code, III. *Scient. Am.* **215**, 55–62.
DARNELL, J. E. JR (1974). The origin of mRNA and the structure of the mammalian chromosome. *Harvey Lect.* **69**, 1–47.

DAVIDSON, J. N. (1977). *The biochemistry of the nucleic acids*, 8th edn. Academic Press, New York.

GEFTER, M. L. (1975). DNA replication. *A. Rev. Biochem.* **44**, 45–78.

KORNBERG, A. (1974). *DNA synthesis*. Freeman, San Francisco.

LEHMAN, I. R. (1974). DNA ligase: structure, mechanism and function. *Science, NY* **186**, 790–7.

LODISH, H. F. (1976). Translational control of protein synthesis. *A. Rev. Biochem.* **45**, 39–65.

RICH, A. and RAJBHANDARY, U. L. (1976). Transfer RNA: molecular structure, sequence and properties. *A. Rev. Biochem.* **45**, 805–60.

TRAVERS, A. A. (1978). *Transcription of DNA*. Carolina Biology Readers, North Carolina.

WATSON, J. B. (1976). *Molecular biology of the gene*, 3rd edn. Benjamin, New York.

WHITE, A., HANDLER, P., SMITH, E. L., HILL, R. L., and LEHMAN, I. R. (1978). Principles of biochemistry, 6th edn, pp. 783–859. McGraw-Hill, New York.

Nucleic acid metabolism

Synthesis. Mammalian tissue cells synthesize both RNA and DNA. With the possible exception of adenine, preformed purines and pyrimidines are not utilized for this purpose, the nitrogenous bases of nucleic acids being built *de novo* from simple precursors. Bases from preformed nucleotides are somewhat more readily incorporated.

Tissue breakdown. RNA is continuously being broken down by enzyme systems within the cell, and has thus to be continuously replaced. Most of the DNA, representing the control centre of the cell, seems to be metabolically stable during the life of the cell. However, the maturation of nucleated red blood cells to form non-nucleated discs involves the destruction of the nucleus, and this contributes a considerable proportion of the purines and pyrimidines freed from cells.

Food nucleic acids. The cellular foodstuffs which are rich in nuclei (liver, kidney, pancreas, yeast) are likewise rich in nucleic acids. These are digested in the duodenum and small intestine, with the enzymic liberation of nucleotides, nucleosides, and free purines and pyrimidines, but little of this is incorporated into tissue nucleic acid or nucleotides.

Freed purines and pyrimidines. The purine bases adenine and guanine are catabolized to uric acid, most of which is excreted [see below].

The pyrimidine bases are completely degraded to urea and simple carbon compounds.

Biosynthesis of purines

The evidence for the biosynthesis of purines is:

1. A growing infant on a milk diet, which is almost purine-free, greatly increases its purine content and the total number of its cell nuclei.

2. A Dalmatian dog fed for a year on a purine- and nucleoprotein-free diet maintained perfect health, and during the time excreted 100 g of uric acid of which not more than 10 g could possibly have come from preformed purine of its tissues.

3. Feeding or injecting isotopically-labelled precursors into animals or human subjects has shown that adenine and guanine (and thence uric acid) can be synthesized from non-purine components of the common metabolic pool. All 9 atoms of the rings have to be contributed by specific sources, as shown in FIGURE IX.30 (a).

4. Experiments with soluble enzyme systems from liver have defined each step of the biosynthetic sequence. An early stage, before any ring-closure, is the introduction of ribose-phosphate;

the final closure is of the pyrimidine ring, giving inosinic acid (inosine-phosphate = hypoxanthine-ribose-phosphate). This nucleotide can be converted reversibly to the nucleotide of adenine, AMP (and thence to ADP and ATP), or irreversibly to the nucleotides of xanthine and guanine.

Biosynthesis of pyrimidines

As with the purines, the ring atoms of the pyrimidines are each contributed from an obligatory source, as shown in FIGURE IX.30 (b). This synthetic route leads to uridylic acid (uracil-ribose-phosphate) from which the other pyrimidines and their nucleotides are derived by substitution. The methyl group of thymine is contributed by methionine.

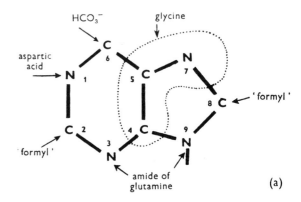

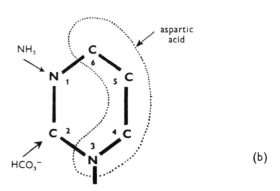

FIG. IX.30. Specific (obligatory) precursors for the biosynthesis of the ring system in (a) Purine derivatives, (b) Pyrimidine derivatives. (Note the different routes to the 6-membered ring in the two structures.)

Metabolism of purines

Purine nucleotides are released during the breakdown of tissue nucleic acids, food nucleic acids, and tissue nucleotides, and are further split by specific enzymes, liberating the free purine bases. The bases can be either reconverted to their respective nucleotides; oxidized to uric acid as a final product; or excreted. The enzyme xanthine oxidase converts xanthine to uric acid.

Catabolism to uric acid. FIGURE IX.23 shows the stages in the oxidation of purines to uric acid. The immediate precursor of uric acid is xanthine, which arises from guanine or hypoxanthine. Adenine does not give rise to hypoxanthine directly, but can do so via the corresponding nucleotides.

Very little is known concerning the metabolic fate of the pyrimidines released by nucleic acid and nucleotide breakdown.

URIC ACID

Formation. In man, uric acid results:

1. Primarily, from the oxidation of the purine moieties of degraded nucleic acids.
2. Secondarily, and in particular in certain gouty individuals, from glycine by a direct pathway not involving nucleic acids.

In birds and some reptiles, the excretion of uric acid constitutes the chief route for the elimination of waste nitrogen from every source.

Fate. In most animals including most breeds of dog uric acid is further oxidized to, and excreted as, allantoin. Strangely enough, the Dalmatian coach-hound is unable to reabsorb uric acid in the renal tubules and thus excretes all the uric acid which is filtered out in the glomeruli; it is, however, able to form allantoin like any other dog if the ureters are tied off to retain the uric acid within the body. In man, most of the uric acid formed from purines is excreted in the urine; it has been shown, however, that a considerable proportion of experimentally administered uric acid is further oxidized to unknown products which do not include allantoin.

Blood uric acid. The normal limits in serum are $0.1-0.4$ mmol l^{-1}. It is raised in gout up to $0.5-0.9$ mmol l^{-1}. Ingestion of purine-containing food has no effect on the blood uric acid in normal people, but raises it in cases of renal insufficiency. It is claimed that uric acid is the first nitrogenous constituent to be retained in nephritis, values of 0.5 mmol l^{-1} occurring before the other nitrogenous constituents are increased. In uraemia [p. 244], values as high as 1.6 mmol l^{-1} may be obtained. In leukaemia considerable breakdown of nuclei of white blood corpuscles occurs and the blood uric acid is raised. Although the kidneys are excreting a great deal of uric acid they are unable to eliminate it as rapidly as it is formed.

Excretion of uric acid. The uric acid in human plasma is mostly present in a freely diffusible form which appears in the glomerular filtrate in a concentration approaching that of plasma urate. Reabsorption of the greater part of the uric acid occurs during passage of the filtrate through the renal tubules. Uricosuric drugs diminish this reabsorption and so increase urinary output.

The total daily output of uric acid (urate) depends partly on the purine content of the diet. On a normal diet the output averages $0.75-1$ g. On a purine-free diet, it falls to about half this, the residue representing the continuing metabolism (endogenous degradation) of tissue nucleic acids.

Uric acid excretion is diminished when the diet has a low purine content, a low protein content, a low calorific value, or consists of fat rather than of carbohydrate. The latter factor causes renal retention of urate and tends to increase the uric acid content of the body.

Gout

Gout is a chronic disorder characterized by:

1. Excess of uric acid in the blood.
2. Deposition of sodium monourate in articular and non-articular structures, producing the so-called tophi.
3. Recurring attacks of acute arthritis. These are due to deposition of microcrystals of monosodium urate in and around the structures of the affected joint [p. 397].

Metabolism of uric acid in gout. The essential abnormality in *primary* gout is increased formation of uric acid from simple carbon and nitrogen compounds without intermediary incorporation into nucleic acids. There is no evidence of decreased destruction of uric acid in the body, nor is there diminished urate excretion by the kidney (except late in the disease when renal failure occurs).

Studies with isotopic ^{15}N uric acid have shown that the amount of freely diffusible uric acid in the body fluids is increased in gout. In normal persons the 'metabolic pool' is $0.7-1.3$ g, of which 50–70 per cent is replaced by new uric acid daily, most of the lost substance appearing in the urine. In gouty subjects the metabolic pool of uric acid is 2–4 g and a value of 31 g has been recorded. The daily turnover is 30–50 per cent. In addition to the diffusible metabolic pool there is in chronic gout a large pool of very slowly diffusible urate in the tophaceous deposits.

In *secondary* gout such as may occasionally occur in polycythaemia, chronic leukaemias, and pernicious anaemia there is an increased breakdown of nucleic acids leading to an excess of the end-product, uric acid.

Drugs in treatment of gout

Acute gout. The most effective drugs are colchicine, phenylbutazone, and indomethacin. Colchicine has no effect on uric acid metabolism and does not relieve pain due to other causes. Its mode of action is still not known. Phenylbutazone increases the renal excretion of urates but it is not known whether this contributes to its beneficial actions in acute gout.

Chronic gout. The best treatment is the prolonged administration of drugs which increase the excretion of uric acid in the urine by diminishing tubular reabsorption of urate. The most effective uricosuric drugs are probenecid (*Benemid*) and salicylates, the former usually being better tolerated. The increased excretion of uric acid lowers the serum uric acid level, reduces the metabolic pool, and may, in the course of months, reduce the size of the tophi. The drug allopurinol inhibits the enzyme xanthine oxidase and thus reduces uric acid synthesis.

REFERENCES

Davidson, J. N. (1977). *The biochemistry of the nucleic acids*, 8th edn. Academic Press, New York.
Henderson, J. F. and Paterson, A. R. P. (1972). *Nucleotide metabolism*. Academic Press, New York.
Murray, A. W. (1971). The biological significance of purine salvage. *A. Rev. Biochem.* **40**, 811–36.
Stanbury, J. B., Wyngaarden, J. B., and Fredrickson, D. S. (eds.) (1972). *The metabolic basis of inherited disease*, 3rd edn, Part VI. *Disorders of purine and pyrimidine metabolism*. McGraw-Hill, New York.

SULPHUR COMPOUNDS IN URINE

The sulphur compounds of urine are derived solely from the sulphur-containing amino acids (methionine, cysteine, and cystine) of the dietary and tissue proteins. The sulphur is excreted in three forms:

1. Inorganic Sulphate. Sulphur-containing amino acids of the amino acid pool that are not used in protein synthesis or other transformations are completely oxidized, the nitrogen appearing as urea and the sulphur as sulphate ions, SO_4^{2-}. The sulphate anions are excreted in the urine, with an equivalent amount of cation (Na^+, K^+, NH_4^+). The level of inorganic sulphate excretion follows that of urea excretion, being high on a high-protein diet and reaching a minimum on a protein deficient diet. The normal range of urinary output is $0.3-3.0$ g of sulphate ions per day. Inorganic sulphate introduced into the body parenterally is excreted without appreciable utilization.

2. Ethereal Sulphate. The urine contains small amounts of or-

ganic sulphate esters, R—O—SO₃H (where R = various aromatic radicals), the so-called ethereal sulphates. These are the forms in which many phenols, R . OH, are 'detoxicated' and excreted in the urine [p. 442]. The conjugation of the phenol with sulphate from amino acids takes place in the liver. The phenols which form ethereal sulphates are: (i) physiological compounds (e.g. oestradiol and oestrone); (ii) non-physiological compounds (e.g. salicylic acid from ingested aspirin; and (iii) phenols formed from aromatic amino acids by the putrefaction of amino acids in the intestine. Thus tryptophan is decomposed to give indole, which is oxidized in the liver to indoxyl and conjugated with SO_4^{2-}. The potassium salt of indoxyl sulphate (called indican) is excreted in the urine.

3. Neutral Sulphur. This term is applied to a heterogenous collection of unoxidized S-containing substances (e.g. cystine, mercaptans) found in the urine in traces; their origins are obscure, but the total neutral S excretion is unaffected by the diet (in the classical terminology, it is endogenous).

METABOLISM IN STARVATION

During starvation, the subject must live on the component tissues of his own body for energy purposes. Even if no physical work is being done, about 8400 kJ will be needed daily. During the first few days the glycogen stores of the liver are called upon, but these at the best only amount to several hundred grams and so are of very limited value. The main source of energy must be the fat reserves and tissue protein. So long as fat is available tissue protein is used sparingly, but some net destruction of protein cannot be avoided.

Death occurs after about 4 weeks, when the body weight is reduced by 50 per cent. The temperature only falls just before the end.

For a discussion of the effects of undernutrition see page 483.

Blood sugar. This is maintained at a steady level almost to the end; it is formed in the liver from amino-acid residues, glycerol (from fats) and lactic acid (from partial dissimilation of muscle glycogen); if hepatectomy is performed, a hypoglycaemia which is rapidly fatal, results.

Use of fat. The fat which is used in starvation is the neutral fat (triglyceride, *élément variable*) which is found in adipose tissue; it is mobilized and taken to the liver, where it is either completely dissimilated or transformed into ketone bodies, which are sent to other tissues for oxidation. The complex lipid (*élément constant*) which forms part of the cell structure is spared till the end.

Ketosis. Owing to lack of carbohydrate, ketogenesis is stimulated and ketones pass into the blood from the liver faster than they are disposed of by the tissues [p. 510]. There is thus a ketosis and ketone bodies appear in the urine. The usual steps are taken to compensate for the tendency to acidaemia, i.e. buffering by means of bicarbonate, increased pulmonary ventilation, and fall of alveolar CO_2 tension; increased acidity of the urine; increased NH_4^+ excretion [p. 229]. As the depot fat is mobilized, there is a slight rise in the blood fat and fatty liver develops.

Use of tissue protein

Tissue protein is treated in starvation like food protein and is hydrolyzed to amino acids but on a larger scale than normally. The tissues are not treated uniformly: brain and heart only lose 3 per cent of their bulk; muscles, liver, and spleen lose 30, 55, and 70 per cent respectively. There is evidence that the breakdown of tissue protein in starvation is controlled, possibly through the action of the adrenal cortex.

The released amino acids enter the 'common pool'.

1. The first call on the pool is to maintain the structure and so preserve the functional efficiency of the essential organs; this would also include formation of enzymes and those hormones which are proteins or amino-acid derivatives.

2. The second call on the pool is to preserve the normal blood sugar level without which brain function fails; as the sugar is steadily withdrawn from the blood by all the organs (and not only by the brain), protein used in this way is acting as a source of energy. The conversion of amino acids to sugar (or the dissimilation of amino acids *in place* of sugar) occurs in the liver and involves preliminary deamination; the NH_3 thus liberated is excreted as urea.

The urinary urea output in starvation is an index of tissue protein consumption. When the fat stores are exhausted, protein alone is available for energy purposes and death rapidly results.

Nitrogen excretion during the first week of starvation averages about 10 g daily (which is far higher than the possible minimum [cf. p. 484]); during the second or third week it may fall to a considerably lower value. The changes in the individual nitrogenous and sulphur-containing constituents are essentially similar to those found on a low protein diet [see TABLE, p. 470]. Just before death the urinary nitrogen output rises steeply ('pre-mortal rise') owing to the rapid destruction of tissue protein.

Sources of energy

An examination of the nitrogenous excretion and of the metabolic rate enables an estimate to be made of the relative amounts of fat and protein which are being combusted. Thus, a starving man on the fifth day of his fast, when his energy output was 8400 kJ, excreted 11.4 g of N. One g of nitrogen in the urine represents the breakdown of 6.25 g of protein. The urinary nitrogen indicates that $6.25 \times 11.4 = 71.5$ g of tissue protein were being broken down, which would yield roughly 1250 kJ (1 g protein = 17 kJ). The rest of the energy, i.e. 7500 kJ, must have been derived from about 190 g of fat (1 g fat = 39 kJ). The fasting body was using protein and fat in the proportion by weight of 71.5:190.

INTEGRATION OF METABOLIC PROCESSES

The elucidation of the metabolic pathways of the three main groups of foodstuffs and body components has led to the following generalization: after initial modifications in the first phases of metabolism, the structures of all carbohydrates, fats, and proteins are no longer distinguishable as to origin, being incorporated into a system of common carbon fragments based on the 'citric acid cycle'.

Metabolic pools

The common system of intermediate dissimilation products is termed the 'common metabolic pool'. It is like a trading market, to which goods are contributed and from which others are drawn. Various reversible and irreversible processes, forming other cycles, govern the operation of the central metabolic system. If one type of fragment is lacking when needed, or is present in excess, then appropriate cycles are set in motion in such a manner as will produce that fragment, or dissimilate it.

Similarly, each distinct reversible metabolic cycle constitutes a 'body pool' for a given compound, e.g. the body glucose pool, the amino-acid pool; the use of isotopically labelled molecules has contributed largely to our knowledge of these metabolic cycles. The size of an individual pool can be measured by an isotopic dilution technique. The amount of substance thus measured is not solely the amount of free substance distributed throughout the body but also includes the amount which could be manufactured reversibly by the operation of other cycles. For example, the 'body glucose pool' would include some portion of the liver glycogen as well as all the free glucose of the body fluids. The meaning of the word 'pool' is thus determined by the measurement techniques that are employed.

Metabolic syntheses

Until recently, it seemed probable that the synthesis of a metabolic storage product (e.g. glycogen or fat) was accomplished by the exact reversal of its dissimilative breakdown, using the same enzymes in reverse. It is now clear that, in a number of instances, entirely different chemical pathways and enzyme systems and subcellular structures are involved, so that dissimilation and synthesis can be controlled separately, and indeed can proceed simultaneously. Examples are:

1. The synthesis of glycogen from glucose 1-phosphate via a uridine diphosphate, while the breakdown of glycogen is a phosphorolysis under the influence of the enzyme phosphorylase.

2. The synthesis of fatty acids from acetate, where the H is provided by a coenzyme carrier different from that which removes H during fatty acid oxidation to active acetate.

Metabolic inter-relationships

The reactions of methionine [p. 471] illustrate how artificial and misleading is the customary division of general metabolism into subsections called carbohydrate, fat, and protein metabolism. Methionine, derived from protein, is responsible for creatine formation, and thus indirectly for carbohydrate utilization in muscle; it is responsible for choline formation, and thus indirectly with preventing undue fat accumulation in the liver; it is indirectly responsible for the formation of acetylcholine (via choline) and of adrenaline, and thus for nervous transmission at cholinergic and adrenergic terminals; by its incorporation (as intact methionine and as the S atom of cysteine) into enzyme- and hormone-proteins it exerts an influence on the metabolism of the whole body.

REFERENCES

BIANCHI, R., MARIANI, G., and McFARLANE, A. S. (eds.) (1976). *Plasma protein turnover*. University Park Press, Baltimore.
GREENBERG, D. M. (ed.) (1975). *Metabolic pathways*, 3rd edn, Vol. VII. *Metabolism of sulfur compounds*. Academic Press, New York.
OLSON, R. E. (ed.) (1975). *Protein–calorie malnutrition*. Academic Press, New York.

PART X
Nutrition

When Lavoisier had correctly interpreted the process of chemical combustion, he realized that the production of energy from the body was a form of the same combustion and could be represented quantitatively by the same chemical reactions. Since then, the physiological functions of the major food components have been determined and a great deal of knowledge has been accumulated on the minor 'accessory food factors'. It has been the aim of nutritional studies to find how food (in its real sense, not as a laboratory material) can best supply the chemicals needed for the optimum physiological functioning of every individual. Some success has been achieved, not least in establishing priorities for communities faced with scarcity of food.

Many problems of nutrition remain—the positive health effects of food, the hazards of excess, the basis of unusual response or demand for specific foods. But these seem of little consequence beside the economic and political problems of putting into practice on a world scale what is already known. While the Western peoples are better fed than ever before, the emergent countries of the world are falling yet nearer to the starvation mark. Indeed, disastrous famines have occurred in the 1970s in parts of Africa and Asia and the growth of the world population increases the risk of further famines.

General references

DAVIDSON, S., PASSMORE, R., BROCK, J. F., and TRUSWELL, R. S. (1975). *Human nutrition and dietetics*, 6th edn. Churchill Livingstone, Edinburgh.
GOODHART, R. S. and SHILLS, M. E. (eds.) (1973). *Modern nutrition in health and disease*, 5th edn. Lea and Febiger, Philadelphia.
HOWARD, A. N. and BAIRD, I. M. (eds.) (1981). *Recent advances in clinical nutrition*. John Libby, London.
PYKE, M. (1975). *Success in nutrition*. Murray, London.
WORTMAN, S. (1976). *Food and agriculture*. Scientific American, San Francisco.

Dietary requirements

ESSENTIALS OF A DIET

An adequate diet must have an energy value sufficient to provide for: the requirements of the basal metabolism, the stimulating action of the foodstuffs, loss in the faeces and urine of unutilized derivatives of the foodstuffs, and the needs of varying degress of muscular and external work. It must have adequate amounts of protein (essential amino acids), fat, carbohydrate, water, and salts (ions), all in suitable proportions, and an ample vitamin content. In children adequate provision must be made for additional tissue formation; the special requirements of menstruation, pregnancy,

lactation, illness and old age must be taken into account. As the food has to be eaten by discriminating human beings it must be well cooked, look attractive, taste nice, and be reasonably varied from day to day. 'Nutrition is a science of test tubes, gastronomy is the art of taste buds' (André Simon) (and, of course, of the olfactory nerve endings).

ENERGY REQUIREMENTS

Available food energy. To determine the energy content that a known weight of food gives when burnt in a small chamber is simple [p. 212]; but to assess the energy that an individual can derive by cooking and eating that food presents problems which have not yet been satisfactorily solved. A variable loss occurs during preparation and cooking; a small proportion escapes digestion and absorption and is eliminated in the faeces (the loss being less on an animal diet than on a vegetable diet with its greater cellulose content).

As a rough dietary guide, the energy yields of the only nutrients that matter as energy providers can be taken as:

carbohydrate (sugar, starch)	17 kJ g⁻¹ (4 kcal g⁻¹)
protein (mixed)	17 kJ g⁻¹ (4 kcal g⁻¹)
fat	37 kJ g⁻¹ (9 kcal g⁻¹)

(1 kcal $\approx$ 4.2 kJ; 1000 kcal $\approx$ 4.2 MJ.)

Requirements in man

When food is freely available a healthy individual assesses his own energy requirements very accurately by means of his appetite. There are wide variations in individual energy requirements even among persons living under similar conditions.

For certain purposes it is necessary to know average requirements among large groups of people such as the Armed Forces, inmates of prisons, or perhaps the whole population in times of scarcity. Assessment of energy requirements has been made by *dietary surveys* or *surveys of energy expenditure*.

A dietary survey is easier to carry out since it merely involves finding out what groups of people in various occupations actually eat, but is is only of value in indicating requirements if food is available in liberal quantities. The drawbacks to such surveys are that people may eat more than they need to satisfy energy requirements and that variations in need with different occupational activities are not actually measured.

Surveys of energy expenditure aim to provide more accurate information on requirements by recording the detailed physical activities of a group of subjects and assessing the energy cost of each activity either by direct measurement of oxygen consumption or from published tables. The sum of all the separate measurements gives the total energy expenditure.

The results of a survey showing the energy expenditures of clerks as compared with miners are given in TABLE X.1. This itemizes the average daily expenditure of energy (kJ) by 10 clerks (average age 28.3 years, weight 64.6 kg) and by 19 miners (average age 33.6 years, weight 65.7 kg) as measured over a whole week by Garry *et al.* (1955):

TABLE X.1

	Clerks	Miners
Asleep and day-time dozing	2100	2100
Activities at work	3700	7300
Non-occupational activities and recreations	5900	5900
Total	11 700	15 300

Factors influencing energy expenditure

Physical activity. The amount of physical activity is the most important factor. TABLE X.2 shows the energy expenditure associated with different types of physical activity.

TABLE X.2
Examples of the energy expenditure of physical activities (After Davidson *et al.* 1975)

Light work at 10–20 kJ min⁻¹		Moderate work at 20–30 kJ min⁻¹
Assembly work	Building industry	General labouring (pick and shovel)
Light industry	Bricklaying	
Electrical industry	Plastering	Agricultural work (non-mechanized)
Carpentry	Painting	
Military drill	Agricultural work (mechanized)	Route march with rifle and pack
Most domestic work with modern appliances	Driving a truck	Ballroom dancing
	Golf	Gardening
Gymnastic exercises	Bowling	Tennis
		Cycling (up to 16 k.p.h.)

Heavy work at 30–40 kJ min⁻¹	Very heavy work at over 40 kJ min⁻¹
Coal mining (hewing and loading)	Lumber work
Football	Furnace men (steel industry)
Country dancing	Swimming (crawl)
	Cross country running
	Hill climbing

Workers in forestry or agriculture, as well as coalminers, may expend more than 17 000 kJ and sedentary workers less than 10 050 kJ daily. The range of daily energy expenditure for the vast majority of people is:

10 000–17 000 kJ daily for men
7150–12 600 kJ daily for women.

Mental activity involves no expenditure of energy, though accompanying increase of muscle tension does.

Age. This affects energy expenditure in two main ways. With advancing age the BMR falls, and the amount of physical activity declines [TABLE X.3].

Body size and weight. Both BMR and the energy expenditure of mechanical work are directly proportional to body weight. However, big people may eat no more than smaller people because they are less physically active.

Climate. People living in a hot climate eat less than in a cold climate. Where the mean annual environmental temperature ex-

ceeds 25 °C the energy intake should be reduced by 5–10 per cent. Since protection against cold is more easily achieved, by clothing and by heating of buildings, the requirements should be increased by only 3 per cent for every 10 °C of mean annual external temperature below the reference temperature of 10 °C.

TABLE X.3
Estimated energy requirements of adults according to age
(Mean annual environmental temperature, 10 °C)
[Davidson and Passmore (1966) p. 32.]

Age (years)	Men weighing 65 kg	Women weighing 55 kg
	kJ day⁻¹	kJ day⁻¹
20–30	13 500	9700
30–40	13 000	9400
40–50	12 600	9080
50–60	11 600	8400
60–70	10 600	7650
70–	9300	6720

Children

The energy intake of children must allow for growth, physical development, and the considerable activity natural to their state.

Infants require about 3000 kJ daily during the first 3 months of life, rising to 4200 kJ daily at 10–12 months.

Children from 1–19 years. The following average requirements have been proposed (Davidson and Passmore 1966):

Age (years)		kJ per day
1– 3	Children	5500
4– 6	Children	7150
7– 9	Children	8820
10–12	Children	10 500
13–15	Boys	13 000
	Girls	10 920
16–19	Boys	15 000
	Girls	10 000

The energy allowances for growing children should not be adjusted to their actual weight. The undersized child may need more energy than the well-developed child, in order to attain the normal adult size. On the other hand over-feeding during early childhood may predispose to obesity in adult life.

Results of inadequate energy intake

Undernutrition and starvation

Undernutrition and starvation lead to wasting of the body with marked loss of adipose tissue and of muscle. The main causes are:

1. Insufficient food in the diet, as in times of famine, which still occur all too frequently in various parts of the world, especially in Asia and Africa.

2. Severe disease of the alimentary tract. Conditions such as the malabsorption syndrome and cancer of the oesophagus prevent absorption of nutrients even if the dietary supply is adequate.

3. Infection and toxaemias. These reduce appetite or interfere with normal metabolism.

In famines the main deficiency may be in energy, but in some cases vitamin deficiencies predominate. The effects of moderate and severe energy deficiency have been studied experimentally in healthy volunteers.

Body changes. The typical changes in body composition produced by starvation leading to 25 per cent loss in body weight are summarized in TABLE X.4.

TABLE X.4
*The changes in body composition of a man who in health
weighed 65 kg and then lost 25 per cent of this weight as a
result of starvation*
[Davidson *et al.* (1975) p. 283]

	In health	After starvation
	(kg)	(kg)
Protein	11.5	8.5
Fat	9	2.5
Carbohydrate	0.5	0.3
Water:		
Extracellular	15	15
Intracellular	25	19
Minerals	4	3.5
	65.0	48.8

The main alterations are the loss of 3 kg of protein, 6.5 kg of fat, and 200 g of carbohydrate. This represents a reserve store of about 300 000 kJ. If such a normal healthy man is deprived of all food (but is provided with water) and takes no exercise, his energy expenditure will approach the basal rate (say 6700 kJ day^{-1}). At this rate of exhaustion he will lose 25 per cent of his body weight in about 45 days.

TABLE X.4 shows that the carbohydrate reserve (glycogen in liver and muscle) is very small. The brain requires 90 g of glucose daily for its metabolic activities. With complete starvation the blood glucose level is maintained by conversion of protein to carbohydrate; the provision of 100 g of carbohydrate daily will reduce endogenous protein breakdown to a minimum and will also prevent ketosis. The increased breakdown of endogenous protein occurs chiefly in muscles and glands. Glucose is formed from the non-nitrogenous part of the liberated amino acids. There is a negative nitrogen balance. The enhanced breakdown of fat in the adipose tissue depots releases free fatty acids (FFA) which can provide energy for muscular contraction and thus help to conserve the blood glucose for the brain which cannot utilize FFA. Intracellular water is decreased but extracellular water is not correspondingly reduced so that a relative excess accumulates, causing eventually famine oedema. Water retention is promoted by several factors. Reduction in plasma protein level reduces the osmotic pressure exerted by blood plasma and allows more fluid to accumulate in the tissue spaces. There is a fall in tissue tension which may be due to loss of fat; famine oedema is greater in old people in whom skin elasticity has already declined. Renal function is disturbed, leading to reduced urinary excretion by day and polyuria at night.

Failure of hormone and enzyme production. Many of the body's hormones and all enzymes are proteins. It is therefore not surprising that when protein breakdown continues for long periods during starvation the synthesis of protein hormones and enzymes is reduced. In girls and women delayed puberty and amenorrhoea are very likely due to reduced gonadotrophic hormone secretion [p. 574] and in men loss of libido and the development of impotence could be similarly caused. Marked atrophy of the thyroid gland is probably due to reduced secretion of thyrotrophic hormone. The terminal diarrhoea after prolonged starvation could be due to reduced formation of digestive enzymes.

Circulation. The marked bradycardia is due in part to thyroid deficiency. Peripheral blood flow and venous pressure are reduced; arterial blood pressure is lowered.

Metabolic rate. The BMR is reduced, partly owing to the smaller mass of active tissues and partly owing to thyroid deficiency. In this way energy amounting to 2500 kJ is saved daily.

Hunger. In people totally deprived of food the feeling of hunger disappears after a few days; in semi-starved people on the other hand the hunger sensation becomes progressively accentuated. It is known that strong contractions of the empty stomach give rise to a sense of hunger (hunger pains), but it is not likely that hunger is always or generally produced in this way. Thus these subjects often complained of hunger immediately after a large meal and when gastric motility and tone were depressed.

Complete starvation. See page 480.

CONSTITUTION OF NORMAL DIET

A normal diet should provide the amount of energy required according to age, body size, physical activity, etc. Energy values range from 7500 to 15 000 kJ. Energy is provided by protein, fat, and carbohydrate.

Protein

Protein is an indispensable constituent of the diet because it is the only source of the amino acids, including the essential amino acids which cannot be synthesized in the body. Amino acids are needed: to build up new tissue during the period of growth or pregnancy and to provide the milk proteins during lactation; to maintain the structure of every tissue cell including its content of protein-containing enzyme systems; to provide the raw materials for the manufacture of certain external secretions (digestive enzymes of alimentary canal) and internal secretions (e.g. those of the anterior pituitary, thyroid, adrenal medulla); and to maintain the normal concentrations of plasma proteins and haemoglobin. Protein also provides 10–15 per cent of the energy in a well-balanced diet [p. 470].

As explained on page 465 there is normally a dynamic equilibrium between the amino acids derived from the food and those derived from the breakdown of the tissues, and also between the amino acids derived from one tissue and those derived from another. But in the course of this endless flux, some amino acids are broken down irreversibly and their nitrogen content is excreted in the urine, mainly as urea. The minimum protein intake in the adult must contain enough of each essential amino acid to make good this steady loss. The individual proteins of animal tissues generally closely resemble those of human tissues in their amino-acid composition; they can consequently be employed more economically for repair and growth; those with such a composition are called proteins of *high biological value*. The individual proteins of vegetable foods frequently have a very different type of amino-acid pattern and thus cannot be so economically built up into human tissues; such proteins are said to be of *low biological value*.

But it must be pointed out that this discussion is rather academic; we do not eat individual animal or individual vegetable proteins; we eat *foodstuffs*, e.g. meat, potato, bread, egg, milk, all of which consist of a mixture of several proteins. The important thing is the amino-acid composition and the biological value of the protein mixture in a single foodstuff, or even more so, in the mixture of foodstuffs found in typical human diet [cf. FIGS. IX.13 and IX.14, p. 466]. Chemical analysis is not the sole criterion of the biological value of a protein, because we know little of the effects of different amino-acid proportions within protein foodstuffs. Although growth rate is largely determined by the absolute amount of the limiting essential amino acid, the efficient utilization of nitrogenous compounds is determined by the relative proportions of many amino acids; immediately after the digestion of an unbalanced diet, all amino acids are rapidly catabolized, even the limiting one. The balance of non-essential amino acids seems of increasing importance as physiological stress increases.

There are two ways of testing the biological value of an individual protein:

1. To see whether it can maintain satisfactory growth and health when it is the sole source of protein supplied.
2. To determine the minimum amount of the protein which must be added to a diet, adequate in other respects, to maintain nitrogenous equilibrium.

Similar determinations can be carried out on a mixture of proteins, a foodstuff or a mixture of foodstuffs. Using these criteria, certain isolated plant proteins, such as gliadin or zein, are very inadequate; the proteins of flour made from wheat endosperm are not satisfactory; on the other hand, the mixed proteins of wheat or maize as found in high-extraction flour, by compensating for one another's deficiencies or quantitative peculiarities, maintain nitrogenous equilibrium at a fairly low level, particularly if they are supplemented by small quantities of other proteins such as those of milk or meat. In any mixed diet, even if wholly of plant origin, the proteins are sure to be sufficiently varied to compensate for any individual inadequacies in amino-acid content, *if only the total amount of protein is sufficient*.

The recommendation for adults is that the protein intake should not be less than 1 g per kg body weight, some of it in the form of animal protein. Pregnant and lactating women need more, e.g. 1.5 to 2 g per kg of total protein and (for safety) a larger proportion of animal protein [p. 487]. Nitrogenous equilibrium and unimpaired health and mental and physical vigour have been maintained for many months on diets which contained no more than 30–50 g protein daily. But the proteins were specially selected and were of high biological value and the experiments were carried out on a very small number of subjects for too short a period. In some experiments nitrogenous equilibrium was maintained on 30–40 g of protein derived exclusively from vegetable sources (cereals, potatoes, and other vegetables and fruit). To provide a margin of safety a minimum of 60 g of protein should be provided in the form of a varied diet.

Proteins of high and low biological value are sometimes called 'first class' and 'second class' proteins respectively. As the best known individual proteins of low biological value are derived from vegetable foods there has been a regrettable tendency to equate the terms 'first' and 'second' class generally with animal and vegetable proteins respectively. Such misuse of the terms should be strenuously avoided. Before vegetable proteins in general are dismissed as second class it should be remembered that many animals build up their muscles (i.e. our highly-prized meat) from the proteins of the humble grass. (Nebuchadnezzar lived for a time on grass and so, it is claimed, have exceptional individuals since his time.) Protein in meat and fish is becoming increasingly expensive. Vegetable protein from various sources, enriched with vitamin B_{12}, can be directly utilized by human beings; when given in this way it is cheaper and more efficient than when used to make animal protein. Defatted soyabean flour, proteins from green leaves, legumes, and even concentrates of plankton and microorganisms could provide more good-quality protein than would be obtained by feeding these materials to animals. Such sources of protein may become very important if the world population continues to rise. Many people who have been accustomed to a diet rich in animal food feel unhappy without it and their efficiency as workers may fall off as a result, but it is hard to assess the relative importance of physiological and psychological factors in such cases.

Protein malnutrition. There is no satisfactory measurement for assessing the state of protein nutrition in a human being. Although the levels of plasma enzymes are often lowered by the loss of liver protein, the main burden of depletion is borne by the muscles.

Kwashiorkor means 'the sickness the older child gets when the next baby is born', and this indicates the serious syndrome of protein deficiency which develops when a child is weaned. It is common in the tropics among children living on diets adequate in calories (from carbohydrate) but deficient in good quality protein. It is characterized by oedema, anaemia, and fatty liver, which must arise from the dietary imbalance rather than a simple deficiency. In countries where kwashiorkor is rife, animal protein is scarce and attempts are being made to prevent the condition by adding small amounts of foods of animal origin (dried skimmed milk or concentrates of fish protein) to high-protein vegetable foods such as soya beans or ground nuts.

Marasmus is total inanition occurring during the first year of life, with severe lack of both energy and protein as well as of all other nutrients. Marasmus is much commoner than kwashiorkor (McLaren 1974).

REFERENCES

LATHAM, M. C. (1974). *Physiol. Rev.* **54**, 541.
McLAREN, D. S. (1974). *Lancet* **ii**, 93.

Fat

The amount of fat consumed varies with the country, economic status, occupation, and the general circumstances. The maximum fat content of the really native diets in Japan before the Second World War was thought to be about 30 g which was then the European minimum. The amount of fat eaten increases steadily with income (in an uncontrolled economy).

Experiments of short duration have been carried out in which health and weight were maintained on diets of high total calorie value containing as little as 10–14 g of fat daily. Although the body can adapt itself to partial or even complete deprivation of fat for short periods, deleterious effects may become apparent after months or years. It is advisable that the normal diet of 13 000 kJ should contain at least 75 g of fat (3000 kJ); the fat content of the diet should always be raised when there is a large increase in the energy expenditure of the body, either in the form of work or because of exposure to cold. Conversely, in middle-aged people leading sedentary lives in affluent countries the fat content of the diet should be reduced.

The significance of fat in the diet depends on several factors. Firstly, it is almost completely absorbed from the alimentary canal. The bulk of the food becomes of importance when the total energy requirements of the body are very large. Weight for weight, fat has double the energy value of starch or sugar. Fat, in addition, is taken without admixture in a pure form, whereas the other foods are all mixed with a considerable proportion of water; when starch is cooked it is swollen up with five to ten times its volume of water. The Swedish and American lumbermen and the Welsh miners obtain a large part of their huge energy intake from fat. Carbohydrates are more subject to fermentative changes in the intestine, with the production of gases and general discomfort.

The animal fats are the most important sources of some of the vitamins, i.e. A and D [pp. 489, 553]. The vegetable fats are an equally effective source of energy, but are deficient in vitamins, except E [p. 489]. When butter is scarce it is essential from the point of view of the health of the community that margarine should have vitamins A and D added to make it equivalent to butter in its vitamin content. This has been enforced in Great Britain. Adequately vitaminized margarine (made from vegetable oils) is equal dietetically in all respects to butter, and is generally cheaper.

Specific evidence for dietary 'essential' fatty acids has not been forthcoming, but there are many who believe that fats containing a high proportion of polyunsaturated fatty acids (e.g. soyabean oil, fish oils, soft margarine) can counteract the cholesterol-depositing tendencies of opulent Western diets. The development of atherosclerosis is associated with high plasma concentrations of cholesterol and triglycerides of saturated fatty acids.

Carbohydrates

These furnish more than 50 per cent of the energy content of most diets, and are a cheap and readily obtained food. If the amount of carbohydrate ingested is greatly reduced in amount ketosis may develop. As both carbohydrate and fat serve chiefly as sources of energy, they can replace one another to a considerable extent, so long as precautions are taken to ensure the minimum amounts of fat specified above and that the change over is not too abrupt. The carbohydrate intake should be sufficient to prevent the need for protein breakdown to provide energy. In adults the daily intake of carbohydrate is commonly in the range 300–500 g (5250–8400 kJ).

Dietary fibre

In addition to protein, fat, carbohydrate, minerals, ions, and vitamins the diet contains unabsorbable plant material called *dietary fibre* which resists digestion in the human gastro-intestinal tract. Dietary fibre consists of a heterogeneous group of polymeric carbohydrate compounds including *celluloses, hemicelluloses,* and *pectins* plus the non-carbohydrate substance *lignin* which is a polymer based on phenylpropane units. Food tables give values for *crude* fibre in the diet: this is the residue of a food after it has been treated with boiling H_2SO_4, NaOH, water, ethanol, and ether. Crude fibre consists of cellulose + lignin and comprises less than 50 per cent of total dietary fibre. Chemical measurement of hemicelluloses and pectins is difficult.

In the UK the total daily dietary fibre intake is 16–28 g, of which 13 per cent occurs in bread and cereals and the rest in vegetables, fruit, and nuts. A vegetarian diet provides about twice this amount of dietary fibre. The chief importance of dietary fibre is that it increases the bulk of the colonic contents partly by its own volume and partly by uptake of water and this in turn stimulates bowel movements so that defecation is facilitated. In rural Africans (Bantus) on an unrefined diet the 'transit time' of a swallowed marker material was 30–36 hours, whereas in UK subjects on a refined diet the 'transit time' was about 80 hours. The unrefined, high-fibre diet also produced more fluid stools, weighing 150–350 g, contrasted with the small hard stools, weighing about 100 g daily in the subjects on a refined diet (Burkitt, Walker, and Painter 1972).

In communities living on a high-residue diet, constipation and colonic diverticulosis do not occur, whereas both are seen frequently in older people taking a low-residue diet. The addition of dietary fibre (e.g. wheat bran) to a low-residue diet relieves constipation and symptoms of diverticulosis.

The colon contains bacteria which may have two effects:

1. They act on dietary fibre to form volatile fatty acids (acetic, propionic, butyric) which are highly ionized at colonic pH and are therefore poorly absorbed. They may act as osmotic or irritant purgatives.

2. Anaerobic colonic clostridia degrade bile salts to carcinogenic substances which might, acting over a long period of time, lead to cancer of the colon. High dietary-fibre intake would dilute the degradation products of bile salts and reduce their formation by hastening the passage of the colonic contents.

The role of dietary fibre needs further prolonged study.

REFERENCES

BURKITT, D. P., WALKER, A. R. P., and PAINTER, N. S. (1972). Dietary fibre and disease. *Br. med. J.* iv, 1408.
CUMMINGS, J. H. (1973). Dietary fibre. (Review.) *Gut* **14**, 69.
EASTWOOD, M. A., FISHER, N., GREENWOOD, C. T., and HUTCHISON, J. B. (1974). Perspective on the bran hypothesis. *Lancet* i, 1029.
PAINTER, N. S. and BURKITT, D. P. (1971). Diverticular disease of the colon. *Br. med. J.* **ii**, 450.

Mineral ions

The mineral constituents of the human body amount to 4.3–4.4 per cent, largely in the skeleton.

1. The only salt commonly consumed as such is sodium chloride; it is also present in small amounts in the food, particularly in milk and in vegetables. The minimum requirements are 1–2 g of NaCl daily; while custom varies considerably, the average intake is about 8–10 g. The effects of NaCl deficiency are considered on page 241.

2. 0.9–1 g of calcium is needed daily; the minimum necessary for the maintenance of a calcium balance is 0.63 g of calcium per 70 kg of body weight. A sufficient calcium supply is very important, especially in children, and is best obtained by the ingestion of liberal quantities of milk and 'fortified' bread.

3. 0.88 g of phosphate (measured as P) is the minimum needed per 70 kg body weight, but there is no risk of phosphate shortage in a diet yielding 12 500 kJ daily unless large amounts of white flour are consumed as the principal cereal.

The sources, function, and metabolism of Ca^{2+} and PO_4^{3-}, and their control by vitamin D and parathyroid secretion, are described on page 546 *et seq.*

4. Sulphur. All proteins contain sulphur in the amino acids methionine, cysteine, and cystine which are the chief dietary source of sulphur. Free sulphydryl (SH) groups account for the activities of coenzyme A and glutathione sulphate occurs in heparin and chondroitin sulphate.

5. Magnesium. A typical British diet contains 200–400 mg of magnesium daily, mainly in cereals and vegetables. Deficiency due to dietary lack does not occur, but may result from excessive loss of magnesium in chronic diarrhoea.

6. The daily intake of iron should not be less than 12 mg, an amount which should should be increased in pregnancy and in lactation [cf. p. 588]. The role of copper is considered on page 43.

7. Iodine in minute traces is essential for the formation of thyroxine the active principle of the thyroid gland [p. 537]. In districts in which the iodine supplies in the drinking water are insufficient, simple goitre frequently develops. This may be prevented, or the condition may be cured, by the deliberate addition of iodine to the public water supply, or preferably by the use of iodized table salt containing added iodate. Many countries now require all domestic salt to be iodized, with personal liberty (or fad) safeguarded by the existence of a non-iodized salt on special order. This compulsory scheme has been recommended for Britain but never adopted.

8. Fluorine. Most human adults ingest 2–3 mg of fluorine daily. The chief source is drinking water, especially if hard. Sea fish and China tea are other sources. Fluoride becomes concentrated in bone and in developing dental enamel. When the water supply contains fluorine in a concentration of 1 part per million the incidence of dental caries in children is significantly reduced. Fluoride may act by reducing the solubility of tooth minerals or by discouraging the growth of acid-forming bacteria. At 3–5 parts per million in water fluoride may produce mottling of teeth and at 10 parts per million chronic poisoning may occur.

There is no doubt that at 1 part per million fluoride safely reduces the incidence of dental decay in children (Royal College of Physicians 1976). Artificial fluoridation of water supplies is, however, strongly opposed by many people, largely on the ground that it interferes with the basic right of any person to choose what he consumes. The introduction of fluoride in toothpaste may largely resolve the problem.

Trace elements include cobalt which is a constituent of vitamin B_{12}, zinc which is present in insulin and in many enzymes, copper which occurs in blood combined with an α-globulin, forming the protein caeruloplasmin, manganese which is concerned in some enzyme systems, selenium which can substitute for vitamin E in certain animal species, and molybdenum which is a component of xanthine oxidase. All these trace elements occur in adequate quantities in human diets and there are no known deficiency diseases attributable to their lack.

REFERENCES

ROYAL COLLEGE OF PHYSICIANS (1976). *Fluoride, teeth and health*. London.

Vitamins

The detailed distribution and function of the vitamins is given in the special sections devoted to them [pp. 488–96 and the References therein].

For practical purposes the following generalizations serve as an adequate guide. Fresh fruit juices and green vegetables ensure an adequate supply of vitamin C. High (85 per cent) extraction flour goes a long way to providing all the necessary vitamin B complex. A plentiful supply of animal fats (milk and milk products, meat fat, and vitaminized margarine) and eggs gives all the necessary vitamin A and D for adults; green vegetables and carrots are important sources of A; in the case of children, cod-liver oil or halibut-liver oil should be added as a supplementary source of these vitamins. Vitamin E or K requirements need not be considered in preparing dietaries.

For the role of vitamin B_{12} in blood formation see pages 39–40.

NUTRITIONAL REQUIREMENTS OF SPECIAL GROUPS

Although all human beings have common nutritional needs, there are variations from one section of the community to another. Requirements change from infancy through childhood to adolescence and adulthood. The needs of the pregnant woman ('feeding two') are not the same as those of a non-pregnant woman. The concept is of the 'physiological group', each group needing special consideration. To the group implied above must be added two others—the aging and aged, and the ill and convalescent.

Pregnancy, lactation, infancy

The body of a baby (weighing 3.5 kg at birth) contains about 500 g of protein, 30 g of Ca, 14 g of P, and 0.4 g of Fe; over two-thirds of the protein is laid down in the last 3 months, and one-third in the last month of pregnancy. In addition the mother lays down new protein in the growing uterus, breasts, and other tissues (perhaps another 500 g). The basal metabolism of the mother at the end of pregnancy has increased by 25 per cent or by about 1500 kJ daily. After birth the baby doubles in weight from 3.5 to 7.0 kg at 3–4 months and increases to 10 kg at one year; each 3.5 kg represents an amount of new tissue equal to that formed throughout pregnancy. 100 ml of human milk contain 1.2 g of protein and 0.03 g of calcium, and yield 225 kJ. At one month after delivery the mother must supply the baby with 7 g of protein, 0.18 g of Ca, and 2400 kJ daily;

at 6 months these amounts have risen to 9.5 g of protein, 0.25 g of Ca and 3700 kJ daily. The milk proteins (lactalbumin and caseinogen) have a specific amino-acid constitution and can only be formed if all the necessary amino acids are supplied in the food in sufficient amounts; the same applies to the new tissues formed by mother and child during pregnancy. It is clear that the pregnant and lactating woman needs a substantially increased protein intake, a considerable part of which should be of animal origin. The recommendations are: for the last 3 months of pregnancy a minimum of 1.5 g of protein per kg (i.e. an extra 30 g) daily; during lactation the allowance should be higher, up to 2 g of protein per kg. The total energy requirements are also increased to the extent indicated.

Calcium requirements are discussed on page 585, vitamin D on page 553, and iron on page 585.

Great stress must be placed on the importance of milk as a 'protective food' especially for children and expectant or nursing mothers, who should drink at least 0.5 litre daily.

The vitamin content of mother's milk (and cow's milk) depends on the maternal diet. It is advisable to supplement both the vitamin C and D intake of the baby and the D intake of the mother. The composition of milk is given on page 598.

The following are recommended for a diet during pregnancy: milk, 0.5–1 litre; green vegetables and an egg once or twice daily; sea fish (for iodine content) every other day; calf's liver once weekly; the welfare food supplements [see below]. The rest of the diet can be, within reason, whatever the woman likes. It should never be forgotten that a pregnant woman has to provide in her diet for the full requirements of the growing fetus. The future of the child depends to a great extent on the adequacy of the maternal diet during pregnancy and lactation.

Welfare foods. The present British scheme for 'welfare foods' offers every expectant mother and every child sufficient supplementation to ensure adequate nutrition, and all virtually free (at least to the individual, but not to the Welfare State, though the results represent a good bargain). The supplements offered are:

orange juice (vitamin C) to expectant mothers and children under 5 years;
cod-liver oil (vitamin D) to children under 5;
tablets containing vitamins A and D, and Fe to expectant and nursing mothers;
cheap dried milk (plus vitamin D) to children under 2;
cheap liquid milk to children under 5.

The aged

Elderly people, after retirement at the customary age of 65 (in the United Kingdom at any rate) become more liable to certain diseases, including infections of the respiratory tract, cardiovascular disorders, and malignant disease. If, however, they remain in good health their dietary requirements differ little from those of the middle-aged. Senescence is an irreversible process, but so long as physical activity is undertaken the energy requirements amount to 8000–9000 kJ with an intake of say 50–55 g of protein, 75 g of fat, and 300 g of carbohydrate, plus sufficient dietary fibre to prevent constipation, diverticular disease, etc. It is particularly important to provide sufficient iron and folate to prevent anaemia, vitamin C to prevent scurvy, and vitamin D to prevent osteomalacia. The cost of a balanced diet may be beyond an old person's means.

The ill

Published tables of nutritional requirements always refer to those who are well. The needs of the sick may differ considerably from those of the healthy; they certainly vary far more widely. For example, surgical patients (especially burn and fracture cases in the post-catabolic phase) have enormous calls for protein, and it is reasonable to suppose that the required proportions of the various essential and non-essential amino acids may differ from those needed for growth in an experimental rat. Unfortunately, appetite is often poorest when the need is greatest.

The nutritional need of those with specific diseases is an important consideration in their medical care, e.g. diabetics.

Appetite and satiety

Appetite, or the relish for food, largely determines the amount of food we eat and hence helps to regulate body weight. In normal persons the appetite equates the energy value of the food eaten and its utilization with remarkable precision. The body weight remains constant over long periods of time in spite of variations in food intake.

In animals feeding is controlled by the activities of the ventromedial satiety centre and the lateral feeding centre of the hypothalamus [p. 344]. Hunger is stimulated by a fall in blood glucose level, a rise in blood free-fatty-acid level, and by a fall in environmental temperature. These factors could promote appetite and initiate feeding in man but customs and habits are more important than hunger in the timing of modern man's mealtimes. These stimuli to eating constitute a quick-acting response to the need for food but other factors are involved in the accurate regulation of body weight over periods of weeks, months, or years. The body normally keeps its content of adipose tissue constant (by mechanisms which are not at all clear) and obesity is due to a failure of this homeostatic process.

Obesity

Although genetic and endocrine factors are contributory, the fundamental cause of obesity is a continued excess of energy intake over output. In young physically active people an increase in energy intake can be compensated for by increased energy output so that body weight remains constant. In most people, as they grow older, the response is less efficient and adipose tissue is deposited in increasing amounts to give the characteristic middle-age spread. It has been observed that overfeeding in early childhood, especially when cow's-milk preparations are bottle-fed, increases the number of adipocytes in fat depots. This not only produces fat children but provides the basis for development of obesity later in life. Adipocytes also multiply at puberty. Obesity predisposes to cardiovascular disease and diabetes, and shortens life.

Obesity in man is commonly associated with the sedentary life and it has been clearly established that obese persons take much less exercise than their non-obese contemporaries. Regular mild or moderate exercise is effective in helping to reduce weight. Energy intake is best cut down by reducing carbohydrate in the diet to 50–100 g a day, by restrictions on sugar, sweets, chocolates, and cakes.

REFERENCES

DAVIDSON, S., PASSMORE, R., BROCK, J. F., and TRUSWELL, R. S. (1975). *Human nutrition and dietetics*, 3rd edn. Churchill Livingstone, Edinburgh.

DRUMMOND, J. C. and WILBRAHAM, A. (1958). *The Englishman's food*. London.

KEYS, A., BROZEK, J., HENSCHEL, A., MICKELSEN, O., and TAYLOR, H. L. (1950). *The biology of human starvation*. Minneapolis.

KON, S. K. and COWIE, A. T. (eds.) (1961). *Milk*. Academic Press, London.

The Lancet (1974). Infant and adult obesity. (Editorial.) *Lancet* i, 17.

McCANCE, R. A. (1962). Food, growth and time. *Lancet* ii, 621.

MAYER, J. (1964). Regulation of food intake. In *Nutrition* (ed. G. H. Beaton and E. W. McHenry) Vol. 1, p. 1. Academic Press, London.

The vitamins

The term 'vitamin' has undergone important changes in meaning since it was first introduced; as a result it carries with it some imprints of all its different meanings. Forty years ago it was believed that the essential constituents of a diet were: protein (in amounts sufficient to maintain nitrogenous equilibrium in the adult and growth in the young), carbohydrate, and fat (these three foodstuffs together must be present in sufficient amounts to yield the full calorie requirements), certain minerals (inorganic ions), and water. But later, when a chemically pure diet of this kind had been prepared and administered, the animals died; natural food, therefore, contains other, non-calorie-providing, but nevertheless essential, constituents for growth, health, and life. In rats, addition of small amounts of milk to a diet which according to the theories then current was adequate (but actually lethal) preserved health and restored growth; the unknown essential factors in milk were called '*accessory food factors*' by Hopkins. In the meantime studies showed that certain other theoretically adequate diets, of which polished rice was the main constituent, produced beriberi; addition of rice 'polishings' to the diet cured the disease. The active principle in the rice polishings was called a vitamine (i.e. an amine essential to life). When it was discovered that few of the 'vitamins' were in fact amines, the word was respelt '*vitamin*'. The term 'vitamin' gradually came to be defined as a substance of unknown chemical composition, probably a complex organic compound, which must be present in the food in minute amounts to enable growth, health, and life to be maintained. The accessory food factors or vitamins were soon divided into:

1. Fat-soluble, i.e. those present in fats and soluble in fat-solvents.
2. Water-soluble.

The fat-soluble were differentiated into vitamins A and D: the water-soluble into B and C. It was soon found that vitamin B was not a single substance but a mixture of several substances; its title was altered to vitamin-B *complex*, and the individual constituents, as they were isolated, were given distinctive names. Rapidly the chemical identity of the vitamins was worked out and many such substances are now known. They are:

Vitamin A
Vitamin D } All fat-soluble
Vitamin E
Vitamin K

Vitamin B complex, i.e.
Thiamine
Nicotinic Acid
Riboflavin
Pyridoxine
Biotin   } All water-soluble
Pantothenic Acid
Folic Acid
Vitamin B_{12}

Vitamin C

The distinctive characteristic of the vitamins is that they are micro-constituents of the diet of high biological activity which cannot be replaced by other normal dietary constituents. They appear to be indispensable for certain specific functions of the body; in most cases their mode of action has been fairly clearly defined at a biochemical level; but in no instance is it possible, in higher animals, to correlate a symptom of deficiency with the diminished function of an enzyme system for which that vitamin is essential. Once an essential portion of the metabolic machinery fails, many possibilities exist for deficiency symptoms which may be far removed from the original biochemical defect. Vitamins do not contribute to the energy of the body; in many cases they mobilize energy.

Sometimes an indispensable vitamin for one species can be synthesized by another species for which the substance in question is then not a vitamin. Thus ascorbic acid is an indispensable vitamin in man; the rat, however, can manage without dietary ascorbic acid. *Both* species need and use ascorbic acid in their tissues, but while man can only get ascorbic acid by taking it in as part of his food, the rat can synthesize it. Another example: the bacteria in the intestine synthesize several members of the vitamin-B group; these vitamins are thus supplied to the body partly by the diet and partly by the intestinal bacteria. If the bacteria are killed off by antibiotics, signs of vitamin-B deficiency may develop because the supplies of dietary vitamin are in themselves inadequate. The generalization is as follows: indispensable chemical compounds or groupings that can be synthesized in the body (the body includes its co-operative intestinal bacteria) need not be provided as such in the food; indispensable chemical compounds or groupings that cannot be synthesized in the body must be provided as such in the food: those organic compounds required only in small amount constitute the vitamins.

Deficiency of a vitamin can arise in two ways: *primary* deficiency, due to inadequate intake of vitamin or its precursor over a prolonged period of time (which may be months or years if previous intake and storage have been satisfactory); or *conditioned* deficiency arising on an adequate diet through other factors which diminish absorption, or prevent release, or increase utilization or excretion.

Vitamin A, vitamins of the B group, and vitamin C are dealt with below. The other vitamins are:

vitamin D (the anti-rachitic vitamin), discussed with parathyroid, page 553;

vitamin E (the antisterility vitamin) occurs in many foodstuffs, particularly wheat embryo. The active agents are fat-soluble tocopherols. The role of vitamin E in man is still uncertain;

vitamin K (the anti-haemorrhagic vitamin), discussed with blood clotting, page 27;

REFERENCES

HARRIS, L. J. (1955). *Vitamins in theory and practice*, 4th edn. London.
MARKS, J. (1968). *The vitamins in health and disease*. Churchill, London.
RECENT RESEARCH ON VITAMINS (1956). *Br. med. Bull.* **12**, No. 1.

Vitamin A (retinol)

All vertebrates require a dietary source of vitamin A or a precursor for the maintenance of vision, epithelia, and skeletal growth.

Chemistry

The vitamin has the formula in FIGURE X.1; a closely-related substance is the hydrocarbon β-*carotene* ($C_{40}H_{56}$) which is readily transformed in the cells of the intestinal wall into vitamin A ($C_{20}H_{30}O$, an alcohol with a terminal $-CH_2OH$ grouping). β-Carotene is converted to two molecules of retinol, but owing to its poor absorption from the alimentary canal it is less active, weight for weight, than retinol.

Sources

Vitamin A is found as such in the fat of milk and therefore in milk products like butter or cream; in eggs; in very large amounts in liver fat, especially in cod-liver oil and in greatest concentration in halibut-liver oil. The vitamin is absent from purely vegetable fats like linseed oil, olive oil, or coconut oil, and consequently from purely vegetable margarines. All margarine in Great Britain, however, has since the last war been reinforced with added vitamin A (and D). Green vegetables and carrots are free from vitamin A but contain variable amounts of carotene. Both the vitamin and carotene are stable and can withstand the ordinary processes of cooking and boiling.

The daily dietary requirement is 300–400 µg of retinol equivalent in children and 750 µg in adults.

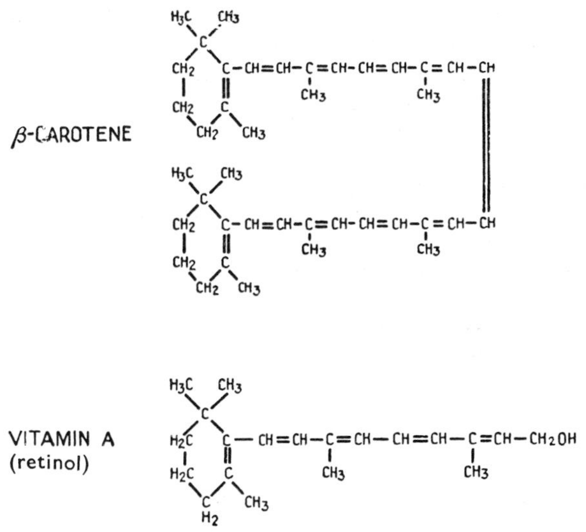

FIG. X.1. Structure of β-carotene and vitamin A.

Absorption

Vitamin A and carotene are absorbed from the small intestine together with fat; the fat acts as a 'carrier' for the fat-soluble vitamin A and fat-soluble carotene.

Transport

Retinol is carried from the gut as retinyl palmitate to the liver where much of it is stored as retinol. Some retinol enters the systemic circulation to become bound to a specific retinol-binding plasma protein.

Relation of vitamin A to vision

See page 381.

Results of vitamin A deficiency

Animal experiments. Vitamin A lack produces striking and characteristic pathological changes in experimental animals. Young rats fed on a diet complete in all respects except for the absence of vitamin A cease to grow, lose weight, and finally die. It should be stressed that lack of any one of the vitamins arrests growth in young animals. Fully grown animals deprived of vitamin A may survive for months, but finally succumb to some intercurrent infection.

Striking pathological changes are found experimentally in the eyes, intestine, and respiratory tract. The outstanding histological feature is the tendency for stratified epithelia to become greatly thickened and for columnar epithelia to be transformed into transitional or stratified epithelia.

1. *Eyes* The lacrimal glands cease to produce tears. The corneal epithelium becomes thickened, dry, and wrinkled and secondarily may undergo necrosis or become infected. Inflammatory processes occur in the conjunctiva and may involve the anterior and posterior chambers of the eye, leading to complete blindness. The eye disorder is referred to as *xerophthalmia*, and is a specific sign of experimental vitamin A deficiency.

2. *Alimentary canal.* The cells of the salivary glands do not secrete, and appear shrunken; the epithelium of the ducts proliferates and the lumen is partially occluded. The mucus-secreting cells of the intestine are atrophied and the tips of the villi are necrosed; masses of bacteria may be found filling the lumina of the glands.

3. The *upper respiratory tract*, particularly the nasal passages, trachea, and bronchi, shows a transformation of the lining cells into a stratified epithelium of flattened cells which undergo extensive keratinization. (Similar changes occur in the vagina and gums.)

4. *Resistance to infection.* As a result of the structural changes in many epithelia and other tissues their local resistance to infection is reduced. From these sites, organisms may pass into the blood stream; bronchopneumonia, enteritis, and inflammation of the eyes may develop and prove fatal. In mice fed on a diet deficient in vitamin A, the introduction of suitably measured doses of mouse-typhoid bacilli may give a mortality rate of 80–100 per cent. When the experiment is repeated on mice fed on the same diet but with the addition of large amounts of vitamin A, the mortality rate is only 10–20 per cent though in all external appearances the two batches of mice may appear equally normal.

5. *Bone.* Vitamin A deficiency in dogs leads to marked overgrowth of certain bones, especially the skull and vertebral column. Owing to the resulting compression of the spinal roots and cranial nerves complex nervous symptoms develop. Vitamin A may be a factor regulating normal bone growth.

Studies in man

Experimental pure vitamin A deficiency. Twenty-three volunteers were kept for up to two years on a diet essentially free from vitamin A or carotene, but the only manifestation was a slight impairment of dark adaptation in three subjects. The very different results in animals and the clinical findings to be described below may be due in part to species differences, the duration of the deficiency and the presence of other conditioning factors, e.g. multiple deficiencies or infections.

Clinical findings. 1. *Xerophthalmia.* The association of this condition in man with vitamin-A deficiency is firmly established. In Denmark, during the First World War, the fats rich in vitamin A were exported, and the populace lived largely on margarine and skimmed milk; several outbreaks of xerophthalmia and bronchopneumonia were reported. Fifteen hundred cases of xerophthalmia occurred in Japanese children who were fed on inadequate diets; the condition was cured by the administration of cod-liver oil.

2. *Skin changes.* Changes in the skin often occur in vitamin A deficiency in man. Thus, in many gaols, asylums, hospitals, and schools in Africa, China, and Sri Lanka the inmates frequently suffer from dryness of the skin and papular eruptions due to destruction of the ducts of the sweat glands and the hair follicles; the condition is called 'toad skin' in Sri Lanka. It usually precedes the development of xerophthalmia, and like the latter condition it is rapidly cured by means of cod-liver oil.

3. *Infections.* There is no clear evidence that vitamin A lack in man lowers resistance to the common infections; excess vitamin A does not raise general immunity above normal.

4. In patients with chronic small intestinal disease impairment of dark-adaptation can occur (Russell *et al.* 1973). Large doses of retinol restore the normal state.

Hypervitaminosis A

Vitamin A shows toxic effects when provided in enormous excess over the normal requirement, and the overzealous or careless administration of high-potency preparations can cause hypervitaminosis A—anorexia, painful swellings over long bones, sparsity of

hair, pruritic rash—disappearing within a week of discontinuing the vitamin supplementation. In rats, hypervitaminosis A during pregnancy has a potent 'teratogenic' effect, causing skeletal deformation of the fetus.

REFERENCES

HAYES, K. C. (1971). *Nutr. Rev.* **29**, 3.
MORTON, R. A. (1957). Vitamin A. In *Lectures on the scientific basis of medicine*, Vol. 5, p. 143. London.
—— (1964). *Proc. Roy. Soc.* **B159**, 510.
RUSSELL, R. M. *et al.* (1973). *Lancet* **ii**, 1161.

Vitamin B group

This group consists of a series of water-soluble organic substances which are found in all cells of all species, from the bacteria, protozoa, and yeasts up to the highest mammalian forms. Most of the members of the group are constituents of fundamental tissue enzyme systems, involved in the oxidation of the foodstuffs, and are therefore indispensable for the normal functioning of all tissues. The best studied members of the group are thiamine, riboflavin, and nicotinic acid, generally found together in foodstuffs but not necessarily in the same proportions. Most members of the vitamin B group can be synthesized by the intestinal bacteria.

Thiamine (aneurin, vitamin B₁)

The molecule (shown in FIG. X.2 as the hydrochloride) contains a pyrimidine and a thiazole (sulphur-containing) ring system. The International Unit is equivalent to 3 μg of thiamine.

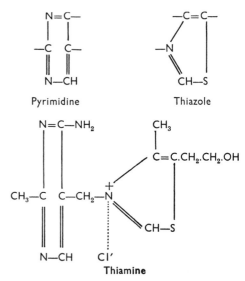

FIG. X.2. Structure of thiamine (vitamin B₁).

Sources. The best sources are cereals, pulses, and yeast. In the cereals thiamine is found mainly in the germ and bran. Wholemeal flour prepared from the entire grain, or flour of a high degree of extraction contains much more thiamine than fine white flour; likewise, so-called polished rice (i.e. rice from which the husk has been removed) is deficient in thiamine. The vitamin is uniformly distributed in the pulses (peas, beans, lentils). It is present in small amounts in meat, milk, and vegetables. The synthetic vitamin is available commercially for the enrichment of white flour.

Requirements. The amount of thiamine needed by the tissues is directly related to body weight, the metabolic rate, and the level of physical activity, and depends on the composition of the diet. As thiamine is specifically employed in carbohydrate utilization the amount required is directly related to the amount of glucose which the diet yields. The daily dietary requirement is 0.5–1.0 mg in children and 1.0–1.5 mg in adults. When the metabolic rate is raised (e.g. in febrile states and in people doing hard manual work), a larger intake is needed. It must always be borne in mind, however, that thiamine is synthesized in unknown and variable amounts by the intestinal bacteria. Signs of thiamine lack can be precipitated on what appears to be an adequate diet by administering antibiotics which kill off the intestinal bacteria. On the other hand, certain bacteria can destroy dietary thiamine.

Role in tissues. In the tissues thiamine is found in the phosphorylated form as diphosphothiamine, which is a coenzyme for reactions which involve decarboxylation of acids. Such decarboxylations (and the reverse process called CO_2 fixation) play an important part in the dissimilation of the metabolites which enter the common metabolic pool. The reactions in which diphosphothiamine is involved are:

1. Pyruvic acid $\leftrightarrows$ Acetyl-Co A + CO_2 [p. 453].
2. Oxaloacetic acid $\leftrightarrows$ Pyruvic acid + CO_2 ['carboxylase' reaction, p. 454].
3. Citric acid $\leftrightarrows$ α-Ketoglutaric acid + CO_2 [Fig. IX.8, p. 454].
4. α-Ketoglutaric acid $\leftrightarrows$ Succinic acid + CO_2.

All these reactions are stages in carbohydrate utilization. It will be recalled that the brain and probably all nervous tissues use blood glucose as their primary source of energy. It is not surprising to find that carbohydrate metabolism (especially in the brain) is deranged in thiamine deficiency.

In thiamine-deficient pigeons pyruvic acid accumulates in the brain, especially in the brain stem, because it cannot be disposed of; the concentration of pyruvic acid also increases in the blood and cerebrospinal fluid. If thiamine is added to a brain slice from such a bird the pyruvate rapidly disappears.

Thiamine deficiency in man

Experimental severe deficiency. The effects of pure thiamine deficiency have been studied in a few women. The extremely unattractive diet that had to be employed consisted of white flour, sugar, tapioca, starch, cheese, butter, hydrogenated fat, tea, and cocoa, with added vitamin C, halibut-liver oil, iron, calcium, and autoclaved brewer's yeast (to supply other members of the B group); the daily thiamine intake was 0.15 mg. After two weeks the first symptoms appeared, consisting of fatigue and loss of appetite; during the second month there was decreased activity, apathy, nausea, decreased food intake, and a fall of blood pressure. During the third month the consumption of food was further decreased (the energy intake was then under 6300 kJ) and the body weight fell slightly. There was dizziness, the heart rate was slow at rest but excessively rapid on exertion, the voltages of the electrocardiogram, especially the T wave, were decreased; uncontrollable vomiting occurred; the motility of the stomach and large intestine was diminished. There was a greatly decreased capacity for work, weakness in the legs, and decreased reflexes. Some of the subjects became very depressed and showed other mental changes; they complained of soreness of the muscles and numbness of the legs, muscular flaccidity, precordial distress, and dyspnoea.

It will be noted that signs of organic changes in the nervous system (anaesthesia, loss of reflexes, and paralysis in legs) made their appearance late; they resemble the neuritic changes found in beriberi. Attention must be drawn to the absence of oedema, cardiac dilatation, or changes in the skin and tongue. As might be expected, the rapidity of development of the symptoms varied to some extent with the degree of muscular activity. The terminal state, especially the nausea, vomiting, and mental changes, resembles the findings in cerebral beriberi.

The symptoms described can probably be attributed solely to thiamine deficiency. The addition of thiamine, without any other change in the diet, produced improvement which was sometimes sudden, but commonly (if organic changes had developed) very gradual.

Biochemical changes. On a normal thiamine intake, the blood thiamine level is 30–90 μg l^{-1}; one-third of any injected thiamine is excreted in the urine. After ingestion or intravenous injection of glucose there is only a very slight rise in the level of blood pyruvic acid (normal 5–10 mg l^{-1}). In severe thiamine deficiency, the blood thiamine level fell to under 30 μg l^{-1}; the excretion of thiamine in the urine progressively declined to zero after the 10th day. In thiamine deficiency carbohydrate metabolism in part stops short at the pyruvic and lactic acid stages; if glucose (0.4 g kg^{-1}) is injected intravenously the blood pyruvic level rises temporarily to an abnormally high level (up to 20–30 mg l^{-1}), as does the blood lactic acid.

Beriberi, the main clinical syndrome of thiamine deficiency is discussed on page 493.

Nicotinic acid amide (nicotinamide, niacin)

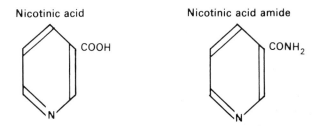

Nicotinic acid   Nicotinic acid amide

Sources and requirements. The richest sources are liver (150 mg kg^{-1}), kidney, and yeast; meat is a valuable source containing 50 mg kg^{-1}: the other important sources are wholemeal flour and green vegetables; milk contains very little, and of the cereals maize contains the least available nicotinic acid. The daily need is about 10–15 mg but some of this can be replaced by excess dietary tryptophane which can be converted metabolically into the vitamin.

Role in tissues. Nicotinamide (which is a pyridine derivative) is found in the tissues as a nucleotide, usually called NAD, formed by the combination of adenine, ribose, phosphate, and nicotinamide [Fig. IX.2, p. 448]. There is also a corresponding NADP.

NAD and NADP are the coenzymes for oxidative enzymes (dehydrogenases); the coenzymes act as intermediate carriers for the hydrogen (2H) released from various substances by the dehydrogenase enzymes [p. 447].

NAD is the coenzyme for the following reactions:

1. Lactic acid $\rightarrow$ Pyruvic acid + 2H.
2. Pyruvic acid + H_2O $\rightarrow$ Acetyl-Co A + CO_2 + 2H (both an oxidation and decarboxylation).
3. β-Hydroxybutyric acid $\rightarrow$ Acetoacetic acid + 2H.

NADP is the coenzyme for the following reactions, among others:

4. Citric acid $\rightarrow$ α-Ketoglutaric acid + CO_2 + 2H [p. 000].
5. Glutamic acid + H_2O $\rightarrow$ α-Ketoglutaric acid + NH_3 + 2H (both an oxidation and deamination).

All the reactions are shown as a loss of 2H from the substrate on the left, the 2H being passed on to the coenzyme, which is thus converted to the reduced form:

$$NAD + 2H \longrightarrow NAD.2H$$

The same enzyme and coenzyme systems are required by the

reactions in reverse, i.e. when reduction of the substrate is taking place and the reduced coenzyme is being reoxidized, e.g.:

$$NAD.2H \longrightarrow NAD + 2H$$
$$2H + Pyruvic\ acid \longrightarrow Lactic\ acid.$$

The oxidation of NAD.2H by oxygen via the electron-transporting cytochromes [p. 448] constitutes a general mechanism for the generation of ATP.

The nicotinamide-containing coenzymes are thus concerned with many of the important energy-producing reactions of metabolism. Nicotinamide (or nicotinic acid) deficiency leads to metabolic disturbances in many tissues; as might be anticipated, the nervous system is gravely involved.

Pellagra, the main clinical syndrome of nicotinamide deficiency, is discussed on page 493.

Riboflavin

Sources and requirements. The main sources are again meat, milk, and wholemeal flour. Loss of riboflavin from food due to exposure to light is more serious than that due to heat; a loss of 85 per cent from milk standing for two hours in sunlight (on the doorstep) should be contrasted with no loss during pasteurization. Clinical studies show that an intake of 0.35–0.5 mg per 4200 total kJ is adequate. On an average diet, a total riboflavin intake of 2 mg daily is regarded as satisfactory, with more for the stress of growth. Riboflavin is synthesized by intestinal bacteria.

Structure. Riboflavin is an orange pigment consisting of a three-ring system (iso-alloxazine) combined with an alcohol derived from ribose (ribitol) [FIG. X.3].

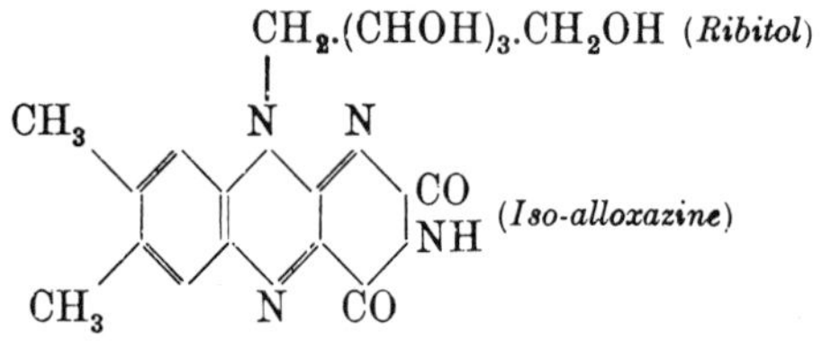

FIG. X.3. Structure of riboflavin.

Role in tissues. In the tissues riboflavin is found as a dinucleotide.

Iso-alloxazine   Adenine
| |
Ribitol—Phosphate—Phosphate—Ribose.

This substance is called flavin-adenine-dinucleotide (FADN), or, for brevity, flavin.

Flavin is the prosthetic group of a number of dehydrogenase enzymes (the flavoproteins) [p. 448]. These enzymes catalyse the release of 2H from various substrates, and their flavin group acts as a temporary hydrogen-carrier. The hydrogen is passed from the flavin to the cytochromes and thence finally to reactions with molecular oxygen to give water. In particular, those oxidation reactions requiring NAD or NADP as initial hydrogen-carriers require flavoproteins as the next hydrogen-carrier of the chain leading to molecular oxygen.

As flavin occupies a key position in reactions leading to the oxidation of hydrogen to water, one would expect that dietary deficiency of riboflavin would lead to dramatic disturbances of function. The actual changes observed clinically are far less striking than might have been anticipated.

Certain other members of the vitamin-B group should be noted, though their role in man is still obscure.

Pyridoxine (vitamin B₆)

Its formula is shown below:

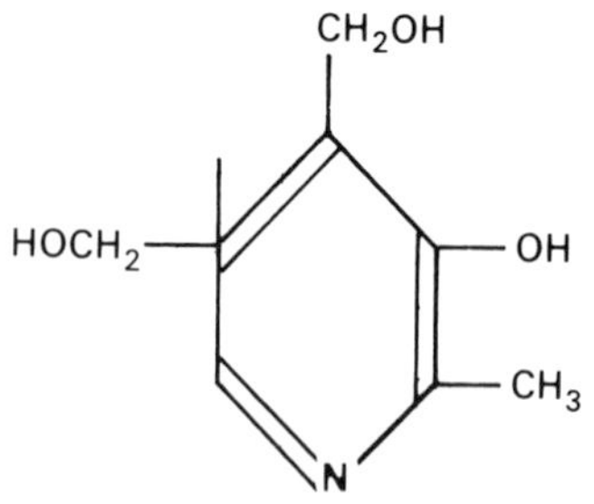

Pyridoxine and its derivatives and their phosphates (the vitamin-B_6 group) act as coenzymes for many of the metabolic reactions of amino acids including transamination reactions [p. 468].

Little is known about the effects of pyridoxine insufficiency in man.

Pantothenic acid

It is concerned (as a dinucleotide, referred to as coenzyme A) with reactions involving the active form of organic acids [p. 447].

Folic acid (pteroyl glutamic acid)

Folic acid is required for the metabolic transference of the 1 C unit of formic acid during biological reactions (e.g. the interconversion of glycine and serine, and the synthesis of the purine ring).

Deficiency in man is expressed clinically in altered haemopoiesis [p. 41]. The deficiency is usually due to non-utilization of naturally-occurring conjugated forms rather than to deficiency of dietary intake.

Cyanocobalamin (vitamin B₁₂)

See page 39 for the relation of this cobalt-containing vitamin to haem synthesis and anaemic conditions.

Clinical vitamin-B-complex deficiencies

As the members of the vitamin-B group are generally found together in foods, isolated deficiency of any single member of the group rarely occurs clinically, though it has been produced experimentally. In the common clinical condition of multiple deficiency the signs of some one single deficiency may predominate and mask the results of the other deficiencies. If the outstanding deficiency is made good the latent deficiencies may become manifest in the appearance of new symptoms.

The principal clinical syndromes resulting from lack of the vitamin-B complex are summarized below:

Beriberi: due primarily to thiamine lack. Three main clinical types are recognized: 1. neuritic (dry); 2. cardiac (wet); 3. cerebral (Wernicke's encephalopathy).

Pellagra: due mainly to lack of nicotinic acid; a high-maize diet is often an aggravating factor.

Ariboflavinosis: the chief changes attributed to riboflavin lack are angular stomatitis, glossitis, scrotal dermatitis, keratitis, defective vision from retrobulbar neuritis, and painful feet.

Mixed syndromes of B₂ Avitaminosis: the term vitamin B_2 is often used to refer to the members of the vitamin-B complex other than thiamine (= vitamin B_1). The varied clinical manifestations are due mainly to lack of nicotinic acid and of riboflavin.

It is interesting to note that when a large population is exposed to a multiple avitaminosis some develop beriberi, some pellagra, some ariboflavinosis, and some develop peculiar disturbances not falling exactly into any of these categories. As will be explained, beriberi and pellagra may involve many organs or systems; in individual cases the disturbances may affect predominantly one or other organ. The reason for these wide individual variations in response to a common dietetic deficiency is quite unknown. It should also be emphasized that clinically vitamin-B-deficient diets are commonly unsatisfactory in other respects; they may be lacking in protein, salts, or other vitamins also. All these aspects of vitamin-B deficiency were revealed by the experiences of British prisoners of war in Singapore during the Second World War.

The clinical findings in beriberi and pellagra are summarized below.

Beriberi

1. Neuritic (dry) form. The spinal cord and peripheral nerves are chiefly involved. Lesions are present in the ventral horn cells, dorsal root ganglia, and the peripheral nerves. There is tenderness of the skin and the deep structures (muscles, bones), wasting and paralysis of muscles, and diminution or loss of deep reflexes; the muscles of respiration may be affected also.

2. Cardiac (wet) form. There is enlargement of the heart owing to an increase in the weight of the right ventricle (resulting from either congestive heart failure, hypertrophy, or oedema). There are signs of circulatory failure: the heart is enlarged, the pulse is rapid and feeble, and cardiac oedema is present which may be extreme and general.

The two syndromes (dry and wet) overlap to a varying degree. The changes in the organs are first reversible and later irreversible. Early treatment with thiamine produces rapid and complete recovery; in chronic cases prolonged treatment with vitamin-B concentrates containing all members of the group produces very gradual and partial recovery.

3. Cerebral beriberi (Wernicke's encephalopathy). This syndrome is due to thiamine lack principally affecting the brain, especially the grey matter round the third ventricle; characteristically, haemorrhages are found in the mammillary bodies. Among the British prisoners in Singapore many cases occurred of which 52 were carefully studied. As might be expected, nearly every one of these patients also had signs of the other forms of beriberi or other vitamin-B deficiencies. The chief findings in this epidemic of cerebral beriberi were as follows:

(i) General. There was loss of appetite, nausea, and vomiting.
(ii) Eye changes. Nystagmus regularly occurred; there was unilateral or bilateral fatigue or palsy of the external recti. The patient had difficulty in reading; he complained that 'everything wavers on looking to the side'; he might see double.
(iii) Mental changes. The patients suffered from sleeplessness and became anxious and depressed; they might cry on being questioned. They lost contact with their environment and seemed to live in a dream-like world. They described their surroundings as somehow different: the brightness and acuteness of life seemed lost; human voices appeared toneless; restless snatches of dreams seemed to become part of reality. There was loss of memory for recent events; in severe cases the preceding 2–3 weeks became blotted out. Administration of thiamine in the early stages of the syndrome caused dramatic improvement within 12–24 hours.

Alcoholic neuropathy. In chronic alcoholism, peripheral nerve changes are common, leading to disturbances of movement and sensation, mainly in the lower limbs. It is suggested that this 'polyneuropathy' is not due to the direct toxic action of alcohol on the nervous system but to a related thiamine deficiency. Chronic alcoholism depresses the appetite and disturbs gastro-intestinal activity, and so may lead to defective intake and absorption of the vitamin. Clinical improvement is claimed to be produced by thiamine therapy, but some observers are not impressed by the results obtained.

Pellagra (= rough skin)

This disease is endemic in the southern states of the United States, Italy, Rumania, and other countries. The findings are summarized below.

1. Gastro-intestinal changes. Glossitis appears early in the disease; the tip and lateral margins of the tongue are red and swollen, and later penetrating ulcers develop. Similar changes take place in the mouth, gums, and pharynx, and follow the same course. There is complaint of a burning sensation in the mouth, oesophagus, and stomach; mucosal changes have been seen in the stomach with the gastroscope. Nausea, vomiting, and diarrhoea are frequent. The urethral and vaginal mucous membranes may show changes similar to those present in the mouth.

2. Skin lesions may develop anywhere, usually on the dorsum of the hands and feet, axillae, elbows, wrists, knees, beneath the breasts, and the perineum. The skin is first red and itchy; later it becomes swollen and tense and vesicles develop; finally desquamation takes place, the underlying skin remaining abnormally thickened and pigmented.

3. Nervous changes. Various forms of mental disorders occur, together with 'polyneuropathy', i.e. symmetrical bilateral involvement of the nerve supply of the lower limbs.

The syndrome can be summarized by the 3 Ds—dermatitis, diarrhoea, and dementia.

The following facts indicate that pellagra is a deficiency disease:

The diet used in pellagrous households generally consists largely of maize, and is poor in meat. Eleven volunteers in a Mississippi prison were given a typical diet consisting of maize meal, white wheat flour, potatoes, salt pork, and syrup; at the end of 6 months 6 had developed pellagra.

Remarkable improvement is obtained by administration of nicotinic acid, or nicotinic-acid amide by any route. In 24–72 hours the redness and swelling of the tongue and mouth and the alimentary symptoms subside. The fiery skin lesions blanch in about 48 hours. Acute mental symptoms disappear; a patient who was previously maniacal may become calm, and confused patients clear. The polyneuropathy, however, is unrelieved by nicotinic acid but is benefited by thiamine. The role of maize in the production of pellagra is not understood; it seems to aggravate in some way the harmful effects of nicotinic-acid deficiency.

The major specific signs of pellagra (gastro-intestinal, skin, and mental changes) are due to deficiency of nicotinic acid. In pellagrous areas many cases of psychosis (without other signs of pellagra) are the result of nicotinic-acid deficiency and are rapidly relieved by this substance. In most clinical cases of pellagra the diet is deficient in many respects (energy, protein, calcium, iron, vitamin A, and the whole B complex). Specific therapy is therefore not a substitute for a general correction of the diet. An adequate intake of tryptophan-containing protein can lessen the demand for dietary nicotinamide.

Pellagra has occurred following antibiotic therapy, probably due to destruction of intestinal bacteria which normally synthesize nicotinamide.

REFERENCES

SMITH and WOODRUFF (1951). Deficiency diseases in Japanese prison camps. *Spec. Rep. Ser. Med. Res. Coun., Lond.* No. 276.
SNELL, E. E. (1953). Metabolic functions of nicotinic acid, riboflavin and vitamin B$_6$. *Physiol. Rev.* **33,** 509.

Vitamin C (ascorbic acid)

Historical

For centuries, scurvy was of common occurrence among seafarers, soldiers, and those without supplies of fresh food. Neither were townsfolk spared during winter months, though the expansion of trade brought a greater variety of food which contributed to its gradual decline. In 1753, Captain Lind published his treatise on scurvy, with experimental work on sailors which established beyond doubt that the disease could be cured (and prevented) by 'summer fruits and salads', especially by lemon juice (not lime juice in the present meaning of the word). (Lind suggested that seamen would be more willing to try this unaccustomed drink if it were mixed with rum.) On his voyage round the world, Captain Cook took a large quantity of sauerkraut, describing it as 'not only a wholesome vegetable food but in my judgment highly antiscorbutic and spoils not by keeping'. Although scurvy was thus early in being recognized as a deficiency disease, and although experimental scurvy was produced in guinea-pigs in 1907, it was not until 1932 that the responsible factor, vitamin C, was isolated. Chemical synthesis followed shortly afterwards.

Chemistry

The chemical name for vitamin C is L-ascorbic acid. Its structure has affinities with the sugars:

ASCORBIC ACID

Vitamin C is almost unique among the vitamins in that only one natural active form is known. Close analogues (synthetically prepared) have little activity.

Properties. The vitamin is water soluble. It is a strong reducing agent, and therefore readily oxidized. Destruction by oxygen is rapid at 100 °C, especially in an alkaline medium. It is therefore generally absent from dried, canned, or preserved foods unless the process is carried out anaerobically. Freezing or 'dehydration' retains the vitamin. During the cooking of vegetables , vitamin C is lost by too much preliminary shredding, too much cooking-water, over-cooking, and by keeping warm for a long time or by re-heating. (There are also culinary reasons why these excesses should be avoided.)

Dietary sources

Ascorbic acid is haphazardly distributed throughout the plant and animal kingdoms. Good dietary sources are:

1. *Fresh fruits* (in order of decreasing vitamin content): blackcurrant, strawberry, orange, lemon, grapefruit; there is little in blackberry, lime, plum, pear, or most varieties of apple. In Britain, concentrated orange juice is available for infants through the Welfare Services; fruit squashes or cordials are not an adequate substitute.

2. *Fresh vegetables* (in decreasing order): brussels sprouts, cauliflower, cabbage, tomato, potato; potatoes (especially old ones) are not particularly rich in vitamin C but are an important

source because of the large quantities eaten; little is available in lettuce, celery, or mushrooms.

The vitamin-C content of human milk is considerably higher than that of cow's milk. A breast-fed baby can derive enough from its mother's milk provided the mother's diet is good, but as cow's milk becomes an increasingly important part of an infant's diet it should be supplemented by fruit juice. Pasteurization destroys the ascorbic acid of cows' milk, but the content is so meagre in any case that this counts for little when weighed against the benefits of pasteurization.

Foods devoid of vitamin C include: fish, eggs, fats and oils, cereal products including bread and nuts.

Synthetic ascorbic acid is available cheaply for therapeutic administration.

Distribution in animals

The vitamin is curiously distributed in animal organs. Concentrations are high in the adrenal and pituitary glands, the walls of the gut, the aqueous and vitreous humours and lens of the eye, and in white blood cells, but low in red blood cells, muscle, and brain.

The output of ascorbic acid in the urine varies with the intake; normally it is barely detectable but rises after the oral administration of daily loading doses of the pure vitamin. The rapidity of response assesses the status of the subject with respect to vitamin C (the 'saturation' test). A 'fully saturated' man contains about 5 g of ascorbic acid, with a plasma level of about 10 mg l^{-1}; a normal diet maintains a plasma level of about half this. Ascorbic acid also occurs in white blood corpuscles, 5–30 μg per 10^8 leucocytes.

Nutritional significance. Mode of action

The only species known to depend on dietary sources for vitamin C are human beings, monkeys, and guinea-pigs. Many other animals contain ascorbic acid in their tissues but synthesize it for themselves. The pathway of this biosynthesis is from D-glucose via D-glucuronic acid lactone, L-gulonolactone, and 2-keto-L-gulonolactone to L-ascorbic acid. L-Gulonolactone oxidase is absent in man and guinea pig.

Deficiency of vitamin C in susceptible species leads to *scurvy*. The biochemical lesions which cause scurvy are unknown, as is the precise role of ascorbic acid in metabolism, though it is probably concerned with the maintenance of connective tissue in general and intercellular ground substance, capillary lining, and bone matrix in particular. Ascorbic acid is required for the formation of hydroxyproline which comprises about 13 per cent of the collagen molecule.

Effects of vitamin-C lack in man. Experimental scurvy

The account which follows is based mainly on the pioneer experiment of Crandon (1940) on himself, and on a later study of 20 British volunteers. The latter were maintained for a control period of 6 weeks on a basal diet completely devoid of vitamin C (it contained less than 1 mg daily). The subjects were then divided into three groups:

1. On the basal diet only (10).
2. Basal diet + 10 mg of C daily (7).
3. Basal diet + 70 mg of C daily (3).

Groups 2 and 3 remained perfectly fit and well. The detailed findings in the completely deprived subjects are considered below.

Blood changes. The average normal vitamin C level in the plasma is 5.5 mg l^{-1}; in the white blood corpuscles the average normal value is 20 μg per 10^8 leucocytes.

1. On a vitamin-C intake of 70 mg daily the blood levels are unaffected.

2. On the vitamin C free diet the plasma vitamin C level fell rapidly and that in the white corpuscles more slowly.

3. On a 10 mg-C intake the plasma changes were the same as in 2 above but the concentration in the white corpuscles was higher than in 2.

Relation between blood vitamin-C level and clinical state. Symptoms appeared after about 5 months in the vitamin-free group, but the intervals between the virtual disappearance of vitamin C from the plasma and from the white cells and the appearance of symptoms were respectively 14 weeks and 3–6 weeks. It should be noted that although vitamin C disappeared from the plasma in the 10 mg-intake group, the subjects remained normal. The virtual disappearance of C from the plasma thus does not necessarily indicate that scurvy is present or imminent; on the other hand a plasma level above 1 mg l^{-1} definitely excludes the diagnosis of scurvy. Scurvy does not occur if the leucocyte content exceeds 5 μg per 10^8 cells.

Clinical changes. 1. Negative findings. Many negative findings must be emphasized. There was no decrease in the red-cell or white-cell count, platelet count, or haemoglobin concentration; vitamin-C lack therefore has no adverse effect on red-marrow activity. There were no general capillary changes; the resistance of the skin capillaries and bleeding time were normal; there was no nose-bleeding; blood or red cells did not appear in the urine, there was no occult blood in the faeces; the microscopic appearances of the capillaries in the nail bed and in the conjunctiva were normal. Body weight was maintained and there was no complaint of general pains or weakness. Susceptibility to infection was not noticeably increased.

2. Skin changes. After 4–6 months there was enlargement and keratosis of the hair follicles in the skin. These changes were marked on the upper arms, back, buttocks, back of thighs, calves, and shins. Microscopic examination showed that the follicles were plugged with horny material in which the hair was coiled or looped; subsequently the local vessels became congested and leaked; the follicles then became red and haemorrhagic.

3. Gums. After 6 months, gum changes developed regularly, consisting of swelling and tiny haemorrhages in the loops of the interdental papillae. In two subjects who had some initial gingivitis gross changes occurred; the gums became very swollen and purplish in colour; later patches of necrosis appeared.

4. Wound healing. Normal healing of a wound (e.g. that produced by a deep skin incision) involves development of new capillaries, proliferation of fibroblasts, and the laying down of reticulin fibres followed later by the appearance of collagen fibres. In severe guinea-pig scurvy all these phases of wound healing are deficient. In Crandon, after 3 months' deprivation of vitamin C, an experimental deep incision in the back healed rapidly and completely within 10 days. A similar incision performed after 6 months' deprivation showed good union of the epidermis after 10 days, but there were no signs of healing in the deeper parts of the wound which was filled with unorganized blood clot; no capillary or fibroblastic proliferation had taken place and there was no laying down of reticulin or collagen fibres. There were considerable variations in the extent of impairment of wound healing in other subjects. The tensile strength of healed wounds was lower than normal. Scars of old wounds that had healed well in the early stages of vitamin-C deprivation become red or livid when clinical scurvy appeared.

5. Neuromuscular co-ordination was slightly impaired and fatigue set in slightly more readily.

6. In three of the 10 British volunteers who were completely deprived of vitamin C grave complications developed which required immediate intensive vitamin-C medication; two developed signs of cardiac ischaemia, i.e. pain in the chest, dyspnoea, and serious electrocardiographic changes.

There thus appears to be stages in the development of experimental scurvy: (i) an initial stage with changes in the vitamin-C content of the blood and no clinical signs; (ii) a stage of clinical scurvy consisting of changes in the skin, gums, and impaired wound healing; (iii) a final stage when life may be in immediate danger. These changes are compared below with the classical manifestations of clinical scurvy.

When 10 mg of C were administered daily to experimental subjects with well-developed scurvy, the skin changes cleared up in 1–2 months and the gums became normal in 3 months. When massive doses of C were given recovery was rapid. Thus the chest pain and dyspnoea in the two subjects mentioned above were relieved in 24 hours. Crandon's experiment was ended by giving him 1 g of vitamin C intravenously daily; a wound inflicted on the final day of his treatment became completely healed in 10 days; the skin petechiae faded in a week and the skin was quite normal in three weeks.

Clinical scurvy

1. Infants. Scurvy can occur in infants who have been fed exclusively on sterilized food, the vitamin-C content of which has been destroyed and not replaced by fruit juice. The characteristic symptoms are haemorrhages from various parts, and anaemia. After being fed on the defective diet for about six months, fretfulness and loss of appetite develop (the long latent period is noteworthy). Haemorrhages occur (i) under the periosteum, e.g. at the lower end of the femur, giving rise to great pain, tenderness, and refusal to use the limb; (ii) from the gums if the teeth have erupted; (iii) from the kidney, intestine, and under the skin. There may be changes in the teeth and bones.

2. Adults. Scurvy is now chiefly encountered, in Western countries at any rate, in old people living alone, or occasionally in those dieting 'on medical advice'. Common signs are small spontaneous bruises on the limbs, extravasation of blood into the tissues, and anaemia, with lethargy and depression, all responding quickly to ascorbic acid. Bleeding gums (one of the classic signs of scurvy) are exceptional in these elderly patients who have shrunken gums and few of their own teeth.

3. Comparison with experimental scurvy. The clinical manifestations are far more severe than those observed in the controlled experiments. It must be remembered, however, that clinically the vitamin deficiency, particularly in adults, may be very prolonged, that multiple vitamin deficiencies may be present which may precipitate or aggravate the symptoms, and infections may develop which increase the rate of disappearance of vitamin C from the body.

It is generally agreed that 'natural' scurvy usually includes anaemia and susceptibility to infection among its features, yet neither of these appeared in the experimental deficiency. The anaemia may be due to blood loss, reduced incorporation of iron into haemoglobin, or to destruction of red cells. Goldberg (1963) injected [^{51}Cr]-labelled normal red cells into scorbutic patients and showed an increased rate of destruction of these cells when compared with their fate in normal people. Thus in scurvy there is haemolytic anaemia. Scorbutic anaemia is corrected by administration of 500 mg of ascorbic acid daily.

In guinea-pig scurvy, the adrenals are depleted of ascorbic acid and of cholesterol; the same happens under various adverse environmental conditions, and it seems possible that the stress of unfavourable physiological surroundings may exacerbate the effects of a poor diet.

Vitamin-C requirements in man

Accumulated experience and the experiments recorded above show that, in adult humans, 10 mg of dietary ascorbic acid is completely protective and curative over long periods. To allow a margin of safety and to provide for individual variations, a daily intake of 30 mg is recommended. This is readily achieved by a normal Western diet—one orange, or half a grapefruit, or a generous helping of lightly-cooked cabbage will furnish the day's requirements. The recommended 30 mg is of ingested vitamin, so that the American aim of 70 mg makes liberal allowances for maltreatment by the cook.

The intake should be the same for children, but increased for adolescents, in pregnancy, and in patients with fever or wounds to heal. Special care must be taken with patients with gastric disorders, who have to be denied the roughage of fruit and vegetables but not the vitamin-C contained therein. Babies, especially if bottle-fed, should receive fresh or concentrated orange or blackcurrant juice from a fortnight onwards.

In recent years there have been suggestions that much larger amounts of ascorbic acid are required in man. The distinguished chemist Linus Pauling (1970) has claimed that 1–2 g of ascorbic acid daily prevent the common cold or reduces its duration and severity. This claim has been contested by some and partly supported by others. Stone (1972) has argued that since all mammals (except guinea-pigs and primates) synthesize ascorbic acid at a rate equivalent to 4 g daily for a 70 kg man, this should be the normal human dietary intake. He claims that although 10–30 mg daily will prevent manifestations of scurvy, much larger quantities (several grams) are needed for optimal health. However, it has not yet been clearly established that prolonged intake of such large amounts of ascorbic acid daily is safe in man, although gorillas take 4.5 g daily without apparent harm!

REFERENCES

GOLDBERG, A. (1963). *Q. Jl Med.* **32**, 51.
MEDICAL RESEARCH COUNCIL (1953). Vitamin C requirements of human adults. *Spec. Rep. Ser. med. Res. Coun., Lond.* No. 208.
PAULING, L. (1970). *Vitamin C and the common cold.* Freeman, San Francisco.
STONE, I. (1972). *The healing factor. 'Vitamin C' against disease.* Grosset and Dunlap, New York.

PART XI
Endocrine glands

Introduction

The endocrine, or ductless, glands are so called because they secrete physiologically active substances called hormones directly into the bloodstream. The word 'endocrine' derives from a Greek word meaning 'I separate within', and is to be contrasted with 'exocrine' which refers to glands whose secretion is delivered to the outside of the body or into the lumen of the alimentary tract or that of the respiratory tract (the salivary glands, gastric glands, and mucous glands, as well as the cutaneous sweat glands, are all exocrine). The word 'hormone', introduced by Starling in 1905, is derived from a Greek word meaning 'I excite or arouse' and was first used in reference to secretin and gastrin [pp. 427 and 431].

The physiological role of the endocrines is to achieve, in Starling's words (1905) 'the chemical correlation of the functions of the body'. The endocrines together with the central and autonomic nervous systems, form the great co-ordinating systems which regulate body functions. They receive information, interpret and integrate it, and initiate responses, via nerve and bloodstream, to arouse the bodily activities most appropriate to the needs of the moment, whilst preserving the constancy of the *milieu intérieur*.

The most important endocrine glands include:

1. The hypothalamus.
2. The anterior pituitary (adenohypophysis). The posterior pituitary (neurohypophysis).
3. The islets of Langerhans in the pancreas.
4. The adrenal cortex and medulla.
5. The thyroid.
6. The parathyroids and kidney.
7. The ovary and testis.

In this part the endocrines 1 to 6 will be discussed. The ovary and testis will be dealt with in Part XII.

Methods of study of endocrines

The functions of endocrine glands have been studied by *clinical investigations on patients* in whom syndromes resulting from under- or oversecretion of hormones have been recognized, by *experimental studies in animals and man*, and by *studies on isolated organs and cell cultures*.

Effects of extirpation. Removal of an endocrine organ shows its importance to the body. The function of an endocrine gland is simply to prevent the physiological and metabolic disturbances which occur after its extirpation. To give examples: removal of the testes before puberty in the male prevents the development of the sex organs and of the secondary sexual characteristics such as growth of hair and breaking of the voice; removal of the thyroid produces myxoedema; removal of the parathyroids produces tetany; removal of the adrenals produces death from adrenocortical deficiency.

Grafts and extracts of endocrine glands have been used to correct the abnormalities produced by removal or destructive diseases of endocrine glands, and the development of chemical techniques has led to the *isolation, identification*, and in many cases, *the synthesis of hormones*.

Properties of a hormone

1. It is synthesized and secreted by living endocrine glandular cells within the body or in cultures of endocrine cells *in vitro*.

2. A hormone is transported by the bloodstream from the endocrine cells to serve as a 'chemical messenger' which acts in specific ways on 'target' cells or organs.

3. A hormone does not provide energy or building materials, but it does have profound regulatory effects on growth, differentiation, and metabolic activities of its target cells by effects on membrane permeability, activation of enzymes, formation of cyclic AMP, etc. The endocrine glandular cells which secrete a hormone and the target cells which respond to it together form a highly specialized system, but it is not known how such co-ordinated specialization is brought about.

4. Hormones belong to different types of chemical structure:

Some are *steroids*, e.g. adrenocortical hormones and male and female sex hormones. These are synthesized from acetate or cholesterol. They are not stored for more than a few minutes in the endocrine cells of synthesis, and for continued action within the body steroid hormones must be continuously synthesized. However, steroid-secreting cells contain large stores of hormone precursor in the form of esterified cholesterol in lipid droplets.

Many hormones are *proteins* or *peptides* and are synthesized by endocrines in the same way as proteins and peptides in general [p. 475]. These hormones may be stored as granules for hours or days and are released by *exocytosis*, a process which depends on the presence of Ca^{2+}, ATP, and ATPase. The anterior-pituitary hormones are glycoproteins or peptides. Hypothalamic hormones, posterior-pituitary hormones, parathyroid hormone, calcitonin, insulin, glucagon, gastrin, secretin, and angiotensin are all peptides. Many of the peptide hormones are formed *in vivo* from larger precursor molecules or *prohormones*.

Thyroid hormones (thyroxine and triiodothyronine) are iodinated tyrosine derivatives, and the adrenal medullary hormones, adrenalin (epinephrine) and noradrenalin (norepinephrine), are *catecholamines* formed also from tyrosine. Thyroxine is stored *extracellularly* in the lumen in the thyroid follicle, attached to the large thyroglobulin molecule, for days or weeks. The catecholamines are also stored as granules in the adrenal medulla in which they are combined with ATP and proteins and are released by exocytosis in response to stimulation of the splanchnic nerve.

5. In the *bloodstream* many hormones, e.g. steroids and thyrox-

ine, are bound to specific plasma carrier proteins. This binding creates a reservoir from which free hormone is released and diffuses to act on the target cells. Catecholamines are not bound to plasma proteins and have a half-life in the blood of a few minutes, whereas thyroxine which is strongly protein bound has a half-life of several days.

6. *Inactivation* of hormones takes place partly in the target organs, and especially in the liver where chemical degradation and conjugation often occur, e.g. insulin, adrenalin, steroids, and thyroxine. The kidney excretes some hormones or their metabolites in detectable quantities in the urine, e.g. steroids, gonadotrophins, and catecholamines.

7. Hormones exert their physiological effects when present in very small concentrations in blood and body fluids. For example, steroid hormones and thyroid hormones act when present in plasma at 10^{-6}–10^{-9} mol l^{-1}, and peptide hormones are even more potent, acting at 10^{-10}–10^{-12} mol l^{-1}. Because of their small concentrations hormones have been difficult to detect in body fluids and special methods have been developed to measure the amounts of hormones in endocrine glands, in gland extracts, and particularly in body fluids.

Hormone assays

In the past, hormone preparations (e.g. insulin, thyroid substance) used for therapeutic purposes required *biological standardization* to ensure constancy of activity. Biological assay of hormone activity in still needed when new synthetic products are being compared with natural hormones. For example, a new anabolic steroid might well have androgenic activity. The new compound would therefore be compared with testosterone by means of dose–response curves obtained by recording the increase in weight of an androgen-sensitive tissue such as the vesiculae seminales in the rat; likewise an analogue of oxytocin can be compared with oxytocin itself for its milk ejection activity by measuring the pressure in the cannulated teat of the lactating rat (Bisset 1974). This and many other biological assay systems have been made so sensitive that they can record the minute concentrations of hormones which occur in circulating blood. With the introduction of various separation techniques such as solvent extraction, chromatography, and molecular sieving, biological assays of steroids, peptides, and catecholamines have become more sensitive, but the development of highly sensitive and accurate *chemical* or *immunochemical* techniques for measuring physiological concentrations of hormones in blood and other body fluids has produced much simpler methods.

Chemical techniques include fluorescence methods for catecholamines, gas-liquid chromatography for steroid hormones, and the use of radioactive isotopes, e.g. ^{131}I and ^{125}I for thyroid hormones.

Competitive radioassay (*saturation analysis*) includes *radioimmunoassay* which was developed in the early 1960s by Berson and Yalow. They noted that in patients with serum antibodies against insulin there was competition between radiolabelled insulin and unlabelled insulin for binding to the specific insulin antibodies. By using serum from animals immunized with insulin (to produce specific insulin antibodies) Berson and Yalow applied this competitive binding procedure to estimate insulin in blood, and later this procedure of immunoassay was widely applied to all the protein and peptide hormones, which can be detected in concentrations of µg l^{-1} or even ng l^{-1}.

Competitive radioassays were first done with antibody as the binding protein, but any naturally occurring protein with high affinity for a hormone can be used instead. For example, specific transport proteins in plasma, such as *transcortin* and *thyroxine-binding globulin* can be used for estimation of cortisol and thyroxine respectively.

The principle involved in competitive radioassay is as follows: the binding sites on antibody or other high affinity protein are saturated with radio(e.g. ^{131}I)-labelled hormone and incubated in the cold with material containing unlabelled (free) hormone. The competition between labelled and unlabelled hormone for the binding sites reduces the proportion of bound radioactive hormone and increases the proportion of free radioactive hormone. As the concentration of unlabelled hormone increases the ratio

$$\frac{\text{bound labelled hormone}}{\text{free labelled hormone}}$$

decreases [FIG. XI.1].

The advantages of competitive radioassay are high sensitivity and high specificity, but it is important to note that radioimmunoassays do not always correlate with biological activity since antibodies can be formed against biologically inactive fragments of peptide chains. There is thus still a need for highly sensitive biological test systems, such as milk ejection in the lactating rat, which can respond to 2.0 pg of intra-arterially injected oxytocin.

Traces of radiolabelled hormone
(●) incubated with excess antisera
or specific binding protein (○)
Little radioactivity remains in
free fraction

Addition of unlabelled hormone
(sample or standard) (○)
increases radioactivity in
supernatant

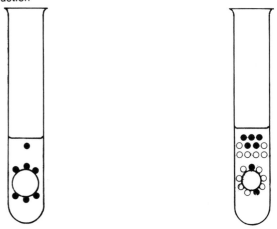

FIG. XI.1. Priniciple of immunochemical assay for measuring hormone in biologic fluids. The amount of radioactivity in the supernatant is a direct function of the amount of hormone in the specimen. (Redrawn from Grodsky, G. M. (1973). In *Review of physiological chemistry*, 14th edn (ed. H. A. Harper) p. 430. Lange Medical, Los Altos, Ca.)

Cytochemical assays. Daly *et al.* (1974) have shown that adrenal steroidogenesis can be detected in slices of guinea-pig adrenal gland, incubated in an ascorbate-enriched culture medium, after the addition of as little as 5 attograms l^{-1} of ACTH (1.0 attogram = 10^{-18} g). This test is therefore much more sensitive than competitive radioassay, and cytochemical techniques have been applied to the measurement of LH, TSH, TSI, and pentagastrin. At present these methods are much more complicated and time-consuming than competitive radioassays but their application could be very useful in measuring the minute basal levels of hormone secretion.

Direct control and regulation of secretion of hormones

In a few instances hormone secretion is regulated by the blood concentration of substances which are directly controlled by the hormones themselves. For example, insulin secretion from β islet cells is promoted by a rise in blood glucose level and glucagon secretion from α-cells by a fall in blood glucose level. These responses help to keep blood glucose levels within narrow limits in spite of variations in carbohydrate intake in the diet, since insulin lowers blood glucose and glucagon raises it [p. 507].

Plasma calcium level is comparably regulated; a fall in plasma calcium promotes secretion of parathyroid hormone (PTH) and a rise reduces PTH secretion and leads to secretion of calcitonin, in each case by direct effects on the appropriate endocrine secretory cells. The contribution of vitamin D metabolites to calcium metabolism must also be considered [p. 553].

Nervous control of endocrine secretions

It is now clear that hormonal secretion from endocrine glands is largely controlled by the central nervous system. Even the endocrine pancreas is partly under nervous control [p. 506].

The evidence for the role of chemical transmission in the autonomic nervous system, in the motor nerves to skeletal muscle and in the central nervous system is discussed on pages 410, 257, and 419. Most of these neurotransmitters are concerned with rapid conveyance of excitation or inhibition over very short distances (nanometres) during very short periods of time (milliseconds). The central nervous system contains neurons which synthesize and release peptides. The highest concentrations of such neurons are found in the hypothalamus where they release peptide hormones which either enter the systemic bloodstream in the neurohypophysis (ADH and oxytocin) or pass to the adenohypophysis via the hypothalamic-adenohypophysial portal vessels (*hypophysiotrophic hormones*) to stimulate or inhibit the secretion of anterior pituitary hormones into systemic blood [p. 518]. Thus we may use the term *neurosecretion* to describe the actions of all neurons which act by release of chemical agents, whether as neurotransmitters or as short range hormones. FIGURE XI.2 illustrates some of the different types of neurosecretory control.

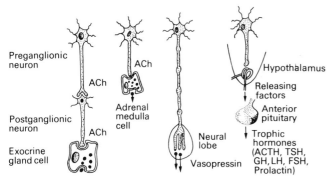

FIG. XI.2. Types of neurochemical transducer systems. *Left*, Secretomotor neurons, postganglionic or preganglionic autonomic fibres make direct synaptic contact with exocrine glandular or hormone-secreting cells. *Right*, Hypothalamic neurosecretory neurons. Neurosecretory neurons of the supraoptic system release ADH (vasopressin) into the systemic bloodstream. Hypothalamic tuberoinfundibular neurons release hypophysiotrophic hormones (releasing factors) into the pituitary portal system to regulate secretion of hormones from the anterior pituitary. *Abbreviations*: ACh, acetylcholine; ACTH, adrenocorticotrophin; TSH, thyroid-stimulating hormone (thyrotrophin); GH, growth hormone (somatotrophin); LH, luteinizing hormone; FSH, follicle-stimulating hormone. (Redrawn from Martin, J. B., Reichlin, S., and Brown, G. M. (1977). *Clinical neuroendocrinology*, p. 5. Davis, Philadelphia.)

The neurons of the supraoptic and paraventricular nuclei are stimulated by cholinergic neurons and inhibited by adrenergic neurons in the hypothalamus. The hypothalamus contains numerous monoaminergic neurons whose terminals are located on the peptidergic hypophysiotrophic neurons which in turn end on the capillary plexus in the median eminence. Since the hypothalamus receives neuronal connections from many regions of the central nervous system it is easy to understand how nervous activity can influence secretion of pituitary hormones [p. 518].

In addition to control by the central nervous system the secretion of anterior pituitary hormones is regulated by feedback effects of the hormones secreted by the target organs (thyroid, adrenal cortex, gonads). The target organ hormones act either on the adenohy-

pophysis or on the hypothalamus, usually by negative feedback. Oestrogens can exert a positive feedback effect [p. 571].

These examples give some idea of the mechanisms whereby the activities of endocrine glands have self-limiting effects on their own rates of secretion.

In some cases many hormones may be concerned in the regulation of a particular activity. For example, the blood glucose level is raised not only by glucagon but also by adrenalin, cortisol (stimulated by ACTH) and growth hormone, while insulin is the only hormone which lowers blood glucose. The actions of glucagon and adrenalin are immediate and are due to breakdown of glycogen in the liver, while those of cortisol and growth hormone are slower in onset, more prolonged, and brought about by quite different mechanisms. The inhibition of insulin secretion by adrenalin is an interesting illustration of the way homeostatic processes may be interconnected.

Modes of action of hormones

Endocrine glands act solely through the hormones which they synthesize and secrete. Now that it is possible to measure the concentration of hormones in body fluids the effects of these physiological concentrations can be studied in intact animals, in human beings, in isolated organs, and in tissue and cell cultures. The *functions* of endocrine glands are to secrete hormones in amounts sufficient to prevent such diseases as myxoedema, diabetes, Addison's disease, tetany, and rickets. The *effects* of hormones were, in the past, often those resulting from large, pharmacological doses. Interest is now directed to attempts to understand the ways in which physiological concentrations of hormones act on cells, subcellular structures, metabolic processes, DNA, RNA, and enzyme activities.

Studies on the actions of hormones in intact animals and human beings can rarely reveal details of mechanisms of action owing to the complexities of regulatory processes mediated by nervous control, compensatory reactions, metabolic degradation of hormones, and our technical inability to control important factors which influence cellular function. On the other hand, whilst detailed mechanisms can be more easily studied in isolated organs, cells, or subcellular fractions, the value of the information obtained may be offset by lack of the normal nervous control and blood supply; in any case 'simple' systems are still complex. However, even approximate answers are worthwhile and it is the convergent and cumulative evidence from different approaches that justifies the putting forward of tentative hypotheses.

The endocrines form a vital co-ordinating system within the body. The co-ordination is seen not only in collaboration between different endocrine glands but even in the actions of one hormone on a sensitive cell or tissue. Thus the growth of breast tissue in girls depends on the combined activity of oestrogen and progesterone, and milk secretion is promoted by prolactin only from breasts previously developed by oestrogen–progesterone actions. At the cellular level glucagon not only promotes glycogenolysis in the liver but simultaneously inhibits glycogen synthesis; insulin has the opposite effects.

Hormones act on cells only after specific combination with receptors, either in the plasma membrane or inside the cells. Numerous effects may result from hormone–receptor combination. Examples are given below:

1. Hormones may increase plasma membrane permeability to glucose and amino acids (e.g. insulin).
2. They may increase ion fluxes (e.g. aldosterone promotes Na^+ influx into distal renal convoluted tubular cells and increased efflux of K^+ and H^+).
3. Antidiuretic hormone increases the permeability of the cells of the distal renal tubules and collecting ducts to water.

4. Hormones may affect cell metabolism by inducing changes in activity of rate-limiting enzymes, e.g. activation of phosphorylase (to be discussed on p. 501).

5. Hormones often promote protein synthesis by actions at different sites: (a) increased entry of amino acids into cells; (b) enhanced DNA directed synthesis of mRNA (transcription); (c) enhanced ribosomal activity (translation)

Actinomycin D blocks DNA-mediated RNA synthesis; it thereby inhibits the synthesis of mRNA, tRNA, and rRNA and hence inhibits ribosomal protein synthesis. Puromycin inhibits protein synthesis at the translation level. However, these inhibitors have other actions on cells and their use as analytic tools is of limited value.

6. Many hormones (e.g. TSH, oestradiol, ACTH) produce multiplication of their target cells (hyperplasia) and often enlargement as well (hypertrophy). The causes of DNA replication and mitosis in these cells are not fully understood, but probably involve de-repression of parts of the DNA molecule.

Hormones and specific receptors

The idea that drugs and hormones combine with specific cellular receptors is now widely accepted, although the problem remains as to how such combination brings about diverse responses. Tepperman (1973) has proposed that a distinction should be made between what he calls the *mobile-receptor model*, which is characterized by specific binding of steroid hormones to cytosol protein and subsequent migration of the hormone–receptor protein complex, (or some derivative of it) into the cell nucleus, and the *fixed-receptor model* in which peptide or catecholamine hormones become bound to receptors on the plasma membrane of the cell and initiate a sequence of intracellular events which constitute a co-ordinated response. In the mobile-receptor model lipid-soluble, small molecular, steroid hormones penetrate the plasma membrane to act within the cell; in the fixed-receptor model the hormones never enter the cell. In non-target body cells steroids enter cells and leave them without producing any action. Only in target cells do steroid hormones bind to cytosol proteins to produce specific actions.

Mobile-receptor model

Oestradiol (oestradiol-17β) was the first steroid hormone to be studied with respect to its detailed mode of action. Gross anatomical and histological evidence showed that it must stimulate protein synthesis in its target organs. Accumulation of DNA, RNA, and protein occur in oestrogen-stimulated tissues and the action of oestradiol on, for example, the uterus is completely prevented by actinomycin D or by puromycin.

Highly radioactive oestradiol, labelled with tritium, has been administered in physiological (μg) amounts to rats. [³H]-Oestradiol-17β is taken up and bound to a specific cytosol receptor protein in oestrogen-sensitive cells in the uterus, vagina, breast, and hypothalamus (other steroid hormones become bound to different cytosol proteins). The oestradiol–receptor protein complex then enters the nucleus and becomes bound directly to the chromosomes, probably to non-histone proteins. One subunit of the oestradiol–receptor protein complex then dissociates and reacts with DNA and stimulates mRNA synthesis, which in turn promotes protein synthesis [Fig. XI.3]. Progesterone, androgenic hormones, and adrenal corticosteroids probably act similarly.

Szego (1974) has shown that [³H]-oestradiol enters lysosomes before it can be detected in the nucleus and that an oestradiol–acid lysosomal protein complex subsequently passes into the nucleus. She suggests that the increased template activity of nuclear chromatin could be due to removal of repressor histones, either by lysosomal proteases or by acidic proteins, with oestradiol playing no direct part in genic derepression. Thus lysosomal constituents could

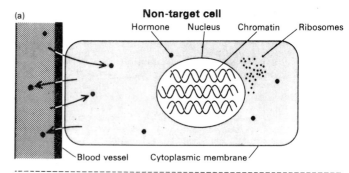

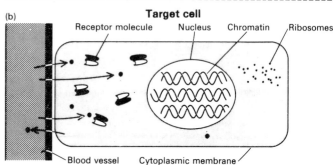

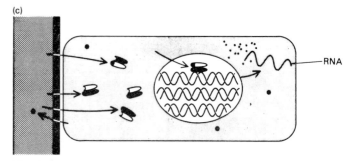

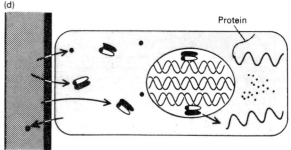

FIG. XI.3. Action of steroid hormones is mediated by receptor molecules found only in cells that respond to the hormone (target cells). In non-target cells (a) the hormone diffuses freely into the cell and out of it, and the concentration remains quite low. In a target cell the hormone is sequestered by receptor molecules (b). Each receptor molecule binds two molecules of hormone, forming a complex that enters the nucleus (c) and becomes attached to the chromatin, the genetic material. The complex stimulates the transcription of particular genes, so that RNA encoding the information in those genes is synthesized. On the organelles called ribosomes the RNA is translated into proteins (d).
(Redrawn from O'Malley, B. W. and Schrader, W. T. (1976). *Scient. Am.* **234**, 32.)

promote both transcriptional and replicative pathways of DNA metabolism, enhancing RNA and protein synthesis as well as inducing mitosis. Lysosomes may have anabolic as well as catabolic effects [p. 5].

Both these proposed mechanisms offer an explanation for the protein synthesis and hyperplasia which steroid hormones induce in their target organs.

Fixed-receptor model

The fixed-receptor model involves hormone binding by a specific receptor on the outer surface of the plasma cell membrane, activation of the enzyme *adenyl cyclase* on the inner surface of the membrane, increased formation of intracellular 3', 5'-cyclic AMP (cAMP) from ATP, and conversion of inactive *protein kinases* to their active forms which phosphorylate many substrates; these reactions are terminated by *phosphodiesterase* which converts cAMP to inactive 5'-AMP. Although the discovery of cAMP by Sutherland in 1957 came from studies on the mode of action of adrenalin on hepatic glycogenolysis, and although subsequently it was shown that many other hormones (e.g. glucagon, ACTH, TSH) act via cAMP it is now realized that cAMP has important regulatory functions in micro-organisms (bacteria, slime mould) in which there is no hormonal control at all. Phylogenetically the cAMP metabolic regulatory process is very ancient and during evolution must have appeared long before hormones and their receptors came on the scene. It is important to appreciate that the cAMP system is a basic regulator of cell metabolism (rather like the Krebs cycle), and that hormones took it over as a going concern when they came to function as conveyors of information from endocrine glands to target organs. This accounts for the widespread occurrence of cAMP as a regulator of intracellular metabolism, and the chief questions which now have to be answered are how is the system regulated so that it can respond specifically to one hormonal stimulus but not to others, and how does the interaction between hormone and surface receptor lead to a highly co-ordinated response of the target cell? To deal with these questions we must discuss more fully the known components of the system, and the biochemical and biological reactions which are influenced by cAMP [see TABLE XI.1 for summary of effects of cAMP].

The sequence of events involved in hormonal activation of the cAMP system is as follows:

The first step is a stereo-specific interaction between the hormone and receptor on the outer surface of the plasma cell membrane. It is quite clear that these receptors are different in the cells of different target organs. For example, ACTH activates adenyl cyclase (and thus cAMP) in cells of the adrenal cortex, and hence promotes cortisol synthesis, whereas it has no effect on liver cells. On the other hand, glucagon activates adenyl cyclase in liver cells and thus promotes glycogenolysis, but has no effect on the cells of the adrenal cortex. A different situation is found in adipose tissue cells, in which cAMP-mediated lipolysis (splitting of triglycerides to free fatty acids and glycerol) is produced by catecholamines, ACTH, TSH, or glucagon, each acting on a different receptor, but all probably activating the same adenyl cyclase on the inner surface of the plasma cell membrane. Thus there is some specificity at the hormone–receptor interaction site, but if different receptors are involved with different hormones how is it that adenyl cyclase is activated in each case?

Transduction of the hormone–receptor signal

All that can be said at present is that some propagated disturbance is conveyed from the hormone–receptor complex on the plasma membrane surface to the adenyl cyclase situated on the inner membrane surface. The role of ionic fluxes, depolarization, and chemical mediators in this process is not certain, but insulin, which increases membrane permeability to glucose, amino acids, and K^+, decreases intracellular cAMP level and prostaglandins may either stimulate or inhibit adenyl cyclase.

Adenyl cyclase occurs in all animal phyla, including micro-organisms in which it is not under hormonal control. The highest concentrations of adenyl cyclase occur in the central nervous sys-tem and retina where it is closely associated with the post-synaptic membrane. Its function is mediated by ATP generation and perhaps also guanosine triphosphate (GTP). Adenyl cyclase is stimulated by low $[Ca^{2+}]$ but inhibited by high $[Ca^{2+}]$; $[Ca^{2+}]$ may be an important regulator of cAMP concentration in the cell.

Phosphodiesterase. Intracellular [cAMP] is the resultant of the rate of cAMP formation by adenyl cyclase and the rate of destruction by phosphodiesterase which catalyses hydrolysis of 3', 5', cyclic AMP to inactive 5'-AMP. Phosphodiesterase is inhibited by methyl xanthines (theophylline, caffeine) which thus tend to increase intracellular cAMP concentration. Insulin appears to enhance phosphodiesterase activity and thus reduces intracellular cAMP concentration.

In certain bacteria the level of intracellular cAMP is determined more by phosphodiesterase than by adenyl cyclase.

Dibutyryl cAMP is a more stable derivative of cAMP which acts in the same way as cAMP and may be used to study the effects of increasing cAMP-like activity within cells.

Protein kinase. The hormone–receptor mechanism leads to selective effects on intracellular [cAMP]. Intracellular cAMP then produces co-ordinated responses within cells by virtue of its ability to phosphorylate enzymes as follows:

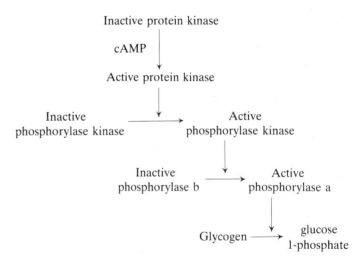

cAMP and glycogen metabolism

cAMP, by activating protein kinase, has two co-ordinated effects on glycogen metabolism.

1. It enhances glycogenolysis to glucose 1-PO$_4$, as shown above.
2. It phosphorylates active glycogen synthetase to form inactive glycogen synthetase.

Thus cAMP simultaneously enhances glycogen breakdown and reduces glycogen synthesis by activating *cAMP-dependent protein kinase*. This enzyme is as widely distributed in mammalian systems as adenyl cyclase and phosphodiesterase. There are probably many protein kinases which have different effects in different cells and many lead to:

1. Permeability changes in membranes.
2. Activation or inactivation of rate-limiting enzymes.
3. Increased or decreased protein synthesis by action on ribosomes.
4. Increased or decreased DNA-mediated RNA synthesis (transcription).
5. Microtubular protein aggregation which promotes glandular secretions.

TABLE XI.1. Effects of 3′, 5′ cyclic AMP. (From Tepperman, J. (1973). In *Metabolic and endocrine physiology*, 3rd edn. Year Book, Chicago.)

Process	Site and activity
Membrane permeability	(1) Ions (synapse, neuromuscular junction, adrenal medulla, retina)
	(2) Water (kidney, toad bladder, toad skin, ADH)
Steroidogenesis	Corpus luteum, adrenal cortex, Leydig cell
Secretory responses	Hypothalamic releasing factors, salivary gland, exocrine pancreas, thyroid, insulin, gastric HCl secretion
Triglyceride and cholesterol ester hydrolysis	Adipose tissue, liver, steroid-producing cells
Inhibition of lipogenesis	Liver, adipose tissue
Movement of intracellular structures	Melanophore dispersion, sperm mobility, cilia (?), cell process maintenance (fibroblasts)
Glycogenolysis stimulation and inhibition	Liver, adipose tissue, muscle
Gluconeogenesis	Liver, kidney
Gene transcription	Micro-organisms (lac operon) Enzyme induction in fetal liver
Protein synthesis and translation	General inhibition of protein synthesis Selective protein synthesis (e.g. adrenal cortex) Stimulation in micro-organisms (tryptophane synthesis)
Motility and aggregation, unicellular organisms	Slime mould aggregation

It has been suggested that cAMP activates protein kinase in the following way [Fig. XI.4]. Protein kinases consist of two combined dissimilar subunits, a regulatory one (R) and a catalytic one (C). cAMP combines with R and leaves C in a free, active state which can phosphorylate enzymes or other proteins.

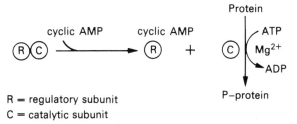

R = regulatory subunit
C = catalytic subunit

FIG. XI.4. The mechanism by which cAMP brings about the activation of protein kinases. The cyclic nucleotide binds to the regulatory subunit of the inactive enzyme causing a dissociation of regulatory from catalytic subunits. The *dissociated* catalytic subunit is catalytically competent, whereas it is not competent when associated with the regulatory subunit. (From Rasmussen, H. (1974). In *Textbook of endocrinology*, 5th edn (ed. R. H. Williams). Saunders, Philadelphia.)

Calcium and cAMP

Although cAMP has been generally regarded as the 'second messenger' in the fixed-receptor model, there is evidence that other second messengers exist and help to control intracellular metabolism. In many systems calcium is necessary for hormone action, and cAMP itself may only act in the presence of calcium. Many peptide hormones promote calcium influx into cells as well as activating adenyl cyclase; cAMP also increases intracellular [Ca²⁺] by promoting efflux of Ca²⁺ from mitochondria. The increased [Ca²⁺] activates or inhibits intracellular enzymes, some of which are phosphorylated proteins formed by cAMP-dependent protein kinases. As Ca²⁺ accumulates in the cell it inhibits adenyl cyclase, thus providing a negative feedback control of cAMP formation.

It is very probable that other factors, besides cAMP and calcium, are involved in the regulation of intracellular activities initiated by hormone–receptor combination in the plasma cell membrane. One such factor is cyclic-guanosine 3′,5′-monophosphate (cGMP) which in some cases antagonises the actions of cAMP. For example, cAMP quickens the rhythmic beating of cultured rat heart cells whereas cGMP slows the rate of beating.

REFERENCES

BISSET, G. W. (1974). In *Handbook of physiology*, Section 7. *Endocrinology*, Vol. IV, Part 1, p. 493. American Physiological Society, Washington, DC.
British Medical Journal (1974). New assays for old. (Editorial.) *Br. med. J.* i, 128.
CATT, K. J. (1971). *An ABC of endocrinology*. The Lancet, London.
DALY, J. R., LOVERIDGE, N., BITENSKY, L., and CHAYEN, J. (1974). *Clin. Endocr.* 3, 311.
GREENGARD, P., ROBISON, G. A., and PAOLETTI, R. (eds.) (1972). *Advances in cyclic nucleotide research*, Vol. 1. *Physiology and pharmacology of cyclic AMP*. Raven Press, New York.
HARRIS, G. W., REED, M., and FAWCETT, C. P. (1966). Hypothalamic releasing factors. *Br. med. Bull.* 22, 266.
LEE, J. and LAYCOCK, J. (1978). *Essential endocrinology*. Oxford University Press.
MARTIN, C. R. (1976). *Textbook of endocrine physiology*. Williams and Wilkins, Baltimore.
MARTIN, J. B., REICHLIN, S., and BROWN, G. M. (1977). *Clinical neuro-endocrinology*. Davis, Philadelphia.
O'MALLEY, B. W. and SCHRADER, W. T. (1976). *Scient. Am.* 234, 32.
RASMUSSEN, H. (1974). Organization and control of endocrine systems .In *Textbook of endocrinology*, 5th edn (ed. R. H. Williams) p. 1. Saunders, Philadelphia.
SMITH, A. D. (1972). *Storage and secretion of hormones*. Scientific Basis of Medicine: Annual Reviews, p. 74. Athlone, London.
SZEGO, C. M. (1974). The lysosome as a mediator of hormone action. *Recent Prog. Horm. Res.* 30, 171.
TEPPERMAN, J. (1973). *Metabolic and endocrine physiology*, 3rd edn. Year Book, Chicago.

The endocrine functions of the pancreas

The exocrine secretion of the pancreas (the pancreatic juice) contains enzymes which promote the digestion of carbohydrate, proteins, and fats to molecules small enough to be absorbed by the intestinal mucosa. These small molecules pass mainly by the portal vein to the liver; some of the digested lipids pass into lymphatics and thence directly into the systemic circulation. The two most important endocrine secretions of the pancreas (insulin and glucagon) also enter the portal vein in relatively high concentrations and are transported direct to the liver where they exert very important actions regulating the metabolism of carbohydrate, fat, and proteins. Thus, the exocrine and endocrine functions of the pancreas are closely interrelated. This view is strongly supported by *embryological* studies which show that the endocrine cells are formed by division of some of the acinar cells to form richly vascularized and innervated islets which have lost direct contact with the lumen of the smaller branches of the pancreatic duct. Before birth, the alpha cells secrete glucagon which prevents fetal hypoglycaemia, but after birth, the normal bihormonal regulatory control by means of insulin from the beta cells, and glucagon, comes into operation and remains active throughout life.

Anatomy and histology

The pancreas is located within the curve of the duodenum. In normal adult man the pancreas weighs 50–75 g, of which 1 g is islet tissue. There are between 0.5 and 1.5 million islets of Langerhans, each being from 75 to 175 µm in diameter.

The islet cells are histologically of three types:

α(alpha)-cells which synthesize and secrete glucagon (15–25 per cent of total)

β(beta)-cells which synthesize and secrete insulin (70–80 per cent)

δ(delta)-cells which may secrete small amounts of gastrin and somatostatin.

The different types of cell are best characterized by the secretory granules which they contain. The granules can be distinguished by their staining properties but even better by electron micrography.

In man, the α-cells are widely distributed in the islets throughout the pancreas. Their secretory granules are round or ovoid, of high electron density, and variable in size. They contain immunoreactive glucagon. α-cells identical with those found in pancreatic islets have also been found in the gastric antrum and in the duodenum. The β-cells contain fewer and larger granules, of less electron density than in α-cells. Many of the granules show as rhomboid crystals surrounded by a clear zone within the limiting membrane. Immunochemical and biochemical studies have shown the β-cell granules to contain insulin and zinc. The δ-cells contain granules of low electron density and immunofluorescence studies suggest the presence of gastrin. The α- and β-cells are separated only by pericapillary spaces, so glucagon could easily be transported to act directly on β-cells and thus promote the secretion of insulin.

The islets are innervated by unmyelinated fibres from both parasympathetic (vagal) and sympathetic nerves whose endings are in close contact with α- and β-cells and can readily influence their secretory activities.

Chemistry of insulin, proinsulin, and glucagon

In 1921, Banting and Best extracted insulin from the pancreas and showed that it reduced blood sugar in dogs. Within a year insulin was sufficiently purified for therapeutic administration to patients with diabetes mellitus, in whom it produced dramatically beneficial effects. Crystalline insulin was prepared by Abel in 1926, the detailed amino-acid structure was elucidated by Sanger in 1952, proinsulin was discovered by Steiner in 1967, and the three-dimensional structure of insulin was revealed by Dorothy Hodgkin in 1969. Insulin and proinsulin can occur as monomers, dimers, or hexamers. β-cell granules contain hexamers incorporating zinc, but the biologically active circulating form of insulin is the monomer.

Proinsulin [Fig. XI.5] is a polypeptide consisting of a chain of 81–86 amino-acid residues, according to species. The molecular weight of the monomer is about 9000. Removal of the connecting peptide from proinsulin produces *insulin* which contains 51 amino-acid residues arranged in two chains, an acidic A-chain containing 21 amino-acid residues, and a basic B-chain containing 30 residues, the chains being joined by two S-S bonds. Thus insulin is a small protein (or a large polypeptide) with a molecular weight of 6000. Its detailed amino-acid composition varies slightly among different animal species. These variations do not affect biological activity (on metabolic processes) but they do account for differences in immunological response. For example, insulin from cattle or sheep is liable to evoke in man the formation of antibodies which cause allergic reactions, whereas pig insulin, which more closely resembles human insulin in structure, is much better tolerated.

Glucagon. Banting and Best detected initial hyperglycaemic activity in their original pancreatic extracts, and in 1923 Murlin called the substance responsible for this activity 'glucagon' (mobilizer of glucose). Many commercial preparations of insulin have been contaminated with glucagon but although α-cells have long been

known to contain glucagon the real physiological significance of this hormone has only recently been revealed (Lefebvre and Unger 1972).

Chemically, glucagon is a straight-chain polypeptide containing 29 amino-acid residues, giving a molecular weight of 2495.

Assay of insulin and glucagon in body fluids

Older methods of biological assay have been replaced by more sensitive and accurate techniques of radioimmunoassay [p. 498] in which the hormone is labelled with ^{131}I or ^{125}I. The amount of radioactive iodine should be not more than one atom of iodine per molecule of hormone, in order to avoid alterations in the biological activity of the hormone. Radioimmunoassay of insulin measures proinsulin as well. In man, the normal fasting plasma *insulin* concentration is 10–50 µiu ml^{-1} (iu = international units). The normal fasting plasma *glucagon* concentration is 100–150 pg ml^{-1}. An increase in blood glucose concentration normally increases the plasma concentration of insulin and reduces that of glucagon.

SYNTHESIS, STORAGE, AND RELEASE OF INSULIN AND GLUCAGON

Biosynthesis of insulin

Isolated islets or cultures of β-cells from the rat pancreas rapidly incorporate radioactive amino acids (e.g. [^{3}H]-leucine), into proinsulin which is synthesized on the membrane-associated ribosomes of the rough endoplasmic reticulum (RER). This translational process is inhibited by specific inhibitors of ribosomal protein synthesis (e.g. puromycin, cycloheximide). Proinsulin migrates in microvesicles from the RER to the Golgi apparatus where conversion to insulin begins to take place by removal of the connecting peptide which has served as a stabilizing 'scaffold' for the alignment of the A- and B-chains until the S-S bonds are formed [Fig. XI.5]. The formation of insulin continues within the storage granules, which therefore contain proinsulin, insulin, and C peptide. Labelled insulin is first detected within 10–15 min and the half-time for conversion of proinsulin to insulin is about one hour. Before discharge of the granule contents by exocytosis most of the proinsulin has been converted to insulin. The blood therefore contains a high proportion of insulin and a little proinsulin; the proportion of C peptide increases as proinsulin is converted to insulin, and since insulin and C peptide have different antigenic properties, measurement of plasma C peptide concentration by immunoassay is an index of plasma insulin concentration.

There is no comparable detailed understanding of the processes involved in the synthesis and release of glucagon.

Both insulin and glucagon enter the portal vein in relatively high concentration and about 50 per cent of both hormones is removed by the liver in a single passage through this organ. Some insulin is excreted by the kidney. Both hormones are inactivated mainly in the liver and kidney and both have a half-life of 5–10 minutes in circulating blood. They persist for much longer in lymph. Insulin is degraded by an enzyme, glutathione-insulin transhydrogenase, which reduces the S-S bonds to SH groups with separation of the A- and B-chains and loss of biological activity. Glucagon is degraded in tissues and in plasma by an aminopeptidase.

Insulin secretion

Insulin, proinsulin, and C peptide circulate in the peripheral blood plasma. The normal basal rate of insulin release from β-cells is 1 iu h^{-1} and the total daily release in man is about 50 iu out of a pancreatic store of 200–250 iu of insulin. Radioimmunoassay measures both insulin and proinsulin, but in peripheral blood the proportion of the latter is small (5–20 per cent); immunoassay can be used to measure C peptide which is split off from proinsulin and is

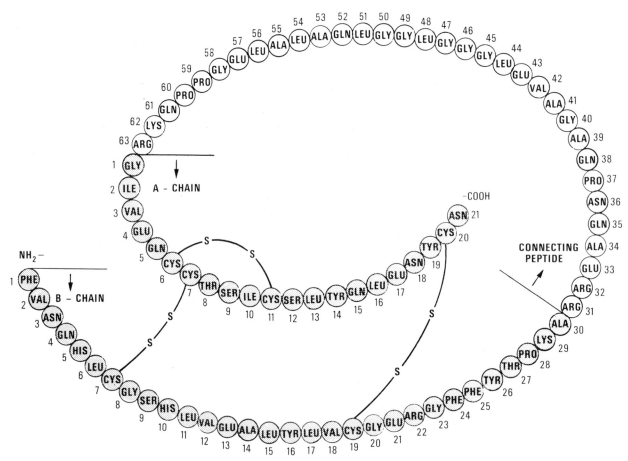

FIG. XI.5. Amino acid sequence of porcine proinsulin. Insulin is formed by removal of the connecting (C) peptide. (From Shaw, W. N. and Chance, R. E. (1968). *Diabetes* **17**, 738.)

proportional to the amount of insulin secreted. In the basal fasting state a normal non-obese man has a plasma insulin concentration of 10–50 µiu ml⁻¹ in peripheral blood plasma which rises 8–10 times after intravenous injection of 20 g of glucose. This response is biphasic with an early rise in blood insulin, commencing at 0.5–1.0 min, reaching a peak at 3–5 min after injection of glucose, and then declining to a level somewhat above the basal. After this *primary* response (first phase), continued infusion of glucose for 2–3 h to raise the blood concentration to over 5.5 mmol l⁻¹ produces a slower, maintained *secondary* increase (second phase) in insulin secretion which increases the plasma insulin concentration to 50–300 µiu ml⁻¹ [FIG. XI.6]. On the basis of experiments performed on the isolated perfused rat pancreas Grodsky (1972) has proposed a two-compartment model to explain the biphasic insulin response. The early primary response is ascribed to release of insulin from a *labile* pool of insulin within the β-islet cells. This labile pool comprises about 2 per cent of the total islet content of insulin. The delayed secondary response is ascribed to release of insulin from a *stable* pool which contains 98 per cent of β-cell content of insulin. If the glucose stimulus is maintained for a long time (e.g. >2 hours) the stable pool will receive some newly synthesized insulin to provide sufficient hormone to deal with its substrate.

Lacy and Malaisse (1973) have offered the following explanation of these observations. Glucose metabolites within the β-cell promote the passage of Ca²⁺ through the plasma cell membrane to raise the intracellular [Ca²⁺]. This causes contraction of microtubules and discharge by exocytosis of the granules already in the tubules. This accounts for the early phase of insulin secretion. During the secondary phase stored granules and newly synthesized granules enter the microtubules to be discharged in their turn by exocytosis.

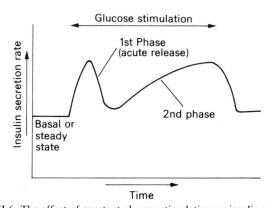

FIG. XI.6. The effect of constant glucose stimulation on insulin secretion. Note the two phases of insulin release and their separation from a basal or steady state prior to glucose stimulation.
(From Williams, R. H. and Porte, D. (1974). The pancreas. In *Textbook of endocrinology*, 5th edn (ed. R. H. Williams) p. 515. Saunders, Philadelphia.)

Regulation of insulin secretion

Basal insulin secretion may be controlled partly by blood glucose, partly by vagal activity and probably by other factors not yet understood. The higher the basal level of insulin secretion the greater is the secretion produced by glucose stimulation.

Substrate stimulation is the most important factor in the control of insulin secretion. In man carbohydrates and proteins are of prime importance, lipids having no significant effects. *Carbohydrates* promote insulin secretion after their degradation during digestion to glucose whose effects we have already discussed. Mannose is also an effective stimulant of insulin release. Both glucose and mannose are metabolized in β-cells. Release of insulin is effected mainly by their metabolites which activate adenyl cyclase and thus promote cyclic AMP formation. Extrusion of insulin granules depends also on Ca^{2+} and ATP. Intracellular micro-electrodes in isolated β-cells record depolarization and action potentials during insulin release by glucose. Fructose is a weak stimulant and probably acts after conversion to glucose. Some sugars, e.g. 2-deoxyglucose, and D-mannoheptulose (from avocado pears) which inhibit phosphorylation, also inhibit insulin secretion. It must be emphasized that glucose not only promotes release of insulin stored in β-cells but also enhances the synthesis of proinsulin, which is converted in these cells to insulin.

Amino acids. Leucine has been shown to release insulin both *in vivo* and from cell cultures of insulin-secreting tumours (the proportion of proinsulin is high in such cases). However, normal β-cells are more effectively stimulated by mixtures of essential amino acids or by basic amino acids such as arginine and lysine. These amino acids stimulate secretion of insulin from isolated islets in the absence of glucose, but the presence of glucose potentiates their effect [FIG. XI.7].

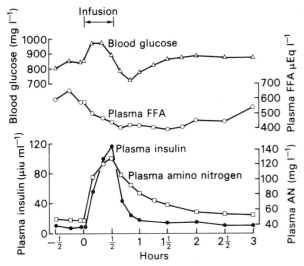

FIG. XI.7. Effect of i.v. administration of 30 g of a mixture of 10 essential amino acids upon mean plasma insulin, amino nitrogen (AN), free fatty acids (FFA), and blood glucose in 35 healthy subjects tested 51 times. (Redrawn from Floyd, J. C. *et al.* (1966). *J. clin Invest.* **45**, 1489.)

The promotion of insulin secretion by breakdown products of protein digestion must have been very appropriate 10 000 years ago when hunting man lived on a diet consisting almost entirely of protein and fat (Fajans and Floyd 1972). Amino acids also stimulate α-cells to secrete glucagon, which, by enhancing glycogenolysis and gluconeogenesis in the liver, would help to prevent insulin-induced hypoglycaemia without, however, reducing the effect of insulin in facilitating the transport of glucose and amino acids into cells.

Hormones

Gastro-intestinal hormones. During digestion of food, several gastro-intestinal hormones are released into the bloodstream to promote secretions of enzymes which degrade the large molecular constituents of foodstuffs to small molecular substances which are absorbed by the intestinal mucosa. These hormones include gastrin, secretin, and pancreozymin [pp. 427 and 431] all of which can either stimulate insulin secretion themselves or enhance the stimulant effects of glucose and amino acids. Glucagon and glucagon-like substances are also released from the gut into the bloodstream during digestion. The result of the release of these gastro-intestinal hormones is an anticipatory release of insulin which precedes the

much greater subsequent secretion of insulin in response to absorbed glucose and amino acids. Increased vagal activity during digestion enhances this gastro-intestinal hormone release as well as directly stimulating β-cell release of insulin.

Glucagon. Glucagon, from α-islet cells, and from the same type of cell in the stomach and duodenum has a powerful stimulant effect on β-islet cells. An immunologically separate 'glucagon-like' substance (*enteroglucagon*) may be released from the gut mucosa. The fact that orally administered glucose produces a greater increase in plasma insulin concentration than intravenous glucose has been attributed to release of enteroglucagon [FIG. XI.8].

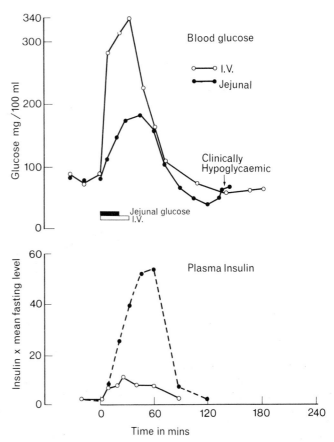

FIG. XI.8. Blood glucose and plasma insulin response to rapid infusion of 60 g glucose intravenously and intrajejunally ($\bullet$—$\bullet$—$\bullet$ = intrajejunal glucose; o—o—o = IV glucose) to normal subject. (McIntyre, N. (1965). *J. clin. Endocr.* **25**, 1317.)

Glucagon stimulates insulin secretion not only by a direct action on β-cells (which may be mediated by the adenyl cyclase–cyclic AMP system), but also indirectly by producing hyperglycaemia. Since hypoglycaemia is the most important stimulus of glucagon secretion the two hormones insulin and glucagon, both of which are secreted into the portal vein in high concentrations, have a co-ordinated function in keeping the blood glucose concentration constant within fairly narrow limits in spite of considerable variations in glucose uptake and utilization. For example, after a large carbohydrate meal insulin counteracts the tendency to hyperglycaemia, whereas during short periods of starvation glucagon helps to prevent hypoglycaemia and ensures an adequate glucose supply to the brain. Insulin and glucagon regulate the output of glucose from the liver by competitive antagonism on the processes concerned with formation and breakdown of glycogen and on those concerned with gluconeogenesis.

Adenohypophysis. Hypophysectomy in rats reduces the insulin content and secretory capacity of the islets. In man, hypophysectomy or panhypopituitarism is associated with reduced plasma insulin responses to glucose or arginine. Administration of growth hormone (GH) restores the insulin responses to normal. In acromegaly hypersecretion of GH enhances the insulin secretion induced by glucose or arginine.

Adrenal cortex. Adrenalectomy reduces glucose-induced insulin secretion and physiological amounts of cortisol restore the response to normal. Large, pharmacological doses of synthetic glucocorticoids (e.g. prednisone which is used to suppress inflammation), greatly enhance glucose-induced insulin secretion, but prolonged treatment may lead to exhaustion of β-islet cells, their failure to respond to glucose, and consequent hyperglycaemia.

Thyroid. Hypothyroidism depresses glucose-induced insulin secretion and thyroid hormones, T_3 and T_4 [p. 537] restore the normal response. Hyperthyroidism depletes the islets of insulin and progressive failure of β-cell function finally produces *metathyroid diabetes*.

Pregnancy enhances glucose-induced insulin secretion, probably owing to secretion of human placental lactogen (HPL).

Oral contraceptives (oestrogen + progesterone-like substances) tend to increase insulin secretion induced by glucose or arginine. Prolonged administration (for more than a year) may reduce insulin secretion in diabetic or potentially diabetic women.

Neural regulation of insulin secretion

The islets of Langerhans are innervated by postganglionic fibres of the parasympathetic (vagus) and sympathetic divisions of the autonomic nervous system.

Vagal stimulation increases the secretion of insulin by the β-cells. This action is mediated by release of ACh at cholinergic nerve endings; stable derivatives of choline, e.g. carbachol, have a similar muscarinic action. Atropine prevents the increased secretion evoked by vagal stimulation but does not alter the basal concentration of insulin in plasma.

Sympathetic nerve stimulation releases NA locally in the islets and splanchnic nerve stimulation releases both Ad and NA from the adrenal medulla into the bloodstream (and hence to the islets). These two catecholamines are the only *endogenous* substances which inhibit glucose- or glucagon-induced insulin secretion. They do so by stimulating α-adrenergic receptors on the β-cells; their inhibitory effect is antagonized by phentolamine, an α-receptor blocking drug, which not only restores the response to exogenous glucose but also raises the basal rate of insulin secretion. NA and Ad also stimulate β-adrenergic receptors, an action tending to increase insulin secretion but usually suppressed by α-receptor stimulation. However, when glucose and Ad are administered together, the suppression of insulin secretion during infusion is followed by an enhanced insulin secretion when the infusion is stopped. This 'rebound' effect is inhibited by the β-receptor blocking drug propranolol. Insulin secretion evoked by arginine or secretin is not inhibited by catecholamines.

Sympathetic nerve stimulation to the islets increases the secretion of glucagon in several species of experimental animal. Thus sympathetic nerve activity will tend to increase the release of glucose from the liver in three ways: 1. by reducing insulin-induced glycogenesis; 2. by increasing the breakdown of glycogen by catecholamines; and 3, most importantly, by release of glucagon which is the most powerful glycogenolytic agent of all.

Central nervous control of islet cell secretions

The autonomic nervous system is everywhere under central ner-

vous control, in man as well in experimental animals. During the period of feeding and absorption of foodstuffs, vagal activity is enhanced by excitation of neurons in the hypothalamus which elicit an increase in insulin secretion and thus promote anabolic processes such as glycogenesis, lipogenesis, and protein synthesis. During starvation hypothalamus-mediated increase in sympathetic activity will, via catecholamines and glucagon, increase catabolic processes to provide energy by release of glucose from liver glycogen, FFA from adipose tissue, and amino acids (for deamination) from muscle; at the same time insulin secretion is reduced.

Some of the important factors which stimulate or inhibit insulin secretion are summarized as follows:

Insulin release	
Stimulated by	Inhibited by
Glucose	Adrenalin
Amino acids	Noradrenalin
Glucagon	D-Mannoheptulose
Sulphonyl ureas	2-Deoxyglucose
Gastrin	Diazoxide
Secretin	Vagotomy
Pancreozymin	Somatostatin (growth hormone-release inhibiting hormone)
Vagal stimulation	

Regulation of glucagon secretion

Since insulin and glucagon seem to act as co-ordinated antagonists in the normal regulation of carbohydrate metabolism it is appropriate to consider here the factors which control the α-cell secretion of glucagon.

Substrate control. A *carbohydrate* meal in normal persons produces a rise in blood glucose concentration, a rise in plasma insulin concentration and a fall in plasma glucagon concentration [FIG. XI.9]. Administration of insulin to produce hypoglycaemia (blood glucose 3 mmol l^{-1}) increases the plasma glucagon concentration. If phloridzin is administered to dogs to produce chronic hypoglycaemia (through excessive renal excretion of glucose) plasma glucagon concentration also increases. Thus the most probable stimulus to glucagon secretion is hypoglycaemia *per se*, and

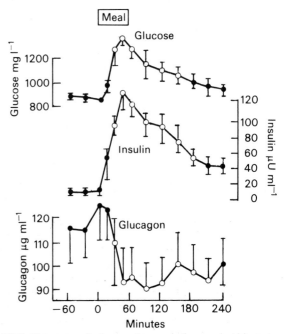

FIG. XI.9. Response of plasma glucagon of normal subjects to a large carbohydrate meal (mean = SEM).
(Redrawn from Unger, R. H. (1970). *New Engl. J. Med.* **283**, 109.)

this has been confirmed by reducing the glucose concentration in the perfusion fluid of the isolated rat pancreas and recording an increased glucagon output.

Protein and amino acids. Oral administration of a protein meal, or intravenous infusion of amino acids, increases the plasma concentration of glucagon as well as that of insulin. Arginine is a potent stimulant of both glucagon and insulin secretion; most glucogenic amino acids increase secretion of both hormones, but alanine probably stimulates only glucagon secretion directly. The physiological significance of this simultaneous stimulation of both α- and β-cells by amino acids is assumed to be the prevention of hypoglycaemia while insulin enhances amino-acid transport into peripheral tissues and promotes protein synthesis. Glucagon, by enhancing glucose output from the liver, ensures an adequate blood glucose supply to the brain.

Fatty acids inhibit glucagon secretion in experimental animals but do not appear to do so in man.

Neural control

Glucagon secretion is increased during exercise in man and this is probably due to increased *sympathetic* activity on the α-islet cells. The enhanced sympathetic activity comes from the hypothalamus. *Vagal stimulation also increases glucagon secretion in man (Bloom 1975).*

Somatostatin (growth hormone-release inhibiting hormone, GH-RIH)

This tetradecapeptide is one of the hypothalamic hormones which passes via the portal vessels to the adenohypophysis [p. 520]. As its name suggests, it inhibits the secretion of growth hormone by the adenohypophysis, but in addition it is a powerful inhibitor of both insulin and glucagon secretion. It acts directly on islet cells and the important implications of its effects on metabolism will be discussed later.

Actions of insulin and glucagon

Since Banting and Best's discovery of insulin in 1921 it has been widely accepted that the disease *diabetes mellitus* is due to insulin deficiency. The original observation of Minkowski and Von Mering in 1889 that pancreatectomy produced a diabetes-like syndrome in dogs was the basis of numerous attempts to isolate from the pancreas a hormone which would restore normal control of the deranged carbohydrate metabolism, and the dramatic effects of insulin administration in reducing the mortality rate of severe ketoacidotic diabetes supported the insulin deficiency hypothesis. However, in 1930 Houssay showed that experimental diabetes in animals could be largely alleviated by hypophysectomy, although the condition of animals without adenohypophysial and pancreatic islet control was unstable. Later it was found that growth hormone was the main diabetogenic pituitary factor, with ACTH as a contributory factor. Recently the role of glucagon has come to be regarded as a fundamental requirement for the production of the diabetic syndrome (Unger 1975) so that in discussing the actions of insulin it is necessary to relate them to the largely antagonistic actions of glucagon. It is also essential to realize that both insulin and glucagon not only promote positively reactions proceeding in one direction, but simultaneously they inhibit metabolic reactions proceeding in the opposite direction. To give examples: insulin produces hypoglycaemia by enhancing the activities of enzymes which promote glycolysis, by stimulating the enzymes which promote glycogen synthesis and, at the same time, inhibiting the enzymatic breakdown of glycogen. Conversely, glucagon causes hyperglycaemia by enhancing the activity of enzymes which promote glycogenolysis and simultaneously inhibiting the activity of glycogen synthetase.

Unger suggests that the α- and β-islet cells form a bihormonal functional unit in which insulin is mainly concerned with anabolic, and glucagon with catabolic, effects on carbohydrate, lipid, and protein metabolism. Their secretions and actions are influenced by other hormones and by autonomic nervous control.

Actions of insulin (and proinsulin)

Studies on normal persons, diabetic patients, intact animals, isolated perfused islets, tissue and cell cultures, using native insulin or labelled (^{131}I, ^{125}I, ^{3}H) insulin or proinsulin have revealed the following information:

1. Insulin is 10–20 times as active as proinsulin on responsive tissues. C peptide is biologically inactive.

2. Insulin does not produce its physiological effects on disrupted cells but it does become bound to separated plasma membranes of responsive cells.

3. All the actions of insulin are brought about by a reversible combination with specific glycoprotein receptors on the surface of the plasma cell membrane. Insulin does not penetrate into the cell yet it exerts profound effects on the intracellular activities of *muscle cells*, *adipocytes*, and *liver cells*. Its biological activity is proportional to the amount bound to the cell surface.

Actions on cell membrane. Insulin acts upon the plasma membrane of muscle and adipose tissue cells to facilitate the transport of glucose, amino acids, K^+ and Mg^{2+}, and inorganic phosphate into the cell. The presence of glucose is not necessary for the insulin-facilitated entry of amino acids and ions into responsive cells. Insulin also facilitates the transport of certain hexoses, e.g. galactose, which are not metabolized in muscle and adipose cells. Glucose itself can cross muscle cell membranes in the absence of insulin, but only if the glucose concentration outside the cells is raised to many times the normal level. In normal persons there is sufficient insulin in plasma to promote glucose uptake with a resting blood [glucose] of 3–5 mmol l^{-1} (50–90 mg per 100 ml). Glucose transport across intestinal epithelial cells, proximal renal tubular cells, erythrocytes, and nerves does not require insulin.

If insulin itself is bound only to the outer surface of the plasma membrane of muscle and adipose cells, how are its effects transmitted to the interior of the cell? Two possibilities have been proposed (Krahl 1972):

(i) Insulin may reduce cAMP formation, particularly in adipose cells. This will reduce the lipolytic response to adrenalin or glucagon and favour anabolic processes leading to synthesis of glycogen and deposition of triglycerides. The attachment of insulin to the cell membrane surface may elicit a propagated disturbance through the membrane which may either inhibit the activity of adenyl cyclase or enhance that of phosphodiesterase, thus reducing cAMP concentration. Both these enzymes are located on the inner surface of the cell membrane.

(ii) The facilitated transport of ions into insulin-responsive cells may influence intracellular enzyme activities. Changes in [K^+] may shift the balance between glycogen synthesis and glycogen breakdown.

Insulin stimulation of protein synthesis in isolated adipose cells does not occur in the absence of Ca^{2+} or Mg^{2+}. Ca^{2+} affects membrane permeability but does not pass into cells; Mg^{2+} activates several intracellular enzymes.

The main actions of insulin are exerted on *muscle, adipose tissue, and liver.* The state of insulin deficiency will be considered under diabetes mellitus, in which other hormones contribute to the metabolic and pathological changes characteristic of this disease.

Muscle and adipose tissue. In these tissues the uptake of glucose is the rate-limiting step for subsequent intracellular glucose metabolism. Thus insulin, by facilitating glucose entry into muscle and adipose cells enhances all aspects of glucose metabolism, that is, glycogen formation, glycolysis, and oxidation (shown by increased O_2 consumption and CO_2 output); the hexose monophosphate shunt (with associated NADPH) and FFA synthesis are all increased. In adipose cells insulin enhances lipid synthesis by providing acetyl-CoA and NADPH for FFA synthesis and glycerol, from α-glycerophosphate, for the formation of triglycerides.

The insulin-facilitated transport of amino acids into muscle is independent of protein synthesis since the enhanced entry still occurs after inhibition of protein synthesis by puromycin. Insulin also facilitates entry of non-metabolized amino acids, e.g. α-aminoisobutyrate.

Insulin acts on several enzymes. It activates hexokinase to convert glucose to intracellular glucose 6-phosphate which is important for two reasons: (i) glucose 6-phosphate cannot pass across the cell membrane, and in muscle there is no phosphatase to release the diffusible glucose (cf. the liver); (ii) glucose 6-phosphate is the starting substance for the various pathways of carbohydrate metabolism mentioned above. Insulin activates glycogen synthetase and simultaneously inhibits enzymic breakdown of glycogen. It enhances the activity of some glycolytic enzymes and inhibits enzymes favouring glucose synthesis. Insulin inhibits lipase in adipose tissue and so reduces the release of FFA. This may be due to an effect on [cAMP] either by inhibiting excitation of adenyl cyclase by other hormones (glucagon, catecholamines, GH, etc.) or by enhancing the activity of phosphodiesterase.

Insulin promotes protein synthesis in muscle by a direct action on ribosomes to increase the translational activity of mRNA. This effect can still occur when RNA synthesis is inhibited by actinomycin D. In experimental diabetes mellitus the polysomes become disaggregated and insulin *in vivo* restores the normal aggregated state.

Liver. The liver cell membrane is freely permeable to glucose so that its extra- and intracellular concentrations are about equal and not dependent on insulin or other hormones. However, insulin does bind specifically and reversibly to the plasma membrane of liver cells and thereby exerts several important actions which are competitively antagonized by glucagon. Some of the actions of insulin on the liver are secondary to its peripheral actions on muscle and adipose tissue, which reduce the amounts of glucose, FFA and amino acids reaching the liver via the bloodstream. However, in the isolated perfused liver insulin acts directly to decrease glucose and urea output from the liver and to increase the uptake of K^+ and phosphate. Insulin may act on a genetic locus in the nucleus which contains the genome for a group of specific enzymes, such as glucokinase, phosphofructokinase, and pyruvate kinase which promote glycolysis, and glycogen synthetase which leads to glycogen formation; insulin represses the enzymes which induce gluconeogenesis. These insulin-induced changes in liver enzyme activities are inhibited by blocking RNA and protein synthesis by actinomycin D and puromycin.

To summarize, the metabolic actions of insulin are:

(i) To promote glucose uptake and catabolism in muscle, adipose tissue and liver.
(ii) To promote synthesis of glycogen.
(iii) To promote synthesis of FFA and triglycerides.
(iv) To promote amino acid uptake, to diminish protein breakdown and to promote protein synthesis.

Actions (i) and (ii) account for the *hypoglycaemia* induced by insulin. Actions (ii), (iii), and (iv) show insulin to be primarily an anabolic hormone with respect to the three main foodstuffs, carbohydrate, fat, and protein.

Interactions between insulin and other hormones

Many other hormones interact with insulin, mostly as antagonists. The effects of insulin in the body can be regulated either by control of insulin secretion itself, or by secretion of insulin antagonists. The most important antagonist is glucagon which participates in many homeostatic processes.

Glucagon acts mostly on the *liver* and *adipose tissue* where it antagonizes the actions of insulin:

1. Glucagon raises blood glucose concentration by enhancing the breakdown of liver glycogen to glucose (*glycogenolysis*) and by promoting *gluconeogenesis* from lactate, pyruvate, glycerol, and amino acids. Glycogenolysis produces a rapid rise in blood glucose within a few minutes; gluconeogenesis produces a slower, more sustained rise in blood glucose lasting for hours or days. Glycogenolysis is mediated by activation of adenyl cyclase in the hepatic cell membrane and subsequent increase in intracellular [cAMP] and activation of protein kinases which in turn activate the phosphorylase responsible for converting glycogen to glucose 6-phosphate, and inactivates glycogen synthetase [p. 451]. Phosphatase in the liver then acts on glucose 6-phosphate to release glucose into the hepatic venous blood. Catecholamines (Ad, NA) act similarly to enhance glycogenolysis but, on a molar basis, they are weaker than glucagon; however, NA released locally at sympathetic nerve terminals might have powerful effects.

Glucagon as a stimulant of glucose output from the liver is, on a molar basis, more potent than insulin as a promoter of glucose retention.

2. Glucagon is a powerful *lipolytic* agent, acting via cAMP to phosphorylate a lipase in adipose tissue which releases FFA and glycerol into the circulation. In the liver FFA in excess are converted to ketone bodies (acetoacetic acid, acetone, and β-hydroxybutyric acid).

Catecholamines antagonize insulin by increasing cAMP formation in the liver, fat and muscle; in the liver this activates phosphorylase, promotes glycogenolysis, and leads to hyperglycaemia. They promote glycogenolysis in muscle and enhance lactate formation; lactate returns via the bloodstream to the liver and is converted to glucose. Catecholamines promote lipolysis in adipose tissue and proteolysis in muscle.

Adrenal medullary tumours (phaeochromocytoma) may cause transient hyperglycaemia and glycosuria.

Diabetes mellitus

Our understanding of the functions of the endocrine pancreas (α- and β-cells) has developed from the recognition and study of the disease *diabetes mellitus*. The word 'diabetes', meaning a 'siphon' or 'running through', was used by Aretaeus the Cappadocian in the second century AD to describe the polyuria; he also noted thirst and emaciation as features of this fatal disease. Susruta in India is said to have referred to diabetes mellitus as 'honey urine' in the fifth century AD. Willis, in 1679, wrote 'those labouring with this disease piss a great deal more than they drink' and went on to say that the urine 'is wonderfully sweet as if it were imbued with Honey or Sugar'. In 1815 the famous French chemist Chevreul (who died at the age of 103!) discovered that the sugar in diabetic urine was *glucose*. By the nineteenth century 'diabetes insipidus' (polyuria with tasteless urine), now ascribed to deficient secretion of antidiuretic hormone from the neurohypophysis [p. 523], was distinguished from diabetes mellitus in which there was in Willis' words 'a running through of sweet urine'. In the 1850s Claude Bernard described the 'internal secretion' of glucose into the blood, from its storage form 'glycogen' in the liver. In 1889 Minkowski and von Mering showed that complete pancreatectomy in dogs produced a condition corresponding to severe diabetes mellitus in man, characterized by polyuria, thirst, emaciation, hunger, glycosuria, hyperglycaemia, and ketonuria, leading to coma and death. After many investigators had attempted to extract an active principle from the pancreas Banting and Best finally succeeded in 1921 in preparing insulin in a form which was effective in overcoming experimental diabetes in pancreatectomized dogs and was also active and highly successful in the treatment of patients with severe diabetes mellitus. This was, and still is, the most outstanding advance in our understanding and control of diabetes mellitus. However, knowledge of hormonal interactions have shown that this disease is not due solely to insulin deficiency but also to the actions of hormones, such as glucagon and GH, which are in many ways antagonistic to insulin.

Features of diabetes mellitus

The most characteristic features of diabetes mellitus are raised blood glucose concentration (hyperglycaemia) and the presence of glucose in the urine (glycosuria). The most typical form is *hereditary idiopathic diabetes mellitus* which occurs in about 1 per cent of the population in the UK and USA. The hereditary aspect is multifactorial, a view which is supported by the very high incidence of the disease in identical twins (where one twin develops the disease the other is almost sure to follow) and in children of parents who are both diabetic. From intensive studies of diabetic patients, their relatives, and population surveys it has been shown that progressive stages in the development of diabetes can be recognized:

1. **Prediabetes (potential diabetes).** These terms are applied to persons with a strong genetic predisposition to diabetes but who do not yet show any abnormality of carbohydrate metabolism. Hypertrophy of islet tissue may produce more insulin than normal and so delay the onset of diabetes.

2. **Latent diabetes** is the term applied to asymptomatic persons who develop glycosuria after stress or glucocorticoid administration. This occurs most commonly in obese subjects in whom plasma insulin concentration is higher than normal after a carbohydrate meal. Excess of fatty tissue creates resistance to the actions of insulin and the normal state can be restored by restriction of carbohydrate intake which reduces body weight by reducing lipogenesis.

3. **Chemical diabetes mellitus** in which there is hyperglycaemia, glycosuria, and symptoms such as thirst, polyuria, and weight loss or gain, which may go unrecognized for a time.

4. **Overt diabetes mellitus** with symptoms takes two main forms with the following features:

Juvenile onset	Maturity onset
(a) Underweight	(a) Normal or overweight
(b) Ketosis, if untreated	(b) Ketosis with severe infection; often absent
(c) Insulin secretion low or absent (a very transient increase may occur at first)	(c) Normal or increased insulin secretion. In advanced cases insulin secretion depressed
(d) Insulin treatment needed	(d) Diet and oral hypoglycaemic drugs effective. Insulin needed only when infection occurs
(e) Patients are sensitive to insulin	(e) Insulin resistant

Secondary diabetes mellitus

Diabetes mellitus may occur with *pancreatitis* leading to fibrosis and destruction of islet tissue and reduced insulin secretion. Excessive secretion of certain *hormones* such as GH and cortisol may produce diabetes (e.g. in acromegaly and Cushing's syndrome).

Aetiology of idiopathic diabetes mellitus

The word 'idiopathic' implies ignorance of causation. Among the multifactorial predisposing factors are:

Heredity, already mentioned. Although only 1 per cent of the population have overt diabetes, 25 per cent are said to be 'potential' diabetics.

Age. The disease is commoner with increasing age.

Obesity. Adipose tissue in obese persons is more resistant to insulin actions than normal adipose tissue. Hence plasma insulin is raised in non-diabetic obese subjects. The increased strain on β-islet cell activity in producing more insulin may eventually lead to exhaustion of β-cells and the onset of diabetes mellitus. Carbohydrate restriction reduces body weight and restores normal sensitivity of adipose tissue to insulin.

Insulin antagonism. Reduction or absence of insulin secretion only produces diabetes mellitus when α-cells are secreting glucagon. Pancreatectomy does not remove the glucagon-secreting α-cells of the stomach and duodenum. In all cases of idiopathic clinical diabetes mellitus and diabetes in animals there is relative or absolute increase in plasma glucagon concentration. Somatostatin [p. 520] suppresses secretion of insulin, glucagon, and GH and prevents the development of experimental diabetes in animals (produced by pancreatectomy, alloxan, or streptozotocin).

Islet destruction. In a few diabetics post-mortem histological studies have revealed lymphocyte infiltration of islet tissue. This has been ascribed speculatively to viral infection, or an autoimmune response. IgG antibodies to pancreatic islet cells have been detected by immunofluorescence in the sera of some patients with insulin-dependent diabetes.

Islet hypertrophy has been reported in the early stages of diabetes. In juvenile cases atrophy quickly follows and in mature type cases atrophy develops much more slowly.

Experimental diabetes mellitus

Various procedures have been introduced to study the diabetic state in animals.

Experimental hereditary obesity in mice is perhaps relevant to the diabetic state which occurs in obese patients. The mice have hyperglycaemia and are very resistant to the actions of insulin; adipocytes and liver cells bind less insulin than normal. However, the outstanding biochemical abnormality in these obese-hyperglycaemic mice is the lack of mobilization of FFA in response to the usual lipolytic stimuli (glucagon, catecholamines, GH, etc.), which derives from the abnormal presence of the enzyme *glycerokinase* in the adipose tissue. This enzyme phosphorylates glycerol to α-glycerophosphate so that any glycerol released from triglyceride in adipocytes will immediately become available for combination with FFA to reform triglyceride within the adipocyte. Thus, there will be no release of FFA and glycerol into the circulation. It remains to be seen how far this model is related to human diabetes.

Pancreatectomy produces a diabetic-like syndrome in experimental animals, due to insulin lack but abolition of exocrine secretion will adversely affect digestion and absorption of foodstuffs.

Chemical destruction of β-islet cells has been achieved by administration of *alloxan* which is also toxic to the kidneys but leaves α-islet cells intact. *Streptozotocin,* an antibiotic which inhibits the growth of certain experimental tumours, is more specifically toxic to β-cells and produces a 'purer' diabetic state than does alloxan. It is important to note that after chemical destruction of β-islet cells the insulin requirement is reduced by subsequent pancreatectomy. This supports the view that the secretion of α-cells contributes to the establishment of the diabetic state.

Anti-insulin serum. Intravenous infusion of anti-insulin serum in rats rapidly induces hyperglycaemia, glycosuria, and ketonuria. The cells become markedly degranulated as insulin is secreted in response to the rising blood glucose concentration. The diabetic state is transient.

Symptoms, signs, and metabolic disturbances in diabetes mellitus

The diabetic syndrome is due to insulin deficiency combined with positive actions of hormones which are normally antagonistic to insulin (glucagon, growth hormone, adrenal glucocorticoids).

Hyperglycaemia, with resting blood glucose rising to 8–11 mmol l^{-1}, and postprandial rises to 22–28 mmol l^{-1}, produces *glycosuria* (when the blood glucose exceeds the normal renal threshold of 8–10 mmol l^{-1}). Hyperglycaemia is not directly harmful to the body but glycosuria produces an osmotic diuresis with loss of water and electrolytes from the body. This *polyuria* causes *thirst* and an increased water intake (polydipsia) as a response to dehydration. Loss of glucose means loss of energy but more important is the electrolyte loss. In acute cases severe loss of water and NaCl may lead to nausea and vomiting which makes dehydration worse. The ensuing haemoconcentration and reduced blood volume leads to circulatory failure, hypotension, reduced renal blood flow, and anuria, ending in coma and death.

In diabetes of more gradual onset loss of fluid may partly account for *loss of weight*, which in juvenile cases may occur in spite of *increased appetite*. Loss of weight in young diabetics may also be due to mobilization of fat and breakdown of proteins, especially in muscle. However, obesity is the main feature in many diabetics of the maturity onset type.

The mobilization of triglycerides from depot fat leads to *hyper-triglyceridaemia* and the liver becomes flooded with fat. In these circumstances it cannot completely oxidize all the FFA; with a high rate of FFA oxidation the process tends to stop at the acetyl CoA stage and the two-carbon units condense to form large amounts of acetoacetic acid and β-hydroxybutyric acid. Acetoacetic acid decarboxylates spontaneously to form acetone. These three substances are called *ketone bodies*.

$$CH_3 - C - CH_2 - COOH$$
Acetoacetic acid

$$CO_2 \quad CH_3 - C - CH_3$$
Acetone
spontaneous

$$CH_3 - CH - CH_2 - COOH$$
β-Hydroxybutyric acid

Ketone bodies are formed only in the liver. In acute insulin deficiency their formation produces a metabolic acidaemia (blood pH may fall to 7.0) which stimulates deep and rapid (Kussmaul) breathing. This eliminates the excess CO_2 released by the buffering action of the H_2CO_3/HCO_3^- system.

As the ketonaemia increases, the ketoacids are excreted in the urine. The urinary excretion of NH_4^+, and later of Na^+, is increased, severe Na^+ and water loss lead to more severe dehydration, which is made even worse by the vomiting induced by acidaemic ketosis. These effects reinforce the osmotic effects of glycosuria and in combination they lead to peripheral circulatory failure, coma, and death [FIG. XI.10].

Insulin lack impairs *protein synthesis* and promotes protein breakdown, especially in muscle. Blood amino-acid concentration rises and in the liver deamination is one of the chief causes of gluconeogenesis. The NH_2 of amino acids is converted to urea and excreted in the urine, giving a negative nitrogen balance. Increased protein catabolism leads to wasting. Protein depletion predisposes to *infection* and the glucose-rich body fluids form a good culture medium for bacteria; hence the liability to furunculosis (boils) and infections of the urinary tract. Increased breakdown of protein and tissue hypoxia due to peripheral circulatory failure promote K^+ loss from ICF to ECF and subsequent excretion of K^+ in the urine.

FIGURE XI.10 summarizes the numerous metabolic and function-

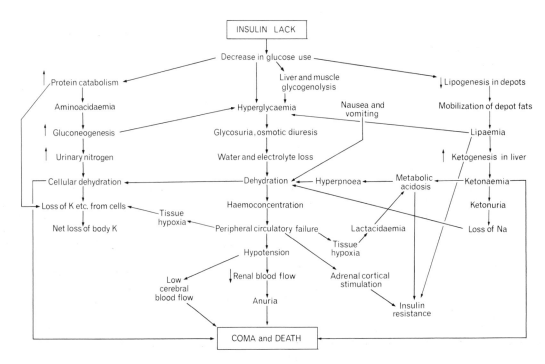

GIG. XI.10. Composite summary of the pathophysiology of diabetic acidosis. Note, particularly, connections among the three general areas of metabolism. (Tepperman, J. (1973). *Metabolic and endocrine physiology*, 3rd edn Wiley, New York.)

al disturbances which occur in diabetic ketoacidosis (Tepperman 1973). The many compensatory attempts to maintain homeostasis finally fail and death ensues.

In diabetic ketoacidosis there is marked resistance to insulin. This might be due to the presence of large amounts of FFA or to relative or absolute increases in the activities of glucagon, GH, or cortisol.

Ketoacidaemic coma must be distinguished, if possible, from the more serious hypoglycaemic coma (Hall *et al.* 1974).

	Ketoacidaemia	Hypoglycaemia
Precipitating factors	Infection, trauma, errors in insulin dosage	In insulin-treated diabetic-missed meal, undue exercise, errors in insulin dosage
Rate of onset	Hours or days	Minutes
Breathing	Deep and rapid; 'Air hunger'	Stertorous
Hydration	Marked dehydration	Normal
Sweating	Absent	Usually marked
CNS	Perhaps diminished reflexes	Various. Often bilateral extensor plantar responses
Urine	Glycosuria and ketonuria marked	No characteristic feature

Chronic complications of diabetes mellitus

In addition to the acute severe complications of insulin deficiency just described there are chronic complications which may occur more frequently in patients who have been inadequately treated (diet, insulin, hypoglycaemic drugs). Poor control of the disease is thought to predispose to the development of degenerative arterial disease (atherosclerosis) in the coronary, cerebral, and peripheral arteries. This has been attributed to the hyperlipaemia and hypercholesterolaemia of the chronic diabetic state. However, in diabetes there is a characteristic vascular lesion, microangiopathy, in which the capillary basement membrane is thicker than normal, and this may be responsible for nephropathy and retinopathy, as well as contributing to neuropathy (degeneration of sensory and motor nerves in the lower part of the body).

There is still argument about how far successful control of hyperglycaemia and glycosuria influences the development of the chronic complications of diabetes. It must be remembered that insulin treatment although highly successful in dealing with most juvenile diabetics, especially in acute ketoacidotic crises, does not control blood glucose concentration during each 24-hour period as effectively as the normal endogenous homeostatic control. Insulin treatment cannot fully replace the precise control effected by the co-ordinated activity of islet α- and β-cells secreting glucagon and insulin respectively into the portal blood, according to the demands of the moment. It is possible that the combined administration of insulin and somatostatin (which suppresses glucagon and GH secretion) could be more effective than insulin alone in controlling all the metabolic disturbances of the diabetic state.

Physiological principles in treatment of diabetes mellitus

1. Diet. In *juvenile* patients, who always need insulin, the diet should be sufficient to promote gain in weight to a level slightly below normal for age and height.

In *maturity onset diabetes* with *obesity* a low energy, low carbohydrate diet is by itself effective in controlling the disease in most patients. Reduction in the amount of body fat increases the sensitivity to endogenous insulin, diminishes the need for excessive secretion of insulin by β-cells and prevents β-cell exhaustion.

2. Insulin is given to control the hyperglycaemia, glycosuria, and ketosis. This is essential in juvenile diabetics who may secrete no insulin at all. Ideally, insulin should be given frequently enough to control the biochemical abnormalities throughout the 24 hours. In practice this is impossible because of speedy degradation of soluble insulin in the body (half-life 15 min). Long acting suspensions of insulin are often used but in neither case can the plasma concentration of insulin be regulated as accurately as that of endogenous insulin in normal persons. Because it is a small protein (or large polypeptide) insulin is destroyed by proteolytic enzymes in the alimentary tract. It is therefore given by injection.

Insulin therapy is intended to restore normal metabolism of carbohydrate, lipids, and protein. Since glucagon and GH produce many of the features of diabetes mellitus, *somatostatin*, which inhibits secretion of both these hormones, is being tested, together with insulin, as a form of mixed hormonal therapy which might control the metabolic disturbances better than insulin alone.

3. Oral hypoglycaemic drugs are of two types:

(a) *Sulphonyl ureas* (tolbutamide, chlorpropamide, glibenclamide). These drugs stimulate insulin secretion from β-cells and reduce hepatic output of glucose; glibenclamide is said also to enhance insulin synthesis. They are of no value in juvenile diabetics but *may* help in non-obese maturity-onset cases.

(b) *Biguanides* (Metformin, Phenformin) unlike the sulphonyl ureas do not lower blood glucose in normal persons and do not stimulate insulin secretion, though they are only effective in the presence of insulin. They increase cellular glucose uptake, reduce glucose output by the liver, and reduce glucose absorption from the gut.

Hypoglycaemia

The normal fasting blood glucose level (specific glucose oxidase method) is $3–5$ mmol l^{-1}. Levels below 2 mmol l^{-1} constitute hypoglycaemia which is a very dangerous and sometimes fatal condition. The most important causes of hypoglycaemia are:

1. Iatrogenic: overdosage of insulin or sulphonylurea drugs.
2. β-Islet cell adenoma or hyperplasia (in infants of diabetic mothers).
3. 'Functional'; delayed (2–4 h) secretion of insulin after oral carbohydrate load.

Symptoms and signs. When blood glucose falls rapidly *adrenalin release* from the adrenal medulla is the earliest compensatory response and produces nervousness, weakness, pallor, anxiety, tachycardia, and headache.

Psychiatric and nervous effects. There is a feeling of great hunger, ascribed to lack of glucose in the cells of the ventromedial satiety centre of the hypothalamus. There is a great sense of fatigue, walking becomes difficult, the patient may become irritable, obstinate, and behave as if intoxicated with alcohol or as if demented.

Vasomotor disturbances (flushing) and profuse sweating may result from central nervous actions (? on the hypothalamus).

Neurological signs include diplopia, aphasia, tremors, and, in severe cases, extensor plantar responses. Finally delirium, convulsions, and coma occur and may, even if not fatal, produce irreversible damage to the central nervous system. Acute hypoglycaemia must be treated with great urgency.

Treatment. The symptoms of hypoglycaemia are rapidly relieved by intravenous administration of glucose.

Natural recovery from hypoglycaemia is brought about by secretion of hormones which raise the blood glucose level to normal. The earliest reactions include stimulation of glucagon and catecholamine secretions. The α-cells are directly stimulated by the fall in blood glucose and by sympathetic nerve stimulation, and the adrenal medulla by sympathetic stimulation [p. 417]. Both hormones increase the hepatic outflow of glucose into the circulation by promoting glycogenolysis and enhanced gluconeogenesis.

REFERENCES

BLOOM, S. R. (1975). *Br. J. hosp. Med.* **13**, 150

FRITZ, I. B. (ed.) (1972) *Insulin action.* Academic Press, New York.

HALL, R., ANDERSON, J., SMART, G. A., and BESSER, M. (1974). *Fundamentals of clinical endocrinology*, 2nd edn, pp. 297–366. Pitman, London.

KRAHL, M. E. (1972). Insulin action at the molecular level. *Diabetes* **21** Suppl. 2, 695–702.

LACY, P. E. and MALAISSE, W. J. (1973). *Recent Prog. Hormone Res.* **29**, 199.

LEFEBVRE, P. J. and UNGER, R. H. (eds.) (1972). *Glucagon.* Pergamon, Oxford.

LORAINE, J. A. and BELL, E. T. (1971). *Hormone assays and their clinical application*, 3rd edn, pp. 227–51. Churchill Livingstone, Edinburgh.

STEINER, D. F. (1972). Proinsulin. *Triangle* **11**, 51–60.

—— and FRANKEL, N. (eds.) (1972). *Handbook of physiology*, Section 7, Vol. 1. American Physiological Society, Washington, DC.

TEPPERMAN, J. (1973). Endocrine function of the pancreas. In *Metabolic and endocrine physiology*, 3rd edn, pp. 167–98. Year Book, Chicago.

UNGER, R. H. and ORCI, L. (1975). Glucagon in diabetes mellitus. *Lancet* **i**, 14–16.

WILLIAMS, R. H. (1974). The pancreas. In *Textbook of endocrinology*, 5th edn (ed. R. H. Williams) pp. 502–626. Saunders, Philadelphia.

Pituitary and endocrine hypothalamus

THE PITUITARY GLAND (HYPOPHYSIS)

The pituitary gland, together with the hypothalamus, forms the most influential endocrine system in the body. The pituitary secretes at least nine hormones some of which are called 'trophic' (or 'tropic'), that is they stimulate the secretion of other endocrine glands, e.g. thyroid, adrenal cortex, and gonads. The word pituitary comes from Latin *pituita,* meaning 'mucus', and was introduced by Galen who thought that the pituitary secreted mucus (or phlegm) into the nasal cavity. The word 'hypophysis' is of Greek origin and means 'outgrowth'.

Embryology. The pituitary gland originates early in embryonic life from two sources. The *adenohypophysis* is formed by an upward evagination of Rathke's pouch (ectoderm) and grows dorsally towards the infundibulum where it meets the *neurohypophysis* which is derived from a downward outgrowth of the infundibular process from the diencephalon. The *pars tuberalis* is a small extension of the adenohypophysis which surrounds the infundibular stem; the posterior wall of Rathke's pouch gives rise to the *pars intermedia* [FIG. XI.11] which lies between the adenohypophysis and neurohypophysis.

Anatomy. The anatomical structure of the human pituitary and its close relationship to the hypothalamus are shown in FIG. XI.11. The pituitary is a complex organ and for our purposes the following nomenclature will be used:

Adenohypophysis = Pars distalis, anterior lobe, or simply anterior pituitary.

Neurohypophysis = Neural lobe, posterior lobe, or simply posterior pituitary.

The median eminence is part of the hypothalamus; the pars intermedia and pars tuberalis are of minor importance.

The pituitary gland lies in the sella turcica of the sphenoid bone at the base of the skull and is difficult of access. In man it normally weights about 0.5 g, the anterior lobe making up about 75 per cent of the whole gland.

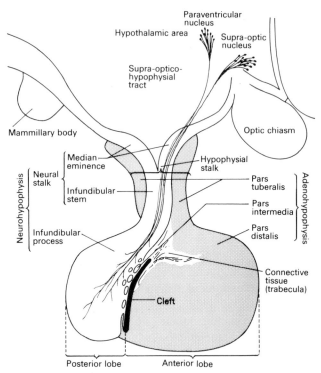

FIG. XI.11. Divisions of the pituitary gland and relationship to hypothalamus.
(From Netter, F. H. (1963). *The pituitary gland.* CIBA Clinical Symposia.)

Nerve supply. Unlike the richly innervated neurohypophysis the adenohypophysis receives no significant nerve supply from any source and its few nerve fibres (of sympathetic origin) supply blood vessels.

Blood supply. The blood supply to the adenohypophysis comes solely from portal vessels. The superior hypophysial arteries, branches of the internal carotid arteries, form a ring round the uppermost part of the pituitary stalk, and divide to produce a dense network of capillary loops supplying the median eminence and upper part of the infundibular stem. The capillary loops drain into veins, the *long portal vessels* which run down the infundibular stem and open into sinusoids supplying about 90 per cent of adenohypophysial tissue. The blood in these long portal vessels has been observed to flow from median eminence to adenohypophysis (Green and Harris 1949). In addition the inferior hypophysial arteries form a second capillary plexus in the lower infundibular stem and give rise to the *short portal vessels* which drain into sinusoids supplying the remaining 10 per cent of the adenohypophysis; the inferior hypophysial arteries mainly supply the neurohypophysis [FIG. XI.12]. The venous drainage of the whole pituitary is via the cavernous sinuses into the jugular vein.

The pituitary gland is under hypothalamic control, the adenohypophysis being influenced by hormones which come from the hypothalamus via the portal vessels, and the neurohypophysis by neurons which convey hormones directly from certain hypothalamic nuclei for storage in the neural lobe. The pituitary and hypothalamus form a single functional unit, but for descriptive purposes the adenohypophysis will be discussed first, then its interrelationship

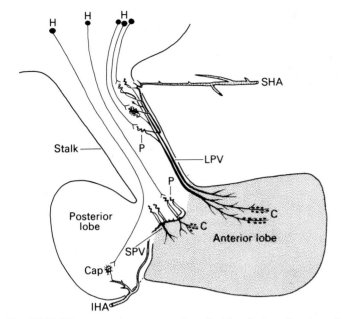

FIG. XI.12. Diagrammatic representation of mid-sagittal section through human hypothalamus and pituitary gland. The Figure shows pathways by which neurohumours secreted by hypothalamic neurons (H) may reach the secreting epithelial cells (C) of the anterior lobe of the pituitary gland. The upper and lower primary capillary beds (P) are the sites at which neurohumours are probably transferred into the portal bloodstream (LPV: long portal vessels; SPV: short portal vessels). Cap: the capillary bed through which the hormones elaborated by many of the neurons of the supraoptic and paraventricular nuclei enter the systemic circulation. SHA, IHA: superior and inferior hypophysial arteries.
(From Daniel, P. M. (1966). *Br. med. Bull.* **22**, 202.)

with the hypothalamus, and the neurohypophysis will be dealt with separately.

The adenohypophysis

The adenohypophysis (anterior pituitary) synthesizes and secretes the following hormones, which are peptides or glycoproteins.

1. Growth hormone (somatotrophin, somatotrophic hormone, GH).
2. Prolactin (lactogenic hormone, galactin, mammotrophin).
3. Corticotrophin (adrenocorticotrophic hormone = ACTH).
4. Thyrotrophin (thyrotrophic hormone, thyroid-stimulating hormone = TSH).
5. Follicle-stimulating hormone (FSH).
6. Luteinizing or interstitial cell-stimulating hormone (LH, ICSH).

TABLE XI.2. *Some important numerical data about most of the adenohypophysial hormones. (After Catt 1971)*

	Hormone	Molecular weight*	Pituitary content (µg)	Secretion rate (µg day^{-1})	Plasma level (µg l^{-1})
Peptides	ACTH	4500	300	10	0.03
	Human GH	21 500	8500	500	1.0–5.0
Glyco-proteins	TSH	31 000	300	110	1.0–2.0
	FSH	30 000	80	30	0.5–1.5
	LH	41 000	35	15	0.5–1.0

*The molecular weights of the glycoproteins are uncertain since subunits of these hormones tend to aggregate.

The functions of the adenohypophysis are numerous. It controls growth, especially of the bones, muscles, and viscera and it influences the metabolism of carbohydrate, protein, and fat. It controls the growth, development, structural integrity, and activity of the adrenal cortex, thyroid, ovary, breast, and testis.

In this section the cytology and the effects of extirpation will be discussed. The relation of the adenohypophysis to the adrenal cortex will be further considered on page 533, to the thyroid on page 543, to the ovary on page 574, to the breast on page 597, and to the testis on page 580.

Cytology of the anterior pituitary

As in other protein or peptide-secreting endocrine glands the cytoplasm of the majority of anterior pituitary cells contains granules which are membrane-surrounded hormone protein or peptide complexes synthesized in the rough endoplasmic reticulum. The adenohypophysis secretes at least six hormones. Three are polypeptides, namely growth hormone, prolactin, and ACTH and three are glycoproteins, namely TSH, FSH, and LH. The cytoplasmic granules are generally regarded as storage forms of the hormones and the question at once arises— are there six different types of cell (or granules) corresponding to the six different hormones? This problem has been studied by many techniques.

1. Histological and histochemical staining procedures are not conclusive because the results are influenced by many factors, especially the fixation fluids. However, histological methods do show that the majority of cells contain stainable granules (*chromophils*) and a minority do not (*chromophobes*). The cells with stainable granules are *acidophils* (i.e. stain red or orange with acidic dyes) or *basophils* (i.e. stain blue or green with basic dyes).
2. Electron microscopy shows very few cells without cytoplasmic granules; granules vary in size, shape, and electron density.
3. Homogenization and fractionation by centrifugation separates different types of granule which can be identified by electron microscopy and assayed biologically.
4. Organ and cell cultures have helped to identify pituitary tumour cells which can be subcultured and their secretory capacity measured.
5. Immunohistological methods are the most accurate for localizing hormone granules which are antigenic in nature.

Cell types in the anterior pituitary

About 75 per cent of adenohypophysial cells are chromophils and 25 per cent chromophobes. The chromophils include:

Growth-hormone cells (somatotrophs, GH cells). These cells contain numerous acidophil granules with a maximal diameter of 350 nm. [FIG. XI.13]. Immunohistological studies confirm staining techniques.
Prolactin cells (mammotrophs, lactotrophs) also contain acidophilic granules but these are larger (600–900 nm diameter) than in GH cells [FIG. XI.13].
Thyrotrophic cells (thyrotrophs, TSH cells) contain angular basophilic granules 100–150 nm in diameter. *Corticotrophic cells* (corticotrophs, ACTH cells) are also basophilic.
Gonadotrophic cells (gonadotrophs) have dense basophilic granules up to 150–200 nm in diameter. Immunohistological studies show that some gonadotrophs contain both FSH and LH. In gonadotroph cell cultures FSH- and LH-forming cells may stain differently.
Basophils contain periodic acid–Schiff (PAS) positive (red) granules of the glycoprotein hormones TSH, FSH, and LH.
Many *chromophobe cells* contain PAS-positive (purple) granules which contain ACTH. Chromophobe cells may be chromophils which have temporarily lost their granules by secretion.

Mitotic activity in adenohypophysial cells is generally low, but removal of the target organs (and hence abolition of hormonal inhibitory feedback) increases the rate of mitosis in the appropriate chromophil cells, e.g. gonadotrophs, corticotrophs, or thyrotrophs. It is also possible that one type of cell can be converted into another type; chromophobes may become chromophils and GH cells may be transformed into prolactin cells (and vice versa).

Granule formation and secretion. The synthesis of granules occurs in a manner similar to that seen in protein-secreting exocrine glands (e.g. pancreas) but the rate of secretion is slower. Hormone synthesis takes place in the polysomes of the rough endoplasmic reticulum (RER); the growing peptide chain enters the lumen of the RER and migrates to the Golgi apparatus where the secretory granules are formed. With glycoproteins the carbohydrate is probably incorporated before granule formation occurs. Sucrose density-gradient centrifugation of adenohypophysial homogenates, followed by electron-microscopic studies and measurement of biological activity, shows that the granules consist of stored hormone; ACTH may be stored in adenohypophysial cells for 20 days and GH for 6 days, depending on the degree of stimulation by hypothalamic hormones. During secretion the lipid membrane of the granule fuses with the cell membrane and by a Ca^{2+}-dependent process the hormone is released into the extracellular fluid (and hence into the blood) by *exocytosis* (emiocytosis) in which the granule membrane is retained within the cell, possibly for further use. Granules which are not secreted are taken up by lysosomes and destroyed by lysosomal enzymes.

Functions of the anterior pituitary

The functions of the adenohypophysis are numerous and are mediated by the secretion of six hormones. In addition the neurohypophysis secretes two hormones and the pars intermedia secretes one hormone.

The functions of the pituitary are simply revealed by studying the long-term effects of removal of the pituitary gland.

Effects of hypophysectomy. The effects of hypophysectomy (removal of the whole pituitary gland) provide important evidence for the role of the pituitary in normal life. Section of the pituitary stalk has only transient effects on pituitary function, because the adenohypophysis is quickly revascularized by regeneration of the portal blood vessels from the median eminence and because the supraoptic and paraventricular nuclei in the hypothalamus can continue to secrete vasopressin and oxytocin respectively into the bloodstream. The results of hypophysectomy have been widely studied in animals, and this operation has been performed in women in the treatment of cancer of the breast and in diabetes mellitus with vascular complications. Grave insufficiency of adenohypophysial function may be associated with pituitary tumours, and occurs in Simmonds' disease (panhypopituitarism) and following severe postpartum haemorrhage leading to extensive pituitary necrosis

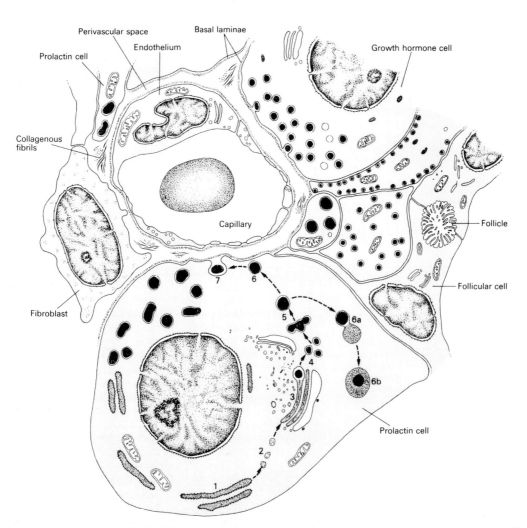

FIG. XI.13. Diagram showing ultrastructural relationship of secretory cells to a capillary in rat hypophysis. Lower prolactin cell illustrates steps in the secretory process: 1, deposition of polypeptides in cisternae of endoplasmic reticulum; 2, transport of polypeptides to Golgi apparatus in vesicles; 3, condensation of polypeptides and granule-formation in Golgi saccules; 4, liberation of membrane-enclosed granules from Golgi apparatus; 5, coalescence of granules; 6, movement of granules to peripheral plasmalemma; and 7, extrusion of the granule by exocytosis. Intracellular disposal of secretion is shown (6a) by fusion of a secretory granule with a dark body containing lysosomal hydrolytic enzymes and digestion of the contents (6b). (From Barker, B. L. (1974). In *Handbook of physiology*, Section 7, Vol. IV, Part 1, p. 71. American Physiological Society, Washington, DC.)

(Sheehan's syndrome). Signs of hypopituitarism are seen when more than 90 per cent of adenohypophysial tissue is destroyed. Hypophysectomy is not fatal provided that glucocorticoids are given.

The consequences of hypophysectomy are partly due to depressed activity of other endocrine glands and partly due to loss of direct pituitary influence on many other organs and tissues.

1. In young animals growth almost ceases. This is due to lack of growth hormone and also to lack of thyroid hormones. In adults no skeletal changes occur.

2. The thyroid atrophies. In man signs of myxoedema develop within 1–6 months. In contrast to primary hypothyroidism plasma thyroid-stimulating hormone (TSH) falls to undetectable levels; serum protein-bound iodine falls and thyroid uptake of ^{131}I decreases, but both these reductions are overcome by administration of TSH. Pituitary hypothyroidism is controlled by giving thyroxine.

3. The adrenal cortex atrophies. Secretion of cortisol is quickly and markedly reduced, and after prolonged pituitary deficiency aldosterone secretion may also be reduced. Prednisone [p. 536] prevents the development of adrenocortical deficiency; ACTH prevents adrenocortical atrophy and maintains normal cortisol secretion.

4. If severe pituitary deficiency occurs before puberty the gonads do not develop and the secondary sexual characters fail to appear. Hypophysectomy in the adult causes gonadal atrophy, with loss of spermatogenesis in the male, and abolition of ovulation and the menstrual cycle in the female. Impotence occurs in the male and sterility in the female. Lack of testicular or ovarian secretion produces the expected effects on secondary sexual characters [pp. 568–77]. There is loss of axillary and pubic hair; in women the vaginal smear becomes post-menopausal in one week and atrophic in four weeks. Urinary gonadotrophin excretion ceases within 2 weeks to 2 months. There are no menopausal symptoms. All these phenomena are due to lack of secretion of FSH and LH (ICSH) from the adenohypophysis and the consequent loss of secretion of testosterone from the testis or oestradiol from the ovary.

5. There is increased sensitivity to insulin; blood glucose falls at the normal rate but recovers very slowly. In diabetics there may be marked improvement in spite of full maintenance doses of adrenal corticosteroids. The improvement is attributed to deficiency of growth hormone.

6. A mild and variable diabetes insipidus develops. The occurrence of permanent, severe diabetes insipidus is only seen when the supraoptic nucleus itself is destroyed, or the median eminence damaged.

Now that it has become clear that the adenohypophysis is controlled by the hypothalamus it is obvious that all the effects here attributed to lack of pituitary hormonal secretions could also be caused by lesions in the hypothalamus or neighbouring parts of the brain [p. 518].

The relation of the adenohypophysis to growth and metabolism

The following adenohypophysial hormones have effects on body growth and metabolism:

1. Growth hormone, which has direct actions on tissues.
2. ACTH, which acts mainly by stimulating secretory activity of the adrenal cortex [p. 533].
3. TSH, which acts by stimulating the thyroid to secrete its hormones T_3 and T_4 [p. 543].

GROWTH HORMONE (GH, SOMATOTROPHIN)

The functional role of the anterior pituitary was first revealed by clinical evidence demonstrating effects on body growth. At the end of the nineteenth century acromegaly and gigantism were shown to be associated with a chromophil tumour of the pituitary. In 1921 Evans and Long enhanced the growth of adult rats by injection of anterior pituitary extract. In the late 1920s P.E. Smith perfected the operation of hypophysectomy in rats and showed the retarding effect of this procedure on skeletal growth, gonadal development, and the appearance of secondary sexual characters.

Chemistry. GH is a protein which shows species differences in structure. Bovine GH is a globulin with a molecular weight of about 45 000. It is active in most laboratory animals but not in monkeys or in man. Human GH (HGH) has a molecular weight of about 21 500 and contains 188 amino-acid residues in a single polypeptide chain. The storage form appears to comprise 191 amino-acid residues and requires activation to produce its physiological effects. Primate GH is active in lower species and, unlike GH from other species, it also has prolactin activity. However, HGH and human prolactin are separate hormones.

The human adenohypophysis contains 5–15 mg of HGH and secretes about 500 μg daily. HGH is synthesized by acidophil adenohypophysial cells distinct from those which synthesize prolactin.

Actions of growth hormone. GH has direct actions on cellular activity and metabolism, especially in muscle, adipose tissue, cartilage, and other connective tissues. These actions have been revealed by hypophysectomy, replacement treatment with GH, and excess GH administration to normal animals.

After *hypophysectomy* in young animals growth ceases. There is marked reduction of biosynthetic processes in many tissues and impairment of carbohydrate metabolism. DNA replication and cell multiplication cease in liver and muscle. Thymidine incorporation into DNA of cartilage and adipose tissue is diminished and synthesis of liver RNA is reduced. Synthesis of glycosaminoglycans (mucopolysaccharides) and collagen in connective tissue is diminished, and sulphate incorporation and hydroxyproline excretion (in urine) are reduced. There is decreased secretion of endogenous insulin in response to a glucose load and increased sensitivity to exogenous insulin, with delayed recovery from insulin-induced hypoglycaemia. The low fasting blood sugar and liver glycogen levels are mainly due to adrenocortical deficiency and can be corrected by glucocorticoid administration.

The administration of GH to hypophysectomized animals restores DNA replication and RNA synthesis in liver, muscle, adipose tissue, and cartilage. The most sensitive and widely used bioassay for GH measures the increased width of tibial cartilage in young hypophysectomized rats. Body growth is resumed with increase in muscle mass, growth of bone, and reduction in body fat. The production of glycosaminoglycans is increased, incorporation of ^{35}S into cartilage enhanced and hydroxyproline excretion raised, all these actions indicating increased collagen synthesis and turnover. Insulin secretion and insulin tolerance return to normal.

In animals with intact pituitary glands GH can increase the growth of bone, liver, kidney, and alimentary tract.

In general, growth hormone increases the number of cells (e.g. in muscle and bone) while insulin increases cytoplasmic growth. Thyroid hormones are required for the full effect of GH on DNA replication. GH has many effects on intermediary metabolism of protein, carbohydrate, and fat.

Protein. GH causes blood urea and amino-acid levels to fall as the result of reduced catabolism of protein and the enhanced uptake of amino acids for protein synthesis in muscle, liver, and connective tissues. GH probably promotes protein synthesis by increasing DNA, RNA, and ribosome activities in responsive cells.

Fat. GH increases the mobilization of fat from adipose tissue and plasma free fatty acid (FFA) level is raised.

Carbohydrate. GH has complex effects on carbohydrate metabolism. Injection of HGH produces an early fall in blood sugar and FFA levels (insulin-like effect) and later it enhances gluconeogenesis, impairs carbohydrate tolerance, and raises blood sugar. In dogs and cats large doses of GH can cause hyperglycaemia and diabetes mellitus, mainly by inhibiting glucose uptake by tissues (and not by exhaustion of insulin secretion). In man large doses of HGH are normally required for diabetogenic effects, but in patients with reduced islet reserve of insulin much smaller doses are effective. In hypophysectomized diabetics quite small doses of HGH can produce marked hyperglycaemia and ketosis. HGH directly antagonizes the peripheral action of insulin (which promotes glucose utilization by skeletal muscle), but it does not affect the inhibitory action of insulin on FFA release.

Ions. HGH in man causes retention of Na^+, Cl^-, Mg^{2+} and Ca^{2+}, the last in spite of increased urinary excretion of Ca^{2+}.

Prolactin-like actions. HGH has prolactin-like properties in some animal species, such as mammogenic and lactogenic actions in rodents and rabbits, stimulation of the pigeon crop sac, and a luteotrophic action on the rodent ovary. HGH can increase lactation in women, but there is no evidence that HGH is the physiological stimulus for milk production. Indeed, lactation can occur in women with isolated HGH deficiency and there is now clear evidence for the existence of human prolactin as a distinct molecular entity.

Mode of action of growth hormone. Growth hormone may not itself act directly on skeletal tissues and the suggestion has been made that GH forms mediators (in the liver and kidney) which have been called 'somatomedins'. One somatomedin is 'sulphation factor' which promotes incorporation of ^{35}S into proteoglycans of growing rat cartilage. These somatomedins can only be isolated from plasma and their importance is not yet established.

Control of secretion of growth hormone

The adenohypophysial content of HGH is very high (5–15 mg per gland) compared with the microgram quantities of the other hormones [see TABLE XI.2, p. 513]. The daily secretion rate is about 500 μg and secretion continues from infancy to old age. It is probable that GH is an important anabolic hormone throughout life.

Plasma HGH levels. In man the secretion of HGH has been studied by measurement of plasma HGH levels, using radioimmunoassay which is much more sensitive than any biological assay system. However, the results of radioimmunoassays must be interpreted with caution since the antigenic portion(s) of the large GH molecule may not correspond with the biologically active regions. The levels of immunoreactive HGH in plasma have been clearly correlated with biological activity only in patients with acromegaly in whom very high plasma HGH levels have been recorded. However, with these reservations, radioimmunoassay studies have provided much valuable information.

Plasma HGH. In the newborn child plasma HGH levels are very high though GH does not seem to be essential for fetal or neonatal growth. Thereafter the basal plasma HGH level is normally 1–5 μg d^{-1}, with intermittent peaks up to 25 μg l^{-1} or more which occur mostly after activity, but also during sleep. The peaks are commoner in women, especially in the middle of the menstrual cycle, and during oestrogen administration.

The following factors are of particular importance in *promoting HGH secretion:*

Insulin-induced hypoglycaemia. A reduction of blood glucose to 50 per cent or less of the normal basal level is the most consistent stimulus of HGH secretion. Physiological fluctuations in blood glucose levels are ineffective. The administration of 2-deoxy-glucose, which inhibits normal glycolysis of glucose, also stimulates HGH secretion.

Fasting and severe malnutrition raise plasma GH in man. The metabolic adaptation to fasting, i.e. reduced glucose utilization and lipolysis, is more dependent on reduced insulin secretion and on the ratio of insulin to HGH than on the absolute level of HGH.

Exercise. Plasma HGH levels rise after exercise, particularly in physically unfit persons in whom lactate production in muscle is greatest.

Amino acid infusion. Intravenous infusion of arginine raises plasma HGH level as much as does insulin-induced hypoglycaemia.

Neural stimuli. Emotional stress, surgical operations, fear, and loud noises may all enhance HGH secretion and, paradoxically, the onset of deep sleep may be associated with large peaks in plasma HGH level.

Inhibition of HGH secretion. The intermittent spikes of HGH secretion which occur in normal resting persons for no known reason are suppressed only transiently by continuous administration of glucose.

Effects of hormones on growth hormone secretion

Thyroid hormones. Hypothyroidism reduces the synthesis, storage, and release of GH and retards growth. In man the HGH secretory response to various stimuli is reduced in hypothyroidism. Thyroid hormone restores body growth rate before the HGH response becomes normal because thyroxine has a direct effect on growth as well as its synergistic action on growth hormone secretion.

Sex hormones. In human beings *oestrogens* stimulate both synthesis and release of HGH. Women have higher basal levels of plasma HGH and more frequent pulses of HGH secretion than men. They also show greater secretory responses to exercise and arginine infusion, especially during the middle of the menstrual cycle. In men oestrogen enhances the plasma HGH response to arginine infusion.

Androgens may also enhance responsiveness to GH in man (and increase GH secretion) but high doses of *progestagens* can suppress the HGH secretion induced by insulin hypoglycaemia.

Corticosteroids inhibit the GH response to insulin hypoglycaemia, but ACTH can stimulate HGH release and responsiveness to hypoglycaemia.

Catecholamines. Infusion of large doses of adrenalin (or cyclic AMP) raises plasma GH level. The effects of adrenergic blockade on the HGH response to insulin hypoglycaemia suggest that stimulation of α-adrenergic receptors promotes GH secretion, while stimulation of β-receptors inhibits secretion.

Hypothalamus and growth hormone

The role of hypothalamic neurohormones in the general control of adenohypophysial hormone secretion is fully discussed on page 518. The hypothalamus influences the secretion of GH by means of two hormones: (i) *Growth hormone-releasing hormone (or factor)*, abbreviated to GH-RH or GH-RF, and (ii) *Growth hormone-release inhibiting hormone (or factor)*, abbreviated to GH-RIH or

GH-RIF (also called *somatostatin*). These two hormones exert antagonistic effects on the GH-secreting acidophils of the adenohypophysis and form the final common paths through which many factors influencing GH secretion ultimately work. For example, insulin-induced hypoglycaemia (blood sugar reduced to < 50 per cent of basal level) raises plasma HGH level to 30–50 μg l⁻¹ by promoting release of GH-RH from the hypothalamus. Infusion of somatostatin prevents the hypoglycaemia-induced secretion of HGH without affecting the increases of prolactin and corticosteroid secretion or the sweating and tachycardia induced by the hypoglycaemia.

Disorders of growth hormone secretion

Hypopituitary dwarfism is due to lack of HGH secretion, which may result from a primary destructive lesion of the adenohypophysis or from a failure of the hypothalamus to secrete GH-RH.

It is important to realize that fetal growth and neonatal growth are largely independent of GH, and that most dwarfed children have normal HGH secretion. However, the rarer hypopituitary dwarfism can now be diagnosed by finding low or absent plasma HGH levels by radioimmunoassay and by demonstrating reduced or absent plasma HGH responses to insulin hypoglycaemia or arginine infusion. The clinical features in such cases include normal birth weight, subsequent severe retardation of growth with a tendency to obesity, low fasting blood sugar, and delayed recovery from insulin hypoglycaemia. Administration of HGH in such cases produces nitrogen retention and a dramatic increase in growth rate (Tanner *et al.* 1971). The best responses occur in children with *isolated HGH deficiency*, and HGH should be given before puberty so that the growth response can be achieved before fusion of the epiphyses of the long bones. If GH deficiency is accompanied by gonadotrophin deficiency it is most important to give HGH alone until the desired amount of growth has occurred, after which gonadotrophin or sex-hormone replacement, which promotes fusion of the epiphyses, may safely be given to stimulate sexual development.

Occasionally severe retardation of growth occurs in association with normal or raised immunoassayable HGH. In such cases there is some alteration in structure of the HGH molecule, with reduction of biological activity and maintenance of normal immunological activity. Large amounts of exogenous HGH promote good growth.

African pygmies show some features suggestive of HGH deficiency; their short stature is due to lack of tissue response to normally secreted amounts of endogenous HGH.

Gigantism and acromegaly. In both gigantism and acromegaly there is usually an acidophil adenoma of the adenohypophysis which secretes excessive amounts of HGH. In some cases there is a primary disorder of the hypothalamus with over-secretion of GH-RH which in turn causes the anterior pituitary to secrete an excess of HGH.

Gigantism occurs when HGH is secreted in excess before fusion of the epiphyses of the long bones which become much longer than usual. Most giants of 2.5 metres or more in height are examples of this disorder.

Acromegaly occurs when there is hypersecretion of HGH after fusion of the epiphyses. The word 'acromegaly' means enlargement of the peripheral regions and is characterized by overgrowth of the bones of the lower half of the face [FIG. XI.14] and by enlargement of the hands and the feet; there is bowing of the spine (kyphosis).

Most patients with acromegaly have much raised plasma levels of HGH, e.g. to 20–250 μg l⁻¹. The plasma of acromegalic patients may also contain growth-promoting peptides distinct from HGH.

Bromocriptine (2-brom-α-ergocriptine), a dopaminergic agonist which suppresses the secretion of prolactin, also inhibits secretion of HGH in patients with acromegaly, with apparently good clinical results. Unlike GH-RIH, which also reduces HGH secretion in acromegaly, bromocriptine does not inhibit secretion of TSH, glucagon, insulin, or gastrin.

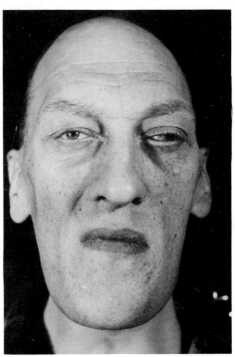

FIG. XI.14. Acromegaly in man aged 42 years. Enlargement of jaw, hands, and feet first noted at 22 years. Diabetes mellitus developed at 24 years. X-rays showed enlarged pituitary fossa.
(Dr J. D. N. Nabarro's case.)

HUMAN PROLACTIN

In most animal species growth hormone (GH) and prolactin (PRL) have been easy to isolate as distinct adenohypophysial hormones. However, in primates separation of these hormones has been only recently achieved (see Friesen and Hwang 1973). There are several clinical and experimental observations which suggest that HGH differs from human PRL (HPRL):

1. Acromegalic patients with raised plasma HGH levels only occasionally have galactorrhoea.
2. Most women who are breast-feeding, or subjects with galactorrhoea, have normal plasma HGH and raised plasma HPRL levels.
3. Some patients with a hereditary form of isolated GH deficiency (ateliotic dwarfism) lactate normally despite the absence of HGH.
4. Section of the pituitary stalk, which does not raise plasma HGH, leads to galactorrhoea in some patients.
5. An extract of pituitary tumour from a patient with galactorrhoea was rich in prolactin but deficient in growth hormone. Preincubation of the extract with antiserum to HGH did not neutralize prolactin activity.

The human prolactin cell

The anterior-pituitary prolactin cell has been identified in tissue cultures of pituitaries from women who died during pregnancy or the post-partum period, and in prolactin-secreting pituitary tumours. The prolactin cell is an acidophil which stains bright red with carmoisine. During pregnancy or the post-partum period PRL cells may constitute over 50 per cent of pituitary acidophils. Immunofluorescent studies have shown that antisera to HGH or human placental lactogen (HPL) do not become localized on pro-lactin cells whereas antiserum to sheep PRL does. Human and sheep PRL share common antigenic determinants not found in HGH or HPL. PRL antiserum has been used to develop a specific radioimmunoassay method for measurement of PRL levels in human body fluids.

Chemistry. Prolactin from ruminants was the first anterior pituit-ary hormone to be isolated but primate prolactin was not isolated till 1971. There were two main reasons for the difficulty in purifying HPRL.

1. HGH constitutes 5–10 per cent of the dry weight of pituitary powder whereas the HPRL content is about 0.1 per cent, a 50–100-fold difference.

2. HGH, unlike GH of other species, has 10–20 per cent of the prolactin activity of the best sheep prolactin.

Thus, if the potency of pure HPRL is 30 iu mg^{-1} (mouse mam-mary gland bioassay) and that of HGH is 5 iu mg^{-1}, then 90 per cent of the total prolactin activity in pituitary homogenates is due to HGH. During pregnancy the number of prolactin cells in the pituit-ary increases dramatically, while the number of HGH cells de-creases. HPRL has many structural features in common with sheep prolactin but differs from HGH. The full amino-acid sequence of HPRL is not yet known but sheep prolactin has a molecular weight of 25 000 and consists of a single chain of 198 amino-acid residues.

Serum prolactin levels in man. Serum HPRL levels have been measured by sensitive *bioassay* (e.g. milk secretion by lobulo-alveolar mammary gland tissue *in vitro*), by *radioimmunoassay*, and by *radioligand assay* (using PRL receptors on plasma mem-branes of mammary gland cells).

The normal serum PRL level in adult females is 6–50 μg l^{-1}, and in adult males it is 6–25 μg l^{-1}. In both sexes the highest values occur 2–3 hours after onset of sleep. Exercise and psychological stress raise serum HPRL levels, in women more than men.

In *pregnancy* serum HPRL levels rise steadily to 50–600 μg l^{-1} at term. *Post-partum* serum HPRL levels fall rapidly to normal except during breast-feeding, when *suckling* raises the serum level to 250 μg l^{-1}.

In *galactorrhoea* due to hypothalamic or pituitary disease, or to drugs (e.g. phenothiazines, reserpine) serum HPRL may rise to 1000 μg l^{-1}.

HYPOTHALAMIC CONTROL OF ANTERIOR PITUITARY SECRETIONS

Stimulation of the hypothalamus causes secretion of several anter-ior pituitary hormones, such as corticotrophin, gonadotrophins, and thyrotrophin. Yet histologically no nerve pathways have been traced from the hypothalamus into the adenohypophysis; the few nerve fibres that are present are probably sympathetic in origin and supply the blood vessels. The link between the hypothalamus and adenohypophysis is provided by a system of portal vessels [p. 512].

The integrity of this hypothalamo-hypophysial portal system is now recognized, largely through the work of G. W. Harris and his colleagues, as essential for the proper functioning of the adeno-hypophysis. If the pituitary stalk is cut through, the target organs of the anterior pituitary, the thyroid, adrenal cortex, and gonads, ini-tially atrophy but recover when the long portal vessels regenerate. However, should regeneration be prevented by a mechanical barrier the atrophy persists. If the adenohypophysis is removed from the sella turcica and placed in the temporal lobe or beneath the capsule of the kidney, it acquires a new vascular supply and survives but the target organs atrophy. However if the transplant is

then replaced beneath the hypothalamus the target organs recover, provided that the portal vessels regenerate. Such findings led Harris to suggest that the terminations of axons of hypothalamic neuro-secretory cells release chemical transmitters, generally peptides, into the capillary plexus of the infundibular stem and these are conveyed by the portal vessels to the anterior pituitary, so regulat-ing its activity. The activity of these hypothalamic neurosecretory cells is in turn regulated by numerous nerve pathways projecting on to the hypothalamus. The transmitters for these pathways are largely monoamines and include catecholamines (NA and DA), serotonin, and histamine.

The substances released by hypothalamic neurons to act on adenohypophysial cells are called as a group *hypothalamic regula-tory hormones*. The individual substances which act predominantly on one type of pituitary cell, are called 'hormones' by some endocri-nologists and 'factors' by others. Some of these peptides have been isolated, structurally identified, and synthesized. Most of these factors (hormones) stimulate pituitary secretion, others inhibit secretion. It is not yet known how many distinct hypothalamic releasing hormones there are.

The hypothalamo-adenohypophysial system serves as a *cascade amplifier* in quantitative terms. A few nanograms of hypothalamic releasing hormone promotes the secretion of micrograms of adeno-hypophysial trophic hormone and this in turn may release milli-grams of hormone from the target organ (cf. cascade amplification in blood clotting, p. 24). As can be seen from these values, the 'amplification' at each stage of this cascade is at least a thousand-fold.

Hypothalamic control of ACTH release

Corticotrophin–releasing hormone (CRH) is stored and pos-sibly synthesized in the medial basal hypothalamus from which it is released into the hypothalamo-hypophysial portal vessels and reaches the adenohypophysis to stimulate ACTH release. This natural hormone (CRH) is only one of a number of natural and synthetic *corticotrophin-releasing factors* (CRF) which can promote ACTH release. The structure of CRH is unknown at present, though it seems certain that it is not adrenalin, noradrenalin, hista-mine, acetylcholine, vasopressin, oxytocin, or a prostaglandin, and it has been proposed that CRH is a short-chain polypeptide related to α-melanocyte-stimulating hormone (MSH) or to vasopressin. Though not the natural hormone, vasopressin is a potent cortico-trophin-releasing factor and may potentiate the response to CRH under physiological conditions. Vasopressin may exert this poten-tiating effect by promoting ACTH synthesis by the adenohypoph-ysis, so that more ACTH is available for release by hypothalamic CRH. The mechanism of action of CRH on the anterior pituitary is still unknown but may involve the activation of cyclic AMP.

Control of CRH release

Stress. Many biological 'stresses' such as exposure to cold, anoxia, severe exercise, trauma, haemorrhage, infections, bacterial toxins, anaesthesia, or hypoglycaemia following insulin administration, lead to an increase in plasma glucocorticoid concentration resulting from release of CRH from the hypothalamus and the consequent release of ACTH from the anterior pituitary. Some of these stimuli, such as trauma or severe exercise, may exert this effect through their emotional accompaniment, and emotions such as fear, anger, and frustration are powerful stimuli for glucocorticoid secretion. The limbic system is known to be involved in emotion and the amygdaloid nuclei, which form part of the limbic system, give rise to a cholinergic pathway which runs to the median eminence and stimulates CRH release in response to stress. Many types of stress produce a barrage of afferent nerve impulses which increase the activity of the reticular activating system, and another cholinergic pathway stimulating CRH release has been shown to reach the median eminence from the reticular formation in the pons and midbrain. Catecholaminergic and serotoninergic nerve pathways also influence CRH secretion.

Glucocorticoid 'feedback'. Injected labelled glucocorticoids are taken up by regions of the brain such as the hippocampus which project to the median eminence. Changes in plasma glucocorticoid levels affect the firing frequency of hypothalamic neurons. Thus an increase in plasma glucocorticoid level may suppress ACTH secretion by reducing CRH release from the hypothalamus as well as by a direct effect on the adenohypophysis. Indeed the application of cortisol to the median eminence leads to a greater suppression of ACTH release than injection into the adenohypophysis itself.

'Circadian' rhythm. There is a *diurnal* or *circadian* rhythm of glucocorticoid secretion and plasma concentration which reach a peak just prior to waking in the morning and decline throughout the hours of daylight and activity, to rise again during darkness and sleep from a low point at about midnight. These changes in glucocorticoid secretion are preceded by corresponding rhythmic variations in the secretion of ACTH and CRH. This rhythm, which can be overridden by the demands of stress, is absent in patients with lesions of the hypothalamus or limbic lobe. Experimental lesions of the anterior hypothalamus, which sever the pathways from the limbic lobe, also abolish the circadian rhythm of glucocorticoid secretion but do not abolish all the responses to stress, suggesting that the circadian rhythm depends on cyclic variations in CRH secretion due to a 'biological clock' in the limbic system. Both serotoninergic and cholinergic pathways have been implicated in this pathway to the hypothalamus.

The activity of this intrinsic 'biological clock' in determining the circadian rhythm of CRH release is supplemented by the effects of the rhythmically alternating cycles of sleep and activity, and the rhythmically alternating cycle of darkness and light during each 24-hour period. If human subjects are exposed to constant light and the length of the sleep/activity cycle varied from 12 to 33 hours, the length of the glucocorticoid cycle is adjusted so that a peak is still attained just before waking. In addition, subjects maintaining a normal sleep/activity cycle, but exposed to abnormal cycles of light and darkness, showed two peaks in plasma glucocorticoid level, one peak corresponding to the transition from sleep to activity and the other coinciding with the change from darkness to light.

Under normal circumstances the transitions from sleep to activity and from dark to light more or less coincide. Moreover these cycles have normally the same 24-hour periodicity as the 'biological clock' so that these several factors contributing to the circadian rhythm mutually reinforce one another. However the abnormalities in the rhythm of CRH, and hence glucocorticoid secretion, when the sleep/activity cycle or the dark/light cycle are experimentally varied, reveal some of the physiologically disruptive effects of modern air-travel; rapid travel across a number of time-zones may disrupt the normal circadian rhythm of adrenocortical activity for up to a week, leading to the impairment of mental and physical performance commonly described as 'jet-lag'.

Hypothalamic control of the release of human growth hormone

The release of growth hormone (HGH, GH, somatotrophin) by the anterior lobe of the pituitary gland is regulated by a *growth hormone-releasing hormone* (GH-RH) derived from neurosecretory cells with endings in the median eminence and upper pituitary stalk. Disease or experimental lesions of the hypothalamus, transplanting the pituitary away from the hypothalamus, or pituitary stalk section, lead to retarded growth which can be restored by administration of GH. Circulating GH concentration can be measured by radioimmunoassay and the level has been shown to be increased by electrical stimulation of the posterior hypothalamus in humans and by stimulation of the median eminence and ventromedial hypothalamic nucleus in the rat. GH is also released by administration of extracts of the ventromedial nucleus or median eminence and GH-RH activity has been demonstrated in hypophysial portal blood.

Hypoglycaemia leads to a rise in plasma GH level which is greatly reduced by hypothalamic lesions. The release of GH-RH which brings about this rise in GH seems to be mediated by 'glucoreceptors' situated in the ventromedial hypothalamic nucleus. Changes in blood glucose level alter the firing rate of neurons in this nucleus while the injection of glucose into the median eminence supresses GH secretion in response to insulin-induced hypoglycaemia.

Emotion and physical stress such as pain, trauma and cold increase plasma GH levels. The response is rapid, within 3–5 minutes, suggesting that nervous pathways to the hypothalamus are responsible for initiating the discharge of GH-RH. GH secretion can be induced by stimulation of the posterior hypothalamus, lateral amygdaloid nucleus, hippocampus, and interpeduncular nucleus, all of which project pathways to regions of the hypothalamus involved in GH-RH release. These structures are also closely related to the limbic system which is intimately involved in emotional behaviour. The release of GH by stimulation of the hippocampus is of additional interest in that plasma GH levels rise during *slow-wave sleep* [p. 339]. Stimulation of the hippocampus also produces the changes of slow-wave sleep in the EEG suggesting that the hippocampus may be involved in the release of GH during deep sleep.

The neurotransmitters for these pathways influencing GH-RH release appear to be catecholamines and serotonin. Injections into the hypothalamus of noradrenalin, L-dopa (a precursor of dopamine and noradrenalin), or serotonin stimulate GH release; injection of isoprenaline inhibits it. It has been proposed that the response to stress and metabolic stimuli such as hypoglycaemia is mediated by noradrenergic pathways since it is reduced by drugs with α-adrenergic blocking effects such as phentolamine and chlorpromazine, and that the slow-wave sleep-induced GH release involves serotoninergic pathways.

Chemical nature of growth hormone-releasing hormone. The chemical identity of GH-RH has not been established, but it appears to be a decapeptide distinct from luteinizing hormone-releasing factor, prolactin-inhibitory factor (PIF), corticotrophin-releasing hormone (CRH), and vasopressin.

Mechanism of action of GH-RH. The current concept of the mechanism of GH release by GH-RH is that binding of GH-RH to a specific receptor site on the membrane of the acidophil cell in the adenohypophysis activates adenyl cyclase leading to the formation of cyclic AMP from adenosine 5′-triphosphate (ATP). This cyclic AMP in turn activates guanyl cyclase to form cyclic guanyl monophosphate (GMP) from guanosine 5′-triphosphate (GTP). The cGMP then stimulates the process by which GH is released from the acidophil cell. Movement of Ca^{2+} into the cell accompanies the action of GH-RH and seems to be an integral part of the GH release process.

Prostaglandins also produce GH release and are found in the hypothalamus but their possible physiological importance as regulators of GH release is unknown.

Growth hormone-release inhibiting hormone (GH-RIH, soma-tostatin)

Certain hypothalamic extracts have been found to contain a hormone which inhibits GH release. This *growth hormone-release inhibiting hormone* or *somatostatin*, which is present in cells of the antrum of the stomach and the pancreatic islets, and in nerve endings in the brain, spinal cord, and autonomic ganglia as well as in the hypothalamus, is a peptide consisting of 14 amino acids (a tetradecapeptide). Somatostatin inhibits GH secretion in normal humans, rats, monkeys, and dogs, and in patients with acromegaly and diabetes. It also inhibits pituitary TSH responses to thyro-trophin-releasing hormone; in addition it inhibits secretion of gastrin and pancreatic secretion of insulin and glucagon. While the physiological importance of somatostatin in the regulation of GH release is unknown it is probable that the hypothalamus exerts a dual excitatory and inhibitory control by varying the balance of GH-RH and somatostatin discharged into its portal circulation.

Hypothalamic control of the release of thyrotrophin (thyrotrophic hormone, thyroid stimulating hormone, TSH)

Conclusive evidence for hypothalamic involvement in the control of thyrotrophin (TSH) secretion was first given by Greer in 1951. The experimental administration of thiouracil produces thyroid enlargement. This is because thiouracil interferes with thyroxine synthesis so that the blood level of thyroxine declines [p. 546]. Since thyroxine normally exercises a feedback inhibition of TSH output by the anterior pituitary, this fall in plasma thyroxine leads to a rise in TSH secretion which results in hypertrophy of the functionally impaired thyroid gland. Greer found that thyroid enlargement following treatment with thiouracil could be prevented by electrolytic lesions placed in the anterior hypothalamus between the supraoptic nucleus and the median eminence. By siting the lesion appropriately it was possible to produce a selective depression of TSH secretion without interfering with the secretion of gonado-trophins or ACTH, indicating that this region of the hypothalamus modifies the release of TSH by producing a specific *thyrotrophin-releasing hormone* (TRH) which reaches the anterior pituitary via the hypothalamic portal vessels.

Conversely, electrical stimulation within an extensive hypothalamic area between the preoptic region and the median eminence depletes the TSH content of the anterior pituitary and causes an increase in circulating TSH level. Another experimental technique for demonstrating the importance of TRH in maintaining normal TSH secretion involves hypophysectomy followed by the implantation of pituitary fragments. If the fragments are transplanted beneath the capsule of the kidney they become revascularized but the activity of the thyroid gland remains at the reduced level seen after hypophysectomy. However, if the implant is placed in the median eminence close to the capillary tufts of the portal system thyroid function is restored to normal.

Control of TRH secretion. The secretion of TRH by the hypothalamus may be controlled to some extent by the level of thyroid hormones in the blood. However, direct depression of pituitary TSH secretion by a raised thyroxine level undoubtedly plays the more important part in controlling thyroid activity. TRH and thyroid hormones act competitively on thyrotrophs of the anterior pituitary. TRH promotes release and synthesis of TSH whereas thyroid hormones depress TSH secretion. Though the hypothalamic region within which electrical stimulation evokes TRH release contains a well-defined system of dopaminergic and noradrenergic

nerve fibres *nervous control* does not seem to play a major role in regulating TRH release. One form of stress which definitely does affect thyroid function is *cold*, especially in the neonate [p. 596] and it is of interest that local cooling of the rostral portion of the preoptic region, which contains neurons sensitive to changes in temperature, leads to an increased release of thyroid hormone in goats and rats. This has led to the suggestion that the discharge of neurons in the preoptic region normally inhibits TRH release and that when metabolism of these neurons is depressed by cooling, this inhibitory discharge is reduced, permitting increased TRH secretion.

There is some evidence for a *circadian pattern* of TRH release. Thyroid hormone levels show a peak at about 4 a.m. and a nadir at about 5 p.m. Whether this is due to the same factors that determine the circadian rhythm of glucocorticoid secretion or whether it is a secondary consequence of the changes in corticosteroid level is not yet known.

Administration of TRH also promotes the release of prolactin from the anterior pituitary but this effect is not thought to be of importance in the physiological control of prolactin secretion.

Chemical nature of TRH. TRH was the first releasing factor to be chemically identified and synthesized. Its structure was shown simultaneously by Guillemin and by Schally and their co-workers in 1969 to be a tripeptide, pyro-Glu-His-Pro-amide. TRH appears to be identical in structure in all mammalian species. The mechanism of TSH release by TRH has not been fully clarified but appears to involve cyclic AMP production and Ca^{2+} uptake by the adenohypophysial cell.

Hypothalamic control of prolactin secretion

Transplantation of the pituitary from the sella turcica to beneath the capsule of the kidney, transection of the pituitary stalk with interposition of a barrier to prevent regeneration of the portal vessels, or electrolytic lesions of the median eminence or pituitary stalk, are all followed by a prolonged *elevation* of prolactin secretion accompanied by atrophy of the ovary, adrenal cortex, and thyroid. These observations suggest that prolactin differs from most other anterior pituitary hormones in that its primary regulation by the hypothalamus is inhibitory, exercised through a *prolactin-inhibiting factor* (PIF) released tonically into the hypothalamo-hypophysial portal vessels. With this restraint removed the pituitary expresses its inherent ability to secrete prolactin at a very rapid rate. The existence of PIF has been demonstrated by blockade of the suckling-induced release of prolactin following administration of hypothalamic extracts, and was elegantly confirmed by Kamberi, Mical, and Porter in 1971. They inserted a microcannula into a hypothalamic portal vessel of a living rat and showed that graded injections of hypothalamic extract produced a graded decrease in the release of prolactin as measured by radioimmunoassay [Fig. XI.15]. They also demonstrated the presence of PIF in samples of blood draining the hypothalamus withdrawn through the cannula [Fig. XI.16]. The *site of PIF production* is the medial basal hypothalamus. Prolactin secretion remains low after this region is surgically isolated from the rest of the hypothalamus except for the median eminence, but rises when the medial basal hypothalamus is destroyed. The secretion of prolactin which is stimulated by suckling has been attributed to inhibition of PIF secretion by neurosecretory cells in this region in response to afferent impulses initiated at the nipple.

The medial basal hypothalamus is rich in catecholamines. Control of the PIF-secreting cells may be exercised by these catecholaminergic neurons since infusions of noradrenalin or dopamine into the third ventricle decrease prolactin release [Fig. XI.16]. The

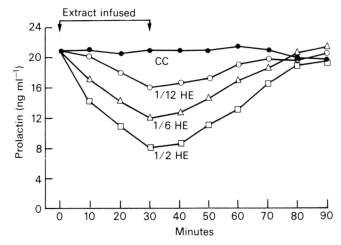

FIG. XI.15. Direct *in vivo* demonstration of existence of hypothalamic prolactin-inhibiting factor in the rat. A microcannula was inserted into a pituitary portal vessel, and the anterior pituitary was infused with an extract of cerebral cortex (CC) or increasing concentrations of hypothalamic extract (HE). One HE is equivalent to the median eminence removed from a single rat. Prolactin concentrations were measured in peripheral plasma. (Adapted from Kamberi, I. A., Mical, R. S., and Porter, J. C. (1971). *Endocrinology* **88**, 1294.)

decreased secretion of PIF in response to suckling could thus be due either to direct inhibition of the PIF-secreting neuron or to inhibition of a tonically discharging catecholaminergic neuron which synapses with it. Treatment with various tranquillizers, which deplete the catecholamine stores of the hypothalamus, has been noted to increase prolactin levels in women and lactation may follow the use of these drugs. Moreover, treatment with L-dopa, which crosses the blood–brain barrier and is converted to dopamine in the hypothalamus, inhibits the secretion of prolactin in women.

The chemical structure of PIF has not been identified but it may be a peptide with a molecular weight of less than 5000. However, dopamine itself directly inhibits prolactin secretion by an action on

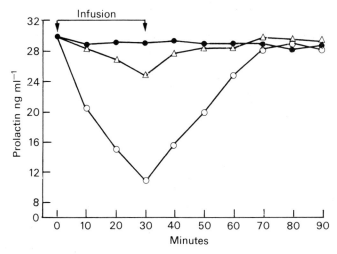

FIG. XI.16. Demonstration of presence of prolactin-inhibiting factor in blood draining the median eminence of the hypothalamus. Graph shows response of anterior pituitary gland of recipient rats (as measured by changes in peripheral plasma levels of prolactin) to portal vessel infusion of plasma from donor rats. Plasma from donor rats was obtained from the femoral vein (●), hypothalamic portal veins of saline-treated rats (△), or hypothalamic portal veins of rats administered dopamine via the third ventricle (○).
(From Kamberi, I. A., Mical, R. S., and Porter, J. C. (1971). *Endocrinology* **89**, 1042.)

the adenohypophysial prolactin-secreting cells; dopamine may be the main PIF.

The production by the hypothalamus of a *prolactin-releasing factor* (PRF) has also been reported.

Hypothalamic control of gonadotrophin secretion

Several of the factors influencing gonadotrophin secretion act through the central nervous system. Examples of particular interest include:

1. Mating. Certain species do not undergo cyclic spontaneous ovulation but ovulate only after coitus or other forms of sexual stimulation. Such 'reflex ovulators' include the rabbit, ferret, and cat. Afferent impulses from the genitalia are generally reponsible for initiating the discharge of luteinizing hormone (LH) which promotes ovulation in these animals, but other sources of afferent stimulation, such as mere exposure to the male, may be sufficient. Ovulation may occur in the pigeon simply in response to the sight of its own reflection in a mirror.

In the rat, a spontaneous cyclic ovulator, coitus is followed by the discharge from the anterior pituitary of prolactin, which is needed to maintain the corpus luteum in this species. Again the link must be by the nervous system since a similar discharge of prolactin follows if the mating is sterile or even if the cervix is mechanically stimulated with a glass rod.

2. Suckling. During lactation, cyclic ovarian function ceases in women and most other mammals. There is inhibition of the discharge of gonadotrophins from the pituitary, which is probably due to an effect on the hypothalamus of afferent impulses originating from the nipple when the infant suckles.

3. Stress. Emotional and physical stress is a well-recognized cause of disturbed menstrual function in women. Amenorrhoea was general among women imprisoned in concentration camps. In rats experimental ether anaesthesia and haemorrhage have been shown to alter plasma levels of FSH and LH.

4. Altered lighting. If rats are exposed constantly to light for prolonged periods the cyclic pattern of gonadotrophin discharge is abolished and the normal oestrous cycle ceases and is replaced by a state of 'constant oestrus'. It has been suggested that the constant light stimulus acts by inhibiting activity of the pineal gland [p. 525]. The menstrual rhythm is frequently disturbed in nurses on night duty or air-hostesses travelling on intercontinental flights.

5. Seasonal factors. Many animals such as the cat and ferret breed only at certain times of the year. A variety of environmental factors may play a part but alterations in the relative length of daylight and darkness are of particular importance. Such observations may give physiological substance to the poet's long-held belief that 'In the spring a young man's fancy lightly turns to thoughts of love'.

Several lines of evidence indicate that the part of the central nervous system most intimately involved in the control of gonadotrophin secretion is the *hypothalamus*.

1. Electrical stimulation. Haterius in 1937 showed that stimulation of the suprachiasmatic region led to ovulation in the anaesthetized rabbit. This has subsequently been confirmed in the rat and the conscious rabbit. However, electrical stimulation of the adenohypophysis itself failed to cause ovulation, indicating that the link between the hypothalamus and the adenohypophysis was not a nervous pathway.

More recently measurement by radioimmunoassay of plasma gonadotrophin levels has supplemented the production of ovulation as an index of gonadotrophin release. Stimulation within a broad area of the hypothalamus, extending from the preoptic region rostrally to the arcuate-median eminence region caudally, leads to elevation of plasma LH concentration and ovulation in the pro-oestrous rat. Plasma FSH concentration is raised by stimulation of an area extending from the anterior hypothalamus to the region of the arcuate nucleus and median eminence. Thus the region from which LH release can be provoked extends slightly more rostrally but otherwise the two areas overlap.

2. Electrical recordings. Recordings of action potentials from units in the preoptic and anterior hypothalamic areas and in the arcuate–median eminence region have revealed increased activity associated with ovulation.

3. Hypothalamic lesions. Destruction of the *median eminence* in the female rat leads to a permanent state of dioestrus with lack of follicular development. The plasma concentrations of FSH and LH fall dramatically within a few hours of the lesion while the concentration of prolactin rises. Following similar lesions in the male rat, atrophy of the testes and accessory sex organs accompanies the fall in LH and FSH and rise in prolactin.

A quite different picture termed *hypothalamic constant oestrus* is produced by lesions placed just above the optic chiasma. The animal appears poised on the brink of ovulation, the ovaries being filled with large follicles which actively secrete oestrogen, but ovulation does not occur because the cyclic burst of LH secretion responsible for ovulation does not occur. Constant oestrus can also be produced by making a knife-cut just in front of the arcuate-median eminence region, showing that a pathway running from the preoptic region to the median eminence is involved in ovulation.

These effects of placing lesions, of electrical stimulation, and of electrical recording at different hypothalamic sites have been interpreted as indicating that there are two groups of neurons associated with gonadotrophin release.

1. A group in the suprachiasmatic region which have long axons running caudally to end in the median eminence around the capillary plexus draining into the hypothalamo-hypophysial portal vessels. At regular intervals determined by a 'hypothalamic clock mechanism', possibly situated in the same region, the circulating ovarian steroids are presumed to exert a positive feedback action on this group of neurons leading to the surge of gonadotrophin release which precedes ovulation. This hypothesis is supported by the increased concentration of gonadotrophin-releasing hormone demonstrated in the blood of women around the time of the ovulatory peak in the LH concentration.

2. A group in the arcuate nucleus which have short axons running to the median eminence. These neurons are responsible for the basal level of gonadotrophin secretion, being under negative feedback control by the level of circulating ovarian steroids. In accordance with this view, changes in the median eminence content of releasing factors have been noted after castration and at puberty.

Hypothalamic gonadotrophin-releasing hormone. (GnRH)
Relatively crude hypothalamic extracts from many mammals, including man, have been shown by both bioassay and radioimmunossay techniques to release LH into the plasma from the adenohypophysis. The existence of a *LH-releasing hormone* (LH-RH) was first demonstrated by McCann *et al.*, in 1960. Since then LH-RH has been shown to be a decapeptide which is active in nanogram doses. It must act directly on the adenohypophysis since it is effective when injected into animals with the median eminence destroyed, when microinjected into the adenohypophysis or perfused into a hypophysial portal vessel, or when added to pituitaries. perfused *in vitro*.

Hypothalamic extracts were shown to have *FSH-releasing activ-*

ity as well by Igarishi and McCann in 1964 and also to promote FSH release from pituitaries incubated *in vitro*. However, there is still uncertainty as to the existence of a specific FSH-releasing factor. In favour of a separate FSH-releasing factor are the observations that stimulation of the preoptic region leads to a large increase in plasma LH concentration but has no effect on FSH, that during the rat oestrous cycle the periods of FSH and LH release overlap but do not coincide, and that when castrated male rats are treated with testosterone there is a dramatic fall in LH but little decrease in FSH. On the other hand, the presence of FSH-releasing activity in the most highly purified preparations of LH-releasing hormone argues against a specific FSH releasing factor. The generally accepted view at present is that there is a single gonadotrophin releasing hormone which releases both LH and FSH from the adenohypophysis and should therefore be termed gonadotrophin-releasing hormone (GnRH) rather than LH-RH and that the relative amounts of FSH and LH released from the anterior pituitary by its action are determined by feedback effects on the adenohypophysis of the varying circulatory levels of the gonadal hormones.

Mechanism of action of gonadotrophin-releasing hormone.

Two principal theories have been proposed to account for the release of gonadotrophins from the adenohypophysis by the gonadotrophin-releasing hormone (GnRH).

1. *The stimulus–secretion coupling hypothesis* of Douglas and Poisner proposes that GnRH alters the membrane permeability of the cell producing the gonadotrophin, leading to depolarization and an uptake of calcium ions which in turn activates the release process. This hypothesis is favoured by the release of FSH and LH from pituitaries incubated *in vitro* on the addition of K^+ (which would be expected to lead to membrane depolarization). This release of gonadotrophins in response to the addition of K^+ or GnRH is depressed if the Ca^{2+} content of the medium is lowered.

2. *The adenyl cyclase hypothesis* proposes that the combination of GnRH with specific receptor sites on the cell membrane activates adenyl cyclase leading to the generation of cyclic AMP from ATP. The cyclic AMP in turn activates release of the gonadotrophin. Evidence favouring this hypothesis includes the activation of adenyl cyclase and the formation of cyclic AMP on addition of hypothalamic extracts to pituitaries incubated *in vitro*, and the potentiation of the response to GnRH by aminophylline. Aminophylline prevents the destruction of cyclic AMP by inhibiting the enzyme phosphodiesterase.

Both mechanisms may in fact be involved, the cyclic AMP generated as a result of the activation of adenyl cyclase having an action on the cell membrane or cytoplasmic organelles, e.g. mitochondria, which leads to Ca^{2+} uptake and activation of the release process. Release of the hormone may involve migration of secretory granules to the cell perimeter where the granule fuses with the cell surface followed by ejection of the core of the granule into the pericapillary space. This process has been observed in growth-hormone release and is probably involved also in the release of gonadotrophins.

Neurotransmitters and release of GnRH. Adrenergic pathways may be involved in GnRH secretion since histochemical fluorescence techniques have demonstrated noradrenergic terminals in the preoptic area and a dopaminergic pathway running from the arcuate nucleus to the median eminence. Injections into the third ventricle of dopamine, noradrenalin, or adrenalin all increase the content of GnRH, and PIF in hypophysial portal blood. Changes in monoamine-oxidase content of the hypothalamus have been observed in the rat during the course of the oestrous cycle.

Extrahypothalamic influences on gonadotrophin release. LH release can be provoked by implants of oestrogen in the *amygdala* suggesting that these nuclei may also participate in the control of gonadotrophin secretion. In several species ovulation has been produced by electrical stimulation of the amygdala. On the other hand the onset of puberty has been shown to be delayed by stimulation, and advanced by lesions, of the amygdala. Lesions have also been claimed to cause a rise in plasma LH. However, some of these conflicting results may turn out to arise from a localization of function within the amygdala.

Stimulation of the *hippocampus* may inhibit spontaneous ovulation or ovulation induced by stimulation of the amygdala or preoptic region, suggesting that the hippocampus is also involved in the regulation of gonadotrophin secretion. This response is abolished by cutting the pathway from the hippocampus to the hypothalamus.

Sexual differentiation of the hypothalamus. Ovarian grafts transplanted into female rats undergo normal cyclic activity, but if the grafts are transplanted into male rats follicles form but do not develop into corpora lutea. This resembles the situation of 'constant oestrus' seen after suprachiasmatic lesions or after dividing the pathway running from the preoptic region to the median eminence. However, ovarian grafts placed in male rats castrated at birth do exhibit cyclic activity. These experiments indicate that the basic pattern of secretion of hypothalamic GnRH is the female cyclic pattern and that 'masculinization' of the hypothalamus is due to the action of testosterone secreted by the infantile testes which results in loss of the ability to produce cyclic variations in gonadotrophin secretion. The preoptic region is probably the site of this action of testosterone since an implant of testosterone in the female produces the picture of 'constant oestrus' seen after destruction of the preoptic region.

The induction of puberty

The mechanism of induction of puberty is still very much of a mystery though a rise in the secretion of gonadal steroids is an important element. Since the immature gonad is quite capable of responding to gonadotrophins and since the immature adenohypophysis is capable of secreting gonadotrophins when appropriately stimulated, the critical factor appears to be the release of GnRH by the hypothalamus. Precocious puberty resulting from lesions of the hypothalamus also implicates the hypothalamus as an important site in this connection.

Perhaps surprisingly, hypothalamic extracts from immature animals do contain releasing factors, and following castration large amounts of gonadotrophins are secreted in immature as well as in mature animals. This has led to the suggestion that before puberty the release of hypothalamic GnRH is more sensitive to the negative feedback action of gonadal steroids. At puberty the threshold for this negative feedback action rises, GnRH is released by the hypothalamus, promotes gonadotrophin secretion by the adenohypophysis, and leads to increased secretion of gonadal steroids. However, the cause of the change in the sensitivity of the hypothalamus to the negative feedback action of the sex steroids remains unknown.

The release of hypothalamic GnRH is also inhibited by melatonin, a hormone produced by the pineal gland [p. 526]. Thus the pineal gland may also participate in regulating the onset of puberty.

REFERENCES

CATT, K. J. (1971). *An ABC of endocrinology*. The Lancet, London.
FRIESEN, H. and HWANG, P. (1973). *A. Rev. Med.* **24**, 251.
GREEN, J. D. and HARRIS, G. W. (1949). *J. Physiol., Lond.* **108**, 359.
KAMBERI, I. A., MICHAL, R. S., and PORTER, J. C. (1971). *Endocrinology* **88**, 1294.
LEE, J. and LAYCOCK, J. (1978). *Essential endocrinology*. Oxford University Press.
MARTIN, J. B., REICHLIN, S., and BROWN, G. M. (1977). *Clinical neuroendocrinology*. Davis, Philadelphia.
MARTINI, L. and BESSER, G. M. (eds.) (1977). *Clinical neuroendocrinology*. Academic Press, London.

THE POSTERIOR PITUITARY

Formation of posterior pituitary hormones

The posterior lobe of the pituitary gland (neurohypophysis) consists of non-myelinated nerve fibres, neuroglia (pituicytes), and blood vessels. There are no 'glandular' cells such as occur in other endocrine glands. The nerve fibres come from the *hypothalamo-hypophysial tract*. This tract takes origin from neurons of the supraoptic and paraventricular nuclei of the hypothalamus and consists of about 100 000 non-myelinated fibres running through the median eminence and pituitary stalk to end in dilated terminations, 0.5–5 μm in diameter, close to the capillaries of the posterior pituitary. The posterior pituitary hormones, *arginine-vasopressin* (ADH) and *oxytocin*, can be extracted from the whole extent of this tract and Harris (1947) demonstrated that electrical stimulation of the tract brought about a reduction in urine flow and an increase in uterine motility. Experimental or pathological lesions involving this hypothalamo-hypophysial pathway produce the condition of *diabetes insipidus* in which large volumes of dilute urine, up to 15 litres, are excreted daily (polyuria). Bargmann (1949) showed that examination under the light microscope after staining by the Gomori technique (acid permanganate oxidation followed by staining with a basic dye, chrome alum haematoxylin) reveals neurosecretory material (NSM), often referred to as Gomori substance, throughout the hypothalamo-hypophysial tract. Under the electron microscope the neurosecretory material is seen to consist of aggregations of membrane bound neurosecretory granules, about 100 nm in diameter [FIG. XI.17], which are believed to contain the polypeptide hormones bound to a carrier protein called *neurophysin*. There are at least two neurophysins; neurophysin I bound to oxytocin and neurophysin II bound to vasopressin.

When ADH secretion is stimulated during dehydration or following injection of hypertonic saline, there is a corresponding depletion of NSM. Following section of the pituitary stalk NSM disappears from the neurohypophysis and accumulates in the hypothalamo-hypophysial fibres proximal to the point of section. Sloper (1966) injected rats with methionine and cysteine labelled

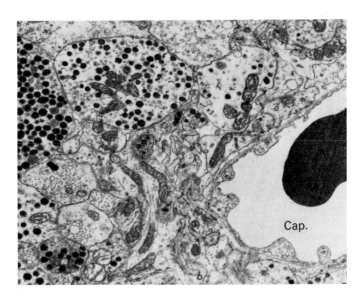

FIG. XI.17. Electron micrograph of rat neurohypophysis, showing neurosecretory granules and small vesicles in the axoplasm of fibres of the hypothalamo-hypophysial tract ending in close relation to a capillary (cap) containing a red blood corpuscle.

with ^{35}S; within minutes the tracer appeared in the cells of the supraoptic and paraventricular nuclei and about 10 hours later tracer could be found in the posterior pituitary. Thus it appears likely that the posterior pituitary hormones are synthesized in the cells of the supraoptic and paraventricular nuclei, vasopressin probably being formed in the supraoptic nucleus and oxytocin in the paraventricular nucleus. The hormones then pass down the axons arising from these nuclei and are stored in their endings in the posterior pituitary from which they are subsequently released by nerve impulses arriving along the hypothalamo-hypophysial pathway. The capillaries in the posterior pituitary have large pores through which the neurosecretory material could pass. The release of hormones from the endings in the posterior pituitary was found by Douglas (1963) to be stimulated by *calcium* and he suggested that depolarization following arrival of the nerve impulse evokes secretion by permitting the inflow of calcium.

Functions of the posterior pituitary

1. The posterior pituitary secretes the antidiuretic hormone (ADH) which controls the secretion of urine by the kidney and thus regulates the water and electrolyte balance of the body fluids [p. 233].

2. It secretes oxytocin which causes ejection of milk from the lactating breast, and probably stimulates the release of lactogenic and galactopoietic factors from the anterior pituitary. It may also have a physiological role in parturition [p. 585].

The nomenclature of the posterior pituitary hormones is confusing. The name 'vasopressin' was given to one hormone when its effects in raising blood pressure in animals, and in causing constriction of precapillary sphincters in man were considered. These effects occur only with large doses; small, physiological doses inhibit diuresis without circulatory effects. The name 'antidiuretic hormone' (ADH) is therefore more appropriate. The name 'oxytocin' was introduced to describe the effect which the other posterior pituitary hormone exerts in 'hastening labour' by its stimulating action on uterine muscle. Its physiological role in lactation is now more firmly established than its role in parturition, so once again the name is unsatisfactory.

Extracts of the posterior lobe of the pituitary contain equal amounts of ADH (arginine-vasopressin) and oxytocin. These hormones have been isolated and identified by du Vigneaud as octapeptides (i.e. peptides containing eight amino-acid residues). They, and many of their analogues with enhanced or novel physiological activities, have been synthesized and are used therapeutically.

Control of release of posterior pituitary hormones

1. Plasma osmotic pressure. Verney (1947) showed that the output of ADH was sensitive to changes in the osmotic pressure of the plasma. Dogs were prepared with exteriorized carotid arteries so that the effect of intracarotid injection of a variety of solutions on an established water diuresis could be studied. Hypertonic solutions of NaCl, Na_2SO_4, or sucrose were found to be highly effective antidiuretic agents, while glucose and urea injection failed to modify the diuresis. To explain these findings Verney suggested that there were 'osmoreceptors' within the hypothalamus which stimulated the secretion of ADH when there was a relative increase in the osmotic pressure of the extracellular fluid. This would account for the antidiuretic affects of NaCl, Na_2SO_4, and sucrose, since the cell membrane is relatively impermeable to sodium ions and sucrose, and for the lack of any effect of glucose or urea to which the cell membrane is freely permeable. The sensitivity to a change in sodium ion concentration lay within physiological limits; a rise in plasma sodium concentration of 3 mmol per litre or only 2 per cent being effective. In further experiments Jewell and Verney (1957)

localized the osmoreceptors to the region of the anterior hypothalamus containing the supraoptic and paraventricular nuclei, since the responses were still obtained after ligation of the external carotid artery and a number of intracranial branches of the internal carotid. More recently the electrical activity of individual neurons in the supraoptic and paraventricular nuclei has been recorded with microelectrodes. Intracarotid injections of hypertonic NaCl produced increases in spike discharge paralleled by a greater rate of release of ADH and oxytocin [FIG. XI.18].

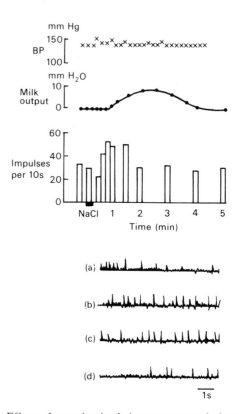

FIG. XI.18. Effects of osmotic stimulation on paraventricular neuron discharges, on milk ejection pressure, and on blood pressure. Time of intracarotid injection of 1 ml of 1 M NaCl is marked by a bar. The number of impulses in paraventricular nucleus neurons during 10s intervals before and after injection is shown in the bar graph. The corresponding milk ejection response, recorded as a mammary duct pressure, is shown in the tracing above this. A calibrating dose of 5 miu. oxytocin given intravenously resulted in a pressure change of 20 mm H_2O. Blood pressure is shown in the top tracing. Actual records of neuron discharges are given below: (a) control period; (b) 30 s after injection; (c) 2 min after; and (d) 5 min after osmotic stimulation. Spikes retouched.
(Redrawn from Brooks, C. McC., Ishikawa, Koizumi, K., and Lu, H.-H. (1966). *J. Physiol., Lond.* **182**, 217–31.)

2. Nervous stimuli. Nerve impulses from many sources may induce secretion of both vasopressin and oxytocin. Electrical stimulation of the central ends of the vagus or ulnar nerves increases the activity of units in the supraoptic and paraventricular nuclei and leads to secretion of posterior pituitary hormones, while any painful stimulus will inhibit diuresis. Verney showed that exercise had an antidiuretic effect in trained dogs but attributed this to the emotional stress involved since the antidiuretic response diminished with repetition of the test. The hypothalamus is intimately involved in the integration of emotional responses [p. 342] and other emotional situations such as fear or anger may also be accompanied by an antidiuretic effect. Oxytocin is released reflexly by stimulation of the nipple in suckling, and by distension of the uterus, and these stimuli also increase the discharge of neurons in the paraventricular nucleus. Coitus may induce milk ejection and uterine contractions. However, neither ADH nor oxytocin is ever secreted in isolation in response to these stimuli but these hormones appear simultaneously in various proportions. Thus while the predominant effect of osmotic stimuli may be antidiuretic, and the response to suckling, uterine distension, and coitus largely a secretion of oxytocin, coitus for example, may also lead to a mild antidiuresis. Similarly, osmotic stimuli lead to a discharge of paraventricular units, though less than of cells in the supraoptic nucleus, and suckling mildly excites supraoptic neurons as well as more markedly exciting paraventricular cells.

3. Chemical transmitters. Intracarotid injection of small amounts of acetylcholine leads to secretion of ADH and oxytocin and an increase in the discharge of cells in the supraoptic and paraventricular nuclei. [FIG. XI.19]. This finding, together with the antidiuretic effect of nicotine and hence of smoking, suggests that the pathways which activate these cells may involve cholinergic neurons. Intracarotid injection of adrenalin inhibits the effects of a following injection of acetylcholine on paraventricular unit activity and on milk ejection and it has long been known that any emotional excitement, which presumably increases sympathetic activity and catecholamine secretion, will interfere with succesful breast-feeding. This inhibitory action of adrenalin on posterior pituitary secretion may help to explain the variability which has been reported in the antidiuretic effect of emotion, since the resulting change in the output of ADH will depend upon the balance between stimulation of the supraoptic and paraventricular nuclei by impulses arriving along nervous pathways, and the inhibitory action of the increased concentration of circulating adrenalin secreted by the adrenal medulla which also forms part of the emotional response. An additional factor in this complex situation is the reduction in urine output due to constriction of the afferent glomerular arterioles by adrenalin and increased sympathetic discharge to the kidneys. It should perhaps be emphasized that the increased frequency of micturition which often accompanies emotional stress is not related to a change in urine production but results from heightened sensitivity of the micturition reflex.

REFERENCES

BARGMANN, W. (1949). *Acta anat.* **8**, 264.
BESSER, G. M. (1974). Hypothalamus as an endocrine organ. *Br. med. J.* **iii**, 560–4; 613–15.
CHARD, D. T. (1975). Posterior pituitary gland. *Clin. endocr.* **4**, 89.
DOUGLAS, W. W. (1963). *Nature, Lond.* **197**, 81.
FORSLING, M. L. (1979). *Antidiuretic hormone*, Vol. 3. Churchill Livingstone, Edinburgh.
HALL, R. (1973). Hypothalamic control of the pituitary. *Br. J. hosp. Med.* **9**, 109–33.
HARRIS, G. W. (1947). *Phil. Trans. R. Soc.* **B232**, 385.
HOCKADAY, T. D. R. (1972). Diabetes insipidus. *Br. med. J.* **ii**, 210–13.
JEWELL, P. A. and VERNEY, E. B. (1957). *Phil. Trans. R. Soc.* **240**, 197.
KNOBIL, E. and SAWYER, W. H. (eds.) (1974). Pituitary gland and its neuroendocrinological control. In *Handbook of physiology*, section 7, vol. 4. American Physiological Society, Washington DC.
MARTINI, L., MOTTA, M., and FRASCHINE, F. (eds.) (1971). *The hypothalamus*. Academic Press, New York.
SLOPER, J. C. (1966). In *The pituitary gland* (ed. G. W. Harris and B. T. Donovan) Vol. 3, p. 131. Butterworth, London.
VERNEY, E. B. (1947). *Proc. R. Soc.* **B135**, 25.

Hypothalamic control of release of melanocyte-stimulating hormone (MSH)

Melanocyte-stimulating hormone (MSH; intermedin), which induces darkening of the skin of fish and amphibia by expanding the melanophores, is produced in the pars intermedia of the pituitary. Its importance in mammals is unknown. The secretion of MSH from the pituitary is regulated by MSH-releasing hormone (MRH) and MSH–release inhibiting hormone (MRIH) which have been extracted from the hypothalamic structures of amphibia and identified as small peptides. Nerve fibres from the hypothalamus reach the pars intermedia through the pituitary stalk. Their endings appear to contain neurosecretory granules and it is thought that MRH and MRIH are transported to the pars intermedia through this nervous pathway rather than through portal vessels.

Little is known of the regulation of MRH and MRIH release. In lower vertebrates blanching and expansion of melanocytes occurs in response to visual and emotional stimuli which are likely to involve the hypothalamus. It has been suggested that melatonin, a powerful melanocyte-constrictor found in the pineal gland, may act by stimulating MRIH secretion.

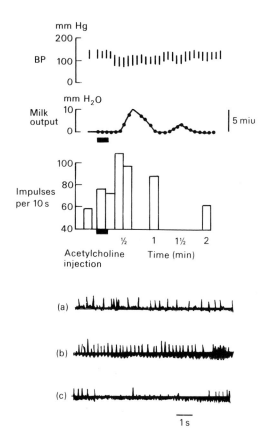

FIG. XI.19. Effect of acetylcholine on neuron activity in paraventricular nucleus, on milk ejection response and on blood pressure. Acetylcholine chloride, 40 µg, was given intra-arterially in 15 s (—). Cellular discharges are shown below: (a) control; (b) 20 s after the injection; and (c) 5 min after the injection. The rates of discharge are plotted in the bar graph. Mammary duct pressure is shown above this. The calibration sign designates maximum pressure change caused by 5 miu of oxytocin given intravenously. Blood pressure recorded from femoral artery is shown at top. Spikes retouched. (Redrawn from Brooks, C. McC., Ishikawa, T., Koizumi, K., and Lu, H.-H. (1966). *J. Physiol., Lond.* **182**, 217–31.)

The pineal gland

The pineal gland has fascinated philosophers for centuries. Descartes (seventeenth century) asserted that it was the seat of the soul. The human pineal weighs about 120 mg and lies between the superior colliculi. Embryologically it develops from the roof of the diencephalon. It has a very rich blood supply.

In lower vertebrates, e.g. amphibians and dogfish, the pineal contains light receptor cells and nerve cells which conduct impulses to the brain. In higher vertebrates, including man, the photoreceptor elements have been completely replaced by parenchymal cells (pinealocytes); the pineal also contains neuroglial cells and numerous postganglionic sympathetic nerve fibres from cells in the superior cervical ganglion. These fibres contain NA and mostly terminate on the pinealocytes. There is an anatomical pathway, via an inferior accessory optic tract and a multisynaptic pathway in the midbrain and medulla, which connects with the superior cervical ganglion so that nerve impulses could be transmitted from the retina to the pineal. The pineal is a neuroendocrine transducer; it forms and secretes a hormone called *melatonin* in response to sympathetic nerve activity and in this respect is analogous to the adrenal medulla, the neurohypophysis and the renin-secreting juxtaglomerular cells of the kidney, all of which secrete hormones in direct response to nervous activity. These organs do not function when transplanted. The pineal is large in young animals and infants; the human pineal begins to involute and calcify during the second decade of life but continues to form melatonin.

Biosynthesis of melatonin in the pineal gland

The following chemical reactions occur in the pineal, the only structure in the body which synthesizes melatonin:

The reactions in the pinealocyte are initiated by β-adrenergic receptor actions of NA released from sympathetic nerve terminals. Consequent activation of adenyl cyclase and cyclic AMP stimulates [^{14}C]-melatonin synthesis from [^{14}C]-tryptophan. Degenerative section of the sympathetic nerves to the pineal stops NA release, activation of adenyl cyclase and the activation of HIOMT; melatonin synthesis is thus abolished.

Melatonin is secreted rapidly from the pineal into the c.s.f. and blood, and by virtue of its high lipid solubility it penetrates into cells in the hypothalamus and midbrain, into peripheral nerve fibres and into the gonads. The human pineal contains melatonin in concentrations of 50–400 μg g^{-1}.

Actions. Melatonin is so named because it contracts the melanophores in amphibian skin (and thus lightens its colour).

In mammals melatonin inhibits oestrus and reduces ovarian or testicular weight. It probably reduces gonadal activity by decreasing the rate of discharge of hypothalamic GnRH and thus reducing the secretion of FSH and LH from the adenohypophysis.

Melatonin also inhibits thyroid and adrenal cortical hormone secretions.

Pinealectomy reduces melatonin secretion to zero and therefore abolishes the normal inhibitory actions of melatonin. The results of pinealectomy in experimental mammals are therefore:

To increase ovarian and uterine weight.
To induce ovulation and oestrus.
To increase weight of the testis, vesiculae seminales, and prostate.
To increase secretions of female and male sex hormones.
To increase thyroid hormone secretion.
To increase adrenal cortical secretions.

Control of melatonin secretion

Darkness stimulates pineal secretion of melatonin, whereas exposure to light inhibits melatonin secretion. These effects are mediated by the retina–optic tract–brain stem– superior cervical ganglion–sympathetic nerve pathway to the pineal.

Physiologically, the pineal may be concerned with regulation of time-dependent effects such as onset of puberty, and rhythmic activities such as hormonal regulation of ovulation. Melatonin also produces slowing of EEG rhythm, sleep and a rise in convulsive threshold.

Tumours consisting of pineal tissue may inhibit the onset of puberty. Tumours which *destroy* the pineal may induce precocious puberty. The pineal could be an important regulatory organ of many kinds of rhythmic activities but much more evidence is needed before the functions of this inaccessible organ can be properly assessed.

It is by no means certain that melatonin is *the* pineal hormone. The antigonadotrophic agents from the pineal may be *polypeptides*, some of which are more potent than melatonin and could be of value as contraceptives (Reiter *et al.* 1975).

REFERENCES

The Lancet (1974). Editorial. *Lancet* **ii**, 1235.
Reiter, R. J. *et al.* (1975). *Lancet* **i**, 741.
Wurtman, R. J. and Cardinali, D. P. (1974). In *Textbook of endocrinology*, 5th edn (ed. R. H. Williams) p. 832. Saunders, Philadephia.

Regulation of growth

Growth is characteristic of living organisms. It includes increases in size and number of cells, leading in human beings to increased height and weight. *Differentiation* of homogeneous cells into cells with specialized functions begins early in embryonic life and continues in parts of the body into old age. In addition to cells, the body contains *intercellular* materials which are important for normal function, e.g. in bone. At the same time as growth occurs, cells in some parts of the body are dying and being shed from infancy onwards, e.g. mucosal cells of the gastro-intestinal tract and cells of the epidermis. Some types of cell, e.g. somatic muscle and nerve cells, cease to divide after 5–6 months of fetal life, though they increase in size (length and weight) after birth, and nerve cells ramify and grow myelin sheaths in postnatal life. Throughout adult life, after height has reached its peak, the capacity for repair after injury, and for growth is retained in certain organs; superficial wounds of the skin may heal completely and if part of the liver is destroyed the remaining cells multiply to restore the normal total number of cells in this organ. In middle and old age although many types of cell continue to multiply, e.g. nails, intestinal epithelium, or enlarge, e.g. adipocytes and the prostate (men), there is in general a tendency for cellular function to decay, e.g. teeth fall out and baldness occurs frequently in men, the reproductive system atrophies after the menopause in women, and by 65–70 the brain in both sexes has lost 20 per cent of the neurons present at birth. However, with passing years there is an increased tendency for the development of new growths (neoplasms), often malignant in character (cancers). Thus growth involves a series of changes which, during embryonic life lead to a spectacular increase in the number of cells so that the fertilized ovum, which is 100 μm in diameter, has developed during 40 weeks into a newborn baby 50 cm long, weighing on average 3.5 kg (3×10^9 times the weight of the ovum).

The processes of embryonic and fetal growth and differentiation belong to the field of embryology and will not be considered here.

After birth the general growth curve shows four distinct phases [FIG. XI.20]:

1. A rapid increase during infancy and especially during the first year (from 3.5 to 10.5 kg).

2. A slower progressive growth from 3 to 12 years of age; during this period boys are slightly taller than girls.

3. A marked acceleration at the time of puberty. In girls the rate of increase in height and weight is greatest at 12–13 years; in boys the adolescent spurt in growth is greatest at 14–15 years. There are considerable individual variations in the age of onset of both puberty and its accompanying growth spurt.

4. Even when the pubertal growth increase has finished, at 18 in girls, and 20 in boys, a small slow growth occurs until 30.

In girls the increase in weight during adolescence is mainly due to increased fat formation, in boys it is mainly due to increased muscular growth (promoted by secretion of testosterone). After adolescence males are on average taller and heavier than females.

The general growth curve applies to the skeleton as a whole, the muscles, and the thoracic and abdominal viscera [FIG. XI.20]. Certain parts of the body have distinctive growth curves. Three main specialized types are recognized:

(i) *Neural type* [FIG. XI.20]. There is a rapid initial increase in size so that the brain, spinal cord, and the organs of special sense, together with the skull, reach 90 per cent of adult size by about 6 years of age. Although nerve cells do not increase in number after birth, some nerve cells increase enormously in mass (up to 200 000 times) by elongation, ramification, and myelination. Nerve cells become rich in RNA which synthesizes the cytoplasm which migrates into nerve fibres.

(ii) *Lymphoid type* [FIG. XI.20]. The lymphoid tissues, including the thymus, tonsils, and lymph nodes throughout the body, grow rapidly in early childhood and reach their maximum size at puberty. After this lymphoid tissue degenerates.

(iii) *Reproductive type* [FIGS. XI.20 and XI.21]. The gonads and the accessory organs of reproduction remain undeveloped till puberty when very rapid growth begins and continues throughout adolescence.

(iv) Certain organs show a different type of growth. The adrenal glands and uterus are relatively large at birth, they then lose weight rapidly and regain their birth weights just before puberty.

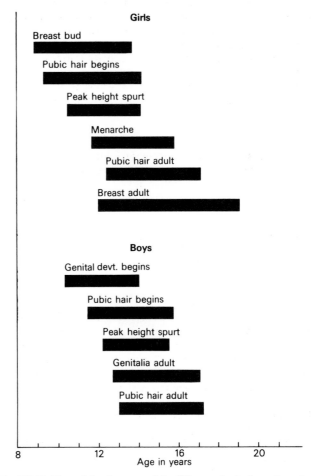

FIG. XI.21. Time of development of the different physical manifestations of puberty in boys and girls.
(Redrawn from Sadow, J. I. D. *et al.* (1980). *Human reproduction.* Croom Helm, London.)

Factors influencing growth and development

Growth and development are mainly controlled by genetic, hormonal, nutritional, and other environmental factors, and by disease.

Genetic control

Genetic factors are very important in relation to growth and stature. Children of tall, heavy parents are likely to have the same stature. Identical twins have the same body growth patterns and the same body proportions when fully grown. Non-identical twins may differ greatly with respect to dental, bone, sexual, and neurological patterns of growth.

Control of body size is complicated and many genes are involved. The form of dwarfism due to failure of growth of the long bones (achondroplasia) is, however, inherited as a Mendelian dominant.

Genetic factors are mainly responsible for certain racial differences in height. The Dinkas of Sudan are among the tallest races

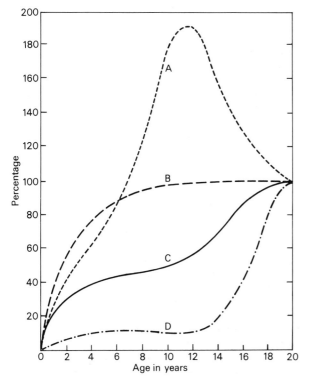

FIG. XI.20. Growth curves of different types of organs. A = lymphoid type; B = neural type; C = general type; D = reproductive type.
(*White House Reports*, 1932.)

in the world and the pygmies of the Congo are one of the smallest. The pygmies secrete adenohypophysial growth hormone in normal amounts and injected GH produces none of the biochemical changes which it arouses in most races. It is assumed that pygmies are small because their tissues do not respond to GH [p. 517].

Although the differences between male and female patterns of growth are hormonally determined, the timing of the adolescent spurt appears to be genetically controlled via the hypothalamus.

Neural control

The adenohypophysis is essential for normal growth and the secretions of this gland are largely controlled by the hypothalamus and related regions of the central nervous system [p. 518].

Peripheral nerves, both motor and sensory, control the integrity of structures which they innervate, possibly by a 'trophic' effect involving release of a chemical (not acetylcholine or other known transmitter) at the nerve terminals.

Hormonal control

Disturbances of growth frequently accompany clinical disorders of the endocrine glands.

Lack of secretion of adenohypophysial GH, and lack of thyroid hormone secretion in early postnatal life lead to dwarfism which may be overcome by administration of GH and thyroid hormones respectively [pp. 517, 540]. Hypothyroidism in childhood has particularly adverse effects on growth of the central nervous system [p. 541]. The hypothalamus controls the secretion of GH and TSH by releasing hypothalamic-hypophysial hormones, so deficiency of hypothalamic function can also produce dwarfism [p. 517]. Conversely, oversecretion of GH before closure of the epiphyses of the long bones leads to gigantism and after epiphysial closure it produces acromegaly [p. 517]. Growth hormone, thyroid hormones, insulin, and androgens all promote protein anabolism in various ways. The adrenal corticosteroid, cortisol, inhibits growth before puberty by antagonizing the actions of GH. It promotes protein catabolism.

The stimulation of growth by GH is effected by proliferation of cartilage cells in the epiphyses. Secretion of testosterone in the male, and oestradiol in the female, at the time of puberty, lead to closure of the epiphyses. Testosterone has anabolic effects which show as an increase in skeletal muscular mass as well as further skeletal growth. The male and female sex hormones promote growth of genitalia and secondary sex characteristics [pp. 577, 568]. The growth of bone requires the co-ordinated activity of parathyroid hormone, calcitonin and 1,25-dihydroxycholecalciferol which control the absorption and distribution of calcium in the body [p. 546].

A further example of growth stimulation by a hormone is the action of erythropoietin on the promotion of red cell production by the bone marrow [p. 37].

Nutrition

The importance of an adequate diet for normal childhood growth and development is vividly revealed in the many parts of the world where near famine conditions prevail. The combined effects of a diet deficient in mass and energy, in proteins, minerals and vitamins, inhibit growth profoundly and, if the lack is sufficiently prolonged, the stunting of growth may be irreversible. With less severe, or shorter periods of undernutrition, restoration of a normal diet leads to a compensatory acceleration in the rate of growth.

Undernutrition can also be caused by malabsorption from the alimentary tract, and if severe enough this may prevent normal growth in children.

Undernutrition affects the growth of different organs and tissues unevenly. Thus dietary deprivation has greater adverse effects on muscle and fat than on the growth of bone, while teeth take precedence over bone; skeletal maturation is less affected than skeletal growth and in the brain total growth is inhibited more than myelination; at puberty undernutrition affects the growth of the genitalia less than that of other organs.

Season and growth. Growth in height is faster in the spring than in the autumn, whereas increase in weight is faster in the autumn than in the spring. The cause of these differences is not yet known but the fact that they are greatest during the adolescent spurt suggests a hormonal basis.

Disease. Ill health from many causes temporarily depresses growth, but during recovery the lost ground is regained, more or less completely. Congenital cardiac defects, especially when associated with hypoxia, often lead to stunting of growth. Surgical correction of the defect restores normal growth.

Exercise. The repeated exercise of skeletal muscles can increase their mass, by producing enlargement of individual fibres. This hypertrophy is favoured by anabolic steroids [p. 580].

Emotional disturbances can decrease the rate of growth in children taking an adequate diet. Removal of the cause restores the normal growth rate.

Old age. Aging or senescence is perhaps difficult to define but two facts are clear:

1. The number, and proportion, of people living beyond the age of 65 is increasing in many countries. Women live longer than men.

2. The number of people reaching the age of 100 is not increasing.

Aging is characterized by cellular degeneration and associated impairment of various functions. Deterioration might be due to loss of non-multiplying cells or to degradative changes in the properties of multiplying cells. Irreparable cell loss in females begins even before birth, since the number of oogonia reaches a peak at 7 months of pregnancy and the number of primary oocytes falls steadily from birth onwards till the ovary atrophies at the time of the menopause [p. 568]. Nerve and muscle cells stop dividing before birth and loss of nerve cells continues throughout life. The special senses (hearing, vision) begin to deteriorate soon after completion of the adolescent spurt in growth, and memory and the capacity to learn are less efficient in early adult life than in childhood. These changes continue and become very obvious in the elderly, in whom loss of memory, impaired control of muscular activity, failure of body temperature control (leading to hypothermia on exposure to cold) and loss of taste and smell are among the numerous manifestations of the loss of 10 000 nerve cells daily.

In addition, old people show loss of subcutaneous fat, reduced elasticity of the skin, rarefaction of bones (osteoporosis) and tendency to fractures which may not heal, stiffness in joints (due to loss of elastic tissue and deposition of collagen), narrowing of intervertebral discs, increased opacity of the lens (senile cataract), and degenerative changes in the cardiovascular system, including coronary and cerebral atherosclerosis. It is often difficult to say whether degenerative changes are physiological, or pathological in nature, but predisposition to fatal infections, malignant disease and cardiovascular catastrophes accounts for the vast majority of deaths in old people.

REFERENCES

BRASEL, J. A. and BLIZZARD, R. M. (1974). The influence of the endocrine glands upon growth and development. In *Textbook of endocrinology*, 5th edn (ed. R. H. Williams) p. 1030. Saunders, Philadelphia.

GREGERMAN, R. L. and BIERMAN, E. L. (1974). Aging and hormones. In *Textbook of endocrinology*, 5th edn (ed. R. H. Williams) p. 1059. Saunders, Philadelphia.

SINCLAIR, D. (1973). *Human growth after birth*, 2nd edn. Oxford University Press.

The adrenal cortex

The adrenal (suprarenal) gland consists of two distinct organs, the outer adrenal cortex and the inner adrenal medulla, which differ in their embryonic development, comparative anatomy, histological structure, and functions. The adrenal medulla is discussed on page 417.

The adrenal cortex is essential to life. It participates in responses to 'stresses' and through its chief hormones, cortisol (hydrocortisone), corticosterone, and aldosterone influences the metabolism of carbohydrate, protein, fat, electrolytes, and water.

Embryology. The adrenal develops from two distinct sources that produce separate organs in fishes, but are combined in mammals. The medulla is derived from ectodermal cells which migrate from the neural crest to their final location inside the adrenal cortex. The cortex is formed during the fourth to sixth week of fetal life from cells of the coelomic mesoderm between the root of the mesentery and the genital ridge. Thus the adrenal cortex is derived from cells in close relation to those which eventually form the ovary or testis. All these cells synthesize steroids, which may be precursors of hormones, or actual hormones which enter the circulation and produce the characteristic effects of ovarian, testicular, or adrenocortical endocrine function.

During fetal life the adrenal cortex is a relatively large organ (at 8 weeks it may be larger than the kidney) and the cells depend on ACTH stimulation. The cortex is very small in the anencephalic fetus (Jost 1975) but can be restored by ACTH administration. The secretion of ACTH from the adenohypophysis is in turn stimulated by corticotrophin-releasing hormone (CRH) from the hypothalamus [p. 518]. The cortisol secreted by the fetal adrenal cortex increases oestrogen production by the placenta. After birth the adrenal cortex atrophies. The three differentiated zones are not formed until the third year of life.

Human anatomy. There are two adrenal glands, which lie behind the peritoneum, one at the top of each kidney. There may also be accessory adrenal glands, consisting of cortex, medulla, or both, at various sites on the posterior abdominal wall. Cortical tissue may sometimes be found close to the ovary or testis, a reminder of their similar embryological derivation. Thus the operation called 'adrenalectomy', involving extirpation of the two main adrenals, may not in fact remove all cortical or medullary activity.

Normally each adult adrenal gland weighs 4–5 g and contains much more cortical than medullary tissue. The weight is less when secretion of ACTH is reduced by pituitary or hypothalamic lesions, and therapeutic administration of ACTH in excess may increase the adrenal weight to four times the normal.

The adrenal medulla receives preganglionic sympathetic nerve fibres but the adrenal cortex has no nerve supply. The adrenals have a rich blood supply; the vein from the left adrenal drains into the renal vein, that from the right adrenal into the inferior vena cava. Thus the adrenal hormones enter the systemic circulation directly, unlike the hormones from the pancreas and gastro-intestinal tract which are conveyed by the portal vein to the liver before they can enter the systemic circulation.

Histology. The cells of the adrenal cortex are arranged in three zones:

1. A narrow outer *zona glomerulosa* which lies directly under the capsule. It consists of relatively small compact cells containing elongated mitochondria.
2. The wider *zona fasciculata* consists of columns of larger cells with spherical mitochondria.
3. The inner *zona reticularis* consists of a network of cells which resemble the cells of the zona fasciculata except that their mitochondria are elongated.

The cells of all three layers have a high lipid content, particularly those of the zona fasciculata; the chief lipid is cholesterol, mainly in the ester form, which is the precursor of the adrenocortical hormones. The cells of the zona glomerulosa secrete *aldosterone;* in man the cells of the zona fasciculata and zona reticularis form a single unit which secretes mainly *cortisol*, some *corticosterone*, *androgenic steroids*, and possibly oestrogens. The adrenal cortex is also rich in ascorbic acid, the function of which is unknown.

Adrenal corticosteroids

Histochemical and analytical chemical techniques, radioactive tracer studies, chromatography, and other isolation procedures have been applied to the detection and estimation of steroids and other cortical constituents, in the cortex itself, in the fluid from perfused adrenals, in adrenal venous blood, in peripheral blood and in the urine. From all these studies, in experimental animals and, where applicable, in man, it is concluded that the physiologically active substances (hormones) secreted by the adrenal cortex are steroids. Of the 30 or more steroids isolated from the adrenal cortex only very few are secreted into the bloodstream, the great majority being intracellular intermediary metabolites. The hormones of the adrenal cortex are usually divided into three main groups.

Cholesterol

Pregnenolone

Corticosterone ('B')

Cortisol ('F')

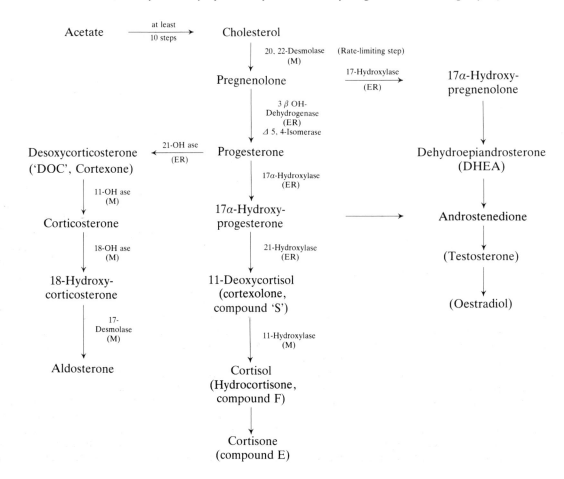

Aldosterone
(aldehyde form)

Aldosterone
(hemi-acetal form)

1. Glucocorticoids. These are so named because they enhance glucose formation. They contain 21 carbon atoms and are called $C_{(21)}$ steroids.

2. Mineralocorticoids. These are $C_{(21)}$ steroids which promote retention of sodium and excretion of potassium by the kidney.

The term 'corticosteroids' includes glucocorticoids and mineralocorticoids.

3. Androgens. These are $C_{(19)}$ steroids with weaker masculinizing activity than testosterone.

Progesterone is formed as an intermediary in the synthesis of both glucocorticoids and mineralocorticoids.

The adult adrenal cortex may form very small amounts of oestrogen.

Biosynthesis of adrenocortical hormones

Cholesterol is the major precursor of all steroid hormones. The adrenal cortex can synthesize cholesterol (from acetyl-CoA) and can also take it up from circulating blood. There is always plenty of cholesterol in the adrenal cortex, mostly within cytoplasmic lipid droplets (especially in the zona fasciculata). When the adrenal cortex is stimulated by ACTH cholesterol is first converted to 20, 22-dihydroxycholesterol; then it is acted upon by the mitochondrial enzyme, 20, 22-desmolase, which forms *pregnenolone*. The conversion of cholesterol to pregnenolone is the rate-limiting step in the biosynthesis of all the adrenocortical steroid hormones.

All the subsequent biosynthetic reactions are effected by enzymes in the endoplasmic reticulum (ER) or in mitochondria (M). The reactions involved in the synthesis of cortisol, corticosterone, aldosterone, and androgens are summarized below.

For relationship with steroid sex hormones see page 573.

Rate of secretion of corticosteroids

The rate of secretion of corticosteroids can be measured in experimental animals by determining:

1. Adrenal gland steroid content.
2. Adrenal vein steroid output.
3. Peripheral blood steroid concentration.
4. Renal excretion of steroids and metabolites.

Methods 3 and 4 are easily applied to man but single estimations of plasma levels are only of value in assessing adrenocortical activity if the corticosteroid concentration is high when it should be low (e.g. at midnight), or low when it should be high (e.g. at 8.0 a.m.). The response to ACTH or other forms of stimulation can be followed by repeated estimations of plasma cortisol levels at suitable intervals.

Renal excretion of corticosteroids and their metabolites (nonconjugated and conjugated with glucuronic acid) can measure steroid output and metabolism over a 24 hour period.

The secretion rate of cortisol can be measured by an isotope dilution method, using [^{14}C]-cortisol. Cortisol secretion in man normally ranges from 5 to 28 mg day^{-1} (mean = 15 mg). Very little cortisol is stored in the adrenal cortex; continuous synthesis and release is required to maintain the physiological concentrations in body fluids which act on target organs.

Blood steroid measurements are made after treating serum or plasma with a lipophilic organic solvent which extracts non-conjugated steroids, whether bound to proteins or not. The techniques for measurement of plasma steroids are as follows:

1. *Plasma 17-hydroxycorticosteroids (17-OHCS)* are measured by a colorimetric method (Porter–Silber reaction) which measures C_{21} steroids and gives values very close to the true plasma cortisol content.

2. *Plasma 11-hydroxycorticosteroids (11-OHCS)* are measured by a fluorimetric method which normally measures cortisol + corticosterone.

3. *Plasma cortisol* can be measured by *protein-binding analysis*. This is the most sensitive technique, which measures cortisol + corticosterone in small samples of blood.

Glucocorticoids in blood. Cortisol is the main glucocorticoid in *human* blood, corticosterone being present in much smaller amounts. In the rat corticosterone predominates.

The *plasma cortisol concentration* shows a marked circadian variation, due to changes in the rates of secretion of ACTH and CRH from the hypothalamus [p. 518]. In normal persons the mean plasma cortisol concentration is maximal between 6.00 and 9.00 a.m., at about 425 nmol l^{-1} (150 µg l^{-1}); the value falls steadily during the day to reach its minimum of about 85 nmol l^{-1} (30 µg l^{-1}) between midnight and 3.00 a.m. [FIG. XI.22].

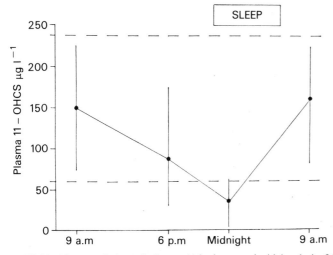

FIG. XI.22. Mean variation of plasma 11-hydroxycorticoid levels in 24 normal subjects. The vertical lines indicate the range of observations. The horizontal dashed lines show the normal range between 9 a.m. and 10 a.m. (After Mattingly, D. (1968). In *Recent advances in medicine*, 15th edn (ed. D. N. Baron, D. N. Compston and A. M. Dawson). Churchill, London.)

In the bloodstream cortisol is normally reversibly bound to a plasma glycoprotein called *transcortin* or *corticosteroid binding globulin* (CBG). At concentrations below 500 nmol l^{-1} 90 per cent of cortisol is bound and the rest is unbound (free) or loosely bound to albumin. With higher plasma cortisol concentrations there is an increased proportion of free and albumin-bound cortisol. Transcortin also binds corticosterone and some of the synthetic corticosteroid derivatives. The half-life of cortisol in the bloodstream is 60–100 min.

Transcortin is synthesized in the liver. In persons on oestrogen therapy (e.g. the contraceptive pill), and in pregnant women, the plasma transcortin concentration may rise to three times the nor-

mal. The unbound free plasma cortisol level at first falls but this stimulates ACTH secretion so that the total plasma cortisol rises to a level which saturates the transcortin and may even increase the amount of free cortisol so that a small degree of hyperadreno-corticism may occur during the third trimester of pregnancy. In cirrhosis of the liver there is decreased transcortin synthesis and in nephrosis there is excessive loss of cortisol-binding proteins in the urine; in either case there is adaptation via the hypothalamus and adenohypophysis so that ACTH secretion is decreased and the free cortisol level returns to normal.

The significance of transcortin is, however, challenged by the observation that certain families may have a genetically determined absence of transcortin, and extremely low plasma cortisol levels. Such persons show no signs of adrenocortical deficiency so it does not seem that transcortin-bound cortisol is necessary to provide a reserve of this hormone.

To summarize, the adrenal cortex, under the control of the hypothalamus and anterior pituitary, normally secretes enough cortisol to satisfy the needs of its target organs regardless of the amount of cortisol which is bound to plasma proteins.

Corticosteroid metabolism and excretion

Cortisol is not broken down in the tissues on which it exerts its physiological actions. A small amount of free (unbound and non-conjugated) cortisol (<100 µg daily) is normally excreted in the urine; in Cushing's syndrome this fraction may increase to 300 µg or more. The main site of cortisol metabolism is the liver where enzymic actions reduce cortisol to dihydrocortisol and tetrahydro-cortisol, both of which are physiologically inactive. Some cortisol is converted to cortisone which is then reduced to tetrahydrocorti-sone. The tetrahydroderivatives of cortisol and cortisone are conju-gated with glucuronic acid by glucuronyl transferase. The glucur-onides are highly water soluble, are not bound to plasma proteins and are rapidly excreted by the kidney.

Aldosterone occurs in plasma in very small concentrations, 60 per cent being bound to albumin. Some aldosterone is converted in the liver to tetrahydroaldosterone glucuronide, and some is conjugated with glucuronic acid in the liver and kidney at the 18 oxo-position. Very little free aldosterone occurs in the urine.

ACTIONS OF ADRENOCORTICAL HORMONES

The adrenal cortex acts by the formation and liberation of the following steroids—cortisol, corticosterone, aldosterone, and androgens, oestrogens and gestagens. Cortisol and corticosterone have been termed *glucocorticoids* because they have striking effects on glucose formation, and aldosterone has been called a *mineralo-corticoid* because its main action is to promote Na^+ retention and K^+ excretion. The functional distinction between a glucocorticoid and a mineralocorticoid is by no means complete. Aldosterone is 500 times more potent than cortisol in causing Na^+ retention in the adrenalectomized dog but is one-third as potent in promoting glu-coneogenesis. However, since in man only about 150 µg of aldos-terone are secreted daily, compared with 20 mg of cortisol, it is clear that aldosterone contributes very little to the regulation of carbohydrate metabolism. On the other hand the daily secretion of cortisol is equivalent to 50 µg of aldosterone which has a significant effect on electrolyte metabolism.

Adrenocortical steroids are lipid soluble and therefore diffuse freely across cell membranes into non-target as well as target organs. In target cells the steroids combine with specific cytoplas-mic receptor proteins and the steroid–receptor complex passes into the cell nucleus where it becomes attached to specific areas of the chromatin. This process derepresses, or activates, certain genes which leads to the formation of new RNA and control of the formation of new proteins, e.g. enzymes. These enzymes modify the activities of the target cells.

Actions of glucocorticoids

Corticosterone is of very little importance in man. It has a weak effect on protein metabolism, favouring conversion to carbohydrate and fat, and it has slight mineralocorticoid activity. It does not inhibit inflammation. Its concentration in human adrenal venous blood is 1/10–1/5 that of cortisol. Corticosterone is important in the rat and mouse.

Cortisol

Cortisol has physiological actions, in daily doses of 25 mg, which allow it to maintain health in a patient with Addison's disease (adrenal insufficiency), and in larger dosage (50–75 mg daily) it has been widely used for the suppression of many forms of inflammation.

Carbohydrate metabolism. Cortisol is necessary for normal carbohydrate metabolism. Cortisol deficiency leads to hypoglycaemia, increased sensitivity to insulin and a marked decrease in liver glycogen. The failure to compensate for the hypoglycaemia is probably due to decreased food intake which is a symptom of adrenal insufficiency. Cortisol excess produces hyperglycaemia, glycosuria, increased resistance to insulin and an increase in liver glycogen. The hyperglycaemia thus induced in a normal animal or person is to some extent counteracted by an increased secretion of insulin. Cortisol accelerates gluconeogenesis in the liver, mainly by favouring glycogen formation from amino acids formed by protein breakdown. The protein in question may come partly from peripheral tissues, e.g. muscle. Cortisol also inhibits glucose uptake by the tissues. The blood pyruvate level rises and allows increased hepatic synthesis of glucose from pyruvate. The increased glucose production due to excess of cortisol thus comes from both nitrogenous and non-nitrogenous sources. Cortisol promotes gluconeogenesis by increasing the activity of the appropriate liver enzymes.

The glycosuria of 'steroid diabetes' is accounted for by hyperglycaemia and increased glomerular filtration rate. There is no evidence of diminished tubular reabsorption of glucose. The release of glucose from glycogen by adrenaline or glucagon depends on the 'permissive' presence of cortisol.

Protein metabolism. Cortisol promotes catabolism of proteins. Normally the breakdown of protein is counterbalanced by anabolic processes, but excess of cortisol causes a negative nitrogen balance accompanied by retardation of growth, wasting of muscles, thinning of the skin, osteoporosis, and reduction in lymphoid tissue. Breakdown of muscle increases creatine excretion, and dissolution of lymphocytes increases uric acid excretion in the urine.

Lipid metabolism. Cortisol excess causes redistribution of fat, with increased deposits on the trunk at the expense of fat in the extremities.

In diabetics cortisol raises plasma lipids and increases ketone body formation but in normal persons these changes are prevented by increased secretion of insulin.

Electrolyte and water metabolism. The effect of cortisol on electrolyte and water excretion is qualitatively similar to that of aldosterone, though it is very much less potent. Thus cortisol promotes retention of Na^+ (and usually of Cl^-) and excretion of K^+ by the kidney. In adrenal cortex insufficiency there is a deficient diuretic response to water drinking so that a water load is not disposed of for 12 or more hours. This is probably due to excessive

transfer of water from the extracellular to the intracellular fluid compartment. Cortisol counteracts this tendency, maintains the extracellular fluid volume, provides an adequate glomerular filtration rate and allows diuresis to occur. Cortisol may antagonize the actions of ADH on the renal tubules and enhance the destruction of ADH by the liver. Both these actions would promote diuresis.

Large amounts of cortisol cause excessive retention of Na^+ and extracellular water leading to oedema and hypertension. K^+ excretion in the urine may be excessive and the resulting hypokalaemia may contribute to muscular weakness and electrocardiographic changes.

Muscle power. Muscular weakness is a feature of adrenocortical insufficiency. It is relieved by cortisol. The exact mechanisms involved are unknown. In excess, cortisol causes muscular weakness by promoting loss of protein and creatine from muscle, associated with oedema and fibrosis (steroid myopathy).

Blood. Adrenocortical insufficiency is usually associated with eosinophilia, lymphocytosis, neutropenia and anaemia, all of which can be corrected by cortisol. Conversely, eosinopenia, lymphopenia, neutrophil leucocytosis and polycythaemia occur in Cushing's syndrome. Excess of cortisol causes lysis of fixed lymphoid tissue of the thymus, spleen and lymph nodes. Cortisol causes lymphopenia by interfering with DNA synthesis, and by promoting destruction of lymphocytes. The eosinophil count is lowered by sequestration of eosinophils in the lungs and spleen and by destruction in the circulating blood. The basophils undergo a similar change. Cortisol increases the platelet count and shortens blood clotting time.

Cardiovascular system. Cortisol can overcome the hypotension associated with adrenocortical insufficiency. It acts partly by restoring the plasma Na^+ and the circulating blood volume, and partly by overcoming myocardial weakness. In excess, cortisol may raise the blood pressure above normal; it increases the production of angiotensinogen which could lead to enhanced formation of angiotensin and this in turn to increased secretion of aldosterone [see pp. 232, 533]. In adrenocortical insufficiency vascular smooth muscle becomes unresponsive to noradrenalin and capillary permeability increases. These changes tend to lead to vascular collapse. The most important circulatory action of cortisol is restoration of the normal sensitivity of vascular smooth muscle to noradrenalin.

Cortisol excess raises blood lipids and the plasma cholesterol level. This leads to atherosclerosis which is a feature of Cushing's syndrome [p. 535].

Central nervous system. Symptoms due to abnormal activities within the central nervous system are seen in association with adrenocortical insufficiency and with cortisol excess. In Addison's disease restlessness, insomnia and inability to concentrate are associated with slowing of electrical discharges in the electroencephalogram. These changes are reversed by cortisol but not by a mineralocorticoid such as DOC. Patients with Cushing's syndrome often show euphoria and are restless and excitable. Cortisol excess lowers the threshold for electrical excitation of the brain and tends to promote fits in epileptic subjects. Mineralocorticoids have the opposite effects.

The thresholds for taste and smell of many chemicals are reduced in adrenocortical insufficiency.

Gastro-intestinal tract. Cortisol increases gastric acidity, and to a smaller degree pepsin production. This action, which may promote peptic ulcer formation, can be blocked by atropine. Cortisol promotes absorption of water-insoluble fats from the intestine.

Bone metabolism. Cortisol excess impedes the development of cartilage, and causes thinning of the epiphysial plate and interruption of growth (in children). There is a defect in synthesis of the protein matrix and decreased deposition of calcium. In addition there is decreased absorption of calcium from the gut (cortisol antagonizes vitamin D) and increased loss of calcium in the urine. It is not surprising that osteoporosis (especially in the vertebrae) is one of the most important hazards of treatment by cortisol and related compounds.

Infection, inflammation, and trauma. There is increased susceptibility to infection both in Addison's disease and in Cushing's syndrome. In very large doses cortisol reduces formation of antibodies by its destructive effect on fixed lymphoid tissue. This action is not seen when the daily dose of cortisol is 60–75 mg, i.e. only 2–3 times the normal physiological output. Cortisol also reduces the tissue response to bacterial invasion and thereby allows infections to spread, e.g. tuberculosis.

Cortisol, in doses exceeding the physiological level, suppresses the vascular and cellular responses to injury and is beneficial in various types of inflammation. It decreases hyperaemia, reduces exudation and diminishes the migration and infiltration of leucocytes at the site of injury. It stabilizes the membranes of the leucocytic granules (lysosomes) and thus prevents the escape of proteases and hydrolytic enzymes into tissue fluid. Damage to neighbouring tissues is thus reduced. These effects are seen in such conditions as rheumatoid arthritis, disseminated lupus erythematosus, periarteritis nodosa, and inflammatory diseases of the eye and skin. Cortisol helps to prevent rejection of transplanted organs (e.g. kidney). Cortisol inhibits hypersensitivity responses to antigen–antibody reactions. It can prevent anaphylactic shock in guinea-pigs and is often very effective in relieving severe bronchial asthma in man, perhaps by reducing oedema of the bronchiolar mucosa. Cortisol does not antagonize the actions of histamine but it inhibits the intracellular synthesis of this substance.

Cortisol has been said to act like 'an asbestos suit against fire', protecting the tissues from damage and preventing their normal responses to injurious agents, but not extinguishing the fire. It does not eradicate the cause of inflammation and when its administration is stopped relapse occurs, unless meanwhile the disease has come to its natural end. Cortisol reduces the formation of granulation tissue and in large doses can delay wound healing, presumably by inhibiting the production of ground substance by collagen-producing cells.

ACTH secretion. Cortisol suppresses secretion of ACTH [p. 519].

Conclusion. Although cortisol is a glucocorticoid its most important physiological functions appear to be control of the distribution of body water and electrolytes, maintenance of blood pressure and glomerular filtration rate and the renal regulation of water excretion.

Actions of mineralocorticoids

Aldosterone

This is the chief mineralocorticoid secreted by the adrenal cortex. It has long been known that the 'amorphous' fraction of adrenal cortex extracts was very effective in keeping adrenalectomized animals alive and in 1952 the work of Simpson and Tait (now Professor and Mrs Tait) at the Middlesex Hospital Medical School led to the isolation of aldosterone from this fraction. It differs from other steroid hormones in having an aldehyde group at C-18 and in solution it exists largely in a tautomeric form, the 11-hemiacetal.

Aldosterone can maintain electrolyte balance and health in adrenalectomized animals and in patients with Addison's disease. It causes retention of sodium and increased urinary excretion of potassium, though it has little effect on water excretion. It acts on the distal renal tubule where it promotes absorption of Na^+ in exchange for K^+ and H^+. In excess aldosterone causes a rise in plasma sodium, a fall in plasma potassium, hypochloraemic alkalosis, increased extracellular fluid volume and hypertension. In excess aldosterone also lowers the sodium content of sweat, saliva and gastro-intestinal secretions. Lack of aldosterone leads to sodium loss, potassium retention, dehydration and circulatory collapse.

At the molecular level aldosterone activates a sodium pump after first combining with a specific cytosol protein ('mobile receptor') which enters the nucleus, acts on nuclear chromatin, increases mRNA formation, and promotes protein synthesis (including enzymes which enhance sodium transport).

In physiological amounts aldosterone has marked effects on sodium and potassium distribution in the body but very little action on carbohydrate or protein metabolism. It has no anti-inflammatory action and it does not inhibit the secretion of ACTH by the anterior pituitary.

Some compounds known as *spironolactones* act as specific aldosterone antagonists and block the sodium–potassium exchange in the distal tubule by competitive enzyme inhibition.

11-*Deoxycorticosterone* (*cortexolone*) has about one-thirtieth of the sodium retaining potency of aldosterone but causes relatively greater potassium loss.

Adrenocortical sex hormones

During fetal life the hyperplastic adrenal cortex (reticularis) produces dehydroepiandrosterone which serves as the main precursor for oestrogen synthesis by the placenta. After birth the involuted adrenal cortex secretes less sex hormones. At puberty in both sexes adrenal androgenic hormone secretion again increases and contributes to increase in muscle mass, sexual hair, and seborrhoea. The development of libido in women depends primarily on the action of adrenal androgens; libido persists after ovariectomy but not after adrenalectomy (even with cortisol substitution therapy).

Adrenal androgens are weak and in the male cannot prevent the hormonal effects of castration.

CONTROL OF ACTIVITY OF ADRENAL CORTEX

The growth of the adrenal cortex, the structural and functional integrity of most of its cells and the secretions of cortisol and corticosterone are controlled by the anterior pituitary through its own secretion of ACTH (adrenocorticotrophic hormone, corticotrophin). The secretion of aldosterone is not under direct pituitary control.

Hypophysectomy causes atrophy of the zona fasciculata and zona reticularis (but less so in the zona glomerulosa which manufactures aldosterone). The cells remain well-filled with lipid precursors of cortisol. In animals the resulting adrenocortical insufficiency is not severe enough to cause death, but in man hypophysectomy produces fatal adrenal cortex deficiency unless appropriate hormone therapy is given.

ACTH

Injection of ACTH causes very rapid secretion of glucocorticoids from the cortex into the adrenal venous blood. Unlike the thyroid and the islet tissue of the pancreas, the adrenal cortex contains only a tiny store of hormones which can be discharged within a matter of seconds. ACTH also causes the cholesterol ester and ascorbic acid concentrations to fall rapidly and the extent of their decline may be used to estimate the degree of cortical stimulation. Once the stores have been depleted further secretion depends on continuous rapid synthesis of the glucocorticoid hormones. Repeated injections of ACTH produce hypertrophy and hyperplasia of the zona fasciculata and the zona reticularis, associated with increased vascularity. ACTH does not promote aldosterone secretion or cause the zona glomerulosa to become depleted of lipid.

ACTH and synthesis of glucocorticoids.
ACTH combines with specific receptors on the surface of the adrenal cortical cell *in vivo* and *in vitro*. This combination, in a Ca^{2+}-containing medium, increases membrane permeability to glucose and certain ions and activates adenyl cyclase. These events lead to:

1. Increase in cAMP concentration which begins almost instantly and reaches a peak at 3 min.
2. Glucocorticoid secretion which begins about 3 min after a 'pulse' of ACTH and reaches a peak at 10 min. *In vitro*, ACTH is active at 10^{-16} mol l^{-1}.
3. A decrease in adrenal ascorbic acid content which is most marked at 1–2 hours.
4. Increase in adrenal size which occurs at 24–48 hours.

With regard to events 1 and 2, the increased entry of glucose into the cell and its subsequent utilization helps in two ways:

(i) Oxidative metabolism of glucose produces ATP which is the precursor for cAMP.

(ii) Via the pentose phosphate pathway $NADPH_2$ is produced. This promotes the hydroxylation reactions at carbon atoms 17, 21, and 11, to produce cortisol.

Cyclic AMP activates protein kinases [p. 501] which phosphorylate at least two proteins, a *lipase* which frees cholesterol from its ester form in lipid droplets, and a *ribosomal protein* which promotes conversion of cholesterol to pregnenolone.

The biological significance of the action of ACTH on adrenal ascorbic acid is unknown. There is no evidence that ascorbic acid is concerned in steroid synthesis.

ACTH has an effect on melanocytes, similar to that produced by *melanocyte-stimulating hormone* (MSH). It is about 1/100 as potent as MSH in dispersing the melanin granules of the melanocytes in frog's skin. In man dispersal of melanin granules cannot occur but MSH stimulates melanin production. The increased pigmentation of Addison's disease is probably due mainly to increased secretion of ACTH, though excessive secretion of MSH from the pars intermedia also occurs [p. 525].

Structure of ACTH.
ACTH is a straight-chain polypeptide composed of 39 amino-acid subunits, with a molecular weight of nearly 4600. A polypeptide containing the first 24 amino acids of ACTH in sequence has been synthesized (tetracosactrin, *Synacthen*) which has all the activity of the natural hormone.

The sequence of the 4–11 amino acids of ACTH is identical with that which makes up β-MSH. It is therefore not surprising that ACTH has melanocyte-stimulating activity.

ACTH is rapidly inactivated in the body and its half-life in plasma after intravenous injection is about 10 minutes. The site of removal or inactivation is not known.

Regulation of glucocorticoid secretion

The secretion of ACTH is largely under nervous control. Information is conveyed from the central nervous system to the adenohypophysis by a chemical agent which is released from the median eminence of the hypothalamus and passes via the hypophysial portal vessels to the anterior pituitary where it promotes the secretion of ACTH into the bloodstream. The chemical mediator of this response is a polypeptide (MW 1000) called *corticotrophin-releasing hormone* (CRH) [p. 518].

Normal glucocorticoid output.
A healthy adult man secretes 5–30 mg of cortisol and 1.5–4.0 mg of corticosterone daily. There is

a distinct *circadian* rhythm in the rate of secretion of cortisol which is revealed by variations in plasma and urinary levels at different times of the day. FIGURE XI.22 shows the diurnal variations in plasma cortisol level (bound and unbound) in normal human subjects. The highest level, about 150 μg l^{-1} (425 nmol l^{-1}), occurs between 6.0 and 9.0 a.m. and the lowest, about 30 μg l^{-1} (80 nmol l^{-1}), occurs near midnight. The morning rise in plasma cortisol mainly occurs during sleep and is therefore not due to the stress of waking up. The changes in plasma cortisol level are preceded by the appropriate changes in plasma ACTH level and the cycle is probably controlled by the 'biological clock' in the limbic lobe. The circadian rhythm is absent in patients with Cushing's syndrome, and in patients with lesions of the hypothalamus or limbic lobe.

Responses to stress.
Many so-called stressful stimuli, which impose strains on the homeostatic mechanisms of the body, promote secretion of ACTH and hence of adrenal glucocorticoids. This response appears to be important for survival, and some types of stressful stimuli can prove lethal in hypophysectomized animals or in adrenalectomized animals given only maintenance doses of glucocorticoid. These stimuli include administration of anaesthetics, insulin (producing hypoglycaemia), infectious diseases, toxins, haemorrhage, traumatic shock, surgical operations, severe burns, exposure to cold, anoxia, muscular exercise, fear, pain, anxiety, anger, frustration, and fatigue.

It is highly probable that physical stimuli such as severe exercise, exposure to cold, burns and trauma act by virtue of their emotional accompaniments. Indeed, strong emotion *per se* causes temporary maximal increases of cortisol secretion, comparable to those seen in patients with Cushing's syndrome. The significance of the tissue metabolic changes induced by cortisol in relation to the psychological factors which promote its release is very difficult to understand.

Glucocorticoid feedback control of ACTH secretion [p. 519].
It is not known precisely how the balance between ACTH and glucocorticoid secretions is held within the normal range but deviations are corrected by some homeostatic mechanism. This is well shown by the effect of removal of one adrenal gland, which leads to hypertrophy of the remaining gland until the total adrenal cortical mass becomes the same as it was originally in the two glands. This hypertrophy only occurs when the pituitary is present and capable of secreting ACTH and GH. When so much adrenal cortical tissue is destroyed that the glucocorticoid inhibition of ACTH secretion is lost, e.g. in Addison's disease, ACTH secretion increases to such an extent that it causes pigmentation.

The effects of cortisol and its analogues on ACTH secretion are clearly shown when these substances are administered in large pharmacological doses to suppress inflammation. Glucocorticoid secretion is greatly depressed and the adrenal cortex atrophies. In these circumstances when steroid therapy is to be terminated the dose should be reduced gradually to allow the adrenal cortex time to recover, and prolonged administration of ACTH will further help to avoid adrenocortical failure.

Control of aldosterone secretion. Renin–angiotensin system.
See page 232.

Adrenal cortex and sympathoadrenal medullary system.
The environmental and emotional disturbances which stimulate adrenocortical secretion also usually activate the sympathoadrenal medullary system, the response in both cases being mediated by the hypothalamus. The release of catecholamines by sympathetic and adrenal medullary stimulation in association with muscular exercise is clearly helpful to the body by promoting redistribution of blood

to active muscles, by increasing cardiac output and coronary blood flow, by converting liver glycogen to glucose, by releasing free fatty acids from adipose tissue and by diminishing fatigue in skeletal muscle. Cortisol enhances some of the actions of catecholamines: it sensitizes arterioles to the constrictor action of noradrenalin; it potentiates the breakdown of glycogen to glucose and the release of FFA from adipose tissue induced by catecholamines; it also promotes protein catabolism. In these ways more energy will become available to the active muscles.

The activation of the hypothalamus–pituitary–adrenal cortex system is nearly as rapid as that of the sympathoadrenal medullary system so interactions between hormones secreted by the two systems can readily occur. There is no evidence that the adrenal medulla controls the secretory activity of the adrenal cortex but the secretion of glucocorticoids by the adrenal enhances the synthesis of phenylethanolamine N-methyltransferase (PNMT) in the adrenal medulla and thereby promotes the conversion of NA to Ad [p. 414]. In addition, ACTH (but not glucocorticoids) increases the formation of tyrosine hydroxylase and dopamine β-hydroxylase, cAMP and ascorbic acid possibly being involved in these effects (Weiner 1975).

DISORDERS OF ADRENOCORTICAL FUNCTION IN MAN

Disorders of the adrenal cortex may cause either defective or excessive secretion of hormones. They have provided evidence of the physiological activities of the adrenal cortex. The clinical syndromes are due to under- or over-secretion of cortisol, aldosterone, and androgens.

Adrenal cortex insufficiency

This can arise in the following ways:

1. Destruction of the adrenal cortex by disease reduces the secretion of all the hormones, though not necessarily equally. Adrenalectomy abolishes all secretion and is lethal unless hormone substitution therapy is given.

2. Diminished secretion of ACTH, due to adenohypophysial or hypothalamic failure, produces atrophy of the two inner zones of the adrenal cortex and reduced secretion of cortisol. Aldosterone secretion is less affected.

3. Congenital failure of cortisol secretion due to defects in the enzymes responsible for its synthesis produces manifestations of cortisol deficiency coupled with secondary hyperplasia of the adrenal cortex and over-secretion of other steroids.

1. Insufficiency may be *acute* (adrenal crisis) or *chronic*.

The acute form occurs after adrenalectomy, after abrupt withdrawal of therapeutically administered glucocorticoids (which cause adrenocortical atrophy), or in patients with reduced basal secretion of cortisol when they are exposed to a sudden stress.

The chronic form is seen in *Addison's disease* in which slow destruction of the adrenal cortex reduces secretion of cortisol and aldosterone. The clinical features are muscular weakness; pigmentation of the skin and buccal mucosa; anorexia, nausea, vomiting and diarrhoea, leading to dehydration and loss of weight; hypotension; mental confusion; decreased ability to withstand stresses due to trauma, infection, etc. If untreated, the disease is fatal. In Addison's disease the morning plasma cortisol level is reduced, sometimes to zero. Administration of ACTH fails to produce the normal rise in plasma cortisol level.

The clinical manifestations are mainly attributable to lack of cortisol and perhaps of aldosterone too, but the pigmentation is due to the increased secretion of ACTH, which occurs when cortisol

secretion falls, and perhaps also to increased secretion of melanocyte-stimulating hormone (MSH).

2. In *anterior pituitary failure* there is deficient secretion of ACTH and hence reduced secretion of glucocorticoids from the zona fasciculata and the zona reticularis. Secretion of aldosterone from the zona glomerulosa is less affected. The main features are those described for cortisol deficiency, especially a reduced capacity to respond to infections and other stressful stimuli. Adrenal insufficiency of pituitary origin differs from that due to adrenal destruction in the following ways:
(i) Electrolyte balance is normal.
(ii) Skin pigmentation is reduced.
(iii) There is evidence of reduced function of other endocrine glands controlled by the adenohypophysis, e.g. thyroid, gonads.

3. *Congenital Adrenal Hyperplasia (Virilism)*. In this condition there are inherited defects in the enzymes responsible for corticosteroid biosynthesis [p. 530]. In the commonest defect lack of 21-*hydroxylase* prevents the conversion of 17-α-OH progesterone to 11-deoxycortisol (compound 'S', cortexolone) and of progesterone to desoxycorticosterone ('DOC', cortexone). The consequence of this block in the pathway for synthesis of cortisol is that 17-OH progesterone accumulates and the urine contains increased amounts of pregnanetriol and *androgenic* 17-ketosteroids. The reduced concentration of cortisol in the blood increases the secretion of ACTH which in turn produces hyperplasia of the adrenal cortex. This leads to overproduction of substances behind the block, the most important of these being androgens. The clinical result is virilism and excessive growth. In boys there is precocious sexual development (*macrogenitosomia praecox*) which may be accompanied by precocious growth of the body as a whole resulting in the stocky, prematurely virile 'infant Hercules' type. In girls there is masculinization. In severe cases genetic female children are born with masculine external genitalia (pseudohermaphroditism). In all cases there may also be signs of cortisol deficiency and in some there is mineralocorticoid deficiency as well.

In 21-hydroxylase deficiency arterial blood pressure is normal, but when there is deficiency of *11-hydroxylase* the accumulation of the mineralocorticoid desoxycorticosterone, and of 11-deoxycortisol produce hypertension as well as virilism; the urine in such cases contains tetrahydro-'S'.

In both types of congenital adrenal hyperplasia the administration of physiological amounts of cortisol will inhibit secretion of ACTH, diminish the hyperplasia and reduce androgen secretion to normal. In this way all the symptoms and signs of disordered adrenal function quickly disappear, and if treatment is started early enough normal sexual development can occur.

Over-activity of the adrenal cortex

There are three main clinical syndromes resulting from oversecretion of adrenocortical hormones, *Cushing's syndrome* due to excessive production of cortisol, primary and secondary *hyperaldosteronism*, and the *acquired adrenogenital syndrome* due to overproduction of androgens or oestrogens. Mixed syndromes can occur.

Cushing's syndrome

The term Cushing's syndrome is applied to the clinical disorder which results from the exposure of body tissues to sustained supraphysiological blood levels of corticosteroids, either endogenous in origin or iatrogenically produced. Essentially Cushing's syndrome is hypercortisolism.

The syndrome may arise from excess of ACTH, or independently of ACTH.

ACTH dependent causes.

1. Bilateral adrenocortical hyperplasia due to increased pituitary

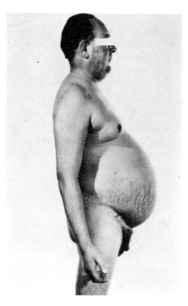
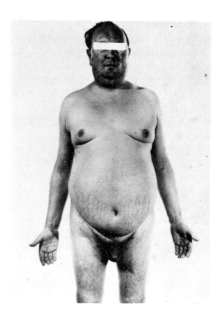

FIG. XI.23. Cushing's disease due to pituitary adenoma. Man aged 39. Red moon-face, thick neck, grossly distended abdomen with purple striations. Wasting of muscles of thighs and pectoral and pelvic girdles. Osteoporosis. Bilateral optic nerve atrophy. Blood pressure 190/120 mm Hg. Plasma corticotrophin concentration = 800 miu 1^{-1} (normal = <10 miu 1^{-1}). Increased urinary hydroxycorticosteroid output. Postmortem: adenocarcinoma of pituitary with chromophobe and chromophil cells present.
(Dr J. D. N. Nabarro's case reported by Shrank, A. B. and Turner, P. (1960). *Br. med. J.* **i**, 849.)

ACTH secretion. This is called *Cushing's disease* (1932) [FIG. XI.23].

2. The *ectopic* ACTH syndrome is due to secretion of ACTH by benign or malignant tumours of non-endocrine origin (e.g. cancer of the lung).

3. Iatrogenic treatment with ACTH or its synthetic analogues.

Non-ACTH dependent causes.

1. Adenomas or carcinoma of the adrenal cortex.
2. Iatrogenic-pharmacological doses of corticosteroids.

In Cushing's disease the excessive ACTH secretion is probably due to excessive secretion of CRH by the hypothalamus. The normal circadian rhythm of ACTH and cortisol secretion is replaced by continuous secretion; the plasma cortisol may be raised to above the normal range but in many cases it is merely maintained, throughout the 24-hour period, at the normal morning value. The excess of ACTH leads to adrenal hyperplasia.

Adrenocortical tumours may secrete large amounts of cortisol, often with various androgens and occasionally aldosterone. The atrophy of the opposite adrenal is due to suppression of ACTH secretion by the high plasma cortisol levels. Secretion by the tumour is not inhibited by cortisol or enhanced by ACTH, i.e. it is autonomous. Extra-adrenal carcinomas, especially of the lung, may secrete large amounts of ACTH, which produces adrenal hyperplasia and high plasma cortisol levels.

Adrenogenital syndrome in adults.

This syndrome in adults is nearly always due to an adrenal tumour. There is a tendency to conversion to the secondary characters of the opposite sex. Thus adult women in whom excess androgen secretion occurs develop male secondary sexual characters (adrenal virilism). The symptoms and signs are deepening of the voice, amenorrhoea, enlargement of the clitoris, growth of hair in the masculine distribution, and marked increase in muscular growth towards the male type. The 17-ketosteroid excretion in the urine is increased from the normal range of 7–12 mg to 30–50 mg daily, indicating androgen production in the body.

Very occasionally adrenal tumours secrete oestrogen to produce feminization in males, with enlargement of the breasts, atrophy of the testes and diminished sexual interest in women.

Aldosteronism (hyperaldosteronism).

Primary aldosteronism (Conn's syndrome) is a condition in which there is prolonged excessive secretion of aldosterone from an adrenocortical adenoma consisting mainly of cells of the glomerulosa type. There is no increase in cortisol secretion. The characteristic features are hypertension, hypokalaemia, hypernatraemia, polyuria, polydipsia, alkalosis and muscular weakness. Urinary aldosterone excretion is increased. Removal of the adenoma usually cures the condition.

Secondary aldosteronism is a condition in which oversecretion of aldosterone is brought about by extra-adrenal factors. It occurs in patients with oedematous states such as congestive heart failure, nephrosis, toxaemia of pregnancy and cirrhosis of the liver. Hypokalaemia may be precipitated by diuretics. Potassium depletion can be prevented by reduction of sodium intake, addition of potassium supplements and the administration of the aldosterone antagonist, spironolactone.

The causal mechanisms of secondary aldosteronism are not yet understood.

Other adrenocortical steroids, synthetic derivatives

Cortisone and cortisol (hydrocortisone) act similarly, the latter being the principal glucocorticoid secreted by the adrenal cortex. Cortisone is converted to cortisol in the body and the opposite change can occur in the liver.

A number of synthetic derivatives of cortisone and cortisol have been made with the aim of increasing anti-inflammatory activity with reduction of unwanted side-actions. However, all that has been achieved is enhancement of glucocorticoid and anti-inflammatory potency with a decrease of Na-retaining activity: *Prednisone* and *prednisolone* are derived from cortisone and cortisol respectively. Their anti-inflammatory potency is about 5 times that of their parent compounds, without any enhancement of mineralocorticoid activity.

Inhibitors of adrenocortical steroid synthesis

A compound called *amphenone* inhibits the synthesis of 17-hydroxycorticoids and of aldosterone but is much too toxic. A related compound called *o,p'*-DDD is better tolerated. It selectively destroys the zona fasciculata and zona reticularis.

Metyrapone (*Metopirone*) inhibits the enzyme 11 β-hydroxylase in the adrenal cortex and thereby interferes with the synthesis of cortisol and corticosterone. The deficiency of glucocorticoids stimulates secretion of ACTH which in these conditions enhances the formation of 11-deoxycortisol and thus leads to increased urinary excretion of 17-hydroxycorticoids. This response is used as a test of pituitary capacity to secrete ACTH.

REFERENCES

BLASCHKO, H., SAYERS, G., and SMITH, A. D. (eds,) (1975). *Handbook of physiology*, Section 7, Vol. IV. *The adrenal gland*. American Physiological Society, Washington, DC.

CHESTER JONES, I. (1957). *The adrenal cortex*. Cambridge University Press.

COPE, C. L. (1965). *Adrenal steroids and disease*. Pitman, London.

DENTON, D. A. (1965). Aldosterone secretion and salt appetite. *Physiol. Rev.* **45**, 245.

EISENSTEIN, A. B. (ed.) (1967). *The adrenal cortex*. Churchill, London.

JOSI, A. (1975). The adrenal cortex. In *Handbook of physiology*, Section 7, Vol. IX. *The adrenal gland* (ed. H. Blaschko, G. Sayers, and A. D. Smith). American Physiological Society, Washington, DC.

LIDDLE, G. W. (1974). The adrenal cortex. In *Textbook of endocrinology*, 5th edn (ed. R. H. Williams) p. 233. Saunders, Philadelphia.

MATTINGLY, D. (1968). Disorders of the adrenal cortex and pituitary gland. In *Recent advances in medicine*, 15th edn (ed. D. N. Baron, N. D. Compston, and A. M. Dawson) p. 125. Churchill. London.

MILLS, J. N. (1966). Human circadian rhythms. *Physiol. Rev.* **46**, 128.

WEINER, N. (1975). In *Handbook of physiology*, Section 7, Vol. VI. *Adrenal cortex* (ed. H. Blaschko, G. Sayers, and A. D. Smith) p. 357. American Physiological Society, Washington, DC.

WOLSTENHOLME, G. E. W. and PORTER, R. (eds.) (1967). *The human adrenal cortex*. Ciba Foundation Study Group, No. 27. Churchill, London.

The thyroid

The name 'thyroid' was introduced by Thomas Wharton in 1656. It is derived from the Greek *thyreos*, a shield.

Embryology. The human thyroid gland begins to develop about 4 weeks after conception when the embryo is 3.5–4.0 mm long. During the first 10–12 weeks fetal growth and development take place without the need for thyroid hormones (except perhaps the minimal amounts which cross the placenta from the maternal circulation). After 12 weeks small amounts of thyroid hormones are formed, and from 20–22 weeks fetal TSH secretion and thyroid hormone secretion increase steadily till the end of pregnancy. Thyroid hormones are required particularly for normal fetal bone maturation and for normal development of the central nervous system.

Anatomy. The normal adult thyroid is the largest endocrine gland, weighing 15–25 g. It consists of two lobes joined by an isthmus and is closely attached to the anterior and lateral aspects of the upper part of the trachea. It has a very rich blood supply and its estimated blood flow of 4–6 ml g^{-1} min^{-1} exceeds even that through the kidney. It has a large lymphatic drainage through which part of the stored large-molecular thyroglobulin may enter the circulation. The thyroid is innervated by the sympathetic system which probably supplies only blood vessels.

Histology. Light microscopy shows the gland to consist of about three million spherical follicles which vary in diameter from 50 to 500 μm. Some 40 follicles are grouped together to form a lobule. The wall of each follicle is lined by a cuboidal epithelium whose height varies with the degree of glandular stimulation. Each follicle contains a clear viscid proteinaceous amber-coloured colloid which normally comprises the greater part of the thyroid mass. The colloid consists mainly of *thyroglobulin*. With increased glandular activity the follicles become smaller, the cells columnar in shape, and the colloid less abundant.

The thyroid is peculiar among endocrine glands in that its product is not stored intracellularly but is secreted through the apex of the follicular cell for extracellular storage in the follicular lumen. Electron microscopy shows that the cells have cytological features of protein-secreting cells combined with structures concerned with absorption of colloid and the release of thyroid hormones from thyroglobulin. The apical end of the follicular cell shows numerous microvilli projecting into the colloid.

Between the follicles there are *parafollicular* or *C cells* which originate in the neural crest and then migrate to the ultimobranchial bodies which fuse with the thyroid in mammals. The C cells secrete *calcitonin* which is discussed on page 553.

The follicular cells secrete the iodine-containing compounds *L-thyroxine* (T_4) and *L-triiodothyronine* (T_3). These are the thyroid hormones (THs).

Formation and secretion of thyroid hormones

Iodine metabolism

Iodine is an essential raw material for the synthesis of thyroid hormones. The richest source of dietary iodine is sea fish, but it also occurs in bread, milk, and vegetables and in areas of the world liable to endemic goitre, iodized salt containing up to 0.05 per cent iodine is generally employed as table salt. Drinking water contains insignificant amounts of iodine. All iodine is converted to iodide in the gut and in this form is completely and rapidly absorbed from the upper gastro-intestinal tract into the blood and ECF. The normal daily dietary intake of iodine is 100–200 μg. Iodine deficiency can develop with a daily intake of less than 50 μg over a long period of time, but if thyroid hormone production falls TSH secretion is increased and may restore normal hormone output with or without enlargement of the thyroid (simple goitre). Some Japanese people, who eat large amounts of sea food (marine fish, shell fish, and seaweed), may assimilate several mg of iodine daily, but in spite of this huge iodine intake thyroid hormonal synthesis is usually kept within normal limits. Thus within the range of 50 μg to several mg of iodine daily the body can adapt thyroid hormone production so as to maintain a euthyroid state. The homeostatic mechanisms include changes in renal excretion of iodine, autoregulatory processes within the thyroid and the remarkable sensitivity of the anterior pituitary thyrotrophic cells to changes in blood concentration of thyroid hormones so that a fall in hormone level promotes secretion of TSH, and vice versa.

The distribution, utilization, and metabolism of iodide in the body have been studied after the administration of tracer amounts of radioactive ^{131}I (half-life 8 days) or ^{132}I (half-life 2.3 hours), which mix with the unlabelled iodide but have no biological actions. About one-third is taken up by the thyroid and about two-thirds excreted by the kidney. The thyroid contains 5000–8000 μg of iodine (95 per cent of the total iodine content of the body), most of which is in the form of iodinated amino acids (iodotyrosines and iodothyronines). The large pool of iodine in the body turns over slowly, at a rate of about 1 per cent daily.

Hormone biosynthesis

The follicular (acinar) cells of the thyroid have two important functions:

1. They synthesize the glycoprotein, thyroglobulin, which is stored as colloid in the follicle.
2. They selectively take up and transport iodide from the blood to the apical end of the cell; hormonal synthesis takes place at or near the interface between cell and colloid.

Thyroglobulin is synthesized on the ribosomes of the endoplasmic reticulum of the acinar cell. Carbohydrate is added to the protein both in the endoplasmic reticulum and in the Golgi complex. The complete, but non-iodinated 19-S glycoprotein (MW 650 000) forms secretion granules which fuse with the plasma membrane of the apical surface of the acinar cell and are discharged into the follicular lumen by exocytosis.

The iodide trap. Plasma iodide concentration is variable and small $(0.5–1.5 \ \mu g \ l^{-1})$. The thyroid can concentrate inorganic iodide so that the ratio of thyroid iodide to plasma iodide (T/P ratio) is 20–50 in normal glands and may rise to several hundreds in stimulated glands. This trapping process is essential for hormone synthesis and is brought about by active transport (iodide pump) which depends on energy provided by aerobic metabolism of the acinar cells. The interior of the acinar cell maintains a negative potential (of about -50 mV) in relation to both interstitial fluid and the luminal colloid. Iodide is pumped into the cell against this negative potential and then diffuses down the electrochemical gradient into the luminal region. The iodide pump depends on the activity of a sodium, potassium-dependent ATPase system. Both the iodide pump and the ATPase system are stimulated by TSH and depressed by hypophysectomy. Thiocyanate, perchlorate, and pertechnetate ions depress iodide transport by competitive inhibition but antithyroid drugs such as carbimazole do not affect the trapping pro-

cess. Other organs such as the salivary glands, gastric mucosa, placenta, and mammary glands can transport iodide against a concentration gradient but the process is not stimulated by TSH and these organs cannot form thyroid hormones.

Oxidation of iodide and organic iodinations. The iodide which enters the thyroid acinar cells is rapidly oxidized by a peroxidase to form iodine $(I^- \rightarrow I_2)$. Iodination of accessible tyrosine residues on the surface of the thyroglobulin molecule then occurs, perhaps assisted by an enzyme, tyrosine iodinase. This process takes place at or near the apical surface of the thyroid follicular cells and leads to the formation of monoiodotyrosine (MIT) and diiodotyrosine (DIT), neither of which has hormonal activity.

Iodothyroinine formation. The iodothyronine hormones, T_3 and T_4 are probably formed by coupling reactions, under oxidative conditions, between adjacent iodotyrosine molecules. The combination of two DIT molecules produces T_4 and the combination of one MIT molecule with one DIT molecule produces T_3. There is no evidence that T_4 is converted to T_3 in the thyroid. Both T_4 and T_3 are in peptide linkage with thyroglobulin.

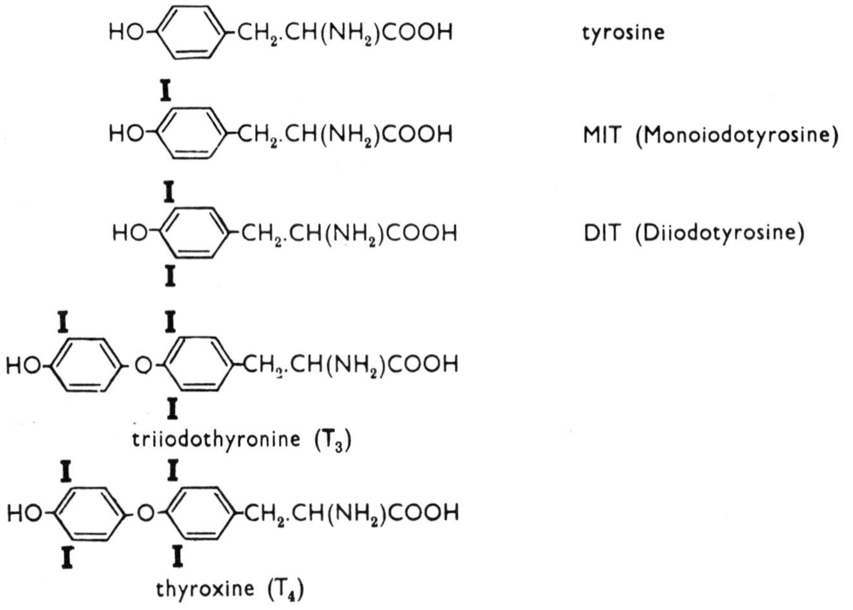

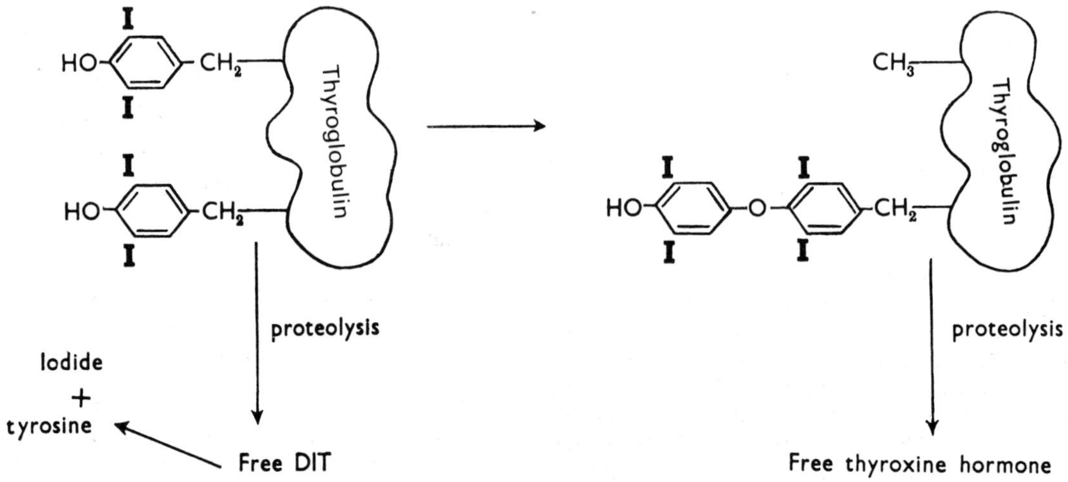

FIG. XI.24. The structures of the tyrosine derivatives concerned in thyroid function. The thyroid hormones are synthesized on the surface of the thyroglobulin molecule.

Storage and release of hormones

The thyroid is unique among endocrine glands by virtue of its large store of hormone in the follicular colloid and the slow rate of turnover of its thyroglobulin-bound T_4 and T_3. The proportions of organic iodine constituents of the human thyroid are as follows:

Iodotyrosines about 60 per cent (DIT and MIT)
T_4 about 35 per cent
T_3 about 5 per cent

Of the total store of 5000–8000 µg of iodine in the thyroid 80–100 µg are secreted daily. Thus if all synthesis of thyroid hormones were suddenly stopped, e.g. by giving an antithyroid drug, physiological levels of thyroid hormones could be maintained in the blood for several weeks.

Release of thyroid hormones is a complex process which is initiated rapidly by the action of TSH. The details of the actions of TSH have been observed *in vivo* in animals and *in vitro* on thyroid slices from experimental animals and man. Within a few minutes TSH promotes the formation of pseudopodia on the apical surface of the follicular cell; these pseudopodia engulf colloid droplets which enter the cell by endocytosis. At the same time lysosomes move from the basal toward the apical end of the cell. The lysosomes fuse with the colloid droplets to form 'phagolysosomes' in which physiologically active proteases are released. These enzymes liberate iodotyrosines and iodothyronines from thyroglobulin, the release being facilitated by reduction of disulphide bonds in the thyroglobulin molecule. The iodothyronines (T_4 and T_3) are released into the cell and traverse the basal membrane into the capillary blood-stream. The iodotyrosines, MIT and DIT are deprived of their iodine by a microsomal enzyme iodotyrosine dehalogenase (deiodinase), found in peripheral tissues as well as in the thyroid, and the released iodine is then largely reclaimed for fresh hormonal synthesis. This enzyme does not release iodine from thyroglobulin-bound MIT or DIT, nor does it deiodinate T_4 or T_3. When the deiodinase is absent (congenital deficiency) the iodotyrosines appear in the blood and are excreted by the kidney. The consequent loss of iodine from the body may produce hypothyroidism by decrease in thyroid hormone synthesis.

Transport, turnover, and metabolism of thyroid hormones (THs)

When the thyroid hormones enter the bloodstream they become firmly but reversibly bound to plasma proteins. The intensity of binding is revealed by the very small proportions of hormones in the free or unbound state, as determined by equilibrium dialysis or gel filtration. Only 0.024 per cent of T_4 and 0.36 per cent of T_3 occur in the free form in plasma (Larsen 1972). In order of decreasing affinity for thyroid hormones the binding hormones are:

Thyroxine (thyronine)-binding globulin (TBG), a glycoprotein with a molecular weight of 60 000. Its concentration in normal plasma is 20 mg l^{-1}, with a half-life of about 5 days. It has a high affinity but low capacity for T_4. The affinity of TBG for T_4 is 2–6 times greater than its affinity for T_3.

Thyroxine-binding pre-albumin (TBPA) has a molecular weight of 50 000 and its plasma concentration is about 250 mg l^{-1} with a half-life of about 2 days. This concentration can bind much T_4 but very little T_3. Thus TBPA has moderate affinity and capacity for T_4. *Albumin* has low affinity but high capacity for T_4 and T_3 binding.

The three plasma proteins together account for the binding capacity for T_4 as follows: TBG 60 per cent, TBPA 30 per cent, albumin 10 per cent though the methods of analysis of binding are subject to artefacts.

The differences in degree of binding of the two hormones T_4 and T_3 help to explain their differences in fractional turnover rate, volume of distribution, serum concentration, and biological activity. T_3 has about four times the biological activity of T_4.

The fractional turnover rate of T_3 is 50–70 per cent daily, whereas that of T_4 is 10 per cent daily. The volume of distribution of T_3 is 35–45 litres, that of T_4 is 10–12 litres. In contrast the total serum concentration of T_4 is about 50 times greater than that of T_3. The total daily production rate of T_4 is 70–90 µg and that of T_3 50–60 µg. It is probable that most of the total daily production of T_3 is due to extrathyroidal conversion of T_4 to T_3. Since T_3 has greater biological activity than T_4 it may be that T_4 owes most of its hormonal effects to deiodination of T_3.

The function of protein binding of thyroid hormones may be to help reduce the proportion of free hormones in the blood. This would reduce hormone loss by kidney and liver, and smooth out the effects of changes in the rate of thyroid hormone secretion, so that the plasma level of free hormones is kept constant.

If free thyroid hormone concentration rises as in thyrotoxicosis, then hyperthyroidism results, whereas if free thyroid hormone concentration falls, as in myxoedema, signs of hypothyroidism develop.

Serum protein-bound iodine (PBI). The term protein-bound iodine (PBI) is also used to designate the iodine which is associated with the proteins precipitated by agents such as trichloracetic and phosphotungstic acids. Most of the iodine in the precipitates consists of T_4 and T_3 so PBI concentration is a useful indicator of thyroid activity.

Metabolism of thyroid hormones (THs)

The conversion of T_4 to T_3

Ever since the discovery of T_3 by Gross and Pitt-Rivers in 1952, the greater physiological potency of this substance as compared with T_4 has raised the question as to whether T_4 is converted to T_3 in the body. Until recently the techniques for separation of T_3 from T_4 (using chromatography and protein-binding analysis) were unsatisfactory but improved methods (gas–liquid chromatography and radioimmunoassay) have shown that most if not all the physiological activity produced by T_4 is due to its conversion to T_3. It seems likely that T_4 is primarily a prohormone and the question now is how much biological activity T_4 has *per se*.

The thyroid gland secretes more T_4 than T_3 and there is no evidence that T_4 is converted to T_3 in the thyroid. There is, however, good evidence that T_4 is deiodinated to T_3 in other tissues and this is very clearly shown in athyreotic patients, in whom administration of T_4 leads to the development of normal or raised levels of T_3 in the serum [FIG. XI.25].

Stimulation of thyroid activity by injection of TSH or TRH raises the serum levels of both T_3 and T_4. In patients with thyrotoxicosis the serum T_3 concentration is raised proportionately more than that of T_4. Enhanced thyroidal secretion of T_3 and greater peripheral conversion of T_4 to T_3 may raise the serum T_3 level 5–10-fold. In some thyrotoxic patients the serum T_4 level is normal while that of T_3 is raised (T_3-thyrotoxicosis). (Studies on *in vitro* cultures of human fibroblasts, liver, and kidney have also shown conversion of T_4 to T_3.)

The physiological effects of T_3, whether on isolated tissues or in whole animals, come on more quickly than those of T_4. T_3 is more rapidly absorbed than T_4, a single dose of 25–100 µg producing a peak serum concentration at 2–4 hours with a return to normal within 24 hours. After administration of a therapeutic dose of T_4 (0.2 mg) to a hypothyroid patient the serum concentrations of T_3 and T_3 rise very slowly over a period of days or weeks.

To sum up:

The thyroid secretes much larger amounts of T_4 than T_3.

T_4 is much more stable in the body than T_3 largely because it is more firmly bound to plasma proteins.

T_4 is an extracellular hormone, while T_3 penetrates tissue fluids and cells readily.

T_3 is quicker acting and more potent than T_4 on isolated tissues, in whole animals and in patients.

T_4 is very largely converted to T_3 in the body. It appears to act as a stable precursor (or prohormone) of T_3 and thus provides a reservoir from which T_3 levels can be continually replenished.

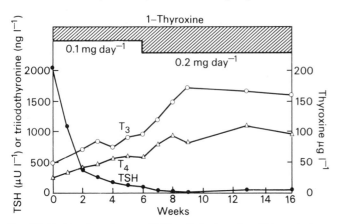

FIG. XI.25. Results of serum TSH, triiodothyronine (T_3), and thyroxine (T_4) measurements in a patient with primary hypothyroidism during treatment with thyroxine.
(From Utiger, R. D. (1974). *A. Rev. Med.* **25**, 299.)

Other metabolic changes in T_3 and T_4

The hormones T_3 and T_4 are metabolized in several ways. About 80 per cent of both hormones is ultimately deiodinated and the remaining 20 per cent is conjugated in the liver, T_4 combining more readily with glucuronic acid and T_3 forming mostly the sulphate ester. The conjugates are secreted in the bile and undergo hydrolysis in the gut where the freed hormone is to some extent reabsorbed, or (if hydrolysis occurs below the region where iodothyronines are absorbed) excreted in the faeces.

Unbound, unconjugated thyroid hormones are destroyed in the body, especially in liver and muscle, by oxidation, deamination, decarboxylation, and splitting of the ether (–O–) linkage as well as deiodination. The deaminated products tetraiodothyroacetic acid (*tetrac* formed from T_4) and triiodothyroacetic acid (*triac* formed from T_3) act much more quickly than their parent compounds to increase O_2 consumption in isolated tissues or in whole animals, but the biological significance of these metabolites is not known.

Actions of thyroid hormones (THs)

If T_4 acts largely after being converted in the body to T_3 it is not surprising that the two substances have the same actions, though T_3 acts more quickly and for a shorter period of time and is, weight for weight, more potent than T_4. The THs have two main types of action:

Metabolic effects such as calorigenesis, regulation of ion and water transport and regulation of intermediary metabolism (of carbohydrates, lipids, and proteins).

Developmental effects such as regulation of growth rate in warm-blooded animals, regulation of metamorphosis in amphibians, and control of protein synthesis (via RNA).

The actions may be studied by administering THs to normal animals in the same way as one might administer a drug, with observations on the responses of the body systems, using physiological, biochemical, and other techniques. The effects of removal of the thyroid gland (thyroidectomy) in experimental animals and the

restoration of normal function by the administration of THs have given most valuable information. Finally, observations on the abnormalities in patients with hypothyroidism (reduced thyroid secretion) or hyperthyroidism (increased thyroid secretion), and their correction by appropriate treatments have provided perhaps the most valuable information of all.

Calorigenesis (heat production)

The outstanding action of THs is to stimulate heat production, as reflected by increased O_2 consumption in whole animals or in isolated tissues *in vitro*. A toxic dose of T_3 in animals can increase O_2 consumption within minutes or 1–2 hours but a small dose acts only after a latent period of 12–24 hours and then lasts for several days. The heart shows the greatest response, with smaller responses from the gastric mucosa, liver, smooth muscle, kidney, and skeletal muscle (in descending order of sensitivity). The spleen, brain, testis, and ovary show no response. These experimental observations in animals correlate well with the long known finding of increased basal metabolic rate (BMR) in hyperthyroidism, with O_2 consumption raised to as much as 50 or 100 per cent above normal, and with decreased BMR (−30 to −40 per cent) in myxoedema. Because of increased heat production, hyperthyroid patients do not tolerate raised external temperatures well and conversely, hypothyroid patients are hypersensitive to cold, as they react less well than normal subjects to a cold environment.

Thyroid hormones produce tachycardia, decreased glycogen content of heart muscle and enhanced sensitivity to the lipolytic action of adrenalin, before an increase in O_2 consumption is observed.

Protein metabolism. In patients with hypothyroidism or in thyroidectomized animals studies with [^{15}N]-glycine show a diminished rate of protein synthesis. Moderate doses of T_4 restore protein synthesis and reduce N excretion in the urine. In cretins or patients with juvenile hypothyroidism normal growth can be restored by administration of T_4. In young thyroidectomized rats it can be shown that T_4 only restores the normal rate of growth when the pituitary is capable of secreting adequate amounts of growth hormone. Thus the protein synthesis needed for growth requires both thyroid hormones and growth hormone.

In hyperthyroidism, whether in thyrotoxicosis or after overdosage with thyroid hormones, there is excessive protein catabolism, bodily wasting and increased N secretion in the urine.

In addition to growth, thyroxine is needed for special tissue *differentiation* and *maturation*, for example in the epiphyseal centres of ossification . Thyroxine also promotes *metamorphosis* in amphibia, e.g. the rapid maturation of tadpoles into dwarf frogs. This effect is not accompanied by increased O_2 consumption.

Carbohydrate metabolism. The effects of thyroid activity on carbohydrate metabolism are many and some are mutually antagonistic. In hyperthyroidism the blood glucose level after fasting is increased; in hypothyroidism it is decreased. THs increase glucose production in the body by increasing glycogenolysis in the liver, heart, and skeletal muscle, by enhancing gluconeogenesis from pyruvate in the liver and by increasing the dietary intake of carbohydrate (increased appetite in hyperthyroidism). In hyperthyroid patients gastric emptying is hastened and this tends to increase the rate of absorption of glucose from the gut. Thyroxine reduces the rate of secretion of insulin and also accelerates its breakdown. By these processes THs can promote or worsen diabetes mellitus.

On the other hand thyroxine increases the rate of glucose uptake and oxidation in the tissues, but not sufficiently to account for the calorigenic action of thyroid hormones (which is more closely related to the increased amounts of circulating free fatty acids).

Mucopolysaccharides. The disturbance of mucopolysaccharide metabolism in hypothyroidism produces a dry, coarse, puffy appearance of the skin which gave rise to the name 'myxoedema' (Ord 1878). Thyroid deficiency leads to excessive accumulation of hyaluronic acid, combined with protein, which causes retention of water and NaCl in the skin and other tissues. Administration of thyroxine increases the breakdown of hyaluronic acid with enhanced urinary excretion of hexosamines produced by depolymerization of the hyaluronic acid–protein complexes. It also causes diuresis and urinary loss of Na.

Lipid metabolism. In myxoedema and cretinism there is a striking rise in serum cholesterol level to 10.5–18 mmol l⁻¹ (400–700 mg per 100 ml) and in hyperthyroidism there is a reduction to about 3.0 mmol l⁻¹ (120 mg per 100 ml) the normal level being 3.6–7.8 mmol l⁻¹ (150–250 mg per 100 ml). THs actually increase cholesterol synthesis, but by enhancing faecal excretion of cholesterol and promoting its conversion to bile acids the end result is a reduction of plasma cholesterol concentration, which is further favoured by increased turnover of the lipoprotein to which cholesterol is bound.

In general THs stimulate all aspects of lipid metabolism, including synthesis, mobilization, and degradation of lipids. Degradation exceeds synthesis so that in hyperthyroidism there is a decrease in the stores of triglycerides and phospholipids as well as cholesterol. Hypothyroidism has the opposite effects.

THs increase lipolysis (and thus release FFA) both by a direct effect via the adenyl cyclase–cyclic AMP system and by sensitizing adipose tissue to other lipolytic agents such as catecholamines, GH, glucocorticoids, and glucagon. Oxidation of the FFA is also increased, and this may account for most of the calorigenic activity of thyroid hormones.

Hepatic synthesis of triglycerides is increased by THs, probably as the result of the greater availability of FFA and glycerol mobilized from adipose tissue. At the same time plasma triglyceride is removed by an increase in lipoprotein lipase.

Vitamin metabolism. Thyroid hormones are necessary for the conversion of β-carotene to vitamin A and for the conversion of vitamin A to retinene. In hypothyroidism a rise in blood carotene level may give a yellowish colour to the skin. Hyperthyroidism may produce vitamin D deficiency.

Thyroid hormones, by stimulating metabolic processes, increase the demand for coenzymes and the vitamins from which they are formed. In hyperthyroidism there is a need for increased supply of water-soluble vitamins such as thiamine, riboflavine, vitamin B_{12}, and ascorbic acid.

The actions of thyroid hormones at the cellular level

The physiological actions of THs cannot be explained in terms of molecular mechanisms. It is important to realise that the same hormone in different doses or in different systems may act in different ways. For example, a large toxic catabolic dose of T_3 causes a very rapid increase (within 2 hours) in the BMR of thyroidectomized rats and at the same time increases mitochondrial respiration, whereas a much smaller anabolic dose acts similarly but only after a latent period of 36 hours. The delayed onset of action of THs, whether concerned with mitochondrial activity or growth, is related to increased rate of RNA synthesis by the nucleus, followed by protein synthesis by ribosomes. Inhibition of RNA synthesis by actinomycin D, or inhibition of protein synthesis by puromycin, prevents the increase in BMR (heat production) and body weight produced by T_3 in thyroidectomized rats. It is suggested that T_3 acts by a time-consuming anabolic process involving protein synthesis in mitochondria. This stimulates O_2 utilization (heat production) without uncoupling oxidative phosphorylation. THs increase the number and size of mitochondria in TH responsive tissues only.

Other cellular actions of thyroid hormones include membrane effects such as:

(a) Increased activity of plasma membrane-bound, Na⁺, K⁺-dependent, ouabain sensitive, ATPase which would cause increased turnover of high energy phosphate and increased substrate oxidations.
(b) Stimulation of adenyl cyclase.
(c) Enhancement of amino-acid transport into responsive cells.

Inorganic ions. The phosphate balance of hyperthyroid patients is negative, that of hypothyroid patients is positive. Hypothyroid patients treated with T_4 promptly excrete large amounts of phosphate in the urine, much of it due to utilization of creatine phosphate in muscles. The phosphaturia is not due to increased parathyroid hormone secretion. Phosphate ions potentiate the effects of THs.

In hyperthyroidism calcium balance is negative, with excessive calcium excretion in urine, faeces, and sweat. The loss of calcium may lead to osteoporosis, and increased urinary hydroxyproline excretion indicates increased resorption of bone. These effects are not due to over-secretion of parathyroid hormone and serum calcium is never much elevated. Absorption of Ca^{2+} from the gut is diminished and does not increase after administration of large doses of vitamin D. In hypothyroidism the calcium pattern is changed in the opposite way with positive Ca balance, etc.

Magnesium metabolism is affected in the opposite way to calcium; in hyperthyroidism Mg balance is positive and in hypothyroidism it is negative.

Administration of T_4 to a myxoedematous patient causes a marked diuresis with excretion of retained NaCl. In normal persons large doses of T_4 cause a smaller diuresis with loss of K, presumably due to cell breakdown. A similar loss of K occurs in hyperthyroidism.

The central nervous system

THs are *essential* for normal activity of the central nervous system though they do not increase oxidative metabolism in nervous tissue. In overt hypothyroidism in adults (myxoedema) there is diminution of mental activity; somnolence; and slowness of thought, speech, and movement. Fatigue, depression, occasional psychotic manifestations, and coma on exposure to cold, due to hypothermia, may also occur. In hyperthyroidism there is increased excitability, irritability, fatigue, anxiety, and emotional instability; fine rhythmic tremor in hands, tongue, or eyeballs may be seen. β-adrenergic receptor blocking drugs, e.g. propranolol, reduce these manifestations.

The central nervous symptoms of abnormal thyroid secretion are accompanied by objective signs.

1. Reflex activity. The achilles tendon reflex (ankle jerk) is slowed and prolonged in myxoedema; it is quickened and shorter in hyperthyroidism. In rats the electroconvulsive shock threshold is raised by thyroidectomy and reduced by THs.

2. The electroencephalogram (EEG). The α-wave frequency is reduced in hypothyroidism and increased in hyperthyroidism.

Growth and development of central nervous system

Cretinism is characterized not only by the central nervous abnormalities seen in myxoedema of adults but also by additional profound effects due to failure of normal growth and development of the central nervous system. Functionally this is shown by the delayed occurrence of the normal stages of the child's development— holding the head up, sitting up, walking, speech—which are reached much later than normal. Mental development is very backward and may always remain subnormal. Anatomically the brain is small and underdeveloped in many ways.

Studies on rats made hypothyroid from birth (by thyroidectomy, by administration of antithyroid drugs or ^{131}I) have revealed the nature of the developmental defects. The weight of the brain is reduced and there is a marked reduction in the cerebral vascular bed, though it is not known whether the latter is the primary cause of the abnormal development. In the cerebral cortex myelination of axons is retarded and the growth and branching of dendrites is reduced, thereby decreasing axodendritic interaction.

From a practical viewpoint it is very important to know how far the nervous abnormalities of hypothyroidism can be reversed by giving THs. If thyroid secretion is defective during the latter part of fetal life or during the first 2 years of postnatal life, it is not always possible to restore normal nervous and mental development, but T_4 should always be given to produce as much benefit as possible. If hypothyroidism develops after the age of 2 years the central nervous manifestations can be reversed by adequate treatment with T_4. Thyroid hormones exert profound effects on the maturation of cerebral tissue during late fetal and early postnatal life. The administration of iodine to pregnant mothers in areas where endemic goitre occurs will help to prevent cretinism, firstly, by promoting fetal synthesis of thyroid hormones and secondly, iodine may promote cerebral development by some process independent of THs.

To sum up: *early diagnosis of cretinism is vital. Failure to commence treatment as soon as possible is a tragic avoidable disaster affecting the patient's whole life.*

Skeletal muscle. Skeletal muscle function is affected in both hypothyroid and hyperthyroid states.

In myxoedema stiffness and aching of muscles are common, with slowness of contraction and relaxation, though muscle strength is normal; in cretinism muscular growth is impaired.

In hyperthyroidism proximal muscular weakness (thyrotoxic myopathy) is common, with undue fatiguability. Biopsy studies show muscular atrophy and abnormal mitochondria, and electromyography reveals continuous muscular activity. The underlying biochemical abnormality appears to be impaired formation of creatine phosphate accompanied by wasteful creatinuria. Since creatine phosphate is required to convert ADP to ATP [p. 470] its reduction must reduce the efficiency of muscular contraction. In some hyperthyroid patients periodic paralysis, with or without hypokalemia, may occur, and myasthenia gravis develops in about 1 per cent of cases.

Cardiovascular system

Hyperthyroidism. In the hyperthyroid state the heart rate is increased, e.g. to 90 during sleep. This effect may be partly due to increased sympathetic activity on β-receptors, via adenyl cyclase–cyclic AMP, since propranolol, which blocks the actions of adrenalin on β-receptors, reduces the tachycardia.

The cardiac output is increased by 50–200 per cent due to increased stroke volume and increased heart rate. The rise in output and velocity of blood flow are greater than the increase in O_2 consumption. Mean arterial blood pressure is slightly reduced and peripheral vascular resistance is much below normal. Myocardial O_2 utilization and coronary blood flow are increased. Cardiac arrhythmias are common and 10 per cent of patients with Graves' disease have atrial fibrillation. Cardiac failure may supervene.

The increased heat production in hyperthyroidism produces a compensatory cutaneous arteriolar dilatation with increased blood flow through the skin to prevent a rise in body temperature. The skin is warm and moist.

Hypothyroidism. The converse changes occur. The heart rate is slow, cardiac output falls e.g. to 2.5 l min^{-1}, stroke volume is decreased, and the velocity of blood flow is much reduced. The heart is enlarged and there may be pericardial effusion containing protein and mucopolysaccharides. The ECG shows low voltage QRS and T waves. Blood pressure is normal. Cutaneous blood flow is reduced and the skin is cold and dry.

Blood

Hypothyroidism. The commonest change is a reduced red cell mass, producing normocytic, normochromic anaemia, with reduced erythroid cellularity of the bone marrow secondary to the reduced O_2 requirements of the body. Iron-deficient anaemia may also occur (in women this is chiefly due to excessive menstrual bleeding). Less commonly the anaemia is macrocytic, usually from vitamin B_{12} deficiency. There may be reduced absorption of vitamin B_{12} from the gut and in some cases autoantibodies against intrinsic factor and gastric parietal cells induce a true pernicious anaemia.

Hyperthyroidism. Thyroid hormones increase red cell mass and cause erythroid hyperplasia proportional to the enhanced O_2 requirements of the body. In hyperthyroid patients the oxyhaemoglobin dissociation curve is shifted to the right, thus releasing more O_2 to the tissues. This effect is due to increased production in the red cell of 2, 3-diphosphoglycerate (2,3-DPG) which directly promotes dissociation of O_2 from oxyhaemoglobin solution [p. 185].

Alimentary tract

Appetite. In hypothyroidism the appetite is reduced: hyperthyroid patients often eat voraciously but lose weight. The mechanisms involved in these appropriate changes in appetite are not known [see p. 344].

Motility. In hypothyroidism gut motility is reduced, constipation is common and may be extreme. In hyperthyroidism gastro-intestinal motility is increased; gastric emptying is hastened and intestinal transit time is shortened, producing diarrhoea. The mode of action of THs in enhancing motility is not known, but it cannot be a potentiation of sympathetic activity since this would reduce motility and cause constipation.

Thyroid and reproductive functions

Puberty. Hypothyroidism in childhood usually delays the onset of puberty, e.g. in untreated cretins. If juvenile hypothyroidism is adequately treated with T_4 puberty can proceed normally.

Hyperthyroidism in girls does not influence the age of menarche or subsequent fertility.

Fertility and ovulation. Hypothyroidism impairs fertility in women. Thyroid hormone treatment restores the fertility to normal, but has no effect on impaired fertility due to causes other than hypothyroidism.

In hypothyroid women there is often an increase in menstrual bleeding, with occurrence of intermenstrual bleeding. The endometrium remains in the proliferative phase, with marked hyperplasia; treatment with progestin produces a secretory type of endometrium and stops the menorrhagia. Hypothyroidism somehow prevents corpus luteum formation (and the subsequent progesterone secretion) and allows persistent oestrogenization of the endometrium. This may be due to interference with the normal ovarian–pituitary feedback processes, perhaps by an action on the hypothalamus.

In hyperthyroidism menstrual bleeding is scanty or absent (amenorrhoea) due to effects of THs on the hypothalamus which would be the opposite of those evoked by hypothyroidism.

Thus alterations of thyroid activity impair fertility in women largely as a result of disordered cyclic ovarian activity.

REGULATION OF THYROID SECRETION

Normal output of hormones

In man the normal metabolic clearance of T_4 (in µg day^{-1}) is about 80 µg. This represents the amount of T_4 secreted by the thyroid. The normal metabolic clearance of T_3 is about 33 µg day $^{-1}$ but since more than two-thirds (say 25 µg) of this is formed by deiodination of T_4 this means that the thyroid secretes only about 8 µg of T_3 daily. The normal serum level of T_4 is about 80 µg l^{-1} and that of T_3 about 120 ng l^{-1}. There are several regulating factors which help to keep secretion rates and serum levels of T_4 and T_3 within normal limits. The most important factors are:

1. TSH *(Thyroid-stimulating hormone, thyrotrophin)* secreted by the anterior pituitary, which in turn is under the control of the hypothalamus through secretion of *thyrotrophin-releasing hormone* (TRH) [p. 520].
2. Intrinsic *autoregulatory processes* within the thyroid which modify the responsiveness to TSH.
3. The *supply of iodine* in the diet [p. 537].

Thyroid-stimulating hormone (TSH)

The synthesis and secretion of thyroid hormones depend almost entirely on the presence of the adenohypophysis. After hypophysectomy the thyroid becomes smaller and less vascular, the follicular cells become thin and flat and the follicular lumen fills with colloid. The amount of thyroid hormone secretion is so much reduced that signs of myxoedema eventually appear. Injection of an anterior pituitary extract restores the atrophic thyroid to its normal state. The active principle is TSH which is a glycoprotein with a molecular weight of 31 000; it is formed and secreted by the adenohypophysial thyrotrophic cells and reaches the thyroid via the bloodstream. Its concentration in serum can be measured by radioimmunassay which gives a mean value of about 2.3 µU ml^{-1}. Studies with radioiodine-labelled TSH in man show that the half-life of TSH in plasma is about 1 hour and the daily secretory rate is 170 mU. In hypothyroidism the serum TSH is raised to more than 20 µU ml^{-1} and plasma half-life and secretory rate of TSH are increased; in most patients with hyperthyroidism serum TSH levels are low or undetectable.

Actions of TSH

TSH acts directly on the thyroid to increase secretion of thyroid hormones, to make the follicular cells cuboidal or columnar in shape, to increase the number of follicular cells (hyperplasia), to reduce the amount of follicular colloid and to increase the vascularity of the gland.

The details of the mode of action have been studied *in vivo* after injection of TSH into rats and in *vitro* after adding TSH to thyroid tissue slices from experimental animals. The *in vivo* experiments have shown that within 5 min of injection there is a marked increase in the formation of pseudopodia and the activity of microvilli at the apex of the thyroid cell and a significant increase in the number of intracellular colloid droplets. As the colloid droplets move into the cell lysomes move from the base of the cell to fuse with the colloid droplets to form phagolysosomes from which proteolytic enzymes are released to split off the thyroid hormones from the thyroglobulin molecule. The T_4 and T_3 so released diffuse from the cell into the

capillary blood and within 10 min of injection of TSH increased concentrations of T_4 and T_3 can be detected in thyroid vein plasma; the iodotyrosines are deiodinated within the cell and the iodine reutilized in hormone synthesis.

TSH not only stimulates the secretion of stored thyroid hormones from the follicular colloid but it also promotes synthesis of fresh thyroid hormones, by increasing iodide transport into the thyroid cells and by enhancing organic binding of iodine to tyrosine and the subsequent coupling to form thyroid hormones on the surface of the thyroglobulin molecules.

Other effects of TSH on thyroid follicular cell activity include increased O_2 consumption, increased glucose and fatty acid utilization, increased CO_2 production, increased phospholipid formation (to help form fresh membrane after endocytosis of colloid droplets), increased RNA and protein synthesis, and, after 48 hours, increased DNA content which precedes cell mitosis.

TSH and cyclic AMP. TSH can act on slices of thyroid tissue, containing intact cells, which shows that its effects are not mediated by neural or vascular pathways. However, TSH has no effect on the contents of ruptured cells and there is much evidence to support the view that it is one of the many hormones which act through the ubiquitous 'second messenger', cyclic AMP. TSH first combines with a specific receptor on the outer surface of the thyroid cell. This activates adenyl cyclase, associated with the cell surface membrane, and this enzyme converts ATP to cyclic AMP, which then combines with a cAMP dependent protein kinase to mediate the actions already described for TSH. Virtually all the effects of TSH on thyroid cell function can be reproduced by cAMP, or its more active dibutyryl derivative (DBC).

Hypothalamic control of TSH secretion [p. 520].

Selective lesions of the anterior hypothalamus just behind the optic chiasma in many species reduce the secretion of TSH and hence produce thyroid atrophy. Conversely, prolonged electrical stimulation of the same part of the hypothalamus through implanted electrodes can increase thyroid secretion. This effect is mediated by the secretion from the median eminence of a *thyrotrophin-releasing hormone* (TRH) which is carried by the hypothalamo-hypophysial portal vessels to stimulate secretion of TSH by the thyrotroph cells of the adenohypophysis [p. 520]. TRH has been isolated from the hypothalamus and shown to be a tripeptide, pyroglutamyl-histidylprolinamide, which can be synthesized. TRH acts on pituitary tissue both *in vivo* and *in vitro* to release TSH, probably via the adenyl cyclase–cAMP system within the thyrotrophs. TRH requires Ca^{2+} and oxidative metabolism in the pituitary for its effects. TRH is rapidly removed from the blood, by degradation in plasma and by renal excretion. The stimulant action of TRH on pituitary thyrotrophs is competitively inhibited by T_4 or T_3. This is the basis for the negative feedback control exerted by thyroid hormones on the secretion of TSH by the adenohypophysis. The deiodination of T_4 to T_3 occurs in many tissues but is greatest in the pituitary, so it seems likely that T_3 is the main hormone producing this negative feedback. There is no clearcut evidence that thyroid hormones influence the secretion of TRH (Burger and Patel 1977).

Thyroid autoregulation

TSH is the chief regulator of thyroid structure and function and is mainly concerned with control of plasma and tissue concentrations of thyroid hormones. However, the thyroid itself possesses autoregulatory mechanisms which maintain the constancy of stored hormones in the thyroid in the face of wide variation in iodine intake. An increased ingestion of iodine does not increase thyroid hormone levels in the plasma or decrease TSH secretion. Conversely iodine deficiency may enhance thyroid hormone secretion without increasing TSH secretion. This autoregulation is probably due to an effect of organic iodine, mainly MIT, excess of which depresses iodide transport into the follicular cell. With iodine deficiency enlargement of the thyroid gland occurs before there is evidence of increased secretion of TSH. In short, the amount of organic iodine (as MIT) affects the sensitivity of the thyroid response to TSH, iodine deficiency increasing the sensitivity and iodine excess decreasing the sensitivity.

Iodine in hyperthyroidism. In hyperfunctioning thyroid glands, e.g. Graves' disease, administration of iodine decreases the rate of thyroid hormone secretion, rapidly lowers the serum T_4 and T_3 levels and relieves the manifestations of thyrotoxicity. At the biochemical level iodine exerts this effect at some stage after the initial effect of cAMP; iodine reduces the proteolytic release of thyroid hormones [p. 546], probably by inhibiting the enzyme glutathione reductase and thereby preventing reduction of S-S bonds in the thyroglobulin molecule. Thus hormone release is reduced, though only for a few weeks. In addition, iodine reduces the thyroid hyperplasia and hypervascularity in patients with Graves' disease. The mode of action in producing this involution of the thyroid is still not understood, but the reduced vascularity may be due to decreased energy metabolism, and subsequent reduction of vasodilator acid metabolites.

Sex and sex hormones on thyroid activity

The effects of THs on gonadal function have been discussed [p. 542]. Conversely the gonads affect thyroid activity. Disorders of thyroid function are commoner in women than in men, e.g. hypothyroidism, Graves' disease, simple goitre during puberty, pregnancy, or at the menopause.

Oestrogens increase the concentration of TBG and of total serum T_4 and T_3 levels. Androgens have the opposite effect.

Pregnancy. During pregnancy the thyroid gland enlarges, thyroid ^{131}I uptake is increased and thyroid iodide clearance raised. These changes are secondary to an increase in renal iodide excretion which produces iodine deficiency. Serum TSH is normal in pregnancy; the placenta secretes a thyroid-stimulating peptide (probably human CG) whose physiological role with respect to the thyroid is uncertain. The total serum T_4 concentration rises and serum T_3 also rises but to a lesser extent. These increases are secondary to the increase in serum TBG concentration and it is doubtful whether there is increased thyroid hormonal activity during pregnancy. The increased BMR (to +20–30 per cent) during the second and third trimesters is due to an increase in total body mass. It is not known how much transfer of thyroid hormones occurs between mother and fetus, but the increase in maternal TBG concentration would hinder transport of maternal THs to a fetus whose thyroid is not functioning properly.

Neonatal thyroid hyperactivity. After delivery the serum TSH concentration of the newborn child increases rapidly to a peak at 30 min after delivery, followed by a rise in serum THs to reach a peak at about 24 hours (Fisher and Odell 1969). The hyperthyroid state of the neonate is attributed to cooling and after 3–5 days the rise in serum TH level exerts a negative feedback reduction of TSH secretion.

Glucocorticoids and stress on thyroid secretion. Both ACTH and adrenal glucocorticoids, in pharmacological doses, decrease TSH secretion by inhibiting endogenous release of TRH.

In man there is no clear evidence that emotional stress causes hyperthyroidism. It is more likely that mental changes are a result of hyperthyroidism than a predisposing cause of it.

Environmental temperature. Exposure of human beings to cold

for several days leads to raised serum T_4 concentration which reaches a peak after three days. Thyroid uptake and clearance rate of ^{131}I are increased. Short exposure to cold does not stimulate thyroid secretion in adults, though, as described above, it does enhance TSH secretion in newborn babies.

Simple goitre

Visible enlargement of the thyroid gland, without signs of abnormal thyroid activity, is called *simple goitre*. The enlargement of the thyroid is due mainly to increased secretion of TSH, which in turn is due to a slight fall in TH level in the blood. The hyperplastic gland will produce more THs and thus prevent hypothyroid manifestations. If, however, TH synthesis still remains low, in spite of great enlargement of the gland, signs of hypothyroidism ensue.

The commonest cause of simple goitre is dietary iodine-lack which is widespread in certain parts of the world, e.g. the Andes, the Himalayas, Zaire, and New Guinea. In such areas women with simple goitre due to dietary iodine deficiency are liable to produce cretinous infants—*endemic cretinism*.

Administration of iodine to the mother during the latter half of pregnancy can prevent cretinism in the offspring, by providing the fetus with sufficient iodine to synthesize its own THs. The addition of small amounts of iodine (100–200 µg daily) in the form of iodized table salt (iodine or iodate) will prevent the development of simple goitre due to dietary iodine deficiency. *Sporadic* cases of simple goitre may be due to relative iodine deficiency, which manifests itself only when additional predisposing factors intensify the body's need for iodine. This is commoner in women than in men and is particularly liable to occur during pregnancy and also at puberty, during menstruation, and at the menopause.

Rare hereditary defects in TH synthesis may involve the iodide trap, the formation of iodotyrosines, the coupling mechanism, TH hormone release, or the deiodination of MIT and DIT. In all these cases TSH secretion will increase and produce simple goitre, but the TH secretion may be so reduced that hypothyroidism develops.

Certain plants and seeds, particularly species of *Brassica* (e.g. cabbage) contain thiocyanates and oxazolidone derivatives which can induce goitres in animals, and also occasionally in man. In addition to these *natural goitrogens*, there are the *antithyroid drugs* which are used in the treatment of thyrotoxicosis [p. 546]. These drugs may reduce THs to a level which stimulates TSH secretion and thus leads to thyroid hyperplasia.

Occasionally, goitre is produced by oversecretion of TSH from the anterior pituitary or of TRH from the hypothalamus.

Hypothyroidism

Hypothyroidism is the condition resulting from reduced circulating levels of free (unbound) T_4 and T_3. In 1891 George Murray successfully treated a myxoedematous patient with sheep's thyroid; this was the first demonstration of substitution therapy for an endocrine glandular deficiency. The administration of thyroid substance was soon shown to be effective in cretins, provided it was given early enough.

Hypothyroidism is not an 'all-or-none' phenomenon but a graded state which results from suboptimal circulating levels of thyroid hormones. The term 'myxoedema' is applied to advanced hypothyroidism with characteristic swelling of skin and subcutaneous tissues. Hypothyroidism may result from removal of too much thyroid tissue during thyroidectomy, damage by radioiodine therapy, autoimmune thyroid disease, marked reduction of iodide in the diet, reduction of TH synthesis, or from TSH deficiency due to hypothalamic-pituitary disease. A completely athyreotic adult patient can live for many years.

Overt hypothyroidism appears in two forms, cretinism in infancy and myxoedema in later life. In adults overt hypothyroidism has the following clinical features: dryness and thickening of the skin, facial puffiness, loss of hair, intolerance of cold and hypothermia, bradycardia, gain of weight, hoarseness of the voice, mental and physical lethargy, intellectual deterioration with poor memory; the severest manifestation is 'myxoedema coma'.

In cretinism the signs of myxoedema are present with the additional features of dwarfism, due to failure of skeletal and muscular growth, gross retardation of mental development and failure of sexual development.

In addition to lowered serum levels of T_4 and T_3 hypothyroidism is characterized by reduced uptake of radioiodine in the region of the thyroid. In secondary hypothyroidism due to pituitary deficiency serum TSH level is reduced but administration of TSH increases uptake of radioiodine by the thyroid. In primary hypothyroidism the serum TSH is abnormally high and is further raised by administration of TRH [FIG. XI.26].

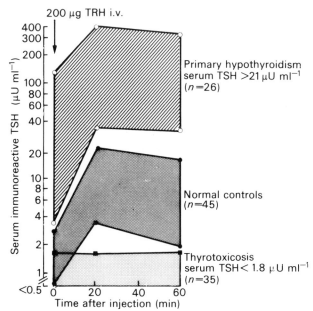

FIG. XI.26. Serum TSH response to TRH in normal controls and in patients with thyrotoxicosis and primary hypothyroidism (values to right of figure refer to levels 20 min after TRH).
(From Hall, R., Anderson, J., Smart, G. A., and Besser, M. (1980). *Fundamental of clinical endocrinology* 3rd edn, p. 115. Pitman, London.)

Hyperthyroidism

Hyperthyroidism is the condition resulting from increased circulating levels of free T_4 and/or T_3. Much the commonest cause is *Graves' disease* (*thyrotoxicosis, exophthalmic goitre*) in which there is moderate general enlargement of the thyroid gland accompanied by signs of excessive thyroid secretion and by exophthalmos.

The outstanding clinical features are summarized as follows; increased *heat production* leads to heat intolerance, with a warm moist skin to increase heat loss; *weight loss* in spite of increased appetite and food intake, diarrhoea; palpitations, due to *tachycardia* (sleeping pulse 90 per min) or *atrial fibrillation*; shortness of breath on exertion, cutaneous vasodilation; undue *fatiguability*, proximal muscle weakness, tremors, *irritability*, and *nervousness*. These effects are due to hypersecretion of THs.

Pathogenesis of Graves' disease

The excessive secretion of thyroid hormones in Graves' disease (hyperthyroidism, thyrotoxicosis) is not due to hypersecretion of TSH. Indeed, serum TSH levels are often undetectable and do not increase after administration of TRH.

Graves' disease is thought to have an autoimmune basis. *Thyroid-stimulating antibodies* (TSAb) are produced by plasma cells derived from B lymphocytes; at the same time the number of T lymphocytes is reduced. The TSAb combine with receptors on human thyroid cell plasma membranes and displace TSH from its binding sites. *In vitro* this combination activates adenyl cyclase to form cAMP; this leads to increased uptake of ^{131}I by thyroid cells and increased synthesis and release of thyroid hormones. Human TSAb appear to occur in the serum of all patients with untreated Graves' disease and thus account for the manifestations of hyperthyroidism. After partial thyroidectomy, HTSAb are found in the serum of less than 20 per cent of patients.

The detection of TSAb in serum helps in the diagnosis of Graves' disease and in the evaluation of various forms of treatment.

Eye signs in Graves' disease. Exophthalmos and lid retraction are two features of Graves' disease which give the eyes a staring look [FIG. XI.27]. Lid retraction may be due to sympathetic overactivity. Exophthalmos means actual protrusion of the eyeball which in Graves' disease is usually symmetrical. The eyeball is pushed forward by increased bulk of the orbital contents—increased fat, muscles enlarged and infiltrated with lymphocytes, increased amounts of mucopolysaccharide, and water. The cause of exophthalmos is unknown. It is not due to excessive thyroid hormone secretion since administration of THs does not induce exophthalmos, though they do produce all the other features of Graves' disease.

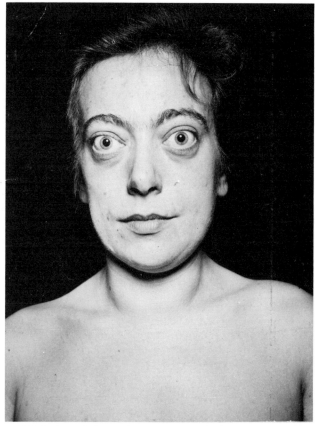

FIG. XI.27. Graves' disease in a woman aged 29. Note enlargement of thyroid and exophthalmos. BMR +65 per cent. Heart rate 120. Increased ^{131}I uptake by thyroid. After 10 days administration of iodine (180 mg daily) BMR fell to +35 per cent. Subtotal thyroidectomy was then performed. (Mr R. Vaughan Hudson's case.)

Antithyroid drugs

Reduction of the hypersecretion of THs in Graves' disease can be achieved by drugs which act in different ways on hormone synthesis and release. All the stages of TH synthesis and release can be depressed.

Potassium perchlorate and thiocyanate inhibit trapping of iodide by the thyroid. The most widely used antithyroid drugs, propyl thiouracil and carbimazole, inhibit oxidation and organic binding of iodine and also the coupling of iodotyrosines to form T_3 and T_4. Propylthiouracil reduces T_3 formation more than T_4 formation in thyrotoxicosis. Iodine or iodide acts mainly by reducing TH release and also reduces the size and vascularity of the gland in preparation for thyroidectomy. The action of iodide lasts for only 1–2 weeks. The therapeutic effectiveness of all these drugs is enhanced by simultaneous administration of β-adrenergic blocking drugs which reduce tachycardia, cardiac arrhythmias, and tremors. Propranolol *per se* may be valuable in preparing patients for thyroidectomy though it does not reduce hormone secretion. Radioiodine (^{131}I) is given to destroy the overactive thyroid tissue in Graves' disease; it is concentrated in the thyroid to such a degree that it selectively damages the secreting cells with minimal damage to other organs.

REFERENCES

Burger, H. G. and Patel, Y. C. (1977). *Clinical neuroendocrinology* (ed. L. Martini and G. M. Besser) p. 67. Academic Press, London.

Catt, K. J. (1971). *An ABC of endocrinology*, p. 82. The Lancet, London.

Dumont, J. E. (1971). *Vitam. Horm.* **29**, 287.

Fischer, D. A. and Odell, W. D. (1969). *J. clin. Invest.* **48**, 1970.

Greer, M. A. and Solomon, D. H. (1974). In *Handbook of physiology*, Section 7, Vol. III. *Thyroid*. American Physiological Society, Washington, D.C.

Hall, R., Anderson, J., Smart, G. A., and Besser, M. (1980). *Fundamentals of clinical endocrinology*, 3rd edn. Pitman, London.

Havard, C. W. H. (1974). *Br. med. J.* **i**, 553.

Hoffenberg, R. (1974). Br. med. J. **iii**, 452–6; 508–10.

Ingbar, S. H. and Woeber, K. A. (1974). The thyroid gland. In *Textbook of endocrinology*, 5th edn (ed. R. H. Williams) p. 95. Saunders, Philadelphia.

Larsen, P. R. (1972). *Metabolism* **21**, 1073.

Levey, G. S. (1971). *Am. J. Med.* **50**, 413.

Sterling, K. (1970). *Recent Prog. Horm. Res.* **26**, 249.

Utiger, R. D. (1974). *A. Rev. Med.* **25**, 289.

Calcium, phosphorus, parathyroid hormone, calcitonin, and vitamin D

Calcium plays a very important role in many body processes and its concentration in body fluids and in cells is maintained within narrow limits by the activities of parathyroid hormone (PTH), vitamin D (which in the body is converted to a hormone, 1,25-dihydroxycholecalciferol ((1,25DHCC)) and perhaps calcitonin (CT). Phosphate and magnesium are closely linked with the actions of calcium.

Summary of roles of calcium in the body

1. About 99 per cent of the body calcium occurs in bones. The rigidity of bones depends on calcification of connective tissue. Some of the calcium in bones is available for calcium homeostasis in body fluids.

2. Calcium promotes excitation–contraction coupling in skeletal, cardiac and smooth muscle [pp. 251, 98, and 409].

3. Calcium stabilizes cell membranes.

4. Calcium is required for release of neurotransmitters from synaptic vesicles.

5. Calcium is required for secretion of granular material from exocrine glands and from endocrines such as the adrenal medulla, β-cells of pancreatic islets (insulin), FSH from adenohypophysis, and ADH from neurohypophysis.

6. Synthesis of nucleic acids and of proteins requires calcium. Calcium promotes mitosis in thymus and bone marrow.

7. Calcium activates or regulates certain enzymes.

8. Calcium is required for several stages of blood clotting [pp. 23–5].

CALCIUM METABOLISM

The average adult human body contains 1000 g (25 mol) of calcium of which 99 per cent is in the skeleton, 4–5 g in the soft tissues (mainly muscle), and 1 g in the extracellular fluids.

Absorption, distribution, and excretion

In a normal human adult the daily diet should contain 0.8–1.0 g of calcium.

Absorption from the intestine

Calcium is absorbed throughout the length of the small intestine, the absorption being greater in the duodenum and proximal jejunum than in the ileum (Willis 1973). The microvilli, on the top of which is the brush border consisting mostly of glycosaminoglycans, effect calcium absorption by active carrier-mediated, energy-dependent transport in the duodenum and passive ionic diffusion and/or facilitated diffusion in the jejunum and ileum. After oral administration Ca absorption is completed within 4 hours.

The rate and extent of Ca absorption depend on many factors, such as age, body requirements, previous dietary Ca intake, the absolute amount of Ca in the gut and the availability of Ca in the gut, including the effects of bile, fatty acids, dietary phosphate, pH of gut contents, PTH, and vitamin D (cholecalciferol).

Age. In young animals Ca transport is greater than in adult animals (e.g. rats). Intestinal Ca transport is increased in late pregnancy and during lactation when Ca requirements are enhanced.

In man, Ca absorption diminishes with age, the decrease beginning at 55–60 years in women and at 60–65 years in men. At still greater ages Ca malabsorption may be evident, due perhaps to deficiency of vitamin D in the diet or to reduced formation in the body of 1,25DHCC.

Previous dietary calcium intake. With a low calcium diet people can increase the proportion of calcium absorbed from the gut and even achieve equilibrium on Ca intakes as low as 100–200 mg daily. Such adaptation is probably due to increased active transport of Ca, which is directly associated with an increased intestinal content of calcium-binding protein (CaBP). This is probably the result of increased 1,25DHCC activity.

These findings in younger adults may not apply to older people whose needs for increased dietary Ca and vitamin D are well established.

Absolute amount of calcium in the gut. The proportion of Ca absorbed from the gut is related inversely to the oral Ca load but the absolute amount of Ca absorbed after oral administration varies directly with the amount given. Thus, after a 1000-mg oral load more Ca is absorbed than after a 500-mg load. As the load increases there is greater saturation of the sites of Ca absorption but the gut has a peak capacity well above the Ca intake from normal dietary loads.

Availability of calcium. For absorption from the gut Ca must be in solution and ionized (Ca^{2+}). Most dietary calcium is bound in complexes which must be broken up before calcium can be ionized and absorbed. An acid environment increases absorption and an alkaline one decreases Ca absorption; thus, Ca absorption is greatest in the duodenum and proximal jejunum where the pH is lower than in distal parts of the gut. Gastric secretion of HCl may be important for Ca absorption.

Phytic and oxalic acids reduce Ca absorption from the gut by forming insoluble calcium salts. Phytic acid (inositol hexaphosphoric acid) also binds Ca as a non-ionizable complex. Phytic acid occurs in large amounts in wholemeal flour and might be expected to reduce Ca absorption considerably. However, many cereals contain a phytase which destroys phytic acid during bread-making, and in many species the intestinal mucosa secretes a phytate-splitting enzyme. The influence of phytate on Ca absorption in man is still controversial, but it is probable that the effect of phytate is less than was formerly supposed. During a short period phytate can produce negative Ca balance but in the long-term adaptation occurs and postive Ca balance is restored in spite of continued intake of phytate.

Phosphate. The mechanisms for Ca and inorganic phosphate absorption are linked and the presence of phosphate in the diet is essential for optimal absorption of Ca. If dietary Ca and phosphate are both high Ca absorption is reduced by precipitation of insoluble Ca phosphate within the gut.

Bile influences Ca absorption in the following ways

1. Bile promotes digestion and absorption of fat. Reduced secretion of bile and bile salts leads to increased amounts of fatty acids in the gut which form insoluble Ca soaps and Ca absorption is diminished.
2. Bile salts increase the solubility of Ca salts.
3. Bile is necessary for optimal absorption of vitamin D.

Parathyroid hormone has a small effect in promoting Ca absorption from the gut.

Cholecalciferol (vitamin D₃) influences Ca absorption after it has been hydroxylated in the body to form the hormone 1,25DHCC. This acts firstly by enhancing transport of Ca across the microvillar membrane and secondly by promoting the synthesis in the gut mucosal cells of a specific Ca-binding transport protein (CaBP). The renal hormone 1,25DHCC is the most important factor promoting Ca absorption in the gut.

Calcium distribution in man

In the intestine the daily dietary intake of 1 g (1000 mg) is supplemented by 600 mg of calcium which enters the gut in various secretions. Of the total 1600 mg, 900 mg is excreted in the faeces and 700 mg is absorbed into the body. This gives a net gain of 100 mg of calcium which enters the pool in plasma and extracellular fluid. The calcium in this pool is constantly being exchanged by absorption from and excretion into the intestine, by glomerular filtration and reabsorption by the kidney, by exchange with intracellular calcium in liver, heart, skeletal muscle, pancreas and nerve, and, most important, by ion exchange with calcium in bone crystals. The extracellular calcium pool exchanges 40–50 times daily but measurements after administration of ^{45}Ca show that this exchange does not lead to any net gain or loss from the pool. The bones contain about 1000 g (1×10^6 mg) of calcium of which 20 g (20 000 mg) is exchanged daily. Only the processes of mineral accretion and bone resorption involve net movement of calcium into or out of bone.

It has been estimated that 10 g (10 000 mg) of calcium is filtered through the glomeruli daily, of which all but 100 mg is reabsorbed, mainly in the proximal tubule. In this way balance is maintained, net output being equal to net intake. The urinary excretion of calcium is largely independent of the diet, and is controlled mainly by the plasma calcium level which in turn is influenced by vitamin D, parathyroid hormone, and calcitonin. When the total plasma calcium level falls below 1.7–2.0 mmol 1^{-1} calcium no longer appears in the urine, and when the plasma calcium exceeds 2.75 mmol 1^{-1} calcium excretion rises above normal but rarely exceeds 500 mg daily, even with marked hypercalcaemia. Renal tubular reabsorption of calcium is increased by parathyroid hormone and is decreased by cortisol and excess of thyroid hormone. Excess of parathyroid hormone and vitamin D increase calcium excretion in the urine but only as a result of hypercalcaemia. Some 90 per cent of the calcium filtered by the glomeruli is in the ionic form, the rest being complexed with citrate, etc., but in the urine the proportion of complexed calcium is much greater due to the relatively high concentrations of phosphate, ascorbate, lactate, glucuronate and amino acids.

In women, calcium is lost from the body during pregnancy and during lactation. The average calcium requirement by the growing fetus is 80 mg daily. The concentration of calcium in fetal blood is higher than that in maternal blood so the placental transfer of calcium must be an active process. During lactation the infant requires 200–400 mg of calcium daily.

Calcium in extracellular fluids

Blood

Calcium in blood is almost entirely in the *plasma* in which it exists in three forms to give a normal *total* plasma Ca concentration of 2.2–2.6 mmol 1^{-1} (8.5–10.5 mg per 100 ml).

1. Protein bound Ca. This form of Ca is loosely bound almost entirely to plasma albumin. This non-diffusible non-ionized Ca comprises about 45 per cent of the total plasma Ca and is physiologically inactive. When the concentration of plasma albumin falls bound Ca concentration also falls, as in nephrosis or liver disease. Conversely, a rise in plasma albumin concentration increases the protein-bound Ca concentration.

2. Complexed calcium is combined with citrate, phosphate, or other anions. This complexed calcium amounts to about 5 per cent of total plasma Ca. It is diffusible, non-ionized, and physiologically inactive.

3. Ionized calcium, which is *diffusible and physiologically active*, comprises about 50 per cent of the total plasma calcium (i.e. 1.25 mmol 1^{-1}). When plasma or serum is dialysed against water or NaCl solution all the calcium, including that 'bound' to protein is removed. The protein-bound fraction constitutes a reserve of calcium to help maintain the constancy of [Ca^{2+}]. The ionized plasma calcium level depends on Ca absorption from the gut (and hence on 1,25DHCC formation) and on the level of secretion of PTH.

Extracellular fluid (ECF) has a calcium concentration about 70 per cent of that in plasma

Plasma calcium homeostasis

The ionized plasma calcium level is normally held constant within narrow limits. After an infusion of a soluble calcium salt, which raises the plasma calcium level by 25 per cent, the normal concentration is restored within 4–6 hours. Conversely, infusion of ethylenediaminetetraacetic acid (EDTA) lowers the plasma calcium, and after the infusion is stopped the plasma calcium level is raised to normal within 4 hours. When infusions of calcium or

EDTA are given to dogs after removal of the thyroid or parathyroid glands the restoration of plasma calcium levels to normal takes over 24 hours in each case. These experiments suggest that PTH (and CT) take part in these homeostatic responses. Vitamin D, or rather its active metabolite 1,25DHCC, also participates in plasma calcium homeostasis. These and other factors are responsible for circadian variation in plasma Ca level (lowest value at 4.00 a.m.) and for changes with age.

Cellular functions of calcium ions

The importance of Ca^{2+} in physiology was first revealed by Sydney Ringer in 1883. He first demonstrated that a solution of NaCl in distilled water would keep a perfused, isolated frog's heart beating for only a few minutes. When his technician, without Ringer's knowledge, used tap water (from London's New River Company) in which to dissolve the NaCl the isolated heart then went on beating for many hours. Ringer found that the tap water contained calcium in amounts which explained this phenomenon and the importance of Ca^{2+} in maintaining physiological activity in nerve and muscle has been appreciated ever since. Later (1888) Ringer also demonstrated the importance of calcium ions in cellular adhesion. Heilbrunn (1927) showed that Ca^{2+} promoted sealing of the surfaces of injured cells and by the late 1930s calcium was known to have various functions, e.g. on cell permeability, on muscle contraction, on gastric secretion, and on blood clotting.

It is now known that calcium is a vital component of cell membranes and of mitochondria, from which it can be released to activate numerous physiological processes. Before discussing these it is necessary to consider the distribution of calcium inside and outside body cells. The distribution of Ca^{2+} is linked with the distribution of H^+ and HPO_4^{2-} so as to form a buffer system for each of these ions. This system is coupled to the CO_2/HCO_3^- system in the control of $[H^+]$ and together they influence the activity of cellular enzymes.

TABLE XI.3. *Estimated ion activities in ECF and in the various fluid phases of a typical mammalian cell (Rasmussen 1974)*

Fluid phase	H^+	Ca^{2+}	HPO_4^{2-}
ECF	40 nmol l^{-1}	500 µmol l^{-1}	200 µmol l^{-1}
Cytosol	150 nmol l^{-1}	250 nmol l^{-1}	100 µmol l^{-1}
Mitochondria (soluble)	12 nmol l^{-1}	20 µmol l^{-1}	500 µmol l^{-1}

The relationship between H^+, Ca^{2+}, and HPO_4^{2-} in an idealized mammalian cell is shown in TABLE XI.3. The points to emphasize are that the $[Ca^{2+}]$ in the cell cytosol is very small indeed compared to the $[Ca^{2+}]$ in ECF and in mitochondria. The pH of ECF is 7.4, that of cytosol 6.8–7.0 and that in mitochondria 7.8–8.0. The asymmetric distribution of these ions, and their translocation across membranes, is the basis of their buffering capacity. The plasma membrane possesses pump mechanisms for extruding Ca^{2+} from the cell while the mitochondria can concentrate large quantities of Ca^{2+} by an energy-dependent process which works at the expense of oxidative phosphorylation (ADP→ATP). The endoplasmic reticulum may also accumulate calcium.

In contrast, the plasma membrane is sufficiently permeable to Mg^{2+} to permit a cytosol concentration of 1.0 mmol l^{-1}. The relatively high extracellular $[Ca^{2+}]$ activates certain extracellular enzymes such as prothrombin, trypsinogen, and amylase, whereas the converse is true of intracellular magnesium-activated enzymes.

Control of intracellular calcium metabolism

Control of the $[Ca^{2+}]$ of the cell cytosol is very important in many types of cell. Changes in cytosol $[Ca^{2+}]$ regulate muscle contraction

[p. 251], exocrine and endocrine secretions (pp. 501, 524), hormone action, and energy metabolism in a wide variety of cells. Cytosol $[Ca^{2+}]$ is regulated and functions in different ways.

In skeletal muscle Ca^{2+} release from sarcoplasmic reticulum (which corresponds to endoplasmic reticulum in other cells) is brought about by depolarization of the T system extension of the sarcolemmal membrane. The released Ca^{2+} activates actomyosin-ATPase and the energy for contraction is obtained from splitting of ATP. Subsequently the sarcoplasmic reticulum sequesters Ca^{2+} and muscular relaxation occurs [p. 252].

The role of calcium in neurotransmitter and cellular extrusion processes (secretions) may be brought about by calcium acting as an ionic bridge between synaptic vesicles or storage granules and the plasma membrane, thus paving the way for exocytosis. However, energy is required for secretory processes and the passage of granules along microtubules and microfilaments may require the splitting of ATP.

The actions of peptide hormones may involve a coupling between events at the plasma membrane with those at the mitochondrial membrane. For example, PTH combines with a receptor site on the cell surface and as a result activates the membrane-bound adenyl cyclase and independently increases membrane permeability so that calcium flows from ECF into the cell cytosol. Adenyl cyclase promotes the conversion of ATP to cAMP which releases calcium from mitochondria to the cytosol. The raised cytosol Ca^{2+} concentration inhibits adenyl cyclase (negative feedback), increases K^+ permeability and in some types of cell increases Na^+ permeability as well.

The regulation of intracellular Mg^{2+} is little understood. Cytosolic $[Mg^{2+}]$ is about 1.0 mmol l^{-1} and the mitochondrial $[Mg^{2+}]$ is about 1.0 mmol l^{-1}; however, mitochondria also contain insoluble $Mg_3(PO_4)_2$. Cellular Mg^{2+} and Ca^{2+} are clearly regulated by independent factors.

Magnesium

The total body content of magnesium is about 25 g (200 mmol) in a 70 kg man. Half of this is present in bone in combination with phosphate and bicarbonate; bone provides a reserve of magnesium for use when there is a shortage elsewhere in the body. Only 1 per cent of body magnesium is in ECF, the plasma concentration being 0.6–1.0 mmol l^{-1}. Erythrocytes contain 2.5 mmol of magnesium per litre and muscle cells about 7.5 mmol l^{-1} Mg, being second only to potassium among intracellular cations. The total concentrations of Mg^{2+} and K^+ in intracellular fluids closely approximate to the supposed ionic composition of the pre-Cambrian oceans. Thus the present-day intracellular environment is very similar to that in which life apparently originated . It is not surprising that both Mg^{2+} and K^+ are important regulators of many enzymes and of protein biosynthesis. Mg^{2+} dependent enzymes and reactions include enzymes with thiamine diphosphate as cofactor, e.g. carboxylase and pyruvate oxidase, enolase, all reactions involving ATP synthesis and hydrolysis, and alkaline phosphatase.

In some respects Mg^{2+} and Ca^{2+} are mutually antagonistic. For example, the action of calcium in promoting release of ACh from synaptic vesicles in cholinergic nerve endings is inhibited by magnesium. The central nervous depression and neuromuscular block produced by high plasma levels of magnesium can be overcome by intravenous injection of calcium salts.

Magnesium deficiency is unlikely to occur on a typical British diet which normally contains 200–400 mg of magnesium daily, mainly in cereals and vegetables. About one-third of this is absorbed from the gut and excreted in the urine. Deficiency can arise from excessive loss of magnesium in chronic diarrhoea or prolonged diuresis. A low plasma magnesium level produces depression, muscular weakness, vertigo, and liability to convulsions. Hypomagnesaemia stimulates secretion of PTH which in turn influences the distribution of magnesium in the body. However, the factors controlling magnesium homeostasis are still largely unknown.

Inorganic phosphate

Inorganic phosphate is intimately related to calcium metabolism and its distribution is influenced by vitamin D and PTH.

The adult human body contains 500–600 g of phosphate (measured as inorganic phosphorus). About 85 per cent of this is in the skeleton and the rest is mainly in the organic and inorganic pools of intracellular phosphate.

Extracellular fluid phosphate

Plasma phosphate. In children the plasma inorganic phosphate level is 1.6–1.9 mmol l^{-1} (5–6 mg per 100 ml) and in adults it is 0.8–1.4 mmol l^{-1} (2.5–4.5 mg per 100 ml). The high level in children is associated with growth rather than sexual maturation. The plasma phosphate level is reduced on a low phosphate diet.

At normal blood pH (7.4) 80 per cent of plasma phosphate occurs as HPO_4^{2-} and the rest as $H_2PO_4^-$.

Relation between plasma calcium and phosphate. In many circumstances the plasma calcium level varies inversely with the plasma inorganic phosphate level, the product $[Ca^{2+}] \times [HPO_4^{2-}]$ being constant. After parathyroidectomy plasma calcium falls and phosphate rises, whereas the opposite happens after administration of PTH. However, in rickets in which absorption of calcium and phosphate is deficient, both calcium and phosphate plasma levels are reduced.

Functions of inorganic phosphate. Inorganic phosphate is incorporated into hydroxyapatite, which gives rigidity to bone and teeth. Phosphates help in the regulation of $[H^+]$ in blood and particularly in urine [p. 227]. In addition phosphate forms a part of such essential organic molecules as the nucleic acids (DNA and RNA), phospholipids, and adenine and guanine nucleotides. Inorganic phosphate is also important in the regulation of glycolysis and energy metabolism. It is strange that so little is known about the mechanisms by which cells obtain an assured supply of phosphate.

Absorption of phosphate. About 70 per cent of ingested phosphate is absorbed from the alimentary tract and 30 per cent is excreted in the faeces. Absorption depends on the presence of sodium and calcium ions and uses metabolic energy. Absorption is enhanced by a low-calcium diet, growth hormone, PTH, vitamin D, and acids; it is decreased by a high calcium diet and by the antacid aluminium hydroxide.

Excretion. With a daily intake of 900 mg of inorganic phosphate the urinary output amounts to 600 mg. In the kidney phosphate undergoes glomerular filtration, reabsorption in the proximal tubule and possibly distal tubular secretion. Urinary excretion of phosphate ceases when the plasma phosphate level falls below 0.65 mmol l^{-1} (2 mg per 100 ml). Urinary excretion of inorganic phosphate is decreased on a low-phosphate diet, by the action of GH, during lactation, and in hypoparathyroidism; excretion is increased on a high phosphate diet, with vitamin D excess and in hyperparathyroidism.

Phosphate deficiency

Severe dietary phosphate deficiency in animals, with a normal intake of vitamin D, causes an abrupt fall in plasma phosphate, disappearance of phosphate from the urine, and the development of rickets. Intracellular phosphate remains unchanged and plasma calcium levels are normal or increased. Thus in phosphate deficiency the maintenance of intracellular phosphate concentration takes precedence over extracellular phosphate concentration, which falls

to such an extent that mineralization of bone matrix ceases and rickets or osteomalacia (depending on age) develops. Matrix formation continues but resorption of bone increases even though PTH secretion is unchanged.

Plasma phosphate deficiency leads to a fall in 2,3-diphosphoglycerate and ATP concentrations in circulating erythrocytes and hence to diminished release of O_2 from HbO_2 in peripheral tissues [p. 185].

Phosphate administration reduces bone resorption, even that induced by PTH. This is shown by reduced excretion of calcium and hydroxyproline in the urine.

BONE

Bone is connective tissue made rigid by orderly deposition of mineral crystals. Bone fibres consist of the protein collagen, encrusted with crystalline mineral, the fibres being set in an amorphous gel of glycosaminoglycans (mucopolysaccharides), called ground substance. Collagen and ground substance together comprise the organic matrix or osteoid.

Chemical composition. Bone consists of 35 per cent mineral salts, chiefly calcium and phosphorus, 20 per cent organic matrix (95 per cent of this is collagen) and 45 per cent water. About 99 per cent of the body calcium is found in bone.

Inorganic constituents. The bone mineral consists almost entirely of hydroxyapatites, of a single crystalline type with a specific lattice structure. Hydroxyapatite consists of Ca^{2+}, (H_3O^+), PO_4^{3-}, and OH^-. The crystals measure $50 \times 25 \times 10$ nm and have a relatively large surface area. They absorb excess phosphate which leads in turn to the binding of water, thus forming the hydration shell on the crystal surface. This allows the exchange of ions between ECF and the surface or interior of the crystal. The structure resembles that of an ion-exchange column and since the total area of the crystals of bone salt in man is of the order of 400 000 m^2 they can be important in metabolic reactions in the body. Some ions such as Na^+ can replace Ca^{2+} at the surface of the crystal, which accounts for the sodium content of bone. Ions inside the crystals have a slow rate of turnover, and some foreign ions, such as lead, strontium, and radium can penetrate to the crystal interior, displace calcium, and remain in bone for years. Anions such as fluoride can replace hydroxyl ions. The removal of the cations can be hastened by chelating agents such as ethylenediaminetetraacetic acid (EDTA). The uptake of H^+, HCO_3^-, and Na^+ enables bone to serve as a site for buffering the pH of ECF.

In addition to crystalline hydroxyapatite, bone, particularly in the young, contains amorphous calcium phosphate $[Ca_3 (PO_4)_2]$. With aging this insoluble calcium phosphate is dissolved and converted to crystalline hydroxyapatite.

Organic constituents. Although inorganic constituents predominate in bone, organic constituents are of vital importance. The calcifiable matrix consists primarily of *collagen*, a fibrous protein rich in glycine, proline, and hydroxyproline. Other, less well defined proteins, and glycosaminoglycans and proteoglycans add to the composition of the bone matrix.

Collagen in all connective tissues is made up of tropocollagen molecules, each of which is a long rod-like cylinder, 300×1.5 nm, with a molecular weight of 300 000. The tropocollagen molecule is composed of three separate monomer units called α-chains which exist as a triple helix (like three strands of a cable), held together by weak intermolecular forces.

There is a very close relationship between the collagenous matrix and the bone mineral. Only the native configuration of collagen can serve as a template to initiate mineral crystal formation from a metastable solution of calcium and phosphate. The relationship between inorganic and organic phases of bone is so intimate that resorption of bone must involve the destruction of both matrix and mineral.

Bone metabolism, growth, and replacement

Bone is continually being remodelled, and bone matrix is always being synthesized, secreted, organized, mineralized, and finally destroyed. Thus bone formation and bone destruction go on continuously and simultaneously throughout life, though the rates change with age and vary in different parts of the skeleton. Bone formation may be endochondral (i.e. takes place after preliminary formation of cartilage), membranous, in which formation occurs without a cartilaginous phase, or endosteal in which bone formation occurs as part of a constant process of remodelling.

The dynamic processes of formation destruction, and remodelling of bone are under cellular control, formation being controlled by single nuclear cells called *osteoblasts* and bone resorption chiefly by multinuclear giant cells called *osteoclasts*. Collagen synthesis begins in the ribosomes of osteoblasts and procollagen is secreted into the extracellular space where it is converted to tropocollagen molecules which undergo considerable polymerization to form microfibrils, which may be centimetres long and 15–130 nm in diameter. During the normal process of mineralization the mineral ions are thought to accumulate within membrane-bound vesicles, derived from the Golgi complex or from the cell membrane.

The process of mineral homeostasis is controlled mainly by PTH, CT, and 1,25DHCC, which also regulate skeletal remodelling. Thyroxine, adrenal glucocorticoids, gonadal hormones, and growth hormone are also important participants in these processes.

Bone cells

There are four types of bone cell; mesenchymal or osteoprogenitor cells, osteoclasts, osteoblasts, and osteocytes. The osteoprogenitor cells are the only ones which undergo mitosis. The remaining cells are formed in the following sequence (Rasmussen 1974): Osteoprogenitor cells→Preosteoclasts→Osteoclasts→Preosteoblasts→ Osteoblasts → Osteocytes. The preosteoclasts may also be converted to osteoblasts directly.

Teeth

Teeth contain three calcified structures, enamel, dentine, and cementum (Irving 1973), which may be compared with bone.

Enamel is the hardest tissue in the body. It contains 95–97 per cent of inorganic matter, 0.2–0.8 per cent organic matter, and 1–4 per cent water. The enamel of dried teeth contains 36 per cent calcium and 16 per cent phosphorus. It is formed by the complex enamel organ, the active cells of which are the ameloblasts. These cells synthesize the organic matrix which consists of a unique protein, very rich in proline but probably containing no hydroxyproline molecules (cf. collagen in bone). The matrix thickens, becomes calcified and then disappears so that enamel of mature teeth consists almost wholly of very long mineral crystals 50–60 nm in width × 25–30 nm in thickness (i.e. much larger than the hydroxyapatite crystals in bone or dentine).

Dentine is much more elastic than enamel. It comprises 61–73 per cent inorganic matter, with an apatite structure, 11–15.5 per cent water, and 20–22.5 per cent organic matter. In dried teeth dentine contains 30 per cent calcium and 13 per cent phosphorus. The dental matrix is laid down by odontoblasts; it consists of collagen + small amounts of glycosaminoglycans.

After full maturation enamel and dentine differ from bone in being metabolically almost inert. They cannot yield calcium or phosphorus to the body for homeostasis.

Erupting teeth, however, are very sensitive to dietary and endocrine factors. Calcium, phosphorus, vitamin D, and PTH are needed for mineralization and vitamins A and C for dentine matrix formation. In animals, hypophysectomy slows the eruption rate and thyroidectomy retards calcification of dentine.

THE PARATHYROID GLANDS

Because of the intimate relationship of the parathyroids to the thyroid the operation of thyroidectomy has always been liable to produce symptoms of parathyroid deficiency, e.g. tetany, due to a fall in plasma $[Ca^{2+}]$.

Recent advances in the study of parathyroid function have followed the development of new techniques in biochemistry, electron microscopy, and immunoassay. It is now realized that calcium, phosphorus, and magnesium metabolism, both intra- and extracellular, are influenced by many other hormones besides PTH. These hormones include 1,25-dihydroxycholecalciferol (1,25-DHCC), formed in the body from vitamin D_3, calcitonin, thyroxine, adrenal glucocorticoids, gonadal hormones, and growth hormone.

Embryology. The two superior parathyroid glands arise from the dorsal portion of the 4th branchial pouch and migrate caudally with the thyroid gland. The two inferior parathyroid glands arise with the thymus from the 3rd branchial pouch; their migration takes them across the path of the superior parathyroids and their final position in relation to the thyroid is less constant, though they are nearly always close to the inferior pole of the thyroid.

Anatomy. In 90 per cent of human beings there are four parathyroid glands (two superior and two inferior), which are small reddish or yellowish-brown bodies measuring about 6 mm long, 3–4 mm wide, and 1–2 mm thick. The glands are situated on the posterior surface of the thyroid, the superior glands near to or embedded in the superior pole of the thyroid. In man the parathyroids are always outside the capsule of the thyroid. The total weight of the parathyroids is about 120 mg. They have a rich blood supply and the blood vessels are innervated by postganglionic sympathetic nerves.

Histology. The parathyroids consist of epithelial cells with numerous wide vascular channels among them. The normal epithelial cells are of two types; the chief cells and the oxyphil cells. In hyperplasia there are also water clear cells.

The chief cells are of two types, the light and the dark. The light chief cell does not secrete parathyroid hormone (PTH). The dark chief cell has a prominent Golgi apparatus and ER, and is rich in secretory granules. It is normally the main source of PTH secretion.

The oxyphil (eosinophil) and water clear cells are derived from chief cells and are capable of secreting PTH.

Functions. The parathyroid glands are *essential for life* and their removal can cause death from asphyxia, resulting from spasm of the laryngeal muscles, thoracic muscles, and diaphragm.

The primary function of the parathyroids is mediated by the secretion of PTH which helps to keep constant the concentration of calcium in intracellular and extracellular fluids, in spite of large variations in calcium intake and excretion. Other hormones are also important in calcium homoeostasis.

Parathyroid hormone (parathormone, PTH)

Chemistry and assay. In man the most abundant form of PTH is a polypeptide chain of 84 amino-acid residues with a molecular weight of 9500. In the parathyroid gland a pro-PTH of higher molecular weight (11 500) has been found (cf. proinsulin in islet cells of pancreas). In peripheral blood smaller biologically active peptides with molecular weights of 7000 and 4500 have been detected by immunological methods. It is not yet clear whether the 84 amino-acid peptide, or its smaller breakdown products, are the more important for mediating parathyroid gland activity.

PTH can be assayed biologically by measuring its calcium-mobilizing activity in thyroparathyroidectomized rats, in which the natural secretions of PTH and calcitonin have been eliminated. Other, more sensitive biological assay methods have been used, but none can measure the small concentrations of PTH in body fluids.

Radioimmunoassay techniques are much more sensitive and can be used to measure the low PTH serum levels of normal persons as well as the higher levels in induced hypocalcaemic patients or in patients with parathyroid adenoma.

Control of parathyroid secretion

The secretion of PTH is controlled by the concentration of Ca^{2+} in the blood which perfuses the parathyroid glands. If the plasma $[Ca^{2+}]$ falls PTH secretion is increased; if the plasma $[Ca^{2+}]$ is raised by infusion of calcium gluconate PTH secretion is reduced [FIG. XI.28]. The possible role of calcitonin in reducing hypercalcaemia will be discussed later, but PTH is the most important hormone involved in calcium homeostasis.

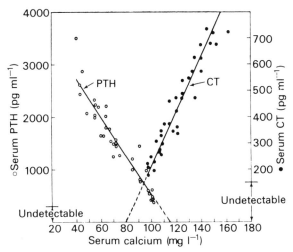

FIG. XI.28. PTH and calcitonin (CT) concentrations in pig serum plotted against serum calcium concentration. The serum calcium concentration was changed radically by the infusion of calcium or EDTA, an agent that chelates calcium.
(From Rasmussen, H. (1974). In *Textbook of endocrinology*, 5th edn (ed. R. H. Williams). Saunders, Philadelphia.)

In patients with severe reduction of plasma $[Mg^{2+}]$ hypocalcaemia develops but PTH secretion is not increased until the Mg^{2+} level is raised towards normal by infusion of a soluble, ionized magnesium salt.

PTH has a short half-life in the circulation—about 20 min. Changes in rate of secretion of PTH in response to fluctuations in plasma $[Ca^{2+}]$ are rapid. PTH is broken down in the liver; fragments of the 84-amino-acid hormone may prolong the biological actions of the hormone.

The administration of vitamin D to a D-deficient person reduces the level of PTH in peripheral blood although the plasma Ca^{2+} concentration remains unchanged or even falls during the first 24–36 hours. This physiological response is attributed to uptake of Ca^{2+} into parathyroid cells (induced by 1,25-DHCC) which is interpreted by these cells as a rise in plasma Ca^{2+} concentration. Thus, 1,25-DHCC changes the set point around which plasma calcium concentration controls PTH secretion.

The plasma inorganic phosphate concentration has no direct effect on PTH secretion but indirectly a raised plasma phosphate may promote PTH secretion by reducing plasma Ca^{2+} concentration.

Actions of PTH

The main organs on which PTH acts are the kidney and bone. Its action on the intestine is possibly mediated by renal synthesis of 1,25-DHCC.

The most important effects of PTH are:

1. Hypercalcaemia and hypophosphataemia.
2. Hyperphosphaturia, hypocalcuria (followed by hypercalcuria).
3. Increased urinary hydroxyproline excretion.
4. Increased resorption of bone; increase in number of osteoclasts and osteoblasts on bone surfaces.
5. Increased conversion 25-HCC to 1,25-DHCC in kidney.
6. Activation of adenyl cyclase in target tissues.

Intestine. Metabolic balance studies in man and experimental animals have shown that, after several hours delay, PTH promotes absorption of calcium from the alimentary tract. This effect may be indirect, due to PTH-induced synthesis of 1,25-DHCC in the kidney.

Bone. PTH has for long been regarded as the most important hormone that promotes bone resorption and thereby raises plasma $[Ca^{2+}]$. The osteolytic effect occurs in two phases, early and late. The early phase can be seen a few minutes after administration of PTH to a parathyroidectomized animal. Plasma $[Ca^{2+}]$ is raised by increased resorptive activity of osteocytes and osteoclasts, and the flow of Ca^{2+} from deep bone to bone surface is enhanced. However, this early phase is small and less important than the late phase, which occurs hours or days after PTH administration and involves the multiplication of osteoprogenitor cells which become osteoclasts. The late phase is accompanied by an early increase in DNA synthesis followed by increases in RNA and protein synthesis. Bone resorption is associated with increased release of lysosomal enzymes so that collagen destruction is added to Ca^{2+} release. However, PTH has a secondary effect of increasing osteoblast formation and thus promoting collagen biosynthesis and fresh bone formation. The increased urinary excretion of hydroxyproline induced by PTH is a reflection of increased synthesis as well as destruction of collagen. Physiologically, these processes of bone destruction and bone formation are harmonized to bring about remodelling of bone which continues throughout life in response to mechanical stresses. PTH is the most important hormone regulating this process but 1,25-DHCC, cortisol, and thyroid hormones also participate.

Kidney. PTH increases urinary excretion of phosphate, sodium, potassium, bicarbonate, and cAMP. It decreases excretion of calcium, magnesium and ammonia. Micropuncture studies have shown that PTH blocks proximal tubular reabsorption of Na^+, Ca^{2+} and HPO_4^{2-} but in the distal tubule it promotes reabsorption of Ca^{2+}. This action of PTH on Ca^{2+} reabsorption is of prime importance in the control of plasma calcium level. In primary hyperparathyroidism urinary calcium excretion increases only because of the increased calcium load in the glomerular filtrate, but the reabsorption process reduces the magnitude of the calcium loss.

PTH promotes the conversion 25-HCC to 1,25-DHCC in the kidney by activating 1-hydroxy cyclase via cAMP.

Mode of action of PTH

PTH activates adenyl cyclase and thus promotes cAMP formation in both bone and kidney, leading to increased urinary excretion of this nucleotide. Exogenous cAMP mimics the effects of PTH very closely both in bone and kidney, but the PTH-induced entry of calcium into bone cells is an independent effect, opposed to the calcium efflux induced by cAMP.

Abnormalities of parathyroid function

Information concerning parathyroid function has been provided by studying the changes which result from reduced secretion (hypoparathyroidism) or excessive secretion (hyperparathyroidism) of PTH.

Hypoparathyroidism

The commonest cause of hypoparathyroidism in man is the removal of, or damage to, the parathyroid glands during thyroidectomy (or extensive operations for laryngeal or oesophageal carcinoma).

The accompanying symptoms are due to reduction of the plasma calcium level. The total plasma calcium may fall to 1–2 mmol l^{-1} (4–8 mg per 100 ml) and the ionized calcium to 0.75 mmol l^{-1} (3 mg per 100 ml), i.e. to about half the normal level. At the same time the plasma inorganic phosphate rises to 2–5 mmol l^{-1} (6–16 mg per 100 ml). Excretion of calcium in the urine usually falls but may be normal and urinary hydroxyproline is reduced. Direct measurement by radio-immunoassay shows very low or undetectable levels of PTH.

The decrease in serum $[Ca^{2+}]$ increases neuromuscular excitability. If nerve fibres are bathed in fluids containing subnormal Ca^{2+} concentrations their excitability is increased. They respond to stimuli which are subthreshold for normal fibres and they may respond to a single stimulus with a repetitive discharge or even generate spontaneous trains of impulses. If motor nerve fibres are involved muscular contractions ensue which are abolished by tubocurarine but not by motor nerve section. The excitability of the central nervous system is also enhanced by hypocalcaemia.

Clinical features of hypoparathyroidism

The characteristic feature of hypoparathyroidism is *tetany*. In an overt form, tetany is accompanied by numbness and tingling of the extremities, a feeling of stiffness in hands and feet, cramps in the extremities, carpopedal spasm, laryngeal stridor (laryngismus stridulus) and generalized convulsions. Death may occur from asphyxia.

Carpopedal spasm may occur spontaneously or it may be brought on by compression of the nerves to a limb by inflation of a sphygmomanometer cuff (Trousseau's sign). In the upper limb the hand generally adopts the 'obstetric' position; the metacarpophalangeal joints are flexed, the fingers are extended, the thumb is drawn on to the palm, the wrist and elbow are flexed. In the lower limb the toes are plantar-flexed and the feet are drawn up.

Laryngeal stridor is due to the sudden contraction of the laryngeal muscles. The glottis is closed, no air can enter the chest and progressive cyanosis develops. After a variable period the spasm relaxes and air enters with a crowing sound.

Facial irritability (Chvostek's sign). Because of the heightened excitability of the nerves to mechanical stimulation, tapping over the facial nerve after its exit from the stylomastoid foramen results in contraction of the facial muscles.

Visceral manifestations of tetany may consist of intestinal colic, biliary colic or bronchospasm. Profuse sweating may also occur. All these effects could be due to increased excitability of autonomic ganglia.

Hypocalcaemia may be accompanied by changes in the electrocardiogram; the ST segment is prolonged and there may be an abnormal T wave.

With prolonged hypoparathyroidism cataracts are commonly seen. The calcium content of the lens is increased in parathyroidectomized animals.

Treatment of parathyroid deficiency

1. The symptoms of tetany can be temporarily relieved by injecting parathyroid hormone which raises the serum calcium, but as this calcium is withdrawn from the bones, such treatment is inadvisable for more than short periods in accidental parathyroidectomy, and is especially undesirable in patients operated on for parathyroid hyperplasia in whom the bones are already dangerously weakened.

2. The serum Ca may be raised temporarily by intravenous injection of soluble calcium salts, e.g. calcium gluconate.

3. The long-term treatment consists in promoting greater absorption of calcium from the intestine. Large doses of calcium (3 g daily) are given by mouth; in addition large doses of vitamin D are given (0.25–1.0 mg daily; 10 000–40 000 iu).

Causes of clinical tetany

Tetany can be produced by hypocalcaemia or by alkalaemia.

Hypocalcaemia. Tetany from hypocalcaemia occurs after parathyroidectomy, in rickets and in osteomalacia, and in renal failure with phosphate retention. It is the fall in *ionized* plasma calcium which affects nerve tissue; alterations in the plasma protein-bound calcium have no effect.

Alkalaemia. Alkalaemia produces tetany although the total calcium concentration in the blood is unchanged. Hyperventilation, profuse vomiting or the excessive ingestion of sodium bicarbonate may cause alkalaemic tetany. It is asserted that alkalaemic states cause an increase of plasma protein ionization and that this 'mops up' Ca^{2+}.

Clinical hyperparathyroidism

This condition may result from diffuse hyperplasia of all the parathyroids or from a localized tumour (adenoma or rarely carcinoma) of one of them. The symptoms of hyperparathyroidism are due to hypercalcaemia, renal stones, and bone disease.

A rise in plasma calcium level to above 3.0 mmol l^{-1} (12 mg per 100 ml) produces weakness, lassitude, loss of musclar tone, thirst, polyuria, anorexia, nausea, vomiting, constipation, and mental symptoms. Polyuria is due to distal renal tubular damage which results in diminished reabsorption of water. This leads to dehydration and thirst.

Renal calculi, consisting of calcium phosphate or oxalate, occur commonly, and may be the only manifestation of hyperparathyroidism. Deposition of calcium in the renal parenchyma (nephrocalcinosis) is also seen.

Nowadays most patients with hyperparathyroidism are diagnosed before evidence of bone disease occurs. The affected bones are painful, there is radiological rarefaction and spontaneous fractures occur, due to excessive resorption of bone (osteitis fibrosa).

Blood. The *plasma calcium* level is raised to 3–5.0 mmol l^{-1} (12–20 mg per 100 ml). The ionized calcium is raised and is responsible for the symptoms. Plasma inorganic phosphate may be reduced to less than 0.8 mmol l^{-1} (2.5 mg per 100 ml) but is often within the normal range of 0.8–1.4 mmol l^{-1} (2.4–4.5 mg per 100 ml). The plasma alkaline phosphatase is increased in those cases of hyperparathyroidism with radiological signs of bone disease.

The PTH level in peripheral venous blood (measured by radioimmunoassay) is usually raised, with still higher values in venous blood from the parathyroid region.

Urine. Excretion of calcium is increased, e.g. to 180 mg daily on a diet containing 120 mg daily or to over 400 mg daily on a normal diet (normal calcium excretion is about 100 mg daily). The hypercalciuria is secondary to hypercalcaemia.

In patients with parathyroid adenoma, without overt bone disease, removal of the tumour causes the plasma calcium level to fall, and the plasma phosphate to rise, so that both reach normal values within a few days. In patients with evident osteitis fibrosa the operation may at first reduce the plasma calcium to a level at which tetany occurs and the plasma phosphate level may fall too. These changes are due to a compensatory increase in osteoblastic activity with increased formation of calcified bone.

Calcitonin (thyrocalcitonin)

Although PTH is the most important hormone in the regulation of extracellular and intracellular $[Ca^{2+}]$ there is also a calcium-lowering hormone, *calcitonin*, which is secreted by the parafollicular or C cells of the thyroid gland (hence the name *thyrocalcitonin*, once used by many). The C-cells have been shown, by a fluorescence technique to be derived from the neural crest, from which also arise the cells which migrate to form the sympathetic ganglia and adrenal medulla. The C-cells migrate first to the ultimobranchial bodies, where they remain in fishes, amphibians, reptiles, and birds. In mammals the C-cells move on to the thyroid during embryonic development, and in man a few C-cells are found in the parathyroids and thymus.

Chemistry. The structure of porcine, bovine, salmon, and human calcitonin has been determined and the human hormone has been synthesized and shown to be biologically active. All types of calcitonin consist of 32 amino-acid residues, but with some differences in amino acid composition. All have a molecular weight of about 3000. Salmon calcitonin is more potent than mammalian calcitonin in man, perhaps because it is more slowly destroyed in the body.

Assay. Calcitonin is estimated by radioimmunoassay. The normal plasma concentration in man is about 200 ng l^{-1} An infusion of calcium raises plasma calcitonin level by 1.5–3-fold. In patients with medullary carcinoma of the thyroid, plasma calcitonin may rise to 500 μg l^{-1}. The half-life of human calcitonin in circulating blood is less than 15 minutes.

Factors affecting secretion of calcitonin. The only known physiological stimulus of calcitonin secretion is a rise in plasma Ca^{2+} level. In man the concentration of plasma calcitonin increases in direct proportion to a rise in plasma calcium concentration above 2.3 mmol l^{-1}.

Actions of calcitonin. Calcitonin inhibits resorption of bone by osteoclasts and osteocytes. It reduces plasma calcium when there is hypercalcaemia, and it reduces urinary excretion of hydroxyproline derived from bone. It increases urinary excretion of Na and phosphate. Calcitonin reduces the calcium concentration in the cytosol of bone cells, partly by increasing calcium efflux, and to a greater extent by increasing calcium uptake by mitochondria.

Physiological role of calcitonin. The role of calcitonin in physiological circumstances is still not certain, but it may help to prevent large rises in plasma calcium level by promoting calcium storage in bone. Its effects are much greater in young than in old animals, and it may play a part in remodelling of bone.

In medullary carcinoma of the thyroid plasma calcitonin concentration may be greatly increased (e.g. to 500 μg l^{-1}) but this in itself produces no clinical manifestations.

There is no clinical syndrome of calcitonin deficiency.

Clinical applications. Calcitonin is of value in treatment of hypercalcaemia due to parathyroid adenoma and vitamin D intoxication. Its major application is in treatment of patients with Paget's disease, in whom overgrowth of bones leads to skeletal deformity and considerable pain. Calcitonin administration relieves pain, restores normal bone structure, and reduces urinary hydroxyproline excretion.

Calcitonin is not of proved value in the treatment of osteoporosis.

VITAMIN D

Rickets

Rickets is a disease characterized mainly by bone deformities in young children. It was recognized by Galen (second century AD) and fully described in the seventeenth century by two English physicians, Whistler and Glisson. The origin of the word 'rickets' is obscure, but Glisson derived from it the artificial Latin word 'rachitis' which is incorporated in the description of vitamin D as 'antirachitic'. During the latter part of the nineteenth century and the early years of the twentieth century there was much argument as to whether rickets was due to a deficient diet or to other environmental influences. The claimed antirachitic effects of cod liver oil, or exposure to sunlight, appeared to confuse the issue. However, in 1919 Mellanby showed that cod liver oil prevented the development of rickets in puppies given a deficient diet, and in the same year Huldschinsky found that rickets could be cured or prevented by exposure of young children to artificial ultraviolet light. The common factor in these seemingly unrelated but clearcut results was the fat-soluble substance, named vitamin D, whose nature and properties will now be discussed.

Vitamin D exists in two forms, both formed by ultraviolet irradiation of steroid precursors:

Vitamin D_2 is formed from ergosterol which occurs in yeast and fungi.

Vitamin D_3 is more important. The precursor substance, 7-dehydrocholesterol, occurs in animal skin. Exposure of a child to ultraviolet light forms enough vitamin D_3 (cholecalciferol) to prevent rickets. Alternatively vitamin D_3 can be obtained from dairy produce (milk, egg-yolk, and butter), the content being greater in summer than winter. Fish liver oils (e.g. cod, halibut) are rich in cholecalciferol.

Vitamin D increases absorption of calcium (and phosphate) from the intestine and corrects the defects of bone development typical of rickets. It also enhances renal tubular reabsorption of Ca and P.

During the past ten years the development of new techniques in organic synthesis, chromatography, and structural identification, together with the use of radioactive vitamin D of very high specific activity, have led to the discovery of vitamin D metabolities in the body after the administration of physiological doses of the vitamin (Holick and De Luca 1974; Kodicek 1974). These more polar products, 25-hydroxycholecalciferol (25-HCC) and 1,25-dihydroxycholecalciferol (1,25-DHCC), particularly the latter, are more active than vitamin D itself. Indeed, 1,25-DHCC, which is finally formed in the kidney and released into the bloodstream, has all the characteristics of a hormone, of which vitamin D and 25OH cholecalciferol can be regarded as precursors. Vitamin D is still an essential dietary component for many people but its status as a hormone precursor means that given an adequate calcium dietary intake, calcium homeostasis in the body is ultimately controlled by hormones.

The sequence of events involved in the formation of 1,25-DHCC is set out overleaf. Starting with cholecalciferol (vitamin D_3) the first polar metabolite is formed by the activity of 25-hydroxylase, a microsomal enzyme in the liver. The compound 25-HCC is transported from the liver bound to an α-globulin of plasma. In the

7–Dehydrocholesterol
(provitamin D)

UV irradiation
of skin →

Cholecalciferol (vitamin D₃)

Hydroxylation by
microsomes in liver

1,25–Dihydroxycholecalciferol
(1,25DHCC)

← Hydroxylation
in kidney
(mitochondria)

25–Hydroxycholecalciferol (25HCC)

kidney 25-HCC is further hydroxylated by 1α-hydroxylase in mitochondria to form 1,25-DHCC. Parathyroidectomy reduces 1α-hydroxylase activity and PTH, AMP and reduction of blood phosphate level all enhance 1 α-hydroxylation (Kodicek 1974).

The hormone 1,25-DHCC acts on intestinal mucosal cells, on bone cells, on kidney cells, and perhaps also on skeletal muscle. In the gut and the kidney it promotes the passage of Ca through cells into the bloodstream. The actions of 1,25-DHCC have been more thoroughly studied on intestinal cells than on other cells. A cytoplasmic receptor protein and a nuclear receptor protein bind 1,25-DHCC and lead to the formation of mRNA which appears to become attached to a polysomal array of 10–11 ribosomes for translation into calcium binding protein (CaBP) in the intestinal mucosal cell. In addition to the formation of CaBP the hormone also increases Ca^{2+} translocation from lumen to cell interior by increasing alkaline phosphatase and a Ca dependent ATPase (?identical with alkaline phosphatase) in the brush border. Ca^{2+} in the cell are mopped up by mitochondria but the cytosolic CaBP has a higher affinity than mitochondria. According to Kodicek the translocation of Ca^{2+} across the serosal surface of the intestinal cell is mediated by an active energy-linked pump mechanism involving ATP and possibly connected with an influx of Na^+. These effects are summarized in FIG. XI.29. Other hydroxylated derivatives of vitamin D have been identified. Holick and DeLuca (1974) have shown that 1,25-DHCC is the vitamin D metabolite which promotes cal-

cium absorption when plasma [Ca] is < 2.4 mmol l⁻¹ (10 mg per 100 ml) but that with higher plasma calcium levels the major circulating metabolite in man is 24,25-DHCC, which is also synthesized in the kidney. Hypocalcaemia promotes secretion of PTH which acts on the kidney to enhance 1 α-hydroxylation of 25-HCC to form 1,25-DHCC. Hypophosphataemia also stimulates 1,25-DHCC production but by a process which is not dependent on PTH. When there is hypercalcaemia and/or hyperphosphataemia the formation

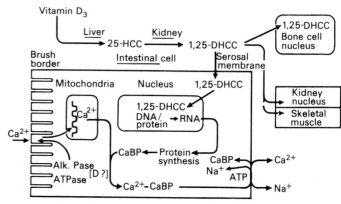

FIG. XI.29. Scheme of vitamin D action (also see text). [From Kodicek, E. (1974). *Lancet* i, 325.]

of 1,25-DHCC is inhibited and the formation of 24,25-DHCC is enhanced. The most important effect of the latter is to promote mineral precipitation on collagen fibrils when plasma is supersaturated with Ca^{2+} and inorganic phosphate.

Thus synthesis of the hormone 1,25-DHCC is regulated by complex feedback processes involving plasma concentrations of calcium, phosphate, and PTH.

Rickets

This disease of children is characterized by bones which are soft from deficient deposition of calcium salts and are therefore easily bent under the weight of the body, so that serious deformities may develop. Further, the process of ossification at the epiphysial line takes place in an abnormal manner. Normally the epiphysial line is a well-defined narrow strip of cartilage 2 mm deep, behind which regular ossification is proceeding. In rickets the epiphysial line forms a wide, irregular band, which can be felt as a marked projection on the surface. Normally the older cartilage cells degenerate and disappear, leaving many spaces into which the blood vessels and osteoblasts of the shaft can penetrate. It rickets this apparently essential preliminary degeneration does not occur and so ossification is retarded. The cartilage cells persist and go on multiplying, giving rise to a characteristic broad irregular cartilaginous zone. In addition, the matrix between the cartilage cells and that of the new bone itself does not become adequately impregnated with lime salts, accounting for the softness of the bones. The level of serum calcium or of phosphate or of both is lowered. Tetany may occur as a complication of rickets owing to the lowered serum calcium. Serum alkaline phosphatase is commonly raised.

Rickets usually sets in about the sixth month of life, and its intensity is directly related to the rapidity of bodily growth. The disease may last for several years with final healing. There is conclusive evidence that the disease is due essentially to lack of vitamin D.

Osteomalacia

Osteomalacia is a form of adult rickets, which is rare in Europe but occurs in the Orient where it is most likely to occur in women after multiple pregnancies and lactations. Many bones become soft and deformed and as in rickets the serum calcium and inorganic phosphate levels are lowered and alkaline phosphatase is raised.

In both rickets and osteomalacia there is an excess of osteoid tissue and a gross defect in mineralization. In both diseases administration of normal amounts of Ca and P in the diet, without vitamin D, are ineffective, but simple rickets and osteomalacia respond to vitamin D (+Ca and P) with rapid mineralization of the affected bones.

Osteoporosis is the commonest generalized disorder of bone, in which there is a reduction in all the constituents of bone. From the age of 40 years there is normally a progressive loss of skeletal mass, leading finally to senile osteoporosis with a liability to fractures (e.g. of the neck of the femur). In women osteoporosis occurs particularly after the menopause, and other predisposing factors are adrenal cortex hyperfunction (or corticosteroid therapy), hyperparathyroidism, hyperthyroidism, prolonged calcium deficiency, and prolonged physical immobilization. There is no simple method of preventing senile osteoporosis but physical exercise and a diet rich in calcium and vitamin D may slow the rate of skeletal rarefaction. Calcitonin is ineffective.

REFERENCES

CUTHBERT, A. W. (ed.) (1970). *Calcium and cellular function.* Macmillan, London.

DE LUCA, H. F. (1976). Parathyroid gland. In *Handbook of physiology*, Vol. VII, Section 7. American Physiological Society, Washington, DC.

HALL, R. *et al.* (1974). Parathyroid glands and calcium metabolism. In *Fundamentals of clinical endocrinology*, 2nd edn, p. 381. Pitman, London.

HEILBRUNN, L. V. (1927). *Arch. exp. Zellforsch* **4**, 246.

HOLICK, M. F. and DELUCA, H. F. (1974). Vitamin D metabolism. *A. Rev. Med.* **25**, 349.

IRVING, J. T. (1973). *Calcium and phosphorus metabolism.* Academic Press, New York.

KODICEK, E. (1974). The story of vitamin D from vitamin to hormone. *Lancet* i, 325.

MALLETTE, L. E. *et al.* (1974). *Medicine, Baltimore* **53**, 127.

RASMUSSEN, H. (1974). Parathyroid hormone, calcitonin and the calciferols. In *Textbook of endocrinology*, 5th edn (ed. R. H. Williams) p. 660. Saunders, Philadelphia.

RINGER, S. (1882). *J. Physiol., Lond* **4**, 29.

SIMKISS, K. (1974). Calcium translocation by cells. *Endeavour* **33**, 119.

TALMAGE, R. V. and MUNSON, P. L. (eds.) (1972). *Calcium, parathyroid hormone and the calcitonins.* Excerpta Medica, Amsterdam.

WASSERMAN, R. H. and CORRADINO, R. A. (1973). Vitamin D, calcium and protein synthesis. *Vitam. Horm.* **31**, 43.

WILLS, M. R. (1973). Intestinal absorption of calcium. *Lancet* ii, 820.

The thymus

The thymus at birth weighs 10–12 g; it increases in weight during childhood and in adolescence it weighs 20–30 g. Thereafter it atrophies progressively until in old age it weighs 3–6 g. In childhood the cortex and medulla are predominant but with increasing age they are replaced by connective tissue and fat.

The mammalian thymus is endodermal in origin, arising from the third branchial pouch. It is composed of a medulla and a cortex. The medulla develops first and consists of reticular-epithelial cells, a few lymphocytes and the concentric corpuscles of Hassall. The cortex is made up of actively multiplying, closely packed lymphocytes and contains no Hassall's corpuscles. The lymphocytes are arranged in packets surrounded by epithelial cells which form a barrier between the capillaries and the lymphocytes. The thymus is thus essentially a lymphoid organ and like other lymphoid tissue it undergoes atrophy in response to adrenal glucocorticoids.

Functions

Extirpation of the thymus from half-grown or fully-grown mammals causes no immediately obvious deficiency syndrome such as follows the removal of the recognized endocrine glands. However removal of the thymus of newborn mice, rats and rabbits has striking effects:

Lymphopenia and atrophy of all lymphoid tissue.

Failure to produce circulating antibody to antigens of bacteria, viruses and red cells.

A fatal wasting disease, partly due to increased susceptibility to infections.

Suppression of delayed hypersensitivity reactions.

Failure to reject foreign tissue transplants.

These effects are not produced by thymectomy in animals more than 2–3 weeks old.

The above described effects of neonatal thymectomy can be prevented by a thymus graft. The normalizing effects of such a graft still occur when the transplanted thymus is placed within a diffusion chamber which has pores large enough to allow the escape of chemical substances but too small to permit the escape of cells. Animals so treated gain weight, produce plasma antibodies, reject foreign skin grafts, and indeed behave like normal animals. It is presumed that the thymic reticular-epithelial tissue secretes a hormone which stimulates lymphopoiesis both within the thymus and in peripheral lymphoid tissue. The thymus thus promotes the development of immunologically competent T-lymphocytes (T = thymic) [p. 57].

It has further been shown that thymectomy in adult mice also leads to a decline in immunological capacity, but only after a lapse of 6–9 months, during which it is assumed that the existing pool of competent lymphocytes becomes gradually depleted. Thus the thymus not only initiates the development of immunologically competent cells in early life but is also responsible for maintenance of an adequate pool of such cells in adult life.

Lymphocyte migration. Labelled lymphocytes in a thymus graft migrate to the host's lymph nodes within a few weeks. Subsequently the graft receives lymphocytes from the bone marrow and after a while these cells also move on to the lymph nodes and spleen.

The lymphoid tissue within the thymus is not immunologically responsive; it lacks germinal centres and plasma cells. The lack of antibody formation may be due to a blood–thymus barrier preventing access of antigen to thymic lymphocytes.

To summarize the immunological role of the thymus:

It provides an environment favourable to lymphopoiesis. It receives precursors from the bone marrow which after development in the thymus pass on to the lymph nodes. By means of a hormone which acts locally in the thymus and also peripherally in lymph nodes it endows lymphocytes with the capacity to respond to antigens.

Thymic hormones. The reticular-epithelial cells secrete a hormone, *thymosin*, which promotes immunocompetence in young T-lymphocytes. It also produces lymphocytosis. It is a peptide with a molecular weight of about 1000.

The thymus also secretes a hormone, thymopoietin (thymin), which inhibits ACh release at motor nerve endings in myasthenia gravis.

Autoimmune diseases. In the 'autoimmune diseases' such as myasthenia gravis, haemolytic anaemia, and Hashimoto's thyroiditis there is hyperplasia of the thymus with the development of follicles with germinal centres like those which occur in normal lymph nodes.

REFERENCES

British Medical Journal (1974). Editorial. *Br. med. J.* **iii**, 75.
MILLER, J. F. A. P. and OSOBA, D. (1967). Immunological function of the thymus. *Physiol. Rev.* **47**, 437.
ROITT, I. M. (1977). *Essential immunology*, 3rd edn. Blackwell, Oxford.
TRANIN, N. (1974). *Physiol. Rev.* **54**, 272.
WOLSTENHOLME, G. E. W. and PORTER, R. (eds.) (1966). *Thymus. Experimental and clinical studies.* Ciba Foundation Symposium, London.

Local hormones

The role of chemical agents as internal secretions or chemical transmitters is well established. Techniques have been developed to reveal the conditions in which these substances are released from the structures in which they are formed. Endocrine glands liberate their secretions into the bloodstream which carries them to the organs and tissues on which they act. Cholinergic and adrenergic nerves release acetylcholine and noradrenalin respectively to influence the structures which they innervate. Living tissues contain many substances which may be activated in certain circumstances to exert profound effects in their immediate neighbourhood. Such substances have been called *local hormones*. It is often difficult to establish the physiological roles of these substances but some of them can undoubtedly contribute to pathological states.

Acetylcholine

In addition to its function as a peripheral cholinergic transmitter [p. 411] acetylcholine is formed in non-nervous tissues which contain choline acetyltransferase and cholinesterase. The functions of locally formed acetylcholine are to promote rhythmic activity in nerve-free smooth muscle (e.g. amniotic membrane), in heart muscle and in cilia of epithelial cells, e.g. in the oesophagus and trachea. All these effects are enhanced by an anticholinesterase such as eserine. Acetylcholine is also synthesized in the placenta (nerve-free) where its function may be to dilate blood vessels.

In the central nervous system cholinergic transmission is accepted in the case of the Renshaw cells [p. 283] and probably occurs at other sites too. However, it is still possible that its presence in nervous tissue is not directly connected with the transmission process.

Histamine

Histamine is found in both plant and animal cells. It is formed from the amino acid, L-histidine, by the action of the enzyme histidine decarboxylase as follows:

$$CH{=}C{-}CH_2{-}CH{-}NH_2 \quad \xrightarrow{-CO_2} \quad CH{=}C{-}CH_2{-}CH_2{-}NH_2$$

histidine
(4-imidazolyl—alanine)

histamine
(4-imidazolyl—ethylamine)

Histamine occurs in nettle stings, wasp and bee venoms in high enough concentrations to cause pain, itch, and the triple response.

In mammalian tissues histamine is widely distributed, but the highest concentrations occur in skin, intestine, and lung, i.e. at surfaces in contact with the outside world. The vascular actions of histamine may help to protect against invading bacteria.

Source of tissue histamine

Tissue histamine is formed by the action of histidine decarboxylase on histidine; the histamine which is formed *in situ* is retained in the tissue.

Bound histamine. The exact form in which histamine is held in the cells is not known, but it appears that most of the histamine in the body occurs in the mast cells of the tissues and the basophil cells and platelets of the blood. In the tissue mast cells histamine is associated with heparin, with which it appears to form a complex.

Within the mast cell histamine is found inside the metachromatic granules which correspond to the mitochondrial fraction separated by centrifugation. The intragranular histamine is readily released when the granule membrane is ruptured by physical or chemical means, and it then becomes physiologically active.

Histamine-forming capacity of tissues. The amounts of histamine which can be extracted from a tissue may be quite unrelated to the rate of formation of histamine in the tissue. Kahlson and his colleagues (1960) have studied the histamine-forming capacity of rapidly growing tissues in embryos and in healing wounds. They have measured the amount of [14C]-histamine produced from [14C]-histidine both *in vitro* and *in vivo* and have shown that although such tissues have a low content of histamine, the substance is being formed and continuously disposed of at a very high rate indeed. Kahlson has produced evidence which suggests that histamine formation may be specifically concerned with initiating and sustaining rapid tissue growth. This work throws quite a new light on the possible physiological role of histamine.

Actions of histamine

Histamine has important actions on the circulation, on smooth muscle, on secretions, and in producing itch or possibly pain.

Circulation. The circulatory actions of histamine vary according to the species. In some species, e.g. rabbit, arteriolar constriction may lead to an initial rise in blood pressure. In all species there is finally a profound fall in blood pressure. Histamine does not depress the heart, and coronary blood flow is increased. In animals the fall in blood pressure is due mainly to marked venular dilatation. Increased vascular permeability allows the escape of plasma into the tissues spaces; this reduces blood volume, produces haemoconcentration, and helps to lower the blood pressure still further.

In man histamine produces arteriolar as well as precapillary sphincter dilatation. Subcutaneous injection of 0.3 mg of histamine base causes general flushing of the skin, a rise of skin temperature of 1–2 °C, a small decline of systolic, a great fall of diastolic, pressure, and a rise of heart rate. Dilation of meningeal vessels stimulates their sensory nerve endings and produces headache.

The triple response to histamine in human skin is described on page 141.

Smooth muscle. Histamine increases the tone of most types of smooth muscle, e.g. intestine and bronchioles. Contraction of the guinea-pig ileum is used in the biological assay of the small concentrations of histamine found in tissues and body fluids.

Secretory actions. Histamine is a powerful stimulant to secretion of HCl by the stomach, and in man a moderate stimulant to secretion of pepsin. It may also stimulate salivary, pancreatic, and intestinal secretions, though to a lesser degree.

Itch and pain. When pricked into the skin, solutions of histamine produce itching, and when injected intradermally or applied to the exposed base of a blister it may produce itching or pain. Concentrations of 10^{-5} g ml^{-1} or higher can produce pain, whereas concentrations down to 10^{-7} g ml^{-1} produce itching only. It is probable that itching results from stimulation of nerve endings just below the epidermis, and pain from stimulation of deeper dermal nerve endings.

Histamine release in urticaria is associated with itching. It is possible that histamine might contribute to the pain and hyperalgesia of severe tissue damage.

Intravenous infusion of histamine induces headache in migranous patients but not in normal persons (Krabbe and Olesen 1980).

Release of histamine. Lewis was the first to suggest that release of histamine or 'H substance' in the skin accounted for the capillary dilatation, wealing, and arteriolar dilatation which make up the 'triple response'. This release was presumed to occur in response to physical stimuli, e.g. firm pressure and application of heat or cold, and also to chemical agents, such as morphine, and in the urticarial reaction which follows administration of certain proteins to allergic subjects.

Anaphylaxis

The actions of histamine resemble those of anaphylactic shock in various animal species. The anaphylactic state is readily induced by injection of foreign proteins such as horse serum and egg albumin into animals such as the guinea-pig, rabbit, or dog. The injected protein, the antigen, provokes the formation of antibodies (IgE) which circulate in the blood stream, some becoming attached to tissues. Two or three weeks after the initial injection of antigen the animal has become sensitized so that an intravenous injection of the specific antigen, which in a normal animal is innocuous, produces a profound and usually fatal anaphylactic shock, the features of which vary among different species. In guinea-pigs obstruction to breathing is the outstanding feature. The lumen of the bronchioles is narrowed by the combined effects of spasm of the smooth muscle, swelling of the mucosa and increased secretion. In rabbits pulmonary arterial constriction, and in dogs constriction of hepatic veins, both lead to profound falls in arterial blood pressure.

The different localizations of the anaphylactic phenomena are due to difference in concentration of fixed tissue IgE. When antigen reacts with tissue antibody histamine is released and will act mainly on immediately neighbouring structures. Histamine release has been demonstrated in the blood and lymph of animals in anaphylactic shock, and by addition of antigen to isolated tissue from sensitized animals.

Thus anaphylactic shock, which occasionally occurs also in man, is mainly due to release of histamine, though heparin (in dogs), and other substances may contribute to the picture. Anti-histamine drugs give good protection against anaphylactic shock.

Certain allergic states in man might be described as local forms of anaphylactic shock. *Urticaria* and *allergic rhinitis* (e.g. hay fever) are almost certainly due to histamine release. Bronchial asthma can be induced by histamine but it is unlikely that attacks are normally due mainly to histamine release. It is true that histamine is released when a specific allergen is added to lung tissue removed from an asthmatic patient but a *'slow reacting substance'* called *SRS-A* (*A* for anaphylaxis) is also released. *SRS-A* consists of peptidolipids derived from arachidonic acid (cf. prostaglandins [p. 559]). Its release in anaphylactic shock is much more prolonged than that of histamine and since it does not show tachyphylaxis it produces long-lasting bronchospasm [p. 60]. Its action is not antagonized by antihistamine drugs, which are of little value in the treatment of bronchial asthma. There are no known antagonists to *SRS-A*.

Prostaglandin F_{2a} may also contribute to the bronchospasm.

Chemical liberators of histamine

Numerous chemical compounds can release histamine from tissues. Among these are drugs such as morphine, pethidine, tubocurarine, and certain diamidines used in trypanosomiasis. A substance called 'Compound 48/80' (a condensation product of *p*-methoxyphenylmethylamine with formaldehyde) is the most powerful known histamine liberator and has been much used in experimental studies on the mechanisms of histamine release. Chemical liberators can release histamine from suspended intracellular particles, but anaphylactic release of histamine only occurs from intact cells. There are many stages involved in the anaphylactic response, beginning with union of antigen with tissue antibody, and ending with release of histamine. One of the intermediate stages is probably enzymic in nature (Mongar and Schild 1962).

Fate of histamine in the body

Small amounts of histamine are excreted via the kidney, partly as free, and partly as conjugated histamine. The conjugated form is N-acetyl histamine.

It is probable that in man the free histamine in urine is released from the tissues, and the acetylated histamine is absorbed from the gut where bacteria bring about conjugation. It can be calculated that 2–3 mg of histamine are released into the blood from the tissues every 24 hours, about 1 per cent appearing unchanged in the urine (Gaddum 1951).

Most of the remaining histamine is oxidized by a widely distributed enzyme known as histaminase or diamine oxidase, the histamine being converted to 4-imidazole acetic acid which has been recovered from the urine as such or as a ribose derivative. The physiological significance of the destruction of histamine by histaminase is attested by the fact that histaminase inhibitors, e.g. aminoguanidine, increase the amount of histamine excreted in the urine and also potentiate the pharmacological actions of histamine. This latter effect is analogous to the potentiating effects of anticholinesterases on the actions of acetylcholine.

In man there is normally very little histaminase in plasma but in pregnancy there is a marked increase, beginning at 2–3 months and reaching a peak at 6–7 months when the histaminase activity may be 1000 times normal. This phenomenon only occurs in the *human* female and its significance is obscure, since there is no reduction in the effects of injected histamine when the plasma histaminase level is very high.

Antihistamine drugs

Drugs may antagonize the actions of histamine in three main ways:

1. By acting in the opposite way to histamine. For example, isoprenaline acts on β-adrenergic receptors to relax bronchiolar muscle and in that way overcomes the increased tone produced by histamine acting on H_1 receptors.

2. H_1-receptor antagonists, e.g. mepyramine and promethazine combine with H_1-receptors and prevent histamine-induced contractions of the smooth muscle of the intestine and bronchi. These drugs often relieve allergic states in which H_1-receptors are activated, e.g. in urticaria and allergic rhinitis, but they are not effective in bronchial asthma because the muscle spasm in this condition is due largely to *SRS-A* [p. 000] and prostaglandin F_{2a}.

3. H_2-receptor antagonists, e.g. cimetidine, inhibit the gastric secretion of HCl evoked by histamine or pentagastrin. Cimetidine may also inhibit the secretion of pepsin and HCl induced by vagal stimulation. Cimetidine relieves pain in patients with duodenal ulcer, probably by reducing gastric secretion, especially at night.

REFERENCES

BLACK, J. W., DUNCAN, W. A. M., DURANT, C. J., GANELLIN, C. R., and PARSONS, E. M. (1973). *Nature, Lond.* **236**, 385.
British Medical Journal (1976). Editorial. *Br. med. J.* i, 789.
CARTER, D. C., FORREST, J. A. H., WERNER, M., HEADING, R. C., PARK, J., and SHEARMAN, D. J. C. (1974). *Br. med. J.* iii, 554.
DALE, H. H. (1948). *Br. med. J.* ii, 281.
GADDUM, J. H. (1951). *Br. med. J.* ii, 987.
HIRSHOWITZ, B. I. (1979). *A Rev. Pharmac.* **19**, 203.
KAHLSON, G. and ROSENGREN, E. (1968). *Physiol. Rev.* **48**, 155.
KRABBE, A. A. and OLESEN, J. (1980). *Pain* **8**, 253.
MONGAR, J. L. and SCHILD, H. O. (1962). *Physiol. Rev.* **42**. 226.
RILEY, J. F. (1959). *The mast cells.* Churchill Livingstone, Edinburgh.
ROCHA E SILVA, M. (1966). In *Handbook of experimental pharmacology*, Vol. XVIII. *Histamines and antihistamines*, Part 1. Histamine. Springer, New York.
SCHAYER, R. W. (1959). Catabolism of histamine in vivo. *Physiol. Rev.* **39**, 116.

5-Hydroxytryptamine (serotonin, enteramine)

5-Hydroxytryptamine (5HT) is another naturally occurring, biologically active amine, though its physiological role is not yet known.

Distribution

5HT is extracted from tissues by acetone and it may be estimated biologically (e.g. on the rat uterus) or by an ultraviolet fluorescence reaction.

Blood. 5HT occurs in the platelets and normally it is barely detec-

table in plasma. Human serum contains about 0.1 μg ml^{-1} of 5HT which is liberated from the platelets during clotting.

Urine. Urine contains 0.1–1.0 mg l^{-1} of 5HT

Alimentary tract. The alimentary tract contains 80–90 per cent of all the 5HT in the body (this explains the term 'enteramine'). It is present almost exclusively in the mucosa, probably in the so-called enterochromaffin cells. It appears that platelets obtain their 5HT while passing through the alimentary tract; there is no evidence that platelets can synthesize 5HT at any stage of their development.

Central nervous system. 5HT occurs in nervous tissue, for example in sympathetic ganglia, brain stem and hypothalamus. The distribution closely resembles that of noradrenalin.

Skin. 5HT occurs in rat's skin, but not in human skin. In the rat it is found in association with histamine and heparin inside mast cells.

5HT also occurs in the skin of certain toads, frogs, and salamanders.

Venoms. 5HT occurs in toad venom in company with bufotenine (N, N-dimethyl-5-hydroxytryptamine) and in high concentration in wasp venom and scorpion venom.

Plant sources. 5HT is found in nettle stings and also in itch powder (cowhage, *Mucuna pruriens*). Bananas are rich in 5HT (and noradrenalin).

Formation and metabolism

The formulae below show how 5HT is formed from tryptophan. The first step consists of oxidation of tryptophan to 5-hydroxytryptophan by an enzyme, tryptophan hydroxylase. This enzyme is inhibited by p-chlorophenylalanine. The next step is decarboxylation of 5-hydroxytryptophan to 5-hydroxytryptamine by means of the enzyme 5-hydroxytryptophan decarboxylase which is present in the alimentary tract, kidney, liver, and nervous system.

Like other naturally occurring amines 5HT is oxidized in the body. This process is brought about by monoamine oxidase and the oxidized product is 5-hydroxyindole acetic acid (5HIAA) which is biologically inert and is excreted in the urine.

Actions

5HT has widespread actions on the circulation, respiration, kidneys, smooth muscle, and on the nervous system.

Circulation. Acting directly, 5HT is a cardiac stimulant and a vasoconstrictor (this is why it is called 'serotonin'). However, the circulatory response in intact animals is modified by many compli-

Tryptophan

5-Hydroxytryptophan (5-HTP)

5-Hydroxytryptamine (5-HT)

5-Hydroxyindole acetic acid (5-HIAA)

cating factors and there are marked species variations. After intravenous injection there is usually a pressor phase, due to peripheral vasoconstriction and cardiac stimulation, but reflex hypotension and bradycardia and peripheral and ganglionic inhibition of vasoconstrictor nerve tone may overcome the tendency to a rise in blood pressure.

In man intravenous injection of 5HT raises systolic and diastolic blood pressure and produces tachycardia. The sites of origin of the reflex circulatory and respiratory effects of 5HT may be carotid chemoreceptors, and afferent endings in the great veins and left atrium. Veins are strongly constricted by 5HT.

Respiration. Reflex apnoea and hyperpnoea can occur, according to the species studied. In man hypernoea is seen.

5HT increases bronchiolar tone by direct action and perhaps also reflexly. In asthmatic subjects it can induce an acute attack of asthma.

Kidney. 5HT has an antidiuretic action which is probably due to afferent glomerular arteriolar constriction. This may be brought about reflexly or by direct action. Ureteric spasm may also stop urine flow temporarily.

Smooth Muscle. 5HT generally increases the tone of smooth muscle, e.g. arterioles, bronchioles, bladder, intestine, uterus, pupil, and nicitating membrane.

It tends to cause evacuation of the bowels. When introduced into the lumen of the intestine 5HT acts reflexly to stimulate peristalsis (Bülbring 1961). The mucosal store of 5HT may be related to peristalsis. A rise of intraluminal pressure releases free 5HT which sensitizes the mucosal pressure receptors so that reflex peristaltic activity is enhanced.

Nerve endings and fibres. Apart from stimulation of chemoreceptors to influence circulation and respiration, 5HT probably acts on the gut by stimulation of cholinergic nerves in the intestinal ganglia.

5HT is also a very potent stimulant of pain nerve endings in human skin, acting in concentrations of 10^{-6}–10^{-8} g ml^{-1}. Release of 5HT from blood platelets after injury, intravascular thrombosis and pulmonary or cardiac infarction could cause pain and reflex respiratory and circulatory effects. 5HT potentiates the algogenic action of bradykinin.

Central actions. 5HT occurs in some nerve terminals in the brain (serotoninergic nerves). When applied by micro-ionotophoresis 5HT excites some neurons and inhibits others. By intravenous injection 5HT has no central nervous actions because it does not cross the blood-brain barrier. Injection of 5HT into the cerebral ventricles induces a lethargic state; 5-hydroxytryptophan, which penetrates the blood–brain barrier and is decarboxylated to form 5HT, induces sleep. These effects, and behavioural responses to reserpine, which depletes brain tissue of 5HT, and to monoamine oxidase inhibitors, which increase the local concentration of 5HT, suggest that serotoninergic neurons may have important physiological effects on mood and behaviour. However, catecholamines, mostly noradrenalin, occur in the same parts of the brain as 5HT and it is difficult to distinguish the roles of adrenergic and serotoninergic neurons. Reserpine and MAO inhibitors have the same effects on noradrenalin as on 5HT.

A serotoninergic pathway which runs caudally from the dorsal raphe nucleus mediates the analgesic action of morphine.

With respect to their effects on body temperature regulation by the hypothalamus 5HT and noradrenalin are antagonists [p. 352]. Feldberg and Myers (1964) showed that injection of 5HT into the lateral ventricle of an unanaesthetized cat raised body temperature, whereas intraventricular injection of noradrenalin lowered body temperature. In the rabbit the converse relationship holds, 5HT lowering, and noradrenalin raising, body temperature. Other species generally behave like the cat or the rabbit and there is evidence that changes in environmental temperature cause the appropriate changes in activity of anterior hypothalamic serotoninergic and adrenergic neurons.

REFERENCES

BRADLEY, P. B. and ELKES, J. (1957). *Brain* **80**, 77.
BÜLBRING, E. (1961). The intrinsic nervous system of the intestine and local effects of 5-hydroxytryptamine. In *Regional neurochemistry* (ed. S. S. Kety and J. Elkes) p. 437. Pergamon, Oxford.
—— and CREMA, A. (1959). *J. Physiol., Lond.* **146**, 18; 29.
FELDBERG, W. and MYERS, R. D. (1964). *J. Physiol., Lond.* **173**, 226.
ROBSON, J. M. and STACEY, R. S. (1962). *Recent advances in pharmacology*, 3rd edn, p. 122. Churchill Livingstone, London.
UDENFRIEND, S., SHORE, P. A., BOGDANSKI, D. F., WEISSBACH, H., and BRODIE, B. B. (1957). *Recent Prog. Horm. Res.* **13**, 1.

Prostaglandins

In the 1930s human semen was found to act on uterine muscle *in vitro* and to lower blood pressure *in vivo*. Because it was thought that the active substance came from the prostate the name *prostaglandin* was introduced by Von Euler in 1937. Actually the seminal vesicles are the chief source of prostaglandins in semen. Subsequently lipid-soluble extracts of many organs were shown to possess smooth muscle-stimulating properties. Bergstrom and his colleagues first isolated prostaglandins E and F (PGE, PGF) in 1960 from sheep prostate and since then the combined use of labelled precursor substances, thin-layer chromatography, gas–liquid chromatography, mass spectrometry, and radioimmunoassay has led to identification of numerous natural prostaglandins, and knowledge of their chemical structure has made possible the synthesis of both naturally-occurring PGs and structural analogues of these compounds.

The immediate precursors of PGs in the body are essential unsaturated fatty acids, e.g. linoleic and arachidonic acids. The PGs formed from these consist of a 20-carbon carboxylic acid containing a cyclopentane ring (a hypothetical structure called prostanoic acid). The commonest prostaglandins (PGA$_1$, PGA$_2$, PGE$_1$, PGE$_2$, PGF$_{1\alpha}$, PGF$_{2\alpha}$) are formed by various types of substitution and degrees of unsaturation in both cyclopentane ring and aliphatic side chains, enzymes being involved at each stage.

The intermediate compounds formed during the biosynthesis of PGs have been shown to have very important actions of their own (Moncada and Vane 1978). The main sequence of biochemical synthetic reactions is as follows:

Arachidonic acid in membrane phospholipids

$\downarrow$ phospholipase A$_2$

Arachidonic acid

$\downarrow$ cyclo-oxygenase (prostaglandin synthetase)

Cyclic endoperoxides

PGG$_2$

Prostacyclin (PGI$_2$) ←—— PGH$_2$ ——→ Thromboxane A$_2$ (TXA$_2$)

$\downarrow$

Thromboxane B$_2$ (TXB$_2$)

PGE$_2$ PGF$_{2\alpha}$

Aspirin, indomethacin, and other non-steroidal anti-inflammatory drugs inhibit cyclo-oxygenase and thereby inhibit the formation of endoperoxides PGG_2 and PGH_2 and their successors. The endoperoxides are unstable (half-life = 5 min) and partly isomerize to form stable prostaglandins such as PGE_2 and $PGF_{2\alpha}$. In addition the cyclic endoperoxides are metabolized to highly unstable compounds, Thromboxane A_2 (half-life = 30 s) and prostacyclin.

The formation of thromboxane A_2 occurs in blood platelets; TXA_2 is a very potent inducer of platelet aggregation and also causes arterial vasoconstriction. TXA_2 is converted spontaneously to the stable biologically inactive compound TXB_2.

Prostacyclin occurs most abundantly in vascular endothelium. It is a highly potent inhibitor of platelet aggregation and is also a vasodilator.

Thus when endothelial damage tends to produce platelet adherence and aggregation [p. 30] by promoting TXA_2 formation in platelets, the extent of these phenomena will be limited by the simultaneous formation of prostacyclin in the damaged endothelium. The antagonistic roles of TXA_2 and prostacyclin in haemostasis and thrombosis remain to be evaluated, as also do the actions of prostaglandins in general. Indeed when one considers that the physiologico-pathological role of the single substance histamine is not fully understood over 50 years after detection of its presence in tissues, it is not surprising that the biological functions of an unknown number of prostaglandins, their precursors and metabolites, have still to be assessed. The main characteristics of prostaglandins will now be discussed.

Metabolism of prostaglandins

PGEs and PGFs are very quickly destroyed in the lungs; it is probable that they are true 'local hormones' acting at the sites of their synthesis and release. PGAs are more stable and their circulation to produce splanchnic vasodilation may help to keep arterial blood pressure within normal limits.

Urinary metabolites of PGEs and PGFs give information about biosynthesis and release of PGEs and PGFs in the body. The activating factors detected in this way include:

1. Cold stress in rats leads to increased excretion of metabolites of PGE_1 and PGE_2.
2. Scalding in the guinea-pig gives evidence of PGE_1 and PGE_2 release.
3. Anaphylactic shock in guinea-pigs is associated with increased $PGF_{1\alpha}$ and $PGF_{2\alpha}$ metabolite excretion.
4. In human pregnancy there is evidence of increased $PGF_{1\alpha}$ and $PGF_{2\alpha}$ release with a further upsurge during labour. Aspirin and indomethacin prevent the increase in urinary metabolite excretion and delay the onset of labour.

Estimation of prostaglandins in tissues. It is important to realize that even gentle handling of tissues promotes synthesis and release of PGs. This applies also to PGs in blood platelets.

Site of action of prostaglandins on cells. The receptors for PGs are located in the plasma membrane; interaction between PGs and their receptors can affect the activity of adenyl cyclase situated in the inner part of the plasma membrane.

Biological actions

The numerous prostaglandins (and their endoperoxides) act on most organs and tissues of the body. The non-vascular actions of prostaglandins are related to cyclic AMP but the vascular actions are independent of cAMP.

Cardiovascular system. PGA_1 and PGA_2 cause peripheral arteriolar dilatation, especially in the splanchnic vascular bed. PGA_2,

isolated from renal medulla and called 'medullin' has an antihypertensive action and also increases renal cortical blood flow which is associated with enhanced urinary excretion of sodium, potassium, and water (Lee 1974).

PGEs produce local vasodilation when given by intra-arterial injection.

Reproductive system. PGEs and PGFs (but not PGAs) act on the female reproductive system:

1. They stimulate contractions of the gravid uterus. $PGF_{2\alpha}$ occurs in amniotic fluid; its concentration in this fluid and in blood rises during labour, suggesting that in women $PGF_{2\alpha}$ may initiate the onset of labour. $PGF_{2\alpha}$ is also found in menstrual fluid where its presence may cause painful uterine contractions (dysmenorrhea).
2. They produce luteolysis and reduced secretion of progesterone in some animal species (e.g. sheep), but this effect has not been clearly demonstrated in women.
3. They promote secretion of hypothalamic gonadotrophin-releasing hormone (GnRH).

Blood platelets. PGE_1 is a very active inhibitor of platelet aggregation. Its action is due to activation of adenyl cyclase, inhibition of phosphodiesterase and subsequent increase in intracellular cAMP concentration.

PGE_2 enhances the ADP-induced aggregation of platelets, acting on the secondary phase. The endoperoxides PGG_2 and PGH_2 also promote platelet aggregation. Aspirin and indomethacin inhibit the conversion of arachidonic acid to endoperoxides.

Platelets normally contain PGE_2 and $PGF_{2\alpha}$. The latter has no effect on aggregation. Thrombin promotes PGE_2 synthesis and release and hence produces platelet aggregation.

Inflammation

Local. PGE and PGA enhance the vascular permeability increase caused by histamine. PGE_1 lowers the threshold of human skin to histamine-induced itch; PGEs also sensitize cutaneous nerve terminals to the pain-producing action of bradykinin (an effect which is inhibited by aspirin). Higher concentrations of PGEs can produce pain directly.

Many types of inflammatory exudate contain prostaglandins which may contribute to the production of itch, pain, vasodilation, increased vascular permeability, and cellular infiltration.

Systemic. Inflammation may produce systemic effects such as headache and fever. Since both these manifestations are relieved by aspirin, prostaglandins may be involved in the production of these symptoms. Feldberg and his colleagues have shown that PGE_1 and PGE_2 produce fever when injected into the cerebral ventricles of the cat, and that bacterial pyrogen causes release of PGE_2 into the CSF. Aspirin inhibits the febrile response to bacterial pyrogen (Feldberg *et al.* 1973).

However, it must be said that PGs are not the sole cause of fever and that their role in both local and systemic manifestations of inflammation requires further study.

Bronchial musculature. The smooth muscle of the bronchi and bronchioles is relaxed by PGEs but contracted by $PGF_{2\alpha}$, which inhibits the bronchodilator action of isoprenaline. $PGF_{2\alpha}$ may contribute to the production of bronchial asthma.

Gastro-intestinal system. Intravenous injection of PGE_1, E_2, or A_1 inhibits the secretion of gastric HCl induced by histamine, pentagastrin, or ingestion of food.

PGEs and $PGF_{2\alpha}$ inhibit the absorption of sodium and water to such an extent that a profuse, watery, cholera-like diarrhoea occurs. PGE₁ and cholera toxin both increase intestinal mucosal adenyl cyclase and cAMP; cholera toxin may act by releasing PGE₁.

PGE and PGF increase intestinal motility *in vitro* and *in vivo*; but it is not known whether they control normal peristalsis.

Metabolic actions. *In vitro*, PGE₁ inhibits the *lipolysis* induced by ACTH, GH, glucagon, and adrenalin, probably by inhibiting adenyl cyclase. The effects of PGs on lipolysis *in vivo* are variable.

Central nervous system. PGs occur in the central nervous system and may function as transmitters or modulators of neuron activity.

Autonomic nervous transmission. PGEs inhibit neurally in-duced release of NA and the responses to NA. They also influence cholinergic neuroeffector junctions, mainly producing stimulation.

The eye. PGE₂ and PGF$_{2\alpha}$ occur in the iris. Their release causes miosis.

To summarize, prostaglandins appear to have important biolo-gical actions but much remains to be learned about their physiolo-gical roles.

REFERENCES

(See the journal *Prostaglandins* for recent work.)
FELDBERG, W., GUPTA, K. P., MILTON, A. S., and WENDLANDT, S. (1973). *J. Physiol., Lond.* **234**, 279.
HEDQVIST, P. (1977). *A. Rev. Pharmac.* **17**, 259.
HORTON, E. W. (1972). *Prostaglandins.* Springer, Berlin.
—— (1979). Prostaglandins and smooth muscle. *Br. med. Bull.* **35** No. 3, 295.
LEE, J. B. (1974). The prostaglandins. In *Textbook of endocrinology*, 5th edn (ed. R. H. Williams) p. 854. Saunders, Philadelphia.
KADOWITZ, P. J., JOINER, P. D., and HYMAN, A. L. (1975). *A. Rev. Phar-mac.* **15**, 285.
MITCHELL, J. R. A. (1981) Prostaglandins in vascular disease: a seminal approach. *Br. med.* **282**, 590.
MONCADA, S. and VANE, J. R. (1978). *Br. med. Bull.* **34**, 129.
SAMUELSSON, B., GRANSTROM, E., GREEN, K., HAMBERG, M., and HAMMER-STROM, S. (1975). *A. Rev. Biochem.* **44**, 669.
VANE, J. R. (1971). *Nature, New Biol.* **231**, 232.

Adenosine derivatives

These occur in all cells and take part in the transfer and storage of energy [p. 446]. They relax smooth muscle, slow the heart and dilate blood vessels. ATP excites ganglion cells, ADP promotes agglutination of platelets and all the adenosine phosphates arouse pain when applied to an exposed blister base in man, in concentra-tions of 10^{-6} and above.

Plasma polypeptides

In addition to local hormones liberated from cells and tissues there are some highly active polypeptides which can be formed from blood plasma, lymph and extracellular fluid.

Angiotensin. This is formed by the action of renin on angiotensi-nogen, an α₂-globulin of plasma [p. 7]. Angiotensin causes vaso-constriction, raises blood pressure, releases catecholamines from the adrenal medulla and promotes secretion of aldosterone from the adrenal cortex.

Plasma kinins. These are polypeptides consisting of 9–11 amino acid residues. *Bradykinin* is a nonapeptide and *kallidin* a decapep-tide, both being formed from kininogen, an α₂-globulin as shown in FIGURE XI.30. It can be seen that the kinin-forming system can be activated in the same way as the blood-clotting system [p. 23]. Contact with foreign surfaces such as glass, kaolin, and urate crys-

tals converts factor XII from its precursor to its active form which then initiates reactions which lead to clotting, kinin formation and some degree of fibrinolysis. Kinin formation is also promoted by tissue injury, trypsin, proteolytic enzymes in snake venoms, and enzymes called kallikreins in pancreas, salivary glands and urine. The kinins formed in plasma are quickly destroyed by the enzyme kininase, which is always present in an active form.

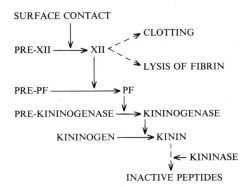

PF = PERMEABILITY FACTORS
PRE-KININOGENASE = PRE-KALLIKREIN
KININOGENASE = KALLIKREIN

FIG. XI.30. The plasma kinin-forming system.

The plasma kinins cause contraction of most types of smooth muscle, e.g. ileum, uterus, bronchiole, but relaxes the rat duode-num. Minute concentrations cause vasodilation and a fall in arterial blood pressure; kinins increase vascular permeability by an action on venules which allows the escape of plasma proteins; they excite sensory nerve endings, causing pain when applied to the skin or injected into an artery or the peritoneum. In high concentrations plasma kinins promote migration of leucocytes from blood to tissue.

Plasma kinins probably contribute to the vascular and sensory phenomena of inflammation; they are formed by antigen–antibody interactions and may thus help to produce bronchoconstriction in bronchial asthma.

Endocrine aspects of carcinoid syndrome

Carcinoid tumours are slow-growing neoplasms of either the en-terochromaffin cells of the alimentary tract or the corresponding cells in other organs such as the pancreas, thyroid, and bronchus.

The most characteristic features of the carcinoid syndrome are episodes of intense flushing of the skin, which spreads downwards from the face and neck, hypertension, abdominal pain, diarrhoea, wheezing (due to bronchoconstriction) and signs of valvular lesions of the right side of the heart. Attacks may be provoked by the taking of alcohol, by exposure to cold and by other circumstances which lead to increased sympathetic activity and secretion of adre-naline. The picture is analogous to that associated with phaeochro-mocytoma and it is clear that the tumour releases one or more humoral agents.

5HT. The first such agent to be suspected was 5HT which is known to be synthesized in normal enterochromaffin tissue, and whose physiological function may be to promote peristalsis. Carcinoid tumours from all sites contain tryptophan-5-hydroxylase which can hydroxylate large amounts of L-tryptophan, to form 5-hydro-xytryptophan (5HTP), which is then decarboxylated by aromatic L-amino acid decarboxylase in the tumour to form 5HT. Intestinal tumours store large amounts of 5HT, e.g. 3 mg g⁻¹. The tumours contain variable amounts of monoamine oxidase (MAO) and may secrete either 5HT itself or its oxidation product, 5HIAA, which is then excreted in the urine. In normal persons, on a diet poor in 5HT, the daily urinary excretion of 5HIAA is 1–10 mg; in patients with carcinoid tumour the daily excretion of 5HIAA ranges from 50 to 1000 mg with only 1–2 mg of 5HT. In tumours from organs derived from the embryonic foregut (pancreas, bronchus, thyroid) there is no decarboxylase and in such cases quite large amounts of 5HTP also appear in the urine.

It is unlikely that 5HT contributes much to the flush. Intravenous injection of 5HT does not reproduce typical attacks of flushing, and there is generally little correlation between free plasma 5HT levels and episodes of flushing. It may help to produce bronchoconstriction, but the symptom for which 5HT is most likely to be responsible is diarrhoea. This view is supported by the fact that the 5HT antagonist, methysergide, relieves the diarrhoea but not the flush or other symptoms. The drug *p*-chlorophenylalanine inhibits tryptophan 5-hydroxylase and thereby reduces the synthesis of 5HT. It too, relieves diarrhoea.

Bradykinin. Intravenous injection of small doses of adrenalin evokes the flush, though clearly not by a direct action, which would produce cutaneous pallor. Oates *et al.* (1964) showed that adrenalin acts by promoting the formation of bradykinin, probably by a direct effect on the enterochromaffin cells leading to release of kallikrein, a tissue kininogenase which acts on kininogen substrate to form bradykinin. Large increases in kallikrein and bradykinin levels in hepatic venous blood, and smaller increases of these substances in brachial artery blood, were recorded during a flush. Intravenous infusion of bradykinin in carcinoid patients produces a flush similar to that seen in spontaneous attacks; it also causes hypotension.

Other substances. Flushing in carcinoid syndrome might also be caused by release of ATP or one of the prostaglandins. Histamine release has been demonstrated in a few cases. Thus, the diarrhoea and abdominal pain of carcinoid syndrome are most probably due to local release of 5HT, and perhaps also bradykinin. The flush is mainly attributable to bradykinin release. The severe headaches which occur in some cases could be due to the actions of 5HT, bradykinin, or prostaglandins. Bronchoconstriction could be brought about by 5HT or bradykinin. The cause of the cardiac lesions in carcinoid syndrome is not known. Carcinoid tumour tissue behaves rather like the venom sacs of poisonous animals in the nature and variety of biologically active substances it manufactures.

The dumping syndrome. This syndrome comprises attacks of nausea, vomiting, and diarrhoea, associated with dizziness and pallor followed by flushing, tachycardia, sweating, and weakness, which occur after a meal in patients who have undergone gastric surgery. The attacks resemble those of carcinoid syndrome but there is no evidence of any abnormality of 5-hydroxyindole metabolism. Zeitlin and Smith (1966) have shown that in attacks experimentally provoked by ingestion of hypertonic glucose solution there was no increase in blood 5HT levels or in urinary excretion of 5HIAA; there was, however, a considerable rise in venous free plasma kinin and a simultaneous fall in kininogen level during the period of flush. Distension of the gut by hypertonic glucose might cause release of kallikrein from enterochromaffin cells.

REFERENCES

GRAHAME-SMITH, D. G. (1972). *The carcinoid syndrome*. Heinemann, London.
The Lancet (1973). Editorial. *Lancet* **ii**, 711.
OATES, J.A. and BUTLER, T. C. (1967). *Adv. Pharmac.* **5**, 109.
ZEITLIN, I J. and SMITH, A. N. (1966). *Lancet* **ii**, 986.

General references on local hormones

BERGSTRÖM, S., CARSON, L. A., and WEEKS, J. R. (1968). The prostaglandins. *Pharmac. Rev.* **20**, 1.
CURTIS, D. R. (1968). Pharmacology and neurochemistry of mammalian central inhibitory processes. In *Proceedings of 4th International Meeting of Neurobiologists*, p. 429. Pergamon, Oxford.
DALE, H. H. (1953). *β*-Iminazolyl-ethylamine (histamine). In *Adventures in physiology*. Pergamon, London.
ERSPAMER, V. (1966). *Hydroxytryptamine and related indolealkylamines*. Springer, Berlin.
GARRATTINI, S. and SHORE, P. A. (eds.) (1968). Biological role of indolealkylamine derivatives. *Adv. Pharmac.* **6**, Part B.
KAHLSON, G. (1965). Histamine. *A. Rev. Pharmac.* **5**, 305.
KEELE, C. A. and ARMSTRONG, D. (1964). *Substances producing pain and itch*. Arnold, London.
LEWIS, G. P. (1968). Pharmacologically active polypeptides. In *Recent Advances in pharmacology* (ed. J. M. Robson and R. S. Stacey) p. 213. Churchill, London.
PEART, W. S. (1965). The renin-angiotensin system. *Pharmac. Rev.* **17**, 143.
ROCHA E SILVA, M. (1966). Histamine and antihistamines. *Handb. exp. Pharmak.* Vol. 18.

PART XII
Physiology of reproduction

The physiological processes involved in human reproduction enable a spermatozoon to penetrate an ovum and produce fertilization, which leads to the development, and finally birth, of a new human being, who is almost always genotypically and phenotypically male or female. To achieve this purpose male and female reproductive systems have anatomical structures whose function is to provide the opportunities for spermatozoa and ova to meet at the right moment in the right place (the Fallopian tube) and for the fertilized ovum to develop to a stage where it moves into the uterus to become embedded in the hormonally prepared endometrium in which the placenta is formed so that the mother can provide the growing fetus with the materials and conditions it needs for nine months of differentiation and growth. After childbirth the mother should continue to provide nourishment by breast-feeding for a further period of 6–9 months.

All this sounds very simple but in fact the growth, function, and regulation of the male and female reproductive systems are very complex. Although much has been discovered about these systems it is clear that much more remains to be learnt, particularly about the processes of control and integration of the activities of components of reproductive systems.

An understanding of the physiology of reproduction is important for the promotion of conception, fetal growth, birth, and early development of healthy children. The need for population control throughout the world has stimulated research on the best ways to prevent conception without harming the mother (or father). Reproduction is a subject with anatomical, physiological, biochemical, genetic, psychological, sociological, economic, political, moral and other aspects. Our emphasis is naturally on physiological aspects.

The organs of reproduction are mainly controlled by hormones secreted by the gonads, the adenohypophysis, the hypothalamus, the placenta, and to a lesser degree by hormones from the adrenal cortex and thyroid. Recent advances in assay techniques—e.g. bioassay, radioimmunoassay, competitive binding assay—have provided much accurate information about blood and body fluid concentrations of follicle-stimulating hormone (FSH), luteinizing hormone (LH), prolactin, gonadal steroids, and placental hormones.

General references

AUSTIN, C. R. and SHORT, R. V. (eds.) (1972–9). *Reproduction in mammals*, 7 Vols. Cambridge University Press.

COHEN, J. (1977). *Reproduction*. Butterworths, London.

HAFEZ, E. S. E. (1978). *Human reproductive physiology*. Ann Arbor Science Publishers, Michigan.

JOHNSON, M. and EVERITT, B. (1980). *Essential reproduction*. Blackwell, Oxford.

ODELL, W. D. and MOYER, D. L. (1971). *Physiology of reproduction*. Mosby, St. Louis.

RHODES, P. (1969). *Reproductive physiology for medical students*. Churchill Livingstone, Edinburgh.

SADOW, J. I. D., GULAMHUSEIN, A. P., MORGAN, M. J., NAFTALIN, N. J., and PETERSEN, S. A. (1980). *Human reproduction*. Croom Helm, London.

SHEARMAN, R. P. (ed.) (1972). *Human reproductive physiology*. Blackwell, Oxford.

SHORT, R. V. (ed.) (1979). Reproduction. *Br. med. Bull.* **35**, No. 2.

Sex determination and sex differentiation

Sexual development in the embryo involves two processes, sex determination and sex differentiation.

Sex determination is a genetic phenomenon and depends on the constitution of the sex chromosomes. What is determined is the *sex genotype* or *genetic sex* which almost invariably accords with the apparent sex of the fully-developed individual. Oogonia and spermatogonia each contain 46 chromosomes. The former contain 44 autosomes + two identical sex chromosomes XX which carry female-determining genes. Spermatogonia contain 44 autosomes + XY where Y is a chromosome carrying male-determining genes. When oocytes undergo reduction division, halving of the chromosome number occurs so that each ovum finally contains 22 + X. When spermatocytes undergo reduction division there are two kinds of resulting secondary spermatocytes (which become spermatozoa). One type contains 22 + X and the other 22 + Y. The ovum may be fertilized by either kind of sperm with the following sex genotypes:

Sperm X + Ovum X = Offspring XX = Female
Sperm Y + Ovum X = Offspring XY = Male.

Thus genetic sex is determined exclusively by the sperm and is quite independent of the ovum. The sex-determining genes on the X and Y chromosomes act on the primitive bipotential gonad to promote its development as a testis or an ovary.

The nuclei of somatic cells in the female contain a visible chromatin mass about 1 μm in diameter which lies against the inner surface of the nuclear membrane [FIG. XII.1]. This *sex chromatin* (Barr body) is thought to result from activation of an X chromosome. The XY chromosome pair of the male forms no detectable mass. Thus the sex genotype can be identified by a cytological test, the most suitable cells being the epithelial cells of the epidermal spinous layer, buccal mucosa or vagina, or the blood leucocytes. Chromosomal sex can be more accurately determined by *karyotyping*.

Sex differentiation in the embryo usually harmonizes with the genetic sex but environmental influences can lead to discordant development so that apparent sex may not accord with genetic sex.

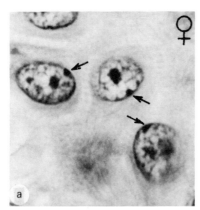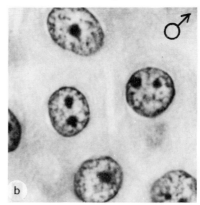

FIG. XII.1. (a) Nuclei in epidermal spinous cell layer of a chromosomal female. The sex chromatin is indicated by the arrows. (b) Spinous cell nuclei, lacking sex chromatin, in a chromosomal male. (H & E×1800.)
(Grumbach, M. M. and Barr, M. L. (1958). *Recent Prog. Horm. Res.* **14**, 255.)

The important feature of sex differentiation is that masculine characteristics of the body have to be imposed by the fetal testicular hormones against a basic feminine trend of the body. Female organogenesis results from the mere absence of testes. Males differentiate very early, females very late (Jost *et al.* 1973).

Sex differentiation takes place in stages:

Gonadogenesis. Primordial germ cells migrate into the urogenital ridge where proliferation of both non-germinal and germinal cells leads to the formation of the primitive gonads which are identical in both sexes up to 42 days of gestation. In the *male*, testicular differentiation begins at 43–50 days with the formation of primitive seminiferous cords. At about 60 days Leydig cells appear and proliferate rapidly in the interstitial spaces between the seminiferous tubules up till about 100 days. In the *female*, the gonad destined to be an ovary continues to show proliferation of the coelomic epithelium and of the primordial germ cells which enlarge gradually and become oogonia. However, it is not till about 80 days that the oogonia undergo meiosis to form oocytes; this marks the point of ovarian differentiation from the undifferentiated gonad. The important point is that in the male the formation of testicular tissue occurs much earlier than ovarian organogenesis. The development of an ovary requires the presence of XX chromosomes; the development of a testis is controlled by the Y chromosome. Thus with XO chromosomal constitution there may be no ovarian tissue and with XXY constitution the Y chromosome causes early testicular development, which prevents subsequent ovarian organogenesis.

Genital ducts and external genitalia. At the eighth week of gestation the Wolffian ducts and Müllerian ducts exist as paired structures in both sexes, and the external genitalia have not yet become differentiated. If the primitive gonad becomes an ovary (or if there is no gonad at all), the Müllerian ducts form the Fallopian tubes, uterus, and upper part of the vagina, the Wolffian ducts disappear, and the external genitalia assume their female form during the third and fourth months. If the primitive gonad becomes a testis the Wolffian ducts form the epididymis, vas deferens, and seminal vesicles, the Müllerian ducts degenerate, and the external genitalia acquire their male characteristics by the fifth month (though the testes do not descend into the scrotum until the eighth month).

In short, the development of male genitalia (and male phenotype) only occurs in the presence of a functioning testis. The development of female genitalia (and female phenotype) occurs in the presence or absence of a functioning ovary. Removal of the fetal testes at an early critical period in experimental animals prevents the formation of male genitalia and results in entirely female development of genitalia; however, castration of male fetuses at a later stage does not affect male sex differentiation [FIG. XII.2].

Modes of action of testis on sex differentiation. The testis influences genital development in three ways:

1. It suppresses growth of the Müllerian duct by a *local* action, i.e. each testis acts only on the Müllerian duct to which it is anatomically closely related. Removal of one testis at an early stage of development allows growth of the Müllerian duct on the same side, whereas regression of duct growth occurs on the side with the intact testis. If a testis is grafted on to an ovary it causes involution of the Müllerian duct on that side.

These phenomena are most simply explained by postulating the release from the fetal testis of a locally active hormone. Testosterone and other androgens (systemically or locally administered) do not suppress Müllerian duct growth (Jost 1972). The influence of the fetal testis on Müllerian duct inhibition has been demonstrated in organ cultures. It is not necessary for the two tissues to be in direct contact for the demonstration of Müllerian inhibition, but the substance from fetal testis does not pass through a dialysis membrane. The active agent is not a steroid but might be a polypeptide or nucleic acid; it is released from fetal testis but not from postnatal testis and is derived from fetal seminiferous tubules, probably the cells of Sertoli.

2. The continued growth of the Wolffian duct, and its derived structures, also requires the presence of fetal testis, but in this case the active agent is testosterone, produced by the Leydig cells. Testosterone released from the fetal testis becomes bound to a high-affinity androgen-binding protein which is probably secreted by the Sertoli cells into the lumen of the seminiferous tubule. Synthesis of this protein is stimulated by FSH. The bound testosterone flows along the Wolffian duct and is released from its binding protein to produce high local concentrations which promote growth of the epididymis, vas deferens, and seminal vesicles. Wolffian duct tissues respond directly to testosterone and do not contain the enzyme which reduces testosterone to dihydrotestosterone.

3. Testosterone and other androgenic hormones are essential for differentiation and growth of *male external genitalia*, but in this case the target organs are stimulated by androgenic hormones distributed via the bloodstream, and testosterone is reduced to dihydrotestosterone which becomes bound to cytosol androgen receptors in the target cells. Male development of the external genitalia occurs only when there is adequate androgenic stimulation during the first 12 weeks of fetal life. The Leydig cells, which secrete testosterone, are most numerous during the third month and early

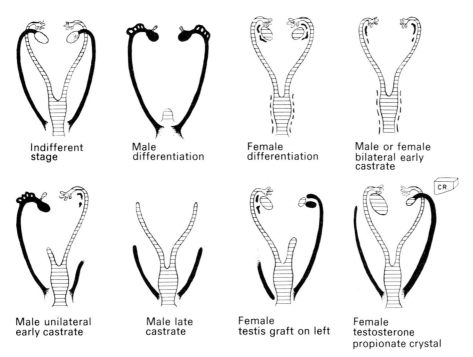

FIG. XII.2. Schematic summary of Jost's experiments with rabbit embryos. The fetal testis plays a decisive role in determining the differentiation of genital ducts. Testosterone stimulates Wolffian development but fails to effect involution of Müllerian structures.
(Grumbach, M. M. and van Wyk, J. J. (1974). Disorders of sex differentiation. In *Textbook of endocrinology* (ed. R. H. Williams) p. 440. Saunders, Philadelphia.)

part of the fourth month of fetal life which coincides with the peak of secretion of human chorionic gonadotrophin (HCG), a placental hormone which acts like pituitary LH to promote development of Leydig cells. The subsequent decline in HCG secretion accounts for the marked decrease in the number of Leydig cells after the fourth month.

If the external genitalia develop for 12 weeks without androgenic stimulation it is impossible to promote normal male external genital development by subsequent administration of androgens. Enlargement of the genital tubercle occurs but fusion of the urethral folds and of the labioscrotal swellings cannot occur after 12 weeks.

Abnormalities of human sex development

Aberrations of sexual development can arise from changes in sex chromosomes or from abnormalities in sex differentiation due to hormonal or environmental causes.

Abnormalities associated with male phenotype

Sex chromosome karyotope 47 XXY (Klinefelter's syndrome). This is typically characterized by the presence of feminine stigmata in an apparent male with very small testes (seminiferous tubule dysgenesis). The sex-chromatin test is positive (genetically female), but the presence of the Y chromosome causes testicular development early in fetal life. This is one of the commonest forms of primary hypogonadism and infertility in the male, with increased prevalence in mentally-retarded men.

Karyotypes XXXY and even XXXXY are known. Testicular development in such cases supports Jost's views on sex differentiation. The presence of more than two X chromosomes leads to severe mental deficiency.

Abnormalities associated with female phenotype

In the absence of the Y chromosome testicular development does not occur and the sex organs develop on female lines.

Ovarian dysgenesis (Turner's syndrome). This is characterized by diminished sexual development, dwarfism, and webbing of the neck in patients with no gonadal tissue. Most patients are chromatin-negative and their karyotype is XO (i.e. only one sex chromosome + 44 autosomes = 45 chromosomes in all; two XX chromosomes are required for a positive sex-chromatin test). The condition is not usually recognized before puberty, the commonest presenting complaint being primary amenorrhoea.

Testicular feminization. Superficially, patients with testicular feminization are normal females, and psychologically completely so. However, after the normal age of puberty, pubic and axillary hair growth is scanty and there is primary amenorrhoea, although breast development is normal. The external genitalia are of female type, but there is no uterus and the vagina ends as a blind pouch. The gonads are testes with immature seminiferous tubules and may lie intraabdominally or in the labia. There is no spermatogenesis but Leydig cells are abundant. The affected persons are genetic males (chromatin-negative somatic cells and XY karyotype). The testes secrete testosterone and oestrogen. The abnormality lies in complete unresponsiveness of the male target organs to testosterone, whilst the breasts, the central nervous system (including the hypothalamus), and external genitalia retain their responsiveness to oestrogens. The failure of development of Müllerian duct derivatives is due to secretion of testicular non-steroidal hormones in early fetal life. This remarkable condition fully supports Jost's evidence and hypotheses on sex differentiation.

True hermaphroditism. In this very rare condition both ovarian and testicular tissue is present, sometimes an ovary on one side and a testis on the other. As would be expected, numerous variations in male and female differentiation can occur affecting both genital duct structures, external genitalia, and the breasts. The sex-chromatin test may be positive or negative. Most true hermaphrodites are whole-body chimeras, consisting of a mixture of 46 XX and 46 XY cells.

Pseudohermaphroditism. Male and female pseudohermaphrodites are persons in whom normal gonadal development has occurred in accordance with their chromosomal sex but with later development of heterosexual characteristics. The commonest type of female hermaphrodite is an individual with ovaries, female ducts, and varying degrees of masculine differentiation of the urogenital sinus and external genitalia. The chromosomal sex is female. The syndrome is usually due to congenital virilizing adrenal hyperplasia [see p. 535].

REFERENCES

GRUMBACH, M M. and VAN WYK, J. J. (1974). Disorders of sex differentiation. In *Textbook of endocrinology*, 5th edn (ed. R. H. Williams) p. 423. Saunders, Philadelphia.
JOST, A. (1972). *Johns Hopkins Med. J.* **130**, 38.
——VIGIER, B., PREPIN, J. and PERCHELLET, J. P. (1973). *Recent Prog. Horm. Res.* **29**, 1.

Female reproductive system

The human female reproductive system consists essentially of two *ovaries* from which ova are released, the *Fallopian tube* (oviduct) in which one or more ova are fertilized by spermatozoa, the *uterus* in which the growing fetus is nourished till childbirth, and the *vagina* and *uterine cervix* through which spermatozoa are introduced into the female reproductive tract as the result of sexual intercourse with the male [FIG. XII.3].

The ovary

There are two ovaries, one on each side, behind and below the Fallopian tubes; each receives a supply of nerves, blood vessels, and lymphatics. The total weight of the adult ovaries is 10–20 g, declining with increasing age. The primary function of the ovary is to produce ova. These are formed from about the tenth week of fetal life from oogonia which multiply freely by mitosis so that at five months of gestation the two ovaries contain 6–7 million germ cells. Formation of oogonia ceases by the seventh month and at term all the oogonia have become primary oocytes, the total number being 2 million oocytes at birth, falling to 300 000 at seven years. The oocytes have entered the prophase of meiotic division by the time of birth. Completion of the first meiotic division does not occur until ovulation (many years later) when, after extrusion of the first polar body, secondary oocytes are formed, which only complete their meiotic division after fertilization by spermatozoa.

The follicle. During the course of maturation the primary oocyte first becomes surrounded by a basal lamina to form a *primordial follicle*; the flat spindle-shaped cells inside the basal lamina become cuboidal and thus form the *primary follicle*. These cuboidal cells later multiply to form *granulosa cells*, around which stromal cells outside the basal lamina differentiate to form the *theca*. The oocyte is immediately surrounded by the *zona pellucida* through which protoplasmic processes pass from the granulosa cells to make contact with the plasma membrane of the oocyte [FIG. XII.4]. The theca is vascularized but the cells within the basal lamina receive no blood supply. After the seventh month of gestation some of the primary follicles show accumulation of fluid and the formation of secondary (Graafian) follicles. However, at birth a large number of primary and secondary follicles have degenerated (*atresia*) and from birth onwards atresia continues until by the menopause there are no follicles left. Two points need to be emphasized:

1. After birth no new oocytes are formed.
2. Only 300–400 follicles proceed to ovulation during the years of female fertile life from menarche to menopause.

By contrast, in the male, spermatogenesis begins at puberty and new spermatozoa are continually being formed till a ripe old age when production gradually ceases.

Premenarchal ovary. From birth to menarche, ovarian weight increases steadily due to increased volume of developing follicles and an increase in stroma. From the age of 8 years oestrogen secretion by the ovary is sufficient to promote an increase in uterine weight.

Postmenarchal ovary. At the menarche, hypothalamic maturation leads to the onset of cyclic ovulation while atresia continues. During each cycle, at approximately monthly intervals, some 10–15 follicles enlarge to become secondary follicles, under the influence of FSH from the adenohypophysis. Fluid accumulation occurs in these follicles but usually only one proceeds to the stage of ovulation in which the ovum, surrounded by the zona pellucida and granulosa cells (cumulus oophorus), is shed into the infundibulum of the Fallopian tube [FIG. XII.5]. The rupture of the follicle is probably due to several factors such as ischaemic necrosis of overlying cells, proteolytic enzyme action, and perhaps also to increased fluid pressure within the Graafian follicle. The reason why only one out of 15 enlarged follicles ovulates each month is unknown; multiple ovulations occur in 1–2 per cent of all cycles.

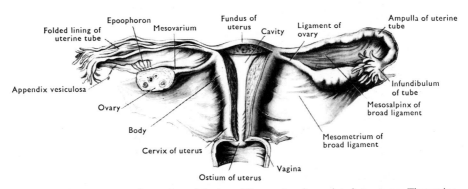

FIG. XII.3 The posterosuperior surface of the broad ligament and associated structures. The vagina, uterus, and left uterine tube have been opened, and left ovary is sectioned parallel to the broad ligament.
(From *Cunningham's manual of practical anatomy*, 13th edn (ed. G. J. Romanes) Vol. 2. Oxford University Press.)

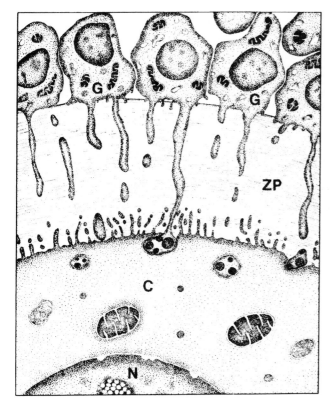

Fig. XII.4. Structure of fully formed zona pellucida (ZP) around an oocyte in a Graafian follicle. Microvilli arising from the oocyte interdigitate with processes from the granulosa cells (G). These processes penetrate into the cytoplasm of the oocyte (C) and may provide nutrients and maternal protein. (N, oocyte nucleus.)
(From Baker, T. G. (1972). Oogenesis and ovulation. In *Reproduction in mammals* (ed. C. R. Austin and R. V. Short) Book 1, p. 25. Cambridge University Press.)

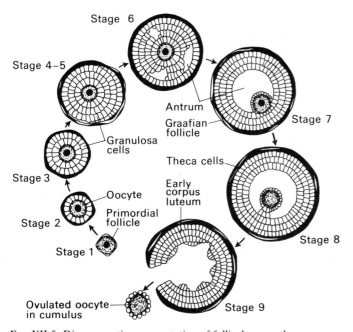

Fig. XII.5. Diagrammatic representation of follicular growth.
(From Baker, T. G. (1972). Oogenesis and ovulation. In *Reproduction in mammals* (ed. C. R. Austin and R. V. Short) Book 1, p. 29. Cambridge University Press.)

Ovulation and corpus luteum formation

Release of the ovum occurs shortly after a high peak of LH and FSH secretion from the adenohypophysis. At the time of ovulation antral fluid escapes and the follicle wall collapses, leading to haemorrhage into the theca interna. After ovulation capillaries from the theca interna invade the rapidly-dividing granulosa layer and the corpus luteum (yellow body) is formed [FIG.XII.6]. This enlarges for 8 or 9 days and if fertilization has not occurred, the corpus luteum regresses and eventually becomes a *corpus albicans*. If pregnancy occurs the corpus luteum continues to grow for several months and begins to degenerate at about the sixth month.

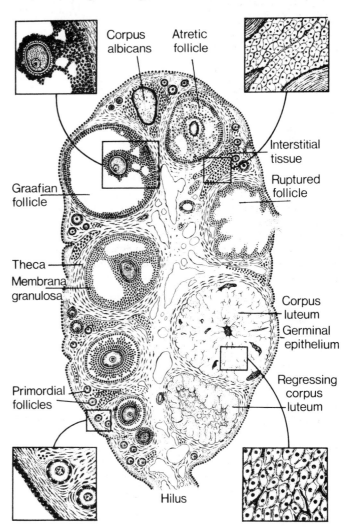

Fig. XII.6. Idealized drawing of the structure of the mammalian ovary showing follicles at various stages in their development and the formation and regression of corpora lutea.
(From Baker, T. G. (1972). Oogenesis and ovulation. In *Reproduction in mammals* (ed. C. R. Austin and R. V. Short) Book 1, p. 40. Cambridge University Press.)

Time of ovulation. An accurate knowledge of the time of ovulation in women is not of academic interest only. It is not known for certain how long an ovum survives after discharge from the ovary or how long the sperm can live after being introduced into the vagina; the evidence suggests that in neither case are they functionally active after an interval of 24–48 hours. For pregnancy to occur it is therefore necessary that coitus should take place within a day or two either side of ovulation, which is presumably the time of maximum fertility. Coitus at other times would tend to be sterile, and it is suggested that the rest of the menstrual cycle constitutes a more or less 'safe period', i.e. pregnancy is unlikely to occur even if no other methods of birth control are employed. This method will, however, fail in its purpose if ovulation in any month is premature or delayed, and such variations occur frequently. There is evidence that ovulation may occur even during the latter part of the menstrual period.

The time of ovulation in women can be simply determined by recording the body temperature before getting up in the morning. In the pre-ovulatory phase the oral temperature is 36.3–36.8 °C and after ovulation the waking temperature increases by 0.3–0.5 °C. The timing of ovulation is of value when it is desired to promote or avoid conception.

The control of the ovarian cycle by the adenohypophysial gonadotrophins (follicle-stimulating hormone, FSH; luteinizing hormone, LH) is discussed on page 571.

REFERENCES

BAKER, T. G. (1972). Oogenesis and ovulation. In *Reproduction in mammals* (ed. C. R. Austin and R. V. Short) Book 1. Cambridge University Press.

Ross, G. T. AND VAN DE WIELE, R. L. (1974). The ovaries. In *Textbook of endocrinology*, 5th edn (ed. R. H. Williams) p. 368. Saunders, Philadelphia.

HORMONAL FUNCTIONS OF THE OVARY

Apart from forming and discharging ova as described above the ovary secretes two hormones, an oestrogenic steroid called *oestradiol*, formed by the cells of the theca interna, and another steroid called *progesterone* formed by the corpus luteum.

These two ovarian hormones regulate the growth, activities and nutrition of the rest of the female reproductive organs:

1. The growth and development of the vagina, uterus, and Fallopian tubes at puberty.
2. The menstrual cycle in women and the oestrous cycle in lower mammals.
3. The appearance and in some cases the persistence of secondary sexual characters.
4. Some of the bodily changes during pregnancy, especially embedding of the blastocyst in the uterus and development of the placenta.
5. Development of the breasts [p. 597].

Puberty

During girlhood, before puberty, oestrogen secretion is too small to cause full development of the reproductive organs. However, the immature ovary can be stimulated by exogenous gonadotrophin so the prepubertal state is not due to ovarian incompetence. The pituitary of immature female animals can secrete gonadotrophins if it is transplanted into the sella turcica of hypophysectomized adult animals. So both ovary and adenohypophysis of immature animals (and, it is presumed, of girls too) are fully capable of responding to physiological stimuli. It seems that before puberty small amounts of FSH and LH are secreted by the adenohypophysis; the ovary secretes small amounts of oestradiol which acts on the hypothalamus to keep the secretion of hypothalamic gonadotrophin-releasing hormone at a low level. At puberty the sensitivity of the hypothalamus to the negative feedback effect of oestradiol decreases and in consequence the hypothalamus secretes more gonadotrophin-releasing hormone, the adenohypophysis secretes more FSH and LH, and the ovaries secrete more oestradiol, progesterone, and androgen which together produce the changes of puberty. It is not known what causes the hypothalamus to become less sensitive to oestrogen inhibition at puberty.

Puberty begins at 10–14 years, and is regarded as ending with the first menstrual period (*menarche*) usually at 13–15 years. At whatever age the menarche appears it is usually some years before regular, adult ovulatory menstrual cycles are established. The main changes which take place at puberty are:

1. The complete ovarian cycle occurs, characterized by ovulation and corpus luteum formation.
2. The uterus and vagina enlarge. The muscle fibres of the uterus increase in number and size; the mucous membrane thickens and the glandular alveoli become larger. All these changes can be produced in immature animals by injection of oestrogen.
3. The breasts begin to appear as a result of outgrowth of ducts from the nipple area, and an increase in the amount of fat, connective tissue, and blood vessels. Such changes as occur before the onset of ovulation are due entirely to oestrogen; during adolescence, owing to the action of both oestrogen and progesterone, glandular alveoli appear and progressively increase in size and numbers with each ovarian cycle.
4. The secondary sexual characters develop; these include the female distribution of fat, giving the characteristic curves to the body, and the appearance of hair in the axilla and on the pubes (in the latter region the upper hair margin is concave upwards). This growth of hair is stimulated by the weak androgens secreted by the ovary and adrenal cortex. The larynx does not enlarge at puberty; hence the female voice remains high pitched.
5. Important psychological changes take place as the girl matures mentally and emotionally through adolescence to young womanhood.

Extirpation of ovary

Before puberty

Little is known about the results of extirpation of the ovary before puberty in girls. Probably puberty does not set in, the menstrual flow does not appear, and the secondary sex characters do not develop. It is obvious that the presence of the ovary is essential for the onset of puberty, though indirectly the anterior pituitary and the hypothalamus are the dominant factors.

In adults

Following the extirpation of the ovary in adults, there is atrophy of the whole genital apparatus—the uterus, the vagina, and the external genital structures. Menstruation ceases permanently. Vasomotor changes are common, e.g. flushing of the skin of the face, neck, and upper chest ('hot flushes'), and a feeling of suffocation. The effects on the breasts are variable: they may increase in size, owing to local accumulation of fat, or they may shrink because the glandular tissue atrophies. Obesity develops from diffuse deposition of fat. Conflicting reports are given concerning the effect on sexual desire, but it is often unaffected; thus in women as in men, though sexual desire is modified by sex hormones it may in large measure be independent of them and be determined by psychological factors. It is also quite certain that sexual desire may persist, sometimes to a heightened degree, in women after ovarian atrophy at the menopause. Complete ovariectomy may result in considerable emotional disturbance, varying from a certain amount of irritability or depression to a condition closely allied to insanity.

Menopause

The menopause (climacteric) is the period of life when menstruation naturally ceases and other phenomena identical with those just described make their appearance. It usually occurs between the ages of 45 and 50, although it may set in earlier or later. The condition is associated with marked changes in the ovaries: they become smaller, the Graafian follicles disappear and are replaced by fibrous tissue; ova, corpora lutea, and the internal secretions of the ovary are no longer formed. These ovarian changes are not due to lack of anterior pituitary hormones, the secretions of follicle-stimulating and luteinizing hormones actually increasing, but to a 'senile' change in the ovary which no longer reacts to the hormones which normally stimulate it.

REFERENCES

DONOVAN, B. T. and VAN DER WERFF TEN BOSCH, J. J. (1965). *Physiology of puberty*. Arnold, London.
ODELL, W. D. and MOYER, D. L. (1971). *Physiology of reproduction*. Mosby, St. Louis.
SHARMAN, A. (1962). The menopause. In *The ovary* (ed. S. Zuckerman, A. M. Mandl, and P. Eckstein) Vol. I, p. 549. Academic Press, London.

The Fallopian tubes (oviducts)

Writing about the transport of gametes in the Fallopian tubes Blandau (1973) says: 'Only this statement can be made with certainty. Mature eggs are ovulated, mature spermatozoa capable of fertilizing them reach the ampulla, and under a suitable environment, union of these cells is accomplished and a new life begins . . . There is no other tubular system in which cells of such different dimensions are transported in opposite directions in a limited period of time.'

Species, hormonal, and other differences abound, and least of all is known about what happens in women. However, with these reservations one can take the evidence obtained from laboratory animals at the physiological, cellular, and biochemical levels and consider how far this can be applied to an understanding of one of the most important events in human life. The events leading to fertilization of the ovum can be described in more detail:

1. Ovulation must occur so that the ovum is free to enter the Fallopian tube [FIGS. XII.7 and XII.8]. The fimbria (from the Latin for 'fringe') clasps and rubs the surface of the ovary at the time of ovulation as the result of contractions of its smooth muscle; the activity of this muscle is appropriately increased by oestrogens and depressed by progesterone. At the time of ovulation the ciliated cells in the mucosa of the infundibulum of the oviduct are maximally developed and most active, also under the influence of oestradiol. The effect of muscular and ciliary activities is to convey the ovum and its surrounding cumulus cells rapidly into the ampulla; the ovum is then held up at the ampullary-isthmic junction for two to three days and, if no fertilization occurs, the ovum degenerates and dies. If mature spermatozoa gain access to the ovum at the right time, fertilization takes place and for two days or so cell divisions occur until the blastocyst is transported into the uterus as the result of relaxation of the sympathetically innervated muscle of the

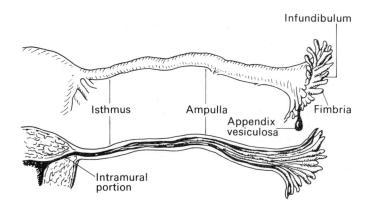

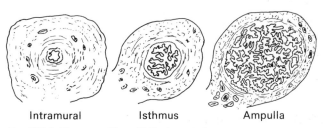

FIG. XII.8. Human oviduct (Fallopian tube), gross and microscopic. (From Odell, W. D. and Moyer, D. L. (1971). *Physiology of reproduction*, p. 39. Mosby, St. Louis.)

isthmus assisted by ciliary action towards the uterus. The muscular relaxation is favoured by progesterone and by prostaglandin E_1.

2. The transport of spermatozoa from the vagina to the ampulla is still very difficult to explain. Spermatozoa must be motile to reach the oviduct but not all motile spermatozoa are capable of fertilizing ova. Cervical mucus is most easily penetrated by spermatozoa at the time of ovulation, but transport from the cervix to the Fallopian tube is not due to any known chemotactic process and must presumably be the result of appropriate ciliary activities and muscular movements in the uterus. The passage through the utero-tubal junction is particularly difficult to explain. It must also be stressed that large numbers of spermatozoa are destroyed by phagocytosis in the uterus and only small numbers reach the tube.

3. The mucosa of the Fallopian tube contains not only ciliated cells but also secretory cells which at the time of ovulation contain granules of glycogen. Glycogen breakdown to *glucose* helps oviducal fluid to provide an environment in which ova and spermatozoa can survive for short periods and the fluid also contains substances needed for cell division up to the stage of blastocyst. Oviducal fluid contains *plasma proteins* which traverse the oestrogen-dilated small blood vessels, specific *mucoproteins* from the secretory cells, *lactate* and *pyruvate* (which provide energy), O_2 and HCO_3^-, which enhance respiration, provide carbon for structural purposes, and help to keep pH in the range 7.5–7.8.

Fertilization of ovum *in vitro*. Although oviducal fluid contains unidentified stimulating substances, it has been possible to induce fertilization of ova by spermatozoa and to promote blastocyst development *in vitro*, using a fluid containing physiological concentrations of Na^+, K^+, Ca^{2+}, and Cl^- together with O_2 and CO_2, in addition to the substances mentioned above to provide energy and structural materials.

The Fallopian tube's unique role as the meeting place for male and female gametes and the provider of nourishment for formation of the blastocyst is such a finely balanced affair that it is not surprising that things sometimes go wrong. For example, the egg and sperms may meet at the wrong time so that fertilization does

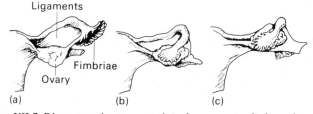

FIG. XII.7. Diagrammatic representation of movements of tube and ovary. (a) Tube extended; no contraction of musculature of ligaments. (b) Contraction has started; tube has curved around upper pole of ovary, and fimbriae are in contact with this pole and one side of ovary. (c) Contraction stronger; fimbriae reach lower ovarian pole and ovary is rotated so as to cause its other side to face ostium abdominale tubae. [From Odell, W. D. and Moyer, D. L. (1971). *Physiology of reproduction*, p. 118. Mosby, St. Louis.]

not occur; if both tubes are blocked by disease, egg and sperms cannot meet at all; fertilization may occur normally but the product of conception may enter the uterus too soon so that implantation cannot take place; alternatively the product of conception may remain in the Fallopian tube and never reach the uterus at all (ectopic gestation), a most dangerous development which may lead to severe life-threatening haemorrhage in the woman.

In some women rendered infertile by Fallopian tube obstruction fertilization has been produced by adding spermatozoa to an isolated ovum *in vitro* and, after a few days, the blastocyst formed outside the body has been inserted into the progestational uterus. Successful implantation has sometimes been achieved and subsequent growth *in utero* has led to the birth of apparently normal 'test tube' babies.

The uterus

The uterus consists of the corpus (body) and the cervix (neck). The Fallopian tubes enter the fundus of the uterus, one on each side, and the cervix opens into the vagina [FIG. XII.3]. The prepubertal uterus is a small organ weighing 10–15 g. The postpubertal uterus which has grown under the influence of oestrogen weighs 30–60 g.

During late pregnancy the uterus enlarges considerably through stretching by the growing fetus and also by hormonal action, the final weight being 800–1000 g.

The cervix differs from the rest of the uterus in having much less muscle and more connective tissue; the mucosa, called the endocervix, contains columnar mucus-secreting epithelium with some ciliated cells. Unlike the endometrium the endocervix is not shed at menstruation. The functions of the cervix are:

1. To allow entry of spermatozoa from the vagina into the uterus and to store viable sperms for 24–48 hours.
2. To allow escape of menstrual debris.
3. To permit passage of the fetus at term.
4. To prevent entry of infectious micro-organisms.

Cervical secretion normally contains about 92 per cent water, NaCl, and glycoproteins of the sialomycin type. Its composition and character change in response to hormonal influences. At the time of ovulation, cervical mucus increases in volume and becomes much more watery (98 per cent water). In this state it is much more easily penetrated by spermatozoa than at any other time. It is possible that spermatozoa may survive in cervical mucus in the glands for some time without undergoing the phagocytosis which quickly destroys them in the uterus and tubes. Spermatozoa may be released from this reservoir to produce fertilization.

During pregnancy, cervical secretion forms a viscous plug which constitutes a barrier against spermatozoa and infectious micro-organisms.

The body of the uterus consists mainly of smooth muscle (*myometrium*) arranged in three layers supplied by sympathetic nerves. Outside the muscle is a serous coat and inside the myometrium is a mucous membrane called the *endometrium* which undergoes characteristic changes during the menstrual cycle.

Two kinds of arterial vessels pass through the myometrium to enter the endometrium:

1. *Spiral* arteries which pursue a very tortuous course and end in capillaries which supply the middle and superficial portions of the endometrium [FIG. XII.9].
2. *Basal* arteries which run only for a short distance and supply the basal portion of the endometrium [FIG. XII.9]

The *endometrium* plays an important role in reproduction. Its secretion nourishes the blastocyst for a few days before implanta-

tion takes place in the hormonally-prepared endometrium. The secretion from the progestational endometrium is rich in glucose, which provides energy, and it also contains enzymes and proteins which stimulate RNA and protein synthesis in the blastocyst.

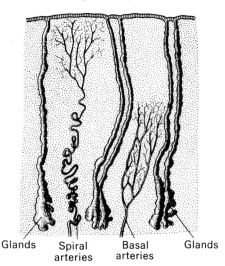

Glands   Spiral   Basal   Glands
arteries   arteries

FIG. XII.9. Blood vessels of uterine mucous membrane. Note uterine glands; spiral arteries supplying inner third of mucosa; straight (basal) arteries supplying basal part of mucosa.
(After Corner, G. W. (1946). *Hormones in human reproduction*. Princeton.)

Oestrous cycles of lower mammals

Certain laboratory animals have been very thoroughly studied with respect to the reproductive cycle in females. However, differences between species make it impossible to generalize on the comparative physiology of the female sex cycle. Some species, such as the dog, have one or two oestrous cycles in the year, when ovulation and receptivity coincide with the male's sexual interest in the female. In others, such as the rabbit, ferret, and cat, ovulation is brought about by a neuroendocrine reflex aroused by stimulation of the vagina during coitus. *Only primates have menstrual cycles*—i.e. recurrent monthly discharge of blood from the female genital canal. In other species there are oestrous cycles of varying duration—e.g. 4–6 days in rat and mouse, 14 days in guinea-pigs, 16 days in sheep, 20 days in cows, and twice a year in bitches, during which there is a special period of desire in the female accompanying the process of ovulation. Only at this time is fruitful coitus possible.

Human menstrual cycle

The menstrual cycle in women is related to ovulation and the secretion of ovarian hormones. Between the menarche and the menopause the endometrium undergoes cyclical changes which are initiated in the hypothalamus and mediated by the pituitary gonadotrophins FSH and LH which act on the ovary to induce release of oestradiol and progesterone respectively [p. 572]. In addition the ovary has inherent genetically determined capacity to secrete its hormones so that they influence the hypothalamico-pituitary activity by feedback control.

The human menstrual cycle is counted from the day on which menstrual bleeding begins. The bleeding typically lasts for about 5 days during which the superficial portion of the endometrium is shed. From day 5 to day 14 the endometrium thickens and proliferates (*proliferative* phase) [FIG. XII.10]; at about day 14 ovulation occurs and at this time there may be a small vaginal discharge of watery cervical fluid. After ovulation the corpus luteum is formed in the ovary and the endometrium undergoes further development (*secretory*, *progestational*, or *luteal* phase), which ends at about 28 days with the onset of menstruation. (The cycle is called menstrual

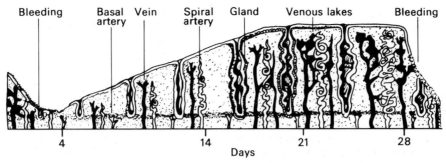

Fig. XII.10. Structural changes of endometrium during the menstrual cycle. (From Odell, W. D. and Moyer, D. L. (1971). *Physiology of reproduction*, p. 35. Mosby, St. Louis.)

from the Latin word 'mensis', a month, i.e. a lunar month of 28 days.) However, the cycle is by no means as regular as the word suggests and cycles lasting a few days more or less than 28 days are quite common.

The proliferative phase. Immediately after bleeding has ended the endometrium is less than 2 mm thick and consists of a ciliated columnar epithelium dipping down into a loose stroma to form simple tubular glands. During days 6 to 14 proliferation occurs in the glands, blood vessels, stroma, and superficial epithelium, until the endometrium becomes about 4 mm thick on day 14 (when ovulation occurs).

Secretory (luteal) phase. About 36 hours after ovulation, when the corpus luteum is well formed, the endometrium becomes still thicker (5 mm) as the glands increase in length and diameter, and become more tortuous and filled with mucus ('saw-toothed' appearance). The stroma cells proliferate and enlarge and resemble those seen in the early placenta. The spiral arteries become more coiled and dilated and the venous lakes become filled with blood; there is exudation of clear and blood-stained fluid from the congested vessels.

Menstruation. If the ovum shed at ovulation is not fertilized, menstruation occurs during the first 5 days (average) of the menstrual cycle. There is bleeding and shedding of the superficial two-thirds of the endometrium which occurs sequentially in different parts of the endometrium so that in about 5 days the whole endometrium has been affected. The immediate cause of endometrial necrosis appears to be spasm of the spiral arteries for several hours; when the vessels relax shedding of necrotic endometrium, leakage of blood and release of mucus make up the debris lost during menstruation. Menstrual blood clots promptly in the uterus but is liquefied by fibrinolysis in the vagina. The total blood loss in normal women varies from 10 to 200 ml (mean 40 ml). Menstrual blood contains prostaglandins.

Hormonal control of the menstrual cycle

The menstrual cycle does not occur after ovariectomy or hypophysectomy in women; failure of menstruation also occurs after hypothalamic lesions which abolish secretion of GnRH [pp. 518–22]. Figure XII.11 shows how the blood concentrations of ovarian steroid hormones and adenohypophysial gonadotrophins vary during the course of the cycle and suggests the following explanation:

1. The hypothalamus acts on the adenohypophysis by release of GnRH [p. 522]. In females hypothalamic control of the adenohypophysis is cyclic but can be influenced by emotional disturbances as well as by feedback effects of oestrogens.

2. Figure XII.11 shows that the blood concentration of FSH rises during menstruation and then declines. This rise in FSH is assumed to promote development of ovarian follicles and at the same time to increase the secretion of oestradiol from the theca interna. The rise in plasma oestradiol concentration is responsible for the proliferative phase in the endometrium. Administration of oestrogen to an ovariectomized woman produces a *proliferative* endometrium. The increase in plasma 17-hydroxyprogesterone concentration is not physiologically significant.

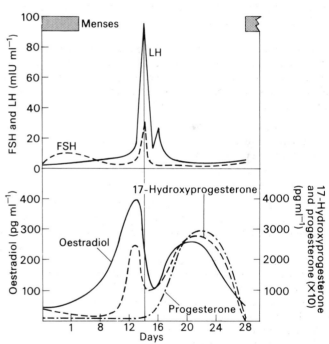

Fig. XII.11. Schematic representation of the fluctuation of serum luteinizing hormone (LH), follicle-stimulating hormone (FSH), progesterone, 17-hydroxyprogesterone, and oestradiol during the normal menstrual cycle in women. Note that both oestradiol and 17-hydroxyprogesterone rise before the LH-FSH ovulation surge. In order to show progesterone and 17-hydroxyprogesterone on the same scale, progesterone concentrations were divided by 10, that is they are actually ten times greater than shown.
(From Odell, W. D. and Moyer, D. L. (1971). *Physiology of reproduction*, p. 66. Mosby, St. Louis.)

3. Plasma oestradiol concentration reaches a peak at 12–13 days. This has a positive feedback effect which induces within 24 hours a sudden surge in LH blood concentration at mid-cycle, which in turn produces ovulation. It will be noted that FSH blood concentration also suddenly rises to a peak, at a lower level, at the same time as the LH peak occurs. The stimulus to ovulation may depend on both LH and FSH (perhaps on the ratio of LH to FSH). The timing of the LH-FSH peak is much more variable than the simplified

FIG. XXI.11 suggests; it may occur at any time between day 11 and day 23 of the menstrual cycle.

4. After ovulation, plasma LH and FSH concentration fall to very low values for the rest of the cycle, but as the corpus luteum is formed plasma progesterone concentration rises markedly and plasma oestradiol also rises during the second half of the cycle. Progesterone, acting on an endometrium primed by oestrogen, produces the secretory phase of endometrial development which is progestational in character, i.e. the endometrium is prepared for implantation of the blastocyst formed from the fertilized ovum. If no fertilization takes place, progesterone secretion (and oestrogen secretion) falls off sharply, the spiral arteries go into spasm, and when they subsequently relax menstruation occurs.

The evidence for the statements just outlined is as follows:

1. In women with absent ovarian function, administration of oestrogen followed by progesterone can produce the proliferative and secretory phases of endometrial development. When ovarian hormone administration is stopped menstruation occurs. Even after oestrogen alone cessation of administration causes bleeding.

2. In women with normal ovaries, menstruation does not occur unless FSH and LH are available in the right amounts at the right time to stimulate ovarian hormone secretions which promote the proliferative and secretory phases of endometrial development. If adenohypophysial function is defective administration of exogenous FSH and LH can mimic the effects of the natural hormones.

3. In women with normal ovaries and adenohypophysis, the menstrual cycle depends on secretion of hypothalamic GnRH [p. 522].

4. In women with amenorrhoea (absence of menstruation) the hormones mentioned above can help to localize the defective link in the hormonal chain.

Reproductive hormones

The reproductive systems of both female and male are wholly dependent on hormones for their differentiation and full development. Chemically the reproductive hormones are different in structure:

Polypeptide hormones

Among these are the *hypothalamic decapeptide* gonadotrophin-releasing hormone (GnRH). This acts on the adenohypophysial cells which synthesize and release FSH and LH.

Prolactin is a long-chain adenohypophysial polypeptide which helps to promote breast development in pregnancy and lactation [pp. 517 and 598].

Oxytocin is a neurohypophysial octapeptide which causes uterine contractions during labour and also promotes milk ejection during breast-feeding.

Glycoproteins

These include *adenohypophysial gonadotrophins* FSH and LH and the placental hormone *chorionic gonadotrophin* (CG). FSH and LH have molecular weights of about 30 000. Sialic acid is an important component of both FSH and LH. FSH and LH each consist of two subunits, one of which is common to the gonadotrophins and to TSH.

Steroid hormones

The ovary, testis, and adrenal cortex secrete lipid-soluble compounds called *steroids*. The basic steroid in the body is cholesterol [FIG. XII.12] which is perhydro-cyclopenteno-phenanthrene. Cholesterol is synthesized from acetate and modifications of its structure produce not only oestrogens, progesterone, testosterone, adrenal corticosteroids [p. 530], but also vitamin-D precursors

[p. 553], bile acids [p. 433], and the steroid component of cardiac glycosides. The synthesis of all steroid hormones involves initial conversion of cholesterol to pregnenolone; the pattern of subsequent routes of biosynthesis depends on the presence of specialized enzyme systems, concentrations of co-factors and precursors and the nature of trophic stimuli. The sequences leading to synthesis of cortisol and aldosterone are set out on page 530 and those leading to the formation of progesterone, testosterone, oestrone, and oestradiol-17β are shown in FIGURE XII.13.

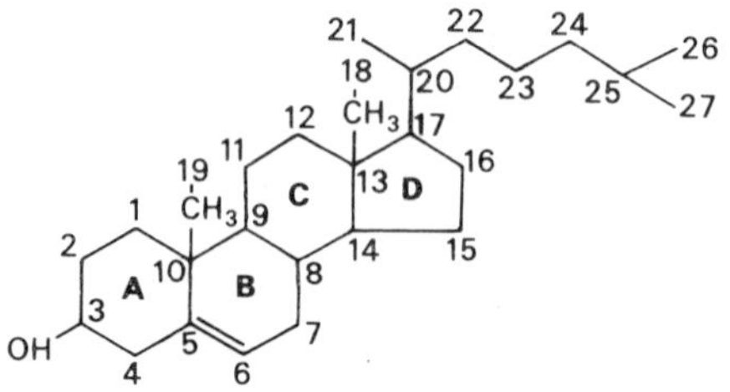

FIG. XII.12. The formula of cholesterol showing the convention for numbering the rings and carbon atoms.
(From Baird, D. T. (1972). Placental hormones. In *Reproduction in mammals* (ed. C. R. Austin and R. V. Short) Book 1. Cambridge University Press.)

Progesterone is secreted in large amounts by the corpus luteum and the placenta and in much smaller amounts by the granulosa cells of the ovarian follicle before ovulation and by the adrenal cortex.

Androgens are formed from pregnenolone and progesterone which are converted first to the weak androgens, dehydroepiandrosterone and androstenedione; these compounds are secreted by the adrenal cortex and ovary. In the Leydig cells of the testis (and to a very small extent in the ovary) these weak androgens are converted to the potent androgen testosterone. Further aspects of androgen metabolism are discussed on page 579.

FIGURE XII.13 shows that the physiologically active oestrogens are synthesized via androstenedione, the weaker *oestrone* being the immediate precursor of the most potent natural oestrogen, *oestradiol-17β*. Oestradiol is secreted mainly by the theca cells of the Graafian follicle and by the corpus luteum. In men a small amount of testosterone is converted somewhere in the body to oestradiol which can be detected in blood and urine.

Oestrogens (oestrus-producing substances)

As oestrus does not occur in women and higher primates the term oestrogen is strictly a misnomer when applied in these species to oestradiol and related substances. However, the term is sanctioned by long usage. The principal and most potent natural oestrogen is *oestradiol-17β*.

In the blood, oestradiol is bound firmly to a β-globulin, the total (free+bound) plasma oestradiol being 183–1830 nmol l^{-1} (500–5000 µg l^{-1}). After radioactive oestradiol is injected into women about 65 per cent can be recovered from urine and 10 per cent from faeces, the fate of the remainder being uncertain. Of the amount detected in the urine 20 per cent occurs as oestradiol and the rest as metabolites such as oestrone and oestriol formed in the liver. Calculation from urinary metabolites suggests a daily production rate of 35–300 µg in different stages of the menstrual cycle and even higher values during pregnancy (15–45 mg daily). Studies with radioactive oestradiol show that although most is bound to plasma β-globulins the hormone is specifically and avidly taken up by target organs such as the uterus, vagina, the anterior hypothalamus, and the adenohypophysis, in which there is a specific binding protein in the cell cytoplasm, or in lysosomes. The bound product rapidly enters the nucleus and via DNA and RNA initiates changes which enhance cell replication and/or protein synthesis [p. 500].

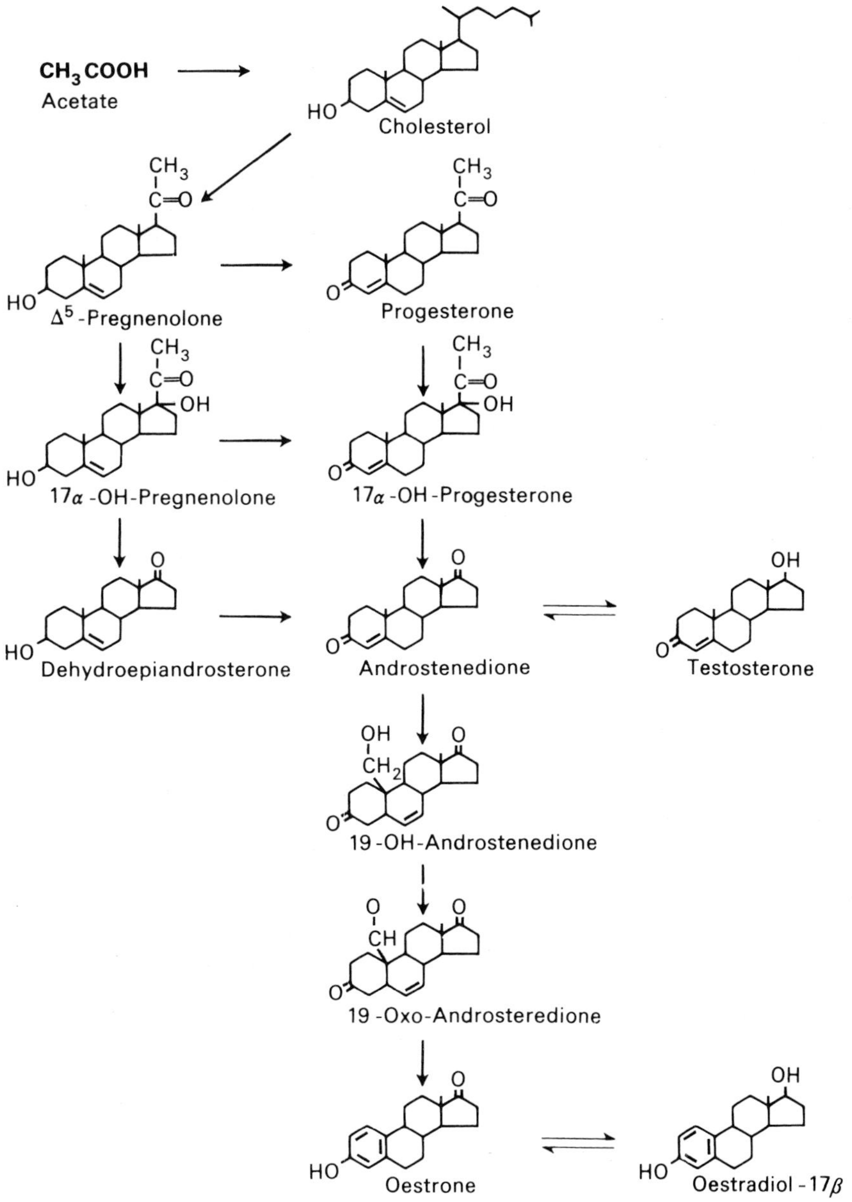

FIG. XII.13. The chief routes of biosynthesis of androgens, oestrogens, and progesterone from acetate and cholesterol.
(From Baird, D. T. (1972). Placental hormones. In *Reproduction in mammals* (ed. C. R. Austin and R. V. Short) Book 1. Cambridge University Press.)

Circulating oestrogens are conjugated in the liver to form water-soluble sulphates and glucuronides which are excreted in the urine.

Actions of oestrogens. Oestradiol is secreted in very small amounts, which have little physiological action, before puberty. At puberty oestradiol is secreted in larger quantities which promote growth of the uterus, vagina, external genitalia, pelvis, and breasts. The growth of the female type of pubic hair and of axillary hair may be caused by local conversion of weak androgens from the ovary and adrenal cortex to testosterone. In immature or ovariectomized women the uterus is small and both myometrium and endometrium underdeveloped. Oestrogen administration promotes mitotic activity in the uterine muscle and endometrium. The muscle fibres enlarge and become more excitable and active; the sensitivity to oxytocin increases. These effects may be due to an action on calcium binding by the myometrial cell membrane. In the endometrium oestrogen stimulates growth of the glandular epithelium, causes hyperaemia, perhaps by histamine release, and increases the content of water, electrolytes, nucleotides, protein, and enzymes. Cervical mucus secretion becomes copious and watery. Oestrogen also stimulates the secretory activity of the cells lining the Fallopian tubes and the motility of the muscle coat and cilia is enhanced.

The vaginal epithelium is particularly sensitive to the action of oestrogen, and examination of vaginal smears can be used to follow the time-course of natural oestrogen secretion or as a bioassay in the measurement of the activity of compounds suspected to have oestrogenic properties. In the vagina of the spayed or immature female rat oestrogens produce an intense wave of mitosis which causes the epithelium to increase in height from 2 or 3 layers of low cuboidal cells to about 10 layers of cells, the most superficial of which show cornification and desquamation. Similar but less clear-cut changes occur in women. Oestrogens increase vaginal secretion and make it acid by promoting the breakdown of glycogen to lactic acid. All these vaginal changes protect against bacterial infection. Oestrogens are necessary for the lubricatory vaginal secretion associated with coitus.

In the skin oestrogen increases the water content and thickness; by antagonizing the actions of androgen it reduces sebaceous-gland secretion and may help to prevent acne.

Oestrogen increases the plasma levels of thyroxine- and cortisol-binding globulins. It also lowers plasma cholesterol level and may therefore help to prevent the development of atherosclerosis.

The implantation of very small amounts of oestrogen in the anterior hypothalamus (amounts too small to have systemic effects) causes oestrus behaviour in ovariectomized cats or rats. However, sexual activity in women is not closely related to oestrogen secretion, and libido is often well marked after the menopause, probably due to secretion of weak androgens from the adrenal cortex and their conversion to testosterone in certain parts of the skin.

Secreted oestrogen influences the menstrual cycle [p. 571] and is important for breast development [p. 568]. It also plays a role in pregnancy and parturition [pp. 581 and 585]. Oestrogen influences the secretion by the anterior pituitary of the gonadotrophic hormones. The mechanism is discussed on page 522.

Mechanism of oestrogen action. Since mitosis and cell growth depend on protein synthesis it is not surprising that inhibitors of protein synthesis, such as actinomycin, prevent oestrogen-induced multiplication of epithelial cells. There is evidence for the existence of a specific oestrogen-binding receptor protein in the cytoplasm, lysosomes, and nucleus in target cells. Attachment of oestrogen to this receptor protein in the nucleus may act on DNA to promote synthesis of new messenger RNA which then stimulates the responses in oestrogen-sensitive cells. According to this view, the oestrogen-receptor complex activates specific gene sites to produce messenger RNA, and oestrogen is not required for any of the subsequent changes in cell response.

Artificial oestrogens. The first artificial oestrogen, *stilboestrol*, was synthesized by Dodds in 1938, but the most potent is *ethinyloestradiol* which is the oestrogen component of the contraceptive pill.

Progesterone is an intermediary in the biosynthesis of cortisol [p. 530], oestradiol, and testosterone [p. 573]. It is secreted into the blood-stream by the corpus luteum and the placenta.

By itself progesterone is a 'non-hormone', i.e. it has no actions on its own. It is not bound specifically to any plasma protein nor any cellular component. Nevertheless, on target organs which have been primed by prior action of oestrogen, progesterone has some pronounced physiological actions:

1. On the oestrogen-stimulated proliferated endometrium it produces the secretory changes which prepare the endometrium for implantation of the fertilized ovum (blastocyst).
2. In some animal species it promotes the growth of alveolar tissue in the breasts. However, in monkeys (? women) oestrogen alone can cause extensive growth of both duct and secretory structures in the breast.
3. In some respects progesterone antagonizes the actions of oestrogen. In many species it decreases the excitability of myometrial cells, reduces spontaneous electrical activity, raises the membrane potential, and decreases the sensitivity of the myometrium to oxytocin. However, in women progesterone may induce slow uterine contractions of high amplitude. Turnbull *et al.* (1974) [FIG. XII.21, p. 582] showed that in women during the last few weeks of pregnancy, right up to the time of labour, the mean level of plasma progesterone fell and that of oestradiol rose (both were measured by highly specific radioimmunoassays). This agrees with the long-known work of Csapo (1961) and will be discussed further on page 585.

It is unlikely that progesterone has effects on other types of smooth muscle in physiological concentrations of 200 µg l^{-1}.

Progesterone inhibits ovulation in women, probably by inhibiting release of LH-releasing factor from the hypothalamus. During pregnancy ovulation is inhibited by progesterone, at first from the corpus luteum, and later from the placenta.

Progesterone raises body temperature slightly, probably due to the formation of progesterone derivatives, mainly etiocholanolone, but also pregnanediol. Progesterone secretion during the luteal phase of the ovarian (and menstrual) cycle probably accounts in this way for the rise in morning body temperature after ovulation.

Relaxin and pelvic ligaments. Relaxin is a polypeptide (molecular weight 8000) which has been isolated from the corporea lutea of many species and its concentration in tissues and blood increases during pregnancy. It causes relaxation of the symphysis pubis, inhibition of uterine contractility, and softening of the cervix. Although it has been identified in the serum of pregnant women its physiological role is difficult to establish.

Therapeutic uses of oestrogens and progesterone

Oestrogens. Stilboestrol and ethinyloestradiol have been used to control menopausal symptoms (flushing, etc.) which occur when ovarian function ceases. Oestrogens are also used in the oral contraceptive pill [p. 575]

Progesterone, or rather its orally absorbed derivatives, are incorporated in the contraceptive pill [p. 575]. Progesterone has been claimed to be of value in pregnant women who have had repeated abortion, either by promoting placenta formation or by reducing uterine contraction. Its value is questionable.

Relationship of hypothalamus and anterior pituitary to ovary

This subject is discussed more fully on pages 521–3, but a few points need to be mentioned in this section:

1. The hypothalamus, through its GnRH, is the final common nervous pathway through which various nervous and psychological factors influence the secretion of the pituitary gonadotrophins, FSH and LH. In women emotional disturbances may cause irregularities of the menstrual cycle and impair fertility.
2. In patients with intact ovaries, but with lack of development of the reproductive organs and secondary sex characteristics, the defect may lie in the adenohypophysis or the hypothalamus. If the pituitary or hypothalamus is defective administration of FSH followed by LH will promote development and normal functions of the organs of reproduction by stimulating the ovary to ovulate and secrete oestradiol and progesterone. In this way many women who have amenorrhoea and are infertile have become pregnant; unfortunately the required dosage of FSH and LH is difficult to predict and in several cases multiple ovulations have been produced with the birth of up to six babies after one course of all-too-successful treatment. A synthetic drug, *clomiphene*, can also induce ovulation, probably by acting on the hypothalamus and thereby promoting LH release from the adenohypophysis.

In some women hypogonadism is due to a defect in the hypothalamus and administration of the decapeptide GnRH can stimulate the normal adenohypophysis to secrete FSH and LH and in this way stimulate ovarian function.

Role of blood oestrogen and progesterone levels

Gonadotrophin secretion by the adenohypophysis is influenced by oestrogens and by progesterone-like compounds (feedback).

The effects of oestrogen vary with the dose. In small doses oestrogen increases the output of FSH, in moderate doses it increases the output of LH and of prolactin, and in large doses it reduces the output of FSH, and also inhibits secretion of prolactin. Progesterone and related substances inhibit the release of LH.

At the menopause when the ovary ceases to secrete oestrogen and progesterone there is increased urinary excretion of FSH, and, to a less extent, of LH. Oestrogen administration then reduces FSH excretion.

The effects of oestrogen and progesterone on gonadotrophin secretion are mediated via centres in the hypothalamus. Lesions in the hypothalamic paraventricular nuclei prevent the atrophy of the gonads induced by administration of oestrogen. Autotransplantation of tiny pieces of ovarian tissue to this region of the hypothalamus suppresses gonadotrophin secretion, presumably by secretion of oestrogen by the graft.

Antifertility drugs. Oral contraceptives

This is an appropriate place to consider the ways in which drugs or homones could interfere with fertilization or prevent the embedding of the fertilized ovum. The problem of birth control is of overwhelming importance in a world where increases in population are outstripping available supplies of food. Fertilization can be prevented in many ways.

If sexual intercourse is confined to the so-called safe period, so that it does not take place a few days before or after the time of ovulation, fertilization should not occur. But it is important to remember that although ovulation most frequently occurs about halfway through the menstrual cycle this is a statistical mean of a distribution which may range from the 7th to the 21st days of the cycle. An individual woman may not always ovulate at the same time in the month. It is possible to detect the time of ovulation by recording the morning mouth temperature which rises by 0.5 °C after ovulation has occurred, but this is not practicable in many parts of the world and attempts to use the rhythm method have not been very satisfactory as a means of preventing pregnancy.

In the male ligation of the vas deferens is a simple, effective procedure but not practicable in a large population. There are drugs which inhibit spermatogenesis but they are too toxic for widespread use. Withdrawal of the penis before orgasm (coitus interruptus), the use of the rubber sheath (condom) by the male, or the diaphragm on the cervix in the female, douches, spermicidal jellies and creams are all less reliable than the intra-uterine device (IUD) which has been widely used in India, and the 'contraceptive pill' used in many Western countries. It is claimed that oral contraceptives, if correctly used, are 100 per cent successful in preventing fertilization.

It was long known that progesterone inhibited ovulation, e.g. during pregnancy, but progesterone itself was inactive when given orally. However, in the early 1950s Gregory Pincus and his colleagues in the United States introduced some new orally active progestagens (gestagens), which, combined with small amounts of oestrogens, have been used extensively during the past 15 years. The common procedure has been to give the pill daily for 21 days from the 5th to the 25th day of the cycle. The gestagens include norethisterone, norethynodrel, and chlormadinone while ethinyloestradiol is the usual oestrogen. The suggested mode of action of the 'classical pill', consisting of gestagen+small dose of oestrogen is:

1. Inhibition of secretion of luteinizing hormone (LH), probably by an action on the hypothalamus. This inhibits ovulation.
2. It renders the cervical mucus hostile to sperm penetration.

3. It may induce endometrial changes which prevent implantation of the blastocyst.

A modification called the 'sequential pill' involves administration of a higher dose of oestrogen alone for 15 days followed by 5 days of oestrogen+gestagen. This inhibits ovulation by suppressing the release of both FSH and LH.

Finally, an oral contraceptive consisting of low dosage of gestagen throughout the whole menstrual cycle controls fertility without inhibiting ovulation. The 'luteal supplementation' may act on cervical mucus or on the endometrium, or perhaps by decreasing the motility of the Fallopian tube.

The safety of oral contraceptives over long periods of time is still being studied. They do seem to increase significantly the risk of thrombo-embolic phenomena but the risks of pregnancy appear to be greater than those of long-term contraceptive administration. They may also produce diabetes mellitus in those predisposed to this disease and they may raise the blood pressure in women with hypertension. However, final judgement will have to wait on even longer periods of observation in women who have taken these substances for 20 or 30 years.

REFERENCES

Diczfalusy, E. (1965). Probable mode of action of oral contraceptives. *Br. med. J.* ii, 1394.
Pincus, G. (1965). *The control of fertility*. Academic Press, New York.
Potts, D. M. (1972). Artificial control of reproduction. In *Reproduction in mammals* (ed. C. R. Austin and R. V. Short) Book 5, p. 43. Cambridge University Press.

Male reproductive system

THE TESTES AND MALE ACCESSORY ORGANS

The male reproductive tract consists of the testes, the epididymes, the vas deferens, vesiculae seminales, prostate, bulbo-urethral (Cowper's) glands, and the penis.

The testis has two distinct but related functions, both of which are under adenohypophysial and hypothalamic control;

1. Production and storage of viable spermatozoa.
2. Synthesis and secretion of the androgenic hormone testosterone.

Structure. The testes are bilateral organs, one usually being larger than the other. In the adult each weighs 10–45 g (mean 25 g) and the weight decreases in old age. The bulk of the testis consists of coiled seminiferous tubules which open into the rete testis; from the rete arise ductuli efferentes, which drain into the epididymis as shown in FIGURE XII.14. The epididymis is attached to the back of the testis and consists of a coiled tube about 7 metres long which continues into the vas (ductus) deferens. The seminiferous tubules and the convoluted ductus epididymis contain smooth muscle.

In addition to the seminiferous tubules the testis contains Leydig (interstitial) cells, together with collagen fibres, a rich blood supply, and a sympathetic nerve supply. The Leydig cells secrete testosterone.

Spermatogenesis is the process by which spermatozoa are formed. It does not begin till puberty and continues throughout adult life, declining in old age. The seminiferous epithelium contains Sertoli cells, each of which extends from the basement mem-

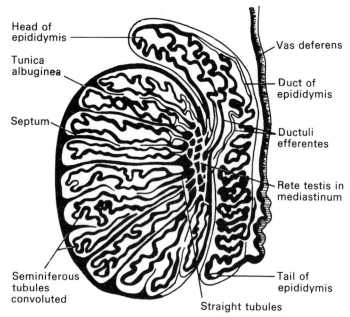

Head of epididymis

Tunica albuginea

Septum

Seminiferous tubules convoluted

Vas deferens

Duct of epididymis

Ductuli efferentes

Rete testis in mediastinum

Tail of epididymis

Straight tubules

FIG. XII.14. Diagram showing the parts of the testis and the epididymis. (Redrawn from Odell, W. D. and Moyer, D. L. (1971). *Physiology of reproduction*, p. 44. Mosby, St. Louis.)

brane to the centre of the lumen of the seminiferous tubule. The developing germ cells lie along the glycogen-containing Sertoli cells and are nourished by them as they grow from spermatogonia, on the basement membrane, to become in turn primary spermatocytes, secondary spermatocytes, spermatids, and finally spermatozoa. Spermatogonia divide by mitosis but the division of primary spermatocytes is by meiosis (maturation division) in which the number of chromosomes is reduced from 46 to 23 in secondary spermatocytes, spermatids, and spermatozoa. Thus sperm and ovum each contain 23 chromosomes and when they unite during fertilization each somatic cell formed subsequently contains 46 chromosomes. A spermatozoon contains either an X or Y sex chromosome and it is this that determines the genetic sex of the embryo [p. 563].

Spermatogenesis does not occur simultaneously in all parts of the testis; at any given moment some areas of seminiferous epithelium are active while others are at rest. Studies on the uptake of [³H]-thymidine into spermatogonia and their successors show that in man spermatogenesis takes 74 days. The sperms in the seminiferous tubules are non-motile and are pushed onwards into the straight tubules, the rete testis, and ductuli efferentes, which are lined by a ciliated epithelium; ciliary activity together with smooth-muscle contraction propels the sperms into the epididymis.

Control of spermatogenesis. Spermatogenesis does not occur after hypophysectomy. FSH alone is not effective but FSH + testosterone is fully active. LH (called interstitial-cell-stimulating hormone—ICSH—in the male) promotes testosterone secretion from Leydig cells; it is probable that testosterone reaches the seminiferous tubules via local lymphatics and that the hormone bound for distant target organs is absorbed directly into the bloodstream.

Spermiogenesis is the term used to describe the conversion of spermatids to spermatozoa. The process takes place within the cytoplasm of the Sertoli cell.

The mature human spermatozoon is 55–65 μm long [FIG. XII.15]. It is described as having a head and a tail. The head which is 4–5 μm long and 3 μm wide consists of a nucleus capped by an acrosome, and the tail comprises the neck, middle piece, principal

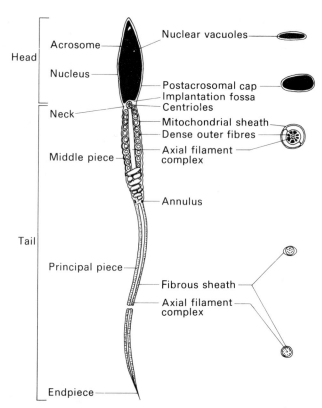

Head

Acrosome

Nucleus

Neck

Middle piece

Tail

Principal piece

Endpiece

Nuclear vacuoles

Postacrosomal cap

Implantation fossa

Centrioles

Mitochondrial sheath

Dense outer fibres

Axial filament complex

Annulus

Fibrous sheath

Axial filament complex

FIG. XII.15. Mature spermatozoon. (From Odell, W. D. and Moyer, D. L. (1971). *Physiology of reproduction*, p. 56. Mosby, St. Louis.)

piece, and endpiece. The head is flattened anteriorly; the nucleus consists of densely staining chromatin material (DNA + basic histones); the acrosome contains mucopolysaccharides and acid phosphatase. The axial filaments originate from the centrioles in the neck and consist of a central pair of fibrils surrounded by two concentric rings of 9 fibrils each. The outer fibrils are probably the contractile components of the spermatozoon. In the middle piece the axial filament is surrounded by a spiral mitochondrial sheath which provides energy for spermatozoal motility.

It is not yet possible to distinguish by morphological, chemical, or immunological methods between sperms bearing X and Y chromosomes, still less to separate them in living semen.

Seminal tract and related glands

The seminal tract consists of the epididymis, the vas (ductus) deferens, its terminal ejaculatory duct which opens into the prostatic urethra, and the penile urethra. The seminal vesicles open into the ejaculatory ducts, the prostatic glands into the prostatic urethra, and the bulbo-urethral glands into the penile urethra [FIG. XII.16].

Epididymis. After the completion of spermiogenesis, spermatozoa in the seminiferous tubular lumen are moved along the tubules to the ductuli efferentes which lead to the convoluted duct of the epididymis. The spermatozoa are then stored in the epididymis mainly in the tail, where they can remain for a month and still be capable of fertilizing an ovum. The wall of the epididymis contains smooth muscle and a secretory columnar epithelium, the secretion of which nourishes spermatozoa and helps them to mature. In the epididymis the spermatozoa are non-motile and they become motile only when they are exposed to oxygen or to a substance which can be metabolized to lactic acid. Epididymal secretion is rich in potassium and has a high K/Na ratio. It has a high concentration of *glycerylphosphoryl-choline*, a potential source of energy.

An enzyme which splits off choline and releases glycerophosphate is found in the endometrium of many animal species; this could enhance sperm motility *in utero*. Epididymal fluid also contains testosterone which may help in the maturation of spermatozoa.

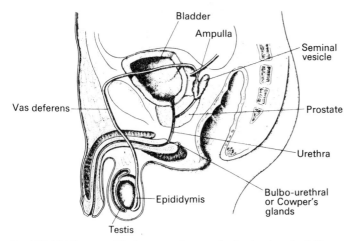

FIG. XII.16. Diagram of the human male reproductive tract, and neighbouring organs. (Littré's glands are very small and not shown here.) (From Austin, C. R. (1972). Fertilization. In *Reproduction in mammals* (ed. C. R. Austin and R. V. Short) Book 1, p. 110. Cambridge University Press.)

Vas deferens (ductus deferens). The tail of the epididymis continues into the vas deferens, a smooth muscular tube lined with a columnar epithelium, which transports spermatozoa into the dilated ampulla. This serves as a secondary storehouse for spermatozoa which will be released at ejaculation.

Vasectomy. Ligation and section of the vas deferens is a simple procedure for sterilization in the male. This may cause some degeneration of the seminiferous tubules but the Leydig cells continue to secrete testosterone and secretions from the seminal vesicles, prostate, and bulbo-urethral glands are unaffected. For a few weeks after vasectomy viable spermatozoa may be released from the ampulla but after two months sexual intercourse may be enjoyed without fear of causing pregnancy.

The vesiculae seminales. These are two lobulated glands, situated between the bladder and rectum, which secrete fluid into the ejaculatory duct. The viscid secretion is rich in potassium, *fructose,* phosphoryl choline, citric acid, and ascorbic acid. Fructose is oxidized by the mitochondria of the middle piece of spermatozoa to provide energy for their movements. The seminal vesicles (and not the prostate) also synthesize and secrete prostaglandins.

Prostate. The glands of the prostate consist of many follicle-like spaces leading into ducts. The epithelium of the follicles secretes the prostatic fluid, which is thin and opalescent and gives the semen its characteristic odour. Between the follicles there is a good deal of muscular tissue.
   Prostatic fluid. This fluid in man is slightly acid in reaction (pH = 6.4). It is rich in calcium and also contains zinc, citric acid, Na⁺ (the main cation), the enzyme fibrinolysin (plasmin), and acid phosphatase.

Bulbo-urethral glands. These form a mucoid secretion which is discharged into the anterior (penile) urethra.

Seminal fluid

1. The semen consists of the products of the seminiferous tubules, the seminal tract (especially the epididymis), and the related glands, i.e. the seminal vesicles, the prostate, and the bulbo-urethral glands. The fluid part is contributed chiefly by the prostate and seminal vesicles. Spermatozoa have very little cytoplasm and depend for their nourishment on the secretions of the accesory sex glands. Semen contains fructose in a concentration which exceeds that of glucose in blood, and it seems likely that the sperm cells use fructose as their metabolic fuel.

2. Human semen is liquid when ejaculated, but soon coagulates *in vitro* or in the vagina; after 15 minutes it undergoes secondary liquefaction. Semen contains fibrinogen and thromboplastin, but no prothrombin or thrombin. Though it is rich in calcium, the excess citrate (from the prostate) must largely remove the calcium from the ionic state. The mechanism of coagulation of semen is obscure, but it presumably involves the conversion of its fibrinogen into fibrin. The secondary liquefaction of the fibrin is due to the specific enzyme, fibrinolysin (plasmin), present in prostatic fluid. At 37 °C, 2 ml of prostatic fluid can liquefy 100–1000 ml of clotted human plasma in 18 hours.

3. Spermatozoa lose their fertilizing capacity before their motility. The maximal duration of fertilizing capacity within the human female reproductive tract (e.g. cervix) is 28–48 hours whereas motility may persist for 48–60 hours.

4. The reaction of semen is alkaline, the acid prostatic fluid being neutralized by the other components; sperms are rapidly immobilized in an acid medium.

Interstitial cells of Leydig

These cells develop from the mesoderm of the embryo; they are abundant in the fourth month of fetal life, fewer in the newborn, and continue to diminish to the end of childhood. Their number again increases at puberty; they remain constant in number during sexual life in man and finally diminish in old age. The interstitial cells are internally secreting cells and are usually arranged round the blood vessels; material with characteristic staining reactions can be traced from the cells into the capillaries. The androgen secreted by these cells is *testosterone.*

It is an interesting and unexpected fact that the testis also secretes *oestrogen.* This is shown in the condition called *testicular feminization* where male genotype is associated with female phenotype [p. 565]. Certain testis tumours (teratomata) form *chorionic gonadotrophin.*

Secretion of testosterone commences at about the age of puberty; androgen, however, appears in the urine earlier. In man, the primates, and the rat, testosterone is secreted continuously; most mammals, however, are seasonal breeders, and in them the secretion of the hormone is correspondingly intermittent. Secretion of testosterone is depressed by undernutrition and especially by vitamin-B deficiency.

Bodily changes at puberty

At puberty the testes increase rapidly in size [FIG. XI.20D, p. 527] and spermatogenesis sets in. The interstitial cells begin to secrete testosterone, and as a result the accessory organs of reproduction (epididymis, vesiculae seminales, prostate, penis) begin to grow and the secondary male sex characters make their appearance. The scrotal skin thickens; there is growth of hair on the face, trunk, and axillae; the pubic hair develops considerably and its upper border is convex upwards. Growth of the larynx occurs and the voice breaks. Considerable muscular development occurs. Occasional erections and discharge of seminal fluid take place. Striking psychological changes also begin to make their appearance at puberty.

The onset of puberty is due to changes which take place in the central nervous system. The hypothalamus begins to secrete its GnRH which acts on the anterior pituitary to release the hormones which promote spermatogenesis and secretion of testosterone [p. 523].

Physiology of coitus

The introduction of sperms into the vagina involves erection of the penis and ejaculation (emission) of the seminal fluid. Both processes are fundamentally reflex in character and can occur in a spinal man following stimulation of the glans penis or related skin areas [p. 363]. In the intact man any or many of the sense organs may constitute a source of appropriate afferent impulses; the response is long-circuited through the brain and involves the activity of the highest cortical levels which can modify the reaction either by way of reinforcement or inhibition. There is no need to stress the enormous importance of psychological influences and especially of emotional states on the act of sexual intercourse. The results of castration show that the reflex arcs are influenced at some point by the internal secretion of the testes. The changes occurring in coitus are considered below.

1. On the efferent side, erection is brought about by the nervi erigentes which relax the muscle coat of the arterioles of the penis and of the spongy tissue of the corpora cavernosa and spongiosa; at the same time the dorsal vein of the penis is compressed. The penis, which in the resting state is small, flabby, and covered with wrinkled skin, becomes thickened, elongated, and rigid and thus well adapted for introduction into the vagina; the angle which the erect penis makes with the trunk follows closely that of the vagina and its length is such that in people of average build the semen is deposited high up in the posterior part of the vagina.

2. Friction between the glans penis and the vaginal mucosa, reinforced by other afferent streams and psychological factors, causes a reflex discharge along the sympathetic to the seminal pathway; the muscle coats of the epididymis, ductus deferens, the seminal vesicles, and the prostate contract, and the sperms accompanied by the secretion of the accessory glands are discharged into the posterior urethra between the internal and external sphincters of the bladder. The semen is thence ejected by the rhythmic contractions of the bulbo- and ischio-cavernosus muscles (supplied by somatic nerves). More prostatic fluid is also secreted, probably owing to parasympathetic stimulation of the glands. During coitus the entire urethra thus takes on a sexual function. It is important to note that the sympathetic nerves which are motor to the seminal tract also close the internal vesical sphincter and thus prevent a reflux of semen into the bladder; the contraction of the sphincter vesicae and the associated inhibition of the detrusor vesicae prevent a simultaneous discharge of urine.

3. The account just given of the innervation of the accessory reproductive organs is supported by sound clinical evidence. Stimulation of the hypogastric (sympathetic) nerves at operation produces ejaculation of semen in man. Ejaculation can no longer occur after bilateral lumbar sympathectomy below L2 or section of the presacral nerve, or after administration of drugs such as guanethidine and methyl-dopa which inhibit the release of noradrenalin from postganglionic sympathetic nerve endings. In these cases penile erection and sensation remain normal. Ganglion-blocking drugs, such as hexamethonium and mecamylamine, inhibit both sympathetic and parasympathetic nerve pathways and therefore reduce both ejaculation and erection. A lesion of all the sacral nerves below S1 which severs the sacral parasympathetic outflow abolishes erection and produces relative anaesthesia of the penis.

4. The 'orgasm' just described in the male should coincide in satisfactory intercourse with appropriate psychological and reflex reactions in the female, consisting in the latter of engorgement of the vulva, relaxation of the adductor muscles of the thighs and of the vaginal orifice, secretion of mucus by the vulval and vaginal

glands, and erection of the clitoris. The vagina becomes distensible, its lining is lubricated, and it becomes easily traversable by the penis. Afferent impulses from the stimulated clitoris may heighten the state of physical excitement in the female and help to promote a complete orgasm. The technique, courtesies, and aesthetics of sexual intercourse are matters of outstanding importance, yet they are never taught by the physiologist and rarely discussed adequately at any stage in the medical curriculum. Sexual relations between civilized men and women are more than a matter of anatomy and physiology.

5. It is possible that coitus may cause secretion of oxytocin from the neurohypophysis. In lactating women coitus sometimes causes a discharge of milk from the breasts. In various species coitus may cause uterine contraction.

Normal sperms are motile and can move at a rate of 1.0 mm min^{-1}, but of the $10^8 - 5 \times 10^8$ sperms released into the vagina at ejaculation less than $10^2(100)$ reach the oviduct. The leading sperms can readily penetrate the watery cervical mucus secreted at the time of ovulation; they pass through the vast uncharted sea of the uterus and reach the oviduct within 30 min, but some spermatozoa can survive in a viable state within the slightly alkaline medium of the cervical mucous glands for up to 48 hours. However, although the fluid medium in the uterus is favourable for sperm metabolism enormous numbers of sperms are destroyed by phagocytosis. Uterine muscular movements and ciliary activity may aid the migration of sperms within the uterus but the real problem is how they get through the utero-tubal junction to reach the ovum at the ampullary-isthmic junction of the Fallopian tube. Only one sperm is usually concerned with fertilization of the ovum; it may be that the acrosome reaction in this one spermatozoon releases enough lytic enzymes locally to enable it to penetrate the cumulus cells and the zona pellucida of the ovum [Fig. XII.17].

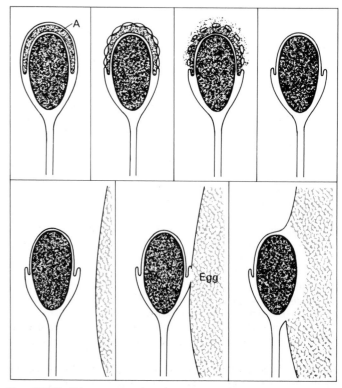

Fig. XII.17. Diagram showing the pattern of the acrosome reaction (*above*), and the first steps in the sperm–egg fusion (*below*). The outline of the spermatozoon represents its plasma membrane; the nucleus is solid black. A, acrosome.
(From Austin, C. R. (1972). In *Reproduction in mammals* (ed. C. R. Austin and R. V. Short) Book 1, p. 117. Cambridge University Press.)

6. The physiological changes which take place in intercourse are by no means restricted to the reproductive organs and adjacent parts. The usual accompaniments of certain kinds of emotional tension are present, e.g. acceleration of the heart (to 150 beats per minute), rise of blood pressure, rapid breathing, flushing of the face, and sweating. Adrenalin is doubtless poured out and it is likely that the anterior pituitary, adrenal cortex, and the thyroid glands are stimulated too.

REFERENCES

Taylor, R. W. (1975). Normal sexual response. *Br. med. J.* ii, 543.

Extirpation of testes

Before puberty

If the testes are removed there is permanent sterility; owing to absence of testosterone the usual pubertal changes do not occur and the accessory organs of reproduction do not develop; the penis, scrotum, vesiculae seminales, and prostate remain small. There is no growth of hair on the face, trunk, or axillae; the pubic hair is of the female type, the outline being concave upwards; the growth of the larynx is arrested. There may be abnormal deposition of fat; accumulations may be found on the buttocks, hips, pubis, and breasts. The muscles are soft and poorly developed. There is some delay in the union of the epiphyses, but no regular tendency to gigantism: some eunuchs are short, others are tall, and on the whole they show the same range of variations as do normal people. The skin is pale and tans poorly when exposed to the sun.

After puberty

Castration after puberty produces changes which vary very much in degree in different subjects. It should be remembered that some of the secondary sexual characters and accessory organs depend on testosterone not only for their development but also for their maintenance; these characters or organs are depressed after castration. Thus the seminal vesicles and prostate always atrophy. Other characters or organs having once developed under the influence of testosterone can persist in its absence; these are unaffected by castration. There is, for example, no alteration in the voice, the penis remains of normal size (it was usually amputated in the case of Eastern eunuchs), and the beard may be unaffected. Common baldness, which is a genetically-determined condition, does not occur in eunuchoid persons. However, if such persons are hereditarily so predisposed administration of androgen causes baldness, though its mode of action is quite obscure. In some cases the general bodily changes resemble those described for prepubertal castration, in others they are not marked. Sexual desire and erection may be absent; but there are many instances in which sexual activity was little impaired, successful coitus (with ejaculation of fluid from the prostate or seminal vesicles) being frequently carried out for as long as twenty-five years after castration; in these cases the pattern of reflex behaviour which was initially induced by testosterone subsequently persisted in spite of its absence. Rats or guinea-pigs may copulate for months after castration and human eunuchs are often quite promiscuous.

There is no evidence that castration damages any essential functions except those related to reproduction; it does not shorten life or produce premature senility. Many castrates have shown the highest intellectual attainments. In some, it is true that a peculiar mental state may result; this is more likely to be due to the psychological trauma produced by castration than to loss of any secretion formed in the testis. If modern psychology is to be believed, the mere subconscious fear of castration may produce serious mental symptoms, although the testes are functioning perfectly normally; it is hardly surprising that the mental results of a real castration are worse. Lucius Apuleius, even after his transformation into an ass, feared castration more than death.

Cryptorchidism. If the testis fails to descend into the scrotum, the seminiferous tubules remain infantile in structure and no development of sperms takes place. If the condition is bilateral the individual is sterile, but as the interstitial cells are structurally normal and continue to secrete, the secondary sexual characters develop normally. The lack of development of the seminiferous tubules in the cryptorchid is attributed to the higher temperature to which the gland is exposed in the abdomen compared with that of the scrotum; if the scrotum is kept artificially warmed, or if the testes are deliberately transferred to the abdomen, spermatogenesis ceases. Inadequate attention has been paid to the effects (if any) on spermatogenesis of wearing the kilt or of residence in a hot country. If the adverse effects of raised temperature on spermatogenesis are as great as suggested in the text one would expect to find a markedly reduced fertility in the tropics; of this there is no evidence.

The differences noted between the effects of cryptorchidism and those of castration demonstrate strikingly that the interstitial cells are responsible for the internal secretion which controls the accessory reproductive organs and the secondary sexual characters.

Testis hormones

The male sex hormone secreted by the testis is *testosterone*, which in the target organs is converted to the more potent *dihydrotestosterone* [Fig. XII.18]. The interstitial cells contain cholesterol ester from which testosterone is synthesized by pathways shown in Figure XII.13. The hormone is converted in the liver into the less potent androsterone and dehydroepiandrosterone, both of which appear in the urine conjugated with glucuronic or sulphuric acids. These conjugates are biologically inactive. Inactivation by the liver accounts for the relative ineffectiveness of oral administration of natural androgens. Methyltestosterone is effective by mouth because it is not inactivated while passing through the liver.

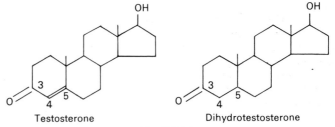

Fig. XII.18.

The male can also synthesize oestrogens, probably in the testis, the pathway for the conversion being illustrated in Figure XII.13 on page 573. Injection of isotopically-labelled testosterone in men shows that it can be converted to oestrone and oestradiol, though the amounts in circulating blood are very much less than in women.

Testosterone circulates in the blood bound to a sex-hormone-binding globulin (probably the same one to which oestradiol is bound). The free, biologically active testosterone is 1–2 per cent of the total blood concentration. The total plasma testosterone concentration in men is 7000 μg l^{-1}, and in women it is 370 μg l^{-1}. The total daily production of testosterone in men is 7 mg; in women it is 350 μg. In women testosterone is formed from androstenedione by enzymatic conversion in the skin as well as in the ovary. The growth of pubic and axillary hair at puberty in girls is due to androgenic stimulation.

In men the concentration of sex hormone-binding globulin rises after the age of 50 years and the proportion of unbound, biologically active testosterone falls. At the same time the secretion of

FSH and LH diminishes. In old age the testes get smaller, spermatogenesis decreases, and the number of Leydig cells declines.

Actions of testosterone and other androgens

Naturally secreted testosterone is responsible for the development of the accessory sex organs and the secondary sexual characters at puberty, and in the case of some of them for their persistence throughout adult life. When administered to an immature male animal testosterone causes precocious development of the accessory sex structures. Testosterone secretion decreases in old age.

Androgens overcome the degenerative changes in the accessory sexual organs resulting from castration. Thus castration produces atrophy of the glandular epithelium of the prostate and seminal vesicles and administration of androgen restores the epithelium to normal. The height of the columnar epithelial-cell lining is proportional to the strength of androgenic stimulation. The penis and scrotum depend on androgen for their development, but much less so for their maintenance.

Testosterone enhances and maintains the motility and fertilizing power of the sperms. In hypophysectomized animals large doses of testosterone stimulate and maintain spermatogenesis in the seminiferous tubules.

Testosterone stimulates the secretion of sebaceous glands and thus gives the skin at puberty an oily appearance, often accompanied by acne vulgaris.

Testosterone is largely responsible for the emotional make-up of the male.

General metabolic effects

Testosterone can be regarded as a specialized growth hormone acting mainly on the accessory organs of reproduction. Like pituitary growth hormone it causes nitrogen retention in the body and increased synthesis and deposition of protein in certain tissues, especially skeletal muscle. This is associated with an increase in muscular strength. Testosterone also causes retention of calcium, phosphorus, sodium, chloride, and water. Attempts have been made to produce related compounds in which *anabolic* activity is enhanced at the expense of androgenic activity. These *anabolic steroids* are 19-nor compounds, i.e. they lack the methyl group in the 19 position. One such compound, 17-ethyl-19-nortestosterone (norethandrolone, *Nilevar*) has anabolic effects in man in one-quarter the dose required to produce androgenic effects. This anabolic action is claimed to be of value in the treatment of premature infants and in patients before and after severe operations.

Testosterone probably acts on its target cells in a manner comparable to that by which oestradiol acts on its target cells, Inside the cells testosterone is reduced to *dihydrotestosterone* which is bound to nuclear chromatin and enhances messenger RNA activity, which in turn stimulates the synthesis of proteins.

Control of testicular activity

Hypothalamus and adenohypophysis

The growth and functions of the testis, like those of the ovary, are controlled by the hypothalamus and the adenohypophysis. The hypothalamic hormone GnRH acts on the anterior pituitary to promote the release of FSH and LH (which in the male has been called the interstitial cell stimulating hormone, ICSH).

1. Hypophysectomy in immature male animals causes the testes to remain infantile; the accessory reproductive organs do not develop owing to lack of testosterone. Hypophysectomy in adults leads to testicular atrophy and the same changes in the accessory reproductive organs and elsewhere as follow castration. In Simmond's disease hypogonadism (depressed spermatogenesis and testosterone secretion) is a common finding.

2. The interstitial-cell-stimulating hormone (ICSH) initiates and sustains the internal secretory activity of the interstitial cells. The secretion of testosterone secondarily causes growth and development of the accessory organs of reproduction. If ICSH is injected into hypophysectomized animals it also causes the return of spermatogenesis; this action is not a direct one on the seminiferous tubules but is due to the release of testosterone which in its turn acts on the tubules. This interpretation is supported by the fact that in hypophysectomized animals and in patients with Simmonds' disease the administration of androgen restores spermatogenesis. In the intact animal, however, the action of administered androgen on the testis is small and variable.

3. It has been supposed that the anterior pituitary (via the hormone known as FSH in the female) directly controls spermatogenesis. This hormone was therefore called 'gametokinetic' to indicate that in both sexes it regulates the formation and maturation of gametes (ovum or sperm). If pure FSH is injected into hypophysectomized animals it does not induce spermatogenesis. If it is given together with ICSH the restoration of spermatogenesis is more complete than with ICSH alone. FSH may thus be an accessory direct stimulating factor on the tubules and may be necessary for the maintenance of an optimal level of spermatogenetic activity.

4. The testis in its turn influences the activity of the anterior pituitary. After castration in animals the basophil cells in the pituitary increase in size and number and there is increased secretion of gonadotrophin. The testis thus normally inhibits the secretion of gonadotrophin. In clinical eunuchoidism due to primary testicular failure there is also excessive gonadotrophin formation, which is relatively immune to androgen treatment but is readily inhibited by oestrogens. (When eunuchoidism is secondary to pituitary insufficiency there is of course decreased secretion of gonadotrophin.)

The source and nature of the testicular inhibitory factor are still under consideration: (i) it might be androgen; but as mentioned above excess gonadotrophin secretion is not inhibited by physiological doses of androgen; (ii) it might be some other unidentified testicular 'factor', derived perhaps from the seminiferous tubules.

It is probably unwise in the present state of knowledge to press too closely analogies between the control of the gonads in the male and female.

5. The anterior pituitary may control the descent of the testis; in some cases of undescended testes the testes have entered the scrotum following treatment with gonadotrophic hormone.

Hormone replacement in male hypogonadism

If androgen deficiency is due to primary testicular failure (castration, disease, underdevelopment) the only effective treatment is administration of testosterone (methyltestosterone is absorbed when given by mouth).

If the testis is normal and the adenohypophysis deficient, injection of FSH and LH stimulates development of Leydig cells, secretion of testosterone, and spermatogenesis. In some patients with hypogonadism the hypothalamus is at fault and the injection of GnRH can stimulate the adenohypophysis to secrete FSH and LH and thus activate normal testicular functions.

Thymus

The thymus normally persists till puberty, when the reproductive organs develop; castration prolongs the period of persistence of the thymus. The sex glands therefore exert a depressant effect on the thymus.

The prostate [see also p. 577]

Carcinoma of the prostate. The growth and maintenance of normal prostatic tissue depends on the presence of testosterone.

The same is largely true for the cells of prostatic cancer which is the best example of a hormone-dependent malignant growth. The cancer *per se* is of unknown origin but most prostatic cancer cells, whether primary in the gland or secondary in bone marrow and lymph nodes, flourish only in the presence of testosterone.

Effect of castration and oestrogens. According to the degree to which the malignant cells are dependent on testosterone, the growth (both primary and secondary) regresses after castration or on administering an oestrogen (e.g. stilboestrol). Adenocarcinoma is much more sensitive to testosterone than undifferentiated carcinoma and is therefore much more susceptible to oestrogen therapy. The oestrogen acts:

1. By inhibiting the release of gonadotrophin by the anterior pituitary and so decreasing natural testosterone secretion.

2. By peripherally competing with and antagonizing the action of testosterone. In successful cases there is clinical improvement, and a fall of serum acid phosphatase sets in after a variable period. Some patients have remained well for as long as 6 years. The administration of androgen makes the patient worse. Generally the dependence of the tumour on androgen is partial, so that after a variable period of oestrogen treatment the growth of the tumour is resumed. Undifferentiated prostatic carcinomas are completely independent of androgen and are therefore unaffected by castration.

REFERENCES

DRILL, V. A. and RIEGEL, B. (1958). Anabolic steroids. *Recent Prog. Horm. Res.* **14**, 29.
HUGGINS, C. (1945). The prostate. *Physiol. Rev.* **25**, 281.
MANN, T. (1964). *The biochemistry of semen and of the male reproductive tract.* Methuen, London.
PAULSEN, C. A. (1974). The testes. In *Textbook of endocrinology* (ed. R. H. Williams) p. 323. Saunders, Philadelphia.
WOLSTENHOLME, G. E. W. and O'CONNOR, M. (eds.) (1967). *Endocrinology of the testis.* Ciba Foundation Colloquia on Endocrinology, Vol. 16. Churchill, London.

Physiology of pregnancy

The physiology of pregnancy is concerned primarily with the nutrition of the growing fetus and with the maternal adaptations needed for this purpose. The maternal changes required for childbirth and lactation are also important. Although it is the reproductive system which is mainly involved, most other systems of the mother's body participate in the adjustments to pregnancy. Many of the changes are due to increased hormonal activity but some physiological adaptations are controlled in other ways.

Fertilization and implantation. Fertilization of the ovum usually takes place in the ampulla of the Fallopian tube [p. 569]. Cell division begins at once to form the blastocyst which at first floats free in the Fallopian tube and then enters the uterus. After about 7 days the blastocyst, by this time consisting of about 200 cells, becomes attached to the progestational endometrium into which it burrows by the lytic action of its trophoblast layer. Maternal recognition of the onset of pregnancy arises from the secretion of chorionic gonadotrophin which causes the corpus luteum to persist.

ENDOCRINOLOGY OF PREGNANCY

Immediately after implantation chorionic gonadotrophin secretion commences and helps to maintain the corpus luteum which conti-

nues to secrete oestrogen and progesterone. As a result ovulation and menstruation are prevented.

Placental hormones

The following hormones are synthesized and secreted by the placenta:

Human chorionic gonadotrophin (HCG) is a glycoprotein which, although structurally slightly different from pituitary LH, acts physiologically like LH. HCG is formed by the syncytiotrophoblastic cells of the placenta and can be detected in serum from about 10 days after ovulation (assuming fertilization has occurred). Its concentration in serum and urine then rises rapidly to reach a peak at 50–60 days after the last menstrual period [FIG. XII.19]. After this the concentration falls to a very much lower level which is maintained till just before labour when it falls to zero. If the fetus dies early HCG disappears from serum and urine. The presence of HCG in the urine forms the basis of all pregnancy diagnosis tests.

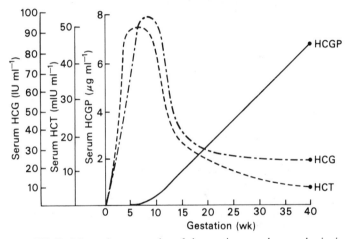

FIG. XII.19. Schematic presentation of changes in serum human chorionic gonadotrophin (HCG), human chorionic growth-hormone prolactin (HCGP), and human chorionic thyrotrophin (HCT).
(From Odell, W. D. and Moyer, D. L. (1971). *Physiology of reproduction*, p. 133. Mosby, St. Louis.)

Human chorionic thyrotrophin (HCT) is a placental substance with properties like those of pituitary TSH. Its concentration in serum follows a curve like that for HCG [FIG XII.19]. The physiological role of HCT is unknown.

Human placental lactogen (HPL), human chorionic somatomammotrophin (HCS), and human chorionic growth-hormone prolactin (HCGP) are different names for a hormone synthesized by placental trophoblast. It resembles pituitary HGH in amino-acid content and molecular weight [p. 515]. Its concentration in serum rises steadily from 10 weeks to term [FIG. XII.19]. The primary function of HPL is to promote growth and development of the breasts during pregnancy. If fetal death occurs late in pregnancy HPL secretion falls.

Oestrogen. During pregnancy oestrogen secretion occurs at first from the corpus luteum [FIG. XII.20] and later from the placenta. Urine and plasma oestrogen concentrations rise steadily to reach a peak at the time of labour [FIG. XII.21]. Urinary oestrogen consists mostly of oestriol with lesser amounts of oestrone and oestradiol, mainly as glucuronides. After delivery oestrogen excretion rapidly declines.

Progesterone secretion, reflected by urinary excretion of pregnanediol glucuronide, rises in parallel with that of oestrogen.

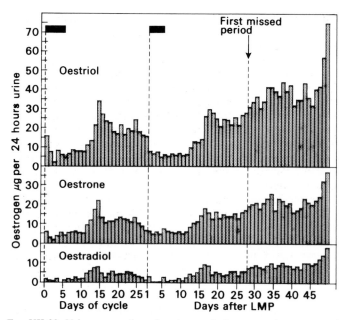

FIG. XII.20. Urinary excretion of oestrogen during a complete menstrual cycle in which conception occurred. ■ = menstrual period.
(From Brown (1956). *Lancet.* **i**, 704.)

However, during the last few weeks of pregnancy the plasma concentration of progesterone falls [FIG. XII.21]. Progesterone is secreted at first by the corpus luteum and later by the placenta.

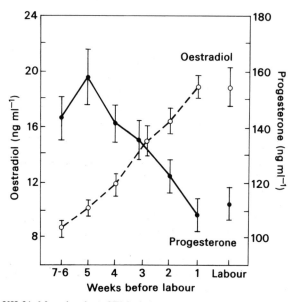

FIG. XII.21. Mean levels ± SEM of plasma progesterone and oestradiol in 33 primigravidae during the seven weeks before the spontaneous onset of labour and in the second stage of labour.
(From Turnbull, A. C., Flint, A. P. F., Jeremy, J. Y., Patten, P. T., Keirse, M. J. N. C., and Anderson, A. B. M. (1974). *Lancet* **i**, 101.)

Human maternal–placental–fetal unit

Both oestrogen and progesterone are required for the initiation and maintenance of pregnancy. In women these hormones are produced mainly by the corpus luteum during the first two months of pregnancy. After this time ovariectomy does not interrupt pregnancy as the maternal–placental–fetal unit takes over the formation of oestrogens and progesterone.

Progesterone. Although the placenta cannot synthesize cholesterol from acetate both mother and fetus can do so and the cholesterol so formed diffuses into the placenta which possesses the enzymes needed to convert cholesterol to progesterone via pregnenolone [FIG. XII.13]. The progesterone can diffuse back into the maternal circulation and so exert its physiological actions; it is converted to pregnanediol which, as the glucuronide, is excreted by the kidney. Progesterone also passes from the placenta into the fetus where it is hydroxylated as described on page 530 to adrenal corticosteroids [FIG. XII.22].

Oestrogens. The formation of oestrogens is more complex. Oestrone and oestradiol are synthesized in the placenta from dehydro-epiandrosterone (DHA) which enters the placenta from maternal and fetal circulations [see FIG. XII.13 for sequence of reactions]. Oestriol, the predominant oestrogen of pregnancy, originates from the fetal adrenal cortex which forms DHA, DHA sulphate, and finally 16-hydroxy-DHA sulphate. The 16-hydroxy-DHA sulphate enters the placenta and is there converted to oestriol which is taken up by the maternal circulation. In the maternal liver oestriol is conjugated with glucuronic acid and the glucuronide is excreted by the kidney. A small amount of oestriol is formed from oestrone and oestradiol.

The reactions just described show how the synthesis of oestrogens and progesterone requires the passage of precursor substances across the barriers between mother, placenta, and fetus. It is therefore apt to speak of a *maternal–placental–fetal* unit.

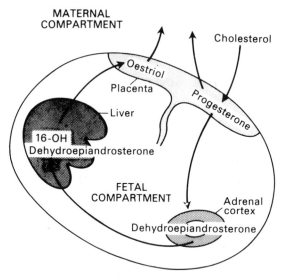

FIG. XII.22. The human feto-placental unit, showing how the mother provides precursor (cholesterol) to the placenta which converts it to progesterone for release into the maternal and fetal circulations. In the fetus, it is further metabolized by the adrenal cortex, the liver, and the placenta to oestriol, which is passed back into the maternal circulation for excretion in the urine.
(From Baird, D. T. (1972). Placental hormones. In *Reproduction in mammals* (ed. C. R. Austin and R. V. Short) Book 1, p. 21. Cambridge University Press.)

Functions. Oestrogens + progesterone promote growth of the uterus early in pregnancy: oestrogens tend to increase myometrial activity and progesterone to decrease it. Hence the rise in plasma oestrogen level and the fall in plasma progesterone level towards the end of pregnancy favour the onset of parturition (Czapo 1972; Turnbull *et al.* 1974).

Pregnancy diagnosis tests

These are all based on the presence of HCG in urine, which can be detected as early as 14 days after conception. The accuracy of these tests is of the order of 99 per cent.

Test	Criterion	Time required
Aschheim–Zondek (A–Z)	Ovulation in immature mice	5 days
Friedman	Ovulation in virgin rabbit	18 hours
Female *Xenopus laevis*	Release of ova	24 hours
Male toads or frogs	Release of sperms	3 hours
Immunological	Inhibition of agglutination of sheep red cells or latex particles coated with HCG	2 hours

Immunological tests. Antibodies to human chorionic gonadotrophin (HCG) can easily be induced in rabbits and the antiserum so produced can be used to detect the presence of HCG in urine or serum from pregnant women by means of complement fixation, haemagglutination, or precipitin tests. The principle of the haemagglutination test is as follows: tanned sheep red cells coated with HCG will agglutinate in HCG antiserum from rabbits. But if urine containing HCG reacts with the antiserum first, the latter's antibodies are used up and so are not available to cause agglutination of the sensitized sheep cells. Thus agglutination means a negative result (i.e. no pregnancy), and failure of agglutination a positive result. The test gives an answer within 2 hours. A comparable test using antigen-coated latex particles instead of haemagglutination has also been developed.

These immunological tests are cheap, reliable, and accurate enough to replace the older biological methods altogether.

REFERENCES

British Medical Journal (1962). *Br. med. J.* **ii,** Annotation 1668.
LORAINE, J. A. and BELL, E. T. (1971). *Hormone assays and their clinical applications*, 3rd edn. Livingstone, Edinburgh.

NUTRITION OF THE FETUS

The role of the placenta in supplying the fetus with the foodstuffs and oxygen essential for its growth and development is described on pages 585-91 in conjunction with the physiology of the fetus itself.

MATERNAL PHYSIOLOGY IN PREGNANCY

The average duration of human pregnancy is about 280 days or 40 weeks when calculated from the first day of the last menstrual period, or 266–270 days when calculated from the time of ovulation (estimated by waking temperature measurements) to the onset of labour.

The youngest age at which childbirth has been reliably reported is probably 4 years 8 months. This is the case of a girl named Lina Medina of Peru who was delivered by caesarean section at a hospital in Lima in 1939. Several American physicians who were in Lima at the time attested to the event and gave their opinion that she was about 5 years old. Actually further investigation suggested that she might have been 5 years 8 months old but the case is little the less remarkable for that. Pregnancy after 47 years is rare and parturition over the age of 52 has not been proved.

The uterus

The most striking change in the maternal organism during pregnancy is the enlargement of the uterus which increases in weight from 50 g in the non-parous state to 1000 g at full term. The enlargement is due mainly to hypertrophy of pre-existing muscle cells, but also to formation of new fibres (hyperplasia) during the early months of pregnancy. The individual muscle fibres grow to become 2–7 times wider and 7–11 times longer than non-parous fibres. There are also increases in the amounts of connective tissue and elastic tissue between the muscle fibres.

During the first two or three months of pregnancy the uterine enlargement is not due to distension from the growing fetus since the same degree of enlargement occurs in extrauterine pregnancy when the ovum is implanted in the Fallopian tube. This initial hypertrophy is attributed to the action of oestrogen, but the subsequent increase in size is due mostly to pressure exerted by the growing products of conception. During the early months the uterine wall becomes thicker than in the non-parous state but later it becomes thinner (5 mm thick) and the fetus can be easily palpated through it.

The middle layer of muscle fibres in the pregnant uterus consists of an interlacing network between which run the blood vessels. When these muscle fibres contract after delivery they constrict the blood vessels like ligatures and thus prevent bleeding from the placental site. If the uterus does not contract firmly enough postpartum haemorrhage will occur, which can be stopped or prevented by administration of the powerful uterine stimulant drug ergometrine.

Changes in body systems and organs during pregnancy

Blood. An outstanding feature of pregnancy is an increase in maternal *blood volume* of about 30 per cent, e.g. from 4 litres to 5.4 litres. This increase is presumed to be an adaptation to meet the demands of an enlarged uterus with its vastly increased blood supply, and to provide increased flows to the skin for the elimination of additional heat and to the kidneys for the excretion of the additional waste products from mother and fetus. The increase in total blood volume comprises increases in both plasma and erthyrocyte volumes, the plasma volume increase being the greater so that the red-cell count, haemoglobin concentration, and haematocrit value tend to decrease. These changes suggest that pregnancy is associated with anaemia, the so-called 'physiological anaemia of pregnancy'. However, this anaemia is only apparent, not real. It is due to the disproportionate increase in plasma volume, and the red cells remain normochromic and normocytic.

Plasma *iron* levels fall but plasma iron-binding capacity increases. There is a great demand for iron during pregnancy which the stores in the body, plus a dietary intake which is adequate in the non-pregnant state, may not always satisfy.

The *bone marrow* becomes hyperplastic during pregnancy, the hyperplasia involving all cell elements.

Plasma proteins. There is an increased plasma fibrinogen level which probably contributes to the increased *erythrocyte sedimentation rate* seen in pregnancy.

Plasma albumin is markedly decreased but α- and β-globulin concentrations are increased. This shows that the diminution in total plasma proteins is not due simply to dilution.

The heart. The heart appears to enlarge during pregnancy, due partly to change in position as the enlarging uterus presses upwards on the diaphragm. It is not certain whether the heart hypertrophies but there is no doubt that the *cardiac output* increases, due partly to increase in stroke volume and partly to increase in heart rate. Measurements with the Hamilton dye method and by the direct Fick method [p. 108] have shown that the cardiac output rises from 4.5 l min^{-1} to a maximum of 6 l min^{-1} before the end of the first trimester of pregnancy and remains at this high level for the rest of pregnancy. Formerly it was stated that after 28–32 weeks the cardiac output fell rather markedly, but recent studies have shown that the fall is only recorded in women lying on their backs, when the large gravid uterus would compress the inferior vena cava and reduce venous return. In women lying on their sides the large gravid uterus does not press on the vena cava and the cardiac output is recorded at 6 l min^{-1} till the end of pregnancy.

The circulation. In normal pregnancy the *arterial blood pressure* shows a slight lowering of systolic and a somewhat greater lowering of diastolic pressure.

The *antecubital venous pressure* is normal but the *femoral* venous pressure is increased due to pressure by the enlarged uterus on the pelvic veins. The raised femoral pressure returns to normal immediately after delivery.

The *blood flow* through the hand and forearm is increased during pregnancy, presumably to dissipate the excess heat generated by increased metabolism as the fetus grows.

The *blood flow* through the *uterus* is very considerably increased [see p. 586].

Respiration. Upward displacement of the diaphragm might be expected to reduce vital capacity, but in fact the diminished height of the pleural cavity is compensated by an increase in width so that vital capacity is unchanged in pregnancy. Tidal volume and pulmonary ventilation are increased, returning to normal during the second week of the puerperium. The increased ventilation may be due to a greater sensitivity of the respiratory centre to CO_2 and results in a lowering of the arterial P_{CO_2} by about 1 kPa, e.g. from 5.3 kPa (40 mm Hg) to 4.3 kPa (33 mm Hg). This may be an effect of the increased progesterone level.

Oxygen consumption is increased by about 15 per cent during pregnancy, mainly to satisfy the needs of the fetus, but also to supply the demands from increased cardiac work, increased respiratory work, and from the added uterine muscle and breast tissue and the placenta.

Digestive tract. In the *stomach* there is hypochlorhydria and decreased motility. Morning nausea and vomiting frequently occur during the early months of pregnancy; the cause is not known. There is loss of tone in the colon and in the bile duct.

Urinary system. Renal blood flow and glomerular filtration rate are increased, probably in parallel with the increase in cardiac output. Increased GFR increases the load of solutes presented for reabsorption. This may account for the glycosuria of pregnancy. The *ureters* are dilated in pregnancy, the right more than the left.

Endocrine glands (apart from ovary and placenta)

The most important endocrine changes in pregnancy, namely the placental production of oestrogen, progesterone, and chorionic gonadotrophin are discussed elsewhere [p. 581].

The thyroid gland may show slight enlargement, with hyperplasia and increased thyroxine output, but since the amount of thyroxine-binding protein in the plasma is increased (by oestrogen) there are no symptoms of hyperthyroidism.

The adrenal cortex enlarges during pregnancy, particularly the zona fasciculata in which cortisol is manufactured. Although cortisol secretion is enhanced and plasma cortisol levels increase (sometimes to values higher than those found in Cushing's syndrome), there are usually no signs of hypercorticism. This is because the excess cortisol is largely bound in a physiologically inactive form to the plasma protein 'transcortin' which is formed in much increased amounts during pregnancy. Nevertheless the well-known regression of rheumatoid arthritis which occurs in pregnancy may be due to increased cortisol secretion.

Nervous system. Mild mental changes are common in pregnancy. Craving for unusual articles of diet and alterations in mood are

characteristic. In a few cases a true psychosis may develop. The causes of these mental changes are not known.

Skin. *Pigmentation* of the nipple and areola in the breast occurs. The linea alba often becomes pigmented too, and occasionally there are brownish patches on the face and neck. These changes disappear after delivery. The explanation of these types of pigmentation is not clear but over-secretion of ACTH or melanocyte stimulating hormone [p. 534] might be responsible.

Metabolic changes in pregnancy

Gain in weight. During pregnancy there is marked increase in body weight averaging about 12.5 kg during the whole 40-week period. This increase is accounted for as follows [Fig. XII.23]:

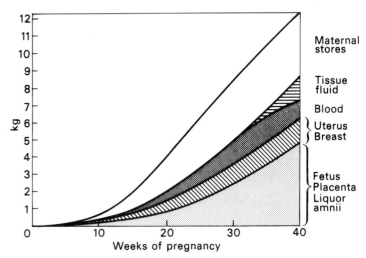

Fig. XII.23. The components of weight gain in normal pregnancy. (Hytten, F. E. and Leitch, I. (1971). *The physiology of human pregnancy*, 2nd edn. Blackwell, Oxford.)

In developing countries where the diet is inadequate the maternal gain in weight may be only 3–6 kg during the whole of pregnancy. The maternal store of fat laid down in the earlier stages of pregnancy is used up in the later stages to nourish both fetus and mother. Even so the weight of the newborn baby may be under 3 kg.

Water metabolism. Increased water retention is one of the most characteristic biochemical changes of later pregnancy. Studies of total body water using D_2O as a tracer show that excess water is retained in the fetus, placenta, and amniotic fluid to the extent of 3.5 litres; increase in blood volume accounts for about 1.3 litres; retention in the breasts, uterus, and other tissues accounts for about 1 litre. During the early days of the puerperium there is a marked diuresis, sweating, and a weight loss of about 2.3 kg.

The retention of water is not completely explained but a fall in plasma protein concentration of 10 g l^{-1} of plasma (particularly of albumin) reduces the colloid osmotic pressure of the plasma by about 20 per cent and would favour retention of water in the tissues. In addition there is retention of sodium to the extent of about 12 g during the 2nd and 3rd trimesters of pregnancy; this is accounted for in the products of conception and in the increased plasma volume. It seems likely that sodium and water retention are due to the steroidal sex hormones.

Protein metabolism. With an adequate diet there is substantial nitrogen retention during pregnancy. The amount of nitrogen retained during pregnancy is greater than that accounted for by retention in the fetus, uterus, and placenta. During lactation, too, there is a positive nitrogen balance, so that more is retained than is needed for the milk.

Carbohydrate metabolism. There is a tendency to glycosuria during pregnancy. Sugar tolerance curves show that this is due to a lowered renal threshold for glucose.

Fat metabolism. In pregnancy there are increases in the blood concentrations of cholesterol, phospholipids, and neutral fat. Adipose tissue depot fat is increased and provides a large energy bank for use in the later stages of pregnancy and during lactation.

Mineral exchange. The *calcium* content of the fetus at term is about 25 g, about two-thirds of the deposition in bone taking place during the last month. Balance studies show that during pregnancy the mother stores about 50 g of calcium and 35–40 g of phosphorus. Only half the calcium goes to the fetus, the rest being stored in maternal tissues, presumably in preparation for lactation.

Iron. The mature human fetus contains 375 mg of iron. This accumulates at the rate of about 0.4 mg per day in the first two-thirds of pregnancy and at about 4 mg per day during the last third. A further 500–700 mg of iron is required by the mother for increased haemoglobin synthesis and a small additional amount of myoglobin formation in the hypertrophied uterus. The pregnant woman requires an extra 1 g of iron which must be provided by food and body reserves. The iron in haemoglobin and myoglobin is retained after delivery but the iron in the newborn child and the iron removed in the blood lost during delivery amounts to 400–500 mg. This net loss represents 1/8–1/10 of the total body iron and must be compensated by the increased dietary intake of iron during pregnancy.

REFERENCES

HYTTEN, F. E. and CHAMBERLAIN, G. V. P. (1980). *Clinical physiology in obstetrics*. Blackwell, Oxford.
RHODES, P. (1969). *Reproductive physiology for medical students*. Churchill Livingstone, London.

Parturition

Uterine contractions of mild, non-painful, intensity can be recorded from the end of the first trimester of pregnancy. Measurements of amniotic-fluid pressure or intra-uterine pressure show increased uterine activity from the 30th week of gestation until during labour the contractions become much greater, more frequent, and painful. The intra-uterine pressure during contraction may rise to 30–50 mm Hg. After delivery of the fetus rhythmical contractions continue for a few days but their frequency and force diminish rapidly.

Initiation and control of parturition

The processes which initiate and control parturition are still not fully understood but the present state of knowledge can be summarized as follows:

1. Direct *nervous control* is not necessary for labour in women. Parturition can still occur after section of the spinal cord in the mid-thoracic region or after cutting the nerve supply (sympathetic) to the uterus.
2. Role of *hormones*. Formerly the onset of labour was ascribed solely to hormonal changes in the mother. There is now evidence to support the view that the fetus initiates parturition by neurohormonal processes (Liggins *et al.* 1973). In sheep and goats parturition is triggered off by release of corticotrophin-releasing factor (from the hypothalamus) which promotes release of ACTH [p. 518] from the adenohypophysis. Destruction of the fetal hypothalamus, or

hypophysectomy, greatly delays the onset of labour. ACTH then acts on the large fetal adrenal cortex to promote secretion of cortisol which passes from the fetus into the placenta where it increases oestrogen secretion and decreases progesterone secretion. As a result the synthesis of prostaglandin $F_{2\alpha}$ is enhanced in the placenta and myometrium. This increases the sensitivity of the myometrium to the uterine stimulant action of oxytocin, so that parturition may be initiated without any increase in oxytocin secretion. However, it has been shown that oxytocin secretion is increased in the later stages of labour, probably by stimuli arising from the distended cervix and vagina. Oxytocin secretion thus ensures expulsion of the fetus and placenta in a reasonable period of time.

It is not yet known how far the studies in sheep and goats apply to women, but the following facts are compatible with these studies:

1. With an anencephalic fetus, which has no hypothalamus and therefore no secretion of ACTH or cortisol, the onset of labour is delayed.
2. During the first stage of labour the amniotic fluid contains PGE_1 and $PGF_{2\alpha}$. Anti-inflammatory drugs, such as aspirin and indomethacin which inhibit prostaglandin synthesis, may delay the onset of labour.
3. The sensitivity of the uterus to oxytocin increases during the latter part of pregnancy and is greatest just before the onset of labour.
4. The concentration of plasma oestrogen increases and that of progesterone decreases during the last few weeks of pregnancy [FIG. XII.21].

REFERENCES

CHALLIS, J. R. G. and THORBURN, G. D. (1975). *Br. med. Bull.* **31**, 57.
CSAPO, A. I. (1972). *J. reprod. Med.* **9**, 400.
GOLDBERG, V. J. and RAMWELL, P. W. (1975). *Physiol. Rev.* **55**, 325.
LIGGINS, G. C., FAIRCLOUGH, R. J., GRIEVES, S. A., KENDALL, J. Z., and KNOX, B. S. (1973). *Recent Prog. Horm. Res.* **29**, 111.
TURNBULL, A. C., FLINT, A. P. F., JEREMY, J. Y., PATTEN, P. T., KIERSE, M. J. N. C., and ANDERSON, A. B. M. (1974). *Lancet* i, 101.

Physiology of the fetus and newborn

THE PLACENTA

The newborn child after 40 weeks gestation weighs about 3.5 kg. Though the fetus grows relatively much faster in the earlier months of pregnancy, the absolute amounts of building materials required increase steadily till birth occurs [FIG. XII.24].

The fertilized human ovum contains very little yolk so the growth of the fetus depends on nutriment derived from the mother. For the first few days after implantation nutritive materials must come from the plasma in the oedematous decidua and from endometrial glandular secretion containing glycogen. Thereafter all the requirements for fetal growth must be obtained from the maternal circulation through the placenta. The rate of placental growth is initially much greater than that of the fetus, but later slows down while that of the fetus continues, so that from about mid-term the weight of the fetus exceeds that of the placenta [FIG. XII.24]. At term the placenta weighs 0.5 kg, about 14 per cent of the weight of the fetus. By the fourth week of pregnancy the placenta has developed to the extent that substances can pass across it from the maternal to the fetal circulation and vice versa. Fetal nutrition involves not only the provision of materials which an animal after birth normally assimilates via the alimentary tract, but also the transfer of O_2 which is supplied after birth by the lungs. In addition the placenta is the organ of excretion for waste products which will be eliminated after birth by the kidney and bowel and it also allows the removal of CO_2.

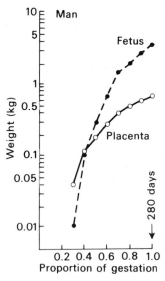

Fig. XII.24. Growth of the human fetus and placenta: the weights are plotted on a logarithmic scale.
(From Dawes, G. S. (1968). *Fetal and neonatal physiology*. Year Book, Chicago.)

Thus for the fetus the placenta combines the functions of the alimentary tract, kidneys, and lungs.

Blood supply of the placenta

There are three aspects of the circulation of the placenta: the relation between the maternal and fetal circulations within the placenta, the uterine circulation which supplies maternal blood to the placenta, and the umbilical circulation which supplies the fetal contribution.

Vascular arrangement within the placenta

In man and the higher sub-human primates the maternal blood vessels in the decidua basalis under the developing embryo become greatly dilated at an early stage of pregnancy. Small finger-like projections from the outer layer of the blastocyst, the chorionic villi, grow into these blood vessels by eroding the decidua. Further erosion of the maternal tissue results in the formation of the *intervillous space*, filled with maternal blood, in which lie the chorionic villi. These, having themselves been invaded by the developing fetal mesoderm, consist of a core of capillaries and connective tissue invested by a layer of trophoblast. The human placenta is thus *haemochorial*, the chorionic villi dipping directly into the maternal blood. The fetal and maternal circulations are thus separated only by three thin layers, the fetal vascular endothelium, the connective tissue of the villus, and the trophoblast.

When it reaches the placenta the umbilical artery divides into main branches which enter the intervillous space and subdivide to supply the chorionic villi. The intervillous space is subdivided by septa forming up to 200 placental lobes or cotyledons each supplied by a main branch of the umbilical artery. The villi are closely packed within the intervillous space which thus comprises a system of communicating crevices about 50 μm wide. The spiral arteries which bring blood from the uterine arteries to the maternal side of the placenta open into the intervillous space all over the surface of the basal decidual plate. The maternal arterial blood thus enters the intervillous space under high pressure and spreads up to the chorion plate before dispersing laterally and downwards past the capillary bed of the fetal villi to reach the basal plate. The blood then drains into the uterine veins [Fig. XII.25].

Uterine blood flow

Uterine blood flow has been measured during pregnancy both directly using the electromagnetic flow-meter and indirectly by methods based on the Fick principle, antipyrine generally being used as the test substance. Most of the observations have been made on larger animals such as sheep, goats, and cows since the vessels in these species are large enough to permit the necessary manipulations. Uterine blood flow is generally expressed in terms of the weight of the uterus plus its contents, i.e. the fetuses, and in these species remains constant at about 200–300 ml min^{-1} kg^{-1} throughout the latter half of pregnancy. A few measurements have been made of uterine flow in women and the values obtained have been in the region of 125–150 ml min^{-1} kg^{-1} indicating a total uterine blood flow near term of 600–750 ml min^{-1}. The lower values obtained in man as opposed to the other species mentioned may have resulted in part from the experimental difficulties inherent in making such measurements in human subjects.

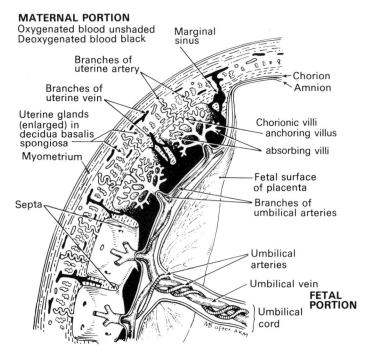

MATERNAL PORTION
Oxygenated blood unshaded
Deoxygenated blood black

Fig. XII.25. Scheme to show the essential features in placental structure which are found after the 60-mm stage. Four cotyledons, including a marginal one, are illustrated; the cotyledons are separated from each other on the maternal side by the septa. They each contain the group of villi which constitute the associated 'fetal' cotyledon. The villi branch freely and there are many adhesions between adjacent ones so giving a partially labyrinthine nature to the intervillous space. The openings of the endometrial arteries and veins into the intervillous space through the basal plate are indicated. About 70 uterine (spiral) arteries supply the mature placenta which has about 200 cotyledons of various sizes.
(After Hamilton, W. J. and Boyd, J. D. (1960). *J. Anat.* **94**, 297–328.)

The expression of uterine flow in terms of the weight of the uterus and its contents may give a misleading impression of the changes in flow to the placenta itself as pregnancy proceeds. In the sheep, placental growth is complete at about the half-way stage of pregnancy but it is only during the second half of pregnancy that the rapid increase in fetal weight occurs. A further factor is that not all the uterine blood flow supplies the placenta. The proportion directed to the placenta has been estimated by injecting radioactively-labelled microspheres of 15–50 μm diameter into the arterial supply. These are trapped in the capillaries and the distribution of the blood flow to the various tissues of the uterus is indicated by their relative radioactivity. About 80–85 per cent of the total uterine blood flow goes to the maternal placenta in the sheep and monkey, the remainder supplying the endometrium and myometrium. Calculation of *placental* blood flow from the values for uterine blood flow during the second half of pregnancy, on the assumptions that placental weight remains constant and that the placenta receives 80 per cent of the total uterine flow, reveals that there is an increase in flow per unit weight of the placenta which parallels the growth of the fetus [FIG. XII.26]. However this increase in placental perfusion rate may be less dramatic in primates in which the placenta continues to grow throughout pregnancy [FIG. XII.24].

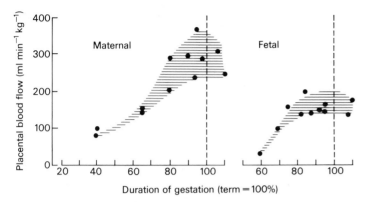

FIG. XII.26. Changes in maternal and fetal placental blood flow during gestation in the sheep. Term in the sheep is 147 days. Note that flow to the placenta from both the maternal and fetal (umbilical) circulations increases greatly during the second half of gestation when fetal growth is most rapid, though there is some tendency for flow to stabilize during the final 20 days of pregnancy.
(Drawn from unpublished data of M. Silver, R. J. Barnes, R. S. Comline, and G. J. Burton, with additional values from Rosenfeld *et al.* (1974) and Mankowski *et al.* (1968).)

Little is known about the way in which maternal flow to the placenta is adjusted to the weight of the developing fetus. The increase in maternal cardiac output and the great reduction in uterine vascular resistance undoubtedly contribute to the increased uterine blood flow but the more precise mechanism remains obscure. One obvious possibility, the levels of blood P_{O_2} and P_{CO_2}, has been discounted. Though the uterine vasculature is responsive to catecholamines, the sympathetic nervous system is unlikely to play an important role since only the non-placental tissue, which receives merely 15 per cent of the uterine blood flow, is innervated. Hormonal influences may play a part, particularly early in pregnancy. Oestrogen has been shown to produce uterine vasodilatation and progesterone a reduction in uterine blood flow; this would be consistent with the observation in the sheep that uterine blood flow (per unit weight) is highest in early pregnancy and falls at about half term when maternal plasma progesterone rises. However, the subsequent increase in placental flow rate as fetal weight rises during the latter part of pregnancy cannot be similarly explained on the basis of the changes in the maternal circulating levels of oestrogen or progesterone during this period.

Umbilical blood flow

Umbilical blood flow has been measured indirectly by the antipyrine-clearance technique and by injecting radioactive microspheres. Values of 170–240 ml min⁻¹ kg⁻¹ fetal weight have been obtained and the relation of flow to fetal weight remains relatively constant at this level throughout the latter half of pregnancy. There have been no direct measurements of umbilical blood flow but flows of 180 ml min⁻¹ kg⁻¹ have been recorded with flow-meters placed around the lower aorta in the fetal lamb. Since some of this flow must be non-placental this represents a value for umbilical flow somewhat lower than that found by indirect measurements.

As mentioned above, fetal weight rises rapidly during the latter half of pregnancy whereas the weight of the placenta is relatively static. Thus if the foregoing figures for umbilical blood flow are referred to placental weight rather than fetal weight the rate of umbilical flow through the placenta is seen to rise sharply during the latter half of gestation [FIG. XII.26]. It can be seen that there is a surprisingly close correlation between the flows to the placenta from the maternal uterine and fetal umbilical circulations.

This increase in total umbilical flow, which is necessary to satisfy the growing demands of the fetus, is probably brought about by the rises in fetal arterial pressure and cardiac output during the latter stages of gestation. Though the umbilical vessels are not innervated and do not appear responsive to changes in the P_{O_2} and P_{CO_2} of the blood perfusing them, adverse circumstances such as hypoxia result in a redistribution of the cardiac output by the autonomic nervous system leading to increased placental flow as well as increases in cerebral and coronary flows.

Transfer of oxygen to the fetus

The oxygen consumption of the fetal lamb at term has been measured by closed-circuit spirometry immediately following caesarian section and has been calculated from the product of umbilical blood flow and the difference in oxygen content of the blood in the umbilical vein and artery. Both methods give values for oxygen consumption of about 7 ml min⁻¹ kg⁻¹. This suggests that the human 3.5 kg fetus, which is similar in weight to the fetal lamb, requires 20–25 ml of oxygen per minute.

Some recently obtained values for P_{O_2} and P_{CO_2} *in the uterine and umbilical arteries and veins* of the sheep are shown in FIGURE XII.27 from which it can be seen that there is a large uterine vein–umbilical vein gradient of 2.3 kPa (17 mm Hg) for oxygen and a much smaller gradient of 0.4 kPa (3 mm Hg) for carbon dioxide. It might seem that this large oxygen gradient can be accounted for by the resistance to the diffusion of oxygen offered by the much thicker and presumably less permeable cell layers which separate the fetal blood in the villous capillaries from the maternal blood in the intervillous space, as compared with the adult lung where only the two thin endothelial layers of the vessels and alveoli separate the blood in the pulmonary capillaries from the oxygen in the air sacs. The much lower CO_2 gradient could be explained by the greater diffusibility of this gas. However, these gradients may give a deceptive indication of the relative efficiency of gas exchange across the placenta. Some 15–20 per cent of the blood in the uterine veins will have traversed non-placental tissue, chiefly the myometrium, while some 5 per cent of the umbilical flow is directed to the fetal membranes. Furthermore *the placenta has itself an appreciable oxygen consumption* which has been put as high as 30–35 ml min⁻¹ kg⁻¹, so that the oxygen usage by the placenta weighing 500 g may be nearly as great as that of the 3 kg fetus. This usage of oxygen by the placenta and non-placental tissues will influence the composition of the uterine and umbilical vein blood which may therefore not be representative of the blood leaving the exchange areas of the placenta. Indeed it is now thought unlikely that the placental membranes constitute a diffusion barrier limiting gas exchange.

The tension of oxygen in the blood supplied to the fetus by the umbilical vein, 4.0–4.7 kPa (30–35 mm Hg), is about one-third of that in the blood returning from the lungs in the adult and Barcroft has referred to this as the 'Mount Everest *in utero*' which every fetus must surmount. Moreover, as shown in FIGURE XII.32, the relatively well oxygenated blood derived from the placenta is progressively diluted as it traverses the fetal circulation so that the fetal tissues are all supplied with blood which is approximately 60 per cent saturated with oxygen.

The human fetus possesses several adaptations which to some extent compensate for the reduced tension at which it receives its supply of oxygen. There is a *shift to the left of the oxygen dissociation curve for fetal blood* as compared with that of adult blood [FIG. XII.28(a)]. As a result, fetal blood can take up much larger volumes

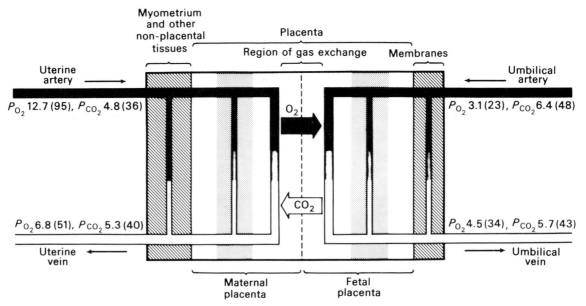

FIG. XII.27. Gaseous exchange at the placenta. Cross-hatched areas represent extraplacental regions supplied by the uterine and umbilical circulations; stippled areas indicate oxygen usage by the placenta. Values for gas tensions are in kPa, with equivalent values in mm Hg in parentheses.
(Values from Comline, R. S. and Silver, M. (1970). *J. Physiol., Lond.* **209**, 587.)

of oxygen than adult blood at low oxygen pressures. Thus the P_{50}, the oxygen partial pressure at which haemoglobin is 50 per cent saturated with oxygen, is 3.6 kPa (27 mm Hg) for adult human blood, while the P_{50} for the blood of the human fetus at term is 2.5 kPa (19 mm Hg).

All the haemoglobin in the early fetus is of the so-called fetal type, *haemoglobin F*; it differs from adult haemoglobin in possessing two γ polypeptide chains in place of the β chains of adult haemoglobin [p. 183], from which it can be distinguished spectrographically and by its characteristic electrophoretic mobility. The adult type begins to appear in the blood at mid-pregnancy when the bone marrow starts to function as a haemopoietic organ. It forms 6 per cent of the total circulating haemoglobin at the 20th week of pregnancy, 20 per cent at birth, 50 per cent at 2 months postnatal, and 90 per cent at 4 months. The last two values suggest that haemoglobin F is not formed after birth and that the corpuscles containing it are destroyed during the first 4 months or so of postnatal life, as would be expected from the known survival time of circulating red cells [p. 44]. However, it is not simply the presence of haemoglobin F which is responsible for the difference in the oxygen dissociation curves of fetal and maternal blood, since solutions of fetal and maternal haemoglobin, separated from their respective cells and dialysed against a common solution, have identical oxygen dissociation curves. The cause of the difference between the dissociation curves of fetal and maternal blood is the presence within the red cells of 2,3-diphosphoglycerate (DPG). DPG competes with oxygen for the binding sites on the haemoglobin molecule and therefore at a given partial pressure of oxygen the percentage saturation of haemoglobin will be reduced in the presence of DPG, i.e. the dissociation curve will be displaced to the right. Other organic phosphates, principally ATP, may have a similar effect. DPG is present in much the same concentration in the red cells of both the human fetus and its mother. However the affinity of fetal haemoglobin for DPG (and for ATP) is considerably less than that of adult haemoglobin. Thus the oxygen dissociation curve for fetal blood is displaced less than that for adult blood, resulting in the fetal blood curve lying to the left of the curve for maternal blood.

Another factor which helps to compensate for the reduced oxygen tension of fetal blood is the *greater concentration of haemoglobin in the fetus*. The maternal haemoglobin concentration falls during pregnancy. This is not due to a reduction in circulating red-cell mass, which actually rises somewhat, but to a proportionately greater expansion of plasma volume. On the other hand in the fetus the haemoglobin concentration during the latter part of pregnancy may exceed normal adult levels and even reach 18–20 g dl^{-1} at term. The extent to which this greater oxygen capacity in the fetus, in conjunction with the shift to the left of the haemoglobin dissociation curve, can increase the oxygen content and thus partially compensate for the reduced oxygen tension in fetal blood, is made evident in FIGURE XII.28 (b). In this figure the oxyhaemoglobin dissociation curves of FIGURE XII.28 (a) are replotted after converting the percentage saturation values on the ordinate to the corresponding oxygen contents of fetal and maternal blood.

A third factor aiding the uptake of oxygen by the fetal blood as it traverses the placenta is the '*double*' *Bohr effect*. The oxyhaemoglobin dissociation curve of fetal blood, like that of adult blood, is displaced to the right by an increase in P_{CO_2} or a reduction in pH [FIG. XII.29 (a)]. This is the normal Bohr effect [p. 185]. In the placenta carbon dioxide is transferred from the fetal to the maternal circulation. Thus while flowing through the placenta the P_{CO_2} of the fetal blood falls and its pH rises, while the P_{CO_2} of the maternal blood rises and its pH falls. Thus the fetal dissociation curve is displaced to the left while the maternal dissociation curve is simultaneously displaced to the right. The result of this 'double' Bohr effect is therefore both to facilitate the unloading of oxygen from the maternal haemoglobin and to augment oxygen uptake by the fetal haemoglobin [FIG. XII.29(b)]. The oxygen-tension values for human maternal and fetal blood given in FIGURE XII.29(b) are all somewhat lower than the corresponding values for sheep blood in FIGURE XII.27. This may represent a species difference but may also reflect the different circumstances under which the samples were obtained. The sheep were unanaesthetized with chronically implanted catheters while the human samples were withdrawn during surgery under anaesthesia.

While the increased haemoglobin concentration and the shift of the oxygen dissociation curve of fetal haemoglobin are undoubtedly advantageous in securing an adequate oxygen supply for the human fetus they may not be as important as has previously been supposed since they are not present in all mammalian species. Moreover the traditional concept that the low P_{O_2} of 4.7–5.3 kPa (35–40 mm Hg) in umbilical-vein blood indicates that the fetus exists poised on the brink of hypoxia may not be correct. Provided

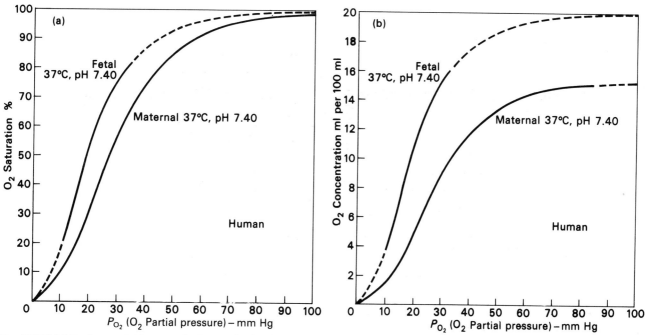

FIG. XII.28 (a) Oxyhaemoglobin curves for human fetal and maternal blood. Abscissa shows oxygen tension; ordinate expresses the percentage saturation of haemoglobin with oxygen. (b) The same oxyhaemoglobin dissociation curves when the percentage saturation values are converted to oxygen concentration. A haemoglobin concentration of 14.9 g dl^{-1} has been taken to represent the fetal value at term (oxygen capacity 20 ml dl^{-1}). For maternal blood the haemoglobin concentration at term is 11.6 g dl^{-1} (oxygen capacity 15.5 ml dl^{-1}). The interrupted lines denote regions of the dissociation curves where the values are less precisely known.

Note: 100 mm Hg represents 13.3 kPa.

(From Metcalfe, J., Bartels, H., and Moll, W. (1967). *Physiol. Rev.* **47**, 782.)

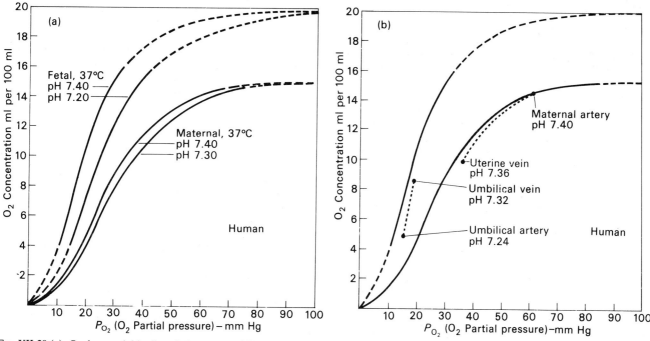

FIG. XII.29 (a). Oxyhaemoglobin dissociation curves of FIG. XII.28(b) showing their displacement due to pH changes of the magnitude observed in the human uterine and umbilical circulations. (b) Representative values reported for oxygen concentrations in blood samples from uterine and umbilical vessels placed on oxyhaemoglobin dissociation curves appropriate to the pH of each blood sample. During placental gas exchange, oxygen tension in maternal blood falls along dotted line on right, departing from the dissociation curve for pH 7.4 as the blood pH falls due to the influx of carbon dioxide and fixed acids from the fetal circulation. Reciprocal changes occur in fetal blood during its passage from the umbilical artery to umbilical vein, the oxygen tension and concentration rising along dotted line on left.

(From Metcalfe, J., Bartels, H., and Moll, W. (1967). *Physiol. Rev.* **47**, 782.)

the blood flow and the oxygen content of the blood supplied to the fetal tissues are large enough, even the comparatively low oxygen saturation of fetal blood is sufficient to supply tissue needs with some margin of safety. It is also possible that a higher fetal P_{O_2}, of say 8.0–10.7 kPa (60–80 mm Hg), might be undesirable because of

toxic effects on tissue growth and development, or because it might lead to premature activation of circulatory changes, such as constriction of the ductus arteriosus or vasodilatation in the lungs, which are normally triggered by the rise in arterial oxygen tension following birth.

Carbon-dioxide excretion by the fetus

The carbon-dioxide tension in umbilical-vein blood is at most only slightly above that in uterine-vein blood [FIG. XII.27]. Owing to the greater diffusibility of carbon dioxide compared with oxygen, the placental membranes offer little hindrance to its passage. Moreover there is a 'double Haldane effect' [p. 187], analogous to the double Bohr effect, whereby the loss of oxygen from the maternal blood passing through the placenta increases its ability to take up carbon dioxide, and the simultaneous oxygenation of the fetal blood helps to promote the unloading of carbon dioxide. The pregnant woman hyperventilates reducing her alveolar P_{CO_2} to 4.0 kPa (30 mm Hg) or less and this also contributes to maintaining the P_{CO_2} in the umbilical vein at approximately the level in the normal adult (5.4 kPa; 40 mm Hg).

Placental transfer of foodstuffs

The 'alimentary' functions of the placenta are necessary for absorption of the materials required for fetal growth. The rate of transfer of nutritive substances from maternal to fetal circulation increases throughout pregnancy, particularly during the last few weeks when the fetal vessels become more numerous and the villous capillaries become thinner.

Passage of substances into the fetal circulation is largely related to their molecular weight. Substances with MW less than 1000 usually cross the placenta readily by simple diffusion and tend to assume equal concentrations on both sides of the placental barrier. Examples are water, sodium, magnesium, chloride, urea, and uric acid. The concentrations of calcium, inorganic phosphorus, and free amino nitrogen are slightly higher in fetal than in maternal blood, raising the possibility of an active transport mechanism in addition to diffusion. The likelihood of an active transport mechanism for amino acids is illustrated by FIGURE XII.30 in which the concentrations of individual amino acids in maternal and fetal plasma are shown. Not only is the overall amino-acid concentration higher in the fetus but the ratio of fetal to maternal concentration varies with the different amino acids. The level of fructose in umbilical cord blood is nearly 50 per cent higher than in maternal blood and it seems likely that the placenta manufactures fructose from glucose though the function of fructose in the fetus remains

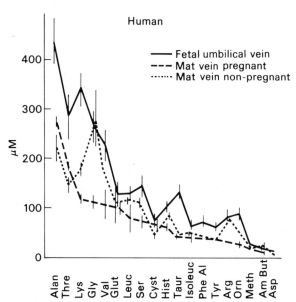

FIG. XII.30. Plasma free amino-acid concentrations (mean ± SE) in the pregnant ewe, and her near-term fetus. The aminograms are arranged in order of the concentrations occurring in the maternal plasma. Fetal: maternal concentration ratios are different for each amino acid.
(From Young, M. and McFadyen, O. (1973). *J. perinat. Med.* **1**, 1.)

obscure. Ascorbic acid is selectively absorbed across the placenta so that at birth its concentration in umbilical cord plasma is 2–4 times higher than that in maternal plasma.

Plasma proteins cross the placenta only in minute amounts. The fetus synthesises its proteins from amino acids transferred across the placenta. Fetal fat arises in two ways, by transfer of fatty acids and cholesterol across the placenta and by synthesis of fat from carbohydrate.

Large molecular size does not preclude passage of substances across the placenta. Antibodies, which are γ-globulins of molecular weight exceeding 100 000 can certainly cross the placenta. For example, antibodies to diphtheria and tetanus toxin can pass from mother to fetus to induce a short-lasting passive immunity in the latter. Even red cells can pass from the fetal to the maternal circulation so it is not surprising that Rh agglutinins can pass in the opposite direction [p. 48].

Drugs and the placenta. Nearly all drugs are of low molecular weight so it is to be expected that the majority can cross the placenta. Central nervous depressant drugs such as morphine, the barbiturates, and general anaesthetics, when given to the mother during labour can enter the fetal circulation in amounts sufficient to depress the breathing of the newborn child. Sulphonamides, penicillin, and other antibiotics pass across in small amounts, as do alcohol and nicotine, and there is increasing evidence that irreparable damage may be caused to the unborn child of an alcoholic mother. It has also been established that the carbon-monoxide concentration in the blood of pregnant women who smoke heavily may be sufficient to produce appreciable carboxyhaemoglobinaemia in the fetus, and the resulting anaemic hypoxia [p. 208] may impair fetal development. The numerous abnormalities of fetal growth following administration of the hypnotic drug thalidomide to mothers in the early stages of pregnancy are a tragic reminder of the potential danger to the fetus from drug therapy during pregnancy.

The amniotic fluid

At a very early period of development a clear fluid collects in the amniotic cavity and surrounds the fetus. At term the volume of fluid is 500–1000 ml. The fluid has a specific gravity of 1007–1025 and contains 98–99 per cent water. The solids consist of half organic and half inorganic matter. One litre of amniotic fluid contains about 5 g of protein, 200 mg of glucose, and also calcium, sodium, potassium, and chloride.

Amniotic fluid is formed to a small extent from fetal pulmonary secretion and fetal urine, but mostly by transudation from maternal blood and/or active transport across the amniotic epithelium. There is a rapid turnover of the fluid, one-third of the volume of water being replaced every hour by water from maternal plasma. Studies with radioactive sodium and potassium show that 13 mmol of sodium and 0.6 mmol of potassium are exchanged each hour. The fluid is removed through drinking by the fetus and by return to the maternal circulation.

The functions of the amniotic fluid are to provide the fetus with fluid to drink, to keep the fetus at an even temperature, to cushion it against injury, and to provide a medium in which it can move easily.

Placental and hormonal influences on fetal growth

Placental size. Several factors, in addition to the obvious ones of the duration of pregnancy, or placental damage, e.g. infarction, have been identified as influencing the size of the fetus. There is a direct relationship between the weight of the infant and the *weight of the placenta* in the final stages of pregnancy. Placental size might limit fetal growth by limiting the transfer of metabolic substrates such as oxygen, carbohydrates, or fats, or of substances, e.g. amino acids, incorporated into the fetal structure, or by reason of inadequate hormone production. A report that babies born during 1955 in Lake County, Colorado, which is at an altitude of 3000 metres where there would be a degree of maternal arterial hypoxaemia, had a median birth weight of 3.07 kg as opposed to 3.29 kg in Denver (altitude 1500 metres) and 3.32 kg at sea-level in Baltimore, lends some support to this hypothesis though further substantiation is needed. Fetal weight is also less in *multiple pregnancies*. Though the size of the placenta is less than normal the reduction in fetal weight is greater than would be expected for this reason alone.

Maternal nutrition does not appear to be a limiting factor to fetal growth in man. In the closing stages of the Second World War, during the winter of 1944 and spring of 1945, the people of Holland endured near starvation, yet there was only a very small reduction in birth weight among Dutch babies born in 1945. On the other hand in the sheep a sharp reduction in food intake during the last third of pregnancy may lead to a fall in fetal weight of 40 per cent. However this effect on fetal weight may be due to the complex metabolic and possibly hormonal changes which the underfed ewe exhibits rather than to a simple reduction in the amount of circulating nutriment available.

Hormonal influences. Fetal production of the *hormones* necessary for development after birth does not appear to influence gross fetal weight. The increase in body weight during gestation in mice, rats, and rabbits is little affected after destruction of the fetal hypophysis by irradiation or intra-uterine decapitation even when accompanied by removal of the thyroid and hypophysectomy of the mother. Birth weight is also normal in human infants who subsequently show signs of congenital hypothyroidism or idiopathic hypopituitary dwarfism within a few months of birth. These findings are surprising since the fetal pituitary produces both TSH and human growth hormone. Exposure of infants to cold during the first 48 hours after birth causes a rise in TSH, and a rise in protein-bound iodine concentration indicative of a normal thyroid response to the TSH secretion [p. 544]. The human fetus has a high plasma level of human growth hormone, several times that in the maternal plasma. This must be derived from the fetal pituitary since isotopically-labelled human growth hormone injected into pregnant women in labour does not appear in the fetal plasma. Moreover in animals fetal hypophysectomy is followed by the disappearance of growth hormone from the fetal plasma.

While intra-uterine hypophysectomy does not affect overall fetal birth weight, hypophysial hormones are required for the normal development of particular organs, especially the changes associated with sexual differentiation and possibly also the development of the brain. Injection of bovine growth hormone into pregnant rats has been shown to increase brain weight, DNA content, and cerebral cortical cell density, and it has been claimed that following such injections the offspring perform better in behavioural tests.

The placental hormone *human placental lactogen* (HPL) [see p. 581], has many properties in common with human growth hormone, including the mobilization of free fatty acids. It is present in the syncytiotrophoblast from the tenth week of pregnancy onwards and is secreted in large quantities into the maternal circulation. However, the relation of HPL to normal fetal growth and development is as yet undetermined.

THE CIRCULATION IN THE FETUS

The umbilical vein, carrying blood from the placenta which is about 80 per cent saturated with oxygen, enters the liver together with the portal vein. After giving off branches which supply blood to the left two-thirds of the liver, the umbilical vein joins the portal sinus, a

branch of the portal vein, to form the ductus venosus. Flow in the portal sinus is from the umbilical vein to the portal vein so that the right one-third of the liver, supplied by branches of the portal vein, receives a mixture of portal venous blood with an oxygen saturation of only 27 per cent and the more highly saturated umbilical-vein blood. The ductus venosus, which in man is about half as wide as the umbilical vein, runs on the under surface of the liver to join the inferior vena cava at its junction with the hepatic veins.

As a result of the admixture of blood from the umbilical vein (80 per cent saturated) with hepatic and systemic venous blood (26 per cent saturated) the inferior vena cava contains blood about 67 per cent saturated with oxygen. The inferior vena cava blood stream entering the heart splits into two on the crista dividens; the majority of the blood passes through the foramen ovale to enter the left atrium, which in the fetus extends dorsally beneath the rest of the heart to join the inferior vena cava at the foramen ovale [Fig. XII.31]. The remainder, together with blood from the superior vena cava, passes into the right atrium where it is joined by blood from the coronary sinus, and thence into the right ventricle.

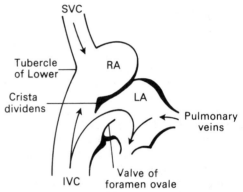

Fig. XII.31. The entry of the great veins into the fetal heart, to illustrate the fact that the foramen ovale lies between the inferior vena cava and left atrium.
(From Dawes, G. S. (1968). *Fetal and neonatal physiology*. Year Book, Chicago.)

The oxygen saturation of the superior vena cava and coronary sinus blood is only 25 per cent. Thus the oxygen saturation of the blood leaving the right ventricle in the pulmonary artery (52 per cent) is appreciably lower than that of the blood leaving the left ventricle through the aorta (62 per cent), since the aortic blood is largely blood derived from the inferior vena cava together with a relatively small contribution which has entered the left ventricle via the pulmonary veins from the lungs [Fig. XII.32].

The two ventricles of the fetal heart are of much the same size and muscular development, unlike the heart of the adult. The pressure in the pulmonary artery exceeds that in the aorta by several mm Hg. Relatively little of the output of the right ventricle is pumped through the lungs since these are collapsed and the vascular resistance offered therein is high. The preponderance of the right ventricular output is directed into the aorta via the ductus arteriosus which joins the pulmonary trunk with the aorta. Thus the two ventricles work in parallel and not in series as in the adult circulation. The left ventricle ejects about 20 per cent more blood per minute than the right. Dawes and his colleagues (1954–8) have measured the flows in the various parts of the fetal circulation of lambs delivered by caesarian section but still attached to the ewe by the umbilical cord. Care was taken to ensure that the lambs were maintained in good physiological condition. Figure XII.33 shows a plan of the fetal circulation with sample figures for the volume flows in different parts of the circuit. Rather more than half the combined output of both ventricles passes through the placenta which offers a lower resistance circuit in parallel with the fetal tissues.

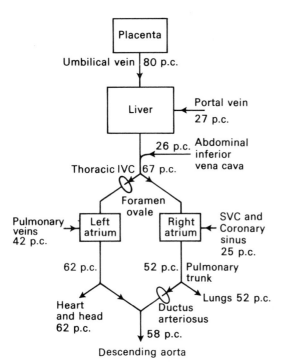

Fig. XII.32. Progressive dilution of well-oxygenated blood (derived from the placenta) by deoxygenated blood from fetal tissues, as it traverses the fetal circulation.
(After G. S. Dawes.)

Development of cardiovascular control mechanisms. The mean arterial blood pressure in the human infant at birth is around 55 mm Hg. Fetal arterial pressure approximately doubles during the last third of gestation and fetal heart rate declines over the same period. These changes are associated with the development of cardiovascular control by the autonomic nervous system and the establishment of the baroreceptor reflexes. In the fetal lamb, from about two-thirds of term onwards, baroreceptor discharges can be detected in the carotid sinus nerve and bradycardia can be elicited by vagal stimulation or by the injection of adrenalin or noradrenalin to increase blood pressure and stimulate the baroreceptors. Conversely haemorrhage leads to a sustained tachycardia. The aortic chemoreceptors are responsive at this stage and are probably responsible for initiating the redistribution of cardiac output during hypoxaemia, so that the brain, heart, and placenta have preference while flow to non-essential regions is reduced. The carotid body chemoreceptors appear to be relatively inactive in the fetus and it may be significant that in the adult the carotid body chemoreceptors

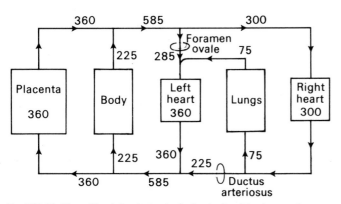

Fig. XII.33. Plan of fetal circulation in the lamb. Combined systemic output of the two ventricles is 585 ml min⁻¹. The left ventricle ejects 360 ml min⁻¹ and the right ventricle 300 ml min⁻¹ (of which 75 ml traverses the lungs and is finally ejected by the left heart).
(After G. S. Dawes.)

have a much greater effect on breathing than do the aortic chemoreceptors, whereas the latter exert a greater effect on the circulation.

Changes in the circulation after birth

The arrest of umbilical blood flow and placental transfusion

If the cord is not tied umbilical blood flow continues for several minutes but at a rapidly decreasing rate. The reason for the reduction in flow is uncertain but compression of the umbilical cord within the uterus and vasoconstriction of the umbilical vessels in response to mechanical stimulation, exposure to the cold air, or the secretion of catecholamines from the infant's adrenal medulla are possible causes.

The normal blood volume at birth is about 90 ml kg⁻¹. If the cord is not clamped until pulsations have ceased and the baby is held at or below the level of the uterus an additional 100 ml of blood can on average be transferred from the placenta to the infant. This amount of blood represents a larger volume of fluid than is generally obtained by breast-feeding in the first 2 days and it contains as much iron as is absorbed from the milk in the first month of life. However if the volume of blood transferred from the placenta is too large the elevation in blood volume, haematocrit, and arterial pressure may embarass the heart and respiration. In a group of infants in which the cord was clamped within 20 seconds of delivery the blood volume was 78 ml kg⁻¹, the haematocrit 47 per cent, and the arterial pressure 44 mm Hg. In a similar group in which the cord was clamped 5 minutes after delivery the corresponding values were blood volume 99 ml kg⁻¹, haematocrit 59 per cent, and arterial pressure 69 mm Hg. Thus only a limited placental transfusion is desirable such as may be produced by leaving the cord untied for no more than one minute. Perhaps more importantly it should be remembered that there is a risk of fetal blood loss to the placenta on delivery if the fetus is raised above the level of the uterus as at caesarian section, or if the cord is partially compressed, blocking venous return from the placenta but not the arterial inflow.

Closure of the foramen ovale

Within a few minutes of delivery the valve of the foramen ovale shuts. There are two reasons for this closure. Firstly, the valve is held open before birth by the pressure and momentum of the blood flowing up the inferior vena cava [FIG. XII.31]. Arrest of the umbilical circulation by tying the cord at delivery causes a decrease in the volume of blood flowing up the inferior vena cava and thus reduces the inferior vena cava–left atrial pressure gradient, favouring closure of the valve. Secondly, arrest of the umbilical circulation also leads to partial asphyxia which initiates breathing movements with a consequent fall in pulmonary vascular resistance (see below). The resulting increase in pulmonary venous return is sufficient to reverse the pressure difference across the foramen ovale and so the valve closes. The valve usually becomes adherent to the edge of the foramen ovale within a few days of birth but in some children anatomical fusion may not be complete for some months or even years. In the majority of these cases the left to right atrial pressure gradient is sufficient to hold the valve shut.

Changes in the pulmonary circulation

The onset of normal inspiration associated with gaseous distension of the previously collapsed lungs is accompanied by a rapid fall in pulmonary arterial pressure together with a sixfold to tenfold increase in pulmonary blood flow. Thus a profound fall in pulmonary vascular resistance must accompany the onset of breathing. In the fetus the pulmonary arterial pressure is slightly higher than that in the aorta and most of the output from the right heart passes through the ductus arteriosus to the aorta. After birth the position is reversed, for the aortic pressure rises and pulmonary artery pressure falls so that blood flow through the ductus, which remains partially open for many hours after birth, occurs from the aorta to the pulmonary trunk, i.e. in the reverse direction to that which takes place in the fetus [FIG. XII.36]. This can be shown most dramatically in the mature fetal lamb by brief occlusion of the ductus arteriosus some minutes after ventilation has been begun. Before ventilation this causes a rise in pulmonary arterial pressure, afterwards it causes a fall [FIG. XII.34].

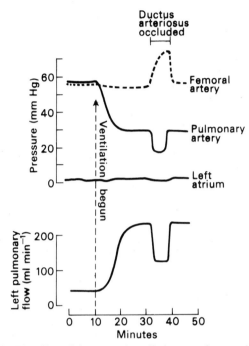

FIG. XII.34. The effect of the onset of ventilation on pulmonary blood flow and pressures in the femoral artery, pulmonary artery and left atrium in a mature fetal lamb. Before ventilation begins pulmonary arterial pressure marginally exceeds that in the femoral artery and pulmonary flow is small. Following the initiation of positive pressure ventilation there is an approximately fivefold increase in pulmonary blood flow and pulmonary artery pressure is halved. Temporary occlusion of the ductus arteriosus now causes a fall in both pulmonary flow and pressure, while femoral pressure rises, indicating that flow in the ductus at this stage was from the aorta to the pulmonary artery.
(Redrawn from the data of Dawes, G. S., Mott, J. C., Widdicombe, J. G., and Wyatt, D. G. (1953). *J. Physiol. Lond.* **121**, 141. (From Dawes, G. S. (1968). *Fetal and neonatal physiology*. Year Book, Chicago.)

Two factors are principally responsible for bringing about the fall in pulmonary vascular resistance when breathing begins. One is the mechanical effects of expansion of the lungs with gas. When the lungs of fetal lambs were ventilated with a gas mixture containing 3 per cent O_2 and 7 per cent CO_2 in N_2 there was considerable pulmonary vasodilatation, as evidenced by the alteration in the pressure-flow curve for the pulmonary circulation [FIG. XII.35], even though there was no significant change in the arterial-blood gas tensions. The mechanism responsible for this lowered pulmonary vascular resistance on gaseous expansion of the lung is the introduction of a gas–fluid interface which leads to a persistent change in the geometry of the alveoli. The other major factor influencing the development of pulmonary vasodilatation is the composition of the gas ventilating the alveoli. If the gas used to ventilate the fetal lungs is changed from a mixture containing 3 per cent O_2 and 7 per cent CO_2 to one containing 21 per cent O_2 and 7 per cent CO_2 there is a further pulmonary vasodilatation [FIG. XII.35] which is enhanced still more by ventilation with air (i.e. 21 per cent O_2 and no CO_2). Thus both hypoxia and hypercapnia contribute to the high pulmonary vascular resistance in the fetus. The aortic chemoreceptors are active at term and activity in the

sympathetic nerves to the lung helps to sustain pulmonary vasoconstrictor tone in the fetus. However, the relief of hypoxia and hypercapnia still results in pulmonary vasodilation after the possibility of a reflex mechanism has been excluded by blocking the sympathetic pathway with hexamethonium, showing that hypoxia and hypercapnia exert their vasoconstrictor effects by a direct local action on the lung.

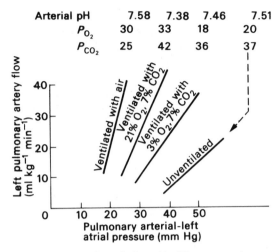

FIG. XII.35. The effect of ventilation with different gas mixtures upon the mean arterial blood gases and pH and pulmonary arterial pressure: flow curves in mature fetal lambs.
(From the data of Cassin, S., Dawes, G. S., Mott, J. C., Ross, B. B., and Strang, L. B. (1964). *J. Physiol. Lond.* **171**, 61.) (From Dawes, G. S. (1968). *Fetal and neonatal physiology*. Year Book, Chicago.)

Closure of the ductus arteriosus

The ductus arteriosus is almost as large as the ascending aorta of the mature fetus and has a thick smooth muscle wall. The ductus constricts rapidly during the first few hours after delivery but final functional closure takes place gradually over the next 1–8 days. For the first hour or so after birth flow continues through the ductus in the fetal direction (i.e. a right-to-left shunt from the pulmonary artery to the aorta) but subsequently flow reverses as the falling pulmonary vascular resistance leads to a progressive reduction in pulmonary artery pressure. This early constriction of the orifice, which remains nevertheless slightly patent, is responsible for the murmur which may be heard by careful auscultation in the immediate neonatal period. The murmur, though continuous, reaches a crescendo with each second heart sound.

Constriction of the ductus arteriosus appears to be due to a direct effect on the smooth muscle wall of the increase in arterial P_{O_2} after birth. This is consistent with the higher incidence of persistent patency of the ductus associated with fetal distress at birth and with delivery at high altitudes. Within a few hours of delivery hypoxia may cause the ductus to dilate again but thereafter the ductus is no longer sensitive to the local O_2 tension. After final occlusion of the ductus by muscular contraction the wall is replaced by fibrous tissue. This permanent sealing of the lumen occurs 2–3 weeks after birth.

The exact mechanism whereby a rise in P_{O_2} constricts the ductus is not yet fully understood. It has been suggested that an increase in the oxygen supply increases the turnover of the cytochrome chain leading to increases in oxidative phosphorylation and the synthesis of ATP. The release of acetylcholine may also have an intermediary role since the constrictor response of the ductus to oxygen is blocked by atropine and acetylcholinesterase and is enhanced by cholinesterase inhibitors. Though of lesser importance than the rise in P_{O_2}, other factors may contribute to the constriction of the ductus arteriosus following birth. These include the release of catecholamines, either from the adrenergic innervation of the ductus itself or from the adrenal medulla; a reduction in the concentration of the dilator prostaglandins PGE_1 and PGE_2, which have been isolated from the ductus, has also been postulated.

FIGURE XII.36 shows the fetal, neonatal, and adult circulations in schematic form and summarizes the effects of these changes in the foramen ovale, pulmonary vasculature, and ductus arteriosus.

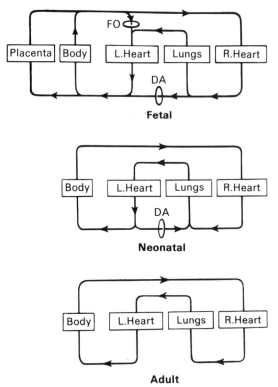

Fetal

Neonatal

Adult

FIG. XII.36. These figures illustrate, in a simplified form, the changes in the circulation at birth. In the fetus both ventricles work in parallel to drive blood from the great veins to the arteries.

Tying the umbilical cord stops the placental circulation, and the consequent reduction in inferior vena caval flow, combined with a great increase in pulmonary blood flow, causes closure of the foramen ovale (FO). These changes occur within a very few minutes of birth and give rise to the *neonatal* circulation in which a large volume of blood flows through the lungs. Because the ductus arteriosus (DA) is still open, this entails a greater left ventricular output.

During the course of the next day or two the lungs become fully functional, and the ductus arteriosus, which constricts considerably soon after birth, finally closes to produce the adult circulation, in which blood is pumped through the lungs and body by the two ventricles operating in series. (After G. S. Dawes.)

(Born, G. V. R. *et al.* (1954). *Symp. quant. Biol.* **19**, 102.)

The *ductus venosus*, which would otherwise constitute a portal–caval shunt bypassing the liver, closes within a few hours of birth. Little is known about the mechanism of closure.

Changes in cardiac muscle

Before birth the wall thickness of the two ventricles is approximately equal though the overall size of the right ventricle may be slightly greater. After birth the left ventricular wall rapidly grows thicker as the systemic arterial pressure rises. Consistent with the fall in pulmonary arterial pressure the thickness of the wall of the right ventricle may be no greater at one year old than at birth. The number of muscle fibres in each ventricle is approximately the same and does not change from birth to adult life, the increase in ventricular size being due to increased length and thickness of the individual fibres. Accompanying the greater development and work load of the left ventricle after birth there is a redistribution of myocardial blood flow, the majority of which becomes directed to the left ventricle.

THE DEVELOPMENT OF BREATHING

Breathing movements *in utero*

If the fetal lamb is delivered by caesarian section into a bath of warm saline and the umbilical cord remains intact, periods of respiratory movements can be observed from 40 days gestation onwards (term is 147 days). Continuous records of fetal intrathoracic pressure obtained with catheters implanted *in utero* have shown that rapid irregular respiratory movements are present in the fetal lamb up to 40 per cent of the time; brief gasps are also seen. The episodes of breathing movements are associated with the signs of rapid eye movement (REM) sleep and show a diurnal variation, becoming gradually more frequent throughout the hours of daylight. The movements are increased by hypercapnia and depressed by hypoxia and hypoglycaemia. During asphyxia they are replaced by a series of gasps. Though the intrapleural pressure changes produced by these respiratory movements are as great or greater than those occurring in breathing after birth, they lead to only small alterations in pulmonary volume, since the lungs of the fetus are filled with fluid with a viscosity and density many times that of air. Thus there is little mixing of the amniotic and lung fluids. It has been suggested by Dawes and his colleagues that the purpose of these breathing movements *in utero* is to exercise and train the respiratory muscles for their function after birth. Ultrasound scanning techniques reveal fetal breathing movements in women as early as 11 weeks gestation. Initially irregular, they gradually become more regular so that by 36 weeks gestation episodes of regular breathing movements at a frequency of 30–70 min^{-1} are present for 55–90 per cent of the time in the human fetus. As in the fetal lamb the breathing movements show a diurnal variation and their incidence diminishes somewhat prior to parturition. Since, also as in the sheep, regular breathing movements in the human fetus are reduced and are supplanted by irregular gasps when the fetus is subjected to hypoxia or asphyxia, monitoring of fetal 'breathing' may prove a useful tool in the assessment of fetal health.

Breathing following birth

Immediately following birth and the tying of the umbilical cord the fetus is generally apnoeic. However within seconds or at most a couple of minutes it begins to make strong gasping efforts. These become more frequent and merge gradually into episodes of regular breathing followed by continuous quiet breathing. With the appreciation that breathing movements are a normal feature of intrauterine life the interesting problem is not so much that of the causes of the onset of breathing but those factors which convert the episodic breathing of the fetus to the continuous breathing of postnatal life. When a baby is born it is removed from a warm moist environment to a cool one where its weight is no longer supported by surrounding fluid. In the process it is squashed and squeezed and handled, albeit gently; it may also be somewhat asphyxiated during delivery and will be totally asphyxiated between the ligation of the umbilical cord and the onset of the first gasps drawing air into the lungs. Exposure to cold, handling, and gravitational stimuli have all been shown to enhance ventilation after birth, but they are not essential for the establishment of maintained regular breathing since lambs delivered into a saline bath at intrauterine temperature will breathe regularly and continuously following ligation of the cord. This suggests that asphyxia may be an important factor in initiating maintained respiration. However asphyxia cannot be a full explanation; once breathing is established after birth the carotid artery P_{O_2} rises from the fetal level of 3.1 kPa (23 mm Hg) to more than 8.1 kPa (60 mm Hg) and the P_{O_2} falls from 6.0 kPa

(45 mm Hg) to 4.7 kPa (35 mm Hg). The arterial chemoreceptors may contribute to this maintenance of breathing even though arterial P_{O_2} has risen and P_{CO_2} has been reduced. Before birth the carotid-body chemoreceptors are relatively inactive despite the low P_{O_2} of the carotid artery blood. The asphyxia induced by tying the umbilical cord leads to a generalized increase in sympathetic nervous activity as evidenced by a rise in arterial blood pressure of up to 40 mm Hg. There is a great increase in the discharge in the cervical sympathetic nerves which supply vasomotor fibres to the carotid body. The resultant reduction in carotid-body blood flow may be sufficient to initiate chemoreceptor activity. The ensuing stimulation of respiration will raise the P_{O_2}, but if the reduction in blood flow is great enough, the oxygen supply to the carotid body may still be diminished despite this rise in P_{O_2} and chemoreceptor stimulation will continue. However, it must be stressed that any such contribution by the chemoreceptors is only one of several factors, including gravitational stress, a change in temperature, and mechanical stimulation, which by their conjoined effects are responsible for the maintained breathing of the newborn.

Surface-active material and the fetal lung fluid

The lungs of the fetus at birth are filled with fluid. This fluid was originally assumed to be aspirated amniotic fluid, but experiments in which fluid continued to drain for 1–2 hours from the cannulated trachea of fetal sheep and goats delivered near term by caesarian section pointed to a pulmonary origin. Moreover the composition of amniotic and lung fluids differs in such a way as to suggest that lung fluid probably originates in the alveoli as a plasma ultrafiltrate. At birth some of the lung fluid is expelled from the mouth as the thorax is compressed in its passage along the birth canal. The remainder is gradually removed by evaporation or through the pulmonary capillaries and lymphatics. Measurements of lymph flow from the lungs following the initiation of ventilation have shown that this can account for about a third of the water and all the protein lost from the lungs of mature lambs within the first 2 hours after delivery.

The entry of air into the lungs with the first breath leads to the formation of an air–liquid interface in the alveoli. There is a surface tension at this interface and because of the Laplace relationship, $P = 2T/r$ (where P is the pressure within a hollow sphere, T the tension in the wall—here the fluid film lining the alveolus, and r the radius), the reduction in alveolar radius during expiration would be expected to lead to an increased alveolar pressure driving more air out of the alveoli and producing their eventual collapse with obliteration of the residual lung volume. This is prevented by the presence in the lung liquid of surface-active material, often referred to as *surfactant*. This material, a complex of proteins and lipids, particularly dipalmitoyl lecithin, forms a thin monomolecular film at the air–fluid interface which reduces the surface tension. During expiration the reduction in the surface area of the fluid lining the alveoli leads to closer packing of the molecules of surfactant bringing about a progressive reduction in surface tension as the size of the alveoli is reduced.

Surfactant is produced by the type II pulmonary alveolar cells or pneumocytes, and osmiophilic lamellar inclusion bodies in these cells may represent surfactant prior to its discharge into the alveolar spaces. The importance of surfactant in the lung at birth is illustrated in Figure XII.37 which is taken from the work of Reynolds and Strang. The continuous line in Figure XII.37(a) shows the pressure–volume curve obtained on inflating with air for the first time the lungs of a fetal lamb near term. An opening pressure of 15 cm H_2O was required before any significant amount of air entered the lung; thereafter air entered readily until full inflation was achieved at a pressure of 40 cm H_2O. On deflation about one-quarter of the air introduced into the lung was retained, i.e. a residual volume had been established. On the second inflation,

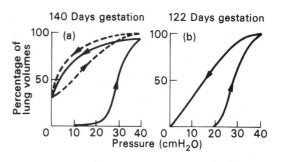

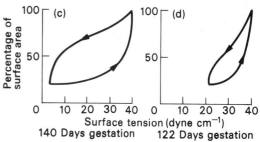

FIG. XII.37. Pressure–volume curves of (a) the first (solid line) and subsequent (interrupted line) pulmonary inflations in a mature fetal lamb of 140 days gestation; (b) in an immature lamb of 122 days gestation, the pressure–volume curves of subsequent inflations superimpose on that of the first; (c) and (d) show the corresponding relationships of surface tension to surface area when a film of lung extract was spread on the surface of liquid in a trough. During compression of the film, the surface tension fell much more in the extract from the mature lung (c), than in the extract from the immature lung (d), indicating the presence of surface active material. (From Dawes, G. S. (1968). *Fetal and neonatal physiology*. Year Book, Chicago.) (Redrawn from Reynolds, E. O. R. and Strang, L. B. (1966). *Br. med. Bull.* **22**, 79.)

shown by the interrupted line, air entered the lung even at the lowest inflation pressures, and at a pressure of 15 cm H_2O the lung was now inflated to three-quarters of its full volume. The comparable pressure–volume curve for the lungs of an immature lamb is shown in Figure XII.37(b). The curve on inflation is similar to that in the near-term lamb but on deflation the lung collapses completely. To produce a second inflation an opening pressure as large as that for the first inflation would be needed, greatly increasing the work of breathing. The lower half of Figure XII.37 shows the relationships between surface area and surface tension when a film of lung extract prepared from the lungs of the two lambs was spread over the surface of liquid in a trough. When the film was compressed by altering the position of a movable barrier dividing the trough, there was a greater fall in surface tension with the extract from the more mature lung indicating the presence of a larger quantity of surfactant. In the human lung there are two chemical pathways for the synthesis of surface-active lecithins. At about 22 weeks of gestation, the methyl-transferase pathway starts to produce palmitoyl myristoyl lecithin. Until about 33 weeks of gestation this is the only surface-active lecithin secreted and the amount synthesized may not be adequate to prevent lung collapse during expiration. From about 33 weeks of gestation onwards, the second synthetic pathway, the choline-phosphotransferase pathway, produces dipalmitoyl lecithin in quantities which rapidly increase until term. In some infants born prematurely the absence of sufficient surfactant is responsible for the *respiratory distress syndrome* or *hyaline membrane disease*. Collapse of the alveoli during expiration leads to reductions in lung compliance, functional residual capacity, and total lung capacity, so that the baby is severely hypoxaemic while the powerful respiratory efforts needed to produce even a limited ventilation rapidly exhaust the infant. In addition pulmonary vasoconstriction, caused by the combined effects of alveolar collapse and hypoxia, raises pulmonary vascular resistance and leads to reopening of the foramen ovale, with right-to-left shunting of blood. In severely affected infants a right-to-left shunt can also occur through the ductus arteriosus. The treatment of the condition has been pioneered by Strang and Reynolds. Oxygen is added to the inspired air to correct the hypoxaemia. Frequent measurements of arterial P_{O_2} are carried out to ensure that the normal arterial oxygen tension is not exceeded, thus avoiding the risks of retrolental fibroplasia and the toxic effects of oxygen on the lung. If necessary, the infant's breathing is assisted with a mechanical ventilator; alternatively lung collapse during expiration may be prevented by making the infant breathe against a continuous inflating pressure.

In specialized intensive-care units, the use of such measures, until with increasing maturity adequate surfactant production develops, has increased to nearly 90 per cent the survival rate of infants born with this illness.

TEMPERATURE REGULATION IN THE NEWBORN

The newborn of all species, including human infants, being much smaller than the corresponding adults, have a greater surface area in relation to their body weight and thus the maintenance of body temperature presents a greater problem. A fall in body temperature can be prevented by reducing heat loss or by increasing heat production. *Cutaneous vasoconstriction*, which reduces heat loss by convection and radiation, has been shown to occur in response to cold exposure even in premature babies. A newborn animal or infant exposed to cold also *reduces its effective surface area* by hunching itself up and tucking in its limbs. Where there are several animals in a litter they will huddle together, further reducing the effective area from which heat is lost. During the first 48 hours after birth, *thyroxine secretion* by the newborn human infant is increased [see p. 544], and this may be related to the colder environment to which the infant is exposed after emerging from the uterus. The newborn animal has a considerable capacity to increase its heat production in the attempt to maintain its body temperature and this is reflected in an increased O_2 consumption in a cold environment.

The temperature at which O_2 consumption at rest or when asleep is minimal is known as the neutral thermal environment and for human infants the thermoneutral zone is from 32.5 °C to 33.5 °C (above this temperature range there is little change in O_2 consumption but sweating is initiated to increase evaporative heat loss). In a neutral thermal environment the O_2 uptake of human infants averages 4.6 ml kg^{-1} min^{-1} and this rises to 15 ml kg^{-1} min^{-1} at 20–25 °C. The corresponding adult values are 3.7 ml kg^{-1} min^{-1} in a neutral thermal environment and 19 ml kg^{-1} min^{-1} in the cold (0–10 °C). Thus the infant can increase its heat production at least as well as can the adult. While the infant is capable of shivering, this is not a prominent feature of its response to cold and the O_2 consumption of newborn rabbits and guinea-pigs still increases markedly on exposure to cold after shivering has been abolished by neuromuscular blockade with gallamine.

The sites of this *non-shivering thermogenesis* are the deposits of *brown adipose tissue* which in the human infant is found as a thin sheet between the shoulder blades, around the neck, behind the sternum, and around the kidney and adrenal gland [FIG. XII.38]. This brown adipose tissue is yellow when full of fat but, due to the high content of cytochrome present in the numerous large mitochondria lying adjacent to the lipid lobules, gradually turns yellowish-brown and then red-brown as the fat is used up. Brown adipose tissue also has a very rich supply of blood vessels and sympathetic nerves. Several lines of evidence have demonstrated that brown adipose tissue is responsible for much of the increased O_2 consumption and heat production on exposure to cold. FIGURE XII.39 shows observations on newborn rabbits in which thermo-

couples were inserted subcutaneously over the interscapular pad of brown adipose tissue and over the sacrospinalis muscle in the lumbar region where there is no brown adipose tissue. A third thermocouple was placed in the colon. At the initial ambient temperature of 35 °C all three temperatures were similar, but 30 minutes after the ambient temperature was reduced to 25 °C the subcutaneous temperature in the lumbar region had fallen by 2 °C and that in the colon by 1 °C. However the temperature over the interscapular brown adipose tissue remained relatively constant. O_2 consumption increased nearly threefold during this period of cold exposure while the rabbit breathed room air, but when the O_2 content of the inspired air was reduced to 5 per cent the metabolic response to cold was abolished and all three temperatures fell and became approximately equal. Excision of the cervical and interscapular deposits of brown adipose tissue in the newborn rabbit reduced the metabolic response to cold by over 80 per cent.

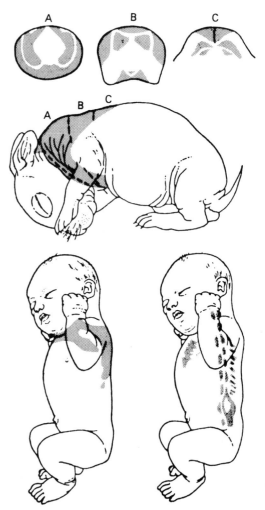

FIG. XII.38. The distribution of brown adipose tissue in the newborn rabbit and human infant.
(From Dawes, G. S. (1968). *Fetal and neonatal physiology*. Year Book, Chicago.) (After Dawkins, K. J. R. and Hull, D. (1965). *Scient. Am.* **213**, 62.)

The increased metabolism of brown adipose tissue in response to cold is under the control of the sympathetic nervous system. Stimulation of the cervical sympathetic nerve provokes a rise in the temperature of the cervical fat pad and sympathectomy abolishes the response to cold. Infusion of noradrenalin is very effective in promoting increased metabolic activity in brown adipose tissue. The metabolic response to cold is abolished by the β-blocker propranolol but α-receptor blockade by phenoxybenzamine has no such effect. It has been suggested that noradrenalin, either infused or liberated by the sympathetic supply to brown adipose tissue, activates a lipase which splits triglyceride. Part of the liberated fatty acid is oxidized, increasing heat production. The remaining fatty acid is re-esterified but the glycerol liberated from the breakdown of triglyceride cannot be used for this. Instead, α-glycerol phosphate is formed from glucose and this is a further exothermic process. Thermogenesis by brown adipose tissue may also result from partial uncoupling of oxidative phosphorylation and stimulation of [Na$^+$ + K$^+$] activated adenosine triphosphatase.

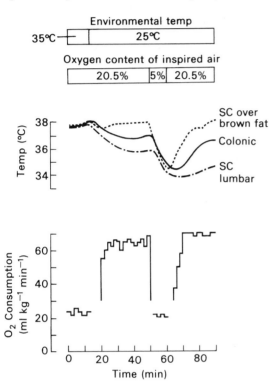

Fig. XII.39. Observations on the effects of cold and hypoxia on a rabbit of 57 g 12 hours after natural birth. The subcutaneous temperature over brown adipose tissue exceeds that elsewhere on cold exposure, provided the supply of oxygen is adequate.
(From Dawes, G. S. (1968). *Fetal and neonatal physiology.* Year Book, Chicago.) (Redrawn from Dawkins, M. J. R. and Hull, D. (1964). *J. Physiol., Lond.* **172**, 216.)

REFERENCES

ALEXANDER, G. (1975). Body temperature control in mammalian young. *Br. med. Bull.* **31**, 62.
BODDY, K. and DAWES, G. S. (1975). Fetal breathing. *Br. med. Bull.* **31**, 3.
COMLINE, R. S., CROSS, K. W., DAWES, G. S. and NATHANIELSZ, P. W. (1973). Foetal and neonatal physiology. Proceedings of the Sir Joseph Barcroft Centenary Symposium held at Cambridge, 25–27 July 1972. Cambridge University Press, London.
—— and SILVER, M. (1974). Recent observations on the undisturbed foetus in utero and its delivery. In *Recent advances in physiology*, No. 9 (ed. R. J. Linden). Churchill Livingstone, Edinburgh.
—— —— (1975). Placental transfer of blood gases. *Br. med. Bull.* **31**, 25.
DAWES, G. S. (1968). *Fetal and neonatal physiology.* Year Book Medical, Chicago.
METCALFE, J., BARTELS, H., and MOLL, W. (1967). Gas exchange in the pregnant uterus. *Physiol. Rev.* **47**, 782.
REYNOLDS, E. O. R. (1975). Management of hyaline membrane disease. *Br. med. Bull.* **31**, 18.
ROBERTSON, N. R. C. (1982). Advances in respiratory distress syndrome. *Br. med. J.* **284**, 917–18.
STRANG, L. B. (1977). *Neonatal respiration.* Blackwell, Oxford.

The mammary glands

STRUCTURE OF MAMMARY GLANDS

The mammary gland consists of a series of ducts, which branch to give rise to terminal tubes; these in turn lead to the alveoli. Covering the external surface of the epithelium of the alveoli and ducts are numerous elongated, branching, longitudinally striated cells which constitute what has been called *myoepithelium.*

The breast arises as an invagination from the surface epithelium which dips down into the underlying connective tissue as solid columns of cells; these gradually become hollowed out to become ducts. At birth the breast is rudimentary, and consists essentially of the tiny nipple from which radiate a few ducts. In the newborn, secretion of a fluid resembling colostrum may occur, probably stimulated by maternal hormones. Little further development occurs until the time of puberty. The changes which occur at puberty in the female vary considerably with the species studied; in many, including the human subject, there is considerable growth and branching of the duct system; in others there may also be formation of glandular tissue. With the recurrence of each sexual (menstrual or oestrous) cycle the gland undergoes further proliferative changes; though this is followed by some degree of regression, on the whole, progressive enlargement takes place, which is due in part to increased deposition of fat. Between each menstrual period (in women) there is hyperaemia of the breasts, increase in the interalveolar stroma, and possibly new formation of alveoli; these changes are transient.

During pregnancy the breasts enlarge greatly and become markedly changed in structure. During the first half of pregnancy there is further duct development, but this is now accompanied by the appearance of many alveoli which form lobules. No milk is secreted by the gland cells at this stage. During the second half of pregnancy the epithelial cells swell and there is gradual initiation of secretory activity with slow accumulation of milk in the alveolar lumina. The further enlargement of the breast which takes place at this stage is not due to an increase in the mass of glandular tissue but to distension of the organ with its secretion. Massage of the breast may squeeze out some of this milk.

Control of breast development

This is due to the complex action of a number of hormones; oestrogen and progesterone are the primary agents responsible for mammary growth, but they appear to work best with the help of the anterior pituitary, adrenal cortex and thyroid glands. The details of the controlling mechanism vary a good deal with the species.

1. Action of oestrogen. The administration of oestrogen to normal or castrated animals, male or female, causes thickening of the nipple and marked growth and branching of the ducts. These results probably account satisfactorily for the duct changes which normally occur at puberty. In most species oestrogen causes little or no glandular development, but in some animals, e.g. cows and goats, oestrogen administration can not only produce alveolar development but even secretion of milk; these latter effects of oestrogen may be mediated via the anterior pituitary. In the human female oestrogen produces only duct development, progesterone being needed for formation of the alveoli.

2. Action of progesterone. Progesterone given alone, when the breast is undeveloped or following its growth under oestrogen treatment, produces no changes. But when given together with oestrogen (i.e. at the same time), marked glandular development occurs, which ultimately may be equivalent to that attained at the end of the first half of normal pregnancy. No secretory changes, however, occur.

3. Action of prolactin. Prolactin acts on a breast that has been caused to grow by oestrogen–progesterone stimulation. It can act directly on mammary epithelial cells to produce localized alveolar hyperplasia. This action is enhanced by growth hormone, cortisol, and thyroxine.

4. Role of placenta. In the pregnant animal the placenta forms both oestrogen and progesterone; in fact it is probable that the placenta is the only source of oestrogen in the pregnant animal, and that none is formed by the ovary itself. Thus, if the ovaries are removed in pregnant mice and the placentae happen to be retained, mammary development proceeds quite normally, indicating that adequate oestrogen and progesterone secretion still take place; the same result occurs if the fetuses as well as the ovaries are removed and the placentae are retained. If, however, the placentae are also aborted following ovariectomy, the breasts rapidly regress. It is clear, therefore, that the placenta is an important organ of internal secretion in relation to breast development during pregnancy; in addition to oestrogen and progesterone the placenta also produces a prolactin-growth-hormone-like factor (placental lactogen).

An intact nerve supply is not essential for the growth of the mammary gland during pregnancy. If the breast is completely transplanted, thus severing all its nervous connections, it may grow during pregnancy, and function, although somewhat inefficiently, after parturition.

Lactation

Lactation consists of two distinct processes.

1. Milk secretion, i.e. the synthesis of milk by the alveolar epithelium and its passage into the lumen of the gland.

2. Milk ejection, i.e. discharge of milk from the breast.

Milk secretion must be considered in two phases:
(i) Initiation of secretion—*lactogenesis*.
Though some secretion is present in the breasts during the latter part of pregnancy, a free flow of milk is established only some days after delivery of the child. The initiation of milk secretion is not under direct nervous control though it is susceptible to psychological influences. Lactogenesis is hormonally controlled in the following way. Relatively *low* circulating levels of oestrogen activate the lactogenic function of the anterior pituitary which is mediated by prolactin. This effect is probably due to reduced secretion of PIF by the hypothalamus [p. 520]. During pregnancy this lactogenic action of oestrogen is inhibited by progesterone. At parturition the rate of progesterone secretion decreases markedly before the decrease in oestrogen occurs, leaving the lactogenic action of the latter unopposed.
(ii) Maintenance of secretion—*galactopoiesis*.
This is also controlled by hormonal factors. Prolactin is important both for initiation and maintenance of lactation in women [p. 517]. In many animal species other hormones are also needed for successful lactation; these include growth hormone, thyroid hormones, insulin, and adrenal and ovarian steroids.

Milk ejection. The discharge of milk from the mammary gland depends not only on the suction exerted by the infant, but also on a contractile mechanism in the breast which expresses milk from the alveoli into the ducts (milk ejection).

It is well known that in the cow the amount of milk present in the cisterns and larger ducts at the commencement of the milking is only a small fraction of the total quantity which can ultimately be collected. It appears that stimulation of the teat produces, after a brief interval, a sudden rise in milk pressure in the udder, and only after this phase has occurred can the full milk yield be obtained. A similar rise of pressure in the ducts occurs in women in response to the stimulus of suckling. Both these effects can also be produced by injection of oxytocin.

The physiological sequence of events is probably as follows: stimulation of the teat or nipple causes nerve impulses to pass (via some unknown pathway) to the paraventricular nucleus and thence along the hypothalamo-hypophysial tract to the neurohypophysis causing the release of oxytocin into the bloodstream. The oxytocin is then carried to the mammary gland where it produces contraction of the myoepithelium surrounding the alveoli, thus expelling their contained milk into the ducts, which are meanwhile kept open by contraction of their longitudinally arranged myoepithelial layer. This sudden outflow of oxytocin into the bloodstream probably also causes the uterine contractions which are known to follow suckling during the puerperium in women. Suckling also promotes secretion of prolactin, presumably by inhibiting the release of PIF from the hypothalamus [p. 520]. Thus suckling can act by neurohumoral mechanisms to cause both secretion and ejection of milk, and it is easy to understand how nervous and psychological factors, acting via the hypothalamus, can influence lactation.

The hypothalamic control of prolactin secretion is discussed on page 520. Briefly, PIF secretion from the hypothalamus inhibits prolactin secretion and therefore the secretion of milk. However, thyrotrophin-releasing hormone not only stimulates secretion of TSH but also prolactin secretion. TRH stimulates prolactin secretion especially in the post-partum period, probably because of the large store of prolactin in the pituitary at this time. TRH also promotes lactation in mothers who have stopped breast-feeding for several days. Since TRH is well absorbed when given by mouth it could be widely used as a galactogogue and could help to make breast-feeding more satisfactory and more fashionable in all parts of the world.

Oxytocin has proved useful clinically in women with engorged painful breasts and a poor flow of milk. It starts off the flow which may then continue well without further treatment.

In the absence of suckling milk accumulates in the breasts and the gland involutes (i.e. regresses).

Lactation is associated with a delay in the return of the menstrual periods and temporary sterility, presumably owing to non-secretion of the gonadotrophins FSH and LH, but women can become pregnant again while nursing.

Milk

Milk is a naturally balanced food, containing about 35 g of first-class protein per litre, mineral salts (especially Ca and P for bone and tooth formation), practically all the vitamins, fat, and soluble carbohydrate. Milk requires only the minimum of supplementation to form a perfect diet, and will itself confer protective supplementation on practically any other dietary intake.

Composition of milk

The fluid secreted during the first three days after parturition is called *colostrum*. It is deep yellow in colour and rich in protein and salts; it is coagulated into solid masses by heat, or even spontaneously. It contains large granular bodies, called colostrum corpuscles, which represent either discharged alveolar cells of the gland, or else leucocytes loaded with fat. These corpuscles are abundant in the first few days, and disappear at the end of the second week.

The milk formed during the first few weeks is called the intermediate or transition milk. Mature milk appears at the end of the first month.

TABLE XII.2 indicates the composition of colostrum, mature human milk, and cow's milk.

The differences between human and cow's milk are very striking; human milk contains considerably less protein, less salts, and more carbohydrate.

	Protein g per cent	Lactose g per cent	Fat g per cent	Ash g per cent	Calcium g per cent
Colostrum (human)	8.5	3.5	2.5	0.37	
Mature human milk 65 kcal/100 ml	1.0–2.0	6.5–8	3.0–5.0	0.18–0.25	0.03
Cows' milk (average) 65 kcal/100 ml	3.5	4.75	3.5	0.75	0.14

1. The protein content of human milk is highest in colostrum (8.5 per cent w/v), and falls during the first few weeks (2.25 per cent) to reach a fairly steady level of about 1.25 per cent; it diminishes rapidly towards the end of lactation. Two proteins are found:

(i) Caseinogen is precipitated by weak acids; it is converted by rennin into calcium caseinate which is insoluble in water, but is easily digested by gastric juices.

(ii) Lactalbumin: resembles serum albumin.

In human milk there are about two parts of lactalbumin to one part of caseinogen. In cow's milk the proportions are very different: the caseinogen is six times in excess of the lactalbumin. Allowing for the difference in the total protein content of the two kinds of milk, it follows that cow's milk contains about six times as much caseinogen as human milk. The caseinogen of cow's milk in the stomach forms large solid masses which are relatively insoluble. Its exact chemical composition, too, is different from that of human caseinogen. When human milk is treated with rennin or dilute acetic acid, fine flocculation occurs.

2. Fat of milk is in the form of minute globules which are emulsified by the dissolved albumin; the fats chiefly present are triolein, tristearin, and tripalmitin. Free fatty acids are only found in minute amounts; cow's milk has about eight times as high a fatty-acid content.

3. The carbohydrate of milk is the disaccharide lactose (glucose-galactose).

4. The ash contains Ca, K, Na, Cl, and P, but only traces of iron: this very low iron content is noteworthy. Human milk contains only 0.03 per cent of Ca (against 0.14 per cent in cow's milk).

5. The vitamin content of milk depends on the maternal diet. Human milk usually contains enough vitamins for the first few months of infancy. Supplements of vitamins A, C, and D are needed later.

Origin of constituents of milk. The specific constituents of milk are elaborated in the gland cells from certain raw materials supplied by the blood. (i) Lactose is derived from the glucose of the plasma. (ii) Proteins of milk come from the plasma amino acids and proteins. Immunoglobulins can pass unchanged from maternal blood to milk. (iii) Fat is formed partly from neutral fat of the blood and partly from acetate.

Conditions affecting composition. Milk is richer in younger women. It is unaffected by the return of menstruation, but is adversely influenced by illness or by emotional disturbances.

Effects of diet. The quantity and composition of milk bear a complicated relation to the diet. Fundamentally a good milk can only be formed from a good diet. A superabundant diet does not increase the total yield or richness of the milk unless the protein content is increased. If the diet is inadequate, it is found that early in lactation the body tissues are used to form milk, which is not reduced much in amount, and weight is lost; late in lactation, however, the yield of milk is reduced. The vitamin content of the milk depends on the amount of these substances in the diet. Alcoholic liquors, like stout, may serve to fatten the mother, but it is very doubtful whether they improve the quality of the milk in any way. Many *drugs* are excreted in milk.

REFERENCES

COWIE, A. T. (1972). Lactation and its hormonal control. In *Reproduction in mammals*, Vol. 3 (ed. C. R. Austin and R. V. Short) p. 106. Cambridge University Press.
—— and FOLLEY, S. J. (1957). In *The neurophypophysis* (ed. H. Heller) p. 183. London.
—— and TINDAL, J. S. (1971). *Physiology of lactation*. Physiological Society Monograph. Edward Arnold, London.
LYONS, W. R., LI, C. H., and JOHNSON, R. E. (1958). Hormonal control of lactation. *Recent Prog. Horm. Res.* **14**, 219.
MARTIN, J. B., REICHLIN, S., and BROWN, G. M. (1977). *Clinical neuroendocrinology*. Davis, Philadelphia.
MARTINI, L. and BESSER, G. M. (eds.) (1977) *Clinical neuroendocrinology*. Academic Press, London.
RICHARDSON, K. C. (1949). *Proc. R. Soc.* **B136**, 30.

Appendix

SI UNITS

The *mass* of a body is the quantity of matter it contains and the basic unit is the kilogram.

The *weight* of a body is the force it exerts on anything which supports it. Normally it exerts this force due to the fact that it itself is being attracted towards the earth by the force of gravity.

Force can be defined as that which changes a body's state of rest or of uniform motion in a straight line. The unit of force is the *newton* (N)—that force which produces an acceleration of 1 metre per second every second (1 m s^{-2}) when it acts on a mass of 1 kg. When the force (F) is expressed in newtons, the mass (m) in kilograms, and the acceleration (a) in metres per second per second:

$$F = ma.$$

It should be noted that 1 newton is equal to 10^5 dynes, where 1 dyne = 1 g × 1 cm s^{-2}.

Work is done when the point of application of a force moves and is measured by the product of the force and the distance moved in the direction of the force. The unit of work is the *joule* (J). One joule of work is done when the point of application of a force of 1 newton (N) moves through a distance of 1 metre in the direction of the force. A kilojoule (kJ) and a megajoule (MJ) are respectively a thousand and a million joules.

In CGS terms the unit of work is 1 erg—equal to a dyne centimetre. As 1 newton = 10^5 dynes, 1 joule = 10^7 ergs.

In mechanics, energy may be divided into potential energy or kinetic energy. When a mass is lifted above the floor, work is done against its weight and this work is stored in the form of gravitational potential energy. 1 Joule is the potential energy lost when a weight of 1 kilogram falls under gravity through 1 metre. The acceleration due to gravitational forces g is 9.81 metres per second per second:

$$g = 9.81 \text{ m s}^{-2}$$

Kinetic energy can be exemplified by a rifle bullet in motion—when it strikes its target it displaces it.

Power is the rate of doing work and is defined in watts (W).

1 watt = 1 joule per second (1 W = 1 J s^{-1}), kilowatts and megawatts represent a thousand and a million watts respectively.

Pressure is the force per unit area and is defined in newtons per square metre (N m^{-2}). The unit is the pascal (Pa)

$$1 \text{ Pa} = 1 \text{ N m}^{-2}$$

Since the days of Torricelli and Pascal (mid- and late-seventeenth century) the barometric pressure due to the atmosphere has been measured in terms of the height of a column of mercury. At sea level this is approximately 760 mm—Pascal himself showed that the barometric pressure at the top of the Puy-de-Dôme in the Auvergne mountains was less than 760 mm Hg.

760 mm Hg = 1 atmosphere = 76 cm Hg. The density of mercury is 13.6 g cm^{-3} and the acceleration due to gravity = 9.81 m s^{-2}.

Thus 1 atm = 76 × 13.6 g cm^{-2}

$$= \frac{76 \times 13.6}{1000} \text{ kg cm}^{-2}$$

$$= \frac{76 \times 13.6 \times 9.81}{1000} \text{ N cm}^{-2}$$

$$= \frac{76 \times 13.6 \times 9.81 \times 100^2}{1000} \text{ N m}^{-2}$$

$$= 101\,400 \text{ N m}^{-2}$$

$$= 101\,400 \text{ Pa}$$

∴ 760 mm Hg = 101.4 kPa

1 mm Hg = 0.133 kPa.

(In passing it might be noted that 10^5 N m^{-2} (=10^5 Pa) is equal to 1 bar or 1000 millibars.)

In medicine it has been customary to express pressures in terms of mm Hg—blood pressure, osmotic pressure, gas pressures, etc. This practice will doubtless continue for a considerable time but from the above arithmetic it is clear that 1 kilopascal equals 7.5 mm Hg—i.e. *whenever the pressure is defined in mm Hg, then it must be multiplied by 4/30 to give the answer in terms of kilopascals.* Thus, 30 mm Hg is 4 kPa, 45 mm Hg is 6 kPa, 90 mm Hg is 12 kPa, etc.

Heat. The unit joule will displace the older term kilocalorie in expressing the energy expenditure—

1 kilocalorie = 4185.5 joules or 4.1855 kJ (kilojoules).

Thus a basal metabolic rate of an adult man formerly expressed as 40 kilocalories per square metre of body surface per hour is 167 kJ m^{-2} h^{-1}.

The unit of quantity is the *mole*. It is the number of atoms ($6.022\,0943 \times 10^{23}$) in 12 grams of carbon—i.e. the atomic weight of carbon. The atomic weight in grams of any element contains this number of atoms. In physiology, most concentrations are defined in millimoles (one thousandth), micromoles (one millionth), or nanomoles (one thousand millionth) according to the substance being considered.

The unit of quantity is the litre which is the cubic decimetre (dm^3). One millilitre (ml) is one thousandth of 1 litre, or 1 cm^3.

Concentrations may be expressed as moles per litre (mol l^{-1}), millimoles per litre (mmol l^{-1}), micromoles per litre (μmol l^{-1}), nanomoles per litre (nmol l^{-1}), etc.

Index

Where more than one page number is given more important sections are indicated by italic type.